## RIGHT TRIANGLE

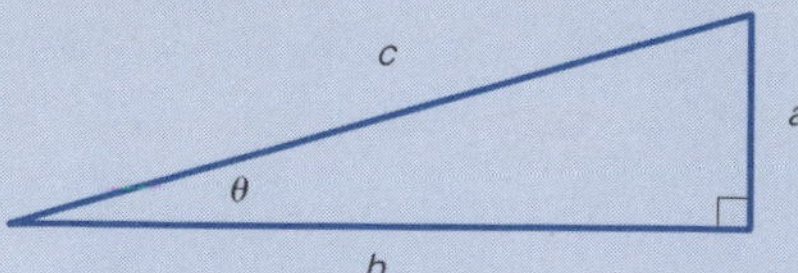


*Pythagorean theorem*

$c^2 = a^2 + b^2$

*Trigonometric functions*

$\sin\theta = \frac{a}{c}$  $\cot\theta = \frac{b}{a}$

$\cos\theta = \frac{b}{c}$  $\sec\theta = \frac{c}{b}$

$\tan\theta = \frac{a}{b}$  $\csc\theta = \frac{c}{a}$

## ANY TRIANGLE

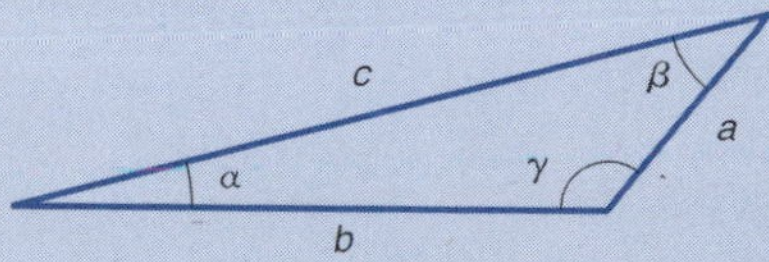


*Law of cosines*

$c^2 = a^2 + b^2 - 2ab\cos\gamma$

*Law of sines*

$$\frac{\sin\alpha}{a} = \frac{\sin\beta}{b} = \frac{\sin\gamma}{c}$$

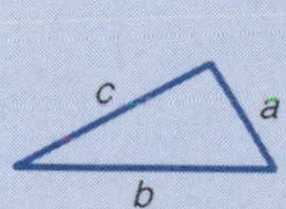


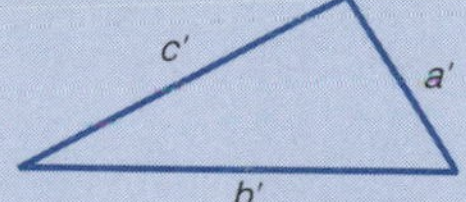


*Ratio of corresponding sides of similar triangles*

$\frac{a}{b} = \frac{a'}{b'}$  $\frac{a}{c} = \frac{a'}{c'}$  $\frac{b}{c} = \frac{b'}{c'}$

# TRIGONOMETRIC FORMULAS

*Fundamental formulas*

$\tan\theta = \frac{\sin\theta}{\cos\theta}$

$\cot\theta = \frac{\cos\theta}{\sin\theta}$

$\sec\theta = \frac{1}{\cos\theta}$

$\csc\theta = \frac{1}{\sin\theta}$

$\sin^2\theta + \cos^2\theta = 1$

$\tan^2\theta + 1 = \sec^2\theta$

$1 + \cot^2\theta = \csc^2\theta$

*Addition formulas*

$\sin(\theta \pm \phi) = \sin\theta\cos\phi \pm \cos\theta\sin\phi$

$\cos(\theta \pm \phi) = \cos\theta\cos\phi \mp \sin\theta\sin\phi$

$\tan(\theta \pm \phi) = \frac{\tan\theta \pm \tan\phi}{1 \mp \tan\theta\tan\phi}$

*Negative angle formulas*

$\sin(-\theta) = -\sin\theta$

$\cos(-\theta) = \cos\theta$

$\tan(-\theta) = -\tan\theta$

*Double angle formulas*

$\sin 2\theta = 2\sin\theta\cos\theta$

$\cos 2\theta = \cos^2\theta - \sin^2\theta = 1 - 2\sin^2\theta$
$= 2\cos^2\theta - 1$

$\tan 2\theta = \frac{2\tan\theta}{1 - \tan^2\theta}$

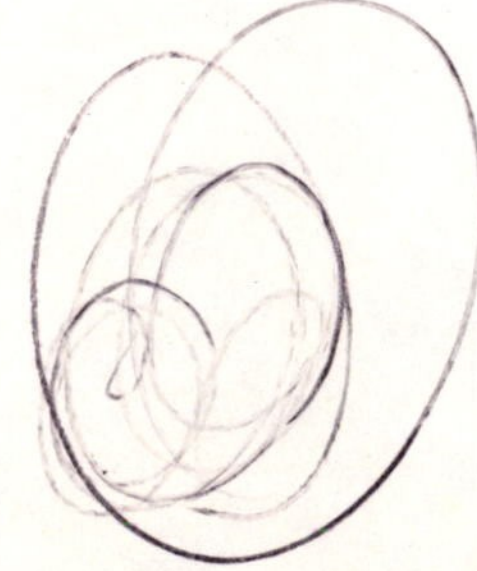

1817

HARPER & ROW, PUBLISHERS, New York
*Cambridge*
*Philadelphia*
*San Francisco*
*Washington*
*London*
*Mexico City*
*São Paulo*
*Singapore*
*Sydney*

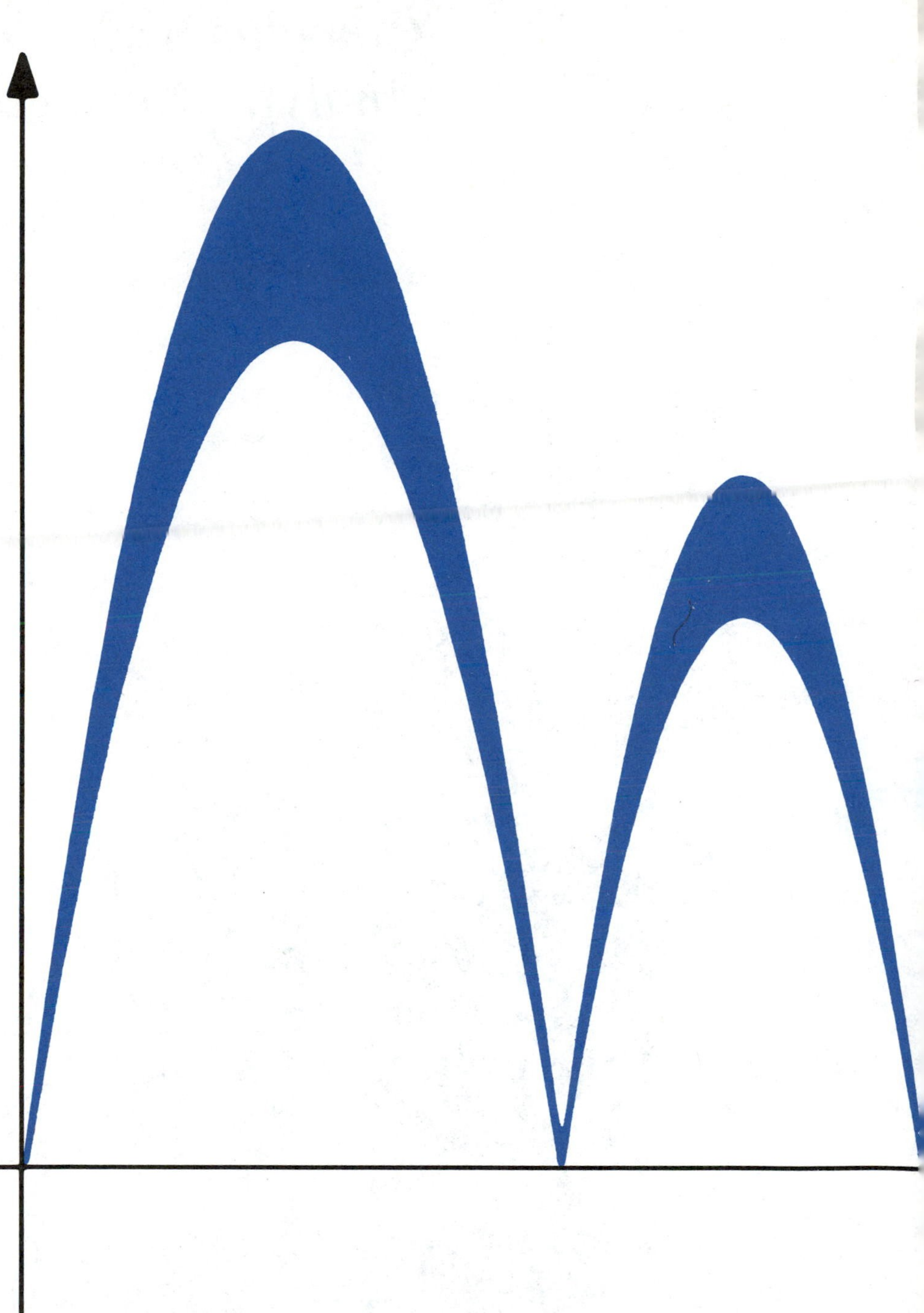

# *Calculus with Analytic Geometry*

# *Calculus with Analytic Geometry*

**RICHARD A. HUNT**
*Purdue University*

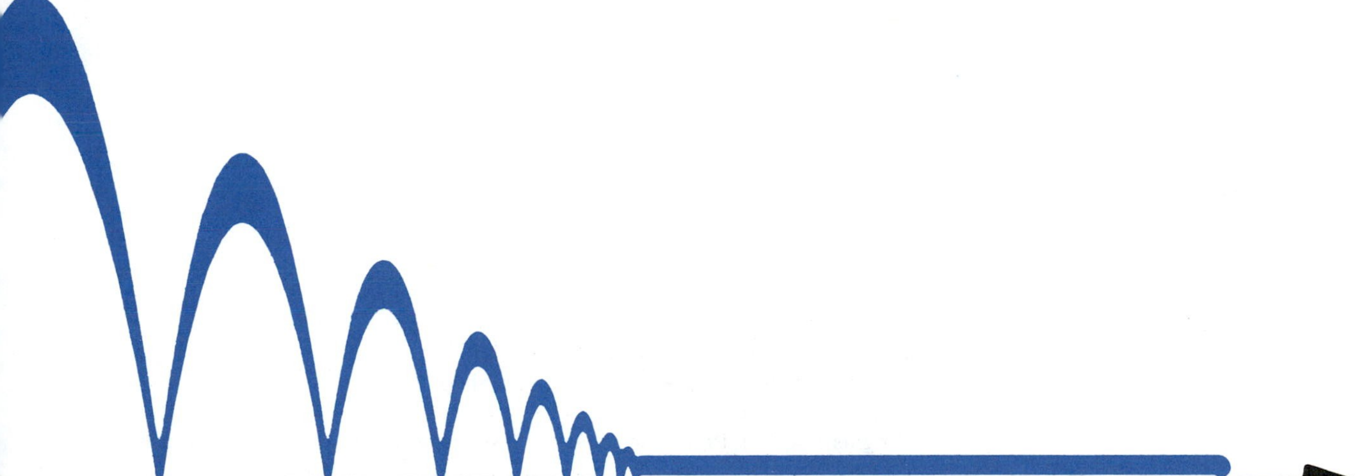

Editor-in-Chief: Judith Rothman
Sponsoring Editors: Ann Trump/Peter Coveney
Development Editors: David R. Esner/Jonathan Haber
Project Editor: Steven Pisano
Text and Cover Design: Betty Binns Graphics
Cover Photo: Joel Gordon
Text Art: ComCom Division of Haddon Craftsmen, Inc.
Production Manager: Kewal Sharma
Compositor: Composition House
Printer and Binder: Kingsport Press
Cover Printer: NEBC

**Calculus with Analytic Geometry**


**Library of Congress Cataloging-in-Publication Data**

Hunt, Richard A., 1937–
Calculus with analytic geometry.

Includes index.
1. Calculus. 2. Geometry, Analytic. I. Title.
QA303.H853 1988 515′.15 87-23612
ISBN 0-06-043036-2

87 88 89 90 9 8 7 6 5 4 3 2 1

# Contents

**CHAPTER TWO**

*Limits, continuity, and differentiability at a point*



**CHAPTER THREE**

*Continuity and differentiability on an interval*



**CHAPTER FOUR**

*Graph sketching and applications of the derivative*

**CHAPTER FIVE**

*The definite integral*



**CHAPTER SIX**

*Applications of the definite integral*



**CHAPTER SEVEN**

*Selected transcendental functions*



**CHAPTER EIGHT**

*Techniques of integration*

**CHAPTER TWELVE**

*Polar coordinates*



**CHAPTER THIRTEEN**

*Three-dimensional space*



**CHAPTER FOURTEEN**

*Vectors*



**CHAPTER FIFTEEN**

*Parametric curves*

**CHAPTER SIXTEEN**

*Differential calculus of functions of several variables*



**CHAPTER SEVENTEEN**

*Multiple integrals*



**CHAPTER EIGHTEEN**

*Topics in vector calculcus*

CONTENTS

# For the instructor

The purpose of this text is to illustrate the concepts, techniques, and applications of calculus. Intended primarily for science and engineering majors, the text contains all the usual topics of a standard three-semester introductory calculus sequence, with some extras.

### *An emphasis on concepts*

The text emphasizes understanding ideas as opposed to blindly memorizing formulas. An important objective is for students to see how the fundamental concepts of calculus, including limits, continuity, and differentiability, relate to the theory. Sometimes the exposition is informal, but the perspective is always that of mathematics. The theory is always mathematically correct, with emphasis on the concepts; only the omission of purely technical details prevents the proofs from being rigorous.

### *The basic approach: laying the groundwork*

The book is distinguished by the manner and extent to which complicated problems are broken down into simpler components that are then synthesized. This occurs both locally, in describing concepts and processes, and globally, as the separate components are studied in advance of a complete problem. As you use this text, you will recognize many places where the groundwork is being set for future work.

The emphasis and approach make the text suitable for a wide range of students. Extra-help sections and drill problems are there for students who need them. Concepts and the more challenging exercises can be emphasized for students with more technical skill and theoretical interest. The manuscript has been used successfully at Purdue both for large lecture classes and for smaller classes of honor students.

### *Bounds of functions*

A significant feature of the approach is the early use, in Section 2.2, of the triangle inequality and other properties of inequalities to find bounds of functions. Bounds of functions then appear in many calculations throughout the text to obtain estimates of error. A systematic study of bounds helps students better understand such problems as approximation by Taylor polynomials and numerical integration. Bounds of functions also provide a foundation for the development of limits. I have found that once students are able to find the bounds of a few representative types of functions they can work competently with the definition of the limit of a function at a point, so this basic concept of calculus can be made accessible to students of varied backgrounds and skills.

### *Multivariable calculus*

The approach makes this text particularly strong in the development of multivariable calculus. I have striven for a text that is as thorough in the third semester as it is in the first. A distinguishing feature is the use of "sketch and describe" problems. In Section 17.1, before multiple integrals are discussed, we introduce problems that teach students to associate the limits of integration of double and triple integrals with regions in the plane and in three-dimensional space. I have found these problems to be effective in helping students with the difficult problem of setting up limits of integration of multiple integrals.

Section 15.1 shows students how to find parametric equations of some standard curves. This allows them to set up and evaluate line integrals in Chapter 18. Continuity and discontinuity of functions of several variables are studied in enough detail in Section 16.2 that students can see why Green's Theorem cannot always be used to evaluate the flux of a vector function across a given closed curve.

### *Development of calculus and noncalculus skills*

Extra help in the basic noncalculus skills required for calculus problems is offered in several chapters. Section 4.5 helps students set up word problems by concentrating on finding equations that relate variables. Section 5.1 contains a development of the skills in analytic geometry that are required for applications of the definite integral. Section 13.2 gives specific instructions on how to sketch three-dimensional objects in the axonometric style of computer graphics and technical drawings. All three-dimensional figures in the text are in this style, and students are encouraged throughout to use sketches in their work. The nearly 2000 figures include over 400 in the answer section so that students can check their sketching skills.

The text contains an outstanding collection of examples to demonstrate the basic techniques and variations needed to solve calculus problems. The examples, which range on each topic from the very simple to the more difficult, are chosen to suggest a wide variety of applications without requiring excessive outside knowledge.

Students are encouraged in these examples to think about problems in general terms. Concepts are illustrated by the simplest examples that make the point, and many hints, suggestions, and warnings are given. The solutions contain discussion of how the general approach to the problem relates to the particular example.

### *Additional features*

While the selection and sequence of topics are essentially standard, the text incorporates several variations in emphasis and order to present the concepts of calculus more effectively:

- Functions and domains are emphasized in extremum problems. The examples and exercises illustrate how practical considerations can restrict the domain of a function that represents a physical quantity and how the domain of a function affects its extreme values.
- Taylor polynomials are introduced in Section 3.6 as a drill for finding higher-order derivatives, along with hints concerning their future use in approximation. Linear and quadratic approximations by Taylor polynomials are studied in Sections 4.8 and 4.9. Estimates of error in the linear and quadratic approximations are given and compared in Section 4.9. Taylor polynomials are reintroduced in Section 10.7 in connection with infinite series.
- Section 7.1 introduces simple forms of l'Hôpital's Rule so they can be used immediately in the development of transcendental functions, which follows. Students should see l'Hôpital's Rule, like calculus itself, as a tool for obtaining information about functions.
- Inverse trigonometric functions illustrate the general theory of inverse functions in Sections 7.2 and 7.3, before the logarithmic and exponential functions are developed. This allows students to see the general theory in connection with familiar functions and provides for an efficient development of transcendental functions.
- Applications of Riemann sums precede introduction of the definite integral in Chapter 5, and students are shown numerical methods of integration before they use the Fundamental Theorem of Calculus to evaluate definite integrals. The aim is to encourage students to think of the definite integral as a sum, an essential idea for understanding its applications.

■ Differential equations are restricted to those that involve only the concepts and techniques of calculus. A short chapter on applications of differential equations follows the chapter on techniques of integration; these differential equations serve as applications of integration. Infinite series solutions of differential equations, placed immediately after the development of power series, emphasize the approximation of functions by power series.

### *Accuracy*

I am well aware of the waste of time and effort caused by errors in a text, especially incorrect answers to the exercises. I have therefore made extensive effort to eliminate errors in the text and to ensure the accuracy of the problems, in both their statement and their solution.

In addition to extensive reviews of the manuscript and extensive class testing at Purdue and elsewhere, two proofreaders each checked galleys against the manuscript, and two mathematics graduate students read page proofs. Of course, I also checked for errors at each stage of the development.

Each exercise has been worked repeatedly. One set of solutions was generated as I class tested the manuscript at Purdue. Another set was provided by a group of four experienced teachers at other universities. A third set of solutions was prepared from page proofs. Finally, I personally worked each exercise for the solutions manuals. Several checkers then compared the answer section of this book with each solution set.

To ensure accuracy, each figure was individually programmed for a Hewlett-Packard two-pen plotter and then loaded on a CAD system, where the colors and screens were separated to produce the finished artwork. The cross-in-a-box symbol printed at the top of each page of this book is a registration mark that allows you to check that the mathematically correct alignment of the computer is preserved in the printed text.

### *Supplements*

This text is supported by a full package of teaching and learning materials.

■ *Instructor's solutions manual* I have prepared a manual for the instructor containing the solutions to all the exercises in the text. You may find it useful for checking the variation of technique required by particular exercises and their degree of difficulty.

■ *Student solutions manual* You may choose to make available for student purchase the *Student Solutions Manual*, containing the solutions to odd-numbered exercises only. Like the *Instructor's Solutions Manual*, it was written by me, to ensure that its style conforms to the examples in the text.

■ *Transparencies* A set of 35 transparencies is available free to instructors. Included are figures from the text of particular complexity or utility in classroom presentation.

■ *Test bank* *CalcTest*, a computerized test-generation system free to adopters, offers instructors a set of core questions for class testing, placement and credit exams, or individual student review. It allows instructors to select a large number of numerical variations on approximately 180 standard questions taken from the text. *CalcTest* will be a valuable teaching resource where many versions of the same test or extensive student practice in a single question is required.

■ *Computer tutorial* *CalcLab*, by Karl E. Petersen of the University of North Carolina, Chapel Hill, is a series of six diskettes, available to students, that review and explore some of the basic concepts and problems of calculus. Its six units cover limits, derivatives, derivatives of trigonometric functions, the shape of curves, related rates, and maxima and minima problems. Each unit of *CalcLab* lets the student produce vivid graphs and simulations. It will help students develop an ability to read problems, model those problems mathematically, visualize and understand relationships between variables, and carry out the steps toward a solution.

## *Acknowledgments*

I have been fortunate to have an excellent group of reviewers. I would like to thank them and other colleagues whose many varied and valuable suggestions improved the text:

Simon Bernau, *Southwestern Missouri State University*
John Brothers, *Indiana University*
Carl Cowen, *Purdue University*
Ron Davis, *Northern Virginia Community College*
Bruce Edwards, *University of Florida*
Garret Etgen, *University of Houston*
Ron Ferguson, *San Antonio College*
Kathy Franklin, *Los Angeles Pierce College*
Charles Friedman, *University of Texas at Austin*
William Fuller, *Purdue University*
August Garver, *University of Missouri, Rolla*
Paul Goodey, *University of Oklahoma*
Richard Hodel, *Duke University*
Edwin Hoefer, *Rochester Institute of Technology*
David Hoff, *Indiana University*
Gail Kaplan, *United States Naval Academy*
B. J. Kirby, *Queen's University*
Paul Kumpel, *State University of New York at Stony Brook*
Stanley Lukawecki, *Clemson University*
Frank Morgan, *Williams College*
Roger Nelsen, *Lewis and Clark College*
Anthony Peressini, *University of Illinois*
William Perry, *Texas A&M University*
Karl Petersen, *University of North Carolina, Chapel Hill*
Thomas Rishel, *Cornell University*
John Scheick, *Ohio State University*

Merriline Smith, *California State Polytechnic Institute*
Roger Smith, *Texas A&M University*
Malcolm Soule, *California State University, Northridge*
Donna Szott, *Allegheny Community College*
Richard Thompson, *University of Arizona*
Eugene Tidmore, *Baylor University*
John Wainwright, *University of Waterloo*
R. O. Wells, *Rice University*

I would like to give special credit to Carl Cowen, who served from beginning to end as my "in-house" consultant at Purdue. I routinely discussed ideas with Carl, and he contributed many original ideas to the text. Special thanks also go to Ed Hoefer, Bill Perry, Gail Kaplan, and Bill Fuller, who supplemented their written reviews with personal discussions, and to the undergraduates here at Purdue whose experiences have helped the book become more pedagogically and mathematically sound.

Bradley Lucier, Purdue University, wrote the BASIC subroutine for generating the figures in the text. John Hunt, my son, and Shuang Zhang programmed many individual figures; my daughter, Julie Wolf, also contributed. Under the direction of David Davis, ComCom transformed the computer programs into the artwork that appears in the book.

Two complete sets of exercise solutions were used to verify the accuracy of the set I personally prepared. One set was produced by David Hart and Ed Hoefer, Rochester Institute of Technology, Mina Wender, The McDonough School, and Karen Zak, United States Naval Academy. The other set was provided by a crew of 20 mathematics graduate students at Purdue. I thank them all.

My association with Harper & Row has been most pleasant. Everyone with whom I have worked on this book has been extremely cooperative. My sincere thanks go to each one. Special thanks to Steven Pisano, who dealt very patiently with my little panics as he coordinated all the pieces of the book, and especially to Ann Trump, who was my first contact with Harper & Row and who gave the book direction for over five years. I greatly appreciate Ann's enthusiasm for the project, her dedication, and the encouragement she provided throughout.

Extra special thanks go to my wife, Ann Hunt, for all she's put up with and done without while I've been laboring over this book these past six years. I couldn't have done it without her understanding and constant loving support.

# *For the student*

This book is written to help you learn calculus. Based on the evaluations of more than a thousand students at Purdue University, I can assure you that you will find it useful and relatively easy to read. The examples are worked out in detail to serve as guides to solving the exercises. Nearly 2000 illustrations, each computer drawn for clarity and accuracy, are included to help you visualize the mathematics. Special attention is given to developing and reviewing the basic skills in algebra, trigonometry, and analytic geometry that you will need to solve calculus problems.

Calculus can be described as a tool for analyzing functions. This analysis gives us information about the physical quantities functions represent and allows us to solve a variety of practical problems that require techniques beyond those of algebra, trigonometry, and analytic geometry.

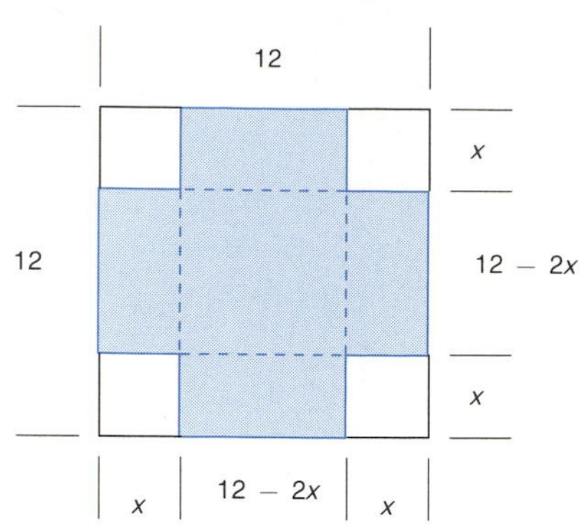


(a)

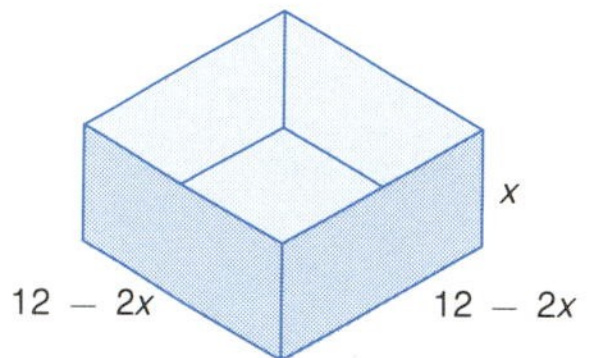


(b)

**FIGURE 1**

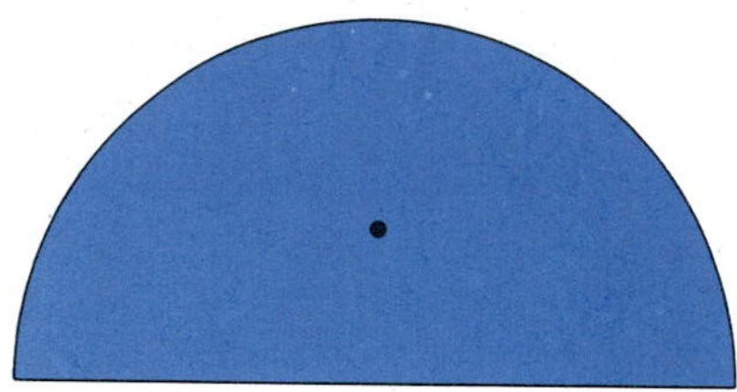

**FIGURE 2**

Suppose, for example, that we want to construct a rectangular box with no top by cutting $x$-inch squares from each corner of a 12-inch square and bending up the sides, as shown in Figure 1. Algebra and geometry can be used to determine that the formula $V = x(12 - 2x)^2$ gives the volume of the box in cubic inches. We then ask which choice of $x$ gives the box with the largest volume. This is not the type of question that is asked in an algebra class, but we can use calculus to answer it *easily*. Many important practical problems involve finding maximum and minimum values of a function, and calculus is well suited to solving problems of this type.

Another fundamental application of calculus is finding approximate values of functions. Approximating functions is inherent in the use of computers and calculators. For example, a calculator display of $\sqrt{2}$ is an approximate value of the function $\sqrt{x}$. Your calculator certainly does not store all values of the function $\sqrt{x}$. Rather, it calculates approximate values.

In your studies, you will use calculus as a tool to solve many types of physical problems. Calculus can be used for such problems as finding the balance point of an object, such as the semicircular plate shown in Figure 2. It can also help us to see that a parachutist will fall no faster than a speed that depends on his or her weight and the effectiveness of the parachute, but not on the height from which the parachutist jumps. Calculus lets us study the bouncing handball on the cover of this book. If the ball always bounces back to three-fourths the height of the previous bounce, it will bounce an infinite number of times. But will it continue bouncing forever? You will discover the answer in Chapter 9. (It has been my experience that students who have not yet studied calculus are split about 50–50 on the answer to this question.)

You will see many applications of calculus in this text, but you should realize that our goal includes more than the examples given here. When you understand the concepts of calculus, you will be able to apply this tool to problems you encounter as you continue to progress in your major area of study.

You can learn calculus by following the same steps you might take in learning to use any new tool. First, you see in general terms what the tool is and what it can do, perhaps by reading the instruction manual and learning about the important parts of the tool. In this step, you study *concepts*. Second, you practice the basic function of the tool until you feel confident with it. You learn *techniques*. Finally, you use the tool to create something useful. You use the tool for *applications*. This text explains and illustrates the concepts, techniques, and applications of calculus. I hope you will become comfortable and skillful with this powerful tool.

*Richard A. Hunt*

**CHAPTER ONE**

# *Some preliminary topics*

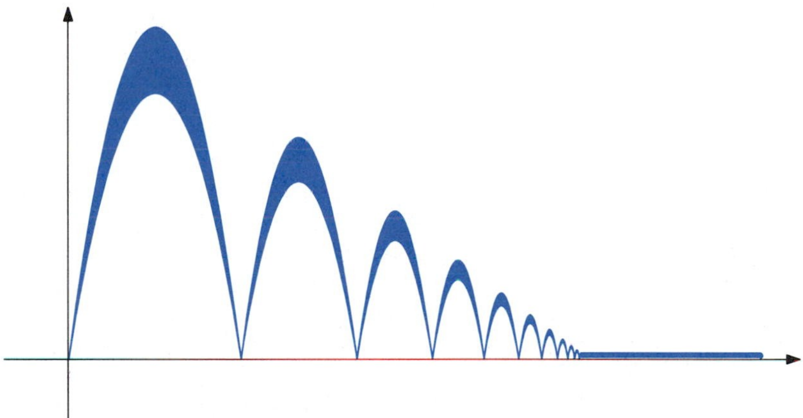

This chapter contains a review of topics of algebra, analytic geometry, and trigonometry that are needed for the study of calculus.

Manipulation of real numbers and algebraic expressions is fundamental to the solution of calculus problems. Our attention is restricted to algebra problems that will appear later as parts of calculus problems. Graphs of lines, parabolas, circles, and other standard curves are reviewed so they will be available to graphically illustrate concepts of calculus. The properties of the trigonometric functions that are reviewed here are used throughout the text.

## 1.1 PROPERTIES OF REAL NUMBERS

In this section we will review solving inequalities, properties of absolute value, and rules of exponents. The development presupposes a working knowledge of the arithmetic properties of the real number system. We also assume familiarity with the association of real numbers with points on a number line, an association that includes the ideas of length, direction, and order. See Figure 1.1.1. We will use the terms "real number" and "point on a line" interchangeably.

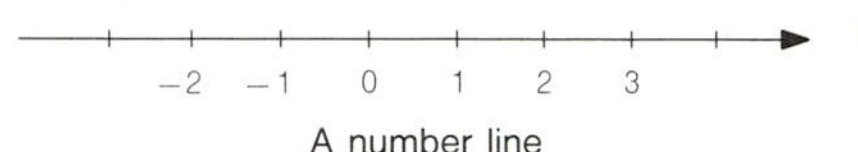


FIGURE 1.1.1

Let us begin with a review of inequalities.

To **solve** an inequality for a variable means to find an equivalent inequality in which the variable is isolated on one side of the inequality; numbers or other known quantities are on the opposite side of the inequality. Recall that the same quantity can be either added to or subtracted from both sides of an inequality without changing the sense (or direction) of the inequality. We can also multiply or divide both sides of an inequality by a positive number without changing the sense. However, multiplication or division by a negative number changes the sense of an inequality.

■ **EXAMPLE 1**

Solve the inequality $2x - 3 < 4 + 5x$.

**SOLUTION**

We gather terms that involve $x$ on the left-hand side of the inequality and constant terms on the right. To do this we add 3 to and subtract $5x$ from both sides of the original inequality and then combine like terms.

$$2x - 3 < 4 + 5x,$$

$$2x - 5x < 4 + 3,$$

$$-3x < 7.$$

Dividing by $-3$ we obtain the solution

$$x > -\frac{7}{3}.$$

Note that the sense of the inequality was changed when we divided by the negative number $-3$. ■

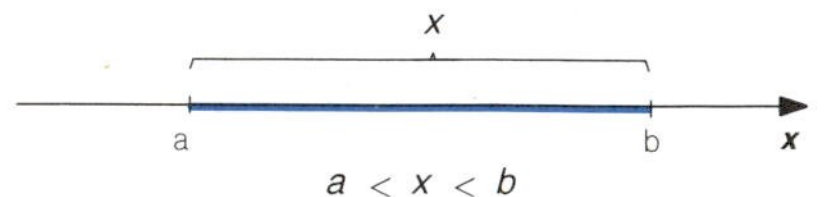


FIGURE 1.1.2

If $a < b$, the **double inequality** $a < x < b$ means $x$ is between $a$ and $b$. The inequalities $a < x$ *and* $x < b$ are *both* true. See Figure 1.1.2.

We can solve certain types of double inequalities by isolating the variables in the middle position. We follow the usual rules of inequalities, except we must remember to add, subtract, multiply, and divide all three terms by the same number. The senses of both inequalities are changed when we multiply or divide by a negative number.

■ **EXAMPLE 2**

Solve the double inequality $-2 < 4 - 3x < 13$.

**SOLUTION**

We first subtract 4 from each of the three terms of the original double inequality to obtain

$$-2 - 4 < -3x < 13 - 4$$

or

$$-6 < -3x < 9.$$

We now divide each term by $-3$, being careful to change the sense of both inequalities, to obtain

$$2 > x > -3.$$

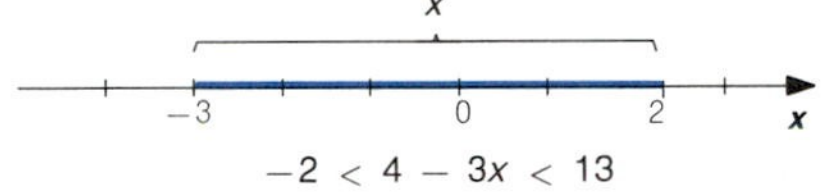


FIGURE 1.1.3

We can rewrite this as $-3 < x < 2$. The solution is sketched in Figure 1.1.3. ■

The **absolute value** of a real number $x$ is

$$|x| = \begin{cases} x, & x \geq 0, \\ -x, & x < 0. \end{cases} \tag{1}$$

The absolute value of a real number is always nonnegative, so $|x| \geq 0$ for any real number $x$.

■ **EXAMPLE 3**

Evaluate (a) $|5|$, (b) $|-3|$, (c) $|1 - \sqrt{2}|$.

**SOLUTION**

(a) Since $5 > 0$, $|5| = 5$.
(b) Since $-3 < 0$, $|-3| = -(-3) = 3$.
(c) We know that $\sqrt{2} > 1$, so $1 - \sqrt{2} < 0$. We then have $|1 - \sqrt{2}| = -(1 - \sqrt{2}) = \sqrt{2} - 1$. ■

For some of our work, it will be useful to express the absolute value of a variable expression in a form that does not involve absolute value notation. Generally, this will require two expressions, one the negative of the other. For example, since $2x - 6 \geq 0$ for $x \geq 3$ and $2x - 6 < 0$ for $x < 3$, we can write

$$|2x - 6| = \begin{cases} 2x - 6, & x \geq 3, \\ -(2x - 6), & x < 3. \end{cases}$$

We have the following properties of absolute value:

$$|-x| = |x|, \quad |xy| = |x| \cdot |y|, \quad \left|\frac{x}{y}\right| = \frac{|x|}{|y|}, \tag{2}$$

and

$$|x + y| \le |x| + |y|. \tag{3}$$

The inequality in (3) is called the **Triangle Inequality**. It can be verified by considering all cases where $x$ and $y$ have like or unlike signs. Equality holds whenever $x$ and $y$ have like signs; inequality holds for unlike signs.

The **distance** between two points $x_0$ and $x_1$ on a number line is $|x_1 - x_0|$. See Figure 1.1.4. *Distance is nonnegative* and

Distance

$x_0$ $x_1$ $x$

FIGURE 1.1.4

$$|x_1 - x_0| = |x_0 - x_1|.$$

The absolute value of $x$ may be interpreted as the distance of $x$ from the origin. It is then easy to see that

$$|x| < d \qquad \text{if and only if } -d < x < d. \tag{4}$$

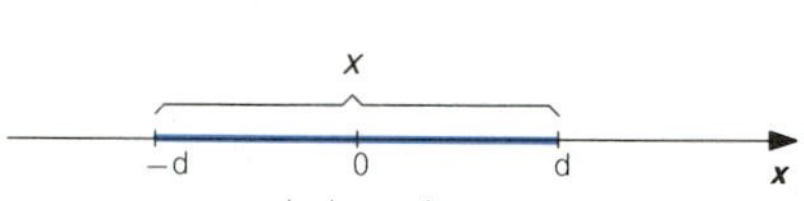


FIGURE 1.1.5

See Figure 1.1.5.

We will need to solve inequalities of the form $|u| < d$, where $u$ is an expression that involves the variable $x$. To do this, we use (4) to write $|u| < d$ as the equivalent double inequality $-d < u < d$, and then we solve the double inequality as before.

■ **EXAMPLE 4**

Solve the inequality $|2x - 3| < 2$.

**SOLUTION**

Using (4), we rewrite $|2x - 3| < 2$ as the double inequality

$$-2 < 2x - 3 < 2.$$

Then

$$-2 + 3 < 2x < 2 + 3,$$

$$1 < 2x < 5,$$

$$\frac{1}{2} < x < \frac{5}{2}.$$

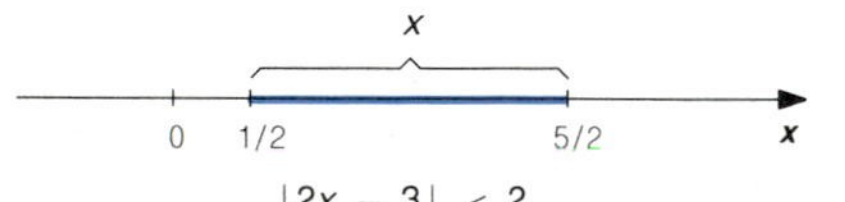


FIGURE 1.1.6

The solution is sketched in Figure 1.1.6. ■

The inequality opposite to that in (4) describes points $x$ that are a distance more than $d$ from the origin.

$$|x| > d \qquad \text{if and only if either } x > d \text{ or } x < -d. \tag{5}$$

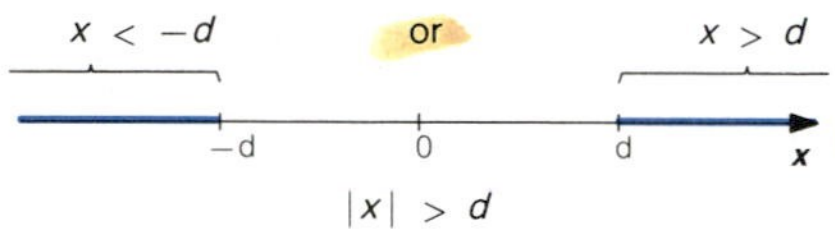

FIGURE 1.1.7

If $d$ is nonnegative, a number $x$ cannot satisfy both $x > d$ and $x < -d$; $|x| > d$ means $x$ is in either the interval $x > d$ or the interval $x < -d$, but not both. See Figure 1.1.7.

Let us now move to a review of exponential expressions. We begin by defining the notation. For any real number $x$ and any positive integer $n$, the $n$th power of $x$ is denoted $x^n$. If $x$ is nonnegative, $x^{1/n}$ or $\sqrt[n]{x}$ will denote the *nonnegative* $n$th root of $x$. (If $n$ is an even integer and $x$ is positive, then $x$ has two real $n$th roots, one positive and one negative. $x^{1/n}$ and $\sqrt[n]{x}$ are taken to be positive if $x$ is positive.) If $x$ is negative and $n$ is odd, then $x$ has only one real $n$th root, denoted either $x^{1/n}$ or $\sqrt[n]{x}$. (A real odd root of a negative number is always negative.) Negative numbers have no real even roots. If $m$ and $n$ are positive integers, $x^{m/n} = (x^m)^{1/n} = (x^{1/n})^m$. $x^0 = 1$ for $x \neq 0$. $0^0$ is undefined. Expressions with negative exponents are defined by $x^{-r} = 1/x^r$.

■ **EXAMPLE 5**

We have

(a) $2^4 = 16$,
(b) $16^{1/4} = \sqrt[4]{16} = 2$,
(c) $-16^{1/4} = -\sqrt[4]{16} = -2$,
(d) $(-16)^{1/4}$ and $\sqrt[4]{-16}$ are undefined,
(e) $2^3 = 8$,
(f) $8^{1/3} = 2$,
(g) $(-2)^3 = -8$,
(h) $(-8)^{1/3} = -2$,
(i) $8^{2/3} = (8^{1/3})^2 = (2)^2 = 4$,
(j) $8^{-1/3} = \dfrac{1}{8^{1/3}} = \dfrac{1}{2}$,
(k) $8^0 = 1$. ■

Since $\sqrt{x^2}$ is the *nonnegative* square root of the nonnegative number $x^2$, we have $\sqrt{x^2} = |x|$. For example, $\sqrt{(2)^2} = \sqrt{4} = 2 = |2|$ and $\sqrt{(-3)^2} = \sqrt{9} = 3 = |-3|$. We can express the distance between points $x_0$ and $x_1$ on a number line as

$$\sqrt{(x_1 - x_0)^2} = |x_1 - x_0|.$$

A working knowledge of the following rules of exponents is required:

$$x^m \cdot x^n = x^{m+n}, \quad (x^m)^n = x^{mn}, \quad (xy)^m = x^m \cdot y^m,$$

$$\left(\frac{x}{y}\right)^m = \frac{x^m}{y^m}, \quad x^{-m} = \frac{1}{x^m}, \quad \frac{x^m}{x^n} = x^{m-n}.$$

■ **EXAMPLE 6**

Express the following in the form $x^r$:

(a) $1/x$,
(b) $x^{1/3} \cdot x^{1/2}$,
(c) $x\sqrt{x}$,
(d) $(x^3)^2$,
(e) $\dfrac{x}{\sqrt{x}}$,
(f) $\dfrac{(x^{-2})^{-3}}{x \cdot x^3}$.

**SOLUTION**

(a) $1/x = x^{-1}$,
(b) $x^{1/3} \cdot x^{1/2} = x^{1/3+1/2} = x^{2/6+3/6} = x^{5/6}$,
(c) $x\sqrt{x} = x^1 x^{1/2} = x^{1+1/2} = x^{3/2}$,
(d) $(x^3)^2 = x^{3(2)} = x^6$,
(e) $\dfrac{x}{\sqrt{x}} = \dfrac{x}{x^{1/2}} = x^{1-1/2} = x^{1/2}$,
(f) $\dfrac{(x^{-2})^{-3}}{x \cdot x^3} = \dfrac{x^{-2(-3)}}{x^{1+3}} = \dfrac{x^6}{x^4} = x^{6-4} = x^2$. ■

The ability to recognize small integer powers of small integers allows us to calculate the exact value of many exponential expressions.

■ **EXAMPLE 7**

We have

(a) $(-3)^3 = -27$,
(b) $(-27)^{1/3} = -3$,
(c) $2^6 = 64$,
(d) $(64)^{1/3} = (2^6)^{1/3} = 2^{6(1/3)} = 2^2 = 4$,
(e) $(25)^{3/2} = ((25)^{1/2})^3 = (5)^3 = 125$,
(f) $\dfrac{(18)^{3/2}}{8^{1/2}} = \dfrac{(2 \cdot 9)^{3/2}}{(2^3)^{1/2}} = \dfrac{2^{3/2} \cdot 9^{3/2}}{2^{3/2}} = (9^{1/2})^3 = (3)^3 = 27$. ■

We can use a calculator to find approximate values of exponential expressions. (When using a calculator, we will ordinarily use the full display, without rounding or truncating.) The symbol $a \approx b$ means $a$ and $b$ are approximately equal.

■ **EXAMPLE 8**

Using a calculator, we have

(a) $10^{1/2} \approx 3.1622777$,
(b) $5^{1.7} \approx 15.425847$,
(c) $3^{-6} \approx 0.00137174$,
(d) $\pi^{1/3} \approx 1.4645919$. ■

We conclude this section by establishing a notation for intervals. This is done in Table 1.1.1. Parentheses indicate that the corresponding endpoints are not included, while square brackets indicate that the corresponding endpoints are included. The *symbols* $\infty$ **(infinity)** and $-\infty$ do not represent real numbers.

Intervals for which both endpoints are real numbers are called **bounded** (or **finite**) **intervals**. Intervals that are not bounded are called **unbounded** (or **infinite**) **intervals**.

Intervals of the form $(a, b)$ are called **open**. Intervals of the form $[a, b]$ are called **closed**.

**TABLE 1.1.1**

| *Interval notation* | *Inequality notation* |
|---|---|
| $x$ in: $(a, b)$ | $a < x < b$ |
| $[a, b]$ | $a \leq x \leq b$ |
| $[a, b)$ | $a \leq x < b$ |
| $(a, b]$ | $a < x \leq b$ |
| $(-\infty, b)$ | $-\infty < x < b$ or simply $x < b$ |
| $(-\infty, b]$ | $-\infty < x \leq b$ or simply $x \leq b$ |
| $(a, \infty)$ | $a < x < \infty$ or simply $x > a$ |
| $[a, \infty)$ | $a \leq x < \infty$ or simply $x \geq a$ |
| $(-\infty, \infty)$ | $-\infty < x < \infty$ or "all real $x$" |

## EXERCISES 1.1

Solve the inequalities in Exercises 1–14.

**1** $4x - 3 < 2x + 5$
**2** $3x + 2 < 10 - x$
**3** $x - 1 > 3x + 4$
**4** $2x + 3 < 5x - 2$
**5** $1 < 2x - 1 < 3$
**6** $-1 < 3x + 5 < 11$
**7** $-4 < 2 - 3x < -1$
**8** $-5 < 3 - 2x < -1$
**9** $|2x - 3| < 1$
**10** $|2x + 1| < 3$
**11** $|3x - 4| < 2$
**12** $|3x + 1| < 1$
**13** $|3 - 2x| < 1$
**14** $|1 - 3x| < 2$

Evaluate the expressions in Exercises 15–22.

**15** $|3 - 4|, |3| + |-4|$
**16** $|5 - 2|, |5| + |-2|$
**17** $|-3 - 2|, |-3| + |-2|$
**18** $|5 + 2|, |5| + |2|$
**19** $|1 - \sqrt{2}| + |2 - \sqrt{2}|$
**20** $|\sqrt{3} - 2| + |\sqrt{3} - 1|$
**21** $\sqrt{(-3)^2}$
**22** $\sqrt{(-2)^2}$

Write the expressions in Exercises 23–32 in the form $x^r$.

**23** $1/x^2$
**24** $1/x^{-1}$
**25** $x^2 \cdot x^3$
**26** $x^{1/3} \cdot x^{1/4}$
**27** $(x^{-3})^2$
**28** $(\sqrt{x})^3$
**29** $1/\sqrt{x}$
**30** $x^2/\sqrt{x}$
**31** $\dfrac{(x^{-1})^2}{x \cdot x^{-4}}$
**32** $\dfrac{x^2 \cdot x^3}{(x^3)^2}$

Evaluate the expressions in Exercises 33–42.

**33** $9^{3/2}$
**34** $(-27)^{2/3}$
**35** $16^{1/4}$
**36** $(-32)^{1/5}$
**37** $(-8)^{2/3}$
**38** $(1/8)^{4/3}$
**39** $10^{1/2} \cdot 125^{-1/6} \cdot 2^{3/2}$
**40** $27^{1/6} \cdot 12^{1/2} \cdot 8^{1/3}$
**41** $27^{4/3}/9$
**42** $9^{2/3}/3^{1/3}$

Use a calculator to find approximate values of the expressions in Exercises 43–48.

**43** $(4.7)^{2/3}$
**44** $(2.4)^{1/4}$
**45** $3^{1.7}$
**46** $2^{1.6}$
**47** $2^{-7}$
**48** $3^{-5}$

Complete the tables in Exercises 49–50.

**49**

| *Interval notation* | *Inequality notation* |
|---|---|
| $(0, 1)$ | |
| $[1, 3]$ | |
| | $-1 < x \leq 1$ |
| | $x > 0$ |
| | $x \leq 1$ |

**50**

| *Interval notation* | *Inequality notation* |
|---|---|
| $(-1, 1)$ | |
| $[0, 2]$ | |
| | $-2 \leq x < 1$ |
| | $x \leq 1$ |
| | $x > -1$ |

## 1.2 ALGEBRA FOR CALCULUS PROBLEMS

The algebra problems in this section are parts of actual calculus problems. The algebra we do here is necessary to complete the solutions of the calculus problems. The better we can handle the algebra, the more time we will have to concentrate on the calculus aspects of the problems.

Calculus requires the simplification of many types of algebraic expressions. The simplest of these involve carrying out multiplications, removing parentheses, and combining like terms.

■ **EXAMPLE 1**

Simplify $(1-2x)(3)+(4+3x)(-2)$.

**SOLUTION**

$$\begin{aligned}(1-2x)(3)+(4+3x)(-2) &= 3-6x-8-6x\\ &= -5-12x.\end{aligned}$$ ■

■ **EXAMPLE 2**

Simplify $(x^2+1)(2x-1)-(x^2-x+1)(2x)$.

**SOLUTION**

$$\begin{aligned}&(x^2+1)(2x-1)-(x^2-x+1)(2x)\\ &\quad=(2x^3-x^2+2x-1)-(2x^3-2x^2+2x)\\ &\quad=2x^3-x^2+2x-1-2x^3+2x^2-2x\\ &\quad=x^2-1.\end{aligned}$$ ■

It is worthwhile to look for factors that are common to all terms. The factored form of the simplified expression is the desired form for most applications.

■ **EXAMPLE 3**

Simplify $(2x-3)^2(4)+(4x-1)(2)(2x-3)(2)$.

**SOLUTION**

$$\begin{aligned}(2x-3)^2(4)+(4x-1)(2)(2x-3)(2) &= 4(2x-3)[(2x-3)+(4x-1)]\\ &= 4(2x-3)(6x-4)\\ &= 4(2x-3)(2)(3x-2)\\ &= 8(2x-3)(3x-2).\end{aligned}$$ ■

■ **EXAMPLE 4**

Simplify

$$\frac{(x+1)^2(3)-(3x-2)(2)(x+1)(1)}{[(x+1)^2]^2}.$$

**SOLUTION**

Factoring $x + 1$ from each term in the numerator, we have

$$\frac{(x+1)^2(3) - (3x-2)(2)(x+1)(1)}{[(x+1)^2]^2}$$

$$= \frac{(x+1)[3(x+1) - 2(3x-2)]}{(x+1)^4}$$

$$= \frac{3x + 3 - 6x + 4}{(x+1)^3}$$

$$= \frac{-3x+7}{(x+1)^3}.$$

■

Cancelling a common variable factor from the numerator and denominator of an expression can give a simplified expression that is defined at a point where the original expression is not defined. We should indicate when this is the case.

**■ EXAMPLE 5**

Simplify

$$\frac{x^2 - x - 6}{x - 3}.$$

**SOLUTION**

$$\frac{x^2 - x - 6}{x - 3} = \frac{(x-3)(x+2)}{x-3} = x + 2, \qquad x \neq 3.$$

The statement $x \neq 3$ indicates that the simplified expression $x + 2$ is equal to the original expression only for $x \neq 3$. The original expression is not defined for $x = 3$, but the simplified expression is defined for all $x$. ■

**■ EXAMPLE 6**

Simplify

$$\frac{\dfrac{1}{x+1} - \dfrac{1}{3}}{x - 2}.$$

**SOLUTION**

$$\frac{\dfrac{1}{x+1} - \dfrac{1}{3}}{x-2} = \frac{\dfrac{1}{x+1}\left(\dfrac{3}{3}\right) - \dfrac{1}{3}\left(\dfrac{x+1}{x+1}\right)}{x-2}$$

$$= \frac{\dfrac{3 - (x+1)}{3(x+1)}}{x-2} = \frac{3 - x - 1}{3(x+1)(x-2)}$$

$$= \frac{-x+2}{3(x+1)(x-2)} = -\frac{(x-2)}{3(x+1)(x-2)}$$

$$= -\frac{1}{3(x+1)}, \qquad x \neq 2.$$ ■

The nature of some calculus formulas gives us expressions involving the sum or difference of powers where the powers differ by one. We must be prepared to handle these expressions. One method is to factor out the term having the smaller exponent. This requires special care when negative exponents are involved.

■ **EXAMPLE 7**

Simplify

$$\left(\frac{3}{2}\right)x^{1/2} - \left(\frac{1}{2}\right)x^{-1/2}.$$

**SOLUTION**

Each term contains a power of $x$. Since $-1/2 < 1/2$, the smaller exponent of $x$ is $-1/2$, so we factor $x^{-1/2}$ from each term. We also factor $1/2$ from each term.

$$\left(\frac{3}{2}\right)x^{1/2} - \left(\frac{1}{2}\right)x^{-1/2} = \left(\frac{3}{2}\right)x^{1-1/2} - \left(\frac{1}{2}\right)x^{-1/2}$$

$$= \frac{x^{-1/2}}{2}(3x^{1} - 1)$$

$$= \frac{3x - 1}{2x^{1/2}}.$$ ■

An alternate method of handling a sum or difference of expressions that involve negative exponents is to write the expressions with positive exponents and obtain a common denominator in order to carry out the addition or subtraction.

■ **EXAMPLE 8**

Simplify

$$(x^3)\left(\frac{1}{3}\right)(1 - x^2)^{-2/3}(-2x) + (1 - x^2)^{1/3}(3x^2).$$

**SOLUTION**

$$(x^3)\left(\frac{1}{3}\right)(1-x^2)^{-2/3}(-2x)+(1-x^2)^{1/3}(3x^2)$$

$$=x^2\left[\frac{-2x^2}{3(1-x^2)^{2/3}}+3(1-x^2)^{1/3}\left(\frac{3(1-x^2)^{2/3}}{3(1-x^2)^{2/3}}\right)\right]$$

$$=x^2\left[\frac{-2x^2+9(1-x^2)}{3(1-x^2)^{2/3}}\right]=x^2\left[\frac{9-11x^2}{3(1-x^2)^{2/3}}\right].$$

■

**■ EXAMPLE 9**

Simplify

$$(x)\left(\frac{1}{2\sqrt{1-2x}}\right)(-2)+(\sqrt{1-2x})(1).$$

**SOLUTION**

$$(x)\left(\frac{1}{2\sqrt{1-2x}}\right)(-2)+(\sqrt{1-2x})(1)$$

$$=\frac{-x}{\sqrt{1-2x}}+\sqrt{1-2x}\left(\frac{\sqrt{1-2x}}{\sqrt{1-2x}}\right)$$

$$=\frac{-x+(1-2x)}{\sqrt{1-2x}}=\frac{1-3x}{\sqrt{1-2x}}.$$

■

Values of the variable for which certain expressions are either zero or undefined are of particular interest for applications of calculus. A quotient is zero when the numerator is zero and the denominator is nonzero. A quotient is undefined when the denominator is zero.

**■ EXAMPLE 10**

Find all values of $x$ for which

$$\frac{(x^2)(-1)-(1-x)(2x)}{(x^2)^2}$$

is (a) zero and (b) undefined.

**SOLUTION**

The expression is undefined when $x=0$. To find where it is zero, we simplify the original expression to obtain

$$\frac{-x^2-2x+2x^2}{x^4}=\frac{x^2-2x}{x^4}=\frac{x(x-2)}{x^4}=\frac{x-2}{x^3}.$$

We see that the numerator of the simplified expression is zero when $x = 2$, so the simplified expression and the original expression are zero when $x = 2$. Even though the numerator of the original expression is zero when $x = 0$, the expression is not zero when $x = 0$ because it is undefined for $x = 0$. ■

**■ EXAMPLE 11**

Find all values of $x$ for which $x^{1/2} - x^{-1/2}$ is (a) zero and (b) undefined.

**SOLUTION**

$$x^{1/2} - x^{-1/2} = x^{1-1/2} - x^{-1/2} = x^{-1/2}(x - 1) = \frac{x - 1}{x^{1/2}}.$$

After simplifying, we see the expression is zero when $x = 1$. We must have $x \geq 0$ in order that $x^{1/2}$ be defined. Since we cannot divide by zero, this expression is defined only for $x > 0$. ■

It is easy to determine values of $x$ for which a polynomial is zero if the polynomial is factored into linear terms. If a quadratic polynomial cannot be easily factored, the **Quadratic Formula** can be used to find values for which it is zero. Recall that the Quadratic Formula tells us that

$$ax^2 + bx + c = 0 \quad \text{has solution} \quad x = \frac{-b \pm \sqrt{b^2 - 4ac}}{2a}.$$

From the Quadratic Formula we see that

$$ax^2 + bx + c \quad \text{has no real zeros if} \quad b^2 - 4ac < 0.$$

**■ EXAMPLE 12**

Find all values of $x$ for which

$$\frac{(x^2 + 1)(2x - 1) - (x^2 - x)(2x)}{(x^2 + 1)^2}$$

is (a) zero and (b) undefined.

**SOLUTION**

Since $x^2 + 1$ is never zero, the expression is defined for all real numbers $x$. The expression is zero when the numerator is zero. That is, when

$$(x^2 + 1)(2x - 1) - (x^2 - x)(2x) = 0.$$

Simplifying this expression, we obtain

$$2x^3 - x^2 + 2x - 1 - 2x^3 + 2x^2 = 0,$$

$$x^2 + 2x - 1 = 0.$$

From the Quadratic Formula, we see this is true for

$$x = \frac{-(2) \pm \sqrt{(2)^2 - 4(1)(-1)}}{2(1)}$$

$$= \frac{-2 \pm \sqrt{8}}{2} = \frac{-2 \pm 2\sqrt{2}}{2} = \frac{2(-1 \pm \sqrt{2})}{2}$$

$$= -1 \pm \sqrt{2}.$$

That is, the expression is zero if either $x = -1 + \sqrt{2}$ or $x = -1 - \sqrt{2}$. ■

In calculus, it is often important to know the sign of a given expression. In Examples 13 and 14, we will illustrate a simple scheme for solving this problem. The scheme is based on the following two facts:

**Most of the expressions we will be dealing with have constant sign on intervals between successive values where they are either zero or undefined.**

**The product or quotient of an odd number of negative factors is negative and the product or quotient of an even number of negative factors is positive.**

### ■ EXAMPLE 13

Find all intervals on which

$$\frac{x(x-2)}{x+1}$$

is positive.

#### SOLUTION

The expression is zero when $x = 0$ and when $x = 2$; it is undefined when $x = -1$. The numbers $-1$, 0, 2 divide the number line into four intervals. The sign of the given expression is constant on each of these intervals. To determine the proper sign in each interval, we list each factor in the left-hand column of a table. We then determine the sign of each factor in each of the intervals. This can be done either by observation or by substitution of a number in the interval for $x$ in the factor. (For example, we may observe that the factor $x + 1$ is negative on intervals to the left of $-1$, where it is zero; $x + 1$ is positive on all intervals to the right of $-1$. Substitution of any number less than $-1$ for $x$ in the expression $x + 1$ gives a negative number. Substitution of any number greater than $-1$ for $x$ in the expression $x + 1$ gives a positive number.) The sign of each factor in each of the four intervals is indicated in the corresponding row. The sign of the given expression in each interval is determined by counting the number of negative signs of factors in the column of the interval. See Table 1.2.1.

TABLE 1.2.1

| | $-\infty$ to $-1$ | $-1$ to $0$ | $0$ to $2$ | $2$ to $\infty$ |
|---|---|---|---|---|
| $x+1$ | $-$ | $+$ | $+$ | $+$ |
| $x$ | $-$ | $-$ | $+$ | $+$ |
| $x-2$ | $-$ | $-$ | $-$ | $+$ |
| $\dfrac{x(x-2)}{x+1}$ | $-$ | $+$ | $-$ | $+$ |

We conclude that $x(x-2)/(x+1)$ is positive for $-1 < x < 0$ and $x > 2$. ■

■ **EXAMPLE 14**

Find all intervals on which

$$\frac{(1-x)^{1/3}}{x^{2/3}}$$

is positive.

**SOLUTION**

The expression is zero when $(1-x)^{1/3} = 0$, which implies $1 - x = 0^3$, or $x = 1$; it is undefined when $x = 0$.

Odd roots of any number have the same sign as the number. Hence, $(1-x)^{1/3}$ has the same sign as $1 - x$. The expression $x^{2/3} = (x^{1/3})^2$ is positive for $x \neq 0$. The corresponding sign table is given in Table 1.2.2.

TABLE 1.2.2

| | $-\infty$ to $0$ | $0$ to $1$ | $1$ to $\infty$ |
|---|---|---|---|
| $x^{2/3}$ | $+$ | $+$ | $+$ |
| $(1-x)^{1/3}$ | $+$ | $+$ | $-$ |
| $\dfrac{(1-x)^{1/3}}{x^{2/3}}$ | $+$ | $+$ | $-$ |

We see from the table that the given expression is positive on the intervals $x < 0$ and $0 < x < 1$. We cannot say that the expression is positive on the interval $x < 1$ because it is not defined for $x = 0$. That is why the factor $x^{2/3}$ and its zero were included in the table even though $x^{2/3}$ does not contribute any negative signs. ■

The method used in Examples 13 and 14 can be used to compare the values of two given expressions.

■ **EXAMPLE 15**

Find all values of $x$ for which

$$x + \frac{1}{x} > 2x.$$

**SOLUTION**

We first rewrite the inequality with zero on one side and then simplify the expression on the other side.

$$x + \frac{1}{x} > 2x,$$

$$-x + \frac{1}{x} > 0,$$

$$\frac{-x^2 + 1}{x} > 0,$$

$$\frac{(1 - x)(1 + x)}{x} > 0.$$

The solution of the inequality is equivalent to the problem of determining values of $x$ for which the expression

$$\frac{(1 - x)(1 + x)}{x}$$

is positive. We can do this by using a table of signs. See Table 1.2.3.

**TABLE 1.2.3**

| | $-\infty$ to $-1$ | $-1$ to $0$ | $0$ to $1$ | $1$ to $\infty$ |
|---|---|---|---|---|
| $1 - x$ | + | + | + | − |
| $1 + x$ | − | + | + | + |
| $x$ | − | − | + | + |
| $\frac{(1 - x)(1 + x)}{x}$ | + | − | + | − |

We see the inequality is true in the intervals $x < -1$ and $0 < x < 1$. ■

## EXERCISES 1.2

Simplify the expressions in Exercises 1–12.

**1** $(2 - x)(3) + (1 + 3x)(-1)$

**2** $(5 - 3x)(2) + (3 + 2x)(-3)$

**3** $(2x + 1)^2(5) + (5x - 1)(2)(2x + 1)(2)$

**4** $\dfrac{(2x - 1)(2)(3x - 2)(3) - (3x - 2)^2(2)}{(2x - 1)^2}$

**5** $\left(-\frac{1}{2}x^{-3/2}\right) - \left(\frac{3}{2}x^{-5/2}\right)$

**6** $\sqrt{x} - \dfrac{1}{\sqrt{x}}$

**7** $(x^2)\left(\frac{1}{3}\right)(x + 1)^{-2/3}(1) + (x + 1)^{1/3}(2x)$

8 $\dfrac{(\sqrt{x^2+1})(1)-(x)\left(\dfrac{1}{2\sqrt{x^2+1}}\right)(2x)}{x^2+1}$

9 $\dfrac{x^2-4}{x-2}$

10 $\dfrac{x^2+3x+2}{x+1}$

11 $\dfrac{\dfrac{1}{x^2}-\dfrac{1}{4}}{x-2}$

12 $\dfrac{\dfrac{1}{x-3}+\dfrac{1}{6}}{x+3}$

In Exercises 13–19, find all values of $x$ for which the given expressions are zero.

13 $x^2-5x-6$

14 $x^2-5x+6$

15 $3x^2-4x+2$

16 $2x^2+3x+2$

17 $(x^2-2x-2)(-1)+(1-x)(2x-2)$

18 $(3x^2-2x+1)(2)+(2x-3)(6x-2)$

19 $(2x-1)(2)(x+2)(1)+(x+2)^2(2)$

In Exercises 20–24, find all values of $x$ for which the given expressions are (a) zero and (b) undefined.

20 $\dfrac{(x-1)(2)(3x+1)(3)-(3x+1)^2(1)}{(x-1)^2}$

21 $x-\dfrac{8}{x^2}$

22 $\sqrt{x-1}-\dfrac{1}{2\sqrt{x-1}}$

23 $\dfrac{\sqrt{2x+1}(3)-(3x+1)\left(\dfrac{1}{2\sqrt{2x+1}}\right)(2)}{2x+1}$

24 $\dfrac{\sqrt{3x-2}(2)-(2x-1)\left(\dfrac{1}{2\sqrt{3x-2}}\right)(3)}{3x-2}$

In Exercises 25–36, find all intervals on which the given expressions are positive.

25 $(x-1)(x+2)$

26 $x^2-3x+2$

27 $\dfrac{x(x+1)}{x-2}$

28 $\dfrac{2x^2-7x+3}{x-3}$

29 $\dfrac{x+1}{x^2-x+1}$

30 $\dfrac{x-2}{x^2-2x+2}$

31 $\dfrac{x}{(x-1)^3}$

32 $\dfrac{x-1}{(x-2)^2}$

33 $\dfrac{(x+1)^{2/3}}{x^{1/3}}$

34 $\dfrac{(x+1)^{1/3}}{x^{2/3}}$

35 $\sqrt{x}-\dfrac{1}{2\sqrt{x}}$

36 $x^{1/3}-x^{-2/3}$

Solve the inequalities in Exercises 37–40.

37 $x^3 < x$

38 $x^3 < x^2$

39 $\dfrac{1}{x} > \dfrac{1}{x-1}$

40 $\dfrac{2}{x+1} > \dfrac{1}{x}$

## 1.3 THE COORDINATE PLANE

We will review the terminology of the coordinate plane and the fundamental concept of the graph of an equation. We will then introduce the idea of a variable point on a graph, where the coordinates of the point are given in terms of a variable. Formulas for the length, slope, and midpoint of a line segment will be reviewed in terms of variable points, as they are used in calculus.

A **rectangular**, or Cartesian, **coordinate system** for points in a plane is determined by any two number lines that are in the plane and are perpendicular to each other. The number lines are called **coordinate axes**. The **coordinates** of a point in the plane are the unique **ordered pair** of numbers determined by the points of intersection of the coordinate axes with lines through the given point and parallel to the axes. Coordinates are written in the form $(a, b)$. The order of the coordinates indicates which of the coordinates is associated with which axis. The coordinate axes can be used to locate the unique point in the plane that corresponds to a given ordered pair.

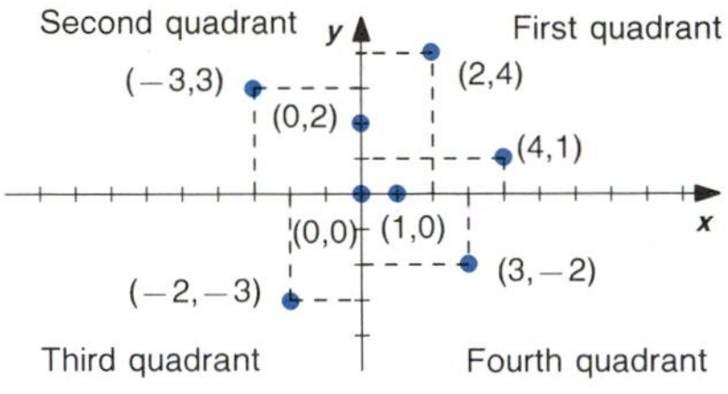


FIGURE 1.3.1

It is customary to draw one coordinate axis horizontal, with positive direction to the right, and the other axis vertical, with positive direction upward. The coordinate axes then divide the plane into four **quadrants**, which are numbered as indicated in Figure 1.3.1. Unless there are specific reasons not to, the same scale is chosen for each axis and the intersection of the axes is zero on both scales. The intersection of the axes then has coordinates $(0, 0)$ and is called the **origin**. The first coordinate of a pair $(a, b)$ corresponds to the horizontal axis, and the second coordinate corresponds to the vertical axis.

In Figure 1.3.1, we have located (or plotted) several points and indicated their coordinates. It is important to be able to locate a point from its coordinates and to determine the coordinates of a point from its position. We will not necessarily distinguish between a point and its coordinates. That is, we may refer to a point $P$ with coordinates $(a, b)$ as either $P$, $(a, b)$, or $P(a, b)$.

We will most often use $x$ as the variable for the horizontal axis and $y$ as the variable for the vertical axis. The first coordinate is then called the $x$-coordinate and the second coordinate is called the $y$-coordinate. The plane is called the $xy$-plane. Other choices of variables are used when that is appropriate for a particular problem.

The **graph** of an equation that relates two variables $x$ and $y$ is the collection of all points $(x, y)$ whose coordinates satisfy the equation.

### ■ EXAMPLE 1

Consider the equation $2x + 3y = 6$. (a) Is $(6, -2)$ on the graph? (b) Is $(-2, 6)$ on the graph?

#### SOLUTION

(a) Substitution of 6 for $x$ and $-2$ for $y$ in the equation gives $2(6) + 3(-2) = 6$, which is true, so $(6, -2)$ is on the graph.

(b) Substitution of $(-2, 6)$ yields $2(-2) + 3(6) = 6$, which is not true, so $(-2, 6)$ is not on the graph. ■

Points on the graph of an equation can be found by choosing a value for one variable and using the equation to solve for the corresponding value or values of the other variable.

Points of intersection of a graph with the coordinate axes are called **intercepts**. Intercepts are often easy to determine and should be plotted. Points $(x, y)$ on the $x$-axis have $y = 0$; points on the $y$-axis have $x = 0$.

### ■ EXAMPLE 2

Find points $(x, y)$ on the graph of $2x + 3y = 6$ that (a) are intercepts, (b) have $x = -3$, (c) have $y = -2$.

#### SOLUTION

(a) Intercepts on the $x$-axis have $y = 0$. When $y = 0$, $2x + 3(0) = 6$ implies $x = 3$. $(3, 0)$ is the $x$-intercept.

Intercepts on the $y$-axis have $x = 0$. When $x = 0$, $2(0) + 3y = 6$ implies $y = 2$. $(0, 2)$ is the $y$-intercept.

(b) When $x = -3$, $2(-3) + 3y = 6$ implies $y = 4$. $(-3, 4)$ is on the graph.

(c) When $y = -2$, $2x + 3(-2) = 6$ implies $x = 6$. $(6, -2)$ is on the graph.

See Figure 1.3.2. ■

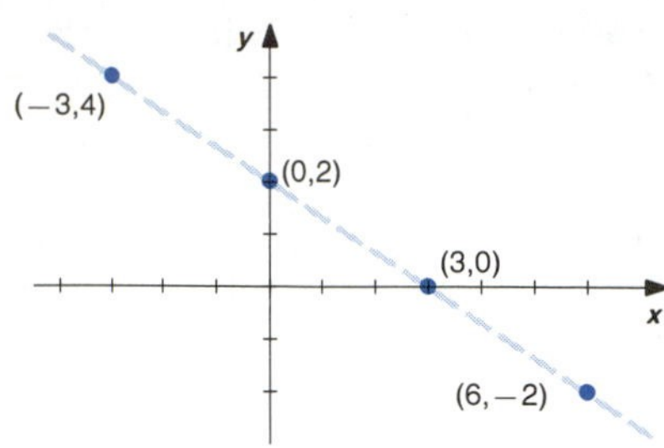


FIGURE 1.3.2

■ **EXAMPLE 3**

Find points $(x, y)$ on the graph of $y = x^2 - 1$ that (a) are intercepts, (b) have $x = 2$, (c) have $x = -2$, (d) have $y = 8$, (e) have $y = -5$.

**SOLUTION**

(a) Intercepts on the $x$-axis have $y = 0$. When $y = 0$, $0 = x^2 - 1$ implies $x = 1$ or $x = -1$. The points $(1, 0)$ and $(-1, 0)$ are $x$-intercepts.

Intercepts on the $y$-axis have $x = 0$. When $x = 0$, $y = (0)^2 - 1$ implies $y = -1$. The point $(0, -1)$ is the $y$-intercept.

(b) When $x = 2$, $y = (2)^2 - 1$ implies $y = 3$. The point $(2, 3)$ is on the graph.

(c) When $x = -2$, $y = (-2)^2 - 1$ implies $y = 3$. The point $(-2, 3)$ is on the graph.

(d) When $y = 8$, $8 = x^2 - 1$ implies $x^2 = 9$. That is, $x = 3$ or $x = -3$. The points $(3, 8)$ and $(-3, 8)$ are both on the graph.

(e) When $y = -5$, $-5 = x^2 - 1$ implies $x^2 = -4$. This equation has no real solution for $x$. There is no point on the graph that has $y$-coordinate $-5$. See Figure 1.3.3. ■

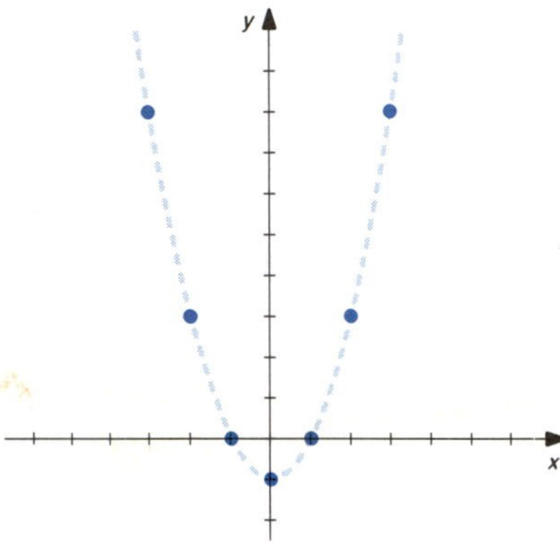


FIGURE 1.3.3

In many calculus problems we will deal with points that have coordinates expressed in terms of a variable. We will refer to such points as **variable points**. A variable point on the graph of an equation in $x$ and $y$ can be expressed in either the variable $x$ or the variable $y$. If $x$ is the variable, then the first coordinate of the variable point is $x$, and the second coordinate is the expression obtained by solving the equation for $y$ in terms of $x$. If $y$ is the variable, the second coordinate is $y$, and the first coordinate is the expression obtained by solving the equation for $x$ in terms of $y$.

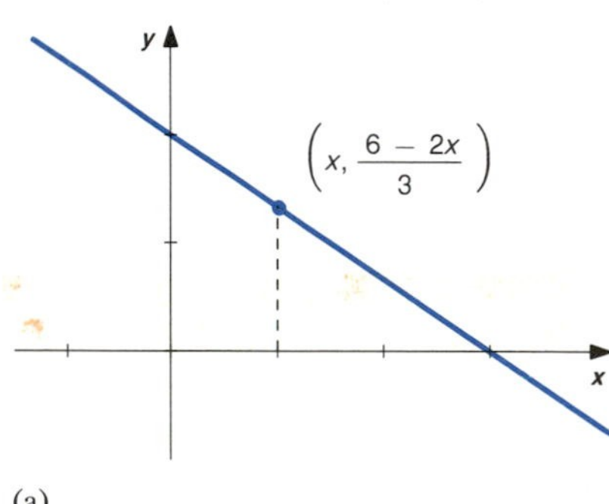


(a)

■ **EXAMPLE 4**

Find the coordinates of a variable point on the graph of $2x + 3y = 6$ in terms of (a) $x$ and (b) $y$.

**SOLUTION**

(a) The point $(x, y)$ is on the graph if $2x + 3y = 6$. Solving this equation for $y$, we have

$$y = \frac{6 - 2x}{3}.$$

$(x, (6 - 2x)/3)$ is a variable point on the graph, in terms of $x$. The points $(x, (6 - 2x)/3)$ vary over the graph as $x$ varies. See Figure 1.3.4a.

(b) $2x + 3y = 6$ implies $x = (6 - 3y)/2$. The point $((6 - 3y)/2, y)$ is a variable point on the graph, in terms of $y$. See Figure 1.3.4b. ■

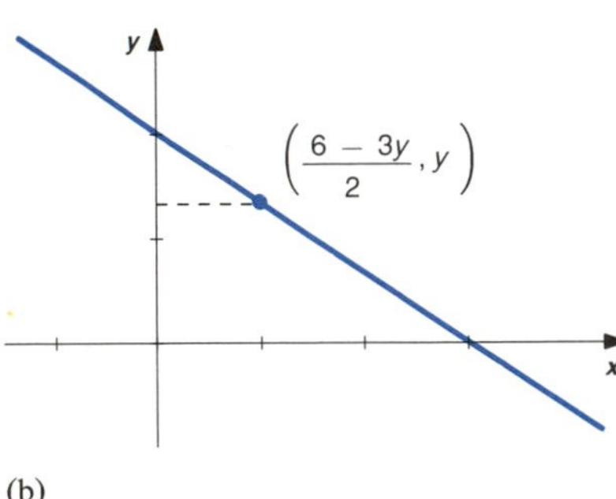


(b)

FIGURE 1.3.4

FIGURE 1.3.5

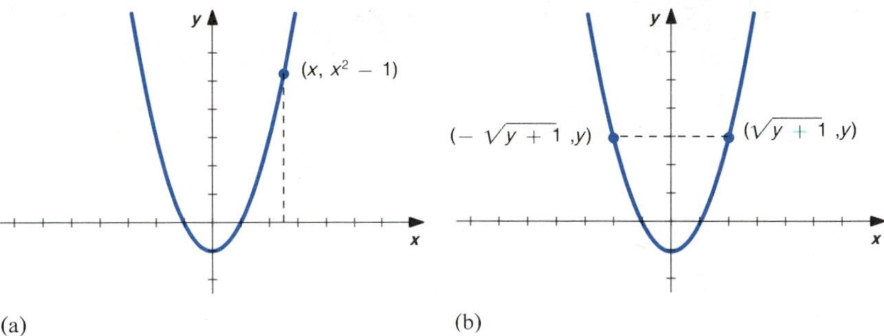


■ **EXAMPLE 5**

Find the coordinates of a variable point on the graph of $y = x^2 - 1$ in terms of (a) $x$ and (b) $y$.

**SOLUTION**

(a) A point $(x, y)$ is on the graph if $y = x^2 - 1$. This means $(x, x^2 - 1)$ is a variable point on the graph, in terms of $x$. See Figure 1.3.5a.

(b) Solving the equation for $x$, we have $x^2 = y + 1$, so $x = \sqrt{y+1}$ or $x = -\sqrt{y+1}$. If $y > -1$, both points $(\sqrt{y+1}, y)$ and $(-\sqrt{y+1}, y)$ are on the graph. When $y = -1$, $x = 0$, so $(0, -1)$ is on the graph. There is no point on the graph with $y < -1$. The points $(\sqrt{y+1}, y)$ and $(-\sqrt{y+1}, y)$ are variable points on the graph for $y \geq -1$. See Figure 1.3.5b. ■

We will use the following formulas.

The **line segment** between two points $P_1(x_1, y_1)$ and $P_2(x_2, y_2)$ has

$$\textbf{Length} = \textbf{Distance between points} = \sqrt{(x_2 - x_1)^2 + (y_2 - y_1)^2},$$

$$\textbf{Slope} = \frac{y_2 - y_1}{x_2 - x_1}, \quad \text{and}$$

$$\textbf{Midpoint} = \left(\frac{x_1 + x_2}{2}, \frac{y_1 + y_2}{2}\right).$$

The distance formula is a consequence of the Pythagorean Theorem. Notice that the points $P_1(x_1, y_1)$, $P_2(x_2, y_2)$, and $(x_2, y_1)$ form a right triangle in Figure 1.3.6. We will use the notation $|P_1P_2|$ to denote the distance between the points $P_1$ and $P_2$. Recall that $\sqrt{(x_2 - x_1)^2} = |x_2 - x_1|$ is the distance between points $x_1$ and $x_2$ on a number line.

We must be careful when using the formula for slope. Choose one point for $P_1$ and the other for $P_2$, but do not interchange $P_1$ and $P_2$ in mid-formula. It may be helpful to remember

$$\text{Slope} = \frac{\text{Rise}}{\text{Run}}.$$

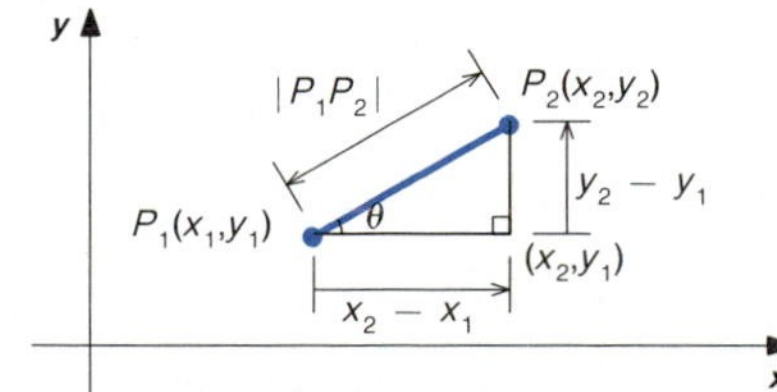


FIGURE 1.3.6

Also, note that the slope is determined by the angle $\theta$ in Figure 1.3.6.

FIGURE 1.3.7

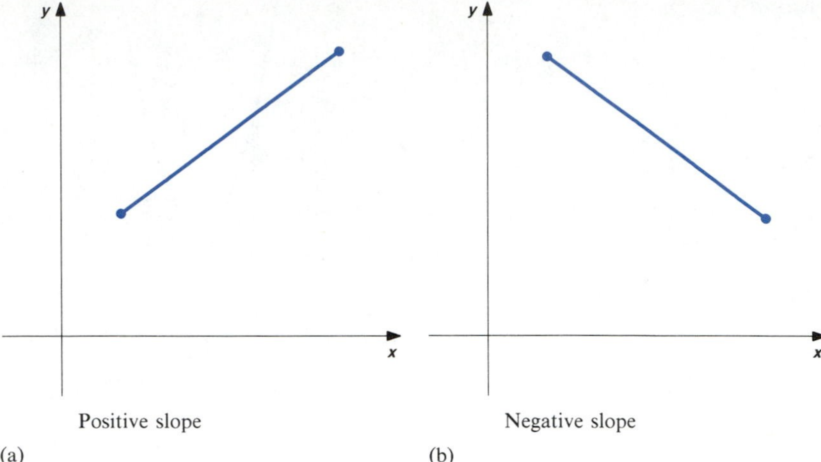

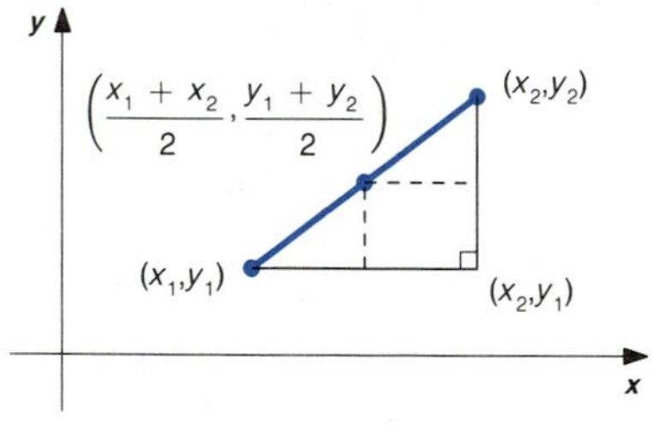

Slope zero

(c)

Slope undefined

(d)

$\left(\frac{x_1 + x_2}{2}, \frac{y_1 + y_2}{2}\right)$ $(x_2, y_2)$ $(x_1, y_1)$ $(x_2, y_1)$

FIGURE 1.3.8

A line segment slopes upward from left to right if the slope is positive. It slopes downward when the slope is negative. Horizontal line segments have slope zero. In case $x_1 = x_2$, the line segment is vertical and the slope is undefined. See Figure 1.3.7.

The coordinates of the midpoint are the averages of the corresponding coordinates of the endpoints. See Figure 1.3.8.

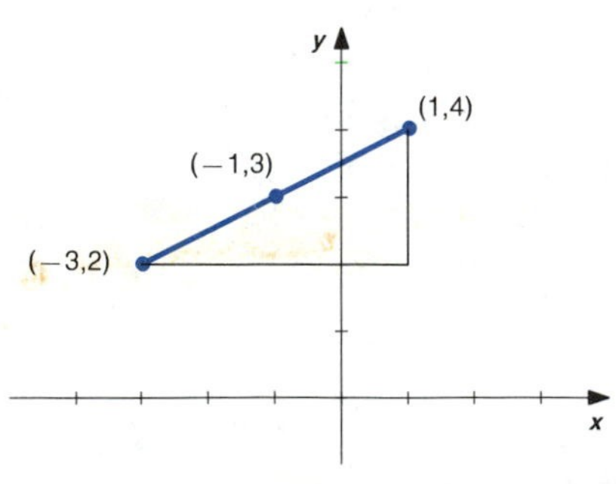


FIGURE 1.3.9

■ **EXAMPLE 6**

Find the length, slope, and midpoint of the line segment between the points $(1, 4)$ and $(-3, 2)$. Simplify.

**SOLUTION**

See Figure 1.3.9.

$$\text{Length} = \sqrt{(-3-1)^2 + (2-4)^2} = \sqrt{(-4)^2 + (-2)^2}$$
$$= \sqrt{20} = \sqrt{4 \cdot 5} = 2\sqrt{5},$$

$$\text{Slope} = \frac{2-4}{-3-1} = \frac{-2}{-4} = \frac{1}{2},$$

$$\text{Midpoint} = \left(\frac{1+(-3)}{2}, \frac{4+2}{2}\right) = (-1, 3).$$ ■

We must be able to use the formulas for length, slope, and midpoint with variable points.

■ **EXAMPLE 7**

Find the length, slope, and midpoint of the line segment between the point (4, 3) and a variable point $(x, y)$ on the graph of $-2x + 3y = 6$, in terms of $x$. Simplify.

**SOLUTION**

Solving the equation $-2x + 3y = 6$ for $y$, we have

$$y = \frac{6+2x}{3}.$$

$(x, (6+2x)/3)$ is a variable point on the graph. Using $P_1 = (4, 3)$ and $P_2 = (x, (6+2x)/3)$, we have

$$\begin{aligned}
\text{Length} &= \sqrt{(x-4)^2 + \left(\frac{6+2x}{3} - 3\right)^2} \\
&= \sqrt{(x-4)^2 + \left(\frac{6+2x-9}{3}\right)^2} \\
&= \sqrt{\frac{9(x-4)^2 + (2x-3)^2}{9}} \\
&= \frac{\sqrt{9x^2 - 72x + 144 + 4x^2 - 12x + 9}}{3} \\
&= \frac{\sqrt{13x^2 - 84x + 153}}{3},
\end{aligned}$$

$$\text{Slope} = \frac{\frac{6+2x}{3} - 3}{x-4} = \frac{\frac{6+2x-9}{3}}{x-4} = \frac{2x-3}{3(x-4)}.$$

When $x = 4$, the line segment is vertical and the slope is undefined.

$$\text{Midpoint} = \left(\frac{4+x}{2}, \frac{3+\frac{6+2x}{3}}{2}\right) = \left(\frac{4+x}{2}, \frac{15+2x}{6}\right).$$

See Figure 1.3.10. ■

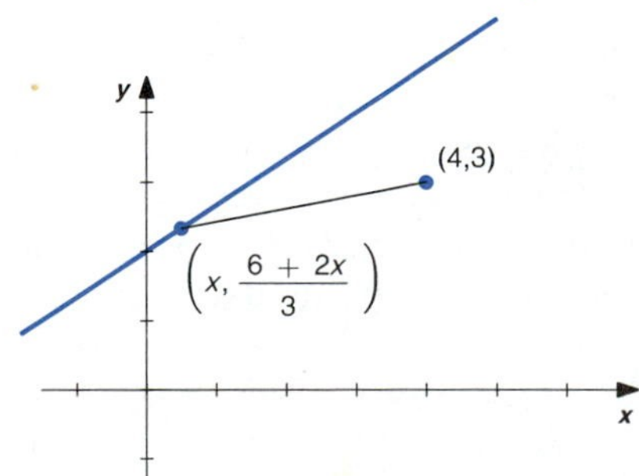


FIGURE 1.3.10

## EXERCISES 1.3

In Exercises 1–6, indicate which of the given points are on the graph of the given equations.

1 $3x - 4y = 7$; $(1, -1)$, $(4, -3)$

2 $2x - 3y = 5$; $(4, -1)$, $(-2, -3)$

3 $y = 2x^2 - 4x + 1$; $(2, 1)$, $(0, 1)$

4 $y = 3x^2 + 6x + 5$; $(-1, 2)$, $(1, 10)$

5 $x^2 + y^2 + 4y - 21 = 0$; $(-3, 2)$, $(2, -3)$

6 $x^2 + y^2 - 6x - 16 = 0$; $(4, -3)$, $(7, -3)$

In Exercises 7–12, plot the intercepts and points with the given $x$- and $y$-coordinates on the graphs of the given equations.

7 $3x + 2y = 6$; $x = -2$, $y = -3$

8 $y = 2x$; $x = 2$, $y = -4$

9 $y = x^2 - 1$; $x = 2$, $y = -2$

10 $y = x^3$; $x = 1, -1$, $y = 8, -8$

11 $x = y^2$; $x = 4, -4$, $y = 1, -1$

12 $x = y(2 - y)$; $x = -3$, $y = 1$

In Exercises 13–18, find the coordinates of a variable point on the graph of the given equation in terms of (a) $x$ and (b) $y$.

13 $y = 2x$

14 $3x + 2y = 6$

15 $y = x^3 - 1$

16 $y = 2(x + 1)^2 - 1$

17 $x = y^2 - 1$

18 $x^2 + y^2 = 4$

In Exercises 19–28, find the length, slope, and midpoint of the line segment between the two given points.

19 $(-1, 2)$, $(3, -1)$

20 $(2, 2)$, $(4, 2)$

21 $(1, -2)$, $(1, 3)$

22 $(3, -1)$, $(3, 2)$

23 $(x, 2x)$, $(3, 0)$

24 $(x, x^2)$, $(0, 1)$

25 $(x, 1 - x)$, $(x, x^2 - 1)$

26 $(x, 1)$, $(x, x^2 + 1)$

27 $(0, y)$, $(4 - y^2, y)$

28 $(-\sqrt{4 - y^2}, y)$, $(\sqrt{4 - y^2}, y)$

In Exercises 29–40, find the slope of the line segment between the given point and a variable point on the graph of the given equation, in terms of $x$.

29 $2x + 3y + 6 = 0$; $(-6, 2)$

30 $3x - 2y + 6 = 0$; $(4, 9)$

31 $y = x^2$; $(0, 0)$

32 $y = x^3$; $(0, 0)$

33 $y = x^3$; $(1, 1)$

34 $y = x^2$; $(2, 4)$

35 $y = x^2 - 3x + 1$; $(2, -1)$

36 $y = x^2 + 2x - 3$; $(1, 0)$

37 $y = \dfrac{1}{x}$; $(1, 1)$

38 $y = \dfrac{1}{x^2}$; $(2, 1/4)$

39 $y = \dfrac{1}{x^2 - 1}$; $(2, 1/3)$

40 $y = \dfrac{1}{2x + 1}$; $(2, 1/5)$

## 1.4 LINES

One of the fundamental applications of calculus is the use of lines to find approximate values of expressions. In this section we will review how to recognize a line from its equation, how to sketch the graph of a line, and how to determine the equation of a line from appropriate data.

Let us consider the graph of an equation of the form

$$ax + by + c = 0, \quad \text{where } a \text{ and } b \text{ not both zero.} \tag{1}$$

The equation in (1) is called the **general equation of a line**. We will see that

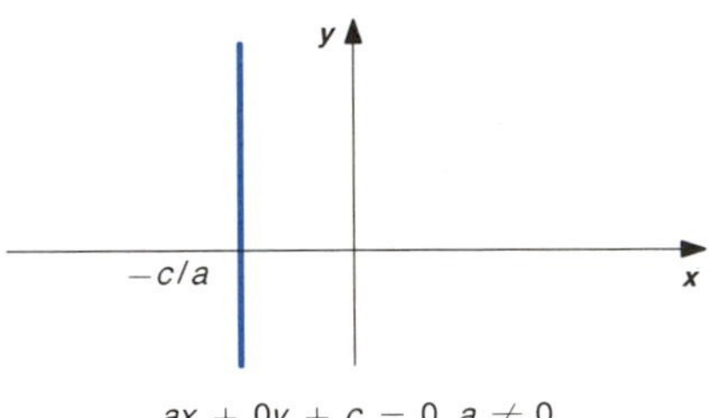

FIGURE 1.4.1

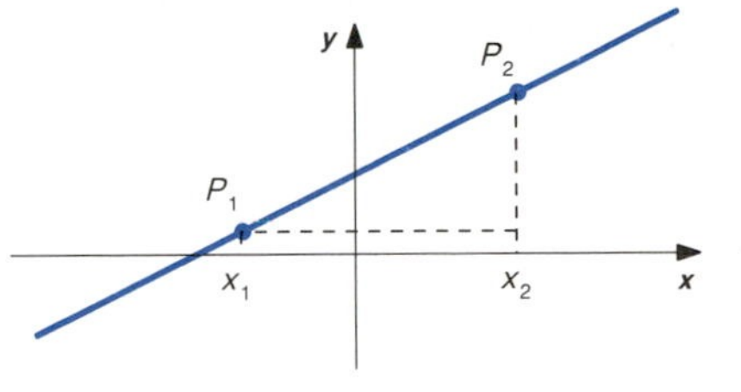

FIGURE 1.4.2

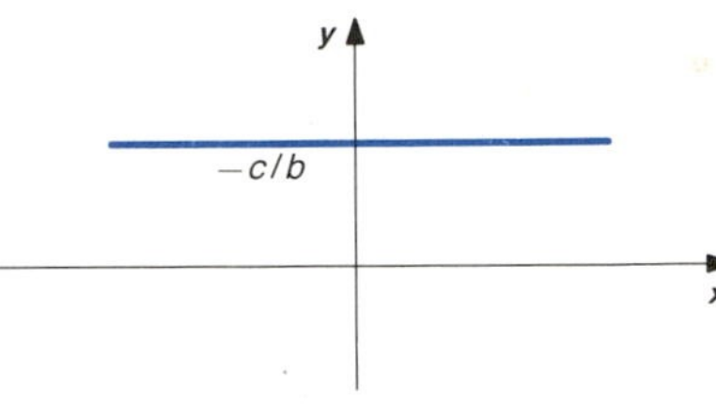

FIGURE 1.4.3

the graph of an equation of this type is a line. Later, we will see that the coordinates of points on any line satisfy an equation of form (1).

In order to show that the graph of an equation of form (1) is a line, we consider the cases $b = 0$ and $b \neq 0$ separately.

If $b = 0$, then $a \neq 0$ and equation (1) reduces to $x = -c/a$. The graph is the vertical line consisting of all points $(x, y)$ with $x = -c/a$. See Figure 1.4.1.

If $b \neq 0$, we solve the equation $ax + by + c = 0$ for $y$ and express a variable point on the graph as $(x, (-c - ax)/b)$. Thus, for each $x$, there is exactly one point $(x, y)$ on the graph. Also, the slopes of the line segments between any two distinct points $P_1(x_1, (-c - ax_1)/b)$ and $P_2(x_2, (-c - ax_2)/b)$ on the graph have a common value. That is,

$$\text{Slope} = \frac{\left(\frac{-c - ax_2}{b}\right) - \left(\frac{-c - ax_1}{b}\right)}{x_2 - x_1} = \frac{-c - ax_2 + c + ax_1}{b(x_2 - x_1)}$$

$$= \frac{-a(x_2 - x_1)}{b(x_2 - x_1)} = -\frac{a}{b}.$$

Since $b \neq 0$, it follows that the graph is a nonvertical line. See Figure 1.4.2. The graph is a horizontal line if $a = 0$. See Figure 1.4.3.

When sketching the graph of an equation of the form (1), we should use the fact that the graph is a line. Since any two points on a line determine the line, we can sketch the graph by plotting only two points and then drawing the line through these points. It is particularly easy to locate the intercepts, and these should be plotted.

■ **EXAMPLE 1**

Graph $2x + 3y = 6$.

**SOLUTION**

First, we recognize that this equation can be written in the form $2x + 3y - 6 = 0$, the general form of the equation of a line. This tells us the graph is a line. The intercepts are easily determined from the original form of the line. When $y = 0$, $2x = 6$, so $x = 3$. When $x = 0$, $3y = 6$, so $y = 2$. We have plotted the points $(3, 0)$ and $(0, 2)$, and then sketched the graph in Figure 1.4.4. ■

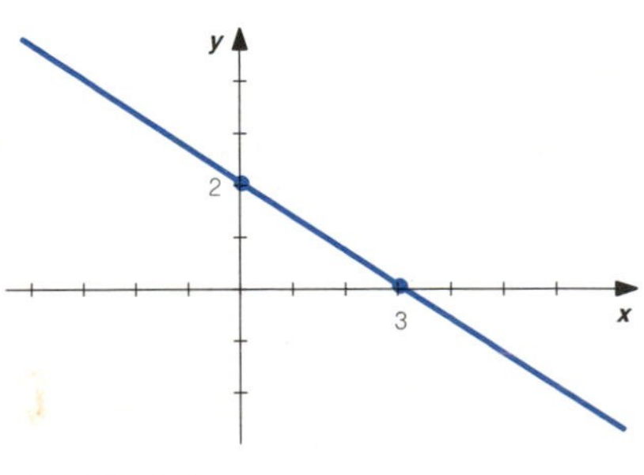

FIGURE 1.4.4

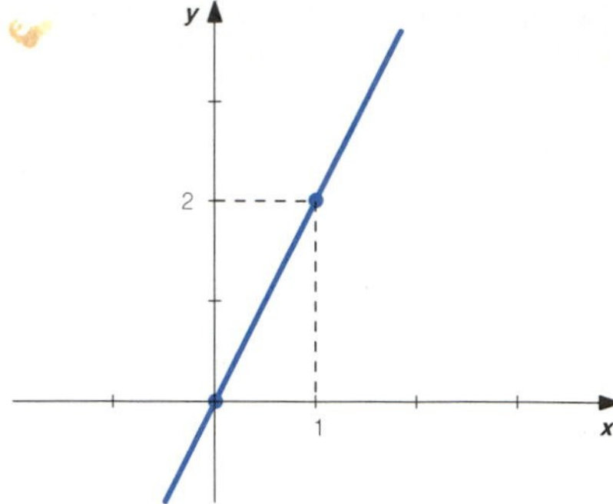

FIGURE 1.4.5

■ **EXAMPLE 2**

Graph $y = 2x$.

**SOLUTION**

We recognize that this is the equation of a line, since it can be written in the general form of the equation of a line. We see that (0, 0) is the only intercept of this line. Any other point will determine the graph. For example, we may choose $x = 1$, so $y = 2$. See Figure 1.4.5. ■

Horizontal and vertical lines are determined by one point.

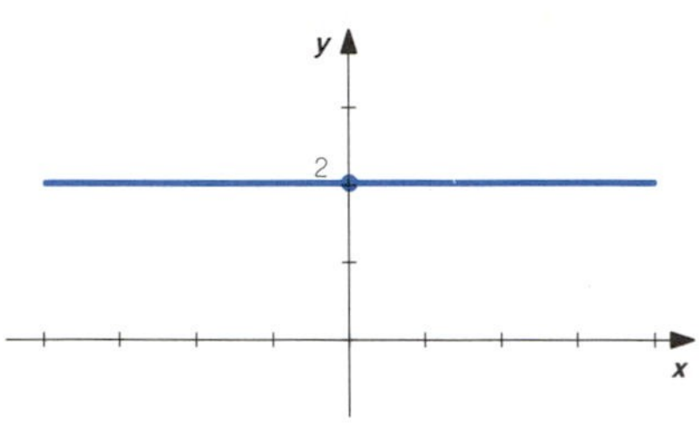

FIGURE 1.4.6

■ **EXAMPLE 3**

Graph $y = 2$.

**SOLUTION**

This is the equation of a line. When $x = 0$, $y = 2$. In fact, the coordinates $(x, 2)$ satisfy the equation for all values of $x$. The graph is a horizontal line. See Figure 1.4.6. ■

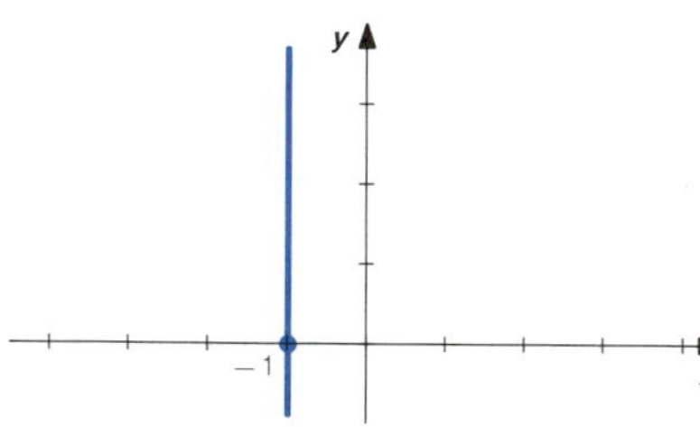

FIGURE 1.4.7

■ **EXAMPLE 4**

Graph $x + 1 = 0$.

**SOLUTION**

This is the equation of a line. When $y = 0$, $x = -1$. The graph contains all points $(-1, y)$ and is a vertical line. See Figure 1.4.7. ■

Let us now consider how to find the equation of a line that satisfies specific conditions. As a consequence, we will see that the coordinates of any line satisfy an equation of the form of (1) $ax + by + c = 0$, where $a$ and $b$ are not both zero.

It is characteristic of a nonvertical line that the line segments between any two points on the line have a common slope, called the **slope of the line**. Note that the slope of the line is determined by the angle $\theta$ in Figure 1.4.8. If $(x_1, y_1)$ is a fixed point on a line with slope $m$, then a point $(x, y)$ is on the line exactly when

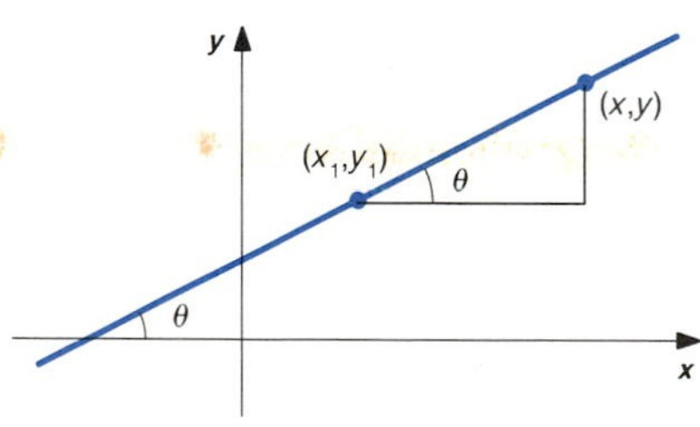

FIGURE 1.4.8

$$\frac{y - y_1}{x - x_1} = m \quad \text{or} \quad y - y_1 = m(x - x_1).$$

This gives us an important characterization of the equation of a nonvertical line.

The **point-slope** form of the equation of the line through the point $(x_1, y_1)$ with slope $m$ is

$$y - y_1 = m(x - x_1).$$

Horizontal lines have slope zero and the point-slope form of the equation of the horizontal line through the point $(x_1, y_1)$ reduces to $y = y_1$.

Slope is not defined for vertical lines, so we cannot use the point–slope form. The equation of the vertical line through the point $(x_1, y_1)$ is $x = x_1$.

At this point we may conclude that the equation of any line, nonvertical or vertical, can be written in the general form of (1) $ax + by + c = 0$, where $a$ and $b$ are not both zero.

Later, we will use calculus to determine the slope of a particular line through a point on a graph. It will then be convenient to use the point–slope form of the equation of a line to *write* the equation of the line.

### ■ EXAMPLE 5

Find an equation of the line through the point $(2, -5)$ with slope 3.

**SOLUTION**

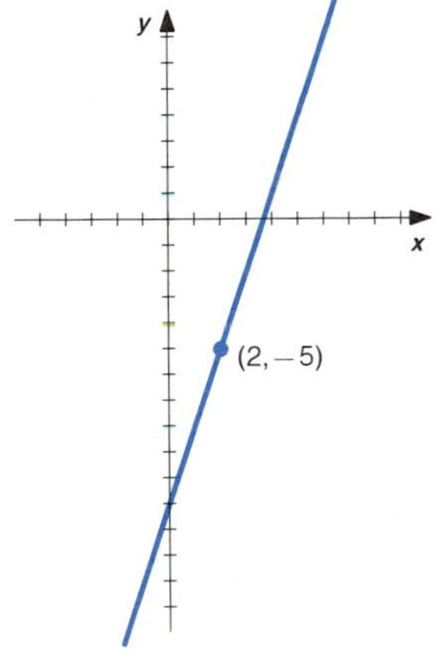


FIGURE 1.4.9

We have the point $(x_1, y_1) = (2, -5)$ and the slope $m = 3$. We substitute directly into the point–slope formula to obtain

$$y - (-5) = 3(x - 2).$$

This equation may be simplified by writing

$$y + 5 = 3x - 6 \quad \text{or} \quad y = 3x - 11.$$

The equation may be expressed in general form as

$$3x - y - 11 = 0.$$

The line is sketched in Figure 1.4.9. ■

### ■ EXAMPLE 6

Find an equation of the line through the points $(2, -1)$ and $(4, 3)$.

**SOLUTION**

We have two points on the line; we need the slope. To find the slope, we use the fact that any two points on the line determine the slope. Hence,

$$\text{Slope} = \frac{3 - (-1)}{4 - 2} = \frac{4}{2} = 2.$$

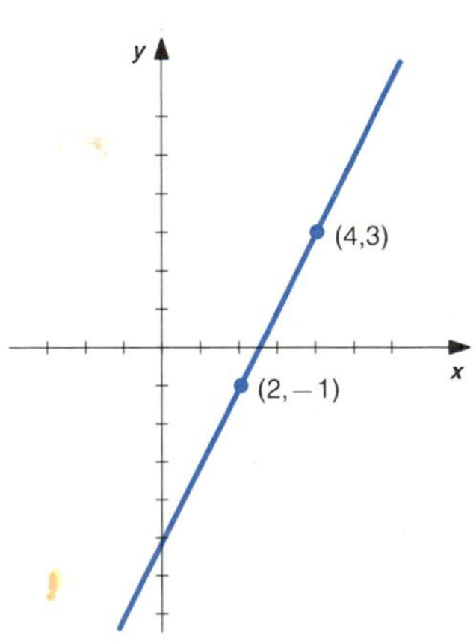


FIGURE 1.4.10

We can now use the point $(4, 3)$ and slope $m = 2$ in the point–slope form and write the desired equation,

$$y - 3 = 2(x - 4).$$

If we used the point $(2, -1)$, we would write the equation

$$y - (-1) = 2(x - 2).$$

Both of the above equations simplify to $2x - y - 5 = 0$. We may use either point to find the equation. The graph is sketched in Figure 1.4.10. ■

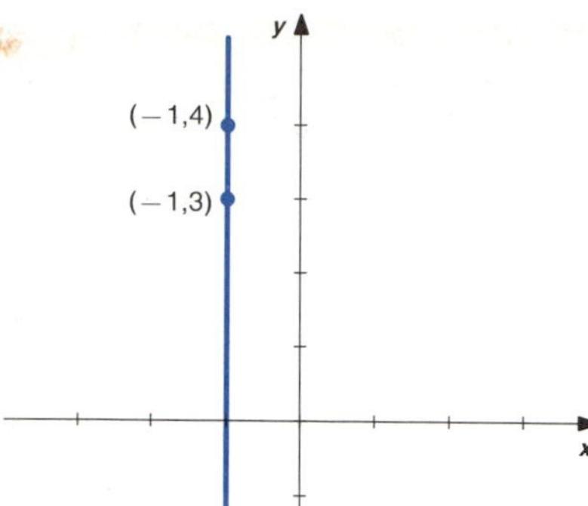

FIGURE 1.4.11

■ **EXAMPLE 7**

Find an equation of the line through the points $(-1, 4)$ and $(-1, 3)$.

**SOLUTION**

We might notice immediately that this is a vertical line with equation $x = -1$. If we did not notice that this is a vertical line, we would proceed as in the previous example to find the slope. That is,

$$\text{Slope} = \frac{4-3}{-1-(-1)} = \frac{1}{0}.$$

The fact that the slope is not defined indicates a vertical line. See Figure 1.4.11. ■

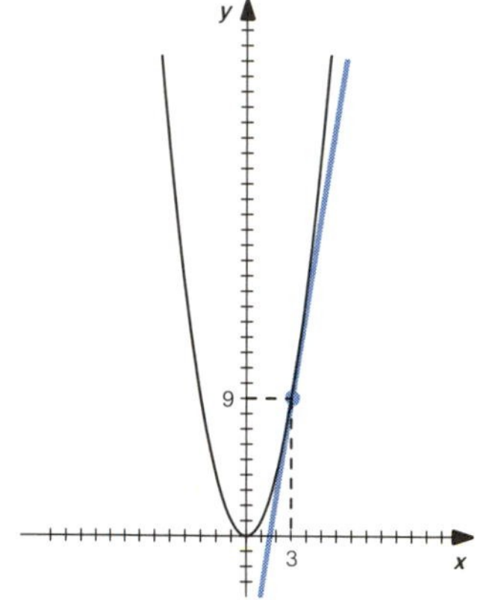

FIGURE 1.4.12

■ **EXAMPLE 8**

Find the equation of the line that has slope $m = 6$ and contains the point on the graph of $y = x^2$ with $x = 3$.

**SOLUTION**

The point on the graph of $y = x^2$ with $x = 3$ has $y = 3^2$. We want an equation of the line through $(3, 9)$ with slope $m = 6$. The point–slope form of the equation is

$$y - 9 = 6(x - 3).$$

Simplifying, we have

$$y - 9 = 6x - 18,$$

$$y = 6x - 9.$$

The graph is sketched in Figure 1.4.12. ■

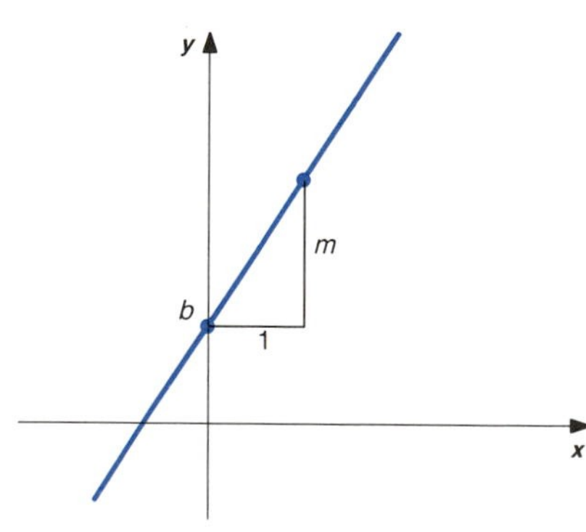

FIGURE 1.4.13

It is not difficult to check that the slope of the line $y = mx + b$ is $m$; its $y$-intercept is $b$. See Figure 1.4.13. This gives us the following characterization.

The **slope–intercept** form of the equation of the line with slope $m$ and $y$-intercept $b$ is

$$\boldsymbol{y = mx + b.}$$

The slope–intercept form can be used to *read* the slope of a line from its equation.

■ **EXAMPLE 9**

Find the slope of the line $3x + 2y - 6 = 0$.

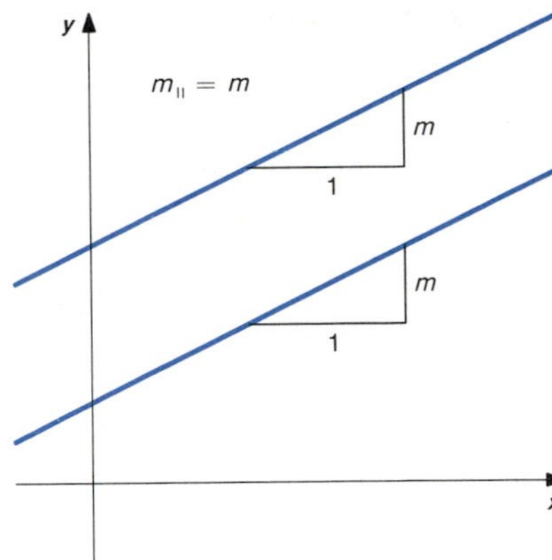

(a)

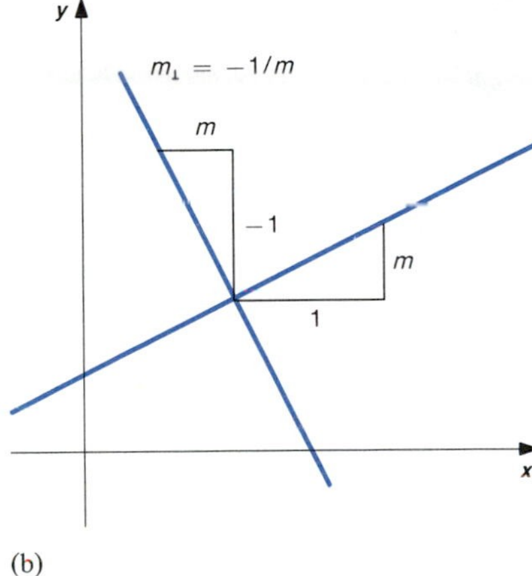

(b)

FIGURE 1.4.14

**SOLUTION**

We first write the equation in slope–intercept form. That is, we solve the equation for $y$.

$$3x + 2y - 6 = 0,$$

$$2y = -3x + 6,$$

$$y = -\frac{3}{2}x + 3.$$

The coefficient of $x$ then gives the slope $m = -3/2$ ■

If a line has nonzero slope $m$, then all parallel lines have slope $m_{\parallel} = m$ and all perpendicular lines have slope $m_{\perp} = -1/m$. (These statements can be proved by showing that the right triangles in Figure 1.4.14 are congruent. Corresponding sides then have equal length, as is indicated by the labeling.) All vertical lines are parallel to each other and are perpendicular to all horizontal lines.

**■ EXAMPLE 10**

Find an equation of the line through the point $(-2, 3)$ and (a) parallel to the line $4x - 3y + 7 = 0$, (b) perpendicular to the line $4x - 3y + 7 = 0$.

**SOLUTION**

We have a point on the lines. We need their slopes. Their slopes are found from the slope of the given line.

$$4x - 3y + 7 = 0,$$

$$-3y = -4x - 7,$$

$$y = \frac{4}{3}x + \frac{7}{3}.$$

We see the given line has slope $m = 4/3$.

(a) Lines parallel to the given line have slope $m_{\parallel} = m = 4/3$. This slope and the point $(-2, 3)$ give the equation

$$y - 3 = \frac{4}{3}(x - (-2)),$$

$$3(y - 3) = 4(x + 2),$$

$$3y - 9 = 4x + 8,$$

$$4x - 3y + 17 = 0.$$

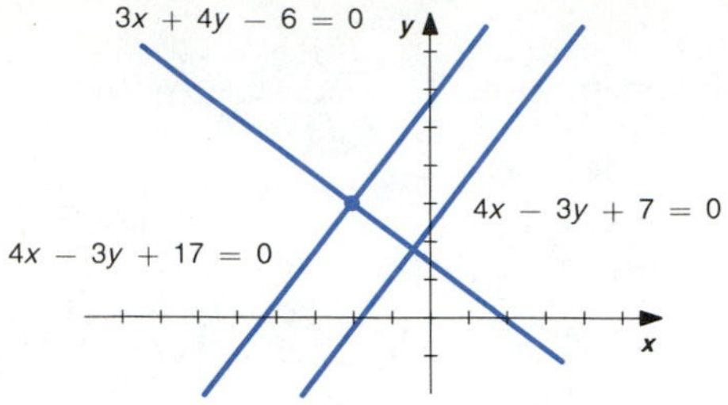

(b) Lines perpendicular to the given line have slope $m_{\perp} = -1/m = -3/4$. We can then write the equation

$$y - 3 = -\frac{3}{4}(x - (-2)),$$

$$4(y - 3) = -3(x + 2),$$

$$4y - 12 = -3x - 6,$$

$$3x + 4y - 6 = 0.$$

The lines are sketched in Figure 1.4.15. ■

FIGURE 1.4.15

If $a$ and $b$ are nonzero, it is not difficult to determine that the equation of the line that contains the points $(a, 0)$ and $(0, b)$ can be written in the form

$$\frac{x}{a} + \frac{y}{b} = 1.$$

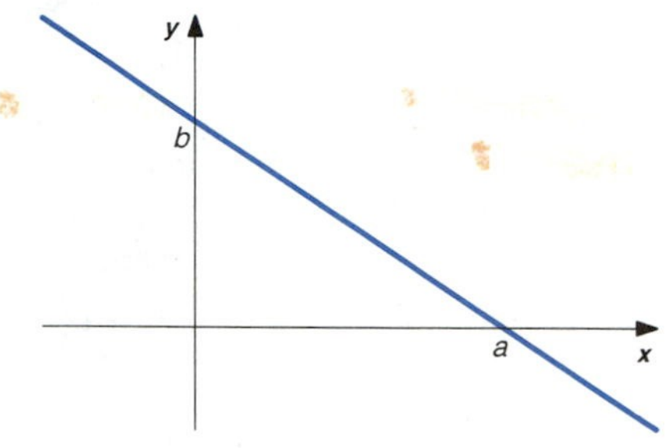

FIGURE 1.4.16

See Figure 1.4.16. This gives us the following characterization.

The **intercept** form of the equation of the line with $x$-intercept $a$ and $y$-intercept $b$ is

$$\mathbf{\frac{x}{a} + \frac{y}{b} = 1.}$$

The intercept form is convenient for writing the equation of a line if we know its intercepts.

■ **EXAMPLE 11**

Find an equation of the line pictured in Figure 1.4.17.

**SOLUTION**

We see from the picture that the $x$-intercept is $a = -2$ and the $y$-intercept is $b = 3$. We then write the equation

$$\frac{x}{-2} + \frac{y}{3} = 1.$$

We can then multiply both sides of the equation by $-6$ to obtain

$$3x - 2y = -6.$$ ■

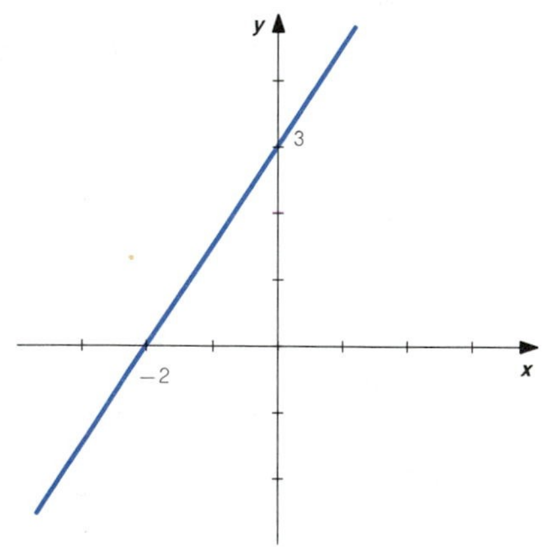

FIGURE 1.4.17

■ **EXAMPLE 12**

Find an equation satisfied by all points $(x, y)$ that are equidistant from the points $(1, 0)$ and $(2, -3)$.

FIGURE 1.4.18

**SOLUTION**

Let $d_1$ be the distance from $(x, y)$ to $(1, 0)$, and let $d_2$ be the distance from $(x, y)$ to $(2, -3)$. See Figure 1.4.18. The desired equation is then

$$d_1 = d_2,$$

$$\sqrt{(x-1)^2 + (y-0)^2} = \sqrt{(x-2)^2 + (y-(-3))^2},$$

$$(x-1)^2 + y^2 = (x-2)^2 + (y+3)^2,$$

$$x^2 - 2x + 1 + y^2 = x^2 - 4x + 4 + y^2 + 6y + 9,$$

$$2x - 6y = 12,$$

$$x - 3y = 6.$$

We should not be surprised that this is the equation of a line. We know from plane geometry that the set of all points that are equidistant from $(1, 0)$ and $(2, -3)$ is the perpendicular bisector of the segment between the points. We could use this fact to obtain the equation of the line. ■

## EXERCISES 1.4

Sketch the graphs of the equations in Exercises 1–14.

1 $2x - 3y = 6$
2 $3x - 2y = 6$
3 $x - 2y = 4$
4 $2x - y = 2$
5 $2x - y = 0$
6 $x - 2y = 0$
7 $x + 3y = 0$
8 $3x + y = 0$
9 $x + 2 = 0$
10 $x - 1 = 0$
11 $x = 0$
12 $y = 0$
13 $y = -1$
14 $y - 1 = 0$

In Exercises 15–44, find an equation of the line that satisfies the given conditions.

15 Through $(2, 3)$, slope $-1$

16 Through $(1, -2)$, slope 3

17 Through $(-1, 2)$, slope $1/3$

18 Through $(2, 4)$, slope $1/2$

19 Through $(2, 1)$ and $(-1, 4)$

20 Through $(-1, 2)$ and $(-3, -1)$

21 Through $(2, 3)$, parallel to the line $2x - y = 4$

22 Through $(-1, 2)$, parallel to the line $6x + 2y = 5$

23 Through $(0, 2)$, perpendicular to the line $x + y = 2$

24 Through $(2, 0)$, perpendicular to the line $x = y$

25 Through $(3, 2)$, parallel to the $x$-axis

26 Through $(-1, 4)$, perpendicular to the $y$-axis

27 Through $(-1, -3)$, perpendicular to the $x$-axis

28 Through $(2, -3)$, parallel to the $y$-axis

29 Slope $-6$, through the point on the graph of $y = x^2$ that has $x = -3$

30 Slope $-6$, through the point on the graph of $y = -x^2$ that has $x = 3$

31 Slope 0, through the point on the graph of $y = x^2 - 2x$ that has $x = 1$

32 Slope 0, through the point on the graph of $y = x^2 + 4x + 2$ that has $x = 2$

33 Through the points on the graph of $y = 1 - x^2$ that have $x = 0$ and $x = 1$

34 Through the points on the graph of $y = x^2 + 4x$ that have $x = -1$ and $x = 1$

35 Through the points on the graph of $y = x^3$ that have $x = 2$ and $x = -1$

36 Through the points on the graph of $y = x^3 - x$ that have $x = 1$ and $x = 2$

**37** Graph is

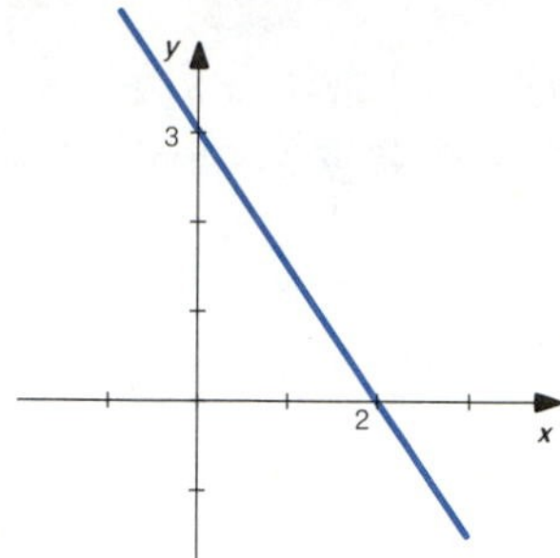


**38** Graph is

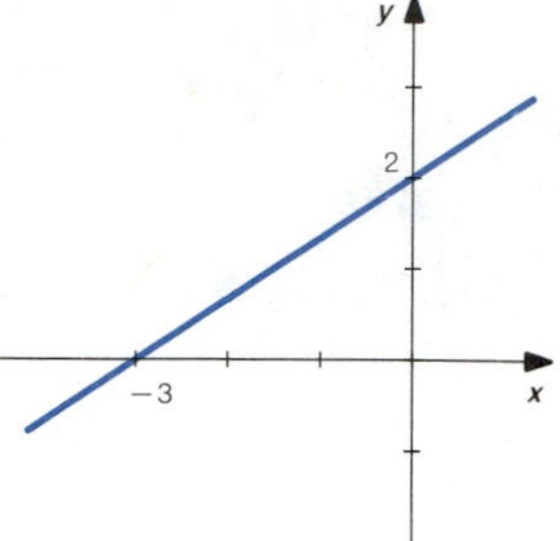


**39** Graph is

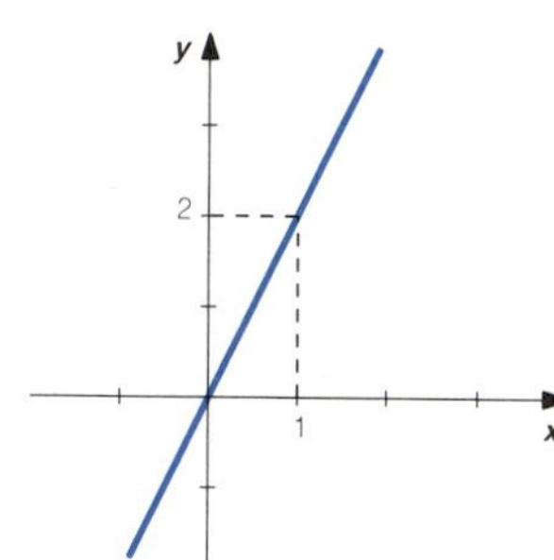


**40** Graph is

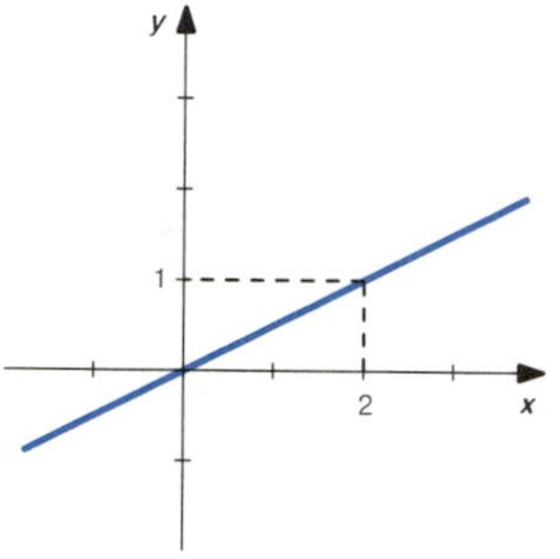


**41** Points on graph are equidistant from $(0, 1)$ and $(2, 0)$

**42** Points on graph are equidistant from $(-2, 1)$ and $(3, -2)$

**43** Points on graph are equidistant from $(-1, 3)$ and $(-1, -1)$

**44** Points on graph are equidistant from $(-1, 2)$ and $(5, 2)$

## 1.5 GRAPHS OF SOME SIMPLE EQUATIONS

In this section we will review graphing techniques for a representative collection of simple equations, so we can use the figures to graphically illustrate calculus concepts. The presentation given here uses the same general ideas that we will use later when using calculus to sketch more complicated curves.

Let us first note that we should not attempt to sketch a graph by plotting a large number of randomly chosen points on the graph. Locating points on a graph is a basic and important idea, but we must realize that any finite number of points on the graph cannot determine the entire graph. Unless we have additional information about the graph, we can never be certain what happens between any two points on the graph. That is, we need to know something of the general character of the graph. Once we know the general character of a graph, the graph can be obtained by plotting only a few key points on the graph and then sketching in the remainder of the graph. For example, if we recognize that an equation is the equation of a line, we know the general character of the graph. We can then use any two points to determine the graph, but we would ordinarily use the intercepts as key points.

Later, we will use calculus to determine the general character and key points of a graph. We will then use the following general procedure.

***Steps for sketching graphs***

- Determine the general character of the graph.
- Determine and plot key points.
- Sketch the remainder of the graph.

For now, we will illustrate the idea with examples for which we can determine the general character and key points from the *form* of the equation, as was done in Section 1.4 to graph lines.

The **standard form** of the equation of the **circle** with center $(h, k)$ and radius $r$ is

$$(x - h)^2 + (y - k)^2 = r^2. \tag{1}$$

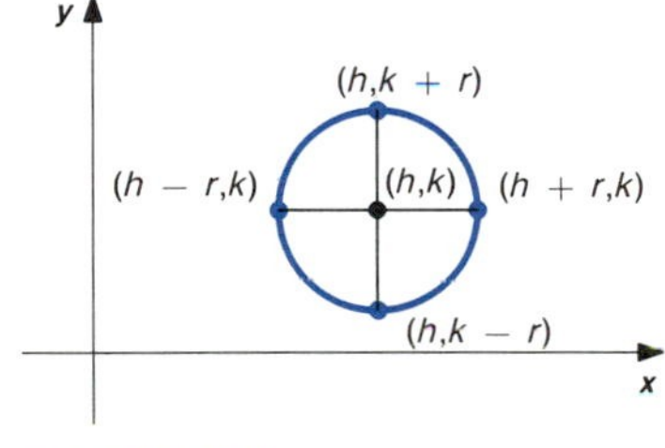


FIGURE 1.5.1

The distance formula tells us that this equation is satisfied by all points $(x, y)$ whose distance from $(h, k)$ is $r$, and that only these points satisfy the equation.

The center of a circle is read easily from the standard form of its equation. To graph a circle, the center should be located and used to plot the four key points $(h + r, k)$, $(h - r, k)$, $(h, k + r)$, and $(h, k - r)$. See Figure 1.5.1.

Equations of the form

$$x^2 + y^2 + Dx + Ey + F = 0 \tag{2}$$

can be put into the standard form of a circle by *completing the squares*. Note that the graph of $(x - h)^2 + (y - k)^2 = 0$ is the single point $(h, k)$. No points satisfy $(x - h)^2 + (y - k)^2 =$ (a negative number).

**■ EXAMPLE 1**

Sketch the graph of the equation $y^2 + 8y = 6x - x^2$.

**SOLUTION**

We must first recognize that the equation can be written in the form of (2), so the graph is a circle, or a single point, or no points satisfy the equation. In particular, we have

$$y^2 + 8y = 6x - x^2, \quad \text{so} \quad x^2 + y^2 - 6x + 8y = 0.$$

To write the equation in standard form, we need to complete the squares. To do this, we group the $x$-terms and the $y$-terms. Planning ahead, we leave spaces to complete the square inside each pair of parentheses. Also, we write blank parentheses on the right for each set of

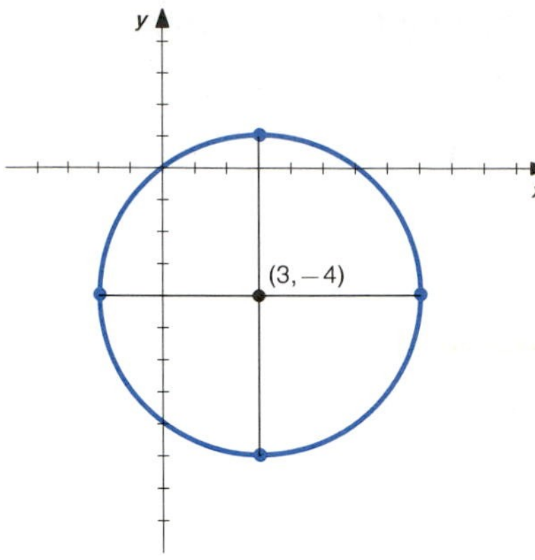


FIGURE 1.5.2

parentheses on the left. We can then fill in equal numbers in corresponding parentheses without destroying equality. We thus obtain

$$(x^2 - 6x \qquad) + (y^2 + 8y \qquad) = 0 + (\ \ ) + (\ \ ).$$

Recalling that the square of one-half the coefficient of the first-order term is added to complete the square, we obtain

$$\left(x^2 - 6x + \left(\frac{-6}{2}\right)^2\right) + \left(y^2 + 8y + \left(\frac{8}{2}\right)^2\right) = 0 + (9) + (16),$$

$$(x^2 - 6x + 9) + (y^2 + 8y + 16) = 25,$$

$$(x - 3)^2 + (y + 4)^2 = 5^2.$$

From the standard form, we see that the circle has center $(3, -4)$ and radius 5. Key points are plotted and the graph is sketched in Figure 1.5.2. ■

The equation

$$\frac{(x - h)^2}{a^2} + \frac{(y - k)^2}{b^2} = 1, \qquad \text{where } a \text{ and } b \text{ are positive and } a \neq b, \qquad (3)$$

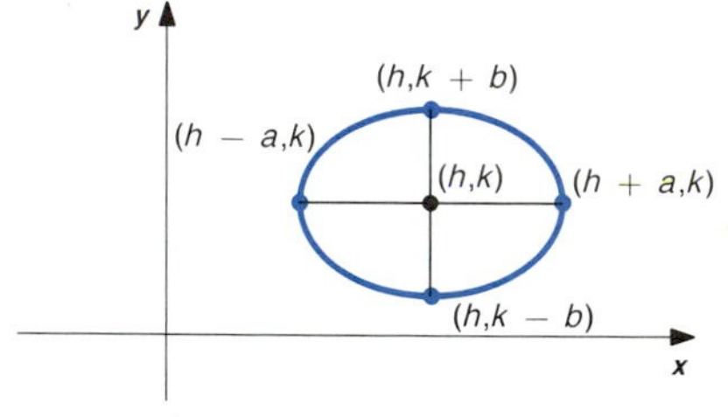


FIGURE 1.5.3

is the **standard form** of the equation of an **ellipse** with center $(h, k)$. The key points $(h - a, k)$ and $(h + a, k)$ are found by moving a distance $a$ to the left and right of the center. The positive number $a$ is the positive square root of the positive number below $(x - h)^2$, and we move $a$ units parallel to the $x$-axis. The key points $(h, k + b)$ and $(h, k - b)$ are found by moving $b$ units up and down from the center. $b$ is the positive square root of the positive number below $(y - k)^2$, and we move $b$ units parallel to the $y$-axis. See Figure 1.5.3.

■ **EXAMPLE 2**

Sketch the graph of the equation

$$\frac{(x - 2)^2}{3^2} + \frac{(y + 1)^2}{4^2} = 1.$$

**SOLUTION**

We first recognize that this is the standard equation of an ellipse with center $(2, -1)$. The key points $(5, -1)$ and $(-1, -1)$ are then plotted by moving three units to the right and left of the center. The key points $(2, 3)$ and $(2, -5)$ are plotted by moving four units up and down from the center. The graph is then sketched through the four key points that have been plotted. See Figure 1.5.4. ■

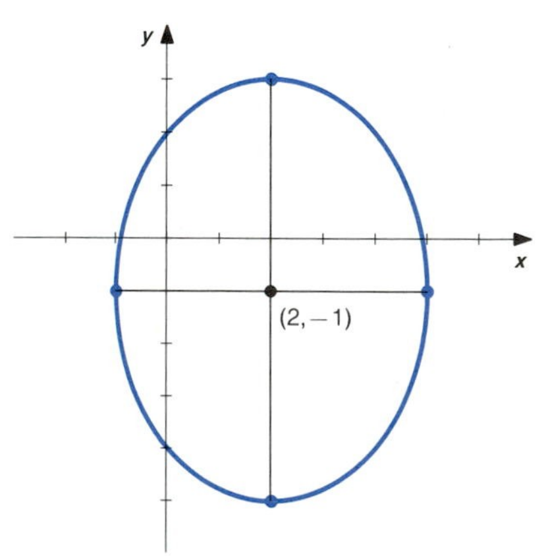


FIGURE 1.5.4

If $A$ and $C$ are nonzero, an equation of the form

$$Ax^2 + Cy^2 + Dx + Ey + F = 0$$

can be written in the form $A(x-h)^2 + C(y-k)^2 = F'$ by completing the squares. If $F'$ is nonzero and $A$, $C$, and $F'$ have the same sign, we can then divide both sides of the equation by $A$ in the case $A = C$ and by $F'$ in the case $A \neq C$ to obtain the standard form of the equation of a circle in the case $A = C$ and the standard equation of an ellipse in the case $A \neq C$.

Let us further illustrate the idea of using the general character and key points to sketch a graph by considering equations of the form

$$y - k = a(x-h)^2, \qquad a \neq 0. \tag{4}$$

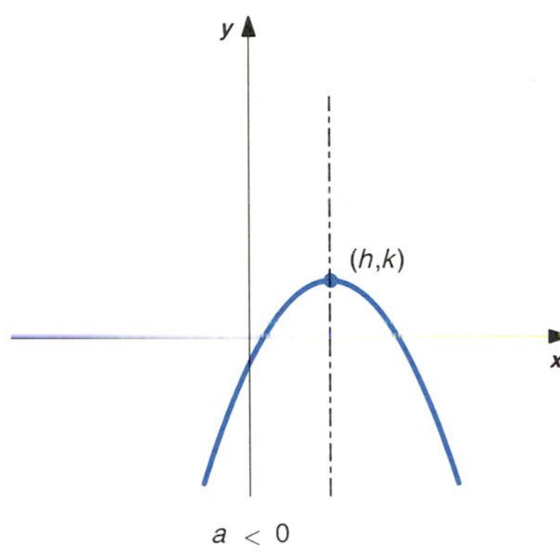


FIGURE 1.5.5

The graph is a **parabola** with **vertex** at the point $(h, k)$. It is symmetric about the vertical line $x = h$. If $a > 0$, then points $(x, y)$ on the graph satisfy $y - k = a(x-h)^2 \geq 0$, so $y \geq k$; the parabola opens upward from the vertex if $a > 0$. If $a < 0$, then $y - k = a(x-h)^2 \leq 0$, so $y \leq k$; the parabola opens downward from the vertex if $a < 0$. See Figure 1.5.5.

An equation of the form

$$y = ax^2 + bx + c, \qquad a \neq 0, \tag{5}$$

can be put into the form of (4) by completing the square. The graph is a parabola that opens upward if $a > 0$ and opens downward if $a < 0$.

To sketch the graph of an equation of the type of (5), we must first recognize that it is a parabola. The sign of the coefficient of $x^2$ then gives us the direction the parabola opens. Next, we determine and plot the intercepts. The graph can then be sketched by drawing a parabola that opens in the proper direction, through the intercepts. If a more accurate sketch is desired, we can determine and plot the vertex before sketching the graph. To find the vertex we complete the square to write the equation in the form of (4). We can read the coordinates of the vertex from the equation in the form of (4).

■ **EXAMPLE 3**

Sketch the graph of the equation $y = 2x - x^2$.

**SOLUTION**

We recognize that the graph is a parabola that opens downward.

When $x = 0$, $y = 2(0) - (0)^2$ implies $y = 0$. The $y$-intercept is $(0, 0)$.

When $y = 0$, we have $0 = 2x - x^2$, or $x(x-2) = 0$. The solutions are $x = 0$ and $x = 2$. The $x$-intercepts are $(0, 0)$ and $(2, 0)$.

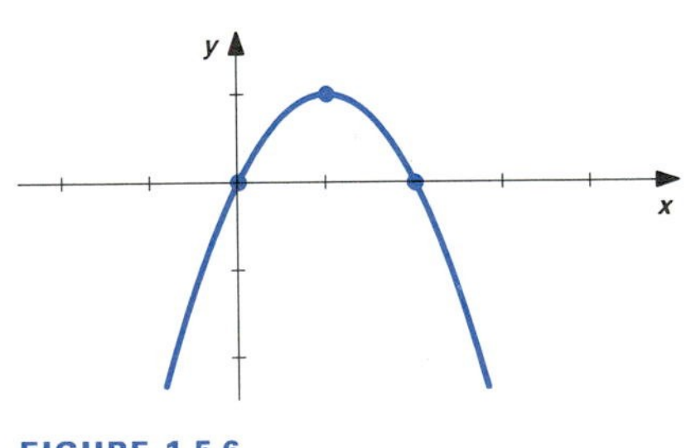


FIGURE 1.5.6

Any sketch of a parabola that opens downward, through the intercepts $(0, 0)$ and $(2, 0)$, would look similar to the actual graph given in Figure 1.5.6. We know from symmetry that the vertex must have $x$-coordinate 1, so its $y$-coordinate is $2(1) - (1)^2 = 1$. We could also determine the vertex by completing the square.

To complete the square, we factor the coefficient of $x^2$ from the $x$-terms and write

$$y - (\qquad) = -(x^2 - 2x + \qquad).$$

We have left space inside the parentheses on the right for the term that will complete the square. Also, we have written blank parentheses on the left to match the parentheses on the right. We have given the pair of parentheses on the left the same coefficient as that on the right, namely, $-1$. We can then retain equality by filling in equal numbers in each of the two pairs of parentheses. We have

$$y - (1) = -\left(x^2 - 2x + \left(\frac{-2}{2}\right)^2\right)$$

$$y - 1 = -(x - 1)^2.$$

From this form of the equation we see that the vertex is $(h, k) = (1, 1)$. ■

■ **EXAMPLE 4**

Sketch the graph of the equation $y = 2x^2 + 4x + 3$.

**SOLUTION**

We recognize that the graph is a parabola that opens upward. When $x = 0$, $y = 3$. The $y$-intercept is $(0, 3)$. When $y = 0$, we have

$$2x^2 + 4x + 3 = 0.$$

This quadratic equation does not factor easily, so we try to solve it by using the Quadratic Formula. That is,

$$x = \frac{-(4) \pm \sqrt{4^2 - 4(2)(3)}}{2(2)}.$$

Since the formula for the solution contains $\sqrt{-8}$, there is no real number solution. This means the graph has no $x$-intercept.

We can determine the vertex and line of symmetry of the parabola by completing the square. Completing the square as before, we have

$$y - 3 + 2(\quad) = 2(x^2 + 2x + \qquad),$$

$$y - 3 + 2(1) = 2\left(x^2 + 2x + \left(\frac{2}{2}\right)^2\right),$$

$$y - 1 = 2(x + 1)^2.$$

We see the vertex is $(h, k) = (-1, 1)$.

In Figure 1.5.7 we have plotted the vertex and $y$-intercept and then used symmetry about the vertical line $x = -1$ to sketch the graph. The $y$-coordinate of the vertex is positive and the parabola opens upward, so there is no $x$-intercept. This agrees with the information we obtained earlier. ■

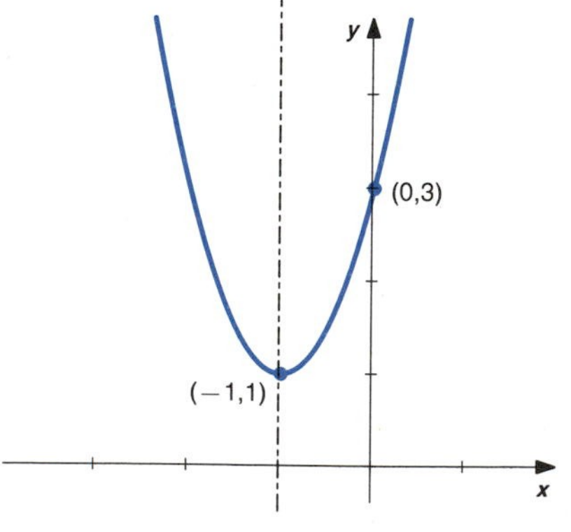


FIGURE 1.5.7

The roles of $x$ and $y$ can be interchanged in the equation of a parabola. That is, the graph of an equation of the form

$$x = ay^2 + by + c, \qquad a \neq 0, \tag{6}$$

is also a parabola. The vertex can be found by completing the square to write the equation in the form

$$x - h = a(y - k)^2. \tag{7}$$

The vertex is then seen to be $(h, k)$. The graph is symmetric about the horizontal line $y = k$. The parabola opens to the right if $a > 0$ and to the left if $a < 0$. See Figure 1.5.8.

$a > 0$

(a)

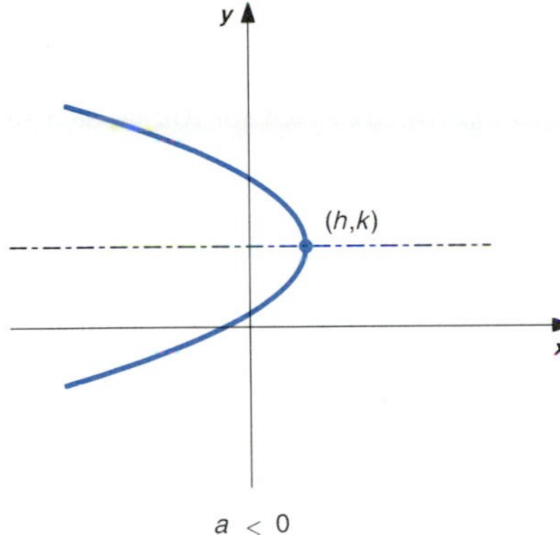


$a < 0$

(b)

FIGURE 1.5.8

■ **EXAMPLE 5**

Sketch the graph of the equation $y^2 - x + 2y - 3 = 0$.

**SOLUTION**

Writing the equation in the form of (6), we have

$$x = y^2 + 2y - 3.$$

From this form we see that the graph is a parabola that opens to the right. When $y = 0$, $x = -3$. The $x$-intercept is $(-3, 0)$. When $x = 0$, we have

$$y^2 + 2y - 3 = 0,$$

$$(y + 3)(y - 1) = 0.$$

The solutions are $y = 1$ and $y = -3$. The $y$-intercepts are $(0, 1)$ and $(0, -3)$.

We could now sketch the graph by drawing a parabola that opens to the right, through the intercepts. If desired, we could complete the square to write the equation in the form

$$x + 4 = (y + 1)^2.$$

This shows that the vertex is $(h, k) = (-4, -1)$. The graph is sketched in Figure 1.5.9. ■

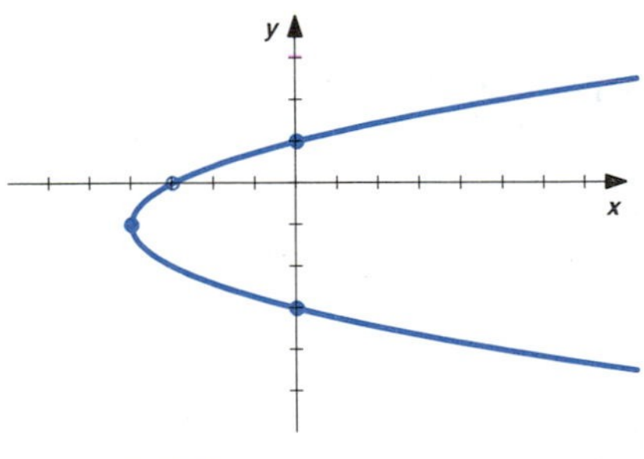


FIGURE 1.5.9

The graphs of some equations that involve a square root may be recognized after the equation is squared. Since squaring an equation can introduce extraneous solutions, we must check that the original equation is satisfied.

■ **EXAMPLE 6**

Sketch the graph of the equation $y = \sqrt{1 - x}$.

**SOLUTION**

Squaring the equation, we obtain

$$y^2 = 1 - x \quad \text{or} \quad x - 1 = -y^2.$$

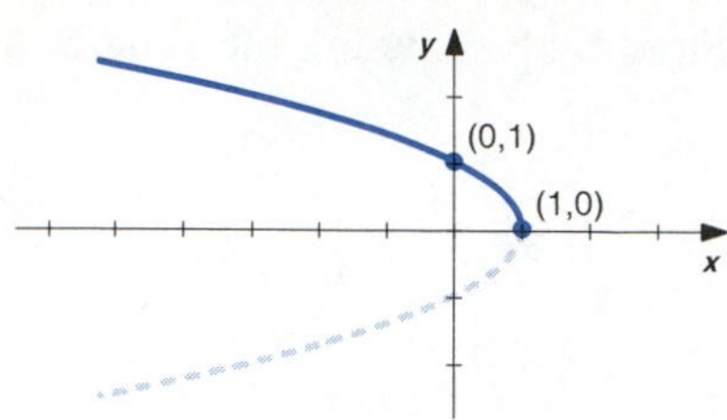

FIGURE 1.5.10

This equation is recognized as the equation of a parabola that opens to the left and has vertex at (1, 0). From the original equation, we have that

$$y = \sqrt{1 - x} \geq 0,$$

since $\sqrt{1-x}$ is the nonnegative square root of the nonnegative number $1 - x$. It follows that the graph of the original equation is the upper half of the parabola.

In Figure 1.5.10 we have plotted the vertex (1, 0) and the intercept (0, 1), and then sketched the graph. ■

■ **EXAMPLE 7**

Sketch the graph of the equation $y = 2 - \sqrt{4 - x^2}$.

**SOLUTION**

Writing the equation with the radical by itself on one side of the equation and then squaring, we have

$$\sqrt{4 - x^2} = 2 - y,$$

$$4 - x^2 = (2 - y)^2,$$

$$4 - x^2 = (y - 2)^2,$$

$$x^2 + (y - 2)^2 = 2^2.$$

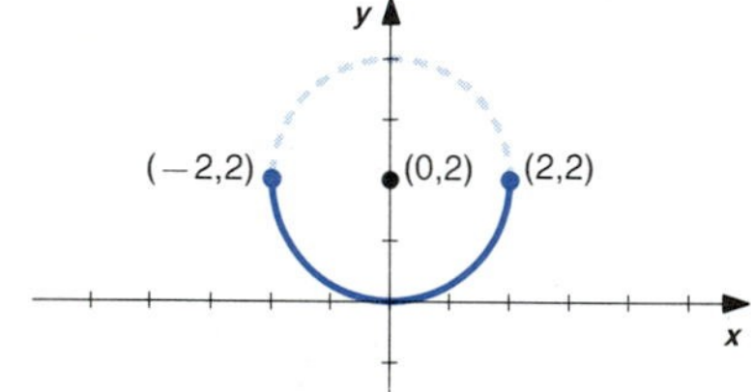

FIGURE 1.5.11

This is the equation of the circle with center at (0, 2) and radius 2. From the original equation, we have

$$y - 2 = -\sqrt{4 - x^2} \leq 0, \quad \text{so} \quad y \leq 2.$$

Points on the graph of the original equation must satisfy $y \leq 2$, so the graph is the lower half of the circle. The graph is sketched in Figure 1.5.11. ■

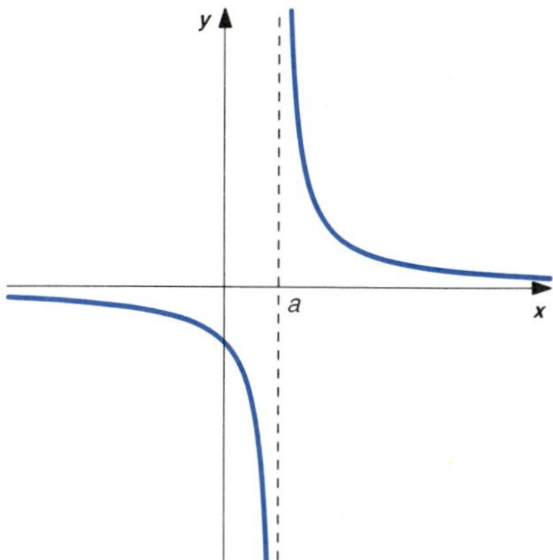

FIGURE 1.5.12

The general character of the graph of

$$y = \frac{1}{x - a}$$

is illustrated in Figure 1.5.12. Note that $|y|$ increases without limit as the corresponding values of $x$ approach $a$ from either side. The line $x = a$ is called a **vertical asymptote** of the graph. For $x < a$, the corresponding values of $y$ are negative and the graph is below the $x$-axis. For $x > a$, the corresponding values of $y$ are positive and the graph is above the $x$-axis. There is no point on the graph that has $x$-coordinate $a$. As $|x|$ becomes large the corresponding values of $y$ approach zero. The line $y = 0$ is called a **horizontal asymptote** of the graph.

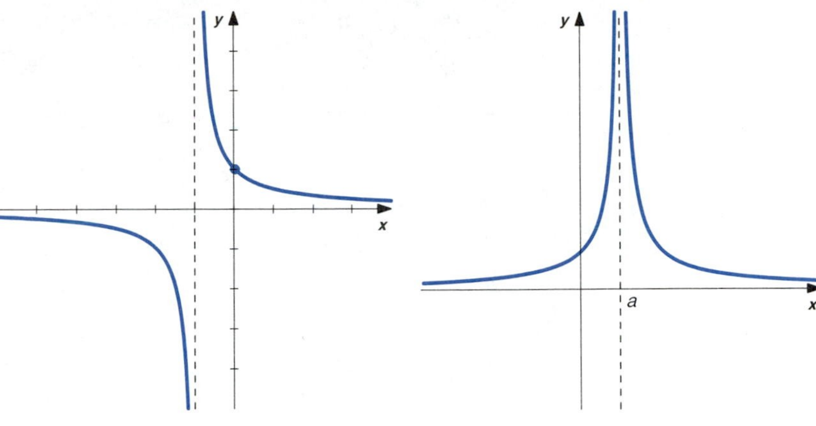


FIGURE 1.5.13

FIGURE 1.5.14

■ **EXAMPLE 8**

Sketch the graph of

$$y = \frac{1}{x+1}.$$

**SOLUTION**

The graph has vertical asymptote $x = -1$ and horizontal asymptote $y = 0$; $y$ is negative for $x < -1$ and positive for $x > -1$. The $y$-intercept (0, 1) is plotted and the graph is sketched in Figure 1.5.13. ■

The general character of the graph of

$$y = \frac{1}{(x-a)^2}$$

is illustrated in Figure 1.5.14. The graph has vertical asymptote $x = a$ and horizontal asymptote $y = 0$. $y$ is positive for $x < a$ and $x > a$. There is no point on the graph with $x$-coordinate $a$.

## EXERCISES 1.5

Sketch the graphs of the equations given in Exercises 1-44. Illustrate the general character of the graphs; only key points and intercepts need be located accurately.

1 $(x-1)^2 + y^2 = 1$

2 $x^2 + (y+1)^2 = 1$

3 $x^2 + y^2 + 4y = 0$

4 $x^2 + y^2 - 6x = 0$

5 $x^2 + y^2 - 2x + 2y = 0$

6 $x^2 + y^2 + 4x - 4y + 6 = 0$

7 $x^2 + y^2 - 3x - y + 9/4 = 0$

8 $x^2 + y^2 + x + 3y + 9/4 = 0$

9 $\frac{x^2}{3^2} + \frac{y^2}{2^2} = 1$

10 $\frac{x^2}{1^2} + \frac{(y-2)^2}{2^2} = 1$

11 $\frac{(x-2)^2}{2^2} + \frac{(y+2)^2}{1^2} = 1$

12 $\frac{(x+1)^2}{1^2} + \frac{(y-3)^2}{3^2} = 1$

13 $4x^2 + 9y^2 + 24x = 36$

14 $x^2 + 4y^2 - 4x - 8y = 0$

15 $y = 2x^2 - 4x + 3$

16 $y = 2x^2 + 4x + 2$

17 $y = x^2 + 1$

18 $y = x^2 + 2x + 2$

19 $y = -x^2 + 4$

20 $y = x^2 - 4$

21 $y = x(x-2)$

22 $y = x(1-x)$

23 $x = y^2$

24 $x = -y^2$

25 $x = -y^2 + 3y - 2$

26 $x = y^2 + 4y + 3$

27 $x = y^2 - y$

28 $x = (y-1)(y-3)$

29 $y = \sqrt{x}$

30 $y = -\sqrt{x}$

31 $x = -\sqrt{y}$

32 $x = \sqrt{y}$

33 $y = \sqrt{x+4}$

34 $y = \sqrt{4-3x}$

35 $x = \sqrt{y-1} + 1$

36 $x = \sqrt{y+1} - 1$

37 $y = -\sqrt{9-x^2}$

38 $y = \sqrt{9-x^2}$

39 $x - \sqrt{2y - y^2} = 0$

40 $x + \sqrt{-2y - y^2} = 0$

41 $y = \frac{1}{x-2}$

42 $y = \frac{1}{x+3}$

43 $y = \frac{1}{(x+1)^2}$

44 $y = \frac{1}{(x-1)^2}$

## 1.6 TRIGONOMETRIC FUNCTIONS

This section contains a review of the basic definitions and properties of the trigonometric functions. We will be dealing with these functions throughout our study of calculus.

We may think of an **angle** as a measure of the rotation from an initial side of the angle to a terminal side. Counterclockwise rotations are associated with positive angles, and clockwise rotations are associated with negative angles. Each rotation is associated with one and only one angle. Unequal angles (such as 45° and 405°) may have the same initial and terminal sides if the angles differ by an integer multiple of 360°. See Figure 1.6.1. An angle that is positioned at the origin with initial side the positive $x$-axis is said to be in **standard position.**

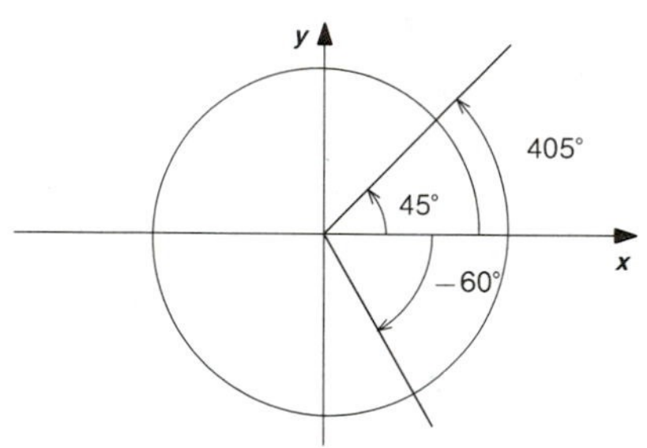

FIGURE 1.6.1

Many of the calculus formulas we will be using are greatly simplified by using **radian measure** of angles. Since we will be using radian measure for calculus problems, it will be useful to know the radian measure of some of the common angles. Recall that the radian measure of an angle is the ratio of the length $s$ of the arc of a circle subtended by the angle and the radius $r$ of the circle. That is,

$$\theta = \frac{s}{r} \text{ radians.}$$

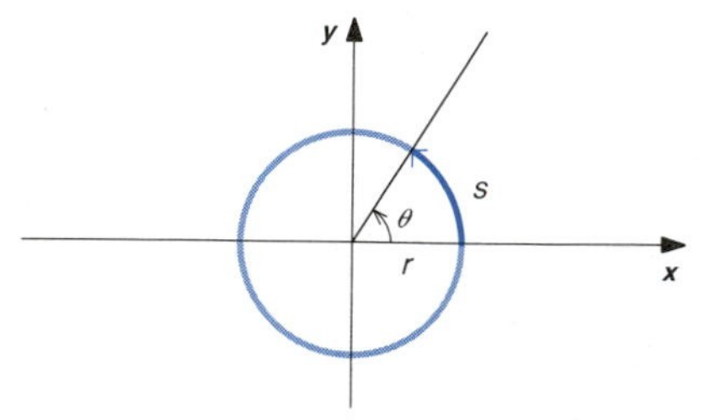

FIGURE 1.6.2

See Figure 1.6.2. This ratio does not depend on the size of the circle. For the purpose of determining the radian measure of an angle, we take the length of an arc of a circle to be positive in the counterclockwise direction and negative in the clockwise direction. This corresponds to the convention established for positive and negative angles.

The radian measure of one revolution is given by

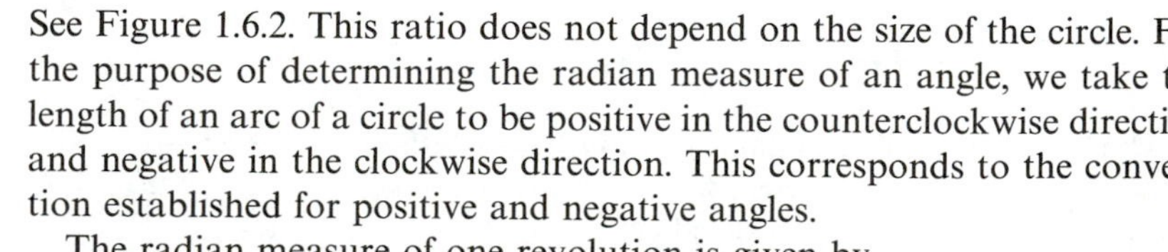

$$\frac{\text{Circumference of circle}}{\text{Radius}} = \frac{2\pi r}{r} = 2\pi.$$

It follows that $360° = 2\pi$ radians (rad), or

$$180° = \pi \text{ rad.}$$

We can change the units of measure of an angle from degrees to radians by multiplying by the unit factor $\pi$ rad/180°. For example,

$$30° = 30°\left(\frac{\pi \text{ rad}}{180°}\right) = \frac{\pi}{6} \text{ rad} \approx 0.52359878 \text{ rad},$$

$$45° = 45°\left(\frac{\pi \text{ rad}}{180°}\right) = \frac{\pi}{4} \text{ rad} \approx 0.78539816 \text{ rad},$$

$$60° = 60°\left(\frac{\pi \text{ rad}}{180°}\right) = \frac{\pi}{3} \text{ rad} \approx 1.0471976 \text{ rad},$$

$$90° = 90°\left(\frac{\pi \text{ rad}}{180°}\right) = \frac{\pi}{2} \text{ rad} \approx 1.5707963 \text{ rad}.$$

See Figure 1.6.3. We can change from radians to degrees by multiplying by the unit factor $180°/\pi$ rad. That is,

$$1 \text{ radian} = (1 \text{ rad})\left(\frac{180°}{\pi \text{ rad}}\right) \approx 57.29578°,$$

$$2 \text{ radians} = (2 \text{ rad})\left(\frac{180°}{\pi \text{ rad}}\right) \approx 114.59156°.$$

See Figure 1.6.4 The radian measure of an angle can be any real number. It is not necessary that radian measure be expressed as a fractional multiple of $\pi$, although the common angles can be expressed conveniently in this form.

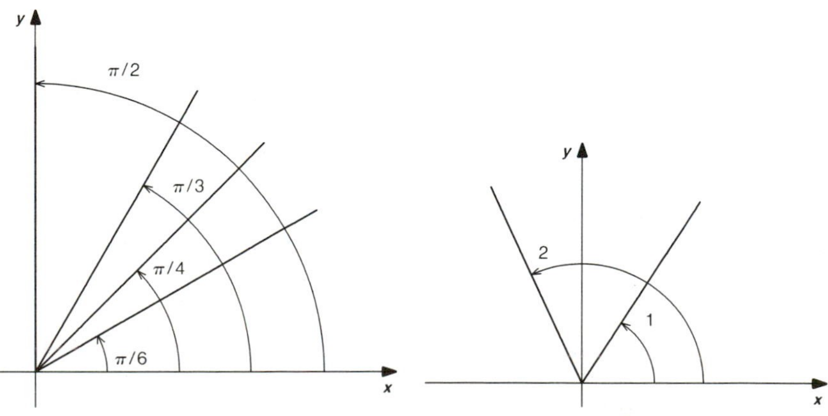


FIGURE 1.6.3

FIGURE 1.6.4

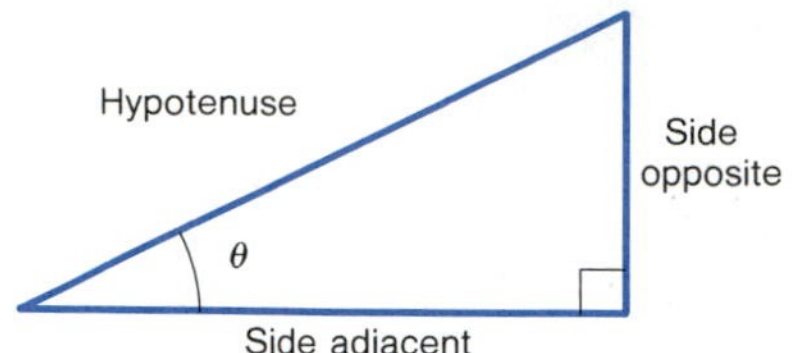


FIGURE 1.6.5

Trigonometric functions of an **acute angle** $\theta$ $(0 < \theta < \pi/2)$ can be described in terms of ratios of lengths of sides of a right triangle. Recall (see Figure 1.6.5) that

$$\sin\theta = \frac{\text{Side opposite}}{\text{Hypotenuse}},$$

$$\cos\theta = \frac{\text{Side adjacent}}{\text{Hypotenuse}},$$

$$\tan\theta = \frac{\text{Side opposite}}{\text{Side adjacent}} = \frac{\sin\theta}{\cos\theta},$$

$$\cot\theta = \frac{\text{Side adjacent}}{\text{Side opposite}} = \frac{\cos\theta}{\sin\theta} = \frac{1}{\tan\theta},$$

$$\sec\theta = \frac{\text{Hypotenuse}}{\text{Side adjacent}} = \frac{1}{\cos\theta},$$

$$\csc\theta = \frac{\text{Hypotenuse}}{\text{Side opposite}} = \frac{1}{\sin\theta}.$$

It is often more convenient to use the exact value of the trigonometric functions of the angles $\pi/6$, $\pi/4$, and $\pi/3$ than to use the multiple-digit approximate values given by a calculator. The values of the trigonometric functions for these common angles are easily read from sketches of representative triangles.

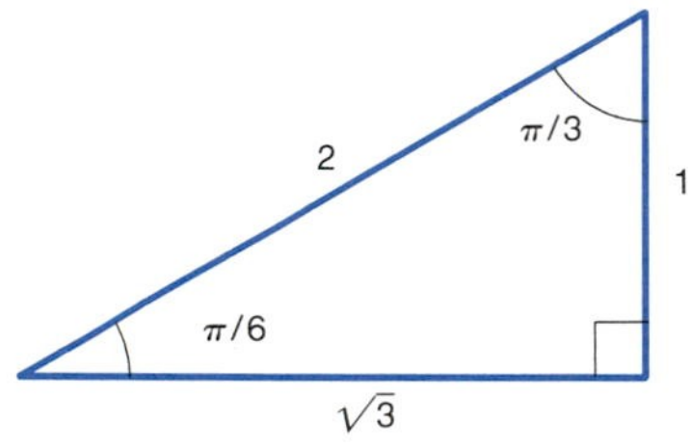


FIGURE 1.6.6

To sketch a $\pi/6$–$\pi/3$–$\pi/2$ triangle, it is only necessary to remember that, since this triangle is half of an equilateral triangle, the side opposite the $\pi/6$ angle is one-half the length of the hypotenuse. That is, $\sin(\pi/6) = 1/2$. If the hypotenuse is 2, then the side opposite the $\pi/6$ angle is 1. The Pythagorean Theorem implies the remaining side has length $\sqrt{3}$. From Figure 1.6.6 we can see that

$$\sin\frac{\pi}{6} = \frac{1}{2}, \qquad \sin\frac{\pi}{3} = \frac{\sqrt{3}}{2},$$

$$\cos\frac{\pi}{6} = \frac{\sqrt{3}}{2}, \qquad \cos\frac{\pi}{3} = \frac{1}{2},$$

$$\tan\frac{\pi}{6} = \frac{1}{\sqrt{3}}, \qquad \tan\frac{\pi}{3} = \sqrt{3},$$

$$\cot\frac{\pi}{6} = \sqrt{3}, \qquad \cot\frac{\pi}{3} = \frac{1}{\sqrt{3}},$$

$$\sec\frac{\pi}{6} = \frac{2}{\sqrt{3}}, \qquad \sec\frac{\pi}{3} = 2,$$

$$\csc\frac{\pi}{6} = 2, \qquad \csc\frac{\pi}{3} = \frac{2}{\sqrt{3}}.$$

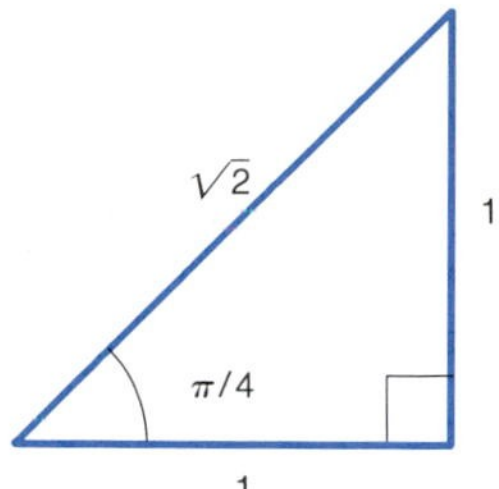

FIGURE 1.6.7

The sides of a $\pi/4$–$\pi/4$–$\pi/2$ triangle are equal, so $\tan(\pi/4) = 1$. If the length of each of its two equal sides is 1, then the hypotenuse is $\sqrt{2}$. We can read the values of the trigonometric functions of $\pi/4$ from Figure 1.6.7. We have

$$\sin\frac{\pi}{4} = \cos\frac{\pi}{4} = \frac{1}{\sqrt{2}},$$

$$\tan\frac{\pi}{4} = \cot\frac{\pi}{4} = 1,$$

$$\sec\frac{\pi}{4} = \csc\frac{\pi}{4} = \sqrt{2}.$$

If the value of one trigonometric function of an acute angle is known, we can determine the values of the other trigonometric functions by drawing a **representative right triangle**.

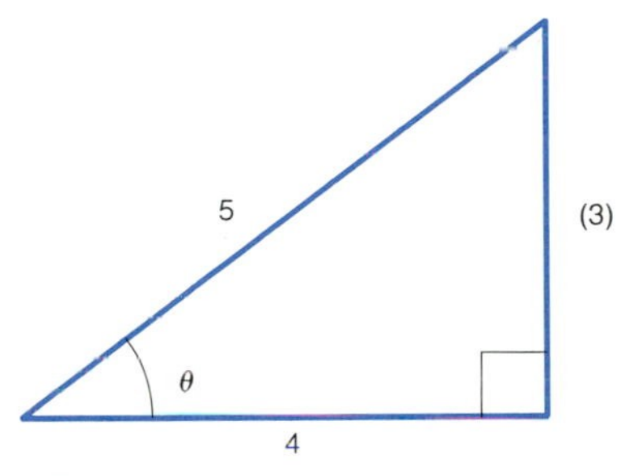

FIGURE 1.6.8

■ **EXAMPLE 1**

Draw a representative right triangle and determine $\tan\theta$ for the acute angle $\theta$ with $\cos\theta = 4/5$.

**SOLUTION**

A representative right triangle is sketched in Figure 1.6.8. The side adjacent to the angle $\theta$ has been labeled 4 and the hypotenuse has been labeled 5. The angle $\theta$ then satisfies $\cos\theta = 4/5$. The Pythagorean Theorem is used to determine that the remaining side has length $\sqrt{5^2 - 4^2} = 3$. We then see that $\tan\theta = 3/4$. ■

We will often encounter problems where the sides and/or angles of a right triangle are variables. The trigonometric functions can then be used to find relations between these variables. It is worth noting that the triangles we will encounter may not appear in a standard or convenient position.

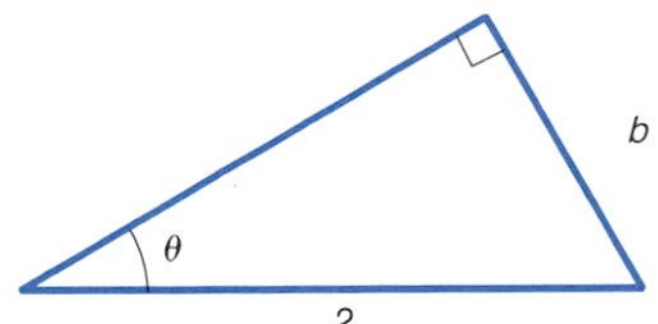

FIGURE 1.6.9

■ **EXAMPLE 2**

(a) Express $b$ in terms of $\theta$ in Figure 1.6.9. (b) Express $d$ in terms of $\phi$ in Figure 1.6.10.

**SOLUTION**

(a) From Figure 1.6.9 we see that $\sin\theta = b/2$, so $b = 2\sin\theta$.
(b) From Figure 1.6.10 we see that $\cot\phi = d/5000$, so $d = 5000\cot\phi$. ■

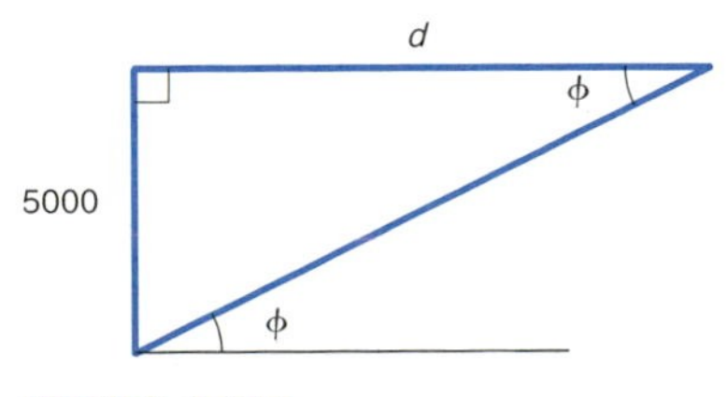

FIGURE 1.6.10

We need to extend the definition of the trigonometric functions to include all angles. Trigonometric functions of an angle that is in standard position can be defined in terms of the coordinates $(x, y)$ of the point of intersection of the terminal side of the angle and the circle $x^2 + y^2 = 1$.

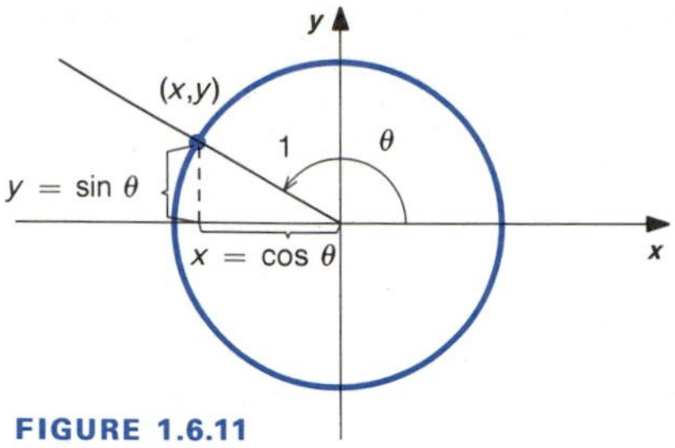

FIGURE 1.6.11

In particular, we define

$$\sin \theta = y \quad \text{and} \quad \cos \theta = x.$$

See Figure 1.6.11. The other trigonometric functions are then defined by

$$\tan \theta = \frac{y}{x}, \quad \cot \theta = \frac{x}{y}, \quad \sec \theta = \frac{1}{x}, \quad \csc \theta = \frac{1}{y}.$$

These definitions agree with the previous definitions of the trigonometric functions of an acute angle in terms of the sides and hypotenuse of a right triangle.

It is clear that the values of the trigonometric functions are equal for angles that differ by integer multiples of $2\pi$.

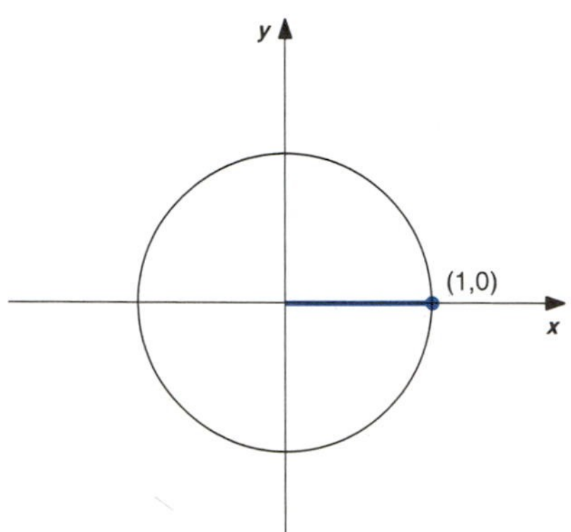

FIGURE 1.6.12

**■ EXAMPLE 3**

Evaluate (a) $\sin 0$, (b) $\sin(\pi/2)$, (c) $\cos \pi$.

**SOLUTION**

In each case we will sketch the angle in standard position and label appropriate coordinates.

(a) From Figure 1.6.12 we see that the terminal side of the angle $\theta = 0$ intersects the unit circle at the point $(x, y) = (1, 0)$. Since the value of the sine is the $y$-coordinate of this point, we have $\sin 0 = 0$.

(b) From Figure 1.6.13 we see that $\sin(\pi/2) = 1$.

(c) From Figure 1.6.14 we see $\cos \pi = -1$. ■

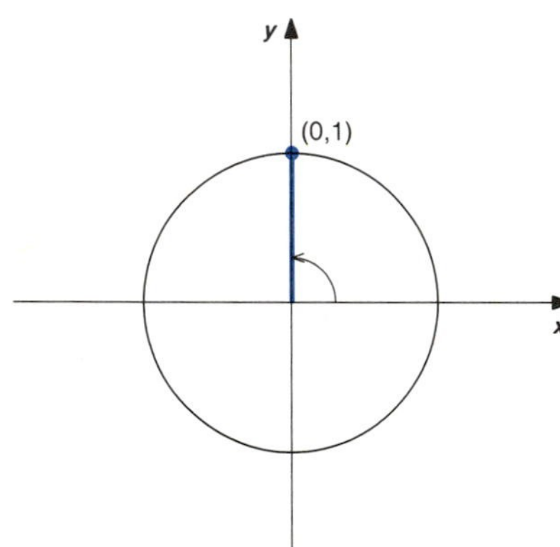

FIGURE 1.6.13

It is possible to use the notion of side opposite, side adjacent, and hypotenuse to determine the trigonometric functions of any angle $\theta$. To do this, we first draw the angle in standard position and then form a right triangle with one side along the $x$-axis, one side vertical, and hypotenuse along the terminal side. This is done by drawing a vertical line segment from a point $(x, y)$ on the terminal side of the angle to the $x$-axis. The base of the triangle is labeled $x$, the height is labeled $y$, and the hypotenuse is labeled $\sqrt{x^2 + y^2}$. The signs of $x$ and $y$ depend on the quadrant that contains the terminal side of $\theta$. The hypotenuse is always labeled positive. See Figure 1.6.15. The terminal side of $\theta$ intersects the unit circle at the point

$$\left(\frac{x}{\sqrt{x^2 + y^2}}, \frac{y}{\sqrt{x^2 + y^2}}\right).$$

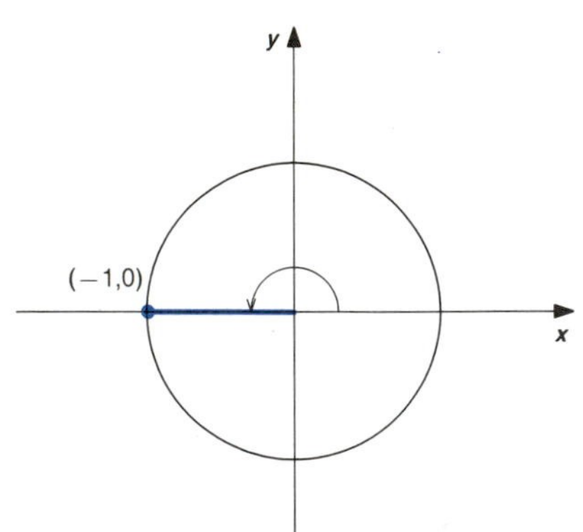

FIGURE 1.6.14

It follows that, for example,

$$\sin \theta = \frac{y}{\sqrt{x^2 + y^2}}.$$

This can be interpreted as the side opposite the acute angle at the origin over the hypotenuse. The other trigonometric functions can also be read from the sketch.

FIGURE 1.6.15

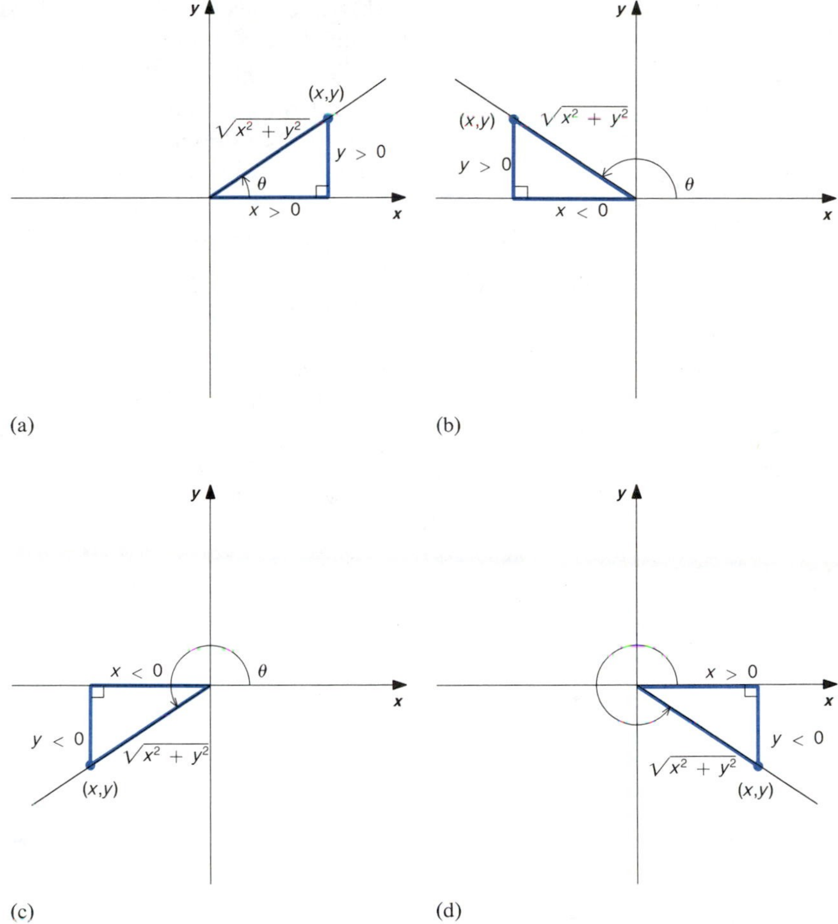


The acute angle between the terminal side and the $x$-axis of an angle in standard position is called the **reference angle** of the angle. Except for sign, the values of the trigonometric functions of an angle are equal to those of its reference angle.

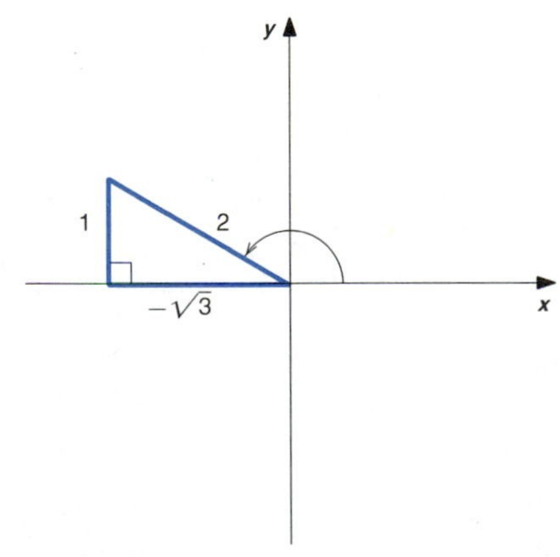


FIGURE 1.6.16

### ■ EXAMPLE 4

Draw representative angles and evaluate (a) $\cos(5\pi/6)$, (b) $\sec(5\pi/4)$, (c) $\tan\theta$, if $\sin\theta = -3/5$ and $\cos\theta < 0$, (d) $\cos\theta$, if $\sin\theta = x/3$ and $-\pi/2 \leq \theta \leq \pi/2$.

#### SOLUTION

(a) An angle of $5\pi/6$ is drawn in standard position in Figure 1.6.16. The terminal side is in the second quadrant. The reference angle is $\pi - 5\pi/6 = \pi/6$. The side opposite the reference angle is labeled 1, the hypotenuse is labeled 2, and the side adjacent is labeled $-\sqrt{3}$. This labeling reflects both the lengths of the sides of a representative $\pi/6$–$\pi/3$–$\pi/2$ triangle and the sign of $x$- and $y$-coordinates in the second quadrant. We can then see that $\cos(5\pi/6) = -\sqrt{3}/2$.

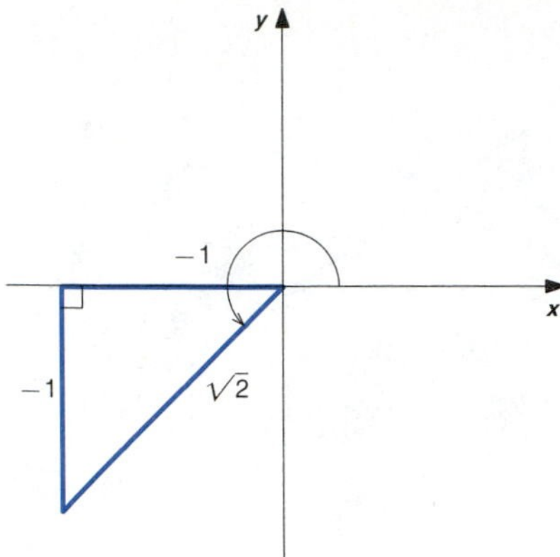

FIGURE 1.6.17

(b) An angle of $5\pi/4$ is drawn in standard position in Figure 1.6.17. The terminal side is in the third quadrant. The reference angle is $5\pi/4 - \pi = \pi/4$. The side opposite the reference angle and the side adjacent are labeled $-1$, and the hypotenuse is labeled $\sqrt{2}$. This labeling reflects both the lengths of the sides of a representative $\pi/4$-$\pi/4$-$\pi/2$ triangle and the sign of $x$- and $y$-coordinates in the third quadrant. We then see that $\sec(5\pi/4) = -\sqrt{2}$.

(c) Representative angles $\theta$ with $\sin\theta = -3/5$ are drawn in Figure 1.6.18. Angles in standard position that satisfy $\sin\theta < 0$ must have terminal sides in either the third or fourth quadrants. The Pythagorean Theorem implies that a right triangle that has hypotenuse 5 and one side 3, must have the other side of length $\sqrt{5^2 - 3^2} = 4$. The angle in Figure 1.6.18b has $\cos\theta = 4/5 > 0$. Since we are given that $\cos\theta < 0$, this angle must be discarded. The angle in Figure 1.6.18a satisfies $\cos\theta = -4/5 < 0$. From Figure 1.6.18a we then see that an angle that satisfies $\sin\theta = -3/5$ and $\cos\theta < 0$, must have $\tan\theta = 3/4$.

FIGURE 1.6.18

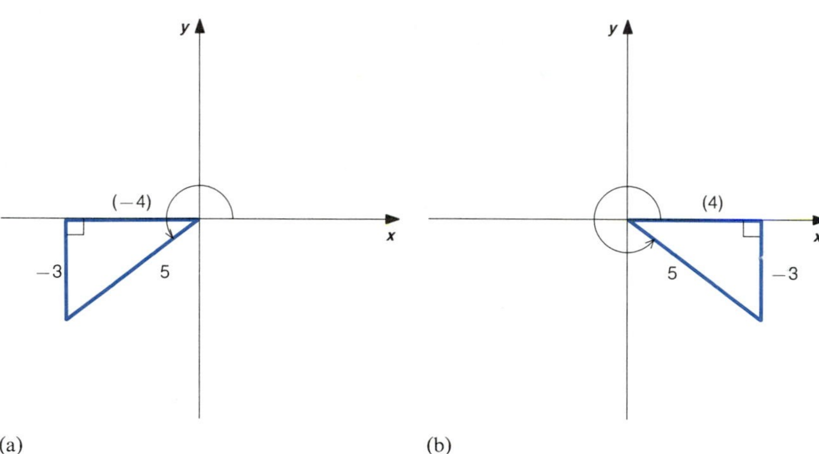


FIGURE 1.6.19

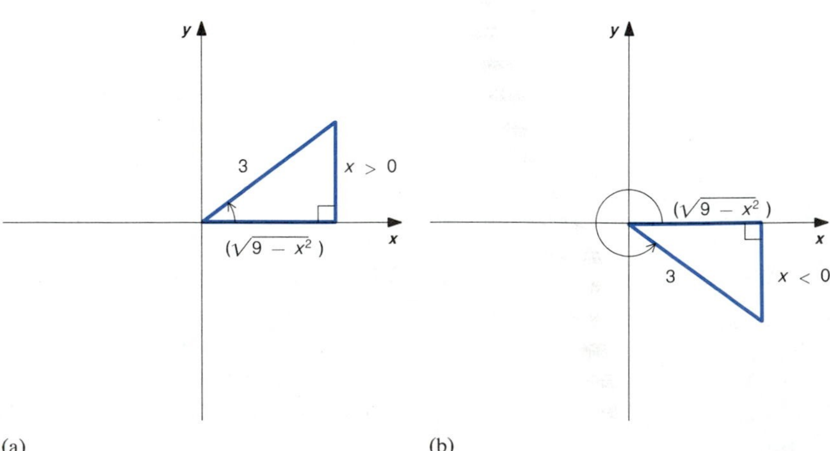

FIGURE 1.6.20

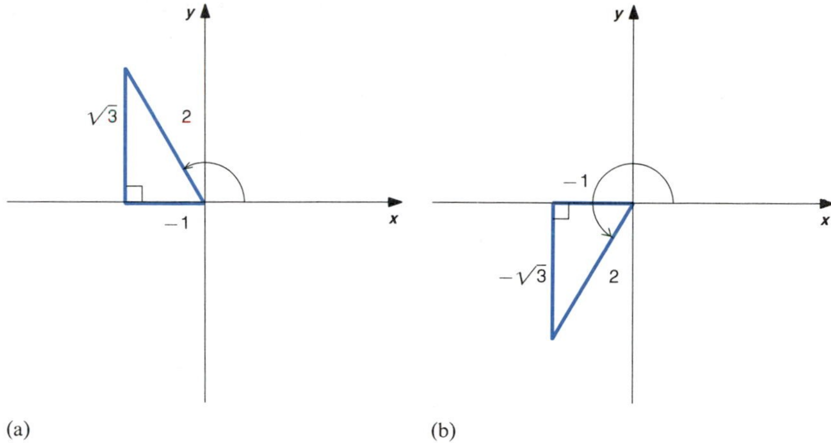


(d) Representative angles with $\sin\theta = x/3$ and $-\pi/2 \leq \theta \leq \pi/2$ are drawn in Figure 1.6.19. The Pythagorean Theorem implies that a right triangle with hypotenuse 3 and one side $x$, must have the other side of length $\sqrt{9-x^2}$. If $x$ is either positive as in Figure 1.6.19a or negative as in Figure 1.6.19b, we have $\cos\theta = \sqrt{9-x^2}/3$. ■

■ **EXAMPLE 5**

Find all $\theta$, $0 \leq \theta \leq 2\pi$, such that $\cos\theta = -1/2$.

**SOLUTION**

Representative angles $\theta$ with $\cos\theta = -1/2$ are drawn in Figure 1.6.20. Angles in standard position that satisfy $\cos\theta < 0$ must have terminal side in either the second or third quadrants. Since the cosine of the reference angle is 1/2, we know that the reference angle is $\pi/3$. From Figure 1.6.20a we then obtain $\theta = \pi - \pi/3 = 2\pi/3$. From Figure 1.6.20b we obtain $\theta = \pi + \pi/3 = 4\pi/3$. (The inverse cosine function on a calculator could be used to determine that $\theta = 2\pi/3 \approx 2.0943951$ is a solution of $\cos\theta = -1/2$. A calculator will not give the value $\theta = 4\pi/3$.) ■

If $k$ is any integer,

$$\sin(k\pi) = 0 \quad \text{and} \quad \cos\left(\frac{\pi}{2} + k\pi\right) = 0.$$

We can use this information to solve certain types of equations.

■ **EXAMPLE 6**

Find all $\theta$, $0 \leq \theta \leq 2\pi$, such that $\cos(2\theta) = 0$.

**SOLUTION**

We know that $\cos(2\theta) = 0$ when $2\theta = \pi/2 + k\pi$, where $k$ is any integer. Solving the equation $2\theta = \pi/2 + k\pi$ for $\theta$, we obtain $\theta = \pi/4 + k\pi/2$. The values $k = 0, 1, 2, 3$ give $\theta = \pi/4, 3\pi/4, 5\pi/4, 7\pi/4$, respectively. These are the solutions of $\cos(2\theta) = 0$ that are between 0 and $2\pi$. See Figure 1.6.21. ■

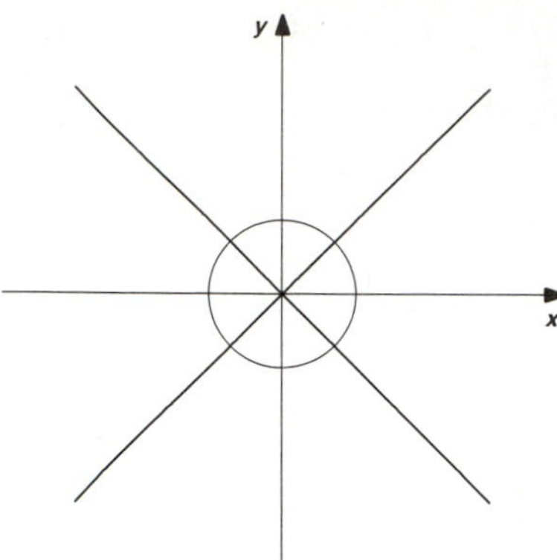


FIGURE 1.6.21

The Law of Cosines and the Law of Sines can be used to find relations between the angles and sides of a triangle. These formulas hold for any triangle, not just for right triangles. See Figure 1.6.22.

**Law of Cosines:** $c^2 = a^2 + b^2 - 2ab\cos\gamma$.

**Law of Sines:** $\dfrac{\sin\alpha}{a} = \dfrac{\sin\beta}{b} = \dfrac{\sin\gamma}{c}$.

If $\gamma = \pi/2$, we have $\cos\gamma = 0$ and the Law of Cosines gives the familiar result of the Pythagorean Theorem. If $\pi/2 < \gamma < \pi$, $\cos\gamma$ is negative and the Law of Cosines gives a value of $c$ that is greater than $\sqrt{a^2 + b^2}$.

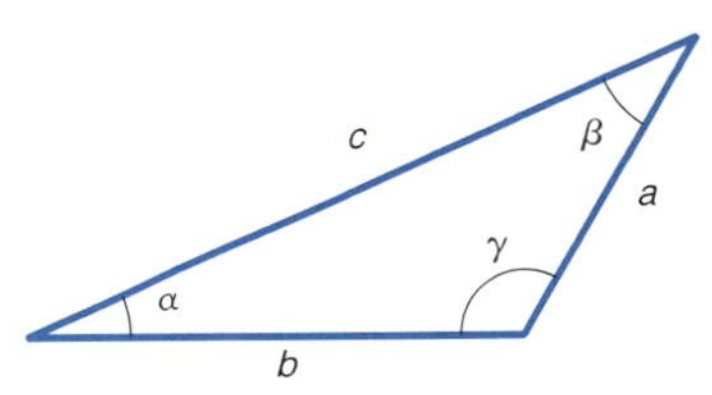


FIGURE 1.6.22

■ **EXAMPLE 7**

(a) Find an equation that relates $b$ and $\beta$ in Figure 1.6.23. (b) Find an equation that relates $c$ and $\alpha$ in Figure 1.6.24.

**SOLUTION**

(a) Side $b$ is the side opposite the angle $\beta$ in the triangle. Since the other two sides of the triangle are known, we can use the law of cosines to obtain the equation

$$b^2 = 10^2 + 10^2 - 2(10)(10)\cos\beta,$$

$$b^2 = 200 - 200\cos\beta.$$

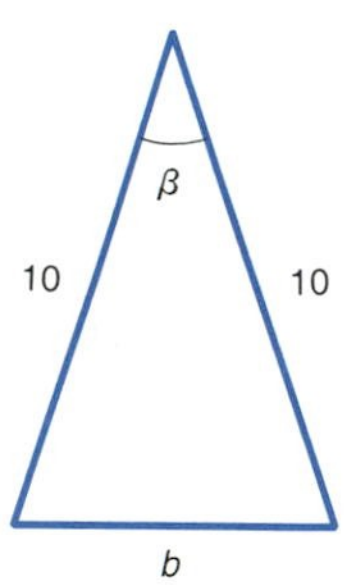


FIGURE 1.6.23

(b) The side opposite the angle $\alpha$ has length 5. The angle opposite the side $c$ is $\pi - \pi/4 = 3\pi/4$. We can then use the Law of Sines to obtain the equation

$$\frac{\sin(3\pi/4)}{c} = \frac{\sin\alpha}{5} \quad \text{or} \quad \frac{1}{\sqrt{2}c} = \frac{\sin\alpha}{5}.$$ ■

Since the radian measure of an angle is a real number that is a ratio of lengths, any real number—positive, negative, or zero—may be considered to be the radian measure of an angle. We may then consider the trigonometric functions to be functions of a real variable. That is, $\sin x$ represents the value of the sine function for the angle that has radian measure $x$. This interpretation is particularly useful for applications of calculus.

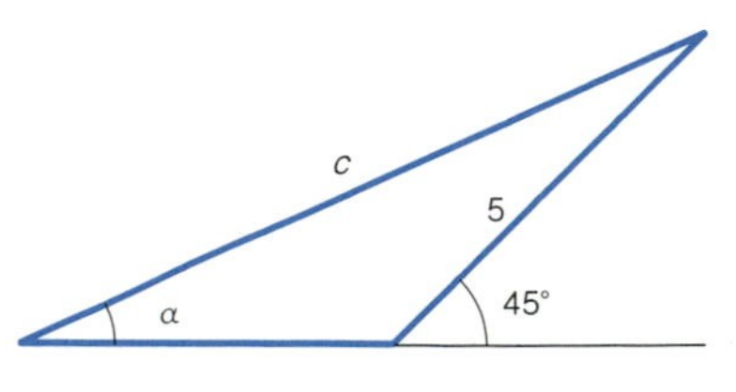


FIGURE 1.6.24

We assume familiarity with the graphs of the trigonometric functions. They are sketched in Figures 1.6.25–1.6.30. The values of $\sin x$, $\cos x$, $\sec x$, and $\csc x$ repeat in intervals of $2\pi$. The functions $\tan x$ and $\cot x$ repeat in intervals of $\pi$.

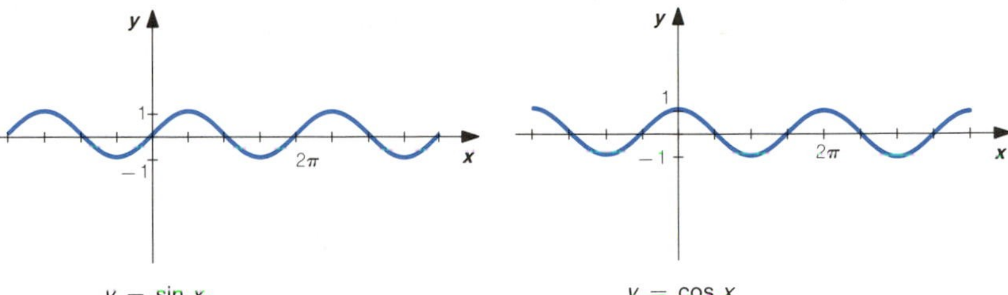


$y = \sin x$

**FIGURE 1.6.25**

$y = \cos x$

**FIGURE 1.6.26**

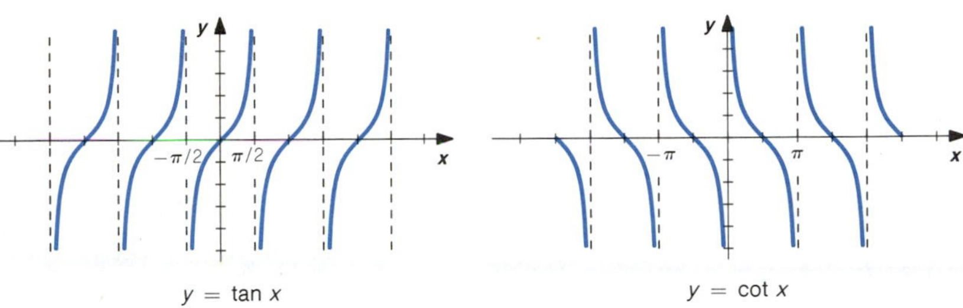


$y = \tan x$

**FIGURE 1.6.27**

$y = \cot x$

**FIGURE 1.6.28**

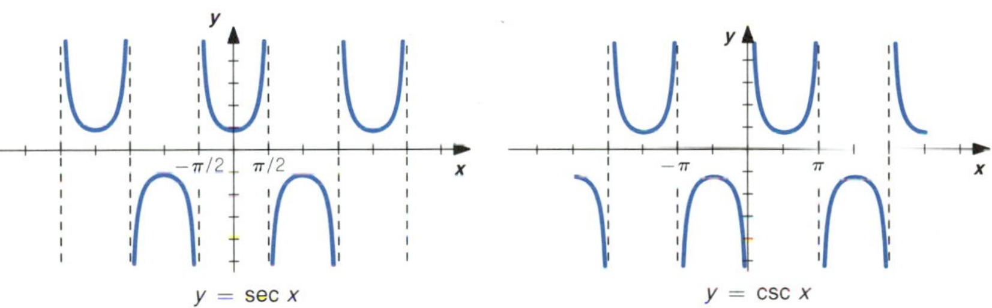


$y = \sec x$

**FIGURE 1.6.29**

$y = \csc x$

**FIGURE 1.6.30**

Let us now see how to use the principles of key points and general character to obtain the graphs of certain modifications of the trigonometric functions.

Functions of the form

$$y = a \cos(bx + \theta)$$

occur in applications from physics and electrical engineering. Let us utilize the fact that the graph has the general character of the cosine curve to sketch its graph.

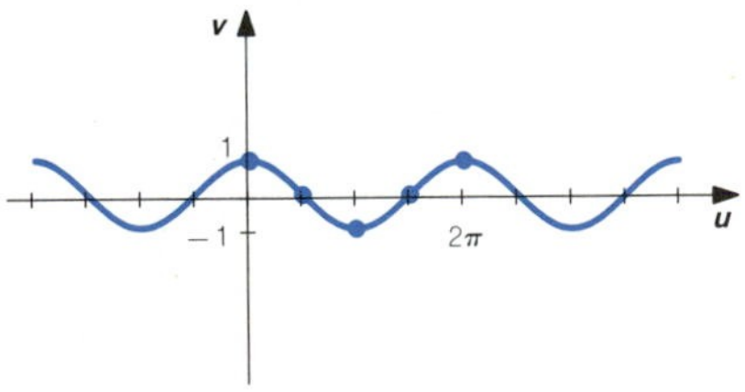

$v = \cos u$

**FIGURE 1.6.31**

■ **EXAMPLE 8**

Sketch the graph of $y = 3 \cos(2x + \pi/2)$.

**SOLUTION**

This function has the same general character as the function $v = \cos u$. From Figure 1.6.31, we see that $v$ decreases from 1 to $-1$ as $u$ goes from 0 to $\pi$; $v$ increases from $-1$ to 1 as $u$ goes from $\pi$ to $2\pi$; and $v = 0$ when $u = \pi/2$ and $u = 3\pi/2$. This general character is repeated as $u$ varies over intervals of length $2\pi$. The coordinates of key points $(u, v)$ on the graph of $v = \cos u$ are listed in the middle columns of Table 1.6.1.

**TABLE 1.6.1**

| $x = (2u - \pi)/4$ | $u$ | $v = \cos u$ | $y = 3v$ |
|---|---|---|---|
| $-\pi/4$ | 0 | 1 | 3 |
| 0 | $\pi/2$ | 0 | 0 |
| $\pi/4$ | $\pi$ | $-1$ | $-3$ |
| $\pi/2$ | $3\pi/2$ | 0 | 0 |
| $3\pi/4$ | $2\pi$ | 1 | 3 |

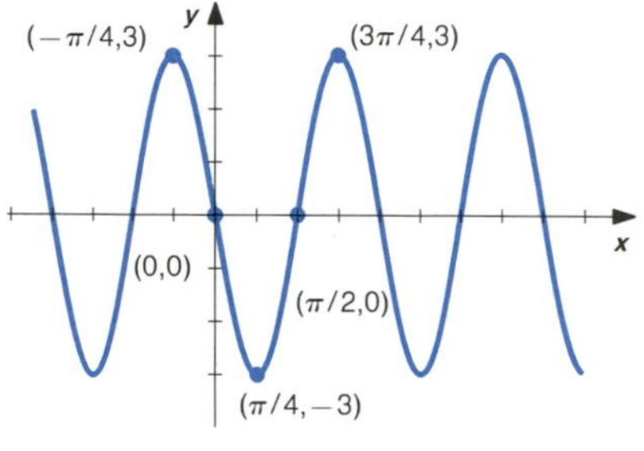

**FIGURE 1.6.32**

If we set $u = 2x + \pi/2$, then we have $x = (2u - \pi)/4$ and $y = 3\cos(2x + \pi/2) = 3\cos u = 3v$. We can then determine values $(x, y)$ that correspond to the key points $(u, v)$. The $(x, y)$ coordinates of key points are listed in the outer columns of Table 1.6.1.

The key points $(x, y)$ in the table are plotted. The graph of $y = 3\cos(2x + \pi/2)$ is then sketched through these points in a way that reflects the general character of the graph. See Figure 1.6.32. The values of $y$ range between $-3$ and 3. Also, the values repeat in intervals of length $\pi$; $\pi$ is the difference of the values of $x$ that correspond to the difference between $u = 0$ and $u = 2\pi$. ■

Let us illustrate the idea of using key points and general character for graphing with a different type of example. The function considered in Example 9 is not of a type that occurs in applications, but it is useful for illustrating several concepts of calculus.

### ■ EXAMPLE 9

Sketch the graph of $y = \cos(2\pi/x)$, $0 < |x| \leq 1$.

#### SOLUTION

Let $u = 2\pi/x$ and $v = \cos u$. Then $x = 2\pi/u$ and $y = v$. We will use the character of the graph of $v = \cos u$ to determine the graph of $y = \cos(2\pi/x)$. To do this we construct a table of values, as was done in Example 8. See Table 1.6.2.

**TABLE 1.6.2**

| $x = 2\pi/u$ | $u$ | $v = \cos u$ | $y = v$ |
|---|---|---|---|
| 1 | $2\pi$ | 1 | 1 |
| 4/5 | $5\pi/2$ | 0 | 0 |
| 2/3 | $3\pi$ | $-1$ | $-1$ |
| 4/7 | $7\pi/2$ | 0 | 0 |
| 1/2 | $4\pi$ | 1 | 1 |
| $1/n$ | $2\pi n$ | 1 | 1 |

Note that $0 < x \leq 1$ implies $u = 2\pi/x \geq 2\pi$. Hence, we begin by considering values of $v$ for $2\pi \leq u \leq 4\pi$.

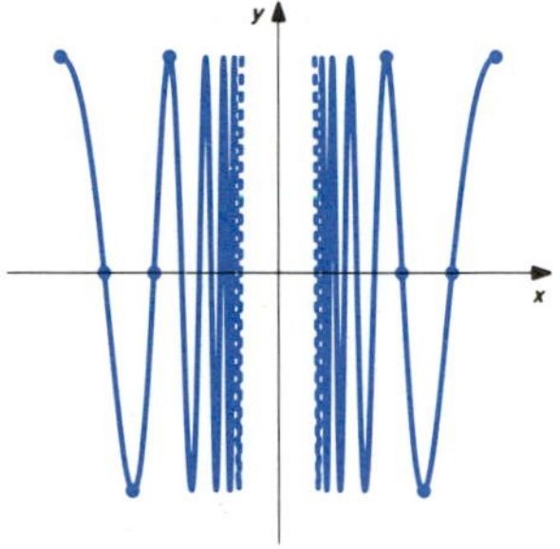


FIGURE 1.6.33

The first five points $(x, y)$ in the table are plotted, and the graph of $y(x)$ is sketched through these points. The curve has the same general character as the cosine curve between these points.

To complete the sketch, we note that the curve repeats the general character of the cosine curve as $u$ varies over intervals of length $2\pi$. That is, between the values $x = 1/n$, $n = 1, 2, 3, \ldots$. The length of these $x$-intervals becomes smaller as $n$ becomes larger and $x$ approaches the origin. Finally, we use the fact that $\cos(-u) = \cos u$ to sketch the graph for $-1 \leq x < 0$. A representative portion of the graph is sketched in Figure 1.6.33. ■

## EXERCISES 1.6

In Exercises 1–8, draw representative right triangles and determine the indicated trigonometric function of the acute angle $\theta$ that satisfies the given condition.

**1** $\cos\theta$, $\sin\theta = 3/5$

**2** $\sin\theta$, $\cos\theta = 5/13$

**3** $\tan\theta$, $\sin\theta = x/2$

**4** $\cos\theta$, $\sin\theta = \sqrt{9 - x^2}/3$

**5** $\sin\theta$, $\tan\theta = x/2$

**6** $\sin\theta$, $\sec\theta = x/2$

In Exercises 7–12, express the indicated variable in terms of a trigonometric function of $\theta$.

**7**

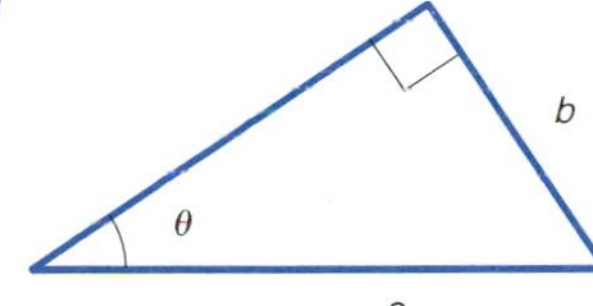

**8**

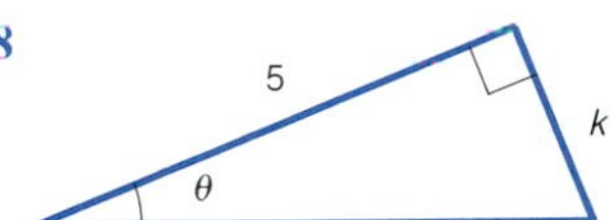

**9**

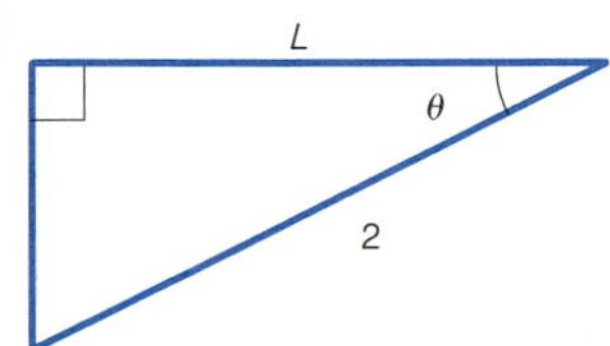

**10**

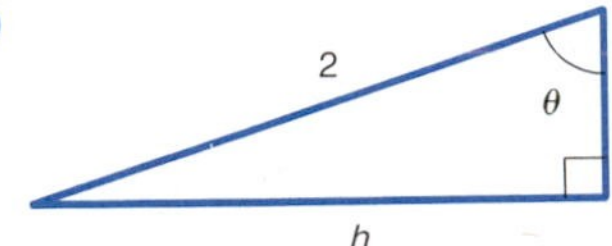

**11**

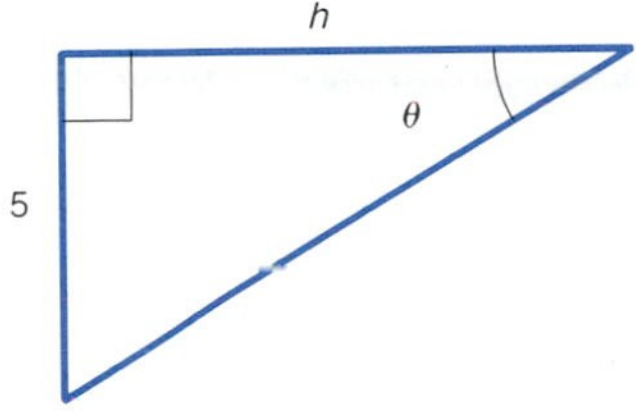

**12**

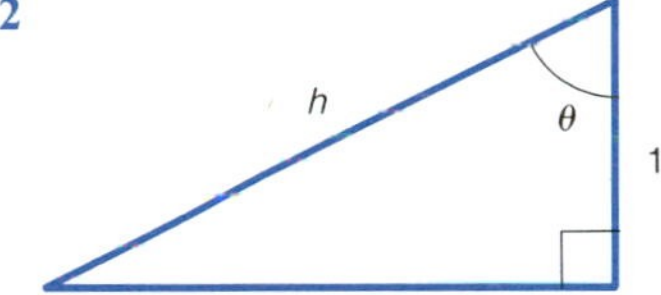


Evaluate the trigonometric functions in Exercises 13–20. (These angles have reference angles of 0, $\pi/6$, $\pi/4$, $\pi/3$, or $\pi/2$, so the exact value answers can be expressed in terms of $\sqrt{2}$ and $\sqrt{3}$.)

**13** (a) $\sin\pi$, (b) $\cos\pi$, (c) $\tan\pi$

**14** (a) $\sin(3\pi/2)$, (b) $\cos(3\pi/2)$, (c) $\tan(3\pi/2)$

**15** (a) $\sin(5\pi/6)$, (b) $\cos(5\pi/6)$, (c) $\tan(5\pi/6)$

**16** (a) $\sin(4\pi/3)$, (b) $\cos(4\pi/3)$, (c) $\tan(4\pi/3)$

**17** (a) $\sin(7\pi/4)$, (b) $\cos(7\pi/4)$, (c) $\tan(7\pi/4)$

**18** (a) $\cot(2\pi/3)$, (b) $\sec(2\pi/3)$, (c) $\csc(2\pi/3)$

**19** (a) $\cot(5\pi/4)$, (b) $\sec(5\pi/4)$, (c) $\csc(5\pi/4)$

**20** (a) $\cot(11\pi/6)$, (b) $\sec(11\pi/6)$, (c) $\csc(11\pi/6)$

In Exercises 21–26, draw representative angles and determine the indicated trigonometric function of the angle $\theta$ that satisfies the given conditions.

**21** $\cos\theta$, if $\sin\theta = -3/5$ and $\cos\theta > 0$

**22** $\sin\theta$, if $\cos\theta = -5/13$ and $\sin\theta < 0$

**23** $\cos\theta$, if $\tan\theta = -1/2$ and $\sin\theta > 0$

**24** $\tan\theta$, if $\cos\theta = -1/3$ and $\sin\theta > 0$

**25** $\cos\theta$, if $\sin\theta = x$ and $-\pi/2 < \theta < \pi/2$

**26** $\sin\theta$, if $\tan\theta = x$ and $-\pi/2 < \theta < \pi/2$

In Exercises 27–34, find all $\theta$, $0 \le \theta \le 2\pi$, that satisfy the given equations. Use radian measure.

**27** $\sin\theta = 1/2$ **28** $\tan\theta = -1$

**29** $\tan\theta = \sqrt{3}$ **30** $\sin\theta = -\sqrt{3}/2$

**31** $\sin(2\theta) = 0$ **32** $\sin(3\theta) = 0$

**33** $\cos(3\theta) = 0$ **34** $\cos(4\theta) = 0$

In Exercises 35–40, use either the Law of Cosines or the Law of Sines to find relations between the variables in the pictured triangles.

**35**

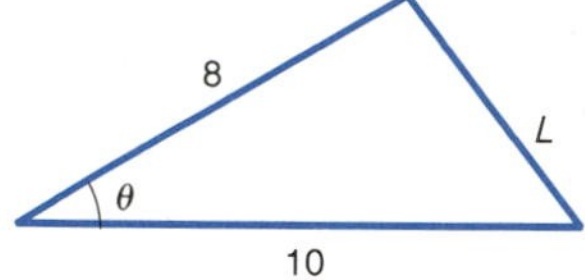


**36**

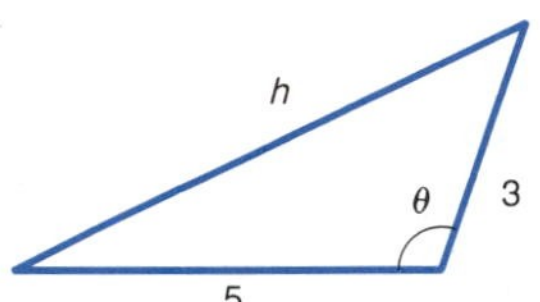


**37**

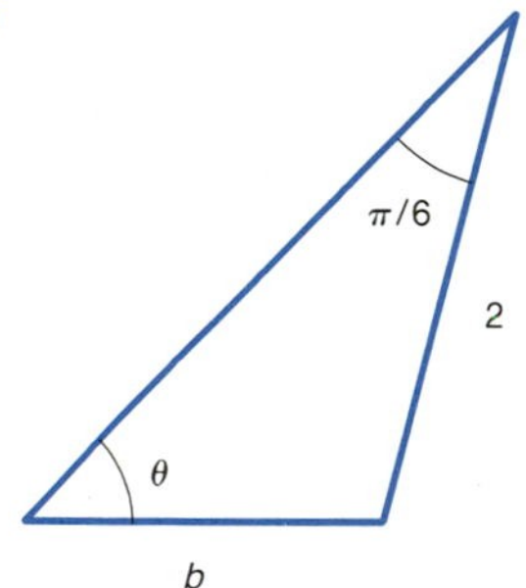


**38**

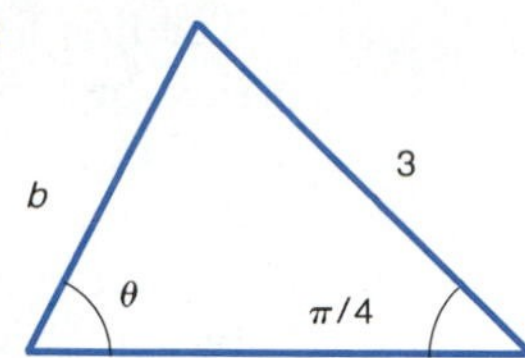


**39**

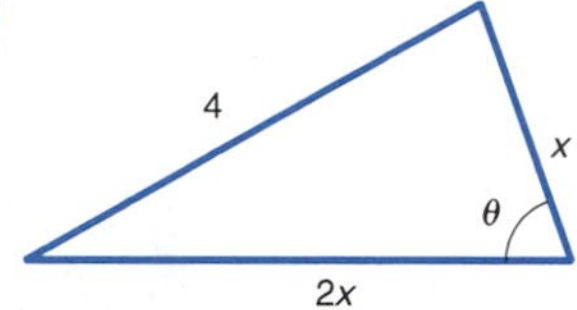


**40**

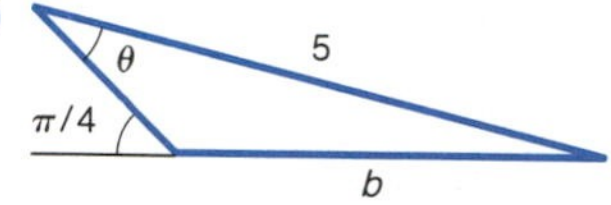


Sketch the graphs of the equations given in Exercises 41–52.

**41** $y = 2\cos\left(x - \frac{\pi}{4}\right)$

**42** $y = 3\cos\left(x + \frac{\pi}{4}\right)$

**43** $y = \cos(x/2)$

**44** $y = 2\cos\left(\frac{x}{2} + \frac{\pi}{2}\right)$

**45** $y = \tan(x/2)$

**46** $y = \sec(2x)$

**47** $y = \cot\left(x - \frac{\pi}{4}\right)$

**48** $y = \csc\left(\frac{x}{2} - \pi\right)$

**49** $y = \sin x$, $-\pi/2 \le x \le \pi/2$

**50** $y = \cos x$, $0 \le x \le \pi$

**51** $y = \tan x$, $-\pi/2 < x < \pi/2$

**52** $y = \sin(1/x)$, $0 < x \le 1/(2\pi)$

## 1.7 TRIGONOMETRIC FORMULAS

The trigonometric formulas that are commonly used in calculus are listed in this section for reference. Verification of these formulas can be found in any standard algebra–trigonometry text and is not given here.

*Fundamental formulas*

$$\tan\theta = \frac{\sin\theta}{\cos\theta}$$

$$\cot\theta = \frac{\cos\theta}{\sin\theta}$$

$$\sec\theta = \frac{1}{\cos\theta}$$

$$\csc\theta = \frac{1}{\sin\theta}$$

$$\sin^2\theta + \cos^2\theta = 1$$

$$\tan^2\theta + 1 = \sec^2\theta$$

$$1 + \cot^2\theta = \csc^2\theta$$

*Addition formulas*

$$\sin(\theta \pm \phi) = \sin\theta\cos\phi \pm \cos\theta\sin\phi$$

$$\cos(\theta \pm \phi) = \cos\theta\cos\phi \mp \sin\theta\sin\phi$$

$$\tan(\theta \pm \phi) = \frac{\tan\theta \pm \tan\phi}{1 \mp \tan\theta\tan\phi}$$

(Note that each of the above equations represents two formulas, one obtained by using the top of the double signs $\pm$ and $\mp$ and the other by using the bottom.)

*Negative angle formulas*

$$\sin(-\theta) = -\sin\theta$$

$$\cos(-\theta) = \cos\theta$$

$$\tan(-\theta) = -\tan\theta$$

***Double angle formulas***

$$\sin 2\theta = 2 \sin \theta \cos \theta$$

$$\cos 2\theta = \cos^2 \theta - \sin^2 \theta = 1 - 2 \sin^2 \theta = 2 \cos^2 \theta - 1$$

$$\tan 2\theta = \frac{2 \tan \theta}{1 - \tan^2 \theta}$$

***Half-angle formulas***

$$\sin^2 \frac{\theta}{2} = \frac{1 - \cos \theta}{2}$$

$$\cos^2 \frac{\theta}{2} = \frac{1 + \cos \theta}{2}$$

$$\tan \frac{\theta}{2} = \frac{1 - \cos \theta}{\sin \theta} = \frac{\sin \theta}{1 + \cos \theta}$$

***Product formulas***

$$\sin \theta \cos \phi = \frac{\sin(\theta + \phi) + \sin(\theta - \phi)}{2}$$

$$\cos \theta \sin \phi = \frac{\sin(\theta + \phi) - \sin(\theta - \phi)}{2}$$

$$\cos \theta \cos \phi = \frac{\cos(\theta - \phi) + \cos(\theta + \phi)}{2}$$

$$\sin \theta \sin \phi = \frac{\cos(\theta - \phi) - \cos(\theta + \phi)}{2}$$

## REVIEW EXERCISES

1 Solve $2x + 3 < 5x - 6$.

2 Solve $|2x - 1| < 3$.

3 Write

$$\frac{x(x^3)^2}{x^5 x^{-2}}$$

in the form $x^r$.

4 Evaluate $(-8)^{-1/3}(18)^{3/2}\sqrt{2}$.

5 Find all values of $x$ for which

$$(3x - 1)(2)(2x - 3)(2) + (2x - 3)^2(3)$$

has value zero.

6 Find all values of $x$ for which $2x^2 - 3x - 3$ has value zero.

7 Simplify

$$\frac{(x^2+1)^{2/3}(1)-(x)(2/3)(x^2+1)^{-1/3}(2x)}{(x^2+1)^{4/3}}.$$

8 Find all values of $x$ for which

$$\sqrt{x-2}-\frac{1}{2\sqrt{x-2}}$$

is (a) zero and (b) undefined.

9 Find all intervals on which $(x-1)(x+2)$ is positive.

10 Find all intervals on which $x(x-2)/(x+1)$ is positive.

11 Find the length, slope, and midpoint of the line segment between the points (2, 3) and (x, 0).

12 Find the slope of the line segment between the point (2, −3) and a variable point $(x, f(x))$ on the graph of $f(x) = 1 - x^2$.

13 Sketch the graph of $2x - 3y + 6 = 0$.

14 Sketch the graph of $2y = 3x$.

15 Sketch the graph of $x^2 + y^2 = 4x$.

16 Sketch the graph of

$$\frac{x^2}{4}+\frac{y^2}{9}=1.$$

17 Sketch the graph of $y = x(x-3)$.

18 Sketch the graph of $y - 1 = 2(x-1)^2$.

19 Sketch the graph of $y = \sqrt{1-x}$.

20 Sketch the graph of

$$y=\frac{1}{x-1}.$$

21 Sketch the graph of

$$y=\frac{1}{(x-1)^2}.$$

22 Find the equation of the line that contains the point (1, −2) and has slope 3.

23 Find the equation of the line that contains the points (4, 3) and (2, −1).

24 Find the equation of the line that contains the point (2, 3) and is perpendicular to the line $2x + 3y = 1$.

25 Find the equation of the line that has graph

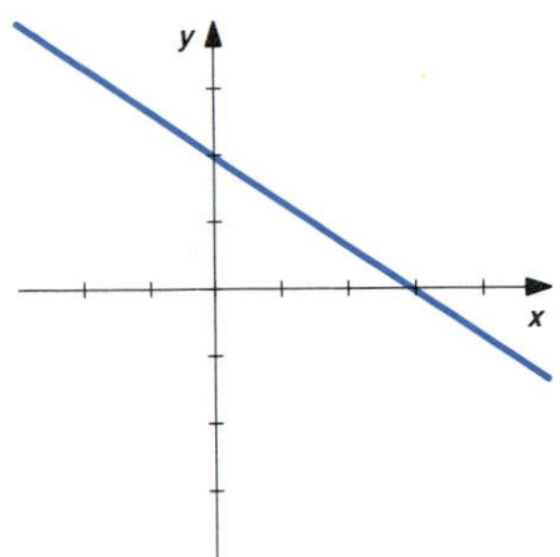


26 Find the equation satisfied by all points $(x, y)$ that are equidistant from (0, 3) and (2, 0).

27 Express $x$ in terms of a trigonometric function of $\theta$.

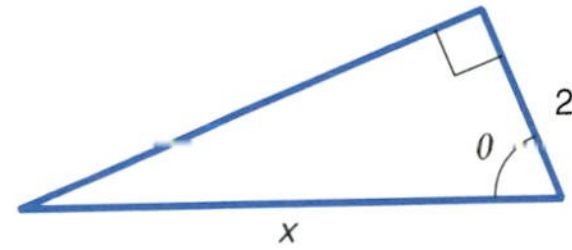


28 Express $x$ in terms of $\cos\theta$.

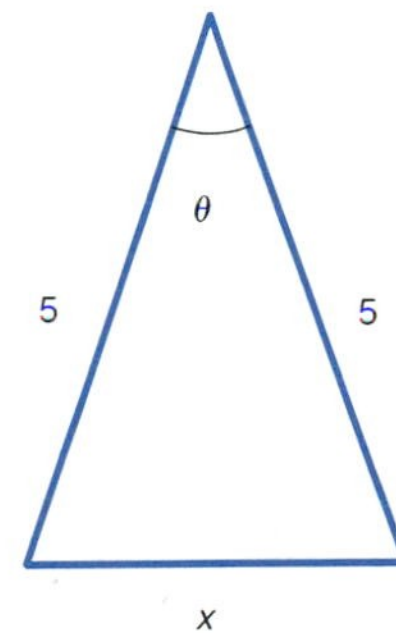


29 Find $\sin\theta$ if $\cos\theta = 1/2$ and $\tan\theta$ is negative.

30 Find all values of $\theta$, $0 \le \theta \le 2\pi$, such that $\sin\theta = -\sqrt{3}/2$.

31 Find all values of $\theta$, $0 \le \theta \le 2\pi$, such that $\cos(2\theta) = 0$.

32 Sketch the graph of $y = 3\sin(2x)$, $0 \le x \le 2\pi$.

33 Sketch the graph of $y = 2\cos(x - \pi/4)$, $0 \le x \le 2\pi$.

## CHAPTER TWO

# *Limits, continuity, and differentiability at a point*

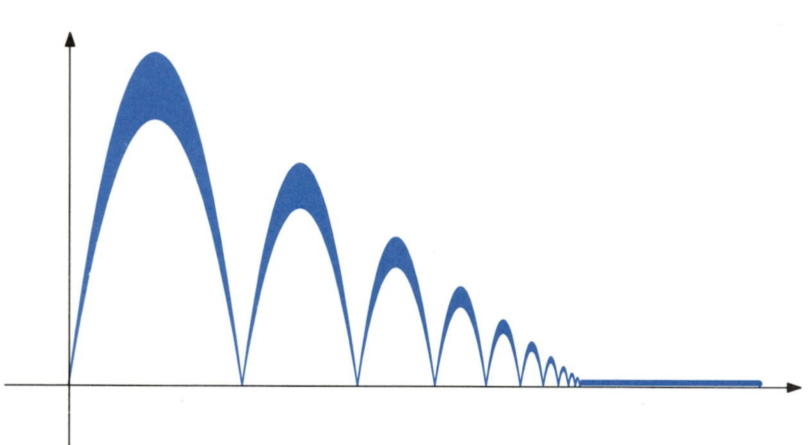

Calculus is used to study functions. By using the techniques of calculus we can obtain information about functions as well as information about the physical quantities represented by functions. In order to use calculus effectively, we must understand the concepts on which it is based. Some of these concepts are introduced in this chapter.

A systematic development of the concept of bounds of functions enables us to obtain information about the size of the values of functions. Such information can be used, for example, to obtain estimates of error in several types of calculations we will be using.

The concept of the limit of a function at a point allows us to describe the behavior of the values of a function near the point. This is one of the key ideas of calculus and distinguishes calculus from other areas of mathematics, such as algebra and trigonometry. The concept of limit is the foundation for the concepts of continuity and differentiability, which are important characteristics of functions. The derivative is one of the central tools in the use of calculus to study functions.

## 2.1 FUNCTIONS

The application of calculus to practical problems depends on expressing physical quantities in terms of functions. Analysis of the functions then gives us information about the physical quantities of interest. In order to use calculus, we must understand the basic ideas of functions.

A formal definition of **function** can be given in terms of what we would ordinarily think of as the **graph** of the function.

*Definition*

A **function with domain** $D$ is a collection of ordered pairs with the property that, for each $x$ in $D$, there is exactly one pair $(x, y)$ in the collection with first element $x$. The set of all second elements $y$ is called the **range** of the function.

Thus, the graph in Figure 2.1.1 represents a function. The graph in Figure 2.1.2 cannot represent a function, because more than one pair $(x, y)$ have the same first element $x$.

A function is specified by a **rule** that determines, for each $x$ in the domain, exactly one corresponding value $y$ in the range. It is customary to use a letter for the name of a function. For example, we may speak of the function $f$. The value $y$ that is determined by a function $f$ for a particular $x$ is denoted $f(x)$—read "$f$ of $x$."

If $f$ is a function and $y = f(x)$, we say $y$ is a function of $x$. The variable $x$ is called the **independent variable**. The variable $y$ is called the **dependent variable**. The value $y = f(x)$ *depends* on the particular choice of $x$ in the domain of $f$. The choice of $x$ determines the corresponding value $y = f(x)$.

Two functions $f$ and $g$ are **equal** if they have the same domain and $f(x) = g(x)$ for every $x$ in that domain.

Let us illustrate the concept of function and graph with an example.

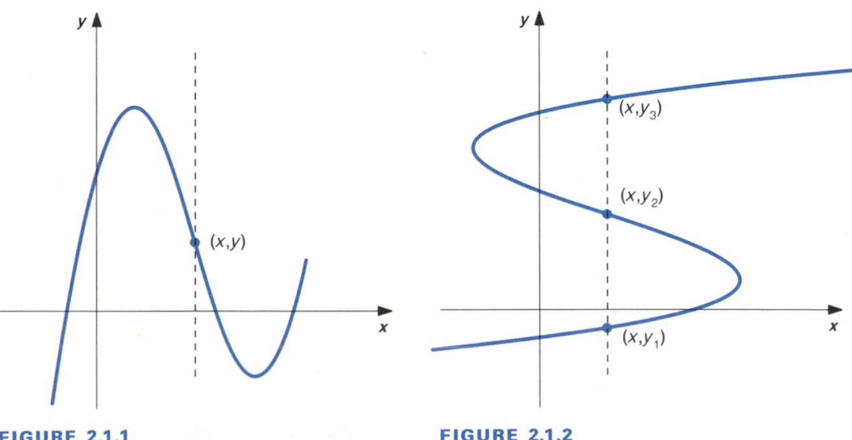


FIGURE 2.1.1

FIGURE 2.1.2

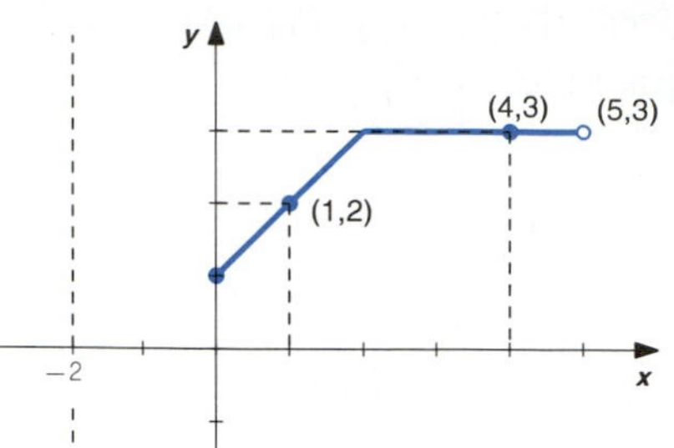

FIGURE 2.1.3

■ **EXAMPLE 1**

The graph in Figure 2.1.3 is the graph of a function $f$. Find the domain and range of $f$. Find $f(-2)$, $f(1)$, and $f(4)$.

**SOLUTION**

The domain of $f$ consists of the set of $x$-coordinates of points $(x, y)$ on the graph. This set is given by the $x$-coordinates of vertical lines that intersect the graph. Since vertical lines intersect the graph if and only if their $x$-coordinates satisfy $0 \leq x < 5$, the domain of $f$ is $0 \leq x < 5$. [The solid dot at $(0, 1)$ emphasizes that the point $(0, 1)$ is on the graph, while the open dot at $(5, 3)$ indicates that point is not on the graph.]

The range of $f$ is given by the set of $y$-coordinates of points $(x, y)$ on the graph. This set is given by the $y$-coordinate of horizontal lines that intersect the graph. Horizontal lines intersect the graph if and only if their $y$-intercepts are in the interval $1 \leq y \leq 3$, so the range of $f$ is $1 \leq y \leq 3$.

There is no point on the graph with $x$-coordinate $-2$. The number $-2$ is not in the domain of $f$; $f(-2)$ is undefined.

The point $(1, 2)$ is on the graph, so $f(1) = 2$.

The point $(4, 3)$ is on the graph, so $f(4) = 3$. ■

Until stated otherwise, the functions we will study will involve only variables that represent real numbers.

We will most often deal with the case where the values of a function are given by a *formula*. Unless stated otherwise, the domain of a function that is given by a formula includes all points where the formula makes sense. The values of the function are obtained by *substitution* into the formula. We will see that:

**Substitution is copying the right thing into the right place.**

Each component of the formula must be copied into its proper position. After the substitution has been carried out, we can perform whatever algebra is necessary to simplify the expression. It is important that substitution and simplification be considered as separate steps.

■ **EXAMPLE 2**

$f(x) = 2x^2 - x - 3$. Find (a) $f(3)$ and (b) $f(x - 2)$.

**SOLUTION**

We first note that the formula $f(x) = 2x^2 - x - 3$ really has nothing to do with $x$. That is, the same function $f$ could be defined with any variable in place of $x$. For example, we could write $f(u) = 2u^2 - u - 3$. The variable serves only as a place holder. In fact, it may be useful to write

$$f(\quad) = 2(\quad)^2 - (\quad) - 3.$$

We can then obtain the value that corresponds to any choice of independent variable simply by *copying* the variable inside each set of parentheses. In each case we will substitute and then simplify.

(a) Substitute: $f(3) = 2(3)^2 - (3) - 3$

Simplify: $= 2(9) - 3 - 3$

$= 12.$

(b) Substitute: $f(x-2) = 2(x-2)^2 - (x-2) - 3$

Simplify: $= 2(x^2 - 4x + 4) - x + 2 - 3$

$= 2x^2 - 8x + 8 - x - 1$

$= 2x^2 - 9x + 7.$

Note that parentheses were useful in both the substitution and simplification stages. ■

The idea of substitution applies to complicated expressions that involve several components. We copy each component into its proper place. Parentheses should be used to organize our work.

■ **EXAMPLE 3**

$f(x) = x^2 - 4x$, $g(x) = 2x - 4$. Find

(a) $f(x) + g(x)$,
(b) $f(x) \cdot g(x)$,
(c) $f(x) - xg(x)$,
(d) $\dfrac{f(x) - f(3)}{x - 3}$,
(e) $f(g(x))$,
(f) $g(f(x))$.

**SOLUTION**

In each case we will first substitute, then simplify:

(a)
$$\begin{aligned} f(x) + g(x) &= (x^2 - 4x) + (2x - 4) \\ &= x^2 - 4x + 2x - 4 \\ &= x^2 - 2x - 4. \end{aligned}$$

(b)
$$\begin{aligned} f(x) \cdot g(x) &= (x^2 - 4x)(2x - 4) \\ &= 2x^3 - 12x^2 + 16x. \end{aligned}$$

(c)
$$\begin{aligned} f(x) - xg(x) &= (x^2 - 4x) - x(2x - 4) \\ &= x^2 - 4x - 2x^2 + 4x \\ &= -x^2. \end{aligned}$$

(d)
$$\begin{aligned} \frac{f(x) - f(3)}{x - 3} &= \frac{(x^2 - 4x) - [(3)^2 - 4(3)]}{x - 3} \\ &= \frac{x^2 - 4x + 3}{x - 3} \\ &= \frac{(x-3)(x-1)}{x - 3} \\ &= x - 1, \qquad x \neq 3. \end{aligned}$$

(We have indicated that $[f(x) - f(3)]/(x - 3) = x - 1$ only for $x \neq 3$.)

(e) $f(g(x))$ is read as "$f$ of $g$ of $x$." Taking one step at a time, we first substitute $2x - 4$ in place of $g(x)$ and then evaluate $f(2x - 4)$.

$$\begin{aligned} f(g(x)) &= f(2x - 4) \\ &= (2x - 4)^2 - 4(2x - 4) \\ &= 4x^2 - 16x + 16 - 8x + 16 \\ &= 4x^2 - 24x + 32. \end{aligned}$$

(f) $$\begin{aligned} g(f(x)) &= g(x^2 - 4x) \\ &= 2(x^2 - 4x) - 4 \\ &= 2x^2 - 8x - 4. \end{aligned}$$ ■

If $f$ and $g$ are functions, the **sum** $f + g$ is the function whose values are given by the formula $(f + g)(x) = f(x) + g(x)$. The domain of $f + g$ consists of those $x$ values for which both $f(x)$ and $g(x)$ are defined. The **product** $fg$, **quotient** $f/g$, and other arithmetic combinations of functions are defined as suggested by the notation. The domains consist of those $x$ for which every component function and the combined arithmetic expression are defined. For example, the domain of $f/g$ consists of those $x$ for which both $f(x)$ and $g(x)$ are defined, except for those $x$ with $g(x) = 0$. The **composite function** $f \circ g$ is defined by the equation

$$f \circ g(x) = f(g(x)).$$

The domain of $f \circ g$ consists of those $x$ in the domain of $g$ for which $g(x)$ is in the domain of $f$. From Examples 3e and 3f we see that it is possible that $f \circ g$ and $g \circ f$ are different functions. The compositions $f \circ g$ and $g \circ f$ are usually unequal.

The domain is an important component of a function. Let us consider some factors that may restrict the domain of a function.

The domain of a function may be restricted by an *explicit statement.*

### ■ EXAMPLE 4

Sketch the graph of the function

$$S(x) = \sin x, \qquad -\frac{\pi}{2} \le x \le \frac{\pi}{2}.$$

Evaluate: (a) $S(\pi/6)$, (b) $S(\pi)$, (c) $S(1)$.

#### SOLUTION

Although $\sin x$ is defined for all $x$, it is stated that the domain of $S$ is restricted to $-\pi/2 \le x \le \pi/2$. The graph of $S$ consists of that part of the graph of $y = \sin x$ with $-\pi/2 \le x \le \pi/2$. See Figure 2.1.4, where we have emphasized that the points $(-\pi/2, -1)$ and $(\pi/2, 1)$ are included in the graph.

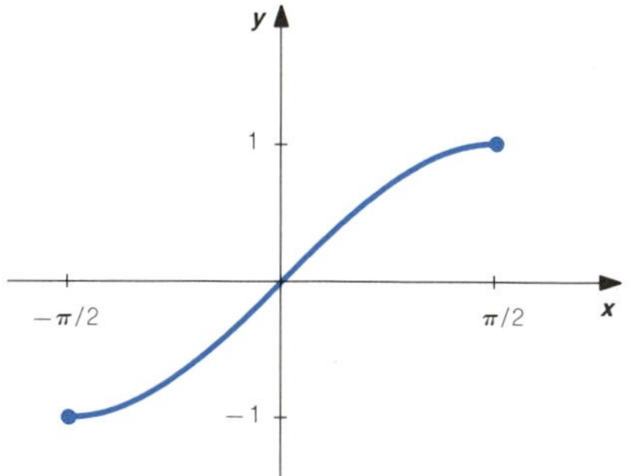


FIGURE 2.1.4

We are told that the values of the function $S$ are given by the formula $\sin x$ only for $-\pi/2 \le x \le \pi/2$. The function $S$ is not defined for other values of $x$.

(a) $S(\pi/6) = \sin(\pi/6) = 1/2$.

(b) $S(\pi)$ is undefined.

(c) $S(1) = \sin 1 \approx 0.84147099$. ■

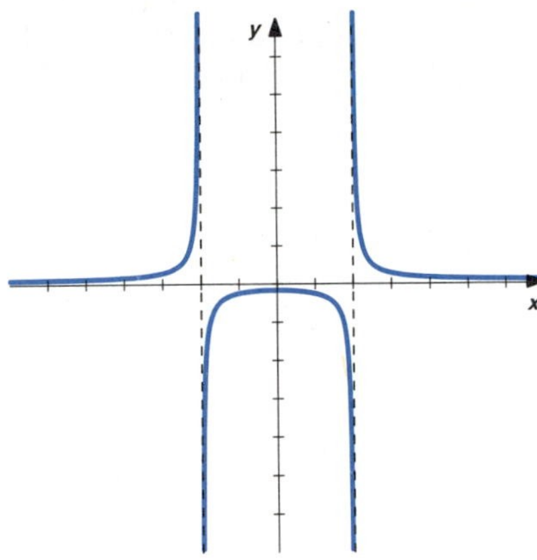


FIGURE 2.1.5

Unless stated otherwise, the domain of a function that is given by a formula includes all points where the formula makes sense. Points where the formula does not make sense are excluded from the domain without comment.

■ **EXAMPLE 5**

Indicate the domains of the functions

$$\text{(a) } f(x) = \frac{1}{x^2 - 4}, \qquad \text{(b) } g(x) = \sqrt{1 - x}, \qquad \text{(c) } h(x) = \tan x.$$

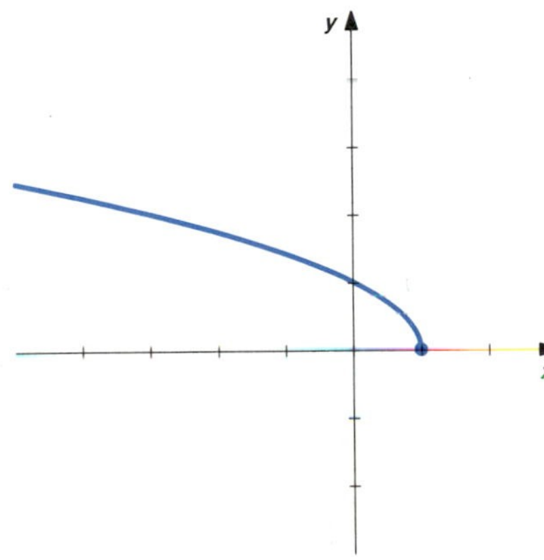


FIGURE 2.1.6

**SOLUTION**

(a) Since we cannot divide by zero, we must avoid $x = \pm 2$. The domain of $f$ consists of all real numbers except $x = 2$ and $x = -2$. The graph of $f$ is given in Figure 2.1.5.

(b) Since the square root of a negative number is undefined, we need $1 - x \ge 0$. Solving this inequality, we find the domain of $g$ consists of all real numbers $x$ with $x \le 1$. See Figure 2.1.6.

(c) The tangent function is undefined at all odd integer multiples of $\pi/2$. The domain of $h$ consists of all real numbers except odd integer multiples of $\pi/2$. See Figure 2.1.7. ■

The domain of a function may be restricted by the interpretation of the *values* of the function as a physical quantity.

■ **EXAMPLE 6**

Express the slope of the line segment between the origin and a variable point on the graph of $y = x^2$, as a function of $x$.

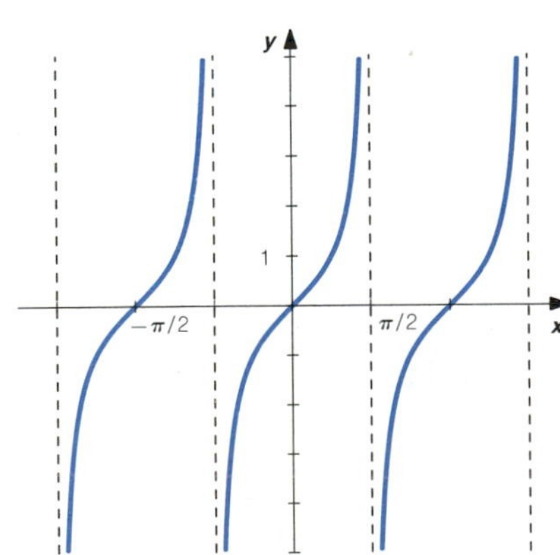


FIGURE 2.1.7

**SOLUTION**

Let $m(x)$ denote the desired slope. A variable point on the graph of $y = x^2$ has coordinates $(x, x^2)$ in terms of $x$. Substitution into the formula for the slope of the line segment between the points $(x, x^2)$ and $(0, 0)$ gives

$$m(x) = \frac{x^2 - 0}{x - 0}.$$

Simplifying, we obtain the function

$$m(x) = x, \qquad x \ne 0.$$

The restriction of the domain of $m$ reflects the fact that $m(x)$ does not represent the slope of the line segment between (0, 0) and $(x, x^2)$ if $x = 0$. When $x = 0$, $(x, x^2)$ and (0, 0) are the same point and there is no line segment between them. See Figure 2.1.8. ■

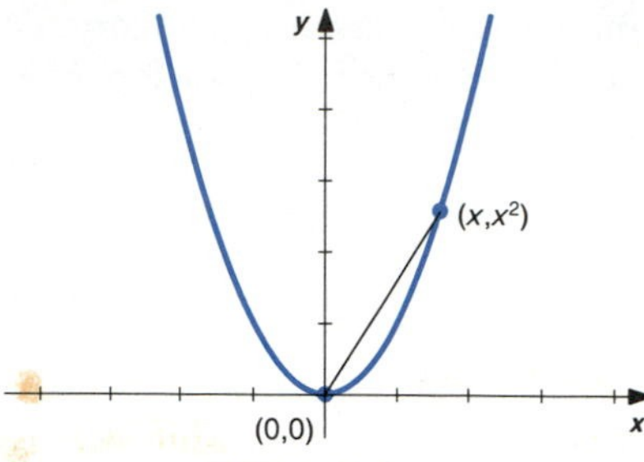

FIGURE 2.1.8

The domain of a function may be restricted by the interpretation of the *independent variable* as a physical quality. In particular:

**Length is nonnegative.**

Variable expressions that represent length must be nonnegative. (Length zero gives degenerate cases of geometric figures such as rectangles, but it will be convenient for future applications of calculus to include these cases. We will need to check that we do not get a degenerate case as the solution of a real problem.)

### ■ EXAMPLE 7

A rectangle has base $x$ and height $2 - x$. Express the area of the rectangle as a function of $x$. Indicate the domain.

FIGURE 2.1.9

#### SOLUTION

The rectangle is sketched and labeled in Figure 2.1.9. The area of a rectangle is given by

Area = (Base)(Height).

Substitution then gives the formula

$$\text{Area} = x(2 - x).$$

The formula makes sense for all $x$, but it will represent the area of a rectangle only if the values of $x$ are restricted. Since $x$ and $2 - x$ represent the length of sides of a rectangle, we must have $x \geq 0$ and $2 - x \geq 0$. These two inequalities imply $0 \leq x \leq 2$. We include the endpoints, $x = 0$ and $x = 2$, in the domain, even though a rectangle with the length of two sides zero is not really a rectangle. Thus, we obtain the function

$$A(x) = x(2 - x), \qquad 0 \leq x \leq 2.$$ ■

A function may be defined by *different formulas* for different values of $x$. It is important to use each formula for the values of $x$ to which it applies.

### ■ EXAMPLE 8

Indicate the domain and sketch the graph of the function

$$f(x) = \begin{cases} 0, & x < 0, \\ x^2, & x \geq 0. \end{cases}$$

Evaluate $f(-1)$ and $f(2)$.

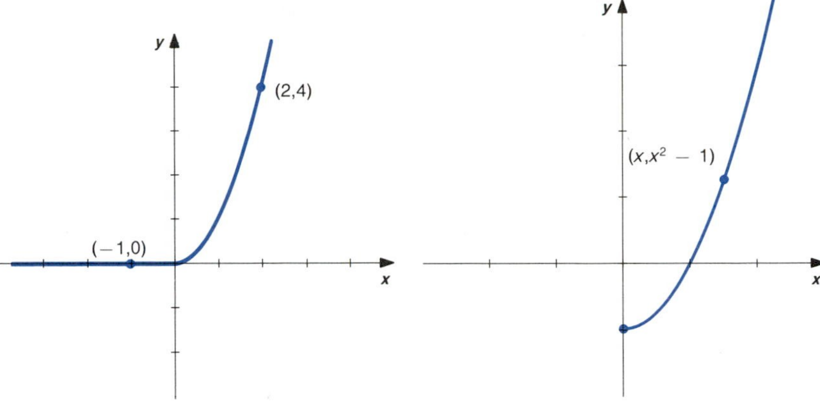


FIGURE 2.1.10

FIGURE 2.1.11

**SOLUTION**

The function is defined for all $x$, so the domain is the set of all real numbers. For $x < 0$, the graph coincides with the graph of the line $y = 0$. For $x \geq 0$, the graph coincides with the parabola $y = x^2$. The graph is sketched in Figure 2.1.10.

We have $-1 < 0$ and $f(x) = 0$ for $x < 0$. Hence, $f(-1) = 0$.

Since $2 \geq 0$ and $f(x) = x^2$ for $x \geq 0$, we have $f(2) = (2)^2 = 4$. ■

An *equation* in $x$ and $y$ determines $y$ as a function of $x$ if and only if the *graph of the equation* determines $y$ as a function of $x$. In many cases we can obtain a formula for a function that is given by an equation in $x$ and $y$ by solving the equation for $y$.

■ **EXAMPLE 9**

Sketch the graphs of the following equations and determine which define $y$ as a function of $x$. Find a formula for the values and indicate the domain and range of each function.

(a) $\sqrt{y+1} = x$, (b) $x^2 + y^2 = 1$.

**SOLUTION**

(a) Solving for $y$, we have

$$\sqrt{y+1} = x,$$

$$y + 1 = x^2,$$

$$y = x^2 - 1.$$

Since the square root function has nonnegative values, we see from the original equation $\sqrt{y+1} = x$ that $x \geq 0$. This means that the graph consists of that part of the parabola $y = x^2 - 1$ with $x \geq 0$. See Figure 2.1.11.

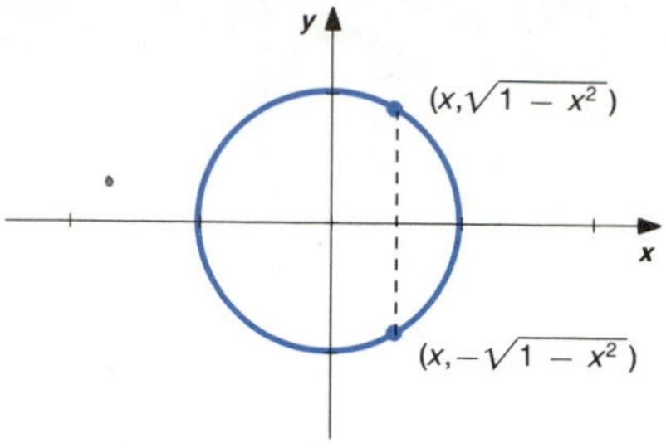


FIGURE 2.1.12

The function is

$$f(x) = x^2 - 1, \qquad x \geq 0.$$

We see that the range consists of all $y \geq -1$.

(b) Solving the equation $x^2 + y^2 = 1$ for $y$, we obtain $y^2 = 1 - x^2$, so $y = \pm\sqrt{1-x^2}$ for $-1 \leq x \leq 1$. For $-1 < x < 1$, both values $y = \sqrt{1-x^2}$ and $y = -\sqrt{1-x^2}$ satisfy the equation. Hence, this equation does not define $y$ as a function of $x$. See Figure 2.1.12. ∎

## EXERCISES 2.1

1 $f(x) = 2x^2 - 3$. Find $f(3)$ and $f(\sqrt{2})$.

2 $f(x) = \dfrac{x}{1-x}$. Find $f(5)$ and $f\left(\dfrac{2}{3}\right)$.

3 $f(x) = 2x - 1$. Find $f\left(\dfrac{1}{x}\right)$ and $\dfrac{1}{f(x)}$.

4 $f(x) = 2x - 3$. Find $f(x^2)$ and $[f(x)]^2$.

5 $f(x) = 3x^2 - 2$. Find $f(x-1)$ and $f(x) - 1$.

6 $f(x) = 3x - 2$. Find $\dfrac{f(x) - f(-3)}{x+3}$.

7 $f(x) = x^2 + x$. Find $\dfrac{f(x) - f(-2)}{x+2}$.

8 $f(x) = \dfrac{x}{x-3}$. Find $\dfrac{f(x) - f(1)}{x-1}$.

In Exercises 9–12, evaluate the given expressions for the functions $f(x) = 3x^2 - 2x + 1$ and $g(x) = 2x - 1$.

9 $f(x) + g(x)$, $f(x) + (x+1)g(x)$

10 $f(x) \cdot g(x)$, $f(x) - [g(x)]^2$

11 $f(g(x))$

12 $g(f(x))$

Sketch the graphs of the functions in Exercises 13–20. Find the values indicated.

13 $f(x) = x^2$, $x \geq 0$; $f(-1)$, $f(0)$, $f(2)$

14 $f(x) = 2x + 4$, $x \geq -2$; $f(-3)$, $f(0)$, $f(2)$

15 $f(x) = \cos x$, $0 \leq x \leq \pi$; $f(-\pi/2)$, $f(\pi/2)$, $f(2\pi/3)$

16 $f(x) = \tan x$, $-\pi/2 < x < \pi/2$; $f(-\pi/2)$, $f(\pi/4)$, $f(3\pi/4)$

17 $f(x) = \begin{cases} 1, & x < 0, \\ 1 - x, & x > 0; \end{cases}$ $f(-2)$, $f(0)$, $f(2)$

18 $f(x) = \begin{cases} x - 1, & x > 1, \\ 1 - x, & x < 1; \end{cases}$ $f(0)$, $f(1)$, $f(3)$

19 $f(x) = \begin{cases} x, & 0 \leq x \leq 2, \\ 5 - x, & x > 2; \end{cases}$ $f(-1)$, $f(2)$, $f(3)$

20 $f(x) = \begin{cases} x^2, & x < 0, \\ x + 1, & 0 \leq x \leq 2; \end{cases}$ $f(-2)$, $f(0)$, $f(3)$

Indicate the domain of the functions in Exercises 21–26.

21 $f(x) = \dfrac{x}{x^2 - 1}$

22 $f(x) = \dfrac{x+1}{x^2 + 5x + 4}$

23 $f(x) = \sqrt{3 - 2x}$

24 $f(x) = \dfrac{1}{\sqrt{x^2 + 3x - 4}}$

25 $f(x) = \sec x$

26 $f(x) = \csc x$

27 Find the slope of the line segment between the point $(1, 2)$ and a variable point on the line $2x + y = 4$, as a function of $x$. Indicate the domain.

28 Find the slope of the line segment between the point $(-2, 4)$ and a variable point on the parabola $y = x^2$, as a function of $x$. Indicate the domain.

29 Find the slope of the line segment between the point on the parabola $y = x^2$ that has $x$-coordinates 1 and a variable point on the graph, as a function of $x$. Indicate the domain.

30 Find the slope of the line segment between the point on the parabola $y = 1 - x^2$ that has $x$-coordinates 2 and a variable point on the graph, as a function of $x$. Indicate the domain.

31 A rectangular box has length $x$, width $2 - x$, and height $3 - x$. Express the volume of the box as a function of $x$. Indicate the domain.

32 Express the total surface area of the six sides of the box in Exercise 31 as a function of $x$. Indicate the domain.

33 A rectangular box has a square base and no top. The length of one side of the square base is $x$. The height of the box is $5 - x$. Express the total surface area of the five sides of the box as a function of $x$. Indicate the domain.

34 Express the volume of the box in Exercise 33 as a function of $x$. Indicate the domain.

Each graph in Exercises 35–38 is the graph of a function. Find the domain, range, and values indicated.

35 $f(-4)$, $f(1)$, $f(3)$

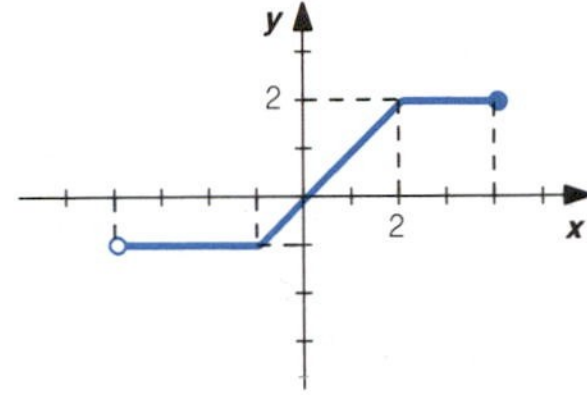


36 $f(-2)$, $f(0)$, $f(2)$

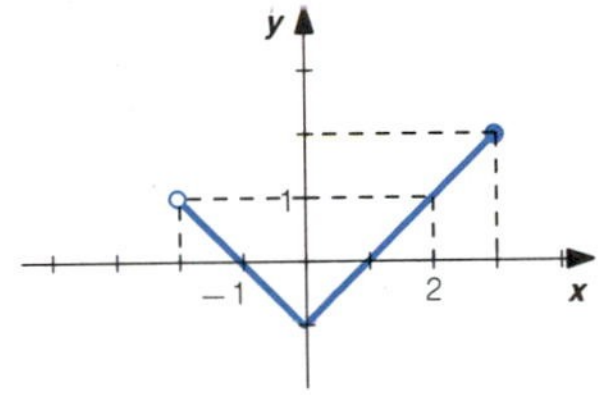


37 $f(-1)$, $f(0)$, $f(2)$

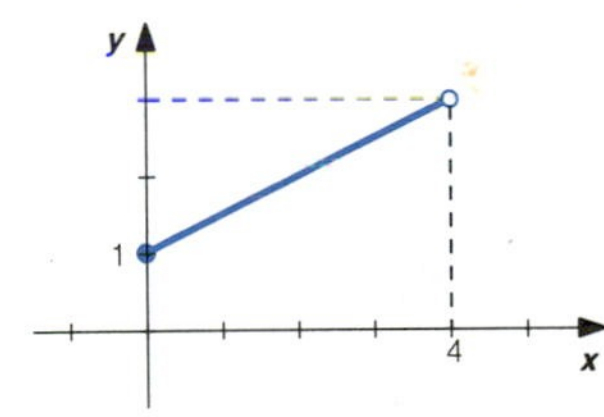


38 $f(1)$, $f(3/2)$, $f(3)$

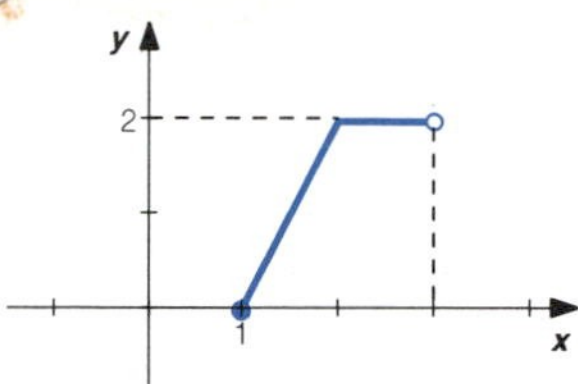


Sketch the graphs of the equations in Exercises 39–48. Determine which of them define $y$ as a function of $x$. Find a formula for the values and indicate the domain and range of each function.

39 $2x + 3y + 6 = 0$

40 $x^2 - 4x + y = 0$

41 $x - y^2 + 1 = 0$

42 $x - \sqrt{y + 1} = 0$

43 $\sqrt{1 - x} + y = 0$

44 $(x - 1)^2 + y^2 = 1$

45 $x^2y = 4$

46 $xy^2 = 4$

47 $x = \sin y$

48 $x = \cos y$

49 Does the equation $4x^2 + 4xy + y^2 = 0$ define $y$ as a function of $x$? Explain.

50 Does the equation $2x^2 + 3xy + y^2 = 0$ define $y$ as a function of $x$? Explain.

51 Find a function of $x$ with a graph that is the upper half of the circle with center the origin and radius 2.

52 Explain why there is no function of $x$ with a graph that is the right half of the circle with center the origin and radius 2.

53 If $f(x) = x^2$ and $g(x) = \sqrt{x}$, explain why $f \circ g \neq x$.

54 If $f(x) = \sqrt{x}$ and $g(x) = x^2$, explain why $f \circ g \neq x$.

55 Verify that $\dfrac{|x| + x}{2} = \begin{cases} x, & \text{if } x \geq 0, \\ 0, & \text{if } x < 0. \end{cases}$

56 Use the absolute value function in a way that is similar to the way it is used in Exercise 55 to write a single formula for the function that has value $f(x)$ if $f(x) \geq g(x)$, and value $g(x)$ if $g(x) > f(x)$. (This function is called the **maximum** of $f$ and $g$.)

## 2.2 USING INEQUALITIES TO FIND BOUNDS OF FUNCTIONS

We continue our study of functions by discussing the concept of bounds of functions. The idea will be introduced through graphs. We will then establish some analytic techniques for finding bounds of certain types of functions. The systematic development of bounds of functions at this time allows us to use the concept in our later work. In particular, we will use the ideas in later sections to facilitate understanding of the concept of the limit of a function. Also, the concept of bounds of functions is central to several types of error analyses that we will be using.

*Definition*

If $|f(x)| \le B$ for all $x$ in a set $S$, then $B$ is a **bound** of $f$ on $S$ and $f$ is said to be **bounded** on $S$. (If $S$ is the entire domain of $f$, we say simply $f$ is **bounded**.)

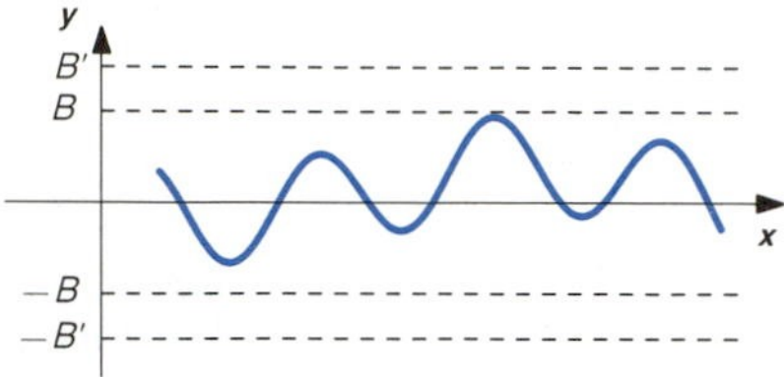

FIGURE 2.2.1

The expression $|f(x)| \le B$ means $-B \le f(x) \le B$. If $f$ is bounded by $B$ on a set $S$, then the graph of $y = f(x)$, $x$ in $S$, lies between (possible touching) the horizontal lines $y = B$ and $y = -B$. If $B$ is a bound of $f$ and $B' \ge B$, then $B'$ is also a bound of $f$, so $f$ does not have a unique bound. See Figure 2.2.1.

A function $f$ is not bounded on a set $S$ if, no matter how large $B$ is chosen, the graph of $y = f(x)$ is either above the line $y = B$ or below the line $y = -B$ for some $x$ in $S$.

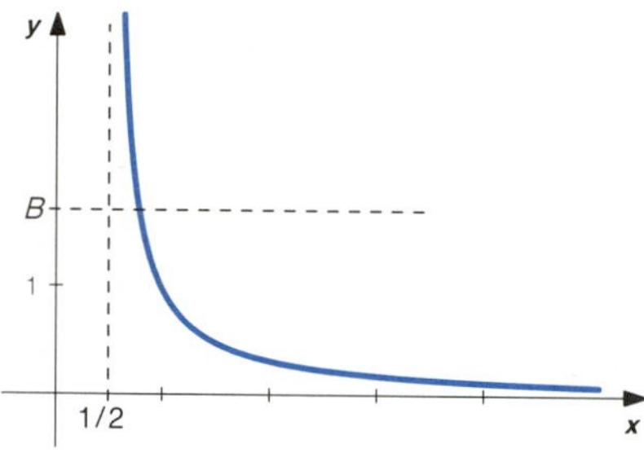

FIGURE 2.2.2

**EXAMPLE 1**

From the graph in Figure 2.2.2 we see that the function $f(x) = 1/(2x - 1)$, $x > 1/2$, is not bounded. Note that no matter how large $B$ is chosen, the graph of $y = f(x)$ is above the line $y = B$ for all values of $x > 1/2$ that are near enough to $1/2$.

We need to be able to determine bounds of functions from the formulas that define the functions. To do this we will use the following properties of absolute value to separate complicated expressions into simpler parts:

$$|f(x) + g(x)| \le |f(x)| + |g(x)|,$$

$$|f(x)g(x)| = |f(x)||g(x)|,$$

$$\left|\frac{f(x)}{g(x)}\right| = \frac{|f(x)|}{|g(x)|}.$$

The **Triangle Inequality** $|f + g| \le |f| + |g|$ allows us to find a bound of a sum by finding bounds of each term separately.

**EXAMPLE 2**

Find a bound of the function

$$f(x) = x^2 - 2x - 1, \qquad |x| \le 2.$$

**SOLUTION**

Using the Triangle Inequality, we have

$$\begin{aligned}|f(x)| = |x^2 - 2x - 1| &\le |x^2| + |-2x| + |-1| \\ &= |x|^2 + |-2||x| + |-1| \\ &= |x|^2 + 2|x| + 1.\end{aligned}$$

Since $|x| \le 2$, we then obtain

$$|f(x)| \le (2)^2 + 2(2) + 1 = 4 + 4 + 1 = 9,$$

so $B = 9$ is a bound of $f$ on the set $|x| \le 2$.

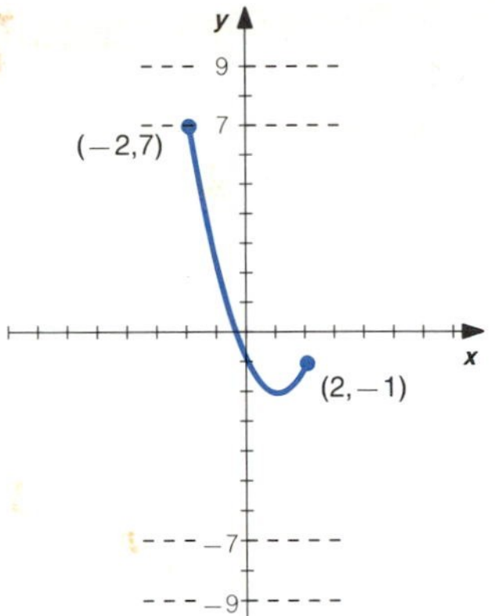


FIGURE 2.2.3

Note that use of the Triangle Inequality results in a *sum of nonnegative terms.* Each term is positive or zero and there are plus signs between each of the terms.

The bound $B = 9$ obtained in Example 2 is not the smallest bound of the function. From the graph in Figure 2.2.3 we see that $B = 7$ is the smallest choice for bounds of $f$. For many problems the exact value of the smallest bound is not important. The important point is that we obtained the bound $B = 9$ in a systematic (and easy) way.

We know that the values of the sine and cosine functions range between $-1$ and 1. Hence,

$$|\sin u| \leq 1 \quad \text{and} \quad |\cos u| \leq 1,$$

where $u$ represents any real number or expression.

■ **EXAMPLE 3**

Find a bound of the function

$$f(x) = 2 \sin x - \cos 3x, \qquad 0 \leq x \leq 2\pi.$$

**SOLUTION**

Using the Triangle Inequality, we have

$$|f(x)| = |2 \sin x - \cos 3x| \leq |2 \sin x| + |-\cos 3x|$$
$$= 2|\sin x| + |\cos 3x| \leq 2(1) + (1) = 3.$$ ■

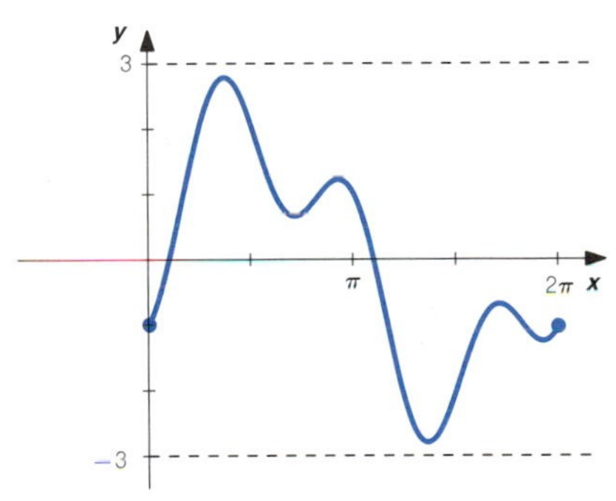


FIGURE 2.2.4

We see from the graph in Figure 2.2.4 that $B = 3$ is not the smallest bound of the function in Example 3. In this example it is somewhat difficult to determine the smallest bound, but it is very easy to establish that $B = 3$ is a bound.

We can use $|fg| = |f||g|$ to find a bound of a product by finding a bound for each factor separately.

■ **EXAMPLE 4**

Find a bound of the function

$$f(x) = x \sin x, \qquad 0 \leq x \leq 2\pi.$$

**SOLUTION**

We see that $f(x)$ is a product of the two factors $x$ and $\sin x$. We have $|x| \leq 2\pi$ for $0 \leq x \leq 2\pi$ and we know that $|\sin x| \leq 1$. It follows that

$$|f(x)| = |x \sin x| = |x||\sin x| \leq (2\pi)(1) = 2\pi.$$

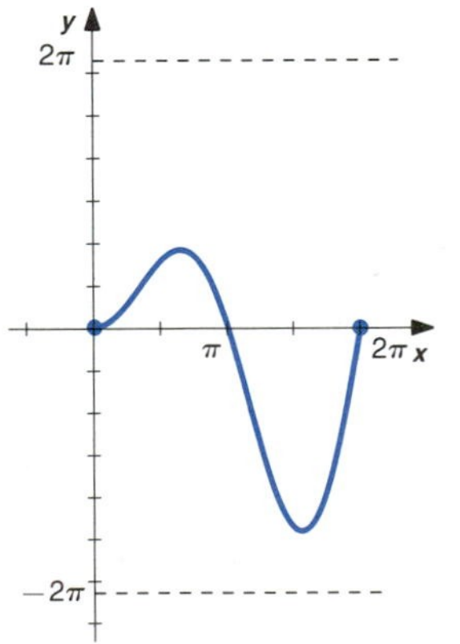


FIGURE 2.2.5

Therefore, $B = 2\pi$ is a bound of $f$ on the set $0 \leq x \leq 2\pi$. See Figure 2.2.5. ■

An expression of the form $1/u(x)$ is large if the values of $u(x)$ in the denominator are close to zero. To show that $1/u(x)$ is bounded, we must show that the values $u(x)$ stay a fixed distance away from zero. That is, we must find a positive number $A$ such that $0 < A \leq |u(x)|$. Then

$$0 < A \leq |u(x)| \quad \textbf{implies} \quad \frac{1}{|u(x)|} \leq \frac{1}{A}.$$

We cannot use the Triangle Inequality to obtain an inequality of the form $A \leq |u(x)|$, because the inequality goes in the wrong direction. *Do not try to use the Triangle Inequality for factors that are in the denominator*.

Generally, it is difficult to obtain bounds of functions that have variable factors in the denominator. We will illustrate how to handle two types of factors in the denominator. Namely:

**(i)** Factors that consist of a sum of nonnegative variable terms plus a positive constant term, and

**(ii)** Linear factors of the form $ax + b$, $a \neq 0$, given a linear inequality that restricts values of $x$.

Techniques for handling these types of factors in the denominator are illustrated in Examples 5 and 6.

**■ EXAMPLE 5**

Find a bound of the function

$$f(x) = \frac{1}{x^4 + 2x^2 + 4}.$$

**SOLUTION**

We need to show that the values of the denominator stay a fixed distance away from zero. We see that the denominator is a sum of nonnegative variable terms and a positive constant. In particular, the variable terms satisfy

$$0 \leq x^4 + 2x^2.$$

Adding the positive constant term to each side of the above inequality, we obtain

$$4 \leq x^4 + 2x^2 + 4.$$

This implies

$$|f(x)| = \frac{1}{x^4 + 2x^2 + 4} \leq \frac{1}{4}.$$

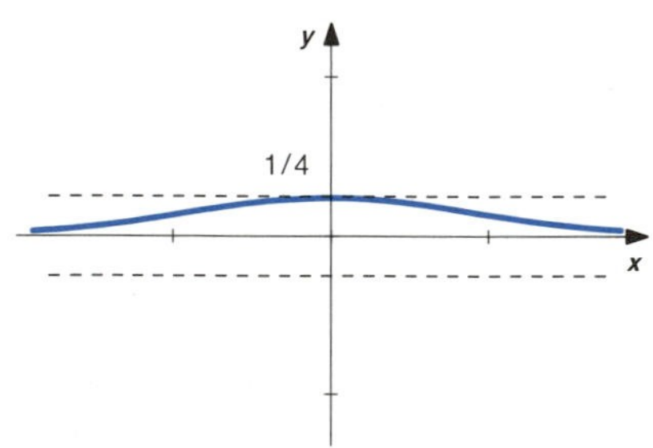


FIGURE 2.2.6

$B = 1/4$ is a bound. See Figure 2.2.6. ■

■ **EXAMPLE 6**

Find a bound of the function

$$f(x) = \frac{1}{2x - 1}, \qquad x > 2.$$

**SOLUTION**

Since the factor $2x - 1$ is in the denominator, we must show that the values of $2x - 1$ stay a fixed distance away from zero for $x > 2$. This means that we need to find a positive number $A$ such that $A \le |2x - 1|$ for $x > 2$. We can do this by "solving" the given inequality $x > 2$ for the expression $2x - 1$. That is, we start with the known information $x > 2$, and use this to determine information about the factor of interest, $2x - 1$. We have

$$x > 2,$$

$$2x > 2(2),$$

$$2x - 1 > 4 - 1.$$

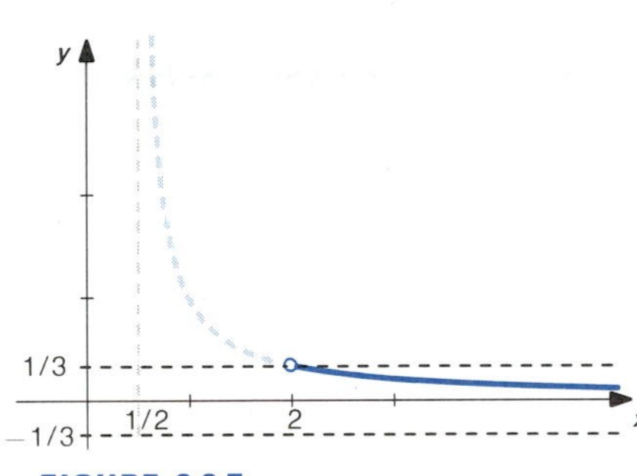

FIGURE 2.2.7

It follows that $|2x - 1| > 3$, so

$$|f(x)| = \frac{1}{|2x - 1|} < \frac{1}{3}.$$

Thc function is bounded by $B = 1/3$ on the set $x > 2$. See Figure 2.2.7. ■

The result of Example 6 should be compared with that of Example 1. In particular, $f(x) = 1/(2x - 1)$ is not bounded on the set $x > 1/2$, but it is bounded on the set $x > 2$. The condition $x > 2$ keeps $x$ a fixed distance away from the point $x = 1/2$, where the denominator becomes zero.

In cases where a factor $u(x)$ of a function satisfies a double inequality such as $A \le |u(x)| \le B$, it is important to know which inequality to use to find a bound of the function.

- **If the factor $u(x)$ is in the numerator, use the inequality**

$$|u(x)| \le B.$$

- **If the factor $u(x)$ is in the denominator, use the inequality $0 < A \le |u(x)|$ to obtain**

$$\frac{1}{|u(x)|} \le \frac{1}{A}.$$

■ **EXAMPLE 7**

Find a bound of

$$f(x) = \frac{x^2 + 4x + 16}{x^2}, \qquad |x - 4| < 2.$$

**SOLUTION**

We write

$$|f(x)| = |x^2 + 4x + 16| \frac{1}{|x|^2}$$

and determine a bound for each of the two factors.

The expression $|x - 4| < 2$ implies $-2 < x - 4 < 2$. Adding 4 to the double inequality, we obtain $2 < x < 6$. This implies

$$2 < |x| < 6.$$

Using $|x| < 6$ and the Triangle Inequality, we can obtain a bound of the first factor:

$$|x^2 + 4x + 16| \leq |x|^2 + 4|x| + 16 < (6)^2 + 4(6) + 16 = 76.$$

To find a bound of the second factor we use $2 < |x|$. This implies

$$4 < |x|^2, \quad \text{so} \quad \frac{1}{|x|^2} < \frac{1}{4}.$$

Combining results we have

$$|f(x)| < (76)\left(\frac{1}{4}\right) = 19.$$

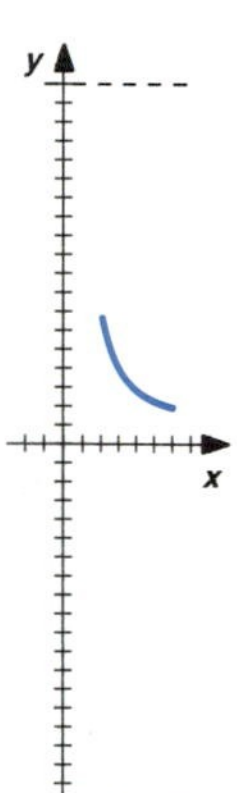

FIGURE 2.2.8

$B = 19$ is a bound of $f$. See Figure 2.2.8. ■

■ **EXAMPLE 8**

Find a bound of

$$f(x) = \frac{(x + 1)^2}{x - 1}, \qquad |x| < 0.2.$$

**SOLUTION**

We need bounds of $(x + 1)^2$ and $1/(x - 1)$. We first rewrite $|x| < 0.2$ as the double inequality

$$-0.2 < x < 0.2.$$

Solving the above double inequality for $x + 1$, we have $0.8 < x + 1 < 1.2$, so $0.8 < |x + 1| < 1.2$. In particular, $|x + 1| < 1.2$ implies $|x + 1|^2 < (1.2)^2 = 1.44$.

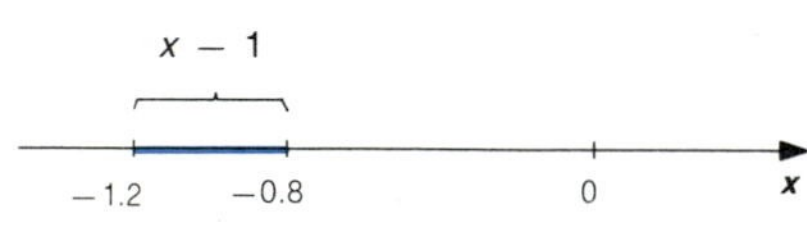

FIGURE 2.2.9

Solving the double inequality $-0.2 < x < 0.2$ for $x - 1$, we have $-1.2 < x - 1 < -0.8$. This implies $0.8 < |x - 1| < 1.2$. (To see the latter implication, make a sketch as in Figure 2.2.9; $|x - 1|$ represents the

FIGURE 2.2.10

distance of $x - 1$ from the origin.) Then

$$0.8 < |x - 1| \quad \text{implies} \quad \frac{1}{|x - 1|} < \frac{1}{0.8}.$$

Combining results we have

$$|f(x)| = |x + 1|^2 \frac{1}{|x - 1|} < (1.44)\left(\frac{1}{0.8}\right) = 1.8.$$

The function is bounded by $B = 1.8$. See Figure 2.2.10. ■

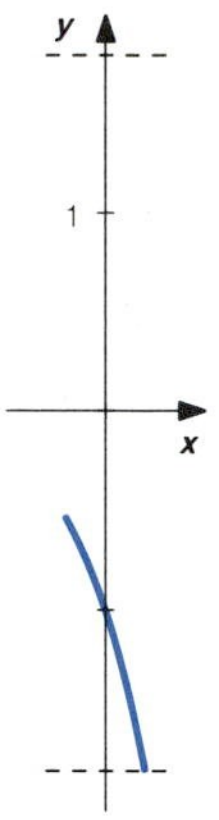

■ **EXAMPLE 9**

$|f(x) - 7| < 2$. Find a bound of (a) $f$ and (b) $1/f$.

**SOLUTION**

$|f(x) - 7| < 2$ implies the double inequality

$$-2 < f(x) - 7 < 2.$$

Adding 7 to the double inequality we obtain

$$5 < f(x) < 9, \quad \text{so} \quad 5 < |f(x)| < 9.$$

(a) $|f(x)| < 9$ implies 9 is a bound of $f$.

(b) $5 < |f(x)|$ implies $\dfrac{1}{|f(x)|} < \dfrac{1}{5}$,

so $1/5$ is a bound of $1/f$. ■

## EXERCISES 2.2

In Exercises 1–40 find bounds of the given functions or indicate they are not bounded.

1 $f(x) = 3x^2 - 4x - 5$, $|x| \le 2$

2 $f(x) = 2x^2 - 3x - 7$, $|x| \le 3$

3 $f(x) = x^3 + 2x - 3$, $|x| \le 2$

4 $f(x) = x^4 - 5x^2 - 8$, $|x| \le 2$

5 $f(x) = 1 - 2x$, all $x$

6 $f(x) = x^2 - 3x - 5$, all $x$

7 $f(x) = 2 \sin x - 3 \cos x$, $0 \le x \le 2\pi$

8 $f(x) = 3 \sin x + 2 \cos x$, $0 \le x \le 2\pi$

9 $f(x) = \sin 2x - \cos 3x$, $0 \le x \le 2\pi$

10 $f(x) = \sin(x/2) + \cos(x/3)$, $0 \le x \le 2\pi$

11 $f(x) = \dfrac{1}{x + 3}$, $x > 2$

12 $f(x) = \dfrac{1}{x + 2}$, $x > 3$

13 $f(x) = \dfrac{1}{x - 2}$, $x > 2$

14 $f(x) = \dfrac{1}{3 - x}$, $x > 3$

15 $f(x) = \dfrac{1}{3x - 2}$, $x > 4$

16 $f(x) = \dfrac{1}{2x - 1}$, $x > 3$

**17** $f(x) = \dfrac{1}{1 - 2x}$, $|x| < 0.3$

**18** $f(x) = \dfrac{1}{1 - 3x}$, $|x| < 0.3$

**19** $f(x) = \dfrac{1}{x}$, $|x - 4| < 1$

**20** $f(x) = \dfrac{1}{x}$, $|x - 5| < 1$

**21** $f(x) = \dfrac{1}{x^2 + 4}$, all $x$

**22** $f(x) = \dfrac{1}{x^2 + 5}$, all $x$

**23** $f(x) = \dfrac{1}{x^4 + 2x^2 + 5}$, all $x$

**24** $f(x) = \dfrac{1}{x^4 + 5x^2 + 2}$, all $x$

**25** $f(x) = x \cos x$, $|x| < 3$

**26** $f(x) = x^2 \sin x$, $|x| < 3$

**27** $f(x) = x^2 \sin^2 x$, $|x| \leq \pi$

**28** $f(x) = x \cos(x^2)$, $|x| \leq \pi$

**29** $f(x) = \dfrac{x}{2x - 6}$, $|x| < 2$

**30** $f(x) = \dfrac{4x}{3x - 11}$, $|x| < 3$

**31** $f(x) = \dfrac{1 + x}{1 - x}$, $|x| < 0.2$

**32** $f(x) = \dfrac{1 - 2x}{1 + 2x}$, $|x| < 0.3$

**33** $f(x) = \dfrac{x + 4}{x^2}$, $|x - 4| < 1$

**34** $f(x) = \dfrac{x + 5}{x^2}$, $|x - 5| < 1$

**35** $f(x) = \dfrac{x^3}{x^2 + 4}$, $|x| < 2$

**36** $f(x) = \dfrac{x^2}{x^2 + 5}$, $|x| < 3$

**37** $f(x) = \dfrac{x^3}{(1 - x)^2}$, $|x| < 0.2$

**38** $f(x) = \dfrac{x^2}{(1 + x)^2}$, $|x| < 0.3$

**39** $f(x) = \dfrac{\sin x + \cos x}{\sin^2 x + 1}$, all $x$

**40** $f(x) = \dfrac{\sin x + \sin 2x}{\tan^2 x + 1}$, $-\pi/2 < x < \pi/2$

**41** $|f(x) - 3| < 1$. Find a bound of $f$.

**42** $|f(x) - 2| < 5$. Find a bound of $f$.

**43** $|g(x) - 2| < 1$. Find a bound of $1/g$.

**44** $|g(x) - 5| < 3$. Find a bound of $1/g$.

**45** $|f(x) - 3| < 0.005$ and $|g(x) - 2| < 0.005$. Find a bound of $|f(x) + g(x) - 5|$.

**46** $|g(x) - M| < M/2$, where $M$ is a fixed positive number. Find a bound of $1/g$ in terms of $M$.

**47** Show that

$$\frac{a^2}{a^2 + b^2} \leq 1.$$

**48** Show that

$$\frac{2|ab|}{a^2 + b^2} \leq 1.$$

[*Hint*: $0 \leq (|a| - |b|)^2$.]

Use the results of Exercises 47 and 48 to verify that the given numbers $B$ are bounds of the functions in Exercises 49–54.

**49** $f(x) = \dfrac{2x^2}{x^2 + 4}$, $B = 2$

**50** $f(x) = \dfrac{18}{x^2 + 9}$, $B = 2$

**51** $f(x) = \dfrac{8x}{x^2 + 4}$, $B = 2$

**52** $f(x) = \dfrac{3x}{x^2 + 9}$, $B = \dfrac{1}{2}$

**53** $f(x) = \dfrac{x^3}{x^2 + 9}$, $|x| \leq 3$, $B = \dfrac{3}{2}$

**54** $f(x) = \dfrac{4x^3}{x^2 + 4}$, $|x| \leq 2$, $B = 4$

## 2.3 THE LIMIT OF A FUNCTION AT A POINT

Limits are the foundation on which calculus is based. To understand calculus, we must understand the ideas behind the concept of limits. We begin our study of the concept of the limit of a function at a point by looking at some specific problems that involve accuracy of measurement. These problems are intended to illustrate the ideas of closeness that are involved in the concept of limits and to develop the technical skill required to work with the definition of limit. Next, the ideas associated with the limit of a function at a point are illustrated graphically. Finally, we will state a formal definition and see how it relates to the ideas.

■ **EXAMPLE 1**

A 120-yd football field is to be marked by using a measuring tape that is approximately 10 yd long. How close to 10 yd must the measuring tape be to insure that the actual length of the field be within 0.3 yd of 120 yd?

**SOLUTION**

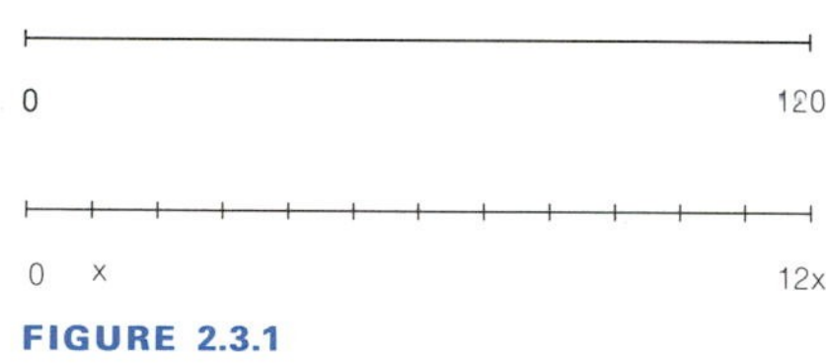


FIGURE 2.3.1

Let $x$ denote the actual length in yards of the measuring tape. Since the length of the tape is essentially 10 yd, 12 lengths of the tape would be used to mark a field 120 yd long. That is, the actual length of the field is $12x$ yd. See Figure 2.3.1. The absolute value of the difference between the actual length and 120 is $|12x - 120|$. We want the actual length to be within 0.3 of 120, or $|12x - 120| < 0.3$. Since

$$|12x - 120| = 12|x - 10|,$$

we want to satisfy the condition $12|x - 10| < 0.3$. Solving the inequality $12|x - 10| < 0.3$ for $|x - 10|$, we have $|x - 10| < 0.3/12 = 0.025$. We conclude that

$$|12x - 120| < 0.3 \qquad \text{whenever } |x - 10| < 0.025.$$

That is, the actual length is within 0.3 yd of 120 yd whenever the length of the measuring tape is within 0.025 yd of 10 yd. (0.025 yd = 0.9 in.) ■

■ **EXAMPLE 2**

How close to 3 must the radius of a circle be to insure that the area of the circle be within 0.01 of $9\pi$? Assume the radius is within 0.5 of 3.

**SOLUTION**

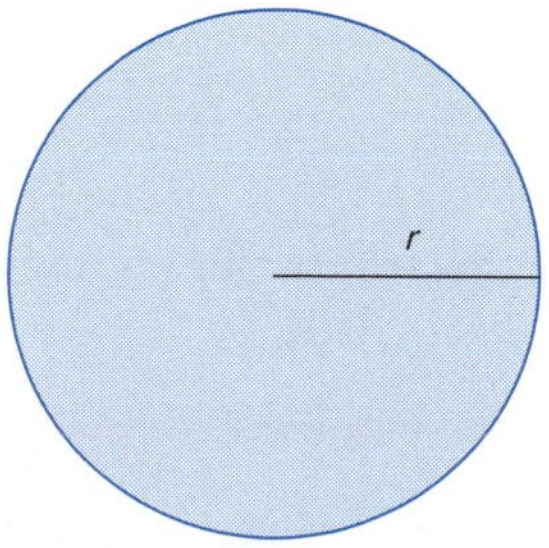

Let $r$ denote the radius of the circle. The area is then given by $\pi r^2$. See Figure 2.3.2. We want $|\pi r^2 - 9\pi| < 0.01$, but it is not necessary to solve this inequality for exact values of $r$. We will illustrate a procedure that reduces the problem to a much simpler inequality.

The first step is to factor $\pi r^2 - 9\pi$. This allows us to express $|\pi r^2 - 9\pi|$ in terms of $|r - 3|$. We have

FIGURE 2.3.2

$$|\pi r^2 - 9\pi| = \pi|r + 3|\,|r - 3|. \tag{1}$$

The next step is to find a bound for the factors other than $|r - 3|$. That is, we need a bound of the factor $r + 3$. Note that $r + 3$ is not bounded unless $r$ is restricted. However, the preliminary assumption that "the radius is within 0.5 of 3" gives us $|r - 3| \leq 0.5$. Thus, we need a bound of the function

$$r + 3, \qquad |r - 3| \leq 0.5.$$

Following the technique of Section 2.2, we write $|r - 3| \leq 0.5$ as the double inequality

$$-0.5 \leq r - 3 \leq 0.5.$$

We can easily solve this double inequality for $r + 3$ by adding 6 to each term. We obtain

$$5.5 \leq r + 3 \leq 6.5, \quad \text{so} \quad 5.5 \leq |r + 3| \leq 6.5.$$

Using the bound of 6.5 for $|r + 3|$ in (1), we obtain

$$|\pi r^2 - 9\pi| \leq \pi(6.5)|r - 3|. \tag{2}$$

It is clear from (2) that $|\pi r^2 - 9\pi| < 0.01$ whenever $\pi(6.5)|r - 3| < 0.01$. It is easy to solve the latter inequality for $|r - 3|$ to obtain $|r - 3| < 0.01/(6.5\pi)$. [Note that $|r - 3| < 0.01/(6.5\pi) \approx 0.00048971$ implies that $|r - 3| < 0.5$, so our preliminary assumption is valid.] We then have

$$|\pi r^2 - 9\pi| < 0.01 \qquad \text{whenever } |r - 3| < 0.01/(6.5\pi).$$

The area of the circle is within 0.01 of $9\pi$ whenever the radius is within $0.01/(6.5\pi)$ of 3. ■

In Example 2 we discovered that $|r - 3| < 0.01/(6.5\pi)$ insures that $|\pi r^2 - 9\pi| < 0.01$. That is, $0.01/(6.5\pi)$ is an "answer" to our problem. The number $0.01/(6.5\pi)$ was determined in a systematic manner by using the preliminary assumption that $|r - 3| < 0.5$. If we used the same procedure with a different preliminary assumption we would get a number different from $0.01/(6.5\pi)$. Thus, it is possible (and true) that $|\pi r^2 - 9\pi|$ is less than 0.01 for values of $r$ that are not within $0.01/(6.5\pi)$ of 3. It would also be a correct solution of our problem to require $r$ to be even closer to 3. For example, we could *round down* the calculator value of $0.01/(6.5\pi) \approx 0.00048971$ and say that

$$|\pi r^2 - 9\pi| < 0.01 \qquad \text{whenever } |r - 3| < 0.00048.$$

That is, if $|r - 3| < 0.00048$, then $|r - 3|$ is also less than the larger number $0.01/(6.5\pi)$, so we know that we must have $|\pi r^2 - 9\pi| < 0.01$.

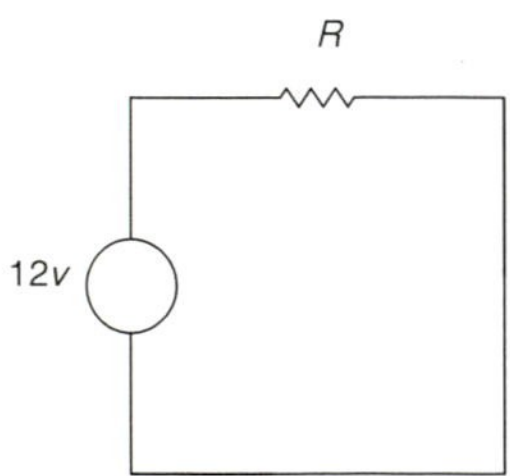


FIGURE 2.3.3

■ **EXAMPLE 3**

How close to 2 Ω (ohms) should the resistance in the 12-V (volt) electrical circuit indicated in Figure 2.3.3 be in order that the current be within 0.05

A (amperes) of 6 A? Assume the resistance is within 0.5 Ω of 2 Ω. $I = V/R$, where $I$ is the current in amperes, $V$ is the voltage in volts, and $R$ is the resistance in ohms.

**SOLUTION**

We are told that $V = 12$, so the current in the circuit is $I = 12/R$. We want to determine $|R - 2|$ so that $|12/R - 6| < 0.05$. Carrying out the arithmetic, we have

$$\left|\frac{12}{R} - 6\right| = \left|\frac{12 - 6R}{R}\right| = \frac{6}{|R|}|2 - R| = \frac{6}{|R|}|R - 2|. \quad (3)$$

We need a bound of the factor $1/R$. We have assumed that the resistance is within 0.5 Ω of 2 Ω. That is,

$$|R - 2| \leq 0.5, \quad \text{so} \quad -0.5 \leq R - 2 \leq 0.5.$$

The double inequality implies

$$1.5 \leq R \leq 2.5, \quad \text{so} \quad 1.5 \leq |R| \leq 2.5.$$

Since the factor $R$ is in the denominator, we use the inequality $1.5 \leq |R|$ to obtain

$$\frac{1}{|R|} \leq \frac{1}{1.5}.$$

Using this bound in (3), we obtain

$$\left|\frac{12}{R} - 6\right| \leq \frac{6}{1.5}|R - 2|.$$

It follows that

$$\left|\frac{12}{R} - 6\right| < 0.05 \qquad \text{whenever } \frac{6}{1.5}|R - 2| < 0.05, \quad \text{or}$$

$$|R - 2| < \frac{(1.5)(0.05)}{6} = 0.0125.$$

(Note that $|R - 2| < 0.0125$ implies that $R$ is within 0.5 of 2, so our preliminary assumption holds.) The current is within 0.05 A of 6 A whenever the resistance is within 0.0125 Ω of 2 Ω. ■

The previous examples illustrate the same general problem.

**GENERAL PROBLEM** Given a positive number $\varepsilon$, find a positive number $\delta$ such that $|f(x) - L| < \varepsilon$ whenever $|x - c| < \delta$.

It is traditional to use the Greek letters $\varepsilon$ (epsilon) and $\delta$ (delta) in the statement of this problem. Think of $\varepsilon$ as a small positive number. The use of $\varepsilon$ in place of a decimal number should not cause any particular difficulty.

There is not a single solution method that works for all functions. However, in many important cases we can use the following.

### *Solution method (one type)*

- **Use algebra and factoring to express $|f(x) - L|$ in terms of $|x - c|$.**
- **Use a preliminary assumption, if necessary, to find bounds of the factors of $|f(x) - L|$ other than $|x - c|$.**
- **Use the results of the above steps to obtain an inequality of the form $|f(x) - L| \leq B|x - c|$. It follows that**

$$|f(x) - L| < \varepsilon \qquad \text{whenever } |x - c| < \varepsilon/B,$$

**so $\delta = \varepsilon/B$ is a solution of the problem.**

■ **EXAMPLE 4**

Find a positive number $\delta$ such that $|x^3 - 1| < 0.5$ whenever $|x - 1| < \delta$. Assume $|x - 1| \leq 1$.

**SOLUTION**

We begin by factoring to express $|x^3 - 1|$ in terms of $|x - 1|$. We obtain

$$|x^3 - 1| = |x^2 + x + 1|\,|x - 1|. \tag{4}$$

We then see that we need a bound of the factor $x^2 + x + 1$. The preliminary assumption $|x - 1| \leq 1$ implies $-1 \leq x - 1 \leq 1$. Solving for $x$ we obtain $0 \leq x \leq 2$, so $0 \leq |x| \leq 2$. Using $|x| \leq 2$ and the Triangle Inequality, we then have

$$|x^2 + x + 1| \leq |x|^2 + |x| + 1 \leq (2)^2 + (2) + 1 = 7.$$

Using this bound in (4), we obtain

$$|x^3 - 1| \leq 7|x - 1|.$$

It follows that

$$|x^3 - 1| < 0.5 \qquad \text{whenever } 7|x - 1| < 0.5.$$

Solving the latter inequality for $|x - 1|$, we obtain

$$|x^3 - 1| < 0.5 \qquad \text{whenever } |x - 1| < 0.5/7,$$

so $\delta = 0.5/7$ is a solution to the problem. See Figure 2.3.4. ■

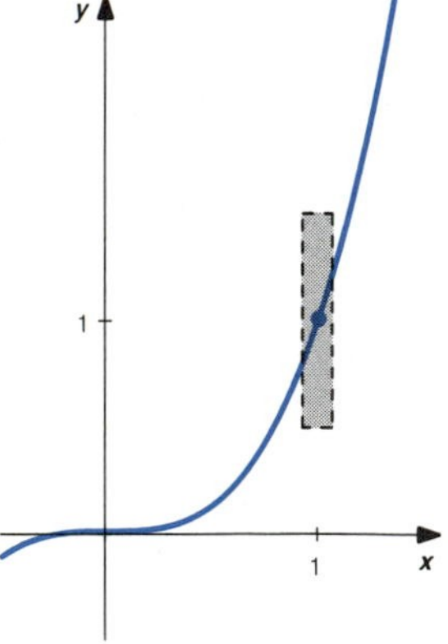

FIGURE 2.3.4

The positive number that corresponds to a given number $\varepsilon$ usually depends on $\varepsilon$, but is not necessarily a constant multiple of $\varepsilon$.

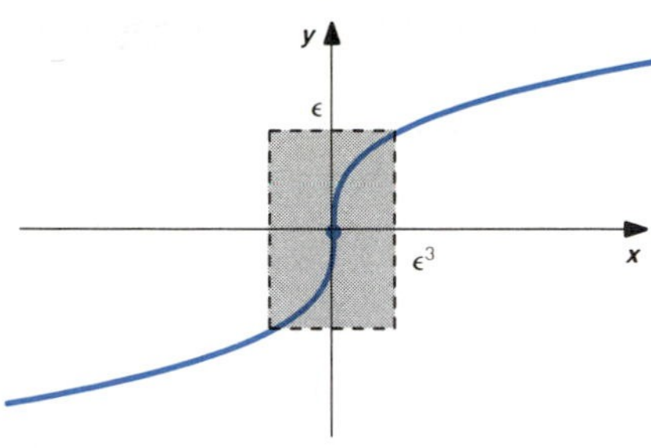


FIGURE 2.3.5

■ **EXAMPLE 5**

For each positive number $\varepsilon$, find a positive number $\delta$ such that $|x^{1/3}| < \varepsilon$ whenever $|x| < \delta$.

**SOLUTION**

In this case we can solve the inequality $|x^{1/3}| < \varepsilon$ for $|x|$ to see that

$$|x^{1/3}| < \varepsilon \qquad \text{whenever } |x| < \varepsilon^3.$$

It follows that $\delta = \varepsilon^3$ is a solution to our problem. See Figure 2.3.5. ■

Solution of the general problem we have been discussing depends on the fact that the values $f(x)$ must be near $L$ for $x$ near $c$. This is related to the concept of a limit of the function at the point $c$. Informally, we say that **the limit of $f(x)$ as $x$ approaches $c$ is $L$** [written $\lim_{x\to c} f(x) = L$] if points $(x, f(x))$ on the graph of $f$ approach the point $(c, L)$ as $x$ approaches $c$ from either side of $c$. We will sometimes write

$$f(x) \to L \quad \text{as} \quad x \to c$$

to indicate that $f(x)$ approaches $L$ as $x$ approaches $c$.

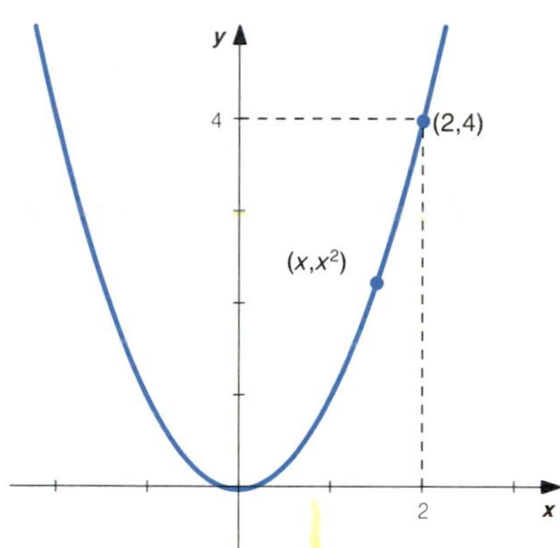


FIGURE 2.3.6

■ **EXAMPLE 6**

From the graph of $f(x) = x^2$ in Figure 2.3.6 we see that the point $(x, x^2)$ approaches the point $(2, 4)$ as $x$ approaches 2, so $\lim_{x\to 2} x^2 = 4$. ■

**The limit of $f(x)$ as $x$ approaches $c$ depends on the values of $f(x)$ as $x$ becomes close to $c$, but we exclude $x = c$.**

It is not necessary that $f$ be defined at $c$ to have a limit at $c$. (We will see in Section 2.6 why it is important that we exclude $x = c$.)

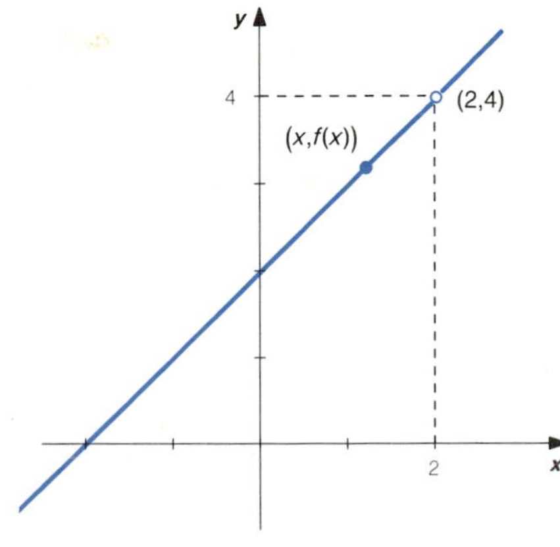


FIGURE 2.3.7

■ **EXAMPLE 7**

From the graph of

$$f(x) = \frac{x^2 - 4}{x - 2}$$

in Figure 2.3.7 we see that

$$\lim_{x\to 2} \frac{x^2 - 4}{x - 2} = 4.$$ ■

Note that

$$f(x) = \frac{x^2 - 4}{x - 2} = \frac{(x-2)(x+2)}{x - 2} = x + 2, \quad x \neq 2.$$

Hence, the graph of $f$ is the same as the graph of the line $y = x + 2$, except $f(x)$ is undefined at $x = 2$.

In Example 7, there may be a reason for writing

$$f(x) = \frac{x^2 - 4}{x - 2}$$

instead of the simpler expression $x + 2$. For example, $f$ may represent the slope of the line segment between the point $(2, 4)$ and a variable point $(x, x^2)$ on the graph of $y = x^2$. If $x = 2$, the points are identical and it does not make sense to speak of the line segment between them. Thus, the fact that $f(x)$ is not defined for $x = 2$ reflects the more important fact that the physical interpretation of $f$ does not make sense for $x = 2$.

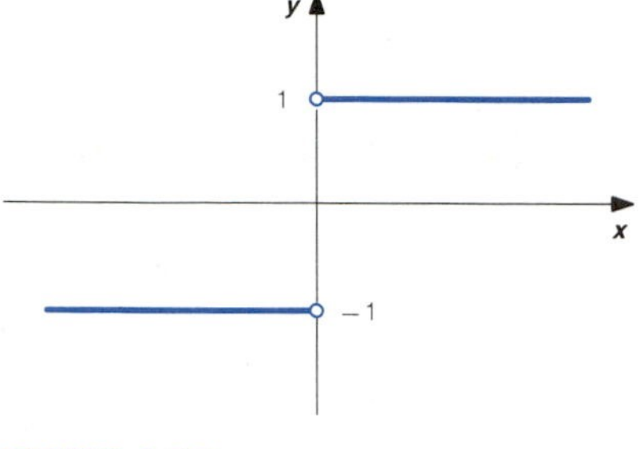

FIGURE 2.3.8

**If $f(x)$ does not approach a *single* number as $x$ approaches $c$, then $\lim_{x \to c} f(x)$ does not exist.**

■ **EXAMPLE 8**

From Figure 2.3.8 we see that

$$\lim_{x \to 0} \frac{|x|}{x} \quad \text{does not exist.}$$

Points on the graph approach $(0, -1)$ as $x$ approaches 0 from the left and they approach $(0, 1)$ as $x$ approaches 0 from the right. Thus, the values $f(x)$ approach both 1 and $-1$ as $x$ approaches 0. ■

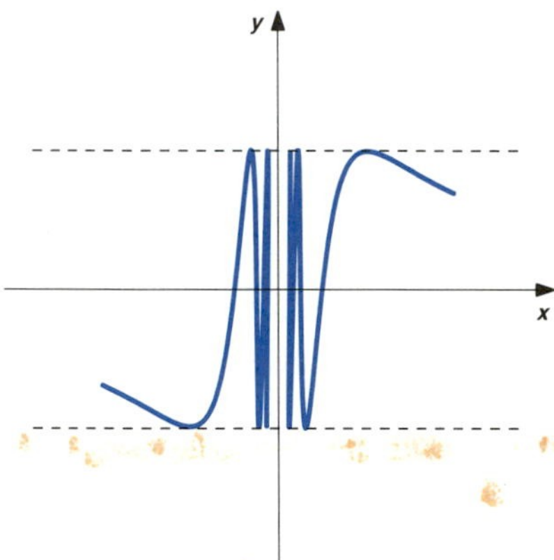

FIGURE 2.3.9

■ **EXAMPLE 9**

From the graph in Figure 2.3.9 we see that

$$\lim_{x \to 0} \sin \frac{1}{x} \quad \text{does not exist.}$$

$\sin(1/x)$ repeatedly assumes all values between 1 and $-1$ as $x$ approaches 0. (A graph of this type was discussed in Section 1.6.) ■

**If $f(x)$ becomes unbounded as $x$ approaches $c$, then $\lim_{x \to c} f(x)$ does not exist.**

If $f(x)$ becomes unbounded as $x$ approaches $c$, then $f(x)$ cannot approach any real number.

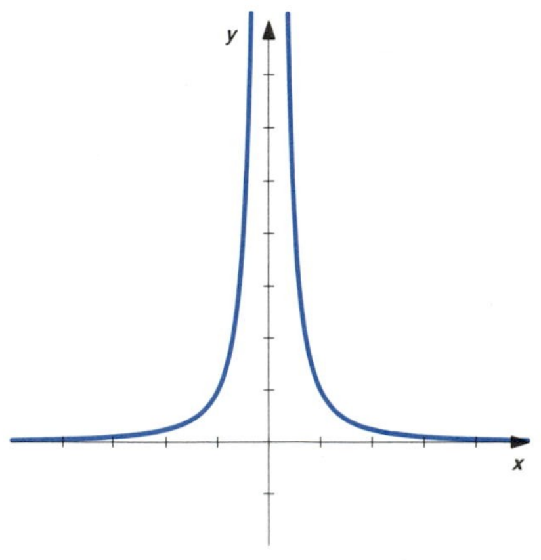

FIGURE 2.3.10

■ **EXAMPLE 10**

From the graph in Figure 2.3.10 we see that

$$\lim_{x \to 0} \frac{1}{x^2} \quad \text{does not exist.}$$

The function $1/x^2$ becomes unbounded as $x$ approaches 0. ■

Let us now state a formal and precise definition of the limit of a function at a point and see how it reflects the ideas we have discussed.

*Definition*

**The limit of $f(x)$ as $x$ approaches $c$ is $L$** [written $\lim_{x \to c} f(x) = L$] if, for each positive number $\varepsilon$, there is a positive number $\delta$ such that

$$|f(x) - L| < \varepsilon \qquad \text{whenever } 0 < |x - c| < \delta.$$

If the condition of the definition is not satisfied for any number $L$, we say $\lim_{x \to c} f(x)$ does not exist. ■

Let us relate the formal definition of limit to the graph of $f$ for $x$ close to $c$. For a given positive number $\varepsilon$, the inequality $|f(x) - L| < \varepsilon$ means that $-\varepsilon < f(x) - L < \varepsilon$, or $L - \varepsilon < f(x) < L + \varepsilon$. Geometrically, this means that the point $(x, f(x))$ on the graph of $f$ is between the horizontal lines $y = L - \varepsilon$ and $y = L + \varepsilon$. The condition of the definition asserts that, for any given positive number $\varepsilon$, there is a positive number $\delta$ such that $|f(x) - L| < \varepsilon$ whenever $0 < |x - c| < \delta$. This means the point $(x, f(x))$ is between the horizontal lines for all $x$ that satisfy $c - \delta < x < c + \delta$, except possibly for $x = c$, which is excluded. As $x$ approaches $c$ from either side, $x$ will eventually be within $\delta$ of $x$, so $(x, f(x))$ eventually stays between the horizontal lines. This means that $(x, f(x))$ can approach only a point or points $(c, y)$ with $L - \varepsilon \leq y \leq L + \varepsilon$ as $x$ approaches $c$. See Figure 2.3.11a. Finally, we use the fact that the condition of the definition holds for each positive number $\varepsilon$ to exclude the possibility that $(x, f(x))$ approaches a point $(c, y)$ with $y \neq L$. That is, if $y \neq L$, we can choose the positive number $\varepsilon$ so small that the point $(c, y)$ is not between the horizontal lines. We conclude that $(x, f(x))$ must approach the single point $(c, L)$ as $x$ approaches $c$, as indicated in Figure 2.3.11b. Thus, $f(x)$ approaches the single number $L$ as $x$ approaches $c$.

It should be noted that the problem of verifying that $\lim_{x \to c} f(x) = L$ is the same general problem that we encountered in Examples 1–5. Namely, given a positive number $\varepsilon$, we must find a positive number $\delta$ such that

$$|f(x) - L| < \varepsilon \qquad \text{whenever } 0 < |x - c| < \delta.$$

Let us see how the formal definition of limit can be used to verify some simple limits.

FIGURE 2.3.11

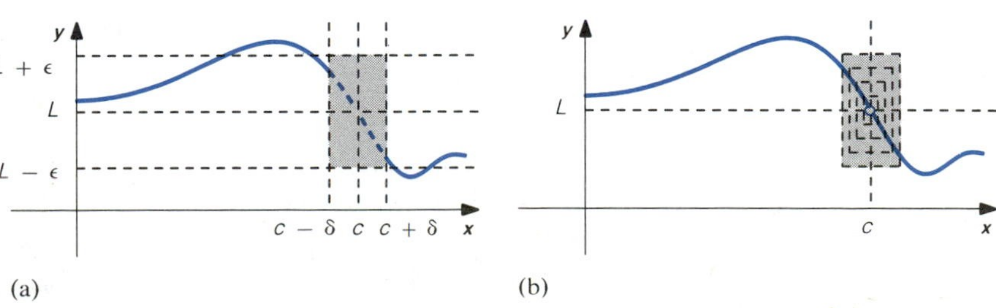

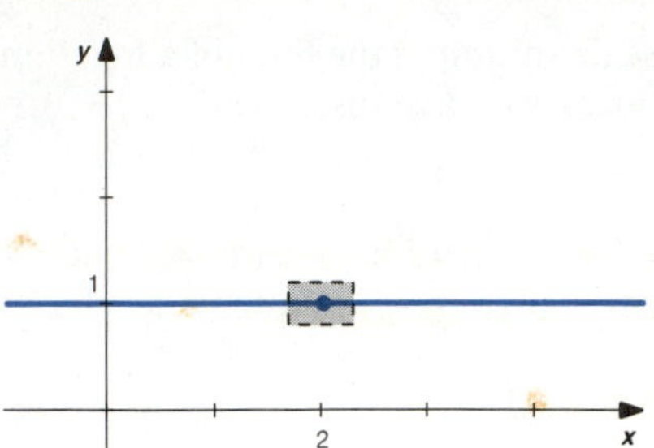


FIGURE 2.3.12

■ **EXAMPLE 11**

Use the definition to verify that $\lim_{x \to 2} 1 = 1$.

**SOLUTION**

If $f(x) = 1$, then for any positive number $\varepsilon$, we have

$$|f(x) - 1| = |1 - 1| = 0 < \varepsilon \qquad \text{whenever } |x - 2| < \delta$$

for any positive number $\delta$. The condition of the definition is satisfied for any choice of $\delta > 0$, so $\lim_{x \to 2} 1 = 1$. (Although any choice of $\delta$ can be used in this example, it is usually true that smaller $\varepsilon$ require smaller $\delta$, so the choice of $\delta$ depends on $\varepsilon$.) See Figure 2.3.12. ■

For any constant function $f(x) = k$ and any point $c$, we can verify as in Example 11 that

$$\lim_{x \to c} k = k. \tag{5}$$

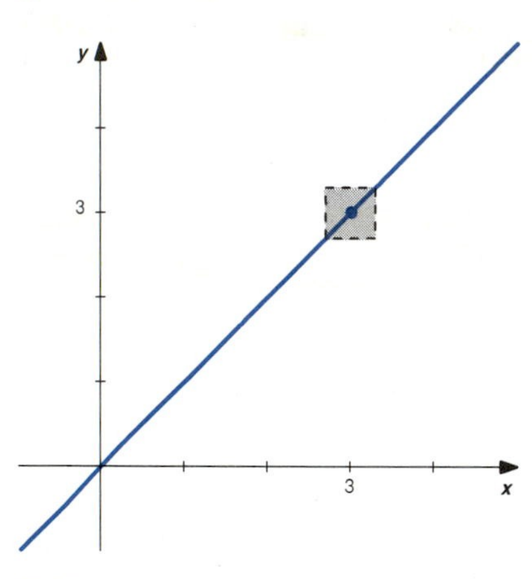


FIGURE 2.3.13

■ **EXAMPLE 12**

Use the definition to verify $\lim_{x \to 3} x = 3$.

**SOLUTION**

If $f(x) = x$, for any positive number $\varepsilon$, we have

$$|f(x) - 3| = |x - 3| < \varepsilon \qquad \text{whenever } |x - 3| < \varepsilon,$$

so the condition of the definition is satisfied with $\delta = \varepsilon$. This shows $\lim_{x \to 3} x = 3$. See Figure 2.3.13. ■

Similarly, for any value of $c$, we have

$$\lim_{x \to c} x = c. \tag{6}$$

■ **EXAMPLE 13**

Use the definition to verify that $\lim_{x \to 4} \sqrt{x} = 2$.

**SOLUTION**

We must show that, for each positive number $\varepsilon$, there is a positive number $\delta$ such that

$$|\sqrt{x} - 2| < \varepsilon \qquad \text{whenever } 0 < |x - 4| < \delta.$$

We make the preliminary assumption that $x \geq 0$.

To verify the condition, we first use the factorization formula

$$x - 4 = (\sqrt{x} - 2)(\sqrt{x} + 2)$$

to express $\sqrt{x} - 2$ in terms of $x - 4$. We obtain

$$|\sqrt{x} - 2| = \frac{|x - 4|}{\sqrt{x} + 2}.$$

(We may think of the above step as rationalization of the numerator.) The next step is to find a bound of the factor $1/(\sqrt{x} + 2)$. The denominator is a sum of a nonnegative variable term $\sqrt{x}$ and a positive constant term 2. Thus, we have

$$0 \leq \sqrt{x},$$

$$2 \leq \sqrt{x} + 2,$$

$$\frac{1}{\sqrt{x} + 2} \leq \frac{1}{2}.$$

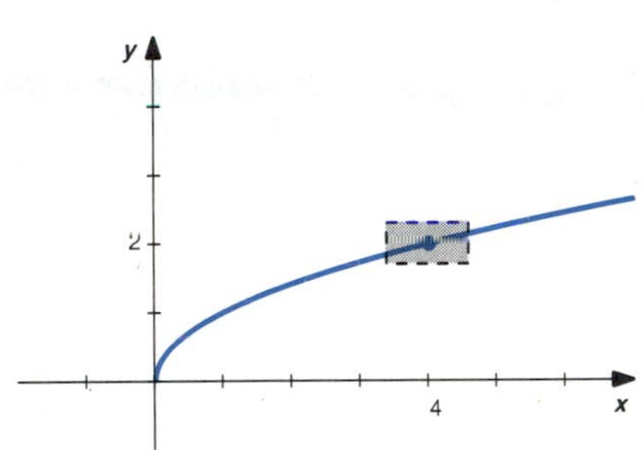


FIGURE 2.3.14

Combining results, we have

$$|\sqrt{x} - 2| \leq \frac{1}{2}|x - 4|.$$

It follows that $|\sqrt{x} - 2| < \varepsilon$ whenever $(1/2)|x - 4| < \varepsilon$, or $|x - 4| < 2\varepsilon$. The condition of the definition is satisfied for $\delta = 2\varepsilon$. see Figure 2.3.14. ■

The ideas used abovc can be used to verify

$$\lim_{x \to c} x^{1/n} = c^{1/n}, \qquad n \text{ a positive integer, } c > 0 \text{ if } n \text{ is even.} \tag{7}$$

For example, if $n = 3$ and $c \neq 0$, we use the factorization formula

$$x - c = (x^{1/3} - c^{1/3})(x^{2/3} + x^{1/3}c^{1/3} + c^{2/3})$$

to express $x^{1/3} - c^{1/3}$ in terms of $x - c$. The case $n = 3$ and $c = 0$ follows from Example 5.

The limit in (7) is restricted to $c > 0$ if $n$ is even, so that $x^{1/n}$ is defined for $x$ on both sides of $c$. No restriction on $c$ is necessary if $n$ is an odd integer, since odd roots are defined for every real number.

It is not always possible to use algebraic methods to verify the condition in the definition of limit. Let us illustrate by verifying the following trigonometric limits.

$$\lim_{\theta \to \theta_0} \cos \theta = \cos \theta_0 \quad \text{and} \quad \lim_{\theta \to \theta_0} \sin \theta = \sin \theta_0. \tag{8}$$

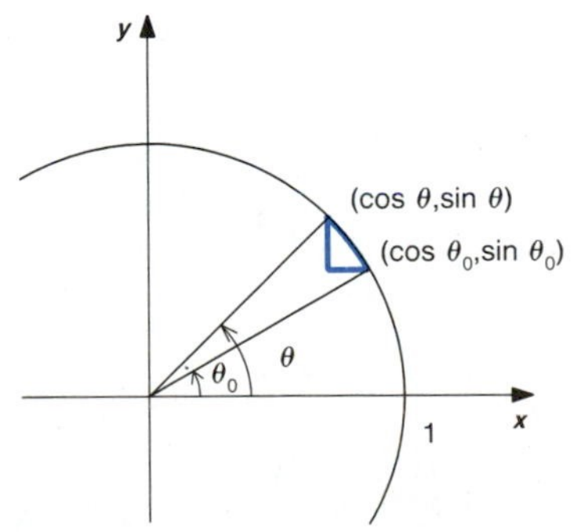


FIGURE 2.3.15

■ *Proof* We will use a geometric argument to verify the condition of the definition for the limits in (8). Recall that the terminal side of an angle of $\theta$ radians in standard position intersects the circle $x^2 + y^2 = 1$ at the point with coordinates $x = \cos \theta$ and $y = \sin \theta$. In Figure 2.3.15 we have labeled the coordinates that correspond to angles $\theta$ and $\theta_0$. It follows from the definition of radian measure (Section 1.6) that the length of the arc of the

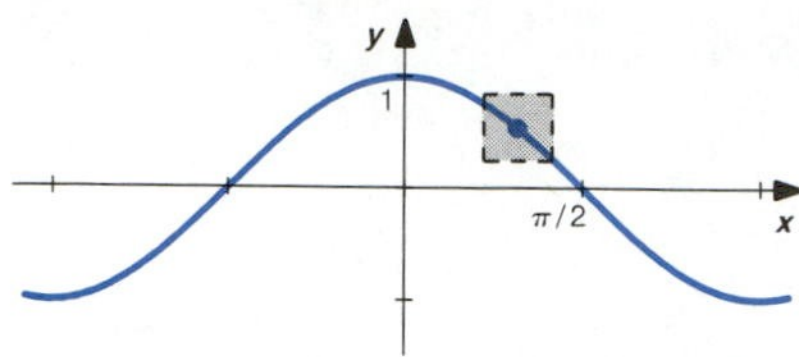
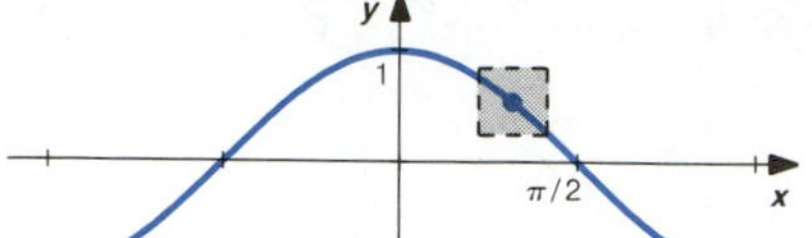


FIGURE 2.3.16

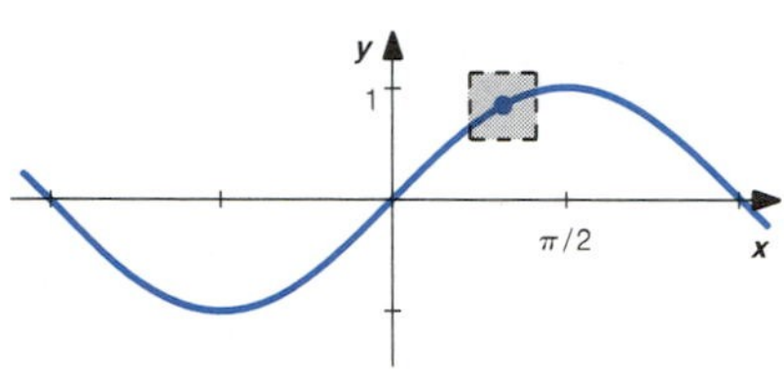


FIGURE 2.3.17

circle between the two points is $|\theta - \theta_0|$. The length of the horizontal side of the "triangle" indicated in Figure 2.3.15 is $|\cos\theta - \cos\theta_0|$, while the length of the vertical side is $|\sin\theta - \sin\theta_0|$. Since the arc is longer than the length of either side, we have

$$|\cos\theta - \cos\theta_0| \le |\theta - \theta_0| \quad \text{and}$$

$$|\sin\theta - \sin\theta_0| \le |\theta - \theta_0|.$$

It is clear from the above equalities that, given any positive number $\varepsilon$, we have

$$|\cos\theta - \cos\theta_0| < \varepsilon \qquad \text{whenever } |\theta - \theta_0| < \varepsilon \quad \text{and}$$

$$|\sin\theta - \sin\theta_0| < \varepsilon \qquad \text{whenever } |\theta - \theta_0| < \varepsilon.$$

In both cases, the condition of the definition is satisfied for $\delta = \varepsilon$. See Figure 2.3.16 and Figure 2.3.17, where we have given our usual graphical interpretation of these results in terms of the graphs of $y = \cos x$ and $y = \sin x$, respectively. ■

## EXERCISES 2.3

For each of the limits in Exercises 1–18, use the given graph to find the value or determine that the limit does not exist.

**1** $\lim\limits_{x\to 0} \dfrac{x^2}{x}$

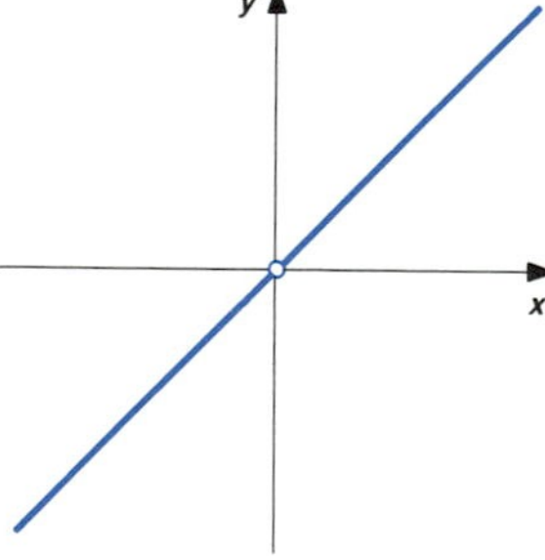


**2** $\lim\limits_{x\to 0} \dfrac{x^3}{x}$

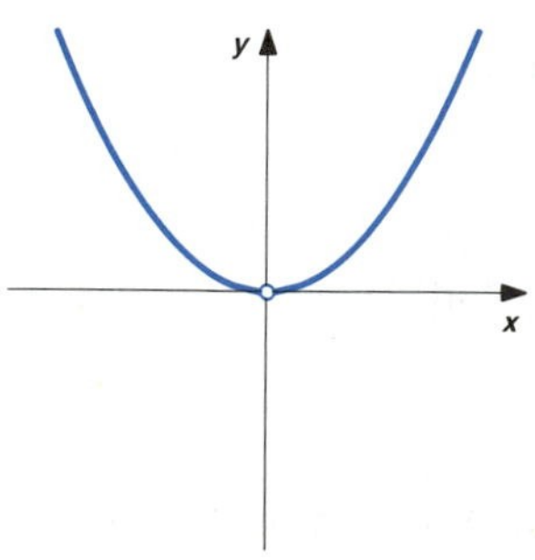


**3** $\lim\limits_{x\to 0} \dfrac{|x| + x}{x}$

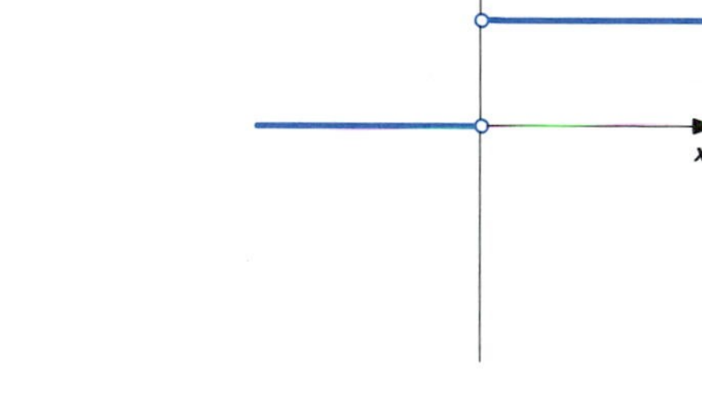


**4** $\lim\limits_{x\to 0} \dfrac{|x| - x}{x}$

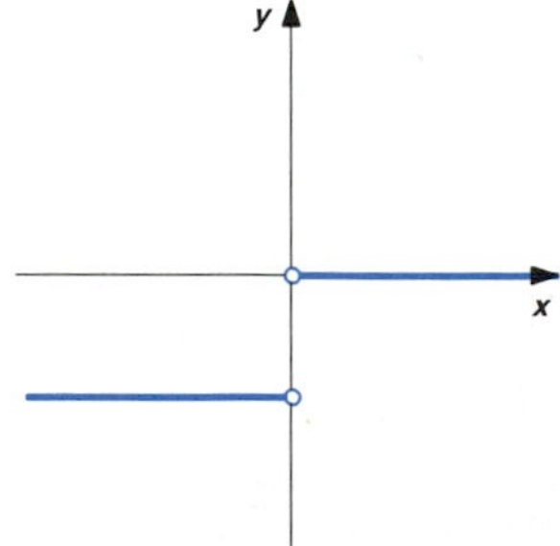

**5** $\lim_{x\to 0} \dfrac{1}{x^{1/3}}$

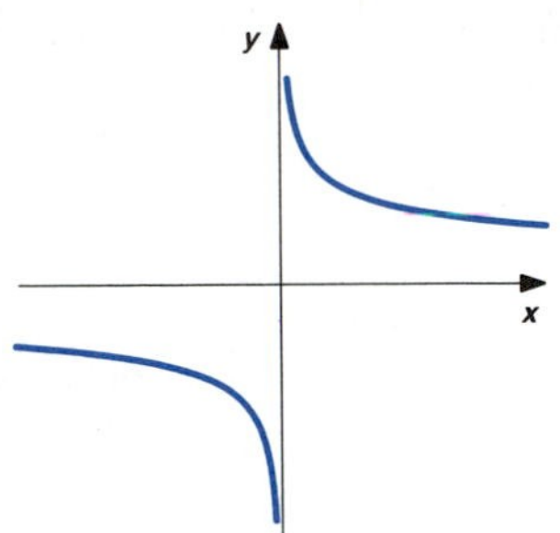


**6** $\lim_{x\to 1} \dfrac{1}{(x-1)^2}$

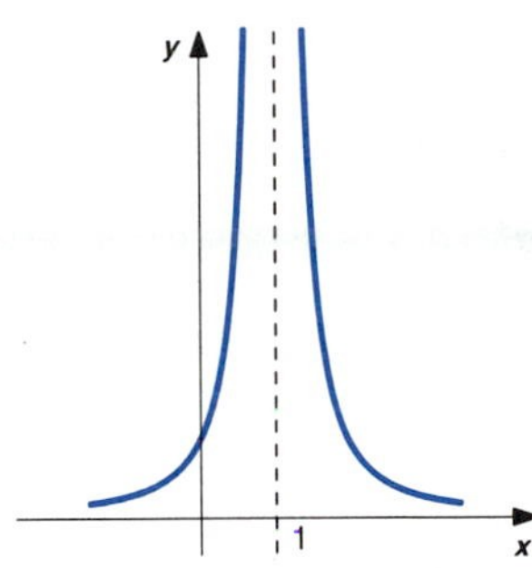


**7** $\lim_{x\to 1} \dfrac{x^2-1}{x-1}$

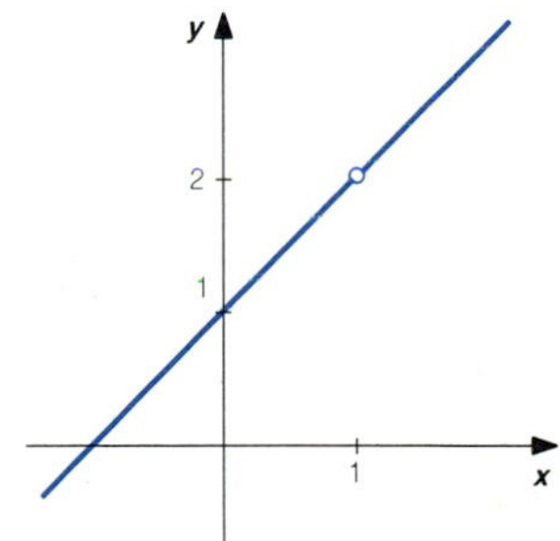


**8** $\lim_{x\to -1} \dfrac{x^2-1}{x+1}$

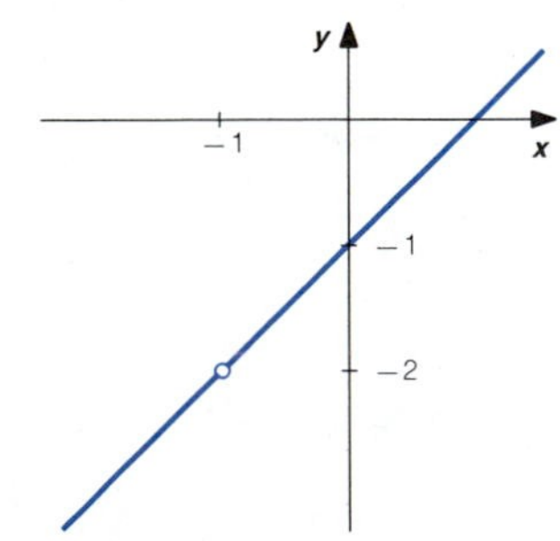


**9** $\lim_{x\to 1} \dfrac{x^3-1}{x-1}$

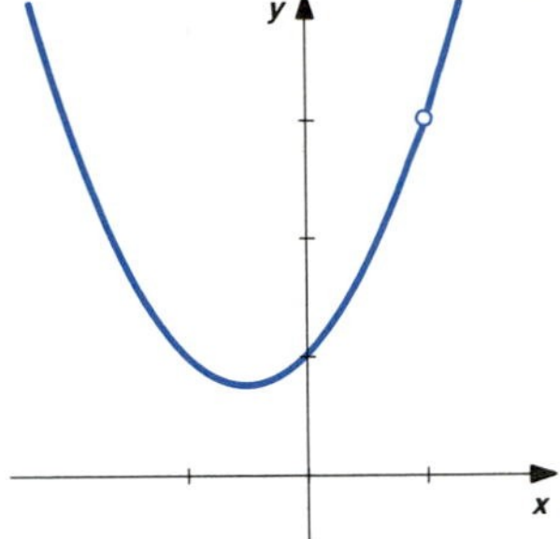


**10** $\lim_{x\to -1} \dfrac{x^3+1}{x+1}$

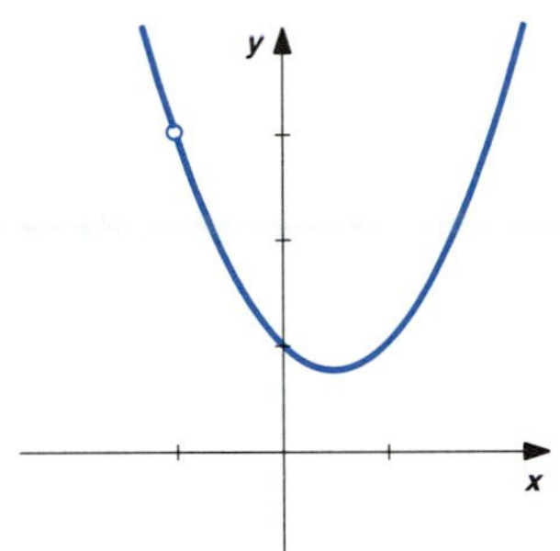


**11** (a) $\lim_{x\to -4} \dfrac{x^2}{4(x+2)}$

(b) 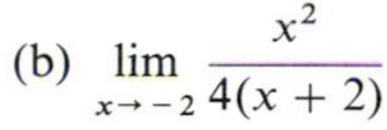 $\lim_{x\to -2} \dfrac{x^2}{4(x+2)}$

(c) 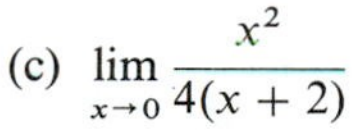 $\lim_{x\to 0} \dfrac{x^2}{4(x+2)}$

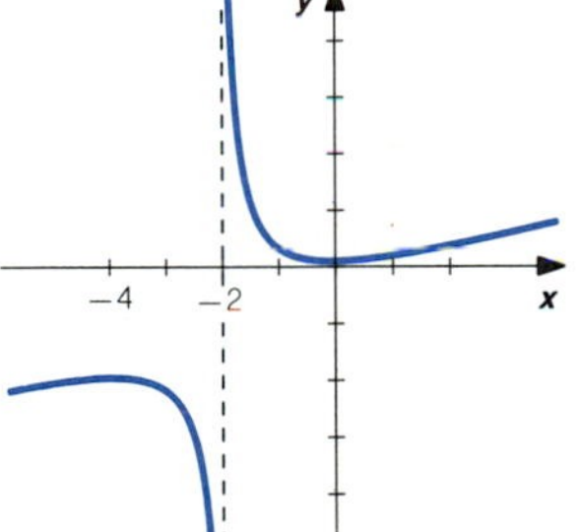


**12** (a) $\lim_{x\to -2} \dfrac{8x}{(x-2)^2}$

(b) 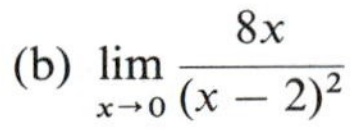 $\lim_{x\to 0} \dfrac{8x}{(x-2)^2}$

(c) 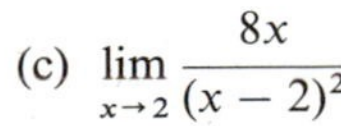 $\lim_{x\to 2} \dfrac{8x}{(x-2)^2}$

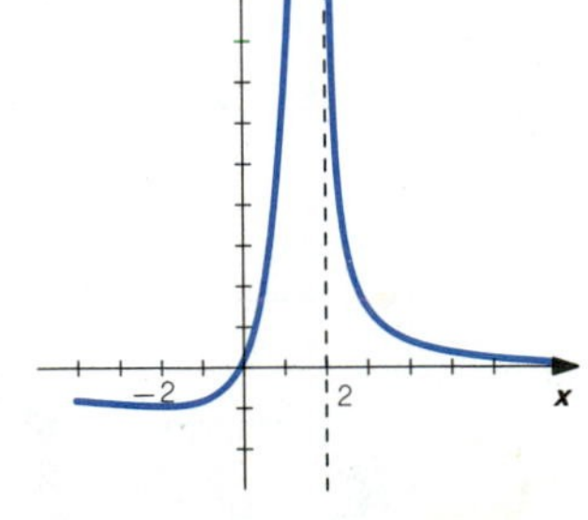

13 $\lim_{x\to 0} \frac{\sin(1/x)}{x}$

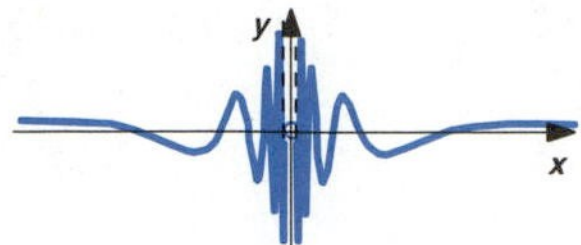


14 $\lim_{x\to 0} (1 - \cos(1/x))$

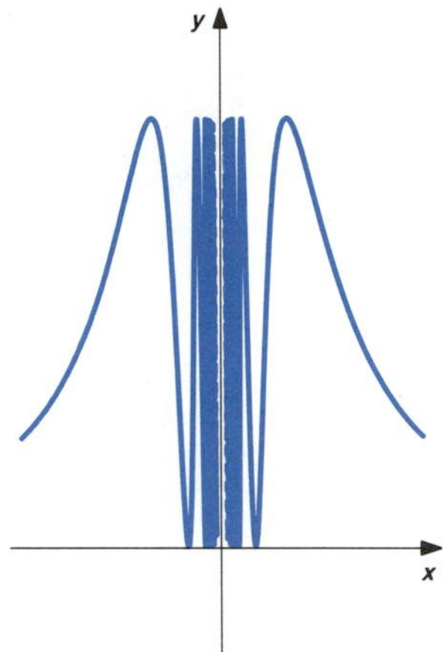


15 $\lim_{x\to 0} x^2\cos^2(1/x)$

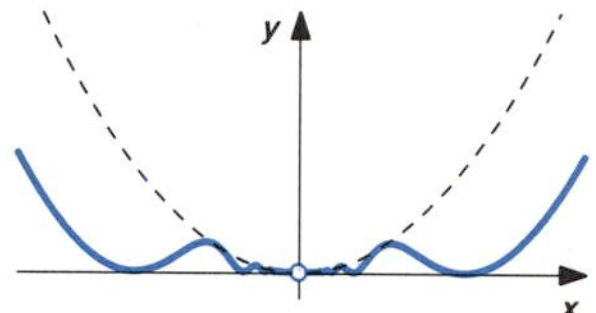


16 $\lim_{x\to 0} x^{1/3}\sin(1/x)$

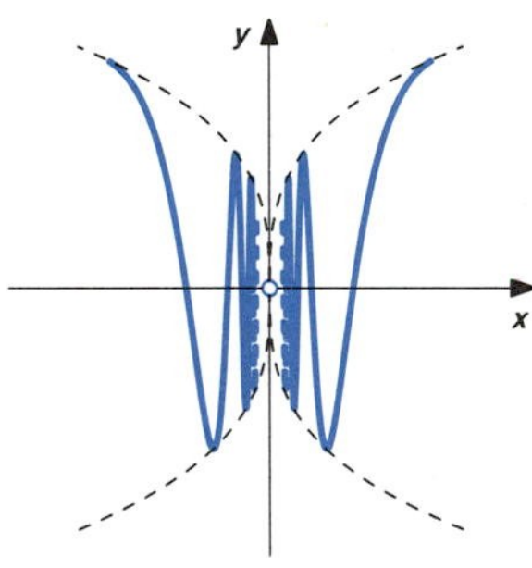


17 $\lim_{x\to 0} \frac{\sin x}{x}$

18 $\lim_{x\to 0} \frac{1 - \cos^2 x}{x}$

19 A 100-m field is to be marked by using a measuring tape that is approximately 10 m long. How close to 10 m must the tape be in order that the actual length of the field be within 0.01 m of 100 m?

20 A 12-ft board is to be marked by using a measuring stick that is approximately 3 ft long. How close to 3 ft must the stick be in order that the actual length of the board be within 0.02 ft of 12 ft?

21 How close to 5 m should each side of a square be in order that the area of the square be within 0.01 $m^2$ of 25 $m^2$? Assume each side is within 0.01 m of 5 m.

22 How close to 4 m should the radius of a circle be in order that the area of the circle be within 0.01 $m^2$ of $16\pi$ $m^2$? Assume the radius is within 0.01 m of 4 m.

23 How close to 2 should each edge of a cube be in order that the volume of the cube be within 0.01 of 8? Assume each edge is within 0.01 of 2.

24 How close to 3 should the radius of a sphere be in order that the volume of the sphere be within 0.01 of $36\pi$? Assume the radius can be measured within 0.01 of 3. $V = 4\pi r^3/3$.

25 How close to 3 Ω must the resistance in a 12-V electrical system be in order that the current be within 0.01 A of 4 A? Assume the resistance is within 1 Ω of 3 Ω. $I = V/R$, where $I$ is the current in amperes, $V$ is the voltage in volts, and $R$ is the resistance in ohms.

26 How close to 4 Ω must the resistance in a 12-V electrical system be in order that the current be within 0.01 A of 3 A? Assume the resistance is within 1 Ω of 4 Ω. $I = V/R$.

In Exercises 27–38, find a positive number $\delta$ such that the given statement is true.

27 For each positive number $\varepsilon$, $|(2x - 3) - 1| < \varepsilon$ whenever $|x - 2| < \delta$.

28 For each positive number $\varepsilon$, $|(4 - 3x) - 7| < \varepsilon$ whenever $|x + 1| < \delta$.

29 For each positive number $\varepsilon$, $|x^3| < \varepsilon$ whenever $|x| < \delta$.

30 For each positive number $\varepsilon$, $|x^5| < \varepsilon$ whenever $|x| < \delta$.

31 For each positive number $\varepsilon$, $|x^{1/3} \sin 3x| < \varepsilon$ whenever $|x| < \delta$.

32 For each positive number $\varepsilon$, $|x^{1/5} \cos 2x| < \varepsilon$ whenever $|x| < \delta$.

33 $|x^2 - 4| < 0.001$ whenever $|x + 2| < \delta$. Assume $|x + 2| \le 1$.

34 $|(x^2 + 5x - 1) + 5| < 0.005$ whenever $|x + 1| < \delta$. Assume $|x + 1| \le 1$.

35 $\left|\frac{1}{x} + \frac{1}{4}\right| < 0.001$ whenever $|x + 4| < \delta$. Assume $|x + 4| \le 1$.

36 $\left|\frac{x}{x+2} - \frac{1}{2}\right| < 0.001$ whenever $|x-2| < \delta$. Assume $|x-2| \le 1$.

37 For each positive number $\varepsilon$, $\left|\frac{x^2-4}{x-2} - 4\right| < \varepsilon$ whenever $0 < |x-2| < \delta$.

38 For each positive number $\varepsilon$, $\left|\frac{2x^2-3x+1}{x-1} - 1\right| < \varepsilon$ whenever $0 < |x-1| < \delta$.

Verify the condition of the formal definition for each of the limits in Exercises 39–53.

39 $\lim_{x\to 3} 2x = 6$

40 $\lim_{x\to 4} (3x-2) = 10$

41 $\lim_{x\to 0} x^{1/3} = 0$

42 $\lim_{x\to 0} x^3 = 0$

43 $\lim_{x\to 16} \sqrt{x} = 4$

44 $\lim_{x\to 9} \sqrt{x} = 3$

45 $\lim_{x\to 27} x^{1/3} = 3$

46 $\lim_{x\to -8} x^{1/3} = -2$

47 $\lim_{x\to \pi} \sin x = 0\ (= \sin \pi)$

48 $\lim_{x\to 0} \cos x = 1\ (= \cos 0)$

49 $\lim_{x\to \pi/4} \sin 2x = 1$

50 $\lim_{x\to \pi/3} \cos 3x = -1$

51 $\lim_{x\to 0} x^2 \cos^2 x = 0$

52 $\lim_{x\to 0} x^{1/3} \sin(1/x) = 0$

53 $\lim_{x\to c} x^{1/3} = c^{1/3}$, $c \ne 0$

## 2.4 EVALUATION OF LIMITS

It is necessary to evaluate limits in order to derive many of the formulas that we will use to solve calculus problems. In this section we will develop a theory that will allow us to evaluate many types of limits easily.

The first step in the theory is to see how *formulas* can be used to evaluate the limit of some specific functions. For example, from statements (5) and (6) of Section 2.3 we have

$$\lim_{x\to c} k = k \quad \text{and} \quad \lim_{x\to c} x = c. \tag{1}$$

Statement (1) tells us the limit as $x$ approaches $c$ of either a constant function $f(x) = k$ or the function $g(x) = x$ is the value of the function at $c$.

**■ EXAMPLE 1**

From statement (1), we have

(a) $\lim_{x\to 5} 3 = 3$, (b) $\lim_{x\to -1} 3 = 3$,
(c) $\lim_{x\to 0} 5 = 5$, (d) $\lim_{x\to -2} x = -2$,
(e) $\lim_{x\to 7} x = 7$, (f) $\lim_{x\to 0} x = 0$. ■

In statements (7) and (8) of Section 2.3, we established that

$$\lim_{x\to c} x^{1/n} = c^{1/n},$$

where $n$ is a positive integer and $c > 0$ if $n$ is even, and that

$$\lim_{x\to c} \cos x = \cos c \quad \text{and} \quad \lim_{x\to c} \sin x = \sin c.$$

These formulas tell us that limits as $x$ approaches $c$ of the sine, cosine, and root functions are the values of the functions at $c$.

**EXAMPLE 2**

We have

(a) $\lim_{x \to \pi/6} \sin x = \sin(\pi/6) = 1/2$,
(b) $\lim_{x \to \pi/2} \cos x = \cos(\pi/2) = 0$,
(c) $\lim_{x \to 16} \sqrt{x} = \sqrt{16} = 4$. ■

The next stage in the theory is to develop formulas that express general rules for evaluating limits. The results we get are based on the following observations.

To verify that $\lim_{x \to c} f(x) = L$, we must show that, for each positive number $\varepsilon$, there is a positive number $\delta$ such that $|f(x) - L| < \varepsilon$ whenever $0 < |x - c| < \delta$. This means that we must show that the expression $|f(x) - L|$ approaches zero as $x$ approaches $c$. The procedure we will use for doing this is to use inequalities and bounds of functions to obtain an inequality of the form $|f(x) - L| \le |z(x)|$, where $z(x)$ is a function that is known to approach zero as $x$ approaches $c$. We will not carry out the details of verifying the $\varepsilon$-$\delta$ definition of limit. It should be clear that the following is true:

**If $|f(x) - L| \le |z(x)|$ and $\lim_{x \to c} z(x) = 0$, then $\lim_{x \to c} f(x) = L$.**

We will also use the following fact:

**The sum of two functions with limit zero has limit zero.**

For example, if $|z_1(x)| < \varepsilon/2$ whenever $0 < |x - c| < \delta_1$ and $|z_2(x)| < \varepsilon/2$ whenever $0 < |x - c| < \delta_2$, then $|z_1(x)| + |z_2(x)| < \varepsilon$ whenever $0 < |x - c| < \delta$, where $\delta$ is the smaller of $\delta_1$ and $\delta_2$. Also, we will use the following:

**The product of a bounded function and a function with limit zero has limit zero.**

The latter statement deserves special attention.

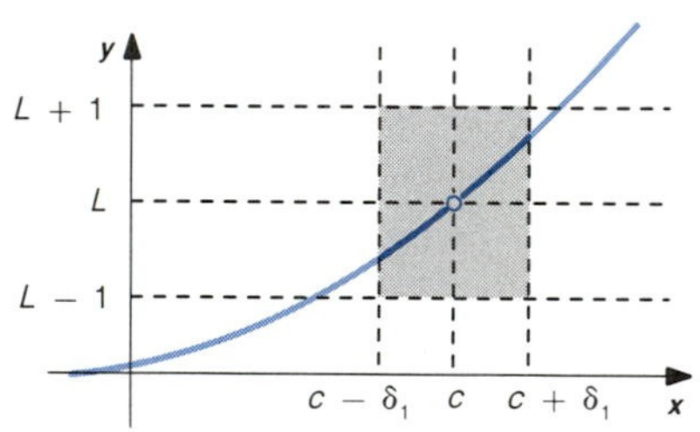


FIGURE 2.4.1

**If $\lim_{x \to c} z(x) = 0$ and the function $b$ is bounded on a set $0 < |x - c| < \delta_1$ for some positive number $\delta_1$, then $\lim_{x \to c} b(x)z(x) = 0$.** (2)

The statement "$b$ is bounded ...," means there is a number $B$ such that $|b(x)| \le B$ whenever $0 < |x - c| < \delta_1$. The product $|b(x)z(x)|$ is then less than $\varepsilon$ whenever $|z(x)|$ is less than $\varepsilon/B$ and $x$ is within $\delta_1$ of $c$, $x \ne c$.

We will need some results that relate limits and bounds of functions.

**If $\lim_{x \to c} f(x) = L$, then $f$ is bounded on a set $0 < |x - c| < \delta_1$ for some $\delta_1 > 0$.**

For example, if $f(x)$ is within one of $L$, then $f$ is bounded. See Figure 2.4.1.

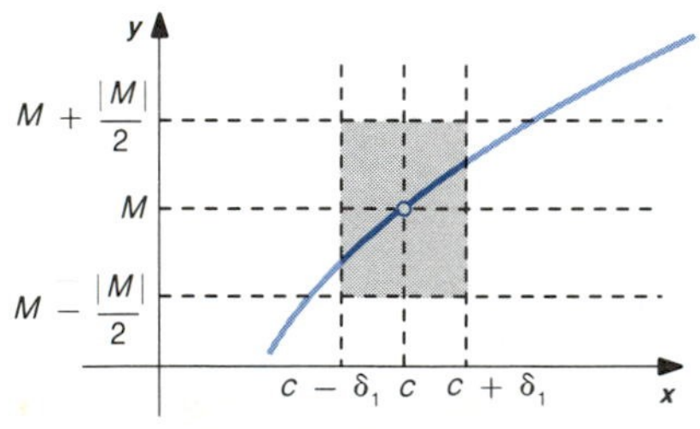


FIGURE 2.4.2

**If $\lim_{x \to c} g(x) = M$ and $M \ne 0$, then $1/g(x)$ is bounded on a set $0 < |x - c| < \delta_1$ for some $\delta_1 > 0$.**

We know that $1/g(x)$ is bounded on sets where $g(x)$ stays away from zero. If $g(x)$ is near $M$ and $M \neq 0$, then $g(x)$ stays away from zero. See Figure 2.4.2.

We now consider the main theorem on limits. This theorem allows us to evaluate the limit of a complicated expression by finding the limits of each of its parts.

*Limit Theorem*

**If $\lim_{x \to c} f(x) = L$ and $\lim_{x \to c} g(x) = M$, then**

**(a)** $\lim_{x \to c} (af(x)) = a \cdot \lim_{x \to c} f(x) = a \cdot L, \quad a$ **any constant,**

**(b)** $\lim_{x \to c} (f(x) + g(x)) = \lim_{x \to c} f(x) + \lim_{x \to c} g(x) = L + M,$

**(c)** $\lim_{x \to c} (f(x) \cdot g(x)) = \lim_{x \to c} f(x) \cdot \lim_{x \to c} g(x) = L \cdot M,$

**(d)** $\lim_{x \to c} \dfrac{1}{g(x)} = \dfrac{1}{\lim_{x \to c} g(x)} = \dfrac{1}{M}, \quad M \neq 0,$

**(e)** $\lim_{x \to c} \dfrac{f(x)}{g(x)} = \dfrac{\lim_{x \to c} f(x)}{\lim_{x \to c} g(x)} = \dfrac{L}{M}, \quad M \neq 0.$

*Proof of the Limit Theorem* If $f(x)$ is near $L$ and $g(x)$ is near $M$, we should expect that $af(x)$ is near $aL$, $f(x) + g(x)$ is near $L + M$, $f(x) \cdot g(x)$ is near $L \cdot M$, $1/g(x)$ is near $1/M$ if $M \neq 0$, and $f(x)/g(x)$ is near $L/M$ if $M \neq 0$. Let us see how to express this in mathematical terms.

We must show that the appropriate differences are small whenever $x$ is close enough to $c$, $x \neq c$. To do this we will express the differences in terms of $|f(x) - L|$ and $|g(x) - M|$. We know that $|f(x) - L|$ and $|g(x) - M|$ both approach zero as $x$ approaches $c$, because $\lim_{x \to c} f(x) = L$ and $\lim_{x \to c} g(x) = M$.

(a) Factoring, we have

$$|a \cdot f(x) - a \cdot L| = |a||f(x) - L|.$$

Since we know that $|f(x) - L|$ approaches zero as $x$ approaches $c$, the product with the constant $|a|$ also approaches zero.

(b) Rearranging the terms and using the Triangle Inequality, we then have

$$\begin{aligned} |(f(x) + g(x)) - (L + M)| &= |(f(x) - L) + (g(x) - M)| \\ &\leq |f(x) - L| + |g(x) - M|. \end{aligned}$$

Since each of the two terms on the right approach zero, the sum also approaches zero.

(c) The algebra trick of adding and subtracting $f(x)M$ and the Triangle Inequality are used to verify this statement.

$$\begin{aligned} |f(x)g(x) - LM| &= |f(x)g(x) - f(x)M + f(x)M - LM| \\ &= |f(x)(g(x) - M) + (f(x) - L)M| \\ &\le |f(x)||g(x) - M| + |f(x) - L||M|. \end{aligned}$$

Since $f$ has a limit at $c$, we know that $f(x)$ is bounded near $c$, $x \neq c$, so the product $|f(x)||g(x) - M| \to 0$ as $x \to c$, as in statement (2). Since $|f(x) - L||M| \to 0$ as $x \to c$, the result follows.

(d) $$\left|\frac{1}{g(x)} - \frac{1}{M}\right| = \frac{|M - g(x)|}{|g(x)M|}.$$

Since $g$ has a nonzero limit, we know that $1/g(x)$ is bounded near $c$, $x \neq c$. It follows that

$$\left|\frac{1}{g(x)} - \frac{1}{M}\right| = \frac{1}{|g(x)|}\frac{1}{|M|}|M - g(x)| \to 0 \qquad \text{as } x \to c.$$

(e) From (d), we have

$$\frac{1}{g(x)} \to \frac{1}{M} \qquad \text{as } x \to c.$$

From (c), applied to the product

$$f(x) \cdot \frac{1}{g(x)},$$

we have

$$\lim_{x \to c} \frac{f(x)}{g(x)} = \lim_{x \to c} f(x) \cdot \frac{1}{g(x)} = L \cdot \frac{1}{M} = \frac{L}{M}, \qquad M \neq 0.$$ ■

The Limit Theorem tells us we can evaluate the limit of a sum by adding the limits of the terms in the sum; the limit of a product can be evaluated by multiplying the limits of the factors; the limit of a quotient can be evaluated by dividing the limit of the numerator by the limit of the denominator, if the limit of the denominator is not zero.

A polynomial $P(x) = a_0 + a_1x + \cdots + a_nx^n$ is a sum of a constant function and terms that are a constant multiple of products of the function $g(x) = x$. The Limit Theorem and statement (1) then imply

$$\begin{aligned} \lim_{x \to c} P(x) &= \lim_{x \to c} (a_0 + a_1x + \cdots + a_nx^n) \\ &= \lim_{x \to c} a_0 + \lim_{x \to c} a_1x + \cdots + \lim_{x \to c} a_nx^n \\ &= a_0 + a_1c + \cdots + a_nc^n \\ &= P(c). \end{aligned}$$

This result and part (e) of the Limit Theorem give the following:

**If $P(x)$ and $Q(x)$ are polynomials, then**

$$\lim_{x \to c} P(x) = P(c) \quad \text{and} \quad \lim_{x \to c} \frac{P(x)}{Q(x)} = \frac{P(c)}{Q(c)}, \qquad Q(c) \neq 0.$$

This tells us that the limit at $c$ of a **rational function** $P/Q$ is the value of the function at $c$, if $P/Q$ is defined at $c$.

■ **EXAMPLE 3**

We have

(a) $\lim_{x \to 2} (x^2 - 3x + 2) = (2)^2 - 3(2) + 2 = 0,$

(b) $\lim_{x \to -3} (x + 5)(x - 3) = (-3 + 5)(-3 - 3) = -12,$

(c) $\lim_{x \to 3} \dfrac{x^2}{x - 5} = \dfrac{3^2}{3 - 5} = -\dfrac{9}{2}.$ ■

Since

$$\tan x = \frac{\sin x}{\cos x}, \quad \cot x = \frac{\cos x}{\sin x}, \quad \sec x = \frac{1}{\cos x}, \quad \text{and} \quad \csc x = \frac{1}{\sin x},$$

we can use the Limit Theorem and what we know about the limits of the sine and cosine functions to conclude the following.

**The limit at $c$ of each of the six trigonometric functions is the value of the function at $c$, as long as the function is defined at $c$.**

■ **EXAMPLE 4**

$$\lim_{x \to \pi/4} \tan x = \tan \frac{\pi}{4} = 1.$$ ■

The Limit Theorem also allows us to evaluate sums, products, and quotients of trigonometric and root functions.

■ **EXAMPLE 5**

$$\lim_{x \to 0} x^{1/3} \cos x = \left(\lim_{x \to 0} x^{1/3}\right)\left(\lim_{x \to 0} \cos x\right)$$
$$= (0)(\cos 0) = 0.$$ ■

■ **EXAMPLE 6**

Evaluate $\lim_{x \to 0} x \cos(1/x)$.

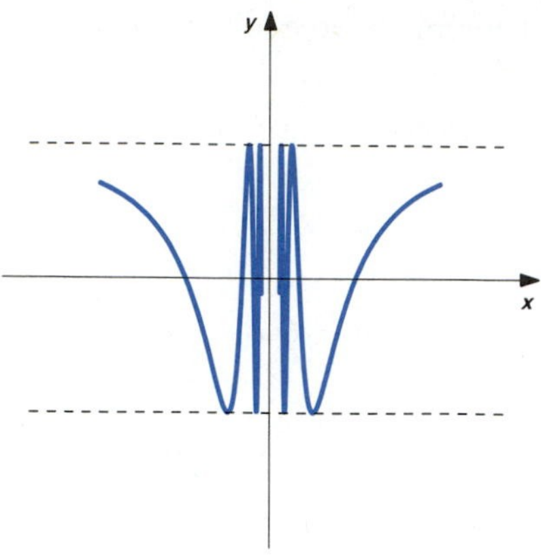


FIGURE 2.4.3

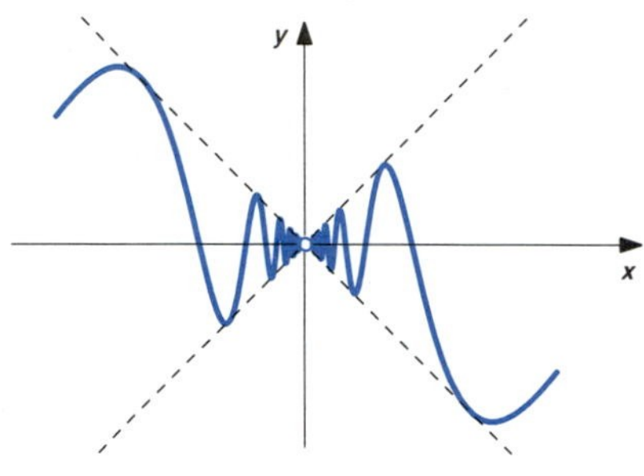


FIGURE 2.4.4

**SOLUTION**

We see this as the limit of the product of $x$ and $\cos(1/x)$. However, we cannot use the Limit Theorem because the factor $\cos(1/x)$ does not have a limit. [The latter statement can be seen from the graph of $\cos(1/x)$ given in Figure 2.4.3.] We do know that $|\cos(1/x)| \leq 1$, $x \neq 0$, so the factor is bounded. Since $\lim_{x\to 0} x = 0$, we can use statement (2), which tells us the product of a bounded function and a function with limit zero has limit zero. The graph of $x\cos(1/x)$ is indicated in Figure 2.4.4. The limit at zero exists, even though the function is not defined at $x = 0$, and we can only indicate the behavior of the graph for $x$ near the origin. ■

In Section 2.6, where the concept of derivative is introduced, we will encounter limits of the form

$$\lim_{x\to c} \frac{f(x) - f(c)}{x - c}.$$

Thus, we are interested in the behavior of the quotient

$$\frac{f(x) - f(c)}{x - c}$$

for $x$ near the point $c$ where the denominator is zero. This important type of limit cannot be evaluated simply by substitution of $c$ for $x$. Some limits of this type can be evaluated by using the following fact.

**If $\lim_{x\to c} g(x)$ exists and there is some positive number $\delta_1$ such that $f(x) = g(x)$ for $0 < |x - c| < \delta_1$, then $\lim_{x\to c} f(x) = \lim_{x\to c} g(x)$.**

This statement is true because limits at $c$ depend only on values of $x$ near $c$, $x \neq c$, and $f(x) = g(x)$ for those $x$. Limits at $c$ do not depend on values at $c$.

■ **EXAMPLE 7**

Evaluate

$$\lim_{x\to 1} \frac{x^2 - 1}{x - 1}.$$

**SOLUTION**

We first note that we cannot evaluate the limit at 1 of $(x^2 - 1)/(x - 1)$ by substitution of 1 for $x$, because the denominator is zero when $x = 1$. However,

$$f(x) = \frac{x^2 - 1}{x - 1} = \frac{(x - 1)(x + 1)}{x - 1} = x + 1, \qquad x \neq 1.$$

That is, if $g(x) = x + 1$, we have $f(x) = g(x)$, $x \neq 1$. Compare the graph of

FIGURE 2.4.5

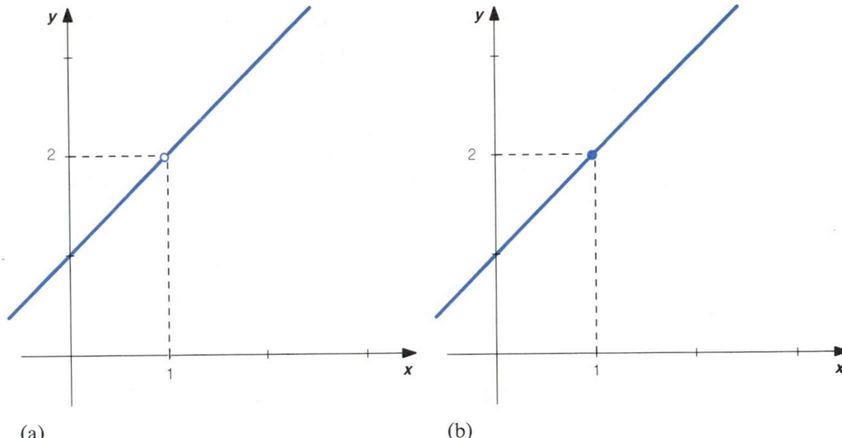


$f$ in Figure 2.4.5a with the graph of $g$ in Figure 2.4.5b. The graphs are the same, except for the point (1, 2). Since $g(x) = x + 1$ is a polynomial, we already know how to evaluate limits of $g$. In particular,

$$\lim_{x \to 1} g(x) = \lim_{x \to 1} (x + 1) = 1 + 1 = 2.$$

Since $f$ and $g$ have the same limit at 1, it follows that $\lim_{x \to 1} f(x) = 2$. ■

The limit of $f(x)$ as $x$ approaches $c$ involves the behavior of $f(x)$ for $x$ on both sides of $c$. The same ideas and theory hold if we restrict attention to $x$ on only one side of $c$.

■ *Definition*

**The limit of $f(x)$ as $x$ approaches $c$ from the right is $L$** (written $\lim_{x \to c^+} f(x) = L$) if, for each $\varepsilon > 0$, there is a $\delta > 0$ such that

$$|f(x) - L| < \varepsilon \quad \text{whenever} \quad c < x < c + \delta.$$

**The limit of $f(x)$ as $x$ approaches $c$ from the left is $L$** (written $\lim_{x \to c^-} f(x) = L$) if, for each $\varepsilon > 0$, there is a $\delta > 0$ such that

$$|f(x) - L| < \varepsilon \quad \text{whenever} \quad c - \delta < x < c.$$ ■

Values of $x$ that are to the right of $c$ are greater than $c$. This is indicated by the plus sign in the symbol $x \to c^+$. The negative sign in the symbol $x \to c^-$ indicates values of $x$ that are to the left of $c$, that is, less than $c$.

The expression $\lim_{x \to c} f(x) = L$ means that $f(x)$ approaches $L$ as $x$ approaches $c$ from either the right or the left. This gives the following relations between a limit and one-sided limits at a point.

**If $\lim_{x \to c} f(x) = L$, then $\lim_{x \to c^+} f(x) = \lim_{x \to c^-} f(x) = L$.**
**If $\lim_{x \to c^+} f(x) = \lim_{x \to c^-} f(x) = L$, then $\lim_{x \to c} f(x) = L$.**
**If $\lim_{x \to c^+} f(x) \neq \lim_{x \to c^-} f(x) = L$, then $\lim_{x \to c} f(x)$ does not exist.**

In the latter case, $f(x)$ does not approach a *single* number $L$ as $x$ approaches $c$.

To apply the theory we have developed to evaluate $\lim_{x\to c} f(x)$, it is necessary that we have a formula for $f(x)$. The formula must hold for $x$ on both sides of $c$, whenever $x$ is near $c$ and $x \neq c$. If different formulas for $f(x)$ hold for $x$ on different sides of $c$, we must use one-sided limits to evaluate $\lim_{x\to c} f(x)$ or show that the limit does not exist. When dealing with functions that are defined by more than one formula, we must be careful to use each formula only where it defines the function.

■ **EXAMPLE 8**

Evaluate (a) $\lim_{x\to 2}|x|$ and (b) $\lim_{x\to 0}|x|$.

**SOLUTION**

We know that

$$|x| = \begin{cases} x & \text{if } x \geq 0, \\ -x & \text{if } x < 0, \end{cases}$$

so $|x|$ is a function that is defined by different formulas for different values of $x$.

(a) The formula $|x| = x$ holds for all $x$ on both sides of 2, $x$ near 2, so we have

$$\lim_{x\to 2}|x| = \lim_{x\to 2} x = 2.$$

(b) Different formulas for $|x|$ hold for $x$ on different sides of 0, so we must look at each one-sided limit separately. The limit from the left at zero involves only $x < 0$. For $x < 0$ we use the formula $|x| = -x$, so

$$\lim_{x\to 0^-} |x| = \lim_{x\to 0^-} (-x) = 0.$$

The limit from the right at zero involves only $x > 0$. For $x > 0$, we use the formula $|x| = x$, so

$$\lim_{x\to 0^+} |x| = \lim_{x\to 0^+} x = 0.$$

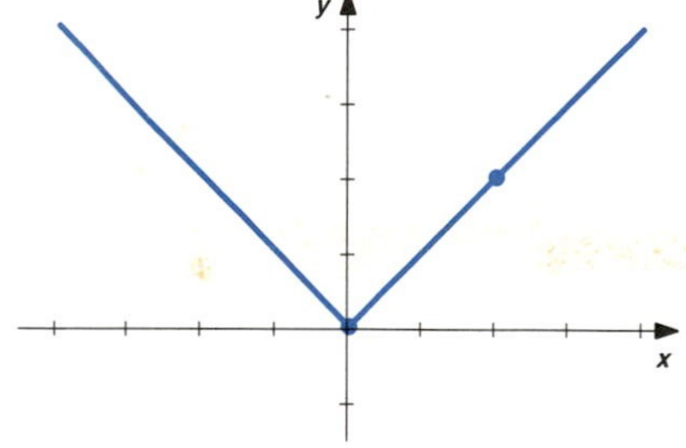


FIGURE 2.4.6

Since both one-sided limits have value zero, $\lim_{x\to 0}|x| = 0$. See Figure 2.4.6. ■

■ **EXAMPLE 9**

Evaluate $\lim_{x\to 0}|x|/x$.

**SOLUTION**

Since $|x| = x$ when $x \geq 0$, and $|x| = -x$ when $x < 0$, we have

$$\frac{|x|}{x} = \begin{cases} 1, & x > 0, \\ -1, & x < 0. \end{cases}$$

The limit from the left at zero involves only $x < 0$, so we use the formula $|x|/x = -1$ to obtain

$$\lim_{x\to 0^-} \frac{|x|}{x} = \lim_{x\to 0^-} (-1) = -1.$$

The limit from the right at zero involves only $x > 1$, so we use the formula $|x|/x = 1$ to obtain

$$\lim_{x\to 0^+} \frac{|x|}{x} = \lim_{x\to 0^+} 1 = 1.$$

Since the one-sided limits have unequal values, $\lim_{x\to 0}|x|/x$ does not exist. See Figure 2.4.7. ■

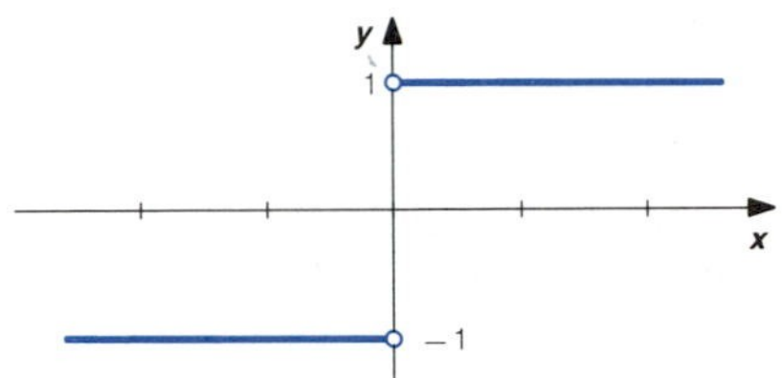

FIGURE 2.4.7

## EXERCISES 2.4

Evaluate the limits in Exercises 1–26.

1 $\lim_{x\to 0} (x^2 - 3x + 4)$

2 $\lim_{x\to 0} (2 - 4x + x^2)$

3 $\lim_{x\to -2} (2x - 1)(x + 4)$

4 $\lim_{x\to -3} (x - 1)(4x + 2)$

5 $\lim_{x\to 8} (2x^{1/3} + 12x^{-2/3})$

6 $\lim_{x\to 4} (\sqrt{x} + 1/\sqrt{x})$

7 $\lim_{x\to 0} (2 \sin x + 3 \cos x)$

8 $\lim_{x\to \pi/4} (3 \sin x + \cos x)$

9 $\lim_{x\to -1} \frac{3x - 1}{x - 1}$

10 $\lim_{x\to -2} \frac{x^2 - 4}{x - 2}$

11 $\lim_{x\to -1} \frac{x^2}{x^2 + 1}$

12 $\lim_{x\to 0} \frac{2x - 1}{x - 2}$

13 $\lim_{x\to \pi/3} \tan x$

14 $\lim_{x\to \pi/6} \sec x$

15 $\lim_{x\to 0} x \sin x$

16 $\lim_{x\to 0} (x + 1)\cos x$

17 $\lim_{x\to 0} x^2 \cos(1/x)$

18 $\lim_{x\to 0} x \cos(1/x^2)$

19 $\lim_{x\to 3} \frac{x^2 - 9}{x - 3}$

20 $\lim_{x\to -5} \frac{x^2 - 25}{x + 5}$

21 $\lim_{x\to 2} \frac{2x^2 - 3x - 2}{x - 2}$

22 $\lim_{x\to 3} \frac{2x^2 + x - 21}{x - 3}$

23 $\lim_{x\to 1} \frac{\frac{1}{x + 1} - \frac{1}{2}}{x - 1}$

24 $\lim_{x\to 2} \frac{\frac{1}{x^2} - \frac{1}{4}}{x - 2}$

25 $\lim_{x\to 2} \frac{\frac{x}{x + 1} - \frac{2}{3}}{x - 2}$

26 $\lim_{x\to 2} \frac{\frac{x}{1 - x} - (-2)}{x - 2}$

In Exercises 27–38 evaluate or determine "does not exist" for each of the limits $\lim_{x\to c^-} f(x)$, $\lim_{x\to c^+} f(x)$, and $\lim_{x\to c} f(x)$, for the given functions $f$ and given numbers $c$.

27 $f(x) = \begin{cases} 0, & x \le 0, \\ \sin x, & x > 0; \end{cases} \quad c = 0$

28 $f(x) = \begin{cases} 1, & x \le 0, \\ \dfrac{1}{1 + x^2}, & x > 0; \end{cases} \quad c = 0$

29 $f(x) = \begin{cases} x + 1, & x < 0, \\ 0, & x = 0, \\ x - 1, & x > 0; \end{cases} \quad c = 0$

30 $f(x) = \begin{cases} 1 - x^2, & x < 0, \\ 0, & x = 0, \\ x^2 - 1, & x > 0; \end{cases} \quad c = 0$

31 $f(x) = \frac{|x - 1|}{x - 1}; c = 1$

32 $f(x) = \frac{|x + 2|}{x + 2}; c = -2$

33 $f(x) = \sqrt{x + 1}; c = -1$

34 $f(x) = \sqrt{1 - x}; c = 1$

35 $f(x) = \begin{cases} x, & x < 0, \\ 2x - x^2, & x > 0; \end{cases} \quad c = 0, c = 1$

**36** $f(x) = \begin{cases} 1 - x, & x < 0, \\ x^2 - 2x + 1, & x > 0; \end{cases}$ $c = 0, c = 1$

**37** $f(x) = \begin{cases} x - 1, & x < 2, \\ \dfrac{x^2}{x - 2}, & x > 2; \end{cases}$ $c = 0, c = 2$

**38** $f(x) = \begin{cases} x + 1, & x < 1, \\ x + 3, & x > 1; \end{cases}$ $c = 0, c = 1$

**39** Show that, if $\lim_{x \to c} [f(x)/g(x)]$ exists and $\lim_{x \to c} g(x) = 0$, then $\lim_{x \to c} f(x) = 0$. (This means that $\lim_{x \to c} [f(x)/g(x)]$ cannot exist if $\lim_{x \to c} g(x) = 0$ and $\lim_{x \to c} f(x) \neq 0$.)

**40** Use the definition of limit to verify that the sum of two functions with limit zero has limit zero. (*Hint*: If $z_1(x)$ and $z_2(x)$ are each less than $\varepsilon/2$ in absolute value, then the absolute value of their sum is less than $\varepsilon$. If $\lim_{x \to c} z(x) = 0$, then for each given positive number $\varepsilon$, there is a positive number $\delta$ such that $|z(x)| < \varepsilon/2$ whenever $0 < |x - c| < \delta$.)

**41** Use the definition of limit to verify that if $\lim_{x \to c} z(x) = 0$ and there is a positive number $\delta_1$ such that the function $b$ is bounded on the set $0 < |x - c| < \delta_1$, then

$$\lim_{x \to c} b(x)z(x) = 0.$$

**42** If $\lim_{x \to c} f(x) = L$, show that there is a positive number $\delta_1$ such that $|f(x)| < 1 + |L|$ whenever $0 < |x - c| < \delta_1$. (*Hint*: $|f(x)| = |f(x) - L + L| \leq |f(x) - L| + |L|$. Use $\varepsilon = 1$ in the definition of $\lim_{x \to c} f(x) = L$.)

**43** If $\lim_{x \to c} g(x) = M$ and $M \neq 0$, show that there is a positive number $\delta_1$ such that $1/|g(x)| < 2/|M|$ whenever $0 < |x - c| < \delta_1$. (*Hint*: $|M| \leq |M - g(x)| + |g(x)|$, so $|g(x)| > |M|/2$ if $|M - g(x)| < |M|/2$.)

## 2.5 CONTINUITY AT A POINT

Continuity of functions is one of the most important concepts of calculus. Continuity is necessary to guarantee a solution to many of the problems we use calculus to solve. To understand when results of calculus can and cannot be applied to solve a problem, we must first have an understanding of continuity. We will see that continuity is closely related to the concept of limit.

We have seen that it is not necessary for a function $f$ to be defined at $c$ in order to have a limit at $c$. However, in many cases the limit of $f$ at $c$ is $f(c)$. This leads to the following definition.

*Definition*

The function $f$ is said to be **continuous at $c$** if $\lim_{x \to c} f(x) = f(c)$.

The definition means that for $f$ to be continuous at $c$,

(i) $f$ must have a limit at $c$,
(ii) $f$ must be defined at $c$, and
(iii) the value of the limit at $c$ must be equal to the value of the function at $c$.

If one (or more) of the above three conditions is not satisfied, $f$ is said to be **discontinuous at $c$**.

If $\lim_{x \to c} f(x)$ exists, but $f$ fails to be continuous at $c$ because either $f(c)$ is not defined or $f(c)$ is defined to be a number other than $\lim_{x \to c} f(x)$, then $f$ is said to have a **removable discontinuity at** $c$. In this case, $f$ can be made continuous at $c$ by defining (or redefining) $f(c)$ to be $\lim_{x \to c} f(x)$. If $\lim_{x \to c} f(x)$ does not exist, then no choice of $f(c)$ can make $f$ continuous at $c$.

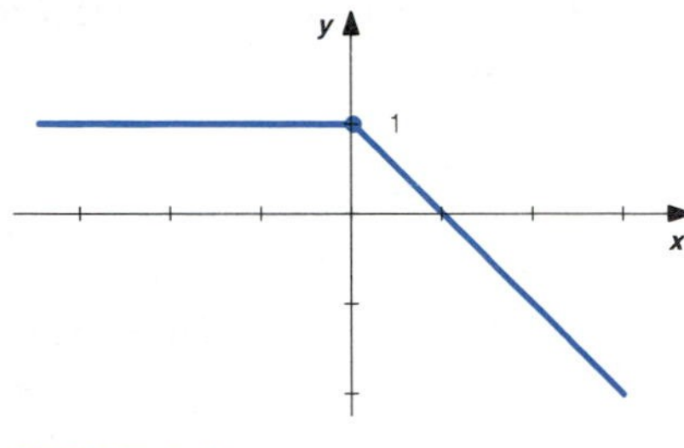

FIGURE 2.5.1

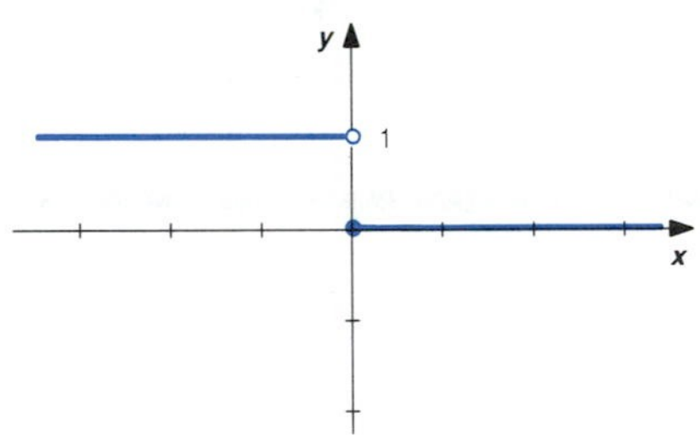

FIGURE 2.5.2

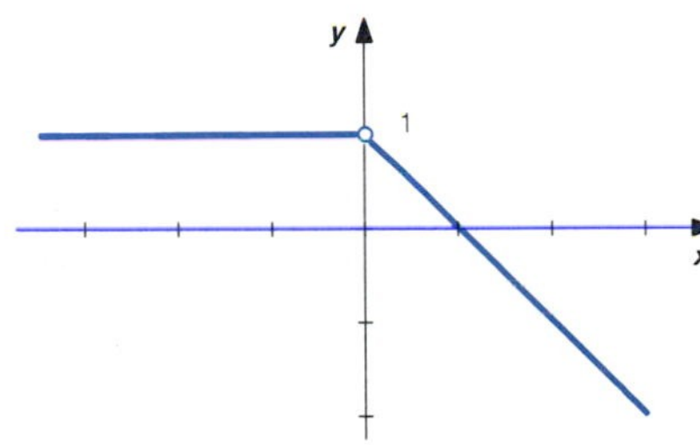

FIGURE 2.5.3

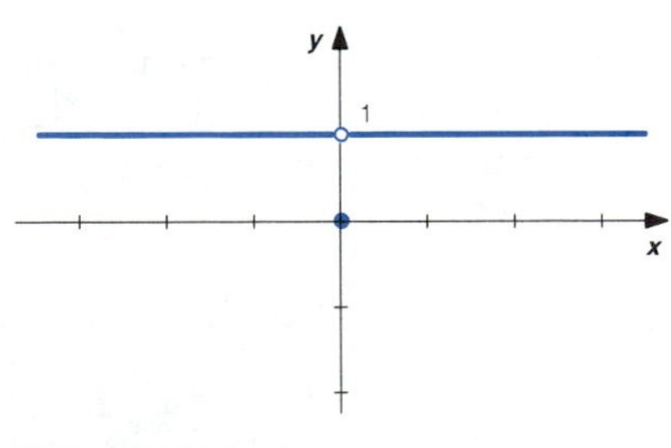

FIGURE 2.5.4

**EXAMPLE 1**

(a) From Figure 2.5.1 we see that

$$f(x) = \begin{cases} 1, & x < 0, \\ 1 - x, & x \geq 0, \end{cases}$$

is continuous at every point.

(b) From Figure 2.5.2 we see that

$$g(x) = \begin{cases} 1, & x < 0, \\ 0, & x \geq 0, \end{cases}$$

is not continuous at zero, because $\lim_{x \to 0} g(x)$ does not exist.

(c) From Figure 2.5.3 we see that

$$h(x) = \begin{cases} 1, & x < 0, \\ 1 - x, & x > 0, \end{cases}$$

is not continuous at zero, because $h$ is not defined at zero. Since $\lim_{x \to 0} h(x)$ exists, the discontinuity at zero is removable. The function can be made continuous at zero by defining $h(0)$ to be 1, the value of the limit at zero.

(d) From Figure 2.5.4 we see that

$$k(x) = \begin{cases} 1, & x \neq 0, \\ 0, & x = 0, \end{cases}$$

is not continuous at zero, because $\lim_{x \to 0} k(x) = 1$, while $k(0) = 0$. Since the limit at zero exists, the function has a removable discontinuity at zero. The function can be made continuous at zero by redefining $k(0)$ to be 1, the value of the limit. ■

It follows from the evaluation of the limits in previous sections that many of the familiar functions we encounter are continuous at each point where they are defined. For example, polynomials are continuous at every point. If $P$ and $Q$ are polynomials, then the rational function $P/Q$ is continuous at all points where $Q$ is not zero. The trigonometric functions are continuous at all points where they are defined. The absolute value function is continuous at every point. If $n$ is any positive integer, the root function $r(x) = x^{1/n}$ is continuous at those $c$ values where it is defined on both sides of $c$.

It follows from the Limit Theorem of Section 2.4 that the sum and product of continuous functions are continuous. The quotient of two continuous functions is continuous at points where the denominator is nonzero.

To investigate the continuity of composite functions $f \circ g(x) = f(g(x))$, we will need the following result on limits:

**If $\lim_{x \to c} g(x) = M$ and $f$ is continuous at $M$, then $\lim_{x \to c} f(g(x)) = f(\lim_{x \to c} g(x)) = f(M)$.** (1)

■ *Proof* We need to show that $|f(g(x)) - f(M)|$ is small whenever $|x - c|$ is small enough, $x \neq c$. Since $f$ is continuous at $M$,

$$|f(g(x)) - f(M)|$$

is small whenever $|g(x) - M|$ is small enough, including when $g(x) = M$. But, $\lim_{x \to c} g(x) = M$ implies $|g(x) - M|$ is small (possibly zero) whenever $|x - c|$ is small enough, $x \neq c$. Combining results, we have (1). ■

■ **EXAMPLE 2**

Evaluate

(a) $\lim\limits_{x \to 2} (3x - 1)^5$,

(b) $\lim\limits_{x \to 3} |1 - 2x - 3x^2|$,

(c) $\lim\limits_{x \to 2} \sqrt{\dfrac{x^2 - 4}{x - 2}}$,

(d) $\lim\limits_{x \to \pi/3} \sin^2 x$,

(e) $\lim\limits_{x \to \pi/6} \cos 2x$.

**SOLUTION**

(a) Let us see how to use statement (1) to evaluate this limit. If $f(x) = x^5$ and $g(x) = 3x - 1$, then

$$(3x - 1)^5 = f(3x - 1) = f(g(x)).$$

We know that $\lim_{x \to 2} g(x) = \lim_{x \to 2}(3x - 1) = 3(2) - 1 = 5$. Also, $f(x) = x^5$ is continuous at 5. Statement (1) then implies

$$\begin{aligned} \lim_{x \to 2} (3x - 1)^5 &= \lim_{x \to 2} f(g(x)) \\ &= f\left(\lim_{x \to 2} g(x)\right) \\ &= f(5) = 5^5. \end{aligned}$$

It is more convenient to use statement (1) without naming the functions "$f$" and "$g$." For example, $(3x - 1)^5$ is the "5th power function of $3x - 1$." Since the 5th power function is continuous at every point, statement (1) tells us that

$$\begin{aligned} \lim_{x \to 2} (3x - 1)^5 &= \left(\lim_{x \to 2}(3x - 1)\right)^5 \\ &= (3(2) - 1)^5 = 5^5. \end{aligned}$$

(b) $|1-2x-3x^2|$ is the "absolute value of $1-2x-3x^2$." Since the absolute value function is continuous at every point, (1) implies that

$$\lim_{x\to 3}|1-2x-3x^2| = \left|\lim_{x\to 3}(1-2x-3x^2)\right|$$
$$= |1-2(3)-3(3)^2| = |-32| = 32.$$

(c) $\sqrt{(x^2-4)/(x-2)}$ is the "square root of $(x^2-4)/(x-2)$." The square root function is continuous at every positive point, so

$$\lim_{x\to 2}\sqrt{\frac{x^2-4}{x-2}} = \sqrt{\lim_{x\to 2}\frac{x^2-4}{x-2}}$$
$$= \sqrt{\lim_{x\to 2}\frac{(x-2)(x+2)}{x-2}}$$
$$= \sqrt{\lim_{x\to 2}(x+2)} = \sqrt{4} = 2.$$

We have used the fact that $(x-2)(x+2)/(x-2) = x+2$ for $x \neq 2$, so the limits at 2 of these two functions are equal.

(d) $\sin^2 x = (\sin x)^2$ is the "square of $\sin x$." The square function is continuous at every point, so we have

$$\lim_{x\to\pi/3}\sin^2 x = \lim_{x\to\pi/3}(\sin x)^2$$
$$= \left(\lim_{x\to\pi/3}\sin x\right)^2$$
$$= \left(\sin\frac{\pi}{3}\right)^2$$
$$= \left(\frac{\sqrt{3}}{2}\right)^2$$
$$= \frac{3}{4}.$$

We have used the fact that the sine function is continuous to evaluate $\lim_{x\to\pi/3}\sin x$.

(e) $\cos 2x$ is the "cosine of $2x$." We have

$$\lim_{x\to\pi/6}\cos 2x = \cos\left(\lim_{x\to\pi/6}2x\right) = \cos 2\left(\frac{\pi}{6}\right) = \cos\frac{\pi}{3} = \frac{1}{2}. \qquad \blacksquare$$

It is a consequence of (1) that "the composition of two continuous functions is continuous." Let us state this as a theorem.

*Theorem*

**If $g$ is continuous at $c$ and $f$ is continuous at $g(c)$, then $f \circ g$ is continuous at $c$.**

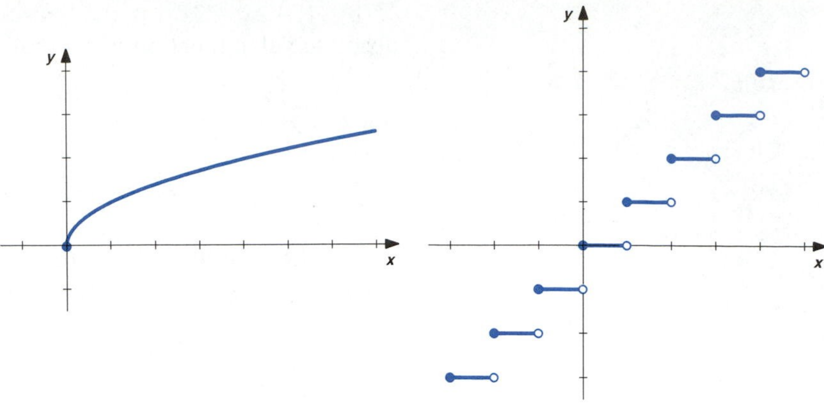


FIGURE 2.5.5

FIGURE 2.5.6

■ *Proof* We first note that $\lim_{x\to c} g(x) = g(c)$, since $g$ is continuous at $c$. Since $f$ is continuous at $g(c)$, (1) implies

$$\lim_{x\to c} f \circ g(x) = \lim_{x\to c} f(g(x)) = f(\lim_{x\to c} g(x)) = f(g(c)) = f \circ g(c). \quad ■$$

The power function $x^{m/n} = (x^{1/n})^m$ is a composition of the polynomial $x^m$ and the root function $x^{1/n}$. It follows that it is continuous at $c$, as long as it is defined on both sides of $c$.

We can use one-sided limits to consider continuity of a function from each side of a point $c$.

■ *Definition*

A function $f$ **is continuous from the left at** $c$ if

$$\lim_{x\to c^-} f(x) = f(c).$$

A function $f$ **is continuous from the right at** $c$ if

$$\lim_{x\to c^+} f(x) = f(c).$$

■

A function $f$ is continuous at $c$ if and only if it is continuous from the left and right at $c$, so $\lim_{x\to c} f(x) = f(c)$.

The square root function is continuous from the right at zero, because $\lim_{x\to 0^+} \sqrt{x} = 0 = \sqrt{0}$. It is not continuous at zero, because it is not defined for negative $x$. See Figure 2.5.5.

The greatest integer less than or equal to a real number $x$ is denoted $[x]$. For example, $[1.95] = 1$, $[3] = 3$, and $[-1.3] = -2$. The **greatest integer function** can be used in computer science to **truncate** decimal numbers. From the graph in Figure 2.5.6 we see that the function is continuous at every noninteger $x$, continuous from the right at every integer, but not continuous from the left at any integer. It is discontinuous at each integer.

## EXERCISES 2.5

In Exercises 1–4 indicate whether the functions whose graphs are given are continuous at each of the points $a$, $b$, and $c$.

**1**

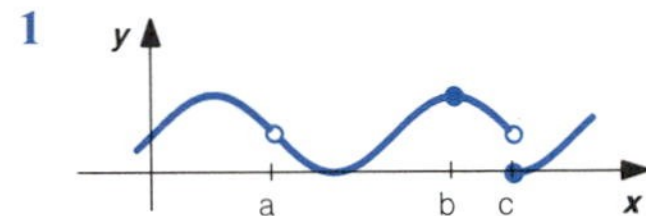


**2**

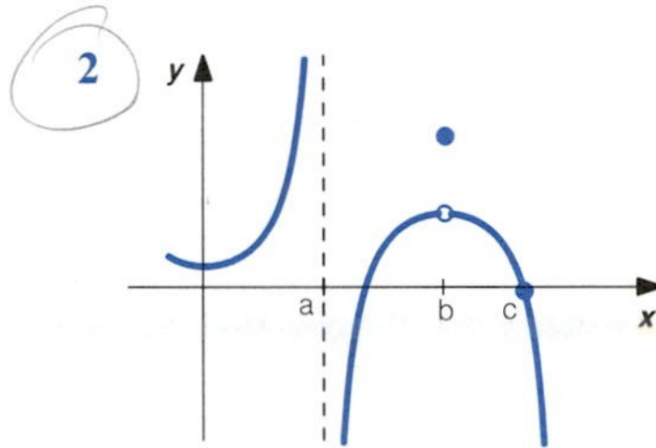


**3**

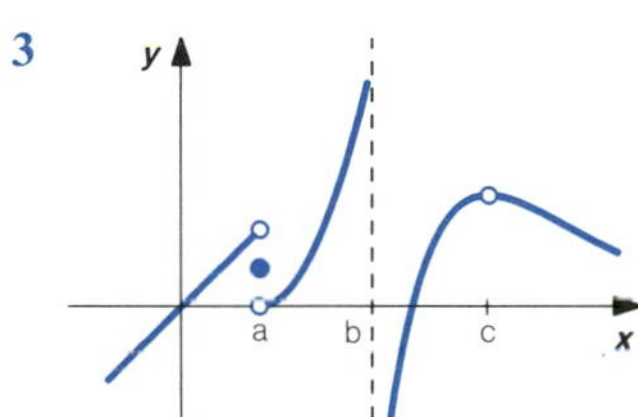


**4**

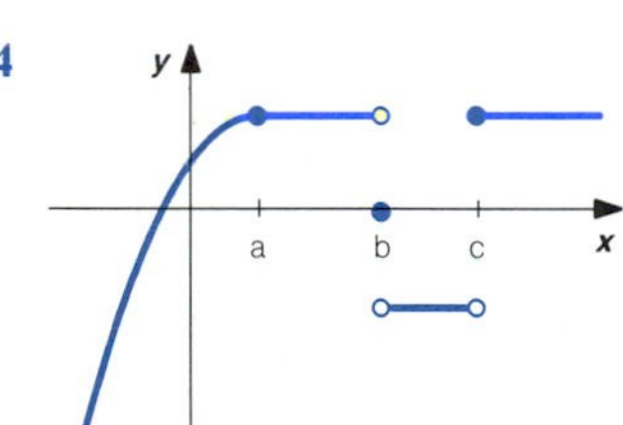


In each of Exercises 5–8, either find a value of $a$ so that $f$ is continuous at 0, or indicate that this is impossible.

**5** $f(x) = \begin{cases} 1 + x, & x < 0 \\ a, & x = 0 \\ 1 - x, & x > 0 \end{cases}$

**6** $f(x) = \begin{cases} 0, & x < 0 \\ a, & x = 0 \\ x \cos x, & x < 0 \end{cases}$

**7** $f(x) = \begin{cases} \dfrac{|x| + x}{2x}, & x \neq 0 \\ a, & x = 0 \end{cases}$

**8** $f(x) = \begin{cases} 0, & x < 0 \\ a, & x = 0 \\ \cos x, & x > 0 \end{cases}$

In Exercises 9–10 find numbers $a$ and $b$ so that $f$ is continuous at every point. (*Hint*: Use the fact that the graph of $y = ax + b$ is a line to sketch the graph of $f$.)

**9** $f(x) = \begin{cases} 2, & x < 0 \\ ax + b, & 0 \leq x \leq 4 \\ 0, & x > 4 \end{cases}$

**10** $f(x) = \begin{cases} x, & x < -1 \\ ax + b, & -1 \leq x \leq 2 \\ 2x - 7, & x > 2 \end{cases}$

Evaluate the limits in Exercises 11–24.

**11** $\lim_{x \to 1} (3x^2 - 7x + 3)^{10}$

**12** $\lim_{x \to -1} \left(\dfrac{10x + 2}{x + 2}\right)^{1/3}$

**13** $\lim_{x \to 0} \left|\dfrac{x + 4}{x - 2}\right|$

**14** $\lim_{x \to 4} \left|\dfrac{x}{2 - x}\right|$

**15** $\lim_{x \to 3} \sqrt{x^2 + 16}$

**16** $\lim_{x \to 5} \sqrt{x^2 + 144}$

**17** $\lim_{x \to 2} (x^2 + 4)^{-1/3}$

**18** $\lim_{x \to 3} |3x - 1|^{2/3}$

**19** $\lim_{x \to 3} \sqrt{\dfrac{x^2 - 4x + 3}{x - 3}}$

**20** $\lim_{x \to 1} \sqrt{\dfrac{x^2 + 2x - 3}{x - 1}}$

**21** $\lim_{x \to \pi/4} \tan 3x$

**22** $\lim_{x \to \pi/3} \sec 2x$

**23** $\lim_{x \to \pi/2} \cos^2(x + \pi/2)$

**24** $\lim_{x \to \pi/2} \sin^2(x + \pi/4)$

In Exercises 25–28 sketch the graphs and determine if the functions are continuous from the left at zero or continuous from the right at zero.

**25** $f(x) = \begin{cases} \dfrac{|x| + x}{2}, & x \neq 0 \\ 0, & x = 0 \end{cases}$

**26** $f(x) = [x] + x$

**27** $f(x) = -[-x]$

**28** $f(x) = \sqrt{x}$

**29** Use the definition of limit to verify that, if $f$ is continuous at $M$ and $\lim_{x \to c} g(x) = M$, then $\lim_{x \to c} f(g(x)) = f(M)$.

**30** Is it true that $\lim_{x \to c} g(x) = 0$ implies $\lim_{x \to c} \sqrt{g(x)} = 0$. Explain.

**31** If

$$f(x) = \begin{cases} 0, & \text{if } x \text{ is rational,} \\ 1, & \text{if } x \text{ is irrational,} \end{cases}$$

show that $f$ is not continuous at any point.

**32** If

$$f(x) = \begin{cases} 0, & \text{if } x \text{ is rational,} \\ x, & \text{if } x \text{ is irrational,} \end{cases}$$

show that $f$ is continuous at 0.

**33** Sketch the graph of the function $f(x) = [10x]/10$.

**34** The function of Exercise 33 truncates a positive decimal number $x$ to one decimal place. Find a function that truncates a positive decimal number to five decimal places.

**35** Sketch the graph of $f(x) = [x + 0.5]$.

**36** The function of Exercise 35 rounds a decimal number $x$ to the nearest integer. Find a function that rounds a decimal number to the nearest hundredth.

## 2.6 THE DERIVATIVE AT A POINT

The derivative is one of the fundamental working tools of calculus. The mathematical idea of a derivative has application to a wide range of physical problems. In this section we will see what a derivative is and how the concept relates to a few of these problems. We must understand the idea of the derivative if we are to understand its applications.

We begin with the mathematical definition of derivative. This gives us a common reference point for the examples that illustrate how the concept of derivative is used.

■ *Definition*

If $\lim_{x \to c}[f(x) - f(c)]/(x - c)$ exists, $f$ is said to be **differentiable at $c$**. The value of the limit is denoted $f'(c)$ (read "$f$ prime of $c$") and is called the derivative of $f$ at $c$. That is, the **derivative of $f$ at $c$** is

$$f'(c) = \lim_{x \to c} \frac{f(x) - f(c)}{x - c},$$

if the limit exists. If the limit does not exist, $f$ is **not differentiable at $c$**. ■

The *mathematical meaning* of the existence of the derivative is that the ratios $[f(x) - f(c)]/(x - c)$ approach a single number $f'(c)$ as $x$ approaches $c$. The derivative *is* the limit of these ratios. The *physical meaning* of the derivative depends on the *interpretation* of the ratios as physical quantities. We will illustrate some of the possible interpretations in this section.

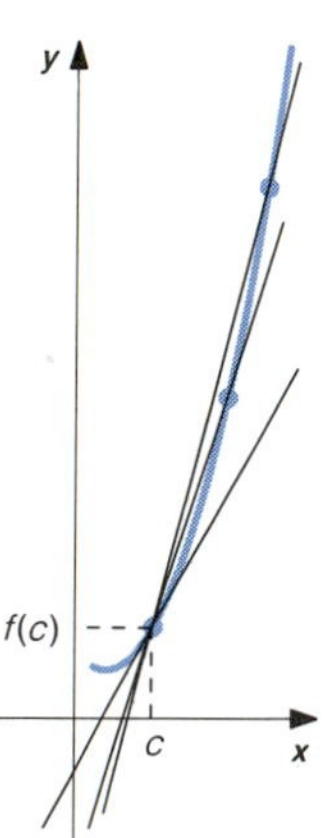


**FIGURE 2.6.1**

### *Slope*

The quotient $[f(x) - f(c)]/(x - c)$, $x \neq c$, may be interpreted as the slope of the line through the fixed point $(c, f(c))$ and a variable point $(x, f(x))$ on the graph of $y = f(x)$. This slope is not defined when $x = c$. However, if $f$ is differentiable at $c$, the slope is close to the number $f'(c)$ whenever $x$ is close enough to $c$, $x \neq c$. See Figure 2.6.1. This leads us to consider the line through $(c, f(c))$ with slope $f'(c)$.

■ *Definition*

If $f$ is differentiable at $c$, the line through the point $(c, f(c))$ with slope $f'(c)$ is said to be **tangent** to the graph of $y = f(x)$ at the point $(c, f(c))$. ■

In later sections we will use the line tangent to the graph of a function to find approximate values of that function. If $f$ is not differentiable at $c$, then there is no line that will serve this purpose. No line has been defined to be tangent to the graph at $(c, f(c))$ in this case.

The point-slope form of the equation of the line through the point $(c, f(c))$ with slope $f'(c)$ is

$$y - f(c) = f'(c)(x - c).$$

This is an **equation of the line tangent to the graph at $(c, f(c))$.**

We should realize that the slope of the line between two *different* points on a graph is a familiar idea, but the idea of a line tangent to a graph needed to be defined.

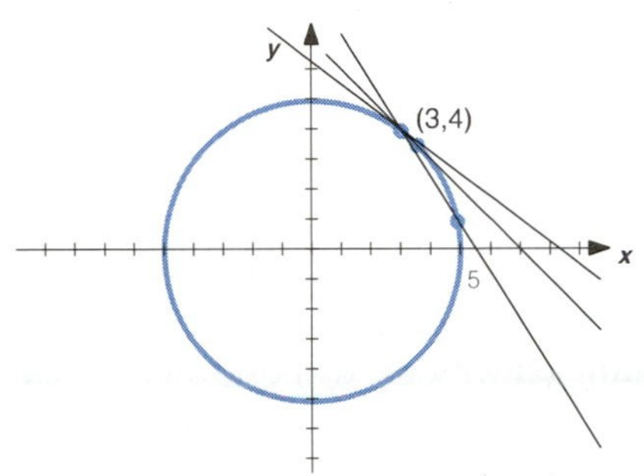


FIGURE 2.6.2

Later in this section we will look more closely at the geometric interpretation of the derivative as a slope. At this time we should note that the idea expressed by the formal definition of the line tangent to a graph agrees with the idea we already have of a line tangent to a circle. For example, we see in Figure 2.6.2 that the slope of the line through the point $(3, 4)$ and another point $(x, \sqrt{25 - x^2})$ on the graph of the circle $x^2 + y^2 = 25$ approaches the slope of the line tangent to the circle at $(3, 4)$ as $x$ approaches 3.

■ **EXAMPLE 1**

(a) Find the slope of the line through the point $(2, 4)$ and a variable point on the graph of $f(x) = x^2$ in terms of $x$. (b) Find the slope of the line tangent to the graph at $(2, 4)$. (c) Find an equation of the line tangent to the graph at $(2, 4)$.

**SOLUTION**

(a) A variable point on the graph has coordinates $(x, x^2)$, in terms of $x$. If $x \neq 2$, the slope of the line through $(2, 4)$ and $(x, x^2)$ is

$$\frac{f(x) - f(2)}{x - 2} = \frac{x^2 - 4}{x - 2}.$$

Simplifying, we obtain

$$\frac{x^2 - 4}{x - 2} = \frac{(x - 2)(x + 2)}{x - 2} = x + 2, \qquad x \neq 2.$$

(b) The slope of the line tangent to the graph at $(2, 4)$ is

$$f'(2) = \lim_{x \to 2} \frac{f(x) - f(2)}{x - 2} = \lim_{x \to 2} \frac{x^2 - 4}{x - 2}.$$

Since the limits at 2 do not depend on the values at 2, we can replace

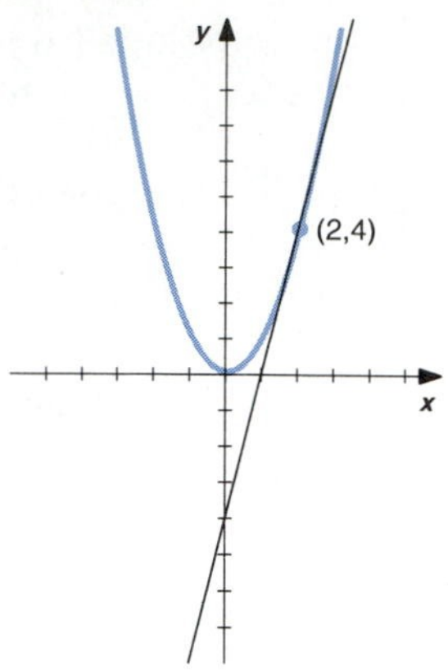

FIGURE 2.6.3

$(x^2 - 4)/(x - 2)$ by the simplified expression $x + 2$ in the above limit. We obtain

$$f'(2) = \lim_{x \to 2} (x + 2) = 2 + 2 = 4.$$

(c) The point-slope form of the equation of the line tangent to the graph at (2, 4) is $y - 4 = 4(x - 2)$. Simplifying, we obtain

$$y = 4x - 4.$$

See Figure 2.6.3. ■

## *Velocity*

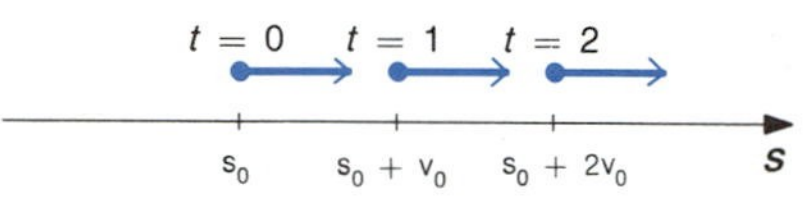

FIGURE 2.6.4

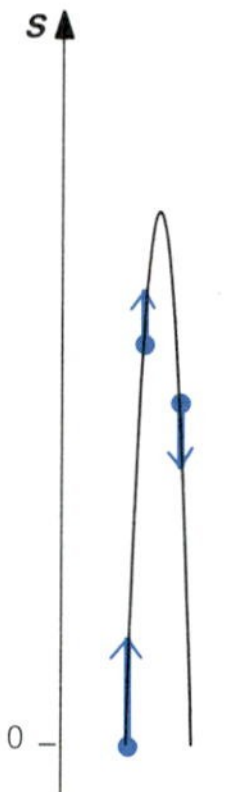

FIGURE 2.6.5

Let us consider a particle moving along a number line. The position of the particle at time $t$ may be given by a function $s(t)$. For example, if a particle moves along a number line with constant velocity $v_0$, its position is given by the formula $s(t) = s_0 + v_0 t$, where $s_0$ is the position of the particle at time $t = 0$. See Figure 2.6.4, where we have used a schematic drawing to indicate the motion; the arrows indicate the direction and speed of the particle. If a particle is thrown upward from ground level with initial velocity $v_0$ (ft/s), its height (in feet) $t$ seconds later is known to be given by the formula $s(t) = -16t^2 + v_0 t$. See Figure 2.6.5, where we have used a schematic drawing to indicate the direction and speed of the particle at various times.

If the position of a particle on a number line at time $t$ is given by $s(t)$, we can determine the average velocity of the particle over any interval of time. That is, the average velocity over the time interval from time $t_0$ to another time $t$ is

$$\text{Average velocity} = \frac{\text{Change in position}}{\text{Length of time}} = \frac{s(t) - s(t_0)}{t - t_0}.$$

If the function $s$ is differentiable at $t_0$, the average velocity of the particle over small intervals of time that either begin at $t_0$ $(t_0 < t)$ or end at $t_0$ $(t < t_0)$ will be near the number $s'(t_0)$. This leads to the following definition.

■ *Definition*

If the position function $s(t)$ is differentiable at $t_0$, the derivative

$$s'(t_0) = \lim_{t \to t_0} \frac{s(t) - s(t_0)}{t - t_0}$$

is called the (instantaneous) **velocity** at time $t_0$. ■

The sign of the velocity indicates the direction of the object on the number line. The **speed** at time $t_0$ is given by $|s'(t_0)|$.

Average velocity over a nonzero interval of time is a familiar idea. However, it does not make sense to consider instantaneous velocity to be the average velocity over a time interval of length zero. Instantaneous velocity is the limit of average velocities.

**■ EXAMPLE 2**

The position of a particle on a number line at time $t$ is given by $s(t) = 1/(t+1)$. (a) Find the average velocity of the particle over the interval of time between 1 and $t$, $t \neq 1$. (b) Find the (instantaneous) velocity at time 1.

**SOLUTION**

(a) For $t \neq 1$, the average velocity over the interval of time between 1 and $t$ is

$$\frac{s(t) - s(1)}{t-1} = \frac{\dfrac{1}{t+1} - \dfrac{1}{1+1}}{t-1}.$$

Simplifying, we have

$$\frac{\dfrac{1}{t+1} - \dfrac{1}{2}}{t-1} = \frac{\dfrac{2-(t+1)}{2(t+1)}}{t-1} = \frac{\dfrac{2-t-1}{2(t+1)}}{t-1}$$

$$= \frac{\dfrac{1-t}{2(t+1)}}{t-1} = \frac{1-t}{2(t+1)(t-1)} = \frac{-(t-1)}{2(t+1)(t-1)}$$

$$= \frac{-1}{2(t+1)}, \qquad t \neq 1.$$

(b) The instantaneous velocity at time 1 is the derivative

$$s'(1) = \lim_{t \to 1} \frac{s(t) - s(1)}{t-1}.$$

Since limits at 1 do not depend on values at 1, we can use the simplified expression for the average velocity in the above limit. We obtain

$$s'(1) = \lim_{t \to 1} \frac{-1}{2(t+1)} = \frac{-1}{2(1+1)} = -\frac{1}{4}.$$

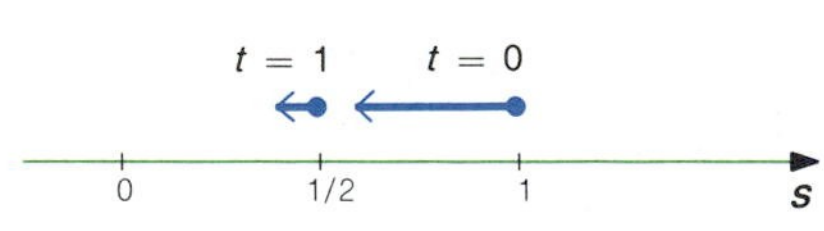

FIGURE 2.6.6

The negative velocity indicates that the particle is moving in the negative direction when $t = 1$. See Figure 2.6.6. ■

## *General rate of change*

Generally, we may interpret the ratio $[f(x) - f(c)]/(x - c)$ as the average rate of change of $f(x)$ with respect to the variable $x$.

*Definition*

If $f$ is differentiable at $c$, the derivative

$$f'(c) = \lim_{x \to c} \frac{f(x) - f(c)}{x - c}$$

is called the (instantaneous) **rate of change of $f$ with respect to the variable $x$ at $c$.** ■

**■ EXAMPLE 3**

Find the rate of change of the volume of a cube with respect to the length of one of its edges when each edge is 10 ft.

**SOLUTION**

It is very important that the length of the edges and the volume of the cube be considered variables. Let $x$ denote the length of each edge of the cube. The volume is then given by $V = x^3$. See Figure 2.6.7.

The rate of change of the volume $V$ with respect to the length of each edge $x$ is defined to be the derivative

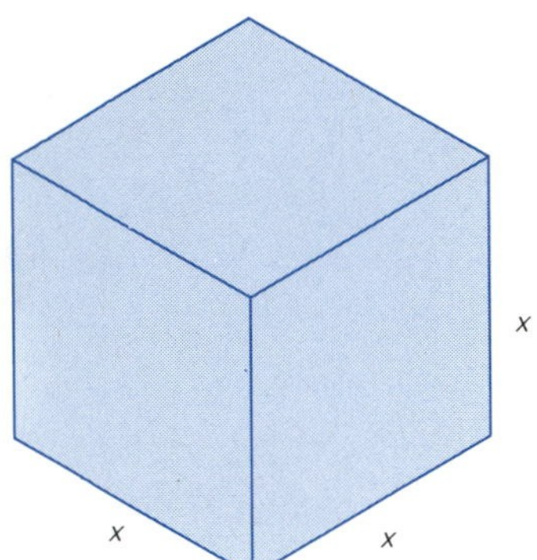


FIGURE 2.6.7

$$\begin{aligned} V'(10) &= \lim_{x \to 10} \frac{V(x) - V(10)}{x - 10} \\ &= \lim_{x \to 10} \frac{x^3 - 10^3}{x - 10}. \end{aligned}$$

Factoring the numerator and simplifying, we have

$$\begin{aligned} V'(10) &= \lim_{x \to 10} \frac{(x - 10)(x^2 + 10x + 10^2)}{x - 10} \\ &= \lim_{x \to 10} (x^2 + 10x + 10^2) \\ &= 10^2 + 10(10) + 10^2 = 300. \end{aligned}$$

The volume of the cube is changing at a rate of 300 ft$^3$ per 1 ft change in the length of its edges at the instant when the edges are 10 ft. ■

**■ EXAMPLE 4**

Find the rate of change of the distance between the points (0, 3) and $(x, 0)$ with respect to $x$ when $x$ is 4.

**SOLUTION**

Note that $(x, 0)$ is a variable point on the $x$-axis. See Figure 2.6.8.

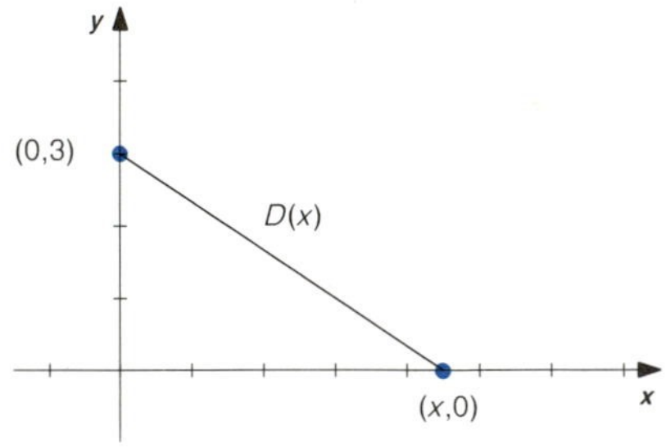

The Distance Formula gives

FIGURE 2.6.8

$$D(x) = \sqrt{(x - 0)^2 + (0 - 3)^2} = \sqrt{x^2 + 9}.$$

The desired rate of change is the derivative

$$D'(4) = \lim_{x\to 4} \frac{\sqrt{x^2+9} - \sqrt{4^2+9}}{x-4} = \lim_{x\to 4} \frac{\sqrt{x^2+9} - 5}{x-4}.$$

Rationalizing the numerator and simplifying, we have

$$\begin{aligned} D'(4) &= \lim_{x\to 4} \frac{\sqrt{x^2+9} - 5}{x-4} \cdot \frac{\sqrt{x^2+9}+5}{\sqrt{x^2+9}+5} \\ &= \lim_{x\to 4} \frac{(x^2+9)-25}{(x-4)(\sqrt{x^2+9}+5)} \\ &= \lim_{x\to 4} \frac{x^2-16}{(x-4)(\sqrt{x^2+9}+5)} \\ &= \lim_{x\to 4} \frac{(x-4)(x+4)}{(x-4)(\sqrt{x^2+9}+5)} \\ &= \lim_{x\to 4} \frac{x+4}{\sqrt{x^2+9}+5} \\ &= \frac{4+4}{\sqrt{4^2+9}+5} = \frac{8}{5+5} = \frac{4}{5}. \end{aligned}$$

The distance is changing at a rate of 4/5 units per unit change in $x$ at the instant that $x = 4$. ■

### *Geometric aspects of differentiability*

If $f$ is continuous at $c$, we already know that points $(x, f(x))$ on the graph of $f$ approach the point $(c, f(c))$ as $x$ approaches $c$. We now show that differentiability implies continuity.

**If $f$ is differentiable at $c$, then $f$ is continuous at $c$.** (1)

■ *Proof* We must show $\lim_{x\to x} f(x) = f(c)$. Using some algebraic manipulation and properties of the limit, we have

$$\begin{aligned} \lim_{x\to c} f(x) &= \lim_{x\to c}\left[\frac{f(x)-f(c)}{x-c}(x-c) + f(c)\right] \\ &= \lim_{x\to c}\frac{f(x)-f(c)}{x-c} \cdot \lim_{x\to c}(x-c) + \lim_{x\to c} f(c) \\ &= f'(c)\cdot(c-c) + f(c) = f(c). \end{aligned}$$

Note that we needed to know that $\lim_{x\to c}[f(x) - f(x)]/(x-c)$ exists in order to use the Limit Theorem. The fact that $f$ is differentiable at $c$ means that this limit exists. ■

It follows from (1) that $f$ *is not differentiable at points where it is not continuous.*

FIGURE 2.6.9

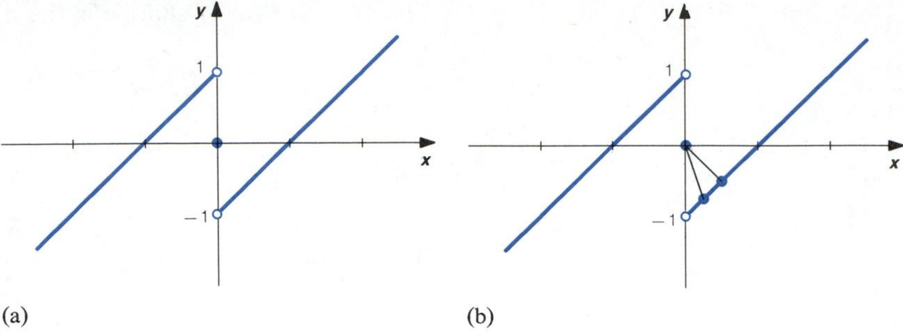


■ **EXAMPLE 5**

Is

$$f(x) = \begin{cases} x + 1, & x < 0, \\ 0, & x = 0, \\ x - 1, & x > 0, \end{cases}$$

differentiable at 0?

**SOLUTION**

We see from the graph in Figure 2.6.9a that $f$ is not continuous at 0, so $f$ is not differentiable at 0. We could also verify directly that

$$f'(0) = \lim_{x \to 0} \frac{f(x) - f(0)}{x - 0}$$

does not exist by investigating the corresponding one-sided limits, neither of which exists. For example, using the formula $f(x) = x - 1$ for $x > 0$, we have

$$\lim_{x \to 0^+} \frac{f(x) - f(0)}{x - 0} = \lim_{x \to 0^+} \frac{(x - 1) - (0)}{x - 0} = \lim_{x \to 0^+} \frac{x - 1}{x}.$$

This limit does not exist, because $(x - 1)/x$ becomes unbounded as $x$ approaches 0 from the right. This corresponds to the fact that the line segment between (0, 0) and a variable point $(x, f(x))$ on the graph becomes vertical as $x$ approaches 0 from the right. See Figure 2.6.9b. ■

**$f$ is not differentiable at $c$ if $[f(x) - f(c)]/(x - c)$ becomes unbounded as $x$ approaches $c$.** (2)

This can happen, even though $f$ is continuous at $c$. In simple cases this means that the graph either passes smoothly through $(c, f(c))$ vertically or has a vertical sharp point at $(c, f(c))$.

■ **EXAMPLE 6**

Is $f(x) = x^{1/3}$ differentiable at 0?

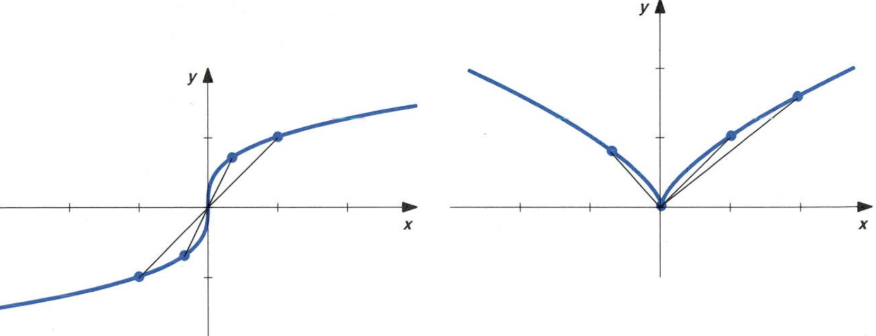

FIGURE 2.6.10

FIGURE 2.6.11

**SOLUTION**

$$f'(0) = \lim_{x\to 0} \frac{f(x) - f(0)}{x - 0} = \lim_{x\to 0} \frac{x^{1/3} - 0}{x} = \lim_{x\to 0} \frac{1}{x^{2/3}}.$$

Since $1/x^{2/3}$ becomes unbounded as $x$ approaches zero, $f'(0)$ does not exist. This means that $f$ is not differentiable at 0. We see from Figure 2.6.10 that the graph passes smoothly through the origin vertically. ■

■ **EXAMPLE 7**

Is $f(x) = x^{2/3}$ differentiable at 0?

**SOLUTION**

$$f'(0) = \lim_{x\to 0} \frac{f(x) - f(0)}{x - 0} = \lim_{x\to 0} \frac{x^{2/3} - 0}{x} = \lim_{x\to 0} \frac{1}{x^{1/3}}.$$

The limit of $1/x^{1/3}$ at zero does not exist, because $1/x^{1/3}$ becomes unbounded as $x$ approaches zero. This means that $f(x) = x^{2/3}$ is not differentiable at zero. From Figure 2.6.11 we see that the slope of the line through the origin and a point $(x, x^{2/3})$ becomes increasing large and positive as $x$ approaches zero from the right. The slope becomes increasing large and negative as $x$ approaches zero from the left. The graph of $f$ has a sharp point at (0, 0). ■

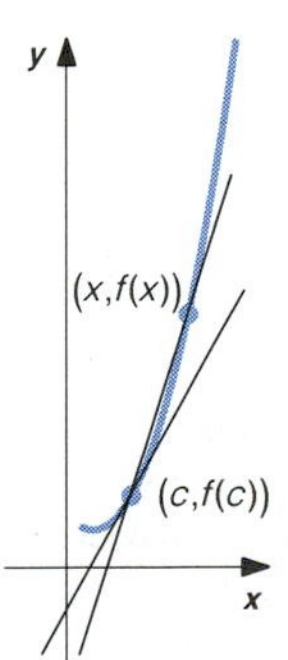


FIGURE 2.6.12

Let us emphasize that a function can be continuous but not differentiable at a point $c$. However, a function cannot be differentiable but not continuous at $c$.

If $f$ is differentiable at $c$, the slope of the line through $(c, f(c))$ and $(x, f(x))$ approaches $f'(c)$, the slope of the line tangent to the graph at $(c, f(c))$, as $x$ approaches $c$. This means the angle between the two lines approaches zero as $x$ approaches $c$. See Figure 2.6.12. In this sense we have the following.

**If $f$ is differentiable at $c$, then the graph of $f$ passes through the point $(c, f(c))$ in the same direction as the line tangent to the graph at that point.** (3)

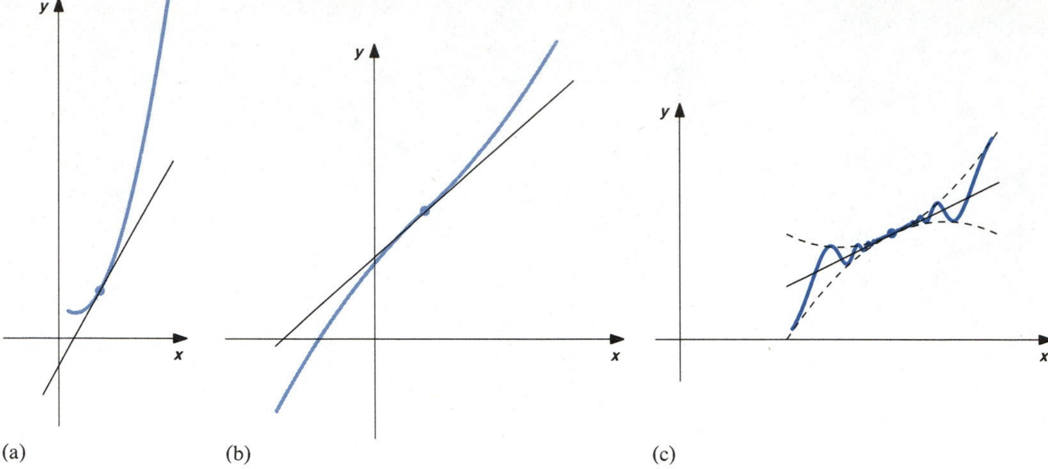


FIGURE 2.6.13

The graph may touch or cross the tangent line at any number of points in addition to the point $(c, f(c))$. See Figure 2.6.13.

**If both one-sided limits exist and**

$$\lim_{x \to c^+} \frac{f(x) - f(c)}{x - c} \neq \lim_{x \to c^-} \frac{f(x) - f(c)}{x - c}, \tag{4}$$

**then $f$ is not differentiable at $c$.**

The graph has a corner at $(c, f(c))$.

■ **EXAMPLE 8**

Is

$$f(x) = \begin{cases} x^2, & x \leq 1, \\ x, & x > 1, \end{cases}$$

differentiable at 1?

**SOLUTION**

One-sided limits or a sketch of the graph can be used to show that $\lim_{x \to 1} f(x) = 1 = f(1)$, so $f$ is continuous at 1.

To evaluate

$$f'(1) = \lim_{x \to 1} \frac{f(x) - f(1)}{x - 1}$$

we must evaluate each one-sided limit, since different formulas for $f(x)$ hold for $x$ on different sides of 1. The limit from the right at 1 involves

$x > 1$, so we use the formula $f(x) = x$ to obtain

$$\lim_{x\to 1^+} \frac{f(x) - f(1)}{x - 1} = \lim_{x\to 1^+} \frac{x - 1}{x - 1} = \lim_{x\to 1^+} 1 = 1.$$

The limit from the left at 1 involves $x < 1$, so we use the formula $f(x) = x^2$ to obtain

$$\lim_{x\to 1^-} \frac{f(x) - f(1)}{x - 1} = \lim_{x\to 1^-} \frac{x^2 - 1}{x - 1}$$

$$= \lim_{x\to 1^-} \frac{(x - 1)(x + 1)}{x - 1} = \lim_{x\to 1^-} (x + 1) = 1 + 1 = 2.$$

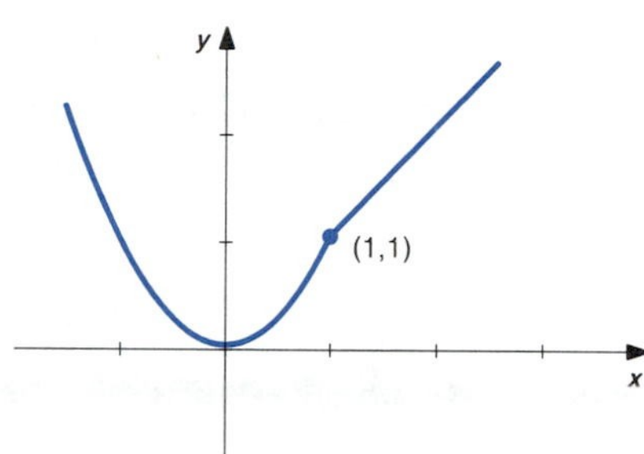

FIGURE 2.6.14

Since the one-sided limits are unequal,

$$f'(1) = \lim_{x\to 1} \frac{f(x) - f(1)}{x - 1}$$

does not exist, so $f$ is not differentiable at 1. In Figure 2.6.14 we see that the graph has a corner at (1, 1). ■

## EXERCISES 2.6

In Exercises 1–8: (a) Find the slope of the line between the point $(c, f(c))$ and a variable point $(x, f(x))$ on the graph of $y = f(x)$ in terms of $x$. (b) Find the slope of the line tangent to the graph at the point $(c, f(c))$. (c) Find the equation of the line tangent to the graph at $(c, f(c))$.

1 $f(x) = x^2 - 3x$; $c = 1$

2 $f(x) = 4x - x^2$; $c = 1$

3 $f(x) = 3x + 4$; $c = -3$

4 $f(x) = 2x + 5$; $c = -3$

5 $f(x) = x^3$; $c = 2$

6 $f(x) = 1 - x^3$; $c = 2$

7 $f(x) = 1/x$; $c = 4$

8 $f(x) = 1/(x + 1)$; $c = 4$

In Exercises 9–16 the position of a particle on a number line at time $t$ is given by $s(t)$. (a) Find the average velocity of the particle as $t$ ranges over the interval of time between $t_0$ and $t$, $t \neq t_0$. (b) Find the (instantaneous) velocity at time $t_0$.

9 $s(t) = t^2 - 2$; $t_0 = 2$

10 $s(t) = 3 - t^2$; $t_0 = 2$

11 $s(t) = 6 - 2t$; $t_0 = 3$

12 $s(t) = 4 + 5t$; $t_0 = 3$

13 $s(t) = \sqrt{t + 1}$; $t_0 = 8$

14 $s(t) = \sqrt{t}$; $t_0 = 4$

15 $s(t) = \dfrac{t}{t + 1}$; $t_0 = 3$

16 $s(t) = \dfrac{t}{t + 2}$; $t_0 = 8$

17 If a particle is thrown vertically from ground level with initial velocity $v_0$ ft/s, its height in feet $t$ seconds later is known to be given by $s(t) = -16t^2 + v_0 t$. Find the average velocity of the particle over the interval of time between 0 and $t$. What is the approximate value of these average velocities for small values of $t$?

18 Use the formula for $s(t)$ given in Exercise 17 to find the average velocity of the particle over the interval of time between $v_0/32$ and $t$. What is the approximate value of these average velocities for values of $t$ near $v_0/32$?

19 Find the rate of change of the area of a circle with respect to its radius when the radius is 3.

20 Find the rate of change of the area of a square with respect to the length of one side when the side is 5.

21 Find the rate of change of the distance between the points $(0, y)$ and $(-3, 0)$ with respect to $y$ when $y$ is $-4$.

22 Find the rate of change of the distance between the points $(0, y)$ and $(12, 0)$ with respect to $y$ when $y$ is 5.

Determine whether the functions given in Exercises 23–38 are differentiable at zero by investigating $\lim_{x\to 0} [f(x) - f(0)]/(x - 0)$. Use one-sided limits where appropriate.

23 $f(x) = |x|$

24 $f(x) = \begin{cases} 1 + x, & x \le 0 \\ 1 - x, & x > 0 \end{cases}$

25 $f(x) = \begin{cases} x^2, & x \le 0 \\ 0, & x > 0 \end{cases}$

**26** $f(x) = \begin{cases} x^2 + x, & x \le 0 \\ x, & x > 0 \end{cases}$

**27** $f(x) = \begin{cases} 1 + x + x^2, & x \le 0 \\ 2 - x - x^2, & x > 0 \end{cases}$

**28** $f(x) = \begin{cases} |x|/x, & x \ne 0 \\ 0, & x = 0 \end{cases}$

**29** $f(x) = x^{4/5}$

**30** $f(x) = x^{4/3}$

**31** $f(x) = x^{5/3}$

**32** $f(x) = x^{3/5}$

**33** $f(x) = |x^2 - 2x|$

**34** $f(x) = |x^3 - 3x^2|$

**35** $f(x) = \begin{cases} x^2 \sin(1/x), & x \ne 0 \\ 0, & x = 0 \end{cases}$

**36** $f(x) = \begin{cases} x \sin(1/x), & x \ne 0 \\ 0, & x = 0 \end{cases}$

**37** $f(x) = \begin{cases} 0, & x \text{ rational} \\ x, & x \text{ irrational} \end{cases}$

**38** $f(x) = \begin{cases} 0, & x \text{ rational} \\ x^2, & x \text{ irrational} \end{cases}$

**39** At what points is the function $f(x) = |x^2 - 3x + 2|$ not differentiable?

**40** At what points is the function $f(x) = |x^3 - 4x^2 + 4x|$ not differentiable?

## REVIEW EXERCISES

**1** $f(x) = 2x^2 - 3x + 1$, $g(x) = 3x - 1$.
Find $f(x - 2) - 2g(x)$. Find $5f(x) - [g(x)]^2$.

**2** $f(x) = 3x^2 - 2x + 5$, $g(x) = 2x - 3$.
Find $[f(x) - f(2)]/(x - 2)$. Find $f(g(x)) - 2g(f(x))$.

**3** $f(x) = \begin{cases} 1 - \cos x, & -2\pi \le x < 0 \\ 0, & x = 0 \\ x^2 - 1, & 0 < x \le 2 \end{cases}$

(a) Sketch the graph of $f$.
(b) Find the domain and range of $f$.
(c) Evaluate $f(-\pi)$, $f(1)$, and $f(3)$.
(d) Evaluate $\lim_{x \to 0^-} f(x)$ and $\lim_{x \to 0^+} f(x)$.
(e) Is $f$ continuous from the left at 0? Is $f$ continuous from the right at 0? Is $f$ continuous at 0?

**4** $f(x) = \begin{cases} x^2 - 1, & -2 \le x \le 0 \\ x - 1, & 0 < x < 3 \end{cases}$

(a) Sketch the graph of $f$.
(b) Find the domain and range of $f$.
(c) Evaluate $f(-1)$, $f(2)$, and $f(4)$.
(d) Is $f$ continuous at 0?
(e) Evaluate

$$\lim_{x \to 0^-} \frac{f(x) - f(0)}{x - 0} \quad \text{and} \quad \lim_{x \to 0^+} \frac{f(x) - f(0)}{x - 0}.$$

(f) Is $f$ differentiable at 0?

**5** Find the domain of the function

$$f(x) = \frac{1}{\sqrt{3 - 2x}}.$$

**6** Find the domain of the function

$$f(x) = \frac{x^2 - 3x - 4}{x - 4}.$$

**7** A rectangular box has length $11 - 2x$, width $8 - 2x$, and height $x$. Express the volume of the box as a function of $x$. Indicate the domain.

**8** A right circular cylinder has radius $r$ and height $2 - r$. Express the total area of the lateral surface, the circular top, and the circular bottom of the cylinder as a function of $r$. Indicate the domain.

Sketch the graphs of the equations in Exercises 9–10 and determine which define $y$ as a function of $x$. Find the domain and range of those that do.

**9** $y^2 = 4x^2$

**10** $x + \sqrt{y} = 1$

Find a bound or indicate unbounded for each of the functions in Exercises 11–18.

**11** $f(x) = 3x^2 - x \cos 2x$; $|x| < 3$

**12** $f(x) = x \sin 3x - x^2 \cos 5x$, $|x| < 2$

**13** $f(x) = \dfrac{x}{2x - 0.8}$, $|x - 1| < 0.2$

**14** $f(x) = \dfrac{x - 1}{x + 1}$, $|x| < 0.2$

**15** $f(x) = \dfrac{x^3}{x^2 + 4}$, $|x - 2| < 2$

16 $f(x) = \dfrac{\cos x}{2 + \sin^2 x}$; $|x| < \pi$

17 $f(x) = \dfrac{6}{x - 2}$, $|x - 3| < 1$

18 $f(x) = \dfrac{1}{1 + \cos x}$, $|x| < \pi$

In Exercises 19–22, find a positive number $\delta$ such that the given statement is true for each positive number $\varepsilon$.

19 $|(2x - 1) - 5| < \varepsilon$ whenever $|x - 3| < \delta$.

20 $\left|\dfrac{x^2 - 1}{x - 1} - 2\right| < \varepsilon$ whenever $0 < |x - 1| < \delta$.

21 $|3x^{1/3}| < \varepsilon$ whenever $|x| < \delta$.

22 $|(x^2 - 3x + 1) - (-1)| < \varepsilon$ whenever $|x - 2| < \delta$. Assume $|x - 2| \leq 1$.

23 A 12-ft length of wood is to be measured by using a measuring stick that is approximately 3 ft long. How close to 3 ft must the measuring stick be to insure that the measured length is within 0.1 ft of 12 ft?

24 How close to 4 Ω should the resistance in a 12-V electrical circuit be in order that the current be within 0.05 A of 3 A? Assume the resistance is within at least 0.5 Ω of 4 Ω. $I = V/R$, where $I$ is the current in amperes, $V$ is the voltage in volts, and $R$ is the resistance in ohms.

25 Evaluate $\lim_{x\to 0} \cos(1/x)$.

26 Evaluate $\lim_{x\to 0} x \cos(1/x)$.

27 Find the slope of the line segment between the point (3, 9) and a variable point $(x, x^2)$ on the graph of $f(x) = x^2$. Find the equation of the line tangent to the graph at the point (3, 9).

28 The position of a particle on a number line at time $t$ is given by $s(t) = -16t^2 + 32t$. Find the average velocity of the particle over the interval of time from 2 and $t$. Find the velocity of the particle at time 2.

29 Use the definition of the derivative to show that $f(x) = x^{1/5}$ is not differentiable at 0. Explain what happens to the slope of the line segment between (0, 0) and $(x, x^{1/5})$ as $x$ approaches 0 from each side. Sketch the graph.

30 Use the definition of the derivative to show that $f(x) = x^{2/5}$ is not differentiable at 0. Explain what happens to the slope of the line segment between (0, 0) and $(x, x^{2/5})$ as $x$ approaches 0 from each side. Sketch the graph.

# CHAPTER THREE

# *Continuity and differentiability on an interval*

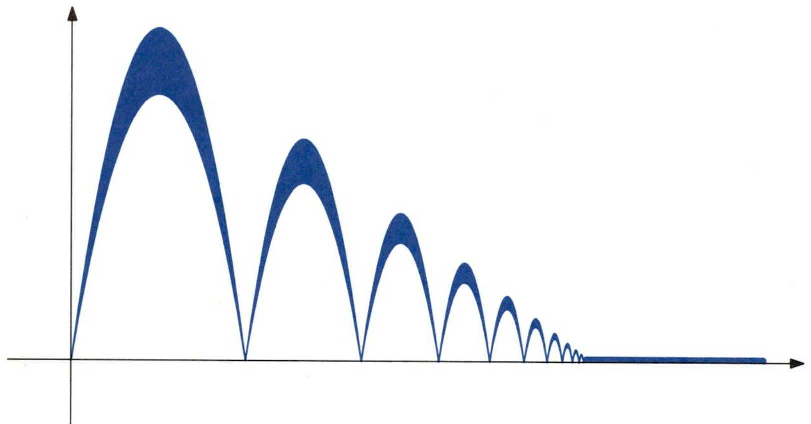

*I*n Chapter 2 we emphasized that the derivative $f'(c)$ is a number that is obtained as the limit of the difference quotient $[f(x) - f(c)]/(x - c)$ as $x$ approaches $c$. This point of view is important for establishing what a derivative is and how it relates to physical quantities. However, in order to use derivatives effectively in practical problems, we need a method of determining values of a derivative that is easier than using the definition as a limit in each individual case. The idea is to use the definition of the derivative to establish general differentiation formulas. We can then use the formulas for future work.

In this chapter, we will develop formulas for the derivatives of power functions and trigonometric functions. Also, we will develop formulas that allow us to determine the derivatives of sums, products, quotients, and compositions of functions. These formulas are essential for the efficient application of the derivative to practical problems.

We will also see how continuity and differentiability at each point of an interval can be used to obtain solutions to some of the fundamental problems of calculus. In particular, we will investigate the problems of finding approximate values of solutions of equations and finding largest and smallest values of functions. We will also see an example of how information about the derivative of a function can be used to obtain other information about the function. These are all problems that we will return to in later chapters.

## 3.1 POWER AND SUM RULES

In this section we will develop formulas for the derivatives of the power functions $f(x) = x^r$, where $r$ is a rational number. We will also see how to differentiate constant multiples of functions and sums of functions.

Before we begin to develop differentiation formulas, let us make the observation that, as $c$ varies over the set where $f$ is differentiable, the values $f'(c)$ determine a *function* $f'$. We will see that it is convenient to use the same independent variable for the functions $f$ and $f'$. For example, if $f$ is a function of $x$, the derivative of $f$ is the function $f'$ that has values $f'(x)$.

The symbols

$$\frac{df}{dx} \quad \text{and} \quad Df$$

are commonly used to denote the derivative $f'$. When the variable of differentiation represents time, the symbol $\dot{f}$ is sometimes used. We will use $f'$ and $df/dx$, but not the others. Both expressions

$$(\text{Formula})' \quad \text{and} \quad \frac{d}{dx}(\text{Formula})$$

represent the derivative of the function whose Formula is indicated. For functions given in the form $y = (\text{Formula})$, we use either $y'$ or $dy/dx$ to indicate the derivative.

Let us now begin a step-by-step development of the derivatives of the power functions. We start by deriving the formula for the differentiation of constant functions:

$$\frac{d}{dx}(k) = 0, \qquad k \text{ a constant.} \tag{1}$$

■ *Proof* To verify formula (1), we will use the definition of the derivative to show that, if $f(x) = k$, a constant function, then $f'(c) = 0$ for every $c$. That is, if $f(x) = k$, then

$$f'(c) = \lim_{x \to c} \frac{f(x) - f(c)}{x - c} = \lim_{x \to c} \frac{k - k}{x - c}$$

$$= \lim_{x \to c} \frac{0}{x - c} = \lim_{x \to c} 0 = 0.$$

[We have used the fact that $0/(x - c) = 0$ for $x \neq c$. Since limits at $c$ do not depend on values at $c$, it follows that $0/(x - c)$ and $0$ have equal limits at $c$.] ■

We next derive a formula for the derivative of power functions $x^n$, where $n$ is a positive integer.

$$\frac{d}{dx}(x^n) = nx^{n-1}, \qquad n \text{ a positive integer.} \tag{2}$$

■ *Proof* We must show that, if $f(x) = x^n$, where $n$ is a positive integer, then $f'(c) = nc^{n-1}$ for every $c$. To do this, we will use the factorization formula

$$x^n - c^n = (x - c)(x^{n-1} + cx^{n-2} + c^2x^{n-3} + \cdots + c^{n-2}x + c^{n-1}),$$

which can be verified by carrying out the multiplication and combining like terms on the right-hand side. We then have

$$\begin{aligned} f'(c) &= \lim_{x\to c} \frac{f(x) - f(c)}{x - c} = \lim_{x\to c} \frac{x^n - c^n}{x - c} \\ &= \lim_{x\to c} \frac{(x - c)(x^{n-1} + cx^{n-2} + c^2x^{n-3} + \cdots + c^{n-2}x + c^{n-1})}{x - c} \\ &= \lim_{x\to c} (x^{n-1} + cx^{n-2} + c^2x^{n-3} + \cdots + c^{n-2}x + c^{n-1}). \end{aligned}$$

Since polynomials are continuous, the latter limit is the value of the polynomial at $c$. We then obtain

$$f'(c) = c^{n-1} + cc^{n-2} + c^2c^{n-3} + \cdots + c^{n-2}c + c^{n-1}.$$

Each of the $n$ terms on the right equals $c^{n-1}$, so

$$f'(c) = nc^{n-1}.$$

■

Formulas (1) and (2) are used to write the derivatives of constant functions and power functions that have exponents that are positive integers.

### ■ EXAMPLE 1

Using (1) and (2), we have

(a) $\frac{d}{dx}(3) = 0,$

(b) $\frac{d}{dx}(-7) = 0,$

(c) $\frac{d}{dx}(x^2) = 2x^{2-1} = 2x^1 = 2x,$

(d) $\dfrac{d}{dx}(x^{15}) = 15x^{15-1} = 15x^{14}$,

(e) $\dfrac{d}{dx}(x) = \dfrac{d}{dx}(x^1) = 1x^{1-1} = 1x^0 = 1 \cdot 1 = 1$. ■

To extend our results for the differentiation of power functions to include rational exponents, we use (2) to verify the following:

If $f(x) = x^{m/n}$, $m$ and $n$ positive integers, then

$$f'(c) = \frac{m}{n} c^{m/n - 1} \tag{3}$$

for every $c > 0$.

■ *Proof* The derivative of $x^{m/n}$ at $c$ is

$$f'(c) = \lim_{x \to c} \frac{x^{m/n} - c^{m/n}}{x - c}.$$

To evaluate this limit, let us set

$$u = x^{1/n} \quad \text{and} \quad d = c^{1/n}.$$

Then $x = u^n$, $c = d^n$, $x^{m/n} = u^m$, and $c^{m/n} = d^m$. If $x \neq c$, then $u \neq d$, and we can write

$$\begin{aligned} \frac{x^{m/n} - c^{m/n}}{x - c} &= \frac{u^m - d^m}{u^n - d^n} \\ &= \frac{u^m - d^m}{u - d} \cdot \frac{u - d}{u^n - d^n} \\ &= \frac{u^m - d^m}{u - d} \cdot \frac{1}{\dfrac{u^n - d^n}{u - d}}. \end{aligned}$$

The first factor above is the difference quotient for the derivative of the function $u^m$ at the point $d$. From (2), we know that this approaches the derivative $md^{m-1}$ as $u$ approaches $d$. The quotient must approach the same value as $x$ approaches $c$, because $u = x^{1/n} \to c^{1/n} = d$ as $x \to c$. The latter fact is a consequence of the continuity at $c$ of the function $x^{1/n}$. Similarly, the quotient in the denominator of the second factor approaches the derivative $nd^{n-1}$ as $x$ approaches $c$. Collecting results, we have

$$f'(c) = md^{m-1} \cdot \frac{1}{nd^{n-1}} = \frac{m}{n} d^{m-n} = \frac{m}{n}(c^{1/n})^{m-n} = \frac{m}{n} c^{m/n - 1}.$$ ■

If $n$ is an odd integer, then the formula for the derivative of $x^{m/n}$ given in (3) also holds for $c < 0$. The above proof holds in this case. If $n$ is odd and

$m/n \geq 1$, the formula holds for $c = 0$. This follows from a straightforward application of the definition of derivative. If $m/n < 1$, $x^{m/n}$ is not differentiable at zero.

**EXAMPLE 2**

From formula (3) we have

$$\text{(a)} \quad \frac{d}{dx}(x^{2/3}) = \frac{2}{3}x^{2/3-1} = \frac{2}{3}x^{-1/3},$$

$$\text{(b)} \quad \frac{d}{dx}(x^{1.7}) = 1.7x^{0.7},$$

$$\text{(c)} \quad \frac{d}{dx}(\sqrt{x}) = \frac{d}{dx}(x^{1/2}) = \frac{1}{2}x^{1/2-1},$$

$$= \frac{1}{2}x^{-1/2} = \frac{1}{2x^{1/2}} = \frac{1}{2\sqrt{x}}.$$

The square root function occurs so often that it is useful to remember that

$$\frac{d}{dx}(\sqrt{x}) = \frac{1}{2\sqrt{x}}.$$

Statement (3) gives us the formula for the derivative of $x^r$, where $r$ is a positive rational number. It is not difficult to remove the requirement that $r$ be positive.

If $f(x) = x^{-r}$, $r$ a positive rational, then

$$f'(c) = -rc^{-r-1} \qquad \text{for every } c > 0. \tag{4}$$

*Proof* We use (3) to prove (4). We have

$$f'(c) = \lim_{x \to c} \frac{f(x) - f(c)}{x - c} = \lim_{x \to c} \frac{x^{-r} - c^{-r}}{x - c}$$

$$= \lim_{x \to c} \frac{\frac{1}{x^r} - \frac{1}{c^r}}{x - c} = \lim_{x \to c} \frac{\frac{c^r - x^r}{x^r \cdot c^r}}{x - c}$$

$$= \lim_{x \to c} \frac{c^r - x^r}{(x - c)x^r c^r}$$

$$= \lim_{x \to c} \left( -\frac{x^r - c^r}{x - c} \cdot \frac{1}{x^r c^r} \right).$$

Formula (3) implies that the limit at $c$ of the first factor above is $rc^{r-1}$. The continuity at $c$ of $x^r$ gives the limit of the second factor. We obtain

$$f'(c) = -rc^{r-1} \cdot \frac{1}{c^r c^r} = -rc^{-r-1}.$$

Let us summarize our results for the derivatives of power functions in terms of the derivative as a function.

*Theorem 1*

$$\frac{d}{dx}(x^r) = rx^{r-1}, \qquad r \text{ a rational number.}$$

We can use Theorem 1 to find the derivative of any power function that has a rational number exponent.

**EXAMPLE 3**

Using Theorem 1, we have

$$\text{(a)}\ \frac{d}{dx}\left(\frac{1}{x}\right) = \frac{d}{dx}(x^{-1}) = (-1)x^{-1-1} = -x^{-2},$$

$$\text{(b)}\ \frac{d}{dx}\left(\frac{1}{\sqrt{x}}\right) = \frac{d}{dx}(x^{-1/2}) = -\frac{1}{2}x^{-3/2}.$$

We will need formulas that tell us how to differentiate various combinations of functions. Our first result of this type is the following:

*Theorem 2*

**If $f$ and $g$ are differentiable at $c$ and $a$ is a constant then $af$ and $f+g$ are differentiable at $c$. Also,**

$$(af)'(c) = af'(c) \quad \text{and} \quad (f+g)'(c) = f'(c) + g'(c).$$

*Proof* Using the definition of the derivative and the Limit Theorem to verify the first statement of Theorem 2, we have

$$\begin{aligned}(af)'(c) &= \lim_{x\to c}\frac{(af)(x)-(af)(c)}{x-c}\\ &= \lim_{x\to c}\frac{af(x)-af(c)}{x-c}\\ &= \lim_{x\to c} a\frac{f(x)-f(c)}{x-c}\\ &= a\left[\lim_{x\to c}\frac{f(x)-f(c)}{x-c}\right] = af'(c).\end{aligned}$$

The second statement of Theorem 2 is verified by rearranging terms and using the Limit Theorem.

$$\begin{aligned}(f+g)'(c) &= \lim_{x\to c}\frac{(f+g)(x)-(f+g)(c)}{x-c}\\ &= \lim_{x\to c}\frac{[f(x)+g(x)]-[f(c)+g(c)]}{x-c}\\ &= \lim_{x\to c}\left[\frac{f(x)-f(c)}{x-c}+\frac{g(x)-g(c)}{x-c}\right]\end{aligned}$$

$$= \lim_{x \to c} \frac{f(x) - f(c)}{x - c} + \lim_{x \to c} \frac{g(x) - g(c)}{x - c}$$

$$= f'(c) + g'(c). \qquad \blacksquare$$

Theorem 2 tells us that the derivative of a constant multiple of a differentiable function is the constant times the derivative of the function, and that the derivative of a sum of differentiable functions is the sum of the derivatives of the functions. The second statement may be extended to apply to the sum of any finite number of terms.

**■ EXAMPLE 4**

Using Theorem 2, we have

(Power function) + (Constant)(Power function) + (Constant function)

$$\text{(a)} \quad \frac{d}{dx}(x^2 - 3x + 7)$$

$$= 2x - 3(1) + 0,$$

(Derivative of power function) + (Constant)(Derivative of power function) + (Derivative of constant function)

$$\text{(b)} \quad \frac{d}{dx}(2x^4 - 3x^3 + x^2 - x + 4) = 2(4x^3) - 3(3x^2) + 2x - 1 + 0$$

$$= 8x^3 - 9x^2 + 2x - 1. \qquad \blacksquare$$

In some cases we can carry out a multiplication or division in order to express $f(x)$ in a form for which we can easily determine its derivative.

**■ EXAMPLE 5**

$$\text{(a)} \quad \frac{d}{dx}(x(x-1)^2) = \frac{d}{dx}(x(x^2 - 2x + 1))$$

$$= \frac{d}{dx}(x^3 - 2x^2 + x)$$

$$= 3x^2 - 2(2x) + 1 = 3x^2 - 4x + 1,$$

$$\text{(b)} \quad \frac{d}{dx}\left(\frac{x^2 - 3x + 5}{x}\right) = \frac{d}{dx}\left(\frac{x^2}{x} - \frac{3x}{x} + \frac{5}{x}\right)$$

$$= \frac{d}{dx}(x - 3 + 5x^{-1})$$

$$= 1 - 0 + 5(-1)x^{-1-1} = 1 - 5x^{-2}. \qquad \blacksquare$$

■ **EXAMPLE 6**

Find the equation of the line tangent to the graph of $f(x) = x^2 - 5x + 1$ at the point on the graph with $x = 2$.

**SOLUTION**

We need the coordinates of the point on the graph that has $x = 2$ and the slope of the line tangent at that point.

$$f(2) = (2)^2 - 5(2) + 1 = -5,$$

so $(2, -5)$ is the point on the graph with $x = 2$.

The slope of the line tangent to the graph of $f$ at $(2, -5)$ is $f'(2)$, the value of the derivative at 2.

$$f(x) = x^2 - 5x + 1, \quad \text{so}$$

$$f'(x) = 2x - 5.$$

Substituting 2 for $x$ in the formula for $f'(x)$, we have

$$f'(2) = 2(2) - 5 = -1.$$

[We now use differentiation formulas to evaluate derivatives. We do not determine $f'(2)$ by evaluating the limit as $x$ approaches 2 of the slope of the line segments between $(2, -5)$ and a variable point on the graph, as we did in Section 2.6.]

The point-slope form of the equation of the line through $(2, -5)$ with slope $-1$ is

$$y - (-5) = (-1)(x - 2).$$

Simplifying, we have

$$y + 5 = -x + 2,$$

$$x + y + 3 = 0.$$

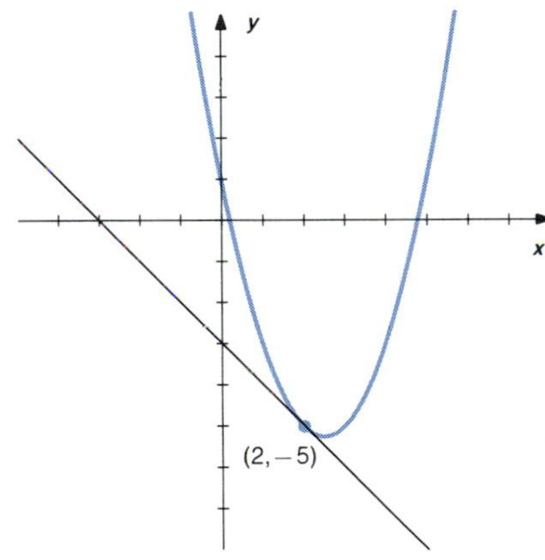

FIGURE 3.1.1

The graph is sketched in Figure 3.1.1. ■

In Example 6 we used the formula for $f(x)$ to determine the $y$-coordinate of the point on the graph with $x = 2$. The formula for $f'(x)$ was used to determine the slope of the line tangent to the graph at $(2, f(2))$. In many problems we will deal with both a function $f$ and its derivative $f'$. Always label your functions so you can tell which function is which. Don't just write the formula for $f'(x)$, write "$f'(x) =$ " followed by the formula.

■ **EXAMPLE 7**

The height in feet above ground level of an object $t$ seconds after it is thrown vertically from ground level with initial velocity 48 ft/s is given by the formula $s(t) = 48t - 16t^2$. The formula is valid until the object returns to ground level. Find the velocity of the object at time $t$. What is the position and velocity of the object when $t = 3$?

**SOLUTION**

The (instantaneous) velocity is the derivative $s'(t)$.

$$s(t) = 48t - 16t^2, \quad \text{so}$$

$$s'(t) = 48(1) - 16(2t)$$
$$= 48 - 32t.$$

When $t = 3$, the position is

$$s(3) = 48(3) - 16(3)^2 = 144 - 144 = 0.$$

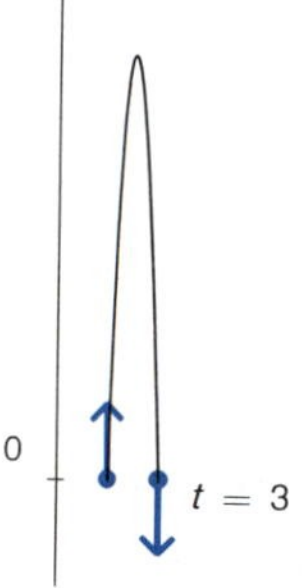

FIGURE 3.1.2

When $t = 3$, the object is 0 ft above ground level. This means that the object returns to ground level 3 s after it is thrown.

When $t = 3$, the velocity is

$$s'(3) = 48 - 32(3) = 48 - 96 = -48.$$

The fact that the velocity at $t = 3$ is negative indicates that the particle is moving in the negative direction (downward) at that time. The object hits the ground at a speed of 48 ft/s, which is the same speed at which it was thrown. The position and velocity of the particle at time $t = 3$ are indicated in Figure 3.1.2. ■

■ **EXAMPLE 8**

$f(x) = x - 3x^{1/3}$. Find all points $(x, f(x))$ on the graph of $f$ for which (a) $f'(x) = 0$ and (b) $f'(x)$ does not exist.

**SOLUTION**

$f(x) = x - 3x^{1/3}$. Applying our rules for differentiation, we have

$$f'(x) = 1 - 3\left(\frac{1}{3}x^{-2/3}\right).$$

Simplifying, we have

$$f'(x) = 1 - x^{-2/3}$$
$$= x^{-2/3}(x^{2/3} - 1)$$
$$= \frac{x^{2/3} - 1}{x^{2/3}}.$$

(a) $f'(x)$ is zero at those values of $x$ for which the numerator is zero. We obtain the equation

$$x^{2/3} - 1 = 0,$$

$$x^{2/3} = 1,$$

$$(x^{2/3})^3 = 1^3,$$

$$x^2 = 1,$$

$$x = 1 \quad \text{or} \quad -1.$$

Therefore, $f'(x) = 0$ when $x = 1$ or $-1$. We have

$$f(1) = 1 - 3(1)^{1/3} = 1 - 3(1) = -2,$$

so $(1, -2)$ is a point on the graph where the derivative is zero. We see that

$$f(-1) = -1 - 3(-1)^{1/3} = -1 - 3(-1) = 2,$$

so $(-1, 2)$ is also a point on the graph where the derivative is zero.

(b) The expression we obtained for $f'(x)$ is undefined for $x = 0$, because we cannot divide by zero. It is also true that $f'(x)$ does not exist for $x = 0$. We have

$$f(0) = 0 - 3(0)^{1/3} = 0,$$

so $(0, 0)$ is a point on the graph where the derivative does not exist.

The graph is sketched in Figure 3.1.3. It has horizontal tangents at $(1, -2)$ and $(-1, 2)$, where the derivative is zero. The graph becomes vertical as it passes through the origin, where the derivative does not exist. ■

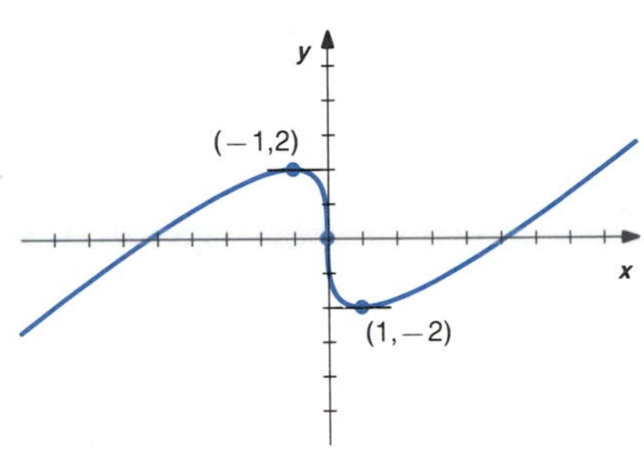


FIGURE 3.1.3

## EXERCISES 3.1

In Exercises 1–20 find the derivatives of the given functions.

1 $f(x) = x^2 - 7x + 3$

2 $f(x) = 3x^2 + 4x - 2$

3 $f(x) = 2x^3 - x + 1$

4 $f(x) = x^4 - 3x^2 + 4$

5 $f(x) = 2x - 3 + x^{-1}$

6 $f(x) = x^2 + x^{-2}$

7 $f(x) = x^{1/2} + x^{-1/2}$

8 $f(x) = x^{1/3} + x^{-1/3}$

9 $f(x) = x^{4/3} - 4x^{1/3}$

10 $f(x) = 6x^{2/3} - 9x^{-1/3}$

11 $f(x) = 2x^{1.3} + 5x^{0.3}$

12 $f(x) = 5x^{-1.6} + 10x^{-0.6}$

13 $f(x) = \sqrt{x}(2x + 1)$

14 $f(x) = \dfrac{3x + 1}{\sqrt{x}}$

15 $f(x) = \dfrac{(x + 1)^2}{4x}$

16 $f(x) = \sqrt{x}(\sqrt{x} - 1)^2$

17 $f(t) = 3(t^2 + 1)^2$

18 $f(t) = (3 - 2t)(1 + t)$

19 $f(t) = \dfrac{1 + \sqrt{t}}{2t}$

20 $f(t) = \dfrac{4 + \sqrt{t}}{3t^2}$

In Exercises 21–26 find the equation of the line tangent to the graph of $y = f(x)$ at the point on the graph with $x = c$.

21 $f(x) = 3x + 1$, $c = -2$

22 $f(x) = 2 - 3x$, $c = 1$

23 $f(x) = x^2 - 2$, $c = 2$

24 $f(x) = 3x^2 - 4x + 1$, $c = 2$

25 $f(x) = 1/x$, $c = -2$

26 $f(x) = 1/x^2$, $c = -2$

In Exercises 27–32 the given function represents the position on a line of a particle at time $t$. Find the velocity of the particle at time $t$. Evaluate the position and velocity at the given time $t_1$.

27 $s(t) = 50t - 16t^2$, $t_1 = 3$

28 $s(t) = 60t - 16t^2$, $t_1 = 2$

29 $s(t) = t + 4t^3$, $t_1 = 1$

30 $s(t) = t^3 - t$, $t_1 = 2$

31 $s(t) = \sqrt{t}$, $t_1 = 4$

32 $s(t) = t^{1/3}$, $t_1 = 8$

In Exercises 33–38 find all points $(x, f(x))$ on the graph of $f$ for which (a) $f'(x) = 0$ and (b) $f'(x)$ does not exist.

**33** $f(x) = x^2 - 6x + 1$

**34** $f(x) = x^3 + x^2 - x - 1$

**35** $f(x) = x + 4x^{-1}$

**36** $f(x) = 2x^{1/3} + x^{-2/3}$

**37** $f(x) = x^{4/3} + 4x^{1/3}$

**38** $f(x) = x^{7/3} - 7x^{1/3}$

**39** Find all points on the graph of $f(x) = x^3 - x$ that have tangent lines with slope 2.

**40** Find all points on the graph of $f(x) = x^2 - 2x$ that have tangent lines parallel to the line $8x - 2y = 5$.

**41** Find all points on the graph of $f(x) = x^2$ that have tangent lines that contain the point (5, 9). [*Hint*: The equation of the line tangent to the graph of $f(x) = x^2$ at a point $(x_0, y_0)$ on the graph is $y - x_0^2 = 2x_0(x - x_0)$. If this line contains the point (5, 9), then $(x, y) = (5, 9)$ must satisfy the equation of the line. Substitute 5 for $x$ and 9 for $y$, and solve the resulting equation for $x_0$.]

**42** Find all points on the graph of $f(x) = x^3$ that have tangent lines that intersect the $x$-axis at the single point (4, 0).

**43** Find all points of intersection of the parabola $y = x^2$ and the line $x + y = 2$. At each point of intersection, find the acute angle of intersection of the line $x + y = 2$ and the line tangent to the parabola at the point. (*Hint*: Use the formula for the tangent of the difference of two angles. The formula is given in Section 1.7.)

**44** Find all points of intersection of the graphs of $y = x^3$ and $y = 3x^2 - 2x$. At each point of intersection, find the acute angle of intersection of the lines tangent to the graphs at the point.

## 3.2 PRODUCT AND QUOTIENT RULES

We develop formulas that are used to find the derivatives of products and quotients.

■ *Theorem 1*

**If $f$ and $g$ are differentiable at $c$, then $fg$ is differentiable at $c$ and**

$$(fg)'(c) = f(c)\cdot g'(c) + g(c)\cdot f'(c).$$

■ *Proof* We will use the algebraic trick of adding and subtracting $f(x)g(c)$ in the numerator to evaluate the limit in the definition of the derivative.

$$\begin{aligned}
(fg)'(c) &= \lim_{x\to c}\frac{(fg)(x) - (fg)(c)}{x - c}\\
&= \lim_{x\to c}\frac{f(x)g(x) - f(c)g(c)}{x - c}\\
&= \lim_{x\to c}\frac{f(x)g(x) - f(x)g(c) + f(x)g(c) - f(c)g(c)}{x - c}\\
&= \lim_{x\to c}\left[f(x)\frac{g(x) - g(c)}{x - c} + g(c)\frac{f(x) - f(c)}{x - c}\right]\\
&= \lim_{x\to c} f(x)\lim_{x\to c}\frac{g(x) - g(c)}{x - c} + g(c)\lim_{x\to c}\frac{f(x) - f(c)}{x - c}\\
&= f(c)\cdot g'(c) + g(c)\cdot f'(c).
\end{aligned}$$

The above limits were evaluated by using the definition of the derivatives of $f$ and $g$, and the continuity of $f$. Since $f$ is differentiable at $c$, $f$ is continuous at $c$, so $\lim_{x\to c} f(x) = f(c)$. ■

The Product Rule does not depend on the names of the functions involved. Think of Theorem 1 as saying

**(Derivative of a product) = (First factor)(Derivative of second) + (Second factor)(Derivative of first).**

When the functions involved are given by formulas, we should write the derivative directly, in one step. It is helpful to use the above form every time you differentiate a product. The use of all four sets of parentheses on the right-hand side helps to organize your work.

■ **EXAMPLE 1**

Using Theorem 1, we have

(a) $$\frac{d}{dx}\Big(\underbrace{(x^2+1)}_{\substack{\uparrow\\ \text{First factor}}}\underbrace{(3x-2)}_{\substack{\uparrow\\ \text{Second factor}}}\Big) = \underbrace{(x^2+1)}_{\substack{\uparrow\\ \text{First factor}}}\ \underbrace{(3)}_{\substack{\uparrow\\ \text{Derivative of second}}} + \underbrace{(3x-2)}_{\substack{\uparrow\\ \text{Second factor}}}\ \underbrace{(2x)}_{\substack{\uparrow\\ \text{Derivative of first}}},$$

(b) $$\frac{d}{dx}((x-2)(x^2+3x+1)) = (x-2)(2x+3) + (x^2+3x+1)(1),$$

(c) $$\frac{d}{dx}((3x-2)(x^2+x^{-2})) = (3x-2)(2x-2x^{-3}) + (x^2+x^{-2})(3).$$

■

The derivatives in Example 1 have not been simplified. Simplifying is an additional step, distinct from using the Product Rule to determine the derivatives. Simplification of expressions that arise from the application of differentiation formulas is discussed in Section 1.2.

■ *Theorem 2*

**If $f$ and $g$ are differentiable at $c$ and $g(c) \neq 0$, then $f/g$ is differentiable at $c$ and**

$$\left(\frac{f}{g}\right)'(c) = \frac{g(c)f'(c) - f(c)g'(c)}{[g(c)]^2}.$$

■ *Proof* The proof of Theorem 2 is similar to that of Theorem 1. We use the fact that $g$ is continuous at $c$, so $g(c) \neq 0$ implies $g(x) \neq 0$ for $x$ near $c$.

$$\left(\frac{f}{g}\right)'(c) = \lim_{x\to c}\frac{\left(\frac{f}{g}\right)(x) - \left(\frac{f}{g}\right)(c)}{x-c} = \lim_{x\to c}\frac{\frac{f(x)}{g(x)} - \frac{f(c)}{g(c)}}{x-c}$$

$$= \lim_{x\to c}\frac{\frac{g(c)f(x) - f(c)g(x)}{g(x)g(c)}}{x-c}$$

$$= \lim_{x \to c} \frac{\dfrac{g(c)f(x) - g(c)f(c) + g(c)f(c) - f(c)g(x)}{g(x)g(c)}}{x - c}$$

$$= \lim_{x \to c} \frac{g(c)[f(x) - f(c)] - f(c)[g(x) - g(c)]}{(x - c)g(x)g(c)}$$

$$= \lim_{x \to c} \frac{\dfrac{g(c)[f(x) - f(c)] - f(c)[g(x) - g(c)]}{x - c}}{g(x)g(c)}$$

$$= \lim_{x \to c} \frac{g(c)\dfrac{f(x) - f(c)}{x - c} - f(c)\dfrac{g(x) - g(c)}{x - c}}{g(x)g(c)}.$$

The Limit Theorem then gives

$$\left(\frac{f}{g}\right)'(c) = \frac{g(c)f'(c) - f(c)g'(c)}{[g(c)]^2}.$$ ■

Think of Theorem 2 as saying

$$\textbf{(Derivative of a quotient)} = \frac{\begin{pmatrix}\text{Bottom}\\ \text{factor}\end{pmatrix}\begin{pmatrix}\text{Derivative}\\ \text{of top}\end{pmatrix} - \begin{pmatrix}\text{Top}\\ \text{factor}\end{pmatrix}\begin{pmatrix}\text{Derivative}\\ \text{of bottom}\end{pmatrix}}{(\text{Bottom factor})^2}.$$

We must be especially careful when using the Quotient Rule. Be sure each term is in its proper place. Using all five sets of parentheses on the right-hand side helps to organize your work.

■ **EXAMPLE 2**

From Theorem 2, we have

(a) $$\frac{d}{dx}\left(\frac{3x - 2}{x^2 + 1}\right) = \frac{(x^2 + 1)\;(3) - (3x - 2)\;(2x)}{(x^2 + 1)^2},$$

(Top factor) ↓ $3x - 2$; (Bottom factor) ↑ $x^2 + 1$; (Bottom factor) ↓ $(x^2+1)$; (Derivative of top) ↓ $(3)$; (Top factor) ↓ $(3x-2)$; (Derivative of bottom) ↓ $(2x)$; (Bottom factor, squared) ↑ $(x^2+1)^2$

(b) $$\frac{d}{dx}\left(\frac{x - 1}{2x + 1}\right) = \frac{(2x + 1)(1) - (x - 1)(2)}{(2x + 1)^2},$$

(c) $$\frac{d}{dx}\left(\frac{x^3}{x^2 + 4}\right) = \frac{(x^2 + 4)(3x^2) - (x^3)(2x)}{(x^2 + 4)^2}.$$ ■

Don't use the Product Rule or the Quotient Rule when one of the factors is a constant. That would give the correct derivative, but it is needlessly complicated. It's easier to use

$$(af)' = a \cdot f', \quad \left(\frac{f}{b}\right)' = \left(\frac{1}{b} \cdot f\right)' = \frac{1}{b} \cdot f', \quad \text{and}$$

$$\left(\frac{a}{x^r}\right)' = (ax^{-r})' = a(-rx^{-r-1}).$$

In Section 3.4 we will learn to differentiate functions of the form $a/g = a \cdot g^{-1}$ without using the Quotient Rule.

**■ EXAMPLE 3**

$$\text{(a)} \ \frac{d}{dx}\left(\frac{4x-3}{7}\right) = \frac{d}{dx}\left[\frac{1}{7}(4x-3)\right] = \frac{1}{7}(4) = \frac{4}{7},$$

$$\text{(b)} \ \frac{d}{dx}\left(\frac{4}{x^2}\right) = \frac{d}{dx}(4x^{-2}) = 4(-2x^{-3}) = -\frac{8}{x^3}. \quad ■$$

In some cases you may find it easier to carry out a multiplication or division instead of using the Product or Quotient Rules. For example, we may write either

$$\begin{aligned}\frac{d}{dx}(x(x^2 - 3x + 2)) &= \frac{d}{dx}(x^3 - 3x^2 + 2x) \\ &= 3x^2 - 6x + 2\end{aligned}$$

or

$$\begin{aligned}\frac{d}{dx}(x(x^2 - 3x + 2)) &= (x)(2x - 3) + (x^2 - 3x + 2)(1) \\ &= 2x^2 - 3x + x^2 - 3x + 2 \\ &= 3x^2 - 6x + 2.\end{aligned}$$

**■ EXAMPLE 4**

Find the equation of the line tangent to the graph of

$$f(x) = (2x^2 - 3x + 1)(3x^2 + 5x - 4)$$

at the point where $x = 2$.

**SOLUTION**

We need the coordinates of the point on the graph where $x = 2$ and the slope of the line tangent at that point.

$$\begin{aligned}f(2) &= [2(2)^2 - 3(2) + 1][3(2)^2 + 5(2) - 4] \\ &= (3)(18) = 54,\end{aligned}$$

so the point (2, 54) is on the graph. The slope of the line tangent to the graph at (2, 54) is the value of the derivative $f'(2)$. It is not convenient to carry out the multiplication of the factors of $f(x)$, so we use the Product Rule to find the derivative.

$$f(x) = (2x^2 - 3x + 1)(3x^2 + 5x - 4), \quad \text{so}$$

$$f'(x) = (2x^2 - 3x + 1)(6x + 5) + (3x^2 + 5x - 4)(4x - 3).$$

Substitution in the formula for $f'$ gives

$$\begin{aligned} f'(2) &= [2(2)^2 - 3(2) + 1][6(2) + 5] + [3(2)^2 + 5(2) - 4][4(2) - 3] \\ &= (3)(17) + (18)(5) = 141. \end{aligned}$$

The point–slope form of the equation of the line through the point (2, 54) with slope 141 is

$$y - 54 = 141(x - 2).$$

Simplifying, we obtain

$$y = 141x - 282 + 54,$$

$$y = 141x - 228.$$

■

### ■ EXAMPLE 5

$f(x) = x^{1/3}/(3x - 2)$. Find all points $(x, f(x))$ on the graph of $f$ for which (a) $f'(x) = 0$ and (b) $f'(x)$ does not exist.

**SOLUTION**

Since

$$f(x) = \frac{x^{1/3}}{3x - 2},$$

the Quotient Rule gives

$$f'(x) = \frac{(3x - 2)\left(\frac{1}{3}x^{-2/3}\right) - (x^{1/3})(3)}{(3x - 2)^2}.$$

To find values of $x$ for which $f'(x) = 0$ and values of $x$ for which $f'(x)$ does not exist, we simplify the expression for $f'(x)$. Each factor in the numerator contains a power of $x$. Factoring the smaller power of $x$ from each term in the numerator, we have

$$f'(x) = \frac{(3x - 2)\left(\frac{1}{3}x^{-2/3}\right) - (x^{1/3})(3)}{(3x - 2)^2}$$

$$= \frac{(3x-2)\left(\frac{x^{-2/3}}{3}\right) - \left(\frac{9x^{1-2/3}}{3}\right)}{(3x-2)^2}$$

$$= \frac{\left(\frac{x^{-2/3}}{3}\right)[(3x-2)-9x]}{(3x-2)^2}$$

$$= \frac{-6x-2}{3x^{2/3}(3x-2)^2}.$$

(a) $f'(x) = 0$ when $-6x - 2 = 0$, or $x = -1/3$. When $x = -1/3$,

$$y = f\left(-\frac{1}{3}\right) = \frac{\left(-\frac{1}{3}\right)^{1/3}}{3\left(-\frac{1}{3}\right)-2} = \frac{-\left(\frac{1}{3}\right)^{1/3}}{-3} = \left(\frac{1}{3}\right)^{4/3} \approx 0.23112043.$$

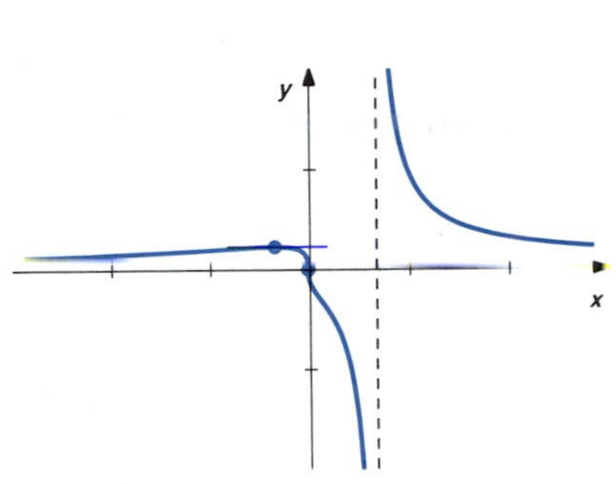


FIGURE 3.2.1

Therefore, $(-1/3, (1/3)^{4/3})$ is a point on the graph with $f'(x) = 0$. The graph has a horizontal tangent at this point.

(b) We see that $f'(x)$ does not exist when $3x^{2/3}(3x-2)^2 = 0$; that is, when $x = 0$ or $x = 2/3$. When $x = 0$, $y = f(0) = 0$. The coordinates $(0, 0)$ represent a point on the graph where $f'(x)$ does not exist. The graph has a vertical tangent at $(0, 0)$. There is no point on the graph of $f$ with $x = 2/3$. See Figure 3.2.1. ■

## EXERCISES 3.2

Use the Product Rule or the Quotient Rule to evaluate the derivatives in Exercises 1–8. Use parentheses to indicate each factor in the formula you are using and then simplify your answers.

1 $\frac{d}{dx}((x^2+1)(2x^2-3x+1))$

2 $\frac{d}{dx}((2x-3)(2x^2+5x-7))$

3 $\frac{d}{dx}((x+x^{-1})(x-x^{-1}))$

4 $\frac{d}{dx}((1+\sqrt{x})(x+\sqrt{x}))$

5 $\frac{d}{dx}\left(\frac{3x}{x^2+1}\right)$

6 $\frac{d}{dx}\left(\frac{2x-7}{x^2+9}\right)$

7 $\frac{d}{dx}\left(\frac{2x-1}{4x+1}\right)$

8 $\frac{d}{dx}\left(\frac{3x+1}{2x-1}\right)$

In Exercises 9–14 find the equation of the line tangent to the graph of $y = f(x)$ at the point on the graph where $x = c$.

9 $f(x) = (x^2+3x+1)(x^3-4x+1)$, $c = 0$

10 $f(x) = (3x-5)(x^3+2x^2+3x+4)$, $c = 0$

11 $f(x) = \frac{x^2+9}{5}$, $c = 4$

12 $f(x) = \frac{x^2+4}{x}$, $c = 2$

13 $f(x) = \frac{x+1}{x-1}$, $c = 2$

14 $f(x) = \frac{x^2-1}{x^2+1}$, $c = 1$

In Exercises 15–18 find all points $(x, f(x))$ on the graph of $f$ for which (a) $f'(x) = 0$ and (b) $f'(x)$ does not exist.

**15** $f(x) = \dfrac{x-1}{x+1}$

**16** $f(x) = \dfrac{x}{x^2+1}$

**17** $f(x) = \dfrac{x^{1/3}}{x^2+5}$

**18** $f(x) = \dfrac{x^{2/3}}{2x+1}$

In Exercises 19–23 use the Product Rule or the Quotient Rule to find formulas for the derivatives of functions in the given form.

**19** $[f(x)]^2$

**20** $[f(x)]^3$

**21** $f(x) \cdot g(x) \cdot h(x)$

**22** $\dfrac{1}{g(x)}$

**23** $\dfrac{1}{[g(x)]^2}$

**24** Use the fact that

$$\frac{d}{dx}(x^r) = rx^{r-1}$$

for positive rational numbers $r$ to verify that

$$\frac{d}{dx}(x^{-r}) = -rx^{-r-1},$$

by applying the Quotient Rule to find the derivative of $1/x^r$.

**25** $F(x) = x \cdot f(x)$, where $f(2) = 3$ and $f'(2) = 5$. Find $F'(2)$.

**26** $F(x) = f(x)/x^2$, where $f(2) = 16$ and $f'(2) = 12$. Find $F'(2)$.

## 3.3 DIFFERENTIATION OF TRIGONOMETRIC FUNCTIONS

In this section we will develop differentiation formulas for the six basic trigonometric functions. Radian measure of angles will be used to develop the formulas. (Properties of trigonometric functions and radian measure are reviewed in Chapter 1.)

We will need the following limit result to verify the formulas for the differentiation of the sine and cosine functions:

$$\lim_{\theta \to 0} \frac{\sin \theta}{\theta} = 1, \qquad \theta \text{ in radians.} \tag{1}$$

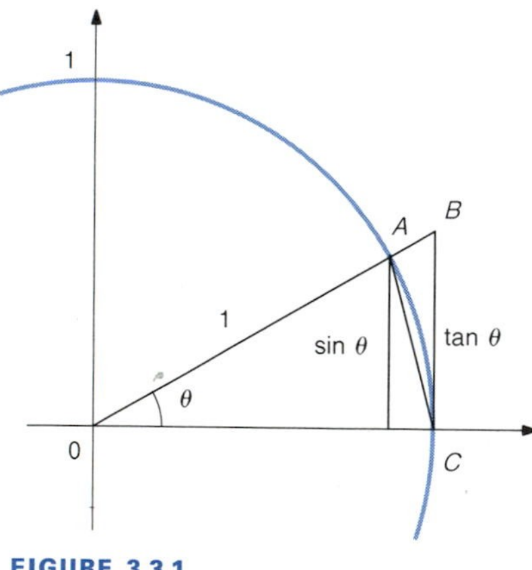


FIGURE 3.3.1

■ *Proof* We see that the limit in (1) cannot be evaluated by factoring and then cancelling $\theta$. We will use a geometric argument to evaluate the limit. We first assume that $0 < \theta < \pi/2$.

From Figure 3.3.1 we see that

$$(\text{Area triangle } OAC) < (\text{Area sector } OAC) < (\text{Area triangle } OBC). \tag{2}$$

Since the area of a sector is proportional to its central angle, we have

$$\frac{(\text{Area of sector } OAC)}{\theta} = \frac{(\text{Area of circle})}{2\pi} = \frac{\pi(1)^2}{2\pi}.$$

It follows that (Area of sector $OAC$) $= \theta/2$. Triangle $OAC$ has height $\sin\theta$ and base 1. Triangle $OBC$ is a right triangle with height $\tan\theta$ and base 1. Substitution into (2) then gives

$$\frac{1}{2}(1)(\sin\theta) < \frac{\theta}{2} < \frac{1}{2}(1)(\tan\theta) \quad \text{or} \quad \frac{\sin\theta}{2} < \frac{\theta}{2} < \frac{\sin\theta}{2\cos\theta}.$$

Since $\sin\theta > 0$ for $0 < \theta < \pi/2$, we can divide the double inequality by $(\sin\theta)/2$ to obtain

$$1 < \frac{\theta}{\sin\theta} < \frac{1}{\cos\theta} \quad \text{or} \quad \cos\theta < \frac{\sin\theta}{\theta} < 1.$$

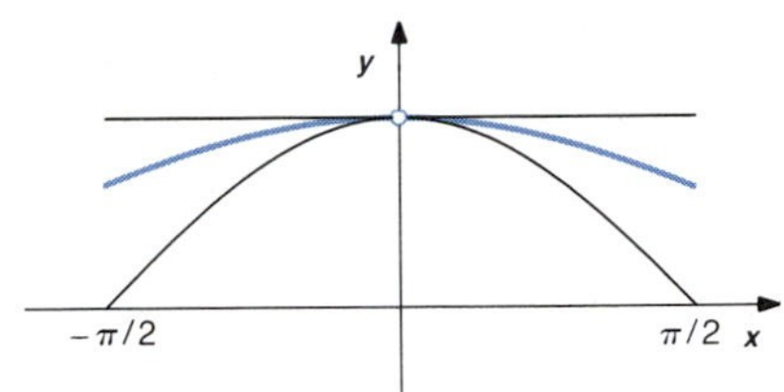


FIGURE 3.3.2

The latter inequality also holds for $-\pi/2 < \theta < 0$, because

$$\frac{\sin(-\theta)}{-\theta} = \frac{\sin\theta}{\theta} \quad \text{and} \quad \cos(-\theta) = \cos\theta.$$

See Figure 3.3.2. Subtracting 1 from each term of the above double inequality, we obtain

$$\cos\theta - 1 < \frac{\sin\theta}{\theta} - 1 < 0, \quad \text{so}$$

$$\left|\frac{\sin\theta}{\theta} - 1\right| < |\cos\theta - 1|, \qquad 0 < |\theta| < \frac{\pi}{2}.$$

Since the cosine function is continuous at zero, we know that

$$\cos\theta - 1 \to \cos 0 - 1 = 0 \qquad \text{as} \quad \theta \to 0.$$

It follows that

$$\lim_{\theta\to 0} \frac{\sin\theta}{\theta} = 1.$$ ■

We will also need

$$\lim_{\theta\to 0} \frac{1-\cos\theta}{\theta} = 0, \qquad \theta \text{ in radians.} \tag{3}$$

■ *Proof* We can use (1) to verify (3). That is,

$$\begin{aligned}
\lim_{\theta\to 0} \frac{1-\cos\theta}{\theta} &= \lim_{\theta\to 0} \left(\frac{1-\cos\theta}{\theta}\right)\left(\frac{1+\cos\theta}{1+\cos\theta}\right) \\
&= \lim_{\theta\to 0} \frac{1-\cos^2\theta}{\theta(1+\cos\theta)} \\
&= \lim_{\theta\to 0} \frac{\sin^2\theta}{\theta(1+\cos\theta)} \\
&= \lim_{\theta\to 0} \left(\frac{\sin\theta}{\theta}\right)\left(\frac{\sin\theta}{1+\cos\theta}\right) \\
&= \lim_{\theta\to 0} \frac{\sin\theta}{\theta} \lim_{\theta\to 0} \frac{\sin\theta}{1+\cos\theta} \\
&= (1)\left(\frac{\sin 0}{1+\cos 0}\right) = 0.
\end{aligned}$$ ■

We can now determine derivatives of the six basic trigonometric functions. We begin with the derivative of the sine function. We will use the limits (1) and (3) that we developed above to evaluate the limit that gives the derivative at a point $c$. *Since the limits in* (1) *and* (3) *were evaluated for* $\theta$ *in radians, radian measure must also be used for the differentiation formulas.*

If $f(x) = \sin x$, then $f'(c) = \cos c$ for all $c$. (4)

■ *Proof* The derivative of $\sin x$ at $c$ is

$$f'(c) = \lim_{x \to c} \frac{\sin x - \sin c}{x - c}.$$

To evaluate this limit we use the trigonometric identity for the sine of a sum. That is,

$$\begin{aligned}\sin x &= \sin[(x - c) + c] \\ &= \sin(x - c) \cdot \cos c + \cos(x - c) \cdot \sin c.\end{aligned}$$

Then

$$\begin{aligned}f'(c) &= \lim_{x \to c} \frac{\sin x - \sin c}{x - c} \\ &= \lim_{x \to c} \frac{\sin(x - c) \cdot \cos c + \cos(x - c) \cdot \sin c - \sin c}{x - c} \\ &= \lim_{x \to c} \left[\left(\frac{\sin(x - c)}{x - c}\right)(\cos c) + \left(\frac{\cos(x - c) - 1}{x - c}\right)(\sin c)\right].\end{aligned}$$

Noting that "$x \to c$" is the same as "$(x - c) \to 0$," we can substitute $\theta$ in place of $x - c$, and then use (1) and (3) to obtain

$$\begin{aligned}f'(c) &= \lim_{\theta \to 0} \left[\left(\frac{\sin \theta}{\theta}\right)(\cos c) + \left(\frac{\cos \theta - 1}{\theta}\right)(\sin c)\right] \\ &= (1)(\cos c) + (0)(\sin c) = \cos c.\end{aligned}$$ ■

Similarly, by using the formula for $\cos[(x - c) + c]$, we could verify the following:

If $f(x) = \cos x$, then $f'(c) = -\sin c$ for all $c$. (5)

Let us express the derivatives of the trigonometric functions as functions. *It is convenient that these six formulas be memorized*:

$$\frac{d}{dx}(\sin x) = \cos x,$$

$$\frac{d}{dx}(\cos x) = -\sin x,$$

$$\frac{d}{dx}(\tan x) = \sec^2 x,$$

$$\frac{d}{dx}(\cot x) = -\csc^2 x,$$

$$\frac{d}{dx}(\sec x) = \sec x \cdot \tan x,$$

$$\frac{d}{dx}(\csc x) = -\csc x \cdot \cot x.$$

The formulas for the derivatives of $\sin x$ and $\cos x$ come from (4) and (5). The other formulas may then be derived by using the Quotient Rule. For example,

$$\begin{aligned}\frac{d}{dx}(\tan x) &= \frac{d}{dx}\left(\frac{\sin x}{\cos x}\right)\\ &= \frac{(\cos x)\left[\frac{d}{dx}(\sin x)\right] - (\sin x)\left[\frac{d}{dx}(\cos x)\right]}{\cos^2 x}\\ &= \frac{(\cos x)(\cos x) - (\sin x)(-\sin x)}{\cos^2 x}\\ &= \frac{\cos^2 x + \sin^2 x}{\cos^2 x} = \frac{1}{\cos^2 x} = \sec^2 x.\end{aligned}$$

Similarly,

$$\begin{aligned}\frac{d}{dx}(\sec x) &= \frac{d}{dx}\left(\frac{1}{\cos x}\right)\\ &= \frac{(\cos x)(0) - (1)(-\sin x)}{\cos^2 x}\\ &= \frac{\sin x}{\cos^2 x} = \left(\frac{1}{\cos x}\right)\cdot\left(\frac{\sin x}{\cos x}\right) = \sec x \cdot \tan x.\end{aligned}$$

(We have used the Quotient Rule even though the numerator is a constant. In Section 3.4 we will learn a more convenient way to find the derivative of such functions.)

Formulas for the derivatives of $\cot x$ and $\csc x$ can be obtained in the same way as those for $\tan x$ and $\sec x$.

We can now differentiate products and quotients that involve trigonometric functions.

**■ EXAMPLE 1**

(a) $\frac{d}{dx}(x^2 \sin x) = (x^2)(\cos x) + (\sin x)(2x)$,

(b) $\dfrac{d}{dx}\left(\dfrac{\cos x}{x}\right) = \dfrac{(x)(-\sin x) - (\cos x)(1)}{x^2}$,

(c) $\dfrac{d}{dx}(\sec x \tan x) = (\sec x)(\sec^2 x) + (\tan x)(\sec x \tan x)$. ■

Differentiation of functions with more that two factors may require various combinations of the Product and Quotient Rules.

■ **EXAMPLE 2**

Find the derivative of

(a) $f(x) = x \sin x \cos x$ and

(b) $g(x) = \dfrac{x \sin x}{1 + \cos x}$.

**SOLUTION**

(a) Let us read $x \sin x \cos x$ as the product of the function $x$ and the function $\sin x \cos x$. That is,

$$f(x) = \underset{\substack{\uparrow \\ \text{(First factor)}}}{[x]}\,\underset{\substack{\uparrow \\ \text{(Second factor)}}}{[\sin x \cos x]}.$$

Then

$$f'(x) = \underbrace{[x]}_{\substack{\uparrow \\ \text{(First factor)}}}\underbrace{[(\sin x)(-\sin x) + (\cos x)(\cos x)]}_{\substack{\uparrow \\ \text{(Derivative of second)}}} + \underbrace{[\sin x \cos x]}_{\substack{\uparrow \\ \text{(Second factor)}}}\ \underbrace{[1]}_{\substack{\uparrow \\ \text{(Derivative of first)}}}.$$

The Product Rule was used to determine the derivative of the second factor. The derivative simplifies to

$$f'(x) = -x \sin^2 x + x \cos^2 x + \sin x \cos x.$$

We can obtain the same derivative by reading $x \sin x \cos x$ as the product of $x \sin x$ and $\cos x$, so

$$f(x) = [x \sin x][\cos x]$$

and

$$f'(x) = [x \sin x][-\sin x] + [\cos x][(x)(\cos x) + (\sin x)(1)].$$

This simplifies to the same expression for $f'(x)$ that we obtained earlier.

(b) We read $g(x)$ as a quotient whose top factor is a product. That is,

$$g(x) = \frac{\overbrace{[x \sin x]}^{\text{(Top factor)}}}{\underbrace{[1 + \cos x]}_{\text{(Bottom factor)}}}.$$

Applying the Quotient Rule, and using the Product Rule to find the derivative of the top factor, we obtain

$$g'(x) = \frac{\overbrace{[1 + \cos x]}^{\text{Bottom factor}}\overbrace{[(x)(\cos x) + (\sin x)(1)]}^{\text{Derivative of top}} - \overbrace{[x \sin x]}^{\text{Top factor}}\overbrace{[-\sin x]}^{\text{Derivative of bottom}}}{\underbrace{(1 + \cos x)^2}_{\text{Bottom factor, squared}}}.$$

Simplifying, we have

$$\begin{aligned} g'(x) &= \frac{x \cos x + \sin x + x \cos^2 x + \cos x \sin x + x \sin^2 x}{(1 + \cos x)^2} \\ &= \frac{x(\cos x + \cos^2 x + \sin^2 x) + (\sin x + \cos x \sin x)}{(1 + \cos x)^2} \\ &= \frac{x(\cos x + 1) + (\sin x)(1 + \cos x)}{(1 + \cos x)^2} \\ &= \frac{(x + \sin x)(1 + \cos x)}{(1 + \cos x)^2} \\ &= \frac{x + \sin x}{1 + \cos x}. \end{aligned}$$

■

**■ EXAMPLE 3**

Find the equation of the line tangent to the graph of $f(x) = 2 \sin x \cos x$ at the point where $x = \pi/3$.

**SOLUTION**

When $x = \pi/3$,

$$y = f\left(\frac{\pi}{3}\right) = 2 \sin \frac{\pi}{3} \cos \frac{\pi}{3} = 2\left(\frac{\sqrt{3}}{2}\right)\left(\frac{1}{2}\right) = \frac{\sqrt{3}}{2},$$

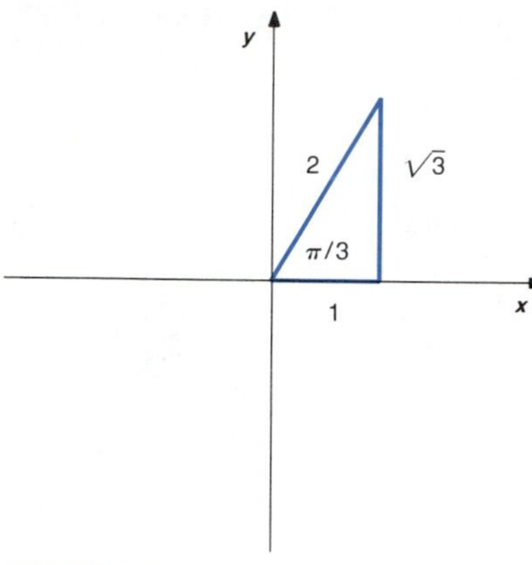

FIGURE 3.3.3

so $(\pi/3, \sqrt{3}/2)$ is the point on the graph where $x = \pi/3$. (The exact values of $\sin(\pi/3)$ and $\cos(\pi/3)$ can be determined from the representative triangle in Figure 3.3.3. A calculator could be used to determine approximate values.) The slope of the line tangent to the graph at the point $(\pi/3, \sqrt{3}/2)$ is the value of the derivative at $x = \pi/3$. Using the Product Rule, we have

$$\begin{aligned} f'(x) &= 2[(\sin x)(-\sin x) + (\cos x)(\cos x)] \\ &= 2[-\sin^2 x + \cos^2 x]. \end{aligned}$$

Hence,

$$\begin{aligned} f'\left(\frac{\pi}{3}\right) &= 2\left[-\sin^2\frac{\pi}{3} + \cos^2\frac{\pi}{3}\right] \\ &= 2\left[-\left(\frac{\sqrt{3}}{2}\right)^2 + \left(\frac{1}{2}\right)^2\right] = -1. \end{aligned}$$

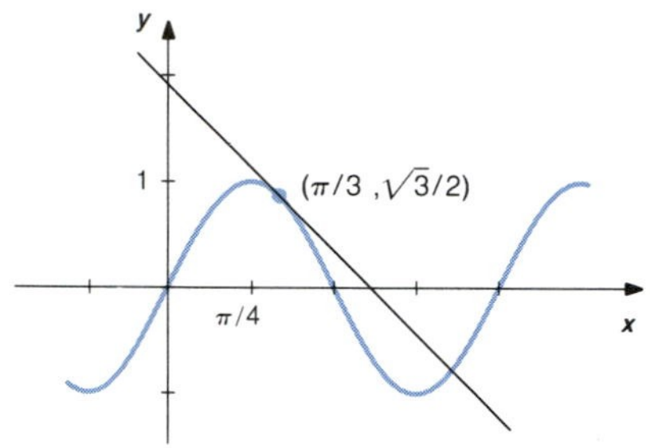

FIGURE 3.3.4

The point–slope form of the equation of the line through the point $(\pi/3, \sqrt{3}/2)$ with slope $-1$ is

$$y - \frac{\sqrt{3}}{2} = (-1)\left(x - \frac{\pi}{3}\right)$$

$$x + y = \frac{\pi}{3} + \frac{\sqrt{3}}{2}.$$

See Figure 3.3.4. ■

■ **EXAMPLE 4**

Find points on the graph of

$$f(x) = \sin x + \cos x, \qquad 0 \le x \le 2\pi,$$

where the tangent line is horizontal.

**SOLUTION**

We know that the slope of the tangent line at the point $(x, f(x))$ is given by the value of the derivative $f'(x)$. Horizontal lines have slope zero, so we need to find values of $x$ for which $f'(x) = 0$.

$$f(x) = \sin x + \cos x, \quad \text{so} \quad f'(x) = \cos x - \sin x.$$

Setting $f'(x) = 0$, we have

$$\cos x - \sin x = 0.$$

To solve this equation, we express it in terms of $\tan x$. We have

$$-\sin x = -\cos x,$$

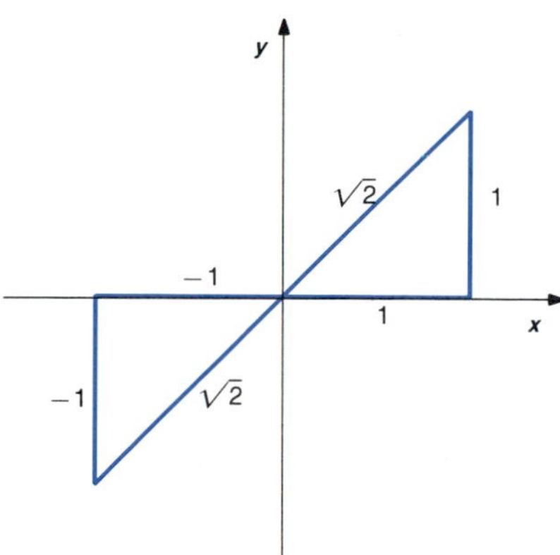


FIGURE 3.3.5

$$\frac{\sin x}{\cos x} = 1,$$

$$\tan x = 1.$$

There are two solutions of the equation $\tan x = 1$ in the interval $0 \le x \le 2\pi$. Both solutions have reference angle $\pi/4$. Representative angles are sketched in Figure 3.3.5. The solutions are $x = \pi/4$ and $x = 5\pi/4$. These are the values of $x$ that correspond to points on the graph with horizontal tangents. The $y$-coordinates are given by the corresponding values $f(x)$.

$$f\left(\frac{\pi}{4}\right) = \sin\frac{\pi}{4} + \cos\frac{\pi}{4} = \frac{1}{\sqrt{2}} + \frac{1}{\sqrt{2}} = \frac{2}{\sqrt{2}} = \sqrt{2} \quad \text{and}$$

$$f\left(\frac{5\pi}{4}\right) = \sin\frac{5\pi}{4} + \cos\frac{5\pi}{4} = \frac{-1}{\sqrt{2}} + \frac{-1}{\sqrt{2}} = -\frac{2}{\sqrt{2}} = -\sqrt{2}.$$

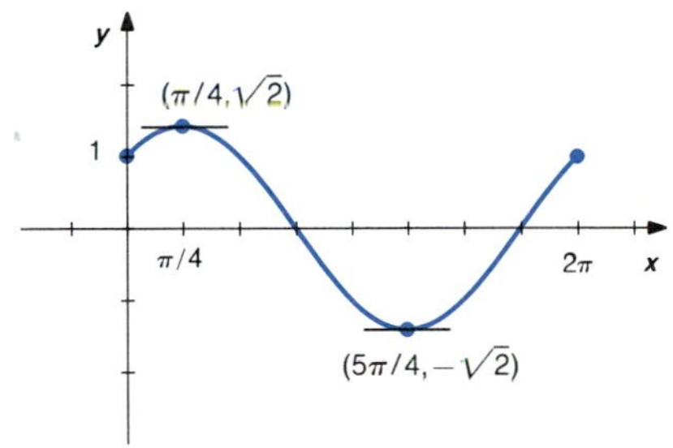


FIGURE 3.3.6

$(\pi/4, \sqrt{2})$ and $(5\pi/4, -\sqrt{2})$ are points on the graph that have horizontal tangents. The graph is sketched in Figure 3.3.6. ■

■ **EXAMPLE 5**

The position of a particle on a number line at time $t$ is given by $s(t) = t - 2\sin t$, $0 \le t \le 2\pi$. Find intervals of time for which the particle is moving in the positive direction.

**SOLUTION**

The particle is moving in the positive direction when the velocity is positive. The velocity is given by the derivative $s'(t)$.

$$s(t) = t - 2\sin t, \quad \text{so} \quad s'(t) = 1 - 2\cos t.$$

We need to find values of $t$, $0 \le t \le 2\pi$, such that

$$s'(t) > 0,$$

$$1 - 2\cos t > 0.$$

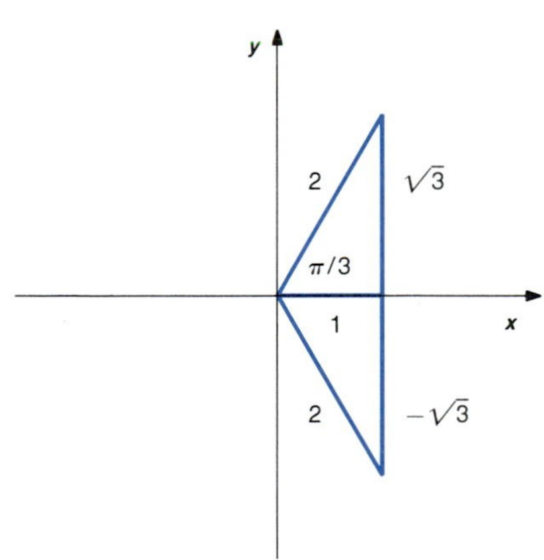


FIGURE 3.3.7

The expression $1 - 2\cos t$ can change sign only at points where $1 - 2\cos t = 0$, or $\cos t = 1/2$. From the representative angles drawn in Figure 3.3.7, we see that $\cos t = 1/2$ for $t = \pi/3$ and $t = 5\pi/3$. The table of signs given in Table 3.3.1 can then be determined by calculating a value of $1 - 2\cos t$ in each interval. We see that $s'(t)$ is positive and the particle is moving in the positive direction for $\pi/3 < t < 5\pi/3$. ■

TABLE 3.3.1

| | $0$ to $\pi/3$ | $\pi/3$ to $5\pi/3$ | $5\pi/3$ to $2\pi$ |
|---|---|---|---|
| $s'(t) = 1 - 2\cos t$ | − | + | − |

We can use the formulas

$$\lim_{\theta\to 0}\frac{\sin\theta}{\theta}=1 \quad\text{and}\quad \lim_{\theta\to 0}\frac{1-\cos\theta}{\theta}=0$$

to evaluate the limits at zero of some other trigonometric expressions. The idea is to write the given expression in terms of $(\sin\theta)/\theta$ and $(1-\cos\theta)/\theta$.

**EXAMPLE 6**

Evaluate

$$\lim_{x\to 0}\frac{x}{\sin 3x}.$$

**SOLUTION**

We have

$$\frac{x}{\sin 3x}=\left(\frac{1}{3}\right)\left(\frac{3x}{\sin 3x}\right)=\left(\frac{1}{3}\right)\left(\frac{1}{\frac{\sin 3x}{3x}}\right)=\left(\frac{1}{3}\right)\left(\frac{1}{\frac{\sin\theta}{\theta}}\right),$$

where $\theta=3x$. Since $\theta=3x\to 0$ as $x\to 0$, we then have

$$\lim_{x\to 0}\frac{x}{\sin 3x}=\lim_{\theta\to 0}\left(\frac{1}{3}\right)\left(\frac{1}{\frac{\sin\theta}{\theta}}\right)=\left(\frac{1}{3}\right)\left(\frac{1}{1}\right)=\frac{1}{3}.$$

■

## EXERCISES 3.3

Find the derivatives in Exercises 1–12. Use parentheses to indicate each factor when using the Product or Quotient Rules, and then simplify your answers.

1 $\frac{d}{dx}(\sin x+\cos x)$

2 $\frac{d}{dx}(2\sin x-\cos x)$

3 $\frac{d}{dx}(\sec x-2\tan x)$

4 $\frac{d}{dx}\left(\frac{2\cot x-\sec x}{3}\right)$

5 $\frac{d}{dx}(3x^{1/3}\cdot\sin x)$

6 $\frac{d}{dx}(4\sec x\cdot\sin x)$

7 $\frac{d}{dx}(2x\sin x)$

8 $\frac{d}{dx}(x^2\sin x)$

9 $\frac{d}{dx}(\tan x\cdot\sin x)$

10 $\frac{d}{dx}\left(\frac{3x^2}{1+\sin x}\right)$

11 $\frac{d}{dx}\left(\frac{\cos x}{x^2+4}\right)$

12 $\frac{d}{dx}\left(\frac{2\cos x}{\sin x+\cos x}\right)$

In Exercises 13–16 find the equation of the line tangent to the graph at the point where $x=c$.

13 $f(x)=x+\sin x$, $c=0$

14 $f(x)=x\cos x$, $c=0$

15 $f(x)=\sin x\cdot\tan x$, $c=\pi/6$

16 $f(x)=\dfrac{\sin x+\cos x}{\sin x-\cos x}$, $c=\pi/2$

In Exercises 17–20 find all points on the graphs of the given functions where the tangent line is horizontal.

17 $f(x)=x+2\sin x$, $0\le x\le 2\pi$

18 $f(x)=x-\sqrt{2}\cos x$, $0\le x\le 2\pi$

**19** $f(x) = \sqrt{3}\sin x + \cos x,\ 0 \le x \le 2\pi$

**20** $f(x) = \sqrt{3}\sin x - \cos x,\ 0 \le x \le 2\pi$

In Exercises 21–24 the position of a particle on a number line at time $t$ is given by the function $s(t)$. Find intervals of time for which the particle is moving in the positive direction.

**21** $s(t) = \sqrt{3}t + 2\cos t,\ 0 \le t \le 2\pi$

**22** $s(t) = \sqrt{3}t + 2\sin t,\ 0 \le t \le 2\pi$

**23** $s(t) = 2t - \tan t,\ 0 \le t < \pi/2$

**24** $s(t) = 2\sec t - \tan t,\ 0 \le t < \pi/2$

**25** Prove that if $f(x) = \cos x$, then $f'(c) = -\sin c$ for all $c$.

**26** Derive the formula for the derivative of $\cot x$ by expressing $\cot x$ in terms of $\sin x$ and $\cos x$ and then using the Quotient Rule.

**27** Derive the formula for the derivative of $\csc x$ by expressing $\csc x$ in terms of $\sin x$ and then using the Quotient Rule.

Use the formulas

$$\lim_{\theta \to 0} \frac{\sin \theta}{\theta} = 1 \quad \text{and} \quad \lim_{\theta \to 0} \frac{1 - \cos \theta}{\theta} = 0$$

to evaluate the limits in Exercises 28–33.

**28** $\displaystyle\lim_{x \to 0} \frac{\tan x}{x}$

**29** $\displaystyle\lim_{x \to 0} \frac{1 - \sec x}{x}$

**30** $\displaystyle\lim_{x \to 0} \frac{\sin(2x)}{x}$

**31** $\displaystyle\lim_{x \to 0} \frac{\tan(4x)}{\sin(2x)}$

**32** $\displaystyle\lim_{x \to 0} x \cot x$

**33** $\displaystyle\lim_{x \to 0} \frac{1 - \cos x}{x^2}$

**34** Show that $x_0 = \tan x_0$ whenever the line tangent to the graph of $y = \sin x$ at the point $(x_0, \sin x_0)$ intersects the origin. See the figure below.

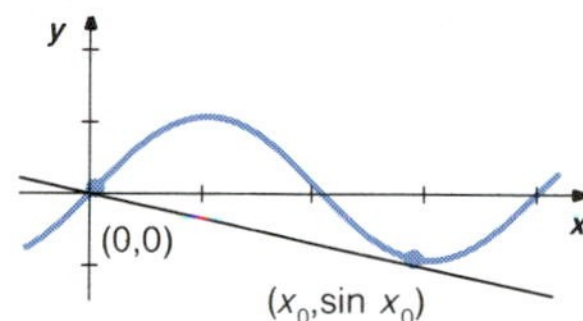


## 3.4 CHAIN RULE

In previous sections we have developed formulas for the differentiation of sums, products, and quotients of functions. In this section we will develop a formula for the differentiation of composite functions and see how to apply the formula to compositions of power functions and the six basic trigonometric functions.

The **Chain Rule** is used to differentiate composite functions.

■ *Theorem (Chain Rule)*

**If $g$ is differentiable at $c$ and $f$ is differentiable at $g(c)$, then the composite function $f \circ g(x) = f(g(x))$ is differentiable at $c$ and**

$$(f \circ g)'(c) = f'(g(c)) \cdot g'(c).$$

■ *Proof* We will evaluate the limit in the definition of the derivative.

$$(f \circ g)'(c) = \lim_{x \to c} \frac{f \circ g(x) - f \circ g(c)}{x - c} = \lim_{x \to c} \frac{f(g(x)) - f(g(c))}{x - c}.$$

We consider the cases $g'(c) \neq 0$ and $g'(c) = 0$ separately.

If $g'(c) \neq 0$, then $g(x) - g(c) \neq 0$ for $x$ near $c$, $x \neq c$. We can then write

$$\frac{f(g(x)) - f(g(c))}{x - c} = \frac{f(g(x)) - f(g(c))}{g(x) - g(c)} \cdot \frac{g(x) - g(c)}{x - c}.$$

Let us set $u = g(x)$ and $d = g(c)$. The first factor on the right above is then

$$\frac{f(u) - f(d)}{u - d}.$$

Since $f$ is differentiable at $d = g(c)$, this ratio approaches $f'(d) = f'(g(c))$ as $u$ approaches $d$. But $u = g(x)$ approaches $d = g(c)$ as $x$ approaches $c$, since $g$ is continuous at $c$. It follows that the first factor approaches $f'(g(c))$ as $x$ approaches $c$. Since the second factor above approaches $g'(c)$ as $x$ approaches $c$, we have $(f \circ g)'(c) = f'(g(c)) \cdot g'(c)$ in case $g'(c) \neq 0$.

If $g'(c) = 0$, then $g(x) - g(c)$ may or may not be zero for $x$ near $c$, $x \neq c$. For points near $c$, $x \neq c$, where $g(x) - g(c) \neq 0$, we can argue as in the case $g'(c) \neq 0$ to show $[f(g(x)) - f(g(c))]/(x - c)$ is near $f'(g(c)) \cdot g'(c)$. When $g(x) - g(c) = 0$, $x \neq c$, we have

$$\frac{f(g(x)) - f(g(c))}{x - c} = f'(g(c)) \cdot g'(c),$$

because both these expressions are zero. It follows that $(f \circ g)'(c) = f'(g(c)) \cdot g'(c)$ in case $g'(c) = 0$, so the result is true in both cases. ■

The Chain Rule may also be expressed in the following form. If $y = y(u)$ is a differentiable function of $u$ and $u = u(x)$ is a differentiable function of $x$, then $y = y(u(x))$ is a differentiable function of $x$ and

$$\frac{dy}{dx} = \frac{dy}{du} \cdot \frac{du}{dx}.$$

This form is easy to remember. It is also suggestive of the proof of the Chain Rule that we used.

In the case $y = u^r$ and $u = u(x)$, the Chain Rule gives

$$\frac{d}{dx}(u^r) = \frac{dy}{dx} = \frac{dy}{du} \cdot \frac{du}{dx} = ru^{r-1} \cdot \frac{du}{dx}.$$

In the formulas

$$\frac{d}{dx}(x^r) = rx^{r-1} \quad \text{and} \quad \frac{d}{du}(u^r) = ru^{r-1},$$

we are differentiating a constant power of the variable of differentiation. We know from Section 3.1 how to differentiate a constant power of the variable of differentiation; the Chain Rule is not required. The Chain Rule is required to differentiate a power of a *function* that is not simply the variable of differentiation. This version of the Chain Rule should be thought of as saying

**Derivative of (Function)$^r$ = $r$(Function)$^{r-1}$(Derivative of function).**

Let us see how to use this formula.

■ **EXAMPLE 1**

Find the following derivatives:

(a) $\dfrac{d}{dx}((x^2 + 7x - 3)^4)$,

(b) $\dfrac{d}{dx}\left(\dfrac{1}{x^2 + 1}\right)$,

(c) $\dfrac{d}{dx}(\sqrt{x^2 + x + 4})$,

(d) $\dfrac{d}{dx}(\sin^2 x)$.

**SOLUTION**

(a) $(x^2 + 7x - 3)^4$ is the 4th power of the function $x^2 + 7x - 3$. We can use the Chain Rule to write the derivative in one step:

$$\frac{d}{dx}\left(\underbrace{(x^2 + 7x - 3)}_{\text{(Function)}}{}^{\overset{(r)}{4}}\right) = \overset{(r)}{4}\underbrace{(x^2 + 7x - 3)}_{\text{(Function)}}{}^{\overset{(r-1)}{3}}\underbrace{(2x + 7)}_{\text{(Derivative of function)}}.$$

(b) Let us write $1/(x^2 + 1)$ as $(x^2 + 1)^{-1}$, the $-1$ power of the function $x^2 + 1$. The Chain Rule then gives the derivative.

$$\frac{d}{dx}\left(\frac{1}{x^2 + 1}\right) = \frac{d}{dx}\left(\underbrace{(x^2 + 1)}_{\text{(Function)}}{}^{\overset{(r)}{-1}}\right) = \overset{(r)}{(-1)}\underbrace{(x^2 + 1)}_{\text{(Function)}}{}^{\overset{(r-1)}{-2}}\underbrace{(2x)}_{\text{(Derivative of function)}}.$$

We could have used the Quotient Rule to differentiate $1/(x^2 + 1)$, but it is much easier to use the Chain Rule. In general, don't use the Quotient Rule to differentiate functions of the form $1/g(x)$. Instead, write $1/g(x)$ as $[g(x)]^{-1}$ and use the Chain Rule.

(c) The square root function can be written as the half-power function. The Chain Rule can then be used to differentiate square roots of functions.

$$\begin{aligned}\frac{d}{dx}(\sqrt{x^2 + x + 4}) &= \frac{d}{dx}((x^2 + x + 4)^{1/2}) \\ &= \frac{1}{2}(x^2 + x + 4)^{-1/2}(2x + 1).\end{aligned}$$

Alternatively, we could use the fact that

$$\frac{d}{du}(\sqrt{u}) = \frac{1}{2\sqrt{u}}$$

and write

$$\frac{d}{dx}(\sqrt{x^2 + x + 4}) = \frac{1}{2\sqrt{x^2 + x + 4}}(2x + 1).$$

(d) We see that $\sin^2 x = (\sin x)^2$ is the square of $\sin x$. The Chain Rule then gives the derivative.

$$\frac{d}{dx}(\sin^2 x) = \frac{d}{dx}((\sin x)^2) = 2(\sin x)(\cos x).$$ ■

It is important to know how to use the Chain Rule for each function we can differentiate. For example, we know that

$$\frac{d}{du}(\sin u) = \cos u.$$

Hence, in the case $y = \sin u$ and $u = u(x)$, the Chain Rule gives

$$\frac{d}{dx}(\sin u) = \frac{dy}{dx} = \frac{dy}{du}\cdot\frac{du}{dx} = \cos u \cdot \frac{du}{dx}.$$

This formula should be thought of as saying

**Derivative of sin(Function) = cos(Function) · (Derivative of Function).**

The versions of the Chain Rule that corresponds to the differentiation formulas we know are listed below.

$$\frac{d}{dx}(u^r) = ru^{r-1}\cdot\frac{du}{dx},$$

$$\frac{d}{dx}(\sin u) = \cos u \cdot \frac{du}{dx}, \qquad \frac{d}{dx}(\cot u) = -\csc^2 u \cdot \frac{du}{dx},$$

$$\frac{d}{dx}(\cos u) = -\sin u \cdot \frac{du}{dx}, \qquad \frac{d}{dx}(\sec u) = \sec u \cdot \tan u \cdot \frac{du}{dx},$$

$$\frac{d}{dx}(\tan u) = \sec^2 u \cdot \frac{du}{dx}, \qquad \frac{d}{dx}(\csc u) = -\csc u \cdot \cot u \cdot \frac{du}{dx}.$$

The above formulas should be read with "function" in place of $u$. In most of the examples we will be dealing with, the "function" will be given by a formula. We should then use the above formulas to write the formulas for the derivatives in one step. It is helpful to use parentheses when doing this.

■ **EXAMPLE 2**

Find the derivatives.

(a) $\frac{d}{dx}(\sin(x^2+1))$,

(b) $\frac{d}{dx}(\tan(2x))$,

(c) $\frac{d}{dx}(\sec\sqrt{x})$.

**SOLUTION**

(a) $\sin(x^2+1)$ is the sine of $x^2+1$. The Chain Rule gives the derivative.

$$\frac{d}{dx}\underbrace{(\sin(x^2+1))}_{\text{(sin(Function))}} = \underbrace{[\cos(x^2+1)]}_{\text{(cos(Function))}}\ \underbrace{[2x]}_{\left(\text{Derivative of function}\right)}.$$

(b) $\tan(2x)$ is the tangent of the function $2x$. The Chain Rule gives the derivative:

$$\frac{d}{dx}\underbrace{(\tan(2x))}_{\text{(tan(Function))}} = \underbrace{[\sec^2(2x)]}_{(\sec^2\text{(Function))}}\ \underbrace{[2]}_{\left(\text{Derivative of function}\right)}.$$

(c) $\sec\sqrt{x}$ is the secant of the square root of $x$. The Chain Rule gives the derivative:

$$\frac{d}{dx}(\sec\sqrt{x}) = [\sec\sqrt{x}\cdot\tan\sqrt{x}]\left[\frac{1}{2\sqrt{x}}\right]. \quad ■$$

Some problems require that we use the Chain Rule more than once, or that we use the Chain Rule with either or both the Product and Quotient Rules.

■ **EXAMPLE 3**

Find the derivatives.

(a) $\frac{d}{dx}\left(x\sin\frac{1}{x}\right)$,

(b) $\frac{d}{dx}\left(\sqrt{\frac{1-x}{2x+1}}\right)$,

(c) $\frac{d}{dx}([1+\sec^2(3x)]^5)$.

**SOLUTION**

(a) We see that $x\sin(1/x)$ is the product of $x$ and $\sin(1/x)$, so the Product Rule is used to find its derivative. Since $\sin(1/x)$ is the sine of the function $1/x$, the Chain Rule is used to differentiate this factor.

$$\frac{d}{dx}\Big(\underbrace{x}_{\text{First factor}}\ \underbrace{\sin\frac{1}{x}}_{\text{Second factor}}\Big) = \underbrace{(x)}_{\text{First factor}}\underbrace{\left[\left(\cos\frac{1}{x}\right)(-x^{-2})\right]}_{\text{Derivative of second}} + \underbrace{\left[\sin\frac{1}{x}\right]}_{\text{Second factor}}\underbrace{(1)}_{\text{Derivative of first}}.$$

(b) The expression

$$\sqrt{\frac{1-x}{2x+1}} = \left(\frac{1-x}{2x+1}\right)^{1/2}$$

is the half-power of the quotient $(1-x)/(2x+1)$. To find the derivative we use the Chain Rule and then the Quotient Rule.

$$\frac{d}{dx}\left(\sqrt{\frac{1-x}{2x-1}}\right) = \frac{d}{dx}\left(\Big(\overbrace{\frac{1-x}{2x+1}}^{\text{(Function)}}\Big)^{\overset{(r)}{1/2}}\right)$$

$$= \underbrace{\frac{1}{2}}_{(r)}\Big(\underbrace{\frac{1-x}{2x+1}}_{\text{(Function)}}\Big)^{\overset{(r-1)}{-1/2}}\underbrace{\left[\frac{(2x+1)(-1)-(1-x)(2)}{(2x+1)^2}\right]}_{\text{(Derivative of function)}}.$$

(c) This example requires the Chain Rule more than once. We will take the derivative one step at a time. We first note that $[1+\sec^2(3x)]^5$ is the 5th power of the function $1+\sec^2(3x)$. The Chain Rule tells us that

$$\frac{d}{dx}([1+\sec^2(3x)]^5) = 5[1+\sec^2(3x)]^4\frac{d}{dx}(1+\sec^2(3x)).$$

We now read $1+\sec^2(3x)$ as the sum $1+[\sec(3x)]^2$. The second term is the square of the function $\sec(3x)$, so we use the Chain Rule to find the derivative of that term. That is,

$$\frac{d}{dx}(1+\sec^2(3x)) = 0 + 2[\sec(3x)]\frac{d}{dx}(\sec(3x)).$$

Since $\sec(3x)$ is the secant of the function $3x$, the Chain Rule is used to find the derivative.

$$\frac{d}{dx}(\sec(3x)) = [\sec(3x)\tan(3x)](3).$$

Combining results we have

$$\frac{d}{dx}([1+\sec^2(3x)]^5)$$
$$= 5[1+\sec^2(3x)]^4\{0+2[\sec(3x)][\sec(3x)\tan(3x)](3)\}.$$

Let us note that the factors in the above derivative were determined in order from left to right. Also, each term is determined by the previous term. Hence, the factors could have been written in order on one line without writing the component derivatives on separate lines and then combining results. That is, the first factor of the derivative that we write is $5[1+\sec^2(3x)]^4$, which is the first term of the derivative of $(\text{Function})^5$, where Function $= 1+\sec^2(3x)$. After we have written this term, we write $0+2\sec(3x)$, which is the derivative of 1 plus the first term of the derivative of $(\text{Function})^2$, where Function $= \sec(3x)$. We then multiply $2\sec(3x)$ by $\sec(3x)\tan(3x)$, which is the first term of the derivative of sec(Function), where Function $= 3x$. $\operatorname{Sec}(3x)\tan(3x)$ is then multiplied by 3, which is the derivative of the function $3x$. The chain stops with the derivative of $3x$, which we know is 3 and does not require the Chain Rule. Pictorially, we have

$$\overbrace{\frac{d}{dx}([1+\sec^2(3x)]^5)}^{\begin{pmatrix}\text{Derivative of}\\(\text{Function})^5\end{pmatrix}}$$

$$= \underbrace{5[1+\sec^2(3x)]^4}_{\begin{pmatrix}\text{First term of derivative}\\\text{of }(\text{Function})^5,\\\text{Function} = 1+\sec^2(3x)\end{pmatrix}}\quad \underbrace{\{0\ +}_{\begin{pmatrix}\text{Derivative}\\\text{of 1}\end{pmatrix}}\quad \underbrace{2[\sec(3x)]}_{\begin{pmatrix}\text{First term}\\\text{of derivative of}\\(\text{Function})^2,\\\text{Function} = \sec(3x)\end{pmatrix}}\quad \underbrace{[\sec(3x)\tan(3x)]}_{\begin{pmatrix}\text{First term}\\\text{of derivative of}\\\text{sec(Function)},\\\text{Function} = 3x\end{pmatrix}}\quad \underbrace{(3)}_{\begin{pmatrix}\text{Derivative of}\\\text{Function} = 3x\end{pmatrix}}\}.$$

■

■ **EXAMPLE 4**

Find all points on the graph of $f(x) = x(1-x)^{1/3}$ for which (a) $f'(x) = 0$ and (b) $f'(x)$ does not exist.

**SOLUTION**

$f(x)$ is the product of $x$ and $(1-x)^{1/3}$, so we use the Product Rule to find $f'(x)$. $(1-x)^{1/3}$ is a power of a function, so we use the Chain Rule to find the derivative of this factor. We have

$$f'(x) = [x]\left[\frac{1}{3}(1-x)^{-2/3}(-1)\right] + [(1-x)^{1/3}][1].$$

We simplify by factoring $(1-x)^{-2/3}/3$ from each term. We have

$$f'(x) = -\frac{x(1-x)^{-2/3}}{3} + \frac{3(1-x)^{1-(2/3)}}{3}$$
$$= \left[\frac{(1-x)^{-2/3}}{3}\right][-x+3(1-x)]$$
$$= \frac{3-4x}{3(1-x)^{2/3}}.$$

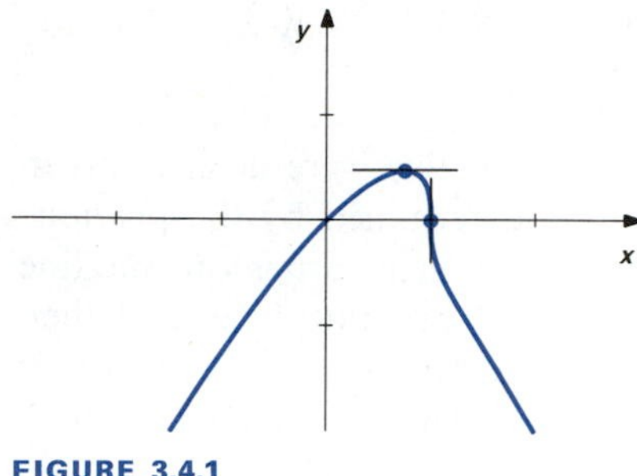

FIGURE 3.4.1

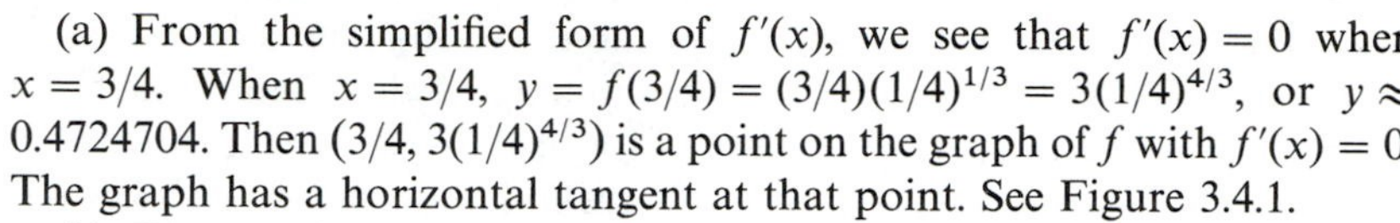

(a) From the simplified form of $f'(x)$, we see that $f'(x) = 0$ when $x = 3/4$. When $x = 3/4$, $y = f(3/4) = (3/4)(1/4)^{1/3} = 3(1/4)^{4/3}$, or $y \approx 0.4724704$. Then $(3/4, 3(1/4)^{4/3})$ is a point on the graph of $f$ with $f'(x) = 0$. The graph has a horizontal tangent at that point. See Figure 3.4.1.

(b) We see that $f'(x)$ does not exist when $x = 1$. When $x = 1$, $y = f(1) = 0$. The point $(1, 0)$ is a point on the graph where $f'(x)$ does not exist. The graph passes through the point $(1, 0)$ vertically. See Figure 3.4.1. ■

The Chain Rule can be used even if one of the functions involved is not completely known. It is only necessary that we know the values of the functions and derivatives at the appropriate points.

■ **EXAMPLE 5**

$V(t) = [s(t)]^3$, where $s(2) = 4$ and $s'(2) = 5$. Find $V'(2)$.

**SOLUTION**

Using the Chain Rule to differentiate the power of the function $s$, we obtain

$$V'(t) = 3[s(t)]^2[s'(t)].$$

When $t = 2$, we have

$$V'(2) = 3[s(2)]^2[s'(2)]$$
$$= 3[4]^2[5] = 240.$$ ■

■ **EXAMPLE 6**

$F(x) = f(3x-1)$, where $f'(5) = 2$. Find $F'(2)$.

**SOLUTION**

The function $F$ is the composition of the function $f$ and the function $3x - 1$. The Chain Rule gives the derivative

$$F'(x) = f'(3x-1)\cdot\frac{d}{dx}(3x-1)$$
$$= f'(3x-1)\cdot(3).$$

Substituting $x = 2$ and $f'(5) = 2$, we obtain

$$\begin{aligned} F'(2) &= f'(3(2) - 1)\cdot(3) \\ &= f'(5)\cdot(3) \\ &= (2)(3) \\ &= 6. \end{aligned}$$

■

## EXERCISES 3.4

Evaluate the derivatives in Exercises 1–22. Use parentheses to indicate each factor and then simplify your answers.

1 $\frac{d}{dx}((2x^2 - 3x + 1)^5)$

2 $\frac{d}{dx}((3x^2 - 4x + 5)^7)$

3 $\frac{d}{dx}((2x^3 - 1)^{-2/3})$

4 $\frac{d}{dx}((x^2 - 2x + 4)^{-1/3})$

5 $\frac{d}{dx}\left(\frac{1}{2\sin x + 3}\right)$

6 $\frac{d}{dx}\left(\frac{1}{3\sec x - 1}\right)$

7 $\frac{d}{dx}(\sin(x^3 - x))$

8 $\frac{d}{dx}(\cos\sqrt{x})$

9 $\frac{d}{dx}(\tan\sqrt{x})$

10 $\frac{d}{dx}(\cos(1 - x))$

11 $\frac{d}{dx}\left(\sec\frac{1}{x}\right)$

12 $\frac{d}{dx}\left(\csc\frac{1}{x^2}\right)$

13 $\frac{d}{dx}(x\sqrt{x^2 + 1})$

14 $\frac{d}{dx}(x^2(1 + 2x)^3)$

15 $\frac{d}{dx}\left(\frac{\sin x}{\sqrt{\cos^2 x + 1}}\right)$

16 $\frac{d}{dx}\left(\frac{\tan x}{\sqrt{\sec^2 x + 1}}\right)$

17 $\frac{d}{dx}(\tan^2(2x))$

18 $\frac{d}{dx}((\sin 2x)(\cos 2x))$

19 $\frac{d}{dx}(\sqrt{x^2 + \sin^2 x})$

20 $\frac{d}{dx}(\sqrt{1 + \cos^2 x})$

21 $\frac{d}{dx}(\sqrt{x + \sqrt{x + x^2}})$

22 $\frac{d}{dx}([1 + (1 + x^2)^2]^2)$

In Exercises 23–30 find the equation of the line tangent to the graph of $y = f(x)$ at the point on the graph with $x = c$.

23 $f(x) = \sqrt{\frac{x^3}{x + 3}}$, $c = 1$

24 $f(x) = \left(\frac{x}{x^2 + 16}\right)^{2/3}$, $c = 4$

25 $f(x) = \frac{1}{\sqrt{x^2 + 9}}$, $c = 4$

26 $f(x) = \frac{1}{3x - 2}$, $c = 2$

27 $f(x) = \cot^2 x$, $c = \pi/4$

28 $f(x) = \csc^2 x$, $c = \pi/3$

29 $f(x) = x - \sin(2x)$, $c = \pi/2$

30 $f(x) = x + \cos(2x)$, $c = \pi/4$

In Exercises 31–36 find all points $(x, f(x))$, on the graph of $f$ for which (a) $f'(x) = 0$ and (b) $f'(x)$ does not exist.

31 $f(x) = x(3x - 4)^{1/3}$

32 $f(x) = x(3x - 20)^{2/3}$

33 $f(x) = \frac{x}{(6x + 4)^{2/3}}$

34 $f(x) = \frac{x}{(3x - 2)^{2/3}}$

35 $f(x) = x - \tan\frac{x}{2}$, $0 < x < \pi$

36 $f(x) = \sin x - \sin^2 x$, $0 \le x \le 2\pi$

37 $F(t) = f(4t + 3)$, where $f'(7) = 5$. Find $F'(1)$.

38 $F(t) = f(t^2 + 1)$, where $f'(5) = 3$. Find $F'(2)$.

39 $F(t) = \sqrt{g(t)}$, where $g(2) = 9$ and $g'(2) = 4$. Find $F'(2)$.

40 $F(t) = (g(t))^{1/3}$, where $g(3) = 8$ and $g'(3) = 6$. Find $F'(3)$.

41 $F(x) = f(1 + x) + f(1 - x)$, where $f$ is differentiable at 1. Find $F'(0)$.

42 $F(x) = f(a + bx) + f(a - bx)$, where $f$ is differentiable at $a$. Find $F'(0)$.

43 Use the formula $|x| = \sqrt{x^2}$ and the Chain Rule to show that

$$\frac{d}{dx}(|x|) = \frac{x}{|x|}, \qquad x \neq 0.$$

It follows that

$$\frac{d}{dx}(|u|) = \frac{u}{|u|}\frac{du}{dx}, \qquad u \neq 0.$$

**44** Find the equation of the line tangent to the graph of $f(x) = |1 - x^2|$ at the point where $x = 2$.

**45** Find the equation of the line tangent to the graph of $f(x) = |\sin(5x)|$ at the point where $x = \pi/4$.

**46** Find the equation of the line tangent to the graph of $f(x) = |(1 - x)^{1/3}|$ at the point where $x = 9$.

In Chapter 7 we will introduce a function $L$ with derivative $L'(x) = 1/x$. Use this formula to evaluate the derivatives in Exercises 47–50.

**47** $\dfrac{d}{dx}(L(x^2))$

**48** $\dfrac{d}{dx}(L(3x))$

**49** $\dfrac{d}{dx}(L(x^2 + 1))$

**50** $\dfrac{d}{dx}(L(\sec x + \tan x))$

**51** In Chapter 7 we will introduce a function $E$ with $E(0) = 1$ and $E'(x) = E(x)$. Use these formulas to show that the line tangent to the graph of $y = E(ax)$ at the point where the graph intersects the $y$-axis has slope $a$.

**52** Find the rate of change of the volume of a weather balloon when its radius is 10 cm if its radius is increasing at a rate of 0.3 cm/s. $V = 4\pi r^3/3$, where $V$ and $h$ are functions of time.

**53** If a balloon is rising vertically from a point on the ground that is 50 ft from a ground-level observer, the height of the balloon and the angle of elevation from the observer to the balloon are related by the equation $h = 50 \tan \theta$, where $h$ and $\theta$ are functions of time. The angle of elevation is observed to be increasing at a rate of 2°/s when the angle of elevation is 45°. Find the rate at which the balloon is rising at that time. (*Hint*: Angles must be expressed in radian measure in order to apply the formulas for the differentiation of the trigonometric functions.)

## 3.5 IMPLICIT DIFFERENTIATION

Recall from Section 2.1 that a graph defines $y$ as a function of $x$ if every vertical line that intersects the graph intersects it at exactly one point. Various parts of a graph may define functions of $x$, even though the entire graph does not. For example, the graph in Figure 3.5.1 does not define $y$ as a function of $x$. However, if we restrict our attention to any one of the rectangles in Figure 3.5.2, the graph does define $y$ as a function of $x$.

*Definition*

A graph **defines $y$ as a function of $x$ in a rectangle with center $(x_0, y_0)$** on the graph if every vertical line that intersects the rectangle intersects the graph at exactly one point $(x, y)$ inside the rectangle. In that case, $y$ is the value of the function at $x$. ■

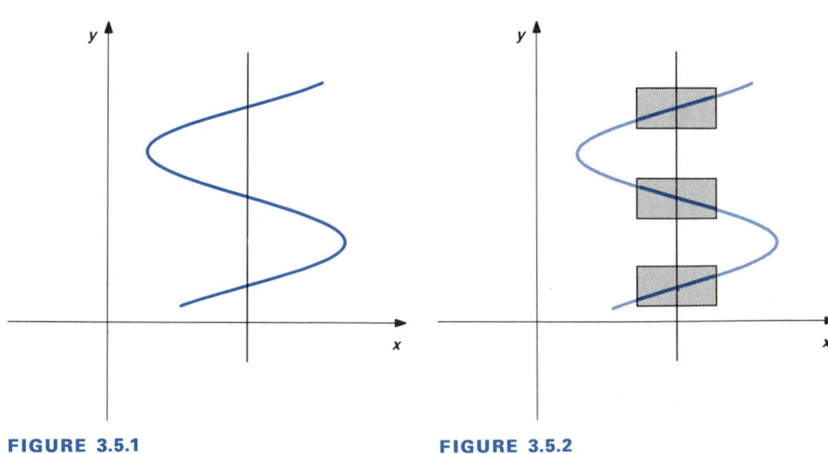


FIGURE 3.5.1 FIGURE 3.5.2

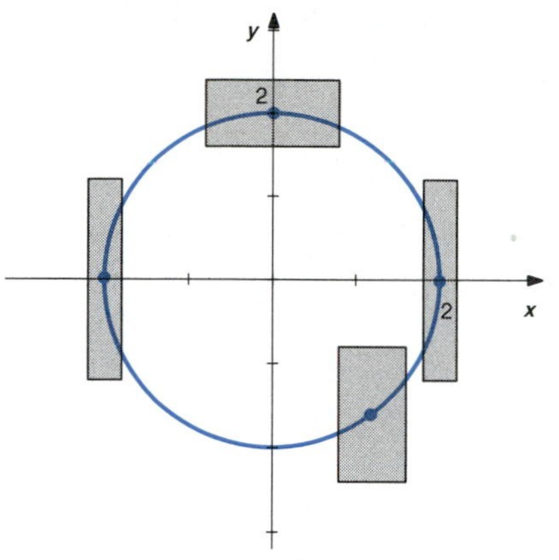


FIGURE 3.5.3

**EXAMPLE 1**

Find all points $(x_0, y_0)$ on the graph of the equation $x^2 + y^2 = 4$ such that the graph defines $y$ as a function of $x$ in some rectangle with center $(x_0, y_0)$.

**SOLUTION**

The graph of the equation $x^2 + y^2 = 4$ is a circle with center $(0, 0)$ and radius 2. We see from Figure 3.5.3 that the graph defines $y$ as a function of $x$ in some rectangle with center $(x_0, y_0)$ at every point $(x_0, y_0)$ on the graph, except for $(2, 0)$ and $(-2, 0)$. Different points $(x_0, y_0)$ may require rectangles of different dimensions to satisfy the condition of the definition. No rectangle with center at either $(2, 0)$ or $(-2, 0)$ can satisfy the desired condition, since some vertical lines that intersect the rectangle would intersect the graph at two points within the rectangle and some vertical lines that intersect the rectangle would not intersect the graph at any point. ■

Equations of the form $y = f(x)$ give $y$ **explicitly** as a function of $x$. Various parts of the graph of an equation in $x$ and $y$ may define $y$ as a function of $x$, even though we cannot solve the equation for $y$ and find a formula for the function. Such functions are said to be given **implicitly** by the equation. When these functions are differentiable, it is possible to find their derivatives without solving the equation for $y$. The method is called **implicit differentiation**.

### *Method of Implicit Differentiation*

- **Assume the equation defines $y$ as a differentiable function of $x$.**
- **Differentiate both sides of the equation with respect to $x$. Remember that $y$ is a function of $x$.**
- **Solve the differentiated equation for $dy/dx$ in terms of $x$ and $y$. (This is always easy to do, because the differentiated equation is linear in $dy/dx$.)**

**EXAMPLE 2**

Use the Method of Implicit Differentiation to find $dy/dx$ for the functions of $x$ defined implicitly by the equation $x^2 + y^2 = 4$.

**SOLUTION**

We assume the equation $x^2 + y^2 = 4$ defines $y$ as a differentiable function of $x$. The equals sign in the equation tells us both sides of the equation represent the same function of $x$. Since the derivatives of equal functions are equal, we can obtain another equation by differentiating both sides of the equation with respect to $x$.

$$x^2 + y^2 = 4,$$

$$\frac{d}{dx}(x^2 + y^2) = \frac{d}{dx}(4),$$

$$\frac{d}{dx}(x^2) + \frac{d}{dx}(y^2) = \frac{d}{dx}(4).$$

We know that

$$\frac{d}{dx}(x^2) = 2x \quad \text{and} \quad \frac{d}{dx}(4) = 0.$$

Since $y$ is a function of $x$, $y^2$ is of the form (Function)$^2$. The Chain Rule then gives

$$\frac{d}{dx}(y^2) = (2y)\frac{dy}{dx}.$$

Substitution in the above formula then gives

$$2x + (2y)\frac{dy}{dx} = 0 \quad \text{or} \quad \frac{dy}{dx} = -\frac{x}{y}.$$

■

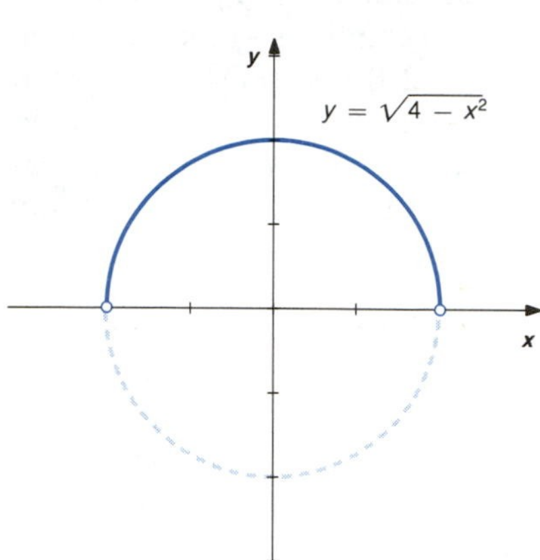


FIGURE 3.5.4

In Example 2 we can find formulas for the implicitly defined functions by solving the equation $x^2 + y^2 = 4$ for $y$. We obtain

$$y = \sqrt{4 - x^2} \quad \text{and} \quad y = -\sqrt{4 - x^2}.$$

The function defined by the upper part of the graph ($y > 0$) is given by the formula $y = \sqrt{4 - x^2}$. See Figure 3.5.4. Using the Chain Rule, we see that the derivative of this function is

$$\frac{dy}{dx} = \frac{1}{2\sqrt{4 - x^2}}(-2x) = -\frac{x}{\sqrt{4 - x^2}}.$$

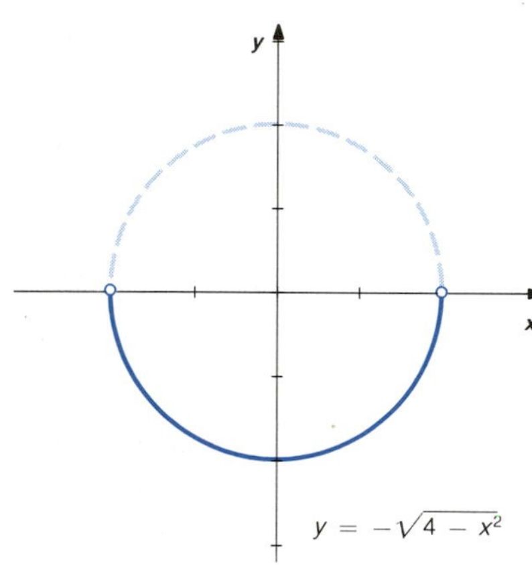


FIGURE 3.5.5

Since $y = \sqrt{4 - x^2}$, we can write $dy/dx = -x/y$. Similarly, the function defined by the lower part of the graph ($y < 0$) is given by the formula $y = -\sqrt{4 - x^2}$. See Figure 3.5.5. The derivative is

$$\frac{dy}{dx} = -\frac{1}{2\sqrt{4 - x^2}}(-2x) = -\frac{x}{-\sqrt{4 - x^2}}.$$

Since $y = -\sqrt{4 - x^2}$, we can write $dy/dx = -x/y$. Thus, the formula $dy/dx = -x/y$ holds in either case, as we had previously discovered by using the (easier) method of Implicit Differentiation.

When we are using the method of Implicit Differentiation it is important to remember that $y$ is a function of $x$ and that we are differentiating with respect to $x$. Hence, powers of $y$ are powers of functions and the Chain Rule is required for their differentiation. We should also be prepared to use combinations of the Chain Rule, Product Rule, and Quotient Rule.

■ **EXAMPLE 3**

Use the Method of Implicit Differentiation to find $dy/dx$ for the functions of $x$ defined implicitly by the equation $x^{2/3} + y^{2/3} = 1$.

**SOLUTION**

We assume $y$ is a differentiable function of $x$. Differentiating the equation

$$x^{2/3} + y^{2/3} = 1$$

with respect to $x$, and using the Chain Rule to find the derivative of the 2/3 power of the function $y$, we have

$$\frac{2}{3}x^{-1/3} + \frac{2}{3}y^{-1/3}\frac{dy}{dx} = 0.$$

(The above expression does not make sense if either $x = 0$ or $y = 0$.) Solving for $dy/dx$, we have

$$\frac{2}{3}y^{-1/3}\frac{dy}{dx} = -\frac{2}{3}x^{-1/3},$$

$$\frac{dy}{dx} = -\frac{x^{-1/3}}{y^{-1/3}},$$

$$\frac{dy}{dx} = -\frac{y^{1/3}}{x^{1/3}} = -\left(\frac{y}{x}\right)^{1/3}.$$

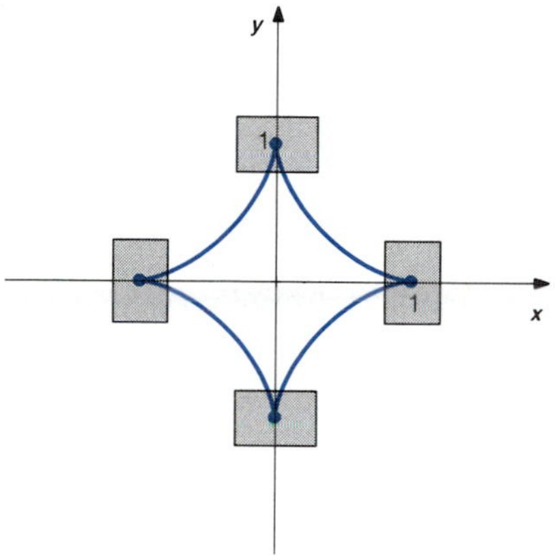

FIGURE 3.5.6

From Figure 3.5.6 we see that the graph of the equation $x^{2/3} + y^{2/3} = 1$ does not define $y$ as a function of $x$ in a rectangle with center at either of the points $(1, 0)$ or $(-1, 0)$ on the graph. The graph defines $y$ as a function of $x$ in rectangles with centers $(0, 1)$ and $(0, -1)$, but the functions are not differentiable at $x = 0$. ■

■ **EXAMPLE 4**

Use the Method of Implicit Differentiation to find $dy/dx$ for the functions of $x$ defined implicitly by the equation $\cos(x + y) = x$.

**SOLUTION**

We assume $y$ is a differentiable function of $x$. Differentiating the equation

$$\cos(x + y) = x$$

with respect to $x$, and using the Chain Rule to differentiate the cosine of the function $x + y$, we have

$$-\sin(x + y)\cdot\left(1 + \frac{dy}{dx}\right) = 1.$$

Solving for $dy/dx$, we have

$$-\sin(x + y) - \sin(x + y)\cdot\frac{dy}{dx} = 1,$$

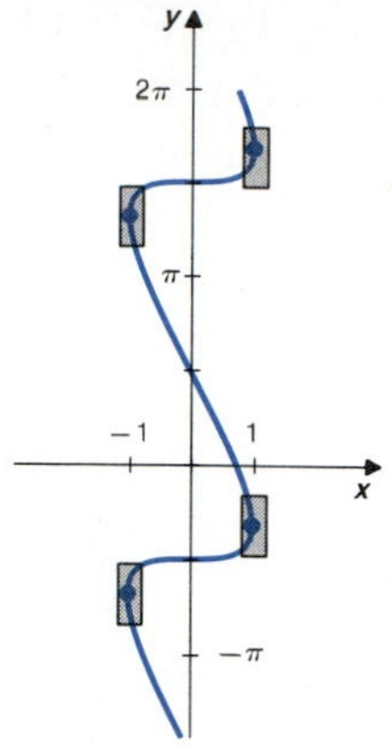

$$-\sin(x+y)\cdot\frac{dy}{dx} = 1 + \sin(x+y),$$

$$\frac{dy}{dx} = \frac{1+\sin(x+y)}{-\sin(x+y)}.$$

We see that the formula for $dy/dx$ does not make sense whenever $\sin(x+y) = 0$—that is, whenever $x + y = k\pi$, $k = 0, \pm 1, \pm 2, \ldots$. For those values of $(x, y)$, the defining equation $\cos(x+y) = x$ becomes $\cos(k\pi) = x$, so $x$ must be 1 when $k$ is even and $-1$ when $k$ is odd. The graph becomes vertical and the equation does not define $y$ as a function of $x$ in rectangles with centers at the corresponding points on the graph. See Figure 3.5.7. ■

FIGURE 3.5.7

For some applications of calculus, we are interested in the direction that is perpendicular to the line that is tangent to the graph of an equation in $x$ and $y$ at a point $(x_0, y_0)$ on the graph. The line that is perpendicular to the tangent line is called the **line normal** to the graph at the point $(x_0, y_0)$. See Figure 3.5.8. We know that the slope of the line tangent to the graph at a point $(x_0, y_0)$ is the value of the derivative $dy/dx$ at the point, which we denote

$$\left.\frac{dy}{dx}\right|_{(x_0, y_0)}.$$

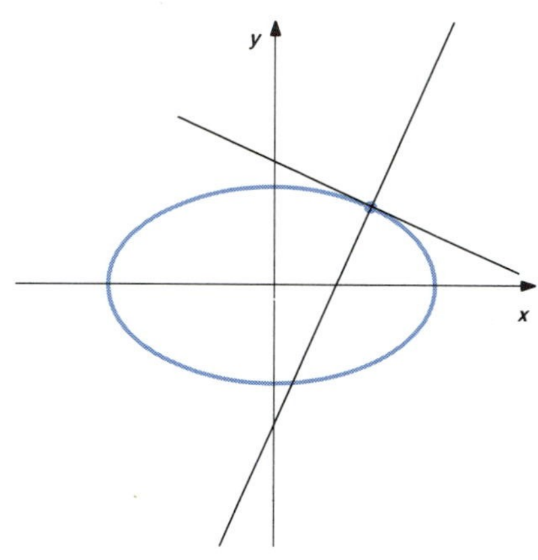

FIGURE 3.5.8

If $dy/dx$ is nonzero at $(x_0, y_0)$, the line normal to the graph at $(x_0, y_0)$ has slope that is the negative reciprocal of $dy/dx$.

**The equation of the line normal to the graph at the point $(x_0, y_0)$ is**

$$y - y_0 = \frac{-1}{\left.\frac{dy}{dx}\right|_{(x_0, y_0)}}(x - x_0).$$

If $dy/dx$ is zero at $(x_0, y_0)$, the normal line is vertical and has equation $x = x_0$.

■ **EXAMPLE 5**

Find the equation of the line normal to the graph of $x^3 + y^3 - 4.5xy = 0$ at the point (2, 1) on the graph.

**SOLUTION**

We need the value of the derivative $dy/dx$ at the point (2, 1). It is not convenient to solve the equation for $y$ in terms of $x$, so we use the Method of Implicit Differentiation to find $dy/dx$. We assume the equation defines $y$ as a differentiable function of $x$ and differentiate the equation

$$x^3 + y^3 - 4.5xy = 0$$

with respect to $x$. Using the Chain Rule to differentiate the 3rd power of the function $y$ and the Product Rule to differentiate the product $xy$, we obtain

$$3x^2 + 3y^2\frac{dy}{dx} - 4.5\left[(x)\left(\frac{dy}{dx}\right) + (y)(1)\right] = 0.$$

Solving for $dy/dx$, we have

$$3x^2 + 3y^2\frac{dy}{dx} - 4.5x\frac{dy}{dx} - 4.5y = 0,$$

$$(3y^2 - 4.5x)\frac{dy}{dx} = 4.5y - 3x^2,$$

$$\frac{dy}{dx} = \frac{4.5y - 3x^2}{3y^2 - 4.5x} = \frac{1.5(3y - 2x^2)}{1.5(2y^2 - 3x)} = \frac{3y - 2x^2}{2y^2 - 3x}.$$

When $(x, y) = (2, 1)$,

$$\frac{dy}{dx} = \frac{3(1) - 2(2)^2}{2(1)^2 - 3(2)} = \frac{-5}{-4} = \frac{5}{4}.$$

The slope of the line normal to the graph at the point (2, 1) is then $-1/(5/4) = -4/5$. The point-slope form of the equation of the line through (2, 1) with slope $-4/5$ is

$$y - 1 = -\frac{4}{5}(x - 2).$$

Simplifying, we obtain

$$5(y - 1) = -4(x - 2),$$

$$5y - 5 = -4x + 8,$$

$$4x + 5y = 13.$$

The graph of the equation and the normal line are sketched in Figure 3.5.9. ■

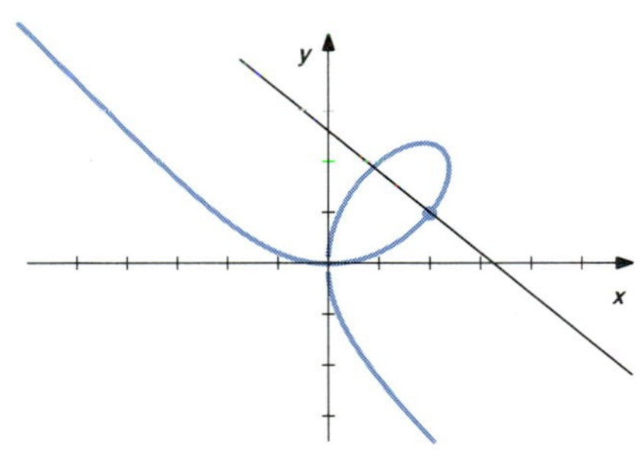


FIGURE 3.5.9

The Method of Implicit Differentiation can be used to find the derivative of any variable with respect to another from an equation that relates the two variables. We must distinguish between the independent variable of differentiation and the dependent variable that is considered to be the function.

■ **EXAMPLE 6**

If a rocket is rising vertically from a point on the ground that is 200 ft from an observer at ground level, the height of the rocket and the angle of elevation from the observer to the rocket are related by the formula $h = 200 \tan \theta$. See Figure 3.5.10. Find $d\theta/dh$, the rate of change of the angle of elevation with respect to the height, when the height is 600 ft.

FIGURE 3.5.10

**SOLUTION**

We first note that we want the derivative of the *function* $\theta$ with respect to the *independent variable* $h$. We assume that the equation $h = 200 \tan \theta$ defines $\theta$ as a differentiable function of $h$ and differentiate the equation with respect to $h$. Since $\tan \theta$ is the tangent of a function, we need to use the Chain Rule to differentiate that term. We have

$$h = 200 \tan \theta,$$

$$1 = 200 \sec^2 \theta \frac{d\theta}{dh},$$

$$\frac{d\theta}{dh} = \frac{\cos^2 \theta}{200}.$$

When $h = 600$, we have

$$600 = 200 \tan \theta, \quad \text{so} \quad \tan \theta = 3.$$

The exact value of $\cos \theta$ can then be determined from the representative angle in Figure 3.5.11, or a calculator can be used to find an approximate value. We obtain $\cos \theta = 1/\sqrt{10}$, so

$$\frac{d\theta}{dh} = \frac{(1/\sqrt{10})^2}{200} = 0.0005 \text{ rad/ft}.$$

Recall that the formulas we use for the differentiation of trigonometric functions hold only if the angles are measured in radians. ■

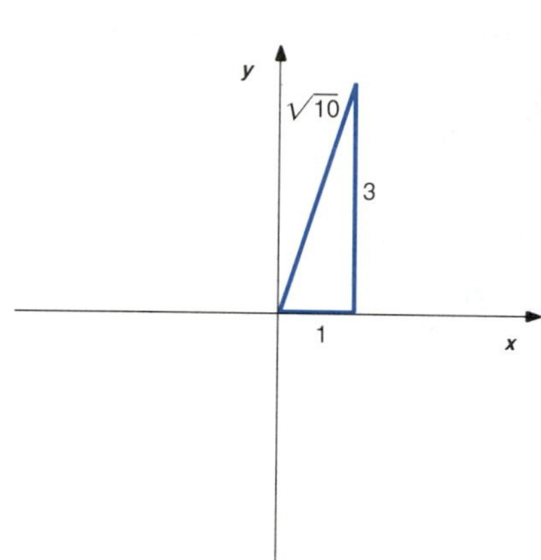


FIGURE 3.5.11

# EXERCISES 3.5

In Exercises 1–8 find all points $(x_0, y_0)$ on the graphs where the graphs define $y$ as a function of $x$ in a rectangle with center $(x_0, y_0)$.

1

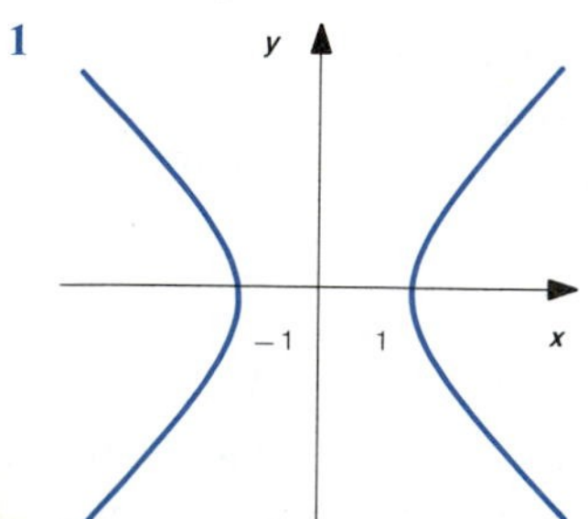


2

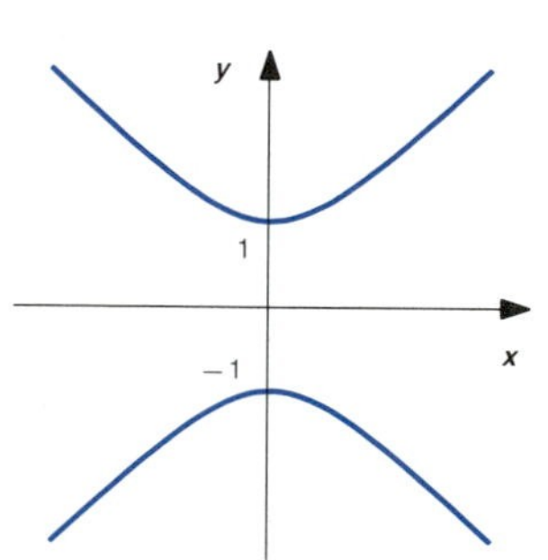

3

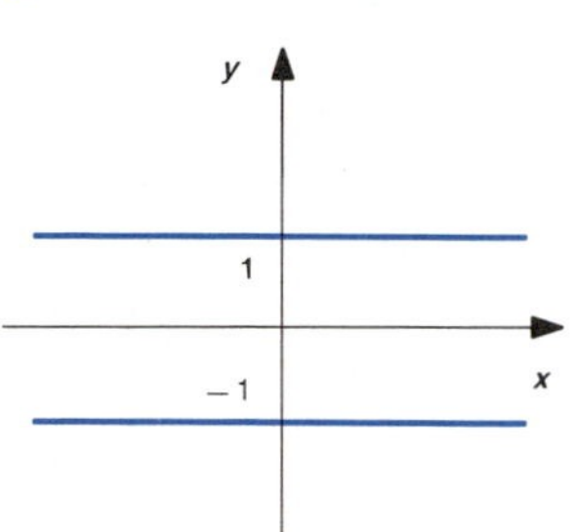


4

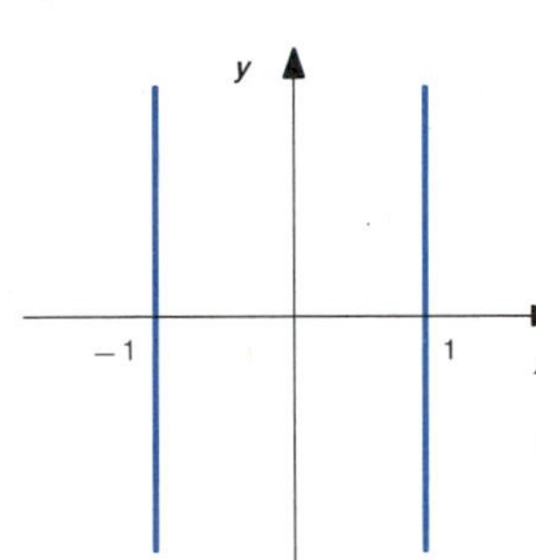


5

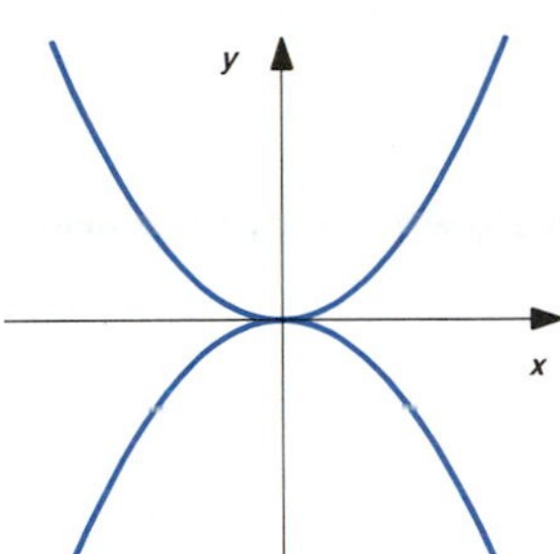


6

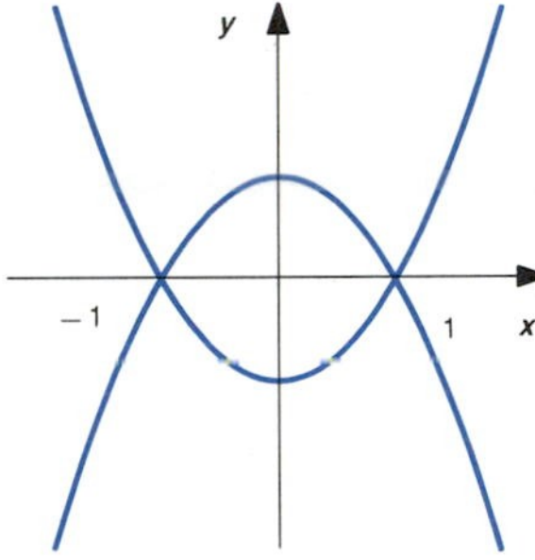


7

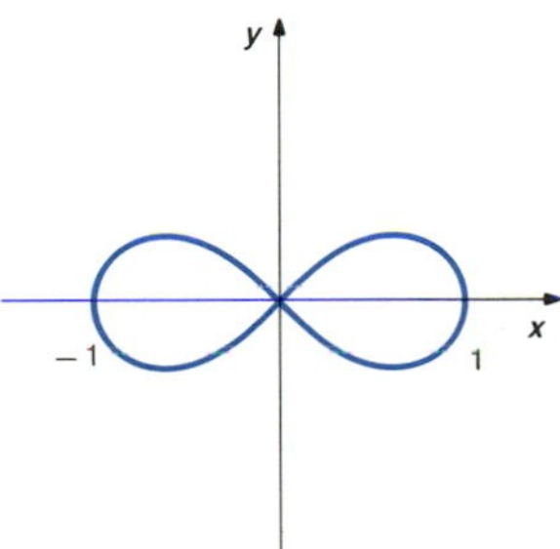


8

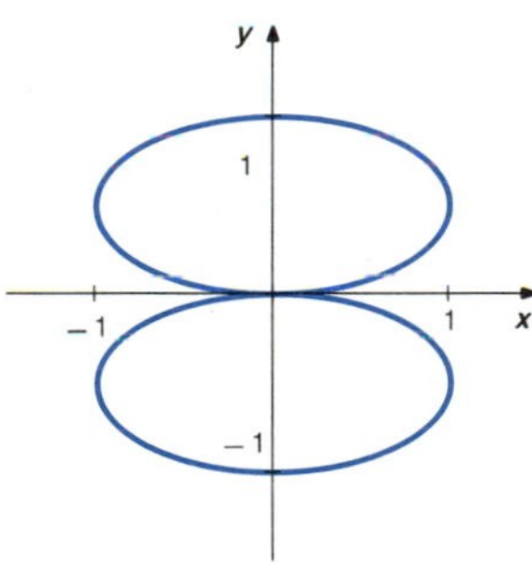


Use the method of Implicit Differentiation to find $dy/dx$ for the functions of $x$ that are defined implicitly by the equations in Exercises 9–22.

9 $x^2 - y^2 = 4$

10 $y^2 - x^2 = 49$

11 $y^3 = x$

12 $y^3 = x^2$

13 $\sqrt{x} + \sqrt{y} = 4$

14 $x^{1/3} + y^{1/3} = 2$

15 $\sin(x + y) = x$

16 $\sin(x + y) = y$

17 $xy^2 - yx^2 = 16$

18 $xy^3 - yx^3 = 4$

19 $x^3 + y^3 - 3xy = 0$

20 $x^4 + 4xy - y^4 = 0$

21 $(\sin x)(\cos y) = y$

22 $(\sin x)(\cos y) = x$

In Exercises 23–28 find the equation of the line tangent to the graph at the given point.

23 $y^2 = 2x^3$, $(2, -4)$

24 $2y^3 = x^2$, $(-4, 2)$

25 $(x + y)^3 = x + y + 6$, $(3, -1)$

26 $\sqrt{x + y} = x + y - 6$, $(5, 4)$

27 $xy^3 - x^3y = 30$, $(2, 3)$

28 $xy^2 + yx^2 = 6$, $(2, -3)$

In Exercises 29–34 find the equation of the line normal to the graph at the given point.

29 $\dfrac{1}{x^2} + \dfrac{1}{y^2} = 20$, $\left(\dfrac{1}{2}, -\dfrac{1}{4}\right)$

30 $\dfrac{1}{x} + \dfrac{1}{y} = 1$, $\left(\dfrac{1}{4}, -\dfrac{1}{3}\right)$

31 $x = \sin y$, $(1/2, \pi/6)$

32 $x = \sec y$, $(2, \pi/3)$

33 $x = \tan y$, $(1, \pi/4)$

34 $x = y - \cos y$, $(\pi/2, \pi/2)$

35 Show that the line normal to the circle $x^2 + y^2 = r^2$ at any point $(x_0, y_0)$ on the circle, passes through the origin.

36 Find the equation of each line that contains the origin and is tangent to the circle $(x - 4)^2 + y^2 = 4$.

37 The volume $V$ and the surface area of a sphere are related by the equation $36\pi V^2 = S^3$. Find $dS/dV$, the rate of change of the surface area with respect to the volume when the volume is $\pi\sqrt{6}$ ft$^3$.

38 The volume $V$ of water in a certain conical tank is given by the equation $V = \pi h^3/3$, where $h$ is the depth of the water above the vertex of the cone. Find $dh/dV$, the rate of change of the depth with respect to the volume when the volume is $9\pi$ ft$^3$.

39 Find and compare $dv/du$ and $du/dv$, if $u = 1 + 2v + 3v^2$.

40 Find and compare $ds/dr$ and $dr/ds$, if $r = 4 - s + 2s^2$.

## 3.6 HIGHER-ORDER DERIVATIVES

Since the derivative of a function is again a function, we may consider successive derivatives of functions. The derivative of a function is then called the first derivative. The derivative of the first derivative is called the second derivative. The derivative of the second derivative is called the third derivative, and so forth. The following notation is used:

| | | | |
|---|---|---|---|
| **First derivative:** | $\dfrac{df}{dx}$ | $f'$ | $f^{(1)}$ |
| **Second derivative:** | $\dfrac{d}{dx}\left(\dfrac{df}{dx}\right) = \dfrac{d^2f}{dx^2}$ | $(f')' = f''$ | $f^{(2)}$ |
| **Third derivative:** | $\dfrac{d}{dx}\left(\dfrac{d^2f}{dx^2}\right) = \dfrac{d^3f}{dx^3}$ | $(f'')' = f'''$ | $f^{(3)}$ |
| $\vdots$ | | | |
| ***n*th derivative:** | $\dfrac{d}{dx}\left(\dfrac{d^{n-1}f}{dx^{n-1}}\right) = \dfrac{d^nf}{dx^n}$ | | $f^{(n)}$ |

Both $d^nf/dx^n$ and $f^{(n)}$ indicate derivatives of order $n$. The function $f$ is sometimes denoted $f^{(0)}$.

The prime notation, $f'$, $f''$, $f'''$ becomes awkward for derivatives of higher order than three, so we will usually use either $d^nf/dx^n$ or $f^{(n)}$ to denote such derivatives.

■ **EXAMPLE 1**

Find the first four derivatives of

(a) $f(x) = 3 - x + 5x^2 - 7x^3$,
(b) $g(x) = 1/x$,
(c) $h(x) = \sin 2x$.

**SOLUTION**

(a) $f(x) = 3 - x + 5x^2 - 7x^3$,

$$f^{(1)}(x) = -1 + 10x - 21x^2,$$

$$f^{(2)}(x) = 10 - 42x,$$

$$f^{(3)}(x) = -42,$$

$$f^{(4)}(x) = 0.$$

(b) $g(x) = x^{-1}$,

$$g^{(1)}(x) = (-1)x^{-2},$$

$$g^{(2)}(x) = (-1)(-2)x^{-3},$$

$$g^{(3)}(x) = (-1)(-2)(-3)x^{-4},$$

$$g^{(4)}(x) = (-1)(-2)(-3)(-4)x^{-5}.$$

(c) Using the Chain Rule, we have

$$h(x) = \sin 2x,$$

$$h^{(1)} = (\cos 2x)(2) = 2\cos 2x,$$

$$h^{(2)}(x) = 2(-\sin 2x)(2) = -4\sin 2x,$$

$$h^{(3)}(x) = -4(\cos 2x)(2) = -8\cos 2x,$$

$$h^{(4)}(x) = -8(-\sin 2x)(2) = 16\sin 2x.$$ ■

**■ EXAMPLE 2**

$f(x) = x/\sqrt{x^2+1}$. Find $f''(x)$.

**SOLUTION**

We must first find $f'$. Using the Quotient Rule, we have

$$f'(x) = \frac{(\sqrt{x^2+1})(1) - (x)\left(\dfrac{1}{2\sqrt{x^2+1}}(2x)\right)}{(\sqrt{x^2+1})^2}.$$

It is worthwhile to simplify $f'(x)$ before calculating $f''(x)$. We have

$$f'(x) = \frac{\sqrt{x^2+1} - \dfrac{x^2}{\sqrt{x^2+1}}}{x^2+1}$$

$$= \frac{\sqrt{x^2+1}\cdot\dfrac{\sqrt{x^2+1}}{\sqrt{x^2+1}} - \dfrac{x^2}{\sqrt{x^2+1}}}{x^2+1}$$

$$= \frac{(x^2+1) - x^2}{(x^2+1)^{3/2}}$$

$$= \frac{1}{(x^2+1)^{3/2}} = (x^2+1)^{-3/2}.$$

The Chain Rule then gives

$$f''(x) = -\frac{3}{2}(x^2+1)^{-5/2}(2x) = \frac{-3x}{(x^2+1)^{5/2}}.$$ ■

**EXAMPLE 3**

$y^2 - 2xy = 1$. Find formulas for $dy/dx$ and $d^2y/dx^2$ in terms of $x$ and $y$.

**SOLUTION**

We assume the equation defines $y$ as a differentiable function of $x$. Differentiating the equation $y^2 - 2xy = 1$ with respect to $x$ and solving the resulting equation for $dy/dx$, we have

$$2y\frac{dy}{dx} - 2\left[(x)\left(\frac{dy}{dx}\right) + (y)(1)\right] = 0,$$

$$2y\frac{dy}{dx} - 2x\frac{dy}{dx} - 2y = 0,$$

$$(2y - 2x)\frac{dy}{dx} = 2y,$$

$$\frac{dy}{dx} = \frac{y}{y - x}.$$

We can obtain $d^2y/dx^2$ by differentiating the above equation with respect to $x$. We must remember that both $y$ and $dy/dx$ are functions of $x$. Using the Quotient Rule, we have

$$\frac{d^2y}{dx^2} = \frac{(y - x)\left(\frac{dy}{dx}\right) - (y)\left(\frac{dy}{dx} - 1\right)}{(y - x)^2}$$

Simplifying, we have

$$\frac{d^2y}{dx^2} = \frac{y\frac{dy}{dx} - x\frac{dy}{dx} - y\frac{dy}{dx} + y}{(y - x)^2}$$

$$= \frac{y - x\frac{dy}{dx}}{(y - x)^2}.$$

Using the formula $dy/dx = y/(y - x)$ found above, we can express the second derivative in terms of $x$ and $y$.

$$\frac{d^2y}{dx^2} = \frac{y - x\left(\frac{y}{y - x}\right)}{(y - x)^2}$$

$$= \frac{y\frac{y - x}{y - x} - \frac{xy}{y - x}}{(y - x)^2}$$

$$= \frac{y(y-x) - xy}{(y-x)^3}$$

$$= \frac{y^2 - 2xy}{(y-x)^3}.$$

Since the original equation is $y^2 - 2xy = 1$, we can further simplify to obtain

$$\frac{d^2y}{dx^2} = \frac{1}{(y-x)^3}.$$ ■

Recall that if the **position** of a particle along a line at time $t$ is given by the function $s(t)$, its **velocity** at time $t$ is given by the formula $v(t) = s'(t)$. We now define the (instantaneous) **acceleration** of the particle to be the rate of change of the velocity with respect to time,

$$a(t) = v'(t) = s''(t).$$

■ **EXAMPLE 4**

The height in feet above ground of an object $t$ seconds after it is thrown vertically from a height of 64 ft with initial velocity 48 ft/s is given by the formula

$$s(t) = -16t^2 + 48t + 64, \qquad t \geq 0.$$

The formula is valid until the object returns to ground level. Find formulas for the velocity $v(t)$ and acceleration $a(t)$ of the object. At what time $t$ is $v(t) = 0$? What is the height when $v(t) = 0$? At what times is the height zero? What is the velocity at the times when the height is zero?

**SOLUTION**

We have

$$s(t) = -16t^2 + 48t + 64,$$

$$v(t) = s'(t) = -32t + 48,$$

$$a(t) = v'(t) = s''(t) = -32.$$

(Note that the acceleration is constant. The negative value indicates the acceleration is downward.)

$$v(t) = 0 \quad \text{when} \quad -32t + 48 = 0, \quad \text{so} \quad t = \frac{48}{32} = \frac{3}{2}.$$

When $t = 3/2$, the height is

$$s\left(\frac{3}{2}\right) = -16\left(\frac{3}{2}\right)^2 + 48\left(\frac{3}{2}\right) + 64 = 100.$$

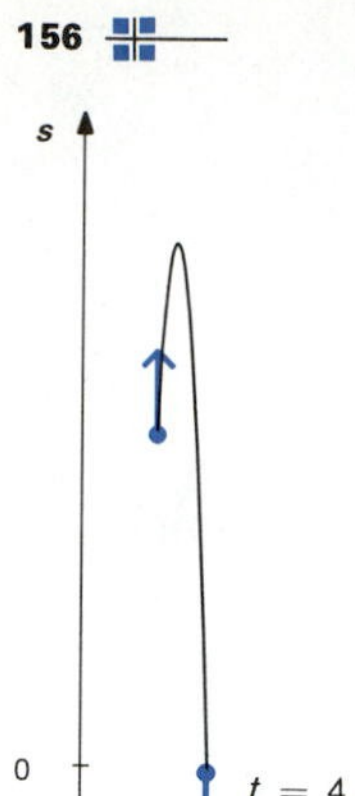


FIGURE 3.6.1

The height is zero when $s(t) = 0$, or

$$-16t^2 + 48t + 64 = 0,$$

$$-16(t^2 - 3t - 4) = 0,$$

$$(t - 4)(t + 1) = 0,$$

$$t = 4, \quad t = -1.$$

When $t = 4$, the velocity is $v(4) = -32(4) + 48 = -80$. The value $t = -1$ is not in the domain of the height function. See Figure 3.6.1. ■

The first and second derivatives have geometric interpretations that we will investigate in Chapter 4. Points on the graphs where either $f'(x) = 0$ or $f''(x) = 0$ will be of special interest.

## ■ EXAMPLE 5

Find all points on the graph of the function $f(x) = 2x^3 - 3x^2 - 12x$ where (a) $f'(x) = 0$ and (b) $f''(x) = 0$.

### SOLUTION

We begin by writing the formula for $f(x)$ and then calculating $f'(x)$ and $f''(x)$. It is important that each of the three functions be labeled properly.

$$f(x) = 2x^3 - 3x^2 - 12x,$$

$$f'(x) = 6x^2 - 6x - 12,$$

$$f''(x) = 12x - 6.$$

(a) We see that $f'(x) = 0$ when

$$6x^2 - 6x - 12 = 0,$$

$$6(x^2 - x - 2) = 0,$$

$$6(x - 2)(x + 1) = 0.$$

The solutions are $x = 2$ and $x = -1$. These are the $x$-coordinates of the points on the graph where $f'(x) = 0$. To find the corresponding $y$-coordinates, we must substitute into the equation for $f(x)$.

When $x = 2$,

$$y = f(2) = 2(2)^3 - 3(2)^2 - 12(2) = -20.$$

When $x = -1$,

$$y = f(-1) = 2(-1)^3 - 3(-1)^2 - 12(-1) = 7.$$

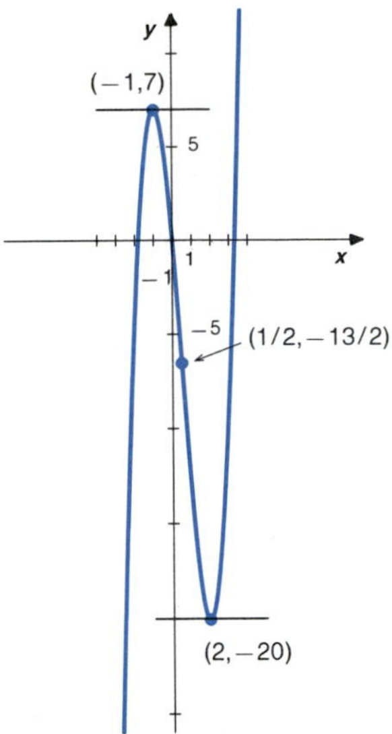


FIGURE 3.6.2

The points $(2, -20)$ and $(-1, 7)$ are the points on the graph where $f'(x) = 0$. The graph has horizontal tangents at these points.

(b) We see that $f''(x) = 0$ when

$$12x - 6 = 0,$$

$$12x = 6,$$

$$x = \frac{1}{2}.$$

The $y$-coordinate of the point on the graph with $x = 1/2$ is

$$y = f\left(\frac{1}{2}\right) = 2\left(\frac{1}{2}\right)^3 - 3\left(\frac{1}{2}\right)^2 - 12\left(\frac{1}{2}\right) = -\frac{26}{4} = -\frac{13}{2}.$$

The point on the graph with $f''(x) = 0$ is $(1/2, -13/2)$. From Figure 3.6.2 we see that the graph opens downward for $x < 1/2$ and opens upward for $x > 1/2$. ■

Let us now see how the values of a function and its derivatives at a single point $c$ can be used to find approximate values of the function for $x$ near $c$. (A sophisticated version of this basic idea is used by your calculator to calculate function values.) We will use **Taylor polynomials** to illustrate the idea. Taylor polynomials will be reintroduced and studied in more detail in later sections.

If $f$ is any function that has $n$ derivatives at $c$, we can form the polynomial

$$P_n(x) = f(c) + f'(c)(x - c) + \frac{f''(c)}{2!}(x - c)^2 + \cdots + \frac{f^{(n)}(c)}{n!}(x - c)^n.$$

$P_n(x)$ is called the **$n$th-order Taylor polynomial of $f$ about $c$.** $P_n(x)$ is expressed in terms of powers of $(x - c)$. If $c \neq 0$, these powers should not be expanded in an attempt to express $P_n(x)$ in terms of powers of $x$. The powers of $(x - c)$ in $P_n(x)$ go from zero to $n$, where the zero power of $(x - c)$ is taken to be one; the coefficient of $(x - c)^k$ is $f^{(k)}(c)/k!$, $0 \leq k \leq n$, where $f^{(0)}(c) = f(c)$, $0! = 1$, and $k! = 1 \cdot 2 \cdots k$ for $k \geq 1$.

The idea of a Taylor polynomial is not entirely new to us. That is, let $P_1(x)$ be the first-order Taylor polynomial of $f$ about $c$; then consider the graph or the equation $y = P_1(x)$. We obtain

$$y = f(c) + f'(c)(x - c),$$

$$y - f(c) = f'(c)(x - c),$$

so the graph of $P_1$, the first-order Taylor polynomial of $f$ at $c$, is the line tangent to the graph of $f$ at the point $(c, f(c))$.

**EXAMPLE 6**

Find the first-, second-, third-, fourth-, fifth-, sixth-, and seventh-order Taylor polynomials of cos $x$ about 0. Compare the values of these Taylor polynomials at $x = 0.2$ with a calculator value of $\cos(0.2)$.

**SOLUTION**

We need the values of $f(x) = \cos x$ and the first seven derivatives at 0:

$$f(x) = \cos x, \qquad f(0) = \cos 0 = 1,$$

$$f'(x) = -\sin x, \qquad f'(0) = -\sin 0 = 0,$$

$$f''(x) = -\cos x, \qquad f''(0) = -\cos 0 = -1,$$

$$f'''(x) = \sin x, \qquad f'''(0) = \sin 0 = 0,$$

$$f^{(4)}(x) = \cos x, \qquad f^{(4)}(0) = \cos 0 = 1,$$

$$f^{(5)}(x) = -\sin x, \qquad f^{(5)}(0) = -\sin 0 = 0,$$

$$f^{(6)}(x) = -\cos x, \qquad f^{(6)}(0) = -\cos 0 = -1,$$

$$f^{(7)}(x) = \sin x, \qquad f^{(7)}(0) = \sin 0 = 0.$$

Substitution of the values in the defining formula gives

$$P_1(x) = f(0) + f'(0)(x - 0) = 1,$$

$$P_2 = f(0) + f'(0)(x - 0) + \frac{f''(0)}{2!}(x - 0)^2 = 1 - \frac{1}{2!}x^2,$$

$$P_3 = f(0) + f'(0)(x - 0) + \frac{f''(0)}{2!}(x - 0)^2 + \frac{f'''(0)}{3!}(x - 0)^3 = 1 - \frac{1}{2!}x^2,$$

$$P_4(x) = P_5(x) = 1 - \frac{1}{2!}x^2 + \frac{1}{4!}x^4,$$

$$P_6(x) = P_7(x) = 1 - \frac{1}{2!}x^2 + \frac{1}{4!}x^4 - \frac{1}{6!}x^6.$$

Evaluating the above Taylor polynomials at $x = 0.2$, we obtain

$$P_1(0.2) = P_2(0.2) = 1,$$

$$P_2(0.2) = P_3(0.2) = 1 - \frac{1}{2}(0.2)^2 = 0.98,$$

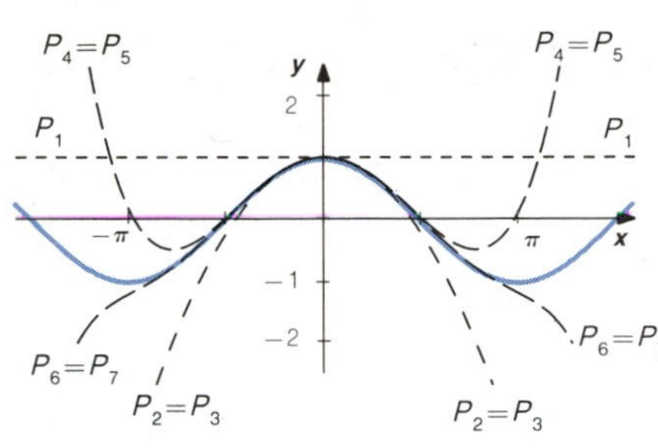


FIGURE 3.6.3

$$P_4(0.2) = P_5(0.2) = 1 - \frac{1}{2}(0.2)^2 + \frac{1}{24}(0.2)^4 \approx 0.98006667,$$

$$P_6(0.2) = P_7(0.2) = 1 - \frac{1}{2}(0.2)^2 + \frac{1}{24}(0.2)^4 - \frac{1}{720}(0.2)^6 \approx 0.98006658.$$

A calculator gives $\cos(0.2) \approx 0.98006658$. (Be sure to use radian measure.) We can see from Figure 3.6.3 how the values of the Taylor polynomials $P_n(x)$ begin to approach $\cos x$ as $n$ increases. ■

## EXERCISES 3.6

Find the first three derivatives of the functions given in Exercises 1–8.

1 $f(x) = 3x^2 - 7x + 2$

2 $f(x) = x^3 - 2x^2 + x - 7$

3 $f(x) = \sqrt{1 - x}$

4 $f(x) = \dfrac{1}{2x + 1}$

5 $f(x) = x^{1/3} + x^{-1/3}$

6 $f(x) = \cos 2x$

7 $f(x) = \tan x$

8 $f(x) = \sec x$

In Exercises 9–14 use the Method of Implicit Differentiation to find formulas for $dy/dx$ and $d^2y/dx^2$ in terms of both $x$ and $y$.

9 $x^2 + y^2 = 4$

10 $y^3 = x^2$

11 $y^2 - 2xy + 4 = 0$

12 $\dfrac{1}{x} + \dfrac{1}{y} = 1$

13 $x = \tan y$

14 $x = \sin y$

In Exercises 15–18 the position of a particle on a line at time $t$ is given by the formula for $s(t)$. Find (a) the velocity $v(t) = s'(t)$ and (b) the acceleration $a(t) = v'(t) = s''(t)$ at time $t$. (c) Determine the position at each time the velocity is zero. (d) Determine the velocity at each time the position is zero.

15 $s(t) = -16t^2 + 32t + 48$

16 $s(t) = \dfrac{t}{(t^3 + 1)^{1/3}}$

17 $s(t) = 1 - 2 \sin t$

18 $s(t) = \sin t - \cos t$

In Exerecises 19–24 find all points on the graph of the given function $f$ where (a) $f'(x) = 0$ and (b) $f''(x) = 0$.

19 $f(x) = 4 + 12x - 3x^2$

20 $f(x) = x^3 - 3x^2$

21 $f(x) = x + \dfrac{1}{x}$

22 $f(x) = \sqrt{x^2 + 1}$

23 $f(x) = x + \cos x,\ 0 \le x \le 2\pi$

24 $f(x) = \sin x + \cos x,\ 0 \le x \le 2\pi$

In Exercises 25–32 find the $n$th-order Taylor polynomial of $f$ about $c$ and compare the values of $f(x_0)$ and $P_n(x_0)$.

25 $f(x) = x^3 - x^2 + x - 1,\ n = 1, 2, 3,\ c = 1,\ x_0 = 1.2$

26 $f(x) = x^3,\ n = 1, 2, 3,\ c = 1,\ x_0 = 1.1$

27 $f(x) = (x - 2)^3,\ n = 1, 2, 3,\ c = 0,\ x_0 = -0.1$

28 $f(x) = \dfrac{1}{1 - x},\ n = 1, 2, 3,\ c = 0,\ x_0 = 0.1$

29 $f(x) = \tan x,\ n = 1, 2, 3,\ c = 0,\ x_0 = 0.2$

30 $f(x) = \sin x,\ n = 2, 4, 6,\ c = 0,\ x_0 = -0.1$

31 $f(x) = \sqrt{x},\ n = 1, 2, 3,\ c = 1,\ x_0 = 0.98$

32 $f(x) = x^{1/3},\ n = 1, 2, 3,\ c = 1,\ x_0 = 1.02$

33 Find a formula for $(fg)''$.

34 Find a formula for $(fg)'''$.

35 Show that

$$\frac{d^2}{dx^2}(f(g(x))) = f'(g(x))g''(x) + f''(g(x))(g'(x))^2.$$

36 If $P(x)$ is a polynomial of the form

$$P(x) = a_0 + a_1(x - c) + a_2(x - c)^2 + \cdots + a_n(x - c)^n,$$

show that

$$a_j = \frac{P^{(j)}(c)}{j!}, \qquad j = 0,1,2\ldots, n.$$

[*Hint*: Use the fact that positive powers of $(x - c)$ are zero when $x = c$ to evaluate $P(c), P'(c), P''(c), \ldots, P^{(n)}(c)$.]

37 If $P(x) = a_0 + a_1x + a_2x^2 + \cdots + a_nx^n$ is a polynomial and $P'(x) = 0$ for all $x$, show that $P(x) = a_0$, a constant. [*Hint*: Use the result of Exercise 36 and the fact that higher-order derivatives of $P$ are derivatives of the constant function $P'(x) = 0$ and, hence, are zero.]

## 3.7 THE INTERMEDIATE VALUE THEOREM

Continuity is the key to the problem of finding approximate values of solutions of equations that cannot be solved by the methods of algebra. The Intermediate Value Theorem provides the theoretical basis for the solution of this problem.

The Intermediate Value Theorem expresses one of several important and useful properties of functions that are continuous at each point of an interval. Let us begin our study by defining the notion of continuity on an interval.

*Definition*

A function $f$ is said to be **continuous on an interval** if (i) $f$ is continuous at each point of the interval that is not an endpoint of the interval; (ii) $f$ is continuous from the right at the left endpoint of the interval, if that point is included in the interval; and (iii) $f$ is continuous from the left at the right endpoint of the interval, if that point is included in the interval. ■

Continuity on an interval depends only on the values of the function at points of the interval. At endpoints that are included in the interval, the function is required to be continuous only from the direction of points inside the interval; no requirements are made at endpoints of the interval that are not included in the interval. If $f$ is continuous on an interval $I$, then $f$ is continuous on any interval contained in $I$.

We can use what we learned in Section 2.5 about continuity at a point to determine continuity on intervals. For example:

- The function $f(x) = x^2 - 7x + 3$ is continuous on any interval, because *polynomials are continuous at every point*. See Figure 3.7.1.
- The function $f(x) = 1/x$ is continuous on any interval that does not contain the origin. In particular, $f$ is continuous on each of the intervals $x > 0$ and $x < 0$. See Figure 3.7.2. We know that *rational functions are continuous at every point where they are defined.*

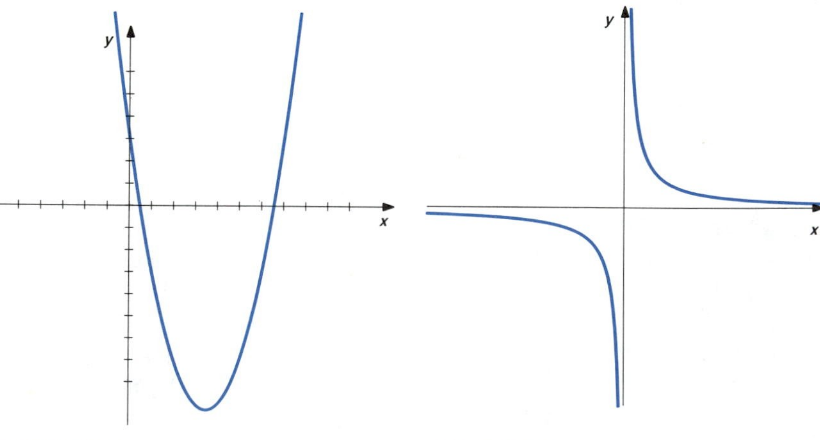


FIGURE 3.7.1

FIGURE 3.7.2

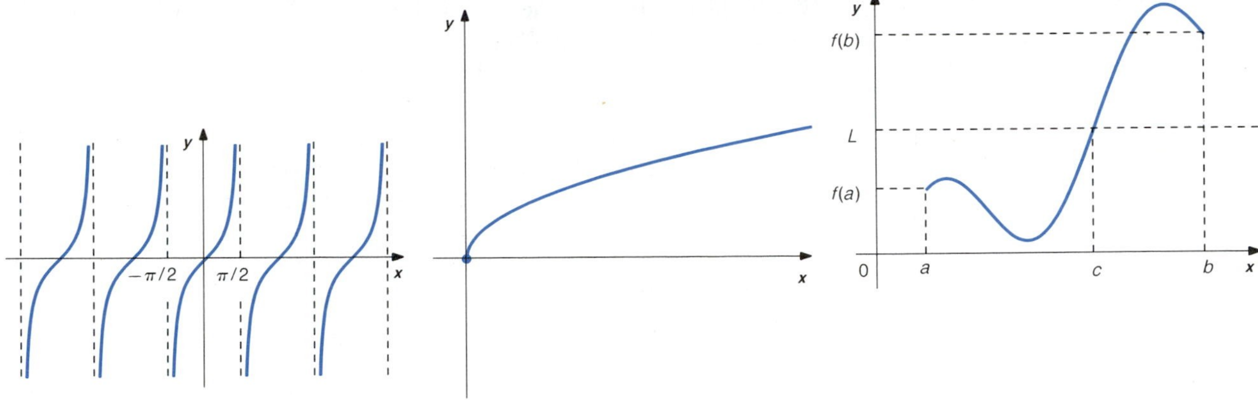


FIGURE 3.7.3

FIGURE 3.7.4

FIGURE 3.7.5

■ The function $f(x) = \tan x$ is continuous on intervals on which it is defined, such as $-\pi/2 < x < \pi/2$. The *trigonometric functions are continuous at every point where they are defined*. See Figure 3.7.3.

■ The function $f(x) = \sqrt{x}$ is continuous at every $x > 0$ and continuous from the right at $x = 0$, so it is continuous on the interval $x \geq 0$ and all of its subintervals. See Figure 3.7.4.

We can now state the Intermediate Value Theorem.

■ *Intermediate Value Theorem*

**If $f$ is continuous on the interval $a \leq x \leq b$, and $L$ is any number between $f(a)$ and $f(b)$, then there is at least one number $c$ such that $a \leq c \leq b$ and $f(c) = L$.**

The Intermediate Value Theorem tells us that the graph of a continuous function cannot jump from one side of a horizontal line $y = L$ to the other without intersecting the line at least once. See Figure 3.7.5.

The Intermediate Value Theorem follows from the following result, which we will use to find approximate values of the solution of equations $f(x) = 0$.

■ *Special case of the Intermediate Value Theorem*

**If $f$ is continuous on the interval $a \leq x \leq b$, and either $f(a) < 0 < f(b)$ or $f(b) < 0 < f(a)$, then there is at least one number $c$ such that $a < c < b$ and $f(c) = 0$.**

■ *Proof* The procedure used to verify this result is called the **bisection method** of locating the zeros of a function. To illustrate how the method works, let us consider the case $f(a) < 0 < f(b)$. We will first locate a particular point $c$ and then show $f(c) = 0$.

We begin by cutting the interval $a \leq x \leq b$ at its midpoint $(a + b)/2$ and selecting one of the two half-intervals: select the left half-interval if

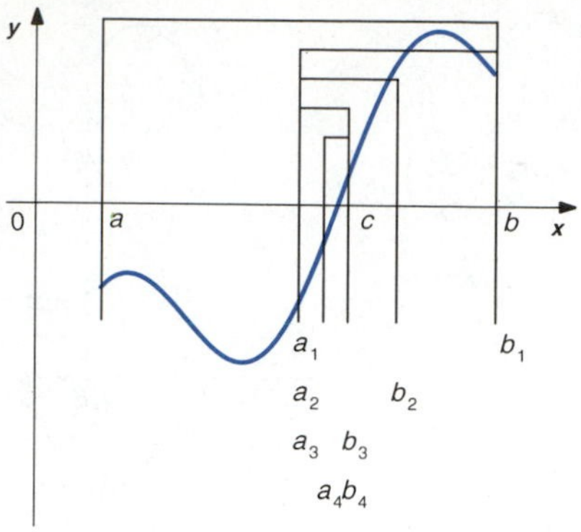

FIGURE 3.7.6

$f((a+b)/2) \geq 0$; select the right half-interval if $f((a+b)/2) < 0$. If $a_1 \leq x \leq b_1$ is the selected half-interval, we will have $f(a_1) < 0 \leq f(b_1)$.

We repeat the procedure of cutting in half and selecting one of the halves, with the interval $a \leq x \leq b$ replaced by the interval $a_1 \leq x \leq b_1$. We obtain a new interval $a_2 \leq x \leq b_2$. See Figure 3.7.6.

Theoretically, this procedure of cutting in half and selecting can be continued indefinitely to obtain an unending sequence of intervals. These intervals are nested, each one inside the previous one, and the lengths of the intervals approach zero. It is intuitively clear that these conditions imply that the intervals shrink down to a single point $c$.

As our selected intervals shrink down to $c$, their endpoints approach $c$. Clearly, $a \leq c \leq b$, so $f$ is continuous at $c$. It follows that the values of $f$ at the endpoints of our selected intervals must approach $f(c)$. Since we have selected the intervals in such a way that the values of $f$ at their left endpoints are less than zero, these values cannot approach any positive number, so $f(c) \leq 0$. Since the values of $f$ at the right endpoints of our selected intervals are greater than or equal to zero, these values cannot approach any negative number, so $f(c) \geq 0$. It follows that $f(c) = 0$ and $a < c < b$. ■

The selection process of the bisection method is easily programmed for use with a computer.

The essential feature of the bisection method is that the selection process produces a sequence of nested intervals, each contained in the previous interval, with the lengths of the intervals approaching zero. This feature can be retained without using bisection strictly. We can modify the selection process if that is convenient for a particular problem. Let us see how the method works with an example.

### ■ EXAMPLE 1

Determine that the equation $\sin x = 1 - x$ has a solution in the interval $0 \leq x \leq \pi/2$, and then find an interval of length 0.1 that contains the solution.

#### SOLUTION

We begin by writing the given equation as an equivalent equation in the form $f(x) = 0$. We have

$$\sin x = 1 - x,$$

$$x - 1 + \sin x = 0.$$

We then see that the solution of the original equation satisfies $f(x) = 0$, where

$$f(x) = x - 1 + \sin x.$$

Evaluating the function $f$ at 0 and $\pi/2$, we obtain

$$f(0) = 0 - 1 + \sin(0) = -1 \quad \text{and} \quad f\left(\frac{\pi}{2}\right) = \frac{\pi}{2} - 1 + \sin\left(\frac{\pi}{2}\right) = \frac{\pi}{2}.$$

Since $f$ is a continuous function that has a negative value at 0 and a positive value at $\pi/2$, the special case of the Intermediate Value Theorem tells us that there is at least one point $c$ between 0 and $\pi/2$ with $f(c) = 0$. This means that $c$ is a solution of our original equation.

To locate the solution $c$ more accurately, we evaluate the function $f$ at some point between 0 and $\pi/2$. Let us try the value at $x = 0.8$, a point that is near the midpoint of the interval between 0 and $\pi/2 \approx 1.57$. We must evaluate the sine of 0.8 *radians*. We have

$$f(0.8) = 0.8 - 1 + \sin(0.8) \approx 0.517.$$

Since $f(0) < 0 < f(0.8)$, we know that there is a solution $c$ between 0 and 0.8. Continuing to cut down the size of the interval that contains a solution, we evaluate $f(0.4)$:

$$f(0.4) = 0.4 - 1 + \sin(0.4) \approx -0.210.$$

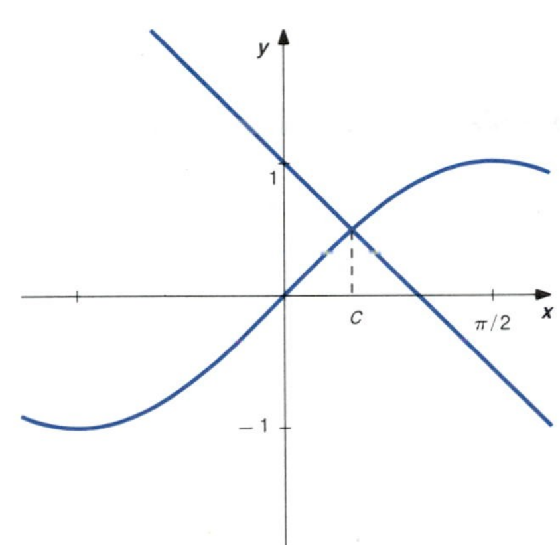


FIGURE 3.7.7

Since $f(0.4) < 0 < f(0.8)$, we know there is a solution $c$ between 0.4 and 0.8. Evaluating $f$ at 0.6, we have

$$f(0.6) = 0.6 - 1 + \sin(0.6) \approx 0.164.$$

We now have $f(0.4) < 0 < f(0.6)$. Evaluating $f$ at 0.5, we obtain

$$f(0.5) = 0.5 - 1 + \sin(0.5) \approx -0.020.$$

Since $f(0.5) < 0 < f(0.6)$, we know that the interval $[0.5, 0.6]$ contains a solution of our equation. See Figure 3.7.7. ■

In Example 1 we determined that the interval $(0.5, 0.6)$ contains a solution of the equation $\sin x = 1 - x$. This means that any point of the interval $[0.5, 0.6]$ can be used as an approximate value of the solution that is accurate to within 0.1 of the exact solution. We could obtain approximate values that are more accurate by continuing to cut down the size of the interval that contains the solution. However, more efficient methods are generally used if accurate approximate values are required. We will discuss one such method in the next section. The bisection method is used to obtain rough approximations of the solutions.

TABLE 3.7.1

| $x$ | $f(x)$ |
|---|---|
| $-3$ | $-67$ |
| $-2$ | $-26$ |
| $-1$ | $-5$ |
| 0 | 2 |
| 1 | 1 |
| 2 | $-2$ |
| 3 | $-1$ |
| 4 | 10 |

■ **EXAMPLE 2**

The equation $x^3 - 4x^2 + 2x + 2 = 0$ has three solutions. Find intervals between successive integers that contain the solutions.

**SOLUTION**

We want to determine the location of the solutions of the equation $f(x) = 0$, where $f(x) = x^3 - 4x^2 + 2x + 2$. The idea is to evaluate $f(x)$ at various integers until we find successive integers for which the sign of $f(x)$ changes. A table of values is given in Table 3.7.1. Note that if $|x|$ is large $f(x)$ will have the same sign as the highest-order term, $x^3$.

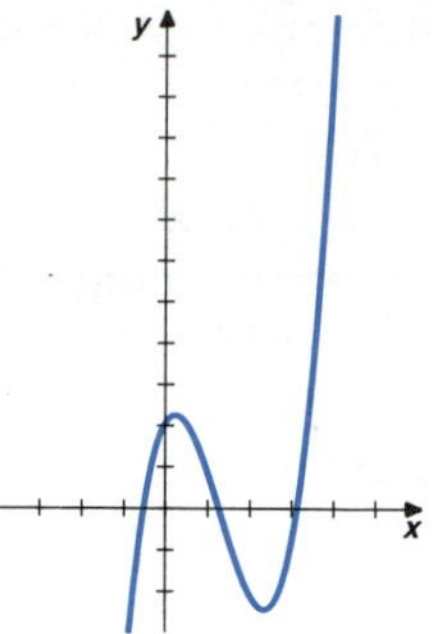


FIGURE 3.7.8

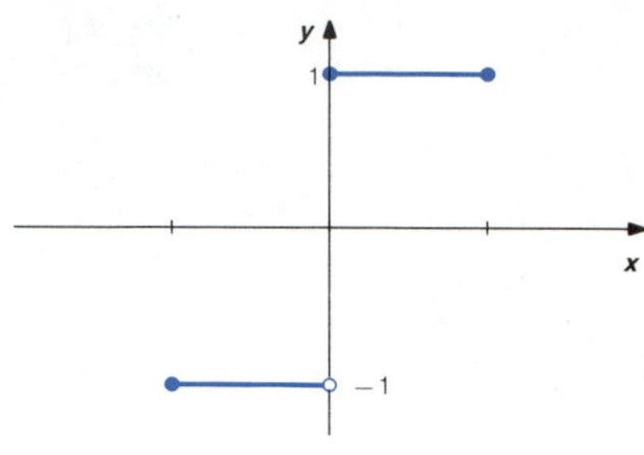


FIGURE 3.7.9

From Table 3.7.1 we see that $f(-1) < 0 < f(0)$, $f(1) > 0 > f(2)$, and $f(3) < 0 < f(4)$, so the intervals $(-1, 0)$, $(1, 2)$, and $(3, 4)$ each contain a solution of the equation $f(x) = 0$. See Figure 3.7.8. ■

Let us emphasize that the Intermediate Value Theorem depends on the continuity of the function on the interval $[a, b]$. If there is even one point between $a$ and $b$ where $f$ is not continuous, the conclusion of the Intermediate Value Theorem may fail. For example, from the graph in Figure 3.7.9, we see that the function

$$f(x) = \begin{cases} -1, & -1 \le x < 0, \\ 1, & 0 \le x \le 1, \end{cases}$$

is not continuous at $x = 0$. We have $f(-1) < 0 < f(1)$, but there is no point $c$ that satisfies $-1 < c < 1$ and $f(c) = 0$. The intervals obtained by the bisection method shrink down to the point of discontinuity, $x = 0$. The graph jumps over the horizontal line $y = 0$ at the point where $x = 0$, where the function is not continuous.

## EXERCISES 3.7

In Exercises 1–8, determine that the given equation has a solution in the given interval and then find an interval of length 0.1 that contains the solution.

1 $x^3 - 4x^2 + x + 3 = 0$, $1 \le x \le 2$

2 $x^3 + 2x^2 - 8x + 2 = 0$, $1 \le x \le 2$

3 $x^4 - 3x^3 + 1 = 0$, $0 \le x \le 1$

4 $2x^4 - 6x + 3 = 0$, $0 \le x \le 1$

5 $x = 2 \sin x$, $\pi/2 \le x \le \pi$

6 $x = \cos x$, $0 \le x \le \pi/2$

7 $\cos x = \tan x$, $0 \le x \le \pi/4$

8 $\sin x = 2 \tan x - 1$, $0 \le x \le \pi/4$

Each of the equations given in Exercises 9–12 has three solutions. Find intervals between successive integers that contain the solutions.

9 $x^3 - 3x + 1 = 0$

10 $x^3 - 12x + 8 = 0$

11 $x^3 - 3x^2 + 1 = 0$

12 $x^3 - 6x^2 + 12 = 0$

13 Show that for any constants $b$, $c$, $d$ the equation $x^3 + bx^2 + cx + d = 0$ has at least one solution.

14 Find a function $f$ such that $f(-1) = -1$ and $f(1) = 1$, where $f$ is continuous on the interval $-1 < x < 1$, but there is no point $c$ between $-1$ and 1 for which $f(c) = 0$.

15 The function $f(x) = |x|/x$ satisfies $f(-1) = -1$ and $f(1) = 1$, but there is no point $c$ between $-1$ and 1 with $f(c) = 0$. Why doesn't the Intermediate Value Theorem apply to this function on the interval $-1 \le x \le 1$?

16 Show that the Intermediate Value Theorem is a consequence of the special case we proved by applying the special case to the function $g(x) = f(x) - L$.

## 3.8 NEWTON'S METHOD

If $f(x_0)$ and $f(x_1)$ have opposite signs and $f$ is continuous on the interval $[x_0, x_1]$, we know from the Intermediate Value Theorem of the previous section that there is at least one $c$ between $x_0$ and $x_1$ such that $f(c) = 0$. That is, $c$ is a solution of the equation $f(x) = 0$. See Figure 3.8.1. We could use the Bisection Method as in Section 3.7 to find approximate values of $c$, but that method is not efficient for problems that require high accuracy. In this section we will investigate a much more efficient method for finding approximate values of solutions of $f(x) = 0$ in the case that $f$ is differentiable. The method is called **Newton's Method.**

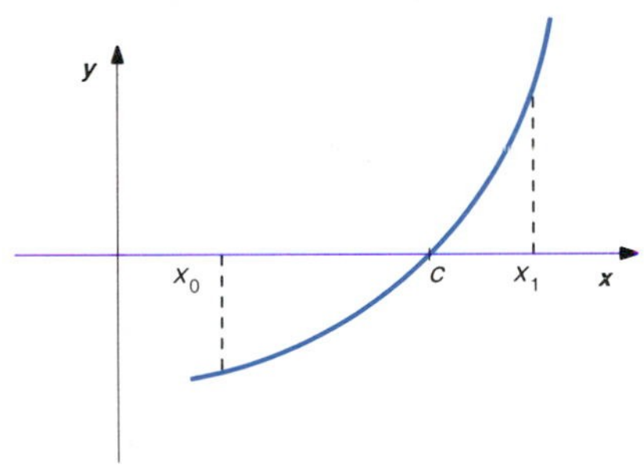


FIGURE 3.8.1

Newton's method is based on the fact that tangent lines can be used to approximate values of a *differentiable* function. For example, suppose $x_1$ is a point that is near the exact solution $c$ of the equation $f(x) = 0$. The equation of the line tangent to the graph of $f$ at the point $(x_1, f(x_1))$ is

$$y - f(x_1) = f'(x_1)(x - x_1).$$

Since the graph of $f$ is near the graph of the tangent line, we should expect that the graph would cross the $x$-axis near the point where the tangent line crosses the $x$-axis. If $f'(x_1) \neq 0$, this line will intersect the $x$-axis at a point $(x_2, 0)$. Substituting these coordinates into the equation of the tangent line and solving for $x_2$, we have

$$0 - f(x_1) = f'(x_1)(x_2 - x_1),$$

$$-f(x_1) = f'(x_1)\cdot x_2 - f'(x_1)\cdot x_1,$$

$$f'(x_1)\cdot x_2 = f'(x_1)\cdot x_1 - f(x_1),$$

$$x_2 = x_1 - \frac{f(x_1)}{f'(x_1)}.$$

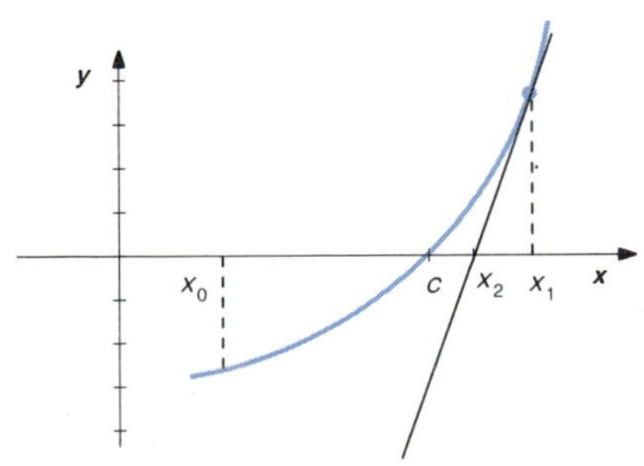


FIGURE 3.8.2

This value of $x_2$ should be better than $x_1$ as an approximate value of the unknown exact solution $c$. See Figure 3.8.2. We could repeat the process with $x_2$ in place of $x_1$ to obtain $x_3 = x_2 - f(x_2)/f'(x_2)$. Generally, we could repeat the process as often as we please and obtain successively more accurate approximate values of the exact solution $c$. Let us summarize:

***Newton's Method for finding approximate values of solutions c of the equation $f(x) = 0$***

**Use values of $f(x)$ to find an interval such that $f(x)$ has opposite signs at the endpoints. Choose $x_1$ to be one endpoint of this interval and then use the formula**

$$x_{n+1} = x_n - \frac{f(x_n)}{f'(x_n)}, \qquad n = 1, 2, 3, \ldots,$$

**to find the successive approximate values $x_2, x_3, x_4, \ldots$. Choose $x_{n+1}$ (appropriately rounded) to be the final approximate value of $c$ if the difference between $x_{n+1}$ and $x_n$ is within the desired accuracy.**

Later in this section we will discuss some conditions that guarantee that the successive approximate values approach the exact solution $c$. The conditions also guarantee that $x_{n+1}$ is of the desired accuracy if the difference between $x_{n+1}$ and $x_n$ is.

■ **EXAMPLE 1**

Use Newton's Method to find an approximate value of the positive solution of the equation $x^2 - 2 = 0$. Obtain accuracy to five decimal places. (The exact solution is $\sqrt{2}$, so we can check the accuracy of the method with a calculator.)

**SOLUTION**

We seek the positive solution of $f(x) = 0$, where

$$f(x) = x^2 - 2.$$

This is the point on the positive $x$-axis where the graph of $f$ crosses the $x$-axis. See Figure 3.8.3.

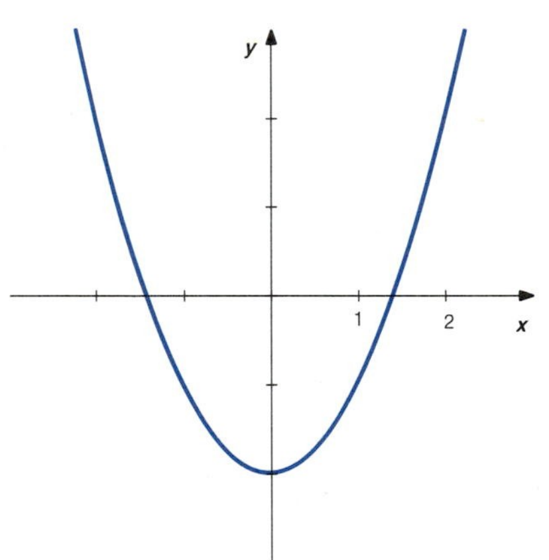


FIGURE 3.8.3

Since $f(1) = 1^2 - 2 = -1 < 0$ and $f(2) = 2^2 - 2 = 2 > 0$, we know there is a solution $c$ between 1 and 2. Let us choose $x_1 = 2$ to be our first approximation of $c$.

We have $f(x) = x^2 - 2$, so $f'(x) = 2x$. Substitution into the formula for successive approximations gives

$$x_{n+1} = x_n - \frac{f(x_n)}{f'(x_n)} = x_n - \frac{x_n^2 - 2}{2x_n}.$$

Simplifying, we have

$$\begin{aligned} x_{n+1} &= x_n - \frac{x_n^2}{2x_n} + \frac{2}{2x_n} \\ &= x_n - \frac{x_n}{2} + \frac{1}{x_n} \\ &= \frac{x_n}{2} + \frac{1}{x_n}. \end{aligned}$$

We have chosen $x_1 = 2$, so the formula (with $n = 1$) gives

$$x_2 = \frac{x_1}{2} + \frac{1}{x_1} = \frac{(2)}{2} + \frac{1}{(2)} = 1.5.$$

Continuing, we obtain

$$x_3 = \frac{x_2}{2} + \frac{1}{x_2} = \frac{(1.5)}{2} + \frac{1}{(1.5)} = 1.4166667,$$

$$x_4 = \frac{x_3}{2} + \frac{1}{x_3} = \frac{(1.4166667)}{2} + \frac{1}{(1.4166667)} = 1.4142157,$$

$$x_5 = \frac{x_4}{2} + \frac{1}{x_4} = \frac{(1.4142157)}{2} + \frac{1}{(1.4142157)} = 1.4142136.$$

(We have used a calculator to evaluate the successive approximations. It is convenient to *store* $x_n$ and then *recall* it as it is needed in the calculation of $x_{n+1}$. A programmable calculator or a computer would be very appropriate for this work.) These calculations are arranged in Table 3.8.1.

**TABLE 3.8.1**

| $n$ | $x_n$ | |
|---|---|---|
| 1 | 2 | ← Our choice |
| 2 | 1.5 | ← Calculated from $x_1 = 2$ |
| 3 | 1.4166667 | ← Calculated from $x_2 = 1.5$ |
| 4 | 1.4142157 | ← Calculated from $x_3 = 1.4166667$ |
| 5 | 1.4142136 | ← Calculated from $x_4 = 1.4142157$ |

The first five decimal digits did not change when $x_5$ was calculated from $x_4$. We use 1.41421 as our approximate value of $\sqrt{2}$. (A calculator gives $\sqrt{2} \approx 1.4142136$, so our approximate value does seem to be accurate to five decimal places.) ■

■ **EXAMPLE 2**

Find an approximate value of the positive solution of $2 \sin x = x$. Obtain accuracy to five decimal places.

**SOLUTION**

We want the positive solution of $f(x) = 0$, where

$$f(x) = 2 \sin x - x.$$

This is the point on the positive $x$-axis where the graph of $f$ crosses the $x$-axis. See Figure 3.8.4. Note that $f(\pi/2) = 2\sin(\pi/2) - \pi/2 = 2(1) - \pi/2 > 0$ and $f(\pi) = 2 \sin \pi - \pi = -\pi < 0$, so there is a solution between $\pi$ and $\pi/2$.

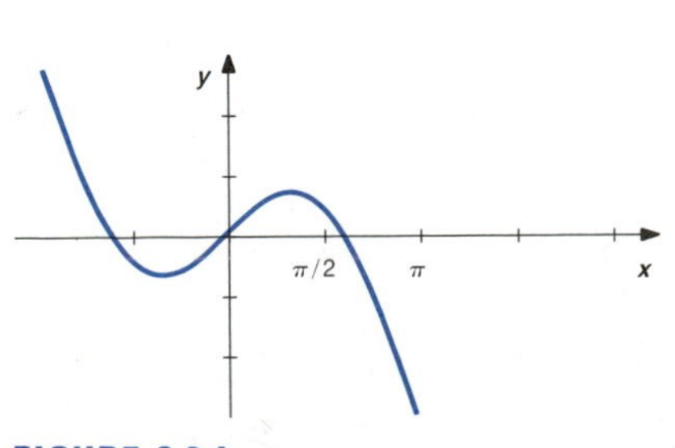

FIGURE 3.8.4

We have $f(x) = 2 \sin x - x$, so $f'(x) = 2 \cos x - 1$. Substituting in the formula for successive approximations, we have

$$x_{n+1} = x_n - \frac{f(x_n)}{f'(x_n)} = x_n - \frac{2 \sin x_n - x_n}{2 \cos x_n - 1}.$$

TABLE 3.8.2

| $n$ | $x_n$ |
|---|---|
| 1 | $\pi$ |
| 2 | 2.0943951 |
| 3 | 1.913223 |
| 4 | 1.8956718 |
| 5 | 1.8954943 |
| 6 | 1.8954943 |

Let us choose $x_1 = \pi$ and calculate the successive approximate values from the above formula. The results are arranged in Table 3.8.2. Don't forget that your calculator must be set for radian measure to calculate the values at $x_n$ of the sine and cosine functions.

Since $x_5$ and $x_6$ agree to five decimal places, we use 1.89549 as our final approximate value. ■

It is clear that we cannot use the formula

$$x_{n+1} = x_n - \frac{f(x_n)}{f'(x_n)}$$

if $f'(x_n) = 0$. Generally, we will want to apply the method on intervals $[x_0, x_1]$ where the derivative $f'$ is either always positive or always negative. Let us see that the behavior of the second derivative $f''$ can also effect the success of Newton's Method.

Suppose we try to use Newton's Method to find the solution of $f(x) = 0$, where $f(x) = x^{1/3}$. We have $f'(x) = (1/3)x^{-2/3}$, so

$$x_{n+1} = x_n - \frac{f(x_n)}{f'(x_n)} = x_n - \frac{x_n^{1/3}}{(1/3)x^{-2/3}} = x_n - 3x_n = -2x_n.$$

This gives

$$x_2 = -2x_1,$$

$$x_3 = -2x_2 = -2(-2x_1) = 4x_1,$$

$$x_4 = -2x_3 = -2(4x_1) = -8x_1,$$

$$\vdots$$

Hence, we see that for any choice of $x_1 \neq 0$, the successive approximations are on the opposite side of zero (the exact solution) and twice as far from zero as the previous approximation. This is illustrated in Figure 3.8.5. The failure of Newton's Method is related to the fact that the graph of $y = x^{1/3}$ opens upward on the interval $x < 0$ and opens downward on the interval $x > 0$. We will see in Chapter 4 that the graph of $f$ opens upward on an interval if $f''(x)$ is positive on the interval; the graph opens downward on an interval if $f''(x)$ is negative on the interval.

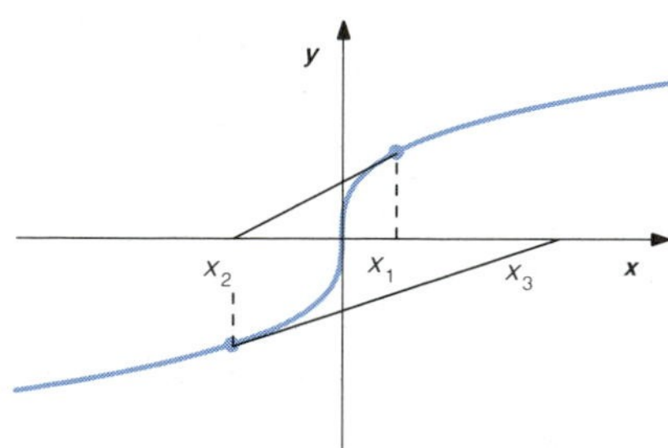


FIGURE 3.8.5

FIGURE 3.8.6

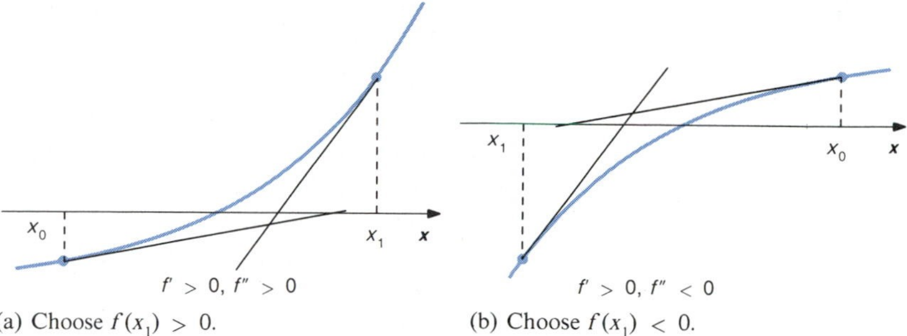


(a) Choose $f(x_1) > 0$. (b) Choose $f(x_1) < 0$.

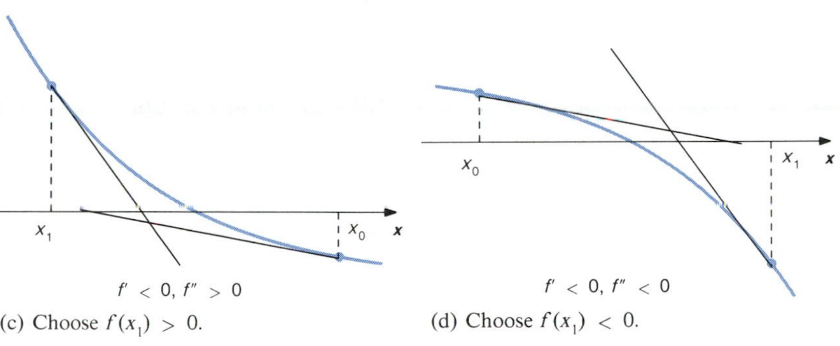


(c) Choose $f(x_1) > 0$. (d) Choose $f(x_1) < 0$.

If $f'(x)$ and $f''(x)$ are both of constant sign on an interval that contains a solution of $f(x) = 0$, then we can guarantee the success of Newton's Method. This is illustrated geometrically in Figure 3.8.6. The tangent linc at the point $(x_1, f(x_1))$ intersects the $x$-axis at a point that is between $x_1$ and the exact solution if the sign of $f(x_1)$ agrees with that of the second derivative $f''$. The tangent line at $(x_0, f(x_0))$ intersects the $x$-axis on the opposite side of the solution if $f(x_0)$ and the sign of $f''$ are opposite. To insure that successive approximations become closer to the exact solution we should **choose $x_1$ so the sign of $f(x_1)$ agrees with that of $f''$**.

The following theorem describes conditions under which the successive approximations of Newton's Method approach the exact solution.

*Theorem*

**Suppose $f'(x)$ and $f''(x)$ are of constant sign with $|f'(x)| \geq m > 0$ and $|f''(x)| \leq M$ for all $x$ between $x_0$ and $x_1$. Assume $f(x_0)$ and $f(x_1)$ are of opposite signs, while $f(x_1)$ and $f''(x_1)$ have the same sign. Then there is a point $c$ between $x_0$ and $x_1$ with $f(c) = 0$. Moreover, $|x_1 - x_0| \leq m/M$ implies that the successive approximate values of $c$ obtained by Newton's Method, $x_1, x_2, \ldots,$ approach the value $c$ and $|x_{n+1} - c| \leq |x_{n+1} - x_n|$.**

We will not verify this theorem.

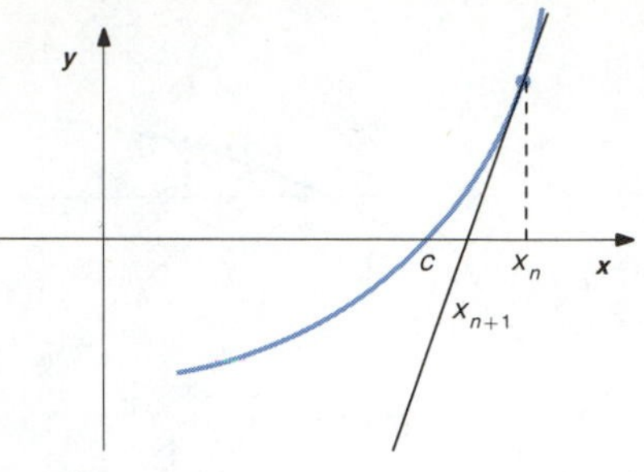

FIGURE 3.8.7

The condition of the theorem that $f'(x)$ is of constant sign implies that $f$ is either increasing on the interval between $x_0$ and $x_1$ or decreasing on that interval. The existence of $f'(x)$ on the interval implies that $f$ is continuous. Since $f(x_0)$ and $f(x_1)$ are of opposite sign, the Intermediate Value Theorem guarantees at least one value of $c$ between $x_0$ and $x_1$ for which $f(c) = 0$. Since $f'(x)$ is of constant sign on the interval, there can be only one such point $c$.

The condition that $f(x_1)$ and $f''(x_1)$ have the same sign implies that successive approximate values of $c$ are all on the same side of $c$. From Figure 3.8.7 we see that the condition $|x_{n+1} - c| \leq |x_{n+1} - x_n|$ in the conclusion of the theorem then implies that the distance between $x_{n+1}$ and $c$ is less than or equal to half the distance between $x_n$ and $c$. This indicates that the approximate values approach $c$. Also, the condition $|x_{n+1} - c| \leq |x_{n+1} - x_n|$ tells us that $x_{n+1}$ is closer to the unknown solution $c$ than to the previous approximate value $x_n$. This means that when the difference between $x_{n+1}$ and $x_n$ is within the desired accuracy, we can use $x_{n+1}$ as our final approximate value of $c$.

When using the above theorem, it is not necessary to evaluate the numbers $x_0$, $m$, and $M$. It is enough to know that such numbers exist. If the other conditions of the theorem are satisfied for some $x_0$, then $x_n$ will eventually be close enough to $c$ so the condition $|x_1 - x_0| \leq m/M$ is true with $x_n$ in place of $x_1$ and some new choice of $x_0$ that is close enough to $c$.

■ **EXAMPLE 3**

Show that the equation $x^3 - 3x + 1 = 0$ has three solutions and find approximate values for each of them. Obtain accuracy to three decimal places.

**SOLUTION**

To show that the equation has three solutions, we need to find three intervals on which the continuous function $f(x) = x^3 - 3x + 1$ changes sign. We can do this by evaluating $f(x)$ at various points until we obtain three changes of sign. For example, we have

$$f(-2) = -1,$$

$$f(-1) = 3,$$

$$f(0) = 1,$$

$$f(1) = -1, \quad \text{and}$$

$$f(2) = 3.$$

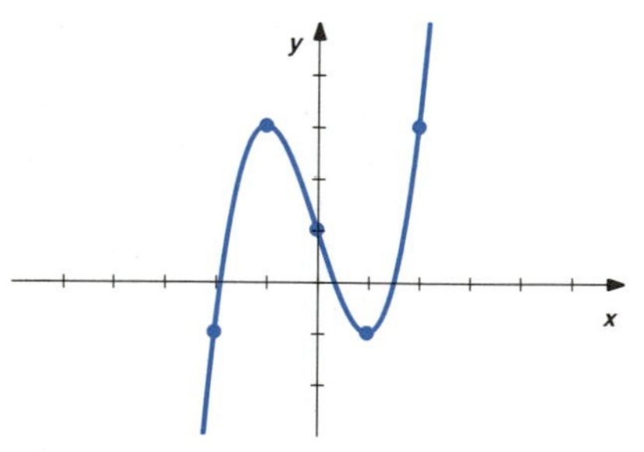

FIGURE 3.8.8

In Figure 3.8.8 we have plotted the corresponding points and then sketched the graph of $f$ through these points. There is a solution of $f(x) = 0$ on each of the intervals $(-2, -1)$, $(0, 1)$, and $(1, 2)$, where the values $f(x)$ change sign.

To apply Newton's Method, we need to find the derivative of $f(x) = x^3 - 3x + 1$. We have

$$f'(x) = 3x^2 - 3.$$

We see that $f'(x) = 0$ when $x = 1, -1$; $f'(x)$ is positive on the intervals $x < -1$ and $x > 1$; $f'(x)$ is negative on the interval $-1 < x < 1$. The second derivative is

$$f''(x) = 6x.$$

We see that $f''(x) = 0$ when $x = 0$; $f''(x) \leq 0$ for $x \leq 0$; $f''(x) \geq 0$ for $x \geq 0$.

We have $f'(x) > 0$ and $f''(x) < 0$ for $x < -1$. We should choose $x_1$ in this interval with $f(x_1) < 0$, so the sign agrees with that of $f''$. Let us choose $x_1 = -2$. Newton's Method will then give us the solution in the interval $(-2, -1)$. ($f'(x)$ stays away from zero on any interval $[x_1, x_0]$ with $x_0 < -1$.)

In the interval $[0, 1)$, we have $f' < 0$ and $f'' > 0$. We want $f(x_1) > 0$, so we choose $x_1 = 0$.

For $x > 1$, we have $f'(x) > 0$ and $f''(x) > 0$. We want $f(x_1) > 0$, so we choose $x_1 = 2$.

The formula for the successive approximations is

$$x_{n+1} = x_n - \frac{f(x_n)}{f'(x_n)} = x_n - \frac{x_n^3 - 3x_n + 1}{3x_n^2 - 3}.$$

Simplifying, we have

$$x_{n+1} = x_n \frac{3x_n^2 - 3}{3x_n^2 - 3} - \frac{x_n^3 - 3x_n + 1}{3x_n^2 - 3}$$

$$= \frac{3x_n^3 - 3x_n - (x_n^3 - 3x_n + 1)}{3x_n^2 - 3}$$

$$= \frac{2x_n^3 - 1}{3x_n^2 - 3}.$$

The results of the calculations are given in Table 3.8.3. The approximate values of the solutions are $-1.879$, 0.347, and 1.532.

**TABLE 3.8.3**

| $n$ | $x_n$ | $x_n$ | $x_n$ |
|---|---|---|---|
| 1 | $-2$ | 0 | 2 |
| 2 | $-1.8888889$ | 0.3333333 | 1.6666667 |
| 3 | $-1.8794516$ | 0.34722222 | 1.5486111 |
| 4 | $-1.8793852$ | 0.34729635 | 1.5323902 |
| | | | 1.532089 |

■

## EXERCISES 3.8

In Exercises 1–10, use Newton's Method to find an approximate value of the equation $f(x) = 0$ in the given interval. Obtain accuracy to four decimal places. (The solutions may be checked by direct calculation on a calculator.)

1 $f(x) = x^2 - 3$, $[1, 2]$

2 $f(x) = x^2 - 5$, $[2, 3]$

3 $f(x) = x^2 - 8$, $[-3, -2]$

4 $f(x) = x^2 - 10$, $[-4, -3]$

5 $f(x) = x^3 - 9$, $[2, 3]$

6 $f(x) = x^3 - 12$, $[2, 3]$

7 $f(x) = x^4 - 8$, $[1, 2]$

8 $f(x) = x^4 - 12$, $[1, 2]$

9 $f(x) = \cos x - 0.4$, $[0, \pi/2]$

10 $f(x) = \sin x - 0.4$, $[0, \pi/2]$

Each of the equations in Exercises 11–20 has one *positive* solution. Use Newton's Method to approximate its value to four decimal places.

11 $x^2 - x - 1 = 0$

12 $x^2 + x - 1 = 0$

13 $3x^2 + 3x - 1 = 0$

14 $3x^2 + 6x - 2 = 0$

15 $x^3 = 1 - 3x$

16 $x^3 = 3 - x$

17 $x = \cos x$

18 $x = 3 \sin x$

19 $x^2 = \cos x$

20 $x^2 = \sin x$

Show that each of the equations in Exercises 21–26 has three solutions. Use Newton's Method to approximate their values to three decimal places.

21 $x^3 - 3x + 1 = 0$

22 $x^3 - 3x - 1 = 0$

23 $x^3 - 3x^2 + 2 = 0$

24 $x^3 - 3x^2 + 3 = 0$

25 $x^3 - 3x^2 - 9x + 10 = 0$

26 $x^3 - 3x^2 - 9x + 8 = 0$

## 3.9 THE EXTREME VALUE THEOREM

Determining the largest and smallest values of a function is a problem for which calculus is particularly well suited. In this section we will see how continuity is related to this problem and how the derivative can be used in its solution. The methods developed here will be used later, when we illustrate how the ideas relate to the solution of physical problems.

We begin with a formal definition of the largest and smallest values of a function on a set.

*Definition*

The value $f(c)$ is called the **maximum value of $f$ on a set $S$** if $c$ is in $S$ and $f(x) \leq f(c)$ for all $x$ in $S$; $f(d)$ is called the **minimum value of $f$ on a set $S$** if $d$ is in $S$ and $f(d) \leq f(x)$ for all $x$ in $S$. Maximum and minimum values of a function on a set are called **extreme values** or **extrema** of the function. ■

The following theorem guarantees the existence of extrema of a function that is continuous on an interval that contains both endpoints.

*Extreme Value Theorem*

**If $f$ is continuous on the interval $a \leq x \leq b$, then there are points $c$ and $d$ in the interval such that $f(c)$ is the maximum value of $f$ on the interval and $f(d)$ is the minimum value of $f$ on the interval.**

We will not verify this theorem. We will illustrate the rule of continuity in the result with some examples. Note that if $f$ is continuous on an interval that contains both endpoints, then the theorem says $f$ has both a maximum value and a minimum value on the interval. If either the interval does not contain both endpoints or $f$ is not continuous at every point of the interval, then $f$ may or may not have maximum or minimum values.

■ **EXAMPLE 1**

Sketch the graphs of the given functions and use the graphs to find their maximum and minimum values; indicate "does not exist" where appropriate.

(a) $f(x) = x^2, \quad -1 \le x \le 2,$

(b) $f(x) = 2 - x, \quad 0 < x \le 2,$

(c) $f(x) = 1/x, \quad x \ge 1,$

(d) $f(x) = \begin{cases} x + 1, & -1 \le x < 0, \\ 0, & x = 0, \\ x - 1, & 0 < x \le 1. \end{cases}$

**SOLUTION**

(a) The function $f(x) = x^2$ is continuous on the interval $-1 \le x \le 2$, and the interval contains both endpoints. Hence, the theorem guarantees maximum and minimum values. From the graph in Figure 3.9.1, we see that $f(2) = 4$ is the maximum value and $f(0) = 0$ is the minimum value.

(b) The theorem does not guarantee either a maximum or minimum value of $f(x) = 2 - x$ on the interval $0 < x \le 2$, because the endpoint 0 is not included in the interval. From Figure 3.9.2, we see that $f(2) = 0$ is the minimum value of $f$; there is no maximum value. Two is not a *value* of $f$, because there is no $x$ in the interval $0 < x \le 2$ with $f(x) = 2$.

(c) The theorem does not guarantee either a maximum or minimum value of $f(x) = 1/x$ on the interval $x \ge 1$, because the interval is infinite. Infinite endpoints are not included in an interval. We see from Figure 3.9.3 that $f$ has a maximum value of $f(1) = 1$; $f$ does not have a minimum value.

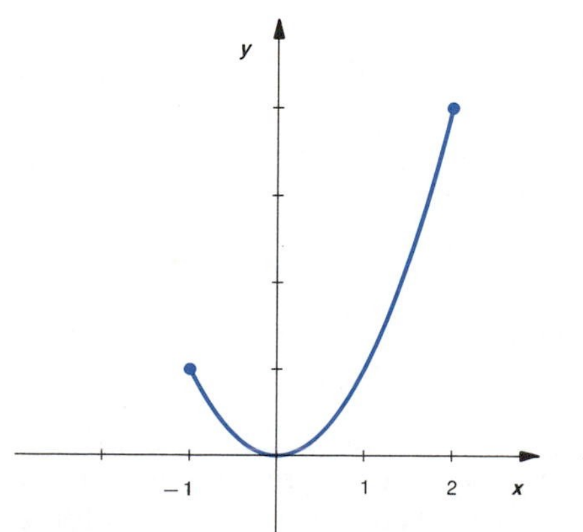

FIGURE 3.9.1

FIGURE 3.9.2

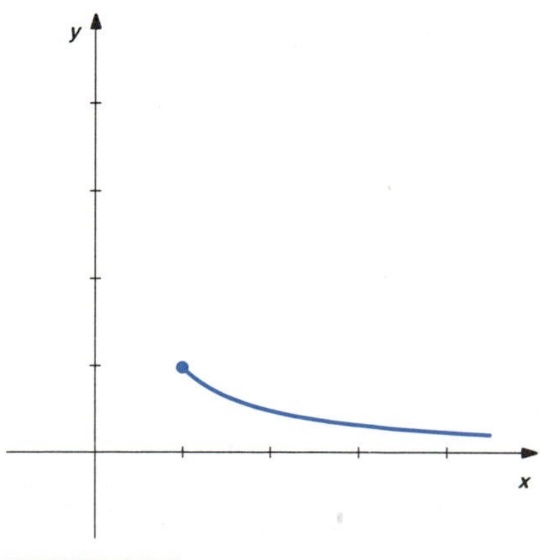

FIGURE 3.9.3

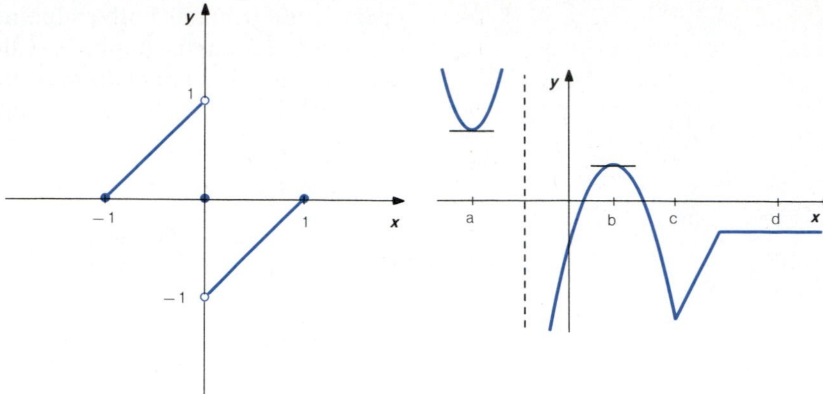


FIGURE 3.9.4

FIGURE 3.9.5

(d) The theorem does not guarantee either a maximum value or a minimum value of $f$ on the interval $-1 \leq x \leq 1$, because the function is discontinuous at $x = 0$, a point on the interval. We see from Figure 3.9.4 that $f$ has neither a maximum value nor a minimum value on the interval; 1 and $-1$ are not values of $f$. ■

The following definition is related to the use of calculus to find extrema.

■ *Definition*

The value $f(c)$ is called a **local maximum** of $f$ if there is some open interval $I$ with center $c$ such that $f(x) \leq f(c)$ for all $x$ in $I$. $f(c)$ is called a **local minimum** of $f$ if there is some open interval $I$ with center $c$ such that $f(c) \leq f(x)$ for all $x$ in $I$. Local maxima and local minima are called **local extrema**. ■

The value $f(c)$ is a local maximum of $f$ if $f(c)$ is greater than or equal to $f(x)$ whenever $x$ is close enough to $c$. A local minimum of $f$ is less than or equal to all values of $f(x)$ for nearby $x$. It is required that $f(x)$ be defined for all $x$ in some interval $|x - c| < d$ in order that $f(c)$ be a local maximum or minimum of $f$.

From the graph in Figure 3.9.5 we see that $f(a)$ and $f(c)$ are local minima of $f$ and that $f(b)$ is a local maximum of $f$. $f(d)$ is both a local minimum and a local maximum of $f$. It is possible that a local minimum is greater than a local maximum.

The following theorem is the basis of using calculus for finding maximum and minimum values of a function.

■ *Theorem 1*

**If $f(c)$ is either a local maximum or minimum of $f$, then either (i) $f'(c) = 0$ or (ii) $f'(c)$ does not exist.**

■ *Proof* Let us verify Theorem 1 in the case where $f(c)$ is a local maximum of $f$, so $f(x) \leq f(c)$ for $x$ near $c$. Then $f(x) - f(c) \leq 0$ for $x$ near $c$. We now look directly at the definition of the derivative of $f$ at $c$.

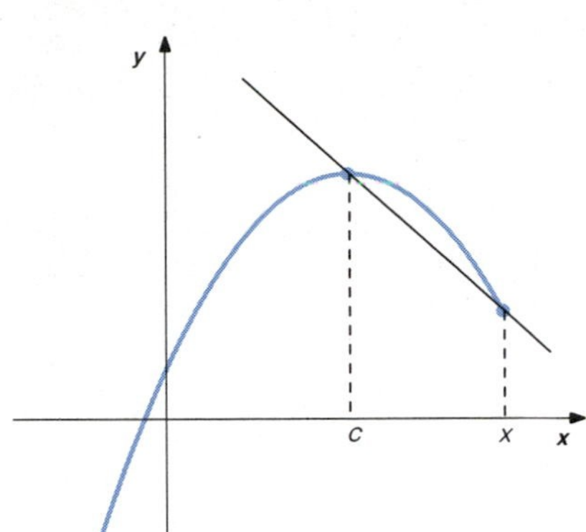


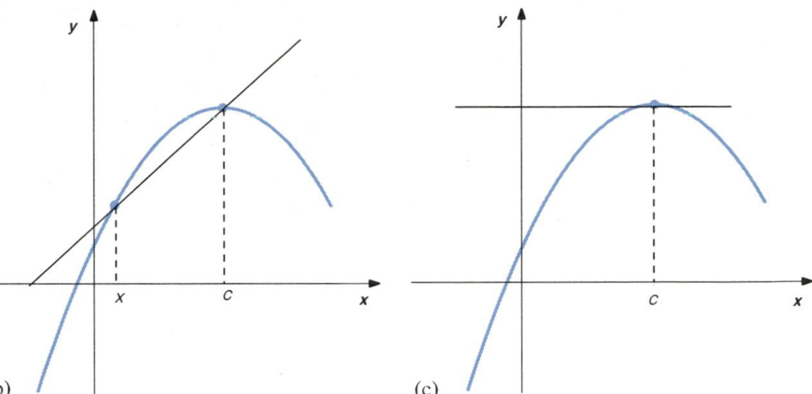


FIGURE 3.9.6

For $x$ near $c$, $x > c$, we have $x - c > 0$, so

$$\frac{f(x) - f(c)}{x - c} \leq 0.$$

This implies that

$$\lim_{x \to c^+} \frac{f(x) - f(c)}{x - c} \leq 0,$$

if that one-sided limit exists.

For $x$ near $c$, $x < c$, we have $x - c < 0$, so

$$\frac{f(x) - f(c)}{x - c} \geq 0.$$

This implies that

$$\lim_{x \to c^-} \frac{f(x) - f(c)}{x - c} \geq 0,$$

if that one-sided limit exists.

If $f'(c)$ exists, then both of the above one-sided limits exist and have the same value. The only possible common value of the one-sided limits is zero, so $f'(c) = 0$. Geometrically, if $f(c)$ is a local maximum of $f$, then the slope of the line through the points $(c, f(c))$ and $(x, f(x))$, $x$ near $c$, $x \neq c$, slopes downward if $x > c$ (Figure 3.9.6a) and upward if $x < c$ (Figure 3.9.6b). If these slopes approach a common value as $x$ approaches $c$, that common value must be $f'(c) = 0$ (Figure 3.9.6c).

It is not difficult to modify the above argument to show that if the derivative exists at a point whose value is a local minimum, then the derivative must equal zero. ■

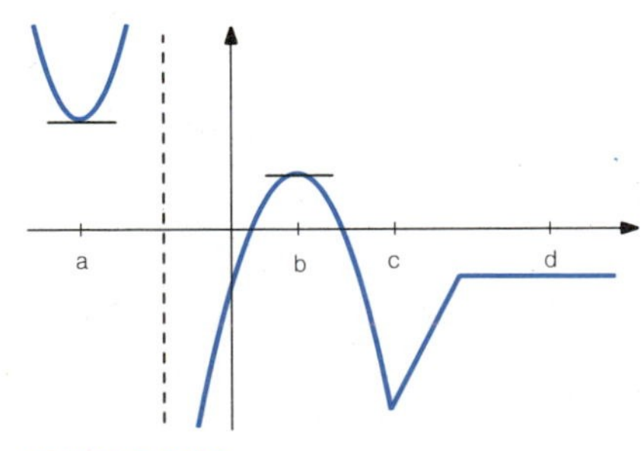


FIGURE 3.9.7

A function can have a local maximum or minimum of $f(c)$ at a point where $f'(c)$ does not exist. In Figure 3.9.7, $f(c)$ is a local minimum, but $f'(c)$ does not exist; $f'(a) = f'(b) = f'(d) = 0$.

If $f$ is continuous on the interval $a \leq x \leq b$, we know from the Extreme Value Theorem that $f$ has a maximum value and a minimum value on the interval. If $f(c)$ is either a maximum or minimum value of $f$ and $a < c < b$, then $f(c)$ must also be a local maximum or minimum, respectively. This means either $f'(c) = 0$ or $f'(c)$ does not exist. Noting that the maximum and minimum values of $f$ on the interval $a \leq x \leq b$ can occur at either of the two endpoints of the interval, we obtain the following result:

*Theorem 2*

**If $f$ is continuous on the closed interval $a \leq x \leq b$, then the maximum and minimum values of $f$ on the interval occur at points $x$ where either**

**(i) $a < x < b$ and $f'(x) = 0$;**

**(ii) $a < x < b$ and $f'(x)$ does not exist; or**

**(iii) $x$ is an endpoint of the interval ($x = a$ or $x = b$).**

Points where either $f'(x) = 0$ or $f'(x)$ does not exist (but $f$ is defined at $x$) are called **critical points** of $f$. Theorem 2 tells us that we can find extrema of a function that is continuous on an interval that contains both endpoints by evaluating $f(x)$ at each critical point inside the interval and at both endpoints of the interval; the largest of these values is the maximum value and the smallest of these values is the minimum value.

### EXAMPLE 2

Find the maximum and minimum values of the function

$$f(x) = x^3 - 12x + 3, 0 \leq x \leq 3.$$

#### SOLUTION

Since $f$ is continuous on the interval $0 \leq x \leq 3$, an interval that contains both endpoints, the Extreme Value Theorem guarantees both maximum and minimum values. To find them we investigate the derivative of $f$.

$$f(x) = x^3 - 12x + 3,$$

$$f'(x) = 3x^2 - 12.$$

(i) $f'(x) = 0$ implies

$$3x^2 - 12 = 0,$$

$$3(x^2 - 4) = 0,$$

$$3(x - 2)(x + 2) = 0.$$

The derivative is zero when $x = 2$. ($x = -2$ is not in the interval $0 \leq x \leq 3$.)

(ii) The derivative exists for all points in the interval.

(iii) The endpoints are $x = 0$ and $x = 3$.

TABLE 3.9.1

| | $x$ | $f(x)$ | |
|---|---|---|---|
| $f' = 0$: | 2 | $-13$ | ← Smallest value of $f$ in table |
| Endpoints: | 0 | 3 | ← Largest value of $f$ in table |
| | 3 | $-6$ | |

The values of $f$ at each $x$ determined above, $x = 2, 0, 3$, are found by substitution in the formula $f(x) = x^3 - 12x + 3$. The values are given in Table 3.9.1. We see that

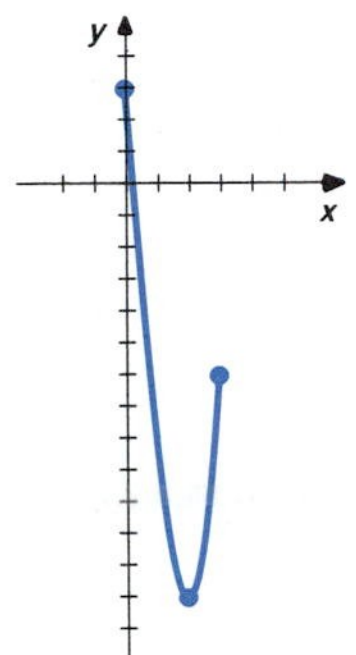


FIGURE 3.9.8

$f(0) = 3$ is the maximum value of $f$;

$f(2) = -13$ is the mimimum value of $f$.

The table of values of $f(x)$ constructed above has values of $f(x)$ for exactly those $x$ in the interval $0 \leq x \leq 3$ for which either (i) $f'(x) = 0$, (ii) $f'(x)$ does not exist, or (iii) $x$ is an endpoint of the interval. These values, and only these values, appear in the table.

The graph of $f(x) = x^3 - 12x + 3$, $0 \leq x \leq 3$, is sketched in Figure 3.9.8. It is not necessary to graph the function to determine its maximum and minimum values. ■

■ **EXAMPLE 3**

Find maximum and minimum values of the function $f(x) = x - 3x^{1/3}$, $-1 \leq x \leq 8$.

**SOLUTION**

Since $f$ is continuous on an interval that contains both endpoints, we know there are maximum and minimum values. We begin by investigating the derivative.

$$f(x) = x - 3x^{1/3},$$

$$\begin{aligned} f'(x) &= 1 - x^{-2/3} \\ &= x^{-2/3}(x^{2/3} - 1) \\ &= \frac{x^{2/3} - 1}{x^{2/3}}. \end{aligned}$$

We have simplified $f'(x)$ to find values of $f$ for which either $f'(x) = 0$ or $f'(x)$ does not exist.

(i) $f'(x) = 0$ whenever $x^{2/3} - 1 = 0$, so

$$x^{2/3} = 1,$$

$$x^2 = 1^3,$$

$$x = 1.$$

($-1$ is an endpoint of the interval.)

TABLE 3.9.2

| | $x$ | $f(x)$ | |
|---|---|---|---|
| $f' = 0$: | 1 | $-2$ | ← Smallest value of $f$ in table |
| $f'$ does not exist: | 0 | 0 | |
| Endpoints: | $-1$ | 2 | ← Largest value of $f$ in table |
| | 8 | 2 | |

(ii) $f'(x)$ does not exist when $x = 0$.
(iii) Endpoints are $x = -1, 8$.

The values of $f(x) = x - 3x^{1/3}$ at the points determined above are given in Table 3.9.2. We see that

$f(-1) = f(8) = 2$ is the maximum value of $f$;

$f(1) = -2$ is the minimum value of $f$.

The graph of $f(x) = x - 3x^{1/3}$, $-1 \le x \le 8$, is shown in Figure3.9.9. ■

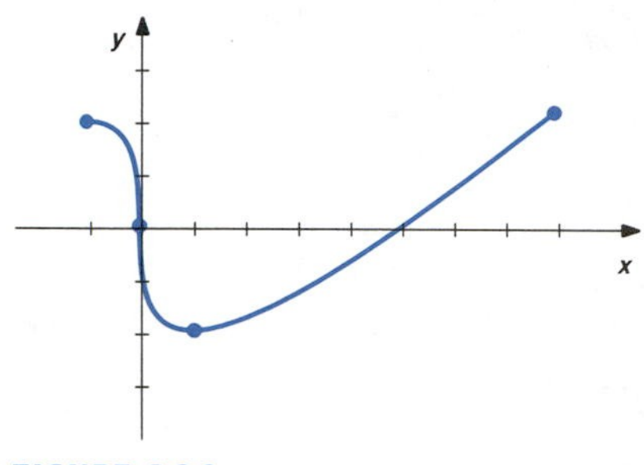

FIGURE 3.9.9

■ **EXAMPLE 4**

The formula $s(t) = -16t^2 + 48t$ gives the height in feet of an object $t$ seconds after it is thrown vertically at a speed of 48 ft/s from ground level. The formula is valid until the object returns to ground level. How high above ground level will the object reach?

FIGURE 3.9.10

**SOLUTION**

Experience suggests that the object leaves ground level at $t = 0$, moves upward to a highest point, and then falls back to ground level, as indicated in Figure 3.9.10. To fit this problem into the mathematical context of this section, we consider the function

$$s(t) = -16t^2 + 48t, \qquad 0 \le t \le t_1,$$

where $t_1$ is the value of $t$ when the object returns to ground level. This means that $s(t_1) = 0$. [It is easy to see that $s(t) = 0$ implies $t = 0, 3$, so $t_1 = 3$.] The greatest height reached by the object is the maximum value of the function $s(t)$. Since $s$ is continuous on the interval $0 \le t \le t_1$, theory guarantees that $s$ will have a maximum value on this interval. We have

$$s(t) = -16t^2 + 48t,$$

$$s'(t) = -32t + 48.$$

(i) $s'(t) = 0$ when $-32t + 48 = 0$, or $t = 1.5$. The height of the object when $t = 1.5$ is

$$s(1.5) = -16(1.5)^2 + 48(1.5) = 36.$$

(ii) $s'(t)$ exists for all $x$.

(iii) $s(t) = 0$ at both endpoints.

We conclude that the maximum height is $s(1.5) = 36$ ft. Note that the velocity is zero when the object is at its greatest height. ■

## EXERCISES 3.9

In Exercises 1–8, sketch the graphs of the given functions and use the graphs to find their maximum and minimum values; indicate "does not exist" where appropriate.

1 $f(x) = x - 2,\ 0 \le x \le 3$

2 $f(x) = x^2 + 1,\ -1 \le x \le 2$

3 $f(x) = 1 - x^2,\ -2 < x < 1$

4 $f(x) = |x - 1| - 1,\ -1 < x < 2$

5 $f(x) = \tan x,\ -\pi/2 < x < \pi/2$

6 $f(x) = \sec x,\ -\pi/2 < x < \pi/2$

7 $f(x) = \begin{cases} x + 1, & -1 \le x < 0 \\ x - 1, & 0 \le x \le 1 \end{cases}$

8 $f(x) = \begin{cases} \dfrac{|x|}{x}, & x \ne 0 \\ 0, & x = 0 \end{cases}$

Find the maximum and minimum values of the functions given in Exercises 9–24.

9 $f(x) = 2x^2 - 8x + 3,\ 0 \le x \le 3$

10 $f(x) = 3x^2 + 6x - 10,\ 0 \le x \le 3$

11 $f(x) = x^3 - 6x^2 + 4,\ -1 \le x \le 6$

12 $f(x) = x^3 - 12x - 8,\ -1 \le x \le 2$

13 $f(x) = 3x^{2/3},\ -1 \le x \le 8$

14 $f(x) = 3x^{1/3},\ -1 \le x \le 8$

15 $f(x) = x(1 - x)^{1/3},\ 0 \le x \le 2$

16 $f(x) = x - 3x^{1/3},\ -1 \le x \le 27$

17 $f(x) = \dfrac{x}{x^2 + 1},\ 0 \le x \le 2$

18 $f(x) = \dfrac{x^2}{x^2 + 1},\ -1 \le x \le 2$

19 $f(x) = \sin \dfrac{3x}{5},\ 0 \le x \le 2\pi$

20 $f(x) = \cos \dfrac{3x}{5},\ 0 \le x \le \pi$

21 $f(x) = \sin x + \cos x,\ 0 \le x \le 2\pi$

22 $f(x) = \sin x + \cos x,\ 0 \le x \le \pi$

23 $f(x) = 4 - |2x - 3|,\ 0 \le x \le 4$

24 $f(x) = |x^3 - 1|,\ -1 \le x \le 2$

25 The formula $s(t) = -16t^2 + 64t$ gives the height in feet of an object $t$ seconds after it is thrown vertically at a speed of 64 ft/s from ground level. The formula is valid until the object returns to ground level. How high above ground level will the object reach?

26 The formula $s(t) = -2.7t^2 + 64t$ gives the height in feet of an object $t$ seconds after it is thrown vertically at a speed of 64 ft/s from the surface of the moon. The formula is valid until the object returns to the surface. How high above the surface of the moon will the object reach?

27 The formula $s(t) = -4.9t^2 + 98t$ gives the height in meters of an object $t$ seconds after it is thrown vertically at a speed of 98 m/s from ground level. The formula is valid until the object returns to ground level. How high above ground level will the object reach?

28 The formula $s(t) = -4.9t^2 + 49t + 6$ gives the height in meters of an object $t$ seconds after it is thrown vertically from a height 6 m above ground level at a speed of 49 m/s. The formula is valid until the object returns to ground level. How high above ground level will the object reach?

29 Find the maximum area of a rectangle that has sides of lengths $x$ and $6 - x$. (Use the fact that length is nonnegative to establish an appropriate interval of definition for the function that gives the area of the rectangle.)

30 Find the maximum volume of a rectangular box that has edges of lengths $x$, $3 - 2x$, and $2 - 2x$.

31 Show that among all rectangles that have lengths of sides $x$ and $5 - 2x$, there is none that has minimum area. (It is convenient to use the fact that length is nonnegative to set up the problem mathematically, but a rectangle that has length of one side zero is not actually a rectangle.)

32 If $f$ is continuous on the interval $a \le x \le b$, differentiable on the interval $a < x < b$, with $f(a) \ge 0$, $f(b) \ge 0$, and $f'(x) \ne 0$ for $a < x < b$, show that $f(x) > 0$ for $a < x < b$. (*Hint*: Can $f$ obtain its minimum value at a point $x$ with $a < x < b$?)

33 Show that $\tan x > x$, $0 < x < \pi/4$.

34 Complete the proof of Theorem 1 by showing that, if $f(c)$ is a local minimum of $f$ and $f'(c)$ exists, then $f'(c) = 0$.

35 **Functions that are derivatives satisfy the Intermediate Value Theorem**, even though they may not be continuous. In particular, if $f'(x)$ is defined for $a \le x \le b$ with $f'(a) < 0$ and $f'(b) > 0$, there must be a point $c$ between $a$ and $b$ with $f'(c) = 0$. Use the Extreme Value Theorem for $f$ and Theorem 1 of this section to verify this special case. [*Hint*: Can $f(a)$ or $f(b)$ be the minimum value of $f$ on the interval $a \le x \le b$?]

## 3.10 THE MEAN VALUE THEOREM

One of the important ideas of calculus is to use information about the derivative of a function to obtain information about the function. The Mean Value Theorem is a basic tool for doing this.

The Mean Value Theorem is a consequence of the following basic result, which we will use to establish several other results.

■ *Rolle's Theorem*

**If $f$ is continuous on the closed interval $a \le x \le b$, $f(a) = f(b) = 0$, and $f$ is differentiable in the open interval $a < x < b$, then there is a point $c$ with $a < c < b$ and $f'(c) = 0$.**

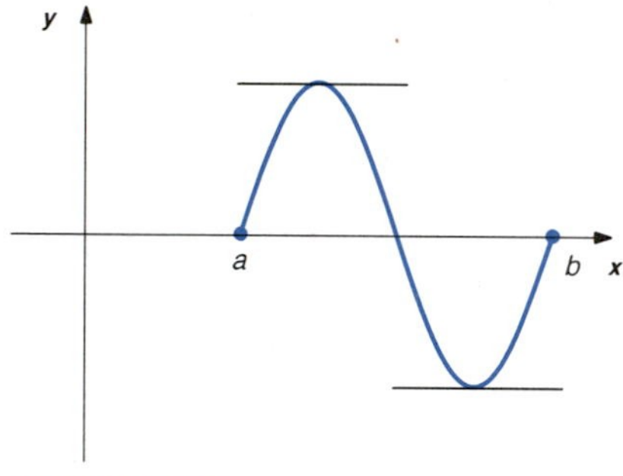


FIGURE 3.10.1

■ *Proof* Since $f$ is continuous on the closed interval $a \le x \le b$, we know that $f$ has both a maximum value and a minimum value on the interval. If both of these are zero, then $f(x) = 0$ for all $a \le x \le b$ and we can choose $c$ to be any point in the interval $a < x < b$. If either the maximum value or the minimum value is $f(c) \ne 0$, then $c$ cannot be $a$ or $b$ [since $f(a) = f(b) = 0$], so $a < c < b$. This means that $f(c)$ is a local maximum or minimum of $f$, so Theorem 1 of Section 3.9 implies either $f'(c) = 0$ or $f'(c)$ does not exist. Since $f$ is differentiable at $c$, we must have $f'(c) = 0$. See Figure 3.10.1. ■

Let us now state the Mean Value Theorem and see how it can be used.

■ *Mean Value Theorem*

**If $f$ is continuous on the closed interval $a \le x \le b$ and differentiable on the open interval $a < x < b$, then there is a point $c$ with $a < c < b$ and**

$$\frac{f(b) - f(a)}{b - a} = f'(c).$$

■ *Proof* To verify the Mean Value Theorem, we apply Rolle's Theorem to the function

$$g(x) = f(x) - f(a) - \left(\frac{f(b) - f(a)}{b - a}\right)(x - a).$$

[The value of $g$ is obtained by subtracting the $y$-coordinate of points on the line through $(a, f(a))$ and $(b, f(b))$ from the corresponding values of $f$.] The function $g$ satisfies the hypotheses of Rolle's Theorem and

$$g'(c) = f'(c) - \frac{f(b) - f(a)}{b - a}.$$

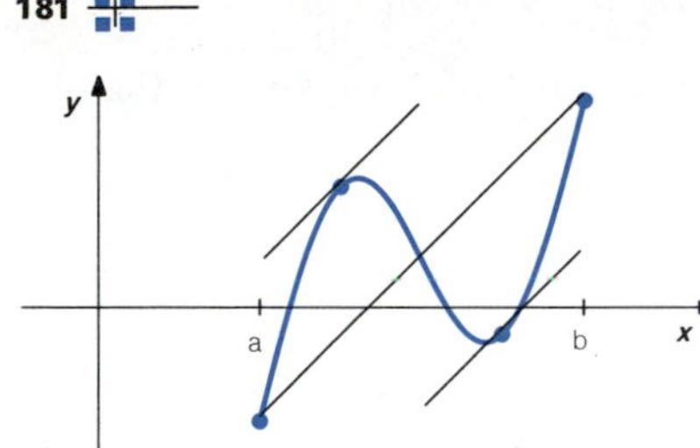


FIGURE 3.10.2



Thus, $g'(c) = 0$ gives the conclusion of the Mean Value Theorem. A representative graph of $f$ is given in Figure 3.10.2. ■

The average (or mean) rate of change of $f$ over the interval $a \le x \le b$ is

$$\frac{f(b) - f(a)}{b - a}.$$

The Mean Value Theorem tells us that, subject to the conditions of the theorem, the mean rate of change of $f$ over the interval is equal to the instantaneous rate of change of $f$ with respect to $x$ at some point $c$, $a < c < b$.

We can use the Mean Value Theorem to obtain useful information about $f$, even though we may not know the particular value of $c$ for which the conclusion holds. In fact, the usefulness of the theorem is that it is not necessary to go to the trouble of finding $c$. It is enough for many purposes to know that such a $c$ exists.

■ **EXAMPLE 1**

Suppose $f$ is continuous on the closed interval $1 \le x \le 3$ and $f'(x) \ge 0$ for $1 < x < 3$. Show that $f(3) \ge f(1)$.

**SOLUTION**

We apply the Mean Value Theorem to the function $f$ on the interval $1 \le x \le 3$ to obtain

$$\frac{f(3) - f(1)}{3 - 1} = f'(c)$$

for some $c$, $1 < c < 3$. Since $f'(x) \ge 0$ for all $x$, $1 < x < 3$, we must have $f'(c) \ge 0$. Then

$$\frac{f(3) - f(1)}{3 - 1} = f'(c) \ge 0,$$

so $f(3) - f(1) \ge 0(3 - 1) = 0$, or $f(3) \ge f(1)$. ■

■ **EXAMPLE 2**

Suppose $f$ is differentiable and $|f'(x) - 2| \le 0.5$ whenever $|x - 2| \le 1$. If $f(2) = 1$, find an interval that is certain to contain the value of $f(2.2)$.

**SOLUTION**

We can apply the Mean Value Theorem to the function $f$ on the interval $2 \le x \le 2.2$ to obtain

$$\frac{f(2.2) - f(2)}{2.2 - 2} = f'(c)$$

for some $c$ between 2 and 2.2. Since $c$ is within 1 of 2, we known that

$$|f'(c) - 2| \leq 0.5,$$

$$-0.5 \leq f'(c) - 2 \leq 0.5,$$

$$1.5 \leq f'(c) \leq 2.5.$$

Combining results, we then have

$$1.5 \leq \frac{f(2.2) - f(2)}{2.2 - 2} \leq 2.5,$$

$$(1.5)(0.2) \leq f(2.2) - 1 \leq (2.5)(0.2),$$

$$1.3 \leq f(2.2) \leq 1.5.$$

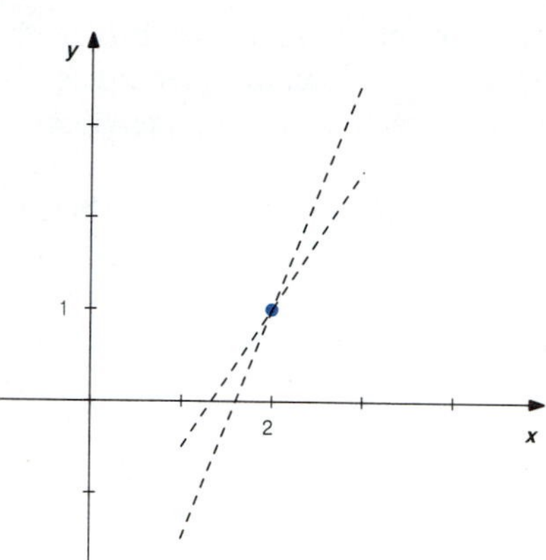

FIGURE 3.10.3

We see that $f(2.2)$ must be in the interval [1.3, 1.5]. Geometrically, the graph of $f$ must be contained between the lines $y = 1 + 1.5(x - 2)$ and $y = 1 + 2.5(x - 2)$ for $|x - 2| \leq 1$. See Figure 3.10.3. ■

We conclude this section with some results that are fundamental to integral calculus, a topic that we will study at great length in later chapters.

■ *Theorem 1*

**If $f'(x) = 0$ for all $x$ in an interval, then $f(x)$ is constant on the interval.**

■ *Proof* Choose any point $x_1$ in the interval. If $x$ is in the interval, the Mean Value Theorem gives

$$\frac{f(x) - f(x_1)}{x - x_1} = f'(c)$$

for some $c$ between $x$ and $x_1$. Then $c$ is in the interval, so $f'(c) = 0$. This gives

$$f(x) - f(x_1) = f'(c) \cdot (x - x_1) = 0,$$

so $f(x) = f(x_1)$. That is, all values of $f(x)$, $x$ in the interval, are equal to the constant $f(x_1)$. ■

It is important in the above theorem that the derivative be zero on an *interval.* A function can have derivative zero at every point in a *set* and not be constant on the set. For example, the function

$$f(x) = \begin{cases} 1, & 0 < x < 1, \\ -1, & -1 < x < 0, \end{cases}$$

has $f'(x) = 0$ for all $x$ in the *set* that consists of the two intervals $(-1, 0)$ and $(0, 1)$. The function $f$ is not constant on the *set*, but it is constant on each of the *intervals* where $f'(x)$ exists and is zero. See Figure 3.10.4.

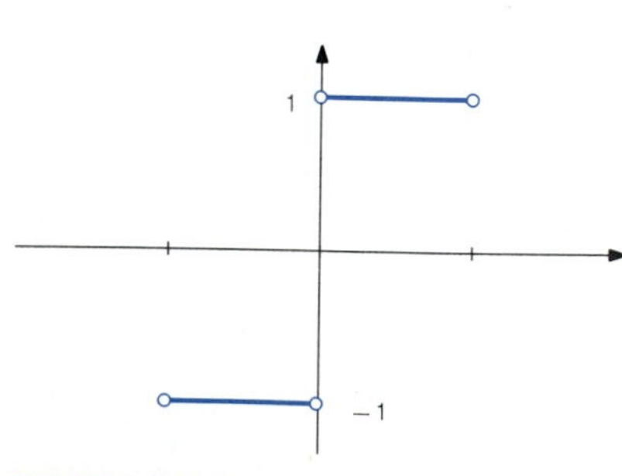

FIGURE 3.10.4

*Theorem 2* **If $f'(x) = g'(x)$ for all $x$ in an interval, then there is some constant $C$ such that $g(x) = f(x) + C$ for $x$ in the interval.**

Theorem 2 can be proved by applying Theorem 1 to the function $g - f$.

## EXERCISES 3.10

1 Suppose $f$ is continuous for $0 \leq x \leq 3$ and differentiable for $0 < x < 3$ with $f'(x) \geq 2$ for $0 < x < 3$. If $f(0) \geq 4$, show $f(3) \geq 10$.

2 Suppose $f$ is continuous for $-1 \leq x \leq 1$ and differentiable for $-1 < x < 1$ with $f'(x) \geq 3$ for $-1 < x < 1$. If $f(1) \leq 0$, show $f(-1) \leq -6$.

3 Suppose $f$ is differentiable with $f'(x) \leq -4$ for all $x$. If $f(-2) \leq 7$, show $f(2) \leq -9$.

4 Suppose $f$ is differentiable with $f'(x) \leq 5$ for all $x$. If $f(4) \geq 18$, show $f(1) \geq 3$.

5 Suppose $f$ is differentiable and $|f'(x) - 5| \leq 0.5$ whenever $|x - 2| \leq 0.4$. If $f(2) - 3$, find an interval that is certain to contain the value of $f(2.3)$.

6 Suppose $f$ is differentiable and $|f'(x) - 5| \leq 0.1$ whenever $|x - 2| \leq 0.4$. If $f(2) = 3$, find an interval that is certain to contain the value of $f(2.3)$.

7 Suppose $f$ is differentiable and $|f'(x) - 3| \leq 0.2$ whenever $|x - 7| \leq 0.4$. If $f(7) = 3$, find an interval that is certain to contain the value of $f(6.9)$.

8 Suppose $f$ is differentiable and $|f'(x) - 1.5| \leq 0.01$ whenever $|x - 3| \leq 0.05$. If $f(3) = 2.25$, find an interval that is certain to contain the value of $f(2.98)$.

9 A motorist tells a police officer that although he drove 84 mi in 1 h 24 min, he never exceeded 55 mph. Explain why the officer (who took calculus) knew that this was impossible. (Assume that his speed was a differentiable function of time.)

10 Suppose $f$ and $g$ are both differentiable for all $x$, $f(0) \geq g(0)$, and $f'(x) \geq g'(x)$ for $x > 0$. Show that $f(x) \geq g(x)$ for $x > 0$. (Apply the Mean Value Theorem to the function $h = f - g$ over the interval $[0, x]$.)

11 The function $f(x) = |x|$, $-1 \leq x \leq 1$, satisfies

$$\frac{f(1) - f(-1)}{1 - (-1)} = 0,$$

but there is no point $c$ in the interval between $-1$ and $1$ where $f'(c) = 0$. Why doesn't the Mean Value Theorem apply to this function on the interval $[-1, 1]$?

12 Find a function $f$ with $f(0) = 0$, $f(1) = 1$, and $f'(x) = 0$ for $0 < x < 1$. Why doesn't the Mean Value Theorem apply to this function on the interval $[0, 1]$?

13 Find the equation of the line through the two points $(a, f(a))$ and $(b, f(b))$. If $L$ denotes the function whose graph is that line, show that the function used in the proof of the Mean Value Theorem is $g = f - L$.

14 Use the Intermediate Value Theorem and Rolle's Theorem to show that the equation $2x^3 + 3x - 1 = 0$ has exactly one solution.

15 Use Rolle's Theorem to show that a polynomial of degree 2 can have at most two distinct zeros. (*Hint*: The derivative must be zero at some point between two distinct zeros of a polynomial. The derivative of a polynomial of degree 2 is a polynomial of degree 1.)

16 If $f''(x) > 0$ for $a < x < b$, show that $f$ can have at most one critical point in the interval. [Recall from Section 3.9 that $c$ is a critical point of $f$ if either $f'(c) = 0$ or $f'(c)$ does not exist.]

17 If $f'(x) = a$ for all $x$ in an interval $I$, show that there is a number $b$ such that $f(x) = ax + b$ for $x$ in $I$. [*Hint*: Consider the derivative of $f(x) - ax$.]

18 Use Theorem 1 to prove Theorem 2.

19 If $f(x) = \sin^2 x$ and $g(x) = -\cos^2 x$, show that $f'(x) = g'(x)$. Find a constant $C$ such that $g(x) = f(x) + C$.

20 If $f(x) = \sec^2 x$ and $g(x) = \tan^2 x$, show that $f'(x) = g'(x)$. Find a constant $C$ such that $g(x) = f(x) + C$.

21 If $f'(x) = \sec^2 x$ and $f(0) = 0$, show that $f(x) = \tan x$, $-\pi/2 < x < \pi/2$. Is it necessarily true that $f(x) = \tan x$ for any values of $x$ that are not in the interval $-\pi/2 < x < \pi/2$?

22 Use the Mean Value Theorem to show that

$$|\sin x - \sin y| \leq |x - y| \qquad \text{for all} \quad x, y.$$

## REVIEW EXERCISES

Find the derivatives in Exercises 1–18.

1 $\dfrac{d}{dx}(x^3 - 2x^2 + x - 4)$

2 $\dfrac{d}{dx}\left(\sqrt{x} + \dfrac{1}{\sqrt{x}}\right)$

3 $\dfrac{d}{dx}(x^2 \cdot \tan x)$

4 $\dfrac{d}{dx}(x \cdot \sin x)$

5 $\dfrac{d}{dx}(\sec x \tan x)$

6 $\dfrac{d}{dx}(\sec^2 x)$

7 $\dfrac{d}{dx}\left(\dfrac{x^2}{\cos x}\right)$

8 $\dfrac{d}{dx}\left(\dfrac{\cos x}{1 + x^2}\right)$

9 $\dfrac{d}{dx}((5x - 1)^{-1/3})$

10 $\dfrac{d}{dx}(\sec \sqrt{x})$

11 $\dfrac{d}{dx}\left(\dfrac{1}{1 - \cos 3x}\right)$

12 $\dfrac{d}{dx}\left(\dfrac{1}{1 - \csc 2x}\right)$

13 $\dfrac{d}{dx}(\sin^2 x)$

14 $\dfrac{d}{dx}(\sin x^2)$

15 $\dfrac{d}{dx}(\csc(\cot x))$

16 $\dfrac{d}{dx}((1 + \sqrt{x^2 + 1})^2)$

17 $\dfrac{d}{dx}(x\sqrt{4x + 1})$

18 $\dfrac{d}{dx}\left(\sqrt{\dfrac{2x + 1}{3x - 1}}\right)$

Use the Method of Implicit Differentiation to find $dy/dx$ for the functions of $x$ defined implicitly by the equations in Exercises 19–22.

19 $x^3 - y^3 = 1$

20 $y^2 + 2xy - x^2 = 0$

21 $x = \cos y$

22 $x = 2y^2 - 3y + 1$

23 Find the equation of the line tangent to the graph of $f(x) = \sqrt{x}$ at the point on the graph with $x = 4$.

24 Find the equation of the line normal to the graph of $x = \tan y$ at the point $(1, \pi/4)$.

25 Find all points on the graph of $f(x) = x(x + 32)^{1/3}$ where the graph has (a) a horizontal tangent and (b) a vertical tangent.

26 Find all points on the graph of $f(x) = x - 2\cos x$, $0 \le x \le 2\pi$, where the graph has a horizontal tangent.

27 The position of a particle on a number line at time $t$ is given by $s(t) = \sqrt{t}/(1 + t)$. Find all intervals of time for which the particle is moving in the positive direction.

28 The distance from the ceiling at time $t$ of a certain weight that is bouncing up and down on a spring attached to the ceiling is given by $s(t) = 2 + \cos 3t$. Find the position of the weight at times when the velocity is zero.

Use the formulas

$$\lim_{\theta \to 0} \frac{\sin \theta}{\theta} = 1 \quad \text{and} \quad \lim_{\theta \to 0} \frac{1 - \cos \theta}{\theta} = 0$$

to evaluate the limits in Exercises 29–32.

29 $\displaystyle\lim_{x \to 0} \frac{\sin 3x}{x}$

30 $\displaystyle\lim_{x \to 0} \frac{\tan 2x}{x}$

31 $\displaystyle\lim_{x \to 0} \frac{1 - \cos^2 x}{x}$

32 $\displaystyle\lim_{x \to 0} \frac{1 - \cos^2 x}{x^2}$

33 A ladder 12 ft long is leaning against a wall. The bottom of the ladder is sliding away from the wall at a rate of 3 ft/s, while the top of the ladder slides down the wall. Let $x$ denote the distance of the bottom of the ladder from the wall and let $y$ denote the height of the top of the ladder. Since the ladder is moving, $x$ and $y$ are functions of time. The functions are related by the equation $x^2 + y^2 = 12^2$. We are given that the rate of change of $x$ with respect to time is $dx/dt = 3$. Find the rate of change of $y$ with respect to time when $x = 3$, 6, and 9. What happens to $dy/dt$ as $x$ approaches 12?

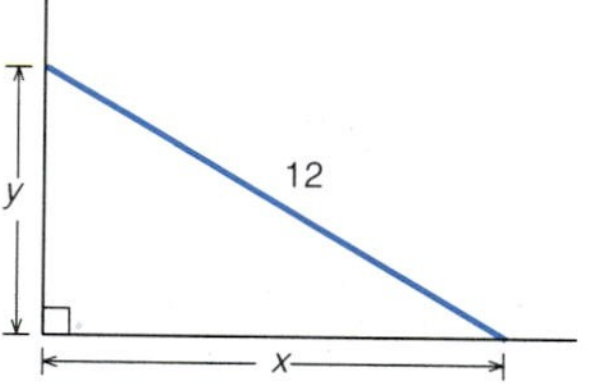


34 If a rocket is rising vertically from a point on the ground that is 60 ft from a ground-level observer, the height of the rocket and the angle of elevation from the observer to the rocket are related by the equation $h = 60 \tan \theta$, where $h$ and $\theta$ are functions of time. The rocket is rising at a rate of 120 ft/s when it is 80 ft high. What is the rate of change of the angle of elevation at this time?

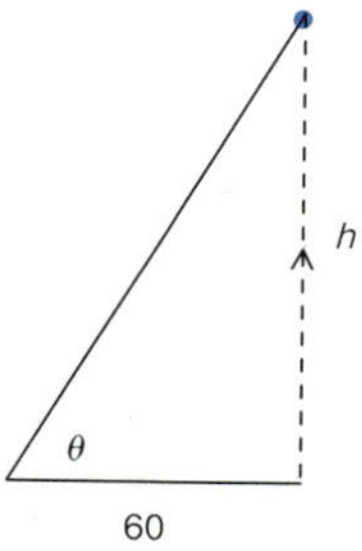

35 In Chapter 7 we will introduce a function $S$ with the derivative

$$S'(x) = \frac{1}{\sqrt{1-x^2}}.$$

Use this formula to evaluate

$$\frac{d}{dx}(S(x/a)), \qquad a \text{ a constant.}$$

36 In Chapter 7 we will introduce a function $T$ with the derivative

$$T'(x) = \frac{1}{1+x^2}.$$

Use this formula to evaluate

$$\frac{d}{dx}(T(x/a)), \qquad a \text{ a constant.}$$

37 Find $dy/dx$ and $d^2y/dx^2$ for the functions of $x$ that are defined implicitly by the equation $x^2 - y^2 = 3$.

38 Assume the equation $x = \tan y$ defines $y$ as a differentiable function of $x$. Show that

$$\frac{dy}{dx} = \frac{1}{1+x^2}.$$

(*Hint*: $\sec^2 y = 1 + \tan^2 y$.)

39 Find the fourth-order Taylor polynomial of $f(x) = \cos(2x)$ about 0.

40 Find the third-order Taylor polynomial of $f(x) = 1/(1+x)$ about 0.

41 Determine that the equation $x^3 + \sqrt{3}x - \sqrt{19} = 0$ has a solution in the interval $1 \le x \le 2$, and then find an interval of length 0.1 that contains the solution.

42 Determine that the equation $x^3 = \cos x$ has a solution in the interval $0 \le x \le \pi/2$, and then find an interval of length 0.1 that contains the solution.

43 The equation $x^3 - 6x - 1 = 0$ has three solutions. Find intervals between successive integers that contain the solutions.

44 The function $f(x) = \csc x$ satisfies $f(-\pi/2) = -1$ and $f(\pi/2) = 1$, but there is no point $c$ between $-\pi/2$ and $\pi/2$ with $f(c) = 0$. Why doesn't the Intermediate Value Theorem apply to this function on the interval $-\pi/2 \le x \le \pi/2$?

45 Use Newton's Method to approximate to four decimal places the solution of the equation $2x = \cos x$.

46 Use Newton's Method to approximate to four decimal places the solution of the equation $x^3 + 3x = 2$.

In Exercises 47–50 sketch the graphs and find the maximum and minimum values of the given functions. Indicate "does not exist" where that is appropriate.

47 $f(x) = 3 - 2x,\ 0 \le x < 2$

48 $f(x) = x^2 - 1,\ -1 < x < 2$

49 $f(x) = 1 - |x|,\ -1 < x < 2$

50 $f(x) = \begin{cases} -1 - x, & -1 \le x < 0 \\ 0, & x = 0 \\ 1 - x, & 0 < x \le 1 \end{cases}$

Find the maximum and minimum values of the functions given in Exercises 51–54.

51 $f(x) = x^3 - x^2 - 2,\ -1 \le x \le 2$

52 $f(x) = \sin \dfrac{2\pi x}{3},\ 0 \le x \le 2$

53 $f(x) = (1 - 3x)^{2/3},\ 0 \le x \le 3$

54 $f(x) = |x^2 - 4x - 5|,\ 0 \le x \le 6$

55 Find the maximum volume of a rectangular box that has edges of lengths $x$, $6 - 2x$, and $6 - 2x$.

56 The perimeter of a triangle that has sides of lengths 1 and 2 with included angle $\theta$ is given by the formula $p(\theta) = 3 + \sqrt{5 - 4\cos\theta}$. Show that among all such triangles there is none for which the perimeter is either maximum or minimum.

57 If $f(2) = 3$ and $f'(x) \ge -1$ for all $x$, show that $f(7) \ge -2$.

58 If $f(2) = 3$ and $|f'(x) - 2| \le 0.2$ for all $x$, show that $3 + 2.2(x-2) \le f(x) \le 3 + 1.8(x-2)$ for $x \le 2$ and $3 + 1.8(x-2) \le f(x) \le 3 + 2.2(x-2)$ for $x \ge 2$, so the graph of $f$ is contained between the two lines $y = 3 + 1.8(x-2)$ and $y = 3 + 2.2(x-2)$.

59 If $f$ is continuous on the interval $x \ge c$ and $f'(x) > 0$ for all $x > c$, show that $f(x) > f(c)$ for all $x > c$.

60 If $f'(x) = \cos x$ and $f(0) = 0$, show that $f(x) = \sin x$.

## CHAPTER FOUR

# *Graph sketching and applications of the derivative*

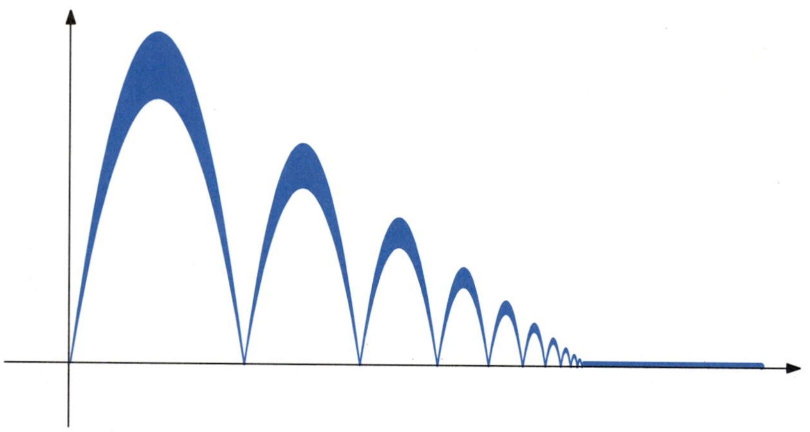

In this chapter we will study some applications of the ideas developed in Chapter 3.

We begin by extending the definition of limit to include infinite-valued limits and limits at infinity. These concepts are then related to the graph of a function.

A sketch of the graph of a function is useful for illustrating properties of the physical quantity that is represented by the function. A sketch is not a detailed drawing. The purpose of a sketch is to indicate key points and general characteristics. We will study key points and general characteristics that are related to the sign of the function, the sign of the derivative, and the sign of the second derivative. In applications, the key points and general characteristics that are illustrated depend on the properties of interest of the physical quantity that the function represents.

To use calculus to solve a practical problem that arises in a real-world situation, it is necessary that the problem be translated into mathematical expressions that involve real variables. This means that physical quantities must be represented by variables and that relations between variables must be established. We will investigate some representative procedures for doing this and then see how to use the results in some problems that involve rates of change and maximum–minimum values.

We continue to investigate the problems of finding approximate values of functions and solutions of equations. We also introduce a new problem: Given a function $f$, can we find a function $F$ such that $F' = f$?

## 4.1 INFINITE-VALUED LIMITS, VERTICAL ASYMPTOTES

Many important characteristics of a function can be illustrated by a sketch of its graph. We begin our study of graph sketching by considering some key points and general characteristics that are directly related to values of the function $f(x)$.

The *domain* is an important part of a function. Our sketch of the graph of a function $f$ should indicate the domain of $f$. The graph should also indicate *intercepts* and the *sign of* $f(x)$, although in some cases it may be difficult to accurately determine the $x$-intercepts. The sign of $f(x)$ tells us whether the graph is above or below the $x$-axis. *Continuity* of a function is an important characteristic. In simple cases, continuity on an interval can be indicated by sketching the graph over the interval "without lifting our pencil," since the graph of a function cannot "jump" on intervals where the function is continuous. It is also useful to note that a function can change sign only at points where it is either zero or discontinuous. Behavior near any *points of discontinuity* should be indicated.

Generally speaking, we wish to sketch the graph of a function in the *simplest way* that reflects the desired key points and general characteristics. For now, this means the graph should reflect the domain, intercepts, sign of $f(x)$, and continuity. "In the simplest way" means with as few wiggles as is possible. In later sections we will learn to use calculus to determine more about the "wiggles."

■ **EXAMPLE 1**

Sketch the graph of the function $f(x) = 2 - \sqrt{x + 1}$.

**SOLUTION**

We first note that $\sqrt{x + 1}$ is defined only for $x + 1 \geq 0$. The domain of $f$ is the interval $x \geq -1$. We should plot the point on the graph with $x = -1$, the left endpoint of the domain. Since $f(-1) = 2 - \sqrt{-1 + 1} = 2$, the point is $(-1, 2)$.

The $x$-intercepts are found by solving the equation $f(x) = 0$. We have

$$2 - \sqrt{x + 1} = 0,$$

$$\sqrt{x + 1} = 2.$$

Squaring both sides, we obtain

$$x + 1 = 4,$$

$$x = 3.$$

Since squaring an equation may introduce extraneous solutions, we must check that $x = 3$ is a solution of the original equation. It is. The point (3, 0) is the only $x$-intercept.

We obtain the $y$-intercept by setting $x = 0$. We have

$$f(0) = 2 - \sqrt{0 + 1} = 1.$$

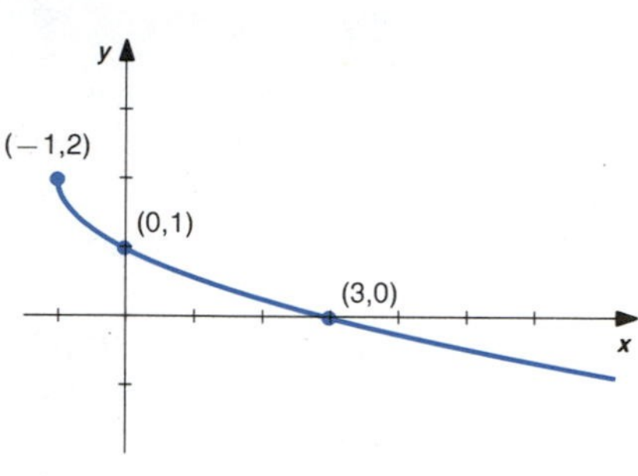


FIGURE 4.1.1

The $y$-intercept is (0, 1).

Since $f$ is continuous on the interval $x \geq -1$, the Intermediate Value Theorem tells us that the sign of $f(x)$ can change only at points where it is zero. In this example, $f(x)$ is positive for $-1 \leq x < 3$ and negative for $x > 3$.

The graph can now be sketched by first plotting the three key points we have found and then drawing a curve through these points in the simplest way that reflects the domain, continuity, and the sign of $f(x)$. The result should look similar to the accurate sketch given in Figure 4.1.1. ■

■ **EXAMPLE 2**

Sketch the graph of the function

$$f(x) = x(x + 1)(x - 2).$$

**SOLUTION**

The function is defined and continuous for all $x$.

The $x$-intercepts are found by solving the equation $f(x) = 0$. We have $x(x + 1)(x - 2) = 0$. The solutions are $x = 0, -1, 2$. The $x$-intercepts are (0, 0), (−1, 0), and (2, 0).

The $y$-intercept is given by $f(0) = 0(0 + 1)(0 - 2) = 0$, but we have already determined that (0, 0) is on the graph.

The sign of $f(x)$ is determined by constructing a table of signs of the factors of $f(x)$, as in Section 1.2. The zeros of the three factors of $f(x)$ divide the number line into four intervals on which the sign of $f(x)$ does not change. Endpoints of these intervals are indicated along the top of Table 4.1.1. The factors of $f(x)$ are listed in the left-hand column, and the sign of each factor in each interval is indicated. The sign of $f(x)$ in each

TABLE 4.1.1

| | $-\infty$ to $-1$ | $-1$ to $0$ | $0$ to $2$ | $2$ to $\infty$ |
|---|---|---|---|---|
| $x$ | − | − | + | + |
| $x + 1$ | − | + | + | + |
| $x - 2$ | − | − | − | + |
| $f(x)$ | − | + | − | + |

interval is then determined by counting the number of factors that are negative in the interval. From Table 4.1.1 we see that $f(x)$ is negative on the intervals $x < -1$ and $0 < x < 2$; $f(x)$ is positive on the intervals $-1 < x < 0$ and $x > 2$.

The graph can now be sketched by plotting the intercepts and then drawing a continuous curve from below the $x$-axis, up through the point $(-1, 0)$, down through the origin, and back up through the point $(2, 0)$. The result should look similar to the accurate sketch given in Figure 4.1.2. ■

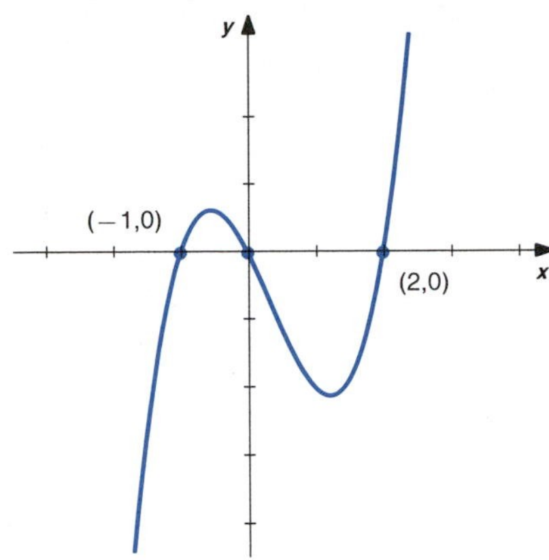

FIGURE 4.1.2

We saw in Chapter 2 that the limit of $f(x)$ as $x$ approaches $c$ is useful for describing a particular behavior of the graph of $f$ for $x$ near $c$. We also know that if $f(x)$ becomes unbounded as $x$ approaches $c$, then $\lim_{x\to c} f(x)$ does not exist. It is important for the concepts of continuity and differentiability that a limit be a real number. For other purposes, it is convenient to extend the idea of limit to include the cases where the values of $f(x)$ approach either infinity or negative infinity as $x$ approaches $c$. The formal definitions follow.

■ *Definition*

We say

$$\lim_{x\to c} f(x) = \infty$$

if, for each positive number $B$, there is a $\delta > 0$ such that $f(x) > B$ whenever $0 < |x - c| < \delta$; we say

$$\lim_{x\to c} f(x) = -\infty$$

if, for each negative number $-B$, there is a $\delta > 0$ such that $f(x) < -B$ whenever $0 < |x - c| < \delta$. See Figure 4.1.3. ■

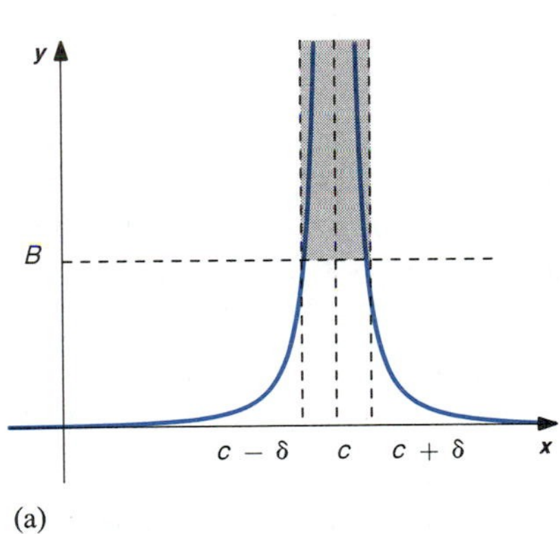

(a)

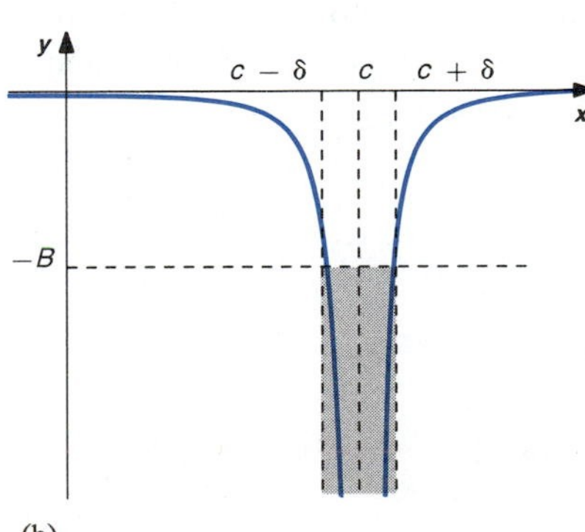

(b)

FIGURE 4.1.3

One-sided limits can also be extended to include the values $\infty$ and $-\infty$.

We say $x = c$ is a **vertical asymptote** of a function $f$ if one of the one-sided limits of $f(x)$ at $c$ is either $\infty$ or $-\infty$. The expression $x = c$ is the equation of a vertical line, so a vertical asymptote is a vertical line. Vertical asymptotes reflect an important general characteristic of a graph and should be indicated in a sketch of the graph.

A rational function $P/Q$ has a vertical asymptote at points $c$ where the denominator is zero and the numerator is not. The sign of $P(x)/Q(x)$ for $x$ on each side of $c$, $x$ near $c$, then determines the behavior of the graph near $x = c$. Some of the possible behaviors of the graph of a function near a vertical asymptote are illustrated in Example 3.

■ **EXAMPLE 3**

Sketch the graphs of the given functions near the vertical asymptotes $x = c$. Indicate the value of the one-sided limits $\lim_{x\to c^-} f(x)$ and $\lim_{x\to c^+} f(x)$.

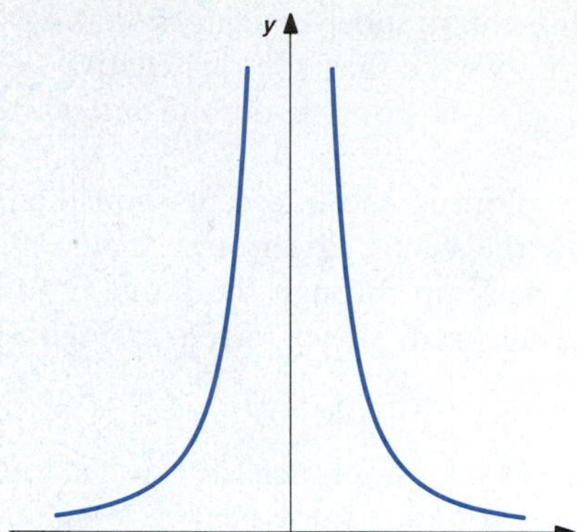

FIGURE 4.1.4

(a) $f(x) = \dfrac{1}{x^2}$,

(b) $f(x) = \dfrac{x}{x-2}$,

(c) $f(x) = \dfrac{x+1}{(x+2)^2}$,

(d) $f(x) = \tan x$, $c = \dfrac{\pi}{2}$.

**SOLUTION**

(a) It is clear that $f(x) = 1/x^2$ is positive for $x \neq 0$ and that

$$\lim_{x\to 0} \frac{1}{x^2} = \infty,$$

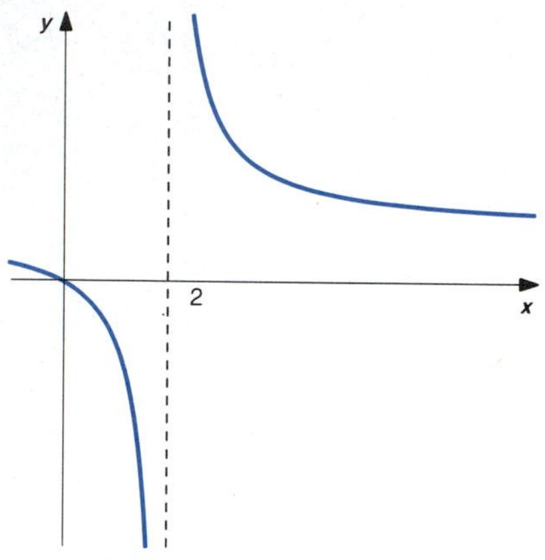

FIGURE 4.1.5

so both one-sided limits of $1/x^2$ at zero are positive infinity. [If you think it would help, you could make a table of values of $f(x)$ for $x$ near zero, $x \neq 0$.] Therefore, $f$ has vertical asymptote $x = 0$. The portion of the graph near $x = 0$ is sketched in Figure 4.1.4.

(b) $f(x) = x/(x-2)$ has vertical asymptote $x = 2$. We can check that $f(x) > 0$ for $x > 2$, and $f(x) < 0$ for $x < 2$, $x$ near 2. It follows that

$$\lim_{x\to 2^-} \frac{x}{x-2} = -\infty \quad \text{and} \quad \lim_{x\to 2^+} \frac{x}{x-2} = \infty.$$

The graph of $f$ for $x$ near 2 is sketched in Figure 4.1.5.

(c) $f(x) = (x+1)/(x+2)^2$ has vertical asymptote $x = -2$. We have $f(x) < 0$ for all $x$ near $-2$, on both sides of $x = -2$. Therefore,

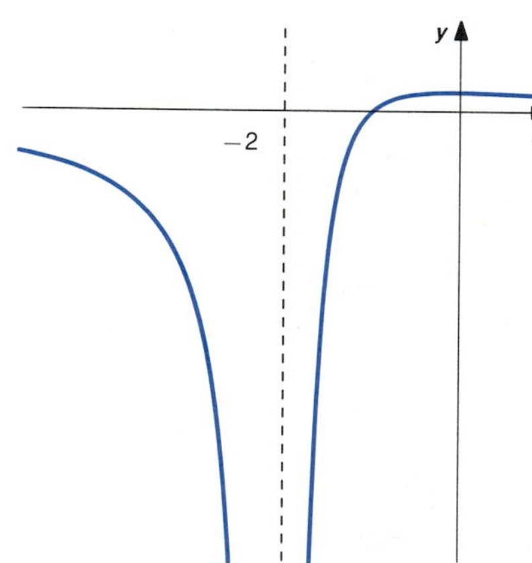

FIGURE 4.1.6

$$\lim_{x\to -2} \frac{x+1}{(x+2)^2} = -\infty$$

and both one-sided limits at $-2$ are negative infinity. See Figure 4.1.6.

(d) $f(x) = \tan x = \sin x/\cos x$ has vertical asymptote $x = \pi/2$, because $\cos x$ approaches $\cos(\pi/2) = 0$ as $x$ approaches $\pi/2$ and $\sin x$ approaches $\sin(\pi/2) = 1 \neq 0$. $f(x)$ is positive in the interval $(0, \pi/2)$, so

$$\lim_{x\to\pi/2^-} \tan x = \infty.$$

The function is negative in the interval $(\pi/2, \pi)$, so

$$\lim_{x\to\pi/2^+} \tan x = -\infty.$$

See Figure 4.1.7. ■

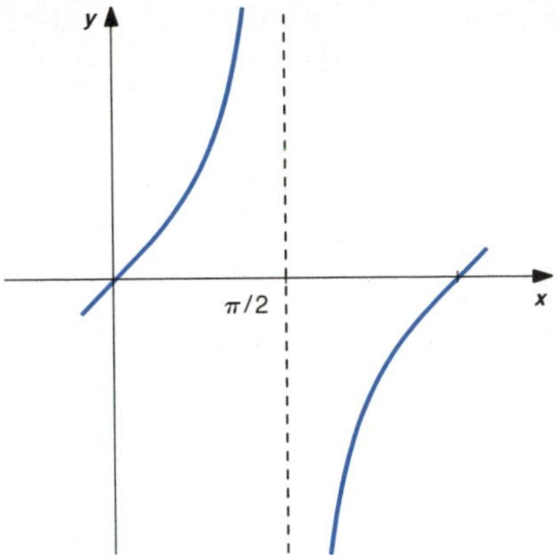

FIGURE 4.1.7

***Warning:*** *The usual rules of limits may not apply to limits that have infinite values.* The problem is that the rules of arithmetic do not apply to $\infty$ and $-\infty$, because they are not real numbers. For example,

$$\lim_{x\to 0}\left(\frac{1}{x^4}-\frac{1}{x^2}\right) \neq \lim_{x\to 0}\frac{1}{x^4} - \lim_{x\to 0}\frac{1}{x^2} = \infty - \infty = (?)$$

and

$$\lim_{x\to 0^+} x\cot x \neq \left(\lim_{x\to 0^+} x\right)\left(\lim_{x\to 0^+} \cot x\right) = 0\cdot\infty = (?).$$

The following results can sometimes be used to evaluate limits of the above types. The results can be verified by using the definition of limits with infinite values.

*Theorem 1*

**If $\lim_{x\to c} f(x) = \lim_{x\to c} g(x) = \infty$, then**

$$\lim_{x\to c}[f(x)+g(x)] = \infty \quad \text{and} \quad \lim_{x\to c} f(x)\cdot g(x) = \infty.$$

**If $\lim_{x\to c} f(x) = L$, $L$ a real number, and $\lim_{x\to c} g(x) = \infty$, then**

$$\lim_{x\to c}[f(x)+g(x)] = \infty,$$

$$\lim_{x\to c} f(x)\cdot g(x) = \begin{cases} \infty, & \text{if } L > 0, \\ -\infty, & \text{if } L < 0, \end{cases}$$

**and**

$$\lim_{x\to c}\frac{f(x)}{g(x)} = 0.$$

Corresponding results hold for limits that have values of negative infinity and also for one-sided limits.

**EXAMPLE 4**

Evaluate the following limits:

(a) $\lim_{x\to 0}\left(\frac{1}{x^2}-\frac{1}{x^4}\right)$,

(b) $\lim_{x\to 0^+} x\cot x$.

**SOLUTION**

(a) Factoring $1/x^4$ from each term, we have

$$\frac{1}{x^2}-\frac{1}{x^4} = \frac{1}{x^4}(x^2-1).$$

Symmetric with respect to $y$-axis
(a)

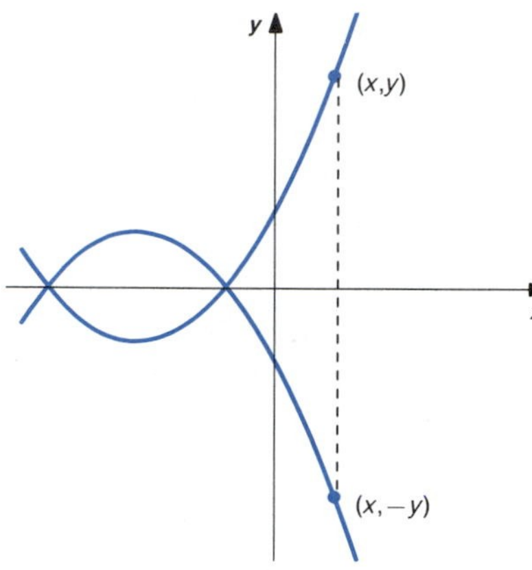


Symmetric with respect to $x$-axis
(b)

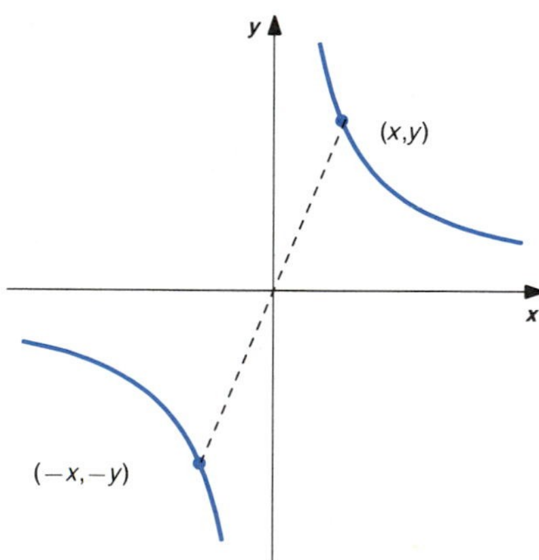


Symmetric with respect to origin
(c)

**FIGURE 4.1.8**

As $x$ approaches zero, the first factor on the right approaches positive infinity and the second factor approaches $-1$. It follows that the product approaches negative infinity. We have

$$\lim_{x\to 0}\left(\frac{1}{x^2}-\frac{1}{x^4}\right)=-\infty.$$

(b) In this example we can rearrange terms and use the fact that

$$\lim_{x\to 0^+}\frac{\sin x}{x}=1.$$

(This limit was used in Section 3.3.) We have

$$\begin{aligned}\lim_{x\to 0^+} x\cot x &= \lim_{x\to 0^+} x\frac{\cos x}{\sin x}\\ &= \lim_{x\to 0^+}\left(\frac{x}{\sin x}\right)(\cos x)\\ &= (1)(1)=1.\end{aligned}$$

■

When sketching graphs, it is sometimes useful to take advantage of certain simple symmetries.

A graph is said to be **symmetric with respect to the $y$-axis** if $(-x, y)$ is on the graph whenever $(x, y)$ is. See Figure 4.1.8a.

A graph is **symmetric with respect to the $x$-axis** if $(x, -y)$ is on the graph whenever $(x, y)$ is. See Figure 4.1.8b.

It is **symmetric with respect to the origin** if $(-x, -y)$ is on the graph whenever $(x, y)$ is. See Figure 4.1.8c.

A function $f$ is called **even** if $f(-x) = f(x)$ for all $x$; $f$ is called **odd** if $f(-x) = -f(x)$ for all $x$.

It is easy to check that the graph of an even function is symmetric with respect to the $y$-axis. The graph of an odd function is symmetric with respect to the origin. Except for the trivial case $f(x) = 0$, the graph of a *function* cannot be symmetric with respect to the $x$-axis.

## ■ EXAMPLE 5

(a) $f(x) = x^3 - x$ is an odd function, because

$$\begin{aligned}f(-x) &= (-x)^3-(-x)\\ &= -(x^3-x) = -f(x).\end{aligned}$$

The graph is symmetric with respect to the origin. See Figure 4.1.9.

(b) $f(x) = x^4 - x^2$ is an even function, because

$$f(-x)=(-x)^4-(-x)^2=x^4-x^2=f(x).$$

The graph is symmetric with respect to the $y$-axis. See Figure 4.1.10.

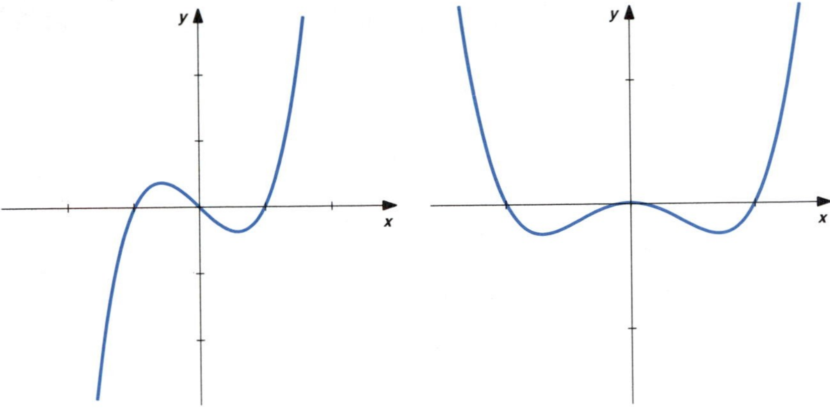


FIGURE 4.1.9

FIGURE 4.1.10

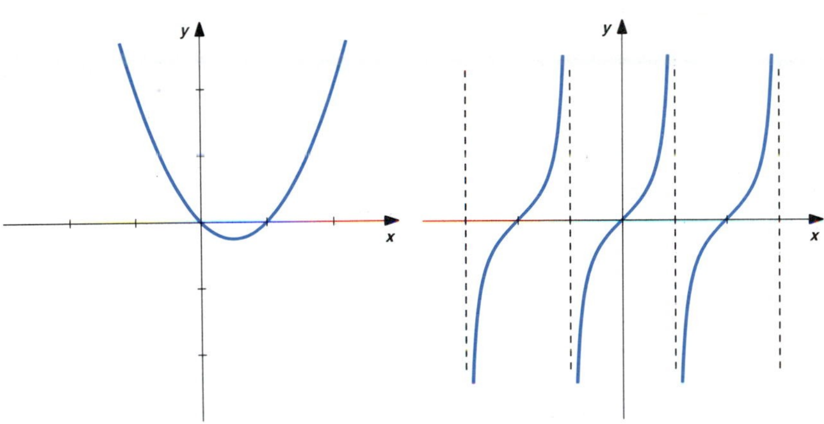


FIGURE 4.1.11

FIGURE 4.1.12

(c) $f(x) = x^2 - x$ is neither an odd function nor an even function. The expression

$$f(-x) = (-x)^2 - (-x) = x^2 + x$$

is different from both $f(x)$ and $-f(x)$. The graph is symmetric with respect to neither the $y$-axis nor the origin. See Figure 4.1.11.

(d) $f(x) = \tan x$ is an odd function, because

$$f(-x) = \tan(-x) = -\tan x = -f(x).$$

The graph is symmetric with respect to the origin. See Figure 4.1.12.

(e) $f(x) = 1/x^2$ is an even function, because

$$f(-x) = \frac{1}{(-x)^2} = \frac{1}{x^2} = f(x).$$

The graph is symmetric with respect to the $y$-axis. See Figure 4.1.13. ■

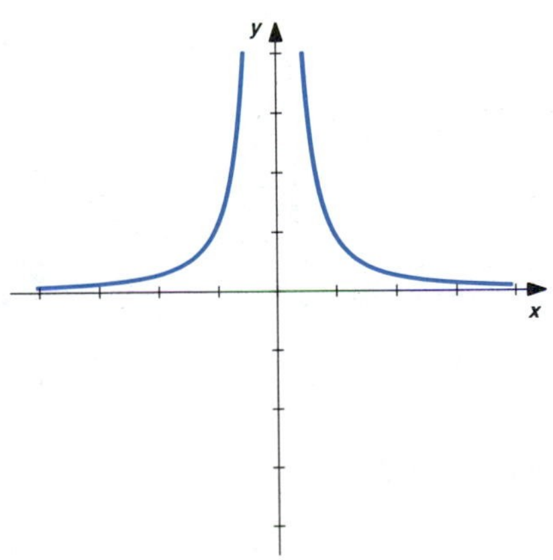


FIGURE 4.1.13

## EXERCISES 4.1

Sketch the graphs of the functions $f$ given in Exercises 1–20 in the simplest way that reflects the domain, intercepts, sign of $f(x)$, and continuity.

1 $f(x) = 2x - 4$

2 $f(x) = 3 - 6x$

3 $f(x) = x(x - 1)$

4 $f(x) = x(x + 2)$

5 $f(x) = 4 + 3x - x^2$

6 $f(x) = 5 + 4x - x^2$

7 $f(x) = x(x^2 - 4)$

8 $f(x) = x^2(x - 1)$

9 $f(x) = x^2(9 - x^2)/4$

10 $f(x) = x^2(x^2 + 1)$

11 $f(x) = x^2(x + 2)$

12 $f(x) = x(x + 1)^2$

13 $f(x) = \sqrt{1 - x}$

14 $f(x) = \sqrt{4 - x}$

15 $f(x) = 1 - \sqrt{x + 4}$

16 $f(x) = 1 - \sqrt{x + 1}$

17 $f(x) = \sqrt{9 - x^2}$

18 $f(x) = -\sqrt{4 - x^2}$

19 $f(x) = \sqrt{x - x^2}$

20 $f(x) = \sqrt{-x - x^2}$

In Exercises 21–34 sketch the graph of the given function $f$ near the vertical asymptote $x = c$. Indicate the value of the one-sided limits $\lim_{x \to c^-} f(x)$ and $\lim_{x \to c^+} f(x)$.

21 $f(x) = \dfrac{1}{x + 1}$

22 $f(x) = \dfrac{1}{x - 2}$

23 $f(x) = \dfrac{x}{x + 1}$

24 $f(x) = \dfrac{x}{x + 2}$

25 $f(x) = \cos x \cot x,\ c = 0$

26 $f(x) = \sin x \tan x,\ c = \pi/2$

27 $f(x) = \sec x \tan x,\ c = \pi/2$

28 $f(x) = \csc x \cot x,\ c = 0$

29 $f(x) = \dfrac{x - 3}{(x - 1)^2}$

30 $f(x) = \dfrac{x + 1}{(x + 3)^2}$

31 $f(x) = \dfrac{x}{1 - x}$

32 $f(x) = \dfrac{x + 1}{x + 2}$

33 $f(x) = \dfrac{\cos x}{x^2}$

34 $f(x) = \dfrac{\sin x}{x^2}$

Evaluate the limits in Exercises 35–42.

35 $\displaystyle\lim_{x \to 0^+} \left(\frac{1}{x^2} - \frac{1}{x}\right)$

36 $\displaystyle\lim_{x \to 0^-} \left(\frac{1}{x^2} + \frac{1}{x^3}\right)$

37 $\displaystyle\lim_{x \to 0^+} \left(\frac{1}{x^3} - \frac{1}{x^2}\right)$

38 $\displaystyle\lim_{x \to 0^-} \left(\frac{1}{x^3} - \frac{1}{x^2}\right)$

39 $\displaystyle\lim_{x \to 0} \frac{\tan(2x)}{x}$

40 $\displaystyle\lim_{x \to 0} x \csc(x/2)$

41 $\displaystyle\lim_{x \to 0} \left(\frac{1}{x} - \frac{\cos^2 x}{x}\right)$

42 $\displaystyle\lim_{x \to 0} \left(\frac{1}{x} - \frac{1}{x \sin x}\right)$

43 Use the definition to show that

$$\lim_{x \to 0} \frac{1}{x^2} = \infty.$$

44 Use the definition to show that

$$\lim_{x \to 3} \frac{1}{(2x - 6)^2} = \infty.$$

45 Write a formal definition of the statement

$$\lim_{x \to c^+} f(x) = \infty$$

and use the definition to show that

$$\lim_{x \to 2^+} \frac{2}{3x - 6} = \infty.$$

46 Write a formal definition of the statement

$$\lim_{x \to c^-} f(x) = -\infty$$

and use the definition to show that

$$\lim_{x \to 2^-} \frac{3}{4x - 8} = -\infty.$$

47 If $1/2 \le f(x) \le 3/2$ for all $x$ and $\lim_{x\to c} g(x) = \infty$, use the definition to show that $\lim_{x\to c} f(x)g(x) = \infty$.

48 If $\lim_{x\to c} g(x) = \infty$, use the definition to show that

$$\lim_{x\to c} \frac{1}{g(x)} = 0.$$

49 If $\lim_{x\to c} g(x) = 0$, explain why it is not necessarily true that

$$\lim_{x\to c} \frac{1}{g(x)} = \infty.$$

50 If $f$ is continuous on the interval $a < x < b$, and $\lim_{x\to a^+} f(x) = \lim_{x\to b^-} f(x) = \infty$, show that $f$ has a minimum value on the interval.

Determine whether the graphs given in Exercises 51–56 appear to be symmetric with respect to the $x$-axis, $y$-axis, or the origin.

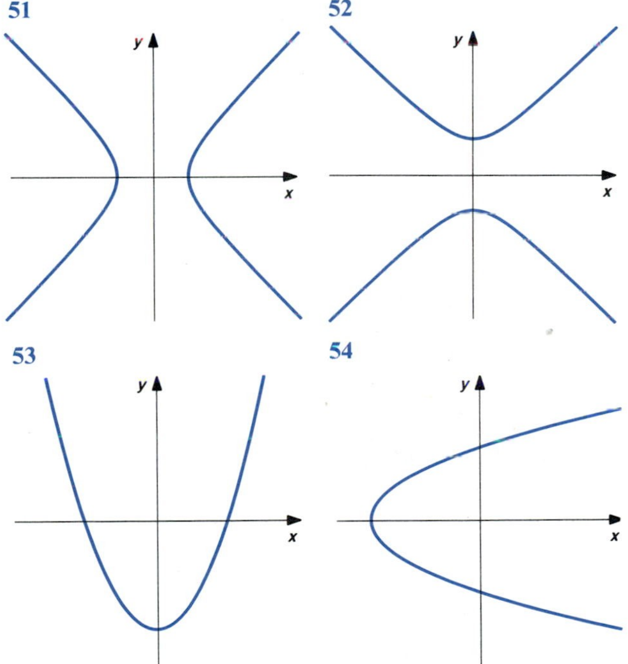


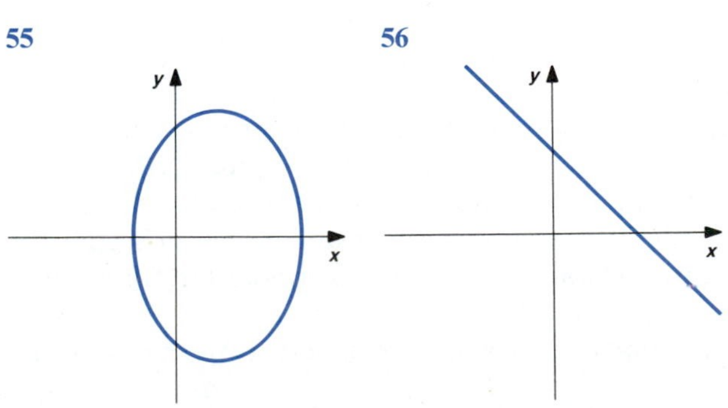


Determine if the functions given in Exercises 57–64 are odd or even.

57 $f(x) = x^4 + 2x^2 + 1$

58 $f(x) = 5x^4 + 3x^2 + 1$

59 $f(x) = 2x^3 - 3x + 1$

60 $f(x) = x^3 + 2x$

61 $f(x) = \dfrac{x^2 + 4}{x^3 - x}$

62 $f(x) = \dfrac{x^3 + 5}{x^2 + 9}$

63 $f(x) = x \sin x$

64 $f(x) = x \cos x$

65 For which integers is $f(x) = x^n$ an even function? For which integers is it an odd function?

66 Is the product or quotient of two even functions odd or even?

67 Is the product or quotient of two odd functions odd or even?

68 Is the product or quotient of an odd function and an even function odd or even?

69 Use the given graph of $f(x)$, $x \ge 0$, to sketch the graphs of $f(-x)$, $x \le 0$, and $f(|x|)$, all $x$. Discuss the symmetry of the graph of $f(|x|)$.

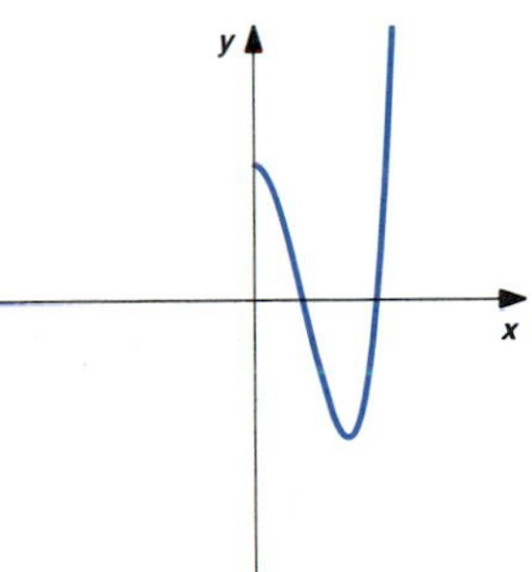


70 Use the given graph of $f(x)$, $x \ge 0$, to sketch the graphs of $f(-x)$, $x \le 0$, and $f(|x|)$, all $x$. Discuss the symmetry of the graph of $f(|x|)$.

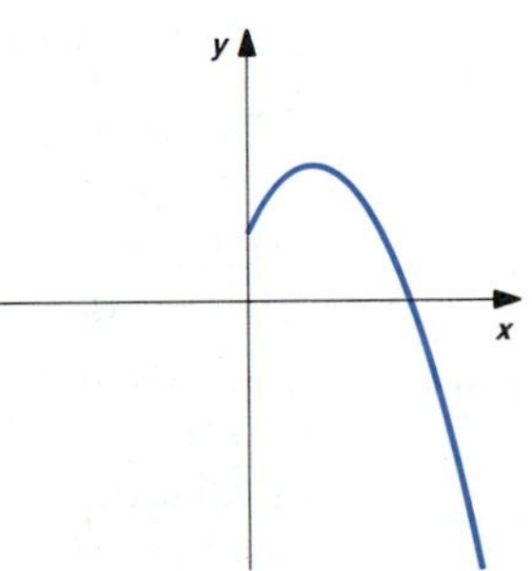

71 Use the given graph of $f(x)$ to sketch the graph of $-f(x)$ and $|f(x)|$. Discuss the symmetry between the graph of $f(x)$ and the graph of $-f(x)$.

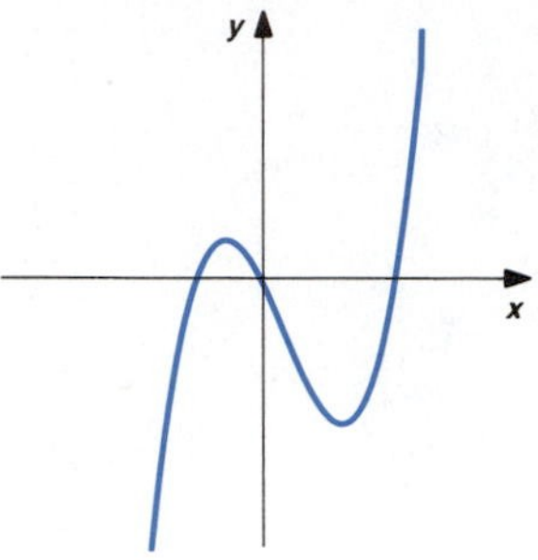


72 Use the given graph of $f(x)$ to sketch the graph of $-f(x)$ and $|f(x)|$. Discuss symmetry between the graph of $f(x)$ and the graph of $-f(x)$.

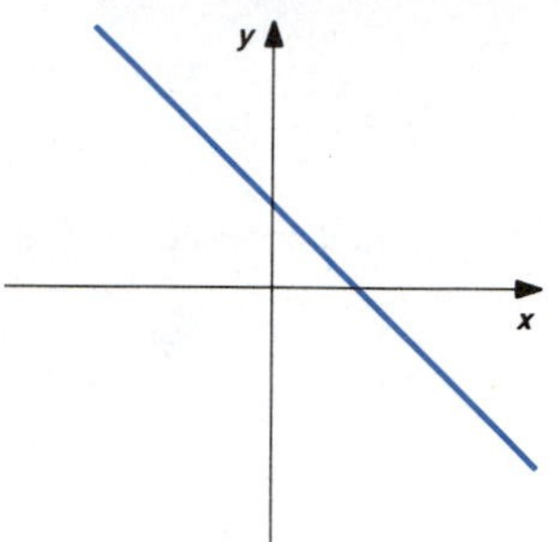


## 4.2 LIMITS AT INFINITY, HORIZONTAL ASYMPTOTES

Knowledge of the behavior of $f(x)$ for $|x|$ large can be used to obtain information about the quantity represented by the function. For example, we may be interested if a certain piece of machinery approaches a steady-state temperature after being run for a long time. Limits at infinity are used to investigate such problems.

We begin our study of the behavior of $f(x)$ for $|x|$ large with the following formal definitions of limits at infinity.

*Definition*

We say

$$\lim_{x\to\infty} f(x) = L$$

if, for each positive number $\varepsilon$, there is a positive number $B$ such that $|f(x) - L| < \varepsilon$ whenever $x > B$. We say

$$\lim_{x\to-\infty} f(x) = L$$

if, for each positive number $\varepsilon$, there is a negative number $-B$ such that $|f(x) - L| < \varepsilon$ whenever $x < -B$. We write $\lim_{x\to\pm\infty} f(x) = L$ if both limits are $L$.

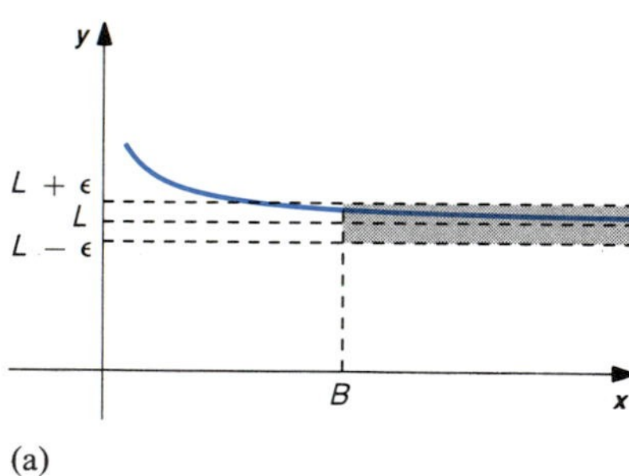


(a)

(b)

FIGURE 4.2.1

See Figure 4.2.1.

The statement $\lim_{x\to\infty} f(x) = L$ means that the values $f(x)$ are near $L$ for all large enough $x$. The statement $\lim_{x\to-\infty} f(x) = L$ means that $f(x)$ is near $L$ for all negative $x$ with large enough absolute value.

The definitions of limits at $\infty$ and $-\infty$ can be modified as in Section 4.1 to include infinite values. The usual rules of limits hold for *finite-valued* limits at $\infty$ and $-\infty$.

Evaluation of many limits at $\infty$ and $-\infty$ is based on the following observation.

**If $r > 0$, then**

$$\lim_{x \to \infty} \frac{1}{x^r} = 0.$$

(1)

**If $x^r$ is defined for $x < 0$, then $r > 0$ implies**

$$\lim_{x \to -\infty} \frac{1}{x^r} = 0.$$

Statement (1) merely reflects the fact that a positive power or root of a large number is large and that dividing a fixed number by a much larger number gives a small number.

We will want to investigate the behavior as $x \to \pm\infty$ of functions that are quotients of functions that involve powers of $x$. The rule for the limit of a quotient cannot be used directly to evaluate

$$\lim_{x \to \pm\infty} \frac{f(x)}{g(x)}$$

in the case where

$$\lim_{x \to \pm\infty} f(x) = \lim_{x \to \pm\infty} g(x) = \infty.$$

To overcome this difficulty, we will *scale the ratio by factoring the highest power of x from each of the numerator and denominator.* We then obtain a single power of $x$ times a quotient of terms that have finite, nonzero limits at infinity. We can then use rules of limits to determine the limit as $x \to \pm\infty$.

■ **EXAMPLE 1**

Evaluate

$$\lim_{x \to \infty} \frac{2x - 3}{x^2 + 7x - 1}.$$

**SOLUTION**

We have a quotient where both the numerator and the denominator approach infinity as $x$ approaches infinity. To investigate the ratio, we scale the ratio by factoring the highest power of $x$ from each of the numerator and denominator. We then obtain

$$\lim_{x \to \infty} \frac{2x - 3}{x^2 + 7x - 1} = \lim_{x \to \infty} \frac{x\left(2 - \dfrac{3}{x}\right)}{x^2\left(1 + \dfrac{7}{x} - \dfrac{1}{x^2}\right)}$$

$$= \lim_{x\to\infty} \left(\frac{1}{x}\right)\left(\frac{2 - \dfrac{3}{x}}{1 + \dfrac{7}{x} - \dfrac{1}{x^2}}\right).$$

The scaling has given us an expression that consists of $1/x$ times a quotient of terms that have finite, nonzero limits at infinity. We can then use rules of limits to evaluate the limit of the expression. Using (1), we obtain

$$\lim_{x\to\infty} \frac{2x - 3}{x^2 + 7x - 1} = \lim_{x\to\infty} \left(\frac{1}{x}\right)\left(\frac{2 - \dfrac{3}{x}}{1 + \dfrac{7}{x} - \dfrac{1}{x^2}}\right) = (0)\,\frac{2 - 0}{1 + 0 - 0} = 0. \quad ■$$

The method of Example 1 can be used to verify the following fact.

**If $P$ and $Q$ are polynomials with the degree of $P$ less than the degree of $Q$, then**

$$\lim_{x\to\pm\infty} \frac{P(x)}{Q(x)} = 0.$$

■ **EXAMPLE 2**

Evaluate

$$\lim_{x\to-\infty} \frac{2x^2 - 1}{3x^2 + x - 4}.$$

**SOLUTION**

As in the previous example, we have a quotient of functions that approach infinity. We scale by factoring the highest power of $x$ from each of the numerator and the denominator. This gives

$$\lim_{x\to-\infty} \frac{2x^2 - 1}{3x^2 + x - 4} = \lim_{x\to-\infty} \frac{x^2\left(2 - \dfrac{1}{x^2}\right)}{x^2\left(3 + \dfrac{1}{x} - \dfrac{4}{x^2}\right)} = \lim_{x\to-\infty} \frac{2 - \dfrac{1}{x^2}}{3 + \dfrac{1}{x} - \dfrac{4}{x^2}} = \frac{2}{3}. \quad ■$$

Example 2 illustrates the following general fact.

**If $P$ and $Q$ are polynomials of equal order, then**

$$\lim_{x\to\pm\infty} \frac{P(x)}{Q(x)}$$

**is the quotient of the coefficient of the highest-order term of $P$ and the coefficient of the highest-order term of $Q$.**

■ **EXAMPLE 3**

Evaluate

$$\lim_{x\to\infty} \frac{x^{-1/3} + 2x^{-1/4}}{3x^{-1/3} - 4x^{-1/4}}.$$

**SOLUTION**

In this example we have a quotient of functions that approach zero. As before, we scale the ratio by factoring the highest power of $x$ from each of the numerator and the denominator. Since $-1/3 < -1/4$, we factor $x^{-1/4}$ from each. We have

$$\lim_{x\to\infty} \frac{x^{-1/3} + 2x^{-1/4}}{3x^{-1/3} - 4x^{-1/4}} = \lim_{x\to\infty} \frac{x^{-1/4}(x^{(-1/3)+(1/4)} + 2)}{x^{-1/4}(3x^{(-1/3)+(1/4)} - 4)}$$

$$= \lim_{x\to\infty} \frac{x^{-1/12} + 2}{3x^{-1/12} - 4}$$

$$= \lim_{x\to\infty} \frac{\left(\dfrac{1}{x^{1/12}}\right) + 2}{\left(\dfrac{3}{x^{1/12}}\right) - 4} = \frac{0+2}{0-4} = -\frac{1}{2}.$$ ■

We cannot use the Limit Theorem to evaluate the limit of a sum of two functions that have infinite limits of opposite sign. In some cases, we can evaluate limits of this type by rewriting the expression as a product of a factor with an infinite limit and a factor with a finite, *nonzero* limit.

■ **EXAMPLE 4**

Evaluate $\lim_{x\to-\infty} (x^3 + 2x^2)$.

**SOLUTION**

We note that $x^3 \to -\infty$ and $2x^2 \to \infty$ as $x \to -\infty$. Rewriting the sum by factoring out the higher-order term, we have

$$x^3 + 2x^2 = x^3\left(1 + \frac{2}{x}\right).$$

We then see that the given expression is the product of $x^3$ and a factor that has limit one as $x$ approaches negative infinity. Since $x^3 \to -\infty$ as $x \to -\infty$, we conclude that

$$\lim_{x\to-\infty} (x^3 + 2x^2) = -\infty.$$ ■

The method of Example 4 can be used to verify the following:

**The limits at infinity of any polynomial are the same as the limits of the highest-order term of the polynomial.**

FIGURE 4.2.2

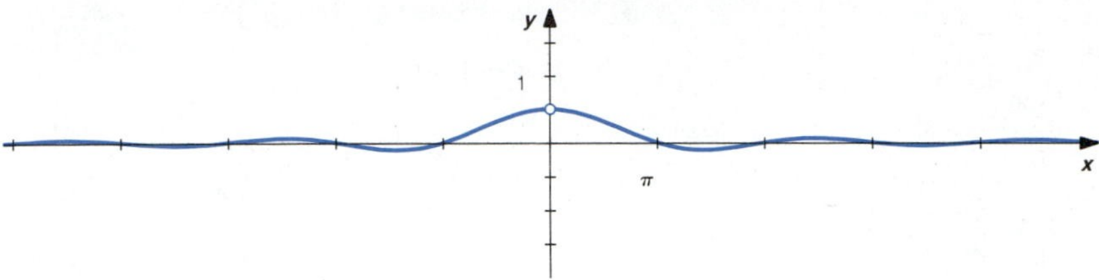


If either $\lim_{x\to\infty} f(x) = L$ or $\lim_{x\to-\infty} f(x) = L$, $y = L$ is called a **horizontal asymptote** of $f$. In that case, points $(x, f(x))$ on the graph $f$ approach the horizontal line $y = L$ as $x$ approaches infinity or negative infinity, respectively. The graph of a function $f$ intersects a horizontal asymptote $y = L$ whenever $f(x) = L$. A graph may touch or cross a horizontal asymptote any number of times, as is illustrated by the graph of $(\sin x)/x$ given in Figure 4.2.2. We see that

$$\lim_{x\to\pm\infty} \frac{\sin x}{x} = 0,$$

so $y = 0$ is a horizontal asymptote of $(\sin x)/x$. The graph crosses the asymptote at each point on the asymptote that has $x$-coordinate a nonzero integer multiple of $\pi$. [$(\sin x)/x$ is undefined for $x = 0$, but

$$\lim_{x\to 0} \frac{\sin x}{x} = 1,$$

so $x = 0$ is not a vertical asymptote.]

Horizontal and vertical asymptotes should be added to our list of general characteristics and key points to be reflected in the graph of a function. This means our graphs should be sketched in the simplest way that reflects the *domain, intercepts, sign of* $f(x)$, *continuity, vertical asymptotes,* and *horizontal asymptotes.* Recall that in the "simplest way" means with as few "wiggles" as possible.

■ **EXAMPLE 5**

Sketch the graph of

$$f(x) = \frac{x}{\sqrt{x^2+1}}.$$

**SOLUTION**

$f$ is defined and continuous for all $x$. The only intercept is $(0, 0)$. It is clear that $f(x)$ is positive for positive $x$ and negative for negative $x$. We may notice that $f(-x) = -f(x)$, so the function is odd and the graph is symmetric with respect to the origin.

We investigate $f(x)$ for $|x|$ large by scaling. Since $\sqrt{x^2} = |x|$, we have

$$f(x) = \frac{x}{\sqrt{x^2+1}} = \frac{x}{\sqrt{x^2\left(1+\dfrac{1}{x^2}\right)}} = \frac{x}{|x|\sqrt{1+\dfrac{1}{x^2}}}.$$

Since

$$\frac{x}{|x|} = \begin{cases} 1, & x > 0, \\ -1, & x < 0, \end{cases}$$

we have

$$f(x) = \begin{cases} \dfrac{1}{\sqrt{1 + \dfrac{1}{x^2}}}, & x > 0, \\[2ex] \dfrac{-1}{\sqrt{1 + \dfrac{1}{x^2}}}, & x < 0. \end{cases}$$

It follows that $\lim_{x \to \infty} f(x) = 1$ and $\lim_{x \to -\infty} f(x) = -1$. The lines $y = 1$ and $y = -1$ are both horizontal asymptotes of $f$; $f(x)$ is near 1 for $x$ positive with large absolute value and $f(x)$ is near $-1$ for negative $x$ with large absolute value. The simplest way to sketch the graph is rising from the asymptote $y = -1$ on the left, through the origin, and rising toward the asymptote $y = 1$ on the right. [We could check that neither of the equations $f(x) = -1$ and $f(x) = 1$ has a solution, so the graph does not touch or cross either asymptote. We might also notice that $\sqrt{x^2 + 1} > |x|$, so $|f(x)| < 1$ for all $x$. This also implies the graph stays strictly between the lines $y = 1$ and $y = -1$.] The graph is sketched in Figure 4.2.3. ■

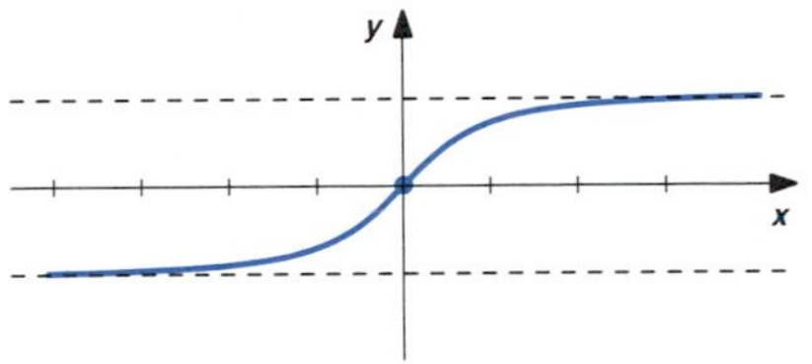


FIGURE 4.2.3

■ **EXAMPLE 6**

Sketch the graph of

$$f(x) = \frac{x^2}{x^2 - 4}.$$

**SOLUTION**

$f(x)$ is defined for all $x$, except for $x = 2$ and $x = -2$. The lines $x = 2$ and $x = -2$ are vertical asymptotes. The function is continuous and positive on the intervals $x < -2$ and $x > 2$. It is continuous and negative for $-2 < x < 2$. $(0, 0)$ is the only intercept. We have $f(-x) = f(x)$, so the function is even and the graph is symmetric with respect to the $y$-axis.

For $|x|$ large, we have

$$f(x) = \frac{x^2}{x^2 - 4} = \frac{x^2}{x^2\left(1 - \dfrac{4}{x^2}\right)} = \frac{1}{1 - \dfrac{4}{x^2}}.$$

It follows that $\lim_{x \to \pm\infty} f(x) = 1$. The line $y = 1$ is a horizontal asymptote of $f$; $f(x)$ is close to 1 for positive and negative $x$ with large absolute value.

To sketch the graph, we plot the intercept, draw the vertical and horizontal asymptotes, determine the behavior near each vertical asymptote and for $|x|$ large, and then sketch the graph in the simplest way that satisfies the conditions. The graph is sketched in Figure 4.2.4. We can

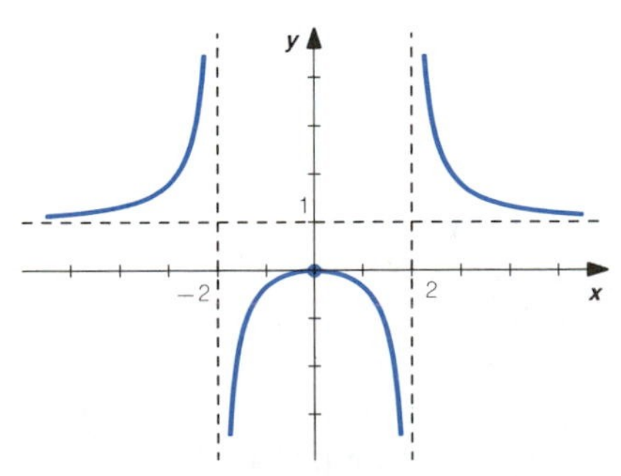


FIGURE 4.2.4

check that the equation $f(x) = 1$ does not have a solution, so this graph does not cross the horizontal asymptote $y = 1$. ■

For many functions $f$ that arise as the solutions of science and engineering problems, it is important to know the behavior of $f(x)$ for $|x|$ large. For example, $f(x)$ might represent the temperature of a particular piece of machinery at time $x$. If $f(x)$ becomes unbounded as $x$ approaches infinity, the machinery will overheat and fail to operate. However, if we knew the rate at which $f(x)$ grows, we could effect design changes in the cooling system to overcome the problem. The study of **asymptotic behavior** allows us to describe and characterize different types of behavior at infinity.

Let us begin by considering the function

$$f(x) = \frac{6x + 3}{2x - 1}.$$

We know that the graph has horizontal asymptote $y = 3$, because

$$\lim_{x \to \pm\infty} \frac{6x + 3}{2x - 1} = \lim_{x \to \pm\infty} \frac{x\left(6 + \dfrac{3}{x}\right)}{x\left(2 - \dfrac{1}{x}\right)} = \lim_{x \to \pm\infty} \frac{6 + \dfrac{3}{x}}{2 - \dfrac{1}{x}} = \frac{6}{2} = 3.$$

To see this from a different point of view, let us carry out the division of $6x + 3$ by $2x - 1$. We obtain

$$\begin{array}{r|l} & \;3 \\ \hline 2x - 1 & 6x + 3 \\ & 6x - 3 \\ \hline & \qquad 6 \end{array}$$

It follows that

$$f(x) = \frac{6x + 3}{2x - 1} = 3 + \frac{6}{2x - 1}.$$

That is, we have $f(x) = A(x) + r(x)$, where $A(x) = 3$ and $r(x) = 6/(2x - 1)$. Since $f(x) - A(x) = r(x)$, and

$$\lim_{x \to \pm\infty} r(x) = \lim_{x \to \pm\infty} \frac{6}{2x - 1} = 0,$$

points $(x, f(x))$ on the graph of $f$ are near corresponding points $(x, A(x))$ on the graph of $A$ whenever $|x|$ is large enough. See Figure 4.2.5.

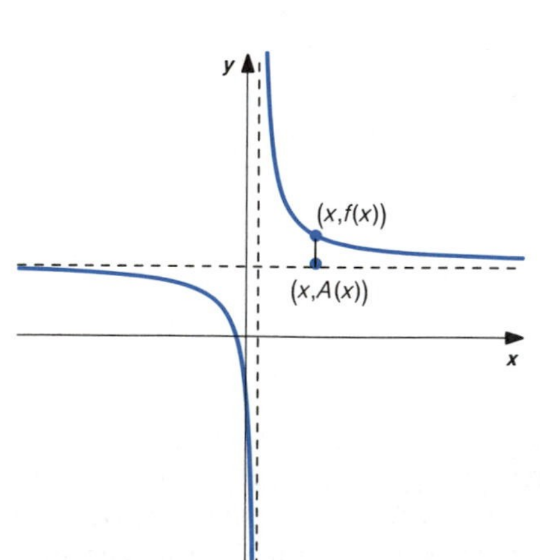

FIGURE 4.2.5

We can use the above ideas to study the behavior of a function $f$ for $|x|$ large, even though $f(x)$ may not have a finite limit as $|x|$ approaches infinity. That is, if $f(x) = A(x) + r(x)$, where $\lim_{x \to \pm\infty} r(x) = 0$, then points $(x, f(x))$ on the graph of $f$ will be close to the points $(x, A(x))$ on the graph of $A$ for $|x|$ large. We can think of the graph of $A$ as an asymptote of $f$. If $A$ is a simple function, this will be useful information. For example, if $A(x) = ax + b$, the graph of $A$ is a line. $A(x) = b$ gives the horizontal asymptote $y = b$. If $a \neq 0$, $A(x) = ax + b$ is called an **oblique asymptote** of $f$.

■ **EXAMPLE 7**

Determine the asymptotic behavior and sketch the graph of

$$f(x) = \frac{(x + 1)(x - 3)}{x + 2}.$$

**SOLUTION**

The function $f$ is defined for all $x$, except for $x = -2$. It is continuous on intervals where it is defined. $f$ has vertical asymptote $x = -2$ and intercepts $(-1, 0)$, $(3, 0)$, and $(0, -3/2)$. The sign of $f(x)$ is determined from the sign chart given in Table 4.2.1. The values $f(x)$ can change sign only at points where it is either zero or undefined.

TABLE 4.2.1

| | $-\infty$ to $-2$ | $-2$ to $-1$ | $-1$ to $3$ | $3$ to $\infty$ |
|---|---|---|---|---|
| $x + 2$ | $-$ | $+$ | $+$ | $+$ |
| $x + 1$ | $-$ | $-$ | $+$ | $+$ |
| $x - 3$ | $-$ | $-$ | $-$ | $+$ |
| $f(x)$ | $-$ | $+$ | $-$ | $+$ |

Let us carry out the division in order to investigate the behavior of $f(x)$ for $|x|$ large. We first expand the numerator to obtain

$$(x + 1)(x - 3) = x^2 - 2x - 3.$$

Then

$$\begin{array}{r|l} & x - 4 \\ x + 2 & x^2 - 2x - 3 \\ & x^2 + 2x \\ \hline & \quad -4x - 3 \\ & \quad -4x - 8 \\ \hline & \qquad\quad 5 \end{array}$$

This gives

$$f(x) = \frac{(x + 1)(x - 3)}{x + 2} = \frac{x^2 - 2x - 3}{x + 2} = x - 4 + \frac{5}{x + 2}.$$

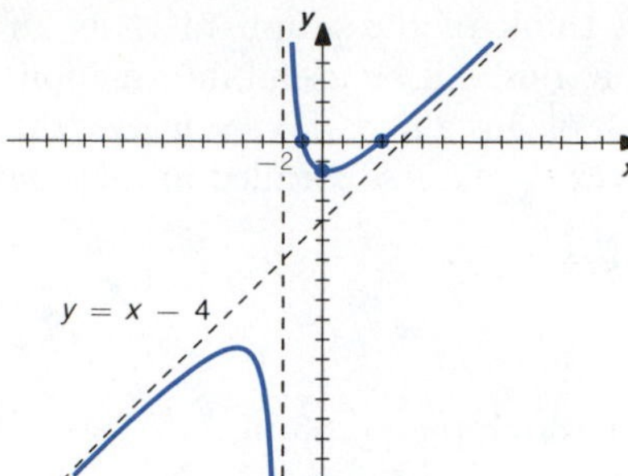

FIGURE 4.2.6

Since

$$\lim_{x \to \pm\infty} \frac{5}{x+2} = 0,$$

we see that points $(x, f(x))$ on the graph of $f$ are near points $(x, x-4)$ on the graph of $A(x) = x - 4$ for $|x|$ large.

The graph is sketched by plotting the intercepts, drawing the vertical asymptote $x = -2$, drawing the oblique asymptote $y = x - 4$, and then sketching the graph in the simplest way that satisfies the conditions we have found. This is done in Figure 4.2.6. Note that

$$f(x) = x - 4 + \frac{5}{x+2} < x - 4 \qquad \text{whenever } x < -2$$

and

$$f(x) = x - 4 + \frac{5}{x+2} > x - 4 \qquad \text{whenever } x > -2.$$

■

■ **EXAMPLE 8**

Determine the asymptotic behavior and sketch the graph of

$$f(x) = \frac{x^3 + 1}{x}.$$

**SOLUTION**

The function is defined for $x \neq 0$ and is continuous on intervals where it is defined. We see $f$ has vertical asymptote $x = 0$ and intercept $(-1, 0)$. We can determine by observation or from a table of signs that $f(x)$ is positive for $x < -1$ and $x > 0$, and that $f(x)$ is negative for $-1 < x < 0$. Writing

$$f(x) = \frac{x^3 + 1}{x} = x^2 + \frac{1}{x},$$

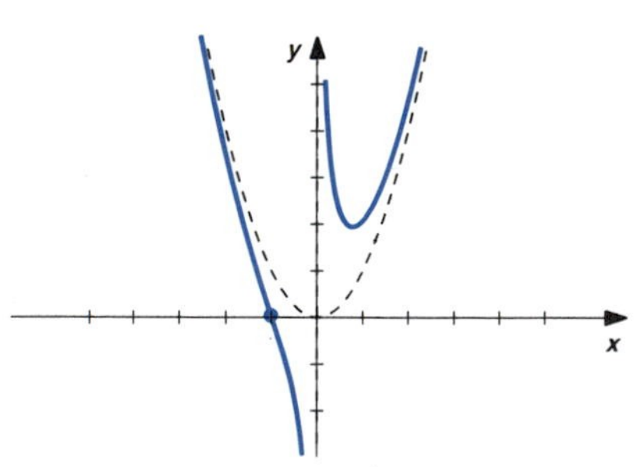

FIGURE 4.2.7

we see that points on the graph of $f$ are near corresponding points on the graph of $A(x) = x^2$ for $|x|$ large. The graph is sketched by plotting the intercept, drawing the vertical asymptote $x = 0$, drawing the asymptotic parabola $y = x^2$, sketching the parts of the graph near the vertical asymptote and for $|x|$ large, and then filling in the graph in the simplest way that satisfies the conditions we have found. This is done in Figure 4.2.7. Note that

$$f(x) = x^2 + \frac{1}{x} < x^2 \qquad \text{whenever } x < 0$$

and

$$f(x) = x^2 + \frac{1}{x} > x^2 \qquad \text{whenever } x > 0.$$

■

FIGURE 4.2.8

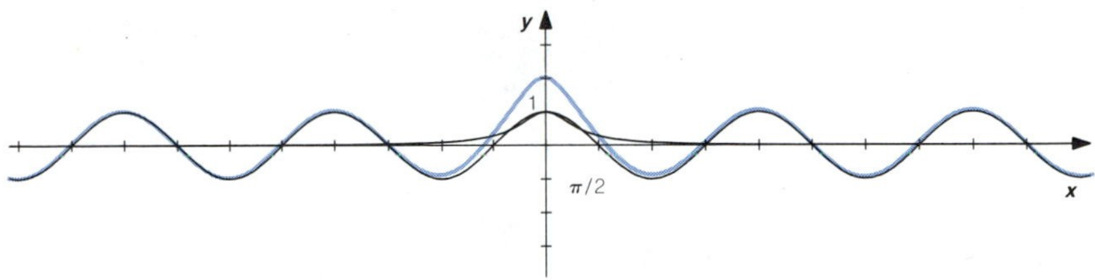


■ EXAMPLE 9

Determine the asymptotic behavior and sketch the graph of

$$i(t) = \frac{1}{t^2 + 1} + \cos t.$$

SOLUTION

The function is defined and continuous for all $t$. The graph can be sketched by adding the values of $y = 1/(t^2 + 1)$ to the graph of $y = \cos t$. This is done in Figure 4.2.8. Note that $1/(t^2 + 1)$ approaches zero as $|t|$ approaches infinity, so the graph of $i(t)$ is asymptotic to $A(t) = \cos t$. Since $f(-x) = f(x)$, the function is even and the graph is symmetric with respect to the $y$-axis. ■

## EXERCISES 4.2

Evaluate the limits in Exercises 1–24.

1 $\lim_{x \to \infty} \dfrac{x^2 + x}{x^3 - x^2}$

2 $\lim_{x \to \infty} \dfrac{(3x - 1)^3}{x^4 + 1}$

3 $\lim_{x \to -\infty} \dfrac{x^2 - 3x + 1}{x^2 + 27}$

4 $\lim_{x \to -\infty} \dfrac{3 - 2x - x^2}{1 - x - 2x^2}$

5 $\lim_{x \to \infty} \dfrac{(x^2 + 1)(x - 3)}{x^3 - 8}$

6 $\lim_{x \to \infty} \dfrac{x^2 - 4x}{(2x + 1)(3x + 1)}$

7 $\lim_{x \to \infty} \sqrt{\dfrac{x^2}{1 + 4x^2}}$

8 $\lim_{x \to -\infty} \sqrt{\dfrac{x}{9x + 4}}$

9 $\lim_{x \to \infty} \dfrac{x^2}{x - 1}$

10 $\lim_{x \to \infty} \dfrac{x^2}{1 - x}$

11 $\lim_{x \to -\infty} \dfrac{x^3}{1 - x}$

12 $\lim_{x \to \infty} \dfrac{1 - x}{1 - x^3}$

13 $\lim_{x \to -\infty} \dfrac{x}{\sqrt{4x^2 + 1}}$

14 $\lim_{x \to \infty} \dfrac{x}{\sqrt{9x^2 + 1}}$

15 $\lim_{x \to \infty} \dfrac{x\sqrt{4x + 1}}{(1 + \sqrt{x})^3}$

16 $\lim_{x \to -\infty} \dfrac{x\sqrt{1 - 4x}}{\sqrt{1 - 9x^3}}$

17 $\lim_{x \to \infty} (x^3 - 4x^2)$

18 $\lim_{x \to \infty} (x - 2x^2)$

19 $\lim_{x \to -\infty} (x - x^2 - x^3 - 4x^4)$

20 $\lim_{x \to -\infty} (2x^5 - 3x^4 + x^3)$

21 $\lim_{x \to \infty} (x^{3/5} - x^{3/4})$

22 $\lim_{x \to \infty} (x - \sqrt{x})$

23 $\lim_{x \to \infty} (\sqrt{x^2 + 4} - x)$

$\left(\textit{Hint}: \text{Multiply by } \dfrac{\sqrt{x^2 + 4} + x}{\sqrt{x^2 + 4} + x}.\right)$

24 $\lim_{x \to \infty} (\sqrt{4x^2 + 1} - 2x)$

$\left(\textit{Hint}: \text{Multiply by } \dfrac{\sqrt{4x^2 + 1} + 2x}{\sqrt{4x^2 + 1} + 2x}.\right)$

In Exercises 25–44 sketch the graphs in the simplest way that reflects the domain, intercepts, sign of $f(x)$, continuity, and behavior near vertical and horizontal asymptotes.

25 $f(x) = \dfrac{x}{x - 1}$

26 $f(x) = \dfrac{x}{x + 1}$

27 $f(x) = \dfrac{x^2}{x^2 - 1}$

28 $f(x) = \dfrac{(x + 1)^2}{x^2 + 2x}$

29 $f(x) = \dfrac{x}{x^2 + x + 1}$

30 $f(x) = \dfrac{x}{x^2 + 1}$

31 $f(x) = \dfrac{x-2}{x^2-4x}$

32 $f(x) = \dfrac{x}{x^2-1}$

33 $f(x) = \dfrac{x^2-4}{x^2-1}$

34 $f(x) = \dfrac{x^2-1}{x^2-4}$

35 $f(x) = \dfrac{x^2-1}{x^3}$

36 $f(x) = \dfrac{x^2-1}{x^4}$

37 $f(x) = \dfrac{1}{\sqrt{x^2+1}}$

38 $f(x) = \dfrac{1}{x^2+1}$

39 $f(x) = \dfrac{|x|}{\sqrt{x^2+1}}$

40 $f(x) = \dfrac{6x}{\sqrt{4x^2+9}}$

41 $f(x) = \dfrac{\sin x}{x^2}$

42 $f(x) = \dfrac{\cos x}{x}$

43 $f(x) = \dfrac{\cos^2 x}{x^2}$

44 $f(x) = \dfrac{\sin^2 x}{x^2}$

Write a formal definition for the statements in Exercises 45–48.

45 $\lim_{x\to\infty} f(x) = \infty$

46 $\lim_{x\to\infty} f(x) = -\infty$

47 $\lim_{x\to-\infty} f(x) = -\infty$

48 $\lim_{x\to-\infty} f(x) = \infty$

49 Use the definition to show that

$$\lim_{x\to\infty} \frac{1}{2x-1} = 0.$$

50 Use the definition to show that

$$\lim_{x\to\infty} \frac{x}{x+1} = 1.$$

51 If $\lim_{x\to\infty} f(x) = L$ and $\lim_{x\to\infty} g(x) = M$, where $L$ and $M$ are real numbers, use the definition to show that

$$\lim_{x\to\infty} [f(x) + g(x)] = L + M.$$

52 Evaluate

$$\lim_{x\to\infty} (\sqrt{x^2+4x} - \sqrt{x^2+1}).$$

In Exercises 53–70 determine the asymptotic behavior and sketch the graphs.

53 $f(x) = x + \dfrac{1}{x}$

54 $f(x) = x - \dfrac{1}{x}$

55 $f(x) = \dfrac{1-4x-x^2}{x+2}$

56 $f(x) = \dfrac{1-2x-x^2}{x+1}$

57 $f(x) = \dfrac{x^2-2x}{x-1}$

58 $f(x) = \dfrac{x^2+4x+5}{x+2}$

59 $f(x) = \dfrac{x^3-1}{x}$

60 $f(x) = \dfrac{x^3-x^2-2x}{x-1}$

61 $f(x) = \dfrac{x^3+8}{x^2}$

62 $f(x) = \dfrac{x^3-1}{x^2}$

63 $f(x) = \dfrac{x^4+1}{x}$

64 $f(x) = \dfrac{x^4-1}{x}$

65 $f(x) = \dfrac{x^4-1}{x^2}$

66 $f(x) = \dfrac{x^4+1}{x^2}$

67 $f(x) = \dfrac{x^4+x^2+4}{x^2}$

68 $f(x) = \dfrac{x^4+x^2-2}{x^2}$

69 $i(t) = \dfrac{t^2}{t^2+1} + \cos t$

70 $i(t) = \dfrac{1}{t} + \sin t$

## 4.3 SIGN OF $f'(x)$, EXTREMA

An important general characteristic of a function involves the rising or falling of the graph as $x$ moves in the positive direction over an interval. Informally, we say $f$ is **increasing** on an interval if the graph rises over the interval, as indicated in Figure 4.3.1a. We say $f$ is **decreasing** if the graph falls over the interval, as illustrated in Figure 4.3.1b. The following formal definition reflects these ideas.

■ *Definition*

A function $f$ is **increasing** on an interval $I$ if $f(x_1) < f(x_2)$ whenever $x_1$ and $x_2$ are in $I$ and $x_1 < x_2$; $f$ is **decreasing** on $I$ if $f(x_1) > f(x_2)$ whenever $x_1$ and $x_2$ are in $I$ and $x_1 < x_2$. ■

FIGURE 4.3.1

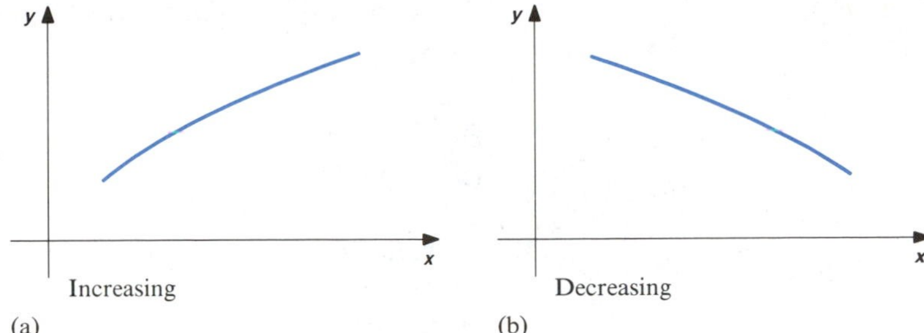


Recall that the slope of the line tangent to the graph of $f$ at $(c, f(c))$ is given by the value of the derivative $f'(c)$. The tangent line is increasing if $f'(c) > 0$ and decreasing if $f'(c) < 0$. The following theorem relates the sign of $f'(x)$ to intervals where the graph of $f$ is either increasing or decreasing.

*Theorem 1*

**If $f'(x) > 0$ for all $x$ in an interval $I$, then $f$ is increasing on $I$. If $f'(x) < 0$ for all $x$ in $I$, then $f$ is decreasing on $I$.**

■ *Proof* Let us verify the first statement of the theorem. Verification of the second statement is similar.

We assume $f'(x) > 0$ for all $x$ in an interval $I$. We must show that $f$ satisfies the condition of the definition of increasing on $I$.

If $x_1$ and $x_2$ are in $I$, then the Mean Value Theorem tells us there is a point $c$ between $x_1$ and $x_2$ such that

$$\frac{f(x_2) - f(x_1)}{x_2 - x_1} = f'(c), \quad \text{or}$$

$$f(x_2) - f(x_1) = f'(c)(x_2 - x_1).$$

Since $c$ is in $I$, we have $f'(c) > 0$. Then $x_1 < x_2$ implies that the product $f'(c)(x_2 - x_1)$ is positive. That is,

$$f(x_2) - f(x_1) = f'(c)(x_2 - x_1) > 0,$$

so $f(x_2) > f(x_1)$ whenever $x_1$ and $x_2$ are in $I$ and $x_1 < x_2$. The condition of the definition is satisfied, so $f$ is increasing on $I$. ■

We should sketch the graph of $f$ increasing on intervals where $f'(x)$ is positive; the graph should be sketched decreasing on intervals where $f'(x)$ is negative.

We will use the fact that $f'(x)$ is of constant sign on the intervals between successive points where either $f'(x) = 0$ or $f'(x)$ does not exist. In some cases the sign of $f'(x)$ in an interval is easily determined by observation of the formula for $f'(x)$. In more complicated examples, the formula for $f'(x)$ should be written in factored form and a table of signs constructed.

Recall from Section 3.9 that values of $x$ in the domain of $f$ for which either

**(i)** $f'(x) = 0$ or

**(ii)** $f'(x)$ does not exist

are called **critical points of *f***. From Theorem 1 of Section 3.9 we see that if $f(c)$ is either a local maximum or local minimum of a function $f$, then $c$ is a critical point of $f$. However, the fact that $c$ is a critical point of $f$ does not guarantee that $f(c)$ is either a local maximum or a local minimum. We will use a sketch of the graph that reflects the sign of $f'(x)$ to determine which critical points correspond to local extrema.

Points on the graph of a function $f$ that correspond to critical points of $f$ should be plotted. We should draw horizontal tangents through points on the graph where $f'(x) = 0$. If $f$ is continuous at $c$ and $|f'(x)|$ approaches infinity as $x$ approaches $c$, the graph of $f$ becomes vertical at the point $(c, f(c))$; we should draw vertical tangents through such points. These horizontal and vertical tangents are then used as guides for sketching the graph.

Let us see how to sketch graphs in a way that reflects the sign of $f'(x)$, as well as the domain, intercepts, sign of $f(x)$, continuity, and asymptotes.

■ **EXAMPLE 1**

Sketch the graph of $f(x) = 1 + 3x - x^3$. Find local extrema.

**SOLUTION**

Let us begin by investigating key points and general characteristics that are directly related to the values of the function $f(x)$. The function is defined and continuous for all $x$. We see that the $y$-intercept is $(0, 1)$, but the $x$-intercepts are not easily determined from the equation $f(x) = 0$. Hence, it is not convenient to determine intervals where $f(x)$ is of constant sign. In order to investigate the behavior of $f(x)$ for $|x|$ large, we write

$$f(x) = 1 + 3x - x^3 = x^3\left(\frac{1}{x^3} + \frac{3}{x^2} - 1\right).$$

The second factor approaches $-1$ as $|x|$ approaches infinity. Since

$$\lim_{x\to-\infty} x^3 = -\infty \quad \text{and} \quad \lim_{x\to\infty} x^3 = \infty,$$

it follows that

$$\lim_{x\to-\infty} f(x) = \infty \quad \text{and} \quad \lim_{x\to\infty} f(x) = -\infty.$$

Let us now consider information related to the derivative $f'(x)$. We have

$f(x) = 1 + 3x - x^3$, so

TABLE 4.3.1

| | $-\infty$ to $-1$ | $-1$ to $1$ | $1$ to $\infty$ |
|---|---|---|---|
| $3(1-x)$ | + | + | − |
| $1+x$ | − | + | + |
| $f'(x)$ | − | + | − |

$$f'(x) = 3 - 3x^2$$
$$= 3(1 - x^2)$$
$$= 3(1 - x)(1 + x).$$

From the factored form of $f'(x)$ we see that $f'(x) = 0$ when $x = 1$ or $x = -1$. The corresponding values of $y$ are

$$f(1) = 1 + 3(1) - (1)^3 = 3 \quad \text{and}$$

$$f(-1) = 1 + 3(-1) - (-1)^3 = -1.$$

The graph has horizontal tangents at $(1, 3)$ and $(-1, -1)$.

The sign of $f'(x)$ can be determined either by direct observation or from the table of signs given in Table 4.3.1. We see that $f'(x)$ is positive on the interval $-1 < x < 1$, so $f$ is increasing on this interval; $f'(x)$ is negative on the intervals $x < -1$ and $x > 1$, so $f$ is decreasing on these intervals.

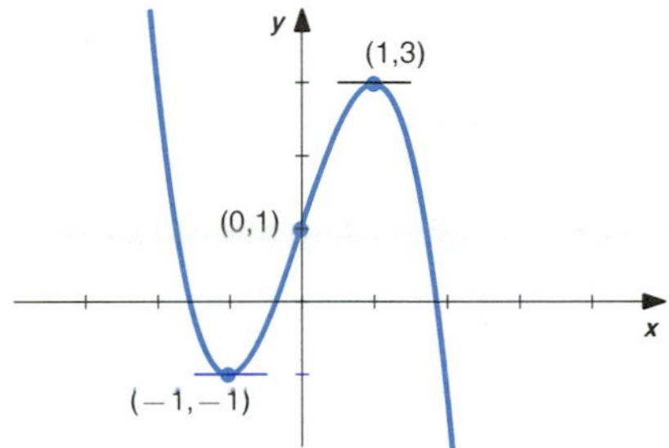

FIGURE 4.3.2

The graph in Figure 4.3.2 reflects the key points and general characteristics that we have found. The graph decreases to the point $(-1, -1)$, where it levels off and begins increasing. It increases through the $y$-intercept at $(0, 1)$ to the point $(1, 3)$, where it levels off and begins decreasing. From the graph we see that there are three $x$-intercepts, one to the left of $-1$, one between $-1$ and 0, and one to the right of 1. We also see that the critical point $x = -1$ gives a local minimum and the critical point $x = 1$ gives a local maximum. ■

■ **EXAMPLE 2**

Sketch the graph of

$$f(x) = \frac{1}{9}(x^4 - 4x^3).$$

Find local extrema.

**SOLUTION**

The function is defined and continuous for all $x$. Factoring $f(x)$, we have

$$f(x) = \frac{x^3}{9}(x - 4).$$

This shows the intercepts are $(0, 0)$ and $(4, 0)$. A sign chart for the function $f(x)$ can be used to see that $f(x) > 0$ for $x < 0$ and $x > 4$, and that $f(x) < 0$ for $0 < x < 4$. Since

$$f(x) = \frac{x^3}{9}(x - 4) = \frac{x^4}{9}\left(1 - \frac{4}{x}\right),$$

we see that $f(x) \to \infty$ as $x \to \pm\infty$.

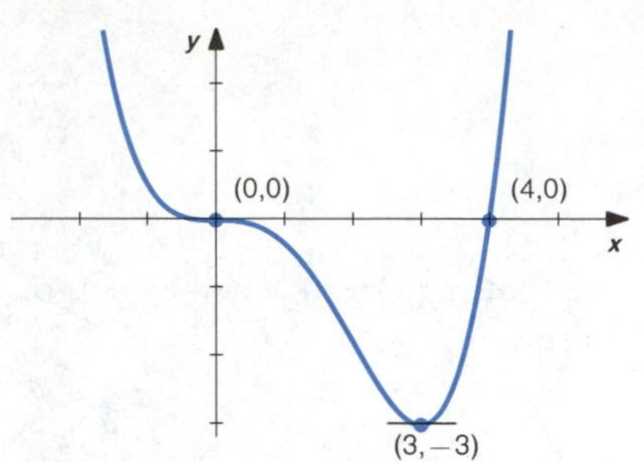


FIGURE 4.3.3

We could now sketch the graph of $f$ in the simplest way that reflects the properties determined above. The result would be similar to the actual graph given in Figure 4.3.3, although we would probably not sketch the "wiggle" at the origin. We would not know for certain on which intervals the graph is *actually* increasing or decreasing, and we would not know the exact location of the local minimum indicated by the sketch. These characteristics are related to the sign of $f'(x)$.

The derivative of

$$f(x) = \frac{1}{9}(x^4 - 4x^3)$$

is

$$f'(x) = \frac{1}{9}(4x^3 - 12x^2)$$

$$= \frac{4x^2}{9}(x - 3).$$

We see that $f'(x) = 0$ when $x = 0$ or $x = 3$. The graph has horizontal tangents at $(0, 0)$ and $(3, -3)$. A table of signs for $f'(x)$ is given in Table 4.3.2.

TABLE 4.3.2

| | $-\infty$ to 0 | 0 to 3 | 3 to $\infty$ |
|---|---|---|---|
| $\frac{4x^2}{9}$ | + | + | + |
| $x - 3$ | − | − | + |
| $f'(x)$ | − | − | + |

Theorem 1 tells us that $f$ is increasing on the interval $x > 3$ and decreasing on each of the intervals $x < 0$ and $0 < x < 3$.

To sketch the graph, we plot the points $(0, 0)$, $(4, 0)$, and $(3, -3)$, and draw horizontal tangents through the points $(0, 0)$ and $(3, -3)$. The graph is then sketched through these points in a way that reflects the sign of $f'(x)$. This is done in Figure 4.3.3. The graph decreases to the point $(0, 0)$, levels off, and continues decreasing to the point $(3, -3)$. Hence, the graph is decreasing on the interval $x < 3$. It levels off at $(3, -3)$ and then increases through the point $(4, 0)$.

From the graph we see that the critical point $x = 3$ gives a local minimum. Also, the value of $f$ at the critical point $x = 0$ is neither a local maximum nor a local minimum. ■

■ **EXAMPLE 3**

Sketch the graph of

$$f(x) = \frac{x}{4 - x^2}.$$

Find local extrema.

**SOLUTION**

The function is defined for all $x$, except $x = 2$ and $x = -2$. It is continuous on intervals where it is defined. The only intercept is $(0, 0)$. The graph has vertical asymptotes $x = 2$ and $x = -2$. We have

$$\lim_{x \to \pm\infty} \frac{x}{4 - x^2} = 0,$$

FIGURE 4.3.4

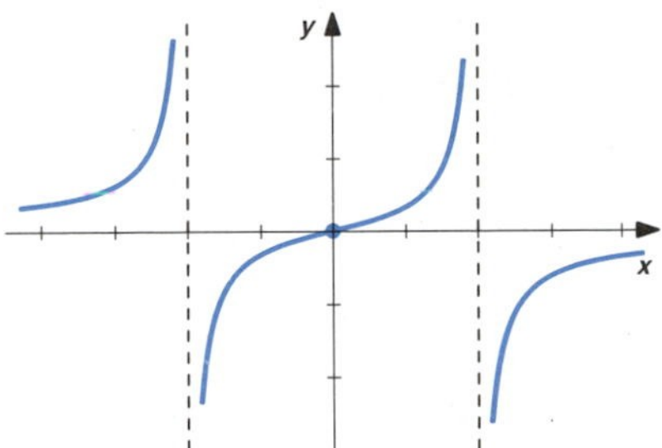


so the graph has horizontal asymptote $y = 0$. A table of signs can be used to determine that $f(x)$ is positive on the intervals $x < -2$ and $0 < x < 2$, and that $f(x)$ is negative on the intervals $-2 < x < 0$ and $x > 2$. We may also notice that $f(-x) = -f(x)$, so the function is odd and the graph is symmetric with respect to the origin.

We could now sketch the graph in the simplest way that reflects the above information. The result would probably look like the graph in Figure 4.3.4. However, we would not be certain about the lack of "wiggles" and local extrema unless we investigate the derivative $f'(x)$:

$$f(x) = \frac{x}{4 - x^2},$$

so the Quotient Rule gives

$$f'(x) = \frac{(4 - x^2)(1) - (x)(-2x)}{(4 - x^2)^2}.$$

Simplifying, we have

$$f'(x) = \frac{4 - x^2 + 2x^2}{(4 - x^2)^2} = \frac{x^2 + 4}{(4 - x^2)^2}.$$

There are no critical points. [$f'(x)$ does not exist for $x = 2$ and $x = -2$, but these points are not in the domain of $f$.] There are no local extrema. We see that $f'(x) > 0$ at every point at which it exists. Theorem 1 tells us $f$ is increasing on each of the intervals $x < -2$, $-2 < x < 2$, and $x > 2$.

We can sketch the graph by plotting the intercept (0, 0), drawing the vertical asymptotes $x = -2$ and $x = 2$, drawing the horizontal asymptote $y = 0$, and then sketching the graph increasing on each of the intervals $x < -2$, $-2 < x < 2$, and $x > 2$. This is done in Figure 4.3.4. We see that $f$ is *not* increasing on the interval $-\infty < x < \infty$. ■

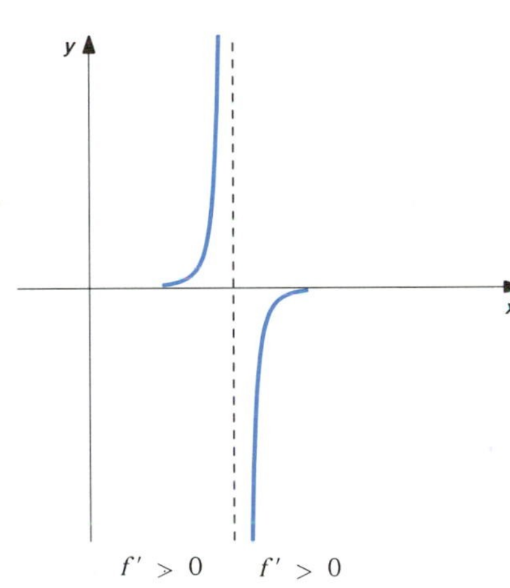


(a)

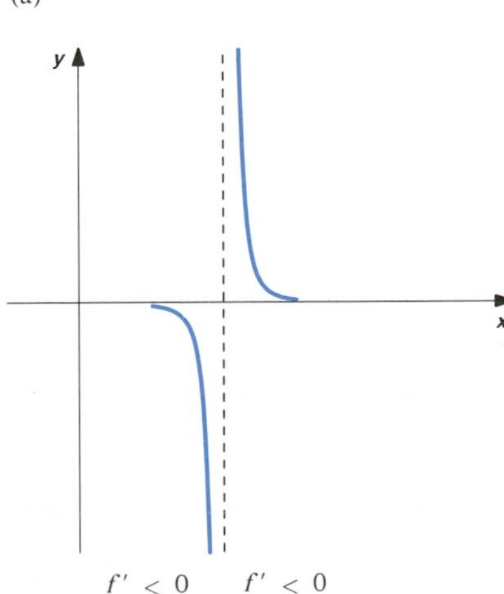


(b)

FIGURE 4.3.5

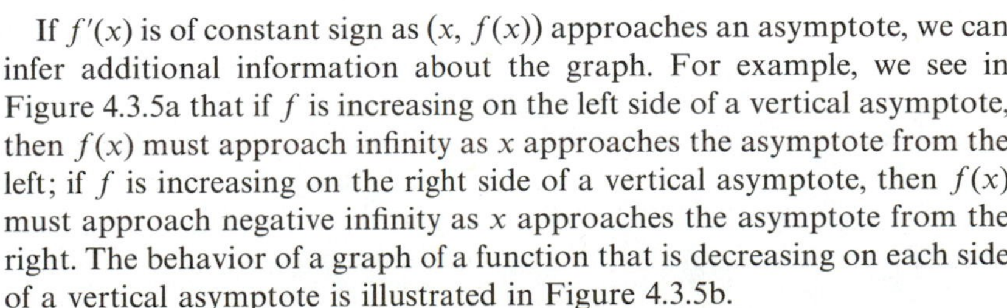

If $f'(x)$ is of constant sign as $(x, f(x))$ approaches an asymptote, we can infer additional information about the graph. For example, we see in Figure 4.3.5a that if $f$ is increasing on the left side of a vertical asymptote, then $f(x)$ must approach infinity as $x$ approaches the asymptote from the left; if $f$ is increasing on the right side of a vertical asymptote, then $f(x)$ must approach negative infinity as $x$ approaches the asymptote from the right. The behavior of a graph of a function that is decreasing on each side of a vertical asymptote is illustrated in Figure 4.3.5b.

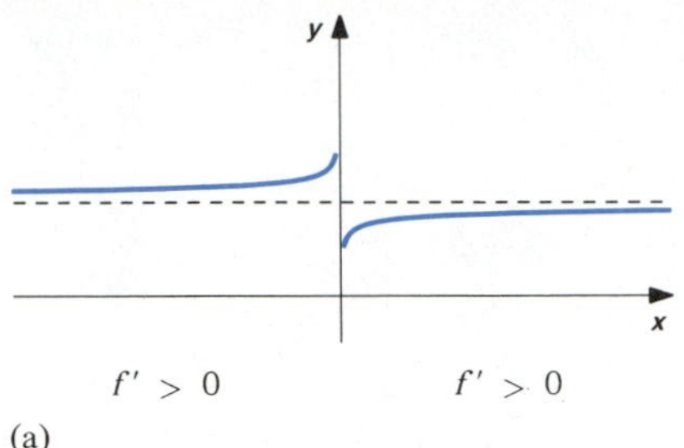


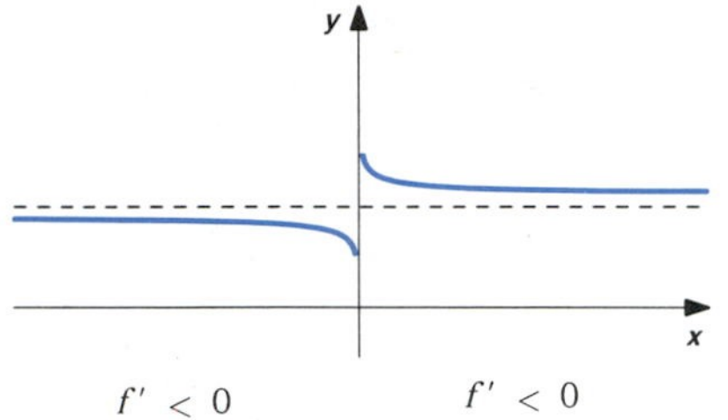


FIGURE 4.3.6

If $f'(x)$ is of constant sign as $x$ approaches either $\infty$ or $-\infty$, and $(x, f(x))$ approaches a horizontal asymptote, the graph must approach the asymptote from only one side of the asymptote. The possibilities are illustrated in Figure 4.3.6. A graph does not cross a horizontal asymptote in infinite intervals where $f'(x)$ is of constant sign. Of course, the graph can cross a horizontal asymptote any number of times if the sign of $f'(x)$ changes.

**EXAMPLE 4**

Sketch the graph of $f(x) = x\sqrt{x+1}$. Find local extrema.

**SOLUTION**

$f(x)$ is defined only for $x + 1 \geq 0$, or $x \geq -1$; it is continuous on the interval $x \geq -1$. The intercepts are $(0, 0)$ and $(-1, 0)$. It is not difficult to check that $f(x)$ is negative for $-1 < x < 0$ and positive for $x > 0$.

The Product Rule is used to find the derivative. Also, the Chain Rule is needed to differentiate the factor that is the square root of the function $x + 1$. We have

$$f(x) = x\sqrt{x+1},$$

$$f'(x) = (x)\left(\frac{1}{2\sqrt{x+1}}\right)(1) + (\sqrt{x+1})(1).$$

To determine the critical points, we simplify $f'(x)$ to obtain

$$f'(x) = \frac{x}{2\sqrt{x+1}} + \sqrt{x+1}\left(\frac{2\sqrt{x+1}}{2\sqrt{x+1}}\right)$$

$$= \frac{x + 2(x+1)}{2\sqrt{x+1}} = \frac{3x+2}{2\sqrt{x+1}}.$$

We see that $f'(x) = 0$ when $x = -2/3$. We have

$$f\left(-\frac{2}{3}\right) = \left(-\frac{2}{3}\right)\sqrt{-\frac{2}{3}+1} = -\left(\frac{2}{3}\right)\sqrt{\frac{1}{3}} = -\frac{2}{3^{3/2}} \approx -0.38.$$

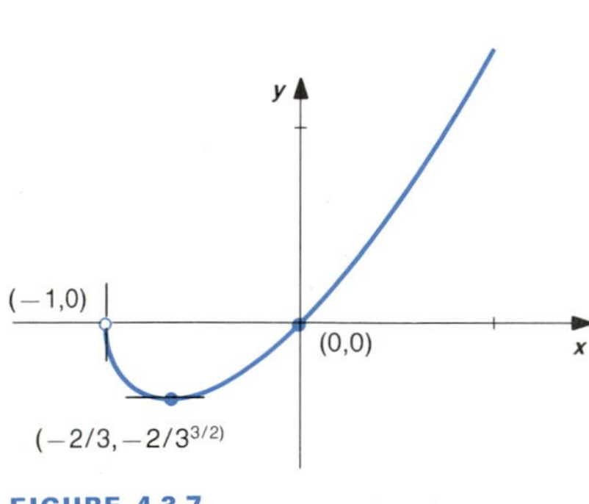


FIGURE 4.3.7

The graph has a horizontal tangent at $(-2/3, -2/3^{3/2})$. We see that $|f'(x)|$ approaches infinity as $x$ approaches $-1$ from the right, so the graph has a vertical tangent at $(-1, 0)$. We have $f'(x) < 0$ for $-1 < x < -2/3$, so the graph is decreasing on this interval. We have $f'(x) > 0$ for $x > -2/3$, so the graph is increasing for $x > -2/3$.

To sketch the graph, we plot the points $(-1, 0)$, $(0, 0)$, and $(-2/3, -2/3^{3/2})$, draw a horizontal tangent through $(-2/3, -2/3^{3/2})$ and a vertical tangent through $(-1, 0)$, and then sketch the graph in a way that reflects the sign of $f'(x)$. This is done in Figure 4.3.7. We see that $f$ has a local minimum at $(-2/3, -2/3^{3/2})$. ■

## EXERCISES 4.3

Sketch the graphs and find local extrema in Exercises 1–56.

1 $f(x) = x^{4/3}$ 2 $f(x) = x^{2/3}$

3 $f(x) = x^3$ 4 $f(x) = x^{1/3}$

5 $f(x) = (10 + 6x - x^2)/6$

6 $f(x) = (2x^2 + 8x - 11)/8$

7 $f(x) = (x^3 - 12x)/8$ 8 $f(x) = 3(x^3 + 2x^2 + x)$

9 $f(x) = 2 + 3x - x^3$ 10 $f(x) = x^3 - 3x^2 + 2$

11 $f(x) = x^3 + x$ 12 $f(x) = 2 - x - x^3$

13 $f(x) = 4x^2 - x^4$ 14 $f(x) = x^4 + x^2$

15 $f(x) = 3x^4 - 4x^3 + 6x^2$ 16 $f(x) = 3x^4 + 4x^3$

17 $f(x) = (12x^5 - 45x^4 + 40x^3)/8$

18 $f(x) = 4x^5 - 5x^4$ 19 $f(x) = x^{7/3} - 7x^{1/3}$

20 $f(x) = (16x^{2/3} - x^{8/3})/16$

21 $f(x) = (x^2 - 1)^{3/5}$ 22 $f(x) = (x^2 - 1)^{2/5}$

23 $f(x) = \sqrt{x}(x - 3)$ 24 $f(x) = x\sqrt{3 - x}$

25 $f(x) = \dfrac{8x}{x^2 + 4}$ 26 $f(x) = \dfrac{1}{x^2 + 1}$

27 $f(x) = \dfrac{x^2}{x^2 + 9}$ 28 $f(x) = \dfrac{4x^3}{x^4 + 3}$

29 $f(x) = \dfrac{6x^{1/3}}{x^2 + 5}$ 30 $f(x) = \dfrac{12x^{2/3}}{x^2 + 8}$

31 $f(x) = x + \dfrac{4}{x}$ 32 $f(x) = x + \dfrac{4}{x^2}$

33 $f(x) = x^2 + \dfrac{1}{x^2}$ 34 $f(x) = x^3 + \dfrac{3}{x}$

35 $f(x) = \dfrac{1}{x^2 - 1}$ 36 $f(x) = \dfrac{x}{x^2 - 1}$

37 $f(x) = \dfrac{8x - 20}{x^2 - 4}$ 38 $f(x) = \dfrac{x^2 - 1}{x^2}$

39 $f(x) = \dfrac{(x^2 - 1)^2}{x^4}$ 40 $f(x) = 4\left(\dfrac{x^2 - 1}{x^4}\right)$

41 $f(x) = x + \sin x,\ 0 \le x \le 2\pi$

42 $f(x) = x + \cos x - 1,\ 0 \le x \le 2\pi$

43 $f(x) = x + \dfrac{1}{2}\sin 2x,\ 0 \le x \le 2\pi$

44 $f(x) = x + 2\sin x,\ 0 \le x \le 2\pi$

45 $f(x) = \sin x + \cos x,\ 0 \le x \le 2\pi$

46 $f(x) = \cos x - \sin x,\ 0 \le x \le 2\pi$

47 $f(x) = \tan x - x,\ -\pi/2 < x < \pi/2$

48 $f(x) = \cot x + x,\ 0 < x < \pi$

49 $f(x) = \tan x - 2\sec x,\ -\pi/2 < x < \pi/2$

50 $f(x) = \tan x + 2\sec x,\ -\pi/2 < x < \pi/2$

51 $f(x) = \dfrac{1}{\sqrt{1 - x^2}}$ 52 $f(x) = \dfrac{1}{\sqrt{2x - x^2}}$

53 $f(x) = \sqrt{4x - x^2}$ 54 $f(x) = -\sqrt{9 - x^2}$

55 $f(x) = \sqrt{x^2 - 1}$ 56 $f(x) = \dfrac{1}{\sqrt{x^2 - 1}}$

In Exercises 57–60 the position of an object along a number line at time $t$ is given by $s(t)$. Find intervals of time for which the object is moving in the positive direction.

57 $s(t) = -16t^2 + 64t + 64,\ t \ge 0$

58 $s(t) = -16t^2 + 128t + 32,\ t \ge 0$

59 $s(t) = \sin t,\ 0 \le t \le 2\pi$

60 $s(t) = \sin 2t,\ 0 \le t \le 2\pi$

61 If $f$ is continuous at $c$ and $\lim_{x\to c^+} f'(x) = \infty$, use the Mean Value Theorem to show that

$$\lim_{x\to c^+} \frac{f(x) - f(c)}{x - c} = \infty.$$

62 If

$$f(x) = \begin{cases} x^2 \sin \dfrac{1}{x^2}, & x \ne 0, \\ 0, & x = 0, \end{cases}$$

show that $f'(x)$ becomes unbounded as $x$ approaches zero, but $f'(0) = 0$.

63 Show that the graph of the function $f(x) = x\sqrt{x + 1}$ of Example 4 approaches the asymptote

$$A(x) = \left(x + \frac{1}{3}\right)^{3/2}$$

as $x$ approaches infinity.

64 Use the Mean Value Theorem to prove the **First Derivative Test for Local Extrema.**

(a) If $f'(c) = 0$, $f'(x) > 0$ for $c - \delta < x < c$, and $f'(x) < 0$ for $c < x < c + \delta$, then $f$ has a local maximum at $(c, f(c))$.

(b) If $f'(c) = 0$, $f'(x) < 0$ for $c - \delta < x < c$, and $f'(x) > 0$ for $c < x < c + \delta$, then $f$ has a local minimum at $(c, f(c))$.

## 4.4 SIGN OF $f''(x)$, CONCAVITY

We continue the study of functions by introducing another general characteristic, that of concavity. Concavity involves the rate of change of the increase or decrease of a function.

In simple terms, a graph is said to be **concave up** if it opens upward. It is said to be **concave down** if it opens downward. The idea of concavity may be expressed more precisely by considering chords drawn between two points on the graph.

*Definition*

The graph of a function $f(x)$ is **concave up** on an interval $I$ if it is below or touching each chord between points on the graph with $x$ in $I$, as indicated in Figure 4.4.1a; it is **concave down** on $I$ if it is above or touching each chord, as in Figure 4.4.1b. ■

The graph of a linear function $f(x) = ax + b$ is the same as the chord between any two points on the line, so linear functions are both concave up and concave down.

To use the concept of concavity in our graph sketching, we will characterize the idea in terms of calculus. We begin with the following result.

FIGURE 4.4.1

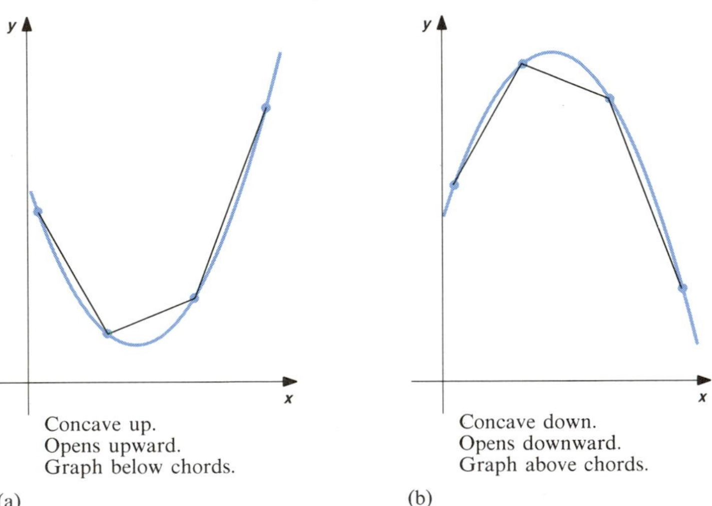


*Lemma 1*

If $f'$ is increasing, then the graph of $f$ is above lines that are tangent to the graph, as indicated in Figure 4.4.2a; if $f'$ is decreasing, the graph is below tangent lines, as indicated in Figure 4.4.2b.

■ *Proof* This result can be established by comparing the slope of the line tangent to the graph of $f$ at $(c, f(c))$ and the slope of the chord between the

FIGURE 4.4.2

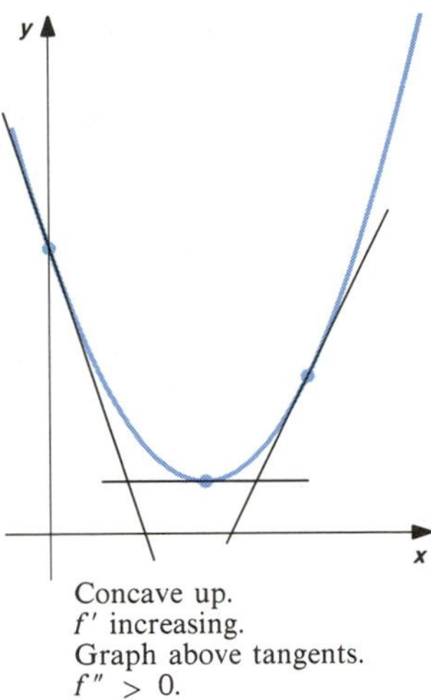


Concave up.
$f'$ increasing.
Graph above tangents.
$f'' > 0$.

(a)

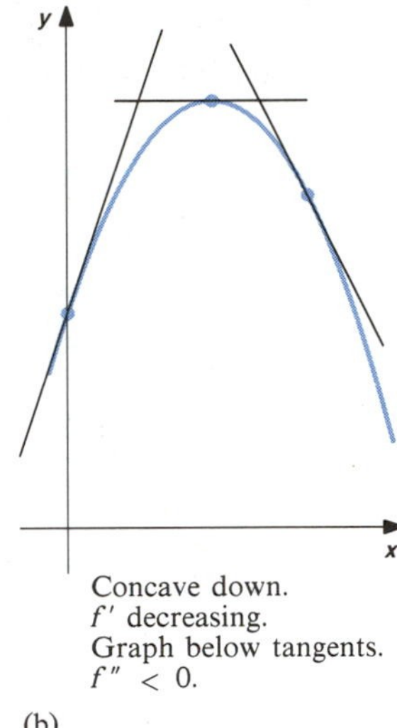


Concave down.
$f'$ decreasing.
Graph below tangents.
$f'' < 0$.

(b)

points $(c, f(c))$ and $(x, f(x))$, $x \neq c$. The tangent line has slope $f'(c)$ and the chord has slope

$$\frac{f(x) - f(c)}{x - c}.$$

We will use the fact that the Mean Value Theorem implies there is some $u$ between $c$ and $x$ such that

$$\frac{f(x) - f(c)}{x - c} = f'(u),$$

so the slope of the chord is $f'(u)$.

The case $c < x$ is illustrated in Figure 4.4.3. Since $u$ is between $c$ and $x$, $c < x$ implies $c < u < x$. Since $f'$ is increasing, $c < u$ implies $f'(c) < f'(u)$. We conclude that the slope of the chord is greater than the slope of the tangent line. The point $(x, f(x))$ is then above the tangent line, as asserted for $f'$ increasing in Lemma 1. The cases for $x < c$ and $f'$ decreasing are similar. ■

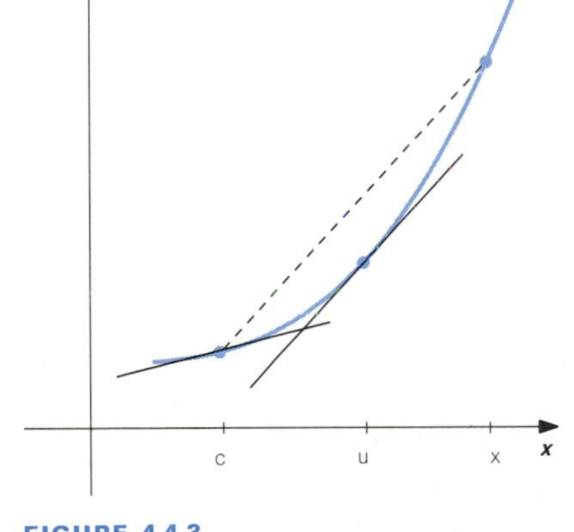


FIGURE 4.4.3

Intuitively, an increasing derivative $f'$ bends the graph upward from its tangent lines and the graph is above its tangent lines. A decreasing derivative $f'$ bends the graph downward, below its tangent lines.

We can use Lemma 1 to obtain the following.

■ *Lemma 2*

The graph of a function $f$ is concave up on intervals where the derivative $f'$ is increasing. It is concave down on intervals where $f'$ is decreasing.

■ *Proof* Let us verify the case where $f'$ is increasing. The other case is similar. We will show that, if $f'$ is increasing and $x_1 < c < x_2$, then $(c, f(c))$

FIGURE 4.4.4

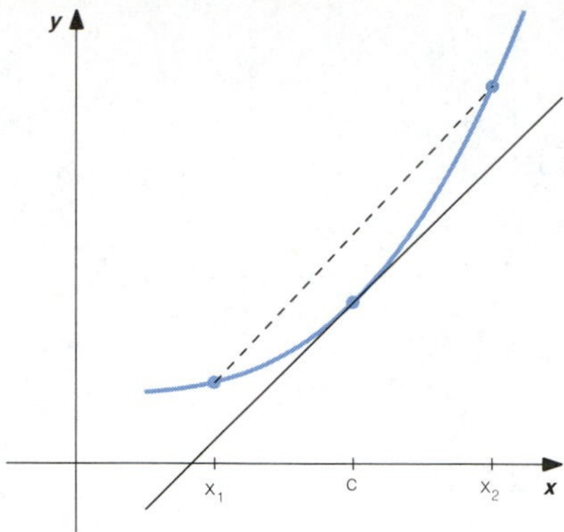


is below the chord between $(x_1, f(x_1))$ and $(x_2, f(x_2))$. From Lemma 1, we know that both points $(x_1, f(x_1))$ and $(x_2, f(x_2))$ must be above the tangent line at $(c, f(c))$. This is only possible if $(c, f(c))$ is below the chord. See Figure 4.4.4. ■

If the second derivative $f'' = (f')'$ is positive, then the derivative $f'$ is increasing. If $f''$ is negative, then $f'$ is decreasing. This gives us the following characterization of concavity.

■ *Theorem*

**If the second derivative $f''(x)$ is positive on an interval $I$, then $f$ is concave up on $I$. If $f''(x)$ is negative on $I$, then $f$ is concave down on $I$.**

The above characterization of concavity is the one we will use for sketching the graphs of functions.

■ **EXAMPLE 1**

Sketch the graphs of $f(x) = x^3$ and $g(x) = x^{1/3}$.

**SOLUTION**

Both functions are defined and continuous for all $x$. Both have intercept (0, 0). Both are negative for $x < 0$ and positive for $x > 0$.

The derivatives are $f'(x) = 3x^2$ and $g'(x) = (1/3)x^{-2/3}$. The graph of $f$ has a horizontal tangent at (0, 0) and the graph of $g$ has a vertical tangent at (0, 0). Both derivatives are positive on the intervals $x < 0$ and $x > 0$. Both functions are increasing on $-\infty < x < \infty$.

The second derivatives are $f''(x) = 6x$ and $g''(x) = -(2/9)x^{-5/3}$. We see that $f''$ is negative for $x < 0$ and positive for $x > 0$. It follows that the graph of $f$ is concave down for $x < 0$ and concave up for $x > 0$. See Figure 4.4.5a. We see that $g''$ is positive for $x < 0$ and negative for $x > 0$. This implies the graph of $g$ is concave up for $x < 0$ and concave down for $x > 0$. See Figure 4.4.5b. ■

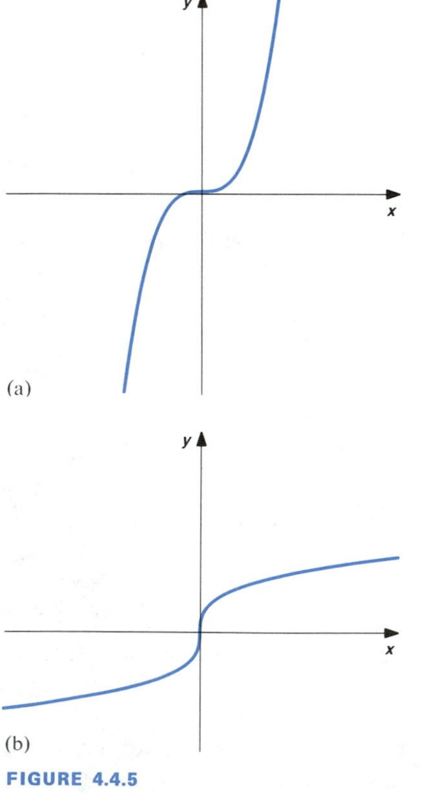


FIGURE 4.4.5

If $c$ is a critical point of $f$ with $f'(c) = 0$ and $f''(c) \neq 0$, the sign of $f''(c)$ can be used to determine if $f$ has a local maximum or minimum at $c$. We have the following result.

*Second Derivative Test for Local Extrema at Critical Points*

**If $f'(c) = 0$ and $f''(c) > 0$, then $f$ has a local minimum at $c$. If $f'(c) = 0$ and $f''(c) < 0$, then $f$ has a local maximum at $c$. If $f'(c) = 0$ and $f''(c) = 0$, $f$ may have either a local maximum, local minimum, or neither at the critical point $c$.**

■ *Proof* Let us verify the first statement. Verification of the second statement is similar. If $f''(c) > 0$, then

$$\lim_{x \to c} \frac{f'(x) - f'(c)}{x - c} = f''(c) > 0.$$

This implies that

$$\frac{f'(x) - f'(c)}{x - c} > 0, \qquad \text{for } x \text{ near } c,\ x \neq c.$$

For $x < c$, we have $x - c < 0$, so multiplication of the above inequality by $x - c$ gives

$$f'(x) - f'(c) < 0.$$

Since $f'(c) = 0$, we have

$$f'(x) < 0, \qquad \text{for } x < c,\ x \text{ near } c.$$

This means that $f$ is decreasing on an interval with right endpoint $c$. Similarly, we can show that

$$f'(x) > 0, \qquad \text{for } x > c,\ x \text{ near } c,$$

so $f$ is increasing on an interval with left endpoint $c$. We conclude that $f(c)$ must be a local minimum. This case is illustrated by the graph of $f(x) = x^2$ in Figure 4.4.6a. The second statement is illustrated by the graph of $f(x) = -x^2$ in Figure 4.4.6b.

FIGURE 4.4.6

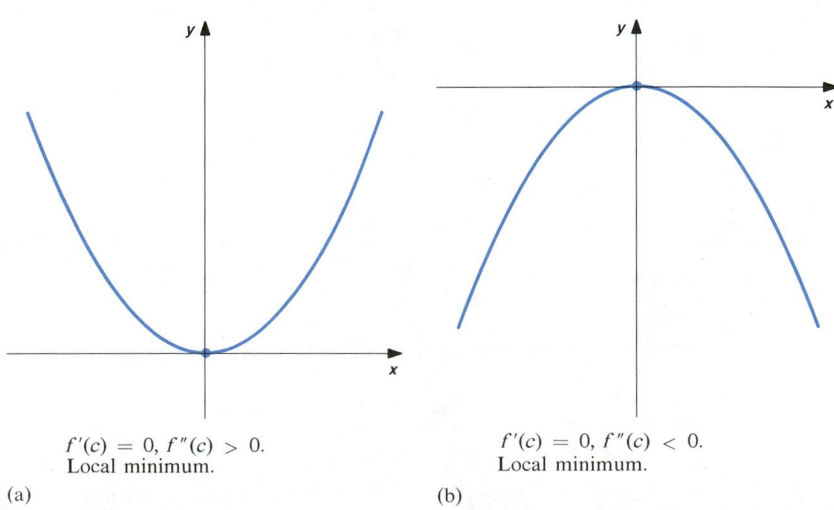


(a) $f'(c) = 0$, $f''(c) > 0$. Local minimum.

(b) $f'(c) = 0$, $f''(c) < 0$. Local minimum.

FIGURE 4.4.7

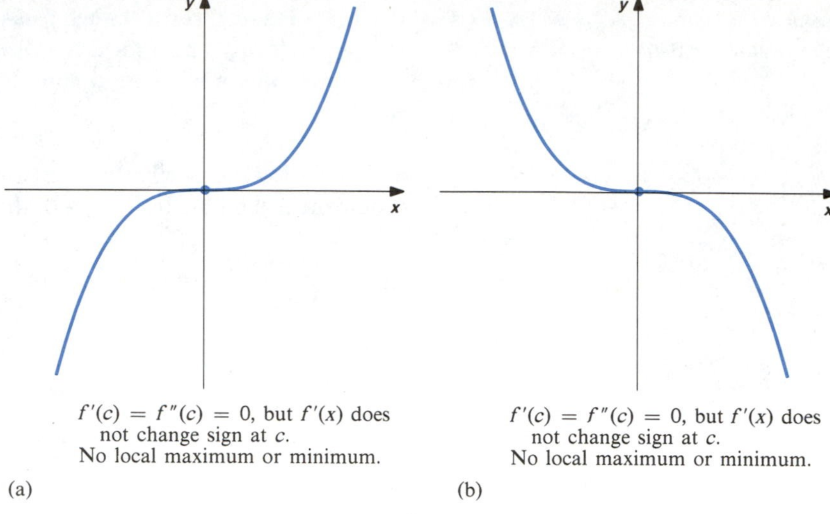


(a) $f'(c) = f''(c) = 0$, but $f'(x)$ does not change sign at $c$. No local maximum or minimum.

(b) $f'(c) = f''(c) = 0$, but $f'(x)$ does not change sign at $c$. No local maximum or minimum.

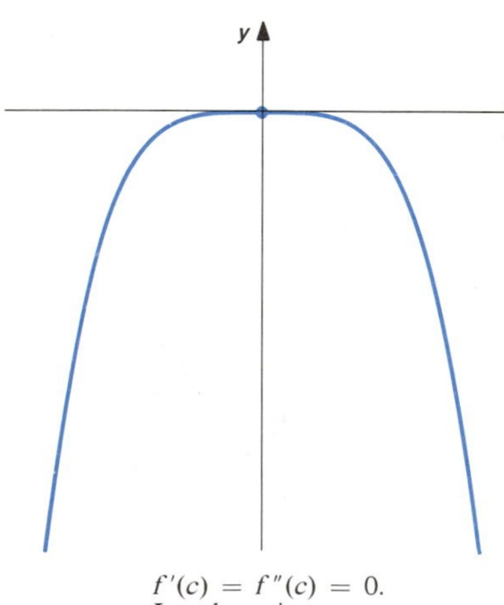


(a) $f'(c) = f''(c) = 0$. Local minimum.

(b) $f'(c) = f''(c) = 0$. Local maximum.

FIGURE 4.4.8

If $f'(c) = f''(c) = 0$, $f'(x)$ may not change sign at $x = c$, so $f$ may not have a local maximum or minimum at the critical point $c$. This is illustrated in Figure 4.4.7 by the graphs of $f(x) = x^3$ and $f(x) = -x^3$.

It is possible that $f$ has a local minimum or maximum value at a point where $f'(c) = f''(c) = 0$. These cases are illustrated in Figure 4.4.8, where we have sketched the graphs of $f(x) = x^4$ and $f(x) = -x^4$. ■

If $f'(c) = f''(c) = 0$, we should use the sign of $f'(x)$ for $x$ near $c$ and a sketch of the graph of $f$ to determine if $f$ has a local extreme value at $c$.

The second derivative $f''(x)$ is of constant sign on intervals between successive points where it is either zero or undefined. Points on the graph where either $f''(x) = 0$ or $f''(x)$ does not exist should be plotted if it is desired that the graph illustrate concavity.

A point on a graph where the graph changes concavity, from concave up to concave down or from concave down to concave up, is called a **point of inflection** of the graph. If $(d, f(d))$ is a point of inflection of the graph of $f$, and $f''(x)$ exists for $x$ near $d$ on each side of $d$, then $f''(x)$ must change sign at $x = d$. This means either

**(i)** $f''(d) = 0$ or

**(ii)** $f''(d)$ does not exist.

The condition that either $f''(d) = 0$ or "$f''(d)$ does not exist" does not guarantee that $(d, f(d))$ is a point of inflection. The condition merely tells us that $(d, f(d))$ is a candidate for a point of inflection. We must check that the concavity changes before we know that it actually is a point of inflection.

At this point it is worthwhile to review the general characteristics of graphs that we have studied. We can obtain information about a graph from the sign of $f(x)$, the sign of $f'(x)$, and the sign of $f''(x)$. It is important that we understand what information is given by each of $f, f'$, and $f''$.

If $f(x)$, $f'(x)$, and $f''(x)$ are each of constant sign for $x$ in an interval $I$ of finite length, then all eight combinations of signs are possible; not all combinations of signs of $f$, $f'$, and $f''$ are possible on infinite intervals or if the graph has asymptotes.

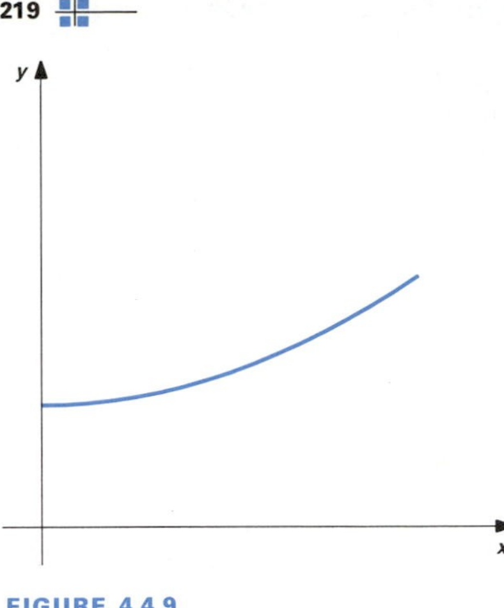


FIGURE 4.4.9

**EXAMPLE 2**

Sketch the graph of a function that satisfies the given conditions or indicate that this is impossible.

(a) $f(x) > 0, f'(x) > 0, f''(x) > 0, 0 < x < 1$.

(b) $f(x) < 0, f'(x) > 0, f''(x) < 0, 0 < x < 1$, vertical asymptote $x = 0$.

(c) $f(x) > 0, f'(x) < 0, f''(x) < 0, 0 < x < 1$, vertical asymptote $x = 0$.

(d) $f(x) > 0, f'(x) < 0, f''(x) < 0, -\infty < x < \infty$.

(e) $f(x) < 0, f'(x) > 0, f''(x) < 0, -\infty < x < \infty$, horizontal asymptote as $x \to \infty$.

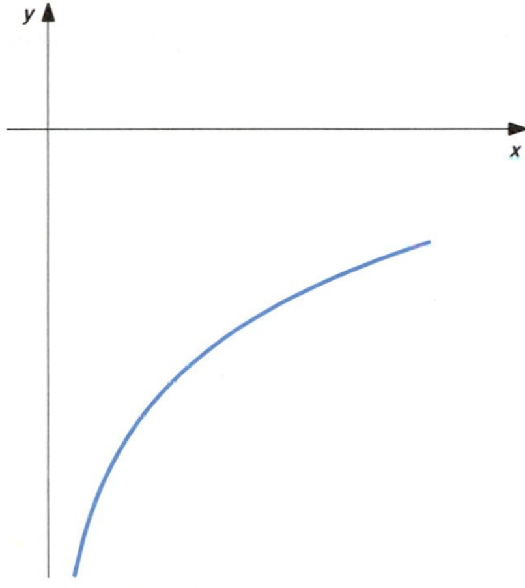


FIGURE 4.4.10

**SOLUTION**

(a) The condition $f > 0$ implies the graph is above the $x$-axis; $f' > 0$ implies the graph is increasing; $f'' > 0$ implies the graph is concave up. See Figure 4.4.9.

(b) The condition $f < 0$ implies the graph is below the $x$-axis; $f' > 0$ implies the graph is increasing; $f'' < 0$ implies the graph is concave down. The graph $f$ can have a vertical asymptote $x = 0$. See Figure 4.4.10.

(c) The condition $f > 0$ implies the graph is above the $x$-axis; $f' < 0$ implies the graph is decreasing; $f'' < 0$ implies the graph is concave down. It is impossible for the graph to have a vertical asymptote $x = 0$. See Figure 4.4.11.

(d) The condition $f > 0$ implies the graph is above the $x$-axis; $f' < 0$ implies the graph is decreasing; $f'' < 0$ implies the graph is concave down. If a function satisfies $f'(x) < 0$ and $f''(x) < 0$ for $-\infty < x < \infty$, then $f(x)$ must approach negative infinity as $x$ approaches infinity. It is impossible that $f(x)$ is positive for all $x$. See Figure 4.4.12.

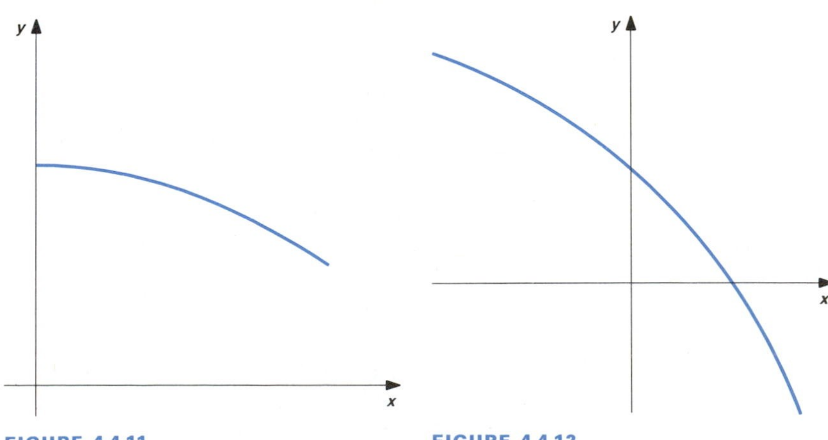


FIGURE 4.4.11

FIGURE 4.4.12

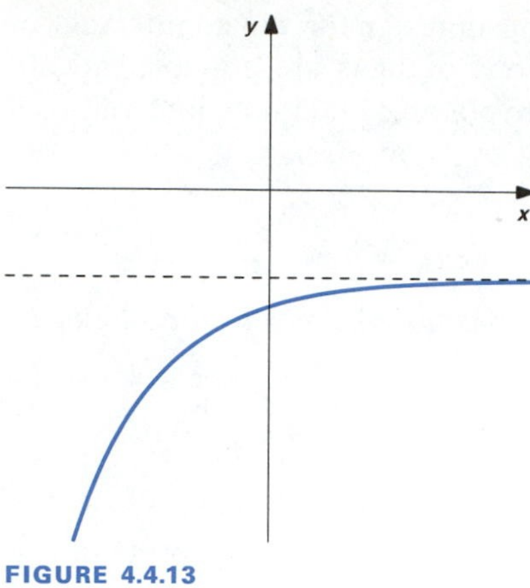


FIGURE 4.4.13

(e) The condition $f < 0$ implies the graph is below the $x$-axis; $f' > 0$ implies the graph is increasing; $f'' < 0$ implies the graph is concave down. The graph can have a horizontal asymptote as $x \to \infty$. In fact, it should seem that a function that is negative and increasing as $x$ approaches infinity should have a horizontal asymptote at infinity. This is true. See Figure 4.4.13. ■

If $f'(x)$ and $f''(x)$ are both of constant sign as $(x, f(x))$ approaches an asymptote, the graph must open away from the asymptote.

We are now ready to sketch the graphs of some specific functions. We have seen that the information we obtain from the values of $f(x)$, $f'(x)$, and $f''(x)$ sometimes overlaps. We should use whatever information is necessary and convenient for illustrating the desired general characteristics. It is not necessary to use all we know for every example, although extra information can be used as a check of our work.

■ **EXAMPLE 3**

Sketch the graph of $f(x) = x^3 - 6x^2 + 9x$. Find local extrema and points of inflection.

**SOLUTION**

The function is defined and continuous for all $x$. Writing $f(x)$ in factored form, we have

$$f(x) = x(x-3)^2.$$

It follows that $f$ has intercepts (0, 0) and (3, 0). We can check that $f(x)$ is negative for $x < 0$ and positive for $0 < x < 3$ and $x > 3$. Figure 4.4.14a illustrates these characteristics.

The derivative of $f(x) = x^3 - 6x^2 + 9x$ is

$$f'(x) = 3x^2 - 12x + 9.$$

Factoring, we have

$$f'(x) = 3(x^2 - 4x + 3) = 3(x-1)(x-3).$$

We see that $f'(x) = 0$ when $x = 1$ and $x = 3$. We have $f(1) = 4$ and $f(3) = 0$. The graph has horizontal tangents at (1, 4) and (3, 0). The first derivative $f'(x)$ is positive and the graph is increasing on the intervals $x < 1$ and $x > 3$. The derivative is negative and the graph is decreasing on the interval $1 < x < 3$. These characteristics are illustrated in Figure 4.4.14b. The function has a local maximum at (1, 4) and a local minimum as (3, 0).

The second derivative is the derivative of

$$f'(x) = 3x^2 - 12x + 9, \quad \text{so}$$

$$f''(x) = 6x - 12.$$

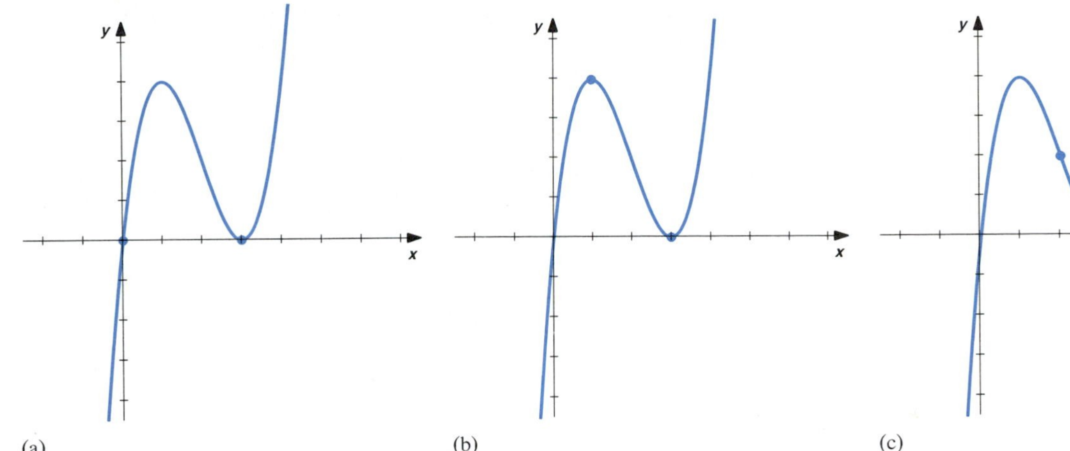


FIGURE 4.4.14

We see that $f''(x) = 0$ when $x = 2$. We have $f(2) = 2$. The second derivative $f''(x)$ is negative and the graph is concave down on the interval $x < 2$; $f''(x)$ is positive and the graph is concave up on the interval $x > 2$. The concavity of the graph is illustrated in Figure 4.4.14c. The concavity changes from negative to positive at the point (2, 2), so (2, 2) is a point of inflection of the graph. Also, $f'(1) = 0$ and $f''(1) = -6 < 0$, so the Second Derivative Test tells us that $f$ has a local maximum at the critical point $c = 1$. Similarly, $f'(3) = 0$ and $f''(3) = 6 > 0$ imply $f$ has a local minimum at the critical point $c = 3$. Of course, we already discovered these facts by looking at the sign of $f'(x)$ for $x$ near $x = 1$ and $x = 3$. ■

■ **EXAMPLE 4**

Sketch the graph of

$$f(x) = \sin x + \cos x, \qquad 0 \leq x \leq 2\pi.$$

Find local extrema and points of inflection.

**SOLUTION**

The function is defined and continuous on the interval $0 \leq x \leq 2\pi$. Points on the graph that correspond to endpoints of the domain are (0, 1) and $(2\pi, 1)$. The $x$-intercepts are given by the equation $f(x) = 0$, or

$$\sin x + \cos x = 0.$$

Solving, we have

$$\sin x = -\cos x,$$

$$\frac{\sin x}{\cos x} = -1,$$

$$\tan x = -1.$$

The solutions that are in the domain are $x = 3\pi/4$ and $x = 7\pi/4$. We can check that $f(x)$ is positive for $0 < x < 3\pi/4$ and $7\pi/4 < x < 2\pi$; $f(x)$ is negative for $3\pi/4 < x < 7\pi/4$.

The derivative of $f(x) = \sin x + \cos x$ is

$$f'(x) = \cos x - \sin x.$$

Setting $f'(x) = 0$, we obtain

$$\cos x - \sin x = 0,$$

$$\sin x = \cos x,$$

$$\tan x = 1.$$

The solutions in the domain are $x = \pi/4$ and $x = 5\pi/4$. The $y$-coordinates of the corresponding points on the graph are given by

$$f\left(\frac{\pi}{4}\right) = \sin\frac{\pi}{4} + \cos\frac{\pi}{4} = \frac{1}{\sqrt{2}} + \frac{1}{\sqrt{2}} = \frac{2}{\sqrt{2}} = \sqrt{2} \quad \text{and}$$

$$f\left(\frac{5\pi}{4}\right) = \sin\frac{5\pi}{4} + \cos\frac{5\pi}{4} = \frac{-1}{\sqrt{2}} + \frac{-1}{\sqrt{2}} = -\sqrt{2}.$$

The graph has horizontal tangents at the points $(\pi/4, \sqrt{2})$ and $(5\pi/4, -\sqrt{2})$. The first derivative $f'(x)$ is positive and the graph is increasing on the intervals $0 < x < \pi/4$ and $5\pi/4 < x < 2\pi$; $f'(x)$ is negative and the graph is decreasing on the interval $\pi/4 < x < 5\pi/4$. The function has a local maximum at $(\pi/4, \sqrt{2})$ and a local minimum at $(5\pi/4, -\sqrt{2})$.

The second derivative is the derivative of

$$f'(x) = \cos x - \sin x, \quad \text{so}$$

$$f''(x) = -\sin x - \cos x.$$

We have $f''(x) = 0$ when $x = 3\pi/4$ and $x = 7\pi/4$. These values happen to be the same as the $x$-intercepts. The second derivative $f''(x)$ is negative and the graph is concave down on the intervals $0 < x < 3\pi/4$ and $7\pi/4 < x < 2\pi$; $f''(x)$ is positive and the graph is concave up on the interval $3\pi/4 < x < 7\pi/4$. The points $(3\pi/4, 0)$ and $(7\pi/4, 0)$ are points of inflection. The graph is sketched in Figure 4.4.15. ■

FIGURE 4.4.15

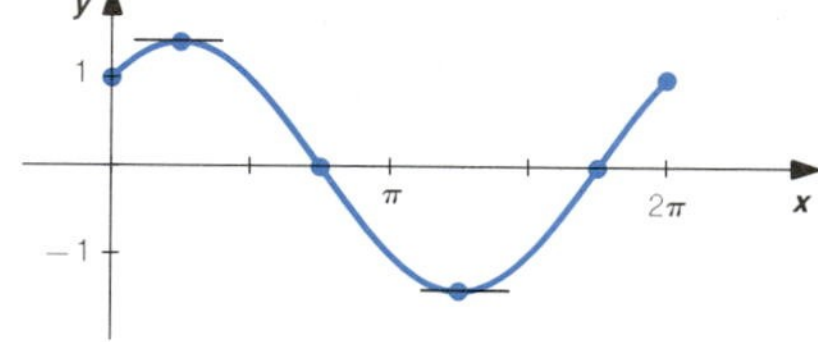

**EXAMPLE 5**

Sketch the graph of

$$f(x) = \frac{5x^2 - 5}{x^3}.$$

Find local extrema and points of inflection.

**SOLUTION**

We see that $f(x)$ is undefined for $x = 0$. The graph has vertical asymptote $x = 0$. The function is continuous for $x < 0$ and $x > 0$. The intercepts are $(1, 0)$ and $(-1, 0)$. The sign of $f(x)$ can be determined from a table of signs. See Table 4.4.1.

TABLE 4.4.1

| | $-\infty$ to $-1$ | $-1$ to $0$ | $0$ to $1$ | $1$ to $\infty$ |
|---|---|---|---|---|
| $5x^2 - 5$ | + | − | − | + |
| $x^3$ | − | − | + | + |
| $f(x)$ | − | + | − | + |

We have

$$\lim_{x \to \pm\infty} \frac{5x^2 - 5}{x^3} = 0,$$

since the denominator is of higher order than the numerator. The graph has horizontal asymptote $y = 0$. We may also notice that $f(-x) = -f(x)$, so $f$ is an odd function and the graph is symmetric with respect to the origin.

The information we have obtained by investigating the values of $f(x)$ can be used to sketch the graph. The graph in Figure 4.4.16 satisfies the conditions we have determined and is as simple as possible. That is, the graph changes from increasing to decreasing or vice versa only when necessary to satisfy the conditions we have determined.

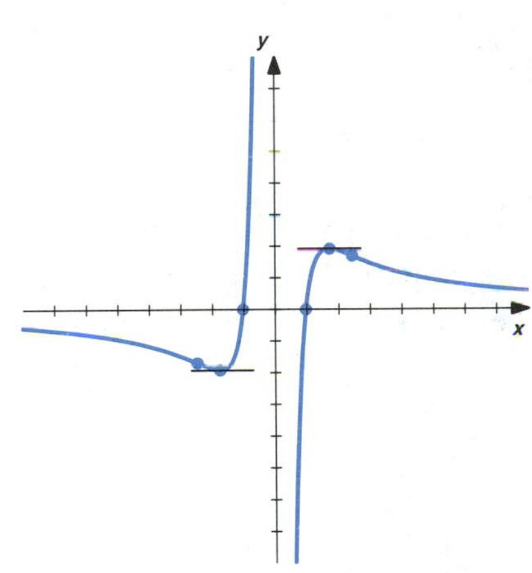


FIGURE 4.4.16

From the sketch in Figure 4.4.16 it appears that $f$ must have a local minimum somewhere in the interval $x < -1$ and a local maximum somewhere in the interval $x > 1$. By comparing the concavity at the local extrema with that as $|x| \to \infty$, we see that there must be a point of inflection in each of the intervals $x < -1$ and $x > 1$. If we want to locate these key points we must investigate the derivatives of $f$. The first derivative can be used to determine intervals on which the graph is either increasing or decreasing. The second derivative can be used to determine concavity.

We can easily carry out the division by $x^3$ to obtain

$$f(x) = 5x^{-1} - 5x^{-3}, \quad \text{so}$$

$$f'(x) = -5x^{-2} + 15x^{-4} \quad \text{and}$$

$$f''(x) = 10x^{-3} - 60x^{-5}.$$

Simplifying the first derivative, we obtain

$$f'(x) = \frac{5(3 - x^2)}{x^4}.$$

We see that $f'(x) = 0$ when $x = \sqrt{3}$ and when $x = -\sqrt{3}$. The corresponding $y$-coordinates are

$$f(\sqrt{3}) = \frac{5(\sqrt{3})^2 - 5}{(\sqrt{3})^3} = \frac{10}{3^{3/2}} \quad \text{and} \quad f(-\sqrt{3}) = -\frac{10}{3^{3/2}}.$$

The function has a local maximum at $(\sqrt{3}, 10/3^{3/2})$ and a local minimum at $(-\sqrt{3}, -10/3^{3/2})$. Investigation of the sign of $f'(x)$ verifies that the graph is increasing on the intervals $-\sqrt{3} < x < 0$ and $0 < x < \sqrt{3}$; $f$ is decreasing on the intervals $x < -\sqrt{3}$ and $x > \sqrt{3}$.

Simplifying the second derivative, we obtain

$$f''(x) = \frac{10(x^2 - 6)}{x^5}.$$

We have $f''(x) = 0$ when $x = \sqrt{6}$ and $x = -\sqrt{6}$. The corresponding $y$-coordinates are

$$f(\sqrt{6}) = \frac{5(\sqrt{6})^2 - 5}{(\sqrt{6})^3} = \frac{25}{6^{3/2}} \quad \text{and} \quad f(-\sqrt{6}) = -\frac{25}{6^{3/2}}.$$

Investigation of the sign of $f''(x)$ shows that the graph is concave up on the intervals $-\sqrt{6} < x < 0$ and $x > \sqrt{6}$. The graph is concave down on the intervals $x < -\sqrt{6}$ and $0 < x < \sqrt{6}$. The graph has points of inflection at $(\sqrt{6}, 25/6^{3/2})$ and $(-\sqrt{6}, -25/6^{3/2})$, since the graph changes concavity at these points. ■

■ **EXAMPLE 6**

Sketch the graph of $f(x) = x^{5/3} + 5x^{2/3}$. Find local extrema and points of inflection.

**SOLUTION**

The function is defined and continuous for all $x$. We have

$$f(x) = x^{5/3} + 5x^{2/3} = x^{2/3}(x + 5),$$

$$f'(x) = \frac{5}{3}x^{2/3} + \frac{10}{3}x^{-1/3} = \frac{5(x + 2)}{3x^{1/3}},$$

$$f''(x) = \frac{10}{9}x^{-1/3} - \frac{10}{9}x^{-4/3} = \frac{10(x - 1)}{9x^{4/3}}.$$

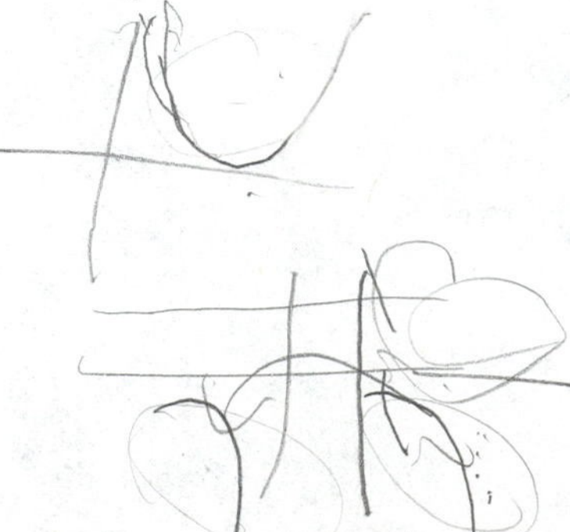

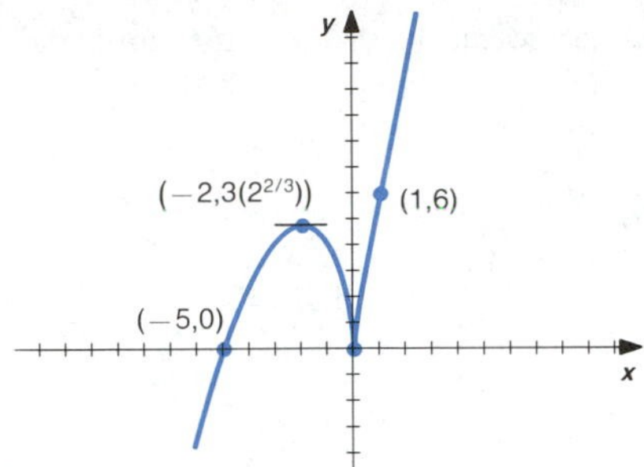


FIGURE 4.4.17

From the factored form of $f(x)$, we see that the graph has intercepts $(0, 0)$ and $(-5, 0)$. $f(x)$ is positive for $-5 < x < 0$ and $x > 0$, and is negative for $x < -5$.

From the factored form of $f'(x)$, we see that $f'(x) = 0$ when $x = -2$, and $f'(x)$ does not exist when $x = 0$. We have $f(-2) = 3 \cdot 2^{2/3}$ and $f(0) = 0$. The graph has a horizontal tangent $[f'(x) = 0]$ at $(-2, 3 \cdot 2^{2/3})$ and a vertical tangent $[|f'(x)| \to \infty]$ at $(0, 0)$. $f$ is increasing $[f'(x) > 0]$ on the intervals $x < -2$ and $x > 0$, decreasing $[f'(x) < 0]$ on the interval $-2 < x < 0$.

We see that $f''(x) = 0$ when $x = 1$, and that $f''(x)$ does not exist when $x = 0$. We have $f(1) = 6$ and $f(0) = 0$. The points $(1, 6)$ and $(0, 0)$ are possible points of inflection. The graph is concave up $[f''(x) > 0]$ on the interval $x > 1$. The graph is concave down $[f''(x) < 0]$ on the intervals $x < 0$ and $0 < x < 1$. The point $(1, 6)$ is a point of inflection, but $(0, 0)$ is not, because the graph does not change concavity at $(0, 0)$. The graph is given in Figure 4.4.17. Notice that it is difficult to discern the concavity for positive $x$ from this accurate sketch. ■

## EXERCISES 4.4

In Exercises 1–16 sketch the graph of a function that satisfies the given conditions or indicate that this is impossible.

1 $f(x) > 0,\ f'(x) > 0,\ f''(x) < 0,\ 0 < x < 1$

2 $f(x) > 0,\ f'(x) < 0,\ f''(x) > 0,\ 0 < x < 1$

3 $f(x) < 0,\ f'(x) > 0,\ f''(x) > 0,\ 0 < x < 1$

4 $f(x) < 0,\ f'(x) < 0,\ f''(x) > 0,\ 0 < x < 1$

5 $f(x) > 0,\ f'(x) > 0,\ f''(x) < 0,\ 0 < x < 1$, vertical asymptote $x = 1$

6 $f(x) > 0,\ f'(x) < 0,\ f''(x) > 0,\ 0 < x < 1$, vertical asymptote $x = 0$

7 $f(x) < 0,\ f'(x) < 0,\ f''(x) < 0,\ 0 < x < 1$, vertical asymptote $x = 1$

8 $f(x) < 0,\ f'(x) < 0,\ f''(x) < 0,\ 0 < x < 1$, vertical asymptote $x = 0$

9 $f(x) > 0,\ f'(x) < 0,\ f''(x) > 0,\ -\infty < x < \infty$

10 $f(x) < 0,\ f'(x) > 0,\ f''(x) > 0,\ -\infty < x < \infty$

11 $f(x) < 0,\ f'(x) < 0,\ f''(x) > 0,\ -\infty < x < \infty$

12 $f(x) < 0,\ f'(x) < 0,\ f''(x) < 0,\ -\infty < x < \infty$

13 $f(x) < 0,\ f'(x) < 0,\ f''(x) < 0,\ -\infty < x < \infty$, horizontal asymptote as $x \to -\infty$

14 $f(x) < 0,\ f'(x) < 0,\ f''(x) < 0,\ -\infty < x < \infty$, horizontal asymptote as $x \to \infty$

15 $f(x) > 0,\ f'(x) < 0,\ f''(x) > 0,\ -\infty < x < \infty$, horizontal asymptote as $x \to -\infty$

16 $f(x) > 0,\ f'(x) < 0,\ f''(x) > 0,\ -\infty < x < \infty$, horizontal asymptote as $x \to \infty$

Sketch the graphs of the functions in Exercises 17–42. Find local extrema and points of inflection.

17 $f(x) = x^3 + 1$

18 $f(x) = (x - 1)^{1/3}$

19 $f(x) = -2 + 3x - x^3$

20 $f(x) = x^3 - 3x^2$

21 $f(x) = x^3 + x$

22 $f(x) = 2 - x - x^3$

23 $f(x) = 64x^2 - 16x$

24 $f(x) = 3x^4 + 4x^3$

25 $f(x) = (6x^5 + 20x^3 - 90x)/32$

26 $f(x) = x^4 - 2x^3$

27 $f(x) = x^{1/3}(x^2 - 7)$

28 $f(x) = x^{2/3}(x^2 - 16)/16$

29 $f(x) = x^{2/3}(x + 40)/4$

30 $f(x) = x(x - 40)^{2/3}/4$

31 $f(x) = \dfrac{8x}{x^2 + 4}$

32 $f(x) = \dfrac{1}{x^2 + 1}$

33 $f(x) = \dfrac{x}{x^2 - 1}$

34 $f(x) = \dfrac{x^2 - 1}{x}$

35 $f(x) = \dfrac{x - 1}{x^2}$

36 $f(x) = \dfrac{x^2 - 1}{x^2}$

37 $f(x) = \dfrac{(x^2 - 1)^2}{x^4}$

38 $f(x) = \dfrac{10(x^2 - 1)}{x^5}$

39 $f(x) = x + \sin 2x,\ 0 \le x \le 2\pi$

40 $f(x) = x - \sin x,\ 0 \le x \le 2\pi$

41 $f(x) = \sin x + \cos x$, $0 \le x \le 2\pi$

42 $f(x) = \cos x - \sin x$, $0 \le x \le 2\pi$

In Exercises 43–46, sketch a short segment of the graph of the given equation that contains the given point on the graph; the sketch should indicate the sign of the slope and the concavity of the graph as it passes through the point.

43 $x = y^3 - y$, $(0, 1)$

44 $x^3 = y^3 + 3y + 1$, $(1, 0)$

45 $x = \tan y$, $(1, \pi/4)$

46 $x = \sin y$, $(1/2, \pi/6)$

## 4.5 FINDING EQUATIONS THAT RELATE VARIABLES

This section is devoted to the basic skill of translating a problem that is described in words into a mathematical expression that involves real variables. No real problems are solved (or even stated) in this section. However, the procedures discussed are essential for the solution of the problems considered in ensuing sections.

The first step in translating a problem from words to equations is to read the problem carefully. Each unknown quantity in the problem is then identified and designated by a variable. If appropriate, a sketch should be made and labeled. Finally, we find equations that relate the variables. We will illustrate some of the ways used to relate variables.

Relations between variables are often obtained from *known formulas* or *given formulas*. In this section we will use the following formulas from geometry:

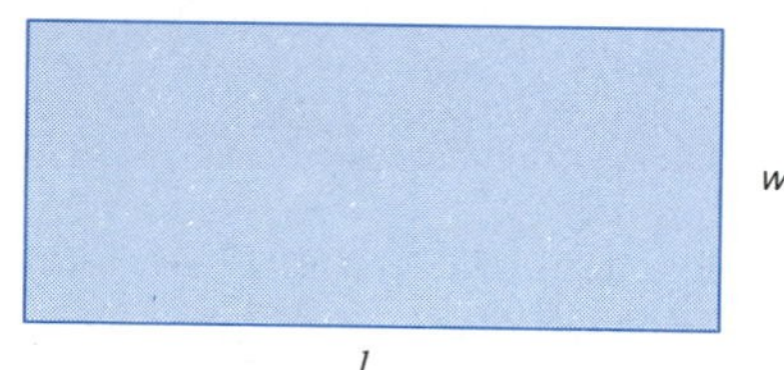

FIGURE 4.5.1

**RECTANGLE** (Figure 4.5.1)

**Area** $= lw$

**Perimeter** $= 2l + 2w$

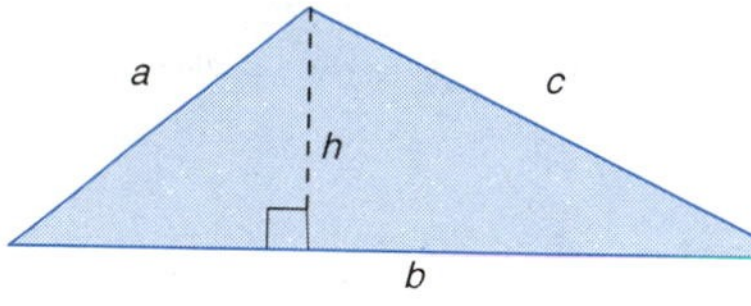

FIGURE 4.5.2

**TRIANGLE** (Figure 4.5.2)

**Area** $= \frac{1}{2} bh$

**Perimeter** $= a + b + c$

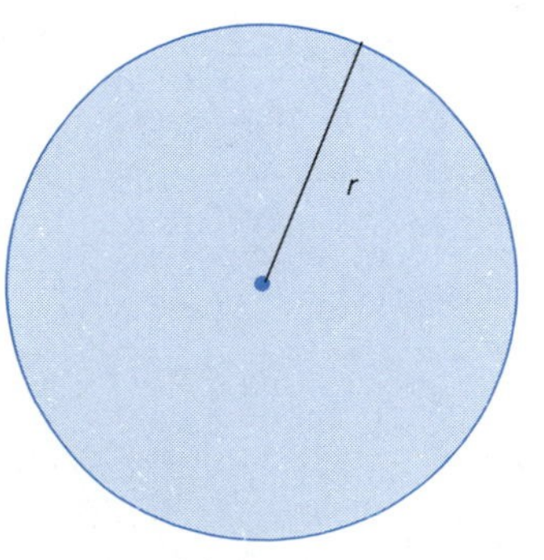

FIGURE 4.5.3

**CIRCLE** (Figure 4.5.3)

**Area** $= \pi r^2$

**Circumference** $= 2\pi r$

**RECTANGULAR BOX** (Figure 4.5.4)

**Volume** $= lwh$

**Areas of faces** $= lw, lh, wh$

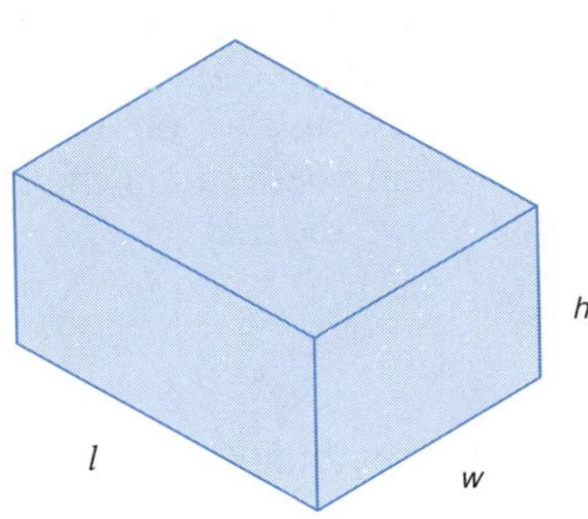


FIGURE 4.5.4

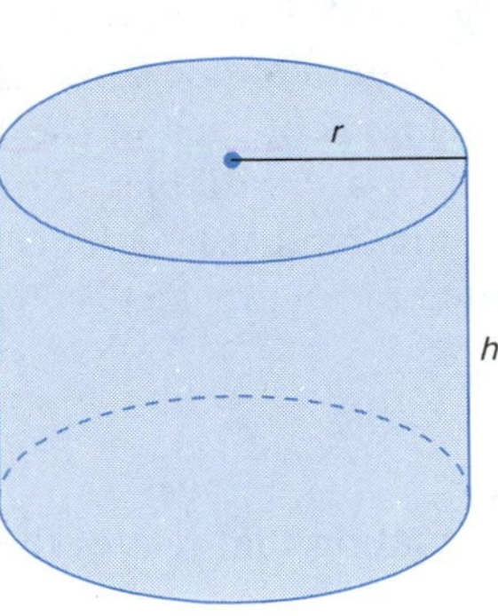


FIGURE 4.5.5

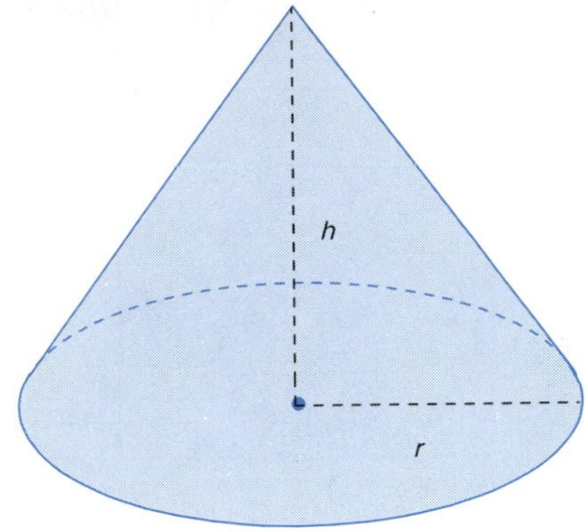


FIGURE 4.5.6

**RIGHT CIRCULAR CYLINDER** (Figure 4.5.5)

**Volume = $\pi r^2 h$**

**Area of curved surface = $2\pi rh$**

**RIGHT CIRCULAR CONE** (Figure 4.5.6)

**Volume = $\frac{1}{3}\pi r^2 h$**

**Area of curved surface = $\pi r\sqrt{r^2 + h^2}$**

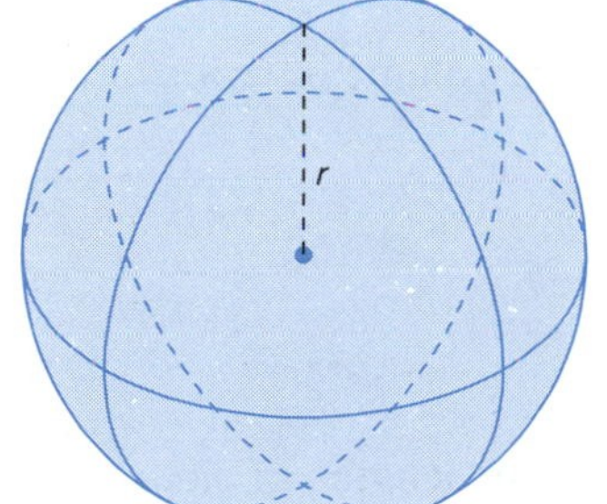


FIGURE 4.5.7

**SPHERE** (Figure 4.5.7)

**Volume = $\frac{4}{3}\pi r^3$**

**Surface area = $4\pi r^2$**

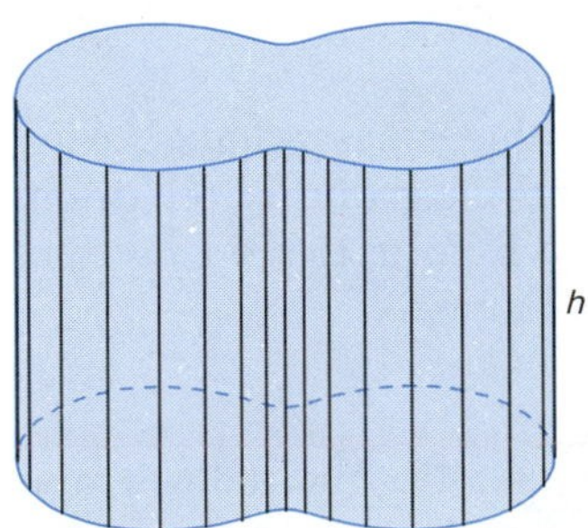


FIGURE 4.5.8

**GENERALIZED RIGHT CYLINDER** (Figure 4.5.8)

**Volume = (Area of base)(Height)**

■ **EXAMPLE 1**

A rectangular box is to be constructed from a 12-in.-square piece of cardboard by cutting equal squares from each corner and then bending up the sides. (a) Find an equation that relates the length of one side of the square base of the box and the height of the box. (b) Express the volume of the box in terms of its height.

(a)

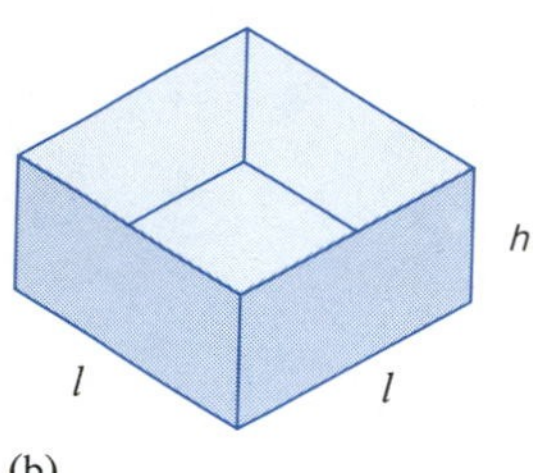

(b)

**FIGURE 4.5.9**

**SOLUTION**

(a) In Figure 4.5.9a we have sketched a 12-in. square and indicated that a square with length of side $h$ inches is to be cut from each corner. A length of $l$ inches remains of each of the 12-in. sides of the original square. The lengths $h$ and $l$ are variables. After the sides are bent up, we obtain a rectangular box that has length of each side of the square base $l$ and height $h$, as indicated in Figure 4.5.9b. From Figure 4.5.9a we see that

$$2h + l = 12.$$

This is the equation that relates the length of each side of the square base and the height of the box.

(b) From Figure 4.5.9b we see that the volume of the box in cubic inches is given by the equation

$$V = hl^2.$$

We want to express the volume in terms of the single variable $h$. To do this, we use the fact from part (a) that the variables $h$ and $l$ are related by the equation

$$2h + l = 12.$$

Solving this equation for $l$ in terms of $h$, we obtain

$$l = 12 - 2h.$$

Substitution in the expression for $V$ then gives the volume in terms of the height. We have

$$V = hl^2,$$

$$V = h(12 - 2h)^2.$$

(The variables $h$ and $l = 12 - 2h$ cannot represent lengths of sides of a rectangular box unless they are nonnegative, since length is nonnegative. Thus, the above expression for $V$ can represent the volume of the box only if $h \geq 0$ and $12 - 2h \geq 0$, or $0 \leq h \leq 6$.) ■

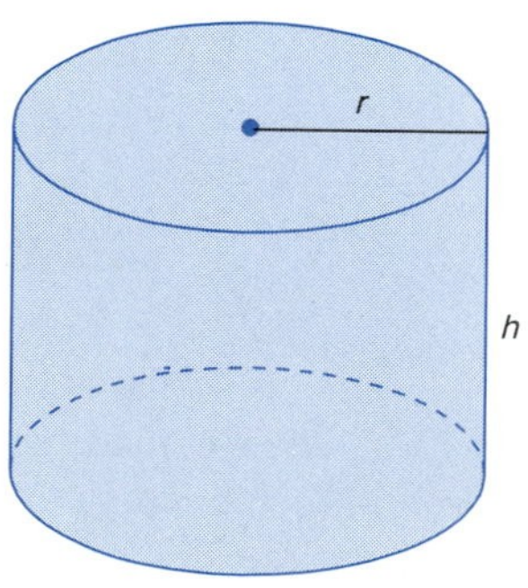

**FIGURE 4.5.10**

■ **EXAMPLE 2**

An oil drum in the shape of a right circular cylinder has volume 5.6 $\text{ft}^3$. (a) Find an equation that relates the height and the radius of the drum. (b) Express the total amount of material needed to form the top, bottom, and curved surface of the drum in terms of its radius.

**SOLUTION**

(a) A sketch of the drum is given in Figure 4.5.10. The radius $r$ and height $h$ have been labeled; these are variables. The volume is not a variable. The radius and height of the cylinder are related because they determine the volume of the cylinder. The equation for the volume of a right circular cylinder tells us the volume is $\pi r^2 h$. We assume that the units of length are feet, so $\pi r^2 h$ gives the volume in cubic feet. Since we know that the volume of the drum is 5.6 ft$^3$, we must have

$$\pi r^2 h = 5.6.$$

This is the equation that relates the variables.

(b) The total amount of material needed to form the drum is the sum of the areas of the top, bottom, and the curved surface. The area of each of the top and bottom is $\pi r^2$ ft$^2$, and the area of the curved surface is $2\pi rh$ ft$^2$. The total amount of material, in terms of the two variables $r$ and $h$, is given by the expression

$$M = 2\pi r^2 + 2\pi rh.$$

Since we want $M$ in terms of the single variable $r$, we use the relation between $r$ and $h$ that was established in part (a) to eliminate the variable $h$. We have

$$\pi r^2 h = 5.6,$$

$$h = \frac{5.6}{\pi r^2}.$$

Substitution then gives

$$M = 2\pi r^2 + 2\pi rh,$$

$$M = 2\pi r^2 + 2\pi r\left(\frac{5.6}{\pi r^2}\right),$$

$$M = 2\pi r^2 + \frac{11.2}{r}.$$

This gives the total amount of material in terms of the radius of the drum. ■

Many practical problems involve right triangles. For these problems it is common to use the *Pythagorean Theorem* to find relations between variables. It is also helpful to be familiar with the definition of the *trigonometric functions* in terms of side opposite, side adjacent, and hypotenuse. See Figure 4.5.11. (The trigonometric functions are reviewed in Section 1.6.)

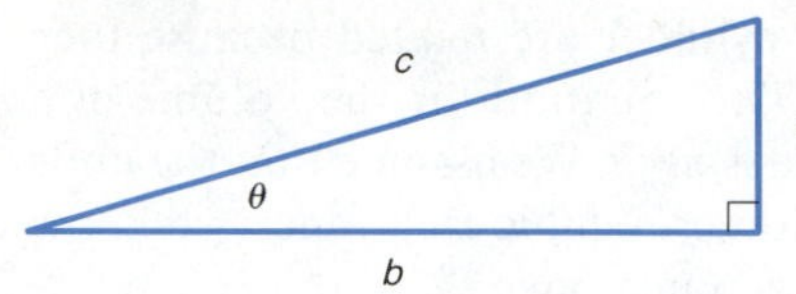

FIGURE 4.5.11

**RIGHT TRIANGLE** (Figure 4.5.11)

*Pythagorean Theorem*

$$c^2 = a^2 + b^2$$

*Trigonometric functions*

$$\sin\theta = \frac{a}{c} \qquad \cot\theta = \frac{b}{a}$$

$$\cos\theta = \frac{b}{c} \qquad \sec\theta = \frac{c}{b}$$

$$\tan\theta = \frac{a}{b} \qquad \csc\theta = \frac{c}{a}$$

■ **EXAMPLE 3**

A taut rope runs from the bow of a small boat to a windlass that is on the edge of a dock. The windlass is 8 ft higher than the point where the rope is attached to the boat. Find an equation that relates the length of the rope between the boat and the windlass and the horizontal distance of the boat from a point directly below the windlass.

**SOLUTION**

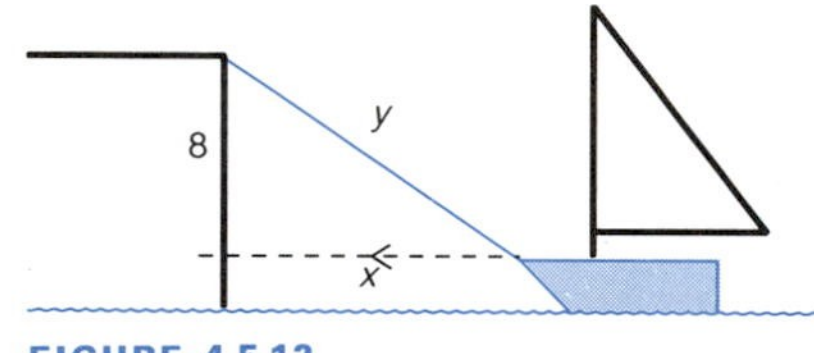

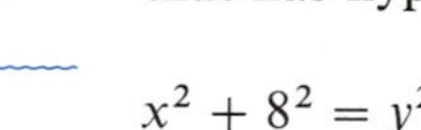
FIGURE 4.5.12

A sketch is given in Figure 4.5.12. The length of rope between the boat and the windlass is labeled $y$, and the horizontal distance of the boat from a point directly below the windlass is labeled $x$; $x$ and $y$ are variables. From the sketch we see that $x$ and 8 are the lengths of the sides of a right triangle that has hypotenuse $y$. The Pythagorean Theorem then gives

$$x^2 + 8^2 = y^2. \qquad ■$$

■ **EXAMPLE 4**

An airplane is flying horizontally at an altitude of 2000 ft in a direction that will take it directly over an observer at ground level. Find an equation that relates the angle of elevation from the observer to the plane and the distance between the observer and the point at ground level directly below the plane.

**SOLUTION**

A sketch is given in Figure 4.5.13. The angle of elevation is labeled $\theta$ and the distance between the observer and the point at ground level directly below the plane is labeled $x$; $\theta$ and $x$ are variables. From Figure 4.5.13 we see that

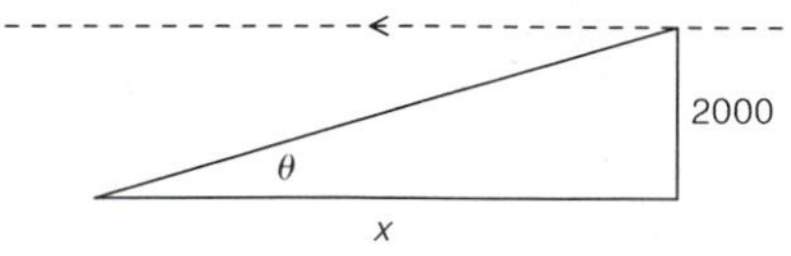

FIGURE 4.5.13

$$\tan\theta = \frac{2000}{x} \quad \text{or} \quad x = 2000\cot\theta. \qquad ■$$

For triangles that are not necessarily right triangles, we can obtain relations by using the *Law of Cosines*, the *Law of Sines*, and *ratios of corresponding sides of similar triangles.*

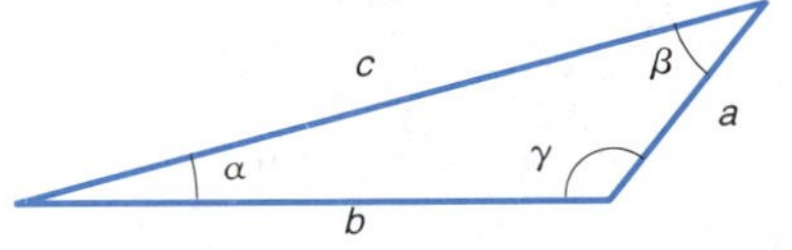


FIGURE 4.5.14

FIGURE 4.5.15

**ANY TRIANGLE** (Figures 4.5.14 and 4.5.15)

*Law of Cosines*

$$c^2 = a^2 + b^2 - 2ab \cos \gamma$$

*Law of Sines*

$$\frac{\sin \alpha}{a} = \frac{\sin \beta}{b} = \frac{\sin \gamma}{c}$$

*Ratios of corresponding sides of similar triangles*

$$\frac{a}{b} = \frac{a'}{b'} \qquad \frac{a}{c} = \frac{a'}{c'} \qquad \frac{b}{c} = \frac{b'}{c'}$$

■ **EXAMPLE 5**

A girl 5 ft tall is walking away from a lamppost that is 15 ft high. Find an equation that relates the length of the girl's shadow and the distance between the girl and the lamppost.

**SOLUTION**

A sketch is given in Figure 4.5.16. The distance between the girl and the lamppost is labeled $d$ and the length of her shadow is labeled $s$; $d$ and $s$ are variables. Using the ratio of corresponding parts of the larger and smaller similar right triangles, we obtain

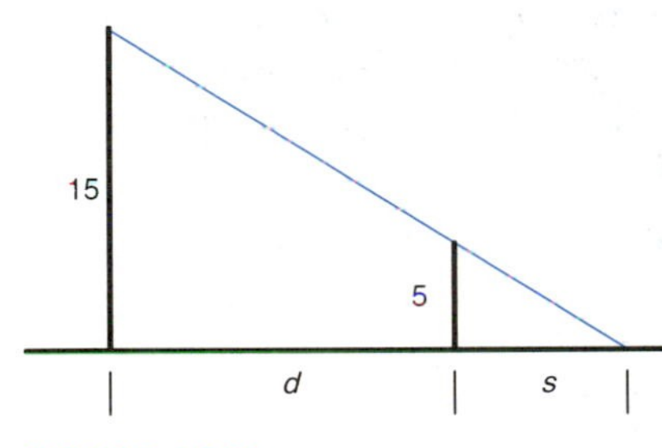


FIGURE 4.5.16

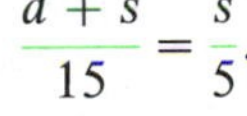

$$\frac{d + s}{15} = \frac{s}{5}.$$

Simplifying, we obtain

$$5d + 5s = 15s,$$

$$5d = 10s,$$

$$d = 2s. \qquad ■$$

*Direct or inverse variation* is a common relation between variables. Either of the statements "$y$ varies directly as $x$" or, more simply, "$y$ varies as $x$" is translated into the equation $y = kx$, where $k$ is a constant. In this case $x$ and $y$ are proportional and $k$ is called the constant of proportionality. The statement "$y$ varies inversely as $x$" is translated into the equation $y = k/x$, where $k$ is a constant. The statement "$y$ varies jointly as $x$ and $z$" means that $y$ varies as the product of $x$ and $z$, so $y = kxz$ for some constant $k$. A variable may simultaneously vary directly as a product of variables and inversely as a product of other variables. For example, the fact that the force $F$ between two masses $m_1$ and $m_2$ due to the attraction of gravity varies directly as the product of the masses and inversely as the square of the distance $d$ between them is indicated by the equation

$$F = \frac{km_1m_2}{d^2}.$$

The value of the gravitational constant $k$ in this equation depends only on the units used to measure mass, distance, and force.

■ **EXAMPLE 6**

**Hooke's Law** asserts that the force required to either stretch or compress a spring from its natural length varies as the displacement. A certain spring is stretched 12 in. beyond its natural length by a 3-lb weight. Express the force in terms of the displacement for this spring.

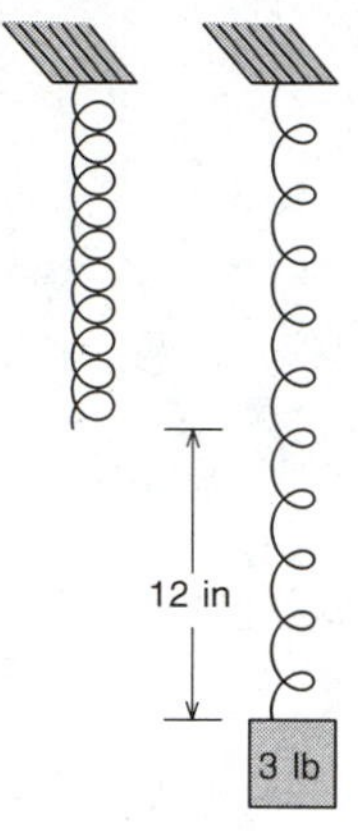


FIGURE 4.5.17

**SOLUTION**

Let $F$ denote the force in pounds that corresponds to a displacement of $x$ in.; $x$ and $F$ are variables. Hooke's Law tells us that

$$F = kx \quad \text{for some constant } k.$$

We are told that a force of 3 lb results in a displacement of 12 in. This means that when the spring is stretched 12 in., the spring force is equal to the force of 3 lb, so $x = 12$ when $F = 3$. See Figure 4.5.17. Substitution of the value 12 for the variable $x$ and the corresponding value 3 for the variable $F$ then allows us to determine the spring constant $k$. We have

$$F = kx,$$

$$3 = k(12),$$

$$k = \frac{1}{4}.$$

Using this value of $k$ we obtain the equation

$$F = \frac{x}{4}.$$ ■

*Linear* relations between variables occur relatively often and we should be prepared to deal with them. Finding the equation of lines in the $xy$-plane is discussed in Section 1.4. The same procedures work for any variables that have a linear relationship. Either variable can play the role of $x$ and either can play the role of $y$, but we must be careful not to switch roles in the middle of the problem.

■ **EXAMPLE 7**

A farmer wishes to plant a grove of apple trees. He knows from experience that each tree will produce approximately 2 bushels of apples if he plants

100 trees per acre. Also, because of crowding, each tree will produce only 1.85 bushels if he plants 110 trees per acre. (a) Find the equation that relates the number of trees per acre, $n$, and the production of apples per tree, $r$, if the relation is assumed to be linear. (b) Express the total bushels per acre produced in terms of the number of trees per acre.

**SOLUTION**

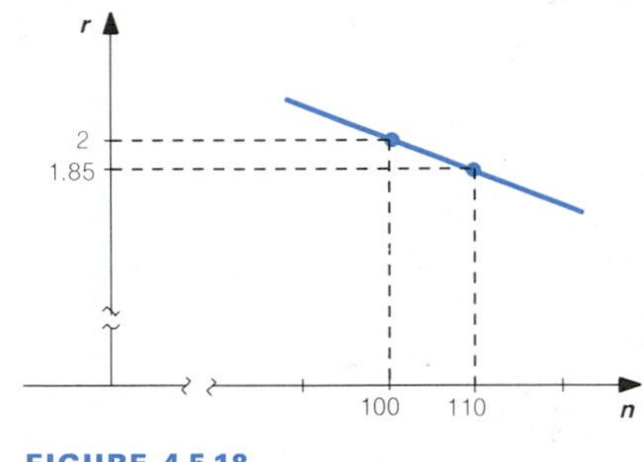


FIGURE 4.5.18

(a) In Figure 4.5.18 we have plotted the points $(n_1, r_1) = (100, 2)$ and $(n_2, r_2) = (110, 1.85)$. We want to find the equation of the line that contains these two points. The slope of the line is given by

$$\frac{r_2 - r_1}{n_2 - n_1} = \frac{1.85 - 2}{110 - 100} = \frac{-0.15}{10} = -0.015.$$

The equation of the line through $(n_1, r_1) = (100, 2)$ with slope $m = -0.015$ is

$$r - r_1 = m(n - n_1),$$

$$r - 2 = -0.015(n - 100),$$

$$r = -0.015n + 3.5.$$

(Be sure that the role of each variable in the calculation of the slope corresponds to the same role the variable plays in the point-slope formula for the equation of the line.)

(b) Let $T$ denote the total bushels per acre of apples produced. This is given by the product of the number of trees per acre and the production per tree, so

$$T = nr.$$

We want to express $T$ in terms of $n$, so we use the linear relation between $n$ and $r$ that was established in part (a) to eliminate the variable $r$. We have

$$r = -0.015n + 3.5.$$

Substitution then gives us

$$T = nr,$$

$$T = n(-0.015n + 3.5).$$

This is the desired equation. ■

Actually, the number of trees per acre planted in Example 7 must be a nonnegative integer. To apply techniques of calculus we will assume that $n$ can be any nonnegative real number. Also, the assumption that the relation between $r$ and $n$ is linear is not likely to hold if either a very large or very small number of trees per acre are planted.

## EXERCISES 4.5

1 A rectangular box with no top is to be constructed by cutting equal squares from each corner of a 8.5 in. × 11 in. piece of cardboard. (a) Find an equation that relates the height and length. (b) Find an equation that relates the height and width. (c) Express the volume of the box in terms of its height.

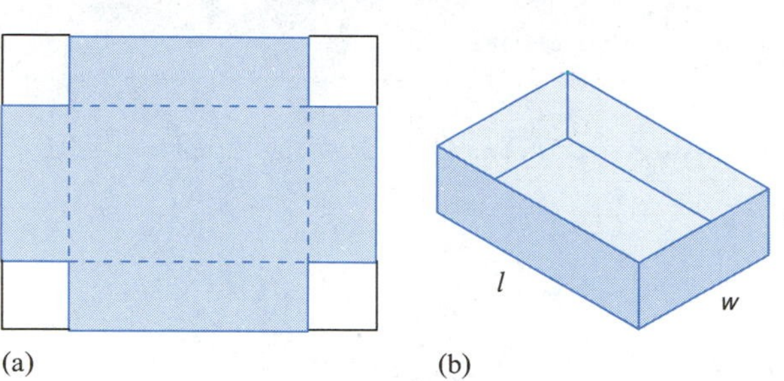


2 A rectangular box with top is to be constructed by cutting and bending a 20 in. × 36 in. piece of cardboard as indicated below. (a) Find an equation that relates the height and length. (b) Find an equation that relates the height and width. (c) Express the volume of the box in terms of its height.

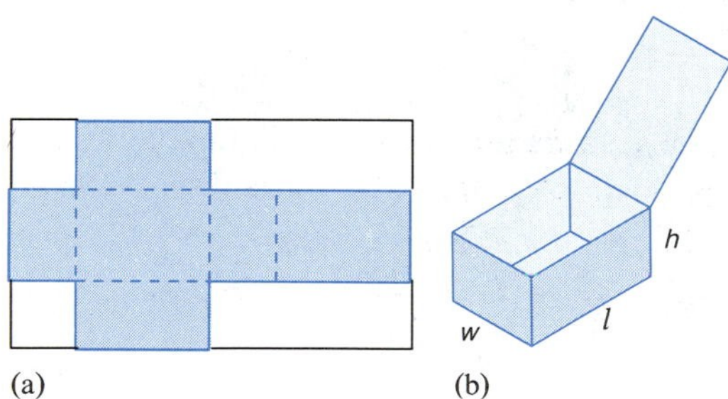


3 A farmer has 600 m of fencing to enclose a rectangular region adjacent to a river. No fencing is required along the river. (a) Find an equation that relates the length and width of the region. (b) Express the area of the region in terms of the length of fence parallel to the river.

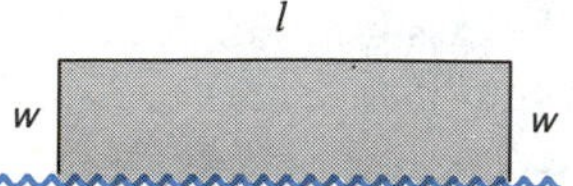


4 A farmer has 800 m of fencing to enclose a rectangular field and divide it in half. (a) Find an equation that relates the length and width of the field. (b) Express the area of the field in terms of the length of fence that divides the field.

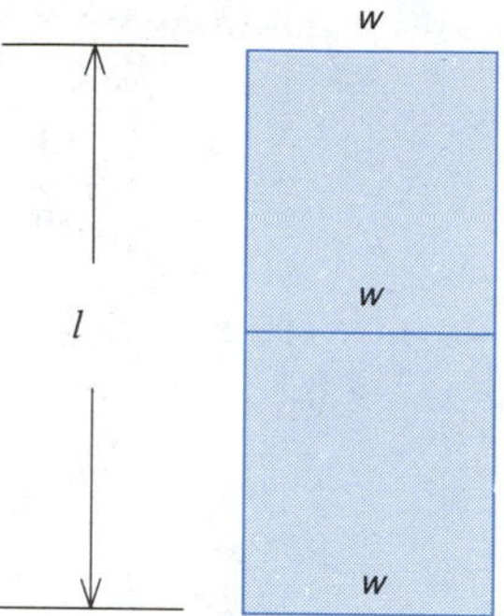


5 (a) Find an equation that relates the height $h$ and the length $l$ of one side of an equilateral triangle. (b) Express the area of the triangle in terms of the common length of the sides.

6 (a) Find an equation that relates the hypotenuse $h$ and the length $l$ of one side of an isosceles right triangle. (b) Express the perimeter of the triangle in terms of the hypotenuse.

7 The area of a rectangular field is to be 20,000 $\text{m}^2$. (a) Find an equation that relates the length and width of the field. (b) Express the perimeter of the field in terms of the length of one of its sides.

8 The diameter of a conical pile of dry, smooth sand is three times its height. Find an equation that relates the volume of sand and the height of the pile.

9 A tank in the shape of a right circular cylinder with two hemispherical ends is to have volume 600 $\text{ft}^3$. (a) Find an equation that relates the radius and height of the cylindrical part. (b) Express the total surface area of the tank in terms of its radius.

10 A window in the shape of a rectangle with a semicircular top is to have total area 16 $\text{ft}^2$. (a) Find an equation that relates the width and the height of the rectangular part.

(b) Express the perimeter of the window in terms of the width of the rectangular part.

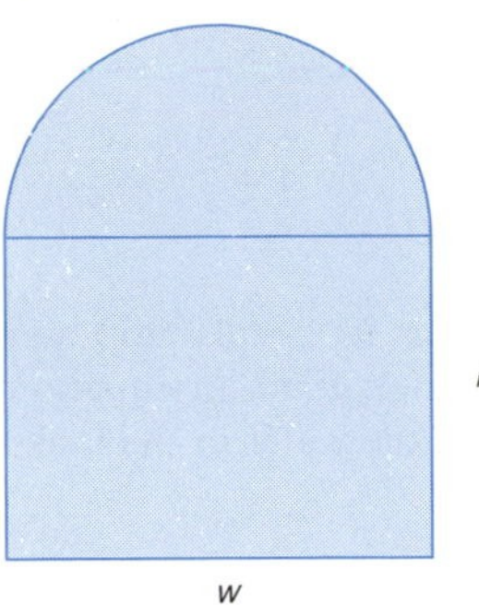


11 The string is taut between a boy and a kite that is 150 ft above ground. Find an equation that relates the length of string and the distance between the boy and the point at ground level that is directly below the kite.

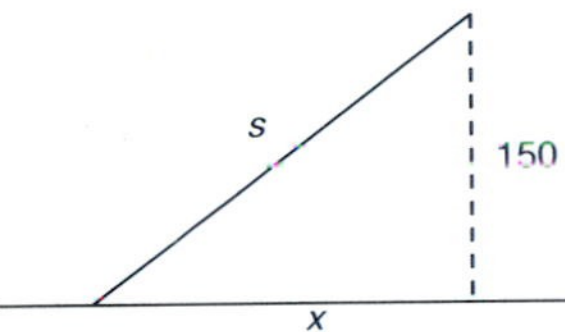


12 A post 12 ft high is 20 ft from a post 16 ft high. A wire is to be run from the top of one post to the ground at a point between the posts and then to the top of the other post. Find an equation that relates the total length of wire required and the distance between the point where the wire touches the ground and the 16-ft-high post.

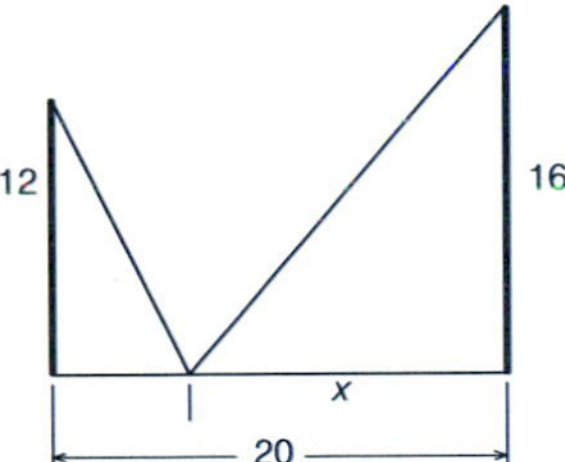


13 A balloon is rising from a point on the ground that is 50 m from an observer. Find an equation that relates the height of the balloon and the angle of elevation from the observer to the balloon.

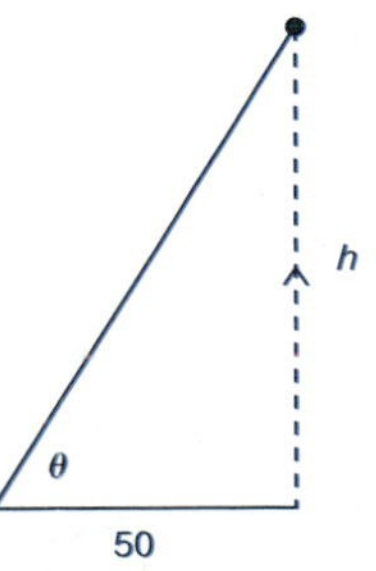


14 A tennis ball is moving in a direction perpendicular to the net, toward a point at the net that is 12 m from a spectator who is sitting in line with the net. Find an equation that relates the distance of the ball from the net and the angle between the net and the line of sight from the spectator to the ball.

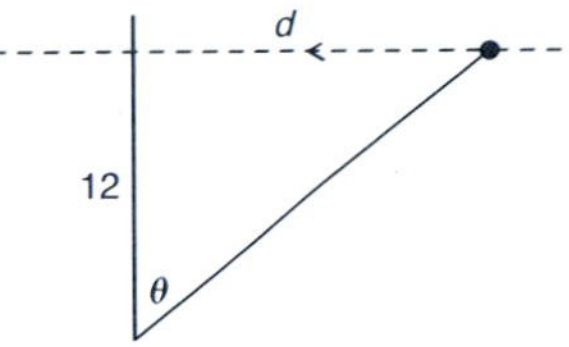


15 A rectangle is inscribed in a right triangle in such a way that one corner of the rectangle is positioned at the right angle of the triangle. The triangle has height 3 and base 4. (a) Find an equation that relates the length and width of the rectangle. (b) Express the area of the rectangle in terms of the length of its side that is along the base of the triangle.

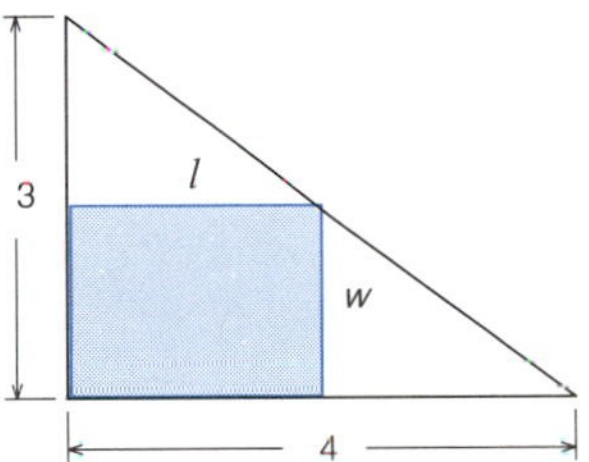


16 A triangular playground is to be constructed at the corner of Lake Street and Elm Steet, which meet at right angles. The playground is to be bounded on two sides by the streets and on the other side by a straight wall that runs between the streets. The wall must be built through a lamppost that is 100 ft from Lake Street and 120 ft away from Elm. (a) Find an equation that relates the lengths of the edges of the playground along the two streets. (b) Express the area of the playground in terms of the length of its edge along Lake Street.

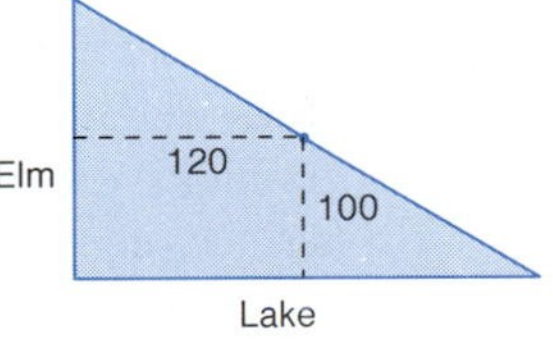


17 Hooke's Law asserts that the force required to either stretch or compress a spring from its natural length varies as the length displaced. A certain spring is stretched 6 in. beyond its natural length by a 0.3-lb weight. Express the force in terms of the displacement for this spring.

18 The force on a particular object due to the gravitational attraction of the earth varies inversely as the square of

the distance of the object from the center of the earth. Find the equation that relates the force due to gravity and the distance from the center of the earth of the object, if the object weighs 150 lb at the surface of the earth, 4000 miles from the center of the earth.

19 The manager of an apartment building estimates that all 80 apartments can be rented if she charges \$300/month, but one apartment becomes vacant for each \$10/month increase in rent. (a) Find an equation that relates the rent and the number of occupied apartments, assuming that the relation is linear. (b) Express the total amount of rent due each month in terms of the monthly rent per unit.

20 A salesman believes he can sell 100 items at \$2 each, but he also believes he can sell an additional 15 items for each \$0.10 he decreases the unit price. (a) Find an equation that relates the unit price and the expected number of items sold, assuming that the relation is linear. (b) Express the total amount of revenue collected in terms of the unit price.

21 A rectangle is inscribed in a semicircle of radius 4 with the base of the rectangle along the diameter of the semicircle. Express the area of the rectangle in terms of its height.

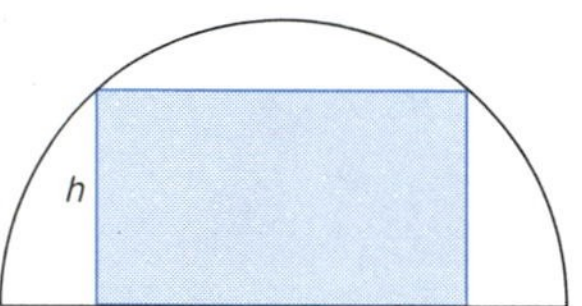


22 A right triangle has hypotenuse 5. Express the area of the triangle in terms of one of its acute angles.

23 A trough 12 ft long is partially filled with water. Each vertical end of the trough is an isosceles triangle that is 3 ft across the top and 2 ft deep. Find an equation that relates the depth and volume of the water in the trough.

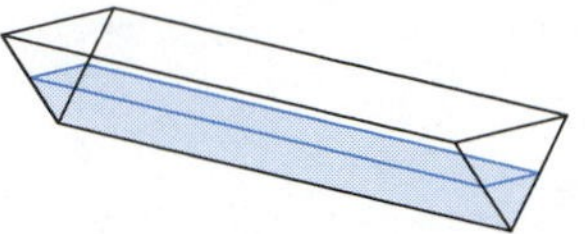

24 A conical tank that is 12 m across its circular top and 15 m deep is partially filled with water. Find an equation that relates the depth and volume of the water in the tank.

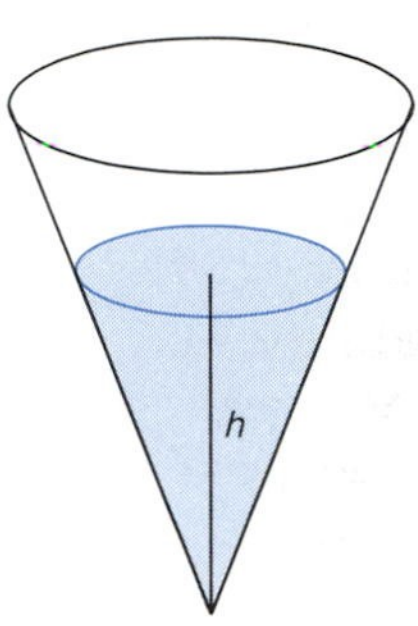


25 The frustrum of a right circular cone has radius of base $b$, radius of top $a$, and height $h$. Express its volume in terms of $a$, $b$, and $h$.

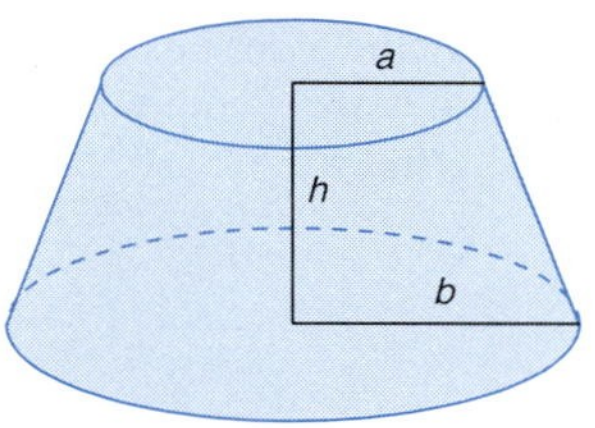


## 4.6 RELATED RATE PROBLEMS

In this section we will study quantities that are functions of time. We will study relations between these quantities and their rates of change. The problems involve determining relations between variables that are described in words, finding relations between their derivatives, and then using known data to determine unknown values and rates of change.

We will use the following procedure.

***Procedure for related rate problems***

- **Carefully read the problem.**
- **Identify and name the variables. Draw a picture, if appropriate.**
- **Determine what you want and what you know.**

- **Find equations that relate the variables.**
- **Differentiate the equations with respect to the time variable.**
- **Substitute the known values of the variables and rates in the original and differentiated equations to solve for what you want.**

The procedure is given at this time to indicate the solution pattern we will be using. It is intended to help you organize your work. You should refer back to this procedure as you go through the examples in this section.

**EXAMPLE 1**

A ladder 10 ft long is leaning against a wall. The bottom of the ladder is sliding away from the wall at a rate of 4 ft/s. Find the rate at which the top of the ladder is moving down the wall when the bottom of the ladder is 8 ft from the wall.

**SOLUTION**

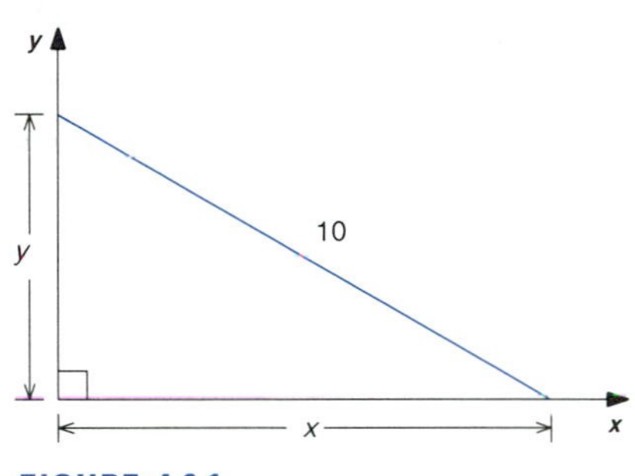

FIGURE 4.6.1

We must first determine the variables. This is done by sketching and labeling a ladder leaning against a wall, as in Figure 4.6.1. The distance of the bottom of the ladder from the wall, $x$, and the height of the top of the ladder, $y$, are variables; they are both changing as time changes. We are interested in the instant when $x = 8$, but we should not label $x = 8$ in our sketch. It is most important to realize that $x$ is a variable and that 8 is one value of this variable. The length of the ladder does not change. We should label the length of the ladder 10.

We next determine what we want and what we know, in terms of the variables. This is done in order to state the problem in mathematical terms. Also, we want to display this information clearly, so it is readily available when needed. We are asked to find the rate of change of $y$ when $x = 8$. The rate of change of $y$ is the derivative of $y$ with respect to time. What we want is displayed as follows:

*We want:* $\dfrac{dy}{dt}$ when $x = 8$.

We are told that the bottom of the ladder is moving away from the wall at a rate of 4 ft/s. Since $x$ is increasing, the rate of change of $x$ with respect to time is positive. We display what we know as follows:

*We know:* $\dfrac{dx}{dt} = +4.$

The next step is to find an equation that relates the variables. From the Pythagorean Theorem, we obtain the equation

$$x^2 + y^2 = 10^2.$$

We now differentiate the equation that relates the variables to obtain a relation between the rates of change. We must remember that $x$ and $y$ are functions of $t$, and that we are differentiating with respect to $t$. Powers of $x$

and $y$ are powers of functions and we must use the Chain Rule for their differentiation. We obtain

$$2x\frac{dx}{dt} + 2y\frac{dy}{dt} = 0.$$

We are now ready to use what we know to determine what we want. When $x = 8$, we know that $dx/dt = 4$. Substitution of these values in the differentiated equation gives

$$2x\frac{dx}{dt} + 2y\frac{dy}{dt} = 0,$$

$$2(8)(4) + 2(y)\left(\frac{dy}{dt}\right) = 0,$$

$$2(y)\left(\frac{dy}{dt}\right) = -2(8)(4).$$

We see that we need to know the value of $y$ when $x = 8$. We determine this by substituting 8 for $x$ in the original equation and then solving for $y$. We obtain

$$x^2 + y^2 = 10^2,$$

$$8^2 + y^2 = 10^2,$$

$$y^2 = 100 - 64,$$

$$y = \sqrt{36} = 6.$$

That is, $y = 6$ when $x = 8$. Using this value of $y$ in the differentiated equation, we obtain

$$2(6)\left(\frac{dy}{dt}\right) = -2(8)(4),$$

$$\frac{dy}{dt} = -\frac{2(8)(4)}{2(6)} = -\frac{16}{3}.$$

The derivative $dy/dt$ is negative. This indicates $y$ is decreasing. The top of the ladder is sliding down the wall at a rate of 16/3 ft/s when the bottom of the ladder is 8 ft from the wall. ■

### ■ EXAMPLE 2

The Universal Gas Law states that $PV = nRT$, where $P$ is the pressure in newtons/m$^2$, $V$ is the volume in cubic meters, $n$ is the number of moles of the gas, $R = 8.314$ joules per mole per degree Kelvin is the universal gas constant, and $T$ is the temperature in degrees Kelvin. The volume of 0.03 mol of a certain gas is increasing at a rate of 0.0025 m$^3$/s while the temperature is decreasing at a rate of 0.07 K/s. Find the rate of change of the pressure when the volume is $0.06\ \mathrm{m}^3$ and the temperature is 295 K. Is the pressure increasing or decreasing at this time?

**SOLUTION**

The variables are $P$, $V$, and $T$.

$$\textit{We want}: \frac{dP}{dt} \quad \text{when} \quad V = 0.06 \quad \text{and} \quad T = 295.$$

We are told that the volume is increasing, so $dV/dt$ is positive. The temperature is decreasing, so $dT/dt$ is negative.

$$\textit{We know}: \frac{dV}{dt} = +0.0025 \quad \text{and} \quad \frac{dT}{dt} = -0.07.$$

Using $n = 0.03$ and $R = 8.314$ we obtain the relation

$$PV = 0.24942T.$$

Using the Product Rule to differentiate the product $PV$, we obtain

$$(P)\left(\frac{dV}{dt}\right) + (V)\left(\frac{dP}{dt}\right) = 0.24942\frac{dT}{dt}.$$

When $V = 0.06$ and $T = 295$, the relation $PV = 0.24942\ T$ gives

$$(P)(0.06) = (0.24942)(295),$$

$$P = \frac{(0.24942)(295)}{0.06} = 1226.315.$$

This is the pressure at the time in which we are interested. The differentiated equation then gives

$$(1226.315)(0.0025) + (0.06)\left(\frac{dP}{dt}\right) = (0.24942)(-0.07),$$

$$\frac{dP}{dt} = \frac{-(0.24942)(0.07) - (1226.315)(0.0025)}{0.06} = -51.387448.$$

When the volume is $0.06\ \mathrm{m}^3$ and the temperature is 295 K, the pressure is decreasing at a rate of $51.387448\ \mathrm{N/m}^2$ per second. (The negative value of $dP/dt$ indicates that the pressure is decreasing.) ■

■ **EXAMPLE 3**

A water trough is 10 ft long. The vertical ends are isosceles triangles, 4 ft across the top with sloping sides of length 3 ft. See Figure 4.6.2a. Water is flowing into the trough at a rate of $2\ \mathrm{ft}^3/\mathrm{min}$. Find the rate of change of the depth of water in the trough when the depth is 1.5 ft.

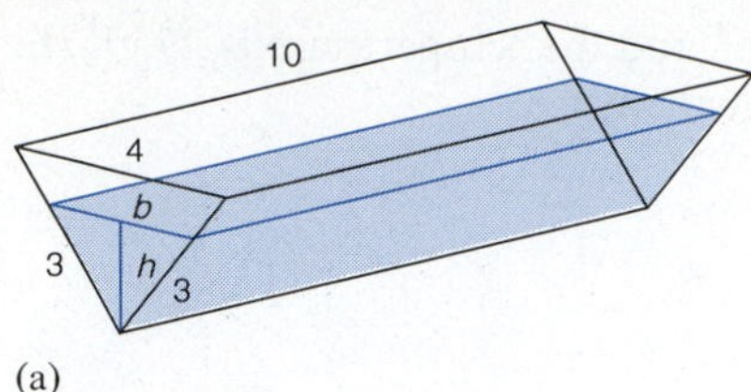


(a)

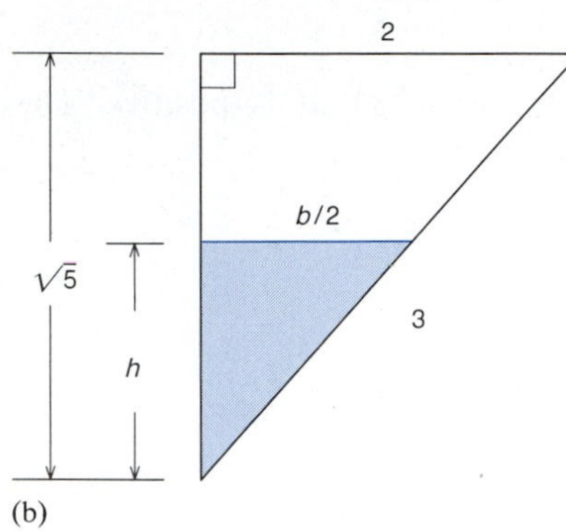


(b)

**FIGURE 4.6.2**

**SOLUTION**

The volume of water in the trough, $V$, and the depth of the water, $h$, are changing as water flows into the trough; $V$ and $h$ are variables.

*We want*: $\dfrac{dh}{dt}$ when $h = 1.5$.

*We know*: $\dfrac{dV}{dt} = +2$.

To find an equation that relates the volume and the depth, it is convenient to introduce an additional variable. In particular, let $b$ denote the distance between the sloping sides of the trough at the uppermost level of the water in the trough. From Figure 4.6.2a we see that the body of water forms a generalized right cylinder with (vertical) triangular base and (horizontal) height 10. The volume of water in the trough is then given by the formula

Volume = (Area of base)(Height).

In this case we have

$$V = \frac{1}{2}(b)(h)(10),$$

$$V = 5bh.$$

Half of the end of the trough forms a right triangle with one side 2 and hypotenuse 3. See Figure 4.6.2b. The Pythagorean Theorem implies that the other side is $\sqrt{3^2 - 2^2} = \sqrt{5}$. We can use the ratio of corresponding sides of similar triangles to obtain the relation

$$\frac{b/2}{h} = \frac{2}{\sqrt{5}},$$

$$b = \frac{4h}{\sqrt{5}}.$$

We now have two equations that relate the variables:

$$V = 5bh \quad \text{and} \quad b = \frac{4h}{\sqrt{5}}.$$

We could differentiate both of these equations and then work with four equations to determine the unknown rate. However, since the information we have and want does not involve the variable $b$, it is easier to use the second equation to eliminate $b$ from the first equation. This gives us a relation between the variables $V$ and $h$, exactly the variables that we are interested in. We obtain

$$V = 5\left(\frac{4h}{\sqrt{5}}\right)h,$$

$$V = 4\sqrt{5}h^2.$$

Differentiating this equation with respect to $t$, we have

$$\frac{dV}{dt} = 4\sqrt{5}\left(2h\frac{dh}{dt}\right).$$

When $h = 1.5$, the differentiated equation gives

$$2 = 4\sqrt{5}(2)(1.5)\left(\frac{dh}{dt}\right),$$

$$\frac{dh}{dt} = \frac{2}{4\sqrt{5}(2)(1.5)} \approx 0.0745 \text{ ft/min.}$$

The level of the water is rising at a rate of approximately 0.0745 ft/min when the depth is 1.5 ft. ■

Since the formulas that we know for differentiating trigonometric functions require that angles be measured in radians, *we must use radian measure in calculus problems that involve differentiation of trigonometric functions.*

**■ EXAMPLE 4**

A balloon is rising vertically from a point on the ground that is 200 m from an observer at ground level. The observer determines that the angle of elevation between the observer and the balloon is increasing at a rate of 0.9°/s when the angle of elevation is 45°. How fast is the balloon rising at this time?

**SOLUTION**

A sketch is given in Figure 4.6.3. The angle of elevation $\theta$ and the height $h$ are variables.

Let us change the given angles from degrees to radians. We have

$$45° = \frac{\pi}{4} \text{ rad} \quad \text{and} \quad 0.9° = (0.9°)\left(\frac{\pi \text{ rad}}{180°}\right) = \frac{\pi}{200} \text{ rad.}$$

We then determine what we want and what we know in the problem.

*We want*: $\dfrac{dh}{dt}$ when $\theta = \dfrac{\pi}{4}$.

*We know*: $\dfrac{d\theta}{dt} = +\dfrac{\pi}{200}$.

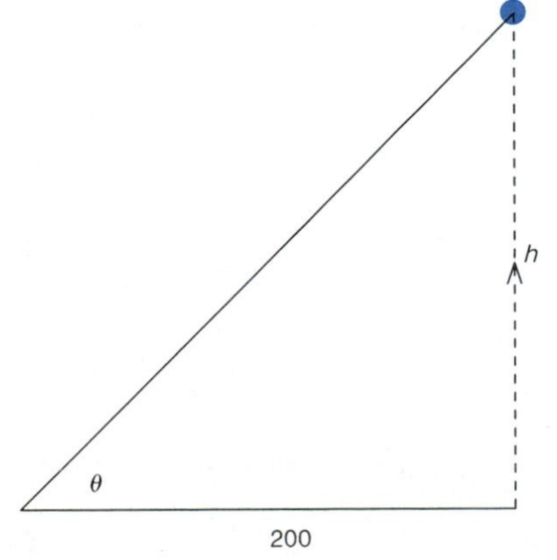

FIGURE 4.6.3

From Figure 4.6.3 we see that $\theta$ and $h$ are related by the equation

$$\tan\theta = \frac{h}{200}, \qquad \text{so} \quad h = 200\tan\theta.$$

We then differentiate the latter equation with respect to $t$ to obtain an equation that involves the rates of change of the variables. Since $\theta$ is a function of time, the Chain Rule is required for the differentiation of $\tan\theta$ with respect to $t$. We obtain

$$\frac{dh}{dt} = 200\sec^2\theta\,\frac{d\theta}{dt}.$$

When $\theta = \pi/4$, we have

$$\begin{aligned}\frac{dh}{dt} &= 200\left(\sec\frac{\pi}{4}\right)^2\left(\frac{\pi}{200}\right)\\ &= 200(\sqrt{2})^2\left(\frac{\pi}{200}\right) = 2\pi.\end{aligned}$$

The balloon is rising at a rate of $2\pi$ m/s $\approx 6.3$ m/s when the angle of elevation is $\pi/4$. ■

■ **EXAMPLE 5**

A connecting rod 60 cm long runs from a point 20 cm from the center of a flywheel to a piston. The piston moves back and forth in line with the center of the flywheel as the flywheel rotates at a rate of 0.5 revolution/s. At what speed is the piston moving when the radial line to the point on the flywheel where the rod is attached makes an angle of 90° with the radial line in the direction of the piston? What is the speed of the piston when the angle is 60°?

**SOLUTION**

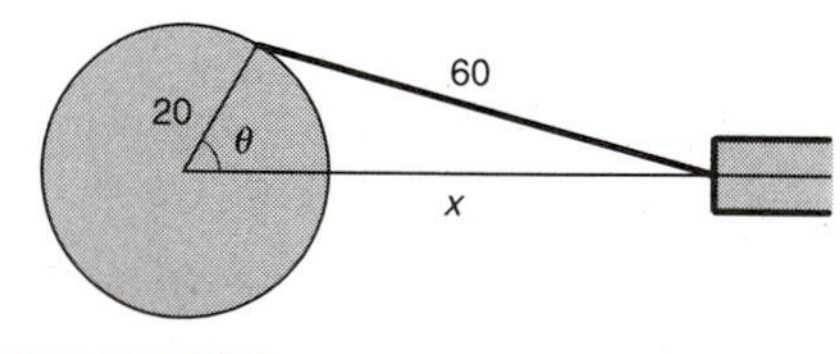


FIGURE 4.6.4

A sketch is given in Figure 4.6.4. The angle $\theta$ and the distance $x$ between the center of the flywheel and the piston are variables.

Let us express the angles in terms of radian measure. We know that

$$90^\circ = \frac{\pi}{2}\text{ rad} \quad\text{and}\quad 60^\circ = \frac{\pi}{3}\text{ rad}.$$

Also, since 1 revolution $= 2\pi$ rad, we have

$$0.5\text{ revolution} = (0.5\text{ revolution})\left(\frac{2\pi\text{ rad}}{1\text{ revolution}}\right) = \pi\text{ rad},$$

so 0.5 revolution/s $= \pi$ rad/s.

*We want*: $\frac{dx}{dt}$ when $\theta = \frac{\pi}{2}$ and when $\theta = \frac{\pi}{3}$.

Let us assume the flywheel is rotating counterclockwise, so $\theta$ is increasing.

*We know*: $\frac{d\theta}{dt} = \pi$ rad/s.

We can use the Law of Cosines to obtain the relation

$$60^2 = 20^2 + x^2 - 2(20)(x)(\cos\theta).$$

To differentiate this equation, we note that $x^2$ is the square of a function, $\cos\theta$ is the cosine of a function, and $x\cos\theta$ is a product of functions. Using the Chain Rule and the Product Rule, we obtain

$$0 = 0 + 2x\frac{dx}{dt} - 40\left[(x)\left(-\sin\theta\frac{d\theta}{dt}\right) + (\cos\theta)\left(\frac{dx}{dt}\right)\right].$$

When $\theta = \pi/2$, we have $\sin\theta = 1$ and $\cos\theta = 0$. The original equation then gives

$$60^2 = 20^2 + x^2 - 2(20)(x)(0),$$

$$x^2 = 60^2 - 20^2,$$

$$x = \sqrt{60^2 - 20^2} = \sqrt{3200} = 40\sqrt{2}.$$

The differentiated equation then gives

$$0 = 2(40\sqrt{2})\left(\frac{dx}{dt}\right) - 40\left[(40\sqrt{2})(-1)(\pi) + (0)\left(\frac{dx}{dt}\right)\right],$$

$$-2(40\sqrt{2})\left(\frac{dx}{dt}\right) = 40(40\sqrt{2})(\pi),$$

$$\frac{dx}{dt} = -20\pi \approx -62.831853.$$

The piston is moving at a speed of $20\pi$ cm/s when $\theta = \pi/2$. The sign of the derivative tells us that $x$ is decreasing, so the piston is moving toward the flywheel at this time.

When $\theta = \pi/3$, we have $\sin\theta = \sqrt{3}/2$ and $\cos\theta = 1/2$. The original equation then gives

$$60^2 = 20^2 + x^2 - 2(20)(x)\left(\frac{1}{2}\right),$$

$$x^2 - 20x - 60^2 + 20^2 = 0,$$

$$x^2 - 20x - 3200 = 0.$$

The Quadratic Formula gives

$$\begin{aligned} x &= \frac{-(-20) \pm \sqrt{(-20)^2 - 4(1)(-3200)}}{2(1)} \\ &= \frac{20 \pm \sqrt{13{,}200}}{2} \\ &= \frac{20 \pm 20\sqrt{33}}{2} \\ &= 10 \pm 10\sqrt{33}. \end{aligned}$$

Since $x$ is positive, we have $x = 10 + 10\sqrt{33}$. The differentiated equation then gives

$$0 = 2(10 + 10\sqrt{33})\left(\frac{dx}{dt}\right) - 40\left[(10 + 10\sqrt{33})\left(-\frac{\sqrt{3}}{2}\right)(\pi) + \left(\frac{1}{2}\right)\left(\frac{dx}{dt}\right)\right].$$

Multiplying both sides of this equation by 2 and then solving for $dx/dt$, we obtain

$$0 = 4(10 + 10\sqrt{33})\left(\frac{dx}{dt}\right) + 40(10 + 10\sqrt{33})(\sqrt{3})(\pi) - 40\left(\frac{dx}{dt}\right),$$

$$(40 - 4(10 + 10\sqrt{33}))\left(\frac{dx}{dt}\right) = 40(10 + 10\sqrt{33})(\sqrt{3})(\pi),$$

$$\begin{aligned} \frac{dx}{dt} &= \frac{40(10 + 10\sqrt{33})(\sqrt{3})(\pi)}{-40\sqrt{33}} = -\frac{(10 + 10\sqrt{33})(\sqrt{3})(\pi)}{\sqrt{33}} \\ &= -\frac{10\pi(1 + \sqrt{33})}{\sqrt{11}} \approx -63.886239. \end{aligned}$$

The piston is moving toward the flywheel at a rate of

$$\frac{10\pi(1 + \sqrt{33})}{\sqrt{11}} \text{ cm/s} \qquad \text{when } \theta = \frac{\pi}{3}.$$ ■

The speed of an object that is moving in a circular path is given by the formula

$$\text{Speed} = (\text{Angular velocity})(\text{Radius of circular path}),$$

where the angular velocity is in terms of radian measure of angle per unit of time. Thus, in Example 5, the end of the rod that is attached to the flywheel is moving at a rate of

$$(\pi \text{ rad/s})(20 \text{ cm}) = 20\pi \text{ cm/s}.$$

This is the same as the speed of the piston when $\theta = \pi/2$. When $\theta = \pi/3$, the speed of the piston is $10\pi(1 + \sqrt{33})/\sqrt{11} \approx 63.886239$, which is greater than $20\pi \approx 62.831853$.

## EXERCISES 4.6

1 The radius of a circle is increasing at a rate of 2 ft/s. Find the rate of change of its (a) circumference and (b) area when the radius is 6 ft.

2 Each side of a square is increasing at a rate of 3 ft/s. Find the rate of change of its (a) perimeter and (b) area when each side is 5 ft.

3 Each edge of a cube is increasing at a rate of 3 cm/s. Find the rate of change of its (a) surface area and (b) volume when each edge is 20 cm.

4 The radius of a sphere is increasing at a rate of 2 cm/s. Find the rate of change of its (a) surface area and (b) volume when the radius is 40 cm.

5 Air is being blown into a spherical balloon at a rate of $2\text{ ft}^3/\text{min}$. Find the rate of change of its radius when the radius is 1.5 ft.

6 Sand is falling into a conical pile at a rate of $2\text{ ft}^3/\text{s}$. The height of the cone is always one-third of the diameter of its base. Find the rate of change of the diameter of the pile when it contains $72\pi\text{ ft}^3$ of sand.

7 A water trough is 8 ft long. The back is vertical and the vertical ends are right triangles that are 3 ft across the top and 2 ft down the back. Water is flowing into the trough at a rate of $5\text{ ft}^3/\text{min}$. Find the rate of change of the depth of the water in the trough when the depth is 1.5 ft.

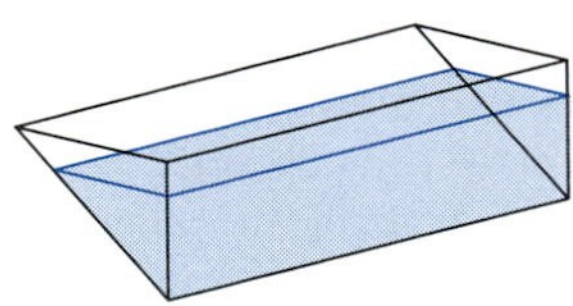

8 A water trough is 12 ft long. The vertical ends are trapezoids that are 4 ft across the top, 1 ft deep, and 2 ft across the bottom. Water is pouring into the trough at a rate of $8\text{ ft}^3/\text{min}$. Find the rate of change of the depth of water in the trough when the depth is 0.2 ft.

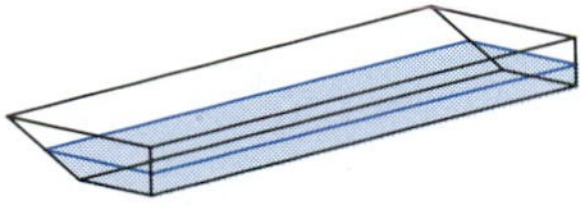

9 A swimming pool has dimensions 10 m × 20 m. The bottom of the pool slants down from a depth of 1 m along one end to a depth of 3 m and then levels off, forming a 10 m × 10 m level square at the other end. Water is being pumped into the pool at a rate of $0.2\text{ m}^3/\text{min}$. Find the rate of change of the depth of the water in the pool when the depth is (a) 0.5 m and (b) 2.5 m.

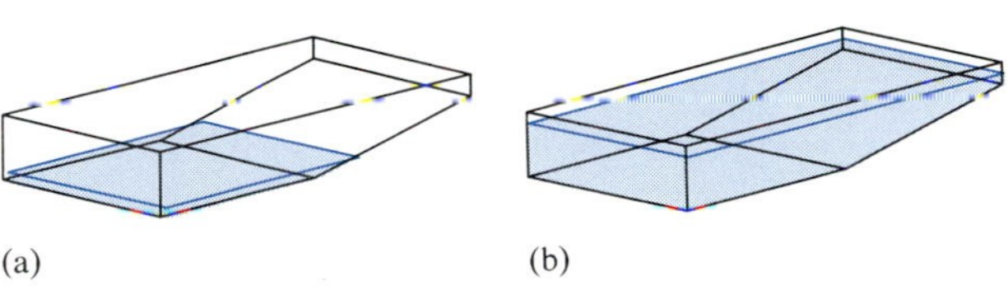

(a) (b)

10 The bottom of a 10 m × 10 m swimming pool slants from a depth of 1.5 m to a depth of 3 m at the opposite end. Water is being pumped into the pool at a rate of $2\text{ m}^3/\text{min}$. Find the rate of change of the depth of water in the pool when the depth is (a) 1 m and (b) 2 m.

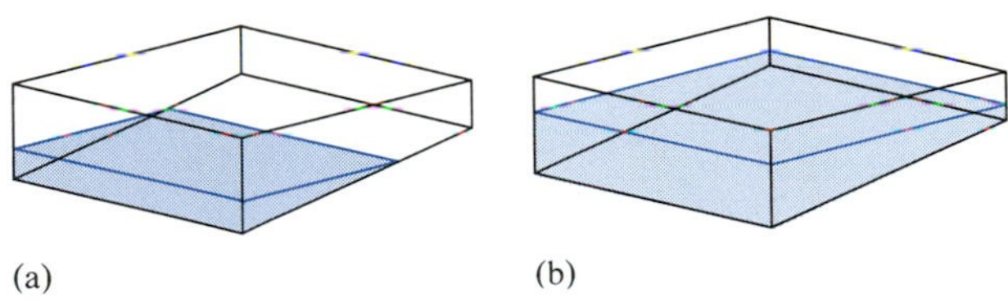

(a) (b)

11 A girl 5 ft tall walks at a rate of 6 ft/s away from a lamppost that is 15 ft high. Find the rate of change of the length of her shadow when she is (a) 10 ft and (b) 20 ft from the lamppost.

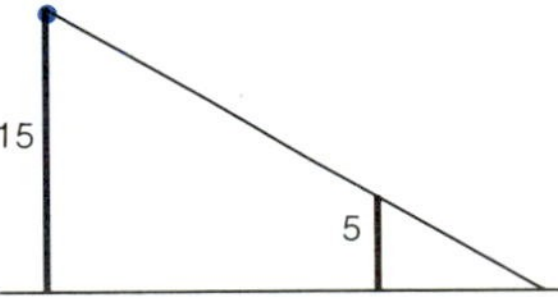


12 A right circular cone is 6 m across the top and 8 m deep. Water is flowing into the cone at a rate of $12\text{ m}^3/\text{min}$. Find

the rate of change of the depth of water in the cone when the depth is 4 m.

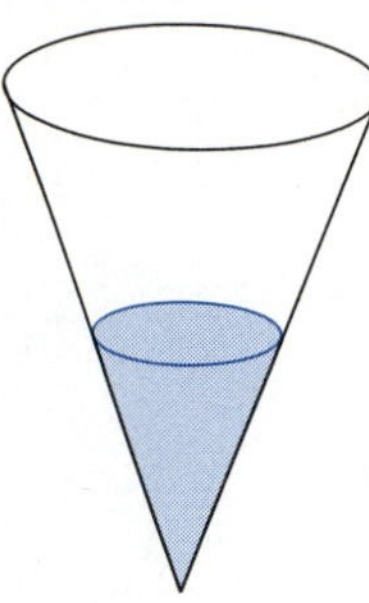

**13** Wind is blowing a kite horizontally at a rate of 6 ft/s, 120 ft above the ground. Find the rate at which the string must be let out to keep the string taut when 130 ft of string is out.

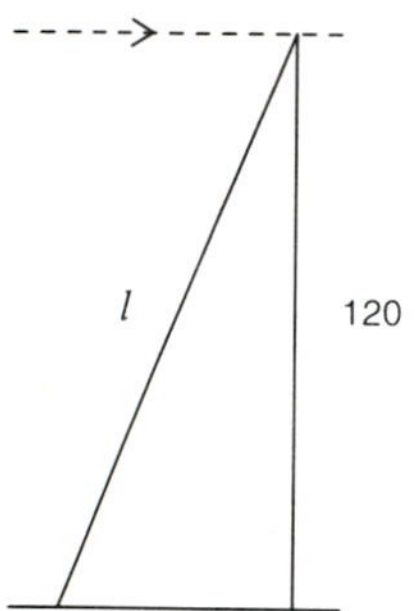


**14** A boat is being towed by a rope that goes from the boat to a pulley on a dock. The pulley is 5 ft above the point on the boat where the rope is attached to the boat. The rope is being pulled in at a rate of 2 ft/s. Find the rate at which the boat is approaching the dock when the boat is 12 ft from the dock.

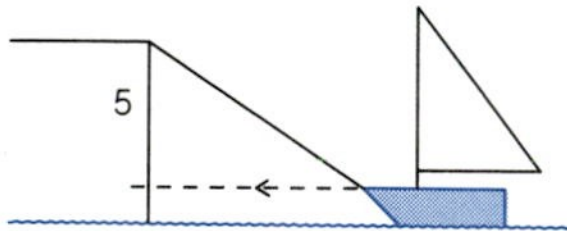


**15** A car is traveling north toward an intersection at a rate of 60 mph while a truck is traveling east away from the intersection at a rate of 50 mph. Find the rate of change of the distance between the car and truck when the car is 4 mi south of the intersection and the truck is 3 mi east of the intersection. Is the distance increasing or decreasing at this time?

**16** Find the rate of change of the distance between the car and truck in Exercise 15 when the car is 3 mi south of the intersection and the truck is 4 mi east of the intersection. Is the distance increasing or decreasing at this time?

**17** Two ships leave from the same point at the same time. The angle between their paths is 60°. One ship travels at 20 mph and the other at 30 mph. Find the rate of change of the distance between the ships 2 h after they separated.

**18** Find the rate of change of the distance between the ships in Exercise 17 2 h after they separated if the angle between their paths is 120°.

**19** A balloon is rising vertically at a rate of 2 ft/s from a point on the ground that is 50 ft from a ground-level observer. Find the rate of change of the angle of elevation between the observer and the balloon when the balloon is 120 ft above ground.

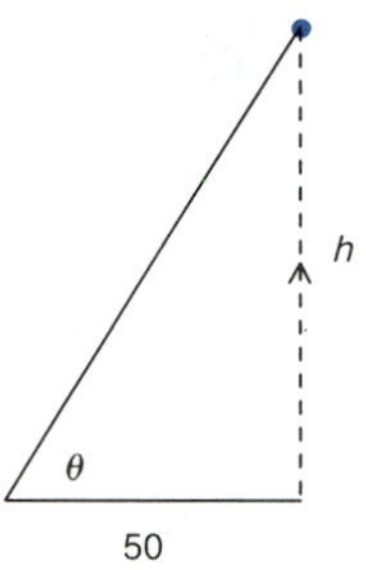


**20** A lighthouse that is 200 ft from a straight shoreline contains a light that is revolving at a rate of 0.2 revolution/s. Find the rate at which the beam from the light is moving along the shore at a point that is 100 ft from the point on the shore nearest the lighthouse.

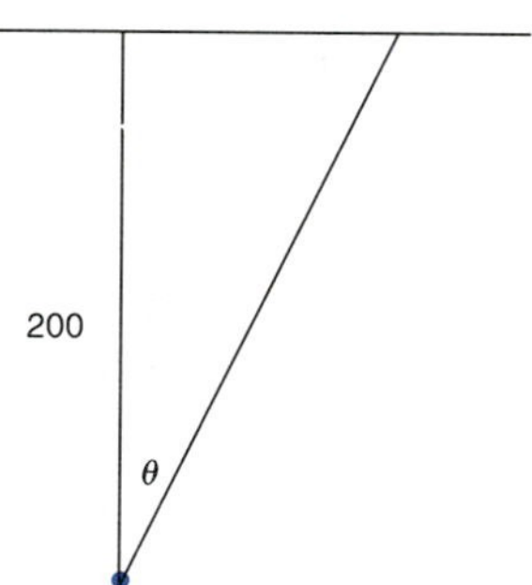


**21** An airplane is flying at 150 ft/s at an altitude of 2000 ft in a direction that will take it directly over an observer at ground level. Find the rate of change of the angle of elevation between the observer and the plane when the plane is directly over a point on the ground that is 2000 ft from the observer.

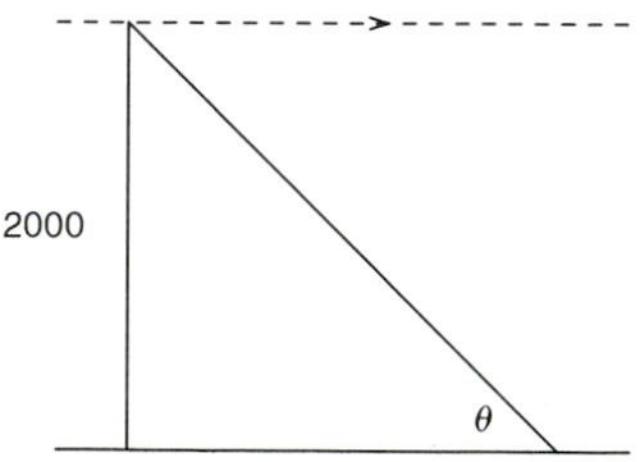

22 An airplane takes off at an angle of 30°, flying at 160 ft/s in a direction that will take it directly over an observer at ground level, 800 ft from the take-off point. Find the rate of change of the angle of elevation between the observer and the plane when the angle of elevation is 60° (a) before and (b) after the plane passes over the observer. (*Hint*: You can use the Law of Sines.)

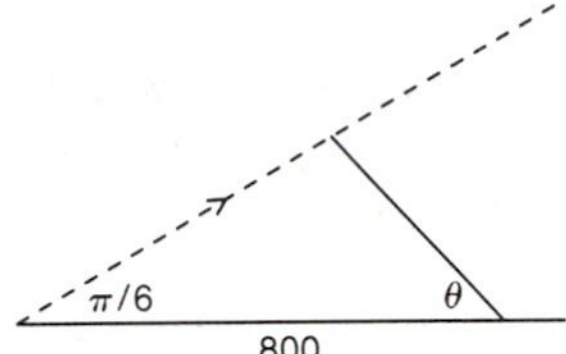


23 A tennis ball is hit at 50 m/s directly toward the net at a point that is 10 m from a spectator sitting in line with the net. Find the rate of change of the angle of the line of sight of the spectator to the ball as the ball passes over the net.

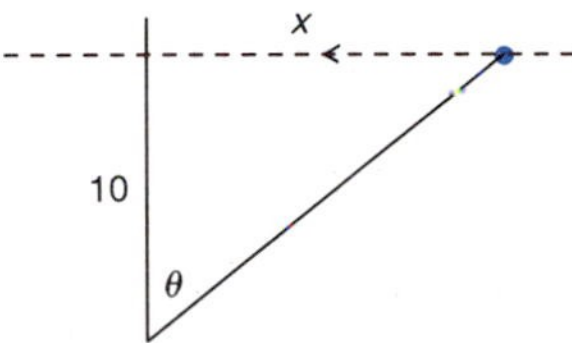


24 A rocket is rising vertically from a point on the ground that is 100 m from an observer at ground level. The observer notes that the angle of elevation is increasing at a rate of 12°/s when the angle of elevation is 60°. Find the speed of the rocket at that instant.

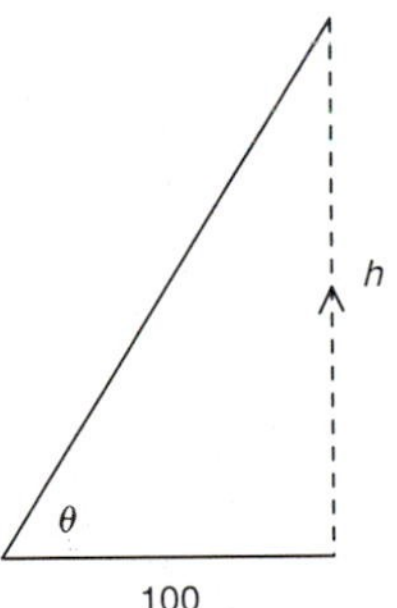


25 A right circular cylinder is being flattened in a manner that does not change its volume. Find the rate of change of its radius when the radius is 3 in. and the height is 4 in., if the height is decreasing at a rate of 0.2 in./s.

26 As a right circular cylinder is being heated, its radius is increasing at a rate of 0.04 m/s and its height is increasing at a rate of 0.15 m/s. Find the rate of change of the volume of the cylinder when the radius is 0.5 m and the height is 0.3 m.

27 Two variable resistors are connected in parallel. The resistance of the first is increasing at a rate of 0.12 Ω/s while the resistance of the second is decreasing at a rate of 0.08 Ω/s. Find the rate of change of the total resistance when the first is 4 Ω and the second is 2 Ω. Is the total resistance increasing or decreasing at this time? The total resistance $R$ of two parallel resistors $R_1$ and $R_2$ is given by the formula $1/R = (1/R_1) + (1/R_2)$.

28 The current in an electrical system is decreasing at a rate of 5 A/s while the voltage remains constant at 9 V. Find the rate of change of the resistance when the current is 18 A. The voltage $V$ in volts, the resistance $R$ in ohms, and the current $I$ in amperes are related by the formula $V = IR$.

29 The pressure of a fixed amount of a certain gas is increasing at a rate of 500 (N/m$^2$)/s while the temperature remains constant. Find the rate of change of the volume of the gas when the pressure is 1600 N/m$^2$ and the volume is 3 m$^3$. Is the volume increasing or decreasing at this time? $PV$ is constant in this case.

30 The pressure and volume of a fixed quantity of air at a fixed temperature are related by the formula $PV^{1.4} =$ constant. Find the rate of change of the pressure at the instant the pressure is 5 lb/in.$^2$ and the volume is 45 in.$^3$, if the volume is increasing at a rate of 3 in.$^3$/s. Is the pressure increasing or decreasing at this time?

31 Water is evaporating from a conical cup at a rate that varies as the area of the surface of the water. Show that the depth of water in the cup decreases at a constant rate which does not depend on the dimensions of the cup.

32 A snowball is melting at a rate that varies as its surface area. Show that its radius decreases at a constant rate.

33 A helicopter is taking off at an angle of 60°, flying at 20 m/s in a straight line away from an observer at ground level, 100 m away from the take-off point. Find the distance between the helicopter and the observer 5 s after take-off. Find the rate of change of the distance at this time.

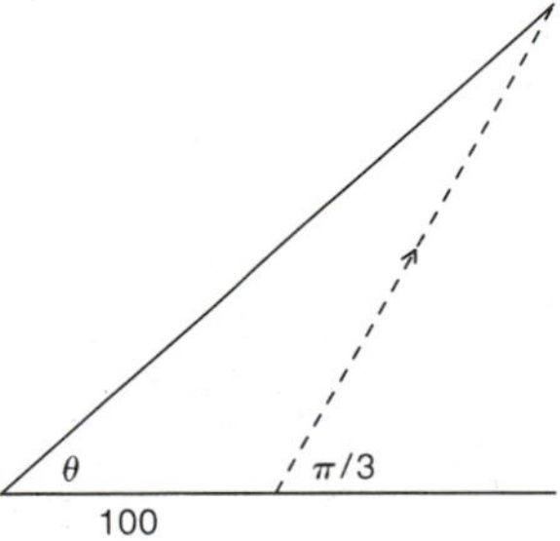


34 For the helicopter in Exercise 33 find the angle of elevation from the observer to the helicopter 5 s after take-off. Find the rate of change of the angle of elevation at this time.

35 A particle is moving along the parabola $y = x^2$ in such a way that its $x$-coordinate is increasing and

$$\left(\frac{dx}{dt}\right)^2 + \left(\frac{dy}{dt}\right)^2 = 1.$$

Find $dx/dt$ and $dy/dt$ when the particle is at the point (2, 4).

36 A particle is moving in the counterclockwise direction around the ellipse

$$\frac{x^2}{10^2} + \frac{y^2}{5^2} = 1$$

in such a way that

$$\left(\frac{dx}{dt}\right)^2 + \left(\frac{dy}{dt}\right)^2 = 1.$$

Find $dx/dt$ and $dy/dt$ when the particle is at the point (6, 4).

37 A baseball scout is using a radar gun to time a pitching prospect. He is sitting in line with the third base line, 60 ft from home plate. The radar gun registered 80 mph when the ball was 50 ft from home plate. How fast was the ball traveling at that time? (*Hint*: The radar gun measures the rate of change of the distance between the ball and the gun.)

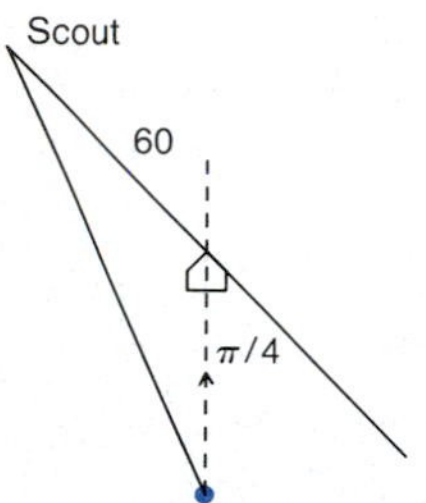


38 A policeman stops a motorist after a radar gun indicates the motorist was driving 65 mph. The motorist claims he couldn't have been going that fast and the radar reading must be higher than his actual speed because the policeman was not in line with the direction of motion of the motorist when the reading was taken. The policeman admits that he was not in line with the direction of motion, but claims that that made the radar reading lower than the actual speed. Who is right and why?

39 A cylindrical beaker has radius $r$ and height $h$. Water is being poured from the beaker by tilting the beaker. Find a formula that relates the rate of change of the volume of the water in the beaker and the rate of change of the angle the beaker makes with the vertical. Assume that the water covers the entire bottom of the beaker. (*Hint*: Use the average depth of the water from the bottom of the beaker to find the volume of water in the beaker.)

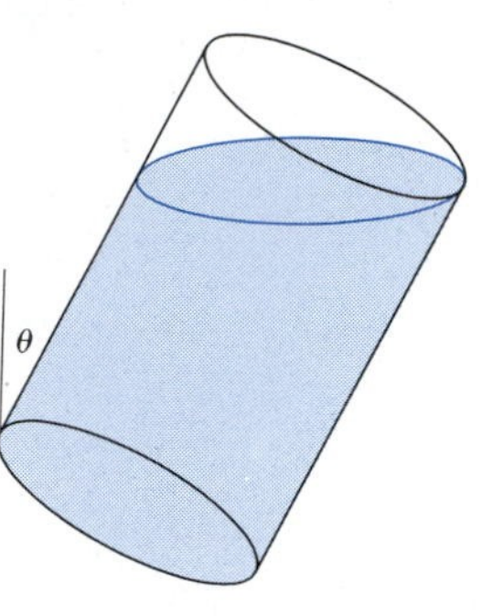


## 4.7 MAXIMUM–MINIMUM VALUE PROBLEMS

Efficient design involves finding maximum and minimum values of quantities subject to some type of constraint. For example, we may want to produce a product in such a way as to maximize profit or minimize cost subject to production and market constraints. To solve this type of problem, we will use the methods of Section 4.5 to express the quantity of interest as a function of a single real variable. We can then use the technique of Section 3.9 to find maximum and minimum values. Let us outline the steps that we will be following in the examples.

***Procedure for maximum–minimum problems***

- **Carefully read the problem.**
- **Identify and name the variables; draw a picture, if appropriate.**
- **Express the maximum-minimum variable of interest in terms of the desired independent variable and any convenient extra variables.**

- **Find equations that relate the independent variable and any extra variables.**
- **Use substitution to eliminate all extra variables from the expression for the maximum–minimum variable.**
- **Determine the domain of the independent variable.**
- **Determine the maximum–minimum values by checking values of the maximum–minimum variable**
  - **(i) at points in the domain where the derivative is zero;**
  - **(ii) at points in the domain where the derivative does not exist; and**
  - **(iii) at (or near) the endpoints of the domain.**

A sketch of the graph of the maximum–minimum variable as a function of the independent variable may be helpful.

The domain is an important part of a function. This is particularly true in maximum–minimum value problems. We have seen in Section 2.1 how the domain of a function may be restricted by the interpretation of variables as physical quantities. Other restrictions on the values of variables and the domain of a function may arise as consequences of physical constraints that exist for a particular problem.

If we obtain the maximum–minimum variable as a function that is continuous on an interval that contains both endpoints, then the Extreme Value Theorem guarantees that the function has both a maximum value and a minimum value. Moreover, the maximum and minimum values must each occur at a point inside the interval where the derivative is zero, a point inside the interval where the derivative does not exist, or an endpoint of the interval. If a function is not continuous at every point of an interval, or if the interval does not contain both endpoints, then the function may or may not have maximum or minimum values. In these cases a sketch can be used to determine if a solution exists.

Most of the functions that occur are continuous on an interval. We try to include the endpoints of the interval, if this is possible. Generally speaking, we include the endpoints of the interval in the domain if the formula we obtain makes sense at the endpoints. Care must be taken when interpreting the physical conditions described at the endpoints. Let us illustrate with a simple geometric example.

### EXAMPLE 1

A triangle is to have one side of length 3 and another side of length 4. (a) Express the area of the triangle as a function of the angle between these sides and find the angle that gives the triangle of maximum area. (b) Express the perimeter of the triangle as a function of the angle between the given sides and find the angle that gives the triangle of minimum perimeter.

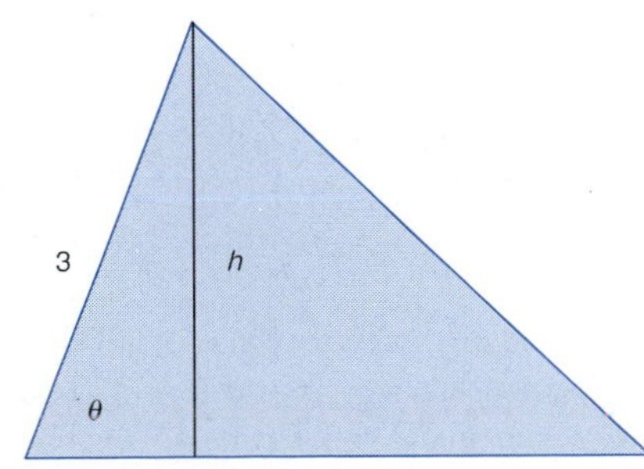

FIGURE 4.7.1

#### SOLUTION

(a) A sketch of the triangle is given in Figure 4.7.1. The maximum–minimum variable of interest is the area of the triangle, which we denote by $A$. The angle between the sides of lengths 3 and 4 is labeled $\theta$; $\theta$ is the desired independent variable. We want to express the area as a function of

$\theta$, but the *easiest* way to describe the area of a triangle is in terms of its base and height. Hence, we introduce the variable $h$, which represents the height of the triangle. We then have

$$A = \frac{1}{2}(4)(h).$$

Simplifying, we obtain

$$A = 2h.$$

Let us now use the relation between $h$ and $\theta$ to express the area in terms of $\theta$. From Figure 4.7.1 we have

$$\sin \theta = \frac{h}{3}, \quad \text{so} \quad h = 3 \sin \theta.$$

Substitution then gives us a formula for the area in terms of the angle. We have

$$A = 2h,$$

$$A = 2(3 \sin \theta),$$

$$A = 6 \sin \theta.$$

The expression $6 \sin \theta$ is defined for all $\theta$, but $\theta$ can represent an angle of a triangle only if $0 < \theta < \pi$. We include the endpoints $\theta = 0$ and $\theta = \pi$ in order to facilitate the mathematical analysis of the problem. This gives us an area function that is continuous on an interval that contains both endpoints. The Extreme Value Theorem then guarantees that the function has a maximum value and we can use the method of Section 3.9 to find its value. If the maximum value of the area function is obtained for a value of $\theta$ that satisfies $0 < \theta < \pi$, that value of $\theta$ gives the triangle of maximum area. The area function is

$$A(\theta) = 6 \sin \theta, \qquad 0 \leq \theta \leq \pi.$$

We have

$$A'(\theta) = 6 \cos \theta.$$

(i) Setting $A'(\theta) = 0$, we obtain

$$6 \cos \theta = 0.$$

**TABLE 4.7.1**

| $\theta$ | $A(\theta)$ |
|---|---|
| $\frac{\pi}{2}$ | 6 |
| 0 | 0 |
| $\pi$ | 0 |

The solution for $0 < \theta < \pi$ is $\theta = \pi/2$. (ii) The derivative exists for all $\theta$ in the interval. (iii) The endpoints of the interval are $\theta = 0$ and $\theta = \pi$. The value of $A(\theta)$ at each $\theta$ determined above is given in Table 4.7.1. The maximum value of the function is the largest value of $A(\theta)$ in the table. We see that the area function has a maximum value of 6 when $\theta = \pi/2$. This

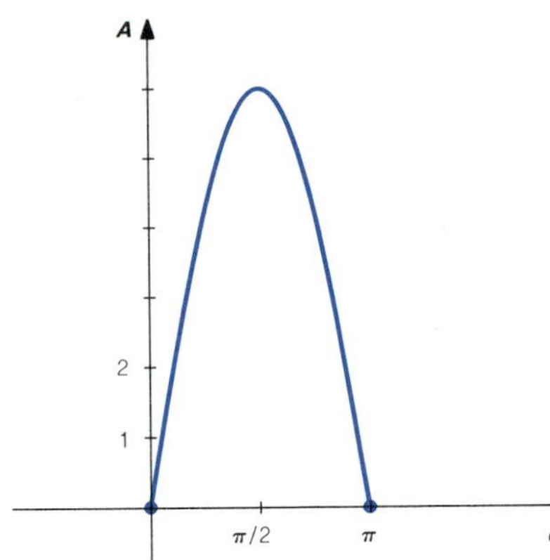

FIGURE 4.7.2

value of $\theta$ gives the triangle of maximum area. The graph of the function $A(\theta)$ is sketched in Figure 4.7.2.

(b) The triangle is sketched in Figure 4.7.3. The maximum–minimum variable is the perimeter $P$. The desired independent variable is the angle $\theta$. The side opposite $\theta$ is labeled $c$. It is easy to express the perimeter of the triangle in terms of the variable $c$. We have

$$P = 4 + 3 + c,$$

$$P = 7 + c.$$

Since we wish to express the perimeter as a function of $\theta$, we need to find an equation that relates the variables $c$ and $\theta$. The Law of Cosines gives us

$$c^2 = 4^2 + 3^2 - 2(3)(4)(\cos\theta).$$

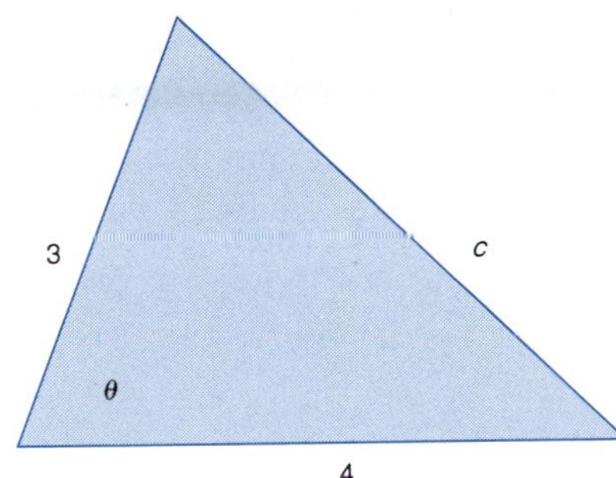

FIGURE 4.7.3

Solving for $c$ in terms of $\theta$, we obtain

$$c = \sqrt{25 - 24\cos\theta}.$$

Substitution in the expression for the perimeter then gives

$$P = 7 + \sqrt{25 - 24\cos\theta}.$$

Consideration of the domain as in part (a) gives us the function

$$P(\theta) = 7 + \sqrt{25 - 24\cos\theta}, \qquad 0 \le \theta \le \pi.$$

Using the Chain Rule to evaluate the derivative, we have

$$P'(\theta) = \frac{1}{2\sqrt{25 - 24\cos\theta}}(-24)(-\sin\theta).$$

There is no value of $\theta$ between 0 and $\pi$ for which $P'(\theta) = 0$, and there is no point in the interval where the derivative does not exist. This means that the minimum value of the function must occur at one of the endpoints of the interval and not at any point inside the interval. We have

$$P(0) = 7 + \sqrt{25 - 24\cos 0} = 7 + \sqrt{1} = 8$$

and

$$P(\pi) = 7 + \sqrt{25 - 24\cos\pi} = 7 + \sqrt{49} = 14.$$

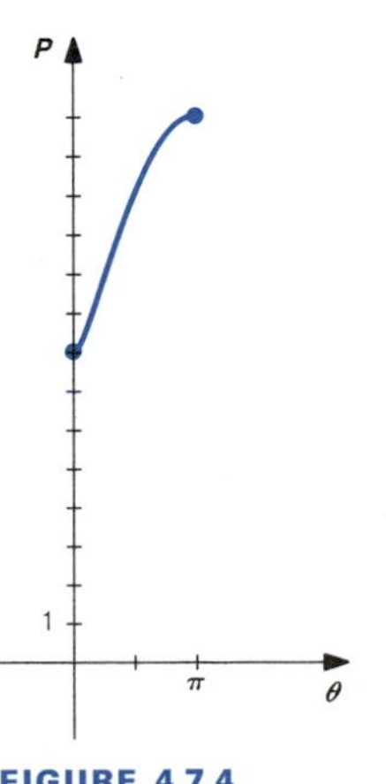

FIGURE 4.7.4

It follows that the minimum value of the perimeter function is $P(0) = 8$. However, when $\theta = 0$ (or $\theta = \pi$) the function does not represent the perimeter of an actual triangle. Among all triangles that have sides of length 3 and 4, there is none that has minimum (or maximum) perimeter. The perimeter approaches 8 as the angle $\theta$ decreases to zero, but none of the triangles has perimeter 8. See Figure 4.7.4. ■

**■ EXAMPLE 2**

A farmer has 200 m of fencing to enclose a rectangular region adjacent to a river. No fencing is required along the river. (a) Express the area of the region as a function of the length of the sides of the region perpendicular to the river. Find the dimensions and area of the region that has maximum area. (b) Repeat part (a) with the constraint that the region cannot extend more than 40 m from the river.

**SOLUTION**

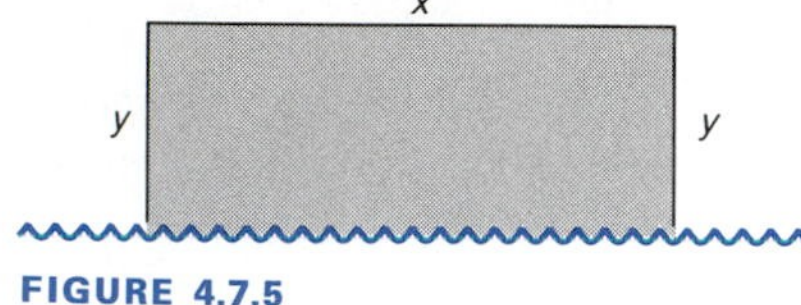


FIGURE 4.7.5

(a) The region is sketched in Figure 4.7.5. The maximum–minimum variable of interest is the area of the region, which we denote by $A$. Note that we have a choice of how much fence to use parallel to the river ($x$) and how much to use perpendicular to the river ($y$), so $x$ and $y$ are variables. Also, $x$ and $y$ are exactly the variables that are needed to obtain an expression for the area, the maximum–minimum variable of interest. As a general rule we will use as many variables as is convenient to write down an expression for the maximum–minimum variable. Any extra variables will be eliminated later, when we find relations between variables. It is very important that finding relations between variables should not be confused with the number one task of expressing the maximum–minimum variable in terms of any convenient variables. We have

$$A = xy.$$

We now find a relation between $x$ and $y$. These variables are related because they determine the amount of fencing, and we are told that the total amount of fencing is 200. From our sketch we see that the total amount of fencing is $x + 2y$, since no fencing is required along the river. Hence, consideration of the total amount of fencing gives the relation

$$x + 2y = 200.$$

The variable $y$ is the desired independent variable, so we solve the above equation for $x$ in terms of $y$. We obtain

$$x = 200 - 2y.$$

Substituting into the expression for the area of the region and then simplifying, we obtain

$$A = xy,$$

$$A = (200 - 2y)y,$$

$$A = 200y - 2y^2.$$

We now need to determine the domain. Since *length is nonnegative*, we must have $x \geq 0$ and $y \geq 0$. Then, substituting $x = 200 - 2y$ into the inequality $x \geq 0$ and solving for $y$, we obtain

$$x \geq 0,$$

$$200 - 2y \geq 0,$$

$$-2y \geq -200,$$

$$y \leq 100.$$

We include the endpoints $y = 0$ and $y = 100$ in the domain, even though these values of $y$ give a "rectangle" with two sides and area zero. This gives

$$A(y) = 200y - 2y^2, \qquad 0 \leq y \leq 100.$$

TABLE 4.7.2

| $y$ | $A(y)$ |
|---|---|
| 50 | 5000 |
| 0 | 0 |
| 100 | 0 |

The area function $A$ is continuous on an interval that contains both endpoints, so we are guaranteed a maximum (and minimum) value. We use the techniques of Section 3.9 to find the maximum value. We have

$$A'(y) = 200 - 4y.$$

(i) The derivative is zero when $y = 50$, which is a point inside the interval of definition. (ii) The derivative exists for all $y$ in the interval. (iii) The endpoints are $y = 0$ and $y = 100$. The formula for $A(y)$ is used to find the corresponding values of the area. The values are given in Table 4.7.2. We see that $A(50) = 5000\ \text{m}^2$ is the maximum area of the region described. When $y = 50$, the formula for $x$ in terms of $y$ gives $x = 200 - 2(50) = 100$. The dimensions of the region that has maximum area are $50\ \text{m} \times 100\ \text{m}$, with the longer side parallel to the river.

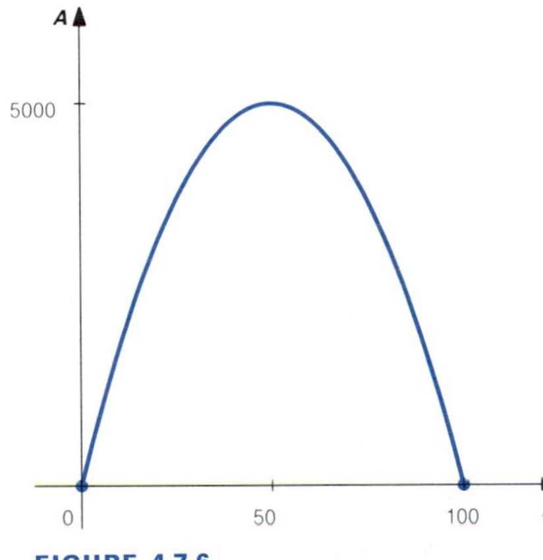

FIGURE 4.7.6

Let us note that the area $A(y)$ is clearly positive for $0 < y < 100$ and $A(0) = A(100) = 0$. It follows that the maximum value of $A(x)$ must occur for some $y$ in the interval $0 < y < 100$. Since the derivative exists for all $y$ in the interval and the derivative is zero at only one point, the value at that point must give the maximum value of $A(y)$. The graph of the function $A(y)$ is sketched in Figure 4.7.6.

(b) This problem is the same as that in part (a), except we are told that the region cannot extend more than 40 m from the river. We obtain the same formula for the area in terms of $y$, but we now have the additional restriction that $y \leq 40$. See Figure 4.7.7. The function is then

x
y 40 y

FIGURE 4.7.7

$$A(y) = 200y - 2y^2, \qquad 0 \leq y \leq 40.$$

The derivative $A'(y) = 200 - 4y$ is not equal to zero for any $y$ between 0 and 40, and the derivative exists at all of these points. It follows that the maximum value of the function must occur at one of the endpoints of the interval. We have

$$A(0) = 200(0) - 2(0)^2 = 0$$

and

$$A(40) = 200(40) - 2(40)^2 = 4800.$$

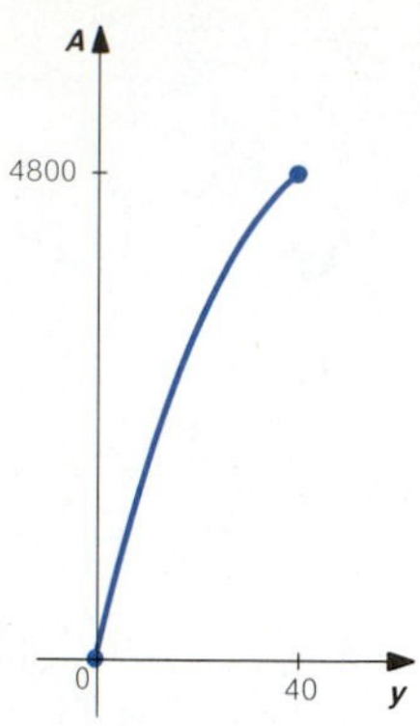

FIGURE 4.7.8

The maximum area of the region subject to the condition that it does not extend more than 40 m from the river is 4800 $m^2$. This maximum value is obtained by constructing the fence with the two sides perpendicular to the river 40 m long and the side parallel to the river 120 m long. The graph of the function $A(y)$ is sketched in Figure 4.7.8. ■

■ **EXAMPLE 3**

A rectangular field is to have area 60,000 $m^2$. Fencing is required to enclose the field and to divide it in half. Express the total amount of fencing required as a function of the length of fence that divides the field. Find the minimum amount of fencing required. What are the outer dimensions of the field that requires the least fencing?

**SOLUTION**

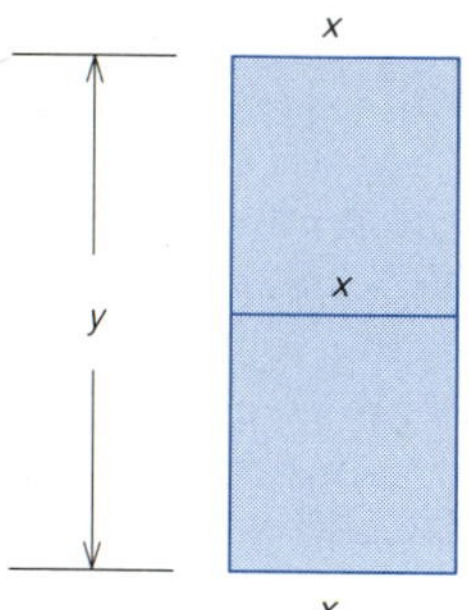

FIGURE 4.7.9

A sketch is given in Figure 4.7.9. The sides of the field are labeled $x$ and $y$; these are variables. The maximum-minimum variable of interest is the total amount of fencing, which is given by

$$F = 3x + 2y.$$

We need to find an equation that relates the variables $x$ and $y$. These variables are related because they determine the area of the field, and the area is given to be 60,000. Consideration of the area of the field gives the relation

$$xy = 60{,}000.$$

Since the desired independent variable is $x$, we solve for $y$ in terms of $x$. We obtain

$$y = \frac{60{,}000}{x}.$$

Substitution in the formula for the total amount of fencing then gives

$$F = 3x + \frac{120{,}000}{x}.$$

Since length is nonnegative, we must have $x \geq 0$ and $y \geq 0$. However, neither $y$ nor the total amount of fencing is defined if $x = 0$. (The area of the field cannot be 60,000 if the length of one side is zero.) There is no difficulty (mathematically) for any positive value of $x$. We obtain the function

$$F(x) = 3x + \frac{120{,}000}{x}, \qquad x > 0.$$

In this problem, we do not have a function that is continuous on an interval that contains both endpoints, so the Extreme Value Theorem does not guarantee a minimum value. However, we can determine the solution by using information about the graph of the function. This involves investigating the derivative and the behavior of the function near the endpoints of the interval of definition. The derivative is

$$F'(x) = 3 - \frac{120{,}000}{x^2}.$$

Simplifying, we obtain

$$F'(x) = \frac{3x^2 - 120{,}000}{x^2}.$$

We see that $F'(x) = 0$ whenever

$$3x^2 - 120{,}000 = 0,$$

$$3x^2 = 120{,}000,$$

$$x^2 = 40{,}000,$$

$$x = 200, \qquad x = -200.$$

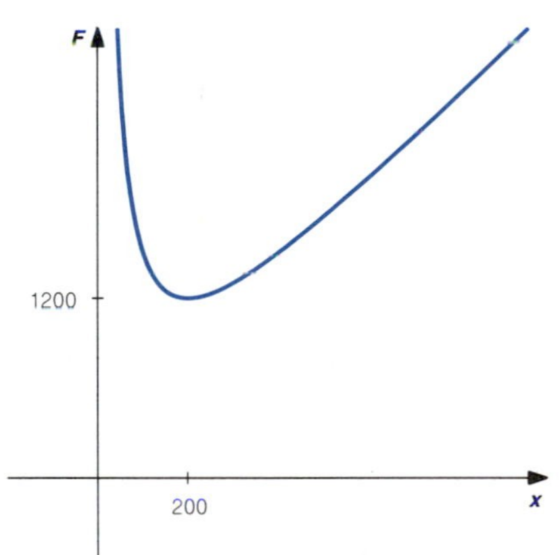

**FIGURE 4.7.10**

The formula for $F(x)$ gives $F(200) = 1200$. The point $(200, 1200)$ is the only critical point in the interval of definition. It is not difficult to see that the values of

$$F(x) = 3x + \frac{120{,}000}{x}$$

approach infinity as $x$ approaches zero and as $x$ approaches infinity. Also, the sign of $F'(x)$ can be used to determine that $F$ is decreasing on the interval $0 < x < 200$ and increasing on the interval $x > 200$. We conclude that $F(200) = 1200$ m is the minimum amount of fencing required. See Figure 4.7.10. When $x = 200$, $y = 60{,}000/200 = 300$. The outer dimensions of the field that requires the least fencing are $200 \text{ m} \times 300 \text{ m}$. ∎

**∎ EXAMPLE 4**

A cabin is located 2 km directly into the woods from a mailbox on a straight road. A store is located on the road, 5 km from the mailbox. A woman wishes to walk from the cabin to the store by cutting diagonally through the woods to a point on the road that is between the mailbox and the store, and then continuing along the road to the store. She can walk 3 km/h through the woods and 4 km/h along the road. Express the total time for her walk as a function of the distance between the mailbox and the point on the road toward which she walks. Find the point on the road toward which she should walk in order to minimize the total time for her walk and find the minimum total time.

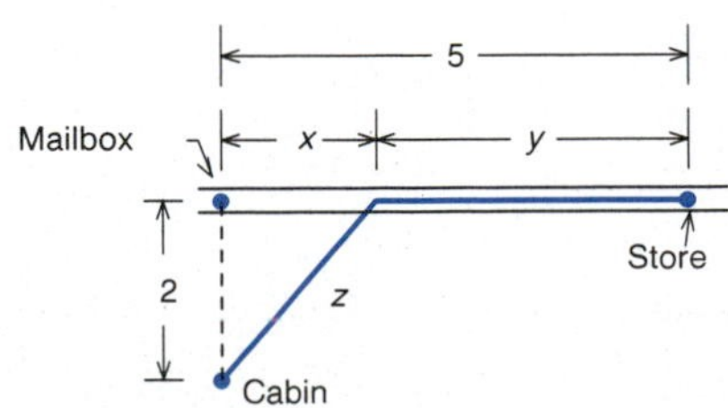

**FIGURE 4.7.11**

**SOLUTION**

A map is sketched in Figure 4.7.11. The variables are the distances labeled $x$, $y$, and $z$. The desired independent variable is $x$.

The maximum–minimum variable of interest is the total time for the walk. We know that

$$\text{Rate} = \frac{\text{Distance}}{\text{Time}}, \quad \text{so} \quad \text{Time} = \frac{\text{Distance}}{\text{Rate}}.$$

The total time $T$ is the sum of time through the woods and the time along the road. Expressing these times in terms of the variables $z$ and $y$, we obtain the expression

$$T = \frac{z}{3} + \frac{y}{4}.$$

Each of the variables $y$ and $z$ can be expressed in terms of the variable $x$. The Pythagorean Theorem gives the relation

$$z^2 = x^2 + 2^2, \quad \text{so} \quad z = \sqrt{x^2 + 4}.$$

Clearly, $x + y = 5$, so $y = 5 - x$.

Substitution then gives

$$T = \frac{\sqrt{x^2 + 4}}{3} + \frac{5 - x}{4}.$$

Since the woman is planning to walk toward a point that is between the mailbox and store, we have $0 \leq x \leq 5$. This gives the function

$$T(x) = \frac{\sqrt{x^2 + 4}}{3} + \frac{5 - x}{4}, \qquad 0 \leq x \leq 5.$$

The derivative is

$$\begin{aligned} T'(x) &= \frac{1}{3}\left(\frac{1}{2\sqrt{x^2 + 4}}\right)(2x) - \frac{1}{4} \\ &= \frac{4x - 3\sqrt{x^2 + 4}}{12\sqrt{x^2 + 4}}. \end{aligned}$$

(i) $T'(x) = 0$ implies

$$\begin{aligned} 4x &= 3\sqrt{x^2 + 4}, \\ 16x^2 &= 9(x^2 + 4), \\ 7x^2 &= 36, \\ x &= \frac{6}{\sqrt{7}}, \quad \text{since} \quad x \geq 0. \end{aligned}$$

Since squaring an equation can give extraneous roots, we must check that we have obtained a solution of the original equation. We have.

$$T\left(\frac{6}{\sqrt{7}}\right) = \frac{\sqrt{(36/7)+4}}{3} + \frac{5-(6/\sqrt{7})}{4} \approx 1.6909586.$$

(ii) $T'(x)$ exists in the interval. (iii) The values at the endpoints are $T(0) = (2/3) + (5/4) \approx 1.9166667$ and $T(5) = \sqrt{29}/3 \approx 1.7950549$. We conclude that the minimum total time is $T(6/\sqrt{7}) \approx 1.6909586$ h, or approximately 1 h 41 min. In order to complete the walk this quickly, the woman should walk toward the point on the road that is $x = 6/\sqrt{7} \approx 2.2677868$ mi, or about $2\frac{1}{4}$ mi from the mailbox. The graph of $T$ is sketched in Figure 4.7.12. ■

FIGURE 4.7.12

In Example 4, the restriction $x \leq 5$ was necessary in the domain of $T(x)$, because we used $5 - x$ to represent a distance. Also, note that $x = 5$ means that the woman walks directly through the woods to the store. For certain combinations of rates and distances this path gives the minimum time. This would be the case in Example 4 if, for example, the store were within 2.26 mi of the mailbox. The endpoint $x = 0$ indicates a path directly toward the road and then along the road to the store. One might verify that the derivative of $T$ at $x = 0$ is negative for any combination of rates and distances, so $x = 0$ can never give a minimum value of $T$.

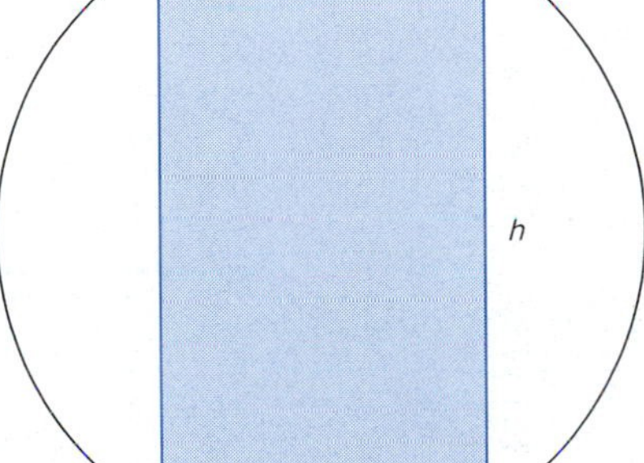

(a)

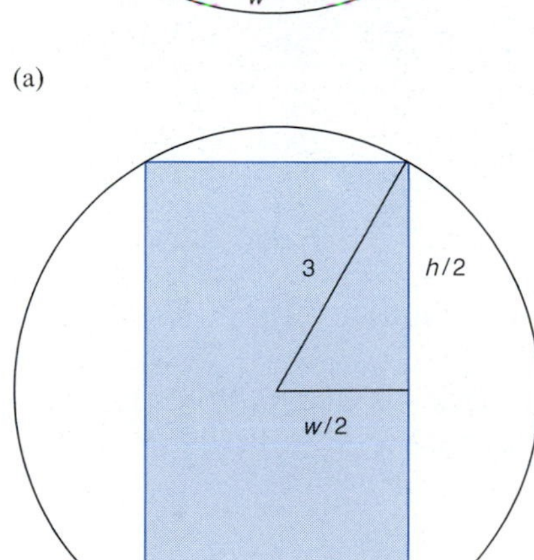


(b)

FIGURE 4.7.13

■ **EXAMPLE 5**

The strength of a rectangular beam varies as the product of the width and the square of the height of a cross section. Find the dimensions of the rectangular beam of maximum strength that can be cut from a cylindrical log with radius 3.

**SOLUTION**

A sketch of a cross section of the beam is given in Figure 4.7.13a. The variables are the width $w$ and the height $h$.

The maximum–minimum variable of interest is the strength of the beam. The statement that the strength of the beam varies as the product of the width and the square of the height translates into the equation

$$S = kwh^2,$$

where $S$ is the strength and $k$ is a constant that depends on type of wood and the units that are used. Even though we cannot determine the value of the constant $k$, $k$ is not a variable in this problem. To express $S$ as a function of a single variable, we need to find an equation that relates the variables $w$ and $h$. To do this, we draw a line from the center of the log to a corner of the beam, as in Figure 4.7.13b. This gives us a right triangle with sides $w/2$ and $h/2$, and hypotenuse 3. The Pythagorean Theorem then gives

$$\left(\frac{w}{2}\right)^2 + \left(\frac{h}{2}\right)^2 = 3^2.$$

We could now choose either $w$ or $h$ to be the independent variable and solve the above equation for the other variable. We see that in either case this will result in an expression that involves a square root. However, since the variable $h$ appears squared in the formula for the strength, we can avoid a square root in the formula for the strength if we choose the independent variable to be $w$. Solving the above equation for $h^2$ in terms of $w$, we obtain

$$w^2 + h^2 = 4(9),$$

$$h^2 = 36 - w^2.$$

Substituting in the expression for the strength and then simplifying, we obtain

$$S = kw(36 - w^2),$$

$$S = k(36w - w^3).$$

It is geometrically clear that the width of the beam cannot be greater than the diameter of the log, so we obtain the function

$$S(w) = k(36w - w^3), \qquad 0 \le w \le 6.$$

The derivative is

$$S'(w) = k(36 - 3w^2).$$

(i) $S'(w) = 0$ when

$$36 - 3w^2 = 0,$$

$$-3w^2 = -36,$$

$$w^2 = 12,$$

$$w = \sqrt{12} = 2\sqrt{3}.$$

(ii) $S'(w)$ exists for all $w$ in the interval of definition. (iii) At the endpoints, we have $S(0) = S(6) = 0$.

Since the function is zero at both endpoints and positive at points between, the value at the single critical number must be the maximum value. When $w = 2\sqrt{3}$, we have

$$h^2 = 36 - w^2,$$

$$h^2 = 36 - 12,$$

$$h = \sqrt{24} = 2\sqrt{6}.$$

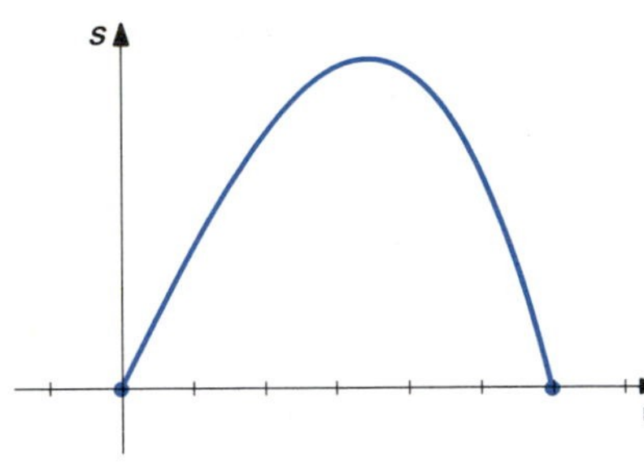


FIGURE 4.7.14

The strongest rectangular beam that can be cut from a cylindrical log with radius 3 has dimensions $2\sqrt{2} \times 2\sqrt{6}$. The shape of the strongest beam does not depend on the constant $k$, although the maximum strength does. The graph of the function $S$ is sketched in Figure 4.7.14. ■

## ■ EXAMPLE 6

A right circular cylinder is inscribed in a right circular cone so that the centerlines of the cylinder and the cone coincide. The cone has height 6 and radius of base 3. Express the volume of the cylinder as a function of its radius. Find the dimensions and volume of the cylinder that has maximum volume.

### SOLUTION

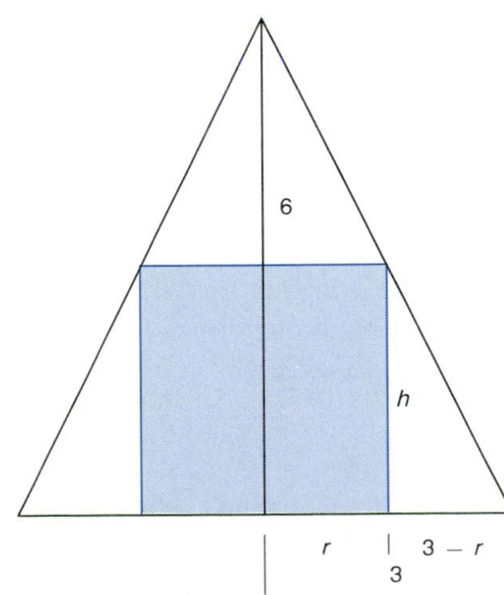


FIGURE 4.7.15

A sketch is given in Figure 4.7.15a. The radius $r$ and the height $h$ of the cylinder are variables. The maximum–minimum variable is the volume of the cylinder. We know that

$$V = \pi r^2 h.$$

To find an equation that relates $r$ and $h$, we sketch a cross section through the common centerline, as indicated in Figure 4.7.15b. Using the ratio of corresponding sides of the smaller and larger similar triangles, we obtain

$$\frac{h}{3 - r} = \frac{6}{3}.$$

The desired independent variable is $r$. Solving for $h$ in terms of $r$, we have

$$h = 2(3 - r),$$

$$h = 6 - 2r.$$

Substituting in the equation for the volume of the cylinder, we obtain

$$V = \pi r^2(6 - 2r),$$

$$V = 2\pi(3r^2 - r^3).$$

Since the radius of the cylinder cannot exceed the radius of the base of the cone, we obtain the function

$$V(r) = 2\pi(3r^2 - r^3), \qquad 0 \le r \le 3.$$

The derivative is

$$V'(r) = 2\pi(6r - 3r^2).$$

We see that $V'(r) = 0$ when

$$6r - 3r^2 = 0,$$

$$3r(2 - r) = 0.$$

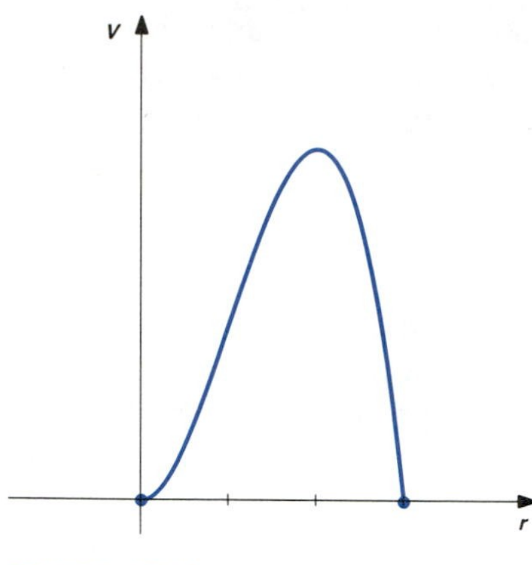

FIGURE 4.7.16

The only solution inside the interval of definition is $r = 2$. The derivative exists at all points in the interval. The function is zero at both endpoints and is positive between, so the maximum volume must occur at the single critical point. The maximum volume of the inscribed cylinder is

$$V(2) = 2\pi(3(2)^2 - (2)^3) = 8\pi.$$

When $r = 2$, the formula for $h$ in terms of $r$ gives

$$h = 6 - 2(2) = 2.$$

The cylinder of maximum volume has height 2 and radius 2. The graph of the function $V$ is sketched in Figure 4.7.16. ■

Many practical maximum-minimum problems involve variables that are restricted to **discrete** values. For example, the number of items manufactured must be a nonnegative integer and the price charged per item must be restricted to values that are not fractions of a cent. Calculus does not apply directly to problems that involve discrete variables. However, we can sometimes get an indication of the solution of the discrete problem by applying calculus to a real-variable **mathematical model** of the problem. That is, we simply assume that each variable ranges over all real numbers in some interval, even though the physical interpretation of the variable makes sense only for discrete values. This allows us to use calculus to find a solution of the problem for the model. Hopefully, the solution of the discrete problem involves values of the variables that are near the values that give the solution of the problem for the model. It is not within the scope of this text to carefully discuss when this is actually the case. We will simply accept the solution of the model problem as an approximate solution of the discrete problem.

### ■ EXAMPLE 7

From experience, a promoter knows that he can sell all 400 tickets to a concert if he charges \$6 per ticket. Also, he knows that for each \$1 increase in price, he will sell approximately 35 fewer tickets. Express the total revenue collected for tickets as a function of the price per ticket. For what price per ticket would he expect to collect the most money from the sale of tickets and how much money would he expect to collect?

#### SOLUTION

The variables in this problem are the price per ticket, $p$, and the number of tickets sold, $n$. In practice, $p$ must be restricted to values that do not involve fractions of a cent and $n$ must be an integer. In order to use calculus, we assume that $p$ and $n$ are real-number variables. The maximum-minimum variable is the total revenue collected, $R$, which is given by the formula

$$R = np.$$

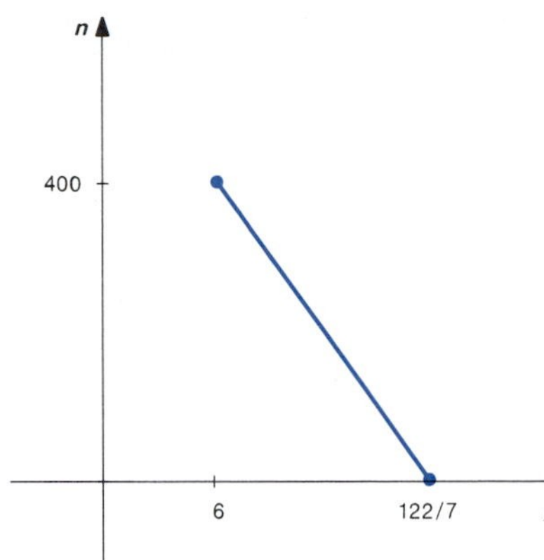

FIGURE 4.7.17

The statement that the number of tickets sold will decrease by 35 for each \$1 increase in ticket price indicates a linear relation between $n$ and $p$. See Figure 4.7.17. Since $n = 400$ when $p = 6$, we know that the line contains the point $(p, n) = (6, 400)$. The slope of the line is

$$\frac{\text{Change in } n}{\text{Change in } p} = \frac{-35}{1} = -35.$$

The equation of the line that contains the point $(p, n) = (6, 400)$ with slope $-35$ is

$$n - 400 = -35(p - 6).$$

The desired independent variable is $p$. Solving for $n$ in terms of $p$, we obtain

$$n = 400 - 35p + 210,$$

$$n = 610 - 35p.$$

Substitution in the formula for the total revenue then gives

$$R = (610 - 35p)(p),$$

$$R = 610p - 35p^2.$$

We assume that $p \geq 6$, since all of the tickets can be sold at this price. Since $n = 610 - 35p$ must be nonnegative, we also have $p \leq 122/7$. This gives the function

$$R(p) = 610p - 35p^2, \qquad 6 \leq p \leq \frac{122}{7}.$$

The derivative is

$$R'(p) = 610 - 70p.$$

We see that $R'(p) = 0$ when $p = 61/7$. This is the only critical point of the function. We have

$$R\left(\frac{61}{7}\right) = 610\left(\frac{61}{7}\right) - 35\left(\frac{61}{7}\right)^2 \approx 2658.$$

The values at the endpoints are $R(6) = 610(6) - 35(6)^2 = 2400$ and $R(122/7) = 0$. The maximum value of the revenue function is given by $p = 61/7 \approx 8.71$. The promoter would expect to collect the most money from tickets for the concert if he charges approximately \$8.71 per ticket. He would expect to collect approximately \$2658. The graph of $R(p)$ is given in Figure 4.7.18. ■

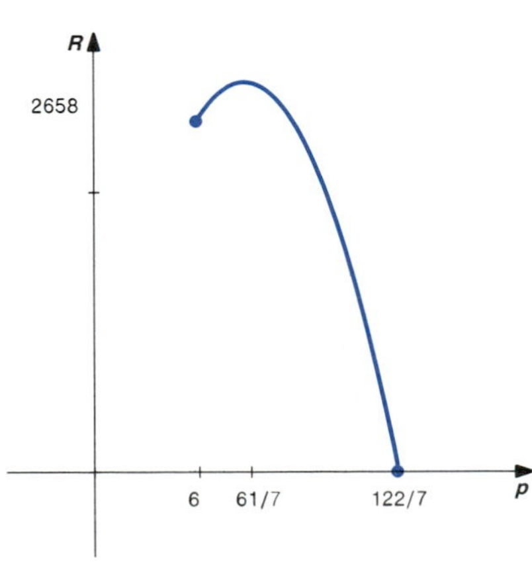

FIGURE 4.7.18

## EXERCISES 4.7

**1** A right triangle has hypotenuse 5. Express the area of the triangle as a function of one of its acute angles. Find the angles and area of the triangle that has maximum area.

**2** A right triangle has hypotenuse 5. Express the perimeter of the triangle as a function of one of its acute angles. Find the angles and perimeter of the triangle that has maximum perimeter.

**3** A farmer has 900 m of fencing. The fencing is to be used to enclose a rectangular field and to divide it in half. Express the area of the field as a function of the length of fencing used to divide the field in half. Find the outer dimensions and area of the field that has maximum area.

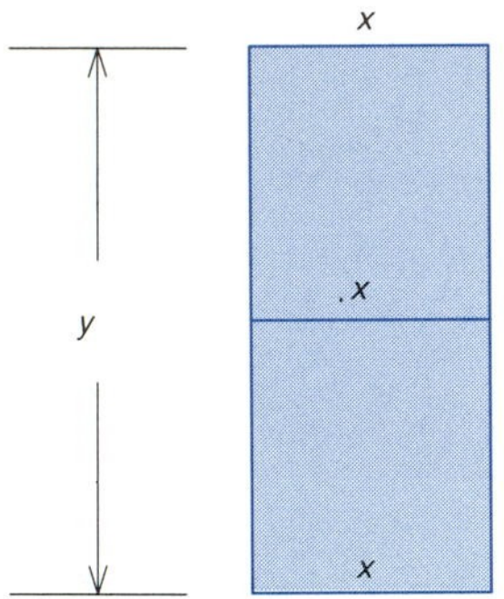


**4** A gutter with vertical sides and no top is to be constructed by bending up equal sides of a rectangular piece of metal that is 10 in. wide. Express the area of the cross section of the gutter as a function of the length turned up on each side. Find the dimensions and area of the cross section that has maximum area.

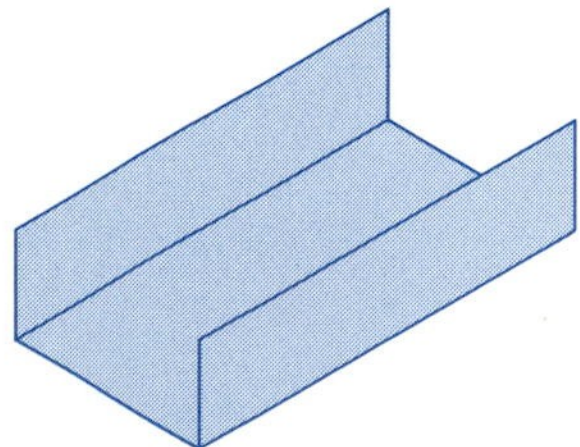

**5** A farmer wishes to enclose a rectangular region adjacent to a river. No fencing is required along the river. The area of the region is to be 15,000 $m^2$. (a) Express the amount of fencing required as a function of the length of fence parallel to the river. Find the dimensions and amount of fencing required for the field that requires the least amount of fencing. (b) Repeat part (a) with the constraint that the field cannot extend more that 150 m along the river.

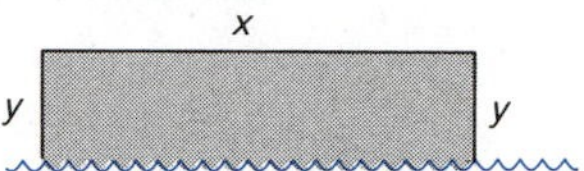


**6** (a) A rectangular pen is to be constructed so that one side of the pen contains all of a 100-ft straight fence that is already standing and also that the pen will have an area of 16,000 $ft^2$. New fencing is not required along the existing fence. Express the amount of new fencing required as a function of the length of the side of the pen that contains the existing fence. Find the dimensions and amount of new fencing required for the pen that requires the least new fencing. (b) Repeat part (a), except assume that one side of the pen must contain all of a 150-ft length of an existing fence.

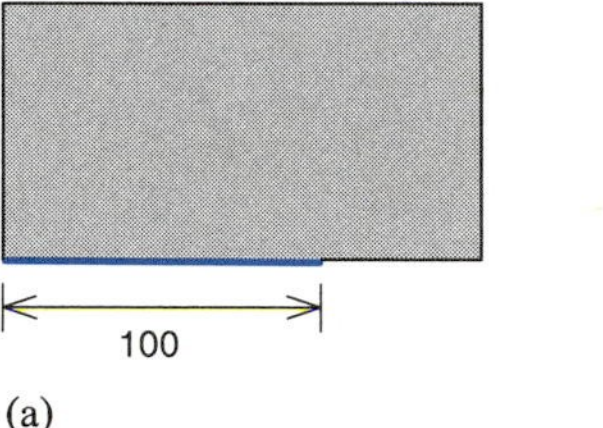

(a)

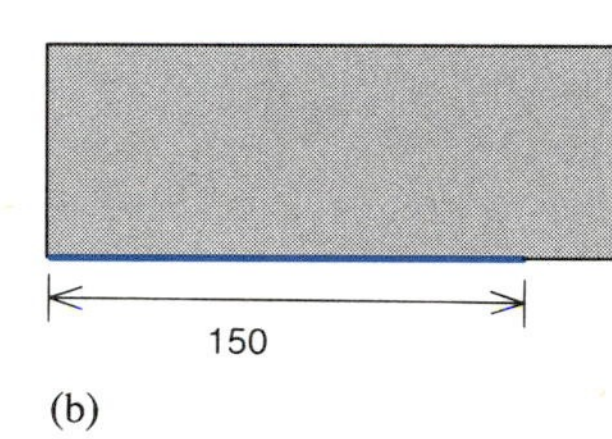

(b)

**7** (a) A 3-m length of wire is to be cut into two pieces. One piece is to be bent to form a circle and the other piece is to be bent to form a square. Find the length of wire that is used for each of the circle and the square, and the total area of the circle and the square that have minimum total area. (b) Can the wire be cut in such a way that the total area of the circle and square are maximum? What configuration gives the maximum area if it is not required that the wire is cut, so all of the wire may be used to form one of either a circle or a square?

**8** Solve Exercise 7 with an equilateral triangle in place of the circle.

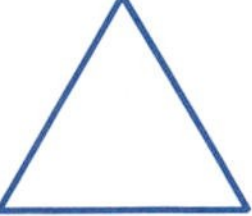

**9** A rectangular box with no top is to be constructed by cutting equal squares from each corner of a 2 ft × 3 ft rectangular piece of material and bending up the sides. Express the volume of the box as a function of its height. Find the dimensions and volume of the box that has maximum volume.

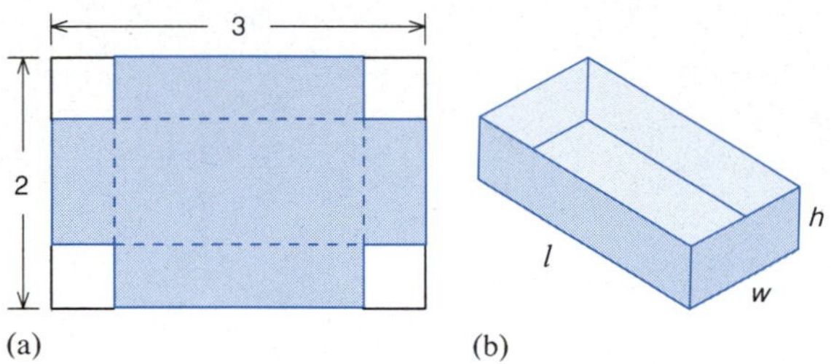

(a) (b)

**10** Material for the top and bottom of a cylindrical container costs $\$2/\text{ft}^2$. Material for the curved part costs $\$1/\text{ft}^2$. The allotment for the total cost of material is \$48. Express the volume of the cylinder as a function of its radius. Find the dimensions and volume of the cylinder that has maximum volume.

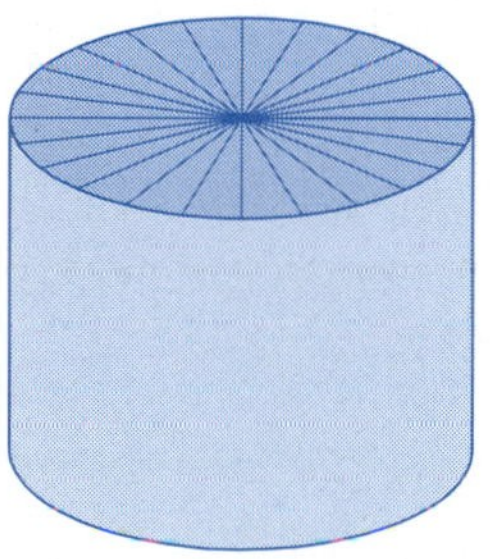

**11** Fencing is required to enclose a rectangular field of area 16,000 $\text{m}^2$ and to divide it in half. Fencing to enclose the field costs \$3/m and the fencing used to divide the field costs \$2/m. Express the total cost of fencing as a function of the length of fence that divides the field. Find the dimensions and total cost of fencing for the field that is least expensive to fence.

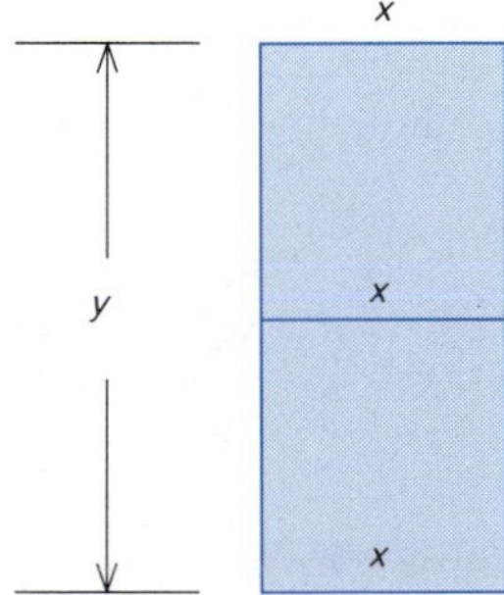


**12** A rectangular field adjacent to a river is to be enclosed. Fencing along the river costs \$5/m and fencing for the other three sides costs \$3/m. The area of the field is to be 9000 $\text{m}^2$. Express the total cost of fencing as a function of the length of fence along the river. Find the dimensions and total cost of fencing for the field that is least expensive to fence.

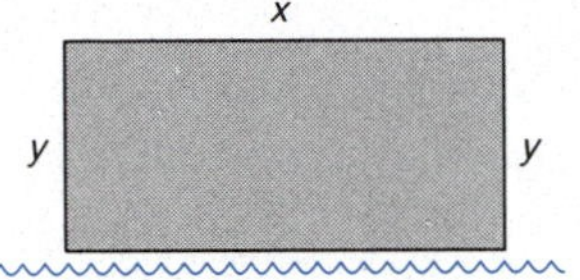


**13** (a) A factory is located 100 m downstream and on the opposite side of a river from a power plant. The river is 60 m wide. A power line is to be run from the power plant, diagonally under the river, and then along the bank to the factory. It costs \$130/m to lay the line under the river and \$50/m along the bank. Express the total cost as a function of the distance between the point across from the power plant and the point where the line emerges from the river. Find the point where the line emerges from the river and the total cost for the cheapest path. (b) Repeat part (a), but with the factory 20 m downstream.

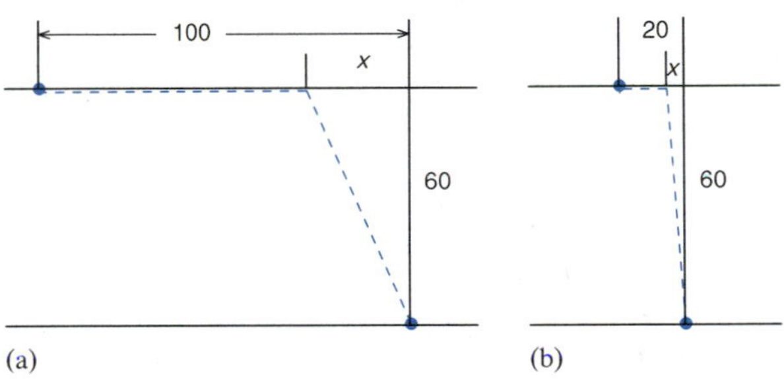

(a) (b)

**14** (a) A man is in a boat, 300 m directly into the ocean from his house on the shore. A lighthouse is on the shore, 200 m from his house. He can row 2 km/h and walk 4 km/h. He plans to row toward a point on the shore between his house and the lighthouse, and then walk to the lighthouse. Find the minimum total time this will take. (b) Repeat part (a), but with the lighthouse 160 m from the man's house.

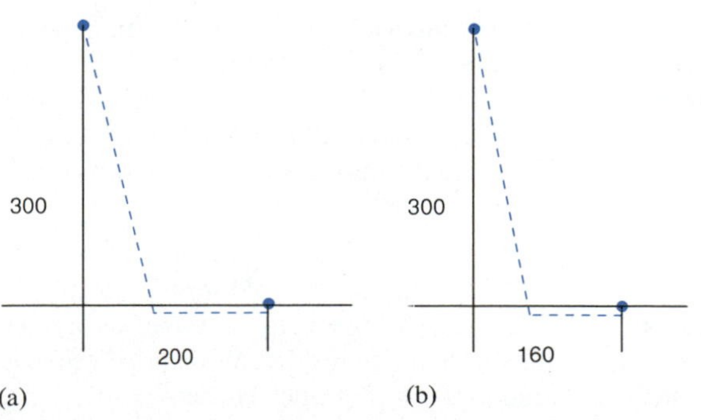

(a) (b)

**15** A post 6 ft high is 21 ft from a post 8 ft high. A line is to be run from the top of one post to the ground and then to the top of the other post. Express the length of line as a function of the distance between the 6-ft post and the point where the line touches the ground. Find the minimum length of line.

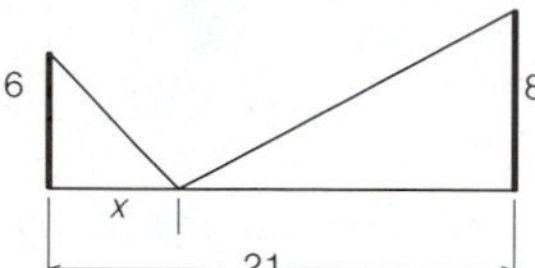


**16** The speed of light through air is $v_0$ ft/s and the speed through water is $v_1$ ft/s. An object is submerged in water 3 ft below a point that is 10 ft from a point where a boy is standing. The boy's eyes are 5 ft above water level. Express the total time in seconds for a light ray to travel from the object to the boy's eyes, as a function of the distance between the boy and the point where the ray emerges from the water. Show that the total time for a ray of light to travel from the object to water level and then through air to the boy's eyes is minimum when

$$\frac{\sin \theta_0}{v_0} = \frac{\sin \theta_1}{v_1}.$$

(Light travels in a straight line through water and air, but bends sharply as it passes from water to air.)

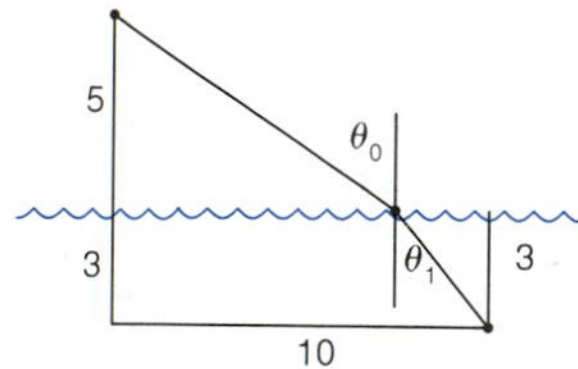


**17** An apartment building with 40 apartments can be completely rented if the rent is set at $200/month. One apartment becomes vacant for every $10 increase in rent. Express the total amount of rent collected as a function of the unit rent. What rent will give the maximum total amount of rent collected per month?

**18** A complete lot of 300 items can be sold at a price of $2 each. One fewer of the items can be sold for each $0.02 the price is increased. Express the total amount of money collected as a function of the unit price. What price will give the maximum total amount of money collected?

**19** A sector of a circle has perimeter 4. Express the area of the sector as a function of the radius. Find the central angle and radius that give the sector of maximum area.

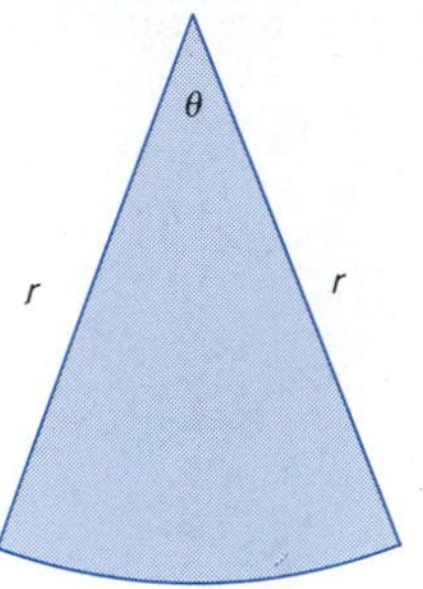


**20** An isosceles triangle has perimeter 4. Express the area of the triangle as a function of the length of the two equal sides. Find the length of the equal sides and the included angle of the triangle that has maximum area.

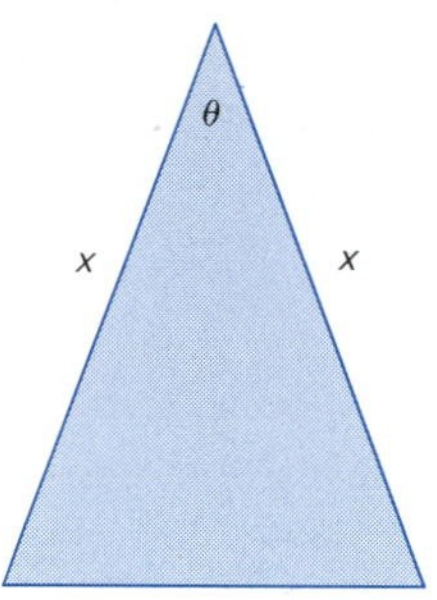


**21** (a) A rectangle is inscribed in a right triangle with the corner of the rectangle at the right angle of the triangle. Show that the area of the rectangle is maximum when a corner of the rectangle is at the midpoint of the hypotenuse of the triangle. (b) A rectangle is inscribed in a triangle with base of the rectangle along the longest side of the triangle. Explain why the rectangle of maximum area has two corners on midpoints of sides of the triangle. [*Hint for part (b)*: Use part (a) to show that any other such rectangle has smaller area.]

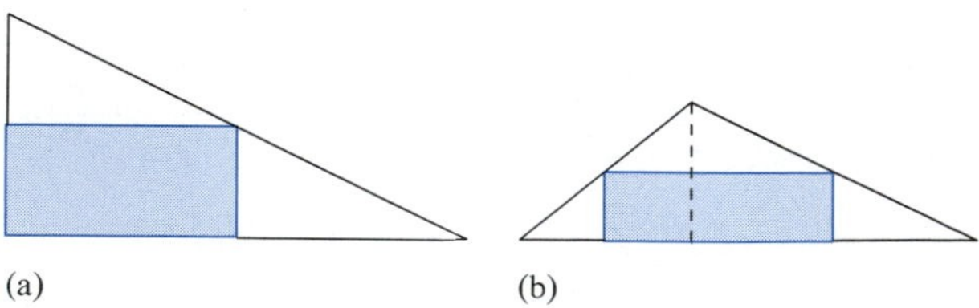


**22** A rectangle is inscribed in the ellipse

$$\frac{x^2}{a^2} + \frac{y^2}{b^2} = 1$$

with the centerlines of the rectangle along the axes of the ellipse. Show that the maximum area of the rectangle is $2ab$.

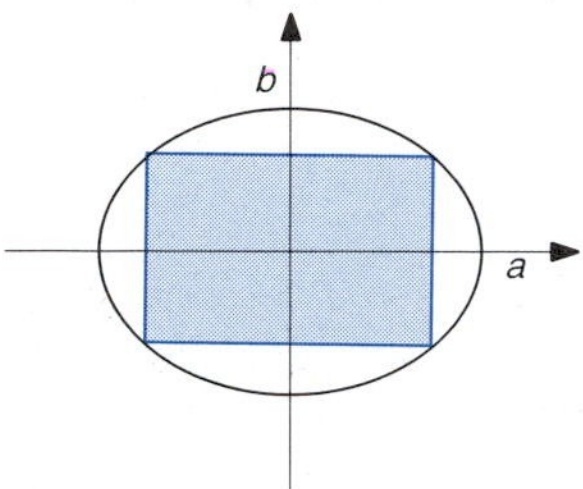


**23** A conical cup is to be made by cutting a sector from a circle of radius 6. Express the volume of the cup as a function of the height of the cup. Find the central angle of the sector that gives the cup of maximum volume.

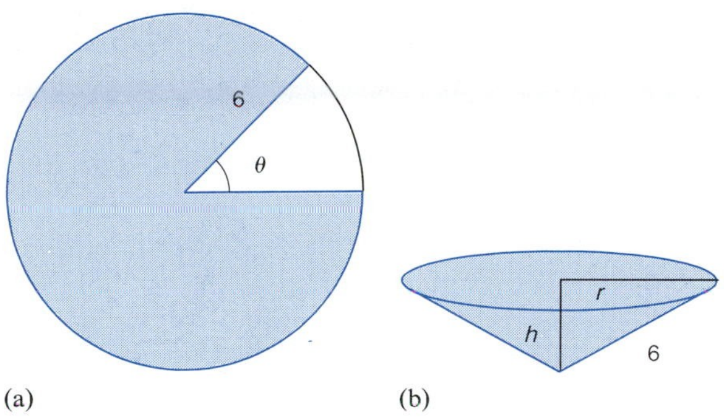


(a) (b)

**24** A conical cup is to be made by cutting a sector from a semicircle of radius 6. Express the volume of the cup as a function of the height of the cup. Find the central angle of the sector that gives the cup of maximum volume. (*Hint*: What is the smallest the height can be?)

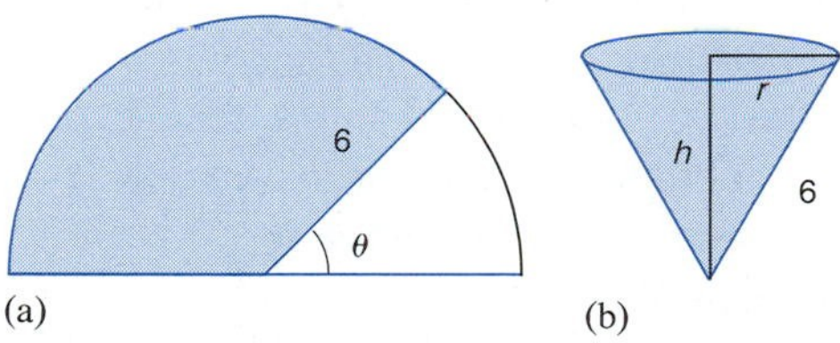


(a) (b)

**25** (a) A gutter is to be formed by bending equal lengths at equal angles from rectangular material that is 12 in. wide. The gutter is to be 3 in. deep. Express the area of a cross section as a function of the angle that each side is turned up. Find the angle that gives the maximum cross-sectional area. (b) Repeat (a) in the case that the gutter must be 5.5 in. deep.

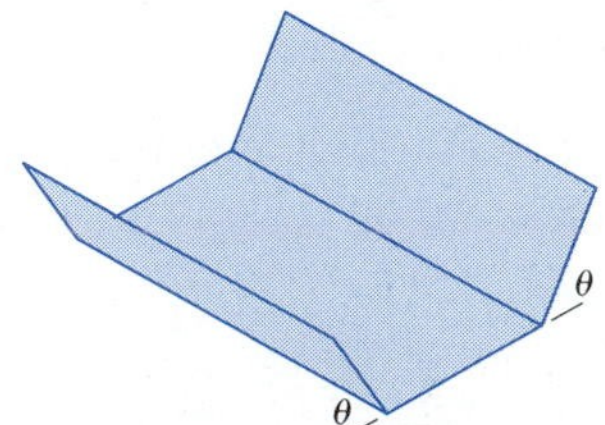


**26** The base of a bin is 6 ft × 10 ft. One of the 6-ft sides is 4 ft tall and hinged at the bottom, so it may be lowered away from the bin. The other three sides are vertical and 4 ft tall; two sides extend in quarter circles beyond the hinged end. Express the volume of the bin to the level of the top of the hinged side as a function of the angle that it makes with the horizontal. Find the angle that gives the maximum volume.

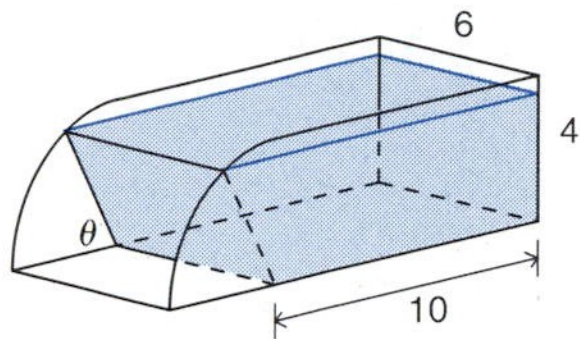


**27** The cross section of a parabolic dome is given by the equation $y = (10{,}000 - x^2)/50$; the floor corresponds to $y = 0$. Find the height at which lights should be installed in order to give the maximum illumination of the floor at its center, if illumination varies directly as the cosine of the angle the rays make with the direction perpendicular to the surface and inversely as the square of the distance between the source and the point of the surface.

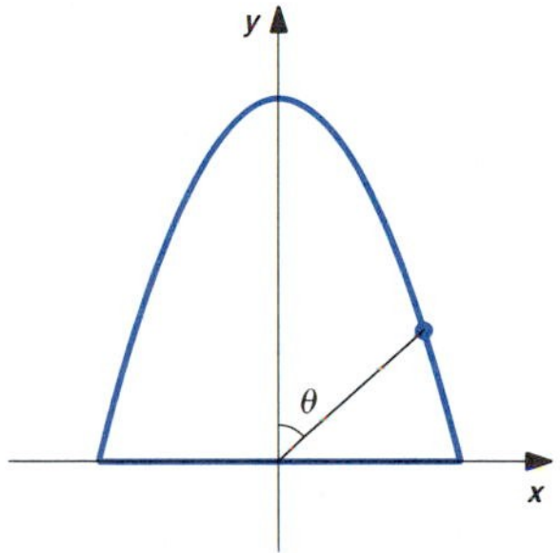


**28** A right circular cylinder with volume $8\pi$ is to be constructed from two equal circles and a rectangle that are cut from a rectangular piece of material in one of the three ways indicated below. Determine the radius and height of the cylinder and the method of cutting that requires the least area of the rectangular piece from which the parts are cut. [*Hint*: $h \geq 2r$ in (a) and $h \geq 4r$ in (b).]

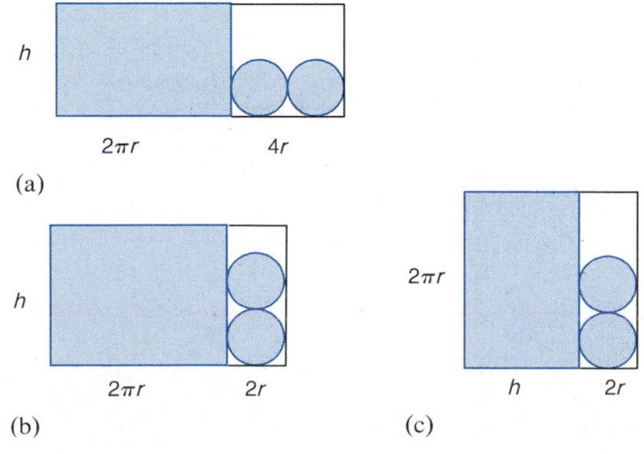


(a)

(b) (c)

**29** Show that the point $(x_1, y_1)$ on the nonhorizontal line $y = mx + b$ that is closest to a point $(x_0, y_0)$ not on the line satisfies

$$\frac{y_1 - y_0}{x_1 - x_0} = -\frac{1}{m}.$$

This means that the line through the points $(x_0, y_0)$ and $(x_1, y_1)$ is perpendicular to the line $y = mx + b$.

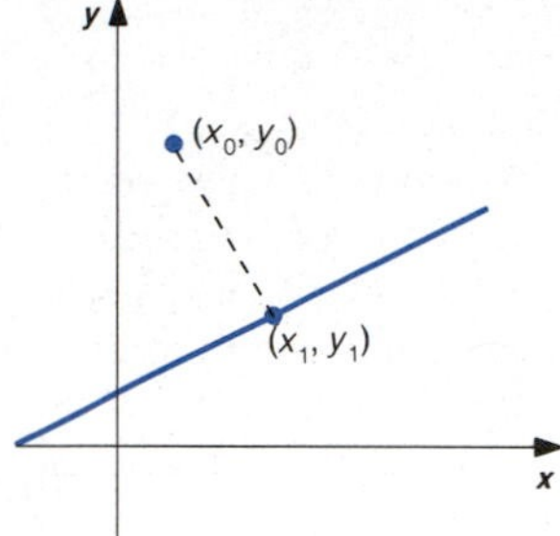


**30** Suppose that the minimum distance between a point $(x_0, y_0)$ and points on the graph of a differentiable function $f$ occurs at the point $(x_1, y_1)$ on the graph. Show that

$$\frac{y_1 - y_0}{x_1 - x_0} = -\frac{1}{f'(x_1)}$$

if $f'(x_1) \neq 0$, and that $x_1 = x_0$ if $f'(x_1) = 0$. In either case, the line through $(x_0, y_0)$ and $(x_1, y_1)$ is perpendicular to the line tangent to the graph of $f$ at the point $(x_1, y_1)$, as long as $(x_0, y_0)$ is not on the graph of $f$.

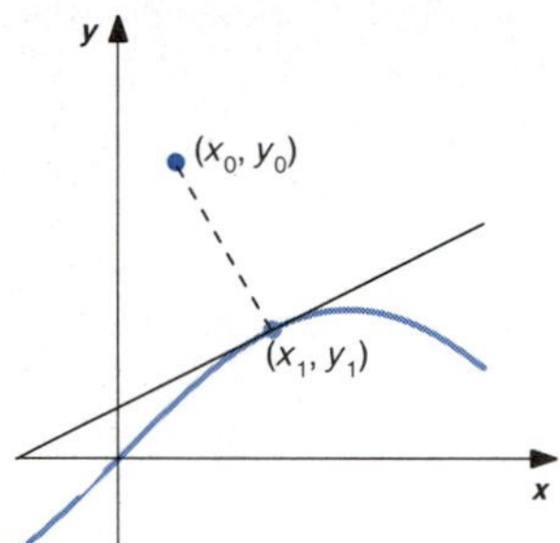


## 4.8 LINEAR APPROXIMATION, DIFFERENTIALS

You are already familiar with applications of the mathematical idea of approximation of values of functions. For example, when you use a calculator to determine the value of $\sqrt{2}$, the value given by your calculator is not the exact value, but an approximate value of the function $\sqrt{x}$ for $x = 2$. In this section, we will illustrate one of the most simple and important types of approximation methods, **linear approximation**. The idea is to use the values of a function and its derivative at a single point to find approximate values of the function at nearby points.

Let us assume that $f$ is differentiable at $c$, and that we are interested in approximate values of $f(x)$ for $x$ near $c$. We know from Section 2.6 that the graph of $f$ passes through the point $(c, f(c))$ in the same direction as the line tangent to the graph at that point. We also note that the tangent line is the graph of the *linear function*

$$L(x) = f(c) + f'(c)(x - c).$$

See Figure 4.8.1.

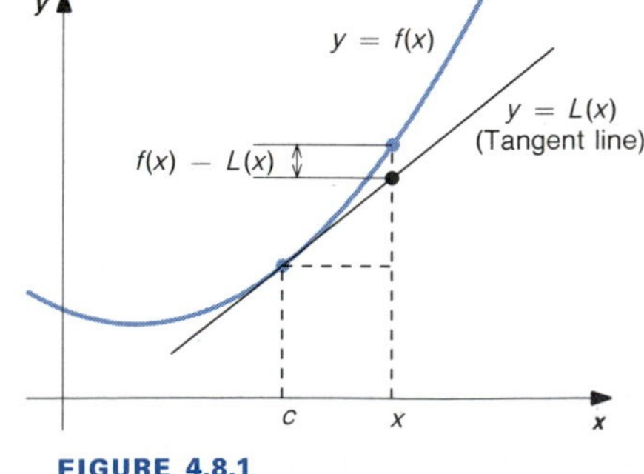


FIGURE 4.8.1

Let us see that the difference $f(x) - L(x)$ becomes small relative to $x - c$ as $x$ approaches $c$. That is, for $x$ near $c$, $x \neq c$, we have

$$\frac{f(x) - L(x)}{x - c} = \frac{f(x) - [f(c) + f'(c)(x - c)]}{x - c}$$

$$= \frac{f(x) - f(c)}{x - c} - \frac{f'(c)(x - c)}{x - c}$$

$$= \frac{f(x) - f(c)}{x - c} - f'(c).$$

Since

$$\lim_{x \to c} \frac{f(x) - f(c)}{x - c} = f'(c),$$

we see that the above ratio approaches zero as $x$ approaches $c$. Thus, if $x$ is near $c$, $x \neq c$, we have

$$\frac{f(x) - L(x)}{x - c} = (\text{A small number}),$$

$$f(x) - L(x) = (\text{A small number})(x - c).$$

If $x$ is near $c$, then $x - c$ is small and the product (A small number)$(x - c)$ is very small. That is, $L(x)$ is very near $f(x)$, and the difference between $f(x)$ and $L(x)$ becomes much smaller than $x - c$ as $x$ approaches $c$. This means that the value $L(x)$ on the tangent line can be used as an approximate value of $f(x)$ for $x$ near $c$. We write

$$\boldsymbol{f(x) \approx f(c) + f'(c)(x - c) \quad \text{whenever } x \approx c.} \tag{1}$$

Let us illustrate how to use (1) to find approximate values of a function $f$. To do this we must know the value $f(c)$ and the value of the derivative $f'(c)$, both at the point $c$. We can then use that information to find approximate values of $f(x)$ for $x$ near $c$.

### ■ EXAMPLE 1

Use a linear approximation of $\sqrt{x}$ for $x$ near 1 to find an approximate value of $\sqrt{1.02}$.

#### SOLUTION

We use (1) with $f(x) = \sqrt{x}$ and $c = 1$. We have $f(1) = \sqrt{1} = 1$ and $f'(x) = 1/(2\sqrt{x})$, so $f'(1) = 1/2 = 0.5$. Substitution in (1) then gives

$$f(x) \approx f(c) + f'(c)(x - c),$$

$$\sqrt{x} \approx 1 + (0.5)(x - 1).$$

Substitution of $x = 1.02$ gives

$$\sqrt{1.02} \approx 1 + (0.5)(1.02 - 1) = 1.01.$$

The approximate value of $\sqrt{1.02}$ obtained by linear approximation is 1.01. (A calculator gives $\sqrt{1.02} \approx 1.0099505$.) The graph of $f(x) = \sqrt{x}$ and the linear function $L(x) = 1 + (0.5)(x - 1)$ are given in Figure 4.8.2. ■

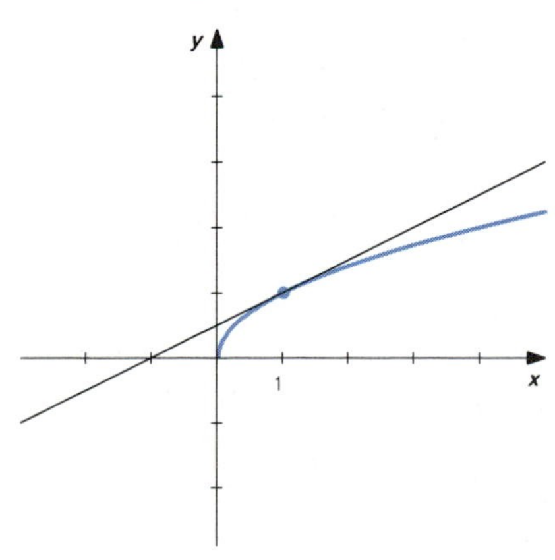


FIGURE 4.8.2

■ **EXAMPLE 2**

Use a linear approximation of $\sin x$ for $x$ near zero to find an approximate value of $\sin(0.2)$.

**SOLUTION**

We are interested in a linear approximation of the function $f(x) = \sin x$ for $x$ near $c = 0$. We have $f(0) = \sin(0) = 0$ and $f'(x) = \cos x$, so $f'(0) = \cos(0) = 1$. Using (1) with $f(x) = \sin x$ and $c = 0$, we have

$$f(x) \approx f(c) + f'(c)(x - c),$$

$$\sin x \approx 0 + (1)(x - 0) = x.$$

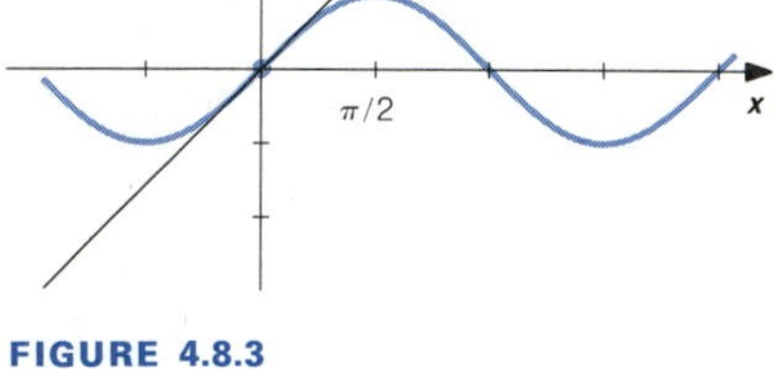


FIGURE 4.8.3

Substitution of 0.2 for $x$ gives

$$\sin(0.2) \approx 0.2.$$

The desired approximate value of $\sin(0.2)$ is 0.2. [A calculator value of the sine of 0.2 rad is $\sin(0.2) \approx 0.19866933$.] The graph of $f(x) = \sin x$ and the linear function $L(x) = x$ are given in Figure 4.8.3. ■

It is not necessary to have an explicit formula for the values of a function in order to use linear approximation. In particular, we can use the method of linear approximation to find approximate values of functions that are defined implicitly.

■ **EXAMPLE 3**

Use the line tangent to the graph of the equation $(x^2 + y^2)^2 = 16xy$ at the point (2, 2) to approximate the value of $y$ near 2 that satisfies the equation when $x = 1.85$.

**SOLUTION**

We assume that the equation $(x^2 + y^2)^2 = 16xy$ defines $y$ as a differentiable function of $x$ near the point (2, 2). We want a linear approximation of this function $y(x)$. Since the graph of this function passes through the point (2, 2), we have $y(2) = 2$. We need the value of $y'(2)$, which we obtain by implicit differentiation. Using the Chain Rule and Product Rule to differentiate both sides of the equation

$$(x^2 + y^2)^2 = 16xy$$

with respect to $x$, we obtain

$$2(x^2 + y^2)(2x + 2yy') = 16[(x)(y') + (y)(1)],$$

$$(x^2 + y^2)(2x) + (x^2 + y^2)(2y)y' = 8xy' + 8y,$$

$$[2y(x^2 + y^2) - 8x)]y' = 8y - 2x(x^2 + y^2),$$

$$y' = \frac{8y - 2x(x^2 + y^2)}{2y(x^2 + y^2) - 8x}.$$

When $(x, y) = (2, 2)$,

$$y'(2) = \frac{8(2) - 2(2)(4 + 4)}{2(2)(4 + 4) - 8(2)} = \frac{-16}{16} = -1.$$

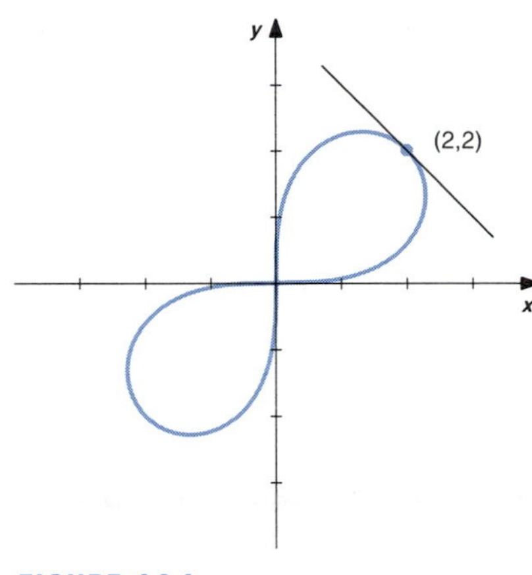

FIGURE 4.8.4

Substitution of $c = 2$ in (1) then gives

$$y(x) \approx y(2) + y'(2)(x - 2)$$
$$= 2 - (x - 2).$$

When $x = 1.85$, we obtain

$$y(1.85) \approx 2 - (1.85 - 2) = 2.15.$$

This is the desired approximate value. The graph of the equation $(x^2 + y^2)^2 = 16xy$ and the line tangent to the graph at (2, 2) are sketched in Figure 4.8.4. ■

In many problems we are interested in the change, or approximate change, of values $f(x)$ that correspond to a change in $x$. If $\boldsymbol{\Delta x = x - c}$ represents a change in $x$, the corresponding change in $f(x)$ is denoted $\boldsymbol{\Delta f = f(x) - f(c)}$. If $f$ is differentiable at $c$, then we know from (1) that

$$f(x) \approx f(c) + f'(c)(x - c), \quad \text{so}$$

$$f(x) - f(c) \approx f'(c)(x - c).$$

Let us rewrite this last statement as

$$\boldsymbol{\Delta f \approx f'(c) \cdot \Delta x.} \tag{2}$$

It is convenient to introduce a notation that suggests the use of the derivative in the linear approximation of the change of values of a function.

■ *Definition*

The **differential** of $f$ at $c$ is $df$, where

$$df = f'(c) \cdot dx, \qquad dx \text{ any number.}$$ ■

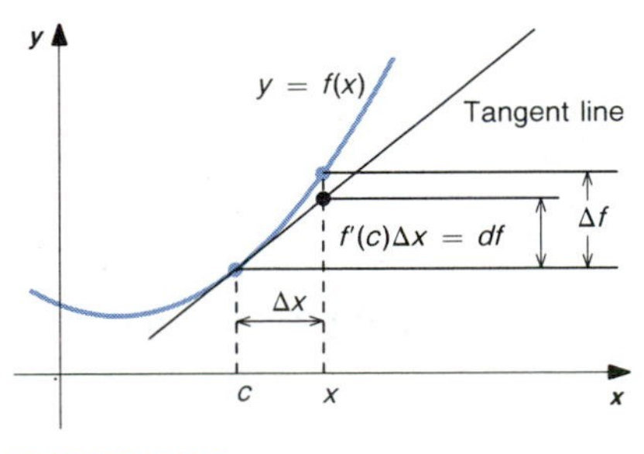

FIGURE 4.8.5

The symbols $dx$ and $df$ represent variables. For a fixed value of $c$, $df$ is a linear function of the variable $dx$. If $dx = \Delta x$, then $df = f'(c) \cdot dx = f'(c) \cdot \Delta x$. The differential $df$ then represents the change of values along the line tangent to the graph of $f$ at the point $(c, f(c))$, so $df$ is a linear approximation of the difference $\Delta f$; $df$ is also called a **differential approximation** of $\Delta f$. See Figure 4.8.5. From (2), we have

$$\boldsymbol{\Delta f \approx f'(c) \cdot \Delta x = f'(c) \cdot dx = df.} \tag{3}$$

It is convenient for some applications of differentials to replace $c$ by the variable $x$, so

$$df = f'(x) \cdot dx.$$

The value of the differential $df$ then depends on the variables $x$ and $dx$.

### EXAMPLE 4

Find the differentials $du$ in terms of the variables $x$ and $dx$.

(a) $u = 3x^2 - 2$,
(b) $u = x^{1/3}$,
(c) $u = \cos 2x$.

#### SOLUTION

In each case $du = u'(x) \cdot dx$.

(a) $u = 3x^2 - 2$ implies $u'(x) = 6x$, so $du = 6x\, dx$.

(b) $u = x^{1/3}$ implies $u'(x) = \dfrac{1}{3}x^{-2/3}$, so $du = \dfrac{1}{3}x^{-2/3}\, dx$.

(c) $u = \cos 2x$ implies $u'(x) = (-\sin 2x)(2)$, so $du = -2\sin 2x\, dx$.

(Note that we used the Chain Rule to find the derivative with respect to $x$ of $\cos 2x$.) ■

### EXAMPLE 5

Find the change in area of a circle due to a change of $dr$ in its radius. Find the differential approximation of the change in area.

#### SOLUTION

The area of a circle in terms of its radius is given by the formula

$$A = A(r) = \pi r^2.$$

The actual change in area of the circle due to change of $dr$ in its radius is

$$\begin{aligned}\Delta A &= A(r + dr) - A(r)\\ &= \pi(r + dr)^2 - \pi r^2\\ &= \pi r^2 + 2\pi r(dr) + \pi(dr)^2 - \pi r^2\\ &= 2\pi r\, dr + \pi(dr)^2.\end{aligned}$$

The differential approximation of the change in area is the differential $dA$. We have $A'(r) = 2\pi r$, so

$$dA = A'(r)\, dr = 2\pi r\, dr.$$

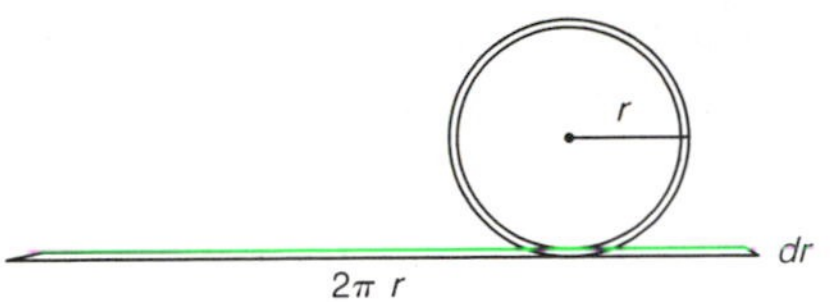


FIGURE 4.8.6

The approximate change in area is given by the circumference of the circle times the change in radius. We can think of $dA$ as the area of the "rectangle" obtained by cutting and unrolling a band of width $dr$ around the circle of radius $r$. See Figure 4.8.6. If $dr$ is small, then

$$\Delta A - dA = 2\pi r\,dr + \pi(dr)^2 - 2\pi r\,dr = \pi(dr)^2$$

is very small and the differential approximation $dA$ is very close to the actual change $\Delta A$.

The difference $\Delta f$ may be interpreted as the **error** in $f$ that corresponds to an error of $\Delta x$ in $x$. The differential $df$ may be used to find the **approximate error** in $f$ due to a **possible error** of $\Delta x$ in $x$. The ratio $df/f$ is the **approximate relative error**.

■ **EXAMPLE 6**

Find the approximate error in the volume of a cube if each edge of the cube is intended to be 2 ft with a possible error of 0.005 ft. What is the approximate relative error?

**SOLUTION**

The volume of a cube with length of each edge $s$ is given by the formula $V = V(s) = s^3$. The approximate error in volume is given by the differential

$$dV = V'(s)\,ds = 3s^2\,ds.$$

We want the value of the differential $dV$ when the intended length of side is $s = 2$ and the possible error in $s$ is $ds = \pm 0.005$. Substitution then gives

$$dV = 3(2)^2(\pm 0.005) = \pm 0.06 \text{ ft}^3.$$

This is the approximate value of the error in volume.

When $s = 2$, we have $V = 2^3$ and the approximate relative error is

$$\frac{dV}{V} = \pm\frac{0.06}{2^3} = \pm 0.0075 = \pm 0.75\%.$$ ■

# EXERCISES 4.8

1 Use a linear approximation of $x^{2/3}$ for $x$ near 8 to find an approximate value of $(7.97)^{2/3}$.

2 Use a linear approximation of $x^{3/4}$ for $x$ near 16 to find an approximate value of $(15.96)^{3/4}$.

3 Use a linear approximation of $x^{17}$ for $x$ near 1 to find an approximate value of $(1.03)^{17}$.

4 Use a linear approximation of $x^{13}$ for $x$ near 1 to find an approximate value of $(1.02)^{13}$.

5 Use a linear approximation of $\cos x$ for $x$ near 0 to find an approximate value of $\cos(0.05)$.

6 Use a linear approximation of $\sin x$ for $x$ near 0 to find an approximate value of $\sin(-0.15)$.

7 Use a linear approximation of $\sec x$ for $x$ near 0 to find an approximate value of $\sec(\pi/12)$.

8 Use a linear approximation of $\tan x$ for $x$ near 0 to find an approximate value of $\tan(\pi/12)$.

9 Use the line tangent to the graph of the equation $x^2 + y^2 = 25$ at the point $(3, 4)$ to approximate the value of $y$ near 4 that satisfies the equation when $x = 2.9$.

10 Use the line tangent to the graph of the equation $x^2 - y^2 = 16$ at the point $(5, 3)$ to approximate the value of $y$ near 3 that satisfies the equation when $x = 5.1$.

11 Use the line tangent to the graph of the equation $x = \tan y$ at the point $(1, \pi/4)$ to approximate the value of $y$ near $\pi/4$ that satisfies the equation when $x = 1.2$.

12 Use the line tangent to the graph of the equation $x = \sin y$ at the point $(0.5, \pi/6)$ to approximate the value of $y$ near $\pi/6$ that satisfies the equation when $x = 0.45$.

13 Use the line tangent to the graph of the equation $(x^2 + y^2)^{3/2} = 2^{3/2}xy$ at the point $(1, 1)$ to approximate the value of $y$ near 1 that satisfies the equation when $x = 1.1$.

14 Use the line tangent to the graph of the equation $(x^2 + y^2)^2 = 3x^2y - y^3$ at the point $(\sqrt{3}/2, 1/2)$ to approximate the value of $y$ near $1/2$ that satisfies the equation when $x = 0.9$.

In Exercises 15–26 find the differentials $du$ in terms of the variables $x$ and $dx$.

15 $u = 3x - 1$

16 $u = 2x + 7$

17 $u = x^2 + 1$

18 $u = x^2 + 2x + 2$

19 $u = x^3 + 4x^2 - 3$

20 $u = x^3 - x$

21 $u = \sqrt{x}$

22 $u = \sqrt{2x + 1}$

23 $u = \cos(3x)$

24 $u = \sin(2x)$

25 $u = \tan x$

26 $u = \sec x$

27 A right circular cylinder has fixed height $h_0$ and variable radius $r$. Find the change in its volume due to a change of $dr$ in its radius. Find the differential approximation of the change in volume.

28 Find the change in volume of a sphere due to a change of $dr$ in its radius. Find the differential approximation of the change in volume.

29 Find the change in the area of a square due to a change of $ds$ in the length of each of its sides. Find the differential approximation of the change in the area.

30 Find the change in the area of an equilateral triangle due to a change of $ds$ in the length of each of its sides. Find the differential approximation of the change in the area.

31 The height of a right circular cylinder is twice the radius of its base. Express the volume of the cylinder in terms of its radius. Use differentials to find the approximate error in volume if the radius is 4 m with a possible error of $\pm 0.03$ m. Find the approximate relative error.

32 The height of a right circular cone is one-half the radius of its base. Express the volume of the cone in terms of its radius. Use differentials to find the approximate error in volume if the radius is 2 m with a possible error of $\pm 0.02$ m. Find the approximate relative error.

33 The height of a tree is determined by measuring the angle of elevation to the top of the tree from a point on the ground that is 40 ft from the base of the tree. Find the height of the tree and the approximate error in the calculation if the angle of elevation is measured to be 45° with a possible error of 1°.

34 The angle of elevation between an observer at ground level and an airplane is measured to be 30° with a possible error of 1° as the plane passes over a point on the ground that is 5000 ft from the observer. Find the height of the plane and the approximate error in the calculation.

## 4.9 QUADRATIC APPROXIMATION, ACCURACY

If $f$ is differentiable at $c$, we have seen that the first-order polynomial function $L(x) = f(c) + f'(c)(x - c)$ can be used to find approximate values of $f(x)$ for $x$ near $c$. We now investigate the possibility of obtaining more accurate approximate values of $f(x)$ by using polynomial functions of higher order. In this section, we introduce a method of quadratic approximation and compare the accuracy of the linear and quadratic methods. Ideas related to higher-order polynomial approximations will be studied in Chapter 10.

The results of this section are based on the following generalizations of the Mean Value Theorem.

*Taylor's formulas*

**Suppose $I$ is an interval that contains $c$. If $f''$ exists on $I$, then for each point $x$ in $I$, there is a point $z$ between $x$ and $c$ such that**

$$f(x) = f(c) + f'(c)(x - c) + \frac{f''(z)}{2}(x - c)^2.$$

**If $f'''$ exists on $I$, then for each point $x$ in $I$, there is a point $z$ between $x$ and $c$ such that**

$$f(x) = f(c) + f'(c) + \frac{f''(c)}{2}(x - c)^2 + \frac{f'''(z)}{6}(x - c)^3.$$

These formulas can be verified in the same manner in which we verified the Mean Value Theorem in Section 3.10. That is, we apply Rolle's Theorem to an auxiliary function $g$. (See Exercises 19 and 20.)

The linear and quadratic approximations of $f(x)$ are given by

$$P_1(x) = f(c) + f'(c)(x - c)$$

and

$$P_2(x) = f(c) + f'(c)(x - c) + \frac{f''(c)}{2}(x - c)^2,$$

respectively. $P_1$ and $P_2$ are the first and second **Taylor polynomials** of $f$ about $c$. (Taylor polynomials were introduced in Section 3.6.) The expressions

$$R_1 = \frac{f''(z)}{2}(x - c)^2 \quad \text{and} \quad R_2 = \frac{f'''(z)}{6}(x - c)^3$$

are called **remainder terms**. Taylor's formulas can be expressed as

$$f(x) = P_1(x) + R_1 \quad \text{and} \quad f(x) = P_2(x) + R_2, \quad \text{or}$$

$$f(x) - P_1(x) = R_1 \quad \text{and} \quad f(x) - P_2(x) = R_2.$$

We then see that the remainder term in Taylor's formula measures the difference between the actual value of the function $f(x)$ and the approximate value given by the corresponding Taylor polynomial. We do not know the value of the remainder term, because we do not know the value of the number $z$ in the formula. However, we can determine a bound of the error in using the Taylor polynomial as an approximate value of $f(x)$ by using the methods of Section 2.2 to obtain a bound of the remainder term that holds for all values of $z$ between $x$ and $c$.

■ **EXAMPLE 1**

Use linear and quadratic approximations of $x^{15}$ for $x$ near 1 to find approximate values of $(0.97)^{15}$. Use the remainder terms in Taylor's formulas to determine the accuracy of the approximations.

**SOLUTION**

We will need formulas for the first three derivatives of the function $f(x) = x^{15}$.

We have

$$f(x) = x^{15},$$

$$f'(x) = 15x^{14},$$

$$f''(x) = 15(14)x^{13},$$

$$f'''(x) = 15(14)(13)x^{12}.$$

When $x = 1$, we obtain

$$f(1) = (1)^{15} = 1,$$

$$f'(1) = 15(1)^{14} = 15,$$

$$f''(1) = 15(14)(1)^{13} = 15(14).$$

The first- and second-order Taylor polynomials about $c = 1$ are

$$P_1(x) = f(1) + f'(1)(x-1) = 1 + 15(x-1) \quad \text{and}$$

$$P_2(x) = f(1) + f'(1)(x-1) + \frac{f''(1)}{2}(x-1)^2$$

$$= 1 + 15(x-1) + \frac{15(14)}{2}(x-1)^2.$$

(Note that we do not expand the powers of $x - c$ in the Taylor polynomial of $f$ about $c$.) The desired linear and quadratic approximations of $(0.97)^{15}$ are

$$P_1(0.97) = 1 + 15(0.97 - 1) = 0.55 \quad \text{and}$$

$$P_2(0.97) = 1 + 15(0.97 - 1) + \frac{15(14)}{2}(0.97 - 1)^2 = 0.6445.$$

To determine the accuracy of the above approximate values, we investigate the remainder terms in Taylor's formulas. The remainder that corresponds to the linear approximation is

$$R_1 = \frac{f''(z)}{2}(x-c)^2, \qquad \text{where } z \text{ is some number between } x \text{ and } c.$$

Using $x = 0.97$, $c = 1$, and the formula for $f''$ from above, we obtain

$$R_1 = \frac{15(14)z^{13}}{2}(0.97 - 1)^2, \qquad \text{where } z \text{ satisfies } 0.97 < z < 1.$$

The inequality $0.97 < z < 1$ implies $0.97 < |z| < 1$. Since $R_1$ involves a positive power of $z$, we use the inequality $|z| < 1$ to obtain

$$\begin{aligned}|R_1| &= \frac{15(14)|z|^{13}}{2}(0.97-1)^2 \\ &< \frac{15(14)(1)^{13}}{2}(0.97-1)^2 \\ &= 0.0945.\end{aligned}$$

It follows that

$$|f(0.97) - P_1(0.97)| = |R_1| < 0.0945,$$

so the linear approximation $P_1(0.97) = 0.55$ is within 0.0945 of the exact value of $f(0.97) = (0.97)^{15}$.

The accuracy of the quadratic approximation can be determined from the remainder $R_2$. We have

$$R_2 = \frac{f'''(z)}{6}(x-c)^3 \qquad \text{for some } z \text{ between } x \text{ and } c.$$

Using $x = 0.97$, $c = 1$, and the formula for $f'''$, we obtain

$$R_2 = \frac{15(14)(13)z^{12}}{6}(0.97-1)^3, \qquad 0.97 < z < 1.$$

As before, $0.97 < z < 1$ implies $|z| < 1$, so

$$\begin{aligned}|R_2| &= \frac{15(14)(13)|z|^{12}}{6}|0.97-1|^3 \\ &< \frac{15(14)(13)(1)^{12}}{6}|0.97-1|^3 \\ &= 0.012285.\end{aligned}$$

We then have

$$|f(0.97) - P_2(0.97)| = |R_2| < 0.012285,$$

so the quadratic approximation $P_2(0.97) = 0.6445$ is within 0.012285 of the actual value of $(0.97)^{15}$. [A calculator gives $(0.97)^{15} \approx 0.63325119$. This is consistent with the results we obtained.] The graphs of $x^{15}$, the linear function $P_1(x)$, and the quadratic function $P_2(x)$ are given in Figure 4.9.1. ■

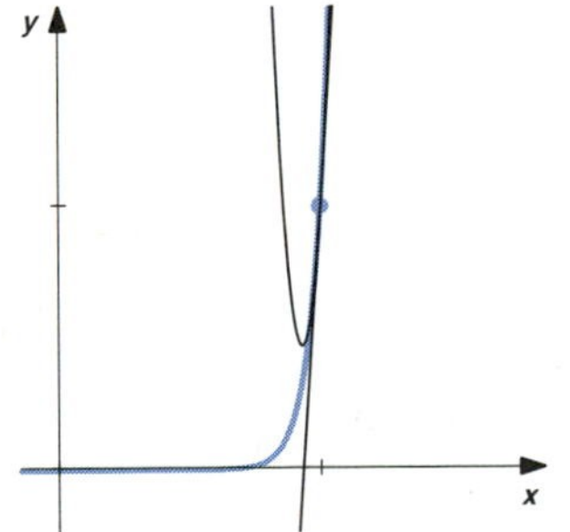


FIGURE 4.9.1

■ **EXAMPLE 2**

Use linear and quadratic approximations of $\cos x$ for $x$ near 0 to find approximate values of $\cos(0.03)$. Use the remainder terms in Taylor's formulas to determine the accuracy of the approximations.

**SOLUTION**

We will need the first three derivatives of $f(x) = \cos x$ and the values of $f$, $f'$, and $f''$ at $c = 0$:

$$f(x) = \cos x, \qquad f(0) = \cos 0 = 1,$$

$$f'(x) = -\sin x, \qquad f'(0) = -\sin 0 = 0,$$

$$f''(x) = -\cos x, \qquad f''(0) = -\cos 0 = -1,$$

$$f'''(x) = \sin x.$$

The first- and second-order Taylor polynomials of $f$ at $c = 0$ are

$$P_1(x) = f(0) + f'(0)(x - 0)$$
$$= 1 + (0)(x - 0) = 1 \quad \text{and}$$

$$P_2(x) = f(0) + f'(0)(x - 0) + \frac{f''(0)}{2}(x - 0)^2$$
$$= 1 + (0)(x) + \frac{-1}{2}(x)^2 = 1 - 0.5x^2.$$

The desired linear and quadratic approximations are

$$P_1(0.03) = 1 \quad \text{and}$$

$$P_2(1.03) = 1 - 0.5(0.03)^2 = 0.99955.$$

Using $x = 0.03$, $c = 0$, and the formula for $f''$ from above, we have

$$R_1 = \frac{f''(z)}{2}(x - c)^2 = \frac{-\cos z}{2}(0.03)^2, \qquad z \text{ between } 0.03 \text{ and } 0.$$

Since $|\cos z| \le 1$ for any value of $z$, we have

$$|R_1| = \frac{|-\cos z|}{2}|0.03|^2 \le \frac{1}{2}(0.03)^2 = 0.00045.$$

We conclude that the linear approximation $P_1(0.03) = 1$ is within 0.00045 of the actual value of $\cos(0.03)$.

The accuracy of the quadratic approximation is determined from $R_2$. Using $x = 0.03$, $c = 0$, and the formula for $f'''$, we obtain

$$R_2 = \frac{f'''(z)}{6}(x - c)^3 = \frac{\sin z}{6}(0.03)^3, \qquad z \text{ between } 0.03 \text{ and } 0.$$

Since $|\sin z| \le 1$, we obtain

$$|R_2| = \frac{|\sin z|}{6}|0.03|^3 \le \frac{1}{6}(0.03)^3 = 0.0000045.$$

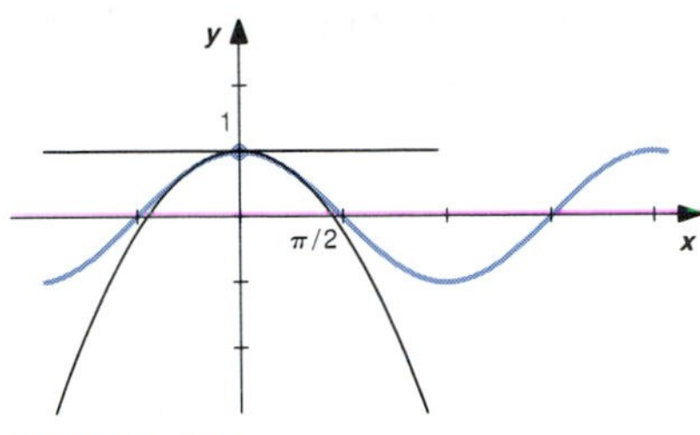


FIGURE 4.9.2

We conclude that the quadratic approximation $P_2(0.03) = 0.99955$ is within 0.0000045 of the actual value of $\cos(0.03)$. [A calculator gives $\cos(0.03) \approx 0.99955003$.] The graphs of $\cos x$, $P_1(x)$, and $P_2(x)$ are given in Figure 4.9.2. ■

The idea in using Taylor polynomials to find approximate values of functions is to use the values of the function and its derivatives at a single point to determine values at nearby points. It is not necessary to have a formula for the function.

■ **EXAMPLE 3**

Suppose that a function $f$ satisfies $f(1) = 0$ and is differentiable for $x > 0$ with $f'(x) = 1/x$. Use linear and quadratic approximations of $f(x)$ for $x$ near 1 to find approximate values of $f(1.2)$. Use the remainder terms in Taylor's formulas to determine the accuracy of the approximations.

**SOLUTION**

We will need the first three derivatives. We are given that

$$f'(x) = x^{-1}.$$

Differentiating, we obtain

$$f''(x) = -x^{-2},$$

$$f'''(x) = 2x^{-3}.$$

We are given that $f(1) = 0$. From the above formulas we see that $f'(1) = (1)^{-1} = 1$ and $f''(1) = -(1)^{-2} = -1$, so

$$\begin{aligned} P_1(x) &= f(1) + f'(1)(x - 1) \\ &= 0 + (1)(x - 1) = x - 1 \end{aligned}$$

and

$$\begin{aligned} P_2(x) &= f(1) + f'(1)(x - 1) + \frac{f''(1)}{2}(x - 1)^2 \\ &= 0 + (x - 1) - \frac{1}{2}(x - 1)^2. \end{aligned}$$

The approximate values of $f(1.2)$ are

$$P_1(1.2) = (1.2 - 1) = 0.2$$

and

$$P_2(1.2) = (1.2 - 1) - \frac{1}{2}(1.2 - 1)^2 = 0.18.$$

Using $x = 1.2$, $c = 1$, and the formula for $f''$, we obtain

$$R_1 = \frac{f''(z)}{2}(x-c)^2 = \frac{-z^{-2}}{2}(1.2-1)^2$$
$$= \left(-\frac{1}{2}\right)\left(\frac{1}{z^2}\right)(0.2)^2, \qquad 1 < z < 1.2.$$

The inequality $1 < z < 1.2$ implies $1 < |z| < 1.2$. Since the factor $z$ is in the denominator, we use the inequality $1 < |z|$ to obtain

$$\frac{1}{|z|} < \frac{1}{1}.$$

We then have

$$|R_1| = \left(\frac{1}{2}\right)\left(\frac{1}{|z|^2}\right)(0.2)^2 < \left(\frac{1}{2}\right)\left(\frac{1}{(1)^2}\right)(0.2)^2 = 0.02.$$

The linear approximation $P_1(1.2) = 0.2$ is within 0.02 of the actual value of $f(1.2)$.

We have

$$R_2 = \frac{f'''(z)}{6}(x-c)^3 = \left(\frac{1}{6}\right)\left(\frac{2}{z^3}\right)(1.2-1)^3, \qquad 1 < z < 1.2.$$

As before, $1 < z < 1.2$ implies

$$\frac{1}{|z|} < \frac{1}{1}, \quad \text{so} \quad |R_2| < \left(\frac{1}{6}\right)\left(\frac{2}{(1)^3}\right)(0.2)^3 \approx 0.00266667.$$

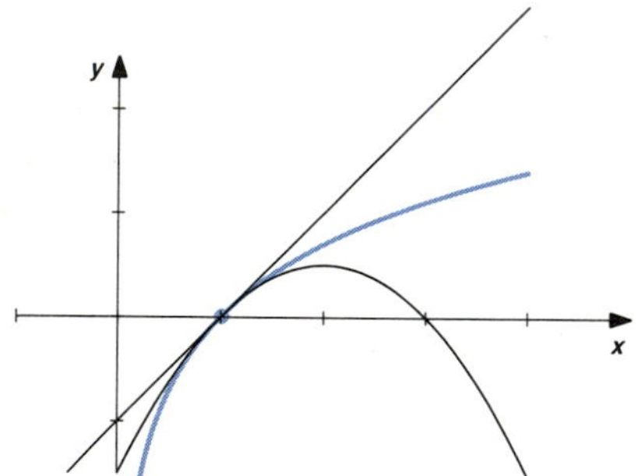

FIGURE 4.9.3

We conclude that the approximate value $P_2(1.2) = 0.18$ is within 0.00266667 of the actual value of $f(0.02)$. The graphs of $f$, $P_1$, and $P_2$ are given in Figure 4.9.3. [The function described in Example 3 is called the **natural logarithmic function** and is denoted $\ln(x)$. We will study this function more in Chapter 7. You may check that the calculator value is $\ln(1.2) \approx 0.18232156$. This is consistent with the results of Example 3.] ■

## EXERCISES 4.9

In Exercises 1–12, use linear and quadratic approximations of the given function near the given point $c$ to find approximate values of the given expression. Use the remainder terms in Taylor's formulas to determine the accuracy of the approximations.

1 $f(x) = x^{20}$, $c = 1$, $(0.97)^{20}$

2 $f(x) = x^{10}$, $c = 1$, $(0.98)^{10}$

3 $f(x) = x^{1/3}$, $c = 8$, $(8.04)^{1/3}$

4 $f(x) = x^{1/4}$, $c = 16$, $(16.05)^{1/4}$

5 $f(x) = \cos x$, $c = 0$, $\cos(0.06)$

6 $f(x) = \sin x$, $c = 0$, $\sin(0.02)$

7 $f(x) = 1/(1-x)$, $c = 0$, $1/(1-0.2)$

8 $f(x) = 1/(1-x)$, $c = 0$, $1/(1-(-0.2))$

9 The function $f$ satisfies $f(0) = 1$, $f'(x) = f(x)$, and $|f(x)| < 3$ for $x \leq 1$; $c = 0$, $f(0.2)$.

10 The function $f$ satisfies $f(0) = 1$, $f'(x) = f(x)$, and $|f(x)| \leq 1$ for $x \leq 0$; $c = 0$, $f(-0.2)$.

11 The function $f$ satisfies $f(0) = 0$ and has derivative $f'(x) = (x+1)^{-1}$; $c = 0$, $f(-0.25)$.

12 The function $f$ satisfies $f(0) = 0$ and has derivative $f'(x) = (1-x)^{-1}$; $c = 0$, $f(0.25)$.

In Exercises 13–16, find linear and quadratic polynomials that approximate the graph of the given equation near the point $(x_0, y_0)$ on the graph. Use the polynomials to find approximate values of the $y$ near $y_0$ that satisfy the equation for the given value of $x$.

13 $x^2 + y^2 = 25$, $(x_0, y_0) = (3, 4)$, $x = 3.1$

14 $x^2 - y^2 = 16$, $(x_0, y_0) = (5, 3)$, $x = 4.9$

15 $x = \sin y$, $(x_0, y_0) = \left(\frac{1}{2}, \frac{\pi}{6}\right)$, $x = 0.55$

16 $x = \tan y$, $(x_0, y_0) = \left(1, \frac{\pi}{4}\right)$, $x = 0.94$

17 Show that

$$|\sin x - x| \leq \frac{|x|^3}{6}.$$

18 Use the fact that $|\sin z| \leq |z|$ to show that

$$\left|\cos x - \left(1 - \frac{x^2}{2}\right)\right| \leq \frac{|x|^4}{6}.$$

19 Suppose $I$ is an interval that contains $c$ and $x$, and that $f''$ exists on $I$. Let

$$g(t) = f(x) - f(t) - f'(t)(x-t) - R_1\frac{(x-t)^2}{(x-c)^2},$$

where $R_1 = f(x) - f(c) - f'(c)(x-c)$. Verify the formula for the remainder term $R_1$ in Taylor's Formula by applying Rolle's Theorem to $g(t)$ on the interval between $x$ and $c$, and showing that $g'(z) = 0$ implies

$$R_1 = \frac{f''(z)}{2}(x-c)^2.$$

20 Suppose $I$ is an interval that contains $c$ and $x$, and that $f'''$ exists on $I$. Let

$$g(t) = f(x) - f(t) - f'(t)(x-t) - \frac{f''(t)}{2}(x-t)^2 - R_2\frac{(x-t)^3}{(x-c)^3},$$

where

$$R_2 = f(x) - f(c) - f'(c)(x-c) - \frac{f''(c)}{2}(x-c)^2.$$

Verify the formula for the remainder term $R_2$ in Taylor's Formula by applying Rolle's Theorem to $g(t)$ on the interval between $x$ and $c$, and showing that $g'(z) = 0$ implies

$$R_2 = \frac{f'''(z)}{6}(x-c)^3.$$

## 4.10 INDEFINITE INTEGRALS

We have seen many applications that involve finding the derivative of a given function. In this section and in subsequent sections we will see that the following problem also has wide application to physical problems.

**Problem:** For a given function $f$, can we find a function (or functions) $F$ such that $F'(x) = f(x)$ for $x$ in an interval?

Let us see that the problem does not have a simple answer. We first observe that if $F$ is any function that satisfies $F'(x) = f(x)$, then for any constant $C$, the function $G(x) = F(x) + C$ also has derivative $G'(x) = F'(x) = f(x)$. Hence, if we can find one solution $F(x)$, then each of the functions $G(x) = F(x) + C$ is also a solution. This leads us to the question: Are there any other, different types of solution? Fortunately, the answer to this question is no. That is, if $F'(x) = f(x)$ and $G'(x) = f(x)$ on an interval, then $F'(x) = G'(x)$ on the interval. Theorem 2 of Section 3.10 then implies there is a constant $C$ such that $G(x) = F(x) + C$ on the interval. The following theorem summarizes these results.

*Theorem 1*

**Suppose $F'(x) = f(x)$ on an interval $I$.**

**(a) If $G(x) = F(x) + C$ for some constant $C$, then $G'(x) = f(x)$ on $I$.**

**(b) If $G'(x) = f(x)$ on $I$, then there is some constant $C$ such that $G(x) = F(x) + C$ on $I$.**

We have seen that the solution to the problem, if there is a solution, consists of a collection of functions $G$ such that $G'(x) = f(x)$ on an interval. This collection of functions is denoted by the *symbol*

$$\int f(x)\, dx.$$

This symbol is called the **indefinite integral** of $f$. The function $f(x)$ is called the **integrand** of the indefinite integral. At this time $\int$ and $dx$ are merely parts of a symbol and have no individual meaning.

If we know of a particular function $F$ such that $F'(x) = f(x)$ on an interval, the collection of functions $G$ such that $G'(x) = f(x)$ on an interval can be described as $F(x) + C$. Thus, we can write

$$\int f(x)\, dx = F(x) + C.$$

A statement of this form is called an **integration formula**. Finding a function $F$ such that $F'(x) = f(x)$ is called evaluating the indefinite integral $\int f(x)\, dx$. The mathematical statement

$$\int f(x)\, dx = F(x) + C \quad \text{means } F'(x) = f(x) \text{ on an interval.}$$

An integration formula $\int f(x)\, dx = F(x) + C$ can be verified by showing that $F'(x) = f(x)$ on an interval.

A function $F$ such that $F'(x) = f(x)$ on an interval is sometimes called an **antiderivative** of $f$.

### EXAMPLE 1

Verify the following integration formulas by differentiation.

(a) $\displaystyle\int x\sqrt{x^2 + 1}\, dx = \frac{1}{3}(x^2 + 1)^{3/2} + C.$

(b) $\displaystyle\int \cos^2 x\, dx = \frac{\cos x \cdot \sin x}{2} + \frac{x}{2} + C.$

(c) $\int x \sin x \, dx = -x \cos x + \sin x + C.$

(d) $\int \sin x \cdot \cos x \, dx = \frac{1}{2} \sin^2 x + C.$

(e) $\int \sin x \cdot \cos x \, dx = -\frac{1}{2} \cos^2 x + C.$

**SOLUTION**

In each case, we must check that the derivative of the right-hand side equals the integrand. (The derivative of the constant $C$ is zero.)

(a) $\frac{d}{dx}\left(\frac{1}{3}(x^2+1)^{3/2}\right) = \left(\frac{1}{3}\right)\left(\frac{3}{2}\right)(x^2+1)^{1/2}(2x) = x\sqrt{x^2+1},$ so

$$\int x\sqrt{x^2+1}\, dx = \frac{1}{3}(x^2+1)^{3/2} + C.$$

(b) $\frac{d}{dx}\left(\frac{\cos x \cdot \sin x}{2} + \frac{x}{2}\right)$

$$= \frac{1}{2}[(\cos x)(\cos x) + (\sin x)(-\sin x) + 1]$$

$$= \frac{1}{2}[\cos^2 x - \sin^2 x + 1]$$

$$= \frac{1}{2}[\cos^2 x - (1 - \cos^2 x) + 1] = \cos^2 x, \quad \text{so}$$

$$\int \cos^2 x \, dx = \frac{\cos x \cdot \sin x}{2} + \frac{x}{2} + C.$$

(We used the fact that $\sin^2 x + \cos^2 x = 1$ to write the derivative in the desired form.)

(c) $\frac{d}{dx}(-x\cos x + \sin x)$

$$= -[(x)(-\sin x) + (\cos x)(1)] + \cos x$$

$$= x \sin x, \quad \text{so}$$

$$\int x \sin x \, dx = -x\cos x + \sin x + C.$$

(d) $\frac{d}{dx}\left(\frac{1}{2}\sin^2 x\right) = \frac{1}{2}(2)(\sin x)(\cos x) = \sin x \cdot \cos x,$ so

$$\int \sin x \cdot \cos x \, dx = \frac{1}{2}\sin^2 x + C.$$

(e) $\frac{d}{dx}\left(-\frac{1}{2}\cos^2 x\right) = -\frac{1}{2}(2)(\cos x)(-\sin x) = \sin x \cdot \cos x,$

so

$$\int \sin x \cdot \cos x \, dx = -\frac{1}{2} \cos^2 x + C.$$

Note that

$$\frac{1}{2} \sin^2 x - \left( -\frac{1}{2} \cos^2 x \right) = \frac{1}{2} (\sin^2 x + \cos^2 x) = \frac{1}{2}.$$

That is, the two different indefinite integrals of $\sin x \cdot \cos x$ that were verified in (d) and (e) differ by a constant, as Theorem 1 says they must. ■

**■ EXAMPLE 2**

Find all $k$ for which $\int \sin 2x \, dx = k \cos 2x + C$.

**SOLUTION**

The expression $\int \sin 2x \, dx = k \cos 2x + C$ means

$$\frac{d}{dx} (k \cos 2x) = \sin 2x.$$

Using the Chain Rule to carry out the differentiation, we obtain the equation

$$k(-\sin 2x)(2) = \sin 2x,$$

$$-2k = 1,$$

$$k = -\frac{1}{2}.$$

The integration formula is true for $k = -1/2$. ■

Note that

$$\int F'(x) \, dx = F(x) + C. \tag{1}$$

Statement (1) is true because the derivative of $F(x)$ on the right is equal to $F'(x)$, the integrand of the indefinite integral.

We have still not addressed the problem of how to evaluate an indefinite integral. Much effort will be given to this question in a later chapter. For now, we will see how to solve the problem for a few basic functions. The idea is to transform a known differentiation formula $F'(x) = f(x)$ into the corresponding integration formula $\int f(x) \, dx = F(x) + C$. For example, we know that

$$\frac{d}{dx} \left( \frac{x^{r+1}}{r+1} \right) = x^r, \quad \text{so} \quad \int x^r \, dx = \frac{x^{r+1}}{r+1} + C.$$

We also know that

$$\frac{d}{dx}(\sin x) = \cos x, \quad \text{so} \quad \int \cos x \, dx = \sin x + C.$$

The integration formulas that correspond to the formulas for the derivatives of power functions and the basic trigonometric functions are listed below:

$$\int x^r \, dx = \frac{x^{r+1}}{r+1} + C, \qquad r \neq -1,$$

$$\int \cos x \, dx = \sin x + C,$$

$$\int \sin x \, dx = -\cos x + C,$$

$$\int \sec^2 x \, dx = \tan x + C,$$

$$\int \csc^2 x \, dx = -\cot x + C,$$

$$\int \sec x \cdot \tan x \, dx = \sec x + C,$$

$$\int \csc x \cdot \cot x \, dx = -\csc x + C.$$

These formulas will occur often enough in our work that they should be memorized. You may also find it helpful to verify these formulas by differentiation each time they are used. You should be very familiar with the corresponding differentiation formulas, so the differentiation can be carried out without any additional writing. Be especially careful with the signs in the trigonometric formulas.

■ **EXAMPLE 3**

Using the first formula above we have

(a) $\int x^2 \, dx = \frac{x^3}{3} + C,$

(b) $\int \sqrt{x} \, dx = \int x^{1/2} \, dx = \frac{x^{(1/2)+1}}{(1/2)+1} + C = \frac{2}{3}x^{3/2} + C,$

(c) $\int \frac{1}{\sqrt{x}} \, dx = \int x^{-1/2} \, dx = \frac{x^{1/2}}{1/2} + C = 2x^{1/2} + C,$

(d) $\int 1 \, dx = \int x^0 \, dx = \frac{x^{0+1}}{0+1} + C = x + C.$ ■

Since $(aF + bG)' = aF' + bG'$, we have

$$\int [af(x) + bg(x)] \, dx = a \int f(x) \, dx + b \int g(x) \, dx. \tag{2}$$

This says that we can evaluate the indefinite integral of a sum by evaluating the indefinite integral of each term. Also, the indefinite integral of a constant multiple of a function is the constant times the indefinite integral of the function. A constant $C$ should be added on the right-hand side when the indefinite integrals are evaluated.

**EXAMPLE 4**

Using (2) we have

(a) $\displaystyle\int (x^2 + 3x - 1)\,dx = \frac{x^3}{3} + 3\frac{x^2}{2} - x + C,$

(b) $\displaystyle\int (\cos x + \sin x)\,dx = \sin x - \cos x + C,$

(c) $\displaystyle\int (2\sec^2 x - \sec x \cdot \tan x)\,dx = 2\tan x - \sec x + C,$

(d) $\displaystyle\int x(6x - 4)\,dx = \int (6x^2 - 4x)\,dx = 6\frac{x^3}{3} - 4\frac{x^2}{2} + C.$

Note that it was necessary to carry out the multiplication before we could use (2) to evaluate the indefinite integral in (d).

(e) $$\int \frac{x+1}{x^{1/3}}\,dx = \int (x^{2/3} + x^{-1/3})\,dx = \frac{x^{5/3}}{5/3} + \frac{x^{2/3}}{2/3} + C$$
$$= \frac{3}{5}x^{5/3} + \frac{3}{2}x^{2/3} + C.$$

It was necessary to carry out the division before we could use (2) to evaluate the indefinite integral in (e). ■

If $G$ is an antiderivative of $f$ and

$$\int f(x)\,dx = F(x) + C,$$

then

$$G(x) = F(x) + C \text{ for some constant } C.$$

We can determine a particular antiderivative $G$ by using additional information to evaluate the constant. $C$

**EXAMPLE 5**

Find a function $G$ such that $G'(x) = \sin x$ and $G(0) = 0$.

**SOLUTION**

We know from (1) that $G$ is an antiderivative of $G'(x) = \sin x$. We have

$$\int \sin x\,dx = -\cos x + C,$$

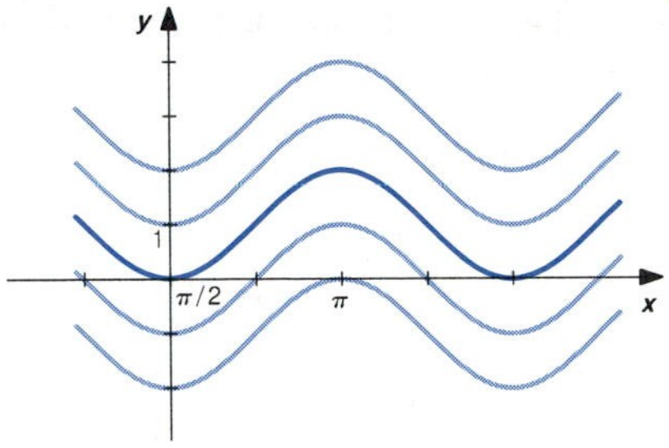

FIGURE 4.10.1

so $G(x) = -\cos x + C$, for some constant $C$. The constant $C$ is evaluated by solving the equation $G(0) = 0$ for $C$. We have

$$G(0) = 0,$$

$$-\cos 0 + C = 0,$$

$$-1 + C = 0,$$

$$C = 1.$$

The desired function is $G(x) = -\cos x + 1$.

The graphs of some of the functions $-\cos x + C$ for different values of $C$ are sketched in Figure 4.10.1. The graph of only one of these functions passes through the point (0, 0). That is the one we found. ■

Let us see how the ideas of an indefinite integral can be used in problems that involve motion along a number line. If the position of an object along a number line at time $t$ is given by $s(t)$, we know the velocity is $v(t) = s'(t)$ and the acceleration is $a(t) = v'(t)$; that is, $s(t)$ is an antiderivative of $v(t)$, so

$$s(t) \quad \text{is given by} \quad \int v(t)\,dt.$$

Since $v(t)$ is an antiderivative of $a(t)$,

$$v(t) \quad \text{is given by} \quad \int a(t)\,dt.$$

If we know the formula for the acceleration $a(t)$ and the velocity at some time $t_0$, we can obtain the formula for the velocity $v(t)$ by using the method of Example 5. Similarly, we can obtain the formula for the position $s(t)$ if we know the formula for the velocity $v(t)$ and the position at some time.

Acceleration due to the gravitational attraction of the earth is essentially constant at points near the surface of the earth. Depending on the units of length used, we will use either

$$g = 32\text{ ft/s}^2 \quad \text{or} \quad g = 9.8\text{ m/s}^2$$

for the magnitude of this acceleration. For problems that involve vertical motion near the surface of the earth, it is generally convenient to choose a vertical coordinate line with positive direction up and origin at ground level. If other forces such as air resistance are neglected, we then have

$$a(t) = -g.$$

The negative value of the acceleration indicates that the acceleration is downward, in the negative direction. We then have

$$\int a(t)\,dt = \int -g\,dt = -gt + C, \quad \text{so}$$

$$v(t) = -gt + C.$$

It is easy to see that $C = v(0)$. The value $v(0)$ is customarily denoted $v_0$ and is called the **initial velocity**. Thus,

$$v(t) = -gt + v_0.$$

We then have

$$\int v(t)\,dt = \int (-gt + v_0)\,dt = -\frac{1}{2}gt^2 + v_0 t + C, \quad \text{so}$$

$$s(t) = -\frac{1}{2}gt^2 + v_0 + C.$$

The value of the above constant $C$ is $s(0)$. The value $s(0)$ is denoted $s_0$ and called the **initial position**. We then have the following formula for the height above ground level of an object that is thrown vertically from an initial height $s_0$ with initial velocity $v_0$.

$$s(t) = -\frac{1}{2}gt^2 + v_0 t + s_0. \tag{3}$$

[Although this formula is easily derived from the formula $a(t) = -g$, some students may just memorize the result. In any case, we are concerned with how to interpret and use the formula.]

### ■ EXAMPLE 6

A ball is thrown vertically upward from a height of 48 ft above ground level at a speed of 32 ft/s. (a) How high above ground will it get? (b) How long after it is thrown will it hit the ground? (c) At what speed will it hit the ground? (d) At what speed will a ball hit the ground if it is thrown downward from a height of 48 ft at 32 ft/s?

#### SOLUTION

(a) We choose $t = 0$ to be the time the ball is thrown. We are told that the initial position is $s_0 = 48$ and the initial velocity is $v_0 = +32$. (The initial velocity is upward, so $v_0$ is positive.) Since the unit of length is feet, we use the value $g = 32$ ft/s$^2$. Formula (3) then gives

$$s(t) = -16t^2 + 32t + 48.$$

The velocity is

$$v(t) = s'(t) = -32t + 32.$$

The formulas for $s(t)$ and $v(t)$ are valid only from the time the ball is thrown ($t = 0$) until it returns to ground level.

The highest above ground that the ball reaches is the maximum value of the function $s(t)$. This occurs when $s'(t) = v(t) = 0$. [We know that the ball will go upward (positive velocity) to its highest point and then start falling (negative velocity). At its highest point the velocity changes from positive to negative, and is zero.] Solving the equation $v(t) = 0$ for $t$, we have

$$v(t) = 0,$$

$$-32t + 32 = 0,$$

$$-32t = -32,$$

$$t = 1.$$

When $t = 1$, the height is

$$s(1) = -16(1)^2 + 32(1) + 48 = 64.$$

The ball reaches a maximum height of 64 ft above ground.

(b) The ball is at ground level when $s(t) = 0$. Using the formula for $s(t)$ from (a), we can solve this equation for $t$. We have

$$s(t) = 0,$$

$$-16t^2 + 32t + 48 = 0,$$

$$-16(t^2 - 2t - 3) = 0,$$

$$-16(t - 3)(t + 1) = 0.$$

The solutions are $t = 3$ and $t = -1$. The ball will hit the ground 3 s after it is thrown.

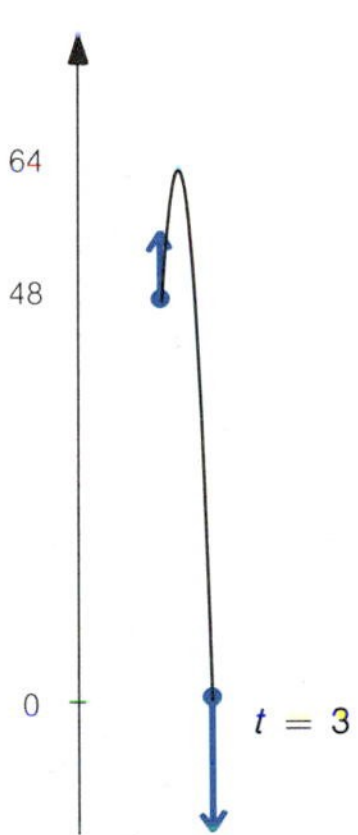


FIGURE 4.10.2

(c) The speed is the absolute value of the velocity. Using the formula from (a), we see that when $t = 3$, the velocity is

$$v(3) = -32(3) + 32 = -64.$$

The ball hits the ground at a speed of 64 ft/s. The negative velocity tells us the ball is falling when it hits the ground. See Figure 4.10.2.

(d) If a ball is thrown downward from a height of 48 ft at 32 ft/s, it has initial position $s_0 = 48$ and initial velocity $v_0 = -32$. (The initial velocity is negative, because the motion of the ball is downward, in the negative direction of our coordinate line.) The position at time $t$ is then

$$s(t) = -16t^2 - 32t + 48.$$

The velocity is

$$v(t) = s'(t) = -32t - 32.$$

The ball will hit the ground when $s(t) = 0$. Solving this equation for $t$, we have

$$s(t) = 0,$$

$$-16t^2 - 32t + 48 = 0,$$

$$-16(t^2 + 2t - 3) = 0,$$

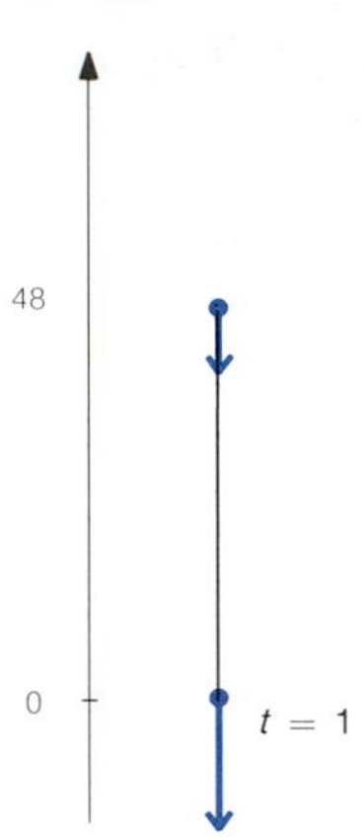


FIGURE 4.10.3

$$-16(t+3)(t-1)=0,$$

$$t=1, \quad t=-3.$$

The ball hits the ground 1 s after it is thrown. When $t = 1$, the velocity is

$$v(1) = -32(1) - 32 = -64.$$

It hits the ground at 64 ft/s. See Figure 4.10.3.

Comparing the answers in Example 6c and 6d, we see that a ball that is thrown directly upward from a height of 48 ft at 32 ft/s hits the ground with the same speed as a ball that is thrown directly downward from the same height at the same speed.

## EXERCISES 4.10

Verify the integration formulas in Exercises 1–8 by differentiation.

**1** $\displaystyle\int \frac{x}{\sqrt{x^2+4}}\,dx = \sqrt{x^2+4} + C$

**2** $\displaystyle\int x\sqrt{x^2+9}\,dx = \frac{1}{3}(x^2+9)^{3/2} + C$

**3** $\displaystyle\int \frac{1}{\sqrt{x}(1+\sqrt{x})^2}\,dx = -\frac{2}{1+\sqrt{x}} + C$

**4** $\displaystyle\int \frac{(1+x^{1/3})^3}{x^{2/3}}\,dx = \frac{3}{4}(1+x^{1/3})^4 + C$

**5** $\displaystyle\int x\sin x\,dx = \sin x - x\cos x + C$

**6** $\displaystyle\int x\cos x\,dx = \cos x + x\sin x + C$

**7** $\displaystyle\int \cos^3 x\,dx = -\frac{1}{3}\sin^3 x + \sin x + C$

**8** $\displaystyle\int \sin^3 x\,dx = \frac{1}{3}\cos^3 x - \cos x + C$

Find all $k$ for which the integration formulas in Exercises 9–16 are valid.

**9** $\displaystyle\int \sqrt{2x+3}\,dx = k(2x+3)^{3/2} + C$

**10** $\displaystyle\int \sqrt{1-x}\,dx = k(1-x)^{3/2} + C$

**11** $\displaystyle\int \cos 3x\,dx = k\sin 3x + C$

**12** $\displaystyle\int \cos\frac{x}{3}\,dx = k\sin\frac{x}{3} + C$

**13** $\displaystyle\int \cos(\pi - x)\,dx = k\sin(\pi - x) + C$

**14** $\displaystyle\int \sin\left(\frac{\pi}{4} - x\right)dx = k\cos\left(\frac{\pi}{4} - x\right) + C$

**15** $\displaystyle\int x^2(3x^3+1)^3\,dx = k(3x^3+1)^4 + C$

**16** $\displaystyle\int \frac{x^2}{(x^3+1)^3}\,dx = \frac{k}{(x^3+1)^2} + C$

Evaluate the indefinite integrals in Exercises 17–28.

**17** $\displaystyle\int (x^2 - 2x + 3)\,dx$

**18** $\displaystyle\int (3x^3 + 2x + 1)\,dx$

**19** $\displaystyle\int (x^{1/2} + x^{-1/2})\,dx$

**20** $\displaystyle\int (x^{1/3} + x^{-1/3})\,dx$

**21** $\displaystyle\int (\sin x + \cos x)\,dx$

**22** $\displaystyle\int (\sec^2 x + \sec x \cdot \tan x)\,dx$

23 $\int \csc x \cot x \, dx$

24 $\int \csc^2 x \, dx$

25 $\int (x + 1)(x - 2) \, dx$

26 $\int x(x^2 + 2) \, dx$

27 $\int \frac{x^2 + 1}{x^2} \, dx$

28 $\int \frac{x - 1}{\sqrt{x}} \, dx$

In Exercises 29–38 find a function $G(x)$ that satisfies the given conditions.

29 $G'(x) = 6x^2 + 2x - 3$, $G(0) = 5$

30 $G'(x) = 3x^2 + 1$, $G(0) = -2$

31 $G'(x) = \sqrt{x}$, $G(1) = 2$

32 $G'(x) = 1/\sqrt{x}$, $G(4) = 0$

33 $G'(x) = \sin x$, $G(0) = 2$

34 $G'(x) = \cos x$, $G(\pi) = -1$

35 $G''(x) = 12x$, $G'(0) = 2$, $G(0) = 3$.

36 $G''(x) = 6x + 4$, $G'(0) = -1$, $G(0) = 4$.

37 $G''(x) = \sin x + \cos x$, $G'(0) = 1$, $G(0) = 1$.

38 $G''(x) = \sin x + \cos x$, $G'(0) = 0$, $G(0) = 0$.

39 A ball is thrown vertically upward from ground level at a speed of 32 ft/s. (a) How high above ground will it get? (b) How long will it take the ball to return to ground level? (c) What will be the velocity of the ball when it hits the ground?

40 A ball is thrown vertically upward from ground level at a speed of 64 ft/s. (a) How high above ground will it get? (b) How long will it take the ball to return to ground level? (c) What will be the velocity of the ball when it hits the ground?

41 A rock is thrown directly downward at a speed of 48 ft/s from the top of a building 64 ft high. (a) How long will it take to hit the ground? (b) At what speed will it hit the ground?

42 A rock is thrown directly downward at a speed of 64 ft/s from the top of a building 80 ft high. (a) How long will it take to hit the ground? (b) At what speed will it hit the ground?

43 A hot air balloon is rising vertically at a rate of 0.7 m/s when a sandbag is released from a height of 42 m. (a) How long will it take the sandbag to hit the ground? (b) At what speed will it hit the ground?

44 A rocket is rising vertically at a rate of 4.9 m/s at a height of 147 m when it releases a spent fuel tank. (a) How long will it take the tank to hit the ground? (b) At what speed will the tank hit the ground?

45 A rocket is accelerating from a position of rest at a rate of $6t$ m/s$^2$. How far will it travel in the interval of time from (a) $t = 0$ to $t = 2$? (b) $t = 2$ to $t = 4$?

46 A racer starts from rest and accelerates at a constant rate of 4 m/s$^2$. How long will it take to travel 400 m?

## REVIEW EXERCISES

In Exercises 1–4, sketch the graphs of the given functions in the simplest way that reflects the domain, intercepts, sign of $f(x)$, and continuity.

1 $f(x) = x^2 - x - 2$

2 $f(x) = x(x - 2)^2$

3 $f(x) = 1 - \sqrt{4 - x}$

4 $f(x) = \sqrt{25 - x^2} - 3$

Evaluate the limits in Exercises 5–12.

5 $\lim_{x \to 2^+} \frac{1 - x}{2 - x}$

6 $\lim_{x \to -3^-} \frac{x}{(x + 3)^2}$

7 $\lim_{x \to \pi^+} \frac{\cos x}{\tan x}$

8 $\lim_{x \to \pi^-} \frac{\sec x}{\sin x}$

9 $\lim_{x \to \infty} \frac{3x - 1}{2x + 1}$

10 $\lim_{x \to \infty} \frac{x^2 + 2x + 5}{x^3 - 1}$

11 $\lim_{x \to -\infty} \frac{6x + 5}{\sqrt{4x^2 + 9}}$

12 $\lim_{x \to -\infty} (1 - 2x^2 - 3x^3)$

Determine the asymptotic behavior and sketch the graphs of the functions given in Exercises 13–16.

13 $f(x) = \frac{-2x}{x + 2}$

14 $f(x) = x + \frac{1}{x^2}$

15 $f(x) = \frac{x^2 + 2x + 1}{x}$

16 $f(x) = \frac{x^3 - x^2 + x}{x - 1}$

Determine if the graphs given in Exercises 17–20 are symmetric with respect to the $x$-axis, $y$-axis, or the origin.

17 18

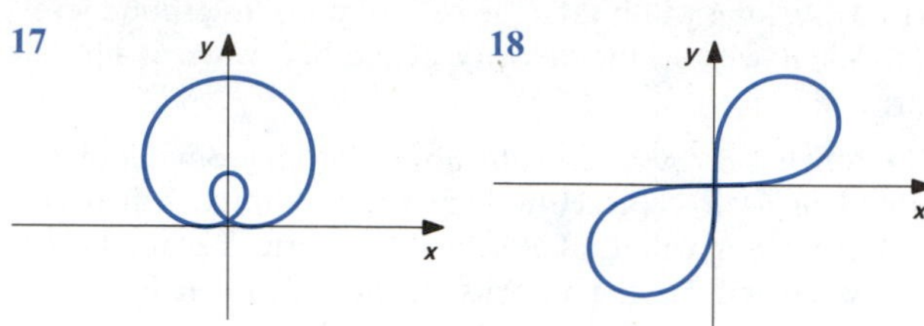

19 20

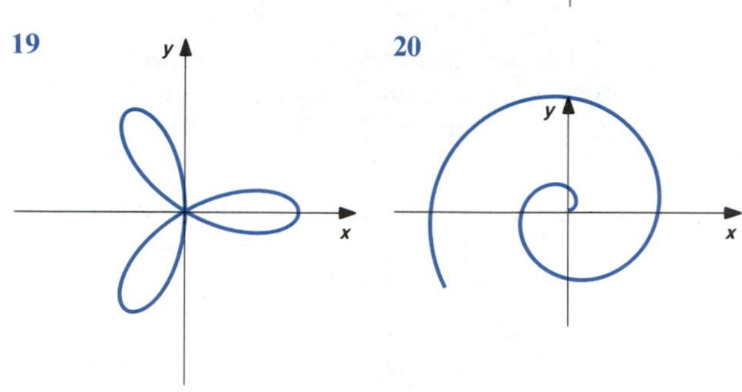

Determine if the functions given in Exercises 21–24 are odd or even.

21 $f(x) = 2x^4 - 1$ 22 $f(x) = 3x^3 + 5x$

23 $f(x) = x^3 - 2x + 1$ 24 $f(x) = x \tan x$

25 Use the given graph of $f(x)$, $x \geq 0$, to sketch the graphs of $f(-x)$, $x \leq 0$, and $f(|x|)$, all $x$. Discuss symmetry of the graph of $f(|x|)$.

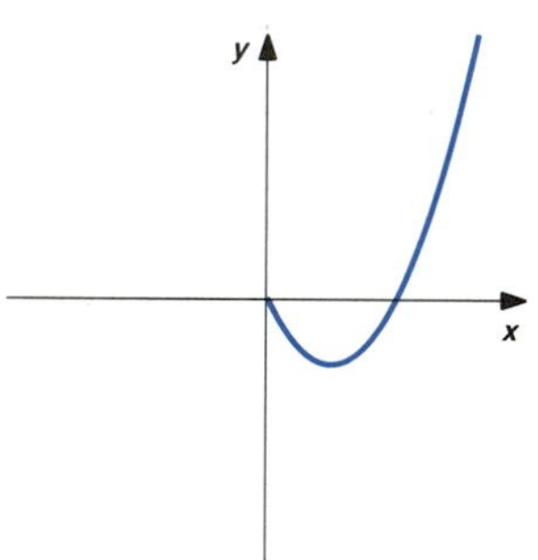

26 Use the given graph of $f(x)$ to sketch the graphs of $-f(x)$ and $|f(x)|$. Discuss symmetry between the graph of $f(x)$ and the graph of $-f(x)$.

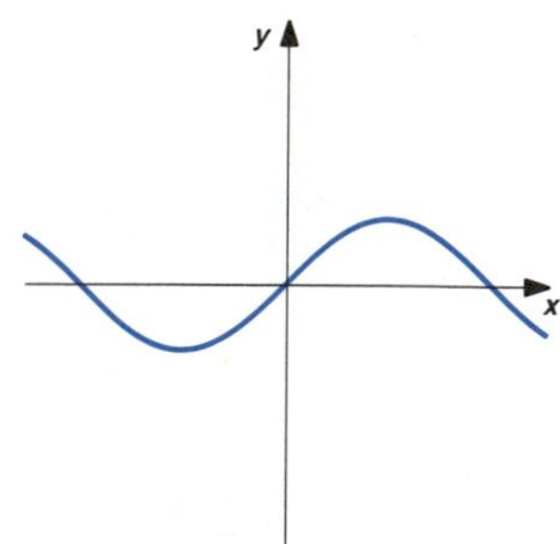

In Exercises 27–30 sketch the graph of a function that will satisfy the given conditions or indicate that this is impossible.

27 $f(x) > 0$, $f'(x) < 0$, $f''(x) < 0$, $0 < x < 1$

28 $f(x) < 0$, $f'(x) > 0$, $f''(x) < 0$, $0 < x < 1$, vertical asymptote $x = 0$

29 $f(x) < 0$, $f'(x) < 0$, $f''(x) > 0$, $0 < x < 1$, vertical asymptote $x = 1$

30 $f(x) < 0$, $f'(x) > 0$, $f''(x) > 0$, $x > 0$

Sketch the graphs of the functions in Exercises 31–40. Find local extrema and points of inflection.

31 $f(x) = x(x^2 - 3)$ 32 $f(x) = (4x^3 - x^4)/5$

33 $f(x) = \dfrac{x}{x+1}$ 34 $f(x) = \dfrac{x^2}{x^2 - 1}$

35 $f(x) = x - \dfrac{4}{x^2}$ 36 $f(x) = \dfrac{20x - 40}{x^3}$

37 $f(x) = 2 \sin x - x$, $0 \leq x \leq 2\pi$

38 $f(x) = \tan x - 2x$, $-\pi/2 < x < \pi/2$

39 $f(x) = (x^2 - 1)^{2/3}$ 40 $f(x) = x\sqrt{x+2}$

41 The radius of a right circular cylinder is increasing at a rate of 3 in./s while the height is decreasing at a rate of 4 in./s. Find the rate of change of the volume of the cylinder when the radius is 6 in. and the height is 8 in. Is the volume increasing or decreasing at this time?

42 The volume of a rectangular box remains constant as the length of each side of its square base increases at a rate of 3 cm/s. Find the rate of change of the height of the box when each side of the base is 5 cm and the height is 7 cm.

43 An airplane is flying 200 ft/s at an altitude of 10,000 ft in a line that will take it directly over a person who is observing the plane from the ground. Find the rate of change of the angle of elevation from the observer to the plane when it is directly over a point on the ground that is 10,000 ft from the observer.

44 An oil tank consists of a right circular cylinder of radius 5 m and height 10 m on top of a cone with radius of top 5 m and height 4 m. Oil is being pumped into the tank at a rate of 50 $m^3$/min. What is the rate of change of the depth of the oil in the tank when the depth at the deepest part is (a) 3 m and (b) 7 m?

45 A farmer has 2000 ft of fencing to enclose a rectangular region adjacent to a river. No fencing is required along the river. (a) Express the area of the region as a function of the length of the fence that is perpendicular to the river. Find the dimensions and area of the region that has maximum area. (b) Repeat part (a) with the constraint that the region cannot extend more than 400 ft from the river.

46 A rectangular box with square base and no top is to have volume of 32 $\text{ft}^3$. (a) Express the total amount of material required for the box as a function of the length of each side of its square base. Find the dimensions and the amount of material required for the box that requires the least material. (b) Repeat part (a) with the restriction that each side of the base must be at least 5 ft.

47 A right triangle is formed in the first quadrant by the $x$-axis, the $y$-axis, and a line through the point (2, 1). Express the area of the triangle as a function of its base. Find the dimensions and area of the triangle of minimum area.

48 A post 2 m high is 3 m from a post 1 m high. A cable is to be run from the top of one post to the ground at a point between the posts, and then to the top of the other post. Express the length of the cable as a function of the distance between the 2-m post and the point where the cable touches the ground. Find the shortest length of cable that can be used.

49 The horizontal distance traveled by a javelin that is thrown with initial velocity 20 m/s while running at 6 m/s is given by the formula

$$D(\theta) = \left(\frac{40}{4.9}\right)(10 \sin \theta \cos \theta + 3 \sin \theta),$$

where $\theta$ is the angle at which it is thrown. Find the angle that gives the longest throw.

50 Use linear and quadratic approximations of $x^{16}$ for $x$ near 1 to find approximate values of $(0.98)^{16}$. Use the remainder terms in Taylor's Formulas to determine the accuracy of the approximations.

51 Use a quadratic approximation of $\cos x$ for $x$ near 0 to find an approximate value of $\cos(-0.15)$. Use the remainder term in Taylor's Formula to determine the accuracy of the approximation.

52 Use the line tangent to the graph of the equation $x = 3y^2 - y^3$ at the point (2, 1) to approximate the value of $y$ near 1 that satisfies the equation when $x = 1.96$.

53 The height of a rectangular box is half the length of one side of its square base. Use differentials to find the approximate error in volume if the length of each side of the base is 2 m with a possible error of $\pm 0.02$ m. Find the approximate relative error.

54 The angle of elevation between an observer at ground level and an airplane is measured to be 20° with a possible error of 1° as the plane passes over a point on the ground that is 1 mi from the observer. Find the height of the plane and the differential approximation of the error in calculation. Find the approximate relative error.

Find all $k$ for which the integration formulas in Exercises 55–58 are valid.

55 $\displaystyle\int \sin 2x\, dx = k \cos 2x + C$

56 $\displaystyle\int \sec^2 \frac{x}{3}\, dx = k \tan \frac{x}{3} + C$

57 $\displaystyle\int \sqrt{5 - 2x}\, dx = k(5 - 2x)^{3/2} + C$

58 $\displaystyle\int \frac{x}{\sqrt{x^2 + 1}}\, dx = k\sqrt{x^2 + 1} + C$

Evaluate the indefinite integrals in Exercises 59–64.

59 $\displaystyle\int (x^3 - x + 2)\, dx$

60 $\displaystyle\int (x^{2/3} - x^{-1/3})\, dx$

61 $\displaystyle\int (\sin x - \cos x)\, dx$

62 $\displaystyle\int (\csc x)(\csc x + \cot x)\, dx$

63 $\displaystyle\int \sqrt{x}(x + 1)\, dx$

64 $\displaystyle\int \frac{x^2 + 1}{x^2}\, dx$

65 Find a function $G(x)$ that satisfies $G'(x) = 6x^2 + 4x - 1$ and $G(1) = 2$.

66 Find a function $G(x)$ that satisfies $G''(x) = 12x^2$, $G'(1) = 3$, and $G(1) = 4$.

67 A ball is thrown vertically upward from ground level at a speed of 64 ft/s. How high above ground will it get? How long will it take to return to ground level?

68 A hot air balloon is rising vertically at a rate of 2 ft/s when a sandbag is released from a height of 1580 ft. How long will it take the sandbag to hit the ground? At what speed will it hit the ground?

## CHAPTER FIVE

# *The definite integral*

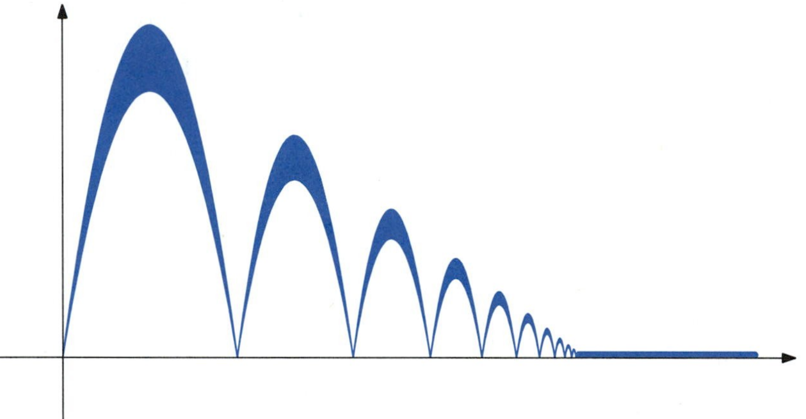

Many physical problems involve a certain type of sum, called a Riemann sum. In this chapter we will see how these types of sums are related to problems of determining the area of regions in the plane and finding the volume of solids. We will also see how the sums are related to problems that involve motion of an object along a number line.

Riemann sums lead us to the concept of the definite integral. A definite integral is a form of a limit of Riemann sums. We develop the concept of the definite integral in a way that is intended to emphasize the relation of Riemann sums to applications of the integral. Approximation techniques are discussed and compared for efficiency of computing. Formulas for the evaluation of definite integrals are developed. Finally, we will see the fundamental relation between the definite integral and the derivative.

## 5.1 ANALYTIC GEOMETRY FOR APPLICATIONS OF THE DEFINITE INTEGRAL

In this section we will illustrate how analytic geometry can be used to describe selected geometric quantities such as length and area in terms of a variable. The skills developed here are necessary in order to apply the concept of the definite integral to physical problems. The examples in this section are presented exactly as they will be used in later sections as parts of calculus problems.

Let us begin by considering the length of a variable vertical line segment. Recall from Section 1.3 that the coordinates of a variable point on the graph of $y = f(x)$ can be expressed in terms of the variable $x$ as $(x, f(x))$. For example, $(x, x^2)$ is a variable point on the graph of $y = x^2$ and $(x, x + 2)$ is a variable point on the graph of $y = x + 2$. Two points are on the same vertical line if they have the same $x$-coordinate. For example, the points $(x, x^2)$ and $(x, x + 2)$ are on the same vertical line for any choice of the variable $x$; the line segment between these points is a variable vertical line segment.

**Always use $x$ as a variable in problems that deal with variable vertical line segments.**

It is not necessary to use the distance formula to obtain the length of a vertical line segment. We should label the coordinates of the endpoints and then use

**(Length of vertical line segment)**
**= ($y$-coordinate of top point) − ($y$-coordinate of bottom point).**

### ■ EXAMPLE 1

Find the length of a variable vertical line segment between $y = x^2$ and the $x$-axis.

#### SOLUTION

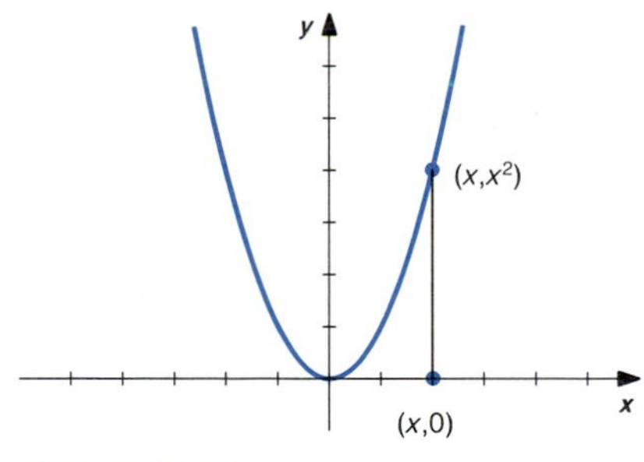


FIGURE 5.1.1

The length of the line segment depends on its position. That is, we will determine length as a function of $x$.

The first step is to make a sketch and label coordinates in terms of the variable $x$. This is done in Figure 5.1.1. The point $(x, x^2)$ is a variable point on the graph of $y = x^2$. The $x$-axis has equation $y = 0$, so $(x, 0)$ is a variable point on the $x$-axis.

The length of a variable vertical line segment is given by the $y$-coordinate of the top point minus the $y$-coordinate of the bottom point. From our sketch we see that $(x, x^2)$ is the top of the line segment and $(x, 0)$ is the bottom, for all values of $x$. We can then write

$$\text{Length} = x^2 - 0 \quad \text{or} \quad L(x) = x^2.$$

The distance formula gives the same result. That is,

$$L(x) = \sqrt{(x - x)^2 + (x^2 - 0)^2} = \sqrt{(x^2)^2} = |x^2| = x^2. \quad ■$$

### ■ EXAMPLE 2

Find the length of a variable vertical line segment between points on the graph of $x + 1 - (y - 1)^2 = 0$.

#### SOLUTION

We wish to express length as a function of $x$. To do this, we need to solve the given equation for $y$ in terms of $x$:

$$x + 1 - (y - 1)^2 = 0,$$

$$(y - 1)^2 = x + 1,$$

$$y - 1 = \pm\sqrt{x + 1},$$

$$y = 1 + \sqrt{x + 1}, \qquad y = 1 - \sqrt{x + 1}.$$

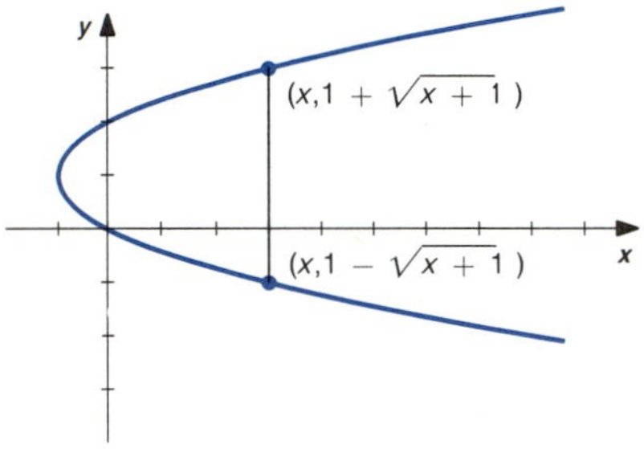

We can now sketch the parabola and label coordinates. This is done in Figure 5.1.2.

We use the sketch to write

$$\text{Length} = (1 + \sqrt{x + 1}) - (1 - \sqrt{x + 1}),$$

$$L(x) = 2\sqrt{x + 1}.$$

FIGURE 5.1.2

The domain of this function is $x \geq -1$. When $x = -1$, the "line segment" has length zero. The vertical line $x = a$ does not intersect the parabola when $a < -1$. ■

Recall from Section 1.3 that the **midpoint** of the line segment between the points $(x_1, y_1)$ and $(x_2, y_2)$ is

$$\left(\frac{x_1 + x_2}{2}, \frac{y_1 + y_2}{2}\right).$$

### ■ EXAMPLE 3

Find the length and midpoint of a variable vertical line segment between $y = x^2$ and $x + y = 2$.

#### SOLUTION

The graphs are sketched and labeled in Figure 5.1.3. We have used the fact that $x + y = 2$ implies $y = 2 - x$.

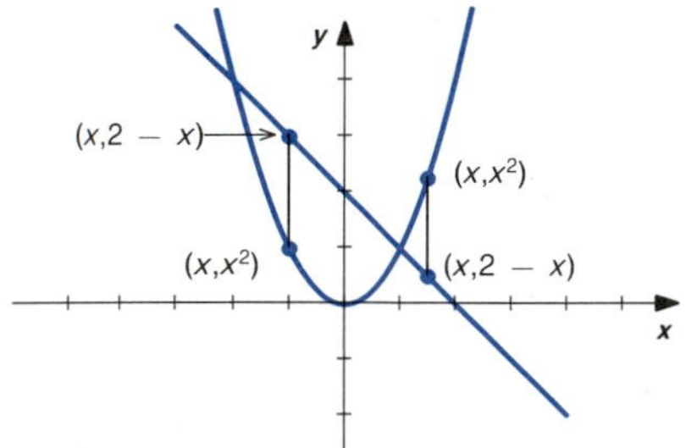

FIGURE 5.1.3

The line is above the parabola between their points of intersection, and the parabola is above the line outside the points of intersection. Hence, *there is not a single variable vertical line segment whose labeling of top and bottom holds for all x.* There are two different types of vertical segments with different labeling. This will be reflected in the fact that two different formulas will be used to express the length—one the negative of the other. To determine intervals where each formula holds, we need the $x$-coordinates of the points of intersection. If a point $(x, y)$ is in the intersection of the graphs, its coordinates will satisfy both equations. This leads to the simultaneous equations

$$y = x^2, \qquad y = 2 - x.$$

We do not need to find pairs $(x, y)$ that satisfy both equations, but only the $x$-coordinates of the solutions.

**We eliminate the undesired variable.**

This is easily done by equating the two expressions for $y$. We obtain

$$x^2 = 2 - x.$$

Solving for $x$, we have

$$x^2 + x - 2 = 0,$$

$$(x + 2)(x - 1) = 0,$$

$$x = -2, \qquad x = 1.$$

The length of a variable vertical line segment is

$$L(x) = \begin{cases} (2 - x) - x^2, & -2 \leq x \leq 1, \\ x^2 - (2 - x), & x < -2 \quad \text{or} \quad x > 1. \end{cases}$$

The midpoint is

$$\left(\frac{x + x}{2}, \frac{x^2 + (2 - x)}{2}\right) = \left(x, \frac{x^2 - x + 2}{2}\right).$$

The two different types of vertical line segments give the same formula for the coordinates of their midpoints. ■

If the distance formula were used to determine the length of the line segment in Example 3, we would have

$$\begin{aligned} L(x) &= \sqrt{(x - x)^2 + [(2 - x) - x^2]^2} \\ &= |(2 - x) - x^2| \\ &= \begin{cases} (2 - x) - x^2, & (2 - x) - x^2 \geq 0, \\ x^2 - (2 - x), & (2 - x) - x^2 < 0. \end{cases} \end{aligned}$$

We would then need to solve the above inequalities to determine intervals of $x$ for which each formula for $L(x)$ is valid. The result would be the same as that obtained in Example 3. (Absolute value signs are not convenient for the calculations we will be making in applications of the definite integral.)

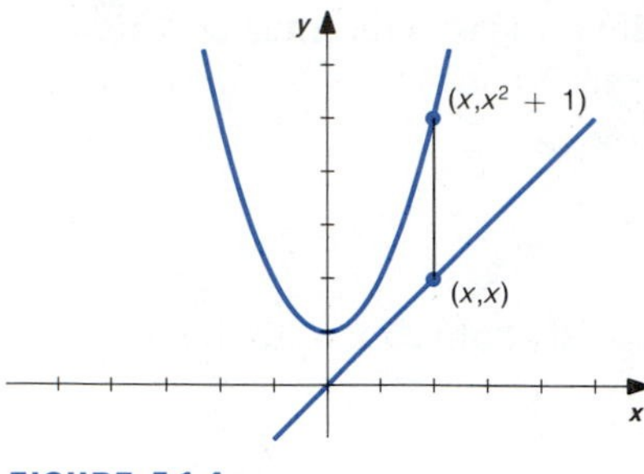

FIGURE 5.1.4

**EXAMPLE 4**

Find the length of a variable vertical line segment between $x - y = 0$ and $x^2 - y + 1 = 0$.

**SOLUTION**

The graphs are sketched and labeled in Figure 5.1.4.

It appears from our sketch that the parabola is always above the line. To be sure of this, we attempt to find the $x$-coordinates of points of intersection. This leads to the equations

$$y = x^2 + 1, \qquad y = x.$$

Equating the two expressions for $y$ and then solving for $x$, we obtain

$$x^2 + 1 = x,$$

$$x^2 - x + 1 = 0.$$

The quadratic formula gives the solutions

$$x = \frac{-(-1) \pm \sqrt{(-1)^2 - 4(1)(1)}}{2(1)}.$$

Since $b^2 - 4ac = (-1)^2 - 4(1)(1) = -3 < 0$, there are no real solutions of this equation. This means the graphs do not intersect. We have length

$$L(x) = (x^2 + 1) - x.$$ ■

The same ideas that we used for vertical line segments also apply to horizontal line segments. Points on a horizontal line have a common $y$-coordinate.

**Always use $y$ as a variable in problems that deal with variable horizontal line segments.**

In this case, we must label the coordinates on our graphs in terms of their $y$-coordinate. Also,

**(Length of horizontal line segment)**
**= ($x$-coordinate of right endpoint) − ($x$-coordinate of left endpoint).**

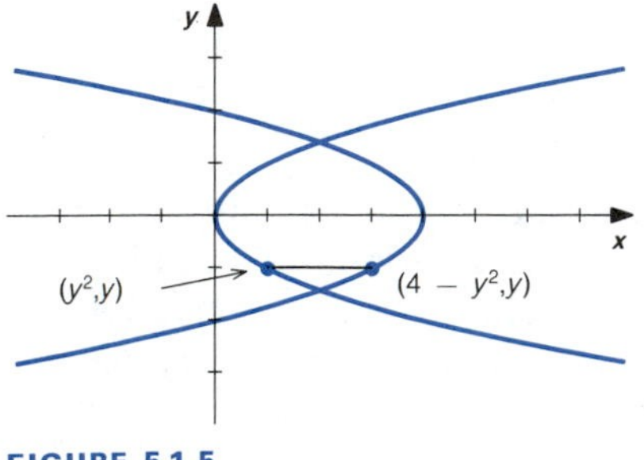

FIGURE 5.1.5

**EXAMPLE 5**

Find the length of a variable horizontal line segment between $x = y^2$ and $x = 4 - y^2$ for values of $y$ that are between the $y$-coordinates of the points of intersection of the graphs.

**SOLUTION**

The graphs are sketched and labeled in Figure 5.1.5.

We need the $y$-coordinates of the points of intersection of the graphs. In this case, we eliminate the variable $x$ from the simultaneous equations

$$x = y^2, \qquad x = 4 - y^2$$

to obtain

$$y^2 = 4 - y^2.$$

Solving for $y$, we have

$$2y^2 = 4,$$

$$y^2 = 2,$$

$$y = \sqrt{2}, \qquad y = -\sqrt{2}.$$

For values of $y$ between the $y$-coordinates of the points of intersection, the parabola $x = 4 - y^2$ is to the right of the parabola $x = y^2$. This means the length is

$$L(y) = (4 - y^2) - y^2, \qquad -\sqrt{2} \leq y \leq \sqrt{2}.$$ ■

We will need to be able to find the area of a variable cross section of certain three-dimensional solids.

**Always use $x$ as a variable in problems that deal with variable cross sections perpendicular to the $x$-axis.**

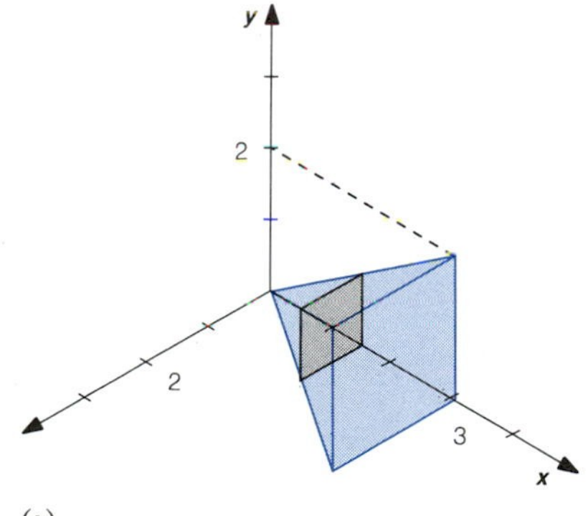


(a)

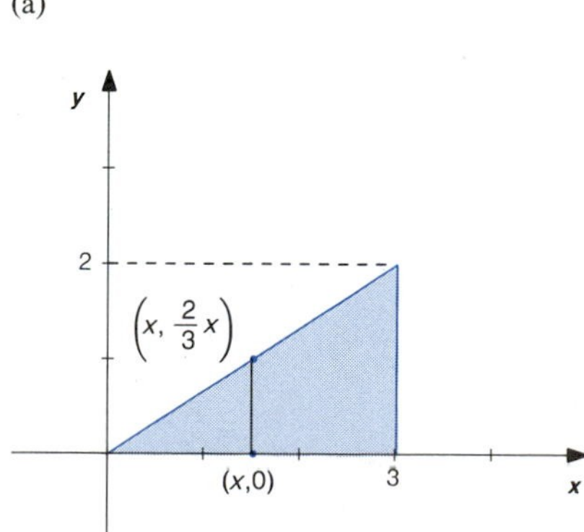


(b)

FIGURE 5.1.6

### ■ EXAMPLE 6

Cross sections perpendicular to the $x$-axis of the prism pictured in Figure 5.1.6a are squares. Find the area of a variable cross section.

#### SOLUTION

We see from Figure 5.1.6a that the length of one side of the square formed by a variable cross section is the length of a variable line segment between the $x$-axis and the line in the $xy$-plane through the origin and the point (3, 2). The slope of this line is

$$\frac{2-0}{3-0} = \frac{2}{3}.$$

The equation of the line through the origin with slope 2/3 is

$$y = \frac{2}{3}x.$$

This line and a variable vertical line segment between it and the $x$-axis are sketched and labeled in Figure 5.1.6b. We see that the length of one side of the square is

$$\text{Length} = \frac{2}{3}x.$$

The area of a variable square cross section is

$$\text{Area} = (\text{Length})^2,$$

$$A(x) = \left(\frac{2}{3}x\right)^2,$$

$$A(x) = \frac{4}{9}x^2, \qquad 0 \le x \le 3. \qquad \blacksquare$$

**Always use $y$ as a variable in problems that deal with variable cross sections perpendicular to the $y$-axis.**

■ **EXAMPLE 7**

Cross sections perpendicular to the $y$-axis of the right circular cone pictured in Figure 5.1.7a are circles. Find the area of a variable cross section.

**SOLUTION**

We need to determine the radius of a variable circular cross section as a function of $y$. This is the length of a variable line segment between the $y$-axis and the line in the $xy$-plane through the points (0, 3) and (1, 0). The slope of this line is $-3$, and the equation is

$$y - 0 = -3(x - 1).$$

Solving for $x$ we have

$$y = -3x + 3,$$

$$3x = 3 - y,$$

$$x = \frac{3 - y}{3}.$$

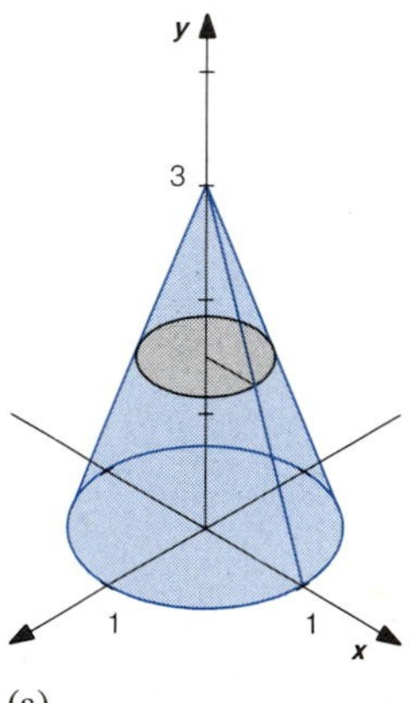

(a)

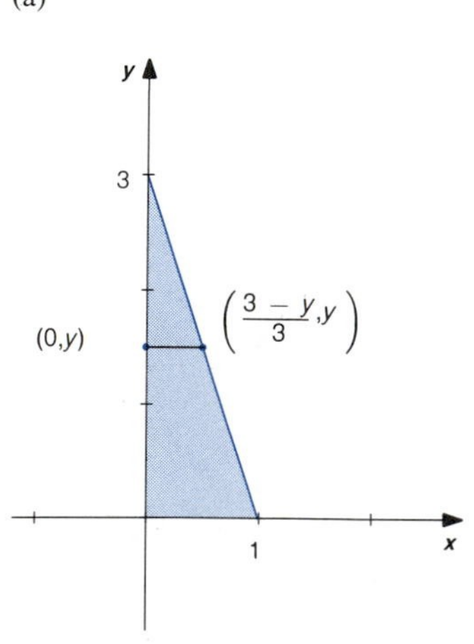

(b)

FIGURE 5.1.7

The line and a variable horizontal line segment are sketched and labeled in Figure 5.1.7b. We see that

$$\text{Radius} = \frac{3 - y}{3}.$$

The area of a circular cross section is

$$\text{Area} = \pi(\text{Radius})^2,$$

$$A(y) = \pi\left(\frac{3 - y}{3}\right)^2, \qquad 0 \le y \le 3. \qquad \blacksquare$$

FIGURE 5.1.8

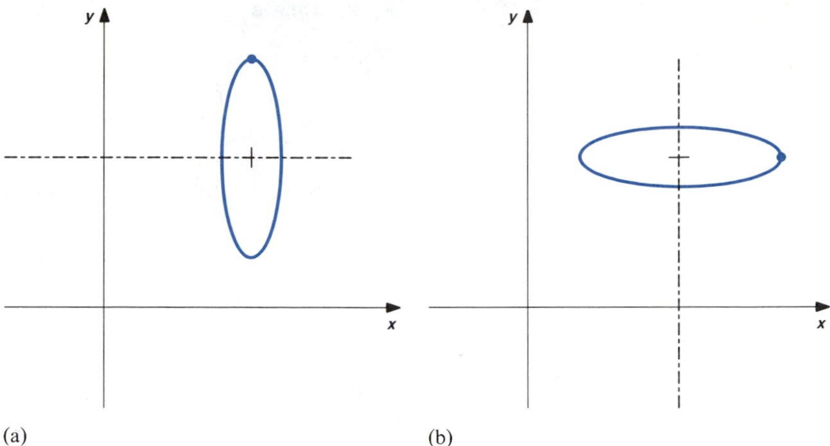


The next type of problem we consider involves revolving a line segment about a line. When a line segment is revolved through space, a surface is traced out. To sketch the surface we must be able to sketch circles obtained by revolving points in the sketch plane about a line in that plane. Figure 5.1.8a illustrates a circle obtained by revolving a point about a line that is parallel to the $x$-axis. Figure 5.1.8b illustrates a circle obtained by revolving a point about a line that is parallel to the $y$-axis. These circles in three-dimensional space are drawn as ellipses in the two-dimensional sketch plane in order to represent an oblique view of the circles.

If a line segment is revolved about a perpendicular line, the resulting surface will be a disk if the segment touches the line (Figure 5.1.9a) and a washer if the segment does not touch the line (Figure 5.1.9b). A line segment that is revolved about a parallel line generates a cylinder (Figure 5.1.9c). The areas of these surfaces are given by the formulas

**Area of disk** $= \pi r^2$,

**Area of washer** $= \pi R^2 - \pi r^2$,

**Lateral surface area of cylinder** $= 2\pi rh$.

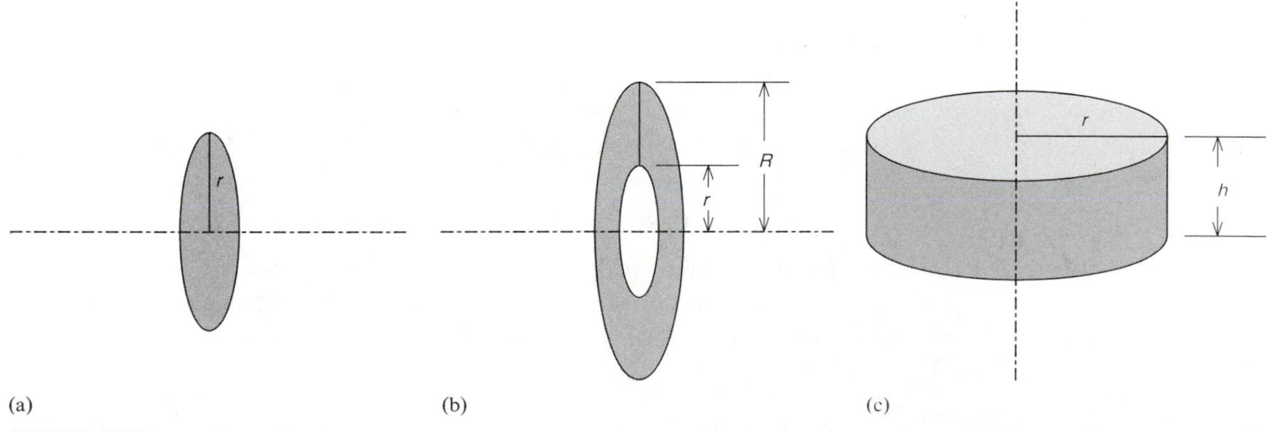


FIGURE 5.1.9

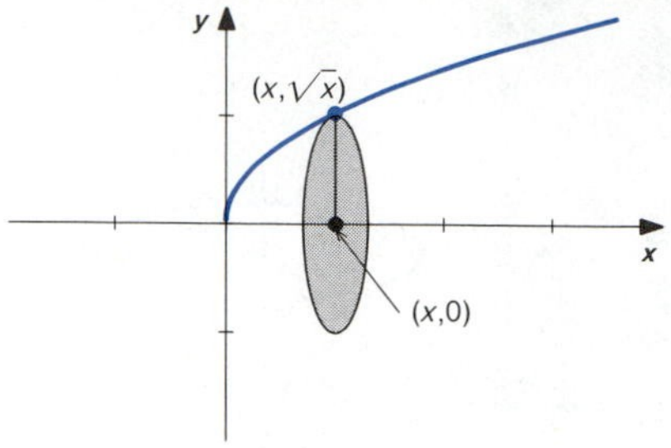

FIGURE 5.1.10

■ **EXAMPLE 8**

Find the area of the disk obtained by revolving a variable vertical line segment between $y = \sqrt{x}$ and the $x$-axis about the $x$-axis.

**SOLUTION**

In Figure 5.1.10 we have sketched the curve, sketched and labeled the endpoints of a variable vertical line segment, and then sketched the circle obtained by revolving the upper endpoint about the $x$-axis. The lower endpoint is on the $x$-axis so that the point remains stationary as the line segment is revolved about the $x$-axis. The radius of the resulting disk is

$$\text{Radius} = \sqrt{x} - 0.$$

The area of the disk is

$$\text{Area} = \pi(\text{Radius})^2,$$

$$A(x) = \pi(\sqrt{x})^2,$$

$$A(x) = \pi x, \qquad x \geq 0.$$

■

■ **EXAMPLE 9**

A variable vertical line segment between $y = \sqrt{x}$ and the $x$-axis is revolved about the line $y = -1$. Find the area of the resulting washer.

**SOLUTION**

The first step is to sketch the curve, draw a variable vertical line segment, and sketch the line $y = -1$. The circles obtained by revolving each endpoint of the segment about the axis of revolution are then sketched, indicating a washer. We then label both endpoints of the variable line segment and the center of the washer in terms of $x$. This is done in Figure 5.1.11. From the sketch, we see that the washer has

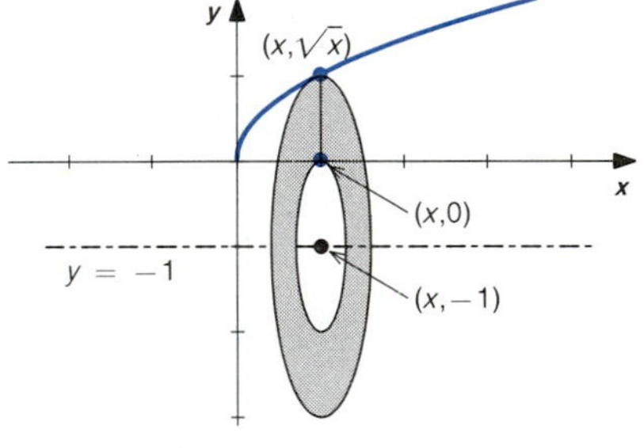


FIGURE 5.1.11

$$\text{Outer radius:} \quad R = \sqrt{x} - (-1) = \sqrt{x} + 1,$$

$$\text{Inner radius:} \quad r = 0 - (-1) = 1.$$

The area is

$$A(x) = \pi(\sqrt{x} + 1)^2 - \pi(1)^2, \qquad x \geq 0.$$

■

■ **EXAMPLE 10**

Find the lateral surface area of the cylinder obtained by revolving a variable vertical line segment between $y = \sqrt{x}$ and the $x$-axis about the $y$-axis.

**SOLUTION**

We first sketch the curve and a variable vertical line segment. Each endpoint of the segment is then revolved about the $y$-axis, forming a circle.

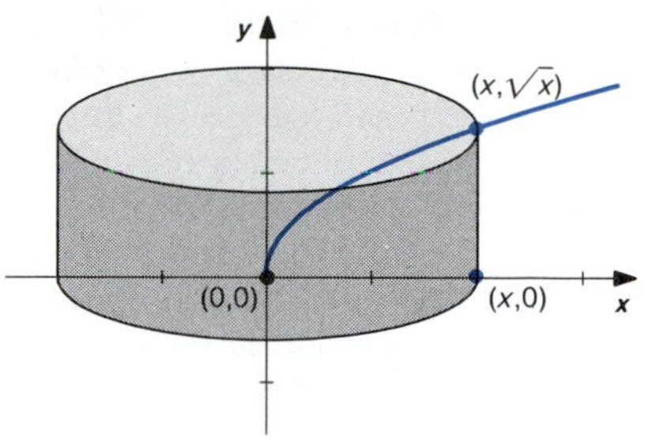


FIGURE 5.1.12

Since the vertical segment is parallel to the axis of revolution, the radii of these two circles are equal. These two circles are then connected by a vertical line segment that indicates the edge of the cylinder generated by revolving the segment. The final step is to label the endpoints of the segment and a point on the axis of revolution. This is done in Figure 5.1.12. From the sketch, we see the cylinder has

Radius: $r = x - 0$,

Height: $h = \sqrt{x} - 0$.

The lateral surface area is

Surface area $= 2\pi(\text{Radius})(\text{Height})$,

$S(x) = 2\pi(x)(\sqrt{x}), \qquad x \geq 0.$ ■

## EXERCISES 5.1

Sketch, label, and find the length of the variable line segments described in Exercises 1–12. Use $x$ as the variable in problems that involve vertical line segments; use $y$ as the variable in problems that involve horizontal line segments. Do not use absolute value signs in formulas for length.

1 Vertical, between $y = x^2$ and $y = -2$

2 Vertical, between $y = -x^2$ and $y = 1$

3 Vertical, between $y = x^2 + 1$ and $y + x^2 = 0$

4 Vertical, between $y = 1 - x^2$ and $x + y = 2$

5 Horizontal, between $x = y^2$ and $x + y = -1$

6 Horizontal, between $x = 1 - y^2$ and $x = 2$

7 Vertical, between points on $(x - 2)^2 + y^2 = 4$

8 Horizontal, between points on $(x - 2)^2 + y^2 = 4$

9 Horizontal, between points on $y = (x - 1)^2 - 1$

10 Vertical, between points on $x = (y + 1)^2 - 1$

11 Vertical, between $y = 2x$ and the $x$-axis

12 Horizontal, between $y = 3x$ and $y = 2x$

Sketch, label, and find the length and midpoint of the line segments described in Exercises 13–20. Use $x$ as the variable in problems that involve vertical line segments; use $y$ as the variable in problems that involve horizontal line segments. Do not use absolute value signs in formulas for length.

13 Vertical, between $y = 2 - x^2$ and $y = x$

14 Vertical, between $y = x^2$ and $y = 4 - x^2$

15 Horizontal, between $x = y^2$ and $2x = 1 + y^2$

16 Horizontal, between $x + y^2 = 1$ and $x - y + 1 = 0$

17 Vertical, between $y = x^2$ and $y = x$, $0 \leq x \leq 1$

18 Horizontal, between $y = x^3$ and $y = x$, $0 \leq y \leq 1$

19 Horizontal, between $y = x^3$ and $y = x^2$, $0 \leq y \leq 1$

20 Vertical, between $y = x^3$ and $y = x^2$, $0 \leq x \leq 1$

The shape of cross sections perpendicular to the $x$-axis is indicated for each of the solids in Exercises 21–26. Express the area of a variable cross section as a function of $x$.

21 Circles

22 Circles

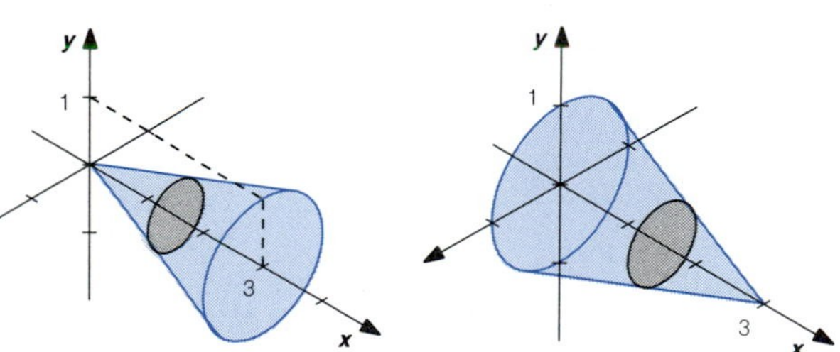


23 Squares

24 Squares

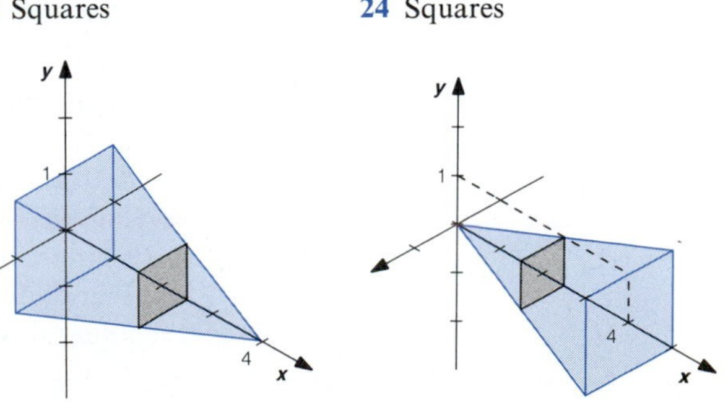

**25** Isosceles right triangles **26** Rectangles

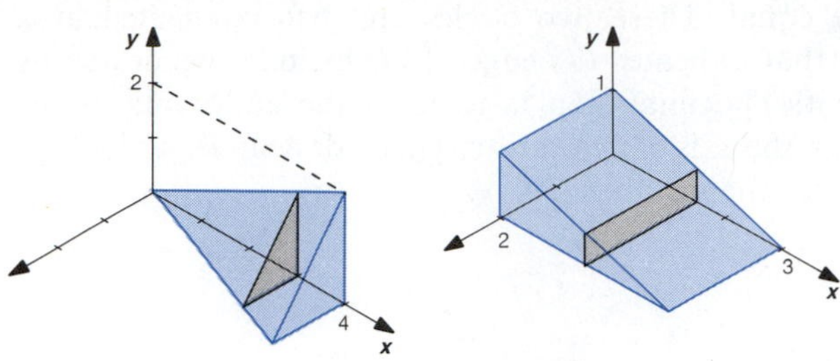


The shape of cross sections perpendicular to the $y$-axis is indicated for each of the solids in Exercises 27–30. Express the area of a variable cross section as a function of $y$.

**27** Squares **28** Squares

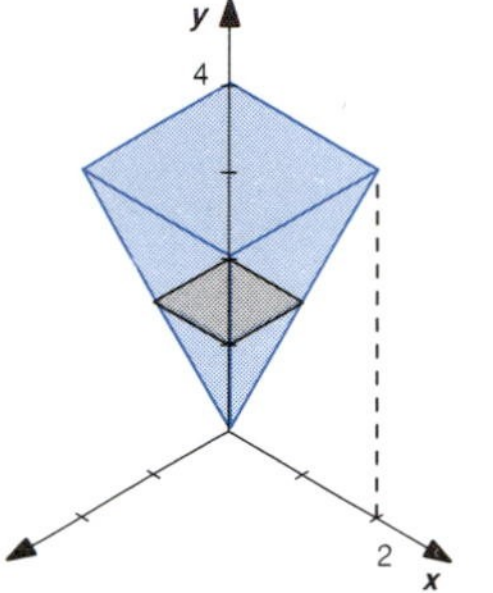


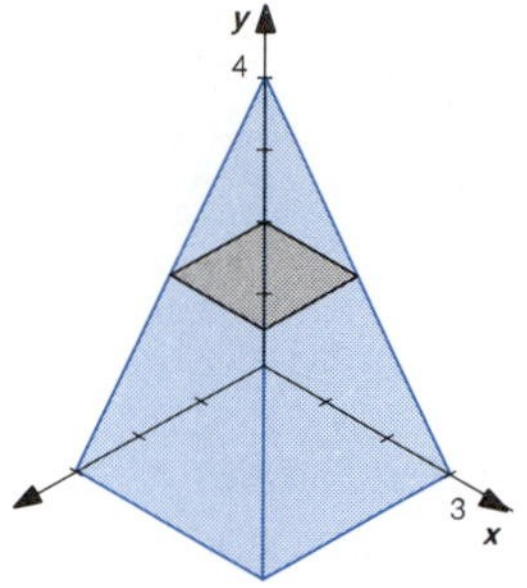


**29** Rectangles **30** Isosceles right triangles

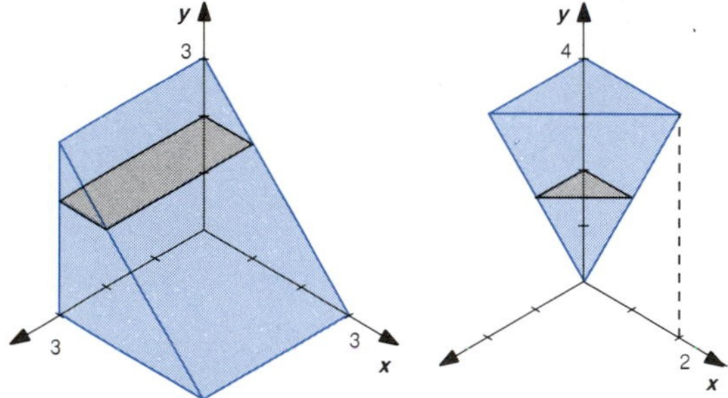


In Exercises 31–44, revolving the given line segment about the indicated line generates either a disk, washer, or cylinder. Sketch and label whichever one is generated in each exercise. Find the area of each disk and washer obtained and find the lateral surface area of each cylinder obtained. Use $x$ as the variable in problems that involve vertical line segments; use $y$ as the variable in problems that involve horizontal line segments. Do not use absolute value signs in formulas for length.

**31** Vertical, between $y = x^2 + 1$ and the $x$-axis; about the $x$-axis

**32** Vertical, between $y = x^2 - 1$ and $y = 1$, $-\sqrt{2} \le x \le \sqrt{2}$; about $y = 1$

**33** Vertical, between $y = x^2 + 1$ and the $x$-axis; about $y = -1$

**34** Vertical, between $y = x^2 + 1$ and $y = 1$; about the $x$-axis

**35** Vertical, between $hy = rx$ and the $x$-axis, $0 \le x \le h$; about the $x$-axis; assume $h$ and $r$ are positive

**36** Horizontal, between $ry = hx$ and $x = r$, $0 \le y \le h$; about the $y$-axis; assume $h$ and $r$ are positive

**37** Vertical, between $(x/r) + (y/h) = 1$ and the $x$-axis, $0 \le x \le r$; about the $y$-axis; assume $h$ and $r$ are positive

**38** Horizontal, between $(x/r) + (y/h) = 1$ and the $y$-axis, $0 \le y \le h$; about the $y$-axis; assume $h$ and $r$ are positive

**39** Horizontal, between $(x/h) + (y/r) = 1$ and the $y$-axis, $0 \le y \le r$; about the $x$-axis; assume $h$ and $r$ are positive

**40** Vertical, between points on $x = y^2 + 1$; about the $y$-axis

**41** Horizontal, between points on $y = x^2$; about the $x$-axis

**42** Horizontal, between $x = \sqrt{r^2 - y^2}$ and the $y$-axis; about the $y$-axis

**43** Horizontal, between $x = y^2$ and $x - y = 2$, $-1 \le y \le 2$; about $x = 4$

**44** Vertical, between points on $x^2 + y^2 = r^2$; about $x = R$, where $0 < r < R$

## 5.2 RIEMANN SUMS

We begin our investigation of the definite integral by introducing Riemann sums. The study of Riemann sums leads to the definition of the definite integral and to an understanding of how definite integrals are used in solving physical problems.

In simple terms, a Riemann sum is a sum of products. Each term is the product of the length of an interval and the value of a function at a point in the interval. We describe Riemann sums in precise mathematical terms to establish the notation and provide a common frame of reference for the examples that follow. Do not expect to understand Riemann sums by reading the mathematical description. It is the set of *examples* that give us an understanding of the idea and how it is used.

A **Riemann sum** of a function $f$ over an interval $[a, b]$ is any sum of the form

$$f(x_1^*)\,\Delta x_1 + f(x_2^*)\,\Delta x_2 + \cdots + f(x_n^*)\,\Delta x_n, \tag{1}$$

where the points $x_0, x_1, \ldots, x_n$ with

$$a = x_0 < x_1 < \cdots < x_n = b$$

divide the interval $[a, b]$ into $n$ subintervals,

$$[x_0, x_1], [x_1, x_2], \ldots, [x_{n-1}, x_n],$$

the length of the subinterval $[x_{j-1}, x_j]$ is

$$\Delta x_j = x_j - x_{j-1},$$

and $x_j^*$ is any point in the interval $[x_{j-1}, x_j]$, so

$$x_{j-1} \le x_j^* \le x_j, \qquad j = 1, \ldots, n.$$

Points $x_0, x_1, \ldots, x_n$ with $a = x_0 < x_1 < \cdots < x_n = b$ are said to form a **partition** of $[a, b]$. The largest of the lengths $\Delta x_j$ is denoted max $\Delta x_j$ and called the **norm** of the partition.

We will always divide the interval $[a, b]$ into $n$ subintervals of equal length. Their common length is then

$$\Delta x_j = \frac{b - a}{n}, \qquad j = 1, \ldots, n,$$

and the points of subdivision are given by the formula

$$x_j = a + (j)\left(\frac{b - a}{n}\right), \qquad j = 0, 1, 2, \ldots, n. \tag{2}$$

A Riemann sum is a *number* that is a sum of products. Although we will calculate the value of some Riemann sums, calculation is only of secondary interest. Our primary goal is to see how Riemann sums are related to some physical problems. We will see that it is the *interpretation* of the products $f(x_j^*)\,\Delta x_j$ that leads to the many applications of the concept. This is analogous to the derivative, where it is the interpretation of the difference quotient that leads to the applications.

### *Application of Riemann sums to area*

Let us consider a region in the plane with the property that a variable vertical line that intersects the region intersects it in a segment of length

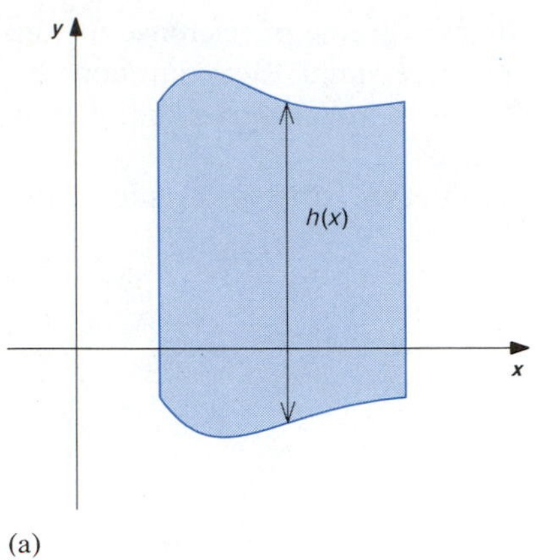


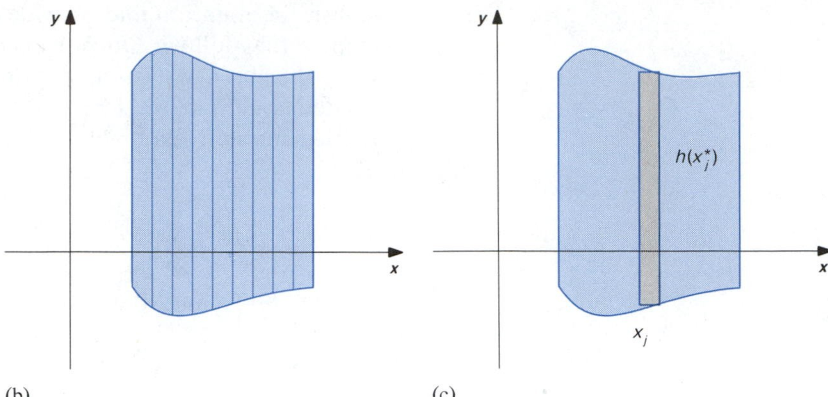


(a) (b) (c)

FIGURE 5.2.1

$h(x)$. Such a region is illustrated in Figure 5.2.1a. To find an approximate value of the area of the region, let us slice the region into $n$ thin vertical strips of width $\Delta x_j$, $j = 1, \ldots, n$, as indicated in Figure 5.2.1b. If $h(x)$ is a continuous function, then the area of each vertical strip is approximately the area of a rectangle with height $h(x_j^*)$ and base $\Delta x_j$, as indicated in Figure 5.2.1c. Also, the error in approximation of the area in each strip becomes *proportionally* smaller than the area in the strip as the width of the strip becomes smaller. This means that the total area of the rectangles becomes closer to the exact area of the region as the approximating slices become thinner. That is, if all $\Delta x_j$ are small enough,

Area of region $\approx$ Sum of areas of rectangles,

$$\textbf{Area} \approx h(x_1^*)\,\Delta x_1 + h(x_2^*)\,\Delta x_2 + \cdots + h(x_n^*)\,\Delta x_n.$$

The sum on the right above is a Riemann sum of the height function $h(x)$. The product $h(x_j^*)\,\Delta x_j$ is interpreted as the area of a rectangle with height $h(x_j^*)$ and base $\Delta x_j$. The Riemann sum is simply the sum of the areas of $n$ rectangles.

If $f(x)$ is nonnegative for $a \leq x \leq b$, the region bounded by

$$y = f(x), \qquad y = 0, \quad x = a, \quad x = b$$

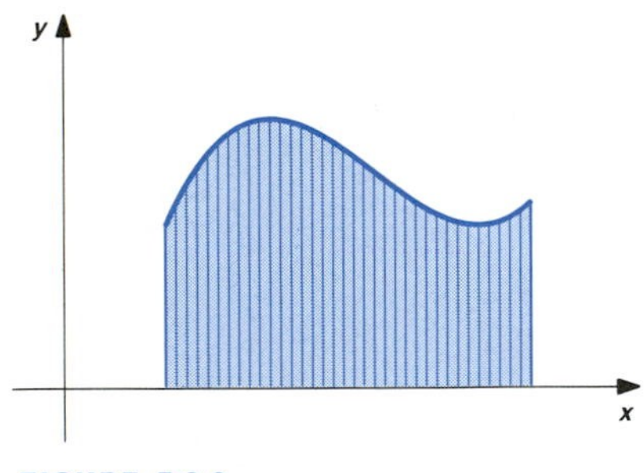


FIGURE 5.2.2

is called the **region under the curve** $y = f(x)$, $a \leq x \leq b$. See Figure 5.2.2. The height function for this region is $f(x)$, so Riemann sums of $f$ give approximate values of the area under the curve.

■ **EXAMPLE 1**

Use Riemann sums with $n = 8$ subintervals of equal length and $x_j^* = x_j$, the right endpoint of the subinterval $[x_{j-1}, x_j]$, $j = 1, \ldots, n$, to find an approximate value of the area under the curve $y = \sqrt{4 - x^2}$, $-2 \leq x \leq 2$.

**SOLUTION**

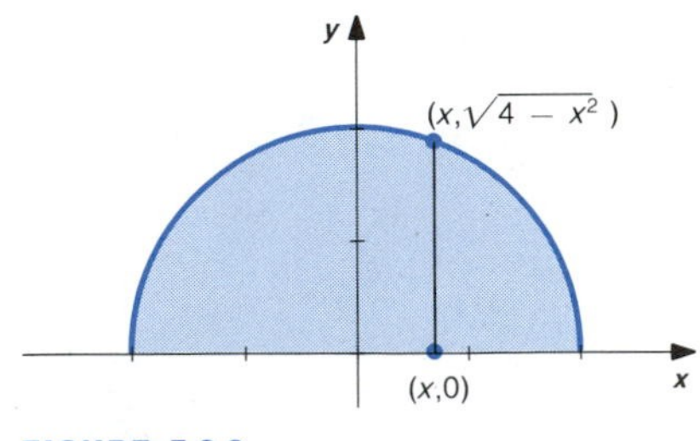


FIGURE 5.2.3

The region is sketched in Figure 5.2.3. The variable height of the region has been indicated by sketching and labeling a variable vertical line segment that intersects the region. We want to evaluate the Riemann sum of $f(x) = \sqrt{4 - x^2}$ over the interval $[-2, 2]$, using $n = 8$ subintervals of equal length

$$\Delta x_j = \frac{b-a}{n} = \frac{2-(-2)}{8} = 0.5,$$

and $x_j^* = x_j$, the right endpoint of each subinterval.

The points $x_0 = -2$, $x_1 = -1.5$, $x_2 = -1$, $x_3 = -0.5$, $x_4 = 0$, $x_5 = 0.5$, $x_6 = 1$, $x_7 = 1.5$, and $x_8 = 2$ divide the interval $[-2, 2]$ into eight subintervals of equal length. The desired Riemann sum is

$$\begin{aligned} R_8 &= f(x_1)\,\Delta x_1 + f(x_2)\,\Delta x_2 + f(x_3)\,\Delta x_3 + f(x_4)\,\Delta x_4 \\ &\quad + f(x_5)\,\Delta x_5 + f(x_6)\,\Delta x_6 + f(x_7)\,\Delta x_7 + f(x_8)\,\Delta x_8 \\ &= \sqrt{4-(-1.5)^2}(0.5) + \sqrt{4-(-1)^2}(0.5) + \sqrt{4-(-0.5)^2}(0.5) \\ &\quad + \sqrt{4-(0)^2}(0.5) + \sqrt{4-(0.5)^2}(0.5) + \sqrt{4-(1)^2}(0.5) \\ &\quad + \sqrt{4-(1.5)^2}(0.5) + \sqrt{4-(2)^2}(0.5) \\ &\approx 5.9914181. \end{aligned}$$

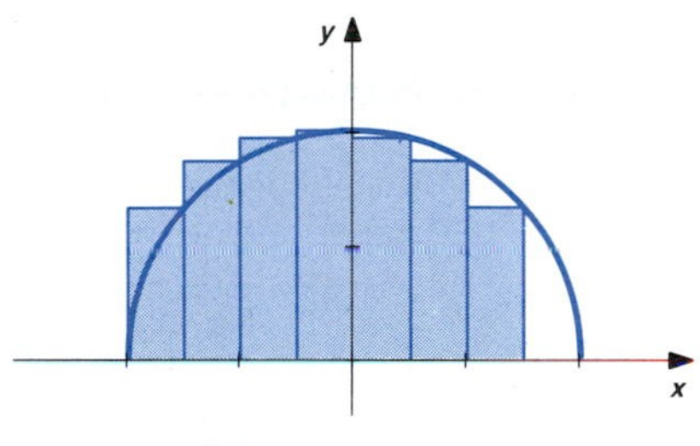


FIGURE 5.2.4

$R_8$ is the area of the rectangles pictured in Figure 5.2.4. There are eight rectangles if we count the "rectangle" on the far right that has height and area zero. (The region under the curve is half of a circle of radius 2. The exact area of the region is $\pi(2)^2/2 = 2\pi \approx 6.2831853$.) ■

■ **EXAMPLE 2**

(a) Use Riemann sums with $n = 5$ subintervals of equal length and $x_j^* = x_j$, the right endpoint of the subinterval $[x_{j-1}, x_j]$, $j = 1, \ldots, n$, to find an approximate value of the area of the triangular region bounded by the lines

$$y = \frac{x}{2}, \quad y = -x, \quad \text{and} \quad x = 1.$$

(b) Repeat (a), except use $n = 10$ subintervals.

**SOLUTION**

The region is sketched in Figure 5.2.5. We have also sketched and labeled a variable vertical line segment that intersects the region. From the sketch we see that

$$\text{Length of a variable vertical line segment} = \frac{x}{2} - (-x).$$

This length gives the height function for this region. We have

$$h(x) = \frac{3x}{2}, \qquad 0 \le x \le 1.$$

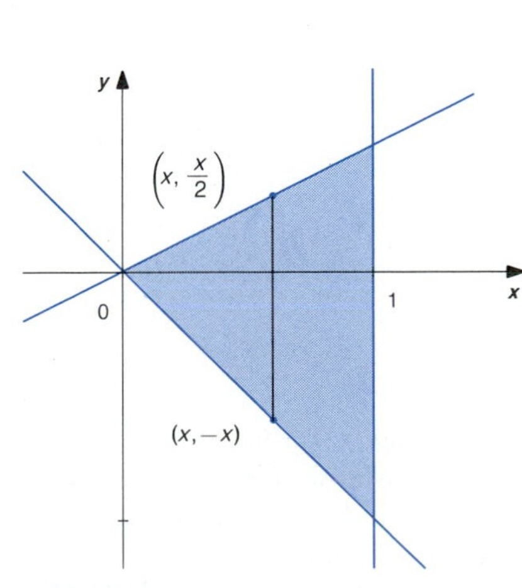


FIGURE 5.2.5

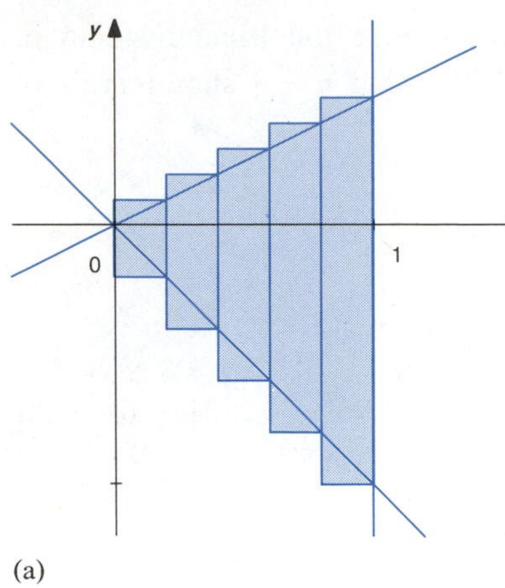


(a)

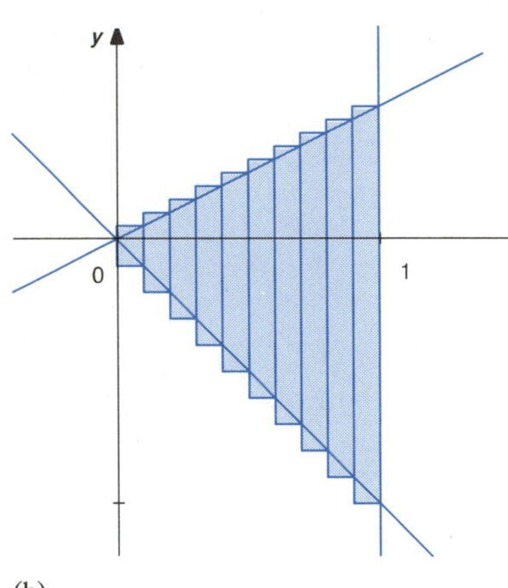


(b)

FIGURE 5.2.6

(a) We want to evaluate the Riemann sum of $h(x) = 3x/2$ over the interval $[0, 1]$, using $n = 5$ subintervals of equal length and $x_j^* = x_j$, the right endpoint of each subinterval.

The points $x_0 = 0$, $x_1 = 0.2$, $x_2 = 0.4$, $x_3 = 0.6$, $x_4 = 0.8$, and $x_5 = 1$ clearly divide the interval $[0, 1]$ into five subintervals of equal length $\Delta x_j = 0.2, j = 1, \ldots, 5$. The desired Riemann sum is

$$\begin{aligned} R_5 &= h(x_1)\,\Delta x_1 + h(x_2)\,\Delta x_2 + h(x_3)\,\Delta x_3 + h(x_4)\,\Delta x_4 + h(x_5)\,\Delta x_5 \\ &= h(0.2)\cdot(0.2) + h(0.4)\cdot(0.2) + h(0.6)\cdot(0.2) \\ &\quad + h(0.8)\cdot(0.2) + h(1)\cdot(0.2) \\ &= \left(\frac{3(0.2)}{2}\right)(0.2) + \left(\frac{3(0.4)}{2}\right)(0.2) + \left(\frac{3(0.6)}{2}\right)(0.2) \\ &\quad + \left(\frac{3(0.8)}{2}\right)(0.2) + \left(\frac{3(1)}{2}\right)(0.2) = 0.9. \end{aligned}$$

$R_5$ is the sum of the areas of the five rectangles pictured in Figure 5.2.6a.

(b) The Riemann sum of $h(x) = 3x/2$ over the interval $[0, 1]$ with ten subintervals of equal length and $x_j^*$ chosen to be the right endpoint of the $j$th subinterval is

$$\begin{aligned} R_{10} &= h(x_1)\,\Delta x_1 + \cdots + h(x_{10})\,\Delta x_{10} \\ &= h(0.1)\cdot(0.1) + h(0.2)\cdot(0.1) + \cdots + h(0.9)\cdot(0.1) + h(1)\cdot(0.1) \\ &= \left(\frac{3(0.1)}{2}\right)(0.1) + \left(\frac{3(0.2)}{2}\right)(0.1) + \left(\frac{3(0.3)}{2}\right)(0.1) + \left(\frac{3(0.4)}{2}\right)(0.1) \\ &\quad + \left(\frac{3(0.5)}{2}\right)(0.1) + \left(\frac{3(0.6)}{2}\right)(0.1) + \left(\frac{3(0.7)}{2}\right)(0.1) \\ &\quad + \left(\frac{3(0.8)}{2}\right)(0.1) + \left(\frac{3(0.9)}{2}\right)(0.1) + \left(\frac{3(1)}{2}\right)(0.1) = 0.825. \end{aligned}$$

$R_{10}$ is the sum of the areas of the ten rectangles pictured in Figure 5.2.6b. Riemann sums are nearer to the exact area of the triangular region when smaller subintervals are used. (The exact area of the triangle is 0.75.) ■

You should be able to use your calculator to evaluate the sum of products in a Riemann sum directly, without preliminary calculations. In any case, avoid rounding the individual terms. Small errors in each of many terms can add up to a significant error in the total.

### *Application of Riemann sums to volume*

Let us consider a solid with the property that variable cross sections perpendicular to the $x$-axis have area $A(x)$, where $A$ is a continuous function. Such a solid is illustrated in Figure 5.2.7a. To approximate the volume of the solid, we divide it into $n$ slices of thickness $\Delta x_j, j = 1, \ldots, n$, by cutting perpendicular to the $x$-axis, as indicated in Figure 5.2.7b. If a

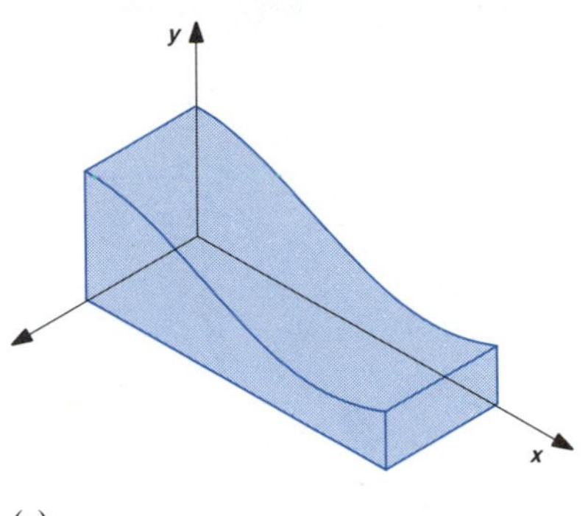


(a)

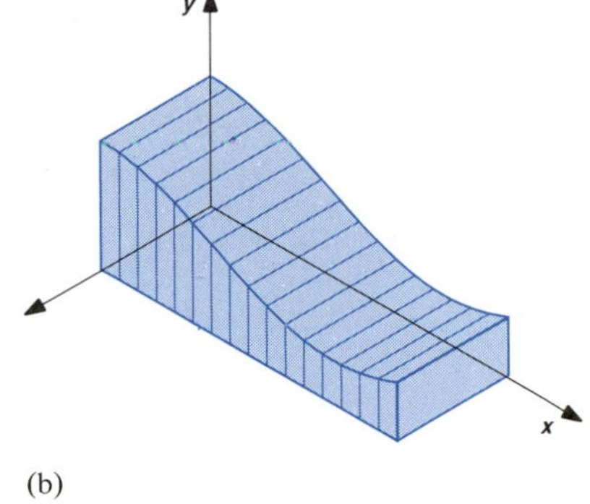


(b)

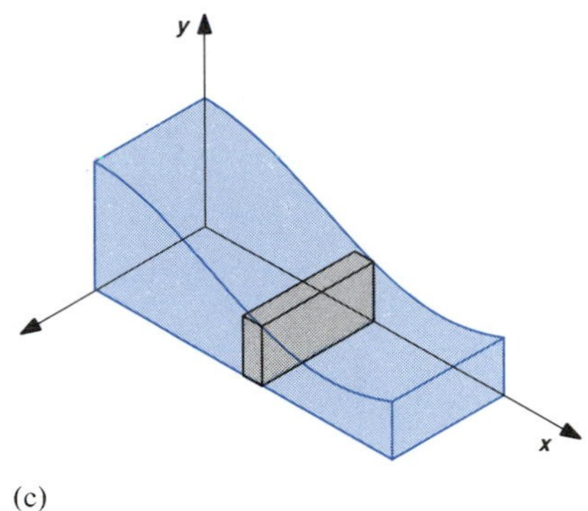


(c)

FIGURE 5.2.7

slice has edges perpendicular to the faces, then cross sections perpendicular to the faces have constant area, and we know

Volume of slice = (Area of cross section)(Thickness).

This is illustrated in Figure 5.2.7c. In case the area of the cross sections of the slice is not constant, we can approximate the volume of the slice by the product

$A(x_j^*)\,\Delta x_j$,

where $A(x_j^*)$ is the area of the cross section at some point within the slice. If $A$ is continuous and the slice is thin enough, the error in the approximate value will be proportionally smaller than the volume of the slice. This means the sum of approximate volumes of the slices will be near the actual volume of the solid. That is, if all $\Delta x_j$ are small enough,

Volume of solid $\approx$ sum of approximate volumes of slices,

$$\textbf{Volume} \approx A(x_1^*)\,\Delta x_1 + A(x_2^*)\,\Delta x_2 + \cdots + A(x_n^*)\,\Delta x_n.$$

The sum is a Riemann sum of the cross-sectional area function $A(x)$. The product $A(x_j^*)\,\Delta x_j$ is interpreted as the volume of a slice that has area of cross section $A(x_j^*)$ and thickness $\Delta x_j$. The Riemann sum is the sum of the volumes of $n$ slices.

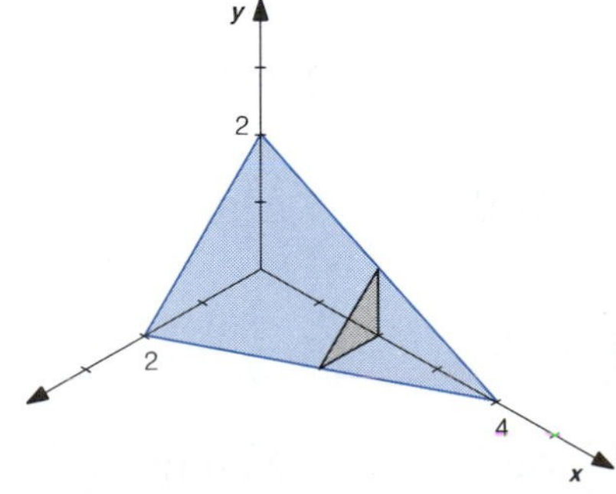


FIGURE 5.2.8

■ **EXAMPLE 3**

Use Riemann sums with $n = 8$ subintervals of equal length and $x_j^* = x_j$, $j = 1, \ldots, n$, to find an approximate value of the prism pictured in Figure 5.2.8. Cross sections perpendicular to the $x$-axis are isosceles right triangles.

**SOLUTION**

We need to determine the area of a variable cross section that is perpendicular to the $x$-axis. These cross sections are isosceles right triangles with base and height given by the length of a variable line segment between the $x$-axis and the line between the points (0, 2) and (4, 0) in the $xy$-plane. This line has slope

$$\frac{0-2}{4-0} = -\frac{1}{2}.$$

The equation of the line through (4, 0) with slope $-1/2$ is

$$y - 0 = -\frac{1}{2}(x-4),$$

$$y = -\frac{1}{2}x + 2.$$

The base $b(x)$ and height $h(x)$ of a variable triangular cross section are then given by

$$b(x) = h(x) = -\frac{1}{2}x + 2.$$

Hence, the area of a cross section is given by

$$\begin{aligned} A(x) &= \frac{1}{2}b(x)h(x) \\ &= \frac{1}{2}\left(-\frac{1}{2}x + 2\right)^2 \\ &= \frac{1}{8}x^2 - x + 2, \qquad 0 \le x \le 4. \end{aligned}$$

The points $x_0 = 0$, $x_1 = 0.5$, $x_2 = 1.0$, $x_3 = 1.5$, $x_4 = 2.0$, $x_5 = 2.5$, $x_6 = 3.0$, $x_7 = 3.5$, and $x_8 = 4.0$ divide the interval [0, 4] into eight subintervals of equal length

$$\Delta x_j = \frac{b-a}{n} = \frac{4-0}{8} = 0.5.$$

The desired Riemann sum is

$$\begin{aligned} &A(0.5)\,\Delta x_1 + A(1.0)\,\Delta x_2 + A(1.5)\,\Delta x_3 + A(2.0)\,\Delta x_4 \\ &\qquad + A(2.5)\,\Delta x_5 + A(3.0)\,\Delta x_6 + A(3.5)\,\Delta x_7 + A(4.0)\,\Delta x_8 \\ &\qquad = \left(\frac{1}{8}(0.5)^2 - (0.5) + 2\right)(0.5) + \left(\frac{1}{8}(1.0)^2 - (1.0) + 2\right)(0.5) \\ &\qquad + \left(\frac{1}{8}(1.5)^2 - (1.5) + 2\right)(0.5) + \left(\frac{1}{8}(2.0)^2 - (2.0) + 2\right)(0.5) \\ &\qquad + \left(\frac{1}{8}(2.5)^2 - (2.5) + 2\right)(0.5) + \left(\frac{1}{8}(3.0)^2 - (3.0) + 2\right)(0.5) \\ &\qquad + \left(\frac{1}{8}(3.5)^2 - (3.5) + 2\right)(0.5) + \left(\frac{1}{8}(4.0)^2 - (4.0) + 2\right)(0.5) \\ &= 2.1875. \end{aligned}$$

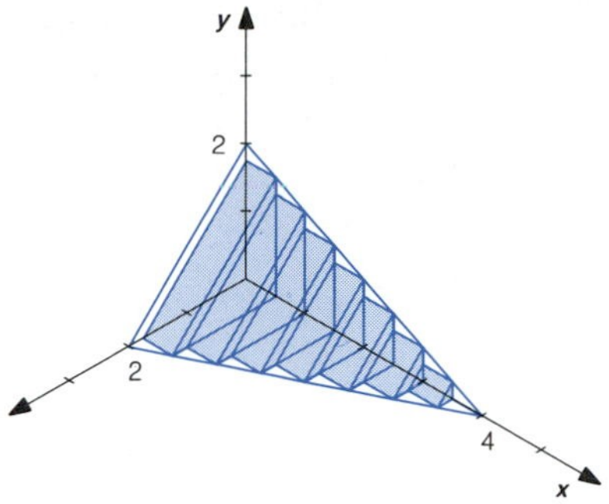

FIGURE 5.2.9

This Riemann sum is the volume of the eight smaller prisms pictured in Figure 5.2.9, where we have counted the "prism" on the far right that has zero volume. ■

### *Application of Riemann sums to velocity and position along a line*

If we travel for 2 h at a rate that varies between 50 and 60 mph, it is clear that we will cover between 100 and 120 mi. That is,

$$\text{Distance} \geq (\text{Minimum velocity})(\text{Length of time}) = (50)(2) = 100$$

and

$$\text{Distance} \leq (\text{Maximum velocity})(\text{Length of time}) = (60)(2) = 120.$$

We could improve the accuracy of our estimate of the distance traveled by breaking the 2 h into smaller intervals of time over which the speed does not vary so much and then adding the results. This idea leads to a Riemann sum.

Suppose the position of an object along a number line at time $t$ is given by $s(t)$. Also, assume that the function $s$ is differentiable, so the (instantaneous) velocity of the object is the derivative $s'(t)$. Let us see how the difference in position of the object at times $t = a$ and $t = b$ is given by a Riemann sum of $s'$ over the time interval $[a, b]$. That is, choose points

$$a = t_0 < t_1 < \cdots < t_{n-1} < t_n = b,$$

so the interval $[a, b]$ is partitioned into $n$ subintervals. Then the difference in position can be written

$$\begin{aligned} s(b) - s(a) &= s(t_n) - s(t_0) \\ &= [s(t_n) - s(t_{n-1})] \\ &\quad + [s(t_{n-1}) - s(t_{n-2})] + \cdots + [s(t_2) - s(t_1)] \\ &\quad + [s(t_1) - s(t_0)]. \end{aligned}$$

Applying the Mean Value Theorem to each of the above terms, we have

$$\begin{aligned} \mathbf{s(b) - s(a)} &= s'(t_n^*) \cdot (t_n - t_{n-1}) \\ &\quad + s'(t_{n-1}^*) \cdot (t_{n-1} - t_{n-2}^*) + \cdots + s'(t_2) \cdot (t_2 - t_1) \\ &\quad + s'(t_1) \cdot (t_1^* - t_0). \end{aligned} \tag{3}$$

The sum in (3) is a Riemann sum of $s'$ over the interval $[a, b]$. (The terms are written in reverse order, but that doesn't matter.) In this application the terms $s'(t_j^*)\,\Delta t_j$ are interpreted as the product of velocity and a length of time. This product gives the distance traveled by an object that travels with average velocity $s'(t_j^*)$ over a length of time $\Delta t_j = t_j - t_{j-1}$. The sign of the term indicates the direction traveled. The algebraic sum of the terms gives the net displacement $s(b) - s(a)$.

We may not know the values of the $t_j^*$'s that give the average velocities over the intervals $[t_{j-1}, t_j]$, $j = 1, \ldots, n$. However, if each $t_j^*$ in (3) is replaced by the $t$ that gives the maximum value of $s'(t)$ in the interval $[t_{j-1}, t_j]$, the corresponding Riemann sum will be greater than or equal to $s(b) - s(a)$. Similarly, if each $t_j^*$ is replaced by a value that gives the minimum of $s'(t)$ in the interval $[t_{j-1}, t_j]$, we obtain a sum that is less than or equal to $s(b) - s(a)$. We can obtain approximate values of $s(b) - s(a)$ by evaluating these two Riemann sums.

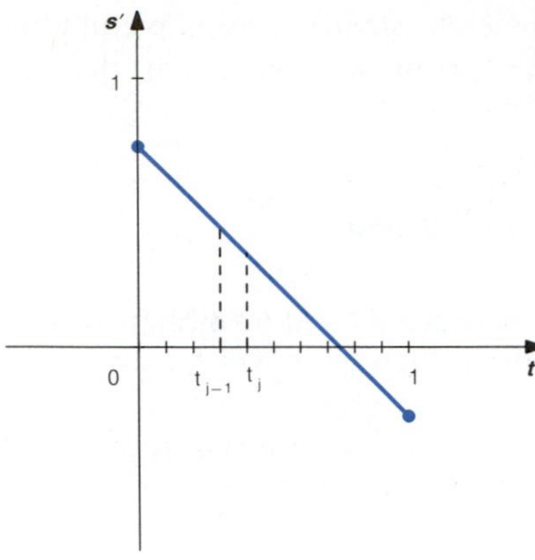


(a)

■ **EXAMPLE 4**

The position of an object along a number line at time $t$ is given by $s(t)$. The formula for $s(t)$ is not known, but it is known that $s(0) = 2$ and $s'(t) = 0.75 - t$. Use Riemann sums of $s'$ over the interval $[0, 1]$ with $n = 10$ subintervals of equal length to find upper and lower bounds of $s(1)$, the position at $t = 1$.

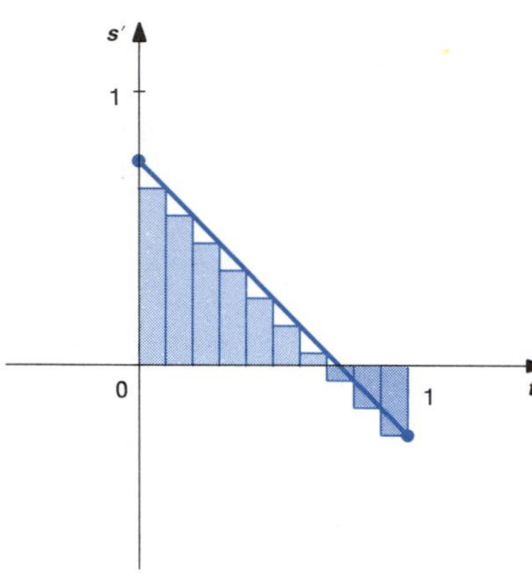


(b)

**SOLUTION**

The points $t_0 = 0, t_1 = 0.1, t_2 = 0.2, \ldots,$ and $t_{10} = 1.0$ clearly divide $[0, 1]$ into ten subintervals of equal length 0.1.

Since the function $s'(t) = 0.75 - t$ is decreasing, the maximum occurs at the left endpoint of each subinterval $t_{j-1}$, and the minimum occurs at the right endpoint $t_j$. See Figure 5.2.10a. That is,

$$s'(t_j) \leq s'(t_j^*) \leq s'(t_{j-1}), \qquad j = 1, \ldots, 10.$$

Therefore, $s(1) - s(0)$ is between the corresponding Riemann sums of $s'$.

The Riemann sum of $s'(t) = 0.75 - t$ over the interval $[0, 1]$ with $n = 10$ subintervals of equal length and $x_j^* = x_j$, the right endpoint of each subinterval is (see Figure 5.2.10b)

$$\begin{aligned} s'(t_1)\,\Delta t_1 &+ s'(t_2)\,\Delta t_2 + \cdots + s'(t_{10})\,\Delta t_{10} \\ &= (0.75 - 0.1)(0.1) + (0.75 - 0.2)(0.1) + (0.75 - 0.3)(0.1) \\ &\quad + (0.75 - 0.4)(0.1) + (0.75 - 0.5)(0.1) + (0.75 - 0.6)(0.1) \\ &\quad + (0.75 - 0.7)(0.1) + (0.75 - 0.8)(0.1) + (0.75 - 0.9)(0.1) \\ &\quad + (0.75 - 1.0)(0.1) = 0.2. \end{aligned}$$

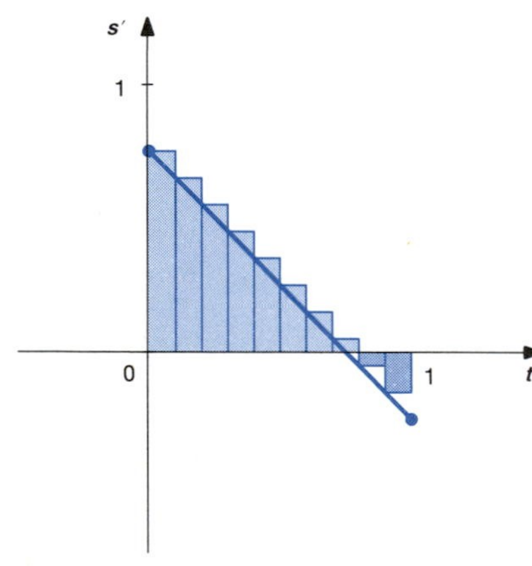


(c)

**FIGURE 5.2.10**

Using $x_j^* = x_{j-1}$, the left endpoint of each subinterval, we obtain (see Figure 5.2.10c)

$$\begin{aligned} s'(t_0)\,\Delta t_1 &+ s'(t_1)\,\Delta t_2 + \cdots + s'(t_9)\,\Delta t_{10} \\ &= (0.75 - 0)(0.1) + (0.75 - 0.1)(0.1) + (0.75 - 0.2)(0.1) \\ &\quad + (0.75 - 0.3)(0.1) + (0.75 - 0.4)(0.1) + (0.75 - 0.5)(0.1) \\ &\quad + (0.75 - 0.6)(0.1) + (0.75 - 0.7)(0.1) + (0.75 - 0.8)(0.1) \\ &\quad + (0.75 - 0.9)(0.1) = 0.3. \end{aligned}$$

We conclude that

$$0.2 \leq s(1) - s(0) \leq 0.3,$$

$$0.2 \leq s(1) - 2 \leq 0.3,$$

$$2.2 \leq s(1) \leq 2.3.$$

■

We can use the methods of Section 4.10 to determine that a function $s(t)$ with $s'(t) = 0.75 - t$ and $s(0) = 2$ must be $s(t) = 2 + 0.75t - 0.5t^2$. It follows that $s(1) = 2.25$. This is consistent with the range for $s(1)$ that we found in Example 4. The connection between the two methods is fundamental to the study of calculus. We will discuss this connection in Section 5.5.

The **net displacement** of an object that travels along a number line is the total distance traveled in the positive direction minus the total distance traveled in the negative direction. The **total distance traveled** is the sum of the distances traveled in each direction. If the position of an object along a number line at time $t$ is given by $s(t)$, we have seen that the net displacement can be approximated by Riemann sums of the velocity $s'$. The total distance traveled by the object can be approximated by Riemann sums of the speed $|s'|$. That is,

$$\text{Total distance traveled} \approx |s'(t_1^*)|\Delta t_1 + |s'(t_2^*)|\Delta t_2 + \cdots + |s'(t_j^*)|\Delta t_n.$$

■ **EXAMPLE 5**

The position of an object along a number line at time $t$ is given by $s(t)$, where $s'(t) = 0.75 - t$. Use Riemann sums of $|s'|$ over the interval $[0, 1]$ with $n = 10$ subintervals of equal length and $t_j^* = t_j$ to find an approximate value of the total distance traveled by the object over the time interval from 0 to 1.

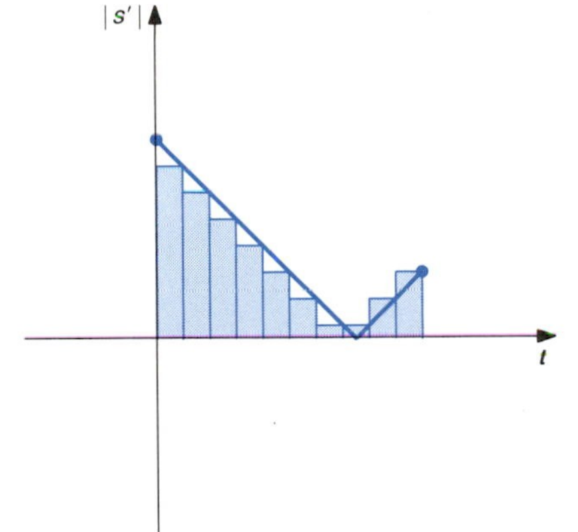


FIGURE 5.2.11

**SOLUTION**

The points $t_0 = 0, t_1 = 0.1, \ldots, t_{10} = 1$ divide $[0, 1]$ into ten subintervals of equal length 0.1. The desired Riemann sum is

$$\begin{aligned}
&|s'(t_1)|\Delta t_1 + |s'(t_2)|\Delta t_2 + \cdots + |s'(t_{10})|\Delta t_{10} \\
&\quad = |0.75 - 0.1|(0.1) + |0.75 - 0.2|(0.1) + |0.75 - 0.3|(0.1) \\
&\qquad + |0.75 - 0.4|(0.1) + |0.75 - 0.5|(0.1) + |0.75 - 0.6|(0.1) \\
&\qquad + |0.75 - 0.7|(0.1) + |0.75 - 0.8|(0.1) + |0.75 - 0.9|(0.1) \\
&\qquad + |0.75 - 1.0|(0.1) = 0.29.
\end{aligned}$$

The graph of $|s'(t)| = |0.75 - t|$ is given in Figure 5.2.11. ■

## EXERCISES 5.2

In Exercises 1–4, use Riemann sums with the given number $n$ of subintervals of equal length and $x_j^* = x_j$, the right endpoint of each subinterval, to find approximate values of the area under the given curves.

**1** $y = x$, $0 \le x \le 1$; $n = 10$

**2** $y = x^2$, $0 \le x \le 2$; $n = 8$

**3** $y = \sin x$, $0 \le x \le \pi$; $n = 6$

**4** $y = \cos x$, $0 \le x \le \pi/2$; $n = 6$

In Exercises 5–10, use Riemann sums with the given number $n$ of subintervals of equal length and $x_j^* = x_j$, the right endpoint of each subinterval, to find approximate values of the area bounded by the given curves.

**5** $y = -2x,\ y = 0,\ x = 0,\ x = 1;\ n = 10$

**6** $y = x + 1,\ x - 2y = 1,\ x = -1,\ x = 1;\ n = 10$

**7** $y = 2x + 3,\ x + y = 0,\ x = -1,\ x = 1;\ n = 10$

**8** $y = 1,\ y = 1 - x^2,\ x = 0,\ x = 2;\ n = 8$

**9** $y = x^2,\ y = x,\ x = 0,\ x = 1;\ n = 5$

**10** $y = x^2 - 1,\ y = x + 1,\ x = -1,\ x = 2;\ n = 6$

The shape of cross sections perpendicular to the $x$-axis is indicated for each of the solids pictured in Exercises 11–20. Use Riemann sums with $n = 10$ subintervals of equal length and $x_j^* = x_j,\ j = 1, \ldots, 10$, to find approximate values of the volumes of the solids.

**11** Circles

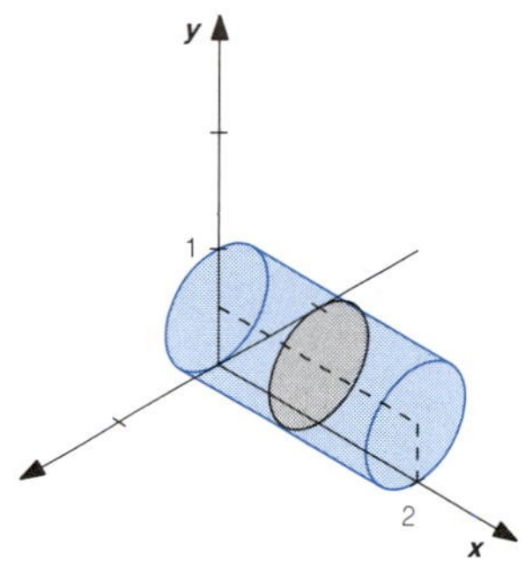


**12** Rectangles

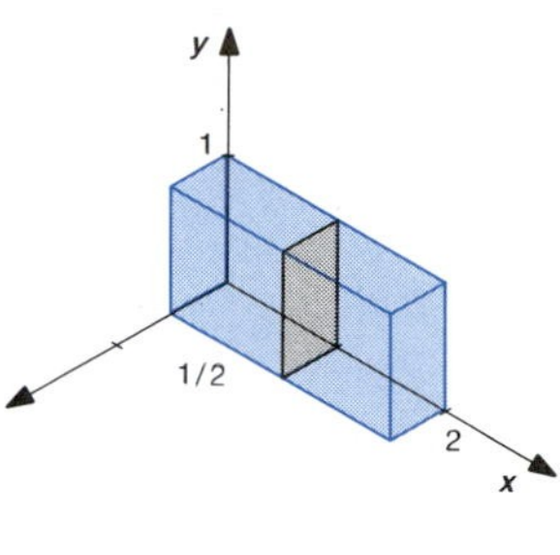


**13** Rectangles

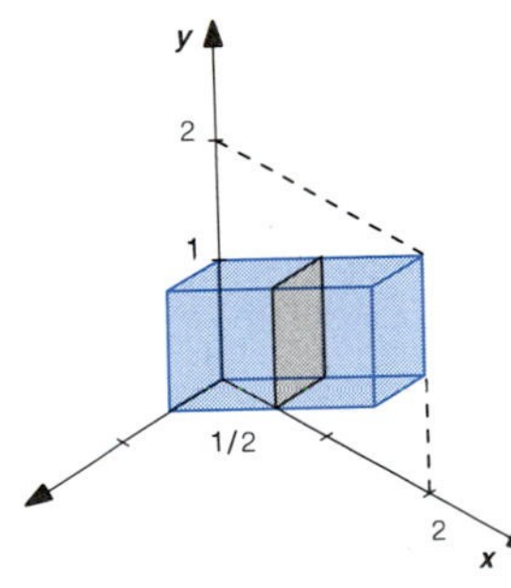


**14** Circles

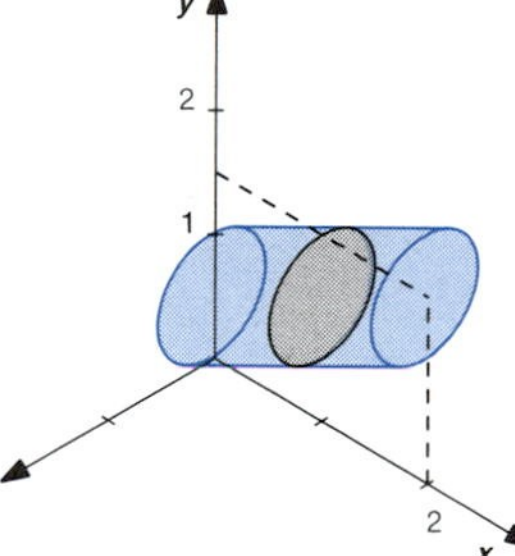


**15** Circles

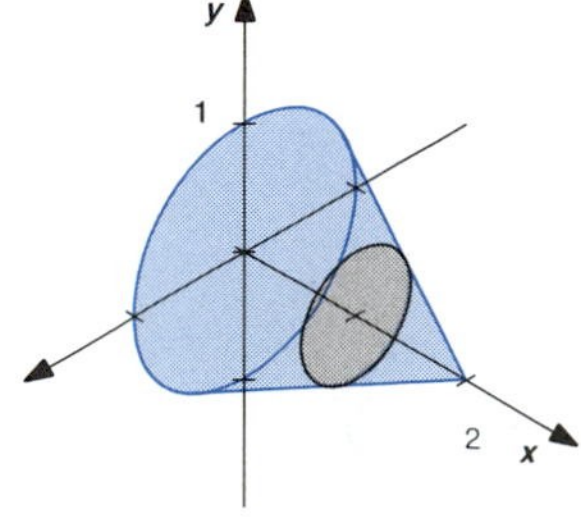


**16** Circles

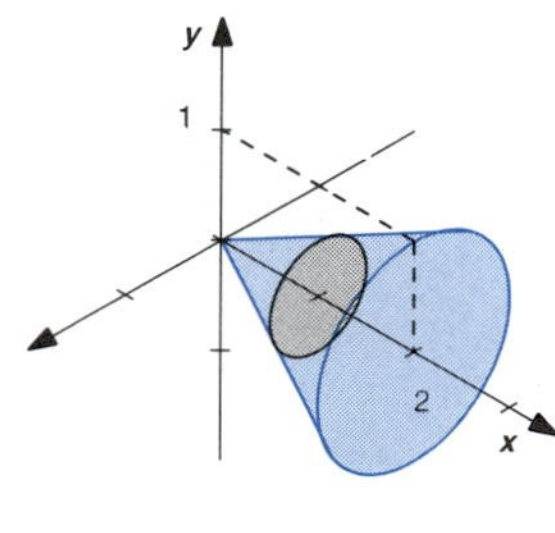


**17** Isosceles right triangles

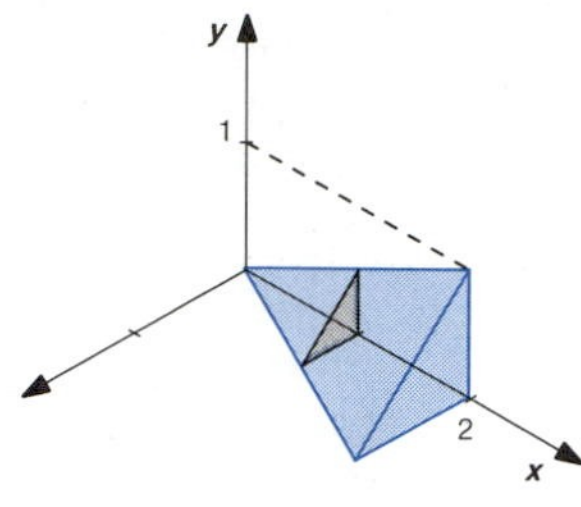


**18** Rectangles

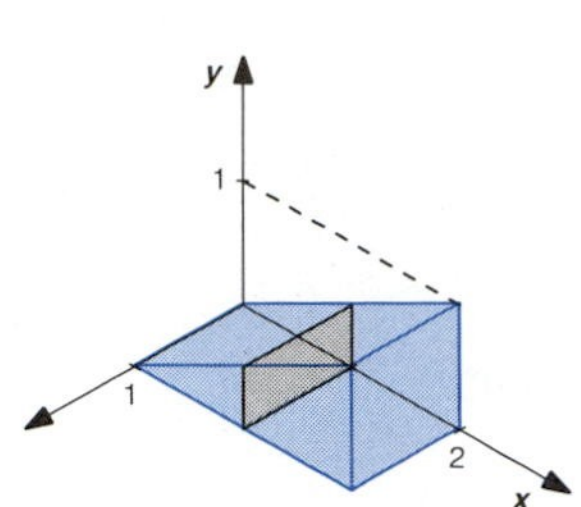


**19** Circles
(Solid is a hemisphere)

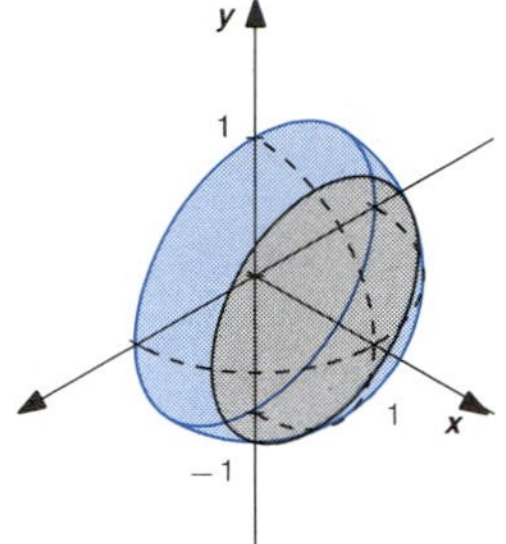


**20** Squares

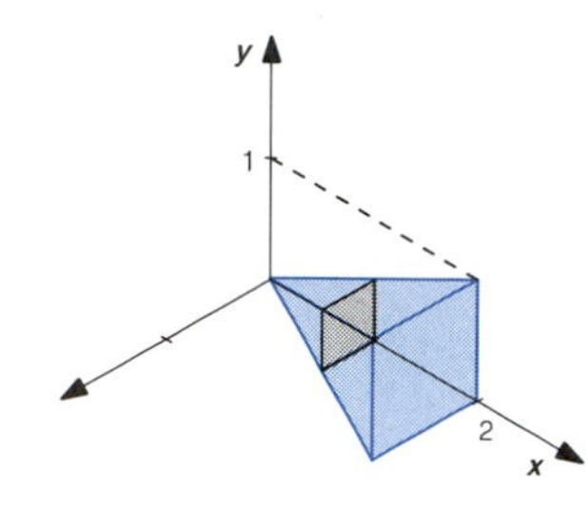


In Exercises 21–28 the initial position $s(0)$, a terminal time $T$, and the instantaneous velocity $s'(t)$ of an object on a number line are given. Use Riemann sums of $s'$ over the interval $[0, T]$ with the given number $n$ of subintervals of equal length to find upper and lower bounds of $s(T)$.

**21** $s'(t) = 2t,\ s(0) = 0,\ T = 1,\ n = 4$

**22** $s'(t) = 2t,\ s(0) = 0,\ T = 2,\ n = 8$

**23** $s'(t) = t^2,\ s(0) = 4,\ T = 2,\ n = 8$

**24** $s'(t) = t^2,\ s(0) = 4,\ T = 1,\ n = 4$

**25** $s'(t) = 32 - 32t,\ s(0) = 0,\ T = 1,\ n = 4$

**26** $s'(t) = 32 - 32t,\ s(0) = 0,\ T = 2,\ n = 8$

**27** $s'(t) = \sqrt{1 + t^3},\ s(0) = 1,\ T = 2,\ n = 8$

**28** $s'(t) = \sqrt{1 + t^3},\ s(0) = 1,\ T = 1,\ n = 4$

In Exercises 29–32 the instantaneous velocity $s'(t)$ of an object on a number line is given. Use Riemann sums of $|s'|$ over the interval $[0, T]$ with the given number $n$ of subintervals of equal length and $x_j^* = x_j$ to find an approximate value of the total distance traveled.

**29** $s'(t) = 49 - 9.8t,\ T = 10,\ n = 10$

**30** $s'(t) = 19.6 - 9.8t,\ T = 6,\ n = 6$

**31** $s'(t) = 2 \sin t,\ T = 2\pi,\ n = 6$

**32** $s'(t) = \cos t,\ T = \pi,\ n = 8$

## 5.3 LIMITS OF RIEMANN SUMS, THE DEFINITE INTEGRAL

We will investigate the behavior of some Riemann sums that involve a large number of subintervals. To do this it is convenient to use $\sum$ (sigma) notation.

The symbol $\sum_{j=1}^{n} a_j$ is defined by the equation

$$\sum_{j=1}^{n} a_j = a + a_2 + \cdots + a_n.$$

Similarly, $\sum_{j=m}^{n} a_j = a_m + a_{m+1} + \cdots + a_n$ for $m \le n$. The values of the terms $a_j$ are often given by a formula. That is, $a_j$ is given as a function of $j$, $a_j = a(j)$.

■ **EXAMPLE 1**

Evaluate

(a) $\sum_{j=1}^{4} 2j$,

(b) $\sum_{j=1}^{5} (2-j)$,

(c) $\sum_{j=1}^{6} 4$,

(d) $\sum_{j=2}^{5} j^2$.

**SOLUTION**

(a) The terms of the sum $\sum_{j=1}^{4} 2j$ are given by the formula $a_j = 2j$, $1 \le j \le 4$. Hence,

$$\sum_{j=1}^{4} 2j = 2(1) + 2(2) + 2(3) + 2(4) = 20.$$

(b) The terms of the sum $\sum_{j=1}^{5} (2-j)$ are given by the formula $a_j = 2 - j$, $1 \le j \le 5$, so

$$\sum_{j=1}^{5} (2-j) = (2-1) + (2-2) + (2-3) + (2-4) + (2-5) = -5.$$

(c) The terms of the sum $\sum_{j=1}^{6} 4$ are given by the formula $a_j = 4$, $1 \le j \le 6$, so the values of $a_j$ are given by a constant function. Then

$$\sum_{j=1}^{6} 4 = 4 + 4 + 4 + 4 + 4 + 4 = 6(4) = 24.$$

(d) The terms of the sum $\sum_{j=2}^{5} j^2$ are given by the formula $a_j = j^2$, $2 \le j \le 5$, so

$$\sum_{j=2}^{5} j^2 = 2^2 + 3^2 + 4^2 + 5^2 = 54.$$

■

We will use the following property of sums.

$$\sum_{j=1}^{n} (ca_j + db_j) = c\left(\sum_{j=1}^{n} a_j\right) + d\left(\sum_{j=1}^{n} b_j\right). \tag{1}$$

■ *Proof* Statement (1) is easily verified by using familiar properties of arithmetic. We have

$$\begin{aligned}\sum_{j=1}^{n} (ca_j + db_j) &= (ca_1 + db_1) + (ca_2 + db_2) + \cdots + (ca_n + db_n)\\ &= (ca_1 + ca_2 + \cdots + ca_n) + (db_1 + db_2 + \cdots + db_n)\\ &= c(a_1 + a_2 + \cdots + a_n) + d(b_1 + b_2 + \cdots + b_n)\\ &= c\left(\sum_{j=1}^{n} a_j\right) + d\left(\sum_{j=1}^{n} b_j\right).\end{aligned}$$

■

The following formulas can be verified by Mathematical Induction:

$$\sum_{j=1}^{n} c = c + c + \cdots + c = nc, \tag{2}$$

$$\sum_{j=1}^{n} j = 1 + 2 + \cdots + n = \frac{n(n+1)}{2}, \tag{3}$$

$$\sum_{j=1}^{n} j^2 = 1^2 + 2^2 + \cdots + n^2 = \frac{n(n+1)(2n+1)}{6}, \tag{4}$$

$$\sum_{j=1}^{n} j^3 = 1^3 + 2^3 + \cdots + n^3 = \left(\frac{n(n+1)}{2}\right)^2. \tag{5}$$

■ **EXAMPLE 2**

Evaluate $\sum_{j=1}^{50} (j^2 - 5j + 3)$.

**SOLUTION**

Using formula (1) and then formulas (2), (3), and (4) with $n = 50$, we have

$$\begin{aligned}\sum_{j=1}^{50} (j^2 - 5j + 3) &= \left(\sum_{j=1}^{50} j^2\right) - 5\left(\sum_{j=1}^{50} j\right) + \left(\sum_{j=1}^{50} 3\right)\\ &= \frac{50(50+1)(2(50)+1)}{6} - 5\left(\frac{50(50+1)}{2}\right) + 50(3)\\ &= 36{,}700.\end{aligned}$$

■

We can use formulas (1) to (5) to investigate the behavior of some Riemann sums that involve a large number of subintervals.

■ **EXAMPLE 3**

(a) Find a simple formula for the Riemann sum of $f(x) = 2x$ over the interval $[0, 1]$, using $n$ subintervals of equal length and $x_j^* = x_j$, $j = 1, \ldots, n$.

(b) Find the limit as $n$ approaches infinity of the Riemann sum found in (a).

**SOLUTION**

(a) From formula (2) of Section 5.2 we have

$$x_j = a + (j)\left(\frac{b-a}{n}\right) = 0 + (j)\left(\frac{1-0}{n}\right) = \frac{j}{n}, \qquad j = 1, \ldots, n.$$

Also,

$$\Delta x_j = \frac{b-a}{n} = \frac{1}{n}, \qquad j = 1, \ldots, n.$$

Using $\sum$ notation we can express the desired Riemann sum as

$$R_n = \sum_{j=1}^{n} f(x_j)\,\Delta x_j = \sum_{j=1}^{n} f\left(\frac{j}{n}\right)\cdot\left(\frac{1}{n}\right) = \sum_{j=1}^{n}\left(2\frac{j}{n}\right)\left(\frac{1}{n}\right).$$

Using property (1) and formula (3) we have

$$R_n = \frac{2}{n^2}\left(\sum_{j=1}^{n} j\right) = \frac{2}{n^2}\left(\frac{n(n+1)}{2}\right) = 1 + \frac{1}{n}.$$

(b) The limit of $R_n$ as $n$ approaches infinity is evaluated by using the techniques of Section 4.2 that were used to evaluate limits as a real variable $x$ approaches infinity. We have

$$\lim_{n\to\infty} R_n = \lim_{n\to\infty}\left(1 + \frac{1}{n}\right) = 1.$$

■

■ **EXAMPLE 4**

(a) Find a simple formula for the Riemann sum of $f(x) = 1 - x^2$ over the interval $[0, 1]$, using $n$ subintervals of equal length and $x_j^* = x_j$, $j = 1, \ldots, n$.

(b) Find the limit as $n$ approaches infinity of the Riemann sum found in (a).

**SOLUTION**

(a) As in Example 3 we have $x_j = j/n$ and $\Delta x_j = 1/n$, $j = 1, \ldots, n$. The desired Riemann sum is

$$R_n = \sum_{j=1}^{n} f(x_j)\,\Delta x_j = \sum_{j=1}^{n} f\left(\frac{j}{n}\right)\cdot\left(\frac{1}{n}\right) = \sum_{j=1}^{n}\left(1 - \left(\frac{j}{n}\right)^2\right)\left(\frac{1}{n}\right).$$

Using property (1) and formulas (2) and (4), we have

$$\begin{aligned}
R_n &= \frac{1}{n}\left(\sum_{j=1}^{n} 1\right) - \frac{1}{n^3}\left(\sum_{j=1}^{n} j^2\right) \\
&= \frac{1}{n}(n) - \frac{1}{n^3}\left(\frac{n(n+1)(2n+1)}{6}\right) \\
&= 1 - \frac{2n^2 + 3n + 1}{6n^2} \\
&= 1 - \frac{1}{3} - \frac{1}{2n} - \frac{1}{6n^2} \\
&= \frac{2}{3} - \frac{1}{2n} - \frac{1}{6n^2}.
\end{aligned}$$

(b) $\lim_{n\to\infty} R_n = \lim_{n\to\infty}\left(\frac{2}{3} - \frac{1}{2n} - \frac{1}{6n^2}\right) = \frac{2}{3}.$ ■

■ **EXAMPLE 5**

(a) Find a simple formula for the Riemann sum of $f(x) = x^3$ over the interval $[0, 2]$, using $n$ subintervals of equal length and $x_j^* = x_j$, $j = 1, \ldots, n$.

(b) Find the limit as $n$ approaches infinity of the Riemann sum found in (a).

**SOLUTION**

(a) From formula (2) of Section 5.2 we have

$$x_j = a + (j)\left(\frac{b-a}{n}\right) = 0 + (j)\left(\frac{2-0}{n}\right) = \frac{2j}{n}, \qquad j = 1, \ldots, n.$$

Also,

$$\Delta x_j = \frac{b-a}{n} = \frac{2}{n}, \qquad j = 1, \ldots, n.$$

The desired Riemann sum is

$$R_n = \sum_{j=1}^{n} f(x_j)\,\Delta x_j = \sum_{j=1}^{n} f\left(\frac{2j}{n}\right)\cdot\left(\frac{2}{n}\right) = \sum_{j=1}^{n}\left(\frac{2j}{n}\right)^3\left(\frac{2}{n}\right) = \frac{16}{n^4}\left(\sum_{j=1}^{n} j^3\right).$$

Using formula (5) we have

$$R_n = \frac{16}{n^4}\left(\frac{n(n+1)}{2}\right)^2 = \frac{16n^2 + 32n + 16}{4n^2} = 4 + \frac{8}{n} + \frac{4}{n^2}.$$

(b) $\lim_{n\to\infty} R_n = \lim_{n\to\infty}\left(4 + \frac{8}{n} + \frac{4}{n^2}\right) = 4.$ ■

Examples 3, 4, and 5 illustrate the idea of a limit of Riemann sums. Let us express this concept in terms of a formal definition.

*Definition*

The function $f$ is said to be **integrable** over the interval $[a, b]$ if there is a number, denoted $\int_a^b f(x)\,dx$, such that, for each $\varepsilon > 0$, there is a positive number $\delta$ such that

$$\left|\sum_{j=1}^{n} f(x_j^*)\,\Delta x_j - \int_a^b f(x)\,dx\right| < \varepsilon$$

whenever $\sum_{j=1}^{n} f(x_j^*)\,\Delta x_j$ is any Riemann sum of $f$ over $[a, b]$ that has partition norm $\max \Delta x_j < \delta$. If it exists, $\int_a^b f(x)\,dx$ is called the **definite integral** of $f$ from $a$ to $b$. The function $f$ is called the **integrand** and the numbers $a$ and $b$ are called **limits of integration.** ■

The value of a definite integral does not depend on the variable used. For example, $\int_a^b f(x)\,dx$ and $\int_a^b f(t)\,dt$ represent the same number. The variable of integration is sometimes called a **dummy variable**.

The number $\int_a^b f(x)\,dx$, if it exists, is a complicated form of a limit of Riemann sums. The notation is suggestive of the Riemann sum $\sum_{j=1}^{n} f(x_j^*)\,\Delta x_j$. That is, the symbol $\int$ corresponds to the summation sign $\sum$; $f(x)$ corresponds to $f(x_j^*)$; and $dx$ corresponds to the length $\Delta x_j$. In applications of the definite integral, we will think of $\int_a^b f(x)\,dx$ as the "sum" of the "products" $f(x)\cdot dx$; $f(x)$ and $dx$ will be interpreted as physical quantities, as though they were components of a Riemann sum.

If $f$ is nonnegative for $a \le x \le b$, we have seen in Section 5.2 that the Riemann sums

$$f(x_1)\,\Delta x_1 + f(x_2)\,\Delta x_2 + \cdots + f(x_n)\,\Delta x_n$$

give approximate values of the area under the curve. It should then seem reasonable that the definite integral $\int_a^b f(x)\,dx$ can be interpreted *geometrically* as the area under the curve. See Figure 5.3.1.

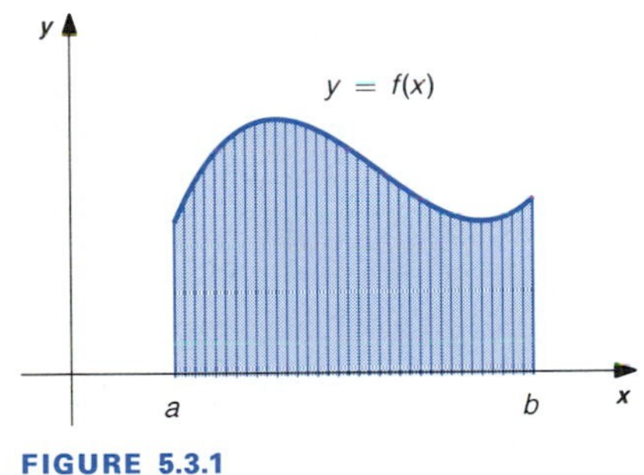

FIGURE 5.3.1

The condition of the definition requires that *all* Riemann sums of $f$ over $[a, b]$ that have all subintervals of small enough length be near the number $\int_a^b f(x)\,dx$. As in the case of the limit of a function at a point, this implies there is at most one number that can satisfy the condition of the definition. Also, as in the case of limits of functions, we are not interested in verifying the condition of the definition. We are interested in using theory to evaluate definite integrals.

The following theorem guarantees the existence of the definite integral for an important class of functions.

*Theorem*

**If $f$ is continuous on the closed interval $[a, b]$, then the definite integral of $f$ from $a$ to $b$ exists.**

We will not verify this theorem, although we note that the continuity of $f$ implies that values $f(x_j^*)$ are nearly equal for all $x_j^*$ in the interval $[x_{j-1}, x_j]$ if $\Delta x_j = x_j - x_{j-1}$ is small.

If we know $\int_a^b f(x)\,dx$ exists, then we can use the condition of the definition to our advantage. For example, we can then find an approximate value of $\int_a^b f(x)\,dx$ by using *any* convenient type of Riemann sum. We are not restricted to any particular choice of points of subdivision or of the points $x_j^*$ in the interval $[x_{j-1}, x_j]$. The Riemann sum used will be a good approximation of the definite integral as long as $[a, b]$ is partitioned into small enough subintervals.

Let us reconsider the results of Example 3. The function $f(x) = 2x$ is continuous on the interval $[0, 1]$, so we know from the above theorem that the definite integral $\int_0^1 2x\,dx$ exists. Also, we know that *any* Riemann sum of $f$ over $[0, 1]$ will be near the number $\int_0^1 2x\,dx$, as long as $[0, 1]$ is partitioned into small enough subintervals. In particular, the Riemann sums $R_n$ with subintervals of equal length and $x_j^* = x_j$ must be near $\int_0^1 2x\,dx$, if $n$ is large enough. The fact that $\lim_{n\to\infty} R_n = 1$ tells us that $R_n$ is near 1 whenever $n$ is large enough. Since all other Riemann sums of $f$ over $[0, 1]$ with small enough subintervals must also be near the same number, we conclude that

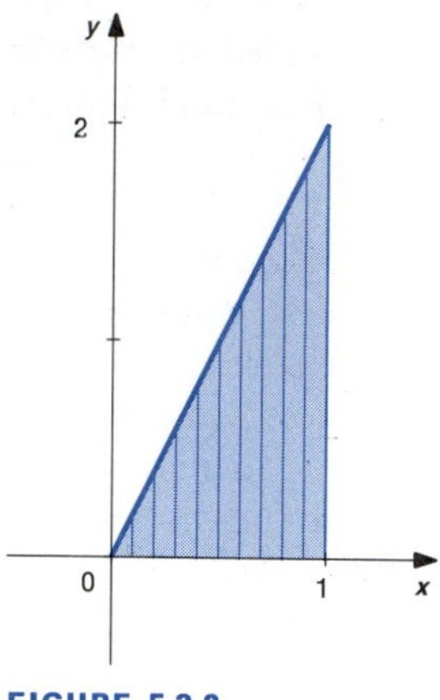

FIGURE 5.3.2

$$\int_0^1 2x\,dx = 1.$$

(The value of 1 is the area under the curve $y = 2x$, $0 \le x \le 1$. See Figure 5.3.2.) Similarly, the results of Examples 4 and 5 tell us

$$\int_0^1 (1 - x^2)\,dx = \frac{2}{3} \quad \text{and} \quad \int_0^2 x^3\,dx = 4, \quad \text{respectively.}$$

The method used in Examples 3–5 is not the most efficient way to evaluate a definite integral. These examples are intended to illustrate the idea that *a definite integral is a limit of Riemann sums.* We will learn better ways to evaluate definite integrals, but we must remember what a definite integral is. The idea of a definite integral as a "sum" is basic to understanding how it is used in applications.

Many properties of the definite integral follow from the corresponding properties of sums.

$$\int_a^b (cf(x) + dg(x))\,dx = c\cdot\int_a^b f(x)\,dx + d\cdot\int_a^b g(x)\,dx. \tag{6}$$

■ *Proof* We assume the three integrals exist. The value of each of them can then be approximated by using the *same subintervals* of $[a, b]$ and the *same points* $x_j^*$. This is true because the definition requires only that $[a, b]$ be partitioned into small enough subintervals. Using property (1) we then have

$$\begin{aligned}\int_a^b (cf(x) + dg(x))\,dx &\approx \sum_{j=1}^n (cf(x_j^*) + dg(x_j^*))\,\Delta x_j \\ &= c\left(\sum_{j=1}^n f(x_j^*)\,\Delta x_j\right) + d\left(\sum_{j=1}^n g(x_j^*)\,\Delta x_j\right) \\ &\approx c\cdot\int_a^b f(x)\,dx + d\cdot\int_a^b g(x)\,dx.\end{aligned}$$ ■

The following formulas define $\int_a^b f(x)\,dx$ in case $b \le a$.

$$\int_a^b f(x)\,dx = -\int_b^a f(x)\,dx \quad \text{and} \quad \int_a^a f(x)\,dx = 0. \tag{7}$$

The formulas in (7) may be considered to be properties of the definite integral. Another important property is

$$\int_a^b f(x)\,dx = \int_a^c f(x)\,dx + \int_c^b f(x)\,dx. \tag{8}$$

■ *Proof* We first assume $a < c < b$. We then choose a partition of $[a, b]$ that includes $c$ as a point of subdivision. Let us say $c = x_m$. Separating the terms of an approximating Riemann sum into those involving subintervals to the left of $c$ and those involving subintervals to the right of $c$, we have

$$\begin{aligned}\int_a^b f(x)\,dx &\approx \sum_{j=1}^{n} f(x_j^*)\,\Delta x_j \\ &= \sum_{j=1}^{m} f(x_j^*)\,\Delta x_j + \sum_{j=m+1}^{n} f(x_j^*)\,\Delta x_j \\ &\approx \int_a^c f(x)\,dx + \int_c^b f(x)\,dx.\end{aligned}$$

Other cases follow from the case $a < c < b$. For example, if $a < b < c$, we know from the above argument that

$$\int_a^c f(x)\,dx = \int_a^b f(x)\,dx + \int_b^c f(x)\,dx, \quad \text{so}$$

$$\int_a^b f(x)\,dx = \int_a^c f(x)\,dx - \int_b^c f(x)\,dx.$$

Property (7) then implies

$$\begin{aligned}\int_a^b f(x)\,dx &= \int_a^c f(x)\,dx - \left[-\int_c^b f(x)\,dx\right] \\ &= \int_a^c f(x)\,dx + \int_c^b f(x)\,dx.\end{aligned}$$ ■

FIGURE 5.3.3

Property (8) is illustrated geometrically in Figure 5.3.3. The area under the curve $y = f(x)$ from $x = a$ to $x = b$ is equal to the sum of the area from $x = a$ to $x = c$ and the area from $x = c$ to $x = b$.

The following four results are sometimes useful:

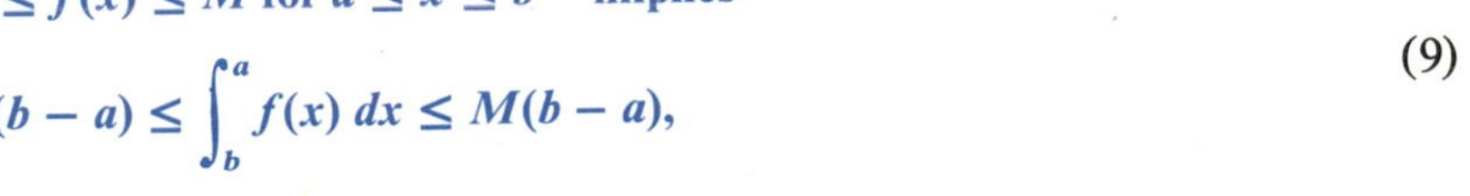

$$m \le f(x) \le M \text{ for } a \le x \le b \quad \text{implies}$$

$$m(b - a) \le \int_b^a f(x)\,dx \le M(b - a), \tag{9}$$

$$f(x) \geq 0 \text{ for } a \leq x \leq b \quad \text{implies} \quad \int_a^b f(x)\,dx \geq 0, \tag{10}$$

$$f(x) \leq g(x) \text{ for } a \leq x \leq b \quad \text{implies} \quad \int_a^b f(x)\,dx \leq \int_a^b g(x)\,dx, \tag{11}$$

$$\left|\int_a^b f(x)\,dx\right| \leq \int_a^b |f(x)|\,dx. \tag{12}$$

■ *Proof of (9)–(12)* Let us verify the right-hand inequality in (9). Verification of the left-hand inequality is similar. We have

$$\int_a^b f(x)\,dx \approx \sum_{j=1}^{n} f(x_j^*)\,\Delta x_j$$
$$\leq \sum_{j=1}^{n} M\,\Delta x_j = M \sum_{j=1}^{n} \Delta x_j = M(b-a).$$

The above inequality is true because $f(x_j^*) \leq M$ for every choice of $x_j^*$ and because $\Delta x_j > 0$. Also, we have used the fact that $\sum_{j=1}^{n} \Delta x_j = b - a$. That is, the sum of the lengths of the subintervals of $[a, b]$ is the length of $[a, b]$.

Property (10) follows from (9) with $m = 0$.

Property (11) is verified in the same manner as (9).

We should think of (12) as saying the absolute value of a sum is less than or equal to the sum of absolute values. Using the Triangle Inequality to verify (12), we have

$$\left|\int_a^b f(x)\,dx\right| \approx \left|\sum_{j=1}^{n} f(x_j^*)\,\Delta x_j\right| \leq \sum_{j=1}^{n} |f(x_j^*)|\,\Delta x_j \approx \int_a^b |f(x)|\,dx. \quad ■$$

### ■ EXAMPLE 6

Show that

$$\frac{36}{13} \leq \int_0^3 \frac{12}{x^2+4}\,dx \leq 9.$$

**SOLUTION**

We need to find upper and lower bounds of the integrand for $0 \leq x \leq 3$. We have $x^2 \geq 0$ so

$$x^2 + 4 \geq 4 \quad \text{and} \quad \frac{12}{x^2+4} \leq \frac{12}{4} = 3.$$

Also, $0 \leq x \leq 3$ implies $x^2 + 4 \leq 3^2 + 4 = 13$, so

$$\frac{12}{x^2+4} \geq \frac{12}{13}.$$

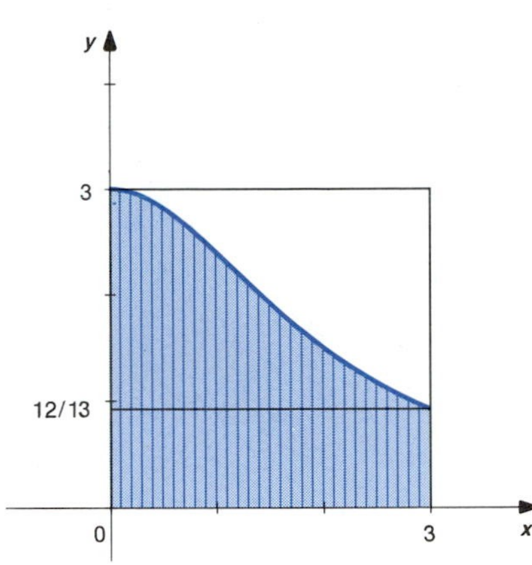


FIGURE 5.3.4

It then follows from (9) that

$$\int_0^3 \frac{12}{x^2+4}\,dx \le 3(3-0) = 9$$

and

$$\int_0^3 \frac{12}{x^2+4}\,dx \ge \frac{12}{13}(3-0) = \frac{36}{13}.$$

Example 6 is illustrated geometrically in Figure 5.3.4. The area between the curve $y = 12/(x^2+4)$ and the $x$-axis between $x = 0$ and $x = 3$ is greater than the area of the rectangle with base 3 and height 12/13, and less than the area of the rectangle with base 3 and height 3. ■

■ **EXAMPLE 7**

Given that $\int_0^7 f(x)\,dx = 3$, $\int_0^4 f(x)\,dx = 5$, and $\int_0^4 g(x)\,dx = 7$, use properties of the definite integral to evaluate

$$\text{(a)} \int_4^7 f(x)\,dx \quad \text{and} \quad \text{(b)} \int_0^4 [6f(x) - 4g(x)]\,dx.$$

**SOLUTION**

(a) Using property (8), we have

$$\int_4^7 f(x)\,dx = \int_0^7 f(x)\,dx - \int_0^4 f(x)\,dx = 3 - 5 = -2.$$

(b) Property (6) implies

$$\int_0^4 [6f(x) - 4g(x)]\,dx = 6\left(\int_0^4 f(x)\,dx\right) - 4\left(\int_0^4 g(x)\,dx\right)$$
$$= 6(5) - 4(7) = 2.$$ ■

The definite integral can be used to define the average value of a function $f(x)$, $a \le x \le b$. To do this, we divide the interval into $n$ subintervals of equal length $\Delta x = (b-a)/n$. Let $x_j^*$ denote any point in the $j$th subinterval, $j = 1, \ldots, n$. The average value of the numbers $f(x_1^*), \ldots, f(x_n^*)$ is

$$\frac{f(x_1^*) + \cdots + f(x_n^*)}{n} = \frac{1}{b-a}[f(x_1^*) + \cdots + f(x_n^*)]\,\Delta x.$$

If the interval is divided into small enough subintervals, the Riemann sum $[f(x_1^*) + \cdots + f(x_n^*)]\,\Delta x$ is nearly equal to $\int_a^b f(x)\,dx$. The following definition should then seem reasonable.

■ *Definition*

The **average value of $f$** over the interval $a \le x \le b$ is

$$f_{\text{avg}} = \frac{1}{b-a}\int_a^b f(x)\,dx.$$ ■

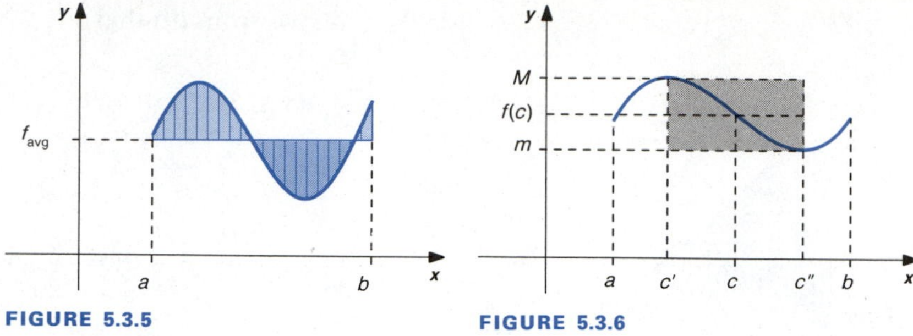


FIGURE 5.3.5

FIGURE 5.3.6

Geometrically, the area of the region above the line $y = f_{avg}$ and below $y = f(x)$ is equal to the area of the region below $y = f_{avg}$ and above $y = f(x)$. See Figure 5.3.5.

We conclude this section with the following.

*Mean Value Theorem for Definite Integrals*

**If $f$ is continuous on $[a, b]$, there is a number $c$ such that $a < c < b$ and**

$$\int_a^b f(x)\,dx = f(c)(b - a).$$

■ *Proof* Since $f$ is continuous on an interval that contains both endpoints, we know from the Extreme Value Theorem of Section 3.9 that $f$ has a maximum value $f(c') = M$ and minimum value $f(c'') = m$ on $[a, b]$. Then $m \le f(x) \le M$ on $[a, b]$, so (9) implies

$$m(b - a) \le \int_a^b f(x)\,dx \le M(b - a), \quad \text{or}$$

$$m \le \frac{1}{b - a}\int_a^b f(x)\,dx \le M.$$

Since

$$\frac{1}{b - a}\int_a^b f(x)\,dx$$

is between the values $f(c')$ and $f(c'')$, the Intermediate Value Theorem of Section 3.7 tells us there is at least one number $c$ that is between $c'$ and $c''$ with

$$f(c) = \frac{1}{b - a}\int_a^b f(x)\,dx, \quad \text{so} \quad \int_a^b f(x)\,dx = f(c)(b - a).$$

Since $c'$ and $c''$ are in the interval $[a, b]$, so is $c$. See Figure 5.3.6. ■

The Mean Value Theorem for Integrals tells us that if $f$ is continuous on an interval $I$, then there is a point $c$ in $I$ such that $f(c)$ is equal to the average value of $f$ on $I$.

# EXERCISES 5.3

Evaluate the sums in Exercises 1–10 either directly or by using Formulas (1)–(5).

1 $\sum_{j=1}^{4} 3j$

2 $\sum_{j=1}^{4} 2j$

3 $\sum_{j=1}^{4} (1-2j)$

4 $\sum_{j=1}^{4} (3-2j)$

5 $\sum_{j=3}^{6} (j^2-j)$

6 $\sum_{j=3}^{6} (j^2-2j)$

7 $\sum_{j=1}^{12} (j^3-6j)$

8 $\sum_{j=1}^{10} (j^3-j^2)$

9 $\sum_{j=1}^{100} (2j^2+4j-3)$

10 $\sum_{j=1}^{120} (3j^2-4j-5)$

In Exercises 11–20, use the method of Examples 3–5 to (a) find a simple formula for the Riemann sum corresponding to the given integrals, using $n$ subintervals of equal length and $x_j^* = x_j, j = 1, \ldots, n$. (b) Evaluate the integrals by finding the limit as $n$ approaches infinity of the Riemann sums found in (a).

11 $\int_0^3 (2x-4)\,dx$

12 $\int_0^2 (4x-2)\,dx$

13 $\int_0^4 (3x^2-8x-4)\,dx$

14 $\int_0^3 (x^2-6x-1)\,dx$

15 $\int_0^2 (6x^3-6x^2-1)\,dx$

16 $\int_0^4 (x^3-6x^2+5)\,dx$

17 $\int_0^b 1\,dx,\ b>0$

18 $\int_0^b x\,dx,\ b>0$

19 $\int_0^b x^2\,dx,\ b>0$

20 $\int_0^b x^3\,dx,\ b>0$

Use the method of Examples 3–5 to evaluate the area under the curves given in Exercises 21–22.

21 $y = 4x+6,\ 0 \le x \le 3$

22 $y = 3x^2+2,\ 0 \le x \le 2$

Use the method of Examples 3–5 to evaluate the average values of the functions given in Exercises 23–24 over the indicated intervals.

23 $f(x) = 6x-4,\ 0 \le x \le 4$

24 $f(x) = 1-x^2,\ 0 \le x \le 3$

Use the properties of the definite integral to evaluate the integrals in Exercises 25–32, given that $\int_5^7 f(x)\,dx = 3$, $\int_1^5 f(x)\,dx = 4$, $\int_1^7 g(x)\,dx = 5$, and $\int_5^7 g(x)\,dx = 6$.

25 $\int_5^1 f(x)\,dx$

26 $\int_7^5 g(x)\,dx$

27 $\int_1^7 f(x)\,dx$

28 $\int_1^5 g(x)\,dx$

29 $\int_5^7 [4f(x)+3g(x)]\,dx$

30 $\int_5^7 [2f(x)-4g(x)]\,dx$

31 $\int_1^5 [3f(x)-5g(x)]\,dx$

32 $\int_1^5 [5f(x)-2g(x)]\,dx$

33 Show that

$$0 \le \int_0^\pi \sin x\,dx \le \pi.$$

34 Show that

$$\left|\int_0^{2\pi} f(x)\sin x\,dx\right| \le \int_0^{2\pi} |f(x)|\,dx.$$

35 If $|f(x)| \le B$ for $a \le x \le b$, show that

$$\int_a^b [f(x)]^2\,dx \le B\int_a^b |f(x)|\,dx.$$

36 Show that

$$\left|\int_0^1 x^k f(x)\,dx\right| \le \int_0^1 |f(x)|\,dx$$

for all nonnegative integers $k$.

37 (a) Use (9) to find upper and lower bounds of

$$\int_0^{1/2} \frac{4}{1+x^2}\,dx \quad\text{and}\quad \int_{1/2}^1 \frac{4}{1+x^2}\,dx.$$

(b) Add the results of (a) to show that

$$2.6 \le \int_0^1 \frac{4}{1+x^2}\,dx \le 3.6.$$

38 (a) Use (9) to find upper and lower bounds of

$$\int_{2j-1}^{2j} \frac{1}{x}\,dx.$$

(b) Add the results of (a) for $j = 1, \ldots, n$ to show that

$$\frac{n}{2} \le \int_1^{2^n} \frac{1}{x}\,dx \le n.$$

39 Show that

$$\int_a^b c\,dx = c(b-a), \qquad c \text{ a constant.}$$

**40** Show that

$$\int_a^b x\,dx = \frac{b^2}{2} - \frac{a^2}{2}.$$

**41** If $f$ and $g$ are continuous with $g > 0$ on the interval $I = [a, b]$, show that there is a point $c$ in $I$ such that

$$\int_a^b f(x)g(x)\,dx = f(c)\int_a^b g(x)\,dx.$$

## 5.4 APPROXIMATE VALUES OF DEFINITE INTEGRALS, NUMERICAL METHODS

We will investigate systematic and efficient ways to calculate Riemann sums that involve a large number of subintervals. The methods are particularly well suited for use with a computer, since the formulas are easily programmed. In Section 5.5 we will learn a method for evaluating certain types of definite integrals without using Riemann sums. However, there are many elementary definite integrals for which the methods of Section 5.5 do not work. We must then rely on the methods of this section or on more sophisticated numerical methods.

We assume $f(x)$ is continuous on the closed interval $[a, b]$, so the definite integral $\int_a^b f(x)\,dx$ exists.

The first step in our systematic construction of Riemann sums of $f$ over $[a, b]$ is to choose points of subdivision of $[a, b]$. It seems reasonable to divide $[a, b]$ into $n$ subintervals of equal length. From statement (2) of Section 5.2 we then have

$$x_j = a + (j)\left(\frac{b-a}{n}\right), \qquad j = 0, 1, \ldots, n.$$

Also,

$$\Delta x_j = \frac{b-a}{n}, \qquad j = 1, \ldots, n.$$

The next step in constructing our Riemann sums is to choose points $x_j^*$ in each of the intervals $[x_{j-1}, x_j], j = 1, \ldots n$. One simple way to do this is to choose $x_j^* = x_j$, so $x_j^*$ is the right endpoint of the subinterval $[x_{j-1}, x_j]$, $j = 1, \ldots n$. Let $R_n$ denote the corresponding Riemann sum. That is,

$$R_n = \left(\frac{b-a}{n}\right)[f(x_1) + f(x_2) + \cdots + f(x_n)].$$

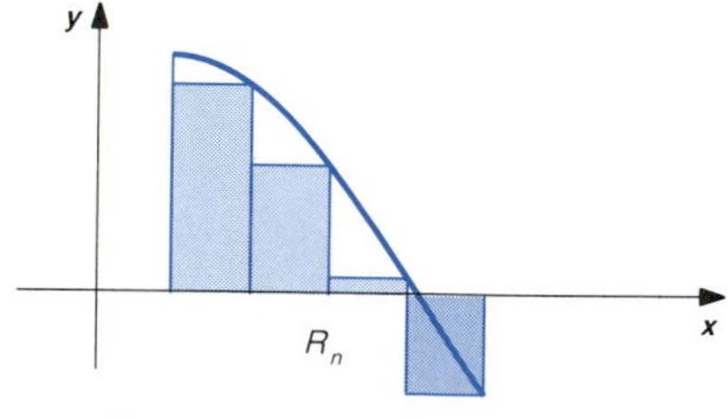

FIGURE 5.4.1

$R_n$ is illustrated geometrically in Figure 5.4.1. Terms of the Riemann sum that correspond to rectangles below the $x$-axis give the *negative* of the area of the corresponding rectangle.

Writing

$$R_n = \left(\frac{b-a}{n}\right)[0 \cdot f(x_0) + 1 \cdot f(x_1) + 1 \cdot f(x_2) + \cdots + 1 \cdot f(x_n)],$$

we see that

$$R_n = B\left(\sum_{j=0}^{n} c_j f(x_j)\right), \tag{1}$$

**where $B = \dfrac{b-a}{n}$, $c_0 = 0$, and $c_1 = c_2 = \cdots = c_n = 1$.**

We will study other types of Riemann sums. They will all be of the form

$$B\left(\sum_{j=0}^{n} c_j f(x_j)\right), \quad \text{where } B, c_0, c_1, \ldots, c_n \text{ are fixed constants.} \tag{2}$$

Different types of Riemann sums give different constants $B, c_0, c_1, \ldots, c_n$. It is quite easy to program a computer to carry out the calculation of sums and products of the form of (2). We will use a calculator and present the calculations in tabular form.

It is useful to know how accurately a particular type of Riemann sum approximates the value of $\int_a^b f(x)\,dx$. The following result gives a bound of the error associated with Riemann sums of type $R_n$.

**If $|f'(x)| \le M$ for $x$ in $[a, b]$, then**

$$\left|\int_a^b f(x)\,dx - R_n\right| \le \frac{M(b-a)^2}{2n}. \tag{3}$$

Statement (3) is verified at the end of this section. For now, let us see how to use the result.

**EXAMPLE 1**

(a) Use $R_{10}$ to find an approximate value of $\int_1^2 (1/x)\,dx$. (b) How large should $n$ be to insure that $|\int_1^2 (1/x)\,dx - R_n| \le 0.0001$?

**SOLUTION**

(a) The calculations are given in Table 5.4.1. The $j$ column runs from 0 to $n = 10$. The $x_j$ column goes from $a = 1$ to $b = 2$ in steps of $(b-a)/n = 0.1$. The $c_j$ column reflects that we are choosing $x_j^*$ to be the right endpoint of each subinterval and the value of $f$ at $x_0$ does not contribute. In the $f(x_j)$ column, we have substituted the value of $x_j$ into the formula for $f(x)$, but we have not elevated each term separately. We can then use a calculator to evaluate the expression $\sum c_j f(x_j)$ directly, without making any preliminary calculations. (This is one way to avoid excessive rounding in preliminary calculations.) The value of $B$ is then multiplied by $\sum c_j f(x_j)$ to obtain the value of the desired Riemann sum. We see that $R_{10} = 0.66877$.

(b) We have $f(x) = x^{-1}$, so $f'(x) = -x^{-2} = -1/x^2$. We need a bound of the factor $1/x^2$. For $1 \le x \le 2$, we have $1 \le |x| \le 2$, so $1^2 \le |x|^2 \le 2^2$. Since the factor $x^2$ is in the denominator, we use the inequality $1^2 \le |x|^2$ to obtain

$$\frac{1}{|x|^2} \le \frac{1}{1^2}.$$

**TABLE 5.4.1**

| $j$ | $x_j$ | $c_j$ | $f(x_j) = 1/x_j$ |
|---|---|---|---|
| 0 | 1 | 0 | 1 |
| 1 | 1.1 | 1 | 1/1.1 |
| 2 | 1.2 | 1 | 1/1.2 |
| 3 | 1.3 | 1 | 1/1.3 |
| 4 | 1.4 | 1 | 1/1.4 |
| 5 | 1.5 | 1 | 1/1.5 |
| 6 | 1.6 | 1 | 1/1.6 |
| 7 | 1.7 | 1 | 1/1.7 |
| 8 | 1.8 | 1 | 1/1.8 |
| 9 | 1.9 | 1 | 1/1.9 |
| 10 | 2 | 1 | 1/2 |

$\sum c_j f(x_j) = 6.68769$

$B = (b-a)/n = 0.1$

$R_{10} = B(\sum c_j f(x_j)) = 0.66877$

We then have

$$|f'(x)| = \frac{1}{|x|^2} \le \frac{1}{1^2} = 1,$$

so we can use $M = 1$ in (3). Substituting into (3), we then obtain

$$\left|\int_1^2 \frac{1}{x}\,dx - R_n\right| \le \frac{M(b-a)^2}{2n} = \frac{(1)(2-1)^2}{2n} = \frac{1}{2n}.$$

It follows that

$$\left|\int_1^2 \frac{1}{x}\,dx - R_n\right| < 0.0001 \quad \text{whenever} \quad \frac{1}{2n} < 0.0001$$

or $\quad n > \dfrac{1}{0.0002} = 5000.$

(If we actually used more than 5000 terms to find an approximate value of the integral, we would need to worry about round-off error in each of the terms. A small error in each of 5000 terms could add up to a significant error.) ■

We can improve the accuracy of the approximate values of a definite integral by choosing the points $x_j^*$ so that $f(x_j^*)$ is between the values $f(x_{j-1})$ and $f(x_j)$. In particular, since $f$ is continuous and the average $(f(x_{j-1}) + f(x_j))/2$ is between the values $f(x_{j-1})$ and $f(x_j)$, the Intermediate Value Theorem tells us there is some point $x_j^*$ between $x_{j-1}$ and $x_j$ such that $f(x_j^*) = (f(x_{j-1}) + f(x_j))/2$. See Figure 5.4.2a. The Riemann sums that correspond to this choice of $x_j^*$ are

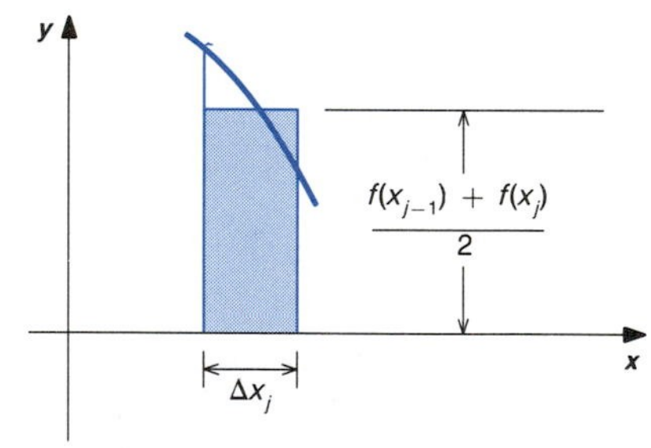

(a)

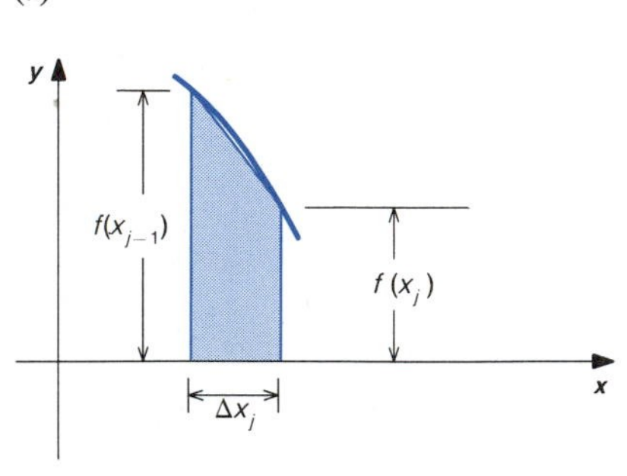

(b)

**FIGURE 5.4.2**

$$\begin{aligned}\sum_{j=1}^{n} f(x_j^*)\,\Delta x_j &= \sum_{j=1}^{n} \left(\frac{f(x_{j-1}) + f(x_j)}{2}\right)\left(\frac{b-a}{n}\right)\\
&= \left(\frac{b-a}{2n}\right)\left(\sum_{j=1}^{n} [f(x_{j-1}) + f(x_j)]\right)\\
&= \left(\frac{b-a}{2n}\right)\Big([f(x_0) + f(x_1)]\\
&\qquad + [f(x_1) + f(x_2)] + \cdots + [f(x_{n-1}) + f(x_n)]\Big)\\
&= \left(\frac{b-a}{2n}\right)[1\cdot f(x_0) + 2\cdot f(x_1) + 2\cdot f(x_2)\\
&\qquad + \cdots + 2\cdot f(x_{n-1}) + 1\cdot f(x_n)].\end{aligned}$$

This Riemann sum is called the $n$th **Trapezoidal approximation** of the definite integral $\int_a^b f(x)\,dx$. [If $f(x) \ge 0$ for $a \le x \le b$, the product

$$\left(\frac{f(x_{j-1}) + f(x_j)}{2}\right)\Delta x_j$$

may be interpreted as the area of the trapezoid pictured in Figure 5.4.2b.] We will denote the $n$th Trapezoidal approximation by $T_n$. The sum $T_n$ involves only values of $f$ at our points of subdivision of $[a, b]$. We can express $T_n$ in the form of (2). In particular,

$$T_n = B\left(\sum_{j=0}^{n} c_j f(x_j)\right), \quad \text{where } B = \frac{b-a}{2n}, \tag{4}$$

$$c_0 = c_n = 1, \text{ and } c_1 = c_2 = \cdots = c_{n-1} = 2.$$

Corresponding to (3) we have the following result for the **Trapezoidal Rule**.

**If $|f''(x)| \le M$ for $a \le x \le b$, then**

$$\left|\int_a^b f(x)\,dx - T_n\right| \le \frac{M(b-a)^3}{12n^2}. \tag{5}$$

We will not verify (5). The ideas are similar to those involved in (3).

## EXAMPLE 2

(a) Use $T_{10}$ to find an approximate value of $\int_1^2 (1/x)\,dx$. (b) How large should $n$ be to insure that $|\int_1^2 (1/x)\,dx - T_n| < 0.0001$?

### SOLUTION

(a) The calculations are carried out in Table 5.4.2. The constants $c_0, \ldots, c_{10}$ and $B$ for the Trapezoidal Rule are given by (4). We see that $T_{10} = 0.69377$.

(b) We can evaluate the second derivative of $f(x) = 1/x$ to obtain

$$f''(x) = \frac{2}{x^3}.$$

We can determine as in Example 1 that

$$\frac{1}{|x|^3} \le \frac{1}{1^3} = 1, \qquad 1 \le x \le 2,$$

so we have

$$|f''(x)| = 2\left(\frac{1}{|x|^3}\right) \le 2(1) = 2.$$

This shows that we can use $M = 2$ in (5). Then

$$\left|\int_1^2 \frac{1}{x}\,dx - T_n\right| \le \frac{M(b-a)^3}{12n^2} = \frac{(2)(2-1)^3}{12n^2} = \frac{1}{6n^2}.$$

TABLE 5.4.2

| $j$ | $x_j$ | $c_j$ | $f(x_j) = 1/x_j$ |
|---|---|---|---|
| 0 | 1 | 1 | 1 |
| 1 | 1.1 | 2 | 1/1.1 |
| 2 | 1.2 | 2 | 1/1.2 |
| 3 | 1.3 | 2 | 1/1.3 |
| 4 | 1.4 | 2 | 1/1.4 |
| 5 | 1.5 | 2 | 1/1.5 |
| 6 | 1.6 | 2 | 1/1.6 |
| 7 | 1.7 | 2 | 1/1.7 |
| 8 | 1.8 | 2 | 1/1.8 |
| 9 | 1.9 | 2 | 1/1.9 |
| 10 | 2 | 1 | 1/2 |

$\sum c_j f(x_j) = 13.8754$

$B = (b-a)/2n = 0.05$

$T_{10} = 0.69377$

It follows that

$$\left|\int_1^2 \frac{1}{x}\,dx - T_n\right| < 0.0001 \quad \text{whenever} \quad \frac{1}{6n^2} < 0.0001$$

or $n > \sqrt{1/0.0006} \approx 40.8$. Since $n$ must be an integer, we should use $n \geq 41$. (Recall from Example 1b that $n > 5000$ is required to insure that $R_n$ is within 0.0001 of the exact value of the integral.) ■

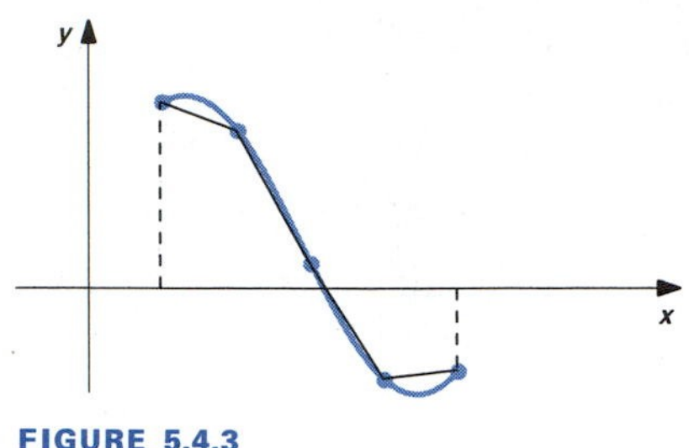

FIGURE 5.4.3

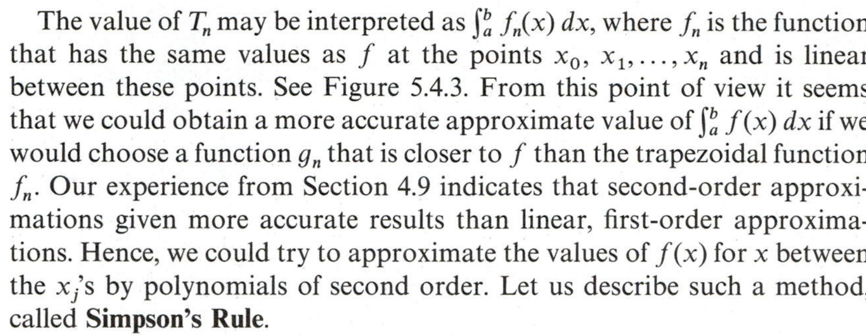

The value of $T_n$ may be interpreted as $\int_a^b f_n(x)\,dx$, where $f_n$ is the function that has the same values as $f$ at the points $x_0, x_1, \ldots, x_n$ and is linear between these points. See Figure 5.4.3. From this point of view it seems that we could obtain a more accurate approximate value of $\int_a^b f(x)\,dx$ if we would choose a function $g_n$ that is closer to $f$ than the trapezoidal function $f_n$. Our experience from Section 4.9 indicates that second-order approximations given more accurate results than linear, first-order approximations. Hence, we could try to approximate the values of $f(x)$ for $x$ between the $x_j$'s by polynomials of second order. Let us describe such a method, called **Simpson's Rule**.

We begin by defining a function $g_n$ for $x_0 \leq x \leq x_2$ by the formula $g_n(x) = c_0 + c_1(x - x_1) + c_2(x - x_1)^2$, where the numbers $c_0$, $c_1$, and $c_2$ are chosen so the graph of $g_n$ passes through the points $(x_0, f(x_0))$, $(x_1, f(x_1))$, and $(x_2, f(x_2))$. The condition $g_n(x_1) = f(x_1)$ clearly implies $c_0 = f(x_1)$. The other two conditions then give us the equations

$$f(x_0) = f(x_1) + c_1(x_0 - x_1) + c_2(x_0 - x_1)^2 \quad \text{and}$$

$$f(x_2) = f(x_1) + c_1(x_2 - x_1) + c_2(x_2 - x_1)^2.$$

Noting that $x_2 - x_1 = x_1 - x_0 = (b - a)/n$, we can solve the above system of two equations for the unknown values of $c_1$ and $c_2$. We obtain

$$c_1 = \frac{f(x_2) - f(x_0)}{2(x_1 - x_0)} = \frac{[f(x_2) - f(x_0)]n}{2(b - a)} \quad \text{and}$$

$$c_2 = \frac{f(x_0) - 2f(x_1) + f(x_2)}{2(x_1 - x_0)^2} = \frac{[f(x_0) - 2f(x_1) + f(x_2)]n^2}{2(b - a)^2}.$$

Using these values of $c_0$, $c_1$, and $c_2$ in the expression for $g_n(x)$, we can show that

$$\int_{x_0}^{x_2} g_n(x)\,dx = \frac{b - a}{3n}[f(x_0) + 4f(x_1) + f(x_2)].$$

(The above integral can be evaluated by the methods of Section 5.3, or by a much easier method that we will learn in Section 5.5.)

We are able to define $g_n$ similarly in each of the intervals $[x_2, x_4]$, $[x_4, x_6], \ldots, [x_{n-2}, x_n]$. This works out evenly only if *n is an even integer*. Evaluating the integral of $g_n$ over each of the above integrals, we obtain

$$\int_a^b g_n(x)\,dx = \int_{x_0}^{x_2} g_n(x)\,dx + \int_{x_2}^{x_4} g_n(x)\,dx + \cdots + \int_{x_{n-2}}^{x_n} g_n(x)\,dx$$

$$= \left(\frac{b-a}{3n}\right)\{[f(x_0) + 4f(x_1) + f(x_2)] + [f(x_2) + 4f(x_3) + f(x_4)] + \cdots + [f(x_{n-2}) + 4f(x_{n-1}) + f(x_n)]\}$$

This expression is denoted $S_n$. Grouping like terms, we obtain

$$S_n = B\left(\sum_{j=0}^{n} c_j f(x_j)\right), \tag{6}$$

**where $B = (b-a)/3n$, $n$ is an even integer, $c_0 = c_n = 1$, $c_1 = c_3 = \cdots = c_{n-1} = 4$ (odd-numbered $c$'s), and $c_2 = c_4 = \cdots = c_{n-2} = 2$ (even-numbered $c$'s, except for $c_0$ and $c_n$).**

A bound of the error in using $S_n$ to find an approximate value of $\int_a^b f(x)\,dx$ is given by the following formula, which we will not verify.

**If $|f^{(4)}(x)| \le M$ for $a \le x \le b$, then**

$$\left|\int_a^b f(x)\,dx - S_n\right| \le \frac{M(b-a)^5}{180n^4}. \tag{7}$$

■ **EXAMPLE 3**

(a) Use $S_{10}$ to find an approximate value of $\int_1^2 (1/x)\,dx$. (b) How large should $n$ be to insure that $|\int_1^2 (1/x)\,dx - S_n| < 0.0001$?

**SOLUTION**

(a) The calculations are carried out in Table 5.4.3. The constants $c_0, \ldots, c_{10}$ and $B$ for Simpson's Rule are given by (6). We see that $S_{10} = 0.69315$.

(b) As in Examples 1 and 2, we can determine that

$$|f^{(4)}(x)| = \frac{24}{|x|^5} \le 24, \qquad 1 \le x \le 2,$$

so we can use $M = 24$ in (7). Then

$$\left|\int_1^2 \frac{1}{x}\,dx - S_n\right| \le \frac{M(b-a)^5}{180n^4} = \frac{(24)(2-1)^5}{180n^4} = \frac{2}{15n^4}.$$

It follows that

$$\left|\int_1^2 \frac{1}{x}\,dx - T_n\right| < 0.0001 \quad \text{whenever} \quad \frac{2}{15n^4} < 0.0001$$

or $n > (2/0.0015)^{1/4} \approx 6.04$. Since $n$ must be an even integer for Simpson's Rule, we should use $n \ge 8$. (Recall that the Trapezodal Rule required $n \ge 41$ to insure an approximate value of $T_n$ within 0.0001 of the integral.) ■

**TABLE 5.4.3**

| $j$ | $x_j$ | $c_j$ | $f(x_j) = 1/x_j$ |
|---|---|---|---|
| 0 | 1 | 1 | 1 |
| 1 | 1.1 | 4 | 1/1.1 |
| 2 | 1.2 | 2 | 1/1.2 |
| 3 | 1.3 | 4 | 1/1.3 |
| 4 | 1.4 | 2 | 1/1.4 |
| 5 | 1.5 | 4 | 1/1.5 |
| 6 | 1.6 | 2 | 1/1.6 |
| 7 | 1.7 | 4 | 1/1.7 |
| 8 | 1.8 | 2 | 1/1.8 |
| 9 | 1.9 | 4 | 1/1.9 |
| 10 | 2 | 1 | 1/2 |

$\sum c_j f(x_j) = 20.79451$

$B = (b-a)/(3n) = 0.03333$

$S_{10} = 0.69315$

Let us conclude this section by proving (3), which gives a bound of the error associated with $R_n$. The proof serves as a reminder of many of the facts we have learned.

■ *Proof of (3)* We will show that if $|f'(x)| \leq M$ for $x$ in $[a, b]$, then

$$\left|\int_a^b f(x)\,dx - R_n\right| \leq \frac{M(b-a)^2}{2n}.$$

We begin by using the fact that the integral over $[a, b]$ is equal to the sum of the integrals over the subintervals. This gives

$$\left|\int_a^b f(x)\,dx - R_n\right| = \left|\sum_{j=1}^n \int_{x_{j-1}}^{x_j} f(x)\,dx - \sum_{j=1}^n f(x_j)\,\Delta x_j\right|.$$

Exercise 39 in Section 5.3 implies

$$\int_{x_{j-1}}^{x_j} c\,dx = c(x_j - x_{j-1}), \qquad c \text{ a constant}$$

Since $f(x_j)$ is a fixed number, we then have

$$f(x_j)\,\Delta x_j = f(x_j)(x_j - x_{j-1}) = \int_{x_{j-1}}^{x_j} f(x_j)\,dx.$$

Substitution of this integral for $f(x_j)\,\Delta x_j$ gives us

$$\left|\sum_{j=1}^n \int_{x_{j-1}}^{x_j} f(x)\,dx - \sum_{j=1}^n f(x_j)\,\Delta x_j\right| = \left|\sum_{j=1}^n \int_{x_{j-1}}^{x_j} f(x)\,dx - \sum_{j=1}^n \int_{x_{j-1}}^{x_j} f(x_j)\,dx\right|.$$

Using properties of sums and integrals we have

$$\begin{aligned}
\left|\sum_{j=1}^n \int_{x_{j-1}}^{x_j} f(x)\,dx - \sum_{j=1}^n \int_{x_{j-1}}^{x_j} f(x_j)\,dx\right| &= \left|\sum_{j=1}^n \left[\int_{x_{j-1}}^{x_j} f(x)\,dx - \int_{x_{j-1}}^{x_j} f(x_j)\,dx\right]\right| \\
&= \left|\sum_{j=1}^n \int_{x_{j-1}}^{x_j} [f(x) - f(x_j)]\,dx\right| \\
&\leq \sum_{j=1}^n \left|\int_{x_{j-1}}^{x_j} [f(x) - f(x_j)]\,dx\right| \\
&\leq \sum_{j=1}^n \int_{x_{j-1}}^{x_j} |f(x) - f(x_j)|\,dx.
\end{aligned}$$

The Mean Value Theorem implies $|f(x) - f(x_j)| = |f'(c_j)(x - x_j)|$ for some $c_j$ between $x$ and $x_j$. Since $|f'(c_j)| \le M$ and $x_j - x > 0$, we have $|f(x) - f(x_j)| \le M(x_j - x)$. We then have

$$\sum_{j=1}^{n} \int_{x_{j-1}}^{x_j} |f(x) - f(x_j)|\, dx \le \sum_{j=1}^{n} \int_{x_{j-1}}^{x_j} M(x_j - x)\, dx.$$

Using the results of Exercises 39 and 40 in Section 5.3 to evaluate the integrals, we obtain

$$\begin{aligned}\sum_{j=1}^{n} \int_{x_{j-1}}^{x_j} & M(x_j - x)\, dx \\ &= \sum_{j=1}^{n} M\left(x_j(x_j - x_{j-1}) - \frac{x_j^2}{2} + \frac{x_{j-1}^2}{2}\right) \\ &= \sum_{j=1}^{n} M\left(x_j^2 - x_j x_{j-1} - \frac{x_j^2}{2} + \frac{x_{j-1}^2}{2}\right) \\ &= \frac{M}{2} \sum_{j=1}^{n} (x_j - x_{j-1})^2 \\ &= \frac{M}{2} \sum_{j=1}^{n} \left(\frac{b-a}{n}\right)^2 = \frac{M}{2}\left(\frac{b-a}{n}\right)^2 (n) = \frac{M(b-a)^2}{2n}.\end{aligned}$$

Collecting results, we have

$$\left| \int_a^b f(x)\, dx - R_n \right| \le \frac{M(b-a)^2}{2n}. \qquad \blacksquare$$

## EXERCISES 5.4

In Exercises 1–8 calculate (a) $R_n$, (b) $T_n$, and (c) $S_n$ for the given functions $f$, the intervals $[a, b]$, and the integers $n$. (d) Use formula (3) to find a bound for the error in (a). (e) Use formula (5) to find a bound for the error in (b). (f) Use formula (7) to find a bound for the error in (c). (g) Find values of $n$ that guarantee $|\int_a^b f(x)\, dx - R_n| < 0.0001$. (h) Find values of $n$ that guarantee $|\int_a^b f(x)\, dx - T_n| < 0.0001$. (i) Find values of $n$ that guarantee $|\int_a^b f(x)\, dx - S_n| < 0.0001$.

**1** $f(x) = 1 - x^2$, $[0, 1]$, $n = 8$

**2** $f(x) = x - x^2$, $[0, 1]$, $n = 8$

**3** $f(x) = x^2$, $[-1, 1]$, $n = 10$

**4** $f(x) = x^3$, $[-1, 1]$, $n = 10$

**5** $f(x) = \sin x$, $[0, \pi]$, $n = 4$

**6** $f(x) = \cos x$, $[0, \pi]$, $n = 4$

**7** $f(x) = x \sin x$, $[0, \pi]$, $n = 6$

**8** $f(x) = x \cos x$, $[0, \pi]$, $n = 6$

In Exercises 9–10 use Simpson's Rule with $n = 8$ to find approximate values of the given integrals.

**9** (a) $\displaystyle\int_1^2 \frac{1}{x}\, dx$; (b) $\displaystyle\int_2^4 \frac{1}{x}\, dx$

**10** (a) $\displaystyle\int_4^8 \frac{1}{x}\, dx$; (b) $\displaystyle\int_8^{16} \frac{1}{x}\, dx$

In Exercises 11–14 use Simpson's Rule with $n = 4$ to find approximate values of the given integrals.

**11** $\displaystyle\int_0^1 \frac{4}{x^2 + 1}\, dx$

**12** $\displaystyle\int_0^{1/\sqrt{2}} 8(\sqrt{1 - x^2} - x)\, dx$

**13** $\int_{-2}^{2} 2\sqrt{4-x^2}\,dx$

**14** $\int_{0}^{4} \sqrt{x}\,dx$

**15** Explain why we cannot use formula (7) to estimate the error in Exercise 13.

**16** Explain why we cannot use formula (7) to estimate the error in Exercise 14.

In Exercises 17–18 the velocity (ft/s) of an object is given at intervals of 1 s over an interval of 10 s. Use Simpson's Rule to find an approximate value of the integral $\int_0^{10} s'(t)\,dt$. (The exact value of this integral is the displacement of the object over the interval of time.)

**17**

| $t$ | 0 | 1 | 2 | 3 | 4 | 5 | 6 | 7 | 8 | 9 | 10 |
|---|---|---|---|---|---|---|---|---|---|---|---|
| $s'(t)$ | 0 | 30 | 65 | 95 | 130 | 160 | 190 | 225 | 255 | 290 | 320 |

**18**

| $t$ | 0 | 1 | 2 | 3 | 4 | 5 | 6 | 7 | 8 | 9 | 10 |
|---|---|---|---|---|---|---|---|---|---|---|---|
| $s'(t)$ | 0 | 1 | 5 | 10 | 15 | 25 | 35 | 50 | 65 | 80 | 100 |

**19** Use Simpson's Rule to find an approximate value that is within 0.05 of the actual value of the area under the curve $y = x^5$, $0 \le x \le 2$.

**20** Use Simpson's Rule to find an approximate value that is within 0.0000001 of the actual value of the area under the curve $y = x^3$, $0 \le x \le 2$.

**21** If $f(x) = c_3x^3 + c_2x^2 + c_1x + c_0$, verify that Simpson's Rule with $n = 2$ gives the exact value of $\int_a^b f(x)\,dx$. (*Hint*: What is the error?)

## 5.5 THE FUNDAMENTAL THEOREM OF CALCULUS

Many (but not all) of the definite integrals we deal with can be evaluated very simply, without calculation of Riemann sums. The idea utilizes the fundamental connection between the concepts of the derivative and the integral.

■ *Theorem 1 (The Fundamental Theorem of Calculus)*

**If $f$ and $F$ are continuous on the interval $a \le x \le b$ and $F'(x) = f(x)$ for $a < x < b$, then**

$$\int_a^b f(x)\,dx = F(b) - F(a).$$

■ *Proof* The proof of this result that we present emphasizes that the definite integral is a limit of Riemann sums. To begin, let us suppose the points $a = x_0 < x_1 < \cdots < x_n = b$ divide the interval $[a, b]$ into $n$ subintervals. Then

$$\begin{aligned} F(b) - F(a) &= F(x_n) - F(x_0) \\ &= [F(x_n) - F(x_{n-1})] + [F(x_{n-1}) - F(x_{n-2})] \\ &\quad + \cdots + [F(x_1) - F(x_0)] \\ &= \sum_{j=1}^{n} [F(x_j) - F(x_{j-1})]. \end{aligned}$$

Applying the Mean Value Theorem on each of the subintervals, we obtain $F(x_j) - F(x_{j-1}) = F'(x_j^*)(x_j - x_{j-1})$ for some $x_j^*$ between $x_j$ and $x_{j-1}$, where $j = 1, \ldots, n$. Since $x_j - x_{j-1} = \Delta x_j$ and $F'(x_j^*) = f(x_j^*)$, we obtain

$$F(b) - F(a) = \sum_{j=1}^{n} f(x_j^*)\,\Delta x_j.$$

We then see that the number $F(b) - F(a)$ can be written as a Riemann sum of $\int_a^b f(x)\,dx$. But all such Riemann sums that have all subintervals small enough must be near $\int_a^b f(x)\,dx$. It follows that the number $F(b) - F(a)$ cannot be different from the number $\int_a^b f(x)\,dx$, so

$$\int_a^b f(x)\,dx = F(b) - F(a).$$

■

■ **EXAMPLE 1**

(a) We know that $f(x) = 2x$ is the derivative of $F(x) = x^2$, so Theorem 1 tells us

$$\int_3^5 2x\,dx = \int_3^5 f(x)\,dx = F(5) - F(3) = 5^2 - 3^2 = 16.$$

(b) The function $f(x) = 2\sin x\cos x$ is the derivative of $F(x) = \sin^2 x$, so Theorem 1 tells us

$$\int_0^{\pi/2} 2\sin x\cos x\,dx = \int_0^{\pi/2} f(x)\,dx = F\left(\frac{\pi}{2}\right) - F(0)$$
$$= \sin^2\frac{\pi}{2} - \sin^2 0 = (1)^2 - (0)^2 = 1.$$

■

To use Theorem 1 to evaluate the definite integral $\int_a^b f(x)\,dx$, we need to know a function $F$ with $F'(x) = f(x)$. The problem of finding a function $F$ with $F' = f$ is the same as the problem of evaluating an indefinite integral, as was studied in Section 4.10. Recall that, for certain functions $f$, we were able to use known differentiation formulas to find functions $F$ such that $F'(x) = f(x)$. Let us recall the basic integration formulas from Section 4.10.

$$\int x^r\,dx = \frac{x^{r+1}}{r+1} + C, \qquad r \neq -1,$$

$$\int \cos x\,dx = \sin x + C,$$

$$\int \sin x\,dx = -\cos x + C,$$

$$\int \sec^2 x\,dx = \tan x + C,$$

$$\int \csc^2 x\,dx = -\cot x + C,$$

$$\int \sec x \cdot \tan x \, dx = \sec x + C,$$

$$\int \csc x \cdot \cot x \, dx = -\csc x + C,$$

$$\int [cf(x) + dg(x)] \, dx = c \cdot \int f(x) \, dx + d \cdot \int g(x) \, dx.$$

The methods we will use to find a function $F$ such that $F'(x) = f(x)$ lead naturally to formulas that involve $x$. That is, we first find a formula for $F(x)$ and then use substitution to find the value of $F(b) - F(a)$. It is convenient to introduce the notation

$$F(x) \Big]_a^b = F(b) - F(a).$$

Recall from Section 4.10 that $\int f(x) \, dx = F(x) + C$ means $F'(x) = f(x)$. Hence, if we assume $f$ and $F$ are continuous on $[a, b]$ Theorem 1 tells us

$$\int f(x) \, dx = F(x) + C \quad \text{implies} \quad \int_a^b f(x) \, dx = F(x) \Big]_a^b = F(b) - F(a).$$

We know that we may have $\int f(x) \, dx = F(x) + C$ and $\int f(x) \, dx = G(x) + C$ for different functions $F$ and $G$. In this case we have $F'(x) = f(x) = G'(x)$, so there is some constant $C_0$ such that $G(x) = F(x) + C_0$. Then

$$G(b) - G(a) = [F(b) + C_0] - [F(a) + C_0] = F(b) - F(a),$$

so we may use either $F$ or $G$ in Theorem 1. Generally, we use the most convenient function.

**■ EXAMPLE 2**

Evaluate the following definite integrals.

(a) $\displaystyle\int_1^4 x^{-1/2} \, dx,$

(b) $\displaystyle\int_0^4 (x^3 - 4x - 3) \, dx,$

(c) $\displaystyle\int_0^{\pi} \sin t \, dt,$

(d) $\displaystyle\int_{-\pi/4}^{\pi/4} \sec^2 \theta \, d\theta.$

**SOLUTION**

In each case we will use the basic integration formulas as in Section 4.10 to find a function $F$ such that $\int_a^b f(x) \, dx = F(x)]_a^b$. Also, recall that the variable of integration is a dummy variable; we may use any real variable in place of $x$.

$$\text{(a)}\quad \int_1^4 x^{-1/2}\,dx = \frac{x^{1/2}}{1/2}\Big]_1^4 = 2x^{1/2}\Big]_1^4$$
$$= 2(4)^{1/2} - 2(1)^{1/2} = 2(2) - 2(1) = 2.$$

$$\text{(b)}\quad \int_0^4 (x^3 - 4x - 3)\,dx = \left(\frac{x^4}{4} - 4\frac{x^2}{2} - 3x\right)\Big]_0^4$$
$$= \left(\frac{(4)^4}{4} - 2(4)^2 - 3(4)\right) - \left(\frac{(0)^4}{4} - 2(0)^2 - 3(0)\right) = 20.$$

$$\text{(c)}\quad \int_0^\pi \sin t\,dt = -\cos t\Big]_0^\pi = (-\cos\pi) - (-\cos 0) = -(-1) + 1 = 2.$$

$$\text{(d)}\quad \int_{-\pi/4}^{\pi/4} \sec^2\theta\,d\theta = \tan\theta\Big]_{-\pi/4}^{\pi/4}$$
$$= \tan\frac{\pi}{4} - \tan\left(-\frac{\pi}{4}\right) = (1) - (-1) = 2. \quad ■$$

A word of caution concerning Theorem 1 is in order. We must be sure that $F'(x) = f(x)$ for all $x$ in the open interval $(a, b)$ and that $F$ and $f$ are continuous on the closed interval $[a, b]$. For example, we know that

$$\int \frac{1}{x^2}\,dx = -\frac{1}{x} + C, \quad \text{because} \quad \frac{d}{dx}\left(-\frac{1}{x}\right) = \frac{1}{x^2},\ x \neq 0.$$

Since $(-1/x)' \neq 1/x^2$ for $x = 0$, we cannot use Theorem 1 to evaluate $\int_a^b (1/x^2)\,dx$ if the interval $[a, b]$ contains the point $x = 0$. In particular,

$$\int_{-1}^1 \frac{1}{x^2}\,dx \underset{(?)}{=} -\frac{1}{x}\Big]_{-1}^1 = \left(-\frac{1}{1}\right) - \left(-\frac{1}{-1}\right) = -2 \text{ is nonsense.}$$

Theorem 1 does not apply. (The integral of the positive function $1/x^2$ over the interval $[-1, 1]$ would not be negative, if it made sense at all.) Even though $F(x) = -1/x$ and $f(x) = 1/x^2$ satisfy $F'(x) = f(x)$ on the interval $0 < x < 1$, we cannot use Theorem 1 to evaluate

$$\int_0^1 \frac{1}{x^2}\,dx$$

because $f(x) = 1/x^2$ and $F(x) = -1/x$ are not continuous on the interval $[0, 1]$. [The fact that $f(0)$ and $F(0)$ are undefined can be corrected by assigning them any values. However, it is impossible to assign values at $x = 0$ that make the functions *continuous* at $x = 0$.]

Let us see how Theorem 1 relates to motion along a line. Suppose the position of an object along a number line at time $t$ is given by $s(t)$. If $s'$ is continuous on the interval $a \leq t \leq b$, Theorem 1 implies

$$s(b) - s(a) = \int_a^b s'(t)\, dt.$$

This formula is consistent with the fact that the **net displacement** of the object over the time interval $[a, b]$ can be approximated by Riemann sums of the velocity. In Section 5.2 we used the formula

$$s(b) - s(a) \approx s'(t_1^*)\, \Delta t_1 + s'(t_2^*)\, \Delta t_2 + \cdots + s'(t_n^*)\, \Delta t_n.$$

In Section 5.2 we also determined that the **total distance traveled** can be approximated by Riemann sums of the speed.

$$\text{Total distance traveled} \approx |s'(t_2^*)|\, \Delta t_1^* + |s'(t_2^*)|\, \Delta t_2 + \cdots + |s'(t_n)|\, \Delta t_n.$$

The following formula should then seem reasonable.

$$\textbf{Total distance traveled} = \int_a^b |s'(t)|\, dt.$$

**■ EXAMPLE 3**

The instantaneous velocity of an object on a number line is given by $s'(t) = 64 - 32t$, $0 \leq t \leq 3$. (a) Find the net displacement of the object. (b) Find the total distance traveled by the object.

**SOLUTION**

(a) The net displacement over the time interval from $t = 0$ to $t = 3$ is

$$\begin{aligned} s(3) - s(0) &= \int_0^3 s'(t)\, dt = \int_0^3 (64 - 32t)\, dt \\ &= 64t - 16t^2 \Big]_0^3 \\ &= [64(3) - 16(3)^2] - [64(0) - 16(0)^2] \\ &= 48. \end{aligned}$$

The net displacement is 48 units in the positive direction.

(b) The total distance traveled is

$$\int_0^3 |s'(t)|\, dt = \int_0^3 |64 - 32t|\, dt.$$

At this point we must recall that the absolute value function is given by two expressions, one the negative of the other. Since our integration formulas require that the same formula holds for the interval of integration, we must determine intervals for which we can express $|64 - 32t|$ by a single formula. It is easy to see that $64 - 32t = 0$ implies $t = 2$, $64 - 32t \geq 0$ if $t \leq 2$, and $64 - 32t \leq 0$ if $t \geq 2$. It follows that

$$|64 - 32t| = \begin{cases} 64 - 32t, & t \leq 2, \\ -(64 - 32t), & t \geq 2. \end{cases}$$

We can now write the integral over the interval [0, 3] as the sum of the integrals over two subintervals that have the property that one formula for $|64 - 32t|$ holds on each subinterval. We obtain

$$\begin{aligned}\int_0^3 |64 - 32t|\,dt &= \int_0^2 |64 - 32t|\,dt + \int_2^3 |64 - 32t|\,dt \\ &= \int_0^2 (64 - 32t)\,dt + \int_2^3 -(64 - 32t)\,dt \\ &= \left[64t - 16t^2\right]_0^2 - \left[64t - 16t^2\right]_2^3 \\ &= \left[\left(64(2) - 16(2)^2\right) - \left(64(0) - 16(0)^2\right)\right] \\ &\quad - \left[\left(64(3) - 16(3)^2\right) - \left(64(2) - 16(2)^2\right)\right] \\ &= [64] - [-16] = 80.\end{aligned}$$

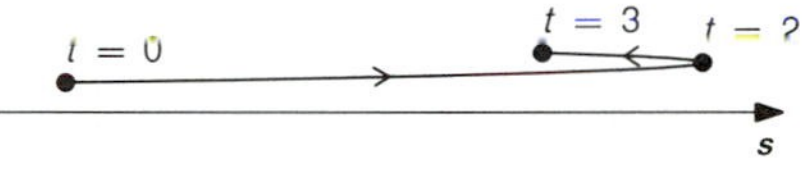

FIGURE 5.5.1

The total distance traveled is 80 units. The object travels 64 units in the positive direction and then 16 units in the negative direction. See Figure 5.5.1. ■

If $f$ is continuous on the interval $[a, b]$, then the function

$$G(x) = \int_a^x f(t)\,dt$$

is defined for $a \le x \le b$. The following theorem tells us that $G(x)$ is differentiable for $a < x < b$, with derivative $f(x)$. The function $\int_a^x f(t)\,dt$ is called an **indefinite integral of $f$.**

■ *Theorem 2 (The Second Fundamental Theorem of Calculus)*

**If $f$ is continuous on the interval $[a, b]$ and $G(x) = \int_a^x f(t)\,dt$, then $G'(x) = f(x)$, $a < x < b$.**

■ *Proof* The proof of Theorem 2 uses the definition of the derivative and properties of the integral. We will show that $G'(c) = f(c)$, $c$ any point in the interval $(a, b)$.

For $x \neq c$, we have

$$\frac{G(x) - G(c)}{x - c} = \frac{1}{x - c}\left(\int_a^x f(t)\,dt - \int_a^c f(t)\,dt\right) = \frac{1}{x - c}\int_c^x f(t)\,dt.$$

The Mean Value Theorem for definite integrals that was given in Section 5.3 implies there is a number $z$ between $c$ and $x$ such that

$$\int_c^x f(t)\,dt = f(z)(x - c).$$

Substitution then gives

$$\frac{G(x) - G(c)}{x - c} = \frac{1}{x - c}[f(z)(x - c)] = f(z).$$

Since $z$ is between $x$ and $c$, $z$ must approach $c$ as $x$ approaches $c$. But $f$ is continuous at $c$, so $f(z)$ must approach $f(c)$ as $z$ approaches $c$. It follows that

$$G'(c) = \lim_{x \to c} \frac{G(x) - G(c)}{x - c} = f(c). \quad \blacksquare$$

If we have a function $F$ such that $F'(x) = f(x)$ on $[a, b]$, we can use Theorem 1 to express the function $G$ of Theorem 2 in terms of $F$. That is,

$$G(x) = \int_a^x f(t)\,dt = F(t)\Big]_a^x = F(x) - F(a).$$

It is then clear that $G'(x) = F'(x) = f(x)$. [The derivative of the constant $F(a)$ is zero.] Theorem 2 tells us how to evaluate the derivative of $\int_a^x f(t)\,dt$ *without* finding a function $F$ with $F'(x) = f(x)$. The derivative is simply the value of the integrand at $x$, which is $f(x)$.

### ■ EXAMPLE 4

From Theorem 2 we have

(a) $\frac{d}{dx}\left(\int_\pi^x \sin t\,dt\right) = \sin x,$

(b) $\frac{d}{dx}\left(\int_1^x \frac{1}{t}\,dt\right) = \frac{1}{x}, \qquad x > 0,$

(c) $\frac{d}{dx}\left(\int_0^x \sin t^2\,dt\right) = \sin x^2.$ ■

In Example 4a we can write $\int_\pi^x \sin t\,dt = -\cos x - 1$, and we then differentiate to obtain the derivative $\sin x$. At this time we cannot evaluate the integral in Example 4b because we do not know a function $F$ with $F'(x) = 1/x$. Such a function will be introduced in Chapter 7, so we will be able to evaluate the integral in Example 4b. There is no combination of elementary functions that has derivative $\sin x^2$, so we must use Theorem 2 in Example 3c.

### ■ EXAMPLE 5

Evaluate

$$\frac{d}{dx}\left(\int_x^0 \frac{1}{4 + t^2}\,dt\right).$$

**SOLUTION**

To use Theorem 2, the indefinite integral must have the variable of differentiation as the upper limit of integration and a constant as the lower limit of integration. The given integral can be written in this form by interchanging the limits of integration. Since interchanging the limits of integration changes the sign of the integral, we have

$$\frac{d}{dx}\left(\int_x^0 \frac{1}{4+t^2}\,dt\right) = \frac{d}{dx}\left(-\int_0^x \frac{1}{4+t^2}\,dt\right) = -\frac{1}{4+x^2}.$$ ■

If the upper limit of integration is a function of the variable of differentiation, we can use the Chain Rule to find the derivative. That is, if

$$y = \int_a^u f(t)\,dt \quad \text{and} \quad u = u(x),$$

then $y$ is a function of $x$ and the Chain Rule gives

$$\frac{dy}{dx} = \frac{dy}{du}\frac{du}{dx} = f(u)\cdot\frac{du}{dx}.$$

We have used the fact that Theorem 2 implies

$$\frac{d}{du}\left(\int_a^u f(t)\,dt\right) = f(u),$$

since the upper limit of integration is the variable of differentiation.

**■ EXAMPLE 6**

Evaluate the derivatives.

(a) $\frac{d}{dx}\left(\int_1^{x^2} \frac{1}{t}\,dt\right)$,

(b) $\frac{d}{dx}\left(\int_0^{\sqrt{x}} \sin t^2\,dt\right)$,

(c) $\frac{d}{dx}\left(\int_x^{x^2} \frac{1}{t}\,dt\right)$, $\quad x > 0$.

**SOLUTION**

(a) We have $y = \int_1^u (1/t)\,dt$, where $u = x^2$. The Chain Rule and Theorem 2 then imply

$$\frac{dy}{dx} = \frac{dy}{du}\frac{du}{dx} \quad \text{and} \quad \frac{dy}{du} = \frac{1}{u} = \frac{1}{x^2},$$

so

$$\frac{d}{dx}\left(\int_1^{x^2} \frac{1}{t}\,dt\right) = \left(\frac{1}{x^2}\right)(2x) = \frac{2}{x}.$$

(b) We have $y = \int_0^u \sin(t^2)\,dt$, where $u = \sqrt{x}$, so

$$\frac{d}{dx}\left(\int_0^{\sqrt{x}} \sin t^2\,dt\right) = (\sin(\sqrt{x})^2)\left(\frac{1}{2\sqrt{x}}\right) = \frac{\sin x}{2\sqrt{x}}.$$

(c) To use Theorem 2, we express the function to be differentiated as a sum of integrals that have a variable expression as the upper limit of integration and a constant as the lower limit. We can then write the derivative of the sum as above. We could use any positive number in place of 1 below:

$$\frac{d}{dx}\left(\int_x^{x^2} \frac{1}{t}\,dt\right) = \frac{d}{dx}\left(\int_x^1 \frac{1}{t}\,dt + \int_1^{x^2} \frac{1}{t}\,dt\right)$$

$$= \frac{d}{dx}\left(-\int_1^x \frac{1}{t}\,dt + \int_1^{x^2} \frac{1}{t}\,dt\right)$$

$$= -\frac{1}{x} + \left(\frac{1}{x^2}\right)(2x) = -\frac{1}{x} + \frac{2}{x} = \frac{1}{x}.$$

■

## EXERCISES 5.5

Use Theorem 1 to evaluate the definite integrals given in Exercises 1–24.

1 $\int_1^4 \sqrt{x}\,dx$

2 $\int_1^8 x^{1/3}\,dx$

3 $\int_1^8 x^{-1/3}\,dx$

4 $\int_1^9 x^{-1/2}\,dx$

5 $\int_1^2 \frac{1}{x^3}\,dx$

6 $\int_1^3 \frac{1}{x^4}\,dx$

7 $\int_0^1 (x^2 + x + 1)\,dx$

8 $\int_0^1 (x^2 - x - 1)\,dx$

9 $\int_1^3 (x^2 - 6x + 2)\,dx$

10 $\int_1^3 (x^2 - 8x + 3)\,dx$

11 $\int_0^{\pi/2} \cos x\,dx$

12 $\int_0^{\pi/2} \sin x\,dx$

13 $\int_0^{\pi/4} \sec^2 x\,dx$

14 $\int_{\pi/6}^{\pi/2} \csc^2 x\,dx$

15 $\int_{\pi/6}^{\pi/3} \csc x \cdot \cot x\,dx$

16 $\int_{\pi/6}^{\pi/3} \sec x \cdot \tan x\,dx$

17 $\int_0^{\pi} \sin x\,dx$

18 $\int_0^{\pi} \cos x\,dx$

19 $\int_0^{2\pi} \cos x\,dx$

20 $\int_0^{2\pi} \sin x\,dx$

21 $\int_{\pi/6}^{\pi/3} \csc^2 x\,dx$

22 $\int_{\pi/6}^{\pi/3} \sec^2 x\,dx$

23 $\int_{-\pi/4}^{\pi/4} \sec x \tan x\,dx$

24 $\int_{\pi/4}^{\pi/2} \csc x \cot x\,dx$

In Exercises 25–30 the instantaneous velocity of an object on a number line is given. Find (a) the net displacement and (b) the total distance traveled by the object over the given time interval.

25 $s'(t) = 32 - 32t,\ 0 \le t \le 3$

26 $s'(t) = 29.4 - 9.8t,\ 0 \le t \le 5$

27 $s'(t) = t - t^3,\ 0 \le t \le 2$

28 $s'(t) = t^2 - 3t + 2,\ 0 \le t \le 3$

29 $s'(t) = \sin t,\ 0 \le t \le 2\pi$

30 $s'(t) = \sin t - \cos t,\ 0 \le t \le \pi$

Use Theorem 2 to evaluate the derivatives in Exercises 31–44.

31 $\frac{d}{dx}\left(\int_1^x \cos t^2\,dt\right)$

32 $\frac{d}{dx}\left(\int_2^x \cos t^2\,dt\right)$

33 $\frac{d}{dx}\left(\int_0^x \sec t^2\,dt\right),\ |x| < \sqrt{\pi/2}$

34 $\frac{d}{dx}\left(\int_0^x \tan t\,dt\right),\ |x| < \frac{\pi}{2},$

35 $\frac{d}{dx}\left(\int_x^1 (1 - t)^5\,dt\right)$

36 $\dfrac{d}{dx}\left(\displaystyle\int_x^8 (10-t)^{-2}\,dt\right)$, $x < 10$,

37 $\dfrac{d}{dx}\left(\displaystyle\int_0^{x^2} \sin t\,dt\right)$

38 $\dfrac{d}{dx}\left(\displaystyle\int_0^{\sqrt{x}} \cos t\,dt\right)$

39 $\dfrac{d}{dx}\left(\displaystyle\int_0^{x+1} (t-1)^4\,dt\right)$

40 $\dfrac{d}{dx}\left(\displaystyle\int_0^{x-1} (t+1)^5\,dt\right)$

41 $\dfrac{d}{dx}\left(\displaystyle\int_1^{2x} \frac{1}{t}\,dt\right)$, $x > 0$

42 $\dfrac{d}{dx}\left(\displaystyle\int_1^{3x} \frac{1}{t}\,dt\right)$, $x > 0$

43 $\dfrac{d}{dx}\left(\displaystyle\int_x^{x^2} \tan t\,dt\right)$, $|x| < \sqrt{\pi/2}$

44 $\dfrac{d}{dx}\left(\displaystyle\int_x^{x^2} \sec t\,dt\right)$, $|x| < \sqrt{\pi/2}$

45 $G(x) = \displaystyle\int_{-1}^{x} \frac{1}{t^2+1}\,dt$. Find $G''(2)$.

46 $G(x) = \displaystyle\int_0^x \sqrt{25-t^2}\,dt$, $-5 < x < 5$. Find $G''(3)$.

## 5.6 SUBSTITUTION AND CHANGE OF VARIABLES

We have seen in Section 5.5 that a formula of the form $\int f(x)\,dx = F(x) + C$ greatly facilitates evaluation of the definite integral $\int_a^b f(x)\,dx$. However, we do not yet have a systematic method of evaluating indefinite integrals, except for our basic formulas. In this section we introduce a method that extends the class of functions for which we can evaluate indefinite integrals. The method is called **substitution**. Substitution is a basic tool for manipulating integrals. It is a type of *algebra* that we need in order to evaluate integrals.

Use of simplifying substitutions to evaluate integrals is based on the Chain Rule for differentiation.

If

$$\frac{d}{du}(F(u)) = f(u) \qquad \text{for } u \text{ in an interval,}$$

we have

$$\int f(u)\,du = F(u) + C. \tag{1}$$

If $u = u(x)$ is a differentiable function of $x$, the Chain Rule gives

$$\frac{d}{dx}\big(F(u(x))\big) = \frac{dF}{du}\frac{du}{dx} = f(u(x)) \cdot u'(x).$$

The integral form of this statement is

$$\int f(u(x))u'(x)\,dx = F(u(x)) + C. \tag{2}$$

The idea of a simplifying substitution is to replace $u(x)$ by $u$ and $u'(x)\,dx$ by $du$ in (2). [Recall from Section 4.8 that $du = u'(x)\,dx$ is the differential of $u(x)$.] This substitution reduces the left side of (2) to the simpler form of the left side of (1). If we can evaluate the integral in (1), we can substitute

$u(x)$ for $u$ in the right side of (1). This gives us the right side of (2). The form of the solution is

$$\int f(u(x))u'(x)\,dx = \int f(u)\,du \qquad \textbf{(Substitute)}$$

$$= F(u) + C \qquad \textbf{(Evaluate)}$$

$$= F(u(x)) + C \qquad \textbf{(Substitute back).}$$

Of course, to successfully use this method, we must be able to evaluate the simplified integral $\int f(u)\,du$. For now, this means that $\int f(u)\,du$ must be one of the following types, or a *constant* multiple of one of them, or a sum of such types.

$$\int u^r\,du = \frac{u^{r+1}}{r+1} + C, r \neq -1, \tag{3}$$

$$\int \cos u\,du = \sin u + C, \tag{4}$$

$$\int \sin u\,du = -\cos u + C, \tag{5}$$

$$\int \sec^2 u\,du = \tan u + C, \tag{6}$$

$$\int \csc^2 u\,du = -\cot u + C, \tag{7}$$

$$\int \sec u \cdot \tan u\,du = \sec u + C, \tag{8}$$

$$\int \csc u \cdot \cot u\,du = -\csc u + C. \tag{9}$$

The above formulas of integration are those that we used in Sections 4.10 and 5.5, except we are now using $u$ for the variable of integration.

To evaluate an integral, we must recognize which of the above formulas applies, choose $u(x)$, evaluate the differential

$$du = u'(x)\,dx$$

and then substitute. If we get a constant multiple of one of the above formulas, we can evaluate the simplified integral. If we cannot reduce a given integral to one of the above types, then we cannot evaluate it at this time. Of course, we can use the formula

$$\int [cf(u) + dg(u)]\,du = c \cdot \int f(u)\,du + d \cdot \int g(u)\,du.$$

to evaluate sums of the above types.

Let us illustrate the method with some examples.

■ **EXAMPLE 1**

Determine which of the integrals reduce to a constant multiple of one of the types (3)–(9). Evaluate those that do.

(a) $\int 2x\sqrt{x^2+1}\,dx,$

(b) $\int x\sqrt{x^2+1}\,dx,$

(c) $\int \sqrt{x^2+1}\,dx,$

(d) $\int x^2\sqrt{x^2+1}\,dx,$

(e) $\int \frac{1}{(1+\sqrt{x})^2}\frac{1}{\sqrt{x}}\,dx,$

(f) $\int \frac{1}{(1+\sqrt{x})^2 x}\,dx,$

(g) $\int x^2 \sin x^3\,dx,$

(h) $\int \cos x^2\,dx,$

(i) $\int \frac{\cos x}{\sqrt{\sin x+1}}\,dx,$

(j) $\int (\cos x)^{1/3}\,dx,$

(k) $\int \frac{\sin x}{\cos x}\,dx,$

(l) $\int (2x-1)^{12}\,dx.$

**SOLUTION**

(a) The integral $\int 2x\sqrt{x^2+1}\,dx$ involves the square root of a function. Let us see if we can reduce the integral to $\int u^{1/2}\,du$. If it is of this form, we must have $u = x^2+1$. Let us try

$$u = x^2+1, \quad \text{so} \quad du = 2x\,dx.$$

Noting that the integrand contains the factor $2x$, we substitute to obtain

$$\int 2x\sqrt{x^2+1}\,dx = \int \underbrace{\sqrt{x^2+1}}_{\sqrt{u}}\,\underbrace{2x\,dx}_{du} = \int \sqrt{u}\,du = \int u^{1/2}\,du.$$

Using (3) to evaluate the simplified integral, we obtain

$$\int u^{1/2}\,du = \frac{u^{3/2}}{3/2} + C.$$

Substituting $x^2 + 1$ for $u$, and combining results, we have

$$\int 2x\sqrt{x^2+1}\,dx = \frac{2}{3}(x^2+1)^{3/2} + C.$$

Recall that we can verify an integration formula $\int f(x)\,dx = F(x) + C$ by showing $F'(x) = f(x)$. Using the Chain Rule to verify the formula obtained above, we have

$$\frac{d}{dx}\left(\frac{2}{3}(x^2+1)^{3/2}\right) = \left(\frac{2}{3}\right)\left(\frac{3}{2}\right)(x^2+1)^{1/2}(2x) = 2x\sqrt{x^2+1},$$

so the formula is correct. (Since we used the integral form of the Chain Rule to evaluate this integral, we should expect that the Chain Rule would be used to verify the formula by differentiation.)

(b) The integral $\int x\sqrt{x^2+1}\,dx$ involves the square root of a function. As in (a), we try to reduce the integral to $\int u^{1/2}\,du$. Let us try

$$u = x^2 + 1, \qquad du = 2x\,dx.$$

We can then write

$$\int x\sqrt{x^2+1}\,dx = \int \underbrace{\sqrt{x^2+1}}_{\sqrt{u}}\left(\frac{1}{2}\right)\underbrace{(2x)\,dx}_{du}$$

We have introduced the factor

$$\left(\frac{1}{2}\right)(2)$$

in the integrand so the exact expression $du = 2x\,dx$ appears in the integrand. For the substitution $u = x^2 + 1$ to simplify the integral, the integrand must contain the *essential factor* $x$, since $u'(x) = (\text{constant})x$. We can modify the essential factor only by a constant. Substitution now gives

$$\int \sqrt{x^2+1}\left(\frac{1}{2}\right)(2x)\,dx = \int u^{1/2}\left(\frac{1}{2}\right)du.$$

This integral is of the form of (3), except for the constant factor 1/2. We evaluate to obtain

$$\int u^{1/2}\left(\frac{1}{2}\right)du = \left(\frac{1}{2}\right)\left(\frac{u^{3/2}}{3/2}\right) + C = \frac{1}{3}u^{3/2} + C.$$

Substituting $x^2 + 1$ for $u$, we then obtain

$$\int x\sqrt{x^2 + 1}\, dx = \frac{1}{3}(x^2 + 1)^{3/2} + C.$$

An alternate method of evaluating the above integral avoids introduction of the factor

$$\left(\frac{1}{2}\right)(2).$$

That is, noting that the integrand contains the factor $x\, dx$, we solve $du = 2x\, dx$ for $x\, dx$. We obtain

$$x\, dx = \frac{1}{2}\, du.$$

Direct substitution of $(1/2)\, du$ for $x\, dx$ gives

$$\int x\sqrt{x^2 + 1}\, dx = \int \sqrt{u}\,\frac{1}{2}\, du.$$

We then continue as before.

(c) The integral $\int \sqrt{x^2 + 1}\, dx$ involves the square root of a function. We try $u = x^2 + 1$, so $du = 2x\, dx$. This substitution will not reduce the integral to the simplified form $\int$ (constant)$u^{1/2}\, du$, because the integrand does not contain the essential factor $x$. Attempts to substitute will lead to either meaningless expressions that involve both $u$ and $x$ or complicated expressions of the variable $u$ if we try to eliminate $x$ by solving $u = x^2 + 1$ for $x$. For example, suppose we try to modify the integrand by the factor

$$\left(\frac{1}{2x}\right)(2x).$$

We would obtain

$$\int \sqrt{x^2 + 1}\, dx = \int \sqrt{x^2 + 1}\left(\frac{1}{2x}\right)(2x)\, dx.$$

We could substitute $u^{1/2}$ for $\sqrt{x^2 + 1}$ and $du$ for $2x\, dx$, but we would still have a factor of $x$ in the denominator. *We cannot simply move a variable factor of the integrand outside of the integral sign*, as we can do with a constant. We could solve the equation $u = x^2 + 1$ for $x$, to obtain $x = \pm(u - 1)^{1/2}$. Substitution of this expression for $x$ gives

$$\int \frac{u^{1/2}}{\pm 2(u - 1)^{1/2}}\, du,$$

which is more complicated than the integral we started with. We cannot reduce this integral to one of the types (3)–(9).

(d) This example is similar to (c). The substitution $u = x^2 + 1$ will not reduce $\int x^2\sqrt{x^2+1}\,dx$ to the form $\int (\text{constant})u^{1/2}\,du$, because the integrand contains an extra factor of $x$. We cannot reduce this integral to one of the types (3)–(9).

(e) The integral

$$\int \frac{1}{(1+\sqrt{x})^2}\frac{1}{\sqrt{x}}\,dx$$

involves a function to the negative two power. Let us try to reduce the integral to a constant multiple of $\int u^{-2}\,du$. We try

$$u = 1+\sqrt{x}, \quad \text{so} \quad du = \frac{1}{2\sqrt{x}}\,dx.$$

We note that the integrand contains the essential factor $1/\sqrt{x}$. We can then write

$$\begin{aligned}\int \frac{1}{(1+\sqrt{x})^2}\frac{1}{\sqrt{x}}\,dx &= \int \frac{1}{(1+\sqrt{x})^2}(2)\frac{1}{2\sqrt{x}}\,dx\\ &= \int u^{-2}(2)\,du\\ &= 2\frac{u^{-1}}{-1}+C\\ &= -2(1+\sqrt{x})^{-1}+C.\end{aligned}$$

(f) This integral involves a function to the negative two power. Let us try the substitution

$$u = 1+\sqrt{x}, \quad \text{so} \quad du = \frac{1}{2\sqrt{x}}\,dx.$$

We can then write

$$\int \frac{1}{(1+\sqrt{x})^2 x}\,dx = \int \frac{1}{(1+\sqrt{2x})^2}\frac{2}{\sqrt{x}}\frac{1}{2\sqrt{x}}\,dx.$$

We can solve $u = 1+\sqrt{x}$ for $\sqrt{x}$ to obtain $\sqrt{x} = u-1$. Substitution of $u$ for $1+\sqrt{x}$, $u-1$ for $\sqrt{x}$, and $du$ for

$$\frac{1}{2\sqrt{x}}\,dx$$

then gives

$$\int \frac{2}{u^2(u-1)}\,du.$$

This integral does not reduce to one of the types (3)–(9).

(g) The integral $\int x^2 \sin x^3\, dx$ involves the sine of a function. Let us set $u = x^3$, so $du = 3x^2\, dx$. Noting that the integrand contains the essential factor $x^2$, we write

$$\begin{aligned}
\int x^2 \sin x^3\, dx &= \int (\sin x^3) \cdot \left(\frac{1}{3}\right)(3x^2)\, dx \\
&= \int (\sin u)\left(\frac{1}{3}\right) du \\
&= \frac{1}{3}(-\cos u) + C \\
&= -\frac{1}{3}\cos x^3 + C.
\end{aligned}$$

(h) The substitution $u = x^2$ does not reduce $\int \cos x^2\, dx$ to $\int \cos u\, du$, because the integrand does not contain the essential factor $x$. $u = x^2$ implies $du = 2x\, dx$. We cannot reduce this integral to one of the types (3)–(9).

(i) The integral

$$\int \frac{\cos x}{\sqrt{\sin x + 1}}\, dx$$

involves the negative one-half power of a function. Let us set

$$u = \sin x + 1, \quad \text{so} \quad du = \cos x\, dx.$$

The integrand contains the essential factor $\cos x$. We have

$$\begin{aligned}
\int \frac{\cos x}{\sqrt{\sin x + 1}}\, dx &= \int (\sin x + 1)^{-1/2} \cos x\, dx \\
&= \int u^{-1/2}\, du \\
&= \frac{u^{1/2}}{1/2} + C \\
&= 2\sqrt{\sin x + 1} + C.
\end{aligned}$$

(j) The integral $\int (\cos x)^{1/3}\, dx$ does not reduce to one of the desired types. The substitution $u = \cos x$ requires the essential factor $\sin x$, since $du = -\sin x\, dx$. We cannot reduce this integral to one of the types seen in (3)–(9).

(k) The integrand of

$$\int \frac{\sin x}{\cos x}\, dx$$

contains the factor $\sin x$. This leads us to try

$$u = \cos x, \quad \text{so} \quad du = -\sin x\, dx.$$

Substitution then gives

$$\int_{\cos x}^{\sin x} dx = \int (\cos x)^{-1}(-1)(-\sin x)\,dx = -\int u^{-1}\,du.$$

We cannot evaluate this integral at this time. (3) does not apply because the exponent is $-1$.

(l) The integral $\int (2x-1)^{12}\,dx$ involves the twelfth power of a function. We set

$$u = 2x - 1, \quad \text{so} \quad du = 2dx.$$

Substitution then gives

$$\begin{aligned}\int (2x-1)^{12}\,dx &= \int (2x-1)^{12}\left(\frac{1}{2}\right)(2)\,dx \\ &= \int u^{12}\left(\frac{1}{2}\right)du \\ &= \frac{1}{2}\frac{u^{13}}{13} + C \\ &= \frac{1}{26}(2x-1)^{13} + C.\end{aligned}$$

■

We can use the method of substitution to evaluate definite integrals.

■ **EXAMPLE 2**

Evaluate $\displaystyle\int_0^3 \frac{x}{\sqrt{x^2+16}}\,dx.$

**SOLUTION**

Using the method of Example 1 to evaluate the indefinite integral, we set

$$u = x^2 + 16, \quad \text{so} \quad du = 2x\,dx.$$

Then

$$\begin{aligned}\int_0^3 \frac{x}{\sqrt{x^2+16}}\,dx &= \int_0^3 (x^2+16)^{-1/2}\left(\frac{1}{2}\right)(2x)\,dx \\ &= \int u^{-1/2}\left(\frac{1}{2}\right)du \\ &= \left(\frac{1}{2}\right)\left(\frac{u^{1/2}}{1/2}\right)\Bigg].\end{aligned}$$

We have not written the limits of integration in the above two steps. It is $x$, not $u$, that varies over the interval $[0, 3]$. Avoid the incorrect statement of equality that would occur if the limits of $x$ were expressed as limits of $u$. Rewrite the limits only after we have substituted back in terms of $x$.

We have

$$\int_0^3 \frac{x}{\sqrt{x^2+16}}\,dx = \sqrt{x^2+16}\,\Big]_0^3 = \sqrt{3^2+16} - \sqrt{0^2+16}$$

$$= 5 - 4 = 1.$$

In some problems it is desirable to change the limits of integration to correspond to a change of the variable of integration. To see how this is done we assume $F'(x) = f(x)$. Theorem 1 of Section 5.5 and (2) then imply

$$\int_a^b f(u(x))u'(x)\,dx = F(u(x))\Big]_a^b = F((u(b)) - F(u(a)),$$

and (1) gives us

$$\int_{u(a)}^{u(b)} f(u)\,du = F(u)\Big]_{u(a)}^{u(b)} = F(u(b)) - F(u(a)).$$

Assuming appropriate differentiability, so that both formulas hold, we conclude that

$$\int_a^b f(u(x))u'(x)\,dx = \int_{u(a)}^{u(b)} f(u)\,du. \tag{10}$$

In formula (10), the limits of integration of the variable $u$ are obtained from the formula for $u(x)$; $x = a$ corresponds to $u = u(a)$; $x = b$ corresponds to $u = u(b)$.

**EXAMPLE 3**

Express $\int_0^2 x(1+2x^2)^{3/2}\,dx$ as a definite integral in terms of the variable $u = 1 + 2x^2$ and evaluate.

**SOLUTION**

We have $u = 1 + 2x^2$, so $du = 4x\,dx$.

$u(0) = 1 + 2(0)^2 = 1$ and $u(2) = 1 + 2(2)^2 = 9$.

Substitution and (10) give

$$\int_0^2 x(1+2x^2)^{3/2}\,dx = \int_0^2 (1+2x^2)^{3/2}\left(\frac{1}{4}\right)(4x)\,dx$$

$$= \int_1^9 u^{3/2}\left(\frac{1}{4}\right)du$$

$$= \left(\frac{1}{4}\right)\frac{u^{5/2}}{5/2}\Big]_1^9$$

$$= \frac{1}{10}(9)^{5/2} - \frac{1}{10}(1)^{5/2} = \frac{1}{10}(243-1) = \frac{121}{5}.$$

We changed the limits of integration to correspond to the variable $u$, and then evaluated the integral in terms of $u$.

## EXERCISES 5.6

Determine which of the integrals in Exercises 1–24 reduce to a constant multiple of one of types in formulas (3)–(9). Evaluate those that do.

**1** $\int (1-x)^{10}\,dx$

**2** $\int (3-2x)^{12}\,dx$

**3** $\int x(x^2+1)^{1/3}\,dx$

**4** $\int \frac{x}{(x^2+1)^{1/3}}\,dx$

**5** $\int \frac{1}{1+\sqrt{x}}\frac{1}{\sqrt{x}}\,dx$

**6** $\int \frac{1}{1+\sqrt{x}}\,dx$

**7** $\int \frac{\sqrt{1+\sqrt{x}}}{\sqrt{x}}\,dx$

**8** $\int x^2(1+x^3)^{1/3}\,dx$

**9** $\int \frac{x+1}{(x^2+2x+3)^2}\,dx$

**10** $\int \frac{x+2}{(x^2+2x+4)^2}\,dx$

**11** $\int \frac{\cos\sqrt{x}}{\sqrt{x}}\,dx$

**12** $\int \frac{\sin\sqrt{x}}{\sqrt{x}}\,dx$

**13** $\int \sin\sqrt{x}\,dx$

**14** $\int \cos x^2\,dx$

**15** $\int x\sec^2(x^2)\,dx$

**16** $\int \sec(x^2)\,dx$

**17** $\int \sec 2x\cdot\tan 2x\,dx$

**18** $\int \sec^2(2x+1)\,dx$

**19** $\int \sqrt{1+\sin x}\cos x\,dx$

**20** $\int \frac{\cos x}{\sqrt{1+\sin x}}\,dx$

**21** $\int 2\sin x\cdot\cos x\,dx$

**22** $\int \cos^2 x\,dx$

**23** $\int \tan^6 x\sec^2 x\,dx$

**24** $\int \tan^4 x\sec^2 x\,dx$

Evaluate the definite integrals in Exercises 25–36.

**25** $\int_0^2 \sqrt{1+4x}\,dx$

**26** $\int_0^1 \frac{1}{\sqrt{1+4x}}\,dx$

**27** $\int_{\pi^3/8}^{\pi^3} \frac{\sin x^{1/3}}{x^{2/3}}\,dx$

**28** $\int_0^{\sqrt{\pi}} x\sin x^2\,dx$

**29** $\int_3^4 \frac{x}{\sqrt{25-x^2}}\,dx$

**30** $\int_3^4 x\sqrt{25-x^2}\,dx$

**31** $\int_1^8 \frac{(1+x^{1/3})^3}{x^{2/3}}\,dx$

**32** $\int_1^8 \frac{(1+x^{2/3})^3}{x^{1/3}}\,dx$

**33** $\int_{-\pi/4}^{\pi/4} \sin 2x\cos 2x\,dx$

**34** $\int_{-\pi/6}^{\pi/6} \sin 3x\cos 3x\,dx$

**35** $\int_0^{\pi/3} \sqrt{4\sec x+1}\sec x\tan x\,dx$

**36** $\int_0^{\pi/4} \sqrt{\tan x}\sec^2 x\,dx$

**37** Express

$$\int_0^{\pi/2} \frac{\cos t}{1+\sin t}\,dt$$

as a definite integral in terms of the variable $u = 1+\sin t$.

**38** Express

$$\int_{-1}^{1} \frac{1}{1+9x^2}\,dx$$

as a definite integral in terms of the variable $u = 3x$.

**39** Show that

$$\int_1^{1/x} \frac{1}{t}\,dt = -\int_1^x \frac{1}{t}\,dt, \qquad x > 0.$$

**40** Show that

$$\int_1^{x^2} \frac{1}{t}\,dt = 2\int_1^x \frac{1}{t}\,dt, \qquad x > 0.$$

## REVIEW EXERCISES

**1** Sketch, label, and find the length and midpoint of a variable vertical line segment between $y = 4 - x^2$ and $x + y = 2$, between points of intersection of the parabola and the line.

**2** Sketch, label, and find the length and midpoint of a variable horizontal line segment between $x = y^2 - 3$ and $x = 5 - y^2$, between points of intersection of the parabolas.

Find the area of a variable cross section of the solids pictured in Exercises 3–4.

3 Cross sections perpendicular to the $x$-axis are semicircles.

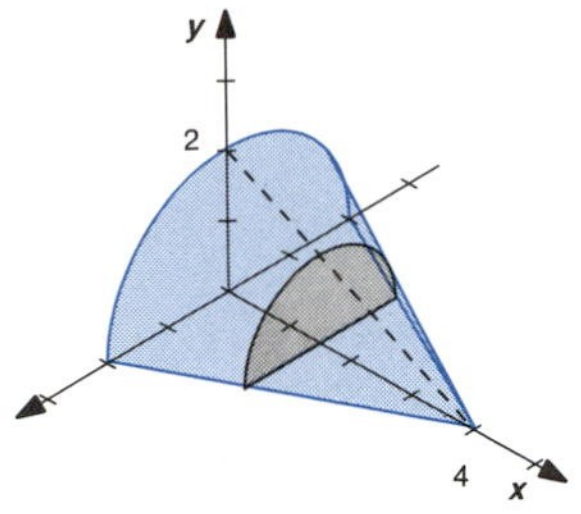


4 Cross sections perpendicular to the $x$-axis are circles.

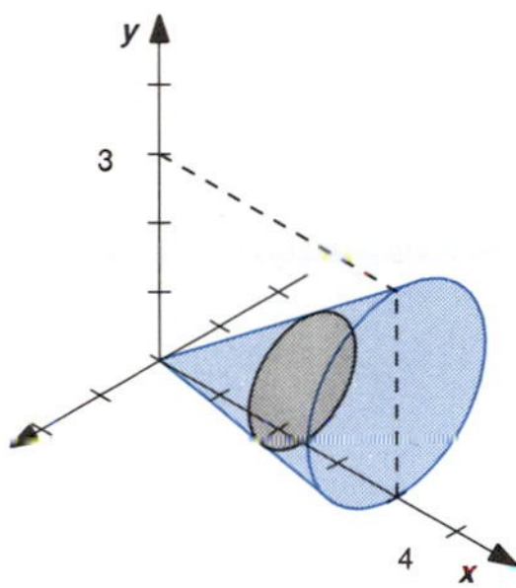


In Exercises 5–6, sketch and label either the disk, washer, or cylinder, whichever one is generated by revolving the given line segment about the indicated line. Find the area of each disk and washer obtained and find the lateral surface area of each cylinder obtained.

5 The variable vertical line segment between $y = \sin x$ and the $x$-axis, $0 \le x \le \pi$: (a) about the $x$-axis, (b) about the $y$-axis, (c) about the line $y = -1$

6 The variable horizontal line segment between $x - y = 1$ and the $y$-axis, $0 \le y \le 1$: (a) about the $x$-axis, (b) about the $y$-axis, (c) about the line $x = 2$

7 Use $R_4$ to find an approximate value of the area of the region between $y = x^2$ and the $x$-axis, $0 \le x \le 2$.

8 Use $R_4$ to find an approximate value of the area of the region between $y = x^2$ and the $x$-axis, $-2 \le x \le 2$.

In Exercises 9–10, use $R_4$ to find an approximate value of the volume of the solids pictured.

9 Cross sections perpendicular to the $x$-axis are squares.

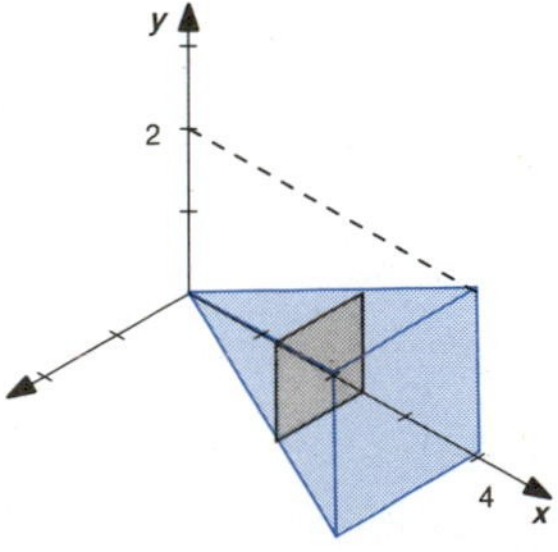


10 Cross sections perpendicular to the $x$-axis are squares.

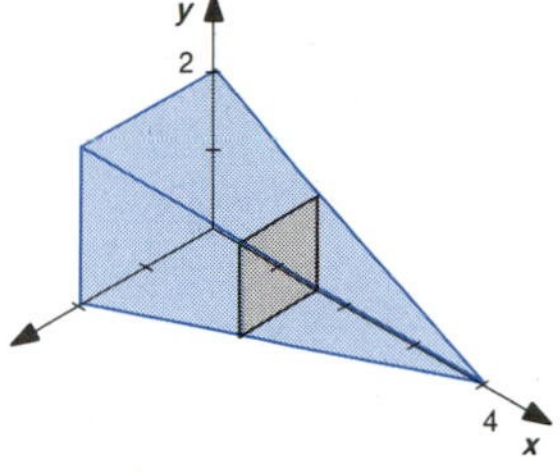


Evaluate the sums in Exercises 11–14.

11 $\sum_{j=1}^{3} (2j - 1)$

12 $\sum_{j=3}^{5} (3j^2 - j)$

13 $\sum_{j=1}^{16} (j^3 + 2j)$

14 $\sum_{j=1}^{100} (2j^2 - 3j + 4)$

In Exercises 15–16, find a simple formula for $R_n$ that corresponds to the given integral and evaluate $\lim_{n \to \infty} R_n$.

15 $\int_0^2 (3x^2 - 4x)\, dx$

16 $\int_0^3 (12x^3 - 5)\, dx$

In Exercises 17–18, the velocity of an object on a number line is given by the formula $v(t)$, $0 \le t \le 1$. Use Riemann sums of $v(t)$ over the interval $[0, 1]$ with six subintervals of equal length to obtain upper and lower bounds for $s(1) - s(0)$, the change in position of the object. [Use the maximum value of $v(t)$ in each subinterval to determine the upper bound; use the minimum value in each subinterval to determine the lower bound.]

17 $v(t) = 2 \sin \pi t$

18 $v(t) = 2 \cos \pi t$

Evaluate the integrals in Exercises 19–20, given that $\int_0^3 f(x)\, dx = 7$, $\int_0^9 f(x)\, dx = 4$, and $\int_0^3 g(x)\, dx = 5$.

19 $\int_0^3 [4g(x) - f(x)]\, dx + \int_3^9 2f(x)\, dx$

20 $\int_0^3 [3f(x) - 2g(x)]\, dx - \int_3^9 f(x)\, dx$

21 Show that

$$\left| \int_0^{2\pi} x \sin x\, dx \right| \le 2\pi^2.$$

22 Show that

$$\left| \int_0^{\pi} x^2 \sin x\, dx \right| \le \frac{\pi^3}{3}.$$

In Exercises 23–24, (a) use $T_4$ to find an approximate value of the given integral. (b) Use $S_4$ to find an approximate value of the given integral. (c) Find a bound of the error in (a). (d) Find a bound of the error in (b). (e) Find values of $n$ that guarantee $T_n$ is within 0.005 of the value of the integral.

(f) Find values of $n$ that guarantee $S_n$ is within 0.005 of the value of the integral.

23 $\displaystyle\int_{-\pi/2}^{\pi/2} \cos x\,dx$ 24 $\displaystyle\int_0^2 x^4\,dx$

Evaluate the indefinite integrals in Exercises 25–30.

25 $\displaystyle\int \sec \pi x \tan \pi x\,dx$ 26 $\displaystyle\int \sqrt{2x+1}\,dx$

27 $\displaystyle\int x\sqrt{x^2+4}\,dx$ 28 $\displaystyle\int \sin^2 x \cos x\,dx$

29 $\displaystyle\int \tan^2 x \sec^2 x\,dx$ 30 $\displaystyle\int \frac{(1+\sqrt{x})^3}{\sqrt{x}}\,dx$

Evaluate the definite integrals in Exercises 31–34.

31 $\displaystyle\int_1^2 (6x^2-4x+1)\,dx$ 32 $\displaystyle\int_{-\pi/4}^{\pi/4} \sec^2 \frac{x}{2}\,dx$

33 $\displaystyle\int_0^4 \frac{x}{(x^2+9)^2}\,dx$ 34 $\displaystyle\int_0^{\pi/2} \sin x \cos x\,dx$

35 Express

$$\int_0^2 \frac{x}{1+4x^2}\,dx$$

as a definite integral in terms of the variable $u = 1 + 4x^2$.

36 Express

$$\int_0^{\pi/4} \frac{\sin x}{1+4\cos^2 x}\,dx$$

as a definite integral in terms of the variable $u = 2\cos x$.

Evaluate the derivatives in Exercises 37–40.

37 $\displaystyle\frac{d}{dx}\left(\int_0^x \sin t^2\,dt\right)$ 38 $\displaystyle\frac{d}{dx}\left(\int_x^1 \frac{1}{t}\,dt\right), \quad x > 0$

39 $\displaystyle\frac{d}{dx}\left(\int_1^{1/x} \frac{1}{t}\,dt\right)$ 40 $\displaystyle\frac{d}{dx}\left(\int_x^{x^3} \frac{1}{t}\,dt\right), \quad x > 0$

CHAPTER SIX

# Applications of the definite integral

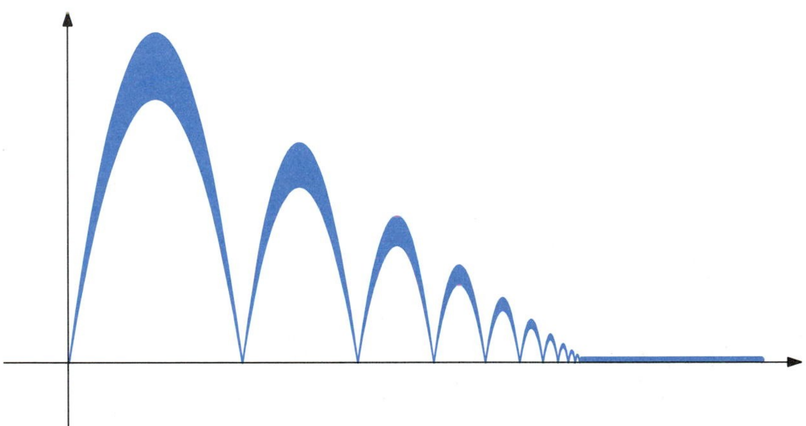

Applications of the definite integral to a variety of physical problems are illustrated. The underlying principle involved is the interpretation of the definite integral as a sum of products, with each component of the integral representing a physical quantity.

## 6.1 AREA OF REGIONS IN THE PLANE

In Chapter 5 we saw how the definite integral is related to the area of certain types of regions in the plane. In this section we will see how definite integrals can be used to find the area of more general regions.

To indicate the idea, let us consider the region bounded by the curves

$$y = f(x), \quad y = g(x), \quad x = a, \quad \text{and} \quad x = b.$$

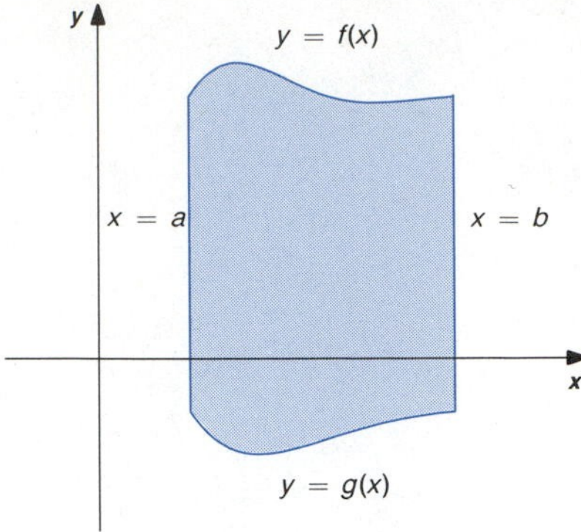


(a)

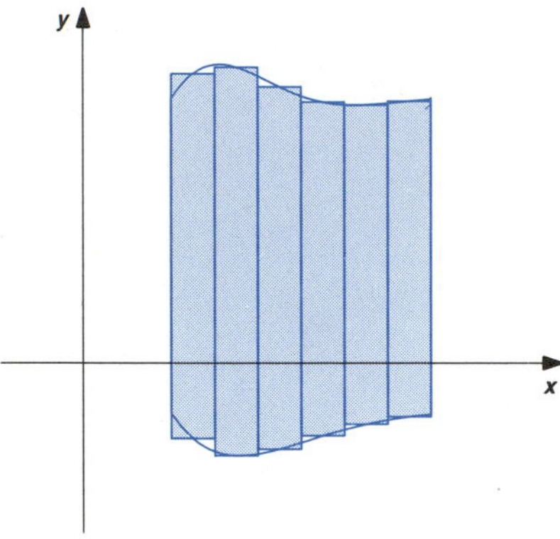


(b)

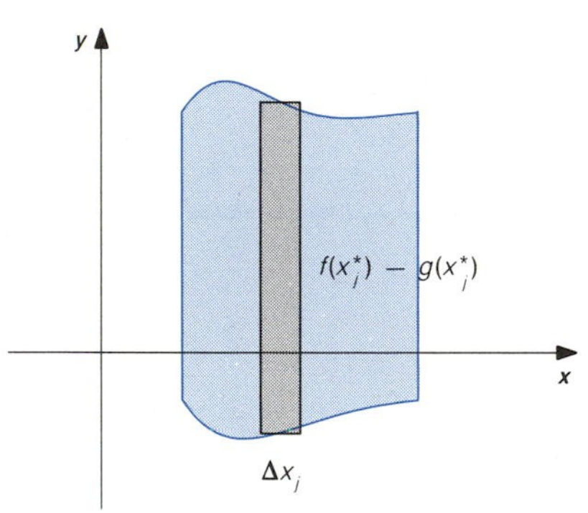


(c)

FIGURE 6.1.1

The region is sketched in Figure 6.1.1a. We assume that $f$ and $g$ are continuous and that $g(x) \leq f(x)$ for $a \leq x \leq b$. It is clear intuitively that the "area" of the region is related to a sum of the areas of rectangles, as pictured in Figure 6.1.1b. (This idea was discussed as an application of Riemann sums in Section 5.2.) The typical rectangle sketched in Figure 6.1.1c has height $f(x_j^*) - g(x_j^*)$ and base $\Delta x_j$, so its area is given by the product $[f(x_j^*) - g(x_j^*)]\,\Delta x_j$. The "area" of the entire region is related to a sum of the areas of such typical rectangles. That is,

Area of region $\approx$ Sum of areas of rectangles,

$$\text{“Area”} \approx \sum_{j=1}^{n} [f(x_j^*) - g(x_j^*)]\,\Delta x_j. \tag{1}$$

We now note that the difference between the area of the rectangle and the "area" bounded by $y = f(x)$, $y = g(x)$, $x = x_{j-1}$, and $x = x_j$ is at most a small proportion of the area of the rectangle if $\Delta x_j$ is small enough. It follows that the sum in (1) is very near the "area" of the region when all $\Delta x_j$'s are small enough. On the other hand, since the sum is a Riemann sum of $f - g$ over the interval $[a, b]$, we know such sums are near the number $\int_a^b [f(x) - g(x)]\,dx$. It should then seem reasonable to *define* the area of the region by the following formula.

$$\textbf{Area} = \int_a^b [f(x) - g(x)]\,dx. \tag{2}$$

This definition of the area of the region is consistent with our intuitive notion of area and gives familiar formulas when applied to simple geometric shapes. When $f(x) \geq 0$ and $g(x) = 0$ for $a \leq x \leq b$, the integral in formula (2) becomes $\int_a^b f(x)\,dx$, which represents the area under the curve $y = f(x)$.

We should not think of (2) as a formula to be memorized. The formula merely reflects the fact that the area is given as a limit of sums of the area of rectangles. We should think of the components of the integral as representing specific physical quantities. That is,

$$\text{Area} = \underbrace{\int_a^b}_{\text{Sum}} \underbrace{\underbrace{[f(x) - g(x)]}_{\text{Length}}\ \underbrace{dx.}_{\text{Width}}}_{\text{Area of rectangle}}$$

To find the area of a region of the type described above.

- **Sketch the region.**
- **Draw a variable vertical rectangular area strip.**
- **Label coordinates of endpoints of the strip in terms of $x$.**
- **Write "area" of strip as (Length)(Width).**
- **Write integral that gives "sum" of areas of strips.**
- **Evaluate the definite integral.**

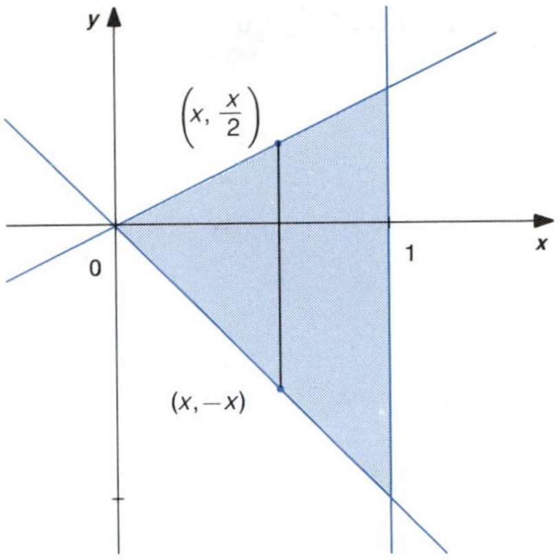

FIGURE 6.1.2

■ **EXAMPLE 1**

Set up and evaluate a definite integral that gives the area of the region bounded by

$$y = \frac{x}{2}, \quad y = -x, \quad x = 0, \quad \text{and} \quad x = 1.$$

**SOLUTION**

The region is sketched and a variable vertical area strip is drawn and labeled in Figure 6.1.2. The variable rectangular strip is drawn as a vertical line segment. The coordinates of the top and bottom of the strip have been labeled in terms of the variable $x$. The strip is typical because the same labeling applies as $x$ varies between $x = 0$ and $x = 1$. We now think of our typical strip as a rectangle with

$$\text{Width} = dx,$$

and

$$\text{Length} = (y\text{-coordinate of top}) - (y\text{-coordinate of bottom})$$
$$= \left(\frac{x}{2}\right) - (-x).$$

(It is clear from Figure 6.1.2 that $x/2 > -x$ for $0 \leq x \leq 1$.)

The area of the region is a limit of the sum of the areas of the rectangles. This is given by the definite integral

$$\text{Area} = \underbrace{\int_0^1}_{\text{Sum}} \underbrace{\underbrace{\left[\left(\frac{x}{2}\right) - (-x)\right]}_{\text{Length}} \underbrace{dx}_{\text{Width}}}_{\text{Area of rectangle}}.$$

Evaluating the integral, we have

$$\text{Area} = \int_0^1 \frac{3}{2} x \, dx = \left(\frac{3}{2}\right)\left(\frac{x^2}{2}\right)\Big]_0^1 = \left(\frac{3}{2}\right)\left(\frac{1^2}{2}\right) - \left(\frac{3}{2}\right)\left(\frac{0^2}{2}\right) = \frac{3}{4}.$$

In this example the region is a triangle with (vertical) base 3/2 and (horizontal) height 1. The usual formula for the area of a triangle gives

$$\text{Area} = \frac{1}{2}(\text{Base})(\text{Height}) = \left(\frac{1}{2}\right)\left(\frac{3}{2}\right)(1) = \frac{3}{4}.$$

This value agrees with that obtained by using the definite integral. ■

Of course, we are interested in using the integral method to find the area of regions where formulas for the area are not available. We will concentrate on setting up area integrals. Evaluation is a separate step. We will not carry out the details of evaluating the integrals for the remaining examples in this section.

### ■ EXAMPLE 2

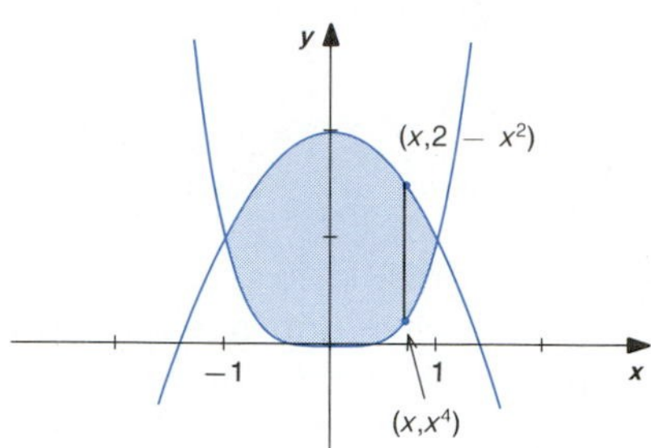


FIGURE 6.1.3

Set up a definite integral that gives the area of the region bounded by $y = x^4$ and $y = 2 - x^2$.

**SOLUTION**

We have sketched the region, drawn a typical rectangular strip, and labeled coordinates in Figure 6.1.3.

The variable $x$ varies between the $x$-coordinates of the points of intersection of the curves $y = x^4$ and $y = 2 - x^2$. These coordinates are found by solving the two equations simultaneously for $x$. That is,

$$y = x^4 \quad \text{and} \quad y = 2 - x^2$$

imply

$$x^4 = 2 - x^2,$$

$$x^4 + x^2 - 2 = 0,$$

$$(x^2 + 2)(x^2 - 1) = 0,$$

$$(x^2 + 2)(x - 1)(x + 1) = 0,$$

so the real number solutions are $x = -1$ and $x = 1$.

We can now write the definite integral that gives the limit of the sum of the areas of the rectangles. We have

$$\text{Area} = \underbrace{\int_{-1}^{1}}_{\text{Sum}} \underbrace{\underbrace{[(2 - x^2) - (x^4)]}_{\text{Length}} \underbrace{dx}_{\text{Width}}}_{\text{Area of rectangle}}.$$

(It is not difficult to verify that the above integral has value 44/15.) ■

### ■ EXAMPLE 3

Set up a definite integral that gives the area of the region bounded by the circle $x^2 + y^2 = r^2$.

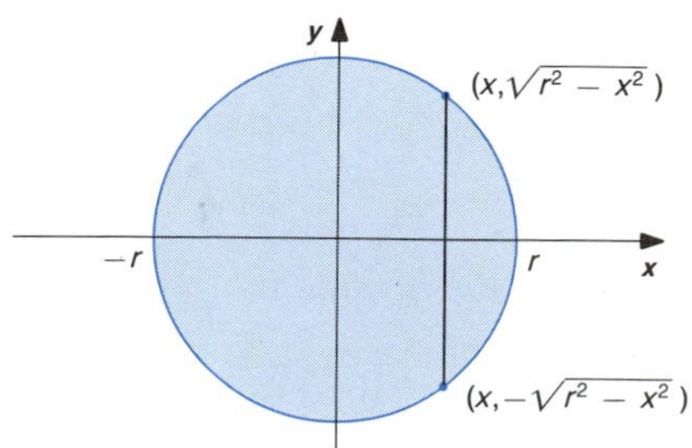


FIGURE 6.1.4

**SOLUTION**

The equation $x^2 + y^2 = r^2$ implies $y = \sqrt{r^2 - x^2}$ or $y = -\sqrt{r^2 - x^2}$. The graph of $y = \sqrt{r^2 - x^2}$ is the upper half of the circle, while $y = -\sqrt{r^2 - x^2}$ gives the lower half. The circle is sketched and a variable vertical rectangular area strip has been drawn and labeled in Figure 6.1.4.

The length of the typical rectangular strip is

$$\sqrt{r^2 - x^2} - (-\sqrt{r^2 - x^2}) = 2\sqrt{r^2 - x^2}.$$

The variable $x$ clearly varies between $x = -r$ and $x = r$. The area of the circle is given by the definite integral

$$\text{Area} = \int_{-r}^{r} 2\sqrt{r^2 - x^2}\, dx.$$

(Later we will learn a technique for evaluating the above integral. Of course, we will discover that its value is $\pi r^2$, the area of a circle of radius $r$.) ■

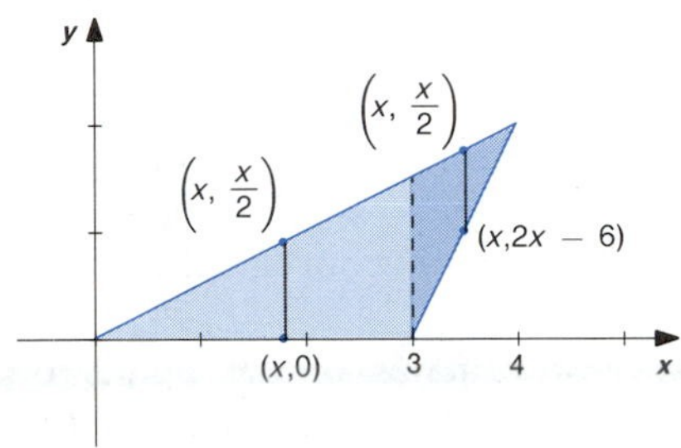

(a)

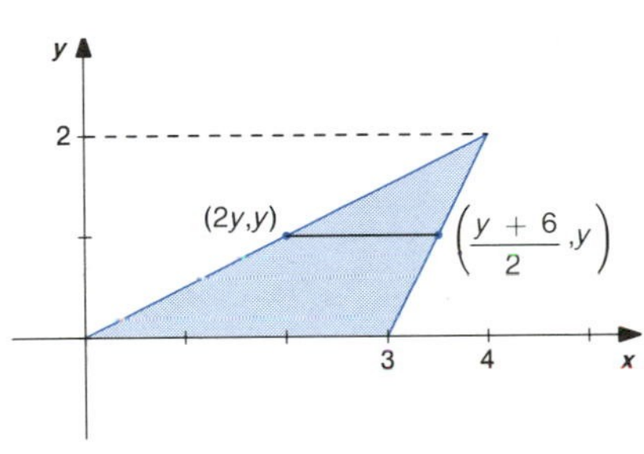

(b)

**FIGURE 6.1.5**

■ **EXAMPLE 4**

Set up an integral or integrals that give the area of the triangular region bounded by $x = 2y$, $y + 6 = 2x$, and $y = 0$.

**SOLUTION**

The region is sketched in Figure 6.1.5a. We see that there is no single variable vertical area strip that is typical for the entire region. Some vertical strips run from $y = 0$ to $y = x/2$ and other vertical strips run from $y = 2x - 6$ to $y = x/2$. This means that two formulas are required to describe the length of a variable vertical area strip, so we must express the area of the region as a sum of two integrals with respect to the variable $x$. The limits of integration are determined by finding the $x$-coordinates of the points of intersection of the boundary lines. It is easy to see that

$x = 2y$ and $y = 0$ imply $x = 0$;

$y + 6 = 2x$ and $y = 0$ imply $x = 3$; and

$x = 2y$ and $y + 6 = 2x$ imply $x = 4$.

We then have

$$\text{Area} = \int_0^3 \left(\frac{x}{2} - 0\right) dx + \int_3^4 \left(\frac{x}{2} - (2x - 6)\right) dx.$$

(Each of the above integrals represents the area of a trianglar portion of the region.)

An alternate solution for this problem is suggested in Figure 6.1.5b. We see that there is a single variable horizontal area strip that is typical for the entire region. In the case of horizontal strips, the variable is $y$ and each component of the corresponding definite integral must be expressed in terms of $y$. We see that $y$ varies between $y = 0$ and the $y$-coordinate of the point of intersection of the lines $x = 2y$ and $x = (y + 6)/2$, which is $y = 2$. The length of a horizontal line segment is

Length = ($x$-coordinate of right endpoint) − ($x$-coordinate of left endpoint).

The area is

$$\text{Area} = \int_0^2 \left(\frac{y + 6}{2} - 2y\right) dy.$$

It is easy to check that both of the above methods give the value of 3 for the area of the triangle. This agrees with the value obtained from the formula

$$\text{Area of triangle} = \frac{1}{2}(\text{Base})(\text{Height}) = \frac{1}{2}(3)(2) = 3. \quad ■$$

**■ EXAMPLE 5**

Set up an integral or integrals that give the area of the region bounded by $x - y = 2$ and $x = y^2 - 1$.

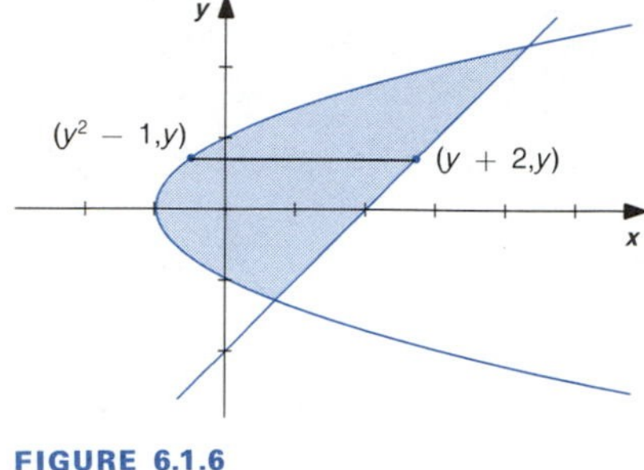

FIGURE 6.1.6

**SOLUTION**

The region is sketched in Figure 6.1.6. There is a single variable horizontal strip that is typical for the entire region. Two integrals would be required if $x$ were chosen as the variable. We should use the more convenient variable strips.

The variable $y$ varies between the $y$-coordinates of the points of intersection of $x = y + 2$ and $x = y^2 - 1$. These values are given by the equation

$$y^2 - 1 = y + 2 \quad \text{or} \quad y^2 - y - 3 = 0.$$

The Quadratic Formula gives the solutions

$$y = \frac{-(-1) \pm \sqrt{(-1)^2 - 4(1)(-3)}}{2(1)}.$$

That is,

$$y = \frac{1 + \sqrt{13}}{2} \quad \text{and} \quad y = \frac{1 - \sqrt{13}}{2}.$$

We have

$$\text{Area} = \int_{(1-\sqrt{13})/2}^{(1+\sqrt{13})/2} [(y + 2) - (y^2 - 1)]\, dy.$$

(The value of this integral is $13\sqrt{13}/6 \approx 7.8120278$.) ■

FIGURE 6.1.7

**■ EXAMPLE 6**

Set up an integral or integrals that give the area of the region bounded by $y = 4x$ and $y = x(x + 1)(x - 2)$.

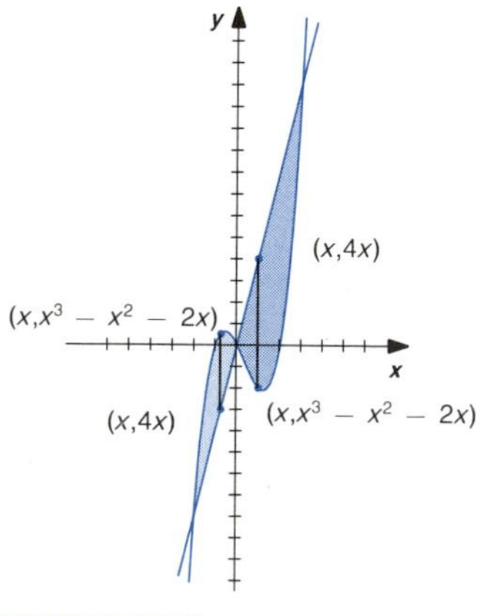

**SOLUTION**

The region is sketched in Figure 6.1.7. We see that the region is composed of two parts, one where the cubic curve is above the line and another where the line is above the cubic curve. We need the $x$-coordinates of the points of intersection of $y = 4x$ and $y = x(x+1)(x-2)$. These are given by

$$x(x+1)(x-2) = 4x,$$

$$x^3 - x^2 - 2x = 4x,$$

$$x^3 - x^2 - 6x = 0,$$

$$x(x^2 - x - 6) = 0,$$

$$x(x-3)(x+2) = 0.$$

The solutions are $x = 0, 3, -2$.

The area is the sum of the areas of the two parts.

$$\text{Area} = \int_{-2}^{0} [(x^3 - x^2 - 2x) - 4x]\, dx + \int_{0}^{3} [4x - (x^3 - x^2 - 2x)]\, dx.$$

(The sum of the two integrals is 253/12.) ■

In Example 6 we could have writtten

$$\text{Area} = \int_{-2}^{3} |(x^3 - x^2 - 2x) - 4x|\, dx,$$

but we could not use our formulas for evaluating definite integrals without determining intervals where the expression inside the absolute value is of constant sign. We have used the sketch in Figure 6.1.7 to help determine the sign of $(x^3 - x^2 - 2x) - 4x$.

## EXERCISES 6.1

Sketch and label the regions bounded by the given lines in Exercises 1–10. Set up and evaluate integrals that give the areas of the regions. Verify the results by using formulas from geometry to find the areas of the regions.

1 $y = 2x,\ y = 0,\ x = 3$

2 $x + y = 4,\ y = 0,\ x = 0$

3 $3x + 2y = 6,\ x - y = 2,\ x = 0$

4 $y = x,\ y = -2x,\ x = 2$

5 $y = 0,\ x - y + 2 = 0,\ 2y + x = 4$

6 $y = 0,\ x + y + 2 = 0,\ 2y - x + 4 = 0$

7 $y = x,\ y = x + 3,\ x = 0,\ x = 2$

8 $y = x + 2,\ y = 0,\ x = 0,\ x = 3$

9 $y = x,\ y = 2,\ y = 0,\ x + y = 5$

10 $y = x,\ y = 2,\ y = 0,\ y = x - 5$

In Exercises 11–24 sketch and label the regions bounded by the given curves. Set up and evaluate integrals that give the areas of the regions.

11 $y = x^2,\ y = 0,\ x = 2$

12 $y = 4 - x^2,\ y = 0$

13 $y = x^2,\ y = x$

14 $y = \sec^2 x,\ y = 0,\ x = 0,\ x = \pi/4$

15 $y = \sin x,\ y = 0,\ x = -\pi/2,\ x = 0$

16 $y = \cos x,\ y = 0,\ x = -\pi/2,\ x = \pi/2$

17 $x = y^2,\ x + y = 2$

18 $x = y^2,\ x - y = 2$

19 $y = \sin x,\ y = 0,\ x = -\pi,\ x = \pi$

20 $y = x^3,\ y = x$

21 $y = \sin x,\ y = 1/2,\ x = 0\ (x \geq 0)$

22 $y = \cos x$, $y = 1/2$, $x = 0$ $(x \geq 0)$

23 $y = x^3/3$, $y = x^2$

24 $y = 32x^{1/3}$, $y = x^2$

In Exercises 25–30 sketch and label the regions bounded by the given curves. Set up integrals that give the areas of the regions.

25 Inside $x^2 + y^2 = 4$ and above $x + y = 2$

26 Inside $x^2 + y^2 = 4$ and above $y = 1$

27 Inside $x^2 + y^2 = 4$ and above $y = -1$

28 Inside $x^2 + y^2 = 4$ and above $2x + y = 2$

29 $y = \sin x$, $y = x$, $x = \pi$

30 $y = \tan x$, $y = x$, $x = \pi/4$

31 Sketch the region bounded by

$$y = 1, \quad y = 0, \quad x = 0, \quad x = 2\pi$$

and the region bounded by

$$y = 1 + \sin^2 x, \quad y = \sin^2 x, \quad x = 0, \quad x = 2\pi.$$

Show that the two regions have equal areas.

32 Sketch the region bounded by

$$x = \sec y, \quad x = 0, \quad y = -\frac{\pi}{4}, \quad y = \frac{\pi}{4}$$

and the region bounded by

$$x = \frac{1}{2}\sec y, \quad x = -\frac{1}{2}\sec y, \quad y = -\frac{\pi}{4}, \quad y = \frac{\pi}{4}.$$

Show that the two regions have equal areas.

33 Show that the length of a variable line segment between two sides of a triangle and parallel to the base depends only on the length of the base and height of the triangle, and on the distance of the segment from the base. Use an area integral to show that triangles that have equal bases and heights have the same area, even though they may not be the same shape.

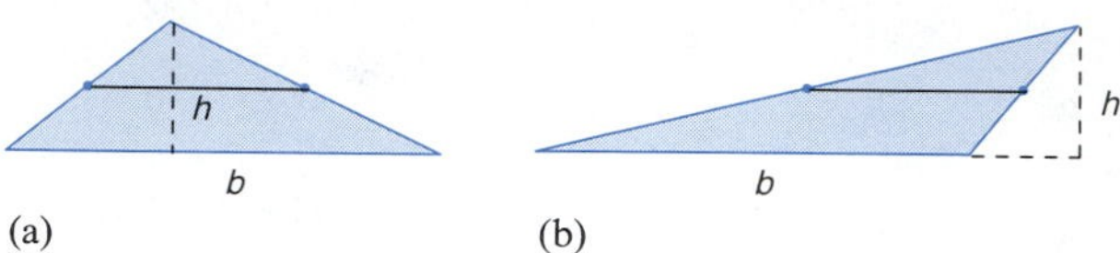


(a) (b)

34 Show that the length of a variable line segment between sides of a parallelogram and parallel to the base depends only on the length of the base. Use an area integral to show that parallelograms that have equal bases and heights have the same area, even though they may not be the same shape.

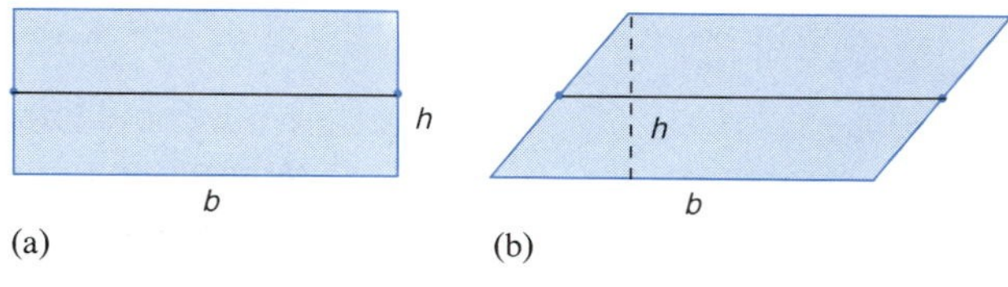


(a) (b)

## 6.2 VOLUMES BY SLICING

The main objective of this section is to strengthen understanding of how the idea of the definite integral as a sum is used in physical problems. The particular problem that we consider involves the volume of three-dimensional solids.

The idea of obtaining the volume of a solid by slicing and adding the volumes of the slices was introduced in Section 5.2 as an application of Riemann sums. Let us review the idea.

Consider a solid object such as is pictured in Figure 6.2.1a. We assume that the area of cross sections perpendicular to the $x$-axis is given by a continuous function $A(x)$, $a \leq x \leq b$. Let us now slice the solid as indicated in Figure 6.2.1b. If the slices are thin enough, the volume of that part of the solid between $x = x_{j-1}$ and $x = x_j$ is approximately $A(x_j^*)\,\Delta x_j$, where $x_j^*$ is any point between $x_j$ and $x_{j-1}$. An approximate value of the volume of the solid is then given by the sum of the approximate volumes of the parts,

Volume of solid $\approx$ Sum of approximate volumes of slices,

$$\text{Volume} \approx \sum_{j=1}^{n} A(x_j^*)\,\Delta x_j.$$

If all slices are thin enough, the above Riemann sum is a good approximation of both the volume of the solid and the value of the definite integral $\int_a^b A(x)\,dx$. Hence, it should seem reasonable that

$$\textbf{Volume} = \int_a^b A(x)\,dx.$$

Each component of this integral should have a definite physical meaning.

$$\text{Volume} = \underbrace{\int_a^b}_{\text{Sum}} \underbrace{\underbrace{A(x)}_{\text{(Area of cross section)}} \quad \underbrace{dx.}_{\text{(Thickness of slice)}}}_{\text{Volume of slice}}$$

(a)

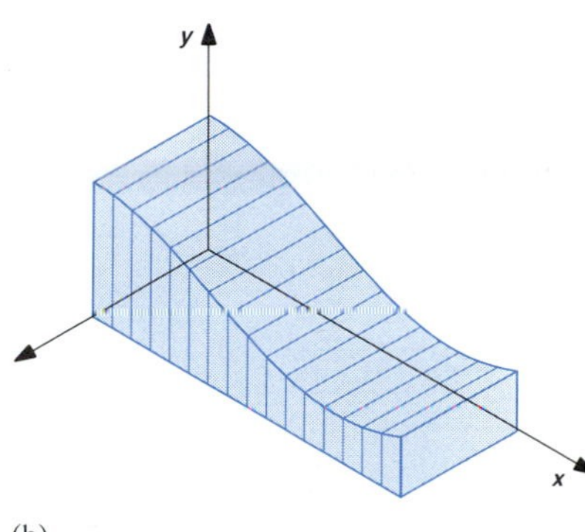


(b)

FIGURE 6.2.1

The limits of integration indicate that the solid consists of slices between $x = a$ and $x = b$.

In order to use the method of slicing to find the volume of the above type of solid,

- **Sketch the solid.**
- **Draw a variable cross section perpendicular to the $x$-axis.**
- **Find the area of the variable cross section in terms of $x$.**
- **Write the "volume" of a slice as (area)(thickness).**
- **Write the integral that gives the "sum" of the slice volumes.**
- **Evaluate the definite integral.**

Let us now use a definite integral to find the exact volume of the prism considered in Example 2 of Section 5.2.

### EXAMPLE 1

Set up and evaluate a definite integral that gives the volume of the prism sketched in Figure 6.2.2. Cross sections perpendicular to the $x$-axis are isosceles right triangles.

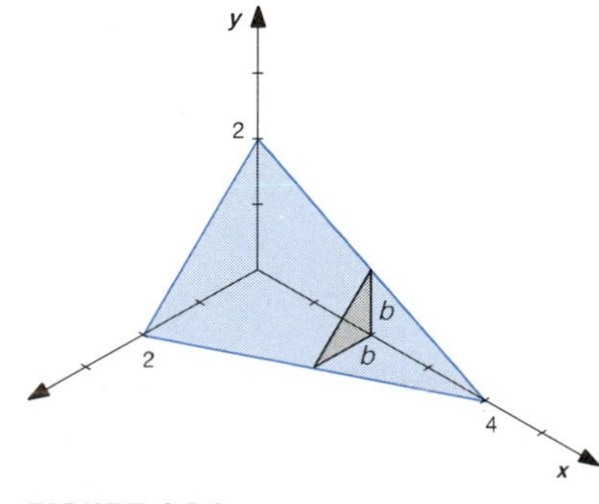


FIGURE 6.2.2

#### SOLUTION

In Figure 6.2.2 we have used $b$ to indicate the equal sides of a variable right triangular cross section. The area of the cross section can then be expressed as

$$A = \frac{1}{2}b^2,$$

where $b$ depends on the position along the $x$-axis. We see that $b$ is the length of the vertical line segment between the $x$-axis and the line through the points (0, 2) and (4, 0) in the $xy$-plane. To express $b$ in terms of the variable $x$, we need the equation of this line. This line has slope $-1/2$. The equation of the line through the point (4, 0) with slope $-1/2$ is

$$y - 0 = -\frac{1}{2}(x - 4).$$

The height (and base) of the triangle at position $x$ are given by the $y$-coordinate of the point on the line. Solving the equation for $y$, we obtain

$$y = \frac{4 - x}{2},$$

so

$$b = \frac{4 - x}{2}.$$

The area of a variable cross section is then

$$A(x) = \frac{1}{2}\left(\frac{4 - x}{2}\right)^2, \qquad 0 \le x \le 4.$$

The volume of the prism is given by the sum of the volumes of the slices between $x = 0$ and $x = 4$. We have

$$\text{Volume} = \underbrace{\int_0^4}_{\text{Sum}} \underbrace{\underbrace{\frac{1}{2}\left(\frac{4 - x}{2}\right)^2}_{\left(\substack{\text{Area of}\\ \text{cross section}}\right)} \underbrace{dx}_{\left(\substack{\text{Thickness}\\ \text{of slice}}\right)}}_{\text{Volume of slice}}.$$

(We are thinking of the integral as a sum of the volumes of a finite number of thin slices. The definite integral gives the limit of such sums.)

Evaluating the definite integral, we obtain

$$\begin{aligned}\text{Volume} &= \int_0^4 \frac{1}{8}(16 - 8x + x^2)\, dx \\ &= \frac{1}{8}\left(16x - 4x^2 + \frac{1}{3}x^3\right)\Big]_0^4 \\ &= \frac{1}{8}\left(16(4) - 4(4)^2 + \frac{1}{3}(4)^3\right) = \frac{8}{3}.\end{aligned}$$

■

The same ideas can be used to find the volume of a solid by slicing perpendicular to the $y$-axis. In this case we use $y$ as the variable of integration, so the area of cross sections must be expressed as a function of $y$.

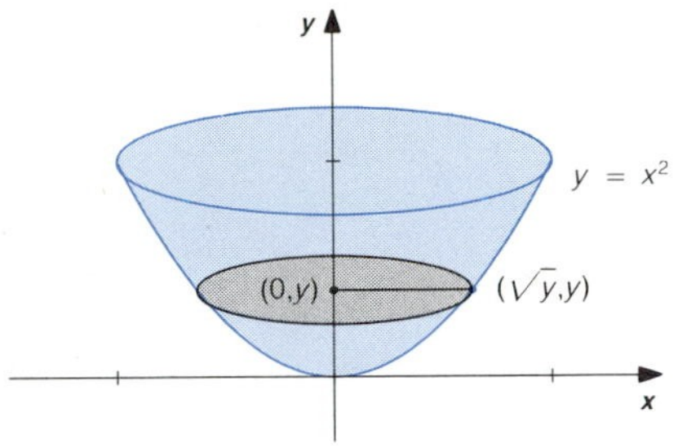

FIGURE 6.2.3

### ■ EXAMPLE 2

Set up an integral that gives the volume of the solid that intersects the $xy$-plane in the region bounded by $y = x^2$ and $y = 1$, if cross sections perpendicular to the $y$-axis are disks with centers on the $y$-axis.

#### SOLUTION

The solid is pictured in Figure 6.2.3. A variable cross section perpendicular to the $y$-axis has been drawn and labeled. The position of a variable slice is given by the variable $y$, so we have labeled the sketch in terms of $y$.

The cross sections perpendicular to the $y$-axis are disks and have area $\pi r^2$, where the radius $r$ depends on the position of the slice. From the sketch we see that the radius $r$, in terms of the variable $y$, is given by $r = \sqrt{y}$. The volume of the solid is

$$\text{Volume} = \underbrace{\int_0^1}_{\text{Sum}} \underbrace{\underbrace{\pi(\sqrt{y})^2}_{\left(\substack{\text{Area of}\\ \text{cross section}}\right)} \; \underbrace{dy.}_{\left(\substack{\text{Thickness}\\ \text{of slice}}\right)}}_{\text{Volumc of slicc}}$$

(The integral has value $\pi/2$.) ■

### ■ EXAMPLE 3

The base of a solid consists of the region bounded by

$$2x + 3y = 6, \quad y = 0, \quad \text{and} \quad x = 0.$$

Set up an integral that gives the volume of the solid if cross sections perpendicular to the $x$-axis are semicircles with diameters on the base.

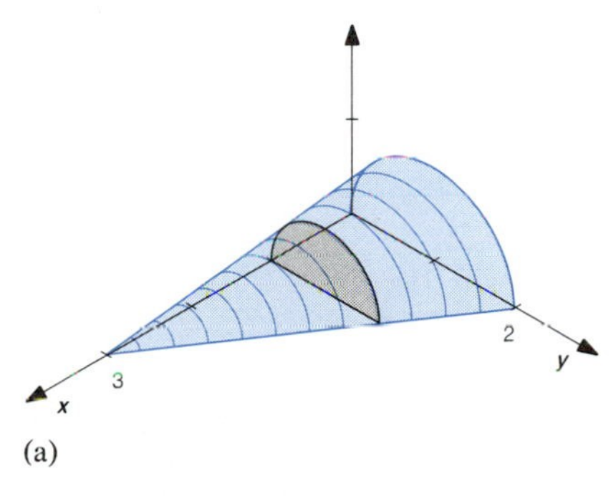

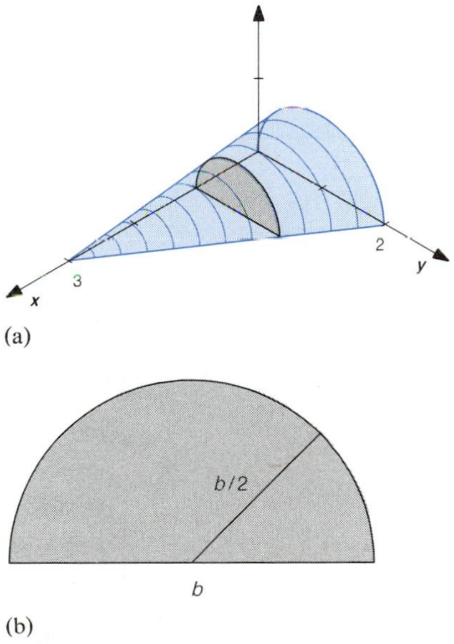

(a)

(b)

FIGURE 6.2.4

#### SOLUTION

The solid is sketched in Figure 6.2.4a.

Let $b$ denote the length of the base of a cross section at the variable position $x$ along the $x$-axis. The cross section is a semicircle with diameter on the base, so $b$ is the length of the diameter of the semicircle. From the sketch of Figure 6.2.4b we see that the area of the cross section is one-half the area of a circle of radius $b/2$, or

$$A = \frac{1}{2}\pi\left(\frac{b}{2}\right)^2 = \frac{\pi}{8}b^2.$$

We need to express $b$ in terms of the variable $x$. From Figure 6.2.4a, we see that $b$ is given by the $y$-coordinate of the point on the line $2x + 3y = 6$. Solving the equation for $y$ in terms of the variable $x$, we have

$$y = \frac{6 - 2x}{3}, \quad \text{so} \quad b = \frac{6 - 2x}{3}.$$

The volume is

$$\text{Volume} = \int_0^3 \frac{\pi}{8}\left(\frac{6-2x}{3}\right)^2 dx.$$

(This integral has value $\pi/2$.) ■

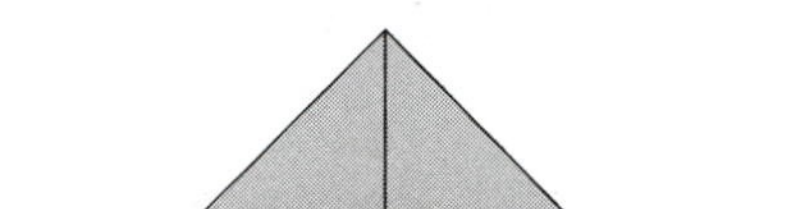

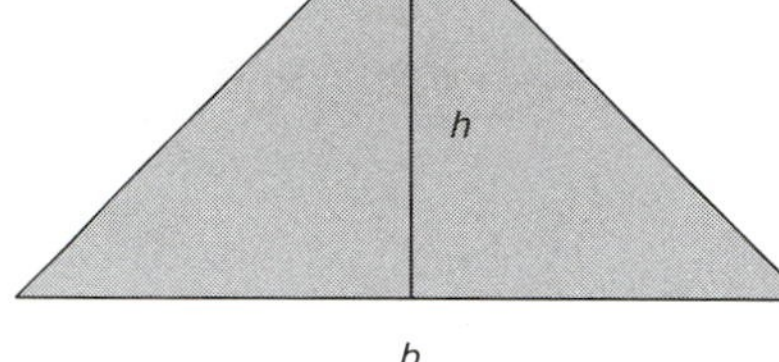


(b)

FIGURE 6.2.5

■ **EXAMPLE 4**

The base of a solid consists of the region inside the circle $x^2 + y^2 = 4$. Set up an integral that gives the volume of the solid if cross sections perpendicular to the $x$-axis are isosceles right triangles with hypotenuses on the base.

**SOLUTION**

The solid is sketched in Figure 6.2.5a.

The area of a variable cross section is given by

$$A = \frac{1}{2}bh,$$

where $b$ is the base (hypotenuse) and $h$ is the height. Since the equal angles in an isosceles right triangle are 45°, we have

$$h = \frac{b}{2}.$$

See Figure 6.2.5b. Then

$$A = \frac{1}{2}bh = \frac{1}{2}b\left(\frac{b}{2}\right) = \frac{b^2}{4}.$$

We need to express $b$ in terms of the variable $x$. The equation $x^2 + y^2 = 4$ implies $y = \sqrt{4 - x^2}$ and $y = -\sqrt{4 - x^2}$, so we see that $b = \sqrt{4 - x^2} - (-\sqrt{4 - x^2}) = 2\sqrt{4 - x^2}$. The volume is

$$\text{Volume} = \int_{-2}^{2} \frac{1}{4}(2\sqrt{4 - x^2})^2\, dx.$$

(This integral has value 32/3.) ■

(a)

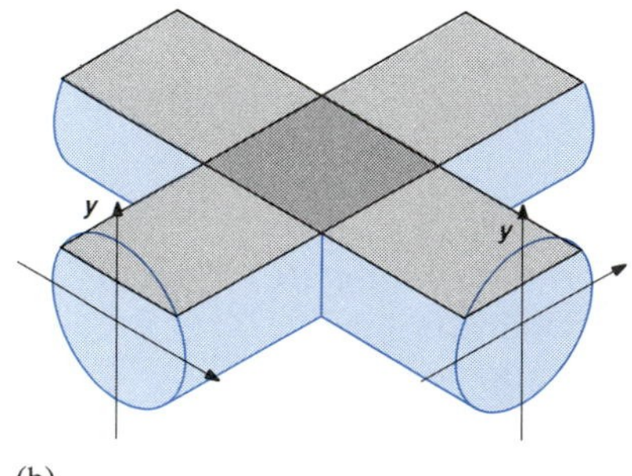


(b)

FIGURE 6.2.6

■ **EXAMPLE 5**

Find the volume of the solid that is bounded by two right circular cylinders of radius $r$, if the axes of the cylinders meet at right angles.

**SOLUTION**

The solid is sketched in Figure 6.2.6a. In Figure 6.2.6b we have sketched two solid right circular cylinders with equal radii that intersect at right angles, sliced along a plane that is parallel to their axes, with the upper portion removed. This shows that the cross section of the intersection of the cylinders in this plane is a square. Using a two-dimensional coordinate system with axes through the center of one of the cylinders and perpendicular to the axis of the cylinder, with $y$ directed upward, we can determine that the length of each side of the square cross section at level $y$ is $2\sqrt{r^2 - y^2}$. The area of the cross section of the intersection of the cylinders at level $y$ is then

$$A(y) = (2\sqrt{r^2 - y^2})^2 = 4(r^2 - y^2).$$

As $y$ varies from $-r$ to $r$, these cross sections generate the entire intersection of the cylinders. We then have

$$\begin{aligned}\text{Volume} &= \int_{-r}^{r} 4(r^2 - y^2)\,dy \\ &= 4r^2y - 4\frac{y^3}{3}\Big]_{-r}^{r} \\ &= \left(4r^2(r) - 4\frac{(r)^3}{3}\right) - \left(4r^2(-r) - 4\frac{(-r)^3}{3}\right) = \frac{16}{3}r^3.\end{aligned}$$

■

It should be clear from the method of slicing that the volumes of two solids are equal if their corresponding cross sections have equal area. This general result is called **Cavalieri's Theorem**.

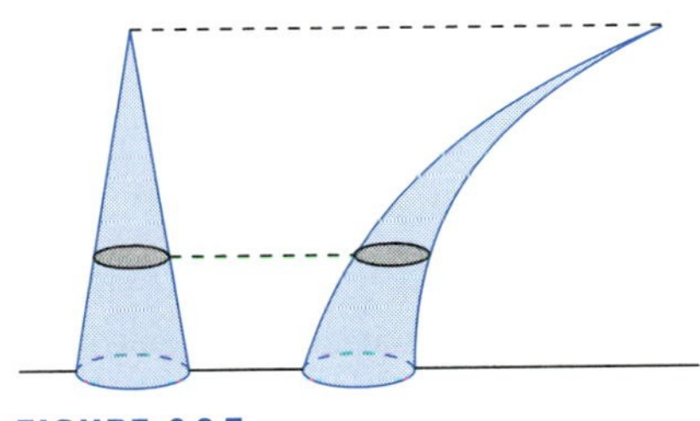

FIGURE 6.2.7

■ EXAMPLE 6

Cavalieri's Theorem implies that the volumes of the solids pictured in Figure 6.2.7 are equal, because corresponding cross sections have equal area. ■

## EXERCISES 6.2

Set up and evaluate integrals that give the volume of the solids pictured in Exercises 1–12.

1 Cross sections are rectangles.

2 Cross sections are right triangles.

3 Cross sections are circles.

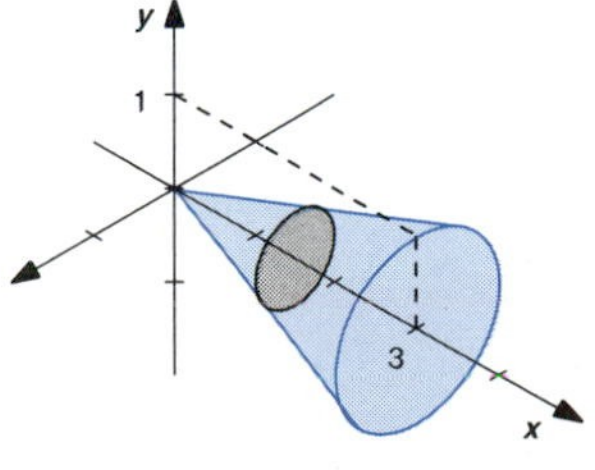


4 Cross sections are squares.

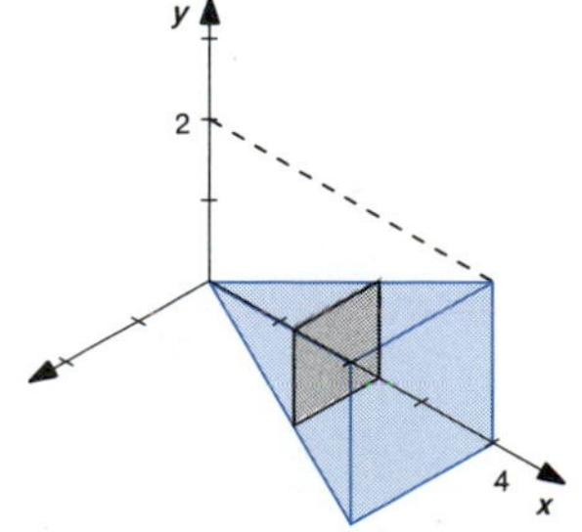

**5** Cross sections are squares.

**6** Cross sections are circles.

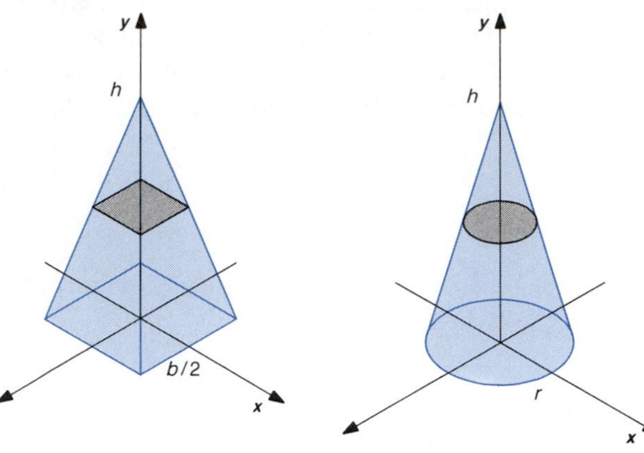


**7** Intersection with the $xy$-plane is the region bounded by

$x^2 - y^2 = 1,\quad y = 1,\quad \text{and}\quad y = -1.$

Cross sections perpendicular to the $y$-axis are circles with centers on the $y$-axis.

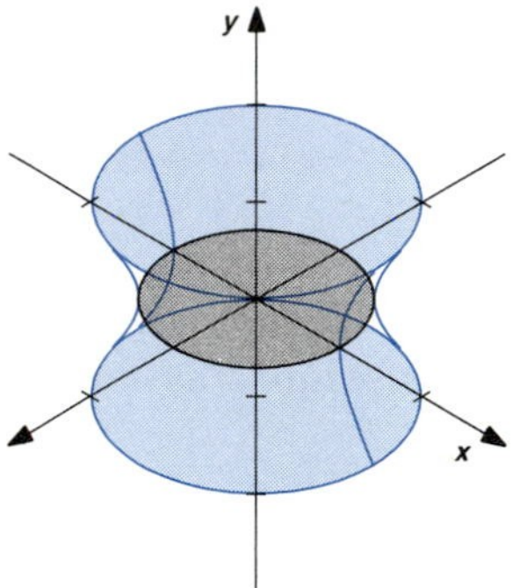


**8** Intersection with the $xy$-plane is the region bounded by

$y = 4 - x^2\quad \text{and}\quad y = 0.$

Cross sections perpendicular to the $y$-axis are circles with centers on the $y$-axis.

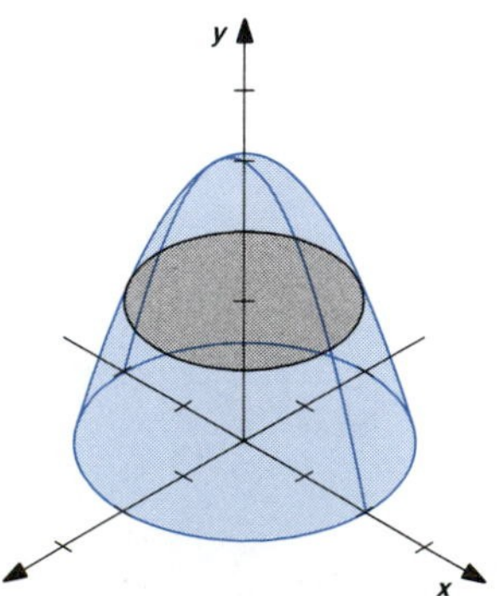


**9** The base is the region in the $xy$-plane bounded by

$x + 3y = 3,\quad x - 3y = 3,\quad \text{and}\quad x = 0.$

Cross sections perpendicular to the $x$-axis are (a) squares, (b) isosceles right triangles with hypotenuse on the base, (c) isosceles right triangles with one side on the base.

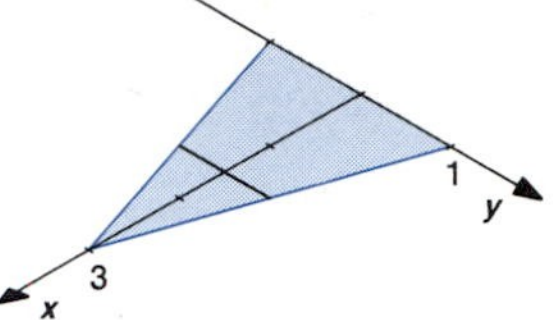


**10** The base consists of the region in the $xy$-plane inside the circle $x^2 + y^2 = 4$. Cross sections perpendicular to the $x$-axis are (a) squares, (b) semicircles with diameter on the base, (c) equilateral triangles with one side on the base.

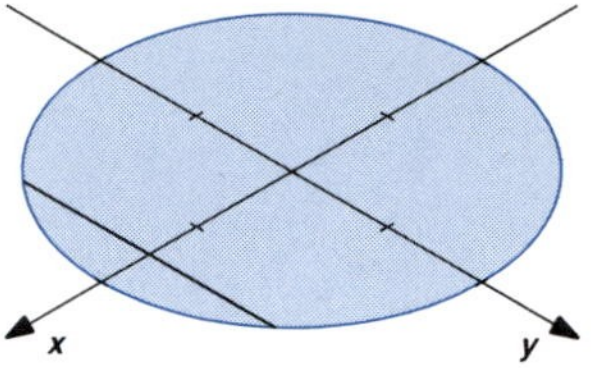


**11** The base is the region in the $xy$-plane bounded by

$x = 4 - y^2\quad \text{and}\quad x = 0.$

Cross sections perpendicular to the $x$-axis are (a) semicircles with diameter on the base, (b) equilateral triangles with one side on the base, (c) rectangles with height half the base.

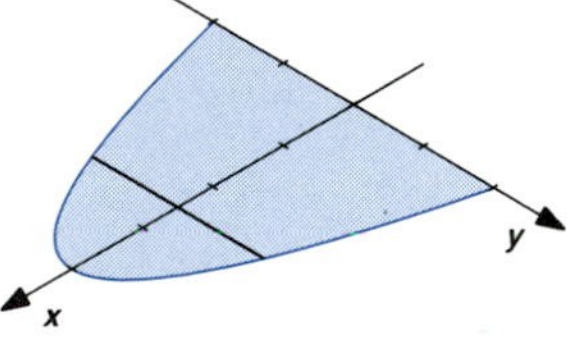


**12** The base is the region in the $xy$-plane bounded by

$y^2 - x^2 = 1,\quad x = 1,\quad \text{and}\quad x = -1.$

Cross sections perpendicular to the $x$-axis are (a) semicircles with diameters on the base, (b) isosceles right triangles with hypotenuses on the base, (c) rectangles with height half the base.

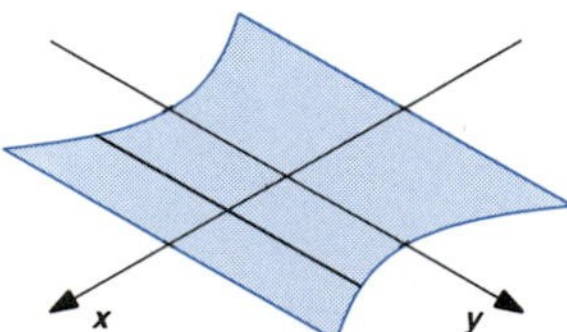

13 A cylindrical hole of radius $r$ is drilled along the axis of a right circular cone with radius of base $2r$ and height $4r$. Set up and evaluate an integral that gives the volume of the resulting solid.

14 A cylindrical hole of radius $r$ is drilled along the diameter of a sphere of radius $2r$. Set up and evaluate an integral that gives the volume of the resulting solid.

15 A wedge is formed from a right circular cylinder of radius 2 in. by cutting the cylinder perpendicular to the axis and at an angle of 30° from the perpendicular plane, in such a way that the edge of the wedge is a diameter of the cylinder. Set up and evaluate an integral that gives the volume of the solid.

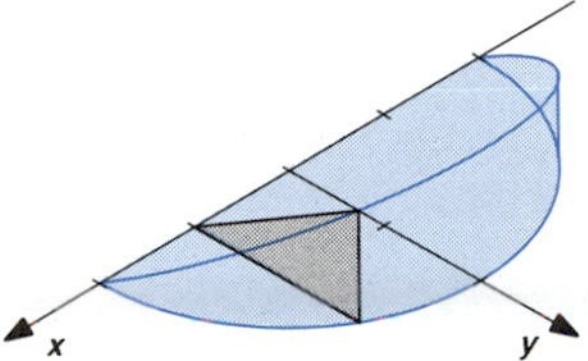


16 A solid is formed by slicing a right circular cylinder of radius 20 cm by two parallel planes that are 7 cm apart, each at an angle $\theta$ with the axis of the cylinder. Set up and evaluate an integral that gives the volume of the solid. (Note that $\int_{-20}^{20} 2\sqrt{20^2 - x^2}\,dx$ is the area of a circle of radius 20.)

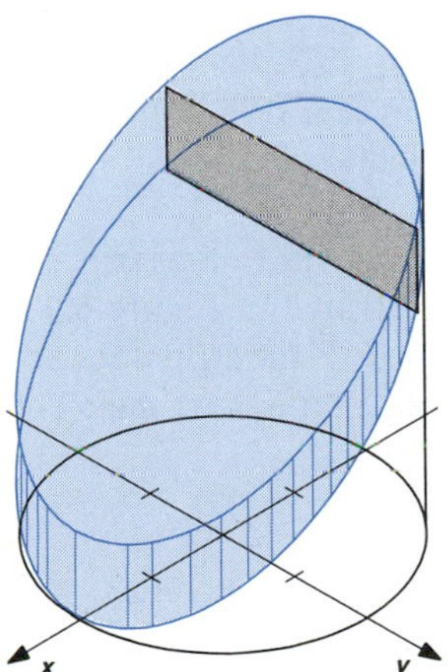


17 A solid intersects the $xy$-plane in the region bounded by

$$y = \sin x, \quad y = -\sin x, \quad x = 0, \quad \text{and} \quad x = \pi.$$

Cross sections perpendicular to the $x$-axis are circles with centers on the $x$-axis. Set up an integral that gives the volume of the solid.

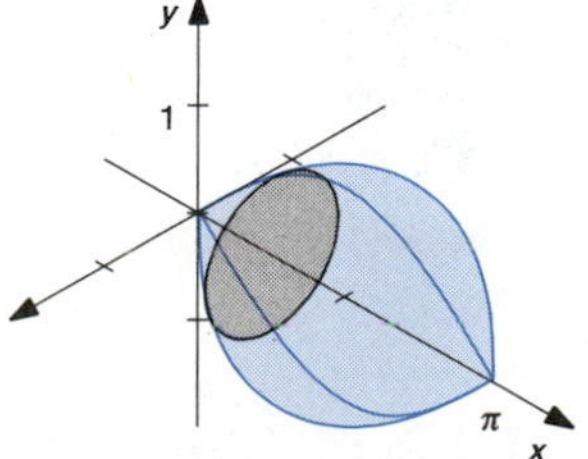


18 A solid intersects the $xy$-plane in the region bounded by

$$y = \cos x, \quad y = -\cos x, \quad x = 0, \quad \text{and} \quad x = \frac{\pi}{2}.$$

Cross sections perpendicular to the $x$-axis are circles with centers on the $x$-axis. Set up an integral that gives the volume of the solid.

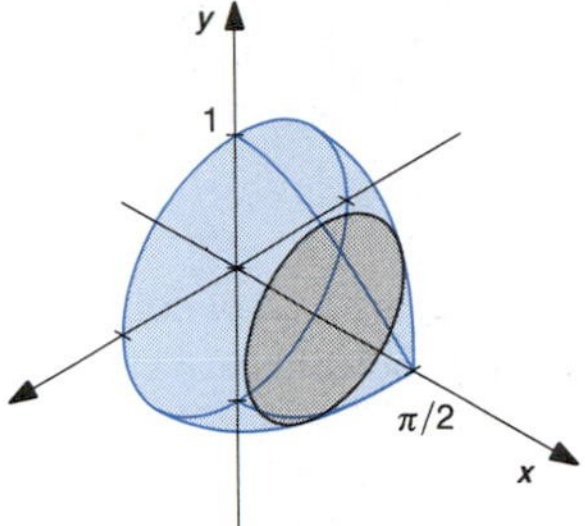


19 Find the volume of the solid that is bounded by two right circular cylinders of radius $r$ if their axes meet at angle $\theta$.

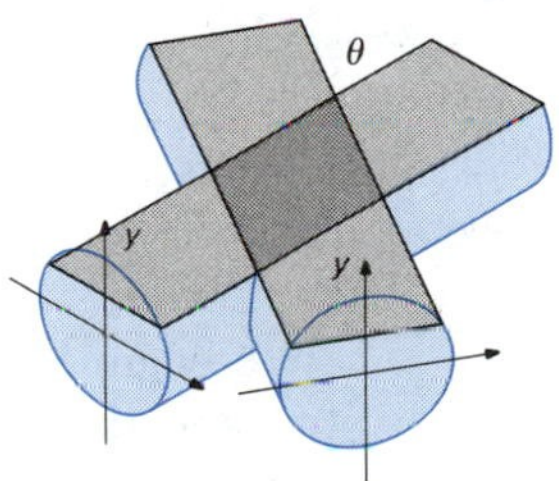


20 The axes of a right circular cylinder of radius $r$ and a right circular cylinder of radius $2r$ meet at a right angle. Set up an integral that gives the volume of the solid bounded by the cylinders.

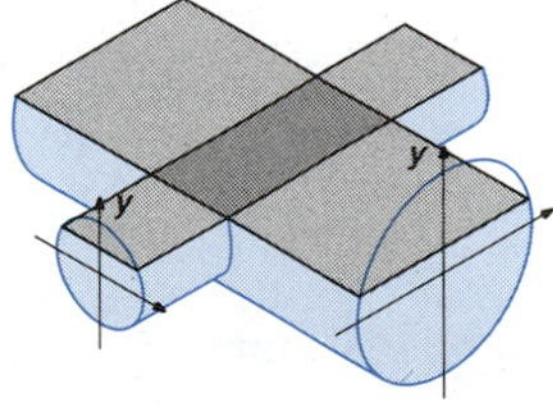


21 The axes of a right circular cylinder of radius $r$ and a square beam with length of sides $r$ meet at a right angle. Set up an integral that gives the volume of that part of the beam that is inside the cylinder.

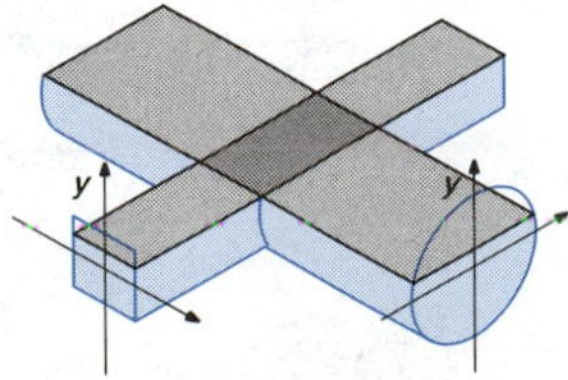

22 A square beam with length of sides $r$ intersects a right circular cylinder of radius $r$ at a right angle and so that the bottom of the beam meets the axis of the cylinder. Set up an integral that gives the volume of that part of the beam that is inside the cylinder.

23 Use integral calculus to explain why a stack of coins has the same total volume whether or not the coins are stacked directly on top of one another.

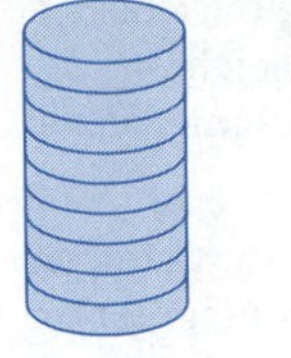

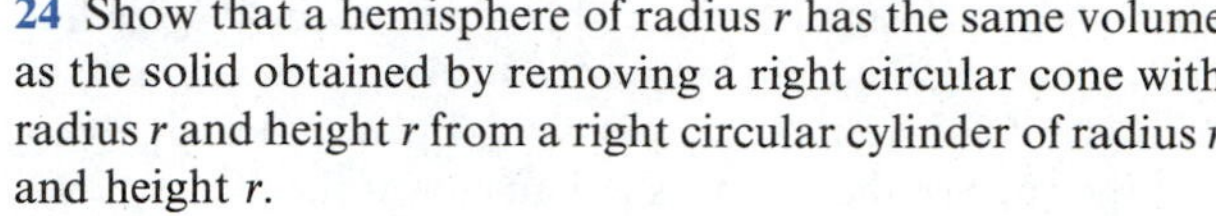

24 Show that a hemisphere of radius $r$ has the same volume as the solid obtained by removing a right circular cone with radius $r$ and height $r$ from a right circular cylinder of radius $r$ and height $r$.

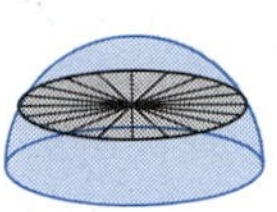

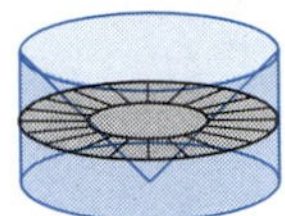

## 6.3 VOLUMES OF SOLIDS OF REVOLUTION

In this section we consider the problem of finding the volume of a solid that is obtained by revolving a region in the plane about a line in the plane that does not pass through the region.

We begin by studying the case where the region is a thin rectangular strip and the axis of revolution is either perpendicular or parallel to the longer side of the strip.

In the case where the axis of revolution is perpendicular to the strip, the solid generated is either a disk or a washer, as illustrated in Figure 6.3.1. We get a disk if the strip touches the axis of revolution. If the strip does not touch the axis of revolution we get a disk with a hole, that is, a washer. The volumes are easily calculated:

**Volume of disk = $\pi$(Radius)$^2$(Thickness);**

and

**Volume of washer = [$\pi$(Outer radius)$^2$ − $\pi$(Inner radius)$^2$][Thickness].**

FIGURE 6.3.1

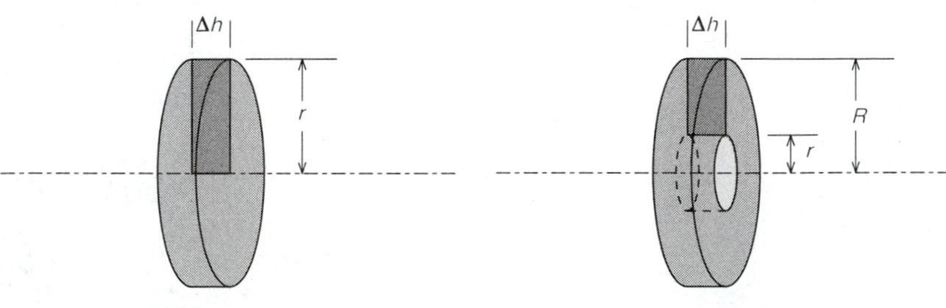

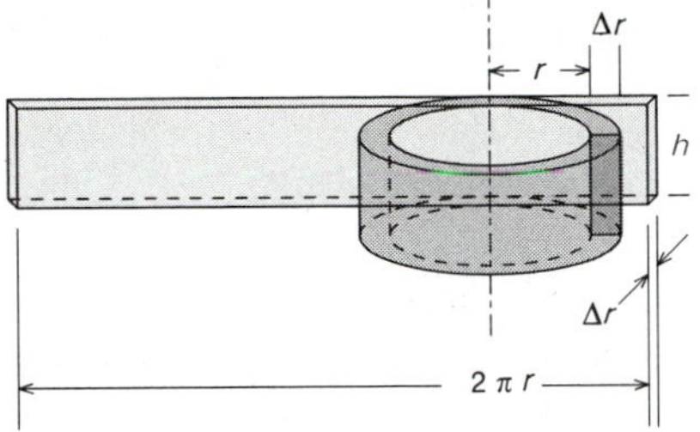


FIGURE 6.3.2

In the case where the axis of revolution is parallel to the longer side of the strip, the solid obtained will be a thin cylindrical shell, as illustrated in Figure 6.3.2. The volume is

$$\begin{aligned} V &= \pi(r + \Delta r)^2 h - \pi r^2 h \\ &= \pi r^2 h + 2\pi r h\, \Delta r + \pi h(\Delta r)^2 - \pi r^2 h \\ &= 2\pi r h\, \Delta r + \pi h(\Delta r)^2. \end{aligned}$$

If $\Delta r$ is small, then $\pi h(\Delta r)^2$ is proportionally much smaller than $2\pi r h\, \Delta r$. We use the formula

**Approximate volume of cylindrical shell**
**$= 2\pi$(Radius)(Height)(Thickness)**

The approximate volume of the cylindrical shell may be thought of as the volume of the rectangular slab obtained by cutting and unrolling the shell. The shell unrolls to a length of $2\pi$(Radius), which is the circumference of a circle.

By using integral calculus we can find the volumes of solids obtained by revolving more general plane regions about a line in the plane. The idea is to divide the area into thin rectangular strips and use the integral to find the limit of the sum of the volumes of the solids obtained from each of the strips. See Figure 6.3.3.

***Solution steps for finding volume of a solid of revolution***

- **Sketch region and axis of revolution.**
- **Choose a convenient variable rectangular area strip.**
- **Sketch the circular path of the endpoints of the strip and identify the solid obtained by revolving the strip as either a disk, washer, or cylindrical shell.**
- **Label appropriate coordinates in terms of the variable that gives the**

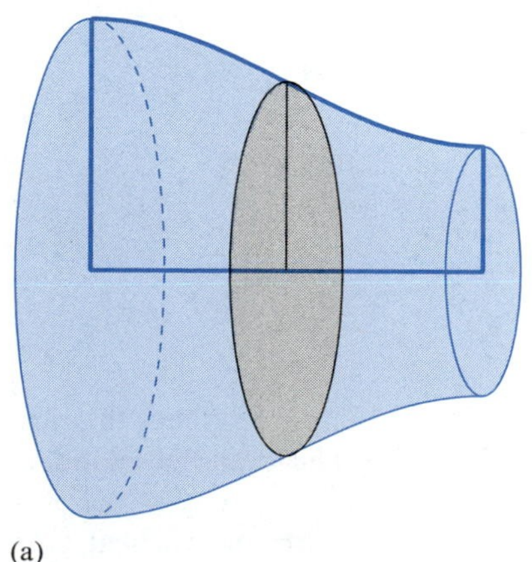

(a)

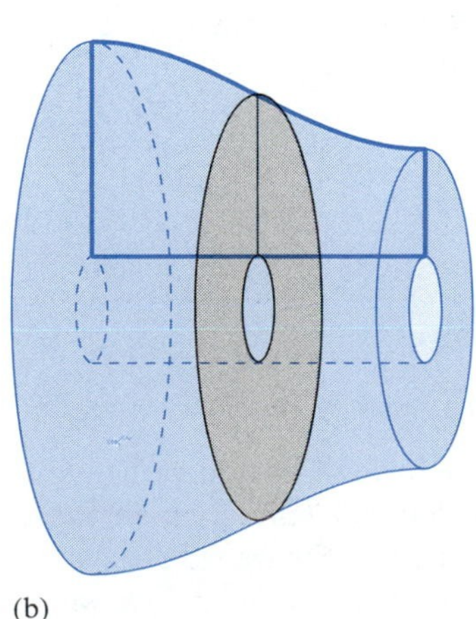

(b)

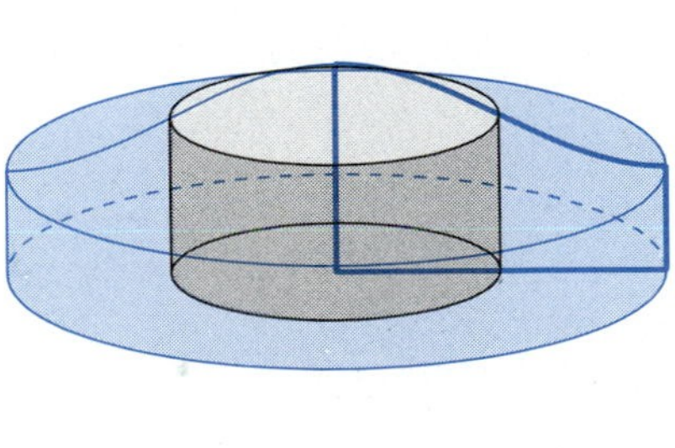

(c)

FIGURE 6.3.3

**position of the variable area strip ($x$ for vertical strips and $y$ for horizontal strips).**

- **Write an integral that gives the limit of sums of the volumes of the disks or washers, or the approximate volumes of the cylindrical shells.**
- **Evaluate the integral.**

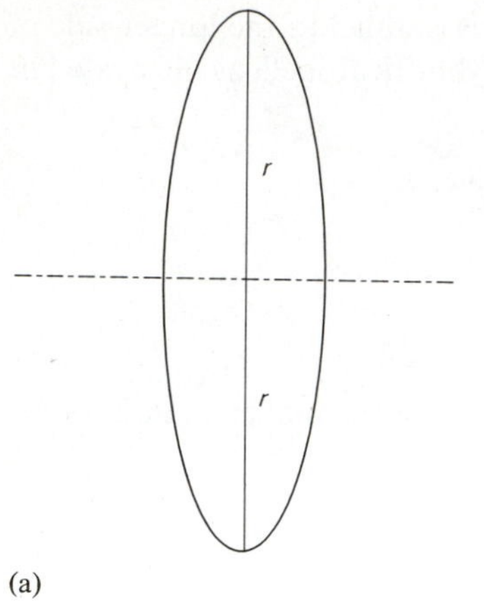


(a)

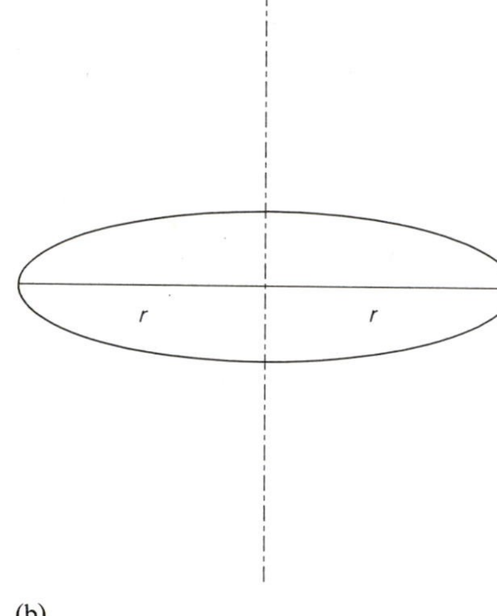


(b)

FIGURE 6.3.4

Recall from Section 5.1 that circles perpendicular to the sketching plane appear as ellipses. See Figure 6.3.4. Some care should be used to sketch your circle through a point that is equidistant and on the opposite side of the axis of revolution from the point being revolved.

■ **EXAMPLE 1**

Consider the region bounded by

$$y = \frac{x}{2}, \quad y = 0, \quad \text{and} \quad x = 2.$$

Set up an integral that gives the volume of the solid obtained by revolving the region about the (a) $x$-axis, (b) line $y = -1$, (c) $y$-axis, (d) line $x = 3$.

**SOLUTION**

We could use either a vertical strip or a horizontal strip, but let us use a vertical strip for each of the four parts of the problem. This means the components of our integrals will be expressed in terms of the variable $x$.

(a) The region, a variable vertical area strip, and the disk obtained by revolving the strip about the $x$-axis are sketched in Figure 6.3.5a. We have labeled coordinates of the endpoints of the area strip in terms of the variable $x$. The lower endpoint of the variable vertical strip is on the axis of revolution and is the center of the disk obtained by revolving the strip. The radius of the disk is the length of the vertical line segment between the points $(x, x/2)$ and $(x, 0)$. The length of a vertical line segment is the difference in the $y$-coordinates of its endpoints. The equation

$$\text{Length} = (y\text{-coordinate of top}) - (y\text{-coordinate of bottom})$$

gives nonnegative length.

The volume is

$$V = \underbrace{\int_0^2}_{\text{Sum}} \underbrace{\underbrace{\pi}_{\pi} \underbrace{\left(\frac{x}{2}\right)^2}_{(\text{Radius})^2} \underbrace{dx.}_{(\text{Thickness})}}_{\text{Volume of disk}}$$

(The value of this integral is $2\pi/3$.)

We have drawn only one typical disk in Figure 6.3.5a. Sketches that are too detailed tend to hide the essential features. The complete solid is sketched in Figure 6.3.5b.

(b) From Figure 6.3.6a we see that revolving a variable vertical area strip about the line $y = -1$ results in a washer. We have labeled the

FIGURE 6.3.5

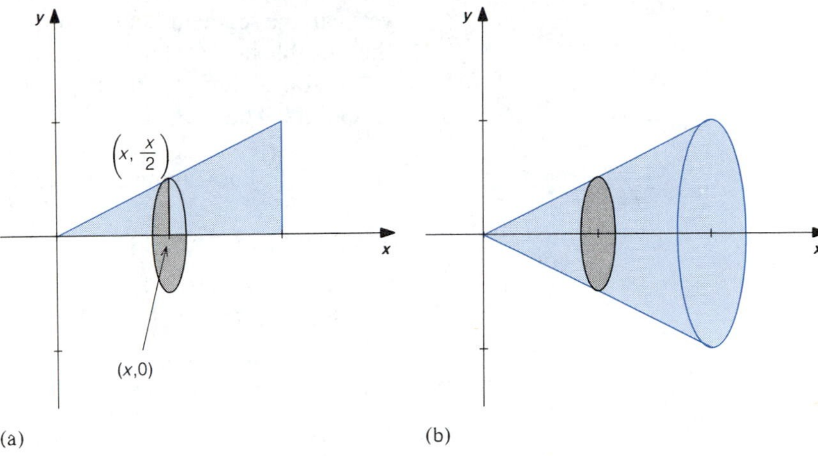


FIGURE 6.3.6

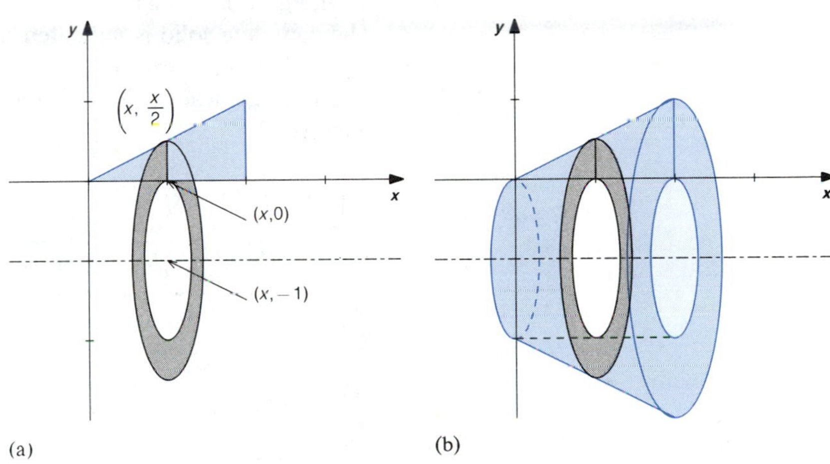


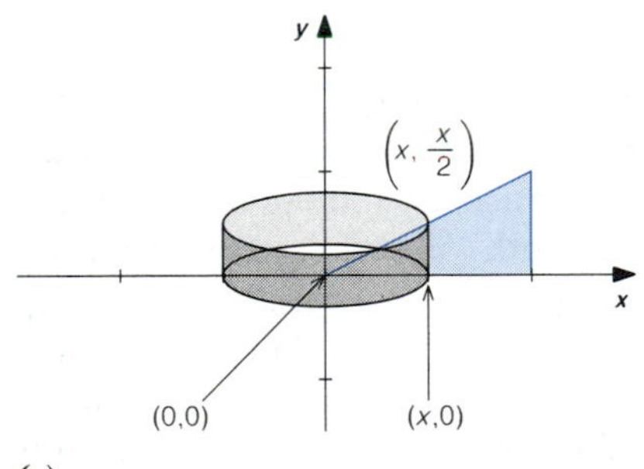


(a)

(b)

FIGURE 6.3.7

endpoints of the area strip and the center of the washer. Labeling coordinates allows us to read the lengths of the outer radius and inner radius from our sketch.

The volume is

$$V = \underbrace{\int_0^2}_{\text{Sum}} \underbrace{\Big[ \underbrace{\pi}_{[\pi} \underbrace{\left(\frac{x}{2} - (-1)\right)^2}_{(\text{Outer radius})^2} - \underbrace{\pi}_{-\,\pi} \underbrace{(0 - (-1))^2}_{(\text{Inner radius})^2]} \Big] \underbrace{dx}_{[\text{Thickness}]}}_{\text{Volume of washer}}.$$

(The value of the integral is $8\pi/3$.)

The complete solid is sketched in Figure 6.3.6b.

(c) Since the $y$-axis is parallel to the variable vertical area strip, we obtain a cylindrical shell. We have labeled the endpoints of the strip and a corresponding point on the axis of revolution in Figure 6.3.7a. The variable is $x$. The height of the cylindrical shell is the length of the variable

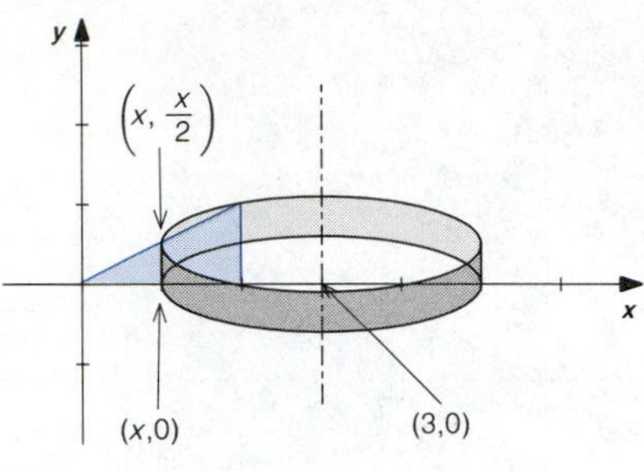

(a)

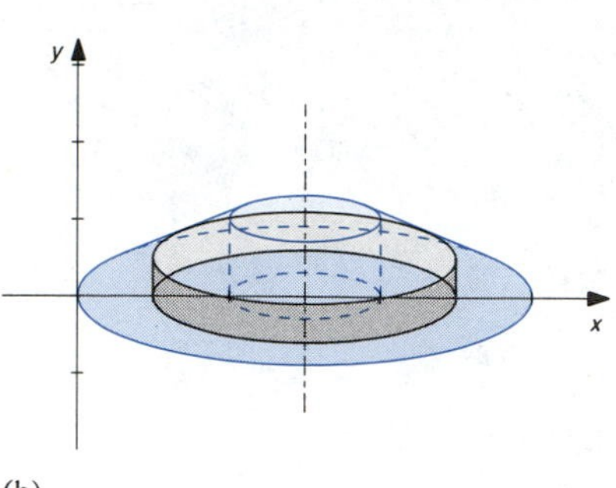

(b)

FIGURE 6.3.8

vertical line segment between $(x, x/2)$ and $(x, 0)$. The radius is the length of the variable horizontal line segment between $(0, 0)$ and $(x, 0)$. The length of a horizontal line segment is the difference of the $x$-coordinates of its endpoints. That is,

Length = ($x$-coordinate of right endpoint) − ($x$-coordinate of left endpoint).

We then have

$$V = \underbrace{\int_0^2}_{\text{Sum}} \underbrace{\underbrace{2\pi}_{2\pi} \underbrace{(x-0)}_{\text{(Radius)}} \underbrace{\left(\frac{x}{2} - 0\right)}_{\text{(Height)}} \underbrace{dx}_{\text{(Thickness)}}}_{\text{Approximate volume of cylindrical shell}}.$$

(This integral has value $8\pi/3$.)

The complete solid is indicated in Figure 6.3.7b.

(d) Revolving a vertical area strip about a vertical line gives a cylindrical shell. A typical shell is sketched and labeled in Figure 6.3.8a. The volume is

$$V = \underbrace{\int_0^2}_{\text{Sum}} \underbrace{\underbrace{2\pi}_{2\pi} \underbrace{(3-x)}_{\text{(Radius)}} \underbrace{\left(\frac{x}{2} - 0\right)}_{\text{(Height)}} \underbrace{dx}_{\text{(Thickness)}}}_{\text{Approximate volume of cylindrical shell}}.$$

(The value of the integral is $10\pi/3$.) ■

The complete solid is indicated in Figure 6.3.8b.

The method of solution we are using does not require that we memorize a different formula for each case that might occur. The only formulas used are that of the area of a circle for disks and washers, and the circumference of a circle for cylindrical shells. Since both of these are well-known formulas, it should not matter whether revolving a variable area strip results in a disk, washer, or cylindrical shell. Just choose the most convenient area strip and use whatever you get. Labeling key coordinates in terms of the appropriate variable allows us to read radii and heights from our sketches. As in all applications of the integral, it is most important to think of the definite integral as a sum. Each component of the integral should be thought of in terms of a specific physical quantity.

■ **EXAMPLE 2**

Consider the region bounded by

$$x = (y-1)^2 \quad \text{and} \quad x = y + 1.$$

Set up an integral that gives the volume of the solid obtained by revolving the region about the (a) $x$-axis, (b) line $y = 3$, (c) $y$-axis, (d) line $x = 4$.

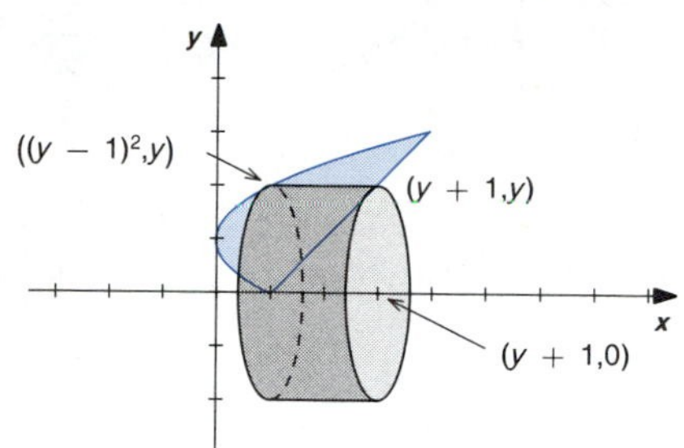


FIGURE 6.3.9

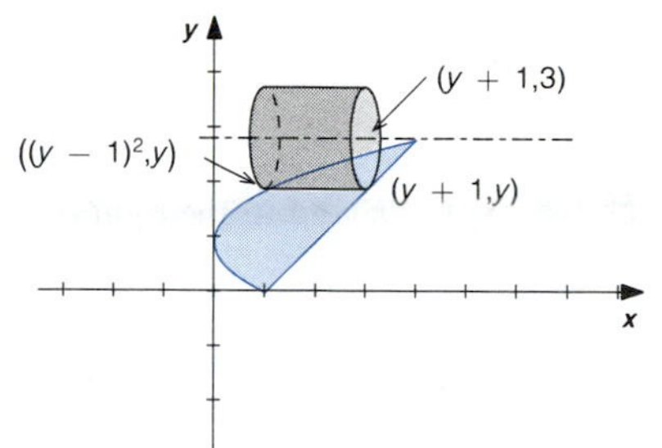


FIGURE 6.3.10

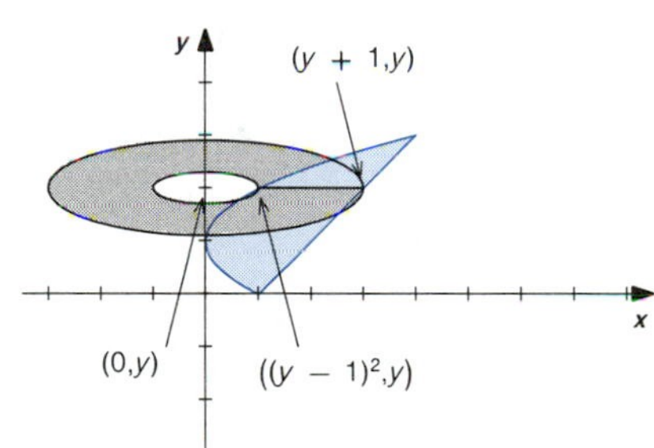


FIGURE 6.3.11

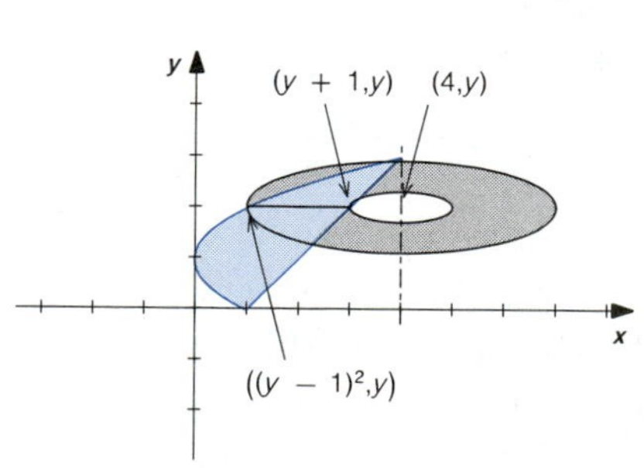


FIGURE 6.3.12

**SOLUTION**

In this problem it is clearly more convenient to use horizontal area strips, so $y$ is the variable. (Note that there is not a single vertical area strip that is typical for the region. One type of vertical strip would go from the bottom to the top of the parabola, while another type of vertical strip would go from the line to the top of the parabola. Two integrals would be required to write the volume as integrals with respect to $x$.)

Appropriate sketches are given in Figures 6.3.9–6.3.12. Note that sketches of the complete solids would be very complicated. The coordinates are labeled in terms of the variable $y$. In each case, the variable $y$ ranges between the $y$-coordinates of the points of intersection of the parabola and the line.

$x = (y-1)^2$ and $x = y + 1$ imply

$$(y-1)^2 = y + 1,$$

$$y^2 - 2y + 1 = y + 1,$$

$$y^2 - 3y = 0,$$

$$y(y-3) = 0.$$

The desired coordinates are $y = 0$ and $y = 3$.

Integrals that give the desired volumes are given below. Use Figures 6.3.9–6.3.12 to identify and check the components of each integral.

(a) Revolving about the $x$-axis (Figure 6.3.9) gives

$$V = \int_0^3 2\pi(y)[(y+1) - (y-1)^2]\,dy.$$

$(V = 27\pi/2.)$

(b) Revolving about the line $y = 3$ (Figure 6.3.10) gives

$$V = \int_0^3 2\pi(3-y)[(y+1) - (y-1)^2]\,dy.$$

$(V = 27\pi/2.)$

(c) Revolving about the $y$-axis (Figure 6.3.11) gives

$$V = \int_0^3 [\pi(y+1)^2 - \pi((y-1)^2)^2]\,dy.$$

$(V = 72\pi/5.)$

(d) Revolving about the line $x = 4$ (Figure 6.3.12) gives

$$V = \int_0^3 [\pi(4-(y-1)^2)^2 - \pi(4-(y+1))^2]\,dy.$$

$(V = 108\pi/5.)$ ■

# EXERCISES 6.3

In Exercises 1–26 set up and evaluate integrals that give the volume of the solids obtained by revolving the regions bounded by the given curves about the given lines.

1 $y = x^2$, $y = 0$, $x = 1$; about (a) $x$-axis, (b) $y = 1$, (c) $y$-axis, (d) $x = 1$

2 $y = x^3$, $y = 0$, $x = 1$; about (a) $x$-axis, (b) $y = 1$, (c) $y$-axis, (d) $x = 1$

3 $y = x^2$, $y = x$; about (a) $x$-axis, (b) $y = 1$, (c) $y$-axis, (d) $x = 1$

4 $y = 1 - x^2$, $x + y = 1$; about (a) $x$-axis, (b) $y = 1$, (c) $y$-axis, (d) $x = 1$

5 $y = 4x - x^2$, $x = y$; about the $x$-axis

6 $y = 4x + x^2$, $x + y = 0$; about the $y$-axis

7 $x = y$, $x + 2y = 3$, $y = 0$; about (a) $x$-axis, (b) $y$-axis

8 $x + y = 1$, $x - 2y = 1$, $y = 1$; about (a) $x$-axis, (b) $y$-axis

9 $y = x^2$, $x + y = 2$, $x = 0$ $(x \geq 0)$; about the $y$-axis

10 $y = 2 - x^2$, $y = 0$, $x = 0$ $(x \geq 0)$; about the $y$-axis

11 $x = 2 - y^2$, $x = 0$; about the $y$-axis

12 $x = y^2 + 1$, $x = 0$, $y = 1$, $y = -1$; about the $y$-axis

13 $y = 1/x$, $y = 0$, $x = 1$, $x = 2$; about (a) $x$-axis, (b) $y$-axis

14 $y = 1/x^3$, $y = 0$, $x = 1$, $x = 2$; about (a) $x$-axis, (b) $y$-axis

15 $y = 2 - x^2$, $x + y = 2$; about (a) $x$-axis, (b) $y$-axis

16 $y = x^2$, $x + y = 2$, $y = 0$; about (a) $x$-axis, (b) $y$-axis

17 $y = 1/x^3$, $x = 0$, $y = 1$, $y = 8$; about (a) $x$-axis, (b) $y$-axis

18 $y = 1/x$, $x = 0$, $y = 1$, $y = 2$; about (a) $x$-axis, (b) $y$-axis

19 $y = \sec x$, $y = 1$, $x = 0$, $x = \pi/4$; about the $x$-axis

20 $y = 2 \sec x$, $y = 1$, $x = 0$, $x = \pi/4$; about the $x$-axis

21 $y = \sin x^2$, $y = 0$, $x = 0$, $x = \sqrt{\pi}$; about the $y$-axis

22 $y = \cos x^2$, $y = 0$, $x = 0$, $x = \sqrt{\pi/2}$; about the $y$-axis

23 $\dfrac{x}{r} + \dfrac{y}{h} = 1$, $y = 0$, $x = 0$; about the $y$-axis. (This gives the volume of a right circular cone with height $h$ and radius of base $r$.)

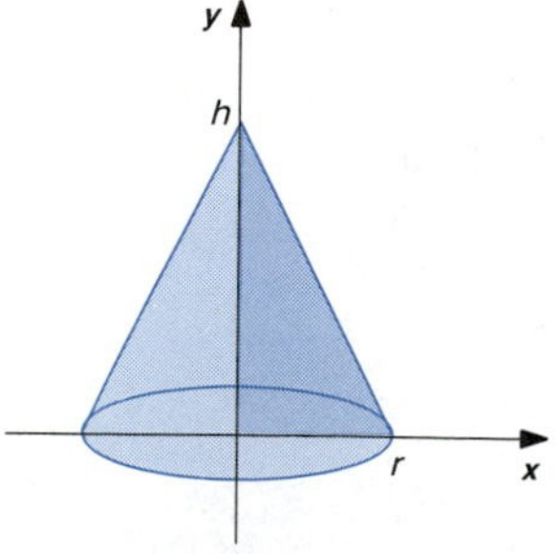


24 $x = \sqrt{r^2 - y^2}$, $x = 0$; about the $y$-axis. (This gives the volume of a sphere of radius $r$.)

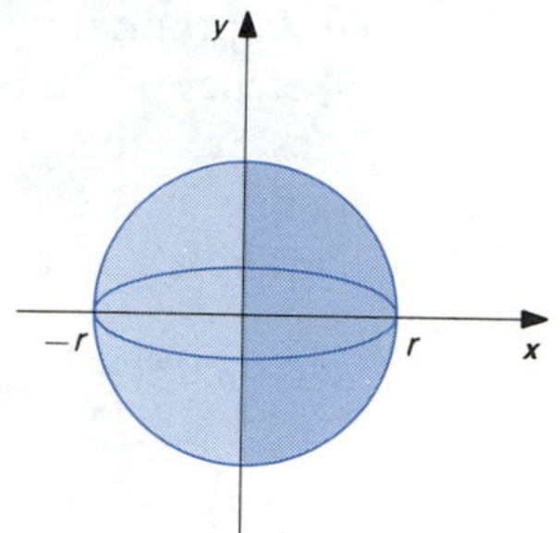


25 Inside $x^2 + y^2 = r^2$, above $y = r - h$, and to the right of $x = 0$; about the $y$-axis. (This gives the volume of a segment of a sphere.)

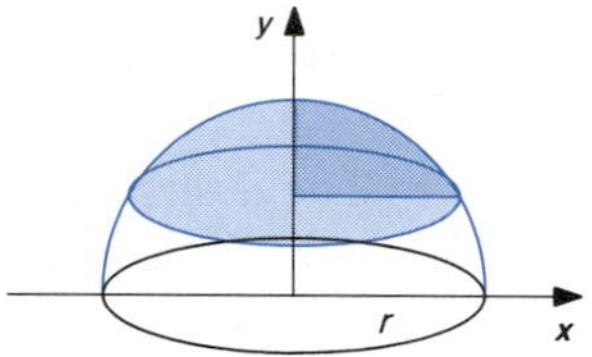


26 $(r - R)y = h(x - R)$, $y = h$, $y = 0$, $x = 0$; about the $y$-axis. (This gives the volume of a truncated cone.)

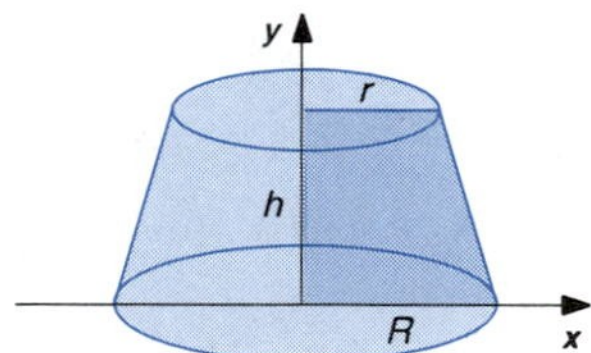


In Exercises 27–34 set up integrals that give the volume of the solids obtained by revolving the given regions about the given lines.

27 Inside $(x - 2)^2 + y^2 = 1$; about the $y$-axis

28 Inside $x^2 + y^2 = 1$; about $x = 2$

29 $y = \sin x$, $y = 0$, $x = 0$, $x = \pi$; about (a) $x$-axis, (b) $y$-axis

30 $y = \cos x$, $y = 0$, $x = 0$, $x = \pi/2$; about (a) $x$-axis, (b) $y$-axis

31 $y = \cos x$, $y = 1$, $x = 0$, $x = \pi/2$; about (a) $x$-axis, (b) $y$-axis

32 $y = \sin x$, $y = x$, $x = 0$, $x = \pi$; about (a) $x$-axis, (b) $y$-axis

33 $y = \tan x$, $y = 0$, $x = 0$, $x = \pi/4$; about (a) $x$-axis, (b) $x = \pi/4$

34 $y = \cot x$, $y = 0$, $x = \pi/4$, $x = \pi/2$; about (a) $x$-axis, (b) $x = \pi/4$

## 6.4 ARC LENGTH AND SURFACE AREA

We are familiar with formulas for the length of a line segment and for the circumference of a circle. We may also know formulas for the area of the surface of a sphere and the lateral surface area of right circular cylinders and cones. In this section we will use integral calculus to find the length of more general curves and the area of more general surfaces.

*Definition*

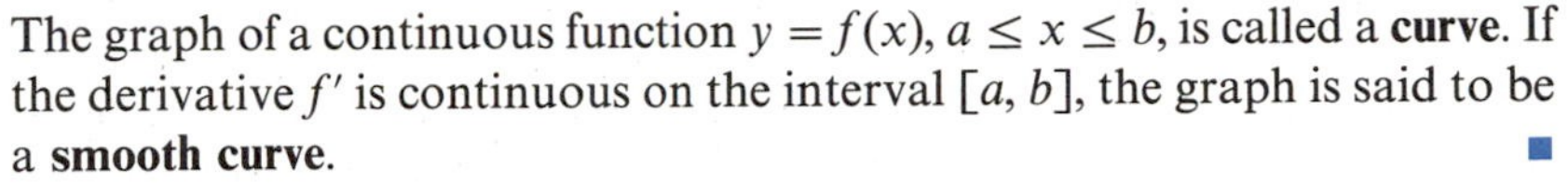

The graph of a continuous function $y = f(x)$, $a \le x \le b$, is called a **curve**. If the derivative $f'$ is continuous on the interval $[a, b]$, the graph is said to be a **smooth curve**.

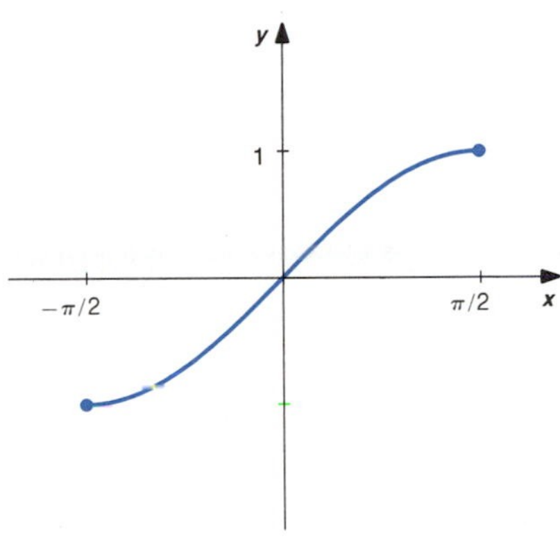


FIGURE 6.4.1

A smooth curve has a tangent line at each point and the slope of the tangent line varies continuously as $x$ varies. A curve is not smooth if it contains a sharp corner. For example, the graph of $y = \sin x$, $-\pi/2 \le x \le \pi/2$, is a smooth curve. See Figure 6.4.1. The graph of $y = |x|$, $-1 \le x \le 1$, is not a smooth curve. See Figure 6.4.2.

FIGURE 6.4.2

Let us consider the **arc length** of a smooth curve that is the graph of $y = f(x)$, $a \le x \le b$. To calculate an approximate value of the arc length of the curve, let us choose points

$$a = x_0 < x_1 < \cdots < x_{n-1} < x_n = b$$

that divide $[a, b]$ into $n$ subintervals. We can then use the length of the line segment between $(x_{j-1}, f(x_{j-1}))$ and $(x_j, f(x_j))$ as an approximate value of the "arc length" of the curve between these points, $j = 1, 2, \ldots, n-1, n$. See Figure 6.4.3. The distance between $(x_{j-1}, f(x_{j-1}))$ and $(x_j, f(x_j))$ is

$$\begin{aligned}\Delta s_j &= \sqrt{(x_j - x_{j-1})^2 + (f(x_j) - f(x_{j-1}))^2}\\ &= \sqrt{(\Delta x_j)^2 + (\Delta y_j)^2}.\end{aligned}$$

Since the curve is smooth, the actual arc length of the curve will be very near the sum of the lengths of these line segments, if $[a, b]$ is divided into small enough subintervals. That is,

$$\text{Arc length} \approx \sum_{j=1}^{n} \Delta s_j = \sum_{j=1}^{n} \sqrt{(\Delta x_j)^2 + (\Delta y_j)^2}.$$

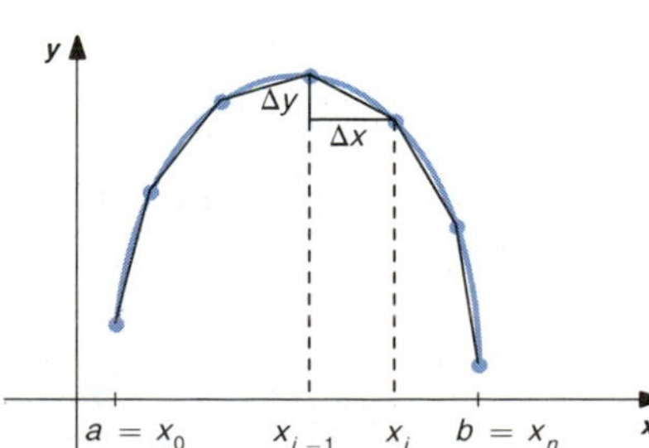


FIGURE 6.4.3

To relate the arc length of the curve to a definite integral, we need to express the expression we have for the approximate arc length as a Riemann sum. To do this, let us use the Mean Value Theorem to obtain

$$\Delta y_j = f(x_j) - f(x_{j-1}) = f'(x_j^*)(x_j - x_{j-1}) = f'(x_j^*)\,\Delta x_j,$$

for some $x_j^*$ between $x_j$ and $x_{j-1}$.

Then

$$\begin{aligned}\Delta s_j &= \sqrt{(\Delta x_j)^2 + (\Delta y_j)^2}\\ &= \sqrt{(\Delta x_j)^2 + (f'(x_j^*)\,\Delta x_j)^2}\\ &= \sqrt{[1 + (f'(x_j^*))^2][\Delta x_j]^2}\\ &= \sqrt{1 + (f'(x_j^*))^2}\,\Delta x_j.\end{aligned}$$

We then have

$$\text{Arc length} \approx \sum_{j=1}^{n} \Delta s_j = \sum_{j=1}^{n} \sqrt{1 + (f'(x_j^*))^2}\,\Delta x_j.$$

We then see that an approximate value of the arc length of the curve is given by a Riemann sum of the continuous function $\sqrt{1 + (f'(x))^2}$ over the interval $[a, b]$. The formula

$$\textbf{Arc length} = \int_a^b \sqrt{1 + (f'(x))^2}\,dx$$

should satisfy our intuitive notion of arc length of the smooth curve $y = f(x)$, $a \le x \le b$. The formula may be considered to be the definition of the arc length of a smooth curve. (It is possible to define arc length in terms of sums of lengths of the line segments we used, without expressing the sums as Riemann sums. Such a definition could be used to calculate the arc length of curves that are not smooth.)

It is useful to think of

$$ds = \sqrt{1 + (f'(x))^2}\,dx$$

as the arc length of a small piece of the smooth curve, located at the point $(x, f(x))$, $a \le x \le b$. The integral

$$\int_a^b \sqrt{1 + (f'(x))^2}\,dx = \int_a^b ds$$

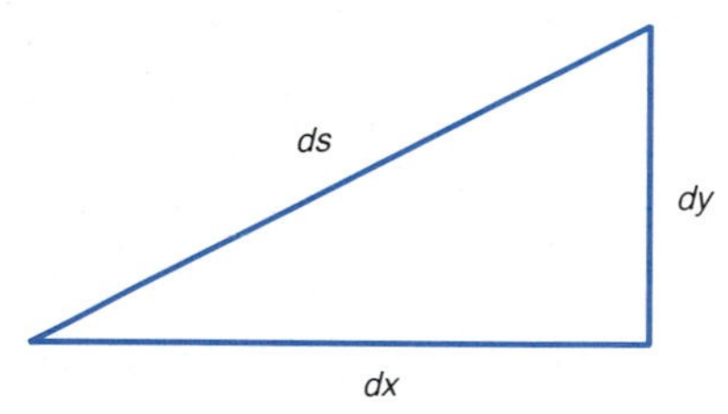

FIGURE 6.4.4

then represents the "sum" of the arc lengths of the pieces of the curve. The formula for $ds$ is suggested by the sketch in Figure 6.4.4. That is,

$$ds = \sqrt{(dx)^2 + (dy)^2} = \sqrt{1 + \left(\frac{dy}{dx}\right)^2}\,dx.$$

We call $ds$ the **differential arc length**.

Let us see that the integral formula for arc length agrees with the distance formula for the length of a line segment.

■ **EXAMPLE 1**

Set up and evaluate an integral that gives the arc length of the curve $y = 2x - 1$, $0 \le x \le 3$.

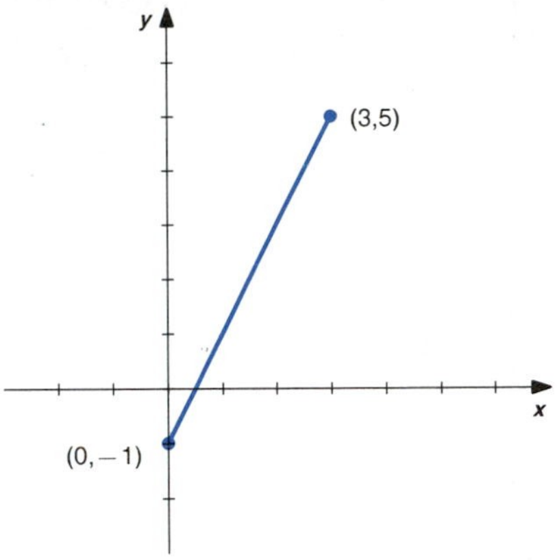

FIGURE 6.4.5

**SOLUTION**

The curve is the line segment between the points $(0, -1)$ and $(3, 5)$. See Figure 6.4.5. The distance formula tells us the length of the segment is

$$s = \sqrt{(3-0)^2 + [5-(-1)]^2} = \sqrt{45} = 3\sqrt{5}.$$

To use the integral formula for arc length, we need the derivative of $y$ with respect to $x$. We have

$$y = 2x - 1, \quad \text{so} \quad \frac{dy}{dx} = 2.$$

Then

$$ds = \sqrt{1 + \left(\frac{dy}{dx}\right)^2}\,dx = \sqrt{1 + (2)^2}\,dx = \sqrt{5}\,dx.$$

The arc length is

$$s = \int ds = \int_0^3 \sqrt{5}\,dx = 3\sqrt{5}.$$

This agrees with the value obtained from the distance formula. ■

■ **EXAMPLE 2**

Set up and evaluate an integral that gives the arc length of the curve $y = x^{3/2}$, $0 \le x \le 4$.

**SOLUTION**

We have

$$y = x^{3/2}, \quad \text{so} \quad \frac{dy}{dx} = \frac{3}{2}x^{1/2}.$$

Then

$$\begin{aligned} ds &= \sqrt{1 + \left(\frac{dy}{dx}\right)^2}\,dx \\ &= \sqrt{1 + \left(\frac{3}{2}x^{1/2}\right)^2}\,dx \\ &= \sqrt{1 + \frac{9}{4}x}\,dx. \end{aligned}$$

The arc length is

$$s = \int ds = \int_0^4 \sqrt{1 + \frac{9}{4}x}\,dx.$$

The definite integral may be evaluated by using the substitution

$$u = 1 + \frac{9}{4}x, \quad \text{so} \quad du = \frac{9}{4}dx.$$

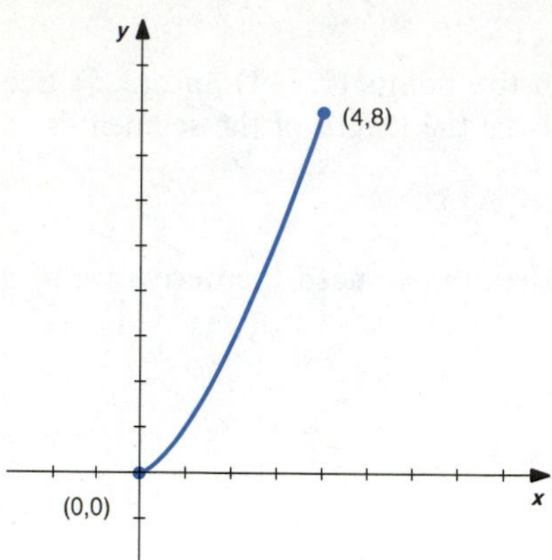


FIGURE 6.4.6

Then

$$\begin{aligned} s &= \int_0^4 \left(1 + \frac{9}{4}x\right)^{1/2} \left(\frac{4}{9}\right)\left(\frac{9}{4}\right) dx \\ &= \int u^{1/2} \frac{4}{9}\, du \\ &= \left(\frac{4}{9}\right)\left(\frac{u^{3/2}}{3/2}\right)\Big] \\ &= \left(\frac{4}{9}\right)\left(\frac{2}{3}\right)\left(1 + \frac{9}{4}x\right)^{3/2}\Big]_0^4 \\ &= \frac{8}{27}\left(1 + \frac{9}{4}(4)\right)^{3/2} - \frac{8}{27}\left(1 + \frac{9}{4}(0)\right)^{3/2} \\ &= \frac{8}{27}(10^{3/2} - 1) \approx 9.0734153. \end{aligned}$$

The curve is sketched in Figure 6.4.6. ■

■ **EXAMPLE 3**

Find the arc length of the curve

$$y = \frac{x^3}{3} + \frac{1}{4x}, \qquad 1 \le x \le 2.$$

**SOLUTION**

We have

$$y = \frac{x^3}{3} + \frac{1}{4x}, \quad \text{so} \quad \frac{dy}{dx} = x^2 - \frac{1}{4x^2}.$$

Then

$$\begin{aligned} ds &= \sqrt{1 + \left(\frac{dy}{dx}\right)^2}\, dx = \sqrt{1 + \left(x^2 - \frac{1}{4x^2}\right)^2}\, dx \\ &= \sqrt{1 + \left(x^4 - 2x^2\frac{1}{4x^2} + \frac{1}{16x^4}\right)}\, dx \\ &= \sqrt{1 + \left(x^4 - \frac{1}{2} + \frac{1}{16x^4}\right)}\, dx \\ &= \sqrt{x^4 + \frac{1}{2} + \frac{1}{16x^4}}\, dx \\ &= \sqrt{\left(x^2 + \frac{1}{4x^2}\right)^2}\, dx \\ &= \left(x^2 + \frac{1}{4x^2}\right) dx. \end{aligned}$$

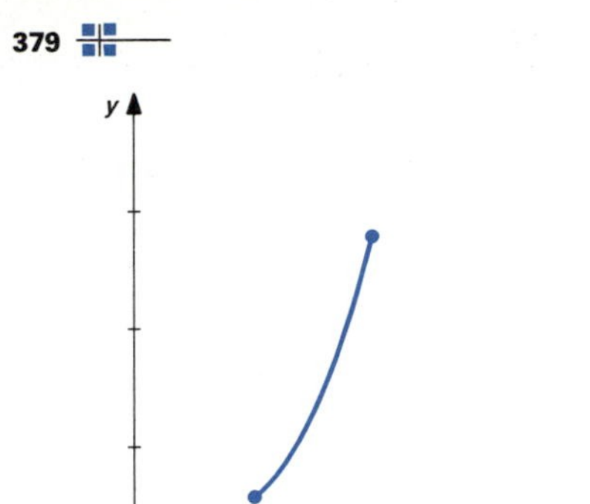

FIGURE 6.4.7

In this example the expression inside the square root turned out to be a perfect square. This greatly simplifies the evaluation of the arc length integral. We have

$$ds = \int ds = \int_1^2 \left(x^2 + \frac{1}{4x^2}\right) dx$$

$$= \left(\frac{x^3}{3} - \frac{1}{4x}\right)\Bigg]_1^2 = \left(\frac{8}{3} - \frac{1}{8}\right) - \left(\frac{1}{3} - \frac{1}{4}\right) = \frac{59}{24}.$$

The curve is sketched in Figure 6.4.7. ■

A problem closely associated with arc length is that of the area of a surface of revolution. Let us consider the surface obtained by revolving the smooth curve

$$y = f(x), \qquad a \le x \le b,$$

about a line. As before, we divide the curve into small pieces that are very nearly line segments of length $\Delta s_j$, $j = 1, \ldots, n$. The result of revolving one of the small pieces about the axis of revolution is a ribbonlike band that is very nearly part of the surface of a right circular cone. We can obtain an approximate value of the area of this band by cutting and unrolling it. We can then think of the band as a rectangle with one length the circumference of the circle formed by the band and the other dimension the *slant height* of the band. See Figure 6.4.8. Noting that the slant height is the arc length of the small piece, we obtain that the approximate area of the band is

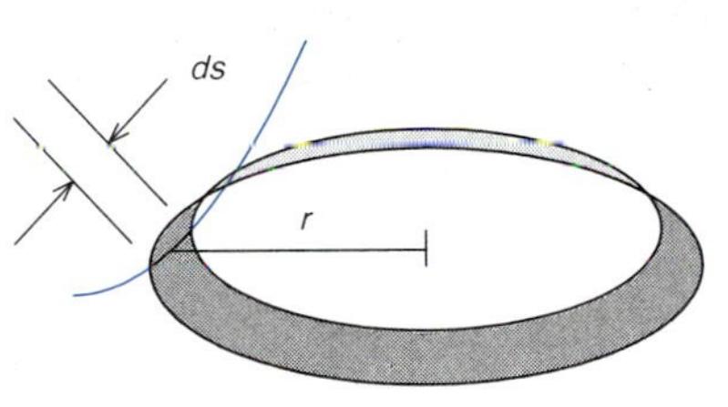

(a)

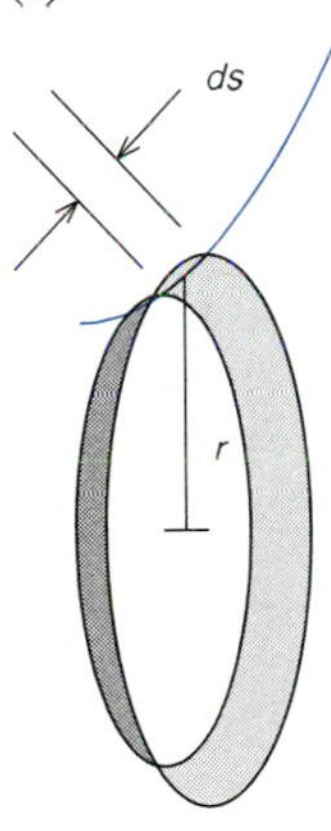

(b)

FIGURE 6.4.8

$$dS = 2\pi r\, ds.$$

(We are using capital $S$ for surface area and lower case $s$ for arc length. Be careful not to confuse them.) It should then seem reasonable that the surface area is given by the following integral:

$$\textbf{Surface area} = S = \int dS = \int_a^b 2\pi r\, ds,$$

where $r$ is expressed in terms of the variable $x$ and $ds = \sqrt{1 + (f'(x))^2}\, dx$ is the differential arc length.

When setting up an integral to determine the area of a surface of revolution, it is useful to draw a typical small piece of the curve and sketch the band obtained by revolving the piece about the axis of revolution. Labeling the location of the piece as $(x, f(x))$ and labeling the center of the band allows us to read the radius of the band in terms of $x$. Don't forget to use $ds$, the slant height of the band.

### ■ EXAMPLE 4

Find the area of the surface obtained by revolving the curve $y = x^2$, $0 \le x \le 1$, about the $y$-axis.

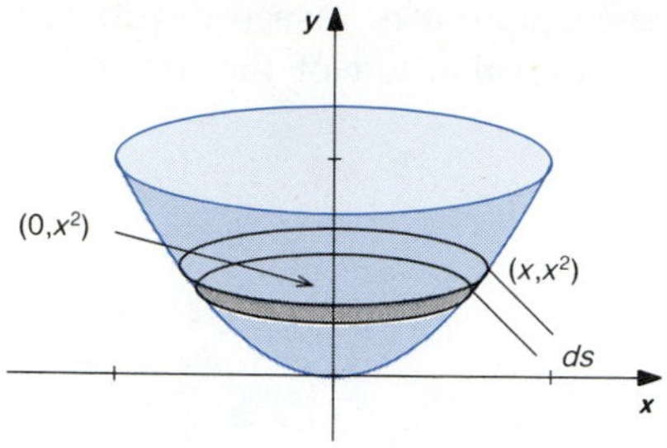

FIGURE 6.4.9

**SOLUTION**

A typical piece of the surface is sketched and labeled in Figure 6.4.9. We see that the radius of the band is $r = x$.

$$y = x^2, \quad \text{so} \quad \frac{dy}{dx} = 2x.$$

Then

$$ds = \sqrt{1 + \left(\frac{dy}{dx}\right)^2}\, dx = \sqrt{1 + 4x^2}\, dx.$$

The surface area is given by

$$S = \underbrace{\int_0^1}_{\text{Sum}} \underbrace{\underbrace{2\pi x}_{\text{(Circumference)}} \underbrace{\sqrt{1 + 4x^2}\, dx}_{\text{(Slant height, } ds)}}_{\text{Area of band}}.$$

This integral involves the square root of a function, so we can try to reduce it to $\int u^{1/2}\, du$. We set $u = 1 + 4x^2$, so $du = 8x\, dx$. Noting that the integrand contains the essential factor $x$, we have

$$\begin{aligned} S &= \int_0^1 2\pi x\sqrt{1 + 4x^2}\, dx \\ &= \int_0^1 2\pi(1 + 4x^2)^{1/2}\left(\frac{1}{8}\right)(8x)\, dx \\ &= \int 2\pi u^{1/2}\left(\frac{1}{8}\right) du \\ &= 2\pi\left(\frac{1}{8}\right)\left(\frac{u^{3/2}}{3/2}\right)\Bigg] \\ &= 2\pi\left(\frac{1}{8}\right)\left(\frac{2}{3}\right)(1 + 4x^2)^{3/2}\Bigg]_0^1 \\ &= \frac{\pi}{6}(1 + 4)^{3/2} - \frac{\pi}{6}(1)^{3/2} = \frac{\pi}{6}(5^{3/2} - 1) \approx 5.3304135. \end{aligned}$$

■

■ **EXAMPLE 5**

Set up an integral that gives the area of the surface obtained by revolving the curve $y = \sin x$, $0 \le x \le \pi$, about the $x$-axis.

**SOLUTION**

$$y = \sin x \quad \text{implies} \quad \frac{dy}{dx} = \cos x.$$

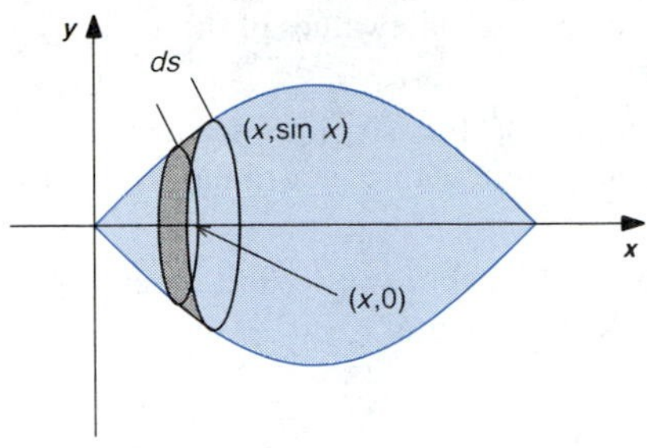

FIGURE 6.4.10

Then

$$ds = \sqrt{1 + \left(\frac{dy}{dx}\right)^2}\, dx = \sqrt{1 + \cos^2 x}\, dx.$$

From Figure 6.4.10 we see that the radius of a typical band is $r = \sin x$. The surface area is

$$S = \underbrace{\int_0^{\pi}}_{\text{Sum}} \underbrace{\underbrace{2\pi \sin x}_{\text{(Circumference)}} \underbrace{\sqrt{1 + \cos^2 x}\, dx}_{\text{(Slant height, } ds)}}_{\text{Area of band}}.$$

(This integral can be evaluated by using techniques we will learn in Chapter 8. It has approximate value 14.423599.) ■

## EXERCISES 6.4

Find the arc lengths of the curves given in Exercises 1–4.

1 $y = \frac{4}{3}(x + 1)^{3/2}, 0 \le x \le 3$

2 $y = \frac{4}{3}x^{3/2} + 1, 0 \le x \le 2$

3 $y = 2x^{2/3}, 1 \le x \le 8$

4 $y = [4 - (x - 8)^{2/3}]^{3/2}, 0 \le x \le 7$

In Exercises 5–8 find the areas of the surfaces obtained by revolving the given curves about the given lines.

5 $y = \frac{1}{3}x^3, 0 \le x \le 1$; about the $x$-axis

6 $y = \frac{1}{2}x^2 + 1, 0 \le x \le 1$; about the $y$-axis

7 $y = \sqrt{25 - x^2}, 0 \le x \le 3$; about the $y$-axis

8 $y = \sqrt{25 - x^2}, 0 \le x \le 3$; about the $x$-axis

In Exercises 9–14, find (a) the arc lengths of the given curves and (b) the areas of the surfaces obtained by revolving the curves about the given line.

9 $y = 4 - 2x, 0 \le x \le 2$; about the $y$-axis

10 $y = 4 - 2x, 0 \le x \le 2$; about the $x$-axis

11 $y = \frac{1}{6}x^3 + \frac{1}{2x}, 1 \le x \le 2$; about the $x$-axis

12 $y = \frac{2x^6 + 1}{8x^2}, 1 \le x \le 3$; about the $x$-axis

13 $y = \frac{1}{24}(x^2 + 16)^{3/2}, 0 \le x \le 3$; about the $y$-axis

14 $y = \frac{2}{27}(x^2 + 9)^{3/2}, 0 \le x \le 4$; about the $y$-axis

15 Find the lateral surface area of a right circular cone with height $h$ and radius of base $r$.

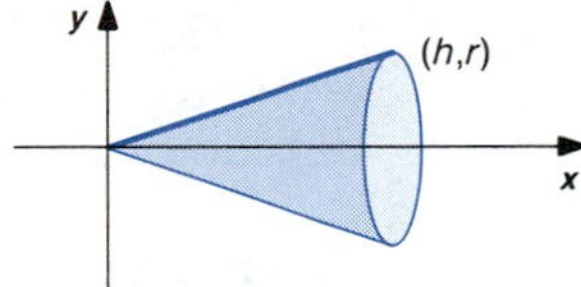


16 Find the lateral surface area of a right circular cylinder with height $h$ and radius $r$.

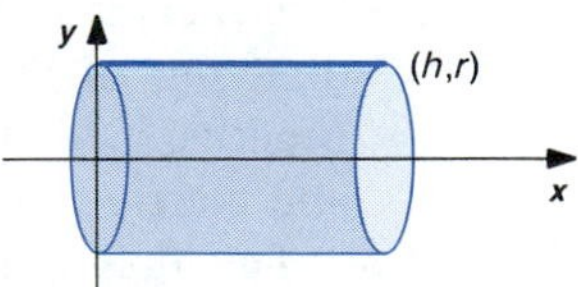


17 (a) Find the area of the surface obtained by revolving the curve $y = \sqrt{r^2 - x^2}, 0 \le x \le x_0$, about the $y$-axis, where

$0 < x_0 < r$. (b) Find the surface area of a hemisphere of radius $r$ by taking the limit as $x_0$ approaches $r$ from the left of the area found in (a).

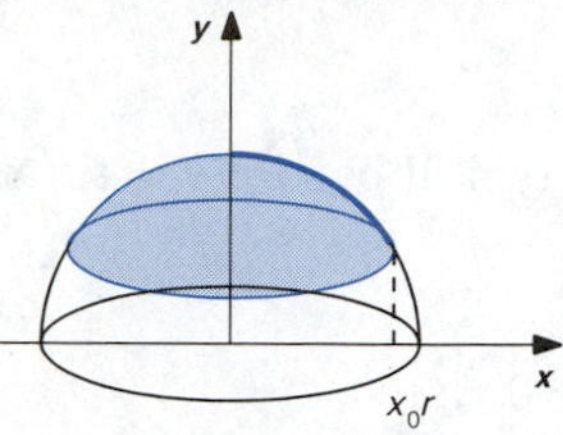


**18** (a) Find the area of the surface obtained by revolving the curve $y = \sqrt{r^2 - x^2}, 0 \leq x \leq x_0$, about the $x$-axis, where $0 < x_0 < r$. (b) Find the surface area of a hemisphere of radius $r$ by taking the limit as $x_0$ approaches $r$ from the left of the area found in (a).

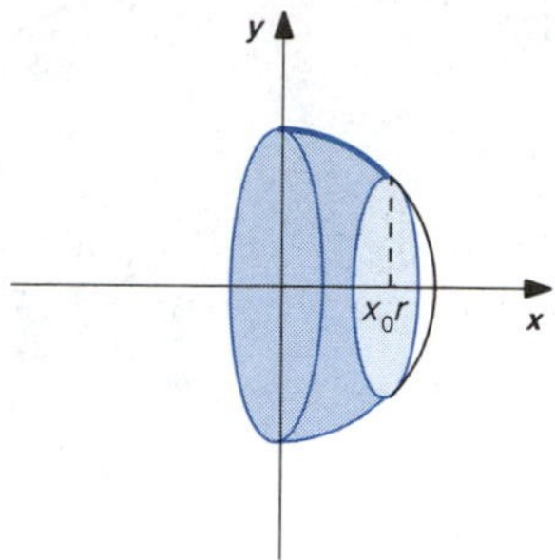


In Exercises 19–23 find approximate values of the arc length of the given curves by:

(a) adding the lengths of the six line segments that connect points on the graph with $x$-coordinates equally spaced;

(b) using Simpson's Rule with $n = 6$ to find an approximate value of the arc length integral.

**19** $y = x^2, 0 \leq x \leq 3$

**20** $y = x^3, 0 \leq x \leq 3$

**21** $y = \sin x, 0 \leq x \leq \pi$

**22** $y = \sin 2x, 0 \leq x \leq \pi$

**23** $y = \sqrt{1 - x^2}, 0 \leq x \leq 1/\sqrt{2}$. (Note that the arc length is one-eighth the circumference of a circle with radius 1.)

**24** If $f'(t)$ is continuous for $a \leq t \leq b$, then the arc length function $s(x) = \int_a^x \sqrt{1 + (f'(t))^2}\, dt$ is defined for $a \leq x \leq b$. Show that the differential of $s$, $ds = s'(x)\, dx$, is what we have called the differential arc length, $ds = \sqrt{1 + (f'(x))^2}\, dx$. Show that $ds = \sec \gamma\, dx$, where $\gamma$ is the angle between the $y$-axis and the line normal to $y = f(x)$ at $(x, f(x))$.

## 6.5 WORK

**Work** is a quantity associated with a force acting on an object as it moves from one position to another.

*Force has magnitude and direction.* In this section we will consider the work done when an object moves along a line, subject to a force in a direction along the line. Work done by a *constant force* is then given by

**Work = (Constant force) · (Distance).**

In this formula, distance should be interpreted as directed distance, with direction indicated by sign. Also, the direction along the line in which a force acts is indicated by its sign. Work done by a force is positive if the direction of the force is in the direction of motion, while the work is negative if the force is opposite to the motion. Units of work in English and metric systems are indicated in Table 6.5.1.

The magnitude of the force on an object due to the gravitational attraction of the earth is the weight of the object. The force is directed toward the center of the earth. The magnitude of the force depends on the distance of the object from the center of the earth, so objects weigh slightly

TABLE 6.5.1

| *System* | *Distance* | *Force* | *Work* |
|---|---|---|---|
| English | feet (ft) | pound (lb) | ft · lb |
| mks (SI) | meter (m) | newton (N) | N · m = joule (J) |
| cgs | centimeter (cm) | dyne | dyne · cm = erg |

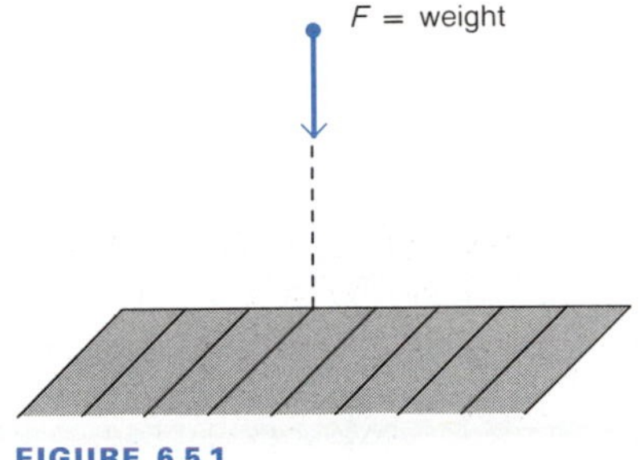

FIGURE 6.5.1

more at sea level than they do at the top of a high mountain. For problems that involve motion near the surface of the earth we assume that the weight of an object does not vary. If the distances involved are relatively small compared to the diameter of the earth, we will consider the earth to be flat and the force downward. See Figure 6.5.1.

■ **EXAMPLE 1**

A 500-lb ball is to be raised 15 ft. The work required is $W = (500) \cdot (15) = 7500$ ft · lb. ■

In Example 1 we calculated the work required to overcome the force due to gravity. This requires an upward force, in the direction of motion, so the work required is positive. The work done by gravity is negative, since the direction of motion is opposite to the force due to gravity. If an object is lowered, the work done by gravity is positive, and the work required to overcome gravity is negative.

Work associated with the force of gravity has the property that it depends only on the initial and terminal heights of the object being moved. For example, if an object is raised and then lowered back to its initial position, the net work done is zero. This is because the work required to overcome gravity is negative when the direction of motion is downward, so it cancels the work required to lift the object. Gravity does not directly offer any resistance to horizontal motion, so no work is required to overcome gravity for horizontal motion. Of course, the force of friction must be overcome for horizontal motion. The force due to friction of a moving object is always in the direction opposite to the motion. Hence, if an object is moved horizontally and then returned to its original position, the work done to overcome moving friction is positive.

To calculate work done in procedures more complicated than that illustrated in Example 1, we use the following basic fact.

**Total work is the sum of the work of each component.**

This allows us to break down a complicated procedure into component parts. Each component must consist of a (nearly) constant force applied over a distance along a straight line. We can then use the basic formula Work = (Constant force) · (Distance) to approximate the component work and add results to determine the total work associated with the procedure. We are particularly interested in cases where a definite integral can be used to evaluate the limit of the sums of component work.

FIGURE 6.5.2

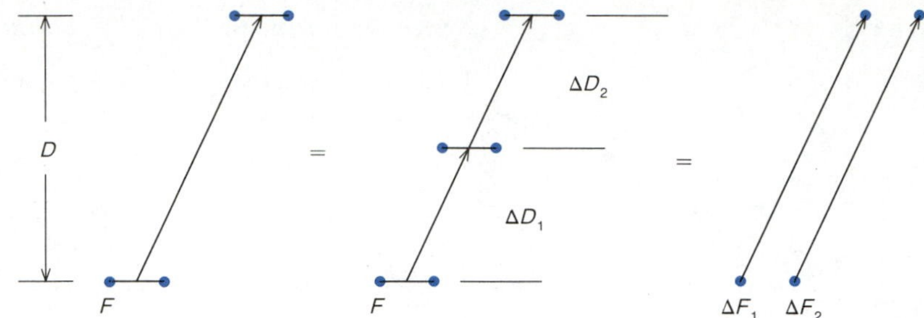


Either the distance or the object being moved may be divided into components. Figure 6.5.2 illustrates three procedures for lifting a weight. The entire weight may be lifted the entire distance ($W = F \cdot D$); the weight may be lifted to an intermediate level and then the rest of the way ($W = F \cdot \Delta D_1 + F \cdot \Delta D_2$); or each component may be lifted separately ($W = \Delta F_1 \cdot D + \Delta F_2 \cdot D$). If $D = \Delta D_1 + \Delta D_2$ and $F = \Delta F_1 + \Delta F_2$, then

$$W = F \cdot D = F \cdot \Delta D_1 + F \cdot \Delta D_2 = \Delta F_1 \cdot D + \Delta F_2 \cdot D.$$

■ **EXAMPLE 2**

The conical tank pictured in Figure 6.5.3 is 8 ft deep and 8 ft across the top. It contains water to a depth of 6 ft. Find the work required to pump the water to the level of the top of the tank. Water weighs 62.4 lb/ft$^3$.

**SOLUTION**

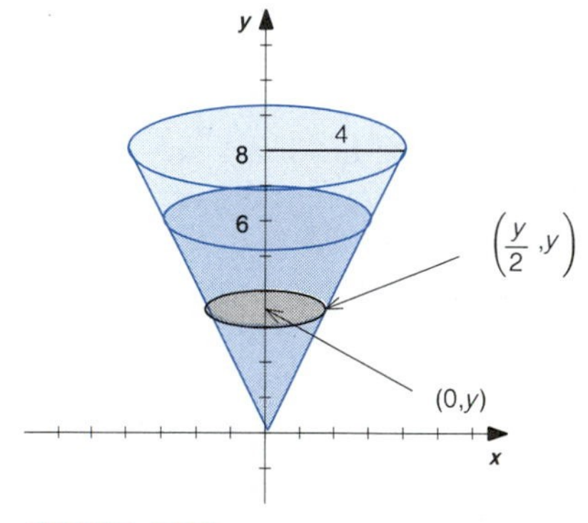


FIGURE 6.5.3

In this problem we need to determine the work required to overcome the force of gravity. This means that the work required depends only on the initial and terminal height of each portion of water. Each thin layer of water at level $y$ must be pumped to the level of the top of the tank. The variable $y$ varies from 0 to 6, since these are the levels of water that we wish to remove. The volume of a thin layer of water at level $y$ and the distance that the water at level $y$ is to be moved depend on the variable $y$.

The cross section of water at level $y$ is a disk. From Figure 6.5.3 we see that the radius of the disk is the $x$-coordinate of the point on the line through the points (0, 0) and (4, 8). This line has slope 2 and equation $y = 2x$, so $x = (1/2)y$. The coordinates of the endpoints of a typical radius have been labeled in Figure 6.5.3 in terms of the variable $y$.

We can express the work required as

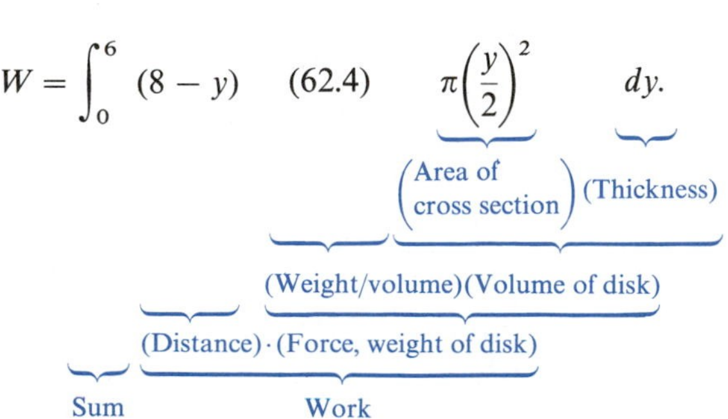

We then have

$$\begin{aligned}
W &= \int_0^6 \frac{(62.4)\pi}{4}(8y^2 - y^3)\,dy \\
&= \frac{(62.4)\pi}{4}\left(8\frac{y^3}{3} - \frac{y^4}{4}\right)\Bigg]_0^6 \\
&= \frac{(62.4)\pi}{4}\left(8\frac{6^3}{3} - \frac{6^4}{4}\right) \\
&= (62.4)(\pi)(63) \approx 12{,}350.229 \text{ ft}\cdot\text{lb}.
\end{aligned}$$

■

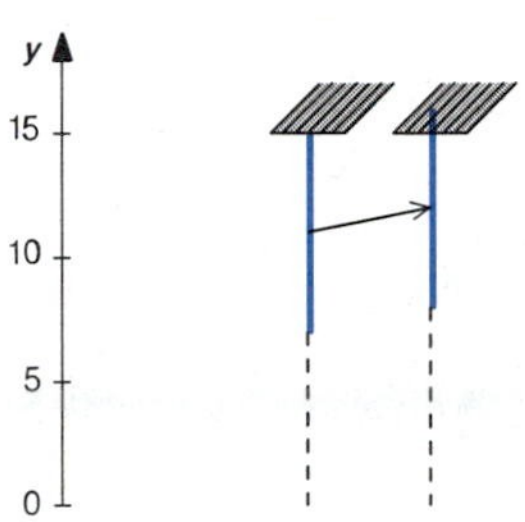

FIGURE 6.5.4

### ■ EXAMPLE 3

A chain 15 ft long and weighing 3 lb/ft is hanging vertically from the top of a building. Find the work required to raise 10 ft of the chain to the level of the top of the building—that is, 5 ft remains hanging.

**SOLUTION**

The work required to overcome gravity depends only on the initial and terminal heights of the components of the chain. Let us choose a vertical coordinate system as in Figure 6.5.4.

In Figure 6.5.4 we have illustrated the procedure of lifting the chain a little at a time. The force is then the weight of the chain left hanging each time it is lifted. The variable $y$ represents the position of the bottom of the chain. $y$ varies from 0 to 10. The work done is

$$W = \underbrace{\int_0^{10}}_{\text{Sum}} \underbrace{\underbrace{\underbrace{\underbrace{3}\ \underbrace{(15-y)}}_{\text{(Weight/length)(Length)}}\ \underbrace{dy.}}_{\text{(Force, weight)}\cdot\text{(Distance)}}}_{\text{Work}}$$

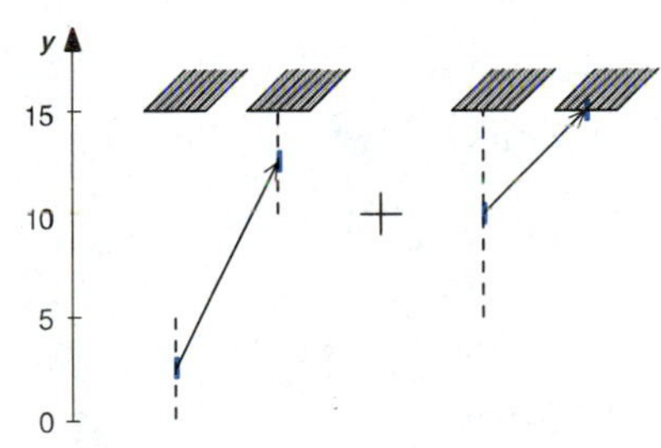

FIGURE 6.5.5

Evaluating this integral gives

$$W = \left(3(15)y - 3\frac{y^2}{2}\right)\Bigg]_0^{10} = (450 - 150) - (0) = 300 \text{ ft}\cdot\text{lb}.$$

It is also possible to adopt the procedure of cutting the chain into small pieces and lifting each piece separately. This is illustrated in Figure 6.5.5. The variable $y$ represents the level of the small piece that is being lifted. Each part of the chain is lifted some distance, so $y$ varies from 0 to 15. For $0 \le y \le 5$, the piece at height $y$ is lifted 10 ft. For $5 \le y \le 15$ the piece is lifted from height $y$ to height 15, a distance of $15 - y$. We can express the work as the sum of two integrals.

$$W = \underbrace{\int_0^{5}}_{\text{Sum}} \underbrace{\underbrace{\underbrace{(10)}\ \underbrace{\underbrace{(3)}\ \underbrace{dy}}_{\text{(Weight/length)(Length)}}}_{\text{(Distance)}\cdot\text{(Force, weight)}}}_{\text{Work}} + \underbrace{\int_5^{15}}_{\text{Sum}} \underbrace{\underbrace{\underbrace{(15-y)}\ \underbrace{\underbrace{(3)}\ \underbrace{dy.}}_{\text{(Weight/length)(Length)}}}_{\text{(Distance)}\cdot\text{(Force, weight)}}}_{\text{Work}}$$

FIGURE 6.5.6

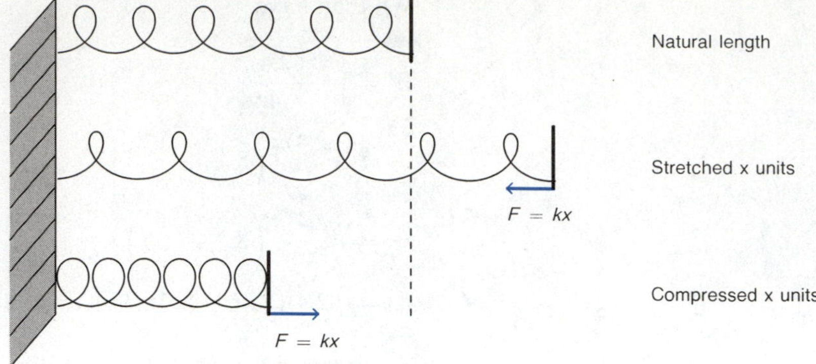


Then

$$W = (10)(3)y\Big]_0^5 + \left(45y - 3\frac{y^2}{2}\right)\Big]_5^{15} = 150 + 150 = 300 \text{ ft}\cdot\text{lb}.$$

The first integral gives the work done in lifting a (3)(5)-lb pièce of chain a distance of 10 ft. That is, we could have thought of lifting the bottom 5 ft of the chain as a single unit.

The two different procedures used in Example 3 require the same amount of work. ■

**Hooke's Law** says the force due to the displacement of a spring from its natural position is directly proportional to the displacement.

$$F = kx, \qquad k \text{ constant.}$$

The force is directed toward the natural position of the spring. See Figure 6.5.6.

■ **EXAMPLE 4**

A 4-lb weight stretches a certain spring 0.8 ft beyond its natural length. How much work is required to stretch the spring from 0.8 ft to 1.3 ft beyond its natural length?

**SOLUTION**

We must first determine the force on the spring. From Hooke's Law we know $F(x) = kx$, where $x$ is the distance the spring is stretched beyond its natural length and $k$ is a constant. We need the value of $k$. Since $x = 0.8$ when $F = 4$, substitution in the equation $F(x) = kx$ gives $4 = k(0.8)$, so $k = 4/0.8 = 5$ and $F(x) = 5x$.

The force $F(x) = 5x$ is not constant as $x$ varies from 0.8 to 1.3. However, the force is nearly constant as $x$ ranges over very small intervals. We can then calculate the work done over each small subinterval of $[0.8, 1.3]$ and use a definite integral to evaluate the limit of the sums. See Figure 6.5.7.

FIGURE 6.5.7

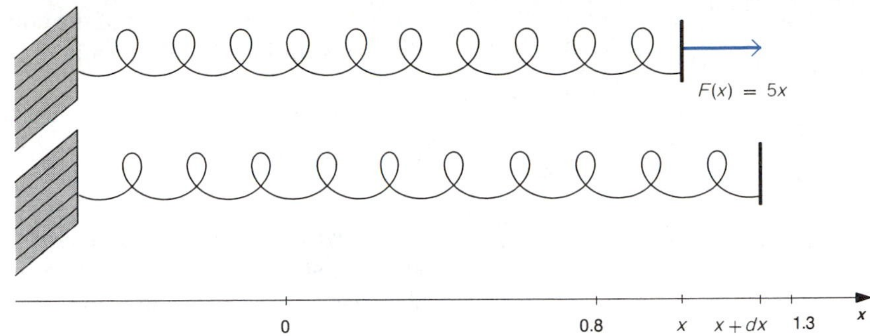


We have

$$W = \underbrace{\int_{0.8}^{1.3}}_{\text{Sum}} \underbrace{\underbrace{5x}_{\text{(Force)}} \cdot \underbrace{dx.}_{\text{(Distance)}}}_{\text{Work}}$$

Evaluating this integral, we have

$$W = \left.\frac{5x^2}{2}\right]_{0.8}^{1.3} = \frac{5}{2}(1.3)^2 - \frac{5}{2}(0.8)^2 = 2.625 \text{ ft} \cdot \text{lb}.$$ ■

The same reasoning as was used in Example 4 shows that the work associated with any continuously varying force $F(x)$ in a direction along the line of motion between $a$ and $b$ is given by

$$W = \int_a^b F(x)\, dx.$$

This integral represents the sum of the components of work given by a force of $F(x)$ over a distance $dx$.

Let us consider a gas that is contained in a cylinder with a movable piston that allows the volume of the gas to vary. See Figure 6.5.8a. The **state** of the gas is determined by its pressure ($P$), volume ($V$), and temperature ($T$). We assume that the pressure and temperature of the gas are the same at every point inside the container, so the gas is in equilibrium. The force due to gas pressure on the piston is

$$F = P \cdot A,$$

where $A$ is the area of the piston. If the force due to gas pressure is different from the exterior force on the piston, the piston will move. If the piston is moved a small distance $dy$, the change in volume of the gas is

$$dV = A\, dy,$$

and the work done by the gas is

$$dW = F \cdot dy = (P \cdot A)\, dy = P \cdot (A\, dy) = P\, dV.$$

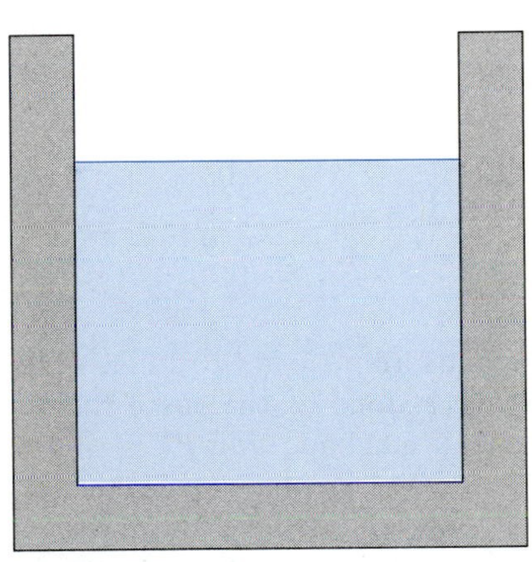

(a)

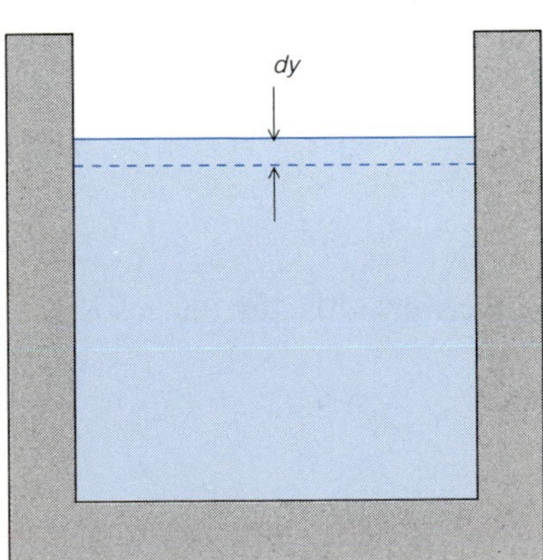


(b)

FIGURE 6.5.8

See Figure 6.5.8b. If the state of the gas changes relatively slowly, so the gas is essentially at equilibrium at each intermediate state, the work done by the gas is given by the sum of the components $dW = P\,dV$. If the volume of the gas is initially $V_i$ and terminally $V_t$, the total work can be expressed as an integral,

$$W = \int_{V_i}^{V_t} P\,dV.$$

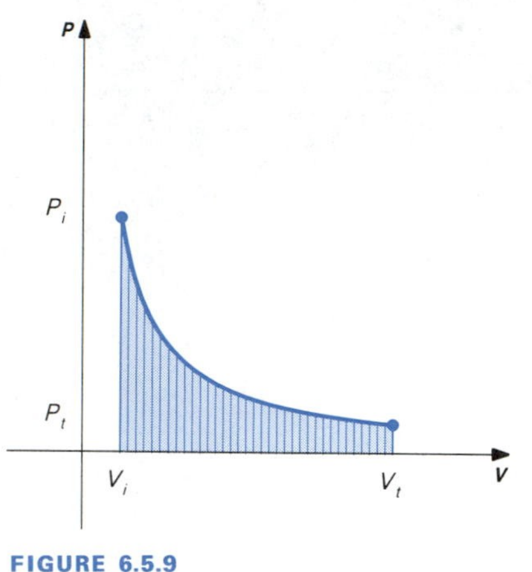


FIGURE 6.5.9

The work done by a gas does not depend only on the initial and terminal states of the gas; it also depends on the intermediate states. An **ideal gas** satisfies the condition

$$\frac{PV}{T} = \text{Constant}$$

($T$ is the absolute temperature in degrees kelvin), so the state of an ideal gas can be determined from its pressure and volume. Changes in the state of a gas can be indicated by a ***PV* diagram**. It is customary to interpret the work done by a gas as the area under the curve in the $PV$ diagram. See Figure 6.5.9.

A process in which an ideal gas changes state without exchanging heat with its surrounding is called an **adiabatic process**. In an adiabatic process, the volume and pressure of an ideal gas are related by an equation of the form

$$PV^{\gamma} = k,$$

where $\gamma$ and $k$ are constants.

### ■ EXAMPLE 5

Air in the cylinder of a diesel engine expands from 1.2 in.$^3$ to 12 in.$^3$. The initial pressure is 600 lb/in.$^2$. Find the work done by the gas if the air behaves as an ideal gas and the expansion is adiabatic with $PV^{1.4} = k$.

**SOLUTION**

We need to determine the value of the constant $k$. Initially, $P_i = 600$ lb/in.$^2$ and $V_i = 1.2$ in.$^3$, so

$$k = (600)(1.2)^{1.4} \approx 774.4707.$$

We then have $PV^{1.4} = 774$, so

$$P = 774V^{-1.4}.$$

The terminal volume is $V_t = 12$ in.$^3$ and the work done by the gas is

$$W = \int_{V_i}^{V_t} P\,dV = \int_{1.2}^{12} 774V^{-1.4}\,dV$$

$$= 774 \left.\frac{V^{-0.4}}{-0.4}\right]_{1.2}^{12} = -1935[(12)^{-0.4} - (1.2)^{-0.4}]$$

$$\approx 1083 \text{ in.} \cdot \text{lb.}$$

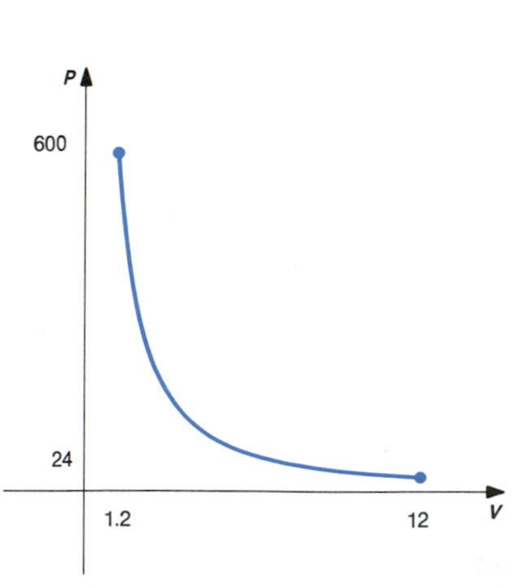


FIGURE 6.5.10

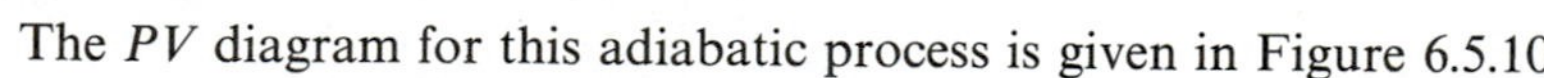

The $PV$ diagram for this adiabatic process is given in Figure 6.5.10. ■

## EXERCISES 6.5

**1** An elevator carries two passengers from the first to the second floor, where an additional passenger enters. Two people get out at the third floor and the remaining person rides to the fourth floor. Find the work required to lift the passengers if each passenger weighs 150 lb and the floors are 12 ft apart.

**2** Eight boxes 1 ft high and weighing 40 lb each are at floor level. Find the work required to stack them one on top of the other.

The containers pictured in Exercises 3–4 are full of water. Find the work required to pump all the water to (a) the level of the top of the containers, (b) a level 2 ft above the top. (Weight = 62.4 lb/ft$^3$.)

**3**

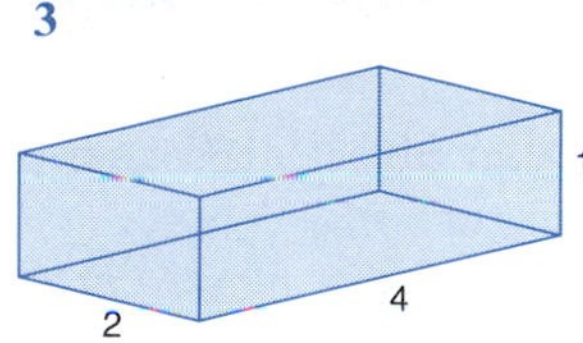


**4**

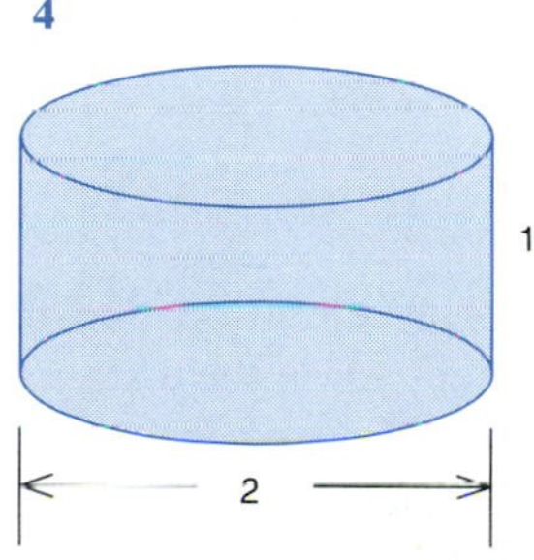


The containers pictured in Exercises 5–8 contain water. (Weight = 62.4 lb/ft$^3$.) Find the work required to pump the water to the level of the top of the containers if (a) they are full, (b) they contain water to a depth of 1 ft.

**5**

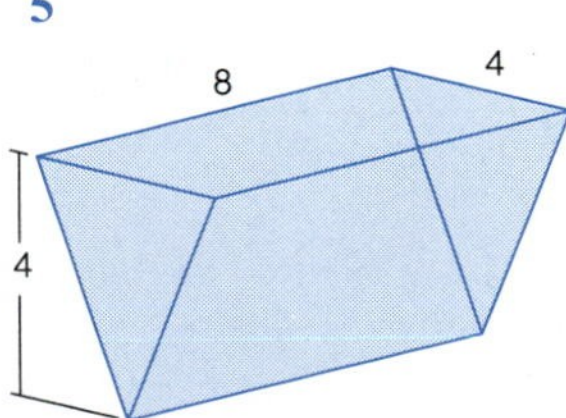


**6**

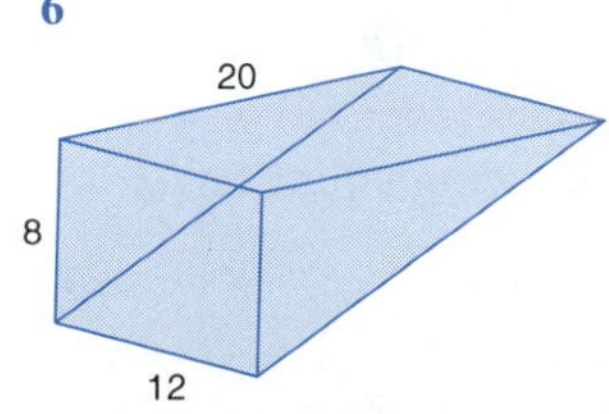


**7**

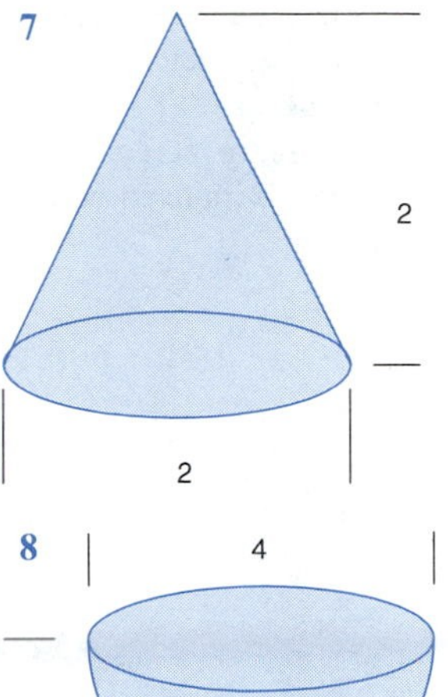


**8**

4

2

**9** A 10-ft chain that weighs 0.8 lb/ft is hanging vertically. (a) Find the work required to raise the entire chain to the level of the top of the chain. (b) Find the work required to raise the chain as in (a) if there is a 20-lb weight attached to the bottom of the chain.

**10** A 30-ft chain that weighs 2 lb/ft is hanging vertically from a crane. (a) Find the work required to roll up 20 ft of the chain, so the bottom is raised 20 ft. (b) Find the work required to raise the chain as in (a) if there is a 200-lb weight attached to the bottom of the chain.

**11** A uniform chain 12 ft long and weighing 4 lb is lying at ground level. Find the work required to raise the chain so it hangs vertically with the bottom of the chain (a) at ground level and (b) 8 ft above ground level.

**12** A uniform chain 10 ft long and weighing 32 lb is lying at ground level. Find the work required to raise the entire chain to a platform that is 4 ft above the ground.

**13** A 2-lb weight stretches a certain spring 1.5 ft. Find the work required to stretch the spring from its natural length to (a) 1.5 ft and (b) 3 ft beyond its natural length.

**14** A 3-lb weight stretches a certain spring 2 ft. Find the work required to stretch the spring from its natural length to (a) 2 ft and (b) 4 ft beyond its natural length.

**15** Two foot-pounds of work are required to stretch a certain spring 1.2 ft beyond its natural length. How far beyond its natural length will a 4-lb weight stretch the spring?

**16** Three foot-pounds of work are required to stretch a certain spring from 2 ft to 4 ft beyond its natural length. How far beyond its natural length will a 5-lb weight stretch the spring?

17 Air in the cylinder of a diesel engine expands from 2 in.$^3$ to 16 in$^3$. The initial pressure is 500 lb/in.$^2$. Find the work done by the gas if the air behaves as an ideal gas and the expansion is adiabatic with $PV^{1.4} = k$.

18 An ideal gas expands from volume 20 in.$^3$ at pressure 15 lb/in$^2$ to volume 30 in.$^3$ at 30 lb/in.$^2$. Sketch the $PV$ diagram and find the work done by the gas (a) if the gas expands with pressure fixed and then pressure increases with fixed volume, and (b) if pressure increases with fixed volume and the gas then expands with pressure fixed.

19 The magnitude of the force due to friction of an object that is moving along a level surface is $F = \mu w$, where $\mu$ is the **coefficient of kinetic friction** and $w$ is the weight of the object. The direction of the force is opposite to the direction of movement. Find the work required to move a 40-lb object 30 ft along a level surface and then return it to its original position if the coefficient of friction is $\mu = 0.5$.

20 A sandbag is being lifted 10 ft at the rate of 2 ft/s. Find the work done if the sandbag initially weighs 80 lb and loses sand at a uniform rate of 3 lb/s as it is being lifted.

## 6.6 MASS, MOMENTS, AND CENTERS OF MASS

The idea of a center of mass or balance point of an object or system of masses involves the mathematical concept of moments.

Moments of a system of masses are defined in terms of mass. Let us distinguish between the mass and the weight of an object. The weight of an object is the force due to the gravitational attraction of the earth. **Newton's Second Law of Motion** tells us that

$$w = ma,$$

where $w$ is the weight, $m$ is the mass, and $a$ is the acceleration due to gravity. The weight and acceleration due to gravity depend on the distance of the object from the center of the earth. The mass depends on the molecular structure and not on the position of the object. For problems that involve motion near the surface of the earth, we assume that acceleration due to gravity has a constant value $g$. We then have

$$w = mg,$$

so the weight of an object is proportional to its mass. We will use either $g = 32$ ft/s$^2$ or $g = 9.8$ m/s$^2$. Units of force, mass, and acceleration in English and metric systems are given in Table 6.6.1.

**TABLE 6.6.1**

| *System* | *Mass* | *Acceleration* | *Force* |
|---|---|---|---|
| English | slug | ft/s$^2$ | slug · ft/s$^2$ = lb |
| mks (SI) | kilogram (kg) | m/s$^2$ | kg · m/s$^2$ = newton (N) |
| cgs | gram (g) | cm/s$^2$ | g · cm/s$^2$ = dyne |

English and metric units are related by the following equations.

$$1 \text{ slug} = 14.57 \text{ kg} \quad \text{and} \quad 1 \text{ N} = 0.225 \text{ lb}.$$

Thus, the *weight* in pounds of a 1-kg *mass* near the surface of the earth is

$$w = mg = (1\text{ kg})(9.8\text{ m/s}^2) = 9.8\text{ N} = (9.8\text{ N})\left(\frac{0.225\text{ lb}}{1\text{ N}}\right) = 2.205\text{ lb}.$$

Let us now define the moment of a system of masses.

*Definition*

The **moment about $x = a$** of a system of masses $m_1, m_2, \ldots, m_n$ located at points $x_1, x_2, \ldots x_n$, respectively, on a number line is

$$M_{x=a} = \sum_{j=1}^{n} (x_j - a)m_j.$$

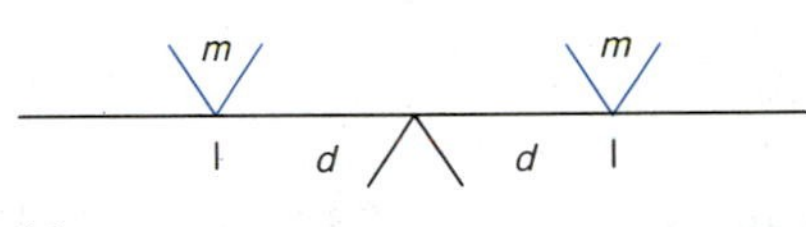


(a)

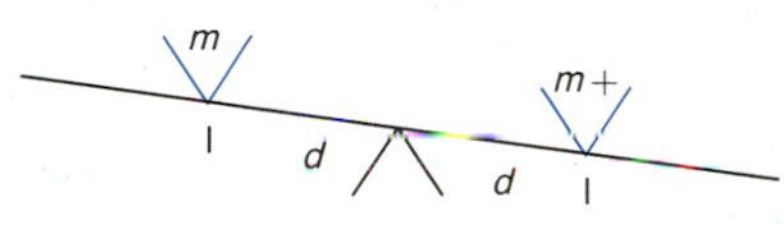


(b)

(c)

FIGURE 6.6.1

We may think of $M_{x=a}$ as a measure of the tendency of the system to rotate about the point $x = a$ due to the force of gravity acting downward on each mass of the system.

The idea of moments and balance points should be consistent with our experience. For example, we know that two equal masses will balance if they are placed equal distances on opposite sides of the fulcrum of a lever (Figure 6.6.1a). If one mass is either increased (Figure 6.6.1b) or moved farther from the fulcrum (Figure 6.6.1c), the lever will rotate down on the side of that mass and up on the opposite side.

It is a rule of physics that a system of masses will balance at the point $x = a$ if $M_{x=a} = 0$. We can use the definition of $M_{x=a}$ to solve the equation for the balance point. Thus, if $x = a$ is the balance point,

$$M_{x=a} = 0,$$

$$\sum_{j=1}^{n} (x_j - a)m_j = 0,$$

$$\sum_{j=1}^{n} x_j m_j - a\sum_{j=1}^{n} m_j = 0,$$

$$a = \frac{\sum_{j=1}^{n} x_j m_j}{\sum_{j=1}^{n} m_j}.$$

We have $\sum_{j=1}^{n} x_j m_j = M_{x=0}$; $M = \sum_{j=1}^{n} m_j$ is the total mass of the system. It is customary to denote the balance point by $\bar{x}$. Then

$$\bar{x} = \frac{\sum_{j=1}^{n} x_j m_j}{\sum_{j=1}^{n} m_j} = \frac{M_{x=0}}{M}.$$

The balance point $\bar{x}$ is called the **center of mass** of the system.

**EXAMPLE 1**

Find the center of mass of the system of masses $m_1 = 5$, $m_2 = 3$, $m_3 = 2$, located at points $x_1 = -4$, $x_2 = 2$, $x_3 = 8$, respectively.

**SOLUTION**

The total mass is

$$M = m_1 + m_2 + m_3 = 5 + 3 + 2 = 10.$$

The moment about $x = 0$ of the system is

$$M_{x=0} = x_1m_1 + x_2m_2 + x_3m_3 = (-4)(5) + (2)(3) + (8)(2) = 2.$$

The center of mass is

$$\bar{x} = \frac{M_{x=0}}{M} = \frac{2}{10} = 0.2.$$

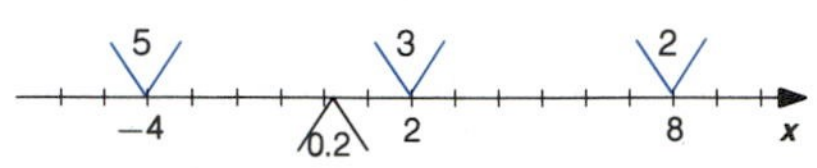


FIGURE 6.6.2

See Figure 6.6.2. ■

The same ideas of moments and centers of mass apply to a system of masses $m_1, m_2, \ldots, m_n$ located at points $(x_1, y_1), (x_2, y_2), \ldots, (x_n, y_n)$ in the plane.

$$M_{x=0} = \sum_{j=1}^{n} x_jm_j \quad \text{and} \quad M_{y=0} = \sum_{j=1}^{n} y_jm_j$$

represent the moments about the *lines* $x = 0$ and $y = 0$, respectively. The center of mass of the system is $(\bar{x}, \bar{y})$, where

$$\bar{x} = \frac{\sum_{j=1}^{n} x_jm_j}{\sum_{j=1}^{n} m_j} = \frac{M_{x=0}}{M} \quad \text{and} \quad \bar{y} = \frac{\sum_{j=1}^{n} y_jm_j}{\sum_{j=1}^{n} m_j} = \frac{M_{y=0}}{M}.$$

It is common to designate $M_{x=0}$ by $M_y$ and $M_{y=0}$ by $M_x$. That is,

$$M_{x=0} = M_{y\text{-axis}} = M_y \quad \text{and} \quad M_{y=0} = M_{x\text{-axis}} = M_x.$$

■ **EXAMPLE 2**

Find the center of mass of the system of masses $m_1 = 6$, $m_2 = 3$, $m_3 = 2$ located at points $(x_1, y_1) = (-4, 2)$, $(x_2, y_2) = (2, 1)$, $(x_3, y_3) = (7, -5)$, respectively.

**SOLUTION**

We have

$$M = m_1 + m_2 + m_3 = 6 + 3 + 2 = 11,$$

$$M_{x=0} = x_1m_1 + x_2m_2 + x_3m_3 = (-4)(6) + (2)(3) + (7)(2) = -4,$$

and

$$M_{y=0} = y_1m_1 + y_2m_2 + y_3m_3 = (2)(6) + (1)(3) + (-5)(2) = 5.$$

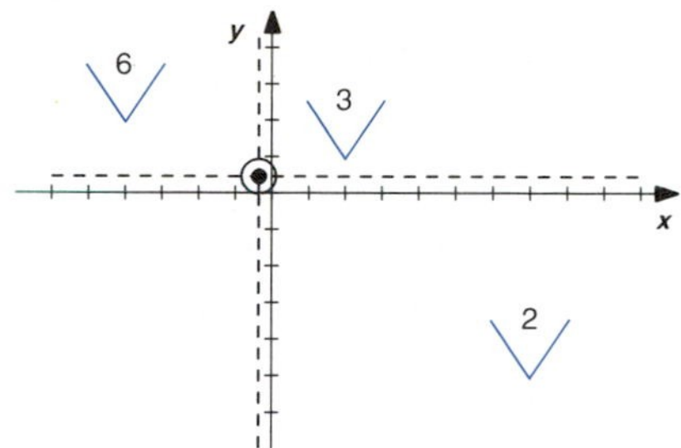

FIGURE 6.6.3

Hence,

$$\bar{x} = \frac{M_{x=0}}{M} = \frac{-4}{11} \quad \text{and} \quad \bar{y} = \frac{M_{y=0}}{M} = \frac{5}{11}.$$

The center of mass is located at $(\bar{x}, \bar{y}) = (-4/11, 5/11)$. See Figure 6.6.3. ■

It is not difficult to see that a system will balance about any line through its center of mass. For example, suppose that $(\bar{x}, \bar{y}) = (0, 0)$, so

$$\sum_{j=1}^{n} x_j m_j = \sum_{j=1}^{n} y_j m_j = 0.$$

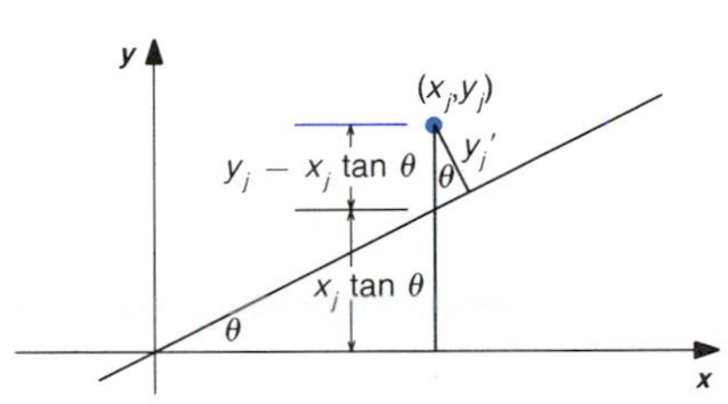

FIGURE 6.6.4

From Figure 6.6.4 we see that the directed distance of a point $(x_j, y_j)$ from the line $l$ through the origin at an angle $\theta$ with the $x$-axis is given by

$$y_j' = (y_j - x_j \tan\theta)(\cos\theta) = y_j \cos\theta - x_j \sin\theta.$$

Hence, the moment about $l$ of the system is

$$M_l = \sum_{j=1}^{n} y_j' m_j = \cos\theta \sum_{j=1}^{n} y_j m_j - \sin\theta \sum_{j=1}^{n} x_j m_j = 0.$$

Actually, the system will balance if supported only at the point that is the center of mass.

Let us now extend the ideas of mass, moment, and center of mass from systems of point masses to solid objects. In this section we will consider thin, flat objects whose shapes are described by regions in the plane. Such an object is called a **lamina**. The **density** of a lamina is its mass per unit area. **Homogeneous** laminas have uniform density, given by a constant $\rho$. The mass is then the product of the density and the area of the region that describes the object:

$$M = \rho A.$$

We will use the following fact.

**The center of mass of a homogeneous rectangular lamina is located at the geometric center of the rectangle.** (1)

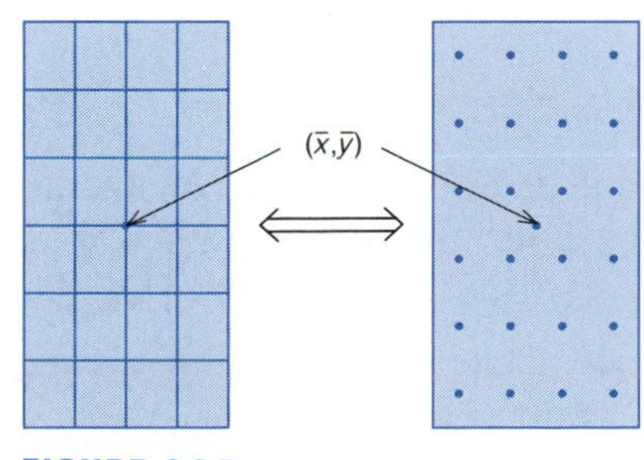

FIGURE 6.6.5

Statement (1) is intuitively clear from the interpretation of center of mass as the balance point. It is worthwhile to use the ideas of calculus to relate the problem to that of a sytstem of point masses. For example, we could divide the rectangle into small pieces of equal size, as illustrated in Figure 6.6.5. Each small piece could then be considered as a point mass and we could determine the center of mass of the system of point masses. Such a procedure would lead to (1).

Similar reasoning gives the following:

**The center of mass of a homogeneous circular lamina is located at the center of the circle.**

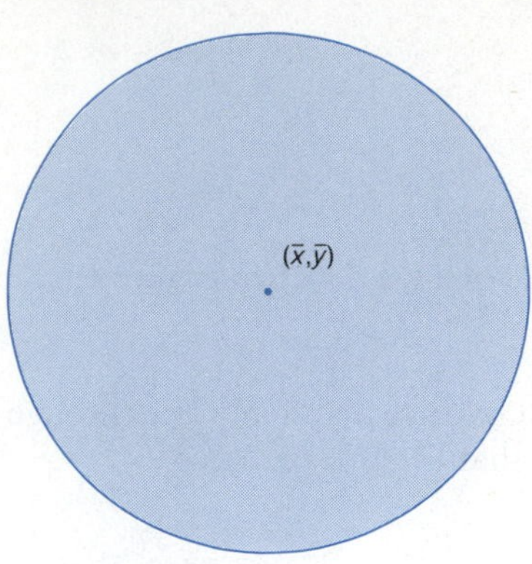


FIGURE 6.6.6

See Figure 6.6.6.

We can also determine the center of mass of a homogeneous triangular lamina.

**The center of mass of a homogeneous triangular lamina is located at the intersection of the medians of the triangle.**

Statement (1) implies that each thin rectangular area strip in Figure 6.6.7 balances about a point very near its center, so the balance point of the triangle must be very near the line formed by the centers of the rectangular strips. The line formed by the centers of the strips is a median of the triangle. Since the strips can be drawn parallel to any side of the triangle, we conclude that the intersection of the medians of the triangular lamina is the balance point. It is useful to recall that the intersection of the medians of a triangle is located at one-third the height from each side. See Figure 6.6.8.

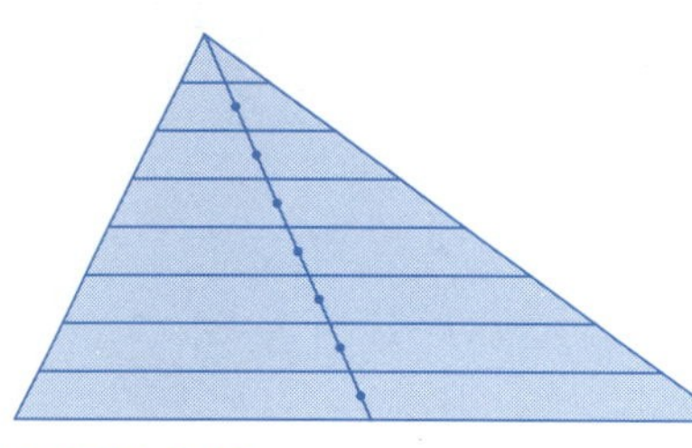

FIGURE 6.6.7

We can rewrite the definition of coordinates of the center of mass of a system in the form

$$M_{x=0} = \bar{x}M \quad \text{and} \quad M_{y=0} = \bar{y}M. \tag{2}$$

We then observe that $\bar{x}M$ and $\bar{y}M$ can be interpreted as the moments about $x = 0$ and $y = 0$, respectively, of a system that consists of a single point mass of $M$ located at the point $(\bar{x}, \bar{y})$. Statement (2) then gives the following fact.

**The moments about $x = 0$ and $y = 0$ of a system of point masses with total mass $M$ are equivalent to the corresponding moments of the system that consists of a single point mass of $M$ located at the center of mass of the original system.**

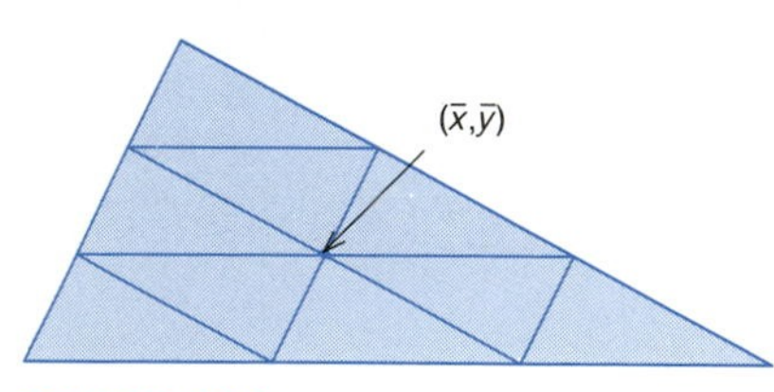


FIGURE 6.6.8

When calculating moments we can replace a system or an object by a point mass located at the center of mass. We can also replace any part of a system by a corresponding point mass at the center of mass of the part.

■ **EXAMPLE 3**

Find the center of mass of the homogeneous lamina pictured in Figure 6.6.9.

**SOLUTION**

Let us obtain an equivalent system of three point masses by replacing the circle, the rectangle, and the triangle by point masses located at their respective centers of mass. The mass of each part is determined by the formula

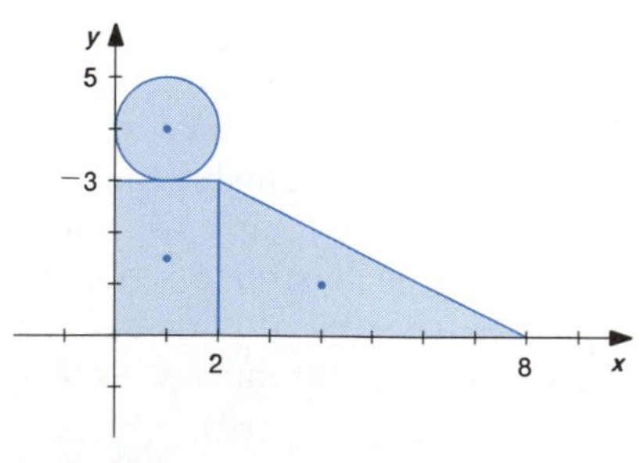


FIGURE 6.6.9

$$\text{Mass} = (\text{Density})(\text{Area})$$

and each center of mass is determined geometrically.

The mass of the circle is $m_1 = \rho\pi(1)^2$, where $\rho$ is the constant density. The center of mass is $(x_1, y_1) = (1, 4)$.

The rectangle has mass $m_2 = \rho(2)(3)$ and center of mass

$$(x_2, y_2) = \left(1, \frac{3}{2}\right).$$

The triangle has mass

$$m_3 = \rho\left(\frac{1}{2}\right)(6)(3)$$

and center of mass $(x_3, y_3) = (4, 1)$.

For the system of three masses, we have

$$M = \pi\rho + 6\rho + 9\rho = (\pi + 15)\rho,$$

$$M_{x=0} = (1)(\pi\rho) + (1)(6\rho) + (4)(9\rho) = (\pi + 42)\rho,$$

and

$$M_{y=0} = (4)(\pi\rho) + \left(\frac{3}{2}\right)(6\rho) + (1)(9\rho) = (4\pi + 18)\rho.$$

Hence,

$$\bar{x} = \frac{M_{x=0}}{M} = \frac{(\pi + 42)\rho}{(\pi + 15)\rho} = \frac{\pi + 42}{\pi + 15} \approx 2.488,$$

$$\bar{y} = \frac{M_{y=0}}{M} = \frac{(4\pi + 18)\rho}{(\pi + 15)\rho} = \frac{4\pi + 18}{\pi + 15} \approx 1.685.$$

From Figure 6.6.9 we see that $(\bar{x}, \bar{y}) \approx (2.488, 1.685)$ seems to be a reasonable position for the balance point. This serves as a rough check on our calculations. ■

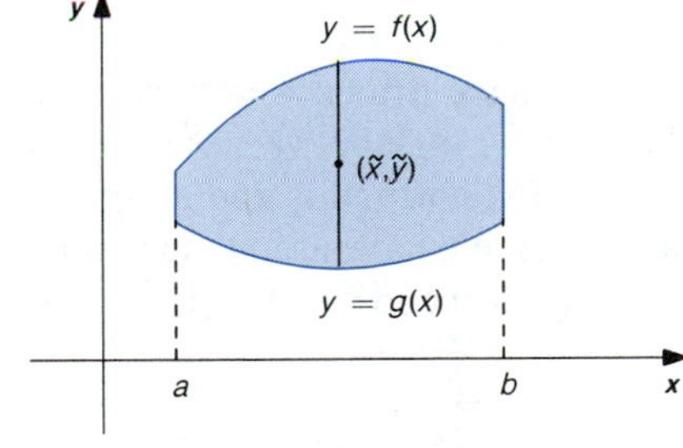


(a)

We can use calculus to determine the center of mass of a homogeneous lamina whose shape is given by equations in $x$ and $y$. See Figure 6.6.10.

(b)

FIGURE 6.6.10

***Solution steps***

- **Sketch the region and draw a variable rectangular area strip.**
- **Label the center of the mass of the variable area strip in terms of the appropriate variable. Denote the center of mass of the rectangle by $(\tilde{x}, \tilde{y})$. [The formula for the midpoint of a line segment can be used to determine $(\tilde{x}, \tilde{y})$.]**
- **Use the integral as a sum to calculate the total mass of the lamina. The mass of a variable rectangle is**

$$dM = \rho\, dA,$$

**where $\rho$ is the density and $dA$ represents the area of the rectangle. The total mass is the limit of sums,**

$$M = \int dM.$$

- **Use the integral as a sum to calculate the moments about $x = 0$ and $y = 0$:**

$$M_{x=0} = \int \tilde{x}\, dM \quad \text{and} \quad M_{y=0} = \int \tilde{y}\, dM.$$

**Note that $\tilde{x}\, dM$ and $\tilde{y}\, dM$ represent the moments about $x = 0$ and $y = 0$, respectively, of the variable rectangle. The integrals give the limits of the sums.**

- **The center of mass is given by**

$$\bar{x} = \frac{\int \tilde{x}\, dM}{\int dM} = \frac{M_{x=0}}{M} \quad \text{and} \quad \bar{y} = \frac{\int \tilde{y}\, dM}{\int dM} = \frac{M_{y=0}}{M}.$$

Let us use the integral formulas to find the center of mass of a homogeneous triangular lamina.

### EXAMPLE 4

Find the center of mass of the homogeneous triangular lamina bounded by $y = 3x$, $y = 0$, and $x = 2$.

#### SOLUTION

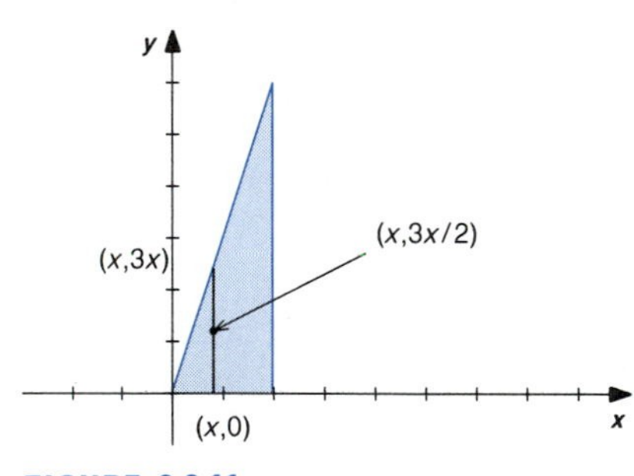


FIGURE 6.6.11

The lamina is sketched in Figure 6.6.11. We have drawn a vertical area strip, so we will use $x$ as our variable. The strips range between $x = 0$ and $x = 2$, so the limits of integration are 0 and 2. The endpoints and midpoint of the strip are labeled in terms of the variable $x$. We have used the formula for the midpoint of the line segment between $(x, 3x)$ and $(x, 0)$ to determine that

$$(\tilde{x}, \tilde{y}) = \left(\frac{x + x}{2}, \frac{3x + 0}{2}\right) = \left(x, \frac{3x}{2}\right).$$

The variable area strip has

Length $= 3x - 0 = 3x$,

Width $= dx$,

Area $= 3x\, dx$,

Mass $= dM = \rho\, dA = \rho 3x\, dx$.

It follows that

$$M = \int dM = \int_0^2 \rho 3x\,dx = 6\rho,$$

$$M_{x=0} = \int \tilde{x}\,dM = \int_0^2 x\rho 3x\,dx = \int_0^2 \rho 3x^2\,dx = 8\rho, \quad \text{and}$$

$$M_{y=0} = \int \tilde{y}\,dM = \int_0^2 \left(\frac{3x}{2}\right)\rho 3x\,dx = 12\rho, \quad \text{so}$$

$$\bar{x} = \frac{M_{x=0}}{M} = \frac{8\rho}{6\rho} = \frac{4}{3} \quad \text{and} \quad \bar{y} = \frac{M_{y=0}}{M} = \frac{12\rho}{6\rho} = 2.$$

We see that the point

$$(\bar{x}, \bar{y}) = \left(\frac{4}{3}, 2\right)$$

is the intersection of the medians of the triangle, as we knew it should be. ■

■ **EXAMPLE 5**

Find the center of mass of the homogeneous lamina bounded by $y = x^2$ and $y = x$.

**SOLUTION**

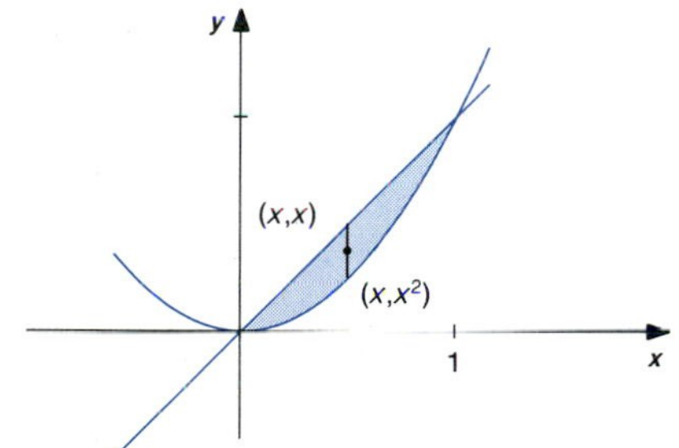

FIGURE 6.6.12

The region is sketched in Figure 6.6.12. We choose a vertical area strip, so the variable is $x$. Note that $x$ ranges between $x = 0$ and $x = 1$, the $x$-coordinates of the points of intersection of $y = x^2$ and $y = x$.

Using the formula for the midpoint of the line segment between $(x, x^2)$ and $(x, x)$, we have

$$(\tilde{x}, \tilde{y}) = \left(\frac{x + x}{2}, \frac{x + x^2}{2}\right) = \left(x, \frac{x + x^2}{2}\right).$$

A variable rectangle has

Length $= x - x^2$,

Width $= dx$,

Area $= dA = (x - x^2)\,dx$,

Mass $= dM = \rho\,dA = \rho(x - x^2)\,dx$.

It follows that

$$M = \int dM = \int_0^1 \rho(x - x^2)\,dx = \frac{\rho}{6},$$

$$M_{x=0} = \int \tilde{x}\,dM = \int_0^1 x\rho(x - x^2)\,dx = \frac{\rho}{12},$$

and

$$M_{y=0} = \int \tilde{y}\, dM = \int_0^1 \left(\frac{x + x^2}{2}\right)\rho(x - x^2)\, dx = \frac{\rho}{2}\int_0^1 (x^2 - x^4)\, dx = \frac{\rho}{15},$$

so

$$\bar{x} = \frac{M_{x=0}}{M} = \frac{\rho/12}{\rho/6} = \frac{1}{2} \quad \text{and} \quad \bar{y} = \frac{M_{y=0}}{M} = \frac{\rho/15}{\rho/6} = \frac{2}{5}.$$

The center of mass is located at $(\bar{x}, \bar{y}) = (1/2, 2/5)$. ■

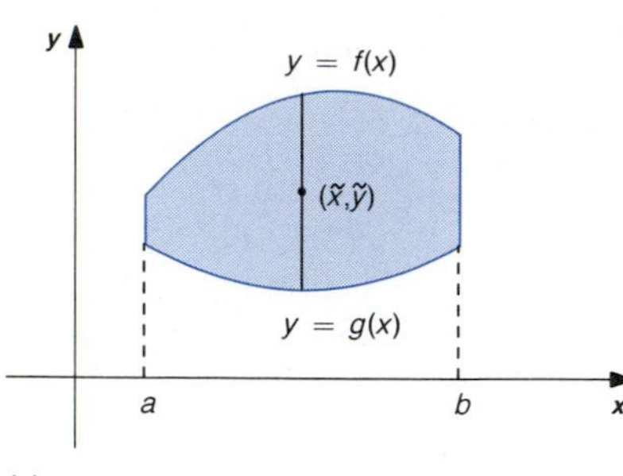

(a)

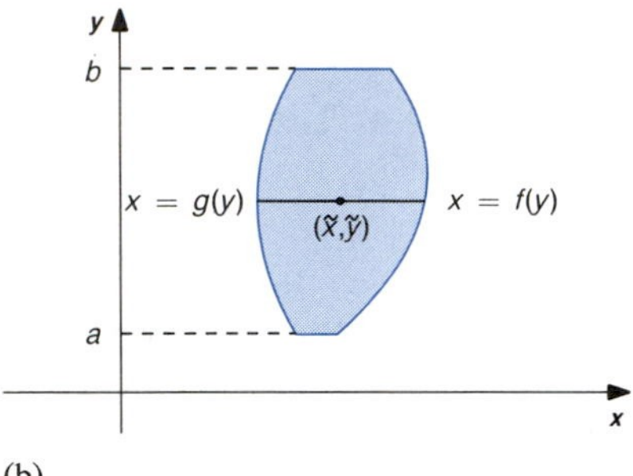

(b)

FIGURE 6.6.13

One may notice that the density $\rho$ of a homogeneous lamina cancels out of the calculation of the center of mass of the lamina. It is a good idea to use $\rho$ in the calculations for homogeneous laminas, even though it does cancel. It is absolutely necessary that $\rho$ be included in calculations that involve variable density.

If the density of a lamina is given by a continuous function of either $x$ or $y$, but not both, we can use definite integrals to evaluate the mass and moments of the lamina. If $\rho = \rho(x)$, then thin vertical rectangles are essentially homogeneous and have center of mass very near their geometric center. We can then calculate mass and moments by dividing the lamina into vertical area strips and using $x$ as the variable of integration. See Figure 6.6.13a. If $\rho = \rho(y)$, then thin horizontal rectangles are essentially homogeneous and we can calculate mass and moments by dividing the lamina into horizontal area strips and using $y$ as the variable of integration. See Figure 6.6.13b.

### ■ EXAMPLE 6

Find the center of mass of the lamina that is bounded by $y = \sqrt{x}$, $y = 1$, and the $y$-axis, if the density is given by $\rho = y$.

#### SOLUTION

The region is sketched in Figure 6.6.14. Since the density is a function of $y$, we must use $y$ as the variable of integration. Thin horizontal rectangles are essentially homogeneous. We have labeled a typical horizontal area strip in terms of the variable $y$. Note that $y = \sqrt{x}$ implies $x = y^2$. A variable horizontal area strip has

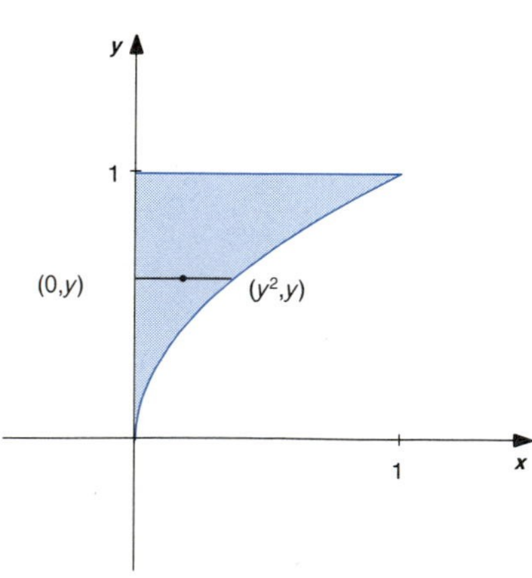

FIGURE 6.6.14

$$\text{Area} = dA = y^2\, dy,$$

$$\text{Mass} = dM = \rho\, dA = (y)(y^2\, dy) = y^3\, dy,$$

and midpoint

$$(\tilde{x}, \tilde{y}) = \left(\frac{y^2}{2}, y\right).$$

The total mass is

$$M = \int dM = \int_0^1 y^3\, dy = \frac{1}{4}.$$

The moments are

$$M_{x=0} = \int \tilde{x}\, dM = \int_0^1 \left(\frac{y^2}{2}\right) y^3\, dy = \int_0^1 \frac{y^5}{2}\, dy = \frac{1}{12}$$

and

$$M_{y=0} = \int \tilde{y}\, dM = \int_0^1 (y) y^3\, dy = \int_0^1 y^4\, dy = \frac{1}{5}, \quad \text{so}$$

$$\bar{x} = \frac{M_{x=0}}{M} = \frac{1/12}{1/4} = \frac{1}{3} \quad \text{and} \quad \bar{y} = \frac{M_{y=0}}{M} = \frac{1/5}{1/4} = \frac{4}{5}.$$

The center of mass is located at

$$(\bar{x}, \bar{y}) = \left(\frac{1}{3}, \frac{4}{5}\right).$$ ■

In the case of a homogeneous lamina defined by a region in the plane, it is customary to refer to the center of mass of the lamina as the **centroid** of the region.

## EXERCISES 6.6

In Exercises 1–4 find $\bar{x}$, the balance point of the given system of masses on a coordinate line.

**1** $m_1 = 4, m_2 = 3; x_1 = -2, x_2 = 4$

**2** $m_1 = 2, m_2 = 5; x_1 = -4, x_2 = 3$

**3** $m_1 = 2, m_2 = 3, m_3 = 5; x_1 = -6, x_2 = -2, x_3 = 4$

**4** $m_1 = 6, m_2 = 4, m_3 = 3; x_1 = -4, x_2 = 2, x_3 = 5$

In Exercises 5–8 find $(\bar{x}, \bar{y})$, the center of mass of the given system of masses in the plane.

**5** $m_1 = 1,\ m_2 = 2,\ m_3 = 3;\ (x_1, y_1) = (-4, 5),\ (x_2, y_2) = (-1, 2),\ (x_3, y_3) = (2, -4)$

**6** $m_1 = 2,\ m_2 = 3,\ m_3 = 4;\ (x_1, y_1) = (-6, 5),\ (x_2, y_2) = (-2, 2),\ (x_3, y_3) = (6, 4)$

**7** $m_1 = 2,\ m_2 = 3,\ m_3 = 3,\ m_4 = 4;\ (x_1, y_1) = (-4, -2),\ (x_2, y_2) = (3, 4),\ (x_3, y_3) = (1, -4),\ (x_4, y_4) = (2, -5)$

**8** $m_1 = 1,\ m_2 = 2,\ m_3 = 3,\ m_4 = 4;\ (x_1, y_1) = (5, 4),\ (x_2, y_2) = (2, 4),\ (x_3, y_3) = (1, -4),\ (x_4, y_4) = (-2, 1)$

Find the center of mass of each homogeneous lamina pictured in Exercises 9–14.

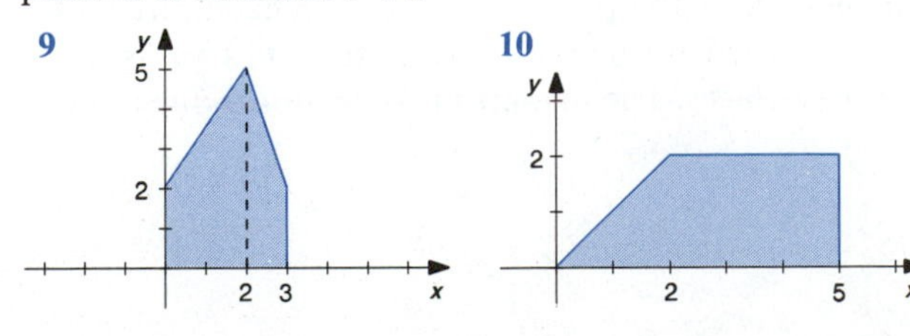


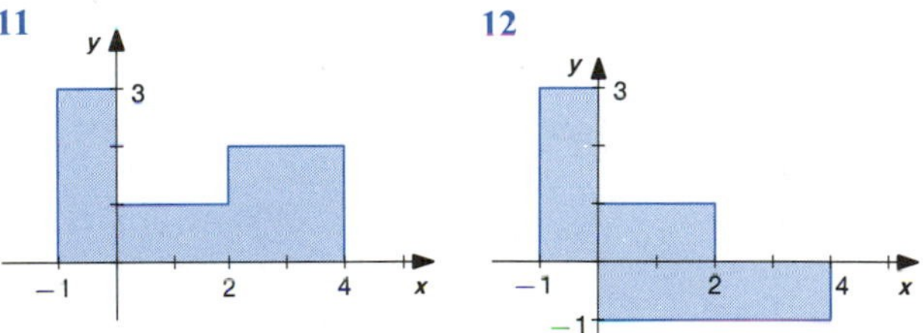


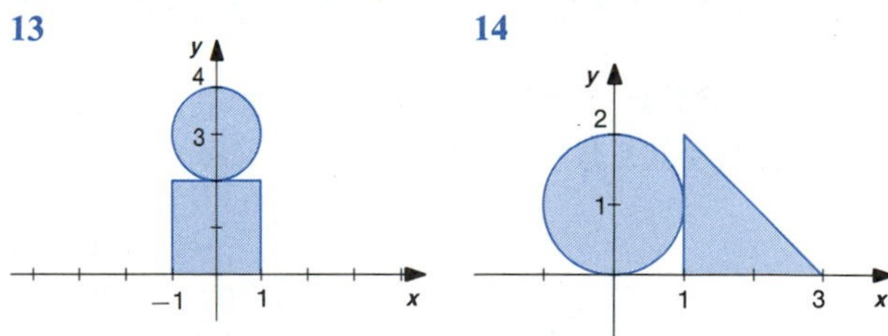


In Exercises 15–24 find the center of mass of the homogeneous lamina bounded by the given curves.

**15** $y = x^2, y = 2x, x = 0, x = 1$

**16** $y = x^2, y = 4x, x = 0, x = 1$

**17** $y = x^2, y = x^2 + 1, x = -1, x = 1$

**18** $y = x^2 + 2, y = 1 - x^2, x = -1, x = 1$

**19** $y = x^2, y = x + 2$

**20** $y = 2 - x^2, y = x$

21 $x = 2y^2$, $x = y^2 + 1$

22 $x = y^2 + 1$, $x = 0$, $y = -1$, $y = 1$

23 $x = 2y$, $x = y^2 + 1$, $y = 0$

24 $x = 1 - y^2$, $x = 2 - 2y$, $y = 0$

25 Verify that the integration method gives the center of mass of the homogeneous triangular lamina pictured to be $(b/3, h/3)$.

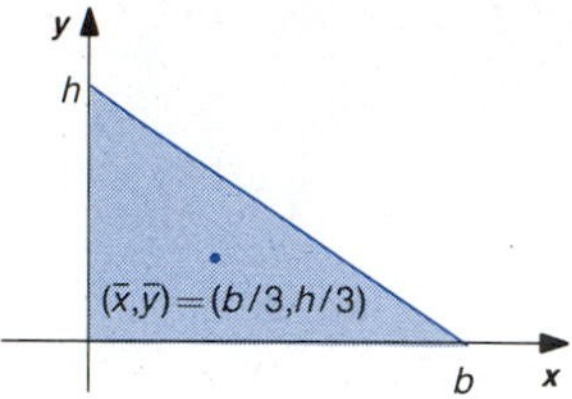


26 Verify that the integration method gives the center of mass of a homogeneous circular lamina to be the center of the circle. ($M = \rho A = \rho\pi r^2$.)

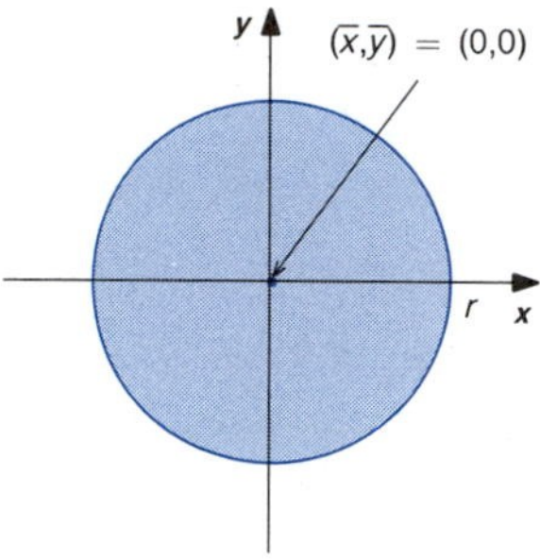


In Exercises 27–34 find the center of mass of the lamina that is bounded by the given curves and has the given density function.

27 $y = 1$, $y = 0$, $x = 0$, $x = 1$; $\rho = x$

28 $y = x$, $y = 0$, $x = 1$; $\rho = x^2$

29 $y = x^3$, $y = 0$, $x = 0$, $x = 1$; $\rho = 2x$

30 $y = x$, $y = x^2$; $\rho = x$

31 $y = x^2$, $y = 0$, $x = 1$ $(x \geq 0)$; $\rho = y$

32 $y = x$, $y = 1$, $x = 0$; $\rho = y$

33 $x + y = 1$, $y = 0$, $x = 0$; $\rho = y^2$

34 $x = 1 - y^2$, $x = 0$; $\rho = y^2$

35 Let $R$ be the region bounded by $y = f(x)$, $y = g(x)$, $x = a$, and $x = b$. Assume $f$ and $g$ are continuous with $g(x) \leq f(x)$ for $a \leq x \leq b$. Let $\bar{x}$ be the $x$-coordinate of the centroid of $R$. Let $A$ denote the area of $R$, and let $V$ denote the volume of the solid obtained by revolving $R$ about a vertical line $x = x_0$, $x_0 \leq a$. Show that $V = 2\pi(\bar{x} - x_0)A$, so that the volume of the solid is the distance traveled by the centroid of the region times the area of the region. This illustrates a general result called **Pappus' Theorem.**

In Exercises 36–38 use the result of Exercise 35 to find the volume of the solids obtained by revolving the regions bounded by the given curves about the given lines.

36 $(x - 1)^2 + y^2 = 1$; about the $y$-axis

37 $y = x$, $y = 0$, $x = 3$; about the $y$-axis

38 $\dfrac{x}{r} + \dfrac{y}{h} = 1$, $y = 0$, $x = 0$ $(r, h > 0)$; about the $y$-axis

## 6.7 FORCE DUE TO FLUID PRESSURE

Fluid exerts force due to fluid pressure on submerged surfaces. We will study forces on flat surfaces. We will see that calculation of force due to fluid pressure on such surfaces involves moment integrals.

Let us begin by considering some basic principles of fluid pressure. Pressure is force per unit area. We will consider problems in which the force is due to the weight of the fluid. We assume the fluid is located near the surface of the earth, so weight is a constant multiple of the mass, $w = mg$. Fluid pressure at a point in a standing body of fluid is then directly proportional to the depth of the point below the level of the fluid surface. **Pascal's Principle** asserts the pressure is equal in each direction.

If a flat surface is submerged horizontally, then the fluid pressure is equal at each point on the surfaces. The magnitude of the force due to fluid pressure is then given by

$$F = \rho g h A,$$

where $\rho$ is the density of the fluid, $h$ is the depth of the surface, and $A$ is the area of the surface. (The density $\rho$ is mass per unit volume and $\rho g$ is the weight per unit volume of the fluid.) The force acts in a direction perpendicular to the surface. Note that $hA$ is the volume of a generalized right cylinder with height $h$ and area of base $A$; $\rho hA$ is the mass and $\rho ghA$ is the weight of that amount of fluid. See Figure 6.7.1a.

It is a consequence of Pascal's Principle that the force due to fluid pressure depends on the depth of the surface below the uppermost fluid level. It does not matter how much fluid is directly above the surface. Thus, if the containers pictured in Figure 6.7.1a and 6.7.1b contain the same type fluid at the same uppermost level and have bases of equal area, then the forces on their bottom surfaces will be equal. This is because the fluid exerts an upward force on the horizontal surface below the fluid level in Figure 6.7.1b, and the surface must exert an equal downward force to contain the fluid. See Figure 6.7.1c. The downward force exerted by the container is transmitted to the total force on the bottom of the container. The downward force exerted by the container is equal to the force exerted by a column of fluid with height at the uppermost fluid level. For example, we know that if a thin straw were inserted as in Figure 6.7.1d, the fluid would be forced up the straw until it reached the uppermost level of the surface.

**FIGURE 6.7.1**

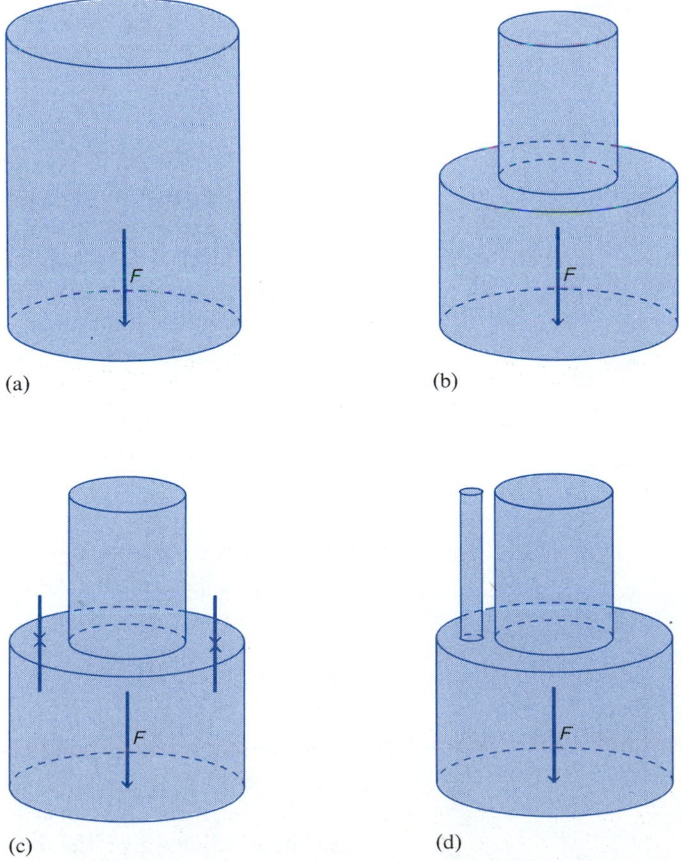

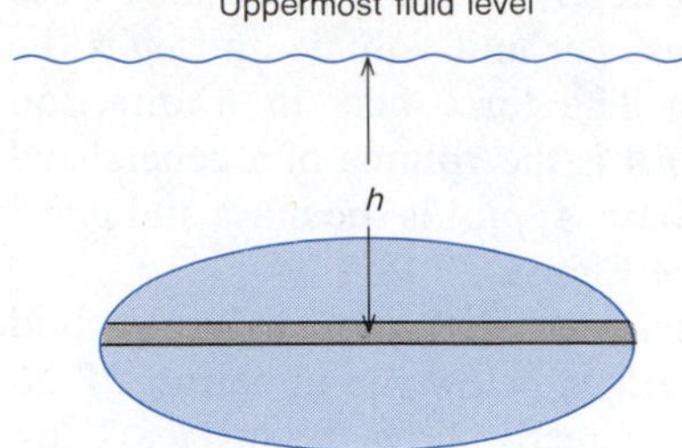


FIGURE 6.7.2

Let us now consider the force on a submerged flat surface that is not necessarily horizontally. We divide the surface into strips so that points on the same strip are at approximately the same level. The force on a strip of area $dA$, located at a depth of approximately $h$, is approximately

$$dF = \rho gh\, dA.$$

See Figure 6.7.2. It is important that each point on the strip be at approximately the same level and

**Depth of strip = (Uppermost fluid level) − (Level of strip).**

The force acts in the direction perpendicular to the surface. (Recall that Pascal's Principle tells us the pressure at a point is equal in every direction.) Since we have assumed that the surface is flat, the forces on different strips act in the same direction. The total force on the surface is the sum of the forces on its parts, which we write as an integral:

$$F = \int \rho gh\, dA.$$

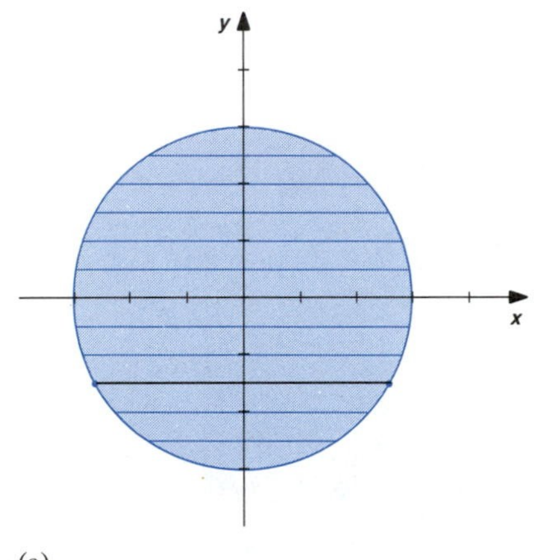


(a)

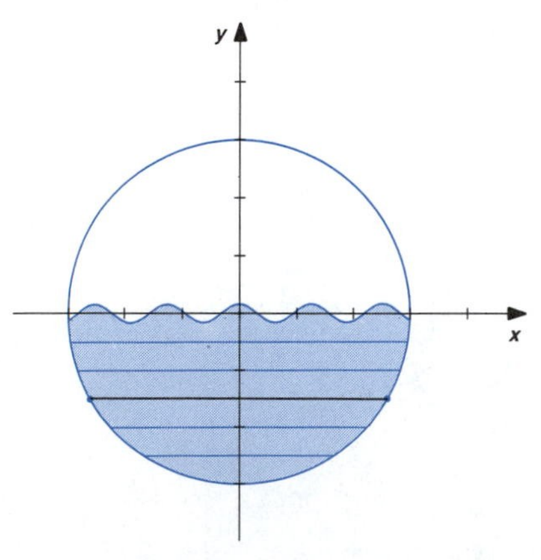


(b)

FIGURE 6.7.3

■ **EXAMPLE 1**

A cylindrical tank is lying on its side. The radius of each circular end is 3 ft. Set up an integral that gives the force on one circular end if the tank is (a) completely full of a fluid with weight $\rho g = 62.4$ lb/ft$^3$ and (b) half full.

**SOLUTION**

Since the tank is lying on its side, the circular ends are vertical. We choose a coordinate system with origin at the center of one end, so that end is described by the equation $x^2 + y^2 = 3^2$. See Figure 6.7.3. The variable $y$ represents the level of a variable area strip. We need to calculate the force on each strip as $y$ varies between the smallest and largest values of the $y$-coordinates of the submerged surface.

(a) If the tank is full, the uppermost fluid level is $y = 3$, so the depth of a variable strip is

$$h = (\text{Uppermost fluid level}) - (\text{Level of strip}) = 3 - y.$$

The variable $y$ ranges from $-3$ to 3. See Figure 6.7.3a. A variable horizontal strip runs from $(-\sqrt{9 - y^2}, y)$ to $(\sqrt{9 - y^2}, y)$. The area of a variable strip is $dA = 2\sqrt{9 - y^2}\, dy$. The total force is

$$F = \int_{-3}^{3} (62.4)(3 - y)2\sqrt{9 - y^2}\, dy.$$

[The above integral has value $(62.4)(27\pi) \approx 5292.9553$.]

(b) If the tank is only half full, the uppermost fluid level is $y = 0$, so $h = 0 - y = -y$. The submerged surface is between levels $y = -3$ and $y = 0$. These values give the limits of intergration. See Figure 6.7.3b.

The total force is

$$F = \int_{-3}^{0} (62.4)(0 - y)2\sqrt{9 - y^2}\, dy.$$

[This integral has value (62.4)(18) = 1123.3.] ■

Let us now see how *force due to fluid pressure is related to a moment integral.* We merely note that

$$F = \int \rho g h \, dA = g\left(\int h\rho \, dA\right).$$

The last integral may be interpreted as the moment integral about the uppermost level of the fluid of the system of masses $\rho \, dA$ at level $h$ below the uppermost fluid level. This observation allows us to use what we know about moment integrals to evaluate forces on submerged flat surfaces. In particular, since the moment of a system of masses is the same as the moment of the system with total mass located at the center of mass, we have

**Force on submerged flat surface**
**= (Depth of centroid of surface)($\rho g$)(Area of surface).**

We can use the above formula to evaluate the integral in Example 1a. The centroid of the circular surface is located at its center, 3 ft below the uppermost fluid level. Thus,

$$F = \bar{y}\rho g A = (3)(62.4)(\pi 3^2) \approx 5292.9553 \text{ lb}.$$

It is not convenient to use this method to evaluate the integral in Example 1b, because we don't know the centroid of a semicircular surface.

■ **EXAMPLE 2**

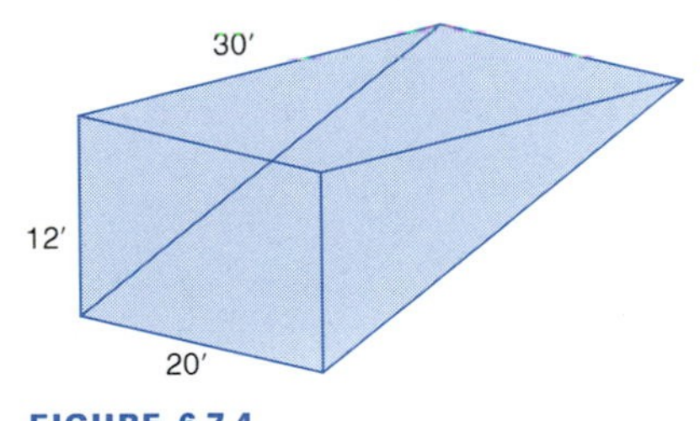


FIGURE 6.7.4

The swimming pool pictured in Figure 6.7.4 is full of water. Find the force due to fluid pressure on (a) the vertical rectangular surface at the deep end, (b) one triangular side, and (c) the slanted bottom of the pool. ($\rho g$ = 62.4 lb/ft$^3$.)

**SOLUTION**

(a) The centroid of the rectangle is located 6 ft below the fluid level. We have

$$F = (\text{Depth of centroid})(\rho g)(\text{Area})$$

$$= (6)(62.4)(20)(12) = 89{,}856 \text{ lb}.$$

(b) The centroid of the triangle is located one-third of the distance down from the horizontal side at the top of the triangle, 4 ft below the fluid level. The force is

$$F = (4)(62.4)\left(\frac{1}{2}\right)(30)(12) = 44{,}928 \text{ lb}.$$

(c) The center of the slanted, rectangular bottom of the pool is 6 ft deep. The length of the slanted side is

$$\sqrt{30^2 + 12^2} = \sqrt{1044}, \quad \text{so}$$

$$F = (6)(62.4)(20)(\sqrt{1044}) \approx 241{,}945 \text{ lb.}$$

## EXERCISES 6.7

**1** The containers pictured are filled with a fluid of weight density $\rho g = 60 \text{ lb/ft}^3$. Find the force due to fluid pressure on the circular bottom surface of each container.

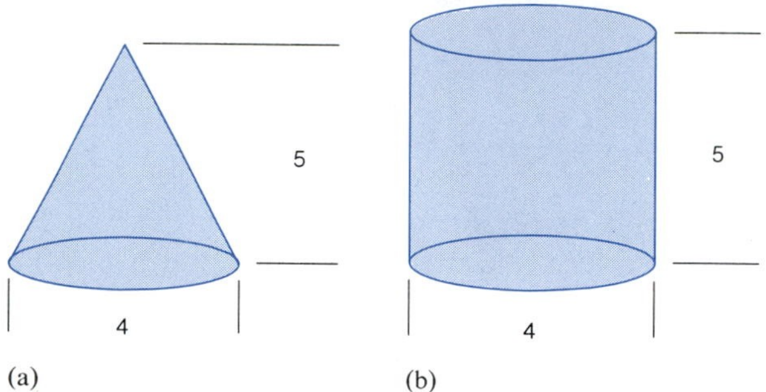


**2** The rectangular tanks pictured are filled with a fluid of weight density $\rho g = 60 \text{ lb/ft}^3$. Find the force due to fluid pressure on the indicated surface of each tank.

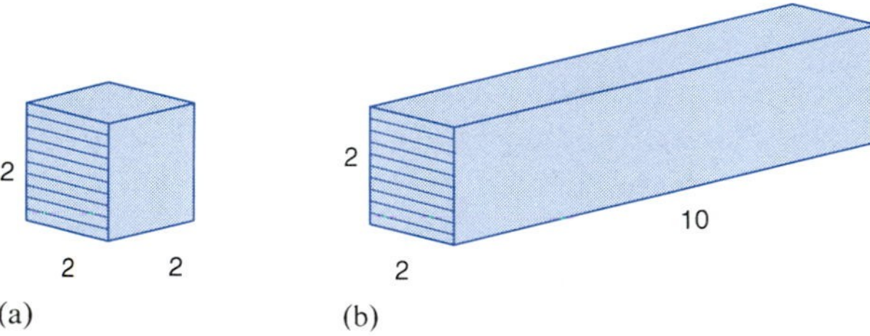


In Exercises 3–6 use the given coordinate system to set up an integral that gives the force due to fluid pressure on the vertically submerged flat surfaces pictured.

**3**

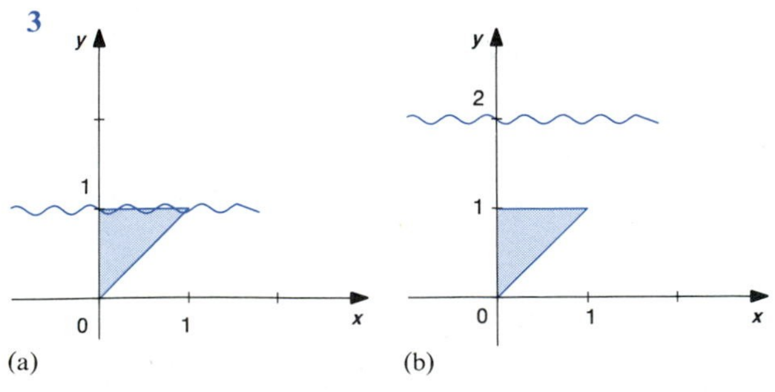


**4**

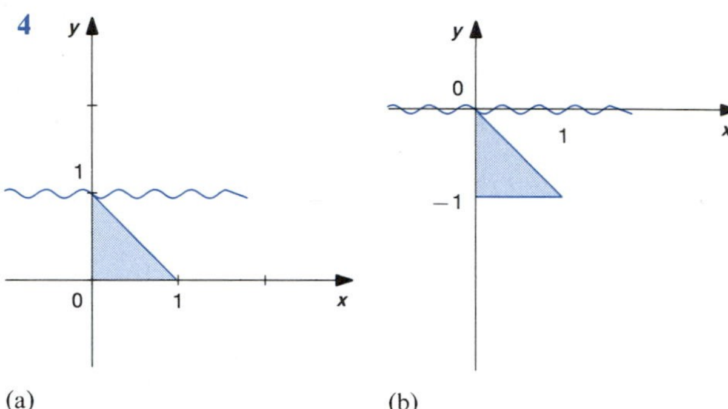


**5**

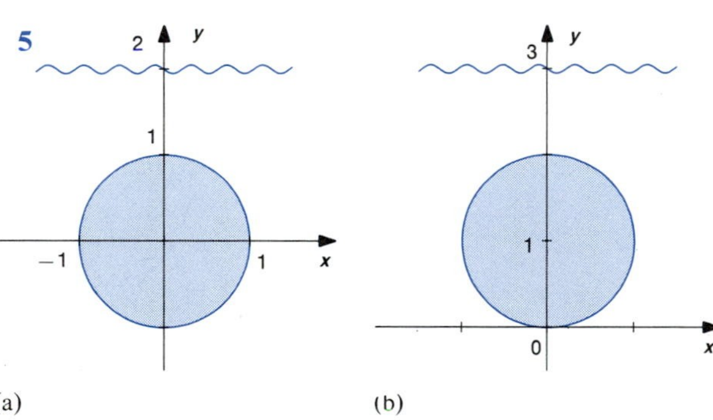


**6**

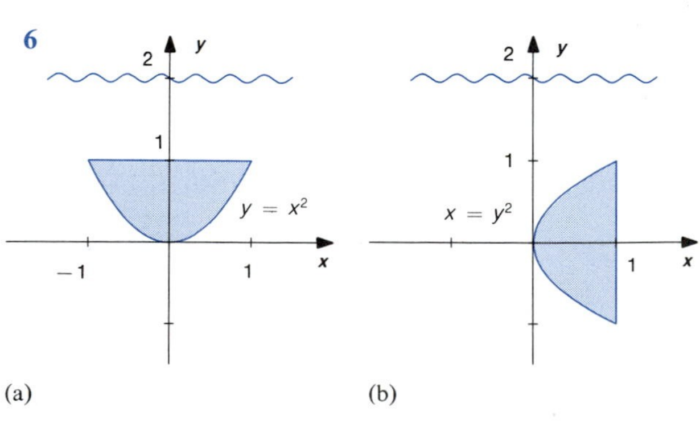


In Exercises 7–10 use the given coordinate system to set up an integral that gives the force due to fluid pressure on the indicated vertical surfaces if the containers have liquid to the level indicated.

**7**

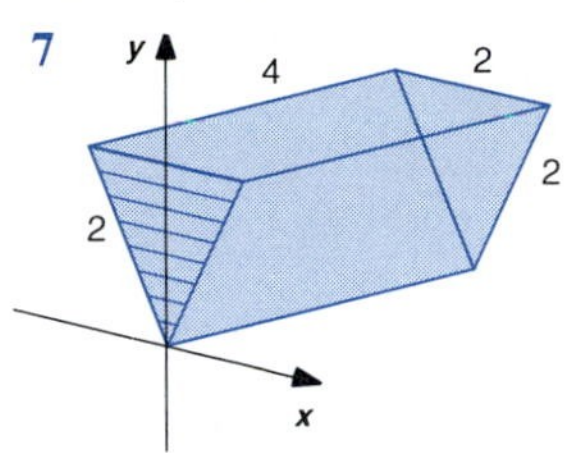


**8**

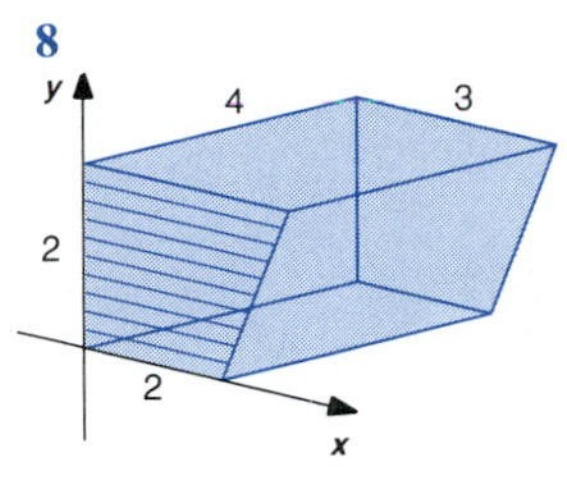


**9**

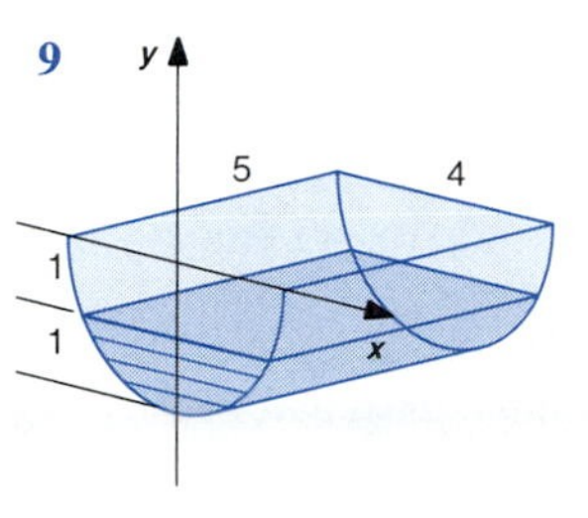


**10**

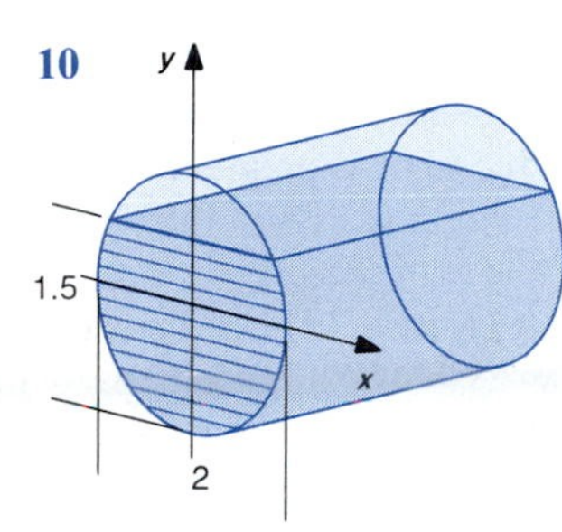


Find the force due to fluid pressure on the vertically submerged flat surfaces pictured in Exercises 11–20.

**11** **12**

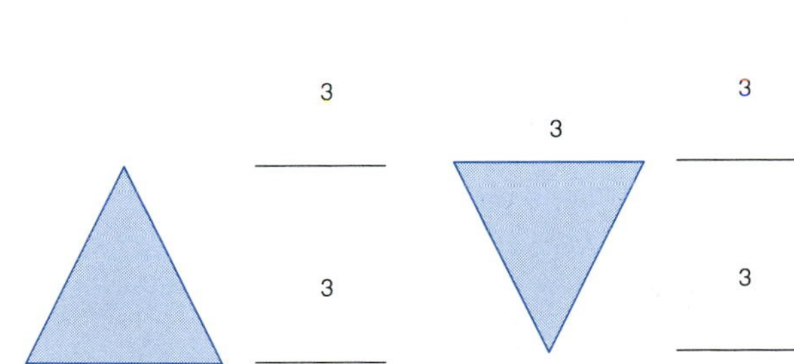


**13**

**14**

**15**

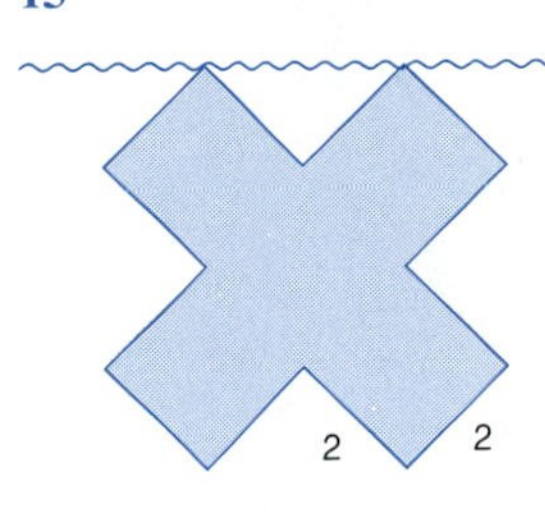


**16**

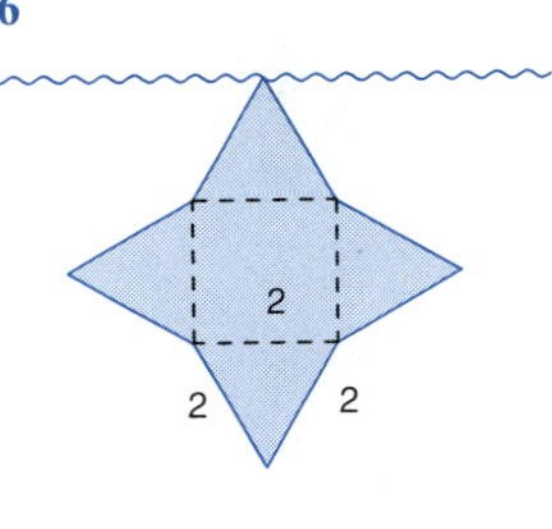


**17** **18**

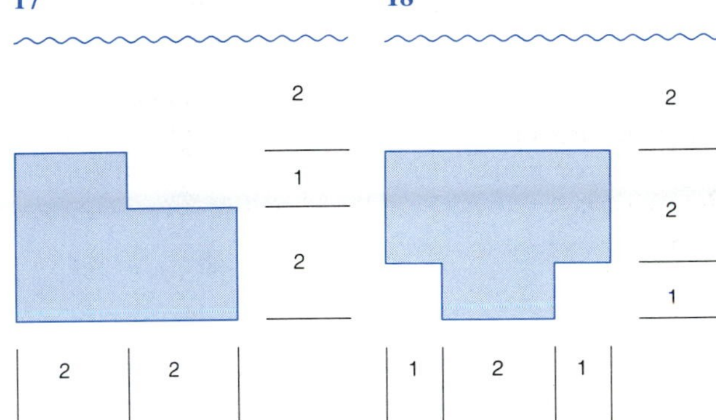


**19**

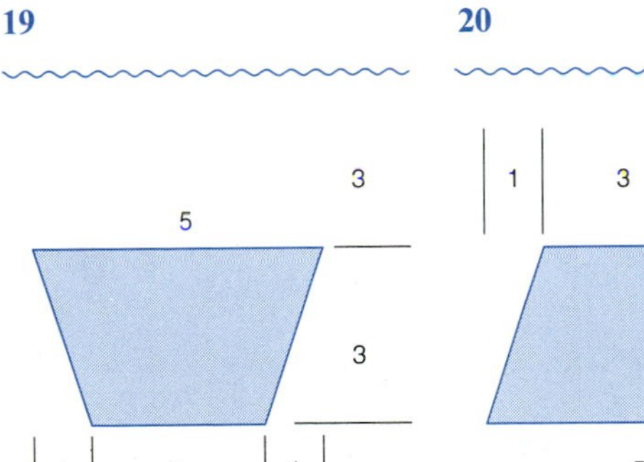


**20**

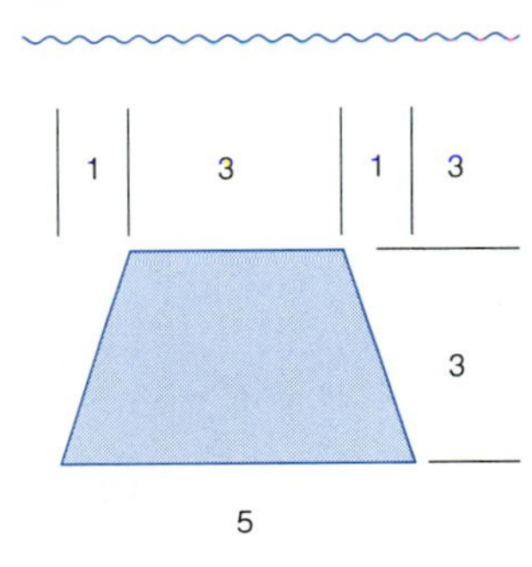


In Exercises 21–24 the containers pictured are filled with fluid. Find the force due to fluid pressure on the indicated flat surfaces.

**21**

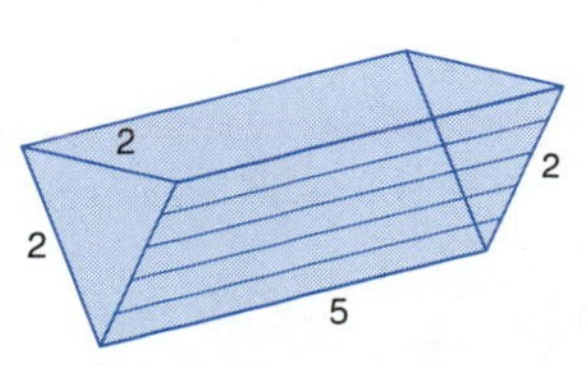


**22**

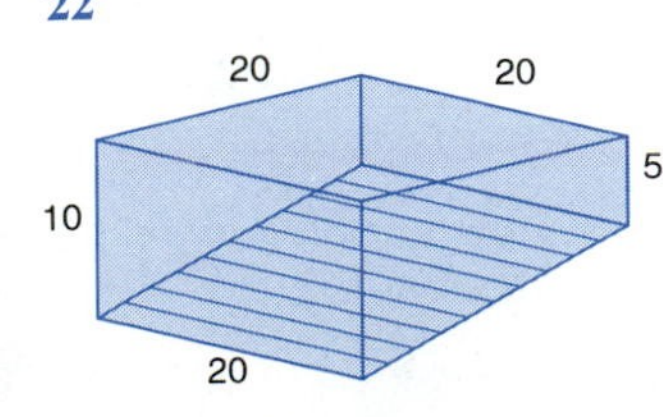

23

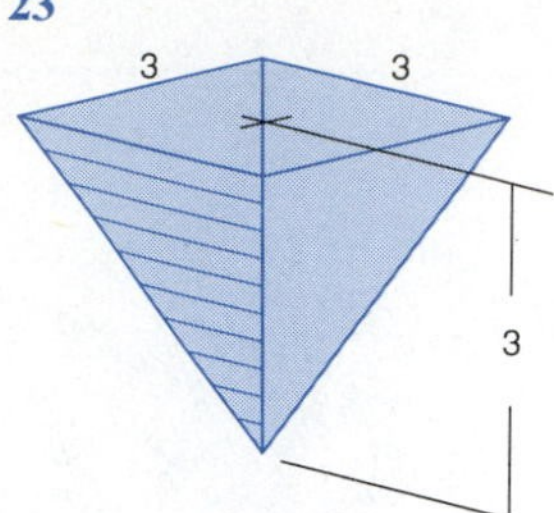


24

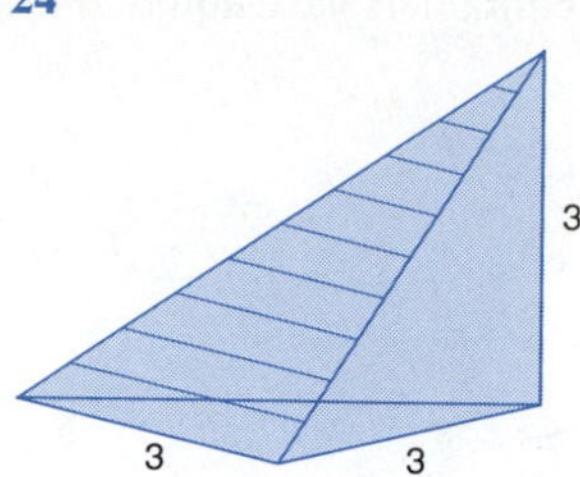


25 A rectangular block of density $\rho_b$ is submerged in a fluid of density $\rho_f$ as pictured. (a) Find the force due to fluid pressure on the top surface of the block. (b) Find the force due to fluid pressure on the bottom surface of the block. (c) Find the force on the block due to gravity. (d) Find the net vertical force on the block due to the forces in (a), (b), and (c). What is the direction of the force if $\rho_b > \rho_f$?

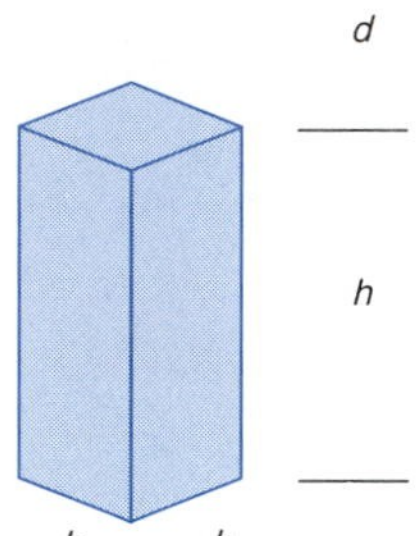


26 A rectangular block of density $\rho_b$ is submerged in a fluid of density $\rho_f$ as pictured. Assume $\rho_b < \rho_f$. (a) Find the force due to fluid pressure on the bottom surface of the block. (b) Find the force on the block due to gravity. (c) Find $d$ so that the net vertical force on the block due to the forces in (a) and (b) is zero. (d) For the value of $d$ found in (c), show that the weight of fluid displaced by the submerged part of the block equals the total weight of the block.

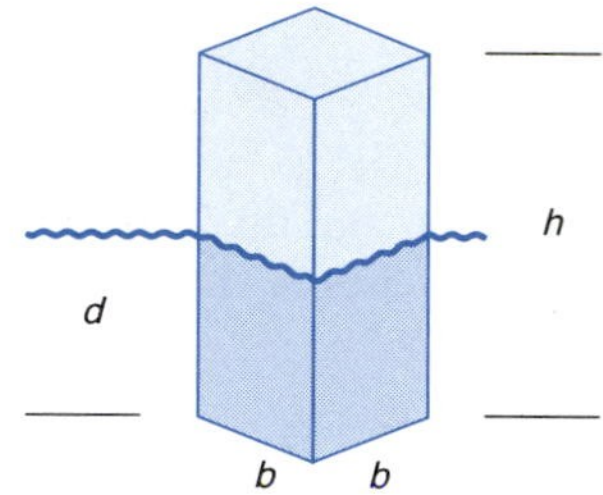


## REVIEW EXERCISES

Set up integrals that give the area of the region bounded by the curves given in Exercises 1–6.

1 $y = x^4$, $y = 8x$

2 $y = x(x - 3)$, $y = x$

3 $y = x$, $x = y(y - 2)$

4 $y = 2x$, $2x + 3y = 6$, $x$-axis

5 $y = 2x$, $y = x/2$, $x + y = 3$

6 $y = x(x - 1)(x + 2)$, $y = 4x$

Set up an integral that gives the volume of each solid pictured in Exercises 7–8.

7 Cross sections perpendicular to the $x$-axis are isosceles right triangles.

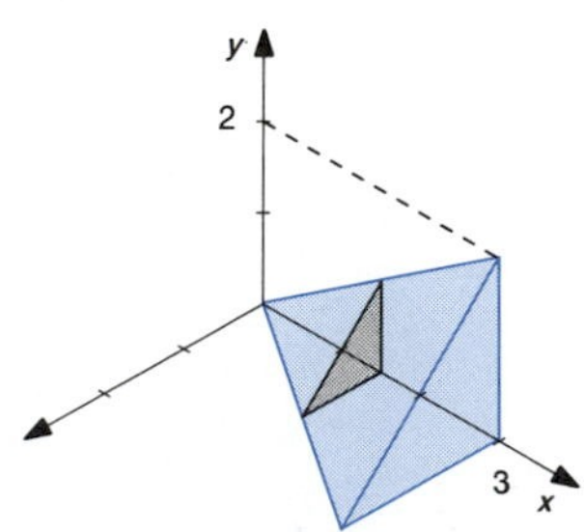

8 Cross sections perpendicular to the $y$-axis are isosceles right triangles.

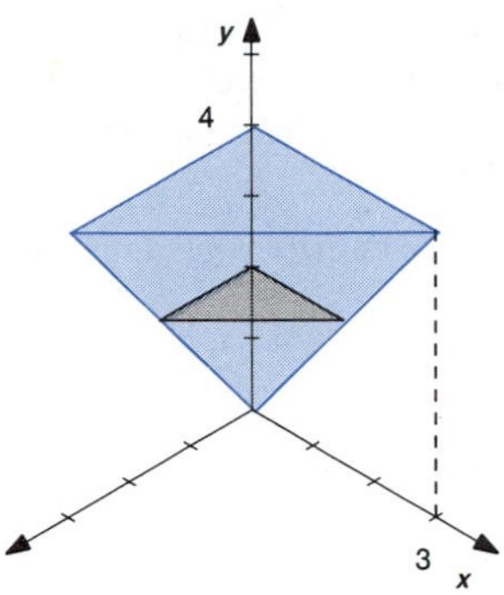


9 Set up an integral that gives the volume of the solid with base the region in the $xy$-plane bounded by $x^2 + y^2 = 1$, if cross sections perpendicular to the $x$-axis are squares.

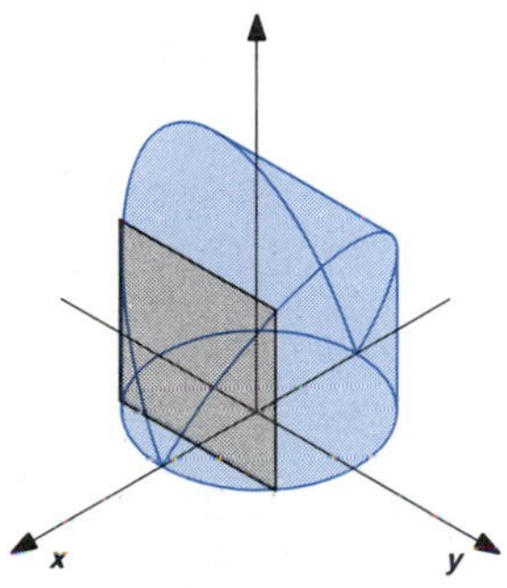


10 Set up an integral that gives the volume of the solid with base the region in the $xy$-plane bounded by $x = 4 - y^2$ and the $y$-axis, if cross sections perpendicular to the $x$-axis are isosceles right triangles with hypotenuses on the base.

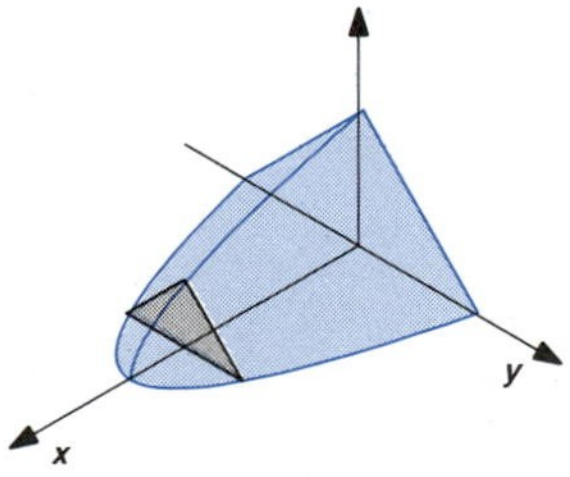


11 Set up an integral that gives the volume of the solid obtained by revolving the region bounded by $y = x^2$ and $y = x$ (a) about the $x$-axis, (b) about the $y$-axis

12 Set up an integral that gives the volume of the solid obtained by revolving the region bounded by $y = \sin x$, the $x$-axis, $x = 0$, and $x = \pi$ (a) about the $x$-axis, (b) about the $y$-axis, (c) about the line $y = 1$.

13 (a) Set up an integral that gives the arc length of the curve $y = \sin x$, $0 \le x \le \pi$. Set up integrals that give the area of the surface obtained by revolving the curve (b) about the $x$-axis, (c) about the $y$-axis.

14 (a) Set up an integral that gives the arc length of the curve $y = x^2, 0 \le x \le 2$. Set up integrals that give the area of the surface obtained by revolving the curve (b) about the $x$-axis, (c) about the $y$-axis.

15 A 2-lb weight stretches a certain spring 2 ft beyond its natural length. Find the work required to stretch the spring from 1 ft to 4 ft beyond its natural length.

16 A 4-lb weight stretches a certain spring 0.5 ft beyond its natural length. Find the work required to stretch the spring from 1 ft to 1.5 ft beyond its natural length.

17 A conical tank is 6 m high and has radius of base 2 m. The vertex of the cone is at the bottom and the tank contains water to a depth of 4 m at the deepest part. Find the work required to pump all the water to the level of the top of the tank. (Density = $1000 \text{ kg/m}^3$, g = $9.8 \text{ m/s}^2$, $\text{kg}\cdot\text{m/s}^2$ = N.)

18 A spherical tank has radius 8 m. The tank contains water to a depth of 5 m at the deepest part. Find the work required to pump all the water to the level of the top of the tank. (Density = $1000 \text{ kg/m}^3$, $g = 9.8 \text{ m/s}^2$, $\text{kg}\cdot\text{m/s}^2$ = N.)

19 A 50-ft chain that weighs 4 lb/ft is hanging vertically. Find the work required to raise 30 ft of the chain to the level of the top, so 20 ft of chain remains hanging.

20 A 15-ft chain that weighs 2 lb/ft is lying at ground level. Find the work required to raise the chain so it hangs vertically with the bottom of the chain 5 ft above ground level.

21 Find the center of mass of the homogeneous lamina pictured.

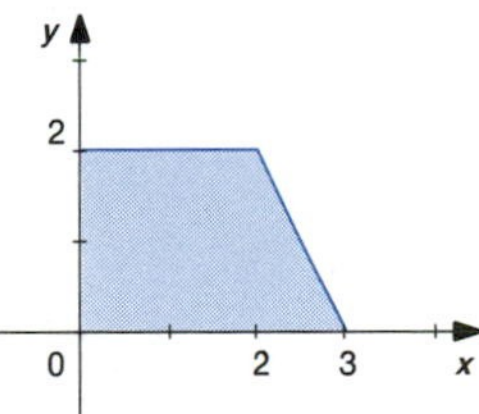


22 Find the center of mass of the homogeneous lamina pictured.

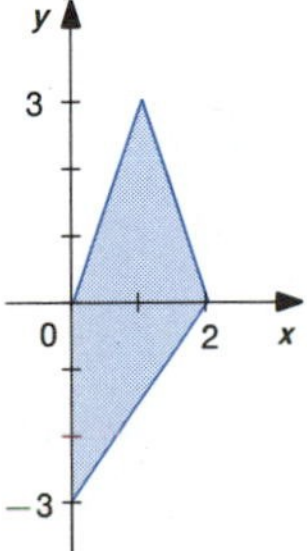

**23** Find the center of mass of the homogeneous lamina inside $x^2 + y^2 = 1$ and above the $x$-axis. (Use a formula for the area of a semicircle.)

**24** Find the center of mass of the homogeneous lamina bounded by $y = x(x - 2)$ and $y = x$.

Find the force due to fluid pressure on the vertically submerged surfaces pictured in Exercises 25–28.

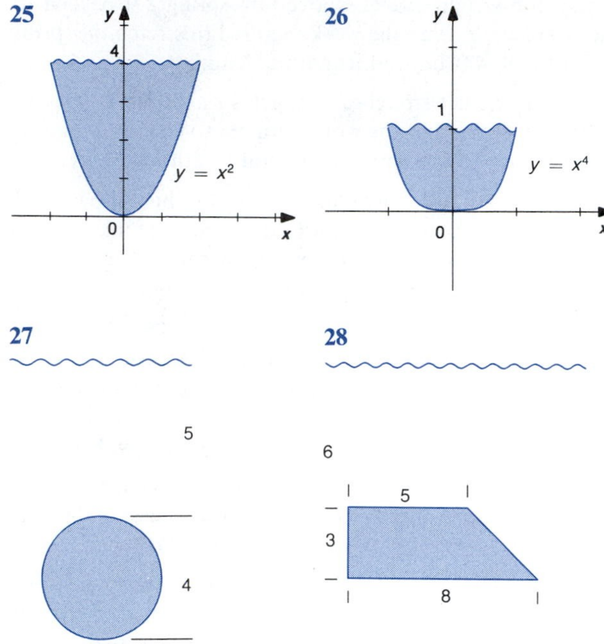


**29** The speed of an object traveling along a number line is the absolute value of its velocity. If the speed is constant, the distance traveled is the product of the speed times the length of time. Use calculus to find the distance traveled if the velocity is $v(t) = 1 - t^2$, $0 \leq t \leq 2$.

**30** A lamina is bounded by $y = x$, $x = 1$, and the $x$-axis. The density per unit area at a point $(x, y)$ is given by $\rho = x$. (a) Find the mass of the lamina. (b) Find the center of mass of the lamina.

**31** The mass of an object of uniform density is given by the formula

$$\text{Mass} = (\text{Density})(\text{Volume}).$$

For objects that may be thought of as lying along a curve, we can speak of density as mass per unit length. Use an integral to find the mass of a beam of length $\pi$ if the density per unit length of the beam at a point $x$ units from one end of the beam is given by $\rho(x) = \sin x$.

**32** For objects that may be thought of as lying along a curve, we can speak of density as mass per unit length. Set up an integral that gives the mass of a rod if the rod is positioned along the curve $y = x^2$, $0 \leq x \leq 1$, and the density per unit length of the rod at points along the rod is given by $\rho = xy$.

**33** The moment of inertia about the $x$-axis of a system of masses $m_1, \ldots, m_n$ located at points $(x_1, y_1), \ldots, (x_n, y_n)$, respectively, in a coordinate plane, is defined to be

$$I_{y=0} = \sum y_j^2 m_j.$$

Use an integral to find the moment of inertia about the $x$-axis of the homogeneous lamina bounded by $x = 0$, $x = h$, $y = 0$, $y = k$.

# CHAPTER SEVEN

# *Selected transcendental functions*

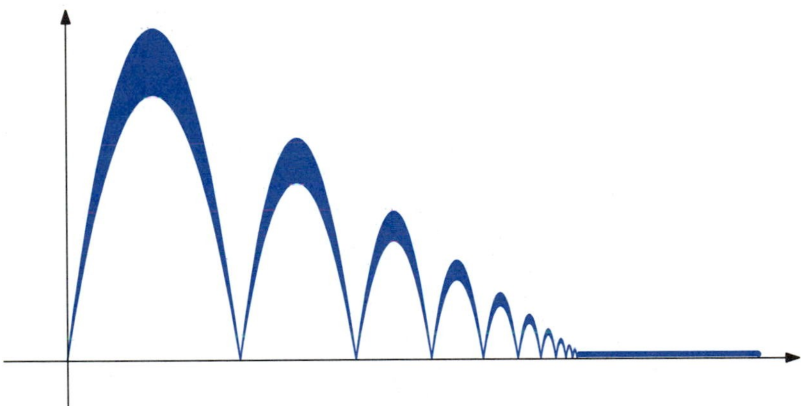

*F*unctions that can be expressed in terms of sums, products, quotients, powers, and roots of polynomials are called **algebraic** functions; all other functions are called **transcendental** functions. The trigonometric functions are examples of transcendental functions. In this chapter, we will develop other transcendental functions that occur in physical problems. The development includes differentiation formulas for each new function. The integration formulas that correspond to the new differentiation formulas allow us to use the Fundamental Theorem of Calculus to evaluate a wider variety of integrals.

## 7.1 L'HÔPITAL'S RULE, INDETERMINATE FORMS 0/0, $\infty/\infty$, $0 \cdot \infty$

In this section we will develop a method that allows us to use derivatives to evaluate ratios of functions. The method is called **l'Hôpital's Rule**. We will use l'Hôpital's Rule to study properties of the transcendental functions that will be introduced in this chapter. It is important that we think of l'Hôpital's Rule as a tool that can be used for the investigation of functions.

L'Hôpital's Rule can be used to compare values of functions. If two functions either both approach zero or both approach infinity at the same point $c$, the limit at $c$ of their ratio can be used to determine the relative rate at which they approach either zero or infinity. For example, in Section 3.3 we used a geometric argument to show that

$$\lim_{x \to 0} \frac{\sin x}{x} = 1.$$

This can be interpreted as saying that the transcendental function $\sin x$ approaches zero at the same rate as the linear function $x$ as $x$ approaches zero.

In some cases, we can evaluate a limit of a ratio or a product by evaluating the limits of each factor separately. For example, to evaluate

$$\lim_{x \to 2} \frac{x^2 + 5}{x + 1},$$

we note that

$$\lim_{x \to 2} (x^2 + 5) = 9 \quad \text{and} \quad \lim_{x \to 2} (x + 1) = 3.$$

Since the limit of the denominator is nonzero, the Limit Theorem tells us

$$\lim_{x \to 2} \frac{x^2 + 5}{x + 1} = \frac{9}{3} = 3.$$

Also, we know that $\lim_{x \to 0} x^2 \sin(1/x) = 0$, because $\lim_{x \to 0} x^2 = 0$ and $\sin(1/x)$ is bounded $[|\sin(1/x)| \leq 1]$ for $x \neq 0$. For certain combinations of factors that have limit either zero or infinity, we cannot determine the limit of an expression by looking directly at the individual factors.

■ *Definition*

If

$$\lim_{x \to c} f(x) = 0 \quad \text{and} \quad \lim_{x \to c} g(x) = 0,$$

$f(x)/g(x)$ is said to have **indeterminate form 0/0** at $x = c$. ■

If $f(x)/g(x)$ has indeterminate form 0/0 at $c$, then $f(x)$ and $g(x)$ both become small as $x$ approaches $c$. The value of the quotient $f(x)/g(x)$ then depends on the relative smallness of $f(x)$ and $g(x)$ and may have any magnitude. The following result can be used to evaluate the limit at $c$ of certain expressions of indeterminate form 0/0.

*L'Hôpital's Rule*

**Suppose that, for some positive number $\delta_0$, $f$ and $g$ are continuous and differentiable for $0 < |x - c| < \delta_0$. If $f(x)/g(x)$ has indeterminate form 0/0, then**

$$\lim_{x \to c} \frac{f'(x)}{g'(x)} = L \quad \text{implies} \quad \lim_{x \to c} \frac{f(x)}{g(x)} = L.$$

**The result also holds for one-sided limits at $c$.**

■ *Proof* Since $f/g$ has indeterminate form 0/0 at $x = c$, we know that

$$\lim_{x \to c} f(x) = \lim_{x \to c} g(x) = 0.$$

We assume that $f(c) = g(c) = 0$. These values agree with the values of the limits of the functions at $c$, so $f$ and $g$ are then continuous at $c$ and at all points in the interval $|x - c| < \delta_0$. (If $f$ and/or $g$ are not zero at $c$, then assign them the value zero. The limits at $c$ do not depend on values at $c$, so we may assign any convenient values.) Let us fix a point $x$ with $|x - c| < \delta_0$. We then apply Rolle's Theorem to the function $h(t) = g(x)f(t) - f(x)g(t)$ in the interval between $x$ and $c$. Note that $x$ is considered to be a constant. The conditions of Rolle's Theorem are satisfied. [The function $h(t)$ is continuous on the closed interval between $x$ and $c$, and is differentiable on the open interval between $x$ and $c$, because $f(t)$ and $g(t)$ are. Clearly $h(x) = 0$. $h(c) = 0$ because we have $f(c) = g(c) = 0$.] The conclusion of Rolle's Theorem gives us a point $z$ that is between $x$ and $c$ with $h'(z) = 0$. That is,

$$g(x)f'(z) - f(x)g'(z) = 0, \quad \text{so} \quad \frac{f(x)}{g(x)} = \frac{f'(z)}{g'(z)}.$$

The algebraic step above requires that $g(x)$ and $g'(z)$ are not zero, which is true if $x$ is near enough to $c$. The existence of the limit of $f'(x)/g'(x)$ as $x \to c$ implies that $f'(x)/g'(x)$ is defined for all $x$ near $c$, $x \neq c$. This means that $g'(z) \neq 0$, if we use a value $x$ that is close enough to $c$. Also, if $g(x) = g(c) = 0$, then Rolle's Theorem would imply $g'$ is zero at a point between $x$ and $c$. Since $g'$ is not zero, we must have $g(x) \neq 0$.

$$\lim_{x \to c} \frac{f'(x)}{g'(x)} = L \quad \text{implies} \quad \frac{f'(z)}{g'(z)} \to L \quad \text{as} \quad z \to c.$$

Since $z$ is between $x$ and $c$, $x \to c$ implies $z \to c$. Hence,

$$\frac{f(x)}{g(x)} = \frac{f'(z)}{g'(z)} \to L \quad \text{as} \quad x \to c.$$

The argument is the same for one-sided limits. ■

If $f$ and $g$ are continuous at $c$ and $f(c) = g(c) = 0$, then $f/g$ has indeterminate form 0/0. This means that we can check that a quotient of continuous functions has indeterminate form 0/0 by substitution of the value $x = c$; $f(c)/g(c) = 0/0$ tells us $f/g$ has indeterminate form 0/0. That $f(c)/g(c)$ is undefined has no bearing on either the existence or the value of the limit of $f(x)/g(x)$ as $x$ approaches $c$.

**EXAMPLE 1**

Evaluate

$$\lim_{x\to 3} \frac{x^2 - 2x - 3}{x - 3}.$$

**SOLUTION**

We could evaluate this limit by factoring, but let us see how to use l'Hôpital's Rule. We have $f(x)/g(x)$, where $f(x) = x^2 - 2x - 3$ and $g(x) = x - 3$. The conditions of l'Hôpital's Rule are satisfied. That is, $f$ and $g$ are continuous and $f(3) = g(3) = 0$, so $f/g$ has indeterminate form 0/0 at $x = 3$. Also,

$$\lim_{x\to 3} \frac{f'(x)}{g'(x)} = \lim_{x\to 3} \frac{2x - 2}{1} = 2(3) - 2 = 4.$$

The quotient $f'/g'$ is the derivative of the numerator of $f/g$ divided by the derivative of the denominator of $f/g$. Do each of the two differentiations separately and do not confuse the process with the rule for differentiating quotients.

L'Hôpital's Rule then tells us

$$\lim_{x\to 3} \frac{x^2 - 2x - 3}{x - 3} = \lim_{x\to 3} \frac{f(x)}{g(x)} = \lim_{x\to 3} \frac{f'(x)}{g'(x)} = 4.$$

We will write the solution in the shorter form

$$\lim_{x\to 3} \frac{x^2 - 2x - 3}{x - 3} \underset{(0/0,\,\text{l'H})}{=} \lim_{x\to 3} \frac{2x - 2}{1} = 4.$$

(0/0, l'H) indicates that we have checked that l'Hôpital's Rule applies to the indeterminate form 0/0 and that equality holds if the limit of $f'/g'$ exists. The existence of the latter limit then justifies the use of l'Hôpital's Rule. If the limit of $f'/g'$ does not exist, then l'Hôpital's Rule does not apply and the equation we have written is not valid. The limit of $f/g$ may exist, even though the limit of $f'/g'$ does not.

The graph of

$$y = \frac{x^2 - 2x - 3}{x - 3}$$

is sketched in Figure 7.1.1. ■

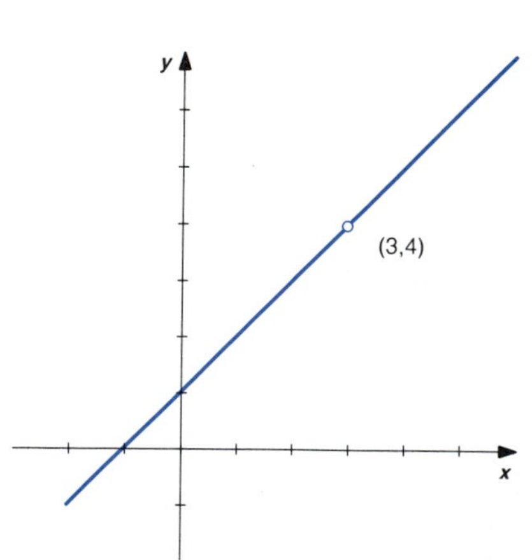


FIGURE 7.1.1

■ **EXAMPLE 2**

Evaluate

$$\lim_{x\to 0} \frac{1-\cos 2x}{x^2}.$$

**SOLUTION**

It is easy to see that $(1 - \cos 2x)/x^2$ has indeterminate form 0/0 at $x = 0$. We can then write the solution in the short form.

$$\lim_{x\to 0} \frac{1-\cos 2x}{x^2} \underset{(0/0,\text{l'H})}{=} \lim_{x\to 0} \frac{-(-\sin 2x)(2)}{2x}$$

$$\underset{(\text{Simplifying})}{=} \lim_{x\to 0} \frac{\sin 2x}{x}$$

$$\underset{(0/0,\text{l'H})}{=} \lim_{x\to 0} \frac{(\cos 2x)(2)}{1} = 2.$$

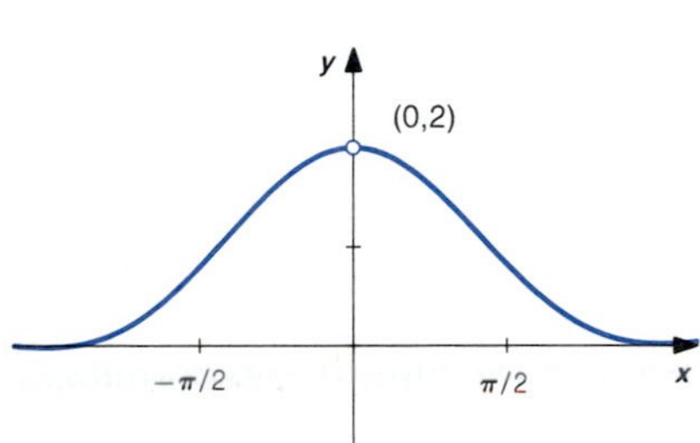


FIGURE 7.1.2

In this example the first application of l'Hôpital's Rule gave us $(\sin 2x)/x$, which also has indeterminate form 0/0. We were able to evaluate the limit after a second application of l'Hôpital's Rule. The graph of $y = (1 - \cos 2x)/x^2$ is sketched in Figure 7.1.2. ■

■ *Definition*

If

$$\lim_{x\to c} f(x) = \pm\infty \quad \text{and} \quad \lim_{x\to c} g(x) = \pm\infty,$$

$f(x)/g(x)$ is said to have **indeterminate form ∞/∞** at $x = c$. ■

If $f(x)/g(x)$ has indeterminate form $\infty/\infty$ at $c$, then $f(x)$ and $g(x)$ both become large as $x$ approaches $c$. The value of the quotient $f(x)/g(x)$ depends on the relative largeness of $f(x)$ and $g(x)$ and may have any value.

L'Hôpital's Rule applies to expressions of indeterminate form $\infty/\infty$ and to limits at infinity. We will not prove l'Hôpital's Rule in these cases, but we will use the results.

■ **EXAMPLE 3**

Evaluate

$$\lim_{x\to\infty} \frac{2x^2-2x+3}{3x^2+1}.$$

**SOLUTION**

$(2x^2 - 2x + 3)/(3x^2 + 1)$ has indeterminate form $\infty/\infty$ at $\infty$. Then

$$\lim_{x\to\infty} \frac{2x^2-2x+3}{3x^2+1} \underset{(\infty/\infty,\text{l'H})}{=} \lim_{x\to\infty} \frac{4x-2}{6x}$$

$$\underset{(\infty/\infty,\text{l'H})}{=} \lim_{x\to\infty} \frac{4}{6} = \frac{2}{3}.$$

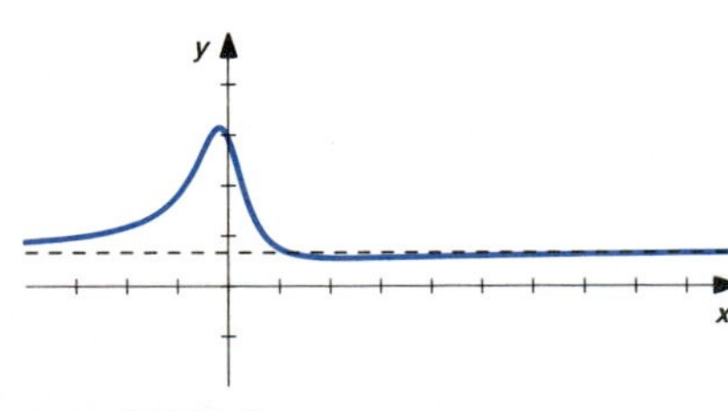


FIGURE 7.1.3

The graph of $y = (2x^2 - 2x + 3)/(3x^2 + 1)$ is sketched in Figure 7.1.3. The line $y = 2/3$ is a horizontal asymptote. ■

■ *Definition*

If

$$\lim_{x\to c} f(x) = 0 \quad \text{and} \quad \lim_{x\to c} g(x) = \pm\infty,$$

the product $fg$ is said to have **indeterminate form $\mathbf{0 \cdot \infty}$** at $c$. ■

By writing either

$$f(x)g(x) = \frac{f(x)}{1/g(x)} \quad \text{or} \quad f(x)g(x) = \frac{g(x)}{1/f(x)} \qquad [\text{if } f(x) \neq 0],$$

we can express $fg$ in indeterminate form $0/0$ or $\infty/\infty$, respectively. We can then apply l'Hôpital's Rule to try to evaluate the limit of the product.

■ **EXAMPLE 4**

Evaluate

$$\lim_{x\to 0} x \csc x.$$

**SOLUTION**

We see that $x \csc x$ has indeterminate form $0 \cdot \infty$ at $x = 0$. Rewriting the product as a quotient, we have

$$\lim_{x\to 0} x \csc x \underset{\text{(Rewriting)}}{=} \lim_{x\to 0} \frac{\csc x}{1/x}$$

$$\underset{(\infty/\infty,\text{ l'H})}{=} \lim_{x\to 0} \frac{-\csc x \cot x}{-1/x^2}.$$

At this point we see that application of l'Hôpital's Rule to the quotient does not seem to be helping, so we abandon this effort and try rewriting the product as a different quotient. We then obtain

$$\lim_{x\to 0} x \csc x \underset{\text{(Rewriting)}}{=} \lim_{x\to 0} \frac{x}{1/\csc x}$$

$$= \lim_{x\to 0} \frac{x}{\sin x}$$

$$\underset{(0/0,\text{ l'H})}{=} \lim_{x\to 0} \frac{1}{\cos x} = 1.$$

The graph of $y = x \csc x$ for $x$ near zero is sketched in Figure 7.1.4. ■

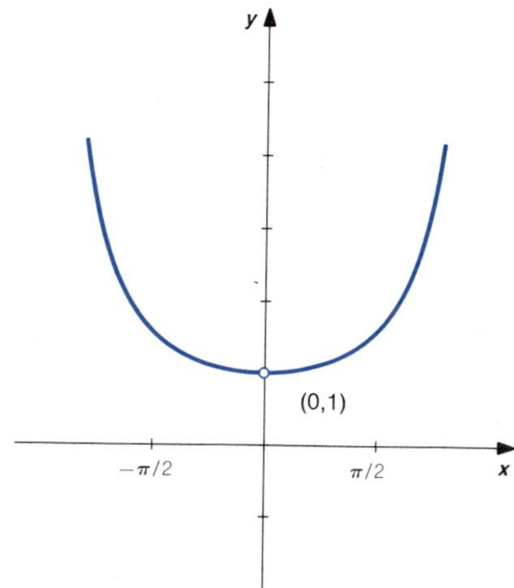


FIGURE 7.1.4

In some cases, expressions of indeterminate form $\infty - \infty$ can be written as a product or quotient to which we may apply l'Hôpital's Rule.

■ **EXAMPLE 5**

Evaluate

$$\lim_{x\to 0}\left(\frac{1}{x}-\frac{1}{\sin x}\right).$$

**SOLUTION**

$$\lim_{x\to 0}\left(\frac{1}{x}-\frac{1}{\sin x}\right) \underset{\text{(Rewriting)}}{=} \lim_{x\to 0}\frac{\sin x - x}{x\sin x}$$

$$\underset{(0/0,\,\text{l'H})}{=} \lim_{x\to 0}\frac{\cos x - 1}{(x)(\cos x)+(\sin x)(1)}$$

$$\underset{(0/0,\,\text{l'H})}{=} \lim_{x\to 0}\frac{-\sin x}{(x)(-\sin x)+(\cos x)(1)+\cos x}$$

$$=\frac{0}{2}=0.$$

The graph of

$$y=\frac{1}{x}-\frac{1}{\sin x}$$

for $x$ near zero is sketched in Figure 7.1.5. ■

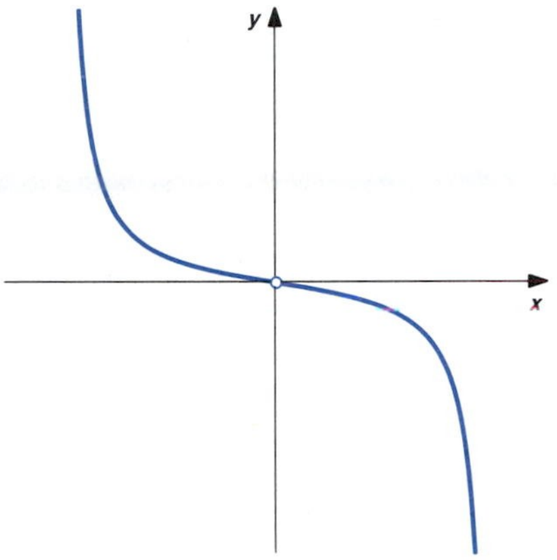

FIGURE 7.1.5

L'Hôpital's Rule is a valuable tool for determining information about functions and the physical quantities that they represent. We should know how and when to use the method but should not use the method blindly. It is often helpful to simplify the expression before applying l'Hôpital's Rule.

■ **EXAMPLE 6**

Evaluate

$$\lim_{x\to 0}\frac{\sin x}{\tan x}.$$

We see that $\sin x/\tan x$ has indeterminate form 0/0. Before applying l'Hôpital's Rule, let us try simplifying. We have $\tan x = \sin x/\cos x$, so $\sin x/\tan x = \cos x$ whenever $\tan x$ is defined and nonzero. In particular, $\sin x/\tan x = \cos x$ for $x$ near 0, $x \neq 0$. We then have

$$\lim_{x\to 0}\frac{\sin x}{\tan x}=\lim_{x\to 0}\cos x = 1.$$

It is not necessary to use l'Hôpital's Rule in this example. The graph of $y = \sin x/\tan x$ is sketched in Figure 7.1.6. ■

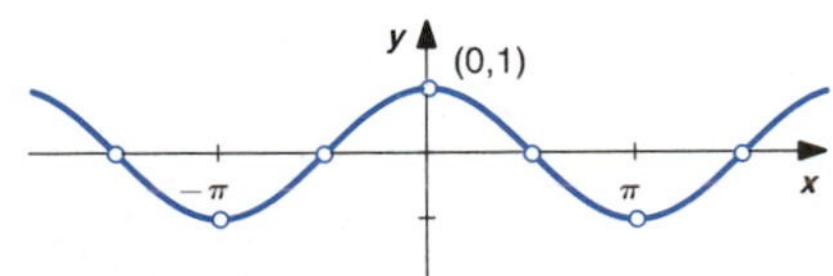


FIGURE 7.1.6

We can simplify the differentiation in some problems by excluding factors that have finite, nonzero limits from the application of l'Hôpital's Rule.

**■ EXAMPLE 7**

Evaluate

$$\lim_{x\to 0} \frac{\sin 3x}{x\sqrt{4-x^2}}.$$

**SOLUTION**

We see that $(\sin 3x)/(x\sqrt{4-x^2})$ has indeterminate form 0/0 at $x = 0$. We also note that the factor $\sqrt{4-x^2}$ has limit 2 as $x \to 0$. Hence, let us write

$$\frac{\sin 3x}{x\sqrt{4-x^2}} = \left(\frac{\sin 3x}{x}\right)\left(\frac{1}{\sqrt{4-x^2}}\right).$$

We then use l'Hôpital's Rule to evaluate the limit of the first expression. We have

$$\lim_{x\to 0} \frac{\sin 3x}{x} \underset{(0/0,\,\text{l'H})}{=} \lim_{x\to 0} \frac{(\cos 3x)(3)}{1} = 3.$$

Then

$$\lim_{x\to 0} \frac{\sin 3x}{x\sqrt{4-x^2}} = \left(\lim_{x\to 0} \frac{\sin 3x}{x}\right)\left(\lim_{x\to 0} \frac{1}{\sqrt{4-x^2}}\right) = (3)\left(\frac{1}{2}\right) = \frac{3}{2}.$$

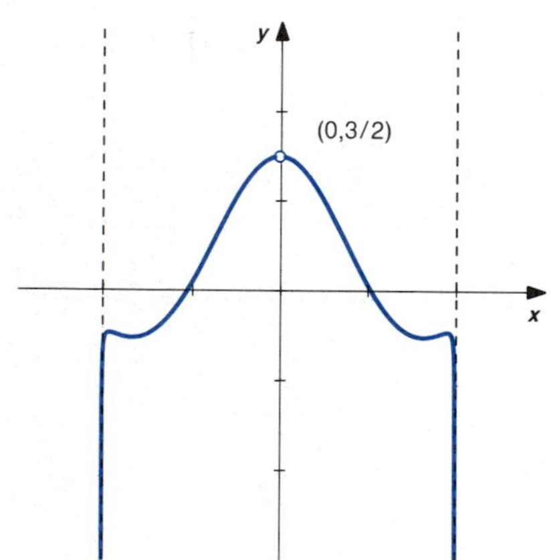


FIGURE 7.1.7

If we applied l'Hôpital's Rule to the given quotient we would have

$$\lim_{x\to 0} \frac{\sin 3x}{x\sqrt{4-x^2}} \underset{(0/0,\,\text{l'H})}{=} \lim_{x\to 0} \frac{3\cos 3x}{(x)\left(\dfrac{1}{2\sqrt{4-x^2}}\right)(-2x) + (\sqrt{4-x^2})(1)}$$

$$= \frac{3}{0+2} = \frac{3}{2}.$$

The graph of $y = (\sin 3x)/(x\sqrt{4-x^2})$ is sketched in Figure 7.1.7. ■

Be sure to check that an expression is of indeterminate form before applying l'Hôpital's Rule.

**■ EXAMPLE 8**

Evaluate

$$\lim_{x\to 0} \frac{\cos 2x}{x^2}.$$

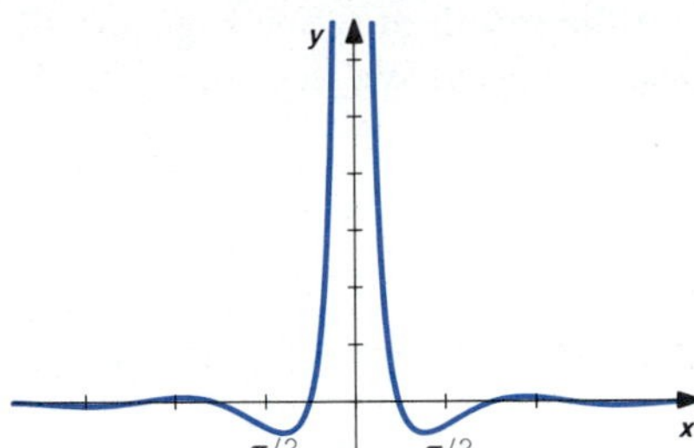

FIGURE 7.1.8

**SOLUTION**

$(\cos 2x)/x^2$ has form $1/0$ at $x = 0$. This is not an indeterminate form. It is not difficult to see that

$$\lim_{x \to 0} \frac{\cos 2x}{x^2} = \infty.$$

It is left to the reader to show that two blind applications of l'Hôpital's Rule give the incorrect value of $-2$ for the limit of $(\cos 2x)/x^2$ as $x$ approaches zero. The graph of $y = (\cos 2x)/x^2$ is sketched in Figure 7.1.8. ■

## EXERCISES 7.1

Evaluate the limits in Exercises 1–34.

**1** $\lim_{x \to 3} \dfrac{x^2 - 9}{x - 3}$

**2** $\lim_{x \to 2} \dfrac{x^2 - 3x + 2}{x - 2}$

**3** $\lim_{x \to \pi/6} \dfrac{\sin x - (1/2)}{x - (\pi/6)}$

**4** $\lim_{x \to \pi/4} \dfrac{\tan x - 1}{x - (\pi/4)}$

**5** $\lim_{x \to 0} \dfrac{\sin 3x}{\cos 2x}$

**6** $\lim_{x \to 0} \dfrac{\cos x}{\sin^2 x}$

**7** $\lim_{x \to 0} \dfrac{\sin 2x}{x}$

**8** $\lim_{x \to 0} \dfrac{\sin 3x}{\sin 2x}$

**9** $\lim_{x \to 0} \dfrac{\sin x - x}{x^3}$

**10** $\lim_{x \to 0} \dfrac{\cos 2x - 1}{x^2}$

**11** $\lim_{x \to 0^+} \dfrac{\cos 2x}{\sin 3x}$

**12** $\lim_{x \to 0^-} \dfrac{\cos 3x}{\sin 2x}$

**13** $\lim_{x \to 0} \dfrac{x^2 + 3x + 2}{x^2 - 1}$

**14** $\lim_{x \to 0} \dfrac{4x^2 - 3x + 5}{x^2 + 2x + 3}$

**15** $\lim_{x \to \infty} \dfrac{3x^2 - 2x + 1}{5x^2 + 4x + 3}$

**16** $\lim_{x \to \infty} \dfrac{2x^2 + 1}{3x^2 - x + 4}$

**17** $\lim_{x \to \infty} \dfrac{x^2 + x + 1}{3x^2 - x + 2}$

**18** $\lim_{x \to \infty} \dfrac{4x^2 - 9}{2x^2 + 3x + 1}$

**19** $\lim_{x \to \infty} \dfrac{2x - 3}{3x^2 - x + 2}$

**20** $\lim_{x \to \infty} \dfrac{x^2 - 2x}{3x - 5}$

**21** $\lim_{x \to \infty} x \sin(1/x)$

**22** $\lim_{x \to \infty} x \sin(1/x^2)$

**23** $\lim_{x \to 0} \sin x \cot x$

**24** $\lim_{x \to 0} \sin x \csc x$

**25** $\lim_{x \to 0} x \cot 2x$

**26** $\lim_{x \to 0} \sin 2x \cot 3x$

**27** $\lim_{x \to 0} \dfrac{\cos^2 x \sin x}{x}$

**28** $\lim_{x \to 0} \dfrac{x\sqrt{x + 4}}{\sin 2x}$

**29** $\lim_{x \to 0} \dfrac{\sin x\sqrt{1 - \sin x}}{x}$

**30** $\lim_{x \to 0} \dfrac{\cos x \sin^2 x}{1 - \cos x}$

**31** $\lim_{x \to \infty} \left(\dfrac{x^2}{x + 1} - \dfrac{x^2}{x - 1}\right)$

**32** $\lim_{x \to \infty} \left(\dfrac{2x^2}{4x + 1} - \dfrac{x}{2}\right)$

**33** $\lim_{x \to 0} \left(\dfrac{1}{x} - \cot x\right)$

**34** $\lim_{x \to 0} (\cot x - \csc x)$

## 7.2 INVERSE FUNCTIONS, INVERSE TRIGONOMETRIC FUNCTIONS

The concept of inverse functions is essential for the development of the transcendental functions that are introduced in this chapter.

Let us begin by expressing the idea of an inverse function in terms of the graph of a function $f$ and the equation $y = f(x)$. Suppose that the graph of $f$ has the property that every horizontal line that intersects the graph intersects it in exactly one point. This means that for each $y$ in the range of $f$, there is exactly one $x$ such that $y = f(x)$. We can then define a function

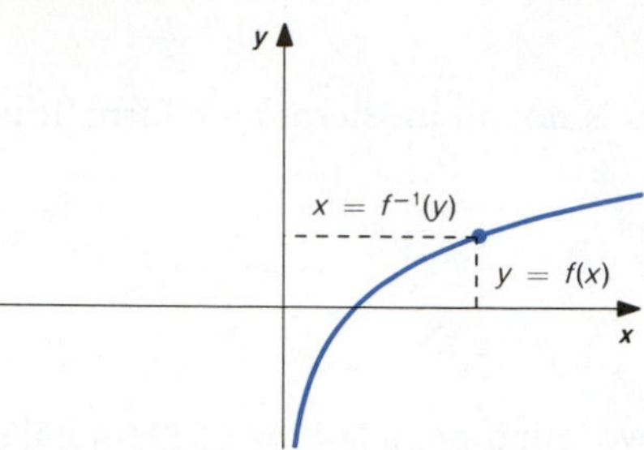

FIGURE 7.2.1

$f^{-1}$ with values $f^{-1}(y) = x$ if $(x, y)$ is on the graph of $f$. See Figure 7.2.1. The function $f^{-1}$ is called the **inverse function** of $f$. We then have

$$y = f(x) \quad \text{if and only if} \quad x = f^{-1}(y). \tag{1}$$

Statement (1) holds for all $x$ in the domain of $f$ and all $y$ in the range of $f$. The domain of $f$ becomes the range of $f^{-1}$ and the range of $f$ becomes the domain of $f^{-1}$.

We can think of the two equations in (1) as different forms of the same equation. In particular, we think of solving the equation $y = f(x)$ for $x$ to obtain $x = f^{-1}(y)$. This gives us a formula for $f^{-1}(y)$. To express the inverse function $f^{-1}$ in terms of the variable $x$, we simply substitute $x$ for $y$ in the formula for $f^{-1}(y)$.

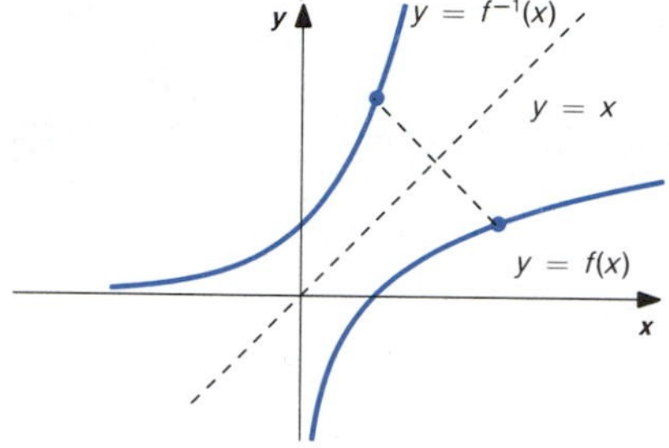

FIGURE 7.2.2

The graphs of $y = f(x)$ and $y = f^{-1}(x)$ are symmetric with respect to the line $y = x$. That is, if $(c, d)$ is on the graph of $y = f(x)$, then $d = f(c)$. This implies that $c = f^{-1}(d)$, so $(d, c)$ is on the graph of $y = f^{-1}(x)$. See Figure 7.2.2.

If $y = f(x)$ and $x = f^{-1}(y)$, then substitution gives us

$$y = f(f^{-1}(y)) \quad \text{and} \quad x = f^{-1}(f(x)). \tag{2}$$

That is, $(f \circ f^{-1})(y) = y$ and $(f^{-1} \circ f)(x) = x$. Statement (2) holds for $x$ in the domain of $f$ (equal to the range of $f^{-1}$) and $y$ in the range of $f$ (equal to the domain of $f^{-1}$).

We can use (2) to solve the equation $y = f(x)$ for $x$. That is, if

$$y = f(x),$$

we can apply the function $f^{-1}$ to both sides of the equation to obtain

$$f^{-1}(y) = f^{-1}(f(x)).$$

Statement (2) then gives

$$f^{-1}(y) = x.$$

In some cases we can find a formula for $f^{-1}$ by using the usual rules of algebra to solve the equation $y = f(x)$ for $x$.

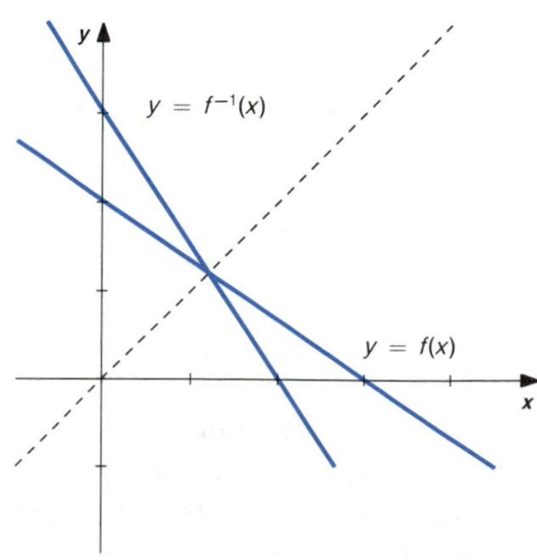

FIGURE 7.2.3

■ **EXAMPLE 1**

Sketch the graph of $f$ and determine whether $f$ has an inverse function $f^{-1}$. If $f^{-1}$ exists, find a formula for $f^{-1}(x)$ and sketch the graph of $y = f^{-1}(x)$.

(a) $f(x) = (6 - 2x)/3$,
(b) $f(x) = x^3 + 1$,
(c) $f(x) = x^2$,
(d) $f(x) = x^2$, $x \geq 0$.

**SOLUTION**

(a) From Figure 7.2.3 we see that the function $f(x) = (6 - 2x)/3$ has an inverse function, since each horizontal line intersects the graph at exactly one point. Substituting in the equation $y = f(x)$ and then solving for $x$, we have

$$y = f(x),$$

$$y = \frac{6 - 2x}{3},$$

$$3y = 6 - 2x,$$

$$2x = 6 - 3y,$$

$$x = \frac{6 - 3y}{2}.$$

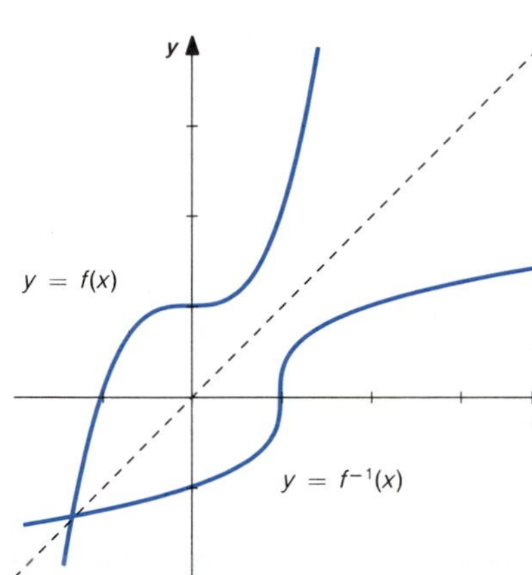


FIGURE 7.2.4

Since $y = f(x)$ implies $x = f^{-1}(y)$, this means

$$f^{-1}(y) = \frac{6 - 3y}{2}.$$

Substituting $x$ for $y$ in this equation, we obtain

$$f^{-1}(x) = \frac{6 - 3x}{2}.$$

The graph of $y = f^{-1}(x)$ is sketched in Figure 7.2.3, along with that of $y = f(x)$.

(b) From Figure 7.2.4 we see that $f(x) = x^3 + 1$ has an inverse function. Solving $y = x^3 + 1$ for $x$, we have

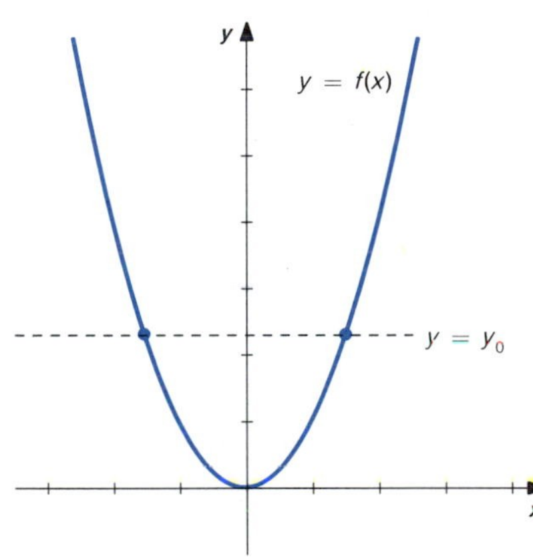


FIGURE 7.2.5

$$y = x^3 + 1,$$

$$x^3 = y - 1,$$

$$x = (y - 1)^{1/3}.$$

This means $f^{-1}(y) = (y - 1)^{1/3}$, or $f^{-1}(x) = (x - 1)^{1/3}$. The graph of $y = f^{-1}(x)$ is sketched in Figure 7.2.4.

(c) $f(x) = x^2$ does not have an inverse function. For any $y_0 > 0$, the horizontal line $y = y_0$ intersects the graph in two points. See Figure 7.2.5. The equation $y = x^2$ does not have a unique solution for $x$ in terms of $y$. For $y > 0$ the equation has two solutions, $x = \sqrt{y}$ and $x = -\sqrt{y}$.

(d) Figure 7.2.6 indicates that $f(x) = x^2$, $x \geq 0$, has an inverse function. The equation $y = x^2$ has the *single* solution $x = \sqrt{y}$ if $x$ is restricted to nonnegative values. We have $f^{-1}(y) = \sqrt{y}$, or $f^{-1}(x) = \sqrt{x}$. The graph of $y = f^{-1}(x)$ is given in Figure 7.2.6. ■

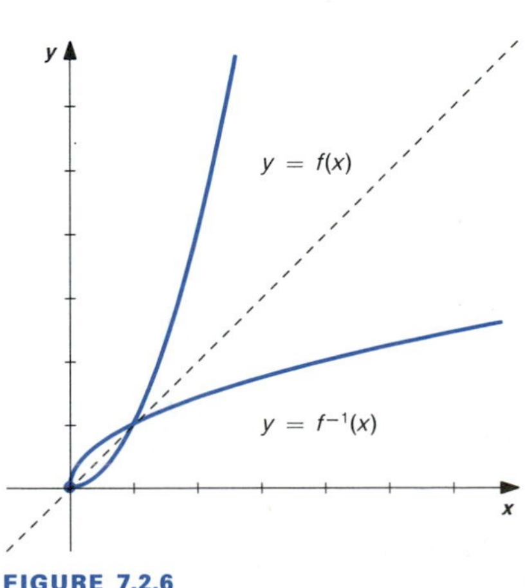


FIGURE 7.2.6

Let us compare Example 1c and 1d. The function $f(x) = x^2$ of Example 1c does not have an inverse function, but the *different function* obtained in Example 1d by restricting the domain to nonnegative values of $x$ does

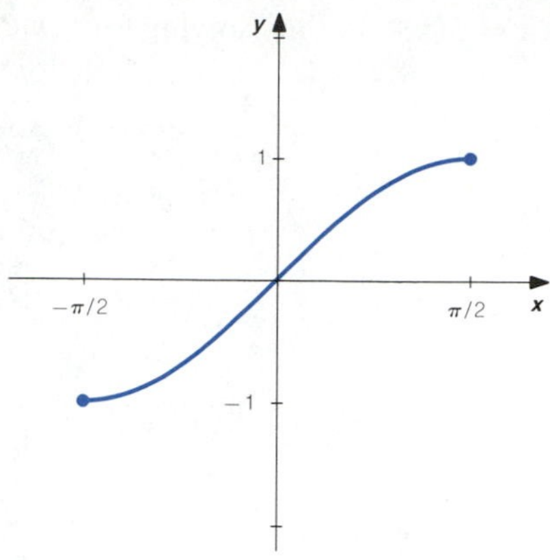

(a)

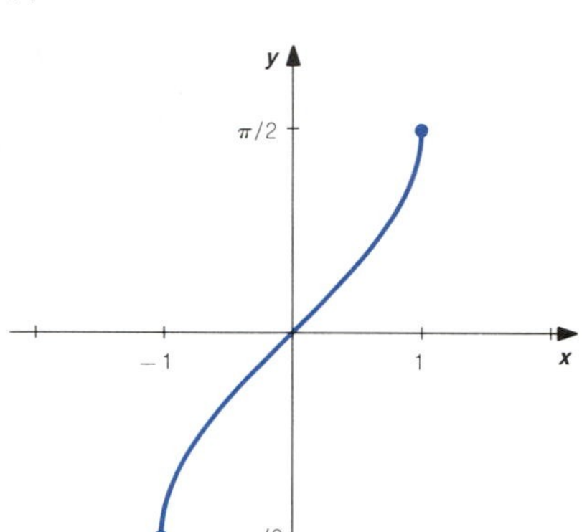

(b)

**FIGURE 7.2.7**

have an inverse. We can use this idea of restricting the domain of a function to define the **inverse trigonometric functions.**

The function $y = \sin x$ does not have an inverse function. However, we can obtain an inverse function by restricting the domain. We want an interval on which $\sin x$ assumes each value between $-1$ and 1 exactly once. The interval $-\pi/2 \le x \le \pi/2$ is most convenient. Thus, the function

$$f(x) = \sin x, \qquad -\frac{\pi}{2} \le x \le \frac{\pi}{2},$$

has an inverse function, which we denote either $\arcsin x$ or $\sin^{-1}x$. See Figure 7.2.7a. (The $-1$ in the notation $\sin^{-1}x$ corresponds to the $-1$ in the notation $f^{-1}$. In particular, this $-1$ is not intended to be an exponent. The negative first power of $\sin x$ should be written $(\sin x)^{-1}$ in order to avoid confusion.) We have

$$y = \sin x, \quad -\frac{\pi}{2} \le x \le \frac{\pi}{2}, \quad \text{if and only if} \quad \sin^{-1}y = x.$$

The graph of $y = \sin^{-1}x$ is given in Figure 7.2.7b. The function $\sin^{-1}x$ is defined for $-1 \le x \le 1$ and has values $-\pi/2 \le y \le \pi/2$.

We cannot use the usual rules of algebra to "solve" the equation $y = \sin x$ for $x$. It follows from the definitions that

$$y = \sin x, \quad -\frac{\pi}{2} \le x \le \frac{\pi}{2}, \quad \text{implies} \quad \sin^{-1}y = x.$$

Similarly,

$$\sin^{-1}y = x \quad \text{implies} \quad y = \sin x.$$

The function $f(x) = \cos x$, $0 \le x \le \pi$, has an inverse function, denoted either $\arccos x$ or $\cos^{-1}x$. See Figure 7.2.8a. We have

$$y = \cos x, \quad 0 \le x \le \pi, \quad \text{if and only if} \quad \cos^{-1}y = x.$$

The graph of $y = \cos^{-1}x$ is given in Figure 7.2.8b. $\text{Cos}^{-1}x$ is defined for $-1 \le x \le 1$ and has values $0 \le y \le \pi$.

The function $f(x) = \tan x$, $-\pi/2 < x < \pi/2$, has an inverse function, denoted either $\arctan x$ or $\tan^{-1}x$. See Figure 7.2.9a. We have

$$y = \tan x, \quad -\frac{\pi}{2} < x < \frac{\pi}{2}, \quad \text{if and only if} \quad \tan^{-1}y = x.$$

The graph of $y = \tan^{-1}x$ is given in Figure 7.2.9b. $\text{Tan}^{-1}x$ is defined for all $x$ and has values $-\pi/2 < y < \pi/2$.

The function $f(x) = \sec x$, $0 \le x \le \pi$, $x \ne \pi/2$, has an inverse function, denoted either $\text{arcsec}\, x$ or $\sec^{-1}x$. See Figure 7.2.10a. We have

$$y = \sec x, \quad 0 \le x \le \pi, \quad x \ne \frac{\pi}{2}, \quad \text{if and only if} \quad \sec^{-1}y = x.$$

FIGURE 7.2.8

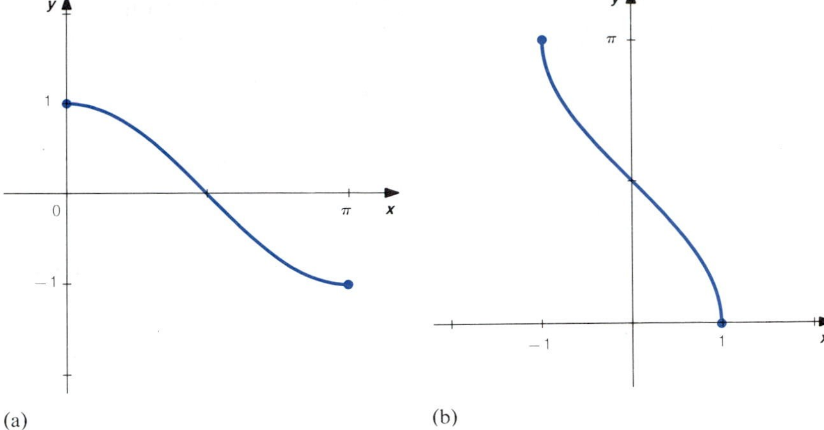


(a) (b)

FIGURE 7.2.9

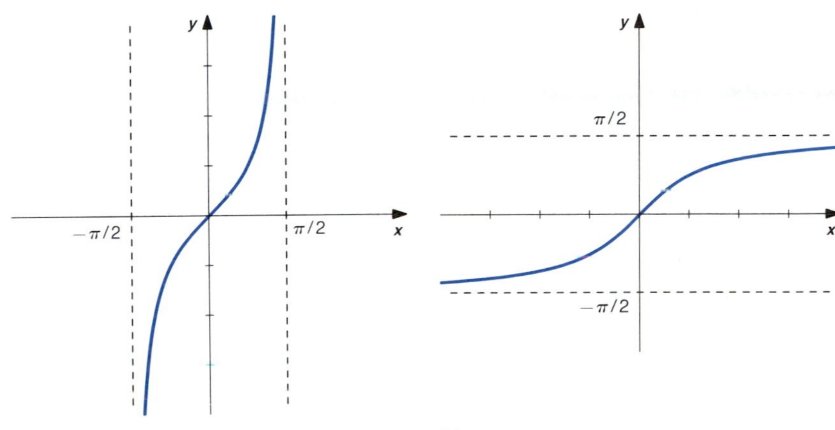


(a) (b)

FIGURE 7.2.10

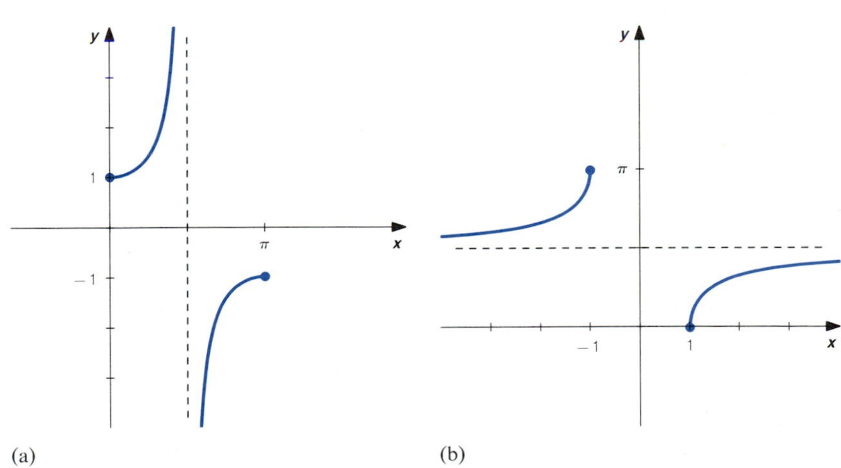


(a) (b)

The graph of $y = \sec^{-1} x$ is given in Figure 7.2.10b. $\mathrm{Sec}^{-1} x$ is defined for $|x| \geq 1$ and has values $0 \leq y \leq \pi$, $y \neq \pi/2$.

Vertical asymptotes of the graph of $y = f(x)$ correspond to horizontal asymptotes of the graph of $y = f^{-1}(x)$. For example, compare the graph in Figure 7.2.10a with that in Figure 7.2.10b. Similarly, horizontal asymptotes of the graph of $y = f(x)$ correspond to vertical asymptotes of the graph of $y = f^{-1}(x)$.

From the graph in Figure 7.2.9b we see that

$$\lim_{x\to\infty} \tan^{-1}x = \frac{\pi}{2} \quad \text{and} \quad \lim_{x\to-\infty} \tan^{-1}x = -\frac{\pi}{2}. \tag{3}$$

From the graph in Figure 7.2.10b we see that

$$\lim_{x\to\infty} \sec^{-1}x = \frac{\pi}{2} \quad \text{and} \quad \lim_{x\to-\infty} \sec^{-1}x = \frac{\pi}{2}. \tag{4}$$

The trigonometric functions and the inverse trigonometric functions are real-valued functions of a real variable. However, it is sometimes convenient to think of a real number as the radian measure of an angle. For example, we may think of $\sin x$ as the sine of the angle that has radian measure $x$. Similarly, we may think of $\sin^{-1}x$ as the angle $\theta$ that satisfies

$$-\frac{\pi}{2} \leq \theta \leq \frac{\pi}{2} \quad \text{and} \quad \sin\theta = x.$$

We can use known values of the trigonometric functions or a calculator to determine values of the inverse trigonometric functions. It is important to remember the range of each of the inverse trigonometric functions.

■ **EXAMPLE 2**

Evaluate

(a) $\sin^{-1}(1)$,
(b) $\cos^{-1}(1/\sqrt{2})$,
(c) $\tan^{-1}(-1)$,
(d) $\sin^{-1}(0.8)$,
(e) $\tan^{-1}(-2)$,
(f) $\sec^{-1}[\sec(-\pi/3)]$,
(g) $\sin[\cos^{-1}(-1/2)]$.

**SOLUTION**

(a) The expression $\sin^{-1}(1)$ represents an angle $\theta$ that satisfies $-\pi/2 \leq \theta \leq \pi/2$ and $\sin\theta = 1$. The angle $\theta = \pi/2$ satisfies these conditions, so $\sin^{-1}(1) = \pi/2$.

(b) The expression $\cos^{-1}(1/\sqrt{2})$ represents an angle $\theta$ with $0 \leq \theta \leq \pi$ and $\cos\theta = 1/\sqrt{2}$. Hence, $\cos^{-1}(1/\sqrt{2}) = \pi/4$.

(c) The expression $\tan^{-1}(-1) = -\pi/4$, because $-\pi/4$ is in the interval $(-\pi/2, \pi/2)$ and $\tan(-\pi/4) = -1$.

(d) The expression $\sin^{-1}(0.8)$ represents an angle $\theta$ with $-\pi/2 \leq \theta \leq \pi/2$, and $\sin\theta = 0.8$. None of the familiar angles $\theta$ satisfies these conditions, so we use a calculator to determine that $\sin^{-1}(0.8) \approx 0.92729522$ (radians).

(e) A calculator gives $\tan^{-1}(-2) \approx -1.1071487$ (radians).

(f) Since $\sec(-\pi/3) = 2$, $\sec^{-1}[\sec(-\pi/3)] = \sec^{-1}[2] = \pi/3$. [Statement (2), which says $f^{-1}(f(x)) = x$, does not apply in this case. $\mathrm{Sec}^{-1}$ is the inverse function of $f(x) = \sec x$, $0 \leq x \leq \pi$, $x \neq \pi/2$. Hence, $x = -\pi/3$ is not in the domain of $f$.]

(g) $\sin[\cos^{-1}(-1/2)] = \sin\theta$, where $\theta = \cos^{-1}(-1/2)$. Then, since $\cos^{-1}(-1/2) = 2\pi/3$, we have $\sin[\cos^{-1}(-1/2)] = \sin[2\pi/3] = \sqrt{3}/2$. ■

Certain combinations of trigonometric functions and inverse trigonometric functions simplify to algebraic expressions. A sketch is often useful for this type of simplification.

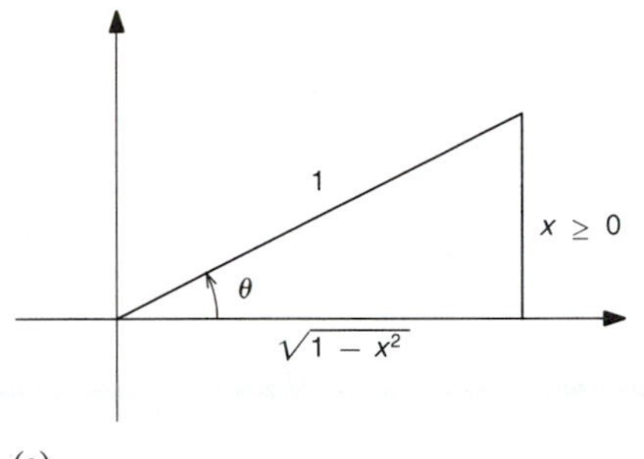


(a)

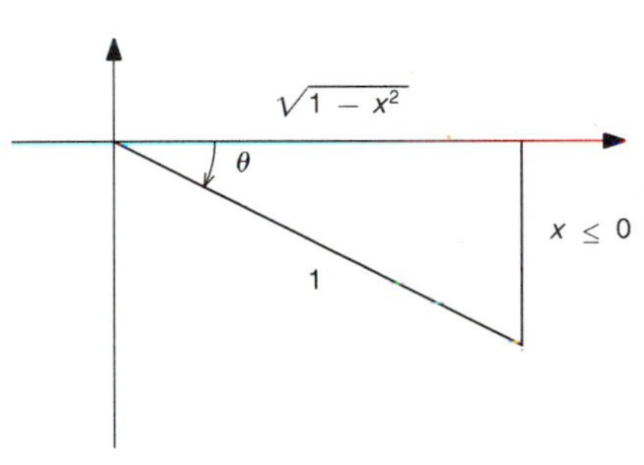


(b)

**FIGURE 7.2.11**

■ **EXAMPLE 3**

Write $\cos(\sin^{-1}x)$ as an algebraic expression.

**SOLUTION**

$\mathrm{Cos}(\sin^{-1}x) = \cos\theta$, where $\theta = \sin^{-1}x$. Thus, $\theta$ satisfies $-\pi/2 \leq \theta \leq \pi/2$ and $\sin\theta = x$. Representative angles with $\sin\theta = x$ and $-\pi/2 \leq \theta \leq \pi/2$ are sketched in Figure 7.2.11. We have used $\sin\theta = x/1$ to label one side and the hypotenuse of each triangle and then used the Pythagorean Theorem to determine that the other side is $\sqrt{1 - x^2}$. We can read from our sketch that $\cos(\sin^{-1}x) = \cos\theta = \sqrt{1 - x^2}$. (Note that $\sqrt{1 - x^2} \geq 0$ and $-\pi/2 \leq \theta \leq \pi/2$ implies $\cos\theta \geq 0$.) ■

We can use the relation between a function and its inverse,

$$y = f(x) \quad \text{if and only if} \quad x = f^{-1}(y),$$

to solve equations. Recall that you can think of these as different forms of the same equation, or you can obtain one from the other by using (2).

■ **EXAMPLE 4**

Solve the equations for $x$. (a) $\tan^{-1}(2x) = -1$, (b) $\sin(2x) = 0.5$, $-\pi/4 \leq x \leq \pi/4$.

**SOLUTION**

(a) $\mathrm{Tan}^{-1}(2x) = -1$ implies $\tan(-1) = 2x$, so $x = \tan(-1)/2$. Using a calculator, we have $x \approx -0.77870386$. Note that $\tan(-1)$ is the tangent of an angle with *radian* measure $-1$.

(b) The equation $\sin(2x) = 0.5$, $-\pi/2 \leq 2x \leq \pi/2$, implies $2x = \sin^{-1}(0.5)$. Hence, $x = \sin^{-1}(0.5)/2 = (\pi/6)/2 = \pi/12$. ■

■ **EXAMPLE 5**

Find all $\theta$, $0 \leq \theta \leq \pi$, such that $\cot(2\theta) = -3/2$.

**SOLUTION**

Let us first rewrite the equation $\cot(2\theta) = -3/2$ as $\tan(2\theta) = -2/3$. $\mathrm{Tan}^{-1}(-2/3)$ is one value of $2\theta$ that satisfies $\tan(2\theta) = -2/3$. Since the

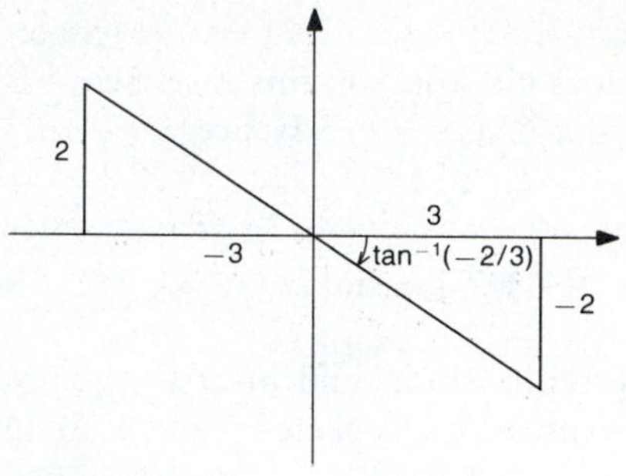

FIGURE 7.2.12

tangent function is periodic with period $\pi$, all angles $2\theta$ that satisfy $\tan(2\theta) = -2/3$ are of the form

$$2\theta = \tan^{-1}(-2/3) + k\pi, \qquad k \text{ an integer.}$$

See Figure 7.2.12. The values of $\theta$ that satisfy $0 \le \theta \le \pi$ and $\tan(2\theta) = -2/3$ are

$$\theta = \frac{\tan^{-1}(-2/3) + \pi}{2} \approx 1.276795$$

and

$$\theta = \frac{\tan^{-1}(-2/3) + 2\pi}{2} \approx 2.8475914.$$

■

Properties of an inverse function $f^{-1}$ can be determined from corresponding properties of $f$. For example,

$$\cos^{-1}x = \frac{\pi}{2} - \sin^{-1}x. \tag{5}$$

■ *Proof* $\text{Cos}^{-1}x$ is the angle $\theta$ that satisfies $0 \le \theta \le \pi$ and $\cos\theta = x$. To verify (5) we show that $\theta = (\pi/2) - \sin^{-1}x$ satisfies these conditions. The first condition follows from the fact that $-\pi/2 \le \sin^{-1}x \le \pi/2$. Since $\cos[(\pi/2) - \phi] = \sin(\phi)$ for any angle $\phi$, we obtain the equation $\cos[(\pi/2) - \sin^{-1}(x)] = \sin(\sin^{-1}x) = x$, so the second condition also holds. ■

We conclude this section by describing the idea of inverse function in mathematical terms and determining some general properties of inverse functions.

■ *Definition*

A function $f$ is said to be **one-to-one** if for each pair $x_1$, $x_2$ in the domain of $f$,

$$x_1 \ne x_2 \quad \text{implies} \quad f(x_1) \ne f(x_2).$$

■

This means that for each $y$ in the range of $f$, there is exactly one $x$ in the domain of $f$ with $y = f(x)$. The inverse function of $f$ can then be defined by the relation

$$y = f(x) \quad \text{if and only if} \quad x = f^{-1}(y).$$

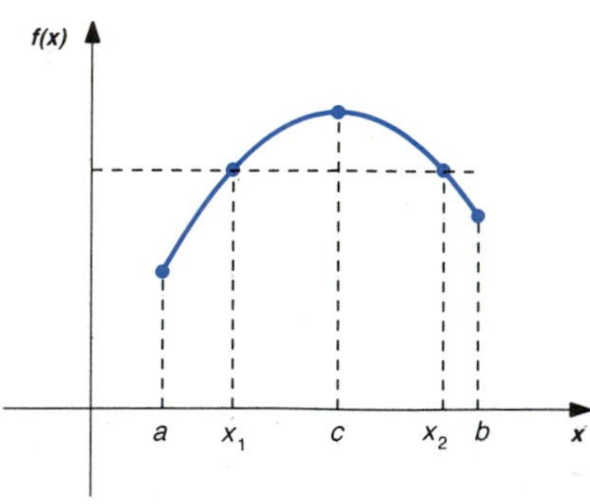

FIGURE 7.2.13

We most often deal with functions that are continuous on an interval. Let us note that a function that is one-to-one and continuous on an interval must be either increasing on the interval or decreasing on the interval. For example, a function with a graph such as that in Figure 7.2.13 cannot be one-to-one, since we could use the Intermediate Value Theorem

FIGURE 7.2.14

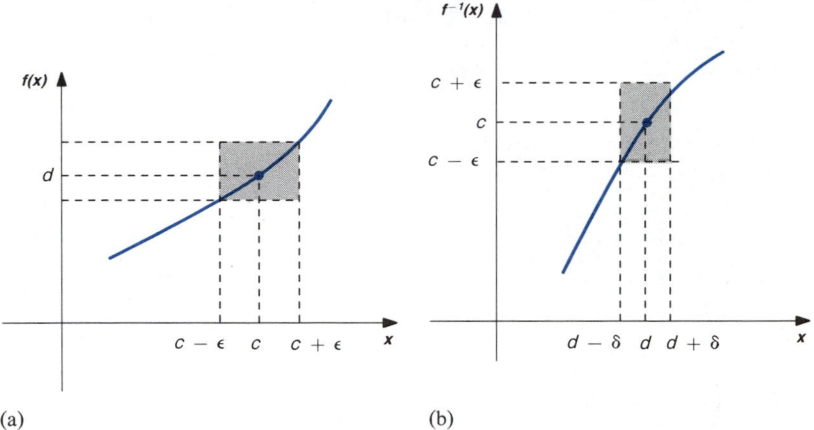


(a) (b)

for Continuous Functions on each of the intervals $[a, c]$ and $[c, d]$ to find points $x_1$ and $x_2$ with $x_1 \neq x_2$ and $f(x_1) = f(x_2)$.

*Theorem 1*

**If $f$ is continuous and one-to-one on an interval $I$, then $f^{-1}$ is continuous at each point $d = f(c)$, $c$ in $I$.**

*Proof* To show that $f^{-1}$ is continuous at $d$ we must verify that, given $\varepsilon > 0$, there is a $\delta > 0$ such that $|f^{-1}(y) - f^{-1}(d)| < \varepsilon$ whenever $|y - d| < \delta$. We know that $f^{-1}$ is defined and that $f$ is either increasing on $I$ or decreasing on $I$. From the sketches in Figure 7.2.14 for the case of $f$ increasing on $I$, it appears that we can use $\delta = \text{minimum}(|f(c - \varepsilon) - f(c)|, |f(c + \varepsilon) - f(c)|)$, where $c = f^{-1}(d)$. The same value of $\delta$ also works in the case of $f$ decreasing. The fact that $\delta$ is positive is a consequence of the fact that $f$ is one-to-one. ■

*Theorem 2*

**If $f$ is differentiable on an interval $I$ with either $f'(x) > 0$ on $I$ or $f'(x) < 0$ on $I$, then $f^{-1}$ is defined and continuous at points $d = f(c)$, $c$ in $I$.**

*Proof* If a function $f$ is differentiable on an interval $I$, then $f$ is continuous on $I$. If $f'(x)$ is of constant sign on $I$, then $f$ is either increasing on $I$ or decreasing on $I$. In either case, $f$ is one-to-one, so Theorem 2 follows from Theorem 1. ■

## EXERCISES 7.2

In Exercises 1–12 sketch the graph of $f$ and determine if $f$ has an inverse function $f^{-1}$. If $f^{-1}$ exists, find a formula for $f^{-1}(x)$ and sketch the graph of $y = f^{-1}(x)$.

**1** $f(x) = 2x + 1$

**2** $f(x) = 1 - \dfrac{x}{2}$

**3** $f(x) = 2(x - 1)^{1/3}$

**4** $f(x) = 1 - x^3$

**5** $f(x) = x^4 - 1$

**6** $f(x) = 2x - x^2$

**7** $f(x) = \sqrt{4 - x^2}$, $0 \le x \le 2$

**8** $f(x) = 2\sqrt{1 - x^2}$, $0 \le x \le 1$

**9** $f(x) = \sqrt{x+1}$ **10** $f(x) = 1 - \sqrt{1-x}$

**11** $f(x) = \dfrac{1}{x-1}$ **12** $f(x) = \dfrac{x}{x+1}$

Evaluate the expressions in Exercises 13–30.

**13** $\sin^{-1}(1/2)$ **14** $\sin^{-1}(\sqrt{3}/2)$

**15** $\cos^{-1}(-1)$ **16** $\cos^{-1}(-1/\sqrt{2})$

**17** $\tan^{-1}(\sqrt{3})$ **18** $\tan^{-1}(1)$

**19** $\sec^{-1}(\sqrt{2})$ **20** $\sec^{-1}(2/\sqrt{3})$

**21** $\sin^{-1}(0.1)$ **22** $\sin^{-1}(0.01)$

**23** $\cos^{-1}(0.2)$ **24** $\cos^{-1}(-0.2)$

**25** $\tan^{-1}(3)$ **26** $\tan^{-1}(0.75)$

**27** $\cos[\sin^{-1}(-1/\sqrt{2})]$ **28** $\sin[\cos^{-1}(\sqrt{3}/2)]$

**29** $\sin[\sin^{-1}(3\pi/4)]$ **30** $\tan^{-1}[\tan(4\pi/3)]$

Write the expressions in Exercises 31–36 as algebraic expressions.

**31** $\tan[\sin^{-1}x]$ **32** $\sin[\tan^{-1}x]$

**33** $\cos[\sin^{-1}(x/2)]$ **34** $\cos[\tan^{-1}(x/3)]$

**35** $\sin[\sec^{-1}(x/3)]$ **36** $\tan[\sec^{-1}(x/2)]$

Solve for $x$ in Exercises 37–40.

**37** $\sin^{-1}(2x) = 0.23$ **38** $\sin^{-1}(x/2) = -0.14$

**39** $\sin(2x) = 0.25,\ -\pi/4 \le x \le \pi/4$

**40** $\tan(2x-1) = -0.5,\ -\pi/2 < 2x - 1 < \pi/2$

Solve for $\theta$ in Exercises 41–44.

**41** $\cos\theta = 1/3,\ 0 \le \theta \le 2\pi$ **42** $\sin\theta = -1/4, 0 \le \theta \le 2\pi$

**43** $\cot(2\theta) = -4,\ 0 \le \theta \le \pi$ **44** $\cot(2\theta) = -3, 0 \le \theta \le \pi$

**45** Verify that $f$ increasing on $[a, b]$ implies $f$ is one-to-one on $[a, b]$.

**46** Verify that $f$ continuous and increasing on $[a, b]$ implies $f^{-1}$ is increasing on $[f(a), f(b)]$.

**47** Find a function $f$ that is one-to-one on $[a, b]$ but is neither increasing on $[a, b]$ nor decreasing on $[a, b]$.

**48** Find a function $f$ that is increasing on $[a, b]$, but $f^{-1}$ is not defined at some points of $[f(a), f(b)]$.

Use the definitions to verify the identities in Exercises 49–52.

**49** $\sin^{-1}(-x) = -\sin^{-1}x$

**50** $\cos^{-1}(-x) = \pi - \cos^{-1}x$

**51** $\sec^{-1}x = \cos^{-1}(1/x)$

**52** $\tan^{-1}x + \tan^{-1}(1/x) = \pi/2,\ x > 0$

**53** A sign 16 ft high is painted on a vertical wall. The bottom of the sign is 9 ft above eye level. Express the angle of sight between the top and bottom of the sign as a function of the distance from the wall.

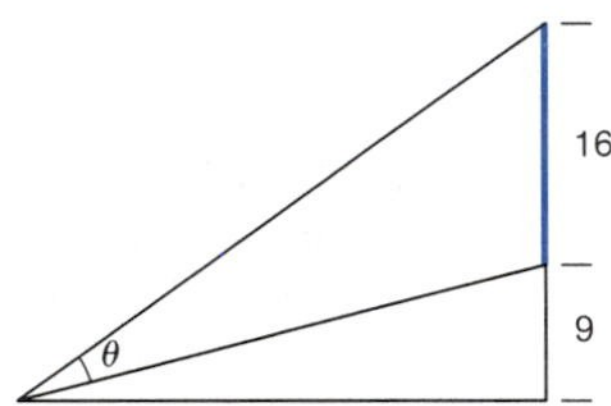


## 7.3 DIFFERENTIATION OF INVERSE FUNCTIONS

To apply the concept of inverse functions to calculus problems, we need to establish differentiation and integration formulas. In this section, we will briefly discuss some general conditions under which an inverse function is differentiable and how the derivative of a function is related to the derivative of its inverse. We will illustrate how to obtain formulas for the derivatives of inverse functions by developing formulas for the derivatives of the inverse trigonometric functions. The corresponding integration formulas allow us to evaluate some new types of integrals.

Let us state the main result as a theorem.

*Theorem 1*

**Suppose $f$ is differentiable on an open interval $I$ with either $f'(x) > 0$ on $I$ or $f'(x) < 0$ on $I$. Then $f$ has an inverse function $f^{-1}$, and for each $c$ in $I$, $f^{-1}$ is differentiable at $d = f(c)$ with**

$$(f^{-1})'(d) = \frac{1}{f'(c)}.$$

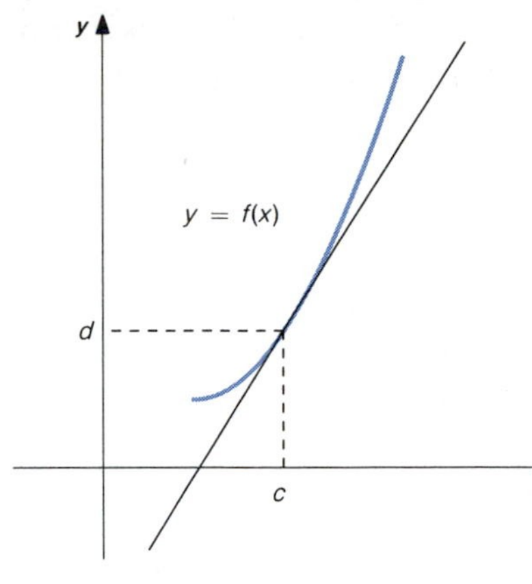


(a)

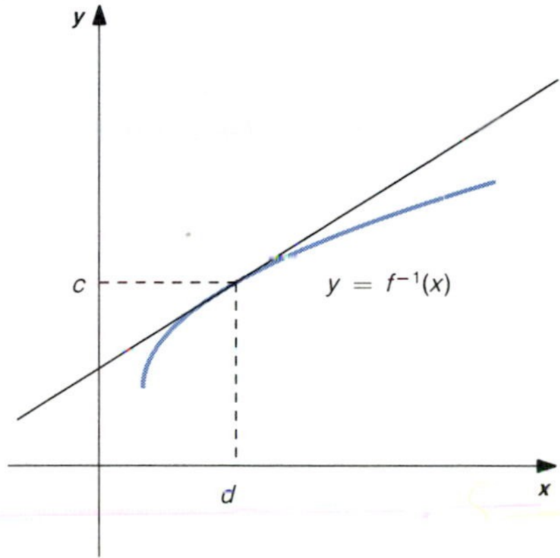


(b)

**FIGURE 7.3.1**

■ *Proof* It follows from Theorem 2 of Section 7.2 that $f^{-1}$ exists and is continuous. To both show that $f^{-1}$ is differentiable and verify the differentiation formula, we must look at the definition of the derivative of $f^{-1}$ at the point $d = f(c)$. We have

$$(f^{-1})'(d) = \lim_{y \to d} \frac{f^{-1}(y) - f^{-1}(d)}{y - d}.$$

We know that $d = f(c)$ implies $f^{-1}(d) = c$. Setting $x = f^{-1}(y)$, so $y = f(x)$, and noting that $y \neq d$ implies $x = f^{-1}(y) \neq f^{-1}(d) = c$, we can rewrite the above quotient as

$$\frac{x - c}{f(x) - f(c)} \quad \text{or} \quad \frac{1}{\dfrac{f(x) - f(c)}{x - c}}.$$

Since $f^{-1}$ is continuous, $x = f^{-1}(y) \to f^{-1}(d) = c$ as $y \to d$. This means

$$\frac{f(x) - f(c)}{x - c} \to f'(c) \quad \text{as} \quad y \to d.$$

Combining results, we obtain the theorem. ■

The relation between the derivative of a function and the derivative of its inverse is illustrated geometrically in Figure 7.3.1. The line tangent to the graph of $y = f(x)$ at $(c, d)$ has slope $f'(c)$, while the line tangent to the graph of $y = f^{-1}(x)$ at $(d, c)$ has slope $(f^{-1})'(d)$. Since the graphs of $y = f(x)$ and $y = f^{-1}(x)$ are symmetric with respect to the line $y = x$, these slopes should be reciprocals, so

$$(f^{-1})'(d) = \frac{1}{f'(c)}, \qquad d = f(c).$$

Let us illustrate how implicit differentiation can be used to find a formula for the derivative of an inverse function.

The function $\sin x$ is differentiable with derivative $\cos x > 0$ for $-\pi/2 < x < \pi/2$. It follows that the inverse function $\sin^{-1}x$ exists and is differentiable for $-1 < x < 1$. To find a formula for the derivative of $\sin^{-1}x$, we set

$$y = \sin^{-1}x, \quad \text{so} \quad \sin y = x.$$

Differentiating the latter equation with respect to $x$, and using the Chain Rule to differentiate the sine of the function $y$, we have

$$\cos y \frac{dy}{dx} = 1, \quad \text{so} \quad \frac{dy}{dx} = \frac{1}{\cos y}.$$

Since $y$ is a function of $x$, it would be convenient to have a formula for the derivative of $y$ in terms of $x$. We need to express $\cos y$ in terms of $x$. We

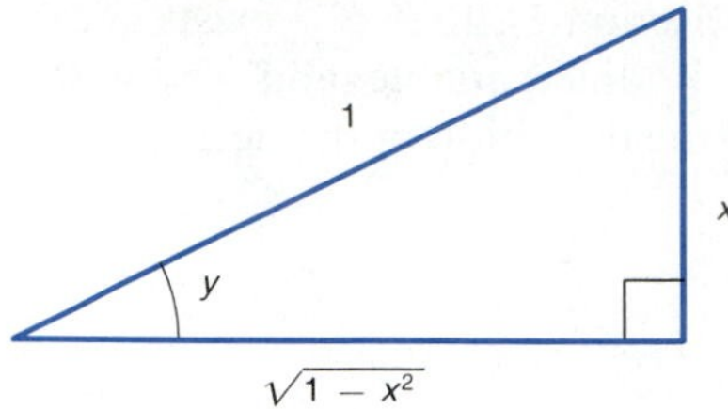


FIGURE 7.3.2

know $\sin y = x$. A representative right triangle with "angle" $y$ is sketched in Figure 7.3.2. We can determine from the sketch that $\cos y = \sqrt{1 - x^2}$. Substitution then gives the formula

$$\frac{d}{dx}(\sin^{-1}x) = \frac{1}{\sqrt{1 - x^2}}.$$

Tan $x$ has a positive derivative for $-\pi/2 < x < \pi/2$, so the inverse function $\tan^{-1}x$ exists and is differentiable. Let

$$y = \tan^{-1}x, \quad \text{so} \quad \tan y = x.$$

Differentiating this equation with respect to $x$, and using the Chain Rule, we obtain

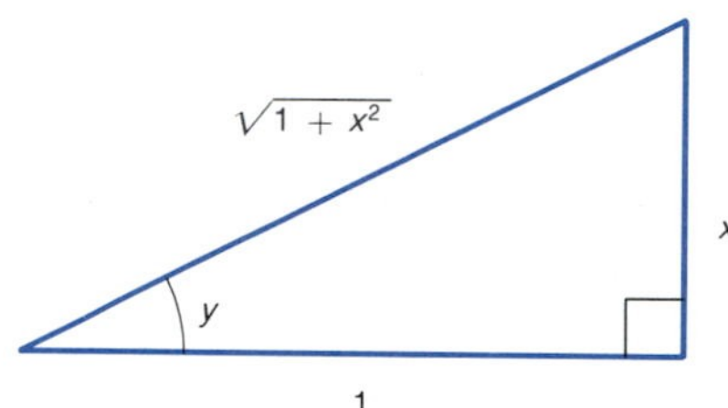


FIGURE 7.3.3

$$\sec^2 y \frac{dy}{dx} = 1, \quad \text{or} \quad \frac{dy}{dx} = \frac{1}{\sec^2 y}.$$

From Figure 7.3.3 we see that $\tan y = x$ implies $\sec y = \sqrt{1 + x^2}$. Substitution then gives

$$\frac{d}{dx}(\tan^{-1}x) = \frac{1}{1 + x^2}.$$

We can use the formula

$$\frac{d}{dx}(\sec x) = \sec x \tan x$$

to verify that

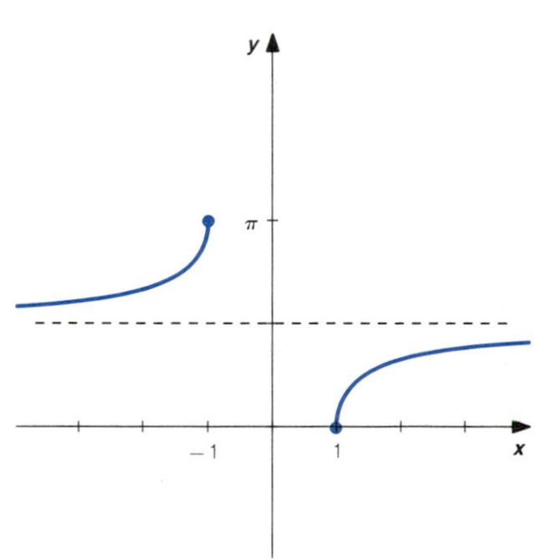


FIGURE 7.3.4

$$\frac{d}{dx}(\sec^{-1}x) = \frac{1}{|x|\sqrt{x^2 - 1}}.$$

We see from the graph in Figure 7.3.4 that the derivative of $\sec^{-1}x$ is positive at all points where it exists.

From Formula (4) of Section 7.2 we know that

$$\cos^{-1}x = \frac{\pi}{2} - \sin^{-1}x, \quad \text{so} \quad \frac{d}{dx}(\cos^{-1}x) = -\frac{d}{dx}(\sin^{-1}x), \quad \text{or}$$

$$\frac{d}{dx}(\cos^{-1}x) = -\frac{1}{\sqrt{1 - x^2}}.$$

We must know the Chain Rule version of each of the new differentiation formulas. For example,

$$\frac{d}{dx}(\sin^{-1}(\text{Function})) = \frac{1}{\sqrt{1 - (\text{Function})^2}}\frac{d}{dx}(\text{Function}).$$

**EXAMPLE 1**

(a) $\dfrac{d}{dx}(\sin^{-1}(x^2)) = \dfrac{1}{\sqrt{1-(x^2)^2}}(2x)$.

(b) $\dfrac{d}{dx}\left(\tan^{-1}\dfrac{x}{2}\right) = \left(\dfrac{1}{1+(x/2)^2}\right)\left(\dfrac{1}{2}\right) = \dfrac{2}{4+x^2}$.

(c) $\dfrac{d}{dx}(\sec^{-1}(|x|)) = \dfrac{1}{\|x\|\sqrt{|x|^2-1}}\dfrac{d}{dx}(|x|)$.

If $x > 1$, then $\|x\| = x$ and $\dfrac{d}{dx}(|x|) = 1$.

If $x < -1$, then $\|x\| = -x$ and $\dfrac{d}{dx}(|x|) = -1$.

In either case the result in (c) simplifies to

$$\frac{d}{dx}(\sec^{-1}(|x|)) = \frac{1}{x\sqrt{x^2-1}}.$$

Let us restate the result of Exercise 1c:

$$\frac{d}{dx}(\sec^{-1}(|x|)) = \frac{1}{x\sqrt{x^2-1}}.$$

**EXAMPLE 2**

A boy is sitting 30 ft from a road, watching cars go by. Find the rate of change of his angle of sight as a car going 30 mph passes the point on the road nearest him.

**SOLUTION**

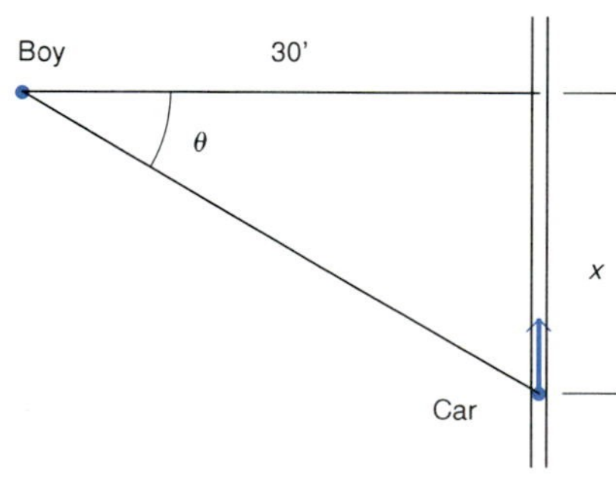


FIGURE 7.3.5

A sketch is given in Figure 7.3.5. The distance of the car from the point on the road nearest the boy has been labeled $x$ and the angle of sight has been labeled $\theta$. It is important that $x$ and $\theta$ are considered to be variables. The units of $x$ should be feet, the units used to indicate the distance of the boy from the street; $\theta$ is measured in radians.

Following the procedure of Section 4.6, we express what we want and what we know in mathematical terms and display the information clearly for future reference.

*We want*: $\dfrac{d\theta}{dt}$ when $x = 0$.

We need to change 30 mph to units of ft/s.

$$30 \text{ mph} = \left(\frac{30 \text{ mi}}{\text{h}}\right)\left(\frac{1 \text{ h}}{60 \text{ min}}\right)\left(\frac{1 \text{ min}}{60 \text{ s}}\right)\left(\frac{5280 \text{ ft}}{1 \text{ mi}}\right) = 44 \text{ ft/s}.$$

*We know:* $\dfrac{dx}{dt} = -44$ ft/s.

($x$ is decreasing, so $dx/dt$ is negative.)

From the sketch, we see that the variables are related by the equation

$$\tan\theta = \frac{x}{30}, \quad \text{with} \quad -\frac{\pi}{2} < \theta < \frac{\pi}{2}, \quad \text{so} \quad \theta = \tan^{-1}\left(\frac{x}{30}\right).$$

Differentiating the latter equation with respect to $t$, we have

$$\frac{d\theta}{dt} = \frac{1}{1 + \left(\dfrac{x}{30}\right)^2}\left(\frac{1}{30}\frac{dx}{dt}\right).$$

When $x = 0$ we have

$$\frac{d\theta}{dt} = \frac{1}{30}(-44) = -1.4666667,$$

so the angle of sight is decreasing at a rate of 1.4666667 rad/s. ■

The following integration formulas may be verified by differentiation. Recall that $\int f(x)\,dx = F(x) + C$ means $F' = f$ on some interval.

$$\int \frac{1}{\sqrt{a^2 - u^2}}\,du = \sin^{-1}\left(\frac{u}{a}\right) + C,$$

$$\int \frac{1}{a^2 + u^2}\,du = \frac{1}{a}\tan^{-1}\left(\frac{u}{a}\right) + C,$$

$$\int \frac{1}{u\sqrt{u^2 - a^2}}\,du = \frac{1}{a}\sec^{-1}\left(\frac{|u|}{a}\right) + C.$$

Note that the integral for $\sin^{-1}$ involves the square root of

$$(\text{Constant})^2 - (\text{Variable})^2.$$

The integral for $\sec^{-1}$ involves the square root of

$$(\text{Variable})^2 - (\text{Constant})^2.$$

■ **EXAMPLE 3**

Evaluate

$$\int \frac{1}{\sqrt{1 - 4x^2}}\,dx.$$

**SOLUTION**

The integral

$$\int \frac{1}{\sqrt{1-4x^2}}\,dx$$

involves the square root of $(\text{Constant})^2 - (\text{Variable})^2$, so it may reduce to a $\sin^{-1}$ integral. We have $4x^2 = (2x)^2$ in place of $u^2$. Let us set $u = 2x$, so that $du = 2\,dx$. Then

$$\begin{aligned}\int \frac{1}{\sqrt{1-4x^2}}\,dx &= \int \frac{(1/2)2}{\sqrt{1-(2x)^2}}\,dx \\ &= \int \frac{1/2}{\sqrt{1-u^2}}\,du \\ &= \frac{1}{2}\sin^{-1}u + C \\ &= \frac{1}{2}\sin^{-1}2x + C.\end{aligned}$$

■

■ **EXAMPLE 4**

Evaluate

$$\int \frac{x}{x^4+16}\,dx.$$

**SOLUTION**

The integral

$$\int \frac{x}{x^4+16}\,dx$$

involves $x^4 + 16 = (x^2)^2 + (4)^2$, so it may reduce to a $\tan^{-1}$ integral. Let us try the substitution $u = x^2$, so $du = 2x\,dx$. Then

$$\begin{aligned}\int \frac{x}{x^4+16}\,dx &= \int \frac{(1/2)\,2x}{(x^2)^2+4^2}\,dx \\ &= \int \frac{1/2}{u^2+4^2}\,du \\ &= \frac{1}{2}\left[\frac{1}{4}\tan^{-1}\left(\frac{u}{4}\right)\right] + C \\ &= \frac{1}{8}\tan^{-1}\left(\frac{x^2}{4}\right) + C.\end{aligned}$$

■

**EXAMPLE 5**

Evaluate

$$\int \frac{1}{x\sqrt{x^4 - 4}}\, dx.$$

**SOLUTION**

The integral

$$\int \frac{1}{x\sqrt{x^4 - 4}}\, dx$$

involves the square root of (Variable)$^2$ − (Constant)$^2$, so it may reduce to a $\sec^{-1}$ integral. Setting $u = x^2$, we have $du = 2x\, dx$, so

$$\begin{aligned}\int \frac{1}{x\sqrt{x^4 - 4}}\, dx &= \int \frac{(1/2)\, 2x}{x^2\sqrt{(x^2)^2 - 2^2}}\, dx \\ &= \int \frac{1/2}{u\sqrt{u^2 - 2^2}}\, du \\ &= \frac{1}{2}\left(\frac{1}{2}\sec^{-1}\frac{|u|}{2}\right) + C \\ &= \frac{1}{4}\sec^{-1}\frac{x^2}{2} + C.\end{aligned}$$

■

**EXAMPLE 6**

Evaluate

$$\int \frac{\tan^{-1}x}{1 + x^2}\, dx.$$

**SOLUTION**

In this integral we recognize

$$\frac{1}{1 + x^2}$$

as the derivative of $\tan^{-1}x$. This suggests the substitution $u = \tan^{-1}x$, so

$$du = \frac{1}{1 + x^2}\, dx.$$

Then

$$\int \frac{\tan^{-1}x}{1 + x^2}\, dx = \int u\, du = \frac{u^2}{2} + C = \frac{1}{2}(\tan^{-1}x)^2 + C.$$

■

■ **EXAMPLE 7**

Find the volume of the solid obtained by revolving the region bounded by

$$y = \frac{3}{\sqrt{x^2+9}}, \qquad y = 0, \quad x = 0, \quad x = 3,$$

about the (a) $x$-axis, (b) $y$-axis.

**SOLUTION**

(a) A typical vertical area strip generates a disk when revolved about the $x$-axis. From the sketch in Figure 7.3.6a we see that the radius of the disk is $3/\sqrt{x^2+9}$. The volume is

$$V = \int_0^3 \pi\left(\frac{3}{\sqrt{x^2+9}}\right)^2 dx = \int_0^3 \frac{9\pi}{x^2+3^2}\,dx = 9\pi\,\frac{1}{3}\tan^{-1}\frac{x}{3}\Big]_0^3$$

$$= 3\pi\tan^{-1}(1) - 3\pi\tan^{-1}(0) = 3\pi\,\frac{\pi}{4} = \frac{3\pi^2}{4}.$$

(b) In this case a typical vertical area strip generates a cylindrical shell with radius $x$ and height $3/\sqrt{x^2+9}$. See Figure 7.3.6b. The volume is

$$V = \int_0^3 2\pi x\left(\frac{3}{\sqrt{x^2+9}}\right) dx = \int_0^3 3\pi(x^2+9)^{-1/2}2x\,dx.$$

Let $u = x^2 + 9$, so $du = 2x\,dx$. Substitution then gives

$$V = \int 3\pi u^{-1/2}\,du = 3\pi\frac{u^{1/2}}{1/2}\Big] = 6\pi\sqrt{x^2+9}\Big]_0^3 = 6\pi\sqrt{18} - 6\pi\sqrt{9}.$$ ■

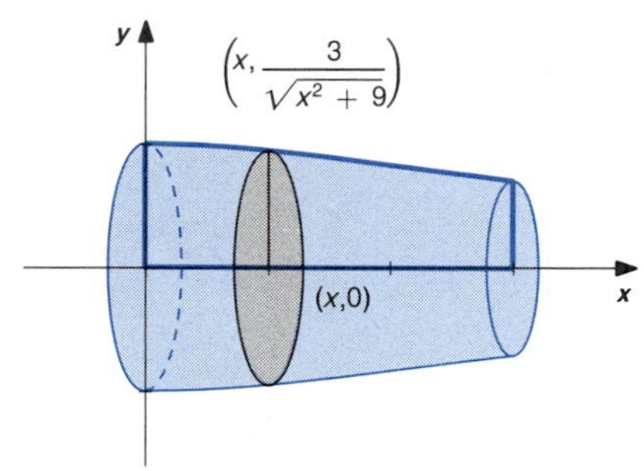


(a)

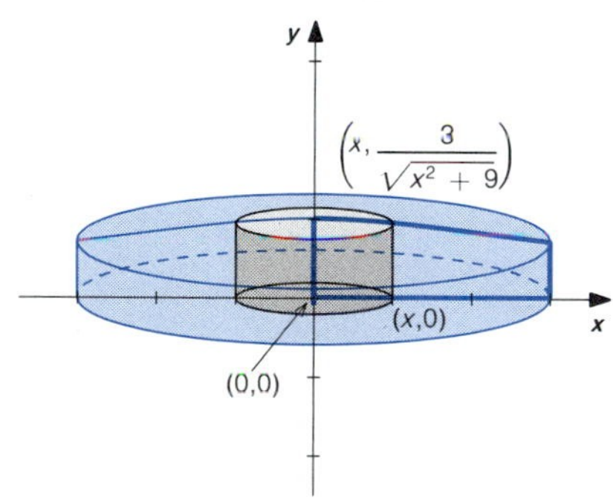


(b)

**FIGURE 7.3.6**

## EXERCISES 7.3

Evaluate the derivatives in Exercises 1–10.

**1** $\frac{d}{dx}\left(\sin^{-1}\frac{x}{3}\right)$

**2** $\frac{d}{dx}(\sin^{-1}2x)$

**3** $\frac{d}{dx}\left(\cos^{-1}\frac{x}{2}\right)$

**4** $\frac{d}{dx}\left(\cos^{-1}\frac{1}{x}\right)$

**5** $\frac{d}{dx}\left(\tan^{-1}\frac{1}{x}\right)$

**6** $\frac{d}{dx}\left(\tan^{-1}\frac{x}{a}\right)$

**7** $\frac{d}{dx}(\sec^{-1}x^2)$

**8** $\frac{d}{dx}(\sec^{-1}2x)$

**9** $\frac{d}{dx}\left(\sin^{-1}\frac{x}{a}\right)$

**10** $\frac{d}{dx}\left(\sec^{-1}\frac{|x|}{a}\right)$

Evaluate the integrals in Exercises 11–20.

**11** $\int \frac{1}{\sqrt{9-x^2}}\,dx$

**12** $\int \frac{1}{\sqrt{4-9x^2}}\,dx$

**13** $\int \frac{1}{9+4x^2}\,dx$

**14** $\int \frac{1}{4+x^2}\,dx$

**15** $\int \frac{1}{x\sqrt{4x^2-1}}\,dx$

**16** $\int \frac{x}{\sqrt{1-x^4}}\,dx$

**17** $\int \frac{x^3}{\sqrt{x^4-1}}\,dx$

**18** $\int \frac{x}{\sqrt{x^2-4}}\,dx$

**19** $\int \frac{\sin^{-1}x}{\sqrt{1-x^2}}\,dx$

**20** $\int \frac{\sqrt{\tan^{-1}x}}{1+x^2}\,dx$

21 Find an equation of the line tangent to the graph of $y = x\tan^{-1}x$ at the point where $x = 1$.

22 Find an equation of the line tangent to the graph of $y = x\sin^{-1}x$ at the point where $x = 1/2$.

23 A balloon is rising vertically at a rate of 6 ft/s from a point on the ground that is 30 ft from an observer. Find the rate of change of the angle of inclination from the observer to the balloon when the balloon is 90 ft above ground.

24 A baseball is thrown by the pitcher toward home plate at a rate of 110 ft/s. At what rate would the batter need to rotate his line of sight to follow the ball as it passes home plate at a point 2 ft from him?

25 Find the area of the region bounded by $y = 1/\sqrt{4 - x^2}$, $y = 0$, $x = -1$, $x = 1$.

26 Find the area of the region bounded by $y = 1/(x^2 + 4)$, $y = 0$, $x = -2$, $x = 2$.

27 Find the volume of the solid obtained by revolving the region bounded by $y = 4/\sqrt{x^2 + 16}$, $y = 0$, $x = 0$, $x = 4\sqrt{3}$ about the (a) $x$-axis, (b) $y$-axis.

28 Find the volume of the solid obtained by revolving the region bounded by $y = 1/(16 - x^2)^{1/4}$, $y = 0$, $x = 0$, $x = 2\sqrt{3}$ about the (a) $x$-axis, (b) $y$-axis.

29 Evaluate $\displaystyle\lim_{x\to\infty} \int_0^x \frac{1}{1 + t^2}\,dt$.

30 Evaluate $\displaystyle\lim_{x\to 1^-} \int_0^x \frac{1}{\sqrt{1 - t^2}}\,dt$.

## 7.4 THE NATURAL LOGARITHMIC FUNCTION

The formal definition of the natural logarithmic function can be related to the area under a curve. See Figure 7.4.1.

■ *Definition*

The **natural logarithmic function** is

$$\ln x = \int_1^x \frac{1}{t}\,dt, \qquad x > 0.$$ ■

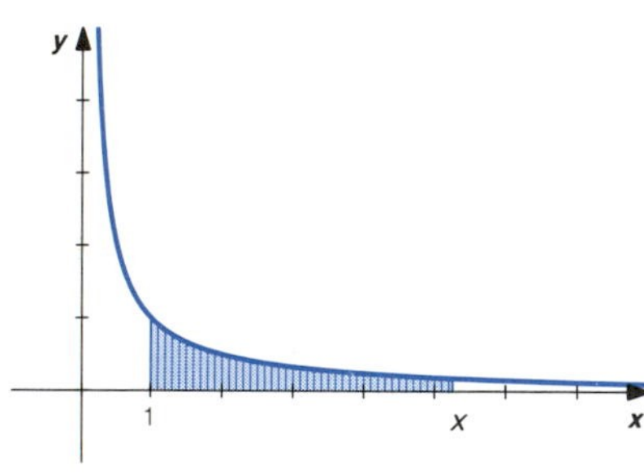

FIGURE 7.4.1

If $x > 0$, then the function $1/t$ is continuous on the closed interval with endpoints 1 and $x$, so the integral that defines $\ln x$ exists. The integral does not exist for $x \le 0$ and $\ln x$ is not defined for $x \le 0$. We have chosen this definition of $\ln x$ to facilitate study of the function from the point of view of calculus. For example, the Fundamental Theorem of Calculus immediately gives us

$$\frac{d}{dx}(\ln x) = \frac{d}{dx}\left(\int_1^x \frac{1}{t}\,dt\right) = \frac{1}{x}.$$

We can also use the integral definition of the logarithmic function to show that $\ln x$ satisfies the familiar properties of logarithms.

$$\ln(ab) = \ln a + \ln b. \tag{1}$$

$$\ln(a^r) = r\ln a, \qquad r \text{ a rational number.} \tag{2}$$

$$\ln\left(\frac{1}{b}\right) = -\ln b. \tag{3}$$

$$\ln\left(\frac{a}{b}\right) = \ln a - \ln b. \tag{4}$$

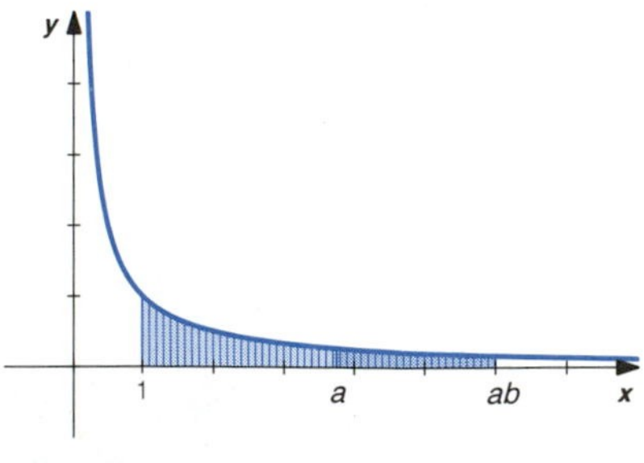

FIGURE 7.4.2

■ *Proof of (1)–(4)* To verify (1) we have (see Figure 7.4.2)

$$\ln(ab) = \int_1^{ab} \frac{1}{t}\,dt = \int_1^a \frac{1}{t}\,dt + \int_a^{ab} \frac{1}{t}\,dt.$$

The substitution $u = (1/a)t$ in the second integral gives

$$\ln(ab) = \int_1^a \frac{1}{t}\,dt + \int_1^b \frac{1}{u}\,du = \ln a + \ln b.$$

Similarly, the substitution $u = t^{1/r}$ can be used to verify (2). Since $1/x = x^{-1}$, (3) is a special case of (2). Equation (4) follows from (1) and (3). ■

The Chain Rule version of the differentiation formula for $\ln x$ is

$$\frac{d}{dx}(\ln(\text{Function})) = \frac{1}{\text{Function}}\frac{d}{dx}(\text{Function}).$$

■ **EXAMPLE 1**

(a) $\dfrac{d}{dx}(\ln(x^2+1)) = \dfrac{1}{x^2+1}(2x)$.

(b) $\dfrac{d}{dx}(\ln(\sec x)) = \dfrac{1}{\sec x}(\sec x \tan x) = \tan x$. ■

It is helpful to use the properties of logarithms to simplify differentiation of the natural logarithmic function of products, quotients, and powers.

■ **EXAMPLE 2**

$$\frac{d}{dx}(\ln\sqrt{x^2+1}) = \frac{d}{dx}\left(\frac{1}{2}\ln(x^2+1)\right) = \frac{1}{2}\left(\frac{1}{x^2+1}\right)(2x) = \frac{x}{x^2+1}.$$ ■

In some cases it is easier to differentiate $\ln|y|$ than $y$. The technique of obtaining the derivative of $y$ by differentiating $\ln|y|$ is called **logarithmic differentiation**.

■ **EXAMPLE 3**

$$y = \frac{x^2(x-1)^3}{x+1}.$$

Use the method of logarithmic differentiation to find $dy/dx$.

**SOLUTION**

Taking logarithms and simplifying, we obtain

$$\ln|y| = 2\ln|x| + 3\ln|x-1| - \ln|x+1|.$$

Using the Chain Rule to differentiate the logarithm of each function, we obtain

$$\frac{1}{y}\frac{dy}{dx} = \frac{2}{x} + \frac{3}{x-1} - \frac{1}{x+1},$$

$$\frac{dy}{dx} = y\left[\frac{2(x-1)(x+1) + 3(x)(x+1) - (x)(x-1)}{x(x-1)(x+1)}\right]$$

$$= \frac{x^2(x-1)^3}{x+1}\left[\frac{4x^2+4x-2}{x(x-1)(x+1)}\right] = \frac{2x(x-1)^2(2x^2+2x-1)}{(x+1)^2}$$ ■

Let us now investigate the graph of ln $x$. We know that

$$\frac{d}{dx}(\ln x) = \frac{1}{x}, \qquad x > 0.$$

It follows that ln $x$ is continuous and increasing on the interval $x > 0$. The second derivative is $-1/x^2$, which is negative, so ln $x$ is concave downward. We have $\ln(1) = 0$, so the graph passes through the point (1, 0). In Example 3 of Section 5.4, we used Simpson's Rule to show that

$$\ln 2 = \int_1^2 \frac{1}{t}\, dt \approx 0.69315,$$

with error less than 0.0001. This allows us to plot the point (2, ln 2). From (2) we have

$$\ln 2^n = n \ln 2.$$

Thus, if $n$ is increased by one, the value of $\ln 2^n$ is increased by ln 2. That is,

$$\ln 2^{n+1} = (n+1)\ln 2 = n \ln 2 + \ln 2 = \ln 2^n + \ln 2.$$

Since ln $x$ is increasing, this gives

$$\lim_{x \to \infty} \ln x = \infty. \tag{5}$$

Similarly, if $n$ is decreased by one, $\ln 2^n$ is decreased by ln 2. Since $2^n \to 0$ as $n \to -\infty$, this implies ln $x$ must have vertical asymptote $x = 0$ and

$$\lim_{x \to 0^+} \ln x = -\infty. \tag{6}$$

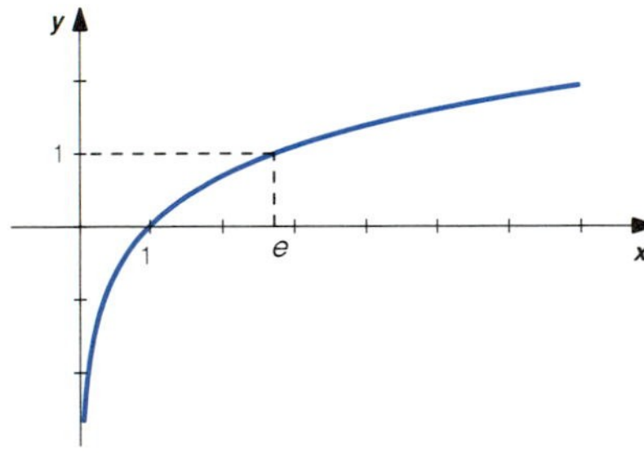

FIGURE 7.4.3

The graph of ln $x$ is given in Figure 7.4.3.

For large values of $x$, ln $x$ is proportionally much smaller than any root $x^r$ of $x$. This can be described by saying that ln $x$ approaches infinity slower than $x^r$ as $x$ approaches infinity. More precisely,

$$\lim_{x \to \infty} \frac{\ln x}{x^r} = 0, \qquad r > 0. \tag{7}$$

Let us verify (7) in the special case $r = 1/2$. Using l'Hôpital's Rule, we have

$$\lim_{x \to \infty} \frac{\ln x}{x^{1/2}} \underset{(\infty/\infty,\, \text{l'H})}{=} \lim_{x \to \infty} \frac{1/x}{1/(2x^{1/2})}$$

$$= \lim_{x \to \infty} 2x^{-1/2} = 0.$$

Similarly, ln $x$ approaches infinity slower than $x^{-r}$ as $x$ approaches zero from the right. That is,

$$\lim_{x \to 0^+} \frac{\ln x}{x^{-r}} = \lim_{x \to 0^+} x^r \ln x = 0, \qquad r > 0. \tag{8}$$

Since $\ln 2 \approx 0.69315 < 1$ and $\ln 4 = \ln 2^2 = 2 \ln 2 > 1$, the Intermediate Value Theorem for Continuous Functions tells us there is a number $e$ with $2 < e < 4$ and

$$\ln e = 1.$$

(See Figure 7.4.3.) It can be shown that the number $e$ is an irrational number with approximate value

$$e \approx 2.7182818.$$

[You should verify with a calculator that $\ln(2.7182818) \approx 1$.]

From (2) we have $\ln(e^r) = r \ln e = r$, so $e^r$ is the unique solution of the equation $\ln x = r$. That is,

$$\ln x = r \quad \text{if and only} \quad \text{if } x = e^r, \qquad r \text{ a rational number.} \tag{9}$$

We know from algebra courses that

$$x = e^r \quad \text{means} \quad \log_e x = r.$$

It then follows from (9) that we must have

$$\ln x = \log_e x.$$

We will discuss logarithms to bases other than $e$ in Section 7.6.

We can use (9) to solve the equation $\ln x = r$ for $x$.

■ **EXAMPLE 4**

Sketch the graph of $f(x) = x \ln x$.

**SOLUTION**

We first note that $f(x)$ is defined only for $x > 0$. We see that $f(1) = 0$, $f(x)$ is positive for $x > 1$, and $f(x)$ is negative for $0 < x < 1$.

Using the product rule, we have

$$\begin{aligned} f'(x) &= \frac{d}{dx}(x \ln x) \\ &= (x)\left(\frac{1}{x}\right) + (\ln x)(1) \\ &= 1 + \ln x. \end{aligned}$$

The equation $f'(x) = 0$ implies $1 + \ln x = 0$, so $\ln x = -1$. Statement (9) then tells us $x = e^{-1}$. We see that $f(e^{-1}) = e^{-1}\ln(e^{-1}) = -e^{-1}$. The graph has a horizontal tangent at $(e^{-1}, -e^{-1})$. (Using a calculator, we have $e^{-1} \approx 0.36787944$.) The first derivative $f'(x)$ is negative and the graph is decreasing for $0 < x < e^{-1}$; $f'(x)$ is positive and the graph is increasing for $x > e^{-1}$; $f(e^{-1}) = -e^{-1}$ is the minimum value of $f(x)$.

We have

$$f''(x) = \frac{1}{x}, \qquad x > 0.$$

The second derivative $f''(x)$ is positive and the graph is concave up for $x > 0$.

It follows from (8) that

$$\lim_{x \to 0^+} f(x) = \lim_{x \to 0^+} x \ln x = 0,$$

but let us see how to use l'Hôpital's Rule to evaluate the limit. We have

$$\lim_{x \to 0^+} x \ln x = \lim_{x \to 0^+} \frac{\ln x}{1/x}$$

$$\underset{(\infty/\infty, \text{l'H})}{=} \lim_{x \to 0^+} \frac{1/x}{-1/x^2}$$

$$= \lim_{x \to 0^+} (-x) = 0.$$

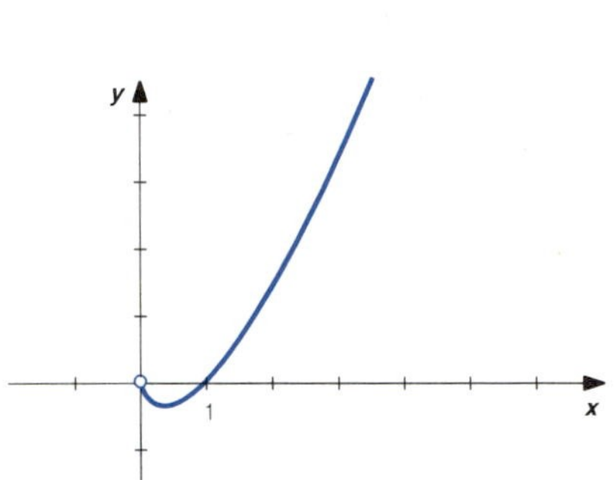

FIGURE 7.4.4

This tells us that the graph approaches the point (0, 0) as $x$ approaches zero from the right. If we note that $f'(x) = 1 + \ln x$ approaches negative infinity as $x$ approaches zero from the right, we could conclude that the graph becomes vertical at the origin. The graph is sketched in Figure 7.4.4. ■

## EXERCISES 7.4

Evaluate the derivatives in Exercises 1–18.

1 $\frac{d}{dx}(\ln(2x + 1))$

2 $\frac{d}{dx}(\ln(3x - 2))$

3 $\frac{d}{dx}(\ln(\sec x + \tan x))$

4 $\frac{d}{dx}(\ln(\csc x))$

5 $\frac{d}{dx}\left(\ln \frac{1}{x^2}\right)$

6 $\frac{d}{dx}\left(\ln \frac{1}{1 + x^2}\right)$

7 $\frac{d}{dx}(\ln(x\sqrt{x^2 + 1}))$

8 $\frac{d}{dx}(\ln\sqrt{x(x^2 + 1)})$

9 $\frac{d}{dx}\left(\ln \frac{2x + 1}{3x - 5}\right)$

10 $\frac{d}{dx}\left(\ln \frac{x}{x + 2}\right)$

11 $\frac{d}{dx}\left(\ln \frac{x}{x^2 + 3}\right)$

12 $\frac{d}{dx}(\ln(\ln x))$

13 $\frac{d}{dx}((\ln x)^2)$

14 $\frac{d}{dx}(\sqrt{\ln x})$

Use logarithmic differentiation to find $dy/dx$ in Exercises 15–18.

15 $y = x(x - 1)^{1/3}(x + 1)^{2/3}$

16 $y = \frac{\sqrt{x}(x + 1)}{x - 1}$

17 $y = \frac{x(x - 1)^{1/3}}{(x + 1)^{2/3}}$

18 $y = \frac{x\sqrt{x^2 + 1}}{(2x - 1)^2}$

Sketch the graphs of the functions given in Exercises 19–24. Indicate relative extrema, points of inflection, the behavior as $x$ approaches zero from the right, and the behavior as $x$ approaches infinity.

19 $y = x^2 \ln x$

20 $y = \sqrt{x} \ln x$

21 $y = \frac{\ln x}{x}$

22 $y = \frac{\ln x}{\sqrt{x}}$

23 $y = \frac{1}{x} + \ln x$

24 $y = x - \ln x$

25 Use the definition to verify that $\ln x^r = r \ln x$, $r$ a rational number.

26 Verify that $\ln(x/y) = \ln(x) - \ln(y)$.

## 7.5 $\int u^{-1}\,du$

To evaluate integrals of the form $\int u^{-1}\,du$, it is convenient to consider the function $\ln|x|$. The graph is sketched in Figure 7.5.1. The function $\ln|x|$ is defined for all $x \neq 0$ and the graph is symmetric about the $y$-axis. The function is increasing on the interval $x > 0$ and decreasing on $x < 0$. It is not difficult to verify that

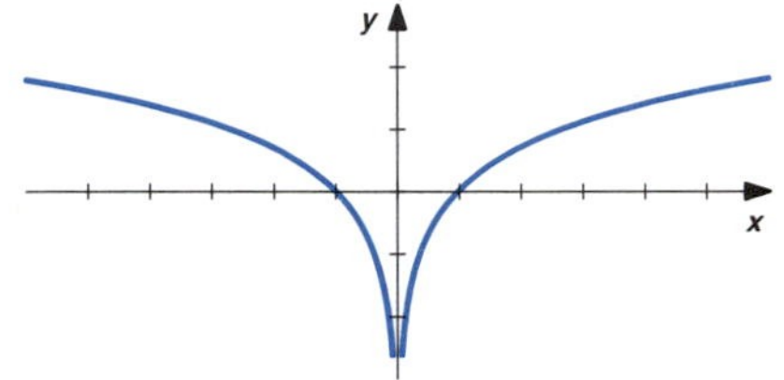


FIGURE 7.5.1

$$\frac{d}{dx}(\ln|x|) = \frac{1}{x}.$$

The corresponding integration formula is

$$\int \frac{1}{u}\,du = \ln|u| + C.$$

This formula is used to integrate quotients whose numerator is a constant multiple of the derivative of its denominator. We will use the method of substitution that was introduced in Section 5.6.

■ **EXAMPLE 1**

Evaluate

$$\int \frac{1}{1-2x}\,dx.$$

**SOLUTION**

We choose $u = 1 - 2x$, so $du = -2\,dx$. Then

$$\begin{aligned}\int \frac{1}{1-2x}\,dx &= \int \frac{1}{1-2x}\left(-\frac{1}{2}\right)(-2)\,dx \\ &= \int \frac{1}{u}\left(-\frac{1}{2}\right)du \\ &= -\frac{1}{2}\ln|u| + C \\ &= -\frac{1}{2}\ln|1-2x| + C.\end{aligned}$$

■

■ **EXAMPLE 2**

Evaluate

$$\int \frac{\cos x}{2 + \sin x}\,dx.$$

**SOLUTION**

Setting $u = 2 + \sin x$, so $du = \cos x\, dx$, we obtain

$$\int \frac{\cos x}{2 + \sin x}\, dx = \int \frac{1}{u}\, du$$
$$= \ln|u| + C$$
$$= \ln(2 + \sin x) + C.$$

■ **EXAMPLE 3**

Evaluate

$$\int \frac{2x - 1}{x + 1}\, dx.$$

**SOLUTION**

We must carry out the division in order to write this integral in a form we can evaluate. We have

$$\begin{array}{r|l} & \;\;2 \\ x + 1 & 2x - 1 \\ & 2x + 2 \\ \hline & \;\;- 3. \end{array}$$

Then

$$\int \frac{2x - 1}{x + 1}\, dx = \int \left(2 - \frac{3}{x + 1}\right) dx = 2x - 3 \int \frac{1}{x + 1}\, dx.$$

Using the substitution $u = x + 1$, $du = dx$, to evaluate

$$\int \frac{1}{x + 1}\, dx,$$

we obtain

$$\int \frac{2x - 1}{x + 1}\, dx = 2x - 3 \ln|x + 1| + C.$$

■ **EXAMPLE 4**

Evaluate $\int \tan x\, dx$.

**SOLUTION**

In this example, we write

$$\tan x = \frac{\sin x}{\cos x}$$

and then recognize that the numerator, $\sin x$, is a constant multiple of the derivative of the denominator, $\cos x$. We set $u = \cos x$, so $du = -\sin x\, dx$.

Substitution then gives

$$\begin{aligned}\int \tan x\,dx = \int \frac{\sin x}{\cos x}\,dx &= \int \frac{1}{\cos x}(-1)(-\sin x)\,dx\\ &= \int \frac{1}{u}(-1)\,du\\ &= -\ln|u| + C\\ &= -\ln|\cos x| + C\\ &= \ln|\cos x|^{-1} + C\\ &= \ln|\sec x| + C.\end{aligned}$$

■

■ **EXAMPLE 5**

Evaluate

$$\int \frac{(\ln x)^2}{x}\,dx.$$

**SOLUTION**

This example is not of the form $\int u^{-1}\,du$, but we recognize that $1/x$ is the derivative of $\ln x$. We then choose

$$u = \ln x, \quad \text{so} \quad du = \frac{1}{x}\,dx.$$

Substitution gives

$$\int \frac{(\ln x)^2}{x}\,dx = \int u^2\,du = \frac{u^3}{3} + C = \frac{1}{3}(\ln x)^3 + C.$$

■

■ **EXAMPLE 6**

Evaluate

$$\int \frac{2x}{\sqrt{x^2+1}}\,dx.$$

**SOLUTION**

In this example, the numerator, $2x$, is not a constant multiple of the derivative of the denominator, $\sqrt{x^2+1}$. The formula $\int u^{-1}\,du$ does not apply. The substitution $u = x^2 + 1$, $du = 2x\,dx$, gives

$$\begin{aligned}\int \frac{2x}{\sqrt{x^2+1}}\,dx &= \int \frac{1}{\sqrt{u}}\,du = \int u^{-1/2}\,du\\ &= \frac{u^{1/2}}{1/2} + C\\ &= 2\sqrt{x^2+1} + C.\end{aligned}$$

■

Do not confuse

$$\int \frac{1}{u}\,du \quad \text{and} \quad \int \frac{1}{u^r}\,du, \qquad r \neq 1.$$

For $r \neq 1$, we have

$$\int \frac{1}{u^r}\,du = \int u^{-r}\,du = \frac{u^{-r+1}}{-r+1} + C.$$

This formula does not make sense for $r = 1$. When $r = 1$, we have

$$\int \frac{1}{u}\,du = \ln|u| + C.$$

We cannot use the formula

$$\int \frac{1}{u}\,du = \ln|u| + C$$

to evaluate a definite integral

$$\int_a^b \frac{1}{u}\,du$$

if $a \leq 0 \leq b$. The Fundamental Theorem does not apply in this case. (The Fundamental Theorem requires that the integrand be continuous on the interval $a \leq u \leq b$. The function $1/u$ becomes unbounded as $u$ approaches zero, so it cannot be assigned any value at zero that makes it continuous on an interval that contains zero.)

■ **EXAMPLE 7**

Find the area of the region bounded by $y = 8/x$, $y = 0$, $x = e$, and $x = e^2$. (Recall that $\ln e = 1$.)

**SOLUTION**

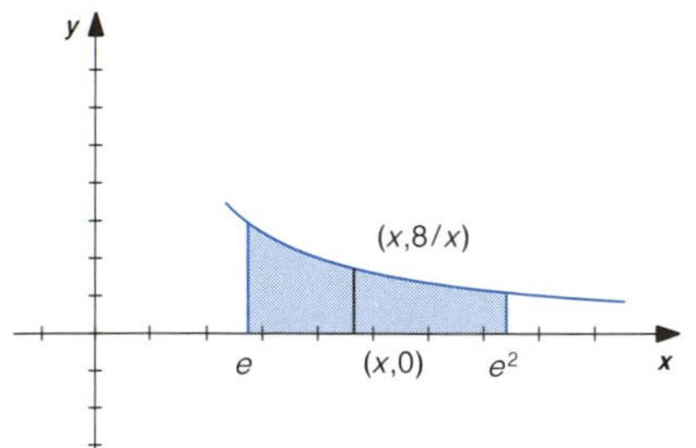


**FIGURE 7.5.2**

The region is sketched in Figure 7.5.2. We have labeled the endpoints of a variable vertical area strip in terms of the variable $x$. We see that a typical strip has height $8/x$ and width $dx$. The integral that gives the "sum" of the areas of the rectangles is

$$\begin{aligned} \text{Area} &= \int_e^{e^2} \frac{8}{x}\,dx = 8\ln|x|\Big]_e^{e^2} \\ &= 8(\ln e^2 - \ln e) = 8(2\ln e - \ln e) = 8(2-1) = 8. \end{aligned}$$

■

■ **EXAMPLE 8**

Find the volume of the solid obtained by revolving the region bounded by $y = 8/(x^2 + 4)$, $y = 0$, $x = 0$, and $x = 1$, about the $y$-axis.

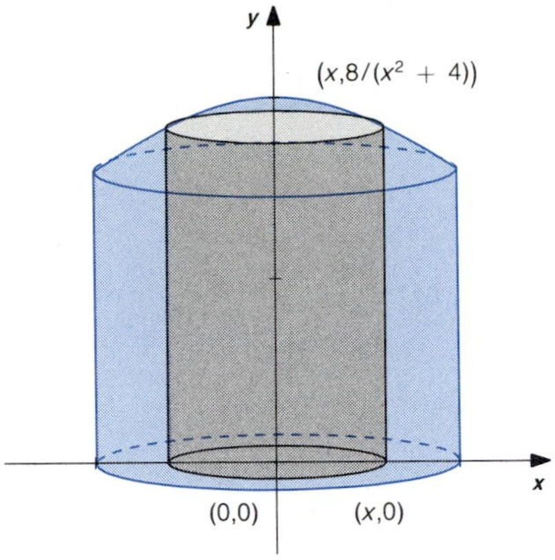

FIGURE 7.5.3

**SOLUTION**

The region is sketched in Figure 7.5.3. A variable vertical area strip generates a cylindrical shell when revolved about the $y$-axis. From the labeling of the sketch we see that the cylinder has a radius $= x$, height $= 8/(x^2 + 4)$, and thickness $dx$. We then have

$$\text{Volume} = \int_0^1 2\pi x\left(\frac{8}{x^2+4}\right) dx.$$

Using the substitution $u = x^2 + 4$, $du = 2x\,dx$, we obtain

$$\begin{aligned}\text{Volume} &= \int_0^1 \frac{8\pi}{x^2+4}\,2x\,dx \\ &= \int \frac{8\pi}{u}\,du \\ &= 8\pi \ln|u|\Big] \\ &= 8\pi \ln(x^2+4)\Big]_0^1 \\ &= 8\pi \ln 5 - 8\pi \ln 4 \\ &= 8\pi \ln\frac{5}{4} \approx 5.6082091.\end{aligned}$$

■

## EXERCISES 7.5

Evaluate the integrals in Exercises 1–20.

1 $\displaystyle\int \frac{1}{2x+1}\,dx$

2 $\displaystyle\int \frac{1}{3x+2}\,dx$

3 $\displaystyle\int \frac{x}{x^2+1}\,dx$

4 $\displaystyle\int \frac{x^2}{x^3+1}\,dx$

5 $\displaystyle\int \frac{x}{x-3}\,dx$

6 $\displaystyle\int \frac{x^2}{x+1}\,dx$

7 $\displaystyle\int \frac{\sin x}{1+\cos x}\,dx$

8 $\displaystyle\int \frac{\sin x - \cos x}{\sin x + \cos x}\,dx$

9 $\displaystyle\int \frac{\sec x \tan x + \sec^2 x}{\sec x + \tan x}\,dx$

10 $\displaystyle\int \cot x\,dx$

11 $\displaystyle\int \frac{1}{x \ln x}\,dx$

12 $\displaystyle\int \frac{1}{x(\ln x)^2}\,dx$

13 $\displaystyle\int \frac{x^2}{x^2+1}\,dx$

14 $\displaystyle\int \frac{x^3}{x^2+1}\,dx$

15 $\displaystyle\int \frac{1}{x^{1/3}(1+x^{2/3})}\,dx$

16 $\displaystyle\int \frac{1}{\sqrt{x}(1+\sqrt{x})}\,dx$

17 $\displaystyle\int \frac{1}{x^2+1}\,dx$

18 $\displaystyle\int \frac{x}{(x^2+1)^2}\,dx$

19 $\displaystyle\int \frac{1}{\sqrt{4-x^2}}\,dx$

20 $\displaystyle\int \frac{x}{\sqrt{4-x^2}}\,dx$

21 Find the area of the region bounded by

$$y = \frac{1}{x}, \qquad y = \frac{1}{x+1}, \qquad x = 1, \quad x = 2.$$

22 Find the area of the region bounded by

$$y = \frac{1}{x}, \qquad y = 0, \quad x = a, \quad x = 2a,$$

where $a$ is a positive constant.

**23** Find the volume of the solid generated by revolving the region bounded by

$$y = \frac{2}{x+1}, \quad y = 0, \quad x = 0, \quad x = 1,$$

about the (a) $x$-axis, (b) $y$-axis.

**24** Find the volume of the solid generated by revolving the region bounded by

$$y = \frac{1}{x}, \quad y = 0, \quad x = 1, \quad x = 2,$$

about the vertical line (a) $x = 1$, (b) $x = 2$.

## 7.6 EXPONENTIAL FUNCTIONS, INDETERMINATE FORMS $0^0$, $1^\infty$, $\infty^0$

We use the natural logarithmic function to define and establish properties of exponential functions.

We know that $\ln x$ has positive derivative for $x > 0$, so $\ln x$ has an inverse function. Let us denote the inverse function of $\ln x$ by $\exp x$. By definition, we have

$$y = \ln x \quad \text{if and only if} \quad x = \exp y.$$

Properties of the **exponential function** $y = \exp x$ can be obtained from what we know of the logarithmic function. The range of $\ln x$ is all real numbers, so the domain of $\exp x$ is all real numbers. The domain of $\ln x$ is all positive real numbers, so the range of $\exp x$ is all positive real numbers. That is, $\exp x$ is defined and positive for all real $x$.

$$\ln 1 = 0 \quad \text{implies} \quad \exp 0 = 1.$$

$$\ln e = 1 \quad \text{implies} \quad \exp 1 = e.$$

$$y = \ln x \to -\infty \quad \text{as} \quad x \to 0^+, \quad \text{so} \quad y = \exp x \to 0 \quad \text{as} \quad x \to -\infty.$$

$$y = \ln x \to \infty \quad \text{as} \quad x \to \infty, \quad \text{so} \quad y = \exp x \to \infty \quad \text{as} \quad x \to \infty.$$

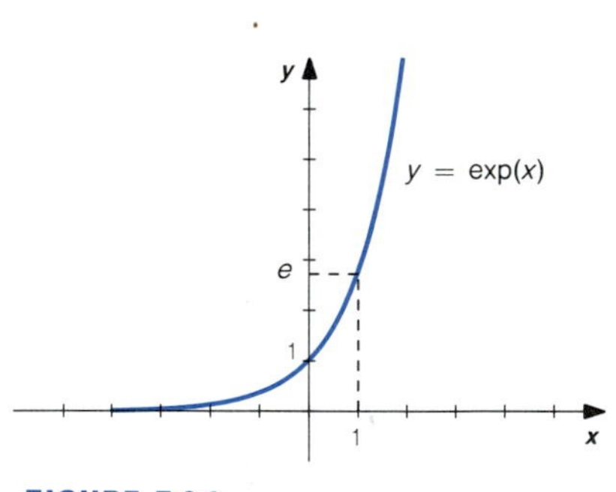

FIGURE 7.6.1

The graph of $y = \exp x$ is sketched in Figure 7.6.1.

For $a > 0$ and $r$ rational, the expression $a^r$ has been defined in terms of powers and roots of $a$; $a^x$ has not been defined for irrational $x$. We will see that the exponential function $\exp x$ can be used to define $a^x$ for all $x$. We first consider the case with base $a = e$, where $\ln e = 1$.

Let us show that

$$e^r = \exp r, \qquad r \text{ rational.}$$

We know that $\exp r$ is the number $z$ that satisfies $\ln z = r$, since $\ln z = r$ is equivalent to $\exp r = z$. But we know that $\ln e^r = r \ln e = r$ for rational numbers $r$, so $\exp r$ must be the number $e^r$.

Since the value of $e^x$ agrees with that of $\exp x$ for rational $x$, we *define* $e^x$ to be $\exp x$ for every value of $x$ and use the simpler notation $e^x$ in place of exp $x$. The relations between $\ln x$ and its inverse function can then be written

$$y = \ln x \quad \text{if and only if} \quad x = e^y,$$

$$\ln e^x = x \quad \text{for all } x,$$

and

$$e^{\ln x} = x, \qquad x > 0.$$

For $a > 0$ and any number $x$, we define $a^x$ by the equation

$$a^x = e^{x \ln a}.$$

If $x$ is rational, this definition agrees with the previous definition of $a^x$ in terms of powers and roots. That is, from Section 7.4 we know that $\ln a^x = x \ln a$, so $a^x = e^{x \ln a}$.

When we showed that $\ln a^r = r \ln a$ in Section 7.4, it was necessary that $r$ be a rational number because $a^r$ was defined only for rational numbers $r$. We now have

$$\ln a^b = \ln e^{b \ln a} = b \ln a \quad \text{for all values of } b.$$

### PROPERTIES OF $e^x$

$$e^a e^b = e^{a+b}.$$

$$(e^a)^b = e^{ab}.$$

$$e^{-a} = \frac{1}{e^a}.$$

$$\frac{e^a}{e^b} = e^{a-b}.$$

These are the familiar rules of exponents. The only difference is that they now make sense for all real values of the exponents; the rules are no longer restricted to rational exponents. These properties follow from corresponding properties of the logarithmic function. For example, $e^{a+b}$ is the number $z$ that satisfies $\ln z = a + b$. Since

$$\ln e^a e^b = \ln e^a + \ln e^b = a \ln e + b \ln e = a + b,$$

we must have $e^{a+b} = e^a e^b$.

We know that $e^x$ is differentiable, since $e^x$ is the inverse function of a function that has positive derivative. Let us obtain a formula for the derivative. We set

$$y = e^x, \quad \text{so} \quad \ln y = x.$$

Using the Chain Rule to differentiate this equation with respect to $x$, we obtain

$$\left(\frac{1}{y}\right)\left(\frac{dy}{dx}\right) = 1, \quad \text{so} \quad \frac{dy}{dx} = y.$$

Since $y = e^x$, substitution gives the formula

$$\frac{d}{dx}(e^x) = e^x.$$

The Chain Rule version of this formula is

$$\frac{d}{dx}(e^{\text{Function}}) = e^{\text{Function}} \frac{d}{dx}(\text{Function}).$$

■ **EXAMPLE 1**

(a) $\frac{d}{dx}(e^{3x}) = e^{3x}(3)$,

(b) $\frac{d}{dx}(e^{x/2}) = e^{x/2}\left(\frac{1}{2}\right)$,

(c) $\frac{d}{dx}(e^{x^2}) = e^{x^2}(2x)$,

(d) $\frac{d}{dx}(xe^x - e^x) = [(x)(e^x) + (e^x)(1)] - e^x = xe^x$. ■

The usual formula for differentiation of a power function holds for all *constant* powers of the variable of differentiation. It is no longer necessary to require that the power be a rational number.

$$\frac{d}{dx}(x^\alpha) = \alpha x^{\alpha - 1}.$$

■ *Proof*

$$\frac{d}{dx}(x^\alpha) = \frac{d}{dx}(e^{\alpha \ln x}) = e^{\alpha \ln x}\left(\alpha \frac{1}{x}\right) = \alpha x^\alpha x^{-1} = \alpha x^{\alpha - 1}.$$ ■

We can find the derivative of a function raised to a power that is a function by writing $f^g = e^{g \ln f}$ and then using the Chain Rule. We could also use the method of logarithmic differentiation.

■ **EXAMPLE 2**

$$\begin{aligned}\frac{d}{dx}(x^x) &= \frac{d}{dx}(e^{x \ln x}) \\ &= e^{x \ln x}\left[(x)\left(\frac{1}{x}\right) + (\ln x)(1)\right] \\ &= x^x[1 + \ln x].\end{aligned}$$ ■

The following integration formula corresponds to the differentiation formula

$$\frac{d}{dx}(e^x) = e^x.$$

$$\int e^u\,du = e^u + C.$$

**EXAMPLE 3**

Evaluate $\int e^{3x}\,dx$.

**SOLUTION**

Setting $u = 3x$, so $du = 3\,dx$, we have

$$\begin{aligned}\int e^{3x}\,dx &= \int e^{3x}\left(\frac{1}{3}\right)3\,dx\\ &= \int e^u\left(\frac{1}{3}\right)du\\ &= \frac{1}{3}e^u + C\\ &= \frac{1}{3}e^{3x} + C.\end{aligned}$$

■

**EXAMPLE 4**

Evaluate $\int xe^{-x^2}\,dx$.

**SOLUTION**

Using $u = -x^2$, so $du = -2x\,dx$, we have

$$\begin{aligned}\int xe^{-x^2}\,dx &= \int e^{-x^2}\left(-\frac{1}{2}\right)(-2x)\,dx\\ &= \int e^u\left(-\frac{1}{2}\right)du\\ &= -\frac{1}{2}e^u + C\\ &= -\frac{1}{2}e^{-x^2} + C.\end{aligned}$$

■

**EXAMPLE 5**

Evaluate $\int \cos x\, e^{\sin x}\,dx$.

**SOLUTION**

Set $u = \sin x$, so $du = \cos x\,dx$. Then

$$\begin{aligned}\int \cos x\, e^{\sin x}\,dx &= \int e^u\,du\\ &= e^u + C\\ &= e^{\sin x} + C.\end{aligned}$$

■

**■ EXAMPLE 6**

Evaluate

$$\int \frac{e^x}{1+e^x}\,dx.$$

**SOLUTION**

In this example, we recognize $e^x$ as the derivative of $1 + e^x$. We then set $u = 1 + e^x$, so $du = e^x\,dx$ and

$$\begin{aligned}\int \frac{e^x}{1+e^x}\,dx &= \int \frac{1}{u}\,du \\ &= \ln|u| + C \\ &= \ln(1+e^x) + C.\end{aligned}$$ ■

The value $e^x$ approaches infinity faster than any constant power of $x$ as $x$ approaches infinity. That is,

$$\lim_{x\to\infty} \frac{x^\alpha}{e^x} = 0. \tag{1}$$

Let us verify this in the case $\alpha = 2$. Using l'Hôpital's Rule, we have

$$\lim_{x\to\infty} \frac{x^2}{e^x} \underset{(\infty/\infty,\,\text{l'H})}{=} \lim_{x\to\infty} \frac{2x}{e^x} \underset{(\infty/\infty,\,\text{l'H})}{=} \lim_{x\to\infty} \frac{2}{e^x} = 0.$$

Similarly,

$$\lim_{x\to-\infty} |x|^\alpha e^x = 0. \tag{2}$$

**■ EXAMPLE 7**

Sketch the graph of $y = xe^{-x}$.

**SOLUTION**

We will investigate the sign of $y$ and the first two derivatives.

$$\begin{aligned} y &= xe^{-x}, \\ y' &= (x)(e^{-x})(-1) + (e^{-x})(1) \\ &= e^{-x}(1-x), \\ y'' &= (e^{-x})(-1) + (1-x)(e^{-x})(-1) \\ &= e^{-x}(x-2). \end{aligned}$$

We know that $e^{-x} > 0$ for all $x$. Hence, $y = 0$ implies $x = 0$. The only intercept is (0, 0). We see that $y > 0$ (graph above $x$-axis) whenever $x > 0$ and $y < 0$ (graph below $x$-axis) whenever $x < 0$.

The equation $y' = 0$ implies $x = 1$ and $y = (1)e^{-(1)} = 1/e$. The graph has a horizontal tangent at $(1, 1/e)$. We see that $y' > 0$ ($y$ increasing) whenever $x < 1$; we also see that $y' < 0$ ($y$ decreasing) whenever $x > 1$. The graph has a local maximum at $(1, 1/e)$.

The equation $y'' = 0$ implies $x = 2$ and $y = (2)e^{-(2)} = 2/e^2$. We see that $y'' > 0$ ($y$ concave upward) whenever $x > 2$; we also see that $y'' < 0$ ($y$ concave downward) whenever $x < 2$. The coordinate $(2, 2/e^2)$ is a point of inflection.

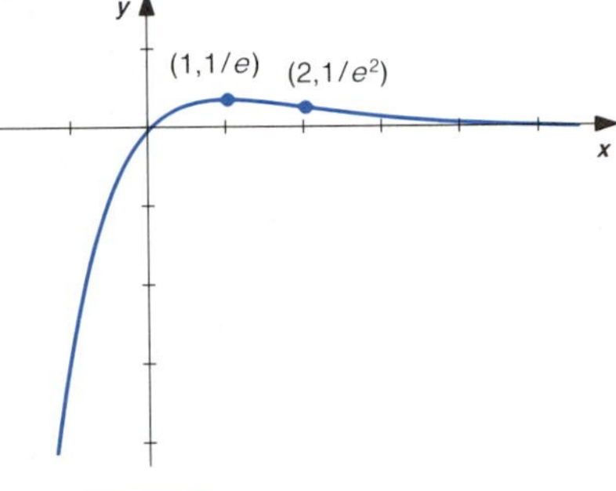


FIGURE 7.6.2

It follows from (1) that $\lim_{x\to\infty} xe^{-x} = 0$, but let us use l'Hôpital's Rule to evaluate the limit. We have

$$\lim_{x\to\infty} xe^{-x} = \lim_{x\to\infty} \frac{x}{e^x} \underset{(\infty/\infty,\text{l'H})}{=} \lim_{x\to\infty} \frac{1}{e^x} = 0.$$

This shows that the graph has horizontal asymptote $y = 0$ as $x$ approaches infinity. As $x$ approaches negative infinity, $xe^{-x}$ approaches negative infinity. The graph is sketched in Figure 7.6.2. ■

Depending on the values of the limits of $f$ and $g$, the function $f^g$ is said to have **indeterminate form $0^0$, $1^\infty$, or $\infty^0$**. Generally, $\ln f^g = g \ln f$ then has indeterminate form $0 \cdot \infty$. For example, if $\lim f(x) = \lim g(x) = 0$, then we have $\ln f^g = g \ln f$, where $\ln f$ approaches negative infinity. Thus, $g \ln f$ has indeterminate form $0 \cdot \infty$.

(NOTE: $1^\infty$ doesn't look like an indeterminate form, because 1 raised to any real number power is 1. However, you can verify on your calculator that $x^y$ can have either very small or very large values if $x$ is near 1, $x \neq 1$, and $y$ is large.)

■ **EXAMPLE 8**

Evaluate

$$\lim_{x\to\infty} \left(1 + \frac{1}{x}\right)^x.$$

**SOLUTION**

The expression

$$\left(1 + \frac{1}{x}\right)^x$$

has indeterminate form $1^\infty$ at infinity. Let

$$L = \lim_{x\to\infty} \left(1 + \frac{1}{x}\right)^x.$$

Since the function ln is continuous, we know that

$$\ln L = \lim_{x\to\infty} \ln\left(1 + \frac{1}{x}\right)^x.$$

Using properties of the function ln, we then have

$$\ln L = \lim_{x\to\infty} x \ln\left(1+\frac{1}{x}\right)$$

$$= \lim_{x\to\infty} \frac{\ln\left(1+\frac{1}{x}\right)}{\frac{1}{x}}$$

$$\underset{(0/0,\,\text{l'H})}{=} \lim_{x\to\infty} \frac{\frac{1}{1+(1/x)}\left(-\frac{1}{x^2}\right)}{-\frac{1}{x^2}}$$

$$\underset{(\text{Simplifying})}{=} \lim_{x\to\infty} \frac{1}{1+\frac{1}{x}} = 1.$$

We now have $\ln L = 1$, but we must remember that $L$ is the value of the limit we want to evaluate.

$$\ln L = 1 \quad \text{implies} \quad L = e^1 = e, \quad \text{so} \quad \lim_{x\to\infty}\left(1+\frac{1}{x}\right)^x = e.$$ ■

Example 8 gives us a way to evaluate approximate values of $e$, the base of the natural logarithmic function. You can use your calculator to verify that

$$\left(1+\frac{1}{x}\right)^x \approx 2.7182818 \quad \text{for large values of } x.$$

For some applications it is convenient to use logarithms and exponentials to a base other than $e$. We have already defined exponentials $a^x$. Namely,

$$a^x = e^{x\ln a}, \qquad a > 0.$$

For $a > 0$, $a \neq 1$, the function $a^x$ has derivative

$$\frac{d}{dx}(a^x) = \frac{d}{dx}(e^{x\ln a}) = e^{x\ln a}(\ln a) = (\ln a)a^x.$$

We see that the derivative is positive if $a > 1$ and negative if $0 < a < 1$. In either case, $a^x$ has an inverse function, which we denote by $\log_a x$. Thus,

$$y = a^x \quad \text{if and only if} \quad x = \log_a y.$$

$\text{Log}_a\, x$ is the familiar function that is discussed in algebra classes. $\text{Log}_{10}\, x$ is called the **common logarithm** of $x$. $\text{Log}_e\, x = \ln x$ is the **natural logarithm** of $x$. Logarithms to any base are related to natural logarithms by the formula

$$\log_a x = \frac{\ln x}{\ln a}.$$

■ *Proof* $\log_a x$ is the number $y$ that satisfies $a^y = x$. Since

$$a^{(\ln x)/(\ln a)} = e^{[(\ln x)/(\ln a)]\ln a} = e^{\ln x} = x,$$

we have

$$\log_a x = \frac{\ln x}{\ln a}.$$ ■

We have

$$\frac{d}{dx}(\log_a x) = \left(\frac{1}{\ln a}\right)\left(\frac{1}{x}\right).$$

The differentiation formula

$$\frac{d}{dx}(a^x) = (\ln a)a^x$$

corresponds to the integration formula

$$\int a^x\,dx = \frac{a^x}{\ln a} + C.$$

The factor $\ln a$ appears in the above differentiation and integration formulas. The formulas are simpler for the base $e$, because $\ln e = 1$. That is why the base $e$ is "natural" for calculus.

## EXERCISES 7.6

Evaluate the derivatives in Exercises 1–16.

1 $\frac{d}{dx}(e^{-x^2/2})$

2 $\frac{d}{dx}(e^{\sqrt{x}})$

3 $\frac{d}{dx}(e^{2x-1})$

4 $\frac{d}{dx}(e^{1-3x})$

5 $\frac{d}{dx}(e^{\tan x})$

6 $\frac{d}{dx}(e^{\sec x})$

7 $\frac{d}{dx}(xe^{-x})$

8 $\frac{d}{dx}((\sin x)(e^{-x}))$

9 $\frac{d}{dx}\left(\frac{e^{-x}}{x}\right)$

10 $\frac{d}{dx}\left(\frac{e^{2x}}{x^2}\right)$

11 $\frac{d}{dx}(\ln(1 + e^x))$

12 $\frac{d}{dx}(\tan^{-1}(e^x))$

13 $\frac{d}{dx}(x^{\cos x})$

14 $\frac{d}{dx}(x^{\ln x})$

15 $\frac{d}{dx}(10^x)$

16 $\frac{d}{dx}\left(\frac{1}{2^x}\right)$

Evaluate the integrals in Exercises 17–28.

17 $\int e^{-x}\,dx$

18 $\int e^{2x}\,dx$

19 $\int \frac{e^{\sqrt{x}}}{\sqrt{x}}\,dx$

20 $\int e^{x^{1/3}}x^{-2/3}\,dx$

21 $\int e^x\sqrt{e^x+1}\,dx$

22 $\int \frac{e^x}{\sqrt{1-e^{2x}}}\,dx$

23 $\int \frac{e^x}{1+e^{2x}}\,dx$

24 $\int \frac{e^x}{(1+e^x)^2}\,dx$

25 $\int \frac{e^{2x}}{1+e^{2x}}\,dx$

26 $\int e^x e^{e^x}\,dx$

27 $\int 2^{-x}\,dx$

28 $\int 10^x\,dx$

Sketch the graphs of the functions in Exercises 29–34. Indicate asymptotes, local extrema, and points of inflection.

29 $y = xe^x$

30 $y = x^2e^x$

31 $y = \frac{e^x - e^{-x}}{2}$

32 $y = \frac{e^x + e^{-x}}{2}$

33 $y = e^x/x$

34 $y = 1/(e^x - 1)$

35 Find the area of the region bounded by $y = e^{-x}$, $y = 0$, $x = 0$, and $x = 1$.

36 Find the volume of the solid obtained by revolving the region bounded by $y = e^x$, $y = 0$, $x = 0$, and $x = 1$, about the $x$-axis.

Evaluate the limits in Exercises 37–46.

37 $\lim_{x\to 0} \frac{e^{2x} - 1 - 2x}{x^2}$

38 $\lim_{x\to\infty} x\ln(1 + e^{-x})$

39 $\lim_{x\to 0^+} x^x$

40 $\lim_{x\to 0^+} x^{1/\ln x}$

41 $\lim_{x\to 0^+} (1 - \sqrt{x})^{1/x}$

42 $\lim_{x\to 0^+} (1 + 2x)^{1/x}$

43 $\lim_{x\to 0^+} x^{1/x}$

44 $\lim_{x\to\infty} x^x$

45 $\lim_{x\to\infty} x^{1/x}$

46 $\lim_{x\to\infty} (1 + 2x)^{1/x}$

Use the corresponding property of $\ln x$ to verify the equations in Exercises 47–48.

47 $(e^x)^y = e^{xy}$

48 $\frac{e^x}{e^y} = e^{x-y}$.

## 7.7 EXPONENTIAL GROWTH AND DECAY

In many natural processes the rate of change of a physical quantity is proportional to the current amount of the quantity. Thus, we are led to find a function $y(t)$ that satisfies an equation of the form

$$y' = ky, \qquad k \text{ a constant.}$$

This equation is called a **differential equation**. A function $y(t)$ that satisfies the equation for $t$ in an interval is called a **solution** of the differential equation. In this section we will solve this differential equation and see how it is related to several scientific problems. We will also consider some other differential equations that are closely related to the differential equation $y' = ky$.

We can solve the differential equation $y' = ky$ by integration. First, let us note that any solution $y$ must be continuous, since it is differentiable. Also, $y'$ must be continuous, since $y' = ky$.

If $y(t)$ is a solution of $y' = ky$, then $y'(t) = ky(t)$ for $t$ in an interval. In case there is some point in the interval for which $y(t)$ is not zero, we can write

$$\frac{y'(t)}{y(t)} = k.$$

Integrating with respect to $t$, we obtain

$$\int \frac{y'(t)}{y(t)}\,dt = \int k\,dt.$$

Using the substitution $y = y(t)$, so that $dy = y'(t)\,dt$, to evaluate the integral on the left, we have

$$\int \frac{1}{y}\,dy = \int k\,dt,$$

$$\ln|y| = kt + c, \qquad c \text{ a constant.}$$

Substitution of $y(t)$ for $y$ then gives

$$\ln|y(t)| = kt + c,$$

$$|y(t)| = e^{kt+c},$$

$$|y(t)| = e^{c}e^{kt}.$$

This shows that $y(t)$ is never zero. Since continuous functions cannot change sign without becoming zero, we must have

$$y(t) = Ce^{kt},$$

where either $C = e^{c}$ or $C = -e^{c}$. If there are no points on the interval where $y(t)$ is nonzero, we have $y(t) = Ce^{kt}$ with $C = 0$. Thus, any solution of $y' = ky$ must be of the form $y(t) = Ce^{kt}$. It is easy to check that any function of this form is a solution of $y' = ky$. That is, $y(t) = Ce^{kt}$ implies $y'(t) = Ce^{kt}(k)$, so $y'(t) = ky(t)$ for all $t$.

We have seen that for every constant $C$, $y(t) = Ce^{kt}$ is a solution of the differential equation $y' = ky$, and that all solutions are of this form. Thus, $Ce^{kt}$ represents all solutions of the differential equation. The collection of functions $Ce^{kt}$ is called the **general solution** of the differential equation. We can find a **particular solution** by using additional information to determine a value of the constant $C$. In particular, if it is required that a solution of the differential equation $y' = ky$ satisfy the condition $y(0) = y_0$, substitution into the equation $y(t) = Ce^{kt}$ gives

$$y(0) = Ce^{k(0)},$$

$$y_0 = C, \quad \text{so} \quad y(t) = y_0e^{kt}.$$

The problem of finding a function $y(t)$ that satisfies the differential equation $y' = ky$ and the **initial condition** $y(0) = y_0$ is called an **initial value problem**. It is convenient to memorize the solution of the above initial value problem, which we restate.

$$y' = ky, \qquad y(0) = y_0 \quad \textbf{has solution} \quad y(t) = y_0e^{kt}.$$

Let us now see how the equation $y' = ky$ and its solution, $y = Ce^{kt}$ occur in some physical problems.

### *Exponential growth*

Under certain conditions, such as enough food and space, bacteria colonies grow at a rate that is directly proportional to the population of the colony. If $P(t)$ denotes the population of the colony at time $t$, this means $P' = kP$, so statement (1) implies

$$P(t) = P_0 e^{kt}. \tag{2}$$

The initial population $P_0$ and the constant $k$ can be determined if we know the population of the colony at two different times.

**■ EXAMPLE 1**

The population of a certain colony of bacteria is known to increase at a rate that is directly proportional to its size. Initially, when $t = 0$, the population was 120. Three hours later the population was 200. What was the population when $t = 2$, two hours after the initial time?

**SOLUTION**

We know from formula (2) that $P(t) = P_0 e^{kt}$ for some constants $P_0$ and $k$. We are given that the initial population is $P_0 = 120$. Also, we know that $P(3) = 200$. That is,

$$120e^{k(3)} = 200.$$

Solving this equation for $k$, we have

$$e^{3k} = \frac{5}{3},$$

$$3k = \ln \frac{5}{3},$$

$$k = \frac{\ln \frac{5}{3}}{3}.$$

We then have

$$P(t) = 120e^{\{[\ln(5/3)]/3\}t}.$$

We were asked to find the population when $t = 2$. This is

$$P(2) = 120e^{\{[\ln(5/3)]/3\}(2)} \approx 169.$$

(We have used a calculator to find the approximate value.) The graph of $P(t)$ is given in Figure 7.7.1. ■

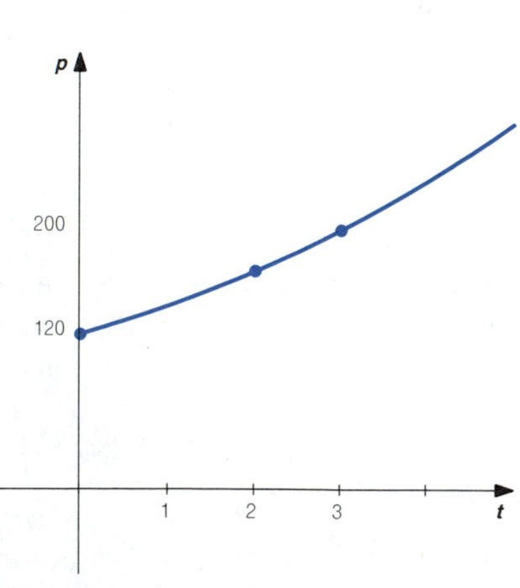


FIGURE 7.7.1

We can use properties of the logarithm to express the formula for the population in Example 1 in a different form. That is,

$$\begin{aligned} P(t) &= 120e^{\{[\ln(5/3)]/3\}t} \\ &= 120e^{(t/3)\ln(5/3)} \\ &= 120e^{\ln[(5/3)^{t/3}]} \\ &= 120\left(\frac{5}{3}\right)^{t/3}. \end{aligned}$$

## *Radioactive decay*

Radioactive material decays at a rate that is directly proportional to the current amount of material. This means $A' = kA$, where $A(t)$ is the amount of material at time $t$. We then know from statement (1) that $A(t) = A_0e^{kt}$ for some constant $k$. (Since the amount is decreasing, $k$ will be negative.)

Suppose that the amount of a particular radioactive substance at time $t$ is given by the equation $A(t) = A_0e^{kt}$. After some time $t_1$, half of the initial amount will have decayed, so half will remain. This means

$$A(t_1) = \frac{1}{2}A_0.$$

Solving this equation for $t_1$, we have

$$A_0e^{kt_1} = \frac{1}{2}A_0,$$

$$e^{kt_1} = 2^{-1},$$

$$kt_1 = \ln(2^{-1}), \quad \text{and}$$

$$t_1 = -\frac{\ln 2}{k}.$$

This shows that the length of time it takes for half of the substance to decay does not depend on the initial amount $A_0$, but only on the constant $k$. The value of $k$ depends on the type of radioactive substance. The length of time it takes for half of a radioactive substance to decay is called the **half-life** of the substance.

If the amount of a radioactive substance at time $t$ is given by $A(t) = A_0e^{kt}$, we have shown that the half-life is given by the formula

$$\text{Half-life} = -\frac{\ln 2}{k}, \quad \text{so} \quad k = -\frac{\ln 2}{\text{half-life}}.$$

We can then write

$$A(t) = A_0e^{-[(\ln 2)/(\text{half-life})]t}. \tag{3}$$

This formula is convenient when dealing with radioactive substances for which the half-life is known.

■ **EXAMPLE 2**

Carbon $^{11}C$ has a half-life of 20 min. How long will it take for 90% of a sample of $^{11}C$ to decay?

**SOLUTION**

From formula (3) with half-life = 20 min, we have $A(t) = A_0 e^{-[(\ln 2)/20]t}$, where $t$ is measured in minutes. When 90% of the sample has decayed, there will be 10% remaining. Thus, we need to determine $t$ such that $A(t) = 0.10A_0$. Solving this equation for $t$ we have

$$A_0 e^{-[(\ln 2)/20]t} = 0.10A_0,$$

$$-\left(\frac{\ln 2}{20}\right)t = \ln 0.10,$$

$$t = \frac{-\ln 0.10}{\frac{\ln 2}{20}} = \frac{\ln\left(\frac{1}{10}\right)^{-1}}{\ln 2}(20) = \frac{\ln 10}{\ln 2}(20) \approx 66.4.$$

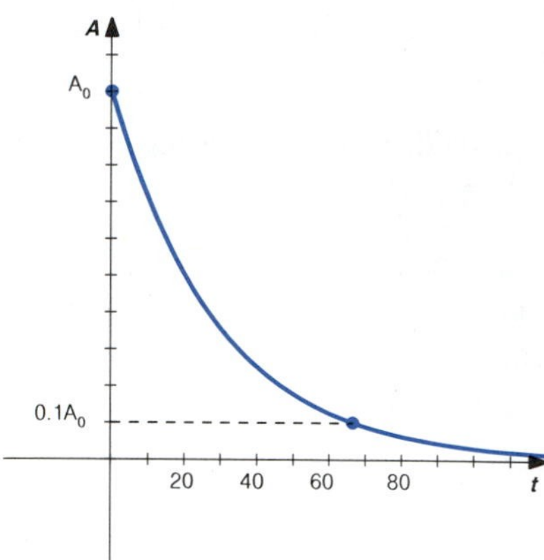

FIGURE 7.7.2

Ninety percent of the sample will decay in approximately 66.4 min. The graph of $A(t)$ is sketched in Figure 7.7.2. ■

## *Carbon dating*

Radiocarbon $^{14}C$ is a radioactive form of carbon that is present in the earth's atmosphere. All living plants and animals absorb radiocarbon as a part of their life processes. Thus, the radiocarbon that decays in a living thing is replaced, so the level remains essentially constant. After the living thing dies, the $^{14}C$ is not replaced and decays with half-life 5750 years. The time of death of an ancient living thing can be determined by measuring the proportion of radiocarbon remaining in a sample.

■ **EXAMPLE 3**

Find the age of a sample that has 70% of the radiocarbon it had while living.

**SOLUTION**

Using formula (3) with half-life = 5750 years, we have

$$A(t) = A_0 e^{-[(\ln 2)/5750]t}.$$

We want to find $t$ so that

$$A(t) = 0.70A_0.$$

Substitution gives

$$A_0 e^{-[(\ln 2)/5750]t} = 0.70 A_0,$$

$$-\left(\frac{\ln 2}{5750}\right)t = \ln 0.70,$$

$$t = -\frac{5750}{\ln 2}\ln 0.70 \approx 2959.$$

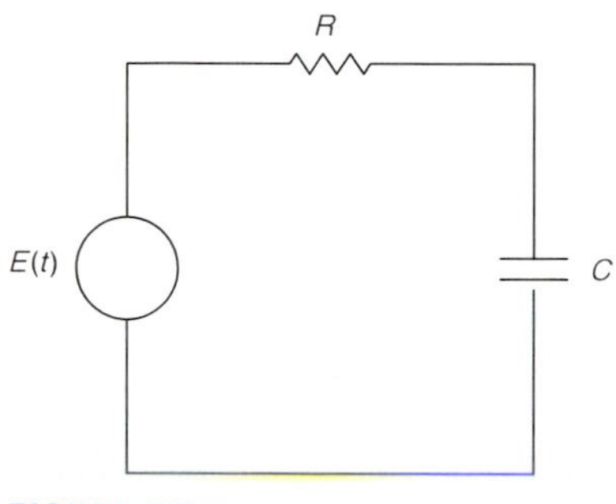

FIGURE 7.7.3

The sample is approximately 2959 years old. ■

### *Electric circuits*

Consider a simple series circuit with resistance $R$ (ohms), capacitance $C$ (farads), and impressed voltage $E(t)$ (volts), as indicated in Figure 7.7.3. Let $i(t)$ (amperes) denote the current at time $t$. One of Kirchhoff's Laws implies that $i(t)$ is a solution of the differential equation

$$Ri' + \frac{1}{C}i = E'(t).$$

If the impressed voltage is constant, we have $E'(t) = 0$, so the equation reduces to

$$i' = -\frac{1}{RC}i.$$

We know from (1) that this equation has solution

$$i(t) = i_0 e^{-t/(RC)}. \tag{4}$$

From Equation (4) we see that $i(t)$ approaches zero as $t$ approaches infinity. That is, the capacitor builds up a charge that approaches the constant value of the impressed voltage and the current decreases to zero over long intervals of time.

■ **EXAMPLE 4**

A simple series circuit consists of a charged capacitance, a 200-Ω resistor, and a constant impressed voltage of 6 V. Find the capacitance if the current in the circuit decreases from 10 A to 5 A in 0.3 s.

**SOLUTION**

The impressed voltage is constant, $E(t) = 6$, so we know the current is given by formula (4). The resistance is $R = 200$, and the capacitance $C$ is unknown. We take the initial current to be $i_0 = 10$, so

$$i(t) = 10e^{-t/(200C)}.$$

We also know that $i(0.3) = 5$. Substitution then gives

$$5 = 10e^{-0.3/(200C)}.$$

Solving for $C$, we obtain

$$e^{0.3/(200C)} = 2,$$

$$\frac{0.3}{200C} = \ln 2,$$

$$C = \frac{0.3}{200 \ln 2} \approx 0.00216404.$$

The capacitance is approximately 0.00216404 F (farads). ■

## *Newton's Law of Cooling*

The rate of change of the temperature of an object is directly proportional to the difference between its temperature and the temperature of its surroundings. (We assume that all parts of the object are the same temperature, and that the temperature of the object does not significantly affect the temperature of the surroundings. These assumptions are reasonable if the object is small and is composed of material that is a good conductor of heat.) Let $T(t)$ denote the temperature of the object and let $K$ denote the constant temperature of the surrounding region. Then Newton's Law of Cooling can be expressed as

$$T' = k(T - K).$$

We can solve this differential equation by integration. If $T(t)$ does not have value $K$, we have

$$T'(t) = k(T(t) - K),$$

$$\frac{1}{T(t) - K} T'(t) = k,$$

$$\int \frac{1}{T(t) - K} T'(t)\, dt = \int k\, dt.$$

Using the substitution

$$u = T(t) - K, \qquad du = T'(t)\, dt$$

to evaluate the integral on the left, we have

$$\ln|T(t) - K| = kt + c,$$

$$|T(t) - K| = e^c e^{kt},$$

$$T(t) - K = Ce^{kt} \qquad (C = \pm e^c).$$

It follows that the temperature is given by

$$\mathbf{T(t) = K + Ce^{kt}.} \tag{5}$$

■ **EXAMPLE 5**

A small object had temperature 200°F when it was dropped into a large pool of water at 80°F. Five minutes after it was put into the water, the temperature of the object was 100°F. What was its temperature 10 min after it was put in the water?

**SOLUTION**

Since the temperature of the surrounding region is $K = 80$, formula (5) tells us that the temperature at time $t$ is given by

$$T(t) = 80 + Ce^{kt}.$$

We know that the initial temperature of the object is $T(0) = 200$. Substitution of $t = 0$ in the equation for $T(t)$ gives $200 = 80 + C$, so $C = 120$ and $T(t) = 80 + 120e^{kt}$. We also know that $T(5) = 100$, so

$$100 = 80 + 120e^{k(5)},$$

$$\frac{1}{6} = e^{5k},$$

$$5k = \ln\frac{1}{6},$$

$$k = \frac{\ln(1/6)}{5}.$$

This gives the formula

$$T(t) = 80 + 120e^{\{[\ln(1/6)]/5\}t}.$$

The temperature 10 min after the object is placed in the water is

$$\begin{aligned} T(10) &= 80 + 120e^{\{[\ln(1/6)]/5\}(10)} \\ &= 80 + 120e^{2\ln(1/6)} \\ &= 80 + 120e^{\ln(1/6)^2} \\ &= 80 + 120e^{\ln(1/36)} \\ &= 80 + 120\left(\frac{1}{36}\right) \approx 83.33°\text{F}. \end{aligned}$$

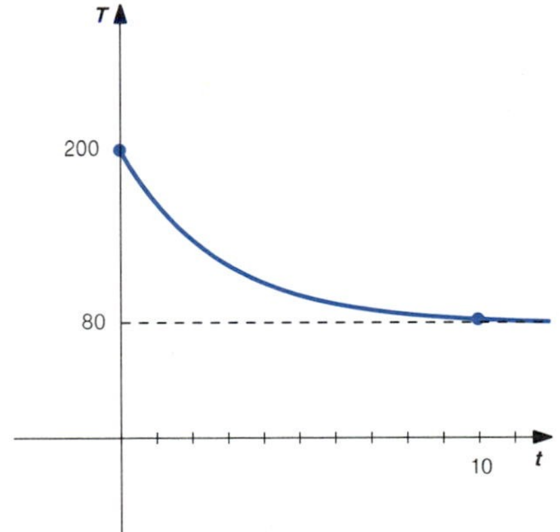


FIGURE 7.7.4

The graph of $T(t)$ is given in Figure 7.7.4. Note that the temperature of the object approaches 80, the temperature of the water. ■

## *Mixing*

Many problems involve the flow of different concentrations of a fluid in and out of a container. Let us illustrate the ideas by an example.

**EXAMPLE 6**

A 100-gal tank is full of brine that has a salt concentration of 2 lb/gal. Brine with salt concentration of 0.5 lb/gal is flowing into the tank at the rate of 4 gal/min. The brines are thoroughly mixed, and the mixture flows out of the tank at a rate of 4 gal/min. Find an equation that gives the salt concentration $t$ min after this process begins.

**SOLUTION**

Let $y(t)$ denote the amount of salt in the tank at time $t$. We will use the fundamental fact that

$$(\text{Amount of salt}) = (\text{Concentration})(\text{Amount of brine}).$$

The amount of brine in the tank is always 100 gal. Thus, the initial amount of salt in the tank is

$$y(0) = 2\left(\frac{\text{lb}}{\text{gal}}\right)(100 \text{ gal}) = 200 \text{ lb}$$

and the concentration of salt in the tank at time $t$ is

$$\frac{\text{Amount of salt}}{\text{Amount of brine}} = \frac{y(t)}{100}\frac{\text{lb}}{\text{gal}}.$$

The rate at which salt is flowing into the tank is the concentration of the brine flowing into the tank times the rate at which it is flowing into the tank. This rate is

$$r_i = \left(0.5\frac{\text{lb}}{\text{gal}}\right)\left(4\frac{\text{gal}}{\text{min}}\right) = 2\frac{\text{lb}}{\text{min}}.$$

Similarly, the rate at which salt is flowing out of the tank is

$$r_o = \left(\frac{y(t)}{100}\frac{\text{lb}}{\text{gal}}\right)\left(4\frac{\text{gal}}{\text{min}}\right) = 0.04y(t)\frac{\text{lb}}{\text{min}}.$$

The rate of change of the amount of salt in the tank at time $t$ is $y'(t)$. Since the rate of change must be the difference between the rate at which the salt is flowing in and out of the tank, we see that $y(t)$ must be the solution of the initial value problem

$$y' = 2 - 0.04y, \qquad y(0) = 200.$$

We can find the general solution of this differential equation by integration. If $y(t)$ is a solution, we have

$$y'(t) = 2 - 0.04y(t),$$

$$y'(t) = -0.04(y(t) - 50),$$

$$\frac{1}{y(t) - 50} y'(t) = -0.04,$$

$$\int \frac{1}{y(t) - 50} y'(t)\, dt = \int -0.04\, dt.$$

Using the substitution

$$y = y(t) - 50, \qquad dy = y'(t)\, dt$$

to evaluate the integral on the left, we have

$$\ln|y(t) - 50| = -0.04t + c,$$

$$|y(t) - 50| = e^c e^{-0.04t},$$

$$y(t) - 50 = Ce^{-0.04t} \qquad (C = \pm e^c),$$

$$y(t) = 50 + Ce^{-0.04t}.$$

Since $y(0) = 200$, we have

$$200 = 50 + C, \quad \text{so} \quad C = 150.$$

The amount of salt is then

$$y(t) = 50 + 150e^{-0.04t}.$$

The concentration is

$$\frac{y(t)}{100} = 0.5 + 1.5e^{-0.04t}.$$

The graph of the concentration as a function of time is given in Figure 7.7.5. Note that the concentration approaches 0.5, the concentration of the brine flowing into the tank. ■

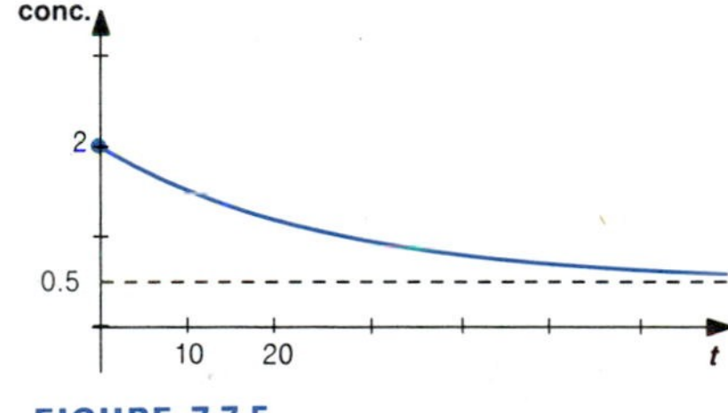

FIGURE 7.7.5

## EXERCISES 7.7

1 A certain bacteria colony increases at a rate proportional to the population. The initial population was 200. After 2 h the population was 400. What was the population 4 h after the initial time?

2 A bacteria colony is observed to increase from 100 to 300 in 2 h. Assuming exponential growth, how long will it take the colony to grow from 300 to 500?

3 Each year the population of a certain town increases by 5%, so $P(t + 1) = (1.05)P(t)$, $t$ in years. Assuming exponential growth, how long will it take the town to grow from 10,000 to 25,000?

4 The population of a colony of bacteria increases by 2% each day, so $P(t + 1) = (1.02)P(t)$, $t$ in days. Assuming exponential growth, what will the population of the colony be 1 week after the population is 900?

5 Five percent of a certain radioactive substance decays in 1 day. What is the half-life of the substance?

6 Carbon $^{11}C$ has half-life 20 min. What percent of a sample of $^{11}C$ remains after 12 min?

7 Find the age of an ancient skull that is measured to have 40% of the $^{14}C$ it contained while living.

8 Determine the age of a sample of wood that contains 25% of the $^{14}C$ it contained while a living tree.

9 A simple series circuit consists of a charged 0.02-F capacitance, a 200-Ω resistor, and no impressed voltage. The initial current is 10 A. Find the current after 0.2 s.

10 A simple series circuit consists of a charged 0.0004-F capacitance, a resistor, and impressed charge of 6 V. Find the resistance if the current in the circuit decreases from 8 A to 2 A in 0.1 s.

11 A thermometer reads 90° when it is moved to a 70° room. Ten minutes later the thermometer reads 80°. What will the thermometer read after an additional 10 min?

12 A small beef roast at a temperature of 50° is put into a 350° oven. After 1 hr the temperature of the roast is 90°. How much longer will it take the roast to reach 150°?

13 Newton's Law of Cooling leads to the formula

$$T(t) = K + [T(0) - K]e^{kt}.$$

Physical considerations imply that $k < 0$ if either $T(0) - K > 0$ or $T(0) - K < 0$. Evaluate $\lim_{t\to\infty} T(t)$ and give a physical interpretation of the meaning of the limit.

14 A thermometer reads 70° when it is first moved outdoors. After being outside 1 min, the thermometer reads 60°. One minute later it reads 55°. Find the outdoor temperature.

15 A 100-gal tank is full of pure water. Brine with a salt concentration of 2 lb/gal is flowing into the tank at a rate of 6 gal/min, while the mixture is flowing out at the same rate. Find a formula for the concentration of salt in the tank.

16 A 30-gal tank contains dye in a concentration of 0.1 oz/gal. Pure water is flowing into the tank at a rate of 2 gal/min, while the mixture is flowing out at the same rate. How long will it take before the concentration in the tank is 0.06 oz/gal?

17 A 50-gal tank contains pure water. For 10 min, brine with salt concentration 2 lb/gal flows into the tank at a rate of 3 gal/min while the mixture flows out at the same rate. For the next 10 min pure water runs into the tank at a rate of 3 gal/min, while the mixture flows out at the same rate. Find the concentration of salt in the tank at the end of the 20-min period.

18 A 500-gal tank contains 200 gal of brine with salt concentration 2 lb/gal. Pure water flows into the tank at a rate of 10 gal/min, while the mixture flows out of the tank at a rate of 5 gal/min. Find the salt concentration in the tank at the time the tank becomes completely filled.

## 7.8 HYPERBOLIC AND INVERSE HYPERBOLIC FUNCTIONS

Certain combinations of the functions $e^x$ and $e^{-x}$ occur often enough in applications that it is convenient to give them a name.

The **hyperbolic sine** is

$$\sinh x = \frac{e^x - e^{-x}}{2}.$$

The **hyperbolic cosine** is

$$\cosh x = \frac{e^x + e^{-x}}{2}.$$

We can use the usual methods to graph $y = \sinh x$. We know that

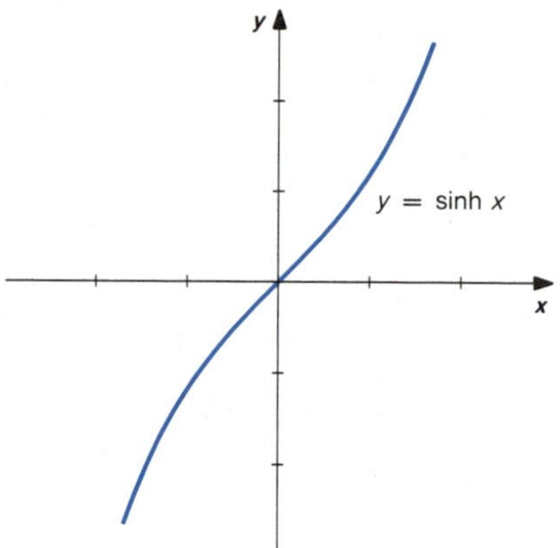


FIGURE 7.8.1

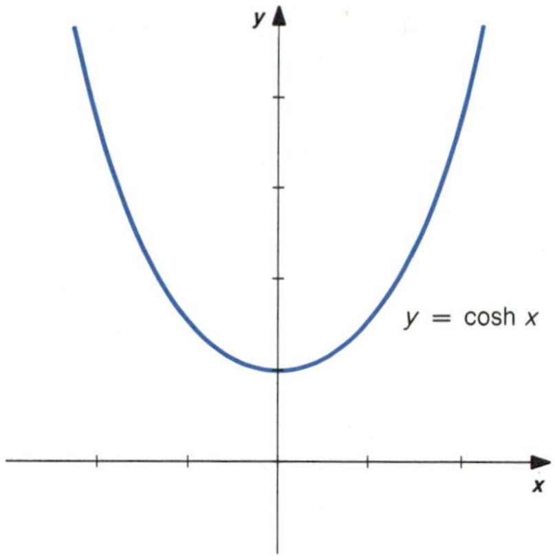

FIGURE 7.8.2

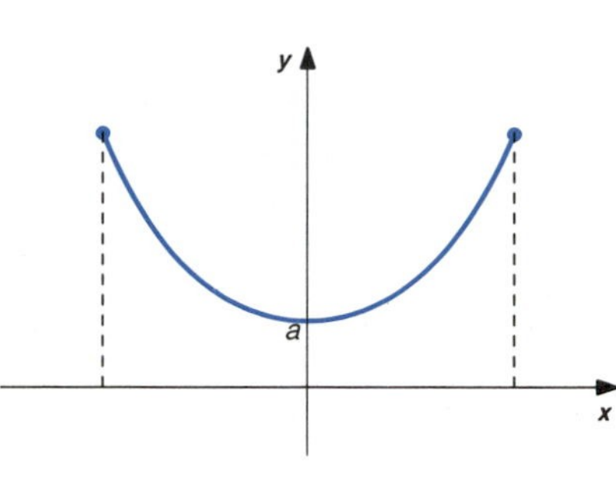


FIGURE 7.8.3

$e^0 = 1$, $e^x > 1 > e^{-x}$ for $x > 0$, and $e^x < 1 < e^{-x}$ for $x < 0$. It follows that $\sinh 0 = 0$, $\sinh x > 0$ for $x > 0$, and $\sinh x < 0$ for $x < 0$.

$$\frac{d}{dx}(\sinh x) = \frac{d}{dx}\left(\frac{e^x - e^{-x}}{2}\right) = \frac{e^x - e^{-x}(-1)}{2} = \frac{e^x + e^{-x}}{2}.$$

The derivative is positive for all $x$, so $\sinh x$ is increasing. The second derivative can be used to show that the graph is concave down for $x < 0$ and concave up for $x > 0$. $(0, 0)$ is a point of inflection. The graph is sketched in Figure 7.8.1.

We can also use the usual methods to analyze the function $\cosh x$. The derivative tells us that $\cosh x$ is decreasing for $x < 0$ and increasing for $x > 0$. The minimum value is $\cosh 0 = 1$. The second derivative tells us that $\cosh x$ is concave up. The graph is sketched in Figure 7.8.2.

It can be shown that a homogeneous flexible cable that is suspended as illustrated in Figure 7.8.3 will hang in the shape of the curve $y = a\cosh(x/a)$. This curve is called a **catenary**.

Sinh $x$ and cosh $x$ are related by the following identity:

$$\cosh^2 x - \sinh^2 x = 1. \tag{1}$$

■ *Proof* The identity is verified by substitution of exponential expressions for $\sinh x$ and $\cosh x$, and then simplifying. We have

$$\begin{aligned}\cosh^2 x - \sinh^2 x &= \left(\frac{e^x + e^{-x}}{2}\right)^2 - \left(\frac{e^x - e^{-x}}{2}\right)^2 \\ &= \frac{(e^x)^2 + 2e^x e^{-x} + (e^{-x})^2}{4} - \frac{(e^x)^2 - 2e^x e^{-x} + (e^{-x})^2}{4} \\ &= \frac{e^{2x} + 2e^{x-x} + e^{-2x}}{4} - \frac{e^{2x} - 2e^{x-x} + e^{-2x}}{4} \\ &= \frac{e^{2x} + 2 + e^{-2x} - e^{2x} + 2 - e^{-2x}}{4} = \frac{4}{4} = 1.\end{aligned}$$

■

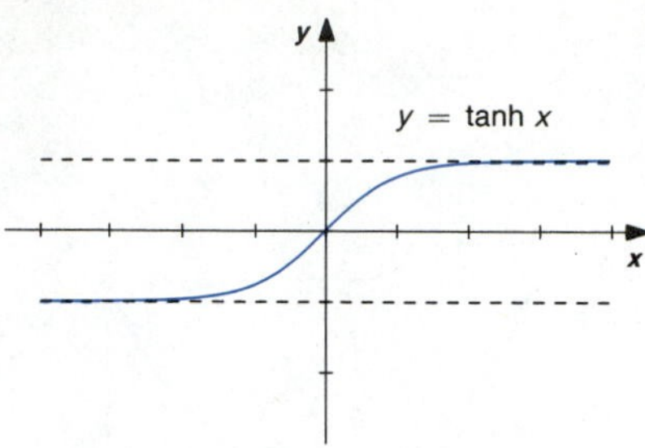


FIGURE 7.8.4

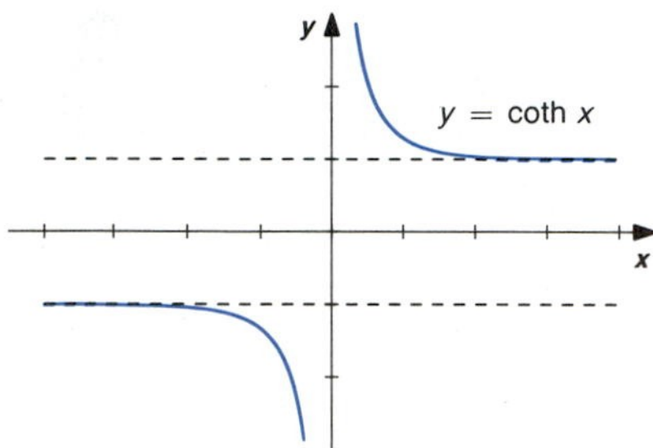


FIGURE 7.8.5

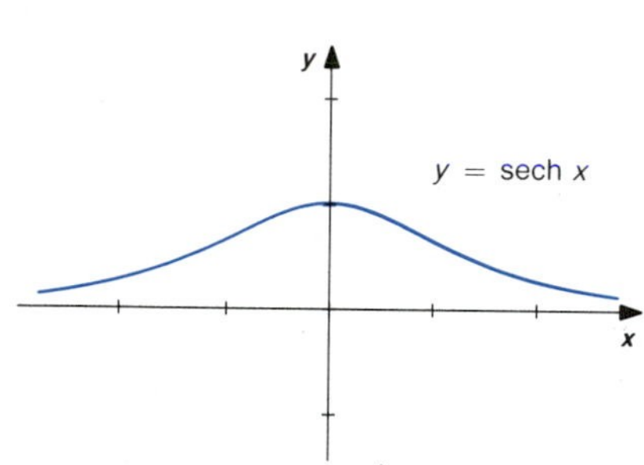


FIGURE 7.8.6

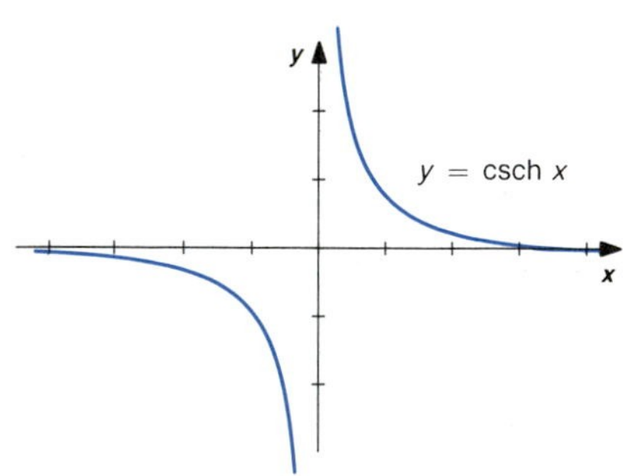


FIGURE 7.8.7

From (1) we see that the coordinates $x = \cosh t$, $y = \sinh t$ satisfy the equation of the hyperbola $x^2 - y^2 = 1$. This suggests the name of the hyperbolic functions. Analogously, the coordinates $x = \sin t$, $y = \cos t$ satisfy the equation of the circle $x^2 + y^2 = 1$. The trigonometric functions are sometimes called the circular functions. We will see that, except for possible changes in sign, many formulas for trigonometric functions also hold for hyperbolic functions.

Corresponding to the trigonometric functions, we define the following:

$$\tanh x = \frac{\sinh x}{\cosh x} = \frac{e^x - e^{-x}}{e^x + e^{-x}},$$

$$\coth x = \frac{\cosh x}{\sinh x} = \frac{e^x + e^{-x}}{e^x - e^{-x}},$$

$$\operatorname{sech} x = \frac{1}{\cosh x} = \frac{2}{e^x + e^{-x}},$$

$$\operatorname{csch} x = \frac{1}{\sinh x} = \frac{2}{e^x - e^{-x}}.$$

The graphs of these functions are given in Figures 7.8.4–7.8.7.

Hyperbolic analogues of many trigonometric identities are easily verified by using the definitions. For example, dividing each side of equation (1) by $\cosh^2 x$, we obtain

$$1 - \tanh^2 x = \operatorname{sech}^2 x. \tag{2}$$

Some other identities are given in the exercises.

It is convenient to express the derivatives of the hyperbolic functions in terms of hyperbolic functions. We have the following:

$$\frac{d}{dx}(\sinh x) = \cosh x,$$

$$\frac{d}{dx}(\cosh x) = \sinh x,$$

$$\frac{d}{dx}(\tanh x) = \operatorname{sech}^2 x,$$

$$\frac{d}{dx}(\coth x) = -\operatorname{csch}^2 x,$$

$$\frac{d}{dx}(\operatorname{sech} x) = -\operatorname{sech} x \tanh x,$$

$$\frac{d}{dx}(\operatorname{csch} x) = -\operatorname{csch} x \coth x.$$

The first two of these formulas are easily verified by using the definitions of $\sinh x$ and $\cosh x$. These two formulas can then be used to derive the others. For example,

$$\frac{d}{dx}(\operatorname{sech} x) = \frac{d}{dx}((\cosh x)^{-1}) = (-1)(\cosh x)^{-2}(\sinh x)$$

$$= -\left(\frac{1}{\cosh x}\right)\left(\frac{\sinh x}{\cosh x}\right) = -\operatorname{sech} x \tanh x.$$

The differentiation formulas for the hyperbolic functions correspond to those of the trigonometric functions, except for the differences in sign.

We should be able to use the Chain Rule version of the differentiation formulas.

■ **EXAMPLE 1**

(a) $\dfrac{d}{dx}\left(a\cosh\left(\dfrac{x}{a}\right)\right) = a\sinh\left(\dfrac{x}{a}\right)\left(\dfrac{1}{a}\right) = \sinh\left(\dfrac{x}{a}\right).$

(b) $\dfrac{d}{dx}(\tanh(x^2)) = \operatorname{sech}^2(x^2)(2x).$ ■

The following integration formulas correspond to the above differentiation formulas:

$$\int \cosh u\, du = \sinh u + C,$$

$$\int \sinh u\, du = \cosh u + C,$$

$$\int \operatorname{sech}^2 u\, du = \tanh u + C,$$

$$\int \operatorname{csch}^2 u\, du = -\coth u + C,$$

$$\int \operatorname{sech} u \tanh u\, du = -\operatorname{sech} u + C,$$

$$\int \operatorname{csch} u \coth u\, du = -\operatorname{csch} u + C.$$

■ **EXAMPLE 2**

Evaluate

$$\int_0^3 \sinh \frac{x}{3}\, dx.$$

**SOLUTION**

Using $u = x/3$, so $du = (1/3)\,dx$, we have

$$\int_0^3 \sinh\frac{x}{3}\,dx = \int_0^3 \sinh\frac{x}{3}(3)\left(\frac{1}{3}\right)dx$$

$$= \int \sinh u\,(3)\,du$$

$$= 3\cosh u\Big]$$

$$= 3\cosh\frac{x}{3}\Big]_0^3$$

$$= 3\cosh\frac{3}{3} - 3\cosh\frac{0}{3}$$

$$= 3\cosh 1 - 3$$

$$= 3\,\frac{e + e^{-1}}{2} - 3 \approx 1.6292419.$$

■

### *Inverse hyperbolic functions*

We know that $\sinh x$ has positive derivative for all $x$. It follows that $\sinh x$ has a differentiable inverse function, $\sinh^{-1} x$. The defining relation is

$$y = \sinh^{-1} x \quad \text{if and only if} \quad x = \sinh y.$$

We can obtain a formula for $\sinh^{-1}x$ by solving the equation $x = \sinh y$ for $y$. That is,

$$x = \sinh y,$$

$$x = \frac{e^y - e^{-y}}{2},$$

$$e^y - 2x - e^{-y} = 0.$$

Multiplying the above equation by $e^y$, we obtain

$$(e^y)^2 - 2x(e^y) - 1 = 0.$$

We then note that we have an equation that is quadratic in $e^y$. The Quadratic Formula gives the solution

$$e^y = \frac{-(-2x) \pm \sqrt{(-2x)^2 - 4(1)(-1)}}{2(1)}.$$

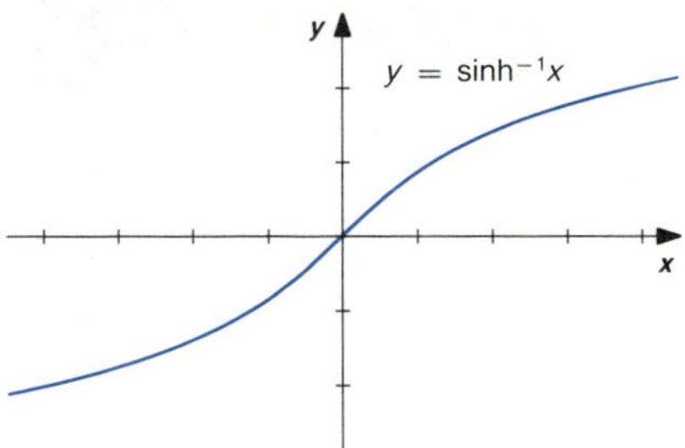

FIGURE 7.8.8

Since $e^y > 0$, we discard the negative solution and simplify to obtain

$$e^y = x + \sqrt{x^2 + 1},$$
$$y = \ln(x + \sqrt{x^2 + 1}).$$

The formula for $\sinh^{-1}x$ is

$$\sinh^{-1}x = \ln(x + \sqrt{x^2 + 1}).$$

We can use the same method to obtain formulas for other inverse hyperbolic functions. In some cases it is necessary to restrict the domain to obtain an inverse function. Let us list the formulas.

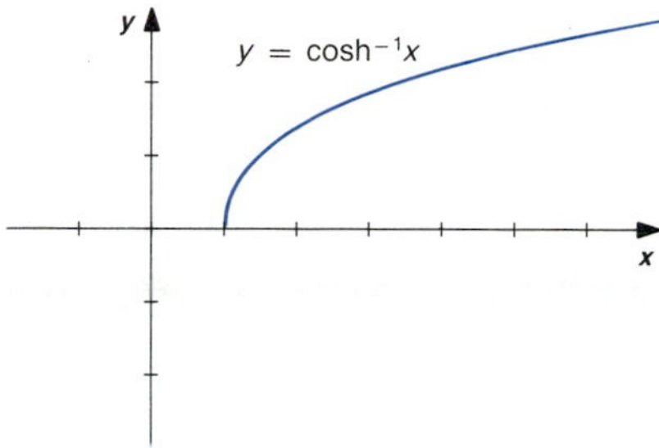

FIGURE 7.8.9

The inverse function of $\sinh x$ is

$$\sinh^{-1}x = \ln(x + \sqrt{x^2 + 1}). \qquad \text{(Figure 7.8.8)}$$

The inverse function of $\cosh x$, $x \geq 0$, is

$$\cosh^{-1}x = \ln(x + \sqrt{x^2 - 1}). \qquad \text{(Figure 7.8.9)}$$

The inverse function of $\tanh x$ is

$$\tanh^{-1}x = \frac{1}{2}\ln\frac{1 + x}{1 - x}. \qquad \text{(Figure 7.8.10)}$$

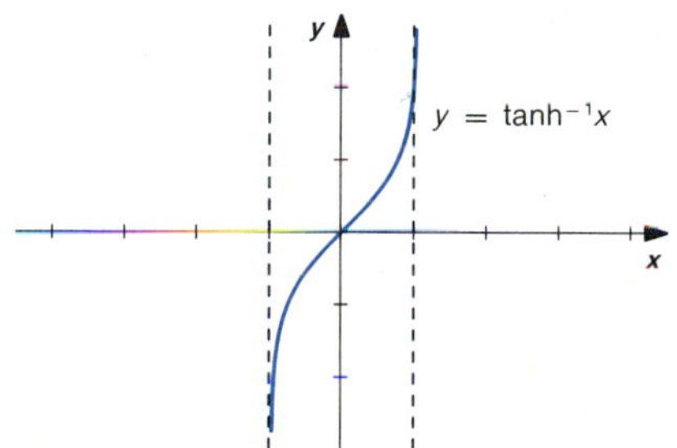

FIGURE 7.8.10

The inverse function of $\coth x$ is

$$\coth^{-1}x = \frac{1}{2}\ln\frac{x + 1}{x - 1}. \qquad \text{(Figure 7.8.11)}$$

The inverse function of $\operatorname{sech} x$, is

$$\operatorname{sech}^{-1}x = \ln\frac{1 + \sqrt{1 - x^2}}{x}. \qquad \text{(Figure 7.8.12)}$$

The inverse function of $\operatorname{csch} x$, is

$$\operatorname{csch}^{-1}x = \ln\left(\frac{1}{x} + \frac{\sqrt{1 + x^2}}{|x|}\right). \qquad \text{(Figure 7.8.13)}$$

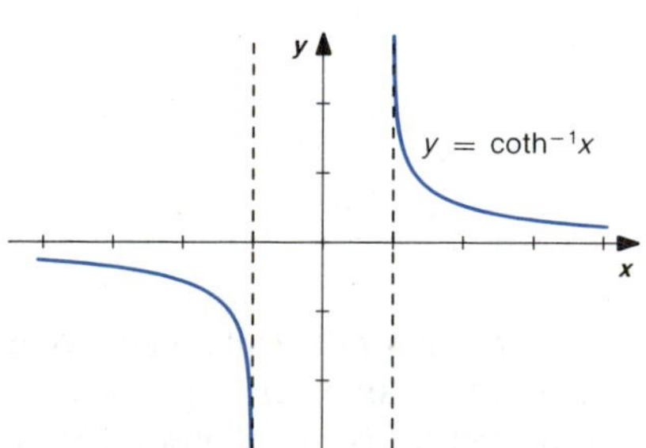

FIGURE 7.8.11

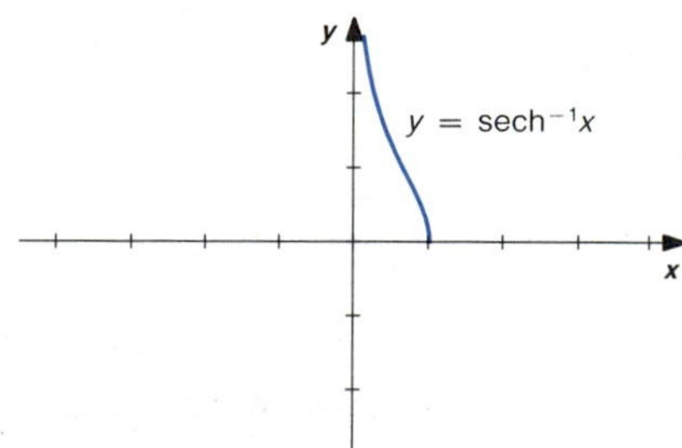

FIGURE 7.8.12

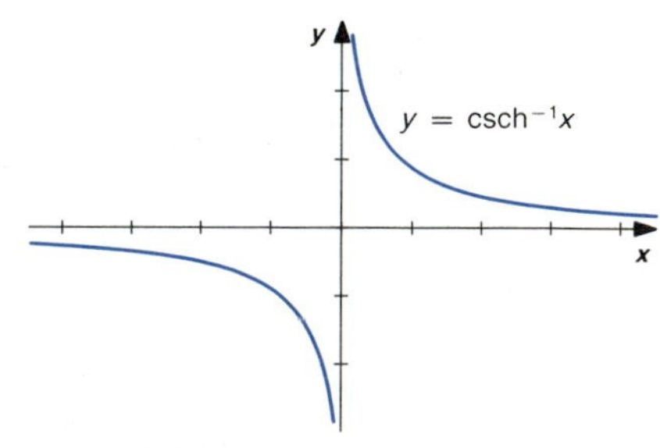

FIGURE 7.8.13

We can use the Chain Rule to obtain the following differentiation formulas:

$$\frac{d}{dx}(\sinh^{-1} x) = \frac{1}{\sqrt{1 + x^2}},$$

$$\frac{d}{dx}(\cosh^{-1} x) = \frac{1}{\sqrt{x^2 - 1}}, \qquad x > 1,$$

$$\frac{d}{dx}(\tanh^{-1} x) = \frac{1}{1 - x^2}, \qquad |x| < 1,$$

$$\frac{d}{dx}(\coth^{-1} x) = \frac{1}{1 - x^2}, \qquad |x| > 1,$$

$$\frac{d}{dx}(\operatorname{sech}^{-1} x) = -\frac{1}{x\sqrt{1 - x^2}}, \qquad 0 < x < 1,$$

$$\frac{d}{dx}(\operatorname{csch}^{-1} x) = -\frac{1}{|x|\sqrt{1 + x^2}}, \qquad x \neq 0.$$

The above differentiation formulas can be used to verify the following integration formulas. We assume that $a > 0$.

$$\int \frac{1}{\sqrt{a^2 + u^2}}\, dx = \sinh^{-1} \frac{u}{a} + C,$$

$$\int \frac{1}{\sqrt{u^2 - a^2}}\, dx = \cosh^{-1} \frac{u}{a} + C,$$

$$\int \frac{1}{a^2 - u^2}\, du = \frac{1}{a} \tanh^{-1} \frac{u}{a} + C, \qquad |u| < a,$$

$$\int \frac{1}{a^2 - u^2}\, du = \frac{1}{a} \coth^{-1} \frac{u}{a} + C, \qquad |u| > a,$$

$$\int \frac{1}{u\sqrt{a^2 - u^2}}\, dx = -\frac{1}{a} \operatorname{sech}^{-1} \frac{|u|}{a} + C,$$

$$\int \frac{1}{u\sqrt{a^2 + u^2}}\, dx = -\frac{1}{a} \operatorname{csch}^{-1} \frac{|u|}{a} + C.$$

To use the above integration formulas to obtain a numerical value of a definite integral, we must be able to evaluate the inverse hyperbolic functions. If these functions are not programmed on your calculator, they can be evaluated by expressing them in terms of the logarithmic function.

**EXAMPLE 3**

Evaluate

$$\int_0^3 \frac{1}{\sqrt{9+x^2}}\,dx.$$

**SOLUTION**

The integration formula gives

$$\int_0^3 \frac{1}{\sqrt{9+x^2}}\,dx = \sinh^{-1}\frac{x}{3}\Big]_0^3 = \sinh^{-1}(1) - \sinh^{-1}(0) = \sinh^{-1}(1)$$

$$= \ln(1+\sqrt{(1)^2+1}) = \ln(1+\sqrt{2}) \approx 0.88137359.$$

■

## EXERCISES 7.8

Verify the identities in Exercises 1–10.

1 $\sinh(-x) = -\sinh x$

2 $\cosh(-x) = \cosh x$

3 $\sinh 2x = 2\sinh x\cosh x$

4 $\cosh 2x = \cosh^2 x + \sinh^2 x$

5 $\cosh(x+y) = \cosh x\cosh y + \sinh x\sinh y$

6 $\sinh(x+y) = \sinh x\cosh y + \cosh x\sinh y$

7 $\tanh(x+y) = \dfrac{\tanh x + \tanh y}{1+\tanh x\tanh y}$

8 $\tanh 2x = \dfrac{2\tanh x}{1+\tanh^2 x}$

9 $\cosh^2\dfrac{x}{2} = \dfrac{1+\cosh x}{2}$

10 $\tanh\dfrac{x}{2} = \dfrac{\sinh x}{1+\cosh x}$

Evaluate the derivatives in Exercises 11–20.

11 $\dfrac{d}{dx}(\sinh x^2)$

12 $\dfrac{d}{dx}(\sinh\sqrt{x})$

13 $\dfrac{d}{dx}(\cosh^2 x)$

14 $\dfrac{d}{dx}(\cosh x^2)$

15 $\dfrac{d}{dx}(\ln(\cosh x))$

16 $\dfrac{d}{dx}(\tan^{-1}(\sinh x))$

17 $\dfrac{d}{dx}(\sinh^{-1} 2x)$

18 $\dfrac{d}{dx}\left(\cosh^{-1}\dfrac{x}{3}\right)$

19 $\dfrac{d}{dx}(\tanh^{-1}\sin x)$

20 $\dfrac{d}{dx}(\operatorname{sech}^{-1}\cos x)$

Evaluate the integrals in Exercises 21–32.

21 $\displaystyle\int \sinh 3x\,dx$

22 $\displaystyle\int \sinh\frac{x}{2}\,dx$

23 $\displaystyle\int x\,\operatorname{sech}^2(x^2)\,dx$

24 $\displaystyle\int \frac{\cosh\sqrt{x}}{\sqrt{x}}\,dx$

25 $\displaystyle\int \tanh x\,dx$

26 $\displaystyle\int \sinh x\cosh x\,dx$

27 $\displaystyle\int \frac{1}{\sqrt{9+4x^2}}\,dx$

28 $\displaystyle\int \frac{1}{\sqrt{9x^2-4}}\,dx$

29 $\displaystyle\int \frac{e^x}{\sqrt{e^{2x}-1}}\,dx$

30 $\displaystyle\int_4^8 \frac{x}{x^4-16}\,dx$

31 $\displaystyle\int_0^2 \frac{1}{9-x^2}\,dx$

32 $\displaystyle\int_3^5 \frac{1}{x^2-4}\,dx$

33 Use the definitions in terms of the exponential function to verify that

$$\frac{d}{dx}(\sinh x) = \cosh x.$$

34 Use the definitions in terms of the exponential function to verify that

$$\frac{d}{dx}(\cosh x) = \sinh x.$$

**35** Use formulas for the derivatives of sinh $x$ and cosh $x$ to verify that

$$\frac{d}{dx}(\tanh x) = \operatorname{sech}^2 x.$$

**36** Use the formula for the derivative of sinh $x$ to verify that

$$\frac{d}{dx}(\operatorname{csch} x) = -\operatorname{csch} x \coth x.$$

**37** Verify that

$$\operatorname{sech}^{-1} x = \ln \frac{1 + \sqrt{1 - x^2}}{x}$$

by solving the equation

$$x = \frac{2}{e^y + e^{-y}}$$

for $y$, $y \geq 0$.

**38** Verify that

$$\tanh^{-1} x = \frac{1}{2} \ln \frac{1 + x}{1 - x}$$

by solving the equation

$$x = \frac{e^y - e^{-y}}{e^y + e^{-y}}$$

for $y$.

**39** Verify that

$$\frac{d}{dx}(\sinh^{-1} x) = \frac{1}{\sqrt{1 + x^2}}$$

(a) by differentiating the logarithmic expression for $\sinh^{-1} x$ and then simplifying and (b) by implicit differentiation of the equation $\sinh y = x$, and then using $\cosh^2 y - \sinh^2 y = 1$ to express $y'$ in terms of $x$.

**40** Verify that

$$\frac{d}{dx}(\tanh^{-1} x) = \frac{1}{1 - x^2}$$

(a) by differentiating the logarithmic expression for $\tanh^{-1} x$ and then simplifying and (b) by implicit differentiation of the equation $\tanh y = x$, and then using $1 - \tanh^2 y = \operatorname{sech}^2 y$ to express $y'$ in terms of $x$.

## REVIEW EXERCISES

Evaluate the limits in Exercises 1–4.

**1** $\displaystyle\lim_{x \to 0} \frac{1 - \cos 4x}{x^2}$ **2** $\displaystyle\lim_{x \to 0} \frac{\sin^{-1} 2x}{x}$

**3** $\displaystyle\lim_{x \to \infty} (1 + 3x)^{1/(\ln x)}$ **4** $\displaystyle\lim_{x \to 0} (1 - x)^{1/x}$

Find the inverse function of the functions given in Exercises 5-8.

**5** $f(x) = 3x - 2$

**6** $f(x) = x^2 - 2x$, $x \geq 1$

**7** $f(x) = \tan(2x) - 1$, $-\pi/4 < x < \pi/4$

**8** $f(x) = 2 \sin(x/2)$, $-\pi \leq x \leq \pi$

**9** Evaluate $\sin^{-1}\left(-\dfrac{\sqrt{3}}{2}\right)$.

**10** Evaluate $\cos^{-1}\left(\cos\left(-\dfrac{\pi}{4}\right)\right)$.

**11** Write $\sin\left(\cos^{-1} \dfrac{x}{2}\right)$ as an algebraic expression.

**12** Write $\sec\left(\tan^{-1} \dfrac{x}{4}\right)$ as an algebraic expression.

Find the derivatives in Exercises 13–20.

**13** $\dfrac{d}{dx}(\sin^{-1} x^2)$ **14** $\dfrac{d}{dx}\left(\tan^{-1} \dfrac{x}{2}\right)$

**15** $\dfrac{d}{dx}\left(\sec^{-1} \dfrac{1}{x}\right)$ **16** $\dfrac{d}{dx}(\ln(x\sqrt{x^2 + 1}))$

**17** $\dfrac{d}{dx}(\ln\sqrt{2 - \sin^2 x})$ **18** $\dfrac{d}{dx}(\ln|\sec x + \tan x|)$

**19** $\dfrac{d}{dx}(e^{-x^2/2})$ **20** $\dfrac{d}{dx}(xe^{-x} + e^{-x})$

Evaluate the definite integrals in Exercises 21–22.

**21** $\displaystyle\int_0^2 \frac{1}{4 + x^2}\,dx$ **22** $\displaystyle\int_0^1 \frac{1}{\sqrt{4 - x^2}}\,dx$

Evaluate the indefinite integrals in Exercises 23–30.

**23** $\displaystyle\int \frac{x}{9 + x^2}\,dx$ **24** $\displaystyle\int \frac{x}{\sqrt{9 - x^2}}\,dx$

25 $\int \frac{1}{x\sqrt{9-x^2}}\,dx$

26 $\int \tan 2x\,dx$

27 $\int \sec^2 x\, e^{\tan x}\,dx$

28 $\int \frac{(\ln x)^3}{x}\,dx$

29 $\int \frac{(\sin^{-1}x)^2}{\sqrt{1-x^2}}\,dx$

30 $\int \frac{e^x - e^{-x}}{e^x + e^{-x}}\,dx$

Sketch the graphs of the equations given in Exercises 31–34.

31 $y = xe^{-x/2}$

32 $y = e^{-x^2}$

33 $y = x \ln x$

34 $y = x - \ln x$

35 A bacteria colony is observed to increase from 200 to 600 in 3 h. Assuming exponential growth, how long will it take the colony to grow from 600 to 1000?

36 Determine the age of a sample of wood that contains 30% of the $^{14}C$ it contained while a living tree. (Radiocarbon $^{14}C$ has a half-life of 5750 years.)

37 A small object had temperature 180°F when it was dropped into a large pool of water at 72°F. Five minutes after it was put into the water, the temperature of the object was 150°F. What was its temperature 10 min after it was put in the water?

38 A 10-gal tank is full of a solution that contains 4 oz of dye. Pure water is running into the tank at a rate of 3 gal/min, while the mixture is flowing out at the same rate. How long will it take before the concentration of die in the tank is less than 0.02 oz/gal?

## CHAPTER EIGHT

# *Techniques of integration*

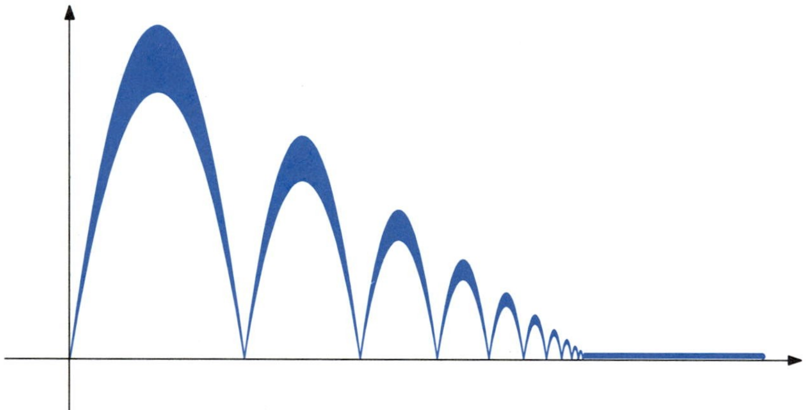

We have seen that the solution of many problems that arise in science and engineering requires the evaluation of integrals. We have also seen that the Fundamental Theorem of Calculus can be used to evaluate an integral, if the appropriate integration formula is either known or can be found in a table of integrals. Computers can also now be used to evaluate integrals.

It is important to be able to algebraically manipulate the form of integrals, even if integration tables or computers are used to evaluate them. In this chapter, we see how substitution, integration by parts, and partial fractions are used to change the form of integrals. We will see how these methods can be used to develop a table of integrals and practice using the formulas to evaluate integrals. In later chapters, we will refer to the table of integrals for the evaluation of integrals.

## 8.1 REVIEW OF KNOWN FORMULAS

Let us review the integration formulas that we know:

$$\int u^{\alpha}\,du = \frac{u^{\alpha+1}}{\alpha+1} + C, \qquad \alpha \neq -1.$$

$$\int \frac{1}{u}\,du = \ln|u| + C.$$

$$\int e^{u}\,du = e^{u} + C.$$

$$\int \cos u\,du = \sin u + C.$$

$$\int \sin u\,du = -\cos u + C.$$

$$\int \sec^2 u\,du = \tan u + C.$$

$$\int \csc^2 u\,du = -\cot u + C.$$

$$\int \sec u \tan u\,du = \sec u + C.$$

$$\int \csc u \cot u\,du = -\csc u + C.$$

$$\int \frac{1}{\sqrt{a^2-u^2}}\,du = \sin^{-1}\frac{u}{a} + C.$$

$$\int \frac{1}{a^2+u^2}\,du = \frac{1}{a}\tan^{-1}\frac{u}{a} + C.$$

$$\int \frac{1}{u\sqrt{u^2-a^2}}\,du = \frac{1}{a}\sec^{-1}\frac{|u|}{a} + C.$$

A simple substitution will often transform a given integral into one of the above basic types. In later sections we will learn more sophisticated methods for evaluating integrals, but we should always look for a simple solution.

Recall that an integration formula can be checked by differentiation. That is, we can verify the formula $\int f(x)\,dx = F(x) + C$ by checking that $F' = f$. In many integration problems it is very easy to differentiate the answer mentally, and we should do so. In practice, as the answers become more complicated, we will not check every problem, but the possibility is still there.

Generally, it is a good idea to write out each step of a substitution, even if it is a simple substitution. However, as you gain experience and confidence, you may carry out simple substitutions without writing the details. When you do this, you should check the result by differentiating mentally.

### EXAMPLE 1

Evaluate $\int e^{x/3}\,dx$.

**SOLUTION**

This integral involves $e$ to a power that is a function, so we try to reduce it to $\int e^u\,du$. We choose $u = x/3$, so $du = (1/3)\,dx$. We will introduce the factor $(3)(1/3)$ in the integrand so the exact expression $du = (1/3)\,dx$ appears in the integrand. We have

$$\begin{aligned}\int e^{x/3}\,dx &= \int e^{x/3}(3)\left(\frac{1}{3}\right)dx\\ &= \int e^u(3)\,du\\ &= 3e^u + C\\ &= 3e^{x/3} + C.\end{aligned}$$

This result is easily checked. Using the Chain Rule, we have

$$\frac{d}{dx}(3e^{x/3}) = 3e^{x/3}\left(\frac{1}{3}\right) = e^{x/3},$$

so the integration result is correct. ■

### EXAMPLE 2

Evaluate $\int \sin(2x + 1)\,dx$.

**SOLUTION**

This integral involves the sine of a function so we try to reduce it to $\int \sin u\,du$. We set $u = 2x + 1$, so $du = 2\,dx$. We then have

$$\begin{aligned}\int \sin(2x+1)\,dx &= \int \sin(2x+1)\left(\frac{1}{2}\right)(2)\,dx\\ &= \int \sin u\left(\frac{1}{2}\right)du\\ &= \frac{1}{2}(-\cos u) + C\\ &= -\frac{1}{2}\cos(2x+1) + C.\end{aligned}$$

Using the Chain Rule to check our work, we have

$$\frac{d}{dx}\left(-\frac{1}{2}\cos(2x+1)\right) = -\frac{1}{2}(-\sin(2x+1))(2) = \sin(2x+1). \qquad \blacksquare$$

### ■ EXAMPLE 3

Evaluate $\int x\sqrt{x^2+1}\,dx$.

**SOLUTION**

This integral involves the square root of a function. Let us try to reduce the integral to $\int u^{1/2}\,du$. We see that the substitution $u = x^2 + 1$, $du = 2x\,dx$ will work because the integrand contains the *essential factor* $x$, a constant multiple of the derivative of $x^2 + 1$. We have

$$\begin{aligned}\int x\sqrt{x^2+1}\,dx &= \int\left(\frac{1}{2}\right)2x(x^2+1)^{1/2}\,dx \\ &= \int \frac{1}{2}u^{1/2}\,du \\ &= \frac{1}{2}\left(\frac{u^{3/2}}{3/2}\right) + C \\ &= \frac{1}{3}(x^2+1)^{3/2} + C.\end{aligned}$$

Checking our work, we have

$$\frac{d}{dx}\left(\frac{1}{3}(x^2+1)^{3/2}\right) = \frac{1}{3}\left(\frac{3}{2}\right)(x^2+1)^{1/2}(2x) = x(x^2+1)^{1/2}. \qquad \blacksquare$$

### ■ EXAMPLE 4

Evaluate

$$\int \frac{x^3}{x^4+1}\,dx.$$

**SOLUTION**

We see that the numerator contains the essential factor $x^3$ necessary for the simplifying substitution $u = x^4 + 1$, $du = 4x^3\,dx$.

$$\begin{aligned}\int \frac{x^3}{x^4+1}\,dx &= \int \frac{(1/4)4x^3}{x^4+1}\,dx \\ &= \int \frac{1}{4}u^{-1}\,du \\ &= \frac{1}{4}\ln|u| + C \\ &= \frac{1}{4}\ln(x^4+1) + C.\end{aligned}$$

Checking this result, we have

$$\frac{d}{dx}\left(\frac{1}{4}\ln(x^4+1)\right)=\frac{1}{4}\left(\frac{1}{x^4+1}\right)(4x^3)=\frac{x^3}{x^4+1}.$$ ■

■ **EXAMPLE 5**

Evaluate

$$\int \frac{e^{\sqrt{x}}}{\sqrt{x}}\,dx.$$

**SOLUTION**

This integral involves $e$ to a power that is a function, so we can try to reduce it to $\int e^u\,du$. Set

$$u=\sqrt{x}, \quad \text{so} \quad du=\frac{1}{2\sqrt{x}}\,dx.$$

The essential factor $1/\sqrt{x}$ makes this substitution work.

$$\begin{aligned}\int \frac{e^{\sqrt{x}}}{\sqrt{x}}\,dx &= \int e^{\sqrt{x}}\frac{2}{2\sqrt{x}}\,dx\\ &= \int 2e^u\,du\\ &= 2e^u + C\\ &= 2e^{\sqrt{x}} + C.\end{aligned}$$

Checking, we have

$$\frac{d}{dx}(2e^{\sqrt{x}})=2e^{\sqrt{x}}\frac{1}{2\sqrt{x}}=\frac{e^{\sqrt{x}}}{\sqrt{x}}.$$ ■

■ **EXAMPLE 6**

Evaluate

$$\int \frac{1}{x\sqrt{4x^2-9}}\,dx.$$

**SOLUTION**

Setting $u = 2x$, so that $du = 2\,dx$, we have

$$\begin{aligned}\int \frac{1}{x\sqrt{4x^2-9}}\,dx &= \int \frac{2}{2x\sqrt{(2x)^2-9}}\,dx\\ &= \int \frac{1}{u\sqrt{u^2-3^2}}\,du\\ &= \frac{1}{3}\sec^{-1}\frac{|u|}{3} + C\\ &= \frac{1}{3}\sec^{-1}\frac{2|x|}{3} + C.\end{aligned}$$

Checking this result, we have

$$\begin{aligned}\frac{d}{dx}\left(\frac{1}{3}\sec^{-1}\frac{2|x|}{3}\right) &= \frac{1}{3}\frac{1}{(2x/3)\sqrt{(2x/3)^2-1}}\frac{2}{3}\\ &= \frac{1}{x\sqrt{4x^2-9}}.\end{aligned}$$

■

In Example 6 it was easier to evaluate the integral algebraically than it was to check the result by differentiation. In such cases we will omit the check.

■ **EXAMPLE 7**

Evaluate $\int \sin^2 x \cos x\,dx$.

**SOLUTION**

We first note that this integral does not match up with any of the formulas we have for the integral of trigonometric functions. We then recognize that the factor $\cos x$ is the derivative of $\sin x$. This means we can try the substitution $u = \sin x$, $du = \cos x\,dx$, which works.

$$\begin{aligned}\int \sin^2 x \cos x\,dx &= \int u^2\,du\\ &= \frac{u^3}{3} + C\\ &= \frac{1}{3}(\sin x)^3 + C.\end{aligned}$$

This result is easy enough to check. We have

$$\frac{d}{dx}\left(\frac{1}{3}(\sin x)^3\right) = \frac{1}{3}(3)(\sin x)^2(\cos x) = \sin^2 x \cos x.$$

■

It should be becoming clear that you must know the formulas for differentiation of the basic functions forwards and backwards. That is, it isn't enough to know that the derivative of sin $x$ is cos $x$. When you see cos $x$, you must recognize that it is the derivative of sin $x$.

**EXAMPLE 8**

Evaluate $\int \tan^4 x \sec^2 x \, dx$.

**SOLUTION**

We recognize that $\sec^2 x$ is the derivative of $\tan x$. This suggests the substitution $u = \tan x$, $du = \sec^2 x \, dx$.

$$\begin{aligned}\int \tan^4 x \sec^2 x \, dx &= \int u^4 \, du \\ &= \frac{u^5}{5} + C \\ &= \frac{1}{5} \tan^5 x + C.\end{aligned}$$

**EXAMPLE 9**

Evaluate $\int \sec^3 x \tan x \, dx$.

**SOLUTION**

The integrand involves powers of sec $x$, and contains the factor sec $x$ tan $x$. This suggests that we try $u = \sec x$, so $du = \sec x \tan x \, dx$. We have

$$\begin{aligned}\int \sec^3 x \tan x \, dx &= \int \sec^2 x \sec x \tan x \, dx \\ &= \int u^2 \, du \\ &= \frac{u^3}{3} + C \\ &= \frac{1}{3} \sec^3 x + C.\end{aligned}$$

In some examples we must complete a square before we substitute.

**EXAMPLE 10**

Evaluate

$$\int \frac{x+1}{x^2 - 6x + 25} \, dx.$$

**SOLUTION**

$$\begin{aligned}\int \frac{x+1}{x^2 - 6x + 25} \, dx &= \int \frac{x+1}{x^2 - 6x + [9] - [9] + 25} \, dx \\ &= \int \frac{x+1}{(x-3)^2 + 16} \, dx.\end{aligned}$$

Setting $u = x - 3$, so that $du = dx$ and $x = u + 3$, we then obtain

$$\begin{aligned}\int \frac{x+1}{x^2 - 6x + 25}\,dx &= \int \frac{(u+3)+1}{u^2+16}\,du \\ &= \int \frac{u}{u^2+16}\,du + \int \frac{4}{u^2+16}\,du \\ &= \int \frac{(1/2)2u}{u^2+16}\,du + \int \frac{4}{u^2+4^2}\,du.\end{aligned}$$

The substitution $v = u^2 + 16$, $dv = 2u\,du$, transforms the first integral above to

$$\int \frac{1/2}{v}\,dv = \frac{1}{2}\ln|v| + C = \frac{1}{2}\ln(u^2 + 16) + C.$$

We then have

$$\begin{aligned}\int \frac{x+1}{x^2 - 6x + 25}\,dx &= \frac{1}{2}\ln(u^2+16) + 4\frac{1}{4}\tan^{-1}\frac{u}{4} + C \\ &= \frac{1}{2}\ln((x-3)^2+16) + \tan^{-1}\frac{x-3}{4} + C \\ &= \frac{1}{2}\ln(x^2 - 6x + 25) + \tan^{-1}\frac{x-3}{4} + C.\end{aligned}$$

■

■ **EXAMPLE 11**

Evaluate

$$\int \frac{1}{\sqrt{x - x^2}}\,dx.$$

**SOLUTION**

$$\begin{aligned}\int \frac{1}{\sqrt{x-x^2}}\,dx &= \int \frac{1}{\sqrt{(1/4) - [x^2 - x + (1/4)]}}\,dx \\ &= \int \frac{1}{\sqrt{(1/2)^2 - [x - (1/2)]^2}}\,dx.\end{aligned}$$

Setting $u = x - (1/2)$, so that $du = dx$, we then have

$$\begin{aligned}\int \frac{1}{\sqrt{x-x^2}}\,dx &= \int \frac{1}{\sqrt{(1/2)^2 - u^2}}\,du \\ &= \sin^{-1}\left(\frac{u}{1/2}\right) + C \\ &= \sin^{-1}\frac{x-(1/2)}{1/2} + C \\ &= \sin^{-1}(2x-1) + C.\end{aligned}$$

■

## EXERCISES 8.1

Evaluate the integrals in Exercises 1–36.

1 $\displaystyle\int \frac{x}{\sqrt{x^2+4}}\,dx$

2 $\displaystyle\int x^3\sqrt{x^4+2}\,dx$

3 $\displaystyle\int \frac{1}{\sqrt{1-9x^2}}\,dx$

4 $\displaystyle\int \frac{x}{\sqrt{1-9x^2}}\,dx$

5 $\displaystyle\int \frac{\cos x}{2-\sin x}\,dx$

6 $\displaystyle\int \frac{e^x-e^{-x}}{e^x+e^{-x}}\,dx$

7 $\displaystyle\int xe^{-x^2}\,dx$

8 $\displaystyle\int \frac{(\ln x)^3}{x}\,dx$

9 $\displaystyle\int \frac{1}{x\ln x}\,dx$

10 $\displaystyle\int e^{\sin x}\cos x\,dx$

11 $\displaystyle\int \sin 2x\,dx$

12 $\displaystyle\int \cos(x/2)\,dx$

13 $\displaystyle\int \frac{\sec(x^{2/3})\tan(x^{2/3})}{x^{1/3}}\,dx$

14 $\displaystyle\int x\sec^2(x^2)\,dx$

15 $\displaystyle\int e^{3x-1}\,dx$

16 $\displaystyle\int \frac{1}{1-2x}\,dx$

17 $\displaystyle\int \frac{1}{(3x-1)^2}\,dx$

18 $\displaystyle\int \frac{x}{(x^2+2)^2}\,dx$

19 $\displaystyle\int \frac{1}{2+9x^2}\,dx$

20 $\displaystyle\int \frac{1}{\sqrt{2-9x^2}}\,dx$

21 $\displaystyle\int \csc^2(2x)\,dx$

22 $\displaystyle\int \frac{\csc\sqrt{x}\cot\sqrt{x}}{\sqrt{x}}\,dx$

23 $\displaystyle\int \sin^4 x\cos x\,dx$

24 $\displaystyle\int \frac{\sin x}{\cos^2 x}\,dx$

25 $\displaystyle\int \tan^5 x\sec^2 x\,dx$

26 $\displaystyle\int \frac{\sec^2 x}{\tan x}\,dx$

27 $\displaystyle\int \sec^5 x\tan x\,dx$

28 $\displaystyle\int \frac{\tan x}{\sec^3 x}\,dx$

29 $\displaystyle\int \frac{e^x}{\sqrt{1-e^{2x}}}\,dx$

30 $\displaystyle\int \frac{e^x}{\sqrt{9-e^{2x}}}\,dx$

31 $\displaystyle\int \frac{1}{\sqrt{4x-x^2}}\,dx$

32 $\displaystyle\int \frac{1}{\sqrt{9x-x^2}}\,dx$

33 $\displaystyle\int \frac{1}{x^2+2x+5}\,dx$

34 $\displaystyle\int \frac{1}{x^2-4x+6}\,dx$

35 $\displaystyle\int \frac{x}{x^2-2x+5}\,dx$

36 $\displaystyle\int \frac{x}{x^2+6x+10}\,dx$

## 8.2 INTEGRATION BY PARTS

The integration version of the formula for the derivative of a product is

$$\int [f(x)g'(x)+g(x)f'(x)]\,dx = f(x)g(x)+C.$$

It is convenient to substitute $u=f$ and $v=g$, and to choose $C=0$. We then write the above equation in the form

$$\int u\,dv + \int v\,du = uv \quad \text{or}$$

$$\boldsymbol{\int u\,dv = uv - \int v\,du.}$$

The latter formula is called the **Integration by Parts Formula.**

Integration by parts can be used to change the form of an integral. In many cases it is easier to evaluate $\int v\,du$ than it is to evaluate $\int u\,dv$.

Depending on the interpretation of the variables, the formula for integration by parts can give useful relations between physical quantities.

To use the integration by parts formula, it is necessary to identify the components $u$ and $dv$ in the given integral. Generally speaking, we want to choose $u$ to be a function that has a derivative that is "nicer" than the function $u$. The functions $x^n$, $\ln x$, and the inverse trigonometric functions are good choices for $u$. We generally choose $dv$ such that $v$ is "no worse" then $dv$. Powers of $x$, exponentials, and the sine and cosine functions are good choices for $dv$. Of course, we must recognize a candidate for $v$ before we choose $dv$. There is no point in choosing a $dv$ for which we can't determine a function $v$.

■ **EXAMPLE 1**

Evaluate $\int x \sin x \, dx$.

**SOLUTION**

We first note that a simple substitution does not reduce this integral to one of the types of Section 8.1. We then see that we could easily evaluate the integral, if we could eliminate the factor $x$ in the integrand. This suggests that we try integration by parts with $u = x$, since differentiation of $x$ gives the constant function 1. That is, $u = x$ implies $du = dx$, so $du$ does not contain the factor $x$ that we were trying to eliminate. The choice of $dv = \sin x \, dx$ (this is the only possible choice once we choose $u = x$, since we have $u \, dv = x \sin x \, dx$) gives us $v = -\cos x$, which is about the same as a factor of $\sin x$. We can use any function $v$ that corresponds to the chosen $dv$. We choose the simplest function $v$ that satisfies $dv = \sin x \, dx$. We have

$$u = x, \qquad dv = \sin x \, dx,$$

$$du = dx, \qquad v = -\cos x.$$

Then

$$\begin{aligned} \int x \sin x \, dx &= \int u \, dv \\ &= uv - \int v \, du \\ &= (x)(-\cos x) - \int (-\cos x) \, dx \\ &= -x \cos x + \sin x + C. \end{aligned}$$

The formula for the differentiation of a product is required to check this integral. We have

$$\begin{aligned} \frac{d}{dx}(-x \cos x + \sin x) &= -[(x)(-\sin x) + (\cos x)(1)] + \cos x \\ &= x \sin x. \end{aligned}$$

■

■ **EXAMPLE 2**

Evaluate $\int x^2e^{2x}\,dx$.

**SOLUTION**

We would like to eliminate the factor $x^2$ in this integral. We can at least lower the power of $x$ by one if we integrate by parts with $u = x^2$. Then

$$u = x^2, \qquad dv = e^{2x}\,dx,$$

$$du = 2x\,dx, \qquad v = \frac{1}{2}e^{2x}.$$

[To find a function $v$ that satisfies $dv = e^{2x}\,dx$, we need to evaluate the integral $\int e^{2x}\,dx$. This involves the substitution $u = 2x$, $du = 2\,dx$, which can be carried out mentally. You can easily check that $v = (1/2)e^{2x}$ satisfies $dv = e^{2x}\,dx$ by using the Chain Rule to differentiate $v$ mentally.] Then

$$\begin{aligned}\int x^2e^{2x}\,dx &= \int u\,dv\\ &= uv - \int v\,du\\ &= (x^2)\left(\frac{1}{2}e^{2x}\right) - \int\frac{1}{2}e^{2x}2x\,dx\\ &= \frac{1}{2}x^2e^{2x} - \int e^{2x}x\,dx.\end{aligned}$$

We integrate by parts once again to eliminate the remaining power of $x$. We use

$$u = x, \qquad dv = e^{2x}\,dx,$$

$$du = dx, \qquad v = \frac{1}{2}e^{2x}.$$

We then have

$$\begin{aligned}\int x^2e^{2x}\,dx &= \frac{1}{2}x^2e^{2x} - \left[\int u\,dv\right]\\ &= \frac{1}{2}x^2e^{2x} - \left[uv - \int v\,du\right]\\ &= \frac{1}{2}x^2e^{2x} - \left[(x)\left(\frac{1}{2}e^{2x}\right) - \int\frac{1}{2}e^{2x}\,dx\right]\\ &= \frac{1}{2}x^2e^{2x} - \frac{1}{2}xe^{2x} + \frac{1}{2}\int e^{2x}\,dx\\ &= \frac{1}{2}x^2e^{2x} - \frac{1}{2}xe^{2x} + \frac{1}{4}e^{2x} + C.\end{aligned}$$

■

**EXAMPLE 3**

Evaluate $\int x \ln x \, dx$.

**SOLUTION**

It wouldn't do much good to eliminate $x$ in this example, because it isn't particularly easy to integrate $\ln x$. Let us try $u = \ln x$. Then $du$ is a power of $x$ and powers of $x$ are generally not too bad to work with. Also, $dv = x\,dx$ gives a $v$ that is a power of $x$. We have

$$u = \ln x, \qquad dv = x\,dx,$$

$$du = \frac{1}{x}\,dx, \qquad v = \frac{x^2}{2}.$$

Then

$$\begin{aligned}\int x \ln x \, dx &= \int u\,dv \\ &= uv - \int v\,du \\ &= (\ln x)\left(\frac{x^2}{2}\right) - \int \left(\frac{x^2}{2}\right)\left(\frac{1}{x}\right) dx = (\ln x)\left(\frac{x^2}{2}\right) - \int \frac{x}{2}\,dx \\ &= \frac{1}{2}x^2 \ln x - \frac{1}{4}x^2 + C.\end{aligned}$$

■

**EXAMPLE 4**

Evaluate $\int \sin^{-1}x \, dx$.

**SOLUTION**

The derivative of $\sin^{-1}x$ involves powers of $x$, so let us try to integrate by parts with $u = \sin^{-1}x$. We must then have $dv = dx$. We have

$$u = \sin^{-1}x, \qquad dv = dx,$$

$$du = \frac{1}{\sqrt{1 - x^2}}\,dx, \qquad v = x.$$

Then

$$\begin{aligned}\int \sin^{-1}x \, dx &= \int u\,dv \\ &= uv - \int v\,du \\ &= (\sin^{-1}x)(x) - \int x \frac{1}{\sqrt{1 - x^2}}\,dx.\end{aligned}$$

The integral above on the right is evaluated by using the substitution $u = 1 - x^2$, so $du = -2x\,dx$. We have

$$\begin{aligned}\int \sin^{-1}x\,dx &= x\sin^{-1}x - \int (1-x^2)^{-1/2}\left(-\frac{1}{2}\right)(-2x)\,dx \\ &= x\sin^{-1}x - \left(-\frac{1}{2}\right)\left(\frac{(1-x^2)^{1/2}}{1/2}\right) + C \\ &= x\sin^{-1}x + \sqrt{1-x^2} + C.\end{aligned}$$

■

We have seen that integration by parts can be used to reduce the power of $x$ in certain integrals. Clever application of integration by parts can also be used to reduce the power of other functions. We will need the following **reduction formulas** for powers of the sine and cosine functions.

$$\int \sin^n x\,dx = -\frac{\sin^{n-1}x\cos x}{n} + \frac{n-1}{n}\int \sin^{n-2}x\,dx. \tag{1}$$

$$\int \cos^n x\,dx = \frac{\cos^{n-1}x\sin x}{n} + \frac{n-1}{n}\int \cos^{n-2}x\,dx. \tag{2}$$

Let us see how to derive (2); formula (1) is derived similarly. We begin by writing $\cos^n x$ as a product and integrating by parts. We will then use the identity

$$\sin^2 x + \cos^2 x = 1.$$

We use

$$u = \cos^{n-1}x, \qquad dv = \cos x\,dx,$$

$$du = (n-1)(\cos^{n-2}x)(-\sin x)\,dx, \qquad v = \sin x.$$

Then

$$\begin{aligned}\int \cos^n x\,dx &= \int \cos^{n-1}x\cos x\,dx \\ &= \int u\,dv \\ &= uv - \int v\,du \\ &= (\cos^{n-1}x)(\sin x) - \int (\sin x)(n-1)(\cos^{n-2}x)(-\sin x)\,dx \\ &= \cos^{n-1}x\sin x + (n-1)\int \cos^{n-2}x\sin^2 x\,dx \\ &= \cos^{n-1}x\sin x + (n-1)\int \cos^{n-2}x(1-\cos^2 x)\,dx \\ &= \cos^{n-1}x\sin x + (n-1)\int \cos^{n-2}x\,dx - (n-1)\int \cos^n x\,dx.\end{aligned}$$

We now have

$$\int \cos^n x\, dx = \cos^{n-1}x \sin x + (n-1)\int \cos^{n-2}x\, dx - (n-1)\int \cos^n x\, dx.$$

This implies

$$n\int \cos^n x\, dx = \cos^{n-1}x \sin x + (n-1)\int \cos^{n-2}x\, dx,$$

so

$$\int \cos^n x\, dx = \frac{\cos^{n-1}x \sin x}{n} + \frac{n-1}{n}\int \cos^{n-2}x\, dx.$$

The following formulas can be derived by setting $n = 2$ in formulas (1) and (2).

$$\int \sin^2 x\, dx = -\frac{\sin x \cos x}{2} + \frac{x}{2} + C. \tag{3}$$

$$\int \cos^2 x\, dx = \frac{\sin x \cos x}{2} + \frac{x}{2} + C. \tag{4}$$

For example, when $n = 2$ in formula (2), we have

$$\int \cos^2 x\, dx = \frac{\sin x \cos^{2-1}x}{2} + \frac{2-1}{2}\int \cos^{2-2}x\, dx$$

$[\cos^{2-2}x = (\cos x)^0 = 1]$

$$= \frac{\sin x \cos x}{2} + \frac{1}{2}\int 1\, dx$$

$$= \frac{\sin x \cos x}{2} + \frac{1}{2}x + C.$$

It is also possible to evaluate $\int \cos^2 x\, dx$ by using the trigonometric identity

$$\cos^2 x = \frac{1 + \cos 2x}{2}.$$

That is,

$$\int \cos^2 x\, dx = \int \frac{1 + \cos 2x}{2}\, dx = \frac{x}{2} + \left(\frac{1}{2}\right)\left(\frac{\sin 2x}{2}\right) + C.$$

The identity $\sin 2x = 2 \sin x \cos x$ can then be used to obtain the same expression obtained by using formula (2). Similarly, we can evaluate $\int \sin^2 x\, dx$ by using either formula (1) or the identity

$$\sin^2 x = \frac{1 - \cos 2x}{2}.$$

In any case, we now consider (3) and (4) to be known formulas to be used in future applications.

**■ EXAMPLE 5**

Evaluate $\int \sin^4 x\, dx$.

**SOLUTION**

We use the appropriate formulas.

$$\begin{aligned}\int \sin^4 x\, dx &\underset{(1)}{=} -\frac{\sin^{4-1}x \cos x}{4} + \frac{4-1}{4}\int \sin^{4-2}x\, dx \\ &= -\frac{\sin^3 x \cos x}{4} + \frac{3}{4}\int \sin^2 x\, dx \\ &\underset{(3)}{=} -\frac{\sin^3 x \cos x}{4} + \frac{3}{4}\left(-\frac{\sin x \cos x}{2} + \frac{x}{2}\right) + C \\ &= -\frac{\sin^3 x \cos x}{4} - \frac{3 \sin x \cos x}{8} + \frac{3x}{8} + C. \end{aligned}$$ ■

The following formulas are useful for evaluating some integrals that will occur in later problems.

$$\int e^{ax} \cos bx\, dx = \frac{e^{ax}}{a^2 + b^2}(b \sin bx + a \cos bx) + C. \tag{5}$$

$$\int e^{ax} \sin bx\, dx = \frac{e^{ax}}{a^2 + b^2}(a \sin bx - b \cos bx) + C. \tag{6}$$

Let us derive (5) in the special case $a = b = 1$. The technique involves integrating by parts twice. The method works because integrating (or differentiating) $\cos x$ twice gives us a negative multiple of $\cos x$, while differentiating (or integrating) $e^x$ twice gives us a positive multiple of $e^x$. The result is that the original integral appears on the opposite side of the equation, multiplied by a factor that is not one. We have

$$u = e^x, \qquad dv = \cos x\, dx,$$

$$du = e^x\, dx, \qquad v = \sin x.$$

Then

$$\begin{aligned}\int e^x \cos x\, dx &= \int u\, dv \\ &= uv - \int v\, du \\ &= e^x \sin x - \int e^x \sin x\, dx.\end{aligned}$$

We now use

$$u = e^x, \qquad dv = \sin x\, dx,$$

$$du = e^x\, dx, \qquad v = -\cos x.$$

We then have

$$\int e^x \cos x\, dx = e^x \sin x - \left(\int u\, dv\right)$$

$$= e^x \sin x - \left(uv - \int v\, du\right)$$

$$= e^x \sin x - \left(e^x(-\cos x) - \int e^x(-\cos x)\, dx\right)$$

$$= e^x \sin x + e^x \cos x - \int e^x \cos x\, dx.$$

We now have

$$\int e^x \cos x\, dx = e^x \sin x + e^x \cos x - \int e^x \cos x\, dx.$$

It follows that

$$2\int e^x \cos x\, dx = e^x \sin x + e^x \cos x, \quad \text{so}$$

$$\int e^x \cos x\, dx = \frac{e^x}{2}(\sin x + \cos x) + C.$$

(We can add the constant $C$ to the right side of the above equation, because the constant has derivative zero.)

It doesn't matter which functions were chosen for $u$ and $dv$ in the first integration by parts above. However, we must be very careful to choose $u$ and $dv$ in the second integration by parts so that one function is differentiated twice while the other function is integrated twice. We would simply undo the previous integration by parts if we integrated the function we just obtained by differentiation and differentiated the function we just obtained by integration.

## EXERCISES 8.2

Use integration by parts to evaluate the integrals in Exercises 1–16.

1 $\int xe^{2x}\, dx$

2 $\int xe^{-x}\, dx$

3 $\int x \sin(x/2)\, dx$

4 $\int x \cos 2x\, dx$

5 $\int x \dfrac{2x}{(1 - x^2)^{3/2}}\, dx$

6 $\int x^2 \dfrac{2x}{(x^2 + 1)^2}\, dx$

7 $\int \sqrt{x}\,\frac{\sin\sqrt{x}}{\sqrt{x}}\,dx$ 8 $\int x^2\cos(x/2)\,dx$

9 $\int x^3e^{-x}\,dx$ 10 $\int x^3\sin x\,dx$

11 $\int \sqrt{x}\ln x\,dx$ 12 $\int x^3\ln x\,dx$

13 $\int \tan^{-1}x\,dx$ 14 $\int \ln x^2\,dx$

15 $\int x\tan^{-1}x\,dx$ 16 $\int x\sec^{-1}|x|\,dx$

Use the technique that was used to verify the special case of formula (5) to evaluate the integrals in Exercises 17–20.

17 $\int e^{2x}\cos x\,dx$ 18 $\int e^x\sin x\,dx$

19 $\int \sin(\ln x)\,dx$ 20 $\int \cos(\ln x)\,dx$

21 Use integration by parts to derive the formula

$$\int e^{ax}\cos bx\,dx = \frac{e^{ax}}{a^2+b^2}(b\sin bx + a\cos bx) + C.$$

22 Use integration by parts to derive the formula

$$\int \sin^n x\,dx = -\frac{\sin^{n-1}x\cos x}{n} + \frac{n-1}{n}\int \sin^{n-2}x\,dx.$$

23 Use integration by parts to derive the formula

$$\int x^n\ln x\,dx = \frac{x^{n+1}\ln x}{n+1} - \frac{x^{n+1}}{(n+1)^2} + C.$$

24 Use integration by parts to derive the formula

$$\int (\ln x)^n\,dx = x(\ln x)^n - n\int (\ln x)^{n-1}\,dx.$$

Use formulas (1)–(6) and the results of Exercises 23 and 24 to evaluate the integrals in Exercises 25–36.

25 $\int \sin^2 x\,dx$ 26 $\int \cos^4 x\,dx$

27 $\int \cos^3 2x\,dx$ 28 $\int \sin^3(x/2)\,dx$

29 $\int e^{-x}\sin(x/2)\,dx$ 30 $\int e^{2x}\sin 3x\,dx$

31 $\int e^{-2x}\cos 5x\,dx$ 32 $\int e^{-2x}\cos 3x\,dx$

33 $\int (\ln x)^2\,dx$ 34 $\int (\ln x)^3\,dx$

35 $\int x^3\ln x\,dx$ 36 $\int x^2\ln x\,dx$

37 Find the volume of the solid obtained by revolving the region bounded by $y = \sin x$, $y = 0$, $0 \le x \le \pi$, about the $x$-axis.

38 Find the volume of the solid obtained by revolving the region bounded by $y = \cos x$, $y = 0$, $0 \le x \le \pi/2$, about the $y$-axis.

## 8.3 TRIGONOMETRIC INTEGRALS

In the previous section we saw how reduction formulas can be used to evaluate powers of the sine and cosine functions. In this section we will develop additional reduction formulas. We will also see that some trigonometric integrals can be evaluated by a simplifying substitution or by using trigonometric identities.

Generally, we will say an integral is *easy* to integrate if a simplifying substitution reduces it to one of the fundamental formulas that we know. Integrals of the form $\int \sin^m x\cos x\,dx$ are *easy* to evaluate by using the simplifying substitution $u = \sin x$, so $du = \cos x\,dx$. Similarly, $\int \cos^n x\sin x\,dx$ is *easy* to integrate if we use the substitution $u = \cos x$, so $du = -\sin x\,dx$. If either $m$ or $n$ is an odd integer, $\int \sin^m x\cos^n x\,dx$ can be written as a sum of integrals of the above easy types. To do this, save one of the odd factors of either sin $x$ or cos $x$ and use $\sin^2 x + \cos^2 x = 1$ to replace extra even powers.

■ **EXAMPLE 1**

Evaluate $\int \sin^3 x \cos x \, dx$.

**SOLUTION**

This is of the *easy* type, since the integrand involves powers of sin $x$ and contains the essential factor cos $x$. Setting $u = \sin x$, so $du = \cos x \, dx$, we have

$$\int \sin^3 x \cos x \, dx = \int u^3 \, du$$

$$= \frac{u^4}{4} + C$$

$$= \frac{\sin^4 x}{4} + C. \quad ■$$

■ **EXAMPLE 2**

Evaluate $\int \sin^3 x \cos^3 x \, dx$.

**SOLUTION**

The power of cos $x$ in the integrand is an odd integer. We then save one factor of cos $x$ and use $\sin^2 x + \cos^2 x = 1$ to express the remaining even power of cos $x$ in terms of sin $x$. We can then use the substitution $u = \sin x$, $du = \cos x \, dx$ to write the integral as a sum of two integrals of easy type. We substitute before writing the integral as a sum.

$$\int \sin^3 x \cos^3 x \, dx = \int \sin^3 x \cos^2 x \cos x \, dx$$

$$= \int \sin^3 x (1 - \sin^2 x) \cos x \, dx$$

$$= \int u^3 (1 - u^2) \, du$$

$$= \int (u^3 - u^5) \, du$$

$$= \frac{u^4}{4} - \frac{u^6}{6} + C$$

$$= \frac{\sin^4 x}{4} - \frac{\sin^6 x}{6} + C.$$

We could also evaluate $\int \sin^3x \cos^3x \, dx$ by saving one factor of $\sin x$ and using the substitution $u = \cos x$, $du = -\sin x \, dx$. This gives

$$\begin{aligned}
\int \sin^3x \cos^3x \, dx &= \int \sin^2x \cos^3x \sin x \, dx \\
&= \int (1 - \cos^2x) \cos^3x(-1)(-\sin x) \, dx \\
&= \int (1 - u^2)u^3(-1) \, du \\
&= \int (u^5 - u^3) \, du \\
&= \frac{u^6}{6} - \frac{u^4}{4} + C \\
&= \frac{\cos^6x}{4} - \frac{\cos^4x}{4} + C.
\end{aligned}$$

The two answers are different, but

$$\begin{aligned}
&\left(\frac{\sin^4x}{4} - \frac{\sin^6x}{6}\right) - \left(\frac{\cos^6x}{6} - \frac{\cos^4x}{4}\right) \\
&\quad = \frac{1}{4}\sin^4x - \frac{1}{6}\sin^6x - \frac{1}{6}(1 - \sin^2x)^3 + \frac{1}{4}(1 - \sin^2x)^2 \\
&\quad = \frac{1}{4}\sin^4x - \frac{1}{6}\sin^6x - \frac{1}{6}(1 - 3\sin^2x + 3\sin^4x - \sin^6x) \\
&\qquad + \frac{1}{4}(1 - 2\sin^2x + \sin^4x) = \frac{1}{12}.
\end{aligned}$$

The two answers differ by a constant, so they have the same derivative, namely, $\sin^3x \cos^3x$. ■

If $m$ and $n$ are both even integers, we cannot write $\int \sin^m x \cos^n x \, dx$ as a sum of the easy types. In that case, we use the following formulas to reduce the powers.

$$\int \sin^n x \, dx = -\frac{\sin^{n-1}x \cos x}{n} + \frac{n-1}{n}\int \sin^{n-2}x \, dx. \tag{1}$$

$$\int \cos^n x \, dx = \frac{\cos^{n-1}x \sin x}{n} + \frac{n-1}{n}\int \cos^{n-2}x \, dx. \tag{2}$$

$$\int \sin^2x \, dx = -\frac{\sin x \cos x}{2} + \frac{x}{2} + C. \tag{3}$$

$$\int \cos^2x \, dx = \frac{\sin x \cos x}{2} + \frac{x}{2} + C. \tag{4}$$

$$\int \sin^m x \cos^n x \, dx = -\frac{\sin^{m-1}x \cos^{n+1}x}{m+n} + \frac{m-1}{m+n}\int \sin^{m-2}x \cos^n x \, dx. \tag{5}$$

$$\int \sin^m x \cos^n x \, dx = \frac{\sin^{m+1}x \cos^{n-1}x}{m+n} + \frac{n-1}{m+n}\int \sin^m x \cos^{n-2}x \, dx. \tag{6}$$

Formulas (1)–(4) were derived in Section 8.2 by using integration by parts. Formulas (5) and (6) can be derived by using the same method. The formulas can be checked by differentiation.

■ **EXAMPLE 3**

Evaluate $\int \sin^2 x \cos^2 x \, dx$.

**SOLUTION**

This integral is not one of the easy types, so we will use the reduction formulas.

$$\begin{aligned}\int \sin^2 x \cos^2 x \, dx &\underset{(5)}{=} -\frac{\sin x \cos^3 x}{4} + \frac{1}{4}\int \cos^2 x \, dx \\ &\underset{(4)}{=} -\frac{\sin x \cos^3 x}{4} + \frac{1}{4}\left(\frac{\sin x \cos x}{2} + \frac{x}{2}\right) + C \\ &= -\frac{\sin x \cos^3 x}{4} + \frac{\sin x \cos x}{8} + \frac{x}{8} + C.\end{aligned}$$

■

We could also evaluate the integral in Example 3 by applying (6) to reduce the power of $\cos x$, and then using (3). In that case we would obtain an answer that contains different powers of $\sin x$ and $\cos x$. Recall that two answers can both be correct if they differ by a constant.

Integrals of even powers of $\sin x$ and $\cos x$ can also be evaluated by using the trigonometric formulas

$$\sin^2 x = \frac{1 - \cos 2x}{2} \quad \text{and} \quad \cos^2 x = \frac{1 + \cos 2x}{2}$$

to reduce the powers. This method results in answers that involve the sine and cosine of multiples of $x$, unless you use additional trigonometric formulas to switch back to $\sin x$ and $\cos x$. This is not convenient for the work we will be doing in Section 8.4, so we will systematically use the reduction formulas.

In some cases we have a choice of writing an integral as a sum of integrals of easy type or using a reduction formula. For example, we can write $\int \cos^3 x \, dx = \int (1 - \sin^2 x) \cos x \, dx$ or we can use (2). Which to use is a matter of personal preference.

Integrals that involve powers of $\sec x$ and $\tan x$ can be evaluated by using the same general principles that we used for powers of $\sin x$ and $\cos x$. Namely:

**Look for an easy substitution.**

If you don't see an easy substitution:

**Use a reduction formula.**

Integrals of the form $\int \tan^n x \sec^2 x \, dx$ are *easy* to evaluate. We use $u = \tan x$, so $du = \sec^2 x \, dx$. Similarly, $\int \sec^{m-1} x \sec x \tan x \, dx$ is *easy* to evaluate. Use $u = \sec x$, so $du = \sec x \tan x\, dx$. If either $m$ is even or $n$ is odd, we can write $\int \sec^m x \tan^n x\, dx$ as sum of integrals of easy type. To do this we use $\tan^2 x + 1 = \sec^2 x$ to replace extra even powers.

## ■ EXAMPLE 4

Evaluate $\int \sec^7 x \tan x\, dx$.

### SOLUTION

The integrand can be written as a power of $\sec x$ times $\sec x \tan x$. We can then use the substitution $u = \sec x$, $du = \sec x \tan x\, dx$ to write the integral in the form of an easy type.

$$\begin{aligned}\int \sec^7 x \tan x\, dx &= \int \sec^6 x \sec x \tan x\, dx \\ &= \int u^6\, du \\ &= \frac{u^7}{7} + C \\ &= \frac{1}{7}\sec^7 x + C.\end{aligned}$$ ■

## ■ EXAMPLE 5

Evaluate $\int \tan^2 x \sec^4 x\, dx$.

### SOLUTION

We save $\sec^2 x$ and use $\sec^2 x = \tan^2 x + 1$ to express the remaining even power of $\sec x$ in terms of $\tan x$. The substitution $u = \tan x$, $du = \sec^2 x\, dx$ then gives

$$\begin{aligned}\int \tan^2 x \sec^4 x\, dx &= \int \tan^2 x\, (\sec^2 x) \sec^2 x\, dx \\ &= \int \tan^2 x\, (\tan^2 x + 1) \sec^2 x\, dx \\ &= \int u^2(u^2 + 1)\, du \\ &= \int (u^4 + u^2)\, du \\ &= \frac{u^5}{5} + \frac{u^3}{3} + C \\ &= \frac{1}{5}\tan^5 x + \frac{1}{3}\tan^3 x + C.\end{aligned}$$ ■

If $m$ is odd and $n$ is even, we cannot write $\int \sec^m x \tan^n x\, dx$ as a sum of integrals of easy type. In this case we need the following results:

$$\int \tan x\, dx = \ln|\sec x| + C. \tag{7}$$

$$\int \sec x\, dx = \ln|\sec x + \tan x| + C. \tag{8}$$

Formula (7) was derived in Example 4 of Section 7.5 by writing $\tan x = \sin x/\cos x$ and then using the substitution $u = \cos x$. Formula (8) may be verified by differentiation. It can be derived by a clever trick:

$$\int \sec x\, dx = \int \sec x \frac{\tan x + \sec x}{\tan x + \sec x}\, dx = \int \frac{\sec x \tan x + \sec^2 x}{\sec x + \tan x}\, dx.$$

The substitution $u = \sec x + \tan x$ then gives (8).

Higher powers of $\sec x$ can be reduced by using

$$\int \sec^m x\, dx = \frac{1}{m-1} \sec^{m-2} x \tan x + \frac{m-2}{m-1} \int \sec^{m-2} x\, dx. \tag{9}$$

Formula (9) can be derived by using a method similar to that used to derive formula (2) of Section 8.2. That is, we write $\sec^m x = \sec^{m-2} x \sec^2 x$, integrate by parts with $u = \sec^{m-2} x$ and $dv = \sec^2 x\, dx$, and use the trigonometric formula $\tan^2 x + 1 = \sec^2 x$. Let us derive (9) in the special case $m = 3$. We have

$$u = \sec x, \qquad dv = \sec^2 x\, dx,$$

$$du = \sec x \tan x\, dx, \qquad v = \tan x.$$

Then

$$\begin{aligned}
\int \sec^3 x\, dx &= \int \sec x \sec^2 x\, dx \\
&= \int u\, dv \\
&= uv - \int v\, du \\
&= (\sec x)(\tan x) - \int (\tan x)(\sec x \tan x)\, dx \\
&= \sec x \tan x - \int \sec x \tan^2 x\, dx \\
&= \sec x \tan x - \int \sec x (\sec^2 x - 1)\, dx \\
&= \sec x \tan x - \int \sec^3 x\, dx + \int \sec x\, dx.
\end{aligned}$$

We now have

$$\int \sec^3 x\, dx = \sec x \tan x - \int \sec^3 x\, dx + \int \sec x\, dx, \quad \text{so}$$

$$2\int \sec^3 x\, dx = \sec x \tan x + \int \sec x\, dx,$$

$$\int \sec^3 x\, dx = \frac{1}{2}\sec x \tan x + \frac{1}{2}\int \sec x\, dx.$$

This agrees with (9) when $m = 3$.

The following formula can be used to reduce powers of $\tan x$:

$$\int \tan^n x\, dx = \frac{1}{n-1}\tan^{n-1} x - \int \tan^{n-2} x\, dx. \tag{10}$$

Formula (10) is not difficult to derive. We have

$$\begin{aligned}\int \tan^n x\, dx &= \int \tan^{n-2} x \tan^2 x\, dx \\ &= \int \tan^{n-2} x\,(\sec^2 x - 1)\, dx \\ &= \int \tan^{n-2} x \sec^2 x\, dx - \int \tan^{n-2} x\, dx.\end{aligned}$$

The first integral above can be evaluated by using the substitution $u = \tan x$, so $du = \sec^2 x\, dx$. We then have

$$\begin{aligned}\int \tan^n x\, dx &= \int u^{n-2}\, du - \int \tan^{n-2} x\, dx \\ &= \frac{u^{n-1}}{n-1} - \int \tan^{n-2} x\, dx, \qquad n \neq 1 \\ &= \frac{1}{n-1}\tan^{n-1} x - \int \tan^{n-2} x\, dx.\end{aligned}$$

If $m$ is odd and $n$ is even (not one of the easy cases), we can use $\tan^2 x + 1 = \sec^2 x$ to express $\sec^m x \tan^n x$ as a sum of powers of $\sec x$, and then reduce the power by using (9). Continue reducing the powers until the integrand is $\sec x$, then use (8).

**■ EXAMPLE 6**

Evaluate $\int \tan^2 x \sec^3 x\, dx$.

**SOLUTION**

This cannot be written as an easy type, so we will use a reduction formula. We first use $\tan^2 x + 1 = \sec^2 x$ to express the integral as a sum of integrals of powers of sec $x$. We will then use a reduction formula to reduce the integral with the *highest* power of sec $x$. We will then combine integrals of like powers of sec $x$ before using another reduction formula.

$$\int \tan^2 x \sec^3 x\,dx = \int (\sec^2 x - 1)\sec^3 x\,dx$$

$$= \left[\int \sec^5 x\,dx\right] - \int \sec^3 x\,dx$$

$$\underset{(9)}{=} \left[\frac{1}{4}\sec^3 x \tan x + \frac{3}{4}\int \sec^3 x\,dx\right] - \int \sec^3 x\,dx$$

$$= \frac{1}{4}\sec^3 x \tan x - \frac{1}{4}\left[\int \sec^3 x\,dx\right]$$

$$\underset{(9)}{=} \frac{1}{4}\sec^3 x \tan x - \frac{1}{4}\left[\frac{1}{2}\sec x \tan x + \frac{1}{2}\int \sec x\,dx\right]$$

$$= \frac{1}{4}\sec^3 x \tan x - \frac{1}{8}\sec x \tan x - \frac{1}{8}\left[\int \sec x\,dx\right]$$

$$\underset{(8)}{=} \frac{1}{4}\sec^3 x \tan x - \frac{1}{8}\sec x \tan x$$

$$- \frac{1}{8}\ln|\sec x + \tan x| + C. \qquad \blacksquare$$

The integrals that we have discussed so far in this section involve many cases of odd and even powers. Rather than trying to blindly memorize all cases, it is recommended that you be familiar with the derivatives and always *look for an easy substitution.* If you can't find an easy substitution, then try a *reduction formula.*

The following trigonometric identities can be used either to evaluate integrals or to derive integration formulas.

$$\sin mx \sin nx = \frac{1}{2}\{\cos[(m-n)x] - \cos[(m+n)x]\},$$

$$\sin mx \cos nx = \frac{1}{2}\{\sin[(m+n)x] + \sin[(m-n)x]\},$$

$$\cos mx \cos nx = \frac{1}{2}\{\cos[(m+n)x] + \cos[(m-n)x]\}.$$

### ■ EXAMPLE 7

Evaluate $\int \sin 2x \sin x\,dx$.

**SOLUTION**

We use the formula for $\sin mx \sin nx$ with $m = 2$ and $n = 1$.

$$\begin{aligned}\int \sin 2x \sin x\,dx &= \int \frac{1}{2}(\cos x - \cos 3x)\,dx \\ &= \frac{1}{2}\left(\sin x - \frac{1}{3}\sin 3x\right) + C \\ &= \frac{1}{2}\sin x - \frac{1}{6}\sin 3x + C.\end{aligned}$$

■

## EXERCISES 8.3

Evaluate the integrals in Exercises 1–30.

**1** $\int \cos^2 x \sin x\,dx$

**2** $\int \sin x \cos x\,dx$

**3** $\int \tan x \sec^2 x\,dx$

**4** $\int \tan^2 x \sec^2 x\,dx$

**5** $\int \sec^5 x \tan x\,dx$

**6** $\int \sec^3 x \tan^3 x\,dx$

**7** $\int \sin^2 2x\,dx$

**8** $\int \tan^2(x/2)\,dx$

**9** $\int \cos^3 x\,dx$

**10** $\int \sec^4 x \tan x\,dx$

**11** $\int \sec^3 2x\,dx$

**12** $\int \cos^4(x/3)\,dx$

**13** $\int \tan^6 x \sec^4 x\,dx$

**14** $\int \sec^6 x\,dx$

**15** $\int \sin^4 x \cos^3 x\,dx$

**16** $\int \cos^2 x \sin^5 x\,dx$

**17** $\int \frac{\sin^3 x}{\cos^2 x}\,dx$

**18** $\int \sec^4 x\,dx$

**19** $\int \cos^4 x \sin^2 x\,dx$

**20** $\int \sec^5 x\,dx$

**21** $\int \frac{1}{\sec^2 x}\,dx$

**22** $\int \frac{1}{\sec^4 x}\,dx$

**23** $\int \tan^3 x\,dx$

**24** $\int \cos^5 x\,dx$

**25** $\int \sin^3 x \cos^2 x\,dx$

**26** $\int \sin^6 x \cos^4 x\,dx$

**27** $\int \sin 3x \sin 2x\,dx$

**28** $\int \sin 5x \cos 3x\,dx$

**29** $\int \cos 4x \sin x\,dx$

**30** $\int \cos 5x \cos 2x\,dx$

**31** If $m$ and $n$ are positive integers, show that

$$\int_0^{2\pi} \sin mx \sin nx\,dx = \begin{cases} \pi, & m = n, \\ 0, & m \neq n. \end{cases}$$

**32** If $m$ and $n$ are positive integers, show that

$$\int_0^{2\pi} \sin mx \cos nx\,dx = 0.$$

## 8.4 TRIGONOMETRIC SUBSTITUTIONS

We will study some substitutions that can be used to transform an integral that contains the square root of a quadratic expression into an integral of trigonometric functions. The methods of Section 8.3 can then be used to evaluate the integral.

The following rules are listed for reference. We will discuss them briefly and then illustrate their use by example.

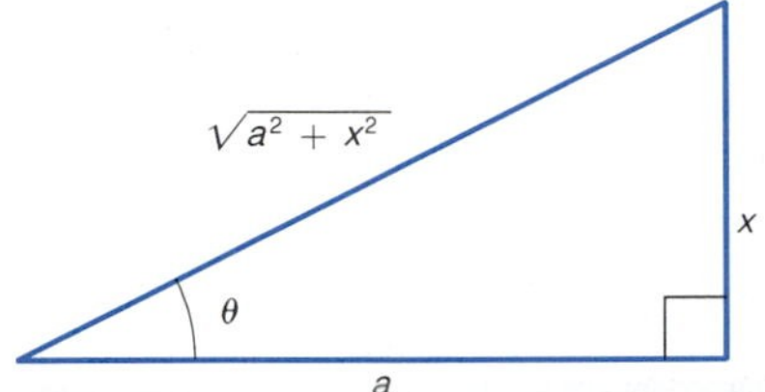


FIGURE 8.4.1

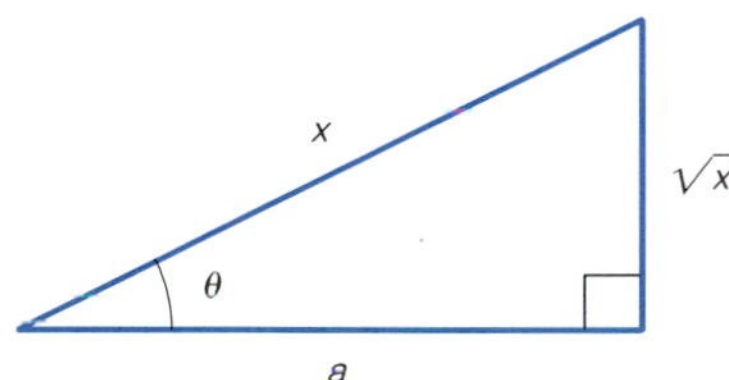


FIGURE 8.4.2

FIGURE 8.4.3

When the integrand contains the factor $\sqrt{a^2 - x^2}$, let $\boldsymbol{x = a \sin \theta}$, where $a > 0$, $-a \leq x \leq a$, and $-\pi/2 \leq \theta \leq \pi/2$. Then $\boldsymbol{dx = a \cos \theta \, d\theta}$, $\boldsymbol{\sqrt{a^2 - x^2} = a \cos \theta}$, and $\boldsymbol{\theta = \sin^{-1}(x/a)}$. See Figure 8.4.1.

When the integrand contains the factor $\sqrt{a^2 + x^2}$, let $\boldsymbol{x = a \tan \theta}$, where $a > 0$, $-\infty < x < \infty$, and $-\pi/2 < \theta < \pi/2$. Then $\boldsymbol{dx = a \sec^2\theta \, d\theta}$, $\boldsymbol{\sqrt{a^2 + x^2} = a \sec \theta}$, and $\boldsymbol{\theta = \tan^{-1}(x/a)}$. See Figure 8.4.2.

When the integrand contains the factor $\sqrt{x^2 - a^2}$, let $\boldsymbol{x = a \sec \theta}$, where $a > 0$, $x \geq a$, and $0 \leq \theta \leq \pi/2$. Then $\boldsymbol{dx = a \sec \theta \tan \theta \, d\theta}$, $\boldsymbol{\sqrt{x^2 - a^2} = a \tan \theta}$, and $\boldsymbol{\theta = \sec^{-1}(x/a)}$. See Figure 8.4.3.

Our usual form of substitution has been to write the new variable as a function of the variable of integration. For the above substitutions, it is more convenient to express the variable of integration in terms of the new variable. For example, we write $x = a \sin \theta$ instead of $\theta = \sin^{-1}(x/a)$. This allows us to easily obtain a formula for $dx$ in terms of trigonometric functions of $\theta$. The formulas for $x$ and $dx$ are then substituted to obtain a trigonometric integral.

The trigonometric substitutions are used to simplify quadratic expressions. The choice of which substitution to use depends on the form of the quadratic expression that we want to reduce.

If $x = a \sin \theta$ ($a > 0$ and $-\pi/2 \leq \theta \leq \pi/2$), then

$$\sqrt{a^2 - x^2} = \sqrt{a^2 - a^2 \sin^2\theta} = \sqrt{a^2 \cos^2\theta} = a \cos \theta,$$

so the substitution $x = a \sin \theta$ simplifies the expression $\sqrt{a^2 - x^2}$.

If $x = a \tan \theta$ ($a > 0$ and $-\pi/2 < \theta < \pi/2$), then

$$\sqrt{a^2 + x^2} = \sqrt{a^2 + a^2 \tan^2\theta} = \sqrt{a^2 \sec^2\theta} = a \sec \theta,$$

so the substitution $x = a \tan \theta$ simplifies the expression $\sqrt{a^2 + x^2}$.

If $x = a \sec \theta$ ($a > 0$ and $0 \leq \theta \leq \pi/2$), then

$$\sqrt{x^2 - a^2} = \sqrt{a^2 \sec^2\theta - a^2} = \sqrt{a^2 \tan^2\theta} = a \tan \theta,$$

so the substitution $x = a \tan \theta$ simplifies the expression $\sqrt{a^2 + x^2}$.

We will use the following procedure:

- **Choose the trigonometric substitution that simplifies the quadratic expression in the integrand.**
- **Evaluate the differential $dx$ and simplify the quadratic expression that is in the integrand.**
- **Substitute.**
- **Evaluate the trigonometric integral.**
- **Use a sketch of a representative right triangle with angle $\theta$ to transform from functions of $\theta$ back to functions of $x$.**

**■ EXAMPLE 1**

Evaluate

$$\int \frac{1}{(1-x^2)^{3/2}}\,dx.$$

**SOLUTION**

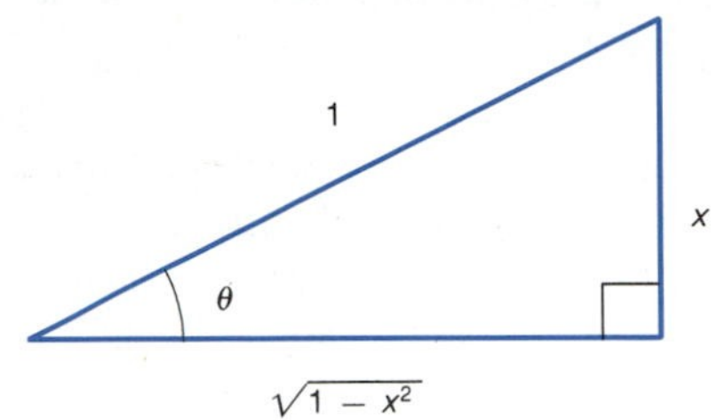

FIGURE 8.4.4

The integrand contains the factor $(1-x^2)^{3/2} = (\sqrt{1-x^2})^3$, so we use the substitution $x = \sin\theta$. Then $dx = \cos\theta\,d\theta$, $\sqrt{1-x^2} = \cos\theta$, and $(1-x^2)^{3/2} = (\cos\theta)^3$. Substitution then gives

$$\int \frac{1}{(1-x^2)^{3/2}}\,dx = \int \frac{1}{(\cos\theta)^3}\cos\theta\,d\theta = \int \sec^2\theta\,d\theta = \tan\theta + C.$$

A representative right triangle with angle $\theta$ that satisfies $\sin\theta = x$ is sketched in Figure 8.4.4. The sketch can be used to express $\tan\theta$ in terms of $x$. We obtain

$$\int \frac{1}{(1-x^2)^{3/2}}\,dx = \frac{x}{\sqrt{1-x^2}} + C.$$ ■

**■ EXAMPLE 2**

Evaluate

$$\int \frac{1}{\sqrt{4+x^2}}\,dx.$$

**SOLUTION**

The integrand contains the factor $\sqrt{4+x^2}$, so we use the substitution $x = 2\tan\theta$. Then $dx = 2\sec^2\theta\,d\theta$ and $\sqrt{4+x^2} = 2\sec\theta$. Substitution yields

$$\int \frac{1}{\sqrt{4+x^2}}\,dx = \int \frac{1}{2\sec\theta}\,2\sec^2\theta\,d\theta$$

$$= \int \sec\theta\,d\theta$$

[From formula (8) of Section 8.3]

$$= \ln|\sec\theta + \tan\theta| + C$$

[From Figure 8.4.5]

$$= \ln\left(\frac{\sqrt{4+x^2}}{2} + \frac{x}{2}\right) + C.$$

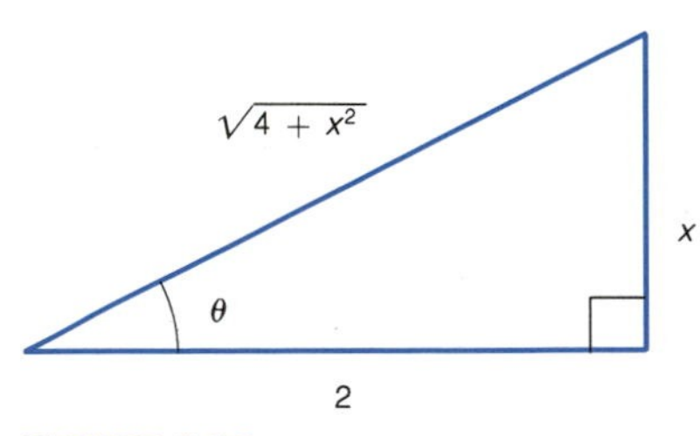

FIGURE 8.4.5

The answer can also be expressed as $\ln(\sqrt{4+x^2} + x) + C$, since

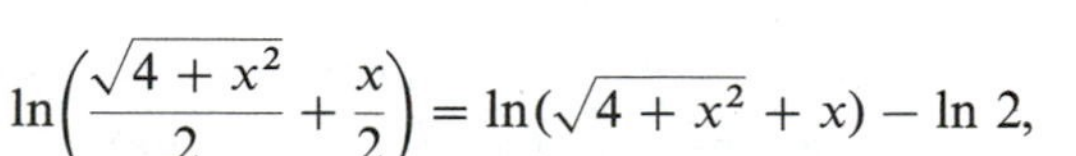

$$\ln\left(\frac{\sqrt{4+x^2}}{2} + \frac{x}{2}\right) = \ln(\sqrt{4+x^2} + x) - \ln 2,$$

so the two answers differ by the constant ln 2. ■

■ **EXAMPLE 3**

Evaluate $\int \frac{1}{(9+x^2)^2}\,dx$.

**SOLUTION**

The integrand contains the factor $(9+x^2)^2 = (\sqrt{9+x^2})^4$, so we use the substitution $x = 3\tan\theta$. Then $dx = 3\sec^2\theta\,d\theta$, $\sqrt{9+x^2} = 3\sec\theta$, and $(9+x^2)^2 = (3\sec\theta)^4$. We substitute to obtain

$$\begin{aligned}
\int \frac{1}{(9+x^2)^2}\,dx &= \int \frac{1}{(3\sec\theta)^4}\,3\sec^2\theta\,d\theta \\
&= \int \frac{1}{27\sec^2\theta}\,d\theta \\
&= \frac{1}{27}\int \cos^2\theta\,d\theta \\
&\qquad\text{[From formula (4) of Section 8.3]} \\
&= \frac{1}{27}\left(\frac{\sin\theta\cos\theta}{2} + \frac{\theta}{2}\right) + C \\
&\qquad\text{[From Figure 8.4.6]} \\
&= \frac{1}{54}\left(\frac{x}{\sqrt{9+x^2}}\right)\left(\frac{3}{\sqrt{9+x^2}}\right) + \frac{1}{54}\tan^{-1}\frac{x}{3} + C \\
&= \frac{x}{18(9+x^2)} + \frac{1}{54}\tan^{-1}\frac{x}{3} + C. \quad ■
\end{aligned}$$

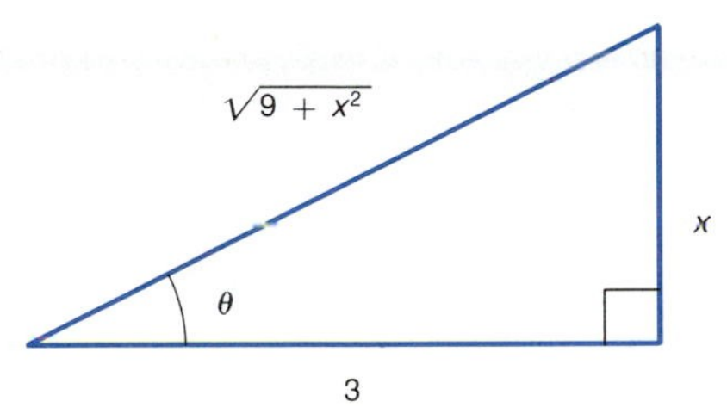

FIGURE 8.4.6

If the quadratic expression contains a linear term $x$, we must complete a square before making a trigonometric substitution.

■ **EXAMPLE 4**

Evaluate $\int \sqrt{x^2 - 2x}\,dx$.

**SOLUTION**

Completing the square, we have

$$x^2 - 2x = x^2 - 2x + [1] - [1] = (x-1)^2 - 1.$$

This suggests that we set $u = x - 1$ and then use the substitution $u = \sec\theta$. Rather than introduce the variable $u = x - 1$, we simply set $x - 1 = \sec\theta$. Then $dx = \sec\theta\tan\theta\,d\theta$ and $\sqrt{(x-1)^2 - 1} = \tan\theta$. We have

$$\begin{aligned}
\int \sqrt{x^2 - 2x}\,dx &= \int \sqrt{(x-1)^2 - 1}\,dx \\
&= \int \tan\theta\sec\theta\tan\theta\,d\theta \\
&= \int \tan^2\theta\sec\theta\,d\theta
\end{aligned}$$

$$= \int (\sec^2\theta - 1) \sec\theta \, d\theta$$

$$= \int \sec^3\theta \, d\theta - \int \sec\theta \, d\theta$$

[Formula (9) of Section 8.3]

$$= \left(\frac{1}{2}\sec\theta\tan\theta + \frac{1}{2}\int \sec\theta \, d\theta\right) - \int \sec\theta \, d\theta$$

$$= \frac{1}{2}\sec\theta\tan\theta - \frac{1}{2}\int \sec\theta \, d\theta$$

[Formula (8) of Section 8.3]

$$= \frac{1}{2}\sec\theta\tan\theta - \frac{1}{2}\ln|\sec\theta + \tan\theta| + C$$

[From Figure 8.4.7]

$$= \frac{1}{2}(x-1)(\sqrt{(x-1)^2 - 1})$$

$$- \frac{1}{2}\ln|(x-1) + \sqrt{(x-1)^2 - 1}| + C$$

$$= \frac{1}{2}(x-1)(\sqrt{x^2 - 2x}) - \frac{1}{2}\ln|x - 1 + \sqrt{x^2 - 2x}| + C. \quad ■$$

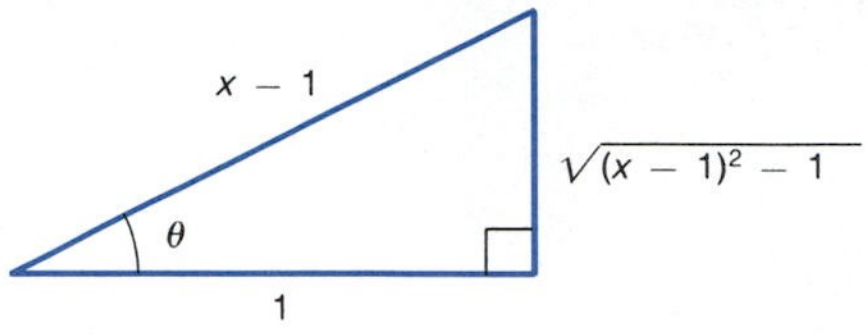

FIGURE 8.4.7

It may be of some interest that we can now evaluate an integral that represents the area of a circle.

■ **EXAMPLE 5**

Find the area of a circle of radius $r$.

**SOLUTION**

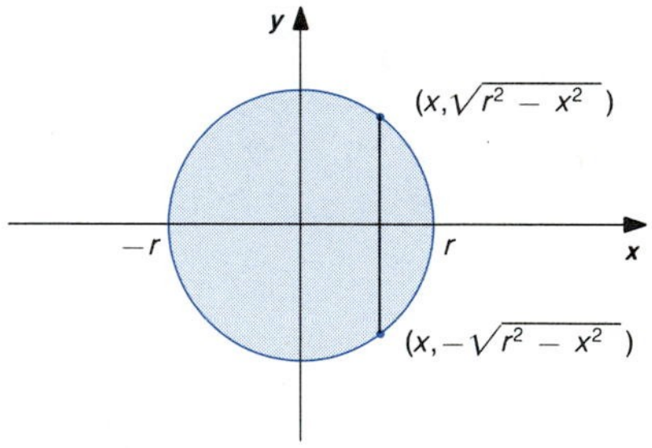

Let us consider the circle $x^2 + y^2 = r^2$. A sketch of the circle is given in Figure 8.4.8. We see that a variable vertical area strip has length $2\sqrt{r^2 - x^2}$. The area of the circle is

$$A = \int_{-r}^{r} 2\sqrt{r^2 - x^2} \, dx.$$

FIGURE 8.4.8

Let $x = r\sin\theta$. Then $dx = r\cos\theta \, d\theta$ and $\sqrt{r^2 - x^2} = r\cos\theta$. Let us change the limits of integration to correspond to the variable $\theta$. When $x = -r$, we have $r\sin\theta = -r$. This implies that $\sin\theta = -1$, so $\theta = \sin^{-1}(-1) = -\pi/2$. Similarly, $x = r$ implies that $\theta = \sin^{-1}(1) = \pi/2$. So

$$A = \int_{-\pi/2}^{\pi/2} 2r\cos\theta \, r\cos\theta \, d\theta = \int_{-\pi/2}^{\pi/2} 2r^2\cos^2\theta \, d\theta$$

[Formula (4) of Section 8.3]

$$= 2r^2\left[\frac{1}{2}\sin\theta\cos\theta + \frac{\theta}{2}\right]_{-\pi/2}^{\pi/2}$$

$$= r^2[\sin(\pi/2)\cos(\pi/2) + (\pi/2)] - r^2[\sin(-\pi/2)\cos(-\pi/2) + (-\pi/2)]$$

$$= \pi r^2. \quad ■$$

Don't forget to look for an easier method before trying a trigonometric substitution. For example, the substitution $x = 2\tan\theta$ could be used to show that

$$\int \frac{1}{x^2+4}\,dx = \frac{1}{2}\tan^{-1}\frac{x}{2} + C,$$

but we should recognize this as an integral for which we already know the formula. Also, we could use $x = \sin\theta$ to show

$$\int x\sqrt{1-x^2}\,dx = -\frac{1}{3}(1-x^2)^{3/2} + C,$$

but it is much easier to use the substitution $u = 1 - x^2$.

## EXERCISES 8.4

Evaluate the integrals in Exercises 1–20.

1 $\int \sqrt{4-x^2}\,dx$

2 $\int \frac{x^2}{\sqrt{4-x^2}}\,dx$

3 $\int \frac{1}{(2-x^2)^{3/2}}\,dx$

4 $\int \frac{x^3}{\sqrt{9-x^2}}\,dx$

5 $\int \frac{1}{(1+9x^2)^2}\,dx$

6 $\int \frac{1}{(1+x^2)^3}\,dx$

7 $\int x^3\sqrt{1-x^2}\,dx$

8 $\int x\sqrt{x^2-4}\,dx$

9 $\int \frac{1}{\sqrt{x^2-3}}\,dx$

10 $\int \frac{1}{\sqrt{4x^2-1}}\,dx$

11 $\int \frac{x^2}{\sqrt{x^2-16}}\,dx$

12 $\int \frac{x^2}{\sqrt{x^2+4}}\,dx$

13 $\int x^2\sqrt{9-x^2}\,dx$

14 $\int \frac{x^2}{(1-x^2)^{3/2}}\,dx$

15 $\int \frac{x}{4+x^2}\,dx$

16 $\int \sqrt{x^2-4}\,dx$

17 $\int \frac{1}{\sqrt{x^2+4x+5}}\,dx$

18 $\int \frac{1}{\sqrt{x^2+4x+3}}\,dx$

19 $\int \frac{1}{9-x^2}\,dx$

20 $\int \sqrt{4x-x^2}\,dx$

21 Find the area of the region bounded by the ellipse

$$\frac{x^2}{a^2} + \frac{y^2}{b^2} = 1.$$

22 Find the area of the region bounded by the hyperbola

$$\frac{x^2}{a^2} - \frac{y^2}{b^2} = 1$$

and the line $x = \sqrt{a^2+b^2}$.

23 Find the volume of the solid that is obtained by revolving the region bounded by $y = x/\sqrt{x^2+1}$, $y = 0$, and $x = 1$, about the $y$-axis.

24 Find the volume of the solid that is obtained by revolving the region bounded by $y = x/\sqrt{x^2+1}$, $y = 0$, and $x = 1$, about the $x$-axis.

25 Find the arc length of the curve $y = x^2$ between $(0, 0)$ and $(2, 4)$.

26 Find the area of the surface obtained by revolving the curve $y = x^2$, $0 \le x \le 2$, about the $x$-axis.

27 Use the substitution $u = a\tan\theta$ and then formula (2) of Section 8.3 to derive the formula

$$\int \frac{1}{(u^2+a^2)^n}\,du = \frac{u}{a^2(2n-2)(u^2+a^2)^{n-1}} + \frac{2n-3}{a^2(2n-2)}\int \frac{1}{(u^2+a^2)^{n-1}}\,du.$$

## 8.5 PARTIAL FRACTIONS

We will illustrate a technique that can be used to evaluate integrals of the form

$$\int \frac{P(x)}{Q(x)}\,dx,$$

where $P$ and $Q$ are polynomials. The idea is to express the quotient $P/Q$ as a sum of terms that can be integrated by using techniques we have previously developed. The particular expression we will describe is called the **partial fraction expansion** of $P/Q$.

The partial fraction expansion of a rational function $P/Q$ contains terms that correspond to the factors of the denominator. It is theoretically possible to factor $Q(x)$ into powers of distinct linear factors $(qx - p)^m$ and powers of distinct irreducible quadratic factors $(ax^2 + bx + c)^n$. (The quadratic $ax^2 + bx + c$ is irreducible if $b^2 - 4ac < 0$. This means the quadratic is never zero. The linear factor $qx - p$ is zero when $x = p/q$.)

If the degree of the numerator is less than the degree of the denominator, the terms of the partial fraction expansion of $P/Q$ are determined by the following rules:

- Each distinct linear or irreducible quadratic factor of $Q$ contributes a term or sum of terms to the partial fraction expansion of $P/Q$. The form and the number of terms that correspond to a factor of $Q$ depend on the form of the factor and the power of the factor in $Q$.
- If $Q(x)$ contains exactly $m$ identical linear factors $qx - p$, the partial fraction expansion of $P/Q$ contains a sum of the form

$$\frac{A_1}{qx - p} + \frac{A_2}{(qx - p)^2} + \cdots + \frac{A_m}{(qx - p)^m}.$$

- If $Q(x)$ contains exactly $n$ identical irreducible quadratic factors $ax^2 + bx + c$, then the partial fraction expansion contains a sum of the form

$$\frac{B_1x + C_1}{ax^2 + bx + c} + \cdots + \frac{B_nx + C_n}{(ax^2 + bx + c)^n}.$$

Each distinct factor of $Q$ contributes a term or terms of the type indicated. In each case, the denominators are powers of the factors, where the powers run from the first power up to the power of the factor in $Q$. The numerators are numbers in the case of linear factors. In the case of irreducible quadratic factors, the numerators are of the form (number)$x$ + (number).

If the degree of the numerators is greater than or equal to the degree of the denominator, we must carry out the division to obtain polynomials $R$ and $S$ with the degree of $R$ less than the degree of $Q$ and

$$\frac{P(x)}{Q(x)} = S(x) + \frac{R(x)}{Q(x)}.$$

Since the degree of $R$ is less than the degree of $Q$, we know the form of its partial fraction expansion. The partial fraction expansion of $P/Q$ is then obtained by replacing $R/Q$ in the above formula by its partial fraction expansion. Don't forget to check the degrees of the numerator and the denominator in order to determine if it is necessary to carry out a division to obtain the partial fraction expansion.

After we have written the partial fraction expansion in the proper form, we must determine the values of the $A$'s, $B$'s, and $C$'s in the expansion.

**EXAMPLE 1**

Evaluate

$$\int \frac{x-3}{x^2-3x+2}\,dx.$$

**SOLUTION**

We first note that

$$\deg(x-3) = 1 < 2 = \deg(x^2-3x+2),$$

so it is not necessary to divide.

Factoring, we have $x^2 - 3x + 2 = (x-1)(x-2)$. We see there are two distinct linear factors, each to the first power. We then know that the partial fraction expansion is of the form

$$\frac{x-3}{(x-1)(x-2)} = \frac{A}{x-1} + \frac{B}{x-2}.$$

Multiplying both sides of the above equation by $(x-1)(x-2)$ to clear fractions, we obtain

$$x - 3 = A(x-2) + B(x-1). \qquad (*)$$

There are unique numbers $A$ and $B$ so that equation (*) is true for all $x$. We must determine these unknown numbers. We will illustrate two methods for doing this.

**METHOD 1** Combine like terms in equation (*) to obtain an equation of the form 0 = (polynomial). We then obtain a linear system of equations in the unknowns $A$ and $B$ by setting all coefficients and the constant term of the polynomial equal to zero. Solution for the unknowns then gives the desired identity. From (*) we have

$$x - 3 = Ax - 2A + Bx - B,$$

$$0 = (A + B - 1)x + (-2A - B + 3).$$

Then

$$\begin{cases} A + B - 1 = 0, & \text{[Coefficient of } x\text{]} \\ -2A - B + 3 = 0. & \text{[Constant term]} \end{cases}$$

Solving for $A$ and $B$, we obtain $A = 2$ and $B = -1$.

**METHOD 2** We obtain a linear system of equations in the unknowns by substituting specific values of $x$ into equation (*). Values of $x$ for which $Q(x) = 0$ give simple equations and these should be chosen before other values of $x$.

Let $x = 1$ in (*).

$$1 - 3 = A(1 - 2) + B(1 - 1) \quad \text{implies} \quad A = 2.$$

Let $x = 2$ in (*).

$$2 - 3 = A(2 - 2) + B(2 - 1) \quad \text{implies} \quad B = -1.$$

Method 1 will work in any case and is easily programmed so the work can be done by a computer. Method 2 is a convenient shortcut in the case of distinct linear factors raised to the first power.

By either method, we obtain

$$\frac{x - 3}{x^2 - 3x + 2} = \frac{2}{x - 1} - \frac{1}{x - 2}.$$

(The above equation can be checked by adding the fractions on the right.) Then

$$\int \frac{x - 3}{x^2 - 3x + 2}\,dx = \int \frac{2}{x - 1}\,dx - \int \frac{1}{x - 2}\,dx$$
$$= 2\ln|x - 1| - \ln|x - 2| + C.$$

■

■ **EXAMPLE 2**

Evaluate

$$\int \frac{x^3 - x - 3\sqrt{2}}{x^2 - 2}\,dx.$$

**SOLUTION**

$\operatorname{Deg}(x^3 - x - 3\sqrt{2}) = 3 \not< 2 = \deg(x^2 - 2)$, so we must divide:

$$\begin{array}{r|l} & \quad x \\ \hline x^2 - 2 & x^3 - \phantom{2}x - 3\sqrt{2} \\ & x^3 - 2x \\ \hline & \qquad x - 3\sqrt{2} \end{array}$$

It follows that

$$\frac{x^3 - x - 3\sqrt{2}}{x^2 - 2} = x + \frac{x - 3\sqrt{2}}{x^2 - 2}.$$

The quadratic $x^2 - 2$ does not have rational zeros, but it is zero when $x = \pm\sqrt{2}$; $x^2 - 2$ is not an irreducible quadratic. Factoring, we have

$$x^2 - 2 = (x - \sqrt{2})(x + \sqrt{2}).$$

(In some cases it could be necessary to use the quadratic formula to find the zeros and "factor" a quadratic expression.) We see that there are two distinct linear factors, each to the first power. We then know that the partial fraction expansion is of the form

$$\frac{x^3 - x - 3\sqrt{2}}{x^2 - 2} = x + \frac{A}{x - \sqrt{2}} + \frac{B}{x + \sqrt{2}}.$$

Clearing fractions, we obtain

$$x^3 - x - 3\sqrt{2} = x(x^2 - 2) + A(x + \sqrt{2}) + B(x - \sqrt{2}). \qquad (*)$$

We need to find the unique numbers $A$ and $B$ that make equation (*) true for all $x$.

**METHOD 1**

$$x^3 - x - 3\sqrt{2} = x^3 - 2x + Ax + \sqrt{2}A + Bx - \sqrt{2}B,$$

$$0 = (A + B - 1)x + (\sqrt{2}A - \sqrt{2}B + 3\sqrt{2}).$$

$$\begin{cases} A + B - 1 = 0, & \text{[Coefficient of } x\text{]} \\ \sqrt{2}A - \sqrt{2}B + 3\sqrt{2} = 0. & \text{[Constant term]} \end{cases}$$

The solution is $(A, B) = (-1, 2)$.

**METHOD 2** Let $x = \sqrt{2}$ in (*):

$$(\sqrt{2})^3 - \sqrt{2} - 3\sqrt{2} = (0) + A(2\sqrt{2}) + B(0) \quad \text{implies} \quad A = -1.$$

Let $x = -\sqrt{2}$ in (*).

$$(-\sqrt{2})^3 + \sqrt{2} - 3\sqrt{2} = (0) + A(0) + B(-2\sqrt{2}) \quad \text{implies} \quad B = 2.$$

By either method we obtain

$$\frac{x^3 - x - 3\sqrt{2}}{x^2 - 2} = x - \frac{1}{x - \sqrt{2}} + \frac{2}{x + \sqrt{2}}.$$

(This equation can be checked by simplifying the expression on the right.) We then have

$$\int \frac{x^3 - x - 3\sqrt{2}}{x^2 - 2}\,dx = \int \left(x - \frac{1}{x - \sqrt{2}} + \frac{2}{x + \sqrt{2}}\right) dx$$

$$= \frac{x^2}{2} - \ln|x - \sqrt{2}| + 2\ln|x + \sqrt{2}| + C. \qquad \blacksquare$$

If you forgot to divide in Example 2 and simply wrote

$$\frac{x^3 - x - 3\sqrt{2}}{x^2 - 2} \underset{(?)}{=} \frac{A}{x - \sqrt{2}} + \frac{B}{x + \sqrt{2}},$$

this incorrect expression could not be an identity for any choice of the unknown numbers $A$ and $B$. Method 1 would lead to an inconsistent system of linear equations. Method 2 would give values of $A$ and $B$, but the resulting equation would be incorrect. Be very careful when using Method 2.

### ■ EXAMPLE 3

Evaluate

$$\int \frac{1}{(x^2 - 1)(x - 1)}\, dx.$$

**SOLUTION**

$\text{Deg}(1) = 0 < 3 = \deg((x^2 - 1)(x - 1))$, so we don't need to divide.
Factoring, we have

$$(x^2 - 1)(x - 1) = (x - 1)(x + 1)(x - 1) = (x - 1)^2(x + 1).$$

In this example we have two distinct linear factors, one to the first power and the other squared. Since the factor $x - 1$ is squared, the partial fraction expansion has one term with denominator $x - 1$ and another term with denominator $(x - 1)^2$. The factor $x + 1$ appears to the first power, so it contributes only one term, with denominator $x + 1$. Since both factors are linear, the numerator of each of the three terms is an unknown number. We have

$$\frac{1}{(x - 1)^2(x + 1)} = \frac{A}{x - 1} + \frac{B}{(x - 1)^2} + \frac{C}{x + 1},$$

$$1 = A(x - 1)(x + 1) + B(x + 1) + C(x - 1)^2. \qquad (*)$$

**METHOD 1**

$$1 = Ax^2 - A + Bx + B + Cx^2 - 2Cx + C,$$

$$0 = (A + C)x^2 + (B - 2C)x + (-A + B + C - 1).$$

$$\begin{cases} A \quad\quad + C \quad = 0, & \text{[Coefficient of } x^2\text{]} \\ \quad\quad B - 2C \quad = 0, & \text{[Coefficient of } x\text{]} \\ -A + B + \ C - 1 = 0. & \text{[Constant term]} \end{cases}$$

The solution is $(A, B, C) = (-1/4, 1/2, 1/4)$.

**METHOD 2** Let $x = 1$ in (*).

$$1 = A(0) + B(2) + C(0) \quad \text{implies} \quad B = 1/2.$$

Let $x = -1$ in (*).

$$1 = A(0) + B(0) + C(-2)^2 \quad \text{implies} \quad C = 1/4.$$

We can now obtain the value of $A$ by substituting the known values of $B$ and $C$, and any convenient value of $x$ into (*). Let us use $x = 0$. Then

$$1 = A(-1)(1) + \frac{1}{2}(1) + \frac{1}{4}(-1)^2,$$

$$1 = -A + \frac{1}{2} + \frac{1}{4},$$

$$A = -\frac{1}{4}.$$

Both methods give

$$\frac{1}{(x^2-1)(x-1)} = \frac{-1/4}{x-1} + \frac{1/2}{(x-1)^2} + \frac{1/4}{x+1}.$$

Then

$$\int \frac{1}{(x^2-1)(x-1)}\,dx = \int \left(\frac{-1/4}{x-1} + \frac{1/2}{(x-1)^2} + \frac{1/4}{x+1}\,dx\right)$$

$$= -\frac{1}{4}\ln|x-1| - \frac{1}{2}\frac{1}{x-1} + \frac{1}{4}\ln|x+1| + C. \quad ■$$

If you incorrectly considered $x^2 - 1$ to be an irreducible quadratic in Example 3 and wrote

$$\frac{1}{(x^2-1)(x-1)} \underset{(?)}{=} \frac{A}{x-1} + \frac{Bx+C}{x^2-1},$$

you could not obtain an identity for any values of $A$, $B$, or $C$. Method 1 would result in an inconsistent system of linear equations. Method 2 would give values of $A$, $B$, and $C$, but an incorrect equation.

### ■ EXAMPLE 4

Evaluate

$$\int \frac{12}{x^4 - x^3 - 2x^2}\,dx.$$

**SOLUTION**

Deg(12) $= 0 < 4 = \deg(x^4 - x^3 - 2x^2)$, so we don't need to divide.
Factoring, we have

$$x^4 - x^3 - 2x^2 = x^2(x-2)(x+1).$$

In this example we have three distinct linear factors, one squared and the others to the first power. Note that $x^2$ is not an irreducible quadratic. The partial fraction expansion is of the form

$$\frac{12}{x^4 - x^3 - 2x^2} = \frac{A}{x} + \frac{B}{x^2} + \frac{C}{x-2} + \frac{D}{x+1},$$

$$\begin{aligned} 12 = {} & A(x)(x-2)(x+1) + B(x-2)(x+1) \\ & + C(x^2)(x+1) + D(x^2)(x-2). \end{aligned} \tag{*}$$

We will use method 1 for this example.

$$12 = A(x)(x^2 - x - 2) + B(x^2 - x - 2) + C(x^2)(x+1) + D(x^2)(x-2),$$

$$12 = Ax^3 - Ax^2 - 2Ax + Bx^2 - Bx - 2B + Cx^3 + Cx^2 + Dx^3 - 2Dx^2,$$

$$\begin{aligned} 0 = {} & (A + C + D)x^3 + (-A + B + C - 2D)x^2 \\ & + (-2A - B)x + (-2B - 12). \end{aligned}$$

$$\begin{cases} A + C + D = 0, & \text{[Coefficient of } x^3\text{]} \\ -A + B + C - 2D = 0, & \text{[Coefficient of } x^2\text{]} \\ -2A - B = 0, & \text{[Coefficient of } x\text{]} \\ -2B - 12 = 0. & \text{[Constant term]} \end{cases}$$

The solution is $(A, B, C, D) = (3, -6, 1, -4)$.
We have

$$\frac{12}{x^4 - x^3 - 2x^2} = \frac{3}{x} + \frac{-6}{x^2} + \frac{1}{x-2} + \frac{-4}{x+1}.$$

Hence,

$$\begin{aligned} \int \frac{12}{x^4 - x^3 - 2x^2}\,dx &= \int \left(\frac{3}{x} + \frac{-6}{x^2} + \frac{1}{x-2} + \frac{-4}{x+1}\right) dx \\ &= 3\ln|x| + \frac{6}{x} + \ln|x-2| - 4\ln|x+1| + C. \end{aligned}$$ ■

■ **EXAMPLE 5**

Evaluate

$$\int \frac{x}{(x^2 + x + 1)(x-1)}\,dx.$$

**SOLUTION**

We do not need to divide in this example.

We have one linear term and one irreducible quadratic in the denominator. Note that $(1)^2 - 4(1)(1) = -3 < 0$, so $x^2 + x + 1$ is irreducible. The numerator of the term that corresponds to the factor $x^2 + x + 1$ is (Unknown number)$x$ + (Unknown number). We write

$$\frac{x}{(x^2 + x + 1)(x - 1)} = \frac{A}{x - 1} + \frac{Bx + C}{x^2 + x + 1},$$

$$x = A(x^2 + x + 1) + (Bx + C)(x - 1), \tag{*}$$

$$x = Ax^2 + Ax + A + Bx^2 - Bx + Cx - C,$$

$$0 = (A + B)x^2 + (A - B + C - 1)x + (A - C).$$

$$\begin{cases} A + B = 0, & \text{[Coefficient of } x^2\text{]} \\ A - B + C - 1 = 0, & \text{[Coefficient of } x\text{]} \\ A - C = 0. & \text{[Constant term]} \end{cases}$$

The solution is $(A, B, C) = (1/3, -1/3, 1/3)$. We obtain

$$\frac{x}{(x^2 + x + 1)(x - 1)} = \frac{1/3}{x - 1} + \frac{(-1/3)x + (1/3)}{x^2 + x + 1}$$
$$= \frac{1}{3}\frac{1}{x - 1} + \frac{1}{3}\frac{1 - x}{x^2 + x + 1}.$$

The integral of the first term on the right is easy. Let us work on evaluating the integral of the second term. We first complete the square.

$$\int \frac{1 - x}{x^2 + x + 1}\,dx = \int \frac{1 - x}{(x^2 + x + (1/4)) - (1/4) + 1}\,dx$$
$$= \int \frac{1 - x}{(x + (1/2))^2 + (\sqrt{3}/2)^2}\,dx.$$

Using $u = x + (1/2)$, so that $du = dx$ and $x = u - (1/2)$, we see that the above integral is

$$\int \frac{1 - (u - (1/2))}{u^2 + (\sqrt{3}/2)^2}\,du = \frac{3}{2}\int \frac{1}{u^2 + (\sqrt{3}/2)^2}\,du - \int \frac{1}{u^2 + (\sqrt{3}/2)^2}\left(\frac{1}{2}\right)(2u)\,du$$

$$= \frac{3}{2}\left(\frac{1}{\sqrt{3}/2}\right)\tan^{-1}\left(\frac{u}{\sqrt{3}/2}\right) - \frac{1}{2}\ln[u^2 + (\sqrt{3}/2)^2] + C$$

$$= \sqrt{3}\tan^{-1}\left(\frac{2(x + (1/2))}{\sqrt{3}}\right) - \frac{1}{2}\ln[(x + (1/2))^2 + (\sqrt{3}/2)^2] + C$$

$$= \sqrt{3}\tan^{-1}\left(\frac{2x + 1}{\sqrt{3}}\right) - \frac{1}{2}\ln(x^2 + x + 1) + C.$$

We then write

$$\int \frac{x}{(x^2+x+1)(x-1)}\,dx = \frac{1}{3}\int \frac{1}{x-1}\,dx + \frac{1}{3}\int \frac{1-x}{x^2+x+1}\,dx$$

$$= \frac{1}{3}\ln|x-1| + \frac{\sqrt{3}}{3}\tan^{-1}\frac{2x+1}{\sqrt{3}} - \frac{1}{6}\ln(x^2+x+1) + C. \qquad \blacksquare$$

■ **EXAMPLE 6**

Evaluate

$$\int \frac{x^3+1}{(x^2+1)^2}\,dx.$$

**SOLUTION**

$\text{Deg}(x^3+1) = 3 < 4 = \deg((x^2+1)^2)$, so we don't need to divide. The term $x^2+1$ is irreducible. We have

$$\frac{x^3+1}{(x^2+1)^2} = \frac{Ax+B}{x^2+1} + \frac{Cx+D}{(x^2+1)^2}. \tag{*}$$

$$x^3+1 = (Ax+B)(x^2+1) + (Cx+D),$$

$$x^3+1 = Ax^3+Bx^2+Ax+B+Cx+D,$$

$$0 = (A-1)x^3+Bx^2+(A+C)x+(B+D-1).$$

$$\begin{cases} A-1=0, \\ B=0, \\ A+C=0, \\ B+D-1=0. \end{cases}$$

The solution is $(A, B, C, D) = (1, 0, -1, 1)$. We have

$$\frac{x^3+1}{(x^2+1)^2} = \frac{x+0}{x^2+1} + \frac{-x+1}{(x^2+1)^2}$$

$$= \frac{x}{x^2+1} - \frac{x}{(x^2+1)^2} + \frac{1}{(x^2+1)^2}.$$

The integral of the first two terms on the right above is easily obtained by using the substitution $u = x^2+1$. For the third term on the right, let us use the substitution $x = \tan\theta$, so

$$dx = \sec^2\theta\,d\theta \quad \text{and} \quad (x^2+1)^2 = (\tan^2\theta+1)^2 = (\sec^2\theta)^2 = \sec^4\theta.$$

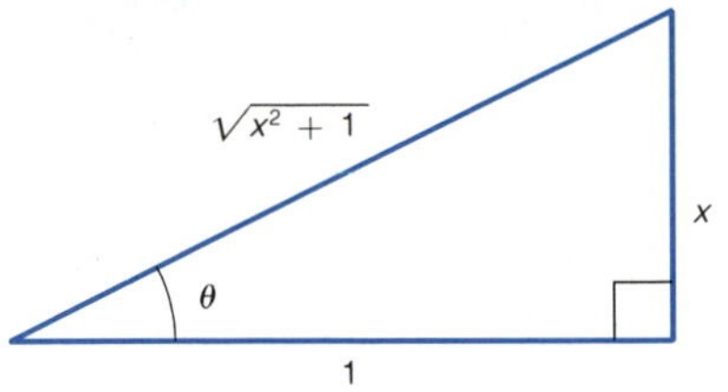

$\sqrt{x^2+1}$, $x$, $\theta$, $1$

FIGURE 8.5.1

Then

$$\int \frac{1}{(x^2+1)^2}\,dx = \int \frac{\sec^2\theta}{\sec^4\theta}\,d\theta = \int \cos^2\theta\,d\theta$$

[Formula (4) of Section 8.3]

$$= \frac{1}{2}\sin\theta\cos\theta + \frac{\theta}{2} + C$$

[From Figure 8.5.1]

$$= \frac{1}{2}\left(\frac{x}{\sqrt{x^2+1}}\right)\left(\frac{1}{\sqrt{x^2+1}}\right) + \frac{1}{2}\tan^{-1}x + C$$

$$= \frac{1}{2}\frac{x}{x^2+1} + \frac{1}{2}\tan^{-1}x + C.$$

Collecting results, we have

$$\int \frac{x^3+1}{(x^2+1)^2}\,dx = \int\left(\frac{x}{x^2+1} - \frac{x}{(x^2+1)^2} + \frac{1}{(x^2+1)^2}\right)dx$$

$$= \int \frac{1}{x^2+1}\left(\frac{1}{2}\right)(2x)\,dx - \int \frac{1}{(x^2+1)^2}\left(\frac{1}{2}\right)(2x)\,dx$$

$$+ \int \frac{1}{(x^2+1)^2}\,dx$$

$$= \frac{1}{2}\ln(x^2+1) + \frac{1}{2}\frac{1}{x^2+1}$$

$$+ \frac{1}{2}\frac{x}{x^2+1} + \frac{1}{2}\tan^{-1}x + C.$$ ■

The following reduction formula can be used in place of the substitution $x = a\tan\theta$ that was used in Example 6.

$$\int \frac{1}{(u^2+a^2)^n}\,du = \frac{u}{a^2(2n-2)(u^2+a^2)^{n-1}}$$

$$+ \frac{2n-3}{a^2(2n-2)}\int \frac{1}{(u^2+a^2)^{n-1}}\,du. \qquad (1)$$

For example, when $a = 1$ and $n = 2$, formula (1) gives

$$\int \frac{1}{(x^2+1)^2}\,dx = \frac{x}{2(x^2+1)} + \frac{1}{2}\int \frac{1}{x^2+1}\,dx$$

$$= \frac{1}{2}\frac{x}{x^2+1} + \frac{1}{2}\tan^{-1}x + C.$$

This agrees with the result we obtained by using the substitution $x = \tan\theta$.

## EXERCISES 8.5

Evaluate the integrals in Exercises 1–36.

1 $\int \frac{1}{x^2 - 5x + 6}\,dx$

2 $\int \frac{2x - 1}{x^2 - x - 2}\,dx$

3 $\int \frac{x - 6}{x^2 - 3x}\,dx$

4 $\int \frac{x - 5}{2x^2 + x - 1}\,dx$

5 $\int \frac{2x^3 - 9x + 10}{x^2 - 4}\,dx$

6 $\int \frac{3x^2 - 33}{x^2 - 9}\,dx$

7 $\int \frac{2x - 3}{(x - 1)^2}\,dx$

8 $\int \frac{3x + 4}{(x + 2)^2}\,dx$

9 $\int \frac{x^2 + 8x - 4}{x^3 - 4x}\,dx$

10 $\int \frac{3x^2 - 3x - 6}{x^3 - x}\,dx$

11 $\int \frac{4x^2 - 11x + 4}{x^3 - 3x^2 + 2x}\,dx$

12 $\int \frac{x + 5}{x^3 - 4x^2 + 3x}\,dx$

13 $\int \frac{x}{(x - 1)^3}\,dx$

14 $\int \frac{x^2}{(x + 1)^3}\,dx$

15 $\int \frac{x^4 - x^3 + x^2 + 2x - 2}{x^3 - x^2}\,dx$

16 $\int \frac{x^5 - 3}{x^3 - x^2}\,dx$

17 $\int \frac{x^2 + 2x + 2}{x^4 + x^3}\,dx$

18 $\int \frac{4}{x^4 - x^3}\,dx$

19 $\int \frac{4}{x^2(x - 2)^2}\,dx$

20 $\int \frac{4x - 4}{x^2(x - 2)^2}\,dx$

21 $\int \frac{x^2}{x^2 - 3}\,dx$

22 $\int \frac{x^4}{x^2 - 2}\,dx$

23 $\int \frac{3x^2 + 2}{x(x^2 + 1)}\,dx$

24 $\int \frac{x^2 - x - 4}{x(x^2 + 4)}\,dx$

25 $\int \frac{4}{x^2 - 1}\,dx$

26 $\int \frac{32}{x^2 - 16}\,dx$

27 $\int \frac{8}{(x^2 + 1)(x^2 + 9)}\,dx$

28 $\int \frac{3x}{(x^2 + 1)(x^2 + 4)}\,dx$

29 $\int \frac{x^3 + 4x - 2}{x^3 - 8}\,dx$

30 $\int \frac{x^3 + x + 2}{x^3 - 1}\,dx$

31 $\int \frac{x^2}{(x^2 + 9)^2}\,dx$

32 $\int \frac{x^3 - 20x + 4}{(x^2 + 4)^2}\,dx$

33 $\int \frac{5}{x(x^2 + 2x + 5)}\,dx$

34 $\int \frac{5}{x(x^2 + 4x + 5)}\,dx$

35 $\int \frac{2x}{x^2 + 2x - 1}\,dx$

36 $\int \frac{2x}{x^2 - 2x - 1}\,dx$

## 8.6 MISCELLANEOUS SUBSTITUTIONS

We will illustrate some general ideas that can sometimes be used to simplify and evaluate indefinite integrals.

Simplification of the denominator is generally useful.

■ **EXAMPLE 1**

Evaluate

$$\int \frac{x}{(2x - 1)^6}\,dx.$$

**SOLUTION**

We could evaluate this integral by the method of partial fractions. However, we will see that the substitution $u = 2x - 1$ allows us to express the integral in a form that is easily integrated.

The equation $u = 2x - 1$ implies $du = 2\,dx$, so

$$dx = \frac{1}{2}\,du \quad \text{and} \quad x = \frac{1}{2}(u + 1).$$

Substitution then gives

$$\begin{aligned}
\int \frac{x}{(2x-1)^6}\,dx &= \int \frac{(1/2)(u+1)}{u^6}\frac{1}{2}\,du \\
&= \int \left(\frac{1}{4}u^{-5} + \frac{1}{4}u^{-6}\right) du \\
&= \frac{1}{4}\left(\frac{u^{-4}}{-4}\right) + \frac{1}{4}\left(\frac{u^{-5}}{-5}\right) + C \\
&= -\frac{1}{16(2x-1)^4} - \frac{1}{20(2x-1)^5} + C \\
&= -\frac{5(2x-1)}{80(2x-1)^5} - \frac{4}{80(2x-1)^5} + C \\
&= \frac{1-10x}{80(2x-1)^5} + C.
\end{aligned}$$

■

**■ EXAMPLE 2**

Evaluate

$$\int \frac{1}{e^x + 1}\,dx.$$

**SOLUTION**

Let us try to simplify the denominator by setting $u = e^x + 1$. Then $du = e^x\,dx$, so

$$dx = \frac{1}{e^x}\,du = \frac{1}{u-1}\,du.$$

Substitution then gives

$$\int \frac{1}{e^x + 1}\,dx = \int \frac{1}{u}\frac{1}{u-1}\,du = \int \frac{1}{u(u-1)}\,du.$$

The partial fraction expansion of the integrand is

$$\frac{1}{u(u-1)} = \frac{A}{u} + \frac{B}{u-1}.$$

It is not difficult to determine that $A = -1$ and $B = 1$.

We then have

$$\begin{aligned}\int \frac{1}{e^x+1}\,dx &= \int \frac{1}{u(u-1)}\,du \\ &= \int\left(-\frac{1}{u}+\frac{1}{u-1}\right)du \\ &= -\ln|u| + \ln|u-1| + C \\ &= -\ln(e^x+1) + \ln(e^x) + C \\ &= x - \ln(e^x+1) + C.\end{aligned}$$

■

We could also use the substitution $u = e^x$ in Example 2. Generally, the substitution $u = e^{ax}$ can be used to transform an integral that involves $e^{ax}$ into an integral that involves powers of $u$.

The substitution $u^n = ax + b$ can sometimes be used to simplify integrals that contain expressions of the form $(ax+b)^{1/n}$.

### ■ EXAMPLE 3

Evaluate $\int x\sqrt{3-x}\,dx$.

**SOLUTION**

Let $u^2 = 3 - x$, so $2u\,du = -dx$. We then have $x = 3 - u^2$, $\sqrt{3-x} = u$, and $dx = -2u\,du$. Substitution then gives

$$\begin{aligned}\int x\sqrt{3-x}\,dx &= \int (3-u^2)(u)(-2u)\,du = \int (2u^4 - 6u^2)\,du \\ &= 2\frac{u^5}{5} - 6\frac{u^3}{3} + C \\ &= \frac{2}{5}(\sqrt{3-x})^5 - 2(\sqrt{3-x})^3 + C \\ &= \frac{2}{5}(3-x)^{5/2} - 2(3-x)^{3/2} + C.\end{aligned}$$

■

In Example 3, we could also use the substitution $u = 3 - x$. This would result in an integrand that is a linear sum of fractional powers of $u$.

### ■ EXAMPLE 4

Evaluate

$$\int \frac{\sqrt{x}}{\sqrt{x}+1}\,dx.$$

**SOLUTION**

Let $u^2 = x$, so $2u\,du = dx$. Then $\sqrt{x} = u$. Substitution gives

$$\int \frac{\sqrt{x}}{\sqrt{x}+1}\,dx = \int \frac{u}{u+1}(2u)\,du = \int \frac{2u^2}{u+1}\,du.$$

Since the degree of the numerator is greater that the degree of the denominator, we carry out the division. We have

$$\begin{array}{r|l} & 2u - 2 \\ \hline u+1 & 2u^2 \\ & 2u^2 + 2u \\ \hline & -2u \\ & -2u - 2 \\ \hline & 2 \end{array}$$

Then

$$\begin{aligned} \int \frac{\sqrt{x}}{\sqrt{x}+1}\,dx &= \int \frac{2u^2}{u+1}\,du \\ &= \int \left(2u - 2 + \frac{2}{u+1}\right) du \\ &= u^2 - 2u + 2\ln|u+1| + C \\ &= x - 2\sqrt{x} + 2\ln(\sqrt{x}+1) + C. \qquad [u = \sqrt{x}] \end{aligned}$$

■

We can remove fractional powers of $x$ from an integrand by using the substitution $u^n = x$, where $n$ is the least common multiple of the denominators of the fractional powers of $x$.

**■ EXAMPLE 5**

Evaluate

$$\int \frac{x^{1/2}}{x^{1/3}+1}\,dx.$$

**SOLUTION**

Let $u^6 = x$, so $6u^5\,du = dx$. Then $x^{1/2} = u^3$ and $x^{1/3} = u^2$. Substitution gives

$$\int \frac{x^{1/2}}{x^{1/3}+1}\,dx = \int \frac{u^3}{u^2+1}(6u^5)\,du = \int 6\,\frac{u^8}{u^2+1}\,du.$$

Carrying out the division, we have

$$\begin{array}{r|l} & u^6 - u^4 + u^2 - 1 \\ \hline u^2+1 & u^8 \\ & u^8 + u^6 \\ \hline & -u^6 \\ & -u^6 - u^4 \\ \hline & u^4 \\ & u^4 + u^2 \\ \hline & -u^2 \\ & -u^2 - 1 \\ \hline & 1 \end{array}$$

Then

$$\int \frac{x^{1/2}}{x^{1/3}+1}\,dx = \int 6\frac{u^8}{u^2+1}\,du$$

$$= \int 6\left(u^6 - u^4 + u^2 - 1 + \frac{1}{u^2+1}\right) du$$

$$= 6\left(\frac{u^7}{7} - \frac{u^5}{5} + \frac{u^3}{3} - u + \tan^{-1} u\right) + C$$

$[u = x^{1/6}]$

$$= \frac{6}{7}x^{7/6} - \frac{6}{5}x^{5/6} + 2x^{1/2} - 6x^{1/6} + 6\tan^{-1}(x^{1/6}) + C. \quad ■$$

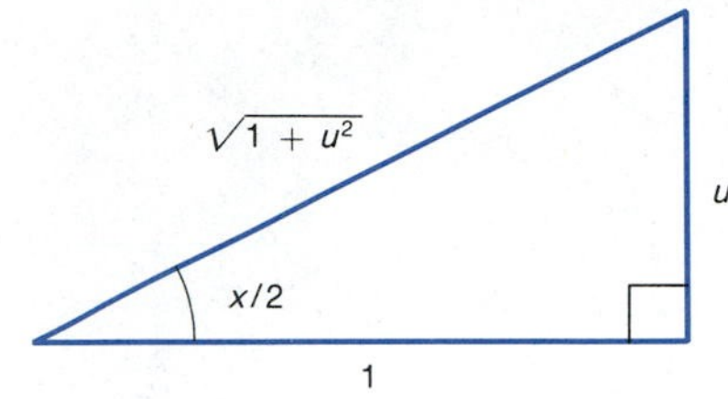


FIGURE 8.6.1

Rational expressions of $\sin x$ and $\cos x$ can be transformed to rational expressions of $u$ by the change of variables $u = \tan(x/2)$. From the representative triangle sketched in Figure 8.6.1 we see that

$$\sin\frac{x}{2} = \frac{u}{\sqrt{1+u^2}} \quad \text{and} \quad \cos\frac{x}{2} = \frac{1}{\sqrt{1+u^2}},$$

so

$$\cos x = 1 - 2\sin^2\frac{x}{2} = 1 - 2\left(\frac{u}{\sqrt{1+u^2}}\right)^2 = \frac{1-u^2}{1+u^2}$$

and

$$\sin x = 2\sin\frac{x}{2}\cos\frac{x}{2} = 2\left(\frac{u}{\sqrt{1+u^2}}\right)\left(\frac{1}{\sqrt{1+u^2}}\right) = \frac{2u}{1+u^2}.$$

The equation $u = \tan(x/2)$, $-\pi < x < \pi$, implies $\tan^{-1}u = x/2$, so $x = 2\tan^{-1}u$. Then

$$dx = \frac{2}{1+u^2}\,du.$$

Summarizing, we see that the substitution

$$u = \tan\frac{x}{2}, \qquad -\pi < x < \pi,$$

gives the equations

$$\cos x = \frac{1-u^2}{1+u^2}, \quad \sin x = \frac{2u}{1+u^2}, \quad \text{and} \quad dx = \frac{2}{1+u^2}\,du.$$

These equations can be used to transform an integral of a rational function of $\sin x$ and $\cos x$ into a rational function of $u$. If we wish to express the answer in terms of sin $x$ and cos $x$, we can use the formula

$$\tan\frac{x}{2} = \frac{\sin x}{1 + \cos x}.$$

**EXAMPLE 6**

Evaluate

$$\int \frac{1}{1 + \sin x}\,dx.$$

**SOLUTION**

Using the formulas for the substitution $u = \tan(x/2)$, we have

$$\begin{aligned}\int \frac{1}{1 + \sin x}\,dx &= \int \frac{1}{1 + \dfrac{2u}{1 + u^2}}\,\frac{2}{1 + u^2}\,du \\ &= \int \frac{2}{(1 + u^2) + (2u)}\,du \\ &= \int \frac{2}{u^2 + 2u + 1}\,du \\ &= \int \frac{2}{(u + 1)^2}\,du \\ &= -\frac{2}{u + 1} + C \\ &= -\frac{2}{\tan(x/2) + 1} + C.\end{aligned}$$

## EXERCISES 8.6

Evaluate the integrals in Exercises 1–32.

1 $\int \frac{x}{\sqrt{2x - 1}}\,dx$

2 $\int x^2\sqrt{x + 2}\,dx$

3 $\int \frac{x}{(3x + 2)^{1/3}}\,dx$

4 $\int \frac{x}{(3x - 1)^{2/3}}\,dx$

5 $\int \frac{1}{1 + \sqrt{x}}\,dx$

6 $\int \frac{x}{1 + \sqrt{x}}\,dx$

7 $\int \frac{1 - \sqrt{x}}{1 + \sqrt{x}}\,dx$

8 $\int x(x + 1)^{1/3}\,dx$

9 $\int \frac{1}{x^{1/2} + x^{1/3}}\,dx$

10 $\int \frac{1}{x^{1/2} + x^{1/4}}\,dx$

11 $\int \frac{x^{1/3}}{1 + x^{2/3}}\,dx$

12 $\int \frac{x^{2/3}}{1 + x^{1/3}}\,dx$

13 $\int \frac{x}{(2x + 5)^{5/2}}\,dx$

14 $\int \frac{x}{(3x - 2)^{7/3}}\,dx$

15 $\int \frac{x}{(2 - x)^4}\,dx$

16 $\int \frac{x^2}{(1 - 2x)^5}\,dx$

17 $\int \frac{e^{2x}}{e^x + 1}\,dx$

18 $\int \frac{1}{e^{2x} + 1}\,dx$

19 $\int e^{2x}\sqrt{1+e^x}\,dx$

20 $\int \frac{e^{2x}}{\sqrt{1+e^x}}\,dx$

21 $\int \frac{1}{1-\sin x}\,dx$

22 $\int \frac{1}{1-\cos x}\,dx$

23 $\int \frac{1}{1+\cos x}\,dx$

24 $\int \frac{1}{1+\sin x+\cos x}\,dx$

25 $\int \frac{1}{5+3\cos x}\,dx$

26 $\int \frac{1}{5+3\sin x}\,dx$

27 $\int \frac{1}{3\sin x+4\cos x}\,dx$

28 $\int \frac{1}{3+5\cos x}\,dx$

29 $\int \frac{1}{\sqrt{1+\sqrt{x}}}\,dx$

30 $\int \sqrt{1+\sqrt{x}}\,dx$

31 $\int e^{\sqrt{x}}\,dx$

32 $\int \sin\sqrt{x}\,dx$

## 8.7 IMPROPER INTEGRALS

Recall that the definite integral was defined as a limit of Riemann sums. The techniques of integration we have learned apply only to integrals for which the appropriate Riemann sums converge. If the Riemann sums do not have a limit, then $\int_a^b f(x)\,dx$ is merely a symbol with no numerical value.

It is a fact that $\int_a^b f(x)\,dx$ does not exist as a limit of Riemann sums unless $f$ is bounded on $[a, b]$. (See Exercise 35 at the end of this section.) Also, we cannot form Riemann sums if the interval of integration is infinite.

■ *Definition*

The integral

$$\int_a^b f(x)\,dx$$

is called **improper** if either:

(i) $f$ is not bounded on $[a, b]$, or

(ii) the interval of integration is infinite. ■

For example:

■ The integral

$$\int_0^1 \frac{1}{\sqrt{1-x^2}}\,dx$$

is improper because $1/\sqrt{1-x^2}$ becomes unbounded as $x$ approaches one, the right endpoint of the interval of integration.

■ The integral

$$\int_0^1 \frac{1}{x}\,dx$$

is improper because $1/x$ becomes unbounded as $x$ approaches zero, the left endpoint of the interval of integration.

- The integral

$$\int_1^\infty \frac{1}{x^2}\,dx$$

is improper because $[1, \infty)$ is an infinite interval of integration.

- The integral

$$\int_{-1}^1 \frac{1}{x^2}\,dx$$

is improper because $1/x^2$ becomes unbounded as $x$ approaches zero, a point inside the interval of integration.

- The integral

$$\int_0^\infty x^{-2/3}\,dx$$

is improper because $x^{-2/3}$ becomes unbounded as $x$ approaches zero and because $[0, \infty)$ is an infinite interval of integration.

So far, an improper integral is only a symbol with no meaning. We will now describe some cases for which certain types of improper integrals may be assigned a numerical value.

Suppose $\int_a^b f(x)\,dx$ is improper only because of the behavior at one endpoint of the interval of integration. For example, suppose that either $f$ becomes unbounded as $x$ approaches $b$ or $b = \infty$, but also that $\int_a^{b'} f(x)\,dx$ exists for every $a < b' < b$. (This means that we can apply the usual techniques of integration to evaluate $\int_a^{b'} f(x)\,dx$.) We then say that

$$\int_a^b f(x)\,dx$$

is **convergent** and

$$\int_a^b f(x)\,dx = \lim_{b' \to b^-} \int_a^{b'} f(x)\,dx,$$

if the limit exists and is finite. Otherwise, we say that $\int_a^b f(x)\,dx$ is **divergent** and no numerical value is assigned to the symbol. Similarly, if $\int_a^b f(x)\,dx$ is improper only at $a$,

$$\int_a^b f(x)\,dx = \lim_{a' \to a^+} \int_{a'}^{b} f(x)\,dx,$$

if the limit exists and is finite.

**EXAMPLE 1**

Evaluate

$$\int_0^1 \frac{1}{\sqrt{1-x^2}}\,dx.$$

**SOLUTION**

The integral is improper only at the right endpoint $x = 1$. For $0 < b < 1$, we have

$$\int_0^b \frac{1}{\sqrt{1-x^2}}\,dx = \sin^{-1}x\Big]_0^b = \sin^{-1}b - \sin^{-1}0 = \sin^{-1}b;$$

$$\lim_{b\to 1^-}\int_0^b \frac{1}{\sqrt{1-x^2}}\,dx = \lim_{b\to 1^-}\sin^{-1}b = \sin^{-1}1 = \frac{\pi}{2}.$$

The integral is convergent with

$$\int_0^1 \frac{1}{\sqrt{1-x^2}}\,dx = \frac{\pi}{2}.$$

■

Note that the Fundamental Theorem of Calculus does not apply directly to

$$\int_0^1 \frac{1}{\sqrt{1-x^2}}\,dx,$$

because the integrand cannot be assigned any value at 1 that would make it continuous on the interval [0, 1].

**EXAMPLE 2**

Evaluate

$$\int_0^1 \frac{1}{x}\,dx.$$

**SOLUTION**

The integral is improper only at the left endpoint $x = 0$. For $0 < a < 1$, we have

$$\int_a^1 \frac{1}{x}\,dx = \ln x\Big]_a^1 = \ln 1 - \ln a = -\ln a.$$

Then

$$\lim_{a\to 0^+}\int_a^1 \frac{1}{x}\,dx = \lim_{a\to 0^+}(-\ln a) = +\infty.$$

Since the limit does not converge to a finite number, we say that

$$\int_0^1 \frac{1}{x}\,dx$$

is divergent. This integral is not assigned a numerical value. ■

■ **EXAMPLE 3**

Evaluate

$$\int_1^\infty \frac{1}{x^2}\,dx.$$

**SOLUTION**

The integral is improper only because the right endpoint is infinite. For $1 < b < \infty$, we have

$$\int_1^b \frac{1}{x^2}\,dx = -\frac{1}{x}\bigg]_1^b = -\frac{1}{b} + \frac{1}{1};$$

$$\lim_{b\to\infty} \int_1^b \frac{1}{x^2}\,dx = \lim_{b\to\infty}\left(-\frac{1}{b} + \frac{1}{1}\right) = 1.$$

Hence, the integral is convergent with

$$\int_1^\infty \frac{1}{x^2}\,dx = 1.$$ ■

Note that we cannot apply the Fundamental Theorem of Calculus directly to the integral in Example 3, because the interval of integration does not contain its right endpoint. Infinity is not a real number.

Suppose $\int_a^b f(x)\,dx$ is improper at other than one endpoint, but we can write the integral as a "sum" of integrals that are each improper at only one endpoint. If *every* integral in the sum is convergent, we say $\int_a^b f(x)\,dx$ is convergent and has a value equal to the sum of the values of the integrals that form the sum. If *any* of the integrals in the sum diverge, we say $\int_a^b f(x)\,dx$ is divergent and no numerical value is assigned to the symbol. The convergence or divergence of an improper integral does not depend on how the interval of integration is divided, as long as we write the integral as a sum of integrals that are improper at only one endpoint.

■ **EXAMPLE 4**

Evaluate

$$\int_{-1}^1 x^{-2}\,dx.$$

**SOLUTION**

This integral is improper at the point $x = 0$, a point in the interval other than an endpoint. We write

$$\int_{-1}^{1} x^{-2}\,dx = \int_{-1}^{0} x^{-2}\,dx + \int_{0}^{1} x^{-2}\,dx.$$

Each of the two integrals in the sum is improper only at one endpoint. We must check that both converge.

Let us check the convergence of

$$\int_{-1}^{0} x^{-2}\,dx.$$

We have

$$\lim_{b\to 0^-} \int_{-1}^{b} x^{-2}\,dx = \lim_{b\to 0^-} \left(-\frac{1}{x}\right)\Bigg]_{-1}^{b} = \lim_{b\to 0^-}\left(-\frac{1}{b} - 1\right) = \infty.$$

Since this integral is divergent, we know that $\int_{-1}^{1} x^{-2}\,dx$ is divergent. It is not necessary to check the convergence of the integral over the interval between zero and one. ■

If you did not notice that the integral in Example 4 is improper, you might write

$$\int_{-1}^{1} x^{-2}\,dx \underset{(?)}{=} -x^{-1}\Big]_{-1}^{1} = -(1) - (-(-1)) = -2,$$

which is an incorrect answer. The Fundamental Theorem of Calculus does not apply to this integral.

■ **EXAMPLE 5**

Evaluate

$$\int_{0}^{\infty} x^{-2/3}\,dx.$$

**SOLUTION**

The integral is improper at $x = 0$ and $\infty$. Let us subdivide the interval of integration at 1, a convenient number, and write

$$\int_{0}^{\infty} x^{-2/3}\,dx = \int_{0}^{1} x^{-2/3}\,dx + \int_{1}^{\infty} x^{-2/3}\,dx.$$

The first integral on the right is improper only at the endpoint $x = 0$, and the second is improper only at $\infty$. We must check the convergence of each separately:

$$\lim_{a\to 0^+} \int_a^1 x^{-2/3}\,dx = \lim_{a\to 0^+} 3x^{1/3}\Big]_a^1 = \lim_{a\to 0^+} (3 - 3a^{1/3}) = 3,$$

so the first integral converges;

$$\lim_{b\to\infty} \int_1^b x^{-2/3}\,dx = \lim_{b\to\infty} 3x^{1/3}\Big]_1^b = \lim_{b\to\infty} (3b^{1/3} - 3) = \infty,$$

so the second integral diverges. Since one of the parts diverges, $\int_0^\infty x^{-2/3}\,dx$ is divergent. ■

**■ EXAMPLE 6**

Evaluate

$$\int_{-\infty}^{\infty} \frac{1}{x^2+1}\,dx.$$

**SOLUTION**

This integral is improper at $\infty$ and $-\infty$. We subdivide the integral at 0, a convenient number, and write

$$\int_{-\infty}^{\infty} \frac{1}{x^2+1}\,dx = \int_{-\infty}^{0} \frac{1}{x^2+1}\,dx + \int_{0}^{\infty} \frac{1}{x^2+1}\,dx.$$

We must check the convergence of each of the two integrals on the right.

$$\lim_{a\to -\infty} \int_a^0 \frac{1}{x^2+1}\,dx = \lim_{a\to -\infty} \tan^{-1}x\Big]_a^0$$

$$= \lim_{a\to -\infty} (\tan^{-1}0 - \tan^{-1}a) = 0 - \left(-\frac{\pi}{2}\right) = \frac{\pi}{2}.$$

$$\lim_{b\to\infty} \int_0^b \frac{1}{x^2+1}\,dx = \lim_{b\to\infty} \tan^{-1}x\Big]_0^b$$

$$= \lim_{b\to\infty} (\tan^{-1}b - \tan^{-1}0) = \frac{\pi}{2}.$$

Both parts are convergent, so the original integral is convergent with

$$\int_{-\infty}^{\infty} \frac{1}{x^2+1}\,dx = \frac{\pi}{2} + \frac{\pi}{2} = \pi.$$ ■

**■ EXAMPLE 7**

Evaluate

$$\int_0^\infty xe^{-x}\,dx.$$

**SOLUTION**

This integral is improper only at infinity. Let us integrate by parts with

$$u = x, \qquad dv = e^{-x}\,dx,$$

$$du = dx, \qquad v = -e^{-x}.$$

For $0 < b < \infty$, we then have

$$\begin{aligned}\int_0^b xe^{-x}\,dx &= \int u\,dv \\ &= uv - \int v\,du \\ &= -xe^{-x}\Big]_0^b + \int_0^b e^{-x}\,dx \\ &= -be^{-b} - e^{-x}\Big]_0^b \\ &= -be^{-b} - e^{-b} + 1.\end{aligned}$$

We need the limit as $b$ approaches infinity of the above expression. Using l'Hôpital's Rule to evaluate the limit of the first term, we have

$$\begin{aligned}\lim_{b\to\infty}\int_0^b xe^{-x}\,dx &= \lim_{b\to\infty}(-be^{-b} - e^{-b} + 1) \\ &= -\left(\lim_{b\to\infty}\frac{b}{e^b}\right) - 0 + 1 \\ &\underset{(\infty/\infty,\,\text{l'H})}{=} -\left(\lim_{b\to\infty}\frac{1}{e^b}\right) + 1 = -(0) + 1 = 1.\end{aligned}$$

The integral is convergent with $\int_0^\infty xe^{-x}\,dx = 1$. ■

It does not make sense in Example 7 to write

$$\int_0^\infty xe^{-x}\,dx \underset{(?)}{=} -xe^{-x} - e^{-x}\Big]_0^\infty \underset{(?)}{=} (-\infty e^{-\infty} - e^{-\infty}) - (0 - 1).$$

You must evaluate a limit to determine the value of an improper integral.

■ **EXAMPLE 8**

Evaluate

$$\int_0^{\infty} \frac{1}{2x^2 + 3x + 1}\,dx.$$

**SOLUTION**

The integral is improper at infinity. We have $2x^2 + 3x + 1 = (2x + 1)(x + 1)$, so the zeros of the denominator are not contained in the interval of integration. The integrand $1/(2x^2 + 3x + 1)$ does not become unbounded on the interval of integration. For $0 < b < \infty$, we have

$$\int_0^{b} \frac{1}{2x^2 + 3x + 1}\,dx = \int_0^{b} \frac{1}{(2x + 1)(x + 1)}\,dx.$$

The partial fraction expansion of the integrand is

$$\frac{1}{(2x + 1)(x + 1)} = \frac{A}{2x + 1} + \frac{B}{x + 1}.$$

It is easy to determine that $A = 2$ and $B = -1$. We then have

$$\begin{aligned}\int_0^{b} \frac{1}{2x^2 + 3x + 1}\,dx &= \int_0^{b} \frac{1}{(2x + 1)(x + 1)}\,dx \\ &= \int_0^{b}\left(\frac{2}{2x + 1} - \frac{1}{x + 1}\right)dx \\ &= \ln|2x + 1| - \ln|x + 1|\Big]_0^b \\ &= \ln\left|\frac{2x + 1}{x + 1}\right|\Bigg]_0^b \\ &= \ln\frac{2b + 1}{b + 1} - \ln 1 \\ &= \ln\frac{2b + 1}{b + 1},\end{aligned}$$

$$\int_0^{\infty} \frac{1}{2x^2 + 3x + 1}\,dx = \lim_{b \to \infty} \ln\frac{2b + 1}{b + 1} = \ln 2.$$

The integral is convergent with value ln 2. ■

In Example 8, we cannot evaluate $\lim_{b\to\infty}[\ln(2b + 1) - \ln(b + 1)]$ as the difference of limits, since each term approaches infinity as $b$ approaches infinity. We need to combine terms to obtain

$$\lim_{b \to \infty} \ln\frac{2b + 1}{b + 2}.$$

## EXERCISES 8.7

Evaluate or indicate divergence for each of the integrals in Exercises 1–32.

1 $\int_0^1 \frac{1}{1-x}\,dx$

2 $\int_0^2 \frac{1}{x-2}\,dx$

3 $\int_0^1 \frac{1}{\sqrt{1-x}}\,dx$

4 $\int_0^2 \frac{1}{(2-x)^{1/3}}\,dx$

5 $\int_0^{\pi/2} \sec^2 x\,dx$

6 $\int_0^{\pi/2} \tan x\,dx$

7 $\int_0^{\pi/2} \frac{\sin x}{1-\cos x}\,dx$

8 $\int_0^1 \frac{1}{\sqrt{x}}\,dx$

9 $\int_0^1 \frac{1}{x(\ln x)^2}\,dx$

10 $\int_0^1 \frac{1}{x\ln x}\,dx$

11 $\int_0^\infty xe^{-x^2}\,dx$

12 $\int_0^\infty \frac{1}{4+x^2}\,dx$

13 $\int_1^\infty \frac{1}{x-1}\,dx$

14 $\int_0^\infty \sin x\,dx$

15 $\int_{-1}^1 \frac{1}{\sqrt{1-x^2}}\,dx$

16 $\int_1^\infty \frac{1}{x\sqrt{x^2-1}}\,dx$

17 $\int_0^\pi \tan x\,dx$

18 $\int_0^\pi \sec x\,dx$

19 $\int_0^\infty \frac{e^{-\sqrt{x}}}{\sqrt{x}}\,dx$

20 $\int_0^\infty \frac{1}{x+1}\,dx$

21 $\int_{-1}^1 x^{-1/3}\,dx$

22 $\int_{-1}^1 x^{-4/3}\,dx$

23 $\int_{-\infty}^\infty \frac{1}{x^2+4x+5}\,dx$

24 $\int_{-\infty}^\infty \frac{x}{x^2+16}\,dx$

25 $\int_{-\infty}^\infty \frac{1}{x^2+4x+3}\,dx$

26 $\int_1^\infty \frac{1}{x^2+x}\,dx$

27 $\int_0^\infty xe^{-x/2}\,dx$

28 $\int_0^1 \ln x\,dx$

29 $\int_1^\infty \frac{1}{x(x^2+1)}\,dx$

30 $\int_0^\infty x^2e^{-x}\,dx$

31 $\int_0^\infty \frac{1}{(x^2+4)^{3/2}}\,dx$

32 $\int_0^\infty e^{-x}\sin x\,dx$

33 Determine the values of $p$ for which $\int_0^1 (1/x^p)\,dx$ is convergent.

34 Determine the values of $p$ for which $\int_1^\infty (1/x^p)\,dx$ is convergent.

35 Let

$$f(x) = \begin{cases} 0, & x = 0, \\ 1/\sqrt{x}, & 0 < x \le 1. \end{cases}$$

Show that for any given number $B$ and for any positive integer $n$, there is a Riemann sum $R$ of $f$ over the interval $[0, 1]$ with $n$ subintervals of equal length and $R > B$. This shows that $\int_0^1 1/\sqrt{x}\,dx$ cannot be the limit of Riemann sums of $f$ over $[0, 1]$. The argument is similar for any function that is unbounded on the interval of integration. [*Hint*: $R \ge f(x_1^*)/n$ for $0 \le x_1^* \le 1/n$.]

36 The velocity at time $t$ of an object that is moving along a straight line is given by the formula

$$v(t) = \frac{1}{1+t^2}, \qquad t \ge 0.$$

What is the total distance the object will travel?

37 The velocity at time $t$ of an object that is moving along a straight line is given by the formula

$$v(t) = \frac{1}{1+t}, \qquad t \ge 0.$$

What is the total distance the object will travel?

38 Show that the arc length of the curve $y = 3x^{2/3}, 0 \le x \le 8$, is given by an improper integral and evaluate the integral.

## REVIEW EXERCISES

Evaluate the integrals in Exercises 1–31.

1 $\int x\sin 2x\,dx$

2 $\int x\sin x^2\,dx$

3 $\int \ln x\,dx$

4 $\int \frac{\ln x}{x}\,dx$

5 $\int x^2e^{-x}\,dx$

6 $\int \sin^4 x\cos x\,dx$

7 $\int \sin^6 x\cos^3 x\,dx$

8 $\int \sin^2 x\,dx$

9 $\int \cos^4 x\,dx$

10 $\int \sec^2 x\tan^8 x\,dx$

11 $\int \sec^4 x\tan^8 x\,dx$

12 $\int \sec^2 x\tan x\,dx$

13 $\int \tan^2 x\,dx$

14 $\int \frac{1}{(9-x^2)^{3/2}}\,dx$

15 $\int x\sqrt{1-x^2}\,dx$

16 $\int \sqrt{4+x^2}\,dx$

17 $\int \frac{1}{\sqrt{9+x^2}}\,dx$

18 $\int \sqrt{x^2-9}\,dx$

19 $\int \frac{1}{(x^2+1)^2}\,dx$

20 $\int \frac{1}{x^2-4}\,dx$

21 $\int \frac{x^3}{x^2-1}\,dx$

22 $\int \frac{x^2}{x^2+2}\,dx$

23 $\int \frac{1}{x^2-3x+2}\,dx$

24 $\int \frac{x}{x^2+6x+10}\,dx$

25 $\int \frac{x}{x^2-4x+8}\,dx$

26 $\int \frac{8x}{(x^2-4)(x+2)}\,dx$

27 $\int \frac{1}{x^3+x^2}\,dx$

28 $\int \frac{x+1}{x^3+x}\,dx$

29 $\int x\sqrt{x+1}\,dx$

30 $\int x^2\sqrt{1-x}\,dx$

31 $\int \frac{1}{1+\sqrt{x}}\,dx$

Evaluate or indicate divergence for each of the integrals in problems 32–38.

32 $\int_0^2 \frac{1}{\sqrt{4-x^2}}\,dx$

33 $\int_0^\infty xe^{-x^2}\,dx$

34 $\int_0^4 \frac{1}{x-1}\,dx$

35 $\int_0^2 \frac{1}{(x-1)^{4/3}}\,dx$

36 $\int_1^\infty xe^{-x}\,dx$

37 $\int_0^1 x\ln(1/x)\,dx$

38 $\int_0^\infty \frac{1}{(x^2+9)^{3/2}}\,dx$

# CHAPTER NINE

# *Applications of integration to differential equations*

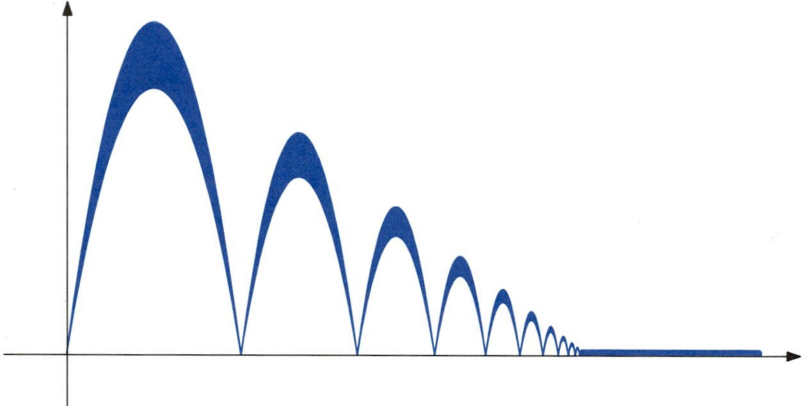

One of the most important applications of calculus is to the solution of differential equations. In Section 7.7 we saw some differential equations that are related to exponential growth and decay. In this chapter we will study a wider variety of differential equations and see how the concept relates to selected applications. Our attention will be restricted to differential equations that can be solved by using the techniques of integration developed in Chapter 8.

## 9.1 INTRODUCTION TO SEPARABLE DIFFERENTIAL EQUATIONS

A central idea in the application of mathematics to practical problems is to find a function that has some prescribed properties. In many problems the desired properties involve rate of change or derivatives of the unknown function. Thus, we are led to an equation that relates an unknown function and one or more of its derivatives to an independent variable. Such equations are called **differential equations**. The following are examples.

$$y' = -32x + 20, \tag{a}$$

$$y'' = -32, \tag{b}$$

$$y' = 2(5 - y), \tag{c}$$

$$yy' = \frac{3}{(4 + x)^2}, \tag{d}$$

$$(y')^2 + 4x^2 = 4, \tag{e}$$

$$y'' + 3y' + 2y = 10 \sin x. \tag{f}$$

In these equations $y$ is considered an unknown function of the independent variable $x$.

The **order** of a differential equation is the order of the highest-order derivative that occurs in the equation. For example, equations (b) and (f) above are second-order differential equations. The others above are first-order differential equations.

A function is said to be a **solution** of a differential equation if the function satisfies the equation for all values of the independent variable in some interval.

■ **EXAMPLE 1**

Show that for any constant $C$, $y(x) = 5 + Ce^{-2x}$ is a solution of the differential equation $y' = 2(5 - y)$.

**SOLUTION**

If $C$ is any constant and $y(x) = 5 + Ce^{-2x}$, then

$$y'(x) = Ce^{-2x}(-2) = -2Ce^{-2x}.$$

Also, substitution gives

$$2(5 - y(x)) = 2(5 - (5 + Ce^{-2x})) = 2(-Ce^{-2x}) = -2Ce^{-2x}.$$

We then see that $y'(x) = 2(5 - y(x))$, so $y(x)$ is a solution of the differential equation. ■

**■ EXAMPLE 2**

Show that for any constants $C_1$ and $C_2$, $y(x) = \sin x - 3\cos x + C_1e^{-x} + C_2e^{-2x}$ is a solution of the differential equation $y'' + 3y' + 2y = 10 \sin x$.

**SOLUTION**

We have

$$y(x) = \sin x - 3\cos x + C_1e^{-x} + C_2e^{-2x},$$

$$y'(x) = \cos x + 3\sin x - C_1e^{-x} - 2C_2e^{-2x}, \quad \text{and}$$

$$y''(x) = -\sin x + 3\cos x + C_1e^{-x} + 4C_2e^{-2x}.$$

Substituting and then combining like terms, we obtain

$$\begin{aligned} y''(x) + 3y'(x) + 2y(x) &= -\sin x + 3\cos x + C_1e^{-x} + 4C_2e^{-2x} \\ &\quad + 3(\cos x + 3\sin x - C_1e^{-x} - 2C_2e^{-2x}) \\ &\quad + 2(\sin x - 3\cos x + C_1e^{-x} + C_2e^{-2x}) \\ &= -\sin x + 3\cos x + C_1e^{-x} + 4C_2e^{-2x} \\ &\quad + 3\cos x + 9\sin x - 3C_1e^{-x} - 6C_2e^{-2x} \\ &\quad + 2\sin x - 6\cos x + 2C_1e^{-x} + 2C_2e^{-2x} \\ &= 10\sin x. \end{aligned}$$

This shows that $y(x)$ is a solution of the differential equation. ■

**■ EXAMPLE 3**

If $C$ is any constant and $y(x)$ is a differentiable function that is given implicitly by the equation

$$y^2 = -\frac{6}{4 + x} + C,$$

show that $y(x)$ is a solution of the differential equation $yy' = 3/(4 + x)^2$.

**SOLUTION**

If $y$ is a differentiable function of $x$, we can differentiate the equation

$$y^2 = -\frac{6}{4 + x} + C$$

with respect to $x$ to obtain

$$2yy' = \frac{6}{(4 + x)^2},$$

so

$$yy' = \frac{3}{(4 + x)^2}.$$

(Remember that $y$ is a function of $x$, so we need to use the Chain Rule to differentiate $y^2$ with respect to $x$.) ■

Some methods of solving differential equations lead to solutions that are given implicitly by an equation in $x$ and $y$, as in Example 3. Solutions in **implicit form** are considered satisfactory. It is not necessary to express $y$ as an explicit function of $x$.

The set of all solutions of a differential equation is called the **general solution** of the equation. For example:

- The function $y(x) = -16x^2 + 20x + C$ is the general solution of the first-order equation $y' = -32x + 20$. That is, $y(x) = -16x^2 + 20x + C$ is a solution for every constant $C$, and every solution can be written in this form for some $C$.
- The function $y(x) = 5 + Ce^{-2x}$ is the general solution of the first-order equation $y' = 2(5 - y)$ in Example 1.
- The function $y(x) = -16x^2 + C_1x + C_2$ is the general solution of the second-order equation $y'' = -32$.
- The function $y(x) = \sin x - 3\cos x + C_1e^{-x} + C_2e^{-2x}$ of Example 2 is the general solution of the second-order equation $y'' + 3y' + 2y = 10 \sin x$.

We should expect that the general solution of a first-order equation involves one arbitrary constant $C$ and the general solution of a second-order equation involves two arbitrary constants. The general solution of an $n$th-order differential equation ordinarily involves $n$ arbitrary constants. Determining the form of the general solution of a differential equation involves concepts that are not ordinarily considered a part of calculus. We will deal with this question only in very limited special cases.

The questions of **existence** and **uniqueness** of the solution of a differential equation are central to the theory. As with the question of the general solution, we will deal with these questions only in very limited special cases.

Generally speaking, in order to have a unique solution to an $n$th-order equation, it is necessary to impose $n$ conditions on the solution. These conditions are used to determine the value of the $n$ unknown constants in the general solution. One way is to prescribe the value of the solution and its first $n - 1$ derivatives at a fixed **initial point** $x_0$. That is, for fixed numbers $x_0, y_0, y_1, \ldots, y_{n-1}$, it is required that

$$y(x_0) = y_0,\ y'(x_0) = y_1, \ldots,\ y^{(n-1)}(x_0) = y_{n-1}.$$

The problem of finding a solution of a differential equation for which the function and some of its derivatives have prescribed values at one point is called an **initial value problem**.

**EXAMPLE 4**

Show that $y(x) = 3e^{2x} + 2e^{-2x}$ is a solution of the initial value problem $y'' = 4y$, $y(0) = 5$, $y'(0) = 2$.

**SOLUTION**

We must show that $y(x)$ satisfies the differential equation and both initial conditions. We have

$$y(x) = 3e^{2x} + 2e^{-2x},$$

$$y'(x) = 3e^{2x}(2) + 2e^{-2x}(-2),$$

$$y''(x) = 3e^{2x}(2)(2) + 2e^{-2x}(-2)(-2)$$

$$= 4(3e^{2x} + 2e^{-2x}).$$

It follows that $y(x)$ satisfies the equation $y'' = 4y$. Since

$$y(0) = 3e^{2(0)} + 2e^{-2(0)} = 5 \quad \text{and} \quad y'(0) = 3e^{2(0)}(2) + 2e^{-2(0)}(-2) = 2,$$

the function $y(x)$ also satisfies the initial conditions. ■

Let us now illustrate how techniques of integration can be used to find the solutions of some simple differential equations.

The most simple differential equations are those of the form

$$y' = f(x).$$

The general solution of this equation can be written as

$$y(x) = \int f(x)\,dx,$$

since the indefinite integral on the right represents the collection of all functions that have derivative $f$. Using techniques of integration to evaluate the indefinite integral, we then obtain a function $F(x)$ such that

$$y(x) = F(x) + C.$$

Initial value problems are solved by using the condition $y(x_0) = y_0$ to determine the value of the constant $C$.

**EXAMPLE 5**

Find the solution of the initial value problem $y' = \sqrt{x + 1}$, $y(0) = 1$.

**SOLUTION**

If $y(x)$ is a solution, we have

$$y'(x) = \sqrt{x + 1}.$$

It follows that

$$y(x) = \int \sqrt{x+1}\, dx,$$

$$y(x) = \frac{2}{3}(x+1)^{3/2} + C.$$

We can check that this is a solution of the given differential equation for any fixed value of $C$. We need to find $C$ so that $y(0) = 1$. Substitution of 0 for $x$ in the formula for $y(x)$ gives us

$$y(0) = \frac{2}{3}(0+1)^{3/2} + C.$$

Setting this expression for $y(0)$ equal to 1, we have

$$\frac{2}{3}(0+1)^{3/2} + C = 1,$$

$$\frac{2}{3} + C = 1,$$

$$C = \frac{1}{3}.$$

The solution of the given initial value problem is

$$y(x) = \frac{2}{3}(x+1)^{3/2} + \frac{1}{3}.$$ ■

In Sections 9.2 and 9.3, we will see some applications of differential equations that can be written in the form

$$M(x) + N(y)y' = 0.$$

Differential equations of this type are called **separable**. [Note that equations of the form $y' = f(x)$ are separable.] Let us see how to find solutions of separable equations by integration.

If $y(x)$ is a solution of the separable differentiable equation $M(x) + N(y)y' = 0$, we have

$$M(x) + N(y(x))y'(x) = 0,$$

$$N(y(x))y'(x) = -M(x).$$

If $M$, $N$, and $y'$ are continuous, we can integrate the above equation with respect to $x$ to obtain

$$\int N(y(x))y'(x)\, dx = -\int M(x)\, dx.$$

Evaluation of the integral on the right gives a function of $x$. We can use the substitution $y = y(x)$, so that $dy = y'(x)\,dx$, to transform the integral on the left to $\int N(y)\,dy$. This integral can then be evaluated as a function of $y$. We then obtain an equation that gives the solution $y$ as an implicit function of $x$. In some cases we can solve for $y$ to obtain an explicit solution.

**■ EXAMPLE 6**

Find the general solution of the separable differential equation $3y^2y' = 2x - 1$.

**SOLUTION**

Let us assume that there is a solution $y(x)$, so

$$3[y(x)]^2y'(x) = 2x - 1.$$

Integrating the above equation with respect to $x$, we obtain

$$\int 3[y(x)]^2y'(x)\,dx = \int (2x - 1)\,dx.$$

We use the substitution $y = y(x)$, $dy = y'(x)\,dx$ in the integral on the left and evaluate both integrals. The constants of integration on the two sides of the equation can be combined and written as a single constant $C$. We have

$$\int 3y^2\,dy = \int (2x - 1)\,dx,$$

$$y^3 = x^2 - x + C,$$

$$[y(x)]^3 = x^2 - x + C.$$

Solving the latter equation for $y(x)$, we obtain

$$y(x) = (x^2 - x + C)^{1/3}.$$

It follows that any solution of the differential equation must be of this form. We can also verify that for any value of the constant $C$, the above $y(x)$ is a solution, so we have found the general solution of the equation. ■

## EXERCISES 9.1

Show that the functions are solutions of the differential equations in Exercises 1–6.

**1** $y' = 2y$, $y(x) = Ce^{2x}$

**2** $y'' - y' - 2y = 0$, $y(x) = C_1e^{-x} + C_2e^{2x}$

**3** $y'' - 2y' + y = 0$, $y(x) = C_1e^x + C_2xe^x$

**4** $y'' - 2y' + 2y = 0$, $y(x) = C_1e^x \cos x + C_2e^x \sin x$

5 $(1 + \ln y)y' = 2x$, $y \ln y = x^2 + C$

6 $yy' = x$, $y^2 - x^2 = C$

Find the general solutions of the separable differential equations in Exercises 7–12.

7 $y' = 6x^2 - 4x + 3$

8 $y' = 6 \sin 2x - 12 \cos 3x$

9 $yy' = x$

10 $y^2y' = 6$

11 $\dfrac{y'}{y} = 2$ (Assume $y > 0$.)

12 $xy' = y$ (Assume $x > 0$ and $y > 0$.)

Solve the initial value problems in Exercises 13–16.

13 $y' + x = 0$, $y(1) = 2$

14 $y^2y' = x^2$, $y(0) = 2$

15 $y' + 3y = 0$, $y(0) = 2$ (Assume $y > 0$.)

16 $y' = 3(y - 1)$, $y(0) = 5$ (Assume $y > 1$.)

## 9.2 FAMILIES OF CURVES, CHEMICAL REACTIONS

In this section we will see how differential equations are related to the study of curves and to certain types of chemical reactions. The examples involve separable differential equations which can be solved by the method of integration that was introduced in Section 9.1.

### *Curves*

We know that the slope of the line tangent to the curve $y = y(x)$ at a point $(x, y)$ on the curve is given by $m = y'(x)$. A curve with a prescribed slope $m$ at each point is a solution of the differential equation $y' = m$. Generally, the prescribed slope $m$ at a point $(x, y)$ can depend on both $x$ and $y$.

■ **EXAMPLE 1**

Find the equation of the curve that has slope $m = 2x$ and contains the point (1, 3).

**SOLUTION**

We want a solution of the differential equation

$$y' = 2x.$$

The general solution of this equation is

$$y = \int 2x\,dx,$$

$$y = x^2 + C.$$

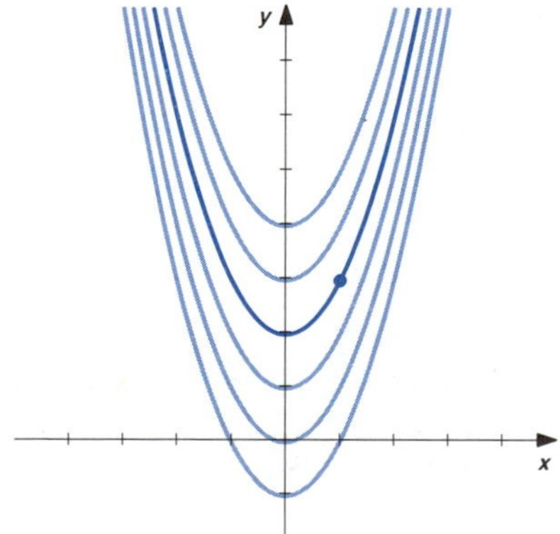

FIGURE 9.2.1

Note that the graph of every solution is a parabola. See Figure 9.2.1. We want the (unique) solution that contains the point (1, 3). If the curve is to contain the point (1, 3), we must have

$$3 = 1^2 + C, \quad \text{so} \quad C = 2.$$

The desired equation is $y = x^2 + 2$. ■

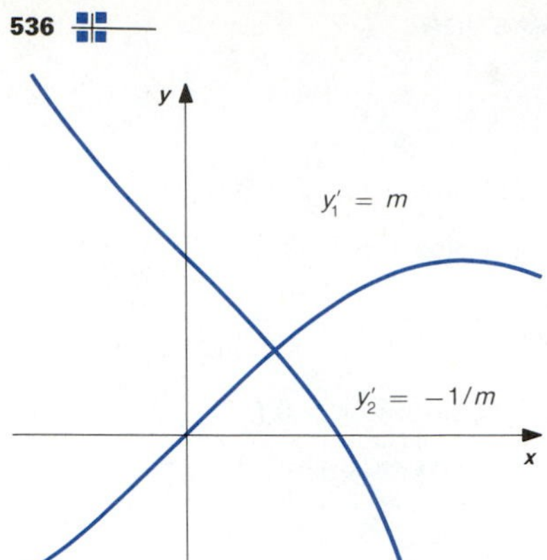

FIGURE 9.2.2

The collection of all curves that form the general solution of a differential equation is called the **family of curves** associated with the differential equation.

Let us consider the families of curves associated with each of the differential equations

$$y' = m(x) \quad \text{and} \quad y' = -\frac{1}{m(x)}.$$

Let $y_1$ be any solution of the differential equation $y' = m(x)$, and $y_2$ be any solution of $y' = -1/m(x)$. The slopes of the lines tangent to the curves at any point of intersection of the two curves, $(x_0, y_1(x_0)) = (x_0, y_2(x_0))$, are $y_1'(x_0) = m(x_0)$ and $y_2'(x_0) = -1/m(x_0)$, respectively. See Figure 9.2.2. Since these slopes are negative reciprocals, *the curves intersect at right angles.*

Two families of curves are called **orthogonal families** if each curve of one family intersects curves of the other family at right angles at every point of intersection.

One example of orthogonal families of curves that occurs in applications involves steady-state heat flow along a thin, flat surface. The heat flows along a family of curves that is orthogonal to the family of isothermal curves where the temperature remains constant.

**■ EXAMPLE 2**

Find the family of curves that is associated with the differential equation $y' = \cos x$, $-\pi/2 < x < \pi/2$. Find an orthogonal family, and sketch three curves from each family.

**SOLUTION**

The general solution of the equation $y' = \cos x$ is

$$y = \int \cos x \, dx,$$

$$y = \sin x + C.$$

Curves of the family of curves associated with the differential equation $y' = \cos x$ have slope $m(x) = \cos x$ at points $(x, y)$ on the curves. The negative reciprocal of this slope is $-1/m(x) = -1/\cos x$. Hence, the general solution of the differential equation

$$y' = -\frac{1}{\cos x}$$

is a family of curves that is orthogonal to the family of curves that satisfy the differential equation $y' = \cos x$. We have

$$y' = -\sec x,$$

$$y = \int -\sec x\, dx,$$

[Formula (8) of Section 8.3]

$$y = -\ln|\sec x + \tan x| + C.$$

To sketch the graph of

$$y = -\ln|\sec x + \tan x| + C, \qquad -\pi/2 < x < \pi/2,$$

we note that $y(0) = -\ln|1 + 0| + C = C$. Also, $y'(x) = -\sec x < 0$, $-\pi/2 < x < \pi/2$, so the function is decreasing in this interval. It is easy to see that $\lim_{x\to\pi/2^-} (\sec x + \tan x) = \infty$, so $y = -\ln|\sec x + \tan x| + C \to -\infty$ as $x \to \pi/2^-$ and the graph has vertical asymptote $x = \pi/2$. Using l'Hôpital's Rule, we have

$$\lim_{x\to-\pi/2^+} (\sec x + \tan x) = \lim_{x\to-\pi/2^+} \left(\frac{1}{\cos x} + \frac{\sin x}{\cos x}\right)$$

$$= \lim_{x\to-\pi/2^+} \frac{1 + \sin x}{\cos x}$$

$$\underset{(0/0,\,\text{l'H})}{=} \lim_{x\to-\pi/2^+} \frac{\cos x}{-\sin x} = 0.$$

Since the logarithmic function becomes unbounded near zero, this shows that the graph has vertical asymptote $x = -\pi/2$. Some of the curves of each family are sketched in Figure 9.2.3. ■

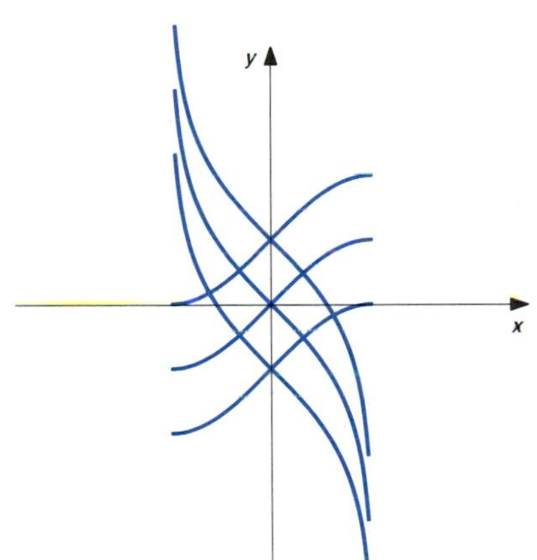


FIGURE 9.2.3

The general solution of the separable equation

$$M(x) + N(y)y' = 0$$

is a family of curves with the property that the slope at a point $(x, y)$ on any of the curves is

$$y' = -\frac{M(x)}{N(y)}.$$

Any orthogonal family of curves will have slope at $(x, y)$ given by the negative reciprocal, so any orthogonal family will satisfy

$$y' = \frac{N(y)}{M(x)}.$$

This is also a separable differential equation.

**■ EXAMPLE 3**

Determine the family of curves associated with the differential equation $x + 2yy' = 0$. Find an orthogonal family and sketch curves in each family.

**SOLUTION**

Let us first find the general solution of the equation

$$x + 2yy' = 0,$$

$$2yy' = -x.$$

Assuming that $y$ is a function of $x$, we integrate with respect to $x$ to obtain

$$\int 2yy'\,dx = -\int x\,dx.$$

Noting that $dy = y'\,dx$, we write

$$\int 2y\,dy = -\int x\,dx.$$

Evaluating the integrals then gives

$$y^2 = -\frac{x^2}{2} + C,$$

$$\frac{x^2}{2} + y^2 = C.$$

For $C > 0$, these curves are ellipses with center at the origin.

The family of curves associated with the equation $x + 2yy' = 0$ has slope $y' = -x/2y$. Any orthogonal family of curves has slope given by the negative reciprocal, or $y' = 2y/x$. Let us find the general solution of this equation. We have

$$\frac{1}{y}y' = \frac{2}{x}.$$

Integrating with respect to $x$ gives

$$\int \frac{1}{y}y'\,dx = \int \frac{2}{x}\,dx,$$

$$\int \frac{1}{y}\,dy = \int \frac{2}{x}\,dx,$$

$$\ln|y| = 2\ln|x| + c,$$

$$|y| = e^{2\ln|x| + c},$$

$$|y| = e^{\ln x^2}e^c$$

$$y = Cx^2,$$

where $C = \pm e^c$. The curves in this family are parabolas.

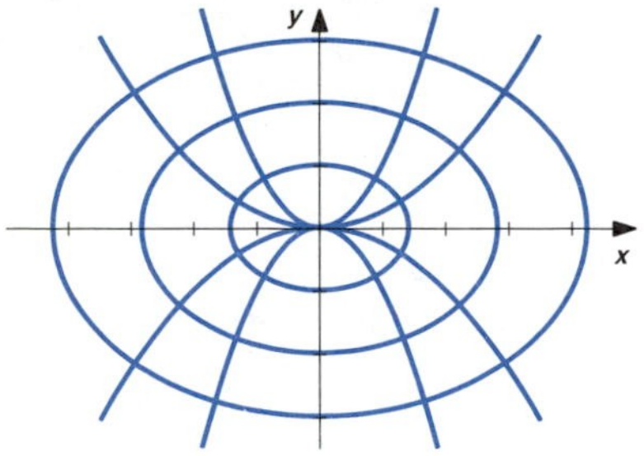

Some curves from each of the orthogonal families are sketched in Figure 9.2.4. ■

FIGURE 9.2.4

## *Chemistry*

Let us consider a simple chemical reaction where a molecule of a substance $A$ combines with a molecule of a substance $B$ to form a molecule of a substance $C$, written

$$A + B \to C.$$

In certain reactions of this type, the reaction depends on the collision of a molecule of substance $A$ with a molecule of substance $B$, and the rate at which molecules of $C$ are formed is proportional to the product of the amount of substance $A$ and the amount of substance $B$ that are present at time $t$. Let $z(t)$ denote the number of molecules of substance $C$ that have been formed after $t$ units of time, so that $z'(t)$ is the rate at which molecules of $C$ are formed. Let $x(t)$ denote the number of molecules of substance $A$ that remains at time $t$, and let $y(t)$ denote the number of molecules of substance $B$ that remain at time $t$. The rate at which molecules of $C$ are formed is proportional to the product of the amounts of $A$ and $B$, a statement that we can express by the equation

$$z'(t) = kx(t)y(t).$$

To translate this statement into a differential equation for $z(t)$, we note that each molecule of $C$ formed reduces the number of molecules of both $A$ and $B$ by one. This means we have $x(t) = x(0) - z(t)$ and $y(t) = y(0) - z(t)$. We can then express $z(t)$ as the solution of the separable differential equation

$$z' = k(x(0) - z)(y(0) - z).$$

It is convenient to express the number of molecules of a substance in terms of **moles** (mol). One mole of a substance is approximately $6.02 \times 10^{23}$ molecules of the substance. The mass in grams of 1 mol of a substance is the **molecular weight** of the substance.

### ■ EXAMPLE 4

Suppose that $A$ and $B$ combine to form $C$ at a rate that is proportional to the product of the number of molecules of $A$ and the number of molecules of $B$. Find the equation for the amount of $C$ formed when 3 mol of $A$ are mixed with 2 mol of $B$.

#### SOLUTION

We want the solution of the initial value problem

$$z' = k(3 - z)(2 - z), \qquad z(0) = 0.$$

Solving the differential equation, we have

$$\frac{1}{(3 - z)(2 - z)} z' = k.$$

Integrating with respect to $t$, we obtain

$$\int \frac{1}{(3 - z)(2 - z)} z' \, dt = \int k \, dt,$$

$$\int \frac{1}{(z - 3)(z - 2)} \, dz = \int k \, dt.$$

To evaluate the integral on the left, we express the integrand in terms of partial fractions. (The method of partial fractions is discussed in Section 8.5.) We have

$$\frac{1}{(z - 3)(z - 2)} = \frac{A}{z - 3} + \frac{B}{z - 2}.$$

Multiplying each side of the above equation by $(z - 3)(z - 2)$, we obtain

$$1 = A(z - 2) + B(z - 3),$$

$$0 = (A + B)z + (-2A - 3B - 1).$$

Then

$$\begin{cases} A + B = 0, & \text{[Coefficient of } z\text{]} \\ -2A - 3B - 1 = 0. & \text{[Constant term]} \end{cases}$$

Solving for $A$ and $B$, we obtain $A = 1$ and $B = -1$, so

$$\frac{1}{(z - 3)(z - 2)} = \frac{1}{z - 3} - \frac{1}{z - 2}.$$

Returning to the solution of our differential equation, we now have

$$\int \left( \frac{1}{z - 3} - \frac{1}{z - 2} \right) dz = \int k \, dt,$$

$$\ln|z - 3| - \ln|z - 2| = kt + c,$$

$$\ln \left| \frac{z - 3}{z - 2} \right| = kt + c,$$

$$\left| \frac{z - 3}{z - 2} \right| = e^{kt + c},$$

$$\frac{z - 3}{z - 2} = Ce^{kt}, \qquad \text{where } C = \pm e^{c}.$$

Since $z(0) = 0$, we have

$$\frac{0-3}{0-2} = C, \quad \text{so } C = \frac{3}{2}$$

and

$$\frac{z-3}{z-2} = \frac{3}{2}e^{kt},$$

$$2(z-3) = 3(z-2)e^{kt},$$

$$(2 - 3e^{kt})z = 6 - 6e^{kt},$$

$$z = \frac{6 - 6e^{kt}}{2 - 3e^{kt}}.$$

FIGURE 9.2.5

Note that when the reaction is complete we would expect that 2 mol of $A$ and 2 mol of $B$ would have combined to form 2 mol of $C$. We would then have 2 mol of $C$, 1 mol of $A$ left over, and no $B$. From the solution we see that $z(t)$ is never 2, but $z(t) \to 2$ as $t \to \infty$. From the graph in Figure 9.2.5 we see that the reaction is nearly complete after a relatively short time. For example, after $4/k$ units of time, $z(4/k) \approx 1.9876386$ mol of $C$ has been formed. ∎

## EXERCISES 9.2

In Exercises 1–8 find the equation of the curve that has the given slope and contains the given point.

**1** $m = 4x - 1$, $(1, 3)$

**2** $m = \sin(x/2)$, $(0, 1)$

**3** $m = \dfrac{1}{1-x}$, $(0, 1)$

**4** $m = e^{2x}$, $(0, 0)$

**5** $m = \dfrac{x}{y^2}$, $(3, 3)$

**6** $m = \dfrac{2x^2}{y}$, $(1, 1)$

**7** $m = 1 + y^2$, $(0, 0)$

**8** $m = \sqrt{1 - y^2}$, $(0, 0)$

In Exercises 9–16 find the family of curves associated with the given differential equation and find an orthogonal family. Sketch three curves from each family.

**9** $y' = 3x^2$

**10** $y' = e^x$

**11** $y' = \dfrac{1}{1+x^2}$

**12** $y' = \sec^2 x$, $-\dfrac{\pi}{2} < x < \dfrac{\pi}{2}$

**13** $y' = 2y$

**14** $y' = y^2$

**15** $xy' = 3y$

**16** $yy' + x = 0$

**17** Suppose that $A$ and $B$ combine to form $C$ at a rate that is proportional to the product of the number of molecules of $A$ and the number of molecules of $B$. Find the equation for the amount of $C$ formed when 3 mol of $A$ is mixed with 4 mol of $B$.

**18** Suppose that $A$ and $B$ combine to form $C$ at a rate that is proportional to the product of the number of molecules of $A$ and the number of molecules of $B$. Find the equation for the amount of $C$ formed when 2 mol of $A$ is mixed with 1 mol of $B$.

19 Suppose that $A$ and $B$ combine to form $C$ at a rate that is proportional to the product of the number of molecules of $A$ and the number of molecules of $B$. Find the equation for the amount of $C$ formed when 1 mol of $A$ is mixed with 1 mol of $B$.

20 The population at time $t$ of a certain community is given approximately by the solution of the initial value problem

$$p' = 200p - 0.01p^2, \qquad p(0) = p_0.$$

Thus, the population increases when $p$ is small and decreases when $p$ is large. Find the solution for (a) $p_0 = 10{,}000$ and (b) $p_0 = 30{,}000$.

21 Find a differential equation satisfied by the family of curves $xy = C\ (x > 0)$ and find a family orthogonal to this family.

## 9.3 MECHANICS, PHASE PLANE EQUATIONS

The examples in this section involve separable differential equations.

### *Mechanics*

Let us consider the motion along a line of a particle with constant mass. We assume the particle is subject to an external force that acts in a direction along the line. **Newton's Second Law of Motion** asserts that the force on the particle is proportional to mass times acceleration. For appropriate systems of units, the constant of proportionality is 1, so

$$F = ma.$$

Some common systems of units are given in Table 9.3.1.

**TABLE 9.3.1**

| *System* | *Length* | *Mass* | *Time* | *Force* |
|---|---|---|---|---|
| English | feet | slugs | seconds | pounds |
| mks(SI) | meters | kilograms | seconds | newtons |
| cgs | centimeters | grams | seconds | dynes |

The weight of an object is the force due to the gravitational attraction of the earth. Thus, Newton's Second Law tells us

$$w = ma,$$

where $w$ is the weight, $m$ is the mass, and $a$ is the acceleration due to gravity. The weight and acceleration due to gravity depend on the distance of the object from the center of the earth. The mass depends on the molecular structure and not on the position of the object.

For problems that involve vertical motion near the surface of the earth, we assume that the magnitude of acceleration due to gravity has a constant value $g$. We will use either

$$g = 32 \text{ ft/s}^2 \quad \text{or} \quad g = 9.8 \text{ m/s}^2,$$

depending on the units. These are approximate values of the acceleration due to gravity at sea level. The weight of an object near the surface of the earth and the mass of the object are then related by the equation

$$w = mg.$$

We know that acceleration is the derivative of the velocity with respect to time, $a = v'$. Hence, we can write Newton's equation as

$$F = mv'.$$

This form of Newton's equation is convenient for the applications we will consider.

The force due to gravity on an object near the surface of the earth is its weight, $w = mg$. If we assume that this is the only force acting on an object, Newton's equation gives $mg = mv'$, or simply

$$v' = g.$$

■ **EXAMPLE 1**

An object is thrown vertically upward with velocity 64 ft/s. Find the maximum height of the object.

**SOLUTION**

We choose a vertical number line with positive direction upward and origin at ground level. Distance is measured in feet, so we choose $g = -32$, where the negative sign indicates the force due to gravity is in the negative, downward direction. The object has initial position $s(0) = 0$ and initial velocity $v(0) = +64$. See Figure 9.3.1.

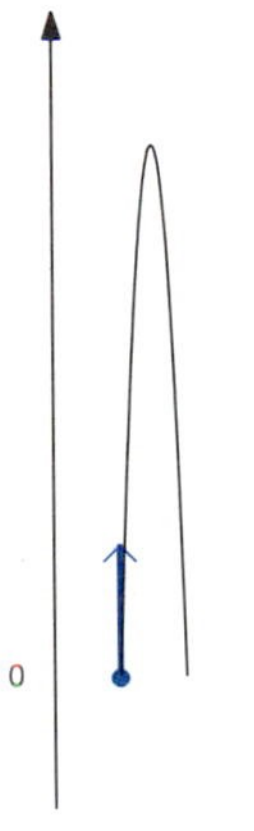

FIGURE 9.3.1

The velocity is the solution of the initial value problem

$$v' = -32, \qquad v(0) = 64.$$

The general solution of the equation is

$$v = \int -32\, dt,$$

$$v = -32t + C_1.$$

When $t = 0$ we have $v(0) = 64$, so $C_1 = 64$ and

$$v = -32t + 64.$$

We know that $s' = v$, where $s(t)$ is the height at time $t$. It follows that the height is the solution of the initial value problem

$$s' = -32t + 64, \qquad s(0) = 0.$$

The general solution is

$$s = \int (-32t + 64)\, dt,$$

$$s = -16t^2 + 64t + C_2.$$

The expression $s(0) = 0$ implies $C_2 = 0$, so

$$s(t) = -16t^2 + 64t.$$

The maximum value of the function $s(t)$ is obtained when the velocity is zero. We have

$$v(t) = -32t + 64,$$

so $v(t) = 0$ implies $t = 2$. When $t = 2$, the height is

$$s(2) = -16(2)^2 + 64(2) = 64.$$

The maximum height is 64 ft. ■

The net force due to two or more forces that are acting in the positive or negative direction along the same line is the algebraic sum of the forces.

■ **EXAMPLE 2**

Determine the vertical velocity and height above ground of a 128-lb parachutist $t$ seconds after jumping from an airplane that is flying slowly and horizontally at an altitude of 5000 ft. Assume that air resistance is eight times the speed and ignore horizontal motion.

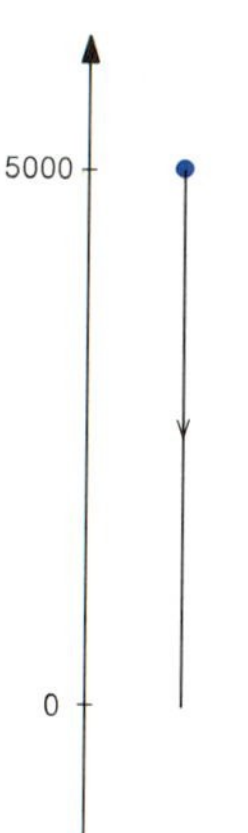


FIGURE 9.3.2

**SOLUTION**

We choose a vertical coordinate line with origin at ground level and positive direction upward. The initial height of the parachutist is $s_0 = 5000$ ft. The initial vertical velocity is $v_0 = 0$. See Figure 9.3.2.

The (total) weight of the parachutist is 128 lb. The force due to gravity is then $-128$, where the negative sign indicates the force is directed downward.

The mass is

$$m = \frac{w}{g} = \frac{128}{32} = 4 \text{ slugs}.$$

The force exerted by the parachute due to air resistance is opposite to the motion and has magnitude $8|v|$, so it is $-8v$. Note that the parachutist is falling, so the velocity is negative and the force $-8v$ is directed upward.

Newton's Second Law of Motion then gives

$$mv' = F,$$

$$4v' = -128 - 8v,$$

$$v' = -2(16 + v).$$

This is a separable differential equation for the velocity as a function of time. Let us solve this equation to determine the velocity. We have

$$\frac{1}{16+v}v' = -2.$$

Integrating with respect to $t$, we obtain

$$\int \frac{1}{16+v}v'\,dt = \int -2\,dt,$$

$$\int \frac{1}{16+v}\,dv = \int -2\,dt,$$

$$\ln|16+v| = -2t + c,$$

$$|16+v| = e^{-2t+c},$$

$$16 + v = Ce^{-2t},$$

where $C = \pm e^c$. Since the initial velocity is zero, we have

$$16 + 0 = C.$$

We then have the equation for the velocity

$$v(t) = -16 + 16e^{-2t}.$$

The height is a solution of the initial value problem

$$s' = v, \qquad s(0) = 5000.$$

We have

$$s' = -16 + 16e^{-2t},$$

$$s = \int (-16 + 16e^{-2t})\,dt,$$

$$s = -16t - 8e^{-2t} + C.$$

Since the initial height is $s_0 = 5000$, we have $5000 = -16(0) - 8 + C$, so $C = 5008$. Thus,

$$s(t) = -16t - 8e^{-t} + 5008.$$ ■

It is interesting to compare the result of Example 2 with the result obtained if air resistance is neglected. This would be a somewhat reasonable assumption without the parachute. In that case we would have the familiar result

$$s(t) = -16t^2 + v_0 t + s_0 = -16t^2 + 5000.$$

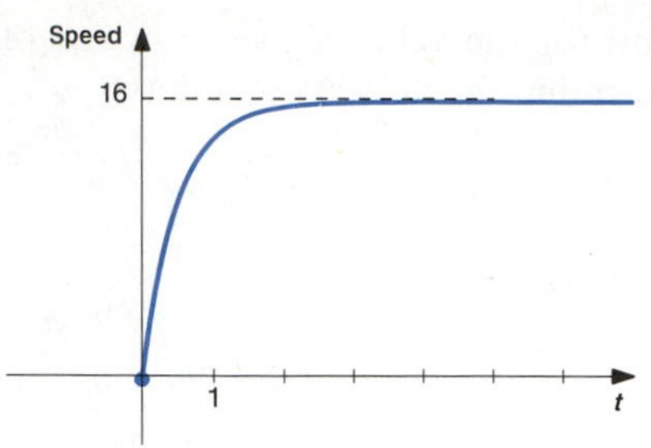


FIGURE 9.3.3

Thus, without a parachute, a person jumping from an altitude of 5000 ft would be expected to hit the ground approximately $\sqrt{5000/16} \approx 17.7$ s later at a speed of $32(17.7) \approx 566$ ft/s. With the greatly increased air resistance due to the parachute, the person in Example 2 is never falling as fast as 16 ft/s, as we can see from the equation

$$v(t) = -16 + 16e^{-2t}.$$

The limiting value of the velocity is called the **terminal velocity**. The speed the parachutist is falling at time $t$, $|v(t)|$, is indicated in Figure 9.3.3.

### *Phase plane equations*

In many problems that involve motions, the external force depends on the position and/or the velocity of the object, but not on time. We can then transform Newton's equation into a differentiable equation for the velocity as a function of the position. To do this, we note that the Chain Rule gives

$$v' = \frac{dv}{dt} = \frac{dv}{dx}\frac{dx}{dt} = \frac{dv}{dx}v.$$

Newton's equation $mv' = F$ can then be written

$$mv\frac{dv}{dx} = F.$$

This form of Newton's equation is called the **phase plane** equation of the motion.

■ **EXAMPLE 3**

A spring is hanging vertically down from a fixed point with a 2-lb weight attached at the bottom. The 2-lb weight stretches the spring 1 ft beyond its natural length. The weight is then pulled downward an additional 1 ft, held motionless at that point, and then released. Find the speed of the weight at the point where the spring is at its natural length.

**SOLUTION**

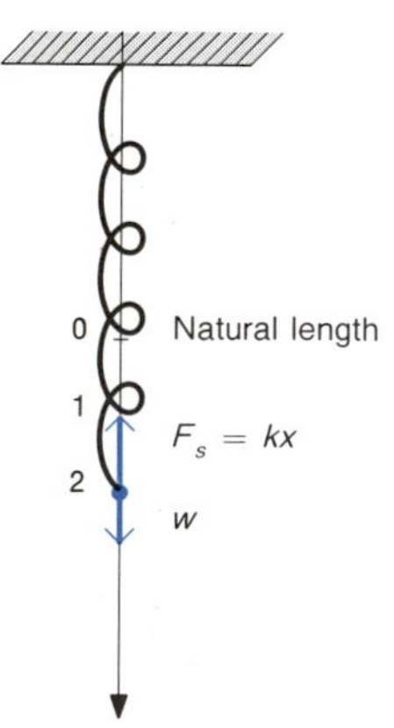


FIGURE 9.3.4

Let us choose a vertical coordinate line with origin at the point where the spring is at its natural length and positive direction downward. See Figure 9.3.4. Since the weight stretches the spring 1 ft and the weight is then pulled downward an additional 1 ft, the initial position of the weight is $x_0 = 2$. Since the weight is held motionless as it is released, its initial velocity is $v_0 = 0$.

Let us determine the force on the weight that is exerted by the spring. **Hooke's Law** tells us that the magnitude of the force is proportional to the displacement from the natural length of the spring. That is,

$$F_s = kx.$$

The force acts in the direction that tends to return the spring to its natural length. The statement that a 2-lb weight stretches the spring 1 ft beyond its natural length tells us that the spring force matches the 2-lb force of gravity when $x$ is 1 ft. When $x$ is positive, the spring force is directed in the negative direction, toward the origin of the coordinate line we have chosen. We conclude that $F_s = -2$ (lb) when $x = 1$ (ft). Substitution then gives

$$-2 = k(1), \quad \text{so} \quad k = -2.$$

The force on the weight due to the spring is then $F_s = -2x$.

The force on the weight due to gravity has magnitude equal to its weight and is directed downward, in the positive direction, so it is equal to 2.

The mass of the weight is

$$m = \frac{w}{g} = \frac{2}{32} = \frac{1}{16}.$$

We assume that the spring force and gravity are the only forces acting on the weight. We see that neither of these forces is a function of time, so we can use the phase plane equation of motion. We have

$$mv\frac{dv}{dx} = F,$$

$$\frac{1}{16}v\frac{dv}{dx} = -2x + 2,$$

$$v\frac{dv}{dx} = -32x + 32.$$

This is a separable differential equation for the velocity as a function of the position. Integrating the equation with respect to $x$, we have

$$\int v\frac{dv}{dx}\,dx = \int (-32x + 32)\,dx,$$

$$\int v\,dv = \int (-32x + 32)\,dx,$$

$$\frac{v^2}{2} = -16x^2 + 32x + C.$$

Substituting the initial values $x_0 = 2$ and $v_0 = 0$, we have

$$0 = -16(2)^2 + 32(2) + C, \quad \text{so} \quad C = 0.$$

Hence, the velocity and position are related by the equation

$$v^2 = -16x^2 + 32x.$$

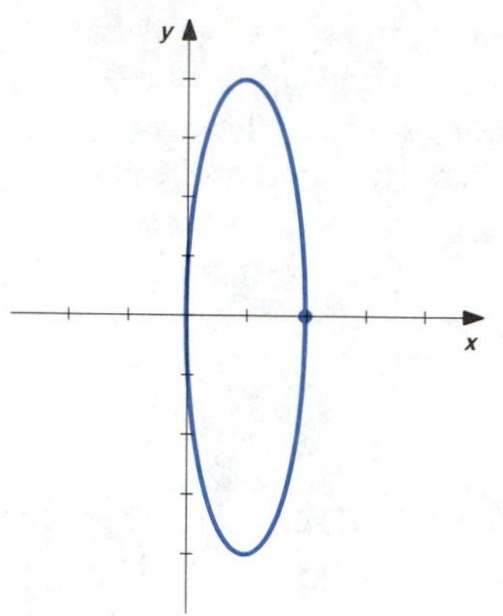

FIGURE 9.3.5

When $x = 0$, the velocity is zero. The graph of the solution in the $xv$-phase plane is given in Figure 9.3.5. The choice of any point on this curve as initial values results in the same motion of the spring. ■

The position of the weight in Example 3 as a function of time is given by the (unique) solution of the second-order initial value problem

$$\frac{d^2x}{dt^2} = -32x + 32, \qquad x(0) = 2, \quad v(0) = 0.$$

We have not learned to solve problems of this type, but we can verify that the solution is

$$x(t) = 1 + \cos\sqrt{32}t.$$

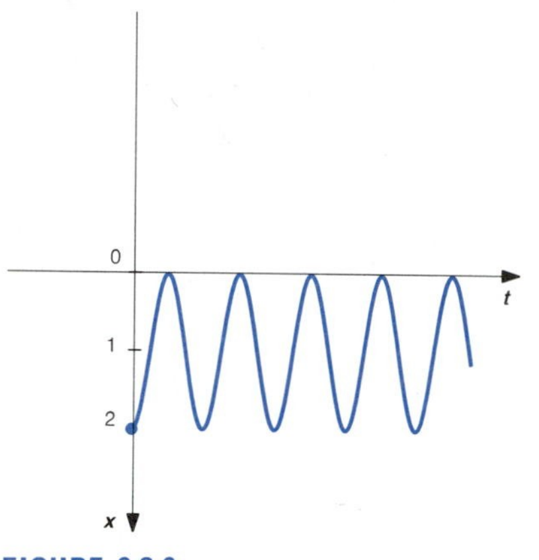

FIGURE 9.3.6

This shows that the weight in Example 3 oscillates between the two points $x = 2$ and $x = 0$, where the velocity is zero. The graph of $x(t)$ is given in Figure 9.3.6.

In a problem that involves the force of gravity acting on an object that does not remain near the surface of the earth, we cannot assume that the force due to gravity is constant. In that case we must use the fact that the force due to the gravitational pull of the earth is proportional to the mass of the object and inversely proportional to the square of the distance of the object from the center of the earth. We can express this force in terms of a vertical coordinate system with origin at the surface of the earth as

$$F = -\frac{R^2gm}{(R + x)^2},$$

where $R$ is the radius of the earth, $g$ is the acceleration due to gravity at sea level, $m$ is the fixed mass of the object, and $x$ is the height of the object above sea level. See Figure 9.3.7. The radius of the earth is approximately $R = 2.1 \times 10^7$ ft.

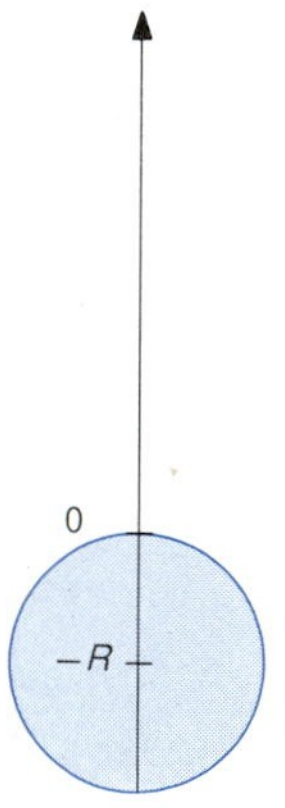

FIGURE 9.3.7

■ **EXAMPLE 4**

A rocket is launched from earth with initial velocity $v_0$. Find a formula for the velocity of the rocket in terms of its height. Assume that gravity is the only force acting on the rocket and that the mass of the rocket is constant. (The assumptions are reasonable if the fuel burns quickly, while the rocket is relatively near the surface of the earth.)

**SOLUTION**

We choose a coordinate line as indicated in Figure 9.3.7. The phase plane equation of the motion is

$$mv\frac{dv}{dx} = -\frac{R^2gm}{(R + x)^2}.$$

This is a separable differential equation. Canceling $m$ from both sides of the equation and then integrating with respect to $x$, we obtain

$$\int v\frac{dv}{dx}\,dx = \int -\frac{R^2g}{(R+x)^2}\,dx,$$

$$\int v\,dv = \int -\frac{R^2g}{(R+x)^2}\,dx,$$

$$\frac{v^2}{2} = \frac{R^2g}{R+x} + C.$$

Since $v = v_0$ when $x = 0$, we have

$$\frac{v_0^2}{2} = Rg + C, \quad \text{so } C = \frac{v_0^2}{2} - Rg.$$

The equation for the velocity in terms of the height is then

$$v^2 = \frac{2R^2g}{R+x} + v_0^2 - 2Rg.$$

The above equation tells us that the velocity will never become zero if $v_0 \geq \sqrt{2Rg} \approx 36{,}660$ ft/s $\approx 25{,}000$ mph. This is the (approximate) initial velocity that a rocket would need to avoid being drawn back to the earth by the earth's gravity. Thus, $\sqrt{2Rg}$ is called the **escape velocity**. Representative graphs of the solution in the $xv$-phase plane for the cases $v_0 \geq \sqrt{2Rg}$ and $v_0 < \sqrt{2Rg}$ are given in Figure 9.3.8. ■

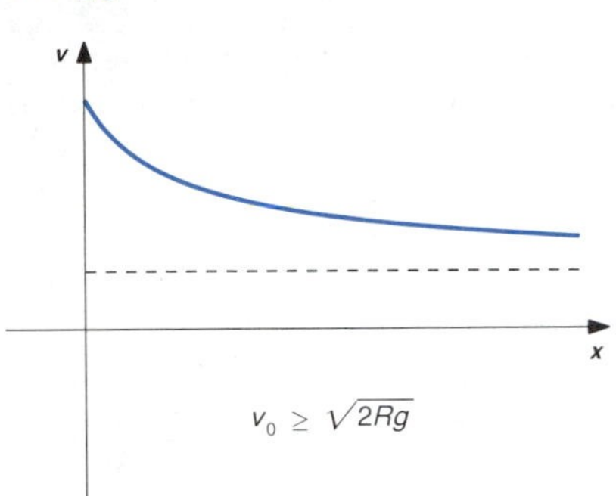


(a)

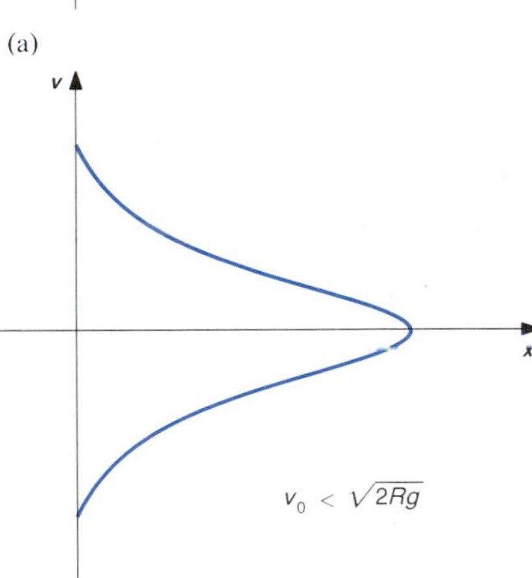


(b)

FIGURE 9.3.8

## EXERCISES 9.3

1 An object is thrown vertically upward from a height of 60 ft with velocity 64 ft/s. How many seconds later will it hit the ground?

2 An object is thrown vertically downward from a height of 60 ft with velocity 64 ft/s. How many seconds later will it hit the ground?

3 A sandbag is released from a balloon that is 480 ft above ground and falling at a rate of 2 ft/s. How many seconds will it take the sandbag to hit the ground?

4 An object is thrown upward from ground level to a maximum height of 120 ft. What was its initial velocity?

5 Determine the velocity and height above ground of a 160-lb parachutist $t$ seconds after jumping from an airplane that is flying slowly and horizontally at 3000 ft. Assume the air resistance is eight times the speed and ignore horizontal motion. Find the terminal velocity of the parachutist.

6 Determine the velocity and height above ground of a 2048-lb package $t$ seconds after it is dropped by parachute from an airplane that is flying slowly and horizontally at 4000 ft. Assume the air resistance is eight times the speed and ignore horizontal motion. Find the terminal velocity of the package.

7 Determine the velocity and height above ground of a 160-lb skydiver $t$ seconds after jumping from an airplane that is flying slowly and horizontally at 3000 ft. Assume the air resistance is four times the speed and ignore horizontal motion. Find the terminal velocity of the skydiver.

8 Determine the velocity and height above ground of a 160-lb skydiver $t$ seconds after jumping from an airplane that is flying slowly and horizontally at 3000 ft. Assume the air resistance is twice the speed and ignore horizontal motion. Find the terminal velocity of the skydiver.

9 Determine the velocity of a 128-lb skydiver $t$ seconds after jumping from an airplane that is flying slowly and horizontally at 3000 ft. Assume the air resistance is twice the square of the speed and ignore horizontal motion. Find the terminal velocity of the skydiver.

10 Determine the vertical velocity of a 288-lb weight $t$ seconds after it is dropped from a height of 2000 ft. Assume the air resistance is twice the square of the velocity. Find the terminal velocity of the weight.

11 A spring is hanging vertically down from a fixed point with a 2-lb weight attached to the bottom. The 2-lb weight stretches the spring 0.5 ft beyond its natural length. The weight is then pulled downward an additional 1 ft, held motionless, and then released. Find the speed of the weight at the point where it originally hung and determine the positions of the weight where the velocity will be zero.

12 A spring is hanging vertically down from a fixed point with a 1-lb weight attached to the bottom. The 1-lb weight stretches the spring 2 ft beyond its natural length. The weight is then pulled downward an additional 1 ft, held motionless, and then released. How far will the weight rise before it falls downward?

13 A spring is hanging vertically down from a fixed point with a 4-lb weight attached to the bottom. The 4-lb weight stretches the spring 1 ft beyond its natural length. The weight is then lifted upward 1 ft so the spring is its natural length, held motionless, and then released. How far will the weight drop before it springs upward?

14 A spring is hanging vertically down from a fixed point with a 3-lb weight attached to the bottom. The 3-lb weight stretches the spring 1 ft beyond its natural length. The weight is then lifted upward 2 ft, so the spring is compressed 1 ft, held motionless, and then released. How far will the weight drop before it springs upward?

15 A force of 1 lb stretches a certain spring 0.3 ft. The spring is positioned horizontally, with one end fixed and a 1-lb weight attached to the free end. The weight is pulled so the spring is stretched 0.5 ft beyond its natural length, held motionless, and then released. Find the speed at which the weight passes the point where the spring is at its natural length. Find the extreme positions of the weight.

16 A force of 2 lb stretches a certain spring 1 ft. The spring is positioned horizontally, with one end fixed and a 2-lb weight attached to the free end. The weight is pushed so the spring is compressed 0.5 ft, held motionless, and then released. Find the speed at which the weight passes the point where the spring is at its natural length. Find the extreme positions of the weight.

17 A force of 3 lb stretches a certain spring 1 ft. The spring is positioned horizontally, with one end fixed and a 3-lb weight attached to the free end. At a certain instant, the weight is at the point where the spring is at its natural length and the weight is moving at a speed of 8 ft/s. Find the extreme positions of the weight.

18 A force of 4 lb stretches a certain spring 1 ft. The spring is positioned horizontally, with one end fixed and a 4-lb weight attached to the free end. At a certain instant, the weight is at the point where the spring is at its natural length and the weight is moving at a speed of 8 ft/s. Find the extreme positions of the weight.

19 A rocket is launched vertically from earth. At the time the engines are turned off, the rocket has an altitude of $R = 2.1 \times 10^7$ ft and a speed of 20,000 ft/s. Assuming that gravity is the only force acting on the rocket, what is the maximum altitude the rocket will reach before it is drawn back toward the earth? $R$ is the radius of the earth.

20 A rocket is launched vertically from earth. At the time the engines are turned off, the rocket has an altitude of $2R$ ft and a speed of 20,000 ft/s. Assuming that gravity is the only force acting on the rocket, what is the maximum altitude the rocket will reach before it is drawn back toward the earth? $R$ is the radius of the earth.

21 What speed at an altitude of $R$ ft would a rocket need to avoid being pulled by gravity back toward the earth? $R$ is the radius of the earth.

22 What speed at an altitude of $2R$ ft would a rocket need to avoid being pulled by gravity back toward the earth? $R$ is the radius of the earth.

23 Use Newton's Method to determine approximately how many seconds the parachutist in Example 2 falls.

## 9.4 FIRST-ORDER LINEAR DIFFERENTIAL EQUATIONS, INTEGRATING FACTORS

Equations of the form

$$y' + p(x)y = q(x)$$

are called **first-order linear** differential equations. If the functions $p$ and $q$ are continuous, we can solve an equation of this type by multiplying the equation by an **integrating factor** $m$, where $m$ is chosen so that

$m(y' + py) = (my)'$. If $\int p$ is any function that has derivative $p(x)$, we will see that

**$e^{\int p}$ is an integrating factor of the equation $y' + p(x)y = q(x)$.**

Let us illustrate how an integrating factor is used to solve the differential equation. Multiplying the equation by $e^{\int p}$, we obtain

$$e^{\int p}[y'(x) + p(x)y(x)] = e^{\int p}\, q(x).$$

Using the fact that

$$\frac{d}{dx}(e^{\int p}) = e^{\int p}\, p,$$

it can be verified that the expression on the left above is the derivative of the product $e^{\int p}y$. The above equation can then be rewritten

$$\frac{d}{dx}(e^{\int p}y) = e^{\int p}\, q, \quad \text{so}$$

$$e^{\int p}y(x) = \int e^{\int p}\, q(x)\, dx.$$

The formula for $y(x)$ is obtained by evaluating the integral on the right and then multiplying the resulting equation by $e^{-\int p}$.

It is not difficult to derive that an integrating factor of the equation $y' + py = q$ is of the form $\pm e^{\int p}$. That is, if $m(y' + py)$ is to be the derivative of the product $my$, we must have

$$(my)' = m(y' + py),$$

$$my' + ym' = my' + mpy,$$

$$ym' = mpy,$$

$$m' = mp.$$

Hence, $m$ must satisfy the separable differential equation

$$\frac{m'}{m} = p.$$

Solving this equation for $m$ by integrating, we obtain

$$\int \frac{m'}{m}\, dx = \int p\, dx,$$

$$\int \frac{1}{m}\, dm = \int p\, dx,$$

$$\ln|m| = \int p\, dx,$$

$$m = \pm e^{\int p\, dx}.$$

Let us now see some examples of the solution of first-order linear differential equations by the method of integrating factors.

**EXAMPLE 1**

Find the solution of

$$y' + \frac{1}{x}y = \frac{1}{x^2}, \qquad x > 0.$$

**SOLUTION**

We have $p(x) = x^{-1}$. Then $\int p = \int x^{-1}\,dx = \ln x$, so an integrating factor is

$$e^{\int p} = e^{\ln x} = x.$$

Multiplying the differential equation by the integrating factor, we obtain

$$x\left(y' + \frac{1}{x}y\right) = x\frac{1}{x^2},$$

$$\frac{d}{dx}(xy) = x^{-1},$$

$$xy = \int x^{-1}\,dx,$$

$$xy = \ln x + C,$$

$$y = \frac{\ln x}{x} + \frac{C}{x}.$$

■

**EXAMPLE 2**

Solve $y' + xy = x$.

**SOLUTION**

In this example the coefficient of $y$ is $x$, so an integrating factor is

$$e^{\int p} = e^{\int x\,dx} = e^{x^2/2}.$$

Then

$$e^{x^2/2}(y' + xy) = e^{x^2/2}x,$$

$$\frac{d}{dx}(e^{x^2/2}y) = e^{x^2/2}x,$$

$$e^{x^2/2}y = \int e^{x^2/2}x\,dx.$$

The integral on the right is easily evaluated by using the substitution $u = x^2/2$, so $du = x\,dx$. We obtain

$$e^{x^2/2}y(x) = e^{x^2/2} + C,$$

$$y(x) = 1 + Ce^{-x^2/2}.$$

■

■ **EXAMPLE 3**

Solve $y' + (\tan x)y = \sec x, \; -\pi/2 < x < \pi/2$.

**SOLUTION**

Using formula (7) of Section 8.3, we have

$$\int p = \int \tan x\,dx = \ln \sec x,$$

so an integrating factor is

$$e^{\int p} = e^{\ln \sec x} = \sec x.$$

Then

$$(\sec x)[y' + (\tan x)y] = \sec^2 x,$$

$$\frac{d}{dx}((\sec x)y) = \sec^2 x,$$

$$(\sec x)y = \int \sec^2 x\,dx,$$

$$(\sec x)y = \tan x + C,$$

$$\frac{1}{\cos x}y = \frac{\sin x}{\cos x} + C,$$

$$y(x) = \sin x + C\cos x.$$

■

■ **EXAMPLE 4**

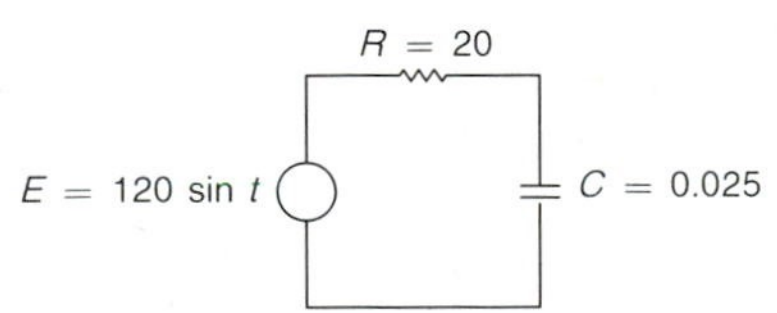


**FIGURE 9.4.1**

A simple series circuit consists of a 0.025-F capacitor, a 20-Ω resistor, and an impressed voltage of $E(t) = 120 \sin t$. See Figure 9.4.1. Kirchhoff's Second Law can be used to show that the current in amperes is a solution of the differential equation

$$20i' + \frac{1}{0.025}i = \frac{d}{dt}(120 \sin t).$$

Find the current in the circuit at time $t$ if the initial current in the circuit is zero.

**SOLUTION**

The current is the solution of the initial value problem

$$20i' + \frac{1}{0.025} i = 120 \cos t, \qquad i(0) = 0.$$

Simplifying the differential equation, we have

$$i' + 2i = 6 \cos t.$$

An appropriate integrating factor for this equation is $e^{2t}$. Then

$$e^{2t}(i' + 2i) = 6e^{2t} \cos t,$$

$$\frac{d}{dt}(e^{2t}i) = 6e^{2t} \cos t,$$

$$e^{2t}i = \int 6e^{2t} \cos t\, dt.$$

Using formula (5) of Section 8.2 to evaluate the integral, we obtain

$$e^{2t}i = 6e^{2t}(0.4 \cos t + 0.2 \sin t) + C,$$

$$i(t) = 2.4 \cos t + 1.2 \sin t + Ce^{-2t}.$$

Since $i(0) = 0$, we have

$$0 = 2.4 + C, \quad \text{so} \quad C = -2.4.$$

Then

$$i(t) = 2.4 \cos t + 1.2 \sin t - 2.4e^{-2t}.$$

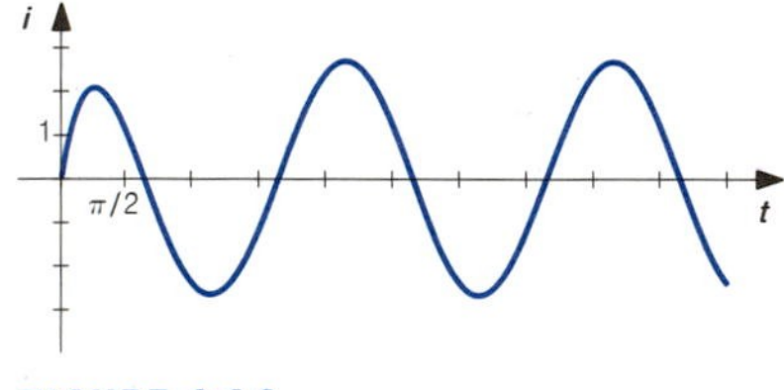

FIGURE 9.4.2

The graph of $i(t)$ is given in Figure 9.4.2. For large values of $t$, we have

$$i(t) \approx 2.4 \cos t + 1.2 \sin t.$$

By using the formula

$$A \cos t + B \sin t = \sqrt{A^2 + B^2} \sin\left(t + \tan^{-1} \frac{A}{B}\right) \qquad (A > 0,\ B > 0),$$

we can determine that

$$2.4 \cos t + 1.2 \sin t = 2.68 \sin(t + 1.107).$$

Thus, the current is essentially a sine curve that is 1.107 rad out of phase with the impressed voltage. ■

## EXERCISES 9.4

Solve the differential equations in Exercises 1–14.

1 $y' + \frac{2}{x}y = x,\ x > 0$

2 $y' + \frac{3}{x}y = 2,\ x > 0$

3 $y' - 2y = e^x$

4 $y' - 2y = e^{2x}$

5 $y' + \left(\frac{1}{x} + 1\right)y = \frac{1}{x},\ x > 0$

6 $y' + y = \frac{1}{1 + e^x}$

7 $y' + \frac{2x}{1 + x^2}y = x$

8 $y' - \frac{2x}{1 + x^2}y = x$

9 $y' - \frac{1}{x}y = \frac{1}{x + 1},\ x > 0$

10 $y' - \frac{1}{x + 1}y = \frac{1}{x},\ x > 0$

11 $y' + (\tan x)y = \tan x,\ -\pi/2 < x < \pi/2$

12 $y' + (\sec x)y = \sec x,\ -\pi/2 < x < \pi/2$

13 $y' + (\sec x)y = 1,\ -\pi/2 < x < \pi/2$

14 $y' - (\tan x)y = \cos x,\ -\pi/2 < x < \pi/2$

15 A circuit consists of a 0.02-F capacitor connected in series to a 60-Ω resistor, with an impressed voltage of 12 V. Kirchhoff's Second Law can be used to show that the current in amperes is a solution of the differential equation

$$60i' + \frac{1}{0.02}i = \frac{d}{dt}(12).$$

Find the current at time $t$ if the initial current is 8 A.

16 A circuit consists of a 0.1-F capacitor connected in series to a 60-Ω resistor, with an impressed voltage of $E(t) = 12 \sin t$. Kirchhoff's Second Law can be used to show that the current in amperes is a solution of the differential equation

$$60i' + \frac{1}{0.1}i = \frac{d}{dt}(12 \sin t).$$

Find the current at time $t$ if the initial current is 8 A.

## REVIEW EXERCISES

1 Find the equation of the curve that has slope $m = 2xy^2$ and contains the point (2, 1).

2 Find the equation of the curve that has slope $m = e^y$ and contains the point (2, 0).

3 Find the family of curves associated with the differential equation $yy' = x$ and find a mutually orthogonal family. Sketch three curves from each family.

4 Find the family of curves associated with the differential equation $yy' = 1$ and find a mutually orthogonal family. Sketch three curves from each family.

5 Suppose that $A$ and $B$ combine to form $C$ at a rate that is proportional to the product of the number of molecules of $A$ and the number of molecules of $B$. Find the equation for the amount of $C$ formed when 1 mol of $A$ is mixed with 3 mol of $B$.

6 Suppose that $A$ and $B$ combine to form $C$ at a rate that is proportional to the product of the number of molecules of $A$ and the number of molecules of $B$. Find the equation for the amount of $C$ formed when 2 mol of $A$ is mixed with 4 mol of $B$.

7 A 256-lb parachutist jumps from an airplane that is flying slowly and horizontally at 3136 ft. (a) Determine the velocity $t$ seconds after jumping and the terminal velocity if the air resistance is eight times the speed. (b) At what speed would the parachutist hit the ground if there were no air resistance? Ignore horizontal motion.

8 Determine the height above ground of a 160-lb skydiver $t$ seconds after jumping from an airplane that is flying slowly and horizontally at 2400 ft. Assume the air resistance is ten times the speed and ignore horizontal motion.

9 A spring is hanging vertically from a fixed point with a 4-lb weight attached to the bottom. The 4-lb weight stretches the spring 0.5 ft beyond its natural length. The weight is then pulled downward an additional 1.5 ft, held motionless at that point, and then released. Determine the position of the weight when the velocity is zero and find the speed of the weight at the point where the spring is at its natural length.

10 A rocket is launched vertically from earth. At the time the engines are turned off, the rocket has an altitude of $R/3$ and a speed of 10,000 ft/s, where the radius of the earth is

$R = 2.1 \times 10^7$ ft. Assuming that gravity is the only force acting on the rocket, determine the maximum altitude it will reach. What speed at an altitude of $R/3$ ft would the rocket need to avoid being pulled by gravity back toward the earth?

Solve the differential equations in Exercises 11–20.

**11** $y' = y^2(x + 1)$

**12** $y' = y^2 - y$

**13** $y' - 2y = 2$

**14** $y' + 4y = e^x$

**15** $y' + y = x$

**16** $y' + y = \sin x$

**17** $y' + \left(\frac{1}{x}\right)y = \frac{1}{x}$, $x > 0$

**18** $y' + \frac{1}{x - 1}y = 1$, $x > 1$

**19** $y' + (\tan x)y = \cos x$, $-\pi/2 < x < \pi/2$

**20** $y' + (\cos x)y = \cos x$

**CHAPTER TEN**

# *Infinite series*

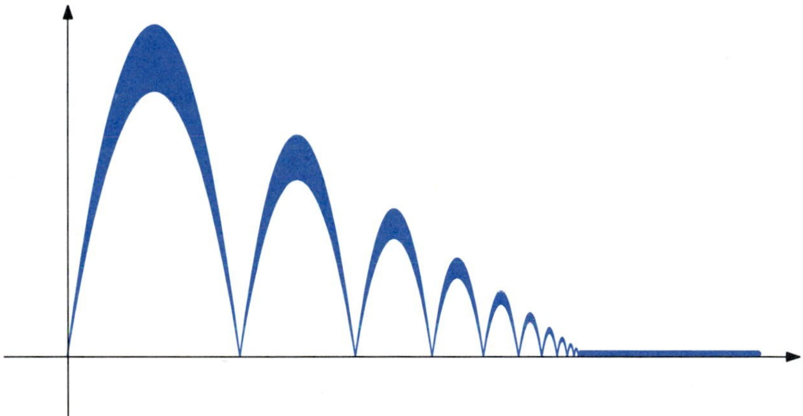

The main objective of this chapter is to study the approximation of functions by Taylor polynomials. Recall that Taylor polynomials were briefly introduced in Section 3.6 and that the approximation of functions by their first- and second-order Taylor polynomials was studied in Sections 4.8 and 4.9. In this chapter we will reintroduce Taylor polynomials and see that higher-order Taylor polynomials can generally be used to obtain approximate values that are as accurate as we desire. We will see how these ideas can be used to obtain approximate values of trigonometric and inverse trigonometric functions, logarithmic and exponential functions, and other transcendental functions.

## 10.1 SEQUENCES

Approximation of functions by their Taylor polynomials depends on the concept of limits of sequences.

*Definition*

A **sequence**, denoted $\{a_n\}_{n\geq 1}$ or $a_1, a_2, a_3, \ldots$, is an ordered set with elements $a_n$ defined for all positive integers $n$; $a_n$ is called the ***n*th term** of the sequence.

We are primarily interested in sequences of numbers. Usually, the numbers $a_n$ are given by a formula. For example, the formula

$$a_n = \frac{1}{n}, \qquad n \geq 1,$$

defines the sequence we express as either

$$\left\{\frac{1}{n}\right\}_{n\geq 1} \quad \text{or} \quad 1, \frac{1}{2}, \frac{1}{3}, \frac{1}{4}, \frac{1}{5}, \frac{1}{6}, \ldots.$$

The sequence $\{a\}_{n\geq 1}$ is the constant sequence $a, a, a, a, \ldots$. Do not confuse the constant *sequence* $\{a\}_{n\geq 1}$ with the *set* that consists of the single element $a$. The sequence $\{a\}_{n\geq 1}$ consists of an infinite number of terms $a$, one for each $n \geq 1$.

The important characteristic of a sequence is that there is a first term, second term, third term, and so on. It is not necessary that the first term of a sequence correspond to $n = 1$ in the formula that defines the terms of the sequence. For example,

$\{2n\}_{n\geq 3}$ is the sequence $6, 8, 10, 12, \ldots$.

The formula

$$a_n = \frac{n}{n+1}, \qquad n \geq 0,$$

gives the sequence

$$0, \frac{1}{2}, \frac{2}{3}, \frac{3}{4}, \ldots.$$

Note that it is theoretically unsound to try to determine a formula for $a_n$ from a partial list of terms. There may be no such formula, or there may be many formulas that give the terms listed. In spite of this, it is customary to indicate some simple sequences by a partial list of terms. For example,

$1, 2, 3, 4, \ldots$ indicates the sequence $\{n\}_{n\geq 1}$,

$\frac{1}{2}, \frac{2}{3}, \frac{3}{4}, \frac{4}{5}, \ldots$ indicates the sequence $\left\{\frac{n}{n+1}\right\}_{n\geq 1}$, and

$1^2, 2^2, 3^2, 4^2, \ldots$ indicates the sequence $\{n^2\}_{n\geq 1}$.

A sequence may be defined inductively or recursively. For example, the equations

$$a_1 = 2, \quad a_{n+1} = \frac{a_n}{2} + \frac{1}{a_n}, \qquad n \geq 1,$$

define a sequence $\{a_n\}_{n\geq 1}$. The equation $a_1 = 2$ implies

$$a_2 = \frac{a_1}{2} + \frac{1}{a_1} = 1.5.$$

Then $a_2 = 1.5$ implies

$$a_3 = \frac{a_2}{2} + \frac{1}{a_2} = \frac{17}{12} \approx 1.4166667.$$

Then $a_3$ determines $a_4 \approx 1.4142157$ and $a_4$ determines $a_5 \approx 1.4142136$. Generally, each term determines the succeeding term. (As $n$ increases, $a_n$ approaches $\sqrt{2}$.)

The idea of a limit of a sequence is important for many problems. Let us state the formal definition.

*Definition*

We say that the real number $L$ is the **limit** of the sequence $\{a_n\}_{n\geq 1}$, written

$$\lim_{n\to\infty} a_n = L,$$

if, for each $\varepsilon > 0$, there is a number $N$ such that $|a_n - L| < \varepsilon$ whenever $n \geq N$. If $\lim_{n\to\infty} a_n = L$, we say the sequence $\{a_n\}_{n\geq 1}$ is **convergent**. If the limit does not exist, we say the sequence is **divergent**. A sequence that approaches infinity is considered divergent. ■

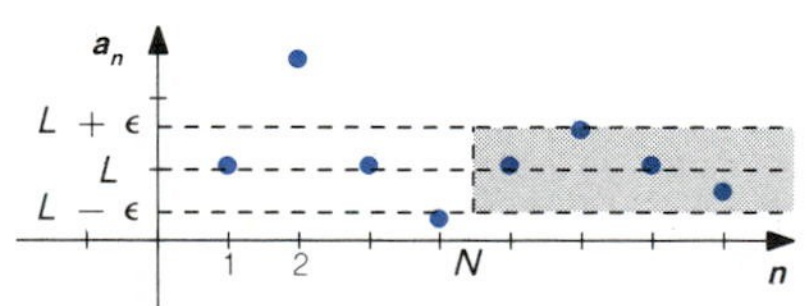


FIGURE 10.1.1

The condition $|a_n - L| < \varepsilon$ whenever $n \geq N$ means that the points $(n, a_n)$ are between the horizontal lines $y = L + \varepsilon$ and $y = L - \varepsilon$ for all $n \geq N$. (The points may or may not be within the lines for $n < N$.) The statement $\lim_{n\to\infty} a_n = L$ implies the points $(n, a_n)$ approach the horizontal asymptote $y = L$ as $n$ approaches infinity. See Figure 10.1.1.

The usual rules for limits of functions apply to limits of sequences. The definition of the limit of a sequence corresponds to that of the limit as $x$ approaches infinity of a function $f(x)$. The following is an immediate consequence of the definitions.

**If $a_n = f(n)$ for all $n$ and $\lim_{x\to\infty} f(x) = L$, then $\lim_{n\to\infty} a_n = L$.**

This means that we can evaluate limits of sequences by treating the formula for $a_n$ as a function of a real variable $n$.

The following basic examples are useful in determining the limit of sequences that are more complicated, but of similar type.

**If $\alpha > 0$, then**

$$\lim_{n\to\infty} \frac{1}{n^\alpha} = 0, \quad \text{and} \quad \lim_{n\to\infty} n^\alpha \quad \text{is divergent.}$$

$$\lim_{n\to\infty} r^n = 0 \quad \text{if} \quad |r| < 1; \ \lim_{n\to\infty} r^n \text{ is divergent if } |r| > 1.$$

(See Exercise 25.)

The ideas and techniques illustrated in the following examples should seem familiar.

■ **EXAMPLE 1**

$$\lim_{n\to\infty} \frac{1}{\sqrt{n}} = 0.$$

See Figure 10.1.2. ■

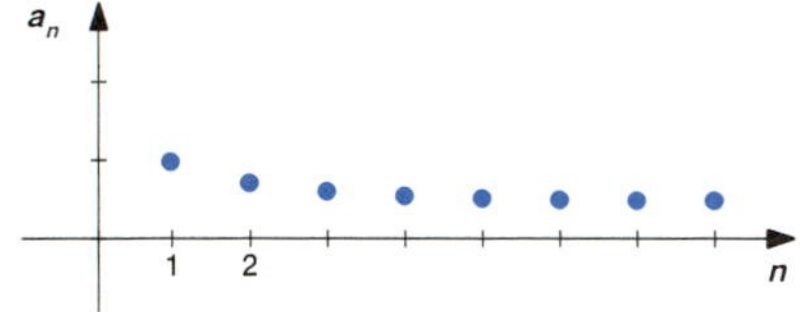


FIGURE 10.1.2

■ **EXAMPLE 2**

$$\lim_{n\to\infty} 2 = 2.$$

See Figure 10.1.3. ■

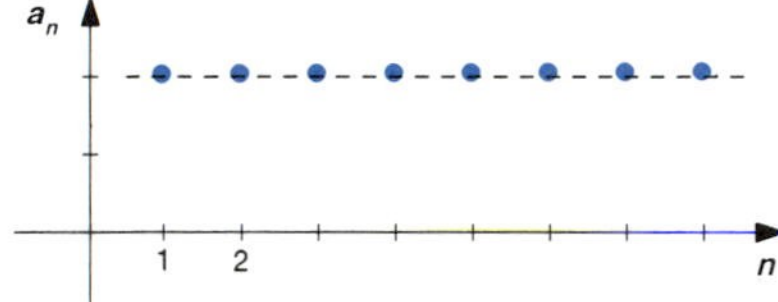


FIGURE 10.1.3

■ **EXAMPLE 3**

$$\lim_{n\to\infty} \left(2 + \frac{1}{n}\right) = 2.$$

See Figure 10.1.4. ■

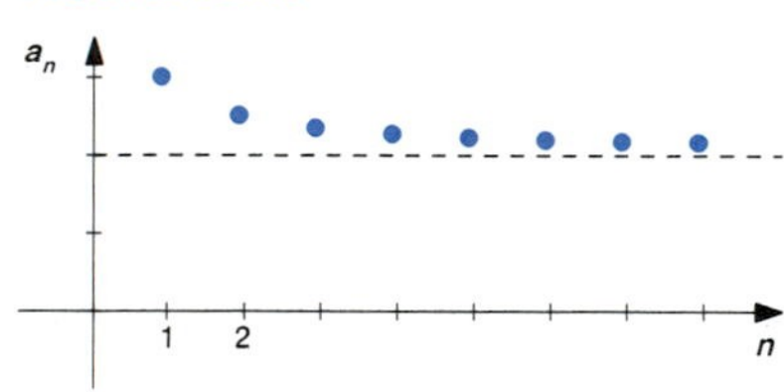


FIGURE 10.1.4

We can determine the limit as $n$ approaches infinity of an algebraic expression of $n$ by factoring the highest power of $n$ from each of the numerator and the denominator.

■ **EXAMPLE 4**

$$\lim_{n\to\infty} \frac{3n-8}{2n-5} = \lim_{n\to\infty} \frac{n\left(3 - \frac{8}{n}\right)}{n\left(2 - \frac{5}{n}\right)}$$

$$= \lim_{n\to\infty} \frac{3 - \frac{8}{n}}{2 - \frac{5}{n}} = \frac{3-0}{2-0} = \frac{3}{2}.$$

See Figure 10.1.5. ■

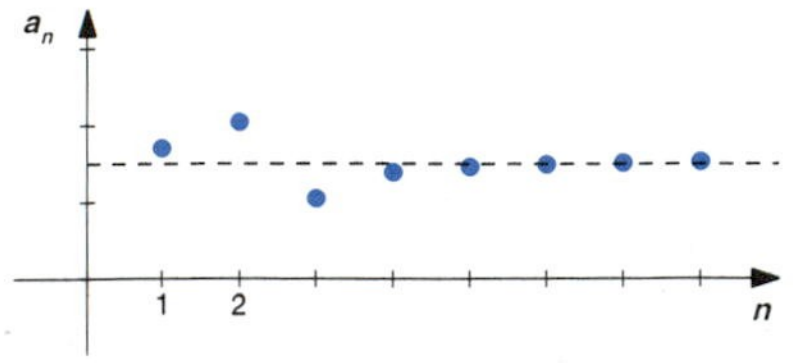

FIGURE 10.1.5

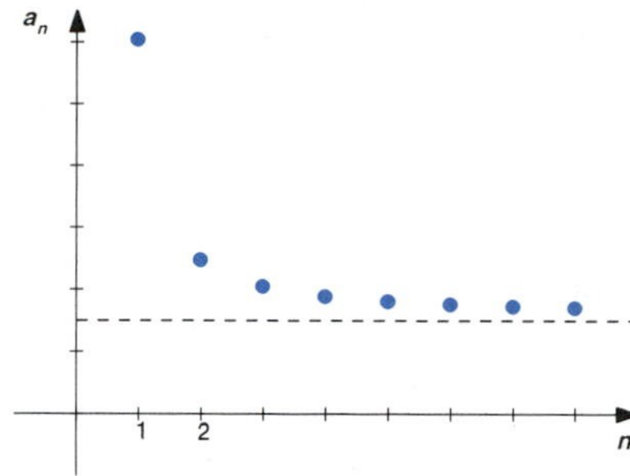


FIGURE 10.1.6

■ **EXAMPLE 5**

$$\lim_{n\to\infty} \frac{3n^2 + 2n + 1}{2n^2 - 1} = \lim_{n\to\infty} \frac{n^2\left(3 + \dfrac{2}{n} + \dfrac{1}{n^2}\right)}{n^2\left(2 - \dfrac{1}{n^2}\right)}$$

$$= \lim_{n\to\infty} \frac{3 + \dfrac{2}{n} + \dfrac{1}{n^2}}{2 - \dfrac{1}{n^2}} = \frac{3}{2}.$$

See Figure 10.1.6. ■

■ **EXAMPLE 6**

$$\lim_{n\to\infty} \frac{\sqrt{4n^2 + 9}}{n} = \lim_{n\to\infty} \frac{\sqrt{n^2\left(4 + \dfrac{9}{n^2}\right)}}{n}$$

$$= \lim_{n\to\infty} \frac{n\sqrt{4 + \dfrac{9}{n^2}}}{n}$$

$$= \lim_{n\to\infty} \sqrt{4 + \frac{9}{n^2}}$$

$$= \sqrt{4} = 2.$$

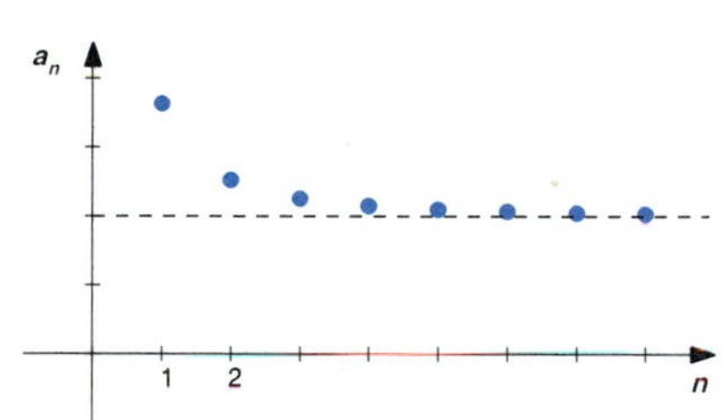


FIGURE 10.1.7

See Figure 10.1.7. ■

■ **EXAMPLE 7**

$$\lim_{n\to\infty} \sin \frac{n\pi}{2} \text{ is divergent.}$$

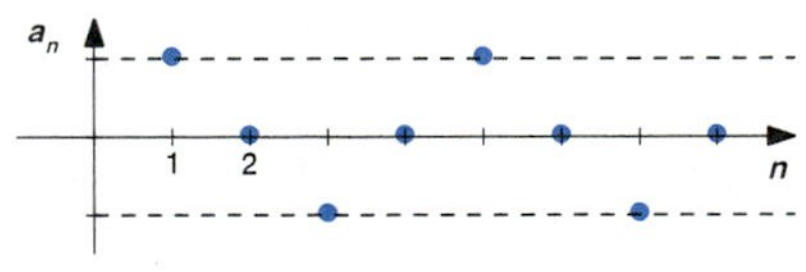


FIGURE 10.1.8

The sequence $\{\sin(n\pi/2)\}_{n\geq 1}$ corresponds to $1, 0, -1, 0, 1, 0, -1, 0, \ldots$. Thus, we see that the terms repeat the three values $-1, 0, 1$ infinitely often; the terms do not approach a *single* number $L$ as $n$ approaches infinity. See Figure 10.1.8. ■

■ **EXAMPLE 8**

$$\lim_{n\to\infty} \frac{\cos n\pi}{\sqrt{n}} = 0.$$

FIGURE 10.1.9

Note that $(\cos n\pi)/\sqrt{n}$ is of the form $a_n b_n$, where $\lim_{n\to\infty} a_n = \lim_{n\to\infty} 1/\sqrt{n} = 0$ and $|b_n| = |\cos n\pi| \leq 1$ for all $n$, so $\{b_n\}_{n\geq 1}$ is a bounded sequence. It follows that the product $a_n b_n$ approaches zero as $n$ approaches infinity. See Figure 10.1.9. ■

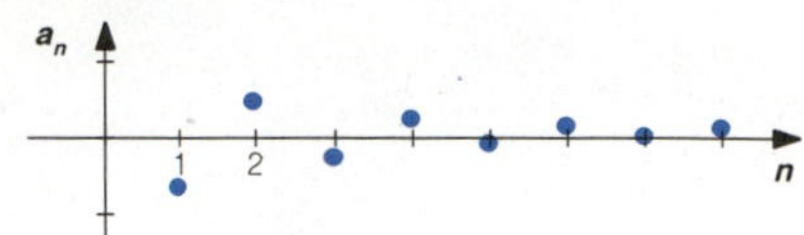


FIGURE 10.1.10

■ **EXAMPLE 9**

$$\lim_{n\to\infty}\left(-\frac{2}{3}\right)^n = 0.$$

This is a consequence of the basic fact that $\lim_{n\to\infty} r^n = 0$ for $|r| < 1$. We have $r = -2/3$ in this example. See Figure 10.1.10. ■

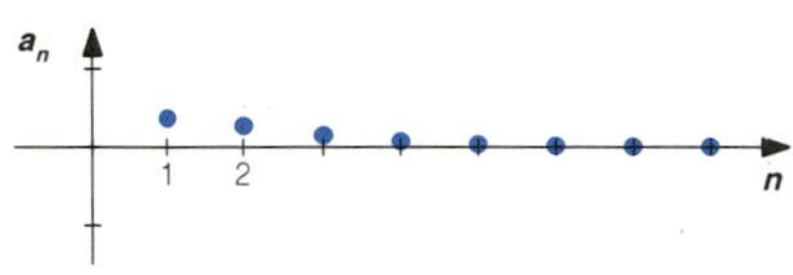
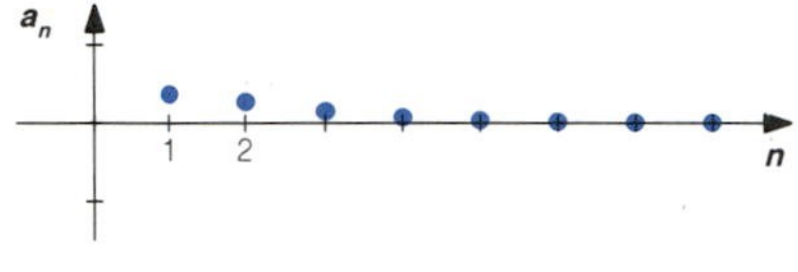


FIGURE 10.1.11

■ **EXAMPLE 10**

$$\lim_{n\to\infty} ne^{-n} = \lim_{n\to\infty}\frac{n}{e^n} \underset{(\infty/\infty,\,\text{l'H})}{=} \lim_{n\to\infty}\frac{1}{e^n} = 0.$$

We have treated $n$ as a real variable and then applied l'Hôpital's Rule. Recall that l'Hôpital's Rule was introduced in Section 7.1. See Figure 10.1.11. ■

We will need the following facts about the convergence of sequences.

**Changing the values of a finite number of terms of a sequencedoes not effect the convergence or divergence of a sequence, nor the limit of the sequence, if it is convergent.** (1)

Statement 1 reflects the fact that the limit of a sequence depends only on $a_n$ for $n$ large, $n \geq N$.

**If $a_n \leq a_{n+1} \leq B$ for all $n$, then $\lim_{n\to\infty} a_n$ exists.** (2)

We will not prove statement 2, but let us see how the Bisection Method can be used to locate a candidate $L$ for the limit of the sequence. We first note that the condition $a_n \leq a_{n+1} \leq B$ for all $n$ implies

$$a_1 \leq a_2 \leq a_3 \leq a_4 \leq \cdots \leq B.$$

This means that all terms $a_n$, $n \geq 1$, are contained in the interval between $a_1$ and $B$. It follows that $L$ must also be in this interval. We then consider the point $B_1 = (a_1 + B)/2$. If there is no $a_n > B_1$, then $L$ must be between $a_1$ and $B_1$. If there is some point $a_n > B_1$, then

$$B_1 \leq a_n \leq a_{n+1} \leq a_{n+2} \leq a_{n+3} \leq a_{n+4} \leq \cdots \leq B,$$

so $L$ must be between $B_1$ and $B$. We continue to bisect as above indefinitely. This gives a sequence of intervals that are nested, each one inside the previous one, and the length of the intervals approaches zero. The intervals shrink down to a single point $L$ and it can be shown that $L$ is the limit of the sequence. See Figure 10.1.12, where we have represented the elements of the sequence as points on a number line.

FIGURE 10.1.12

A sequence that satisfies $a_n \leq a_{n+1}$ for all $n$, is said to be **nondecreasing**. Similarly, we say a sequence is **nonincreasing** if $a_n \geq a_{n+1}$ for all $n$. A sequence is called **monotone** if the sequence is either nondecreasing or nonincreasing. Statement 2 says that a bounded, nondecreasing sequence is convergent.

## EXERCISES 10.1

Evaluate the limits in Exercises 1–24.

**1** $\lim_{n\to\infty} \frac{3n-1}{2n+1}$

**2** $\lim_{n\to\infty} \frac{\sqrt{n}}{2n-3}$

**3** $\lim_{n\to\infty} \frac{n}{n^2+1}$

**4** $\lim_{n\to\infty} \frac{1-n^2}{1+n^2}$

**5** $\lim_{n\to\infty} \frac{(2n-1)^2}{3n^2-n+7}$

**6** $\lim_{n\to\infty} \frac{2n^2+3n+1}{(2n+3)^2}$

**7** $\lim_{n\to\infty} \frac{n}{\sqrt{4n^2+9}}$

**8** $\lim_{n\to\infty} \frac{n^2}{\sqrt{9n^2+4}}$

**9** $\lim_{n\to\infty} (-1)^n$

**10** $\lim_{n\to\infty} \frac{1+(-1)^n}{2}$

**11** $\lim_{n\to\infty} \sin(\pi/n)$

**12** $\lim_{n\to\infty} \cos(\pi/n)$

**13** $\lim_{n\to\infty} n \cos n\pi$

**14** $\lim_{n\to\infty} e^{-n} \sin(n\pi/2)$

**15** $\lim_{n\to\infty} (-1)^n e^{1/n}$

**16** $\lim_{n\to\infty} \frac{\cos n\pi}{n}$

**17** $\lim_{n\to\infty} 2^{-n} \cos(n\pi)$

**18** $\lim_{n\to\infty} \frac{(-1)^n}{\ln n}$

**19** $\lim_{n\to\infty} \frac{e^{1/n}}{n}$

**20** $\lim_{n\to\infty} \frac{n(-1)^n}{n+1}$

**21** $\lim_{n\to\infty} n\, 2^{-n}$

**22** $\lim_{n\to\infty} n \sin(1/n)$

**23** $\lim_{n\to\infty} \frac{2^n}{3^n+1}$

**24** $\lim_{n\to\infty} \frac{3^n}{2^n+1}$

**25** Show that $\lim_{n\to\infty} r^n = 0$ if $|r| < 1$, and $\lim_{n\to\infty} r^n$ is divergent if $|r| > 1$.

**26** Show that $\lim_{r\to\infty} r^n = 1$ if $r = 1$, and that $\lim_{r\to\infty} r^n$ does not exist if $r = -1$.

## 10.2 INFINITE SERIES

Let us consider the following question.

*If a ball always bounces back to half its previous height, will it bounce forever?*

We know that the height of a ball $t$ seconds after it is dropped from an initial height of $s_0$ feet is given by

$$s(t) = s_0 - 16t^2.$$

Hence, it will first hit ground level after $t_0$ seconds, where

$$s(t_0) = 0,$$

$$s_0 - 16t_0^2 = 0,$$

$$t_0 = \frac{\sqrt{s_0}}{4}.$$

After the first bounce, the ball will bounce to a height $s_1 = s_0/2$ and return to ground level after an additional time interval of length

$$t_1 = 2\frac{\sqrt{s_1}}{4} = \frac{\sqrt{s_0/2}}{2}.$$

It will then bounce to a height $s_2 = s_1/2 = s_0/2^2$ and return to ground level after an additional time interval of length

$$t_2 = 2\frac{\sqrt{s_2}}{4} = \frac{\sqrt{s_0/2^2}}{2}.$$

Generally, after the $n$th bounce, the ball will bounce to a height of $s_n = s_0/2^n$ and return to ground level in a time interval of length

$$t_n = \frac{\sqrt{s_0/2^n}}{2}.$$

Thus, we see that the ball bounces an infinite number of times and the total time that the ball bounces should be the "sum"

$$\begin{aligned} T &= t_0 + t_1 + t_2 + t_3 + \cdots \\ &= \frac{\sqrt{s_0}}{4} + \frac{\sqrt{s_0/2}}{2} + \frac{\sqrt{s_0/2^2}}{2} + \frac{\sqrt{s_0/2^3}}{2} + \cdots. \end{aligned}$$

In this section we will investigate the meaning of such "sums" of an infinite number of terms and then return to answer the question about the bouncing ball.

We need the following definitions.

*Definition*

The *symbol*

$$\sum_{n=1}^{\infty} a_n \quad \text{or} \quad a_1 + a_2 + a_3 + \cdots$$

denotes an **infinite series**. Infinite series are sometimes called infinite sums or, simply, series. The ***n*th partial sum** of a series is the number

$$S_n = \sum_{j=1}^{n} a_j = a_1 + a_2 + \cdots + a_n,$$

the sum of the first $n$ terms of the series. If the sequence of partial sums $\{S_n\}_{n \geq 1}$ converges to a number $S$,

$$\lim_{n \to \infty} S_n = S,$$

we say the series is **convergent** with **sum** $S$ and write

$$\sum_{n=1}^{\infty} a_n = S.$$

If the sequence of partial sums is divergent, we say the series is **divergent** and it is not assigned a value. ■

Do not confuse the *series* $\sum_{n=1}^{\infty} a_n$ with the *sequence* $\{a_n\}_{n\geq 1}$. The dictionary definitions of series and sequence are similar, but they are not the same mathematically. In particular, a sequence is convergent if the terms $a_n$ approach a finite limit. However, convergence of a series requires that the sequence of partial sums $S_n = a_1 + a_2 + \cdots + a_n$ approach a finite limit. The fact that the sum of a series is the limit of partial sums is suggested by the notation $a_1 + a_2 + a_3 + \cdots$. The Greek letter $\sum$ (sigma) is used as an abbreviation of *s*um in the notation $\sum_{n=1}^{\infty} a_n$.

The statement $a_1 + a_2 + a_3 + \cdots = S$ does not mean that $S$ is the algebraic sum of an infinite number of terms, although the notation seems to suggest this. We cannot perform the operation of adding an infinite number of terms. We can add any finite number of terms and, hence, determine $S_n$ for any given value of $n$. The statement $a_1 + a_2 + a_3 + \cdots = S$ means that $S$ is the limit of the partial sums $S_n$.

*Convergent* series have some of the properties of finite sums. That is, if $\sum_{n=1}^{\infty} a_n$ and $\sum_{n=1}^{\infty} b_n$ are convergent, then

$$\sum_{n=1}^{\infty} (ca_n + db_n) \quad \text{is convergent}$$

and

$$\sum_{n=1}^{\infty} (ca_n + db_n) = c \sum_{n=1}^{\infty} a_n + d \sum_{n=1}^{\infty} b_n.$$

This property follows from the corresponding property of the finite sums $S_n$ and the Limit Theorem.

We will use the partial sums $S_n$ as approximate values of the sum of a convergent series.

■ **EXAMPLE 1**

Find the first four partial sums of the series

$$\sum_{n=1}^{\infty} \frac{(-1)^{n+1}}{(2n-1)!}(0.5)^{2n-1}.$$

Compare with a calculator value of sin 0.5. (In later sections we will see that the partial sums of the series are Taylor polynomials of sin $x$ about $c = 0$, evaluated at $x = 0.5$, and that the series is convergent with sum sin 0.5. Since the Taylor polynomials involve derivatives of trigonometric functions, we must use radian measure.)

**SOLUTION**

We see that

$$a_n = \frac{(-1)^{n+1}}{(2n-1)!}(0.5)^{2n-1}.$$

Recall that $n! = (1)(2)(3)\cdots(n)$. Then

$$S_1 = a_1 = \frac{1}{1!}(0.5) = 0.5,$$

$$S_2 = a_1 + a_2 = \frac{1}{1!}(0.5) - \frac{1}{3!}(0.5)^3 = 0.47916667,$$

$$S_3 = a_1 + a_2 + a_3 = \frac{1}{1!}(0.5) - \frac{1}{3!}(0.5)^3 + \frac{1}{5!}(0.5)^5 = 0.47942708,$$

$$\begin{aligned} S_4 &= a_1 + a_2 + a_3 + a_4 \\ &= \frac{1}{1!}(0.5) - \frac{1}{3!}(0.5)^3 + \frac{1}{5!}(0.5)^5 - \frac{1}{7!}(0.5)^7 = 0.47942553. \end{aligned}$$

A calculator value of the sine of 0.5 *radians* is $\sin(0.5) \approx 0.47942554$, which is very close to the value $S_4$. ■

The following formula relates the $n$th term and the $n$th partial sum of a series.

$$S_n = S_{n-1} + a_n, \qquad n > 1. \tag{1}$$

That is,

$$S_{n-1} + a_n = (a_1 + a_2 + \cdots + a_{n-1}) + a_n = S_n.$$

It is usually not an easy task to find a formula for $S_n$, even if we have a nice formula for $a_n$. One special case for which we can obtain a formula for the partial sums is that of geometric series.

The series

$$\sum_{n=0}^{\infty} r^n = 1 + r + r^2 + r^3 + r^4 + \cdots$$

is called a **geometric series**. The $n$th partial sum is

$$S_n = 1 + r + r^2 + \cdots + r^{n-1}.$$

Multiplying the above equation by $r$, we obtain

$$rS_n = r + r^2 + \cdots + r^{n-1} + r^n.$$

Then

$$S_n - rS_n = 1 + r + r^2 + \cdots + r^{n-1} - r - r^2 - \cdots - r^{n-1} - r^n$$
$$= 1 - r^n,$$

so

$$(1 - r)S_n = 1 - r^n.$$

If $r \neq 1$, we then divide by $1 - r$ to obtain the formula

$$\mathbf{S_n = 1 + r + r^2 + \cdots + r^{n-1} = \frac{1 - r^n}{1 - r}, \qquad r \neq 1.} \tag{2}$$

If $|r| < 1$, we know that $\lim_{n\to\infty} r^n = 0$, so

$$\lim_{n\to\infty} S_n = \lim_{n\to\infty} \frac{1 - r^n}{1 - r} = \frac{1}{1 - r}.$$

This means that the geometric series is convergent with sum $1/(1 - r)$ if $|r| < 1$. If $|r| \geq 1$, the partial sums of the geometric series do not converge. Note that $S_n = n$ if $r = 1$. Summarizing, we have

$$\mathbf{\sum_{n=0}^{\infty} r^n = 1 + r + r^2 + r^3 + \cdots = \frac{1}{1 - r}, \qquad |r| < 1.} \tag{3}$$

**The series is divergent if $|r| \geq 1$.**

Series of the form $\sum_{n=n_0}^{\infty} ar^n$ are also geometric series. We can obtain partial sums and the sum (if $|r| < 1$) of such series by factoring the common first term from every term of the series and then using formulas (2) and (3).

■ **EXAMPLE 2**

Find the formula for the $n$th partial sum $S_n$ and evaluate

$$\sum_{n=1}^{\infty} 8\left(-\frac{1}{3}\right)^n.$$

**SOLUTION**

We recognize that this is a geometric series with $r = -1/3$. The series converges because $r = -1/3$ satisfies $|r| < 1$. Let us write out the first few terms and then factor the common first term. We have

$$\sum_{n=1}^{\infty} 8\left(-\frac{1}{3}\right)^n = 8\left(-\frac{1}{3}\right) + 8\left(-\frac{1}{3}\right)^2 + 8\left(-\frac{1}{3}\right)^3 + \cdots$$
$$= 8\left(-\frac{1}{3}\right)\left[1 + \left(-\frac{1}{3}\right) + \left(-\frac{1}{3}\right)^2 + \cdots\right].$$

We can now use formula (2) to obtain a formula for $S_n$. In particular, the $n$th partial sum of the above series is the product of the factor $8(-1/3)$ and the $n$th partial sum of the series inside the square brackets. The series inside the brackets is of the form $1 + r + r^2 + r^3 + \cdots$, so formula (2) tells us that it has an $n$th partial sum

$$\frac{1 - r^n}{1 - r}, \qquad \text{with } r = -\frac{1}{3}$$

in this case.

We then obtain

$$\begin{aligned} S_n &= 8\left(-\frac{1}{3}\right)\left[1 + \left(-\frac{1}{3}\right) + \cdots + \left(-\frac{1}{3}\right)^{n-1}\right] \\ &= 8\left(-\frac{1}{3}\right)\left[\frac{1 - \left(-\frac{1}{3}\right)^n}{1 - \left(-\frac{1}{3}\right)}\right] \qquad \text{[Formula (2) with } r = -1/3] \\ &= -2\left[1 - \left(-\frac{1}{3}\right)^n\right]. \end{aligned}$$

We can then obtain the sum as the limit of the partial sums. That is,

$$S = \lim_{n \to \infty} S_n = \lim_{n \to \infty} \left\{-2\left[1 - \left(-\frac{1}{3}\right)^n\right]\right\} = -2. \qquad \blacksquare$$

If we are interested in only the sum and not the partial sums of a convergent geometric series, we can use formula (3).

## ■ EXAMPLE 3

Express the infinitely repeating decimal number 0.77777... as a ratio of two integers.

### SOLUTION

We have

$$\begin{aligned} 0.77777\ldots &= 0.7 + 0.07 + 0.007 + 0.0007 + 0.00007 + \cdots \\ &= 0.7(1 + 0.1 + 0.01 + 0.001 + 0.0001 + \cdots) \\ &= 0.7(1 + 0.1 + (0.1)^2 + (0.1)^3 + (0.1)^4 + \cdots). \end{aligned}$$

The expression inside the parentheses is a convergent geometric series with $r = 0.1$. Substitution in formula (3) then gives

$$0.77777\ldots = 0.7\left(\frac{1}{1 - 0.1}\right) = 0.7\left(\frac{1}{0.9}\right) = \frac{7}{9}. \qquad \blacksquare$$

Let us now return to the question of the bouncing ball. We determined that the total time that the ball bounces is

$$T = \frac{\sqrt{s_0}}{4} + \frac{\sqrt{s_0/2}}{2} + \frac{\sqrt{s_0/2^2}}{2} + \frac{\sqrt{s_0/2^3}}{2} + \frac{\sqrt{s_0/2^4}}{2} + \cdots$$

$$= \frac{\sqrt{s_0}}{4} + \frac{\sqrt{s_0/2}}{2}(1 + \sqrt{1/2} + \sqrt{1/2^2} + \sqrt{1/2^3} + \cdots)$$

$$= \frac{\sqrt{s_0}}{4} + \frac{\sqrt{s_0/2}}{2}(1 + \sqrt{1/2} + (\sqrt{1/2})^2 + (\sqrt{1/2})^3 + \cdots).$$

We see that the series inside the parentheses is a convergent geometric series with $r = \sqrt{1/2}$, so the ball will not bounce forever, even though it bounces an infinite number of times. We can use formula (3) to determine the total time. We have

$$T = \frac{\sqrt{s_0}}{4} + \frac{\sqrt{s_0/2}}{2}\left(\frac{1}{1 - \sqrt{1/2}}\right).$$

Formula (3) is very important. You should know this formula and how to use it.

We can find simple formulas for the $n$th partial sums of certain types of series because of cancellation between the terms. Let us illustrate with some examples.

■ **EXAMPLE 4**

Find a formula for the $n$th partial sum $S_n$ and evaluate the series

$$\sum_{n=1}^{\infty} \left(\frac{1}{n} - \frac{1}{n+1}\right).$$

**SOLUTION**

Let us write a few partial sums and see if a pattern can be determined.

$$S_1 = a_1 = \frac{1}{1} - \frac{1}{1+1}$$

$$= 1 - \frac{1}{2},$$

$$S_2 = a_1 + a_2$$

$$= \left(1 - \frac{1}{2}\right) + \left(\frac{1}{2} - \frac{1}{2+1}\right)$$

$$= 1 + \left(-\frac{1}{2} + \frac{1}{2}\right) - \frac{1}{3}$$

$$= 1 - \frac{1}{3},$$

$$S_3 = a_1 + a_2 + a_3$$
$$= \left(1 - \frac{1}{2}\right) + \left(\frac{1}{2} - \frac{1}{3}\right) + \left(\frac{1}{3} - \frac{1}{4}\right)$$
$$= 1 + \left(-\frac{1}{2} + \frac{1}{2}\right) + \left(-\frac{1}{3} + \frac{1}{3}\right) - \frac{1}{4}$$
$$= 1 - \frac{1}{4}.$$

Generally, we have

$$S_n = a_1 + a_2 + a_3 + a_4 + \cdots + a_n$$
$$= \left(1 - \frac{1}{2}\right) + \left(\frac{1}{2} - \frac{1}{3}\right) + \left(\frac{1}{3} - \frac{1}{4}\right) + \left(\frac{1}{4} - \frac{1}{5}\right) + \cdots + \left(\frac{1}{n} - \frac{1}{n+1}\right)$$
$$= 1 + \left(-\frac{1}{2} + \frac{1}{2}\right) + \left(-\frac{1}{3} + \frac{1}{3}\right) + \left(-\frac{1}{4} + \frac{1}{4}\right)$$
$$+ \left(-\frac{1}{5} + \frac{1}{5}\right) + \cdots + \left(-\frac{1}{n} + \frac{1}{n}\right) - \frac{1}{n+1}$$
$$= 1 - \frac{1}{n+1}.$$

(We have verified this formula for $n = 1$, 2, and 3. It can be proved formally for all positive integers $n$ by using mathematical induction.)

The sum of the series is the limit of the partial sums. We have

$$S = \lim_{n\to\infty} S_n = \lim_{n\to\infty} \left(1 - \frac{1}{n+1}\right) = 1.$$

The series is convergent with

$$\sum_{n=1}^{\infty} \left(\frac{1}{n} - \frac{1}{n+1}\right) = 1.$$

■

**■ EXAMPLE 5**

Find a formula for the $n$th partial sum $S_n$ and evaluate the series

$$\sum_{n=1}^{\infty} \left(\frac{n}{n+1} - \frac{n+2}{n+3}\right).$$

**SOLUTION**

We have

$$S_1 = a_1 = \frac{1}{1+1} - \frac{1+2}{1+3} = \frac{1}{2} - \frac{3}{4},$$

$$S_2 = S_1 + a_2 = \left(\frac{1}{2} - \frac{3}{4}\right) + \left(\frac{2}{3} - \frac{4}{5}\right) = \frac{1}{2} + \frac{2}{3} - \frac{3}{4} - \frac{4}{5},$$

$$S_3 = S_2 + a_3 = \left(\frac{1}{2} + \frac{2}{3} - \frac{3}{4} - \frac{4}{5}\right) + \left(\frac{3}{4} - \frac{5}{6}\right) = \frac{1}{2} + \frac{2}{3} - \frac{4}{5} - \frac{5}{6},$$

$$S_4 = S_3 + a_4 = \left(\frac{1}{2} + \frac{2}{3} - \frac{4}{5} - \frac{5}{6}\right) + \left(\frac{4}{5} - \frac{6}{7}\right) = \frac{1}{2} + \frac{2}{3} - \frac{5}{6} - \frac{6}{7},$$

$$S_5 = S_4 + a_5 = \left(\frac{1}{2} + \frac{2}{3} - \frac{5}{6} - \frac{6}{7}\right) + \left(\frac{5}{6} - \frac{7}{8}\right) = \frac{1}{2} + \frac{2}{3} - \frac{6}{7} - \frac{7}{8}.$$

Generally, we have

$$S_n = S_{n-1} + a_n = \frac{1}{2} + \frac{2}{3} - \frac{n+1}{n+2} - \frac{n+2}{n+3}, \qquad n \geq 2.$$

(This formula can be proved formally by mathematical induction.) The series is convergent with sum

$$S = \lim_{n\to\infty} S_n = \lim_{n\to\infty} \left(\frac{1}{2} + \frac{2}{3} - \frac{n+1}{n+2} - \frac{n+2}{n+3}\right) = \frac{1}{2} + \frac{2}{3} - 1 - 1 = -\frac{5}{6}. \quad \blacksquare$$

Series of the form

$$\sum_{n=1}^{\infty} (b_n - b_{n+k})$$

are called **telescoping series**. The series

$$\sum_{n=1}^{\infty} \left(\frac{1}{n} - \frac{1}{n+1}\right)$$

of Example 4 is a telescoping series with $b_n = 1/n$ and $k = 1$. The series

$$\sum_{n=1}^{\infty} \left(\frac{n}{n+1} - \frac{n+2}{n+3}\right)$$

of Example 5 is a telescoping series with $b_n = n/(n+1)$ and $k = 2$. Generally, the telescoping series

$$\sum_{n=1}^{\infty} (b_n - b_{n+k})$$

has partial sums

$$S_n = (b_1 + b_2 + \cdots + b_k) - (b_{n+1} + b_{n+2} + \cdots + b_{n+k}), \qquad n \geq k. \quad (4)$$

For example, in the case $k = 1$, we have

$$\sum_{n=1}^{\infty} (b_n - b_{n+1}).$$

Then

$$\begin{aligned} S_n &= (b_1 - b_2) + (b_2 - b_3) + (b_3 - b_4) + \cdots + (b_n - b_{n+1}) \\ &= b_1 + (-b_2 + b_2) + (-b_3 + b_3) + (-b_4 + b_4) + \cdots \\ &\quad + (-b_n + b_n) - b_{n+1} \\ &= b_1 - b_{n+1}. \end{aligned}$$

This agrees with formula (4) when $k = 1$. For each fixed value of $k$, mathematical induction on $n$ can be used to verify that the partial sums are given by formula (4).

Let us note that Examples 2, 3, and 4 are intended to reinforce the fact that *the sum of a series is the limit of the partial sums.* In most practical problems it is not necessary to find a formula for the partial sums of a series. Numerical values of a finite number of the partial sums are calculated and used as approximate values of the sum, as in Example 1. However, it is important that we know that the limit of partial sums exists. It doesn't make sense to try to approximate the limit of the sequence of partial sums if this sequence does not have a limit.

## EXERCISES 10.2

Find the first four partial sums of the series and compare with a calculator value of the given sum in Exercises 1–8.

**1** $\sum_{n=1}^{\infty} \frac{1}{n}(0.5)^n = \ln 2$

**2** $\sum_{n=1}^{\infty} \frac{1}{2n-1}(0.6)^{2n-1} = \ln 2$

**3** $\sum_{n=1}^{\infty} \frac{1}{2n-1}(0.8)^{2n-1} = \ln 3$

**4** $\sum_{n=1}^{\infty} \frac{1}{n}\left(\frac{2}{3}\right)^n = \ln 3$

**5** $\sum_{n=1}^{\infty} \frac{1}{(n-1)!}2^{n-1} = e^2$ $\quad (0! = 1)$

**6** $\sum_{n=1}^{\infty} \frac{1}{(n-1)!}(-1)^{n-1} = e^{-1}$ $\quad (0! = 1)$

**7** $\sum_{n=1}^{\infty} \frac{(-1)^{n+1}}{2n-1}(0.5)^{2n-1} = \tan^{-1} 0.5$

**8** $\sum_{n=1}^{\infty} \frac{(-1)^{n+1}}{(2n-2)!}(0.4)^{2n-2} = \cos 0.4$ $\quad$ (cosine of 0.4 rad)

Find a formula for the $n$th partial sum $S_n$ and evaluate the series that are convergent in Exercises 9–22.

**9** $\sum_{n=0}^{\infty} 8(0.2)^n$

**10** $\sum_{n=0}^{\infty} 6(-0.2)^n$

**11** $\sum_{n=1}^{\infty} 5(-1.1)^n$

**12** $\sum_{n=1}^{\infty} (9.9)^n$

**13** $\sum_{n=0}^{\infty} \frac{3}{2^n}$

**14** $\sum_{n=0}^{\infty} \frac{(-1)^n}{3^n}$

**15** $\sum_{n=1}^{\infty} \left(\frac{n}{n+1} - \frac{n+1}{n+2}\right)$

**16** $\sum_{n=1}^{\infty} \left(\frac{n}{2n-1} - \frac{n+1}{2n+1}\right)$

17 $\sum_{n=1}^{\infty} \left(\frac{n}{2^n} - \frac{n+1}{2^{n+1}}\right)$

18 $\sum_{n=1}^{\infty} \left(\frac{2n}{3n-2} - \frac{2n+4}{3n+4}\right)$

19 $\sum_{n=1}^{\infty} \left(\frac{1}{n} - \frac{1}{n+2}\right)$

20 $\sum_{n=1}^{\infty} \left(\frac{n^2}{n+1} - \frac{(n+2)^2}{n+3}\right)$

21 $\sum_{n=1}^{\infty} \frac{1}{n^2+5n+6}$

(*Hint*: Use partial fractions.)

22 $\sum_{n=1}^{\infty} \frac{1}{n^2+4n+3}$

(*Hint*: Use partial fractions.)

23 Express the infinitely repeating decimal number 0.55555... as a geometric series and then use formula (3) to express the number as a ratio of two integers.

24 Express the infinitely repeating decimal number 0.252525... as a geometric series and then use formula (3) to express the number as a ratio of two integers.

25 A handball bounces back to two-thirds the height from which it is dropped. Find the total distance traveled by a handball that is dropped from a height of 6 ft.

26 Verify formula (4) in the case $k = 2$.

27 Show that $\sum_{n=0}^{\infty} r^n$ diverges for $r = 1$ and $r = -1$.

## 10.3 CONVERGENCE TESTS

In this section we will develop some tests that are used to answer the question:

$$\textit{Is } \sum_{n=1}^{\infty} a_n \textit{ convergent or divergent?}$$

This is a reasonable question to ask *before* we start trying to find approximate values of the series. If the series is divergent, it does not make sense to try to find approximate values of the sum. If we know the series is convergent, then we know that the partial sums of the series can be used for approximate values of the sum.

The following preliminary observations should be kept in mind.

**The initial terms, any number of them, do not affect convergence or divergence of a series. (Of course, the initial terms affect the *value* of a convergent series.)** (1)

**If $c \neq 0$, then $\sum_{n=1}^{\infty} a_n$ and $\sum_{n=1}^{\infty} ca_n$ are either both convergent or both divergent.** (2)

We will use the following fact about the terms of a convergent series:

If $\sum_{n=1}^{\infty} a_n$ is convergent, then $\lim_{n\to\infty} a_n = 0$. (3)

■ *Proof* If $\lim_{n\to\infty} S_n = S$, then

$$\lim_{n\to\infty} a_n = \lim_{n\to\infty} (S_n - S_{n-1}) = S - S = 0.$$ ■

Statement (3) cannot be used as a test for convergence, but we can restate it to obtain a logically equivalent **test for divergence**.

**If $\lim_{n\to\infty} a_n \neq 0$, then $\sum_{n=1}^{\infty} a_n$ is divergent.** (4)

■ **EXAMPLE 1**

We see that

$$\sum_{n=1}^{\infty} \frac{n}{n+1}$$

is divergent, because

$$\lim_{n\to\infty} \frac{n}{n+1} = 1 \neq 0.$$

■

Note that statement (4) is only a test for divergence of a series. *Statement (4) cannot be used as a test for convergence.* It is very important to note that statement (3) does not say that a series is convergent if its $n$th term approaches zero as $n$ approaches infinity. A series may diverge, even if its $n$th term approaches zero. For example, $\lim_{n\to\infty}(1/n) = 0$, but

**$\sum_{n=1}^{\infty} \frac{1}{n}$ is divergent.** (5)

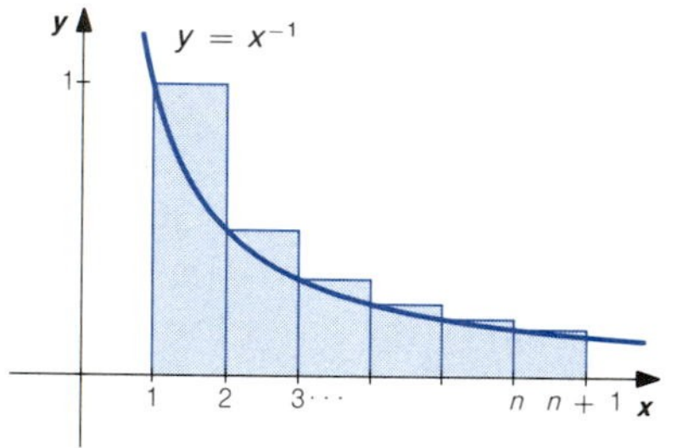


FIGURE 10.3.1

■ *Proof* We can verify statement (5) by showing that the partial sums of the series become unbounded as $n$ approaches infinity. To do this, we first note that any positive number $a_j$ can be interpreted as the area of a rectangle with height $a_j$ and base one. Using Figure 10.3.1 to compare the sum of the areas of the rectangles with the corresponding area under the curve $y = 1/x$, we see that

$$S_n = 1 + \frac{1}{2} + \frac{1}{3} + \frac{1}{4} + \cdots + \frac{1}{n} > \int_1^{n+1} \frac{1}{x}\,dx = \ln(n+1) \to \infty$$

as $n \to \infty$. ■

Let us compare the amount of effort required to show divergence of the series in Example 1 and in statement (5). In Example 1 we used available theory in the form of statement (4). We could then determine divergence by applying a rather simple test that involved looking at just the $n$th term of the series. In statement (5) we did not have the appropriate theory available, so we needed to estimate the size of $S_n$ to show divergence. It should seem easier to use a general theory.

We will now extend our theoretical results concerning the convergence and divergence of series. For the remainder of this section we will consider only series that have nonnegative terms. The following result is the basis of the theory of convergence of series with nonnegative terms.

*Theorem 1*

**Let $S_n$ denote the $n$th partial sum of a series**

$$\sum_{n=1}^{\infty} a_n$$

**with nonnegative terms $a_n \geq 0$. Then either $S_n$ diverges to infinity and $\sum_{n=1}^{\infty} a_n$ is divergent or there is some number $B$ such that $S_n \leq B$ for all $n$ and $\sum_{n=1}^{\infty} a_n$ is convergent.**

■ *Proof* We first note that the partial sums of a series with *nonnegative* terms form a nondecreasing sequence. That is, $a_{n+1} \geq 0$ for all $n$, so $S_{n+1} = S_n + a_{n+1} \geq S_n$ for all $n$. It follows that the sequence of partial sums $S_n$ either increases to infinity or there is some number $B$ such that $S_n \leq B$ for all $n$. See Figure 10.3.2, where we have represented the values $S_n$ as points on a number line. Of course, the series is divergent if $S_n$ increases to infinity. If the partial sums are bounded from above, convergence follows from statement (2) of Section 10.2, which tells us that a *bounded, nondecreasing* sequence is convergent. ■

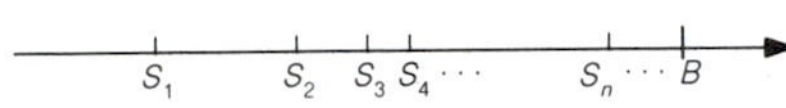


FIGURE 10.3.2

Let us see how Theorem 1 can be used.

$$\sum_{n=1}^{\infty} \frac{1}{n^p} \text{ is convergent if } p > 1. \tag{6}$$

■ *Proof* According to the theorem, it is enough to show there is a number $B$ such that $S_n \leq B$ for all $n$. To do this, we use Figure 10.3.3 to compare the second through $n$th terms of the partial sum $S_n$ with the corresponding areas under the curve $y = x^{-p}$. We see that

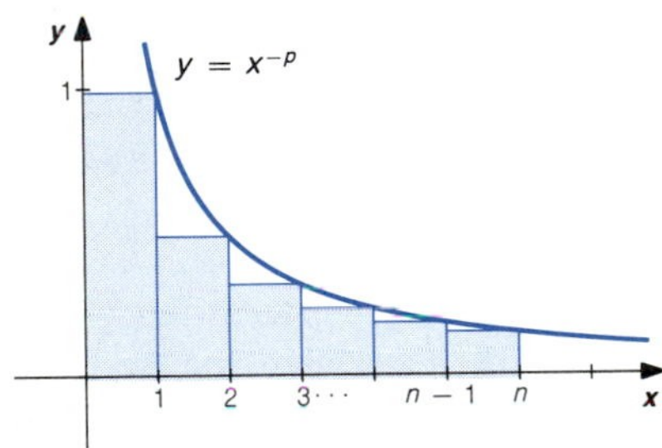


FIGURE 10.3.3

$$\begin{aligned} S_n &= \frac{1}{1^p} + \frac{1}{2^p} + \frac{1}{3^p} + \cdots + \frac{1}{n^p} \\ &\leq 1 + \int_1^2 \frac{1}{x^p}\,dx + \int_2^3 \frac{1}{x^p}\,dx + \cdots + \int_{n-1}^n \frac{1}{x^p}\,dx \\ &= 1 + \int_1^n \frac{1}{x^p}\,dx \\ &= 1 + \left[\frac{1}{1-p}x^{1-p}\right]_1^n \\ &= 1 + \frac{n^{1-p}}{1-p} - \frac{1}{1-p} \\ &= 1 + \frac{1}{p-1} - \frac{n^{1-p}}{p-1}. \end{aligned}$$

Since $p > 1$ implies

$$-\frac{n^{1-p}}{p-1} < 0,$$

we see that

$$S_n < 1 + \frac{1}{p-1} = \frac{p}{p-1}.$$

We have shown that

$$S_n \leq \frac{p}{p-1} \qquad \text{for all } n,$$

so Theorem 1 tells us the series is convergent. ■

We also have the following

$$\sum_{n=1}^{\infty} \frac{1}{n^p} \text{ is divergent if } p < 1. \tag{7}$$

■ *Proof* To verify statement (7), we use Figure 10.3.4 to obtain

$$S_n = \frac{1}{1^p} + \frac{1}{2^p} + \frac{1}{3^p} + \cdots + \frac{1}{n^p}$$

$$\geq \int_1^{n+1} x^{-p}\,dx = \frac{1}{1-p}((n+1)^{1-p} - 1)).$$

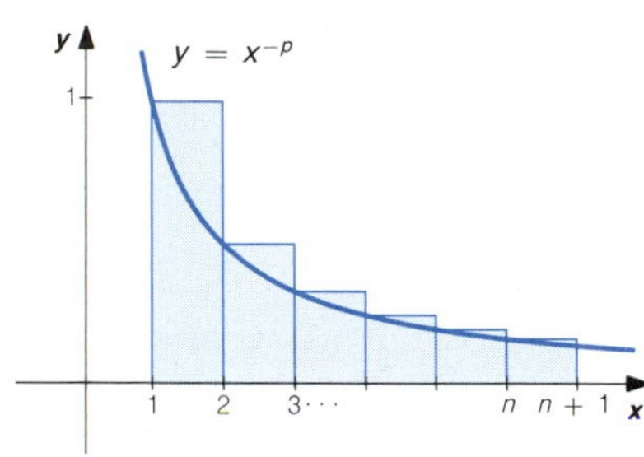

FIGURE 10.3.4

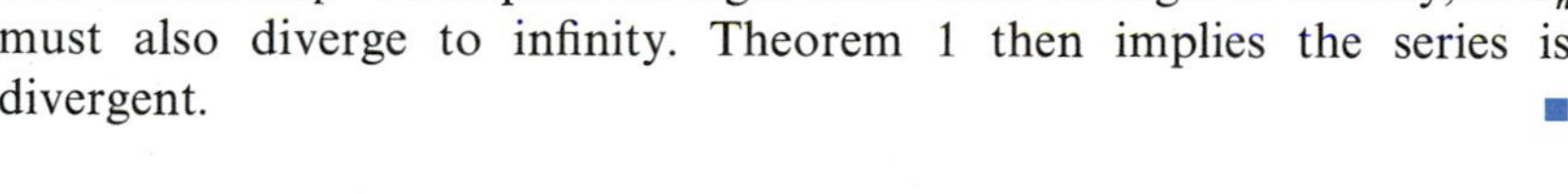
The condition $p < 1$ implies the right-hand term diverges to infinity, so $S_n$ must also diverge to infinity. Theorem 1 then implies the series is divergent. ■

At this point let us repeat the warning given in Section 10.2 not to confuse the *series* $\sum_{n=1}^{\infty} a_n$ with the *sequence* $\{a_n\}_{n \geq 1}$. In particular, the sequences

$$\left\{\frac{1}{n}\right\}_{n \geq 1} \quad \text{and} \quad \left\{\frac{1}{n^2}\right\}_{n \geq 1}$$

are both convergent with limit zero. However, the series

$$\sum_{n=1}^{\infty} \frac{1}{n}$$

is divergent and the series

$$\sum_{n=1}^{\infty} \frac{1}{n^2}$$

is convergent. That is, we have

$$1 + \frac{1}{2} + \frac{1}{3} + \frac{1}{4} + \cdots + \frac{1}{n} > \int_1^{n+1} \frac{1}{x}\,dx = \ln(n+1) \to \infty \qquad \text{as } n \to \infty.$$

Thus, even though the $n$th term of the series $\sum_{n=1}^{\infty}(1/n)$ approaches zero as $n$ approaches infinity, $1/n$ does not approach zero *fast enough* to keep the partial sums from becoming unbounded. On the other hand, we have

$$1+\frac{1}{2^2}+\frac{1}{3^2}+\cdots+\frac{1}{n^2}<1+\int_1^n \frac{1}{x^2}\,dx=2-\frac{1}{n}<2.$$

Thus, $1/n^2$ approaches zero *fast enough* that the partial sums of the series $\sum_{n=1}^{\infty}(1/n^2)$ remain bounded and the series is convergent.

Let us summarize statements (5), (6), and (7). The ***p*-series**

$$\sum_{n=1}^{\infty}\frac{1}{n^p} \text{ is convergent if } p>1 \text{ and divergent if } p\le 1.$$

We can then use statement (2) to determine convergence or divergence of constant multiples of $p$-series.

## ■ EXAMPLE 2

Determine convergence or divergence of the series

$$\sum_{n=1}^{\infty}\frac{1}{2\sqrt{n}}.$$

### SOLUTION

We have

$$a_n=\frac{1}{2\sqrt{n}}=\frac{1}{2}\frac{1}{n^{1/2}}=c\frac{1}{n^p}, \quad \text{where} \quad c=\frac{1}{2}\neq 0 \quad \text{and} \quad p=\frac{1}{2}<1.$$

A nonzero multiple of a divergent $p$-series is divergent, so

$$\sum_{n=1}^{\infty}\frac{1}{2\sqrt{n}}$$

is divergent. ■

## ■ EXAMPLE 3

Determine convergence or divergence of the series

$$\sum_{n=1}^{\infty}\frac{100}{n^{4/3}}.$$

### SOLUTION

We have

$$a_n=\frac{100}{n^{4/3}}=c\frac{1}{n^p}, \quad \text{where} \quad c=100\neq 0 \quad \text{and} \quad p=\frac{4}{3}>1.$$

A multiple of a convergent $p$-series is convergent, so

$$\sum_{n=1}^{\infty} \frac{100}{n^{4/3}}$$

is convergent. ■

We could use the partial sums of the convergent series in Example 3 to obtain approximate values of its sum. It does not make sense to use the partial sums of the divergent series in Example 2 to approximate its sum, since divergent series do not have a sum.

The following theorem can be used to determine convergence or divergence of a series with nonnegative terms $\sum a_n$ by comparing the size of its terms to the terms of a series with nonnegative terms $\sum b_n$ for which convergence or divergence is known.

■ *Comparison Test*

**If $0 \leq a_n \leq b_n$ for all $n$, and $\sum_{n=1}^{\infty} b_n$ is convergent, then $\sum_{n=1}^{\infty} a_n$ is convergent. If $a_n \geq b_n \geq 0$ for all $n$, and $\sum_{n=1}^{\infty} b_n$ is divergent, then $\sum_{n=1}^{\infty} a_n$ is divergent.**

■ *Proof* The comparison test is an immediate consequence of Theorem 1. That is, $a_n \leq b_n$ for all $n$ implies

$$a_1 + a_2 + \cdots + a_n \leq b_1 + b_2 + \cdots + b_n.$$

If $\sum_{n=1}^{\infty} b_n$ is convergent, the partial sums on the right remain bounded. It follows that the partial sums on the left also are bounded, so Theorem 1 implies $\sum_{n=1}^{\infty} a_n$ is convergent. (Note that we are dealing with series that have nonnegative terms, so Theorem 1 applies.) Similarly, $a_n \geq b_n$ for all $n$ implies

$$a_1 + a_2 + \cdots + a_n \geq b_1 + b_2 + \cdots + b_n.$$

If $\sum_{n=1}^{\infty} b_n$ is divergent, then the partial sums on the right become unbounded. This implies the partial sums on the left also become unbounded, so $\sum_{n=1}^{\infty} a_n$ is divergent. ■

The following is a much more useful version of the Comparison Test.

■ *Limit Version of Comparison Test*

**Suppose $a_n$ and $b_n$ are both positive for all $n$.**

**If $0 \leq \lim_{n \to \infty} \frac{a_n}{b_n} < \infty$, then**

**$\sum_{n=1}^{\infty} b_n$ convergent implies $\sum_{n=1}^{\infty} a_n$ convergent.**

**If $0 < \lim_{n \to \infty} \frac{a_n}{b_n} \leq \infty$, then**

**$\sum_{n=1}^{\infty} b_n$ divergent implies $\sum_{n=1}^{\infty} a_n$ divergent.**

■ *Proof* We note that

$$0 \leq \lim_{n\to\infty} \frac{a_n}{b_n} < \infty$$

implies there are numbers $N$ and $c > 0$ such that $a_n/b_n < c$, so $a_n < cb_n$ for all $n \geq N$. Since neither the initial terms nor multiplying the terms by a nonzero constant affects convergence or divergence of a series, the conclusion then follows from the Comparison Test. Similarly,

$$0 < \lim_{n\to\infty} \frac{a_n}{b_n} \leq \infty$$

implies there are numbers $N$ and $c > 0$ such that $a_n/b_n > c$, so $a_n > cb_n$ for all $n \geq N$, and the second statement follows. ■

Note that if $a_n$ and $b_n$ are positive and the limit of $a_n/b_n$ as $n$ approaches infinity is *finite and nonzero*, then

$$\sum_{n=1}^{\infty} b_n \text{ convergent implies } \sum_{n=1}^{\infty} a_n \text{ convergent}$$

and

$$\sum_{n=1}^{\infty} b_n \text{ divergent implies } \sum_{n=1}^{\infty} a_n \text{ divergent.}$$

The limit version of the Comparison Test tells us that if $a_n$ and $b_n$ are positive and the limit of $a_n/b_n$ as $n$ approaches infinity is finite and nonzero, then $\sum_{n=1}^{\infty} a_n$ and $\sum_{n=1}^{\infty} b_n$ either both converge or both diverge.

We can now use the results for the convergence or divergence of $p$-series, along with the Limit Version of the Comparison Test, to determine *easily* the convergence or divergence of series of positive terms that are algebraic expressions of $n$. To determine which $p$-series to use for comparison, we factor the highest power of $n$ from each of the numerator and the denominator.

**■ EXAMPLE 4**

Determine convergence or divergence of the series

$$\sum_{n=1}^{\infty} \frac{n}{3n^2 + 1}.$$

**SOLUTION**

Since the terms of this series are algebraic expressions of $n$, we should compare its terms to a $p$-series. We have

$$a_n = \frac{n}{3n^2 + 1} = \frac{n}{n^2(3 + 1/n^2)} = \frac{1}{n}\left(\frac{1}{3 + 1/n^2}\right).$$

Since for large $n$, $1/(3 + 1/n^2) \approx 1/3$, this suggests we compare with the series

$$\sum_{n=1}^{\infty} \frac{1}{n}.$$

We have

$$\lim_{n\to\infty} \frac{a_n}{1/n} = \lim_{n\to\infty} \frac{1}{3 + (1/n^2)} = \frac{1}{3}.$$

Since

$$\sum_{n=1}^{\infty} \frac{1}{n}$$

is divergent, the Limit Version of the Comparison Test tells us that so is

$$\sum_{n=1}^{\infty} \frac{n}{3n^2 + 1}.$$

■

**■ EXAMPLE 5**

Determine convergence or divergence of the series

$$\sum_{n=1}^{\infty} \frac{3\sqrt{n}}{2n^2 - 1}.$$

**SOLUTION**

The terms are algebraic expressions of $n$, so we compare with a $p$-series. We have

$$a_n = \frac{3\sqrt{n}}{2n^2 - 1} = \frac{3\sqrt{n}}{n^2(2 - (1/n^2))} = \frac{3}{n^{3/2}(2 - (1/n^2))}.$$

This suggests we compare with

$$\sum_{n=1}^{\infty} \frac{1}{n^{3/2}}.$$

We have

$$\lim_{n\to\infty} \frac{a_n}{1/n^{3/2}} = \lim_{n\to\infty} \frac{3}{2 - (1/n^2)} = \frac{3}{2}.$$

Since

$$\sum_{n=1}^{\infty} \frac{1}{n^{3/2}}$$

is convergent, so is

$$\sum_{n=1}^{\infty} \frac{3\sqrt{n}}{2n^2 - 1}.$$ ■

It is useful for the purpose of comparison to recall from Chapter 7 the following relations between the sizes of the logarithmic, exponential, and power functions.

$$\lim_{n \to \infty} \frac{\ln n}{n^{\alpha}} = 0, \qquad \alpha > 0,$$

$$\lim_{n \to \infty} \frac{n^{\alpha}}{e^n} = 0.$$

The first limit tells us that $\ln n$ becomes smaller than any positive power of $n$ as $n$ approaches infinity; the second limit tells us that any power of $n$ becomes smaller than $e^n$ as $n$ approaches infinity.

■ **EXAMPLE 6**

Determine convergence or divergence of the series

$$\sum_{n=1}^{\infty} \frac{\ln n}{n^2}.$$

**SOLUTION**

Let us compare the terms

$$a_n = \frac{\ln n}{n^2}$$

with those of a $p$-series. Our first choice might be $p = 2$. We have

$$\lim_{n \to \infty} \frac{(\ln n)/n^2}{1/n^2} = \lim_{n \to \infty} \left(\frac{\ln n}{n^2}\right)\left(\frac{n^2}{1}\right) = \lim_{n \to \infty} \ln n = \infty.$$

Since $\sum_{n=1}^{\infty} 1/n^2$ is convergent and the ratio $((\ln n)/n^2)/(1/n^2)$ approaches infinity, the Limit Version of the Comparison Test tells us nothing about the convergence or divergence of the series $\sum_{n=1}^{\infty} (\ln n)/n^2$.

We might next try $p = 1$. Then

$$\lim_{n \to \infty} \frac{(\ln n)/n^2}{1/n} = \lim_{n \to \infty} \left(\frac{\ln n}{n^2}\right)\left(\frac{n}{1}\right) = \lim_{n \to \infty} \frac{\ln n}{n} = 0.$$

Since $\sum_{n=1}^{\infty} 1/n$ is divergent and the ratio $((\ln n)/n^2)/(1/n)$ approaches zero, the Limit Version of the Comparison Test tells us nothing about the convergence or divergence of the series $\sum_{n=1}^{\infty} (\ln n)/n^2$.

Let us try $p = 1.5$. Then

$$\lim_{n\to\infty} \frac{(\ln n)/n^2}{1/n^{1.5}} = \lim_{n\to\infty} \left(\frac{\ln n}{n^2}\right)\left(\frac{n^{1.5}}{1}\right) = \lim_{n\to\infty} \frac{\ln n}{n^{0.5}} = 0.$$

Since $\sum_{n=1}^{\infty} 1/n^{1.5}$ is convergent and the ratio $((\ln n)/n^2)/(1/n^{1.5})$ approaches zero, the Limit Version of the Comparison Test tells us that the series $\sum_{n=1}^{\infty} (\ln n)/n^2$ is also convergent. (We could reach the same conclusion by comparison with any $p$-series, $1 < p < 2$.) ■

Recall that we used integrals to determine convergence or divergence of the $p$-series in statements (5), (6), and (7). The same ideas can be used to verify the following.

■ *Integral Test*

**If $f$ is nonnegative, nonincreasing, and continuous on the interval $x \geq N$, with $a_n = f(n)$ for $n \geq N$, then the series $\sum_{n=1}^{\infty} a_n$ and the improper integral $\int_N^{\infty} f(x)\,dx$ are either both convergent or both divergent. Moreover, if the improper integral is convergent, then**

$$0 \leq S - S_N \leq \int_N^{\infty} f(x)\,dx.$$

The inequality

$$0 \leq S - S_N \leq \int_N^{\infty} f(x)\,dx$$

gives a bound of the error in using $S_N$ as an approximate value of $S$. Let us verify the right-hand inequality. We assume that the conditions of the test are satisfied and that the series and the improper integral are both convergent. Using the essential facts that $f$ is nonnegative and *nonincreasing*, we see from Figure 10.3.5 that

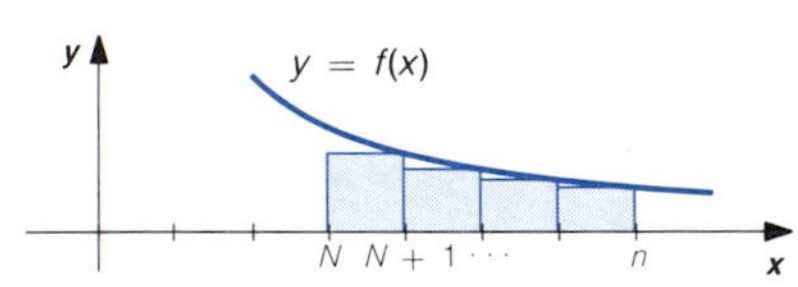


FIGURE 10.3.5

$$S_n - S_N = a_{N+1} + a_{N+2} + \cdots + a_n \leq \int_N^n f(x)\,dx, \qquad n > N.$$

Then

$$S - S_N = \lim_{n\to\infty} (S_n - S_N) \leq \lim_{n\to\infty} \int_N^n f(x)\,dx = \int_N^{\infty} f(x)\,dx.$$

■ **EXAMPLE 7**

Estimate

$$S = \sum_{n=1}^{\infty} \frac{1}{n^3}$$

by a partial sum $S_N$ with $0 \leq S - S_N \leq 0.01$.

**SOLUTION**

The series is a convergent $p$-series, so we know that the sum $S$ is a real number. We will use the integral test to determine the accuracy of using $S_N$ as an approximate value of $S$. The conditions of the Integral Test apply with $f(x) = x^{-3}$. We then have

$$0 \le S - S_N \le \int_N^\infty x^{-3}\,dx = \lim_{n\to\infty} \int_N^n x^{-3}\,dx = \lim_{n\to\infty} \left(\frac{1}{2N^2} - \frac{1}{2n^2}\right) = \frac{1}{2N^2}.$$

It follows that

$$0 \le S - S_N \le 0.001 \quad \text{whenever} \quad \frac{1}{2N^2} < 0.01 \quad \text{or} \quad N^2 > 50.$$

The smallest integer $N$ to satisfy this condition is $N = 8$. The desired approximate value of $S$ is

$$S_8 = \frac{1}{1^3} + \frac{1}{2^3} + \frac{1}{3^3} + \frac{1}{4^3} + \frac{1}{5^3} + \frac{1}{6^3} + \frac{1}{7^3} + \frac{1}{8^3} \approx 1.1951602.$$

(We have used a calculator to evaluate $S_8$. Note that we need to worry about roundoff error if $N$ is large.) ■

Let us see how we can combine the Integral Test and the Limit Version of the Comparison Test to determine convergence or divergence of a series with nonnegative terms.

■ **EXAMPLE 8**

Determine convergence or divergence of the series

$$\sum_{n=2}^\infty \frac{n}{(n^2 - 5)(\ln n)^2}.$$

**SOLUTION**

The first step is to determine a simpler series that converges or diverges as does the given series. To do this, we factor $n^2$ from the quadratic expression in the denominator to obtain

$$a_n = \frac{n}{(n^2 - 5)(\ln n)^2} = \frac{n}{n^2(1 - (5/n^2))(\ln n)^2}$$
$$= \frac{1}{n(1 - (5/n^2))(\ln n)^2}.$$

This indicates that we could compare $\sum a_n$ with the series $\sum b_n$, where

$$b_n = \frac{1}{n(\ln n)^2}.$$

That is,

$$\lim_{n\to\infty} \frac{a_n}{b_n} = 1,$$

so the sequences are either both convergent or both divergent. We can use the integral test to determine the convergence or divergence of $\sum b_n$. We have

$$b_n = f(n), \quad \text{where} \quad f(x) = \frac{1}{x(\ln x)^2}.$$

The function $f$ is nonnegative, nonincreasing, and continuous on the interval $x \geq 2$, so the conditions of the Integral Test are satisfied. Using the substitution

$$u = \ln x, \qquad du = \frac{1}{x}\,dx,$$

we have

$$\int_2^n \frac{1}{x(\ln x)^2}\,dx = \int u^{-2}\,du = -u^{-1}\Big] = -\frac{1}{\ln x}\Big]_2^n$$
$$= \frac{1}{\ln 2} - \frac{1}{\ln n} \to \frac{1}{\ln 2} \qquad \text{as } n \to \infty,$$

so the improper integral is convergent. It follows that $\sum_{n=1}^{\infty} b_n$ and, hence, $\sum_{n=1}^{\infty} a_n$ are convergent. ■

We cannot apply the Integral Test unless we can integrate the corresponding function $f$. However, we can sometimes use the Limit Version of the Comparison Test to replace the given series with a series for which the corresponding function is easier to integrate, as was done in Example 8.

## EXERCISES 10.3

Determine convergence or divergence for each of the series in Exercises 1–32.

1 $\sum_{n=1}^{\infty} \frac{7}{n^{2/3}}$

2 $\sum_{n=1}^{\infty} \frac{3}{n^{3/2}}$

3 $\sum_{n=1}^{\infty} \frac{1}{2n-1}$

4 $\sum_{n=1}^{\infty} \frac{n}{n^2+4}$

5 $\sum_{n=1}^{\infty} n^{-1.01}$

6 $\sum_{n=1}^{\infty} n^{-0.99}$

7 $\sum_{n=1}^{\infty} \frac{n}{\sqrt{n^2+1}}$

8 $\sum_{n=1}^{\infty} \frac{1}{n\sqrt{n^2+4}}$

9 $\sum_{n=1}^{\infty} ne^{-n^2}$

10 $\sum_{n=1}^{\infty} \frac{2n}{3n+1}$

11 $\sum_{n=2}^{\infty} \frac{1}{n \ln n}$

12 $\sum_{n=2}^{\infty} \frac{1}{n^3-1}$

13 $\sum_{n=1}^{\infty} \frac{1}{\sqrt{2n-1}}$

14 $\sum_{n=1}^{\infty} \frac{1}{\sqrt{n^4+16}}$

15 $\sum_{n=1}^{\infty} \sin(1/n^2)$

16 $\sum_{n=1}^{\infty} \cos(1/n^2)$

17 $\sum_{n=1}^{\infty} \frac{\ln n}{n^3}$

18 $\sum_{n=1}^{\infty} \frac{1}{1+\sqrt{n}}$

19 $\sum_{n=1}^{\infty} \frac{2n+5}{n^2+3n+2}$

20 $\sum_{n=1}^{\infty} \frac{2n^2+3}{n^4+2n^2+4}$

21 $\sum_{n=2}^{\infty} \frac{1}{n(\ln n)^2}$

22 $\sum_{n=1}^{\infty} e^{-n}$

23 $\sum_{n=1}^{\infty} e^{-1/n}$

24 $\sum_{n=1}^{\infty} \frac{1}{n(n+1)(n+2)}$

25 $\sum_{n=1}^{\infty} ne^{-n}$

26 $\sum_{n=1}^{\infty} \frac{e^n}{1+e^{2n}}$

27 $\sum_{n=1}^{\infty} \frac{n}{(n+1)e^n}$

28 $\sum_{n=2}^{\infty} \frac{n}{(n^2+1)\ln(2n+1)}$

29 $\sum_{n=1}^{\infty} \frac{\ln n}{ne^n}$

30 $\sum_{n=1}^{\infty} n^3 e^{-n}$

31 $\sum_{n=1}^{\infty} \frac{\cos^2 n}{n^2}$

32 $\sum_{n=1}^{\infty} \frac{e^{1/n}}{n^3}$

33 Estimate

$$S = \sum_{n=1}^{\infty} n^{-4}$$

by a partial sum $S_N$ with $0 \le S - S_N \le 0.01$.

34 Estimate

$$S = \sum_{n=1}^{\infty} e^{-n}$$

by a partial sum $S_N$ with $0 \le S - S_N \le 0.01$.

35 Determine the smallest value of $N$ for which the partial sum $S_N$ of

$$S = \sum_{n=1}^{\infty} \frac{1}{n^2+1}$$

satisfies $0 \le S - S_N \le 0.001$.

36 Determine the smallest value of $N$ for which the partial sum $S_N$ of

$$S = \sum_{n=1}^{\infty} \frac{1}{(n+1)[\ln(n+1)]^2}$$

satisfies $0 \le S - S_N \le 0.1$.

37 Determine values of $p$ for which

$$\sum_{n=2}^{\infty} \frac{1}{n(\ln n)^p}$$

is convergent.

## 10.4 MORE CONVERGENCE TESTS

We develop additional tests that aid us in determining the convergence or divergence of infinite series.

*Ratio Test*

**If $a_n > 0$ for all $n$, then**

$$\sum_{n=1}^{\infty} a_n \text{ is convergent if } \lim_{n\to\infty} \frac{a_{n+1}}{a_n} < 1;$$

$$\sum_{n=1}^{\infty} a_n \text{ is divergent if } \lim_{n\to\infty} \frac{a_{n+1}}{a_n} > 1.$$

■ *Proof* The proof of the Ratio Test involves comparison with a geometric series. For example,

$$\lim_{n\to\infty} \frac{a_{n+1}}{a_n} = \rho < 1$$

implies that the ratio approaches $\rho$ as $n$ approaches infinity. This means there are numbers $r$ and $N$ such that

$$\rho < r < 1 \quad \text{and} \quad \frac{a_{n+1}}{a_n} < r \qquad \text{for } n \geq N.$$

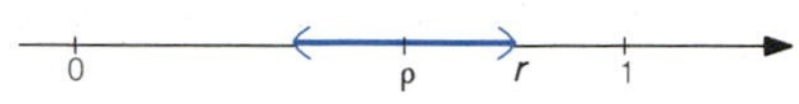


FIGURE 10.4.1

See Figure 10.4.1.

We have

$$\frac{a_{n+1}}{a_n} < r \qquad \text{for all } n \geq N,$$

so $a_{n+1} < ra_n$ for all $n \geq N$. It follows that

$$a_{N+1} < a_N r,$$
$$a_{N+2} < a_{N+1} r < a_N r^2,$$
$$a_{N+3} < a_{N+2} r < a_N r^3,$$
$$\vdots$$
$$a_{N+k} < a_N r^k.$$

Since

$$\sum_{k=1}^{\infty} a_N r^k$$

is a convergent geometric series for $0 < r < 1$, the Comparison Test implies

$$\sum_{k=1}^{\infty} a_{N+k}$$

is convergent, so

$$\sum_{n=1}^{\infty} a_n = a_1 + a_2 + \cdots + a_N + \sum_{k=1}^{\infty} a_{N+k}$$

is also convergent.

If $a_n > 0$ for all $n$, then

$$\lim_{n\to\infty} \frac{a_{n+1}}{a_n} > 1$$

implies $a_n$ does not approach zero as $n$ approaches infinity, so the series

$$\sum_{n=1}^{\infty} a_n$$

must be divergent. ■

The Ratio Test gives no information if

$$\lim_{n\to\infty} \frac{a_{n+1}}{a_n} = 1.$$

The series

$$\sum_{n=1}^{\infty} a_n$$

may converge or diverge in this case. For example, both of the series

$$\sum_{n=1}^{\infty} n^{-2} \quad \text{and} \quad \sum_{n=1}^{\infty} n^{-1/2}$$

have

$$\lim_{n\to\infty} \frac{a_{n+1}}{a_n} = 1.$$

The first series is convergent and the second is divergent. *Generally, the Ratio Test is not useful for series with terms that consist of only algebraic expressions of n.*

**The Ratio Test is very useful for series with terms that contain factorials and/or $n$th powers of constants.**

■ **EXAMPLE 1**

Determine convergence or divergence of the series

$$\sum_{n=1}^{\infty} \frac{2^n}{n!}.$$

**SOLUTION**

We have

$$a_n = \frac{2^n}{n!}, \quad \text{so} \quad a_{n+1} = \frac{2^{n+1}}{(n+1)!}.$$

Since

$$(n+1)! = (1)(2)\cdots(n)(n+1) = n!(n+1),$$

we have

$$\frac{a_{n+1}}{a_n} = a_{n+1}\frac{1}{a_n} = \frac{2^{n+1}}{(n+1)!}\frac{n!}{2^n} = \frac{2^{n+1}}{n!(n+1)}\frac{n!}{2^n}$$

$$= \frac{2}{n+1} \to 0 < 1 \qquad \text{as } n \to \infty.$$

The Ratio Test then implies convergence. ■

■ **EXAMPLE 2**

Determine convergence or divergence of the series

$$\sum_{n=1}^{\infty} n^2 e^{-n}.$$

**SOLUTION**

We have $a_n = n^2 e^{-n}$, so $a_{n+1} = (n+1)^2 e^{-(n+1)}$. Then

$$\frac{a_{n+1}}{a_n} = \frac{(n+1)^2 e^{-(n+1)}}{n^2 e^{-n}} = \left(\frac{n+1}{n}\right)^2 \frac{1}{e} \to \frac{1}{e} < 1 \qquad \text{as } n \to \infty.$$

Convergence follows from the Ratio Test. ■

■ **EXAMPLE 3**

Determine convergence or divergence of the series

$$\sum_{n=1}^{\infty} \frac{5^n}{(2n-1)!}.$$

**SOLUTION**

We have

$$a_n = \frac{5^n}{(2n-1)!}, \quad \text{so} \quad a_{n+1} = \frac{5^{(n+1)}}{(2(n+1)-1)!} = \frac{5^{n+1}}{(2n+1)!}.$$

We have

$$(2n+1)! = (1)(2)\cdots(2n-1)(2n)(2n+1) = (2n-1)!(2n)(2n+1).$$

Then

$$\frac{a_{n+1}}{a_n} = \frac{5^{n+1}}{(2n+1)!}\frac{(2n-1)!}{5^n} = \frac{5^{n+1}}{(2n-1)!(2n)(2n+1)}\frac{(2n-1)!}{5^n}$$

$$= \frac{5}{2n(2n+1)} \to 0 < 1 \qquad \text{as } n \to \infty.$$

The Ratio Test then implies convergence. ■

**■ EXAMPLE 4**

Determine convergence or divergence of the series

$$\sum_{n=1}^{\infty} \frac{2^n}{n^2}.$$

**SOLUTION**

We have

$$a_n = \frac{2^n}{n^2}, \quad \text{so} \quad a_{n+1} = \frac{2^{n+1}}{(n+1)^2}.$$

Then

$$\frac{a_{n+1}}{a_n} = \frac{2^{n+1}}{(n+1)^2}\frac{n^2}{2^n} = \frac{2n^2}{(n+1)^2} \to 2 > 1 \qquad \text{as } n \to \infty.$$

The Ratio Test implies divergence. ■

■ *Root Test*

**If $a_n \geq 0$ for all $n$, then**

$$\sum_{n=1}^{\infty} a_n \text{ is convergent if } \lim_{n \to \infty} (a_n)^{1/n} < 1;$$

$$\sum_{n=1}^{\infty} a_n \text{ is divergent if } \lim_{n \to \infty} (a_n)^{1/n} > 1.$$

The proof of the Root Test is similar to that of the Ratio Test. Note that $0 \leq (a_n)^{1/n} < r$ implies $a_n < r^n$. As in the case of the Ratio Test, we can conclude nothing from the Root Test when the limit of the $n$th root of $a_n$ is one.

**The Root Test is useful for series with terms that involve factors raised to the *n*th power.**

**■ EXAMPLE 5**

Determine convergence or divergence of the series

$$\sum_{n=1}^{\infty} \frac{(\ln n)^{2n}}{n^n}.$$

**SOLUTION**

We have

$$a_n = \frac{(\ln n)^{2n}}{n^n}, \quad \text{so} \quad (a_n)^{1/n} = \frac{(\ln n)^2}{n}.$$

Then

$$\lim_{n\to\infty} (a_n)^{1/n} = \lim_{n\to\infty} \frac{(\ln n)^2}{n} \underset{(\infty/\infty,\,\text{l'H})}{=} \lim_{n\to\infty} \frac{2(\ln n)(1/n)}{1}$$

$$= \lim_{n\to\infty} \frac{2\ln n}{n}$$

$$\underset{(\infty/\infty,\,\text{l'H})}{=} \lim_{n\to\infty} \frac{2/n}{1} = 0.$$

The Root Test then implies convergence. ■

We will encounter series where successive terms are of opposite sign. Such series are called **alternating series**. The following test is useful for determining convergence for some of these series.

■ *Alternating Series Test*

**If $\sum_{n=1}^{\infty} a_n$ is an alternating series with**

**(i) $|a_{n+1}| < |a_n|$ for all $n$ and**
**(ii) $\lim_{n\to\infty} |a_n| = 0$,**

**then the series is convergent and $|S - S_N| < |a_{N+1}|$.**

■ *Proof* The conditions of the Alternating Series Test imply that $S_{N+k}$ is between $S_N$ and $S_{N+1} = S_N + a_{N+1}$ for all $k > 1$. That is, adding $a_{N+1}$ to $S_N$ gives $S_{N+1}$. Since $a_{N+2}$ has sign opposite to that of $a_{N+1}$, adding $a_{N+2}$ to $S_{N+1}$ will take us back toward $S_N$. Since $|a_{N+2}| < |a_{N+1}|$, we will not go back as far as $S_N$. Adding $a_{N+3}$ will take us toward, but not past, $S_{N+1}$, and so forth. See Figure 10.4.2, where we have represented the partial sums as points on a number line. It is geometrically evident that the condition that $\lim_{n\to\infty} |a_n| = 0$ then implies $S_n$ converges to a point $S$ within $|a_{N+1}|$ of $S_N$. ■

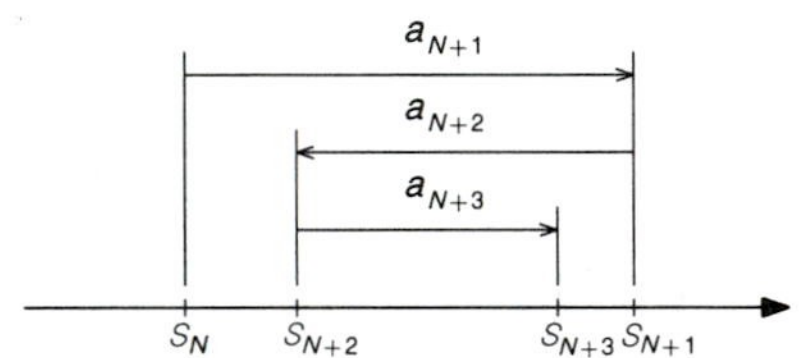


FIGURE 10.4.2

■ EXAMPLE 6

Determine convergence or divergence of the series

$$\sum_{n=1}^{\infty} \frac{(-1)^n}{n^{1/2}}.$$

SOLUTION

We first note that successive terms have opposite signs, so the series is alternating. The values $|a_n| = n^{-1/2}$ satisfy

(i) $(n+1)^{-1/2} < n^{-1/2}$ and
(ii) $\lim_{n\to\infty} n^{-1/2} = 0$.

The Alternating Series Test then implies convergence. ■

**EXAMPLE 7**

Determine convergence or divergence of the series

$$\sum_{n=1}^{\infty} \frac{(-1)^n(\ln n)}{n}.$$

**SOLUTION**

This is an alternating series. We have

$$|a_n| = \frac{\ln n}{n}.$$

To check condition (i), we set

$$f(x) = \frac{\ln x}{x}.$$

Then

$$f'(x) = \frac{(x)(1/x) - (\ln x)(1)}{x^2}.$$

We see that $f'(x) < 0$, so $f$ is decreasing, for $1 - \ln x < 0$, or $x > e$. It follows that

$$\text{(i)}\quad \frac{\ln(n+1)}{n+1} < \frac{\ln n}{n}, \quad \text{at least for } n \geq 3 > e, \text{ which is enough.}$$

(If $\sum_{n=3}^{\infty} a_n$ converges, so does $\sum_{n=1}^{\infty} a_n$. We can always ignore a finite number of terms in order to determine convergence or divergence of a series.)

$$\text{(ii)}\quad \lim_{n\to\infty} \frac{\ln n}{n} \underset{(\infty/\infty,\,\text{l'H})}{=} \lim_{n\to\infty} \frac{1/n}{1} = 0.$$

Convergence then follows from the Alternating Series Test. ■

If the condition $\lim_{n\to\infty} |a_n| = 0$ is not satisfied, we know that the series $\sum_{n=1}^{\infty} a_n$ must be divergent, since the terms of a convergent series must approach zero as $n$ approaches infinity.

**EXAMPLE 8**

Determine convergence or divergence of the series

$$\sum_{n=1}^{\infty} \frac{n(-1)^n}{n+1}.$$

**SOLUTION**

This is an alternating series. We have

$$|a_n| = \frac{n}{n+1} \to 1 \neq 0 \qquad \text{as } n \to \infty.$$

Since the terms do not approach zero, the series must diverge. ■

■ **EXAMPLE 9**

Approximate

$$\sum_{n=0}^{\infty} \frac{(-1)^n}{(2n+1)!}$$

with error less than 0.001.

**SOLUTION**

We first note that the series satisfies the conditions of the Alternating Series Test. This tells us the series convergences to a number $S$ and that

$$|S - S_N| < |a_{N+1}|.$$

This means the sum of the first $N$ terms of the series will be within 0.001 of $S$ if the next term in the series has absolute value less than 0.001. Writing out a few terms of the series, we have

$$S = 1 - \frac{1}{3!} + \frac{1}{5!} - \frac{1}{7!} + \frac{1}{9!} - \frac{1}{11!} + \cdots.$$

Checking the absolute value of each term in the series, we see that $1/7! \approx 0.00019841$ is the first term that is smaller than the acceptable error of 0.001. We then have

$$S \approx S_3 = 1 - \frac{1}{3!} + \frac{1}{5!} \approx 0.84166667.$$

The error is less than 0.001, because the next term in the series has absolute value $1/7! \approx 0.00019841 < 0.001$. (We will see in Section 10.6 that $S$ is the sine of an angle with radian measure 1. A calculator gives $S = \sin 1 \approx 0.84147099$.) ■

## EXERCISES 10.4

Determine convergence or divergence for each of the series in Exercises 1–30.

**1** $\sum_{n=1}^{\infty} \frac{n}{2^n}$

**2** $\sum_{n=1}^{\infty} \frac{3^n}{n!}$

**3** $\sum_{n=1}^{\infty} \frac{(-1)^n}{n+1}$

**4** $\sum_{n=1}^{\infty} \frac{(-1)^n n}{n+1}$

**5** $\sum_{n=1}^{\infty} \frac{e^n}{n}$

**6** $\sum_{n=1}^{\infty} \frac{1}{n^n}$

7 $\sum_{n=1}^{\infty} \frac{n^2 2^n}{(2n-1)!}$

8 $\sum_{n=1}^{\infty} \frac{3^{2n}}{(2n-1)!}$

9 $\sum_{n=1}^{\infty} \frac{(-1)^n}{n^{1/3}}$

10 $\sum_{n=2}^{\infty} \frac{(-1)^n}{\ln n}$

11 $\sum_{n=2}^{\infty} \frac{1}{(\ln n)^n}$

12 $\sum_{n=1}^{\infty} \left(\frac{n}{2n-1}\right)^{2n}$

13 $\sum_{n=1}^{\infty} \frac{(-1)^n}{\sqrt{1+\ln n}}$

14 $\sum_{n=1}^{\infty} \frac{(-1)^n}{\sqrt{n+1}}$

15 $\sum_{n=1}^{\infty} \frac{n!}{2^n}$

16 $\sum_{n=1}^{\infty} \frac{(-e)^n}{1+e^n}$

17 $\sum_{n=1}^{\infty} \frac{2n+1}{3n^3+n}$

18 $\sum_{n=1}^{\infty} \frac{2n}{n^2+1}$

19 $\sum_{n=2}^{\infty} \frac{1}{\sqrt{n}\ln n}$

20 $\sum_{n=2}^{\infty} \frac{\ln n}{n}$

21 $\sum_{n=1}^{\infty} \frac{e^n}{(n+1)^n}$

22 $\sum_{n=1}^{\infty} \frac{(-1)^n n}{e^n}$

23 $\sum_{n=1}^{\infty} \frac{(1)(3)\cdots(2n-1)}{(3)(6)\cdots(3n)}$

24 $\sum_{n=1}^{\infty} \frac{(2)(4)\cdots(2n)}{(2)(5)\cdots(3n-1)}$

25 $\sum_{n=1}^{\infty} \frac{3^n}{[1+\cos(1/n)]^n}$

26 $\sum_{n=1}^{\infty} \frac{2n}{n+1}$

27 $\sum_{n=1}^{\infty} \cos n\pi \sin(\pi/n)$

28 $\sum_{n=1}^{\infty} \frac{\cos n\pi}{1+\sqrt{n}}$

29 $\sum_{n=1}^{\infty} \frac{n!n^n}{(2n)!}$

30 $\sum_{n=1}^{\infty} \frac{n!n^{2n}}{(2n)!}$

Approximate the values of the series in Exercises 31–34 with error less than 0.01.

31 $\sum_{n=1}^{\infty} \frac{(-1)^n}{n!}$

32 $\sum_{n=1}^{\infty} \frac{(-1/2)^n}{n}$

33 $\sum_{n=1}^{\infty} \frac{(-1)^{n+1}}{(2n-1)!}\left(\frac{\pi}{4}\right)^{2n-1}$

34 $\sum_{n=1}^{\infty} \frac{(-1)^{n+1}}{(2n-2)!}\left(\frac{\pi}{3}\right)^{2n-2}$

## 10.5 ABSOLUTE CONVERGENCE, POWER SERIES

The concept of absolute convergence is useful in dealing with series that have terms of variable sign.

*Definition*

A series $\sum_{n=1}^{\infty} a_n$ is said to be **absolutely convergent** if the series $\sum_{n=1}^{\infty} |a_n|$ is convergent. ■

**EXAMPLE 1**

Show that

$$\sum_{n=1}^{\infty} (-1)^n n^{-2}$$

is absolutely convergent.

**SOLUTION**

We have $a_n = (-1)^n n^{-2}$, so

$$|a_n| = n^{-2} = \frac{1}{n^p}, \qquad p = 2.$$

We see that $\sum_{n=1}^{\infty} |a_n|$ is a convergent $p$-series, so $\sum_{n=1}^{\infty} a_n$ is absolutely convergent. ■

Note that the Alternating Series Test shows that the series in Example 1 is convergent. The following general result is true.

■ *Theorem*

**If $\sum_{n=1}^{\infty} a_n$ is absolutely convergent, then it is convergent.**

■ *Proof* Let $b_n = |a_n| - a_n$. Note that $b_n = 0$ if $a_n \geq 0$, and $b_n = 2|a_n|$ if $a_n < 0$. In either case we have $0 \leq b_n \leq 2|a_n|$. The Comparison Test for series with nonnegative terms then shows that $\sum_{n=1}^{\infty} b_n$ is convergent, since $\sum_{n=1}^{\infty} |a_n|$ is convergent. It follows that

$$\sum_{n=1}^{\infty} a_n = \sum_{n=1}^{\infty} (|a_n| - b_n)$$

is convergent. ■

Let us consider the above theorem from a different point of view. If $S_n$ denotes the $n$th partial sum of the series $\sum_{n=1}^{\infty} a_n$, then the triangle inequality implies that a difference of partial sums satisfies

$$\begin{aligned} |S_n - S_N| &= |a_{N+1} + a_{N+2} + \cdots + a_n| \\ &\leq |a_{N+1}| + |a_{N+2}| + \cdots + |a_n|, \qquad n > N. \end{aligned}$$

The sum on the right is the corresponding difference of partial sums of the series $\sum_{n=1}^{\infty} |a_n|$. Thus, the partial sums of $\sum_{n=1}^{\infty} a_n$ differ the most if the terms are all of the same sign. Differences of sign tend to make the partial sums closer to each other, and make convergence more likely.

To determine convergence of a series, we should first check for absolute convergence. We can use the tests we have developed for convergence of series with nonnegative terms to test absolute convergence. If it is absolutely convergent, we know it is also convergent. Absolute convergence tells more about a series than does convergence. A series may be convergent, even though it is not absolutely convergent.

■ *Definition*

We say $\sum_{n=1}^{\infty} a_n$ is **conditionally convergent** if the series is convergent, but not absolutely convergent. ■

(The expression "conditionally convergent" indicates that the convergence of the series depends on the order of the terms. It is true that a series that is absolutely convergent is convergent and has the same sum for every arrangement of its terms. We will not use nor verify this fact.)

■ **EXAMPLE 2**

Show that

$$\sum_{n=1}^{\infty} (-1)^n n^{-1/2}$$

is conditionally convergent.

**SOLUTION**

We have $a_n = (-1)^n n^{-1/2}$, so $|a_n| = n^{-1/2}$. Then $\sum_{n=1}^{\infty} |a_n|$ is a divergent $p$-series, so $\sum_{n=1}^{\infty} a_n$ is not absolutely convergent. We then note that the terms $a_n$ alternate sign, $|a_{n+1}| \leq |a_n|$, and $|a_n| \to 0$ as $n \to \infty$. The Alternating Series Test then shows that $\sum_{n=1}^{\infty} a_n$ is convergent. Since the series is convergent, but not absolutely convergent, it is conditionally convergent. ■

The idea of absolute convergence plays an important role in the study of series of functions. We are particularly interested in developing properties and applications of series with terms that consist of constant multiples of powers of $x - c$, where $x$ is a variable and $c$ represents a fixed number.

■ *Definition*

A series of the form

$$\sum_{n=0}^{\infty} a_n(x - c)^n$$

is called a **power series** about $c$. ■

In the context of power series, we adopt the convention that

$$(c - c)^0 = 1.$$

For each value of the variable $x$, the power series

$$\sum_{n=0}^{\infty} a_n(x - c)^n$$

is a series of real numbers. We can then determine convergence or divergence by using any of the tests we have previously developed.

A power series is a function of $x$, with domain the set of all $x$ for which the series converges. We will see that the domain of a power series is an interval.

The series

$$\sum_{n=0}^{\infty} a_n(x - c)^n$$

converges to $a_0$ when $x = c$, because the partial sums

$$S_n = a_0(c - c)^0 + a_1(c - c)^1 + a_2(c - c)^2 + \cdots + a_{n-1}(c - c)^{n-1} = a_0$$

for all $n$. [Recall that $(c - c)^0 = 1$ by convention.] We will see that it is possible that the series converges only for $x = c$. If the series converges at a point $x_1 \neq c$, we can use the following fact to obtain additional information about the set of convergence.

*Lemma 1* **If a power series**

$$\sum_{n=0}^{\infty} a_n(x-c)^n$$

**converges at a point $x_1$, then the series is absolutely convergent for $|x-c| < |x_1 - c|$.**

■ *Proof* We first note that convergence of the series $\sum_{n=0}^{\infty} a_n(x_1-c)^n$ implies

$$\lim_{n\to\infty} a_n(x_1-c)^n = 0.$$

This means there is a number $N$ such that $|a_n(x_1-c)^n| < 1$ for $n \geq N$. We then have

$$\begin{aligned}|a_n(x-c)^n| &= |a_n(x-c)^n|\left(\frac{|x_1-c|}{|x_1-c|}\right)^n \\ &= |a_n(x_1-c)^n|\left(\frac{|x-c|}{|x_1-c|}\right)^n < \left(\frac{|x-c|}{|x_1-c|}\right)^n \qquad \text{for } n \geq N.\end{aligned}$$

If $|x-c| < |x_1-c|$, we have

$$\frac{|x-c|}{|x_1-c|} = r < 1,$$

and the absolute convergence of $\sum_{n=0}^{\infty} a_n(x-c)^n$ is a consequence of comparison of

$$\sum_{n=0}^{\infty} |a_n(x-c)^n|$$

with the convergent geometric series

$$\sum_{n=0}^{\infty} r^n.$$

See Figure 10.5.1. ■

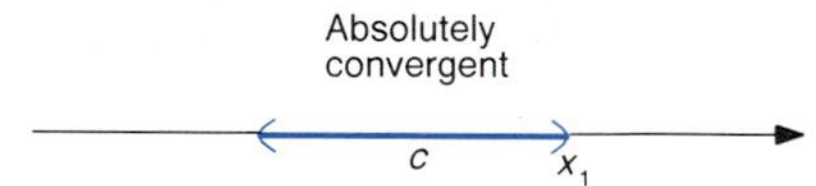

FIGURE 10.5.1

The following statement is an immediate consequence of Lemma 1.

*Lemma 2* **If a power series**

$$\sum_{n=0}^{\infty} a_n(x-c)^n$$

**does not converge absolutely at a point $x_2$, then the series is divergent for $|x-c| > |x_2-c|$.**

Lemmas 1 and 2 can be used to verify a fundamental fact of power series.

■ *Theorem*

**A power series $\sum_{n=0}^{\infty} a_n(x-c)^n$ either**

**(i) converges only for $x = c$,**
**(ii) converges absolutely for all $x$, or**
**(iii) there is a number R such that the series converges absolutely for $|x - c| < R$ and diverges for $|x - c| > R$.**

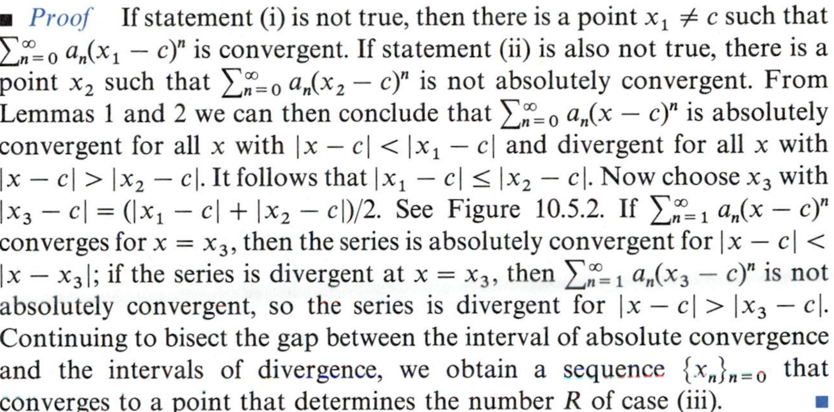

■ *Proof* If statement (i) is not true, then there is a point $x_1 \neq c$ such that $\sum_{n=0}^{\infty} a_n(x_1 - c)^n$ is convergent. If statement (ii) is also not true, there is a point $x_2$ such that $\sum_{n=0}^{\infty} a_n(x_2 - c)^n$ is not absolutely convergent. From Lemmas 1 and 2 we can then conclude that $\sum_{n=0}^{\infty} a_n(x-c)^n$ is absolutely convergent for all $x$ with $|x - c| < |x_1 - c|$ and divergent for all $x$ with $|x - c| > |x_2 - c|$. It follows that $|x_1 - c| \leq |x_2 - c|$. Now choose $x_3$ with $|x_3 - c| = (|x_1 - c| + |x_2 - c|)/2$. See Figure 10.5.2. If $\sum_{n=1}^{\infty} a_n(x-c)^n$ converges for $x = x_3$, then the series is absolutely convergent for $|x - c| < |x - x_3|$; if the series is divergent at $x = x_3$, then $\sum_{n=1}^{\infty} a_n(x_3 - c)^n$ is not absolutely convergent, so the series is divergent for $|x - c| > |x_3 - c|$. Continuing to bisect the gap between the interval of absolute convergence and the intervals of divergence, we obtain a sequence $\{x_n\}_{n=0}$ that converges to a point that determines the number $R$ of case (iii). ■

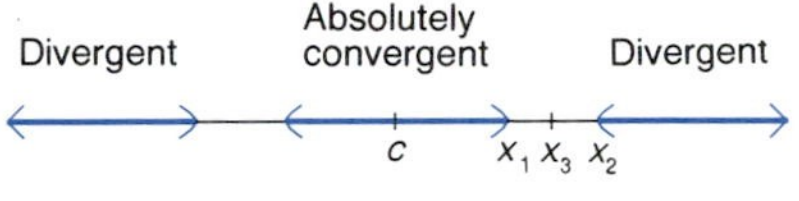


**FIGURE 10.5.2**

The number $R$ in case (iii) of the previous theorem is called the **radius of convergence** of the power series. In case (i), we say the radius of convergence is zero. In case (ii), we say the radius of convergence is infinity. The **interval of convergence** is the set of all $x$ for which a power series is convergent. The interval of convergence contains $|x - c| < R$, and possibly one or both endpoints of that interval.

■ **EXAMPLE 3**

Find the interval of convergence of $\sum_{n=1}^{\infty} nx^n$.

**SOLUTION**

The series is a power series with $a_n = n$ and $c = 0$. Let $u_n = n|x|^n$, so $u_{n+1} = (n+1)|x|^{n+1}$. Then

$$\frac{u_{n+1}}{u_n} = \frac{(n+1)|x|^{n+1}}{n|x|^n} = \frac{n+1}{n}|x| \to |x| \qquad \text{as } n \to \infty.$$

The limit is less than one whenever $|x| < 1$. The Ratio Test then shows $\sum_{n=1}^{\infty} u_n$ is convergent for $|x| < 1$, so $\sum_{n=1}^{\infty} nx^n$ is absolutely convergent for $|x| < 1$. Similarly, the series is divergent for $|x| > 1$. This means the radius of convergence is $R = 1$.

We know that the series is convergent for $-1 < x < 1$. We need to check convergence at the endpoints of this interval. When $x = 1$, we have $\sum_{n=1}^{\infty} n$. Since the $n$th term of this series does not approach zero as $n$ approaches infinity, we know this series must diverge. Similarly, the series is divergent when $x = -1$. The interval of convergence is $-1 < x < 1$. ■

**■ EXAMPLE 4**

Find the interval of convergence of

$$\sum_{n=1}^{\infty} \frac{x^n}{n}.$$

**SOLUTION**

Let

$$u_n = \frac{|x|^n}{n}, \quad \text{so} \quad u_{n+1} = \frac{|x|^{n+1}}{n+1}.$$

Then

$$\frac{u_{n+1}}{u_n} = \frac{|x|^{n+1}}{n+1} \frac{n}{|x|^n} = \frac{n}{n+1}|x| \to |x| \qquad \text{as } n \to \infty.$$

The limit is less than one whenever $|x| < 1$ and greater than one whenever $|x| > 1$. This implies the radius of convergence is $R = 1$.

We need to check convergence at the endpoints of the interval $-1 < x < 1$. When $x = 1$, we have

$$\sum_{n=1}^{\infty} \frac{1}{n}.$$

This is a divergent $p$-series. When $x = -1$, we have

$$\sum_{n=1}^{\infty} \frac{(-1)^n}{n}.$$

The Alternating Series Test shows that this series is convergent. The interval of convergence is $-1 \leq x < 1$. ■

**■ EXAMPLE 5**

Find the interval of convergence of

$$\sum_{n=1}^{\infty} \frac{x^n}{n^2}.$$

**SOLUTION**

Let

$$u_n = \frac{|x|^n}{n^2}, \quad \text{so} \quad u_{n+1} = \frac{|x|^{n+1}}{(n+1)^2}.$$

Then

$$\frac{u_{n+1}}{u_n} = \frac{|x|^{n+1}}{(n+1)^2} \frac{n^2}{|x|^n} = \frac{n^2}{(n+1)^2}|x| \to |x| \qquad \text{as } n \to \infty.$$

As before, this shows the radius of convergence is $R = 1$. It is easy to see that the series is absolutely convergent at both endpoints. The interval of convergence is $-1 \le x \le 1$. ■

■ **EXAMPLE 6**

Find the interval of convergence of

$$\sum_{n=0}^{\infty} n!x^n.$$

**SOLUTION**

Let

$$u_n = n!|x|^n, \quad \text{so} \quad u_{n+1} = (n+1)!|x|^{n+1}.$$

Then

$$\frac{u_{n+1}}{u_n} = \frac{(n+1)!|x|^{n+1}}{n!|x|^n} = \frac{n!(n+1)|x|^{n+1}}{n!|x|^n} = (n+1)|x| \to \infty \qquad \text{as } n \to \infty$$

for every $x \neq 0$. This series converges only for $x = 0$. The radius of convergence is zero. ■

■ **EXAMPLE 7**

Find the interval of convergence of

$$\sum_{n=1}^{\infty} \frac{x^n}{n!}.$$

**SOLUTION**

Let

$$u_n = \frac{|x|^n}{n!}, \quad \text{so} \quad u_{n+1} = \frac{|x|^{n+1}}{(n+1)!}.$$

Then

$$\frac{u_{n+1}}{u_n} = \frac{|x|^{n+1}}{(n+1)!}\,\frac{n!}{|x|^n} = \frac{|x|}{n+1} \to 0 \qquad \text{as } n \to \infty \text{ for each value of } x.$$

The series converges absolutely for all $x$. The radius of convergence is infinity. ■

Since the $n$th term of a convergent series approaches zero as $n$ approaches infinity, we can conclude from Example 7 that

$$\lim_{n \to \infty} \frac{x^n}{n!} = 0 \qquad \text{for all } x. \tag{1}$$

We will need to use (1) in Section 10.7.

■ **EXAMPLE 8**

Find the interval of convergence of

$$\sum_{n=1}^{\infty} \frac{(-1)^n(x-2)^n}{n4^n}.$$

**SOLUTION**

Let

$$u_n = \frac{|x-2|^n}{n4^n}, \quad \text{so} \quad u_{n+1} = \frac{|x-2|^{n+1}}{(n+1)4^{n+1}}.$$

Then

$$\frac{u_{n+1}}{u_n} = \frac{|x-2|^{n+1}}{(n+1)4^{n+1}} \frac{n4^n}{|x-2|^n} = \frac{n|x-2|}{(n+1)4} \to \frac{|x-2|}{4} \qquad \text{as } n \to \infty.$$

The Ratio Test gives convergence for

$$\frac{|x-2|}{4} < 1$$

and divergence for

$$\frac{|x-2|}{4} > 1.$$

Solving the first inequality above, we have

$$|x-2| < 4,$$

$$-4 < x-2 < 4,$$

$$-2 < x < 6.$$

When $x = -2$, the series is

$$\sum_{n=1}^{\infty} \frac{(-1)^n(-2-2)^n}{n4^n} = \sum_{n=1}^{\infty} \frac{1}{n},$$

which is a divergent $p$-series. When $x = 6$, we have

$$\sum_{n=1}^{\infty} \frac{(-1)^n(6-2)^n}{n4^n} = \sum_{n=1}^{\infty} \frac{(-1)^n}{n}.$$

The Alternating Series Test shows that this series is convergent. The interval of convergence is $-2 < x \leq 6$. ■

**■ EXAMPLE 9**

Find the interval of convergence of

$$\sum_{n=1}^{\infty} \frac{(x-2)^{2n}}{4^n}.$$

**SOLUTION**

Let

$$u_n = \frac{|x-2|^{2n}}{4^n}, \quad \text{so} \quad u_{n+1} = \frac{|x-2|^{2(n+1)}}{4^{n+1}}.$$

$$\frac{u_{n+1}}{u_n} = \frac{|x-2|^{2(n+1)}}{4^{n+1}} \frac{4^n}{|x-2|^{2n}}$$

$$= \frac{|x-2|^2}{4} \to \frac{|x-2|^2}{4} \qquad \text{as } n \to \infty.$$

The Ratio Test gives convergence for

$$\frac{|x-2|^2}{4} < 1$$

and divergence for

$$\frac{|x-2|^2}{4} > 1.$$

Solving the first inequality above, we have

$$|x-2|^2 < 4,$$

$$|x-2| < 2,$$

$$-2 < x-2 < 2,$$

$$0 < x < 4.$$

When either $x = 0$ or $x = 4$, the series is $\sum_{n=1}^{\infty} 1$, which is clearly divergent. The interval of convergence is $0 < x < 4$. ■

**■ EXAMPLE 10**

Find the interval of convergence of

$$\sum_{n=1}^{\infty} \frac{n!}{(1)(4)(7)\cdots(3n-2)} x^n.$$

**SOLUTION**

Let

$$u_n = \frac{n!}{(1)(4)(7)\cdots(3n-2)}|x|^n, \quad \text{so}$$

$$u_{n+1} = \frac{(n+1)!}{(1)(4)(7)\cdots(3n-2)(3(n+1)-2)}|x|^{n+1}$$
$$= \frac{(n+1)!}{(1)(4)(7)\cdots(3n-2)(3n+1)}|x|^{n+1}.$$

$$\frac{u_{n+1}}{u_n} = \frac{(n+1)!|x|^{n+1}}{(1)(4)(7)\cdots(3n-2)(3n+1)} \frac{(1)(4)(7)\cdots(3n-2)}{n!|x|^n}$$
$$= \frac{n+1}{3n+1}|x| \to \frac{|x|}{3} \qquad \text{as } n \to \infty.$$

We obtain absolute convergence for $|x| < 3$ and divergence for $|x| > 3$. To check convergence at the points $x = 3$ and $x = -3$, we note that

$$\frac{n!3^n}{(1)(4)(7)\cdots(3n-2)} = \frac{(1)(2)(3)\cdots(n)3^n}{(1)(4)(7)\cdots(3n-2)}$$
$$= \frac{(3)(6)(9)\cdots(3n)}{(1)(4)(7)\cdots(3n-2)}$$
$$= \left(\frac{3}{1}\right)\left(\frac{6}{4}\right)\left(\frac{9}{7}\right)\cdots\left(\frac{3n}{3n-2}\right) > 1.$$

This shows that the $n$th term of the series with either $x = 3$ or $x = -3$ does not approach zero, so the series must diverge for these values. The interval of convergence is $-3 < x < 3$. ■

## EXERCISES 10.5

Determine either absolute convergence, conditional convergence, or divergence for each of the series in Exercises 1–10.

1 $\sum_{n=0}^{\infty} \frac{(-1)^n}{1+n^2}$

2 $\sum_{n=0}^{\infty} \frac{(-1)^n n}{1+n^2}$

3 $\sum_{n=0}^{\infty} \frac{(-1)^n n^2}{1+n^2}$

4 $\sum_{n=0}^{\infty} \frac{n(-2)^n}{1+3^n}$

5 $\sum_{n=1}^{\infty} \frac{(-1)^n}{n(1+2^n)}$

6 $\sum_{n=1}^{\infty} \frac{(-3)^n}{n(1+2^n)}$

7 $\sum_{n=1}^{\infty} \frac{(-1)^n \ln n}{n^2}$

8 $\sum_{n=1}^{\infty} \frac{(-1)^n \ln n}{n}$

9 $\sum_{n=2}^{\infty} \frac{(-1)^n}{\ln n}$

10 $\sum_{n=2}^{\infty} \frac{(-1)^n}{n(\ln n)^2}$

Find the radius of convergence and the interval of convergence of the series in Exercises 11–30.

11 $\sum_{n=0}^{\infty} (x-1)^n$

12 $\sum_{n=0}^{\infty} n(x-1)^{n-1}$

13 $\sum_{n=0}^{\infty} \frac{(x-1)^{n+1}}{n+1}$

14 $\sum_{n=0}^{\infty} \frac{(x+1)^{n+2}}{(n+1)(n+2)}$

15 $\sum_{n=0}^{\infty} \frac{x^n}{1+2^n}$

16 $\sum_{n=0}^{\infty} \frac{x^{2n}}{1+2^n}$

17 $\sum_{n=2}^{\infty} (\ln n)x^n$

18 $\sum_{n=2}^{\infty} \frac{x^n}{\ln n}$

19 $\sum_{n=1}^{\infty} \frac{n(x-2)^n}{2^n}$

20 $\sum_{n=1}^{\infty} \frac{3^n(x-1)^n}{n}$

21 $\sum_{n=1}^{\infty} \frac{(-1)^n(x-3)^n}{n^2 3^n}$

22 $\sum_{n=0}^{\infty} \frac{(-1)^n(x+2)^n}{4^n}$

23 $\sum_{n=0}^{\infty} \frac{n!}{2^n} x^n$

24 $\sum_{n=1}^{\infty} n^n x^n$

25 $\sum_{n=0}^{\infty} \frac{(-1)^n x^{2n}}{(2n)!}$

26 $\sum_{n=0}^{\infty} \frac{(-1)^n x^{2n+1}}{(2n+1)!}$

27 $\sum_{n=1}^{\infty} \frac{(1)(3)\cdots(2n-1)}{(n+2)!} x^n$

28 $\sum_{n=1}^{\infty} \frac{(2)(5)\cdots(3n-1)}{(n+2)!} x^n$

29 $\sum_{n=1}^{\infty} \frac{(3)(5)\cdots(2n+1)}{n!} x^n$

30 $\sum_{n=1}^{\infty} \frac{(4)(7)\cdots(3n+1)}{n!} x^n$

## 10.6 POWER SERIES REPRESENTATION OF FUNCTIONS

We will see that functions that can be represented as power series can be treated as polynomials. This means, within their interval of convergence, we can add, subtract, multiply, divide, differentiate, and integrate term by term.

We already know of one function that can be represented as a power series. Namely, in Section 10.2 we showed that the series

$$\sum_{n=0}^{\infty} r^n$$

is convergent with sum

$$\frac{1}{1-r} \qquad \text{if } |r| < 1.$$

This gives the following power series representation of the function $1/(1-x)$.

$$\frac{1}{1-x} = 1 + x + x^2 + x^3 + \cdots, \qquad |x| < 1. \tag{1}$$

We can obtain power series representations of other functions by *substitution.*

■ **EXAMPLE 1**

Show that

$$\frac{1}{1+x} = 1 - x + x^2 - x^3 + \cdots, \qquad |x| < 1.$$

**SOLUTION**

Rewriting and then substituting $-x$ for $x$ in (1), we obtain

$$\frac{1}{1+x} = \frac{1}{1-(-x)} = 1 + (-x) + (-x)^2 + (-x)^3 + \cdots$$
$$= 1 - x + x^2 - x^3 + \cdots, \qquad |-x| < 1.$$

Similarly, substitution of $-x^2$ for $x$ in (1) gives

$$\frac{1}{1+x^2} = 1 - x^2 + x^4 - x^6 + \cdots, \qquad |x| < 1. \tag{2}$$

The following theorem tells us how to perform the operations of addition, subtraction, and multiplication by a constant.

*Theorem 1*

**Suppose $\sum_{n=0}^{\infty} a_n(x-c)^n$ has radius of convergence $R_1$ and $\sum_{n=0}^{\infty} b_n(x-c)^n$ has radius of convergence $R_2$. Let $R = \text{minimum}(R_1, R_2)$. Then**

$$d\left(\sum_{n=0}^{\infty} a_n(x-c)^n\right) + e\left(\sum_{n=0}^{\infty} b_n(x-c)^n\right)$$
$$= \sum_{n=0}^{\infty} (da_n + eb_n)(x-c)^n, \qquad |x| < R.$$

Theorem 1 means we can carry out the operations on the left by combining terms with like powers of $x - c$, as if the series were finite sums. Replacing each of the two series on the left with their $n$th partial sums, and then combining terms with like powers of $x - c$, gives the $n$th partial sum of the series on the right. Theorem 1 is then an immediate consequence of the Limit Theorem.

*Theorem 2*

**Suppose $\sum_{n=0}^{\infty} a_n(x-c)^n$ has radius of convergence $R_1$ and $\sum_{n=0}^{\infty} b_n(x-c)^n$ has radius of convergence $R_2$. Let $R = \text{minimum}(R_1, R_2)$.**

**If $c_n = a_0b_n + a_1b_{n-1} + a_2b_{n-2} + \cdots + a_{n-1}b_1 + a_nb_0$, then**

$$\left(\sum_{n=0}^{\infty} a_n(x-c)^n\right)\left(\sum_{n=0}^{\infty} b_n(x-c)^n\right) = \sum_{n=0}^{\infty} c_n(x-c)^n, \qquad |x| < R.$$

Theorem 2 tell us that, within both intervals of convergence, we can multiply two power series as if they were finite sums. Note that multiplication of the partial sums

$$a_0 + a_1(x-c) + a_2(x-c)^2 + \cdots + a_N(x-c)^N \quad \text{and}$$
$$b_0 + b_1(x-c) + b_2(x-c)^2 + \cdots + b_N(x-c)^N$$

and then combining terms with like powers of $x - c$ gives a polynomial in $x - c$ with coefficients $c_n$ for $0 \le n \le N$, plus some higher-order terms. The fact that the sum of the extra, higher-order, terms approaches zero as $N$ approaches infinity depends on the absolute convergence of power series at points interior to their interval of convergence. We will not verify this.

**■ EXAMPLE 2**

Find a power series representation about $c = 0$ of the function

$$\frac{x}{x^2 - 3x + 2}.$$

**SOLUTION**

Let us first find the partial fraction expansion of the function. We have

$$\frac{x}{x^2 - 3x + 2} = \frac{x}{(x-1)(x-2)} = \frac{A}{x-1} + \frac{B}{x-2}.$$

It is not difficult to determine that $A = -1$ and $B = 2$, so

$$\frac{x}{x^2 - 3x + 2} = \frac{-1}{x-1} + \frac{2}{x-2}.$$

Rewriting and then using (1), we obtain

$$\begin{aligned}
\frac{x}{x^2 - 3x + 2} &= \frac{1}{1-x} - \frac{1}{1-\dfrac{x}{2}} \\
&= (1 + x + x^2 + x^3 + \cdots) \\
&\quad - \left(1 + \left(\frac{x}{2}\right) + \left(\frac{x}{2}\right)^2 + \left(\frac{x}{2}\right)^3 + \cdots\right) \\
&= (1 - 1) + \left(1 - \frac{1}{2}\right)x \\
&\quad + \left(1 - \left(\frac{1}{2}\right)^2\right)x^2 + \left(1 - \left(\frac{1}{2}\right)^3\right)x^3 + \cdots \\
&= \frac{1}{2}x + \frac{3}{4}x^2 + \frac{7}{8}x^3 + \cdots, \qquad |x| < 1.
\end{aligned}$$

As an alternate solution of this problem, let us rewrite as a product and then multiply the power series. That is,

$$\begin{aligned}
\frac{x}{x^2 - 3x + 2} &= \left(\frac{x}{2}\right)\left(\frac{1}{1-x}\right)\left(\frac{1}{1-\dfrac{x}{2}}\right) \\
&= \left(\frac{x}{2}\right)(1 + x + x^2 + \cdots)\left(1 + \frac{x}{2} + \left(\frac{x}{2}\right)^2 + \cdots\right) \\
&= \left(\frac{x}{2}\right)\left(1 + \left(1 + \frac{1}{2}\right)x + \left(1 + \frac{1}{2} + \frac{1}{4}\right)x^2 + \cdots\right) \\
&= \frac{x}{2} + \left(\frac{1}{2}\right)\left(\frac{3}{2}\right)x^2 + \left(\frac{1}{2}\right)\left(\frac{7}{4}\right)x^3 + \cdots \\
&= \frac{1}{2}x + \frac{3}{4}x^2 + \frac{7}{8}x^3 + \cdots, \qquad |x| < 1.
\end{aligned}$$

We can obtain the same power series by division. That is,

$$\begin{array}{r|l} & (1/2)x + (3/4)x^2 + (7/8)x^3 + \cdots \\ \hline 2 - 3x + x^2 & x \\ & x - (3/2)x^2 + (1/2)x^3 \\ \hline & (3/2)x^2 - (1/2)x^3 \\ & (3/2)x^2 - (9/4)x^3 + (3/4)x^4 \\ \hline & (7/4)x^3 - (3/4)x^4 \\ & (7/4)x^3 - (21/8)x^4 + (7/8)x^5 \\ \hline & \vdots \end{array}$$

Note the order of the terms of the above divisor. Because we want the powers of $x$ in the power series to increase, we will always arrange the divisor and dividend in increasing order of terms, with space left for each missing term. ■

Generally, the radius of convergence of a power series that is obtained by division cannot be described in terms of the radii of convergence of the dividend and the divisor. It seems reasonable that we would need to avoid values of $x$ for which the denominator is zero and the numerator is nonzero.

The next result tells us that functions that have a power series representation can be differentiated and integrated term by term.

■ *Theorem 3*

**Let $R$ be the radius of convergence of $\sum_{n=0}^{\infty} a_n(x-c)^n$. Then $|x| < R$ implies**

$$\frac{d}{dx}(a_0 + a_1(x-c) + a_2(x-c)^2 + a_3(x-c)^3 + a_4(x-c)^4 + \cdots)$$
$$= a_1 + 2a_2(x-c) + 3a_3(x-c)^2 + 4a_4(x-c)^3 + \cdots$$

**and**

$$\int (a_0 + a_1(x-c) + a_2(x-c)^2 + a_3(x-c)^3 + a_4(x-c)^4 + \cdots)\,dx$$
$$= C + a_0(x-c) + \frac{a_1}{2}(x-c)^2 + \frac{a_2}{3}(x-c)^3 + \frac{a_3}{4}(x-c)^4 + \frac{a_4}{5}(x-c)^5 + \cdots.$$

We will not verify this result.

■ **EXAMPLE 3**

$$f(x) = \sum_{n=1}^{\infty} \frac{1}{n} x^n.$$

Evaluate the third derivative of $f$ at zero.

**SOLUTION**

Let us write out a few terms of the series. We have

$$f(x) = x + \frac{1}{2}x^2 + \frac{1}{3}x^3 + \frac{1}{4}x^4 + \frac{1}{5}x^5 + \cdots.$$

We then differentiate term by term to obtain

$$f'(x) = 1 + x + x^2 + x^3 + x^4 + \cdots,$$

$$f''(x) = 1 + 2x + 3x^2 + 4x^3 + \cdots,$$

$$f'''(x) = 2 + 6x + 12x^2 + \cdots.$$

Substitution then gives

$$f'''(0) = 2 + 6(0) + 12(0)^2 + \cdots = 2.$$ ■

Let us see how to use Theorem 2 to obtain power series representations of additional types of functions.

■ **EXAMPLE 4**

Find a power series representation about $c = 0$ of

(a) $\dfrac{-2x}{(1 + x^2)^2}$,

(b) $\ln\left(\dfrac{1}{x - 1}\right)$.

**SOLUTION**

(a) In this case we notice that the given function is the derivative of a function with a known power series representation, namely $(1 + x^2)^{-1}$. Using (2), we then have

$$\begin{aligned}\frac{-2x}{(1 + x^2)^2} &= \frac{d}{dx}\left(\frac{1}{1 + x^2}\right) = \frac{d}{dx}(1 - x^2 + x^4 - x^6 + \cdots)\\ &= -2x + 4x^3 - 6x^5 + \cdots, \qquad |x| < 1.\end{aligned}$$

(b) In this case we express the function as an integral of a power series and then use Theorem 2.

$$\begin{aligned}\ln\frac{1}{1 - x} &= \int_0^x \frac{1}{1 - t}\,dt\\ &= \int_0^x (1 + t + t^2 + t^3 + \cdots)\,dt\\ &= x + \frac{x^2}{2} + \frac{x^3}{3} + \frac{x^4}{4} + \cdots, \qquad |x| < 1.\end{aligned}$$ ■

Let us restate the result of Example 4b for future reference.

$$\ln \frac{1}{1-x} = x + \frac{1}{2}x^2 + \frac{1}{3}x^3 + \frac{1}{4}x^4 + \cdots, \qquad |x| < 1. \tag{3}$$

If a function $f$ can be expressed as a power series, then the partial sums of the power series can be used as approximate values of $f(x)$. Let us illustrate this important idea.

■ **EXAMPLE 5**

Use partial sums of the power series expansion of

$$\ln \frac{1}{1-x}$$

with one, two, three, and four nonzero terms to obtain approximate values of ln 2. Compare with a calculator value of ln 2.

**SOLUTION**

From (3) we see that the partial sums of the power series for

$$\ln \frac{1}{1-x}$$

that have one, two, three, and four nonzero terms are

$$P_1(x) = x,$$

$$P_2(x) = x + \frac{1}{2}x^2,$$

$$P_3(x) = x + \frac{1}{2}x^2 + \frac{1}{3}x^3, \quad \text{and}$$

$$P_4(x) = x + \frac{1}{2}x^2 + \frac{1}{3}x^3 + \frac{1}{4}x^4, \quad \text{respectively.}$$

To use these formulas, we need to find $x$ such that

$$\frac{1}{1-x} = 2.$$

Solving for $x$, we have

$$1 = 2(1 - x),$$

$$1 = 2 - 2x,$$

$$2x = 1,$$

$$x = 0.5.$$

We then obtain

$$P_1(0.5) = 0.5,$$

$$P_2(0.5) = 0.5 + \frac{1}{2}(0.5)^2 = 0.625,$$

$$P_3(0.5) = 0.5 + \frac{1}{2}(0.5)^2 + \frac{1}{3}(0.5)^3 = 0.66666667,$$

$$P_4(0.5) = 0.5 + \frac{1}{2}(0.5)^2 + \frac{1}{3}(0.5)^3 + \frac{1}{4}(0.5)^4 = 0.68229167.$$

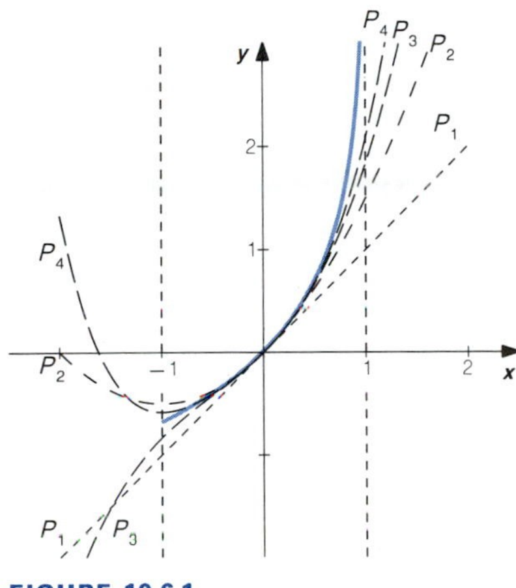

FIGURE 10.6.1

A calculator gives $\ln 2 \approx 0.69314718$.

The graphs of

$$\ln \frac{1}{1-x}, \qquad |x| < 1,$$

and the approximating partial sums of the series are given in Figure 10.6.1. Note that the series for $\ln(1/(1-x))$ converges for $x = -1$, by the Alternating Series Test. It appears from the graph that the limit is

$$\ln \frac{1}{1-(-1)} = \ln 0.5.$$ ■

We can use the power series of

$$\ln \frac{1}{1-x}$$

to obtain approximate values of $\ln u$ for

$$u = \frac{1}{1-x}, \qquad |x| < 1.$$

Solving the inequality $|x| < 1$ for the expression that gives $u$, we obtain

$$-1 < x < 1,$$

$$1 > -x > -1,$$

$$2 > 1 - x > 0,$$

$$\frac{1}{1-x} > \frac{1}{2}.$$

It follows that we can use this method to approximate values of $\ln u$ only for $u > 0.5$. The following result allows us to use a single power series to calculate approximate values of $\ln u$ for all $u > 0$.

$$\ln \sqrt{\frac{1+x}{1-x}} = x + \frac{x^3}{3} + \frac{x^5}{5} + \frac{x^7}{7} + \frac{x^9}{9} + \cdots, \qquad |x| < 1. \tag{4}$$

■ *Proof* We note that

$$\begin{aligned} \ln \sqrt{\frac{1+x}{1-x}} &= \frac{1}{2}(\ln(1+x) - \ln(1-x)) \\ &= \int_0^x \left(\frac{1}{2}\frac{1}{1+t} - \frac{1}{2}\frac{-1}{1-t}\right) dt \\ &= \int_0^x \frac{1}{1-t^2}\, dt. \end{aligned}$$

Using (1) with $t^2$ in place of $x$ to expand $1/(1-t^2)$, and then integrating, we obtain

$$\begin{aligned} \ln \sqrt{\frac{1+x}{1-x}} &= \int_0^x (1 + t^2 + (t^2)^2 + (t^2)^3 + (t^2)^4 + \cdots)\, dt \\ &= \int_0^x (1 + t^2 + t^4 + t^6 + t^8 + \cdots)\, dt \\ &= x + \frac{x^3}{3} + \frac{x^5}{5} + \frac{x^7}{7} + \frac{x^9}{9} + \cdots. \end{aligned}$$ ■

To use (4) to find an approximate value of $\ln u$, we solve

$$u = \sqrt{\frac{1+x}{1-x}}$$

for $x$. We obtain

$$u^2 = \frac{1+x}{1-x},$$

$$u^2(1-x) = 1 + x,$$

$$-(u^2+1)x = -u^2 + 1,$$

$$x = \frac{u^2-1}{u^2+1}.$$

It is not difficult to verify that each positive number $u$ corresponds to a value of $x$ that satisfies $|x| < 1$.

■ **EXAMPLE 6**

Use partial sums of the series expansion of

$$\ln\sqrt{\frac{1+x}{1-x}}$$

with one, two, three, and four nonzero terms to approximate values of ln 2. Compare with a calculator value of ln 2.

**SOLUTION**

We want

$$\sqrt{\frac{1+x}{1-x}} = 2, \quad \text{so} \quad x = \frac{2^2-1}{2^2+1} = 0.6.$$

The partial sum of the series in (4) that contains $n$ nonzero terms is

$$P_{2n}(x) = x + \frac{x^3}{3} + \frac{x^5}{5} + \frac{x^7}{7} + \cdots + \frac{x^{2n-1}}{2n-1}, \qquad n \geq 1.$$

[This notation is consistent with that used in the next section. Note that $P_{2n}(x)$ contains all terms of the series that have order less than or equal to $2n$.] The desired partial sums are $P_{2(1)} = P_2$, $P_{2(2)} = P_4$, $P_{2(3)} = P_6$, and $P_{2(4)} = P_8$, evaluated at $x = 0.6$. We have

$$P_2(0.6) = 0.6,$$

$$P_4(0.6) = 0.6 + \frac{(0.6)^3}{3} = 0.672,$$

$$P_6(0.6) = 0.6 + \frac{(0.6)^3}{3} + \frac{(0.6)^5}{5} = 0.687552,$$

$$P_8(0.6) = 0.6 + \frac{(0.6)^3}{3} + \frac{(0.6)^5}{5} + \frac{(0.6)^7}{7} = 0.69155109.$$

A calculator gives $\ln 2 = 0.69314718$.

Note that the approximate value obtained by using four nonzero terms in Example 6 is closer to the exact value of ln 2 than the corresponding approximate value in Example 5. The graphs of

$$\ln\sqrt{\frac{1-x}{1+x}}$$

and some of the approximating partial sums are given in Figure 10.6.2. ■

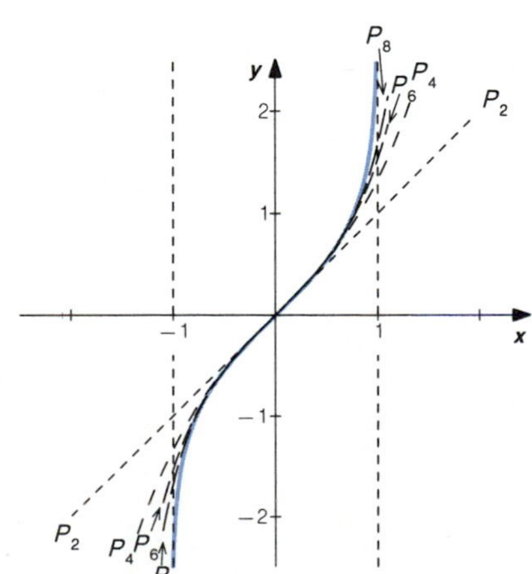

FIGURE 10.6.2

$$\tan^{-1}x = x - \frac{x^3}{3} + \frac{x^5}{5} - \frac{x^7}{7} + \cdots, \qquad |x| < 1. \tag{5}$$

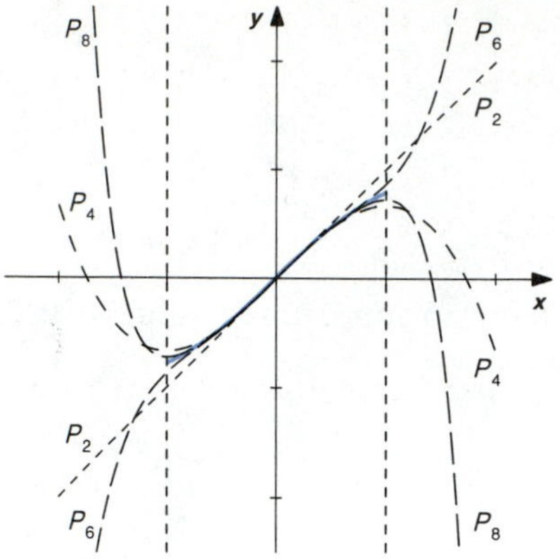

FIGURE 10.6.3

■ *Proof* To prove (5), we integrate the power series of (2). That is,

$$\tan^{-1}x = \int_0^x \frac{1}{1+t^2}\,dt$$

$$= \int_0^x (1 - t^2 + t^4 - t^6 + \cdots)\,dt$$

$$= x - \frac{x^3}{3} + \frac{x^5}{5} - \frac{x^7}{7} + \cdots, \qquad |x| < 1. \qquad ■$$

The graph of $\tan^{-1}x$ and some of the approximating partial sums are given in Figure 10.6.3.

■ **EXAMPLE 7**

Find a power series representation about $c = 0$ of

$$f(x) = \begin{cases} \dfrac{\tan^{-1}x}{x}, & x \neq 0, \\ 1, & x = 0. \end{cases}$$

**SOLUTION**

From (5), we have

$$f(x) = \frac{x - \dfrac{x^3}{3} + \dfrac{x^5}{5} - \dfrac{x^7}{7} + \cdots}{x}$$

$$= \frac{x\left(1 - \dfrac{x^2}{3} + \dfrac{x^4}{5} - \dfrac{x^6}{7} + \cdots\right)}{x}$$

$$= 1 - \frac{x^2}{3} + \frac{x^4}{5} - \frac{x^6}{7} + \cdots,$$

for $0 < |x| < 1$. The value of the power series also agrees with that of $f$ when $x = 0$. ■

■ **EXAMPLE 8**

Approximate

$$\int_0^{0.5} \frac{\tan^{-1}x}{x}\,dx$$

to within 0.001.

**SOLUTION**

Using the power series representation we obtained in Example 7, we have

$$\int_0^{0.5} \frac{\tan^{-1}x}{x}\,dx = \int_0^{0.5}\left(1 - \frac{x^2}{3} + \frac{x^4}{5} - \frac{x^6}{7} + \cdots\right)dx$$

$$= 0.5 - \frac{1}{9}(0.5)^3 + \frac{1}{25}(0.5)^5 - \frac{1}{49}(0.5)^7 + \cdots.$$

We see that the series alternates sign, and that the absolute value of the terms *decrease* to zero. The Alternating Series Test then tells us that the difference in the exact value of the series and a partial sum is less than the absolute value of the first term omitted from the partial sum. It follows that

$$\int_0^{0.5} \frac{\tan^{-1}x}{x}\,dx \approx 0.5 - \frac{1}{9}(0.5)^3 + \frac{1}{25}(0.5)^5 \approx 0.48736111$$

with error less than

$$\frac{1}{49}(0.5)^7 \approx 0.00015944 < 0.001.$$ ■

## EXERCISES 10.6

Find power series representations about $c = 0$ for the functions in Exercises 1–8.

1 (a) $\dfrac{1}{1-x^2}$, (b) $\dfrac{x}{1-x^2}$, (c) $\dfrac{x}{(1-x^2)^2}$

2 (a) $\dfrac{1}{2x-3}$, (b) $\dfrac{x}{2x-3}$, (c) $\dfrac{1}{(2x-3)^2}$

3 (a) $\ln(1+x)$, (b) $x\ln(1+x)$, (c) $\dfrac{\ln(1+2x)}{x}$

4 (a) $\tan^{-1}2x$, (b) $\dfrac{\tan^{-1}2x}{x}$, (c) $\dfrac{\tan^{-1}2x - 2x}{x^3}$

5 (a) $\dfrac{1}{x^2+2x-3}$, (b) $\dfrac{x^2}{x^2+2x-3}$

6 (a) $\dfrac{1}{2x^2-x-3}$, (b) $\dfrac{x}{2x^2-x-3}$

7 (a) $\dfrac{1+x}{1-x}$, (b) $\dfrac{x^2+1}{x-1}$

8 (a) $(x-1)\tan^{-1}x$, (b) $(x^2-1)\tan^{-1}x$

Use partial sums of the series in formula (4) with one, two, three, and four terms to find approximate values of the numbers in Exercises 9–12. Compare with calculator values.

9 $\ln 0.25$

10 $\ln 0.2$

11 $\ln 5$

12 $\ln 4$

Approximate the integrals in Exercises 13–18 to within 0.001.

13 $\displaystyle\int_0^{0.2} \ln(1+x)\,dx$

14 $\displaystyle\int_0^{0.3} \tan^{-1}x\,dx$

15 $\displaystyle\int_0^{0.25} x\tan^{-1}x\,dx$

16 $\displaystyle\int_0^{0.5} x^2\ln(1+x)\,dx$

17 $\displaystyle\int_0^{0.5} \frac{x}{1+x^2}\,dx$

18 $\displaystyle\int_0^{0.2} \frac{x^2}{1+x^2}\,dx$

19 $f(x) = \sum_{n=0}^{\infty} 2^n x^n$. Find $f''(0)$.

20 $f(x) = \sum_{n=1}^{\infty} nx^n$. Find $f''(0)$.

21 $f(x) = \sum_{n=0}^{\infty} \frac{1}{n!} x^{2n}$. Find $f'''(0)$.

22 $f(x) = \sum_{n=0}^{\infty} \frac{1}{(2n+1)!} x^{2n+1}$. Find $f'''(0)$.

23 $f(x) = \sum_{n=0}^{\infty} \frac{x^n}{n!}$. Show that $f'(x) = f(x)$.

24 $f(x) = 1 + kx + \frac{k(k-1)}{2!}x^2 + \frac{k(k-1)(k-2)}{3!}x^3 + \frac{k(k-1)(k-2)(k-3)}{4!}x^4 + \cdots, \quad |x| < 1.$

Show that $(1 + x)f'(x) = kf(x)$.

25 Show that

$$\frac{1}{x} = 1 - (x-1) + (x-1)^2 - (x-1)^3 + (x-1)^4 - \cdots, \qquad |x - 1| < 1.$$

$\left[\textit{Hint}: \frac{1}{x} = \frac{1}{1 + (x-1)}.\right]$

Use the result of Exercise 25 to find power series representations about $c = 1$ for the functions in Exercises 26–27.

26 $\ln x$

27 $1/x^2$

## 10.7 TAYLOR POLYNOMIALS WITH REMAINDER, TAYLOR SERIES

In the two previous sections we studied properties of power series and found power series representations of selected functions. Let us now consider the question:

*Given a function f, can we find a power series representaion of f?*

Theorem 1 provides us with some important information about the answer to this question.

*Theorem 1*

**If $f(x) = \sum_{n=0}^{\infty} a_n(x-c)^n$, $|x - c| < R$, $R > 0$, then**

$$a_n = \frac{f^{(n)}(c)}{n!}, \qquad n \geq 0.$$

*Proof* To verify this result, we evaluate the derivatives of $f$ at the point $c$. We will illustrate the idea by carrying out the calculations for the first few terms. Differentiating the series term by term, we have

$$f(x) = a_0 + a_1(x-c) + a_2(x-c)^2 + a_3(x-c)^3 + a_4(x-c)^4 + \cdots,$$

$$f'(x) = a_1 + a_2(2)(x-c) + a_3(3)(x-c)^2 + a_4(4)(x-c)^3 + \cdots,$$

$$f''(x) = a_2(2) + a_3(3)(2)(x-c) + a_4(4)(3)(x-c)^2 + \cdots,$$

$$f'''(x) = a_3(3)(2) + a_4(4)(3)(2)(x-c) + \cdots.$$

Setting $x = c$ in the first of the above formulas, we obtain

$$f(c) = a_0, \quad \text{so} \quad a_0 = \frac{f^{(0)}(c)}{0!},$$

since $f^{(0)}(c) = f(c)$ and $0! = 1$. Then, setting $x = c$ in each of the other three formulas above we obtain

$$f'(c) = a_1, \quad \text{so} \quad a_1 = \frac{f^{(1)}(c)}{1!},$$

$$f''(c) = a_2(2), \quad \text{so} \quad a_2 = \frac{f^{(2)}(c)}{2!},$$

$$f'''(c) = a_3(3)(2), \quad \text{so} \quad a_3 = \frac{f^{(3)}(c)}{3!}.$$

This verifies the formula for $a_n$, $0 \le n \le 3$. Mathematical induction could be used to verify the formula for all $n$. ■

Recall from Section 3.6 that the polynomial

$$P_n(x) = f(c) + f'(c)(x - c) + \frac{f''(c)}{2!}(x - c)^2 + \cdots + \frac{f^{(n)}(c)}{n!}(x - c)^n$$

is called the $n$th **Taylor polynomial** of $f$ about $c$. The series

$$\sum_{n=0}^{\infty} \frac{f^{(n)}(c)}{n!}(x - c)^n$$

is called the **Taylor series** of $f$ about $c$. If $c = 0$, the series is also called the **Maclaurin series** of $f$. Theorem 1 tells us, if $f$ does have a power series representation, then the series must be the Taylor series of $f$. For example, the power series representations

$$\frac{1}{1 - x} = 1 + x + x^2 + x^3 + \cdots, \qquad |x| < 1,$$

$$\ln\left(\frac{1}{1 - x}\right) = x + \frac{x^2}{2} + \frac{x^3}{3} + \frac{x^4}{4} + \cdots, \qquad |x| < 1,$$

$$\ln\sqrt{\frac{1 + x}{1 - x}} = x + \frac{x^3}{3} + \frac{x^5}{5} + \frac{x^7}{7} + \cdots, \qquad |x| < 1,$$

$$\tan^{-1}x = x - \frac{x^3}{3} + \frac{x^5}{5} - \frac{x^7}{7} + \cdots, \qquad |x| < 1, \quad \text{and}$$

$$\frac{\tan^{-1}x}{x} = 1 - \frac{x^2}{3} + \frac{x^4}{5} - \frac{x^6}{7} + \cdots, \qquad |x| < 1,$$

which were established in Section 10.5, are all Taylor series.

Let us now return to the original question. For a given function $f$, if $f$ has a power series representation, it must be the Taylor series of $f$. If $f$ has derivatives of all orders at $c$, we can form the Taylor series of $f$,

$$\sum_{n=0}^{\infty} \frac{f^{(n)}(c)}{n!}(x - c)^n.$$

Then either

(a) the series converges only for $x = c$, or
(b) the series has a positive interval of convergence $I$.

In case (b), either

$$\text{(i)}\ f(x) = \sum_{n=0}^{\infty} \frac{f^{(n)}(c)}{n!}(x-c)^n \qquad \text{for all } x \text{ in } I,$$

so the Taylor series of $f$ is a power series representation of $f$, or

$$\text{(ii)}\ f(x) \neq \sum_{n=0}^{\infty} \frac{f^{(n)}(c)}{n!}(x-c)^n \qquad \text{for some } x \text{ in } I.$$

In the latter case, $f$ does not have a power series representation on $I$ since any power series representation of $f$ on $I$ must be the Taylor series of $f$.

An example of case (b)(ii) is the function

$$f(x) = \begin{cases} e^{-1/x^2}, & x \neq 0, \\ 0, & x = 0. \end{cases}$$

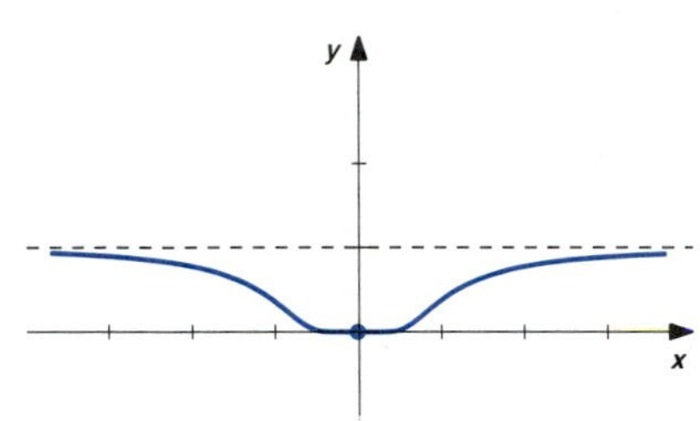

FIGURE 10.7.1

We can use the definition of the derivative as the limit of a difference quotient to show that $f^{(n)}(0) = 0$ for all $n \geq 1$. See Exercise 27 at the end of this section. Hence, the Maclaurin series of $f$ converges to zero for all $x$. This is not equal to $f(x)$, except for $x = 0$. The graph of $f$ is given in Figure 10.7.1.

To investigate if the Taylor series of a function converges to the function, we define $R_n(x)$ to be the **remainder** when $P_n(x)$ is subtracted from $f(x)$. That is,

$$\begin{aligned} R_n(x) &= f(x) - P_n(x) \\ &= f(x) - \left( f(c) + f'(c)(x-c) + \frac{f''(c)}{2!}(x-c)^2 + \cdots \right. \\ &\qquad \left. + \frac{f^{(n)}(c)}{n!}(x-c)^n \right). \end{aligned}$$

Since $R_n(x)$ is the difference between $f(x)$ and $P_n(x)$, $P_n(x)$ approaches $f(x)$ as $n$ approaches infinity if and only if $R_n(x)$ approaches zero as $n$ approaches infinity. That is,

$$f(x) = \sum_{n=0}^{\infty} \frac{f^{(n)}(c)}{n!}(x-c)^n$$

**at exactly those points $x$ for which $R_n(x) \to 0$ as $n \to \infty$.**

We can find a simple formula for $R_n(x)$ in case $f(x) = 1/(1-x)$. The $n$th-order Taylor polynomial of $1/(1-x)$ about $c = 0$ is

$$P_n(x) = 1 + x + x^2 + x^3 + \cdots + x^n.$$

We can then use formula (2) of Section 10.2 to obtain

$$\frac{1}{1-x} - (1 + x + x^2 + x^3 + \cdots + x^n) = \frac{1}{1-x} - \frac{1-x^{n+1}}{1-x} = \frac{x^{n+1}}{1-x}.$$

It follows that the $n$th remainder is

$$R_n(x) = f(x) - P_n(x) = \frac{x^{n+1}}{1-x}.$$

Thus, we see that $R_n(x) \to 0$ as $n \to \infty$ for $|x| < 1$, as expected, since

$$\frac{1}{1-x} = 1 + x + x^2 + x^3 + \cdots, \qquad |x| < 1.$$

If $f$ is any function that has derivatives of all orders in some open interval that contains $c$, the following theorem provides us with a concise formula for the remainder term $R_n(x)$.

■ *Theorem 2 (Taylor's Formula with remainder)*

**If $f, f', f'', \ldots, f^{(n)}$, and $f^{(n+1)}$ are continuous on the closed interval between $x$ and $c$, then there is some point $z$ between $x$ and $c$ such that**

$$f(x) = f(c) + f'(c)(x-c) + \frac{f''(c)}{2!}(x-c)^2 + \cdots + \frac{f^{(n)}(c)}{n!}(x-c)^n + \frac{f^{(n+1)}(z)}{(n+1)!}(x-c)^{(n+1)}.$$

**Hence,**

$$R_n(x) = \frac{f^{(n+1)}(z)}{(n+1)!}(x-c)^{n+1} \quad \text{for some } z \text{ between } x \text{ and } c.$$

**The remainder can also be expressed in integral form as**

$$R_n(x) = \int_c^x \frac{f^{(n+1)}(t)}{n!}(x-t)^n\, dt.$$

■ *Proof* Theorem 2 can be verified by applying Rolle's Theorem to an auxilary function $g$. This is the same idea that was used to verify the Mean Value Theorem in Section 3.10 and Taylor's Formulas in Section 4.9. We will only outline the steps here. Let

$$g(t) = f(x) - f(t) - \frac{f'(t)}{1!}(x-t) - \frac{f''(t)}{2!}(x-t)^2 - \cdots - \frac{f^{(n)}(t)}{n!}(x-t)^n - R_n(x)\frac{(x-t)^{n+1}}{(x-c)^{n+1}},$$

where $R_n(x) = f(x) - P_n(x)$. It is clear that $g(x) = 0$. Also, $g(c) = 0$.

Differentiating $g(t)$ with respect to $t$, and noting the cancellation of terms that occurs, we obtain

$$g'(t) = -\frac{f^{(n+1)}(t)}{n!}(x-t)^n + R_n(x)(n+1)\frac{(x-t)^n}{(x-c)^{n+1}}.$$

We can then apply Rolle's Theorem to obtain a point $z$ between $x$ and $c$ such that $g'(z) = 0$. Substitution of the formula for $g'$ in the equation $g'(z) = 0$, and solving for $R_n(x)$ gives the first form of the remainder. To obtain the integral form of the remainder, we note that $\int_c^x g'(t)\,dt = g(x) - g(c) = 0$. The integral form is obtained by substituting the formula for $g'$ in the equation $\int_c^x g'(t)\,dt = 0$. ■

■ **EXAMPLE 1**

Show that

$$e^x = 1 + x + \frac{x^2}{2!} + \frac{x^3}{3!} + \cdots \qquad \text{for all } x.$$

**SOLUTION**

We first calculate the coefficients of the Maclaurin series of $f(x) = e^x$. We have $f^{(n)}(x) = e^x$, so $f^{(n)}(0) = 1$ for all $n \geq 0$. This shows that

$$\frac{f^{(n)}(c)}{n!} = \frac{1}{n!},$$

so the Maclaurin series of $e^x$ is

$$1 + x + \frac{x^2}{2!} + \frac{x^3}{3!} + \cdots.$$

We must show that this series converges to $e^x$ for all $x$. We will show that $R_n(x) \to 0$ as $n \to \infty$ for all $x$. We have

$$R_n(x) = \frac{f^{(n+1)}(z)}{(n+1)!}x^{n+1} = \frac{e^z}{(n+1)!}x^{n+1}$$

for some $z$ between $x$ and 0. If $x \geq 0$, then $0 \leq z \leq x$, so $e^z \leq e^x$. If $x < 0$, then $x < z < 0$, so $e^z < e^0 = 1$. We then have

$$|R_n(x)| \leq \begin{cases} \dfrac{e^x}{(n+1)!}|x|^{n+1}, & x \geq 0, \\[2mm] \dfrac{1}{(n+1)!}|x|^{n+1}, & x < 0. \end{cases}$$

Each of the expressions on the right above approaches zero as $n$ ap-

FIGURE 10.7.2

proaches infinity for every fixed value of $x$. This is a consequence, for example, of the fact that

$$\sum_{n=1}^{\infty} \frac{|x|^{n+1}}{(n+1)!}$$

converges, as was noted in statement (1) of Section 10.5. It follows that $R_n(x)$ approaches zero as $n$ approaches infinity, so the Maclaurin series of $e^x$ converges to $e^x$ for all $x$. The graph of $e^x$ and some of the approximating Taylor polynomials are sketched in Figure 10.7.2. ■

■ **EXAMPLE 2**

Show that

$$\sin x = x - \frac{x^3}{3!} + \frac{x^5}{5!} - \frac{x^7}{7!} + \cdots \qquad \text{for all } x.$$

**SOLUTION**

Calculations for determining the Maclaurin series of $\sin x$ are given in Table 10.7.1.

TABLE 10.7.1

| $n$ | $f^{(n)}(x)$ | $f^{(n)}(0)$ | $f^{(n)}(0)/n!$ |
|---|---|---|---|
| 0 | $\sin x$ | 0 | 0 |
| 1 | $\cos x$ | 1 | $\frac{1}{1!}$ |
| 2 | $-\sin x$ | 0 | 0 |
| 3 | $-\cos x$ | $-1$ | $-\frac{1}{3!}$ |
| 4 | $\sin x$ | 0 | 0 |
| 5 | $\cos x$ | 1 | $\frac{1}{5!}$ |
| ⋮ | | | |

We see that the Maclaurin series of $\sin x$ is

$$x - \frac{x^3}{3!} + \frac{x^5}{5!} - \frac{x^7}{7!} + \cdots.$$

The series will converge to $\sin x$ if $R_n(x)$ approaches zero as $n$ approaches infinity. We see that $f^{(n+1)}(z)$ is either $\pm\sin z$ or $\pm\cos z$. Since the sine and cosine functions are both bounded by one, we have

$$|R_n(x)| = \left|\frac{f^{n+1}(z)}{(n+1)!}x^{n+1}\right| \le \frac{|x|^{n+1}}{(n+1)!}.$$

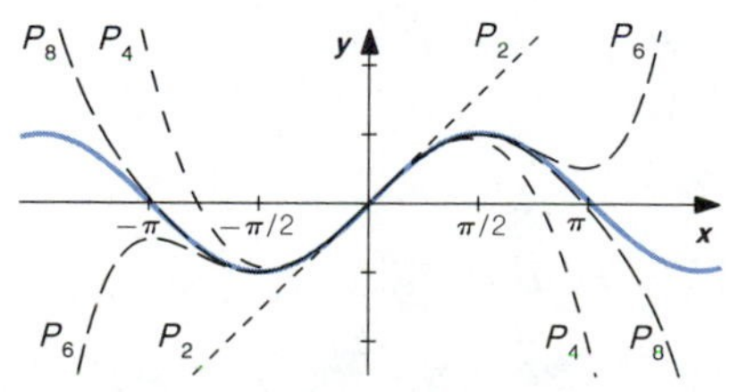

FIGURE 10.7.3

The expression on the right above and, hence, $R_n(x)$ do approach zero as $n$ approaches infinity for any fixed $x$, as was indicated in Example 1. This means that the Maclaurin series of $\sin x$ converges to $\sin x$ for all $x$. The graph of $\sin x$ and some of the approximating Taylor polynomials are sketched in Figure 10.7.3. ■

■ **EXAMPLE 3**

Find the Taylor series of $\sin x$ about $\pi/4$ and show that it converges to $\sin x$ for all $x$.

**SOLUTION**

Calculations for determining the Taylor series of $\sin x$ about $\pi/4$ are given in Table 10.7.2.

**TABLE 10.7.2**

| $n$ | $f^{(n)}(x)$ | $f^{(n)}(\pi/4)$ | $f^{(n)}(\pi/4)/n!$ |
|---|---|---|---|
| 0 | $\sin x$ | $\frac{1}{\sqrt{2}}$ | $\frac{1}{\sqrt{2}}$ |
| 1 | $\cos x$ | $\frac{1}{\sqrt{2}}$ | $\frac{1}{\sqrt{2}}$ |
| 2 | $-\sin x$ | $-\frac{1}{\sqrt{2}}$ | $-\frac{1}{2\sqrt{2}}$ |
| 3 | $-\cos x$ | $-\frac{1}{\sqrt{2}}$ | $-\frac{1}{6\sqrt{2}}$ |
| 4 | $\sin x$ | $\frac{1}{\sqrt{2}}$ | $\frac{1}{24\sqrt{2}}$ |
| 5 | $\cos x$ | $\frac{1}{\sqrt{2}}$ | $\frac{1}{120\sqrt{2}}$ |
| $\vdots$ | | | |

The Taylor series of $\sin x$ about $\pi/4$ is

$$\frac{1}{\sqrt{2}}+\frac{1}{\sqrt{2}}\left(x-\frac{\pi}{4}\right)-\frac{1}{2\sqrt{2}}\left(x-\frac{\pi}{4}\right)^2-\frac{1}{6\sqrt{2}}\left(x-\frac{\pi}{4}\right)^3$$
$$+\frac{1}{24\sqrt{2}}\left(x-\frac{\pi}{4}\right)^4+\cdots.$$

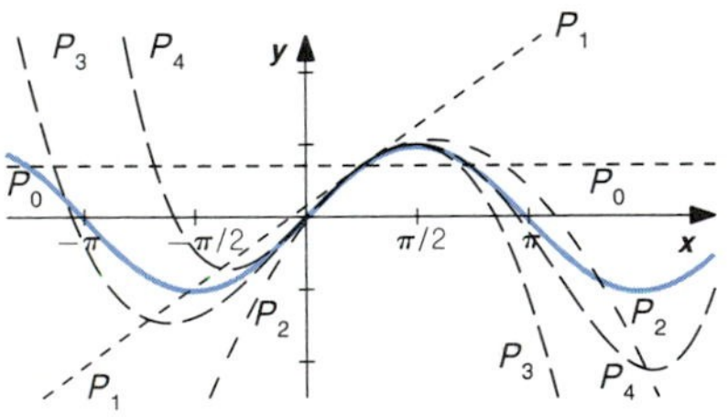

FIGURE 10.7.4

The series will converge to $\sin x$ if $R_n(x)$ approaches zero as $n$ approaches infinity. Since $f^{(n+1)}(z)$ is either $\pm\sin z$ or $\pm\cos z$, we can verify, as in Example 2, that $R_n(x)$ approaches zero as $n$ approaches infinity for all $x$. The function $\sin x$ and some of the Taylor polynomials about $\pi/4$ are sketched in Figure 10.7.4. ■

It is not always necessary to evaluate the derivatives $f^{(n)}(c)$ in order to find a Taylor series. Theorem 1 tells us that if we arrive at a power series representation of $f$ by any (legitimate) method, the power series is the Taylor series of $f$.

■ **EXAMPLE 4**

Show that

$$\cos x = 1-\frac{x^2}{2!}+\frac{x^4}{4!}-\frac{x^6}{6!}+\cdots \qquad \text{for all } x.$$

**SOLUTION**

We have

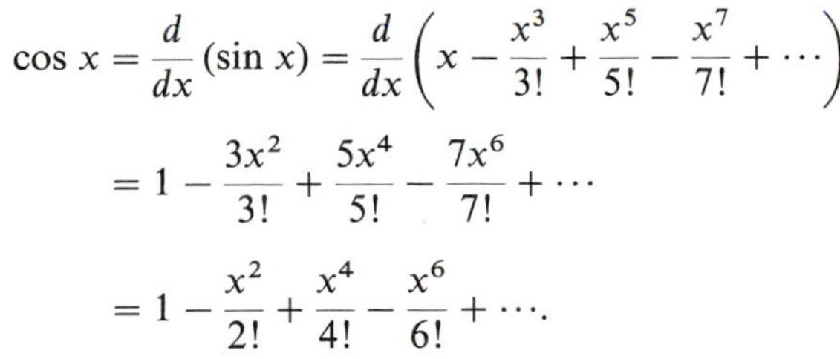

$$\cos x = \frac{d}{dx}(\sin x) = \frac{d}{dx}\left(x - \frac{x^3}{3!} + \frac{x^5}{5!} - \frac{x^7}{7!} + \cdots\right)$$

$$= 1 - \frac{3x^2}{3!} + \frac{5x^4}{5!} - \frac{7x^6}{7!} + \cdots$$

$$= 1 - \frac{x^2}{2!} + \frac{x^4}{4!} - \frac{x^6}{6!} + \cdots.$$

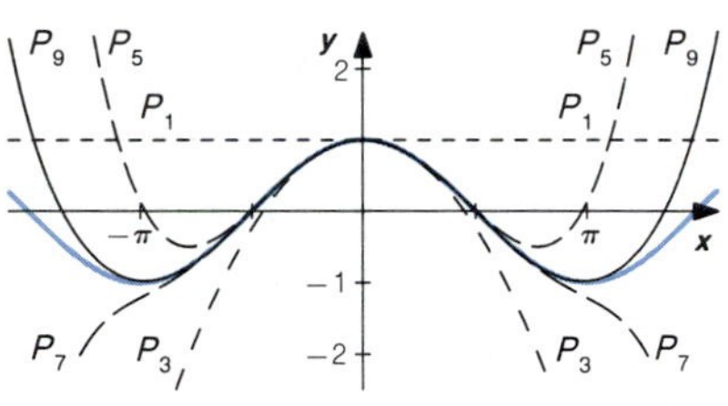


FIGURE 10.7.5

This is the Maclaurin series of $\cos x$. We can see from Figure 10.7.5 how the Taylor polynomials $P_n$ begin to approach $\cos x$ as $n$ increases. ■

■ **EXAMPLE 5**

Show that

$$\frac{\sin x}{x} = 1 - \frac{x^2}{3!} + \frac{x^4}{5!} - \frac{x^6}{7!} + \cdots.$$

**SOLUTION**

We assume that $(\sin x)/x$ is defined to be 1 at $x = 0$. [$\lim_{x\to 0}(\sin x)/x = 1$.] For $x \neq 0$, Example 2 gives

$$\frac{\sin x}{x} = \frac{x - \dfrac{x^3}{3!} + \dfrac{x^5}{5!} - \dfrac{x^7}{7!} + \cdots}{x}$$

$$= \frac{x\left(1 - \dfrac{x^2}{3!} + \dfrac{x^4}{5!} - \dfrac{x^6}{7!} + \cdots\right)}{x}$$

$$= 1 - \frac{x^2}{3!} + \frac{x^4}{5!} - \frac{x^6}{7!} + \cdots.$$

Since the value of the series agrees with the (assigned) value of $(\sin x)/x$ at $x = 0$, the above formula holds for all $x$. The series is the Maclaurin series of the function. ■

■ **EXAMPLE 6**

Approximate

$$\int_0^{0.5} \sin x^2\, dx$$

to within 0.001.

**SOLUTION**

Substituting $x^2$ for $x$ in the series established in Example 2, we have

$$\int_0^{0.5} \sin x^2\,dx = \int_0^{0.5}\left((x^2) - \frac{(x^2)^3}{3!} + \frac{(x^2)^5}{5!} - \frac{(x^2)^7}{7!} + \frac{(x^2)^9}{9!} - \cdots\right)dx$$

$$= \int_0^{0.5}\left(x^2 - \frac{x^6}{6} + \frac{x^{10}}{120} - \cdots\right)dx$$

$$= \frac{(0.5)^3}{3} - \frac{(0.5)^7}{42} + \frac{(0.5)^{11}}{1320} - \cdots.$$

The Alternating Series Test applies and tells us that the error in using a partial sum to approximate the sum is less than the absolute value of the first term omitted. Since

$$\frac{(0.5)^7}{42} = 0.00018601 < 0.001,$$

we have

$$\int_0^{0.5} \sin x^2\,dx \approx \frac{(0.5)^3}{3} = 0.04166667,$$

with error less than 0.001. ■

Many calculations that involve power series involve alternating series and we can use the Alternating Series Test to determine the accuracy of the calculations. In case we do not have an alternating series, we can use estimates of the remainder term to determine accuracy.

In Section 7.4 we showed that $2 < e < 4$. We can now use Taylor series to greatly improve this preliminary estimate of the value of $e$.

■ **EXAMPLE 7**

Assuming that you know that $e < 4$, use the Maclaurin series of $e^x$ to approximate the value of $e$ to within 0.0001.

**SOLUTION**

Note that $e = e^1$. We use partial sums of the Maclaurin series of $e^x$ with $x = 1$ to obtain approximate values of $e$. We have

$$e^x = 1 + x + \frac{1}{2!}x^2 + \frac{1}{3!}x^3 + \frac{1}{4!}x^4 + \cdots.$$

The partial sum that contains $n$ nonzero terms is

$$P_{n-1}(x) = 1 + x + \frac{1}{2!}x^2 + \frac{1}{3!}x^3 + \frac{1}{4!}x^4 + \cdots + \frac{1}{(n-1)!}x^{n-1}.$$

We know from the solution of Example 1 that

$$|R_{n-1}(x)| \le \frac{e^x}{n!}|x|^n, \qquad x \ge 0.$$

**TABLE 10.7.3**

| $n$ | $4/n!$ |
|---|---|
| 2 | 2 |
| 3 | 0.66666667 |
| 4 | 0.16666667 |
| 5 | 0.03333333 |
| 6 | 0.00555556 |
| 7 | 0.00079365 |
| 8 | 0.00009921 < 0.0001 |

Using $x = 1$ and the fact that $e < 4$, we have

$$|R_{n-1}(1)| \le \frac{4^1}{n!}|1|^n = \frac{4}{n!}.$$

In Table 10.7.3 we have tabulated values of $4/n!$ in order to find an integer $n$ that gives the desired accuracy.

We see that we need $n = 8$ nonzero terms of the Maclaurin series of $e^x$ with $x = 1$. We have

$$e \approx P_7(1) = 1 + \frac{1}{1!} + \frac{1}{2!} + \frac{1}{3!} + \frac{1}{4!} + \frac{1}{5!} + \frac{1}{6!} + \frac{1}{7!} = 2.718254.$$

This gives an approximate value of $e$ with error less than $0.00009921 < 0.0001$. (A calculator gives $e \approx 2.7182818$.) ■

## EXERCISES 10.7

Use values of the derivatives at $c$ to find the Taylor series about $c$ of the functions in Exercises 1–6.

1 $f(x) = x^3 + 3x - 1$, $c = 1$

2 $f(x) = x^4 - x^2 + 1$, $c = -1$

3 $f(x) = (x - 2)^3$, $c = 0$

4 $f(x) = x^3 - 9x^2 + 27x - 27$, $c = 3$

5 $f(x) = \sinh x$, $c = 0$

6 $f(x) = \cosh x$, $c = 0$

Use established formulas to find the Maclaurin series of the functions in Exercises 7-14.

7 $f(x) = \sin 2x$

8 $f(x) = \cos(x/2)$

9 $f(x) = e^{-x^2}$

10 $f(x) = xe^{-x}$

11 $f(x) = \cos^2 x$

12 $f(x) = \cos x^2$

13 $f(x) = \dfrac{1 - \cos x}{x^2}$

14 $f(x) = \dfrac{1 - e^x}{x}$

Use the fact that $f(x) = \sum_{n=0}^{\infty} a_n x^n$ for some $x \neq 0$ implies $a_n = f^{(n)}(0)/n!$ to evaluate the derivatives in Exercises 15–18 at $x = 0$.

15 $\dfrac{d^5}{dx^5}(x^2 \sin 2x)$

16 $\dfrac{d^5}{dx^5}\left(\dfrac{x - \sin x}{x^3}\right)$

17 $\dfrac{d^{100}}{dx^{100}}(x \cos x)$

18 $\dfrac{d^{99}}{dx^{99}}(e^{x^2})$

Use Taylor series to approximate the expressions in Exercises 19–26 to within 0.001.

19 $\sin 0.125$

20 $\cos 0.25$

21 $1/\sqrt{e}$

22 $e^{-2}$

23 $\displaystyle\int_0^{0.5} \frac{\sin x}{x}\,dx$

24 $\displaystyle\int_0^{0.2} \frac{1 - \cos x}{x^2}\,dx$

25 $\displaystyle\int_0^{0.2} e^{-x^2}\,dx$

26 $\displaystyle\int_0^{0.3} x \sin x\,dx$

27 $f(x) = \begin{cases} e^{-1/x^2}, & x \neq 0, \\ 0, & x = 0. \end{cases}$

(a) Show that

$$\lim_{x \to 0} \frac{f(x) - f(0)}{x - 0} = 0.$$

This shows $f'(0) = 0$.

(b) Show that

$$\lim_{x \to 0} \frac{f'(x) - f'(0)}{x - 0} = 0.$$

This shows $f''(0) = 0$. [Mathematical induction could be used to show that $f^{(n)}(0) = 0$ for all $n$.]

28 $f(x) = \begin{cases} 1 - x, & x \le 1, \\ 0, & x > 1. \end{cases}$

Find the Maclaurin series of $f$. For what values of $x$ does the series converge? For what values of $x$ does the series converge to $f(x)$?

## 10.8 BINOMIAL SERIES

The Maclaurin series of $(1 + x)^k$ is called the **binomial series**. We have

$$(1 + x)^k = 1 + kx + \frac{k(k-1)}{2!}x^2 + \frac{k(k-1)(k-2)}{3!}x^3 + \frac{k(k-1)(k-2)(k-3)}{4!}x^4 + \cdots, \qquad |x| < 1. \qquad (1)$$

To verify formula (1), we first evaluate the coefficients of the Maclaurin series of $(1 + x)^k$. This is done in Table 10.8.1.

TABLE 10.8.1

| $n$ | $f^{(n)}(x)$ | $f^{(n)}(0)$ | $f^{(n)}(0)/n!$ |
|---|---|---|---|
| 0 | $(1 + x)^k$ | 1 | 1 |
| 1 | $k(1 + x)^{k-1}$ | $k$ | $k$ |
| 2 | $k(k-1)(1 + x)^{k-2}$ | $k(k-1)$ | $k(k-1)/2!$ |
| 3 | $k(k-1)(k-2)(1 + x)^{k-3}$ | $k(k-1)(k-2)$ | $k(k-1)(k-2)/3!$ |
| $\vdots$ | | | |

We see that $f^{(n)}(0)/n! = k(k-1)(k-2)\cdots(k-n+1)/n!$, $n > 0$. This shows that the series is the Maclaurin series of $(1 + x)^k$. The Ratio Test can be used to determine that the series is convergent for $|x| < 1$, but we must verify that the series has sum $(1 + x)^k$. This can be accomplished by showing that the integral form of the remainder $R_n(x)$ approaches zero as $n$ approaches infinity. Later in this section we will illustrate how this is done by considering the special case $k = -1/2$. For now, let us see how the formula is used.

If $k$ is a positive integer, formula (1) contains only a finite number of nonzero terms. In this case (1) reduces to the usual binomial formula and is true for all $x$. Substitution of $k = 2$ and $k = 3$ in formula (1) gives

$$(1 + x)^2 = 1 + 2x + x^2 + 0x^3 + 0x^3 + 0x^4 + \cdots = 1 + 2x + x^2$$

and

$$(1 + x)^3 = 1 + 3x + 3x^2 + x^3 + 0x^4 + 0x^5 + \cdots = 1 + 3x + 3x^2 + x^3,$$

respectively.

■ **EXAMPLE 1**

Show that

$$(1 + x)^{-2} = 1 - 2x + 3x^2 - 4x^3 + \cdots, \qquad |x| < 1.$$

**SOLUTION**

This is immediate from (1) with $k = -2$. ■

**EXAMPLE 2**

Show that

$$\sqrt{1-x} = 1 - \frac{1}{2}x - \frac{1}{8}x^2 - \frac{1}{16}x^3 - \cdots, \qquad |x| < 1.$$

**SOLUTION**

This follows from (1) with $k = 1/2$ and $x$ replaced by $-x$. ■

**EXAMPLE 3**

Show that

$$\sqrt{x} = 1 + \frac{1}{2}(x-1) - \frac{1}{8}(x-1)^2 + \frac{1}{16}(x-1)^3 - \cdots, \qquad |x-1| < 1.$$

**SOLUTION**

Write $\sqrt{x} = \sqrt{1 + (x-1)}$ and then substitute in (1) with $k = 1/2$ and $x - 1$ in place of $x$. ■

**EXAMPLE 4**

Show that

$$\frac{1}{\sqrt{1-x^2}} = 1 + \frac{1}{2}x^2 + \frac{3}{8}x^4 + \frac{5}{16}x^6 + \cdots, \qquad |x| < 1.$$

**SOLUTION**

Use (1) with $k = -1/2$ and $x$ replaced by $-x^2$. ■

**EXAMPLE 5**

Show that

$$\sin^{-1}x = x + \frac{1}{6}x^3 + \frac{3}{40}x^5 + \frac{5}{112}x^7 + \cdots, \qquad |x| < 1.$$

**SOLUTION**

Using the results of Example 4, we have

$$\begin{aligned}\sin^{-1}x &= \int_0^x \frac{1}{\sqrt{1-t^2}}\,dt \\ &= \int_0^x \left(1 + \frac{1}{2}t^2 + \frac{3}{8}t^4 + \frac{5}{16}t^6 + \cdots\right) dt \\ &= x + \frac{1}{6}x^3 + \frac{3}{40}x^5 + \frac{5}{112}x^7 + \cdots, \qquad |x| < 1.\end{aligned}$$

The graph of $\sin^{-1}x$ and some of the approximating Taylor polynomials are given in Figure 10.8.1. ■

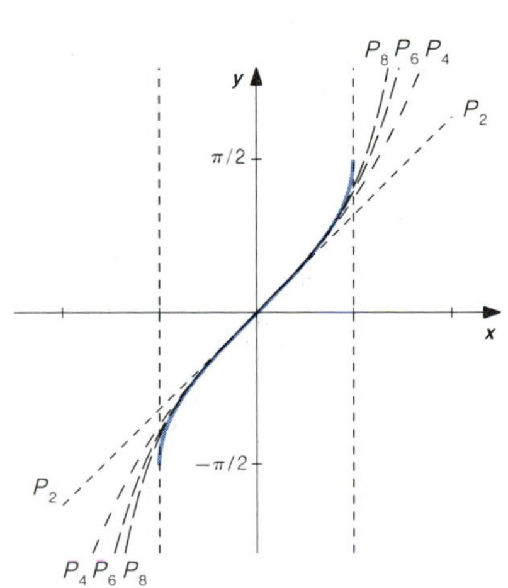

FIGURE 10.8.1

■ **EXAMPLE 6**

Use the first four nonzero terms of the Maclaurin series of $\sin^{-1}x$ to find an approximate value of $\sin^{-1}0.3$. Compare with the calculator value of $\sin^{-1} 0.3$.

**SOLUTION**

From Example 5 we see that the desired Taylor polynomial of $\sin^{-1}x$ is

$$P_8(x) = x + \frac{1}{6}x^3 + \frac{3}{40}x^5 + \frac{5}{112}x^7.$$

Substituting 0.3 for $x$, we obtain

$$P_8(0.3) = 0.3 + \frac{1}{6}(0.3)^3 + \frac{3}{40}(0.3)^5 + \frac{5}{112}(0.3)^7 = 0.30469201.$$

A calculator gives $\sin^{-1} 0.3 \approx 0.30469265$ rad. ■

Let us now return to the verification of formula (1) for the binomial series in the case $k = -1/2$. We will show that the integral form of the remainder approaches zero as $n$ approaches infinity for $|x| < 1$. Recall from Theorem 2 of Section 10.7 that the integral form of the remainder is

$$R_n(x) = \int_0^x \frac{f^{(n+1)}(t)}{n!}(x-t)^n\,dt.$$

For $f(t) = (1+t)^{-1/2}$, we have

$$\begin{aligned} f^{(n+1)}(t) &= \left(-\frac{1}{2}\right)\left(-\frac{3}{2}\right)\left(-\frac{5}{2}\right) \\ &\qquad \cdots\left(-\frac{2n-1}{2}\right)\left(-\frac{2n+1}{2}\right)(1+t)^{-(2n+3)/2} \\ &= (-1)^{n+1}\frac{(1)(3)(5)\cdots(2n-1)(2n+1)}{2^{n+1}}(1+t)^{-n-3/2}. \end{aligned}$$

Substitution then gives

$$\begin{aligned} R_n(x) &= \int_0^x (-1)^{n+1}\frac{(1)(3)(5)\cdots(2n-1)(2n+1)}{2^{n+1}n!}(1+t)^{-n-3/2}(x-t)^n\,dt \\ &= \int_0^x (-1)^{n+1}\frac{(1)(3)(5)\cdots(2n-1)(2n+1)}{(2)(4)(6)\cdots(2n)(2)}(1+t)^{-3/2}\left(\frac{x-t}{1+t}\right)^n dt. \end{aligned}$$

We can now obtain a bound of $R_n(x)$ by finding a bound of each of the factors in the integrand. We have

$$\left|(-1)^{n+1}\frac{(1)(3)(5)\cdots(2n-1)(2n+1)}{(2)(4)(6)\cdots(2n)(2)}\right|$$
$$=\left(\frac{1}{2}\right)\left(\frac{3}{4}\right)\left(\frac{5}{6}\right)\cdots\left(\frac{2n-1}{2n}\right)\left(\frac{2n+1}{2}\right) < (1)(1)(1)\cdots(1)\left(\frac{2n+1}{2}\right)=\frac{2n+1}{2}.$$

The conditions $|x| < 1$, $t$ between $x$ and 0, imply

$$|1+t|^{-3/2} \le \begin{cases} 1, & 0 \le x < 1, \\ (1-|x|)^{-3/2}, & -1 < x < 0, \end{cases}$$

(the negative power of $|1 + t|$ is largest when $1 + t$ is nearest zero) and

$$\left|\frac{x-t}{1+t}\right|^n \le |x|^n$$

(find the maximum value of $|(x - t)/(1 + t)|$ as a function of $t$ for each of the cases $0 \le t \le x < 1$ and $-1 < x \le t < 0$).

We then have

$$|R_n(x)| \le \int_0^{|x|} \frac{2n+1}{2}(1-|x|)^{-3/2}|x|^n\,dt$$
$$= \frac{2n+1}{2}(1-|x|)^{-3/2}|x|^{n+1},$$

where the factor $(1 - |x|)^{-3/2}$ is not needed if $x$ is positive. We can show $(2n + 1)|x|^{n+1}$ approaches zero as $n$ approaches infinity. It follows that $R_n(x)$ also approaches zero for $|x| < 1$.

## EXERCISES 10.8

Use established series to find the Maclaurin series of the functions in Exercises 1-12.

1 $f(x) = \sqrt{1+x}$

2 $f(x) = \dfrac{1}{\sqrt{1+x}}$

3 $f(x) = \dfrac{1}{(1-x)^{1/3}}$

4 $f(x) = (1+x)^4$

5 $f(x) = \sqrt{4+x^2}$

6 $f(x) = \dfrac{1}{\sqrt{9+x^2}}$

7 $f(x) = \dfrac{1}{(x+2)^3}$

8 $f(x) = \dfrac{x}{(x-2)^2}$

9 $f(x) = \sqrt{2x+1}$

10 $f(x) = (3x-1)^{1/3}$

11 $f(x) = \dfrac{\sin^{-1}x}{x}$

12 $f(x) = \sin^{-1}2x$

Use the first four nonzero terms of the Maclaurin series to approximate the expressions in Exercises 13-18.

13 $\sin^{-1}0.15$

14 $\sin^{-1}0.25$

15 $\displaystyle\int_0^{0.5} \frac{x^2}{\sqrt{1+x^4}}\,dx$

16 $\displaystyle\int_0^{0.3} \frac{1}{\sqrt{1+x^4}}\,dx$

17 $\displaystyle\int_0^{0.2} \sin^{-1}x\,dx$

18 $\displaystyle\int_0^{1} \sqrt{4+x^2}\,dx$

## 10.9 APPROXIMATION OF SOLUTIONS OF DIFFERENTIAL EQUATIONS

In this section we will see how infinite series can be used to find polynomial approximations of solutions of differential equations. The following theorem is the theoretical basis for doing this.

*Theorem*

**If $p_0, p_1, \ldots, p_n$, and $q$ are polynomials, and $p_n(c) \neq 0$, then the initial value problem**

$$p_n y^{(n)} + \cdots + p_1 y' + p_0 y = q,\ y(c) = y_0, \ldots, y^{(n-1)}(c) = y_{n-1},$$

**has an infinite series solution**

$$y(x) = \sum_{n=0}^{\infty} a_n (x-c)^n$$

**in some open interval that contains $c$.**

We will not verify this theorem.

We have seen that a small number of terms can sometimes give accurate approximate values of the series

$$\sum_{n=0}^{\infty} a_n (x-c)^n \qquad \text{for } x \text{ near } c.$$

This suggests that the partial sums of an infinite series solution of an initial value problem can be used to give approximate values of the solution. Let us illustrate a systematic method for finding the coefficients of infinite series solutions of differential equations.

### EXAMPLE 1

Use the first four nonzero terms of the infinite series solution of the initial value problem

$$y' - 2y = 0, \qquad y(0) = 3,$$

to obtain a polynomial approximation of the solution.

#### SOLUTION

Note that the initial condition involves $c = 0$. Theory guarantees that there is a solution of the form

$$y(x) = a_0 + a_1 x + a_2 x^2 + a_3 x^3 + a_4 x^4 + \cdots.$$

We then differentiate term by term to obtain

$$y'(x) = a_1 + 2a_2 x + 3a_3 x^2 + 4a_4 x^3 + \cdots.$$

Substitution into the equation $y' - 2y = 0$ then gives

$$[a_1 + 2a_2 x + 3a_3 x^2 + 4a_4 x^3 + \cdots]$$
$$- 2[a_0 + a_1 x + a_2 x^2 + a_3 x^3 + a_4 x^4 + \cdots] = 0.$$

Combining like terms, we obtain

$$(a_1 - 2a_0) + (2a_2 - 2a_1)x + (3a_3 - 2a_2)x^2 + (4a_4 - 2a_3)x^3 + \cdots = 0.$$

Just as in the case of a polynomial, the fact that the above equation holds for all $x$ in an open interval implies that all coefficients of the series are zero. Setting each coefficient of the series equal to zero, we obtain the equations

$$\begin{cases} a_1 - 2a_0 = 0, \\ 2a_2 - 2a_1 = 0, \\ 3a_3 - 2a_2 = 0, \\ 4a_4 - 2a_3 = 0, \\ \quad\vdots \end{cases}$$

Substitution of $x = 0$ in the equation

$$y(x) = a_0 + a_1 x + a_2 x^2 + a_3 x^3 + a_4 x^4 + \cdots$$

gives $y(0) = a_0$. The initial condition $y(0) = 3$ then implies

$$a_0 = 3.$$

The value of $a_0$ determines the value of the remaining coefficients. That is, substitution of $a_0$ in the first equation in the above system of equations gives

$$a_1 = 2a_0 = 2(3) = 6.$$

Substitution of this value in the second equation above gives

$$a_2 = \frac{2a_1}{2} = \frac{2(6)}{2} = 6.$$

Continuing, we obtain

$$a_3 = \frac{2a_2}{3} = \frac{2(6)}{3} = 4.$$

This gives us the first four nonzero terms of the infinite series solution. We have

$$y(x) = 3 + 6x + 6x^2 + 4x^3 + \cdots.$$

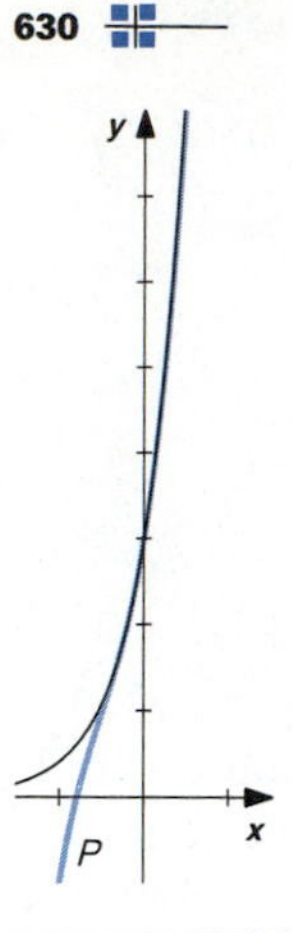

FIGURE 10.9.1

The desired polynomial approximation of the solution is

$$P(x) = 3 + 6x + 6x^2 + 4x^3.$$

We know from Section 7.7 that the solution of the initial value problem is $y(x) = 3e^{2x}$. It can be verified that we have obtained the first four nonzero terms of the Maclaurin series of $3e^{2x}$. The graphs of the solution $y(x)$ and the polynomial approximation $P(x)$ are given in Figure 10.9.1. ■

### ■ EXAMPLE 2

Use the first three nonzero terms of the infinite series solution of the initial value problem

$$y'' + y = 0, \qquad y(0) = 1, \qquad y'(0) = 0,$$

to obtain a polynomial approximation of the solution.

#### SOLUTION

Theory guarantees a solution of the form

$$y(x) = a_0 + a_1x + a_2x^2 + a_3x^3 + a_4x^4 + a_5x^5 + \cdots.$$

We then differentiate term-by-term to obtain

$$y'(x) = a_1 + 2a_2x + 3a_3x^2 + 4a_4x^3 + 5a_5x^4 + \cdots,$$

$$y''(x) = 2a_2 + 3(2)a_3x + 4(3)a_4x^2 + 5(4)a_5x^3 + \cdots.$$

Substitution into the equation $y'' + y = 0$ then gives

$$[2a_2 + 3(2)a_3x + 4(3)a_4x^2 + 5(4)a_5x^3 + \cdots] \\ + [a_0 + a_1x + a_2x^2 + a_3x^3 + a_4x^4 + a_5x^5 + \cdots] = 0.$$

As before, we combine like terms and then set each coefficient equal to zero to obtain a system of equations.

$$(2a_2 + a_0) + (6a_3 + a_1)x + (12a_4 + a_2)x^2 + (20a_5 + a_3)x^3 + \cdots = 0.$$

$$\begin{cases} 2a_2 + a_0 = 0, \\ 6a_3 + a_1 = 0, \\ 12a_4 + a_2 = 0, \\ 20a_5 + a_3 = 0, \\ \quad\vdots \end{cases}$$

The initial conditions determine the values of $a_0$ and $a_1$. We obtain the value of $a_0$ by substituting $c = 0$ into the equation

$$y(x) = a_0 + a_1x + a_2x^2 + a_3x^3 + a_4x^4 + a_5x^5 + \cdots.$$

This gives $y(0) = a_0$. The initial condition $y(0) = 1$ then implies

$$a_0 = 1.$$

Substitution of $x = 0$ in the equation

$$y'(x) = a_1 + 2a_2x + 3a_3x^2 + 4a_4x^3 + 5a_5x^4 + \cdots$$

gives $y'(0) = a_1$. The initial condition $y'(0) = 0$ then implies

$$a_1 = 0.$$

The values of $a_0$ and $a_1$ determine all other coefficients of the solution. Substitution in the first equation of the above system of equations gives

$$a_2 = -\frac{a_0}{2} = -\frac{1}{2}.$$

Substitution in the second equation above gives

$$a_3 = -\frac{a_1}{6} = \frac{0}{6} = 0.$$

Continuing, we obtain

$$a_4 = -\frac{a_2}{12} = -\frac{-1/2}{12} = \frac{1}{24}.$$

This gives the first three nonzero terms of the infinite series solution. We have

$$y(x) = 1 - \frac{1}{2}x^2 + \frac{1}{24}x^4 + \cdots$$

and the desired polynomial approximation of the solution is

$$P(x) = 1 - \frac{1}{2}x^2 + \frac{1}{24}x^4.$$

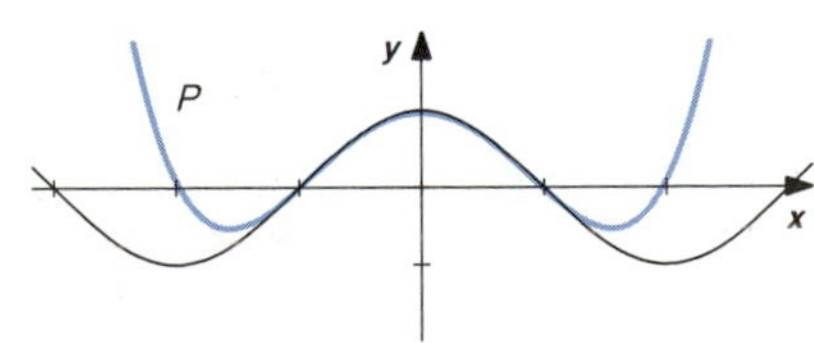

FIGURE 10.9.2

It can be verified that $y(x) = \cos x$ is the solution of the given initial value problem. Note that we have obtained the first three nonzero terms of the Maclaurin series of $\cos x$. The graphs of the solution $y = \cos x$ and the polynomial approximation $P(x)$ are given in Figure 10.9.2. ■

**■ EXAMPLE 3**

Use the first four nonzero terms of the infinite series solution of the initial value problem

$$y'' - 2y' + y = 0, \qquad y(0) = 0, \qquad y'(0) = 1,$$

to obtain a polynomial approximation of the solution.

**SOLUTION**

Theory guarantees a solution of the form

$$y(x) = a_0 + a_1 x + a_2 x^2 + a_3 x^3 + a_4 x^4 + a_5 x^5 + \cdots.$$

We then differentiate term by term to obtain

$$y'(x) = a_1 + 2a_2 x + 3a_3 x^2 + 4a_4 x^3 + 5a_5 x^4 + \cdots,$$

$$y''(x) = 2a_2 + 3(2)a_3 x + 4(3)a_4 x^2 + 5(4)a_5 x^3 + \cdots.$$

Substitution into $y'' - 2y' + y = 0$ then gives

$$\begin{aligned}[2a_2 &+ 3(2)a_3 x + 4(3)a_4 x^2 + 5(4)a_5 x^3 + \cdots] \\ &- 2[a_1 + 2a_2 x + 3a_3 x^2 + 4a_4 x^3 + 5a_5 x^4 + \cdots] \\ &+ [a_0 + a_1 x + a_2 x^2 + a_3 x^3 + a_4 x^4 + a_5 x^5 + \cdots] = 0.\end{aligned}$$

Combining like terms and then setting each coefficient equal to zero, we obtain

$$\begin{aligned}(2a_2 - 2a_1 + a_0) &+ (6a_3 - 4a_2 + a_1)x + (12a_4 - 6a_3 + a_2)x^2 \\ &+ (20a_5 - 8a_4 + a_3)x^3 + \cdots = 0.\end{aligned}$$

$$\begin{cases} 2a_2 - 2a_1 + a_0 = 0, \\ 6a_3 - 4a_2 + a_1 = 0, \\ 12a_4 - 6a_3 + a_2 = 0, \\ 20a_5 - 8a_4 + a^3 = 0, \\ \qquad \vdots \end{cases}$$

The initial conditions determine the values of $a_0$ and $a_1$. Since $y(0) = 0$, we have

$$a_0 = 0.$$

The equation $y'(0) = 1$ implies

$$a_1 = 1.$$

The values of $a_0$ and $a_1$ then determine the other coefficients. Substitution in the above equations then gives

$$a_2 = \frac{2a_1 - a_0}{2} = 1,$$

$$a_3 = \frac{4a_2 - a_1}{6} = \frac{3}{6} = \frac{1}{2},$$

$$a_4 = \frac{6a_3 - a_2}{12} = \frac{2}{12} = \frac{1}{6}.$$

This gives the first four nonzero terms of the infinite series solution. We have

$$y(x) = x + x^2 + \frac{1}{2}x^3 + \frac{1}{6}x^4 + \cdots$$

and the desired polynomial approximation is

$$P(x) = x + x^2 + \frac{1}{2}x^3 + \frac{1}{6}x^4.$$

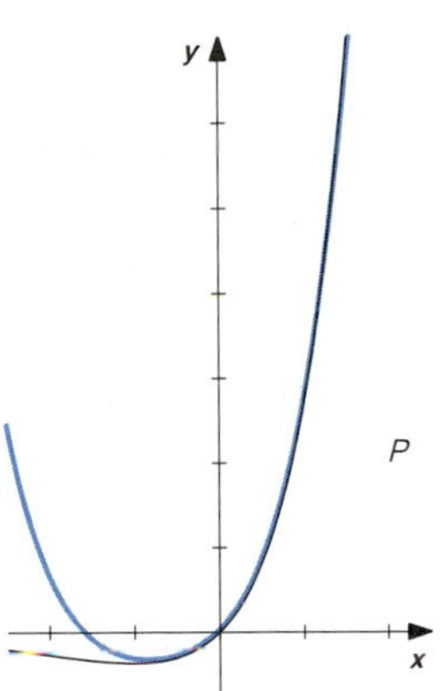

FIGURE 10.9.3

It can be verified that $y(x) = xe^x$ is the solution of the initial value problem. We have obtained the first four nonzero terms of the Maclaurin series of $xe^x$. The solution and approximate solution are sketched in Figure 10.9.3. ■

■ **EXAMPLE 4**

Use the first six nonzero terms of the infinite series solution of the initial value problem

$$y'' + (x^2 + 2)y = 4 - 3x, \qquad y(0) = 1, \qquad y'(0) = 2,$$

to obtain a polynomial approximation of the solution.

**SOLUTION**

Theory guarantees a solution of the form

$$y(x) = a_0 + a_1x + a_2x^2 + a_3x^3 + a_4x^4 + a_5x^5 + \cdots.$$

We then differentiate term by term to obtain

$$y'(x) = a_1 + 2a_2x + 3a_3x^2 + 4a_4x^3 + 5a_5x^4 + \cdots,$$

$$y''(x) = 2a_2 + 3(2)a_3x + 4(3)a_4x^2 + 5(4)a_5x^3 + \cdots.$$

Substitution into $y'' + (x^2 + 2)y = 4 - 3x$ then gives

$$\begin{aligned}&[2a_2 + 3(2)a_3x + 4(3)a_4x^2 + 5(4)a_5x^3 + \cdots]\\ &\qquad + (x^2 + 2)[a_0 + a_1x + a_2x^2 + a_3x^2 + a_4x^4 + a_5x^5 + \cdots]\\ &\quad = 4 - 3x,\end{aligned}$$

$$\begin{aligned}&[2a_2 + 3(2)a_3x + 4(3)a_4x^2 + 5(4)a_5x^3 + \cdots]\\ &\qquad + x^2[a_0 + a_1x + a_2x^2 + a_3x^3 + a_4x^4 + a_5x^5 + \cdots]\\ &\qquad + 2[a_0 + a_1x + a_2x^2 + a_3x^3 + a_4x^4 + a_5x^5 + \cdots]\\ &\quad = 4 - 3x.\end{aligned}$$

Combining like terms, we have

$$\begin{aligned}&(2a_2 + 2a_0 - 4) + (6a_3 + 2a_1 + 3)x + (12a_4 + a_0 + 2a_2)x^2\\ &\quad + (20a_5 + a_1 + 2a_3)x^3 + \cdots = 0.\end{aligned}$$

Setting each coefficient equal to zero then gives

$$\begin{cases} 2a_2 + 2a_0 + -4 = 0,\\ 6a_3 + 2a_1 + 3 = 0,\\ 12a_4 + a_0 + 2a_2 = 0,\\ 20a_5 + a_1 + 2a_3 = 0,\\ \qquad \vdots \end{cases}$$

The initial conditions determine $a_0$ and $a_1$. Since $y(0) = 1$, we have

$$a_0 = 1.$$

The equation $y'(0) = 2$ implies

$$a_1 = 2.$$

The values of $a_0$ and $a_1$ determine the other coefficients. Substituting in the above equations, we have

$$a_2 = \frac{-2a_0 + 4}{2} = 1,$$

$$a_3 = \frac{-2a_1 - 3}{6} = -\frac{7}{6},$$

$$a_4 = \frac{-2a_2 - a_0}{12} = -\frac{3}{12} = -\frac{1}{4},$$

$$a_5 = \frac{-2a_3 - a_1}{20} = \frac{1/3}{20} = \frac{1}{60}.$$

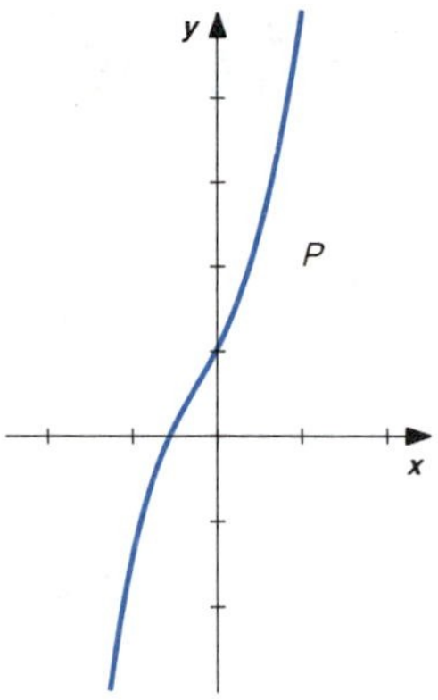

FIGURE 10.9.4

This gives the first six nonzero terms of the infinite series solution. We have

$$y(x) = 1 + 2x + x^2 - \frac{7}{6}x^3 - \frac{1}{4}x^4 + \frac{1}{60}x^5 + \cdots$$

and the desired polynomial approximation is

$$P(x) = 1 + 2x + x^2 - \frac{7}{6}x^3 - \frac{1}{4}x^4 + \frac{1}{60}x^5.$$

The graph of the polynomial approximation of the solution is given in Figure 10.9.4. The solution is not a familiar function. ■

We have seen in Examples 1 through 4 that $n$ initial conditions determine the first $n$ coefficients of an infinite series solution of an $n$th-order initial value problem. Each successive coefficient is determined by previous coefficients, so it is possible to obtain a **recursive formula** that gives coefficients in terms of previous coefficients. The recursive formula could be used to write a computer program to generate partial sums of the infinite series solution.

## EXERCISES 10.9

Use the first four nonzero terms of the infinite series solutions of the initial value problems to obtain polynomial approximations of the solutions in Exercises 1–16.

1 $y' - y = 0$, $y(0) = 2$

2 $y' + y = 0$, $y(0) = 3$

3 $y'' - y = 0$, $y(0) = 1$, $y'(0) = 1$

4 $y'' - y = 0$, $y(0) = 1$, $y'(0) = -1$

5 $y'' + y = 0$, $y(0) = 0$, $y'(0) = 1$

6 $y'' + 4y = 0$, $y(0) = 3$, $y'(0) = 0$

7 $y'' - y = x^2$, $y(0) = 1$, $y'(0) = 2$

8 $y'' + y = x^2$, $y(0) = 1$, $y'(0) = 2$

9 $y'' - 3y' + 2y = 0$, $y(0) = 2$, $y'(0) = -1$

10 $y'' + 4y' + 4y = 0$, $y(0) = 3$, $y'(0) = 2$

11 $y'' + y' + 2y = 0$, $y(0) = 2$, $y'(0) = 1$

12 $y'' - y' + y = 0$, $y(0) = -2$, $y'(0) = 3$

13 $(1 - x)y' - y = 0$, $y(0) = 1$

14 $2(1 + x)y' - y = 0$, $y(0) = 1$

15 $y'' + y' + y = x + 2$, $y(0) = 3$, $y'(0) = 1$

16 $y'' - 2y' + y = 2x - 1$, $y(0) = 1$, $y'(0) = 4$

## REVIEW EXERCISES

Evaluate or indicate "does not exist" for the limits in Exercises 1–8.

1 $\lim_{n\to\infty} \dfrac{3n + 5}{2n - 1}$

2 $\lim_{n\to\infty} \dfrac{\sqrt{9n^2 - 4n + 16}}{2n}$

3 $\lim_{n\to\infty} \dfrac{6n^2 + n - 3}{2n^2 - 3n + 4}$

4 $\lim_{n\to\infty} \dfrac{(-1)^n n}{2n + 1}$

5 $\lim_{n\to\infty} \dfrac{\cos n\pi}{n}$

6 $\lim_{n\to\infty} \dfrac{n^2}{n + 2}$

7 $\lim_{n\to\infty} ne^{-\sqrt{n}}$

8 $\lim_{n\to\infty} \frac{\ln n}{n}$

Find a simple formula for the $n$th partial sum $S_n$ and evaluate the sum $S$ for the series in Exercises 9–14.

9 $\sum_{n=1}^{\infty}\left(\frac{2n}{n+1}-\frac{2n+4}{n+3}\right)$

10 $\sum_{n=1}^{\infty} 2\left(\frac{1}{3}\right)^n$

11 $\sum_{n=1}^{\infty} \frac{2}{n(n+2)}$

12 $\sum_{n=1}^{\infty}\left(\frac{n}{2n-1}-\frac{n+1}{2n+1}\right)$

13 $\sum_{n=1}^{\infty} 4\left(\frac{3}{4}\right)^n$

14 $\sum_{n=0}^{\infty} 9\left(-\frac{1}{2}\right)^n$

Indicate convergent or divergent for the series in Exercises 15–28.

15 $\sum_{n=1}^{\infty} \frac{n}{n^2+3}$

16 $\sum_{n=1}^{\infty} \frac{n}{2n+1}$

17 $\sum_{n=1}^{\infty} (2n+7)e^{-n^2}$

18 $\sum_{n=1}^{\infty} \frac{2^n}{n^2}$

19 $\sum_{n=1}^{\infty} \frac{3^n}{n^n}$

20 $\sum_{n=1}^{\infty} \frac{(-1)^n}{n^{1/3}}$

21 $\sum_{n=1}^{\infty} \frac{\sqrt{n^3+2n}}{n^3}$

22 $\sum_{n=1}^{\infty} \frac{1}{n^2 \ln(n+1)}$

23 $\sum_{n=1}^{\infty} \frac{1}{n(\ln(n+1))^2}$

24 $\sum_{n=1}^{\infty} \left(\frac{n}{2n+1}\right)^n$

25 $\sum_{n=1}^{\infty} \frac{2n^2+7}{n^3+2}$

26 $\sum_{n=1}^{\infty} \frac{(-1)^n n}{2n-3}$

27 $\sum_{n=1}^{\infty} \frac{(1)(3)\cdots(2n-1)}{(3)(6)\cdots(3n)}$

28 $\sum_{n=1}^{\infty} \frac{(-1)^n n}{n^2+1}$

Indicate absolutely convergent, conditionally convergent, or divergent for the series in Exercises 29–32.

29 $\sum_{n=1}^{\infty} \frac{(-1)^n}{n^{2/3}}$

30 $\sum_{n=1}^{\infty} \frac{(-1)^n}{n^{3/2}}$

31 $\sum_{n=1}^{\infty} \frac{(-1)^n}{n^{1/3}}$

32 $\sum_{n=1}^{\infty} \frac{(-1)^n n}{\sqrt{n^2+1}}$

Find the interval of convergence of the series in Exercises 33–36.

33 $\sum_{n=1}^{\infty} \frac{1}{n}(x-1)^n$

34 $\sum_{n=1}^{\infty} \frac{1}{n^2} x^n$

35 $\sum_{n=1}^{\infty} n(x-2)^n$

36 $\sum_{n=1}^{\infty} \frac{(-1)^n n}{n^2+1}(x-1)^n$

Use an estimate of $|S - S_n|$ to find $n$ so that $S_n$ is within 0.001 of $S$ in Exercises 37–40.

37 $\sum_{n=1}^{\infty} n^{-3}$

38 $\sum_{n=1}^{\infty} n^{-4}$

39 $\sum_{n=1}^{\infty} \frac{(-1)^n}{n^2}$

40 $\sum_{n=1}^{\infty} \frac{(-1)^n}{n^3+1}$

Find the first three nonzero terms of the Maclaurin series of the functions in Exercises 41–52.

41 $\frac{\cos x}{1+x}$

42 $\sin 2x$

43 $\frac{x}{1+x^2}$

44 $\ln(1+x^2)$

45 $(2-x)^{-3/2}$

46 $\frac{1}{1+x^2}$

47 $\frac{-2x}{(1+x^2)^2}$

48 $(1-x)^{3/2}$

49 $\frac{1}{x-2}$

50 $\frac{1-\cos x}{x^2}$

51 $\cos^2 x$

52 $(3+x)^{1/3}$

Approximate the values of the integrals in Exercises 53–56 to within 0.0001.

53 $\int_0^{0.3} \cos x^2 \, dx$

54 $\int_0^{0.2} e^{-x^2} \, dx$

55 $\int_0^{0.2} \frac{\sin x}{x} \, dx$

56 $\int_0^{0.1} \frac{1-\cos 2x}{x^2} \, dx$

Use the first four nonzero terms of the infinite series solutions of the initial value problems to obtain polynomial approximations of the solutions in Exercises 57–60.

57 $y' - y = x$, $y(0) = 2$

58 $y' + 2y = 1 - 3x$, $y(0) = 3$

59 $y'' + xy' + x^2 y = 0$, $y(0) = 1$, $y'(0) = 2$

60 $y'' - 2xy' + x^2 y = x$, $y(0) = 2$, $y'(0) = 3$

## CHAPTER ELEVEN

# *Topics in analytic geometry*

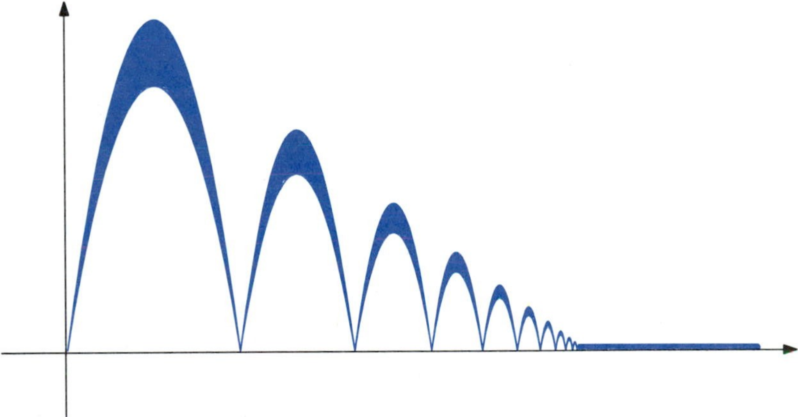

In this chapter we will study parabolas, ellipses, and hyperbolas from the point of view of analytic geometry. That is, we will see how to obtain information about these curves from their equations, and how to determine the equation of a parabola, ellipse, or hyperbola that has prescribed properties.

Let us note that ellipses, parabolas, and hyperbolas occur in many scientific and engineering applications. For example, natural laws result in nearly elliptical orbits of planets and comets; parabolas and hyperbolas are used in the design of a variety of light and sound reflectors.

## 11.1 PARABOLAS

*Definition*

A **parabola** is the set of all points $P$ in a plane that are equidistant from a fixed line (**directrix**) and a fixed point (**focus**) not on the line. ■

The line through the focus, perpendicular to the directrix, is called the **axis** of the parabola. The point on the axis that is equidistant from the focus and the directrix is called the **vertex** of the parabola. The vertex is the intersection of the parabola and the axis. See Figure 11.1.1.

Let us establish an equation that is satisfied by points $(x, y)$ on the parabola with horizontal directrix $y = k - p$ and focus $(h, k + p)$. From Figure 11.1.2 we see that the parabola has vertical axis $x = h$ and vertex $(h, k)$. The parabola opens upward, as in Figure 11.1.2, if $p > 0$; the parabola opens downward if $p < 0$. We see from Figure 11.1.2 that $(x, y)$ is on the parabola whenever

$$d_1 = d_2,$$

$$\sqrt{(x-h)^2 + (y-(k+p))^2} = |y-(k-p)|,$$

$$\sqrt{(x-h)^2 + ((y-k)-p)^2} = |(y-k)+p)|,$$

$$(x-h)^2 + (y-k)^2 - 2p(y-k) + p^2 = (y-k)^2 + 2p(y-k) + p^2,$$

$$(x-h)^2 = 4p(y-k).$$

The equation

$$(x-h)^2 = 4p(y-k)$$

is the **standard form** of the equation of the parabola with vertex $(h, k)$, axis $x = h$, directrix $y = k - p$, and focus $(h, k + p)$. The parabola opens upward if $p > 0$ and downward if $p < 0$.

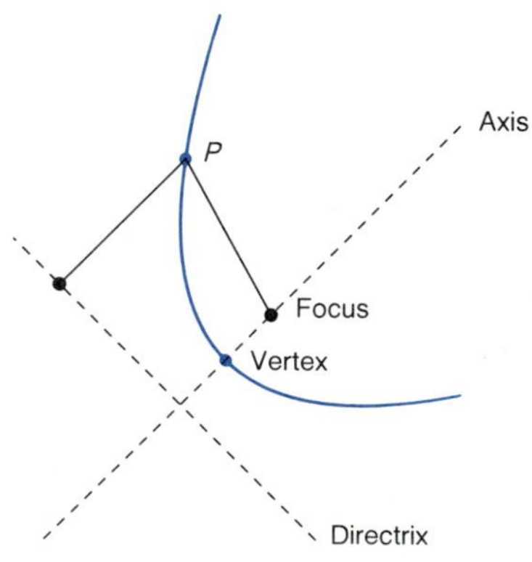


FIGURE 11.1.1

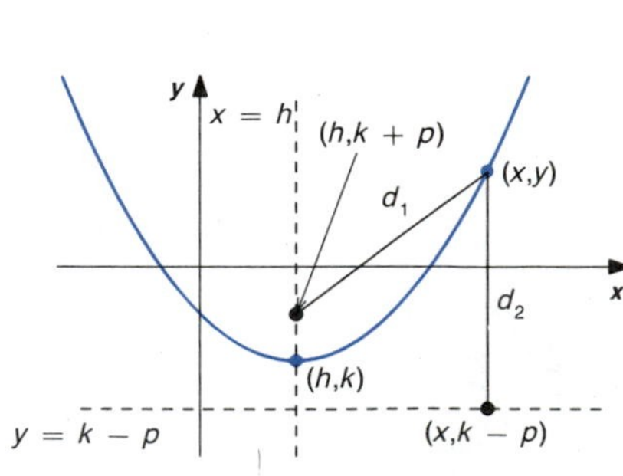


FIGURE 11.1.2

FIGURE 11.1.3

Similarly,

$$(y - k)^2 = 4p(x - h)$$

is the **standard form** of the equation of the parabola with vertex $(h, k)$, axis $y = k$, directrix $x = h - p$, and focus $(h + p, k)$. The parabola opens to the right if $p > 0$ and to the left if $p < 0$. See Figure 11.1.3.

The numbers $h$, $k$, and $p$ can be determined from the above standard forms of the equation of a parabola. The numbers $h$ and $k$ give the location of the vertex $(h, k)$. The focus and directrix are then located by moving a distance of $|p|$ in the appropriate direction from the vertex. The direction depends on the sign of $p$ and which of the two standard types the parabola is. It is easy to sketch the parabola after the vertex, focus, and directrix have been located.

■ **EXAMPLE 1**

Find the vertex, focus, and directrix, and then sketch the parabola $y = x^2 - 1$.

**SOLUTION**

Rewriting the equation in standard form, we have

$$x^2 = y + 1 \quad \text{or}$$

$$(x - 0)^2 = 4\left(\frac{1}{4}\right)(y - (-1)).$$

Comparing this to the standard form

$$(x - h)^2 = 4p(y - k),$$

we see that $h = 0$, $k = -1$, and $p = 1/4$. It follows that the parabola has vertex $(0, -1)$, focus $(0, -1 + 1/4) = (0, -3/4)$, and directrix $y = -1 - 1/4$, or $y = -5/4$. See Figure 11.1.4. ■

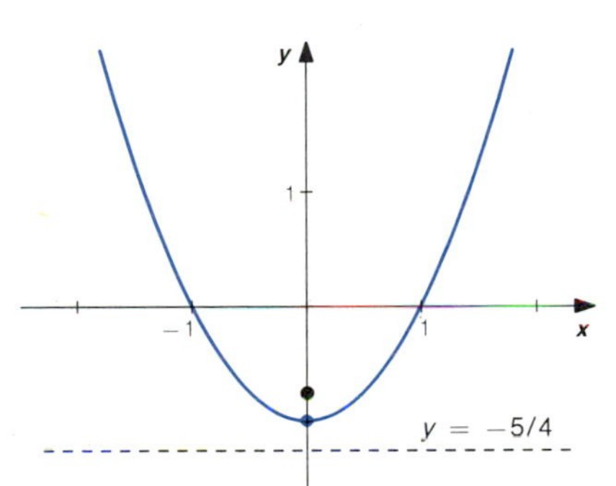

FIGURE 11.1.4

In some cases it is necessary to complete a square to write the equation of a parabola in standard form.

■ **EXAMPLE 2**

Find the vertex, focus, and directrix, and then sketch the parabola $y^2 + 6y = 8x - 1$.

**SOLUTION**

Rewriting the equation in standard form, we have

$$y^2 + 6y = 8x - 1,$$

$$y^2 + 6y + (9) = 8x - 1 + (9),$$

$$(y + 3)^2 = 4(2)(x + 1).$$

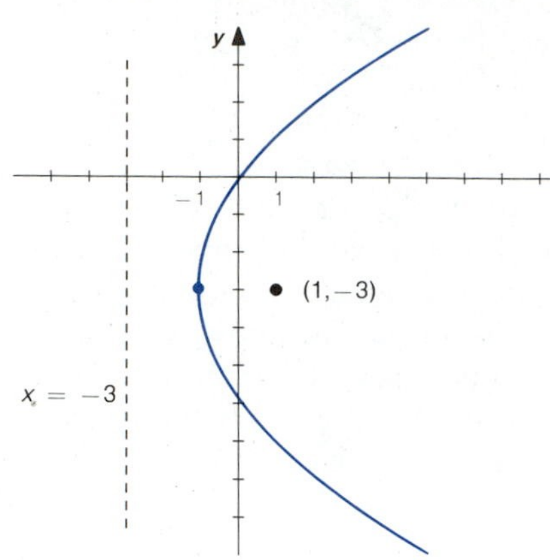


FIGURE 11.1.5

Comparing this to the standard form

$$(y-k)^2 = 4p(x-h),$$

we see that $h = -1$, $k = -3$, and $p = 2$. The parabola has vertex $(-1, -3)$, focus $(-1+2, -3) = (1, -3)$, and directrix $x = -1 - 2$, or $x = -3$. See Figure 11.1.5. ■

If we are given any two of the focus, directrix, or vertex of a parabola that has either vertical or horizontal axis, we can use a sketch to determine the values needed to write the standard form of the equation of the parabola.

### ■ EXAMPLE 3

Find an equation of the parabola with focus (2, 1) and directrix $y = 3$.

**SOLUTION**

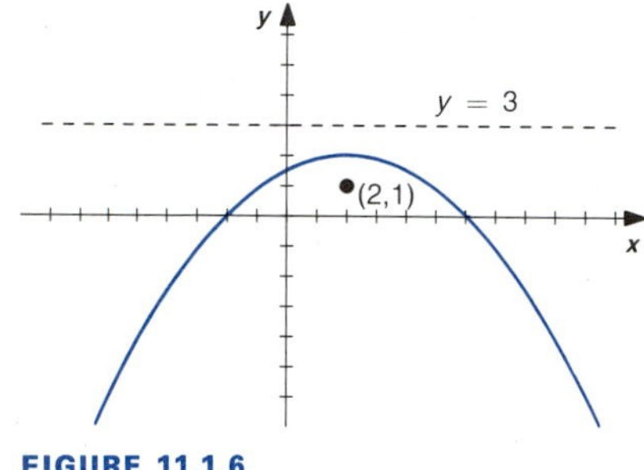


FIGURE 11.1.6

The parabola is sketeched in Figure 11.1.6. The axis of the parabola is the vertical line through the focus, $x = 2$. The axis intersects the directrix at the point (2, 3). The vertex is at the midpoint of the line segment between (2, 1) and (2, 3), so $(h, k) = (2, 2)$. We know that $|p|$ is the distance between the focus and the vertex, so $|p| = 1$. Since the parabola opens downward, we have $p = -1$. The standard equation of the parabola is then

$$(x-2)^2 = -4(y-2).$$ ■

### ■ EXAMPLE 4

Find an equation of the parabola that has horizontal axis, vertex (2, 4), and passes through the origin.

**SOLUTION**

Since the parabola has horizontal axis, we know the standard form is

$$(y-k)^2 = 4p(x-h).$$

We are given that the vertex is $(h, k) = (2, 4)$, so we have

$$(y-4)^2 = 4p(x-2).$$

To determine the value of the unknown number $p$, we use the fact that a point is on the graph of an equation if the coordinates of the point satisfy the equation. If the parabola passes through the origin, the coordinates (0, 0) must satisfy the equation of the parabola. Substitution then gives

$$(0-4)^2 = 4p(0-2),$$

$$16 = -8p,$$

$$p = -2.$$

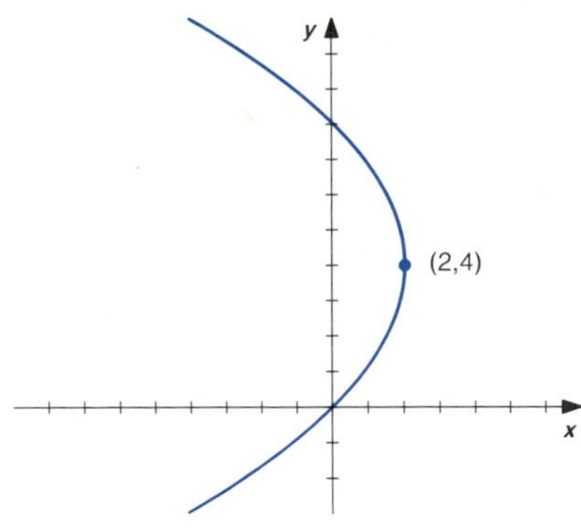

FIGURE 11.1.7

We can then write the equation in standard form as

$$(y-4)^2 = -8(x-2).$$

The graph is sketched in Figure 11.1.7. ■

There is exactly one parabola with vertical axis that contains three given points that are not collinear and have distinct $x$-coordinates. We can determine the equation of the parabola by substituting the coordinates of each of the given points for $(x, y)$ in the equation

$$y = Ax^2 + Bx + C.$$

This gives us three linear equations in the three unknowns $A$, $B$, and $C$, which we solve. The equations have a unique solution if the three given points have distinct $x$-coordinates, and $A \neq 0$ if the points are not collinear. Whenever $A \neq 0$, the graph of $y = Ax^2 + Bx + C$ is a parabola. To find the equation of the parabola with horizontal axis that contains three given points, we follow the same procedure, but use the equation

$$x = Ay^2 + By + C.$$

### ■ EXAMPLE 5

Find the equation of the parabola that has vertical axis and contains the points (0, 0), (2, 0), and (3, 6).

#### SOLUTION

The equation of the parabola is of the form $y = Ax^2 + Bx + C$, where $A$, $B$, and $C$ are unknown numbers. Substitution of the $(x, y)$ coordinates of each of the three given points into the equation gives

$$\begin{cases} 0 = A(0)^2 + B(0) + C, & [(x, y) = (0, 0)] \\ 0 = A(2)^2 + B(2) + C, & [(x, y) = (2, 0)] \\ 6 = A(3)^2 + B(3) + C. & [(x, y) = (3, 6)] \end{cases}$$

The resulting system of linear equations,

$$\begin{cases} 0 = C, \\ 0 = 4A + 2B + C, \\ 6 = 9A + 3B + C, \end{cases}$$

has solution $(A, B, C) = (2, -4, 0)$. The equation is then

$$y = 2x^2 - 4x.$$

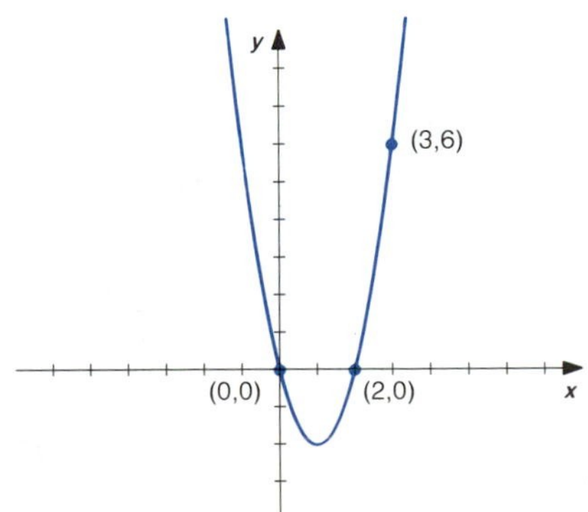

FIGURE 11.1.8

The graph is sketched in Figure 11.1.8. ■

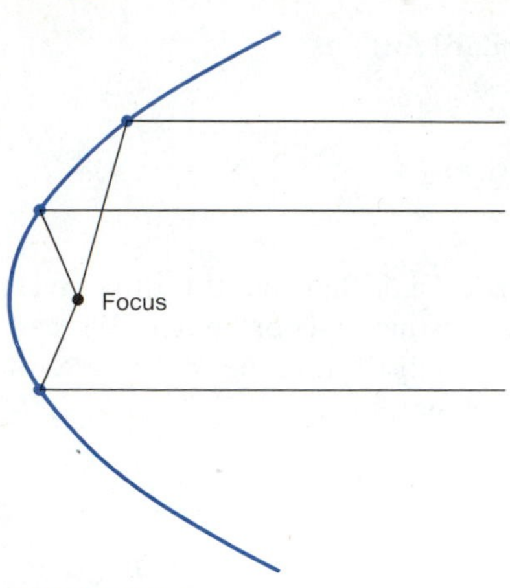


FIGURE 11.1.9

Parabolas have the property that lines parallel to the axis reflect through the focus. See Figure 11.1.9.

The reflection property of parabolas is used in applications. For example, a light source that is positioned at the focus of a parabolic reflector produces a beam of parallel light rays, or a spotlight. Parabolic reflectors are also used in the listening devices you may have seen being used during television broadcasts. The reflectors concentrate the sound waves coming from one direction at the focus.

## EXERCISES 11.1

Find the vertex, focus, and directrix, and then sketch each of the parabolas in Exercises 1–10.

1 $4y = 4 - x^2$

2 $8y = 16 - x^2$

3 $8x = 16 - y^2$

4 $4x = 4 - y^2$

5 $y = x^2 - 3x$

6 $2y = 2x - x^2$

7 $8x = 3y^2 + 6y - 5$

8 $8x = 13 - 6y - 3y^2$

9 $4y = 3x^2 - 6x + 7$

10 $5y = 4x^2 + 8x - 11$

Find an equation of the parabola with the properties described in each of Exercises 11–20.

11 Focus (0, 0); directrix $x = -2$

12 Focus (0, 0); directrix $y = 4$

13 Focus (4, 0); vertex (4, 2)

14 Focus $(-2, 2)$; vertex (0, 2)

15 Vertex $(0, -1)$; directrix $y = -3$

16 Vertex (1, 1); directrix the $y$-axis

17 Vertex $(6, -4)$; contains (0, 0); horizontal axis

18 Vertex $(2, -1)$; contains $(0, -4)$; vertical axis

19 Contains $(-1, 0)$, (0, 3), and (3, 0); vertical axis

20 Contains (0, 3), (0, 1), and (3, 0); horizontal axis

21 The parabola $4y = x^2 - 4$ is sketched below.

(a) Find the equation of the line tangent to the parabola at the point (4, 3).

(b) Find the $y$-intercept of the tangent line of (a).

(c) Verify that the focus of the parabola is the origin.

(d) Show that the distance between the focus and (4, 3) is equal to the distance between the focus and the $y$-intercept of (b).

(e) Use (d) to find a relation between the angles $\alpha$ and $\beta$. (This verifies the reflection property for this parabola at this point on the parabola.)

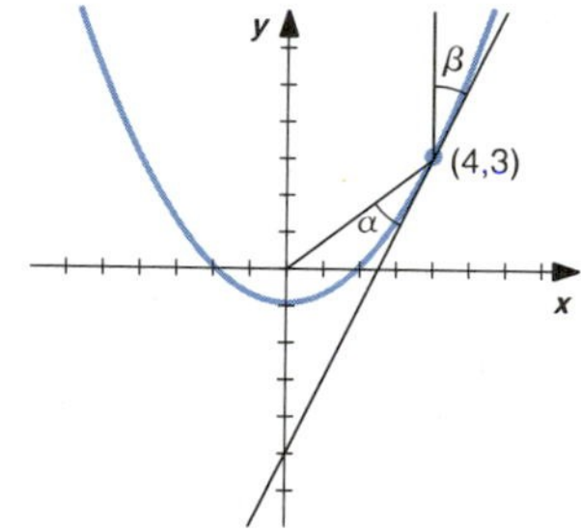


22 The parabola $x^2 = 4p(y + p)$, $p > 0$, is sketched below.

(a) Find the equation of the line tangent to the parabola at a point $(x_0, y_0)$ on the parabola, $x_0 \neq 0$.

(b) Find the $y$-intercept of the tangent line of (a).

(c) Verify that the focus of the parabola is the origin.

(d) Show that the distance between the focus and $(x_0, y_0)$ is equal to the distance between the focus and the $y$-intercept of (b).

(e) Use (d) to find a relation between the angles $\alpha$ and $\beta$. (This verifies the reflection property for parabolas.)

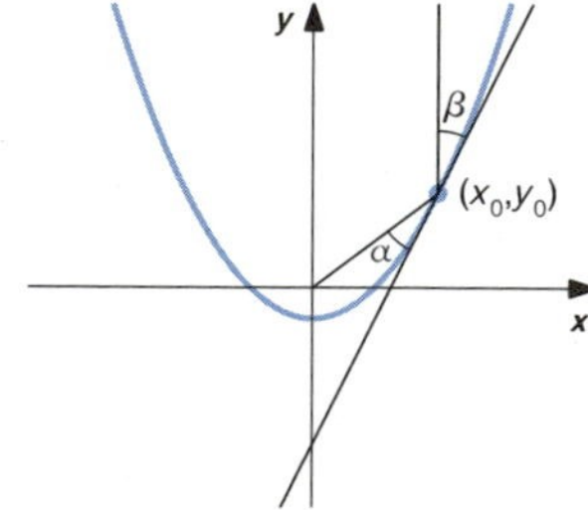

## 11.2 ELLIPSES

*Definition*

An **ellipse** is the set of all points $P$ in a plane such that the sum of the distances between $P$ and two distinct fixed points (**foci**) is constant and greater than the distance between the fixed points.

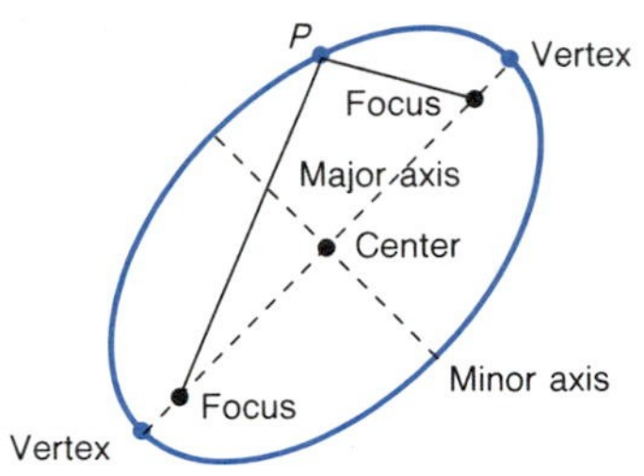


FIGURE 11.2.1

The midpoint of the line segment between the two foci is called the **center** of the ellipse. The two points of intersection of the ellipse with the line through the foci are called **vertices**. The line segment between the vertices is called the **major axis**. The line segment through the center, perpendicular to the major axis, with endpoints on the ellipse is called the **minor axis**. See Figure 11.2.1.

Let us determine an equation of an ellipse that has horizontal major axis. In Figure 11.2.2 we have labeled the center $(h, k)$, foci $(h + c, k)$ and $(h - c, k)$, and vertices $(h + a, k)$ and $(h - a, k)$, where $a > c > 0$. The sum of the distances from one vertex to each of the foci is $2a$. (Note that we would not obtain an ellipse unless $a > c$.) The sum of the distances from any point on the ellipse to each of the foci must also be $2a$. We then see from Figure 11.2.2 that a point $(x, y)$ is on the ellipse whenever

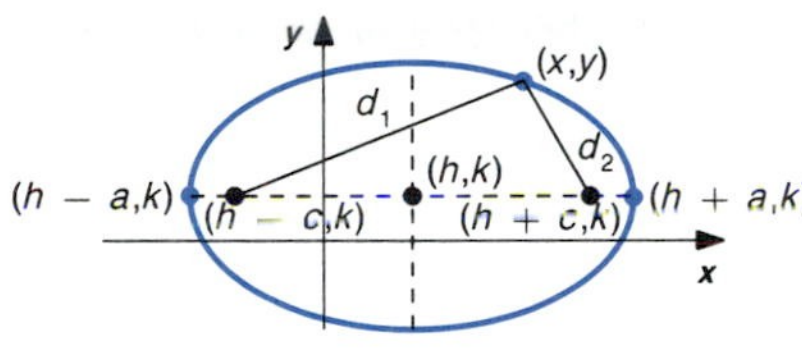


FIGURE 11.2.2

$$d_1 + d_2 = 2a,$$

$$\sqrt{(x-(h-c))^2+(y-k)^2}+\sqrt{(x-(h+c))^2+(y-k)^2}=2a,$$

$$\sqrt{((x-h)+c)^2+(y-k)^2}=2a-\sqrt{((x-h)-c)^2+(y-k)^2},$$

$$((x-h)+c)^2+(y-k)^2=4a^2-4a\sqrt{((x-h)-c)^2+(y-k)^2}$$
$$+((x-h)-c)^2+(y-k)^2,$$

$$(x-h)^2+2c(x-h)+c^2+(y-k)^2$$
$$=4a^2-4a\sqrt{((x-h)-c)^2+(y-k)^2}$$
$$+(x-h)^2-2c(x-h)+c^2+(y-k)^2,$$

$$a\sqrt{((x-h)-c)^2+(y-k)^2}=a^2-c(x-h),$$

$$a^2[((x-h)-c)^2+(y-k)^2]=(a^2-c(x-h))^2,$$

$$a^2(x-h)^2-2a^2c(x-h)+a^2c^2+a^2(y-k)^2$$
$$=a^4-2a^2c(x-h)+c^2(x-h)^2,$$

$$(a^2-c^2)(x-h)^2+a^2(y-k)^2=a^2(a^2-c^2),$$

$$\frac{(x-h)^2}{a^2}+\frac{(y-k)^2}{b^2}=1,$$

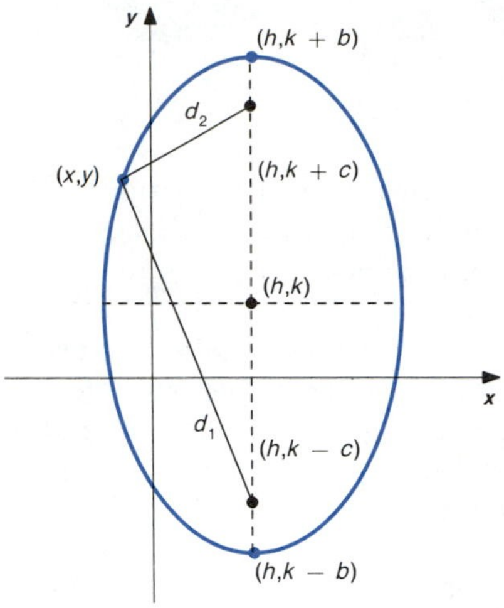

FIGURE 11.2.3

where $b^2 = a^2 - c^2$, $b > 0$. (Note that $a > b$.) Summarizing, we have the following:

The **standard form** of the ellipse with horizontal major axis, center $(h, k)$, vertices $(h - a, k)$ and $(h + a, k)$, foci $(h - c, k)$, and $(h + c, k)$ is

$$\frac{(x-h)^2}{a^2} + \frac{(y-k)^2}{b^2} = 1,$$

where $a > c > 0$, $a > b > 0$, and $b^2 = a^2 - c^2$.

We obtain an equation of an ellipse with vertical major axis in a similar manner, by using Figure 11.2.3.

The **standard form** of the ellipse with vertical major axis, center $(h, k)$, vertices $(h, k - b)$ and $(h, k + b)$, foci $(h, k - c)$, and $(h, k + c)$ is

$$\frac{(x-h)^2}{a^2} + \frac{(y-k)^2}{b^2} = 1,$$

where $b > c > 0$, $b > a > 0$, and $a^2 = b^2 - c^2$.

The standard equation of an ellipse can be used to determine all important points of the ellipse. From the equation

$$\frac{(x-h)^2}{a^2} + \frac{(y-k)^2}{b^2} = 1,$$

we see that the center has coordinates $(h, k)$. The axes are then found by moving a distance of $a$ to the right and left of the center, and a distance of $b$ up and down from the center. Note that $a$ is the square root of the number below $(x - h)^2$ and we move $a$ units parallel to the $x$-axis; $b$ is the square root of the number below $(y - k)^2$ and we move $b$ units parallel to the $y$-axis. If $a > b$, the ellipse has horizontal major axis, vertices $(h - a, k)$ and $(h + a, k)$, foci $(h - c, k)$ and $(h + c, k)$, where $c$ is determined from the equation $c^2 = a^2 - b^2$. If $b > a$, the ellipse has vertical major axis, vertices $(h, k - b)$ and $(h, k + b)$, foci $(h, k - c)$ and $(h, k + c)$, where $c$ is determined from the equation $c^2 = b^2 - a^2$. The graph is a circle if $a = b$.

### ■ EXAMPLE 1

Find the center, vertices, and foci of the ellipse $2x^2 + y^2 = 8$. Sketch the graph.

#### SOLUTION

Writing the equation in standard form, we have

$$2x^2 + y^2 = 8,$$

$$\frac{x^2}{4} + \frac{y^2}{8} = 1,$$

$$\frac{x^2}{2^2} + \frac{y^2}{(\sqrt{8})^2} = 1.$$

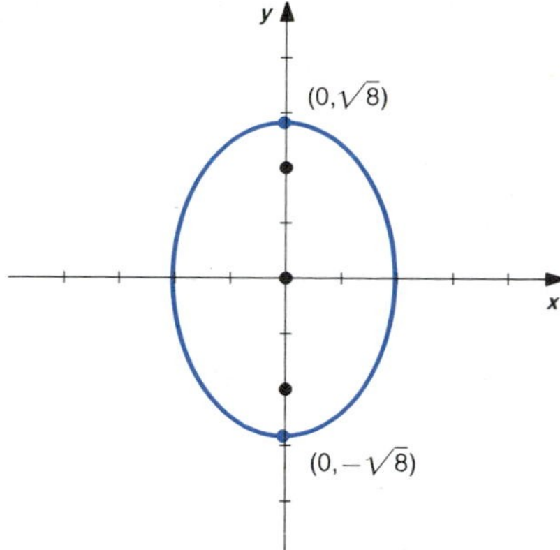

FIGURE 11.2.4

We see that the ellipse has center $(h, k) = (0, 0)$, $a = 2$, and $b = \sqrt{8}$. Since $b > a$, the ellipse has vertical axis. The vertices are $(0, \sqrt{8})$ and $(0, -\sqrt{8})$. We have $c^2 = 8 - 4 = 4$, so $c = \pm 2$. The foci are $(0, 2)$ and $(0, -2)$. The graph is sketched in Figure 11.2.4.

■ **EXAMPLE 2**

Find the center, vertices, and foci of the ellipse

$$4x^2 + 9y^2 - 24x + 36y + 36 = 0.$$

Sketch the graph.

**SOLUTION**

We first complete the squares.

$$4x^2 - 24x \qquad + 9y^2 + 36y \qquad = -36,$$

$$4(x^2 - 6x + [9]) + 9(y^2 + 4y + [4]) = -36 + 4[9] + 9[4],$$

$$4(x-3)^2 + 9(y+2)^2 = 36,$$

$$\frac{(x-3)^2}{9} + \frac{(y+2)^2}{4} = 1,$$

$$\frac{(x-3)^2}{3^2} + \frac{(y+2)^2}{2^2} = 1.$$

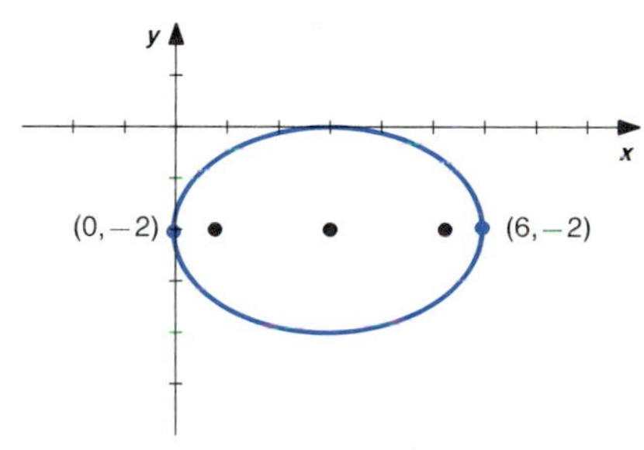

FIGURE 11.2.5

We see that the center is $(h, k) = (3, -2)$. Since $3^2 > 2^2$, the ellipse has horizontal major axis and vertices $(0, -2)$ and $(6, -2)$. We have $c^2 = 3^2 - 2^2$, so $c = \sqrt{5}$. The foci are $(3 - \sqrt{5}, -2)$ and $(3 + \sqrt{5}, -2)$. The graph is sketched in Figure 11.2.5. ■

We can use properties of symmetry and the relation between $a$, $b$, and $c$ to determine the equation of ellipses with prescribed characteristics.

■ **EXAMPLE 3**

Find an equation of the ellipse that has vertices at $(0, 0)$ and $(0, 10)$, and a focus at $(0, 1)$.

**SOLUTION**

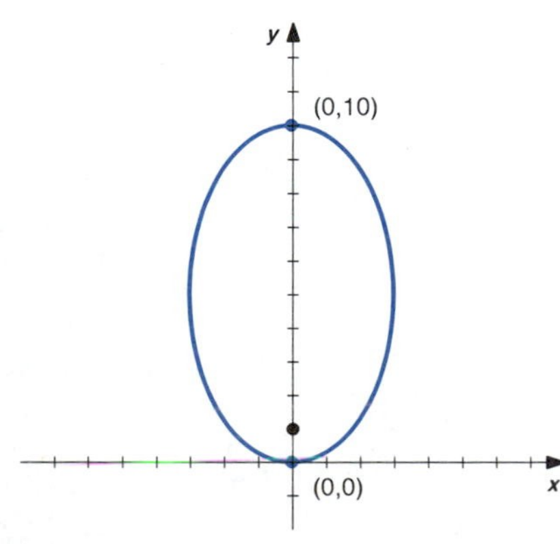

FIGURE 11.2.6

The ellipse is sketched in Figure 11.2.6.

We see that the center is the midpoint of the line segment between the vertices $(0, 0)$ and $(0, 10)$, so $(h, k) = (0, 5)$. The distance between the center and each focus is $c = 4$ and the distance between the center and each vertex is $b = 5$. We then obtain $a^2 = 5^2 - 4^2 = 9$, so $a = 3$. We then write the equation in standard form as

$$\frac{(x-0)^2}{3^2} + \frac{(y-5)^2}{5^2} = 1.$$ ■

*Definition*

The **eccentricity** of an ellipse is

$$e = \frac{\text{Distance between center and each focus}}{\text{Distance between center and each vertex}}.$$

The distance between the center and each vertex is the larger of $a$ and $b$, and $0 < e < 1$. The ellipse is long and thin if the eccentricity is near one. The shape approaches that of a circle as the eccentricity approaches zero.

**EXAMPLE 4**

Sketch the graphs of the ellipses with center at the origin, vertices $(-5, 0)$ and $(5, 0)$, and eccentricity (a) 0.6, (b) 0.8.

**SOLUTION**

(a) The distance between the center and each vertex is $a = 5$. Then $e = c/a$ implies $0.6 = c/5$, so $c = 3$. Then $b^2 = 5^2 - 3^2 = 16$, so $b = 4$. The equation is

$$\frac{x^2}{5^2} + \frac{y^2}{4^2} = 1.$$

(b) We have $a = 5$. Then $e = c/a$ implies $0.8 = c/5$, so $c = 4$. We have $b^2 = 5^2 - 4^2 = 9$, so $b = 3$. The equation is

$$\frac{x^2}{5^2} + \frac{y^2}{3^2} = 1.$$

The graphs are sketched in Figure 11.2.7.

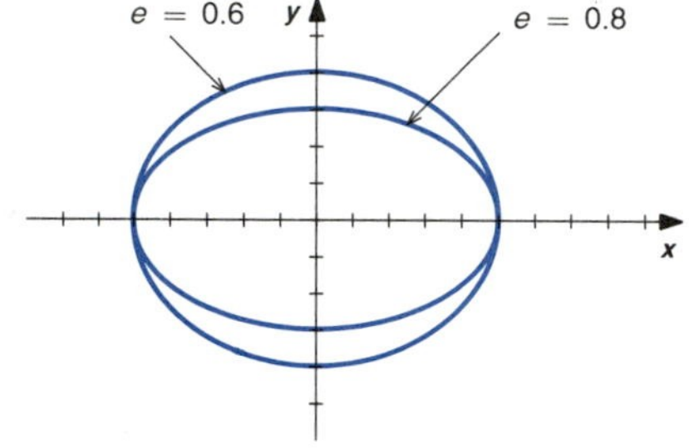

FIGURE 11.2.7

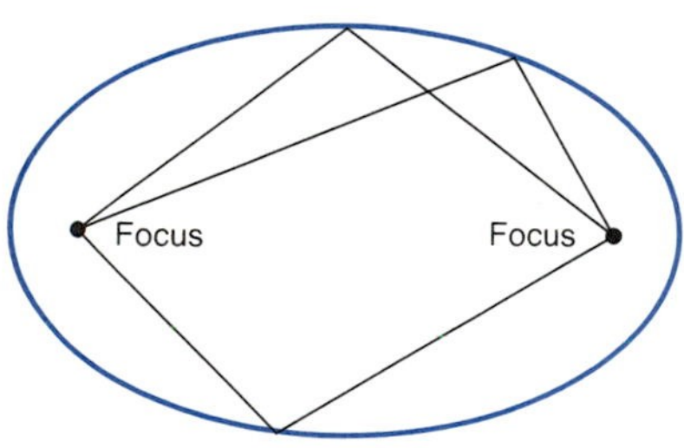

FIGURE 11.2.8

Sound waves that are emitted in all directions from one focus of an ellipse will reflect off the ellipse toward the other focus. The waves will travel the same time along each path. See Figure 11.2.8. In a room that has an ellipsoidal ceiling, words whispered at one focus can be heard quite distinctly by a listener at the other focus, but not by listeners at other points of the room.

## EXERCISES 11.2

Find the center, vertices, and foci, and then sketch the ellipses given in Exercises 1–10.

1 $25x^2 + 9y^2 = 225$

2 $25x^2 + 169y^2 = 4225$

3 $x^2 - 4x + 4y^2 = 0$

4 $4x^2 + y^2 + 4y = 0$

5 $100x^2 + 36y^2 - 180y = 0$

6 $4x^2 - 12x + 36y^2 = 0$

7 $4x^2 - 12x + 9y^2 - 18y = -9$

8 $4x^2 - 8x + y^2 - 4y = -7$

9 $3x^2 - 6x + y^2 - 2y = 0$

10 $x^2 + 4x + 5y^2 + 10y = 0$

Find an equation of the ellipse that satisfies the conditions in each of Exercises 11–18.

11 Center (0, 0); vertex (0, 13); focus (0, 12)

12 Center (0, 0); vertex (5, 0); focus (3, 0)

13 Vertices (−5, 3) and (5, 3); focus (4, 3)

14 Vertex (0, 0); foci (0, 2) and (0, 8)

15 Vertices (1, −4) and (1, 4); contains (2, 0)

16 Vertices (0, 2) and (6, 2); contains (3, 0)

17 Center (0, 0); vertex (4, 0); contains (2, 1)

18 Center (0, 0); vertex (0, 2); contains (1, 1)

19 Find equations of all ellipses that have focus at the origin, vertex (5, 0), and eccentricity 0.5.

20 Find equations of all ellipses that have focus at the origin, vertex (0, 2), and eccentricity 0.6.

21 Describe the set of all $(x, y)$ that satisfy the condition that the sum of the distances between $(x, y)$ and each of the points $(-c, 0)$ and $(c, 0)$ is $2c$, $c > 0$.

22 Describe the set of all $(x, y)$ that satisfy the condition that the sum of the distances between $(x, y)$ and each of the points $(-c, 0)$ and $(c, 0)$ is $2d$, $c > d > 0$.

23 Find the equation of all $(x, y)$ that satisfy the condition that the distance between $(x, y)$ and the origin is one-half the distance of $(x, y)$ from the line $x = -3$.

24 Find the equation of all $(x, y)$ that satisfy the condition that the distance between $(x, y)$ and the origin is $e$ times the distance of $(x, y)$ from the line $x = d$, $0 < e < 1$, $d \neq 0$.

25 The ellipse $16x^2 + 25y^2 = 400$ is sketched below.

(a) Find the slope of the line tangent to the ellipse at the point (4, 12/5).

(b) Find the foci of the ellipse.

(c) Find the slopes of the lines through (4, 12/5) and each focus.

(d) Find the $\tan \alpha$ and $\tan \beta$. [Note that the tangent of the angle between lines that have slopes $m_1$ and $m_2$ is $(m_2 - m_1)/(1 + m_1 m_2)$.]

(e) Find a relation between the angles $\alpha$ and $\beta$. (This verifies the reflection property of this ellipse at this point on the ellipse.)

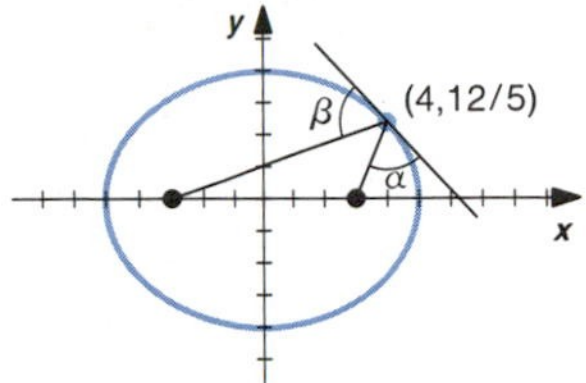


26 The ellipse $b^2x^2 + a^2y^2 = a^2b^2$, $a > b > 0$, is sketched below.

(a) Find the slope of the line tangent to the ellipse at a point $(x_0, y_0)$ on the ellipse, $y_0 \neq 0$.

(b) Find the foci of the ellipse.

(c) Find the slopes of the lines through $(x_0, y_0)$ and each focus.

(d) Find the $\tan \alpha$ and $\tan \beta$. [Note that the tangent of the angle between lines that have slopes $m_1$ and $m_2$ is $(m_2 - m_1)/(1 + m_1 m_2)$.]

(e) Find a relation between the angles $\alpha$ and $\beta$. (This verifies the reflection property for ellipses.)

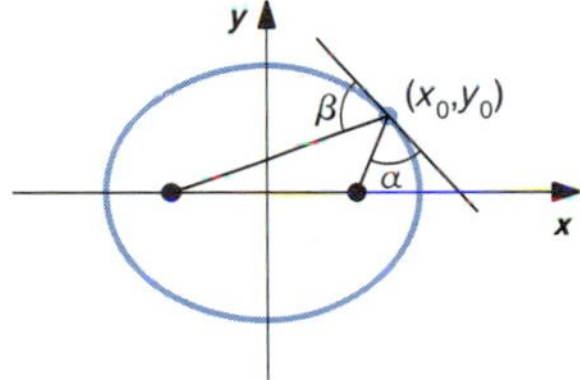


## 11.3 HYPERBOLAS

*Definition*

A **hyperbola** is the set of all points $P$ in a plane such that the difference between distances from $P$ and two distinct fixed points (**foci**) is constant and less than the distance between the fixed points. ■

The midpoint of the line segment between the two foci is called the **center** of the hyperbola. The line through the foci is called the **transverse axis**. The line through the center, perpendicular to the transverse axis, is called the **conjugate axis**. The two points of intersection of the transverse axis and the

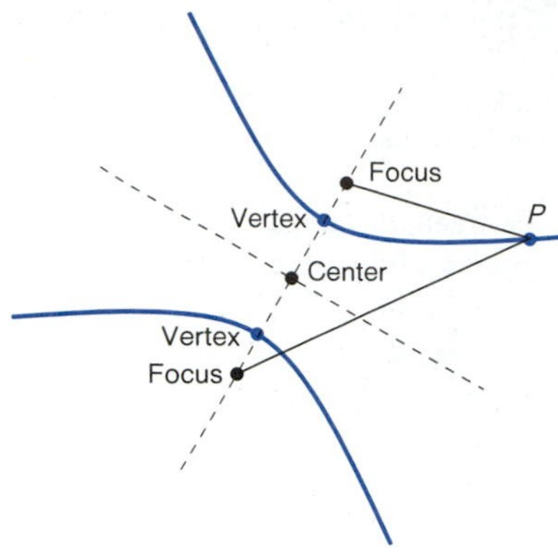


FIGURE 11.3.1

hyperbola are called **vertices**. A hyperbola consists of two separate parts called **branches**. See Figure 11.3.1.

In Figure 11.3.2 we have labeled the center $(h, k)$, vertices $(h - a, k)$ and $(h + a, k)$, and foci $(h - c, k)$ and $(h + c, k)$ of a hyperbola with horizontal transverse axis $y = k$, where $c > a > 0$. The difference of the distances between either vertex and each of the foci is $\pm 2a$. (Note that we would not obtain a hyperbola unless $a < c$.) A point $(x, y)$ is on the hyperbola whenever

$$d_2 - d_1 = \pm 2a.$$

We can now follow the procedure that was used in Section 11.2 to find the equation of an ellipse to find the equation of the hyperbola. [Be sure to isolate one radical on one side of the equation before squaring. You will see that the term $\pm a$ appears squared in the final formula, and $(\pm a)^2 = a^2$. Since $c > a$, $a^2 - c^2$ is negative. We set $-b^2 = a^2 - c^2$, $b > 0$.] Let us summarize the results.

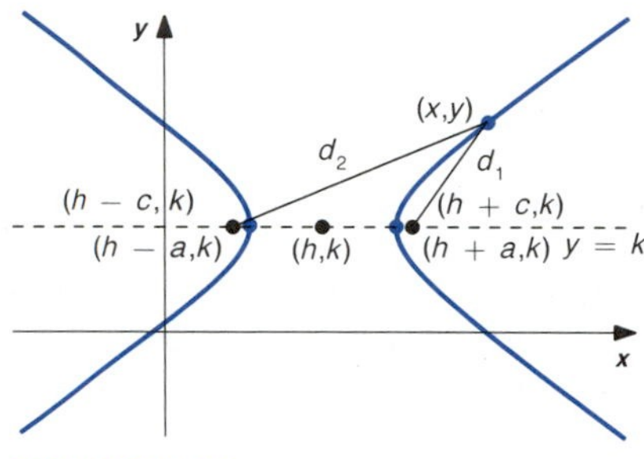


FIGURE 11.3.2

The **standard form** of the equation of a hyperbola with center $(h, k)$, vertices $(h - a, k)$ and $(h + a, k)$, and foci $(h - c, k)$ and $(h + c, k)$ is

$$\frac{(x - h)^2}{a^2} - \frac{(y - k)^2}{b^2} = 1,$$

where $c^2 = a^2 + b^2$, $c > a > 0$, $c > b > 0$.

We can use Figure 11.3.3 to establish the equation of a hyperbola that has vertical transverse axis.

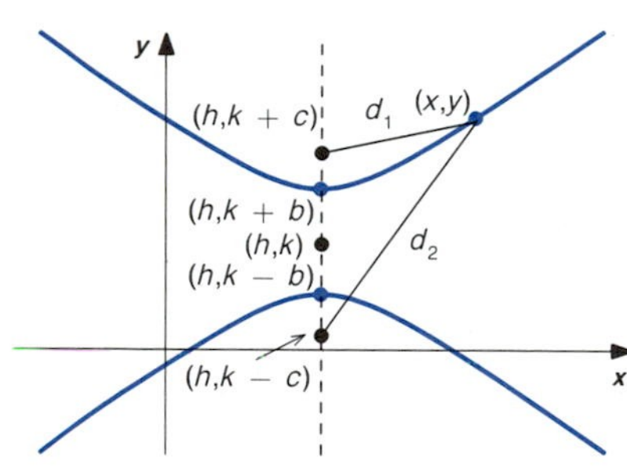


FIGURE 11.3.3

The **standard form** of the equation of a hyperbola with center $(h, k)$, vertices $(h, k - b)$ and $(h, k + b)$, and foci $(h, k - c)$ and $(h, k + c)$ is

$$-\frac{(x - h)^2}{a^2} + \frac{(y - k)^2}{b^2} = 1,$$

where $c^2 = a^2 + b^2$, $c > a > 0$, $c > b > 0$.

The hyperbolas given by each of the above standard forms have **asymptotes**

$$\frac{y - k}{b} = \frac{x - h}{a} \quad \text{and} \quad \frac{y - k}{b} = -\frac{x - h}{a}.$$

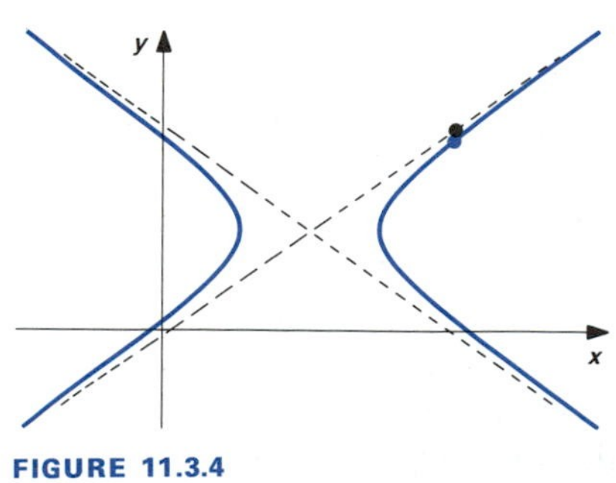


FIGURE 11.3.4

This means that the graphs of the hyperbolas approach the asymptotic lines as $|x|$ and $|y|$ approach infinity. For example, in Figure 11.3.4 the distance between the asymptotic line $y = k + (b/a)(x - h)$ and a corresponding point on the upper right portion of the hyperbola

$$\frac{(x - h)^2}{a^2} - \frac{(y - k)^2}{b^2} = 1$$

satisfies

$$\lim_{x\to\infty}\left[\left(k+\frac{b}{a}(x-h)\right)-\left(k+\frac{b}{a}\sqrt{(x-h)^2-a^2}\right)\right]$$

$$=\lim_{x\to\infty}\frac{b}{a}[(x-h)-\sqrt{(x-h)^2-a^2}]$$

$$=\lim_{x\to\infty}\frac{b}{a}[(x-h)-\sqrt{(x-h)^2-a^2}]\left[\frac{(x-h)+\sqrt{(x-h)^2-a^2}}{(x-h)+\sqrt{(x-h)^2-a^2}}\right]$$

$$=\lim_{x\to\infty}\frac{b}{a}\frac{a^2}{(x-h)+\sqrt{(x-h)^2-a^2}}=0.$$

Note that the asymptotes of a hyperbola can be obtained by replacing the "1" in the general equation by "0." For example,

$$\frac{(x-h)^2}{a^2}-\frac{(y-k)^2}{b^2}=0 \quad \text{implies} \quad \frac{y-k}{b}=\pm\frac{x-h}{a}.$$

It is easy to sketch a hyperbola given in standard form. We first determine the numbers $h$ and $k$ that give the coordinates of the center. The numbers $a$ and $b$ are then used to locate the four points $(h-a, k)$, $(h+a, k)$, $(h, k-b)$, and $(h, k+b)$. A rectangle with sides parallel to the coordinate axes is then drawn through these four points. The asymptotes, which we should draw, are the lines that pass through opposite corners of this rectangle. We then determine which pair of points, either $(h-a, k)$ and $(h+a, k)$, or $(h, k-b)$ and $(h, k+b)$, are the vertices of the hyperbola. (The coordinates of the vertices will satisfy the equation of the hyperbola, the other pair of coordinates will not.) We can then sketch the graph through the vertices, approaching the asymptotes. If required, we can determine $c$ from the equation $c^2 = a^2 + b^2$, and then locate the foci.

### ■ EXAMPLE 1

Find the center, vertices, foci, and asymptoes of the hyperbola $x^2 - 4y^2 = 4$. Sketch the graph.

#### SOLUTION

We have

$$\frac{x^2}{2^2}-\frac{y^2}{1^2}=1.$$

The center is $(0, 0)$. We see that $a = 2$ and $b = 1$. The four points through which we draw a rectangle are $(-2, 0)$, $(2, 0)$, $(0, -1)$, and $(0, 1)$. The coordinates $(-2, 0)$ and $(2, 0)$ satisfy the equation, so the vertices

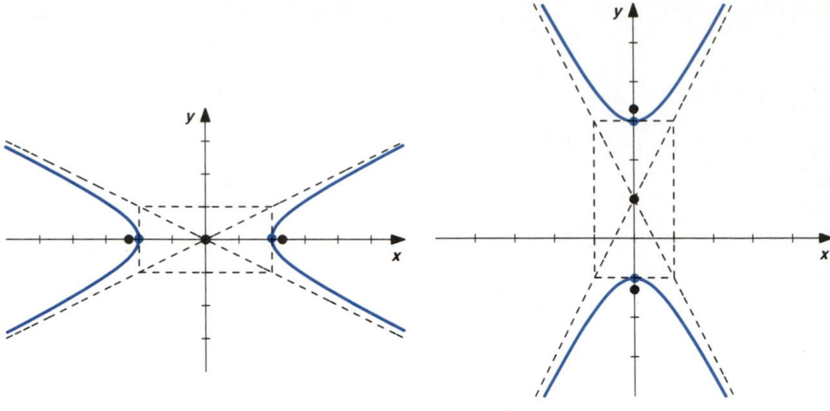


FIGURE 11.3.5

FIGURE 11.3.6

are $(-2, 0)$ and $(2, 0)$. We have $c^2 = 2^2 + 1^2$, so $c = \pm\sqrt{5}$. The foci are $(-\sqrt{5}, 0)$ and $(\sqrt{5}, 0)$. The asymptotes are

$$\frac{x^2}{2^2} - \frac{y^2}{1^2} = 0 \quad \text{or} \quad \frac{x}{2} = \pm y.$$

The graph is sketched in Figure 11.3.5. ■

■ **EXAMPLE 2**

Find the center, vertices, foci, and asymptotes of the hyperbola $4x^2 - (y-1)^2 + 4 = 0$. Sketch the graph.

**SOLUTION**

The equation $4x^2 - (y-1)^2 + 4 = 0$ implies

$$-\frac{x^2}{1^2} + \frac{(y-1)^2}{2^2} = 1.$$

The center is $(0, 1)$. We see that $a = 1$ and $b = 2$. The four points through which we draw a rectangle are $(-1, 1)$, $(1, 1)$, $(0, -1)$, and $(0, 3)$. The coordinates of the points $(0, -1)$ and $(0, 3)$ satisfy the equation and give the vertices of the hyperbola. We have $c^2 = 1^2 + 2^2$, so $c = \pm\sqrt{5}$. The foci are $(0, 1 - \sqrt{5})$ and $(0, 1 + \sqrt{5})$. The asymptotes are

$$-\frac{x^2}{1^2} + \frac{(y-1)^2}{2^2} = 0 \quad \text{or} \quad \frac{y-1}{2} = \pm x.$$

The graph is sketched in Figure 11.3.6. ■

■ **EXAMPLE 3**

Find an equation of the hyperbola that has foci $(0, -5)$ and $(0, 5)$, and one vertex $(0, 4)$.

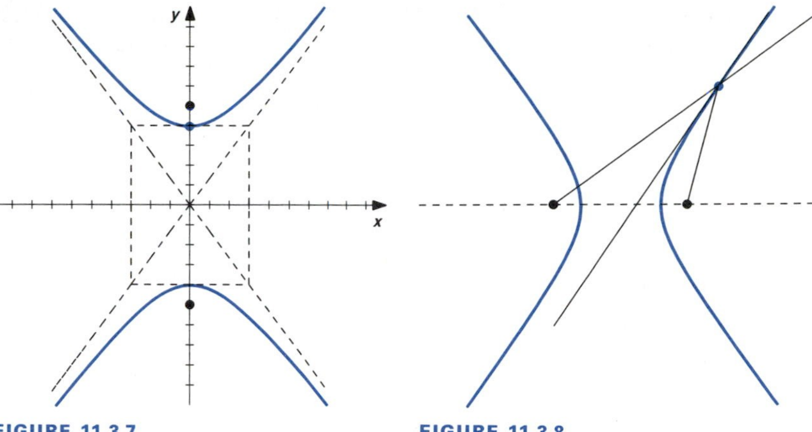


FIGURE 11.3.7

FIGURE 11.3.8

**SOLUTION**

We know that the center is the midpoint of the line segment between the foci, so $(h, k) = (0, 0)$. We have $c = 5$ and $b = 4$, so $5^2 = a^2 + 4^2$ and $a = 3$. We then have

$$-\frac{x^2}{3^2} + \frac{y^2}{4^2} = 1.$$

A sketch is given in Figure 11.3.7. ■

■ *Definition*

The **eccentricity** of a hyperbola is

$$e = \frac{\text{Distance between center and each focus}}{\text{Distance between center and each vertex}}.$$ ■

Note that the eccentricity of a hyperbola is greater than one.

Hyperbolas have the reflection property illustrated in Figure 11.3.8. Light rays emitted from one focus reflect off a hyperbolic reflector in the direction of the line from the other focus through the point of reflection. This spreads the light and produces a floodlight effect.

## EXERCISES 11.3

Find the center, vertices, foci, and asymptotes, and then sketch the hyperbolas given in Exercises 1–10.

1 $25x^2 - 144y^2 = 3600$

2 $25x^2 - 144y^2 + 3600 = 0$

3 $9x^2 - 16y^2 + 144 = 0$

4 $9x^2 - 16y^2 = 144$

5 $x^2 - 2x - y^2 = 0$

6 $x^2 - 4y^2 - 16y = 12$

7 $4x^2 + 8x - y^2 + 4y = 1$

8 $x^2 - 2x - y^2 - 2y + 1 = 0$

9 $9x^2 - 36x - 16y^2 + 32y = 16$

10 $16x^2 + 48x - 9y^2 - 36y = 36$

Find an equation of the hyperbola that satisfies the conditions in each of Exercises 11–18.

11 Center (0, 0); vertex (0, 12); focus (0, 13)

12 Center (0, 0); vertex (4, 0); focus (5, 0)

13 Vertices $(-4, 3)$ and (4, 3); focus (5, 3)

14 Vertex (0, 0); foci $(0, -2)$ and (0, 8)

15 Vertices $(1, -3)$ and (1, 3); contains (0, 5)

16 Vertices (0, 2) and (6, 2); contains $(-2, 0)$

17 Center (0, 0); vertex (0, 4); contains (1, 6)

18 Center (0, 0); vertex (3, 0); contains (4, 5)

19 Find equations of all hyperbolas that have focus at the origin, vertex (2, 0), and eccentricity 4.

20 Find equations of all ellipses that have focus at the origin, vertex (0, 3), and eccentricity 3.

21 Describe the set of all $(x, y)$ that satisfy the condition that the difference of the distances between $(x, y)$ and each of the points $(-c, 0)$ and $(c, 0)$ is $2c$, $c > 0$.

22 Describe the set of all $(x, y)$ that satisfy the condition that the difference of the distances between $(x, y)$ and each of the points $(-c, 0)$ and $(c, 0)$ is $2d$, $d > c > 0$.

23 Find an equation of all $(x, y)$ that satisfy the condition that the distance between $(x, y)$ and the origin is twice the distance of $(x, y)$ from the line $x = -3$.

24 Find an equation of all $(x, y)$ that satisfy the condition that the distance between $(x, y)$ and the origin is $e$ times the distance of $(x, y)$ from the line $x = d$, $e > 1$, $d \neq 0$.

25 The hyperbola $16x^2 - 9y^2 = 144$ is sketched below.

(a) Find the slope of the line tangent to the hyperbola at the point (15/4, 3).

(b) Find the foci of the hyperbola.

(c) Find the slopes of the lines through (15/4, 3) and each focus.

(d) Find $\tan \alpha$ and $\tan \beta$. [Note that the tangent of the angle between lines that have slopes $m_1$ and $m_2$ is $(m_2 - m_1)/(1 + m_1 m_2)$.]

(e) Find a relation between the angles $\alpha$ and $\beta$. (This verifies the reflection property of this hyperbola at this point on the hyperbola.)

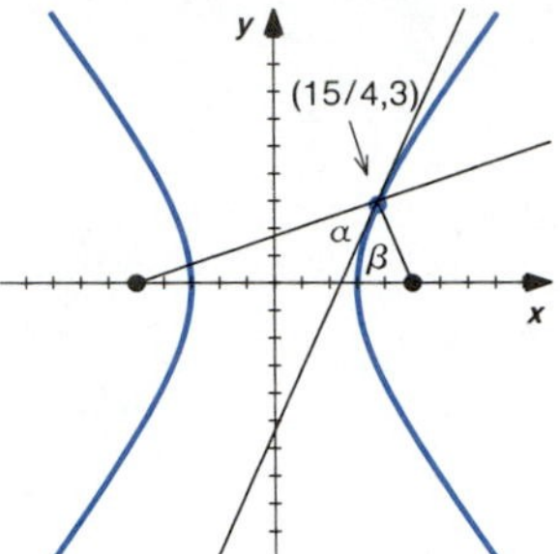


26 The hyperbola $b^2x^2 - a^2y^2 = a^2b^2$, $a > 0$, $b > 0$, is sketched below.

(a) Find the slopes of the lines tangent to the hyperbola at a point $(x_0, y_0)$ on the hyperbola, $y_0 \neq 0$.

(b) Find the foci of the hyperbola.

(c) Find the slopes of the lines through $(x_0, y_0)$ and each focus.

(d) Find $\tan \alpha$ and $\tan \beta$. [Note that the tangent of the angle between lines that have slopes $m_1$ and $m_2$ is $(m_2 - m_1)/(1 + m_1 m_2)$.]

(e) Find a relation between the angles $\alpha$ and $\beta$. (This verifies the reflection property for hyperbolas.)

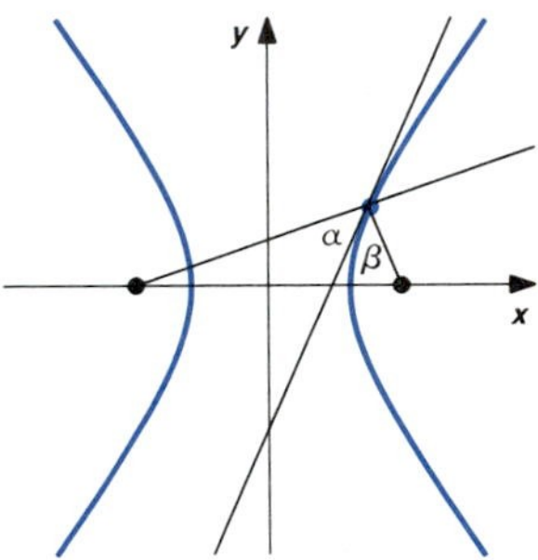


27 Use Figure 11.3.2 to verify that the equation $d_2 - d_1 = \pm 2a$ leads to the standard form of the equation of the hyperbola.

## 11.4 TRANSLATION AND ROTATION

We will see two different ways to change from a given coordinate system to a new coordinate system. The new coordinates will be used to simplify and analyze equations given in terms of the original coordinates.

### *Translation*

The relation between the coordinates of a point with respect to a rectangular coordinate system $(x, y)$ and a rectangular coordinate system

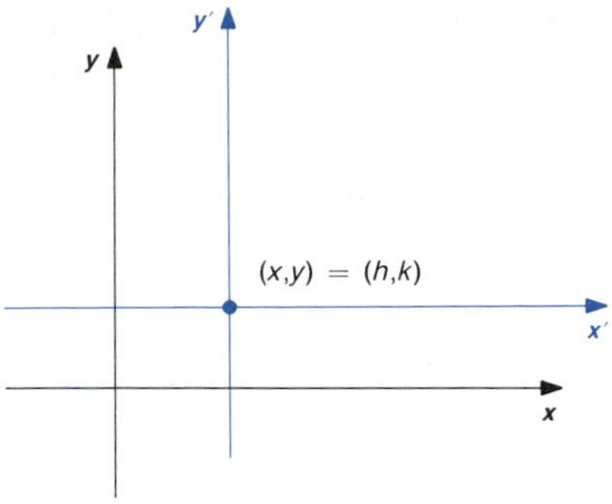

FIGURE 11.4.1

$(x', y')$ obtained by **translation** of the origin to the point $(x, y) = (h, k)$ is given by the equations

$$\begin{cases} \boldsymbol{x = x' + h,} \\ \boldsymbol{y = y' + k.} \end{cases}$$

This translation is illustrated in Figure 11.4.1. Note that the translated $x'$-axis is parallel to the original $x$-axis; the translated $y'$-axis is parallel to the original $y$-axis. From the equation $x = x' + h$, we see that $x' = 0$ corresponds to $x = h$. The equation $y = y' + k$ tells us that $y' = 0$ corresponds to $y = k$. That is, the origin of the translated system, $(x', y') = (0, 0)$, corresponds to the point $(x, y) = (h, k)$.

The equations of translation,

$$\begin{cases} \boldsymbol{x = x' + h,} \\ \boldsymbol{y = y' + k,} \end{cases}$$

give the original $(x, y)$ coordinates in terms of the new coordinates $(x', y')$. This form is convenient for substitution in an equation that involves $x$ and $y$ in order to obtain an equation in $x'$ and $y'$. The translated coordinates $(x', y')$ can be expressed in terms of the original coordinates by the equivalent system of equations

$$\begin{cases} x' = x - h, \\ y' = y - k. \end{cases}$$

■ **EXAMPLE 1**

Express the equation $3x^2 - 12x - 8y = 0$ in terms of the $(x', y')$ coordinates given by the translation

$$\begin{cases} x = x' + 2, \\ y = y' - \dfrac{3}{2}. \end{cases}$$

Sketch the graph and both sets of coordinate axes.

**SOLUTION**

Substitution of the equations of translation gives

$$3x^2 - 12x - 8y = 0,$$

$$3(x' + 2)^2 - 12(x' + 2) - 8\left(y' - \frac{3}{2}\right) = 0,$$

$$3x'^2 + 12x' + 12 - 12x' - 24 - 8y' + 12 = 0,$$

$$3x'^2 - 8y' = 0.$$

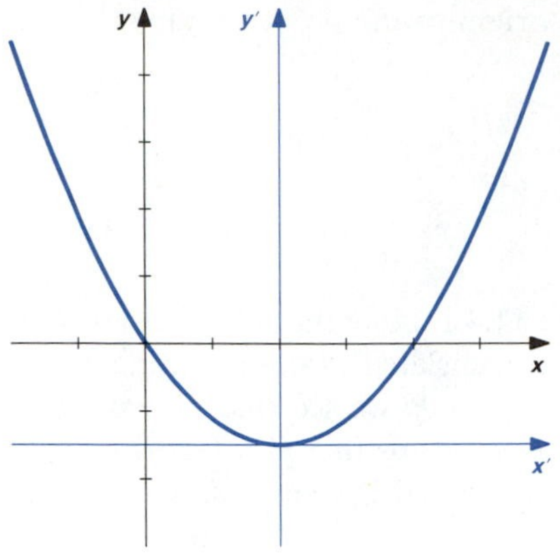


FIGURE 11.4.2

We see that the graph is a parabola with vertex at $(x', y') = (0, 0)$. The equations of translation tell us that $(x', y') = (0, 0)$ corresponds to $(x, y) = (2, -3/2)$. The $(x', y')$ coordinate axes have been drawn with origin at the point $(x, y) = (2, -3/2)$ and the graph is sketched in Figure 11.4.2. ■

**■ EXAMPLE 2**

Find the equations of translation that translate the origin to the center of the ellipse $x^2 - 4x + 4y^2 = 0$. Express the equation in terms of the translated coordinates $(x', y')$. Sketch the graph and both sets of coordinate axes.

**SOLUTION**

The center of the ellipse is found by completing the square and writing the equation in standard form. We have

$$x^2 - 4x + [4] + 4y^2 = [4],$$

$$(x - 2)^2 + 4y^2 = 4,$$

$$\frac{(x - 2)^2}{2^2} + \frac{y^2}{1^2} = 1.$$

We see that the ellipse has center $(h, k) = (2, 0)$. The desired equations of translation are then

$$\begin{cases} x = x' + 2, \\ y = y'. \end{cases}$$

[It is a good idea to check that $(x', y') = (0, 0)$ corresponds to the proper $(x, y)$ coordinates, $(x, y) = (2, 0)$ in this case.] In terms of the $(x', y')$ coordinates, the equation is

$$\frac{x'^2}{2^2} + \frac{y'^2}{1^2} = 1.$$

The $(x', y')$ coordinate axes have been drawn with origin at $(x, y) = (2, 0)$ and the graph sketched in Figure 11.4.3. ■

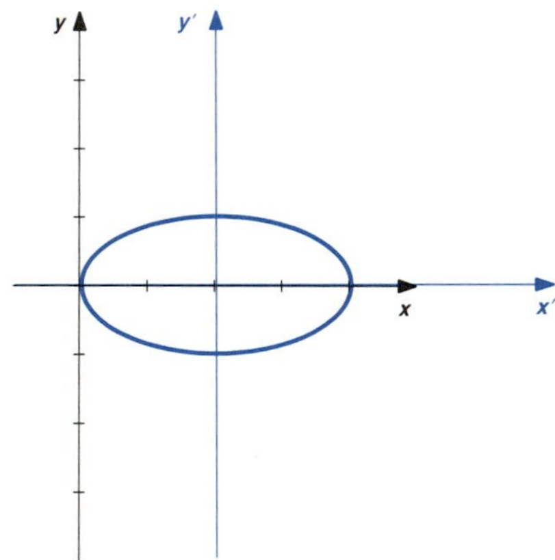


FIGURE 11.4.3

## *Rotation*

We want to establish a new coordinate system by rotating the $x$ and $y$ axes about the origin by an angle $\theta$. This is illustrated in Figure 11.4.4.

A point $P$ in the plane has coordinates $(x, y)$ in the original coordinate system, and coordinates $(x', y')$ in the new, rotated system. We want to determine the relation between the two coordinate systems. From Figure 11.4.4, we have

$$x' = d \cos \alpha \quad \text{and} \quad y' = d \sin \alpha.$$

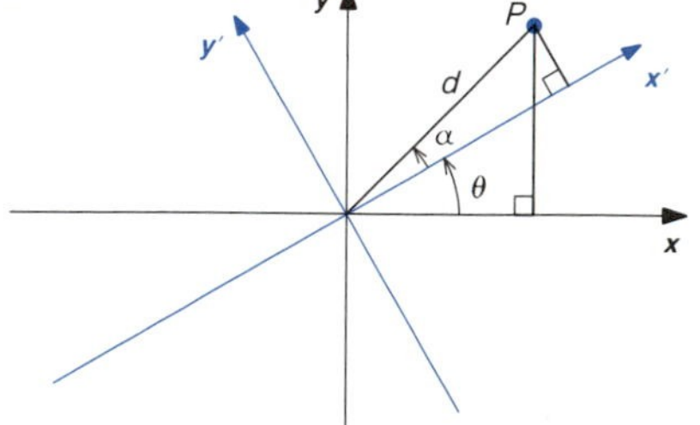


FIGURE 11.4.4

Also,

$$\begin{aligned} x &= d\cos(\alpha + \theta) \\ &= d[\cos\alpha\cos\theta - \sin\alpha\sin\theta] \\ &= x'\cos\theta - y'\sin\theta, \end{aligned}$$

and

$$\begin{aligned} y &= d\sin(\alpha + \theta) \\ &= d[\cos\alpha\sin\theta + \sin\alpha\cos\theta] \\ &= x'\sin\theta + y'\cos\theta. \end{aligned}$$

Summarizing, we have

$$\begin{cases} x = x'\cos\theta - y'\sin\theta, \\ y = x'\sin\theta + y'\cos\theta. \end{cases}$$

Substitution of the above values of $x$ and $y$ in an equation gives an equivalent equation in terms of the new coordinates that are obtained by rotation of the $(x, y)$ axes by an angle of $\theta$.

The equations of rotation can be solved for $x'$ and $y'$ in terms of $x$ and $y$. We obtain

$$\begin{cases} x' = x\cos\theta + y\sin\theta, \\ y' = -x\sin\theta + y\cos\theta \end{cases}$$

[Note that we can interpret these equations as corresponding to a rotation of the $(x', y')$ axes by an angle of $-\theta$.]

■ **EXAMPLE 3**

Express the equation $2x - 3y + 6 = 0$ in terms of the $(x', y')$ coordinates given by a rotation of $\theta = \tan^{-1}(2/3)$. Sketch the graph and both sets of coordinate axes.

**SOLUTION**

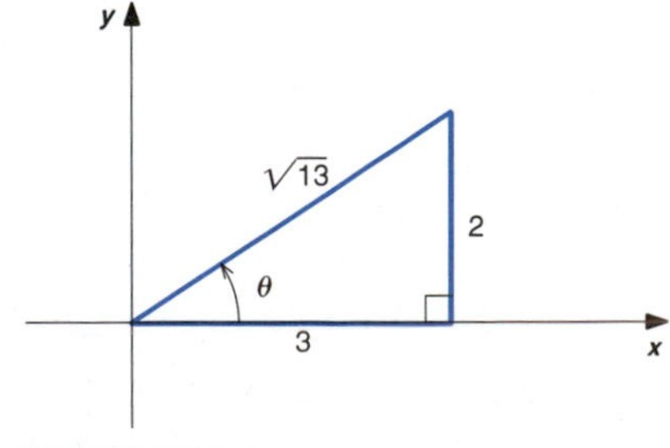


FIGURE 11.4.5

A representative sketch of the angle $\theta = \tan^{-1}(2/3)$ is given in Figure 11.4.5. We see that $\sin\theta = 2/\sqrt{13}$ and $\cos\theta = 3/\sqrt{13}$. Substitution into the equations of rotation then gives

$$x = x'\cos\theta - y'\sin\theta = \frac{3x' - 2y'}{\sqrt{13}},$$

$$y = x'\sin\theta + y'\cos\theta = \frac{2x' + 3y'}{\sqrt{13}}.$$

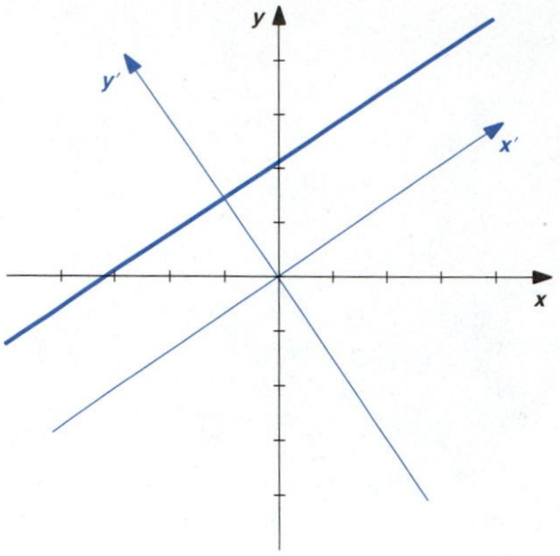

We then have

$$2x - 3y + 6 = 0,$$

$$2\left(\frac{3x' - 2y'}{\sqrt{13}}\right) - 3\left(\frac{2x' + 3y'}{\sqrt{13}}\right) + 6 = 0,$$

$$6x' - 4y' - 6x' - 9y' + 6\sqrt{13} = 0,$$

$$y' = \frac{6\sqrt{13}}{13}.$$

FIGURE 11.4.6

The graph and both sets of axes are sketched in Figure 11.4.6. Note that the line is parallel to the $x'$-axis. This is true because the angle of rotation $\theta$ is equal to the angle between the line and the $x$-axis. ■

■ **EXAMPLE 4**

Express the equation $xy - x + y = 0$ in terms of coordinates $(x', y')$ that correspond to a rotation of $\theta$. Find $0 < \theta < \pi/2$ such that the new equation contains no $x'y'$ term. Sketch the graph and both sets of axes.

**SOLUTION**

Substitution gives

$$xy - x + y = 0,$$

$$(x' \cos\theta - y' \sin\theta)(x' \sin\theta + y' \cos\theta)$$
$$-(x' \cos\theta - y' \sin\theta) + (x' \sin\theta + y' \cos\theta) = 0,$$

$$(\cos\theta \sin\theta)x'^2 + (\cos^2\theta - \sin^2\theta)x'y' - (\sin\theta \cos\theta)y'^2$$
$$+ (\sin\theta - \cos\theta)x' + (\sin\theta + \cos\theta)y' = 0.$$

We see that the new equation will contain no $x'y'$ term if $\theta$ is chosen so that

$$\cos^2\theta - \sin^2\theta = 0.$$

This gives $\tan\theta = \pm 1$. This equation and the condition that $0 < \theta < \pi/2$ are satisfied by the angle $\theta = \pi/4$. Since $\sin(\pi/4) = \cos(\pi/4) = 1/\sqrt{2}$, substitution of $\theta = \pi/4$ into the above equation in $(x', y')$ gives

$$\left(\frac{1}{\sqrt{2}}\frac{1}{\sqrt{2}}\right)x'^2 + \left(\frac{1}{2} - \frac{1}{2}\right)x'y' - \left(\frac{1}{\sqrt{2}}\frac{1}{\sqrt{2}}\right)y'^2 + \left(\frac{1}{\sqrt{2}} - \frac{1}{\sqrt{2}}\right)x'$$
$$+ \left(\frac{1}{\sqrt{2}} + \frac{1}{\sqrt{2}}\right)y' = 0,$$

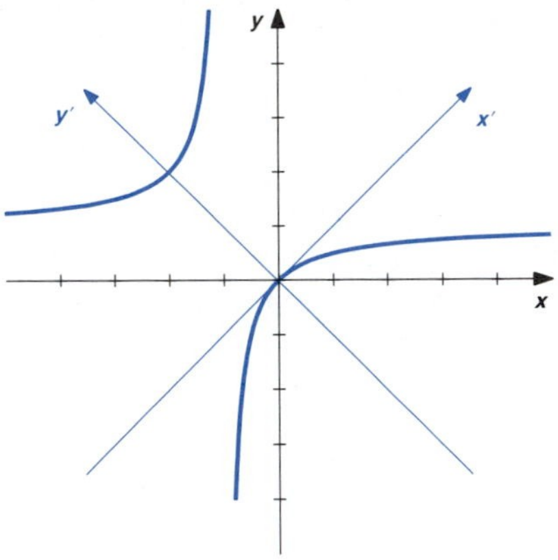

FIGURE 11.4.7

$$x'^2 - y'^2 + \frac{4}{\sqrt{2}}y' = 0,$$

$$x'^2 - y'^2 + 2\sqrt{2}y' = 0.$$

Completing the square to write this equation in standard form, we obtain

$$x'^2 - (y'^2 - 2\sqrt{2}y' + [2]) = -([2]),$$

$$x'^2 - (y' - \sqrt{2})^2 = -2,$$

$$-\frac{x'^2}{(\sqrt{2})^2} + \frac{(y' - \sqrt{2})^2}{(\sqrt{2})^2} = 1.$$

We see that this is the equation of a hyperbola with center $(x', y') = (0, \sqrt{2})$. The vertices are $(x', y') = (0, 0)$ and $(x', y') = (0, 2\sqrt{2})$. The graph and both sets of axes are sketched in Figure 11.4.7. ■

■ **EXAMPLE 5**

Find the $(x, y)$ coordinates of the center, vertices, and foci of the hyperbola $xy - x + y = 0$ of Example 4. Find the equations of the asymptotes in terms of $(x, y)$.

**SOLUTION**

From the equation obtained in Example 4,

$$-\frac{x'^2}{(\sqrt{2})^2} + \frac{(y' - \sqrt{2})^2}{(\sqrt{2})^2} = 1,$$

we see that the center is $(x', y') = (0, \sqrt{2})$. Since the angle of rotation is $\theta = \pi/4$, the corresponding $(x, y)$ coordinates are

$$x = x' \cos\theta - y' \sin\theta = (0)\left(\frac{1}{\sqrt{2}}\right) - (\sqrt{2})\left(\frac{1}{\sqrt{2}}\right) = -1,$$

$$y = x' \sin\theta + y' \cos\theta = (0)\left(\frac{1}{\sqrt{2}}\right) + (\sqrt{2})\left(\frac{1}{\sqrt{2}}\right) = 1.$$

The center is $(x, y) = (-1, 1)$. Similarly, the vertices $(x', y') = (0, 0)$ and $(x', y') = (0, 2\sqrt{2})$ correspond to $(x, y) = (0, 0)$ and $(x, y) = (-2, 2)$, respectively.

We need the $(x', y')$ coordinates of the foci. From the $x'$, $y'$ equation of the hyperbola, we see that $c^2 = a^2 + b^2 = 2 + 2 = 4$, so $c = 2$. The foci are $(x', y') = (0, \sqrt{2} - 2)$ and $(x', y') = (0, \sqrt{2} + 2)$. These correspond to $(x, y) = (\sqrt{2} - 1, -\sqrt{2} + 1)$ and $(x, y) = (-\sqrt{2} - 1, \sqrt{2} + 1)$, respectively.

The equations of the asymptotes are

$$-\frac{x'^2}{(\sqrt{2})^2} + \frac{(y' - \sqrt{2})^2}{(\sqrt{2})^2} = 0,$$

or $y' = x' + \sqrt{2}$ and $y' = -x' + \sqrt{2}$. Using the equations $x' = x \cos \theta + y \sin \theta$, $y' = -x \sin \theta + y \cos \theta$, $\theta = \pi/4$, and substituting, we have

$$y' = x' + \sqrt{2},$$

$$(-x \sin \theta + y \cos \theta) = (x \cos \theta + y \sin \theta) + \sqrt{2},$$

$$-\frac{x}{\sqrt{2}} + \frac{y}{\sqrt{2}} = \frac{x}{\sqrt{2}} + \frac{y}{\sqrt{2}} + \sqrt{2},$$

$$-\frac{2x}{\sqrt{2}} = \sqrt{2},$$

$$x = -1.$$

Similarly, the equation $y' = -x' + \sqrt{2}$ can be shown to have $(x, y)$ equation $y = 1$. ■

The equation

$$Ax^2 + Bxy + Cy^2 + Dx + Ey + F = 0$$

($A$, $B$, $C$ not all three zero) is called the **general second-degree equation**. Rotation of axes by an angle $\theta$ transforms this equation to the equation

$$A'x'^2 + B'x'y' + C'y'^2 + D'x' + E'y' + F' = 0,$$

where

$$A' = A \cos^2 \theta + B \sin \theta \cos \theta + C \sin^2 \theta,$$

$$B' = B(\cos^2 \theta - \sin^2 \theta) + 2(C - A)\sin \theta \cos \theta,$$

$$C' = A \sin^2 \theta - B \sin \theta \cos \theta + C \cos^2 \theta,$$

$$D' = D \cos \theta + E \sin \theta,$$

$$E' = -D \sin \theta + E \cos \theta,$$

$$F' = F.$$

In particular, we see that we can obtain an equation that has no $x'y'$ term if we choose $\theta$ such that

$$B(\cos^2 \theta - \sin^2 \theta) + 2(C - A)\sin \theta \cos \theta = 0.$$

We can use trigonometric identities for double angles to solve this equation for $\theta$. That is,

$$B \cos 2\theta + (C - A) \sin 2\theta = 0,$$

$$\cot 2\theta = \frac{A - C}{B}.$$

If $B \neq 0$, this equation will be satisfied by an angle $0 < \theta < \pi/2$. If $B = 0$, the original equation has no $xy$ term.

It is easy to analyze the equation in terms of the $(x', y')$ coordinates, if that equation contains no $x'y'$ term. That is, $A'x'^2 + B'x'y' + C'y^2 + D'x' + E'y' + F' = 0$ is an ellipse (including the degenerate case of a single point) whenever $B' = 0$ and $A'$ and $C'$ have the same sign. In this case $B'^2 - 4A'C' < 0$. The graph is a hyperbola (or the degenerate case of intersecting lines) whenever $B' = 0$ and $A'$ and $C'$ have opposite sign, so $B'^2 - 4A'C' > 0$. The equation will be a parabola (or the degenerate case of a line) whenever $B' = 0$ and one of $A'$ and $C'$ is zero, so $B'^2 - 4A'C' = 0$.

The expression

$$B^2 - 4AC$$

is called the **discriminant** of the equation

$$Ax^2 + Bxy + Cy^2 + Dx + Ey + F = 0.$$

It can be shown that the equation

$$A'x'^2 + B'x'y' + C'y^2 + D'x' + E'y' + F' = 0$$

that is obtained by any rotation satisfies

$$B'^2 - 4A'C' = B^2 - 4AC.$$

We can then conclude the following **discriminant test** from the discussion above.

**If**

$$Ax^2 + Bxy + Cy^2 + Dx + Ey + F = 0,$$

**then**

**$B^2 - 4AC < 0$ implies the graph is an ellipse,**

**$B^2 - 4AC > 0$ implies the graph is a hyperbola, and**

**$B^2 - 4AC = 0$ implies the graph is a parabola.**

(Each case includes degenerate cases.)

## EXERCISES 11.4

Express the equations in Exercises 1–4 in terms of the $(x', y')$ coordinates of the given translations. Sketch the graphs and both sets of axes.

1 $x = y^2 - 1$; $x = x' - 1$, $y = y'$

2 $x + y^2 + 1 = 0$; $x = x' - 1$, $y = y'$

3 $y = 2x - x^2$; $x = x' + 1$, $y = y' + 1$

4 $y = x^2 + 4x$; $x = x' - 2$, $y = y' - 4$

Find the equations of translation that transform the origin to the center of the graphs of Exercises 5–8. Sketch the graphs and both sets of axes.

5 $x^2 - 4x + 4y^2 - 8y + 7 = 0$

6 $x^2 + 9y^2 - 18y + 8 = 0$

7 $x^2 - y^2 + 2y = 0$

8 $4x^2 - 12x - 9y^2 + 18y + 9 = 0$

Express the equations in Exercises 9–12 in terms of the $(x', y')$ coordinates obtained by rotation by the given angle $\theta$. Sketch the graphs and both sets of axes.

9 $x - y + 2 = 0$; $\theta = \pi/4$

10 $\sqrt{3}x + y = 3$; $\theta = \pi/6$

11 $2x + y = 2$; $\theta = \tan^{-1}(1/2)$

12 $3x - 4y + 12 = 0$; $\theta = \tan^{-1}(3/4)$

In Exercises 13–18, find $0 < \theta < \pi/2$ so rotation by $\theta$ gives an equation with no $x'y'$ term. Express the equations in terms of the $(x', y')$ coordinates obtained by that rotation. Sketch the graphs and both sets of axes.

13 $xy + 1 = 0$

14 $xy = x + 1$

15 $x^2 - 2xy + y^2 - \sqrt{2}x - \sqrt{2}y = 0$

16 $x^2 + 2\sqrt{3}xy + 3y^2 + 2\sqrt{3}x - 2y = 0$

17 $7x^2 - 6\sqrt{3}xy + 13y^2 = 16$

18 $31x^2 + 10\sqrt{3}xy + 21y^2 = 144$

19 Find $-\pi/2 < \theta < \pi/2$ so rotation by $\theta$ changes the equation $4x + 3y = 12$ into an equation with no $x'$ term. Express the equation in terms of the $(x', y')$ coordinates obtained by that rotation. Sketch the graph and both sets of axes.

20 Find $-\pi/2 < \theta < \pi/2$ so rotation by $\theta$ changes the equation $ax + by = c$, $b \neq 0$, into an equation with no $x'$ term. Express the equation in terms of the $(x', y')$ coordinates obtained by that rotation.

21 Verify that rotation by any angle $\theta$ changes the equation

$$Ax^2 + Bxy + Cy^2 + Dx + Ey + F = 0$$

to an equation

$$A'x'^2 + B'x'y' + C'y'^2 + D'x' + E'y' + F' = 0$$

with $B'^2 - 4A'C' = B^2 - 4AC$.

22 Use the discriminant test to identify the graphs in Exercises 13–18.

## REVIEW EXERCISES

1 Find an equation of the ellipse with center $(4, 0)$, one focus $(0, 0)$, and contains $(4, 3)$.

2 Find an equation of the ellipse with foci $(-3, 0)$ and $(3, 0)$, and contains $(0, 2)$.

3 Find an equation of the hyperbola with foci $(-1, 0)$ and $(9, 0)$, and eccentricity 1.25.

4 Find an equation of the hyperbola with vertices $(0, 0)$ and $(-2, 0)$, and one focus $(1, 0)$.

5 Find an equation of the parabola with directrix $x = -2$ and focus $(2, 0)$.

6 Find an equation of the parabola with vertical axis that has vertex $(0, -2)$ and contains the point $(-1, -1)$.

Sketch the graphs of the equations in Exercises 7–12.

7 $x^2 = 16 - 4y^2$

8 $4x^2 + y^2 - 4y = 0$

9 $y^2 = 2 + x$

10 $x^2 = 8 - 8y$

11 $\frac{y^2}{4} - \frac{x^2}{1} = 1$

12 $4x^2 - y^2 - 8x = 0$

In Exercises 13–14, find $-\pi/2 < \theta < \pi/2$ so the equations

$$\begin{cases} x = x' \cos \theta - y' \sin \theta, \\ y = x' \sin \theta + y' \cos \theta, \end{cases}$$

transform the given equation to an equation in $(x', y')$ with no $x'$ term. Express the equation in terms of the corresponding $(x', y')$ coordinates. Sketch both sets of axes and the graph of the equation.

13 $y = \sqrt{3}(x + 1)$

14 $x + y = 2$

In Exercises 15–16, find $0 < \theta < \pi/2$ so the equations

$$\begin{cases} x = x' \cos \theta - y' \sin \theta, \\ y = x' \sin \theta + y' \cos \theta. \end{cases}$$

transform the given equation to an equation in $(x', y')$ with no $x'y'$ term. Express the equation in terms of the corresponding $(x', y')$ coordinates. Sketch both sets of axes and the graph of the equation.

15 $xy = y + 1$

16 $2x^2 + \sqrt{3}xy + y^2 + 2x - 2\sqrt{3}y = 0$

17 Find the coordinates of the vertices of the ellipse $3x^2 - 2xy + 3y^2 = 4$.

18 Find the coordinates of the vertices of the hyperbola $3x^2 + 10xy + 3y^2 = 8$.

**CHAPTER TWELVE**

# *Polar coordinates*

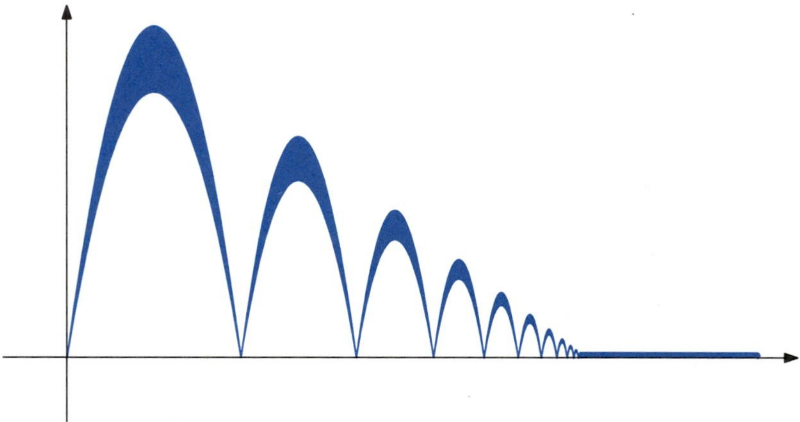

Rectangular coordinates give the position of a point in a plane relative to two perpendicular lines in the plane. There are other ways to describe the position of a point in a plane. In this chapter we will study one such system called the polar coordinate system. Polar coordinates are useful for a variety of problems. The study of planetary motion is one example of a problem for which polar coordinates are particularly well suited.

## 12.1 POLAR COORDINATE SYSTEM

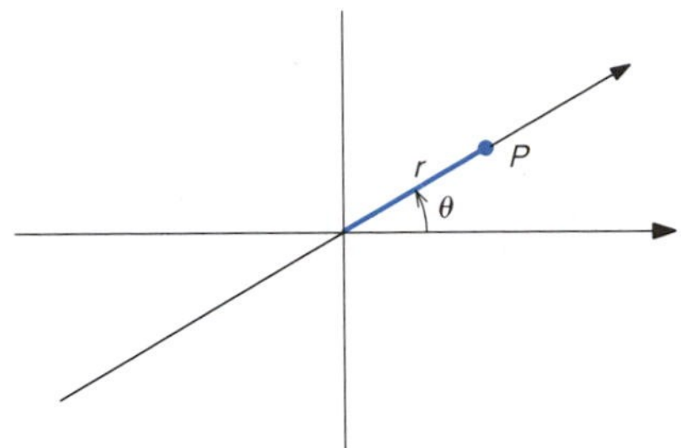

FIGURE 12.1.1

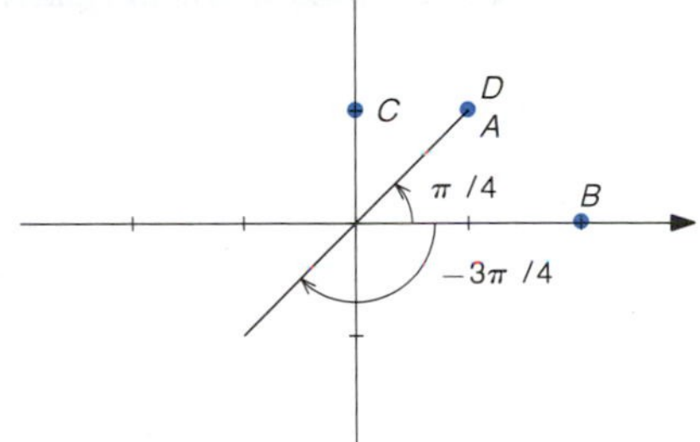

FIGURE 12.1.2

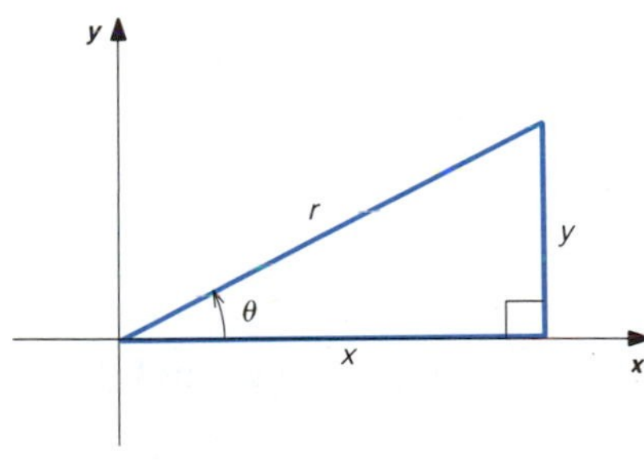

FIGURE 12.1.3

In this section we will introduce the polar coordinate system and investigate polar equations of some familiar curves.

Consider a plane with a fixed number line (**polar axis**) with origin (**pole**) $O$. We will choose the polar axis to be horizontal with positive direction to the right, as the $x$-axis is usually located. The **polar coordinates** $(r, \theta)$ are associated with the point $P$ in the plane that is located by moving a *directed distance* $r$ along the number line that is obtained by rotating the polar axis an angle $\theta$ about the pole. As usual, positive angles correspond to counterclockwise rotations. See Figure 12.1.1. The coordinates $(r, \theta)$ are called polar coordinates of the point $P$.

■ **EXAMPLE 1**

Points with polar coordinates $A(\sqrt{2}, \pi/4)$, $B(2, 0)$, $C(1, \pi/2)$, and $D(-\sqrt{2}, -3\pi/4)$ are plotted in Figure 12.1.2. ■

We see in Example 1 that the points $A$ and $D$ have different polar coordinates, but correspond to the same point. Generally, we have

$$P(r, \theta) = P(r, \theta + 2\pi n), \qquad n \text{ any integer,}$$

and

$$P(r, \theta) = P(-r, \theta + f).$$

Polar coordinates determine rectangular coordinates. From Figure 12.1.3 we see that if the polar axis and the $x$-axis coincide, then

$$x = r \cos \theta, \quad y = r \sin \theta.$$

■ **EXAMPLE 2**

Find rectangular coordinates of the points with polar coordinates $A(\sqrt{2}, \pi/4)$, $B(2, 0)$, $C(1, \pi/2)$, and $D(-\sqrt{2}, -3\pi/4)$.

**SOLUTION**

Using $x = r \cos \theta$ and $y = r \sin \theta$, we have

$$A(x, y) = \left(\sqrt{2} \cos \frac{\pi}{4}, \sqrt{2} \sin \frac{\pi}{4}\right) = (1, 1),$$

$$B(x, y) = (2 \cos 0, 2 \sin 0) = (2, 0),$$

$$C(x, y) = \left(1 \cos \frac{\pi}{2}, 1 \sin \frac{\pi}{2}\right) = (0, 1), \quad \text{and}$$

$$D(x, y) = \left(-\sqrt{2} \cos\left(-\frac{3\pi}{4}\right), -\sqrt{2} \sin\left(-\frac{3\pi}{4}\right)\right) = (1, 1).$$

These points were plotted by using polar coordinates in Figure 12.1.2. ■

To change rectangular coordinates to polar coordinates, we use a sketch and the equations

$$r^2 = x^2 + y^2,$$

$$\tan\theta = \frac{y}{x}.$$

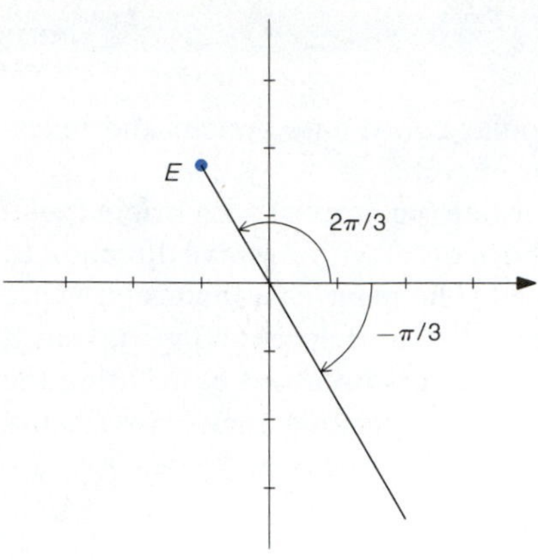

FIGURE 12.1.4

■ **EXAMPLE 3**

Find two sets of polar coordinates for each of the points $E(x, y) = (-1, \sqrt{3})$ and $F(x, y) = (0, 1)$.

**SOLUTION**

We see that $E(x, y) = (-1, \sqrt{3})$ has $r^2 = x^2 + y^2 = 4$, so $r = \pm 2$. We also have $\tan\theta = y/x = -\sqrt{3}/1$. This equation is satisfied by $\theta = -\pi/3 + n\pi$, $n$ any integer. From the sketch in Figure 12.1.4 we see that two sets of polar coordinates for $E$ are $E(r, \theta) = (2, 2\pi/3)$ and $(-2, -\pi/3)$.

$F(x, y) = (0, 1)$ has $r^2 = 1$ and $\theta = \pm\pi/2$. From Figure 12.1.5 we see that two sets of polar coordinates for $F$ are $F(r, \theta) = (1, \pi/2)$ and $(-1, -\pi/2)$. ■

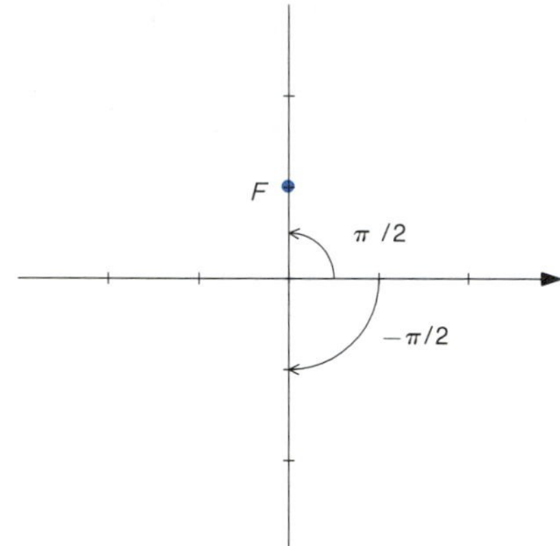

FIGURE 12.1.5

The graph of a polar equation in $r$ and $\theta$ is the set of all points $P$ that correspond to polar coordinates $(r, \theta)$ that satisfy the equation.

■ **EXAMPLE 4**

The graph of the polar equation $r = 2$ is given in Figure 12.1.6. Note that $(r, \theta) = (2, \theta)$ satisfies the equation for any choice of $\theta$. Also, note that the polar equation $r = -2$ has the same graph as $r = 2$, since $P(2, \theta) = P(-2, \theta + \pi)$. Both polar equations $r = 2$ and $r = -2$ satisfy $r^2 = 4$. The corresponding equation in rectangular coordinates is $x^2 + y^2 = 4$. ■

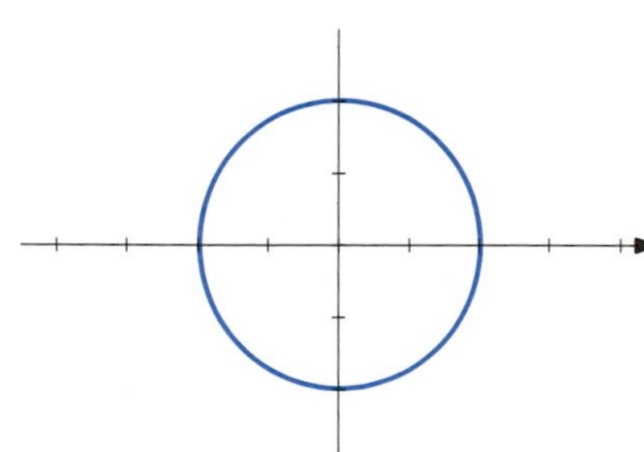
FIGURE 12.1.6

**The graph of $r = r_0$ is a circle with center at the origin and radius $|r_0|$.**

■ **EXAMPLE 5**

The graph of $\theta = \pi/4$ is given in Figure 12.1.7. Note that $P(r, \pi/4)$ is on the graph for all choices of $r$. ■

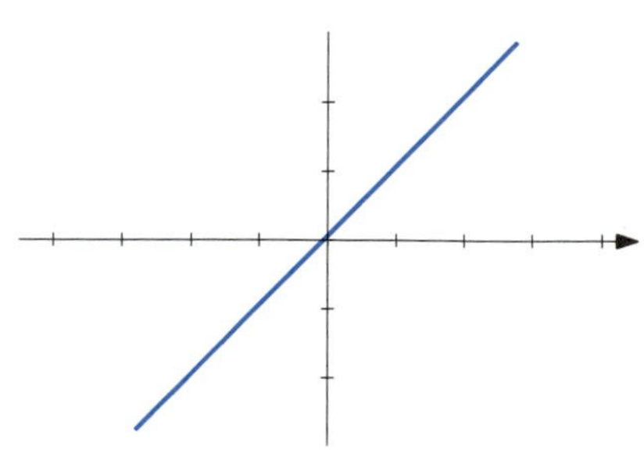
FIGURE 12.1.7

**The graph of $\theta = \theta_0$ is a line through the pole at angle $\theta_0$.**

■ **EXAMPLE 6**

The graph of

$$r = \frac{1}{\cos\theta + \sin\theta}$$

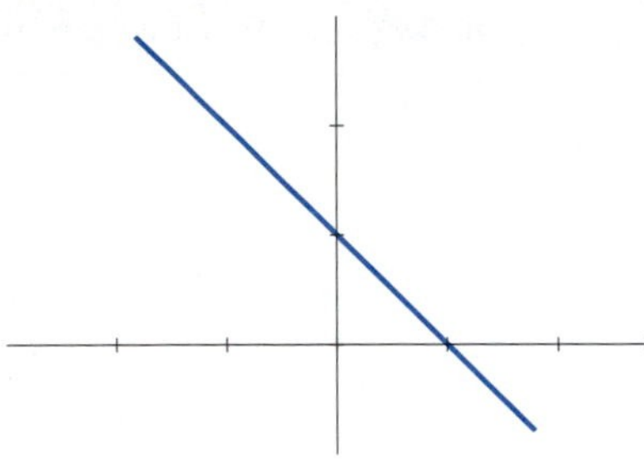

FIGURE 12.1.8

is sketched in Figure 12.1.8. We could graph this equation either by plotting lots of points, or by noticing that

$$r = \frac{1}{\cos\theta + \sin\theta}$$

implies $r\cos\theta + r\sin\theta = 1$. The polar equation $r\cos\theta + r\sin\theta = 1$ corresponds to the rectangular equation $x + y = 1$, which is the equation of a line. If we recognize the original polar equation is a line, we can determine the graph by plotting any two polar points. For example, $\theta = 0$ implies

$$r = \frac{1}{\cos 0 + \sin 0} = 1,$$

so $(r, \theta) = (1, 0)$ is a polar point on the graph. Similarly, we can determine that the polar point $(r, \theta) = (1, \pi/2)$ is on the graph. It is not necessary to change from polar to rectangular coordinates to sketch the graph. ■

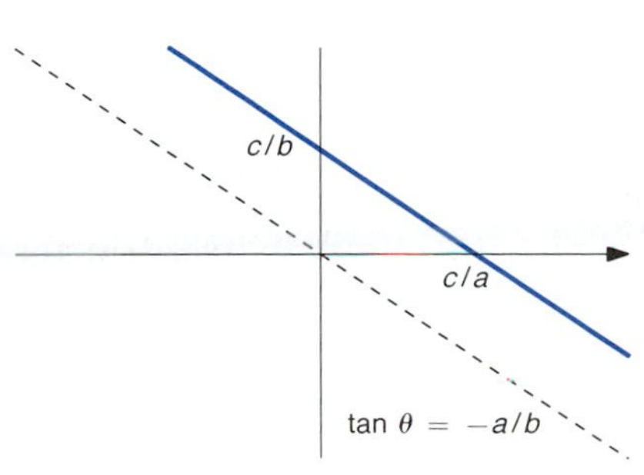


FIGURE 12.1.9

**The graph of $r = \dfrac{c}{a\cos\theta + b\sin\theta}$ is a line.**

We see that $r \to \pm\infty$ as $a\cos\theta + b\sin\theta \to 0$ and that there is no point on the graph with $a\cos\theta + b\sin\theta = 0$. Tan $\theta = -a/b$ is the slope of the line. See Figure 12.1.9.

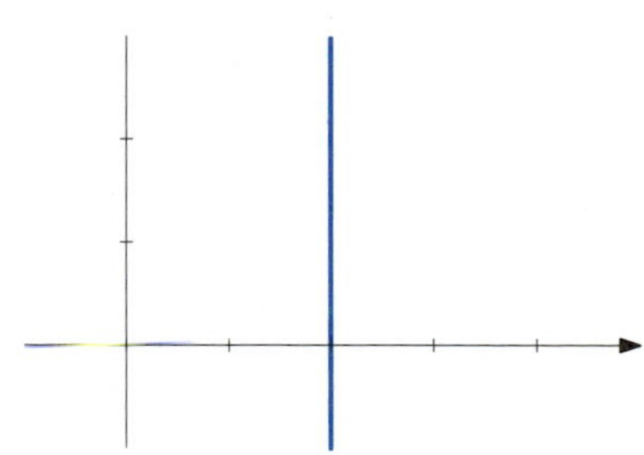

FIGURE 12.1.10

■ **EXAMPLE 7**

The graph of $r = 2\sec\theta$ is given in Figure 12.1.10. We can rewrite the equation $r = 2\sec\theta$ as $r = 2/\cos\theta$. This is the polar form of the equation of a line, as we can see by substituting $a = 1$, $b = 0$, and $c = 2$ in the above general equation. By writing $r\cos\theta = 2$, we see that the rectangular form of the equation is $x = 2$. ■

■ **EXAMPLE 8**

The graph of $r = 2\sin\theta$ is given in Figure 12.1.11. We could plot points or recognize that $r = 2\sin\theta$ implies $r^2 = 2r\sin\theta$. Changing to rectangular coordinates, we then obtain $x^2 + y^2 = 2y$. Completing the square then gives $x^2 + (y - 1)^2 = 1$. ■

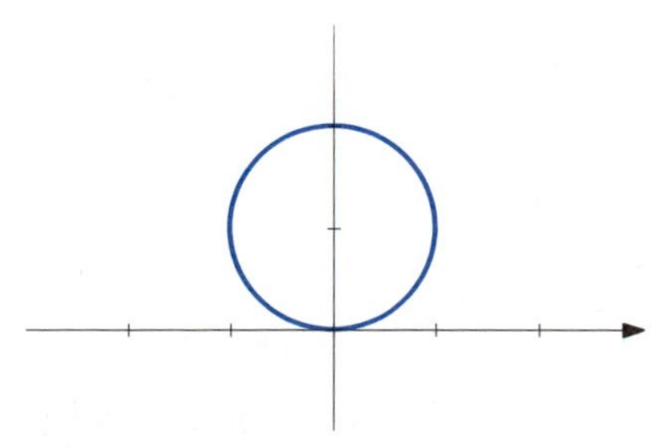

FIGURE 12.1.11

**The graph of $r = 2a\cos\theta$ is a circle through the pole with center on the polar axis. The graph of $r = 2a\sin\theta$ is a circle through the pole with center on the line $\theta = \pm\pi/2$.**

Let us reconsider Example 8 and see how to obtain the graph of $r = 2\sin\theta$ without changing to rectangular coordinates. The idea is to plot a few key points and then draw a smooth curve through the points, in the direction of increasing $\theta$. Coordinates of some key points are given in Table 12.1.1.

TABLE 12.1.1

| $\theta$ | $0$ | $\pi/2$ | $\pi$ | $3\pi/2$ | $2\pi$ |
|---|---|---|---|---|---|
| $r = 2 \sin \theta$ | $0$ | $2$ | $0$ | $-2$ | $0$ |

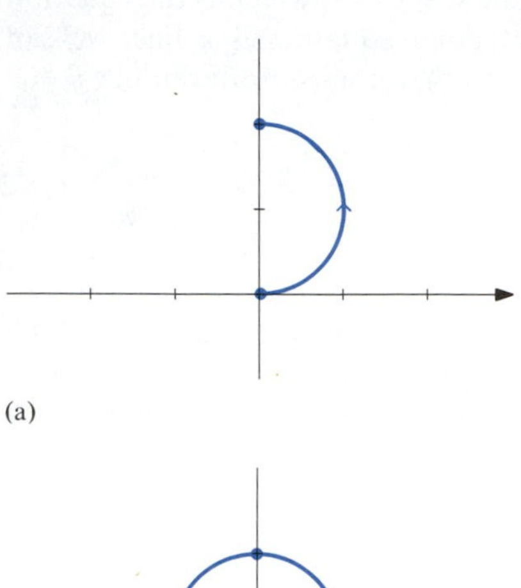

(a)

(b)

FIGURE 12.1.12

Each of the points determined in the table is plotted. We then note that $r = 0$ when $\theta = 0$, and the value of $r$ increases to 2 as $\theta$ increases from 0 to $\pi/2$. We use this information to sketch that part of the curve that corresponds to $0 \leq \theta \leq \pi/2$. The curve should be traced in the direction of increasing $\theta$, from the pole to the point $(2, \pi/2)$. This step is illustrated in Figure 12.1.12a.

Continuing, we note that $r$ decreases from 2 to zero as $\theta$ ranges from $\pi/2$ to $\pi$. This means that the graph goes from the point $(2, \pi/2)$ to the pole as $\theta$ goes from $\pi/2$ to $\pi$. That part of the curve that corresponds to $\pi/2 \leq \theta \leq \pi$ has been added in Figure 12.1.12b. Again, the curve is traced in the direction of increasing $\theta$. As $\theta$ varies from $\pi$ to $2\pi$, $r$ is negative and the graph retraces the circle. Finally, we note that if we know the graph of $r = 2 \sin \theta$ is a circle through the pole, we can use the above method to obtain a very accurate sketch. Also, note that the point $(2, \pi/2)$ then determines the orientation and scale of the circle.

As usual, we will try to sketch graphs by using general characteristics and key points. Key points of polar graphs should include points with $\theta = 0, \pi/2, \pi$, and $3\pi/2$. Values of $\theta$ for which either $r = 0$ or $r$ is undefined should also be noted. The sign of $r$ is an important characteristic. Use the sign of $r$ to *trace the curve in the direction of increasing* $\theta$, between the key points that have been plotted. Of course, sketching any graph is easier if you know the general shape in advance, so you need only to determine orientation and scale. Always check the equation to see if it is a familiar type.

**■ EXAMPLE 9**

Sketch the graph of the polar equations

(a) $r = \dfrac{6}{2 - \sin \theta}$,

(b) $r = \dfrac{2}{1 + \cos \theta}$,

(c) $r = \dfrac{3}{1 - 2 \sin \theta}$.

**SOLUTION**

(a) Table 12.1.2 contains some key values of $r$. The corresponding points are plotted and a smooth curve is then drawn through them, as in

TABLE 12.1.2

| $\theta$ | 0 | $\pi/2$ | $\pi$ | $3\pi/2$ | $2\pi$ |
|---|---|---|---|---|---|
| $r = 6/(2 - \sin \theta)$ | 3 | 6 | 3 | 2 | 3 |

TABLE 12.1.3

| $\theta$ | 0 | $\pi/2$ | $\pi$ | $3\pi/2$ | $2\pi$ |
|---|---|---|---|---|---|
| $r = 2/(1 + \cos \theta)$ | 1 | 2 | * | 2 | 1 |

TABLE 12.1.4

| $\theta$ | 0 | $\pi/6$ | $\pi/2$ | $5\pi/6$ | $\pi$ | $3\pi/2$ | $2\pi$ |
|---|---|---|---|---|---|---|---|
| $r = 3/(1 - 2 \sin \theta)$ | 3 | * | $-3$ | * | 3 | 1 | 3 |

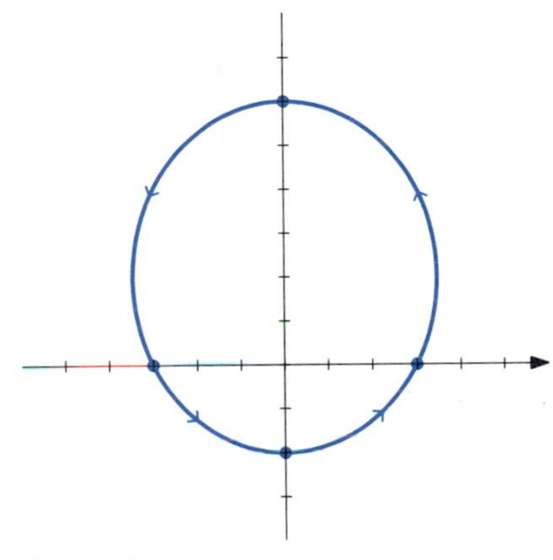

FIGURE 12.1.13

Figure 12.1.13. Starting at the point $(r, \theta) = (3, 0)$, the curve is traced in the direction of increasing $\theta$, as indicated by the arrows. We have used the fact that $r$ is always positive.

(b) Key values of $r$ are given in Table 12.1.3. The key points are plotted and that part of the graph that corresponds to $0 \leq \theta < \pi$ is sketched in Figure 12.1.14a. Note that $r$ approaches infinity as $\theta$ approaches $\pi$ from below; $r$ is undefined when $\theta = \pi$. As $\theta$ increases from $\pi$ to $2\pi$, the values of $r$ decrease from infinity to one. That part of the graph has been added in Figure 12.1.14b.

(c) In this example, we see that $r = 3/(1 - 2 \sin \theta)$ becomes unbounded as $\theta$ approaches $\pi/6$ and $5\pi/6$, where $1 - 2 \sin \theta = 0$; $r$ is undefined at these values of $\theta$. We add these values of $\theta$ to our list of key points in Table 12.1.4.

FIGURE 12.1.14

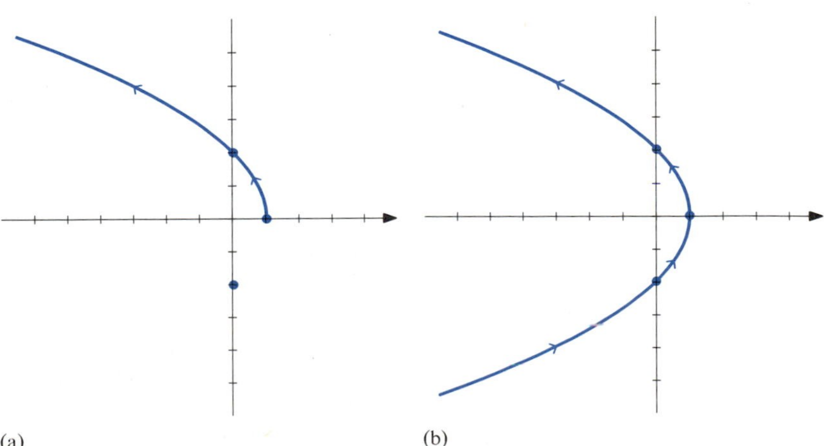

As $\theta$ varies from 0 to $\pi/6$, $r$ increases from 3 to infinity. This part of the graph is sketched in Figure 12.1.15a. As $\theta$ varies from $\pi/6$ to $5\pi/6$, we see that $r$ is negative. This means that the part of the graph that corresponds to $\pi/6 < \theta < 5\pi/6$ will be drawn in the sector opposite to that between $\pi/6$ and $5\pi/6$. As $\theta$ increases from $\pi/6$, $r$ increases from negative infinity to $-3$ when $\theta = \pi/2$, and then decreases to negative infinity as $\theta$ approaches $5\pi/6$ from below. This gives the lower part of the graph that has been added in Figure 12.1.15b. As $\theta$ increases from $5\pi/6$, $r$ decreases from infinity to 3 when $\theta = \pi$, continues decreasing to 1 when $\theta = 3\pi/2$, and then increases to 3 as $\theta$ approaches $2\pi$. This part of the graph has been added in Figure 12.1.15c. ■

(a)

You may suspect from looking at the sketches that the graphs of the equations in Example 9 are conics. They are. For example,

$$r = \frac{2}{1 + \cos\theta} \quad \text{implies}$$

$$r + r\cos\theta = 2,$$

$$r = 2 - r\cos\theta,$$

$$r^2 = (2 - r\cos\theta)^2,$$

$$x^2 + y^2 = (2 - x)^2,$$

$$x^2 + y^2 = 4 - 4x + x^2,$$

$$y^2 = 4 - 4x, \quad \text{a parabola.}$$

(b)

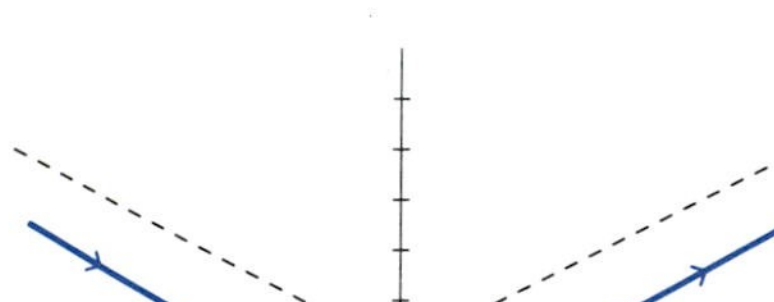

**The graphs of**

$$r = \frac{c}{1 \pm e\sin\theta} \quad \text{and} \quad r = \frac{c}{1 \pm e\cos\theta}$$

**are conics with a focus at the pole and axis on one of the coordinate axes.**

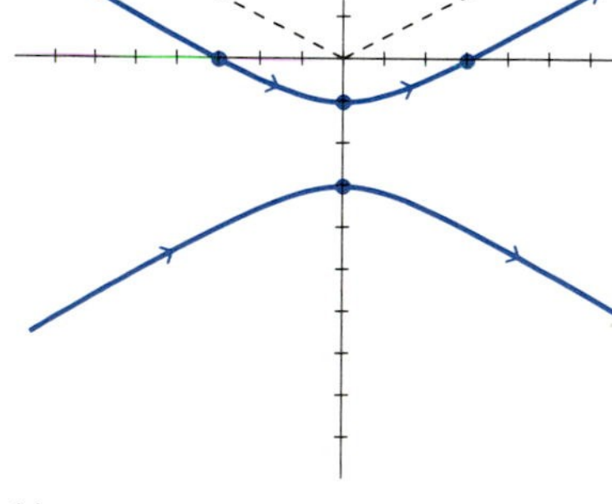

(c)

FIGURE 12.1.15

The graph is an ellipse with eccentricity $e$ if $0 < e < 1$, a parabola if $e = 1$, and a hyperbola with eccentricity $e$ if $e > 1$. It is a circle if $e = 0$.

If we know in advance that the graph is a conic, then the pattern made by the points on the coordinate axes tell which conic it is, as well as give the orientation and scale.

We have seen how to change polar equations of lines and certain circles and conics to rectangular equations. The general idea in changing from polar to rectangular coordinates is to manipulate the polar equation so that it contains only variable expressions of the form $r^2$, $r\cos\theta$, and $r\sin\theta$. The procedure differed in each of the cases studied. We should not expect to be able to change every polar equation to a rectangular equation, although this may be possible with great effort. On the other hand, the equations $x = r\cos\theta$ and $y = r\sin\theta$ can be used to change easily any rectangular equation to a polar equation.

# EXERCISES 12.1

Plot the points with the given polar coordinates in Exercises 1-2.

1 $A(1, 0)$, $B(2, \pi/2)$, $C(-\sqrt{2}, \pi/4)$, $D(-2, 3\pi/2)$, $E(1, 1)$

2 $A(2, 0)$, $B(1, \pi/2)$, $C(-2, \pi)$, $D(2, 7\pi/6)$, $E(\pi/2, 2)$

Find the rectangular coordinates of the points with the given polar coordinates in Exercises 3-4.

3 $A(-1, \pi)$, $B(2, \pi/3)$, $C(3, \pi/2)$, $D(2, 3\pi/4)$

4 $A(-2, 0)$, $B(\sqrt{2}, \pi/4)$, $C(2, -\pi/6)$, $D(-3, \pi/2)$

Find two sets of polar coordinates of the points with the rectangular coordinates given in Exercises 5-6.

5 $A(1, 0)$, $B(0, 2)$, $C(-1, \sqrt{3})$, $D(2, -2)$

6 $A(2, 0)$, $B(0, 1)$, $C(-\sqrt{2}, \sqrt{2})$, $D(-\sqrt{3}, 1)$

Change the polar equations in Exercises 7-14 to equivalent equations in rectangular coordinates.

7 $r = 2$

8 $\theta = \dfrac{\pi}{3}$

9 $r = \sec\theta$

10 $r = \dfrac{6}{2\cos\theta + 3\sin\theta}$

11 $r = 4\cos\theta$

12 $r = \dfrac{6}{2 + \cos\theta}$

13 $r = \dfrac{2}{1 - \sin\theta}$

14 $r = \dfrac{3}{1 + 2\cos\theta}$

Change the rectangular equations given in Exercises 15-22 to polar equations.

15 $x^2 + y^2 = 9$

16 $y + x = 0$

17 $y = 3$

18 $y = 2x - 1$

19 $x^2 + 6x + y^2 = 0$

20 $x^2 + y^2 - 4y = 0$

21 $y^2 = 4x + 4$

22 $16x^2 - 96x + 25y^2 = 256$

Identify and sketch the graph of the polar equations in Exercises 23-40.

23 $r = 3$

24 $r = -2$

25 $\theta = \pi/2$

26 $\theta = \pi/3$

27 $r = \dfrac{2}{\cos\theta + 2\sin\theta}$

28 $r = \dfrac{2}{2\cos\theta - \sin\theta}$

29 $r = \csc\theta$

30 $r = -\sec\theta$

31 $r = \sin\theta$

32 $r = \cos\theta$

33 $r = -3\cos\theta$

34 $r = -2\sin\theta$

35 $r = \dfrac{3}{1 + \sin\theta}$

36 $r = \dfrac{6}{2 - \sin\theta}$

37 $r = \dfrac{6}{2 + \cos\theta}$

38 $r = \dfrac{3}{1 - \cos\theta}$

39 $r = \dfrac{2}{1 - 2\sin\theta}$

40 $r = \dfrac{2}{1 + 2\cos\theta}$

41 Find a formula for the distance between the points with polar coordinates $(r_0, \theta_0)$ and $(r_1, \theta_1)$.

42 Show that the equation $r = a\sin\theta + b\cos\theta$ is the polar equation of a circle if $(a, b) \neq (0, 0)$.

43 Find a polar equation of points that satisfy the condition that the distance of the point from the pole is equal to the distance of the point from the line $x = -d$, $d \neq 0$.

44 Find a polar equation of points that satisfy the condition that the distance of the point from the pole is twice the distance of the point from the line $x = -d$, $d \neq 0$.

45 Find a polar equation of points that satisfy the condition that the distance of the point from the pole is half the distance of the point from the line $x = -d$, $d \neq 0$.

46 Verify that

$$r = \frac{c}{1 + e\sin\theta}$$

is an ellipse with one focus at the pole and eccentricity $e$ if $0 < e < 1$.

47 Verify that

$$r = \frac{c}{1 - e\cos\theta}$$

is a hyperbola with one focus at the pole and eccentricity $e$ if $e > 1$.

48 Verify that

$$r = \frac{c}{1 - \sin\theta}$$

is a parabola with focus at the pole.

## 12.2 GRAPHS OF POLAR EQUATIONS

Section 12.1 was intended to increase confidence in handling polar equations by practicing with familiar curves. In this section we will look at some curves that are more convenient to study in polar coordinates than in rectangular coordinates. We will study graphs of polar equations of the form $r = r(\theta)$, where $r$ is a continuous function of $\theta$. Analysis of the graph will be based on the sign of $r(\theta)$. [This corresponds to graphing a rectangular equation $y = f(x)$ by using the sign of $f(x)$, as was done in Section 4.1.]

The following result is very useful for sketching the graphs of polar equations.

*Theorem*

**Suppose that $r$ is a continuous function of $\theta$ and $r(\theta_0) = 0$, but $r(\theta) \neq 0$ for $\theta$ near $\theta_0$, $\theta \neq \theta_0$. Then the graph of the polar equation $r = r(\theta)$ approaches and leaves the pole at the angle of the line $\theta = \theta_0$ as $\theta$ passes through $\theta_0$. That is, the curve is tangent to the line $\theta = \theta_0$ at the pole.**

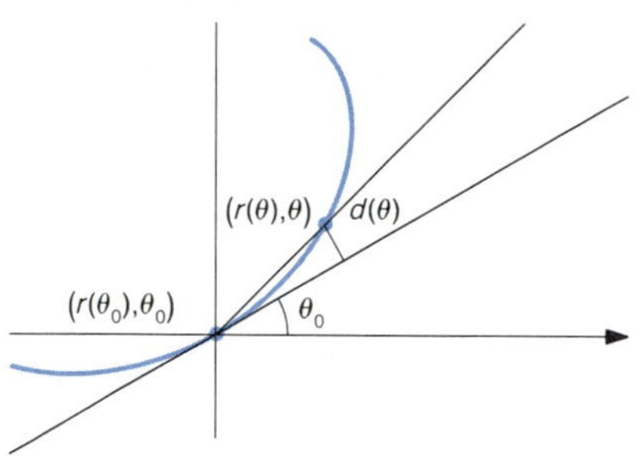


FIGURE 12.2.1

■ *Proof* We first note that $r(\theta) \to r(\theta_0) = 0$ as $\theta \to \theta_0$ because $r(\theta)$ is continuous. This implies the point $(r(\theta), \theta)$ approaches the pole as $\theta$ approaches $\theta_0$. In Figure 12.2.1 we have set $d(\theta)$ equal to the distance from the point $(r(\theta), \theta)$ to the line $\theta = \theta_0$. The distance of the point from the pole is $r(\theta)$. To show that the point approaches the line tangentially, we must verify that the ratio $d(\theta)/r(\theta) \to 0$ as $\theta \to \theta_0$. From Figure 12.2.1 we see that

$$\sin(\theta - \theta_0) = \frac{d(\theta)}{r(\theta)}.$$

Since sine is a continuous function, $\sin(\theta - \theta_0) \to 0$ as $\theta \to \theta_0$. We conclude that

$$\lim_{\theta \to \theta_0} \frac{d(\theta)}{r(\theta)} = 0,$$

so the graph is tangent to the line $\theta = \theta_0$ as $\theta$ approaches $\theta_0$. ■

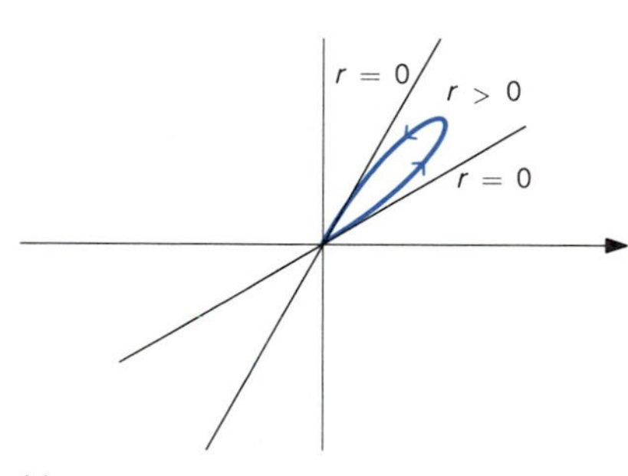


(a)



(b)

FIGURE 12.2.2

A continuous function $r(\theta)$ does not change sign between successive zeros. The graph of $r = r(\theta)$ will make a **positive loop** between successive zeros of $r$ where $r(\theta)$ is positive. This is illustrated in Figure 12.2.2a. The graph will make a **negative loop** between successive zeros of $r$ where $r(\theta)$ is negative. See Figure 12.2.2b. Note that a negative loop is drawn in the sector opposite to that of the corresponding $\theta$.

When sketching the graph of a polar equation, we should determine those values of $\theta_0$ for which $r(\theta_0) = 0$ and lightly draw the lines $\theta = \theta_0$ through the pole. This marks the angles at which the graph approaches the pole. By determining the sign of $r(\theta)$ between successive zeros, we can determine whether to sketch a positive loop or a negative loop.

■ **EXAMPLE 1**

Sketch the graph of the polar equation $r = 1 - \cos\theta$.

**SOLUTION**

We determine key values of $r(\theta)$ from key values of $\cos\theta$. See Table 12.2.1.

**TABLE 12.2.1**

| $\theta$ | $0$ | $\pi/2$ | $\pi$ | $3\pi/2$ | $2\pi$ |
|---|---|---|---|---|---|
| $r$ | $0$ | $1$ | $2$ | $1$ | $0$ |

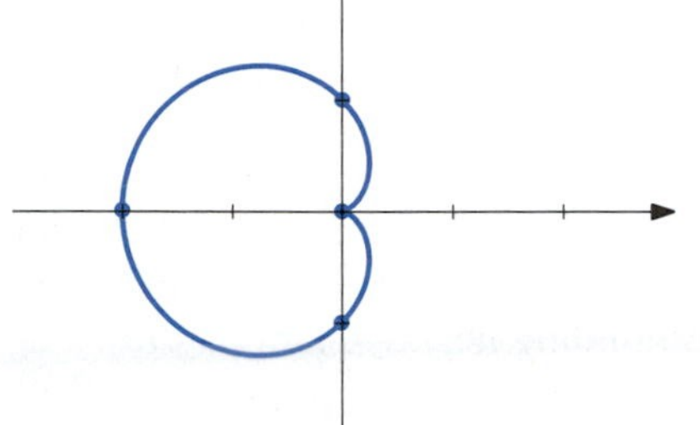

**FIGURE 12.2.3**

We see that $r(0) = r(2\pi) = 0$ and that $r(\theta)$ is positive for $0 < \theta < 2\pi$. The graph makes a positive loop between $\theta = 0$ and $\theta = 2\pi$. The graph is sketched in Figure 12.2.3. Note the graph leaves the pole tangent to the line $\theta = 0$ as $\theta$ increases from 0. The graph approaches the pole tangent to the line $\theta = 2\pi$ as $\theta$ increases toward $2\pi$. ■

■ **EXAMPLE 2**

Sketch the graph of the polar equation $r = 1 + 2\sin\theta$.

**SOLUTION**

The equation $r = 0$ implies $1 + 2\sin\theta = 0$, or $\sin\theta = -1/2$. The solutions $\theta = 7\pi/6$ and $\theta = 11\pi/6$ are included in our table of key values. See Table 12.2.2.

**TABLE 12.2.2**

| $\theta$ | $0$ | $\pi/2$ | $\pi$ | $7\pi/6$ | $3\pi/2$ | $11\pi/6$ | $2\pi$ |
|---|---|---|---|---|---|---|---|
| $r$ | $1$ | $3$ | $1$ | $0$ | $-1$ | $0$ | $1$ |

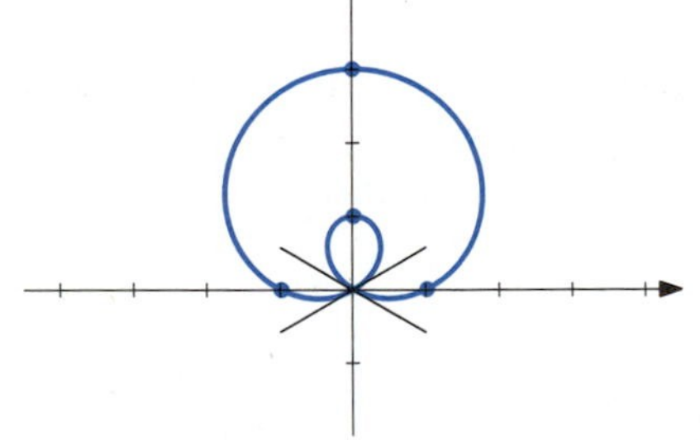

**FIGURE 12.2.4**

The graph is sketched in Figure 12.2.4. Note that the graph passes through the pole at the angle of the lines $\theta = 7\pi/6$ and $\theta = 11\pi/6$. The graph makes a negative loop between $\theta = 7\pi/6$ and $\theta = 11\pi/6$. This is the smaller, inner loop. Note that the line $\theta = 11\pi/6$ is the same as the line $\theta = -\pi/6$. We can then describe the larger, outer loop as a positive loop between $\theta = -\pi/6$ and $\theta = 7\pi/6$. It is important for the integration problems that we will do in later sections that loops be described in terms of increasing $\theta$. ■

The graph of an equation of the form either

$$r = a + b\sin\theta \quad \text{or} \quad r = a + b\cos\theta$$

is called a **limaçon**. If $|a| = |b|$, the graph is also called a **cardioid**. The graph will have an inner loop if $|b| > |a|$. The case $|a| > |b|$ requires more detailed analysis than we are using in this section. We will study this case in Section 12.3.

**TABLE 12.2.3**

| $\theta$ | 0 | $\pi/4$ | $\pi/2$ | $3\pi/4$ | $\pi$ | $5\pi/4$ | $3\pi/2$ | $7\pi/4$ | $2\pi$ |
|---|---|---|---|---|---|---|---|---|---|
| $r$ | 0 | 2 | 0 | $-2$ | 0 | 2 | 0 | $-2$ | 0 |

$0 \leq \theta \leq \pi$

(a)

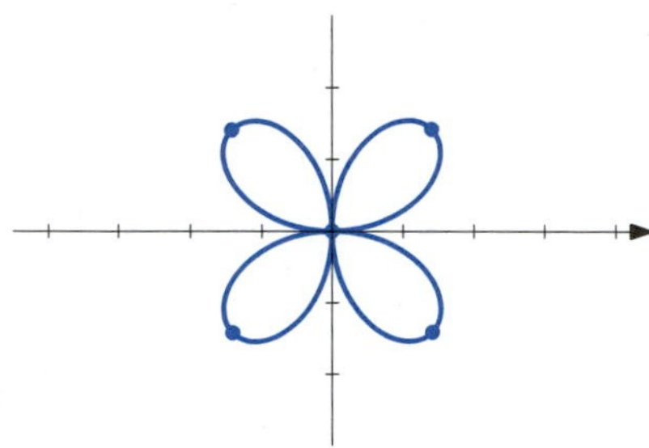

(b)

**FIGURE 12.2.5**

### ■ EXAMPLE 3

Sketch the graph of the polar equation $r = 2 \sin 2\theta$.

**SOLUTION**

Table 12.2.3 includes the values of $\theta$ between 0 and $2\pi$ where $\sin 2\theta$ is 0, 1, and $-1$. We see that values of $r(\theta)$ repeat in intervals of length $\pi$.

The graph approaches the pole at the angle of each of the coordinate axes. The graph makes a positive loop between $\theta = 0$ and $\theta = \pi/2$, a negative between $\theta = \pi/2$ and $\theta = \pi$, a positive loop between $\theta = \pi$ and $\theta = 3\pi/2$, and then a negative loop between $\theta = 3\pi/2$ and $\theta = 2\pi$. In Figure 12.2.5a we have sketched the graph as $\theta$ varies between 0 and $\pi$. As $\theta$ varies between $\pi/2$ and $\pi$, the corresponding negative loop is drawn in the opposite sector, the fourth quadrant. The complete graph is sketched in Figure 12.2.5b. ■

### ■ EXAMPLE 4

Sketch the graph of the polar equation $r = \cos 3\theta$.

**SOLUTION**

Values of $r(\theta)$ repeat in intervals of length $2\pi/3$. Table 12.2.4 includes values where $r = 0$, 1, or $-1$, for $\theta$ between 0 and $\pi$.

**TABLE 12.2.4**

| $\theta$ | 0 | $\pi/6$ | $\pi/3$ | $\pi/2$ | $2\pi/3$ | $5\pi/6$ | $\pi$ |
|---|---|---|---|---|---|---|---|
| $r$ | 1 | 0 | $-1$ | 0 | 1 | 0 | $-1$ |

The graph completes part of a positive loop as $\theta$ varies between 0 and $\pi/6$, and makes a negative loop between $\theta = \pi/6$ and $\theta = \pi/2$. This part of the graph is sketched in Figure 12.2.6a. The graph makes a positive loop

**FIGURE 12.2.6**

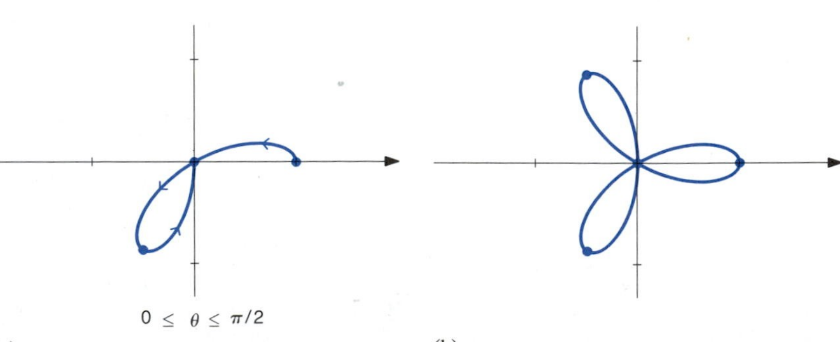

$0 \leq \theta \leq \pi/2$

(a) (b)

between $\theta = \pi/2$ and $\theta = 5\pi/6$, and makes half of a negative loop between $\theta = 5\pi/6$ and $\theta = \pi$. Note that this completes the graph, as illustrated in Figure 12.2.6b. The graph is retraced as $\theta$ varies between $\pi$ and $2\pi$. ■

Graphs of equations of the form either

$$r = a \sin n\theta \quad \text{or} \quad r = a \cos n\theta,$$

$n$ an integer, are called **rose curves**. The zeros divide $[0, 2\pi]$ into $2n$ intervals of length $\pi/n$. The graph alternates positive and negative loops as $\theta$ varies over these intervals. There will be $2n$ **leaves** if $n$ is even and each leaf is traversed once as $\theta$ increases from 0 to $2\pi$. If $n$ is odd, there will be $n$ leaves and each leaf is traversed twice as $\theta$ increases from 0 to $2\pi$.

■ **EXAMPLE 5**

Sketch the graph of the polar equation $r^2 = 2 \sin 2\theta$.

**SOLUTION**

Note that $r^2 = 2 \sin 2\theta$ implies either

$$r = +\sqrt{2 \sin 2\theta} \quad \text{or} \quad r = -\sqrt{2 \sin 2\theta}.$$

Since these two polar equations have the same graphs, let us restrict our attention to $r = \sqrt{2 \sin 2\theta}$. Values of $r(\theta)$ repeat in intervals of $\pi$. Table 12.2.5 contains key values of $r(\theta)$ for $\theta$ between 0 and $\pi$.

TABLE 12.2.5

| $\theta$ | 0 | $\pi/4$ | $\pi/2$ | $3\pi/4$ | $\pi$ |
|---|---|---|---|---|---|
| $r$ | 0 | $\sqrt{2}$ | 0 | * | 0 |

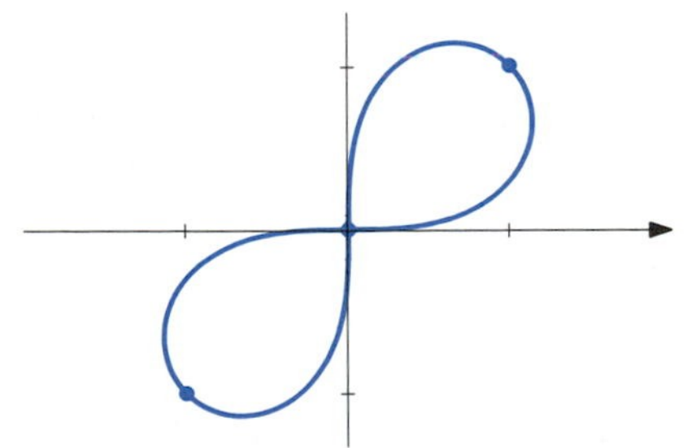

FIGURE 12.2.7

The graph is sketched in Figure 12.2.7. The loop in the first quadrant is formed by $r = \sqrt{2 \sin 2\theta}$ as $r$ varies between 0 and $\pi/2$. The loop in the third quadrant is formed by $r = \sqrt{2 \sin 2\theta}$ as $\theta$ varies between $\pi$ and $3\pi/2$. (The loops would simply be interchanged if we had used $r = -\sqrt{2 \sin 2\theta}$.) There are no points on the graph that correspond to values of $\theta$ between either $\pi/2$ and $\pi$ or $3\pi/2$ and $2\pi$, because $\sin 2\theta$ is negative for those values of $\theta$ and there is no real number $r$ that satisfies $r^2 < 0$. ■

The graph of an equation of the form either

$$r^2 = a \sin 2\theta \quad \text{or} \quad r^2 = a \cos 2\theta$$

is called a **lemniscate**.

■ **EXAMPLE 6**

Sketch the graph of the polar equation $r = \theta/2$, $\theta \geq 0$.

TABLE 12.2.6

| $\theta$ | 0 | $\pi/2$ | $\pi$ | $3\pi/2$ | $2\pi$ |
|---|---|---|---|---|---|
| $r$ | 0 | $\pi/4$ | $\pi/2$ | $3\pi/4$ | $\pi$ |

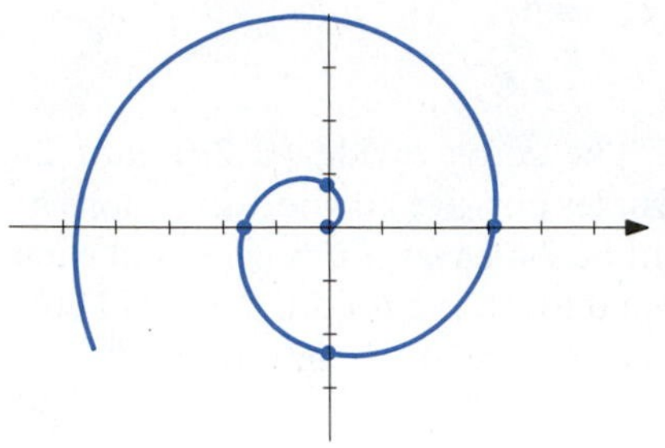

FIGURE 12.2.8

**SOLUTION**

Some key points of the graph are given in Table 12.2.6.

We see that the graph leaves the pole tangent to the line $\theta = 0$ and that $r(\theta)$ increases as $\theta$ increases. The graph is sketched in Figure 12.2.8. ■

Graphs of equations of the form

$$r = a\theta$$

are called **spirals**.

## EXERCISES 12.2

Sketch the graphs of the polar equations given in Exercises 1–28.

1 $r = 1 + \cos\theta$

2 $r = 2 - 2\cos\theta$

3 $r = 2 - 2\sin\theta$

4 $r = 1 + \sin\theta$

5 $r = 1 - 2\cos\theta$

6 $r = 1 + 2\cos\theta$

7 $r = 1 - \sqrt{2}\sin\theta$

8 $r = \sqrt{3} + 2\sin\theta$

9 $r = 2\cos 2\theta$

10 $r = \sin 2\theta$

11 $r = -\cos 3\theta$

12 $r = -\sin 3\theta$

13 $r = \sin 4\theta$

14 $r = \cos 4\theta$

15 $r^2 = 4\cos 2\theta$

16 $r^2 = \sin 2\theta$

17 $r^2 = -9\sin 2\theta$

18 $r^2 = \cos 2\theta$

19 $r = \theta/\pi,\ \theta \geq 0$

20 $r = \theta/4,\ \theta \geq 0$

21 $r^2 = \sin 3\theta$

22 $r^2 = \sin 4\theta$

23 $r = \sin(\theta/2)$

24 $r = \sin(\theta/3)$

25 $r^2 = \theta$

26 $r = \cos^2\theta$

27 $r = \begin{cases} \sin(4\theta/3),\ 0 \leq \theta \leq 3\pi/4 \\ \sin(4\theta/3 - 4\pi/3),\ \pi \leq \theta \leq 7\pi/4 \\ \sin(4\theta),\ 3\pi/4 \leq \theta \leq \pi,\ 7\pi/4 \leq \theta \leq 2\pi \end{cases}$

28 $r = \ln\theta$

29 Find polar coordinates of the point nearest the origin where the graph of $r = \ln\theta$ intersects itself.

## 12.3 LINES TANGENT TO POLAR CURVES

We have seen that the line $\theta = \theta_0$ is tangent to the graph of the polar curve $r = r(\theta)$ at the pole if $r(\theta_0) = 0$. In this section we will investigate lines tangent to polar curves at points other than the pole.

If $r(\theta)$ is differentiable at $\theta_0$ and $r(\theta_0) \neq 0$, it should seem reasonable that the graph of $r = r(\theta)$ has a tangent line at the point $(r(\theta_0), \theta_0)$. Let us verify this fact and develop a formula for the slope of the tangent line. We will see that it is convenient to express the slope of a polar curve in terms of

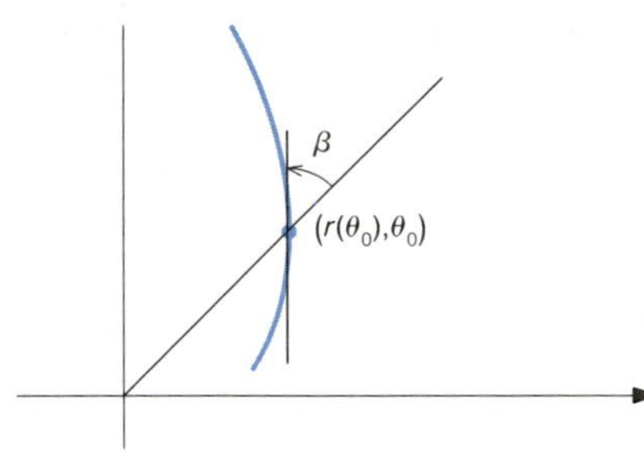

FIGURE 12.3.1

the angle $\beta$ illustrated in Figure 12.3.1; $\beta$ is the angle from the radial line $\theta = \theta_0$ to the line tangent to the graph at $(r(\theta_0), \theta_0)$.

Let $\beta'$ denote the angle from the radial line $\theta = \theta_0$ to the line through the points $(r(\theta_0), \theta_0)$ and $(r(\theta), \theta)$, as illustrated in Figure 12.3.2. Note that the circular arc from $(r(\theta), \theta)$ to $(r(\theta), \theta_0)$ is orthogonal to the radial line $\theta = \theta_0$. Also, if $\theta$ is near $\theta_0$, $\theta \neq \theta_0$, the arc is nearly straight. It follows that

$$\tan \beta' \approx \frac{r(\theta)(\theta - \theta_0)}{r(\theta) - r(\theta_0)} = \frac{r(\theta)}{\dfrac{r(\theta) - r(\theta_0)}{\theta - \theta_0}} \approx \frac{r(\theta_0)}{r'(\theta_0)},$$

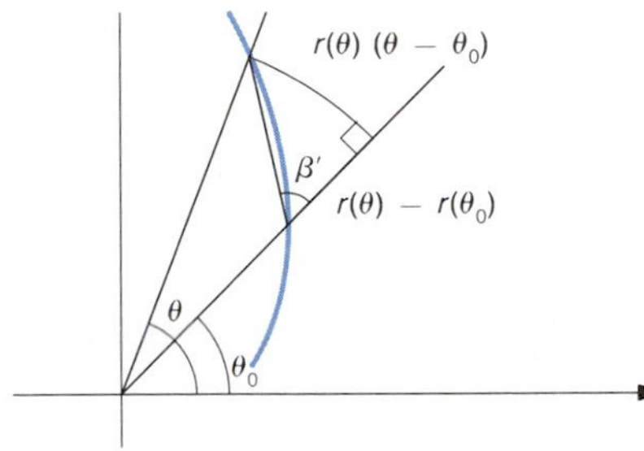

FIGURE 12.3.2

where we have assumed that $r$ is differentiable at $\theta_0$ with $r(\theta_0) \neq 0$ and $r'(\theta_0) \neq 0$. As $\theta$ approaches $\theta_0$, the angles $\beta'$ approach the value $\beta$. We obtain the formula

$$\tan \beta = \frac{r(\theta_0)}{r'(\theta_0)}.$$

If $r(\theta_0) \neq 0$ and $r'(\theta_0) = 0$, we choose $\beta = \pm\pi/2$. In case $r(\theta_0) = 0$, we know the graph approaches and leaves the pole tangent to the line $\theta = \theta_0$, so we choose $\beta = 0$.

We can use values of $\tan \beta$ to improve the accuracy of our graphs.

■ **EXAMPLE 1**

Determine the angle $\beta$ between the graph of $r = 1 - \cos \theta$ and each of the radial lines $\theta = 0, \pi/2, \pi, 3\pi/2$, and $2\pi$. Sketch the graph.

**SOLUTION**

We include the values of $\tan \beta = r/r'$ along with key values of $r$ in Table 12.3.1. The equation $r = 1 - \cos \theta$ implies $r' = \sin \theta$, so we have $\tan \beta = (1 - \cos \theta)/\sin \theta$.

TABLE 12.3.1

| $\theta$ | $0$ | $\pi/2$ | $\pi$ | $3\pi/2$ | $2\pi$ |
|---|---|---|---|---|---|
| $r$ | 0 | 1 | 2 | 1 | 0 |
| $r'$ | 0 | 1 | 0 | $-1$ | 0 |
| $\tan \beta$ | * | 1 | * | $-1$ | * |
| $\beta$ | 0 | $\pi/4$ | $\pm\pi/2$ | $-\pi/4$ | 0 |

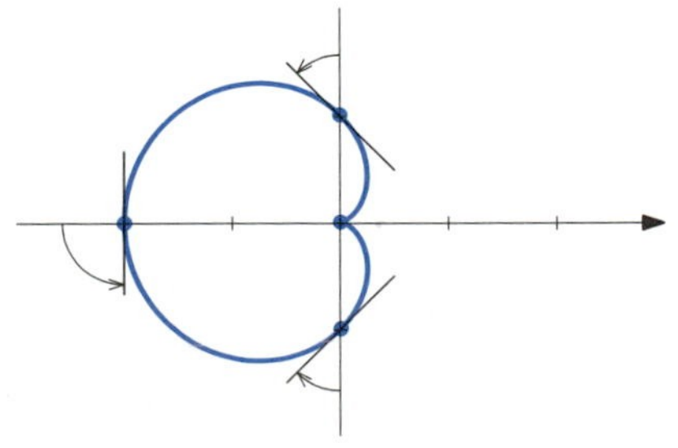
FIGURE 12.3.3

In Figure 12.3.3 we have plotted the key points, drawn the tangents to the graph at the key points, and then sketched the graph through the points at the angle of the tangents. ■

It is not difficult to obtain the polar equation of a line tangent to a polar curve. To do this, let $\beta$ denote the angle associated with the line tangent to the curve $r = r(\theta)$ at the point $(r(\theta_0), \theta_0)$, so $\tan \beta = r(\theta_0)/r'(\theta_0)$. A point $(r, \theta)$ is on the tangent line whenever the angle from the radial line $\theta = \theta_0$

FIGURE 12.3.4

to the line through $(r, \theta)$ and $(r(\theta_0), \theta_0)$ is $\beta$. We can then apply the Law of Sines to the triangle in Figure 12.3.4 to obtain

$$\frac{r}{\sin(\pi - \beta)} = \frac{r(\theta_0)}{\sin(\beta + \theta_0 - \theta)}.$$

Using the trigonometric identity for the sine of a difference to simplify, we obtain the equation

$$\boldsymbol{r = \frac{r(\theta_0)\sin\beta}{\sin(\beta + \theta_0)\cos\theta - \cos(\beta + \theta_0)\sin\theta}}.$$

■ **EXAMPLE 2**

Find a polar equation of the line tangent to $r = 2\cos\theta$ at the point where $\theta = \pi/3$.

**SOLUTION**

We need to determine the angle $\beta$ at the point where $\theta_0 = \pi/3$. We have $r = 2\cos\theta$ and $r' = -2\sin\theta$. When $\theta = \pi/3$, $r = 2(1/2) = 1$ and $r' = -2(\sqrt{3}/2) = -\sqrt{3}$. Then $\tan\beta = r/r' = -1/\sqrt{3}$, so $\beta = -\pi/6$ and $\theta_0 + \beta = \pi/6$. Substitution then gives

$$r = \frac{r(\theta_0)\sin\beta}{\sin(\beta + \theta_0)\cos\theta - \cos(\beta + \theta_0)\sin\theta},$$

$$r = \frac{(1)(\sin(-\pi/3))}{\sin(\pi/6)\cos\theta - \cos(\pi/6)\sin\theta},$$

$$r = \frac{-\sqrt{3}/2}{(1/2)\cos\theta - (\sqrt{3}/2)\sin\theta},$$

$$r = \frac{-\sqrt{3}}{\cos\theta - \sqrt{3}\sin\theta}.$$

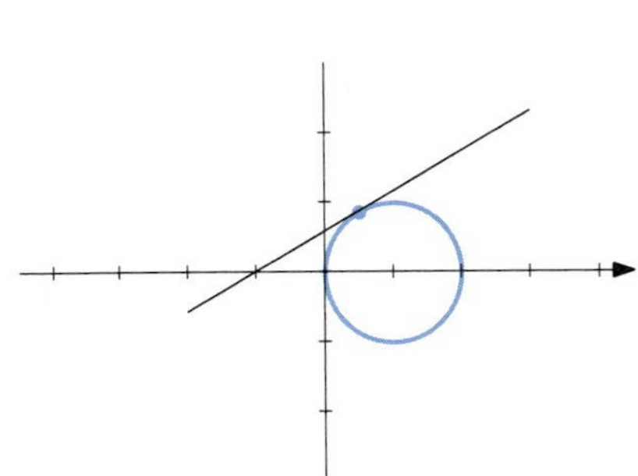

FIGURE 12.3.5

The graph is sketched in Figure 12.3.5. ■

The slope of the line tangent to the polar curve $r = r(\theta)$ at a point $(r(\theta), \theta)$ is $\tan\alpha$, as illustrated in Figure 12.3.6. Since $\alpha = \theta + \beta$, we have

$$\tan\alpha = \tan(\theta + \beta) = \frac{\tan\theta + \tan\beta}{1 - \tan\theta\tan\beta}$$

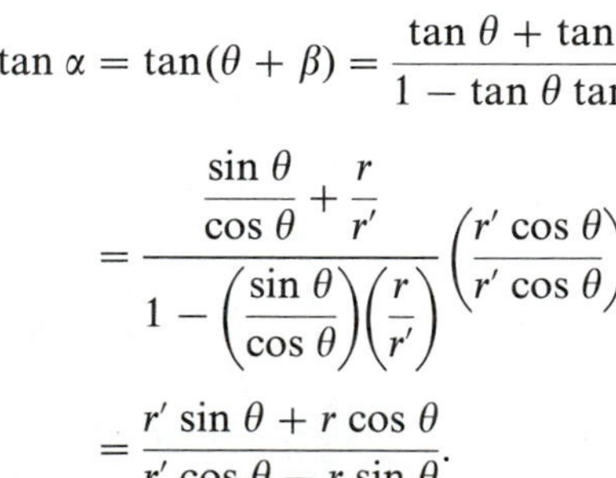

$$= \frac{\dfrac{\sin\theta}{\cos\theta} + \dfrac{r}{r'}}{1 - \left(\dfrac{\sin\theta}{\cos\theta}\right)\left(\dfrac{r}{r'}\right)}\left(\frac{r'\cos\theta}{r'\cos\theta}\right)$$

$$= \frac{r'\sin\theta + r\cos\theta}{r'\cos\theta - r\sin\theta}.$$

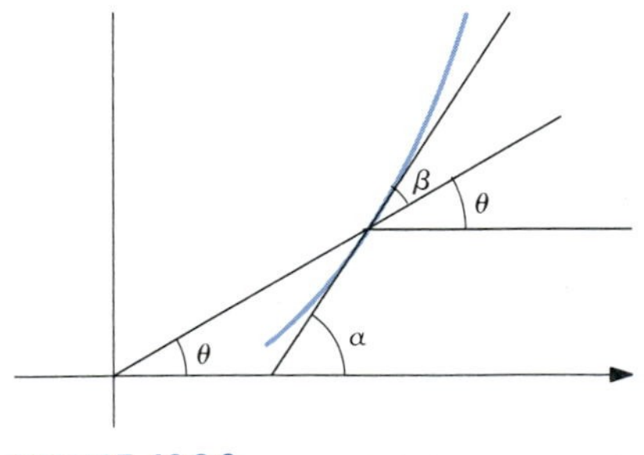


FIGURE 12.3.6

Summarizing, we have

$$\text{Slope} = \frac{r' \sin\theta + r\cos\theta}{r'\cos\theta - r\sin\theta}.$$

There are horizontal tangents where $r' \sin\theta + r\cos\theta = 0$, and vertical tangents where $r'\cos\theta - r\sin\theta = 0$, unless both expressions are zero.

### ■ EXAMPLE 3

Sketch the graph of the polar curve $r = 3 - 2\cos\theta$. Find all points on the graph that have vertical tangent lines.

#### SOLUTION

The graph will have vertical tangents whenever

$$r'\cos\theta - r\sin\theta = 0.$$

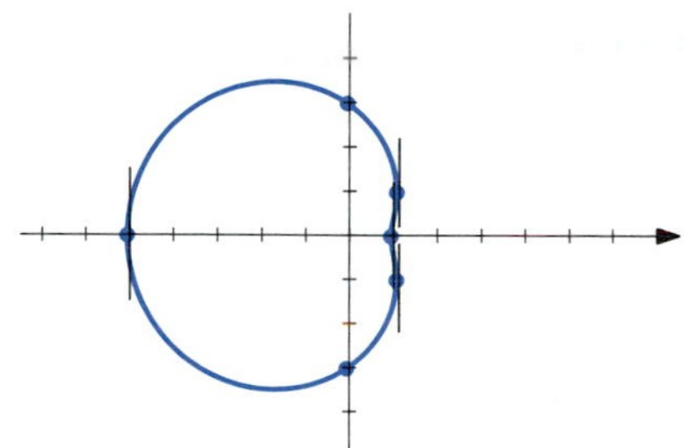

FIGURE 12.3.7

We have $r = 3 - 2\cos\theta$, so $r' = 2\sin\theta$. Substitution then gives

$$(2\sin\theta)(\cos\theta) - (3 - 2\cos\theta)(\sin\theta) = 0,$$

$$(\sin\theta)(4\cos\theta - 3) = 0.$$

We see then that $\sin\theta = 0$ gives the values $\theta = 0$ and $\theta = \pi$, and that $4\cos\theta - 3 = 0$ has solution $\cos^{-1}(3/4)$ in the first quadrant and solution $-\cos^{-1}(3/4)$ in the fourth quadrant. These values of $\theta$ are included in Table 12.3.2.

TABLE 12.3.2

| $\theta$ | $\cos^{-1}(3/4)$ | $0$ | $\pi/2$ | $\pi$ | $3\pi/2$ | $-\cos^{-1}(3/4)$ | $2\pi$ |
|---|---|---|---|---|---|---|---|
| $r$ | $3/2$ | $1$ | $3$ | $5$ | $3$ | $3/2$ | $1$ |

The graph is sketched in Figure 12.3.7. ■

### ■ EXAMPLE 4

Sketch the graph of the polar curve $r = 3 - \cos\theta$. Find all points on the graph that have vertical tangent lines.

#### SOLUTION

The graph will have vertical tangents whenever

$$r'\cos\theta - r\sin\theta = 0.$$

Since $r' = \sin\theta$, substitution gives

$$(\sin\theta)(\cos\theta) - (3 - \cos\theta)(\sin\theta) = 0,$$

$$(\sin\theta)(2\cos\theta - 3) = 0.$$

TABLE 12.3.3

| $\theta$ | 0 | $\pi/2$ | $\pi$ | $3\pi/2$ | $2\pi$ |
|---|---|---|---|---|---|
| $r$ | 2 | 3 | 4 | 3 | 2 |

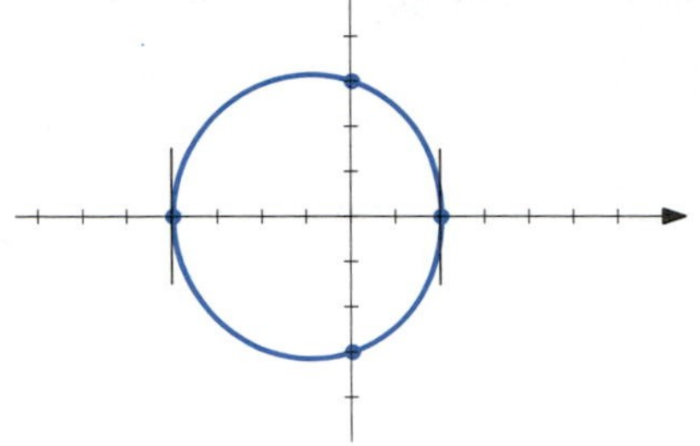

FIGURE 12.3.8

The equation $\sin \theta = 0$ gives the values $\theta = 0$ and $\theta = \pi$, and $2 \cos \theta - 3 = 0$ implies $\cos \theta = 3/2$. This equation has no solution, since the cosine function is bounded by one. Some key values are given in Table 12.3.3. The graph is sketched in Figure 12.3.8. ■

Let us compare the graphs in Figure 12.3.3, 12.3.7, and 12.3.8. Each is the graph of a limaçon, with equation

$$r = a + b \cos \theta, \qquad |a| \geq |b|.$$

It can be shown that the graph has a cusp (sharp point) at the pole when $|a| = |b|$ (Figure 12.3.3). The graph has a smooth indentation at $(a + b, 0)$ for $|b| < |a| < 2|b|$ (Figure 12.3.7). The indentation disappears for $|a| \geq 2|b|$ (Figure 12.3.8). The limaçon has an inner loop if $|a| < |b|$, as we have seen in the previous section.

## EXERCISES 12.3

Determine the angle $\beta$ between the graphs given in Exercises 1-6 and each of the radial lines $\theta = 0, \pi/2, \pi, 3\pi/2$, and $2\pi$. Sketch the graphs.

**1** $r = 1 + \sin \theta$

**2** $r = 1 + 2 \sin \theta$

**3** $r = \dfrac{1}{\cos \theta + 2 \sin \theta}$

**4** $r = \dfrac{6}{2 + \sin \theta}$

**5** $r = \dfrac{2}{1 + \sin \theta}$

**6** $r = \dfrac{3}{1 - 2 \sin \theta}$

Find the polar equation of the line tangent to the given graphs at the point with the given value of $\theta_0$ in Exercises 7-10. Sketch the graphs.

**7** $r = 2, \theta_0 = \pi/3$

**8** $r = 2 \cos \theta, \theta_0 = \pi/6$

**9** $r = \sin 3\theta, \theta_0 = \pi/2$

**10** $r = 2 \cos 2\theta, \theta = 0$

Sketch the graphs in Exercises 11-14. Find all points on the graphs that have vertical tangent lines.

**11** $r = 1 + \cos \theta$

**12** $r = \sqrt{2} + \cos \theta$

**13** $r = 2 + \cos \theta$

**14** $r = 3 + \cos \theta$

Sketch the graphs in Exercises 15-18. Find all points on the graph that have horizontal tangent lines.

**15** $r = 1 + \cos \theta$

**16** $r = 2 \cos \theta$

**17** $r = \dfrac{6}{2 + \sin \theta}$

**18** $r = \dfrac{2}{1 + \sin \theta}$

**19** $r = a - b \cos \theta, a > b > 0$.

(a) Show that the graph has four distinct points with vertical tangent lines whenever $a < 2b$. Sketch the graph.

(b) Show that the graph has only two distinct points with vertical tangent lines whenever $a \geq 2b$. Sketch the graph.

## 12.4 AREA OF REGIONS BOUNDED BY POLAR CURVES

Consider the region in the plane bounded by the polar curves

$$r = r(\theta), \quad \theta = a, \quad \text{and} \quad \theta = b.$$

We assume that $r$ is continuous for $a \leq \theta \leq b$. To find the area of the region, we divide it into sectors by slicing it along radial lines $\theta = \theta_j$ from the pole to the curve $r = r(\theta)$, as illustrated in Figure 12.4.1. The area of each sector is approximately $(1/2)r^2\Delta\theta$, the area of a sector of a circle with radius $r$ and central angle $\Delta\theta$. See Figure 12.4.2. The area of the region should be the limit of sums of the areas of the sectors. That is,

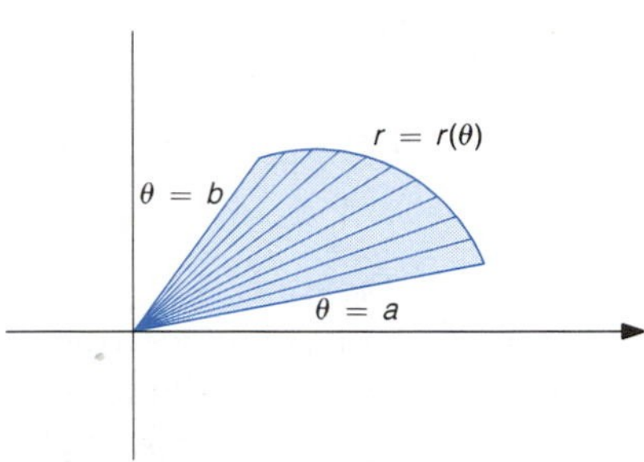

FIGURE 12.4.1

$$\textbf{Area} = \int_a^b \frac{1}{2} [r(\theta)]^2 \, d\theta.$$

Be careful to use the above area formula only for regions to which it applies. The integral represents a limit of sums of the areas of sectors. To apply to a region, radial lines must divide the region into sectors. For any fixed $\theta$ between $a$ and a *larger* number $b$, radial lines at angle $\theta$ must intersect the region from the pole to the point $(r(\theta), \theta)$ on the boundary curve. We also note that, since the area of a sector depends on the *square* of $r(\theta)$, the area formula applies to regions that have $r(\theta) < 0$.

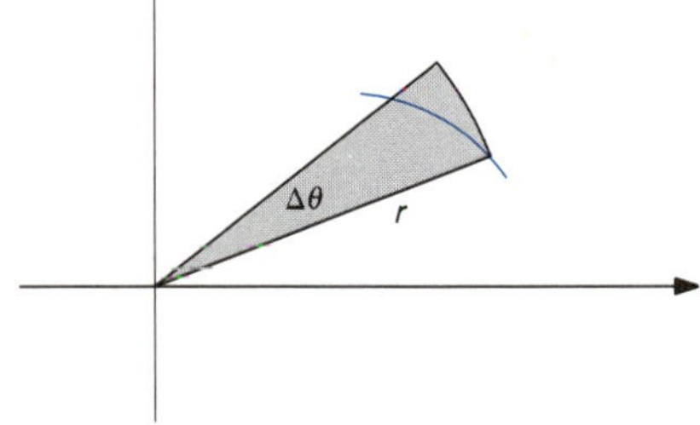

FIGURE 12.4.2

■ **EXAMPLE 1**

Find the area of the region enclosed by $r = 2 \sin \theta$.

**SOLUTION**

The region and a typical area sector are sketched in Figure 12.4.3. The sector runs from the pole to the curve $r = 2 \sin \theta$. These sectors will cover the region once as $\theta$ ranges from $\theta = 0$ to $\theta = \pi$. The area is

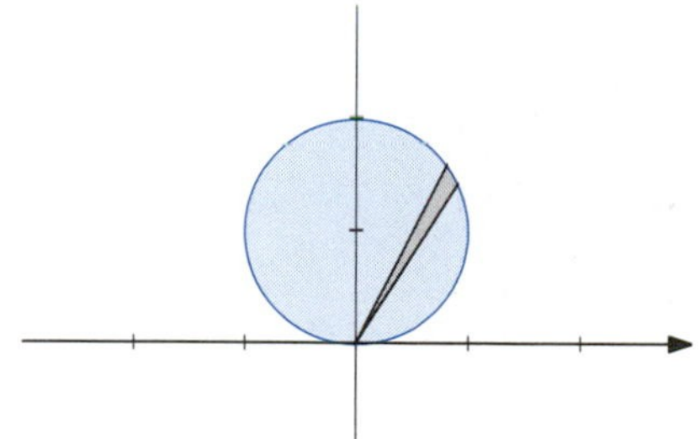
FIGURE 12.4.3

$$\text{Area} = \int_0^\pi \frac{1}{2} [2 \sin \theta]^2 \, d\theta = 2 \int_0^\pi \sin^2\theta \, d\theta$$

[Formula (3) of Section 8.3]

$$= 2\left[-\frac{\sin \theta \cos \theta}{2} + \frac{\theta}{2}\right]_0^\pi$$

$$= [0 + \pi] - [0 + 0] = \pi.$$

■

■ **EXAMPLE 2**

Find the area of the region bounded by the lines $\theta = 0$ and $\theta = 2\pi/3$, and that part of the spiral $r = e^\theta$ with $0 \leq \theta \leq 2\pi/3$.

**SOLUTION**

The region and a typical area sector are pictured in Figure 12.4.4. The sectors run from the pole to $r = e^\theta$. The region is covered once as the

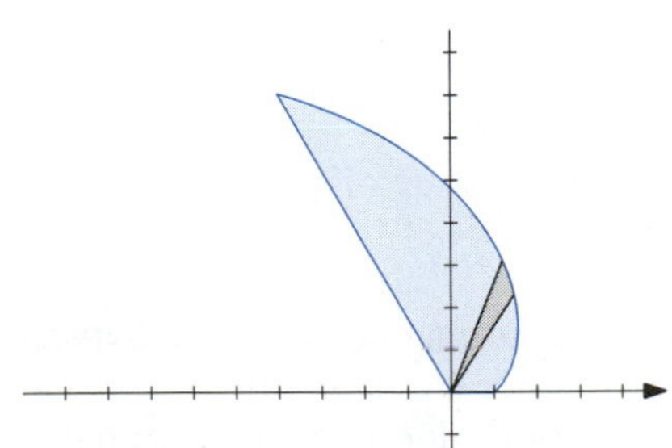
FIGURE 12.4.4

sectors range from $\theta = 0$ to $\theta = 2\pi/3$. The area is

$$\text{Area} = \int_0^{2\pi/3} \frac{1}{2}[e^\theta]^2\, d\theta = \frac{1}{2}\int_0^{2\pi/3} e^{2\theta}\, d\theta$$

$$= \frac{1}{2}\frac{e^{2\theta}}{2}\Big]_0^{2\pi/3} = \frac{1}{4}[e^{4\pi/3} - 1].$$ ■

■ **EXAMPLE 3**

Find the area of the region bounded by the inner loop of $r = 1 + \sqrt{2}\cos\theta$.

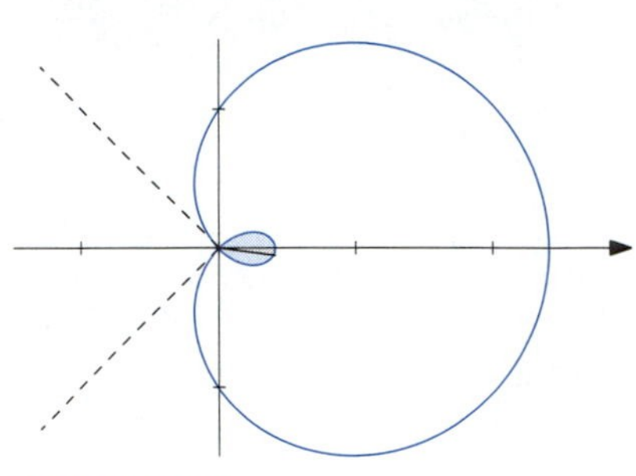

FIGURE 12.4.5

**SOLUTION**

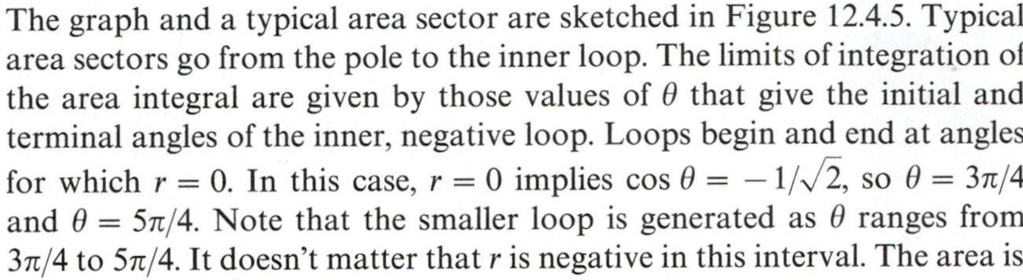

The graph and a typical area sector are sketched in Figure 12.4.5. Typical area sectors go from the pole to the inner loop. The limits of integration of the area integral are given by those values of $\theta$ that give the initial and terminal angles of the inner, negative loop. Loops begin and end at angles for which $r = 0$. In this case, $r = 0$ implies $\cos\theta = -1/\sqrt{2}$, so $\theta = 3\pi/4$ and $\theta = 5\pi/4$. Note that the smaller loop is generated as $\theta$ ranges from $3\pi/4$ to $5\pi/4$. It doesn't matter that $r$ is negative in this interval. The area is

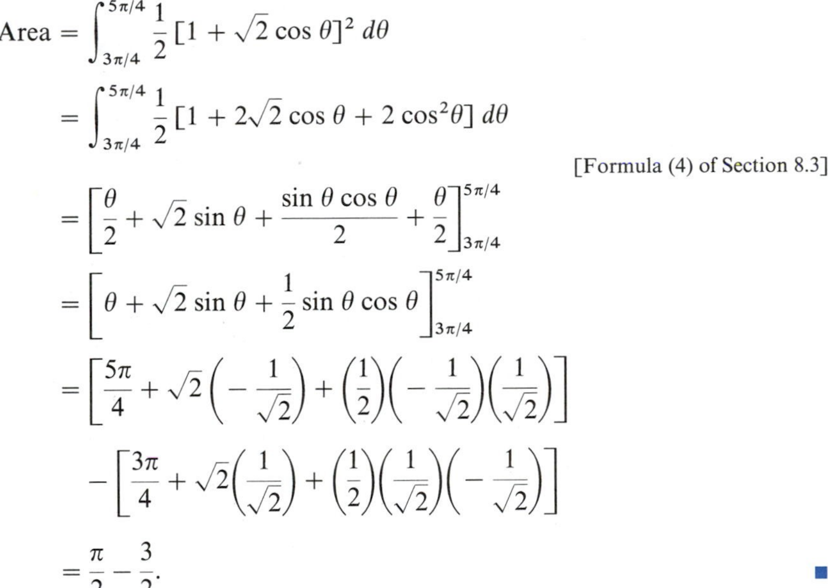

$$\text{Area} = \int_{3\pi/4}^{5\pi/4} \frac{1}{2}[1 + \sqrt{2}\cos\theta]^2\, d\theta$$

$$= \int_{3\pi/4}^{5\pi/4} \frac{1}{2}[1 + 2\sqrt{2}\cos\theta + 2\cos^2\theta]\, d\theta$$

[Formula (4) of Section 8.3]

$$= \left[\frac{\theta}{2} + \sqrt{2}\sin\theta + \frac{\sin\theta\cos\theta}{2} + \frac{\theta}{2}\right]_{3\pi/4}^{5\pi/4}$$

$$= \left[\theta + \sqrt{2}\sin\theta + \frac{1}{2}\sin\theta\cos\theta\right]_{3\pi/4}^{5\pi/4}$$

$$= \left[\frac{5\pi}{4} + \sqrt{2}\left(-\frac{1}{\sqrt{2}}\right) + \left(\frac{1}{2}\right)\left(-\frac{1}{\sqrt{2}}\right)\left(\frac{1}{\sqrt{2}}\right)\right]$$

$$- \left[\frac{3\pi}{4} + \sqrt{2}\left(\frac{1}{\sqrt{2}}\right) + \left(\frac{1}{2}\right)\left(\frac{1}{\sqrt{2}}\right)\left(-\frac{1}{\sqrt{2}}\right)\right]$$

$$= \frac{\pi}{2} - \frac{3}{2}.$$ ■

■ **EXAMPLE 4**

Find the area of the region bounded by the outer loop of $r = 1 + \sqrt{2}\cos\theta$.

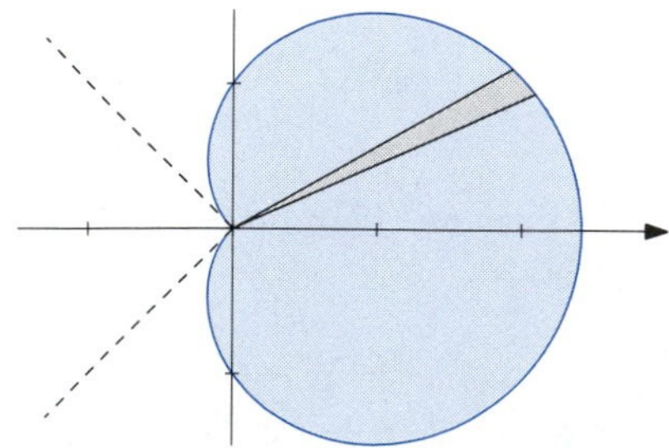

FIGURE 12.4.6

**SOLUTION**

The graph and typical area sector are sketched in Figure 12.4.6. Typical area sectors go from the pole to the outer loop. We need values of $\theta$ that give the initial and terminal angles of the outer loop. As in Example 3, $r = 0$ implies $\cos\theta = -1/\sqrt{2}$, but this time we choose the solutions

$\theta = -3\pi/4$ and $\theta = 3\pi/4$. The larger loop is generated as $\theta$ increases from $-3\pi/4$ to the *larger* value $3\pi/4$. The area is

$$\text{Area} = \int_{-3\pi/4}^{3\pi/4} \frac{1}{2}[1 + \sqrt{2}\cos\theta]^2\, d\theta$$

[As in Example 3]

$$= \left[\theta + \sqrt{2}\sin\theta + \frac{1}{2}\sin\theta\cos\theta\right]_{-3\pi/4}^{3\pi/4}$$

$$= \left[\frac{3\pi}{4} + \sqrt{2}\left(\frac{1}{\sqrt{2}}\right) + \left(\frac{1}{2}\right)\left(\frac{1}{\sqrt{2}}\right)\left(-\frac{1}{\sqrt{2}}\right)\right]$$

$$-\left[-\frac{3\pi}{4} + \sqrt{2}\left(-\frac{1}{\sqrt{2}}\right) + \left(\frac{1}{2}\right)\left(-\frac{1}{\sqrt{2}}\right)\left(-\frac{1}{\sqrt{2}}\right)\right]$$

$$= \frac{3\pi}{2} + \frac{3}{2}.$$

■

Note that we cannot describe the outer loop of Example 4 by saying that $\theta$ ranges (backward) from $5\pi/4$ to $3\pi/2$. Evaluation of the corresponding area integral,

$$\int_{5\pi/4}^{3\pi/4} \frac{1}{2}[1 + \sqrt{2}\cos\theta]^2\, d\theta \quad \left(= -\int_{3\pi/4}^{5\pi/4} \frac{1}{2}[1 + \sqrt{2}\cos\theta]^2\, d\theta\right),$$

would give the negative of the area of the smaller loop. It is very important that the limits of integration go from one number to a larger number, and that they describe those area sectors that cover the region exactly once.

■ **EXAMPLE 5**

Find the area of the region enclosed by $r^2 = 9\sin 2\theta$.

**SOLUTION**

Let us first note that the equation $r^2 = 9\sin 2\theta$ is not really of the form $r = r(\theta)$ that is required to apply our area formula. However, $r^2 = 9\sin\theta$ implies $r = \sqrt{9\sin 2\theta}$ or $r = -\sqrt{9\sin 2\theta}$, and each of the latter equations have the same graph as the original equation. Let us work with the equation $r = \sqrt{9\sin 2\theta}$, so $r(\theta) \geq 0$ whenever $\theta$ is in either the first or third quadrants. We see that $r(\theta)$ is undefined for $\theta$ in either the second or fourth quadrants. We have $[r(\theta)]^2 = 9\sin 2\theta$ whenever $r(\theta)$ is defined. The region is sketched in Figure 12.4.7. *Using symmetry*, we see that the entire area is twice the area of one loop. The region in the first quadrant is covered by sectors from the pole to the curve, as $\theta$ varies between 0 and $\pi/2$. Note that 0 and $\pi/2$ are successive zeros of $r$. We have

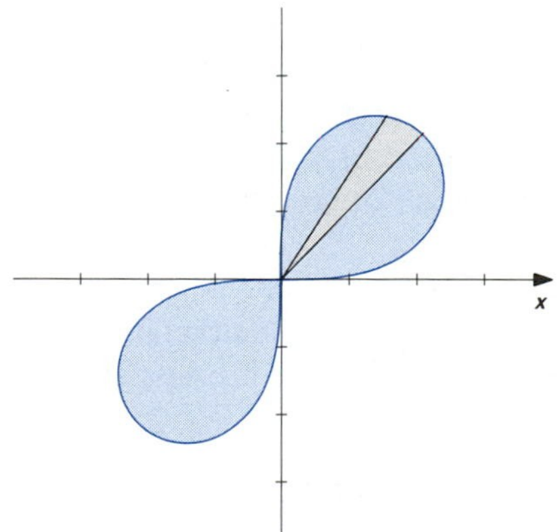


**FIGURE 12.4.7**

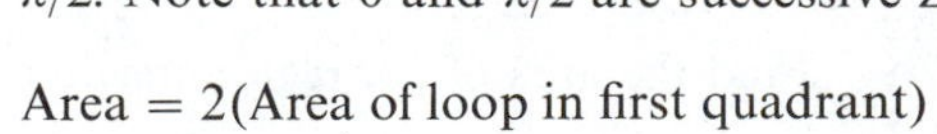

Area = 2(Area of loop in first quadrant)

$$\text{Area} = 2\int_0^{\pi/2} \frac{1}{2} 9\sin 2\theta\, d\theta = -\frac{9}{2}\cos 2\theta\Big]_0^{\pi/2} = -\frac{9}{2}(\cos\pi - \cos 0)$$

$$= -\frac{9}{2}(-1 - 1) = 9.$$

■

Note that we cannot express the area of the region in Example 5 as $\int_0^{2\pi}(1/2)9 \sin 2\theta \, d\theta$. Evaluation of this integral gives zero. Even though there are no points on the graph of $r = \sqrt{9 \sin 2\theta}$ in the second and fourth quadrants, the expression $9 \sin 2\theta$ is well defined and can be integrated over the corresponding values of $\theta$. The reason the integral does not give the proper area is that $9 \sin 2\theta \neq (\sqrt{9 \sin 2\theta})^2$ for $\theta$ in the second and fourth quadrants, because we cannot take the positive square root of a negative number. Thus, the integral does not represent the area bounded by $r = \sqrt{9 \sin 2\theta}$, $\theta = 0$, and $\theta = 2\pi$.

We can use the area formula we have developed to find the area of a region between two polar curves. As in the case of rectangular coordinates, this involves finding the points of intersection of the curves.

The problem of finding all points of intersection of two polar curves deserves a word of caution. The difficulty arises because different coordinate pairs $(r, \theta)$ can represent the same point. If a point has a coordinate pair $(r, \theta)$ that satisfies both equations, it will correspond to a point of intersection. However, there may be points of intersection for which no coordinate pair of the point satisfies both of the given equations. It may be necessary to replace the given equations by different equations that have the same graph. For example, from Figure 12.4.8 we see that the graphs of $r = 2$ and $\theta = \pi/3$ have two points of intersection. The point of intersection in the first quadrant has a coordinate pair $(2, \pi/3)$ that satisfies both equations. The coordinates of this point are given by the simultaneous solution of the equations $r = 2$, $\theta = \pi/3$. The point of intersection in the third quadrant does not have a coordinate pair that satisfies both given equations. The coordinates of this point are not given by a simultaneous solution of the equations $r = 2$, $\theta = \pi/3$. However, if we replace the equation $r = 2$ by the equivalent equation $r = -2$, we can obtain the point of intersection in the third quadrant as the simultaneous solution of the equations $r = -2$, $\theta = \pi/3$. Note that the polar equations $r = 2$ and $r = -2$ have the same graph.

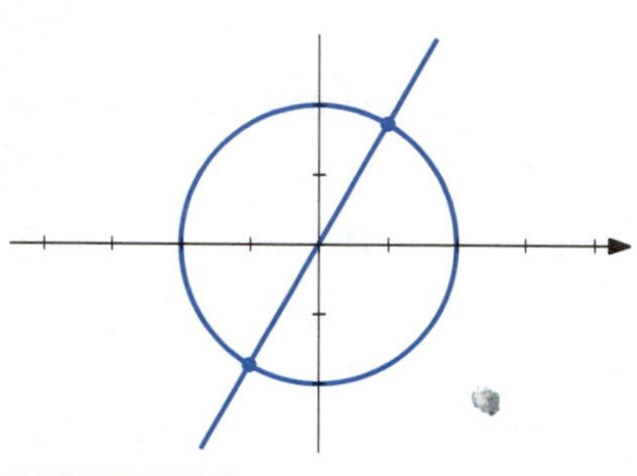

FIGURE 12.4.8

To find all points of intersection of two polar curves it is a good idea to *sketch both curves* and use the sketch to roughly locate and determine the number of points of intersection. Then use simultaneous solutions of the two given equations to find coordinate pairs that satisfy both equations. If this does not give all points of intersection, substitute equivalent equations for the given equations.

Let us now return to the problem of area. We will see that the area of regions between two polar curves can sometimes be determined as a difference of areas of regions of the type to which our area formula applies.

**■ EXAMPLE 6**

Find the area of the region inside $r = 1 + \sin \theta$ and outside $r = 1$.

**SOLUTION**

The region is sketched in Figure 12.4.9. The points of intersection of the graphs are easily seen to be given by the values $\theta = 0$ and $\theta = \pi$. Radial lines at angles $\theta$, $\theta$ between the points of intersection, intersect the region from $(1, \theta)$ to $(1 + \sin \theta, \theta)$. The area of the region is the difference of the

FIGURE 12.4.9

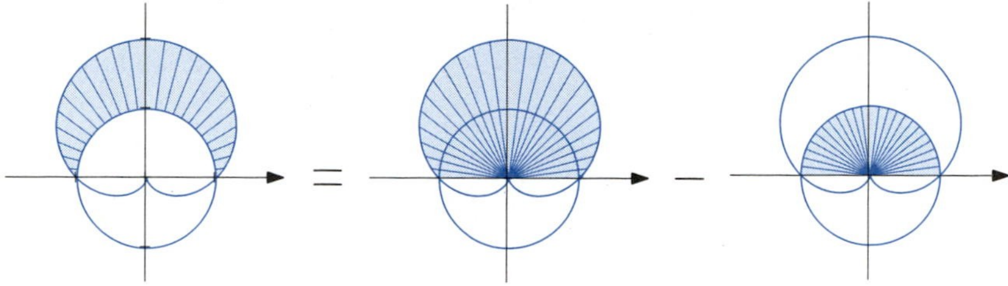

FIGURE 12.4.10

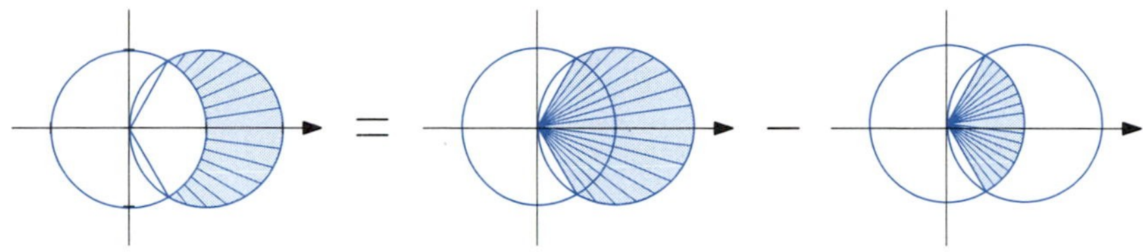

areas of the regions bounded by $r = 1 + \sin\theta$ and $r = 1$, between these values of $\theta$. That is,

$$\begin{aligned}
\text{Area} &= \int_0^\pi \frac{1}{2}(1 + \sin\theta)^2\,d\theta - \int_0^\pi \frac{1}{2}(1)^2\,d\theta \\
&= \int_0^\pi \frac{1}{2}[(1 + \sin\theta)^2 - (1)^2]\,d\theta \\
&= \int_0^\pi \frac{1}{2}[1 + 2\sin\theta + \sin^2\theta - 1]\,d\theta \\
&= \int_0^\pi \left(\sin\theta + \frac{1}{2}\sin^2\theta\right) d\theta.
\end{aligned}$$

[Formula (3) of Section 8.3]

$$\begin{aligned}
&= -\cos\theta + \frac{1}{2}\left(-\frac{\sin\theta\cos\theta}{2} + \frac{\theta}{2}\right)\Bigg]_0^\pi \\
&= \left(1 + \frac{\pi}{4}\right) - (-1) = 2 + \frac{\pi}{4}.
\end{aligned}$$

■

■ **EXAMPLE 7**

Find the area of the region inside $r = 2\cos\theta$ and outside $r = 1$.

**SOLUTION**

The region is sketched in Figure 12.4.10. Radial lines at angles $\theta$, $\theta$ between the points of intersection, intersect the region from $(1, \theta)$ to $(2\cos\theta, \theta)$. We need the points of intersection of the curves. The equations $r = 2\cos\theta$ and

$r = 1$ imply $\cos\theta = 1/2$, so $\theta = -\pi/3$ and $\theta = \pi/3$ are the desired coordinates. The area of the region is the difference of the areas of the regions bounded by $r = 2\cos\theta$ and $r = 1$, between these values of $\theta$. We have

$$\text{Area} = \int_{-\pi/3}^{\pi/3} \frac{1}{2}(2\cos\theta)^2\,d\theta - \int_{-\pi/3}^{\pi/3} \frac{1}{2}(1)^2\,d\theta$$

$$= \int_{-\pi/3}^{\pi/3} \left(2\cos^2\theta - \frac{1}{2}\right) d\theta$$

[Formula (4) of Section 8.3]

$$= 2\left(\frac{\sin\theta\cos\theta}{2} + \frac{\theta}{2}\right) - \frac{\theta}{2}\Bigg]_{-\pi/3}^{\pi/3}$$

$$= \sin\theta\cos\theta + \frac{\theta}{2}\Bigg]_{-\pi/3}^{\pi/3}$$

$$= \left(\left(\frac{\sqrt{3}}{2}\right)\left(\frac{1}{2}\right) + \frac{\pi}{6}\right) - \left(\left(-\frac{\sqrt{3}}{2}\right)\left(\frac{1}{2}\right) - \frac{\pi}{6}\right) = \frac{\sqrt{3}}{2} + \frac{\pi}{3}.$$ ■

■ **EXAMPLE 8**

Find the area of the region between the inner and outer loops of $r = 1 + \sqrt{2}\cos\theta$.

**SOLUTION**

The region is sketched in Figure 12.4.11. The desired area is the area of the outer loop minus the area of the inner loop. Using the results of Examples 3 and 4, we have

$$\text{Area} = (\text{Area of outer loop}) - (\text{Area of inner loop})$$

$$= \left(\int_{-3\pi/4}^{3\pi/4} \frac{1}{2}[1 + \sqrt{2}\cos\theta]^2\,d\theta\right)$$

$$- \left(\int_{3\pi/4}^{5\pi/4} \frac{1}{2}[1 + \sqrt{2}\cos\theta]^2\,d\theta\right)$$

$$= \left(\frac{3\pi}{2} + \frac{3}{2}\right) - \left(\frac{\pi}{2} - \frac{3}{2}\right) = \pi + 3.$$

Note that the integrals that give the areas of these two loops have different limits of integration. We cannot combine the integrals in this example. ■

■ **EXAMPLE 9**

Find the area of the region in the first quadrant that is inside both $r = 1$ and $r = 2\cos\theta$.

**SOLUTION**

The region is sketched in Figure 12.4.12. We see that there are two distinct types of radial lines associated with the region. One type runs from the pole to $r = 1$, and the other runs from the pole to $r = 2\cos\theta$. This means that we can express the area as a sum of the areas of two regions to which

FIGURE 12.4.11

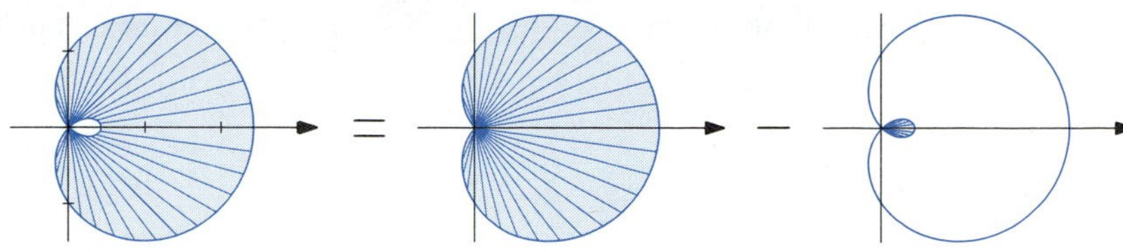

FIGURE 12.4.12

our formula applies. We need the point of intersection. The equations $r = 1$ and $r = 2\cos\theta$ imply $2\cos\theta = 1$. The solution in the first quadrant is $\theta = \pi/3$. The area is

$$\begin{aligned}
\text{Area} &= \int_0^{\pi/3} \frac{1}{2}(1)^2\, d\theta + \int_{\pi/3}^{\pi/2} \frac{1}{2}(2\cos\theta)^2\, d\theta \\
&= \int_0^{\pi/3} \frac{1}{2}\, d\theta + \int_{\pi/3}^{\pi/2} 2\cos^2\theta\, d\theta \qquad \text{[Formula (4) of Section 8.3]} \\
&= \frac{\pi}{6} + \Big[\sin\theta\cos\theta + \theta\Big]_{\pi/3}^{\pi/2} \\
&= \frac{\pi}{6} + \left[\left((1)(0) + \frac{\pi}{2}\right) - \left(\left(\frac{\sqrt{3}}{2}\right)\left(\frac{1}{2}\right) + \frac{\pi}{3}\right)\right] \\
&= \frac{\pi}{3} - \frac{\sqrt{3}}{4}.
\end{aligned}$$

■

## EXERCISES 12.4

Find the area of the regions described in Exercises 1–24.

1 Bounded by $r = 2\sec\theta$, $\theta = 0$, $\theta = \pi/3$

2 Bounded by $r = 2\csc\theta$, $\theta = \pi/4$, $\theta = \pi/2$

3 Bounded by $r = \theta$, $\theta = 0$, $\theta = \pi$

4 Bounded by $r = e^{-\theta/2}$, $\theta = 0$, $\theta = \pi$

5 Enclosed by $r = 2 - 2\sin\theta$

6 Enclosed by $r = 3 - 2\sin\theta$

7 Enclosed by $r^2 = \sin\theta$

8 Enclosed by $r^2 = \cos 2\theta$

9 Enclosed by $r = \sin 2\theta$

10 Enclosed by $r = \cos 3\theta$

11 The smaller loop of $r = 1 - 2\cos\theta$

12 The larger loop of $r = 1 - 2\cos\theta$

13 Inside $r = 2$ and to the right of $r = \sec\theta$

14 Inside $r = 1$ and outside $r = 1 - \cos\theta$

15 Bounded by $r = 2\theta$, $r = \theta$, $\theta = 0$, $\theta = \pi$

16 Inside $r = 2\cos\theta$ and outside $r = \cos\theta$

17 Inside $r = 2\sin 2\theta$ and outside $r = 1$

18 Inside $r^2 = 2\cos 2\theta$ and outside $r = 1$

19 Between loops of $r = 2\cos\theta - 1$

20 Between loops of $r = 2\sin\theta + \sqrt{3}$

21 Inside $r = 2$ and above $r = -\csc\theta$

22 In the first quadrant, inside both $r = 1$ and $r^2 = 2\cos\theta$

23 Inside both $r^2 = \sin\theta$ and $r = \sqrt{2}\sin\theta$

24 Inside both $r = \cos\theta$ and $r = \sqrt{3}\sin\theta$

## 12.5 ARC LENGTH AND SURFACES OF REVOLUTION

Let us consider the length of a polar curve

$$r = r(\theta), \qquad \theta_0 \le \theta \le \theta_1.$$

We assume that $r$ has a continuous derivative. The length of the curve between two points on the curve that are near each other is approximately the length of the line segment between the points. From Figure 12.5.1 we see that

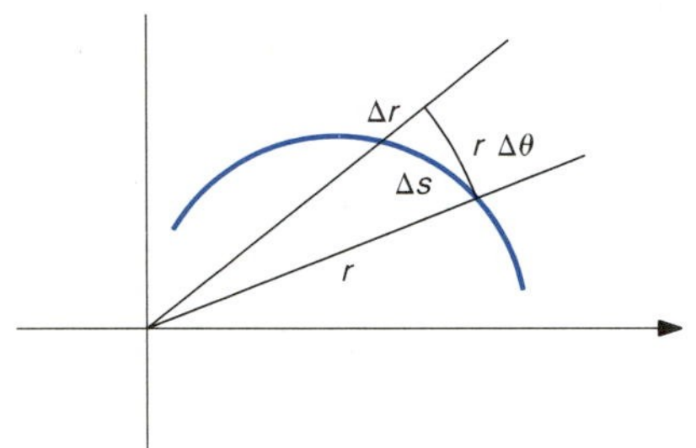

FIGURE 12.5.1

$$\Delta s \approx \sqrt{(r\Delta\theta)^2 + (\Delta r)^2} = \sqrt{r^2 + \left(\frac{\Delta r}{\Delta\theta}\right)^2}\,\Delta\theta.$$

The differential form of this statement is

$$ds = \sqrt{r^2 + r'^2}\, d\theta.$$

It should seem reasonable that the length is given by the formula

$$s = \int ds = \int_{\theta_0}^{\theta_1} \sqrt{[r(\theta)]^2 + [r'(\theta)]^2}\, d\theta.$$

■ **EXAMPLE 1**

Find the length of the polar curve $r = 1 + \cos\theta$.

**SOLUTION**

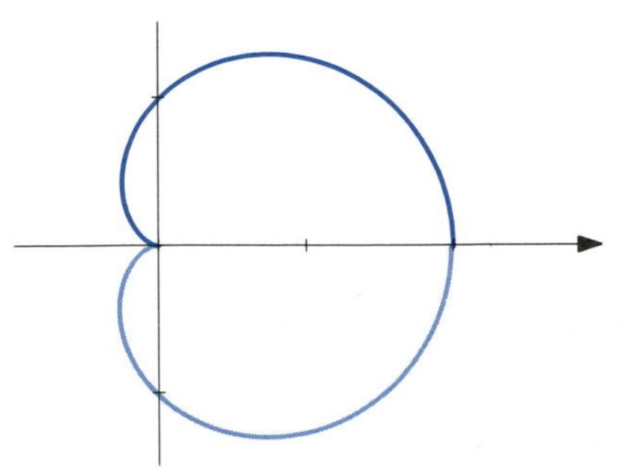
FIGURE 12.5.2

The curve is sketched in Figure 12.5.2. We see that the curve is traced exactly once as $\theta$ varies from 0 to $2\pi$. Using symmetry, we also see that half the length is traced as $\theta$ varies between 0 and $\pi$. This means we can express the length as

$$s = 2\int_0^{\pi} ds.$$

Later, we will see why it is advantageous to express the length as an integral over the interval $[0, \pi]$, instead of integrating over $[0, 2\pi]$.

We have $r = 1 + \cos\theta$, so $r' = -\sin\theta$. Then

$$\begin{aligned} ds &= \sqrt{r^2 + r'^2}\,d\theta \\ &= \sqrt{(1+\cos\theta)^2 + (-\sin\theta)^2}\,d\theta \\ &= \sqrt{1 + 2\cos\theta + \cos^2\theta + \sin^2\theta}\,d\theta \end{aligned}$$

The arc length is

$$s = 2\int_0^{\pi} ds = 2\int_0^{\pi} \sqrt{2\cos\theta + 2}\,d\theta.$$

The integral can be evaluated by using the trigonometric identity

$$1 + \cos\theta = 2\cos^2\frac{\theta}{2}.$$

Then

$$\sqrt{2 + 2\cos\theta} = \sqrt{4\cos^2\frac{\theta}{2}} = 2\left|\cos\frac{\theta}{2}\right|.$$

For $0 \le \theta \le \pi$, we have $\cos(\theta/2) \ge 0$, so $|\cos(\theta/2)| = \cos(\theta/2)$. {Note that $|\cos(\theta/2)| = -\cos(\theta/2)$ for $\pi \le \theta \le 2\pi$, since $\cos(\theta/2)$ is negative for $\theta$ in that range. We have avoided dealing with the two different formulas for $|\cos(\theta/2)|$ by expressing the total length of the curve as an integral over the interval $[0, \pi]$}. We then have

$$s = 2\int_0^{\pi} 2\cos\frac{\theta}{2}\,d\theta = 4(2)\sin\frac{\theta}{2}\Big]_0^{\pi} = 8. \qquad \blacksquare$$

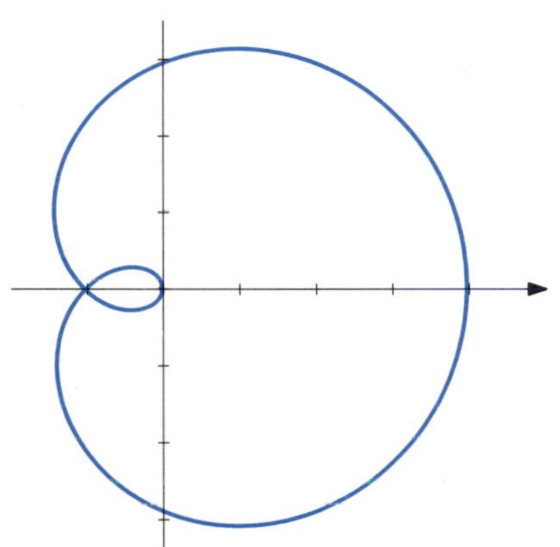

FIGURE 12.5.3

■ **EXAMPLE 2**

Find the length of the polar curve $r = 4\sin^4(\theta/4)$.

**SOLUTION**

Some key values are given in Table 12.5.1.

TABLE 12.5.1

| $\theta$ | 0 | $\pi/2$ | $\pi$ | $3\pi/2$ | $2\pi$ | $5\pi/2$ | $3\pi$ | $7\pi/2$ | $4\pi$ |
|---|---|---|---|---|---|---|---|---|---|
| $r$ | 0 | 0.086 | 1 | 2.914 | 4 | 2.914 | 1 | 0.086 | 0 |

The curve is sketched in Figure 12.5.3. The curve is traced once as $\theta$ varies between 0 and $4\pi$. The equation $r = 4\sin^4(\theta/4)$ implies

$$r' = 4(4)\left(\sin\frac{\theta}{4}\right)^3\left(\cos\frac{\theta}{4}\right)\left(\frac{1}{4}\right) = 4\sin^3\frac{\theta}{4}\cos\frac{\theta}{4}.$$

Then

$$ds = \sqrt{\left(4\sin^4\frac{\theta}{4}\right)^2 + \left(4\sin^3\frac{\theta}{4}\cos\frac{\theta}{4}\right)^2}\,d\theta$$

$$= \sqrt{4^2\left(\sin^3\frac{\theta}{4}\right)^2\left(\sin^2\frac{\theta}{4} + \cos^2\frac{\theta}{4}\right)}\,d\theta$$

$$= 4\sin^3\frac{\theta}{4}\,d\theta,$$

as long as $\sin(\theta/4)$ is nonnegative on the interval of integration. In particular, this is true for $\theta$ between 0 and $4\pi$.

The arc length is

$$s = \int ds = \int_0^{4\pi} 4\sin^3\frac{\theta}{4}\,d\theta$$

$$= 4\int_0^{4\pi}\left(1 - \cos^2\frac{\theta}{4}\right)(-4)\left(-\frac{1}{4}\right)\sin\frac{\theta}{4}\,d\theta$$

$[u = \cos(\theta/4),\ du = -1/4\sin(\theta/4)\,d\theta]$

$$= 4\int(1 - u^2)(-4)\,du$$

$$= 4(-4)\left[u - \frac{u^3}{3}\right]$$

$$= 4(-4)\left[\cos\frac{\theta}{4} - \frac{\cos^3(\theta/4)}{3}\right]_0^{4\pi}$$

$$= -16\left[-1 - \frac{(-1)^3}{3}\right] + 16\left[1 - \frac{1}{3}\right]$$

$$= -16\left(-\frac{2}{3}\right) + 16\left(\frac{2}{3}\right) = \frac{64}{3}.$$

■

In Section 6.4 we developed a formula for the area of a surface of revolution that is obtained by revolving a curve in a plane about a line in the plane. We have

**Surface area = $S = \int dS = \int 2\pi(\text{radius})\,ds$,**

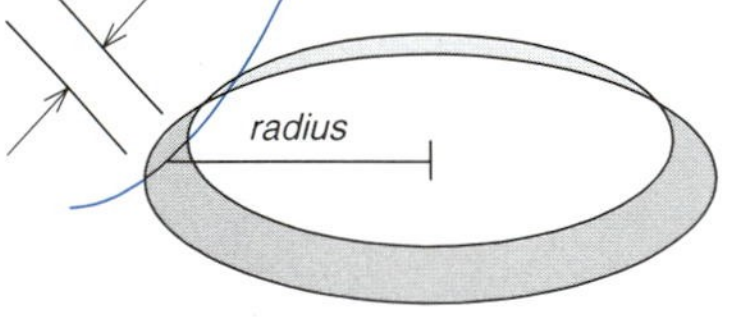


FIGURE 12.5.4

where $2\pi(\text{radius})\,ds$ represents the area of the ribbonlike band obtained by revolving a small piece of the curve of length $ds$ about the axis of revolution. See Figure 12.5.4. The integral represents the sum of the areas from each part of the curve. We can apply the same ideas to obtain the area of a surface obtained by revolving a polar curve $r = r(\theta)$ about a line in the plane. We use the polar form of the differential arc length $ds$ and express the radius in terms of $\theta$.

■ **EXAMPLE 3**

Find the area of the surface obtained by revolving the polar curve $r = 1 + \cos\theta$ about the polar axis.

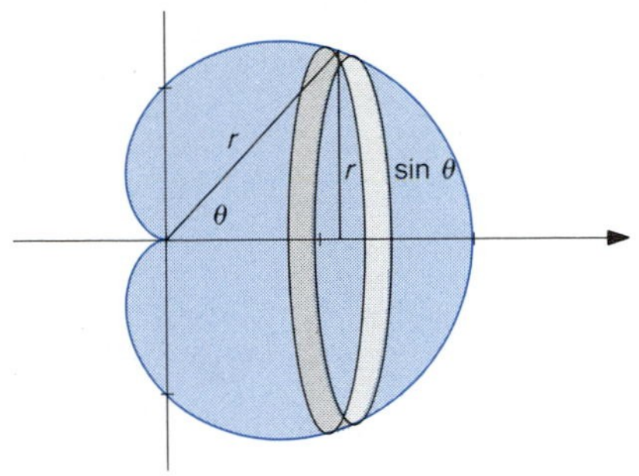


FIGURE 12.5.5

**SOLUTION**

The curve and a typical piece of the surface are sketched in Figure 12.5.5. Note that the entire surface is obtained by one complete revolution about the polar axis of that part of the curve above the polar axis, with $0 \le \theta \le \pi$. The radius of a typical band is

$$\text{Radius} = r(\theta)\sin\theta = (1 + \cos\theta)\sin\theta.$$

The equation $r = 1 + \cos\theta$ implies $r' = -\sin\theta$. As in Example 1, we obtain

$$ds = \sqrt{2(1 + \cos\theta)}\, d\theta.$$

The surface area is

$$\begin{aligned}
S &= \int 2\pi(\text{radius})\, ds \\
&= \int_0^\pi 2\pi(1 + \cos\theta)\sin\theta\sqrt{2(1 + \cos\theta)}\, d\theta \\
&= \int_0^\pi 2\sqrt{2}\pi(1 + \cos\theta)^{3/2}(-1)(-\sin\theta)\, d\theta \\
&\qquad [u = 1 + \cos\theta,\ du = -\sin\theta\, d\theta] \\
&= \int 2\sqrt{2}\pi u^{3/2}(-1)\, du \\
&= -2\sqrt{2}\pi\frac{u^{5/2}}{5/2} \\
&= -2\sqrt{2}\pi\frac{(1 + \cos\theta)^{5/2}}{5/2}\Bigg]_0^\pi \\
&= \frac{4\sqrt{2}\pi(2)^{5/2}}{5} = \frac{32\pi}{5}.
\end{aligned}$$

■

■ **EXAMPLE 4**

Find the area of the surface obtained by revolving the polar curve $r = 2e^{\theta/2}$, $0 \le \theta \le \pi/2$, about the line $\theta = \pi/2$.

**SOLUTION**

The graph and a typical piece of the surface are sketched in Figure 12.5.6. We see that the radius of a typical band is

$$\text{Radius} = r(\theta)\cos\theta = 2e^{\theta/2}\cos\theta.$$

The equation $r = 2e^{\theta/2}$ implies $r' = 2e^{\theta/2}(1/2) = e^{\theta/2}$. Then

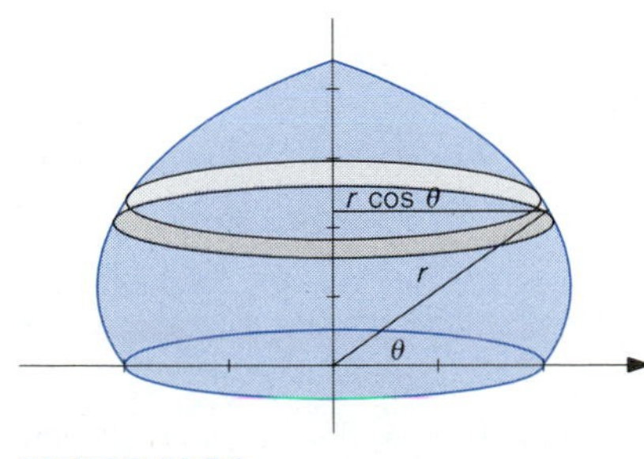


FIGURE 12.5.6

$$\begin{aligned}
ds &= \sqrt{(2e^{\theta/2})^2 + (e^{\theta/2})^2}\, d\theta \\
&= \sqrt{5e^\theta}\, d\theta \\
&= \sqrt{5}e^{\theta/2}\, d\theta.
\end{aligned}$$

The surface area is

$$S = \int 2\pi(\text{radius})\, ds = \int_0^{\pi/2} 2\pi(2e^{\theta/2}\cos\theta)\sqrt{5}e^{\theta/2}\, d\theta$$

$$= 4\pi\sqrt{5}\int_0^{\pi/2} e^{\theta}\cos\theta\, d\theta$$

[Formula (5) of Section 8.2]

$$= 4\pi\sqrt{5}\left[\frac{e^{\theta}\cos\theta}{2} + \frac{e^{\theta}\sin\theta}{2}\right]_0^{\pi/2}$$

$$= 2\pi\sqrt{5}(e^{\pi/2} - 1).$$

## EXERCISES 12.5

Find the arc length of the polar curves given in Exercises 1–8.

1 $r = 4\cos\theta$

2 $r = 1 + \cos\theta$ [*Hint*: $1 + \cos\theta = 2\cos^2(\theta/2)$].

3 $r = e^{\theta/2}$, $0 \le \theta \le 2\pi$

4 $r = e^{-\theta/2}$, $0 \le \theta \le 2\pi$

5 $r = \sin^2(\theta/2)$

6 $r = \theta^2$, $0 \le \theta \le 2\pi$

7 $r = \tan^2\theta\sec\theta$, $0 \le \theta \le \pi/6$

8 $r = \sin^3(\theta/3)$

Find the surface area of the surface of revolution obtained by revolving the curves given in Examples 9–12 about the polar axis.

9 $r = \dfrac{2}{1 + \cos\theta}$, $0 \le \theta \le \pi/2$

10 $r = e^{-\theta/2}$, $0 \le \theta \le \pi$

11 $r^2 = \sin 2\theta$

12 $r = \sin^2(\theta/2)$ [*Hint*: $2\sin(\theta/2)\cos(\theta/2) = \sin\theta$.]

Find the surface area of the surface of revolution obtained by revolving the curves given in Examples 13–16 about the line $\theta = \pi/2$.

13 $r = 2\cos\theta$

14 $r = 1 + \sin\theta$

15 $r = e^{-\theta/2}$, $-\pi/2 \le \theta \le \pi/2$

16 $r = r_0$, $-\pi/2 \le \theta \le \pi/2$

## REVIEW EXERCISES

Sketch the graphs of the polar equations in Exercises 1–16.

1 $r = 2\cos\theta$

2 $r = -2$

3 $r = \dfrac{2}{\cos\theta + \sin\theta}$

4 $r = \dfrac{2}{2\sin\theta - \cos\theta}$

5 $r = \dfrac{1}{1 - \sin\theta}$

6 $r = \dfrac{2}{1 + \cos\theta}$

7 $r = \dfrac{3}{1 - 2\cos\theta}$

8 $r = \dfrac{3}{2 - \sin\theta}$

9 $r = \sin 2\theta$

10 $r = \cos 3\theta$

11 $r^2 = \sin 2\theta$

12 $r = e^{\theta}$

13 $r = 1 + 2\sin\theta$

14 $r = 1 - \cos\theta$

15 $r = 2 + \cos\theta$

16 $r = 3 - 2\sin\theta$

17 Find the angle between the line $\theta = \pi/2$ and the line tangent to the polar curve $r = 2 - 3\cos\theta$ at the point $(2, \pi/2)$.

18 Find the angle between the line $\theta = \pi/2$ and the line tangent to the polar curve $r = e^{2\theta}$ at the point $(e^{\pi}, \pi/2)$.

19 Find the angle between the lines tangent to the polar curves $r = 1 - \cos\theta$ and $r = -3\cos\theta$ at the point $(3/2, 2\pi/3)$.

20 Find the angle between lines tangent to the polar curves $r = 2\cos\theta$ and $r = 2\sin\theta$ at each point of intersection.

Set up integrals that give the areas of the regions indicated in Exercises 21–24.

21 Inside $r = 2 - 2\cos\theta$

22 Inside $r^2 = \sin 2\theta$

23 Inside $r = 2 - 2\cos\theta$ and outside $r = 3$

24 Inside $r = 2\sin\theta$ and outside $r = 1$

Set up integrals that give the length of the polar curves given in Exercises 25–26.

25 $r = 2 - \sin\theta$

26 $r = \theta$, $0 \le \theta \le 2\pi$

In Exercises 27–28, set up integrals that give the area of the surface obtained by revolving the given curve about the given line.

27 $r^2 = \cos 2\theta$, about the polar axis

28 $r = \sin\theta$, about the line $\theta = \pi/2$

**CHAPTER THIRTEEN**

# *Three-dimensional space*

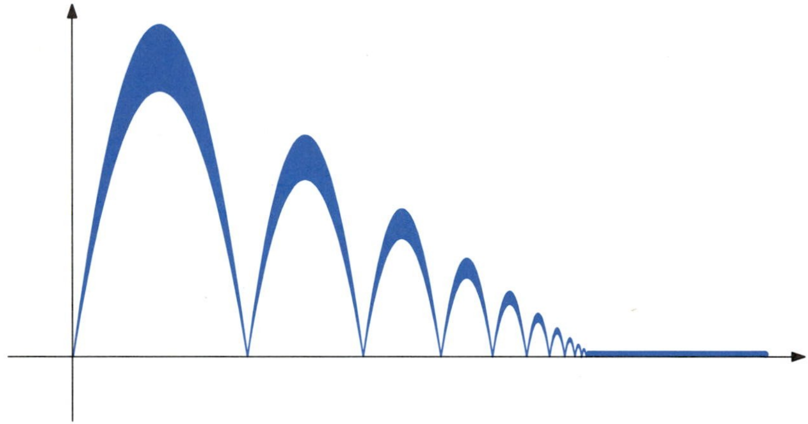

We will introduce a rectangular coordinate system that is used to describe points in three-dimensional space and give some detailed suggestions for sketching curves and surfaces in three-dimensional space. Our sketches will be done in a style that is commonly used in computer graphics and technical drawings. In later chapters, we will see that the skill in sketching three-dimensional figures that is developed here is very useful for understanding calculus of several variables.

# 13.1 RECTANGULAR COORDINATE SYSTEM

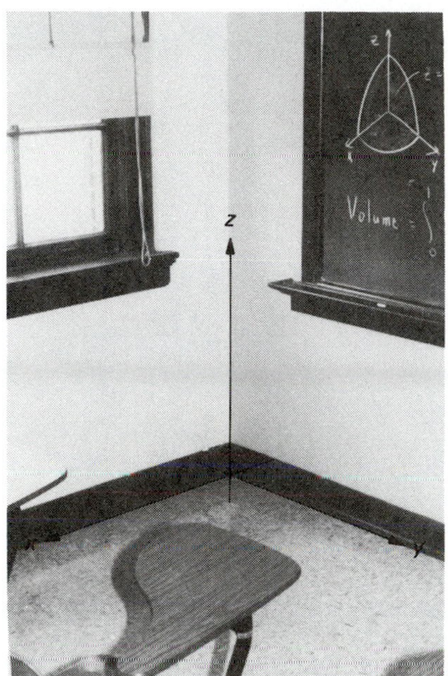


FIGURE 13.1.1

The basic concept of a **three-dimensional rectangular coordinate system** can be illustrated by viewing a rectangular room. We choose a corner where two walls intersect the floor to be the origin. The intersection of the floor with the wall to the left as you face the corner forms the $x$-axis, the intersection of the floor with the wall to the right forms the $y$-axis, and the intersection of the two walls forms the $z$-axis. Any point in the room can be located by specifying its coordinates with respect to these three coordinate axes. That is, the ordered triple of coordinates $(x, y, z)$ is associated with the point found by moving from the origin $x$ units parallel to the $x$-axis, $y$ units parallel to the $y$-axis, and $z$ units parallel to the $z$-axis. See Figure 13.1.1, where we have superimposed the coordinate system on a photograph of the corner of a room.

To sketch objects in three-dimensional space we need to establish a point from which the object is viewed. We will ordinarily orient the coordinate axes as viewed from a point that is an equal distance from each axis, as illustrated in Figure 13.1.1. This view is symmetric with respect to the three coordinate axes and is the most common view used in computer graphics and other technical drawings. We can represent it pictorially by drawing the axes at angles of 120° and using the same scale on each axis. The $z$-axis is drawn with positive direction upward, the positive $x$-axis is drawn to the left and downward, while the positive $y$-axis is drawn to the right and downward. See Figure 13.1.2. The point with coordinates $(x, y, z)$ is plotted by moving from the origin $x$ units parallel to the $x$-axis, $y$ units parallel to the $y$-axis, and $z$ units parallel to the $z$-axis. See Figure 13.1.3. The axes are oriented to form a **right-hand rectangular coordinate system**. This means that, if the $z$-axis is grasped with the *right hand* so the fingers wrap from the direction of the positive $x$-axis to the direction of the positive $y$-axis, the thumb will point in the direction of the positive $z$-axis. See Figure 13.1.4.

The coordinate planes divide three-dimensional space into eight parts, called **octants**. The **first octant** is that portion that has all three coordinates positive.

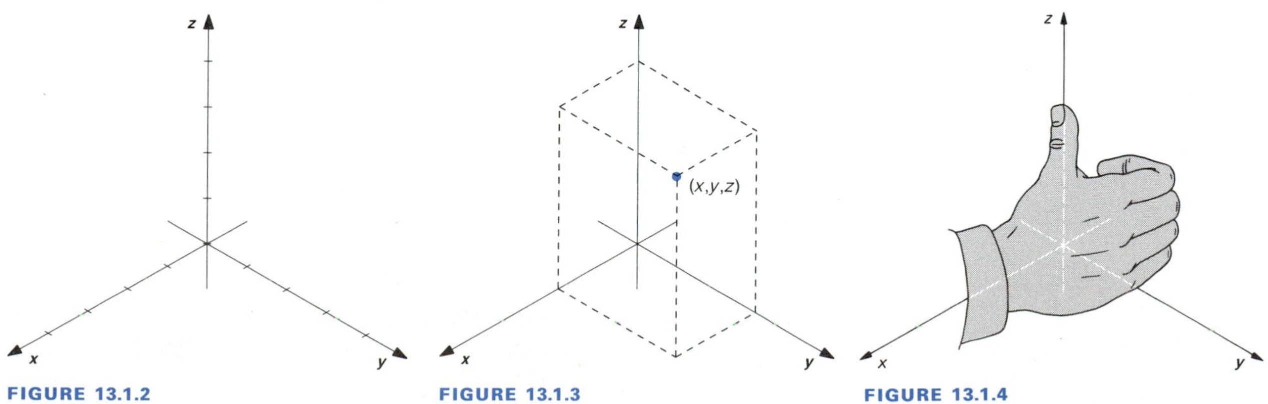


FIGURE 13.1.2 FIGURE 13.1.3 FIGURE 13.1.4

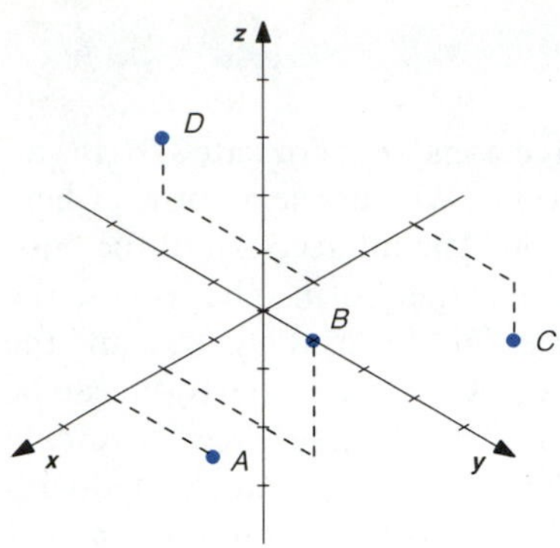

FIGURE 13.1.5

■ **EXAMPLE 1**

Plot the points $A(3, 2, 0)$, $B(2, 3, 2)$, $C(-3, 2, -1)$, and $D(-1, -3, 1)$. Which points are in the first octant?

**SOLUTION**

The points are plotted in Figure 13.1.5. Only $B$ is in the first octant. ■

The **distance** between two points $(x_1, y_1, z_1)$ and $(x_2, y_2, z_2)$ is (see Figure 13.1.6.)

$$d = \sqrt{(x_2 - x_1)^2 + (y_2 - y_1)^2 + (z_2 - z_1)^2}.$$

The **midpoint** $M$ of the line segment between the points is (see Figure 13.1.6.)

$$M = \left(\frac{x_1 + x_2}{2}, \frac{y_1 + y_2}{2}, \frac{z_1 + z_2}{2}\right).$$

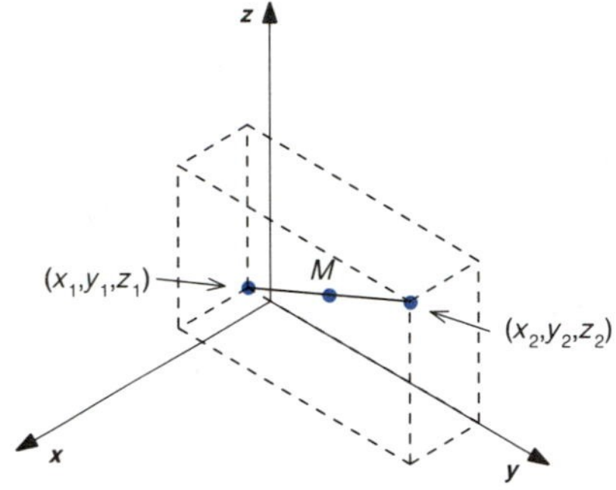

FIGURE 13.1.6

■ **EXAMPLE 2**

Sketch the line segment between the points $(1, -1, 0)$ and $(-1, 3, 4)$. Find the length and midpoint of the segment.

**SOLUTION**

The points are plotted and the segment is sketched in Figure 13.1.7. The length is

$$d = \sqrt{((-1) - 1)^2 + (3 - (-1))^2 + (4 - 0)^2}$$
$$= \sqrt{4 + 16 + 16} = \sqrt{36} = 6.$$

The midpoint is

$$M = \left(\frac{1 - 1}{2}, \frac{-1 + 3}{2}, \frac{0 + 4}{2}\right) = (0, 1, 2).$$ ■

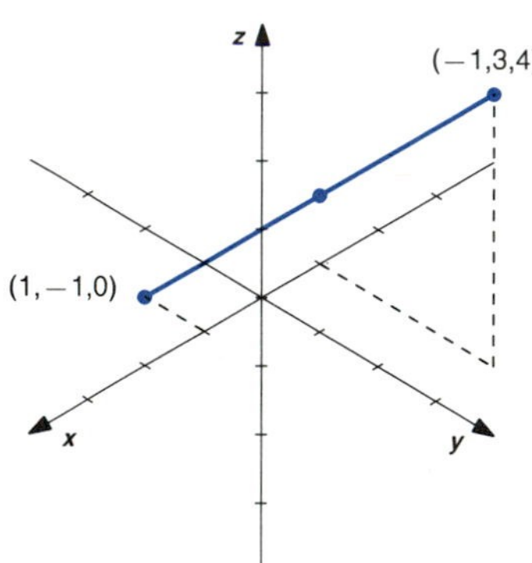

FIGURE 13.1.7

A **sphere** with center $(h, k, l)$ and radius $r$ is the set of all points $(x, y, z)$ in three-dimensional space that are at a distance $r$ from $(h, k, l)$. It follows from the distance formula that the equation of the sphere with center $(h, k, l)$ and radius $r$ is

$$(x - h)^2 + (y - k)^2 + (z - l)^2 = r^2.$$

This is called the **standard equation** of the sphere.

■ **EXAMPLE 3**

Find the standard equation of the sphere that passes through the points $(2, 3, -1)$ and $(-2, 1, 3)$, and has center at the midpoint of the line segment between the points.

**SOLUTION**

We need the coordinates of the center and the length of the radius to write the equation of the sphere.

The midpoint of the line segment between the points is

$$M = \left(\frac{2-2}{2}, \frac{3+1}{2}, \frac{-1+3}{2}\right) = (0, 2, 1).$$

This is the center of the sphere.

The radius is the distance between the center and any point on the sphere. We know that $(2, 3, -1)$ is on the sphere, so

$$r = \sqrt{(2-0)^2 + (3-2)^2 + (-1-1)^2} = 3.$$

The standard equation of the sphere is

$$x^2 + (y-2)^2 + (z-1)^2 = 3^2.$$

The sphere is sketched in Figure 13.1.8. ■

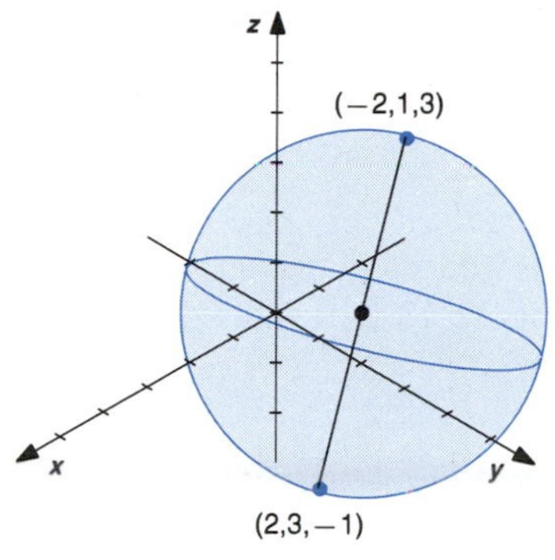


FIGURE 13.1.8

An equation of the form

$$x^2 + y^2 + z^2 + ax + by + cz + d = 0$$

can be written in the form

$$(x-h)^2 + (y-k)^2 + (z-l)^2 = D$$

by completing the squares. If $D > 0$, this is the standard form of the equation of a sphere. If $D = 0$, the graph will be a single point; the equation will have no points on its graph if $D < 0$.

**■ EXAMPLE 4**

Find the center and radius of the sphere

$$x^2 + y^2 + z^2 + 4x - 6z = 0.$$

**SOLUTION**

The center and radius are easily read from the standard form of the sphere. Completing the squares, we have

$$x^2 + y^2 + z^2 + 4x - 6z = 0,$$

$$x^2 + 4x + [4] + y^2 + z^2 - 6z + [9] = 0 + [4] + [9],$$

$$(x+2)^2 + y^2 + (z-3)^2 = (\sqrt{13})^2.$$

The center is $(-2, 0, 3)$ and the radius is $\sqrt{13}$. The sphere is sketched in Figure 13.1.9. ■

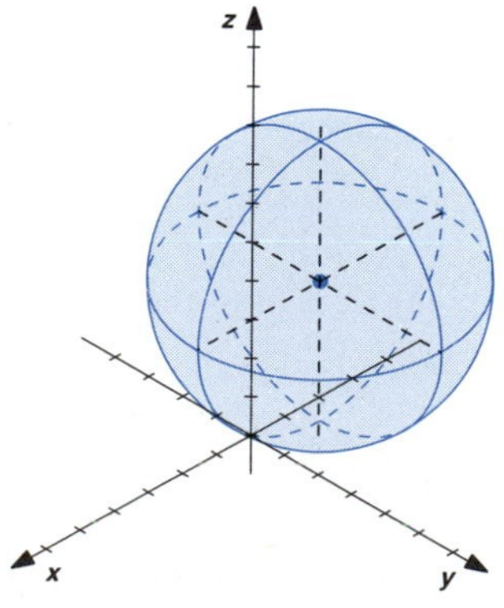


FIGURE 13.1.9

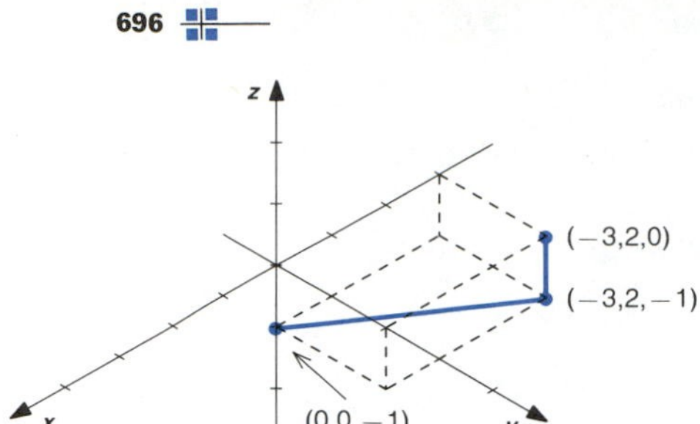


FIGURE 13.1.10

■ **EXAMPLE 5**

Find the distance from the point $(-3, 2, -1)$ to the $z$-axis. Find the distance of the point $(-3, 2, -1)$ to the $xy$-plane.

**SOLUTION**

The distance from the point $(-3, 2, -1)$ to the $z$-axis is the distance between $(-3, 2, -1)$ and the point on the $z$-axis that has the same $z$-coordinate, $(0, 0, -1)$. This is the point on the $z$-axis that is closest to $(-3, 2, -1)$. See Figure 13.1.10. We have

$$\sqrt{(-3-0)^2 + (2-0)^2 + (-1-(-1))^2} = \sqrt{13}.$$

The distance from the point $(-3, 2, -1)$ to the $xy$-plane is the distance between $(-3, 2, -1)$ and $(-3, 2, 0)$. See Figure 13.1.10. This distance is one. ■

## EXERCISES 13.1

**1** Plot the points $A(2, 2, 0)$, $B(1, 2, 1)$, $C(-1, -2, 1)$, and $D(2, -1, 1)$. Which points are in the first octant?

**2** Plot the points $A(1, 3, 1)$, $B(0, 1, 2)$, $C(-3, -3, -1)$, and $D(3, -2, 1)$. Which points are in the first octant?

**3** Sketch the line segment between the points $(1, -1, 0)$ and $(-3, 3, -2)$. Find the length and midpoint of the segment.

**4** Sketch the line segment between the points $(2, 0, 3)$ and $(0, 4, 1)$. Find the length and midpoint of the segment.

**5** Sketch the points $A(2, 0, 0)$, $B(0, 0, 3)$, $C(1, 2, 2)$, and $D(3, 2, -1)$. Show that the midpoint of the line segment between $A$ and $C$ coincides with the midpoint of the line segment between $B$ and $D$, so the four points form a parallelogram.

**6** Sketch the points $A(2, 1, 0)$, $B(0, 0, 3)$, $C(3, 1, 3)$ and show that they form an isosceles triangle. Find the midpoint of its longest side.

**7** Find all values of $a$ so the points $A(0, 0, a)$, $B(3, 0, 0)$, and $C(1, 2, 1)$ form a right triangle with right angle at $C$. (If the lengths of the sides of a triangle satisfy $d_1^2 = d_2^2 + d_3^2$, the Law of Cosines implies the angle opposite the side of length $d_1$ is a right angle.)

**8** Find an equation satisfied by all points $(x, y, z)$ that are equidistant from the point $(1, 3, 2)$ and the origin.

Find the standard form of the equation of the sphere described in each of Exercises 9–14.

**9** The points $(2, 0, 0)$ and $(0, 4, 2)$ form the endpoints of a diameter.

**10** Center $(1, 2, 0)$; contains $(0, 0, 2)$

**11** Center $(4, 3, 12)$; contains the origin

**12** Center $(1, 2, 3)$; tangent to the $xy$-plane

**13** Center $(1, -2, 1)$; tangent to the $z$-axis

**14** Radius 2; center in the first octant; tangent to all three coordinate planes

**15** Find the center and radius of the sphere

$$x^2 + y^2 + z^2 - 2x + 4y = 0.$$

**16** Find the center and radius of the sphere

$$x^2 + y^2 + z^2 + 6y + 2z + 7 = 0.$$

**17** Find the distance from the point $(2, 3, 2)$ to the $z$-axis.

**18** Find the distance from the point $(-1, -2, 1)$ to the $x$-axis.

**19** Find the distance from the point $(2, -4, 3)$ to the $xy$-plane.

**20** Find the distance from the point $(-3, 4, 12)$ to the $yz$-plane.

## 13.2 SKETCHING

The ability to sketch at least a few types of three-dimensional figures is very useful for the understanding of many of the concepts that we will study. In this and in the next section we will offer some hints for sketching and go over some of the more common examples.

The most common type of drawings that are used to represent objects in three-dimensional space in computer graphics and other technical drawings are **axonometric drawings**. In axonometric drawings, lines that are parallel to an axis are drawn parallel to and with the same scale as the axis. The direction and scale at which each axis is drawn depends on the point from which an object is viewed. Computers can be programmed to calculate appropriate directions and scales for any view and change views in a way that makes it appear that an object is rotating slowly.

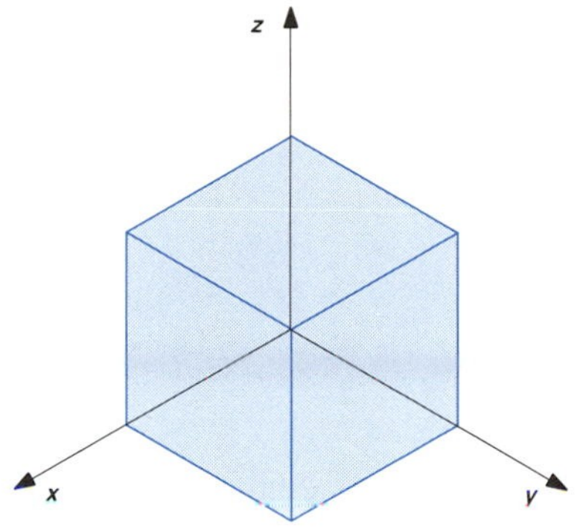


FIGURE 13.2.1

The easiest type of axonometric drawing to make manually, and the most common type used in technical drawing, is an **isometric drawing**. In isometric drawings axes are drawn at angles of 120°, and equal scales are used on the three axes. This corresponds to viewing the origin from a point that is an equal distance from each of the coordinate axes. An isometric drawing of a cube with one corner at the origin and edges along the axes is given in Figure 13.2.1.

As a general rule, if a sketch does not look good in the usual isometric sketch, then we should change the view. For example, we can represent the view from a point that is lower than usual by drawing the $x$-axis and the $y$-axis at an angle greater than 120° and using a relatively large scale on the $z$-axis, as is illustrated in Figure 13.2.2a. A view from a point that is higher than usual is obtained by drawing the $x$-axis and $y$-axis at an angle less than 120° and using a relatively small scale on the $z$-axis, as in Figure 13.2.2b.

We are primarily interested in the graphs of equations. Generally speaking, we should expect the graph of one equation in three variables to be a surface in three-dimensional space. The graph of two simultaneous equations in three variables is the intersection of two surfaces and is generally a curve in three-dimensional space. For example, a line (curve) in space may be considered to be the intersection of two planes (surfaces).

FIGURE 13.2.2

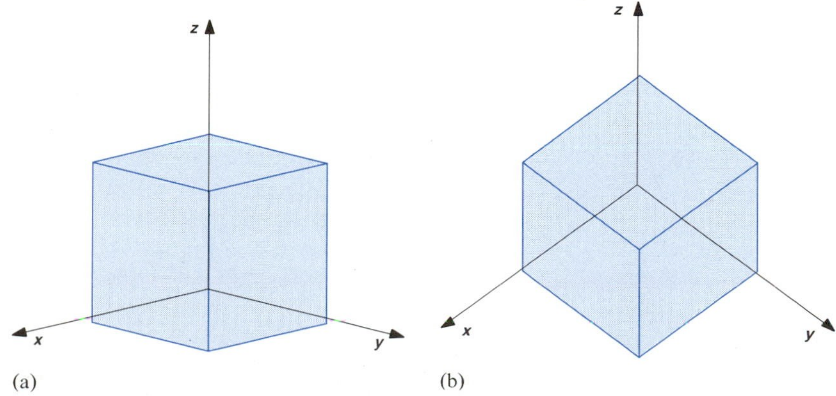

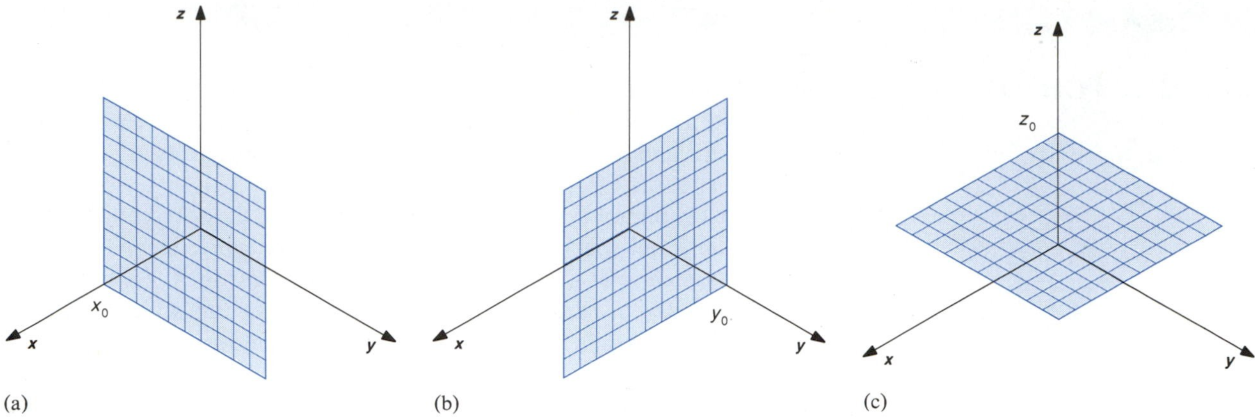


(a) (b) (c)

**FIGURE 13.2.3**

We begin our study of sketching by noting the following:

- $x = x_0$ is the equation of the plane that contains $(x_0, 0, 0)$ and is parallel to the $yz$-plane. See Figure 13.2.3a.
- $y = y_0$ is the equation of the plane that contains $(0, y_0, 0)$ and is parallel to the $xz$-plane. See Figure 13.2.3b.
- $z = z_0$ is the equation of the plane that contains $(0, 0, z_0)$ and is parallel to the $xy$-plane. See Figure 13.2.3c.

Let us see how to sketch curves that are contained in a plane that is parallel to one of the coordinate planes. This is a basic skill that is necessary for sketching three-dimensional figures.

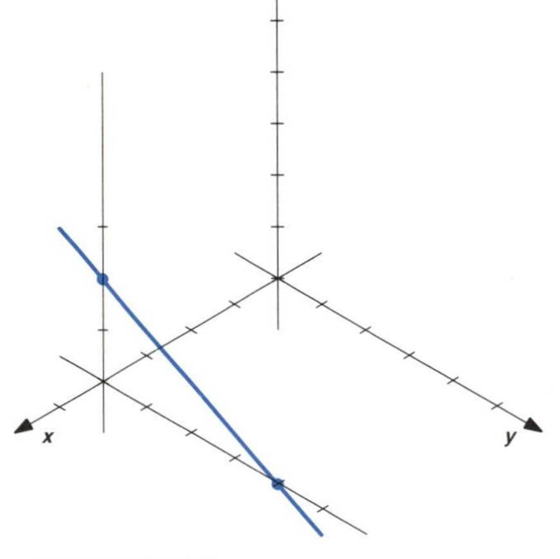


**FIGURE 13.2.4**

### ■ EXAMPLE 1

Sketch the curve $x = 4$, $y + 2z = 4$.

**SOLUTION**

The curve consists of all points with coordinates $(x, y, z)$ that satisfy both equations. The equation $x = 4$ tells us that the graph is in the plane $x = 4$. We begin by sketching scaled lines through $(4, 0, 0)$, parallel to each of the $y$-axis and the $z$-axis. This gives us a coordinate system for the $y$ and $z$ variables in the plane $x = 4$. We then graph the equation $y + 2z = 4$, considered to be a function of the two variables $y$ and $z$, in this coordinate plane. The graph is a line in the plane $x = 4$. It is easily sketched by plotting the intercepts with the coordinate axes we have drawn in the plane $x = 4$. See Figure 13.2.4. ■

### ■ EXAMPLE 2

Sketch the curve $y = 4$, $9x^2 + 4z^2 = 9y$.

**SOLUTION**

The equation $y = 4$ tells us the curve is contained in the plane $y = 4$. We establish a coordinate system for the variables $x$ and $z$ in this plane by drawing scaled lines through the point (0, 4, 0), parallel to each of the $x$-axis and the $z$-axis. We then plot the graph of the equation $9x^2 + 4z^2 = 9y$ in this plane. Since $y = 4$, we can substitute 4 for $y$ in the equation $9x^2 + 4z^2 = 9y$ to obtain $9x^2 + 4z^2 = 9(4)$, or

$$\frac{x^2}{2^2} + \frac{z^2}{3^2} = 1.$$

We then see that the graph is an ellipse in the plane $y = 4$. It is sketched, as usual, by plotting and then sketching through the four points on the axes, except we use the coordinate plane we have established in the plane $y = 4$. See Figure 13.2.5. ■

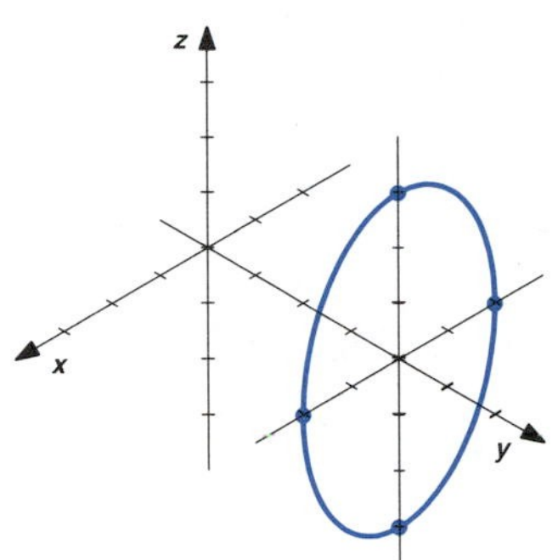


FIGURE 13.2.5

■ **EXAMPLE 3**

Sketch the curve $z = 2$, $y = x^2 - z^2$.

**SOLUTION**

The equation $z = 2$ indicates the curve is in the plane $z = 2$. A coordinate system for the variables $x$ and $y$ in this plane is established by drawing scaled lines through $z = 2$, parallel to each of the $x$-axis and the $y$-axis. We want to sketch the graph of $y = x^2 - z^2$ in this plane. Since $z = 2$, we can substitute 2 for $z$ to obtain an equation in the variables $x$ and $y$. We obtain $y = x^2 - 2^2$, or $y = x^2 - 4$. The graph is a parabola in the plane $z = 2$. It is sketched by finding its intercepts on the coordinate axes we have drawn in the plane $z = 2$, and then sketching the parabola through the intercepts. See Figure 13.2.6. ■

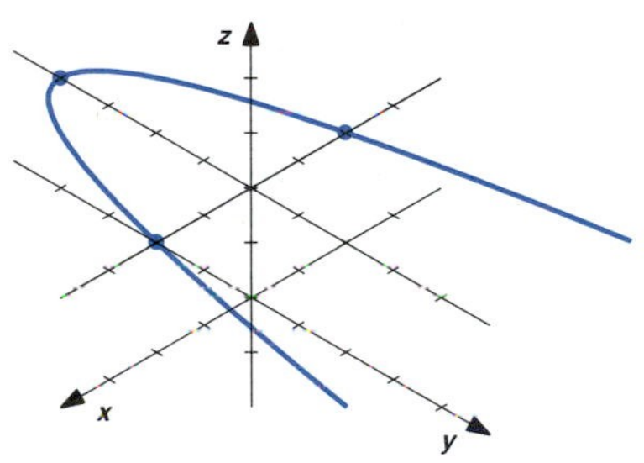


FIGURE 13.2.6

Let us now turn to the problem of sketching surfaces. Generally, it is difficult to sketch a three-dimensional surface on a two-dimensional surface such as a piece of paper. As with all sketches, the idea is to illustrate the general character. A standard method for representing a surface is to draw a number of curves that are the intersection of the surface with planes that are parallel to a coordinate plane. The number of such curves that are drawn depends on the complexity of the surface. An example of a surface that is represented by many curves is given in Figure 13.2.7. This is a familiar type of computer-generated illustration, much more detailed than the sketches we will be doing.

The intersection of a surface with a plane is called a **trace** of the surface in the plane. We will use a selection of traces in planes parallel to the coordinate planes to represent three-dimensional surfaces.

Generally, we should sketch the traces in each coordinate plane. However, the sketches should be kept as simple as possible. If you draw too many curves, the general character of the surface may be lost in the complexity of your sketch. Many surfaces can be represented very nicely by drawing only a trace in one coordinate plane and a typical trace in a perpendicular plane. We will use traces to give our sketches boundary edges, rather than letting the surfaces just fade away.

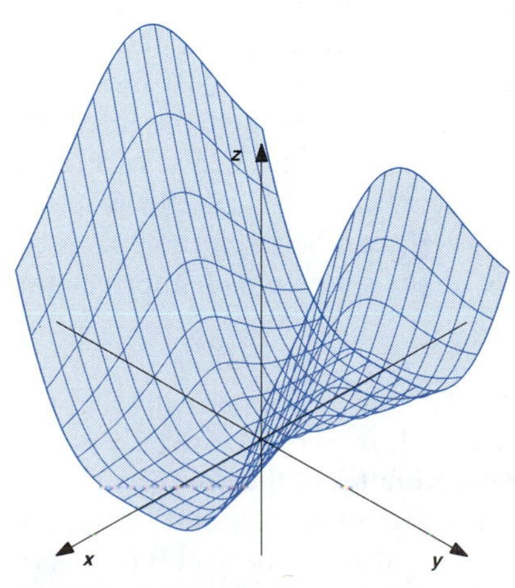


FIGURE 13.2.7

Points of intersection of a surface with each of the coordinate axes are called **intercepts**. Intercepts are often useful in determining the traces of the surface in the coordinate planes.

The most simple surfaces in three-dimensional space are planes. The **general equation** of a plane in three-dimensional space is

$$ax + by + cz + d = 0, \qquad \text{where } (a, b, c) \neq (0, 0, 0).$$

We will derive this equation and study the character of planes in the next chapter. At this time we are interested only in sketching the graph. To sketch a plane, we should determine and plot the intercepts. The intercepts are then used to sketch the coordinate traces. The coordinate traces of a plane are lines through the intercepts.

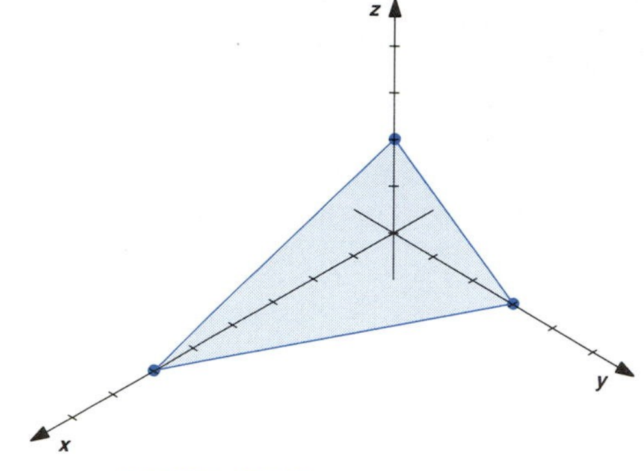

FIGURE 13.2.8

■ EXAMPLE 4

Sketch the plane $x + 2y + 3z = 6$.

SOLUTION

The $x$-intercept is found by setting $y = z = 0$ in the equation. This gives $x + 2(0) + 3(0) = 6$, so $(6, 0, 0)$ is the $x$-intercept. The other intercepts are easily seen to be $(0, 3, 0)$ and $(0, 0, 2)$. The three intercepts are plotted and used to sketch the traces. The traces are the lines through the intercepts in each of the coordinate planes. The segments of the traces between the intercepts are used to form the boundary edges of our sketch. See Figure 13.2.8. ■

■ EXAMPLE 5

Sketch the graph of $x + y = 2$ (in three-dimensional space).

SOLUTION

We recognize that the graph is a plane (not a line), because we know we are dealing with three-dimensional space. The $xy$-trace is the line $z = 0$, $x + y = 2$, which is easily drawn by plotting the intercepts on the $x$-axis and the $y$-axis. The $yz$-trace is given by the equations $x = 0$, $x + y = 2$. Substituting 0 for $x$ in the equation $x + y = 2$, we obtain $y = 2$. The graph is a vertical line in the $yz$-plane, through the intercept $(0, 2, 0)$. Similarly, we determine that the $xz$-trace is the vertical line through the intercept $(2, 0, 0)$. The coordinate traces are sketched in Figure 13.2.9a. In Figure 13.2.9b we have drawn the boundary curve $z = 3$, $x + y = 2$ to give the sketch an upper boundary edge. This line is drawn between the points on our sketch where the coordinate traces of the plane $z = 3$ intersect the coordinate traces of the plane $x + y = 2$. ■

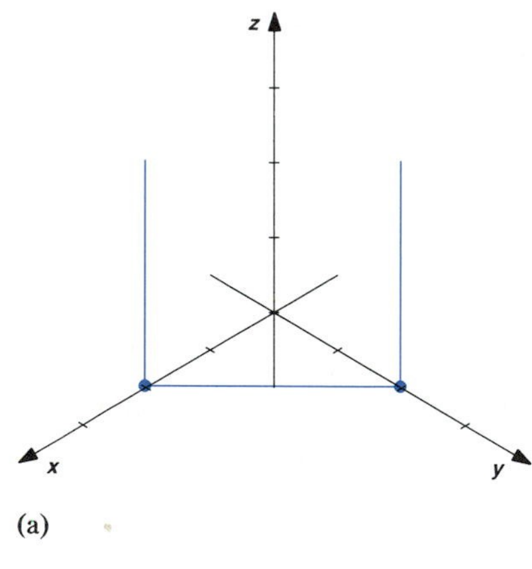

(a)

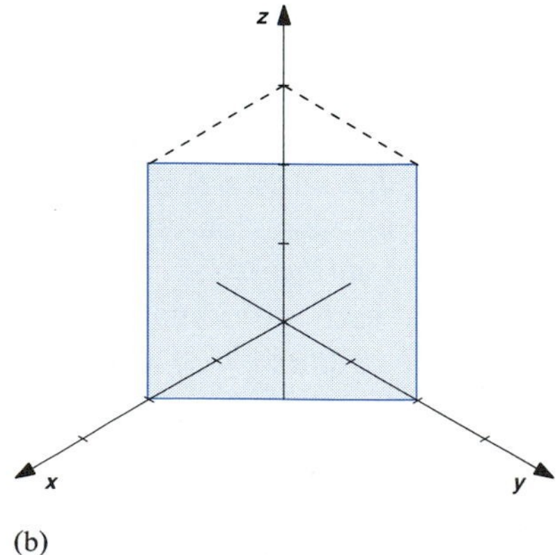

(b)

FIGURE 13.2.9

Note that the equation $x + y = 2$ of Example 5 does not involve the variable $z$. This means that whenever the coordinates of a point $(x_0, y_0, z_0)$ satisfy the equation $x + y = 2$, then the coordinates of the point $(x_0, y_0, z)$ satisfy the equation for every value of $z$. In particular, if a point $(x_0, y_0, 0)$ is on the $xy$-trace, then the point $(x_0, y_0, z)$ is also on the graph for every value of $z$. As $z$ varies over all real numbers, the points $(x_0, y_0, z)$ form a

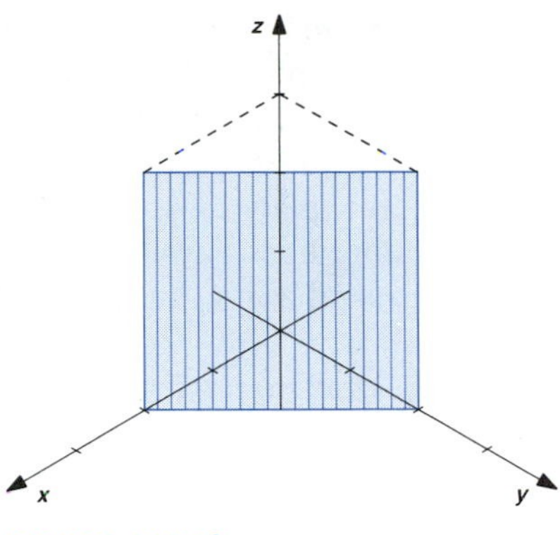


FIGURE 13.2.10

line through $(x_0, y_0, 0)$, parallel to the $z$-axis. The entire plane is generated by these lines as $(x_0, y_0, 0)$ varies over the $xy$-trace. See Figure 13.2.10. These ideas allow us to use the $xy$-trace to sketch the $yz$-trace and the $xz$-trace, without dealing directly with the equations of the latter traces. That is, the $yz$-trace is obtained by drawing a line through the $y$-intercept $(0, 2, 0)$, parallel to the $z$-axis; the $xz$-trace is drawn through the $x$-intercept $(2, 0, 0)$, parallel to the $z$-axis.

**EXAMPLE 6**

Sketch the plane $y = z$.

**SOLUTION**

The graph is a plane in three-dimensional space. The $yz$-trace is the line $x = 0$, $y = z$. Since the equation $y = z$ does not involve the variable $x$, we can use the sketch of the $yz$-trace to determine the other coordinate traces. That is, whenever a point $(0, y_0, z_0)$ is on the $yz$-trace, then the point $(x, y_0, z_0)$ is on the graph for every value of $x$. These points generate a line through $(0, y_0, z_0)$, parallel to the $x$-axis. In particular, since the $yz$-trace contains the point $(0, 0, 0)$, we know that the graph contains the line through $(0, 0, 0)$, parallel to the $x$-axis. This line is the $x$-axis, which is the trace of the plane in the $xy$-plane and in the $xz$-plane. (When a coordinate trace contains a coordinate axis, be sure to mark the axis in such a way that will remind you that it is a trace.) The coordinate traces are sketched in Figure 13.2.11a. We see that the traces form a single line, so the isometric view of the plane is parallel to the plane and the plane appears as a line. In Figure 13.2.11b we have redrawn the traces as viewed from a point that is higher than usual, so the $x$-axis and the $y$-axis are drawn at an angle less than 120° and a relatively small scale is used on the $z$-axis. In Figure 13.2.11c we have used the intersection of the plane $y = z$ with each of the planes $x = 6$ and $y = 3$ to give the sketch boundary edges and have indicated that the plane $y = z$ is generated by lines through the $yz$-trace, parallel to the $x$-axis. ■

FIGURE 13.2.11

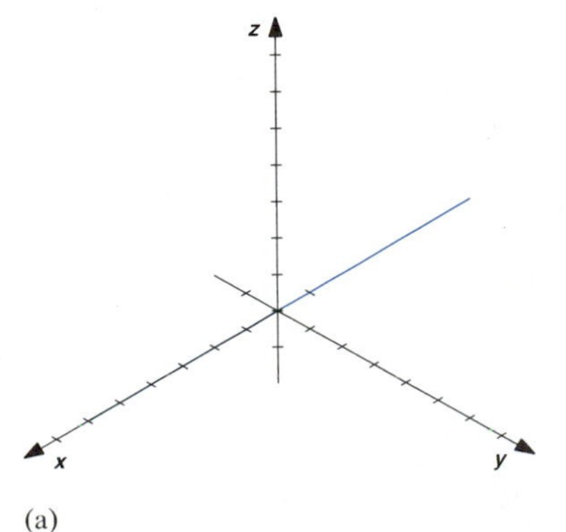


(a)

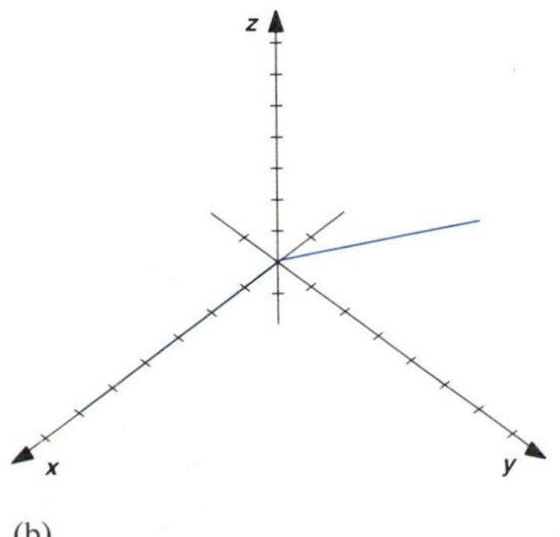


(b)

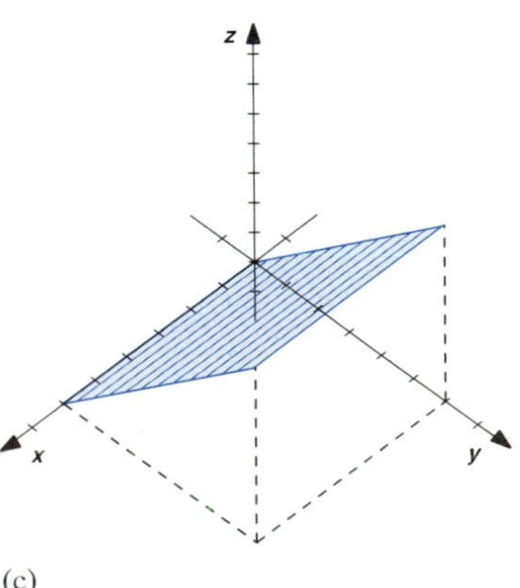


(c)

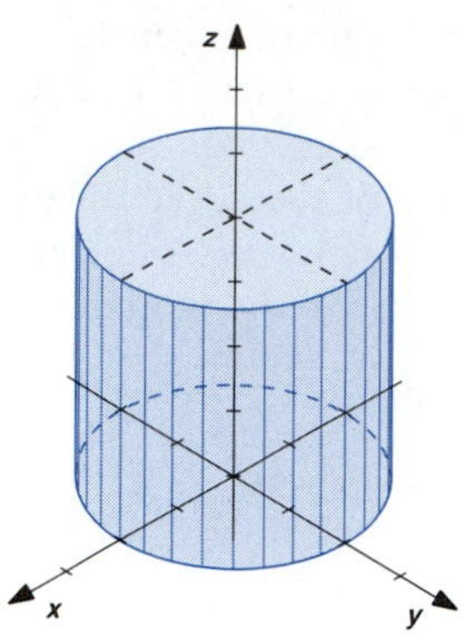


FIGURE 13.2.12

A surface in three-dimensional space that is generated by a collection of parallel lines is called a **generalized cylinder**. The graph in three-dimensional space of an equation that has one (or more) of the three variables missing will be a generalized cylinder with generating lines parallel to the axis of the missing variable(s). For example, the plane $x + y = 2$ of Example 5 is a generalized cylinder with generating lines parallel to the $z$-axis. The plane $y = z$ of Example 6 is a generalized cylinder with generating lines parallel to the $x$-axis.

A generalized cylinder can be sketched by drawing a few of the generating lines between two traces that are parallel to each other and perpendicular to the generating lines.

### ■ EXAMPLE 7

Sketch the graph of $x^2 + y^2 = 2^2$ (in three-dimensional space).

#### SOLUTION

The graph is sketched in Figure 13.2.12. The graph is a (generalized) cylinder in three-dimensional space. The generating lines are parallel to the $z$-axis and perpendicular to the $xy$-plane. The $xy$-trace is the circle $z = 0$, $x^2 + y^2 = 2^2$. We have sketched the $xy$-trace and the parallel trace $z = 4$, $x^2 + y^2 = 4$, and drawn some of the generating lines between these traces. ■

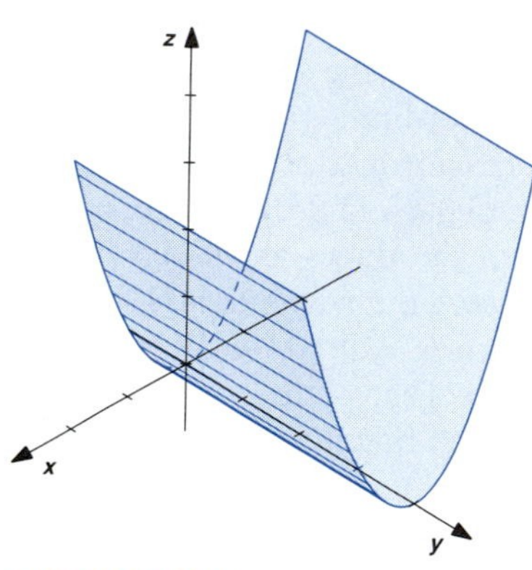


FIGURE 13.2.13

### ■ EXAMPLE 8

Sketch the graph of $z = x^2$.

#### SOLUTION

We assume that we are interested in a graph in three-dimensional space, so the graph is a cylinder with generating lines parallel to the $y$-axis. See Figure 13.2.13. We have drawn the trace $y = 0$, $z = x^2$ (a parabola), the parallel trace $y = 4$, $z = x^2$, and some of the generating lines between these traces. ■

### ■ EXAMPLE 9

Sketch the graph of $z = \sin y$.

FIGURE 13.2.14

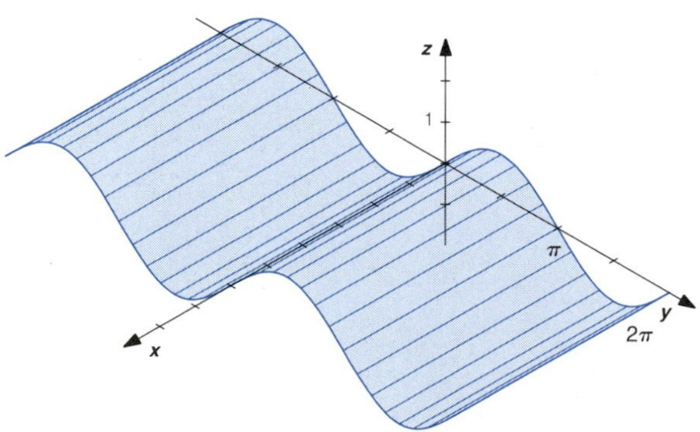

**SOLUTION**

The graph is sketched in Figure 13.2.14. The surface is a generalized cylinder with generating lines parallel to the $x$-axis. We have drawn the trace $x = 0$, $z = \sin y$, the parallel trace $x = 6$, $z = \sin y$, and some of the generating lines between these traces. ■

## EXERCISES 13.2

Sketch the curves with equations given in Exercises 1-10.

1 $x = -2$, $3z - 2y = 6$

2 $x = 3$, $z = 1 - y^2$

3 $y = 2$, $x^2 + z^2 = 4$

4 $y = 2$, $2x + 3z = 6$

5 $z = 2$, $x^2 + y^2 = 2z$

6 $z = 4$, $z = x + 2y$

7 $y = 2$, $z = y^2 - x^2$

8 $z = 0$, $4x^2 + 4z^2 = y^2$

9 $z = 1$, $z = y^2 - x^2$

10 $y = 0$, $x^2 - y^2 - z^2 = 1$

Sketch the surfaces with equations given in Exercises 11-30.

11 $2x + y + 3z = 6$

12 $2x + 3y + 3z = 6$

13 $3x + 2z = 6$

14 $3y + 4z = 12$

15 $z = x + 2y$

16 $z = 2x + 2y$

17 $2x - 2y + z = 4$

18 $3x + 2y - 2z = 6$

19 $y = 2x$

20 $z = 2x$

21 $x^2 + y^2 - 2y = 0$

22 $y^2 + z^2 = 4$

23 $4x^2 + z^2 = 4$

24 $4x^2 + y^2 - 8x = 0$

25 $z = 4 - y^2$

26 $z = y^2 - 1$

27 $y = \sin x$

28 $z = \cos y$

29 $z = |y|$

30 $z = -y/|y|$

## 13.3 QUADRIC SURFACES

We will be dealing with surfaces that have equations of the form

$$Ax^2 + By^2 + Cz^2 + Dx + Ey + Fz + G = 0,$$

where $A$, $B$, and $C$ are not all equal to zero. Surfaces of this type are called **quadric surfaces**. Many of these surfaces can be sketched by drawing only one coordinate trace and a typical trace in a perpendicular plane. We should look particularly for typical traces that are circles or ellipses. These occur quite often in quadric surfaces, they are relatively easy to draw, and they give a good indication of the character of the surface.

Be sure you can draw circles in three-dimensional space. Figure 13.3.1 contains sketches of circles parallel to each of the coordinate planes. Each of these circles appears as an ellipse. Circles are easily sketched by locating the center, drawing diameters that are parallel to coordinate axes, marking off the length of the radius along the diameters that have been drawn, and then sketching an ellipse through the marks. Tangents to the circles at the points we have marked are drawn parallel to a coordinate axis.

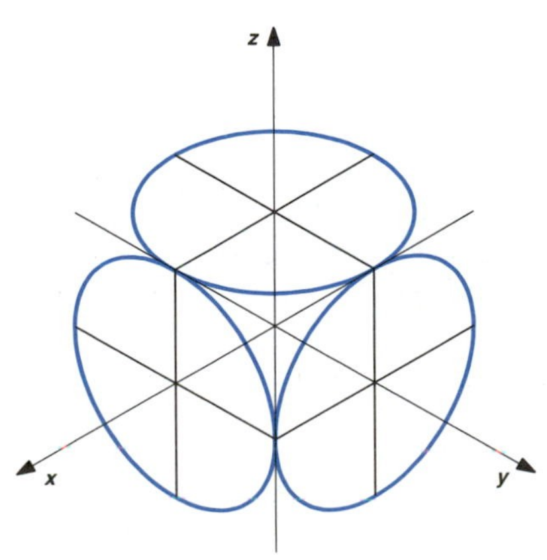


FIGURE 13.3.1

■ **EXAMPLE 1**

Sketch the graph of $z = x^2 + y^2$.

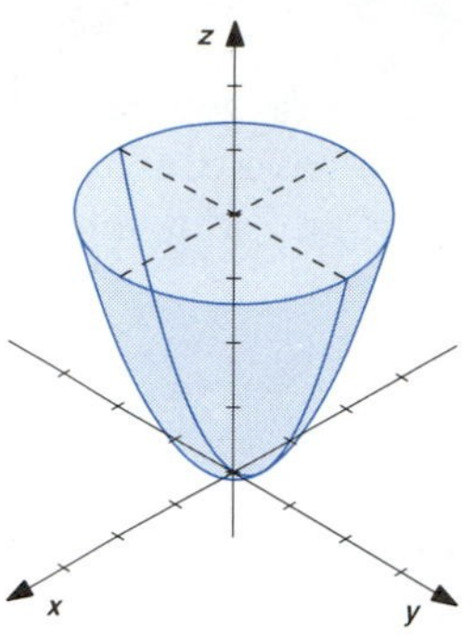


FIGURE 13.3.2

**SOLUTION**

The graph is sketched in Figure 13.3.2. We begin our sketch by looking at the $yz$-trace. This is given by the equations

$$x = 0, \qquad z = x^2 + y^2.$$

Substituting 0 for $x$ in the second equation, we obtain $z = y^2$. The $yz$-trace is then seen to be a parabola, which we sketch. We then notice that traces of the surface in planes $z = z_0$ have equations

$$z = z_0, \qquad z_0 = x^2 + y^2.$$

These are circles that are parallel to the $xy$-plane, with centers on the $z$-axis. We have sketched the trace in the plane $z = 4$ to indicate a typical circle and to give our sketch a top edge. The radius of the circle is determined by the intersection of the coordinate traces of the plane $z = 4$ and the parabola in the $yz$-plane. The sketch is given side edges to complete the picture. The side edges of the sketch do not coincide with either the $yz$-trace or the $xz$-trace. ■

Surfaces of the type illustrated in Figure 13.3.2 are called **paraboloids**.

■ **EXAMPLE 2**

Sketch the graph of $4x^2 - y^2 + 4z^2 = 0$.

**SOLUTION**

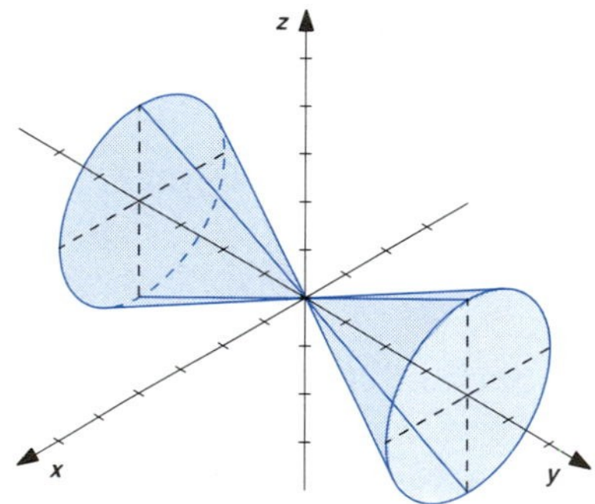


FIGURE 13.3.3

The graph is sketched in Figure 13.3.3. The first step is to determine the $yz$-trace. The $yz$-trace is given by the equations

$$x = 0, \qquad 4x^2 - y^2 + 4z^2 = 0.$$

Simplifying by substituting 0 for $x$ in the second equation, we obtain $-y^2 + 4z^2 = 0$, so $y = 2z$ or $y = -2z$. The graph of the $yz$-trace is then seen to be two lines, which we sketch. The intersection of the surface with the plane $y = y_0$, $y_0$ a constant, is given by the equations

$$y = y_0, \qquad x^2 + z^2 = \frac{y_0^2}{4}.$$

These curves are circles that are parallel to the $xz$-plane, with centers on the $y$-axis. We draw the circles in the planes $y = -4$ and $y = 4$ to represent typical traces and give our sketch side edges. The $yz$-trace we have drawn determines the radius of these circles. The sketch is completed by drawing top and bottom edges. ■

The graph in Figure 13.3.3 is a **cone**.

■ **EXAMPLE 3**

Sketch the graph of $16x^2 + 9y^2 + 36z^2 = 144$.

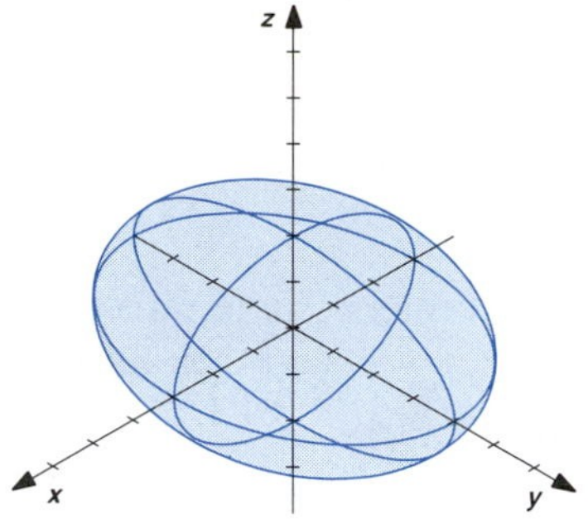


FIGURE 13.3.4

**SOLUTION**

The graph is sketched in Figure 13.3.4. It is easy to determine the equations of the three coordinate traces. We obtain

$$x = 0, \qquad \frac{y^2}{4^2} + \frac{z^2}{2^2} = 1,$$

$$y = 0, \qquad \frac{x^2}{3^2} + \frac{z^2}{2^2} = 1,$$

$$z = 0, \qquad \frac{x^2}{3^2} + \frac{y^2}{4^2} = 1.$$

The coordinate traces are ellipses, which we draw. The sketch is then given an outer edge. ■

The graph in Figure 13.3.4 is called an **ellipsoid**.

■ **EXAMPLE 4**

Sketch the graph of $4x^2 - y^2 + 4z^2 + 4 = 0$.

**SOLUTION**

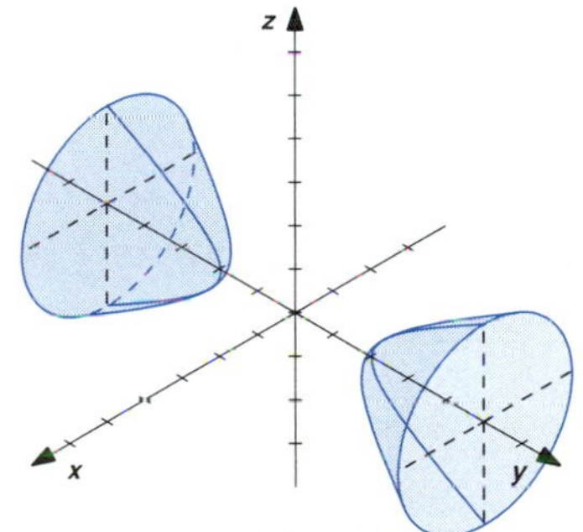


FIGURE 13.3.5

The graph is sketched in Figure 13.3.5. The $yz$-trace is given by the equations

$$x = 0, \qquad \frac{y^2}{2^2} - \frac{z^2}{1^2} = 1.$$

This is a hyperbola, which we draw. If $y_0^2 > 4$, the traces

$$y = y_0, \qquad x^2 + z^2 = y_0^2 - 4$$

are circles that are parallel to the $xz$-plane, with centers on the $y$-axis. The traces are single points for $y_0 = \pm 2$, and empty for $-2 < y_0 < 2$. We have drawn the circles in the planes $y = 5$ and $y = -5$, and then filled in the top and bottom edges. ■

The surface in Figure 13.3.5 is called a **hyperboloid of two sheets**.

■ **EXAMPLE 5**

Sketch the graph of $4x^2 + 4y^2 - z^2 = 4$.

**SOLUTION**

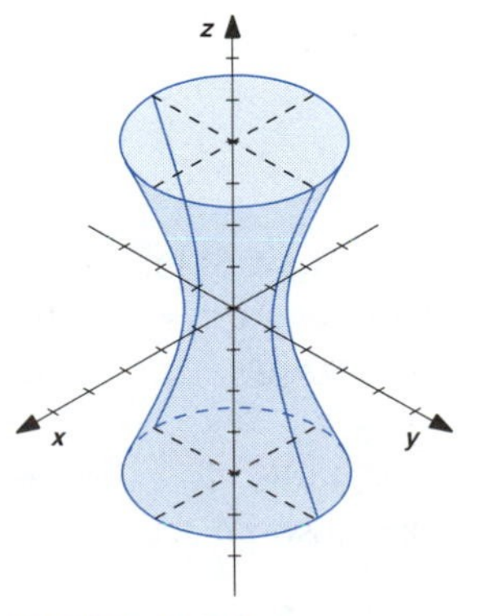


FIGURE 13.3.6

The graph is sketched in Figure 13.3.6. The $yz$-trace is given by the equations

$$x = 0, \qquad \frac{y^2}{1^2} - \frac{z^2}{2^2} = 1.$$

This is a hyperbola, which we draw. The traces

$$z = z_0, \qquad x^2 + y^2 = 1 + \frac{z_0^2}{4}$$

are circles that are parallel to the $xy$-plane, with centers on the $z$-axis. We have drawn the circles in the planes $z = 4$ and $z = -4$, and then filled in the side edges. ■

The surface in Figure 13.3.6 is called a **hyperboloid of one sheet**.

It is not necessary to memorize the form of the equations to distinguish between a hyperboloid of one sheet and a hyperboloid of two sheets. In either case, we first sketch a hyperbolic trace, determine the coordinate plane to which the circular traces are parallel, and then sketch the circular traces. If these traces are drawn correctly, the completed sketch will have the correct number of sheets.

■ **EXAMPLE 6**

Sketch the graph of $4z = y^2 - 4x^2$.

**SOLUTION**

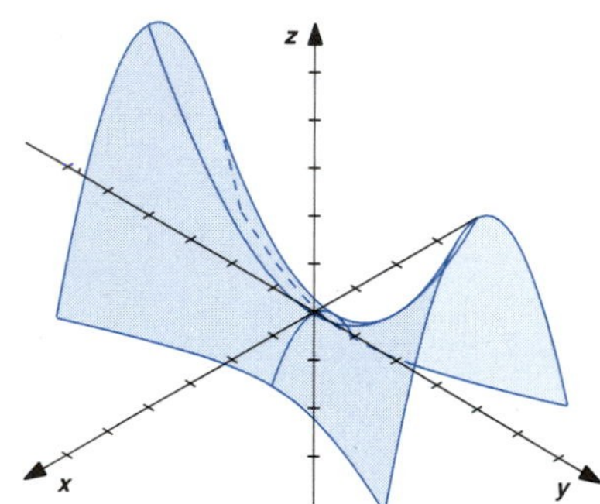


FIGURE 13.3.7

The graph is sketched in Figure 13.3.7. The $yz$-trace is given by

$$x = 0, \qquad 4z = y^2,$$

which is a parabola that opens upward. This is drawn. The $xz$-trace is given by the equations

$$y = 0, \qquad z = -x^2,$$

a parabola that opens downward. This trace is drawn. The traces given by

$$y = y_0, \qquad 4z = y_0^2 - 4x^2,$$

are also parabolas. We sketch the parabolas in the planes $y = 4$ and $y = -4$ for side edges of our sketch. The traces given by

$$z = z_0, \qquad 4z_0 = y^2 - 4x^2$$

are hyperbolas for $z_0 \neq 0$. When $z_0 = 0$, the $xy$-trace is a pair of lines that intersect at the origin. We sketch the hyperbola in the plane $z = -1$ to give our sketch a bottom edge. Finally, we fill in the top edge. ■

The surface in Figure 13.3.7, which resembles a saddle, is called a **hyperbolic paraboloid**. These are difficult to draw because no trace is a circle or ellipse.

(a)



(b)

**FIGURE 13.3.8**

■ **EXAMPLE 7**

Sketch the graph of $x = \sqrt{y^2 + z^2}$.

**SOLUTION**

The $yz$-trace consists of the single point at the origin. The $xz$-trace is given by

$$y = 0, \qquad x = \sqrt{z^2}.$$

Since $\sqrt{z^2} = |z|$, the second equation can be written $x = |z|$. We have sketched the $xz$-trace in Figure 13.3.8a. The traces of the surface in planes $x = x_0$, $x_0 > 0$, are given by

$$x = x_0, \qquad y^2 + z^2 = x_0^2,$$

which are circles that are parallel to the $yz$-plane, with centers on the $x$-axis. We have drawn the circle in the plane $x = 4$ and added the edge lines in Figure 13.3.8a. At this point we notice that the sketch does not seem to indicate the general character of the surface very well. The graph should look like a single-napped cone. We can obtain a better picture by moving the point of view toward the $yz$-plane, so the $y$-axis and the $z$-axis are drawn at an angle greater than 120° and a relatively large scale is used on the $x$-axis. This is done in Figure 13.3.8b. ■

## EXERCISES 13.3

Sketch the graphs of the equations given in Exercises 1–16.

1 $x^2 + z^2 = y$

2 $x^2 + y^2 + z = 4$

3 $4x^2 + 4y^2 + z^2 = 4$

4 $x^2 + y^2 + z^2 = 9$

5 $x^2 - 9y^2 - 9z^2 = 0$

6 $4x^2 + 4y^2 = z^2$

7 $z = \sqrt{3x^2 + 3y^2}$

8 $z = -2\sqrt{x^2 + y^2}$

9 $z = 4 - 2\sqrt{x^2 + y^2}$

10 $3y^2 + 3z^2 = x$

11 $4x^2 + 4y^2 - z^2 + 4 = 0$

12 $x^2 - 4y^2 - 4z^2 = 4$

13 $9x^2 - y^2 + 9z^2 = 9$

14 $x^2 - 9y^2 - 9z^2 + 9 = 0$

15 $8z = x^2 - 8y^2$

16 $z = y^2 - 2x^2$

## REVIEW EXERCISES

1 Sketch the line segment between the points $(3, -2, 1)$ and $(-1, 2, 3)$. Find the length and midpoint of the segment.

2 Sketch the line segment between the points $(2, 3, -1)$ and $(-3, -2, 2)$. Find the length and midpoint of the segment.

3 Find the standard equation of the sphere that has center $(2, 0, 1)$ and contains the point $(0, 2, 2)$.

4 Find the standard equation of the sphere that has center $(3, 2, 4)$ and is tangent to the $xy$-plane.

5 Find the center and radius of the sphere

$$x^2 + y^2 + z^2 + 4x - 6y = 0.$$

6 Find the center and radius of the sphere

$$x^2 + y^2 + z^2 + 2x - y - 3z = 0.$$

Sketch the curves with equations given in Exercises 7–12.

7 $x = 2,\ 2z + 3y = 6$

8 $x = -1,\ 4y^2 + 9z^2 = 36$

9 $y = -2,\ z = x^2 - y^2$

10 $y = 1,\ x^2 - z^2 = y$

11 $z = -1,\ x^2 + 4y^2 = 4z^2$

12 $z = -2,\ x + 2y + z + 2 = 0$

Sketch the surfaces with equations given in Exercises 13–28.

13 $3x + 2y + 6z = 6$

14 $3y + 2z = 6$

15 $z = 2x + 2y$

16 $z = 2y$

17 $x^2 + y^2 - 4x = 0$

18 $x^2 + 4z^2 = 4$

19 $z = y^2$

20 $z^2 = 4y^2$

21 $z = x^2 + y^2$

22 $z = 4 - x^2 - y^2$

23 $y^2 = 4x^2 + 4z^2$

24 $9x^2 + 4y^2 + 36z^2 = 36$

25 $z^2 = 4x^2 + 4y^2 + 4$

26 $y^2 = 4x^2 + 4z^2 + 4$

27 $z = y^2 - x^2$

28 $z = x^2 - y^2$

CHAPTER FOURTEEN

# *Vectors*

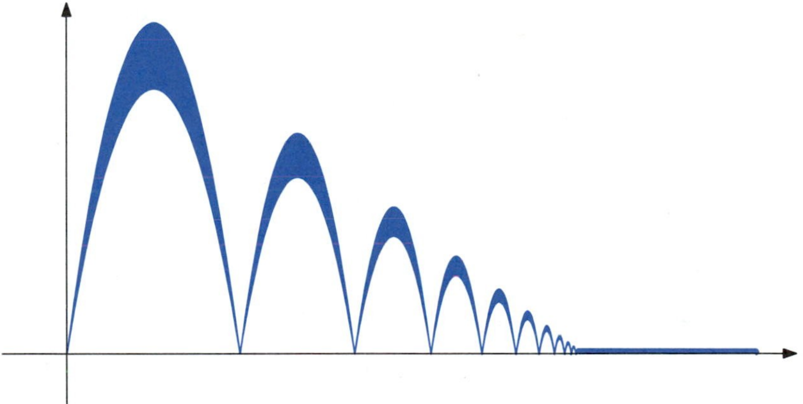

Many physical quantities involve direction as well as magnitude. For example, the change in position of an object that is moving in a coordinate plane or three-dimensional space involves the speed and the direction of movement. Also, the effect of a force on an object depends on the magnitude and the direction of the force. The mathematical objects that are used to study problems that involve magnitude and direction are called **vectors**. In this chapter we will introduce two-dimensional and three-dimensional vectors. We will study geometric interpretations of vectors and see how vectors are used in a few applications.

## 14.1 VECTORS IN THE PLANE

We define a mathematical system that embodies the concepts of length and direction as well as the arithmetical operations we will need for applications.

*Definition*

An ordered pair of real numbers $\langle a, b\rangle$ is called a **two-dimensional vector**. The numbers $a$ and $b$ are called **components** of the vector. Two vectors are **equal** if and only if corresponding components are equal. That is,

$$\langle a, b\rangle = \langle c, d\rangle \quad \text{means} \quad a = c \quad \text{and} \quad b = d.$$

The **sum** of two vectors is

$$\langle a, b\rangle + \langle c, d\rangle = \langle a + c, b + d\rangle$$

and the **product** of a scalar (number) and a vector is

$$c\langle a, b\rangle = \langle ca, cb\rangle.$$

Addition of two vectors and multiplying a vector by a scalar involve the operations of addition and multiplication of real numbers, carried out component by component. The sum of two vectors is a vector and the product of a scalar and a vector is a vector.

We will use boldface to distinguish vectors from scalars. For example, $c\mathbf{U}$ denotes the product of the scalar $c$ and the vector $\mathbf{U}$. Another way to distinguish vectors is to draw an arrow over them.

The **zero vector** is

$$\mathbf{0} = \langle 0, 0\rangle.$$

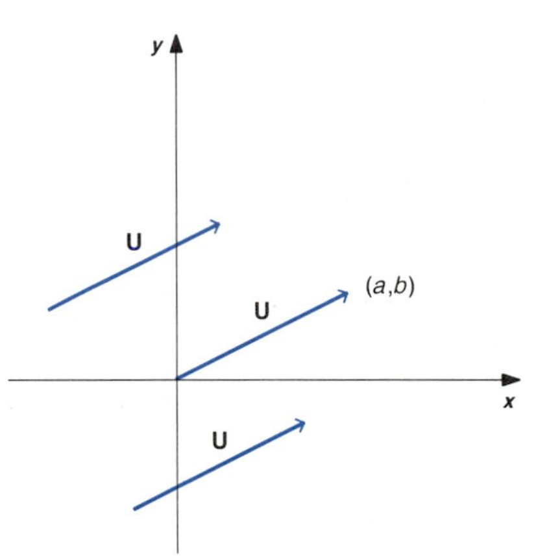

FIGURE 14.1.1

The **negative** of the vector $\mathbf{U} = \langle a, b\rangle$ is

$$-\mathbf{U} = -\langle a, b\rangle = \langle -a, -b\rangle.$$

The vector $\mathbf{U} = \langle a, b\rangle$ can be represented geometrically as an **arrow** from the origin to the point $(a, b)$, or by any other arrow that has the same length and direction. We sketch a vector by sketching any one of its representative arrows. See Figure 14.1.1. Two vectors are equal if and only if each representative arrow of one of the vectors is also a representative arrow of the other. We will not necessarily distinguish between a vector and a representative arrow of the vector. Thus, we may speak of the length and direction of a two-dimensional vector when we mean the length and direction of a representive arrow.

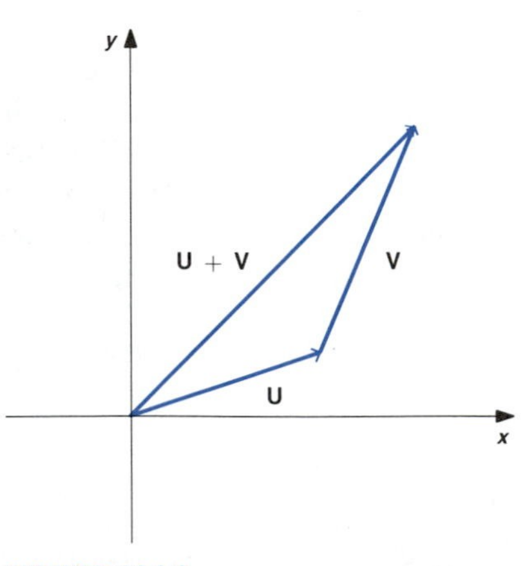

FIGURE 14.1.2

Addition of vectors and multiplication of a vector by a scalar have geometric interpretations.

The sum of $\mathbf{U}$ and $\mathbf{V}$ can be determined geometrically by moving the initial point of $\mathbf{V}$ to the terminal point of $\mathbf{U}$. The sum $\mathbf{U} + \mathbf{V}$ is then the vector from the initial point of $\mathbf{U}$ to the terminal point of $\mathbf{V}$. See Figure 14.1.2.

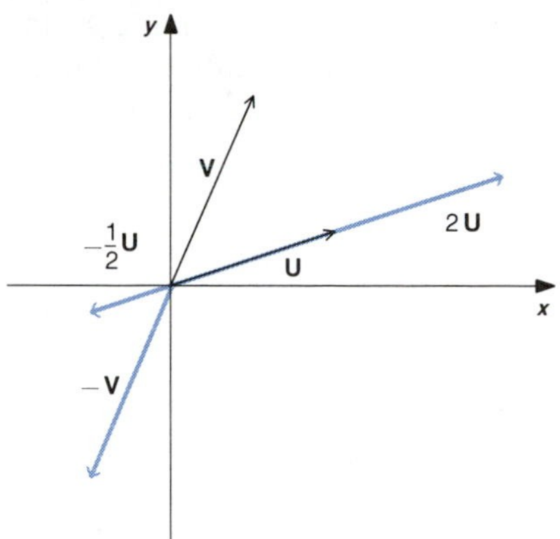


FIGURE 14.1.3

If $c > 0$, $c\mathbf{U}$ is a vector in the same direction as $\mathbf{U}$ with length $c$ times the length of $\mathbf{U}$. If $c < 0$, $c\mathbf{U}$ is in the direction opposite to $\mathbf{U}$ with length $|c|$ times the length of $\mathbf{U}$. The vector $-\mathbf{V} = (-1)\mathbf{V}$ is in the direction opposite to $\mathbf{V}$ with the same length as $\mathbf{V}$. See Figure 14.1.3.

Operations on vectors have the following properties which correspond to properties of addition and multiplication of real numbers.

$$\mathbf{U} + (\mathbf{V} + \mathbf{W}) = (\mathbf{U} + \mathbf{V}) + \mathbf{W},$$

$$\mathbf{U} + \mathbf{V} = \mathbf{V} + \mathbf{U},$$

$$\mathbf{U} + \mathbf{0} = \mathbf{0} + \mathbf{U} = \mathbf{U},$$

$$\mathbf{U} + (-\mathbf{U}) = (-\mathbf{U}) + \mathbf{U} = \mathbf{0},$$

$$c(\mathbf{U} + \mathbf{V}) = c\mathbf{U} + c\mathbf{V},$$

$$(c + d)\mathbf{U} = c\mathbf{U} + d\mathbf{U},$$

$$(cd)\mathbf{U} = c(d\mathbf{U}),$$

$$1\mathbf{U} = \mathbf{U}.$$

These properties are shared by many mathematical systems. They are characteristic properties of a **vector space**.

It is convenient to define two special two-dimensional vectors,

$$\mathbf{i} = \langle 1, 0\rangle \quad \text{and} \quad \mathbf{j} = \langle 0, 1\rangle.$$

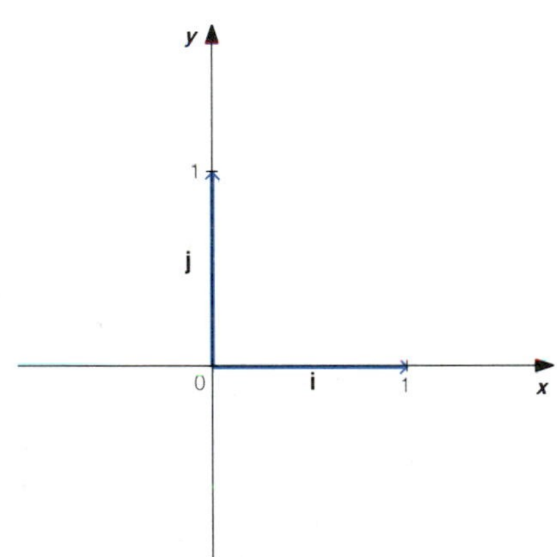


FIGURE 14.1.4

See Figure 14.1.4.

The vector $\langle a, b\rangle$ can be expressed in terms of the vectors $\mathbf{i}$ and $\mathbf{j}$. We have

$$\langle a, b\rangle = a\mathbf{i} + b\mathbf{j}. \tag{1}$$

■ *Proof*

$$\begin{aligned}\langle a, b\rangle &= \langle a, 0\rangle + \langle 0, b\rangle \\ &= a\langle 1, 0\rangle + b\langle 0, 1\rangle \\ &= a\mathbf{i} + b\mathbf{j}.\end{aligned}$$

■

Addition and scalar multiplication of vectors can be expressed in terms of the $\mathbf{i}$ and $\mathbf{j}$ notation as follows.

$$(a\mathbf{i} + b\mathbf{j}) + (c\mathbf{i} + d\mathbf{j}) = (a + c)\mathbf{i} + (b + d)\mathbf{j} \tag{2}$$

and

$$c(a\mathbf{i} + b\mathbf{j}) = ca\mathbf{i} + cb\mathbf{j}. \tag{3}$$

This means we can perform the operations of vector addition and scalar multiplication by using the usual rules of algebra and combining "like" terms. The vectors **i** and **j** act as place holders for the components.

**■ EXAMPLE 1**

Evaluate
(a) $3(2\mathbf{i} - \mathbf{j})$,
(b) $(2\mathbf{i} + 3\mathbf{j}) + (\mathbf{i} - 2\mathbf{j})$,
(c) $5(\mathbf{i} + 2\mathbf{j}) - 2(2\mathbf{i} + 7\mathbf{j})$.

**SOLUTION**

(a) $3(2\mathbf{i} - \mathbf{j}) = 6\mathbf{i} - 3\mathbf{j}$.
(b) $(2\mathbf{i} + 3\mathbf{j}) + (\mathbf{i} - 2\mathbf{j}) = 2\mathbf{i} + 3\mathbf{j} + \mathbf{i} - 2\mathbf{j} = 3\mathbf{i} + \mathbf{j}$.
(c) $5(\mathbf{i} + 2\mathbf{j}) - 2(2\mathbf{i} + 7\mathbf{j}) = 5\mathbf{i} + 10\mathbf{j} - 4\mathbf{i} - 14\mathbf{j} = \mathbf{i} - 4\mathbf{j}$. ■

The length of a two-dimensional vector is defined to agree with the length of any of its representative arrows.

■ *Definition*

The **length** of a two-dimensional vector is

$$|a\mathbf{i} + b\mathbf{j}| = \sqrt{a^2 + b^2}.$$ ■

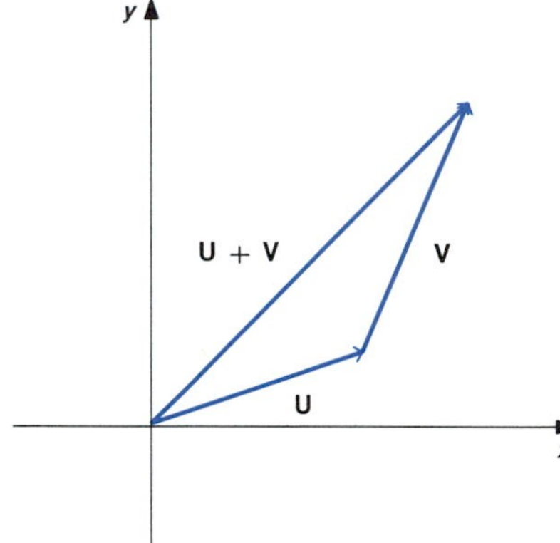

FIGURE 14.1.5

The length of vectors has the following two properties.

$$|c\mathbf{U}| = |c||\mathbf{U}|,$$

$$|\mathbf{U} + \mathbf{V}| \leq |\mathbf{U}| + |\mathbf{V}|.$$

These properties correspond to properties of the absolute value of real numbers. (Note that $|c|$ is the absolute value of the real number $c$.) The first property can be verified directly from the definitions. The second property is evident from the geometric interpretation of the sum of two vectors. That is, the length of one side of a triangle is less than the sum of the lengths of the other two sides. See Figure 14.1.5. Equality occurs whenever **U** and **V** are in the same direction.

**■ EXAMPLE 2**

Evaluate $|3\mathbf{i} - 4\mathbf{j}|$.

**SOLUTION**

$$|3\mathbf{i} - 4\mathbf{j}| = \sqrt{3^2 + (-4)^2} = 5.$$

See Figure 14.1.6. ■

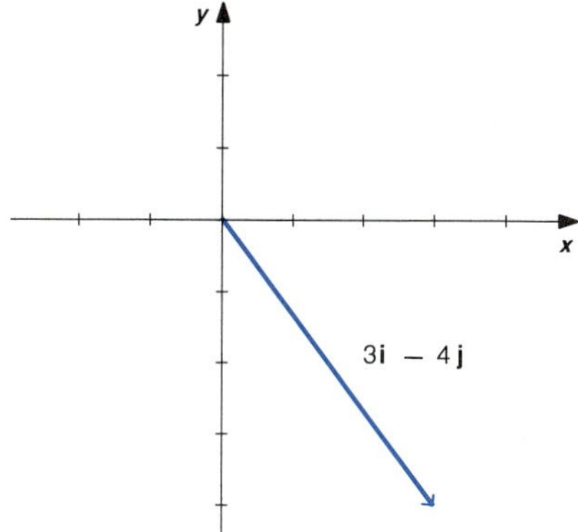

FIGURE 14.1.6

To use vectors to solve geometric problems, the vectors need to be expressed in the form $a\mathbf{i} + b\mathbf{j}$. We can use the following two observations to do this.

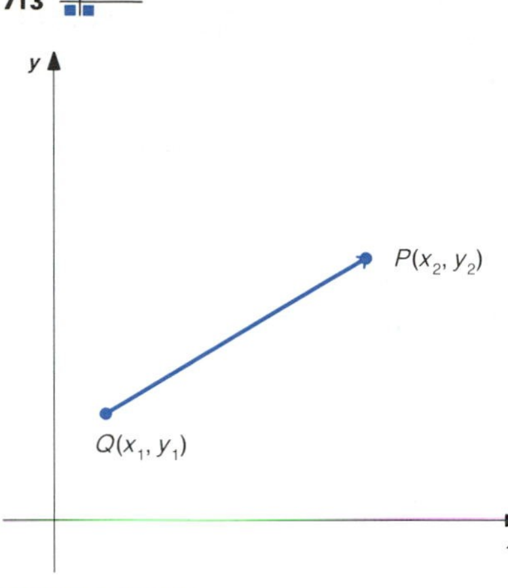


FIGURE 14.1.7

If $Q(x_1, y_1)$ and $P(x_2, y_2)$ are points in a coordinate plane, the vector from $Q$ to $P$ is

$$\mathbf{QP} = (x_2 - x_1)\mathbf{i} + (y_2 - y_1)\mathbf{j}.$$

See Figure 14.1.7.

The vector

$$r \cos \theta \mathbf{i} + r \sin \theta \mathbf{j}$$

has length $r$ and direction at an angle $\theta$ with the positive $x$-axis. See Figure 14.1.8.

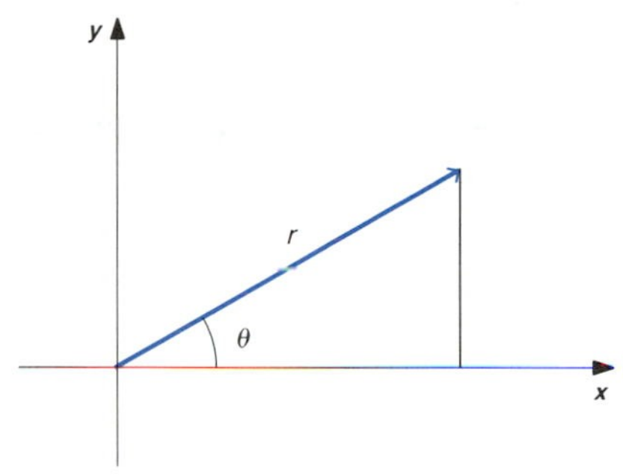


FIGURE 14.1.8

An important application of vectors is to problems that involve force. That is, a force has magnitude and direction, and can be associated with a vector. If two or more forces are applied simultaneously to an object, the **resultant force** is the vector sum of the corresponding vectors.

The zero vector is the resultant of two forces that are of equal magnitude and opposite direction. The effect of several forces can be counteracted by a single force with magnitude that of the resultant force and direction opposite to the resultant force.

■ **EXAMPLE 3**

A person with a broken femur is in **Russell's traction**, as indicated in Figure 14.1.9a. The magnitude of each of the three forces acting on the leg due to the pulley system is 5 lb, the same as the weight at the end of the rope. Find the magnitude and direction of the resultant force on the femur due to the traction.

**SOLUTION**

The vertical force on the sling at the knee can be represented by the vector 5**j**. Each of the two horizontal forces at the foot can be represented by the vector 5**i**. The resultant force is then

$$(5\mathbf{j}) + (5\mathbf{i}) + (5\mathbf{i}) = 10\mathbf{i} + 5\mathbf{j}.$$

The resultant vector has length

$$\sqrt{10^2 + 5^2} = \sqrt{125} = 5\sqrt{3}.$$

The resultant force has magnitude $5\sqrt{3}$ lb.

From Figure 14.1.9b we see that the angle that the resultant vector makes with the horizontal is $\theta = \tan^{-1}(5/10) \approx 26.6°$. This is the direction of the resultant force. ■

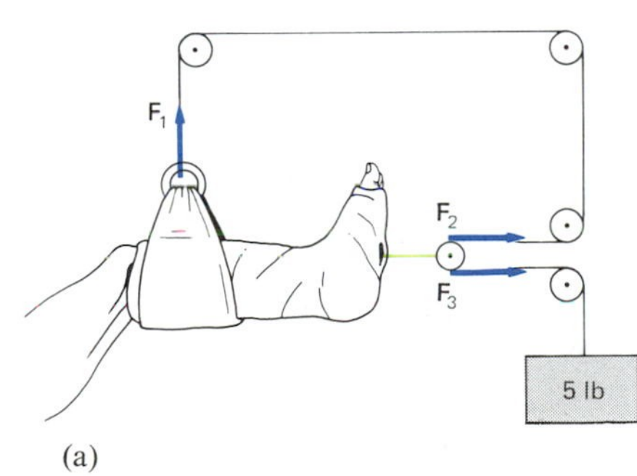


(a)

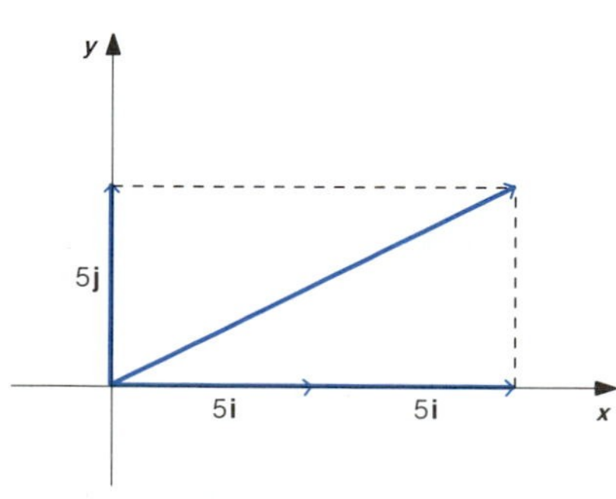


(b)

FIGURE 14.1.9

The vector $\mathbf{U} - \mathbf{V}$ can be determined geometrically. If the vectors $\mathbf{U}$ and $\mathbf{V}$ are drawn with the same initial point, then $\mathbf{U} - \mathbf{V}$ is the vector from the terminal point of $\mathbf{V}$ to the terminal point of $\mathbf{U}$. Be careful to draw the difference vector in the correct direction. See Figure 14.1.10.

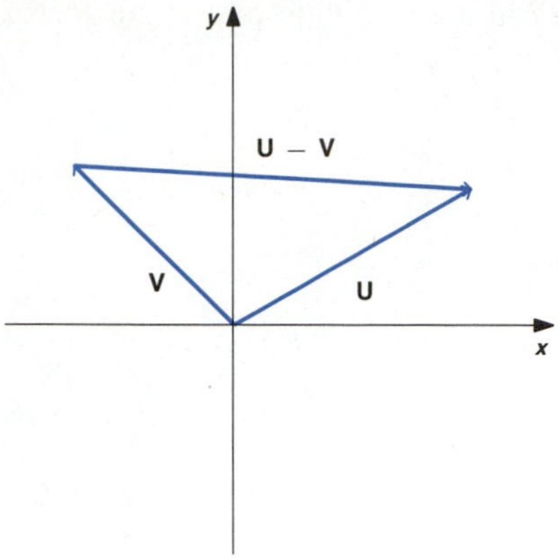

Two vectors $\mathbf{U}$ and $\mathbf{V}$ that are drawn with the same initial point determine a triangle with sides $\mathbf{U}$ and $\mathbf{V}$. The third side is formed by the vector $\mathbf{U} - \mathbf{V}$, as in Figure 14.1.10.



FIGURE 14.1.10

We can use the triangle formed by the vectors $\mathbf{U}$, $\mathbf{V}$, and $\mathbf{U} - \mathbf{V}$ to express the angle between $\mathbf{U}$ and $\mathbf{V}$ in terms of the components of $\mathbf{U}$ and $\mathbf{V}$. If the angle $\theta$ between two nonzero vectors $\mathbf{U} = a\mathbf{i} + b\mathbf{j}$ and $\mathbf{V} = c\mathbf{i} + d\mathbf{j}$ satisfies $0 < \theta < \pi$, the Law of Cosines (Section 1.6) gives

$$|\mathbf{U} - \mathbf{V}|^2 = |\mathbf{U}|^2 + |\mathbf{V}|^2 - 2|\mathbf{U}||\mathbf{V}|\cos\theta.$$

This equality is also true in case either $\mathbf{U} = \mathbf{0}$, $\mathbf{V} = \mathbf{0}$, $\theta = 0$, or $\theta = \pi$. Expressing the squares of the lengths of the vectors in the above equality in terms of their components and then simplifying, we have

$$(a - c)^2 + (b - d)^2 = a^2 + b^2 + c^2 + d^2 - 2|\mathbf{U}||\mathbf{V}|\cos\theta,$$

$$a^2 - 2ac + c^2 + b^2 - 2bd + d^2 = a^2 + b^2 + c^2 + d^2 - 2|\mathbf{U}||\mathbf{V}|\cos\theta,$$

$$2|\mathbf{U}||\mathbf{V}|\cos\theta = 2ac + 2bd,$$

$$|\mathbf{U}||\mathbf{V}|\cos\theta = ac + bd. \tag{4}$$

We see that the formula for the angle between the vectors $a\mathbf{i} + b\mathbf{j}$ and $c\mathbf{i} + d\mathbf{j}$ involves the expression $ac + bd$. This suggests the following definition.

■
*Definition*

The **dot product** of two two-dimensional vectors is

$$(a\mathbf{i} + b\mathbf{j})\cdot(c\mathbf{i} + d\mathbf{j}) = ac + bd.$$ ■

The dot product is the sum of the products of corresponding components. The dot product of two vectors is a *scalar*.

We can express equation (4) in terms of the dot product as

$$|\mathbf{U}||\mathbf{V}|\cos\theta = \mathbf{U}\cdot\mathbf{V}.$$

This gives the following.

**The angle between two nonzero vectors U and V is**

$$\theta = \cos^{-1}\left(\frac{\mathbf{U}\cdot\mathbf{V}}{|\mathbf{U}||\mathbf{V}|}\right).$$

Two vectors are said to be **orthogonal** if their dot product is zero. Two nonzero vectors are orthogonal if and only if they are perpendicular. The zero vector is orthogonal to every vector.

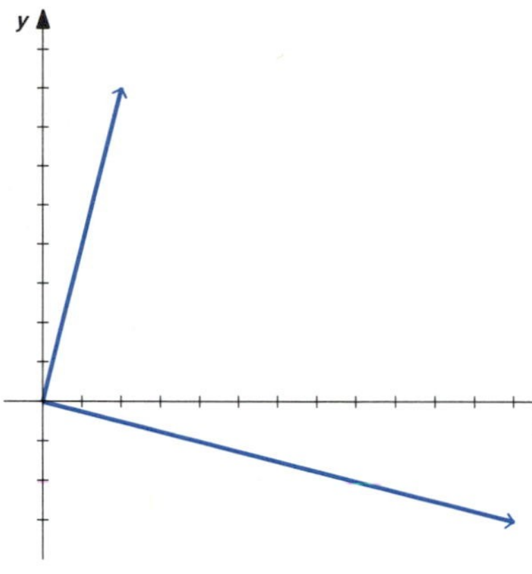

FIGURE 14.1.11

Two vectors are **parallel** if one of the vectors is a scalar multiple of the other. The vectors may have the same direction ($\theta = 0$) or opposite directions ($\theta = \pi$).

Since $|\cos\theta| \le 1$, $\mathbf{U}\cdot\mathbf{V} = |\mathbf{U}||\mathbf{V}|\cos\theta$ implies

$$|\mathbf{U}\cdot\mathbf{V}| \le |\mathbf{U}||\mathbf{V}|.$$

■ **EXAMPLE 4**

Sketch the vectors and find the angle between the vectors.

(a) $12\mathbf{i} - 3\mathbf{j}$, $2\mathbf{i} + 8\mathbf{j}$,

(b) $\mathbf{i} - \mathbf{j}$, $\mathbf{j}$,

(c) $2\mathbf{i} + \mathbf{j}$, $-4\mathbf{i} - 2\mathbf{j}$.

**SOLUTION**

(a) The vectors are sketched in Figure 14.1.11.

$$(12\mathbf{i} - 3\mathbf{j})\cdot(2\mathbf{i} + 8\mathbf{j}) = (12)(2) + (-3)(8) = 0.$$

Hence, $\cos\theta = 0$, so $\theta = \pi/2$. The vectors are orthogonal.

(b) The vectors are sketched in Figure 14.1.12.

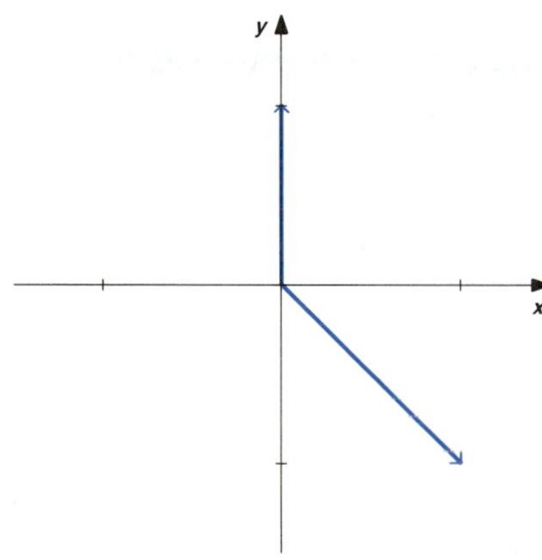

FIGURE 14.1.12

$$(\mathbf{i} - \mathbf{j})\cdot\mathbf{j} = (\mathbf{i} - \mathbf{j})\cdot(0\mathbf{i} + \mathbf{j}) = (1)(0) + (-1)(1) = -1.$$

We have $|\mathbf{i} - \mathbf{j}| = \sqrt{2}$ and $|\mathbf{j}| = 1$. Hence,

$$\cos\theta = \frac{(\mathbf{i} - \mathbf{j})\cdot\mathbf{j}}{|\mathbf{i} - \mathbf{j}||\mathbf{j}|} = \frac{-1}{(\sqrt{2})(1)}, \quad \text{so} \quad \theta = \frac{3\pi}{4}.$$

(c) The vectors are sketched in Figure 14.1.13.

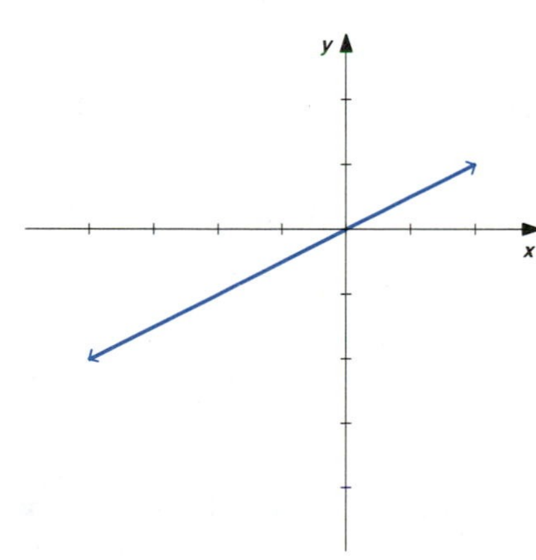

FIGURE 14.1.13

$$(2\mathbf{i} + \mathbf{j})\cdot(-4\mathbf{i} - 2\mathbf{j}) = (2)(-4) + (1)(-2) = -10.$$

We have $|2\mathbf{i} + \mathbf{j}| = \sqrt{5}$ and $|-4\mathbf{i} - 2\mathbf{j}| = \sqrt{20}$. Hence,

$$\cos\theta = \frac{(2\mathbf{i} + \mathbf{j})\cdot(-4\mathbf{i} - 2\mathbf{j})}{|2\mathbf{i} + \mathbf{j}||-4\mathbf{i} - 2\mathbf{j}|} = \frac{-10}{(\sqrt{5})(\sqrt{20})} = -1, \quad \text{so} \quad \theta = \pi.$$

The vectors are parallel and in the opposite direction. ■

The dot product has the following properties.

$$(\mathbf{U} + \mathbf{V})\cdot\mathbf{W} = \mathbf{U}\cdot\mathbf{W} + \mathbf{V}\cdot\mathbf{W},$$

$$(c\mathbf{U})\cdot\mathbf{V} = c(\mathbf{U}\cdot\mathbf{V}) = \mathbf{U}\cdot(c\mathbf{V}),$$

$$\mathbf{U}\cdot\mathbf{V} = \mathbf{V}\cdot\mathbf{U},$$

$$\mathbf{U}\cdot\mathbf{U} > 0 \quad \text{unless} \quad \mathbf{U} = \mathbf{0}.$$

These properties correspond to those of a product of real numbers.

We have

$$(a\mathbf{i} + b\mathbf{j}) \cdot (a\mathbf{i} + b\mathbf{j}) = |a\mathbf{i} + b\mathbf{j}|^2 = a^2 + b^2.$$

That is,

$$\mathbf{U} \cdot \mathbf{U} = |\mathbf{U}|^2. \tag{5}$$

It follows from the definition that

$$\mathbf{i} \cdot \mathbf{j} = (1\mathbf{i} + 0\mathbf{j}) \cdot (0\mathbf{i} + 1\mathbf{j}) = (1)(0) + (0)(1) = 0.$$

In the same way, we can verify that

$$\mathbf{i} \cdot \mathbf{j} = \mathbf{j} \cdot \mathbf{i} = 0 \quad \text{and} \quad \mathbf{i} \cdot \mathbf{i} = \mathbf{j} \cdot \mathbf{j} = 1. \tag{6}$$

We can treat each vector factor $a\mathbf{i} + b\mathbf{j}$ in a dot product as a sum and use the properties of the dot product to remove parentheses. We can then use the formulas in equation (6) to evaluate the dot product.

## ■ EXAMPLE 5

Evaluate

(a) $(\mathbf{i} + 2\mathbf{j}) \cdot (-3\mathbf{j})$,

(b) $(2\mathbf{i} - \mathbf{j}) \cdot (\mathbf{i} + 2\mathbf{j})$.

### SOLUTION

From the definition of dot product, we have

(a) $(\mathbf{i} + 2\mathbf{j}) \cdot (-3\mathbf{j}) = (1\mathbf{i} + 2\mathbf{j}) \cdot (0\mathbf{i} - 3\mathbf{j}) = (1)(0) + (2)(-3) = -6$ and

(b) $(2\mathbf{i} - \mathbf{j}) \cdot (\mathbf{i} + 2\mathbf{j}) = (2)(1) + (-1)(2) = 0$.

Considering these expressions as dot products of sums and using equation (6), we have

(a) $(\mathbf{i} + 2\mathbf{j}) \cdot (-3\mathbf{j}) = -3\mathbf{i} \cdot \mathbf{j} - 6\mathbf{j} \cdot \mathbf{j} = -3(0) - 6(1) = -6$ and

(b) $(2\mathbf{i} - \mathbf{j}) \cdot (\mathbf{i} + 2\mathbf{j}) = 2\mathbf{i} \cdot \mathbf{i} + 4\mathbf{i} \cdot \mathbf{j} - \mathbf{j} \cdot \mathbf{i} - 2\mathbf{j} \cdot \mathbf{j} = 2 + 0 - 0 - 2 = 0$. ■

The vector $a\mathbf{i} + b\mathbf{j}$ is called a **unit vector** if it has length one. That is, $a\mathbf{i} + b\mathbf{j}$ is a unit vector if

$$|a\mathbf{i} + b\mathbf{j}| = \sqrt{a^2 + b^2} = 1.$$

**If $\mathbf{V} \neq \mathbf{0}$, $\mathbf{V}/|\mathbf{V}|$ is the unit vector in the direction of $\mathbf{V}$.**

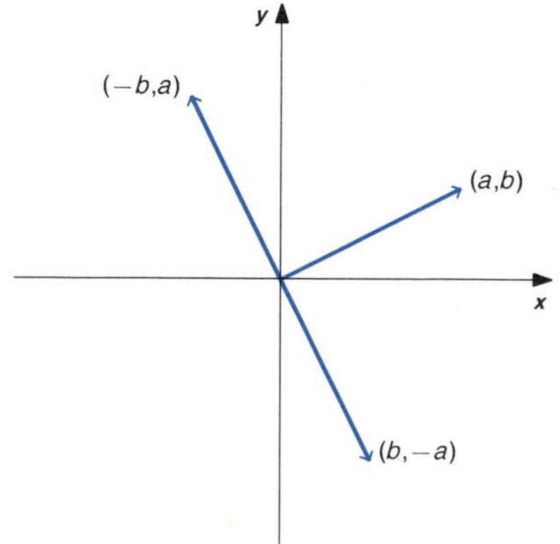

FIGURE 14.1.14

In some problems it is desired to find a vector that is orthogonal to a given vector $a\mathbf{i} + b\mathbf{j}$. We have

$$(a\mathbf{i} + b\mathbf{j}) \cdot (b\mathbf{i} - a\mathbf{j}) = (a)(b) + (b)(-a) = 0,$$

so the vectors $a\mathbf{i} + b\mathbf{j}$ and $b\mathbf{i} - a\mathbf{j}$ are orthogonal. Any scalar multiple of $b\mathbf{i} - a\mathbf{j}$ is also orthogonal to $a\mathbf{i} + b\mathbf{j}$. In particular, the vector $-b\mathbf{i} + a\mathbf{i}$ is also orthogonal to $a\mathbf{i} + b\mathbf{j}$. See Figure 14.1.14.

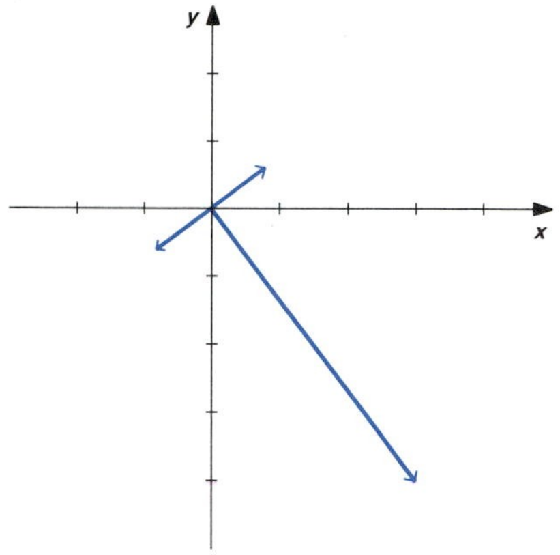

**■ EXAMPLE 6**

Find two unit vectors that are orthogonal to $3\mathbf{i} - 4\mathbf{j}$.

**SOLUTION**

Interchanging components and changing the sign of one component, we obtain

$$4\mathbf{i} + 3\mathbf{j} \quad \text{and} \quad -4\mathbf{i} - 3\mathbf{j}.$$

Both of these vectors are orthogonal to $3\mathbf{i} - 4\mathbf{j}$. Both vectors have length $\sqrt{4^2 + 3^2} = 5$. Unit vectors in the direction of each of the two vectors are obtained by dividing the vectors by their length. This gives the vectors

$$0.8\mathbf{i} + 0.6\mathbf{j} \quad \text{and} \quad -0.8\mathbf{i} - 0.6\mathbf{j}.$$

These two unit vectors are orthogonal to $3\mathbf{i} - 4\mathbf{j}$. See Figure 14.1.15. ■

FIGURE 14.1.15

■ *Definition*

The **work** associated with a constant force **F** that is applied to an object as the object moves in a straight line from a point **Q** to a point **P** is

$$W = \mathbf{F} \cdot \mathbf{QP}.$$ ■

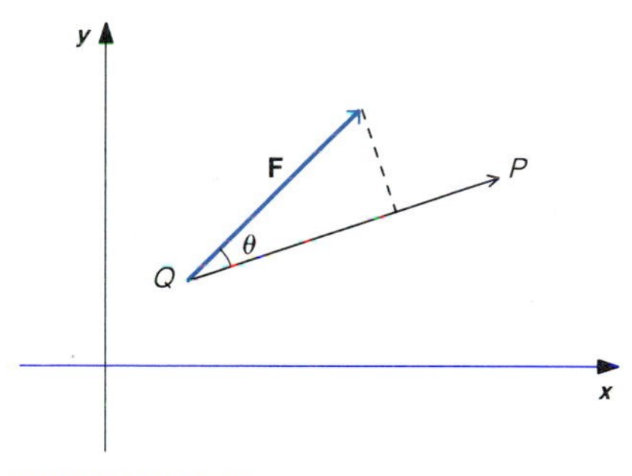


FIGURE 14.1.16

We have

$$W = \mathbf{F} \cdot \mathbf{QP} = |\mathbf{F}||\mathbf{QP}|\cos\theta,$$

where $\theta$ is the angle between the force and the direction of motion. See Figure 14.1.16. If the force is parallel to the direction of motion, then $\cos\theta = \pm 1$ and the work is plus or minus the product of the magnitude of the force and the distance moved. This agrees with the definition of work associated with a force that acts in the direction of motion, as defined in Section 6.5.

**■ EXAMPLE 7**

A horizontal force of 30 lb is applied to a box as the box moves a distance of 40 ft along an incline of 15°. Find the work associated with the force.

**SOLUTION**

A sketch is given in Figure 14.1.17. We can represent the force by the vector

$$\mathbf{F} = 30\mathbf{i}.$$

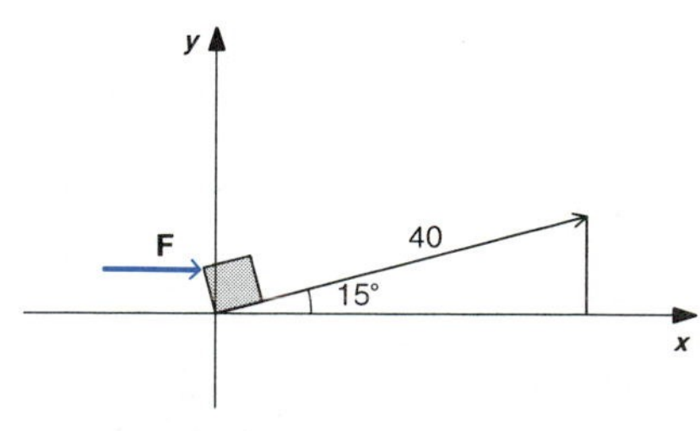


FIGURE 14.1.17

The motion is represented by the vector

$$\mathbf{QP} = (40\cos 15^\circ)\mathbf{i} + (40\sin 15^\circ)\mathbf{j}.$$

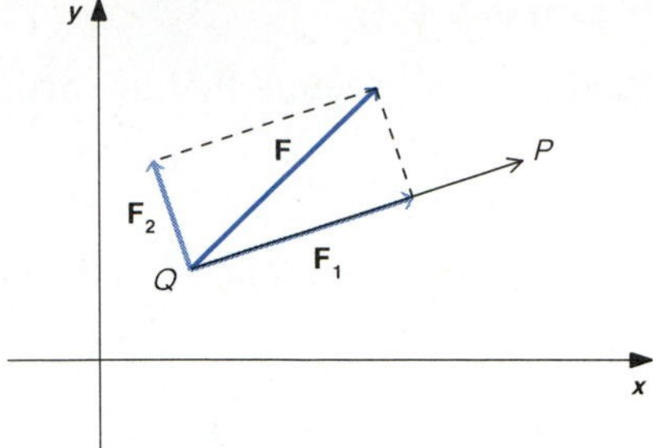

FIGURE 14.1.18

The work is then

$$\begin{aligned} W = \mathbf{F}\cdot\mathbf{QP} &= (30\mathbf{i})\cdot[(40\cos 15^\circ)\mathbf{i} + (40\sin 15^\circ)\mathbf{j}] \\ &= (30)(40\cos 15^\circ) + (0)(40\sin 15^\circ) \\ &= 1200\cos 15^\circ \approx 1159 \text{ ft-lb.} \end{aligned}$$

■

Suppose that a force **F** is applied as an object moves in a straight line from a point $Q$ to a point $P$. Further, suppose that we can write $\mathbf{F} = \mathbf{F}_1 + \mathbf{F}_2$, where $\mathbf{F}_1$ is parallel to **QP** and $\mathbf{F}_2$ is orthogonal to **QP**. See Figure 14.1.18. The work associated with **F** is then

$$W = \mathbf{F}\cdot\mathbf{QP} = (\mathbf{F}_1 + \mathbf{F}_2)\cdot\mathbf{QP} = \mathbf{F}_1\cdot\mathbf{QP} + \mathbf{F}_2\cdot\mathbf{PQ} = \mathbf{F}_1\cdot\mathbf{QP},$$

since $\mathbf{F}_2\cdot\mathbf{QP} = 0$. It follows that work associated with **F** is the same as the work associated with $\mathbf{F}_1$, which is in the direction of motion. Zero work is associated with $\mathbf{F}_2$, which is orthogonal to the direction of motion. This leads us to consider the following problem.

### ■ PROBLEM

Given a vector **U** and a nonzero vector **V**, can we find vectors $\mathbf{U}_1$ and $\mathbf{U}_2$ such that $\mathbf{U} = \mathbf{U}_1 + \mathbf{U}_2$, $\mathbf{U}_1$ is parallel to **V**, and $\mathbf{U}_2$ is orthogonal to **V**?

#### SOLUTION

$\mathbf{U}_1$ parallel to **V** means $\mathbf{U}_1 = c\mathbf{V}$ for some scalar $c$; $\mathbf{U}_2$ orthogonal to **V** means $\mathbf{U}_2\cdot\mathbf{V} = 0$. If

$$\mathbf{U} = \mathbf{U}_1 + \mathbf{U}_2, \quad \mathbf{U}_1 = c\mathbf{V}, \quad \text{and} \quad \mathbf{U}_2\cdot\mathbf{V} = 0,$$

then

$$\begin{aligned} \mathbf{U}\cdot\mathbf{V} &= (\mathbf{U}_1 + \mathbf{U}_2)\cdot\mathbf{V} \\ &= (c\mathbf{V} + \mathbf{U}_2)\cdot\mathbf{V} \\ &= (c\mathbf{V})\cdot\mathbf{V} + \mathbf{U}_2\cdot\mathbf{V} \\ &= c(\mathbf{V}\cdot\mathbf{V}) + 0. \end{aligned}$$

It follows that

$$c = \frac{\mathbf{U}\cdot\mathbf{V}}{\mathbf{V}\cdot\mathbf{V}}.$$

If we set

$$\mathbf{U}_1 = \frac{\mathbf{U}\cdot\mathbf{V}}{\mathbf{V}\cdot\mathbf{V}}\mathbf{V} \quad \text{and} \quad \mathbf{U}_2 = \mathbf{U} - \mathbf{U}_1 = \mathbf{U} - \frac{\mathbf{U}\cdot\mathbf{V}}{\mathbf{V}\cdot\mathbf{V}}\mathbf{V},$$

then $\mathbf{U} = \mathbf{U}_1 + \mathbf{U}_2$, $\mathbf{U}_1$ is parallel to $\mathbf{V}$, and

$$\mathbf{U}_2\cdot\mathbf{V} = \left(\mathbf{U} - \frac{\mathbf{U}\cdot\mathbf{V}}{\mathbf{V}\cdot\mathbf{V}}\mathbf{V}\right)\cdot\mathbf{V} = \mathbf{U}\cdot\mathbf{V} - \frac{\mathbf{U}\cdot\mathbf{V}}{\mathbf{V}\cdot\mathbf{V}}\mathbf{V}\cdot\mathbf{V} = 0,$$

so $\mathbf{U}_2$ is orthogonal to $\mathbf{V}$. ■

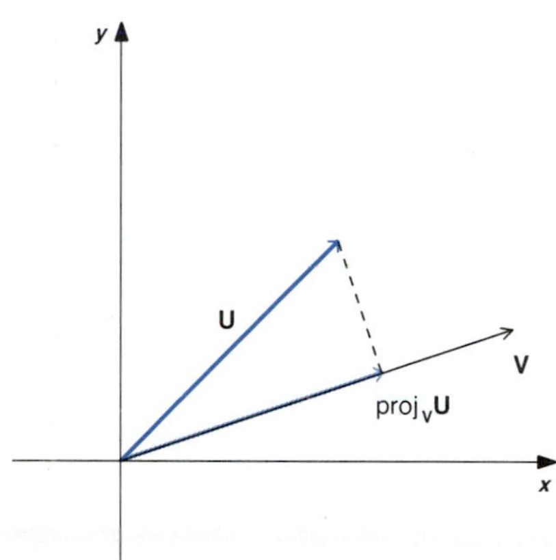


FIGURE 14.1.19

The vector

$$\text{proj}_\mathbf{V}\mathbf{U} = \frac{\mathbf{U}\cdot\mathbf{V}}{\mathbf{V}\cdot\mathbf{V}}\mathbf{V}$$

is called the **projection of U in the direction of V**. See Figure 14.1.19. We have

$$(\text{proj}_\mathbf{V}\mathbf{U})\cdot\mathbf{V} = \left(\frac{\mathbf{U}\cdot\mathbf{V}}{\mathbf{V}\cdot\mathbf{V}}\mathbf{V}\right)\cdot\mathbf{V} = \frac{\mathbf{U}\cdot\mathbf{V}}{\mathbf{V}\cdot\mathbf{V}}(\mathbf{V}\cdot\mathbf{V}) = \mathbf{U}\cdot\mathbf{V}. \tag{7}$$

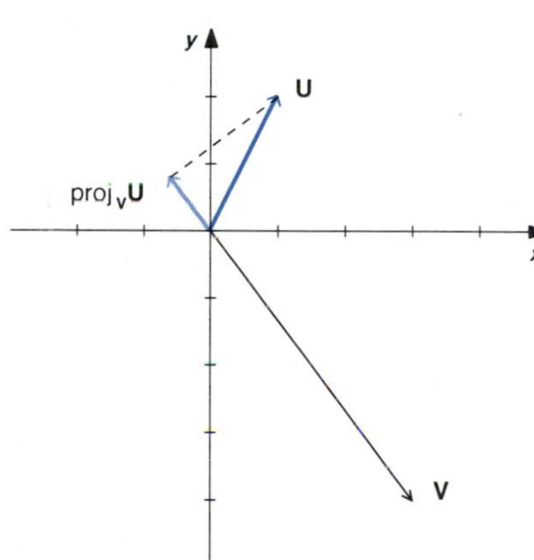


FIGURE 14.1.20

Also,

$$\mathbf{U} - \text{proj}_\mathbf{V}\mathbf{U} \quad \text{is orthogonal to } \mathbf{V}. \tag{8}$$

■ **EXAMPLE 8**

Find the projection of $\mathbf{U} = \mathbf{i} + 2\mathbf{j}$ in the direction of $\mathbf{V} = 3\mathbf{i} - 4\mathbf{j}$. Find vectors $\mathbf{U}_1$ and $\mathbf{U}_2$ such that $\mathbf{U} = \mathbf{U}_1 + \mathbf{U}_2$, $\mathbf{U}_1$ is parallel to $\mathbf{V}$, and $\mathbf{U}_2$ is orthogonal to $\mathbf{V}$.

**SOLUTION**

We have

$$\mathbf{U}\cdot\mathbf{V} = (\mathbf{i} + 2\mathbf{j})\cdot(3\mathbf{i} - 4\mathbf{j}) = (1)(3) + (2)(-4) = 3 - 8 = -5 \quad \text{and}$$

$$\mathbf{V}\cdot\mathbf{V} = (3\mathbf{i} - 4\mathbf{j})\cdot(3\mathbf{i} - 4\mathbf{j}) = (3)(3) + (-4)(-4) = 9 + 16 = 25, \quad \text{so}$$

$$\text{proj}_\mathbf{V}\mathbf{U} = \frac{\mathbf{U}\cdot\mathbf{V}}{\mathbf{V}\cdot\mathbf{V}}\mathbf{V} = \frac{-5}{25}(3\mathbf{i} - 4\mathbf{j}) = -0.6\mathbf{i} + 0.8\mathbf{j}.$$

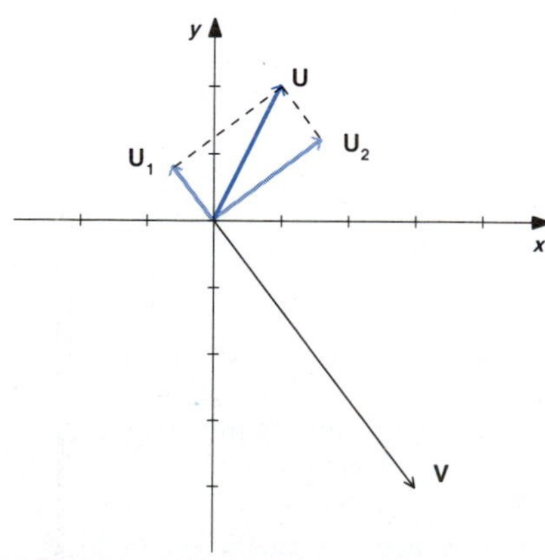


FIGURE 14.1.21

See Figure 14.1.20. The vector $\mathbf{U}_1 = \text{proj}_\mathbf{V}\mathbf{U} = -0.6\mathbf{i} + 0.8\mathbf{j}$ is parallel to $\mathbf{V}$, $\mathbf{U}_2 = \mathbf{U} - \text{proj}_\mathbf{V}\mathbf{U} = (\mathbf{i} + 2\mathbf{j}) - (-0.6\mathbf{i} + 0.8\mathbf{j}) = 1.6\mathbf{i} + 1.2\mathbf{j}$ is orthogonal to $\mathbf{V}$ by (8), and $\mathbf{U} = \mathbf{U}_1 + \mathbf{U}_2$. See Figure 14.1.21. ■

## EXERCISES 14.1

Evaluate the expressions in Exercises 1–12.

1 $-(2\mathbf{i} - 3\mathbf{j})$

2 $3(3\mathbf{i} - 2\mathbf{j})$

3 $2(\mathbf{i} + 2\mathbf{j}) - 3(2\mathbf{i} - 3\mathbf{j})$

4 $4(3\mathbf{i} + \mathbf{j}) - 5(2\mathbf{i} + 2\mathbf{j})$

5 $|12\mathbf{i} - 5\mathbf{j}|$

6 $|4\mathbf{i} - 3\mathbf{j}|$

7 $|-2\mathbf{i} + 3\mathbf{j}|$

8 $|4\mathbf{i} - \mathbf{j}|$

9 $(2\mathbf{i} + 3\mathbf{j})\cdot(4\mathbf{i})$

10 $(4\mathbf{i} - 3\mathbf{j})\cdot(2\mathbf{i} + 2\mathbf{j})$

11 $(5\mathbf{i} - 3\mathbf{j})\cdot(2\mathbf{i} + 3\mathbf{j})$

12 $(4\mathbf{i} - 3\mathbf{j})\cdot(4\mathbf{i} - 3\mathbf{j})$

Evaluate $\mathbf{U} + \mathbf{V}$, $\mathbf{U} - \mathbf{V}$, $2\mathbf{U}$, and $-\mathbf{V}$, for the vectors given in Exercises 13–16. Sketch each vector.

13 $\mathbf{U} = \mathbf{i} + \mathbf{j}$, $\mathbf{V} = \mathbf{i} - \mathbf{j}$

14 $\mathbf{U} = \mathbf{i} + 3\mathbf{j}$, $\mathbf{V} = 2\mathbf{i} - \mathbf{j}$

15 $\mathbf{U} = 2\mathbf{i} - 3\mathbf{j}$, $\mathbf{V} = \mathbf{i} + 2\mathbf{j}$

16 $\mathbf{U} = 3\mathbf{i} - 2\mathbf{j}$, $\mathbf{V} = 2\mathbf{i} + 3\mathbf{j}$

Find the angle between the vectors in Exercises 17–24.

17 $3\mathbf{i} + 4\mathbf{j}$, $4\mathbf{i} - 3\mathbf{j}$

18 $\mathbf{i} + \mathbf{j}$, $\mathbf{i} - \mathbf{j}$

19 $2\mathbf{i} - \mathbf{j}$, $4\mathbf{i} - 2\mathbf{j}$

20 $\mathbf{i} - 2\mathbf{j}$, $-\mathbf{i} + 2\mathbf{j}$

21 $\sqrt{3}\mathbf{i} + \mathbf{j}$, $-\sqrt{3}\mathbf{i} - \mathbf{j}$

22 $-\mathbf{i} + \mathbf{j}$, $-\mathbf{i} - \mathbf{j}$

23 $\mathbf{i} + 2\mathbf{j}$, $\mathbf{i} + 3\mathbf{j}$

24 $2\mathbf{i} - 3\mathbf{j}$, $\mathbf{i} + 2\mathbf{j}$

Find two unit vectors that are orthogonal to the vectors given in Exercises 25–28.

25 $5\mathbf{i} - 12\mathbf{j}$

26 $2\mathbf{i} + \mathbf{j}$

27 $3\mathbf{i}$

28 $4\mathbf{i} - 3\mathbf{j}$

In Exercises 29–32, find the projection of $\mathbf{U}$ in the direction of $\mathbf{V}$. Find vectors $\mathbf{U}_1$ and $\mathbf{U}_2$ such that $\mathbf{U} = \mathbf{U}_1 + \mathbf{U}_2$, $\mathbf{U}_1$ is parallel to $\mathbf{V}$, and $\mathbf{U}_2$ is orthogonal to $\mathbf{V}$.

29 $\mathbf{U} = 2\mathbf{i} - \mathbf{j}$, $\mathbf{V} = 3\mathbf{i} + 4\mathbf{j}$

30 $\mathbf{U} = -\mathbf{i} + 2\mathbf{j}$, $\mathbf{V} = 12\mathbf{i} + 5\mathbf{j}$

31 $\mathbf{U} = 3\mathbf{i} + 2\mathbf{j}$, $\mathbf{V} = -\mathbf{i} + \mathbf{j}$

32 $\mathbf{U} = 2\mathbf{i} + 3\mathbf{j}$, $\mathbf{V} = 3\mathbf{i} - 2\mathbf{j}$

Find the resultant of the forces given in Exercises 33–34.

33 $\mathbf{F}_1 = -2\mathbf{j}$, $\mathbf{F}_2 = \mathbf{i} + 3\mathbf{j}$

34 $\mathbf{F}_1 = -\mathbf{i} - 2\mathbf{j}$, $\mathbf{F}_2 = -\mathbf{i} - \mathbf{j}$

Find a single force that counteracts the resultant of the forces given in each of Exercises 35–36.

35 $\mathbf{F}_1 = 2\mathbf{i}$, $\mathbf{F}_2 = \mathbf{i} + 3\mathbf{j}$

36 $\mathbf{F}_1 = 5\mathbf{i} - 4\mathbf{j}$, $\mathbf{F}_2 = 3\mathbf{i} + 4\mathbf{j}$

Find the projection of the force vector $\mathbf{F}$ in the direction of $\mathbf{V}$ in Exercises 37–38. Write $\mathbf{F}$ as a sum of a force in the direction of $\mathbf{V}$ and a force that is orthogonal to $\mathbf{V}$.

37 $\mathbf{F} = \mathbf{i} + \mathbf{j}$, $\mathbf{V} = -\mathbf{i} + 2\mathbf{j}$

38 $\mathbf{F} = 4\mathbf{i}$, $\mathbf{V} = \mathbf{i} + \mathbf{j}$

39 A force of 18 lb is applied at an angle of 30° from the horizontal to pull a wagon 80 ft along a flat road. Find the work associated with the force.

40 A boat is pulled 50 ft along a 20° incline by two men who are pulling with a force of 60 lb each on a rope that is parallel to the incline, while two women are lifting the boat with a vertical force of 45 lb each. Find the total work done.

41 A boy and a girl are pulling a wagon along a straight line. The girl is pulling with a force of 8 lb on a horizontal rope that makes an angle of 30° with the direction of motion of the wagon. The boy is pulling on a horizontal rope that makes an angle of 45° with the direction of motion. What is the magnitude of the force applied by the boy?

42 A 120-lb person with a broken femur is in **Buck's traction**, as indicated in the figure below. At what angle $\theta$ should the bed be tilted so that the force due to the person's weight has component parallel to the bed that counteracts the force from the traction, if the force due to the 5-lb weight has a magnitude of 5 lb and acts in the direction parallel to the bed?

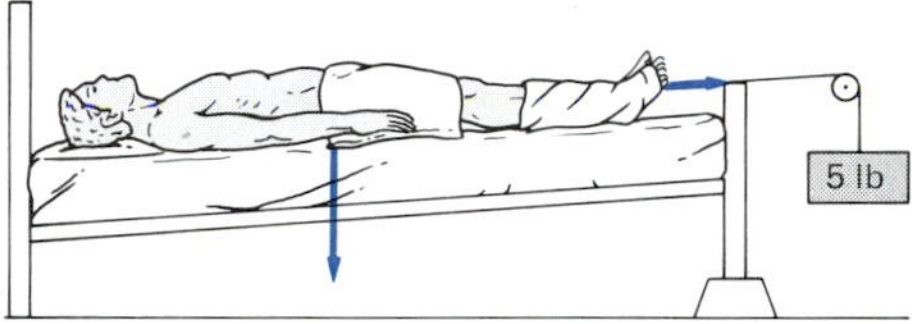


43 A man can row 1.2 m/s in still water. He wishes to row directly across a stream that is 45 m wide. At what angle upstream from directly across the stream should he head if the current of the stream is 0.3 m/s? How long will it take him to cross the stream? See the figure below.

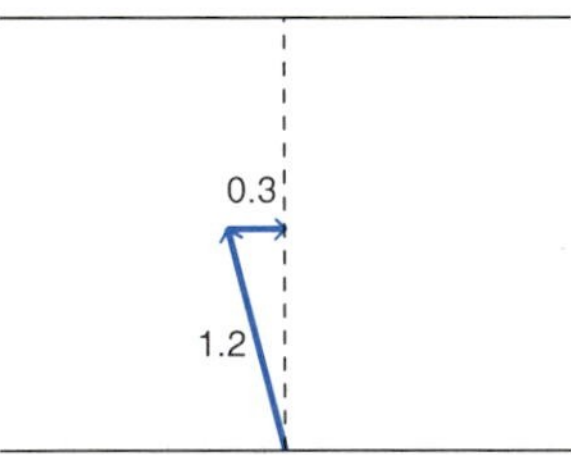


44 If an object is moving along a flat surface, the force due to **kinetic friction** $\mathbf{F}_\mathbf{k}$ acts opposite to the direction of motion and has magnitude that is proportional to the magnitude of the force on the object in the direction orthogonal to the level surface. That is, $|\mathbf{F}_\mathbf{k}| = \mu_k|\mathbf{N}|$. See part (a) of the figure below. The number $\mu_k$ is called the **coefficient of kinetic friction**. Find the work required to overcome kinetic friction

if a horizontal force of 40 lb is applied to a 50-lb box as the box moves 10 ft up an incline of 12°, if the coefficient of kinetic friction is 0.1. See part (b) of the figure below. What is the work required to overcome kinetic friction if the 40-lb force is applied parallel to the incline? See part (c) of the figure below.

Use the technique of the proof of equation (1) to prove the formulas in Exercises 45–46.

45 $(a\mathbf{i} + b\mathbf{j}) + (c\mathbf{i} + d\mathbf{j}) = (a + c)\mathbf{i} + (b + d)\mathbf{j}$

46 $c(a\mathbf{i} + b\mathbf{j}) = ca\mathbf{i} + cb\mathbf{j}$

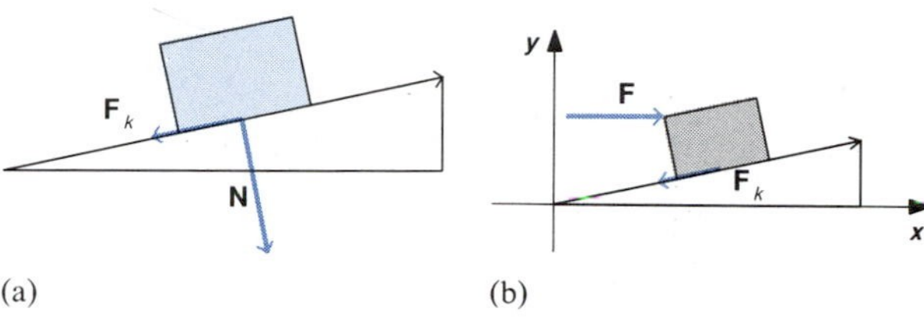


(a)

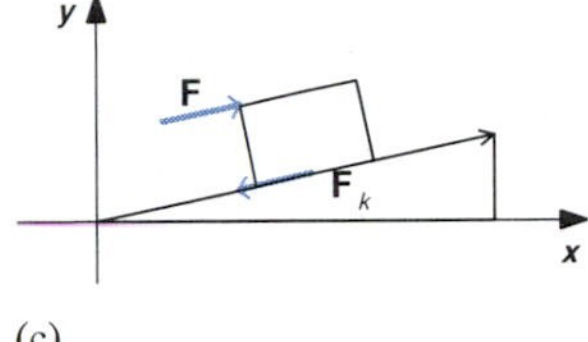


(b)

y
F
F k
x

(c)

## 14.2 VECTORS IN THREE-DIMENSIONAL SPACE, CROSS PRODUCT

The definitions and basic properties of three-dimensional vectors are completely analogous to those of two-dimensional vectors.

**Three-dimensional vectors** are defined to be ordered triples of real numbers $\langle a, b, c\rangle$. The equation

$$\langle a_1, b_1, c_1\rangle = \langle a_2, b_2, c_2\rangle \quad \text{means} \quad a_1 = a_2, \quad b_1 = b_2, \quad \text{and} \quad c_1 = c_2.$$

We define the **sum** of two vectors by

$$\langle a_1, b_1, c_1\rangle + \langle a_2, b_2, c_2\rangle = \langle a_1 + a_2, b_1 + b_2, c_1 + c_2\rangle.$$

**Multiplication by a scalar** is defined by

$$d\langle a, b, c\rangle = \langle da, db, dc\rangle.$$

The **zero vector** is

$$\mathbf{0} = \langle 0, 0, 0\rangle,$$

and the **negative** of $\mathbf{U} = \langle a, b, c\rangle$ is

$$-\mathbf{U} = -\langle a, b, c\rangle = \langle -a, -b, -c\rangle.$$

These definitions can be used to verify that three-dimensional vectors satisfy the characteristic properties of a **vector space** given in Section 14.1.

We define the special three-dimensional vectors

$$\mathbf{i} = \langle 1, 0, 0\rangle, \quad \mathbf{j} = \langle 0, 1, 0\rangle, \quad \text{and} \quad \mathbf{k} = \langle 0, 0, 1\rangle.$$

See Figure 14.2.1.

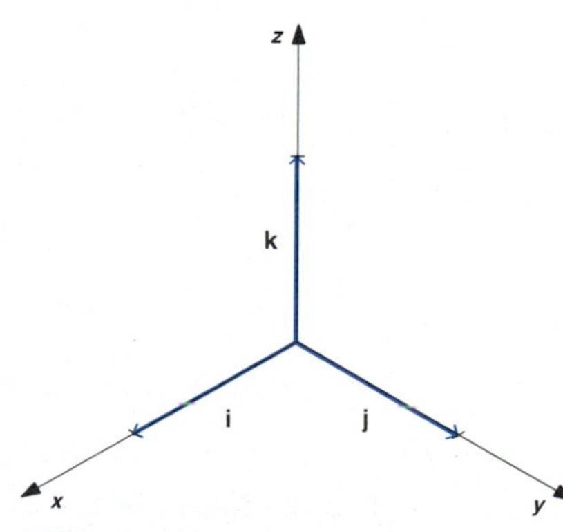


FIGURE 14.2.1

We can use the definitions to verify that

$$\langle a, b, c\rangle = a\mathbf{i} + b\mathbf{j} + c\mathbf{k}.$$

Sums of vectors and multiplication of vectors by scalars are then evaluated as in the case of two-dimensional vectors.

■ **EXAMPLE 1**

Evaluate $2(4\mathbf{i} - 5\mathbf{k}) - 3(\mathbf{i} + 2\mathbf{j} - 4\mathbf{k})$.

**SOLUTION**

$$\begin{aligned} 2(4\mathbf{i} - 5\mathbf{k}) - 3(\mathbf{i} + 2\mathbf{j} - 4\mathbf{k}) &= 8\mathbf{i} - 10\mathbf{k} - 3\mathbf{i} - 6\mathbf{j} + 12\mathbf{k} \\ &= 5\mathbf{i} - 6\mathbf{j} + 2\mathbf{k}. \end{aligned}$$ ■

The **dot product** of the two three-dimensional vectors is

$$(a_1\mathbf{i} + b_1\mathbf{j} + c_1\mathbf{k})\cdot(a_2\mathbf{i} + b_2\mathbf{j} + c_2\mathbf{k}) = a_1a_2 + b_1b_2 + c_1c_2.$$

The dot product of three-dimensional vectors has the same properties as the dot product of two-dimensional vectors. Recall that these properties are analogous to the product of real numbers. We have

$$\mathbf{i}\cdot\mathbf{j} = \mathbf{j}\cdot\mathbf{i} = \mathbf{i}\cdot\mathbf{k} = \mathbf{k}\cdot\mathbf{i} = \mathbf{j}\cdot\mathbf{k} = \mathbf{k}\cdot\mathbf{j} = 0 \quad \text{and} \quad \mathbf{i}\cdot\mathbf{i} = \mathbf{j}\cdot\mathbf{j} = \mathbf{k}\cdot\mathbf{k} = 1.$$

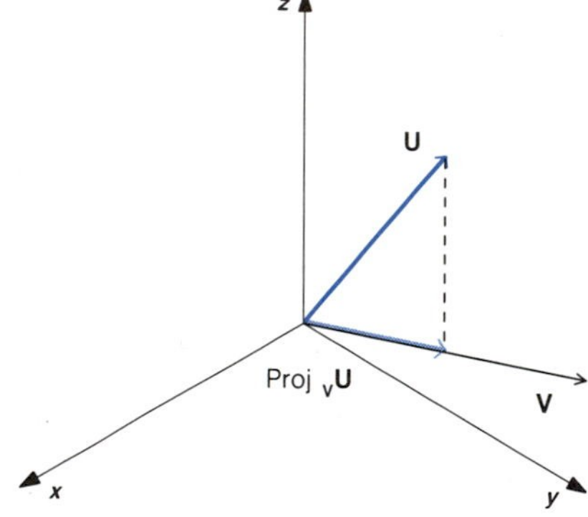

FIGURE 14.2.2

The **length** of a three-dimensional vector is

$$|a\mathbf{i} + b\mathbf{j} + c\mathbf{k}| = \sqrt{a^2 + b^2 + c^2}.$$

The **angle between two nonzero vectors** is

$$\theta = \cos^{-1}\frac{\mathbf{U}\cdot\mathbf{V}}{|\mathbf{U}||\mathbf{V}|}.$$

If $\mathbf{V} \neq \mathbf{0}$, the **projection of U in the direction of V** is the vector

$$\text{proj}_{\mathbf{V}}\mathbf{U} = \frac{\mathbf{U}\cdot\mathbf{V}}{\mathbf{V}\cdot\mathbf{V}}\mathbf{V}.$$

$\mathbf{U} - \text{proj}_{\mathbf{V}}\mathbf{U}$ **is orthogonal to V.**

See Figure 14.2.2.

The **arrow** from a point $Q(x_1, y_1, z_1)$ to a point $P(x_2, y_2, z_2)$ in three-dimensional space is associated with the vector

$$\mathbf{QP} = (x_2 - x_1)\mathbf{i} + (y_2 - y_1)\mathbf{j} + (z_2 - z_1)\mathbf{k}.$$

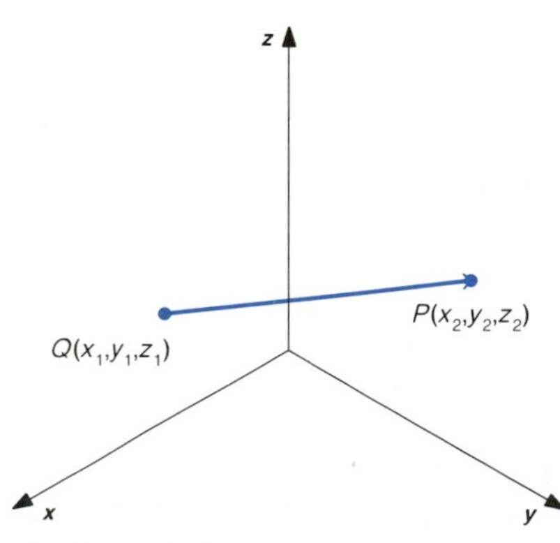

FIGURE 14.2.3

See Figure 14.2.3. The vector $\mathbf{U} = a\mathbf{i} + b\mathbf{j} + c\mathbf{k}$ is associated with the arrow from the origin to the point $(a, b, c)$, and all other arrows that have

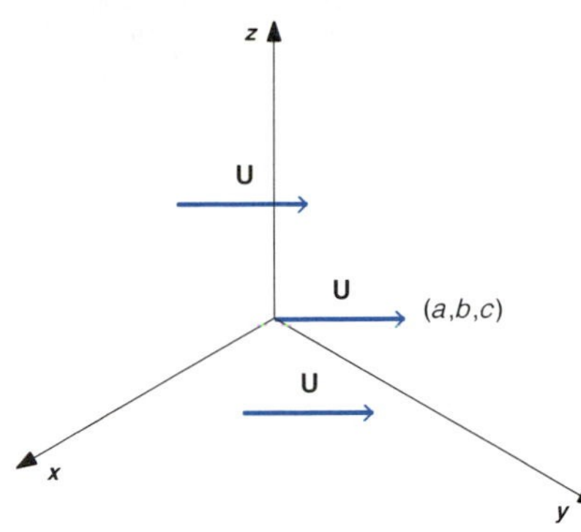


FIGURE 14.2.4

the same length and direction. See Figure 14.2.4. As in the case of two-dimensional vectors, three-dimensional vectors can be added, subtracted, or multiplied by a scalar geometrically. Length of a vector and the angle between vectors correspond to the geometric interpretation.

**EXAMPLE 2**

Sketch the vectors $\mathbf{U} = 2\mathbf{i} + \mathbf{j} + 2\mathbf{k}$ and $\mathbf{V} = 3\mathbf{j} + 4\mathbf{k}$. Find the angle between $\mathbf{U}$ and $\mathbf{V}$. Find the projection of $\mathbf{U}$ in the direction of $\mathbf{V}$. Find vectors $\mathbf{U}_1$ and $\mathbf{U}_2$ such that $\mathbf{U} = \mathbf{U}_1 + \mathbf{U}_2$, $\mathbf{U}_1$ is parallel to $\mathbf{V}$, and $\mathbf{U}_2$ is orthogonal to $\mathbf{V}$.

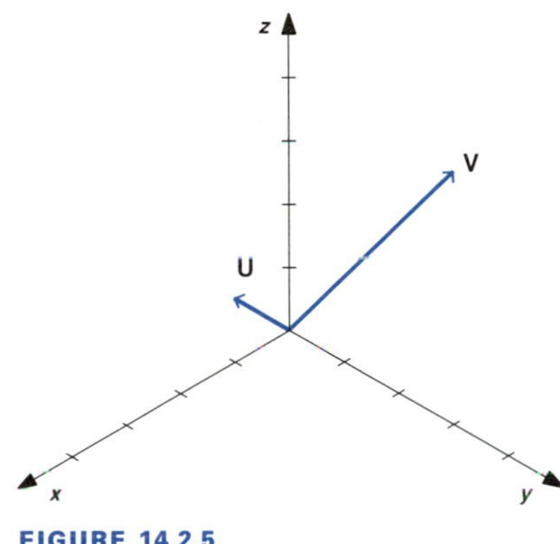


FIGURE 14.2.5

**SOLUTION**

The vectors $\mathbf{U}$ and $\mathbf{V}$ are sketched in Figure 14.2.5. We have

$$\mathbf{U}\cdot\mathbf{U} = |\mathbf{U}|^2 = |2\mathbf{i} + \mathbf{j} + 2\mathbf{k}|^2 = 2^2 + 1^2 + 2^2 = 9,$$

$$\mathbf{V}\cdot\mathbf{V} = |\mathbf{V}|^2 = |3\mathbf{j} + 4\mathbf{k}|^2 = 0^2 + 3^2 + 4^2 = 25,$$

$$\mathbf{U}\cdot\mathbf{V} = (2\mathbf{i} + \mathbf{j} + 2\mathbf{k})\cdot(3\mathbf{j} + 4\mathbf{k}) = (2)(0) + (1)(3) + (2)(4) = 11.$$

The angle $\theta$ between $\mathbf{U}$ and $\mathbf{V}$ satisfies

$$\cos\theta = \frac{\mathbf{U}\cdot\mathbf{V}}{|\mathbf{U}||\mathbf{V}|} = \frac{11}{(3)(5)},$$

so the angle between the vectors is $\theta = \cos^{-1}(11/15) \approx 0.7475843$ rad (calculator value).

The projection of $\mathbf{U}$ in the direction of $\mathbf{V}$ is the vector

$$\text{proj}_{\mathbf{V}}\mathbf{U} = \frac{\mathbf{U}\cdot\mathbf{V}}{\mathbf{V}\cdot\mathbf{V}}\mathbf{V} = \left(\frac{11}{25}\right)(3\mathbf{j} + 4\mathbf{k}) = \frac{33}{25}\mathbf{j} + \frac{44}{25}\mathbf{k}.$$

$$\mathbf{U}_1 = \text{proj}_{\mathbf{V}}\mathbf{U} = \frac{33}{25}\mathbf{j} + \frac{44}{25}\mathbf{k} \quad \text{is parallel to } \mathbf{V},$$

$$\begin{aligned}\mathbf{U}_2 &= \mathbf{U} - \text{proj}_{\mathbf{V}}\mathbf{U}\\ &= (2\mathbf{i} + \mathbf{j} + 2\mathbf{k}) - \left(\frac{33}{25}\mathbf{j} + \frac{44}{25}\mathbf{k}\right)\\ &= 2\mathbf{i} - \frac{8}{25}\mathbf{j} + \frac{6}{25}\mathbf{k}\end{aligned}$$

is orthogonal to $\mathbf{V}$, and $\mathbf{U} = \mathbf{U}_1 + \mathbf{U}_2$. ■

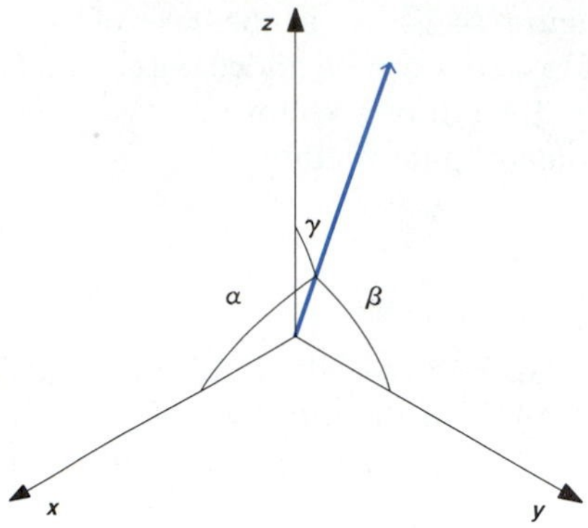

The angles $\alpha$, $\beta$, $\gamma$ between a vector **U** and the vectors **i**, **j**, **k**, respectively, are called **direction angles** of **U**. See Figure 14.2.6. The cosines of the angles are called **direction cosines of U**. If $\mathbf{U} = a\mathbf{i} + b\mathbf{j} + c\mathbf{k}$, then

FIGURE 14.2.6

$$\cos\alpha = \frac{\mathbf{U}\cdot\mathbf{i}}{|\mathbf{U}||\mathbf{i}|} = \frac{a}{|\mathbf{U}|},$$

$$\cos\beta = \frac{\mathbf{U}\cdot\mathbf{j}}{|\mathbf{U}||\mathbf{j}|} = \frac{b}{|\mathbf{U}|}, \quad \text{and}$$

$$\cos\gamma = \frac{\mathbf{U}\cdot\mathbf{k}}{|\mathbf{U}||\mathbf{k}|} = \frac{c}{|\mathbf{U}|}.$$

We see that $\cos\alpha\mathbf{i} + \cos\beta\mathbf{j} + \cos\gamma\mathbf{k} = U/|U|$, so $\cos\alpha\mathbf{i} + \cos\beta\mathbf{j} + \cos\gamma\mathbf{k}$ is the unit vector in the direction of **U**.

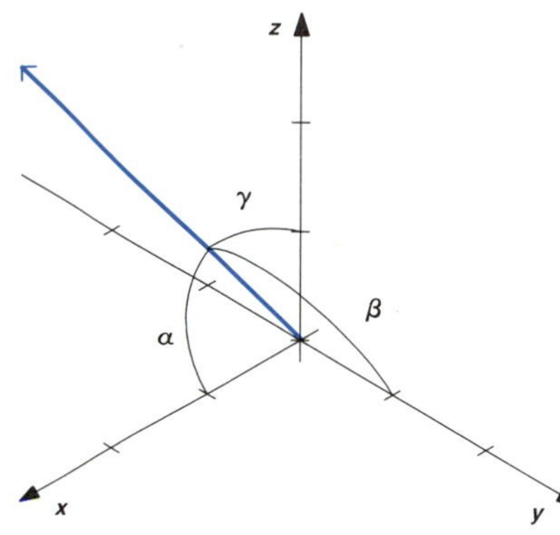


FIGURE 14.2.7

■ **EXAMPLE 3**

Find the direction cosines of the vector $\mathbf{i} - 2\mathbf{j} + 2\mathbf{k}$.

**SOLUTION**

We have

$$|\mathbf{i} - 2\mathbf{j} + 2\mathbf{k}| = \sqrt{1^2 + (-2)^2 + 2^2} = 3,$$

so $\cos\alpha = 1/3$, $\cos\beta = -2/3$, and $\cos\gamma = 2/3$. See Figure 14.2.7. ■

Many physical problems involve a particular type of product of three-dimensional vectors, called a **vector product** or **cross product**. For example, torque, angular momentum, and the force on a charged particle as it moves through a magnetic field can be described in terms of a cross product of vectors.

■ *Definition*

The **cross product** of two three-dimensional vectors is

$$(a_1\mathbf{i} + b_1\mathbf{j} + c_1\mathbf{k}) \times (a_2\mathbf{i} + b_2\mathbf{j} + c_2\mathbf{k})$$
$$= (b_1c_2 - b_2c_1)\mathbf{i} - (a_1c_2 - a_2c_1)\mathbf{j} + (a_1b_2 - a_2b_1)\mathbf{k}.$$

*The cross product of two three-dimensional vectors is a three-dimensional vector.* ■

**U × V is orthogonal to both U and V.** (1)

■ *Proof* We need to show that $\mathbf{U}\cdot(\mathbf{U}\times\mathbf{V}) = 0$ and $\mathbf{V}\cdot(\mathbf{U}\times\mathbf{V}) = 0$. If

$$\mathbf{U} = a_1\mathbf{i} + b_1\mathbf{j} + c_1\mathbf{k} \quad \text{and} \quad \mathbf{V} = a_2\mathbf{i} + b_2\mathbf{j} + c_2\mathbf{k},$$

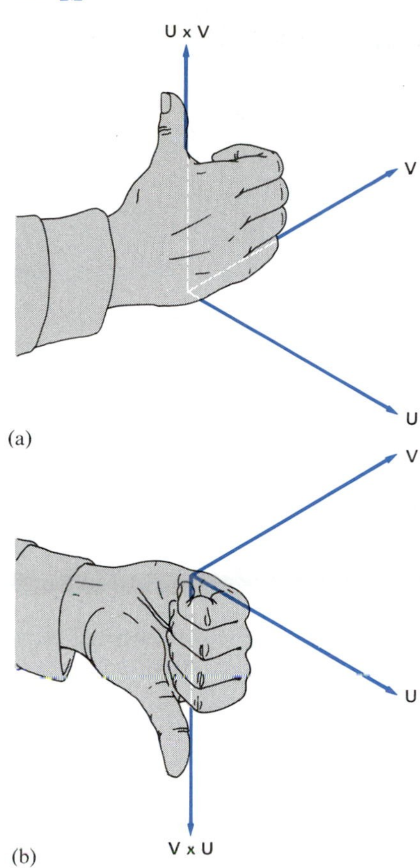

FIGURE 14.2.8

then

$$\mathbf{U}\cdot(\mathbf{U}\times\mathbf{V}) = (a_1)(b_1c_2 - b_2c_1) - (b_1)(a_1c_2 - a_2c_1) + (c_1)(a_1b_2 - a_2b_1)$$
$$= a_1b_1c_2 - a_1b_2c_1 - a_1b_1c_2 + a_2b_1c_1 + a_1b_2c_1 - a_2b_1c_1$$
$$= 0.$$

Verification that $\mathbf{V}\cdot(\mathbf{U}\times\mathbf{V}) = 0$ is similar. ■

We know from (1) that $\mathbf{U}\times\mathbf{V}$ is orthogonal to both **U** and **V**. It is also true that the vectors **U**, **V**, and $\mathbf{U}\times\mathbf{V}$ (in that order) form a right-hand system. That is, if the right hand is placed on the plane formed by **U** and **V** in such a way that the fingers curl from **U** toward **V** in the direction that the angle from **U** to **V** is smaller, between 0 and $\pi$, then the thumb indicates the direction of $\mathbf{U}\times\mathbf{V}$. See Figure 14.2.8a. Note that the direction of $\mathbf{V}\times\mathbf{U}$ is obtained by wrapping the fingers of the right hand from **V** toward **U** in the direction that the angle from **V** to **U** is smaller; this angle is in the direction opposite to the smaller angle from **U** to **V**. This indicates the direction of $\mathbf{V}\times\mathbf{U}$ is opposite to that of $\mathbf{U}\times\mathbf{V}$. See Figure 14.2.8b.

We can verify the following equation by using the method of the proof of equation (1).

$$|\mathbf{U}\times\mathbf{V}|^2 = |\mathbf{U}|^2|\mathbf{V}|^2 - (\mathbf{U}\cdot\mathbf{V})^2. \tag{2}$$

If $\theta$ is the angle between **U** and **V**, we can use equation (2) and the fact that $\mathbf{U}\cdot\mathbf{V} = |\mathbf{U}||\mathbf{V}|\cos\theta$ to express the length of $\mathbf{U}\times\mathbf{V}$ in terms of $\theta$ and the lengths of **U** and **V**. We have

$$|\mathbf{U}\times\mathbf{V}|^2 = |\mathbf{U}|^2|\mathbf{V}|^2 - (\mathbf{U}\cdot\mathbf{V})^2$$
$$= |\mathbf{U}|^2|\mathbf{V}|^2 - (|\mathbf{U}||\mathbf{V}|\cos\theta)^2$$
$$= |\mathbf{U}|^2|\mathbf{V}|^2(1-\cos^2\theta)$$
$$= |\mathbf{U}|^2|\mathbf{V}|^2\sin^2\theta.$$

Taking the square root of each side of the above equation gives

$$|\mathbf{U}\times\mathbf{V}| = |\mathbf{U}||\mathbf{V}|\sin\theta. \tag{3}$$

From Figure 14.2.9 we see that the area of the parallelogram formed by two nonzero vectors **U** and **V** is

$$(\text{Base})(\text{Height}) = (|\mathbf{U}|)(|\mathbf{V}|\sin\theta).$$

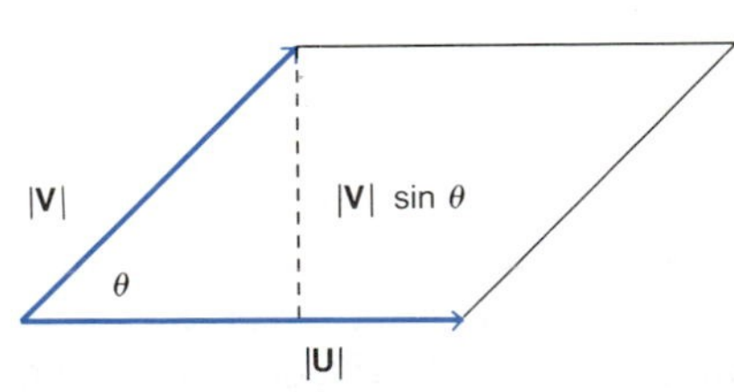

FIGURE 14.2.9

Collecting results, we have

$$(\text{Area of parallelogram formed by } \mathbf{U} \text{ and } \mathbf{V}) = |\mathbf{U}\times\mathbf{V}|. \tag{4}$$

It is straightforward, but tedious, to use the definition of cross product to verify the following properties.

$$\mathbf{U} \times (\mathbf{V} + \mathbf{W}) = \mathbf{U} \times \mathbf{V} + \mathbf{U} \times \mathbf{W},$$

$$(\mathbf{U} + \mathbf{V}) \times \mathbf{W} = \mathbf{U} \times \mathbf{W} + \mathbf{V} \times \mathbf{W},$$

$$(c\mathbf{U}) \times \mathbf{V} = c(\mathbf{U} \times \mathbf{V}) = \mathbf{U} \times (c\mathbf{V}),$$

$$\mathbf{0} \times \mathbf{U} = \mathbf{U} \times \mathbf{0} = \mathbf{0},$$

$$\mathbf{U} \times \mathbf{U} = \mathbf{0},$$

$$\mathbf{U} \times \mathbf{V} = -\mathbf{V} \times \mathbf{U}.$$

The first four properties above correspond to usual properties of multiplication. The properties $\mathbf{U} \times \mathbf{U} = \mathbf{0}$ and $\mathbf{U} \times \mathbf{V} = -\mathbf{V} \times \mathbf{U}$ do not correspond to multiplication of real numbers. *The order of the cross product of three-dimensional vectors is important.*

There is a relatively easy way to remember the definition of cross product. This involves the idea of the **determinant** of a **matrix**. A matrix is simply a rectangular array of numbers.

■ *Definition*

The **determinant** of a $2 \times 2$ matrix is

$$\begin{vmatrix} a & b \\ c & d \end{vmatrix} = ad - cb.$$ ■

The determinant of a $3 \times 3$ matrix can be expressed in terms of determinants of $2 \times 2$ matrices.

■ *Definition*

The **determinant** of a $3 \times 3$ matrix is

$$\begin{vmatrix} a_0 & b_0 & c_0 \\ a_1 & b_1 & c_1 \\ a_2 & b_2 & c_2 \end{vmatrix} = \begin{vmatrix} b_1 & c_1 \\ b_2 & c_2 \end{vmatrix} a_0 - \begin{vmatrix} a_1 & c_1 \\ a_2 & c_2 \end{vmatrix} b_0 + \begin{vmatrix} a_1 & b_1 \\ a_2 & b_2 \end{vmatrix} c_0.$$ ■

The expression on the right above is called a **row expansion** of the $3 \times 3$ determinant on the left. Note that the terms $a_0$, $b_0$, and $c_0$ on the right side of the above equation are the elements of the first **row** of the determinant on the left. The $2 \times 2$ determinant that is multiplied by $a_0$ on the right is obtained by deleting the first row and first column of the determinant on the left, which are the row and column that contain $a_0$. That is,

$$\begin{vmatrix} \not{a}_0 & \not{b}_0 & \not{c}_0 \\ \not{a}_1 & b_1 & c_1 \\ \not{a}_2 & b_2 & c_2 \end{vmatrix} = \begin{vmatrix} b_1 & c_1 \\ b_2 & c_2 \end{vmatrix}.$$

The determinant coefficient of $b_0$ on the right is the *negative* of the $2 \times 2$ determinant obtained by deleting the first row and second column of the determinant on the left. These are the row and column that contain $b_0$.

Deletion of the first row and the third column on the left, the row and column that contain $c_0$, gives the determinant coefficient of $c_0$ on the right.

We can use the idea of determinants to express cross products as

$$(a_1\mathbf{i} + b_1\mathbf{j} + c_1\mathbf{k}) \times (a_2\mathbf{i} + b_2\mathbf{j} + c_2\mathbf{k})$$

$$= \begin{vmatrix} \mathbf{i} & \mathbf{j} & \mathbf{k} \\ a_1 & b_1 & c_1 \\ a_2 & b_2 & c_2 \end{vmatrix} = \begin{vmatrix} b_1 & c_1 \\ b_2 & c_2 \end{vmatrix}\mathbf{i} - \begin{vmatrix} a_1 & c_1 \\ a_2 & c_2 \end{vmatrix}\mathbf{j} + \begin{vmatrix} a_1 & b_1 \\ a_2 & b_2 \end{vmatrix}\mathbf{k}.$$

Note that the above "$3 \times 3$ determinant" is not really a determinant, because the terms are not numbers. However, the same rules apply formally.

Remember that the order of the factors is important in a cross product. Be sure to use the correct order when using the above formula to evaluate a cross product. In particular, note that the components of the first factor form the middle row of the $3 \times 3$ determinant, and the components of the second factor form the bottom row. Also, don't forget the negative sign that precedes the **j** term in the expansion. Finally, note that it is very helpful to use parentheses when evaluating a cross product.

### ■ EXAMPLE 4

$\mathbf{U} = \mathbf{i} - \mathbf{j}$ and $\mathbf{V} = \mathbf{j} + 2\mathbf{k}$.

Evaluate

(a) $\mathbf{U} \times \mathbf{V}$,
(b) $\mathbf{V} \times \mathbf{U}$,
(c) $\mathbf{U} \cdot (\mathbf{U} \times \mathbf{V})$ and $\mathbf{V} \cdot (\mathbf{U} \times \mathbf{V})$,
(d) $|\mathbf{U} \times \mathbf{V}|$,
(e) the angle $\theta$ between **U** and **V** and $|\mathbf{U}||\mathbf{V}|\sin\theta$.

**SOLUTION**

(a) $\mathbf{U} \times \mathbf{V} = (\mathbf{i} - \mathbf{j}) \times (\mathbf{j} + 2\mathbf{k})$

$$= \begin{vmatrix} \mathbf{i} & \mathbf{j} & \mathbf{k} \\ 1 & -1 & 0 \\ 0 & 1 & 2 \end{vmatrix} = \begin{vmatrix} -1 & 0 \\ 1 & 2 \end{vmatrix}\mathbf{i} - \begin{vmatrix} 1 & 0 \\ 0 & 2 \end{vmatrix}\mathbf{j} + \begin{vmatrix} 1 & -1 \\ 0 & 1 \end{vmatrix}\mathbf{k}$$

$$= [(-1)(2) - (1)(0)]\mathbf{i} - [(1)(2) - (0)(0)]\mathbf{j} + [(1)(1) - (0)(-1)]\mathbf{k}$$

$$= -2\mathbf{i} - 2\mathbf{j} + \mathbf{k}.$$

(b) $\mathbf{V} \times \mathbf{U} = (\mathbf{j} + 2\mathbf{k}) \times (\mathbf{i} - \mathbf{j})$

$$= \begin{vmatrix} \mathbf{i} & \mathbf{j} & \mathbf{k} \\ 0 & 1 & 2 \\ 1 & -1 & 0 \end{vmatrix} = \begin{vmatrix} 1 & 2 \\ -1 & 0 \end{vmatrix}\mathbf{i} - \begin{vmatrix} 0 & 2 \\ 1 & 0 \end{vmatrix}\mathbf{j} + \begin{vmatrix} 0 & 1 \\ 1 & -1 \end{vmatrix}\mathbf{k}$$

$$= [(1)(0) - (-1)(2)]\mathbf{i} - [(0)(0) - (1)(2)]\mathbf{j} + [(0)(-1) - (1)(1)]\mathbf{k}$$

$$= 2\mathbf{i} + 2\mathbf{j} - \mathbf{k}.$$

($\mathbf{U} \times \mathbf{V} = -\mathbf{V} \times \mathbf{U}$, as is true in general.)

(c) $\mathbf{U}\cdot(\mathbf{U}\times\mathbf{V}) = (\mathbf{i}-\mathbf{j})\cdot(-2\mathbf{i}-2\mathbf{j}+\mathbf{k})$

$$= (1)(-2) + (-1)(-2) + (0)(1) = 0 \quad \text{and}$$

$$\mathbf{V}\cdot(\mathbf{U}\times\mathbf{V}) = (\mathbf{j}+2\mathbf{k})\cdot(-2\mathbf{i}-2\mathbf{j}+\mathbf{k})$$

$$= (0)(-2) + (1)(-2) + (2)(1) = 0.$$

($\mathbf{U}\times\mathbf{V}$ is orthogonal to both $\mathbf{U}$ and $\mathbf{V}$, as is true in general.)

(d) $|\mathbf{U}\times\mathbf{V}| = |-2\mathbf{i}-2\mathbf{j}+\mathbf{k}| = \sqrt{(-2)^2+(-2)^2+(1)^2} = 3.$

(e) We have

$$|\mathbf{U}| = |\mathbf{i}-\mathbf{j}| = \sqrt{(1)^2+(-1)^2} = \sqrt{2},$$

$$|\mathbf{V}| = |\mathbf{j}+2\mathbf{k}| = \sqrt{(1)^2+(2)^2} = \sqrt{5},$$

$$\mathbf{U}\cdot\mathbf{V} = (\mathbf{i}-\mathbf{j})\cdot(\mathbf{j}+2\mathbf{k}) = (1)(0) + (-1)(1) + (0)(2) = -1.$$

The angle between $\mathbf{U}$ and $\mathbf{V}$ is

$$\theta = \cos^{-1}\left(\frac{\mathbf{U}\cdot\mathbf{V}}{|\mathbf{U}||\mathbf{V}|}\right) = \cos^{-1}\left(\frac{-1}{\sqrt{10}}\right).$$

From Figure 14.2.10 we see that $\sin\theta = 3/\sqrt{10}$, so

$$|\mathbf{U}||\mathbf{V}|\sin\theta = (\sqrt{2})(\sqrt{5})\left(\frac{3}{\sqrt{10}}\right) = 3.$$

($|\mathbf{U}\times\mathbf{V}| = |\mathbf{U}||\mathbf{V}|\sin\theta$, as is true in general.) ■

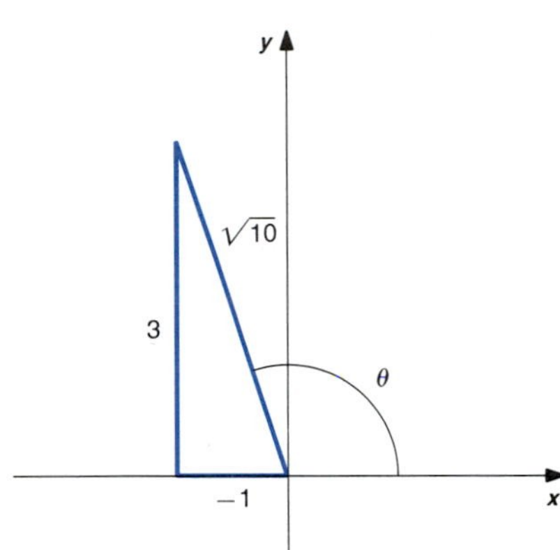


FIGURE 14.2.10

## ■ EXAMPLE 5

Evaluate $\mathbf{i}\times(\mathbf{i}\times\mathbf{j})$ and $(\mathbf{i}\times\mathbf{i})\times\mathbf{j}$.

### SOLUTION

We begin the evaluation of $\mathbf{i}\times(\mathbf{i}\times\mathbf{j})$ by evaluating the vector inside the parentheses. We have

$$\mathbf{i}\times\mathbf{j} = \begin{vmatrix} \mathbf{i} & \mathbf{j} & \mathbf{k} \\ 1 & 0 & 0 \\ 0 & 1 & 0 \end{vmatrix} = \begin{vmatrix} 0 & 0 \\ 1 & 0 \end{vmatrix}\mathbf{i} - \begin{vmatrix} 1 & 0 \\ 0 & 0 \end{vmatrix}\mathbf{j} + \begin{vmatrix} 1 & 0 \\ 0 & 1 \end{vmatrix}\mathbf{k}$$

$$= [(0)(0)-(1)(0)]\mathbf{i} - [(1)(0)-(0)(0)]\mathbf{j} + [(1)(1)-(0)(0)]\mathbf{k}$$

$$= \mathbf{k}.$$

Then

$$\mathbf{i} \times (\mathbf{i} \times \mathbf{j}) = \mathbf{i} \times (\mathbf{k}) = \begin{vmatrix} \mathbf{i} & \mathbf{j} & \mathbf{k} \\ 1 & 0 & 0 \\ 0 & 0 & 1 \end{vmatrix}$$

$$= \begin{vmatrix} 0 & 0 \\ 0 & 1 \end{vmatrix}\mathbf{i} - \begin{vmatrix} 1 & 0 \\ 0 & 1 \end{vmatrix}\mathbf{j} + \begin{vmatrix} 1 & 0 \\ 0 & 0 \end{vmatrix}\mathbf{k}$$

$$= [(0)(1) - (0)(0)]\mathbf{i} - [(1)(1) - (0)(0)]\mathbf{j} + [(1)(0) - (0)(0)]\mathbf{k}$$

$$= -\mathbf{j}.$$

In order to evaluate $(\mathbf{i} \times \mathbf{i}) \times \mathbf{j}$, we again evaluate the vector inside the parentheses first. Since the cross product of any vector with itself is the zero vector, and the cross product of the zero vector with any vector is the zero vector, we have

$$(\mathbf{i} \times \mathbf{i}) \times \mathbf{j} = (\mathbf{0}) \times \mathbf{j} = \mathbf{0}.$$ ■

From Example 5 we see that $\mathbf{U} \times (\mathbf{V} \times \mathbf{W})$ is generally *not* equal to $(\mathbf{U} \times \mathbf{V}) \times \mathbf{W}$. *Since the Associative Law does not hold for cross products, we must be very careful with parentheses (as well as order) when dealing with expressions that involve cross products.* In particular, we cannot express either $\mathbf{U} \times (\mathbf{V} \times \mathbf{W})$ or $(\mathbf{U} \times \mathbf{V}) \times \mathbf{W}$ as $\mathbf{U} \times \mathbf{V} \times \mathbf{W}$; $\mathbf{U} \times \mathbf{V} \times \mathbf{W}$ is undefined.

One use of the cross product is to find a vector that is orthogonal to two given vectors.

### ■ EXAMPLE 6

Find two unit vectors that are orthogonal to both $2\mathbf{i} - \mathbf{j} + 2\mathbf{k}$ and $\mathbf{i} + 2\mathbf{j} - 3\mathbf{k}$.

#### SOLUTION

The cross product of the vectors (in either order) will be orthogonal to both vectors. The unit vector in the direction of the cross product and the negative of that vector are both unit vectors that are orthogonal to the given vectors. See Figure 14.2.11. We have

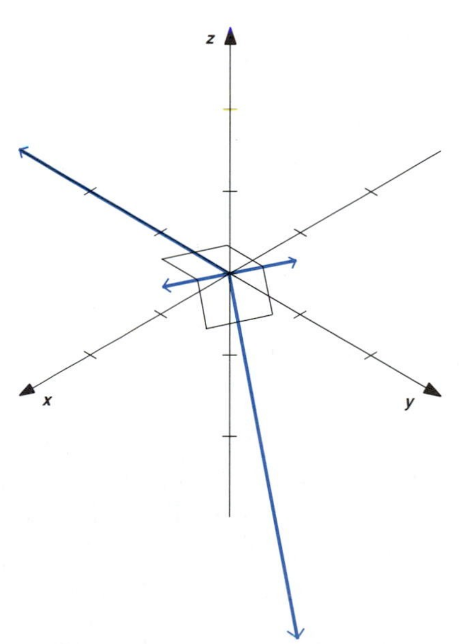


FIGURE 14.2.11

$$(2\mathbf{i} - \mathbf{j} + 2\mathbf{k}) \times (\mathbf{i} + 2\mathbf{j} - 3\mathbf{k})$$

$$= \begin{vmatrix} \mathbf{i} & \mathbf{j} & \mathbf{k} \\ 2 & -1 & 2 \\ 1 & 2 & -3 \end{vmatrix} = \begin{vmatrix} -1 & 2 \\ 2 & -3 \end{vmatrix}\mathbf{i} - \begin{vmatrix} 2 & 2 \\ 1 & -3 \end{vmatrix}\mathbf{j} + \begin{vmatrix} 2 & -1 \\ 1 & 2 \end{vmatrix}\mathbf{k}$$

$$= [(-1)(-3) - (2)(2)]\mathbf{i} - [(2)(-3) - (1)(2)]\mathbf{j} + [(2)(2) - (1)(-1)]\mathbf{k}.$$

$$= -\mathbf{i} + 8\mathbf{j} + 5\mathbf{k}.$$

$$|-\mathbf{i} + 8\mathbf{j} + 5\mathbf{k}| = \sqrt{(-1)^2 + 8^2 + 5^2} = \sqrt{90}.$$

The desired unit vectors are

$$\frac{-\mathbf{i} + 8\mathbf{j} + 5\mathbf{k}}{\sqrt{90}} \quad \text{and} \quad \frac{\mathbf{i} - 8\mathbf{j} - 5\mathbf{k}}{\sqrt{90}}. \qquad \blacksquare$$

We have seen in equation (4) that

(Area of parallelogram formed by **U** and **V**) = $|\mathbf{U} \times \mathbf{V}|$.

It follows that

$$\text{(Area of triangle formed by } \mathbf{U} \text{ and } \mathbf{V}) = \frac{1}{2}|\mathbf{U} \times \mathbf{V}|. \tag{5}$$

See Figure 14.2.12.

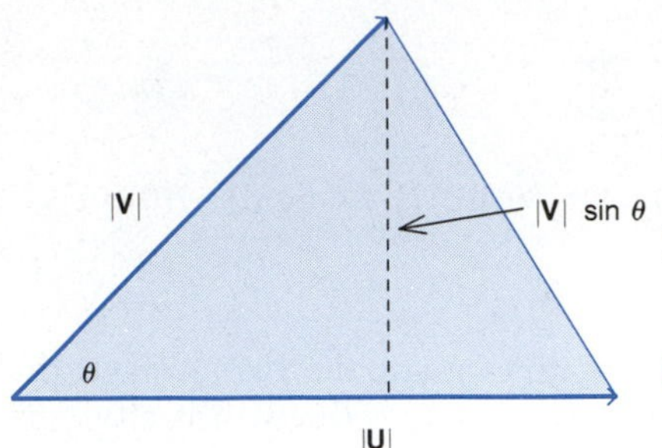


FIGURE 14.2.12

■ **EXAMPLE 7**

Find the area of the triangle with vertices at the origin and the points $P(3, 2, -1)$ and $Q(-2, 1, 1)$.

**SOLUTION**

The triangle is sketched in Figure 14.2.13. We see that the triangle is formed by the vectors from the origin to the points $P(3, 2, -1)$ and $Q(-2, 1, 1)$, which are

$$3\mathbf{i} + 2\mathbf{j} - \mathbf{k} \quad \text{and} \quad -2\mathbf{i} + \mathbf{j} + \mathbf{k},$$

respectively. We have

$$(3\mathbf{i} + 2\mathbf{j} - \mathbf{k}) \times (-2\mathbf{i} + \mathbf{j} + \mathbf{k})$$

$$= \begin{vmatrix} \mathbf{i} & \mathbf{j} & \mathbf{k} \\ 3 & 2 & -1 \\ -2 & 1 & 1 \end{vmatrix} = \begin{vmatrix} 2 & -1 \\ 1 & 1 \end{vmatrix}\mathbf{i} - \begin{vmatrix} 3 & -1 \\ -2 & 1 \end{vmatrix}\mathbf{j} + \begin{vmatrix} 3 & 2 \\ -2 & 1 \end{vmatrix}\mathbf{k}$$

$$= (2 + 1)\mathbf{i} - (3 - 2)\mathbf{j} + (3 + 4)\mathbf{k}$$

$$= 3\mathbf{i} - \mathbf{j} + 7\mathbf{k}.$$

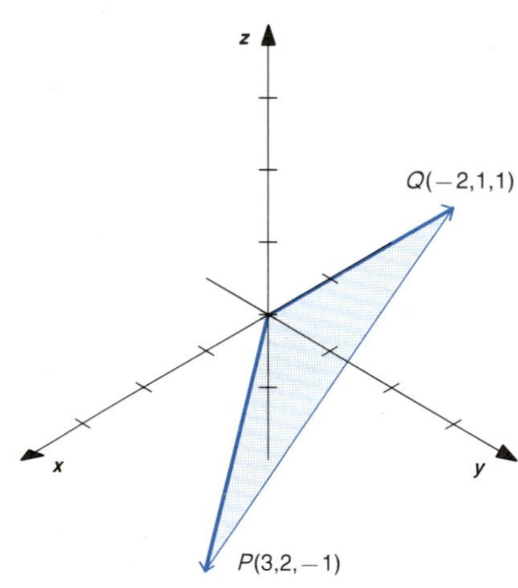


FIGURE 14.2.13

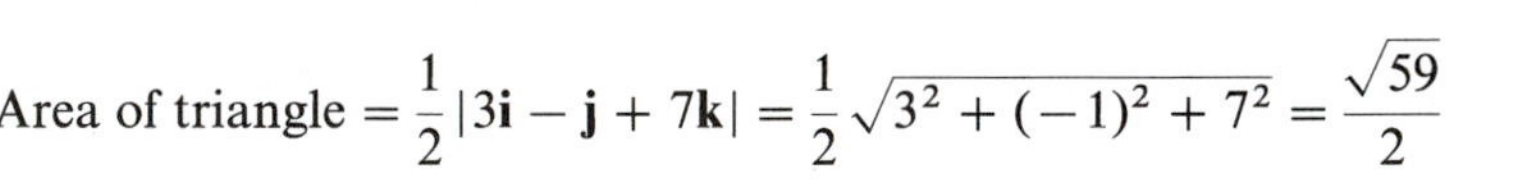

$$\text{Area of triangle} = \frac{1}{2}|3\mathbf{i} - \mathbf{j} + 7\mathbf{k}| = \frac{1}{2}\sqrt{3^2 + (-1)^2 + 7^2} = \frac{\sqrt{59}}{2} \qquad \blacksquare$$

From Figure 14.2.14 we see that the vectors **U**, **V**, and **W** form a parallelepiped with area of base $|\mathbf{U} \times \mathbf{V}|$ and height $|\mathbf{W}||\cos\phi|$, since $\mathbf{U} \times \mathbf{V}$ is perpendicular to the base formed by **U** and **V**. The volume of the parallelepiped is then

$$\text{(Area of base)(Height)} = (|\mathbf{U} \times \mathbf{V}|)(|\mathbf{W}||\cos\phi|)$$
$$= |(\mathbf{U} \times \mathbf{V}) \cdot \mathbf{W}|.$$

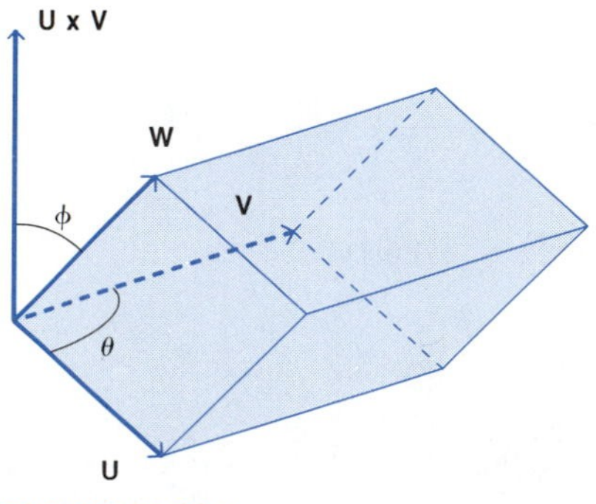


FIGURE 14.2.14

That is,

**(Volume of parallelepiped formed by U, V, and W) = $|(\mathbf{U} \times \mathbf{V}) \cdot \mathbf{W}|$.**

The expression on the right above can be evaluated by using determinants. That is, let

$$\mathbf{U} = a_1\mathbf{i} + b_1\mathbf{j} + c_1\mathbf{k}, \quad \mathbf{V} = a_2\mathbf{i} + b_2\mathbf{j} + c_2\mathbf{k}, \quad \text{and}$$
$$\mathbf{W} = a_0\mathbf{i} + b_0\mathbf{j} + c_0\mathbf{k}.$$

Then

$$\begin{aligned}(\mathbf{U} \times \mathbf{V}) \cdot \mathbf{W} &= (b_1c_2 - b_2c_1)a_0 - (a_1c_2 - a_2c_1)b_0 + (a_1b_2 - a_2b_1)c_0 \\ &= \begin{vmatrix} b_1 & c_1 \\ b_2 & c_2 \end{vmatrix} a_0 - \begin{vmatrix} a_1 & c_1 \\ a_2 & c_2 \end{vmatrix} b_0 + \begin{vmatrix} a_1 & b_1 \\ a_2 & b_2 \end{vmatrix} c_0 \\ &= \begin{vmatrix} a_0 & b_0 & c_0 \\ a_1 & b_1 & c_1 \\ a_2 & b_2 & c_2 \end{vmatrix}\end{aligned}$$

Finally, note that the volume of the parallelepiped does not depend on which two vectors are considered to form the base. This means the vectors can be taken in any order. The scalar

$$(\mathbf{U} \times \mathbf{V}) \cdot \mathbf{W}$$

is called the **triple product** of **U**, **V**, and **W**. We have

$$(\mathbf{U} \times \mathbf{V}) \cdot \mathbf{W} = \mathbf{U} \cdot (\mathbf{V} \times \mathbf{W}). \tag{7}$$

■ **EXAMPLE 8**

Find the volume of the parallelepiped formed by **i** + 2**j**, 3**j**, and **i** + 2**j** + 3**k**.

**SOLUTION**

We begin by evaluating the determinant that corresponds to the triple product of the three given vectors. We have

$$\begin{vmatrix} 1 & 2 & 0 \\ 0 & 3 & 0 \\ 1 & 2 & 3 \end{vmatrix} = \begin{vmatrix} 3 & 0 \\ 2 & 3 \end{vmatrix}(1) - \begin{vmatrix} 0 & 0 \\ 1 & 3 \end{vmatrix}(2) + \begin{vmatrix} 0 & 3 \\ 1 & 2 \end{vmatrix}(0)$$
$$= (9)(1) - (0)(2) + (-3)(0) = 9.$$

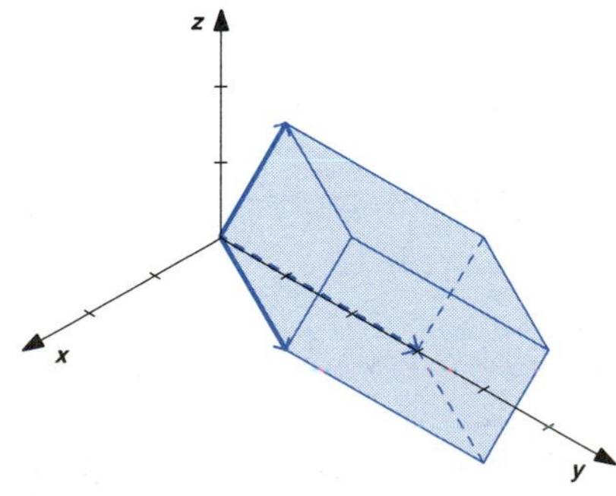

FIGURE 14.2.15

The volume is then $|9| = 9$. The parallelepiped is sketched in Figure 14.2.15. (In this example, one side of the parallelepiped is contained in the $xy$-plane and has area 3. The height from that side is 3, so we can determine geometrically that the volume is 9.) ■

## EXERCISES 14.2

Evaluate the expressions in Examples 1-20.

1 $-2(3\mathbf{i} - 3\mathbf{j} + \mathbf{k})$

2 $(\mathbf{i} - 2\mathbf{j} + 3\mathbf{k}) + (\mathbf{i} + 3\mathbf{j} - 2\mathbf{k})$

3 $2(3\mathbf{i} + \mathbf{j} - 2\mathbf{k}) - 3(2\mathbf{i} + \mathbf{k})$

4 $3(4\mathbf{i} - 2\mathbf{j} + \mathbf{k}) - 2(6\mathbf{i} + \mathbf{j} + \mathbf{k})$

5 $|3\mathbf{i} - 4\mathbf{j} + 5\mathbf{k}|$

6 $|12\mathbf{i} - 4\mathbf{j} - 3\mathbf{k}|$

7 $|\mathbf{i} - 2\mathbf{j} + 2\mathbf{k}|$

8 $|2\mathbf{i} + \mathbf{j} - \mathbf{k}|$

9 $(2\mathbf{i} - 3\mathbf{j} + \mathbf{k}) \cdot (\mathbf{i} - \mathbf{j} - \mathbf{k})$

10 $(2\mathbf{i} + 3\mathbf{j} + 2\mathbf{k}) \cdot (\mathbf{i} + \mathbf{j} + \mathbf{k})$

11 $(\mathbf{i} + 4\mathbf{j}) \cdot (3\mathbf{i} - \mathbf{k})$

12 $(3\mathbf{i} + 4\mathbf{j}) \cdot (3\mathbf{j} + 4\mathbf{k})$

13 $(\mathbf{i} + 2\mathbf{j} + 3\mathbf{k}) \times \mathbf{i}$

14 $(4\mathbf{i} - 5\mathbf{j} - 6\mathbf{k}) \times (\mathbf{i} + \mathbf{j})$

15 $(3\mathbf{i} + 4\mathbf{j} - 5\mathbf{k}) \times (\mathbf{i} + \mathbf{j} + \mathbf{k})$

16 $(5\mathbf{i} + 2\mathbf{j} + 3\mathbf{k}) \times (\mathbf{i} - \mathbf{j} - \mathbf{k})$

17 $\mathbf{i} \times (\mathbf{i} + \mathbf{k})$

18 $(\mathbf{i} + \mathbf{k}) \times \mathbf{i}$

19 $[(\mathbf{i} - \mathbf{j}) \times (\mathbf{i} + \mathbf{k})] \cdot \mathbf{i}$

20 $[(2\mathbf{i} + \mathbf{j}) \times (\mathbf{i} + 2\mathbf{k})] \cdot \mathbf{k}$

In Exercises 21-24, sketch the vectors **U** and **V**, find the angle between **U** and **V**, and find the projection of **U** in the direction of **V**.

21 $\mathbf{U} = \mathbf{i} + \mathbf{j} + \mathbf{k}$, $\mathbf{V} = \mathbf{i} + \mathbf{j}$

22 $\mathbf{U} = 2\mathbf{i} + 2\mathbf{j} + \mathbf{k}$, $\mathbf{V} = 2\mathbf{i} + 2\mathbf{j} - \mathbf{k}$

23 $\mathbf{U} = 2\mathbf{i} - 3\mathbf{j}$, $\mathbf{V} = 3\mathbf{i} + 4\mathbf{k}$

24 $\mathbf{U} = 2\mathbf{i} - \mathbf{j} + 2\mathbf{k}$, $\mathbf{V} = 2\mathbf{i} + 2\mathbf{j} - \mathbf{k}$

Find the direction cosines of the vectors in Exercises 25-28.

25 $4\mathbf{i} - 5\mathbf{k}$

26 $\mathbf{i} - 2\mathbf{j} - 2\mathbf{k}$

27 $3\mathbf{i} + 2\mathbf{j} + 3\mathbf{k}$

28 $3\mathbf{i} - 12\mathbf{j} + 4\mathbf{k}$

Find the two unit vectors that are orthogonal to the vectors given in each of Exercises 29-32.

29 $\mathbf{i} + \mathbf{j}$, $\mathbf{j} + \mathbf{k}$

30 $\mathbf{i} + 2\mathbf{j} + 2\mathbf{k}$, $\mathbf{i} + \mathbf{j} - \mathbf{k}$

31 $2\mathbf{i} - 2\mathbf{j} + 3\mathbf{k}$, $\mathbf{i} + \mathbf{j}$

32 $3\mathbf{i} + 4\mathbf{j} + 2\mathbf{k}$, $2\mathbf{i} + 3\mathbf{j} - 4\mathbf{k}$

Find the area of the parallelogram formed by the vectors given in Exercises 33-36.

33 $\mathbf{i}$, $\mathbf{j} + \mathbf{k}$

34 $6\mathbf{i} + 4\mathbf{j}$, $\mathbf{i} - \mathbf{k}$

35 $2\mathbf{i} + 3\mathbf{j} + \mathbf{k}$, $\mathbf{j} + \mathbf{k}$

36 $3\mathbf{i} + 4\mathbf{j}$, $3\mathbf{j} + 4\mathbf{k}$

Find the volume of the parallelepiped formed by the vectors given in Exercises 37-40.

37 $\mathbf{i}$, $\mathbf{i} + \mathbf{j}$, $2\mathbf{k}$

38 $\mathbf{i} + \mathbf{j}$, $\mathbf{i} - \mathbf{j}$, $\mathbf{i} + \mathbf{j} + 2\mathbf{k}$

39 $2\mathbf{i} - 2\mathbf{j} + \mathbf{k}$, $2\mathbf{i} + 2\mathbf{j} + \mathbf{k}$, $3\mathbf{i}$

40 $3\mathbf{i} - 4\mathbf{j} + 2\mathbf{k}$, $4\mathbf{i} + 3\mathbf{j} + 2\mathbf{k}$, $\mathbf{i} + \mathbf{j}$

Find the area of the triangle formed by the three points given in Exercises 41-44.

41 (0, 0, 0), (1, 2, 0), (−1, 3, 2)

42 (0, 0, 0), (−1, 1, 1), (1, 2, 1)

43 (4, 0, 0), (0, 3, 0), (0, 0, 4)

44 (1, 1, 1), (0, 3, 0), (2, 3, 1)

In Exercises 45-48, express **F** as a sum of two forces, one parallel to **V** and one orthogonal to **V**.

45 $\mathbf{F} = 3\mathbf{k}$, $\mathbf{V} = \mathbf{i} + 2\mathbf{j} + 2\mathbf{k}$

46 $\mathbf{F} = 4\mathbf{i} - \mathbf{k}$, $\mathbf{V} = 2\mathbf{i} + 3\mathbf{j}$

47 $\mathbf{F} = 2\mathbf{i} - \mathbf{j} + \mathbf{k}$, $\mathbf{V} = \mathbf{i}$

48 $\mathbf{F} = -\mathbf{i} + \mathbf{j} + 2\mathbf{k}$, $\mathbf{V} = \mathbf{j}$

## 14.3 PLANES

A **plane** is determined by one point on the plane and a nonzero vector **N** that is normal to the plane. A point $P(x, y, z)$ is on the plane that contains the point $P_0(x_0, y_0, z_0)$ and has normal vector $\mathbf{N} = a\mathbf{i} + b\mathbf{j} + c\mathbf{k}$ if and only if the vector $\mathbf{P_0P} = (x - x_0)\mathbf{i} + (y - y_0)\mathbf{j} + (z - z_0)\mathbf{k}$ is orthogonal to **N**. See Figure 14.3.1. Expressing the orthogonality in terms

FIGURE 14.3.1

of the dot product, we obtain the **vector form** of an equation of the plane,

$$\mathbf{N}\cdot\mathbf{P_0P} = 0.$$

Evaluating the dot product, we have

$$(a\mathbf{i} + b\mathbf{j} + c\mathbf{k})\cdot((x - x_0)\mathbf{i} + (y - y_0)\mathbf{j} + (z - z_0)\mathbf{k}) = 0,$$

$$a(x - x_0) + b(y - y_0) + c(z - z_0) = 0. \qquad (1)$$

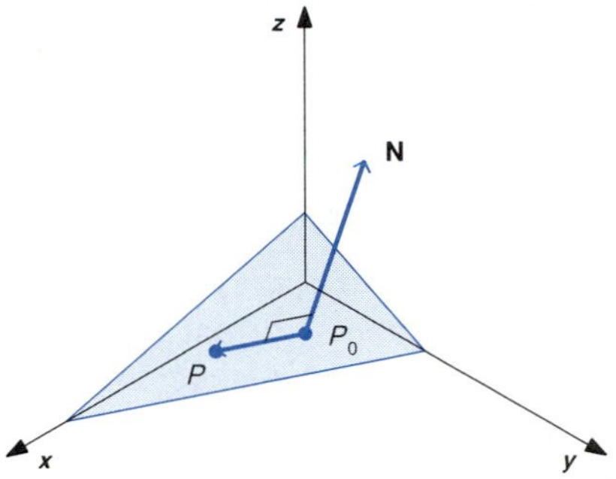

Equation (1) is used to *write* an equation of the plane that contains the point $P_0(x_0, y_0, z_0)$ and has normal vector $\mathbf{N} = a\mathbf{i} + b\mathbf{j} + c\mathbf{k}$. For example, an equation of the plane that contains the point $(2, -1, 3)$ and has normal vector $\mathbf{N} = \mathbf{i} + 2\mathbf{j} - 5\mathbf{k}$ is

$$(1)(x - 2) + (2)(y - (-1)) + (-5)(z - 3) = 0.$$

Let us simplify the equation we obtained for the plane that contains the point $P_0(x_0, y_0, z_0)$ and has normal vector $\mathbf{N} = a\mathbf{i} + b\mathbf{j} + c\mathbf{k}$. We have

$$a(x - x_0) + b(y - y_0) + c(z - z_0) = 0,$$

$$ax - ax_0 + by - by_0 + cz - cz_0 = 0,$$

$$ax + by + cz + (-ax_0 - by_0 - cz_0) = 0,$$

$$ax + by + cz + d = 0,$$

where $d = -ax_0 - by_0 - cz_0$. On the other hand, suppose we start with an equation of the form

$$ax + by + cz + d = 0,$$

where $a$, $b$, and $c$ are not all equal to zero. We then find some point $P_0(x_0, y_0, z_0)$ that satisfies the equation, so $ax_0 + by_0 + cz_0 + d = 0$. We can then reverse the above steps to obtain

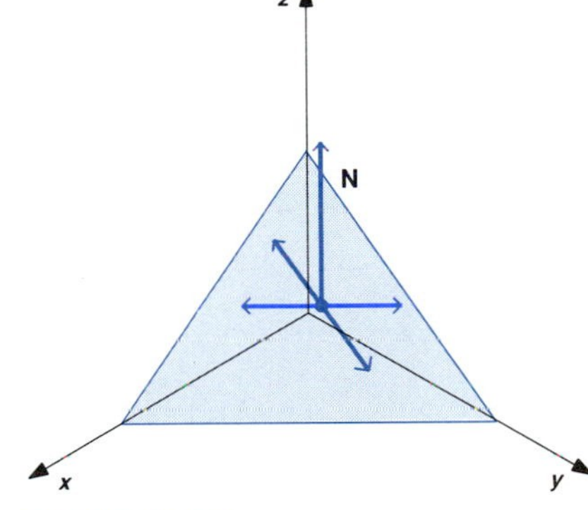

FIGURE 14.3.2

$$a(x - x_0) + b(y - y_0) + c(z - z_0) = 0.$$

We conclude that

$$ax + by + cz + d = 0 \qquad (2)$$

is an equation of a plane that has normal vector $\mathbf{N} = a\mathbf{i} + b\mathbf{j} + c\mathbf{k}$, if $a$, $b$, and $c$ are not all equal to zero. Equation (2) is called the **general equation of a plane**. Note that we can *read* the components of a normal vector from the general equation of a plane. For example, the plane $x - 3y + 2z = 0$ has normal vector $\mathbf{N} = i - 3\mathbf{j} + 2k$.

Any nonzero multiple of a normal vector of a plane is also a normal vector of the plane. Also, a normal vector of a plane is orthogonal to any vector in the plane. See Figure 14.3.2. Also, see Figure 14.3.3

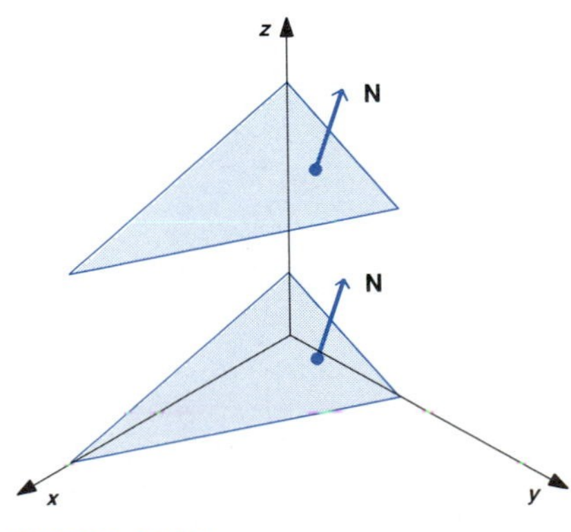

FIGURE 14.3.3

**Parallel planes have parallel normal vectors.**

■ **EXAMPLE 1**

Find an equation of the plane that contains the point $(1, -2, 3)$ and is parallel to the plane $z = 2x + 3y - 4$.

**SOLUTION**

To write an equation of a plane, we need a point on the plane and a normal vector **N**. We have the point $(1, -2, 3)$. To determine a normal vector, we use the fact that a normal vector of a plane is also a normal vector of all parallel planes. This means we can choose **N** to be a vector that is normal to the plane $z = 2x + 3y - 4$. Writing this equation in standard form $-2x - 3y + z + 4 = 0$, we see that the vector $-2\mathbf{i} - 3\mathbf{j} + \mathbf{k}$ is a normal vector. An equation of the plane that contains the point $(1, -2, 3)$ and has normal vector $\mathbf{N} = -2\mathbf{i} - 3\mathbf{j} + \mathbf{k}$ is

$$-2(x - 1) - 3(y + 2) + (z - 3) = 0.$$

This equation simplifies to

$$-2x - 3y + z - 7 = 0 \quad \text{or} \quad z = 2x + 3y + 7.$$ ■

■ **EXAMPLE 2**

Find an equation of the plane that contains the three points $A(1, 1, 1)$, $B(2, 4, 3)$, and $C(3, 2, 1)$.

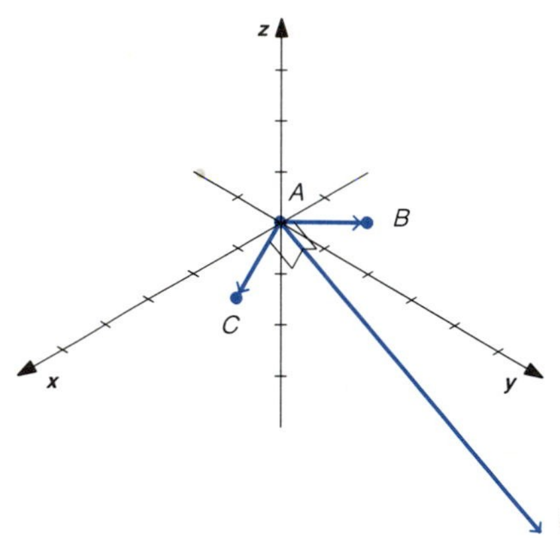


FIGURE 14.3.4

**SOLUTION**

We need a point $P_0$ on the plane and a normal vector **N**. We are given three points on the plane. Let us choose the point $A(1, 1, 1)$ to be $P_0$. To find a normal vector, we first note that vectors from any one of the points $A$, $B$, $C$, to any other of these points are in the plane that contains the points. Let us consider

$$\mathbf{AB} = (2 - 1)\mathbf{i} + (4 - 1)\mathbf{j} + (3 - 1)\mathbf{k} = \mathbf{i} + 3\mathbf{j} + 2\mathbf{k} \quad \text{and}$$

$$\mathbf{AC} = (3 - 1)\mathbf{i} + (2 - 1)\mathbf{j} + (1 - 1)\mathbf{k} = 2\mathbf{i} + \mathbf{j}.$$

See Figure 14.3.4. Since a normal vector of a plane is orthogonal to any vector in the plane, we want a normal vector that is orthogonal to both of the vectors **AB** and **AC**. Since the cross product of two vectors is orthogonal to both factors, this condition will be satisfied by the vector

$$\mathbf{N} = \mathbf{AB} \times \mathbf{AC} = \begin{vmatrix} \mathbf{i} & \mathbf{j} & \mathbf{k} \\ 1 & 3 & 2 \\ 2 & 1 & 0 \end{vmatrix} = \begin{vmatrix} 3 & 2 \\ 1 & 0 \end{vmatrix}\mathbf{i} - \begin{vmatrix} 1 & 2 \\ 2 & 0 \end{vmatrix}\mathbf{j} + \begin{vmatrix} 1 & 3 \\ 2 & 1 \end{vmatrix}\mathbf{k}$$

$$= -2\mathbf{i} + 4\mathbf{j} - 5\mathbf{k}.$$

An equation of the plane that contains the point $(1, 1, 1)$ and has normal vector $\mathbf{N} = -2\mathbf{i} + 4\mathbf{j} - 5\mathbf{k}$ is

$$-2(x - 1) + 4(y - 1) - 5(z - 1) = 0,$$

$$-2x + 2 + 4y - 4 - 5z + 5 = 0,$$

$$-2x + 4y - 5z + 3 = 0.$$

We would get an equivalent equation if any of the points $A$, $B$, $C$ were used as the point $P_0$ and any two vectors that connect two different pairs of points $A$, $B$, $C$ were used to determine a normal vector. ■

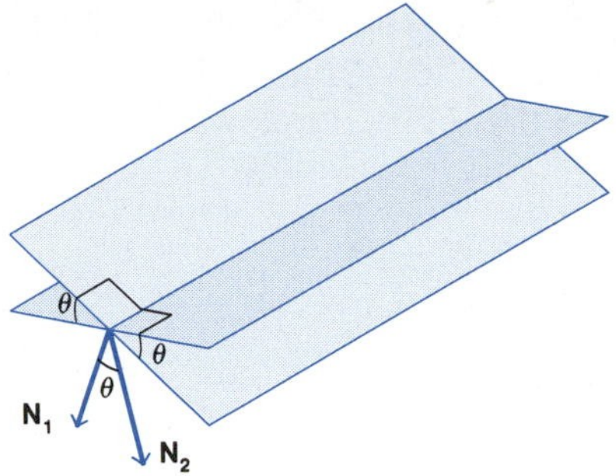

FIGURE 14.3.5

The **angle of intersection** of two intersecting planes with normal vectors $\mathbf{N}_1$ and $\mathbf{N}_2$ is defined to be

$$\theta = \cos^{-1}\left(\frac{|\mathbf{N}_1 \cdot \mathbf{N}_2|}{|\mathbf{N}_1||\mathbf{N}_2|}\right).$$

This is the angle between the normal vectors, if the normal vectors are chosen so the angle between them is acute. Taking the absolute value of the dot product of $\mathbf{N}_1$ and $\mathbf{N}_2$ insures $0 \leq \theta \leq \pi/2$. Geometrically, the angle between two planes is the smaller of the two angles between the lines formed by a cross section that is perpendicular to their line of intersection. See Figure 14.3.5.

■ **EXAMPLE 3**

Find the angle of intersection of the plane $x + y - z = 0$ and the plane $x - 3y + z - 1 = 0$.

**SOLUTION**

The plane $x + y - z = 0$ has normal vector $\mathbf{N}_1 = \mathbf{i} + \mathbf{j} - \mathbf{k}$. The plane $x - 3y + z - 1 = 0$ has normal vector $\mathbf{N}_2 = \mathbf{i} - 3j + \mathbf{k}$. Then

$$\frac{|\mathbf{N}_1 \cdot \mathbf{N}_2|}{|\mathbf{N}_1||\mathbf{N}_2|} = \frac{|(1)(1) + (1)(-3) + (-1)(1)|}{\sqrt{1^2 + 1^2 + (-1)^2}\sqrt{1^2 + (-3)^2 + 1^2}} = \frac{|-3|}{\sqrt{3}\sqrt{11}} = \frac{\sqrt{3}}{\sqrt{11}}.$$

The angle between the planes is

$$\theta = \cos^{-1}\left(\frac{\sqrt{3}}{\sqrt{11}}\right) \approx 1.0213291 \text{ rad.}$$ ■

The distance from a point $P_0$ to a plane is the distance from $P_0$ to the point in the plane that is closest to $P_0$. If $P_2$ is the point on the plane with the property that $\mathbf{P}_2\mathbf{P}_0$ is perpendicular to the plane, then the distance between $P_2$ and $P_0$ is smaller than the distance between $P_0$ and any other point on the plane, so

$$(\text{Distance of } P_0 \text{ to plane}) = |\mathbf{P}_2\mathbf{P}_0|.$$

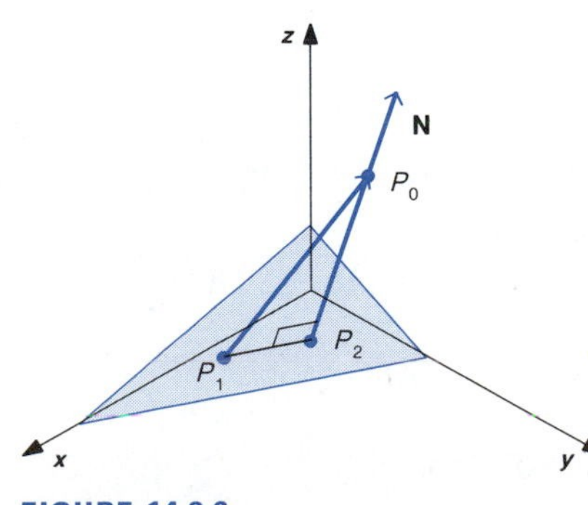

FIGURE 14.3.6

See Figure 14.3.6. Let us see how to use the equation of the plane to evaluate the distance $|\mathbf{P}_2\mathbf{P}_0|$. The equation of the plane determines a normal vector $\mathbf{N}$. We can also use the equation of the plane to find the coordinates of a point $P_1$ on the plane by choosing any convenient values (such as zero) for two of the variables $x$, $y$, $z$, and then solving the equation of the plane for the other variable. From Figure 14.3.6 we see that the distance between $P_0$ and $P_2$ is the length of the projection of $\mathbf{P_1P_0}$ in the direction of $\mathbf{P_2P_0}$. Since $\mathbf{P_2P_0}$ is perpendicular to the plane, the direction of $\mathbf{P_2P_0}$ is either the same as or opposite that of $\mathbf{N}$. It follows that

$$\textbf{(Distance of } P_0 \textbf{ to plane)} = |\mathbf{P_2P_0}|$$

$$= |\text{proj}_{\mathbf{N}}\mathbf{P_1P_0}| = \left|\frac{\mathbf{P_1P_0}\cdot\mathbf{N}}{\mathbf{N}\cdot\mathbf{N}}\mathbf{N}\right| = \frac{|\mathbf{P_1P_0}\cdot\mathbf{N}|}{\mathbf{N}\cdot\mathbf{N}}|\mathbf{N}|$$

$$= \frac{|\mathbf{P_1P_0}\cdot\mathbf{N}|}{|\mathbf{N}|^2}|\mathbf{N}| = \frac{|\mathbf{P_1P_0}\cdot\mathbf{N}|}{|\mathbf{N}|}.$$

If desired, we can find the coordinates of the point $P_2$ in the plane that is closest to the point $P_0$ by subtracting the components of $\text{proj}_{\mathbf{N}}\mathbf{P_1P_0}$ from the coordinates of $P_0$.

### ■ EXAMPLE 4

Find the distance from the point $P_0(2, 5, 4)$ to the plane $x + 2y + 2z = 2$. Find the point on the plane that is closest to $P_0$.

#### SOLUTION

We need the coordinates of a point $P_1$ on the plane and a normal vector of the plane. From the equation of the plane we see that $\mathbf{N} = \mathbf{i} + 2\mathbf{j} + 2\mathbf{k}$ is a normal vector. Setting $y = z = 0$ in the equation of the plane and then solving for $x$, we obtain that the point $P_1(2, 0, 0)$ is on the plane. We have

$$\mathbf{P_1P_0} = (2 - 2)\mathbf{i} + (5 - 0)\mathbf{j} + (4 - 0)\mathbf{k} = 5\mathbf{j} + 4\mathbf{k},$$

$$\mathbf{P_1P_0}\cdot\mathbf{N} = (5\mathbf{j} + 4\mathbf{k})\cdot(\mathbf{i} + 2\mathbf{j} + 2\mathbf{k}) = (0)(1) + (5)(2) + (4)(2) = 18,$$

$$\mathbf{N}\cdot\mathbf{N} = |\mathbf{N}|^2 = (1)^2 + (2)^2 + (2)^2 = 9.$$

The distance from $P_0(2, 5, 4)$ to the plane is then

$$\frac{|\mathbf{P_1P_0}\cdot\mathbf{N}|}{|\mathbf{N}|} = \frac{18}{\sqrt{9}} = 6.$$

The projection of $\mathbf{P_1P_0}$ in the direction of $\mathbf{N}$ is

$$\text{proj}_{\mathbf{N}}\mathbf{P_1P_0} = \frac{\mathbf{P_1P_0}\cdot\mathbf{N}}{\mathbf{N}\cdot\mathbf{N}}\mathbf{N} = \frac{18}{9}(\mathbf{i} + 2\mathbf{j} + 2\mathbf{k}) = 2\mathbf{i} + 4\mathbf{j} + 4\mathbf{k}.$$

Let $\mathbf{P_0}$ denote the vector from the origin to the point $P_0$. Then

$$\mathbf{P_0} - \text{proj}_{\mathbf{N}}\mathbf{P_1P_0} = (2\mathbf{i} + 5\mathbf{j} + 4\mathbf{k}) - (2\mathbf{i} + 4\mathbf{j} + 4\mathbf{k}) = \mathbf{j},$$

so the point on the plane that is closest to $P_0$ is $(0, 1, 0)$. ■

(a)

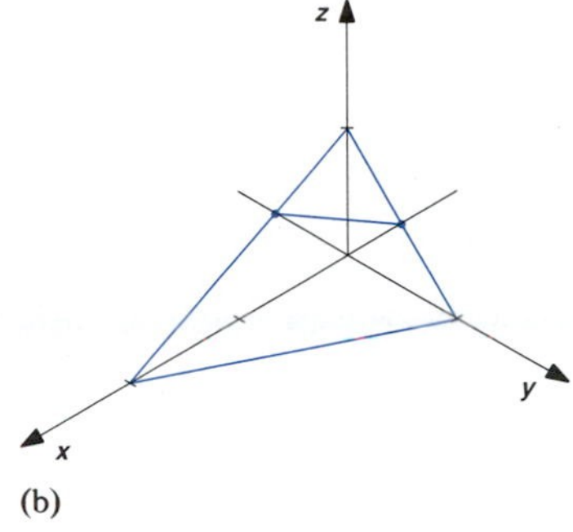


(b)

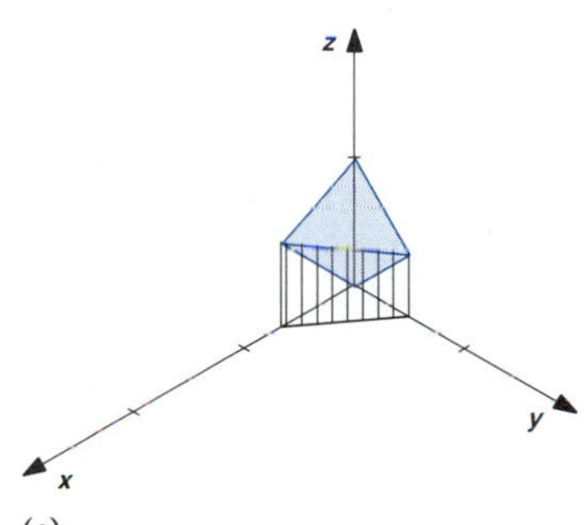


(c)

FIGURE 14.3.7

■ **EXAMPLE 5**

Find a vertical plane that contains the line of intersection of the planes $x + 2y + 2z = 2$ and $x + y - z = 0$. Sketch all three planes and the line of intersection.

**SOLUTION**

Sketches of the planes are made by determining their coordinate traces. Coordinate traces of the plane $x + 2y + 2z = 2$ are sketched by determining the intercepts and then drawing the lines between the intercepts. The plane $x + y - z = 0$ is indicated by sketching the trace $x = 0$, $y = z$, and the trace $y = 0$, $x = z$. See Figure 14.3.7a.

To sketch the line of intersection of the two planes, we note that the intersection of the $yz$-traces and the intersection of the $xz$-traces give two points on the line of intersection. Since the intersection of two planes is a straight line, we can sketch the line of intersection by drawing a straight line through the points we have determined from the two coordinate traces. See Figure 14.3.7b.

To find a vertical plane that contains the line of intersection, we note that points in the line of intersection satisfy both of the equations

$$x + 2y + 2z = 2 \quad \text{and} \quad x + y - z = 0.$$

If these equations are algebraically combined to obtain a new equation, then points on the intersection will satisfy the new equation. This means that the graph of the new equation contains the line of intersection. We want the new equation to be the equation of a vertical plane. A plane is vertical if its equation contains no $z$ term. Hence, we want to combine the two equations in such a way that the new equation is linear and contains no $z$ term. That is, if

$$x + 2y + 2z = 2 \quad \text{and} \quad x + y - z = 0,$$

the second equation gives $z = x + y$. Substitution into the first equation then gives

$$x + 2y + 2(x + y) = 2,$$

$$x + 2y + 2x + 2y = 2,$$

$$3x + 4y = 2.$$

This is the equation of a vertical plane that contains the line of intersection of the two given planes. We have indicated this plane in Figure 14.3.7c by drawing vertical lines between the line of intersection and the $xy$-plane. ■

The intersection of the $xy$-plane with the vertical generalized cylinder that contains a curve is called the **projection** of the curve onto the $xy$-plane.

In Chapter 17 we will learn a procedure to describe certain types of regions in three-dimensional space. An important part of this procedure involves finding a vertical generalized cylinder that contains the intersection of two given surfaces. We will use the procedure introduced in Example 5. Namely,

**To sketch the intersection of two surfaces,**

- **Sketch the coordinate traces of each surface,**
- **Find points of intersection of the coordinate traces,**
- **Connect the points of intersection you have found with a continuous curve. (We assume that the surfaces intersect in a single continuous curve.)**

**To find a vertical general cylinder that contains the intersection of two given surfaces, we combine the equations of the surfaces in such a way that eliminates $z$ from the new equation.**

If the equations of two given surfaces are combined algebraically, the graph of the new equation is a surface that contains the intersection of the two surfaces. If the new equation contains no $z$ term, its graph is a vertical generalized cylinder. That is, the graph is a generalized cylinder with generating lines parallel to the $z$-axis. The intersection of this generalized cylinder with the $xy$-plane is the projection of the curve of intersection onto the $xy$-plane.

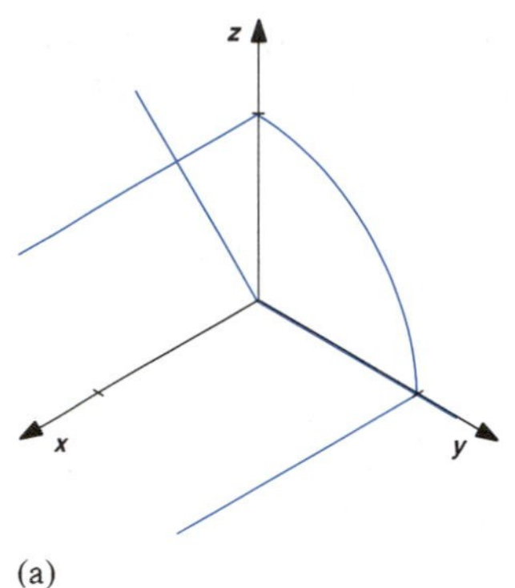

(a)

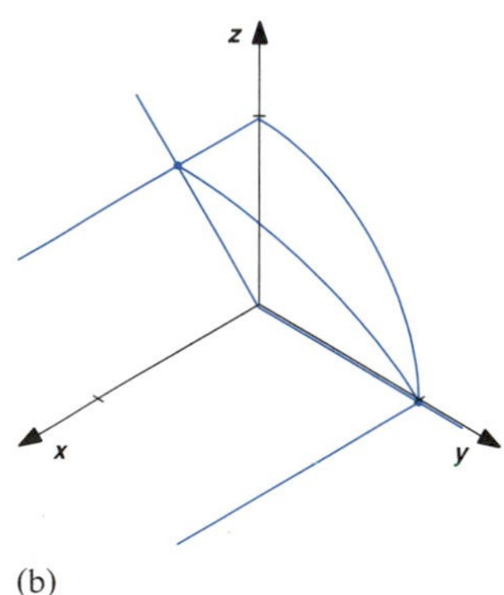

(b)

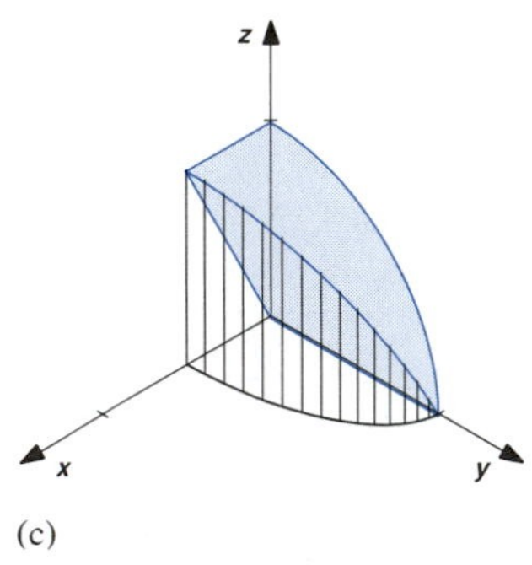

(c)

**FIGURE 14.3.8**

### EXAMPLE 6

Sketch the part in the first octant of the cylinder $y^2 + z^2 = 1$, the plane $z = 2x$, and their curve of intersection. Find a vertical generalized cylinder that contains the curve of intersection. Sketch the generalized cylinder and the projection of the curve of intersection onto the $xy$-plane.

#### SOLUTION

The surface $y^2 + z^2 = 1$ is a cylinder with generating lines parallel to the $x$-axis. The $yz$-trace is a circle, which we sketch. The other traces are then obtained by drawing lines parallel to the $x$-axis from the intersections of the $yz$-trace with each of the $y$-axis and $z$-axis.

The $yz$-trace of the plane $z = 2x$ is the $y$-axis. The $xz$-trace is the line $y = 0$, $z = 2x$.

The traces of the cylinder and plane are sketched in Figure 14.3.8a. In Figure 14.3.8b, we have indicated intersections of the traces in the $yz$-plane and the $xz$-plane, and then drawn the curve of intersection between these two points of intersection.

To find a vertical cylinder that contains the curve of intersection, we combine the equations

$$y^2 + z^2 = 1, \qquad z = 2x,$$

by substituting $2x$ for $z$ in the first equation to obtain

$$y^2 + 4x^2 = 1.$$

This is the equation of a vertical elliptical cylinder. This cylinder and the projection of the curve of intersection onto the $xy$-plane have been added to the sketch in Figure 14.3.8c. ■

## EXERCISES 14.3

Find an equation of the plane that satisfies the conditions given in Exercises 1–12.

1 Contains (2, 3, 1), normal vector $4\mathbf{i} + 3\mathbf{j} - 2\mathbf{k}$

2 Contains (3, −1, 2), normal vector $\mathbf{i} - \mathbf{j} + 2\mathbf{k}$

3 Contains (1, −1, 0), parallel to $x - 3y + 2z = 0$

4 Contains (4, 3, −2), parallel to $2x - y + 3z = 2$

5 Contains (5, 1, 3), parallel to $xy$-plane

6 Contains (2, −2, 1), parallel to $yz$-plane

7 Contains (1, −2, 1), (2, 2, 2), and (0, 0, 0)

8 Contains (3, 2, 0), (5, 2, 0), and (4, 0, 0)

9 Contains (4, 6, 5), (3, 2, 7), and (2, 1, 4)

10 Contains (1, 0, −1), (−1, 1, 0), and (0, 1, 1)

11 Contains (2, 1, −2), contains vectors in the direction of $\mathbf{i} + 2\mathbf{k}$ and $\mathbf{j} - 3\mathbf{k}$

12 Contains (3, −1, 2), contains vectors in the direction of $\mathbf{i} - 2\mathbf{k}$ and $\mathbf{j}$

Find the angle of intersection of the planes in Exercises 13–16.

13 $x - y + z = 1$, $2x + y - z = 3$

14 $x = z$, $x + y + z = 0$

15 $2x + y - z = 2$, $3x + y - z = 3$

16 $3x - 2y = z$, $2x + 3y = z$

Find the distance of the given point from the plane, and the point in the plane that is closest to the given point in each of Exercises 17–20.

17 (0, 0, 0), $2x + 3y + 2z = 6$

18 (1, 2, 0), $3x - y + 2z = 6$

19 (−1, 1, 2), $x + y = 2$

20 (3, −2, 2), $2x + 2y + z = 6$

Find the distance between the parallel planes given in Exercises 21–22.

21 $2x - 3y + z = 6$, $2x - 3y + z = 2$

22 $x + 2y - z = 0$, $x + 2y - z = 4$

In Exercises 23–32 sketch the part in the first octant of the given surfaces and their curve of intersection. Find a vertical generalized cylinder that contains the intersection of the surfaces. Sketch the generalized cylinder and the projection of the curve of intersection onto the $xy$-plane.

23 $x + y + z = 3$, $z = 2x$

24 $x + z = 3$, $y + z = 3$

25 $x + 2y + 2z = 4$, $y = 2z$

26 $z = 2x + 2y$, $z = 3$

27 $z = 2x + 2y$, $x + z = 3$

28 $z = 3x$, $x + z = 3$

29 $z = 3x + 3y$, $z = 4 - x^2 - y^2$

30 $2y = z$, $x^2 + z^2 = 9$

31 $x + 2y + 2z = 6$, $x^2 + y^2 = 4$

32 $x^2 + y^2 = 9$, $x^2 + z^2 = 9$

## 14.4 LINES IN THREE-DIMENSIONAL SPACE

A **line** in three-dimensional space is determined by a point on the line and a direction vector. A point $P(x, y, z)$ is on the line that contains the point $P_0(x_0, y_0, z_0)$ and has direction $\mathbf{V} = a\mathbf{i} + b\mathbf{j} + c\mathbf{k}$ if and only if the vector $\mathbf{P_0P} = (x - x_0)\mathbf{i} + (y - y_0)\mathbf{j} + (z - z_0)\mathbf{k}$ is a multiple of $\mathbf{V}$. See Figure 14.4.1.

That is,

$$\mathbf{P_0P} = t\mathbf{V}, \qquad t \text{ any real number.}$$

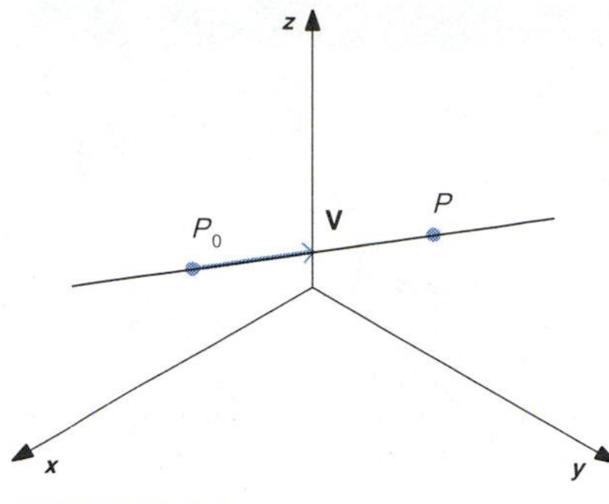


FIGURE 14.4.1

This is the **vector form** of the equation of the line. Let us express this equation in terms of the components of the vectors. This gives

$$(x - x_0)\mathbf{i} + (y - y_0)\mathbf{j} + (z - z_0)\mathbf{k} = t(a\mathbf{i} + b\mathbf{j} + c\mathbf{k}),$$

$$(x - x_0)\mathbf{i} + (y - y_0)\mathbf{j} + (z - z_0)\mathbf{k} = at\mathbf{i} + bt\mathbf{j} + ct\mathbf{k}.$$

Equating corresponding components of the vectors on each side of the above equation, we obtain the equations

$$x - x_0 = at, \quad y - y_0 = bt, \quad z - z_0 = ct, \quad \text{or}$$

$$x = x_0 + at, \quad y = y_0 + bt, \quad z = z_0 + ct.$$

These equations are called **parametric equations** of the line that contains the point $(x_0, y_0, z_0)$ and has direction vector $\mathbf{V} = a\mathbf{i} + b\mathbf{j} + c\mathbf{k}$. The variable $t$ is called a **parameter**.

We will deal with lines and other curves that are expressed in parametric form in Chapter 15. For now, let us find a way of expressing a line as an intersection of planes. If none of $a$, $b$, or $c$ is zero, we can solve each of the three parametric equations of a line for the parameter $t$. Thus,

$$t = \frac{x - x_0}{a}, \quad t = \frac{y - y_0}{b}, \quad t = \frac{z - z_0}{c}.$$

We can then express the equations of the line as

$$\frac{x - x_0}{a} = \frac{y - y_0}{b} = \frac{z - z_0}{c}.$$

This is called the **symmetric form** of the equations of the line that contains the point $(x_0, y_0, z_0)$ and has direction vector $\mathbf{V} = a\mathbf{i} + b\mathbf{j} + c\mathbf{k}$. If one or two of $a$, $b$, or $c$ are zero, the corresponding variable is constant. We should replace the equation that involves the corresponding variable by the equation that expresses the constant value. For example, the equations of the line through the point (1, 3, 2) with direction vector $\mathbf{V} = \mathbf{j} - 2\mathbf{k}$ should be written as

$$x = 1, \quad \frac{y - 3}{1} = \frac{z - 2}{-2}, \quad \text{instead of} \quad \frac{x - 1}{0} = \frac{y - 3}{1} = \frac{z - 2}{-2}.$$

Symmetric equations of a line can be considered as two equations. For example, the symmetric equations

$$\frac{x - x_0}{a} = \frac{y - y_0}{b} = \frac{z - z_0}{c}$$

give the two equations

$$\frac{x - x_0}{a} = \frac{y - y_0}{b} \quad \text{and} \quad \frac{y - y_0}{b} = \frac{z - z_0}{c}.$$

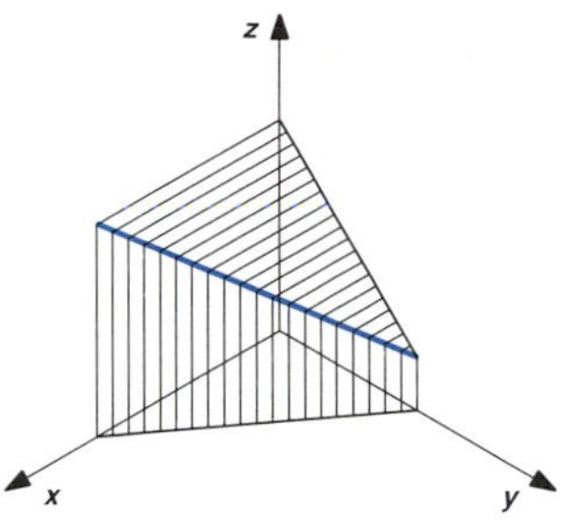

FIGURE 14.4.2

The other equation that relates $x$ and $z$ is a consequence of these two equations. These equations are both equations of planes. We can think of a line in three-dimensional space as the intersection of two planes. Also, the equation

$$\frac{x - x_0}{a} = \frac{y - y_0}{b}$$

is a vertical generalized cylinder that contains the line. This plane gives the projection of the line onto the $xy$-plane. Similarly, the equation

$$\frac{y - y_0}{b} = \frac{z - z_0}{c}$$

gives the projection of the line onto the $yz$-plane. See Figure 14.4.2.

If desired, we can obtain parametric equations of a line with symmetric equations

$$\frac{x - x_0}{a} = \frac{y - y_0}{b} = \frac{z - z_0}{c}$$

by setting each of the three equal expressions equal to a parameter $t$. We obtain

$$\frac{x - x_0}{a} = t, \quad \frac{y - y_0}{b} = t, \quad \frac{z - z_0}{c} = t, \quad \text{so}$$

$$x = x_0 + at, \quad y = y_0 + bt, \quad z = z_0 + ct.$$

Equations of a line that correspond to different points $P_0$ on the line and different (but parallel) direction vectors $\mathbf{V}$ may appear very different, even though they represent the same line.

We can *read* a direction vector from either the parametric or symmetric equations of a line. For example, the line with symmetric equations

$$\frac{x - 1}{2} = \frac{y + 3}{-1} = \frac{z}{3}$$

has direction vector $2\mathbf{i} - \mathbf{j} + 3\mathbf{k}$; the line with parametric equations

$$x = -1 + 2t, \quad y = 3 + 5t, \quad z = -1 - 4t,$$

has direction vector $2\mathbf{i} + 5\mathbf{j} - 4\mathbf{k}$. Also:

**Parallel lines have parallel direction vectors.**

### ■ EXAMPLE 1

Find symmetric equations and parametric equations of the line that contains the point (3, 5, 2) and is parallel to the line

$$\frac{x}{-2} = \frac{y}{1} = \frac{z}{3}.$$

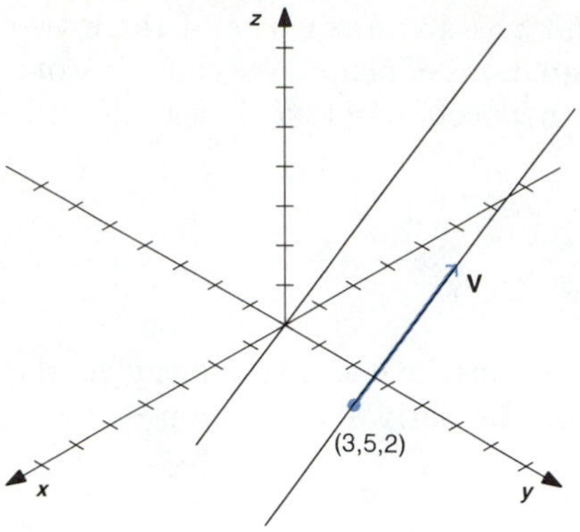

FIGURE 14.4.3

**SOLUTION**

We need a point on the line and a direction vector **V**. We have the point (3, 5, 2). Parallel lines have parallel direction vectors. A direction vector of the given parallel line is $\mathbf{V} = -2\mathbf{i} + \mathbf{j} + 3\mathbf{k}$. Using **V** for the direction vector, we obtain the symmetric equations

$$\frac{x-3}{-2} = \frac{y-5}{1} = \frac{z-2}{3}.$$

Setting each of the above expressions equal to $t$, we obtain

$$\frac{x-3}{-2} = t, \quad \frac{y-5}{1} = t, \quad \frac{z-2}{3} = t, \quad \text{or}$$

$$x = 3 - 2t, \quad y = 5 + t, \quad z = 2 + 3t.$$

These are parametric equations of the line. See Figue 14.4.3. ■

■ **EXAMPLE 2**

Find symmetric equations of the line that contains the point (3, 2, 1) and is perpendicular to the plane

$$x + 2y - 2z = 2.$$

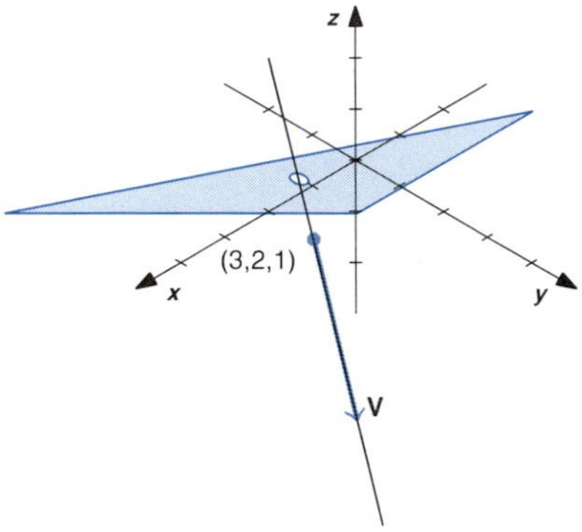

FIGURE 14.4.4

**SOLUTION**

We have a point on the line. We need a direction vector. If the line is to be perpendicular to the plane, then any direction vector of the line must be parallel to any normal vector of the plane. See Figure 14.4.4. This means we can use a normal vector of the plane as a direction vector of the line. A vector normal to the plane $x + 2y - 2z = 2$ is $\mathbf{N} = \mathbf{i} + 2\mathbf{j} - 2\mathbf{k}$. Symmetric equations of the line that contains the point (3, 2, 1) and has direction vector $\mathbf{i} + 2\mathbf{j} - 2\mathbf{k}$ are

$$\frac{x-3}{1} = \frac{y-2}{2} = \frac{z-1}{-2}.$$ ■

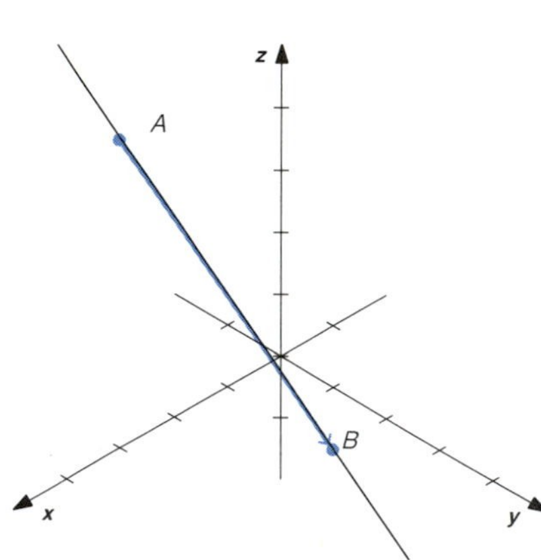


FIGURE 14.4.5

■ **EXAMPLE 3**

Find symmetric equations of the line that contains the two points $A(1, -2, 3)$ and $B(3, 4, 2)$.

**SOLUTION**

We need a point on the line and a direction vector. Let us use the point $A(1, -2, 3)$. We need a direction vector that is parallel to the vector **AB**. See Figure 14.4.5. Choose

$$\mathbf{V} = \mathbf{AB} = (3-1)\mathbf{i} + (4+2)\mathbf{j} + (2-3)\mathbf{k} = 2\mathbf{i} + 6\mathbf{j} - \mathbf{k}.$$

The equations are

$$\frac{x-1}{2} = \frac{y+2}{6} = \frac{z-3}{-1}.$$

■

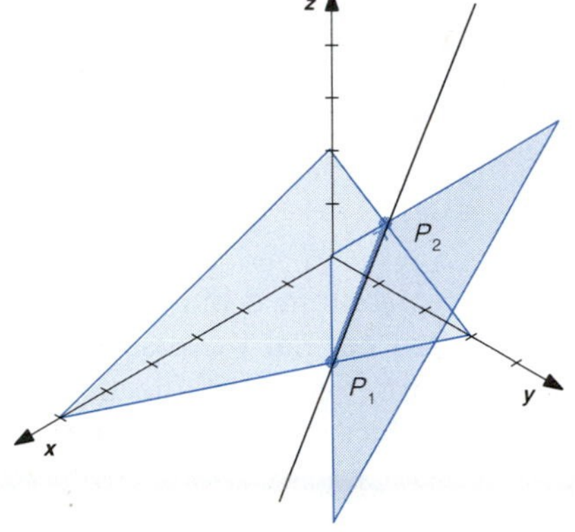


FIGURE 14.4.6

■ **EXAMPLE 4**

Find symmetric equations of the line of intersection of the two planes $x - y + z = 0$ and $x + 2y + 3z = 6$.

**SOLUTION**

We need a point on the line and a direction vector. We also note that any two points on the line can be used to determine a direction vector. See Figure 14.4.6.

Coordinates of points on the line of intersection of the planes are solutions of the simultaneous equations

$$\begin{cases} x - y + z = 0, \\ x + 2y + 3z = 6. \end{cases}$$

We can find a solution of these equations by choosing any convenient value for one of the variables and then solving the equations for the other two variables. For example, let $z = 0$. We then have

$$\begin{cases} x - y = 0, \\ x + 2y = 6. \end{cases}$$

The solutions are $x = 2$ and $y = 2$. The point $P_1 = (2, 2, 0)$ is on the line of intersection. Setting $x = 0$ in the equations of the planes, we obtain

$$\begin{cases} -y + z = 0, \\ 2y + 3z = 6. \end{cases}$$

Solution of these equations gives $y = z = 6/5$. The point $P_2 = (0, 6/5, 6/5)$ is on the line.

To find a direction vector, we choose the vector between the two points on the line that we have found. That is,

$$\begin{aligned} \mathbf{V} = \mathbf{P}_1\mathbf{P}_2 &= (0-2)\mathbf{i} + \left(\frac{6}{5} - 2\right)\mathbf{j} + \left(\frac{6}{5} - 0\right)\mathbf{k} \\ &= -2\mathbf{i} - \frac{4}{5}\mathbf{j} + \frac{6}{5}\mathbf{k}. \end{aligned}$$

Symmetric equations of the line that contains the point (2, 2, 0) and has direction vector $-2\mathbf{i} - (4/5)\mathbf{j} + (6/5)\mathbf{k}$ are

$$\frac{x-2}{-2} = \frac{y-2}{-4/5} = \frac{z-0}{6/5}.$$

■

Any nonzero multiple of a direction vector of a line is also a direction vector of the line. For example, we can obtain a direction vector with integer components in Example 4 by using the direction vector

$$\frac{5}{2}\mathbf{V} = \left(\frac{5}{2}\right)\left(-2\mathbf{i} - \frac{4}{5}\mathbf{j} + \frac{6}{5}\mathbf{k}\right) = -5\mathbf{i} - 2\mathbf{j} + 3\mathbf{k}.$$

The equations of the line can then be written

$$\frac{x-2}{-5} = \frac{y-2}{-2} = \frac{z-0}{3}.$$

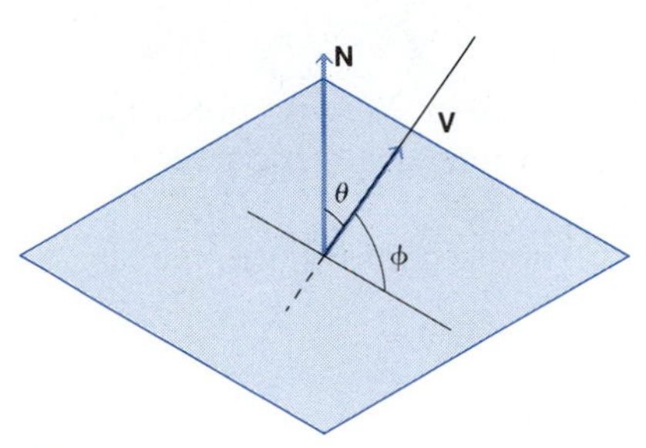


FIGURE 14.4.7

Since the line of intersection of two planes is contained in each plane, any direction vector of the line is orthogonal to normal vectors of each plane. This means that we can find a direction vector of the line of intersection of two planes by taking a cross product of normal vectors of the planes.

Suppose a line with direction vector **V** intersects a plane that has normal vector **N**. See Figure 14.4.7. The acute angle $\theta$ between **V** and **N** satisfies

$$\cos\theta = \frac{|\mathbf{V}\cdot\mathbf{N}|}{|\mathbf{V}||\mathbf{N}|}.$$

(Taking absolute values of the inner product gives $0 \leq \theta \leq \pi/2$.) The **angle of intersection** of the line and plane is defined to be

$$\phi = \frac{\pi}{2} - \theta, \quad \text{so}$$

$$\sin\phi = \frac{|\mathbf{V}\cdot\mathbf{N}|}{|\mathbf{V}||\mathbf{N}|}.$$

The angle $\phi$ is the smallest angle between **V** and any vector in the plane.

Coordinates of the intersection of a line and a plane are given by the simultaneous solution of three linear equations in three variables $x$, $y$, and $z$. Two of these equations are given by symmetric equations of the line and the other equation is the equation of the plane. The equations will have exactly one solution unless the line is parallel to the plane. We should check that the solution satisfies the symmetric equations of the line and the equation of the plane.

■ **EXAMPLE 5**

Find the point of intersection and the angle of intersection of the line

$$\frac{x-1}{3} = \frac{y+1}{1} = \frac{z-2}{-2}$$

and the plane $2x - 3y + z = 6$.

**SOLUTION**

The coordinates of the point of intersection are given by the simultaneous solution of the equations

$$\frac{x-1}{3}=\frac{y+1}{1},\quad \frac{y+1}{1}=\frac{z-2}{-2},\quad 2x-3y+z=6.$$

Clearing fractions and simplifying, we obtain

$$\begin{cases} x-3y=4, \\ -2y-z=0, \\ 2x-3y+z=6. \end{cases}$$

The solution is $(x, y, z) = (-2, -2, 4)$. We check that

$$\frac{(-2)-1}{3}=\frac{(-2)+1}{1}=\frac{(4)-2}{-2}=-1 \quad \text{and} \quad 2(-2)-3(-2)+(4)=6,$$

so the point is on both the line and the plane.

A direction vector of the line is $\mathbf{V} = 3\mathbf{i} + \mathbf{j} - 2\mathbf{k}$. A normal vector of the plane is $\mathbf{N} = 2\mathbf{i} - 3\mathbf{j} + \mathbf{k}$. Then

$$\begin{aligned} \sin\phi &= \frac{|\mathbf{V}\cdot\mathbf{N}|}{|\mathbf{V}||\mathbf{N}|} \\ &= \frac{|(3)(2)+(1)(-3)+(-2)(1)|}{\sqrt{3^2+1^2+(-2)^2}\sqrt{2^2+(-3)^2+1^2}} \\ &= \frac{1}{14}. \end{aligned}$$

The angle of intersection of the line and the plane is

$$\phi = \sin^{-1}\frac{1}{14} \approx 0.07 \text{ rad.}$$

■

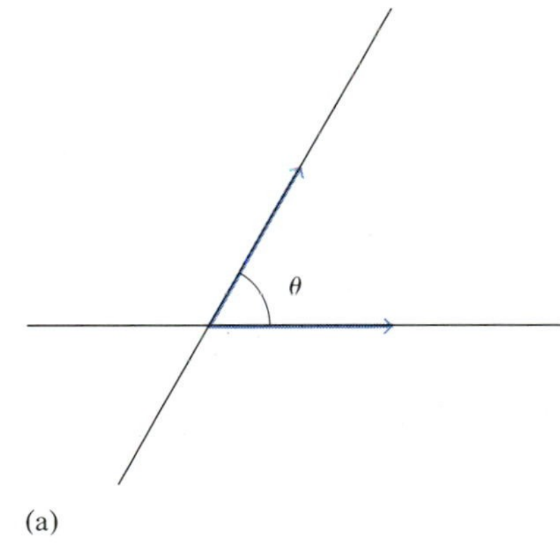


(a)

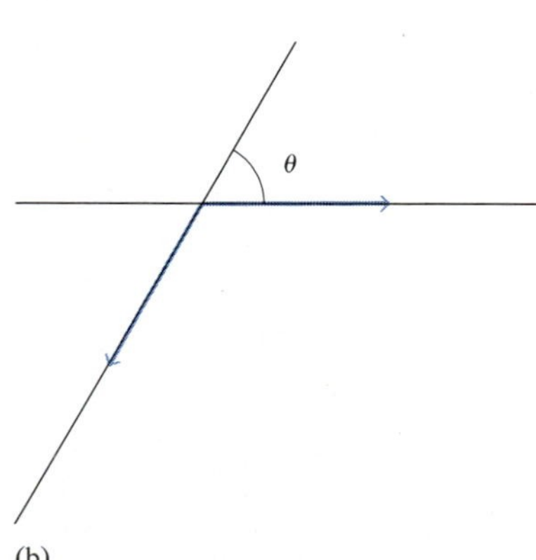


(b)

**FIGURE 14.4.8**

The **angle of intersection** of two intersecting lines with direction vectors $\mathbf{V}_1$ and $\mathbf{V}_2$ satisfies

$$\cos\theta = \frac{|\mathbf{V}_1\cdot\mathbf{V}_2|}{|\mathbf{V}_1||\mathbf{V}_2|}.$$

See Figure 14.4.8.

Symmetric equations of a line give two equations in the variables $x$, $y$, and $z$. Thus, the coordinates of the point of intersection of two lines must satisfy four equations. When finding the point of intersection of two lines, be sure to check that the coordinates you find actually satisfy symmetric equations of both lines.

**■ EXAMPLE 6**

Find the point of intersection and the angle between the lines

$$\frac{x+1}{2} = \frac{y-2}{1} = \frac{z-4}{-2} \quad \text{and} \quad \frac{x-3}{-1} = \frac{y-1}{1} = \frac{z-0}{1}.$$

**SOLUTION**

The coordinates of the point of intersection, if the lines actually intersect, are given by the equations

$$\frac{x+1}{2} = \frac{y-2}{1}, \quad \frac{y-2}{1} = \frac{z-4}{-2}, \quad \frac{x-3}{-1} = \frac{y-1}{1}, \quad \frac{y-1}{1} = \frac{z-0}{1}.$$

We can rewrite this system in the form

$$x+1 = 2y-4, \quad -2y+4 = z-4, \quad x-3 = -y+1, \quad y-1 = z$$

or

$$\begin{cases} x - 2y \phantom{+z} = -5, \\ \phantom{x-} 2y + z = 8, \\ x + \phantom{2}y \phantom{+z} = 4, \\ \phantom{x+2} y - z = 1. \end{cases}$$

It is not difficult to determine that the last three equations have the simultaneous solution $x = 1$, $y = 3$, $z = 2$. We then verify that these values satisfy the given equations of both lines. Namely,

$$\frac{(1)+1}{2} = \frac{(3)-2}{1} = \frac{(2)-4}{-2} = 1 \quad \text{and} \quad \frac{(1)-3}{-1} = \frac{(3)-1}{1} = \frac{(2)-0}{1} = 2.$$

We conclude that (1, 3, 2) is the point of intersection of the lines.

The lines have direction vectors

$$\mathbf{V}_1 = 2\mathbf{i} + \mathbf{j} - 2\mathbf{k} \quad \text{and} \quad \mathbf{V}_2 = -\mathbf{i} + \mathbf{j} + \mathbf{k}.$$

The angle between the lines satisfies

$$\begin{aligned} \cos\theta &= \frac{|\mathbf{V}_1 \cdot \mathbf{V}_2|}{|\mathbf{V}_1||\mathbf{V}_2|} \\ &= \frac{|(2\mathbf{i} + \mathbf{j} - 2\mathbf{k}) \cdot (-\mathbf{i} + \mathbf{j} + \mathbf{k})|}{|2\mathbf{i} + \mathbf{j} - 2\mathbf{k}||-\mathbf{i} + \mathbf{j} + \mathbf{k}|} \\ &= \frac{|-2+1-2|}{\sqrt{4+1+4}\sqrt{1+1+1}} \\ &= \frac{3}{3\sqrt{3}} = \frac{1}{\sqrt{3}}. \end{aligned}$$

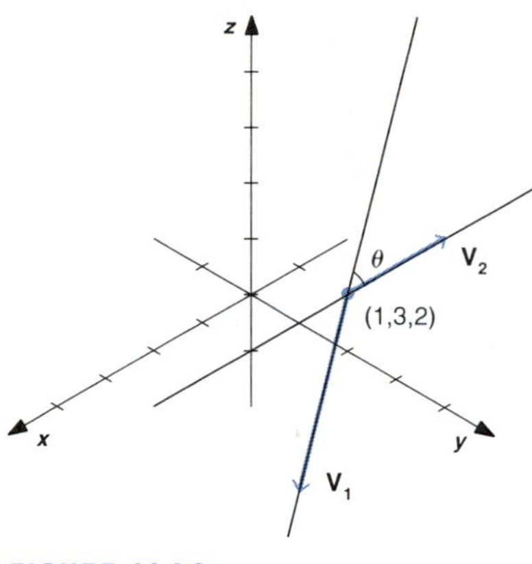

FIGURE 14.4.9

Then $\theta = \cos^{-1}(1/\sqrt{3}) \approx 0.9553166$ rad. See Figure 14.4.9. ■

■ **EXAMPLE 7**

Find the equation of the plane that contains the line

$$\frac{x-1}{1} = \frac{y-4}{-2} = \frac{z-5}{-1}$$

and is parallel to the line

$$\frac{x}{1} = \frac{y}{2} = \frac{z}{-1}.$$

Show that the two lines do not intersect.

**SOLUTION**

To write the equation of a plane, we need a point on the plane and a normal vector. Since the plane is to contain the line

$$\frac{x-1}{1} = \frac{y-4}{-2} = \frac{z-5}{-1},$$

the plane must contain the point (1, 4, 5), which is on the line, and the normal of the plane must be orthogonal to the direction vector of the line, $\mathbf{V}_1 = \mathbf{i} - 2\mathbf{j} - \mathbf{k}$. Since the plane is to be parallel to the line

$$\frac{x}{1} = \frac{y}{2} = \frac{z}{-1},$$

the normal to the plane must also be orthogonal to a direction vector of that line, $\mathbf{V}_2 = \mathbf{i} + 2\mathbf{j} - \mathbf{k}$. See Figure 14.4.10. Since the cross product is orthogonal to both factors, we can use $\mathbf{V}_1 \times \mathbf{V}_2$ as a normal vector of the plane.

FIGURE 14.4.10

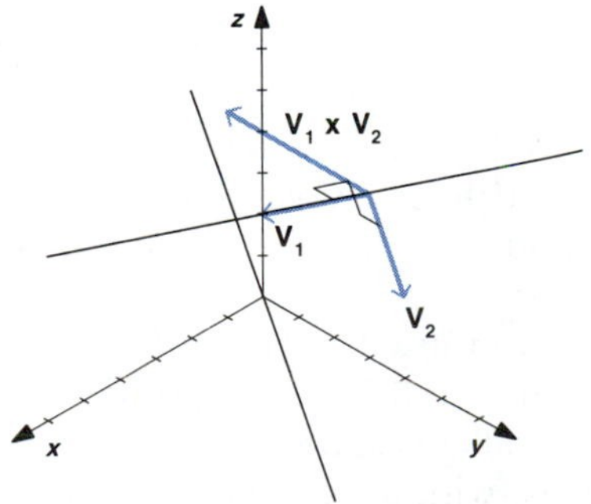

We have

$$\mathbf{V}_1 \times \mathbf{V}_2 = \begin{vmatrix} \mathbf{i} & \mathbf{j} & \mathbf{k} \\ 1 & -2 & -1 \\ 1 & 2 & -1 \end{vmatrix} = \begin{vmatrix} -2 & -1 \\ 2 & -1 \end{vmatrix}\mathbf{i} - \begin{vmatrix} 1 & -1 \\ 1 & -1 \end{vmatrix}\mathbf{j} + \begin{vmatrix} 1 & -2 \\ 1 & 2 \end{vmatrix}\mathbf{k}$$

$$= 4\mathbf{i} + 4\mathbf{k}.$$

The equation of the plane that contains the point (1, 4, 5) with normal vector $4\mathbf{i} + 4\mathbf{k}$ is

$$4(x-1) + 0(y-4) + 4(z-5) = 0,$$

$$4x - 4 + 4z - 20 = 0,$$

$$4x + 4z = 24,$$

$$x + z = 6.$$

Let us now verify that the two lines do not intersect. The coordinate of a point of intersection would satisfy the equations

$$\frac{x-1}{1} = \frac{y-4}{-2}, \quad \frac{y-4}{-2} = \frac{z-5}{-1}, \quad \frac{x}{1} = \frac{y}{2}, \quad \frac{y}{2} = \frac{z}{-1}.$$

Rewriting, we have

$$-2x + 2 = y - 4, \quad y - 4 = 2z - 10, \quad 2x = y, \quad y = -2z,$$

or

$$\begin{cases} 2x + y = 6, \\ y - 2z = -6, \\ 2x - y = 0, \\ y + 2z = 0. \end{cases}$$

The first three of these equations have the unique simultaneous solution $x = 3/2$, $y = 3$, $z = 9/2$. This means that any point on the intersection of the two lines must have these coordinates. Since these values do not satisfy the fourth equation, nor the symmetric equations of both lines, we conclude that the lines do not intersect. ■

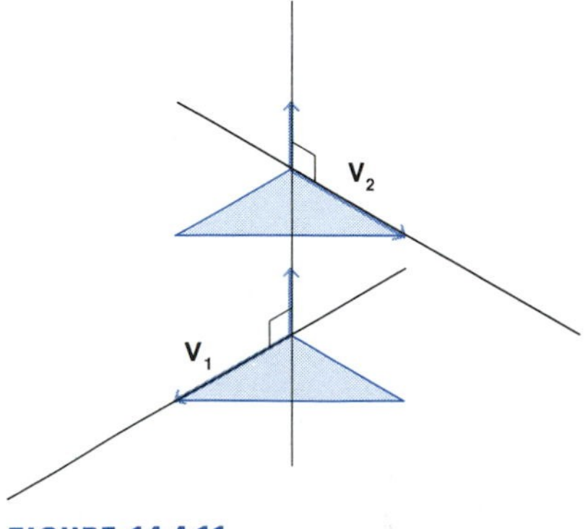

FIGURE 14.4.11

We can use the ideas of Example 7 to find parallel planes that contain two nonparallel lines. Suppose the directions of two lines are given by the nonparallel vectors $\mathbf{V}_1$ and $\mathbf{V}_2$. See Figure 14.4.11. A plane that contains the line with direction $\mathbf{V}_1$ has normal vector that is orthogonal to $\mathbf{V}_1$. A plane that contains the line with direction $\mathbf{V}_2$ has normal vector orthogo-

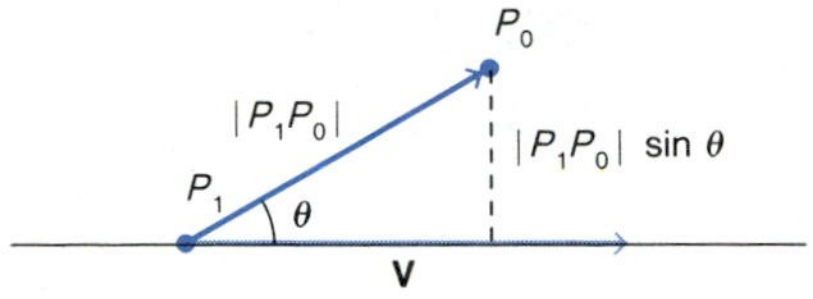

FIGURE 14.4.12

nal to $\mathbf{V}_2$. Parallel planes have parallel normal vectors, so we can choose both normal vectors to be any vector that is orthogonal to both $\mathbf{V}_1$ and $\mathbf{V}_2$. The cross product $\mathbf{N} = \mathbf{V}_1 \times \mathbf{V}_2$ is such a vector.

Let us see how to find the distance of a point $P_0$ from a line in three-dimensional space. We first choose any point $P_1$ that is on the line. From Figure 14.4.12 we see that the distance of the point from the line is

$$d = |\mathbf{P}_1\mathbf{P}_0|\sin\theta,$$

where $\theta$ is the angle between the vector $\mathbf{P}_1\mathbf{P}_0$ and a direction vector $\mathbf{V}$ of the line. From Section 14.2 we know that

$$|\mathbf{P}_1\mathbf{P}_0 \times \mathbf{V}| = |\mathbf{P}_1\mathbf{P}_0||\mathbf{V}|\sin\theta.$$

Substitution into the formula for $d$ then gives

$$(\text{Distance of } \mathbf{P}_0 \text{ to line}) = |\mathbf{P}_1\mathbf{P}_0|\frac{|\mathbf{P}_1\mathbf{P}_0 \times \mathbf{V}|}{|\mathbf{P}_1\mathbf{P}_0||\mathbf{V}|} = \frac{|\mathbf{P}_1\mathbf{P}_0 \times \mathbf{V}|}{|\mathbf{V}|}.$$

■ **EXAMPLE 8**

An airplane approaches the end of a runway from the east with an angle of descent of 20°. A tower that is 50 ft high is located 400 ft south and 800 ft east of the end of the runway. How close does the airplane come to a warning light that is at the top of the tower?

**SOLUTION**

A coordinate system is chosen as indicated in Figure 14.4.13. We want the distance of the point $P_0(400, 800, 50)$ from the line of descent of the airplane. The point $P_1(0, 0, 0)$ is on the line and the line has direction vector

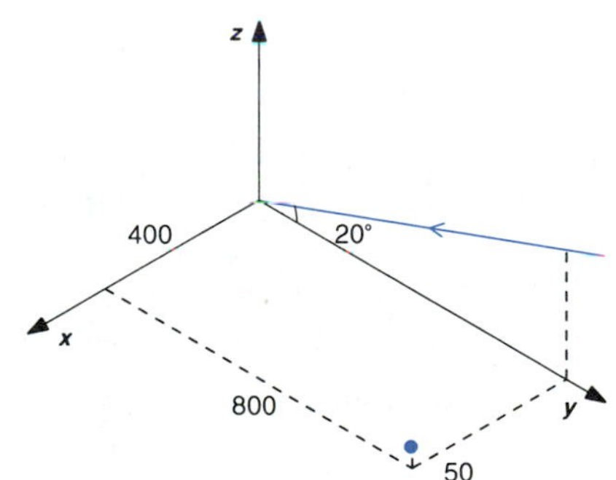


FIGURE 14.4.13

$$\mathbf{V} = \cos 20^\circ\,\mathbf{j} + \sin 20^\circ\,\mathbf{k}.$$

We then have

$$\mathbf{P}_1\mathbf{P}_0 = 400\mathbf{i} + 800\mathbf{j} + 50\mathbf{k}$$

and

$$\mathbf{P}_1\mathbf{P}_0 \times \mathbf{V} = \begin{vmatrix} \mathbf{i} & \mathbf{j} & \mathbf{k} \\ 400 & 800 & 50 \\ 0 & \cos 20^\circ & \sin 20^\circ \end{vmatrix}$$

$$= \begin{vmatrix} 800 & 50 \\ \cos 20^\circ & \sin 20^\circ \end{vmatrix}\mathbf{i} - \begin{vmatrix} 400 & 50 \\ 0 & \sin 20^\circ \end{vmatrix}\mathbf{j} + \begin{vmatrix} 400 & 800 \\ 0 & \cos 20^\circ \end{vmatrix}\mathbf{k}$$

$$= (800\sin 20^\circ - 50\cos 20^\circ)\mathbf{i} - 400\sin 20^\circ\mathbf{j} + 400\cos 20^\circ\mathbf{k}.$$

The distance of $P_0$ from the line is

$$d = \frac{|\mathbf{P_1P_0} \times \mathbf{V}|}{|\mathbf{V}|}$$

$$= \frac{|(800 \sin 20° - 50 \cos 20°)\mathbf{i} - 400 \sin 20°\mathbf{j} + 400 \cos 20°\mathbf{k}|}{|\cos 20°\mathbf{j} + \sin 20°\mathbf{k}|}$$

$$= \frac{\sqrt{(800 \sin 20° - 50 \cos 20°)^2 + (-400 \sin 20°)^2 + (400 \cos 20°)^2}}{\sqrt{\cos^2 20° + \sin^2 20°}}$$

$$\approx 459.74105.$$

The airplane is never closer than approximately 460 ft from the warning light. ■

## EXERCISES 14.4

Find symmetric equations and parametric equations of the lines that satisfy the conditions given in Exercises 1–14.

**1** Contains $(1, -1, 2)$, direction vector $2\mathbf{i} - 2\mathbf{j} + 3\mathbf{k}$

**2** Contains $(2, 3, -1)$, direction vector $3\mathbf{i} + \mathbf{j} + 2\mathbf{k}$

**3** Contains $(2, 3, -1)$, parallel to $\dfrac{x-1}{2} = \dfrac{y-3}{-1} = \dfrac{z+1}{3}$

**4** Contains $(3, 1, -2)$, parallel to $\dfrac{x+1}{3} = \dfrac{y-2}{2} = \dfrac{z-4}{-2}$

**5** Contains $(2, 4, -3)$, parallel to $x$-axis

**6** Contains $(-3, 2, -1)$, parallel to $y$-axis

**7** Contains $(1, -2, 3)$, perpendicular to $2x - 3y + z = 6$

**8** Contains $(3, 2, -2)$, perpendicular to $x + 2y = 2$

**9** Contains $(2, 2, 1)$ and $(1, 1, 3)$

**10** Contains $(1, -2, 1)$ and $(3, 1, 2)$

**11** Contains $(5, 3, 7)$ and $(2, 4, 3)$

**12** Contains $(3, 2, 5)$ and $(3, 2, 1)$

**13** Intersection of $x + y - z = 0$ and $2x + 2y - z = 2$

**14** Intersection of $3x + y + 2z = 6$ and $2x - 2y + 3z = 6$

Find the point of intersection and the angle of intersection of the line and plane given in each of Exercises 15–18.

**15** $3x - 3y + 2z = 6$, $\dfrac{x-1}{2} = \dfrac{y-1}{2} = \dfrac{z+1}{-1}$

**16** $2x + 2y + z = 4$, $\dfrac{x}{2} = \dfrac{y+2}{1} = \dfrac{z-1}{1}$

**17** $x + y + z = 1$, $\dfrac{x-2}{3} = \dfrac{y+1}{-2} = \dfrac{z-3}{2}$

**18** $3x + y + 2z = 13$, $\dfrac{x}{2} = \dfrac{y}{1} = \dfrac{z}{3}$

Determine if the lines in Exercises 19–22 intersect. If they do, find the point of intersection and the angle of intersection.

**19** $\dfrac{x-2}{2} = \dfrac{y-3}{2} = \dfrac{z-3}{1}$, $\dfrac{x+2}{2} = \dfrac{y-2}{-1} = \dfrac{z}{2}$

**20** $\dfrac{x}{1} = \dfrac{y}{1} = \dfrac{z}{2}$, $\dfrac{x+2}{2} = \dfrac{y+2}{2} = \dfrac{z-2}{1}$

**21** $\dfrac{x}{1} = \dfrac{y+1}{1} = \dfrac{z+2}{2}$, $\dfrac{x-1}{1} = \dfrac{y-1}{1} = \dfrac{z-2}{2}$

**22** $\dfrac{x-2}{2} = \dfrac{y-2}{2} = \dfrac{z+1}{-1}$, $\dfrac{x-4}{2} = \dfrac{y+4}{-2} = \dfrac{z}{1}$

In Exercises 23–26, find the distance from the point to the line.

**23** $(0, 0, 0)$, $\dfrac{x-1}{3} = \dfrac{y-3}{2} = \dfrac{z-2}{4}$

**24** $(1, -2, -1)$, $\dfrac{x-2}{2} = \dfrac{y+2}{-3} = \dfrac{z-1}{1}$

**25** $(2, 3, -1)$, $\dfrac{x+1}{1} = \dfrac{y+2}{1} = \dfrac{z-2}{-1}$

**26** $(3, -2, 2)$, $\dfrac{x+2}{3} = \dfrac{y-1}{-1} = \dfrac{z+1}{2}$

**27** Find an equation of the plane that contains

$$\frac{x-1}{2} = \frac{y-1}{2} = \frac{z-2}{-1}$$

and is parallel to

$$\frac{x}{1} = \frac{y}{2} = \frac{z}{1}.$$

**28** Find an equation of the plane that contains

$$\frac{x+1}{2} = \frac{y-2}{-2} = \frac{z-3}{1}$$

and is parallel to

$$\frac{x}{3} = \frac{y}{2} = \frac{z}{4}.$$

In each of Exercises 29–30, determine a vector **N** that is normal to the plane that contains the first given line and is parallel to the second given line. Evaluate the length of the projection of a vector from a point on one line to a point on the other line in the direction of **N**. This is the (shortest) distance between the lines.

**29** $\frac{x-1}{2} = \frac{y-1}{1} = \frac{z-1}{1}$ and $\frac{x-2}{1} = \frac{y-1}{-2} = \frac{z+1}{2}$

**30** $\frac{x-2}{1} = \frac{y+1}{1} = \frac{z+2}{-2}$ and $\frac{x-1}{2} = \frac{y-2}{-2} = \frac{z+1}{1}$

In Exercises 31–36, determine if there is a plane that contains the two given lines. If there is such a plane, find its equation.

**31** $\frac{x-2}{1} = \frac{y-0}{3} = \frac{z-1}{-1}, \frac{x-1}{3} = \frac{y-1}{1} = \frac{z-1}{2}$

**32** $\frac{x-2}{1} = \frac{y-0}{3} = \frac{z-1}{-1}, \frac{x-3}{3} = \frac{y-3}{1} = \frac{z-0}{2}$

**33** $\frac{x+1}{2} = \frac{y-1}{-1} = \frac{z+2}{2}, \frac{x-1}{-2} = \frac{y-0}{3} = \frac{z-0}{1}$

**34** $\frac{x+1}{2} = \frac{y-1}{-1} = \frac{z+2}{2}, \frac{x-1}{-2} = \frac{y-2}{3} = \frac{z-0}{1}$

**35** $\frac{x-2}{1} = \frac{y-0}{3} = \frac{z-1}{-1}, \frac{x-1}{1} = \frac{y-1}{3} = \frac{z+1}{-1}$

**36** $\frac{x+1}{2} = \frac{y-1}{-1} = \frac{z+2}{2}, \frac{x-1}{2} = \frac{y-2}{-1} = \frac{z-0}{2}$

**37** A 5-ft-square target is leaning with its bottom edge on the floor. The bottom edge is 3 ft from the west wall and one side is against the north wall. An arrow is shot to the center of the target from a point that is 21.5 ft from the west wall, 8.5 ft from the north wall, and 4 ft above the floor. Find the angle between the arrow and the target. See figure below.

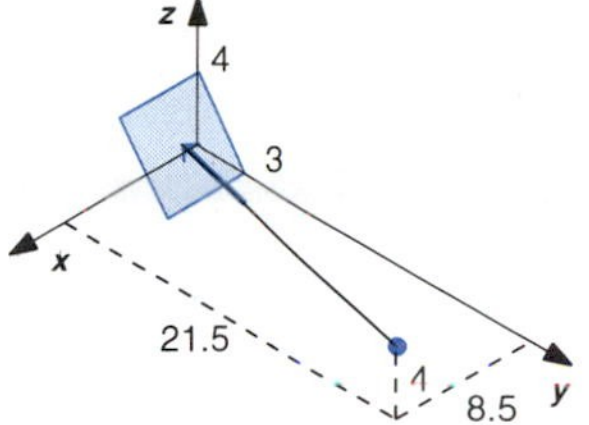


**38** An airplane that is flying a straight line course at an altitude of 2000 ft passes over a point on the ground that is 8000 ft due south of a radio station at ground level and passes over another point that is 8000 ft due east of the station. What is the closest that the airplane comes to the station? See figure below.

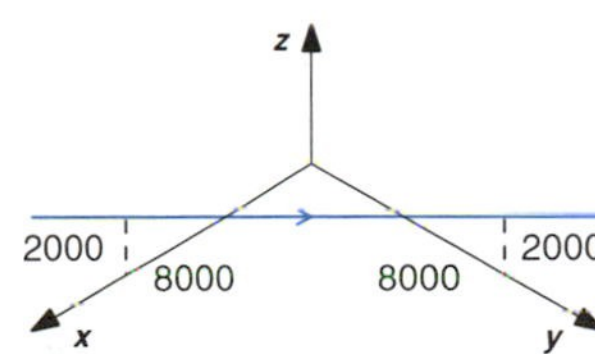


## 14.5 VECTOR-VALUED FUNCTIONS

In this section we will study **vector-valued functions** of a real variable $t$ and see they can be used to study the motion of objects in a coordinate plane or three-dimensional space. Vector-valued simply means that the values $\mathbf{R}(t)$ are vectors. The vectors $\mathbf{R}(t)$ can be either two-dimensional or three-dimensional vectors.

A vector-valued function that has values that are two-dimensional vectors can be expressed in terms of two real-valued **component functions**, $f$ and $g$. That is,

$$\mathbf{R}(t) = f(t)\mathbf{i} + g(t)\mathbf{j},$$

where **i** and **j** are the usual two-dimensional vectors. Similarly, a vector-valued function that has values that are three-dimensional vectors can be expressed in terms of three real-valued component functions,

$$\mathbf{R}(t) = f(t)\mathbf{i} + g(t)\mathbf{j} + h(t)\mathbf{k},$$

where **i**, **j**, and **k** are the usual three-dimensional vectors.

We can represent the position of an object in a coordinate plane at time $t$ by the **position vector**

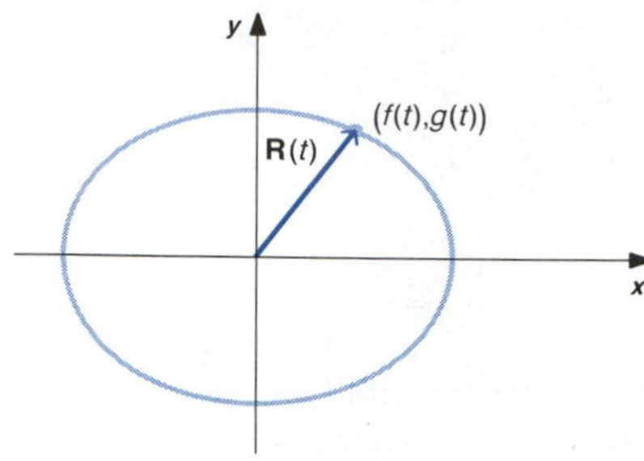


FIGURE 14.5.1

$$\mathbf{R}(t) = f(t)\mathbf{i} + g(t)\mathbf{j}.$$

That is, the coordinates of the position at time $t$ are $(f(t), g(t))$. See Figure 14.5.1. Similarly, we can represent the position of an object in three-dimensional space at time $t$ by a three-dimensional vector-valued function

$$\mathbf{R}(t) = f(t)\mathbf{i} + g(t)\mathbf{j} + h(t)\mathbf{k}.$$

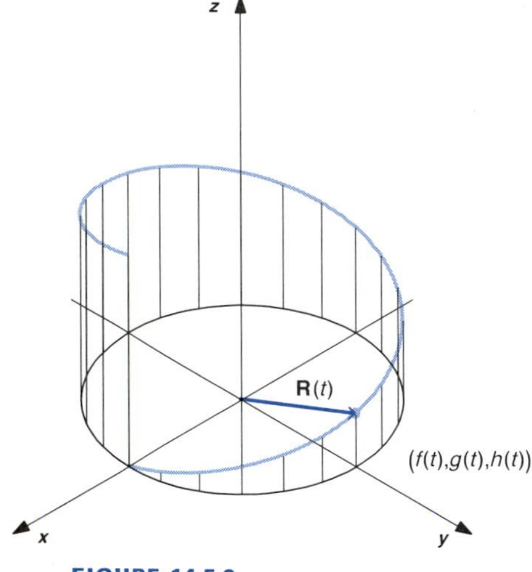


FIGURE 14.5.2

The position at time $t$ then has coordinates $(f(t), g(t), h(t))$. See Figure 14.5.2. If the component functions are continuous, then the coordinates trace a curve as $t$ varies over an interval of time. We will study properties of the curve and how properties of the curve relate to motion along the curve in Chapter 15.

To study the motion of objects in a coordinate plane or three-dimensional space, we need to develop calculus of vector-valued functions. The development involves the same ideas as calculus of real-valued functions. As in the case of real-valued functions, the concepts of differentiation and integration of vector-valued functions depend on limits.

Much of the calculus of vector-valued functions is completely parallel to that of real-valued functions. We merely use the length of a vector in place of the absolute value of a real number. For example, a vector-valued function $\mathbf{R}(t)$ is said to be **bounded** on an interval $I$ if there is a number $B$ such that $|\mathbf{R}(t)| \leq B$ for all $t$ in $I$. $\mathbf{R}(t)$ is close to $\mathbf{L}$ (within $\varepsilon$) whenever $|\mathbf{R}(t) - \mathbf{L}|$ is small (less than $\varepsilon$). The definition of the limit of a vector-valued function can then be stated in the same terms as the limit of a real-valued function. The formal definition follows.

■ *Definition*

We say that a vector $\mathbf{L}$ is the **limit of R(*t*) as *t* approaches *c*** (written $\lim_{t\to c}\mathbf{R}(t) = \mathbf{L}$) if, for each $\varepsilon > 0$, there is a $\delta > 0$ such that

$$|\mathbf{R}(t) - \mathbf{L}| < \varepsilon \quad \text{whenever} \quad 0 < |t - c| < \delta.$$ ■

The properties

$$|c(t)\mathbf{R}(t)| = |c(t)||\mathbf{R}(t)| \quad \text{and} \quad |\mathbf{U}(t) + \mathbf{V}(t)| \leq |\mathbf{U}(t)| + |\mathbf{V}(t)|$$

can be used to verify the Limit Theorem for vector-valued functions. That is, the limit of a sum of two vector-valued functions is the sum of the limits, and the limit of the product of a real-valued function and a vector-valued function is the product of the limits.

It is important that we understand the geometric meaning of the limit of a vector-valued function. Geometrically, $\lim_{t\to c}\mathbf{R}(t) = \mathbf{L}$ means that the

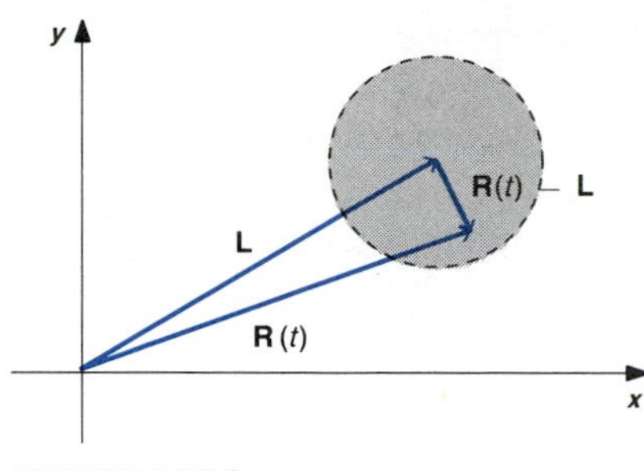

FIGURE 14.5.3

length of $\mathbf{R}(t)$ approaches the length of $\mathbf{L}$ and, if $\mathbf{L} \neq \mathbf{0}$, the direction of $\mathbf{R}(t)$ approaches the direction of $\mathbf{L}$ as $t$ approaches $c$. See Figure 14.5.3.

To evaluate the limit of vector-valued functions, we use the fact that $\mathbf{R}(t)$ approaches $\mathbf{L}$ if and only if each component of $\mathbf{R}(t)$ approaches the corresponding component of $\mathbf{L}$. That is, for two-dimensional vector-valued functions,

$$\lim_{t\to c} (f(t)\mathbf{i} + g(t)\mathbf{j}) = \left(\lim_{t\to c} f(t)\right)\mathbf{i} + \left(\lim_{t\to c} g(t)\right)\mathbf{j}.$$

Thus, *we can evaluate limits of vector-valued functions component by component*. This statement also holds for three-dimensional vector-valued functions.

■ **EXAMPLE 1**

$\mathbf{R}(t) = t^2\mathbf{i} - 3t\mathbf{j}$. Evaluate

$$\lim_{t\to 2} \frac{\mathbf{R}(t) - \mathbf{R}(2)}{t - 2}.$$

**SOLUTION**

We first note that the expression

$$\frac{\mathbf{R}(t) - \mathbf{R}(2)}{t - 2}$$

is well defined for $t \neq 2$. That is, we can divide the vector $\mathbf{R}(t) - \mathbf{R}(2)$ by the real number $t - 2$, since this is equivalent to multiplying the vector by the real number $1/(t - 2)$. Let us now write the expression in terms of components and simplify. For $t \neq 2$, we have

$$\begin{aligned}\frac{\mathbf{R}(t) - \mathbf{R}(2)}{t - 2} &= \frac{[t^2\mathbf{i} - 3t\mathbf{j}] - [2^2\mathbf{i} - 3(2)\mathbf{j}]}{t - 2} \\ &= \frac{t^2 - 2^2}{t - 2}\mathbf{i} + \frac{-3t + 3(2)}{t - 2}\mathbf{j} \\ &= \frac{(t - 2)(t + 2)}{t - 2}\mathbf{i} - \frac{3(t - 2)}{t - 2}\mathbf{j} \\ &= (t + 2)\mathbf{i} - 3\mathbf{j}.\end{aligned}$$

Then

$$\lim_{t\to 2} \frac{\mathbf{R}(t) - \mathbf{R}(2)}{t - 2} = \lim_{t\to 2} [(t + 2)\mathbf{i} - 3\mathbf{j}] = 4\mathbf{i} - 3\mathbf{j}.$$ ■

The **derivative** of a vector-valued function is defined to be

$$\mathbf{R}'(c) = \lim_{t\to c} \frac{\mathbf{R}(t) - \mathbf{R}(c)}{t - c}.$$

$\mathbf{R}'$ is a vector-valued function. *We can evaluate the derivative of vector-valued functions component by component.* Each component is a real-valued function and we can use all the formulas that we already know to find the derivatives. Both $\mathbf{i}$ and $\mathbf{j}$ are *vector* constants. Don't forget that scalar constants have derivative zero. Vector constants have derivative equal to the *zero vector*.

■ **EXAMPLE 2**

$\mathbf{R}(t) = (1 - 2t)\mathbf{i} + (t^2 - t)\mathbf{j}$. Find $\mathbf{R}'(t)$ and $\mathbf{R}''(t)$.

**SOLUTION**

$$\mathbf{R}(t) = (1 - 2t)\mathbf{i} + (t^2 - t)\mathbf{j},$$

$$\mathbf{R}'(t) = -2\mathbf{i} + (2t - 1)\mathbf{j},$$

$$\mathbf{R}''(t) = \mathbf{0} + 2\mathbf{j} = 2\mathbf{j}.$$

■

The following rules of differentiation hold for vector-valued functions $\mathbf{U}$ and $\mathbf{V}$. Here $f$ denotes a scalar function.

$$(\mathbf{U} + \mathbf{V})' = \mathbf{U}' + \mathbf{V}', \tag{1}$$

$$(f\mathbf{V})' = f\mathbf{V}' + f'\mathbf{V}, \tag{2}$$

$$(\mathbf{U} \cdot \mathbf{V})' = \mathbf{U} \cdot \mathbf{V}' + \mathbf{U}' \cdot \mathbf{V}, \tag{3}$$

$$(\mathbf{U} \times \mathbf{V})' = \mathbf{U} \times \mathbf{V}' + \mathbf{U}' \times \mathbf{V}. \tag{4}$$

These formulas can be verified by expressing the functions in terms of their components and then differentiating component-by-component. Don't forget that we must be careful with the order of a cross product.

**If $\mathbf{U}(t)$ is a vector with constant length, then $\mathbf{U}$ and $\mathbf{U}'$ are orthogonal.** (5)

■ *Proof* Since $|\mathbf{U}|^2 = \mathbf{U} \cdot \mathbf{U}$, we can use formula (3) to obtain

$$\frac{d}{dt}(|\mathbf{U}|^2) = \frac{d}{dt}(\mathbf{U} \cdot \mathbf{U}) = \mathbf{U} \cdot \mathbf{U}' + \mathbf{U}' \cdot \mathbf{U} = 2\mathbf{U} \cdot \mathbf{U}'.$$

On the other hand, if $|\mathbf{U}|$ is constant, then $|\mathbf{U}|^2$ is also constant, so

$$\frac{d}{dt}(|\mathbf{U}|^2) = 0.$$

Collecting results, we have $2\,\mathbf{U} \cdot \mathbf{U}' = 0$, so $\mathbf{U} \cdot \mathbf{U}' = 0$. This shows that $\mathbf{U}$ and $\mathbf{U}'$ are orthogonal. ■

If $\mathbf{R}(t)$ is defined for $a \le t \le b$, we can form a Riemann sum $\sum_{j=1}^{n} \mathbf{R}(t_j^*)\,\Delta t_j$. Each term of the sum is the product of a scalar $\Delta t_j$ and a vector $\mathbf{R}(t_j^*)$. Hence, the sum is a vector. In case $\mathbf{R}(t)$ is continuous, the Riemann sums that have all $\Delta t_j$'s small enough will be close to a single vector $\int_a^b \mathbf{R}(t)\,dt$. We can evaluate vector-valued integrals component by component. In case of two-dimensional vectors, we have

$$\int_a^b [f(t)\mathbf{i} + g(t)\mathbf{j}]\,dt = \left(\int_a^b f(t)\,dt\right)\mathbf{i} + \left(\int_a^b g(t)\,dt\right)\mathbf{j}$$

$$\int [f(t)\mathbf{i} + g(t)\mathbf{j}]\,dt = \left(\int f(t)\,dt\right)\mathbf{i} + \left(\int g(t)\,dt\right)\mathbf{j} + \mathbf{C}.$$

Note that the constant in the indefinite integral is a vector.

■ **EXAMPLE 3**

Find $\mathbf{R}(t)$ if $\mathbf{R}'(t) = 6t\mathbf{i} + \mathbf{j}$ and $\mathbf{R}(0) = \mathbf{i} - \mathbf{j}$.

**SOLUTION**

We have

$$\mathbf{R}(t) = \int \mathbf{R}'(t)\,dt$$

$$= \int (6t\mathbf{i} + \mathbf{j})\,dt$$

$$= 3t^2\mathbf{i} + t\mathbf{j} + \mathbf{C}.$$

We have $\mathbf{R}(0) = \mathbf{i} - \mathbf{j}$, so substitution gives

$$\mathbf{i} - \mathbf{j} = \mathbf{0} + \mathbf{0} + \mathbf{C}.$$

Hence, $\mathbf{C} = \mathbf{i} - \mathbf{j}$ and

$$\mathbf{R}(t) = 3t^2\mathbf{i} + t\mathbf{j} + \mathbf{i} - \mathbf{j}.$$ ■

Let us now see how we can use calculus of vector-valued functions to study motion of objects in a coordinate plane or three-dimensional space.

If $\mathbf{R}(t)$ is the position of an object at time $t$, the change in position over the interval of time from $t$ to $t + \Delta t$ is

$$\Delta\mathbf{R} = \mathbf{R}(t + \Delta t) - \mathbf{R}(t).$$

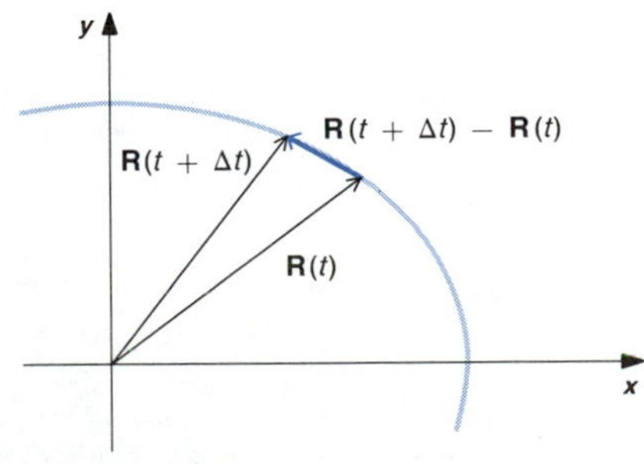


FIGURE 14.5.4

The change in position divided by the length of the time interval is

$$\frac{\Delta\mathbf{R}}{\Delta t} = \frac{\mathbf{R}(t + \Delta t) - \mathbf{R}(t)}{\Delta t}.$$

See Figure 14.5.4. The limit of the above ratio as $\Delta t \to 0$ is the derivative $\mathbf{R}'$, which is the **velocity** of the object,

$$\mathbf{V} = \mathbf{R}'.$$

Similarly, the **acceleration** is given by

$$\mathbf{A} = \mathbf{V}' = \mathbf{R}''.$$

The velocity and acceleration of an object that is moving either in a coordinate plane or in three-dimensional space are vectors.

**The speed of the object is the scalar value $|\mathbf{V}|$.**

■ **EXAMPLE 4**

$$\mathbf{R} = \cos\frac{t}{2}\,\mathbf{i} + \sin\frac{t}{2}\,\mathbf{j}.$$

Find the velocity, speed, and acceleration.

**SOLUTION**

$$\mathbf{R} = \cos\frac{t}{2}\,\mathbf{i} + \sin\frac{t}{2}\,\mathbf{j},$$

$$\mathbf{V} = \mathbf{R}' = -\frac{1}{2}\sin\frac{t}{2}\,\mathbf{i} + \frac{1}{2}\cos\frac{t}{2}\,\mathbf{j},$$

$$\mathbf{A} = \mathbf{V}' = \mathbf{R}'' = -\frac{1}{4}\cos\frac{t}{2}\,\mathbf{i} - \frac{1}{4}\sin\frac{t}{2}\,\mathbf{j},$$

$$\text{Speed} = |\mathbf{V}| = \sqrt{\left(-\frac{1}{2}\sin\frac{t}{2}\right)^2 + \left(\frac{1}{2}\cos\frac{t}{2}\right)^2} = \frac{1}{2}.$$ ■

Note that the $(x, y)$ coordinates of the object in Example 4 satisfy $x^2 + y^2 = 1$. This means that the path of the object is along the circle with center at the origin and radius one. Representative velocity and acceleration vectors are sketched in Figure 14.5.5. Note that the velocity vector is

FIGURE 14.5.5

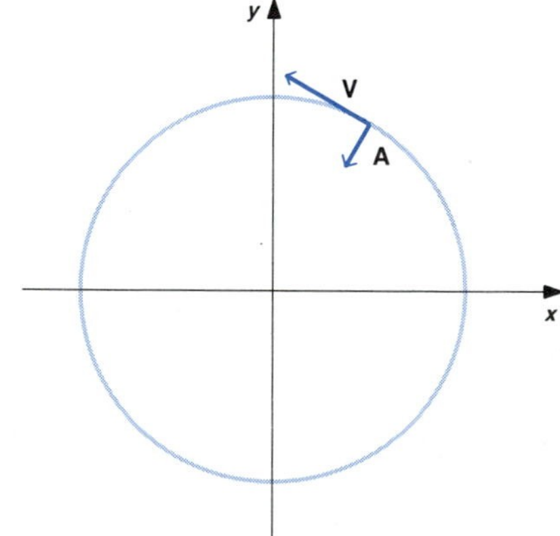

tangent to the circle and the acceleration vector always points toward the origin. Since **V** is of constant length, we know from equation (5) that **V** and $\mathbf{A} = \mathbf{V}'$ must be orthogonal.

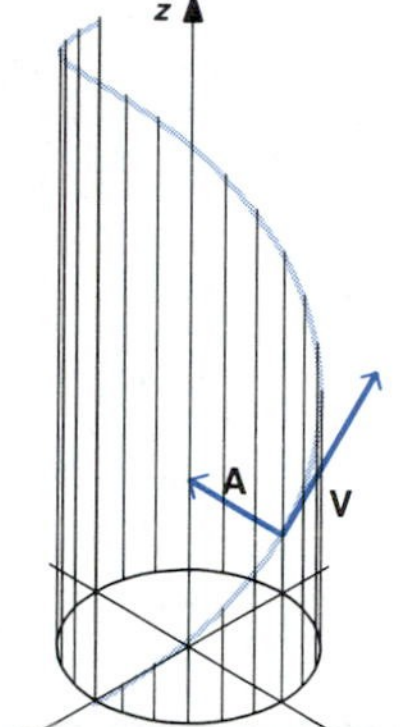


FIGURE 14.5.6

**EXAMPLE 5**

$\mathbf{R} = \cos t\mathbf{i} + \sin t\mathbf{j} + t\mathbf{k}$. Find the velocity and acceleration.

**SOLUTION**

$$\mathbf{R} = \cos t\mathbf{i} + \sin t\mathbf{j} + t\mathbf{k},$$

$$\mathbf{V} = \mathbf{R}' = -\sin t\mathbf{i} + \cos t\mathbf{j} + \mathbf{k},$$

$$\mathbf{A} = \mathbf{V}' = \mathbf{R}'' = -\cos t\mathbf{i} - \sin t\mathbf{j}.$$

The path of the object and representative velocity and acceleration vectors are sketched in Figure 14.5.6. ■

Since

$$\mathbf{V} = \int \mathbf{V}'\,dt = \int \mathbf{A}\,dt \quad \text{and} \quad \mathbf{R} = \int \mathbf{R}'\,dt = \int \mathbf{V}\,dt,$$

we can determine the velocity and the position from the acceleration by integration, if initial values are known.

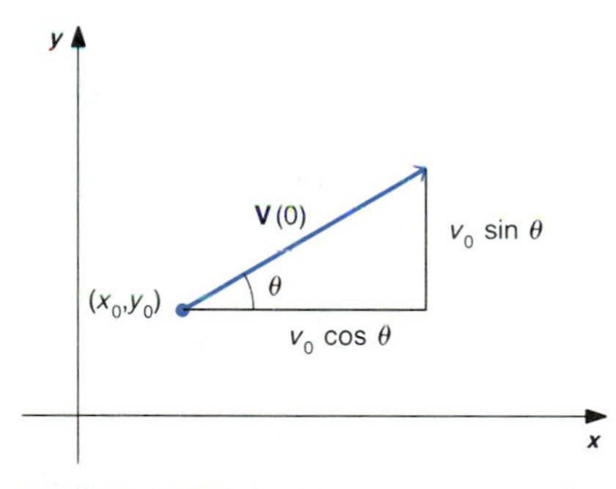


FIGURE 14.5.7

Let us determine the equation of motion of an object that is thrown near the surface of the earth. We assume that the distances involved are small relative to the radius of the earth. We choose a two-dimensional coordinate system with $y$-axis vertical and origin at ground level. We use the vector-valued function $\mathbf{R}(t)$ to represent the position of the object at time $t$. The initial position of the object is $\mathbf{R}(0) = x_0\mathbf{i} + y_0\mathbf{j}$. The initial velocity is $\mathbf{V}(0) = v_0 \cos\theta\mathbf{i} + v_0 \sin\theta\mathbf{j}$, where $\theta$ is the angle of elevation of the initial velocity and $v_0$ is the initial speed. See Figure 14.5.7. We neglect air resistance, so the only force on the object is due to gravity. The acceleration due to gravity has constant magnitude and is directed downward. That is, $\mathbf{A} = -g\mathbf{j}$, where $g = 32\ \text{ft/s}^2$ or $9.8\ \text{m/s}^2$. We can integrate the acceleration vector twice and substitute the initial values of the velocity and position to obtain

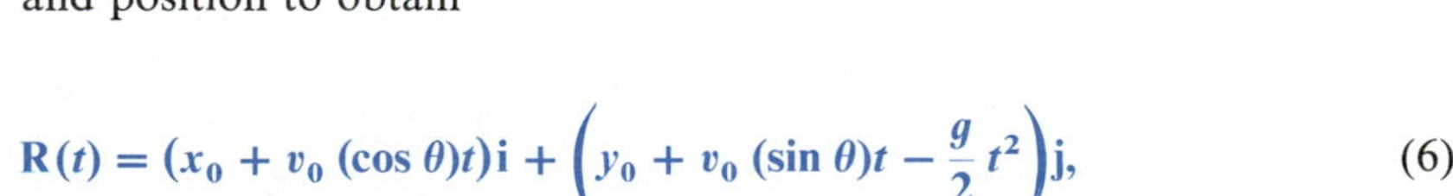

$$\mathbf{R}(t) = (x_0 + v_0(\cos\theta)t)\mathbf{i} + \left(y_0 + v_0(\sin\theta)t - \frac{g}{2}t^2\right)\mathbf{j}, \tag{6}$$

It is important to recall that the position of the object at time $t$ is given by the coordinates $(x, y)$, where

$$x = x_0 + v_0(\cos\theta)t \quad \text{and} \quad y = y_0 + v_0(\sin\theta)t - \frac{g}{2}t^2.$$

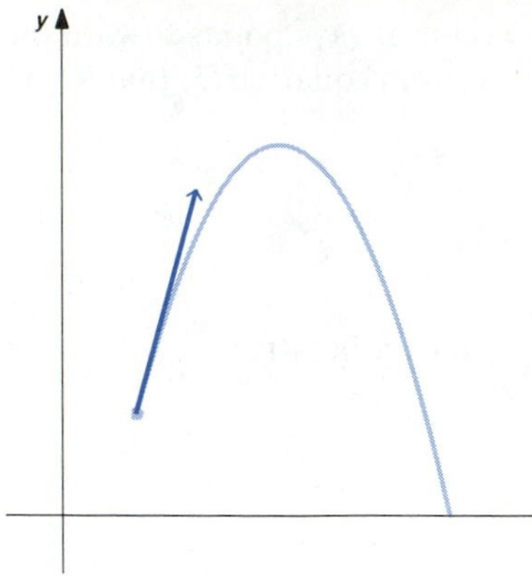

FIGURE 14.5.8

By solving the first of the above equations for $t$, substituting into the second equation, and completing the square, we obtain the equation

$$y - \left(y_0 + \frac{v_0^2 \sin^2\theta}{2g}\right) = \frac{-g}{2v_0^2 \cos^2\theta}\left[x - \left(x_0 + \frac{v_0^2 \sin\theta\cos\theta}{g}\right)\right]^2.$$

This is the equation of a parabola with vertex

$$\left(x_0 + \frac{v_0^2 \sin\theta\cos\theta}{g}, y_0 + \frac{v_0^2 \sin^2\theta}{2g}\right).$$

A representative path is sketched in Figure 14.5.8. Note that the $y$-coordinate of the vertex gives the maximum height of the object. This is the value of $y$ when

$$\frac{dy}{dt} = 0.$$

■ **EXAMPLE 5**

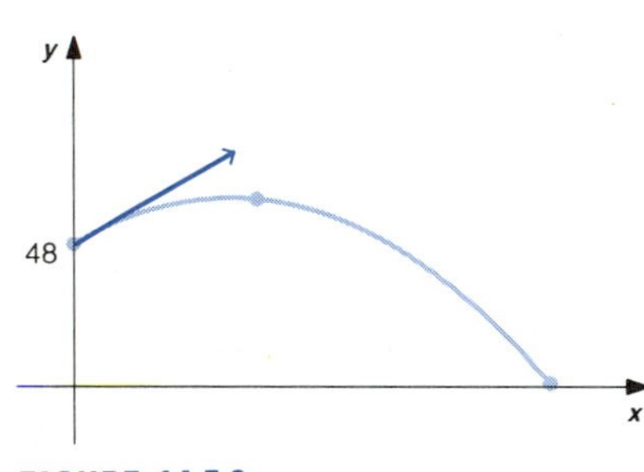

FIGURE 14.5.9

A ball is thrown with a speed of 64 ft/s from a height of 48 ft at an angle of 30° from horizontal. How long will it take the ball to hit the ground? Where will the ball hit the ground? At what speed will the ball hit the ground? What is the maximum height the ball will attain?

**SOLUTION**

A coordinate system is sketched in Figure 14.5.9. The initial position is $(x_0, y_0) = (0,48)$.

We are given that the initial speed is $v_0 = 64$ and that $\theta = 30°$. The units are feet and seconds, so we use $g = 32$ ft/s$^2$. Subtitution of the data into the equation of motion, equation (6), then gives

$$\begin{aligned}\mathbf{R}(t) &= (0 + 64\cos 30°\, t)\mathbf{i} + (48 + 64(\sin 30°)t - 16t^2)\mathbf{j}\\ &= (32\sqrt{3}\, t)\mathbf{i} + (48 + 32t - 16t^2)\mathbf{j}.\end{aligned}$$

The ball will hit the ground when the $\mathbf{j}$ component of the position vector is 0. This corresponds to $y = 0$. That is, when

$$48 + 32t - 16t^2 = 0,$$

$$-16(t^2 - 2t - 3) = 0,$$

$$-16(t - 3)(t + 1) = 0.$$

This implies $t = 3$, since we want $t > 0$. When $t = 3$, the position vector is

$$\mathbf{R}(3) = (32\sqrt{3})(3)\mathbf{i} + (0)\mathbf{j} = 96\sqrt{3}\mathbf{i}.$$

The ball will hit the ground at a point that is $96\sqrt{3} \approx 166.3$ ft from the point on the ground that is directly below where it was thrown.

We have

$$\mathbf{R}(t) = (32\sqrt{3}t)\mathbf{i} + (48 + 32t - 16t^2)\mathbf{j},$$
$$\mathbf{V}(t) = \mathbf{R}'(t) = 32\sqrt{3}\mathbf{i} + (32 - 32t)\mathbf{j}.$$

When $t = 3$, we have

$$\mathbf{V}(3) = 32\sqrt{3}\mathbf{i} - 64\mathbf{j},$$

so the speed when the ball hits the ground is

$$|\mathbf{V}(3)| = \sqrt{(32\sqrt{3})^2 + (-64)^2} = \sqrt{7168} \approx 84.66 \text{ ft/s}.$$

The maximum height attained is the $y$-coordinate of the vertex of the parabolic path,

$$y_0 + \frac{v_0^2 \sin^2\theta}{2g} = 48 + \frac{64^2(1/2)^2}{64} = 64 \text{ ft.}$$ ■

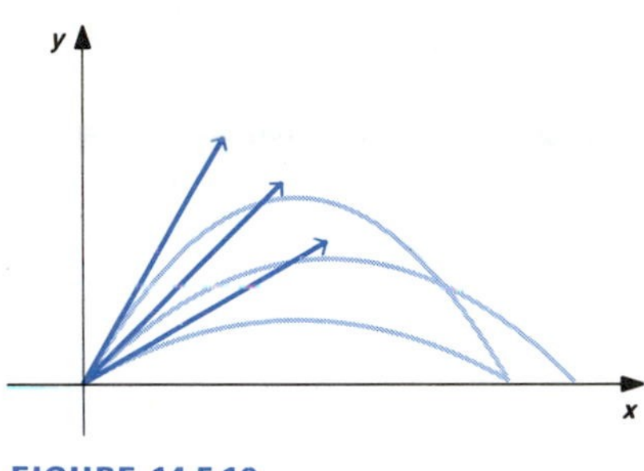

FIGURE 14.5.10

Suppose that a projectile is launched at ground level with fixed initial speed $v_0$. Let us determine the angle of elevation that gives the maximum horizontal distance $L$ the projectile travels before returning to ground level. See Figure 14.5.10. The initial position is taken to be $x_0 = y_0 = 0$. The equation of motion for a particular choice of $\theta$ is

$$\mathbf{R}(t) = (v_0(\cos\theta)t)\mathbf{i} + \left(v_0(\sin\theta)t - \frac{g}{2}t^2\right)\mathbf{j}.$$

We see from the sketch that the horizontal distance $L$ occurs at that time when the $y$-coordinate of the position vector is zero. Thus, at some time $t$ we must have

$$v_0(\cos\theta)t = L \quad \text{and}$$

$$v_0(\sin\theta)t - \frac{g}{2}t^2 = 0.$$

The distance $L$ and the hitting time $t$ both depend on the angle of elevation $\theta$. We can eliminate $t$ from the two equations above and obtain an equation that relates $L$ and $\theta$. That is, the second equation gives

$$t = \frac{2v_0 \sin\theta}{g}.$$

Substitution in the first equation then gives

$$L = v_0 \cos\theta\left(\frac{2v_0 \sin\theta}{g}\right), \quad \text{or}$$

$$L = \frac{2v_0^2}{g}\sin\theta\cos\theta, \qquad 0 \le \theta \le \frac{\pi}{2}.$$

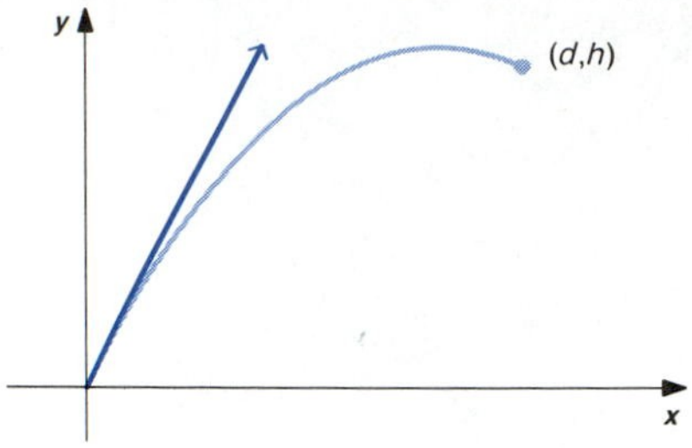

We now have the length $L$ as a continuous function of the elevation angle $\theta$. We can then determine the maximum distance and the corresponding angle $\theta$ as usual. Namely, check values where the derivative is zero and at endpoints. We find that the maximum occurs for $\theta = \pi/4$.

A projectile is to hit a point that is $d$ units horizontally and $h$ units above the launch point. See Figure 14.5.11. Let us determine the launch angle that gives the minimum initial speed.

FIGURE 14.5.11

We choose initial position $x_0 = y_0 = 0$. The position vector is then

$$\mathbf{R}(t) = (v_0 (\cos \theta)t)\mathbf{i} + \left(v_0 (\sin \theta)t - \frac{g}{2} t^2\right)\mathbf{j}.$$

The path of the projectile is required to pass through the point $(d, h)$. This means, for some value of $t$, we must have

$$d = v_0 (\cos \theta)t \quad \text{and}$$

$$h = v_0 (\sin \theta)t - \frac{g}{2} t^2.$$

Note that $t$, $v_0$, and $\theta$ are unknown variables in the above two equations. We can obtain an equation that relates $\theta$ to $v_0$ by solving the first equation for $t$ and then substituting into the second equation. We have

$$t = \frac{d}{v_0 \cos \theta},$$

and then

$$h = v_0 \sin \theta\left(\frac{d}{v_0 \cos \theta}\right) - \frac{g}{2}\left(\frac{d}{v_0 \cos \theta}\right)^2,$$

$$2hv_0^2 \cos^2\theta = 2dv_0^2 \sin \theta \cos \theta - gd^2,$$

$$2 \cos \theta(h \cos \theta - d \sin \theta)v_0^2 = -gd^2,$$

$$v_0^2 = \frac{gd^2}{2 \cos \theta(d \sin \theta - h \cos \theta)}.$$

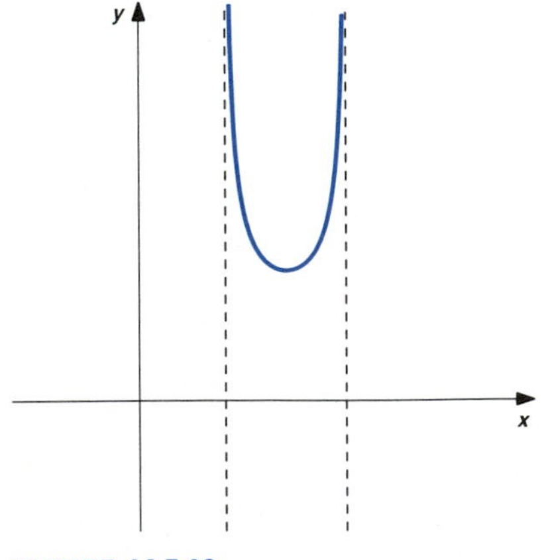

FIGURE 14.5.12

A representative sketch of the graph of $v_0(\theta)$, $\tan^{-1}(h/d) < \theta < \pi/2$, is given in Figure 14.5.12. (Note that $g > 0$. Since $v_0^2$ must be positive, we must have $d \sin \theta - h \cos \theta > 0$. This implies $\tan \theta > h/d$.) The minimum value of $v_0$ occurs where

$$\frac{dv_0}{d\theta} = 0.$$

With some work, we can see that this condition gives the equation

$$\tan\theta = \frac{h + \sqrt{h^2 + d^2}}{d}.$$

There is exactly one value of $\theta$ in the interval $\tan^{-1}(h/d) < \theta < \pi/2$ that satisfies this equation. This is the launch angle that gives the minimum initial speed.

It is possible to verify that the $x$-coordinate of the vertex of the parabolic path of a projectile that is launched at the above angle $\theta$ with the corresponding initial speed $v_0$ is less than $d$. Hence, the projectile is falling as it approaches the point $(d, h)$, as was indicated in Figure 14.5.11.

Suppose the position of an object of mass $m$ at time $t$ is given by $\mathbf{r}(t)$ and its velocity is $\mathbf{v}(t)$.

**The linear momentum of the object is $\mathbf{p} = m\mathbf{v}$.**

**The angular momentum about the origin is $\mathbf{L} = \mathbf{r} \times \mathbf{p}$.**

If the motion of an object is due to a resultant exterior force $\mathbf{F}$, then Newton's Second Law says

$$\mathbf{F} = \frac{d\mathbf{p}}{dt}.$$

That is, the exterior force is equal to the rate of change of the linear momentum of the object. The rate of change of the angular momentum about the origin of the object is

$$\frac{d\mathbf{L}}{dt} = \frac{d}{dt}(\mathbf{r} \times \mathbf{p}).$$

Using property (4), we obtain

$$\begin{aligned}\frac{d\mathbf{L}}{dt} &= \mathbf{r} \times \frac{d\mathbf{p}}{dt} + \frac{d\mathbf{r}}{dt} \times \mathbf{p} \\ &= \mathbf{r} \times \mathbf{F} + \mathbf{v} \times (m\mathbf{v}) \\ &= \mathbf{r} \times \mathbf{F},\end{aligned}$$

since $\mathbf{v} \times (m\mathbf{v}) = m(\mathbf{v} \times \mathbf{v}) = m\mathbf{0} = \mathbf{0}$. For any vector function $\mathbf{F}$, we define the **torque about the origin** to be

$$\boldsymbol{\tau} = \mathbf{r} \times \mathbf{F}.$$

(The units of torque are length times force. Torque is not a force.) We have shown that the rate of change of the angular momentum about the origin of an object is the torque about the origin of the resultant exterior force on the object.

## EXERCISES 14.5

Evaluate the expressions in Exercises 1-18.

1 $\lim_{t\to 2} (t^2\mathbf{i} - 3t\mathbf{j})$

2 $\lim_{t\to \pi/2} (\sin t\mathbf{i} + \cos t\mathbf{j})$

3 $\lim_{t\to 3} \left(\frac{t^2-9}{t-3}\mathbf{i} + \frac{2t-6}{t-3}\mathbf{j}\right)$

4 $\lim_{t\to 0} \left(\frac{\sin 2t}{t}\mathbf{i} + \frac{1-\cos^2 2t}{t}\mathbf{j}\right)$

5 $\frac{d}{dt}(t^2\mathbf{i} - 3t\mathbf{j})$

6 $\frac{d}{dt}(t\mathbf{i} + t^2\mathbf{j} + t^3\mathbf{k})$

7 $\frac{d}{dt}(t\mathbf{i} + (e^{-t}\sin t)\mathbf{j} + (e^{-t}\cos t)\mathbf{k})$

8 $\frac{d}{dt}(te^{-t}\mathbf{i} + e^{-3t}\mathbf{j} + e^{-t}\mathbf{k})$

9 $\frac{d}{dt}(\tan t\mathbf{i} + \sec t\mathbf{j})$

10 $\frac{d}{dt}(\tan^{-1} t\mathbf{i} + \ln t\mathbf{j})$

11 $\int (2t\mathbf{i} + 6t^2\mathbf{j})\, dt$

12 $\int (t^2\mathbf{i} + t^3\mathbf{j} + t^4\mathbf{k})dt$

13 $\int (\sin t\mathbf{i} + \cos t\mathbf{j})dt$

14 $\int (2 \sin 2t\mathbf{i} + 2\cos 2t\mathbf{j})dt$

15 $\int (\sec^2 2t\mathbf{i} + (\sec 2t \tan 2t)\mathbf{j})\, dt$

16 $\int (\sin^2 t\mathbf{i} + \cos^2 t\mathbf{j})\, dt$

17 $\int \left(\frac{t}{t^2+4}\mathbf{i} + \frac{1}{t^2+4}\mathbf{j}\right) dt$

18 $\int \left(\frac{t}{t^2-4}\mathbf{i} + \frac{1}{t^2-4}\mathbf{j}\right) dt$

In Exercises 19-22 find the velocity, acceleration, and speed of an object with the given position vector.

19 $\mathbf{R}(t) = t\mathbf{i} + t^2\mathbf{j}$

20 $\mathbf{R}(t) = 2t^2\mathbf{i} - 3t^2\mathbf{j}$

21 $\mathbf{R}(t) = \cos 2t\mathbf{i} + \sin 2t\mathbf{j}$

22 $\mathbf{R}(t) = \cos 3t\mathbf{i} + \sin 3t\mathbf{j} + 3t\mathbf{k}$

In Exercises 23-26 find the position vector for an object with the given acceleration, initial position, and initial velocity.

23 $\mathbf{A}(t) = -32\mathbf{j}$, $R_0 = \mathbf{0}$, $\mathbf{V}_0 = \mathbf{i} + 2\mathbf{j}$

24 $\mathbf{A}(t) = -32\mathbf{k}$, $\mathbf{R}_0 = \mathbf{0}$, $\mathbf{V}_0 = \mathbf{i} + \mathbf{j} + \mathbf{k}$

25 $\mathbf{A}(t) = 12t\mathbf{j}$, $\mathbf{R}_0 = 3\mathbf{j}$, $\mathbf{V}_0 = 4\mathbf{i}$

26 $\mathbf{A}(t) = 12t^2\mathbf{i} + 6t\mathbf{j} + \mathbf{k}$, $\mathbf{R}_0 = 2\mathbf{k}$, $\mathbf{v}_0 = \mathbf{i} + 2\mathbf{j}$

27 A projectile is launched with initial speed 64 ft/s from ground level at an angle of 60° from the horizontal. How high is it after 3 s? What is the maximum height the projectile will reach?

28 A projectile is launched with initial speed 64 ft/s from ground level at an angle of 30° from the horizontal. How far will it travel horizontally?

29 A projectile is launched horizontally from a point 64 ft above ground level at an initial speed of 80 ft/s. How long will it take to reach ground level? How long would it take for an object that is dropped from a height of 64 ft to reach the ground?

30 A projectile is to be launched from ground level at an angle of 45° from the horizontal. Find the initial speed with which it will reach a maximum height of 144 ft.

31 A projectile is to pass through a point that is 120 ft horizontally and 90 ft above the launch point. What is the minimum initial velocity that must be used?

32 Let $\mathbf{U}(t) = f_1(t)\mathbf{i} + g_1(t)\mathbf{j}$ and $\mathbf{V}(t) = f_2(t)\mathbf{i} + g_2(t)\mathbf{j}$.

(a) Evaluate $(\mathbf{U}\cdot\mathbf{V})'$ by expressing the dot product in terms of the components and then differentiating.

(b) Express $\mathbf{U}\cdot\mathbf{V}' + \mathbf{U}'\cdot\mathbf{V}$ in terms of components.

(c) Show that the expressions in (a) and (b) are equal. This verifies the formula $(\mathbf{U}\cdot\mathbf{V})' = \mathbf{U}\cdot\mathbf{V}' + \mathbf{U}'\cdot\mathbf{V}$.

33 If the variables $v_0$ and $\theta$ are related by the equation

$$v_0^2 = \frac{gd^2}{2\cos\theta(d\sin\theta - h\cos\theta)}, \quad \tan^{-1}(h/d) < \theta < \pi/2,$$

show that

$$\frac{dv_0}{d\theta} = 0 \quad \text{implies} \quad \tan\theta = \frac{h + \sqrt{h^2 + d^2}}{d}.$$

34 If

$$v_0^2 = \frac{gd^2}{2\cos\theta(d\sin\theta - h\cos\theta)}, \quad \tan^{-1}(h/d) < \theta < \pi/2,$$

and

$$\tan\theta = \frac{h + \sqrt{h^2 + d^2}}{d},$$

show that the $x$-coordinate of the vertex of the parabolic path of a projectile with initial position $\mathbf{0}$ is less than $d$.

35 A ball is thrown at an angle of $\theta$ with the horizontal from ground level on the side of a hill in a direction where the hill makes an angle of $\alpha$ with the horizontal. See below. (Take $\alpha$ to be positive if the ball is thrown up the hill and negative if the ball is thrown down the hill.) For a fixed initial speed $v_0$, show that the maximum distance along the hill that the ball travels before hitting the ground is obtained when the ball is thrown at an angle $\theta$ that satisfies $(\tan 2\theta)(\tan\alpha) = -1$.

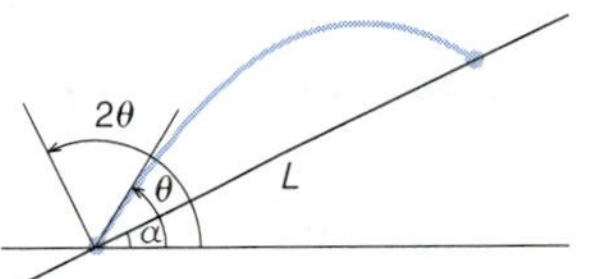


## REVIEW EXERCISES

1 $\mathbf{U} = 2\mathbf{i} + 6\mathbf{j}$, $\mathbf{V} = \mathbf{i} - 3\mathbf{j}$. Find
(a) $\mathbf{U} - 2\mathbf{V}$
(b) $|\mathbf{U}|$
(c) $|\mathbf{V}|$
(d) $\mathbf{U} \cdot \mathbf{V}$
(e) the angle between $\mathbf{U}$ and $\mathbf{V}$
(f) a unit vector in the direction of $\mathbf{V}$
(g) $\text{proj}_{\mathbf{V}} \mathbf{U}$

2 $\mathbf{U} = \mathbf{i} + 2\mathbf{j} + 2\mathbf{k}$, $\mathbf{V} = 3\mathbf{i} + 4\mathbf{j} - 12\mathbf{k}$. Find
(a) $2\mathbf{U} - \mathbf{V}$
(b) $|\mathbf{U}|$
(c) $|\mathbf{V}|$
(d) $\mathbf{U} \cdot \mathbf{V}$
(e) the angle between $\mathbf{U}$ and $\mathbf{V}$
(f) direction cosines of $\mathbf{V}$
(g) $\mathbf{U} \times \mathbf{V}$

3 Find the projection of the force $\mathbf{F} = 5\mathbf{i} + 2\mathbf{j}$ in the direction of the vector $\mathbf{V} = 3\mathbf{i} + 4\mathbf{j}$.

4 Write $\mathbf{F} = 2\mathbf{i} + \mathbf{j}$ as a sum of vectors $\mathbf{F}_1$ and $\mathbf{F}_2$, where $\mathbf{F}_1$ is parallel to $\mathbf{V} = \mathbf{i} + 3\mathbf{j}$ and $\mathbf{F}_2$ is orthogonal to $\mathbf{V}$.

5 Find the area of the parallelogram formed by the vectors $2\mathbf{i} + \mathbf{j}$ and $-\mathbf{i} + \mathbf{j}$.

6 Find the volume of the parallelepiped formed by the vectors $2\mathbf{i} + 2\mathbf{j} - \mathbf{k}$, $-\mathbf{i} + 2\mathbf{j}$, and $\mathbf{i} + \mathbf{j} + \mathbf{k}$.

7 Find the area of the triangle formed by the three points (6, 4, 3), (3, 1, 1), and (4, 2, 2).

8 Find the area of the parallelogram formed by the points (0, 0), (1, 2), (3, −1), and (4, 1).

In Exercises 9–12, find an equation of the plane that satisfies the given conditions.

9 Contains the three points (3, 0, 3), (5, 4, 1), (1, 1, 1)

10 Contains (1, −2, 2) and is parallel to the plane $x - 2y + 3z = 4$

11 Contains (−1, 2, 3) and is perpendicular to the line

$$\frac{x}{3} = \frac{y}{4} = \frac{z}{-5}$$

12 Contains the line

$$\frac{x}{3} = \frac{y}{-1} = \frac{z}{2}$$

and is parallel to the line

$$\frac{x-4}{5} = \frac{y+2}{3} = \frac{z-1}{-4}$$

In Exercises 13–16, find symmetric equations and parametric equations of the line that satisfies the given conditions.

13 Contains (7, 5, 3) and (4, 1, 2)

14 Contains (3, 5, −2) and is parallel to the line

$$\frac{x}{-2} = \frac{y}{1} = \frac{z}{3}$$

15 Contains $(1, -1, 2)$ and is perpendicular to the plane $2x - y + 3z = 1$

16 Line of intersection of both the planes $x + y = 2$ and $x + z = 2$

17 Find the angle between the lines

$$\frac{x}{2} = \frac{y}{-2} = \frac{z}{1} \quad \text{and} \quad \frac{x}{-1} = \frac{y}{-2} = \frac{z}{2}.$$

18 Find the angle between the planes $3x + 4y = 0$ and $2x - y - 2z = 0$.

19 Find the point of intersection and the angle of intersection of the line

$$\frac{x}{2} = \frac{y}{-3} = \frac{z}{1}$$

and the plane $2x + 3y + z = 6$.

20 Find the point of intersection and the angle of intersection of the line

$$\frac{x-2}{3} = \frac{y+1}{4} = \frac{z+3}{-2}$$

and the plane $z = x + 2y$.

21 Find the distance from the point $(3, 1, -2)$ to the plane $z = 2x - 3y$.

22 Find the distance from the point $(1, 3, -2)$ to the line

$$\frac{x}{4} = \frac{y}{-2} = \frac{z}{3}.$$

In Exercises 23–24, the position of an object at time $t$ is given by **R**. Find the velocity, speed, and acceleration at the time indicated.

23 $\mathbf{R} = 2 \cos t\mathbf{i} + 3 \sin t\mathbf{j}$, when $t = \pi/3$

24 $\mathbf{R} = \sin^2 t\mathbf{i} + \cos t\mathbf{j}$, when $t = \pi/4$

25 Find **R** if $\mathbf{R}' = 16t^3\mathbf{i} - 15t^2\mathbf{j}$, $\mathbf{R}(0) = 6\mathbf{i}$.

26 Find **R** if $\mathbf{R}' = 3t^{1/2}\mathbf{i} + t^{-1/2}\mathbf{j}$, $\mathbf{R}(0) = \mathbf{i} + \mathbf{j}$.

27 A golf ball is hit from ground level at 90 ft/s at an angle of 40° from the horizontal. How far will it travel horizontally before returning to ground level?

28 A ball is thrown at an angle of 45° to a point that is 100 ft horizontally and 40 ft vertically away. Determine the initial speed.

**CHAPTER FIFTEEN**

# *Parametric curves*

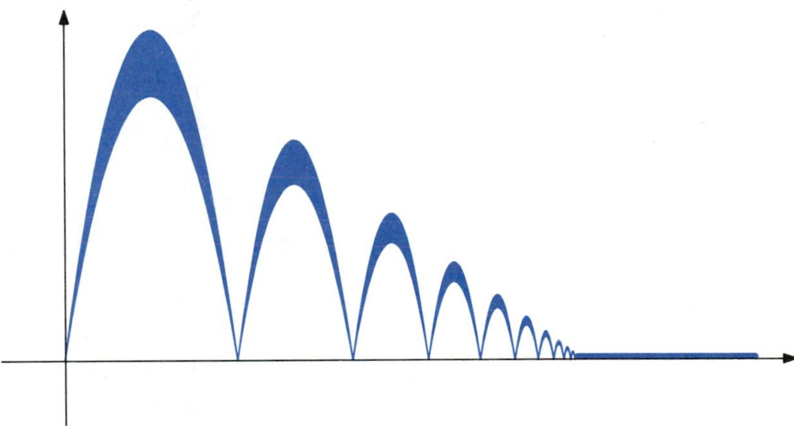

*I*n this chapter we will investigate properties of curves and see how these properties are related to the motion of objects along the curves. That is, we will investigate direction, length, and the sharpness of turning of curves and see how these properties of curves relate to the velocity, acceleration, and speed of an object that is moving along the curve. In Chapter 18 we will use properties of curves that are developed here to study several problems that involve the measurement of physical quantities at points along a curve.

## 15.1 PARAMETRIC EQUATIONS

We have seen that many curves in the $xy$-plane can be described as the graph of an equation in $x$ and $y$. For some applications of science and engineering it is more convenient to express each coordinate of a point on a curve as a continuous function of a real variable rather than determine the coordinates from an equation in $x$ and $y$.

If $x$ and $y$ are continuous functions of $t$ for $t$ in an interval $I$, we can describe a curve $C$ in the $xy$-plane by writing

$$C: \quad x = x(t), \quad y = y(t), \qquad t \text{ in } I.$$

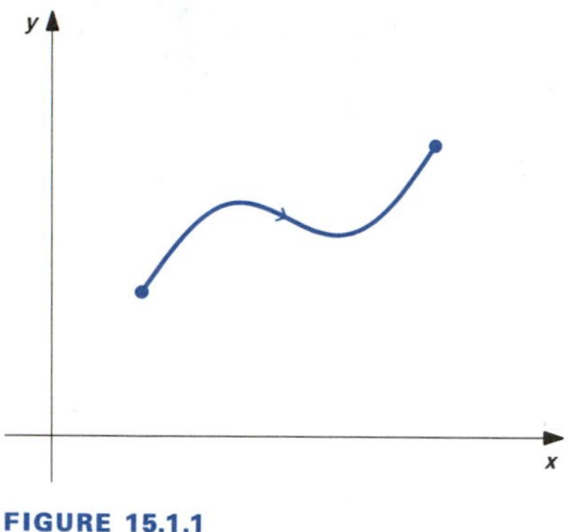

FIGURE 15.1.1

The equations $x = x(t)$, $y = y(t)$ are called **parametric equations** and the real variable $t$ is called a **parameter** of the curve. The **graph** of the **parametric curve** $C$ is the set of all points $(x(t), y(t))$, $t$ in $I$. Note that each value of the parameter $t$ in $I$ determines a point $(x(t), y(t))$ on the graph of the curve; the graph is traced as the parameter varies over $I$. The **direction** of the curve is along the graph, in the direction of increasing $t$. If the interval $I$ contains its left endpoint $a$, the point $(x(a), y(a))$ is called the **initial point** of the curve; if $I$ contains its right endpoint $b$, $(x(b), y(b))$ is called the **terminal point** of the curve. See Figure 15.1.1.

A basic method of sketching a parametric curve is to draw line segments between an ordered set of points on the curve, where the points are ordered according to increasing values of the parameter.

■ **EXAMPLE 1**

Sketch the curve

$$C: \quad x = \cos t - \cos 2t, \quad y = \sin t - \sin 2t, \qquad 0 \leq t \leq 2\pi,$$

by connecting points that correspond to increments of $\Delta t = \pi/4$ with line segments. Sketch a smooth curve through the points that have been plotted.

**SOLUTION**

The values of the coordinates that correspond to values of $t$ in increments of $\Delta t = \pi/4$ from $t = 0$ to $t = 2\pi$ are given in Table 15.1.1. The graph is sketched by first plotting the initial point and then repeating the process of plotting the next point and drawing a line segment from the previous point to the new point. This process is continued until the terminal point is reached. The sketch is given in Figure 15.1.2, where we have superimposed a smooth curve. ■

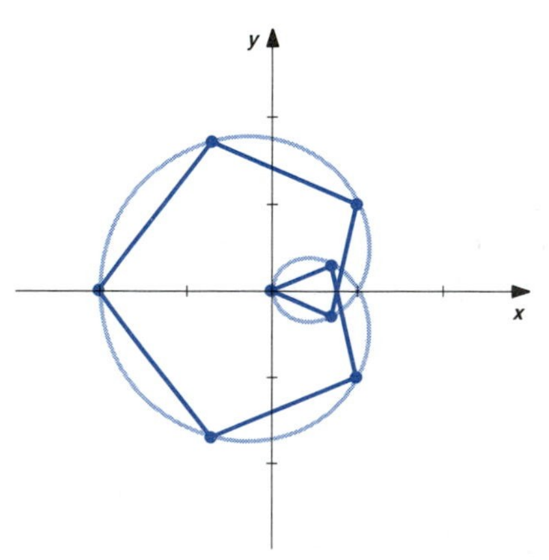

FIGURE 15.1.2

The graphing technique of Example 1 is well suited for computer graphics. It is easy to program a computer to repeat the process of calculating coordinates of a point from parametric equations and then drawing a line segment from the previous point to the new point. All of the figures in this text were computer generated in this way. The curves appear

TABLE 15.1.1

| $t$ | $x$ | $y$ |
|---|---|---|
| $0$ | $0$ | $0$ |
| $\frac{\pi}{4}$ | $\frac{1}{\sqrt{2}} \approx 0.7$ | $\frac{1}{\sqrt{2}} - 1 \approx -0.3$ |
| $\frac{\pi}{2}$ | $1$ | $1$ |
| $\frac{3\pi}{4}$ | $-\frac{1}{\sqrt{2}} \approx -0.7$ | $\frac{1}{\sqrt{2}} + 1 \approx 1.7$ |
| $\pi$ | $-2$ | $0$ |
| $\frac{5\pi}{4}$ | $-\frac{1}{\sqrt{2}} \approx -0.7$ | $-\frac{1}{\sqrt{2}} - 1 \approx -1.3$ |
| $\frac{3\pi}{2}$ | $1$ | $-1$ |
| $\frac{7\pi}{4}$ | $-\frac{1}{\sqrt{2}} \approx -0.7$ | $-\frac{1}{\sqrt{2}} + 1 \approx 0.3$ |
| $2\pi$ | $0$ | $0$ |

quite smooth if the increment $\Delta t$ is small enough, as indicated by the smooth curve in Figure 15.1.2.

In some cases we can identify the graph of a plane parametric curve by eliminating the parameter to obtain a familiar equation in $x$ and $y$ that is satisfied by points on the curve. The curve can then be described without reference to a parameter by indicating its initial and terminal points on the graph of the equation in $x$ and $y$. The graph of the parametric curve may not include all points on the graph of the equation obtained by eliminating the parameter.

■ **EXAMPLE 2**

Identify, sketch, and describe the curve

$$C: \quad x = 2t - 1, \quad y = t + 1, \qquad 0 \le t \le 2.$$

**SOLUTION**

In this example we can eliminate the parameter by solving the second equation for $t$ in terms of $y$ and then substituting into the first equation. We have

$$y = t + 1 \quad \text{implies} \quad t = y - 1$$

and then

$$x = 2t - 1 \quad \text{implies} \quad x = 2(y - 1) - 1.$$

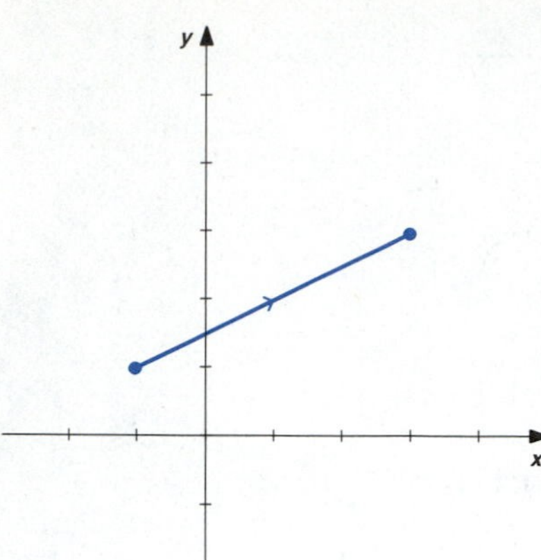


FIGURE 15.1.3

Points of the curve satisfy the equation

$x = 2(y - 1) - 1,$

$x = 2y - 2 - 1,$

$x - 2y + 3 = 0.$

This is the equation of a line. The curve consists of the line segment from the initial point $(-1, 1)$, where $t = 0$, to the terminal point $(3, 3)$, where $t = 2$. See Figure 15.1.3. ■

■ **EXAMPLE 3**

Identify, sketch, and describe the curve

$$C: \quad x = 2t^3 + 1, \quad y = t^3 + 2, \qquad -1 \le t \le 1.$$

**SOLUTION**

We see that

$y = t^3 + 2$ implies $t^3 = y - 2,$

and then

$x = 2t^3 + 1$ implies $x = 2(y - 2) + 1,$ so

$x = 2y - 4 + 1,$

$x - 2y + 3 = 0.$

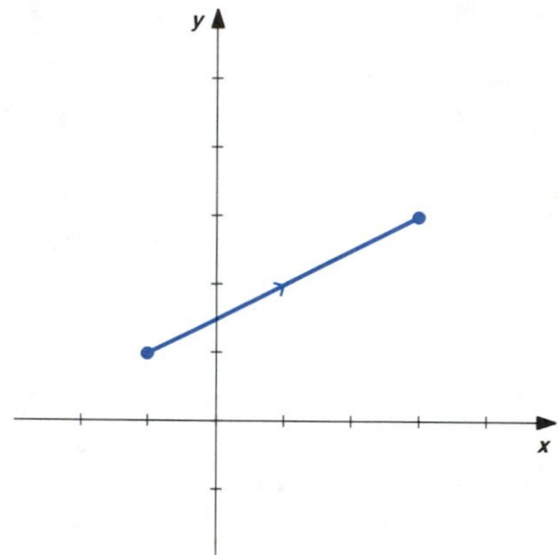


FIGURE 15.1.4

The curve consists of the segment of this line from the initial point $(-1, 1)$, where $t = -1$, to the terminal point $(3, 3)$, where $t = 1$. See Figure 15.1.4. ■

The parametric descriptions of the curves in Examples 2 and 3 are different, but the curves have the same graph and the same direction.

■ **EXAMPLE 4**

Identify, sketch, and describe the curve

$$C: \quad x = 2 \sin^2 t, \quad y = 2 \sin t, \qquad 0 \le t \le \pi/2.$$

**SOLUTION**

We eliminate $t$ by noting that $\sin t = y/2$, so

$x = 2(\sin t)^2,$

$$x = 2\left(\frac{y}{2}\right)^2,$$

$$x = \frac{y^2}{2}.$$

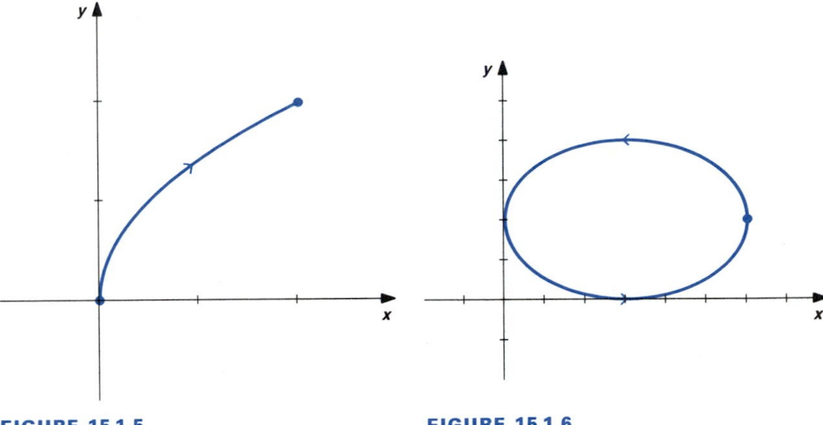


FIGURE 15.1.5

FIGURE 15.1.6

This is the equation of a parabola. The curve consists of that portion of the parabola from the initial point (0, 0) to the terminal point (2, 2). The graph is sketched in Figure 15.1.5. ■

■ **EXAMPLE 5**

Identify, sketch, and describe the curve

$$C\colon\quad x = 3 + 3\cos t, \quad y = 2 + 2\sin t, \qquad 0 \le t \le 2\pi.$$

**SOLUTION**

In this case we eliminate $t$ by using the identity $\sin^2 t + \cos^2 t = 1$. We have

$$\sin t = \frac{y-2}{2} \quad \text{and} \quad \cos t = \frac{x-3}{3},$$

so

$$\frac{(y-2)^2}{2^2} + \frac{(x-3)^2}{3^2} = 1.$$

This is the equation of an ellipse. The curve goes from the initial point (6, 2), counterclockwise once around the ellipse, and returns to the initial point. The sketch is given in Figure 15.1.6. ■

To solve problems that involve the measurement of physical quantities at points along a curve, we will need to find parametric equations for a collection of curves that are described geometrically or in terms of equations in $x$ and $y$. The following rules are useful.

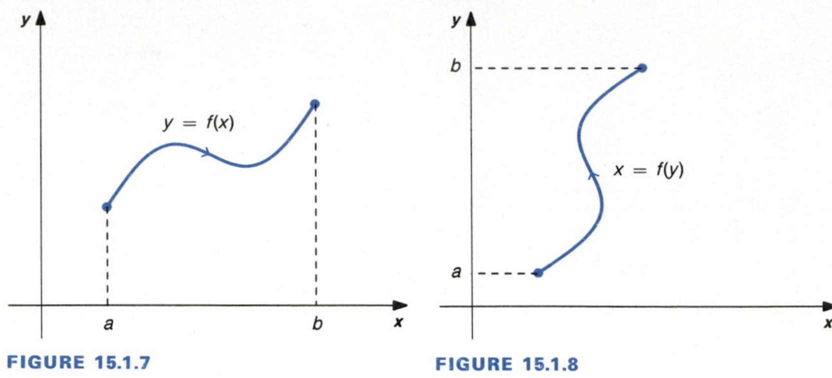


FIGURE 15.1.7

FIGURE 15.1.8

**The curve along the graph of $y = f(x)$ in the direction of increasing $x$ from $(a, f(a))$ to $(b, f(b))$ can be expressed as** (1)

$$C: \quad x = t, \quad y = f(t), \qquad a \le t \le b.$$

See Figure 15.1.7.

**The curve along the graph of $x = f(y)$ in the direction of increasing $y$ from $(f(a), a)$ to $(f(b), b)$ can be expressed as** (2)

$$C: x = f(t), \quad y = t, \qquad a \le t \le b.$$

See Figure 15.1.8.

### EXAMPLE 6

Find parametric equations of the curve from $(0, 0)$ to $(2, 2)$ along the parabola $x = y^2/2$.

#### SOLUTION

The equation $x = y^2/2$ gives $x$ as a function of $y$, and the direction of the curve is in the direction of increasing $y$. See Figure 15.1.9. Formula (2) then gives the parametric representation

$$C: \quad x = \frac{t^2}{2}, \quad y = t, \qquad 0 \le t \le 2. \qquad \blacksquare$$

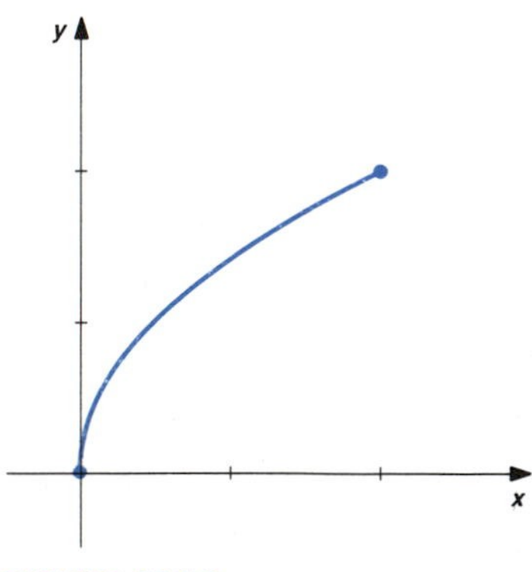


FIGURE 15.1.9

The curves in Examples 4 and 6 have the same graph and the same direction, but the parametric equations that describe the curves are different. Examples 2 and 3 also illustrated that different parametric equations can describe the same curve. There is not a unique way to parametrize a curve. We should try to choose the parameter in a systematic way that gives simple equations. Note that the parametric equations in Example 6 appear more natural than the equations used in Example 4.

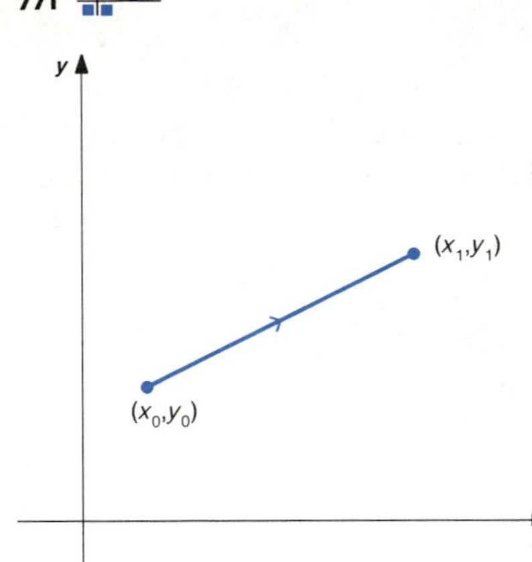


FIGURE 15.1.10

We can find parametric equations of a line segment in the $xy$-plane by using the following formula, which is easily verified. See Figure 15.1.10.

**The line segment from $(x_0, y_0)$ to $(x_1, y_1)$ is given by** (3)

$$C: \quad x = x_0 + (x_1 - x_0)t, \quad y = y_0 + (y_1 - y_0)t, \qquad 0 \leq t \leq 1.$$

[The equations in (3) should be compared with the parametric equations of a line that were given in Section 14.4.]

### ■ EXAMPLE 7

Find parametric equations of the line segment from $(1, -2)$ to $(-3, 1)$.

**SOLUTION**

From formula (3), we obtain

$$C: \quad x = 1 + (-3 - 1)t, \quad y = -2 + (1 - (-2))t, \qquad 0 \leq t \leq 1.$$

Simplifying, we have

$$C: \quad x = 1 - 4t, \quad y = -2 + 3t, \qquad 0 \leq t \leq 1.$$

The graph is sketched in Figure 15.1.11. ■

**Parametric equations of the curve that goes from the point $(h + r, k)$ counterclockwise once around the circle $(x - h)^2 + (y - k)^2 = r^2$ are** (4)

$$C: \quad x = h + r \cos t, \quad y = k + r \sin t, \qquad 0 \leq t \leq 2\pi.$$

See Figure 15.1.12.

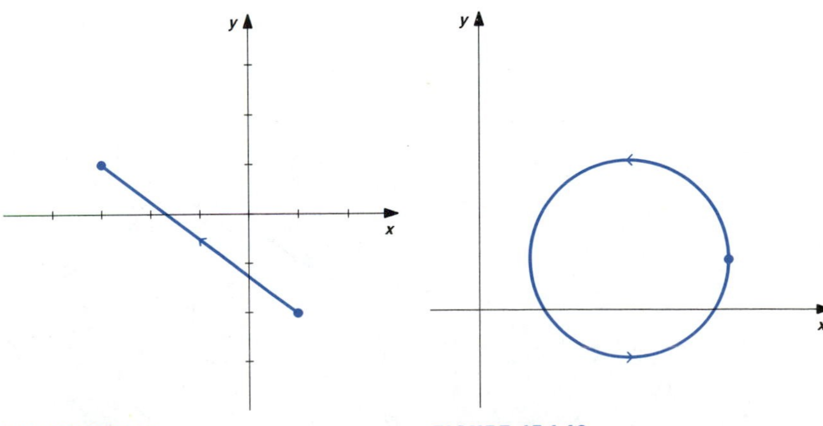


FIGURE 15.1.11

FIGURE 15.1.12

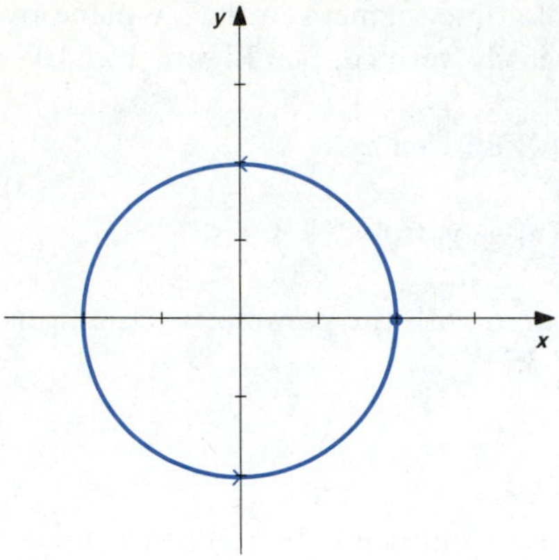

FIGURE 15.1.13

■ **EXAMPLE 8**

Find parametric equations of the curve that goes from the point (2, 0) counterclockwise once around the circle with center at the origin and radius 2.

**SOLUTION**

Using Formula (4) with $(h, k) = (0, 0)$ and $r = 2$, we obtain

$$C: \quad x = 2 \cos t, \quad y = 2 \sin t, \qquad 0 \le t \le 2\pi.$$

See Figure 15.1.13. ■

**Parametric equations of the curve that goes from the point $(h + a, k)$ counterclockwise once around the ellipse**

$$\frac{(x-h)^2}{a^2} + \frac{(y-k)^2}{b^2} = 1 \tag{5}$$

**are**

$$C: \quad x = h + a \cos t, \quad y = k + b \sin t, \qquad 0 \le t \le 2\pi.$$

See Figure 15.1.14.

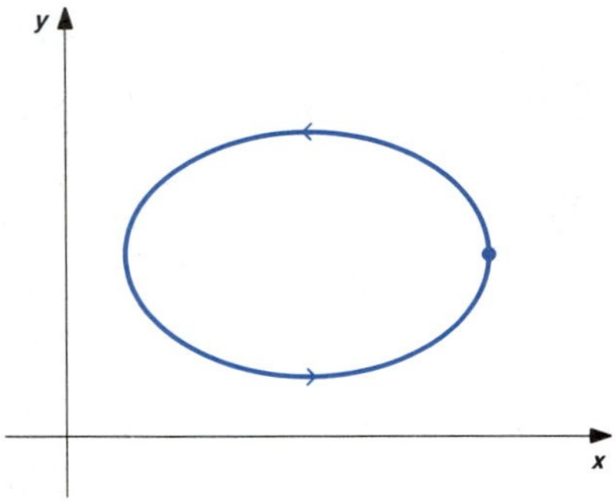


FIGURE 15.1.14

**Parametric equations of the hyperbola**

$$\frac{(x-h)^2}{a^2} - \frac{(y-k)^2}{b^2} = 1 \tag{6}$$

**are**

$$C: \quad x = h + a \sec t, \quad y = k + b \tan t.$$

Once branch is traced for $-\pi/2 < t < \pi/2$; the other branch is traced for $\pi/2 < t < 3\pi/2$. See Figure 15.1.15.

FIGURE 15.1.15

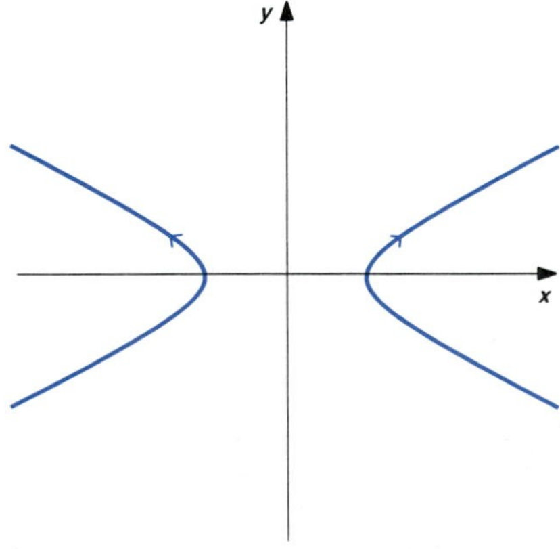

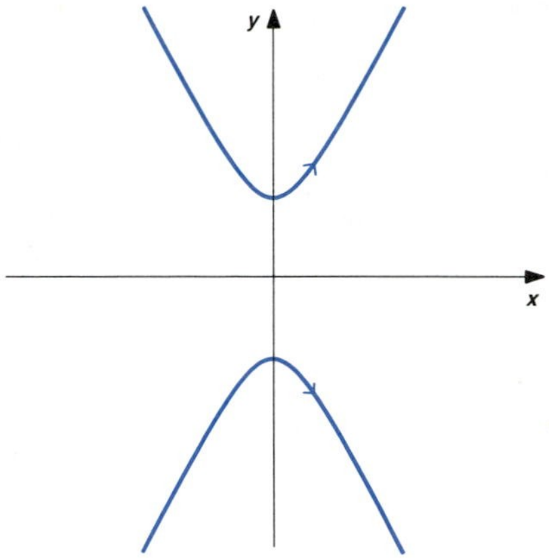

FIGURE 15.1.16

**Parametric equations of the hyperbola**

$$-\frac{(x-h)^2}{a^2}+\frac{(y-k)^2}{b^2}=1 \tag{7}$$

**are**

$$C:\quad x = h + a \tan t, \quad y = k + b \sec t.$$

One branch is traced for $-\pi/2 < t < \pi/2$; the other branch is traced for $\pi/2 < t < 3\pi/2$. See Figure 15.1.16.

In some cases we can use geometry and trigonometry to determine parametric equations of a curve.

■ **EXAMPLE 9**

Find parametric equation of the curve traced by a point on the circumference of a wheel of radius $r$ as the wheel rolls along a level surface.

**SOLUTION**

We choose a coordinate system as indicated in Figure 15.1.17a, so that the initial position of the point on the circumference is (0, 0). As the wheel rolls, the position of the point depends upon the change in position of the center ($t$) and the angle of rotation about the center ($\theta$). See Figure 15.1.17b. We see that

$$x = t - r\sin\theta \quad \text{and} \quad y = r - r\cos\theta.$$

We can obtain a relation between $t$ and $\theta$ by noting that the horizontal distance $t$ is equal to the length along the circumference subtended by $\theta$. This gives

$$r\theta = t.$$

FIGURE 15.1.17

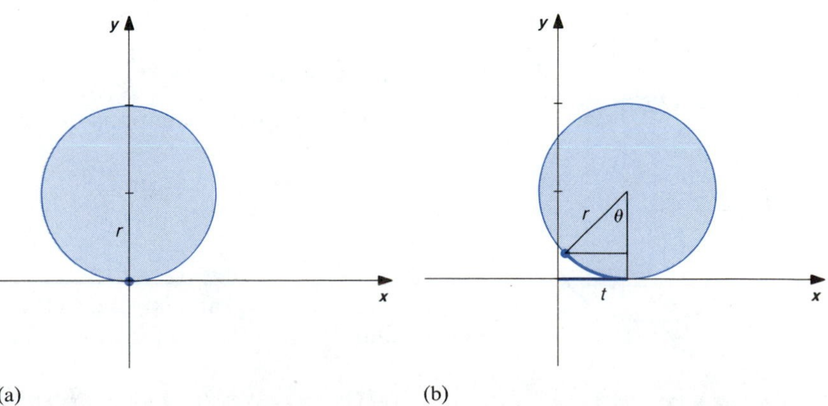

FIGURE 15.1.18

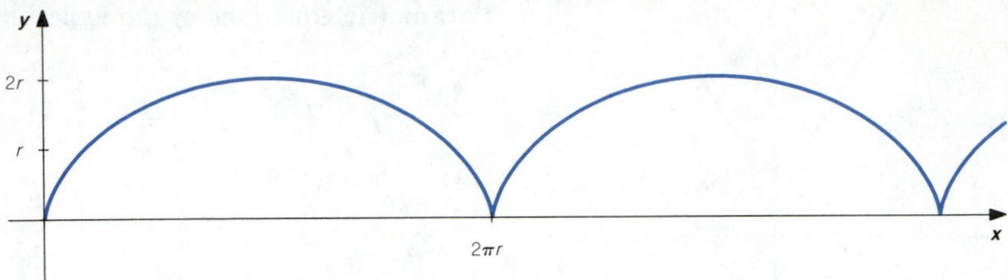


We can solve this equation for $\theta$ in terms of $t$ and then substitute in the above equations for $x$ and $y$ to obtain

$$C: \quad x = t - r \sin\frac{t}{r}, \quad y = r - r\cos\frac{t}{r}.$$

This curve is called a **cycloid**. A sketch is given in Figure 15.1.18. ■

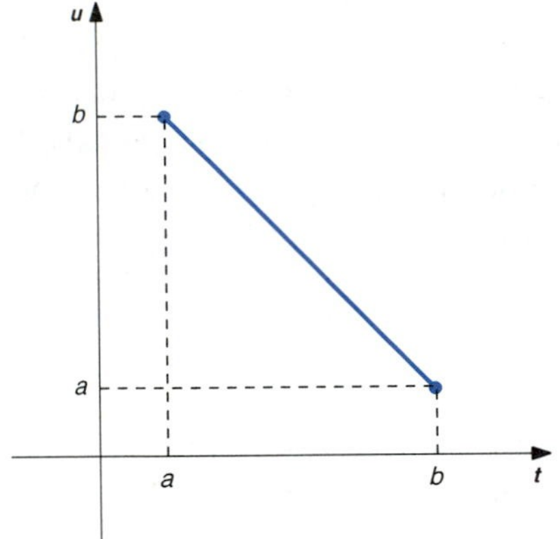


FIGURE 15.1.19

As $t$ varies from $a$ to $b$, the variable $u = a + b - t$ varies in the opposite direction, from $b$ to $a$. See Figure 15.1.19. We can use this observation to find parametric equations of the curve $-C$ that has the same graph, but direction opposite to that of a given curve $C$. We merely solve the equation $u = a + b - t$ for $t$ in terms of $u$ to obtain $t = a + b - u$, and then substitute $a + b - u$ for $t$ in the parametric equations of the given curve $C$. We have chosen $u$ so that the parameters $t$ and $u$ vary over the same interval.

**The curve that has the same graph, but direction opposite of**

$$C: \quad x = f(t), \quad y = g(t), \qquad a \le t \le b, \tag{8}$$

**can be expressed as**

$$-C: \quad x = f(a + b - u), \quad y = g(a + b - u), \qquad a \le u \le b.$$

The graphs of $C$ and $-C$ are identical, but the directions are opposite.

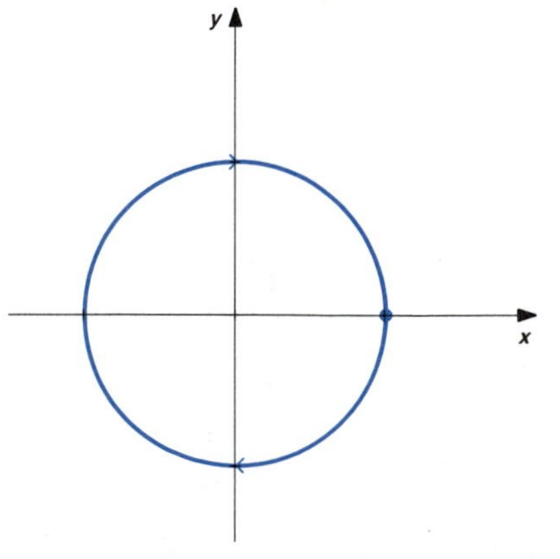


FIGURE 15.1.20

■ **EXAMPLE 10**

Find parametric equations of the curve $-C$, where

$$C: \quad x = \cos t, \quad y = \sin t, \qquad 0 \le t \le 2\pi.$$

**SOLUTION**

We have $a = 0$ and $b = 2\pi$, so the formula $t = a + b - u$ gives $t = 2\pi - u$. Substituting $2\pi - u$ for $t$ in the parametric equations of $C$, we obtain

$$-C: \quad x = \cos(2\pi - u), \quad y = \sin(2\pi - u), \qquad 0 \le u \le 2\pi.$$

The parametric equations simplify to give

$$-C: \quad x = \cos u, \quad y = -\sin u, \qquad 0 \le u \le 2\pi.$$

The curve $-C$ goes clockwise from (1, 0) once around the circle $x^2 + y^2 = 1$. See Figure 15.1.20. ■

### ■ EXAMPLE 11

Find parametric equations of the curve that goes from (2, 4) to (0, 0) along the parabola $y = x^2$.

#### SOLUTION

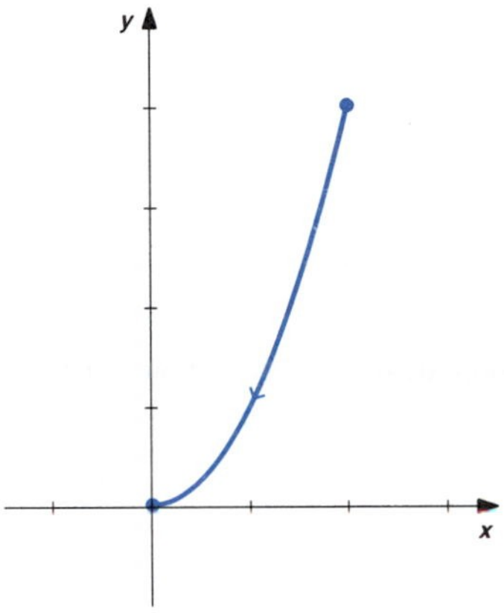

The curve is sketched in Figure 15.1.21. We can use formula (1) to determine parametric equations for the curve that goes along the parabola in the direction of increasing $x$, from (0, 0) to (2, 4). Denoting that curve $-C$, we have

$$-C: \quad x = t, \quad y = t^2, \qquad 0 \le t \le 2.$$

FIGURE 15.1.21

The curve $-C$ is in the direction opposite to the curve we want. We can then use formula (8) to obtain the desired equations. We substitute $2 - u$ for $t$ in the parametric equations of $-C$ to obtain

$$C: \quad x = 2 - u, \quad y = (2 - u)^2, \qquad 0 \le u \le 2.$$ ■

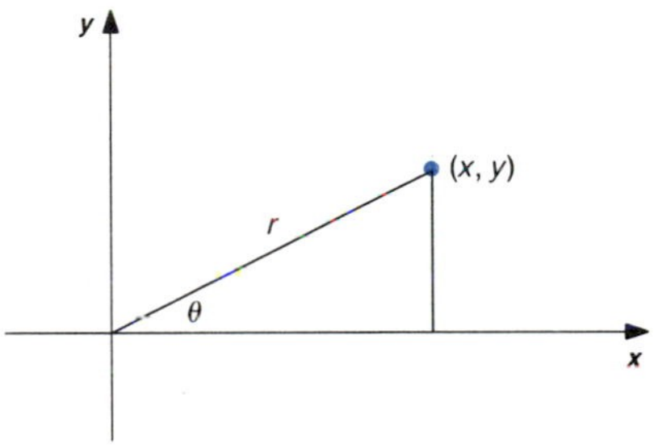

The $(x, y)$ coordinates of a point in the plane that has polar coordinates $(r, \theta)$ are given by the equations

$$x = r \cos \theta, \qquad y = r \sin \theta.$$

See Figure 15.1.22. It follows that

FIGURE 15.1.22

**The polar curve $r = r(\theta)$ in the direction of increasing $\theta$ can be expressed as** (9)

$$C: \quad x = r(t) \cos t, \quad y = r(t) \sin t.$$

### ■ EXAMPLE 12

Find parametric equations of the curve that goes in the direction of increasing $\theta$ from the point $(r, \theta) = (1, 0)$ once around the polar curve $r = \cos 3\theta$.

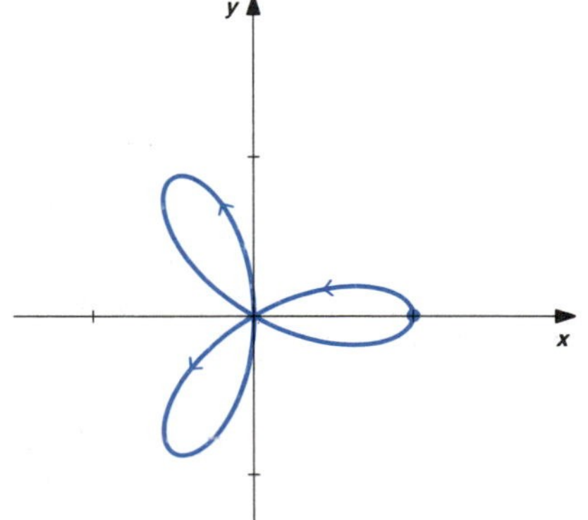

FIGURE 15.1.23

#### SOLUTION

The graph is sketched in Figure 15.1.23. Using the polar graphing techniques of Section 12.2, we can determine that the curve is traced once as $\theta$ varies from 0 to $\pi$. Also, the given initial point corresponds to $\theta = 0$. From formula (9) we then obtain

$$C: \quad x = (\cos 3t)(\cos t), \quad y = (\cos 3t)(\sin t), \qquad 0 \le t \le \pi.$$ ■

# EXERCISES 15.1

Sketch the curves in Exercises 1-8 by connecting points that correspond to the given increments with line segments. Sketch a smooth curve through the points.

1 $C: x = \cos t,\ y = \sin t,\ 0 \leq t \leq 2\pi,\ \Delta t = \pi/4$

2 $C: x = 2\cos t,\ y = \sin t,\ 0 \leq t \leq 2\pi,\ \Delta t = \pi/4$

3 $C: x = 2\cos^2 t,\ y = \sin 2t,\ -\pi/2 \leq t \leq \pi/2,\ \Delta t = \pi/6$

4 $C: x = \sin 2t,\ y = \sin^2 t,\ 0 \leq t \leq \pi,\ \Delta t = \pi/6$

5 $C: x = \dfrac{2\cos t}{\cos t + \sin t},\ y = \dfrac{2\sin t}{\cos t + \sin t},\ -\pi/6 \leq t \leq 2\pi/3,$ $\Delta t = \pi/6$

6 $C: x = 2,\ y = 2\tan t,\ -\pi/3 \leq t \leq \pi/3,\ \Delta t = \pi/6$

7 $C: x = t,\ y = t^2,\ -2 \leq t \leq 2,\ \Delta t = 1$

8 $C: x = t^2,\ y = t,\ -2 \leq t \leq 2,\ \Delta t = 1$

Identify, sketch, and describe the curves given below in Exercises 9-20.

9 $C: x = -3 + 6t,\ y = -1 + 4t,\ 0 \leq t \leq 1$

10 $C: x = 3 - 3t,\ y = -2t,\ 0 \leq t \leq 1$

11 $C: x = t,\ y = 2t - t^2,\ 0 \leq t \leq 2$

12 $C: x = t^2,\ y = t,\ 0 \leq t \leq 2$

13 $C: x = t,\ y = \cos t,\ 0 \leq t \leq 2\pi$

14 $C: x = \cos t,\ y = t,\ 0 \leq t \leq \pi$

15 $C: x = \sin t,\ y = t,\ -\pi/2 \leq t \leq \pi/2$

16 $C: x = t,\ y = \sin t,\ -\pi/2 \leq t \leq \pi/2$

17 $C: x = \cos t,\ y = \sin t,\ 0 \leq t \leq \pi$

18 $C: x = \cos t,\ y = -\sin t,\ 0 \leq t \leq \pi$

19 $C: x = \dfrac{2t}{1 + t^2},\ y = \dfrac{1 - t^2}{1 + t^2},\ -\infty < t < \infty$

20 $C: x = \tan t,\ y = \sec t,\ -\pi/2 < t < \pi/2$

Find parametric equations of the curves described below in Exercises 21-40.

21 Along $y = x^3$, from $(-1, -1)$ to $(2, 8)$

22 Along $y = x^2 - 2x$, from $(0, 0)$ to $(3, 3)$

23 Along $x = y^2$, from $(1, -1)$ to $(4, 2)$

24 Along $x = y^3 - y$, from $(-6, -2)$ to $(6, 2)$

25 Along the line segment from $(0, 0)$ to $(3, 2)$

26 Along the line segment from $(0, 3)$ to $(2, 0)$

27 Along the line segment from $(2, 1)$ to $(-2, 0)$

28 Along the line segment from $(1, 3)$ to $(-1, 1)$

29 Once around the circle $x^2 + y^2 = 4$, counterclockwise from $(2, 0)$

30 Once around the circle $x^2 + (y - 1)^2 = 1$, counterclockwise from $(1, 1)$

31 Once around the ellipse $9x^2 + 4y^2 = 36$, counterclockwise from $(2, 0)$

32 Once around the ellipse $x^2 + 4y^2 = 4$, counterclockwise from $(2, 0)$

33 Along the branch of the hyperbola $x^2 - 4y^2 = 4$ with $x > 0$, in the direction of increasing $y$.

34 Along the branch of the hyperbola $x^2 - 4y^2 = 4$ with $x < 0$, in the direction of increasing $y$

35 Along the branch of the hyperbola $4x^2 - 9y^2 + 36 = 0$ with $y > 0$, in the direction of increasing $x$

36 Along the branch of the hyperbola $4x^2 - 9y^2 + 36 = 0$ with $y < 0$, in the direction of increasing $x$

37 Along $y = x^2$, from $(2, 4)$ to $(-2, 4)$

38 Along $y = x^3$, from $(2, 8)$ to $(0, 0)$

39 Once around the circle $x^2 + y^2 = 1$, clockwise from $(1, 0)$

40 Along $x = 1 - y^2$, from $(0, 1)$ to $(-3, -2)$

Find parametric equations of the polar curves given in Exercises 41-44.

41 $r = 1 - \cos\theta,\ 0 \leq \theta \leq 2\pi$

42 $r = 1 + 2\cos\theta,\ 0 \leq \theta \leq 2\pi$

43 $r = \theta/2,\ 0 \leq \theta \leq 2\pi$

44 $r = \sin 3\theta,\ 0 \leq \theta \leq \pi$

45 Find parametric equations of the curve traced by a point that is a distance $d$ from the center of a wheel of radius $r$ as the wheel rolls along a level surface.

46 Assume that $f$ is continuous on the interval $a \leq x \leq b$, and that $a \leq c < d \leq b$ implies $f(c) < f(d)$. This implies that $f$ has an inverse function $f^{-1}$. Use values of $f$ to determine parametric equations that describe the graph of $y = f^{-1}(x)$ in the direction of increasing $x$.

We can find parametric equations for the general second-degree curve

$$Ax^2 + Bxy + Cy^2 + Dx + Ey + F = 0$$

by using the equations of rotation

$$x = x' \cos \theta - y' \sin \theta, \quad y = x' \sin \theta + y' \cos \theta$$

with $\cot 2\theta = (A - C)/B$ to obtain an equation that contains no $x'y'$ term. We then use the methods of this section to parametrize the $x'y'$ curve. The equations of rotation then give parametric equations for the variables $x$ and $y$.

**47** Find parametric equations of $3x^2 + 2xy + 3y^2 = 4$.

**48** Find parametric equations of $x^2 + 4xy + y^2 = 3$.

**49** One end of a 36-ft rope is attached to the side of a circular silo with a radius of 12-ft, and the other end is attached to a goat. The region where the goat can graze is indicated below. The boundary of the region in the first and fourth quadrants is a semicircular arc. In the second and third quadrants the boundary is generated by the endpoint of the rope as it wraps part way around the silo and then follows a straight tangential path from the silo. (a) Express the coordinates of that part of the boundary curve in the second quadrant in terms of the angle $\theta$ subtended by the arc of the silo between the point where the rope is attached to the point where it is tangent to the silo. (b) Use the fact that the area in the second quadrant where the goat can graze is a limit of sums of the areas of the "trangles" indicated to express that area as a definite integral with respect to $\theta$. (c) Find the total area of the region where the goat can graze.

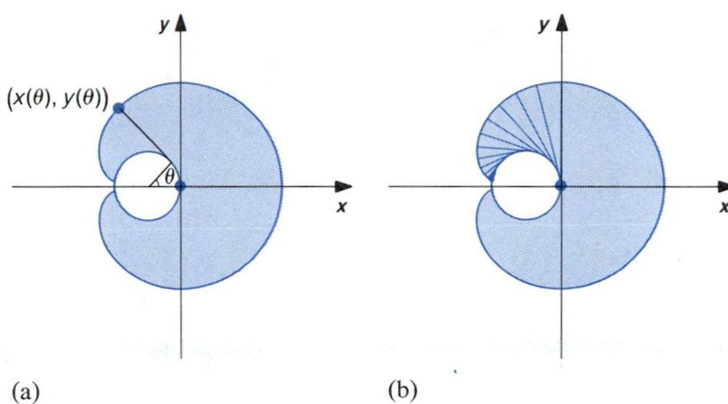


## 15.2 SLOPE, CONCAVITY, ARC LENGTH, CHANGE OF PARAMETER

If $y$ is a function of $x$, we know that

$$\frac{dy}{dx} \quad \text{and} \quad \frac{d^2y}{dx^2}$$

give information about the slope and concavity of the graph of $y$. If the parametric equations of a curve determine $y$ as a function of $x$, we can use what we know about the geometric interpretation of the first and second derivatives of a function to obtain information about the parametric curve.

As in the case of functions that are defined implicitly by an equation in $x$ and $y$, parametric equations of a curve may define $y$ as a function of $x$ only for various parts of the curve. See Figure 15.2.1.

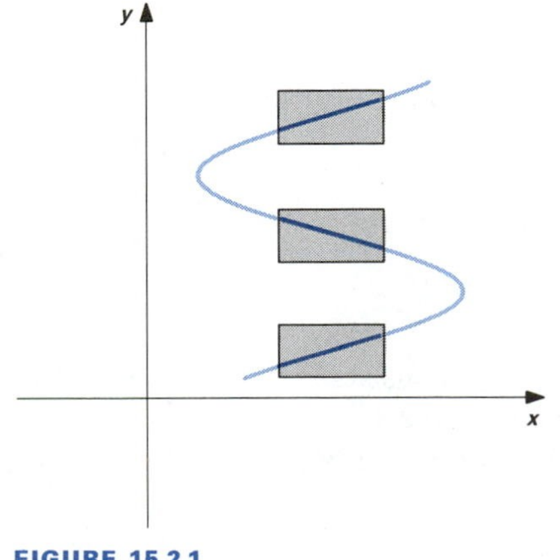


FIGURE 15.2.1

Let us consider the curve

$$C: \ x = x(t), \quad y = y(t), \qquad a \leq t \leq b,$$

where $x$ and $y$ are differentiable functions of $t$. We first note that the sign of $dx/dt$ determines the direction of $C$. That is, if $dx/dt > 0$ for $t$ in a subinterval $I$ of $[a, b]$, then $x$ is increasing as $t$ increases so the curve is in

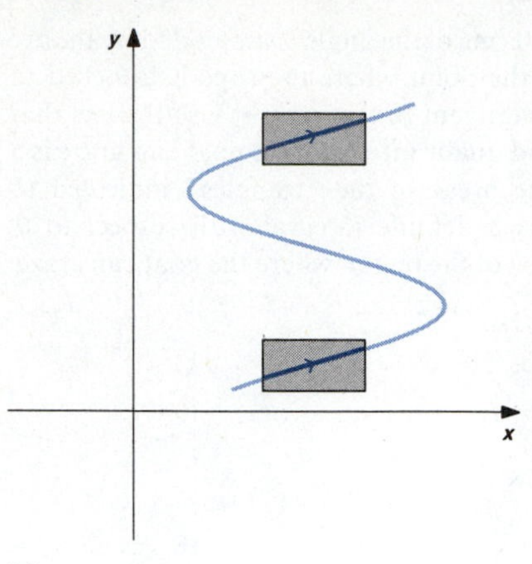
(a)

(b)

FIGURE 15.2.2

the direction of increasing $x$ for $t$ in $I$. See Figure 15.2.2a. If $dx/dt < 0$ for $t$ in $I$, the curve is in the direction of decreasing $x$ for $t$ in $I$. See Figure 15.2.2b.

We assume that $dx/dt$ is either always positive or always negative as $t$ varies over a subinterval $I$ of $[a, b]$. This condition guarantees that the function $x(t)$, $t$ in $I$, has an inverse function. From Section 7.3 we know that the equation $x = x(t)$, $t$ in $I$, defines $t = t(x)$ as a differentiable function of $x$ with

$$\frac{dt}{dx} = \frac{1}{\dfrac{dx}{dt}}.$$

For $t$ in $I$, the equations

$$y = y(t) \quad \text{and} \quad t = t(x)$$

define $y$ as a differentiable function of $x$ and the Chain Rule gives

$$\frac{dy}{dx} = \frac{dy}{dt}\frac{dt}{dx}.$$

Combining results, we obtain

$$\frac{dy}{dx} = \frac{\dfrac{dy}{dt}}{\dfrac{dx}{dt}}. \tag{1}$$

This formula gives $dy/dx$ as a function of the parameter $t$. Its value is the slope of the line tangent to the curve at the point $(x(t), y(t))$.

**■ EXAMPLE 1**

Sketch the curve

$$C: \ x = t^2 - 1, \quad y = t, \qquad -2 \le t \le 3.$$

Find the equation of the line tangent to the curve at the point where $t = 2$.

**SOLUTION**

Substituting $y$ for $t$ in the first equation, we obtain

$$x = y^2 - 1.$$

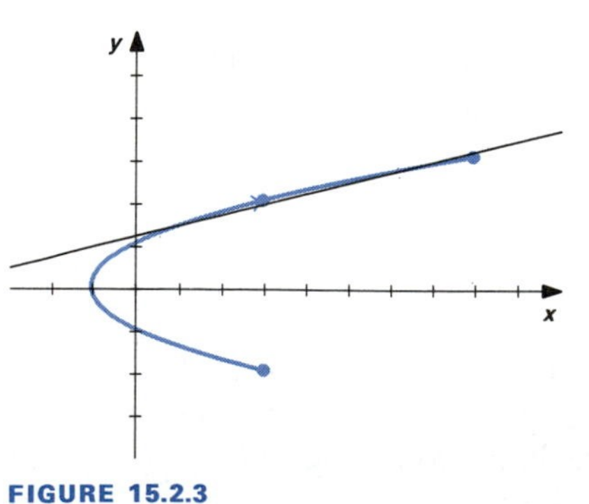
FIGURE 15.2.3

This is the equation of a parabola. The curve consists of that portion of the parabola from $(3, -2)$ to $(8, 3)$. See Figure 15.2.3.

From the equations $x = t^2 - 1$ and $y = t$, we obtain

$$\frac{dx}{dt} = 2t \quad \text{and} \quad \frac{dy}{dt} = 1.$$

Formula (1) then implies

$$\frac{dy}{dx} = \frac{\dfrac{dy}{dt}}{\dfrac{dx}{dt}} = \frac{1}{2t}.$$

When $t = 2$, we have

$$x = 2^2 - 1 = 3, \quad y = 2, \quad \text{and} \quad \frac{dy}{dx} = \frac{1}{4}.$$

The equation of the line through the point (3, 2) with slope 1/4 is

$$y - 2 = \frac{1}{4}(x - 3),$$

$$4y - 8 = x - 3,$$

$$x - 4y + 5 = 0.$$

■

In Example 1 we have $dx/dt = 2t$, so $dx/dt = 0$ when $t = 0$. Formula (1) does not apply when $t = 0$. From Figure 15.2.3 we see that the curve does not determine $y$ as a function of $x$ as the curve passes through the point $(-1, 0)$, which corresponds to $t = 0$.

Assuming appropriate differentiability, we can calculate the second derivative of $y$ with respect to $x$ as follows. We first calculate $dy/dx$ as a function of $t$ and then differentiate this function with respect to $t$ to obtain

$$\frac{d}{dt}\left(\frac{dy}{dx}\right).$$

Then

$$\frac{d^2y}{dx^2} = \frac{d}{dx}\left(\frac{dy}{dx}\right) = \frac{\dfrac{d}{dt}\left(\dfrac{dy}{dx}\right)}{\dfrac{dx}{dt}}. \tag{2}$$

This formula gives $d^2y/dx^2$ in terms of the parameter $t$. The second derivative gives us information about the concavity of the graph.

**■ EXAMPLE 2**

Evaluate $dy/dx$ and $d^2y/dx^2$ at the point on the curve

$$C\colon \; x = e^t \cos t, \quad y = e^t \sin t,$$

where $t = \pi/2$. Sketch an arc of the curve as it passes through that point.

Indicate slope, concavity, and the direction of the curve as it passes through the point.

**SOLUTION**

We have

$$x = e^t \cos t,$$

$$y = e^t \sin t,$$

$$\frac{dx}{dt} = (e^t)(-\sin t) + (\cos t)(e^t) = e^t(\cos t - \sin t), \quad \text{and}$$

$$\frac{dy}{dt} = (e^t)(\cos t) + (\sin t)(e^t) = e^t(\cos t + \sin t).$$

It then follows from formula (1) that

$$\frac{dy}{dx} = \frac{\dfrac{dy}{dt}}{\dfrac{dx}{dt}} = \frac{\cos t + \sin t}{\cos t - \sin t}, \quad \text{so}$$

$$\frac{d}{dt}\left(\frac{dy}{dx}\right) = \frac{(\cos t - \sin t)(-\sin t + \cos t) - (\cos t + \sin t)(-\sin t - \cos t)}{(\cos t - \sin t)^2}.$$

Substitution of $t = \pi/2$ then gives $x = 0$, $y = e^{\pi/2}$, and $dy/dx = -1$. When $t = \pi/2$, we also have

$$\frac{dx}{dt} = -e^{\pi/2} \quad \text{and} \quad \frac{d}{dt}\left(\frac{dy}{dx}\right) = 2,$$

so formula (2) implies

$$\frac{d^2y}{dx^2} = \frac{\dfrac{d}{dt}\left(\dfrac{dy}{dx}\right)}{\dfrac{dx}{dt}} = \frac{2}{-e^{\pi/2}} = -2e^{-\pi/2}.$$

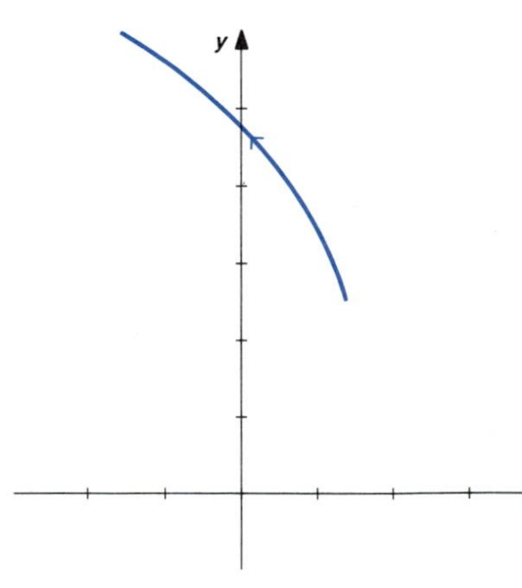

FIGURE 15.2.4

We see that the graph of the curve has slope $-1$ as it passes through the point $(0, e^{\pi/2})$ and it is concave down near that point. Also, since $dx/dt$ is negative, the $x$-coordinates of points on the curve are decreasing as $t$ increases. These features are indicated in the segment of the graph sketched in Figure 15.2.4. (The entire graph is a spiral that spirals counterclockwise away from the origin as $t$ increases.) ■

Let us consider the length of the parametric curve

$$C\colon\ x = x(t), \quad y = y(t), \qquad a \le t \le b.$$

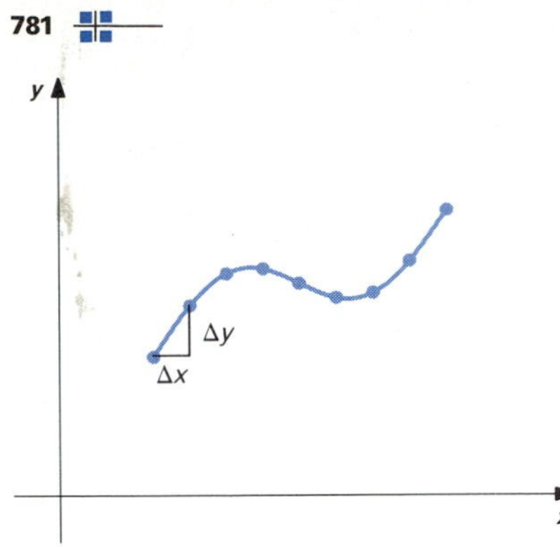


FIGURE 15.2.5

We assume the derivatives $dx/dt$ and $dy/dt$ are continuous. We divide the interval $[a, b]$ into subintervals of length $\Delta t$ and draw a polygonal path through points on the curve that correspond to the points of subdivision. See Figure 15.2.5. If $\Delta t$ is small enough, the length of the curve should be approximately

$$\sum \sqrt{(\Delta x)^2 + (\Delta y)^2} = \sum \sqrt{\left(\frac{\Delta x}{\Delta t}\right)^2 + \left(\frac{\Delta y}{\Delta t}\right)^2}\, \Delta t.$$

The following definition should then seem reasonable.

*Definition*

The **arc length** of the curve $C$: $x = x(t)$, $y = y(t)$, $a \le t \le b$, is

$$\textbf{Arc length} = \int ds = \int_a^b \sqrt{\left(\frac{dx}{dt}\right)^2 + \left(\frac{dy}{dt}\right)^2}\, dt.$$

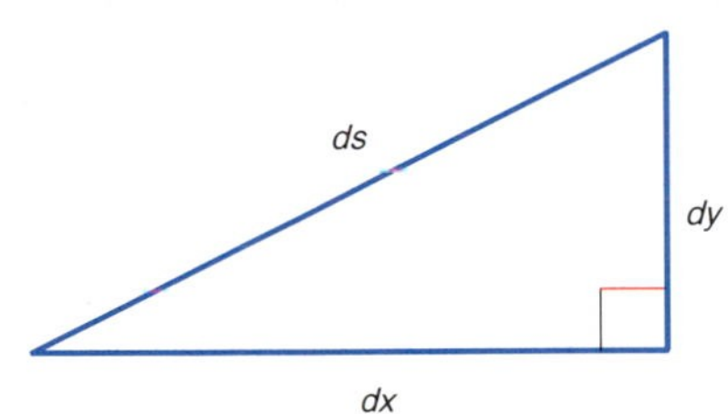


FIGURE 15.2.6

The expression

$$ds = \sqrt{(dx)^2 + (dy)^2} = \sqrt{\left(\frac{dx}{dt}\right)^2 + \left(\frac{dy}{dt}\right)^2}\, dt$$

is the **differential arc length.** See Figure 15.2.6.

### EXAMPLE 3

Find the length of the curve

$$C\colon\ x = 3\cos t, \quad y = 3\sin t, \qquad 0 \le t \le 2\pi.$$

**SOLUTION**

The curve traces the circle $x^2 + y^2 = 3^2$ once, counterclockwise from the point $(3, 0)$. See Figure 15.2.7. We have

$$\frac{dx}{dt} = -3\sin t \quad \text{and} \quad \frac{dy}{dt} = 3\cos t,$$

so

$$ds = \sqrt{(-3\sin t)^2 + (3\cos t)^2}\, dt = \sqrt{9(\sin^2 t + \cos^2 t)}\, dt = 3\, dt.$$

The arc length is

$$s = \int ds = \int_0^{2\pi} 3\, dt = 6\pi.$$

This is the circumference of a circle of radius 3, as we know.

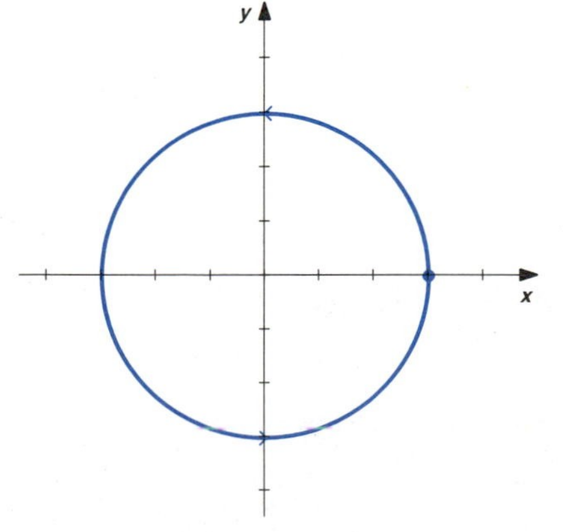


FIGURE 15.2.7

**■ EXAMPLE 4**

Find the length of the curve

$$C:\ x = t^2, \quad y = t^3, \qquad 0 \le t \le \sqrt{5}.$$

**SOLUTION**

We have

$$\frac{dx}{dt} = 2t \quad \text{and} \quad \frac{dy}{dt} = 3t^2.$$

Hence,

$$ds = \sqrt{(2t)^2 + (3t^2)^2}\, dt = \sqrt{4t^2 + 9t^4}\, dt = t\sqrt{4 + 9t^2}\, dt, \qquad t \ge 0.$$

$$\text{Arc length} = \int ds = \int_0^{\sqrt{5}} t\sqrt{4 + 9t^2}\, dt.$$

Let us express this integral in terms of the variable $u = 4 + 9t^2$. Then $du = 18t\, dt$. Also, $u = 49$ when $t = \sqrt{5}$, and $u = 4$ when $t = 0$. The arc length is then

$$= \int_4^{49} \frac{1}{18} u^{1/2}\, du$$

$$= \frac{1}{18} \frac{u^{3/2}}{3/2}\Big]_4^{49}$$

$$= \frac{1}{27}[(49)^{3/2} - (4)^{3/2}]$$

$$= \frac{1}{27}(343 - 8) = \frac{335}{27}.$$

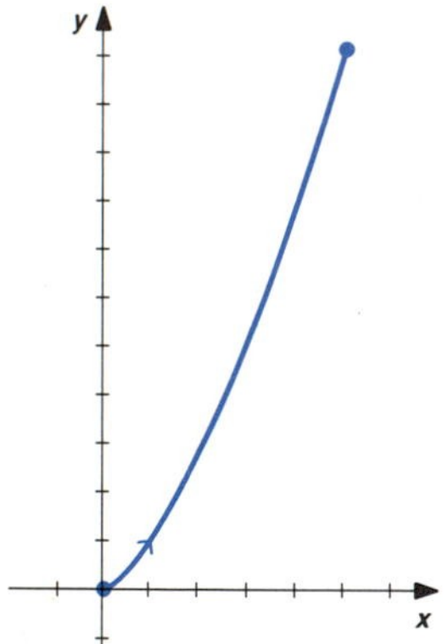


FIGURE 15.2.8

The curve is sketched in Figure 15.2.8. ■

If a curve is the graph of a differentiable function, $y = f(x)$, $a \le x \le b$, we can write

$$C:\ x = x, \quad y = f(x), \qquad a \le x \le b.$$

In the above equations we are using $x$ as the parameter that determines the coordinates $(x, y)$ of points on the curve. The differential arc length is then

$$ds = \sqrt{(dx)^2 + (dy)^2} = \sqrt{\left(\frac{dx}{dx}\right)^2 + \left(\frac{dy}{dx}\right)^2}\, dx = \sqrt{1 + [f'(x)]^2}\, dx.$$

This is the same formula we derived in Section 6.4.

We can find the arc length of a polar curve $r = r(\theta)$ by writing

$$C\colon\ x = r(\theta)\cos\theta,\quad y = r(\theta)\sin\theta,\qquad a \le \theta \le b.$$

Here we are using $\theta$ as the parameter that determines the rectangular coordinates $(x, y)$ of points on the curve. Then

$$\begin{aligned}\left(\frac{dx}{d\theta}\right)^2 + \left(\frac{dy}{d\theta}\right)^2 &= (-r\sin\theta + r'\cos\theta)^2 + (r\cos\theta + r'\sin\theta)^2\\ &= r^2\sin^2\theta - 2rr'\sin\theta\cos\theta + r'^2\cos^2\theta\\ &\quad + r^2\cos^2\theta + 2rr'\sin\theta\cos\theta + r'^2\sin^2\theta\\ &= r^2 + (r')^2.\end{aligned}$$

Hence, we obtain the formula

$$ds = \sqrt{r^2 + \left(\frac{dr}{d\theta}\right)^2}\,d\theta.$$

This is the formula that we derived in Section 12.5 for the arc length of polar curves $r = r(\theta)$. Note that the formula applies to curves that are given in terms of polar coordinates and that it is not necessary to write the curve in parametric form in order to use the formula. If the rectangular coordinates of a curve are given by parametric equations, we use the formula $ds = \sqrt{(dx)^2 + (dy)^2}$, even if the parameter happens to be $\theta$.

■ **EXAMPLE 5**

Find the length of the polar spiral

$$r = \theta^2,\qquad 0 \le \theta \le \pi.$$

**SOLUTION**

We use the formula for the differential arc length of a polar curve. We have

$$r = \theta^2,\quad\text{so}\quad \frac{dr}{d\theta} = 2\theta.$$

Then

$$ds = \sqrt{r^2 + \left(\frac{dr}{d\theta}\right)^2}\,d\theta = \sqrt{(\theta^2)^2 + (2\theta)^2}\,d\theta = \theta\sqrt{\theta^2 + 4}\,d\theta,\qquad \theta \ge 0.$$

$$\begin{aligned}\text{Arc length} &= \int_0^\pi \theta\sqrt{\theta^2 + 4}\,d\theta \qquad [u = \theta^2 + 4,\ du = 2\theta\,d\theta]\\ &= \frac{1}{3}(\theta^2 + 4)^{3/2}\Big]_0^\pi\\ &= \frac{1}{3}[(\pi^2 + 4)^{3/2} - 8].\end{aligned}$$

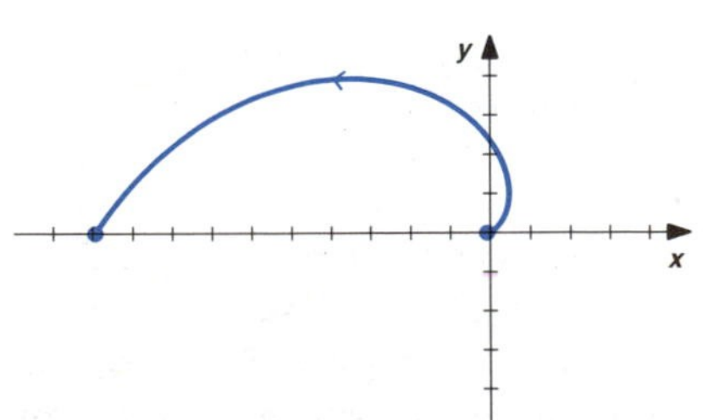


FIGURE 15.2.9

The curve is sketched in Figure 15.2.9. ■

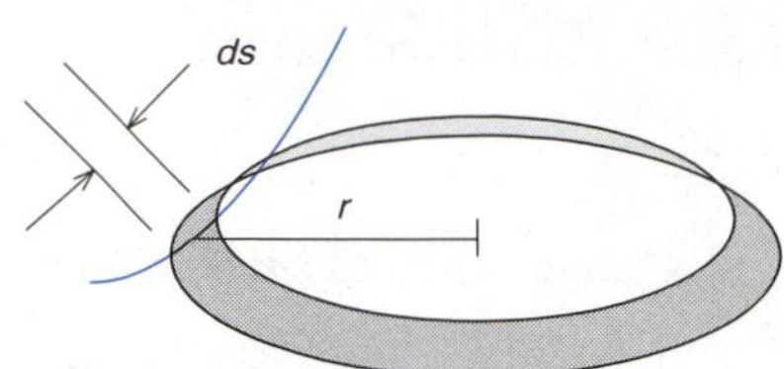


FIGURE 15.2.10

Recall that differential arc length $ds$ is used to determine the area of a surface obtained by revolving a plane curve about a line in the plane. See Figure 15.2.10. We have

$$\textbf{Surface area} = \int 2\pi r\, ds,$$

where $r$ is the radius of the band obtained when a small variable piece of the curve is revolved. If the curve is given in parametric form, we can evaluate the surface area by expressing $r$ and $ds$ in terms of the parameter and then evaluating the integral.

■ **EXAMPLE 6**

Set up an integral that gives the area of the surface obtained by revolving the half-ellipse

$$C: x = 3 \cos t, \quad y = 2 \sin t, \quad 0 \le t \le \pi,$$

about the $x$-axis.

**SOLUTION**

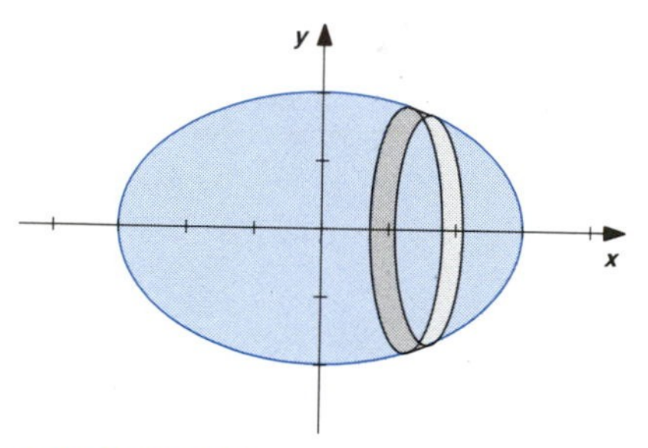


FIGURE 15.2.11

The curve is sketched in Figure 15.2.11. We have also sketched the band that is obtained when a typical piece of arc is revolved about the $x$-axis. The area of a typical band is

$$dS = 2\pi(\text{Radius})(\text{Slant height}) = 2\pi y\, ds.$$

We need to express $y$ and the differential arc length $ds$ in terms of the parameter $t$, and then use a definite integral to give us the limit of the sums of areas of the bands. We have $x = 3 \cos t$ and $y = 2 \sin t$, so

$$\frac{dx}{dt} = -3 \sin t \quad \text{and} \quad \frac{dy}{dt} = 2 \cos t.$$

Then

$$ds = \sqrt{(-3 \sin t)^2 + (2 \cos t)^2}\, dt = \sqrt{9 \sin^2 t + 4 \cos^2 t}\, dt.$$

Since $y = 2 \sin t$, we have

$$\text{Surface area} = \int_0^{\pi} 2\pi\, 2 \sin t \sqrt{9 \sin^2 t + 4 \cos^2 t}\, dt.$$

[This integral could be evaluated by writing $9 \sin^2 t + 4 \cos^2 t = 9 - 5 \cos^2 t$ and then using the substitution $u = (\sqrt{5}/3) \cos t$. We would then need a trigonometric substitution or an integration formula. Also, the methods of Section 5.4 could be used to obtain an approximate value of the integral.] ■

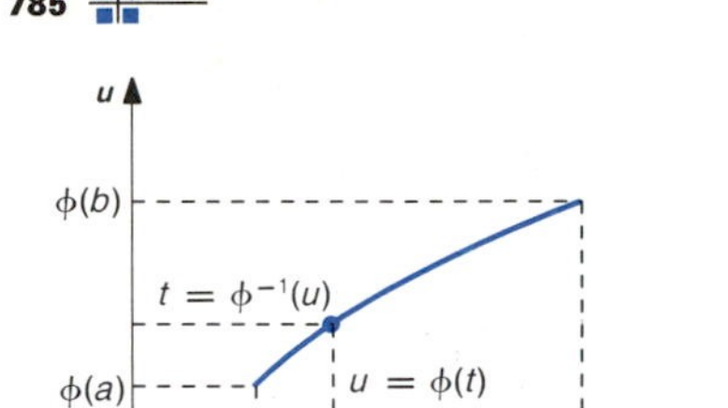

FIGURE 15.2.12

We have seen that a curve that is described geometrically or in terms of an equation in $x$ and $y$ can be parametrized in different ways. Let us now systematically study how to change the parametrization of a curve

$$C: \ x = f(t), \quad y = g(t), \qquad a \leq t \leq b.$$

Suppose $u = \phi(t)$, where $\phi'$ exists and is positive for $a \leq t \leq b$, so $\phi$ has an inverse function $\phi^{-1}$ with

$$u = \phi(t) \quad \text{if and only if} \quad t = \phi^{-1}(u).$$

See Figure 15.2.12. As $u$ varies from $\phi(a)$ to $\phi(b)$, the variable $t = \phi^{-1}(u)$ assumes each value from $a$ to $b$; these values of $t$ are assumed exactly once and no other values of $t$ are assumed for $\phi(a) \leq u \leq \phi(b)$. It follows that the curve

$$C_\phi: \ x = f(\phi^{-1}(u)), \quad y = g(\phi^{-1}(u)), \qquad \phi(a) \leq u \leq \phi(b),$$

has the same graph and the same direction as $C$. In this sense, we consider $C$ and $C_\phi$ to be different parametrizations of the same curve.

Let us see how to use arc length to obtain a parametrization of a curve

$$C: \ x = f(t), \quad y = g(t), \qquad a \leq t \leq b.$$

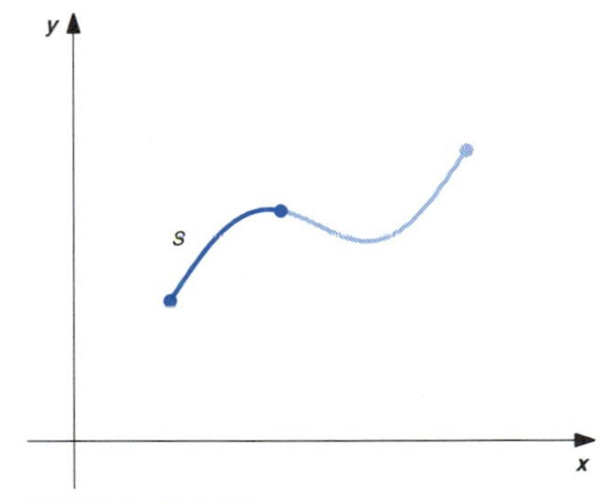

FIGURE 15.2.13

We assume that $f$ and $g$ have continuous derivatives and that

$$\left(\frac{dx}{dt}\right)^2 + \left(\frac{dy}{dt}\right)^2 \neq 0 \qquad \text{for } a \leq t \leq b.$$

We then define the **arc length function**

$$\phi(t) = \int_a^t \sqrt{\left(\frac{dx}{dt}\right)^2 + \left(\frac{dy}{dt}\right)^2}\, d\tau, \qquad a \leq t \leq b,$$

where the derivatives of $x$ and $y$ with respect to the parameter $t$ are expressed in terms of the dummy variable of integration $\tau$. The variable $s = \phi(t)$ represents the length of that portion of the curve between $(f(a), g(a))$ and $(f(t), g(t))$. See Figure 15.2.13. The Fundamental Theorem of Calculus tells us that $\phi$ is differentiable with

$$\phi' = \sqrt{\left(\frac{dx}{dt}\right)^2 + \left(\frac{dy}{dt}\right)^2} > 0.$$

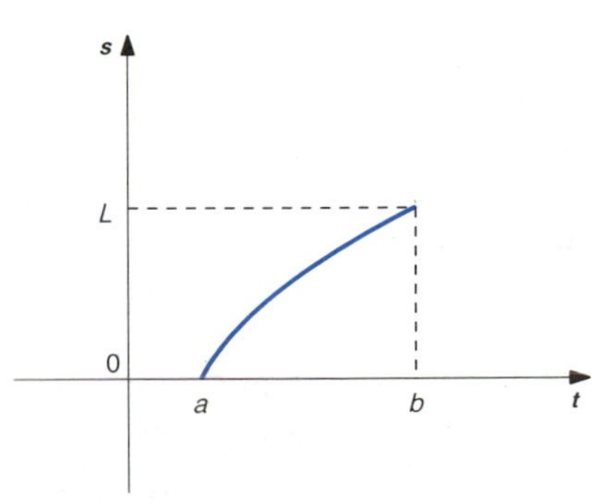

FIGURE 15.2.14

See Figure 15.2.14. Note that $\phi(a) = 0$ and $\phi(b) = L$, where $L$ is the total arc length of the curve. The function $\phi$ has an inverse function $\phi^{-1}$ with

$$s = \phi(t) \quad \text{if and only if} \quad t = \phi^{-1}(s).$$

The curve

$$C_\phi\colon\ x = f(\phi^{-1}(s)),\quad y = g(\phi^{-1}(s)),\qquad 0 \le s \le L,$$

is called the **parametrization of $C$ in terms of arc length**. We shall see in the next section that this parametrization is quite important for the study of the properties of curves.

In order to parametrize a curve in terms of arc length, we need to determine the arc length function and then find its inverse. This is theoretically possible for any curve as described above, but in practice we can carry out the details and obtain formulas in only a few cases.

### ■ EXAMPLE 7

Find the parametrization in terms of arc length of the curve

$$C\colon\ x = 4 - 4t,\quad y = 3t,\qquad 0 \le t \le 1.$$

#### SOLUTION

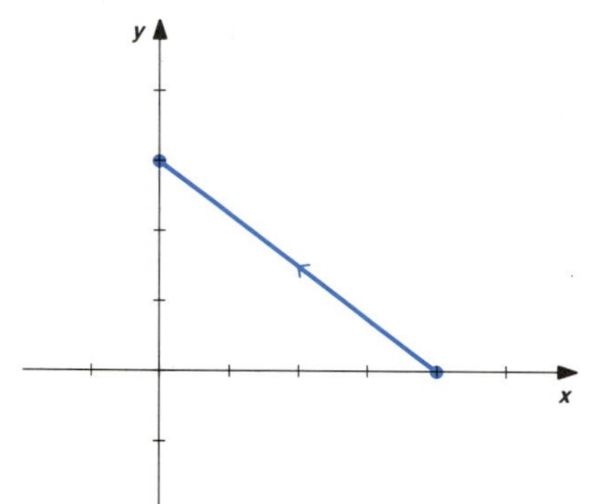

FIGURE 15.2.15

This curve is the line segment from (4, 0) to (0, 3). See Figure 15.2.15. We see that

$$\frac{dx}{dt} = -4 \quad\text{and}\quad \frac{dy}{dt} = 3,$$

so the arc length function is

$$s = \int_0^t \sqrt{(-4)^2 + (3)^2}\, d\tau = 5t.$$

This gives the relation

$$s = 5t.$$

Solving this equation for $t$, we obtain $t$ in terms of the arc length $s$,

$$t = \frac{s}{5}.$$

Substitution into the given parametric equations of $C$ gives

$$x = 4 - 4\left(\frac{s}{5}\right) = 4 - 0.8s \quad\text{and}\quad y = 3\left(\frac{s}{5}\right) = 0.6s.$$

The total length of the curve is given by setting $t = 1$ in the equation $s = 5t$, so $L = 5$. The parametrization of $C$ in terms of arc length is

$$C\colon\quad x = 4 - 0.8s,\quad y = 0.6s,\qquad 0 \le s \le 5.$$

■

■ **EXAMPLE 8**

Find the parametrization in terms of arc length of the curve

$$C:\ x = 3\cos t,\quad y = 3\sin t,\qquad 0 \le t \le 2\pi.$$

**SOLUTION**

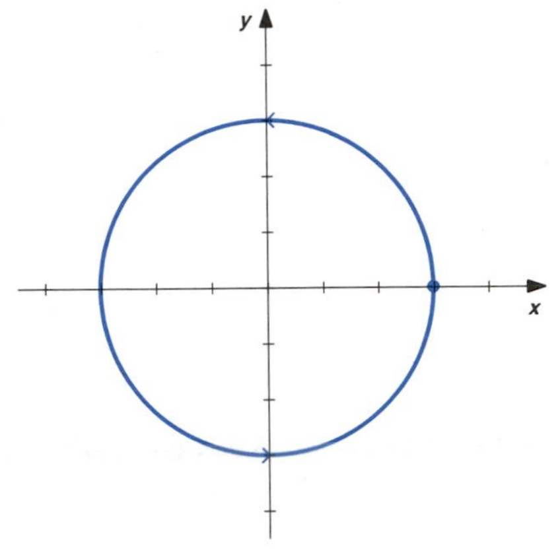


FIGURE 15.2.16

The curve is a circle with center (0, 0) and radius 3. See Figure 15.2.16. We have

$$\frac{dx}{dt} = -3\sin t \quad \text{and} \quad \frac{dy}{dt} = 3\cos t.$$

Expressing these derivatives in terms of the dummy variable $\tau$, we have

$$\begin{aligned} s &= \int_0^t \sqrt{(-3\sin\tau)^2 + (3\cos\tau)^2}\, d\tau \\ &= \int_0^t \sqrt{(3)^2(\sin^2\tau + \cos^2\tau)}\, d\tau \\ &= \int_0^t 3\, d\tau = 3t. \end{aligned}$$

We now have

$$s = 3t.$$

Solving this equation for $t$, we obtain $t$ in terms of the arc length $s$,

$$t = \frac{s}{3}.$$

When $t = 2\pi$, the equation $s = 3t$ gives the total arc length $6\pi$. The parametrization of $C$ in terms of arc length is

$$C:\ x = 3\cos\frac{s}{3},\quad y = 3\sin\frac{s}{3},\qquad 0 \le s \le 6\pi.$$ ■

■ **EXAMPLE 9**

Find the parametrization in terms of arc length of the curve

$$C:\ x = 3t,\quad y = 2t^{3/2},\qquad 0 \le t \le 3.$$

**SOLUTION**

We have

$$\frac{dx}{dt} = 3 \quad \text{and} \quad \frac{dy}{dt} = 3t^{1/2},$$

so the arc length function is

$$\begin{aligned} s &= \int_0^t \sqrt{(3)^2 + (3\tau^{1/2})^2}\, d\tau \\ &= \int_0^t \sqrt{9 + 9\tau}\, d\tau \\ &= \int_0^t 3(1 + \tau)^{1/2}\, d\tau \\ &= 3\frac{(1+\tau)^{3/2}}{3/2}\Big]_0^t \\ &= 2(1 + t)^{3/2} - 2. \end{aligned}$$

We now have

$$s = 2(1 + t)^{3/2} - 2.$$

When $t = 3$, this equation gives the total arc length,

$$L = 2(1 + 3)^{3/2} - 2 = 14.$$

Solving the equation for $t$ in terms of the arc length $s$, we have

$$s = 2(1 + t)^{3/2} - 2,$$

$$s + 2 = 2(1 + t)^{3/2},$$

$$(1 + t)^{3/2} = \frac{s + 2}{2},$$

$$1 + t = \left(\frac{s + 2}{2}\right)^{2/3},$$

$$t = \left(\frac{s + 2}{2}\right)^{2/3} - 1.$$

The parametrization in terms of arc length is

$$C\colon\ x = 3\left[\left(\frac{s + 2}{2}\right)^{2/3} - 1\right],\quad y = 2\left[\left(\frac{s + 2}{2}\right)^{2/3} - 1\right]^{3/2},\qquad 0 \le s \le 14.$$

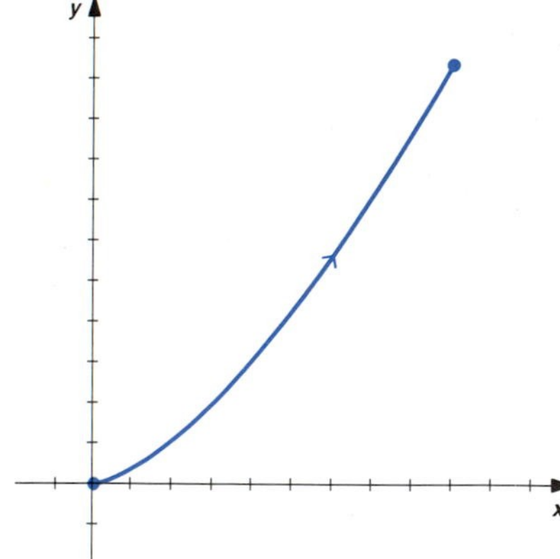


FIGURE 15.2.17

The graph is sketched in Figure 15.2.17. ■

## EXERCISES 15.2

In each of Exercises 1–4, find the equation of the line tangent to the curve at the indicated point, and sketch the curve and the tangent line.

**1** $C\colon x = t^2,\ y = t^3,\ -1 \le t \le 2;\ t = 1$

**2** $C\colon x = t^3,\ y = t^2,\ -1 \le t \le 2;\ t = 1$

**3** $C: x = \sin t,\ y = t,\ -\pi/2 \le t \le \pi/2;\ t = \pi/6$

**4** $C: x = \cos t,\ y = t,\ 0 \le t \le \pi;\ t = \pi/4$

In Exercises 5–10 evaluate $dy/dx$ and $d^2y/dx^2$ at the indicated point. Sketch an arc of the curve as it passes through that point. Indicate slope, concavity, and the direction of the curve as it passes through the point.

**5** $C: x = t^2,\ y = t^3;\ t = 1$

**6** $C: x = 1 + \cos t,\ y = \sin t;\ t = \pi/6$

**7** $C: x = t^2 + t,\ y = t^2 + 2t - 1;\ t = 1$

**8** $C: x = t - t^2,\ y = t^2 + t;\ t = 1$

**9** $C: x = \ln t,\ y = t^2;\ t = 1$

**10** $C: x = t \cos t,\ y = t \sin t;\ t = 3\pi/2$

Find the length of the curves in Exercises 11–20.

**11** $C: x = 4t,\ y = 3 - 3t,\ 0 \le t \le 1$

**12** $C: x = 12t,\ y = 5t,\ 0 \le t \le 1$

**13** $C: x = t^3,\ y = t^2,\ 0 \le t \le 2$

**14** $C: x = t^2/2,\ y = t^3/3,\ 0 \le t \le 1$

**15** $C: x = \cos t,\ y = 1 + \sin t,\ 0 \le t \le 2\pi$

**16** $C: x = 2 \cos t,\ y = 2 \sin t,\ 0 \le t \le \pi$

**17** $C: x = t^2 \cos t,\ y = t^2 \sin t,\ 0 \le t \le 2\pi$

**18** $C: x = e^{2t} \cos t,\ y = e^{2t} \sin t,\ 0 \le t \le 2\pi$

**19** $C: x = 2 \ln t,\ y = t + (1/t),\ 1 \le t \le 4$

**20** $C: x = 2e^t,\ y = (1/3)e^{3t} + e^{-t},\ 0 \le t \le 1$

Find the length of the polar curves in Exercises 21–22.

**21** $r = 2 \cos \theta,\ 0 \le \theta \le \pi$

**22** $r = e^{-\theta},\ 0 \le \theta \le 2\pi$

In Exercises 23–30 find the area of the surface obtained when the given parametric curve is revolved about the indicated line.

**23** $C: x = 3t,\ y = 4t,\ 0 \le t \le 1$; about the $y$-axis

**24** $C: x = rt,\ y = h - ht,\ 0 \le t \le 1$; about the $y$-axis

**25** $C: x = t^2,\ y = t,\ 0 \le t \le 2$; about the $x$-axis

**26** $C: x = t,\ y = t^3/3,\ 0 \le t \le 3$; about the $x$-axis

**27** $C: x = 2 \cos t,\ y = 2 \sin t,\ 0 \le t \le \pi$; about the $x$-axis

**28** $C: x = r \cos t,\ y = r \sin t,\ 0 \le t \le \pi$; about the $x$-axis

**29** $C: x = 3 + \cos t,\ y = \sin t,\ 0 \le t \le 2\pi$; about the $y$-axis

**30** $C: x = R + r \cos t,\ y = r \sin t,\ 0 \le t \le 2\pi$; about the $y$-axis; $R > r > 0$

Find the parametrization in terms of arc length of the curves in Exercises 31–36.

**31** $C: x = 4t,\ y = 3t,\ 0 \le t \le 1$

**32** $C: x = at,\ y = b - bt,\ 0 \le t \le 1$

**33** $C: x = 3 \cos 2t,\ y = 3 \sin 2t,\ 0 \le t \le \pi$

**34** $C: x = r \cos \omega t,\ y = r \sin \omega t,\ 0 \le t \le 2\pi/\omega$

**35** $C: x = 2t^3/3,\ y = t^2,\ 0 \le t \le 3$

**36** $C: x = t^2/2,\ y = t^3/3,\ 0 \le t \le 1$

Find the parametrization in terms of arc length of the polar curves in Exercises 37–38.

**37** $r = e^{\theta},\ 0 \le \theta \le 2\pi$

**38** $r = \theta^2,\ 0 \le \theta \le 2\pi$

## 15.3 UNIT TANGENT AND NORMAL VECTORS, CURVATURE

Given a curve in parametric form,

$$C:\ x = x(t), \quad y = y(t), \qquad a \le t \le b,$$

we can express the curve in **vector form** by writing

$$\mathbf{R}(t) = x(t)\mathbf{i} + y(t)\mathbf{j}, \qquad a \le t \le b.$$

We will see that the vector representation of $C$ is convenient for studying the change of direction of a curve.

Let us think of the vector function

$$\mathbf{R}(t) = x(t)\mathbf{i} + y(t)\mathbf{j}, \qquad a \le t \le b,$$

as representing the position of an object along a curve. See Figure 15.3.1.

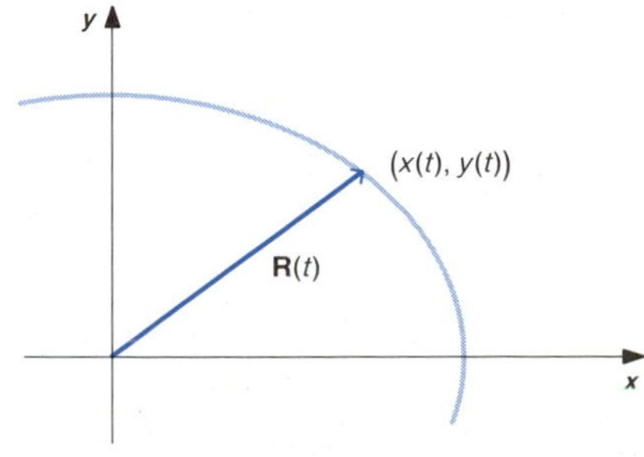

FIGURE 15.3.1

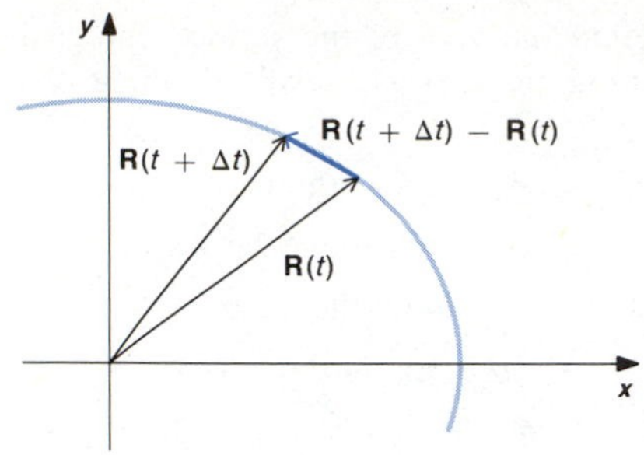

FIGURE 15.3.2

We assume that $x$ and $y$ are differentiable and $(x')^2 + (y')^2 \neq 0$ for $a \leq t \leq b$. This means the derivative

$$\mathbf{R}'(t) = x'(t)\mathbf{i} + y'(t)\mathbf{j}$$

exists and is nonzero. For $\Delta t \neq 0$, the vector from $\mathbf{R}(t)$ to $\mathbf{R}(t + \Delta)$ is $\mathbf{R}(t + \Delta) - \mathbf{R}(t)$. See Figure 15.3.2. As $\Delta t \to 0$, the direction of the vector

$$\frac{\mathbf{R}(t + \Delta t) - \mathbf{R}(t)}{\Delta t}$$

approaches the direction of the *nonzero* vector $\mathbf{R}'(t)$. Thus, $\mathbf{R}'(t)$ is tangent to the curve at position $\mathbf{R}(t)$. Also, $\mathbf{R}'(t)$ points in the direction of increasing $t$. The unit vector in the direction of $\mathbf{R}'(t)$ is

$$\mathbf{T}(t) = \frac{\mathbf{R}'(t)}{|\mathbf{R}'(t)|}.$$

Since $\mathbf{T}$ is a positive multiple of $\mathbf{R}'$, $\mathbf{T}$ is also a tangent to the curve and points in the direction of increasing $t$. The vector $\mathbf{T}$ is called the **unit tangent vector** of the curve.

Since $\mathbf{T}(t)$ is a vector with constant length one, it follows from statement (5) of Section 14.5 that $\mathbf{T}$ and $\mathbf{T}'$ are orthogonal. If $\mathbf{T}' \neq \mathbf{0}$, we define the (principal) **unit normal vector** to the curve to be

$$\mathbf{N} = \frac{\mathbf{T}'}{|\mathbf{T}'|}.$$

Since $\mathbf{N}$ is a multiple of $\mathbf{T}'$,

**T and N are orthogonal.**

We will see in Section 15.4 that the unit normal vector $\mathbf{N}$ does not depend on the direction the curve is traced; $\mathbf{N}$ is always orthogonal to the curve and points in the direction the curve is bending. See Figures 15.3.3.

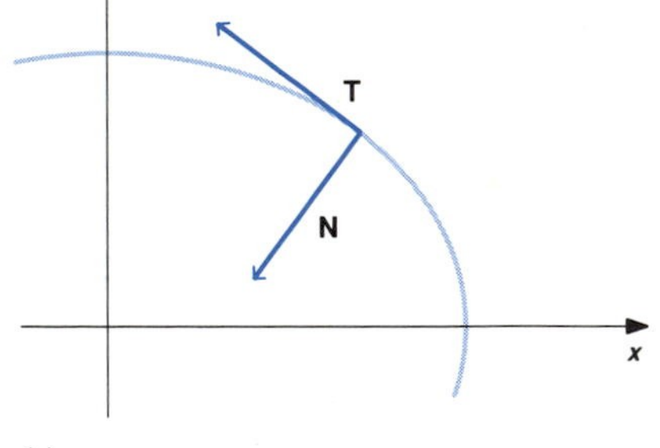

(a)

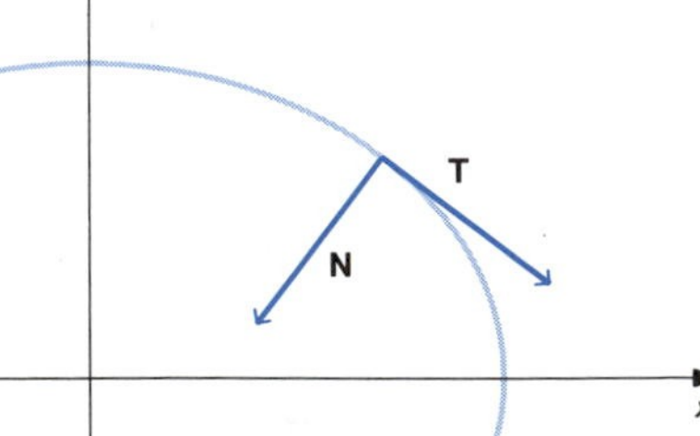

(b)

FIGURE 15.3.3

■ **EXAMPLE 1**

Find the unit tangent and normal vectors of the curve

$$C\colon\ x = 3\cos 2t, \quad y = 3\sin 2t, \qquad 0 \leq t \leq 2\pi.$$

**SOLUTION**

We express the curve in vector form by writing

$$\mathbf{R}(t) = (3\cos 2t)\mathbf{i} + (3\sin 2t)\mathbf{j}.$$

The derivative is

$$\mathbf{R}'(t) = (-6\sin 2t)\mathbf{i} + (6\cos 2t)\mathbf{j}$$

and the length of the derivative is

$$|\mathbf{R}'(t)| = \sqrt{(-6\sin 2t)^2 + (6\cos 2t)^2}$$
$$= \sqrt{36(\sin^2 2t + \cos^2 2t)} = 6.$$

The unit tangent vector is

$$\mathbf{T}(t) = \frac{\mathbf{R}'(t)}{|\mathbf{R}'(t)|} = \frac{(-6\sin 2t)\mathbf{i} + (6\cos 2t)\mathbf{j}}{6} = (-\sin 2t)\mathbf{i} + (\cos 2t)\mathbf{j}.$$

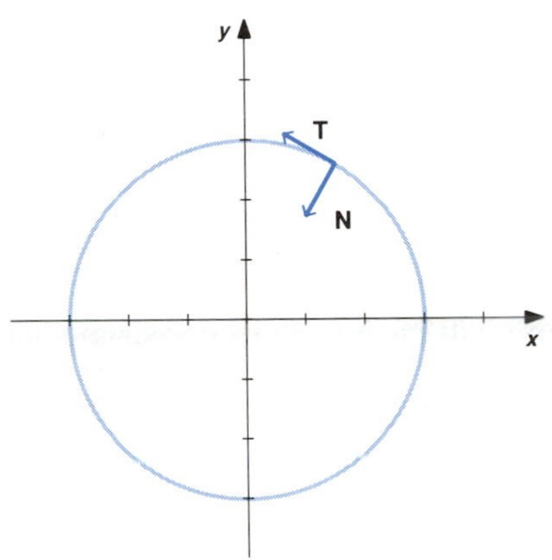

FIGURE 15.3.4

We have

$$\mathbf{T}'(t) = (-2\cos 2t)\mathbf{i} + (-2\sin 2t)\mathbf{j},$$

so

$$|\mathbf{T}'(t)| = \sqrt{(-2\cos 2t)^2 + (-2\sin 2t)^2} = 2 \quad \text{and}$$

$$\mathbf{N} = \frac{\mathbf{T}'}{|\mathbf{T}'|} = (-\cos 2t)\mathbf{i} + (-\sin 2t)\mathbf{j}.$$

The graph of the curve and typical unit tangent and normal vectors are sketched in Figure 15.3.4. Note that the normal vector points toward the center of the circle. ■

The curve

$$C\colon\ x = x(t), \quad y = y(t), \qquad a \le t \le b,$$

is called **smooth** if $x$ and $y$ have continuous derivatives and

$$\left(\frac{dx}{dt}\right)^2 + \left(\frac{dy}{dt}\right)^2 \neq 0 \qquad \text{for } a \le t \le b.$$

If $C$ is a smooth curve, the unit tangent vector $\mathbf{T}$ exists at every point on the curve and the direction of $\mathbf{T}$ varies continuously. This means the graph of a smooth curve cannot have any sharp corners for $a < t < b$, so the curve in Figure 15.3.5 cannot be the graph of a smooth curve. If the initial

FIGURE 15.3.5

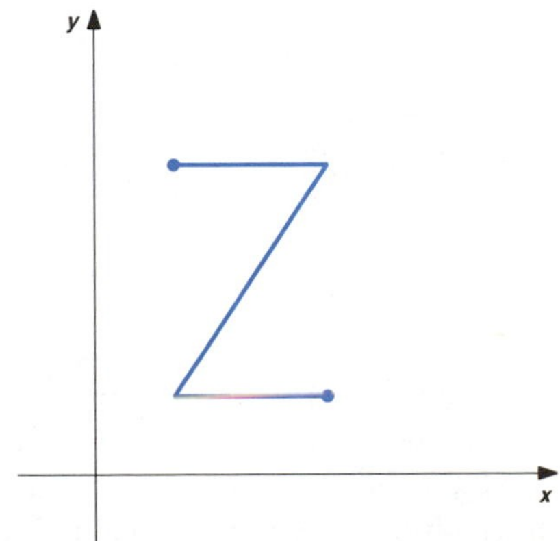

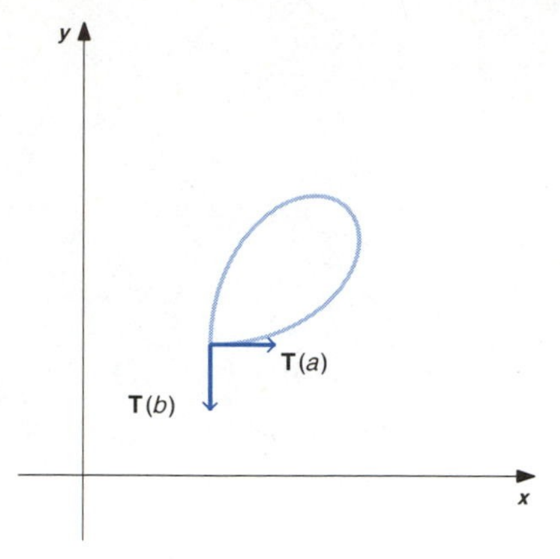


FIGURE 15.3.6

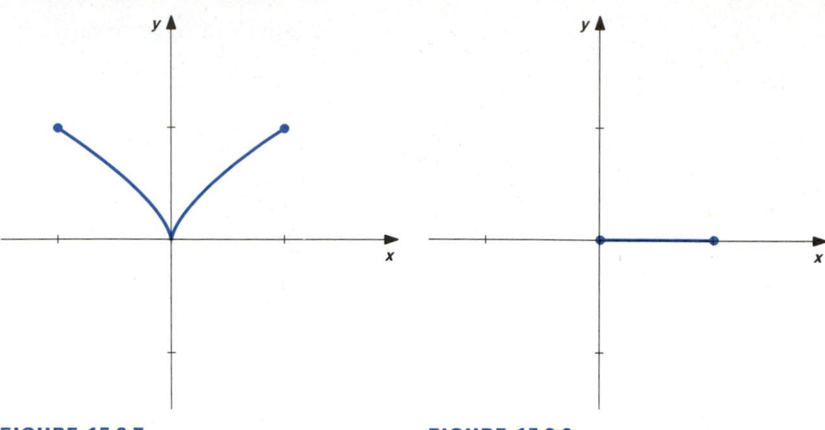


FIGURE 15.3.7

FIGURE 15.3.8

and terminal points coincide, the graph will have a sharp corner at that point unless $\mathbf{T}(a) = \mathbf{T}(b)$. See Figure 15.3.6.

The graph of a curve can have a sharp corner, even if the coordinate functions have continuous derivatives. For example, consider the curve

$$C\colon\ x = t^3, \quad y = t^2, \qquad -1 \le t \le 1.$$

The derivatives $dx/dt$ and $dy/dt$ are continuous, but the graph has a corner at the point (0, 0) where $t = 0$. Note that $(dx/dt)^2 + (dy/dt)^2 = 0$ at that point. See Figure 15.3.7.

The condition $(dx/dt)^2 + (dy/dt)^2 \neq 0$ insures that a smooth curve cannot abruptly reverse direction and double back on itself. For example, consider

$$C\colon\ x = \sin t, \quad y = 0, \qquad 0 \le t \le 3\pi/2.$$

The graph of the curve is the line segment between (0, 0) and (1, 0), which appears "smooth." See Figure 15.3.8. However, the curve goes from (0, 0) to (1, 0) as $t$ varies from 0 to $\pi/2$, doubles back on itself to (0, 0) as $t$ varies from $\pi/2$ to $\pi$, and then returns to (1, 0) as $t$ varies from $\pi$ to $3\pi/2$. We have

$$\left(\frac{dx}{dt}\right)^2 + \left(\frac{dy}{dt}\right)^2 = 0 \qquad \text{for } t = \pi/2 \text{ and } t = 3\pi/2,$$

so the curve is not smooth. The curve

$$C'\colon\ x = t, \quad y = 0, \qquad 0 \le t \le 1,$$

has the same graph as $C$ and is a smooth curve. $C'$ goes directly from the point (0, 0) to the point (1, 0).

If the terminal point of a curve $C_1$ coincides with the initial point of a curve $C_2$, we denote by $C_1 + C_2$ the curve that goes from the initial point of $C_1$, along $C_1$ to the terminal point of $C_1$, which is the initial point of $C_2$,

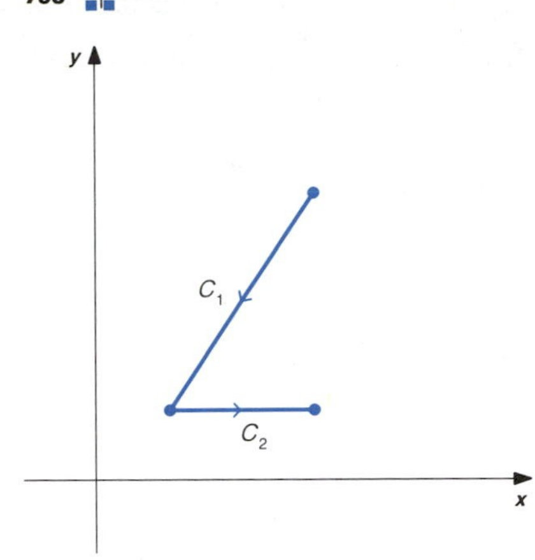


FIGURE 15.3.9

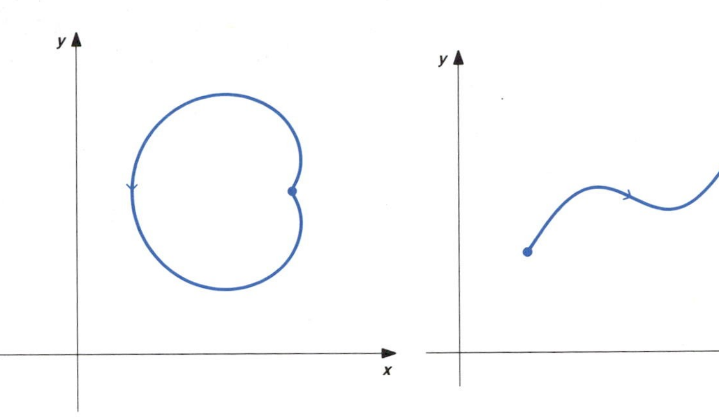


FIGURE 15.3.10

FIGURE 15.3.11

and then along $C_2$ to the terminal point of $C_2$. See Figure 15.3.9. A curve that is the sum of a finite number of smooth curves is called **piecewise smooth**.

Although a smooth curve cannot abruptly reverse direction and double back on itself, it can intersect itself. If the initial point of a curve coincides with its terminal point, we say the curve is **closed**. See Figure 15.3.10. If a curve does not intersect itself except possibly at its endpoints, the curve is said to be **simple**. See Figure 15.3.11. A simple curve may or may not be closed; a closed curve may or may not be simple.

■ **EXAMPLE 2**

Consider the curve $C$ that goes from the point (2, 0) counterclockwise around the circle $x^2 + y^2 = 4$ to the point $(-2, 0)$ and then follows the $x$-axis to (2, 0). Determine if $C$ is either simple or closed. Determine if $C$ can be expressed as either a smooth or piecewise smooth parametric curve, and find a parametric representation.

**SOLUTION**

The curve is sketched in Figure 15.3.12. $C$ is simple because it does not intersect itself except at its endpoints. $C$ is closed because its initial point coincides with its terminal point. $C$ cannot be expressed as a smooth parametric curve because of the corner at $(-2, 0)$. We have $C = C_1 + C_2$, where

$$C_1\colon\ x = 2\cos t,\quad y = 2\sin t,\qquad 0 \le t \le \pi,\quad \text{and}$$

$$C_2\colon\ x = t,\quad y = 0,\qquad -2 \le t \le 2.$$

$C_1$ and $C_2$ are smooth, so $C$ is piecewise smooth. ■

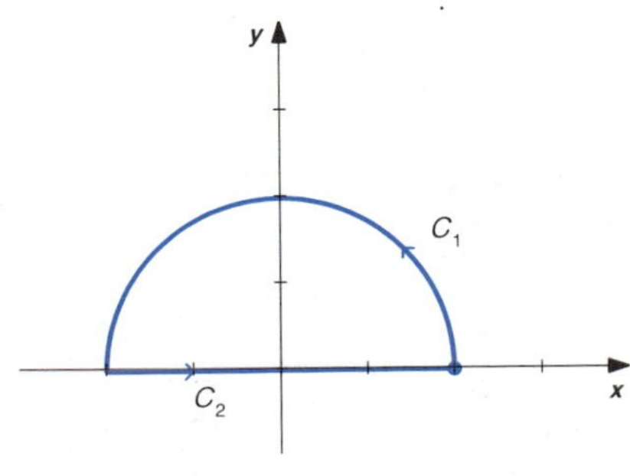


FIGURE 15.3.12

Given a smooth curve

$$C\colon\ x = x(t),\quad y = y(t),\qquad a \le t \le b,$$

we have seen in Section 15.2 that it is theoretically possible to express the curve in terms of arc length $s$, where

$$s = \int_a^t \sqrt{\left(\frac{dx}{dt}\right)^2 + \left(\frac{dy}{dt}\right)^2}\, d\tau \quad \text{and} \quad \frac{ds}{dt} = \sqrt{\left(\frac{dx}{dt}\right)^2 + \left(\frac{dy}{dt}\right)^2}.$$

The vector representation of $C$ is

$\mathbf{R} = x\mathbf{i} + y\mathbf{j}$, so

$\mathbf{R}' = x'\mathbf{i} + y'\mathbf{j}$, and

$$|\mathbf{R}'| = \sqrt{(x')^2 + (y')^2} = \frac{ds}{dt}.$$

It follows that

$$\mathbf{T} = \frac{\mathbf{R}'}{|\mathbf{R}'|} = \frac{x'}{|\mathbf{R}'|}\mathbf{i} + \frac{y'}{|\mathbf{R}'|}\mathbf{j} = \frac{\frac{dx}{dt}}{\frac{ds}{dt}}\mathbf{i} + \frac{\frac{dy}{dt}}{\frac{ds}{dt}}\mathbf{j} = \frac{dx}{ds}\mathbf{i} + \frac{dy}{ds}\mathbf{j}.$$

That is,

$$\mathbf{T} = \frac{d\mathbf{R}}{ds},$$

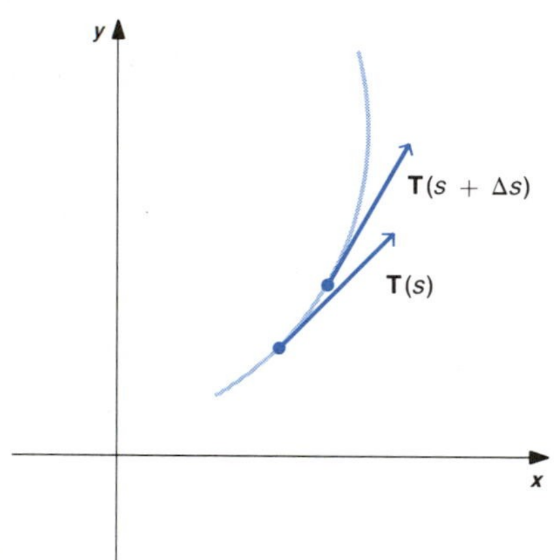


(a)

so the unit tangent vector is the rate of change of the position with respect to arc length. Let us see that $|d\mathbf{T}/ds|$ measures the rate of change of the direction of $\mathbf{T}$. In Figure 15.3.13a we have sketched unit tangent vectors at points on the curve that correspond to $s$ and $s + \Delta s$, $\Delta s \neq 0$. From Figure 15.3.13b we see that, since $\mathbf{T}(s + \Delta s)$ and $\mathbf{T}(s)$ are unit vectors, the length of the vector $\mathbf{T}(s + \Delta s) - \mathbf{T}(s)$ is approximately the radian measure of the angle between the vectors, for $\Delta s$ small. That is,

$$\left|\frac{\mathbf{T}(s + \Delta s) - \mathbf{T}(s)}{\Delta s}\right| \approx \left|\frac{\Delta\theta}{\Delta s}\right|.$$

It follows that $|d\mathbf{T}/ds|$ is the instantaneous rate of change of the direction of $\mathbf{T}$ with respect to arc length.

$$\kappa = \left|\frac{d\mathbf{T}}{ds}\right|$$

is called the **curvature** of the curve. We can calculate the curvature from given parametric equations of the curve by using the formula

$$\kappa = \left|\frac{d\mathbf{T}}{ds}\right| = \frac{\left|\frac{d\mathbf{T}}{dt}\right|}{\frac{ds}{dt}} = \frac{|\mathbf{T}'|}{|\mathbf{R}'|}.$$

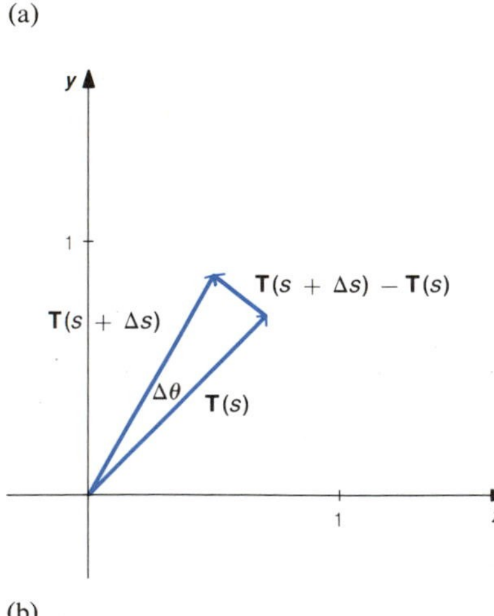


(b)

FIGURE 15.3.13

■ **EXAMPLE 3**

Find the curvature of the circle

$$(x - h)^2 + (y - k)^2 = r^2.$$

**SOLUTION**

We can express the circle in vector form as

$$\mathbf{R}(t) = (h + r \cos t)\mathbf{i} + (k + r \sin t)\mathbf{j}.$$

It follows that

$$\mathbf{R}'(t) = (-r \sin t)\mathbf{i} + (r \cos t)\mathbf{j},$$

$$|\mathbf{R}'(t)| = r,$$

$$\mathbf{T}(t) = (-\sin t)\mathbf{i} + (\cos t)\mathbf{i},$$

$$\mathbf{T}'(t) = (-\cos t)\mathbf{i} + (-\sin t)\mathbf{i}, \quad \text{and}$$

$$|\mathbf{T}'(t)| = 1.$$

The curvature is then

$$\kappa = \frac{|\mathbf{T}'|}{|\mathbf{R}'|} = \frac{1}{r}.$$ ■

We define the **radius of curvature** at a point on a curve where the curvature is nonzero to be

$$\rho = \frac{1}{\kappa}.$$

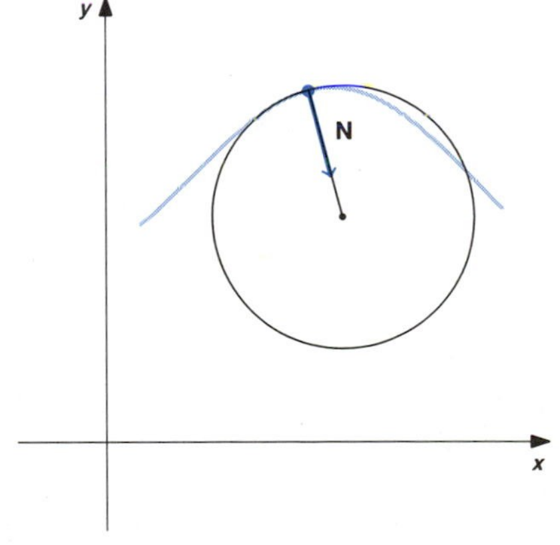

FIGURE 15.3.14

The **circle of curvature** is the circle that has radius $\rho$ and center at a distance of $\rho$ along the normal vector **N** from the point on the curve. The center of the circle of curvature is called the **center of curvature**. See Figure 15.3.14. It follows from Example 3 that the circle of curvature of a circle coincides with the circle. A curve with curvature $\kappa$ at a point is bending the same amount as a circle with radius $1/\kappa$ as the curve passes through the point.

■ **EXAMPLE 4**

Find the unit tangent vector, the unit normal vector, and the curvature of the curve $y = x^2$ at the origin. Find the radius of curvature and the center of curvature at the origin. Sketch the curve and the circle of curvature at the origin.

SOLUTION

We can represent the curve in vector form as

$$\mathbf{R} = t\mathbf{i} + t^2\mathbf{j}.$$

The origin corresponds to $t = 0$. We have

$$\mathbf{R}' = \mathbf{i} + 2t\mathbf{j},$$

$$|\mathbf{R}'| = \sqrt{(1)^2 + (2t)^2} = \sqrt{1 + 4t^2},$$

$$\mathbf{T} = \frac{\mathbf{i} + 2t\mathbf{j}}{\sqrt{1 + 4t^2}} = \frac{1}{\sqrt{1 + 4t^2}}\mathbf{i} + \frac{2t}{\sqrt{1 + 4t^2}}\mathbf{j},$$

$$\mathbf{T}' = -\frac{1}{2}(1 + 4t^2)^{-3/2}(8t)\mathbf{i} + \frac{(1 + 4t^2)^{1/2}(2) - (2t)(1/2)(1 + 4t^2)^{-1/2}(8t)}{1 + 4t^2}\mathbf{j}.$$

It is important that we do not substitute $t = 0$ into the expression for $\mathbf{T}$ before we differentiate to obtain $\mathbf{T}'$. Since we will not be taking any more derivatives, we can now substitute the value $t = 0$ and simplify. We obtain

$$\mathbf{T}' = 0\mathbf{i} + 2\mathbf{j}, \quad \text{so}$$

$$|\mathbf{T}'| = 2 \quad \text{and}$$

$$\mathbf{N} = \mathbf{j}.$$

When $t = 0$, we also have

$$T = \mathbf{i} \quad \text{and} \quad |\mathbf{R}'| = 1, \quad \text{so}$$

$$\kappa = \frac{|\mathbf{T}'|}{|\mathbf{R}'|} = \frac{2}{1} = 2.$$

The radius of curvature is

$$\rho = \frac{1}{\kappa} = \frac{1}{2}.$$

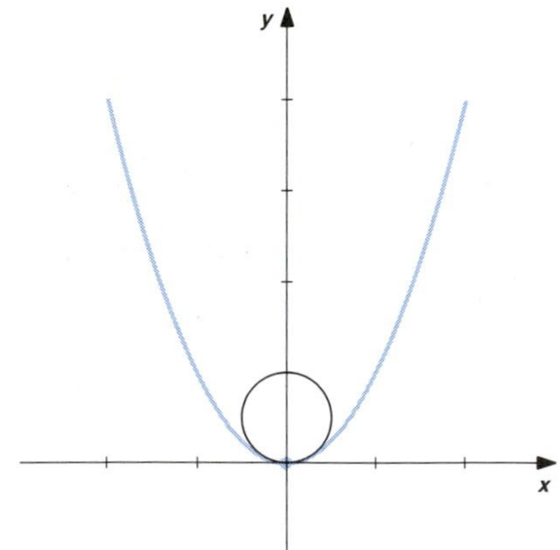

FIGURE 15.3.15

The center of curvature is located at a distance of $\rho = 1/2$ from the origin in the direction of the unit normal vector $\mathbf{N} = \mathbf{j}$. This is the point $(0, 1/2)$. The graph of $y = x^2$ and the circle of curvature are sketched in Figure 15.3.15. ■

■ **EXAMPLE 5**

Find the unit tangent vector, the unit normal vector, and the curvature of the curve $y = \sin x$.

**SOLUTION**

We have

$$\mathbf{R} = t\mathbf{i} + (\sin t)\mathbf{j},$$

$$\mathbf{R}' = \mathbf{i} + (\cos t)\mathbf{j},$$

$$|\mathbf{R}'| = \sqrt{1 + \cos^2 t},$$

$$\mathbf{T} = \frac{1}{(1 + \cos^2 t)^{1/2}}\,\mathbf{i} + \frac{\cos t}{(1 + \cos^2 t)^{1/2}}\,\mathbf{j},$$

$$\mathbf{T}' = -\frac{1}{2}(1 + \cos^2 t)^{-3/2}(2\cos t)(-\sin t)\,\mathbf{i}$$

$$+\frac{(1 + \cos^2 t)^{1/2}(-\sin t) - (\cos t)(1/2)(1 + \cos^2 t)^{-1/2}(2\cos t)(-\sin t)}{1 + \cos^2 t}\,\mathbf{j}$$

$$= \frac{\sin t \cos t}{(1 + \cos^2 t)^{3/2}}\,\mathbf{i} - \frac{\sin t}{(1 + \cos^2 t)^{3/2}}\,\mathbf{j},$$

$$|\mathbf{T}'| = \frac{|\sin t|}{1 + \cos^2 t}.$$

(The absolute value comes from taking the nonnegative square root of $\sin^2 t$.)

$$\mathbf{N} = \left(\frac{\sin t}{|\sin t|}\right)\left(\frac{\cos t}{\sqrt{1 + \cos^2 t}}\,\mathbf{i} - \frac{1}{\sqrt{1 + \cos^2 t}}\,\mathbf{j}\right),$$

$$\kappa = \frac{|\sin t|}{(1 + \cos^2 t)^{3/2}}.$$

Notice that the direction of the unit normal vector **N** corresponds to the concavity of the curve. The unit normal is undefined where $y = 0$. These points are points of inflection of the curve. See Figure 15.3.16. ■

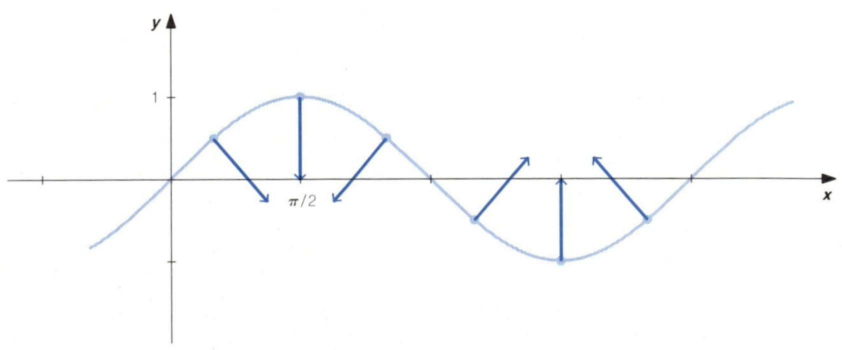


FIGURE 15.3.16

Let us develop a formula for the curvature in terms of the parametric equations of a curve,

$$C\colon\ x = x(t), \quad y = y(t), \qquad a \le t \le b.$$

The vector form of the curve is

$$\mathbf{R} = x\mathbf{i} + y\mathbf{j}.$$

The curvature is $\kappa = |\mathbf{T}'|/|\mathbf{R}'|$. We have

$$\mathbf{R}' = x'\mathbf{i} + y'\mathbf{j} \quad \text{and} \quad |\mathbf{R}'| = \sqrt{x'^2 + y'^2},$$

$$T' = \frac{d}{dt}(\mathbf{T}) = \frac{d}{dt}\left(\frac{\mathbf{R}'}{|\mathbf{R}'|}\right) = \frac{d}{dt}\left(\frac{x'}{\sqrt{x'^2 + y'^2}}\mathbf{i} + \frac{y'}{\sqrt{x'^2 + y'^2}}\mathbf{j}\right).$$

Carrying out the differentiation and then simplifying, we obtain

$$\mathbf{T}' = \frac{y'(y'x'' - x'y'')}{(x'^2 + y'^2)^{3/2}}\mathbf{i} + \frac{x'(x'y'' - y'x'')}{(x'^2 + y'^2)^{3/2}}\mathbf{j}.$$

Calculating the length of this vector, we obtain

$$|\mathbf{T}'| = \frac{|x'y'' - y'x''|}{x'^2 + y'^2}.$$

Dividing by $|\mathbf{R}'|$, we obtain the formula

$$\kappa = \frac{|x'y'' - y'x''|}{(x'^2 + y'^2)^{3/2}}.$$

This formula is convenient for finding the curvature of a curve when we do not need to calculate the unit normal vector.

### ■ EXAMPLE 6

Find a formula for the curvature of the ellipse

$$C\colon\ x = 3\cos t, \quad y = 2\sin t, \qquad 0 \le t \le 2\pi.$$

Find maximum and minimum values of the curvature and find points on the ellipse where these values occur.

**SOLUTION**

We have $x' = -3\sin t$, $x'' = -3\cos t$, $y' = 2\cos t$, and $y'' = -2\sin t$. From the formula we then obtain

$$\begin{aligned}\kappa &= \frac{|(-3\sin t)(-2\sin t) - (2\cos t)(-3\cos t)|}{((-3\sin t)^2 + (2\cos t)^2)^{3/2}} \\ &= \frac{6(\sin^2 t + \cos^2 t)}{(9\sin^2 t + 4\cos^2 t)^{3/2}} \\ &= 6(9\sin^2 t + 4\cos^2 t)^{-3/2}.\end{aligned}$$

Let us use the derivative of $\kappa$ to find the maximum and minimum values. We have

$$\frac{d\kappa}{dt} = 6\left(-\frac{3}{2}\right)(9\sin^2 t + 4\cos^2 t)^{-5/2}[(18)(\sin t)(\cos t) + (8)(\cos t)(-\sin t)]$$

$$= \frac{-90(\sin t)(\cos t)}{(9\sin^2 t + 4\cos^2 t)^{5/2}}.$$

The derivative is zero whenever $(\sin t)(\cos t) = 0$. That is, whenever $t = 0$, $\pi/2$, $\pi$, $3\pi/2$, and $2\pi$. The maximum curvature is $\kappa(0) = \kappa(\pi) = \kappa(2\pi) = 6(4)^{-3/2} = 3/4$, at the vertices of the ellipse. The minimum curvature is $\kappa(\pi/2) = \kappa(3\pi/2) = 6(9)^{-3/2} = 2/9$, at the endpoints of the minor axis of the ellipse.

In this example, we could have determined the maximum and minimum value of the curvature without using a derivative. For example, we know that $\sin^2 t + \cos^2 t = 1$, so

$$\kappa = 6(9\sin^2 t + 4\cos^2 t)^{-3/2}$$
$$= 6(9\sin^2 t + 4(1 - \sin^2 t))^{-3/2}$$
$$= 6(5\sin^2 t + 4)^{-3/2}.$$

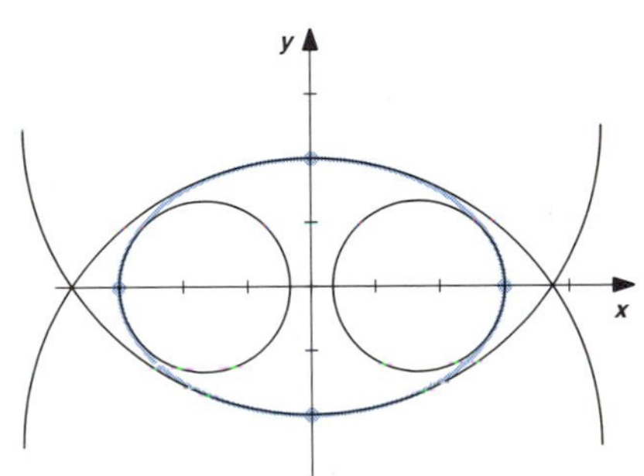

FIGURE 15.3.17

Since $4 \le 5\sin^2 t + 4 \le 9$, we see that

$$6(4)^{-3/2} \le \kappa \le 6(9)^{-3/2}.$$

The ellipse and the circles of curvature at these points are sketched in Figure 15.3.17. ■

If a curve is the graph of a function $y = y(x)$, we can parameterize the curve by writing

$$C\colon\ x = x, \quad y = y(x).$$

In the above equations we are using $x$ as the parameter. We have $x' = 1$, $x'' = 0$, so the formula for curvature reduces to

$$\kappa = \frac{|y''|}{(1 + y'^2)^{3/2}}, \quad \text{if} \quad y = y(x).$$

■ **EXAMPLE 7**

Find the maximum curvature and the point on the curve $y = x^3/9$ where the maximum occurs.

**SOLUTION**

We have $y = x^3/9$, $y' = x^2/3$, and $y'' = 2x/3$. The formula then gives us

$$\kappa = \frac{|2x/3|}{(1 + x^4/9)^{3/2}} = \frac{18|x|}{(9 + x^4)^{3/2}}.$$

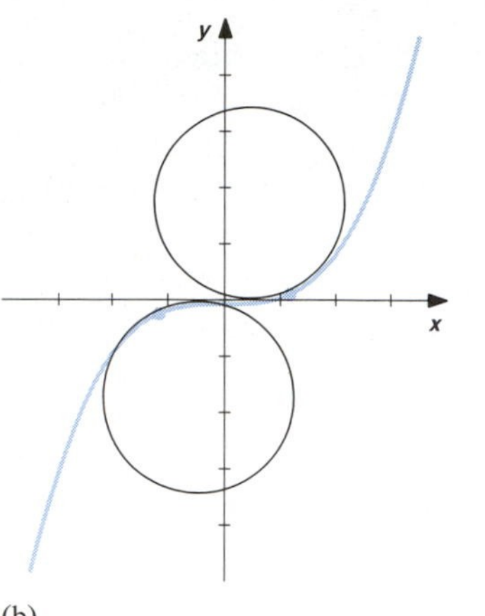


FIGURE 15.3.18

We see that the curvature is zero at the origin, is symmetric about the $y$-axis, and approaches zero as $x \to \infty$. See Figure 15.3.18a. For $x > 0$, we have

$$\kappa = \frac{18x}{(9 + x^4)^{3/2}} \quad \text{and}$$

$$\frac{d\kappa}{dx} = \frac{(9 + x^4)^{3/2}(18) - (18x)(3/2)(9 + x^4)^{1/2}(4x^3)}{(9 + x^4)^3}$$

$$= \frac{18(9 + x^4) - 18(6)x^4}{(9 + x^4)^{5/2}}$$

$$= \frac{18(9 - 5x^4)}{(9 + x^4)^{5/2}}.$$

Setting the derivative equal to zero, we obtain

$$9 - 5x^4 = 0,$$

$$x = (1.8)^{1/4}.$$

When $x = (1.8)^{1/4}$, $y = (1.8)^{3/4}/9$. The maximum curvature of

$$\kappa = \frac{18(1.8)^{1/4}}{(9 + 1.8)^{3/2}} \approx 0.58742848$$

occurs at the points $((1.8)^{1/4}, (1.8)^{3/4}/9)$ and $(-(1.8)^{1/4}, -(1.8)^{3/4}/9)$. See Figure 15.3.18b. ■

## EXERCISES 15.3

Determine which of the curves in Exercises 1–8 are simple and which are closed. Determine which can be expressed as either smooth or piecewise smooth parametric curves, and find parametric representations.

1 From $(-1, 1)$, along $y = x^{2/3}$ to $(1, 1)$

2 From $(-1, -1)$, along $y = x^{1/3}$ to $(1, 1)$

3 From $(-1, -1)$, along $y = x^3$ to $(1, 1)$, along $y = x$ to $(-1, -1)$

4 From $(0, 0)$, along $y = x^2$ to $(1, 1)$, along $y = x$ to $(0, 0)$

5 From $(1, 0)$, counterclockwise twice around $x^2 + y^2 = 1$

6 From $(1, 0)$, counterclockwise one-and-one-half times around $x^2 + y^2 = 1$

7 From $(1, 0)$, counterclockwise once around $x^2 + y^2 = 1$

8 From $(-1, 0)$, along the $x$-axis to $(1, 0)$, counterclockwise once around $x^2 + y^2 = 1$

9 A curve goes once around the polar curve $r = 1 - \cos\theta$. For which initial points is the curve smooth?

10 A smooth curve that is given by a function $f(x)$ can have a unit tangent vector that is vertical. Can $f$ be differentiable at such a point?

Find the unit tangent vector, the unit normal vector, and the curvature of the curves given in Exercises 11–16.

11 $C$: $x = 2\cos t$, $y = 2\sin t$

12 $C$: $x = 1 - 2t$, $y = 2 + 2t$

13 $C$: $x = t$, $y = \ln t$

14 $C$: $x = t^2/2$, $y = t^3/3$, $t > 0$

15 $C: x = e^t\cos t,\ y = e^t\sin t$

16 $C: x = 2t,\ y = e^t + e^{-t}$

For the curves given in Exercises 17–20, find the unit tangent vector, the unit normal vector, the curvature, the radius of curvature, and the center of curvature at the indicated point. Sketch the graph of the curve and sketch the circle of curvature at the indicated point.

17 $C: x = t,\ y = \cos t;\ t = 0$

18 $C: x = e^t,\ y = t;\ t = 0$

19 $C: x = 4\cos t,\ y = 3\sin t;\ t = 0,\ \pi/2$

20 $C: x = t\cos t,\ y = t\sin t;\ t = 0$

Find a formula for the curvature of the curves given in Exercises 21–28.

21 $y = 1/x,\ x > 0$

22 $y = 1/x^2$

23 $y = \ln(\sec x),\ -\pi/2 < x < \pi/2$

24 $y = \ln(\sec x + \tan x),\ 0 \leq x < \pi/2$

25 $C: x = 2\cos t,\ y = \sin t$

26 $C: x = e^t + e^{-t},\ y = e^t - e^{-t}$

27 $C: x = \sec t,\ y = \tan t$

28 $C: x = t - r\cos(t/r),\ y = r - r\sin(t/r)$

29 Find the maximum curvature of the curve $y = \ln x$, and find the point on the curve where the curvature is maximum.

30 Find the maximum curvature of the curve $y = e^x$, and find the point on the curve where the curvature is maximum.

31 Find the maximum curvature of the curve $C: x = 2t$, $y = e^t + e^{-t}$ and find the point on the curve where the curvature is maximum.

32 Show that the curvature of a polar curve $r = r(\theta)$ is given by the formula

$$\kappa = \frac{|r^2 + 2(r')^2 - rr''|}{(r^2 + (r')^2)^{3/2}},$$

where the derivatives are with respect to $\theta$.

Use the formula of Exercise 32 to find a formula for the curvature of the polar curves given in Exercises 33–35.

33 $r = 2 - \cos\theta$

34 $r = 1 + \sin\theta$

35 $r = 1 - \sin\theta$

36 Use the formula of Exercise 32 to show that the curvature of a polar curve $r = r(\theta)$ at the pole is given by

$$\kappa = \frac{2}{|r'|}.$$

Use the formula of Exercise 36 to find the curvature at the pole of the polar curves given in Exercises 37–39.

37 $r = 2\cos 2\theta$

38 $r = 2\sin 3\theta$

39 $r = \sin 4\theta$

40 The curve from $(-1, -1)$, along $y = -\sqrt{1 - (x + 1)^2}$ to $(0, 0)$, and then along $y = \sqrt{4 - (x - 2)^2}$ to $(2, 2)$ is smooth, because the unit tangent vector varies continuously. Use the fact that the curve consists of two circular arcs to discuss the curvature. (Abrupt changes in curvature along the cross section of an airfoil could seriously affect its aerodynamic properties.)

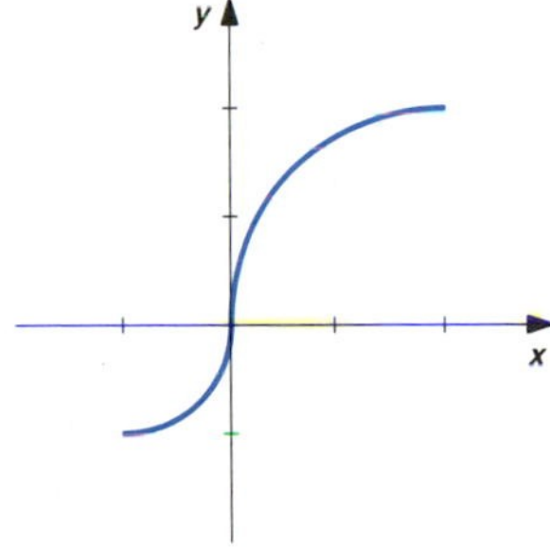


## 15.4 CURVES IN THREE-DIMENSIONAL SPACE, COMPONENTS OF ACCELERATION

The concepts we developed for curves in the plane also apply to space curves.

The parametric form of a curve in three-dimensional space is

$$C: x = x(t), \quad y = y(t), \quad z = z(t),$$

where $x$, $y$, and $z$ are continuous functions of $t$.

We can sketch parametric curves in three-dimensional space by connecting an ordered set of points on the curve. To illustrate the three-dimensional character of the curve, it is useful to indicate the relation between the curve and its projection onto one of the coordinate planes. For example, if we combine the equations for $x$ and $y$ in such a way that eliminates $t$, we will obtain the equation of the vertical generalized cylinder that contains $C$. The intersection of this cylinder with the $xy$-plane is the projection of $C$ onto the $xy$-plane. Drawing the projection and some of the vertical lines between the projection and the curve helps orient the curve in the sketch.

### EXAMPLE 1

Sketch the graph of the curve

$$C: \ x = 3 \cos t, \quad y = 3 \sin t, \quad z = t, \qquad 0 \leq t \leq 2\pi.$$

Find the equation of the vertical generalized cylinder that contains $C$; sketch the projection of $C$ onto the $xy$-plane and draw some of the vertical lines between the curve and its projection.

**SOLUTION**

Values of the coordinates of points on the curve that correspond to increments of $\Delta t = \pi/2$ are given in Table 15.4.1. From the equations $x = 3 \cos t$ and $y = 3 \sin t$, we see that

$$x^2 + y^2 = 3^2.$$

TABLE 15.4.1

| $t$ | $x$ | $y$ | $z$ |
|---|---|---|---|
| 0 | 3 | 0 | 0 |
| $\frac{\pi}{2}$ | 0 | 3 | $\frac{\pi}{2}$ |
| $\pi$ | $-3$ | 0 | $\pi$ |
| $\frac{3\pi}{2}$ | 0 | $-3$ | $\frac{3\pi}{2}$ |
| $2\pi$ | 3 | 0 | $2\pi$ |

This is the vertical generalized cylinder that contains the curve; it is a right circular cylinder with radius 3 and axis along the $z$-axis. The projection onto the $xy$-plane is the circle with center at the origin and radius 3. The $z$-coordinate increases as the curve winds counterclockwise around the cylinder. The curve is called a **helix**. The graph is sketched in Figure 15.4.1. ■

FIGURE 15.4.1

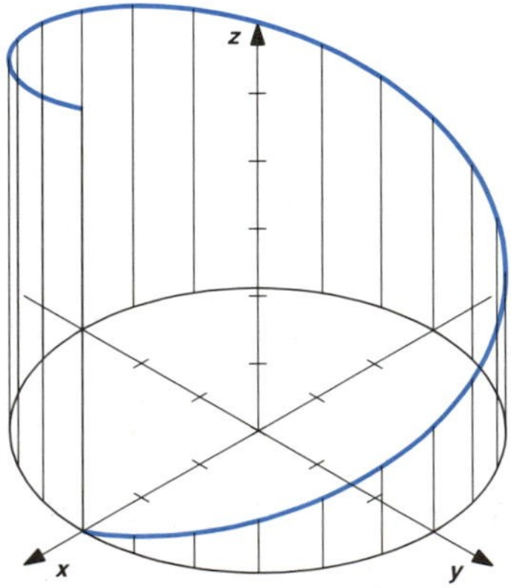

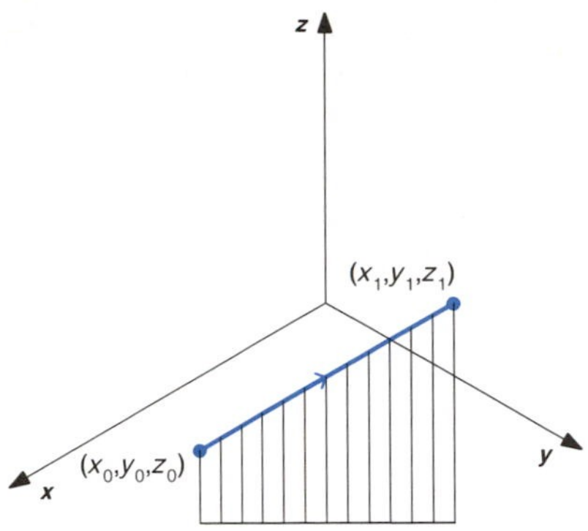

FIGURE 15.4.2

In Section 14.4 we observed that parametric equations of the line in three-dimensional space that contains the point $(x_0, y_0, z_0)$ and has direction vector $\mathbf{V} = a\mathbf{i} + b\mathbf{j} + c\mathbf{k}$ are

$$x = x_0 + at, \quad y = y_0 + bt, \quad z = z_0 + ct.$$

We also know that the line segment from $(x_0, y_0, z_0)$ to $(x_1, y_1\ z_1)$ has direction $\mathbf{V} = (x_1 - x_0)\mathbf{i} + (y_1 - y_0)\mathbf{j} + (z_1 - z_0)\mathbf{k}$. See Figure 15.4.2. It follows that

**The line segment from $(x_0, y_0, z_0)$ to $(x_1, y_1, z_1)$ is given by**

$$C\colon\ x = x_0 + (x_1 - x_0)t, \quad y = y_0 + (y_1 - y_0)t,$$
$$z = z_0 + (z_1 - z_0)t, \qquad 0 \le t \le 1.$$

This formula corresponds to the formula for the parametric equations of the line segment between two points in the plane.

We can find parametric equations for the curve of intersection of a quadric surface

$$Ax^2 + By^2 + Cz^2 + Dx + Ey + Fz + G = 0$$

and a plane $ax + by + cz + d = 0$ by using the equation of the plane to eliminate one variable from the quadric equation, finding parametric equations for the resulting second-order equation in two variables, and then using the linear equation to obtain the other variable in terms of the parameter.

■ **EXAMPLE 2**

Find parametric equations of the curve of intersection of the ellipsoid $4x^2 + y^2 + 4z^2 = 4$ and the plane $x + y = 1$. Sketch the curve.

**SOLUTION**

Let us solve the linear equation for $y$ and then substitute in the quadric equation to eliminate $y$.

$$x + y = 1 \quad \text{implies} \quad y = 1 - x.$$

Substitution then gives

$$4x^2 + y^2 + 4z^2 = 4,$$

$$4x^2 + (1 - x)^2 + 4z^2 = 4,$$

$$4x^2 + 1 - 2x + x^2 + 4z^2 = 4,$$

$$5x^2 - 2x + 4z^2 = 3.$$

This is the equation of an ellipse in the variables $x$ and $z$. We can find parametric equations by completing the square and expressing the equation in the standard form of an ellipse. We have

$$5\left(x^2 - \frac{2}{5}x + \frac{1}{25}\right) + 4z^2 = 3 + 5\left(\frac{1}{25}\right),$$

$$5\left(x - \frac{1}{5}\right)^2 + 4z^2 = \frac{16}{5},$$

$$\frac{\left(x - \frac{1}{5}\right)^2}{\left(\frac{4}{5}\right)^2} + \frac{z^2}{\left(\frac{2}{\sqrt{5}}\right)^2} = 1.$$

From formula (5) of Section 15.1, we then obtain the following parametric equations:

$$x = \frac{1}{5} + \frac{4}{5}\cos t, \quad z = \frac{2}{\sqrt{5}}\sin t, \qquad 0 \le t \le 2\pi.$$

Since $y = 1 - x$, we also have

$$y = 1 - \left(\frac{1}{5} + \frac{4}{5}\cos t\right) = \frac{4}{5} - \frac{4}{5}\cos t.$$

Combining results, we obtain the following parametric representation:

$$C\colon\ x = \frac{1}{5} + \frac{4}{5}\cos t, \quad y = \frac{4}{5} - \frac{4}{5}\cos t, \quad z = \frac{2}{\sqrt{5}}\sin t, \qquad 0 \le t \le 2\pi.$$

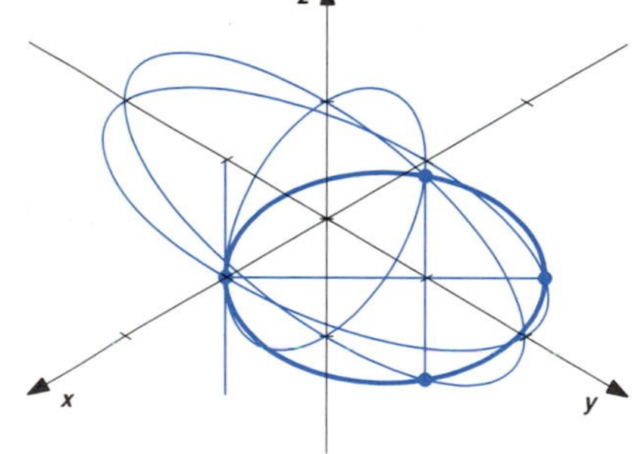

FIGURE 15.4.3

To sketch the curve, we first sketch the traces of the ellipsoid and the plane in each coordinate plane, determine the points of intersection of the traces in each coordinate plane, and then draw the curve of intersection through these points. This is done in Figure 15.4.3. Note that you are looking for points of intersection of traces that are in the same coordinate plane. You must be very careful to recognize which traces are in which coordinate plane. It is also a great help to have some idea of the general character of the curve of intersection of an ellipsoid and a plane. ■

If the coordinate functions have continuous derivatives, we define the **differential arc length** of the curve

$$C\colon\ x = x(t), \quad y = y(t), \quad z = z(t), \qquad a \le t \le b,$$

to be

$$ds = \sqrt{(dx)^2 + (dy)^2 + (dz)^2} = \sqrt{\left(\frac{dx}{dt}\right)^2 + \left(\frac{dy}{dt}\right)^2 + \left(\frac{dz}{dt}\right)^2}\, dt.$$

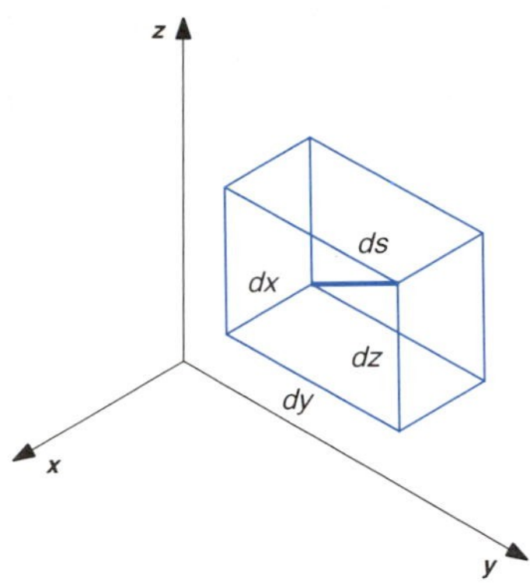


FIGURE 15.4.4

See Figure 15.4.4. The **length** of the curve is

$$L = \int ds = \int_a^b \sqrt{\left(\frac{dx}{dt}\right)^2 + \left(\frac{dy}{dt}\right)^2 + \left(\frac{dz}{dt}\right)^2}\, dt.$$

The **arc length function** is

$$s = \int_a^t \sqrt{\left(\frac{dx}{dt}\right)^2 + \left(\frac{dy}{dt}\right)^2 + \left(\frac{dz}{dt}\right)^2}\, d\tau.$$

If the coordinate functions are continuous with

$$\left(\frac{dx}{dt}\right)^2 + \left(\frac{dy}{dt}\right)^2 + \left(\frac{dz}{dt}\right)^2 \neq 0 \qquad \text{for } a \leq t \leq b,$$

so the curve is smooth, we can use the inverse of the arc length function to parametrize the curve in terms of arc length.

### ■ EXAMPLE 3

Find the length of the helix

$$C: \quad x = 3\cos t, \quad y = 3\sin t, \quad z = t, \qquad 0 \leq t \leq 2\pi.$$

Find the parametrization of the curve in terms of arc length.

**SOLUTION**

We have

$$\frac{dx}{dt} = -3\sin t, \quad \frac{dy}{dt} = 3\cos t, \quad \text{and} \quad \frac{dz}{dt} = 1,$$

so

$$\left(\frac{dx}{dt}\right)^2 + \left(\frac{dy}{dt}\right)^2 + \left(\frac{dz}{dt}\right)^2 = (-3\sin t)^2 + (3\cos t)^2 + (1)^2$$
$$= 9(\sin^2 t + \cos^2 t) + 1 = 9 + 1 = 10.$$

It follows that the total length of the curve is

$$L = \int_0^{2\pi} \sqrt{10}\, d\tau = 2\pi\sqrt{10}$$

and the arc length function is

$$s = \int_0^t \sqrt{10}\, dt = t\sqrt{10}, \qquad 0 \leq t \leq 2\pi.$$

Solving the equation $s = t\sqrt{10}$ for $t$, we obtain

$$t = \frac{s}{\sqrt{10}}.$$

Substitution then gives the parametric representation of $C$ in terms of arc length,

$$C\colon\ x = 3\cos\frac{s}{\sqrt{10}},\quad y = 3\sin\frac{s}{\sqrt{10}},\quad z = \frac{s}{\sqrt{10}},\qquad 0 \le s \le 2\pi\sqrt{10}.\qquad \blacksquare$$

The space curve

$$C\colon\ x = x(t),\quad y = y(t),\quad z = z(t),$$

can be expressed in **vector form** as

$$\mathbf{R} = x(t)\mathbf{i} + y(t)\mathbf{j} + z(t)\mathbf{k}.$$

We then have

$$|\mathbf{R}'| = \sqrt{\left(\frac{dx}{dt}\right)^2 + \left(\frac{dy}{dt}\right)^2 + \left(\frac{dz}{dt}\right)^2} = \frac{ds}{dt}.$$

If $C$ is a smooth curve, we can define the unit tangent vector, the unit normal vector, and the curvature in terms of the arc length parametrization of $C$. The definitions correspond to those of a curve in the plane. As before, we can carry out the calculations in terms of any given parameterization.

The unit tangent vector is $\mathbf{T} = \dfrac{d\mathbf{R}}{ds} = \dfrac{\mathbf{R}'}{|\mathbf{R}'|}$.

The principal unit normal vector is $\mathbf{N} = \dfrac{\dfrac{d\mathbf{T}}{ds}}{\left|\dfrac{d\mathbf{T}}{ds}\right|} = \dfrac{\mathbf{T}'}{|\mathbf{T}'|}$.

(Note that the principal unit normal vector is not defined if $\mathbf{T}' = \mathbf{0}$.)

The curvature is $\kappa = \left|\dfrac{d\mathbf{T}}{ds}\right| = \dfrac{\left|\dfrac{d\mathbf{T}}{dt}\right|}{\dfrac{ds}{dt}} = \dfrac{|\mathbf{T}'|}{|\mathbf{R}'|}$.

For curves in three-dimensional space we define the **unit binormal** to be

$$\mathbf{B} = \mathbf{T} \times \mathbf{N}.$$

The vectors **T**, **N**, and **B** (in that order) form a right-handed rectangular coordinate system at the corresponding point on the curve. See Figure 15.4.5.

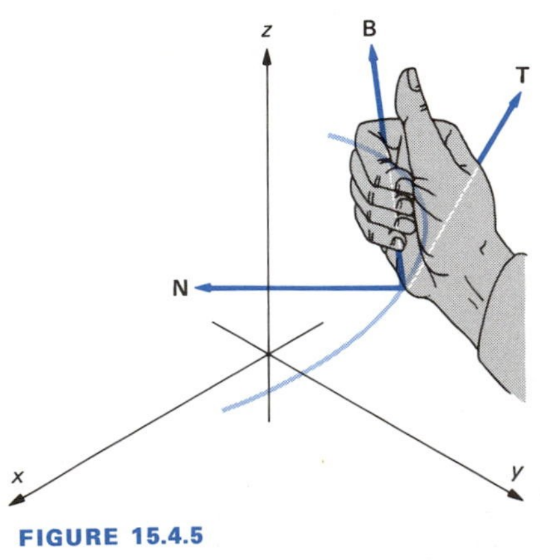

FIGURE 15.4.5

**■ EXAMPLE 4**

Find the unit tangent vector, the principal unit normal vector, the unit binormal, and the curvature of the helix

$$C\colon\ x = 3\cos t,\quad y = 3\sin t,\quad z = t,\qquad 0 \le t \le 2\pi.$$

**SOLUTION**

Using the vector form of the curve, we have

$$\mathbf{R} = (3\cos t)\mathbf{i} + (3\sin t)\mathbf{j} + t\mathbf{k},\quad \text{so}$$

$$\mathbf{R}' = (-3\sin t)\mathbf{i} + (3\cos t)\mathbf{j} + \mathbf{k}\quad \text{and}$$

$$|\mathbf{R}'| = \sqrt{(-3\sin t)^2 + (3\cos t)^2 + (1)^2} = \sqrt{10}.$$

The unit tangent vector is

$$\mathbf{T} = \frac{\mathbf{R}'}{|\mathbf{R}'|} = \left(\frac{-3\sin t}{\sqrt{10}}\right)\mathbf{i} + \left(\frac{3\cos t}{\sqrt{10}}\right)\mathbf{j} + \left(\frac{1}{\sqrt{10}}\right)\mathbf{k}.$$

We have

$$\mathbf{T}' = \left(\frac{-3\cos t}{\sqrt{10}}\right)\mathbf{i} + \left(\frac{-3\sin t}{\sqrt{10}}\right)\mathbf{j},\quad \text{so}$$

$$|\mathbf{T}'| = \frac{3}{\sqrt{10}}.$$

The principal unit normal vector is

$$\mathbf{N} = \frac{\mathbf{T}'}{|\mathbf{T}'|} = (-\cos t)\mathbf{i} + (-\sin t)\mathbf{j}.$$

The binormal is

$$\mathbf{B} = \mathbf{T} \times \mathbf{N} = \begin{vmatrix} \mathbf{i} & \mathbf{j} & \mathbf{k} \\ \dfrac{-3\sin t}{\sqrt{10}} & \dfrac{3\cos t}{\sqrt{10}} & \dfrac{1}{\sqrt{10}} \\ -\cos t & -\sin t & 0 \end{vmatrix}$$

$$= \begin{vmatrix} \dfrac{3\cos t}{\sqrt{10}} & \dfrac{1}{\sqrt{10}} \\ -\sin t & 0 \end{vmatrix}\mathbf{i} - \begin{vmatrix} \dfrac{-3\sin t}{\sqrt{10}} & \dfrac{1}{\sqrt{10}} \\ -\cos t & 0 \end{vmatrix}\mathbf{j} + \begin{vmatrix} \dfrac{-3\sin t}{\sqrt{10}} & \dfrac{3\cos t}{\sqrt{10}} \\ -\cos t & -\sin t \end{vmatrix}\mathbf{k}$$

$$= \frac{\sin t}{\sqrt{10}}\mathbf{i} - \frac{\cos t}{\sqrt{10}}\mathbf{j} + \frac{3}{\sqrt{10}}\mathbf{k}.$$

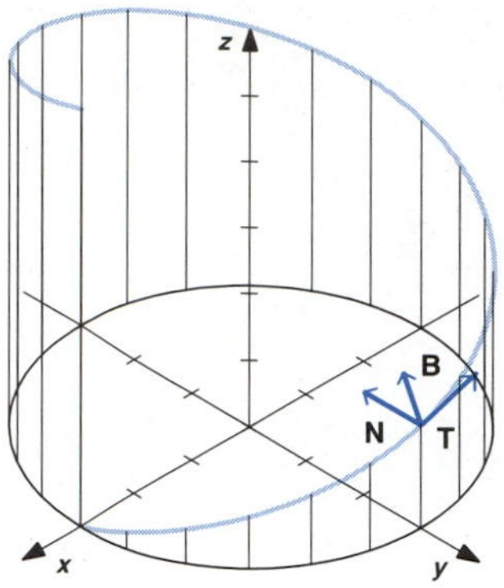

FIGURE 15.4.6

The curvature is

$$\kappa = \frac{|\mathbf{T}'|}{|\mathbf{R}'|} = \frac{\dfrac{3}{\sqrt{10}}}{\sqrt{10}} = 0.3.$$

The curve and representative vectors **T**, **N**, and **B** are sketched in Figure 15.4.6. ■

Let us now see how to use the properties of a curve to study the motion of an object along the curve. In particular, let

$$\mathbf{R}(t) = x(t)\mathbf{i} + y(t)\mathbf{j} + z(t)\mathbf{k}$$

represent the position vector of an object at time $t$. We can also think of **R** as the vector form of the curve traced by the object. We want to relate the velocity and acceleration of the object to the unit tangent and normal vectors of the curve.

The velocity of the object is $\mathbf{V} = \mathbf{R}'$ and the speed is

$$|\mathbf{V}| = |\mathbf{R}'| = \sqrt{\left(\frac{dx}{dt}\right)^2 + \left(\frac{dy}{dt}\right)^2 + \left(\frac{dz}{dt}\right)^2} = \frac{ds}{dt}.$$

We have

$$\mathbf{V} = \mathbf{R}' = |\mathbf{R}'|\mathbf{T} = \frac{ds}{dt}\mathbf{T},$$

so the velocity is in the direction of the unit tangent vector. Also, the acceleration is

$$\mathbf{A} = \mathbf{V}' = \frac{d}{dt}(\mathbf{V}) = \frac{d}{dt}\left(\frac{ds}{dt}\mathbf{T}\right) = \frac{d^2s}{dt^2}\mathbf{T} + \frac{ds}{dt}\mathbf{T}' = \frac{d^2s}{dt^2}\mathbf{T} + \kappa\left(\frac{ds}{dt}\right)^2\mathbf{N},$$

where we have used

$$\mathbf{T}' = |\mathbf{T}'|\mathbf{N} \quad \text{and} \quad |\mathbf{T}'| = \kappa\frac{ds}{dt}, \quad \text{so} \quad \mathbf{T}' = \kappa\frac{ds}{dt}\mathbf{N}.$$

This shows that the acceleration of the object is in the plane determined by the vectors **T** and **N**.

$a_{\mathbf{T}} = \dfrac{d^2s}{dt^2}$ is called the **tangential component of acceleration.**

$a_{\mathbf{N}} = \kappa\left(\dfrac{ds}{dt}\right)^2$ is called the **normal component of acceleration.**

The normal component of acceleration is nonnegative, so *the principal normal vector points in the direction the curve is bending.* Also, the magnitude of the force associated with the normal component of acceleration varies jointly as the curvature and the square of the speed. This suggests that a car should slow down in order to navigate a sharp turn, where the curvature is large and the turning radius is small.

The tangential and normal components of acceleration can be calculated from $\mathbf{R}'$ and $\mathbf{R}''$, without calculating $\mathbf{T}$ and $\mathbf{N}$. Since $\mathbf{A} = \mathbf{R}''$, we have

$$\mathbf{R}'' = \frac{d^2s}{dt^2}\mathbf{T} + \kappa\left(\frac{ds}{dt}\right)^2\mathbf{N}.$$

Taking the dot product of each side of this equation with $\mathbf{T}$, we obtain

$$\mathbf{T}\cdot\mathbf{R}'' = \mathbf{T}\cdot\left(\frac{d^2s}{dt^2}\mathbf{T} + \kappa\left(\frac{ds}{dt}\right)^2\mathbf{N}\right) = \frac{d^2s}{dt^2}\mathbf{T}\cdot\mathbf{T} + \kappa\left(\frac{ds}{dt}\right)^2\mathbf{T}\cdot\mathbf{N} = \frac{d^2s}{dt^2}.$$

We also have

$$\mathbf{T}\cdot\mathbf{R}'' = \left(\frac{\mathbf{R}'}{|\mathbf{R}'|}\right)\cdot\mathbf{R}'' = \frac{\mathbf{R}'\cdot\mathbf{R}''}{|\mathbf{R}'|}.$$

We conclude that

$$a_{\mathbf{T}} = \frac{d^2s}{dt^2} = \frac{\mathbf{R}'\cdot\mathbf{R}''}{|\mathbf{R}'|}.$$

Similarly, we have

$$\mathbf{T}\times\mathbf{R}'' = \mathbf{T}\times\left(\frac{d^2s}{dt^2}\mathbf{T} + \kappa\left(\frac{ds}{dt}\right)^2\mathbf{N}\right) = \frac{d^2s}{d^2t}\mathbf{T}\times\mathbf{T} + \kappa\left(\frac{ds}{dt}\right)^2\mathbf{T}\times\mathbf{N}$$

$$= \kappa\left(\frac{ds}{dt}\right)^2\mathbf{B}.$$

Since

$$\mathbf{T}\times\mathbf{R}'' = \left(\frac{\mathbf{R}'}{|\mathbf{R}'|}\right)\times\mathbf{R}'' = \frac{\mathbf{R}'\times\mathbf{R}''}{|\mathbf{R}'|} \quad \text{and} \quad |\mathbf{B}| = 1,$$

we conclude that

$$a_{\mathbf{N}} = \kappa\left(\frac{ds}{dt}\right)^2 = \frac{|\mathbf{R}'\times\mathbf{R}''|}{|\mathbf{R}'|}.$$

Note that the above formula for $a_{\mathbf{N}}$ involves the *length* of the vectors $\mathbf{R}'\times\mathbf{R}''$ and $\mathbf{R}'$. The components of acceleration are *numbers*, not vectors.

■ **EXAMPLE 5**

The position of an object at time $t$ is given by

$$\mathbf{R} = t^2\mathbf{i} + t\mathbf{j} + \frac{t^2}{2}\mathbf{k}.$$

Find the tangential and normal components of acceleration.

**SOLUTION**

$$\mathbf{R} = t^2\mathbf{i} + t\mathbf{j} + \frac{t^2}{2}\mathbf{k},$$

$$\mathbf{R}' = 2t\mathbf{i} + \mathbf{j} + t\mathbf{k},$$

$$\mathbf{R}'' = 2\mathbf{i} + 0\mathbf{j} + \mathbf{k},$$

$$|\mathbf{R}'| = \sqrt{(2t)^2 + (1)^1 + (t)^2} = \sqrt{5t^2 + 1},$$

$$\mathbf{R}' \cdot \mathbf{R}'' = (2t)(2) + (1)(0) + (t)(1) = 5t,$$

$$\begin{aligned} \mathbf{R}' \times \mathbf{R}'' &= \begin{vmatrix} \mathbf{i} & \mathbf{j} & \mathbf{k} \\ 2t & 1 & t \\ 2 & 0 & 1 \end{vmatrix} \\ &= \begin{vmatrix} 1 & t \\ 0 & 1 \end{vmatrix}\mathbf{i} - \begin{vmatrix} 2t & t \\ 2 & 1 \end{vmatrix}\mathbf{j} + \begin{vmatrix} 2t & 1 \\ 2 & 0 \end{vmatrix}\mathbf{k} \\ &= \mathbf{i} - 0\mathbf{j} - 2\mathbf{k}, \end{aligned}$$

$$|\mathbf{R}' \times \mathbf{R}''| = \sqrt{(1)^2 + (0)^2 + (-2)^2} = \sqrt{5}.$$

The tangential component of acceleration is

$$a_{\mathbf{T}} = \frac{\mathbf{R}' \cdot \mathbf{R}''}{|\mathbf{R}'|} = \frac{5t}{\sqrt{5t^2 + 1}}.$$

The normal component of acceleration is

$$a_{\mathbf{N}} = \frac{|\mathbf{R}' \times \mathbf{R}''|}{|\mathbf{R}'|} = \frac{\sqrt{5}}{\sqrt{5t^2 + 1}}.$$

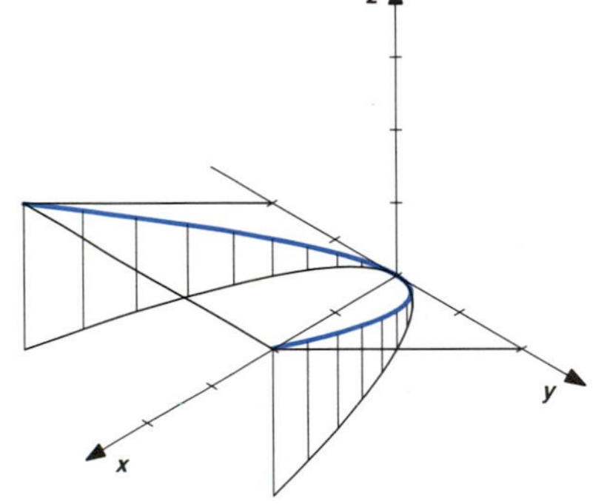

FIGURE 15.4.7

The graph of the curve $\mathbf{R} = t^2\mathbf{i} + t\mathbf{j} + (t^2/2)\mathbf{k}$ is sketched in Figure 15.4.7. The curve is the intersection of the parabolic cylinder $x = y^2$ and the plane $x = 2z$. ■

We have

$$\begin{aligned} |\mathbf{R}''|^2 &= \mathbf{R}'' \cdot \mathbf{R}'' \\ &= (a_{\mathbf{T}}\mathbf{T} + a_{\mathbf{N}}\mathbf{N}) \cdot (a_{\mathbf{T}}\mathbf{T} + a_{\mathbf{N}}\mathbf{N}) \\ &= a_{\mathbf{T}}^2\mathbf{T} \cdot \mathbf{T} + 2a_{\mathbf{T}}a_{\mathbf{N}}\mathbf{T} \cdot \mathbf{N} + a_{\mathbf{N}}^2\mathbf{N} \cdot \mathbf{N} \\ &= a_{\mathbf{T}}^2 + a_{\mathbf{N}}^2. \end{aligned}$$

This gives the formula

$$\boldsymbol{a_{\mathbf{N}} = \sqrt{|\mathbf{R}''|^2 - a_{\mathbf{T}}^2}.}$$

We can use this formula to calculate the normal component of acceleration without evaluating the cross product $\mathbf{R}' \times \mathbf{R}''$.

The formulas that we derived for space curves may also be applied to plane curves, since we may think of a plane curve as a space curve with $z$-coordinate zero.

### ■ EXAMPLE 6

The position of an object in the plane at time $t$ is given by

$$\mathbf{R} = t \cos t\mathbf{i} + t \sin t\mathbf{j}.$$

Find the tangential and normal components of acceleration.

**SOLUTION**

$$\mathbf{R} = t \cos t\mathbf{i} + t \sin t\mathbf{j},$$

$$\begin{aligned}\mathbf{R}' &= [(t)(-\sin t) + (\cos t)(1)]\mathbf{i} + [(t)(\cos t) + (\sin t)(1)]\mathbf{j}\\ &= (-t \sin t + \cos t)\mathbf{i} + (t \cos t + \sin t)\mathbf{j},\end{aligned}$$

$$\begin{aligned}\mathbf{R}'' &= [-((t)(\cos t) + (\sin t)(1)) - \sin t]\mathbf{i}\\ &\quad + [(t)(-\sin t) + (\cos t)(1) + \cos t]\mathbf{j}\\ &= (-t \cos t - 2 \sin t)\mathbf{i} + (-t \sin t + 2 \cos t)\mathbf{j},\end{aligned}$$

$$\begin{aligned}|\mathbf{R}'|^2 &= (-t \sin t + \cos t)^2 + (t \cos t + \sin t)^2\\ &= t^2 \sin^2 t - 2t(\sin t)(\cos t) + \cos^2 t\\ &\quad + t^2 \cos^2 t + 2t(\sin t)(\cos t) + \sin^2 t\\ &= t^2 + 1,\end{aligned}$$

$$\begin{aligned}|\mathbf{R}''|^2 &= (-t \cos t - 2 \sin t)^2 + (-t \sin t + 2 \cos t)^2\\ &= t^2 \cos^2 t + 4t(\sin t)(\cos t) + 4 \sin^2 t\\ &\quad + t^2 \sin^2 t - 4t(\sin t)(\cos t) + 4 \cos^2 t\\ &= t^2 + 4,\end{aligned}$$

$$\begin{aligned}\mathbf{R}' \cdot \mathbf{R}'' &= (-t \sin t + \cos t)(-t \cos t - 2 \sin t)\\ &\quad + (t \cos t + \sin t)(-t \sin t + 2 \cos t)\\ &= t^2(\sin t)(\cos t) - t \cos^2 t + 2t \sin^2 t - 2(\sin t)(\cos t)\\ &\quad - t^2(\sin t)(\cos t) - t \sin^2 t + 2t \cos^2 t + 2(\sin t)(\cos t)\\ &= -t + 2t = t.\end{aligned}$$

The tangential component of acceleration is

$$a_{\mathbf{T}} = \frac{\mathbf{R}' \cdot \mathbf{R}''}{|\mathbf{R}'|} = \frac{t}{\sqrt{t^2 + 1}}.$$

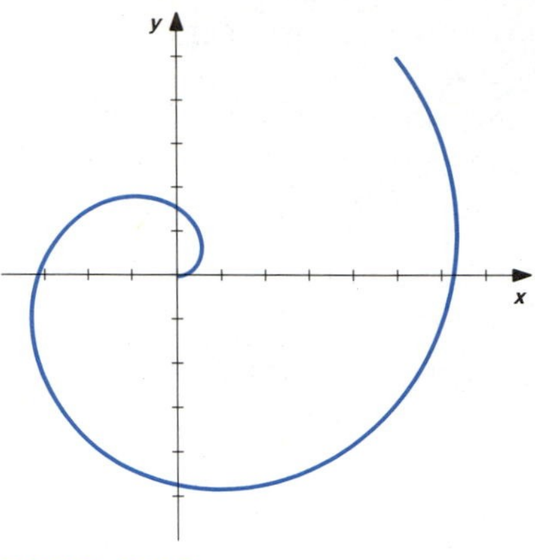

FIGURE 15.4.8

We then have

$$\begin{aligned}a_{\mathbf{N}}^2 &= |\mathbf{R}''|^2 - a_{\mathbf{T}}^2\\ &= t^2 + 4 - \frac{t^2}{t^2+1}\\ &= \frac{(t^2+4)(t^2+1) - t^2}{t^2+1}\\ &= \frac{t^4 + 4t^2 + 4}{t^2+1}\\ &= \frac{(t^2+2)^2}{t^2+1}.\end{aligned}$$

The normal component of acceleration is

$$a_{\mathbf{N}} = \frac{t^2+2}{\sqrt{t^2+1}}.$$

A sketch of the curve $\mathbf{R} = t\cos t\mathbf{i} + t\sin t\mathbf{j}$ is given in Figure 15.4.8. ■

Since

$$a_{\mathbf{N}} = \kappa\left(\frac{ds}{dt}\right)^2 = \frac{|\mathbf{R}' \times \mathbf{R}''|}{|\mathbf{R}'|} \quad \text{and} \quad \frac{ds}{dt} = |\mathbf{R}'|,$$

we have the formula

$$\kappa = \frac{|\mathbf{R}' \times \mathbf{R}''|}{|\mathbf{R}'|^3}.$$

This formula gives the curvature in terms of $\mathbf{R}'$ and $\mathbf{R}''$. If

$$C\colon\ x = x(t), \quad y = y(t), \quad z = 0,$$

so $C$ is a curve in the $xy$-plane, the formula becomes

$$\kappa = \frac{|x'y'' - y'x''|}{(x'^2 + y'^2)^{3/2}}.$$

This is the same formula we used in Section 15.3.

## EXERCISES 15.4

Sketch the graphs of the parametric curves given below in Exercises 1-8.

1 $C\colon x = t,\ y = t,\ z = t^2$

2 $C\colon x = t^2,\ y = t,\ z = t^2$

3 $C\colon x = \cos t,\ y = \sin t,\ z = t/2$

4 $C\colon x = \cos t,\ y = -\cos t,\ z = \sin t$

5 $C\colon x = t^2,\ y = t,\ z = t^3$

6 $C\colon x = -t^3,\ y = t,\ z = t^3$

7 $C\colon x = \dfrac{\cos t}{t+1},\ y = t,\ z = \dfrac{\sin t}{t+1},\ t \geq 0$

8 $C\colon x = t^2 \cos t,\ y = t^2 \sin t,\ z = t$

Find parametric equations of the curves described below in Exercises 9-14 and sketch the curves.

9 Line segment from $(2, 1, -1)$ to $(5, -2, 3)$

10 Line segment from $(0, 2, 1)$ to $(4, 3, 5)$

11 Intersection of $x^2 + y^2 = 1$ and $x + z = 1$

12 Intersection of $x^2 + z^2 = 4$ and $x + y = 0$

13 Intersection of $x^2 + y^2 + z^2 = 4$ and $x + z = 0$

14 Intersection of $z = x^2 + y^2$ and $2x + 2y + z = 2$

Find the unit tangent vector, the principal unit normal vector, the unit binormal, and the curvature of the curves in Exercises 15-20.

15 $C\colon x = e^t \cos t,\ y = e^t \sin t,\ z = e^t$

16 $C\colon x = \cos t,\ y = \sqrt{2} \sin t,\ z = \cos t$

17 $C\colon x = t,\ y = t^2/2,\ z = t$; where $t = 0$

18 $C\colon x = e^t + e^{-t},\ y = \sqrt{2}\,t,\ z = \sqrt{2}\,t$; where $t = 0$

19 $C\colon x = \sqrt{2} \ln t,\ y = 1/t,\ z = t$; where $t = 1$

20 $C\colon x = e^t,\ y = e^{-t},\ z = \sqrt{2}\,t$; where $t = 0$

Find the tangential and normal components of acceleration for the motion described by the position vectors given in Exercises 21-28.

21 $\mathbf{R} = 3 \cos 2t\mathbf{i} + 3 \sin 2t\mathbf{j}$

22 $\mathbf{R} = 2 \cos t\mathbf{i} + 3 \sin t\mathbf{j}$

23 $\mathbf{R} = t\mathbf{i} + t^2\mathbf{j}$

24 $\mathbf{R} = t^3\mathbf{i} + t^2\mathbf{j}$

25 $\mathbf{R} = \cos t\mathbf{i} + \sin t\mathbf{j} + t\mathbf{k}$

26 $\mathbf{R} = \ln t\mathbf{i} + t\mathbf{j} + \ln t\mathbf{k}$

27 $\mathbf{R} = e^{-t} \cos t\mathbf{i} + e^{-t} \sin t\mathbf{j} + t\mathbf{k}$; where $t = 0$

28 $\mathbf{R} = t \cos t\mathbf{i} + t \sin t\mathbf{j} + t\mathbf{k}$; where $t = 0$

Find the curvature of the curves given in Exercises 29-30.

29 $C\colon x = t,\ y = t^2,\ z = t^3$; where $t = 1$

30 $C\colon x = t^2,\ y = t^2,\ z = t^3$; where $t = 1$

31 The position of an object at time $t$ is given by

$$\mathbf{R} = t^2\mathbf{i} + t^3\mathbf{j} + t\mathbf{k}.$$

Find the equation of the plane determined by the vectors $\mathbf{T}$ and $\mathbf{N}$ at the point on the curve where $t = 1$. (*Hint*: The vectors $\mathbf{R}'$ and $\mathbf{R}''$ are in the plane.)

## REVIEW EXERCISES

In Exercises 1-4 find the $xy$-equation of each given curve. Sketch the graph of each curve; show initial point, terminal point, and direction.

1 $C\colon x = -\cos t,\ y = \sin t,\ 0 \leq t \leq \pi$

2 $C\colon x = 1/\sqrt{t-1},\ y = 1/(t-1),\ 5/4 \leq t \leq 5$

3 $C\colon x = t + 1,\ y = 3 - 2t,\ 0 \leq t \leq 1$

4 $C\colon x = 2 \sin t,\ y = 3 \cos t,\ 0 \leq t \leq 2\pi$

Sketch the curves in Exercises 5-6.

5 $C\colon x = 2 \cos t,\ y = 2 \sin t,\ z = t/2,\ 0 \leq t \leq 2\pi$

6 $C\colon x = t,\ y = t^2,\ z = t/2,\ -2 \leq t \leq 2$

In Exercises 7-12 find parametric equations of the indicated curves.

7 Along the line segment from $(2, -1, 3)$ to $(-1, 3, 1)$

8 From $(2, 0)$, counterclockwise around $9x^2 + 4y^2 = 36$

9 Along $x = y^2$ from (4, 2) to (0, 0)

10 The intersection of the ellipsoid $4x^2 + 36y^2 + 9z^2 = 36$ and the plane $z + y = 1$

11 From the origin, counterclockwise once around the polar curve $r = 1 + \cos\theta$

12 From the origin, counterclockwise once around the polar curve $r = \sin\theta$

13 Find parametric equations of the line tangent to the curve $C: x = te^t$, $y = e^t$, at the point where $t = 1$.

14 Find parametric equations of the line tangent to the curve $C: x = 2t^2$, $y = t^3$, $z = t^2/2$, at the point where $t = 1$.

Evaluate $dy/dx$ and $d^2y/dx^2$ at the indicated point for the curves given in Examples 15–16. Sketch an arc of each curve as it passes through the points; show slope, concavity, and direction.

15 $C: x = 1 - t^2$, $y = t^4$; $t = 1$

16 $C: x = 4\cos t$, $y = 2 - 2\sin t$; $t = 3\pi/4$

17 Find the length of the curve $C: x = t^3$, $y = t^2$, $0 \le t \le 1$.

18 Set up an integral that gives the length of the curve $C: x = 2\cos t$, $y = 3\sin t$, $0 \le t \le 2\pi$.

In Exercises 19–20 set up an integral that gives the area of the surface obtained when the given curve is rotated about the indicated line.

19 $C: x = t + \sin t$, $y = 1 - \cos t$, $0 \le t \le 2\pi$; rotated about the $x$-axis

20 $C: x = e^t + e^{-t}$, $y = e^t - e^{-t}$, $0 \le t \le 1$; rotated about the $y$-axis

Find the parametrization in terms of arc length of the curves given in Exercises 21–24.

21 $C: x = 1 - 3t$, $y = 3 + 4t$, $0 \le t \le 1$

22 $C: x = -2\cos 3t$, $y = 2 - 2\sin 3t$, $0 \le t \le \pi/3$

23 $C: x = t^3/3$, $y = t^2/2$, $0 \le t \le 2$

24 The polar curve $r = e^{2\theta}$, $0 \le \theta \le \pi$

Determine which of the curves in Exercises 25–26 are simple and which are closed. Determine which can be expressed as either smooth or piecewise smooth parametric curves, and find parametric representations.

25 From (0, 0), along $y = x^2$ to (1, 1), then along $y = x$ to $(-1, -1)$

26 From (1, 0), along the $x$-axis to $(-1, 0)$, then along the circle $x^2 + y^2 = 1$ through $(0, -1)$ to (1, 0)

27 $C: x = 2\cos t$, $y = 3\sin t$, $0 \le t \le 2\pi$. Find the unit tangent vector, the unit normal vector, the curvature, the radius of curvature, and the center of curvature at the point where $t = 0$. Sketch the curve and the circle of curvature.

28 $\mathbf{R} = \sqrt{5}\,t\mathbf{i} + [(t^2/2) + 2t]\mathbf{j}$. Find the unit tangent vector, the unit normal vector, and the curvature at the point where $t = 0$.

29 $\mathbf{R} = t\mathbf{i} + (2/3)t^{3/2}\mathbf{j}$. Find $\mathbf{T}$, $\mathbf{N}$, and $\kappa$ at the point where $t = 3$.

30 $\mathbf{R} = \cos t\mathbf{i} + \sin t\mathbf{j} + t\mathbf{k}$. Find $\mathbf{T}$, $\mathbf{N}$, $\mathbf{B}$, and $\kappa$ at the point where $t = 0$.

Find formulas for the curvature of the curves given in Exercises 31–34.

31 $C: x = t^2$, $y = t^3/3$

32 $C: x = t^2$, $y = t$, $z = t^3/3$

33 $y = \cos x$

34 $y = x^3 - 1$

Find the components of acceleration and the curvature for the curves given in Exercises 35–36.

35 $\mathbf{R} = t^2\mathbf{i} + t\mathbf{j} + t^2\mathbf{k}$

36 $\mathbf{R} = e^{2t}\mathbf{i} + t^2\mathbf{j}$

## CHAPTER SIXTEEN

# *Differential calculus of functions of several variables*

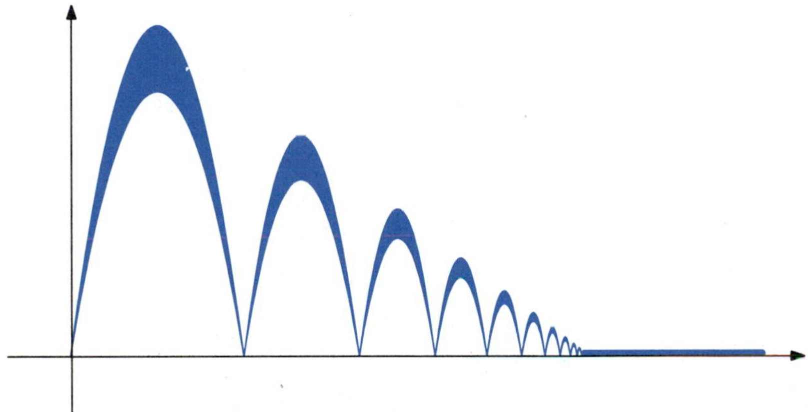

Physical quantities that depend on more than one real variable lead to the study of functions of several variables. As in the case of functions of one real variable, the calculus of functions of several variables depends on the concepts of limits, continuity, and the derivative. We will introduce these concepts and then see how they can be used to find approximate values and extreme values of functions of several variables.

## 16.1 FUNCTIONS OF SEVERAL VARIABLES

The concepts and notation of real-valued functions of several variables are similar to those of real-valued functions of one variable. For example, $f(x, y)$ signifies that $f$ is a function of the variables $x$ and $y$. Similarly, $g(x, y, z)$ indicates that $g$ is a function of the three variables $x$, $y$, and $z$. As in the case of one variable, the values of functions in several variables will often be given by a formula. For example, $f(x, y) = x^2 + y^2$ or $g(x, y, z) = y + 1$. Unless stated otherwise, the domain of a function that is given by a formula includes all points where the formula makes sense. This is the same convention that was used for functions of one variable.

**■ EXAMPLE 1**

Find the domain of the function

$$f(x, y) = \frac{x^2 + y^2}{x - y}.$$

Evaluate $f(2, 0)$, $f(0, -3)$, and $f(2a, a)$ for $a \neq 0$.

**SOLUTION**

The function is defined for all $(x, y)$ except for $(x, y)$ on the *line* $y = x$ where the denominator is zero. See Figure 16.1.1. Substitution into the formula gives

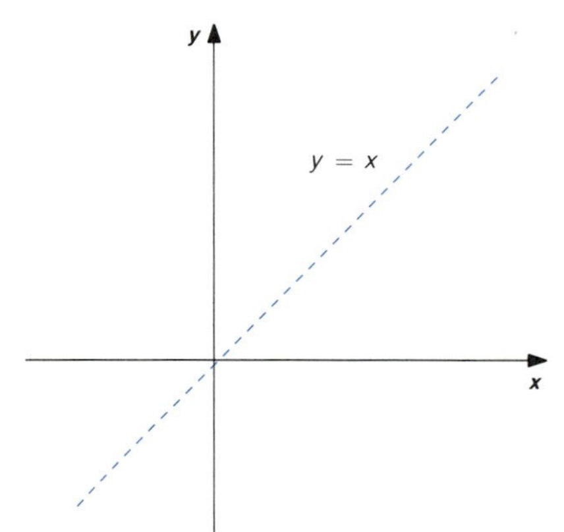


FIGURE 16.1.1

$$f(2, 0) = \frac{(2)^2 + (0)^2}{(2) - (0)} = 2,$$

$$f(0, -3) = \frac{(0)^2 + (-3)^2}{(0) - (-3)} = 3,$$

$$f(2a, a) = \frac{(2a)^2 + (a)^2}{(2a) - (a)} = 5a, \qquad a \neq 0.$$

The condition $a \neq 0$ implies $a \neq 2a$, so $(2a, a)$ is not on the line $y = x$ and $f(2a, a)$ is defined. ■

**■ EXAMPLE 2**

Find the domain of the function $f(x, y) = \sqrt{4 - x^2 - 4y^2}$. Evaluate $f(1, 1/2)$ and $f(1, 0)$.

**SOLUTION**

The function is defined for all $(x, y)$ that satisfy $4 - x^2 - 4y^2 \geq 0$, or $x^2 + 4y^2 \leq 4$. The domain consists of all points on or inside the ellipse

$$\frac{x^2}{2^2} + \frac{y^2}{1^2} = 1.$$

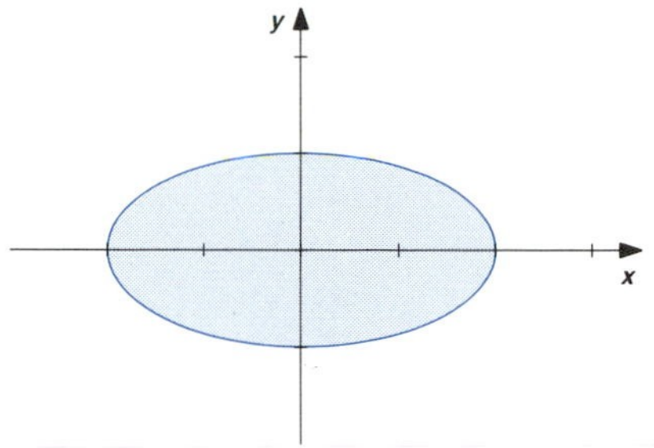


FIGURE 16.1.2

See Figure 16.1.2. Substitution into the formula gives

$$f\left(1, \frac{1}{2}\right) = \sqrt{4 - (1)^2 - 4\left(\frac{1}{2}\right)^2} = \sqrt{2},$$

$$f(1, 0) = \sqrt{4 - (1)^2 - 4(0)^2} = \sqrt{3}.$$

The **graph of a function $f(x, y)$** of two variables $x$ and $y$ is the set of all points $(x, y, z)$ in three-dimensional space such that $z = f(x, y)$. Generally, we would expect the graph of a function of two variables to be a surface in three-dimensional space. Graphing surfaces in three-dimensional space was discussed in Chapter 13. It may be useful to review Chapter 13 at this time.

For a fixed value of $z_0$, the set of all points $(x, y)$ in the plane that satisfy the equation $f(x, y) = z_0$ is called a **level curve** of $f$. Sketching and labeling a selection of level curves is one way to indicate the values and general character of a function of two variables.

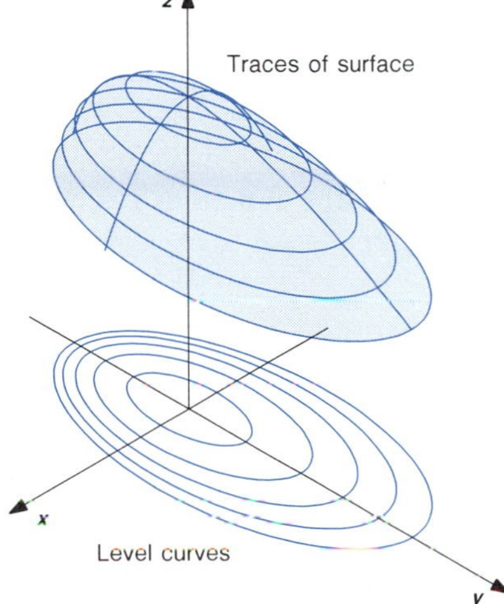


FIGURE 16.1.3

The graph of $z = f(x, y)$ is a surface in three-dimensional space. The level curve $f(x, y) = z_0$ is a curve in the $xy$-plane. The level curve $f(x, y) = z_0$ is related to the curve in three-dimensional space with equations $z = z_0$, $z = f(x, y)$, which is the trace of the surface $z = f(x, y)$ in the plane $z = z_0$. The level curve $f(x, y) = z_0$ is the projection of the trace $z = z_0$, $z = f(x, y)$ onto the $xy$-plane. The graph of $z = f(x, y)$ can be constructed from level curves by using the level curves $f(x, y) = z_0$ for several values of $z_0$ to sketch the traces $z = z_0$, $z = f(x, y)$. See Figure 16.1.3.

Level curves are used on topographic maps to indicate the function whose value at a point is the elevation above sea level of the point. These level curves are called **contour curves**. All points on a particular contour curve are at the same elevation. The elevation changes more rapidly—the hill is steeper—where the contour curves are closer together. See Figure 16.1.4.

FIGURE 16.1.4

FIGURE 16.1.5

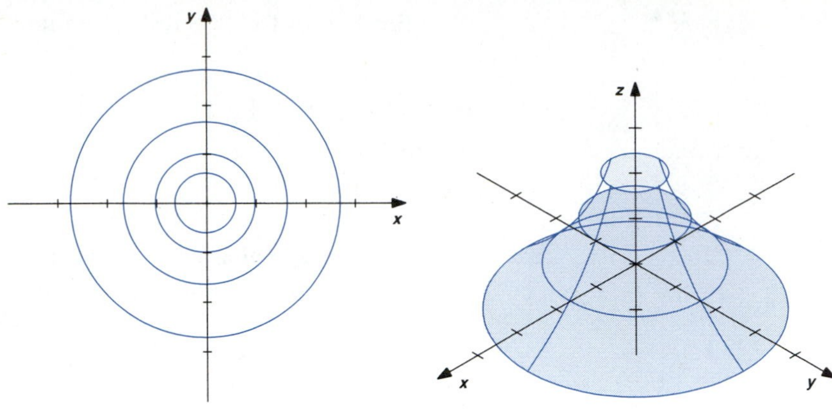


(a) (b)

**EXAMPLE 3**

Characterize and sketch several level curves of the function $f(x, y) = 1 - \ln(x^2 + y^2)$. Sketch the graph of the function.

**SOLUTION**

The level curve that corresponds to a number $z_0$ is the set of all $(x, y)$ that satisfy the equation

$$f(x, y) = z_0, \quad \text{or}$$

$$1 - \ln(x^2 + y^2) = z_0.$$

Rewriting the equation, we obtain

$$\ln(x^2 + y^2) = 1 - z_0,$$

$$x^2 + y^2 = e^{(1 - z_0)}.$$

This is the equation of a circle with center at the origin and radius $e^{(1 - z_0)/2}$. The level curves that correspond to $z_0 = -1, 0, 1,$ and $2$ are sketched in Figure 16.1.5a. [These level curves become closer as they move toward the origin. Since the $z$-increments between the level curves are equal, this indicates that the values of the function are changing more rapidly at $(x, y)$ as $(x, y)$ moves toward the origin. This should be compared with the fact that a hill is steeper where the contour curves are closer together.] The graph of $f$ is sketched in Figure 16.1.5b, where the traces that correspond to $z_0 = -1, 0, 1,$ and $2$ have been indicated. Note that the values $f(x, y)$ increase more rapidly as $(x, y)$ moves toward the origin. ■

**EXAMPLE 4**

Characterize and sketch several level curves of the function

$$f(x, y) = \frac{y^2 - x^2}{2}.$$

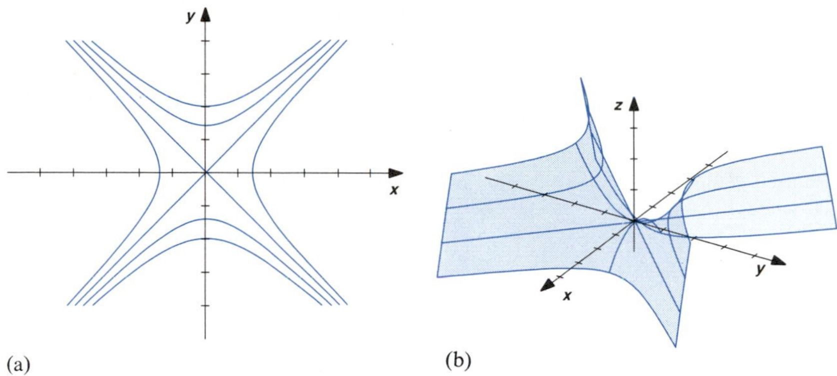


FIGURE 16.1.6

Sketch the graph of the function.

**SOLUTION**

The level curves that correspond to $f(x, y) = z_0$ are

$$y^2 - x^2 = 2z_0.$$

For $z_0 \neq 0$, this is the equation of a hyperbola with center at the origin and asymptotes $y = \pm x$; the vertices are $(\pm\sqrt{-2z_0}, 0)$ if $z_0 < 0$ and $(0, \pm\sqrt{2z_0})$ if $z_0 > 0$. The level curve that corresponds to $z_0 = 0$ consists of the two lines $y = x$ and $y = -x$. Level curves that correspond to $z_0 = -1$, 0, 1, and 2 are given in Figure 16.1.6a. The graph of $f$ is given in Figure 16.1.6b, where the traces that correspond to $z_0 = -1, 0, 1,$ and 2 have been indicated. ■

The graph of a function $f(x, y, z)$ of three variables is the set of all points $(x, y, z, w)$ such that $w = f(x, y, z)$. It doesn't seem likely that we would be able to effectively sketch a graph of points in four-dimensional space. It is possible, however, to indicate the **level surfaces** $f(x, y, z) = w_0$ of a function of three variables as surfaces in three-dimensional space.

### ■ EXAMPLE 5

Characterize and sketch several level surfaces of the function $f(x, y, z) = 2x + 3y + 6z$.

**SOLUTION**

The level surface that corresponds to $w_0 = f(x, y, z)$ is

$$w_0 = 2x + 3y + 6z.$$

This is the equation of a plane. Several of the level surfaces are sketched in Figure 16.1.7. ■

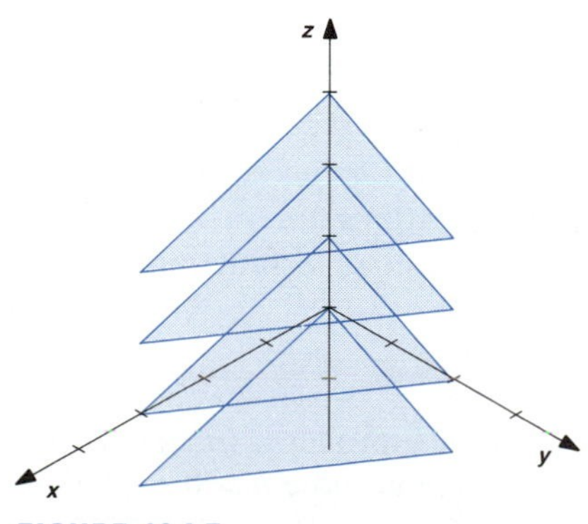


FIGURE 16.1.7

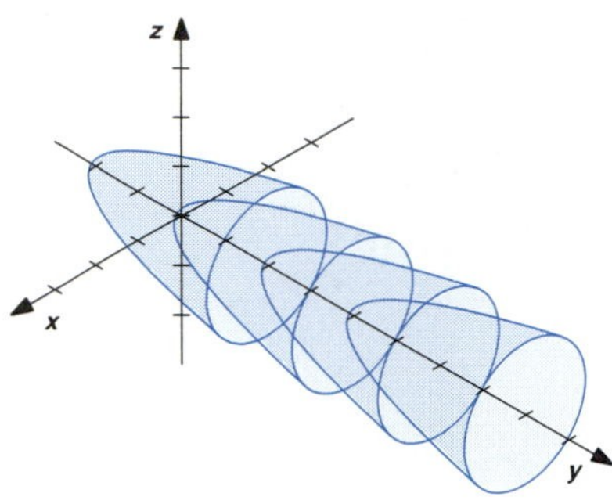


FIGURE 16.1.8

■ **EXAMPLE 6**

Characterize and sketch several level surfaces of the function $f(x, y, z) = y - 2x^2 - 2z^2$.

**SOLUTION**

The level surface that corresponds to $w_0 = f(x, y, z)$ is

$$w_0 = y - 2x^2 - 2z^2.$$

This is the equation of a paraboloid with vertex at the point $(0, w_0, 0)$ and axis along the $y$-axis. Several of the level surfaces are sketched in Figure 16.1.8. ■

## EXERCISES 16.1

Find the domain and indicated values of the functions in Exercises 1–6.

1 $f(x, y) = \dfrac{\sin(x + y)}{x}$; $f(2, -2)$, $f(\pi, \pi/2)$

2 $f(x, y) = \dfrac{2x - y}{x + 2y}$; $f(2, 3)$, $f(3, 2)$

3 $f(x, y) = \dfrac{x + y}{x^2 + y^2}$; $f(0, 1)$, $f(2, 0)$, $f(a, a)$, $a \neq 0$

4 $f(x, y) = \dfrac{x + y}{x^2 - y^2}$; $f(0, 1)$, $f(0, 2)$, $f(1, a)$, $a \neq \pm 1$

5 $f(x, y) = \sqrt{9 - x^2 - y^2}$; $f(2, 1)$, $f(-2, 1)$

6 $f(x, y, z) = \dfrac{1}{\sqrt{36 - x^2 - 4y^2 - 9z^2}}$; $f(0, 2, 1)$, $f(4, -1, 1)$

Characterize and sketch several level curves of the functions given in Exercises 7–14. Sketch the graphs of the functions.

7 $f(x, y) = 3 - x - y$

8 $f(x, y) = 2x + 2y$

9 $f(x, y) = e^{-x^2 - y^2}$

10 $f(x, y) = \sqrt{9 - x^2 - y^2}$

11 $f(x, y) = x^2 + y^2$

12 $f(x, y) = 2\sqrt{x^2 + y^2}$

13 $f(x, y) = \dfrac{\sqrt{x^2 - y^2}}{2}$

14 $f(x, y) = \dfrac{\sqrt{y^2 - x^2}}{2}$

Characterize and sketch several level surfaces of the functions given in Exercises 15–20.

15 $f(x, y, z) = 6x + 3y + 2z$

16 $f(x, y, z) = 2x + 3y$

17 $f(x, y, z) = x^2 + y^2 + z$

18 $f(x, y, z) = x^2 + y^2 + z^2$

19 $f(x, y, z) = z - 2\sqrt{x^2 + y^2}$

20 $f(x, y, z) = z - x^2 - y^2$

## 16.2 LIMITS AND CONTINUITY

The ideas of limits and continuity are important for the development of calculus. Let us briefly discuss these ideas for functions of several variables. We begin with the formal definition of the limit of a function of two variables.

■ *Definition*

We say the function $f(x, y)$ has limit $L$ as $(x, y)$ approaches $(a, b)$ (written $\lim_{(x, y) \to (a, b)} f(x, y) = L$) if, for each $\varepsilon > 0$, there is some $\delta > 0$ such that $|f(x, y) - L| < \varepsilon$ whenever $0 < \sqrt{(x - a)^2 + (y - b)^2} < \delta$. ■

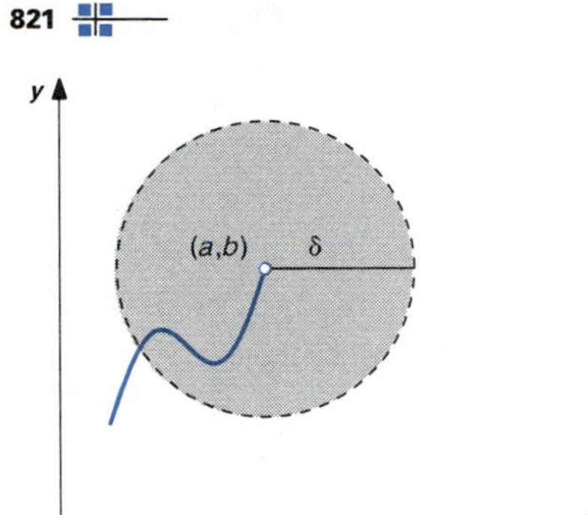


FIGURE 16.2.1

The definition of $\lim_{(x,y)\to(a,b)} f(x, y) = L$ requires that for each $\varepsilon > 0$, there is a positive number $\delta$ such that the values $f(x, y)$ are within $\varepsilon$ of $L$ for all $(x, y)$ that are within $\delta$ of $(a, b)$, excluding the point $(a, b)$. This means that the values $f(x, y)$ must approach a value or values that are within $\varepsilon$ of $L$ as $(x, y)$ approaches $(a, b)$ along any curve. See Figure 16.2.1. The fact that $\varepsilon$ can be any positive number excludes the possibility that $f(x, y)$ approaches any number other than $L$. It follows that $f(x, y)$ must approach the single number $L$ as $(x, y)$ approaches $(a, b)$ along any path.

The formal definition of the limit of a function of two variables corresponds to the definition of the limit of a function of one variable, except that here the distance of $(x, y)$ from $(a, b)$ is given by $\sqrt{(x-a)^2 + (y-b)^2}$, while the distance between numbers $x$ and $a$ on the line is given by $|x - a|$. Since $|x - a| = \sqrt{(x-a)^2}$, the definitions are completely analogous. The definition of limit for functions of three variables is similar. We merely use the formula for the distance between two points in three-dimensional space. For functions of more than three variables, we use the obvious extensions of the formulas for distance. All of the theory we developed for the evaluation of functions of one variable also holds for functions of several variables. In particular, the Limit Theorem for the evaluation of the limit of sums, products, and quotients holds for functions of several variables. We also have the following result for the composition of functions. This result is proved in the same way as the corresponding one-variable result was proved earlier in Section 2.5.

**If $f(t)$ is continuous at $M$ and $\lim_{(x,y)\to(a,b)} g(x, y) = M$, then**

$$\lim_{(x,y)\to(a,b)} f(g(x, y)) = f(M).$$

The Limit Theorem and the above result for the limit of composite functions can be used to evaluate the limits of many common functions of several variables. We first use the definition of limit to prove that

$$\lim_{(x,y)\to(a,b)} x = a \quad \text{and} \quad \lim_{(x,y)\to(a,b)} y = b.$$

The Limit Theorem then tells us that we can evaluate the limit of sums, products, and quotients of $x$ and $y$ by substitution, as long as we do not divide by zero. The above-stated result for composite functions tells us that the limit of a continuous function of a real variable of the above type functions can be evaluated by substitution.

■ **EXAMPLE 1**

Evaluate

$$\lim_{(x,y)\to(0,0)} \cos xy.$$

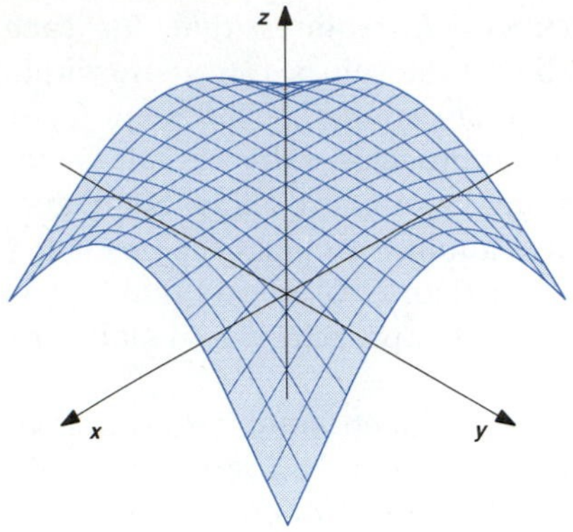

FIGURE 16.2.2

**SOLUTION**

We know that the product $xy$ approaches zero as $(x, y)$ approaches $(0, 0)$. Also, the cosine function is a continuous function of one variable, so

$$\lim_{(x,y)\to(0,0)} \cos xy = \cos 0 = 1.$$

The graph is sketched in Figure 16.2.2. ■

**■ EXAMPLE 2**

Evaluate

$$\lim_{(x,y)\to(2,1)} \frac{x^2}{x^2+y^2}.$$

**SOLUTION**

Substitution gives

$$\lim_{(x,y)\to(2,1)} \frac{x^2}{x^2+y^2} = \frac{(2)^2}{(2)^2+(1)^2} = 0.8.$$

The graph is sketched in Figure 16.2.3. ■

The definition of limit of $f(x, y)$ at $(a, b)$ requires that $f(x, y)$ must approach a *single* number as $(x, y)$ approaches $(a, b)$ along any curve. In particular,

**If $f(x, y)$ approaches different numbers as $(x, y)$ approaches $(a, b)$ along different curves, then $\lim_{(x,y)\to(a,b)} f(x, y)$ does not exist.**

We can show that a limit does not exist at a point by showing that $f(x, y)$ approaches different values as $(x, y)$ approaches the point along two different curves.

FIGURE 16.2.3

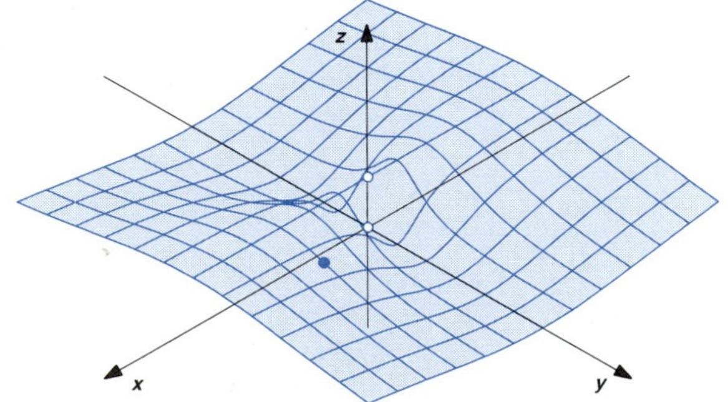

**EXAMPLE 3**

Show that

$$\lim_{(x, y)\to(0, 0)} \frac{x^2 - y^2}{x^2 + y^2}$$

does not exist.

**SOLUTION**

The expression

$$\frac{x^2 - y^2}{x^2 + y^2}$$

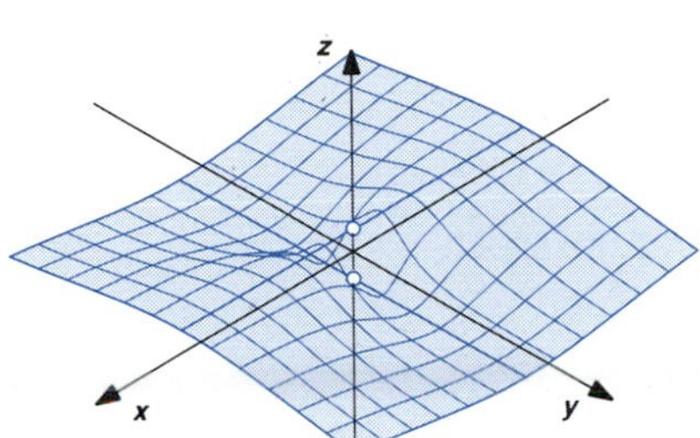


FIGURE 16.2.4

has value 1 at all points $(x, 0)$ on the $x$-axis except the origin, so the values approach 1 as $(x, y) = (x, 0)$ approaches the origin along the $x$-axis. Similarly, the values approach $-1$ as $(x, y) = (0, y)$ approaches the origin along the $y$-axis. Since the values approach different numbers as $(x, y)$ approaches the origin along these curves, the limit at the origin does not exist. See Figure 16.2.4. ■

**EXAMPLE 4**

Suppose

$$f(x, y) = \begin{cases} 1, & (x, y) \text{ on } C, \\ 0, & \text{otherwise,} \end{cases}$$

where $C$ is the parametric curve

$$C\colon\ x = t \cos t, \quad y = t \sin t, \qquad 0 < t \leq 2\pi.$$

Show that $\lim_{(x, y)\to(0, 0)} f(x, y)$ does not exist.

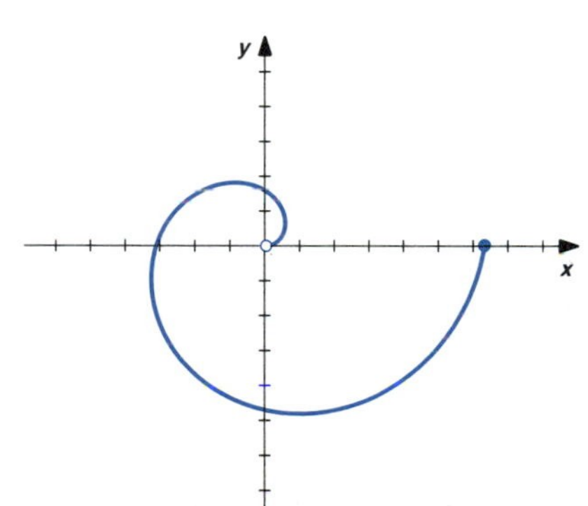


FIGURE 16.2.5

**SOLUTION**

From Figure 16.2.5a we see that any straight-line path to the origin intersects $C$ exactly once before it gets to the origin. Since the values of $f(x, y)$ are 0, except on $C$, it follows that $f(x, y)$ approaches 0 as $(x, y)$ approaches the origin along any straight line. Since $f(x, y)$ approaches 1 as $(x, y)$ approaches the origin along $C$, we conclude that the limit of $f(x, y)$ does not exist at the origin. The graph of $f$ is given in Figure 16.2.5b. ■

**EXAMPLE 5**

Show that

$$\lim_{(x, y)\to(2, 1)} \frac{x^2 - 4}{x - 2}$$

does not exist.

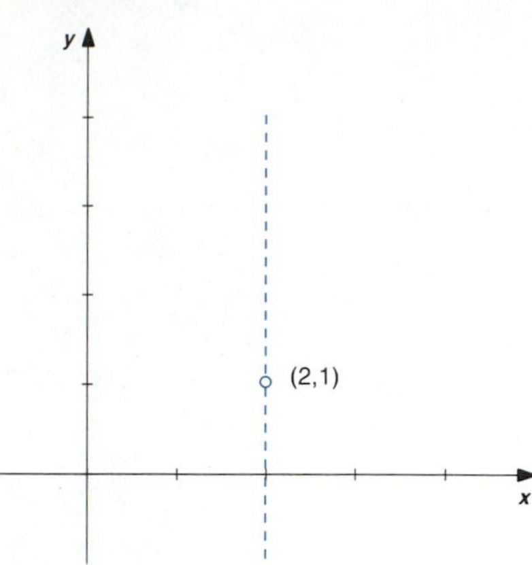

FIGURE 16.2.6

SOLUTION

The expression

$$\frac{x^2 - 4}{x - 2}$$

is not defined on the *line* $x = 2$, so it cannot have a limit as $(x, y)$ approaches $(2, 1)$ along this line. A function $f(x, y)$ cannot have a limit at a point unless the function is defined in some disk with center at the point, excluding only the point itself. See Figure 16.2.6. ■

■ EXAMPLE 6

Show that $\lim_{(x, y)\to(0, 0)} \ln (x^2 + y^2)$ does not exist.

SOLUTION

As $(x, y)$ approaches $(0, 0)$ along any curve, $x^2 + y^2$ approaches zero, so $\ln(x^2 + y^2)$ becomes unbounded. A function does not have a real-valued limit at a point if it becomes unbounded as $(x, y)$ approaches the point. See Figure 16.2.7. ■

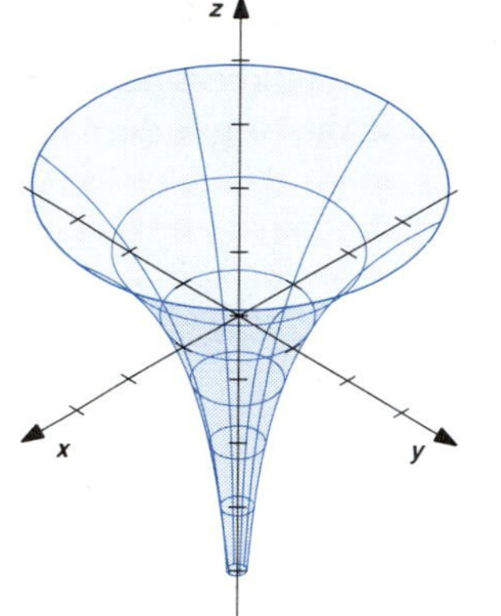

FIGURE 16.2.7

It is possible that a function of two variables has a limit at a point where the denominator is zero. To illustrate this, we will use the following inequalities.

$$|x| \le \sqrt{x^2 + y^2},$$

$$|y| \le \sqrt{x^2 + y^2}.$$

These inequalities are geometrically evident from Figure 16.2.8.

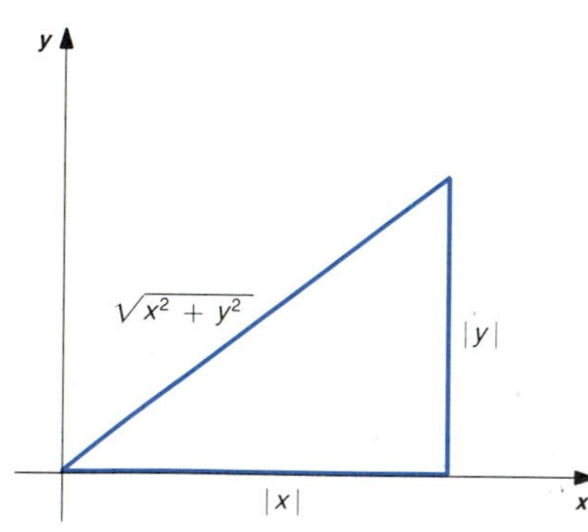

FIGURE 16.2.8

■ EXAMPLE 7

Show that

$$\lim_{(x, y)\to(0, 0)} \frac{y^4}{x^2 + y^2} = 0.$$

SOLUTION

The expression

$$\frac{y^4}{x^2 + y^2}$$

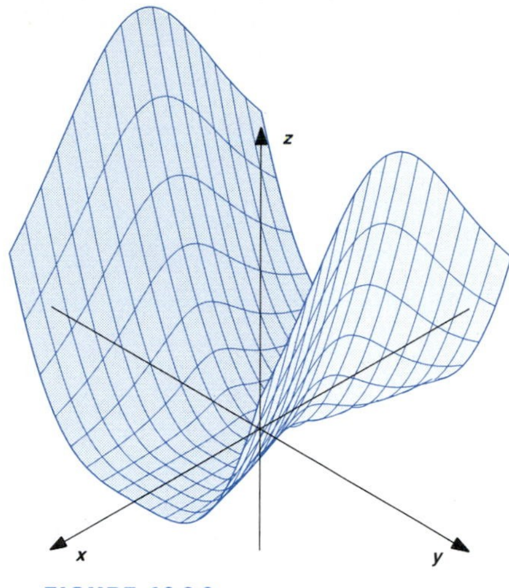


FIGURE 16.2.9

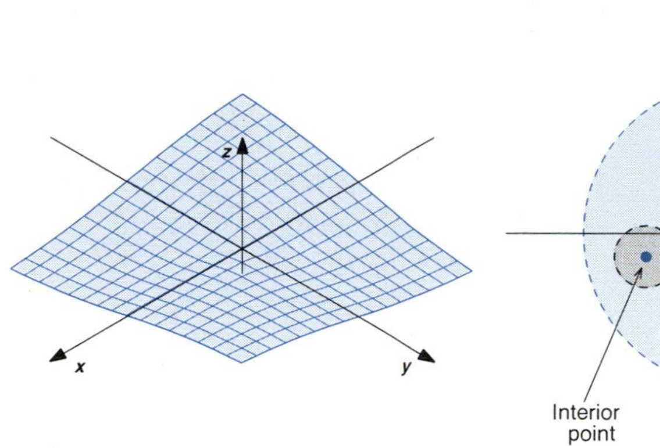


FIGURE 16.2.10

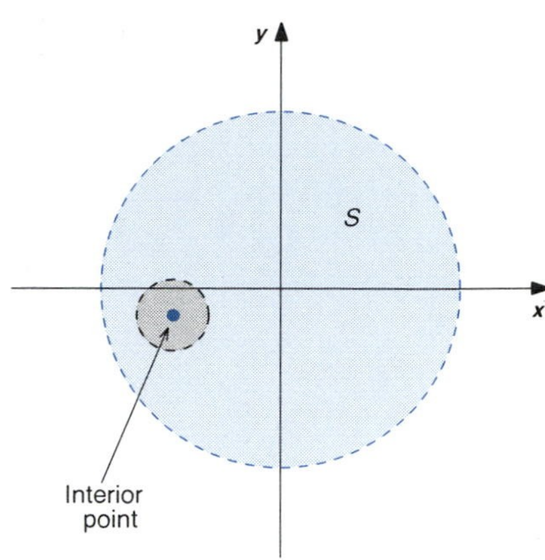


FIGURE 16.2.11

is defined for all $(x, y)$, except for $(x, y) = (0, 0)$. Since $y^4 \leq (\sqrt{x^2 + y^2})^4 = (x^2 + y^2)^2$, we have

$$\left| \frac{y^4}{x^2 + y^2} \right| \leq \frac{(x^2 + y^2)^2}{x^2 + y^2} = x^2 + y^2 \to 0 \qquad \text{as } (x, y) \to (0, 0).$$

The graph is sketched in Figure 16.2.9. ■

■ **EXAMPLE 8**

Show that

$$\lim_{(x, y) \to (0, 0)} \frac{xy}{\sqrt{x^2 + y^2}} = 0.$$

**SOLUTION**

Since $|xy| \leq (\sqrt{x^2 + y^2})^2$, we have

$$\left| \frac{xy}{\sqrt{x^2 + y^2}} \right| \leq \frac{(\sqrt{x^2 + y^2})^2}{\sqrt{x^2 + y^2}} = \sqrt{x^2 + y^2} \to 0 \qquad \text{as } (x, y) \to (0, 0).$$

The graph is sketched in Figure 16.2.10. ■

■ *Definition*

A point is called an **interior point** of a set $S$ if there is some $\delta > 0$ such that all points that are within $\delta$ of the point are in $S$. A set $S$ is called **open** if all points in $S$ are interior points of $S$. ■

The set of points $(x, y)$ that satisfy $x^2 + y^2 < 1$ is an example of an open set. See Figure 16.2.11.

FIGURE 16.2.12

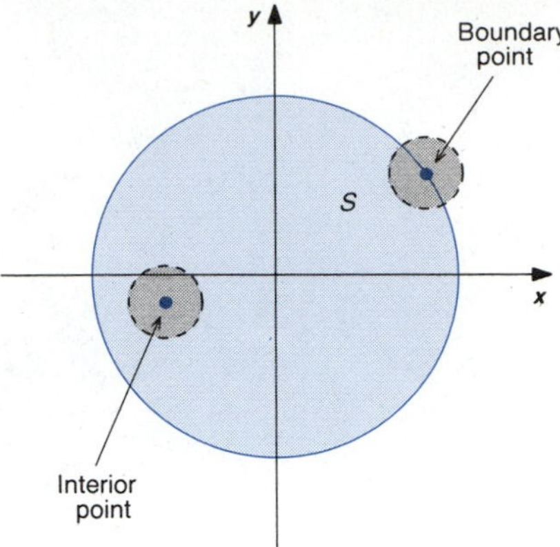


*Definition*

A point is called a **boundary point** of a set $S$ if, for each $\delta > 0$, there is a point in $S$ within $\delta$ of the point and a point not in $S$ that is within $\delta$ of the point. A set $S$ is called **closed** if it contains all its boundary points.

The set of points $(x, y)$ that satisfy $x^2 + y^2 \leq 1$ is an example of a closed region. The boundary of the region is the set of all points on the circle $x^2 + y^2 = 1$, and the interior of the region is the set of points that are inside the circle. See Figure 16.2.12.

Every point of a set must be either an interior point of $S$ or a boundary point of $S$; no point of $S$ can be both an interior point and a boundary point of $S$.

*Definition*

A function $f(x, y)$ is **continuous at a point** $(a, b)$ if, for each positive number $\varepsilon$, there is some $\delta > 0$ such that $|f(x, y) - f(a, b)| < \varepsilon$ whenever $(x, y)$ is in the domain of $f$ and $\sqrt{(x-a)^2 + (y-b)^2} < \delta$. A function $f$ is said to be **continuous on a set** if it is continuous at every point of the set.

Continuity of functions of three or more variables is defined similarly.

The definition of continuity at a point requires that the function be defined at the point. If $(a, b)$ is an *interior point* of the domain of $f$, the definition can be stated in the form

$$\lim_{(x, y) \to (a, b)} f(x, y) = f(a, b).$$

The fact that the definition excludes consideration of points that are not in the domain of $f$ allows us to consider continuity at boundary points of the domain.

Continuity of functions of several variables at a point can be established by using what we know about limits. We can also verify the following.

**If $f(t)$ is continuous on an interval $I$, and $g(x, y)$ is continuous on a set $S$ with values in $I$ for all $(x, y)$ in $S$, then $f(g(x, y))$ is continuous on $S$.**

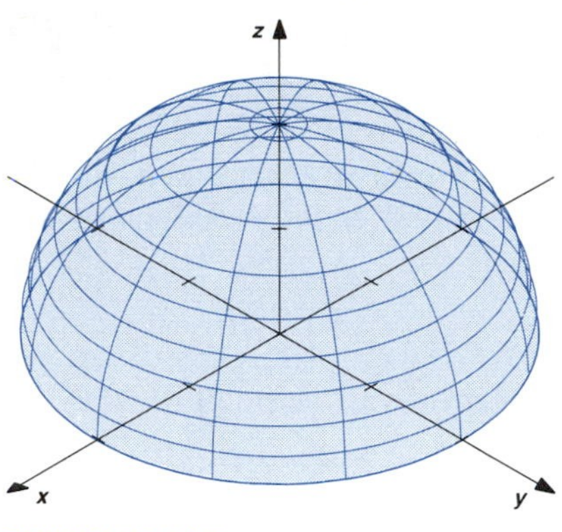


FIGURE 16.2.13

■ **EXAMPLE 9**

Determine the set where $f(x, y) = \sqrt{4 - x^2 - y^2}$ is continuous. Find all boundary points of the set where $f$ is continuous and determine if that set is either open or closed.

We first note that $\sqrt{4 - x^2 - y^2}$ is defined only for $4 - x^2 - y^2 \geq 0$, so $x^2 + y^2 \leq 4$. The function $4 - x^2 - y^2$, $x^2 + y^2 \leq 4$, is a continuous function of $(x, y)$ with nonnegative values. Since the square root function is continuous on the interval $[0, \infty)$, we conclude that $f$ is continuous on its entire domain.

The boundary of the set of points $(x, y)$ that satisfy $x^2 + y^2 \leq 4$ consists of those points on the circle $x^2 + y^2 = 4$. Since the set of points with $x^2 + y^2 \leq 4$ contains all of its boundary points, it is a closed set.

The graph is sketched in Figure 16.2.13. ■

## EXERCISES 16.2

For each of the limits in Exercises 1–22, evaluate the limit or show the limit does not exist.

**1** $\lim_{(x,y)\to(4,2)} (2x - 3y + 5)$

**2** $\lim_{(x,y)\to(3,\pi/2)} \sin xy$

**3** $\lim_{(x,y)\to(0,0)} e^{x^2+y^2}$

**4** $\lim_{(x,y)\to(3,-2)} \ln(x + y)$

**5** $\lim_{(x,y)\to(0,0)} \cot(x^2 + y^2)$

**6** $\lim_{(x,y)\to(3,-2)} \ln\sqrt{x^2 + y^2}$

**7** $\lim_{(x,y)\to(1,-1)} \dfrac{x^2 - 2xy + y^2}{x - y}$

**8** $\lim_{(x,y)\to(2,2)} \dfrac{x^2 - y^2}{x - y}$

**9** $\lim_{(x,y)\to(3,1)} \dfrac{x^2 - y^2}{x - y}$

**10** $\lim_{(x,y)\to(0,-2)} \dfrac{x^2 - y^2}{x - y}$

**11** $\lim_{(x,y)\to(0,0)} \dfrac{x^2}{x^2 + y^2}$

**12** $\lim_{(x,y)\to(0,0)} \dfrac{y^3}{x^2 + y^2}$

**13** $\lim_{(x,y)\to(0,0)} \dfrac{xy^2}{x^2 + y^2}$

**14** $\lim_{(x,y)\to(0,0)} \dfrac{xy}{x^2 + y^2}$

**15** $\lim_{(x,y)\to(0,0)} \dfrac{x^2 + 2y^2}{x^2 + y^2}$

**16** $\lim_{(x,y)\to(0,0)} \dfrac{x^3 + y^3}{x^2 + y^2}$

**17** $\lim_{(x,y)\to(0,0)} \dfrac{x^3 + y}{x^2 + y^2}$

**18** $\lim_{(x,y)\to(0,0)} \dfrac{x^3 + 2x^2y}{x^2 + y^2}$

**19** $\lim_{(x,y)\to(0,0)} f(x, y)$, where $f(x, y) = \begin{cases} x, & y = x^2 \\ 0, & y \neq x^2 \end{cases}$

**20** $\lim_{(x,y)\to(0,0)} f(x, y)$, where $f(x, y) = \begin{cases} x, & y = x^3 \\ y, & y \neq x^3 \end{cases}$

**21** $\lim_{(x,y)\to(0,0)} f(x, y)$, where $f(x, y) = \begin{cases} 1, & y = x^2 + 1 \\ 0, & y \neq x^2 + 1 \end{cases}$

**22** $\lim_{(x,y)\to(0,0)} f(x, y)$, where

$$f(x, y) = \begin{cases} 1 - x^2 - y^2, & x^2 + y^2 < 1 \\ 0, & x^2 + y^2 \geq 1 \end{cases}$$

**23** Determine the set where the function $f(x, y) = \sqrt{1 - x^2 - y^2}$ is continuous. Find the boundary points of that set and determine if the set is either open or closed.

**24** Determine the set where the function

$$f(x, y) = \ln(1 - x^2 - y^2)$$

is continuous. Find the boundary points of that set and determine if the set is either open or closed.

**25** Determine the set where the function of Exercise 19 is continuous. Find the boundary points of that set and determine if the set is either open or closed.

**26** Determine the set where the function of Exercise 20 is continuous. Find the boundary points of that set and determine if the set is either open or closed.

**27** Determine the set where the function of Exercise 21 is continuous. Find the boundary points of that set and determine if the set is either open or closed.

**28** Determine the set where the function of Exercise 22 is continuous. Find the boundary points of that set and determine if the set is either open or closed.

**29** Use the definition of limit to show $\lim_{(x,y)\to(a,b)} x = a$.

**30** Use the definition of limit to show $\lim_{(x,y)\to(a,b)} y = b$.

## 16.3 PARTIAL DERIVATIVES

We begin our study of differentiation of functions of several variables by defining **partial derivatives**. At this point we are mostly concerned with the notation and the evaluation of partial derivatives. In later sections we will see how they are used in applications.

*Definition*

The **partial derivative of $f(x, y)$ with respect to $x$** at the point $(a, b)$ is

$$\frac{\partial f}{\partial x} = \lim_{x \to a} \frac{f(x, b) - f(a, b)}{x - a}, \qquad \text{if the limit exists.}$$

The **partial derivative with respect to $y$** at $(a, b)$ is

$$\frac{\partial f}{\partial y} = \lim_{y \to b} \frac{f(a, y) - f(a, b)}{y - b}.$$

The partial derivatives of $f(x, y)$ with respect to $x$ and $y$ at $(a, b)$ may be interpreted geometrically as the slopes of the curves of intersection of the surface $z = f(x, y)$ with the planes $y = b$ and $x = a$, respectively. See Figure 16.3.1.

Generally, a function has a partial derivative that corresponds to each of its variables. For example, a function $f(x, y, z)$ of three variables has three partial derivatives, $\partial f/\partial x$, $\partial f/\partial y$, and $\partial f/\partial z$, defined by

$$\frac{\partial f}{\partial x} = \lim_{x \to a} \frac{f(x, b, c) - f(a, b, c)}{x - a},$$

$$\frac{\partial f}{\partial y} = \lim_{y \to b} \frac{f(a, y, c) - f(a, b, c)}{y - b}, \quad \text{and}$$

$$\frac{\partial f}{\partial z} = \lim_{z \to c} \frac{f(a, b, z) - f(a, b, c)}{z - c}.$$

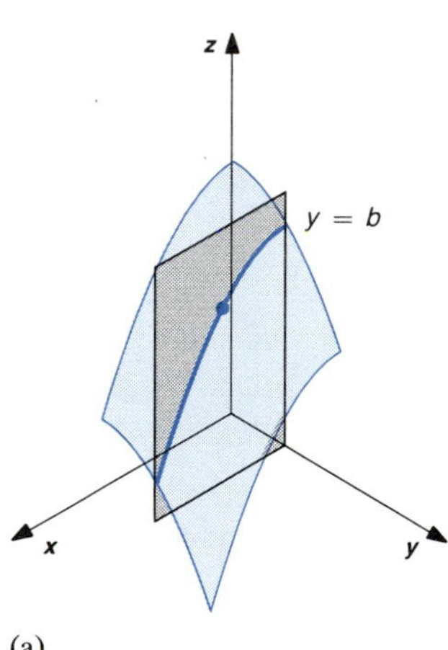


(a)

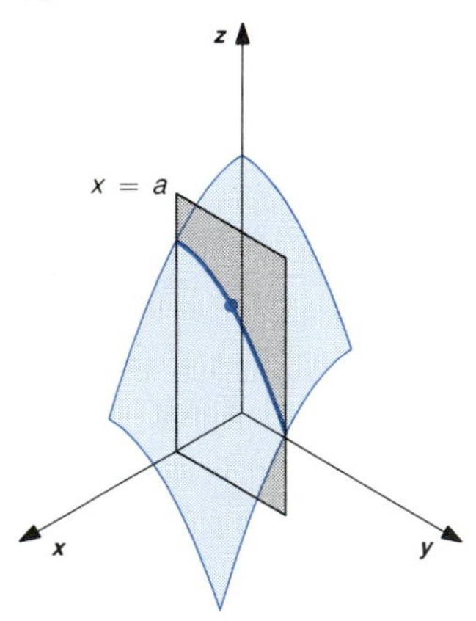


(b)

**FIGURE 16.3.1**

The notation $f_x$ is sometimes used to denote the partial derivative of $f$ with respect to the variable $x$. If $w = f(x, \ldots)$, we may also write $\partial w/\partial x$ in place of $\partial f/\partial x$ or $w_x$ in place of $f_x$.

As in the case of the derivative of a function of one real variable, we have defined partial derivatives at a *point* in the domain of the function. As this point varies we obtain a *function.* The partial derivatives of $f$ are functions of the same variables as the original function. The domain of each partial derivative consists of those points in the domain of $f$ where the partial derivative exists.

Each partial derivative is defined in terms of a limit of a single real variable. Partial derivatives are ordinary derivatives of the functions of one variable that are obtained by fixing the values of all but one variable of the function. This means that we can use all of the rules of differentiation we developed for functions of one variable to evaluate partial derivatives.

It is only necessary to treat all variables, except the variable of partial differentiation, as constants. If the function is given by a formula, the rules of differentiation give us formulas for the partial derivatives. The value of a partial derivative at a point can then be determined by substitution.

■ **EXAMPLE 1**

$f(x, y) = x^2y^3$. Evaluate $f_x(3, 5)$ and $f_y(3, 5)$.

**SOLUTION**

We first find formulas for $f_x(x, y)$ and $f_y(x, y)$. To find $f_x(x, y)$ we consider $y$ to be constant and differentiate $f(x, y) = x^2y^3$ with respect to $x$. The expression $x^2y^3$ is then read as $x^2$ times the "constant" $y^3$. The derivative with respect to $x$ is then the derivative of $x^2$ times the "constant" $y^3$. We obtain

$$f_x(x, y) = (2x)(y^3) = 2xy^3.$$

The partial derivative with respect to $y$ is found by considering $x$ to be a constant and differentiating with respect to $y$. In this case, $x^2y^3$ is read as the "constant" $x^2$ times $y^3$. The derivative with respect to $y$ is the "constant" $x^2$ times the derivative of $y^3$. This gives

$$f_y(x, y) = (x^2)(3y^2) = 3x^2y^2.$$

Substitution into the above formulas gives

$$f_x(3, 5) = 2(3)(5)^3 = 750 \quad \text{and}$$

$$f_y(3, 5) = 3(3)^2(5)^2 = 675.$$

■

■ **EXAMPLE 2**

$f(x, y) = x^2 - 2xy + 3y^2$. Evaluate $f_x$ and $f_y$.

**SOLUTION**

To evaluate the partial derivative $f_x$, we consider $y$ to be a constant and differentiate with respect to the variable $x$. We have

$$f(x, y) = x^2 - 2xy + 3y^2, \quad \text{so}$$

$$f_x(x, y) = 2x - 2y + 0 = 2x - 2y.$$

The partial derivative with respect to $y$ is found by considering $x$ to be a constant and differentiating with respect to $y$. We have

$$f(x, y) = x^2 - 2xy + 3y^2, \quad \text{so}$$

$$f_y(x, y) = 0 - 2x + 3(2y) = -2x + 6y.$$

■

**■ EXAMPLE 3**

$f(x, y) = xe^{xy}$. Find $f_x$ and $f_y$.

**SOLUTION**

If $f(x, y) = xe^{xy}$, then

$$f_x = (x)(e^{xy}y) + (e^{xy})(1) = e^{xy}(xy + 1) \quad \text{and}$$

$$f_y = x(e^{xy}x) = x^2e^{xy}.$$

Note that the Product Rule was used to find $f_x$ and the Chain Rule was used to differentiate $e^{(\text{Function})}$ in both formulas. ■

**■ EXAMPLE 4**

$f(x, y, z) = ze^{x^2y^3}$, Find $f_x$, $f_y$, and $f_z$.

**SOLUTION**

The partial derivative with respect to $x$ is found by considering the variables $y$ and $z$ to be constants and then differentiating $f(x, y, z) = ze^{x^2y^3}$ with respect to $x$. We obtain

$$f_x = z(e^{x^2y^3})(2xy^3).$$

If $x$ and $z$ are considered to be constants, differentiation of $f(x, y, z) = ze^{x^2y^3}$ with respect to $y$ gives

$$f_y = z(e^{x^2y^3})(3x^2y^2).$$

The partial derivative with respect to $z$ is obtained by considering $x$ and $y$ to be constants while we differentiate $f(x, y, z) = ze^{x^2y^3}$ with respect to $z$. This gives

$$f_z = e^{x^2y^3}.$$

■

**■ EXAMPLE 5**

$f(x, y, z) = \sin(x - y)$. Find $f_x$, $f_y$, and $f_z$.

**SOLUTION**

If $f(x, y, z) = \sin(x - y)$, then

$$f_x = (\cos(x - y))(1 - 0) = \cos(x - y),$$

$$f_y = (\cos(x - y))(0 - 1) = -\cos(x - y),$$

$$f_z = 0.$$

■

Since partial derivatives are functions, we may take partial derivatives of the partial derivatives. Notation for such higher-order partial derivatives is indicated below.

$$\frac{\partial}{\partial x}\left(\frac{\partial f}{\partial x}\right) = \frac{\partial^2 f}{\partial x^2} = (f_x)_x = f_{xx},$$

$$\frac{\partial}{\partial y}\left(\frac{\partial f}{\partial x}\right) = \frac{\partial^2 f}{\partial y \partial x} = (f_x)_y = f_{xy},$$

$$\frac{\partial}{\partial x}\left(\frac{\partial f}{\partial y}\right) = \frac{\partial^2 f}{\partial x \partial y} = (f_y)_x = f_{yx},$$

$$\frac{\partial}{\partial y}\left(\frac{\partial f}{\partial y}\right) = \frac{\partial^2 f}{\partial y^2} = (f_y)_y = f_{yy}.$$

The partial derivatives

$$\frac{\partial^2 f}{\partial y \partial x} = \frac{\partial}{\partial y}\left(\frac{\partial f}{\partial x}\right) \quad \text{and} \quad f_{xy} = (f_x)_y$$

are evaluated by first taking the partial derivative with respect to $x$ and then taking the partial derivative with respect to $y$. The order is reversed to evaluate

$$\frac{\partial^2 f}{\partial x \partial y} = \frac{\partial}{\partial x}\left(\frac{\partial f}{\partial y}\right) \quad \text{and} \quad f_{yx} = (f_y)_x.$$

The functions $f_x$ and $f_y$ are called **first-order** partial derivatives of $f$; $f_{xx}$, $f_{xy}$, $f_{yx}$, and $f_{yy}$ are called **second-order** partial derivatives of $f$.

### ■ EXAMPLE 6

Find all first- and second-order partial derivatives of the function $f(x, y) = x \sin y$.

#### SOLUTION

Since $f$ is a function of two variables, there are two first-order derivatives. We have $f(x, y) = x \sin y$, so

$$f_x = \sin y,$$

$$f_y = x \cos y.$$

The second-order derivatives are obtained by taking partial derivatives of each of the first-order partials. We obtain

$$f_{xx} = (f_x)_x = \frac{\partial}{\partial x}(\sin y) = 0,$$

$$f_{xy} = (f_x)_y = \frac{\partial}{\partial y}(\sin y) = \cos y,$$

$$f_{yx} = (f_y)_x = \frac{\partial}{\partial x}(x \cos y) = \cos y,$$

$$f_{yy} = (f_y)_y = \frac{\partial}{\partial y}(x \cos y) = -x \sin y.$$

■

For most of the functions we deal with it is true that corresponding **mixed partials** such as $f_{xy}$ and $f_{yx}$ are equal, even though the order of differentiation is different. The following theorem tells us that the order in which the partial derivatives are taken does not matter if all of the partial derivatives involved are continuous. If the partial derivatives are not continuous, it is possible that $f_{xy} \neq f_{yx}$. (See Exercise 31.)

■ *Theorem*

**If $f(x, y)$ is defined in an open region $R$, and the functions $f, f_x, f_y, f_{xy}$, and $f_{yx}$ are all continuous in $R$, then $f_{xy} = f_{yx}$ in $R$.**

We will not verify this result.

## EXERCISES 16.3

**1** $f(x, y) = x^2 - y^2$. Evaluate $f_x(2, 3)$ and $f_y(2, 3)$.

**2** $f(x, y) = x^2 + 2xy + y^2$. Evaluate $f_x(1, 2)$ and $f_y(1, 2)$.

**3** $f(x, y) = x^3y^2 - xy^2$. Evaluate $f_x(-1, 3)$ and $f_y(-1, 3)$.

**4** $f(x, y) = x^3 - 3x^2y^2 + 3xy^2 - y^3$. Evaluate $f_x(3, 2)$ and $f_y(3, 2)$.

Find all first-order partial derivatives of the functions in Exercises 5–24.

**5** $f(x, y) = 2x^2y - xy^3 + 3x - y + 7$

**6** $f(x, y) = x^3y^2 - 4xy^4 - x^2 + 2y - 5$

**7** $f(x, y) = \dfrac{x}{x + y}$

**8** $f(x, y) = \dfrac{2x - y}{3x + 5y}$

**9** $f(x, y) = x\sqrt{x^2 + y^2}$

**10** $f(x, y) = y \sin(x + y)$

**11** $f(x, y) = \ln(x^2 + y^2)$

**12** $f(x, y) = e^{-x^2 - y^2}$

**13** $f(x, y) = \tan^{-1}(y/x)$

**14** $f(x, y) = \sin^{-1} xy$

**15** $f(x, y) = \tan(2x - y)$

**16** $f(x, y) = \cos 2xy$

**17** $f(x, y) = \sin^2(x^2y^2)$

**18** $f(x, y) = 1/\sqrt{x^2 + y^2}$

**19** $f(x, y, z) = xy + yz + xz$

**20** $f(x, y, z) = \sqrt{x^2 + y^2 + z^2}$

**21** $f(x, y, z) = \dfrac{x - y}{x + z}$

**22** $f(x, y, z) = 2\sqrt{xyz}$

**23** $f(x, y, z) = e^{2x - y}$

**24** $f(x, y, z) = z \ln yz$

Find all first- and second-order partial derivatives of the functions in Exercises 25–30.

**25** $f(x, y) = x^3 + x^2y - 3xy^2 + y^3$

**26** $f(x, y) = 3x^2 - 2xy + 5y^2 - x + 4y + 1$

27 $f(x, y) = x \ln y$

28 $f(x, y) = e^{xy}$

29 $f(x, y) = \sin xy$

30 $f(x, y) = 1 - \sqrt{x^2 + y^2}$

31 $f(x, y) = \begin{cases} \dfrac{xy^3 - x^3y}{x^2 + y^2}, & (x, y) \neq (0, 0), \\ 0, & (x, y) = (0, 0). \end{cases}$

Show that $f_{xy}(0, 0) \neq f_{yx}(0, 0)$. [*Hint*: Show that $f_x(0, y) = y$, so $f_{xy}(0, 0) = 1$. Show that $f_y(x, 0) = -x$, so $f_{yx}(0, 0) = -1$.]

## 16.4 LINEAR APPROXIMATION, DIFFERENTIABILITY

In this section we will investigate the possibility of using a linear function to find approximate values of a function $f(x, y)$ for $(x, y)$ near a point $(a, b)$. The ideas extend to functions of any number of variables and lead to the concept of differentiability of functions of several variables.

If the function $f(x)$ of a single variable is differentiable at $c$, we have seen in Section 4.8 that the linear function

$$L(x) = f(c) + f'(c)(x - c)$$

can be used to find approximate values of $f(x)$ for $x$ near $c$. The function $L$ is the linear function with the property that its value and the value of its derivative at $c$ agree with those of $f(x)$. This suggests that, for functions $f(x, y)$ of two variables, we consider the linear function

$$L(x, y) = f(a, b) + f_x(a, b)(x - a) + f_y(a, b)(y - b).$$

The values of this function and the values of its two first partial derivatives at $(a, b)$ agree with those of $f(x, y)$.

The problem of determining when the linear function $L(x, y)$ gives good approximate values of $f(x, y)$ is more complicated than in the case of functions of a single variable. First, we should note that the partial derivatives $f_x(a, b)$ and $f_y(a, b)$ depend only on the behavior of $f(x, y)$ for $(x, y)$ on the two lines $x = a$ and $y = b$ in the $xy$-plane. The values of $f(x, y)$ at other points do not have any effect on $L$. Figure 16.4.1 contains a sketch of the graph of the function

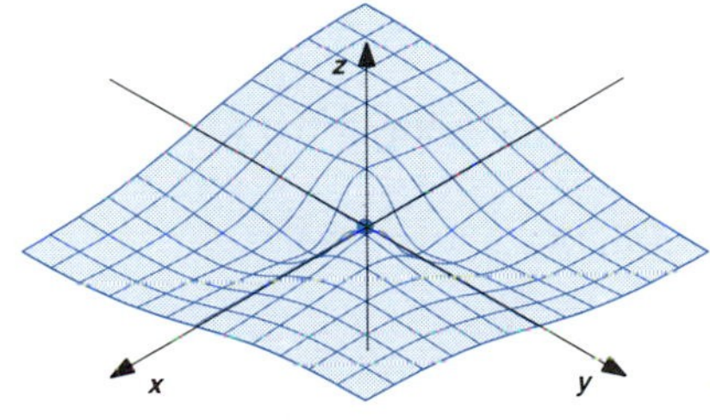

FIGURE 16.4.1

$$f(x, y) = \begin{cases} \dfrac{xy}{x^2 + y^2}, & (x, y) \neq (0, 0), \\ 0, & (x, y) = (0, 0). \end{cases}$$

We have $f(0, 0) = f_x(0, 0) = f_y(0, 0) = 0$, so the linear "approximation" of $f(x, y)$ near the origin is $L(x, y) = 0$. We do have $L(x, y) = f(x, y) = 0$ for $(x, y)$ on either the $x$-axis or the $y$-axis, but zero is not a good approximation of $f(x, y)$ for points that are not on the axes. For example, $f(x, x) = 1/2$ for $x \neq 0$, so the value of $f(x, y)$ is $1/2$ as $(x, y)$ approaches the origin along the line $y = x$. This example also shows that existence of the partial derivatives at a point does not even imply that the function is continuous at the point.

The following theorem gives us conditions that guarantee that the values of $L(x, y)$ and $f(x, y)$ are close to each other for $(x, y)$ near $(a, b)$.

*Theorem*

**If $f(x, y)$ has partial derivatives $f_x$ and $f_y$ in a disk with center $(a, b)$, and $f_x$ and $f_y$ are continuous at $(a, b)$, then there are functions $\varepsilon_1$ and $\varepsilon_2$ of $x$ and $y$ such that $\varepsilon_1 \to 0$ and $\varepsilon_2 \to 0$ as $(x, y) \to (a, b)$, and**

$$f(x, y) - f(a, b) = f_x(a, b)(x - a) + f_y(a, b)(y - b) + \varepsilon_1 \cdot (x - a) + \varepsilon_2 \cdot (y - b).$$

*Proof* Let us write

$$f(x, y) - f(a, b) = f(x, y) - f(x, b) + f(x, b) - f(a, b).$$

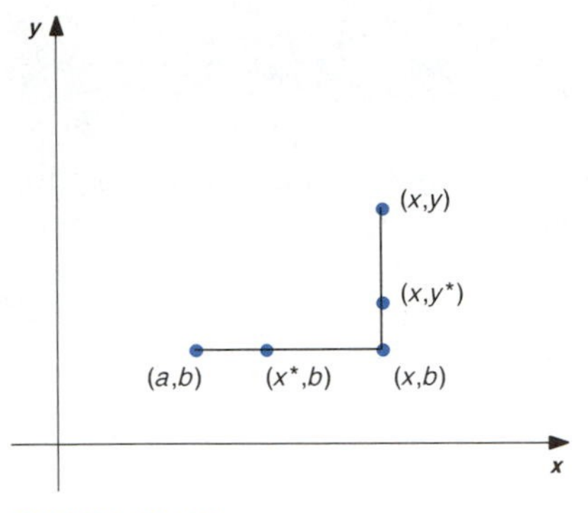

FIGURE 16.4.2

Since $f_y$ is the derivative of the function of $y$ that we obtain by fixing the value of $x$ in $f(x, y)$, the Mean Value Theorem applied to that function of $y$ gives $f(x, y) - f(x, b) = f_y(x, y^*)(y - b)$ for some $y^*$ between $y$ and $b$. Similarly, $f(x, b) - f(a, b) = f_x(x^*, b)(x - a)$ for some $x^*$ between $x$ and $a$. See Figure 16.4.2. Combining results, we obtain

$$\begin{aligned} f(x, y) - f(a, b) &= f_y(x, y^*)(y - b) + f_x(x^*, b)(x - a) \\ &= f_x(a, b)(x - a) + f_y(a, b)(y - b) \\ &\quad + [f_x(x^*, b) - f_x(a, b)](x - a) \\ &\quad + [f_y(x, y^*) - f_y(a, b)](y - b). \end{aligned}$$

The conclusion of the theorem is satisfied if we choose $\varepsilon_1 = f_x(x^*, b) - f_x(a, b)$ and $\varepsilon_2 = f_y(x, y^*) - f_y(a, b)$. The continuity of $f_x$ and $f_y$ at $(a, b)$ implies that $\varepsilon_1$ and $\varepsilon_2$ approach zero as $(x, y)$ approaches $(a, b)$. ■

Let us rewrite the conclusion of the above theorem in the form

$$f(x, y) - [f(a, b) + f_x(a, b)(x - a) + f_y(a, b)(y - b)] = \varepsilon_1 \cdot (x - a) + \varepsilon_2 \cdot (y - b), \quad \text{or}$$

$$f(x, y) - L(x, y) = \varepsilon_1 \cdot (x - a) + \varepsilon_2 \cdot (y - b).$$

This shows that the values of $f(x, y)$ and $L(x, y)$ are close whenever $(x, y)$ is close to $(a, b)$. In fact, we have

$$\left|\frac{f(x, y) - L(x, y)}{\sqrt{(x - a)^2 + (y - b)^2}}\right| = \left|\frac{\varepsilon_1 \cdot (x - a)}{\sqrt{(x - a)^2 + (y - b)^2}} + \frac{\varepsilon_2 \cdot (y - b)}{\sqrt{(x - a)^2 + (y - b)^2}}\right|$$
$$\leq |\varepsilon_1| + |\varepsilon_2| \to 0 \qquad \text{as } (x, y) \to (a, b),$$

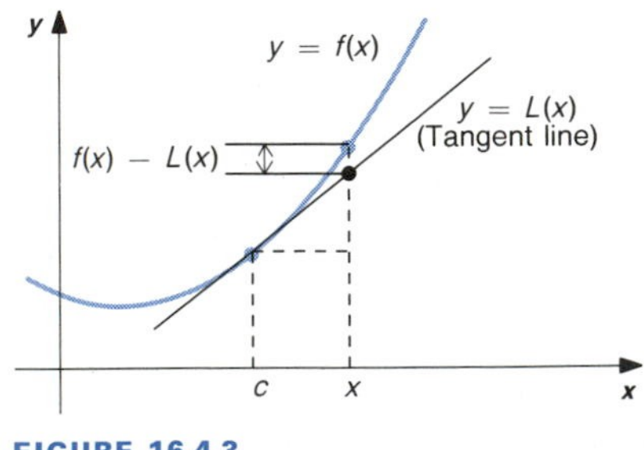

FIGURE 16.4.3

so the ratio of the difference $f(x, y) - L(x, y)$ and the distance between $(x, y)$ and $(a, b)$ approaches zero as $(x, y)$ approaches $(a, b)$. This corresponds to the behavior of a function of one variable near a point where the function is differentiable. That is, $y = L(x)$ is the line tangent to the graph of $f(x)$ at $(c, f(c))$. The differentiability of $f(x)$ at $c$ implies that the ratio of the difference $f(x) - L(x)$ and the distance between $x$ and $c$ approaches zero as $x$ approaches $c$. See Figure 16.4.3.

For functions of two variables, we take the condition of the conclusion of the above theorem to be the definition of differentiable. It is convenient to express the condition in terms of delta notation. For a fixed point $(a, b)$, let

$$\Delta x = x - a, \quad \Delta y = y - b, \quad \text{and} \quad \Delta f = f(x, y) - f(a, b).$$

*Definition*

A function $f(x, y)$ is said to be **differentiable** at $(a, b)$ if the partial derivatives $f_x(a, b)$ and $f_y(a, b)$ exist, and if there are functions $\varepsilon_1$ and $\varepsilon_2$ such that $\varepsilon_1$ and $\varepsilon_2$ approach zero as $(\Delta x, \Delta y)$ approaches $(0, 0)$, and

$$\Delta f = f_x(a, b)\,\Delta x + f_y(a, b)\,\Delta y + \varepsilon_1\,\Delta x + \varepsilon_2\,\Delta y.$$

We shall see that the above form of the definition of differentiability is convenient for *theory*. In particular, we will use the definition to develop formulas of differentiation. The definition is not convenient for determining if a given function is differentiable. We will not need to verify the condition of the definition for individual functions. The previous theorem tells us that *a function that has first partial derivatives at all points near $(a, b)$ is differentiable at $(a, b)$ if the partials are continuous there.*

The *form* of the condition of differentiability extends easily to any number of variables. The only difference is in the geometric interpretation of the condition. The theory is the same. For example, a function $f(x, y, z)$ is defined to be **differentiable** at $(a, b, c)$ if all first partial derivatives of $f$ exist at $(a, b, c)$, and there are functions $\varepsilon_1$, $\varepsilon_2$, and $\varepsilon_3$ such that $\varepsilon_1$, $\varepsilon_2$, and $\varepsilon_3$ approach zero as $(\Delta x, \Delta y, \Delta z)$ approaches $(0, 0, 0)$, and

$$\Delta f = f_x(a, b, c)\Delta x + f_y(a, b, c)\Delta y + f_z(a, b, c)\Delta z + \varepsilon_1\Delta x + \varepsilon_2\Delta y + \varepsilon_3\Delta z.$$

A function of three variables is differentiable at $(a, b, c)$ if it has first partial derivatives at all points near $(a, b, c)$ and these derivatives are continuous at $(a, b, c)$. We say a function is **differentiable on a set** if it is differentiable at each point of the set.

It follows from the definition of differentiability that *a function that is differentiable at a point is continuous at the point.* Continuity requires only that $\Delta f \to 0$ as $(\Delta x_1, \ldots, \Delta x_n) \to (0, \ldots, 0)$.

As in the case of a single variable, it is convenient to describe linear approximation in terms of differentials. If a function $f(x, y)$ has partial derivatives $f_x(a, b)$ and $f_y(a, b)$, we define the **differential** of $f$ at $(a, b)$ to be

$$df = f_x(a, b)\,dx + f_y(a, b)\,dy,$$

where $dx$ and $dy$ are real-valued variables. The differential of a function of any number of variables is defined similarly.

For some problems it is convenient to replace $(a, b)$ by $(x, y)$ in the definition of differentials.

■ **EXAMPLE 1**

$f(x, y) = x^2y$. Express the differential $df$ in terms of the variables $x$, $y$, $dx$, and $dy$.

**SOLUTION**

$$df = f_x(x, y)\,dx + f_y(x, y)\,dy = 2xy\,dx + x^2\,dy. \quad ■$$

The idea of differential may be used with functions of any number of variables.

### ■ EXAMPLE 2

$f(x, y, z) = z\,\sin(x^2 + y^2)$. Express the differential $df$ in terms of the variables $x$, $y$, $z$, $dx$, $dy$, and $dz$.

**SOLUTION**

$$\begin{aligned} df &= f_x(x, y, z)\,dx + f_y(x, y, z)\,dy + f_z(x, y, z)\,dz \\ &= z\cos(x^2 + y^2)(2x)\,dx + z\cos(x^2 + y^2)(2y)\,dy + \sin(x^2 + y^2)\,dz. \quad ■ \end{aligned}$$

We can use differentials of functions of several variables in the same way we used differentials of a single variable. We have

$$\Delta f = f(x, y) - f(a, b), \qquad \Delta x = x - a, \quad \Delta y = y - b.$$

If $f(x, y)$ is differentiable at $(a, b)$ and we set $dx = \Delta x$ and $dy = \Delta y$, then we have

$$\Delta f \approx df$$

for small values of the increments. Also,

$$f(x, y) \approx f(a, b) + df.$$

The expression $\Delta f$ represents the change in $f$ due to a change in $x$ and $y$; $df$ represents the approximate change in $f$. The differential $df$ is sometimes interpreted as the approximate error in $f$ due to possible errors of $dx$ and $dy$ in the variables $x$ and $y$, respectively. The ideas are the same for functions of any number of variables.

### ■ EXAMPLE 3

$f(x, y) = 1/\sqrt{x^2 + y^2}$. Calculate $f(3, 4)$, $f(3.2, 3.9)$, and $\Delta f = f(3.2, 3.9) - f(3, 4)$. Find $df$, the differential approximation of $\Delta f$.

**SOLUTION**

$$f(3, 4) = \frac{1}{\sqrt{3^2 + 4^2}} = \frac{1}{5} = 0.2,$$

$$f(3.2, 3.9) = \frac{1}{\sqrt{(3.2)^2 + (3.9)^2}} \approx 0.19822394 \quad \text{(calculator value)},$$

$$\Delta f = f(3.2, 3.9) - f(3, 4) \approx -0.00177606.$$

The differential $df$ is calculated with $(x, y) = (3.2, 3.9)$ and $(a, b) = (3, 4)$, so $dx = x - a = 0.2$ and $dy = y - b = -0.1$.

$$f_x(x, y) = -\frac{1}{2}(x^2 + y^2)^{-3/2}(2x) = -\frac{x}{(x^2 + y^2)^{3/2}} \quad \text{and}$$

$$f_y(x, y) = -\frac{1}{2}(x^2 + y^2)^{-3/2}(2y) = -\frac{y}{(x^2 + y^2)^{3/2}},$$

so $f_x(3, 4) = -0.024$ and $f_y(3, 4) = -0.032$. Then

$$df = (-0.024)(0.2) + (-0.032)(-0.1) = -0.0016.$$

■

### ■ EXAMPLE 4

Two adjacent sides and the included angle of a triangle are measured to be 4 ft, 5 ft, and 30°, respectively. Find the area of the triangle if the measurements are assumed to be exact. Use differentials to find the approximate maximum error in the area of the triangle due to possible errors of 0.2 ft in the measurements of the sides and an error of 5° in the measurement of the angle.

#### SOLUTION

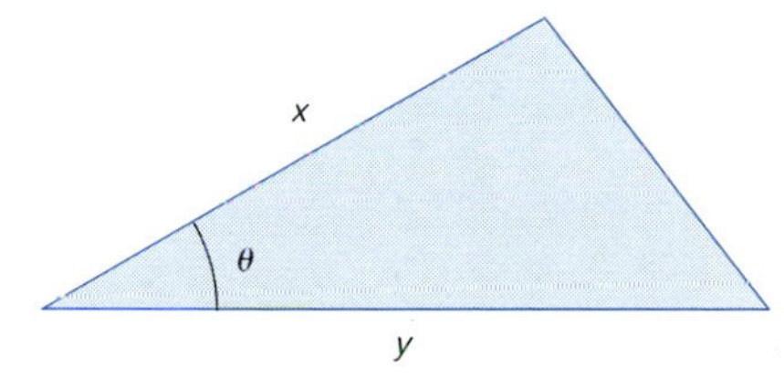

FIGURE 16.4.4

A sketch of the triangle is given in Figure 16.4.4. The variables $x$ and $y$ represent the lengths of the sides in feet and the variable $\theta$ represents the angle. Note that we cannot study the effect of changes in measurements of the sides and angle on the area unless we consider the values of the measurements to be variables.

The area of the triangle in square feet is

$$A(x, y, \theta) = \frac{1}{2} xy \sin \theta.$$

Since we are going to be differentiating a trigonometric function in this problem, we measure angles in *radians*. (Recall that the formulas for the differentiation of the trigonometric functions were derived for radian measure.) We have

$$30° = \frac{\pi}{6} \text{ rad} \quad \text{and} \quad 5° = 5\left(\frac{\pi}{180}\right) = \frac{\pi}{36} \text{ rad}.$$

If the measurements are exact, the area is

$$A(4, 5, \pi/6) = \frac{1}{2}(4)(5)\left(\frac{1}{2}\right) = 5 \text{ ft}^2.$$

The desired approximate value of the error in area is given by the differential $dA$, with the partial derivatives evaluated at $(x, y, \theta) = (4, 5, \pi/6)$, $dx = \pm 0.2$, $dy = \pm 0.2$, and $d\theta = \pm\pi/36$. We have

$$A_x(x, y, \theta) = \frac{1}{2} y \sin\theta, \quad \text{so} \quad A_x(4, 5, \pi/6) = 1.25,$$

$$A_y(x, y, \theta) = \frac{1}{2} x \sin\theta, \quad \text{so} \quad A_y(4, 5, \pi/6) = 1,$$

$$A_\theta(x, y, \theta) = \frac{1}{2} xy \cos\theta, \quad \text{so} \quad A_\theta(4, 5, \pi/6) = 5\sqrt{3}.$$

Then

$$\begin{aligned} dA &= A_x dx + A_y dy + A_\theta d\theta \\ &= (1.25)(\pm 0.2) + (1)(\pm 0.2) + (5\sqrt{3})(\pm\pi/36). \end{aligned}$$

The *maximum* possible error would occur when the errors that correspond to each variable have the same sign. The approximate value of the maximum possible error is the sum of the absolute values of each of the terms in the differential. Thus, we obtain

$$|(1.25)(0.2)| + |(1)(0.2)| + |(5\sqrt{3})(\pi/36)| \approx 1.2057497 \text{ ft}^2.$$ ■

■ **EXAMPLE 5**

The magnitude of the force due to gravitational attraction between two masses $m_1$ and $m_2$ that are a distance $D$ apart is

$$F = k\frac{m_1 m_2}{D^2},$$

where $k$ is a constant. Find the approximate maximum percent error in $F$ due to possible errors of 5 percent in measurement of the masses and a possible error of 4 percent in the measurement in the distance.

**SOLUTION**

The magnitude of the force is a function of the three variables $m_1$, $m_2$, and $D$. We have

$$F_{m_1} = \frac{km_2}{D^2}, \quad F_{m_2} = \frac{km_1}{D^2}, \quad \text{and} \quad F_D = -\frac{2km_1 m_2}{D^3}.$$

The errors in measurements are $dm_1 = \pm 0.05m_1$, $dm_2 = \pm 0.05m_2$, and $dD = \pm 0.04D$. The differential gives the approximate error in $F$. We have

$$\begin{aligned} dF &= F_{m_1} dm_1 + F_{m_2} dm_2 + F_D dD \\ &= \left(\frac{km_2}{D^2}\right)(\pm 0.05m_1) + \left(\frac{km_1}{D^2}\right)(\pm 0.05m_2) + \left(\frac{-2km_1 m_2}{D^3}\right)(\pm 0.04D) \\ &= \pm 0.05F \pm 0.05F \mp 0.08F. \end{aligned}$$

The approximate maximum error is the sum of the absolute values of the terms of the differential,

$$0.05F + 0.05F + 0.08F = 0.18F, \quad \text{or} \quad 18 \text{ percent of } F. \qquad \blacksquare$$

It is important that we realize that the idea of differentials, or linear approximation, is to obtain information about a function from information about its partial derivatives. Thus, we may find approximate values of a function for which we may not have a formula, but for which values of the partial derivatives have been determined experimentally.

### ■ EXAMPLE 6

It is desired to know the effect on the total resistance of a complicated electric circuit due to changes in the resistance of two resistors in the circuit. It is determined experimentally that keeping the second resistor fixed and changing the first by $\Delta R_1$ results in a change of $0.2\, \Delta R_1$ in the resistance of the circuit. Changing the second resistor by $\Delta R_2$ while the first is kept fixed gives a change of $0.15\, \Delta R_2$ in the resistance of the circuit. What is the approximate change in the resistance of the circuit if the first resistor is increased by 10 ohms while the second resistor is decreased by 4 ohms?

#### SOLUTION

The resistance of the circuit $R$ is given by an unknown function of the resistance of the two variable resistors $R_1$ and $R_2$,

$$R = R(R_1, R_2).$$

For fixed values of $R_2$, the rate of change of $R$ with respect to $R_1$ is the partial derivative of $R$ with respect to $R_1$. The experimental data indicate that

$$\frac{\partial R}{\partial R_1} \approx \frac{\Delta R}{\Delta R_1} = \frac{0.2\, \Delta R_1}{\Delta R_1} = 0.2,$$

so we take $\partial R/\partial R_1 = 0.2$. Similarly,

$$\frac{\partial R}{\partial R_2} \approx \frac{\Delta R}{\Delta R_2} = \frac{0.15\, \Delta R_2}{\Delta R_2} = 0.15,$$

so we take $\partial R/\partial R_1 = 0.15$. Using $dR_1 = +10$ and $dR_2 = -4$, we have

$$dR = \frac{\partial R}{\partial R_1}\, dR_1 + \frac{\partial R}{\partial R_2}\, dR_2 = (0.2)(10) + (0.15)(-4) = 1.4.$$

The resistance of the circuit will increase by approximately 1.4 ohms. ■

## EXERCISES 16.4

Express the differential $df$ in terms of the variables $x$, $y$, $dx$, and $dy$ for each of the functions given in Exercises 1–2.

1 $f(x, y) = 3x^2 - 5xy + y^2$

2 $f(x, y) = xy^2 - x + 2y - 7$

Express the differential $df$ in terms of the variables $x$, $y$, $z$, $dx$, $dy$, and $dz$ for each of the functions given in Exercises 3–4.

3 $f(x, y, z) = x^2y^3z$

4 $f(x, y, z) = xy^2/z$

5 $f(x, y) = \sqrt{x^2 + y^2}$. Calculate $f(4, 3)$, $f(3.9, 3.1)$, and $\Delta f = f(3.9, 3.1) - f(4, 3)$. Find $df$, the differential approximation of $\Delta f$.

6 $f(x, y) = x^2y$. Calculate $f(2, 3)$, $f(1.9, 3.1)$, and $\Delta f = f(1.9, 3.1) - f(2, 3)$. Find $df$, the differential approximation of $\Delta f$.

7 $f(x, y, z) = xyz$. Calculate $f(2, 3, 1)$, $f(2.1, 3.1, 0.9)$, and $\Delta f = f(2.1, 3.1, 0.9) - f(2, 3, 1)$. Find $df$, the differential approximation of $\Delta f$.

8 $f(x, y, z) = 2x + y - z$. Calculate $f(1, 2, 3)$, $f(1.2, 1.9, 3.1)$, and $\Delta f = f(1.2, 1.9, 3.1) - f(1, 2, 3)$. Find $df$, the differential approximation of $\Delta f$.

9 The angle of elevation to the top of a vertical flagpole from a point measured to be 100 ft horizontally from the base of the flagpole is measured to be 30°. Find the height of the flagpole if the measurements are assumed to be exact. Use differentials to find the approximate maximum error in the height due to possible errors of 2 ft in the measurement of the distance and 1° in the measurement of the angle.

10 Two adjacent sides and the included angle of a triangle are measured to be 2 ft, 3 ft, and 60°. Find the area of the triangle if the measurements are assumed to be exact. Use differentials to find the approximate maximum error in the area due to possible errors of 0.15 ft in the measurement of the sides and an error 2° in the measurement of the angle.

11 Find the total surface area of a right circular cylinder that has radius 2 m and height 3 m. Use differentials to find the approximate maximum error in the surface area due to possible errors of 0.1 m in the measurement of the radius and 0.15 m in the measurement of the height. ($S = 2\pi r^2 + 2\pi rh$.)

12 Find the volume of a right circular cone that has radius of base 3 m and height 6 m. Use differentials to find the approximate maximum error in the volume due to possible errors of 0.2 m in the measurement of the radius and 0.4 m in the measurement of the height. ($V = \pi r^2h/3$.)

13 Find the approximate maximum percent error in the volume of a rectangular box due to possible errors of 3 percent in the measurement of each of its length, width, and height.

14 Find the approximate maximum percent error in the volume of a rectangular box with square base due to possible errors of 5 percent in the measurement of the length of each side of its square base and its height.

15 Find the approximate maximum percent error in the total surface area of the six sides of the rectangular box of Exercise 13.

16 Find the approximate maximum percent error in the total surface area of the six sides of the rectangular box of Exercise 14.

17 It has been determined experimentally that changing one resistor in a circuit by $\Delta R_1$ with other variables in the circuit fixed results in a change of $0.3\,\Delta R_1$ in the resistance of the circuit. Changing a second resistor by $\Delta R_2$ with other variables fixed gives a change of $0.5\,\Delta R_2$ in the resistance of the circuit. What is the approximate change in the resistance of the circuit if the first resistor is increased by 25 Ω while the second resistor is decreased by 12 Ω?

## 16.5 THE CHAIN RULE

If $z = f(x, y)$ is a function of $x$ and $y$, and if $x$ and $y$ are functions of $t$, then $z = f(x(t), y(t))$ is a function of $t$. We wish to develop a formula for the derivative of $z$ with respect to $t$. We assume that $f$ is a differentiable function of $(x, y)$ at $(x(c), y(c))$ and that $x$ and $y$ are differentiable functions of $t$ at $c$.

The derivative of $z$ with respect to $t$ at a point $c$ is

$$\frac{dz}{dt} = \lim_{t \to c} \frac{f(x(t), y(t)) - f(x(c), y(c))}{t - c}.$$

Since $f(x, y)$ is differentiable at $(x(c), y(c))$, we know there are functions $\varepsilon_1$ and $\varepsilon_2$ that approach zero as $(x, y)$ approaches $(x(c), y(c))$ and such that

$$\begin{aligned} f(x(t), y(t)) - f(x(c), y(c)) = {} & f_x(x(c), y(c))(x(t) - x(c)) \\ & + f_y(x(c), y(c))(y(t) - y(c)) \\ & + \varepsilon_1 \cdot (x(t) - x(c)) + \varepsilon_2 \cdot (y(t) - y(c)). \end{aligned}$$

Dividing the above equation by $t - c$, taking the limit as $t$ approaches $c$, and noting that $(x(t), y(t)) \to (x(c), y(c))$ as $t \to c$, we see that

$$\frac{dz}{dt} = f_x(x(c), y(c)) \frac{dx}{dt}(c) + f_y(x(c), y(c)) \frac{dy}{dt}(c).$$

It is useful to write this result in the following form.

If $z = z(x, y)$, $x = x(t)$, and $y = y(t)$, then

$$\frac{dz}{dt} = \frac{\partial z}{\partial x}\frac{dx}{dt} + \frac{\partial z}{\partial y}\frac{dy}{dt}.$$

**EXAMPLE 1**

$z = \ln(y/x)$, $x = \cos t$, $y = \sin t$. Find $dz/dt$.

**SOLUTION**

We first write the formula that applies and then carefully substitute for each term. [Note that $z = \ln(y/x) = \ln y - \ln x$, so the partial derivatives are easy to evaluate.]

$$\begin{aligned} \frac{dz}{dt} &= \frac{\partial z}{\partial x}\frac{dx}{dt} + \frac{\partial z}{\partial y}\frac{dy}{dt} \\ &= \left(-\frac{1}{x}\right)(-\sin t) + \left(\frac{1}{y}\right)(\cos t). \end{aligned}$$

If we wanted to express the derivative in Example 1 in terms of $t$, we could substitute to obtain

$$\frac{dz}{dt} = \left(-\frac{1}{\cos t}\right)(-\sin t) + \left(\frac{1}{\sin t}\right)(\cos t) = \tan t + \cot t.$$

We could also substitute before differentiating to obtain

$$z = \ln\left(\frac{\sin t}{\cos t}\right) = \ln(\sin t) - \ln(\cos t).$$

The Chain Rule for functions of one variable would then give us

$$\frac{dz}{dt} = \left(\frac{1}{\sin t}\right)(\cos t) - \left(\frac{1}{\cos t}\right)(-\sin t).$$

This agrees with what we obtained previously.

Let us now consider the case $z = z(x, y)$, where $x$ and $y$ are functions of the two variables $s$ and $t$. Then $z = z(x(s, t), y(s, t))$ is a function of $s$ and $t$, so we may consider formulas for the two partial derivatives $\partial z/\partial s$ and $\partial z/\partial t$. Since partial derivatives are ordinary derivatives of the function obtained by holding all variables but the variable of partial differentiation constant, we should be able to use the formula that we derived previously. Thus, we have the following results.

If $z = z(x, y)$, $x = x(s, t)$, and $y = y(s, t)$, then (assuming appropriate differentiability)

$$\frac{\partial z}{\partial s} = \frac{\partial z}{\partial x}\frac{\partial x}{\partial s} + \frac{\partial z}{\partial y}\frac{\partial y}{\partial s} \quad \text{and} \quad \frac{\partial z}{\partial t} = \frac{\partial z}{\partial x}\frac{\partial x}{\partial t} + \frac{\partial z}{\partial y}\frac{\partial y}{\partial t}.$$

We have used "$\partial$" to indicate partial derivatives of a function of more than one variable; "$d$" indicates differentiation of a function of a single variable.

■ **EXAMPLE 2**

$w = x^2 - y^2$, $x = r \cos \theta$, $y = r \sin \theta$. Find $\partial w/\partial r$ and $\partial w/\partial \theta$.

**SOLUTION**

We write the appropriate formula and substitute for each term.

$$\begin{aligned}\frac{\partial w}{\partial r} &= \frac{\partial w}{\partial x}\frac{\partial x}{\partial r} + \frac{\partial w}{\partial y}\frac{\partial y}{\partial r}\\ &= (2x)(\cos \theta) + (-2y)(\sin \theta).\end{aligned}$$

Similarly,

$$\begin{aligned}\frac{\partial w}{\partial \theta} &= \frac{\partial w}{\partial x}\frac{\partial x}{\partial \theta} + \frac{\partial w}{\partial y}\frac{\partial y}{\partial \theta}\\ &= (2x)(-r \sin \theta) + (-2y)\,(r \cos \theta).\end{aligned}$$ ■

The above examples should serve as a hint that there are many versions of the Chain Rule for functions of several variables. At this point it is worthwhile to consider the general case, so that we will have a common frame of reference for the discussion of individual cases.

Suppose $z = z(x_1, \ldots, x_m)$, where

$$x_1 = x_1(t_1, \ldots, t_n), \ldots, x_m = x_m(t_1, \ldots, t_n).$$

Then $z$ is a function of $(t_1, \ldots, t_n)$. Assuming appropriate differentiability, we have

$$\frac{\partial z}{\partial t_1} = \frac{\partial z}{\partial x_1}\frac{\partial x_1}{\partial t_1} + \cdots + \frac{\partial z}{\partial x_m}\frac{\partial x_m}{\partial t_1}.$$

$$\vdots$$

$$\frac{\partial z}{\partial t_n} = \frac{\partial z}{\partial x_1}\frac{\partial x_1}{\partial t_n} + \cdots + \frac{\partial z}{\partial x_m}\frac{\partial x_m}{\partial t_n}.$$

We can say that $z$ is a function of $m$ *dependent* variables $x_1, \ldots, x_m$, and that each dependent variable is a function of $n$ *independent* variables $t_1, \ldots, t_n$. The Chain Rule then gives a formula for the partial derivative of $z$ with respect to each of the $n$ independent variables. Each formula contains $m$ terms, one for each dependent variable. Each term contains two factors. The first factor in each term is a partial derivative of $z$ with respect to a dependent variable and the second factor is the partial derivative of that dependent variable with respect to the variable of partial differentiation, an independent variable.

When using the Chain Rule, it is a good idea to organize your work in a systematic way. Determine how many variables there are of each type. This tells how many formulas there will be and how many terms each formula will contain. Also, write out each formula so you will have a guide for substitution.

■ **EXAMPLE 3**

$w = x^2 + y^2 + z^2$, $x = t + 1$, $y = 2t + 1$, $z = 3t + 1$. Find $dw/dt$.

**SOLUTION**

We see that $w$ is a function of three dependent variables $x$, $y$, and $z$. Each of $x$, $y$, and $z$ is a function of the single independent variable $t$. Thus, the Chain Rule gives us one formula, for the derivative of $w$ with respect to $t$. The formula has three terms, one for each of the three dependent variables. We write the appropriate formula and then substitute for each term.

$$\frac{dw}{dt} = \frac{\partial w}{\partial x}\frac{dx}{dt} + \frac{\partial w}{\partial y}\frac{dy}{dt} + \frac{\partial w}{\partial z}\frac{dz}{dt}$$

$$= (2x)(1) + (2y)(2) + (2z)(3).$$

Note that we used $d$'s in place of $\partial$'s to indicate when we were taking the derivative of a function of a single variable, not a partial derivative of a function of more than one variable. ■

We can use the Chain Rule even if we do not have a formula for each function involved. It is only necessary that we know the appropriate values.

**■ EXAMPLE 4**

Suppose $w = F(t^2 + 2t + 1, t^3 - t + 2)$, where $F(u, v)$ is differentiable with $F_u(1, 2) = 3$ and $F_v(1, 2) = 4$. Find $dw/dt$ when $t = 0$.

**SOLUTION**

We have

$$w = F(u, v), \quad u = t^2 + 2t + 1, \quad v = t^3 - t + 2.$$

The Chain Rule gives

$$\frac{dw}{dt} = \frac{\partial F}{\partial u}\frac{du}{dt} + \frac{\partial F}{\partial v}\frac{dv}{dt}.$$

When $t = 0$, we have

$$u = (0)^2 + 2(0) + 1 = 1 \quad \text{and} \quad v = (0)^3 - (0) + 2 = 2.$$

The partial derivatives of $F$ are evaluated at $(u, v) = (1, 2)$ and we are given that $F_u(1, 2) = 3$ and $F_v(1, 2) = 4$. The derivatives of $u$ and $v$ are evaluated at $t = 0$. We have

$$\frac{du}{dt} = 2t + 2 \quad \text{and} \quad \frac{dv}{dt} = 3t^2 - 1.$$

When $t = 0$,

$$\frac{du}{dt} = 2 \quad \text{and} \quad \frac{dv}{dt} = -1.$$

Substitution of the values into the formula for $dw/dt$ gives

$$\frac{dw}{dt} = (3)(2) + (4)(-1) = 2.$$

■

**■ EXAMPLE 5**

Suppose $w = x^2 - 3xy + y^2$, where $x(t)$ and $y(t)$ are differentiable with

$$x(2) = -1, \quad y(2) = 2, \quad \frac{dx}{dt}(2) = 3, \quad \text{and} \quad \frac{dy}{dt}(2) = -2.$$

Find $dw/dt$ when $t = 2$.

**SOLUTION**

We have

$$w = x^2 - 3xy + y^2, \quad x = x(t), \quad y = y(t).$$

The Chain Rule gives

$$\begin{aligned}\frac{dw}{dt} &= \frac{\partial w}{\partial x}\frac{dx}{dt} + \frac{\partial w}{\partial y}\frac{dy}{dt} \\ &= (2x - 3y)\left(\frac{dx}{dt}\right) + (-3x + 2y)\left(\frac{dy}{dt}\right).\end{aligned}$$

When $t = 2$, we are given that $x = -1$, $y = 2$, $dx/dt = 3$, and $dy/dt = -2$. Substitution then gives

$$\frac{dw}{dt} = [2(-1) - 3(2)](3) + [-3(-1) + 2(2)](-2) = -38.$$ ■

Suppose $z = f(x, y)$ and $y = y(x)$, so $z$ is a function of $x$. Let us find a formula for $dz/dx$. We can fit this problem into our general scheme by writing

$$z = f(x, y), \quad x = x, \quad y = y(x).$$

We then consider $z$ to be a function of two dependent variables $x$ and $y$. These variables are functions of the single independent variable $x$. Thus, we have one formula that has two terms. Namely,

$$\frac{dz}{dx} = \frac{\partial f}{\partial x}\frac{dx}{dx} + \frac{\partial f}{\partial y}\frac{dy}{dx}.$$

Since $dx/dx = 1$, we obtain the formula

$$\frac{dz}{dx} = \frac{\partial f}{\partial x} + \frac{\partial f}{\partial y}\frac{dy}{dx}, \qquad \text{where } z = f(x, y) \text{ and } y = y(x). \tag{1}$$

Let us suppose the equation $F(x, y) = 0$ defines $y$ (implicitly) as a differentiable function of $x$. If $F$ is a differentiable function of $(x, y)$, we can obtain a formula for the derivative of $y$ with respect to $x$ in terms of partial derivatives of $F$. Differentiating each side of the equation $F(x, y) = 0$ with respect to $x$, and using formula (1), we obtain

$$\frac{\partial F}{\partial x} + \frac{\partial F}{\partial y}\frac{dy}{dx} = 0.$$

If $\partial F/\partial y \neq 0$, we then have

$$\boldsymbol{\frac{dy}{dx} = -\frac{\dfrac{\partial F}{\partial x}}{\dfrac{\partial F}{\partial y}} = -\frac{F_x}{F_y}, \qquad \text{where } F(x, y) = 0 \text{ and } y = y(x).} \tag{2}$$

If $y$ is given explicitly by an equation $y = f(x)$, we can write the equation in the form $F(x, y) = 0$, where $F(x, y) = f(x) - y$. Equation (2) then gives

$$\frac{dy}{dx} = -\frac{F_x}{F_y} = -\frac{f'(x)}{-1} = f'(x),$$

as it should.

■ **EXAMPLE 6**

Assume the equation $x^3 + y^3 = x - y$ defines $y$ as a differentiable function of $x$. Find $dy/dx$.

**SOLUTION**

Let us write $F(x, y) = x^3 + y^3 - x + y$. Our equation can then be expressed as $F(x, y) = 0$. We can then use the formula

$$\frac{dy}{dx} = -\frac{F_x}{F_y}$$

to obtain

$$\frac{dy}{dx} = -\frac{3x^2 - 1}{3y^2 + 1}.$$

[This result could also be obtained by the one variable method of implicit differentiation. That is, we could consider $y$ to be a function of $x$, differentiate the equation $x^3 + y^3 = x - y$ with respect to $x$, and then solve the resulting equation for $dy/dx$. Formula (2) avoids the algebra required for the one variable method.] ■

If the equation $F(x, y, z) = 0$ defines $z$ as a function of $x$ and $y$, we may express the partial derivatives of $z$ with respect to $x$ and $y$ in terms of partial derivatives of $F$. We assume $F$ is a differentiable function of $(x, y, z)$ and that the partial derivatives of $z$ exist. We consider $F(x, y, z)$ to be a function of the three dependent variables $x$, $y$, and $z$. These variables are functions of the two independent variables $x$ and $y$. That is, $x = x$, $y = y$, and $z = z(x, y)$. There are then two formulas for partial derivatives of $F$. Each formula has three terms. Using the Chain Rule, we take partial derivatives of both sides of the equation $F(x, y, z) = 0$ to obtain

$$\frac{\partial F}{\partial x}\frac{\partial x}{\partial x} + \frac{\partial F}{\partial y}\frac{\partial y}{\partial x} + \frac{\partial F}{\partial z}\frac{\partial z}{\partial x} = 0 \quad \text{and}$$

$$\frac{\partial F}{\partial x}\frac{\partial x}{\partial y} + \frac{\partial F}{\partial y}\frac{\partial y}{\partial y} + \frac{\partial F}{\partial z}\frac{\partial z}{\partial y} = 0.$$

The variable $y$ is constant while we are taking the partial derivative with respect to $x$, so $\partial y/\partial x = 0$. Since $\partial x/\partial x = 1$, the first formula above gives us

$$\frac{\partial F}{\partial x} + \frac{\partial F}{\partial z}\frac{\partial z}{\partial x} = 0.$$

Similarly, the second formula above gives us

$$\frac{\partial F}{\partial y} + \frac{\partial F}{\partial z}\frac{\partial z}{\partial y} = 0.$$

Summarizing, if $\partial F/\partial z \neq 0$ we have

$$\frac{\partial z}{\partial x} = -\frac{\dfrac{\partial F}{\partial x}}{\dfrac{\partial F}{\partial z}} = -\frac{F_x}{F_z}, \qquad \frac{\partial z}{\partial y} = -\frac{\dfrac{\partial F}{\partial y}}{\dfrac{\partial F}{\partial z}} = -\frac{F_y}{F_z}, \tag{3}$$

$F(x, y, z) = 0, \quad z = z(x, y).$

■ **EXAMPLE 7**

$yz + xz + xy = 1$. Find $\partial z/\partial x$, $\partial z/\partial y$, and $\partial^2 z/\partial x^2$.

**SOLUTION**

Our equation can be expressed as $F(x, y, z) = 0$, where $F(x, y, z) = yz + xz + xy - 1$. We then have

$$\frac{\partial z}{\partial x} = -\frac{F_x}{F_z} = -\frac{z + y}{y + x},$$

$$\frac{\partial z}{\partial y} = -\frac{F_y}{F_z} = -\frac{z + x}{y + x}.$$

The second partial $\partial^2 z/\partial x^2$ is obtained by taking the partial derivative of $\partial z/\partial x$ with respect to $x$. To do this, we must remember that $y$ is considered to be constant and that $z$ is then considered a function of $x$. Using the Quotient Rule, we have

$$\frac{\partial^2 z}{\partial x^2} = \frac{\partial}{\partial x}\left(\frac{\partial z}{\partial x}\right) = \frac{\partial}{\partial x}\left(-\frac{z + y}{y + x}\right)$$

$$= -\frac{(y + x)(z_x + 0) - (z + y)(0 + 1)}{(y + x)^2}.$$

We now substitute the formula for $z_x$ that we obtained above to obtain

$$\frac{\partial^2 z}{\partial x^2} = -\frac{(y + x)\left(-\dfrac{z + y}{y + x}\right) - (z + y)}{(y + x)^2}$$

$$= -\frac{-z - y - z - y}{(y + x)^2} = \frac{2z + 2y}{(y + x)^2}.$$ ■

FIGURE 16.5.1

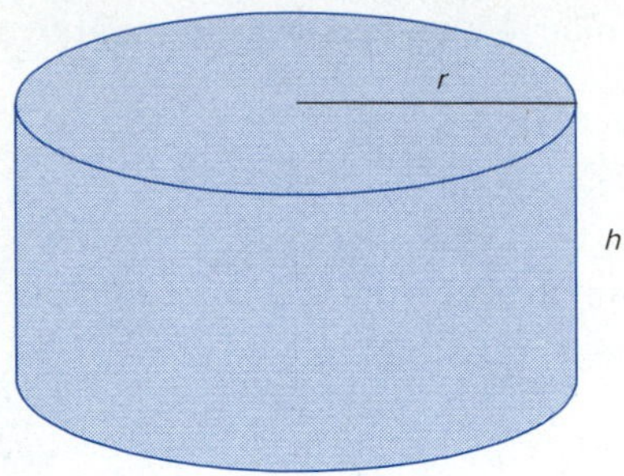


### ■ EXAMPLE 8

The radius of a right circular cylinder is increasing at a rate of 6 ft/s while the height is decreasing at a rate of 4 ft/s. Find the rate of change of the volume of the cylinder when the radius is 12 ft and the height is 10 ft.

#### SOLUTION

The cylinder is pictured in Figure 16.5.1. The radius $r$, the height $h$, and the volume $V$ are variables. Expressing the problem in terms of the variables, we write

*We want*: $\dfrac{dV}{dt}$ when $r = 12$ and $h = 10$.

*We know*: $\dfrac{dr}{dt} = +6$ and $\dfrac{dh}{dt} = -4$.

The volume of the cylinder is given by the formula

$$V = \pi r^2 h.$$

Since $r$ and $h$ are functions of time, the Chain Rule tells us

$$\begin{aligned}\frac{dV}{dt} &= \frac{\partial V}{\partial h}\frac{dh}{dt} + \frac{\partial V}{\partial r}\frac{dr}{dt}\\ &= (\pi r^2)\left(\frac{dh}{dt}\right) + (\pi h 2r)\left(\frac{dr}{dt}\right).\end{aligned}$$

When $r = 12$ and $h = 10$, we have

$$\begin{aligned}\frac{dV}{dt} &= (\pi)(12)^2(-4) + (\pi)(10)(2)(12)(6)\\ &= 864\pi.\end{aligned}$$

The volume is increasing at a rate of $864\pi$ ft$^3$/s. ■

We could use the Product Rule and the Chain Rule for differentiation of functions of a single variable to differentiate both sides of the equation

$$V = \pi r^2 h$$

with respect to $t$ in Example 8. This gives

$$\frac{dV}{dt} = (\pi r^2)\left(\frac{dh}{dt}\right) + (h)\left(\pi 2r\,\frac{dr}{dt}\right),$$

which agrees with what we obtained in Example 8 by using the Chain Rule for functions of several variables.

## EXERCISES 16.5

Use the appropriate version of the Chain Rule to evaluate the derivatives in Exercises 1–12. (Do not eliminate variables by substitution.)

**1** $z = x^2y^3$, $x = 2t - 1$, $y = 3t + 4$. Find $dz/dt$.

**2** $z = \sqrt{xy}$, $x = t^2 + 1$, $y = t^4 + 1$. Find $dz/dt$.

**3** $z = \ln(x^2 + y^2)$, $x = \cos t$, $y = \sin t$. Find $dz/dt$.

**4** $z = e^{2xy}$, $x = \ln t$, $y = t$. Find $dz/dt$.

**5** $z = x^2 + y^2$, $x = r\cos\theta$, $y = r\sin\theta$. Find $\partial z/\partial r$ and $\partial z/\partial\theta$.

**6** $z = x^2y$, $x = r\cos\theta$, $y = r\sin\theta$. Find $\partial z/\partial r$ and $\partial z/\partial\theta$.

**7** $z = r^2\cos\theta$, $r = \sqrt{x^2 + y^2}$, $\theta = \tan^{-1}(y/x)$. Find $\partial z/\partial x$ and $\partial z/\partial y$.

**8** $z = r\sin\theta$, $r = \sqrt{x^2 + y^2}$, $\theta = \tan^{-1}(y/x)$. Find $\partial z/\partial x$ and $\partial z/\partial y$.

**9** $w = xy^2/z$, $x = t - 1$, $y = 3t + 2$, $z = 2t + 3$. Find $dw/dt$.

**10** $w = xy + xz + yz$, $x = t^2$, $y = t^3$, $z = 2t$. Find $dw/dt$.

**11** $w = u/v$, $u = x^2 - y^2$, $v = 2xy$. Find $\partial w/\partial x$ and $\partial w/\partial y$.

**12** $w = \ln(u + v)$, $u = x^2 + y^2$, $v = 2xy$. Find $\partial w/\partial x$ and $\partial w/\partial y$.

**13** $x^3 + xy + y^3 = 0$. Find $dy/dx$.

**14** $x^2 + xy^2 + y^4 = 0$. Find $dy/dx$.

**15** $e^{xy} + e^{-xy} = 4$. Find $dy/dx$.

**16** $\ln(x^2 + y^2) = 2xy$. Find $dy/dx$.

**17** $e^{xyz} - x^2 - y^2 - z^2 = 1$. Find $\partial z/\partial x$ and $\partial z/\partial y$.

**18** $yx^2 + y^2z + xz^2 = 1$. Find $\partial z/\partial x$ and $\partial z/\partial y$.

**19** $x + yz + xz^2 = 0$. Find $\partial z/\partial x$, $\partial z/\partial y$, and $\partial^2 z/\partial x^2$.

**20** $xy - yz + xz = 1$. Find $\partial z/\partial x$, $\partial z/\partial y$, and $\partial^2 z/\partial x\partial y$.

**21** $w = F(t^2, t^3)$, where $F(u, v)$ is differentiable with $F_u(1, 1) = 5$ and $F_v(1, 1) = 7$. Find $dw/dt$ when $t = 1$.

**22** $w = F(3t + 2, 2t + 1)$, where $F(u, v)$ is differentiable with $F_u(2, 1) = -3$ and $F_v(2, 1) = 5$. Find $dw/dt$ when $t = 0$.

**23** $w = u^2/v$, where $u(t)$ and $v(t)$ are differentiable with $u(1) = 3$, $v(1) = 2$, $u'(1) = 5$, and $v'(1) = -4$. Find $dw/dt$ when $t = 1$.

**24** $w = u^2v^3$, where $u(t)$ and $v(t)$ are differentiable with $u(2) = 2$, $v(2) = 3$, $u'(2) = -1$, and $v'(2) = 5$. Find $dw/dt$ when $t = 2$.

**25** $y = F(u, v)$, $u = u(x)$, $v = v(x)$. If $F(u, v) = uv$, show that the formula for $dy/dx$ given by the Chain Rule is the Product Rule.

**26** $y = F(u, v)$, $u = u(x)$, $v = v(x)$. If $F(u, v) = u/v$, show that the formula for $dy/dx$ given by the Chain Rule is the Quotient Rule.

**27** The length and width of a rectangular box are increasing at a rate of 1 ft/s while the height is decreasing at a rate of 1 ft/s. (a) What is the rate of change of the volume of the box when the length is 3 ft, the width is 2 ft, and the height is 1 ft? (b) What is the rate of change of the total area of the six sides of the box at the same instant?

**28** Two sides of a triangle are each increasing at a rate of 2 cm/s while the included angle is decreasing at a rate of 5°/s. What is the rate of change of the area of the triangle when one of the sides is 30 cm, the other side is 20 cm, and the included angle is 30°?

## 16.6 DIRECTIONAL DERIVATIVES AND THE GRADIENT

The first partial derivatives $f_x(a, b)$ and $f_y(a, b)$ may be considered to be the rate of change of $f(x, y)$ at the point $(a, b)$ in the direction of the $x$-axis and $y$-axis, respectively. We will extend this idea to include the rate of change of $f$ in any direction.

Suppose $\mathbf{U} = u_1\mathbf{i} + u_2\mathbf{j}$ is a unit vector, so $u_1^2 + u_2^2 = 1$. We use the vector $\mathbf{U}$ to determine the direction and scale of a number line through $(a, b)$. This line has parametric equations

$$x = a + u_1 t, \qquad y = b + u_2 t.$$

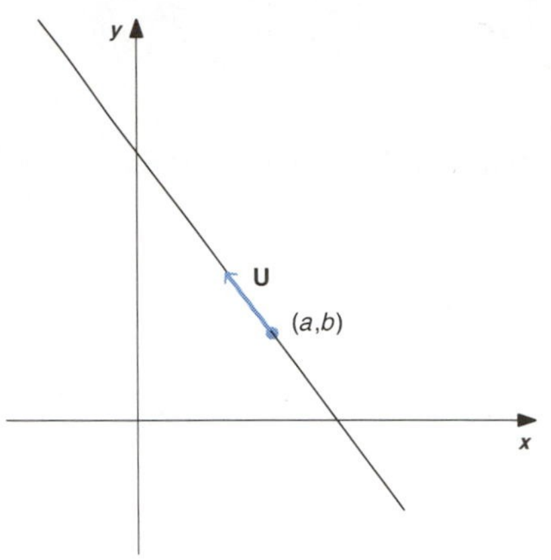

FIGURE 16.6.1

The origin of this number line, where $t = 0$, is at $(a, b)$. The positive direction of the line is given by the direction of $\mathbf{U}$. The fact that $\mathbf{U}$ is a unit vector implies that distance along the line in units of $t$ agrees with distance in terms of $(x, y)$ coordinates. That is, this is a parametrization of the line in terms of arc length. See Figure 16.6.1.

We now consider the function of the real variable $t$ defined by the equations

$$z = f(x, y), \quad x = a + u_1 t, \quad y = b + u_2 t.$$

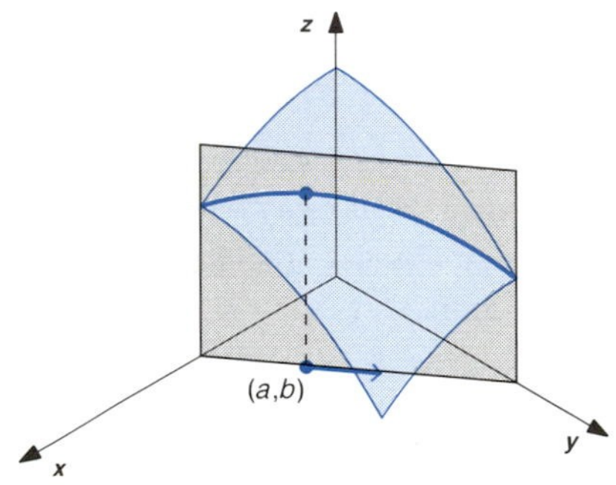

FIGURE 16.6.2

The derivative of $z$ with respect to $t$ at $t = 0$, if it exists, is called the **directional derivative** of $f$ in the direction of the unit vector $\mathbf{U}$ at $(a, b)$, and is denoted $D_{\mathbf{U}}f(a, b)$. $D_{\mathbf{U}}f(a, b)$ may be interpreted as the rate of change of $f$ in the direction of $\mathbf{U}$ at $(a, b)$. See Figure 16.6.2.

If $f$ is a differentiable function of $(x, y)$ and

$$z = f(x, y), \quad x = a + u_1 t, \quad y = b + u_2 t,$$

then $z$ is a differentiable function of $t$ and the Chain Rule gives

$$\begin{aligned}\frac{dz}{dt} &= \frac{\partial f}{\partial x}\frac{dx}{dt} + \frac{\partial f}{\partial y}\frac{dy}{dt} \\ &= f_x(x, y)u_1 + f_y(x, y)u_2.\end{aligned}$$

When $t = 0$, we have $x = a$, $y = b$, and

$$D_{\mathbf{U}}(a, b) = f_x(a, b)u_1 + f_y(a, b)u_2.$$

We can express this equation as

$$D_{\mathbf{U}}f = \nabla f \cdot U, \tag{1}$$

where $|\mathbf{U}| = 1$ and

$$\nabla f = f_x\mathbf{i} + f_y\mathbf{j}.$$

The vector $\nabla f = f_x\mathbf{i} + f_y\mathbf{j}$ is called the **gradient** of the function $f(x, y)$. The gradient of $f(x, y)$ is a vector-valued function of $(x, y)$. **Grad** $f$ is also used to denote the gradient. Ordinarily, we will use formula (1) to evaluate directional derivatives.

We have

$$D_{\mathbf{i}}f = \nabla f \cdot \mathbf{i} = f_x \quad \text{and} \quad D_{\mathbf{j}}f = \nabla f \cdot j = f_y,$$

so the directional derivatives of $f$ in the directions of the coordinate axes are partial derivatives of $f$.

**■ EXAMPLE 1**

$f(x, y) = x^2 + y^2$. Evaluate the directional derivative of $f$ at (3, 2) in the direction of the vector $4\mathbf{i} - 3\mathbf{j}$.

**SOLUTION**

We have

$$\nabla f = f_x\mathbf{i} + f_y\mathbf{j} = 2x\mathbf{i} + 2y\mathbf{j}.$$

When $(x, y) = (3, 2)$,

$$\nabla f = 2(3)\mathbf{i} + 2(2)\mathbf{j} = 6\mathbf{i} + 4\mathbf{j}.$$

We need a *unit* vector in the direction of $4\mathbf{i} - 3\mathbf{j}$. We have

$$|4\mathbf{i} - 3\mathbf{j}| = \sqrt{(4)^2 + (-3)^2} = 5,$$

so we take

$$\mathbf{U} = \frac{4\mathbf{i} - 3\mathbf{j}}{|4\mathbf{i} - 3\mathbf{j}|} = \frac{4}{5}\mathbf{i} - \frac{3}{5}\mathbf{j}.$$

Then

$$\begin{aligned} D_{\mathbf{U}}f(3, 2) &= \nabla f \cdot \mathbf{U} \\ &= (6\mathbf{i} + 4\mathbf{j}) \cdot \left(\frac{4}{5}\mathbf{i} - \frac{3}{5}\mathbf{j}\right) \\ &= (6)\left(\frac{4}{5}\right) + (4)\left(-\frac{3}{5}\right) = \frac{12}{5}. \end{aligned}$$

■

We know that the dot product of two vectors can be expressed in terms of the angle between the vectors. We can use this fact to obtain information about the directional derivatives of a function $f(x, y)$. If $\mathbf{U}$ is a unit vector, we have

$$D_{\mathbf{U}}f = \nabla f \cdot \mathbf{U} = |\nabla F||\mathbf{U}|\cos\theta = |\nabla f|\cos\theta,$$

where $\theta$ is the angle between $\nabla f$ and $\mathbf{U}$. Since $-1 \le \cos\theta \le 1$, we have

$$-|\nabla f| \le D_{\mathbf{U}}f \le |\nabla f|.$$

The maximum value of the directional derivative is $|\nabla f|$, obtained when $\theta = 0$. When $\theta = 0$ the angle between $\nabla f$ and $\mathbf{U}$ is zero, so $\mathbf{U}$ is a unit vector in the direction of the gradient. The minimum value of the directional derivative is $-|\nabla f|$, obtained when $\mathbf{U}$ is a unit vector in the direction opposite to that of the gradient. This gives the following.

*Theorem 1*

**If $f(x, y)$ is differentiable, the directional derivative has maximum value $|\nabla f|$ in the direction of $\nabla f$; the directional derivative has minimum value $-|\nabla f|$ in the direction of $-\nabla f$.**

**EXAMPLE 2**

$f(x, y) = xy^2 - 9y$. Find a unit vector $\mathbf{U}$ such that the rate of change of $f$ at (3, 2) in the direction of $\mathbf{U}$ is maximum. Evaluate the rate of change of $f$ at (3, 2) in the direction of that vector $\mathbf{U}$.

**SOLUTION**

Theorem 1 tells us that the maximum rate of change of $f$ is in the direction of $\nabla f$, and that the maximum rate of change is $|\nabla f|$. We have

$$f(x, y) = xy^2 - 9y, \quad \text{so}$$

$$\nabla f = f_x\mathbf{i} + f_y\mathbf{j} = y^2\mathbf{i} + (2xy - 9)\mathbf{j}.$$

When $(x, y) = (3, 2)$,

$$\nabla f = (2)^2\mathbf{i} + (2(3)(2) - 9)\mathbf{j} = 4\mathbf{i} + 3\mathbf{j} \quad \text{and}$$

$$|\nabla f| = \sqrt{4^2 + 3^2} = 5.$$

The unit vector in the direction of $\nabla f$ is

$$\mathbf{U} = \frac{\nabla f}{|\nabla f|} = \frac{4\mathbf{i} + 3\mathbf{j}}{5} = 0.8\mathbf{i} + 0.6\mathbf{j}.$$

The rate of change of $f$ at (3, 2) is maximum in the direction of $\mathbf{U}$ and the maximum rate of change is

$$D_{\mathbf{U}}f = |\nabla f| = 5.$$

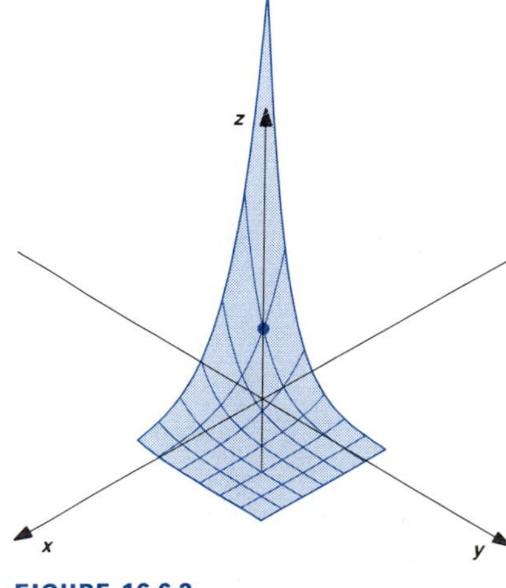

FIGURE 16.6.3

**EXAMPLE 3**

A marble is set on the surface $z = e^{-2x-3y}$ at the point (0, 0, 1). Along what direction in the $xy$-plane will the marble roll?

**SOLUTION**

The surface is pictured in Figure 16.6.3. We assume that the marble will roll in the direction in which the surface is falling away at the fastest rate. That is, the marble will roll in the direction in which the rate of change of $z$

is minimum. This is the direction opposite to $\nabla z$, namely, in the direction of $-\nabla z$. We have

$$z_x = e^{-2x-3y}(-2) \quad \text{and} \quad z_y = e^{-2x-3y}(-3).$$

When $x = 0$ and $y = 0$, the gradient of $z$ is

$$\nabla z = z_x\mathbf{i} + z_y\mathbf{j} - 2\mathbf{i} - 3\mathbf{j}.$$

The marble will roll along the surface in the direction in the $xy$-plane of the vector

$$-\nabla z = z_x\mathbf{i} + z_y\mathbf{j} = 2\mathbf{i} + 3\mathbf{j}.$$ ■

The ideas of gradient and directional derivatives extend to any number of variables. For example, we define the gradient of a function $F$ of three variables to be the vector

$$\nabla F(x, y, z) = F_x(x, y, z)\mathbf{i} + F_y(x, y, z)\mathbf{j} + F_z(x, y, z)\mathbf{k}.$$

Generally, the gradient of a function of $n$ variables is a vector that has $n$ components, each component being a partial derivative of $F$. The gradient of $F$ is a vector-valued function of the same variables as is $F$.

If $F$ is a differentiable function of any number of variables, the directional derivative of $F$ in the direction of a unit vector $\mathbf{U}$ is given by the dot product

$$D_{\mathbf{U}}F = \nabla F \cdot \mathbf{U}, \qquad |\mathbf{U}| = 1.$$

$\mathbf{U}$ must have the same number of components as the gradient of $F$.

The analogue of Theorem 1 is true for any number of variables.

*Theorem 2*

**If $F(x, \ldots)$ is differentiable, the directional derivative has maximum value $|\nabla F|$ in the direction of $\nabla F$; the directional derivative has minimum value $-|\nabla F|$ in the direction of $-\nabla F$.**

### ■ EXAMPLE 4

$F(x, y, z) = x^2 + y^2 + xyz$. Find a unit vector $\mathbf{U}$ such that the rate of change of $F$ at $(2, 3, -1)$ in the direction of $\mathbf{U}$ is maximum. Evaluate $D_{\mathbf{U}}F$ for that vector $\mathbf{U}$.

#### SOLUTION

We know that the value of the directional derivative is maximum in the direction of the gradient. We have

$$\nabla F = F_x\mathbf{i} + F_y\mathbf{j} + F_z\mathbf{k} = (2x + yz)\mathbf{i} + (2y + xz)\mathbf{j} + (xy)\mathbf{k}.$$

When $(x, y, z) = (2, 3 -1)$ the gradient is

$$\nabla F = \mathbf{i} + 4\mathbf{j} + 6\mathbf{k}.$$

FIGURE 16.6.4

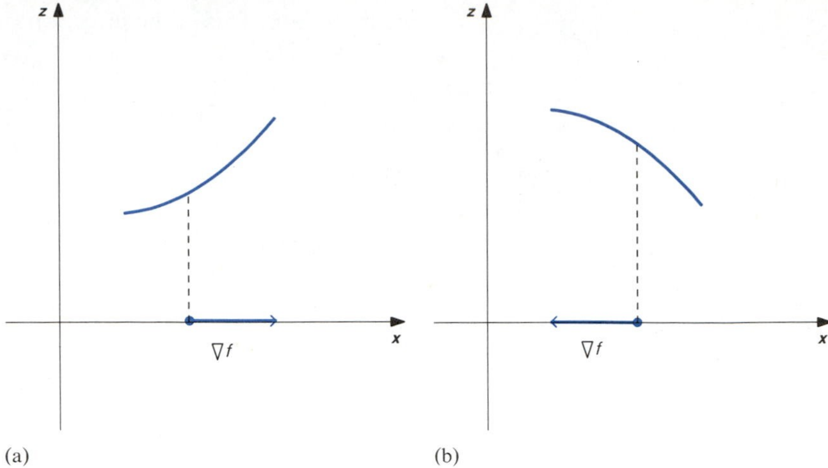


A unit vector in the direction of the gradient at (2, 3, −1) is

$$\mathbf{U} = \frac{\nabla F}{|\nabla F|} = \frac{\mathbf{i} + 4\mathbf{j} + 6\mathbf{k}}{\sqrt{53}}.$$

The maximum rate of increase of $F$ at (2, 3, −1) is

$$D_{\mathbf{U}}F = |\nabla F| = \sqrt{53}.$$ ■

In case $z = f(x)$ is a function of one variable, the gradient of $f$ is a vector with one component. Namely, $\nabla f = f'\mathbf{i}$. This vector points in the positive direction when $f'$ is positive and in the negative direction when $f'$ is negative. In each case, the gradient points in the direction in which $f$ is increasing, as indicated in Theorem 2. See Figure 16.6.4.

## EXERCISES 16.6

In Exercises 1–6 evaluate the directional derivative of the given function at the given point in the direction of the indicated vector.

1 $f(x, y) = xy^2$; (3, 2); $4\mathbf{i} + 3\mathbf{j}$

2 $f(x, y) = x^2 - 4xy + y^2$; (−1, 2): $3\mathbf{i} - 4\mathbf{j}$

3 $f(x, y) = e^{xy}$; (0, 2); $\mathbf{i}$

4 $f(x, y) = \ln(2x + 3y)$; (−1, 1); $3\mathbf{i} + 2\mathbf{j}$

5 $f(x, y, z) = x^2y^3z$; (2, −1, 3); $\mathbf{i} - 2\mathbf{j} - 2\mathbf{k}$

6 $f(x, y, z) = x^2 + y^2 + z^2$; (1, −1, 2); $2\mathbf{i} - 2\mathbf{j} + \mathbf{k}$

In Exercises 7–14, find a unit vector $\mathbf{U}$ such that the rate of change of $f$ in the direction of $\mathbf{U}$ at the given point is maximum and evaluate $D_{\mathbf{U}}f$ at that point.

7 $f(x, y) = 2x - x^2$; (3, 1)

8 $f(x, y) = 8 - y^3$; (2, −1)

9 $f(x, y) = x^2 - 3xy - y^2$; (1, 2)

10 $f(x, y) = x^2y + 4xy$; (1, −1)

11 $f(x, y, z) = e^{xyz}$; (2, 3, 0)

12 $f(x, y, z) = \sin xyz$; (3, 5, π)

13 $f(x, y, z) = x^2 + y^2 + z^2 - xyz$; $(2, -1, 3)$

14 $f(x, y, z) = x^2yz + xy^2z + xyz^2$; $(-1, 2, 3)$

In Exercises 15–18 find two unit vectors $\mathbf{U}$ such that the rate of change of $f$ in the direction of $\mathbf{U}$ at the indicated point is zero.

15 $f(x, y) = 3x - 4y$; $(2, -3)$

16 $f(x, y) = x^2 - 3xy + y^2$; $(3, 2)$

17 $f(x, y) = x^2y^3$; $(2, -1)$

18 $f(x, y) = \sqrt{x^2 + y^2}$; $(3, 4)$

19 If $D_{(\mathbf{i}-\mathbf{j})/\sqrt{2}} f(a, b) = 1$ and $D_{(\mathbf{i}+\mathbf{j})/\sqrt{2}} f(a, b) = 3$, what is $\nabla f(a, b)$?

20 If $D_{(\mathbf{i}+\mathbf{j})/\sqrt{2}} f(a, b) = 2/\sqrt{2}$ and $D_{(3\mathbf{i}+4\mathbf{j})/5} f(a, b) = 2$, what is $\nabla f(a, b)$?

## 16.7 NORMAL LINES AND TANGENT PLANES

There is an important relation between level curves of the function $z = f(x, y)$ and the gradient of $f$. We will see that this relation is a consequence of the following basic result.

*Theorem 1*

**Suppose $f(x, y)$ is differentiable at $(a, b)$ and**

$$C: \quad x = x(s), \quad y = y(s),$$

**is a smooth curve with $(x(s_0), y(s_0)) = (a, b)$. We assume that $C$ is parametrized in terms of arc length, so the unit tangent vector is**

$$\mathbf{T} = \frac{dx}{ds}\mathbf{i} + \frac{dy}{ds}\mathbf{j}.$$

**Then**

$$\lim_{s \to s_0} \frac{f(x(s), y(s)) - f(a, b)}{s - s_0} = D_{\mathbf{T}} f(a, b).$$

■ *Proof* Since $f(x, y)$ is differentiable at $(a, b)$, we have

$$f(x(s), y(s)) - f(a, b) = f_x(a, b)(x(s) - a) + f_y(a, b)(y(s) - b) + \varepsilon_1 \cdot (x(s) - a) + \varepsilon_2 \cdot (y(s) - b),$$

where $\varepsilon_1$ and $\varepsilon_2$ approach zero as $(x(s), y(s))$ approaches $(a, b)$. Since $a = x(s_0)$ and $b = y(s_0)$, we have that $\varepsilon_1$ and $\varepsilon_2$ approach zero as $s$ approaches $s_0$. Also,

$$\lim_{s \to s_0} \frac{x(s) - a}{s - s_0} = \frac{dx}{ds} \quad \text{and}$$

$$\lim_{s \to s_0} \frac{y(s) - b}{s - s_0} = \frac{dy}{ds}.$$

FIGURE 16.7.1

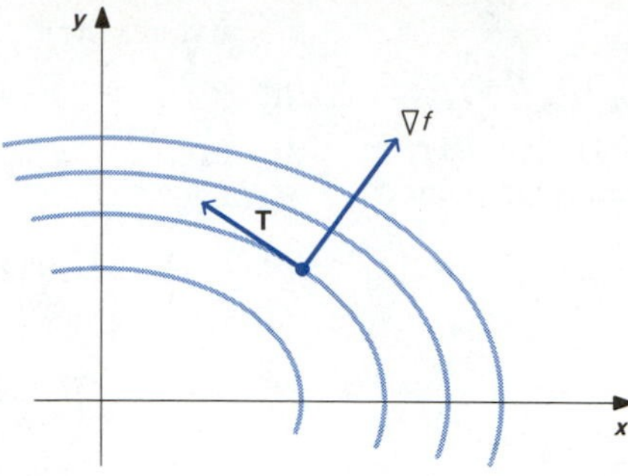


Dividing the above expression by $s - s_0$ and then taking the limit as $s$ approaches $s_0$ then gives

$$\lim_{s \to s_0} \frac{f(x(s), y(s)) - f(a, b)}{s - s_0} = f_x(a, b)\frac{dx}{ds} + f_y(a, b)\frac{dy}{ds}$$

$$= D_{\mathbf{T}} f(a, b). \qquad \blacksquare$$

If the curve $C$ in Theorem 1 is a level curve of $z = f(x, y)$, so $f(x(s), y(s)) = z_0 = f(a, b)$ for all values of $s$, the limit in Theorem 1 is zero. Theorem 1 then implies $D_{\mathbf{T}}f(a, b) = \nabla f \cdot T = 0$. See Figure 16.7.1. This gives the following result.

*Theorem 2*

**If $f(x, y)$ is differentiable at $(a, b)$, then:**

**(a) The directional derivative of $f$ at $(a, b)$ in the direction of the level curve of $f(x, y)$ through $(a, b)$ is zero;**

**(b) The gradient $\nabla f$ is orthogonal to the level curve of $f(x, y)$ through $(a, b)$.**

**EXAMPLE 1**

$f(x, y) = xy$. Sketch the level curves $f(x, y) = z_0$, for $z_0 = 1, 2, 3$. Evaluate $\nabla f(1, 2)$ and sketch that vector with initial position at (1, 2). Evaluate the directional derivative of $f$ at (1, 2) in the direction of a unit vector in the direction of $\nabla f$. Sketch the unit vector $\mathbf{U} = (1/\sqrt{5})\mathbf{i} + (-2/\sqrt{5})\mathbf{j}$ with initial position at (1, 2) and evaluate $D_{\mathbf{U}}f(1, 2)$.

**SOLUTION**

We have $f(x, y) = xy$, so $f_x = y$ and $f_y = x$. The gradient is $\nabla f = y\mathbf{i} + x\mathbf{j}$. When $(x, y) = (1, 2)$, we have $\nabla f(1, 2) = 2\mathbf{i} + \mathbf{j}$.

A unit vector in the direction of $\nabla f(1, 2)$ is given by

$$\mathbf{V} = \frac{\nabla f}{|\nabla f|} = \frac{2\mathbf{i} + \mathbf{j}}{\sqrt{5}} = \frac{2}{\sqrt{5}}\mathbf{i} + \frac{1}{\sqrt{5}}\mathbf{j}.$$

The directional derivative of $f$ in the direction of a unit vector in the direction of $\nabla f$ at (1, 2) is

$$D_{\mathbf{V}} f = \nabla f \cdot \mathbf{V} = (2)\left(\frac{2}{\sqrt{5}}\right) + (1)\left(\frac{1}{\sqrt{5}}\right) = \frac{5}{\sqrt{5}} = \sqrt{5}.$$

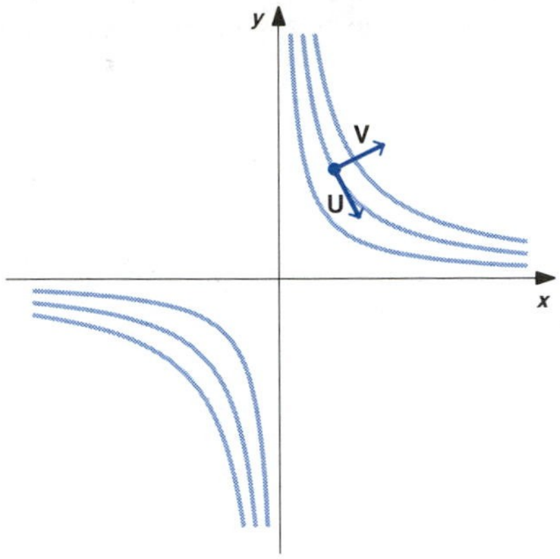

FIGURE 16.7.2

The directional derivative of $f$ in the direction of the unit vector $\mathbf{U} = (1/\sqrt{5}, -2/\sqrt{5})$ is

$$D_{\mathbf{U}}f = \nabla f \cdot \mathbf{U} = (2)\left(\frac{1}{\sqrt{5}}\right) + (1)\left(\frac{-2}{\sqrt{5}}\right) = 0.$$

The three level curves and the vectors $\nabla f(1, 2)$ and $\mathbf{U}$ are sketched in Figure 16.7.2. The level curves are hyperbolas. We see that the vector $\mathbf{U}$ is tangent to the level curve $xy = 2$ at $(1, 2)$; the directional derivative of $f$ in the direction of the level curve is zero. We also see that the gradient of $f$ is perpendicular to the level curve $xy = 2$ at $(1, 2)$. Also, the gradient points toward the level curve $xy = 3$, where the values of $f(x, y)$ are larger than the value $f(1, 2) = 2$. We know from Section 16.6 that $\nabla f$ points in the direction of greatest increase of $f$. ■

Since we can think of a curve that is defined by an equation $f(x, y) = 0$ as a level curve of $f(x, y)$, Theorem 2b implies the following.

■ *Theorem 3*

**If $f(x, y)$ is differentiable, the gradient $\nabla f$ is orthogonal to the curve $f(x, y) = 0$.**

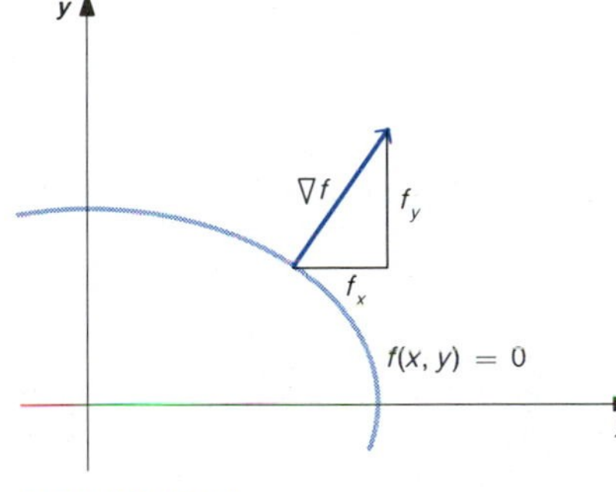


FIGURE 16.7.3

If $\nabla f = f_x\mathbf{i} + f_y\mathbf{j}$ is orthogonal to the curve $f(x, y) = 0$ at a point $(a, b)$ on the curve, then the line that is normal to the curve at $(a, b)$ has slope $f_y/f_x$. See Figure 16.7.3. The line that is tangent to the curve at $(a, b)$ has slope that is the negative reciprocal $-f_x/f_y$. We have seen in Section 16.5 that if the equation $f(x, y) = 0$ defines $y$ as a differentiable function of $x$, then $dy/dx = -f_x/f_y$.

■ **EXAMPLE 2**

Find the equation of the line that is normal to the curve $x^2 = y^2 + 5$ at the point $(3, 2)$.

**SOLUTION**

We write the equation of the curve in the form $f(x, y) = 0$, where

$$f(x, y) = x^2 - y^2 - 5.$$

We then know from Theorem 3 that the gradient of $f$ at $(3, 2)$ is orthogonal to the curve at that point. We have

$$\nabla f = f_x\mathbf{i} + f_y\mathbf{j} = 2x\mathbf{i} - 2y\mathbf{j}.$$

When $(x, y) = (3, 2)$,

$$\nabla f = 2(3)\mathbf{i} - 2(2)\mathbf{j} = 6\mathbf{i} - 4\mathbf{j}.$$

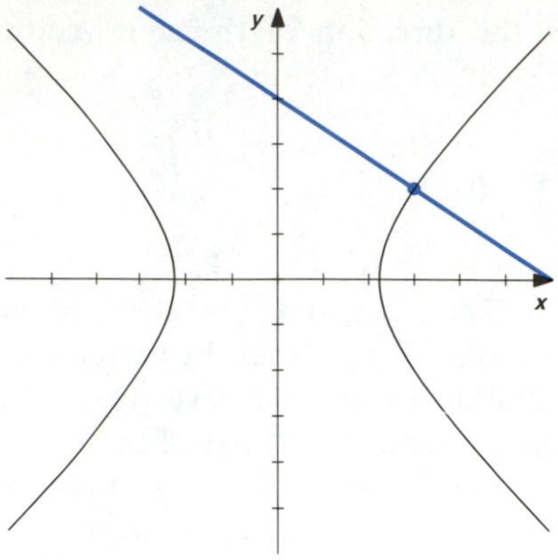

The direction of the gradient gives the slope of the line that is normal to the curve at (3, 2). This slope is

$$m_{\perp} = \frac{-4}{6} = -\frac{2}{3}.$$

The equation of the line through the point (3, 2) with slope $-2/3$ is

$$y - 2 = -\frac{2}{3}(x - 3),$$

$$3(y - 2) = -2(x - 3),$$

$$3y - 6 = -2x + 6,$$

$$2x + 3y - 12 = 0.$$

This is the equation of the line that is normal to the curve at the point (3, 2). See Figure 16.7.4. ■

FIGURE 16.7.4

The analogue of Theorem 2 holds for any number of variables and gives the following result for functions of three variables.

■ *Theorem 4*

**If $F(x, y, z)$ is differentiable at $(a, b, c)$, then:**

**(a) The directional derivative of $F$ in the direction of any vector in the plane tangent to the level surface of $F(x, y, z)$ at $(a, b, c)$ is zero;**

**(b) The gradient $\nabla F$ is orthogonal to the level surface of $F(x, y, z)$ at $(a, b, c)$.**

Since we can think of the surface $F(x, y, z) = 0$ as a level surface of $F(x, y, z)$, Theorem 4 gives the following result.

■ *Theorem 5*

**If $F(x, y, z)$ is differentiable, the gradient $\nabla F$ is orthogonal to the surface $F(x, y, z) = 0$.**

■ **EXAMPLE 3**

Find the equation of the plane tangent to the surface $z^2 = x^2 + y^2$ at the point (3, 4, 5).

**SOLUTION**

Let us express the surface as $F(x, y, z) = 0$, where

$$F(x, y, z) = z^2 - x^2 - y^2.$$

The surface has normal vector

$$\nabla F = F_x\mathbf{i} + F_y\mathbf{j} + F_z\mathbf{k} = -2x\mathbf{i} - 2y\mathbf{j} + 2z\mathbf{k}.$$

When $(x, y, z) = (3, 4, 5)$, we have

$$\nabla F = -6\mathbf{i} - 8\mathbf{j} + 10\mathbf{k}.$$

Since $\nabla F$ is normal to the surface at the point (3, 4, 5), it is also normal to the tangent plane at that point. It follows that an equation of the tangent plane is

$$(-6)(x - 3) + (-8)(y - 4) + (10)(z - 5) = 0.$$

Simplifying the equation, we obtain

$$-3(x - 3) - 4(y - 4) + 5(z - 5) = 0,$$

$$-3x + 9 - 4y + 16 + 5z - 25 = 0,$$

$$3x + 4y - 5z = 0.$$

■

### ■ EXAMPLE 4

Find symmetric equations and parametric equations of the line that is normal to the surface $z^2 = x^2 + y^2 + 4$ at the point $(2, -1, 3)$.

**SOLUTION**

The equation of the surface can be written as $F(x, y, z) = 0$, where

$$F(x, y, z) = z^2 - x^2 - y^2 - 4.$$

The gradient of $F$ is then a vector that is normal to the surface, so the gradient is in the direction of the normal line. We have

$$\nabla F = -2x\mathbf{i} - 2y\mathbf{j} + 2z\mathbf{k}.$$

When $(x, y, z) = (2, -1, 3)$ we have

$$\nabla F = -4\mathbf{i} + 2\mathbf{j} + 6\mathbf{k}.$$

Symmetric equations of the normal line are

$$\frac{x - 2}{-4} = \frac{y + 1}{2} = \frac{z - 3}{6}.$$

We can obtain parametric equations of the line by setting each term in the symmetric equations equal to a parameter $t$. This gives

$$x = 2 - 4t, \quad y = -1 + 2t, \quad z = 3 + 6t.$$

■

A surface with equation $z = f(x, y)$ can be written in the form $F(x, y, z) = 0$, with $F(x, y, z) = z - f(x, y)$. A normal vector to the surface $z = f(x, y)$ is then given by

$$\nabla F = -f_x\mathbf{i} - f_y\mathbf{j} + \mathbf{k}.$$

Let us restate this result as a theorem.

*Theorem 6*

**If $f(x, y)$ is differentiable at $(a, b)$, then the vector $-f_x(a, b)\mathbf{i} - f_y(a, b)\mathbf{j} + \mathbf{k}$ is normal to the surface $z = f(x, y)$ at the point $(a, b, f(a, b))$.**

Do not confuse the normal vector to the surface $z = f(x, y)$ with the gradient of $f(x, y)$. The gradient $\nabla f = f_x\mathbf{i} + f_y\mathbf{j}$ is a vector in the $xy$-plane that is orthogonal to the level curves of $f(x, y)$. The vector $-f_x\mathbf{i} - f_y\mathbf{j} + \mathbf{k}$ that is normal to the surface $z = f(x, y)$ is a vector in three-dimensional space.

**EXAMPLE 5**

Find an equation of the plane that is tangent to the surface $z = x^3 + y^3$ at the point $(2, -1, 7)$.

**SOLUTION**

The equation of the surface is of the form $z = f(x, y)$, where $f(x, y) = x^3 + y^3$. The tangent plane has normal vector

$$\mathbf{N} = -f_x\mathbf{i} - f_y\mathbf{j} + \mathbf{k} = -(3x^2)\mathbf{i} - (3y^2)\mathbf{j} + \mathbf{k}.$$

When $(x, y, z) = (2, -1, 7)$,

$$\mathbf{N} = -3(2)^2\mathbf{i} - 3(-1)^2\mathbf{j} + \mathbf{k} = -12\mathbf{i} - 3\mathbf{j} + \mathbf{k}.$$

An equation of the plane through the point $(2, -1, 7)$ with normal vector $-12\mathbf{i} - 3\mathbf{j} + \mathbf{k}$ is

$$(-12)(x - 2) + (-3)(y - (-1)) + (1)(z - 7) = 0,$$

$$-12x + 24 - 3y - 3 + z - 7 = 0,$$

$$z = 12x + 3y - 14.$$

## EXERCISES 16.7

In Exercises 1–4, sketch the level curves $f(x, y) = z_0$; evaluate $\nabla f$ and $D_{\mathbf{U}}f$ at the indicated point; sketch $\nabla f$ and $\mathbf{U}$ with initial position at the point indicated.

1 $f(x, y) = y/x^2$; $z_0 = 1, 2, 3$; $\mathbf{U} = (\mathbf{i} + 4\mathbf{j})/\sqrt{17}$; $(1, 2)$

2 $f(x, y) = (4 - x)/y^2$; $z_0 = 1, 2, 3$; $\mathbf{U} = (4\mathbf{i} + \mathbf{j})/\sqrt{17}$; $(2, -1)$

3 $f(x, y) = \ln(x^2 + y^2)$; $z_0 = -1, 0, 1$; $\mathbf{U} = (-\mathbf{i} + \mathbf{j})/\sqrt{2}$; $(1/\sqrt{2}, 1/\sqrt{2})$

4 $f(x, y) = \ln xy$; $z_0 = -1, 0, 1$; $\mathbf{U} = (\mathbf{i} - \mathbf{j})/\sqrt{2}$; $(1, 1)$

In Exercises 5–8, find an equation of the line that is normal to given curves at the indicated point.

5 $x^2 + y^2 = 25$; $(3, -4)$

6 $xy = 8$; $(2, 4)$

7 $x^2 - 3xy + 2y^2 = 0$; $(2, 1)$

8 $x = ye^y$; $(0, 0)$

In Exercises 9–14, find an equation of the plane that is tangent to the given surface at the indicated point.

9 $z^2 = x^2 + y^2$; $(3, -4, -5)$

10 $z^2 = xyz + 3$; $(-1, 2, -3)$

11 $ze^z = x^2 + y^2$; $(e, -e, 2)$

12 $z^2 + z - 2xy^2 = 0$; $(3, 1, 2)$

13 $z = xy$; $(-2, 3, -6)$

14 $z = \cos(x^2 + y^2)$; $(0, \sqrt{\pi}, -1)$

In Exercises 15–20 find symmetric equations and parametric equations of the line that is normal to the given surface at the indicated point.

15 $z = x^2 + y^2$; $(-1, 2, 5)$

16 $z^2 = 2x - 3y$; $(3, -1, -3)$

17 $e^{xyz} = 2x + y$; $(2, -3, 0)$

18 $6e^{xz} = yx - yz$; $(2, 3, 0)$

19 $z = xe^y$; $(0, 0, 0)$

20 $z = (\sin x)(\cos y)$; $(0, 0, 0)$

21 Find a vector that is tangent to the curve of intersection of the surfaces $z = xy$ and $x^2 + y^2 + z^2 = 9$ at the point $(1, 2, 2)$.

22 Find a vector that is tangent to the curve of intersection of the cylinders $y^2 + z^2 = 1$ and $x^2 + z^2 = 1$ at the point $(1/2, 1/2, \sqrt{3}/2)$.

## 16.8 EXTREMA

The theory of extrema for functions of several variables is similar to the theory for functions of one variable. For example, we say that $f(c_1, \ldots)$ is the (absolute) **maximum** value of $f$ in a region $R$ if $(c_1, \ldots)$ is in $R$ and $f(c_1, \ldots) \geq f(x_1, \ldots)$ for all $(x_1, \ldots)$ in $R$; $f(c_1, \ldots)$ is the (absolute) **minimum** value of $f$ in $R$ if $(c_1, \ldots)$ is in $R$ and $f(c_1, \ldots) \leq f(x_1, \ldots)$ for all $(x_1, \ldots)$ in $R$. The value $f(c_1, \ldots)$ is a **local** maximum or minimum value of $f$ if there is some $\delta > 0$ such that $f(c_1, \ldots)$ is a maximum or minimum, respectively, of $f$ in the set $(x_1 - c_1)^2 + \cdots + (x_n - c_n)^2 < \delta^2$.

A function $f(x_1, \ldots)$ cannot have a local extreme value at a point where $f$ is either increasing or decreasing in any direction. It follows that $f$ can have a local extreme value only at points where either

(i) all first partial derivatives of $f$ are zero, or
(ii) some first partial derivative of $f$ does not exist.

Points $(c_1, \ldots)$ in the interior of the domain of $f$ where either (i) or (ii) hold are called **critical points** of $f$. As in the case of one variable, values at critical points are not necessarily local extrema. We can sometimes use the graph of $f$ to determine if the value at a critical point is a local maximum, a local minimum, or neither of these.

### ■ EXAMPLE 1

Find critical points and local extrema of

$$f(x, y) = x^2 - 6x + y^2 - 4y.$$

#### SOLUTION

This function is defined with first partial derivatives for all $(x, y)$. The first partial derivatives are

$$f_x(x, y) = 2x - 6,$$

$$f_y(x, y) = 2y - 4.$$

Setting each first partial derivative equal to zero, we obtain the following system of two equations.

$$\begin{cases} 2x - 6 = 0, \\ 2y - 4 = 0. \end{cases}$$

The solution of this system is $(x, y) = (3, 2)$. This is the only critical point of the function. The corresponding value is

$$f(3, 2) = (3)^2 - 6(3) + (2)^2 - 4(2) = -13.$$

In this example we can complete squares to write

$$\begin{aligned} f(x, y) &= x^2 - 6x + (9) + y^2 - 4y + (4) - (9) - (4) \\ &= (x - 3)^2 + (y - 2)^2 - 13. \end{aligned}$$

We then see that the graph is a paraboloid with vertex at $(3, 2, -13)$ and that $f(3, 2) = -13$ is a local minimum which is also the absolute minimum value of the function. See Figure 16.8.1. ■

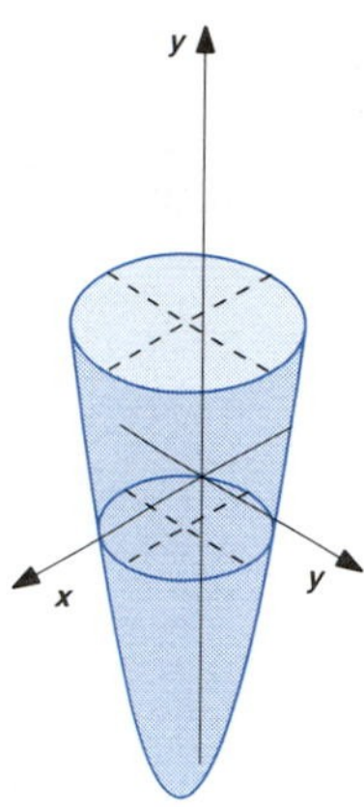

FIGURE 16.8.1

■ **EXAMPLE 2**

Find critical points and extrema of

$$f(x, y) = 4 - 2\sqrt{x^2 + y^2}.$$

**SOLUTION**

For $(x, y) \neq (0, 0)$, we have

$$f_x(x, y) = -2\left(\frac{1}{2}\right)(x^2 + y^2)^{-1/2}(2x),$$

$$f_y(x, y) = -2\left(\frac{1}{2}\right)(x^2 + y^2)^{-1/2}(2y).$$

These partial derivatives are not *simultaneously* zero for $(x, y) \neq (0, 0)$. Since

$$f(x, 0) = 4 - 2|x| \quad \text{and} \quad f(0, y) = 4 - |y|,$$

neither partial derivative exists at $(0, 0)$, so $(0, 0)$ is a critical point. The corresponding value is

$$f(0, 0) = 4.$$

The graph is a cone with vertex at the point $(0, 0, 4)$. The value $f(0, 0) = 4$ is a local maximum which is also the absolute maximum of the function. See Figure 16.8.2. ■

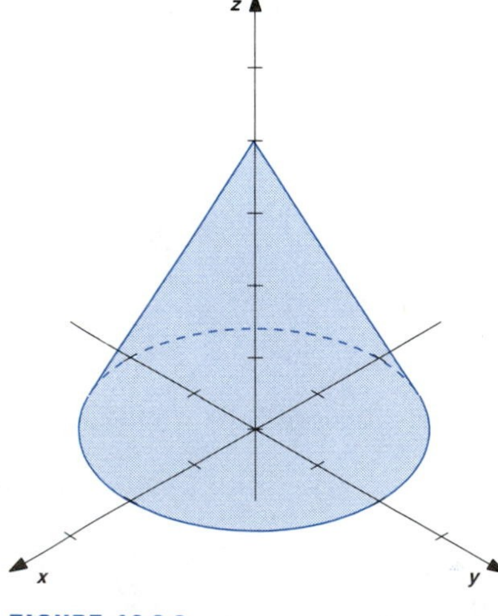

FIGURE 16.8.2

■ **EXAMPLE 3**

Find critical points and local extrema of

$$f(x, y) = \frac{x^2 - y^2}{4}.$$

**SOLUTION**

The function is defined with partial derivatives for all $(x, y)$. We have

$$f_x(x, y) = \frac{2x}{4},$$

$$f_y(x, y) = \frac{-2y}{4}.$$

Setting each first partial derivative equal to zero, we obtain

$$x = 0 \quad \text{and} \quad y = 0.$$

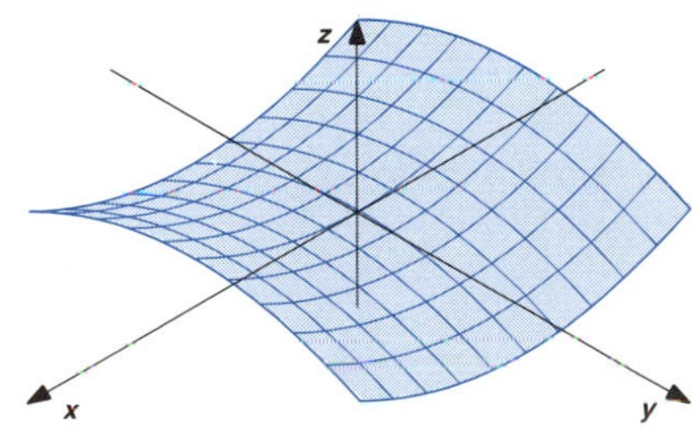

FIGURE 16.8.3

The point $(0, 0)$ is the only critical point. The corresponding value is

$$f(0, 0) = 0.$$

The graph of $f$ is a hyperbolic paraboloid. The value $f(0, 0) = 0$ is neither a maximum nor minimum value; the graph is concave up at the origin in the direction of the $x$-axis; the graph is concave down at the origin in the direction of the $y$-axis. See Figure 16.8.3. ■

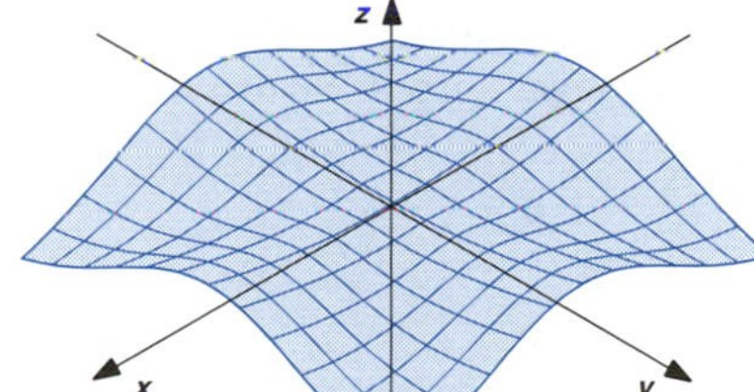

FIGURE 16.8.4

If the value of a function at a critical point is neither a maximum nor minimum value, the corresponding point on the graph is called a **saddle point**.

For functions of a single real variable, we can use the Second Derivative Test to determine if a local extreme value is a local maximum or a local minimum. Let us see what the corresponding result is for functions of two variables. First, we note that it is not enough to require that both second partial derivatives at a critical point are nonzero with the same sign. For example, it is possible that $f(x, y)$ is continuous with $f_x(0, 0) = f_y(0, 0) = 0$, $f_{xx}(0, 0) > 0$, $f_{yy}(0, 0) > 0$, and $f(0, 0)$ not a local minimum. This is illustrated in Figure 16.8.4. The second partials being positive implies that the traces of the surface in the planes $x = 0$ and $y = 0$ are concave up at the origin. We would seem to have a better chance of having a local minimum if the graph were concave up in *every* direction at $(0, 0)$. This is illustrated in Figure 16.8.5.

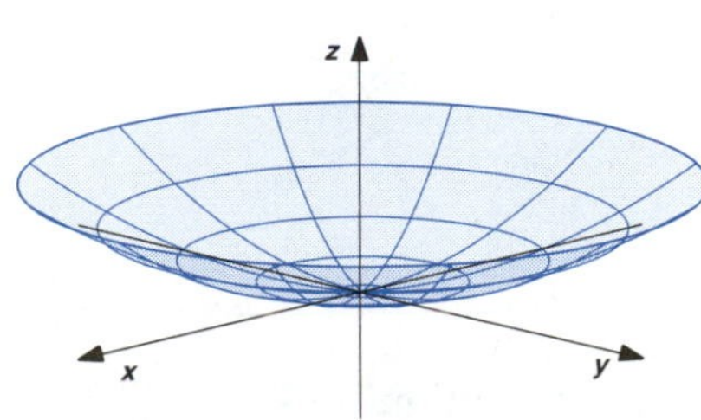

FIGURE 16.8.5

The idea illustrated in Figure 16.8.5 can be used to develop a version of the Second Derivative Test for Local Extrema of functions of two variables. That is, we calculate the second derivative with respect to $t$ of the function

$$z = f(x, y),\ x = a + u_1 t,\ y = b + u_2 t,$$

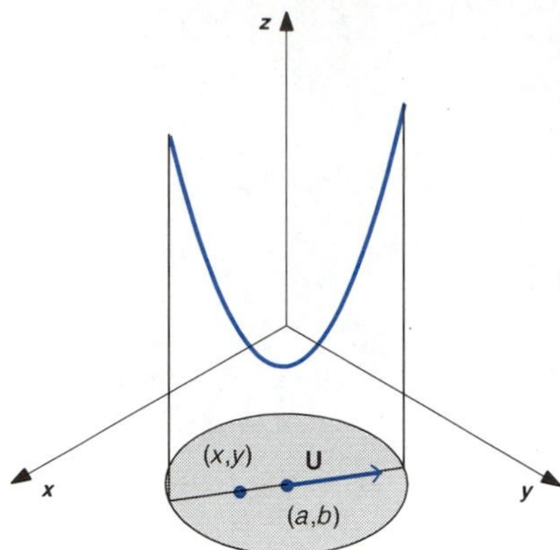


FIGURE 16.8.6

where $u_1^2 + u_2^2 = 1$. The sign of $z''(t)$ gives the concavity of $f$ at $(a, b)$ in the direction of the vector $u_1\mathbf{i} + u_2\mathbf{j}$. See Figure 16.8.6. Assuming appropriate differentiability, we obtain

$$z'(t) = f_x(x, y)u_1 + f_y(x, y)u_2 \quad \text{and}$$

$$z''(t) = [f_{xx}(x, y)u_1 + f_{xy}(x, y)u_2]u_1 + [f_{yx}(x, y)u_1 + f_{yy}(x, y)u_2]u_2.$$

Assuming the second partial derivatives are continuous, so that $f_{xy} = f_{yx}$, we have

$$z''(t) = f_{xx}(x, y)u_1^2 + 2f_{xy}(x, y)u_1u_2 + f_{yy}(x, y)u_2^2.$$

We want $z''(t)$ to be of constant sign for all values of $t$, $u_1$, and $u_2$ for which $(x, y) = (a + u_1t, b + u_2t)$ is in a disk with center $(a, b)$. If $(a, b)$ is a critical point of $f$, so that $f_x(a, b) = f_y(a, b) = 0$, we then have

$$z'(0) = f_x(a, b)u_1 + f_y(a, b)u_2 = 0$$

and two applications of the Mean Value Theorem give

$$\begin{aligned} f(x, y) - f(a, b) &= z(t) - z(0) \\ &= z'(t_1)(t - 0) \\ &= [z'(t_1) - z'(0)]t \\ &= z''(t_2)t^2, \end{aligned}$$

where either $0 < t_2 < t_1 < t$ or $t < t_1 < t_2 < 0$. It follows that $f(x, y) - f(a, b)$ is of constant sign as $(x, y)$ varies over a disk with center $(a, b)$, so $f(a, b)$ is either a local maximum or a local minimum, depending on the constant sign of $z''$. To determine when $z''(t)$ is of constant sign, we note that a quadratic expression $Ax^2 + Bx + C$ is never zero if the discriminant $B^2 - 4AC$ is negative. We can use this fact to conclude that $z''(t)$ is not zero for any choice of $(u_1, u_2)$ with $u_1^2 + u_2^2 = 1$ if

$$[2f_{xy}(x, y)]^2 - 4f_{xx}(x, y)f_{yy}(x, y) < 0, \quad \text{or}$$

$$f_{xx}(x, y)f_{yy}(x, y) - [f_{xy}(x, y)]^2 > 0.$$

Since we have assumed that the second partial derivatives of $f$ are continuous, if we have

$$f_{xx}(a, b)f_{yy}(a, b) - [f_{xy}(a, b)]^2 > 0,$$

then we also have

$$f_{xx}(x, y)f_{yy}(x, y) - [f_{xy}(x, y)]^2 > 0$$

for all $(x, y)$ in a disk with center $(a, b)$. The condition

$$f_{xx}(a, b)f_{yy}(a, b) - [f_{xy}(a, b)]^2 > 0$$

implies that $f_{xx}(a, b)$ and $f_{yy}(a, b)$ are of the same sign and that the sign of $z''(t)$ agrees with them both. Also, if

$$f_{xx}(a, b)f_{yy}(a, b) - [f_{xy}(a, b)]^2 < 0,$$

then the concavity of $f$ is upward in one direction and downward in another direction, so $f(a, b)$ must be a saddle point. This gives the following theorem.

*Theorem 1*

**If $f(x, y)$ has second partial derivatives that are continuous in a disk with center $(a, b)$, and $f_x(a, b) = f_y(a, b) = 0$, then:**

**(i) $f$ has a local maximum at $(a, b)$ if**

$$f_{xx}(a, b)f_{yy}(a, b) - [f_{xy}(a, b)]^2 > 0 \quad \text{and} \quad f_{xx}(a, b) < 0.$$

**(ii) $f$ has a local minimum at $(a, b)$ if**

$$f_{xx}(a, b)f_{yy}(a, b) - [f_{xy}(a, b)]^2 > 0 \quad \text{and} \quad f_{xx}(a, b) > 0.$$

**(iii) $f$ has a saddle point at $(a, b)$ if**

$$f_{xx}(a, b)f_{yy}(a, b) - [f_{xy}(a, b)]^2 < 0.$$

The theorem does not give information for all possible cases. If $f_{xx}f_{yy} - (f_{xy})^2 = 0$, other tests are required to determine if a value at a critical point is either a local maximum, a local minimum, or a saddle point.

If $f_{xy} = f_{yx}$ the expression $f_{xx}f_{yy} - (f_{xy})^2$ is the value of the determinant

$$\begin{vmatrix} f_{xx} & f_{xy} \\ f_{yx} & f_{yy} \end{vmatrix}.$$

This is called the **Hessian** of the function $f$. The extension of the above theorem to functions of more that two variables involves corresponding determinants. We leave this to more advanced texts.

### EXAMPLE 4

Find all critical points of the function

$$f(x, y) = \frac{1}{3}x^3 - x + xy^2.$$

Find the value at each critical point and characterize it as a local maximum, local minimum, or saddle point.

**SOLUTION**

We have

$$f_x(x, y) = x^2 - 1 + y^2,$$

$$f_y(x, y) = 2xy.$$

Setting the first partial derivatives equal to zero, we obtain

$$x^2 - 1 + y^2 = 0, \tag{1}$$

$$2xy = 0. \tag{2}$$

Equation (2) implies either $x = 0$ or $y = 0$. If $x = 0$, substitution into equation (1) gives

$$(0)^2 - 1 + y^2 = 0,$$

$$y^2 = 1,$$

$$y = 1, \ -1.$$

This gives the critical points (0, 1) and (0, −1). Similarly, if $y = 0$, equation (1) implies $x = 1, -1$ and we obtain the critical points (1, 0) and (−1, 0).

Let us use the second derivatives to determine if each critical point corresponds to a local maximum, a local minimum, or a saddle point. We have

$$f_x = x^2 - 1 + y^2, \qquad f_y = 2xy,$$

$$f_{xx} = 2x, \qquad f_{yy} = 2x,$$

$$f_{xy} = 2y,$$

$$f_{xx}f_{yy} - (f_{xy})^2 = (2x)(2x) - (2y)^2 = 4x^2 - 4y^2.$$

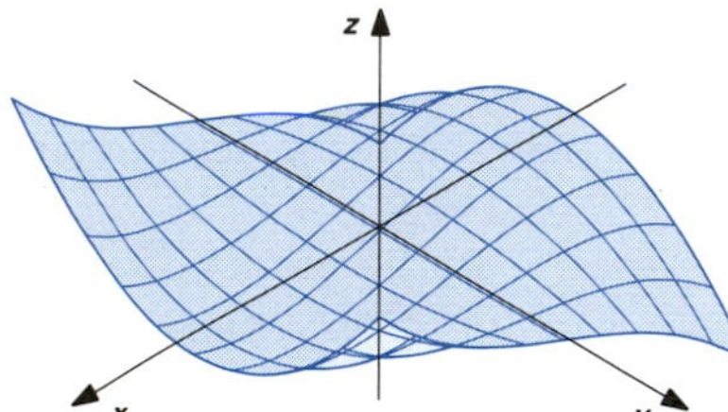


FIGURE 16.8.7

The values of $f$, $f_{xx}f_{yy} - (f_{xy})^2$, and $f_{xx}$ at each critical point are given in Table 16.8.1. We see that (0, 1) and (0, −1) correspond to saddle points; $f(1, 0) = -2/3$ is a local minimum, and $f(-1, 0) = 2/3$ is a local maximum. The graph is sketched in Figure 16.8.7. ■

TABLE 16.8.1

| $(x, y)$ | $f$ | $f_{xx}f_{yy} - (f_{xy})^2$ | $f_{xx}$ | Type |
|---|---|---|---|---|
| (0, 1) | 0 | $-4 < 0$ | — | Saddle point |
| (0, −1) | 0 | $-4 < 0$ | — | Saddle point |
| (1, 0) | −2/3 | $4 > 0$ | $2 > 0$ | Local minimum |
| (−1, 0) | 2/3 | $4 > 0$ | $-2 < 0$ | Local maximum |

Many problems involve finding maximum or minimum values of a function of several variables subject to an equation of constraint that relates the variables.

■ **EXAMPLE 5**

Find the minimum value of $f(x, y) = x^2 - 6x + y^2 - 4y$ subject to the condition that $(x, y)$ is on the line $x + y = 7$.

**SOLUTION**

The values of $f(x, y)$ that we want to consider are indicated in Figure 16.8.8a.

We can use one-variable techniques to find the desired maximum value. That is, we use the condition $x + y = 7$ to express $y$ in terms of $x$. In particular, $y = 7 - x$. We can then substitute for $y$ in the expression for $f(x, y)$ to obtain a function of $x$,

$$\begin{aligned} z &= f(x, 7 - x) \\ &= x^2 - 6x + (7 - x)^2 - 4(7 - x) \\ &= x^2 - 6x + 49 - 14x + x^2 - 28 + 4x \\ &= 2x^2 - 16x + 21. \end{aligned}$$

This is a quadratic function of $x$ that has a minimum at the point where $dz/dx = 0$. We have

$$\frac{dz}{dx} = 4x - 16,$$

$$\frac{dz}{dx} = 0 \quad \text{implies} \quad 4x - 16 = 0, \quad \text{so} \quad x = 4.$$

When $x = 4$, we have $y = 7 - x = 3$. The desired minimum value is

$$f(4, 3) = (4)^2 - 6(4) + (3)^2 - 4(3) = -11.$$ ■

(a)

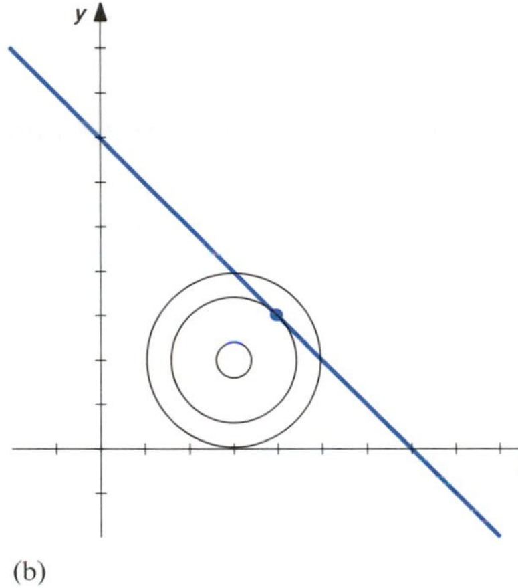


(b)

**FIGURE 16.8.8**

In Figure 16.8.8b we have sketched some level curves of the function $f(x, y) = x^2 - 6x + y^2 - 4y$ in Example 5. Note that a level curve is tangent to the line $x + y = 7$ at the point (4, 3), where the minimum value of the function for $(x, y)$ on the line

We know from the Extreme Value Theorem in Section 3.9 that a function of a single real variable that is continuous on a closed and bounded interval has maximum and minimum values on the interval, and that these values occur at either critical points or endpoints of the interval. Analogous results are true for functions of several variables. Recall from Section 16.2 that a region $R$ in $n$-dimensional space is **closed** if it contains all its boundary points. The region $R$ is said to be **bounded** if there is a number $B$ such that $x_1^2 + \cdots + x_n^2 < B^2$ for all $(x_1, \ldots, x_n)$ in $R$. We then have the following result for functions of several variables.

■ *Theorem 2*

**If $f$ is continuous on a closed, bounded region, then $f$ has a maximum value and a minimum value on the region. These extrema occur either**

**(i) where all first partial derivatives of $f$ are zero;**

**(ii) where some first partial derivative of $f$ does not exist; or**

**(iii) on the boundary of the region.**

We will not verify this result. Note that in order to find boundary extrema we must find extrema subject to the constraint of the equation that determines the boundary. Parts of the boundary that are given by different equations must be treated separately.

**■ EXAMPLE 6**

Find maximum and minimum values of

$$f(x, y) = x^2 - 6x + y^2 - 4y$$

on the closed region bounded by the $x$-axis, the $y$-axis, and the line $x + y = 7$.

**SOLUTION**

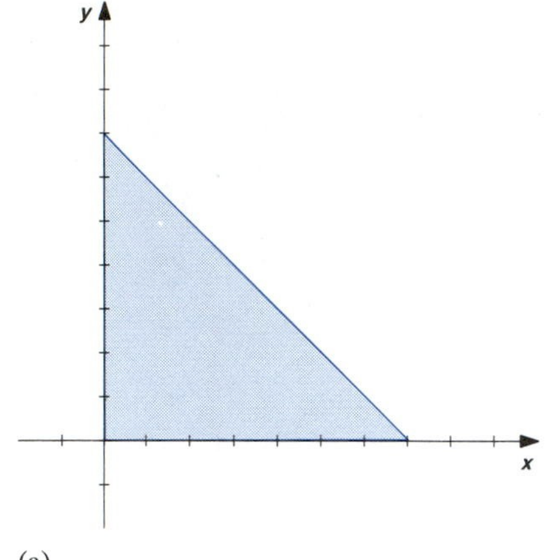


(a)

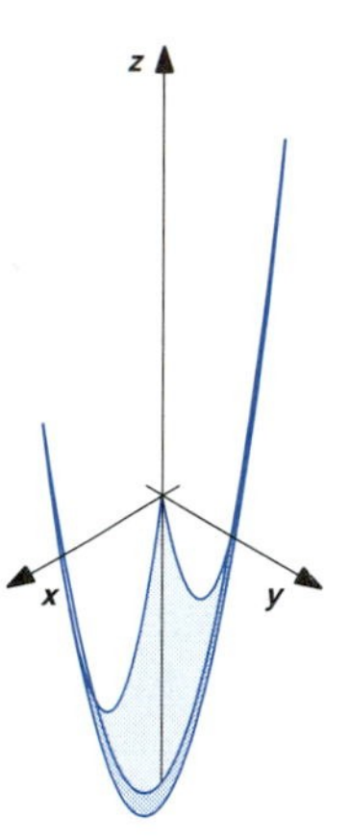


(b)

**FIGURE 16.8.9**

The region bounded by the $x$-axis, the $y$-axis, and the line $x + y = 7$ is sketched in Figure 16.8.9a. The graph of $f$ is sketched in Figure 16.8.9b. The continuous function has maximum and minimum values on the closed and bounded region.

In Example 1 we saw that the first partial derivatives of $f$ are both zero at the point (3, 2). This is an interior point of the domain. There are no points where the first partial derivatives of the function do not exist.

The boundary of the region consists of the lines $y = 0$, $x = 0$, and $x + y = 7$. Each part must be treated separately.

We need extrema of $f(x, y)$ for $(x, y)$ on the line $x + y = 7$, between the points (0, 7) and (7, 0). As in Example 5, we obtain a function of $x$ by substituting $7 - x$ for $y$. However, we must now restrict the domain to those values of $x$ that correspond to points on the line segment between (0 ,7) and (7, 0). We obtain

$$f(x, 7 - x) = 2x^2 - 16x + 21, \qquad 0 \leq x \leq 7.$$

Setting the derivative with respect to $x$ equal to zero gives the point (4, 3). We must also consider the endpoints (0, 7) and (7, 0).

Consideration of extrema on the boundary of the region along $y = 0$ leads to the function

$$f(x, 0) = x^2 - 6x, \qquad 0 \leq x \leq 7.$$

Setting the derivative with respect to $x$ equal to zero gives the point (3, 0). The endpoints are (0, 0) and (7, 0).

Consideration of extrema on the boundary of the region along $x = 0$ leads to the function

$$f(0, y) = y^2 - 4y, \qquad 0 \leq y \leq 7.$$

Setting the derivative with respect to $y$ equal to zero gives the point (0, 2). The endpoints are (0, 0) and (0, 7).

**TABLE 16.8.2**

| $(x, y)$ | $f(x, y)$ |
|---|---|
| (3, 2) | −13 |
| (4, 3) | −11 |
| (3, 0) | −9 |
| (0, 2) | −4 |
| (0, 0) | 0 |
| (7, 0) | 7 |
| (0, 7) | 21 |

All candidates for the maximum and minimum values are listed in Table 16.8.2. We see that the minimum value is $f(3, 2) = -13$; the maximum value is $f(0, 7) = 21$. ■

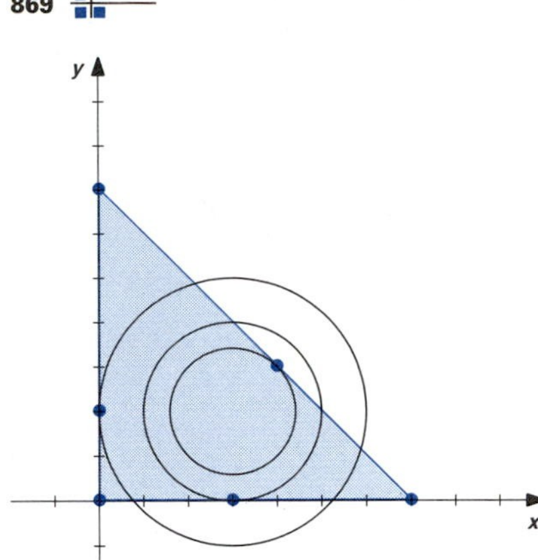


FIGURE 16.8.10

In Figure 16.8.10 we have sketched the domain and some level curves of the function of Example 6. Note that the candidates for extrema on the boundary are exactly corner points of the boundary and points on the boundary where a level curve is tangent to the boundary.

### ■ EXAMPLE 7

Find the minimum distance between the point $(1, 2, 0)$ and the cone $z = 2\sqrt{x^2 + y^2}$.

**SOLUTION**

A sketch is given in Figure 16.8.11. It is more convenient to work with the square of the distance between two points than with the distance. The minimum value of the square corresponds to the minimum value of the distance. The square of the distance between the points $(1, 2, 0)$ and $(x, y, z)$ is

$$D = (x - 1)^2 + (y - 2)^2 + z^2.$$

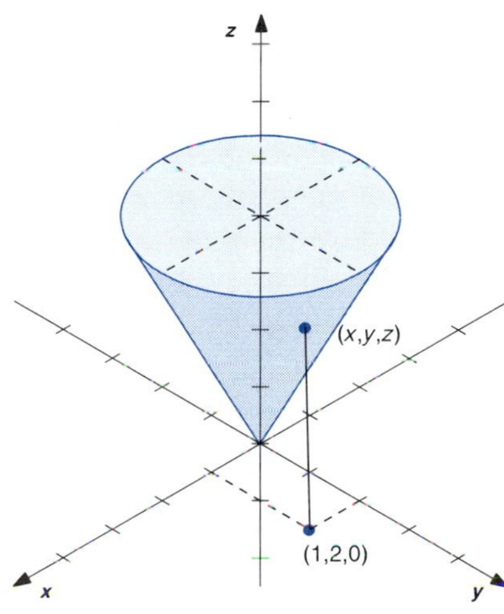


FIGURE 16.8.11

A point $(x, y, z)$ is on the cone if the coordinates of the point satisfy the equation of the cone. Thus, the equation of the cone is a constraint for this problem. Using the equation of the cone to eliminate $z$ from the expression for $D$, we obtain

$$D(x, y) = (x - 1)^2 + (y - 2)^2 + 4(x^2 + y^2).$$

Setting the two partial derivatives of $D(x, y)$ equal to zero, we obtain the system

$$\begin{cases} 2(x - 1) + 4(2x) = 0, \\ 2(y - 2) + 4(2y) = 0. \end{cases}$$

The solution of the system is $x = 0.2$, $y = 0.4$. The corresponding value of $z$ is $z = 2\sqrt{(0.2)^2 + (0.4)^2} = 2\sqrt{0.2} \approx 0.89442719$. It is clear from the geometry that these values give the minimum distance. We have

$$D(0.2, 0.4) = (0.2 - 1)^2 + (0.4 - 2)^2 + 4((0.2)^2 + (0.4)^2) = 4.$$

The minimum distance is then $\sqrt{D} = 2$. ■

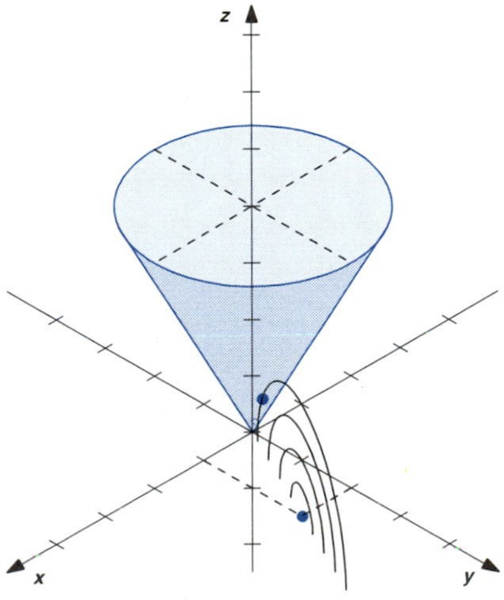

Note that the level surfaces of the function

$$D = (x - 1)^2 + (y - 2)^2 + z^2$$

are spheres with center $(1, 2, 0)$. It seems that the point on the cone $z = 4\sqrt{x^2 + y^2}$ that is closest to $(1, 2, 0)$ should be the first point on the cone that is touched as the level surfaces expand from $(1, 2, 0)$, and that the sphere should be tangent to the cone at that point. This is true. See Figure 16.8.12.

FIGURE 16.8.12

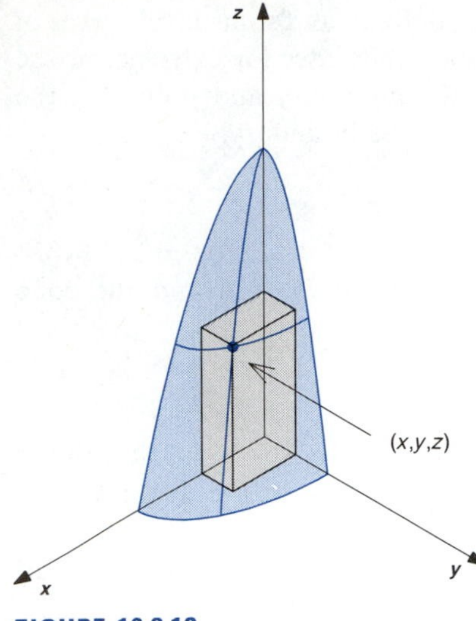

FIGURE 16.8.13

■ **EXAMPLE 8**

Find the dimensions and volume of the rectangular box of maximum volume that is in the first octant with one corner at the origin and the opposite corner on the paraboloid $z = 4 - x^2 - 4y^2$.

**SOLUTION**

The box is sketched in Figure 16.8.13. The volume of the box is

$$V = xyz.$$

The point $(x, y, z)$ is on the paraboloid if the coordinates of the point satisfy the equation of the paraboloid. The equation of the paraboloid is a constraint for the problem. We use the equation of constraint to eliminate $z$ from the expression for the volume. We obtain

$$\begin{aligned} V(x, y) &= xy(4 - x^2 - 4y^2) \\ &= 4xy - x^3y - 4xy^3. \end{aligned}$$

The domain of $V(x, y)$ is the closed region in the first quadrant bounded by $x = 0$, $y = 0$, and $x^2 + 4y^2 = 4$. These correspond to each of $x$, $y$, and $z$ being zero. Since $V(x, y)$ is continuous on a closed and bounded region, we know there is a maximum value. The volume is positive in the interior of the region and zero on the boundary, so the maximum must occur at a critical point. Setting the partial derivatives of $V(x, y)$ equal to zero, we obtain the system

$$\begin{cases} 4y - 3x^2y - 4y^3 = 0, \\ 4x - x^3 - 12xy^2 = 0, \end{cases}$$

or

$$\begin{cases} y(4 - 3x^2 - 4y^2) = 0, \\ x(4 - x^2 - 12y^2) = 0. \end{cases}$$

We may disregard cases where either $x = 0$ or $y = 0$, since these values correspond to boundary points. The system then simplifies to

$$\begin{cases} 4 - 3x^2 - 4y^2 = 0, \\ 4 - x^2 - 12y^2 = 0. \end{cases}$$

The difference of the two equations then gives

$$-2x^2 + 8y^2 = 0, \quad \text{so} \quad x = 2y.$$

Substitution then gives

$$4 - 3(2y)^2 - 4y^2 = 0, \quad \text{so} \quad y = \frac{1}{2}.$$

Then $x = 2y = 2(1/2) = 1$ and $z = 4 - (1)^2 - 4(1/2)^2 = 2$. The box of maximum volume has dimensions $1 \times 1/2 \times 2$ and volume

$$V\left(1, \frac{1}{2}\right) = 4(1)\left(\frac{1}{2}\right) - (1)^3\left(\frac{1}{2}\right) - 4(1)\left(\frac{1}{2}\right)^3 = 1.$$ ■

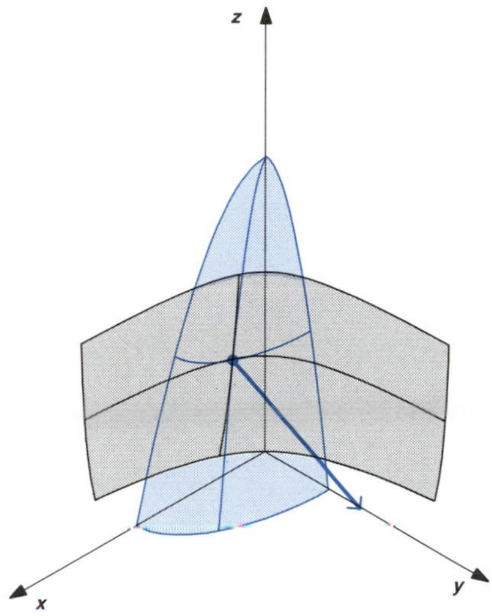

FIGURE 16.8.14

In Example 8 we found the maximum value of $V = xyz$ subject to the constraint $z = 4 - x^2 - 4y^2$. Let us see that the level surface of $V = xyz$ is tangent to the paraboloid $z = 4 - x^2 - 4y^2$ at the point $(1, 1/2, 2)$ that gives the maximum value. In particular, the gradient of $V = xyz$ at $(1, 1/2, 2)$ is

$$\nabla V = V_x\mathbf{i} + V_y\mathbf{j} + V_z\mathbf{k} = yz\mathbf{i} + xz\mathbf{j} + xy\mathbf{k} = \mathbf{i} + 2\mathbf{j} + \frac{1}{2}\mathbf{k}.$$

The gradient of $V$ is normal to the level surface of $V$. Let us write the equation of constraint as $g(x, y) = 0$, where $g(x, y) = x^2 + 4y^2 + z - 4$. The gradient of $g$ is then normal to the paraboloid. We have

$$\nabla g = g_x\mathbf{i} + g_y\mathbf{j} + g_z\mathbf{k} = 2x\mathbf{i} + 8y\mathbf{j} + \mathbf{k} = 2\mathbf{i} + 4\mathbf{j} + \mathbf{k}.$$

Since $\nabla V = \nabla g/2$, the normal vectors of the surfaces are parallel and the surfaces are tangent. See Figure 16.8.14.

## EXERCISES 16.8

Find all critical points of the functions given in Exercises 1–10. Find the value at each critical point and characterize it as a local maximum, local minimum, or saddle point.

**1** $f(x, y) = x^3 - 3x + y^2$

**2** $f(x, y) = x^3 - 12x + y^2 + 8y$

**3** $f(x, y) = x^3 - 3x + y^3 - 3y$

**4** $f(x, y) = x^3 + 3x^2 + y^3 - 3y^2$

**5** $f(x, y) = 2x^3 + 6xy + 3y^2$

**6** $f(x, y) = 2x^3 - 6xy + y^2$

**7** $f(x, y) = x^2 + xy + y^2 + x + 2y$

**8** $f(x, y) = x^3 + y^3 - 3xy$

**9** $f(x, y) = \sqrt{x^2 + y^2}$

**10** $f(x, y) = \sqrt{x^2 + y^2 - 2x + 4y + 5}$

In Exercises 11–14, find extrema of the given functions subject to the indicated conditions; determine if each value is a maximum or minimum value subject to the condition.

**11** $f(x, y) = xy$, $y = 6 - 3x$

**12** $f(x, y) = xy$, $y = x - 4$

**13** $f(x, y) = x^2 - y^2$, $y = 3 - 2x$

**14** $f(x, y) = x^2 - y^2$, $2y = x + 3$

In Exercises 15–20, find the maximum and minimum values of the given functions in the indicated regions.

**15** $f(x, y) = x^3 - 3x + y^2$, $x^2 - 2x + y^2 \le 0$

**16** $f(x, y) = xy + 3$, $x^2 + 4y^2 \le 8$

**17** $f(x, y) = x^2 - 3y + y^3$, $x^2 + 4y^2 \le 36$

**18** $f(x, y) = 9x^2 + y^3 - 12y$, $x^2 + y^2 - 3y \le 0$

**19** $f(x, y) = x^2 + 2x + y^2 - 2y$, region bounded by $x = 0$, $x = 3$, $y = 0$, and $y = 3$

**20** $f(x, y) = 4x - x^2 + 4y - y^2$, region bounded by $x = 0$, $x = 3$, $y = 0$, and $y = 1$

**21** Find the point on the line $y = 2x + 3$ that is nearest the origin.

**22** Find the point or points on the parabola $2y = x^2 - 1$ that are nearest the origin.

23 Find the point or points on the parabola $y = x^2 - 1$ that are nearest the origin.

24 Find the point or points on the paraboloid $z = x^2 + (y^2/4) - 1$ that are nearest the origin.

25 Find the dimensions and volume of the rectangular box of maximum volume that is in the first octant with one corner at the origin and the opposite corner on the plane $x + 2y + z = 3$.

26 Find the dimensions and volume of the rectangular box of maximum volume that is in the first octant with one corner at the origin and the opposite corner on the paraboloid $x^2 + 2y^2 + z = 8$.

27 It is assumed that experimentally determined points $(x_1, y_1), (x_2, y_2), \ldots, (x_n, y_n)$, lie near some line $y = \alpha x + \beta$ and it is desired to find the line that "best" fits the data. One way to do this is to choose the line that minimizes the sum of the squares of the vertical distances of the data points from the line. That is, choose $\alpha$ and $\beta$ such that

$$(\alpha x_1 + \beta - y_1)^2 + (\alpha x_2 + \beta - y_2)^2 + \cdots + (\alpha x_n + \beta - y_n)^2$$

is a minimum. This line is called the **least squares regression line** for the data. Find formulas for $\alpha$ and $\beta$ in terms of $n$ and the coordinates of the data points for this line.

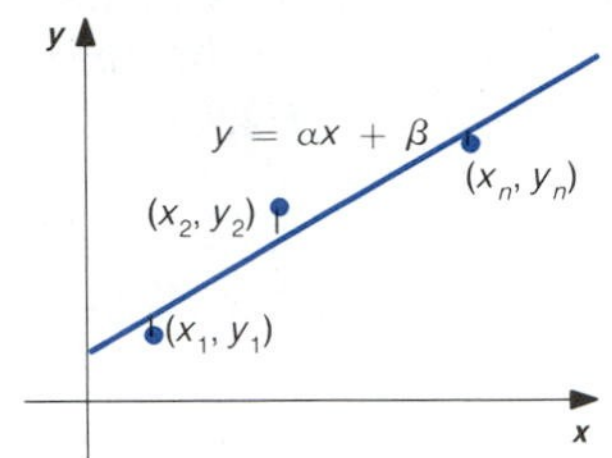


## 16.9. LAGRANGE MULTIPLIERS

The method of substitution used in Section 16.8 for finding extrema subject to constraints requires that certain variables be chosen as independent variables and other variables eliminated. In many problems it is difficult or impossible to solve the equations of restraint to obtain explicit formulas for the variables that are to be eliminated. Also, we must determine the domain of the chosen independent variables. In this section we will discuss a different method of finding extrema of a function subject to constraints. The method allows us to treat all of the original independent variables in a similar manner, without choosing some to be considered as independent variables and eliminating others.

Let us consider the problem of finding $(x, y)$ that give extrema of $f(x, y)$, subject to the constraint $g(x, y) = 0$. That is, we want the extrema of $f(x, y)$ for $(x, y)$ on the curve given by the equation $g(x, y) = 0$. If the directional derivative of $f$ in the direction of the tangent to the curve $g(x, y) = 0$ is not zero at a point on the curve, then the values of $f(x, y)$ are either increasing or decreasing as $(x, y)$ moves along the curve through the point. This means that extreme values of $f(x, y)$ on the curve $g(x, y) = 0$ can occur only at points where the directional derivative of $f$ in the direction of the curve are zero, so the gradient of $f$ must be orthogonal to the curve at those points. Since the plane vectors $\nabla f$ and $\nabla g$ are both orthogonal to the curve $g(x, y) = 0$, $\nabla f$ must be parallel to $\nabla g$. If $\nabla g \neq \mathbf{0}$, this means that $\nabla f$ must be a scalar multiple of $\nabla g$. We can also conclude that the level curves of $f(x, y)$ are tangent to the curve $g(x, y) = 0$ at the points on the curve where the extrema of $f(x, y)$ on the curve occur. See Figure 16.9.1.

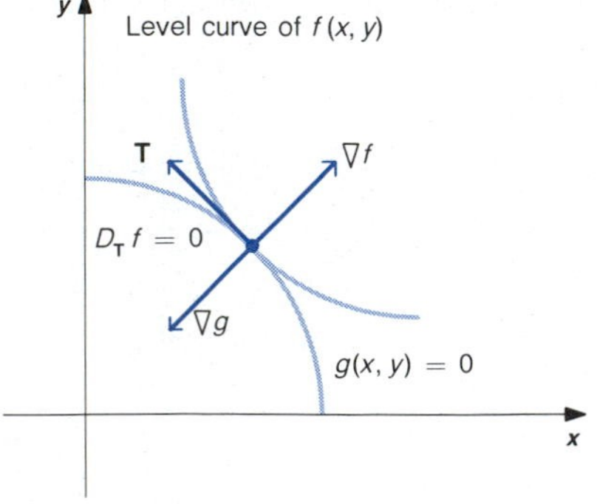


FIGURE 16.9.1

Let us summarize the above results in the following form.

*Theorem 1*

**If $f(x, y)$ and $g(x, y)$ are differentiable and $\nabla g \neq 0$ at points where $g(x, y) = 0$, then extrema of $f(x, y)$ subject to the constraint $g(x, y) = 0$ can occur only at points where there is a number $\lambda$ such that $\nabla f = \lambda \nabla g$.**

The Greek letter $\lambda$ (lambda) introduced above is called a **Lagrange multiplier**. To use the Lagrange multiplier, we note that the vector equation $\nabla f = \lambda \nabla g$ and the equation of constraint give the three equations

$$f_x(x, y) = \lambda g_x(x, y),$$

$$f_y(x, y) = \lambda g_y(x, y),$$

$$g(x, y) = 0.$$

These equations can be obtained by defining the function

$$w = w(x, y, \lambda) = f(x, y) - \lambda g(x, y)$$

and then setting the first partial derivatives of $w$ with respect to $x$, $y$, and $\lambda$ equal to zero. This is the format we will use.

### EXAMPLE 1

Find extrema of $f(x, y) = 4x + 3y$ for $(x, y)$ on the circle $x^2 + y^2 = 25$.

**SOLUTION**

The equation of constraint can be written as $g(x, y) = 0$, where $g(x, y) = x^2 + y^2 - 25$. Setting

$$\begin{aligned} w &= f(x, y) - \lambda g(x, y) \\ &= 4x + 3y - \lambda(x^2 + y^2 - 25), \end{aligned}$$

we have

$$w_x = 4 - \lambda(2x),$$

$$w_y = 3 - \lambda(2y),$$

$$w_\lambda = -(x^2 + y^2 - 25).$$

Setting each of the three partial derivatives equal to zero, we obtain the system of equations

$$\begin{cases} 4 - \lambda(2x) = 0, \\ 3 - \lambda(2y) = 0, \\ x^2 + y^2 - 25 = 0. \end{cases}$$

We want the $(x, y)$ values of the solution of this system. In this example, we can solve the first equation for $\lambda$ and then substitute into the second equation in order to obtain $y$ as a function of $x$. That is,

$$4 - \lambda(2x) = 0, \qquad \text{so} \quad x \neq 0 \quad \text{and} \quad \lambda = \frac{2}{x}.$$

Then $3 - \lambda(2y) = 0$ implies

$$3 - \left(\frac{2}{x}\right)(2y) = 0,$$

$$3x - 4y = 0,$$

$$y = \frac{3x}{4}.$$

We now substitute into the third equation and solve for $x$:

$$x^2 + y^2 - 25 = 0,$$

$$x^2 + \left(\frac{3x}{4}\right)^2 = 25,$$

$$16x^2 + 9x^2 = 25(16),$$

$$25x^2 = 25(16),$$

$$x^2 = 16,$$

$$x = 4, \ -4.$$

When $x = 4$, $\quad y = \dfrac{3(4)}{4} = 3.$

When $x = -4$, $\quad y = \dfrac{3(-4)}{4} = -3.$

The $(x, y)$ values that give solutions of our system of equations are $(4, 3)$ and $(-4, -3)$. These are the only candidates to give extrema of $f(x, y)$ on the circle. Since $f$ is continuous on the circle, which is a closed and bounded set, $f$ has a maximum value and a minimum value on the circle. The value $f(4, 3) = 25$ is the maximum value and the value $f(-4, -3) = -25$ is the minimum value. From Figure 16.9.2 we see that the level curves of $f(x, y) = 4x + 3y$ are tangent to the circle $x^2 + y^2 = 25$ at the points $(4, 3)$ and $(-4, -3)$. ■

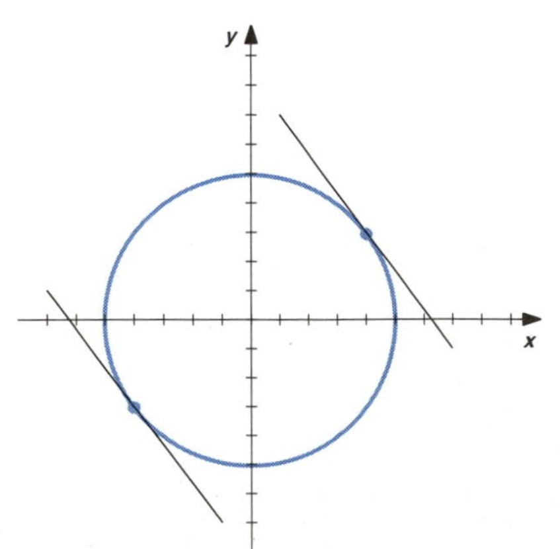


FIGURE 16.9.2

Let us now consider extrema of $f(x, y, z)$, subject to the constraint $g(x, y, z) = 0$. Extrema can occur only at points $(x, y, z)$ on the surface $g(x, y, z) = 0$ where the directional derivative of $f$ is zero in every direction of the plane tangent to the surface. This means that the gradient of $f$ is orthogonal to the surface $g(x, y, z) = 0$, so the gradient of $f$ is parallel to the gradient of $g$, and the level surface of $f$ is tangent to the surface

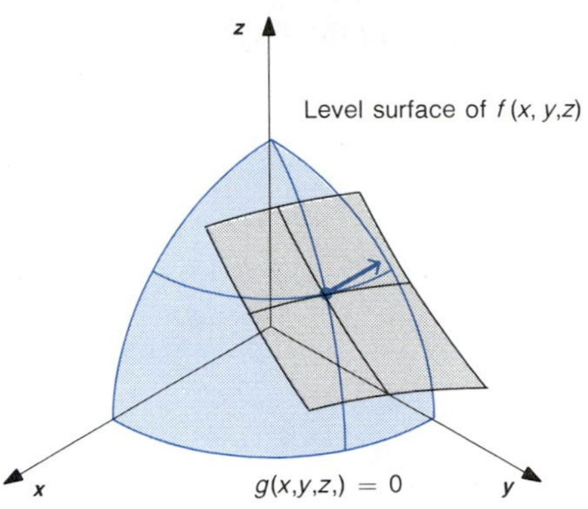

FIGURE 16.9.3

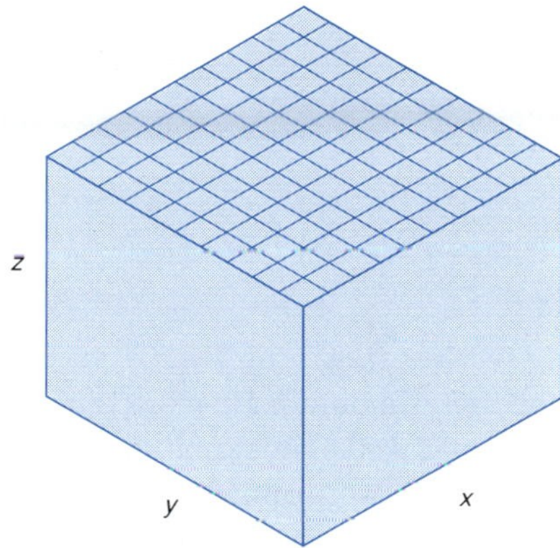

FIGURE 16.9.4

$g(x, y, z) = 0$. See Figure 16.9.3. The analogue of Theorem 1 holds for functions of three variables and the procedure for determining the extrema is similar. We define

$$w = w(x, y, z, \lambda) = f(x, y, z) - \lambda g(x, y, z)$$

and set all four first partial derivatives of $w$ equal to zero.

■ **EXAMPLE 2**

Material for the top of a rectangular box costs $\$3/\text{ft}^2$. Material for the bottom and four sides costs $\$5/\text{ft}^2$. What is the largest volume the box can have if the total cost of material for the box is \$96?

**SOLUTION**

The box is sketched in Figure 16.9.4. The variables are the length, width, height, and the volume of the box. The volume is the max–min variable. We first express the max–min variable in terms of the other variables in the problem.

$$V = xyz.$$

The total cost gives a constraint. We have

$$\begin{aligned} \text{Total cost} &= (\text{Cost of top}) + (\text{Cost of bottom and sides}) \\ &= 3(\text{Area of top}) + 5(\text{Area of bottom plus area of sides}) \\ &= 3xy + 5(xy + 2xz + 2yz). \end{aligned}$$

If the total cost is to be \$96, we obtain the equation of constraint

$$96 = 3xy + 5(xy + 2xz + 2yz),$$

which we rewrite as

$$8xy + 10xz + 10yz - 96 = 0, \quad \text{or}$$

$$4xy + 5xz + 5yz - 48 = 0.$$

Let us now use the method of Lagrange multipliers to find the maximum volume subject to the constraint of total cost. We set

$$w = xyz - \lambda(4xy + 5xz + 5yz - 48).$$

The partial derivatives of $w$ are

$$w_x = yz - \lambda(4y + 5z),$$

$$w_y = xz - \lambda(4x + 5z),$$

$$w_z = xy - \lambda(5x + 5y),$$

$$w_\lambda = -(4xy + 5xz + 5yz - 48).$$

Setting each partial derivative equal to zero, we obtain a system of four equations:

$$\begin{cases} yz - \lambda(4y + 5z) = 0, \\ xz - \lambda(4x + 5z) = 0, \\ xy - \lambda(5x + 5y) = 0, \\ 4xy + 5xz + 5yz - 48 = 0. \end{cases}$$

To solve this system of equations, we multiply the first three equations by $x$, $y$, and $z$, respectively. This gives us

$$\begin{cases} xyz - \lambda(4xy + 5xz) = 0, \\ xyz - \lambda(4xy + 5yz) = 0, \\ xyz - \lambda(5xz + 5yz) = 0, \\ 4xy + 5xz + 5yz - 48 = 0. \end{cases}$$

Subtracting the second equation from the first, we obtain $5\lambda(y - x)z = 0$. This implies that either $\lambda = 0$, $y - x = 0$, or $z = 0$. The equation $z = 0$ clearly does not give the maximum volume, so we disregard that case. Similarly, if $\lambda = 0$, then one of $x$, $y$, or $z$ must be zero, so we disregard that case. We conclude that the maximum value occurs when $y - x = 0$, so $y = x$. Subtracting the third equation from the second, we obtain $\lambda(5z - 4y)x = 0$. As before, we conclude that

$$z = \frac{4y}{5}, \quad \text{so} \quad z = \frac{4x}{5}.$$

Substituting into the fourth equation above, we have

$$4x(x) + 5x\left(\frac{4x}{5}\right) + 5(x)\left(\frac{4x}{5}\right) - 48 = 0,$$

$$12x^2 - 48 = 0,$$

$$x^2 = 4,$$

$$x = 2 \qquad (\text{we disregard } x = -2).$$

$$x = 2 \quad \text{implies} \quad y = 2 \quad \text{and} \quad z = \frac{4(2)}{5} = \frac{8}{5}.$$

The maximum volume is

$$V\left(2, 2, \frac{8}{5}\right) = (2)(2)\left(\frac{8}{5}\right) = \frac{32}{5} = 6.4 \text{ ft}^3.$$

■ **EXAMPLE 3**

A rectangular box with no top is to have volume 8 ft$^3$. Material for the bottom of the box costs twice as much as material for the sides. Find the dimensions of the box that has minimum total cost of materials.

**SOLUTION**

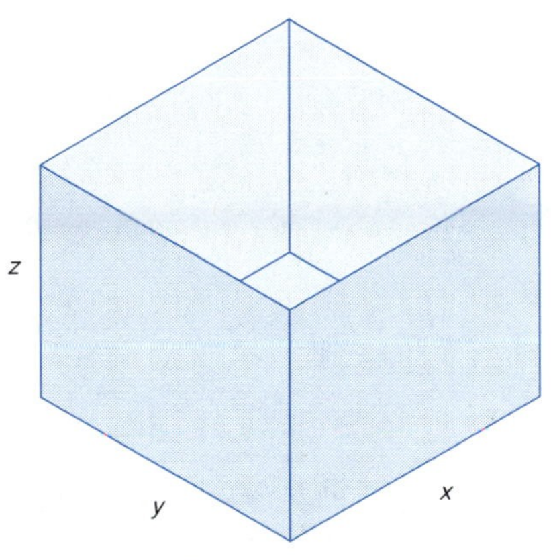


FIGURE 16.9.5

The box is pictured in Figure 16.9.5. The variables are $x$, $y$, $z$, and $C$, the total cost of material.

Let $k$ denote the cost per unit area of material for the sides of the box. The cost per unit area of material for the bottom of the box is then $2k$. The total cost of material is

$$C = 2kxy + k(2yz + 2xz).$$

The cost is subject to the constraint that the volume is 8. That is,

$$xyz = 8 \quad \text{or} \quad xyz - 8 = 0.$$

Let us use the method of Lagrange multipliers to find the least expensive box. We have

$$w = w(x, y, z, \lambda) = 2kxy + 2kyz + 2kxz - \lambda(xyz - 8).$$

The partial derivatives are

$$w_x = 2ky + 2kz - \lambda yz,$$

$$w_y = 2kx + 2kz - \lambda xz,$$

$$w_z = 2ky + 2kx - \lambda xy,$$

$$w_\lambda = -(xyz - 8).$$

Setting each partial derivative equal to zero, we obtain the system

$$\begin{cases} 2ky + 2kz - \lambda yz = 0, \\ 2kx + 2kz - \lambda xz = 0, \\ 2ky + 2kx - \lambda xy = 0, \\ xyz - 8 = 0. \end{cases}$$

Multiplying the first three equations by $x$, $y$, and $z$, respectively, we obtain the system

$$\begin{cases} 2kxy + 2kxz - \lambda xyz = 0, \\ 2kxy + 2kyz - \lambda xyz = 0, \\ 2kyz + 2kxz - \lambda xyz = 0, \\ xyz - 8 = 0. \end{cases}$$

The difference of the first two equations gives

$$2kz(x - y) = 0, \quad \text{so } x = y. \qquad \text{(We disregard } z = 0.\text{)}$$

The difference of the second and third equations gives

$$2kx(y - z) = 0, \quad \text{so } z = y. \qquad \text{(We disregard } x = 0.\text{)}$$

Since $x = y = z$, the fourth equation gives $x^3 = 8$, so $x = 2$. The least expensive box has length of each edge 2. ■

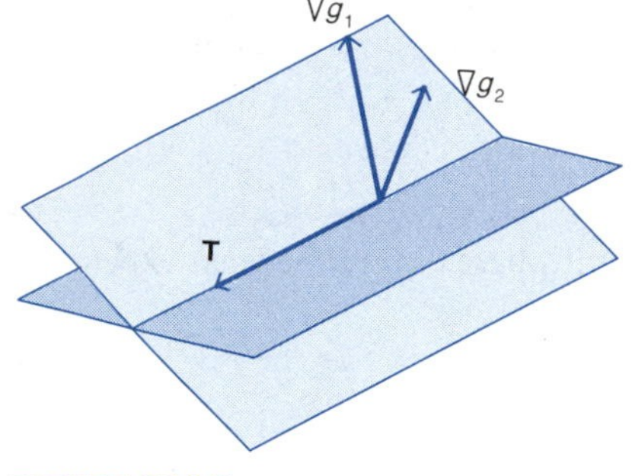


FIGURE 16.9.6

Let us consider the problem of finding extrema of $f(x, y, z)$, subject to $g_1(x, y, z) = 0$ and $g_2(x, y, z) = 0$. That is, we want extreme values of $f(x, y, z)$ for $(x, y, z)$ on the curve of intersection of the surfaces $g_1(x, y, z) = 0$ and $g_2(x, y, z) = 0$. Extrema can occur only where the directional derivative of $f$ in the direction of the curve is zero. If $\nabla g_1 \neq \mathbf{0}$ and $\nabla g_2 \neq \mathbf{0}$, this means that the gradient of $f$ must be in the plane determined by $\nabla g_1$ and $\nabla g_2$. See Figure 16.9.6. It follows that there are numbers $\lambda_1$ and $\lambda_2$ such that

$$\nabla f = \lambda_1 \nabla g_1 + \lambda_2 \nabla g_2.$$

We can then find the desired extrema by defining

$$w = w(x, y, z, \lambda_1, \lambda_2) = f(x, y, z) - \lambda_1 g_1(x, y, z) - \lambda_2 g_2(x, y, z)$$

and setting all five first partial derivatives of $w$ equal to zero.

The same ideas can be used for problems that involve any number of variables and any number of constraints.

If $f$, $g_1, \ldots,$ and $g_n$ are differentiable, and $\nabla g_j \neq \mathbf{0}$ at points where $g_j = 0$, $j = 1, \ldots, n$, then extrema of the function

$$f(x_1, \ldots, x_m)$$

subject to the constraints

$$g_1(x_1, \ldots, x_m) = 0, \ldots, g_n(x_1, \ldots, x_m) = 0$$

occur at points that are solutions of the system of $m + n$ equations obtained by setting all first partial derivatives of the function

$$\begin{aligned} w &= w(x_1, \ldots, x_m, \lambda_1, \ldots, \lambda_n) \\ &= f(x_1, \ldots, x_m) - \lambda_1 g_1(x_1, \ldots, x_m) - \cdots - \lambda_n g_n(x_1, \ldots, x_m) \end{aligned}$$

equal to zero.

Note that we introduce one Lagrange multiplier for each equation of constraint.

### ■ EXAMPLE 4

Find points on the curve of intersection of the cone $z^2 = 4x^2 + 4y^2$ and the plane $4x + 3y + 4z = 7$ that are nearest to and farthest from the origin.

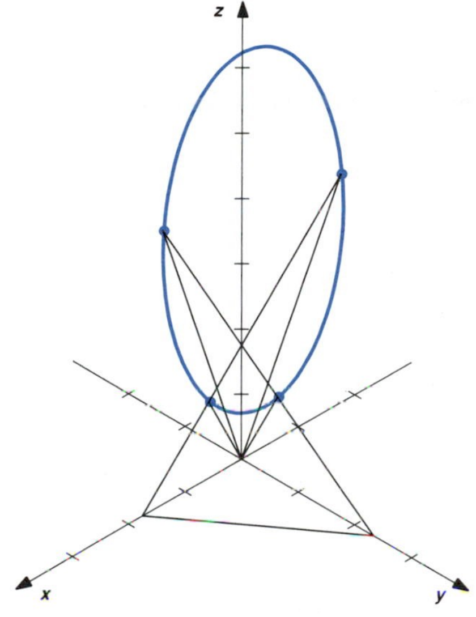


FIGURE 16.9.7

#### SOLUTION

The curve is sketched in Figure 16.9.7. Note that the curve of intersection of the cone with the plane is closed. Since the distance from the origin is a continuous function, we know that there is a point on the curve that gives the maximum distance and a point on the curve that gives the minimum distance. We want the extrema of

$$D(x, y, z) = x^2 + y^2 + z^2,$$

subject to the constraints

$$4x^2 + 4y^2 - z^2 = 0 \quad \text{and} \quad 4x + 3y + 4z - 7 = 0.$$

Note that $D(x, y, z)$ is the *square* of the function that gives the distance of the point $(x, y, z)$ from the origin. This function is easier to work with and has extrema at the same points as the distance function.

There are two constraints in this problem, so we introduce two Lagrange multipliers. We define

$$\begin{aligned} w(x, y, z, \lambda_1, \lambda_2) = x^2 + y^2 + z^2 - \lambda_1(4x^2 + 4y^2 - z^2) \\ - \lambda_2(4x + 3y + 4z - 7). \end{aligned}$$

The partial derivatives are

$$w_x = 2x - \lambda_1(8x) - \lambda_2(4),$$

$$w_y = 2y - \lambda_1(8y) - \lambda_2(3),$$

$$w_z = 2z - \lambda_1(-2z) - \lambda_2(4),$$

$$w_{\lambda_1} = -(4x^2 + 4y^2 - z^2),$$

$$w_{\lambda_2} = -(4x + 3y + 4z - 7).$$

Setting each partial derivative equal to zero, we obtain the system

$$\begin{cases} 2x - \lambda_1(8x) - \lambda_2(4) = 0, & (1) \\ 2y - \lambda_1(8y) - \lambda_2(3) = 0, & (2) \\ 2z - \lambda_1(-2z) - \lambda_2(4) = 0, & (3) \\ 4x^2 + 4y^2 - z^2 = 0, & (4) \\ 4x + 3y + 4z - 7 = 0. & (5) \end{cases}$$

We now multiply the first three equations by $x$, $y$, $z$, respectively, to obtain

$$\begin{cases} 2x^2 - \lambda_1(8x^2) - \lambda_2(4x) = 0, \\ 2y^2 - \lambda_1(8y^2) - \lambda_2(3y) = 0, \\ 2z^2 - \lambda_1(-2z^2) - \lambda_2(4z) = 0. \end{cases}$$

Adding these three equations gives

$$2(x^2 + y^2 + z^2) - \lambda_1(8x^2 + 8y^2 - 2z^2) - \lambda_2(4x + 3y + 4z) = 0.$$

Using equations (4) and (5) then gives

$$2(x^2 + y^2 + z^2) - \lambda_1(0) - \lambda_2(7) = 0, \quad \text{so}$$

$$\lambda_2 = \frac{2(x^2 + y^2 + z^2)}{7}. \tag{6}$$

We can obtain an equation that relates $x$, $y$, and $\lambda_1$ by multiplying equation (1) by 3 and equation (2) by 4 and substracting the resulting equations. We obtain

$$\begin{cases} 6x - 24\lambda_1 x - 12\lambda_2 = 0, \\ 8y - 32\lambda_1 y - 12\lambda_2 = 0, \end{cases}$$

and

$$6x - 8y - \lambda_1(24x - 32y) = 0, \quad \text{so}$$

$$2(3x - 4y) - 8(3x - 4y)\lambda_1 = 0,$$

$$2(3x - 4y)(1 - 4\lambda_1) = 0.$$

It follows that either $3x - 4y = 0$ or $1 - 4\lambda_1 = 0$. If $\lambda_1 = 1/4$, equation (1) implies

$$2x - \left(\frac{1}{4}\right)(8x) - 4\lambda_2 = 0, \quad \text{so} \quad \lambda_2 = 0.$$

It then follows from equation (6) that $x^2 + y^2 + z^2 = 0$, so $(x, y, z) = (0, 0, 0)$. However, the point $(0, 0, 0)$ is not in the plane $4x + 3y + 4z = 7$, so this point cannot give an extreme value for our problem. The remaining case is $3x - 4y = 0$, so

$$y = \frac{3x}{4}.$$

Equation (4) then gives

$$z^2 = 4x^2 + 4\left(\frac{3x}{4}\right)^2 = \frac{25x^2}{4}, \quad \text{so}$$

$$z = \frac{5x}{2} \quad \text{or} \quad z = -\frac{5x}{2}.$$

If $z = 5x/2$, equation (5) gives

$$4x + 3\left(\frac{3x}{4}\right) + 4\left(\frac{5x}{2}\right) = 7,$$

$$\frac{65x}{4} = 7,$$

$$x = \frac{28}{65}.$$

When $x = 28/65$,

$$y = \frac{3(28/65)}{4} = \frac{21}{65} \quad \text{and} \quad z = \frac{5(28/65)}{2} = \frac{70}{65}.$$

If $z = -5x/2$, equation (5) gives

$$4x + 3\left(\frac{3x}{4}\right) + 4\left(-\frac{5x}{2}\right) = 7,$$

$$-\frac{15x}{4} = 7,$$

$$x = -\frac{28}{15}.$$

When $x = -28/15$,

$$y = \frac{3(-28/15)}{4} = -\frac{21}{15} \quad \text{and} \quad z = -\frac{5(-28/15)}{2} = \frac{70}{15}.$$

The maximum and minimum values of the distance between points on the curve and the origin can occur only at

$$\left(\frac{28}{65}, \frac{21}{65}, \frac{70}{65}\right) \quad \text{or} \quad \left(-\frac{28}{15}, -\frac{21}{15}, \frac{70}{15}\right).$$

The distance of the point (28/65, 21/65, 70/65) from the origin is

$$\sqrt{\left(\frac{28}{65}\right)^2 + \left(\frac{21}{65}\right)^2 + \left(\frac{70}{65}\right)^2} \approx 1.2040366;$$

this is the point on the curve that is nearest to the origin.

The distance of the point $(-28/15, -21/15, 70/15)$ from the origin is

$$\sqrt{\left(-\frac{28}{15}\right)^2 + \left(-\frac{21}{15}\right)^2 + \left(\frac{70}{15}\right)^2} \approx 5.2174919;$$

this is the point on the curve that is farthest from the origin. ■

## ■ EXAMPLE 5

Find the maximum and minimum values of

$$f(x, y, z) = x^2 - 2x + y^2 - 4y + z^2 - 4z$$

for $(x, y, z)$ that are on or inside the sphere $x^2 + y^2 + z^2 = 36$.

### SOLUTION

Since $f$ is continuous on the closed and bounded region, we know that there are maximum and minimum values of $f$ in the region. The maximum and minimum values can occur either at a critical point in the interior of the region or on the boundary of the region.

Let us begin by determining the critical points of $f$. The first partial derivatives of $f$ are

$$f_x = 2x - 2,$$

$$f_y = 2y - 4,$$

$$f_z = 2z - 4.$$

Setting each partial derivative of $f$ equal to zero, we obtain

$$\begin{cases} 2x - 2 = 0, \\ 2y - 4 = 0, \\ 2z - 4 = 0. \end{cases}$$

This system of equations has solution (1, 2, 2), which is a point inside the circle $x^2 + y^2 + z^2 = 36$.

We also need to determine extrema of $f$ on the boundary of the sphere. That is, we need extrema of

$$x^2 - 2x + y^2 - 4y + z^2 - 4z,$$

subject to the constraint

$$x^2 + y^2 + z^2 - 36 = 0.$$

Using the method of Lagrange multipliers, we define

$$\begin{aligned} w &= w(x, y, z, \lambda) \\ &= x^2 - 2x + y^2 - 4y + z^2 - 4z - \lambda(x^2 + y^2 + z^2 - 36). \end{aligned}$$

The partial derivatives of $w$ are

$$w_x = 2x - 2 - \lambda(2x),$$

$$w_y = 2y - 4 - \lambda(2y),$$

$$w_z = 2z - 4 - \lambda(2z),$$

$$w_\lambda = -(x^2 + y^2 + z^2 - 36).$$

Setting each of the partial derivatives equal to zero gives the system of equations

$$\begin{cases} x - 1 - \lambda x = 0, \\ y - 2 - \lambda y = 0, \\ z - 2 - \lambda z = 0, \\ x^2 + y^2 + z^2 = 36. \end{cases}$$

$$x - 1 - \lambda x = 0 \quad \text{implies} \quad x \neq 0 \quad \text{and} \quad \lambda = \frac{x-1}{x}.$$

$$y - 2 - \lambda y = 0 \quad \text{implies} \quad y \neq 0 \quad \text{and} \quad \lambda = \frac{y-2}{y}.$$

It follows that

$$\frac{y-2}{y} = \frac{x-1}{x},$$

$$x(y - 2) = (x - 1)y,$$

$$xy - 2x = xy - y,$$

$$y = 2x.$$

Similarly,

$$z - 2 - \lambda z = 0 \quad \text{implies} \quad z \neq 0, \quad \lambda = \frac{z-2}{z}, \quad \text{and} \quad z = 2x.$$

Substitution into the fourth equation of our system then gives

$$x^2 + (2x)^2 + (2x)^2 = 36,$$

$$9x^2 = 36,$$

$$x^2 = 4,$$

$$x = 2, -2.$$

When $x = 2$, $y = 4$ and $z = 4$. When $x = -2$, $y = -4$ and $z = -4$.

The values of $f(x, y)$ at each of the candidates for extrema are given in Table 16.9.1. We see that $f(1, 2, 2) = -9$ is the minimum value and $f(-2, -4, -4) = 72$ is the maximum value. ■

**TABLE 16.9.1**

| $(x, y, z)$ | $f(x, y, z)$ |
|---|---|
| $(1, 2, 2)$ | $-9$ |
| $(2, 4, 4)$ | $0$ |
| $(-2, -4, -4)$ | $72$ |

## EXERCISES 16.9

In Exercises 1–14, find all extrema of the given functions subject to the indicated constraints.

**1** $f(x, y) = 3x - 4y$, $x^2 + y^2 = 25$

**2** $f(x, y) = 12x + 5y$, $x^2 + y^2 = 169$

**3** $f(x, y) = xy$, $x^2 + 3y^2 = 6$

**4** $f(x, y) = (x^2/4) - y^2$, $x^2 + y^2 = 1$

**5** $f(x, y, z) = x + 2y - 2z$, $x^2 + y^2 + z^2 = 9$

**6** $f(x, y, z) = 3x + 4y + 12z$, $x^2 + y^2 + z^2 = 169$

**7** $f(x, y, z) = xyz$, $xy + xz + 3yz = 4$

**8** $f(x, y, z) = xyz$, $xy + 3xz + 4yz = 9$

**9** $f(x, y, z) = 2xy + 4xz + yz$, $xyz = 1$

**10** $f(x, y, z) = 9xy + 3xz + yz$, $xyz = 1$

**11** $f(x, y, z) = xyz$, $x^2 + y^2 + z^2 = 4$, $x + y = 2$

**12** $f(x, y, z) = x + y + z$, $x^2 + y^2 + z^2 = 1$, $x + 2y - 2z = 0$

**13** $f(x, y, z) = 2x - 3y + z$, $x^2 + y^2 = 1$, $y^2 + z^2 = 1$

**14** $f(x, y, z) = xyz$, $x^2 + y^2 = 1$, $y^2 + z^2 = 1$

In Exercises 15–20, find the maximum and minimum values of the given functions in the indicated regions.

**15** $f(x, y) = -x^2 + 4x - y^2 + 2y$, $x^2 + y^2 \leq 20$

**16** $f(x, y) = x^2 - 12x + y^2 - 9y$, $x^2 + y^2 \leq 25$

**17** $f(x, y, z) = x^2 - 4x + y^2 + 2y + z^2 - 6z$, $x^2 + y^2 + z^2 \leq 56$

**18** $f(x, y, z) = x^2 - 2x + y^2 - 4y + z^2 - 4z$, $x^2 + y^2 + z^2 \leq 9$

**19** $f(x, y, z) = 2xy$, $x^2 + y^2 + z^2 \leq 1$

**20** $f(x, y, z) = 2xy + 2z^2$, $x^2 + y^2 + z^2 \leq 1$

**21** A rectangular box with no top is to have volume 32. Find the dimensions and material required for the box that requires the least material for the four sides and bottom.

**22** A rectangular box with no top is to have volume 12 ft$^3$. Material for the bottom of the box costs \$3/ft$^2$. Material for the four sides costs \$1/ft$^2$. Find the dimensions and cost of the box that has minimum total cost of materials.

23 Material for the top of a rectangular box costs \$2/ft$^2$. Material for the bottom and four sides costs \$1/ft$^2$. Find the dimensions and volume of the box with largest volume that can be made with \$9 for cost of material.

24 Find the point on the plane $x + 2y - z = 12$ that is nearest the origin.

25 Show that the volume of a rectangular box with fixed total surface area of the six sides is largest for a cube.

26 Show that the area of a triangle with fixed perimeter is largest for an equilateral triangle.

27 A trough is to be made from a rectangular piece of metal that is 12 in. wide. Equal lengths from each side along the length are to be folded up the same angle to form slanted sides. Find the length that should be folded from each end and the angle that the sides should be bent so the cross section area of the trough is maximum.

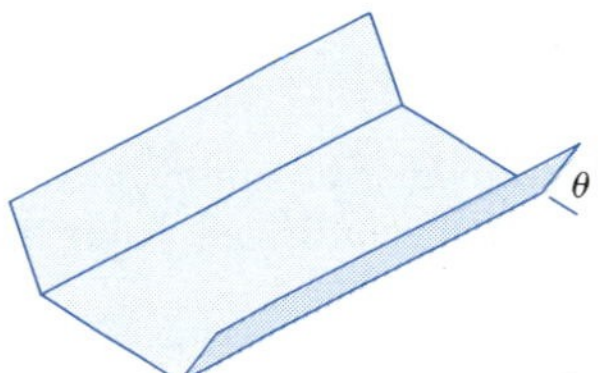


28 A trough with cross-sectional area 4 ft$^2$ is to be made from a rectangular piece of metal by bending equal lengths from each side along the length. Each side is to be folded up the same angle. Find the length that should be folded from each end and the angle that the sides should be bent so the material required is smallest.

## REVIEW EXERCISES

Find the domain of the functions in Exercises 1–2.

1 $f(x, y) = \dfrac{x^2 - y^2}{x - y}$

2 $f(x, y) = \sqrt{1 - \ln(x^2 + y^2)}$

Characterize and sketch several level curves of the functions given in Exercises 3–4. Sketch the graph of the functions.

3 $f(x, y) = x^2 + 2y$

4 $f(x, y) = \dfrac{x^2 + y^2}{y}$

Characterize and sketch several level surfaces of the functions in Exercises 5–6.

5 $f(x, y, z) = 2\sqrt{x^2 + z^2} - y$

6 $f(x, y, z) = z + x^2 + y^2$

For each of the limits in Exercises 7–10, evaluate the limit or show that the limit does not exist.

7 $\displaystyle\lim_{(x,y)\to(0,0)} \frac{x^2}{x^2 + y^2}$

8 $\displaystyle\lim_{(x,y)\to(0,0)} \frac{2xy}{\sqrt{x^2 + y^2}}$

9 $\displaystyle\lim_{(x,y)\to(0,0)} \frac{2x^2 - y^2}{x^2 + y^2}$

10 $\displaystyle\lim_{(x,y)\to(0,0)} \frac{x^3}{x^2 + y^2}$

Find all first- and second-order partial derivatives of the functions in Exercises 11–14.

11 $f(x, y) = x^2 - 2xy + 3y^2$

12 $f(x, y) = \sin xy$

13 $f(x, y) = \ln\sqrt{x^2 + y^2}$

14 $f(x, y) = \tan^{-1}(x/y)$

15 The magnitude of the force due to gravity of an object with mass $m$ at a distance $d$ from the center of the earth is $F = km/d^2$, where $k$ is a constant. Find the approximate maximum percent error in $F$ due to possible errors of 0.2 percent in the measurement of the mass and possible error of 1.5 percent in measurement of the distance.

16 Find the approximate maximum percent error in the current $I$ due to possible errors of 5 percent in measurement of the voltage $V$ and 3 percent in measurement of the resistance $R$. ($I = V/R$.)

17 $z = x^2/y$, $x = \tan t$, $y = \cos^2 t$. Find $dz/dt$.

18 $z = \cos(x^2 - y^2)$, $x = t^2 - 3t + 4$, $y = t^3 - 1$. Find $dz/dt$.

19 $z = x^2 - y^2$, $x = r\cos\theta$, $y = r\sin\theta$. Find $\partial z/\partial r$ and $\partial z/\partial\theta$.

20 $xz^2 + yz = 1$. Find $\partial z/\partial x$ and $\partial z/\partial y$.

21 $x^2y + y^2z + z^2x = 0$. Find $\partial z/\partial x$ and $\partial z/\partial y$.

22 $x^2 + y^2 + z^2 = 4$. Find $\dfrac{\partial^2 z}{\partial y\,\partial x}$.

23 $z^2 = x^2 + y^2$. Find $\dfrac{\partial^2 z}{\partial y\,\partial x}$.

24 The radius of a right circular cylinder is increasing at a rate of 2 ft/s while the height is decreasing at a rate of 3 ft/s. Find the rate of change of the (a) volume and (b) the total surface area of the cylinder when the radius is 6 ft and the height is 4 ft.

Find an equation of the plane tangent to the given surface at the indicated point in each of Exercises 25–26.

**25** $z = x^2 + y^2$, at (2, 3, 13)

**26** $z^2 = x^2 + y^2$, at (3, 4, −5)

Find symmetric equations and parametric equations of the line normal to the given surface at the indicated point in each of Exercises 27–28.

**27** $z^2 = x^2 - y^2$, at (5, −4, 3)

**28** $z = xy$, at (3, 2, 6)

For the functions given in Exercises 29–31, sketch the level curves $f(x, y) = 2$, 3, and 4. Evaluate $\nabla f$ at the point (2, 1). Find a unit vector **U** such that $D_{\mathbf{U}}f(2, 1) = 0$. Sketch $\nabla f$ and **U** at (2, 1).

**29** $f(x, y) = x + y$

**30** $f(x, y) = x^2 - y^2$

**31** $f(x, y) = x^2 - y$

**32** Find the rate of change of $f(x, y, z) = x^2 - xz + yz + y^2$ at the point (2, −1, 3) in the direction of $\mathbf{i} - 2\mathbf{j} + 2\mathbf{k}$.

In Exercises 33–34 find a unit vector **U** such that the rate of change of the given function at (2, 3) in the direction of **U** is maximum and evaluate that maximum value.

**33** $f(x, y) = xy^2 - yx^2$

**34** $f(x, y) = x^2y^2 - xy$

Find all extrema of the functions given in Exercises 35–38.

**35** $f(x, y) = x^2 + 2xy + 3y^2$

**36** $f(x, y) = x^3 + 3xy - y^3$

**37** $f(x, y, z) = x^2 + y^2 + z^2$, $x - y + z = 1$ (minimum)

**38** $f(x, y, z) = x^2 + y^2 + z^2$, $x - y = 1$ (minimum)

**39** A rectangular box with no top has volume 32 ft$^3$. Find the dimensions and surface area of the box that has surface area a minimum.

**40** A rectangular box with no top has surface area 48 ft$^2$. Find the dimensions and volume of the box that has volume a maximum.

CHAPTER SEVENTEEN

# *Multiple integrals*

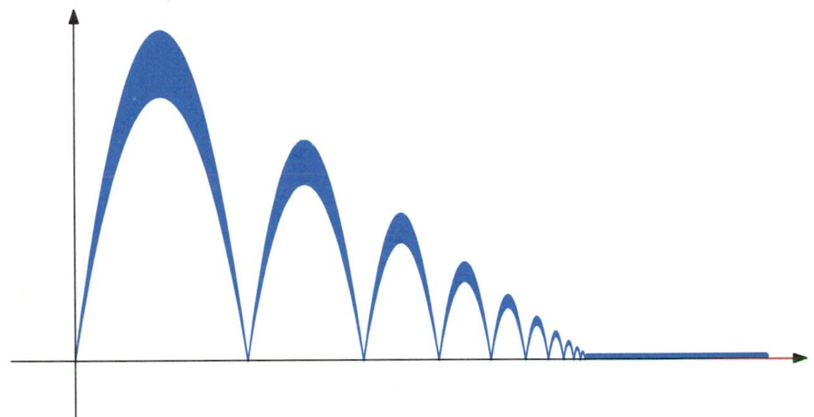

We have seen that definite integrals of functions of one real variable have applications to many physical problems. In this chapter we will study analogous results that involve functions of several variables.

## 17.1 REGIONS IN THE PLANE AND IN THREE-DIMENSIONAL SPACE

In later sections we will study integrals over certain types of regions in the plane and in three-dimensional space. It is essential that we are able to describe these regions in the way that we will develop in this section.

Let us consider the region in the plane bounded by

$$y = g_1(x), \quad y = g_2(x), \quad x = a, \quad \text{and} \quad x = b,$$

FIGURE 17.1.1

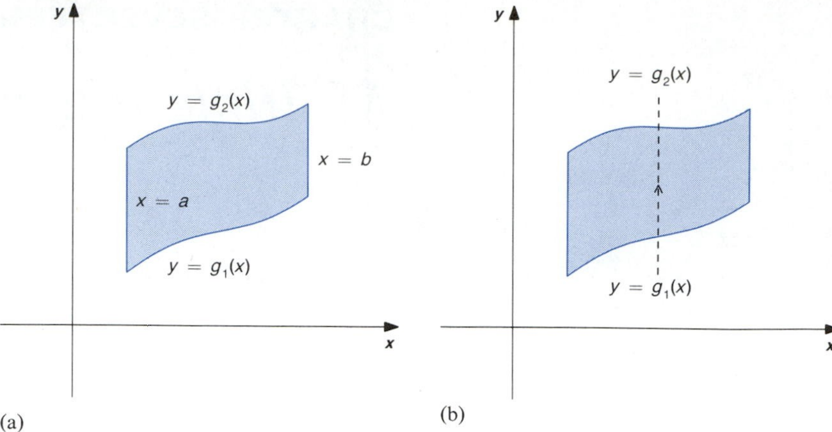


FIGURE 17.1.2

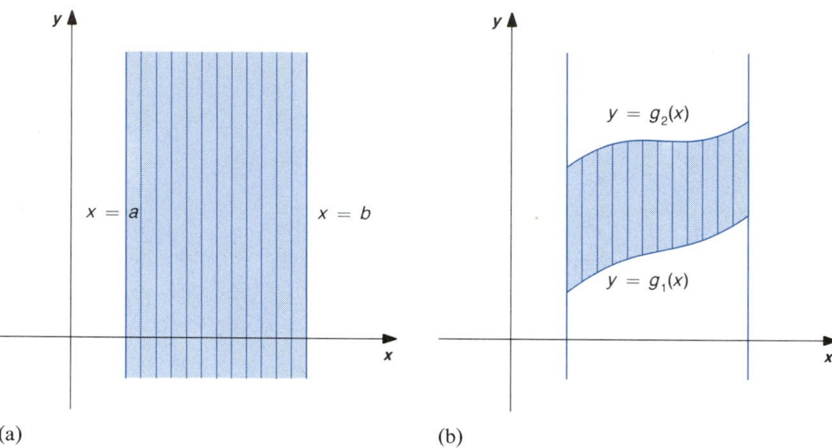


where $g_1$ and $g_2$ are continuous with $g_1(x) \leq g_2(x)$ for $a \leq x \leq b$. A typical region of this type is sketched and labeled in Figure 17.1.1a. An arrow that is shot vertically from below the region follows a trajectory with $x$ fixed, enters the region at a point on the curve $y = g_1(x)$, and leaves the region at a point on the curve $y = g_2(x)$. See Figure 17.1.1b. We will call such regions **vertically simple**, or **simple in the direction of the $y$-axis**, if each of $g_1(x)$ and $g_2(x)$ is given by a *single formula* that holds for $x$ between $a$ and $b$. We can then **describe** the region by writing:

For $x$ fixed, $y$ varies from $y = g_1(x)$ to $y = g_2(x)$.
$x$ varies from $x = a$ to $x = b$.

For $x$ fixed, $y$ varies from the *lower* curve $y = g_1(x)$ to the *upper* curve $y = g_2(x)$; $x$ varies from the *smaller* number $x = a$ to the *larger* number $x = b$.

The description of the region determines the region. The statement "$x$ varies from $x = a$ to $x = b$" tells us that the region is bounded on the sides by the two vertical lines, $x = a$ and $x = b$. See Figure 17.1.2a. The statement "For $x$ fixed, $y$ varies from $y = g_1(x)$ to $y = g_2(x)$" tells us that the region is bounded on the bottom by $y = g_1(x)$ and on the top by $y = g_2(x)$. See Figure 17.1.2b.

■ **EXAMPLE 1**

Sketch and describe the region in the plane bounded by $y = x^2$, $y = -1$, $x = 0$, and $x = 2$.

**SOLUTION**

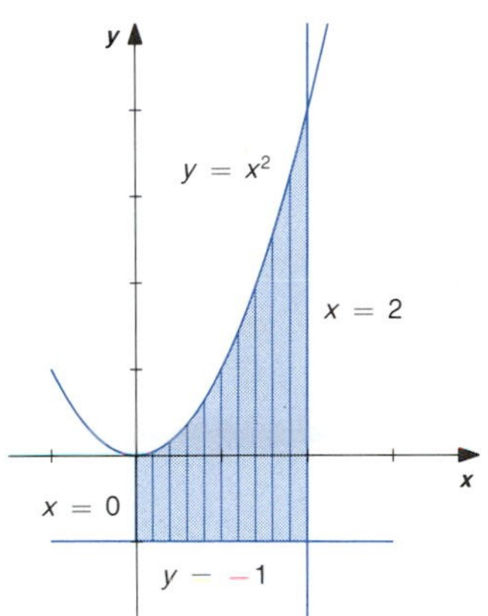


FIGURE 17.1.3

Each of the four curves is sketched and labeled in Figure 17.1.3. We see that the region bounded by the curves is vertically simple. From the sketch, we see that

For $x$ fixed, $y$ varies from $y = -1$ to $y = x^2$.
$x$ varies from $x = 0$ to $x = 2$. ■

In Example 1 we were given the four equations that determined the top, bottom, and two sides of the region. The two equations ($y = -1$ and $y = x^2$) that involve the variable $y$ give the top and bottom of the region; the two equations ($x = 0$ and $x = 2$) that do not involve $y$ give the sides of the region. The top and bottom of any vertically simple region must be given by equations that involve $y$, one equation for the top and one equation for the bottom. Since the equations of the side boundaries do not involve $y$, it follows that we must be given exactly two equations that involve $y$ in order to describe such a region. If a side boundary ($x =$ constant) is determined by the intersection of the top and bottom, the equation of that side is generally not given. We must determine the equation by finding the $x$-coordinate of the point of intersection of the top and bottom.

■ **EXAMPLE 2**

Sketch and describe the region in the plane bounded by $y = x$, $x + y = 2$, and the $y$-axis.

**SOLUTION**

The region is sketched and labeled in Figure 17.1.4. Note that $x + y = 2$ implies $y = 2 - x$. We see from the sketch that the region is vertically simple. Also:

For $x$ fixed, $y$ varies from $y = x$ to $y = 2 - x$.

Since the top and bottom intersect to form the right-side boundary, we must combine the equations of the top and bottom to find the $x$-coordinate of the intersection:

$y = x$ and $y = 2 - x$ imply $x = 2 - x$, so $x = 1$.

The region is bounded on the right by the vertical line $x = 1$. The region is bounded on the left by the $y$-axis, which has equation $x = 0$. Hence:

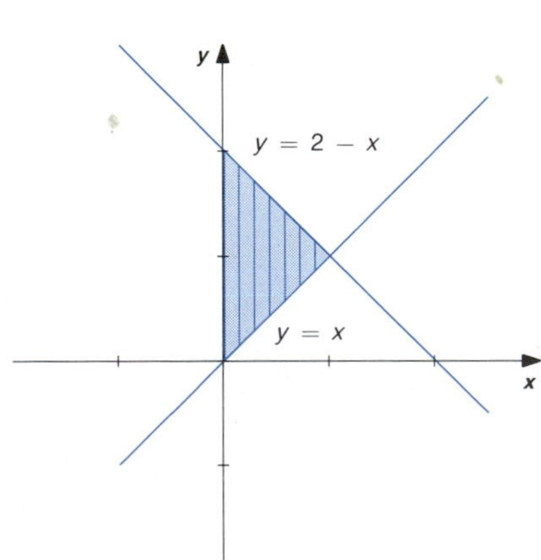


FIGURE 17.1.4

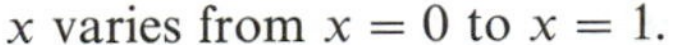

$x$ varies from $x = 0$ to $x = 1$. ■

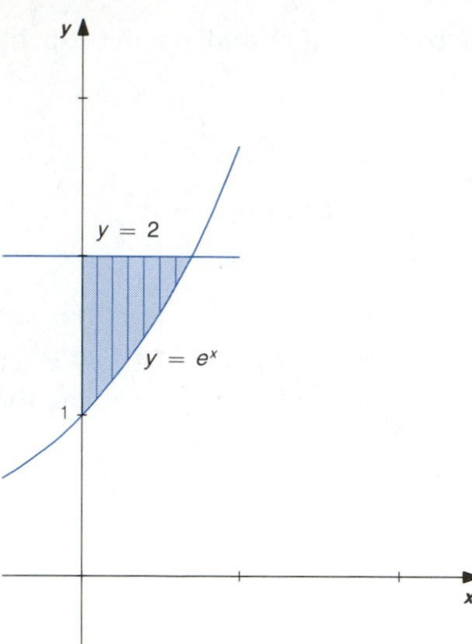


FIGURE 17.1.5

■ **EXAMPLE 3**

Sketch and describe the region in the plane bounded by $y = e^x$, $y = 2$, and the $y$-axis.

**SOLUTION**

The region is sketched and labeled in Figure 17.1.5. The region is vertically simple.

For $x$ fixed, $y$ varies from $y = e^x$ to $y = 2$.

The top and bottom intersect to form the right-side boundary. Solving the equations of the top and bottom for $x$, we have

$y = e^x$ and $y = 2$ imply $e^x = 2$, so $x = \ln 2$.

$x$ varies from $x = 0$ to $x = \ln 2$. ■

A region in the plane bounded by

$$x = g_1(y), \quad x = g_2(y), \quad y = a, \quad \text{and} \quad y = b,$$

where $g_1$ and $g_2$ are continuous with $g_1(y) \leq g_2(y)$ for $a \leq y \leq b$, is illustrated in Figure 17.1.6a. An arrow that is shot horizontally from the left of the region follows a trajectory with $y$ fixed, enters the region at a point on the curve $x = g_1(y)$, and leaves the region at a point on the curve $x = g_2(y)$. See Figure 17.1.6b. Such a region is called **horizontally simple**, or **simple in the direction of the $x$-axis**, if each of $g_1(y)$ and $g_2(y)$ is given by a single formula as $y$ varies between $a$ and $b$. We can then describe the region by writing:

For $y$ fixed, $x$ varies from $x = g_1(y)$ to $x = g_2(x)$.
$y$ varies $y = a$ to $y = b$.

The description of a horizontally simple region includes exactly two equations that involve the variable $x$.

The region of Example 1—bounded by $y = x^2$, $y = -1$, $x = 0$, and $x = 2$—is not horizontally simple. An arrow shot horizontally from the left

FIGURE 17.1.6

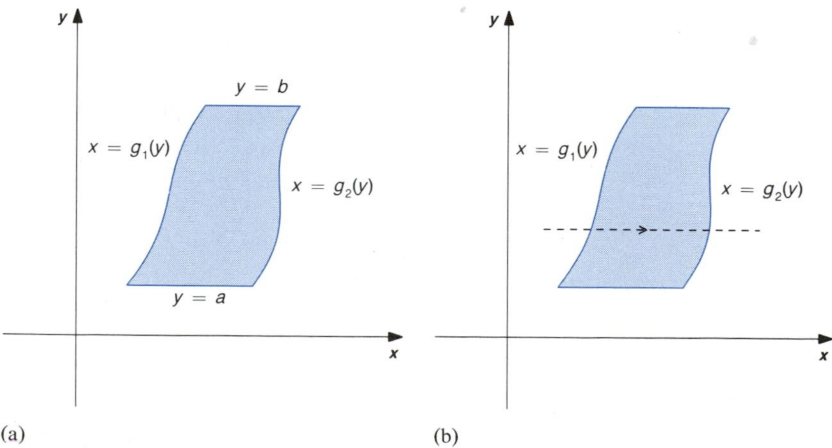

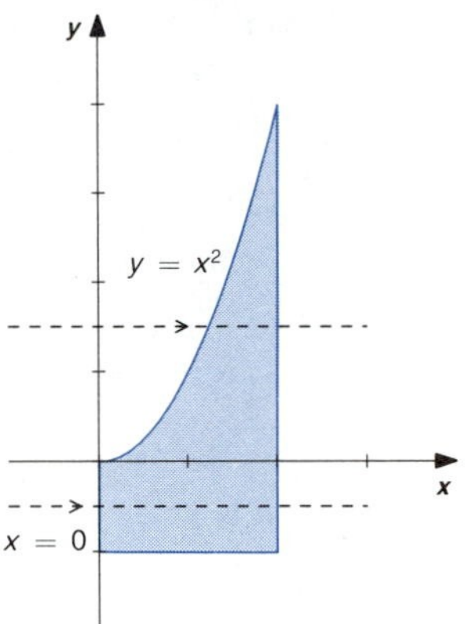


FIGURE 17.1.7

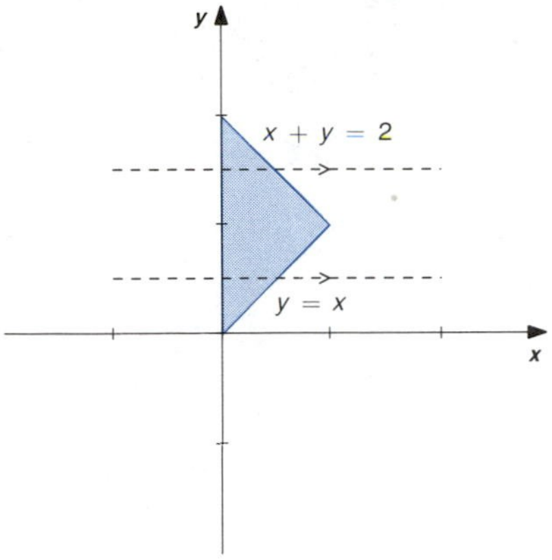

FIGURE 17.1.8

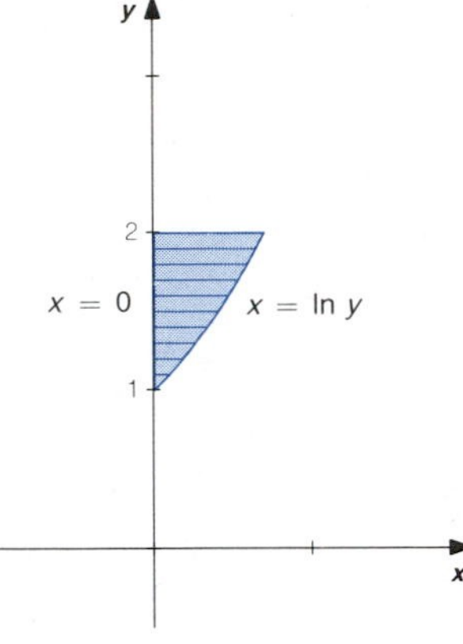


FIGURE 17.1.9

of the region could enter the region at a point on either the parabola $y = x^2$ or the vertical line $x = 0$. See Figure 17.1.7.

The region of Example 2—bounded by $y = x$, $x + y = 2$, and the $y$-axis—is not horizontally simple. An arrow that is shot horizontally from the left of the region could leave the region at a point on either the line $y = x$ or the line $x + y = 2$. See Figure 17.1.8.

The region of Example 3, bounded by $y = e^x$, $y = 2$, and the $y$-axis, is horizontally simple. See Figure 17.1.9. We see that the equation $y = e^x$ can be written $x = \ln y$. The $y$-coordinate of the intersection of $x = 0$ and $x = \ln y$ is 1. We can then describe the region by writing:

For $y$ fixed, $x$ varies from $x = 0$ to $x = \ln y$.
$y$ varies from $y = 1$ to $y = 2$.

**■ EXAMPLE 4**

Sketch and describe the region in the plane bounded by $x = 2y^2 - 1$ and $x = y^2$.

**SOLUTION**

The region is sketched in Figure 17.1.10. The region is not vertically simple, because an arrow shot vertically from a point below the region with positive $x$-coordinate would enter and leave the region twice. The region is horizontally simple. We have:

For $y$ fixed, $x$ varies from $x = 2y^2 - 1$ to $x = y^2$.

The sides intersect to form the top and bottom boundaries.

$x = 2y^2 - 1$ and $x = y^2$ imply

$2y^2 - 1 = y^2,$

$y^2 = 1,$

$y = 1, \quad y = -1.$

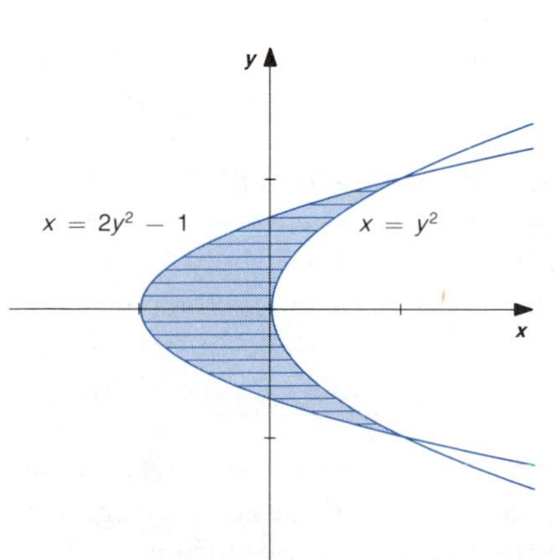


FIGURE 17.1.10

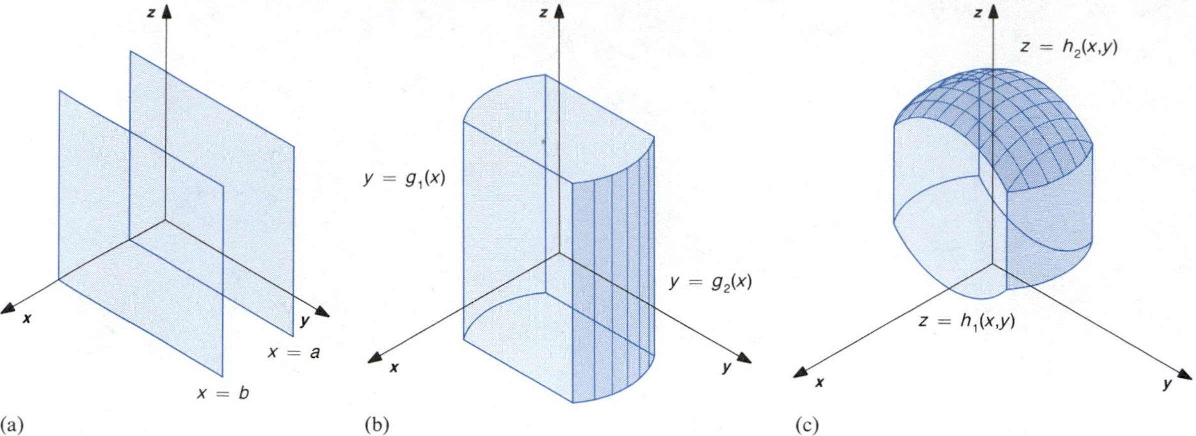

FIGURE 17.1.11

We conclude that

$y$ varies from $y = -1$ to $y = 1$. ■

The same ideas that were used to describe regions in the plane can be used to describe regions in three-dimensional space. We will deal with regions in three-dimensional space that can be described by writing:

For $x$ and $y$ fixed, $z$ varies from $z = h_1(x, y)$ to $z = h_2(x, y)$.
For $x$ fixed, $y$ varies from $y = g_1(x)$ to $y = g_2(x)$.
$x$ varies from $x = a$ to $x = b$.

The statement "$x$ varies from $x = a$ to $x = b$" tells us the region is bounded in the **front** by the plane $x = b$ and is bounded in the **back** by the plane $x = a$. See Figure 17.1.11a. "For $x$ fixed, $y$ varies from $y = g_1(x)$ to $y = g_2(x)$" tells us the region is bounded on the **right side** by the vertical generalized cylinder $y = g_2(x)$ and is bounded on the **left side** by the vertical generalized cylinder $y = g_1(x)$. See Figure 17.1.11b. The statement "For $x$ and $y$ fixed, $z$ varies from $z = h_1(x, y)$ to $z = h_2(x, y)$" tells us that the region is bounded on the **top** by the surface $z = h_2(x, y)$ and is bounded on the **bottom** by the surface $z = h_1(x, y)$. See Figure 17.1.11c.

The statements "For $x$ fixed, $y$ varies from $y = g_1(x)$ to $y = g_2(x)$" and "$x$ varies from $x = a$ to $x = b$," from the description of a region in three-dimensional space, correspond to the description of a region in the $xy$-plane that is simple in the direction of the $y$-axis. This region in the $xy$-plane is the **projection** of the region in three-dimensional space onto the $xy$-plane. See Figure 17.1.12. An arrow that is shot vertically from below the projection follows a trajectory with $x$ and $y$ fixed, enters the region at a point on the bottom surface $z = h_1(x, y)$, travels parallel to the vertical surfaces $y = g_1(x)$, $y = g_2(x)$, $x = a$, and $x = b$, and leaves the region at a point on the top surface $z = h_2(x, y)$.

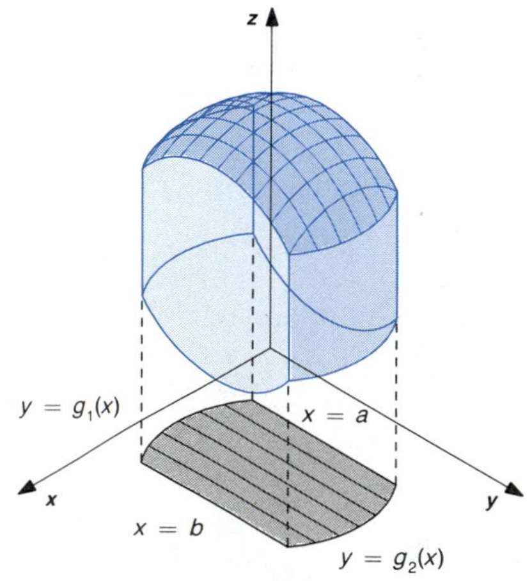

FIGURE 17.1.12

As in the case of regions in the plane, it may be necessary to combine given equations to find all the formulas needed to describe a region in three-dimensional space. We will use the following procedure.

FIGURE 17.1.13

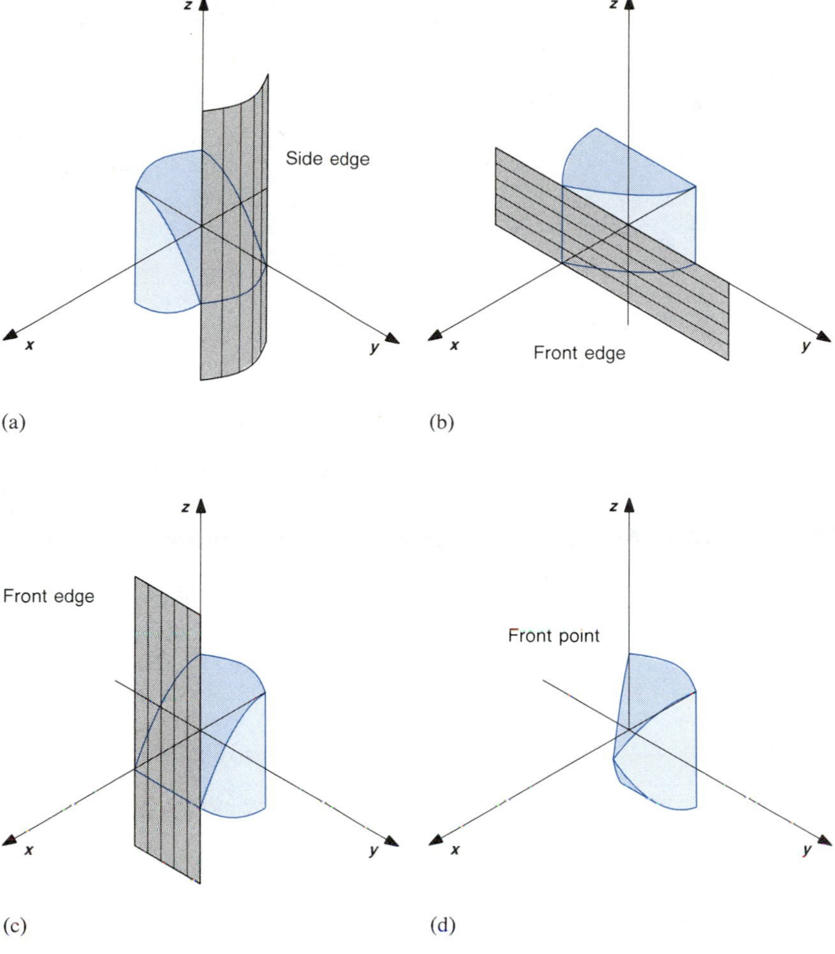


### *Steps to sketch and describe a region in three-dimensional space*

- **Sketch coordinate traces of each surface.**
- **Find points of intersection of traces in each coordinate plane.**
- **Sketch region; draw all boundary curves.**
- **Identify *top* and *bottom surfaces*. The equations of the top and bottom must involve the variable $z$. We must be given exactly two equations that involve the variable $z$.**
- **Identify *side cylinders*. The equations of the side cylinders must involve $y$ and cannot involve $z$. If top and bottom surfaces intersect to form a *side edge* (Figure 17.1.13a), use the equations of the top and bottom to eliminate $z$. This gives the equation of a cylinder that contains the side edge and is used as the side cylinder.**
- **Identify *front* and *back planes*. If either the sides intersect (Figure 17.1.13b), the top and bottom intersect (Figure 17.1.13c), or both (Figure 17.1.13d), there will be either a *front* or *back edge* or a *front* or *back point*. The equations can then be used to find the $x$-coordinate of the intersection. This gives the front or back plane.**

FIGURE 17.1.14

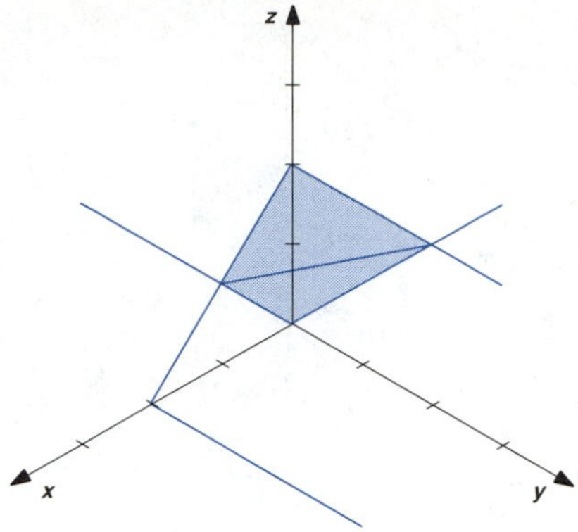


■ **EXAMPLE 5**

Sketch and describe the region in the first octant bounded by $x + z = 2$ and $x + y - z = 0$.

**SOLUTION**

The region is sketched in Figure 17.1.14. The points of intersection of the planes in the $yz$-plane and the $xz$-plane have been connected to indicate a boundary edge. The fact that the region is in the first octant tells us that $x = 0$, $y = 0$, and $z = 0$ are candidates for boundary surfaces of the region.

For $x$ and $y$ fixed, $z$ varies from $z = x + y$ to $z = 2 - x$.

The top and bottom intersect to form a right side edge. We see that

$z = x + y$ and $z = 2 - x$ imply $x + y = 2 - x$, so $y = 2 - 2x$.

This equation gives the right-side cylinder.

For $x$ fixed, $y$ varies from $y = 0$ to $y = 2 - 2x$.

The sides, top, and bottom intersect to form a front point. We use the equations of the sides to find the $x$-coordinate of the intersection:

$y = 0$ and $y = 2 - 2x$ imply $0 = 2 - 2x$, so $x = 1$.

$x$ varies from $x = 0$ to $x = 1$.

FIGURE 17.1.15

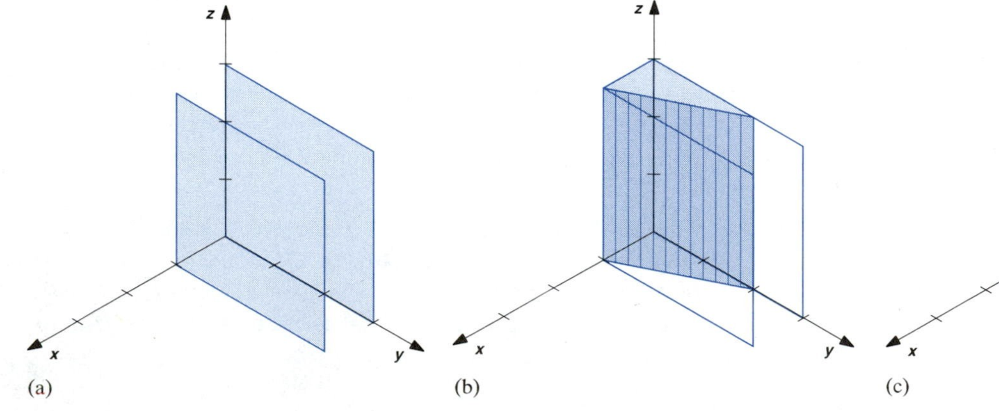

Figure 17.1.15 illustrates how the description of the region determines the region. Note that the right-side cylinder $y = 2 - 2x$ contains the line of intersection of the top and the bottom planes and gives the projection of that line onto the $xy$-plane. ■

**■ EXAMPLE 6**

Sketch and describe the region that is inside the cylinder $x^2 + y^2 = 1$ and between the planes $z = 0$ and $y + z = 2$.

**SOLUTION**

The region is sketched in Figure 17.1.16.

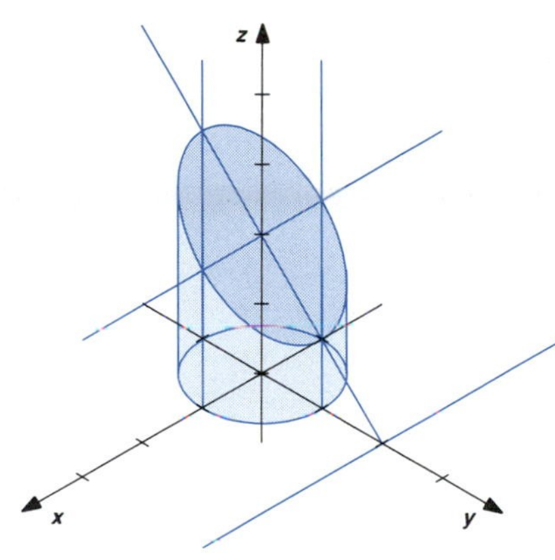


FIGURE 17.1.16

For $x$ and $y$ fixed, $z$ varies from $z = 0$ to $z = 2 - y$.

The side cylinders are given by the equation $x^2 + y^2 = 1$, which gives $y = \sqrt{1 - x^2}$ and $y = -\sqrt{1 - x^2}$.

For $x$ fixed, $y$ varies from $y = -\sqrt{1 - x^2}$ to $y = \sqrt{1 - x^2}$.

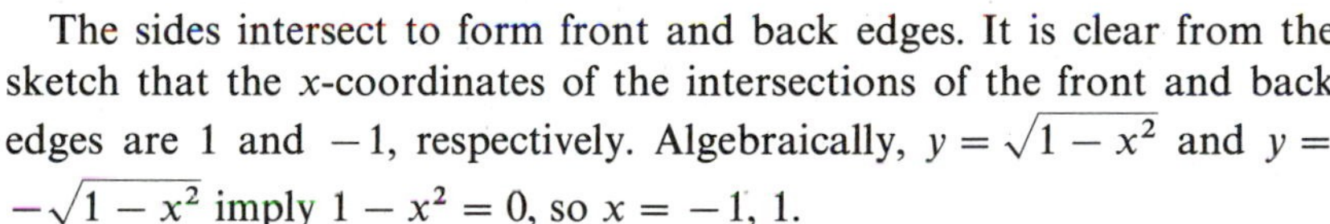

The sides intersect to form front and back edges. It is clear from the sketch that the $x$-coordinates of the intersections of the front and back edges are 1 and $-1$, respectively. Algebraically, $y = \sqrt{1 - x^2}$ and $y = -\sqrt{1 - x^2}$ imply $1 - x^2 = 0$, so $x = -1, 1$.

$x$ varies from $x = -1$ to $x = 1$.

Figure 17.1.17 illustrates how the description of the region determines the region. The projection of the region onto the $xy$-plane is given by the equation $x^2 + y^2 = 1$. ■

FIGURE 17.1.17

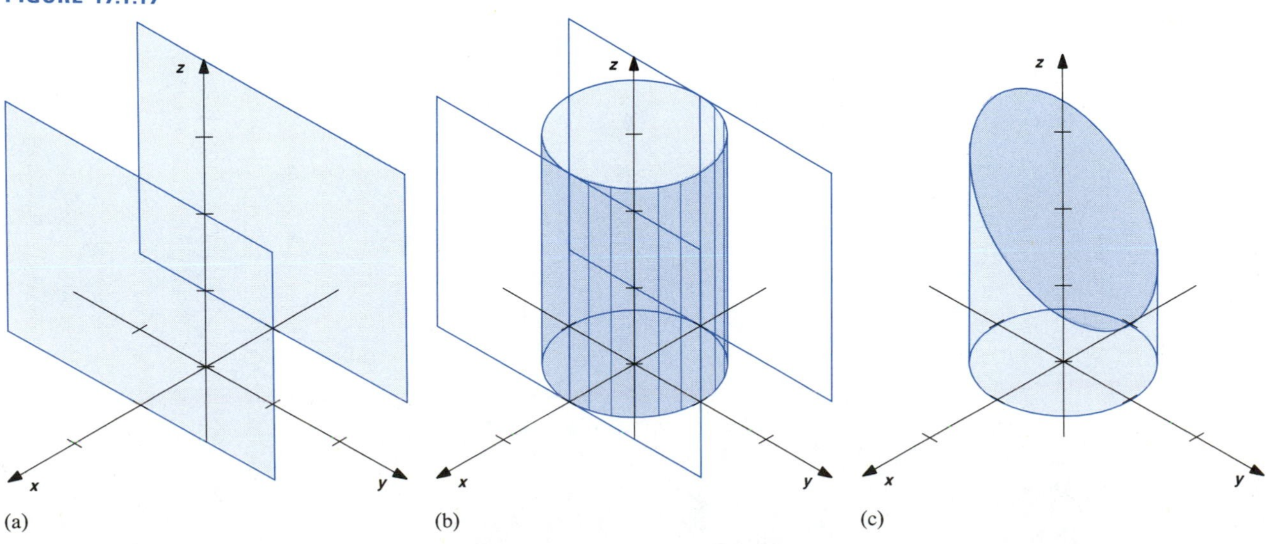

FIGURE 17.1.18

■ **EXAMPLE 7**

Sketch and describe the region in the first octant that is inside both the cone $4x^2 + 4y^2 = z^2$ and the sphere $x^2 + y^2 + z^2 = 4$.

**SOLUTION**

The region is sketched in Figure 17.1.18.

For $x$ and $y$ fixed, $z$ varies from $z = \sqrt{4x^2 + 4y^2}$ to $z = \sqrt{4 - x^2 - y^2}$.

The top and bottom intersect to form a right-side edge.

$$z = \sqrt{4x^2 + 4y^2} \quad \text{and} \quad z = \sqrt{4 - x^2 - y^2} \quad \text{imply}$$

$$4x^2 + 4y^2 = 4 - x^2 - y^2,$$

$$5y^2 = 4 - 5x^2,$$

$$y = \sqrt{\frac{4 - 5x^2}{5}}.$$

For $x$ fixed, $y$ varies from $y = 0$ to $y = \sqrt{\dfrac{4 - 5x^2}{5}}$.

The sides intersect to form a front point.

$$y = \sqrt{\frac{4 - 5x^2}{5}} \text{ and } y = 0 \text{ imply } \frac{4 - 5x^2}{5} = 0, \text{ so } x = \sqrt{\frac{4}{5}}.$$

$x$ varies from $x = 0$ to $x = \sqrt{\dfrac{4}{5}}$.

Figure 17.1.19 illustrates how the description determines the region. The vertical cylinder $y = \sqrt{(4 - 5x^2)/5}$ contains the curve of intersection of the cone and the sphere and gives the projection of that curve onto the $xy$-plane. ■

FIGURE 17.1.19

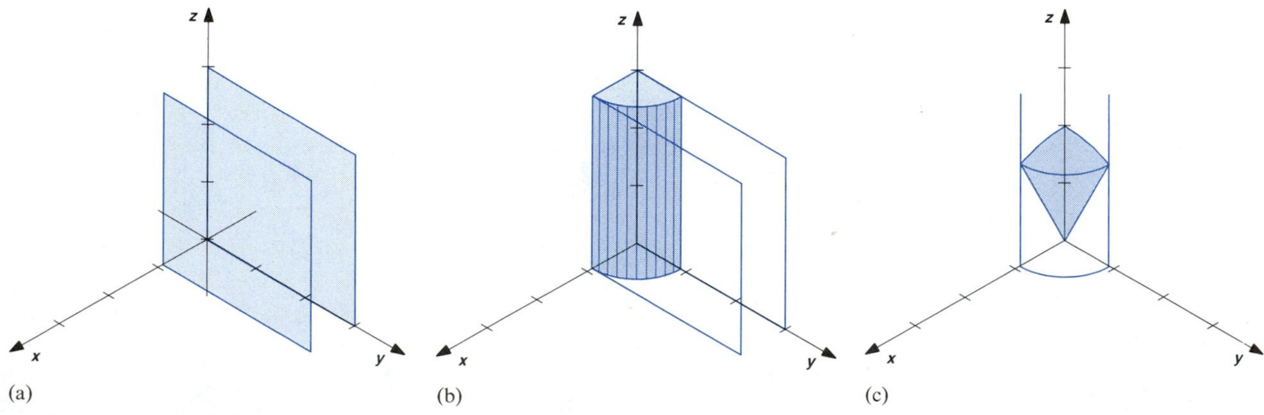

FIGURE 17.1.20

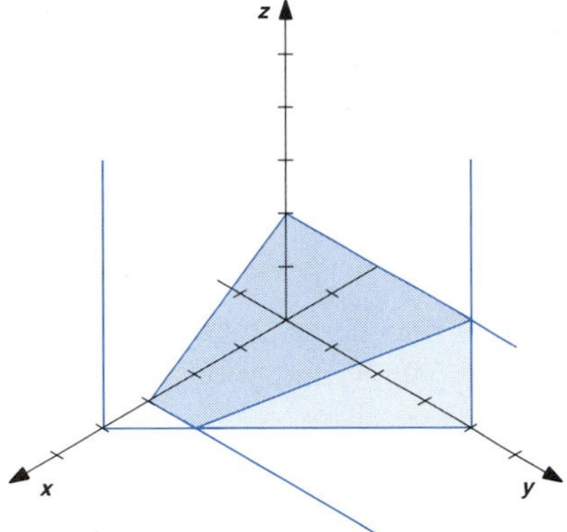


■ **EXAMPLE 8**

Sketch and describe the region in the first octant bounded by the planes $2x + 3z = 6$ and $x + y = 4$.

**SOLUTION**

The region is sketched in Figure 17.1.20.

For $x$ and $y$ fixed, $z$ varies from $z = 0$ to $z = \frac{2}{3}(3 - x)$.

For $x$ fixed, $y$ varies from $y = 0$ to $y = 4 - x$.

The top and bottom intersect to form a front edge. We need the $x$-coordinate of that edge:

$z = \frac{2}{3}(3 - x)$ and $z = 0$ imply $\frac{2}{3}(3 - x) = 0$, so $x = 3$.

$x$ varies from $x = 0$ to $x = 3$.

See Figure 17.1.21. The projection of the region onto the $xy$-plane is given by the equations $y = 0$, $y = 4 - x$, $x = 0$, and $x = 3$. ■

FIGURE 17.1.21

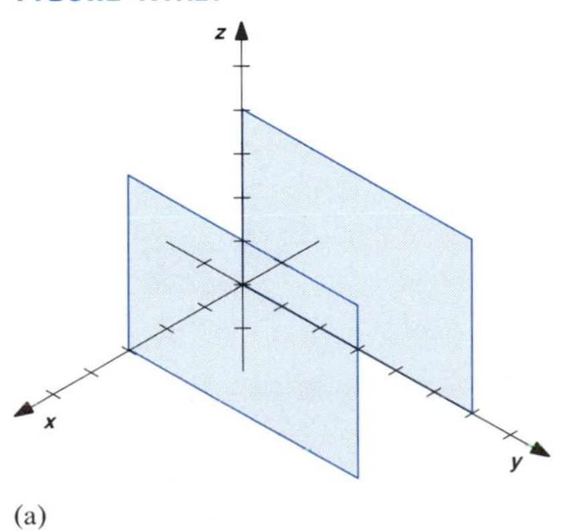


(a)

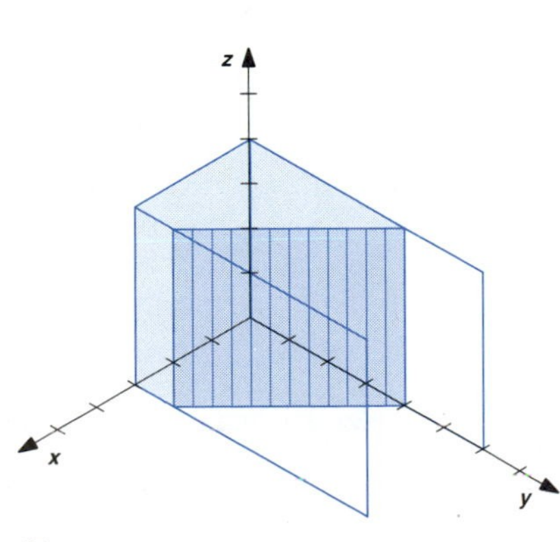


(b)

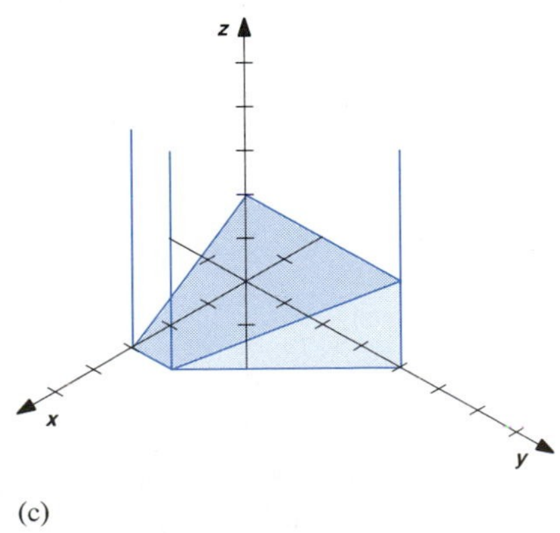


(c)

To sketch regions bounded by more than two surfaces that are not coordinate planes, it is a good idea to sketch the traces and curve of intersection of two of the surfaces before adding the traces of additional surfaces.

### ■ EXAMPLE 9

Sketch and describe the region in the first octant bounded by the planes $y = 2z$, $y = 4$, and $y = 2x$.

**SOLUTION**

Let us begin by sketching the traces and intersection of the planes $y = 2z$ and $y = 4$. This is done in Figure 17.1.22a. Note that neither of these two equations involves the variable $x$, so the line of intersection of the planes extends through the intersection of the traces in the $yz$-plane, parallel to the $x$-axis.

In Figure 17.1.22b we have added the coordinate traces of the plane $y = 2x$ and the lines of intersection of this plane with the planes sketched in Figure 17.1.22a. Note that the intersection of the planes $y = 2x$ and $y = 4$ is a vertical line through the intersection of their traces in the $xy$-plane. The point of intersection of this vertical line with the line of intersection of the planes $y = 2z$ and $y = 4$ gives a point that is in the intersection of $y = 2z$ and $y = 2x$. Since the origin is also in the intersection of $y = 2z$ and $y = 2x$, the boundary of the region contains the line between the latter two points. (You should expect to get a region of the type pictured in Figure 17.1.22b if the wedge pictured in Figure 17.1.22a is cut along the vertical plane $y = 2x$.)

From Figure 17.1.22b we see that

For $x$ and $y$ fixed, $z$ varies from $z = 0$ to $z = \dfrac{y}{2}$.

For $x$ fixed, $y$ varies from $y = 2x$ to $y = 4$.

FIGURE 17.1.22

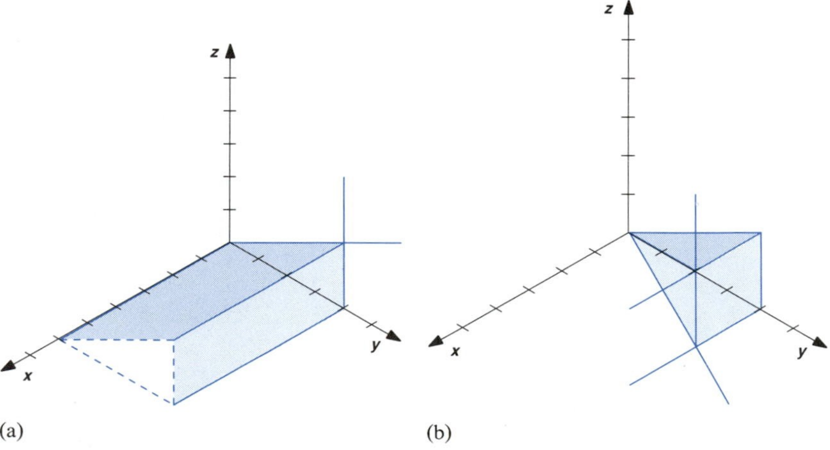

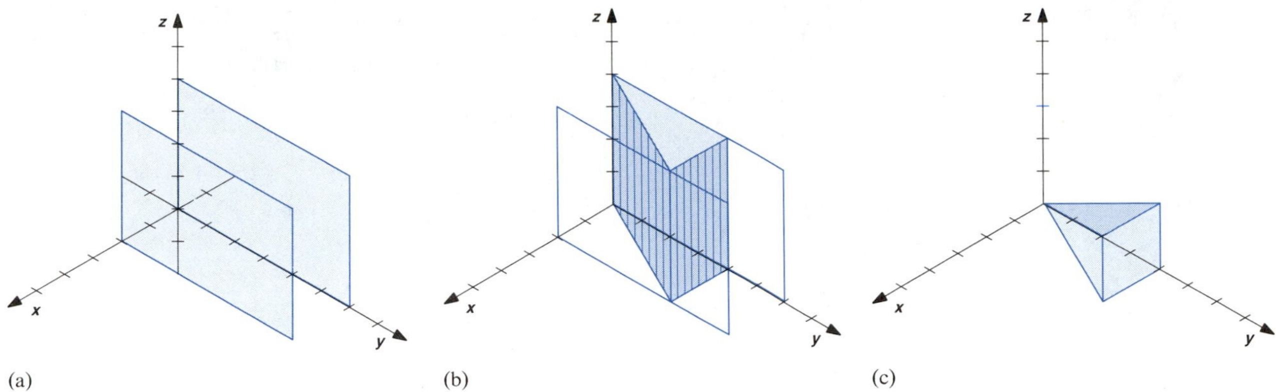


FIGURE 17.1.23

The front plane is given by the $x$-coordinate of the point of intersection of the side cylinders.

$y = 2x$ and $y = 4$ imply $x = 2$.

$x$ varies from $x = 0$ to $x = 2$.

See Figure 17.1.23. The projection of the region onto the $xy$-plane is given by the equations $y = 2x$, $y = 4$, and $x = 0$. ■

We know that certain regions in the plane can be described with either $x$ or $y$ fixed in the first step. Similarly, certain regions in three-dimensional space can be described by fixing the variables $x$, $y$, and $z$ in one or more of six possible orders. We will deal with such regions in Section 17.5.

## EXERCISES 17.1

In Exercises 1–14 sketch and describe the regions in the plane bounded by the given equations.

1 $y = x^2 + 1$, $y = x$, $x = 0$, $x = 1$

2 $y = x^3$, $y = -1$, $x = 1$

3 $y = x$, $y = 1$, $x = 0$

4 $y = x$, $y = 0$, $x = 1$

5 $x + 2y = 2$, $y = 0$, $x = 0$

6 $y = 4 - x^2$, $y = 0$, $x = 0$ $(x \geq 0)$

7 $y = \sin x$, $y = \cos x$, $x = 0$ $(x \geq 0)$

8 $y = \cos x$, $y = 0$, $x = 0$ $(x \geq 0)$

9 $y = x$, $x = 2 - y^2$

10 $y = 1$, $y = 0$, $x + y^2 = 0$, $x = 1 + y^2$

11 $y = 1$, $y = 0$, $x = 0$, $y = \ln x$

12 $y = 1$, $y = 0$, $y = e^x - 1$, $x = 2$

13 $x^2 + y^2 = 4$, $y = 0$ $(y \geq 0)$

14 $x^2 + y^2 = 4$, $x = 0$ $(x \geq 0)$

In Exercises 15–32 sketch and describe the regions in three-dimensional space bounded by the given equations.

15 Above $z = x + y$, below $x + y + z = 2$, in the first octant

16 $y + z = 4$, $z = 2x + 2y$, in the first octant

17 Below $x + y + z = 3$, above $z = x$, in the first octant

18 Below $x + 2y + z = 6$, above $z = 2y$, in the first octant

19 Inside $x^2 + y^2 = 1$, below $x + y + z = 2$, above $z = 0$

20 Inside $x^2 + y^2 + z^2 = 4$, above $z = 1$

21 $y + z = 3$, $x + y = 2$, in the first octant

22 Inside $z = \sqrt{2x^2 + 2y^2}$, below $x + z = 1$, in the first octant

23 Inside $x^2 + y^2 = 4$ and $y^2 + z^2 = 4$, in the first octant

24 Inside $x^2 + y^2 + z^2 = 4$, above $z = \sqrt{3}$, in the first octant

25 Inside $x^2 + y^2 = 1$, outside $z^2 = 4x^2 + 4y^2$

26 $x^2 + z^2 = 1$, $x + y = 3$, in the first octant

27 $y = 3$, $z = 2$, $z = 2x$, in the first octant

28 $y = 3$, $2x + 3z = 6$, in the first octant

29 $x = 2$, $y = 2$, $z = 2$, in the first octant

30 $x = 2$, $y + 2z = 4$, between $y = 0$ and $y = 2$, in the first octant

31 $x = 2$, above $y = 2z$, below $y + 2z = 4$, in the first octant

32 $x + y = 2$, above $y = 2z$, below $y + 2z = 4$, in the first octant

## 17.2 DOUBLE INTEGRALS

We will discuss the concept of the double integral of a bounded function over a region in a rectangular coordinate plane and see how double integrals relate to the area of regions in the plane.

Let us consider a function $f(x, y)$ that is bounded for $(x, y)$ in a region $R$. We assume that $R$ is a bounded region. That is, $R$ is contained in some disk. To define the double integral of $f$ over $R$, we need to consider some analogue of the Riemann sums that were used to define the integral of a function of one variable over an interval. The idea is to cover the region with rectangles $R_j$ that have sides parallel to the coordinate axes, choose a point $(x_j, y_j)$ in each rectangle, and take the sum of the products of $f(x_j, y_j)$ times the area of the corresponding rectangle $R_j$. See Figure 17.2.1. One difficulty is what to do with rectangles that are part in and part out of the region. A nice way out of this difficulty is to consider only regions for which it doesn't matter what we do with the "boundary rectangles." Thus, we assume that $R$ satisfies the following condition.

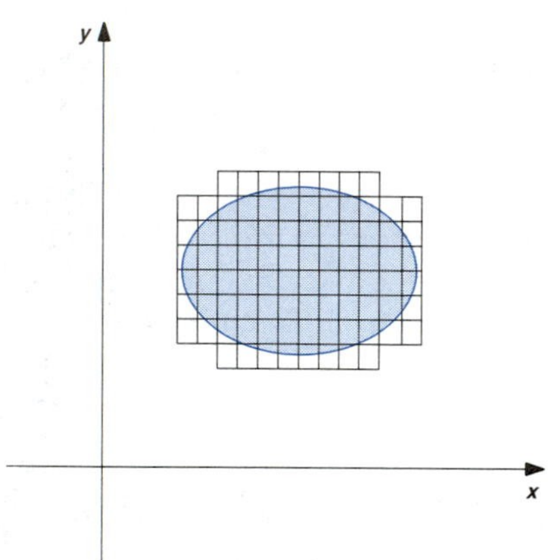

FIGURE 17.2.1

The region $R$ can be completely covered with rectangles in such a way that the total area of all rectangles that are part in and part out of the region is smaller than any prescribed positive number. (1)

It can be proved that regions that are bounded by a finite number of smooth curves satisfy this condition. Experience with using the definite integral to calculate the area of vertically simple and horizontally simple regions in the plane should lead us to believe that regions of these types satisfy condition (1). This is illustrated in Figure 17.2.2. If the top and bottom of the region are given by functions that are continuous for $a \leq x \leq b$, it seems reasonable that the total area of the "boundary rectangles" approaches zero as the width and height of the rectangles approach zero. The sum of the products of the area of each of these rectangles times the value of a *bounded* function $f(x, y)$ also approaches zero.

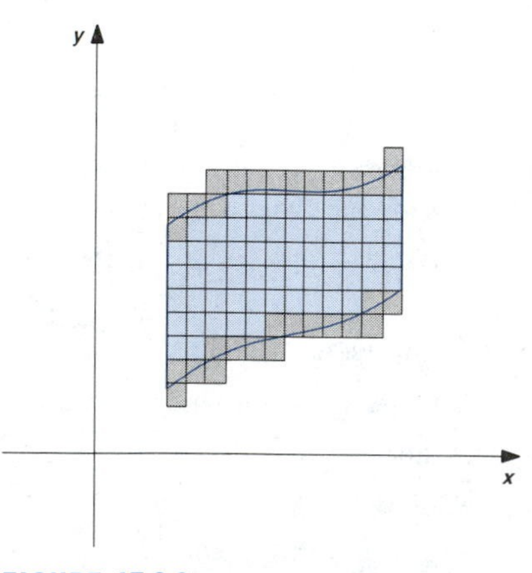

FIGURE 17.2.2

For regions $R$ that have property (1), a **Riemann sum** of a bounded function $f(x, y)$ over $R$ is any sum of the form

$$\sum f(x_j, y_j)\Delta A_j,$$

where $R$ is covered by a grid of rectangles $R_j$ that have sides parallel to the axes, $(x_j, y_j)$ is any point in $R_j$, $\Delta A_j$ is the area of $R_j$, and the sum is over all rectangles that are inside $R$.

We are interested in the case when the Riemann sums of $f$ over $R$ approach a single number as the lengths of sides of the rectangles approach zero. This idea is expressed by the following formal definition.

*Definition*

A function $f$ is **integrable** over the region $R$ if there is a number [denoted $\iint_R f(x, y)\,dA$] such that, for each $\varepsilon > 0$, there is a positive number $\delta$ such that

$$\left|\sum f(x_j, y_j)\Delta A_j - \iint_R f(x, y)\,dA\right| < \varepsilon$$

whenever the lengths of the sides of all rectangles in the Riemann sum are less than $\delta$.

There can be at most one number $\iint_R f(x, y)\,dA$ that satisfies the condition of the definition. If it exists, that number is called the **double integral** of $f$ over $R$. It can be proved that the integral exists if $R$ satisfies condition (1) and $f$ is continuous on the set that consists of $R$ and all boundary points of $R$. In particular, the integral exists if $R$ is bounded by a finite number of smooth curves and $f$ is continuous.

As in the case of a single variable, we will establish methods of evaluating double integrals without dealing directly with Riemann sums, but the idea of the integral as a sum is very important for understanding applications of the concept. We should think of a double integral as a sum of products. Applications of the concept arise from the interpretation of the terms of the products as physical quantities.

The usual properties of the integral are true for double integrals of *integrable* functions. For example, if each of the functions involved are integrable, then

$$\iint_R (f + g)\,dA = \iint_R f\,dA + \iint_R g\,dA,$$

$$\iint_R kf\,dA = k\left(\iint_R f\,dA\right), \quad k \text{ a constant, and}$$

$$\iint_R f\,dA = \iint_{R'} f\,dA + \iint_{R''} f\,dA,$$

where $R$ is the union of the disjoint regions $R'$ and $R''$. See Figure 17.2.3.

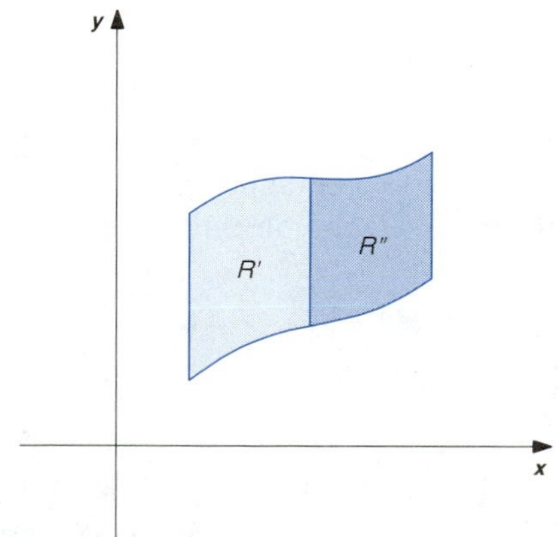

FIGURE 17.2.3

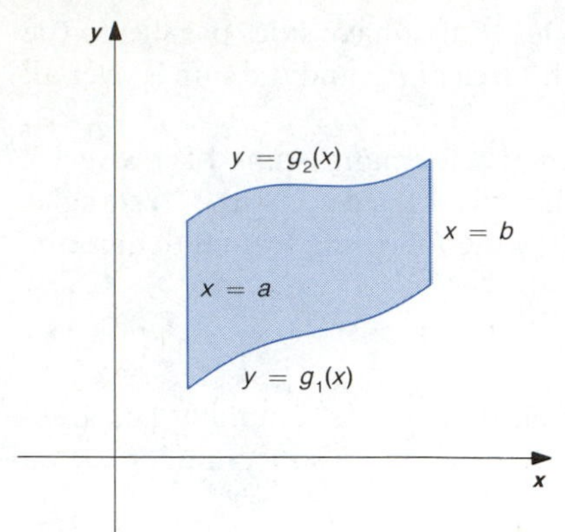


(a)

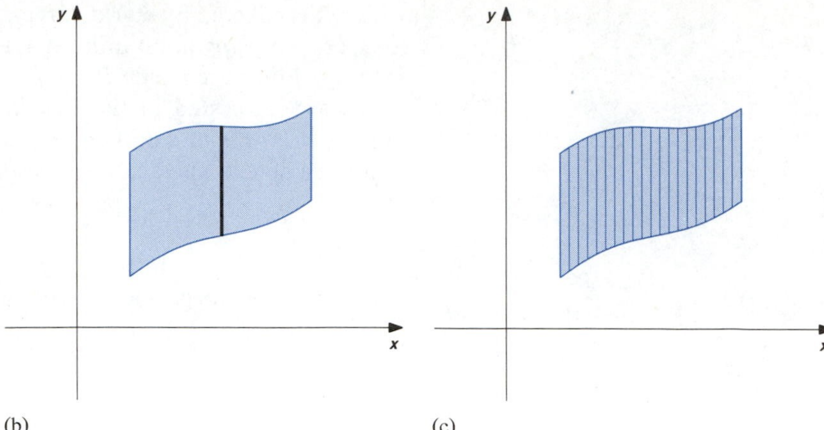


(b) (c)

**FIGURE 17.2.4**

Let us consider a vertically simple region that can be described by the following statements:

For $x$ fixed, $y$ varies from $y = g_1(x)$ to $y = g_2(x)$.
$x$ varies from $x = a$ to $x = b$.

We assume the boundary functions are continuous with $g_1(x) \leq g_2(x)$ for $a \leq x \leq b$. See Figure 17.2.4a. As $y$ varies from $y = g_1(x)$ to $y = g_2(x)$ with $x$ fixed, the points $(x, y)$ form a vertical line segment between the top and bottom of the region. See Figure 17.2.4b. As $x$ varies from $x = a$ to $x = b$, these line segments sweep out the entire region. See Figure 17.2.4c.

We need to develop a formula for the evaluation of the double integral of a function $f(x, y)$ over a vertically simple region $R$ as described above. We begin by covering the region with a grid of rectangles $R_{ij}$ that have sides parallel to the coordinate axes and area $\Delta A_{ij} = \Delta x_i \, \Delta y_j$. This is illustrated in Figure 17.2.5a. We evaluate $f$ at the lower left-hand corner of each rectangle that is contained in the region to obtain the Riemann sum

$$\sum_{i,j} f(x_i, y_j) \, \Delta A_{ij}.$$

The terms that involve the same $x_i$ correspond to a column of rectangles. See Figure 17.2.5b. We group and add the terms that involve the same $x_i$ to obtain

$$\sum_i \left[ \sum_j f(x_i, y_j) \, \Delta y_j \right] \Delta x_i.$$

We see that the sum inside the brackets above is a Riemann sum with respect to $y$ of the function of one variable $f(x_i, y)$ obtained by fixing the value $x_i$; for $x_i$ fixed, $y$ varies from essentially $y = g_1(x_i)$ to $y = g_2(x_i)$. If

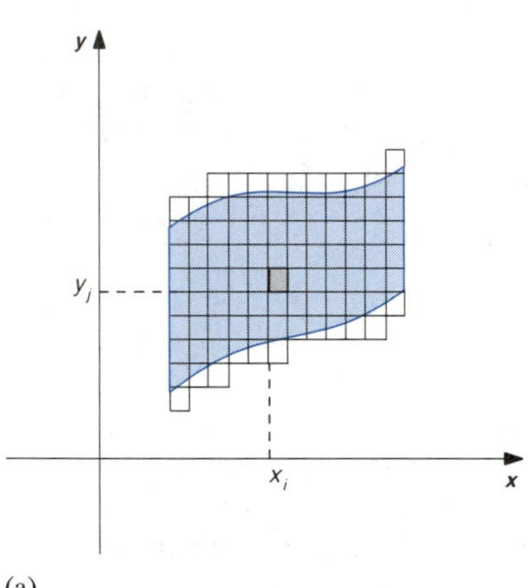


(a)

(b)

**FIGURE 17.2.5**

every $\Delta y_j$ is small enough, the sum inside the brackets is approximately

$$\int_{g_1(x_i)}^{g_2(x_i)} f(x_i, y)\, dy,$$

so our double Riemann sum is approximately

$$\sum_i \left[\int_{g_1(x_i)}^{g_2(x_i)} f(x_i, y)\, dy\right] \Delta x_i.$$

This is a Riemann sum with respect to $x$ of the function of $x$ defined by

$$\int_{g_1(x)}^{g_2(x)} f(x, y)\, dy.$$

The variable $x$ varies from $x = a$ to $x = b$. If every $\Delta x_i$ is small enough, this Riemann sum is approximately

$$\int_a^b \left[\int_{g_1(x)}^{g_2(x)} f(x, y)\, dy\right] dx.$$

Combining results, we obtain that the double integral of $f$ over a vertically simple region $R$ that can be described by the statements "For $x$ fixed, $y$ varies from $y = g_1(x)$ to $y = g_2(x)$" and "$x$ varies from $x = a$ to $x = b$" satisfies

$$\iint_R f(x, y)\, dA = \int_a^b \left[\int_{g_1(x)}^{g_2(x)} f(x, y)\, dy\right] dx.$$

The integral on the right is called an **iterated integral**. It is evaluated by considering $x$ to be constant (fixed) while we integrate the inner integral with respect to $y$ from $g_1(x)$ to $g_2(x)$. This gives a function of $x$, which we then integrate with respect to $x$ from $a$ to $b$. Note how *the description of the region corresponds to the limits of integration and the evaluation of the iterated integral.* In particular, the statement, "For $x$ fixed, $y$ varies from $y = g_1(x)$ to $y = g_2(x)$," gives the limits of integration of the inner integral and suggests that $x$ is fixed while that integral is evaluated. The statement "$x$ varies from $x = a$ to $x = b$" gives the limits of integration of the outer integral.

It is common to omit the brackets around the inner integral of an iterated integral.

## ■ EXAMPLE 1

Evaluate

$$\int_0^1 \int_{\sqrt{x}}^1 24x^2 y\, dy\, dx.$$

**SOLUTION**

We begin by evaluating the inner integral. The variable $x$ is considered to be constant while we integrate with respect to $y$. We obtain

$$\int_0^1 \int_{\sqrt{x}}^1 24x^2y\,dy\,dx = \int_0^1 \left[\int_{\sqrt{x}}^1 24x^2y\,dy\right] dx$$
$$= \int_0^1 \left[24x^2\frac{y^2}{2}\right]_{y=\sqrt{x}}^{y=1} dx.$$

[We have written $y = 1$ and $y = \sqrt{x}$ to remind us that $y$ is the variable of integration and we must substitute for $y$ in the expression $24x^2(y^2/2)$.] We now have

$$\int_0^1 \int_{\sqrt{x}}^1 24x^2y\,dy\,dx = \int_0^1 [12x^2(1)^2 - 12x^2(\sqrt{x})^2]\,dx$$
$$= \int_0^1 [12x^2 - 12x^3]\,dx$$
$$= 12\frac{x^3}{3} - 12\frac{x^4}{4}\Big]_0^1$$
$$= [4(1)^3 - 3(1)^4] - [4(0)^3 - 3(0)^4] = 1. \qquad ■$$

Let us consider a horizontally simple region $R$ that can be described by the following statements:

For $y$ fixed, $x$ varies from $x = g_1(y)$ to $x = g_2(y)$.
$y$ varies from $y = a$ to $y = b$.

See Figure 17.2.6a. As $x$ varies from $x = g_1(y)$ to $x = g_2(y)$ with $y$ fixed, the points $(x, y)$ form a horizontal line segment between the sides of the region. See Figure 17.2.6b. As $y$ varies from $y = a$ to $y = b$, these line segments sweep out the entire region. See Figure 17.2.6c. In a manner that is

FIGURE 17.2.6

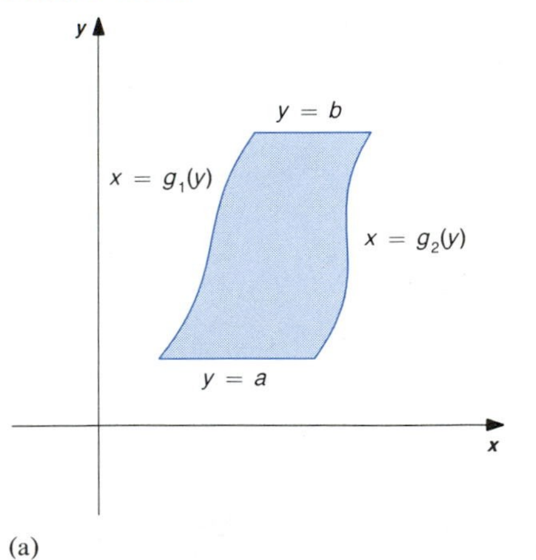


(a)

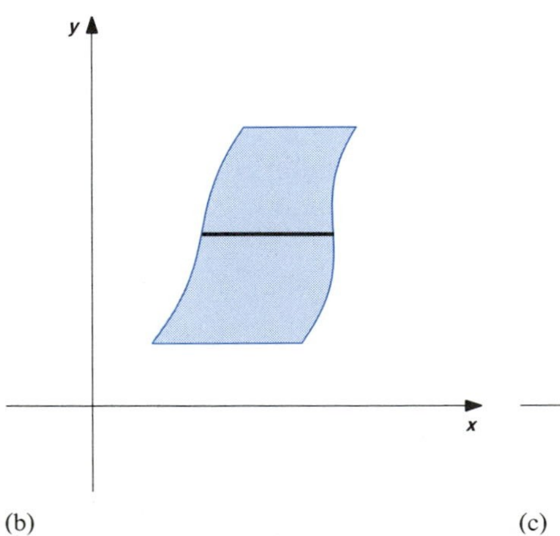


(b)

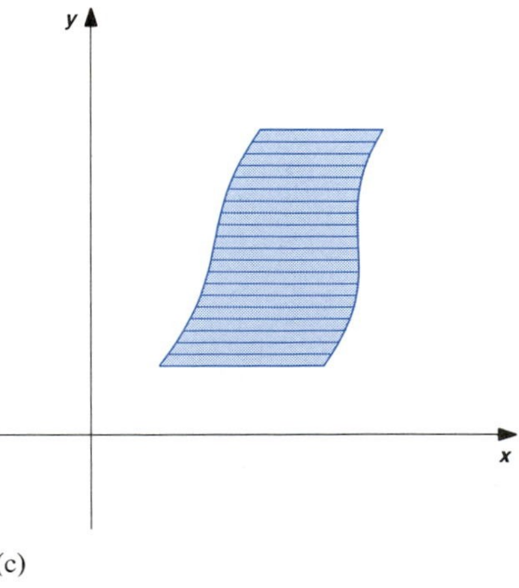


(c)

completely analogous to the case of a vertically simple region, we then obtain

$$\iint_R f(x, y)\, dA = \int_a^b \left[ \int_{g_1(y)}^{g_2(y)} f(x, y)\, dx \right] dy.$$

Iterated integrals of this type are evaluated by treating $y$ as a constant while evaluating the inner integral with respect to $x$, and then integrating with respect to $y$.

### ■ EXAMPLE 2

Evaluate

$$\int_0^\pi \int_0^y \sin x\, dx\, dy.$$

**SOLUTION**

The inner integral is with respect to $x$. We have

$$\begin{aligned}
\int_0^\pi \int_0^y \sin x\, dx\, dy &= \int_0^\pi \left[ \int_0^y \sin x\, dx \right] dy \\
&= \int_0^\pi \left[ -\cos x \right]_{x=0}^{x=y} dy \\
&= \int_0^\pi [(-\cos y) - (-\cos 0)]\, dy \\
&= \int_0^\pi [1 - \cos y]\, dy \\
&= y - \sin y \Big]_0^\pi \\
&= (\pi - \sin \pi) - (0 - \sin 0) = \pi.
\end{aligned}$$

■

Evaluation of iterated integrals should not cause any particular new difficulty. Be sure to evaluate the inner integral first, and remember which is the variable of integration and which variable is considered constant.

If $R$ is either a vertically simple region or a horizontally simple region, we can use the corresponding iterated integral to evaluate the double integral over $R$ of an integrable function. If $R$ is both vertically simple and horizontally simple, we can use an iterated integral with either of the two orders of integration. The two iterated integrals will have the same value, since they represent two different ways of evaluating the same double integral.

We have seen how the limits of integration of an iterated integral correspond to the description of the region. We can determine the limits of integration of an iterated integral over a region by exactly the same method that we used in Section 17.1 to describe the region.

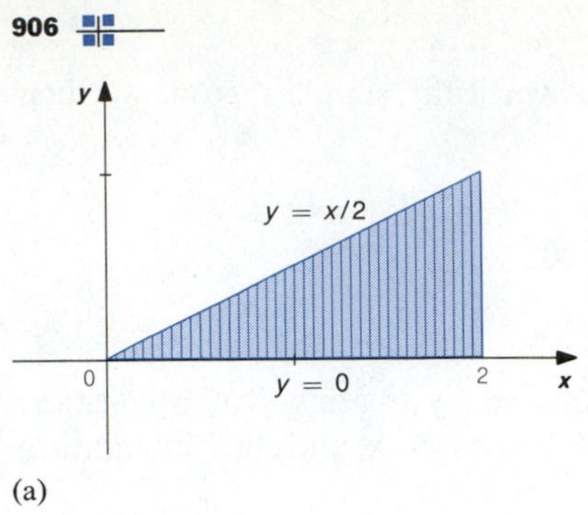

(a)

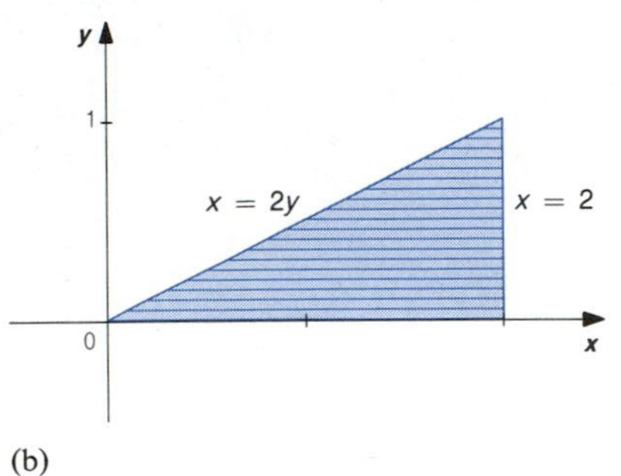

(b)

**FIGURE 17.2.7**

**EXAMPLE 3**

Evaluate $\iint_R (6x + 24y)\, dA$, where $R$ is the region bounded by $y = x/2$, $y = 0$, and $x = 2$.

**SOLUTION**

The region is sketched in Figure 17.2.7a. We see that the region is vertically simple and can be described by the statements

For $x$ fixed, $y$ varies from $y = 0$ to $y = x/2$.
$x$ varies from $x = 0$ to $x = 2$.

The corresponding iterated double integral is

$$\int_0^2 \int_0^{x/2} (6x + 24y)\, dy\, dx.$$

The inner integral is evaluated by treating $x$ as a constant while integrating with respect to $y$. We have

$$\begin{aligned}\int_0^2 \int_0^{x/2} (6x + 24y)\, dy\, dx &= \int_0^2 \Big[ 6xy + 12y^2 \Big]_{y=0}^{y=x/2} dx \\ &= \int_0^2 \left\{ \left[ 6x\left(\frac{x}{2}\right) + 12\left(\frac{x}{2}\right)^2 \right] - [6x(0) + 12(0)^2] \right\} dx \\ &= \int_0^2 6x^2\, dx = 2x^3 \Big]_0^2 = 16.\end{aligned}$$

From the sketch in Figure 17.2.7b, we see that the region is also horizontally simple and can be described by the following statements:

For $y$ fixed, $x$ varies from $x = 2y$ to $x = 2$.
$y$ varies from $y = 0$ to $y = 1$.

The corresponding iterated integral is

$$\begin{aligned}\int_0^1 \int_{2y}^{2} (6x + 24y)\, dx\, dy &= \int_0^1 \Big[ 3x^2 + 24yx \Big]_{x=2y}^{x=2} dy \\ &= \int_0^1 \{[3(2)^2 + 24y(2)] - [3(2y)^2 + 24y(2y)]\}\, dy \\ &= \int_0^1 (12 + 48y - 60y^2)\, dy \\ &= 12y + 24y^2 - 20y^3 \Big]_0^1 = 12 + 24 - 20 = 16.\end{aligned}$$

As expected, the two iterated integrals are equal. ■

**EXAMPLE 4**

Evaluate $\iint_R 4x\, dA$, where $R$ is the region bounded by $y = \ln x$, $y = 0$, and $x = 2$.

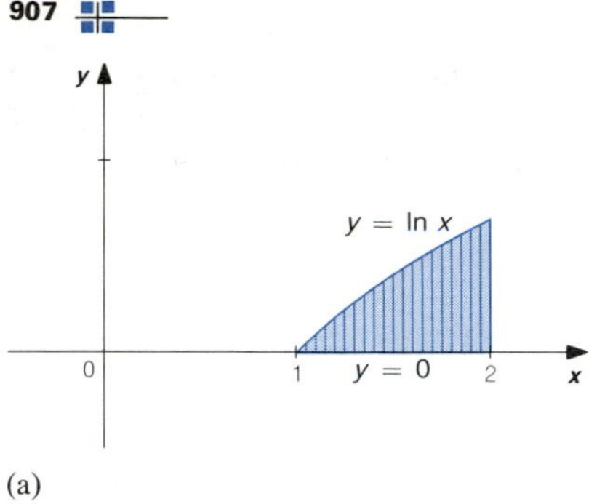


(a)

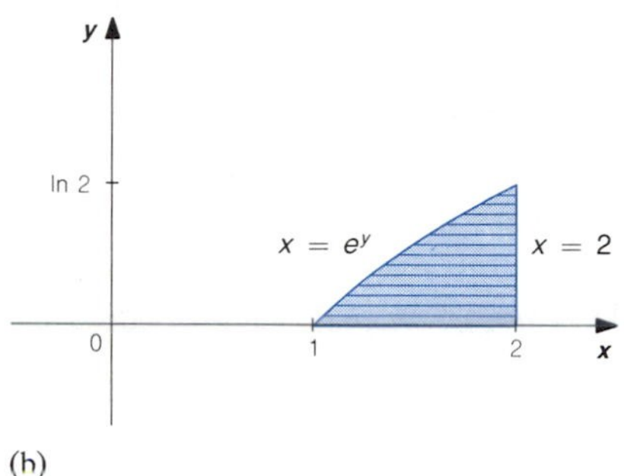


(b)

FIGURE 17.2.8

**SOLUTION**

The region is sketched in Figure 17.2.8a. We see that the region is vertically simple and can be described by the following statements:

For $x$ fixed, $y$ varies from $y = 0$ to $y = \ln x$.
$x$ varies from $x = 1$ to $x = 2$.

The corresponding iterated integral is

$$\int_1^2 \int_0^{\ln x} 4x \, dy \, dx = \int_1^2 4xy \Big]_{y=0}^{y=\ln x} dx$$
$$= \int_1^2 4x \ln x \, dx.$$

Using the table of integrals (or integrating by parts), we obtain

$$\int_1^2 \int_0^{\ln x} 4x \, dy \, dx = 2x^2 \ln x - x^2 \Big]_1^2$$
$$= (8 \ln 2 - 4) - (0 - 1)$$
$$= 8 \ln 2 - 3.$$

From the sketch in Figure 17.2.8b we see that the region is also horizontally simple. We can express the equation $y = \ln x$ as $x = e^y$. Also:

$x = e^y$ and $x = 2$ imply $e^y = 2$, so $y = \ln 2$.

We can then describe the region by the following statements:

For $y$ fixed, $x$ varies from $x = e^y$ to $x = 2$.
$y$ varies from $y = 0$ to $y = \ln 2$.

The corresponding iterated integral is

$$\int_0^{\ln 2} \int_{e^y}^{2} 4x \, dx \, dy = \int_0^{\ln 2} 2x^2 \Big]_{x=e^y}^{x=2} dy$$
$$= \int_0^{\ln 2} [2(2)^2 - 2(e^y)^2] \, dy$$
$$= \int_0^{\ln 2} (8 - 2e^{2y}) \, dy$$
$$= 8y - e^{2y} \Big]_0^{\ln 2}$$
$$= (8 \ln 2 - e^{2 \ln 2}) - (0 - e^0)$$
$$= 8 \ln 2 - 4 + 1 = 8 \ln 2 - 3.$$

We see that the integration is more elementary in the above iterated integral than in the integral with the opposite order of integration. ■

Let us see how to use a double integral to find the area of a region. We cover the region $R$ with a grid of rectangles that have sides parallel to the

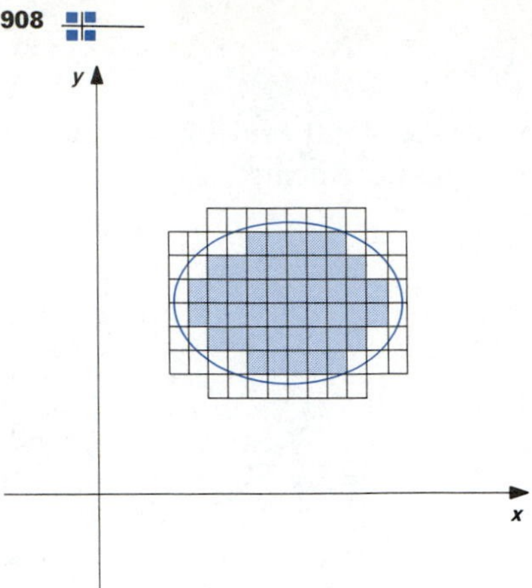


FIGURE 17.2.9

axes and form the sum of the areas of the rectangles contained in $R$. See Figure 17.2.9. If the total area of the boundary rectangles in the grid of rectangles that cover $R$ is small, then the sum of the areas of the rectangles inside $R$ is approximately the area of $R$. This suggests that the area of the region is

$$\textbf{Area} = \iint_R dA.$$

**■ EXAMPLE 5**

Use double integrals to find the area of the region bounded by $x = y - y^2$ and $x + y = 0$.

**SOLUTION**

The region is sketched in Figure 17.2.10. From the sketch, we see that the region is not vertically simple, but it is horizontally simple. We can then describe the region by writing:

For $y$ fixed, $x$ varies from $x = -y$ to $x = y - y^2$.

$y$ varies between the $y$-coordinates of the intersection of the two boundary curves. The equations

$x = y - y^2$ and $x = -y$ imply $-y = y - y^2$, so $y^2 - 2y = 0$.

The solutions are $y = 0$ and $y = 2$, so

$y$ varies from $y = 0$ to $y = 2$

$$\begin{aligned}\text{Area} &= \int_0^2 \int_{-y}^{y-y^2} dx\, dy \\ &= \int_0^2 x\Big]_{x=-y}^{x=y-y^2} dy \\ &= \int_0^2 [(y - y^2) - (-y)]\, dy.\end{aligned}$$

[The above integral represents the sum of the areas of rectangles with horizontal length $(y - y^2) - (-y)$ and width $dy$. This corresponds to the integral obtained by the single variable method we used to determine area.]

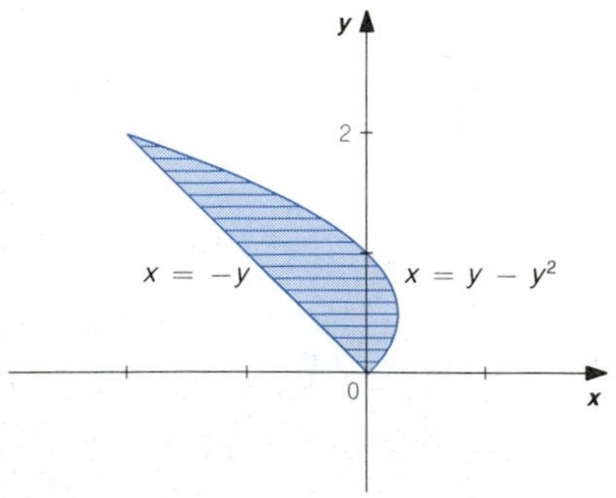

$$\begin{aligned}\text{Area} &= \int_0^2 (2y - y^2)\, dy \\ &= y^2 - \frac{y^3}{3}\Big]_0^2 \\ &= \left(4 - \frac{8}{3}\right) - (0) = \frac{4}{3}.\end{aligned}$$

■

FIGURE 17.2.10

If a region is neither vertically simple nor horizontally simple, it may require more than one iterated integral to evaluate a double integral over the region.

■ **EXAMPLE 6**

Use double integrals to find the area of the region in the first quadrant bounded by $y = 2x - 4$ and $8y = 16 + x^2$.

**SOLUTION**

From the sketch in Figure 17.2.11a we see that the region is neither vertically simple nor horizontally simple. However, we can express the region as the union of two disjoint regions that are vertically simple, as indicated in Figure 17.2.11b. (The region could also be expressed as a union of disjoint horizontally simple regions.) We need the $x$-coordinate of the point of intersection of $y = 2x - 4$ and $8y = 16 + x^2$.

$y = 2x - 4$ and $8y = 16 + x^2$ imply $16 + x^2 = 8(2x - 4)$, so

$16 + x^2 = 16x - 32,$

$x^2 - 16x + 48 = 0,$

$(x - 4)(x - 12) = 0,$

$x = 4, \quad x = 12.$

The solution we want is $x = 4$.

The total area is then the sum of the areas of the parts. We have

$$\begin{aligned}\text{Area} &= \int_0^2 \int_0^{(16+x^2)/8} dy\,dx + \int_2^4 \int_{2x-4}^{(16+x^2)/8} dy\,dx \\ &= \int_0^2 \left(2 + \frac{x^2}{8}\right) dx + \int_2^4 \left[\left(2 + \frac{x^2}{8}\right) - (2x - 4)\right] dx \\ &= \int_0^2 \left(2 + \frac{x^2}{8}\right) dx + \int_2^4 \left(\frac{x^2}{8} - 2x + 6\right) dx \\ &= \left[2x + \frac{x^3}{24}\right]_0^2 + \left[\frac{x^3}{24} - x^2 + 6x\right]_2^4 \\ &= \left[\left(4 + \frac{1}{3}\right) - (0)\right] + \left[\left(\frac{8}{3} - 16 + 24\right) - \left(\frac{1}{3} - 4 + 12\right)\right] = \frac{20}{3}. \quad ■\end{aligned}$$

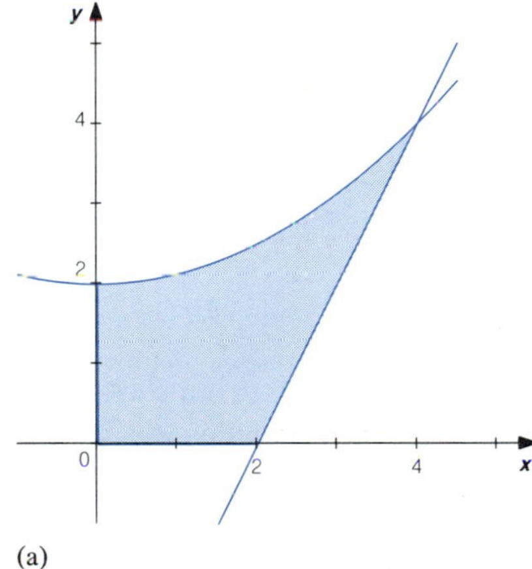

(a)

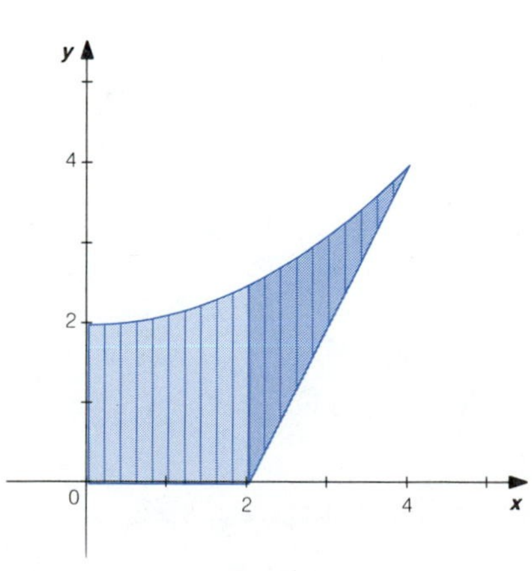

(b)

FIGURE 17.2.11

It is useful to be able to change the order of integration of an iterated integral. For example, a scientific problem may lead to an iterated double integral that is easier to evaluate with the opposite order of integration. We will use the following steps to change the order of integration.

***To change the order of integration of an iterated double integral***

- **Use the given limits of integration to determine the equations of the boundary of the corresponding region.**
- **Sketch the region.**
- **Use the sketch and boundary equations to determine the desired limits of integration.**

Do not try do go directly from the limits of integration for one order of integration to the limits of integration for the opposite order. Always go from the limits of integration to a sketch of the region and then to the other limits of integration.

■ **EXAMPLE 7**

Express

$$\int_0^4 \int_{x/2}^2 e^{y^2}\, dy\, dx$$

as an iterated integral with the opposite order of integration and evaluate.

**SOLUTION**

From the given limits of integration we see that:

For $x$ fixed, $y$ varies from $y = x/2$ to $y = 2$.
$x$ varies from $x = 0$ to $x = 4$.

The corresponding region is sketched in Figure 17.2.12. From the sketch, we see that the region can also be described by the following statements:

For $y$ fixed, $x$ varies from $x = 0$ to $x = 2y$.
$y$ varies from $y = 0$ to $y = 2$.

The corresponding iterated integral is

$$\int_0^2 \int_0^{2y} e^{y^2}\, dx\, dy.$$

FIGURE 17.2.12

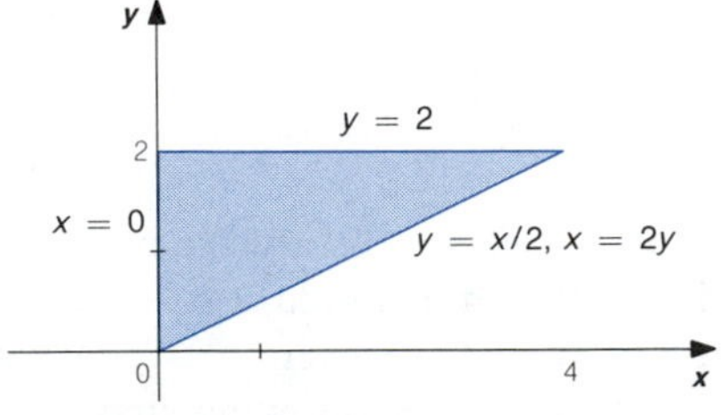

This integral has the same value as the given integral, since both iterated integrals represent the same double integral.

We cannot use the Fundamental Theorem of Calculus to evaluate the original iterated integral, because the indefinite integral $\int e^{y^2}\,dy$ cannot be expressed in terms of any of the usual functions. However, we do have

$$\int_0^2 \int_0^{2y} e^{y^2}\,dx\,dy = \int_0^2 xe^{y^2}\Big]_{x=0}^{x=2y} dy$$

$$= \int_0^2 2y\,e^{y^2}\,dy \qquad [u = y^2,\ du = 2y\,dy]$$

$$= e^{y^2}\Big]_0^2 = e^4 - 1.$$

■

It is worthwhile to consider different methods of evaluating a double integral, since some methods may be significantly easier than others.

## EXERCISES 17.2

Evaluate the integrals in Exercises 1–16.

1 $\int_0^1 \int_0^x xy\,dy\,dx$

2 $\int_0^1 \int_0^{\sqrt{x}} xy\,dy\,dx$

3 $\int_0^1 \int_{\sqrt{x}}^1 y\,dy\,dx$

4 $\int_0^1 \int_x^1 x\,dy\,dx$

5 $\int_0^{\pi/2} \int_0^y \cos x \sin y\,dx\,dy$

6 $\int_0^{\pi/2} \int_0^{\sin y} x \sin y\,dx\,dy$

7 $\int_0^{\pi/4} \int_0^y \sec x \tan x\,dx\,dy$

8 $\int_0^{\pi/4} \int_0^y \sec^2 x\,dx\,dy$

9 $\iint_R x^2\,dA$, where $R$ is the region bounded by $x + y = 4$, $y = 0$, $x = 0$, and $x = 2$

10 $\iint_R x^2\,dA$, where $R$ is the region bounded by $y = 3 - x^2$ and $y = 0$

11 $\iint_R \sqrt{y}\,dA$, where $R$ is the region bounded by $y = x^2$, $y = 4$, and $x = 0$ $(x \geq 0)$

12 $\iint_R \cos y\,dA$, where $R$ is the region bounded by $y = x/2$, $y = 0$, and $x = \pi/2$

13 $\iint_R e^{x/2}\,dA$, where $R$ is the region bounded by $x + y = 4$, $y = 0$, $y = 2$, and $x = 0$

14 $\iint_R 2xy\,dA$, where $R$ is the region bounded by $x + y = 0$, $y = x/2$, and $x = 2$

15 $\iint_R 2xy\,dA$, where $R$ is the region bounded by $y = 2x$, $y = 2$, and $x = 0$

16 $\iint_R x^2\,dA$, where $R$ is the region bounded by $y = \sin^{-1} x$, $y = \pi/2$, and $x = 0$

Use double integrals to find the area of the regions bounded by the equations given in Exercises 17–24.

17 $y = \cos x$, $y = 1/2$ $(-\pi/3 \leq x \leq \pi/3)$

18 $y = \sin x$, $y = 2x/\pi$ $(0 \leq x \leq \pi/2)$

19 $x = 4 - y^2$, $x = 0$

20 $x = y^2$, $x - y = 2$

21 $y = x$, $y = x - 1$, $y = 0$, $x = 2$

22 $y - 1 = x/2$, $y + 2 = 2x$, $x = 0$, $y = 0$

23 $y = x(x - 2)(x + 1)$, $y = 0$

24 $y = \cos x$, $y = 1 - (2x/\pi)$, $x = 0$, $x = 2\pi/3$

Express the integrals in Exercises 25–32 as iterated double integrals with order of integration opposite to that given and evaluate.

25 $\int_0^2 \int_{x^2}^{2x} xy\,dy\,dx$

26 $\int_0^3 \int_{2(3-x)/3}^2 (x + y)\,dy\,dx$

**27** $\int_0^{\pi/2} \int_{\sin x}^{1} \cos x \, dy \, dx$

**28** $\int_0^{\pi/2} \int_0^{\cos x} \sin x \, dy \, dx$

**29** $\int_{1/2}^{1} \int_{\sin^{-1} x}^{\pi/2} \cot y \, dy \, dx$

**30** $\int_0^{1} \int_{\tan^{-1} x}^{\pi/4} \cos y \, dy \, dx$

**31** $\int_1^{e} \int_{\ln x}^{1} x \, dy \, dx$

**32** $\int_0^{1} \int_x^{1} e^{y^2} \, dy \, dx$

**33** Why does the double integral of $e^{xy}$ over $R$ exist, where $R$ is the region that consists of points $(x, y)$ with $0 \le x \le 1$ and $0 \le y \le 1$? We have

$$\iint_R e^{xy} \, dA = \int_0^1 \int_0^1 e^{xy} \, dy \, dx.$$

Evaluate $\int_0^1 e^{xy} \, dy$ for $0 < x \le 1$ and for $x = 0$. Is $\int_0^1 e^{xy} \, dy$ a continuous function of $x$?

## 17.3 APPLICATIONS OF DOUBLE INTEGRALS

We will see how to use double integrals to find both the volume of regions and the area of surfaces in three-dimensional space. We will also see how to find the mass, moments, and centers of mass of laminas that have variable density.

### *Volume*

Consider a region in three-dimensional space that is generated by vertical line segments between the surfaces $z = h_1(x, y)$ and $z = h_2(x, y)$ as $(x, y)$ varies over a region $R$ in the $xy$-plane. We assume that $h_1$ and $h_2$ are continuous with $h_1(x, y) \le h_2(x, y)$ for $(x, y)$ in $R$. $R$ is the projection of the region in space onto the $xy$-plane. See Figure 17.3.1a. Let us divide the region $R$ in the $xy$-plane into rectangles with area $\Delta A$. That part of the region in three-dimensional space that is above each rectangle is essentially a rectangular column with volume

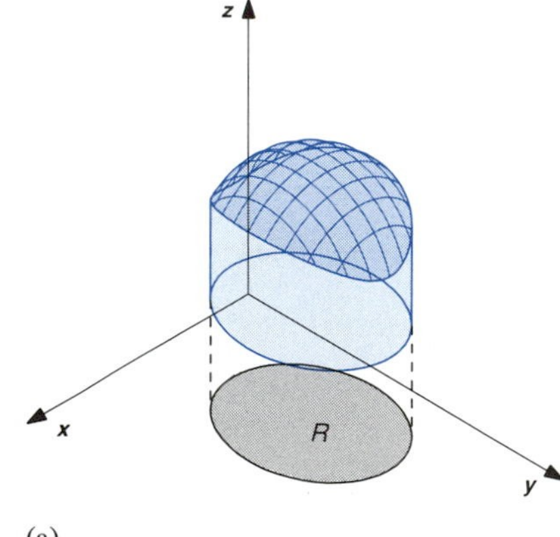

(a)

$$\Delta V \approx (\text{Height})(\text{Area of base}) \approx [h_2(x, y) - h_1(x, y)] \, \Delta A.$$

See Figure 17.3.1b. The height of the column depends on the $(x, y)$ coordinates of its position. The total volume of the region is the sum of the volumes of its parts, so

$$\text{Volume} = \sum(\text{Volume of columns}) \approx \sum[h_2(x, y) - h_1(x, y)] \, \Delta A.$$

The latter sum is a Riemann sum of the height function $h_2 - h_1$ over the projection of the region in space onto the $xy$-plane. It should seem reasonable that the volume of the region in space should be given by the corresponding double integral,

$$\text{Volume} = \iint_R (h_2 - h_1) \, dA.$$

(b)

FIGURE 17.3.1

If the region $R$ is simple in the direction of either the $x$-axis or the $y$-axis, we can express this volume as an iterated double integral.

FIGURE 17.3.2

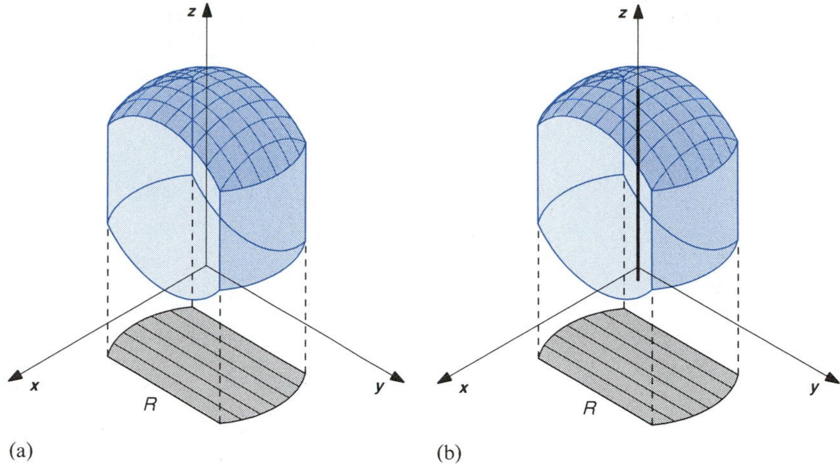


Suppose a region in three-dimensional space can be described by the statements

For $x$ and $y$ fixed, $z$ varies from $z = h_1(x, y)$ to $z = h_2(x, y)$.
For $x$ fixed, $y$ varies from $y = g_1(x)$ to $y = g_2(x)$.
$x$ varies from $x = a$ to $x = b$.

The surfaces $z = h_1(x, y)$ and $z = h_2(x, y)$ form the bottom and top of the region. The surfaces $y = g_1(x)$, $y = g_2(x)$, $x = a$, and $x = b$ form the vertical sides of the region and give the projection of the region onto the $xy$-plane. The projection of the region onto the $xy$-plane is a region $R$ of the $xy$-plane that is simple in the direction of the $y$-axis. See Figure 17.3.2a. As $z$ varies from $z = h_1(x, y)$ to $z = h_2(x, y)$ with $x$ and $y$ fixed, the points $(x, y, z)$ form a vertical line segment between the top and bottom of the region. See Figure 17.3.2b. As $(x, y)$ varies over the projection $R$, these line segments generate the entire region in space. Since the double integral over $R$ can be expressed as an iterated integral, we have

$$\textbf{Volume} = \iint_R (h_2 - h_1)\, dA = \int_a^b \int_{g_1(x)}^{g_2(x)} [h_2(x, y) - h_1(x, y)]\, dy\, dx.$$

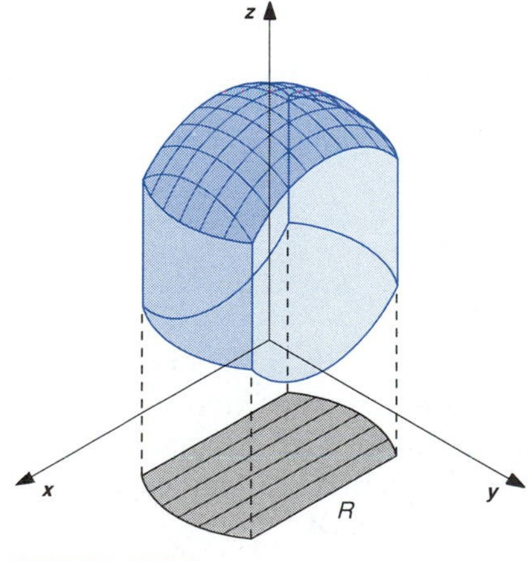

FIGURE 17.3.3

Note that the equations used in the description of the region give the integrand and the limits of integration of the volume integral. Ordinarily, the equations of all six boundary surfaces are not given. We will use the procedure of Section 17.1 to determine missing equations. The same ideas apply if the roles of $x$ and $y$ are interchanged, so the projection of the region onto the $xy$-plane is a region of the $xy$-plane that is simple in the direction of the $x$-axis. See Figure 17.3.3.

## ■ EXAMPLE 1

Use a double integral to find the volume of the region in the first octant bounded by $x + 2y + 3z = 6$.

FIGURE 17.3.4

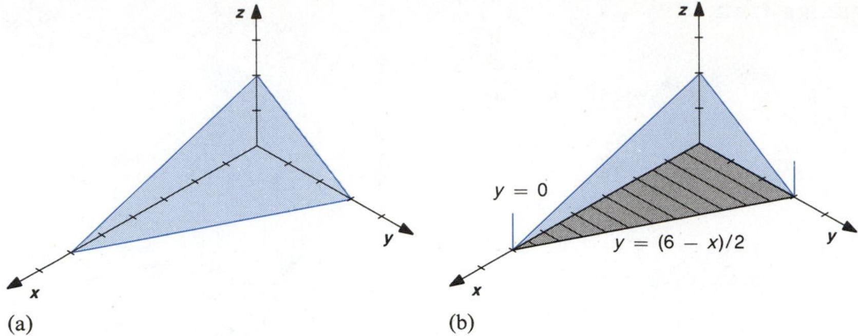


(a) (b)

**SOLUTION**

The region is sketched in Figure 17.3.4a. Let us use the method of Section 17.1 to describe the region. We see that:

For $x$ and $y$ fixed, $z$ varies from $z = 0$ to $z = \dfrac{6 - x - 2y}{3}$.

The top and bottom intersect to form a right-side edge. The equation of the vertical generalized cylinder that contains this edge is obtained by eliminating $z$ from the equations of the top and bottom.

$z = 0$ and $z = \dfrac{6 - x - 2y}{3}$ imply

$$6 - x - 2y = 0,$$

$$y = \frac{6 - x}{2}.$$

For $x$ fixed, $y$ varies from $y = 0$ to $y = \dfrac{6 - x}{2}$.

The sides intersect to form a front point.

$y = 0$ and $y = \dfrac{6 - x}{2}$ imply $0 = 6 - x$, so $x = 6$.

$x$ varies from $x = 0$ to $x = 6$.

The volume of the region is the double integral of the height function over the projection of the region onto the $xy$-plane. We can now use the description of the region to set up an iterated double integral that gives the volume of the region. The height of a variable vertical column is given by the equations of the top and bottom of the region. We have

$$\text{Height} = (\text{Top}) - (\text{Bottom}) = \frac{6 - x - 2y}{3}.$$

The equations that describe the two sides and the front and back boundaries of the region in three-dimensional space give a description of the projection of the region in space onto the $xy$-plane. This description corresponds to a region of the $xy$-plane that is simple in the direction of the $y$-axis. See Figure 17.3.4b. We then have

$$
\begin{aligned}
\text{Volume} &= \int_0^6 \int_0^{(6-x)/2} \frac{6 - x - 2y}{3}\, dy\, dx \\
&= \int_0^6 \int_0^{(6-x)/2} \left(2 - \frac{x}{3} - \frac{2y}{3}\right) dy\, dx \\
&= \int_0^6 \left[2y - \frac{xy}{3} - \frac{y^2}{3}\right]_{y=0}^{y=(6-x)/2} dx \\
&= \int_0^6 \left[(6 - x) - \frac{x(6-x)}{6} - \frac{(6-x)^2}{12}\right] dx \\
&= \int_0^6 \left(6 - x - x + \frac{x^2}{6} - 3 + x - \frac{x^2}{12}\right) dx \\
&= \int_0^6 \left(\frac{x^2}{12} - x + 3\right) dx \\
&= \frac{x^3}{36} - \frac{x^2}{2} + 3x\Big]_0^6 \\
&= 6 - 18 + 18 = 6.
\end{aligned}
$$

■

FIGURE 17.3.5

The projection of the region of Example 1 onto the $xy$-plane is a region of the $xy$-plane that is simple in the direction of the $x$-axis as well as simple in the direction of the $y$-axis. See Figure 17.3.5. The projection can be described by the following statements:

For $y$ fixed, $x$ varies from $x = 0$ to $x = 6 - 2y$.
$y$ varies from $y = 0$ to $y = 3$.

The volume of the region can then be expressed as

$$\text{Volume} = \int_0^3 \int_0^{6-2y} \frac{6 - x - 2y}{3}\, dx\, dy.$$

This integral has value 6.

(a)

■ **EXAMPLE 2**

Use a double integral to find the volume of the region in the first octant bounded by $y + z = 2$ and $x + y - z = 0$.

**SOLUTION**

The region is sketched in Figure 17.3.6a. Let us describe the region. We see that:

For $x$ and $y$ fixed, $z$ varies from $z = x + y$ to $z = 2 - y$.

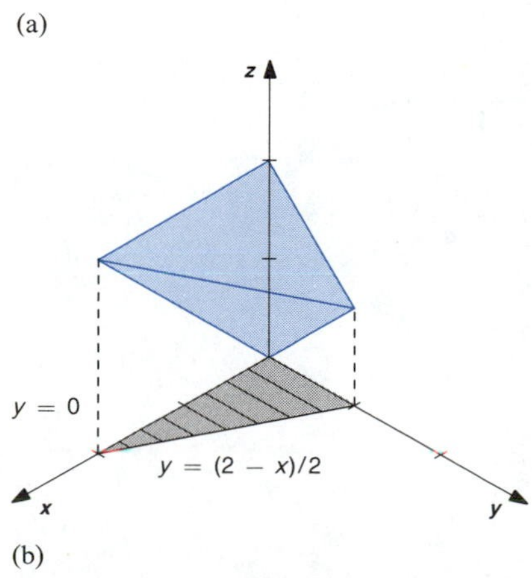


(b)

FIGURE 17.3.6

The top and bottom intersect to form a right-side edge.

$z = x + y$ and $z = 2 - y$ imply

$x + y = 2 - y,$

$2y = 2 - x,$

$y = \dfrac{2 - x}{2}.$

For $x$ fixed, $y$ varies from $y = 0$ to $y = \dfrac{2 - x}{2}$.

The sides intersect to form a front point.

$y = 0$ and $y = \dfrac{2 - x}{2}$ imply $0 = 2 - x$, so $x = 2$.
$x$ varies from $x = 0$ to $x = 2$.

The volume of the region is the double integral of the height function over the projection of the region onto the $xy$-plane. The height of a variable vertical column is

Height = (Top) − (Bottom) = $(2 - y) - (x + y) = 2 - x - 2y.$

The equations of the sides, front, and back of the region in three-dimensional space give limits of integration for an iterated double integral over the projection of the region onto the $xy$-plane. See Figure 17.3.6b. We have

$$\begin{aligned}
\text{Volume} &= \int_0^2 \int_0^{(2-x)/2} (2 - x - 2y)\, dy\, dx \\
&= \int_0^2 \Big[ 2y - xy - y^2 \Big]_{y=0}^{y=(2-x)/2} dx \\
&= \int_0^2 \left[ (2 - x) - \frac{x(2 - x)}{2} - \frac{(2 - x)^2}{4} \right] dx \\
&= \int_0^2 \left( 2 - x - x + \frac{x^2}{2} - 1 + x - \frac{x^2}{4} \right) dx \\
&= \int_0^2 \left( 1 - x + \frac{x^2}{4} \right) dx \\
&= x - \frac{x^2}{2} + \frac{x^3}{12} \Big]_0^2 \\
&= \left( 2 - 2 + \frac{8}{12} \right) - (0) = \frac{2}{3}.
\end{aligned}$$

■

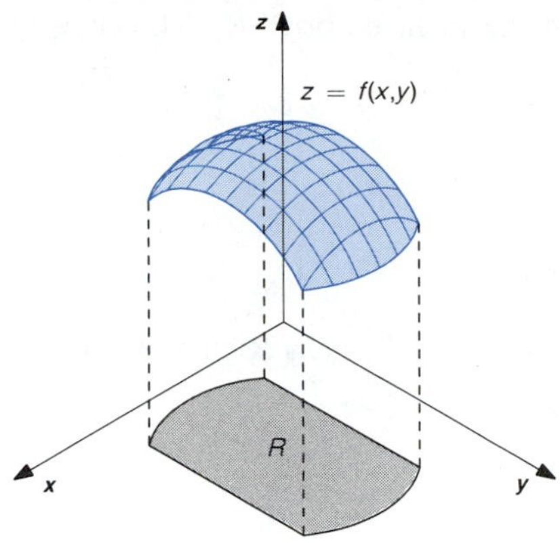

FIGURE 17.3.7

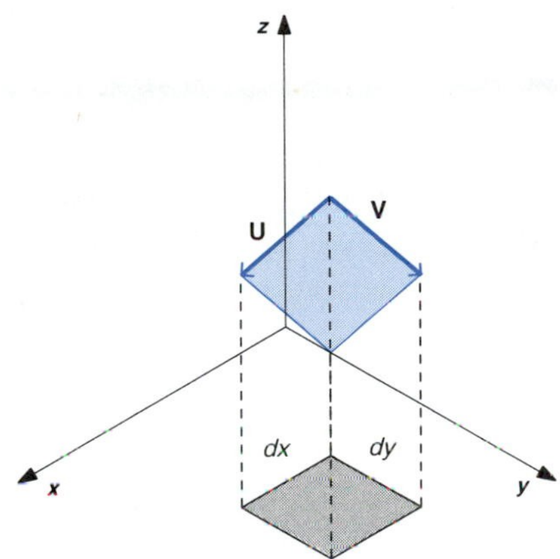

FIGURE 17.3.8

### *Surface area*

Consider a surface given by

$$z = f(x, y), \qquad (x, y) \text{ in the region } R.$$

See Figure 17.3.7. We assume that $f$ has continuous first partial derivatives for $(x, y)$ in $R$ and on the boundary of $R$, so $f$ is differentiable and the surface has a tangent plane at each point. The area of that part of the surface above a small rectangle $R_j$ with one corner at $(x_j, y_j)$ and opposite corner at $(x_j + dx, y_j + dy)$ is approximately the area of the part above $R_j$ of the plane tangent to the surface at $(x_j, y_j)$. See Figure 17.3.8. Let us use vectors to find the area of that part of the tangent plane.

The vector

$$\mathbf{R} = x\mathbf{i} + y\mathbf{j} + f(x, y)\mathbf{k}$$

gives the position of the point on the surface $z = f(x, y)$ that is above the point $(x, y)$ in the $xy$-plane. The differential change in position due to changes of $dx$ in $x$ and $dy$ in $y$ is

$$d\mathbf{R} = dx\mathbf{i} + dy\mathbf{j} + \left(\frac{\partial f}{\partial x}\,dx + \frac{\partial f}{\partial y}\,dy\right)\mathbf{k}.$$

The vector

$$\mathbf{U} = dx\mathbf{i} + f_x(x_j, y_j)(dx)\mathbf{k}$$

corresponds to a change in the $xy$-plane from $(x_j, y_j)$ to $(x_j + dx, y_j)$. The vector

$$\mathbf{V} = dy\mathbf{j} + f_y(x_j, y_j)(dy)\mathbf{k}$$

corresponds to a change in the $xy$-plane from $(x_j, y_j)$ to $(x_j, y_j + dy)$. See Figure 17.3.8. The area of that part of the surface $z = f(x, y)$ above the rectangle $R_j$ is essentially the area of the parallelogram formed by $\mathbf{U}$ and $\mathbf{V}$. From Section 14.2 we know that this area is given by $|\mathbf{U} \times \mathbf{V}|$. We have

$$\begin{aligned}
\mathbf{U} \times \mathbf{V} &= \begin{vmatrix} \mathbf{i} & \mathbf{j} & \mathbf{k} \\ dx & 0 & f_x(x_j, y_j)(dx) \\ 0 & dy & f_y(x_j, y_j)(dy) \end{vmatrix} \\
&= \begin{vmatrix} 0 & f_x(x_j, y_j)(dx) \\ dy & f_y(x_j, y_j)(dy) \end{vmatrix}\mathbf{i} - \begin{vmatrix} dx & f_x(x_j, y_j)(dx) \\ 0 & f_y(x_j, y_j)(dy) \end{vmatrix}\mathbf{j} + \begin{vmatrix} dx & 0 \\ 0 & dy \end{vmatrix}\mathbf{k} \\
&= -f_x(x_j, y_j)(dx)(dy)\mathbf{i} - f_y(x_j, y_j)(dx)(dy)\mathbf{j} + (dx)(dy)\mathbf{k}.
\end{aligned}$$

The area of that part of the tangent plane above the rectangle $R_j$ is then

$$|\mathbf{U} \times \mathbf{V}| = \sqrt{[f_x(x_j, y_j)]^2 + [f_y(x_j, y_j)]^2 + 1}\;dx\,dy.$$

This area is approximately the area of the surface above $R_j$. That is,

$$\Delta S_j \approx |\mathbf{U} \times \mathbf{V}| = \sqrt{[f_x(x_j, y_j)]^2 + [f_y(x_j, y_j)]^2 + 1}\, dx\, dy.$$

It follows that the area of the surface over the region $R$ is

$$S = \sum \Delta S_j \approx \sum \sqrt{[f_x(x_j, y_j)]^2 + [f_y(x_j, y_j)]^2 + 1}\, dx\, dy,$$

where the sum is taken over all rectangles in $R$. This is a Riemann sum, so it should seem reasonable to define

$$S = \iint_R \sqrt{[f_x(x, y)]^2 + [f_y(x, y)]^2 + 1}\, dA.$$

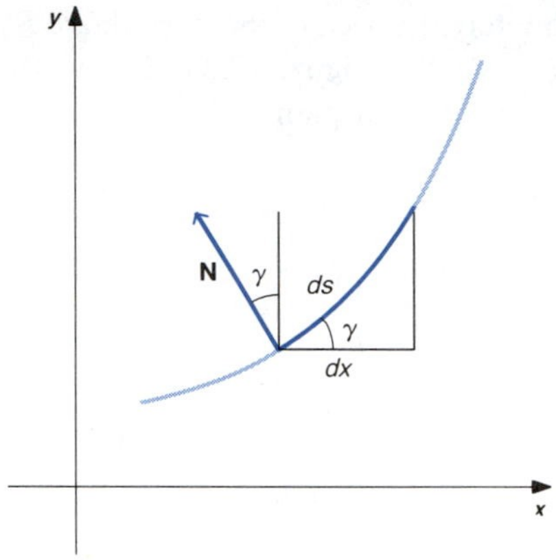

The formula for surface area given previously should be compared with the formula

$$s = \int_a^b \sqrt{[f'(x)]^2 + 1}\, dx$$

for the arc length of the curve $y = f(x)$, $a \le x \le b$. From Figure 17.3.9 we see that the ratio of differential arc length of the curve $y = f(x)$ and length along the $x$-axis is given by

FIGURE 17.3.9

$$\sec \gamma = \frac{ds}{dx} = \sqrt{(f')^2 + 1},$$

where $\gamma$ is the angle between a normal to the curve and the vertical. Let us see that the ratio of differential surface area of the surface $z = f(x, y)$ and area in the $xy$-plane is

$$\sec \gamma = \sqrt{(f_x)^2 + (f_y)^2 + 1},$$

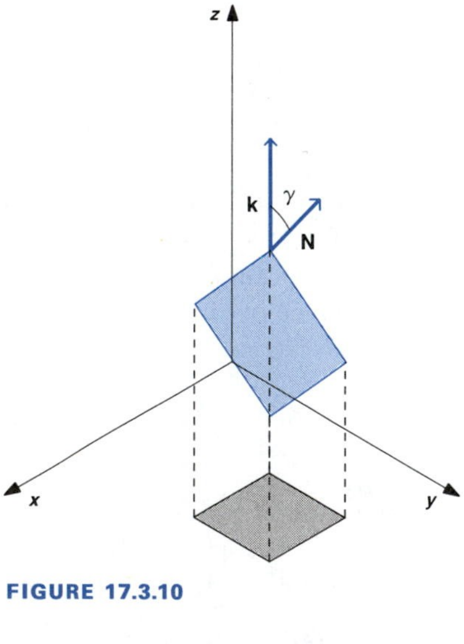

where $\gamma$ is the angle between a normal to the surface and the vertical. See Figure 17.3.10. That is, the vector

$$\mathbf{N} = -f_x\mathbf{i} - f_y\mathbf{j} + \mathbf{k}$$

is normal to the surface $z = f(x, y)$. [$\mathbf{N} = \nabla F$, where $F(x, y, z) = z - f(x, y)$.] If $\gamma$ is the angle between $\mathbf{N}$ and the vertical vector $\mathbf{k}$, we have

$$\cos \gamma = \frac{\mathbf{N} \cdot \mathbf{k}}{|\mathbf{N}||\mathbf{k}|} = \frac{1}{\sqrt{(f_x)^2 + (f_y)^2 + 1}}, \quad \text{so}$$

FIGURE 17.3.10

$$\sec \gamma = \sqrt{(f_x)^2 + (f_y)^2 + 1}.$$

■ **EXAMPLE 3**

Find the surface area of that part of the surface

$$z = \frac{4}{3} - \frac{2}{3}x^{3/2} - \frac{2}{3}y^{3/2}$$

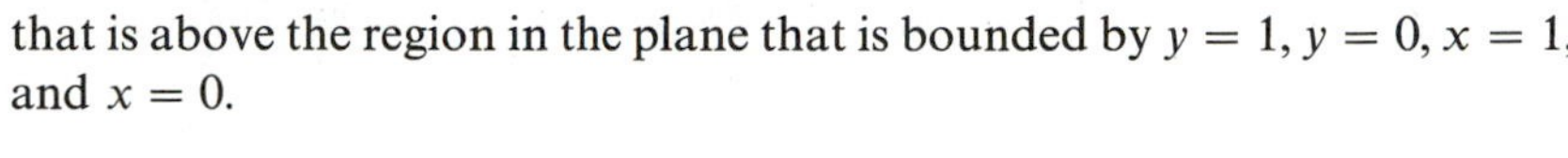

that is above the region in the plane that is bounded by $y = 1$, $y = 0$, $x = 1$, and $x = 0$.

**SOLUTION**

The surface is sketched in Figure 17.3.11. We need the partial derivatives of $z$:

$$z_x = -x^{1/2} \quad \text{and} \quad z_y = -y^{1/2}.$$

Then

$$\begin{aligned}
S &= \iint_R \sqrt{z_x^2 + z_y^2 + 1}\, dA \\
&= \int_0^1 \int_0^1 \sqrt{x + y + 1}\, dy\, dx \\
&= \int_0^1 \frac{2}{3}(x + y + 1)^{3/2}\Big]_{y=0}^{y=1} dx \\
&= \int_0^1 \left[\frac{2}{3}(x + 2)^{3/2} - \frac{2}{3}(x + 1)^{3/2}\right] dx \\
&= \frac{4}{15}(x + 2)^{5/2} - \frac{4}{15}(x + 1)^{5/2}\Big]_0^1 \\
&= \left(\frac{4}{15}3^{5/2} - \frac{4}{15}2^{5/2}\right) - \left(\frac{4}{15}2^{5/2} - \frac{4}{15}1^{5/2}\right) \\
&= \frac{4}{15}[3^{5/2} - 2(2^{5/2}) + 1] \\
&= \frac{4}{15}(3^{5/2} - 2^{7/2} + 1) \approx 1.4066.
\end{aligned}$$

■

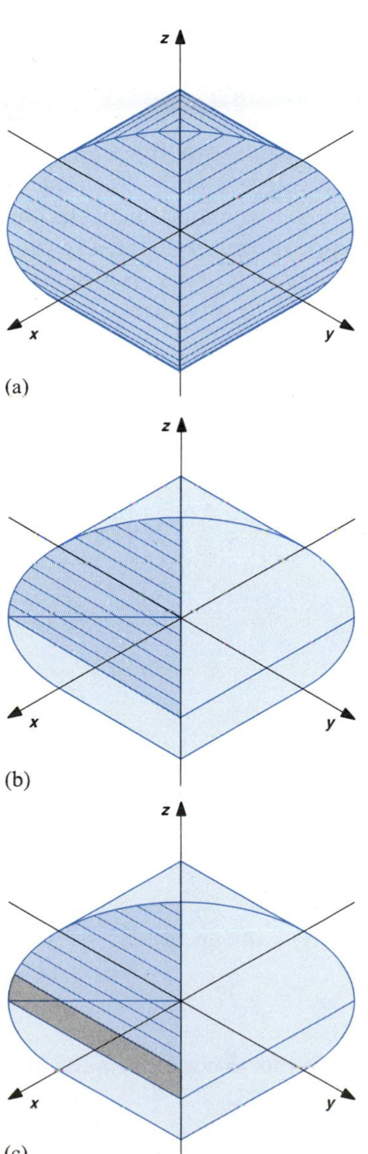



FIGURE 17.3.11

(a)

(b)

(c)

FIGURE 17.3.12

■ **EXAMPLE 4**

Find the surface area of the solid that is bounded by the cylinders $x^2 + z^2 = r^2$ and $y^2 + z^2 = r^2$.

**SOLUTION**

The solid is sketched in Figure 17.3.12a. We see that the total surface area is eight times the area of that part of the surface indicated in Figure 17.3.12b. We need equations of the region in the $xy$-plane below that part of the surface. To find these equations, we note that coordinates of points on the intersection of the cylinders satisfy both

$$x^2 + z^2 = r^2 \quad \text{and} \quad y^2 + z^2 = r^2, \quad \text{so} \quad x^2 = y^2.$$

The total surface area is eight times that part of the surface $z = \sqrt{r^2 - x^2}$ above the region of the $xy$-plane bounded by $y = -x$, $y = x$, and $x = r$.

We have

$$z_x = \frac{-x}{\sqrt{r^2 - x^2}} \quad \text{and} \quad z_y = 0, \quad \text{so}$$

$$\sqrt{z_x^2 + z_y^2 + 1} = \sqrt{\frac{x^2}{r^2 - x^2} + 1}$$
$$= \frac{r}{\sqrt{r^2 - x^2}}.$$

The surface area should be

$$S = 8 \int_0^r \int_{-x}^{x} \frac{r}{\sqrt{r^2 - x^2}}\, dy\, dx.$$

However, we see that the integrand becomes unbounded as $(x, y)$ approaches the line $x = r$, so the formula we developed does not apply directly. We can overcome this difficulty by treating the integral as an improper integral. That is, we evaluate the integral with the limit of integration $r$ replaced by a number $b$ with $0 < b < r$ and then take the limit as $b$ approaches $r$ from the left. This means we omit part of the surface in the evaluation of the surface area; the surface area of the part omitted approaches 0 as $b$ approaches $r$. See Figure 17.3.12c. We have

$$8 \int_0^b \int_{-x}^{x} \frac{r}{\sqrt{r^2 - x^2}}\, dy\, dx$$
$$= 8 \int_0^b \frac{r2x}{\sqrt{r^2 - x^2}}\, dx$$

$[u = r^2 - x^2,\ du = -2x\, dx]$

$$= -16r\sqrt{r^2 - x^2}\Big]_0^b$$
$$= -16r\sqrt{r^2 - b^2} + 16r^2.$$

$$\lim_{b \to r^-} \left(-16r\sqrt{r^2 - b^2} + 16r^2\right) = 16r^2,$$

so the surface area is $16r^2$. ■

### *Mass, moments, and center of mass*

The **mass** of an object that has constant density is given by

Mass = (Density)(Volume).

If the density varies continuously, we can divide the object into small parts with nearly constant density. Then

Total mass = $\sum$ (Mass of parts).

If an object is thin and flat (i.e., a lamina), we may think of the object as occupying a region in a plane, rather than a region in space. We may then describe the density of the lamina as density per unit area. Dividing the region into small parts and adding the masses leads to a double integral. That is, let

$\rho(x, y) =$ Density per unit area at $(x, y)$.

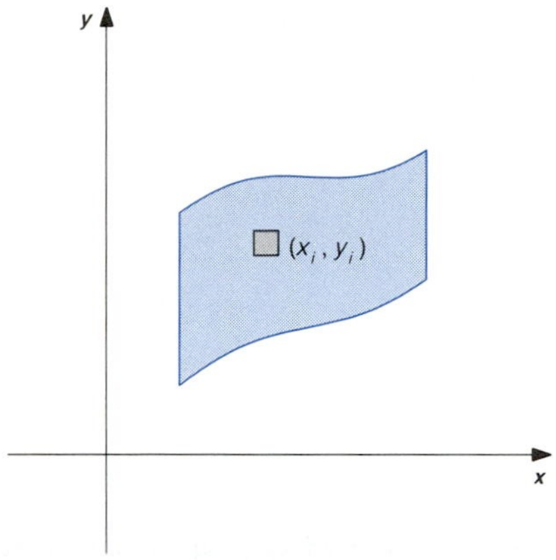

The mass of a rectangular piece of the lamina located at $(x_i, y_j)$ is approximately the density per unit area at the point $(x_i, y_j)$ times the area of the rectangle. That is,

$$\Delta M \approx \rho(x_i, y_j)\, \Delta A_{ij}.$$

See Figure 17.3.13. The differential mass is

$$dM = \rho(x, y)\, dA$$

FIGURE 17.3.13

and total mass is given by

$$M = \iint_R dM = \iint_R \rho(x, y)\, dA.$$

■ **EXAMPLE 5**

Find the mass of the lamina bounded by

$$y = \sqrt{4 - x^2} \quad \text{and} \quad y = 0$$

if the density is given by $\rho(x, y) = y^2$.

**SOLUTION**

From the sketch in Figure 17.3.14 we see that:

For $x$ fixed, $y$ varies from $y = 0$ to $y = \sqrt{4 - x^2}$.
$x$ varies from $x = -2$ to $x = 2$.

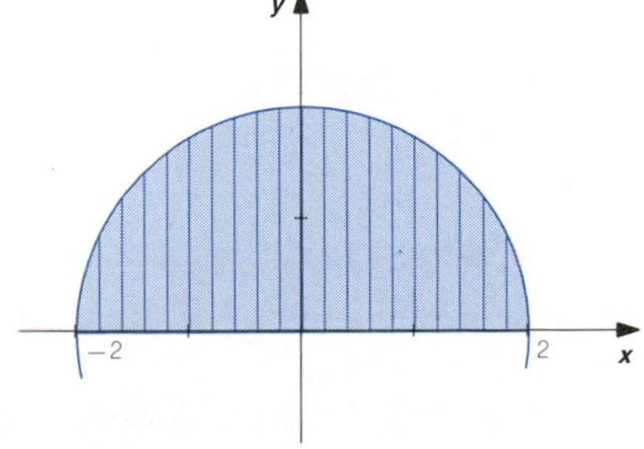


FIGURE 17.3.14

The mass of the lamina is

$$M = \iint_R dM = \iint_R \rho\, dA$$

$$= \int_{-2}^{2} \int_{0}^{\sqrt{4-x^2}} y^2\, dy\, dx$$

$$= \int_{-2}^{2} \frac{y^3}{3}\Bigg]_{y=0}^{y=\sqrt{4-x^2}} dx$$

$$= \int_{-2}^{2} \frac{1}{3}(4 - x^2)^{3/2}\, dx.$$

Using the table of integrals to evaluate the integral, we obtain

$$M = \frac{1}{3}\left[-\frac{x}{8}(2x^2 - 5(4))\sqrt{4 - x^2} + \frac{3(16)}{8}\sin^{-1}\left(\frac{x}{2}\right)\right]_{-2}^{2}$$

$$= \frac{1}{3}\left[-\frac{2}{8}(2(2)^2 - 5(4))\sqrt{4 - 2^2} + \frac{3(16)}{8}\sin^{-1}\left(\frac{2}{2}\right)\right]$$

$$- \frac{1}{3}\left[-\frac{-2}{8}(2(-2)^2 - 5(4))\sqrt{4 - (-2)^2} + \frac{3(16)}{8}\sin^{-1}\left(\frac{-2}{2}\right)\right]$$

$$= 2\sin^{-1}(1) - 2\sin^{-1}(-1) = 2\left(\frac{\pi}{2}\right) - 2\left(-\frac{\pi}{2}\right) = 2\pi.$$

■

The **moment** of a point mass in the plane about one of the coordinate axes in the plane is the product of the mass times the directed distance of the mass from the axis. The total moment of a system of point masses is the sum of the moments of each mass. This leads to the following formulas for moments of a lamina about the axes.

$$M_{y=0} = \iint_R y\, dM, \qquad M_{x=0} = \iint_R x\, dM.$$

These moments are related to the tendency of the lamina to rotate about each coordinate axis.

The **center of mass** (balance point) of a lamina is $(\bar{x}, \bar{y})$, where

$$\bar{x} = \frac{M_{x=0}}{M} = \frac{\iint_R x\, dM}{\iint_R dM} \quad \text{and} \quad \bar{y} = \frac{M_{y=0}}{M} = \frac{\iint_R y\, dM}{\iint_R dM}.$$

■ **EXAMPLE 6**

Find the center of mass of the lamina bounded by

$y = x^2$ and $y = 2x$ if the density is given by $\rho(x, y) = 840xy$.

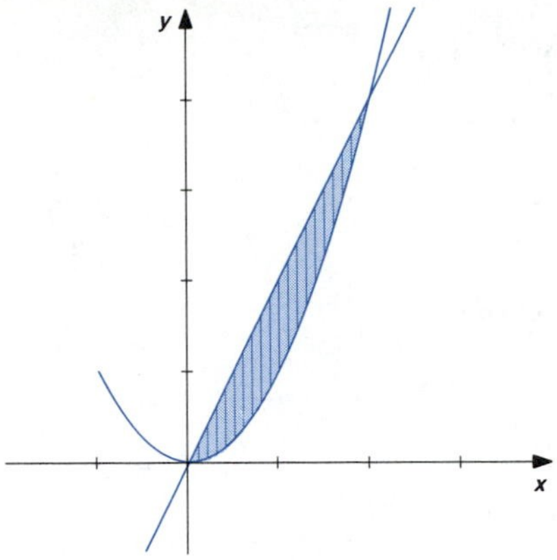

FIGURE 17.3.15

**SOLUTION**

The lamina is sketched in Figure 17.3.15. We need to determine the limits of integration for iterated integrals over the region. We see:

For $x$ fixed, $y$ goes from $y = x^2$ to $y = 2x$.

We need the $x$-coordinate of each point of intersection of the top and bottom curves.

$y = x^2$ and $y = 2x$ imply $x^2 = 2x$, so

$x^2 - 2x = 0,$

$x(x - 2) = 0,$

$x = 0, \quad x = 2.$

Hence,

$x$ goes from $x = 0$ to $x = 2$.

The mass of a small piece of the lamina is

$dM = \rho dA = 840xy\, dy\, dx.$

We then have

$$M = \iint_R dM = \int_0^2 \int_{x^2}^{2x} 840xy\, dy\, dx$$

$$= \int_0^2 420xy^2 \Big]_{y=x^2}^{y=2x} dx$$

$$= \int_0^2 (1680x^3 - 420x^5)\, dx$$

$$= 420x^4 - 70x^6 \Big]_0^2$$

$$= (6720 - 4480) - (0) = 2240,$$

$$M_{x=0} = \iint_R x\, dM = \int_0^2 \int_{x^2}^{2x} 840x^2y\, dy\, dx$$

$$= \int_0^2 420x^2y^2 \Big]_{y=x^2}^{y=2x} dx$$

$$= \int_0^2 (1680x^4 - 420x^6)\, dx$$

$$= 336x^5 - 60x^7 \Big]_0^2$$

$$= (10{,}752 - 7680) - (0) = 3072,$$

$$M_{y=0} = \iint_R y\,dM = \int_0^2 \int_{x^2}^{2x} 840xy^2\,dy\,dx$$

$$= \int_0^2 280xy^3 \Big]_{y=x^2}^{y=2x} dx$$

$$= \int_0^2 (2240x^4 - 280x^7)\,dx$$

$$= 448x^5 - 35x^8 \Big]_0^2$$

$$= (14{,}336 - 8960) - (0) = 5376,$$

$$\bar{x} = \frac{M_{x=0}}{M} = \frac{3072}{2240} = \frac{48}{35} \approx 1.37 \quad \text{and} \quad \bar{y} = \frac{M_{y=0}}{M} = \frac{5376}{2240} = \frac{12}{5} = 2.4.$$

The center of mass is at the point (48/35, 12/5). ■

## EXERCISES 17.3

Use double integrals to find the volume of the regions bounded by the surfaces given in Exercises 1-6.

1 $x + z = 3$, $y = 3$, in the first octant

2 $x + 2y + 3z = 6$, in the first octant

3 Below $2x + 3y + 3z = 6$, above $z = x + y$, in the first octant

4 Below $x + z = 4$, above $z = x + 2y$, in the first octant

5 $z = 4 - x^2 - y^2$, $z = 0$

6 Inside $x^2 + y^2 = 1$, below $z = 1 - xy$, in the first octant

Find the area of each surface given in Exercises 7–12.

7 $z = 2\sqrt{x^2 + y^2}$; $0 \le x \le 2$, $0 \le y \le 3$

8 $x + 3y + 2z = 6$; in the first octant

9 $z = x + (2/3)y^{3/2}$; $0 \le x \le 1$, $0 \le y \le 1$

10 $z = \sqrt{2xy}$; $0 \le x \le 1$, $0 \le y \le 1$

11 $x^2 + z^2 = r^2$; $y = 0$, $y = h$, in the first octant

12 A sphere of radius $r$

Find the mass of each lamina indicated in Exercises 13–18.

13 $y + 2x = 4$, $y = 0$, $x = 0$; $\rho(x, y) = x$

14 $y = 4 - x^2$, $y = 0$; $\rho(x, y) = y$

15 $x + y = 2$, $x = 0$ between $y = 0$ and $y = 1$; $\rho(x, y) = 2 - y$

16 $x = y^2$, $x + y = 2$, $y = 0$, in the first quadrant; $\rho(x, y) = xy$

17 $y = \sin x$, $y = 0$, between $x = 0$ and $x = \pi$; $\rho(x, y) = y$

18 $y = \cos x$, $y = 0$, $x = 0$ $(x \ge 0)$; $\rho(x, y) = x$

Find the center of mass of each lamina indicated in Exercises 19–24.

19 $y = 0$, $y = 4$, $x = 0$, $x = 2$; $\rho(x, y) = 2y$

20 $x + y = 3$, $y = 0$, $x = 0$; $\rho(x, y) = 6xy$

21 $x = \sqrt{1 - y^2}$, $x = 0$; $\rho(x, y) = x$

22 $x^2 + y^2 = 4$, in the first quadrant; $\rho(x, y) = xy$

23 $x^2 + y^2 = 4$, $y = 0$, $y = x$, in the first quadrant; $\rho(x, y) = xy$

24 $x^2 + y^2 = 4$, in the first quadrant; $\rho(x, y) = x^2 + y^2$

## 17.4 DOUBLE INTEGRALS IN POLAR COORDINATES

Many problems involve regions and functions that can be conveniently described in terms of polar coordinates. We develop double integrals in terms of polar coordinates for these problems.

For a given plane region $R$, let us use a grid of radial lines $\theta =$ constant and circular arcs $r =$ constant to divide $R$ into "polar rectangles," as indicated in Figure 17.4.1a. It is not difficult to determine that the area of the typical polar rectangle picture in Figure 17.4.1b is

$$\left(r + \frac{\Delta r}{2}\right) \Delta r \, \Delta\theta,$$

which is essentially $r \, \Delta r \, \Delta\theta$ when $\Delta r$ and $\Delta\theta$ are small. Note that the sides of the polar rectangle are orthogonal. For small values of $\Delta r$ and $\Delta\theta$, the polar rectangle is very nearly an actual rectangle with sides of length $\Delta r$ and $r\Delta\theta$; we can think of $r \, \Delta r \, \Delta\theta$ as the area of this rectangle. We have

$$\Delta A \approx r \, \Delta r \, \Delta\theta.$$

Using the notation of differentials, we have

$$dA = r \, dr \, d\theta.$$

If each polar rectangle is small, the sum of the products of the value of a function $f$ at a point in each polar rectangle times the area of the polar rectangle should be approximately $\iint_R f \, dA$. (Recall that we defined the double integral in terms of rectangles with sides parallel to the axes. To be completely rigorous, it would be necessary to show that we arrive at the same value of the double integral by using polar rectangles. It should seem reasonable that this is true.)

Let us now consider a region $R$ that can be described in terms of polar coordinates by the following statements:

For $\theta$ fixed, $r$ goes from $r = r_1(\theta)$ to $r = r_2(\theta)$.
$\theta$ goes from $\theta = \theta_1$ to $\theta = \theta_2$.

FIGURE 17.4.1

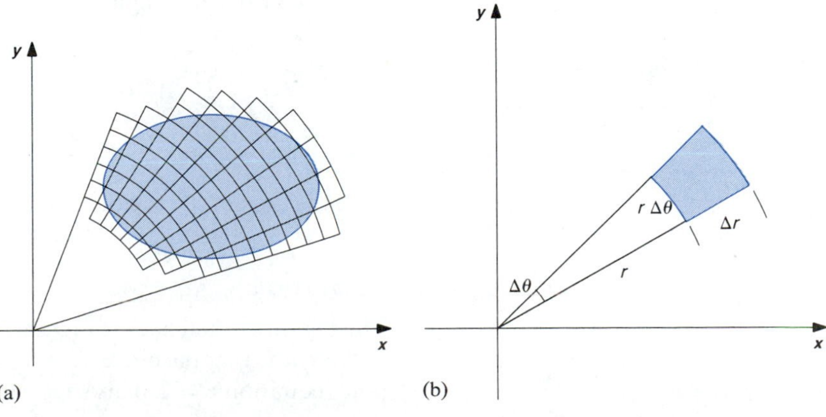

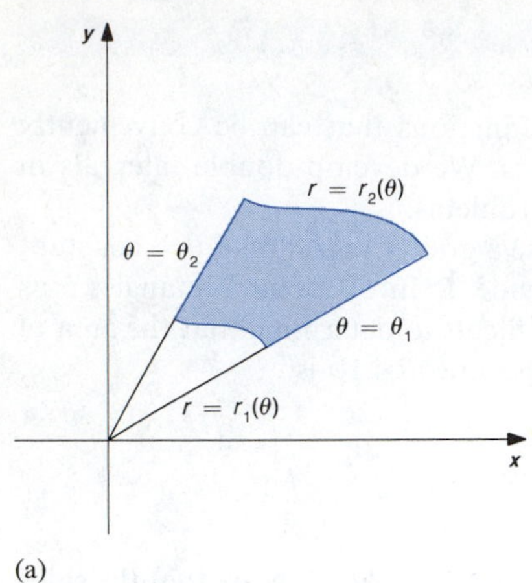


(a)

(b)

(c)

FIGURE 17.4.2

(We assume $0 \le r_1 \le r_2$, $\theta_1 \le \theta_2$.) Such a region is sketched in Figure 17.4.2a. An arrow that is shot from the pole follows a trajectory with $\theta$ fixed, enters the region at a point on the curve $r = r_1(\theta)$, and leaves the region at a point on $r = r_2(\theta)$. As $r$ varies from $r = r_1(\theta)$ to $r = r_2(\theta)$ with $\theta$ fixed, the points $(r, \theta)$ form a line segment along the ray from the origin at angle $\theta$. See Figure 17.4.2b. As $\theta$ varies from $\theta_1$ to $\theta_2$, these line segments sweep out the entire region. See Figure 17.4.2c. The description of a region in terms of polar coordinates determines the limits of integration of an iterated integral in a way that is analogous to rectangular coordinates. Using the formula for the differential area $dA$ in terms of polar coordinates, we have

$$\iint_R f\,dA = \int_{\theta_1}^{\theta_2} \int_{r_1(\theta)}^{r_2(\theta)} f(r, \theta)\, r\, dr\, d\theta.$$

Note that the integrand must be expressed in terms of polar coordinates. Recall that we can use the equations

$$x = r \cos \theta \quad \text{and} \quad y = r \sin \theta$$

to change from rectangular to polar coordinates. Also, we have

$$x^2 + y^2 = r^2.$$

■ **EXAMPLE 1**

Find the area of the region inside the circle $x^2 + y^2 = 4$ and above the line $y = 1$.

**SOLUTION**

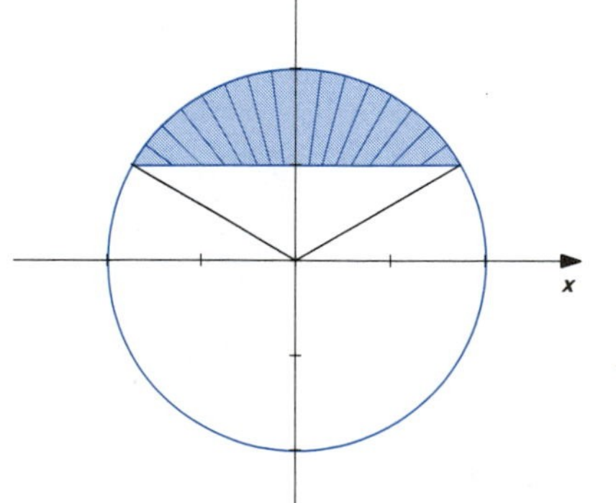


FIGURE 17.4.3

The region is sketched in Figure 17.4.3.

We know that the circle $x^2 + y^2 = 4$ can be expressed by the very simple polar equation $r = 2$. This suggests that we might want to try to use polar coordinates in this problem. The line $y = 1$ can be expressed as a polar equation by writing

$$r \sin \theta = 1, \quad \text{so} \quad r = \csc \theta.$$

From Figure 17.4.3 we see that an arrow that is shot from the pole follows a trajectory with $\theta$ fixed, enters the region at a point on the line $r = \csc \theta$, and leaves at a point on the circle $r = 2$. The intersection of the line and the circle satisfies

$$r = 2 \quad \text{and} \quad r = \csc \theta, \quad \text{so} \quad \csc \theta = 2.$$

The $\theta$-coordinates of the points of intersection are $\theta = \pi/6$ and $\theta = 5\pi/6$. The area is then given by

$$\begin{aligned} A &= \iint_R dA \\ &= \int_{\pi/6}^{5\pi/6} \int_{\csc \theta}^{2} r\, dr\, d\theta \\ &= \int_{\pi/6}^{5\pi/6} \left. \frac{r^2}{2} \right]_{r=\csc\theta}^{r=2} d\theta \\ &= \int_{\pi/6}^{5\pi/6} \frac{1}{2}(4 - \csc^2\theta)\, d\theta \\ &= \left. \frac{1}{2}(4\theta + \cot \theta) \right]_{\pi/6}^{5\pi/6} \\ &= \frac{1}{2}\left(4\frac{5\pi}{6} - \sqrt{3}\right) - \frac{1}{2}\left(4\frac{\pi}{6} + \sqrt{3}\right) \\ &= \frac{4\pi}{3} - \sqrt{3}. \end{aligned}$$

■

We can use an iterated polar integral to evaluate any double integral that occurs in applications, as long as the region can be described in terms of polar coordinates. We should be alert for regions and integrands that can be conveniently expressed in terms of polar coordinates.

### ■ EXAMPLE 2

Find the center of mass of a homogeneous half-disk of radius 1.

#### SOLUTION

We need to choose a coordinate system. It is a good idea to center the disk at the origin and to use polar coordinates. The half-disk is sketched in Figure 17.4.4. It occupies the region inside the (polar) circle $r = 1$ for $0 \le \theta \le \pi$. An arrow that is shot from the pole follows a trajectory with fixed $\theta$, enters the region at the pole $r = 0$, and leaves at a point on the circle $r = 1$.

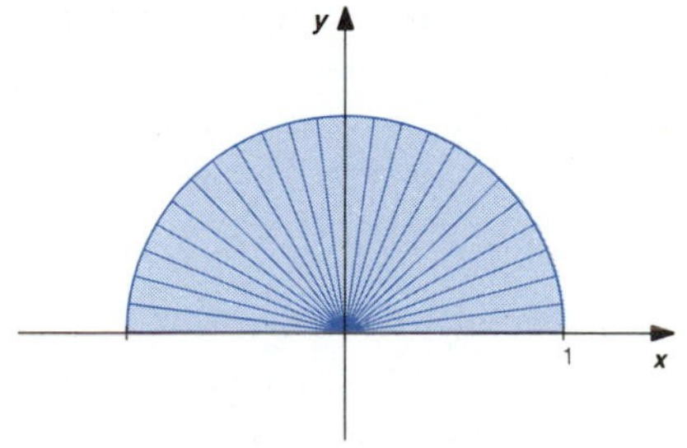

FIGURE 17.4.4

Let $k$ denote the density of the lamina. We then set up the integrals for the mass, $\bar{x}$ and $\bar{y}$ as usual, except we use iterated polar integrals to evaluate the integrals. (Don't forget that we must use $dA = r\,dr\,d\theta$ when using polar coordinates to evaluate $\iint_R f\,dA$.) The mass is

$$M = \iint_R k\,dA = \int_0^\pi \int_0^1 kr\,dr\,d\theta = \int_0^\pi \frac{kr^2}{2}\Bigg]_{r=0}^{r=1} d\theta = \int_0^\pi \frac{k}{2}\,d\theta = \frac{k\pi}{2}.$$

The moment about the $x$-axis is

$$M_{y=0} = \iint_R yk\,dA = \int_0^\pi \int_0^1 r\sin\theta\,kr\,dr\,d\theta.$$

(We used $y = r\sin\theta$ to express $y$ in terms of polar coordinates.)

$$= \int_0^\pi \int_0^1 kr^2 \sin\theta\,dr\,d\theta$$

$$= \int_0^\pi \left[k\left(\frac{r^3}{3}\right)\sin\theta\right]_{r=0}^{r=1} d\theta$$

$$= \int_0^\pi \frac{k}{3}\sin\theta\,d\theta$$

$$= \frac{k}{3}(-\cos\theta)\Bigg]_{\theta=0}^{\theta=\pi} = \frac{2k}{3}.$$

We then have

$$\bar{y} = \frac{M_{y=0}}{M} = \frac{2k/3}{k\pi/2} = \frac{4}{3\pi}.$$

By symmetry, we know that the center of mass is on the $y$-axis, so $\bar{x} = 0$. The center of mass is at the point $(0, 4/(3\pi))$. ■

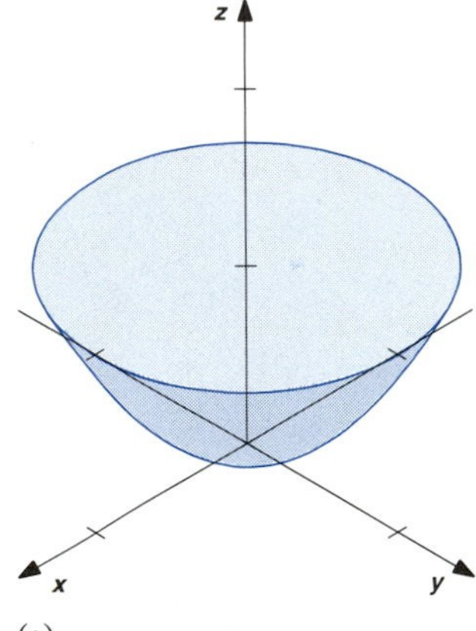

(a)

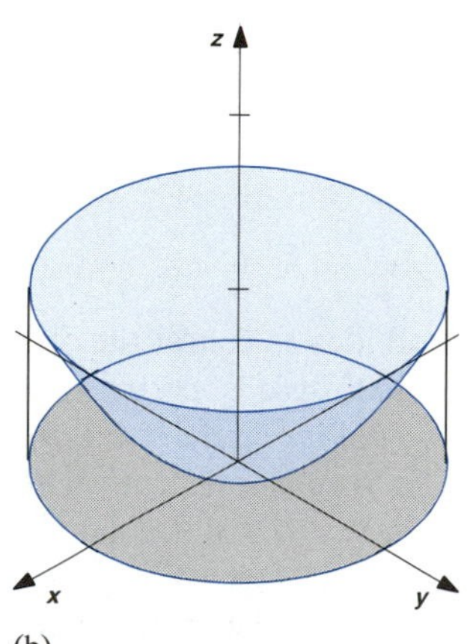

(b)

**FIGURE 17.4.5**

■ **EXAMPLE 3**

Find the surface area of the part of the paraboloid $z = x^2 + y^2$ that is below the plane $z = 1$.

**SOLUTION**

The surface is sketched in Figure 17.4.5a. The intersection of the paraboloid and the plane is contained in the cylinder $x^2 + y^2 = 1$. The intersection of this cylinder with the $xy$-plane gives the region $R$ in the $xy$-plane over which we integrate. See Figure 17.4.5b. The surface area of the part of the paraboloid $z = x^2 + y^2$ above the region $R$ is

$$S = \iint_R \sqrt{z_x^2 + z_y^2 + 1}\,dA$$

$$= \iint_R \sqrt{(2x)^2 + (2y)^2 + 1}\,dA.$$

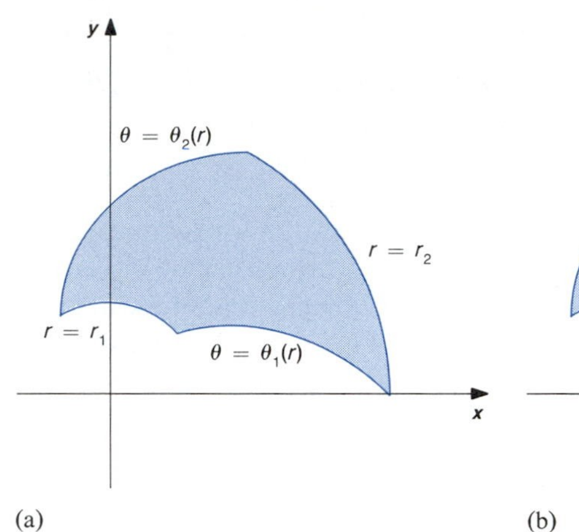


(a)

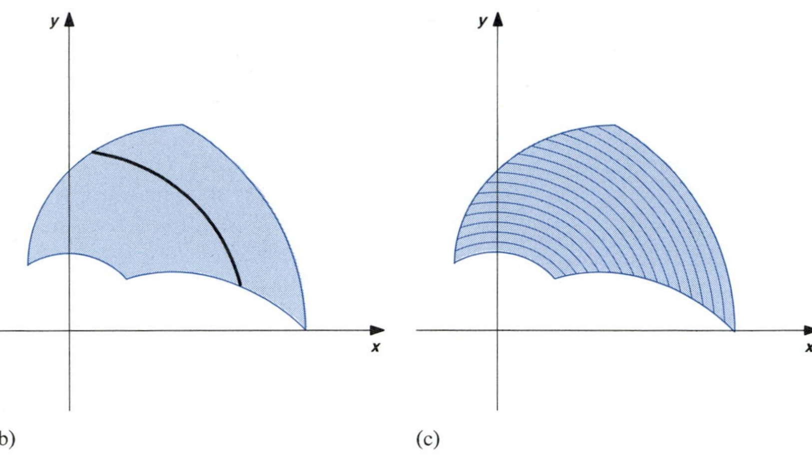


(b) (c)

**FIGURE 17.4.6**

At this point we notice that both the region of integration and the function that is being integrated are very well suited for polar coordinates. The region is bounded by the (polar) circle $r = 1$. The expression $x^2 + y^2$ in the integrand can be replaced by the polar expression $r^2$. We then have

$$S = \int_0^{2\pi} \int_0^1 \sqrt{4r^2 + 1}\, r\, dr\, d\theta \qquad [u = 4r^2 + 1,\ du = 8r\, dr]$$

$$= \int_0^{2\pi} \frac{1}{12}(4r^2 + 1)^{3/2}\Big]_{r=0}^{r=1} d\theta$$

$$= \int_0^{2\pi} \frac{1}{12}(5^{3/2} - 1)\, d\theta = \frac{\pi}{6}(5^{3/2} - 1) \approx 5.3304135. \qquad \blacksquare$$

Let us now consider a region $R$ that can be described in terms of polar coordinates by the statements:

For $r$ fixed, $\theta$ goes from $\theta = \theta_1(r)$ to $\theta = \theta_2(r)$.
$r$ goes from $r = r_1$ to $r = r_2$.

(We assume $\theta_1 \le \theta_2$, $0 \le r_1 \le r_2$.) Such a region is sketched in Figure 17.4.6a. Each counterclockwise circular trajectory that intersects the region has $r$ fixed, enters the region at a point on the curve $\theta = \theta_1(r)$, and leaves the region at a point on $\theta = \theta_2(r)$. As $\theta$ varies from $\theta = \theta_1(r)$ to $\theta = \theta_2(r)$ with $r$ fixed, the points $(r, \theta)$ form an arc of a circle with center at the origin and radius $r$. See Figure 17.4.6b. As $r$ varies from $r = r_1$ to $r = r_2$, these arcs sweep out the entire region. See Figure 17.4.6c. As before, we have

$$\iint_R f\, dA = \int_{r_1}^{r_2} \int_{\theta_1(r)}^{\theta_2(r)} f(r, \theta)\, r\, d\theta\, dr.$$

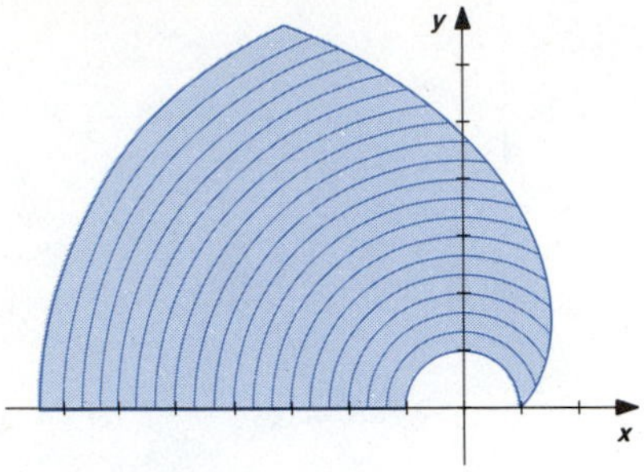

FIGURE 17.4.7

■ **EXAMPLE 4**

Find the area of the region bounded by the polar curves $r = e^{\theta}$ $(0 \le \theta \le 2)$, $r = 1$ $(0 \le \theta \le \pi)$, $r = e^2$ $(2 \le \theta \le \pi)$, and the polar axis.

**SOLUTION**

The region is sketched in Figure 17.4.7. We see that an arrow that is shot from the pole could leave the region at a point on either of the curves $r = e^{\theta}$ or $r = e^2$. We cannot express the area of the region as a single iterated integral with the $r$-integration inside. However, counterclockwise trajectories with fixed $r$ enter the region at a point on $r = e^{\theta}$ and leave at a point on $\theta = \pi$. Since

$$r = e^{\theta} \quad \text{implies} \quad \theta = \ln r,$$

we can describe the region with the following statements:

For $r$ fixed, $\theta$ varies from $\theta = \ln r$ to $\theta = \pi$.
$r$ varies from $r = 1$ to $r = e^2$.

The area of the region is given by the corresponding iterated polar integral. We have

$$\text{Area} = \int_1^{e^2} \int_{\ln r}^{\pi} r \, d\theta \, dr$$
$$= \int_1^{e^2} (\pi r - r \ln r) \, dr.$$

Using the table of integrals (or integration by parts) to evaluate the indefinite integral of $r \ln r$, we obtain

$$\text{Area} = \frac{\pi r^2}{2} - \frac{r^2 \ln r}{2} + \frac{r^2}{4}\Bigg]_1^{e^2}$$
$$= \left(\frac{\pi e^4}{2} - \frac{e^4 \ln e^2}{2} + \frac{e^4}{4}\right) - \left(\frac{\pi}{2} - \frac{\ln 1}{2} + \frac{1}{4}\right)$$
$$= \frac{\pi e^4}{2} - \frac{3e^4}{4} - \frac{\pi}{2} - \frac{1}{4} \approx 42.993165.$$

■

## EXERCISES 17.4

Find the area of the regions indicated in Exercises 1–4.

1 $x^2 + y^2 = 2$, above $y = 1$

2 Between $x^2 + y^2 = 9$ and $x^2 + y^2 = 4$

3 $x^2 + y^2 = 9$, $2y = x$, $x$-axis, in the first quadrant

4 Inside $(x - 1)^2 + y^2 = 1$ and outside $x^2 + y^2 = 1$

5 Find the volume of a sphere of radius $R$.

6 Find the volume of a right circular cone with height $H$ and radius of base $R$.

7 Find the mass of a circular lamina with radius $R$ and density at a point proportional to the square of the distance of the point to the center.

8 Find the mass of a circular lamina with radius $R$ and density at a point proportional to the distance of the point to the center.

Find the center of mass of each lamina described below in Exercises 9–12.

9 $y = \sqrt{1 - x^2}$, $y = 0$; $\rho(x, y) = ky$, $k$ a constant

10 $x^2 + y^2 = 4$, in the first quadrant; $\rho(x, y) = \sqrt{x^2 + y^2}$

11 Between $x^2 + y^2 = 4$ and $x^2 + y^2 = 1$, in the first quadrant; $\rho(x, y) = k$

12 Polar equations $r = r_0$, $0 \le \theta \le \theta_0$, homogeneous

13 Find the area of the curved surface of a right circular cone with height $H$ and radius of base $R$.

14 Find the area of that part of a plane that is inside a right circular cylinder with radius $R$ if the plane intersects the cylinder at an angle $\phi_0$, so the angle between the axis of the cylinder and the normal to the plane is $\pi/2 - \phi_0$.

15 Find the area of that part of the surface of a sphere of radius $R$ that is inside a right circular cone with vertex at the center of the sphere if each line along the cone through its vertex makes an angle of $\phi_0$ with the axis of the cone.

16 Find the surface area of a sphere of radius $R$.

Find the area of the regions bounded by the polar curves given in Exercises 17–20.

17 $r = \theta\ (0 \le \theta \le 2)$, $r = 2\theta\ (0 \le \theta \le 1)$, $r = 2\ (1 \le \theta \le 2)$

18 $r = e^{\theta}\ (0 \le \theta \le 2)$, $r = e^{2\theta}\ (0 \le \theta \le 1)$, $r = e^2\ (1 \le \theta \le 2)$

19 $\theta = \pi/(1 + r^2)\ (\pi/4 \le \theta \le \pi)$, $r = \sqrt{3}\ (0 \le \theta \le \pi/4)$, and the polar axis

20 $r = \ln(\pi/2\theta)\ (\pi/6 \le \theta \le \pi/2)$, $r = \ln 3\ (0 \le \theta < \pi/6)$, and the polar axis

21 Use the formula for the area of a sector of a circle to find the area of the polar rectangle bounded by the polar curves $r = r_0$, $r = r_0 + \Delta r$, $\theta = \theta_0$, $\theta = \theta_0 + \Delta\theta$.

22 An important formula in statistics involves the integral $I = \int_0^{\infty} e^{-x^2}\,dx$. In this exercise we will illustrate a way of evaluating this integral. (The Fundamental Theorem of Calculus cannot be used to evaluate $\int e^{-x^2}\,dx$.)

(a) Use polar coordinates to evaluate

$$\iint_{D_R} e^{-x^2-y^2}\,dA,$$

where $D_R$ is the region in the first quadrant bounded by $x^2 + y^2 = R^2$.

(b) Show that

$$I_R^2 = \iint_{S_R} e^{-x^2-y^2}\,dA,$$

where $I_R = \int_0^R e^{-x^2}\,dx = \int_0^R e^{-y^2}\,dy$ and $S_R$ is the region given by $0 \le x \le R$ and $0 \le y \le R$.

(c) Show that

$$\iint_{D_R} e^{-x^2-y^2}\,dA \le I_R^2 \le \iint_{D_{\sqrt{2}R}} e^{-x^2-y^2}\,dA.$$

(d) Evaluate

$$I = \lim_{R\to\infty} I_R.$$

## 17.5 TRIPLE INTEGRALS

Triple integrals involve the same ideas as double integrals. A region $R$ in three-dimensional space is divided into small rectangular prisms with sides parallel to the coordinate axes. See Figure 17.5.1. A Riemann sum over $R$ of a bounded function $f(x, y, z)$ is the sum of products of $f$ evaluated at a point in each prism times the volume of the prism,

$$\sum f(x_j, y_j, z_j)\,\Delta V_j.$$

The sum is taken over all prisms that are inside the region. As we did in the case of double integrals, we must avoid regions that do not have a nice

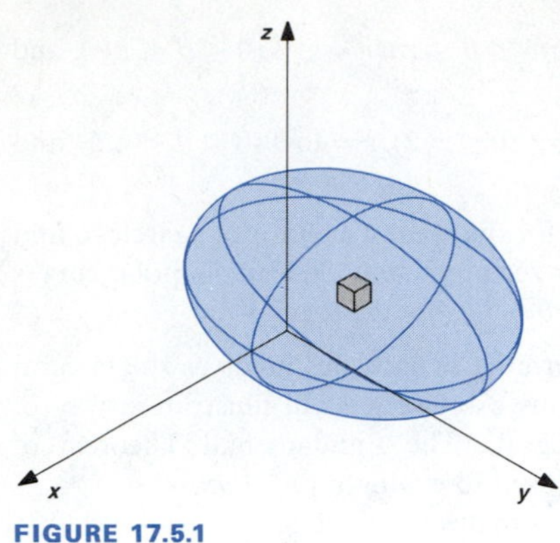

FIGURE 17.5.1

enough boundary. In particular, we require that we can cover the region in such a way that the total volume of all rectangular prisms that touch the boundary is less than any prescribed positive number. If $f$ is continuous at all points of $R$ and the boundary of $R$, then the Riemann sums of $f$ over $R$ approach a single number, denoted

$$\iiint_R f(x, y, z)\, dV,$$

as the lengths of sides of the rectangular prisms approach zero. As usual, applications of the idea depend on the interpretation of the factors as physical quantities.

Let us consider a region $R$ that can be described by the following statements:

For $x$ and $y$ fixed, $z$ varies from $z = h_1(x, y)$ to $z = h_2(x, y)$.
For $x$ fixed, $y$ varies from $y = g_1(x)$ to $y = g_2(x)$.
$x$ varies from $x = a$ to $x = b$.

We assume the boundary functions are continuous with $h_1(x, y) \leq h_2(x, y)$ and $g_1(x) \leq g_2(x)$. The surfaces $z = h_1(x, y)$ and $z = h_2(x, y)$ form the bottom and top of the region, the vertical generalized cylinders $y = g_1(x)$ and $y = g_2(x)$ form the sides of the region, and the vertical planes $x = a$ and $x = b$ form the back and front boundaries of the region. See Figure 17.5.2a. An arrow that is shot vertically from below the region follows a trajectory with $x$ and $y$ fixed, enters the region at a point on the surface $z = h_1(x, y)$, travels parallel to the vertical surfaces $y = g_1(x)$, $y = g_2(x)$, $x = a$, and $x = b$, and leaves the region at a point on the surface $z = h_2(x, y)$. The intersection of the surfaces $y = g_1(x)$, $y = g_2(x)$, $x = a$, and $x = b$ with the $xy$-plane is the projection of the region onto the $xy$-plane, which is a region of the $xy$-plane that is simple in the direction of the $y$-axis. See Figure 17.5.2b. As $z$ varies from $z = h_1(x, y)$ to $z = h_2(x, y)$

FIGURE 17.5.2

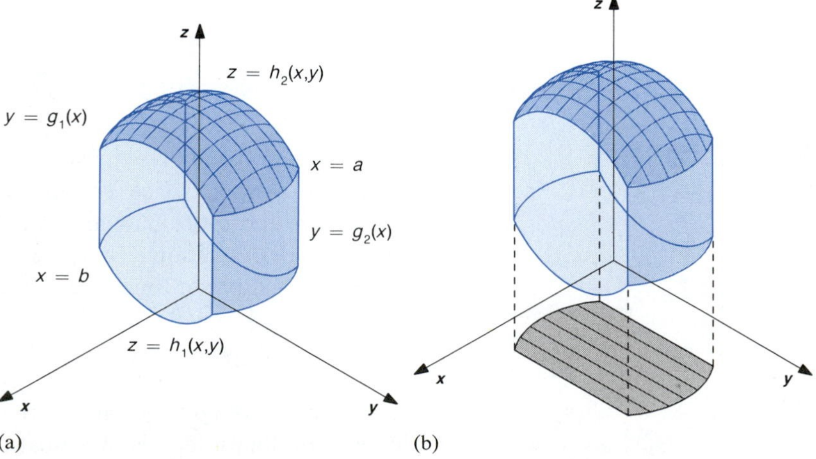

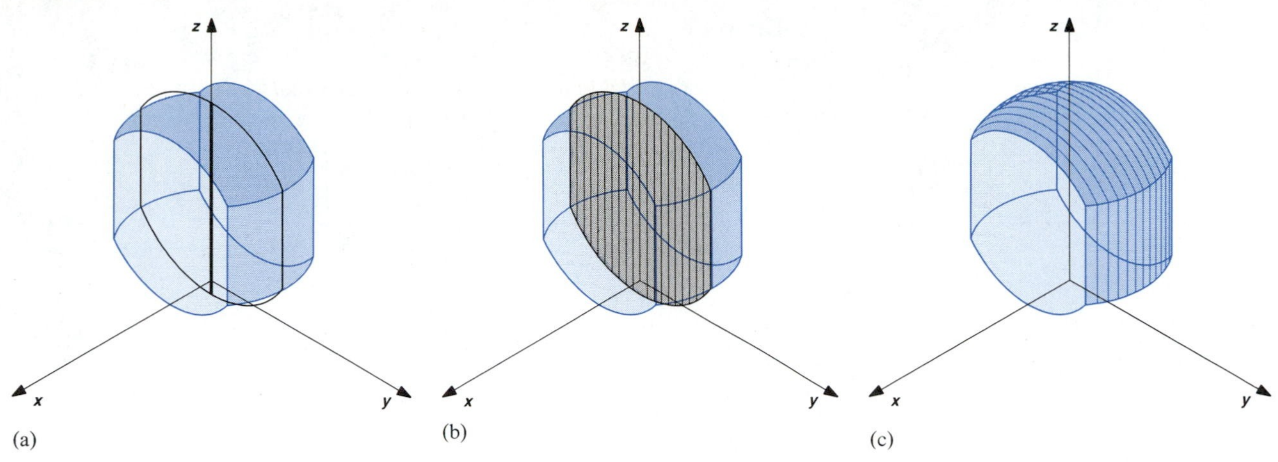


FIGURE 17.5.3

with $x$ and $y$ fixed, the points $(x, y, z)$ form a vertical line segment between the top and the bottom of the region. See Figure 17.5.3a. As $y$ then varies from $y = g_1(x)$ to $y = g_2(x)$ with $x$ fixed, these line segments generate a cross section of the region. See Figure 17.5.3b. As $x$ varies from $x = a$ to $x = b$, the cross sections sweep out the entire region. See Figure 17.5.3c. In a manner similar to that used for double integrals, we can express a triple integral over $R$ as an **iterated triple integral**. We obtain

$$\iiint_R f(x, y, z)\, dV = \int_a^b \int_{g_1(x)}^{g_2(x)} \int_{h_1(x, y)}^{h_2(x, y)} f(x, y, z)\, dz\, dy\, dx.$$

Note how the limits of integration of the iterated triple integral correspond to the description of the region. We will use the method of Section 17.1 to determine the limits of integration.

The iterated triple integral

$$\int_a^b \int_{g_1(x)}^{g_2(x)} \int_{h_1(x, y)}^{h_2(x, y)} f(x, y, z)\, dz\, dy\, dx$$

is evaluated by first integrating the inner integral with respect to $z$ while $x$ and $y$ are treated as constants. The next integration is carried out by treating $x$ as a constant while we integrate with respect to $y$. We then integrate with respect to $x$.

■ **EXAMPLE 1**

Evaluate the iterated integral

$$\int_0^1 \int_0^x \int_0^{xy} 128xyz\, dz\, dy\, dx.$$

**SOLUTION**

$$\begin{aligned}\int_0^1\int_0^x\int_0^{xy} 128xyz\,dz\,dy\,dx &= \int_0^1\int_0^x\left(\int_0^{xy} 128xyz\,dz\right)dy\,dx\\ &= \int_0^1\int_0^x\left(64xyz^2\Big]_{z=0}^{z=xy}\right)dy\,dx\\ &= \int_0^1\left(\int_0^x 64x^3y^3\,dy\right)dx\\ &= \int_0^1\left(16x^3y^4\Big]_{y=0}^{y=x}\right)dx\\ &= \int_0^1 16x^7\,dx\\ &= 2x^8\Big]_0^1 = 2.\end{aligned}$$

■

If a region $R$ is divided into small rectangular prisms, the sum of the volumes of the prisms contained in $R$ should be approximately the volume of the region. This suggests that

$$\textbf{Volume} = \iiint_R dV.$$

■ **EXAMPLE 2**

Use a triple integral to find the volume of the region in the first octant that is bounded by $z = xy$, $z = 0$, $y = 0$, $y = 1$, $x = 0$, and $x = 1$.

**SOLUTION**

The region is sketched in Figure 17.5.4a. We have drawn the $x$-axis and $y$-axis at an angle less than 120° and have used a relatively small scale on the $z$-axis to give a perspective that looks down on the top surface of the region. It is clear that an arrow that is shot vertically from below the region follows a trajectory with $x$ and $y$ fixed, enters the region at a point

FIGURE 17.5.4

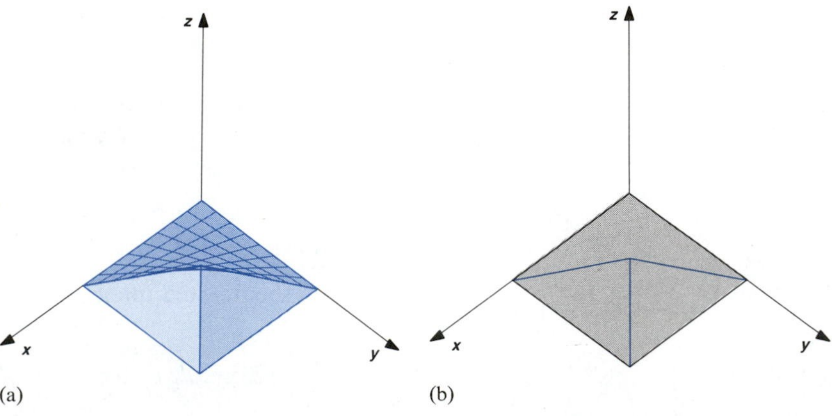

on the surface $z = 0$, and leaves the region at a point on the surface $z = xy$. Also, the projection of the region onto the $xy$-plane is given by the vertical planes $y = 0$, $y = 1$, $x = 0$, and $x = 1$. See Figure 17.5.4b. We can describe the region by the following statements:

For $x$ and $y$ fixed, $z$ varies from $z = 0$ to $z = xy$.
For $x$ fixed, $y$ varies from $y = 0$ to $y = 1$.
$x$ varies from $x = 0$ to $x = 1$.

The volume is

$$\begin{aligned} V &= \int_0^1 \int_0^1 \int_0^{xy} dz\, dy\, dx \\ &= \int_0^1 \int_0^1 xy\, dy\, dx \\ &= \int_0^1 x \frac{y^2}{2} \Big]_{y=0}^{y=1} dx \\ &= \int_0^1 \frac{x}{2}\, dx \\ &= \frac{x^2}{4}\Big]_0^1 = \frac{1}{4}. \end{aligned}$$

■

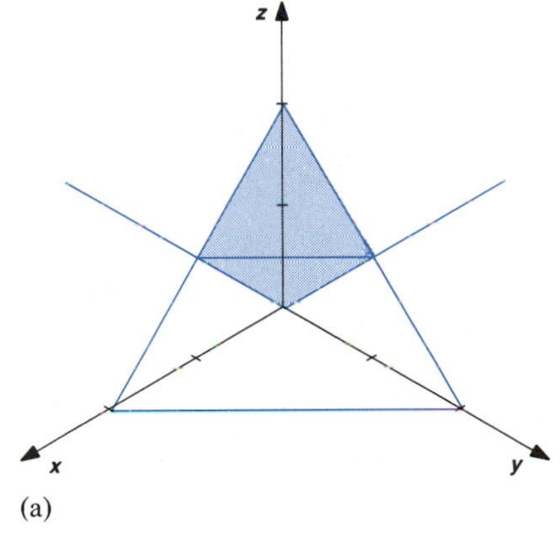

(a)

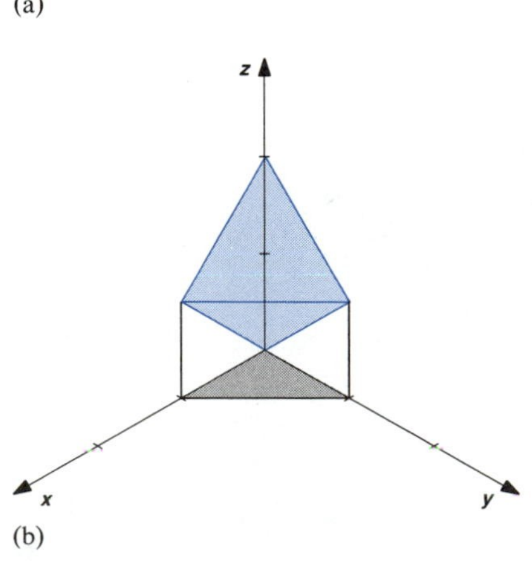

(b)

**FIGURE 17.5.5**

■ **EXAMPLE 3**

Use a triple integral to find the volume of the region in the first octant that is below the plane $x + y + z = 2$ and above the plane $z = x + y$.

**SOLUTION**

The region is sketched in Figure 17.5.5a. Note that the points of intersection of the planes in the $xz$- and $yz$-planes have been connected with a straight line. An arrow that is shot parallel to the $z$-axis from below the region will enter the region at a point on the surface $z = x + y$ and leave the region at a point on the surface $z = 2 - x - y$. The projection of the region onto the $xy$-plane is given by the vertical planes $x = 0$, $y = 0$, and the vertical plane that contains the line of intersection of the top and bottom of the surface. See Figure 17.5.5b. We have:

For $x$ and $y$ fixed, $z$ goes from $z = x + y$ to $z = 2 - x - y$.

The top and bottom intersect to form a right-side edge. $z = 2 - x - y$ and $z = x + y$ imply $x + y = 2 - x - y$, so $2x + 2y = 2$, or $y = 1 - x$.

For $x$ fixed, $y$ goes from $y = 0$ to $y = 1 - x$.

The sides intersect to form a front point:

$y = 1 - x$ and $y = 0$ imply $0 = 1 - x$, so $x = 1$.

$x$ goes from $x = 0$ to $x = 1$.

The volume is

$$V = \int_0^1 \int_0^{1-x} \int_{x+y}^{2-x-y} dz\, dy\, dx$$

$$= \int_0^1 \int_0^{1-x} [(2 - x - y) - (x + y)]\, dy\, dx$$

$$= \int_0^1 \int_0^{1-x} (2 - 2x - 2y)\, dy\, dx$$

$$= \int_0^1 \left[ 2(1 - x)y - y^2 \right]_{y=0}^{y=1-x} dx$$

$$= \int_0^1 [2(1 - x)^2 - (1 - x)^2]\, dx$$

$$= \int_0^1 (1 - x)^2\, dx$$

$$= -\frac{(1 - x)^3}{3}\Big]_0^1$$

$$= \frac{1}{3}.$$ ■

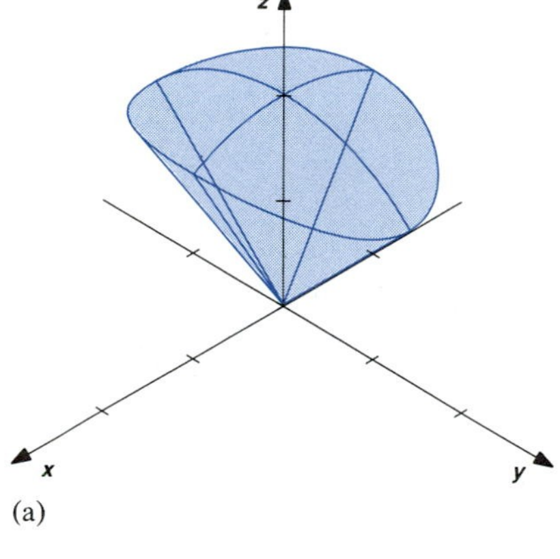


(a)

(b)

**FIGURE 17.5.6**

■ **EXAMPLE 4**

Set up an iterated triple integral that gives the volume of the region inside both the elliptical cone $z = \sqrt{3x^2 + y^2}$ and the sphere $x^2 + y^2 + z^2 = 4$.

**SOLUTION**

The region is sketched in Figure 17.5.6a. An arrow that is shot parallel to the $z$-axis from below the region will enter the region at a point on the cone and leave the region at a point on the sphere. The vertical generalized cylinder that contains the curve of intersection of the cone and cylinder gives the projection of the region onto the $xy$-plane. See Figure 17.5.6b.

The top of the region is formed by the sphere. The bottom is formed by the cone.

$x^2 + y^2 + z^2 = 4$, $z^2 = 3x^2 + y^2$ imply $4x^2 + 2y^2 = 4$, so $y^2 = 2 - 2x^2$. The sides of the region are contained in the cylinders $y = -\sqrt{2 - 2x^2}$ and $y = \sqrt{2 - 2x^2}$.

$y = -\sqrt{2 - 2x^2}$ and $y = \sqrt{2 - 2x^2}$ imply $2 - 2x^2 = 0$, so the $x$-coordinates of the front and back points are $x = 1$ and $x = -1$.

The volume is

$$V = \int_{-1}^1 \int_{-\sqrt{2-2x^2}}^{\sqrt{2-2x^2}} \int_{\sqrt{3x^2+y^2}}^{\sqrt{4-x^2-y^2}} dz\, dy\, dx.$$ ■

## 17.5 TRIPLE INTEGRALS

Triple integrals can be expressed as iterated integrals with six possible orders of integration, if the region is appropriate. Triple integrals over some regions cannot be expressed as a single iterated integral for some of the possible orders. To express a triple integral as an iterated integral with a particular variable on the inside, it is necessary that the region satisfy the condition that arrows that follow a trajectory parallel to the axis of that variable must all enter the region through the same surface and they must all leave the region through the same surface. Also, it must be possible to describe the projection of the region onto the plane of the other two variables in terms of these two variables.

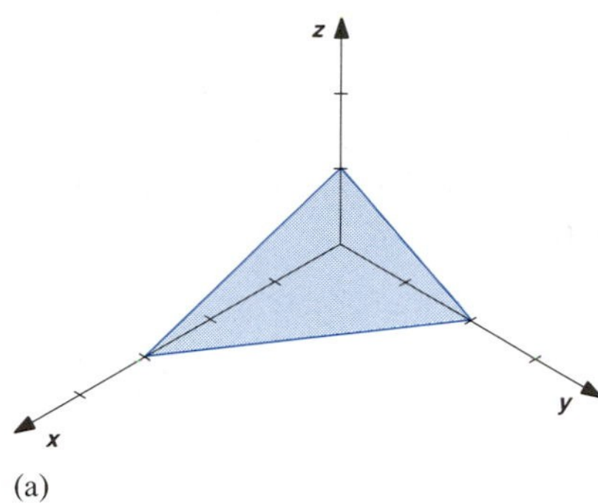

(a)

### ■ EXAMPLE 5

Express the volume of the region in the first octant bounded by the plane $2x + 3y + 6z = 6$ as a single iterated integral for every order of integration for which this is possible.

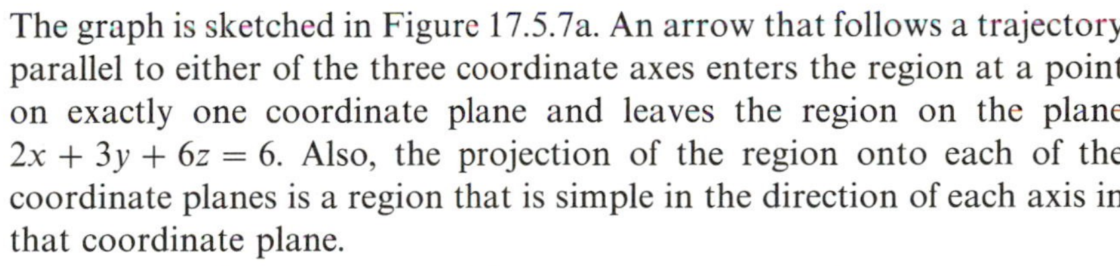

(b)

#### SOLUTION

The graph is sketched in Figure 17.5.7a. An arrow that follows a trajectory parallel to either of the three coordinate axes enters the region at a point on exactly one coordinate plane and leaves the region on the plane $2x + 3y + 6z = 6$. Also, the projection of the region onto each of the coordinate planes is a region that is simple in the direction of each axis in that coordinate plane.

Let us consider an iterated triple integral that has the integration with respect to $x$ on the inside, to be done first. We note that an arrow that is shot parallel to the $x$-axis from behind the region enters the region at a point on the plane $x = 0$ and leaves the region at a point on the plane $x = (6 - 3y - 6z)/2$. We can then say:

For $y$ and $z$ fixed, $x$ varies from $x = 0$ to $x = \dfrac{6 - 3y - 6z}{2}$.

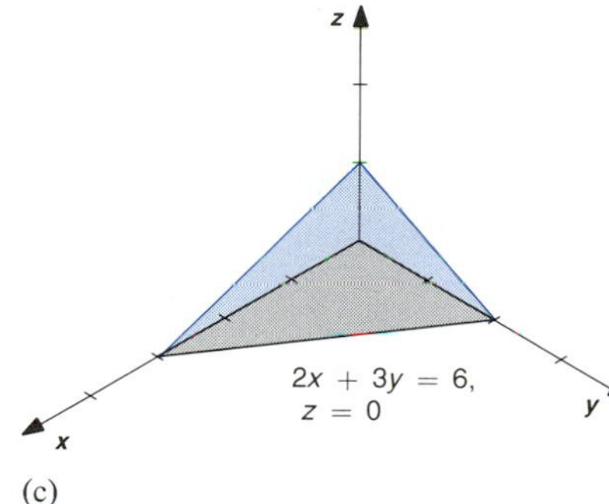

(c)

$x = 0$ and $x = \dfrac{6 - 3y - 6z}{2}$ imply

$$0 = \frac{6 - 3y - 6z}{2},$$

$$0 = 6 - 3y - 6z,$$

$$y + 2z = 2.$$

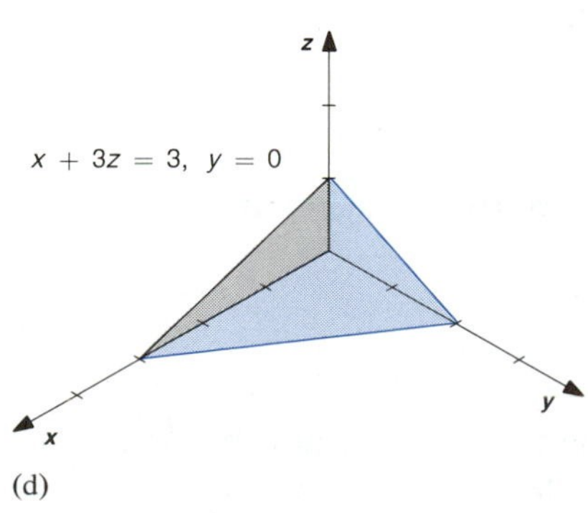

(d)

**FIGURE 17.5.7**

The projection of the region onto the $yz$-plane is given by the planes $y = 0$, $z = 0$, and $y + 2z = 2$. The projection is a region of the $yz$-plane that is simple in the direction of the $z$-axis. See Figure 17.5.7b. The projection onto the $yz$-plane can be then described by the following statements:

For $y$ fixed, $z$ varies from $z = 0$ to $z = (2 - y)/2$.
$y$ varies from $z = 0$ to $y = 2$.

The volume integral that corresponds to this description is

$$\int_0^2 \int_0^{(2-y)/2} \int_0^{(6-3y-6z)/2} dx\, dz\, dy.$$

Since the projection of the region onto the $yz$-plane is also a region of the $yz$-plane that is simple in the direction of the $y$-axis, we can describe this projection by the following statements:

For $z$ fixed, $y$ varies from $y = 0$ to $y = 2 - 2z$.
$z$ varies from $z = 0$ to $z = 1$.

The corresponding volume integral is

$$\int_0^1 \int_0^{2-2z} \int_0^{(6-3y-6z)/2} dx\, dy\, dz.$$

The volume integrals that correspond to an arrow that is shot parallel to the $z$-axis and involve the projection of the region onto the $xy$-plane are

$$\int_0^3 \int_0^{(6-2x)/3} \int_0^{(6-2x-3y)/6} dz\, dy\, dx$$

and

$$\int_0^2 \int_0^{(6-3y)/2} \int_0^{(6-2x-3y)/6} dz\, dx\, dy.$$

See Figure 17.5.7c. Corresponding to an arrow that is shot parallel to the $y$-axis and the projection of the region onto the $xz$-plane, we have

$$\int_0^1 \int_0^{3-3z} \int_0^{(6-2x-6z)/3} dy\, dx\, dz$$

and

$$\int_0^3 \int_0^{(3-x)/3} \int_0^{(6-2x-6z)/3} dy\, dz\, dx.$$

See Figure 17.5.7d. (Each of the six iterated triple integrals has the value 1, which is the volume of the region.) ■

### ■ EXAMPLE 6

Express the volume of the region in the first octant that is below both of the planes $2x + 3y + 3z = 6$ and $x + 2z = 2$ as a single iterated triple integral for every order of integration for which this is possible.

FIGURE 17.5.8

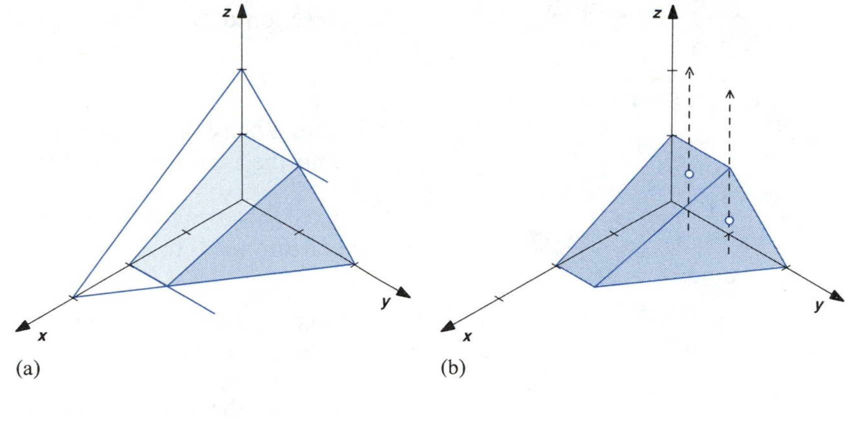


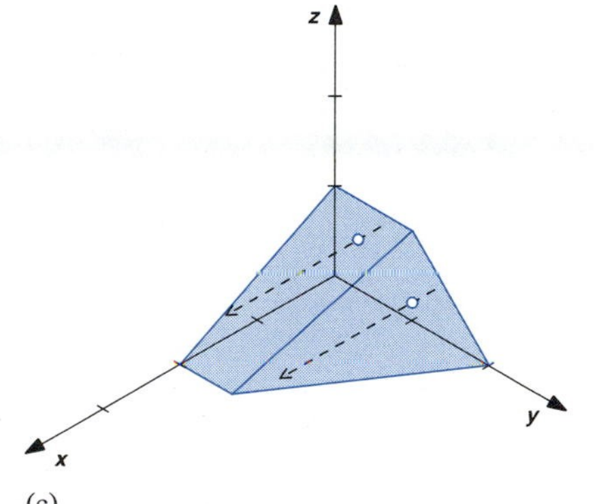


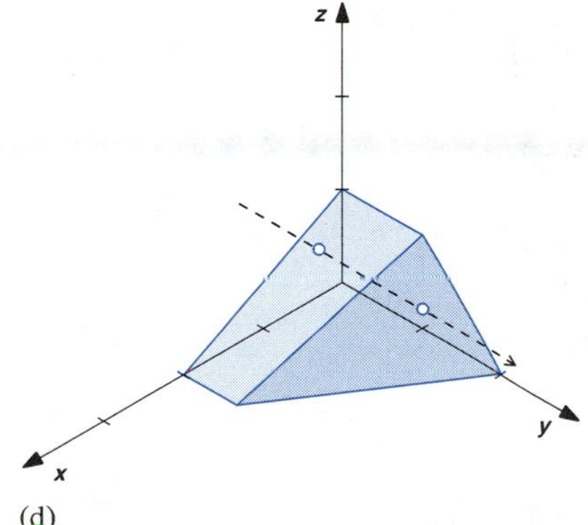


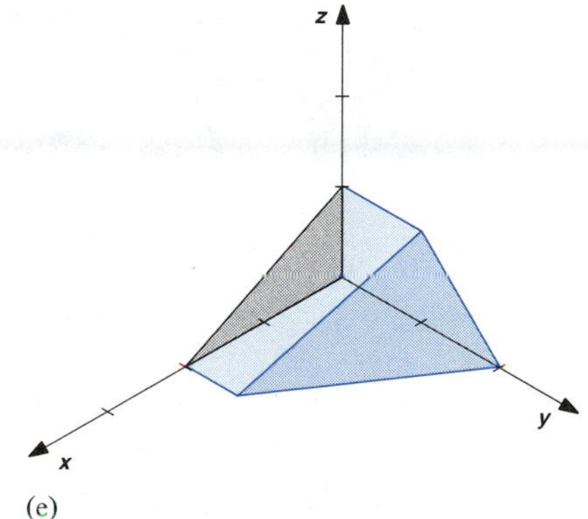


**SOLUTION**

The region is sketched in Figure 17.5.8a.

From Figure 17.5.8b we see that an arrow that is shot parallel to the $z$-axis from below the region could leave the region at a point on either the plane $2x + 3y + 3z = 6$ or the plane $x + 2z = 2$. Hence, there is not a single equation that describes the top of the region, and we cannot express the volume of this region as a single iterated triple integral that has the integration with respect to $z$ on the inside, to be done first.

Similarly, from Figure 17.5.8c we see that an arrow that is shot parallel to the $x$-axis from behind the region could leave the region at a point on either of two planes. This means we cannot express the volume as a single iterated triple integral that has the integration with respect to $x$ on the inside.

An arrow that is shot parallel to the $y$-axis from the left of the region will enter the region at a point on the plane $y = 0$; travel parallel to the planes $x = 0$, $z = 0$; and $x + 2z = 2$; and leave the region at a point on the plane $2x + 3y + 3z = 6$. See Figure 17.5.8d. We can say:

For $x$ and $z$ fixed, $y$ varies from $y = 0$ to $y = \dfrac{6 - 2x - 3z}{3}$.

The projection of the region onto the $xz$-plane is given by the equations

$$x = 0, \quad z = 0, \quad \text{and} \quad x + 2z = 2.$$

See Figure 17.5.8e. This projection is simple in the direction of both the $x$-axis and the $z$-axis. We can then say either of the following:

For $x$ fixed, $z$ varies from $z = 0$ to $z = (2 - x)/2$.
$x$ varies from $x = 0$ to $x = 2$.

For $z$ fixed, $x$ varies from $x = 0$ to $x = 2 - 2z$.
$z$ varies from $z = 0$ to $z = 1$.

The volume of the region is then given by the corresponding two integrals,

$$\int_0^2 \int_0^{(2-x)/2} \int_0^{(6-2x-3z)/3} dy\, dz\, dx$$

and

$$\int_0^1 \int_0^{2-2z} \int_0^{(6-2x-3z)/3} dy\, dx\, dz. \quad \blacksquare$$

## EXERCISES 17.5

Evaluate the integrals in Exercises 1–10.

1 $\int_0^3 \int_0^2 \int_0^1 xyz\, dz\, dy\, dx$

2 $\int_0^1 \int_0^1 \int_0^1 x^2 y e^{xyz}\, dz\, dy\, dx$

3 $\int_0^1 \int_0^x \int_0^y xyz\, dz\, dy\, dx$

4 $\int_0^1 \int_0^x \int_0^{xy} z\, dz\, dy\, dx$

5 $\int_0^\pi \int_0^x \int_0^{xy} x^2 \sin z\, dz\, dy\, dx$

6 $\int_0^r \int_0^{\sqrt{r^2-x^2}} \int_0^{\sqrt{r^2-x^2-y^2}} xy\, dz\, dy\, dx$

7 $\int_0^1 \int_0^z \int_0^y z e^{y^2}\, dx\, dy\, dz$

8 $\int_0^1 \int_0^{\sqrt{z}} \int_0^{\sqrt{yz}} x\, dx\, dy\, dz$

9 $\int_1^e \int_0^y \int_0^{1/y} x^2 z\, dx\, dz\, dy$

10 $\int_1^2 \int_0^{y^2} \int_0^{z^2} yz\, dx\, dz\, dy$

Use triple integrals to find the volume of each region described in Exercises 11–16.

11 $2x + y + 2z = 4$, in the first octant

12 $z = 2x + 2y$, $z = 4$, in the first octant

13 Below $x + 2y + 2z = 4$, above $z = 2x$, in the first octant

14 Below $x + z = 2$, above $z = x + y$, in the first octant

15 Inside both $x^2 + y^2 = r^2$ and $x^2 + z^2 = r^2$

16 Inside $x^2 + y^2 = 4$, below $x + y + z = 4$, in the first octant

Set up an iterated triple integral that gives the volume of each region indicated in Exercises 17–20.

17 Inside both $x^2 + y^2 = r^2$ and $x^2 + z^2 = R^2$, $0 < r < R$

18 Inside both $x^2 + y^2 = r^2$ and $x^2 + y^2 + z^2 = R^2$, $0 < r < R$

19 Inside both $x^2 + y^2 = 1$ and $(y + 4)^2 = 4x^2 + 4z^2$

20 Inside $x^2 + y^2 = 1$, between $z = 4 - \sqrt{x^2 + y^2}$ and $z = 0$

In Exercises 21–26 express the volume of each of the given regions as a single iterated triple integral for every order of integration for which this is possible.

**21** Below both $z = 2x$ and $x + 2y + 3z = 6$, in the first octant

**22** $x + z = 4$, $z = 2$, $y = 4$, $x = 0$, $y = 0$, $z = 0$

**23** $x + y + z = 4$, $z = 1$, $x = 0$, $y = 0$, $z = 0$

**24** $x + z = 3$, $y + z = 3$, $z = 1$, $x = 0$, $y = 0$, $z = 0$

**25** Below both $x + y + z = 6$ and $x + z = 4$, above $z = 2x$, in the first octant

**26** Below both $y + z = 4$ and $y = 3z$, $y = x$, $x = 0$, $z = 0$, in the first octant

## 17.6 APPLICATIONS OF TRIPLE INTEGRALS

Triple integrals are particularly well suited for the evaluation of the mass, moments, and center of mass of solid objects.

The **differential mass** of a small solid with density $\rho$ and volume $dV$ is

$$dM = \rho\, dV.$$

The **mass** of an object that occupies a region $R$ in three-dimensional space and has density at $(x, y, z)$ given by a continuous density function $\rho(x, y, z)$ is

$$M = \iiint_R dM = \iiint_R \rho\, dV.$$

■ **EXAMPLE 1**

Find the mass of the solid in the first octant that is bounded by $z = y/2$, $x = 3$, $y = 4$, if the density at a point $(x, y, z)$ is given by $\rho(x, y, z) = 4y$.

**SOLUTION**

The solid is sketched in Figure 17.6.1. The mass is

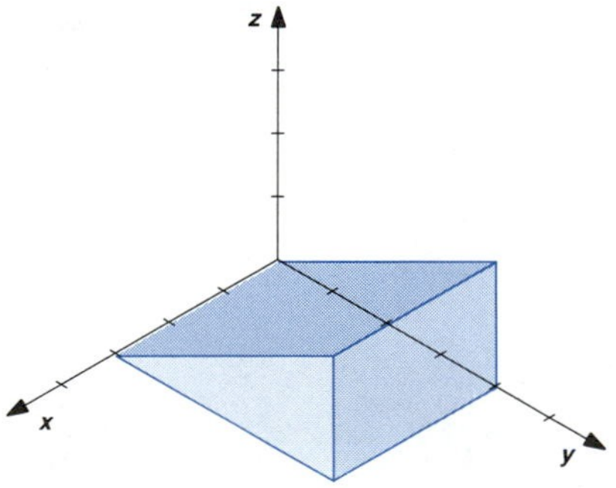

$$\begin{aligned} M &= \iiint_R dM = \iiint_R \rho\, dV \\ &= \int_0^3 \int_0^4 \int_0^{y/2} 4y\, dz\, dy\, dx \\ &= \int_0^3 \int_0^4 4yz\Big]_{z=0}^{z=y/2} dy\, dx \\ &= \int_0^3 \int_0^4 2y^2\, dy\, dx \\ &= \int_0^3 \frac{2y^3}{3}\Big]_{y=0}^{y=4} dx \\ &= \int_0^3 \frac{128}{3}\, dx = 128. \end{aligned}$$

■

FIGURE 17.6.1

First **moments** of an object about the coordinate planes are

$$M_{x=0} = \iiint_R x\,dM, \quad M_{y=0} = \iiint_R y\,dM, \quad \text{and} \quad M_{z=0} = \iiint_R z\,dM.$$

Coordinates of the **center of mass** are

$$\bar{x} = \frac{M_{x=0}}{M} = \frac{\iiint_R x\,dM}{\iiint_R dM},$$

$$\bar{y} = \frac{M_{y=0}}{M} = \frac{\iiint_R y\,dM}{\iiint_R dM}, \quad \text{and}$$

$$\bar{z} = \frac{M_{z=0}}{M} = \frac{\iiint_R z\,dM}{\iiint_R dM}.$$

**■ EXAMPLE 2**

Find the center of mass of a homogeneous solid that is bounded by $z = 4 - x^2 - y^2$, $z = 0$, $y = -1$, $y = 1$, $x = -1$, and $x = 1$.

**SOLUTION**

The solid is sketched in Figure 17.6.2. The density is constant, $\rho(x, y, z) = k$. The mass is

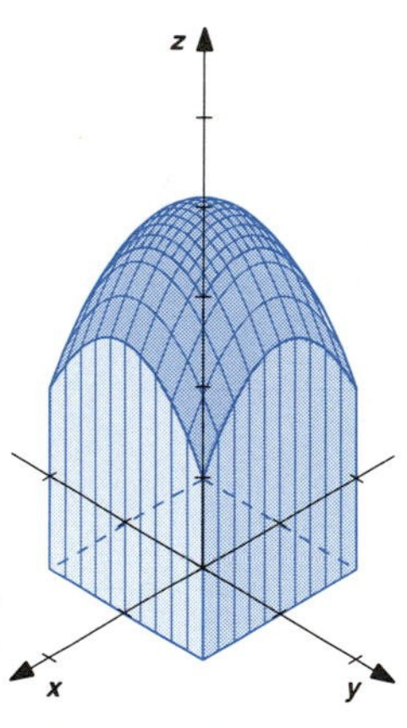

FIGURE 17.6.2

$$M = \iiint_R dM = \iiint_R \rho\,dV$$

$$= \int_{-1}^{1}\int_{-1}^{1}\int_{0}^{4-x^2-y^2} k\,dz\,dy\,dx$$

$$= \int_{-1}^{1}\int_{-1}^{1} k(4 - x^2 - y^2)\,dy\,dx$$

$$= \int_{-1}^{1} k\left[4y - x^2y - \frac{1}{3}y^3\right]_{y=-1}^{y=1} dx$$

$$= \int_{-1}^{1} k\left[\left(4 - x^2 - \frac{1}{3}\right) - \left(-4 + x^2 + \frac{1}{3}\right)\right] dx$$

$$= \int_{-1}^{1} k\left(\frac{22}{3} - 2x^2\right) dx$$

$$= k\left[\frac{22}{3}x - 2\frac{x^3}{3}\right]_{-1}^{1}$$

$$= k\left[\left(\frac{22}{3} - \frac{2}{3}\right) - \left(-\frac{22}{3} + \frac{2}{3}\right)\right] = \frac{40}{3}k.$$

The moment about the $xy$-plane is

$$M_{z=0} = \iiint\limits_R z\,dM = \iiint\limits_R z\,\rho\,dV$$

$$= \int_{-1}^{1}\int_{-1}^{1}\int_{0}^{4-x^2-y^2} kz\,dz\,dy\,dx$$

$$= \int_{-1}^{1}\int_{-1}^{1} k\frac{z^2}{2}\Big]_{z=0}^{z=4-x^2-y^2} dy\,dx$$

$$= \int_{-1}^{1}\int_{-1}^{1} \frac{k}{2}(4-x^2-y^2)^2\,dy\,dx$$

$$= \int_{-1}^{1}\int_{-1}^{1} \frac{k}{2}(16-8x^2-8y^2+x^4+2x^2y^2+y^4)\,dy\,dx$$

$$= \int_{-1}^{1} \frac{k}{2}\left[16y-8x^2y-\frac{8y^3}{3}+x^4y+\frac{2x^2y^3}{3}+\frac{y^5}{5}\right]_{y=-1}^{y=1} dx$$

$$= \int_{-1}^{1} k\left(16-8x^2-\frac{8}{3}+x^4+\frac{2x^2}{3}+\frac{1}{5}\right) dx$$

$$= \int_{-1}^{1} k\left(\frac{203}{15}-\frac{22}{3}x^2+x^4\right) dx$$

$$= k\left[\frac{203}{15}x-\frac{22}{9}x^3+\frac{x^5}{5}\right]_{-1}^{1}$$

$$= 2k\left(\frac{203}{15}-\frac{22}{9}+\frac{1}{5}\right) = \frac{1016}{45}k.$$

Then

$$\bar{z} = \frac{M_{z=0}}{M} = \frac{\frac{1016}{45}k}{\frac{40}{3}k} = \left(\frac{1016}{45}\right)\left(\frac{3}{40}\right) = \frac{127}{75} \approx 1.6933333.$$

Symmetry implies that the center of mass is on the $z$-axis, so $\bar{x} = 0$ and $\bar{y} = 0$. The center of mass is at the point

$$(\bar{x}, \bar{y}, \bar{z}) = \left(0, 0, \frac{127}{75}\right). \qquad \blacksquare$$

Let us see how the study of **kinetic energy** (K.E.) leads to a type of moment integral that is different from first moments. The kinetic energy of an object with mass $m$ that is moving with speed $v$ is

$$\text{K.E.} = \frac{1}{2}mv^2.$$

Suppose the object is moving in a circular path with radius of rotation $r$ and constant angular velocity $\omega$. The speed is then

$$v = \omega r$$

and the kinetic energy is

$$\text{K.E.} = \frac{1}{2} m(\omega r)^2 = \frac{\omega^2}{2} r^2 m.$$

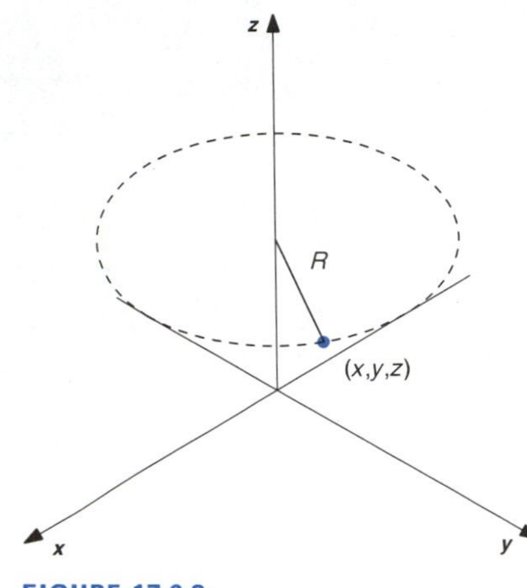

FIGURE 17.6.3

If the object is rotating about the $z$-axis, the radius of rotation is

$$r = \sqrt{x^2 + y^2}$$

and the kinetic energy is

$$\text{K.E.} = \frac{\omega^2}{2} (x^2 + y^2)m.$$

See Figure 17.6.3. Let us define the **moment of inertia** about the $z$-axis of the object to be

$$I_{z\text{-axis}} = (x^2 + y^2)m.$$

The moment of inertia about the $z$-axis of the object is then proportional to the kinetic energy of the object as it rotates about the $z$-axis with constant angular speed.

The moment of inertia about the $z$-axis of a system of masses is the sum of the moments of inertia of each part. The moment of inertia about the $z$-axis of a solid object that occupies a region $R$ in space and has density function $\rho(x, y, z)$ is defined to be the triple integral

$$I_{z\text{-axis}} = \iiint_R (x^2 + y^2)\, dM = \iiint_R (x^2 + y^2)\rho(x, y, z)\, dV.$$

The moments of inertia of a solid about the other coordinate axes are

$$I_{x\text{-axis}} = \iiint_R (y^2 + z^2)\rho\, dV \quad \text{and}$$

$$I_{y\text{-axis}} = \iiint_R (x^2 + z^2)\rho\, dV.$$

Moments of inertia are also called **second moments**.

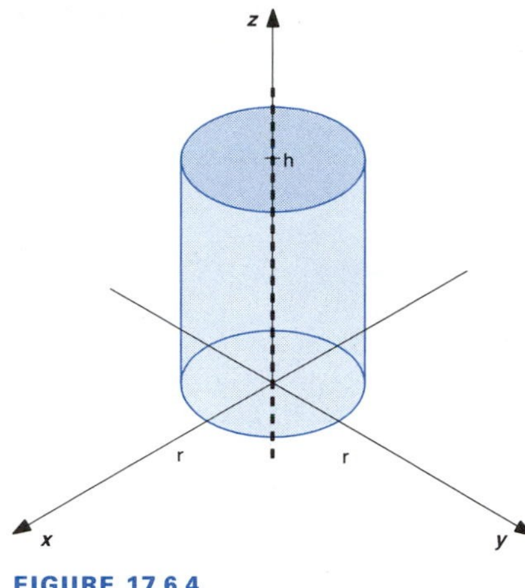


FIGURE 17.6.4

■ **EXAMPLE 3**

Find the moment of inertia of a homogeneous right circular cylinder with radius $r$ and height $h$ about the axis of the cylinder. Express the answer in terms of the mass of the cylinder.

**SOLUTION**

The first step is to choose a coordinate system. Let us choose the axis of the cylinder along the $z$-axis so the desired moment of inertia is $I_{z\text{-axis}}$. See Figure 17.6.4. The density of a homogeneous object is constant. Let $k$ denote the constant density. We then have

$$\begin{aligned}
I_{z\text{-axis}} &= \iiint_R (x^2+y^2)\,dM = \iiint_R (x^2+y^2)\rho\,dV \\
&= \int_{-r}^{r}\int_{-\sqrt{r^2-x^2}}^{\sqrt{r^2-x^2}}\int_0^h k(x^2+y^2)\,dz\,dy\,dx \\
&= \int_{-r}^{r}\int_{-\sqrt{r^2-x^2}}^{\sqrt{r^2-x^2}} kh(x^2+y^2)\,dy\,dx \\
&= \int_{-r}^{r} kh\left[x^2y+\frac{y^3}{3}\right]_{y=-\sqrt{r^2-x^2}}^{y=\sqrt{r^2-x^2}} dx \\
&= \int_{-r}^{r} kh\left[2x^2\sqrt{r^2-x^2}+\frac{2}{3}(r^2-x^2)^{3/2}\right] dx \\
&= 2kh\int_{-r}^{r} x^2\sqrt{r^2-x^2}\,dx + \frac{2kh}{3}\int_{-r}^{r}(r^2-x^2)^{3/2}\,dx.
\end{aligned}$$

The above two integrals are evaluated by using the table of integrals. We obtain

$$\begin{aligned}
I_{z\text{-axis}} &= 2kh\left[\frac{x}{8}(2x^2-r^2)\sqrt{r^2-x^2}+\frac{r^4}{8}\sin^{-1}\left(\frac{x}{r}\right)\right]_{-r}^{r} \\
&\quad + \frac{2kh}{3}\left[-\frac{x}{8}(2x^2-5r^2)\sqrt{r^2-x^2}+\frac{3r^4}{8}\sin^{-1}\left(\frac{x}{r}\right)\right]_{-r}^{r} \\
&= 2kh\left(\frac{r^4\pi}{8}\right)+\frac{2kh}{3}\left(\frac{3r^4\pi}{8}\right) = k\frac{\pi h r^4}{2}.
\end{aligned}$$

We can express the moment of inertia of the cylinder in terms of its radius, height, and mass by noting that the total mass of the cylinder is

$$m = k(\text{Volume}) = k\pi r^2 h.$$

We can then express the density in terms of $m$, $r$ and $h$. That is,

$$k = \frac{m}{\pi r^2 h}.$$

Substitution then gives

$$I_{z\text{-axis}} = \frac{m}{\pi r^2 h}\frac{\pi h r^4}{2} = \frac{mr^2}{2}.$$

■

■ **EXAMPLE 4**

Find the moment of inertia of a homogeneous right circular cylinder with radius $r$ and height $h$ about a line that is perpendicular to the axis of the cylinder and through the axis at one end of the cylinder. Express the answer in terms of the mass of the cylinder.

**SOLUTION**

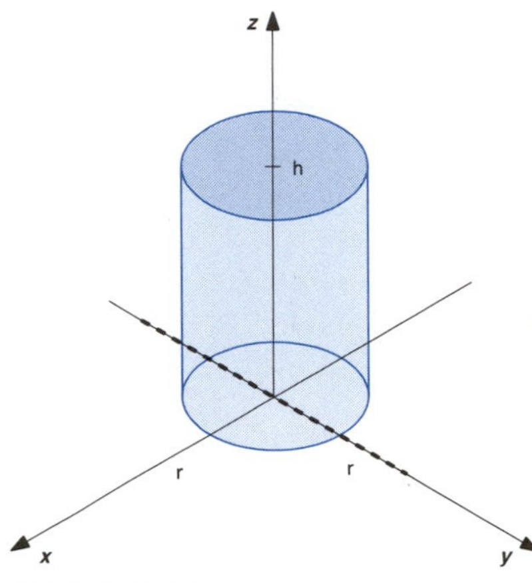

FIGURE 17.6.5

Let us position the cylinder with its axis along the $z$-axis. See Figure 17.6.5.

We want the moment of inertia about a line through the origin in the $xy$-plane. Let us choose $I_{y\text{-axis}}$. Let $k$ denote the constant density. We have

$$I_{y\text{-axis}} = \iiint_R (x^2 + z^2)\,dM = \iiint_R (x^2 + z^2)\rho\,dV$$

$$= \int_{-r}^{r}\int_{-\sqrt{r^2-x^2}}^{\sqrt{r^2-x^2}}\int_0^h k(x^2 + z^2)\,dz\,dy\,dx$$

$$= \int_{-r}^{r}\int_{-\sqrt{r^2-x^2}}^{\sqrt{r^2-x^2}} k\left[x^2 z + \frac{z^3}{3}\right]_{z=0}^{z=h} dy\,dx$$

$$= \int_{-r}^{r}\int_{-\sqrt{r^2-x^2}}^{\sqrt{r^2-x^2}} k\left(hx^2 + \frac{h^3}{3}\right) dy\,dx$$

$$= \int_{-r}^{r} k\left(hx^2 + \frac{h^3}{3}\right)y\Big]_{y=-\sqrt{r^2-x^2}}^{y=\sqrt{r^2-x^2}} dx$$

$$= \int_{-r}^{r} k\left(2hx^2\sqrt{r^2 - x^2} + \frac{2h^3}{3}\sqrt{r^2 - x^2}\right) dx$$

$$= 2kh\int_{-r}^{r} x^2\sqrt{r^2 - x^2}\,dx + \frac{2kh^3}{3}\int_{-r}^{r}\sqrt{r^2 - x^2}\,dx$$

[Table of integrals]

$$= 2kh\left[\frac{x}{8}(2x^2 - r^2)\sqrt{r^2 - x^2} + \frac{r^4}{8}\sin^{-1}\left(\frac{x}{r}\right)\right]_{-r}^{r}$$

$$+ \frac{2kh^3}{3}\left[\frac{x}{2}\sqrt{r^2 - x^2} + \frac{r^2}{2}\sin^{-1}\left(\frac{x}{r}\right)\right]_{-r}^{r}$$

$$= 2kh\left(\frac{r^4\pi}{8}\right) + \frac{2kh^3}{3}\left(\frac{r^2\pi}{2}\right).$$

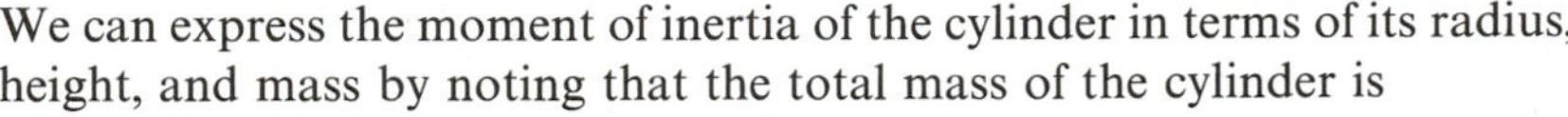

We can express the moment of inertia of the cylinder in terms of its radius, height, and mass by noting that the total mass of the cylinder is

$$m = k(\text{Volume}) = k\pi r^2 h.$$

We can then express the density in terms of $m$, $r$ and $h$. That is,

$$k = \frac{m}{\pi r^2 h}.$$

Substitution then gives

$$I_{y\text{-axis}} = \frac{m}{\pi r^2 h} \frac{hr^4 \pi}{4} + \frac{m}{\pi r^2 h} \frac{h^3 r^2 \pi}{3} = m\left(\frac{r^2}{4} + \frac{h^2}{3}\right). \quad ■$$

Let us see how moments of inertia can be related to the **angular momentum** of a solid. We assume that an object is rotating about the $z$-axis with constant angular velocity $\omega$. The angular momentum about the origin of a small piece of the object with mass $dm$ is

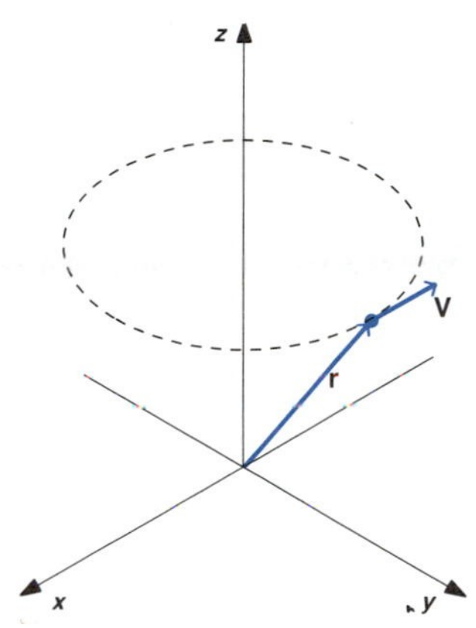

FIGURE 17.6.6

$$d\mathbf{L} = \mathbf{r} \times \mathbf{v}\ dm,$$

where $\mathbf{r} = x\mathbf{i} + y\mathbf{j} + z\mathbf{k}$ is the position vector of the object and $\mathbf{v}$ is its velocity. If the object is rotating counterclockwise about the $z$-axis, we have

$$\mathbf{v} = -\omega y\mathbf{i} + \omega x\mathbf{j}.$$

Note that the speed $|\mathbf{v}|$ is $\omega$ times the radius of rotation and $\mathbf{v}$ is tangent to the circle of rotation. See Figure 17.6.6. Then

$$\mathbf{r} \times \mathbf{v} = (x\mathbf{i} + y\mathbf{j} + z\mathbf{k}) \times (-\omega y\mathbf{i} + \omega x\mathbf{j}) = -\omega xz\mathbf{i} - \omega yz\mathbf{j} + \omega(x^2 + y^2)\mathbf{k}.$$

The total angular momentum is the vector sum of the angular momentum of each of the parts,

$$\mathbf{L} = \iiint d\mathbf{L} = \iiint \mathbf{r} \times \mathbf{v}\ dm.$$

If the object is *symmetric about the z-axis*, the $\mathbf{i}$ and $\mathbf{j}$ components of $\mathbf{r} \times \mathbf{v}$ of diametric points will cancel each other and the angular momentum will become

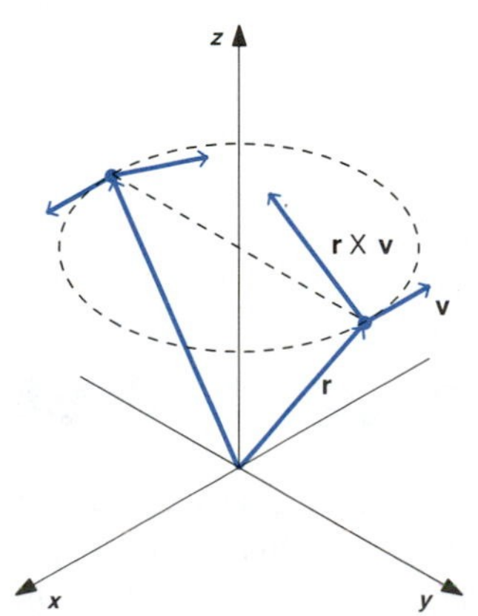

FIGURE 17.6.7

$$\mathbf{L} = \iiint \omega(x^2 + y^2)\mathbf{k}\ dm = \omega I_{z\text{-axis}}\mathbf{k}.$$

See Figure 17.6.7. We know from Section 14.5 that $d\mathbf{L}/dt = \mathbf{0}$ if there is no external force acting on the object, so $\mathbf{L}$ is constant. This means that

$$\omega I_{z\text{-axis}} = \text{constant}.$$

It follows that the angular velocity $\omega$ will increase if the moment of inertia is made smaller. For example, a figure skater can increase the angular velocity of a spin by drawing his or her arms and legs closer to the axis of rotation.

## EXERCISES 17.6

Find the mass of the solids described in Exercises 1–4.

**1** Bounded by $z = 2 - y$, $x = 3$, in the first octant; $\rho(x, y, z) = z$

**2** Bounded by $x + 2y + 2z = 6$, in the first octant; $\rho(x, y, z) = 6 - x$

**3** Bounded by $x = 4$, $y = 6$, $z = 3$, in the first octant; $\rho(x, y, z) = x^2 + y^2 + z^2$

**4** Bounded by $x^2 + y^2 = 4$, $z = 1$, $z = 0$; $\rho(x, y, z) = z$

Find the center of mass of each solid described in Exercises 5–8.

**5** Bounded by $z = 2x + 2y$, $x = 1$, $y = 1$, in the first octant; $\rho(x, y, z) = 6$

**6** Bounded by $x + y + z = 1$, in the first octant; $\rho(x, y, z) = 1$

**7** Bounded by $x = 3$, $y = 2$, $z = 1$, in the first octant; $\rho(x, y, z) = 1 + xyz$

**8** $x^2 + y^2 = 1$, $z = 4$, $z = 0$; $\rho(x, y, z) = z^2$

**9** Find the moment of inertia of a homogeneous right circular cylinder with radius $R$ and height $H$ about its axis. Express the answer in terms of the mass of the cylinder.

**10** Find the moment of inertia of a homogeneous cube with length of each edge $S$ about a line through the center of opposite sides. Express the answer in terms of the mass of the cube.

**11** Find the moment of inertia of a homogeneous cube with length of each edge $S$ about a line along one edge. Express the answer in terms of the mass of the cube.

**12** Set up an iterated triple integral that gives the moment of inertia of a homogeneous solid cone with radius of base $r$ and height $h$, about its axis. $\rho(x, y, z) = 3m/(\pi r^2 h)$.

**13** A figure skater is spinning at a rate of three revolutions per second. Assuming that there are no external forces acting on the skater, find the rate of spin after she decreases the moment of inertia of her body about the axis of rotation by 20 percent by drawing her arms and legs closer to the axis of rotation.

The moment of inertia of an object about any line in three-dimensional space is defined to be the integral of the product of the density function at each point of the object and the square of the distance of the point from the line. From Section 14.4 we know that the distance $d$ of a point $P$ from a line that contains the point $P_0$ and has direction vector $\mathbf{V}$ is given by the formula

$$d = \frac{|\mathbf{PP}_0 \times \mathbf{V}|}{|\mathbf{V}|}.$$

**14** Find the moment of inertia of a homogeneous cube with length of each edge $S$ about a line through opposite corners. Express the answer in terms of the mass of the cube. (*Hint*: Position the cube in the first octant with one corner at the origin. Evaluate the moment of inertia about the line through the origin with direction vector $\mathbf{i} + \mathbf{j} + \mathbf{k}$.)

**15** Set up an iterated triple integral that gives the moment of inertia of the homogeneous solid cone with radius of base $r$ and height $h$, about a line through its vertex along its curved surface. $\rho(x, y, z) = 3m/(\pi r^2 h)$. (*Hint*: Position the cone with vertex at the origin and axis along the positive $z$-axis. Set up the integral that gives the moment of inertia about the line through the origin with direction vector $r\mathbf{j} + h\mathbf{k}$.)

**16** Let $(\bar{x}, \bar{y}, \bar{z})$ be the center of mass of a solid. The first moments of the solid about $x = \bar{x}$ and $y = \bar{y}$ then satisfy $M_{x=\bar{x}} = M_{y=\bar{y}} = 0$. Let $l$ denote the vertical line through the center of mass of the solid, and let $M$ denote the mass of the solid. Show that $I_{x\text{-axis}} = I_l + (\bar{x}^2 + \bar{y}^2)M$, so that the second moment about the $z$-axis is the second moment about the parallel line through the center of mass, plus the square of the distance between the lines times the mass. This illustrates the **Parallel Axis Theorem** for second moments.

## 17.7 CYLINDRICAL AND SPHERICAL COORDINATES

We have seen that it is convenient to express some double integrals in terms of polar coordinates. In Section 17.8 we will see that many triple integrals are considerably easier to evaluate if they are expressed in coordinates other than rectangular. In this section we will introduce cylindrical and spherical coordinate systems and use them to sketch and

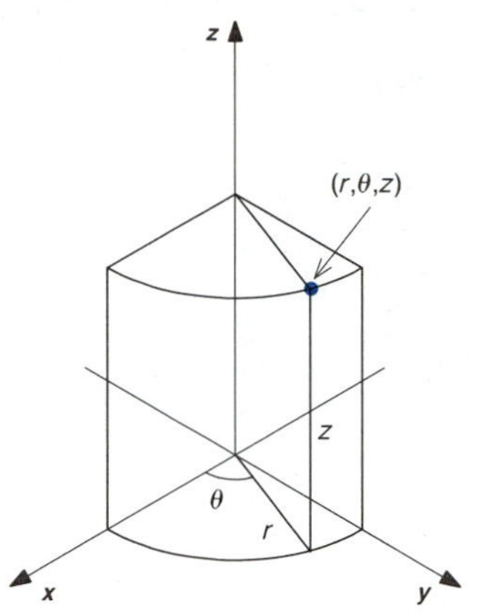


FIGURE 17.7.1

describe regions in three-dimensional space. As in the case of rectangular coordinates, the description of the regions will be used to determine limits of integration of iterated triple integrals.

## *Cylindrical coordinates* $(r, \theta, z)$

The point in three-dimensional space that is associated with the cylindrical coordinates $(r, \theta, z)$ is illustrated in Figure 17.7.1. The cylindrical $z$-coordinate is the same as the $z$-coordinate in rectangular coordinates; $r$ and $\theta$ correspond to polar coordinates of the projection of the point onto the $xy$-plane. Cylindrical and rectangular coordinates are related by the equations

$$x = r \cos \theta,$$

$$y = r \sin \theta,$$

$$z = z.$$

We also have

$$x^2 + y^2 = r^2.$$

The $r$-coordinate is generally taken to be nonnegative.

The three basic surfaces associated with cylindrical coordinates are

- $r = r_0$, a right circular cylinder (Figure 17.7.2a);
- $\theta = \theta_0$ $(r \geq 0)$, a half-plane with edge along the $z$-axis (Figure 17.7.2b);
- $z = z_0$, a horizontal plane (Figure 17.7.2c).

FIGURE 17.7.2

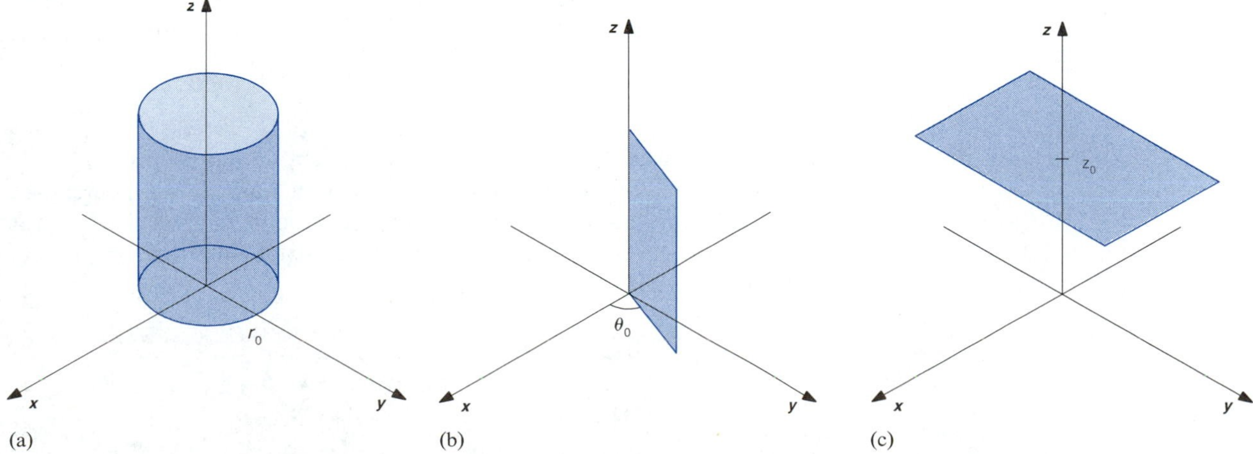

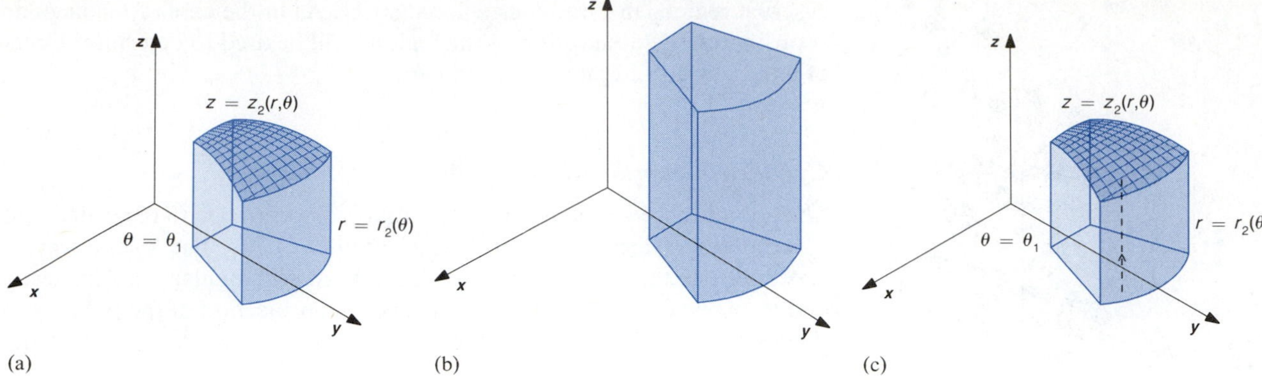


(a) (b) (c)

**FIGURE 17.7.3**

Let us consider a region in three-dimensional space that can be described by the following statements:

For $r$ and $\theta$ fixed, $z$ varies from $z = z_1(r, \theta)$ to $z = z_2(r, \theta)$.
For $\theta$ fixed, $r$ varies from $r = r_1(\theta)$ to $r = r_2(\theta)$.
$\theta$ varies from $\theta = \theta_1$ to $\theta = \theta_2$.

We assume the boundary functions are continuous with $z_1(r, \theta) \leq z_2(r, \theta)$, $0 \leq r_1(\theta) \leq r_2(\theta)$, and $\theta_1 \leq \theta_2$. See Figure 17.7.3a. The vertical generalized cylinders $r = r_1(\theta)$, $r = r_2(\theta)$, $\theta = \theta_1$, and $\theta = \theta_2$ form the vertical side boundaries of the region and give the projection of the region onto the $xy$-plane. See Figure 17.7.3b. The surfaces $z = z_1(r, \theta)$ and $z = z_2(r, \theta)$ form the bottom and top of the region. An arrow that is shot vertically from below the region follows a trajectory with $r$ and $\theta$ fixed; travels parallel to the vertical surfaces $r = r_1(\theta)$, $r = r_2(\theta)$, $\theta = \theta_1$, and $\theta = \theta_2$; enters the region at a point on the bottom surface $z = z_1(r, \theta)$; and leaves the region at a point on the top surface $z = z_2(r, \theta)$. See Figure 17.7.3c.

### ■ EXAMPLE 1

Sketch and use cylindrical coordinates to describe the region in the first octant that is inside the cylinder $x^2 + y^2 = 2$ and below the plane $x + y + z = 4$.

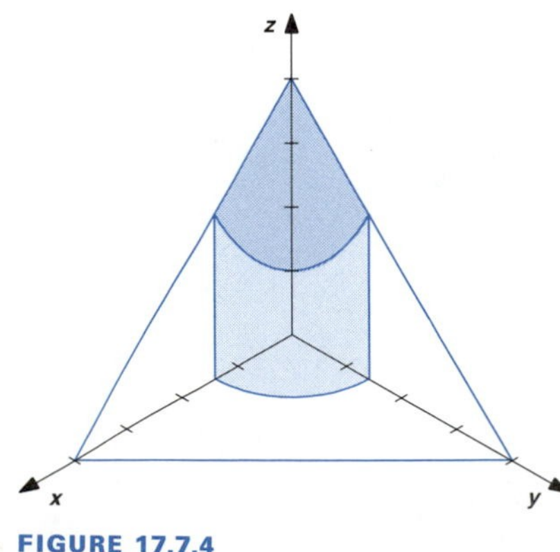


**FIGURE 17.7.4**

#### SOLUTION

The region is sketched in Figure 17.7.4. An arrow that is shot vertically from below the region follows a trajectory with $r$ and $\theta$ fixed, enters the region at a point on the plane $z = 0$ and leaves the region at a point on the plane $x + y + z = 4$. We need to express the given equations in terms of cylindrical coordinates.

$x^2 + y^2 = 2$ implies $(r \cos \theta)^2 + (r \sin \theta)^2 = 2$, so $r = \sqrt{2}$.
$x + y + z = 4$ implies $r \cos \theta + r \sin \theta + z = 4$, so
$z = 4 - r \cos \theta - r \sin \theta$.

For $r$ and $\theta$ fixed, $z$ varies from $z = 0$ to $z = 4 - r\cos\theta - r\sin\theta$.
For $\theta$ fixed, $r$ varies from $r = 0$ to $r = \sqrt{2}$.
$\theta$ varies from $\theta = 0$ to $\theta = \pi/2$. ■

■ **EXAMPLE 2**

Sketch and use cylindrical coordinates to describe the region in the first octant that is inside the sphere $x^2 + y^2 + z^2 = 4$ and above the plane $z = 1$.

**SOLUTION**

The region is sketched in Figure 17.7.5a. An arrow that is shot vertically will enter the region at a point on the plane and leave the region at a point on the sphere. We need to express the equation of the sphere in terms of cylindrical coordinates.

$$x^2 + y^2 + z^2 = 4 \quad \text{implies} \quad (r\cos\theta)^2 + (r\sin\theta)^2 + z^2 = 4.$$

Since $z \geq 0$, we obtain $z = \sqrt{4 - r^2}$. This is the top of the region. $z = 1$ is the bottom of the region.

The top and bottom intersect to form a side boundary.

$$z = \sqrt{4 - r^2} \quad \text{and} \quad z = 1 \quad \text{imply} \quad 1^2 = 4 - r^2, \quad \text{so} \quad r = \sqrt{3}.$$

This is the equation of the vertical cylinder that contains the intersection of the top and bottom. See Figure 17.7.5b.

For $r$ and $\theta$ fixed, $z$ varies from $z = 1$ to $z = \sqrt{4 - r^2}$.
For $\theta$ fixed, $r$ varies from $r = 0$ to $r = \sqrt{3}$.
$\theta$ varies from $\theta = 0$ to $\theta = \pi/2$. ■

FIGURE 17.7.5

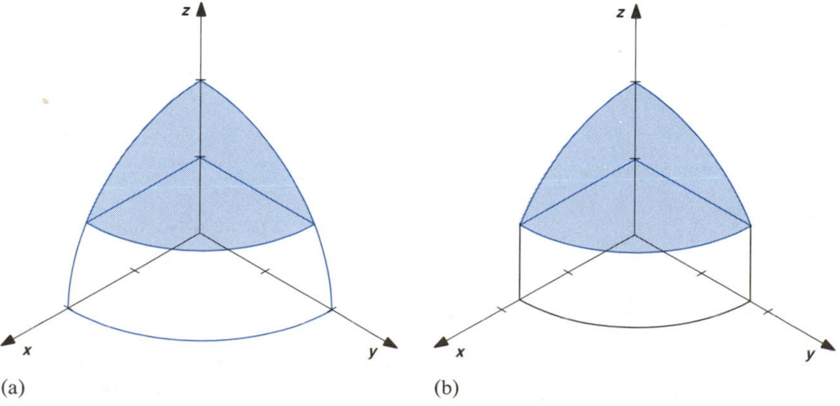

FIGURE 17.7.6

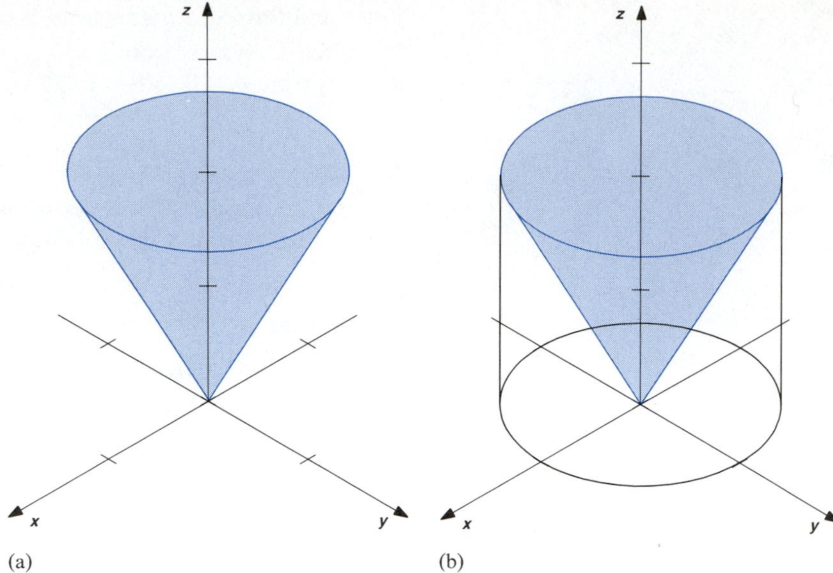


(a) (b)

FIGURE 17.7.7

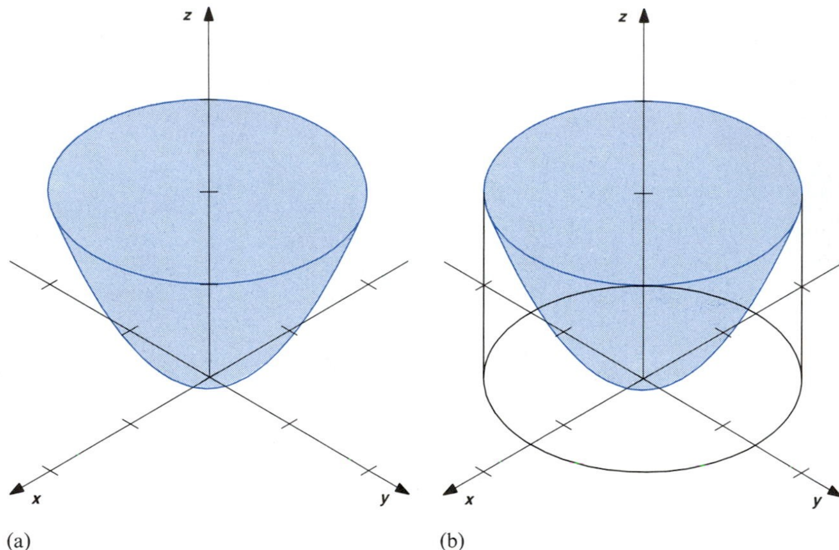


(a) (b)

**EXAMPLE 3**

Sketch and use cylindrical coordinates to describe the region that is inside the cone $z = 2\sqrt{x^2 + y^2}$ and below the plane $z = 2$.

**SOLUTION**

The region is sketched in Figure 17.7.6a. An arrow that is shot vertically will enter the region at a point on the cone and leave the region at a point on the plane $z = 2$. The cylindrical equation of the cone is $z = 2r$. The intersection of the surfaces $z = 2r$ and $z = 2$ is contained in the cylinder $r = 1$. See Figure 17.7.6b.

For $r$ and $\theta$ fixed, $z$ varies from $z = 2r$ to $z = 2$.
For $\theta$ fixed, $r$ varies from $r = 0$ to $r = 1$.
$\theta$ varies from $\theta = 0$ to $\theta = 2\pi$. ■

■ **EXAMPLE 4**

Sketch and use cylindrical coordinates to describe the region that is inside the paraboloid $z = x^2 + y^2$ and below the plane $z = 2$.

**SOLUTION**

The region is sketched in Figure 17.7.7a. An arrow shot vertically enters the region on the paraboloid and leaves on the plane. The cylindrical equation of the paraboloid is $z = r^2$. The intersection of the surfaces $z = r^2$ and $z = 2$ is contained in the cylinder $r = \sqrt{2}$. See Figure 17.7.7b.

For $r$ and $\theta$ fixed, $z$ varies from $z = r^2$ to $z = 2$.
For $\theta$ fixed, $r$ varies from $r = 0$ to $r = \sqrt{2}$.
$\theta$ varies from $\theta = 0$ to $\theta = 2\pi$. ■

### *Spherical coordinates* $(\rho, \theta, \phi)$

The point in three-dimensional space that is associated with the spherical coordinates $(\rho, \theta, \phi)$ is illustrated in Figure 17.7.8. The spherical coordinate $\rho$ is the distance of the point from the origin; $\theta$ corresponds to $\theta$ in cylindrical coordinates; $\phi$ is the angle between the positive $z$-axis and the ray from the origin to the point; $\phi = 0$ is in the direction of the positive

FIGURE 17.7.8

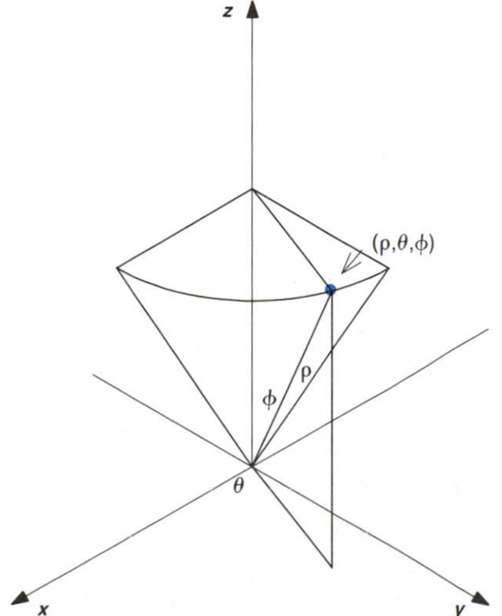

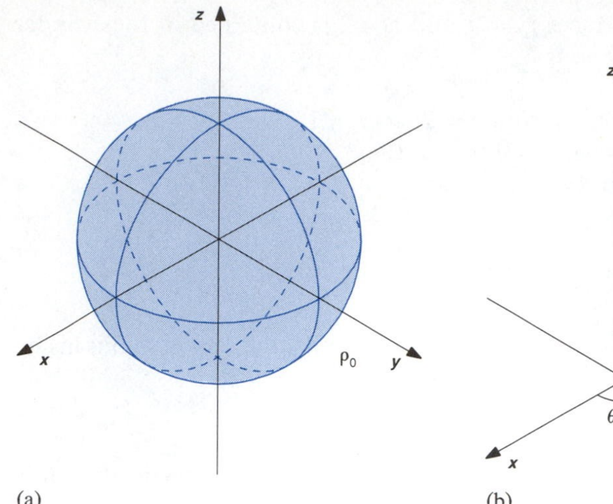


(a)

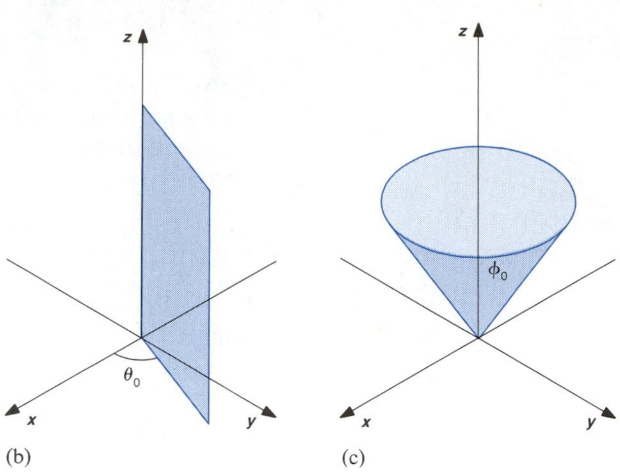


(b) (c)

**FIGURE 17.7.9**

$z$-axis, and $\phi = \pi$ is in the direction of the negative $z$-axis. Spherical and rectangular coordinates are related by the equations

$$x = \rho \sin \phi \cos \theta,$$

$$y = \rho \sin \phi \sin \theta,$$

$$z = \rho \cos \phi.$$

We also have

$$x^2 + y^2 + z^2 = \rho^2.$$

The $\rho$-coordinate is taken to be nonnegative and $0 \le \phi \le \pi$.

The three basic surfaces associated with spherical coordinates are

- $\rho = \rho_0$, a sphere (Figure 17.7.9a);
- $\theta = \theta_0$, a half-plane with edge along the $z$-axis (Figure 17.7.9b);
- $\phi = \phi_0$, a cone (Figure 17.7.9c).

(The graph of $\phi = \pi/2$ is the $xy$-plane.)

A region of the type illustrated in Figure 17.7.10a may be described by the following statements:

For $\theta$ and $\phi$ fixed, $\rho$ varies from $\rho = \rho_1(\theta, \phi)$ to $\rho = \rho_2(\theta, \phi)$.
For $\theta$ fixed, $\phi$ varies from $\phi = \phi_1(\theta)$ to $\phi = \phi_2(\theta)$.
$\theta$ varies from $\theta = \theta_1$ to $\theta = \theta_2$.

The boundary surfaces $\phi = \phi_1(\theta)$, $\phi = \phi_2(\theta)$, $\theta = \theta_1$, and $\theta = \theta_2$ are generated by rays from the origin. See Figure 17.7.10b. An arrow that is shot from the origin along a trajectory with fixed $\theta$ and $\phi$ will enter the region at a point on the surface $\rho = \rho_1(\theta, \phi)$ and leave the region at a point on the surface $\rho = \rho_1(\theta, \phi)$. See Figure 17.7.10c.

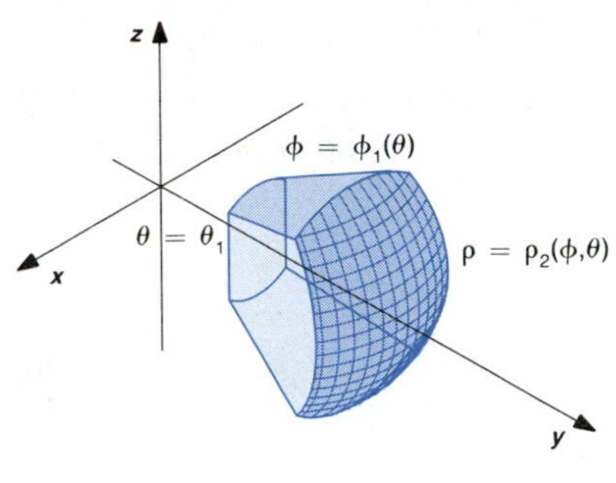


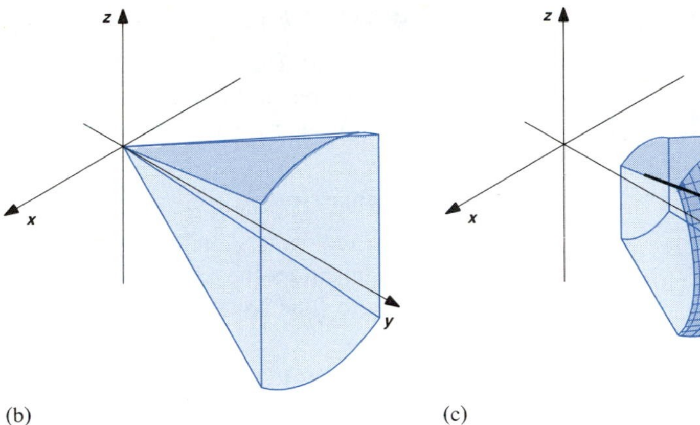


(a) (b) (c)

FIGURE 17.7.10

### ■ EXAMPLE 5

Sketch and use spherical coordinates to describe the region that is inside both the cone $z = \sqrt{x^2 + y^2}$ and the sphere $x^2 + y^2 + z^2 = 4$.

**SOLUTION**

The region is sketched in Figure 17.7.11. An arrow that is shot from the origin enters the region at the origin and leaves the region at a point on the sphere. The spherical equation of the sphere is $\rho = 2$ and the spherical equation of the cone is $\phi = \pi/4$. That is,

FIGURE 17.7.11

$x^2 + y^2 + z^2 = 4$ implies

$$\rho^2 \sin^2 \phi \cos^2 \theta + \rho^2 \sin^2 \phi \sin^2 \theta + \rho^2 \cos^2 \phi = 4,$$

$$\rho^2 \sin^2 \phi(\cos^2 \theta + \sin^2 \theta) + \rho^2 \cos^2 \phi = 4,$$

$$\rho^2 \sin^2 \phi + \rho^2 \cos^2 \phi = 4,$$

$$\rho^2 = 4,$$

$$\rho = 2;$$

$z^2 = x^2 + y^2$ implies

$$\rho^2 \cos^2 \phi = \rho^2 \sin^2 \phi \cos^2 \theta + \rho^2 \sin^2 \phi \sin^2 \theta,$$

$$\cos^2 \phi = \sin^2 \phi,$$

$$\tan^2 \phi = 1,$$

$$\tan \phi = 1,$$

[$\tan \phi > 0$ in this example, since $z > 0$.]

$$\phi = \frac{\pi}{4}.$$

For $\theta$ and $\phi$ fixed, $\rho$ varies from $\rho = 0$ to $\rho = 2$.
For $\theta$ fixed, $\phi$ varies from $\phi = 0$ to $\phi = \pi/4$.
$\theta$ varies from $\theta = 0$ to $\theta = 2\pi$. ■

### ■ EXAMPLE 6

Sketch and use spherical coordinates to describe the region in the first octant that is between the planes $y = 0$ and $y = x$, and inside the sphere $x^2 + y^2 + z^2 = 1$.

FIGURE 17.7.12

#### SOLUTION

The region is sketched in Figure 17.7.12. An arrow that is shot from the origin enters the region at the origin and leaves the region at a point on the sphere. The spherical equation of the half-plane $y = 0$ $(x \geq 0)$ is $\theta = 0$.

$y = x$ implies

$$\rho \sin \phi \sin \theta = \rho \sin \phi \cos \theta,$$

$$\sin \theta = \cos \theta,$$

$$\tan \theta = 1,$$

$$\theta = \frac{\pi}{4}.$$

The $xy$-plane has spherical equation $\phi = \pi/2$.

For $\theta$ and $\phi$ fixed, $\rho$ varies from $\rho = 0$ to $\rho = 1$.
For $\theta$ fixed, $\phi$ varies from $\phi = 0$ to $\phi = \pi/2$.
$r$ varies from $\theta = 0$ to $\theta = \pi/4$. ■

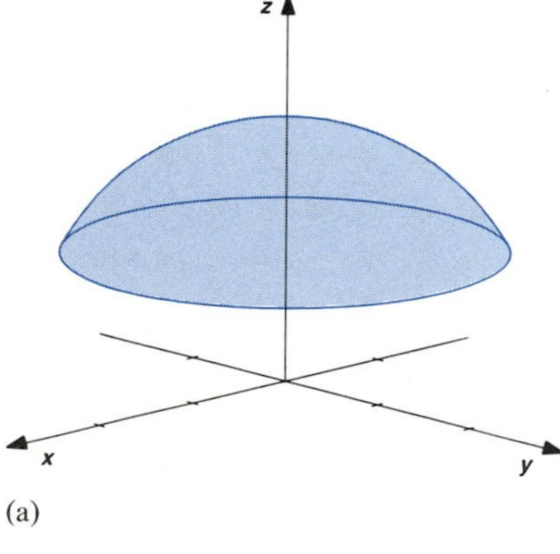


(a)

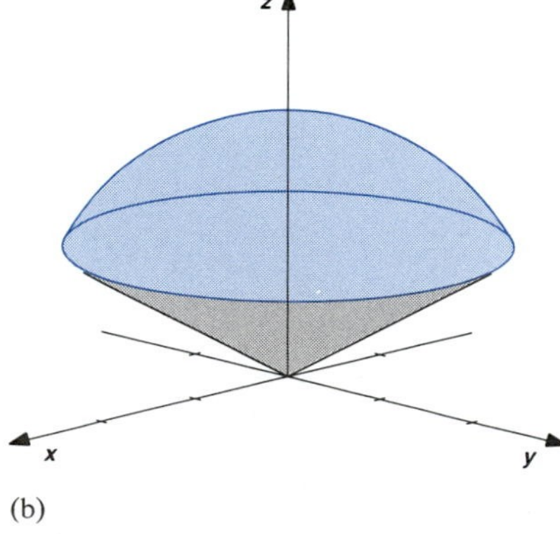


(b)

FIGURE 17.7.13

### ■ EXAMPLE 7

Sketch and use spherical coordinates to describe the region that is inside the sphere $x^2 + y^2 + z^2 = 4$ and above the plane $z = 1$.

#### SOLUTION

The region is sketched in Figure 17.7.13a. An arrow that is shot from the origin enters the region at a point on the plane and leaves the region at a point on the sphere. We need spherical equations of the plane $z = 1$.

$z = 1$ implies

$$\rho \cos \phi = 1, \quad \text{so} \quad \rho = \sec \phi.$$

Points of intersection of the sphere $\rho = 2$ and the plane $\rho = \sec \phi$ satisfy $\sec \phi = 2$. It follows that the intersection is contained in the cone $\phi = \pi/3$. See Figure 17.7.13b.

For $\theta$ and $\phi$ fixed, $\rho$ varies from $\rho = \sec \phi$ to $\rho = 2$.
For $\theta$ fixed, $\phi$ varies from $\phi = 0$ to $\phi = \pi/3$.
$\theta$ varies from $\theta = 0$ to $\theta = 2\pi$. ■

**EXAMPLE 8**

Sketch and use spherical coordinates to describe the region that is inside the sphere $x^2 + y^2 + z^2 = 4$ and outside the cylinder $x^2 + y^2 = 1$.

**SOLUTION**

The region is sketched in Figure 17.7.14a. An arrow shot from the origin enters the region at a point on the cylinder and leaves the region at a point on the sphere. We need spherical equations of the cylinder.

$x^2 + y^2 = 1$ implies

$$\rho^2 \sin^2\phi \cos^2\theta + \rho^2 \sin^2\phi \sin^2\theta = 1,$$

$$\rho^2 \sin^2\phi(\cos^2\theta + \sin^2\theta) = 1,$$

$$\rho^2 \sin^2\phi = 1,$$

$$\rho^2 = \csc^2\phi,$$

$$\rho = \csc \phi.$$

Points of intersection of the sphere $\rho = 2$ and the cylinder $\rho = \csc \phi$ satisfy $\csc \phi = 2$. It follows that the intersection is contained between the two cones, $\phi = \pi/6$ and $\phi = 5\pi/6$. See Figure 17.7.14b.

For $\theta$ and $\phi$ fixed, $\rho$ varies from $\rho = \csc \phi$ to $\rho = 2$.
For $\theta$ fixed, $\phi$ varies from $\phi = \pi/6$ to $\phi = 5\pi/6$.
$\theta$ varies from $\theta = 0$ to $\theta = 2\pi$. ■

FIGURE 17.7.14

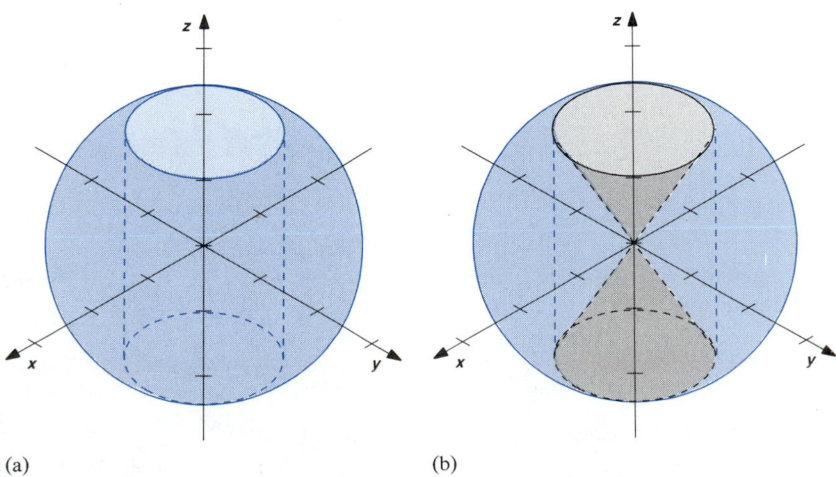

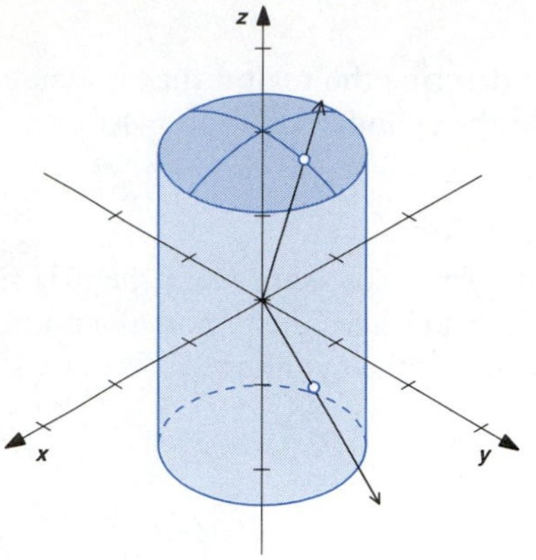

FIGURE 17.7.15

The region that is inside both the sphere $x^2 + y^2 + z^2 = 4$ and the cylinder $x^2 + y^2 = 1$ is not the type of region that we can describe in terms of spherical coordinates with one set of equations, because an arrow that is shot from the origin could leave the region at a point on either the sphere or the cylinder. See Figure 17.7.15. The region is easy to describe in terms of cylindrical coordinates.

Be careful when using the inverse tangent function to find the spherical equation of a cone $z = -h\sqrt{x^2 + y^2}/r$, where $h$ and $r$ are positive. See Figure 17.7.16a. Substitution gives

$$\rho \cos \phi = -\frac{h(\rho \sin \phi)}{r}, \quad \text{so} \quad \tan \phi = -\frac{r}{h}.$$

Since the value of $\tan^{-1}(-r/h)$ is between $-\pi/2$ and 0, we have

$$\phi = \tan^{-1}\left(-\frac{r}{h}\right) + \pi.$$

See Figure 17.7.16b.

FIGURE 17.7.16

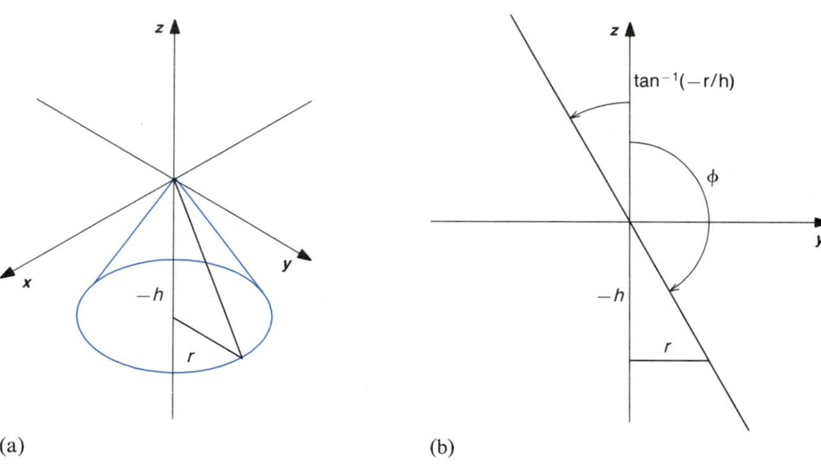


## EXERCISES 17.7

Plot the points with the given cylindrical coordinates $(r, \theta, z)$ in Exercises 1–2.

**1** $A(1, 0, 0)$, $B(2, \pi/4, 2)$, $C(3, \pi, 1)$, $D(3, -\pi/2, -1)$

**2** $A(2, \pi/2, 0)$, $B(2, \pi/3, 0)$, $C(2, 0, -1)$, $D(3, -\pi/4, 2)$

Plot the points with the given spherical coordinates $(\rho, \theta, \phi)$ in Exercises 3–4.

**3** $A(2, 0, \pi)$, $B(2, \pi/2, \pi/4)$, $C(3, -\pi/4, \pi/2)$, $D(3, \pi/2, 3\pi/4)$

**4** $A(3, 0, 0)$, $B(3, \pi/2, \pi/6)$, $C(2, \pi/4, \pi/2)$, $D(3, -\pi/2, 5\pi/6)$

Find the rectangular coordinates of the points with the given cylindrical coordinates $(r, \theta, z)$ in Exercises 5–6.

**5** $A(0, \pi, 2)$, $B(2, \pi/4, -1)$, $C(2, \pi/3, 3)$

**6** $A(\pi, 0, \pi)$, $B(4, -\pi/4, 2)$, $C(2, 2\pi/3, 1)$

Find rectangular coordinates of the points with the given spherical coordinates $(\rho, \theta, \phi)$ in Exercises 7–8.

**7** $A(8, \pi/2, \pi/4)$, $B(1, \pi/4, \pi/2)$, $C(2, \pi/3, 2\pi/3)$

**8** $A(4, \pi/2, \pi/3)$, $B(3, \pi, \pi/2)$, $C(2, \pi/4, 3\pi/4)$

Find cylindrical coordinates of the points with the given rectangular coordinates $(x, y, z)$ in Exercises 9–10.

**9** $A(2, 2, 2)$, $B(0, 0, -1)$, $C(4, -4\sqrt{3}, 1)$

**10** $A(1, -1, 1)$, $B(0, 2, -1)$, $C(-2, 0, 2)$

Find spherical coordinates of the points with the given rectangular coordinates $(x, y, z)$ in Exercises 11–12.

**11** $A(0, 0, -2)$, $B(2, 2, 0)$, $C(0, 2, 2)$

**12** $A(0, 0, 1)$, $B(0, 2, -2)$, $C(2, 0, 2)$

Sketch and use cylindrical coordinates to describe the regions given in Exercises 13–24.

**13** Inside $x^2 + y^2 = 4$, below $z = 2x + 2y$, in the first octant

**14** Inside both $x^2 + y^2 = 1$ and $x^2 + y^2 + z^2 = 4$

**15** Inside $x^2 + y^2 = 9$, below $z = \sqrt{x^2 + y^2}$, above $z = 0$

**16** Inside $x^2 + y^2 = 4$, below $x + z = 4$, in the first octant

**17** $z = 4 - x^2 - y^2$, $z = 0$

**18** $z = 4 - 2\sqrt{x^2 + y^2}$, $z = 0$

**19** $z = 2 - e^{x^2+y^2}$, $z = 0$

**20** $z = \cos\sqrt{x^2 + y^2}$, $z = 0$ $[x^2 + y^2 \le (\pi/2)^2]$

**21** Inside $x^2 + y^2 + z^2 = 16$, $x + y = 4$ $(x + y \le 4)$, in the first octant

**22** $x + y + z = 3$, in the first octant

**23** $x^2 + y^2 = 2x$, $z = 4$, $z = 0$

**24** $x^2 + y^2 = 4y$, $y + z = 4$, $z = 0$

Sketch and use spherical coordinates to describe the regions given in Exercises 25–36.

**25** $z = \sqrt{9 - x^2 - y^2}$, $z = 0$

**26** Inside $x^2 + y^2 + z^2 = 4$, above $z = -\sqrt{x^2 + y^2}$

**27** Inside $x^2 + y^2 + z^2 = 1$, in the first octant

**28** Above $z = \sqrt{x^2 + y^2}$, below $z = 4$

**29** Inside $x^2 + y^2 = 16$, below $z = \sqrt{x^2 + y^2}$, above $z = 0$

**30** Inside $z = 4 - 2\sqrt{x^2 + y^2}$, outside $x^2 + y^2 = 1$, above $z = 0$

**31** Inside $x^2 + y^2 + z^2 = 25$, above $z = 3$

**32** Inside $x^2 + y^2 + z^2 = 25$, below $z = -3$

**33** $x + y + z = 3$, in the first octant

**34** Inside $x^2 + y^2 + z^2 = 1$, between $y = \sqrt{3}x$ and $y = x/\sqrt{3}$, in the first octant

**35** Inside $z = 2\sqrt{x^2 + y^2}$, below $x + z = 2$

**36** Inside $z = -3\sqrt{x^2 + y^2}$, above $z = -4$

## 17.8 MULTIPLE INTEGRALS IN CYLINDRICAL AND SPHERICAL COORDINATES

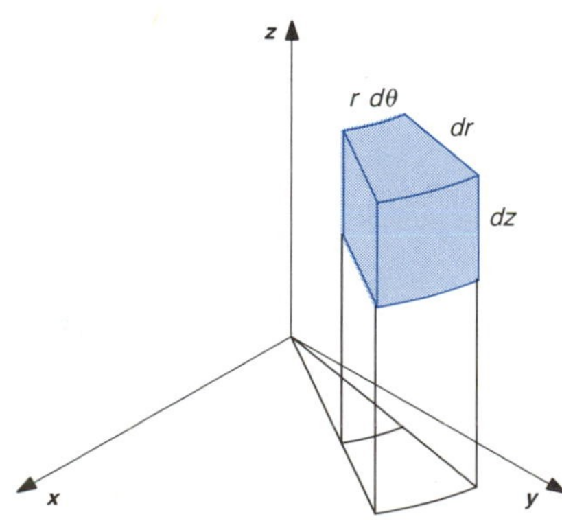


FIGURE 17.8.1

### *Cylindrical coordinates*

Let us see how to evaluate a triple integral

$$\iiint_R f\, dV$$

in terms of cylindrical coordinates. We begin by dividing the region $R$ into prisms that are bounded by the surfaces

$$r = \text{constant}, \quad \theta = \text{constant}, \quad \text{and} \quad z = \text{constant}.$$

See Figure 17.8.1. Each of these prisms has eight edges that are line segments and four edges that are arcs of circles. The edges are orthogonal. For small increments, the four edges that are arcs of circles are nearly straight and the prisms are nearly rectangular. The differential volume of a small prism is

$$dV = r\, dz\, dr\, d\theta.$$

If each prism is small, the sum of products of the value of $f$ at some point in each prism times the differential volume of the prism should be approximately $\iiint_R f\,dV$.

Suppose the region $R$ can be described in terms of cylindrical coordinates by the statements:

For $r$ and $\theta$ fixed, $z$ varies from $z = z_1(r, \theta)$ to $z = z_2(r, \theta)$.
For $\theta$ fixed, $r$ varies from $r = r_1(\theta)$ to $r = r_2(\theta)$.
$\theta$ varies from $\theta = \theta_1$ to $\theta = \theta_2$.

The surfaces $z = z_1(r, \theta)$ and $z = z_2(r, \theta)$ form the bottom and top of the region. The vertical surfaces $r = r_1(\theta)$, $r = r_2(\theta)$, $\theta = \theta_1$, and $\theta = \theta_2$ form the side boundaries of the region. See Figure 17.8.2a. As $z$ varies from $z = z_1(r, \theta)$ to $z = z_2(r, \theta)$, the points $(r, \theta, z)$ form a vertical line segment between the top and bottom of the region. See Figure 17.8.2b. As $r$ varies from $r = r_1(\theta)$ to $r = r_2(\theta)$ with $\theta$ fixed, these line segments generate a cross section of the region in the half-plane with edge along the $z$-axis and in the direction of $\theta$. See Figure 17.8.2c. As $\theta$ varies from $\theta = \theta_1$ to $\theta = \theta_2$,

FIGURE 17.8.2

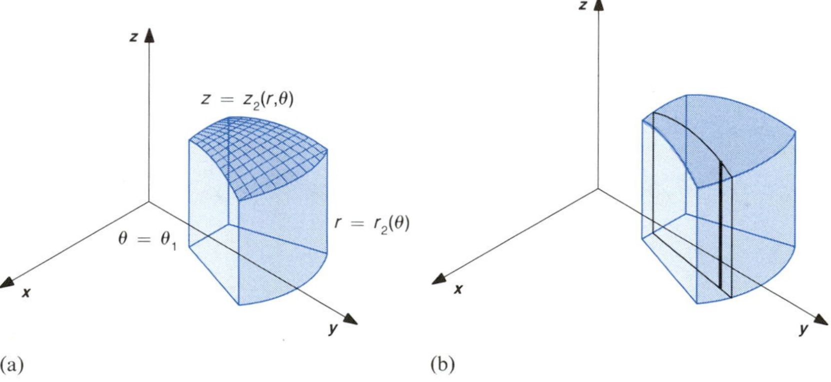


(a) (b)

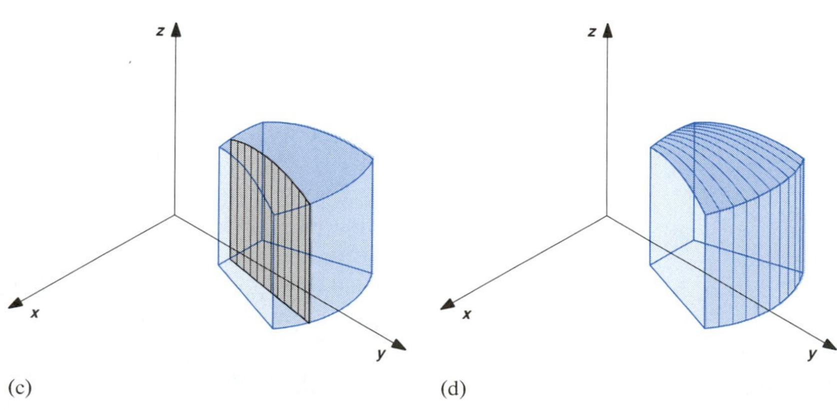


(c) (d)

the cross sections sweep out the entire region. See Figure 17.8.2d. Using the formula above for the differential volume $dV$, we then have

$$\iiint_R f\,dV = \int_{\theta_1}^{\theta_2} \int_{r_1(\theta)}^{r_2(\theta)} \int_{z_1(r,\theta)}^{z_2(r,\theta)} f(r, \theta, z) r\,dz\,dr\,d\theta.$$

Note that the integrand must be expressed in terms of cylindrical coordinates. Recall that we can use the equations

$$x = r\cos\theta,$$

$$y = r\sin\theta,$$

$$z = z,$$

to change from rectangular coordinates to cylindrical coordinates. It is useful to remember that

$$x^2 + y^2 = r^2.$$

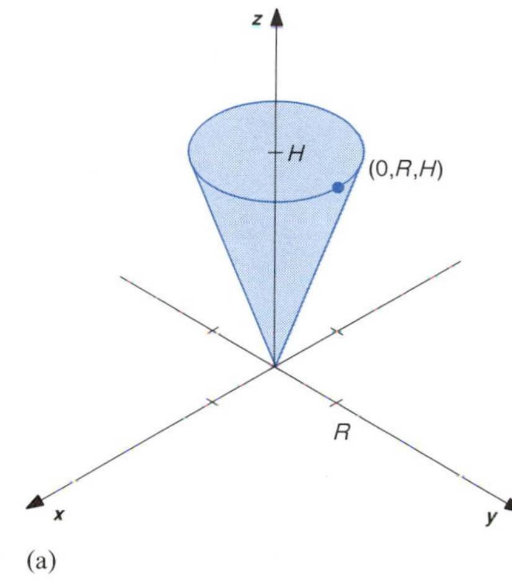

(a)

We may use cylindrical coordinates to evaluate a triple integral whenever the region and the integrand are appropriate. The appearance of the expression $x^2 + y^2$ in either an equation of the boundary of the region or in the integrand is an indication that cylindrical coordinates might be appropriate.

**EXAMPLE 1**

Find the volume of a cone that has height $H$ and radius of base $R$.

**SOLUTION**

Let us position the cone as indicated in Figure 17.8.3a. The equation of the cone in rectangular coordinates is then $z = H\sqrt{x^2 + y^2}/R$. We want the volume of that part of the cone that is below $z = H$.

$$z = H \quad \text{and} \quad z = H\sqrt{x^2 + y^2}/R \quad \text{imply} \quad x^2 + y^2 = R^2.$$

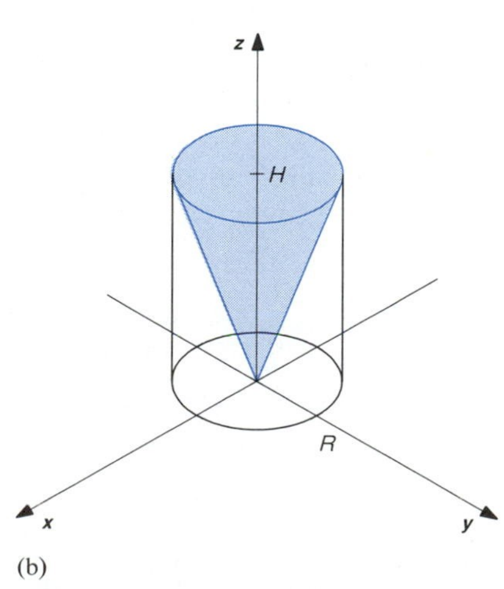

(b)

FIGURE 17.8.3

See Figure 17.8.3b. We see that the region can be described easily in terms of cylindrical coordinates. The cone has cylindrical equation $z = Hr/R$. Using $dV = r\,dz\,dr\,d\theta$, we then have

$$V = \iiint_R dV = \int_0^{2\pi} \int_0^R \int_{Hr/R}^H r\,dz\,dr\,d\theta$$

$$= \int_0^{2\pi} \int_0^R \left(Hr - \frac{Hr^2}{R}\right) dr\,d\theta$$

$$= \int_0^{2\pi} \left(\frac{HR^2}{2} - \frac{HR^3}{3R}\right) d\theta = \left(\frac{HR^2}{6}\right)(2\pi) = \frac{\pi R^2 H}{3}. \quad \blacksquare$$

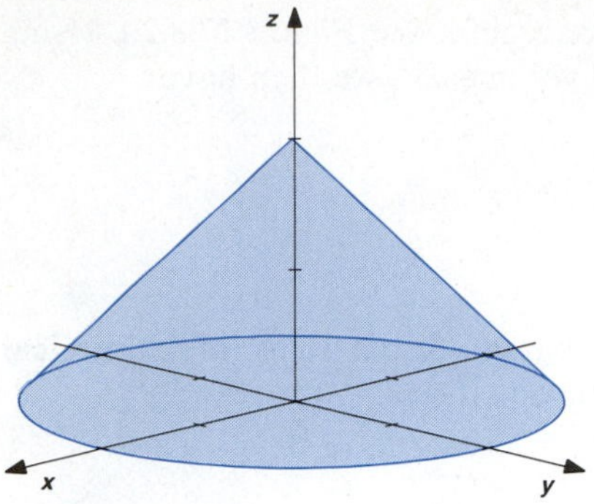

FIGURE 17.8.4

■ **EXAMPLE 2**

Find the mass of a pile of snow that is in the shape of the cone $z = 2 - \sqrt{x^2 + y^2}$ if the density of the snow is given by $\rho(x, y, z) = 2 - z$.

**SOLUTION**

The snow pile is pictured in Figure 17.8.4. We can avoid the square root in the limits of integration of the iterated integral that describes the region by using cylindrical coordinates. The cone is given by the cylindrical equation $z = 2 - r$. The mass is

$$\iiint_R dM = \iiint_R \rho\, dV = \int_0^{2\pi} \int_0^2 \int_0^{2-r} (2 - z)r\, dz\, dr\, d\theta$$

$$= \int_0^{2\pi} \int_0^2 \left(2z - \frac{z^2}{2}\right) r \Bigg]_{z=0}^{z=2-r} dr\, d\theta$$

$$= \int_0^{2\pi} \int_0^2 \left[2(2 - r) - \frac{(2 - r)^2}{2}\right] r\, dr\, d\theta$$

$$= \int_0^{2\pi} \int_0^2 \left[4 - 2r - 2 + 2r - \frac{r^2}{2}\right] r\, dr\, d\theta$$

$$= \int_0^{2\pi} \int_0^2 \left[2r - \frac{r^3}{2}\right] dr\, d\theta$$

$$= \int_0^{2\pi} \left[r^2 - \frac{r^4}{8}\right]_{r=0}^{r=2} d\theta$$

$$= \int_0^{2\pi} [4 - 2]\, d\theta = 4\pi.$$ ■

■ **EXAMPLE 3**

Find the moment of inertia of a homogeneous right circular cylinder with radius $R$ and height $H$ about a line along its curved surface, parallel to its axis. Express the answer in terms of the mass of the cylinder.

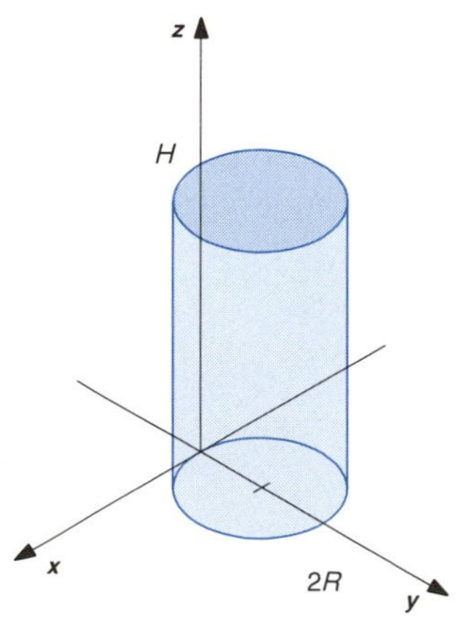

FIGURE 17.8.5

**SOLUTION**

Let us position the cylinder as in Figure 17.8.5. The desired moment of inertia is given by $I_{z\text{-axis}}$. The equation of the cylinder in rectangular coordinates is

$$x^2 + (y - R)^2 = R^2.$$

Simplifying, we obtain

$$x^2 + y^2 - 2Ry + R^2 = R^2,$$

$$x^2 + y^2 = 2Ry.$$

In terms of cylindrical coordinates, the equation is

$$r^2 = 2Rr \sin \theta,$$

$$r = 2R \sin \theta.$$

The cylinder is generated as $\theta$ varies from 0 to $\pi$. The values of $r$ are also nonnegative for $\theta$ in that interval. The density is $\rho(x, y, z) = k$. We then have

$$\begin{aligned} I_{z\text{-axis}} &= \iiint_R (x^2 + y^2)\rho \, dV \\ &= \int_0^{\pi} \int_0^{2R\sin\theta} \int_0^{H} (r^2)(k)r \, dz \, dr \, d\theta \\ &= \int_0^{\pi} \int_0^{2R\sin\theta} kHr^3 \, dr \, d\theta \\ &= \int_0^{\pi} kH \frac{r^4}{4}\Bigg]_{r=0}^{r=2R\sin\theta} d\theta \\ &= \int_0^{\pi} 4kHR^4 \sin^4 \theta \, d\theta. \end{aligned}$$

Using the table of integrals to evaluate the integral, we obtain

$$\begin{aligned} I_{z\text{-axis}} &= 4kHR^4 \left\{ -\frac{\sin^3 \theta \cos \theta}{4}\Bigg]_{\theta=0}^{\theta=\pi} + \frac{3}{4}\int_0^{\pi} \sin^2 \theta \, d\theta \right\} \\ &= 3kHR^4 \int_0^{\pi} \sin^2 \theta \, d\theta \\ &= 3kHR^4 \left[ -\frac{\sin \theta \cos \theta}{2} + \frac{\theta}{2} \right]_0^{\pi} = k\frac{3\pi HR^4}{2}. \end{aligned}$$

We want to express the constant $k$ in terms of the mass of the cylinder. The mass is

$$M = k(\text{Volume}) = k\pi R^2 H$$

Then

$$k = \frac{M}{\pi R^2 H}, \quad \text{so}$$

$$I_{z\text{-axis}} = \frac{M}{\pi R^2 H} \frac{3\pi HR^4}{2} = \frac{3MR^2}{2}.$$

(In Example 3 of Section 17.6 we showed that the moment of inertia of the cylinder about its axis is $MR^2/2$.) ■

FIGURE 17.8.6

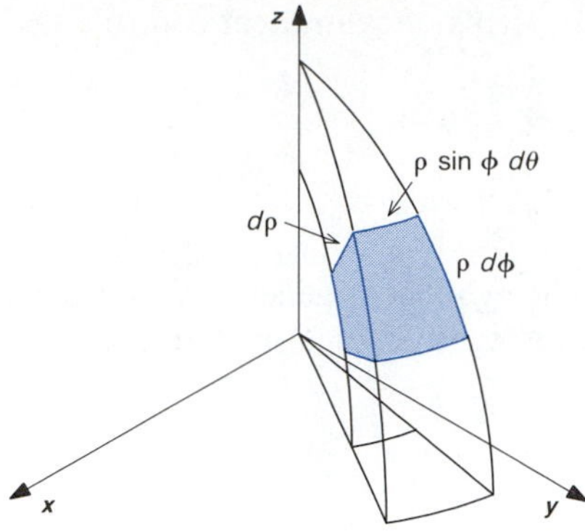


### *Spherical coordinates*

To express a triple integral in terms of spherical coordinates, we divide the region $R$ along surfaces

$$\rho = \text{constant}, \quad \phi = \text{constant}, \quad \text{and} \quad \theta = \text{constant}.$$

For small increments, the resulting prisms are essentially rectangular. See Figure 17.8.6. The differential volume is then

$$dV = \rho^2 \sin \phi \; d\rho \; d\phi \; d\theta.$$

The sum of the products of $f$ evaluated at a point in each prism times the differential volume of the prism should be approximately

$$\iiint_R f \, dV.$$

Suppose the region $R$ can be described in terms of spherical coordinates by the following statements:

For $\theta$ and $\phi$ fixed, $\rho$ varies from $\rho = \rho_1(\theta, \phi)$ to $\rho = \rho_2(\theta, \phi)$.
For $\theta$ fixed, $\phi$ varies from $\phi = \phi_1(\theta)$ to $\phi = \phi_2(\theta)$.
$\theta$ varies from $\theta = \theta_1$ to $\theta = \theta_2$.

The surfaces $\phi = \phi_1(\theta)$, $\phi = \phi_2(\theta)$, $\theta = \theta_1$, and $\theta = \theta_2$ are generated by rays from the origin. An arrow that is shot from the origin and follows a trajectory with $\theta$ and $\phi$ fixed enters the region at a point on the surface $\rho = \rho_1(\theta, \phi)$ and leaves the region at a point on the surface $\rho = \rho_2(\theta, \phi)$. See Figure 17.8.7a. As $\rho$ varies from $\rho = \rho_1(\theta, \phi)$ to $\rho = \rho_2(\theta, \phi)$ with $\phi$ and $\theta$ fixed, the points $(\rho, \theta, \phi)$ form a line segment along a ray from the origin. See Figure 17.8.7b. As $\phi$ varies from $\phi = \phi_1(\theta)$ to $\phi = \phi(\theta)$ with $\theta$ fixed, these line segments generate a cross section of the region in the half-plane with edge along the $z$-axis and in the direction of $\theta$. See Figure 17.8.7c. As $\theta$ varies from $\theta = \theta_1$ to $\theta = \theta_2$, the cross sections sweep out the

FIGURE 17.8.7

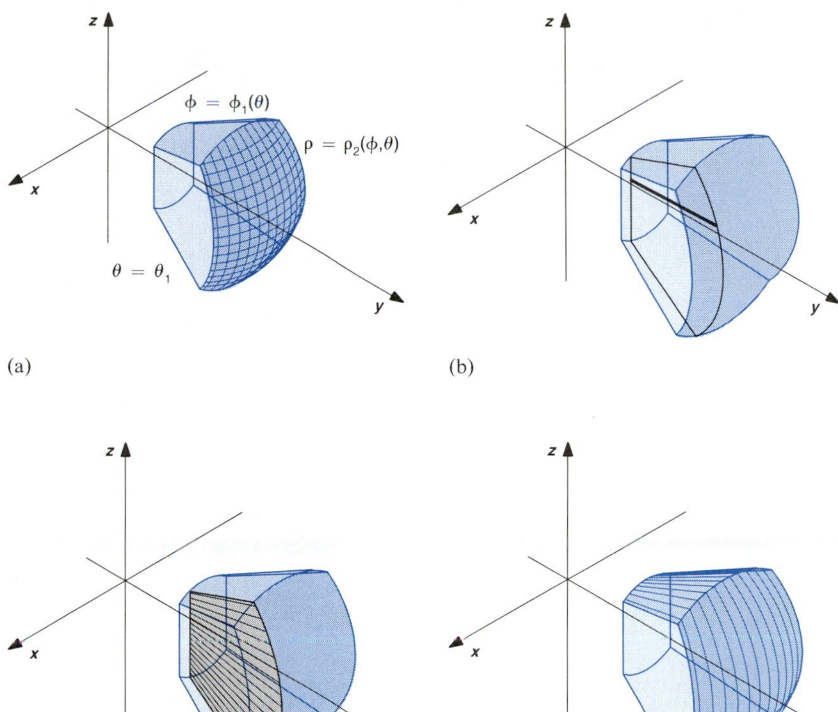


entire region. See Figure 17.8.7d. Using the formula for the differential volume that we obtained above, we then have

$$\iiint_R f\,dV = \int_{\theta_1}^{\theta_2}\int_{\phi_1(\theta)}^{\phi_2(\theta)}\int_{\rho_1(\theta,\phi)}^{\rho_2(\theta,\phi)} f(\rho,\theta,\phi)\rho^2 \sin\phi \, d\rho \, d\phi \, d\theta.$$

Note that the integrand must be expressed in terms of spherical coordinates. Recall that we can use the equations

$$x = \rho \sin\phi \cos\theta,$$

$$y = \rho \sin\phi \sin\theta,$$

$$z = \rho \cos\phi,$$

to change from rectangular to spherical coordinates. Also,

$$x^2 + y^2 + z^2 = \rho^2.$$

We should consider using spherical coordinates to evaluate a triple integral if the integrand is a function of $x^2 + y^2 + z^2$ or if the boundary of the region consists of spheres and/or cones. Recall that the spherical equation $\phi = \phi_0$ is a cone, except when $\phi_0 = \pi/2$ it is the $xy$-plane. See Figure 17.8.8.

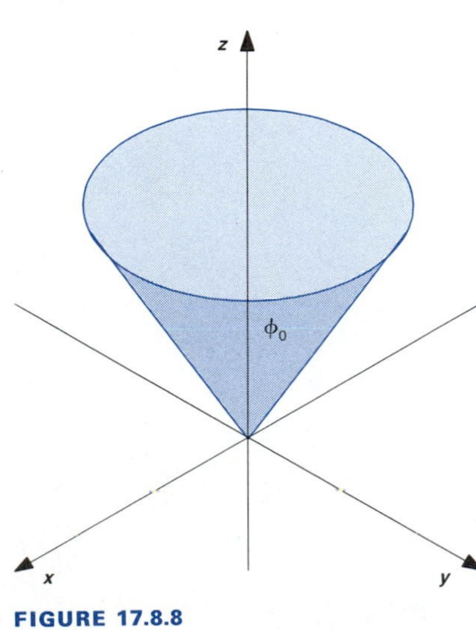


FIGURE 17.8.8

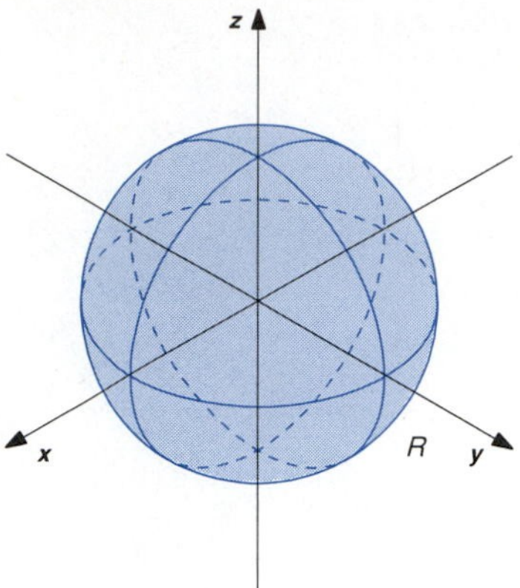


**FIGURE 17.8.9**

■ **EXAMPLE 4**

Find the volume of a sphere of radius $R$.

**SOLUTION**

Let us position the sphere with center at the origin. See Figure 17.8.9. The sphere then has rectangular equation $x^2 + y^2 + z^2 = R^2$. The region bounded by the sphere is described easily in terms of spherical coordinates. The spherical equation of the sphere is $\rho = R$. Using $dV = \rho^2 \sin\phi \, d\rho \, d\phi \, d\theta$, we have

$$V = \iiint dV = \int_0^{2\pi} \int_0^{\pi} \int_0^{R} \rho^2 \sin\phi \, d\rho \, d\phi \, d\theta$$

$$= \int_0^{2\pi} \int_0^{\pi} \frac{R^3}{3} \sin\phi \, d\phi \, d\theta$$

$$= \int_0^{2\pi} \frac{R^3}{3}(-\cos\pi + \cos 0) \, d\theta$$

$$= \frac{R^3}{3}(2)(2\pi) = \frac{4\pi R^3}{3}.$$ ■

■ **EXAMPLE 5**

Find the center of mass of the solid inside both the sphere $x^2 + y^2 + z^2 = 1$ and the cone $z = \sqrt{x^2 + y^2}$ if the density at a point is equal to the distance of the point from the origin.

**SOLUTION**

The solid is pictured in Figure 17.8.10. The spherical equation of the sphere is $\rho = 1$. We can obtain the spherical equation of the cone either by noting that the cone has vertex angle $\phi = \pi/4$ or by arriving at this conclusion by substituting into the rectangular equation. That is,

$$z = \sqrt{x^2 + y^2},$$

$$\rho \cos\phi = \sqrt{(\rho \sin\phi \cos\theta)^2 + (\rho \sin\phi \sin\theta)^2},$$

$$\rho \cos\phi = \sqrt{\rho^2 \sin^2\phi(\cos^2\theta + \sin^2\theta)},$$

$$\rho \cos\phi = |\rho \sin\phi|,$$

$$\rho \cos\phi = \rho \sin\phi, \qquad [\sin\phi \geq 0, \text{ because } 0 \leq \phi \leq \pi]$$

$$1 = \tan\phi,$$

$$\phi = \frac{\pi}{4}.$$

**FIGURE 17.8.10**

The density $\rho$ is equal to the spherical coordinate $\rho$ in this problem. The mass of the solid is

$$\begin{aligned}
M &= \iiint dM = \iiint \rho\, dV \\
&= \int_0^{2\pi}\int_0^{\pi/4}\int_0^1 \rho\rho^2 \sin\phi\, d\rho\, d\phi\, d\theta \\
&= \int_0^{2\pi}\int_0^{\pi/4}\int_0^1 \rho^3 \sin\phi\, d\rho\, d\phi\, d\theta \\
&= \int_0^{2\pi}\int_0^{\pi/4} \frac{\rho^4}{4}\sin\phi\Big]_{\rho=0}^{\rho=1} d\phi\, d\theta \\
&= \int_0^{2\pi}\int_0^{\pi/4} \frac{1}{4}\sin\phi\, d\phi\, d\theta \\
&= \int_0^{2\pi} \frac{1}{4}\Big[-\cos\phi\Big]_{\phi=0}^{\phi=\pi/4} d\theta \\
&= \int_0^{2\pi} \frac{1}{4}\left(-\frac{\sqrt{2}}{2}+1\right) d\theta \\
&= \left(\frac{1}{4}\right)\left(\frac{2-\sqrt{2}}{2}\right)(2\pi) = \frac{(2-\sqrt{2})\pi}{4}.
\end{aligned}$$

We also have

$$\begin{aligned}
M_{z=0} &= \iiint_R z\, dM = \iiint_R z\,\rho\, dV \\
&= \int_0^{2\pi}\int_0^{\pi/4}\int_0^1 \rho\cos\phi\,\rho\rho^2 \sin\phi\, d\rho\, d\phi\, d\theta \\
&= \int_0^{2\pi}\int_0^{\pi/4}\int_0^1 \rho^4 \cos\phi\sin\phi\, d\rho\, d\phi\, d\theta \\
&= \int_0^{2\pi}\int_0^{\pi/4} \frac{\rho^5}{5}\cos\phi\sin\phi\Big]_{\rho=0}^{\rho=1} d\phi\, d\theta \\
&= \int_0^{2\pi}\int_0^{\pi/4} \frac{1}{5}\cos\phi\sin\phi\, d\phi\, d\theta \qquad [u=\sin\phi,\ du=\cos\phi\, d\phi] \\
&= \int_0^{2\pi} \frac{1}{5}\left(\frac{\sin^2\phi}{2}\right)\Big]_{\phi=0}^{\phi=\pi/4} d\theta \\
&= \int_0^{2\pi} \frac{1}{20}\, d\theta = \frac{\pi}{10}.
\end{aligned}$$

$$\bar{z} = \frac{M_{z=0}}{M} = \frac{\dfrac{\pi}{10}}{\dfrac{(2-\sqrt{2})\pi}{4}} = \frac{2}{5(2-\sqrt{2})} \approx 0.68284271.$$

It follows from the symmetry of the region and the density function that $\bar{x} = \bar{y} = 0$. The center of mass is at the point

$$\left(0, 0, \frac{2}{5(2 - \sqrt{2})}\right).$$

■

■ **EXAMPLE 6**

Find the moment of inertia of a homogeneous hemisphere of radius $r$ about a diameter that is perpendicular to its base. Express the answer in terms of the mass of the hemisphere.

**SOLUTION**

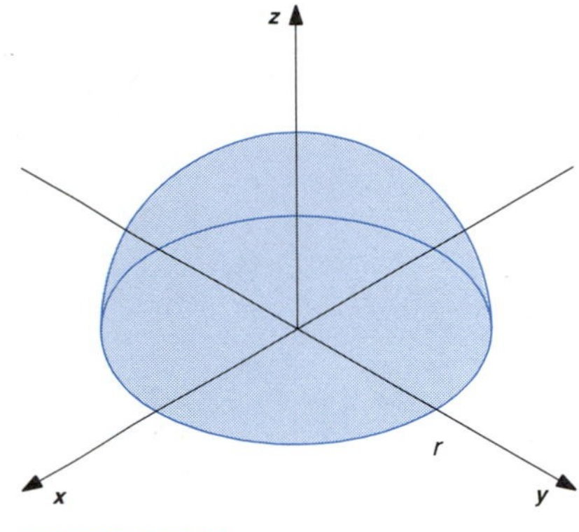


FIGURE 17.8.11

Let us position the hemisphere with center at the origin and base in the $xy$-plane. We want the moment of inertia about the $z$-axis. See Figure 17.8.11. Let $k$ denote the constant density of the sphere. We have

$$\begin{aligned}
I_{z\text{-axis}} &= \iiint_R (x^2 + y^2)k\, dV \\
&= \int_0^{2\pi} \int_0^{\pi/2} \int_0^r (\rho^2 \sin^2\phi)k\rho^2 \sin\phi\, d\rho\, d\phi\, d\theta \\
&= \int_0^{2\pi} \int_0^{\pi/2} \int_0^r k\rho^4 \sin^3\phi\, d\rho\, d\phi\, d\theta \\
&= \int_0^{2\pi} \int_0^{\pi/2} k\frac{\rho^5}{5} \sin^3\phi \Big]_{\rho=0}^{\rho=r} d\phi\, d\theta \\
&= \int_0^{2\pi} \int_0^{\pi/2} k\frac{r^5}{5} \sin^3\phi\, d\phi\, d\theta.
\end{aligned}$$

This integral can be evaluated either by writing $\sin^3\phi = (1 - \cos^2\phi)\sin\phi$ and using the substitution $u = \cos\phi$, $du = -\sin\phi\, d\phi$, or by using the table of integrals. Using the first method, we obtain

$$\begin{aligned}
I_{z\text{-axis}} &= \int_0^{2\pi} k\frac{r^5}{5}\left[\frac{\cos^3\phi}{3} - \cos\phi\right]_{\phi=0}^{\phi=\pi/2} d\theta \\
&= \int_0^{2\pi} k\frac{r^5}{5}\left(\frac{2}{3}\right) d\theta = k\frac{2r^5}{15}(2\pi) = k\frac{4\pi r^5}{15}.
\end{aligned}$$

We can express the moment of inertia in terms of the mass of the hemisphere by noting that the mass is

$$M = \int_0^{2\pi} \int_0^{\pi/2} \int_0^r k\,\rho^2 \sin\phi\, d\rho\, d\phi\, d\theta = k\frac{2\pi r^3}{3}$$

$M = k\dfrac{2\pi r^3}{3}$ implies $k = \dfrac{3M}{2\pi r^3}$. Then

$$I_{z\text{-axis}} = \frac{3M}{2\pi r^3}\frac{4\pi r^5}{15} = \frac{2Mr^2}{5}.$$

■

## EXERCISES 17.8

Find the volume of the regions indicated in Exercises 1–8.

1 $z = H - H\sqrt{x^2 + y^2}/R$, $z = 0$

2 $z = 4 - x^2 - y^2$, $z = 0$

3 Inside both $x^2 + y^2 + z^2 = 16$ and $x^2 + y^2 = 4$

4 Inside $x^2 + (y - 1)^2 = 1$, between $y + z = 4$ and $z = 0$

5 A hemisphere of radius $R$

6 Inside both $x^2 + y^2 + z^2 = 4$ and $z = \sqrt{3x^2 + 3y^2}$

7 Inside $x^2 + y^2 + z^2 = 4$, above $z = 1$

8 Inside the "apple" that has spherical equation $\rho = 1 - \cos\phi$

9 Find the center of mass of a homogeneous right circular cone with height $H$ and radius of base $R$.

10 Find the center of mass of a homogeneous hemisphere with radius $R$.

11 Find the center of mass of the cylinder bounded by $x^2 + y^2 = R^2$, $z = H$, $z = 0$, if the density is $\rho(x, y, z) = 1 + z\sqrt{x^2 + y^2}$.

12 Find the center of mass of the hemisphere bounded by $z = \sqrt{R^2 - x^2 - y^2}$ and $z = 0$, if the density is $\rho(x, y, z) = 1 + \sqrt{x^2 + y^2 + z^2}$.

13 Find the moment of inertia of a homogeneous spherical shell with outer radius $R$ and inner radius $r$, about a diameter. Express your answer in terms of the mass of the shell.

14 Find the moment of inertia of a homogeneous cylindrical shell with outer radius $R$, inner radius $r$, and height $H$, about its axis. Express your answer in terms of mass.

15 Find the moment of inertia about the $z$-axis of a homogeneous solid that is bounded by $z = x^2 + y^2$ and $z = 1$. Express your answer in terms of the mass of the solid.

16 Find the moment of inertia of a wedge cut from a homogeneous sphere with radius $R$ by two planes that intersect at an angle of $\theta_0$ along the diameter, about the diameter. Express your answer in terms of the mass of the wedge.

17 (a) Find the moment of inertia of a homogeneous right circular cylinder with radius $R$ and height $H$, about a line that is parallel to the axis of the cylinder and at a distance of $D$ from the axis. Express your answer in terms of the mass of the cylinder. (b) For what value of $D$ is the moment of inertia in (a) minimum?

18 (a) Find the moment of inertia of a homogeneous sphere with radius $R$, about a line that is at a distance of $D$ from the center of the sphere. Express your answer in terms of the mass of the sphere. (b) For what value of $D$ is the moment of inertia in (a) minimum?

## 17.9 TRANSFORMATION OF COORDINATES

We have seen in previous sections that it is useful to express certain integrals in terms of either polar, cylindrical, or spherical coordinates. In the next section we will extend the ideas used to include general change of variables in multiple integrals. In preparation, we will introduce the concept of transformation of coordinates in this section.

Let us consider coordinates $(u, v)$ that are related to coordinates $(x, y)$ by the equations

$$u = u(x, y), \qquad v = v(x, y). \tag{1}$$

These equations associate a point $(x, y)$ in the $xy$-coordinate plane with the point $(u(x, y), v(x, y))$ in the rectangular $uv$-coordinate plane; regions

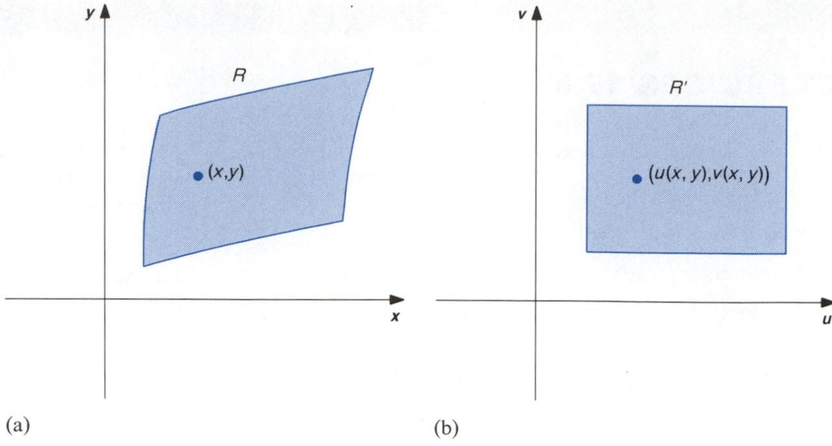


FIGURE 17.9.1

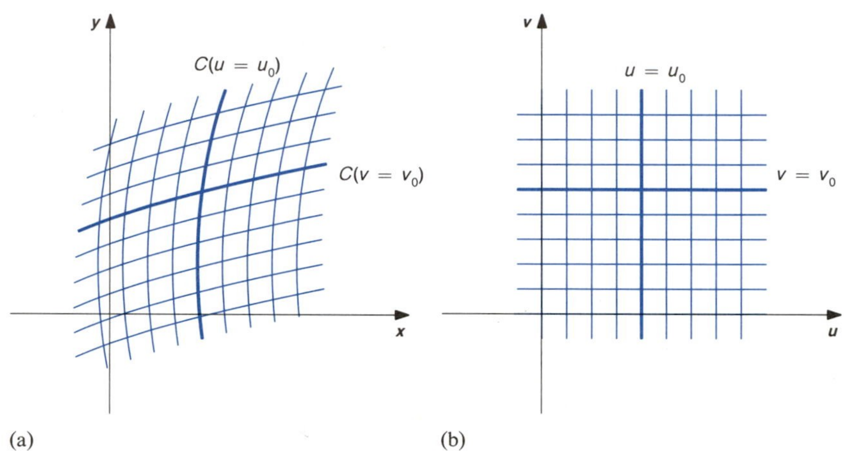


FIGURE 17.9.2

$R$ in the $xy$-plane are associated with regions $R'$ in the $uv$-plane. See Figure 17.9.1. If each point $(u, v)$ in a region $R'$ of the $uv$-plane corresponds to exactly one point $(x, y)$ in a region $R$ of the $xy$-plane, we say the transformation is **one-to-one** from $R$ **onto** $R'$. In this case the equations

$$u = u(x, y), \quad v = v(x, y), \quad (x, y) \text{ in } R,$$

can theoretically be solved for $x$ and $y$ in terms of $u$ and $v$ to obtain the equations

$$x = x(u, v), \quad y = y(u, v), \quad (u, v) \text{ in } R'. \tag{2}$$

Generally, we assume that the transformation of coordinates is one-to-one from a region $R$ of the $xy$-plane onto a region $R'$ of the $uv$-plane. We will then express the transformation in the form of the equations in (2), since these equations are equivalent to those in (1) but are more convenient for the applications of Section 17.10.

Suppose the coordinates $(x, y)$ and $(u, v)$ are related by the equations

$$x = x(u, v), \quad y = y(u, v).$$

As $u$ varies with $v$ fixed, the points $(u, v)$ trace a horizontal line in the $uv$-plane while the corresponding points $(x(u, v), y(u, v))$ trace a curve $C(v = v_0)$ in the $xy$-plane; as $v$ varies with $u$ fixed, the points $(u, v)$ trace a vertical line in the $uv$-plane while the corresponding points $(x(u, v), y(u, v))$ trace a curve $C(u = u_0)$ in the $xy$-plane. See Figure 17.9.2.

**EXAMPLE 1**

Coordinates in the $xy$-plane are related to coordinates in the $uv$-plane by the equations

$$x = uv, \quad y = \frac{v}{u}, \qquad u \text{ and } v \text{ positive.}$$

Sketch the lines $u = 1/\sqrt{2}$, $u = 1$, $u = \sqrt{2}$, $v = 1$, $v = 2$, $v = 3$ in the first quadrant of the $uv$-plane and sketch the corresponding curves in the $xy$-plane.

**SOLUTION**

The curve in the $xy$-plane that corresponds to $u = 1/\sqrt{2}$ satisfies the equations

$$x = \left(\frac{1}{\sqrt{2}}\right)v, \quad y = \frac{v}{\left(\frac{1}{\sqrt{2}}\right)}, \quad \text{or}$$

$$x = \frac{v}{\sqrt{2}}, \quad y = \sqrt{2}v.$$

Solving the first equation for $v$ and then substituting into the second equation, we have

$$v = \sqrt{2}x, \quad \text{so} \quad y = \sqrt{2}(\sqrt{2}x), \quad \text{or} \quad y = 2x.$$

$$u = \frac{1}{\sqrt{2}} \quad \text{corresponds to} \quad y = 2x.$$

Similarly,

$$u = 1 \quad \text{corresponds to} \quad y = x, \quad \text{and}$$

$$u = \sqrt{2} \quad \text{corresponds to} \quad y = \frac{x}{2}.$$

FIGURE 17.9.3

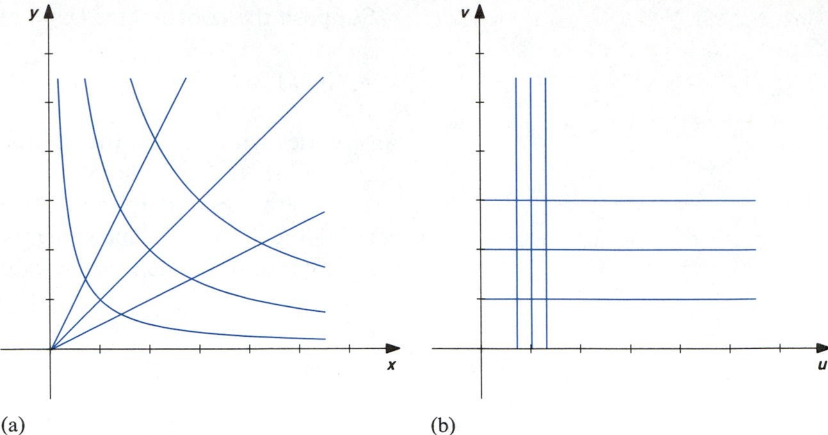


(a) (b)

The curve in the $xy$-plane that corresponds to $v = 3$ satisfies the equations

$$x = u(3), \quad y = \frac{3}{u}.$$

These equations imply

$$u = \frac{x}{3}, \quad \text{so} \quad y = \frac{3}{(x/3)}, \quad \text{or} \quad y = \frac{9}{x}.$$

$$v = 3 \quad \text{corresponds to} \quad y = \frac{9}{x}.$$

Similarly,

$$v = 2 \quad \text{corresponds to} \quad y = \frac{4}{x}, \quad \text{and}$$

$$v = 1 \quad \text{corresponds to} \quad y = \frac{1}{x}.$$

Since $u$ and $v$ are positive, the equations $x = uv$ and $y = v/u$ imply that the $x$ and $y$ are also positive. The restrictions of the curves to the first quadrant are sketched in Figure 17.9.3. ■

■ **EXAMPLE 2**

Coordinates in the $xy$-plane are related to coordinates in the $uv$-plane by the equations

$$x = u^2 - v^2, \quad y = 2uv, \qquad u \text{ and } v \text{ positive.}$$

Sketch the lines $u = 1/2$, $u = 1$, $u = 3/2$, $v = 1/2$, $v = 1$, $v = 3/2$ in the first quadrant of the $uv$-plane and sketch the corresponding curves in the $xy$-plane.

**SOLUTION**

The curve in the $xy$-plane that corresponds to $u = 1/2$ satisfies the equations

$$x = \left(\frac{1}{2}\right)^2 - v^2, \quad y = 2\left(\frac{1}{2}\right)v.$$

Solving the second equation for $v$ and substituting into the first equation, we obtain

$$v = y \quad \text{and} \quad x = \frac{1}{4} - y^2.$$

$$u = \frac{1}{2} \quad \text{corresponds to} \quad x = \frac{1}{4} - y^2.$$

Similarly,

$$u = 1 \quad \text{corresponds to} \quad x = 1 - \left(\frac{y}{2}\right)^2,$$

$$u = \frac{3}{2} \quad \text{corresponds to} \quad x = \frac{9}{4} - \left(\frac{y}{3}\right)^2,$$

$$v = \frac{1}{2} \quad \text{corresponds to} \quad x = y^2 - \frac{1}{4},$$

$$v = 1 \quad \text{corresponds to} \quad x = \left(\frac{y}{2}\right)^2 - 1, \quad \text{and}$$

$$v = \frac{3}{2} \quad \text{corresponds to} \quad x = \left(\frac{y}{3}\right)^2 - \frac{9}{4}.$$

Since $u$ and $v$ are positive, $y = 2uv$ is also positive; $x = u^2 - v^2$ can have any value. The graphs are sketched in Figure 17.9.4. ■

FIGURE 17.9.4

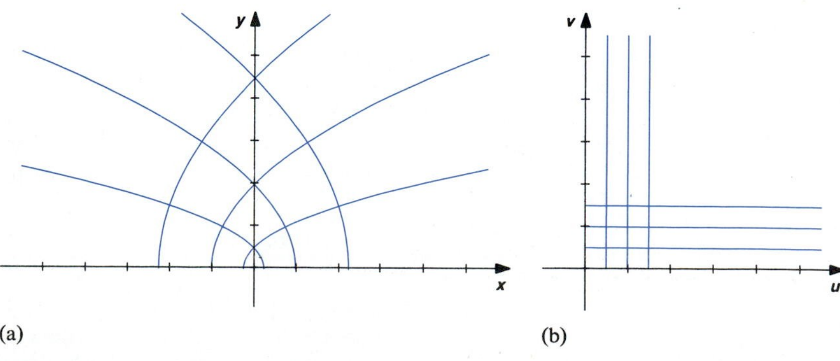

FIGURE 17.9.5

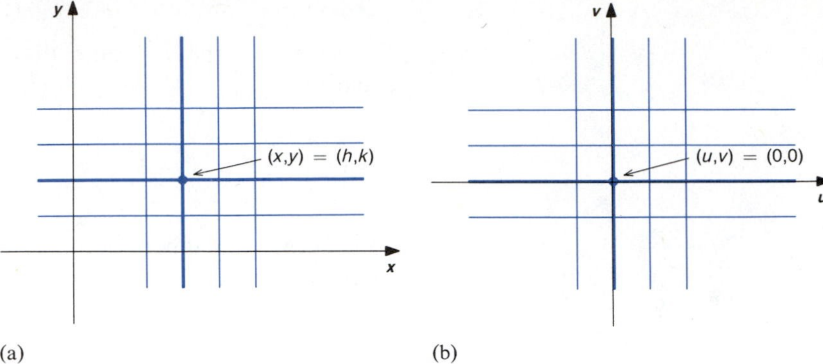


FIGURE 17.9.6

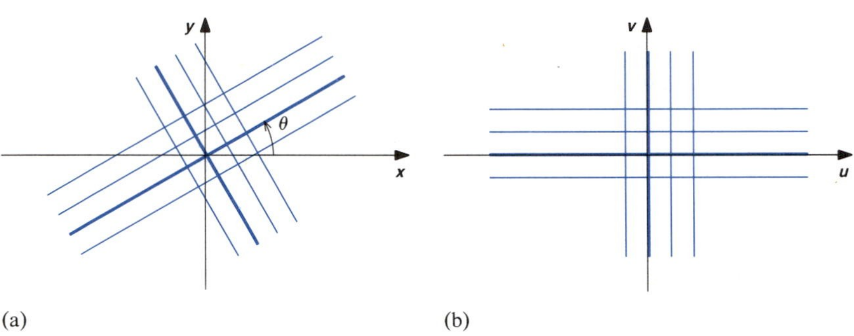


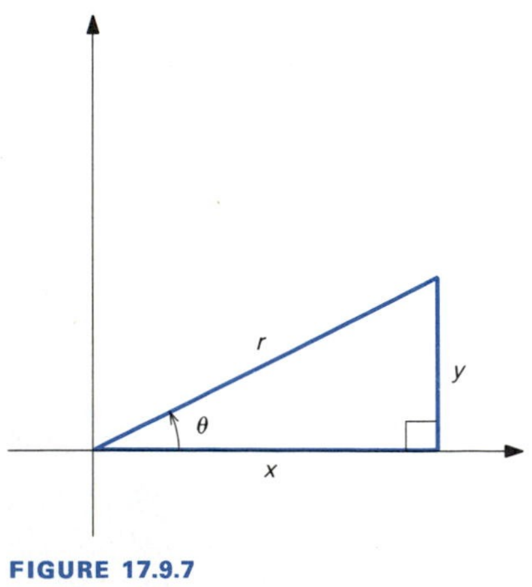


FIGURE 17.9.7

Let us now review some of the coordinate transformations that we have studied previously.

The equations

$$x = u + h \quad \text{and} \quad y = v + k$$

correspond to a **translation** of the $xy$-axes. See Figure 17.9.5. The equations

$$x = (\cos\theta)u - (\sin\theta)v \quad \text{and} \quad y = (\sin\theta)u + (\cos\theta)v$$

correspond to a **rotation** of the $xy$-axis by the angle $\theta$. See Figure 17.9.6. Equations related to translation and rotation of axes were studied in Section 11.4.

Polar coordinates $(r, \theta)$ in the $xy$-plane are related to rectangular coordinates $(x, y)$ by the equations

$$x = r\cos\theta, \quad y = r\sin\theta, \qquad r > 0, \quad 0 \le \theta < 2\pi.$$

See Figure 17.9.7. The line $r = r_0$ in the *rectangular* $r\theta$-plane corresponds to the circle $x^2 + y^2 = r_0^2$ in the $xy$-plane; the line $\theta = \theta_0$ in the *rectangular* $r\theta$-plane corresponds to the ray in the $xy$-plane from the origin that makes an angle of $\theta_0$ with the positive $x$-axis. See Figure 17.9.8. (We have avoided the origin of the $xy$-plane and points in the $r\theta$-plane with $r = 0$ in order that the transformation be one-to-one.)

Let us introduce some other types of transformations of coordinates. The transformation

$$x = au, \qquad y = bv$$

corresponds to a linear **change of scale** of the $xy$-axes. The ellipse

$$\frac{x^2}{a^2} + \frac{y^2}{b^2} = r^2$$

in the $xy$-plane corresponds to the circle $u^2 + v^2 = r^2$ in the $uv$-plane. See Figure 17.9.9.

FIGURE 17.9.8

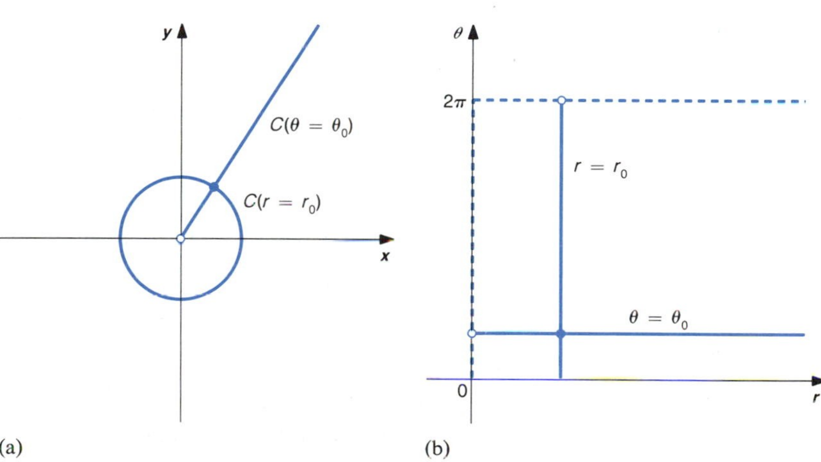


FIGURE 17.9.9

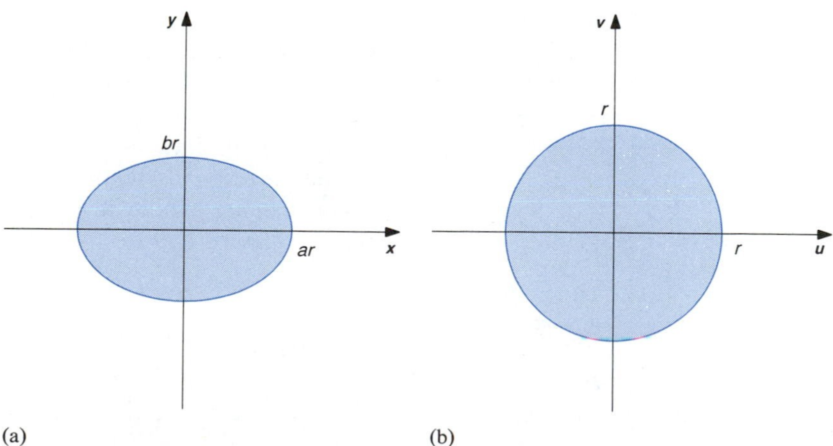

A transformation that is given by equations of the form

$$x = au + bv, \qquad y = cu + dv,$$

where $a$, $b$, $c$, and $d$ are constants, is called a **linear transformation**. The rotation transformation

$$x = (\cos\theta)u - (\sin\theta)v, \qquad y = (\sin\theta)u + (\cos\theta)v$$

and the change of scale transformation

$$x = au, \qquad y = bv$$

are linear transformations. The translation transformation

$$x = u + h, \qquad y = v + k$$

is not a linear transformation unless $(h, k) = (0, 0)$.

If

$$ad - bc \neq 0, \tag{3}$$

we can solve the equations

$$x = au + bv \quad \text{and} \quad y = cu + dv$$

for $u$, $v$ in terms of $x$, $y$ to obtain

$$u = \frac{dx - by}{ad - bc}, \qquad v = \frac{-cx + ay}{ad - bc}.$$

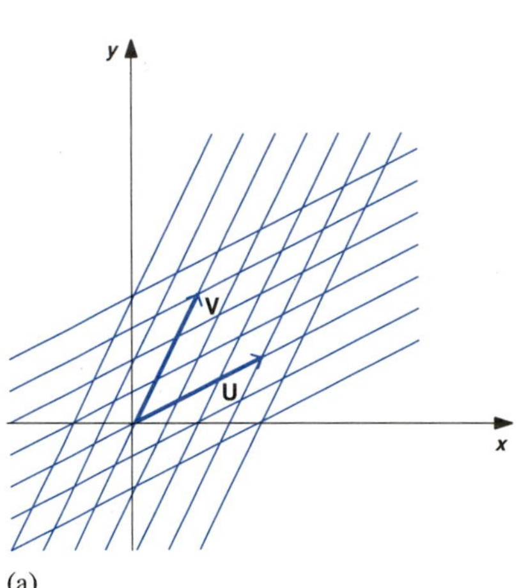

(a)

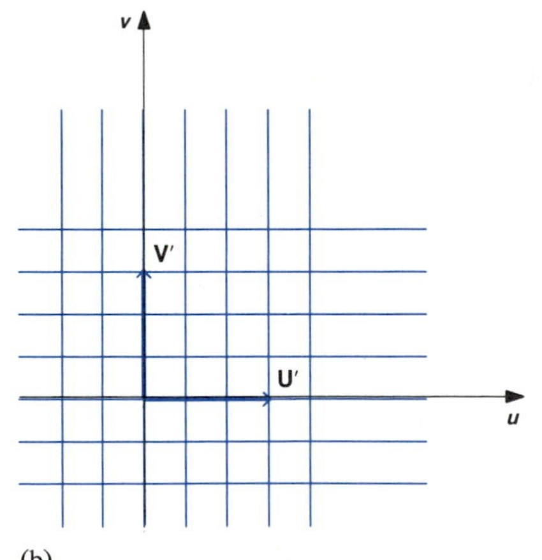

(b)

FIGURE 17.9.10

Each point $(u, v)$ in the $uv$-plane corresponds to exactly one point $(x, y)$ in the $xy$-plane. The equations then define a one-to-one transformation of the entire $xy$-plane onto the entire $uv$-plane.

If condition (3) is satisfied, the linear transformation

$$x = au + bv, \qquad y = cu + dv$$

transforms parallel lines in the $xy$-plane into parallel lines in the $uv$-plane. The vector $\mathbf{U} = a\mathbf{i} + c\mathbf{j}$ in the $xy$-plane corresponds to the vector $\mathbf{U}' = \mathbf{i}$ in the $uv$-plane; lines in the $xy$-plane that are parallel to $\mathbf{U}$ are transformed into horizontal lines in the $uv$-plane. The vector $\mathbf{V} = b\mathbf{i} + d\mathbf{j}$ in the $xy$-plane corresponds to the vector $\mathbf{V}' = \mathbf{j}$ in the $uv$-plane; lines in the $xy$-plane that are parallel to $\mathbf{V}$ are transformed into vertical lines in the $uv$-plane. See Figure 17.9.10.

### ■ EXAMPLE 3

Sketch the region $R$ in the $xy$-plane that is bounded by the lines $y = -x/4$, $y = 2x$, and $x + y = 3$. Sketch the corresponding region $R'$ in the $uv$-plane

FIGURE 17.9.11

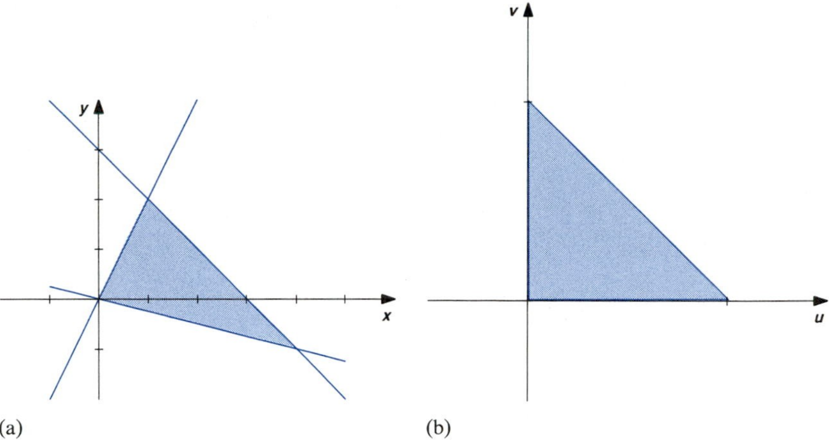


if coordinates in the $xy$-plane are related to coordinates in the $uv$-plane by the linear transformation

$$x = 4u + v, \qquad y = -u + 2v.$$

**SOLUTION**

The region $R$ in the $xy$-plane is sketched in Figure 17.9.11a. We can obtain the boundary equations of $R'$ by substitution into the boundary equations of $R$. The equation $y = -x/4$ implies

$$-u + 2v = -\frac{4u + v}{4},$$

$$-4u + 8v = -4u - v,$$

$$9v = 0,$$

$$v = 0.$$

It follows that

$$y = -\frac{x}{4} \quad \text{corresponds to} \quad v = 0.$$

(This linear transformation transforms lines in the $xy$-plane with direction vector $\mathbf{U} = 4\mathbf{i} - \mathbf{j}$ into horizontal lines in the $uv$-plane.) Similarly,

$$y = 2x \quad \text{corresponds to} \quad u = 0.$$

(The transformation transforms lines in the $xy$-plane with direction vector $\mathbf{V} = \mathbf{i} + 2\mathbf{j}$ into vertical lines in the $uv$-plane.) Also,

$$x + y = 3 \quad \text{corresponds to} \quad u + v = 1.$$

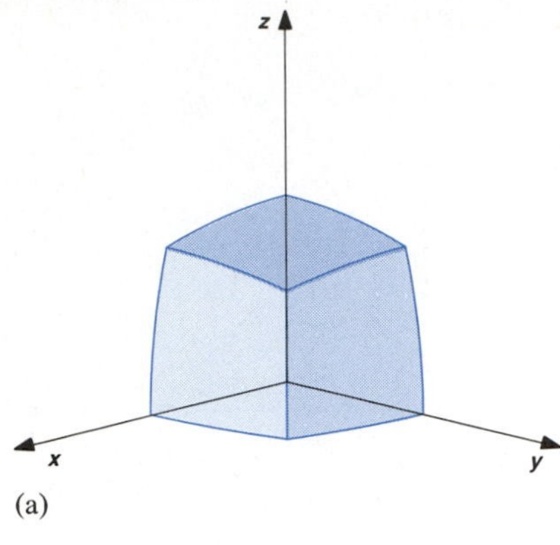

(a)

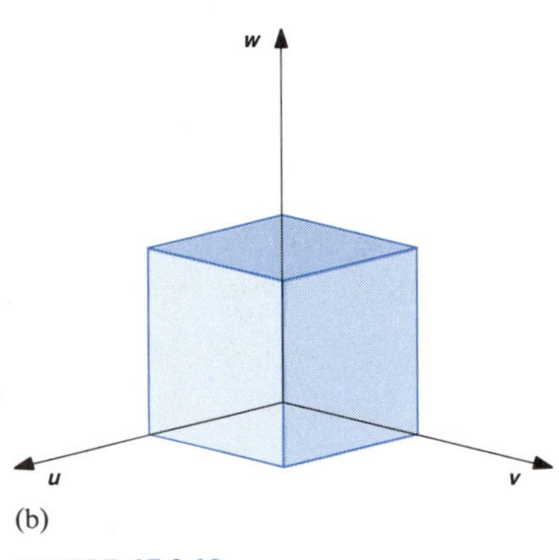

(b)

FIGURE 17.9.12

The region $R'$ in the $uv$-plane is sketched in Figure 17.9.11b. Note that a double integral over $R'$ with respect to the variables $u$ and $v$ could be evaluated by a single iterated double integral with either order of integration. But double integrals over the region $R$ with respect to $x$ and $y$ cannot be evaluated by a single iterated double integral for either order of integration. ■

Transformations of variables in three-dimensional space are similar to transformations of variables in the plane. The equations

$$x = x(u, v, w), \quad y = y(u, v, w), \quad z = z(u, v, w)$$

associate a point $(u, v, w)$ in rectangular $uvw$-space with the point $(x(u, v, w), y(u, v, w), z(u, v, w))$ in $xyz$-space. As one of the variables $u, v, w$ varies with the other two fixed, the points $(u, v, w)$ trace a line in $uvw$-space while the corresponding points $(x(u, v, w), y(u, v, w), z(u, v, w))$ trace a curve in $xyz$-space. See Figure 17.9.12.

We are familiar with the transformation from rectangular coordinates $(x, y, z)$ to cylindrical coordinates $(r, \theta, z)$ given by the equations

$$x = r \cos \theta, \quad y = r \sin \theta, \quad z = z.$$

A typical rectangular box with sides parallel to the coordinate planes in $r\theta z$-space and the corresponding region in $xyz$-space are indicated in Figure 17.9.13.

The transformation from rectangular coordinates $(x, y, z)$ to spherical coordinates $(\rho, \theta, \phi)$ is given by the equations

$$x = \rho \sin \phi \cos \theta, \quad y = \rho \sin \phi \sin \theta, \quad z = \rho \cos \phi.$$

A typical rectangular box with sides parallel to the coordinate planes in $\rho\theta\phi$-space and the corresponding region in $xyz$-space are indicated in Figure 17.9.14.

The **change of scale** transformation in three-dimensional space is given by the equations

$$x = au, \quad y = bv, \quad \text{and} \quad z = cw.$$

FIGURE 17.9.13

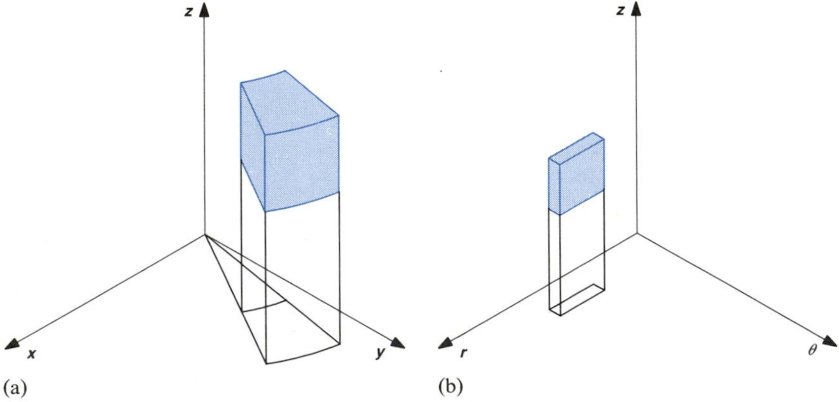

FIGURE 17.9.14

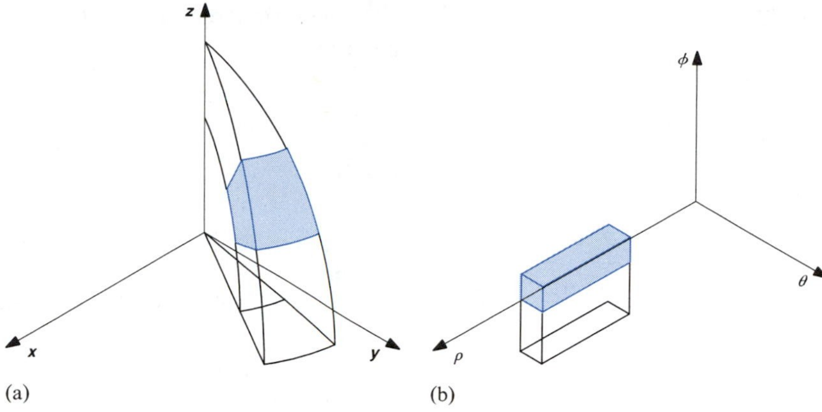


FIGURE 17.9.15

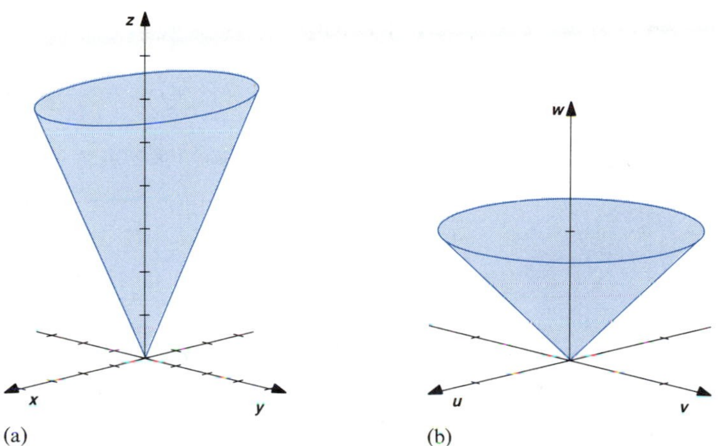


**EXAMPLE 4**

Sketch the region $R$ in $xyz$-space bounded by the elliptical cone $z = \sqrt{4x^2 + 9y^2}$ and the plane $z = 6$. Sketch the region $R'$ in $uvw$-space that corresponds to $R$ if the coordinates are related by the equations

$$x = 3u, \quad y = 2v, \quad z = 6w.$$

**SOLUTION**

The region $R$ is sketched in Figure 17.9.15a. The equation $z = \sqrt{4x^2 + 9y^2}$ corresponds to the equation

$$6w = \sqrt{4(3u)^2 + 9(2v)^2},$$

$$6w = \sqrt{36u^2 + 36v^2},$$

$$w = \sqrt{u^2 + v^2}.$$

This is a right circular cone. The equation $z = 6$ corresponds to the equation

$$6w = 6,$$

$$w = 1.$$

The region $R'$ in $uvw$-space bounded by $w = \sqrt{u^2 + v^2}$ and $w = 1$ is sketched in Figure 17.9.15b. Note that $R'$ can be described easily in terms of either cylindrical or spherical coordinates. ■

A **linear transformation** of coordinates in three-dimensional space is given by equations of the form

$$\begin{cases} x = u_1 u + v_1 v + w_1 w, \\ y = u_2 u + v_2 v + w_2 w, \\ z = u_3 u + v_3 v + w_3 w, \end{cases}$$

where the coefficients of $u$, $v$, and $w$ are constants. If the determinant

$$\begin{vmatrix} u_1 & v_1 & w_1 \\ u_2 & v_2 & w_2 \\ u_3 & v_3 & w_3 \end{vmatrix}$$

is nonzero, the transformation equations can be solved for $u$, $v$, $w$ in terms of $x$, $y$, $z$. In that case, each point in $xyz$-space corresponds to exactly one point in $uvw$-space. The transformation then transforms planes in $xyz$-space into planes in $uvw$-space and lines in $xyz$-space into lines in $uvw$-space. Also, the vectors

$$\mathbf{U} = u_1\mathbf{i} + u_2\mathbf{j} + u_3\mathbf{k},$$

$$\mathbf{V} = v_1\mathbf{i} + v_2\mathbf{j} + v_3\mathbf{k},$$

$$\mathbf{W} = w_1\mathbf{i} + w_2\mathbf{j} + w_3\mathbf{k},$$

in $xyz$-space are transformed into the vectors $\mathbf{i}$, $\mathbf{j}$, $\mathbf{k}$ in $uvw$-space. Lines in the $xyz$-plane with direction vectors $\mathbf{U}$, $\mathbf{V}$, $\mathbf{W}$ are transformed into lines in the $uvw$-plane that are parallel to the $u$-axis, $v$-axis, $w$-axis, respectively. See Figure 17.9.16.

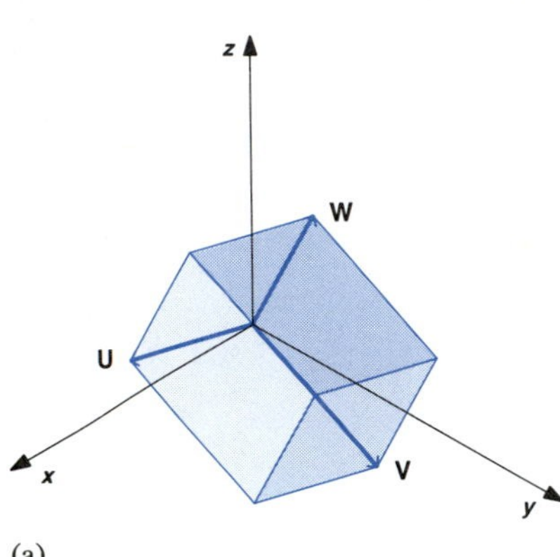


(a)

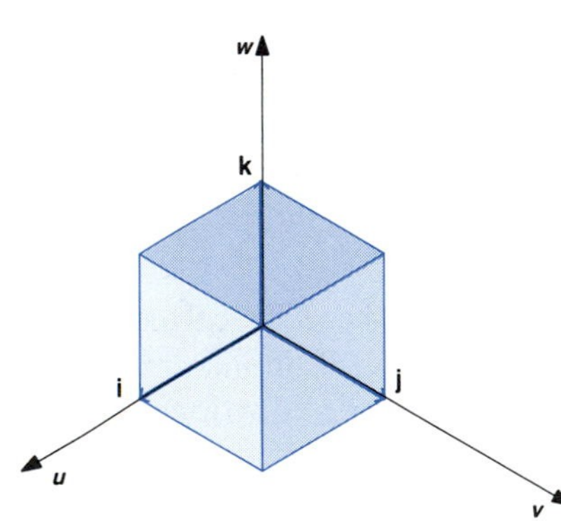


(b)

FIGURE 17.9.16

■ **EXAMPLE 5**

Sketch the region $R$ in $xyz$-space bounded by the planes $3z = 3x + y$, $z = x + y$, $x = 0$, and $x + y + z = 4$. Sketch the corresponding region $R'$ in $uvw$-space if coordinates in $xyz$-space are related to coordinates in $uvw$-space by the linear transformation

$$x = 2u, \quad y = 3v + 2w, \quad z = 2u + v + 2w.$$

FIGURE 17.9.17

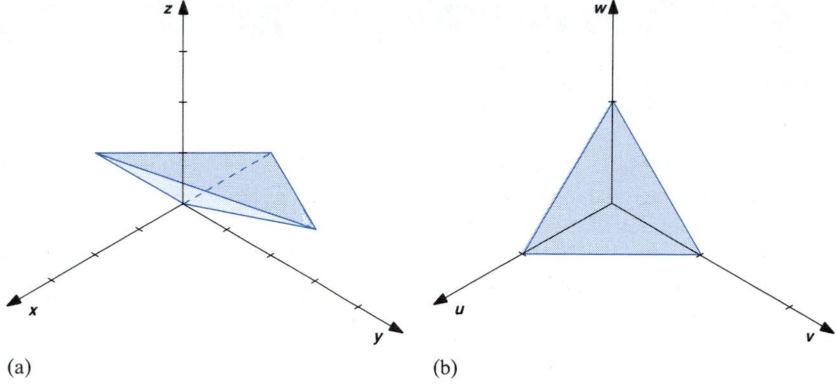


**SOLUTION**

The region $R$ in $xyz$-space is sketched in Figure 17.9.17a. We can obtain the boundary equations of $R'$ by substitution into the boundary equations of $R$.

$3z = 3x + y$ implies

$3(2u + v + 2w) = 3(2u) + (3v + 2w),$

$6u + 3v + 6w = 6u + 3v + 2w,$

$4w = 0,$

$w = 0.$

It follows that

$3z = 3x + y$ corresponds to $w = 0.$

Similarly,

$z = x + y$ corresponds to $v = 0,$

$x = 0$ corresponds to $u = 0,$ and

$x + y + z = 4$ corresponds to $u + v + w = 1.$

The region $R'$ in $uvw$-space is sketched in Figure 17.9.17b. Note that a triple integral over $R'$ with respect to the variables $u$, $v$, and $w$ could be evaluated by using a single iterated triple integral with any order of integration. Triple integrals over the region $R$ with respect to the variables $x$, $y$, and $z$ cannot be evaluated by a single iterated triple integral for any of the six possible orders of integration. ■

## EXERCISES 17.9

**1** Coordinates in the $xy$-plane are related to coordinates in the $uv$-plane by the equations $x = 2u - v$, $y = u + 2v$. Find the $(x, y)$ coordinates that correspond to (a) $(u, v) = (1, 0)$ and (b) $(u, v) = (0, 1)$. Find the coordinates $(u, v)$ that correspond to (c) $(x, y) = (1, 0)$ and (d) $(x, y) = (0, 1)$. (e) Solve the equations for $(u, v)$ in terms of $(x, y)$.

**2** Coordinates in the $xy$-plane are related to coordinates in the $uv$-plane by the equations $x = 3u + 2v$, $y = u - 4v$. Find the $(x, y)$ coordinates that correspond to (a) $(u, v) = (1, 0)$ and (b) $(u, v) = (0, 1)$. Find the $(u, v)$ coordinates that correspond to (c) $(x, y) = (1, 0)$ and (d) $(x, y) = (0, 1)$. (e) Solve the equations for $(u, v)$ in terms of $(x, y)$.

**3** Coordinates in $xyz$-space are related to coordinates in $uvw$-space by the equations $x = u - v$, $y = v - 2w$, $z = u + w$. Find the $(x, y, z)$ coordinates that correspond to (a) $(u, v, w) = (1, 0, 0)$, (b) $(u, v, w) = (0, 1, 0)$, and (c) $(u, v, w) = (0, 0, 1)$. Find the $(u, v, w)$ coordinates that correspond to (d) $(x, y, z) = (1, 0, 0)$, (e) $(x, y, z) = (0, 1, 0)$, and (f) $(x, y, z) = (0, 0, 1)$. (g) Solve the equations for $(u, v, w)$ in terms of $(x, y, z)$.

**4** Coordinates in $xyz$-space are related to coordinates in $uvw$-space by the equations $x = u + v + w$, $y = u + w$, $z = u + v$. Find the $(x, y, z)$ coordinates that correspond to (a) $(u, v, w) = (1, 0, 0)$, (b) $(u, v, w) = (0, 1, 0)$, and (c) $(u, v, w) = (0, 0, 1)$. Find the $(u, v, w)$ coordinates that correspond to (d) $(x, y, z) = (1, 0, 0)$, (e) $(x, y, z) = (0, 1, 0)$, and (f) $(x, y, z) = (0, 0, 1)$. (g) Solve the equations for $(u, v, w)$ in terms of $(x, y, z)$.

Coordinates in the $xy$-plane are related to coordinates in the $uv$-plane by the equations given in Exercises 5–10. Sketch the lines $u = u_0$ and $v = v_0$ in the $uv$-plane and sketch the corresponding curves in the $xy$-plane for the given values of $u_0$ and $v_0$.

**5** $x = u + 2v$, $y = 2u - v$; $u_0 = -1, 0, 1$; $v_0 = -1, 0, 1$

**6** $x = 3u + v$, $y = u - 2v$; $u_0 = -1, 0, 1$; $v_0 = -1, 0, 1$

**7** $x = 2u$, $y = 3v$; $u_0 = -1, 0, 1$; $v_0 = -1, 0, 1$

**8** $x = v \cos u$, $y = v \sin u$, $v > 0$, $0 \le u < 2\pi$; $u_0 = 0, \pi/4, 3\pi/4$; $v = 1, 2, 3$

**9** $x = u^2 - v^2$, $y = u^2 + v^2$, $u$ and $v$ nonnegative; $u_0 = 0, 1, 2$; $v_0 = 0, 1, 2$

**10** $x = u^2 - v^2$, $y = 2uv$, $u$ and $v$ positive; $u_0 = 1, 2, 3$; $v_0 = 1, 2, 3$

Coordinates in the $xy$-plane are related to coordinates in the $uv$-plane by the equations given in Exercises 11–16. Sketch the graph of the given $xy$-equation. Find the corresponding $uv$-equation and sketch its graph in the $uv$-plane.

**11** $x = 3u + 3v$, $y = u + 2v$; $2x - 3y = 6$

**12** $x = u - v$, $y = u + v$; $x - y = 1$

**13** $x = u + 1$, $y = v - 2$; $x^2 + y^2 - 2x + 4y = 6$

**14** $x = u - 2$, $y = v$; $x^2 - y^2 + 4x = 5$

**15** $x = (u - v)/\sqrt{2}$, $y = (u + v)/\sqrt{2}$; $xy = 1$

**16** $x = (u + v)/\sqrt{2}$, $y = (-u + v)/\sqrt{2}$; $xy = 1$

Coordinates in the $xy$-plane are related to coordinates in the $uv$-plane by the equations given in Exercises 17–22. Sketch the graph of the region in the $xy$-plane bounded by the given equations and sketch the graph of the corresponding region $R'$ in the $uv$-plane.

**17** $x = 2u$, $y = 3v$; $9x^2 + 4y^2 = 36$

**18** $x = 4u$, $y = 3v$; $9x^2 + 16y^2 = 144$

**19** $x = 3u + 2v$, $y = u + 3v$; $y = x/3$, $2y = 3x$, $2x + y = 7$

**20** $x = 2u + v$, $y = u + 2v$; $y = x/2$, $y = 2x$, $x + y = 3$

**21** $x = (u - v)/\sqrt{2}$; $y = (u + v)/\sqrt{2}$; $x + y = 2$, $x = 0$, $y = 0$

**22** $x = (u + \sqrt{3}v)/2$, $y = (-\sqrt{3}u + v)/2$; $\sqrt{3}y = x$, $y = -\sqrt{3}x$, $x = 1$

Coordinates in $xyz$-space are related to coordinates in $uvw$-space by the equations given in Exercises 23–28. Sketch the graph of the region in $xyz$-space that is bounded by the given equations and sketch the graph of the corresponding region $R'$ in $uvw$-space.

**23** $x = 2u$, $y = 3v$, $z = 4w$; $36z = 9x^2 + 4y^2$, $z = 4$

**24** $x = 2u$, $y = 6v$, $z = 3w$; $y = \sqrt{9x^2 + 4z^2}$, $y = 6$

**25** $x = 4u$, $y = 3v + w$, $z = v + 3w$; $y = 3z$, $z = 3y$, $x = 0$, $x + y + z = 4$

**26** $x = 3u + v$, $y = u + 3v$, $z = 4w$; $y = 3x$, $x = 3y$, $z = 0$, $x + y + z = 4$

**27** $x = 2u + 3v$, $y = 2v + 3w$, $z = 3u + 2w$; $6y = 4x + 9z$, $6z = 9x + 4y$, $6x = 9y + 4z$, $x + y + z = 5$

**28** $x = u + 2v$, $y = v + 2w$, $z = 2u + w$, $2y = x + 4z$, $2z = 4x + y$, $2x = 4y + z$, $x + y + z = 3$

## 17.10 CHANGE OF VARIABLES IN MULTIPLE INTEGRALS

In this section we will see how to use transformations of coordinates to change variables in double and triple integrals.

Let us consider rectangular coordinates $(x, y)$ that are related to the coordinates $(u, v)$ by the equations

$$x = x(u, v), \quad y = y(u, v). \tag{1}$$

We assume that $x$ and $y$ have continuous first partial derivatives with respect to $u$ and $v$, so $x$ and $y$ are differentiable. Equations (1) then imply

$$dx = \frac{\partial x}{\partial u}\, du + \frac{\partial x}{\partial v}\, dv, \qquad dy = \frac{dy}{\partial u}\, du + \frac{\partial y}{\partial v}\, dv. \tag{2}$$

We also assume

$$\frac{\partial x}{\partial u}\frac{\partial y}{\partial v} - \frac{\partial y}{\partial u}\frac{\partial x}{\partial v} \neq 0. \tag{3}$$

If condition (3) is satisfied, the system of equations in (2) can be solved for $(du, dv)$ in terms of $(dx, dy)$, so differential changes in $x$ and $y$ are related to specific differential changes in $u$ and $v$. It can then be proved that the system of equations in (1) can theoretically be solved for $u$ and $v$ in terms of $x$ and $y$, at least for small changes in the variables.

We want to establish a formula for expressing the double integral of a function $f(x, y)$ over a region $R$ in the $xy$-plane in terms of the coordinates $(u, v)$. To do this, let $R'$ be a region in the $uv$-plane that has the property that every point $(x, y)$ in $R$ corresponds to exactly one point $(u, v)$ in $R'$. We then divide $R'$ into rectangles with horizontal and vertical lines in the $uv$-plane. The corresponding curves $C(u = u_0)$ and $C(v = v_0)$ divide $R$ into curvilinear rectangles. See Figure 17.10.1. If the increments are small, these

FIGURE 17.10.1

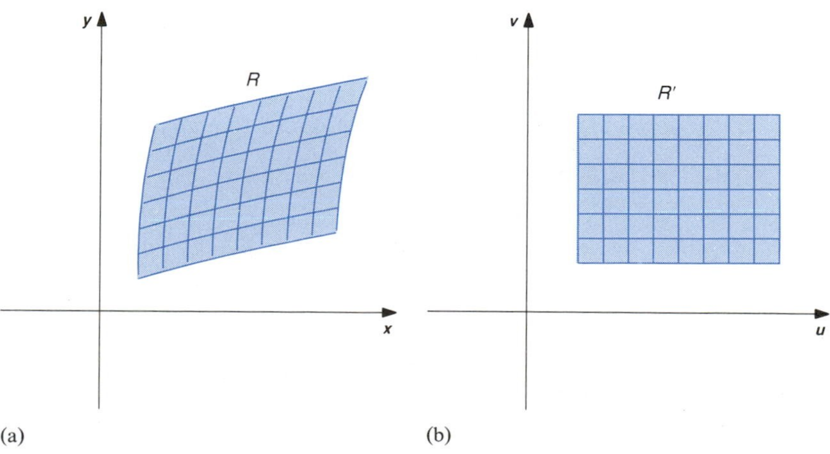

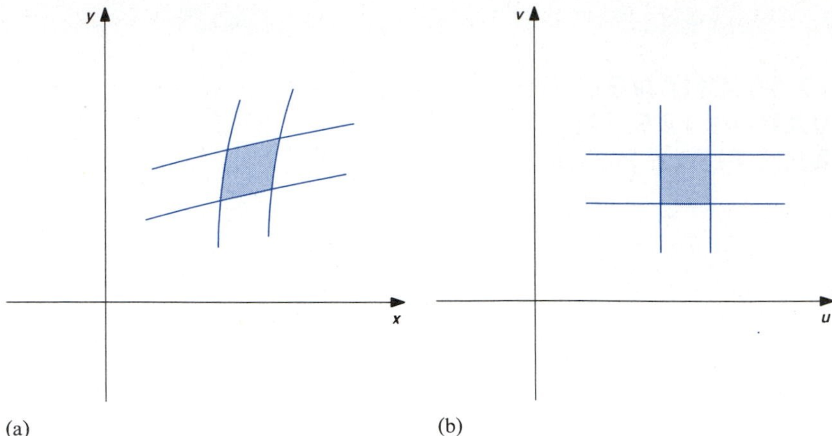


FIGURE 17.10.2

curvilinear rectangles are essentially parallelograms. See Figure 17.10.2. We need to express the differential approximation $dA(x, y)$ of the area of a curvilinear rectangle in terms of the differential area $dA(u, v) = du\,dv$ of the corresponding rectangle in the $uv$-plane. Let us use vectors to do this.

The vector

$$\mathbf{R} = x(u, v)\mathbf{i} + y(u, v)\mathbf{j}$$

gives the position of the point $(x, y)$ in the $xy$-plane that corresponds to the point $(u, v)$ in the $uv$-plane. The differential change in position corresponding to changes of $du$ in $u$ and $dv$ in $v$ is

$$d\mathbf{R} = \left(\frac{\partial x}{\partial u}\,du + \frac{\partial x}{\partial v}\,dv\right)\mathbf{i} + \left(\frac{\partial y}{\partial u}\,du + \frac{\partial y}{\partial v}\,dv\right)\mathbf{j}.$$

The vector

$$\mathbf{U} = \left(\frac{\partial x}{\partial u}\,du\right)\mathbf{i} + \left(\frac{\partial y}{\partial u}\,du\right)\mathbf{j}$$

corresponds to a change of $du$ in $u$ with $v$ fixed; $\mathbf{U}$ is tangent to the curve $C(v = v_0)$. The vector

$$\mathbf{V} = \left(\frac{\partial x}{\partial v}\,dv\right)\mathbf{i} + \left(\frac{\partial y}{\partial v}\,dv\right)\mathbf{j}$$

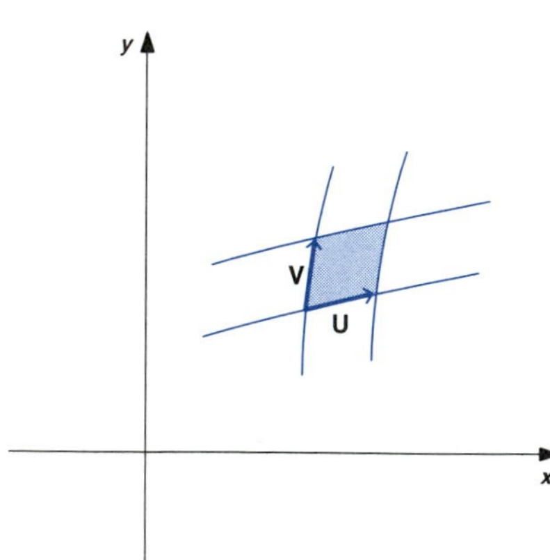


FIGURE 17.10.3

**corresponds to a change of** $dv$ in $v$ with $u$ fixed; $\mathbf{V}$ is tangent to the curve $C(u = u_0)$. The area of the parallelogram formed by $\mathbf{U}$ and $\mathbf{V}$ is the differential approximation of the curvilinear rectangle. See Figure 17.10.3. From statement (4) of Section 14.2 we know that the area of the parallelogram formed by **U** and **V** is the length of the vector **U** × **V**. We have

$$\mathbf{U}\times\mathbf{V} = \begin{vmatrix} \mathbf{i} & \mathbf{j} & \mathbf{k} \\ \dfrac{\partial x}{\partial u}\,du & \dfrac{\partial y}{\partial u}\,du & 0 \\ \dfrac{\partial x}{\partial v}\,dv & \dfrac{\partial y}{\partial v}\,dv & 0 \end{vmatrix}$$

$$= \begin{vmatrix} \dfrac{\partial y}{\partial u}\,du & 0 \\ \dfrac{\partial y}{\partial v}\,dv & 0 \end{vmatrix}\mathbf{i} - \begin{vmatrix} \dfrac{\partial x}{\partial u}\,du & 0 \\ \dfrac{\partial x}{\partial v}\,dv & 0 \end{vmatrix}\mathbf{j} + \begin{vmatrix} \dfrac{\partial x}{\partial u}\,du & \dfrac{\partial y}{\partial u}\,du \\ \dfrac{\partial x}{\partial v}\,dv & \dfrac{dy}{\partial v}\,dv \end{vmatrix}\mathbf{k}$$

$$= \left(\frac{\partial x}{\partial u}\frac{\partial y}{\partial v} - \frac{\partial x}{\partial v}\frac{\partial y}{\partial u}\right)(du\,dv)\mathbf{k}.$$

Since $du\,dv = dA(u, v)$, we then have

$$dA(x, y) = |\mathbf{U}\times\mathbf{V}| = \left|\frac{\partial x}{\partial u}\frac{\partial y}{\partial v} - \frac{\partial x}{\partial v}\frac{\partial y}{\partial u}\right| dA(u, v).$$

Let us define

$$\frac{\partial(x, y)}{\partial(u, v)} = \begin{vmatrix} \dfrac{\partial x}{\partial u} & \dfrac{\partial x}{\partial v} \\ \dfrac{\partial y}{\partial u} & \dfrac{\partial y}{\partial v} \end{vmatrix} = \frac{\partial x}{\partial u}\frac{\partial y}{\partial v} - \frac{\partial y}{\partial u}\frac{\partial x}{\partial v}.$$

This is called the **Jacobian** of $x$ and $y$ with respect to $u$ and $v$. [Condition (3) says that the Jacobian is not zero.] We can express the relation between the differential area of a curvilinear rectangular in the $xy$-plane and the differential area of the corresponding rectangle in the $uv$-plane as

$$dA(x, y) = \left|\frac{\partial(x, y)}{\partial(u, v)}\right| dA(u, v). \tag{4}$$

It should then seem reasonable that

$$\iint_R f(x, y)\,dA(x, y) = \iint_{R'} f(x(u, v), y(u, v))\left|\frac{\partial(x, y)}{\partial(u, v)}\right| dA(u, v). \tag{5}$$

This is the formula that relates double integrals that correspond to the change of variables $x = x(u, v)$, $y = y(u, v)$.

We know that rectangular coordinates $(x, y)$ are related to polar coordinates by the equations

$$x = r\cos\theta, \quad y = r\sin\theta.$$

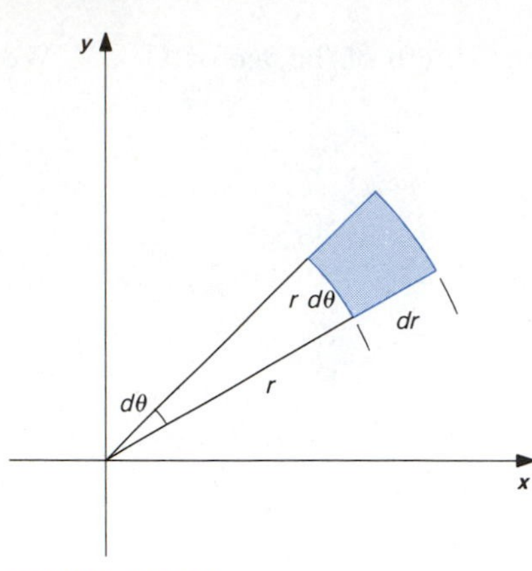


FIGURE 17.10.4

The Jacobian is

$$\frac{\partial(x, y)}{\partial(r, \theta)} = \begin{vmatrix} \frac{\partial x}{\partial r} & \frac{\partial x}{\partial \theta} \\ \frac{\partial y}{\partial r} & \frac{\partial y}{\partial \theta} \end{vmatrix} = \frac{\partial x}{\partial r}\frac{\partial y}{\partial \theta} - \frac{\partial y}{\partial r}\frac{\partial x}{\partial \theta}$$

$$= (\cos \theta)(r \cos \theta) - (\sin \theta)(-r \sin \theta)$$

$$= r(\cos^2\theta + \sin^2\theta) = r.$$

Formula (4) then gives

$$dA(x, y) = \left|\frac{\partial(x, y)}{\partial(r, \theta)}\right| dA(r, \theta) = r\, dr\, d\theta,$$

which is the same formula we derived in Section 17.4. This should not be surprising, since the procedure to derive formula (4) corresponds to that used in Section 17.4. See Figure 17.10.4. Also, note that we used this change of variables in cases where the region $R$ in the $xy$-plane could be conveniently described in terms of polar coordinates. Generally, we apply formula (5) to regions $R$ that can be described easily in terms of the variables $(u, v)$.

Let us compare formula (5) with the formula from Section 5.6 for change of variables in a definite integral of a single variable. We assume that $x = x(u)$ is differentiable and that $dx/du$ is continuous and nonzero, so $dx/du$ has constant sign. Then

$$dx = \frac{dx}{du}\, du,$$

$u = a$ corresponds to $x = x(a)$, and $u = b$ corresponds to $x = x(b)$, so

$$\int_{x(a)}^{x(b)} f(x)\, dx = \int_a^b f(x(u))\frac{dx}{du}\, du. \tag{6}$$

We can rewrite (6) in the form

$$\int_I f(x)\, dx = \int_{I'} f(x(u))\left|\frac{dx}{du}\right| du,$$

where $I' = [a, b]$ and $I$ is either $[x(a), x(b)]$ or $[x(b), x(a)]$, depending on whether $dx/du$ is positive or negative for $a \le u \le b$. We then see that the derivative $dx/du$ in (6) corresponds to the Jacobian in (5). Also, we have expressed the limits of integration of the integral with respect to $x$ in terms of the variable $u$. This is consistent with the way in which we will use formula (5).

■ **EXAMPLE 1**

Use the linear transformation

$$x = 2u + 3v, \qquad y = 2u - v$$

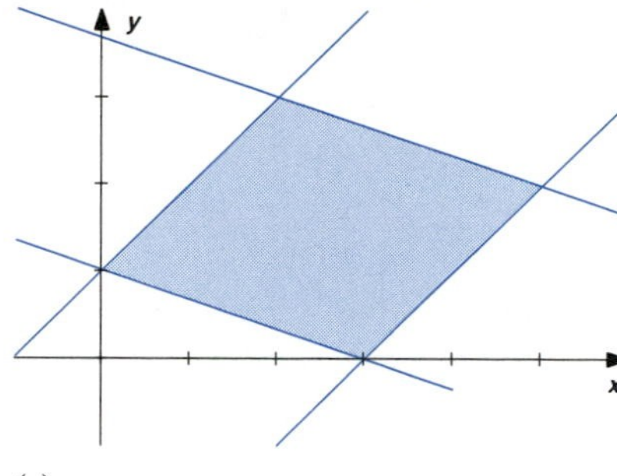

(a)

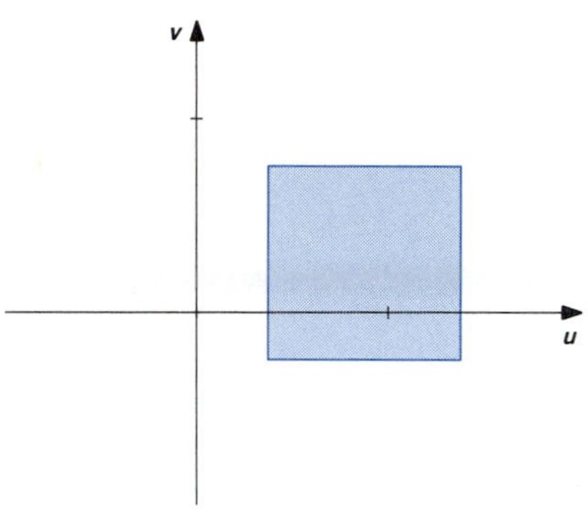

(b)

FIGURE 17.10.5

to evaluate $\iint_R (x + 3y)\, dA$, where $R$ is the parallelogram bounded by the lines

$$y = x - 3, \quad y = x + 1, \quad x + 3y = 3, \quad \text{and} \quad x + 3y = 11.$$

**SOLUTION**

The region $R$ in the $xy$-plane is sketched in Figure 17.10.5a. We determine the boundary equations of the corresponding region in the $uv$-plane by substitution. From the equation $y = x - 3$, we obtain the corresponding equation

$$2u - v = (2u + 3v) - 3,$$

$$v = \frac{3}{4}.$$

This is a horizontal line in the $uv$-plane. The line $y = x + 1$, which is parallel to $y = x - 3$, corresponds to the horizontal line $v = -1/4$.

The equation $x + 3y = 3$ corresponds to

$$(2u + 3v) + 3(2u - v) = 3,$$

$$8u = 3,$$

$$u = \frac{3}{8}.$$

This is a vertical line in the $uv$-plane. The line $x + 3y = 11$ corresponds to the vertical line $u = 11/8$.

The region $R'$ in the $uv$-plane that corresponds to the region $R$ in the $xy$-plane is sketched in Figure 17.10.5b.

The Jacobian of the transformation is

$$\frac{\partial(x, y)}{\partial(u, v)} = \begin{vmatrix} \dfrac{\partial x}{\partial u} & \dfrac{\partial x}{\partial v} \\ \dfrac{\partial y}{\partial u} & \dfrac{\partial y}{\partial v} \end{vmatrix} = \begin{vmatrix} 2 & 3 \\ 2 & -1 \end{vmatrix} = (2)(-1) - (2)(3) = -8.$$

We then have

$$\begin{aligned} \iint_R (x + 3y)\, dA &= \iint_{R'} [(2u + 3v) + 3(2u - v)]|-8|\, dA(u, v) \\ &= \int_{3/8}^{11/8} \int_{-1/4}^{3/4} 64u\, dv\, du \\ &= \int_{3/8}^{11/8} 64u\left[\frac{3}{4} - \left(-\frac{1}{4}\right)\right] du \\ &= 32u^2\Big]_{3/8}^{11/8} = 32\left(\frac{11}{8}\right)^2 - 32\left(\frac{3}{8}\right)^2 = 56. \end{aligned}$$

■

FIGURE 17.10.6

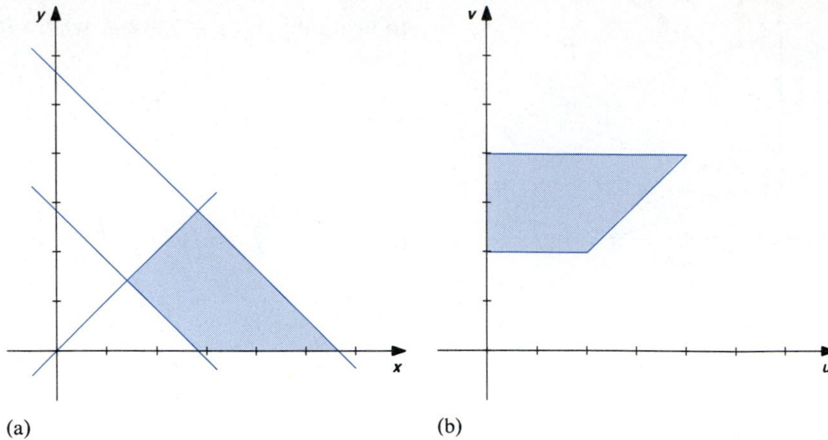


■ EXAMPLE 2

Use the rotation transformation

$$x = \frac{u + v}{\sqrt{2}}, \qquad y = \frac{-u + v}{\sqrt{2}}$$

to evaluate

$$\iint_R \frac{x - y}{x + y}\, dA,$$

where $R$ is the region bounded by the lines $x + y = 2\sqrt{2}$, $x + y = 4\sqrt{2}$, $x = y$, and the $x$-axis.

SOLUTION

The region $R$ is sketched in Figure 17.10.6a. The line $x + y = 2\sqrt{2}$ corresponds to

$$\left(\frac{u + v}{\sqrt{2}}\right) + \left(\frac{-u + v}{\sqrt{2}}\right) = 2\sqrt{2},$$

$$u + v - u + v = 4,$$

$$v = 2.$$

$x + y = 4\sqrt{2}$ corresponds to $v = 4$.

$x = y$ corresponds to $u = 0$.

The $x$-axis has equation $y = 0$. This corresponds to $u = v$. The region $R'$ in the $uv$-plane that corresponds to the region $R$ in the $xy$-plane is sketched in

Figure 17.10.6b. Note that the region is horizontally simple. The Jacobian is

$$\frac{\partial(x, y)}{\partial(u, v)} = \begin{vmatrix} \dfrac{\partial x}{\partial u} & \dfrac{\partial x}{\partial v} \\ \dfrac{\partial y}{\partial u} & \dfrac{\partial y}{\partial v} \end{vmatrix} = \begin{vmatrix} \dfrac{1}{\sqrt{2}} & \dfrac{1}{\sqrt{2}} \\ -\dfrac{1}{\sqrt{2}} & \dfrac{1}{\sqrt{2}} \end{vmatrix}$$

$$= \left(\frac{1}{\sqrt{2}}\right)\left(\frac{1}{\sqrt{2}}\right) - \left(-\frac{1}{\sqrt{2}}\right)\left(\frac{1}{\sqrt{2}}\right) = 1.$$

We then have

$$\iint_R \frac{x - y}{x + y}\, dA(x, y) = \iint_{R'} \frac{\left(\dfrac{u + v}{\sqrt{2}}\right) - \left(\dfrac{-u + v}{\sqrt{2}}\right)}{\left(\dfrac{u + v}{\sqrt{2}}\right) + \left(\dfrac{-u + v}{\sqrt{2}}\right)}\, dA(u, v)$$

$$= \int_2^4 \int_0^v \frac{u}{v}\, du\, dv$$

$$= \int_2^4 \frac{u^2}{2v}\bigg]_{u=0}^{u=v} dv$$

$$= \int_2^4 \frac{v}{2}\, dv$$

$$= \frac{v^2}{4}\bigg]_2^4 = 4 - 1 = 3.$$

■

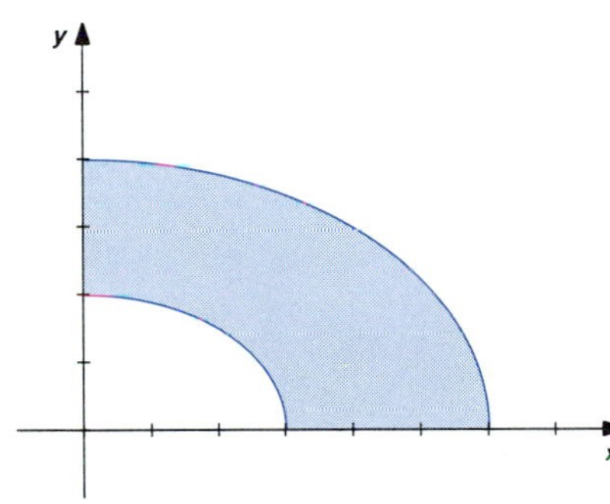

(a)

(b)

FIGURE 17.10.7

■ **EXAMPLE 3**

Use the change of scale transformation

$$x = 3u, \qquad y = 2v$$

to evaluate the area of the region $R$ in the first quadrant that is between the ellipses

$$\frac{x^2}{3^2} + \frac{y^2}{2^2} = 1 \quad \text{and} \quad \frac{x^2}{6^2} + \frac{y^2}{4^2} = 1.$$

**SOLUTION**

The region $R$ is sketched in Figure 17.10.7a. The ellipse

$$\frac{x^2}{3^2} + \frac{y^2}{2^2} = 1$$

corresponds to the circle $u^2 + v^2 = 1$; the ellipse

$$\frac{x^2}{6^2} + \frac{y^2}{4^2} = 1$$

corresponds to the circle $u^2 + v^2 = 4$; the $x$-axis corresponds to the $u$-axis; and the $y$-axis corresponds to the $v$-axis. The region $R'$ in the $uv$-plane that corresponds to the region $R$ in the $xy$-plane is sketched in Figure 17.10.7b. The Jacobian of the transformation is

$$\frac{\partial(x, y)}{\partial(u, v)} = \begin{vmatrix} \dfrac{\partial x}{\partial u} & \dfrac{\partial x}{\partial v} \\ \dfrac{\partial y}{\partial u} & \dfrac{\partial y}{\partial v} \end{vmatrix} = \begin{vmatrix} 3 & 0 \\ 0 & 2 \end{vmatrix} = (3)(2) - (0)(0) = 6.$$

We then have

$$\text{Area} = \iint_R dA(x, y) = \iint_{R'} |6| \, dA(u, v).$$

Let us use polar coordinates in the $uv$-plane to evaluate the latter integral. We have

$$u = r \cos \theta, \quad v = r \sin \theta, \quad \text{and}$$

$$dA(u, v) = r \, dA(r, \theta) = r \, dr \, d\theta.$$

We then have

$$\begin{aligned} \text{Area} &= \iint_{R'} 6 \, dA(u, v) \\ &= \int_0^{\pi/2} \int_1^2 6r \, dr \, d\theta \\ &= \int_0^{\pi/2} 3r^2 \bigg]_{r=1}^{r=2} d\theta \\ &= \int_0^{\pi/2} 9 \, d\theta = \frac{9\pi}{2}. \end{aligned}$$

■

The change of scale transformation

$$x = au, \qquad y = bv$$

followed by the change to polar coordinates in the $uv$-plane

$$u = r \cos \theta, \qquad v = r \sin \theta$$

can be expressed by the equations

$$x = ar \cos \theta, \qquad y = br \sin \theta.$$

The Jacobian of the latter transformation is the product of the Jacobians of the two previous transformations.

FIGURE 17.10.8

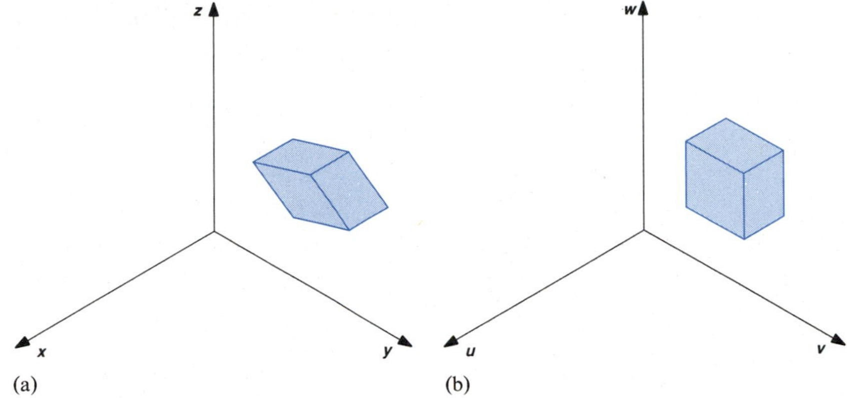


The ideas involved in change of variables in triple integrals are similar to those for double integrals. Equations

$$x = x(u, v, w), \quad y = y(u, v, w), \quad z = z(u, v, w)$$

relate points in a region $R$ of $xyz$-space to points in a region $R'$ in $uvw$-space. We assume that $x$, $y$, and $z$ have continuous first partial derivatives. The **Jacobian** of $x$, $y$, and $z$ with respect to $u$, $v$, and $w$ is defined to be the determinant

$$\frac{\partial(x, y, z)}{\partial(u, v, w)} = \begin{vmatrix} \dfrac{\partial x}{\partial u} & \dfrac{\partial x}{\partial v} & \dfrac{\partial x}{\partial w} \\ \dfrac{\partial y}{\partial u} & \dfrac{\partial y}{\partial v} & \dfrac{\partial y}{\partial w} \\ \dfrac{\partial z}{\partial u} & \dfrac{\partial z}{\partial v} & \dfrac{\partial z}{\partial w} \end{vmatrix}.$$

We assume the Jacobian is nonzero, so differential changes in $x$, $y$, and $z$ correspond to specific differential changes in $u$, $v$, and $w$. We assume that every point $(x, y, z)$ in $R$ corresponds to exactly one point $(u, v, w)$ in $R'$. Small rectangular prisms in $R'$ with differential volume $dV(u, v, w) = du\,dv\,dw$ correspond to curvilinear prisms in $R'$ that are essentially parallelepipeds. See Figure 17.10.8. The differential volume $dV(x, y, z)$ of the curvilinear prisms is the volume of the parallelepiped determined by the vectors

$$\mathbf{U} = \left(\frac{\partial x}{\partial u}\,du\right)\mathbf{i} + \left(\frac{\partial y}{\partial u}\,du\right)\mathbf{j} + \left(\frac{\partial z}{\partial u}\,du\right)\mathbf{k},$$

$$\mathbf{V} = \left(\frac{\partial x}{\partial v}\,dv\right)\mathbf{i} + \left(\frac{\partial y}{\partial v}\,dv\right)\mathbf{j} + \left(\frac{\partial z}{\partial v}\,dv\right)\mathbf{k}, \quad \text{and}$$

$$\mathbf{W} = \left(\frac{\partial x}{\partial w}\,dw\right)\mathbf{i} + \left(\frac{\partial y}{\partial w}\,dw\right)\mathbf{j} + \left(\frac{\partial z}{\partial w}\,dw\right)\mathbf{k}.$$

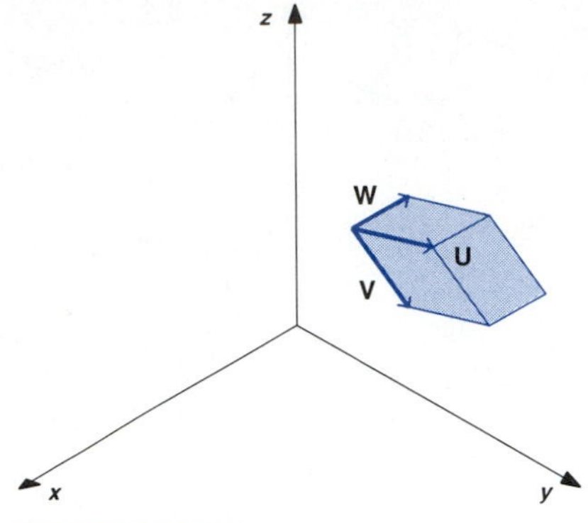

FIGURE 17.10.9

See Figure 17.10.9. From Section 14.2 we know that the volume of the parallelepiped determined by **U**, **V**, and **W** is the absolute value of the determinant

$$\begin{vmatrix} \frac{\partial x}{\partial u}\,du & \frac{\partial x}{\partial v}\,dv & \frac{\partial x}{\partial w}\,dw \\ \frac{\partial y}{\partial u}\,du & \frac{\partial y}{\partial v}\,dv & \frac{\partial y}{\partial w}\,dw \\ \frac{\partial z}{\partial u}\,du & \frac{\partial z}{\partial v}\,dv & \frac{\partial z}{\partial w}\,dw \end{vmatrix}.$$

The value of this determinant can be expressed as

$$\frac{\partial(x, y, z)}{\partial(u, v, w)}\,du\,dv\,dw, \quad \text{so}$$

$$dV(x, y, z) = \left|\frac{\partial(x, y, z)}{\partial(u, v, w)}\right| dV(u, v, w), \quad \text{and}$$

$$\iiint_R f(x, y, z)\,dV(x, y, z)$$

$$= \iiint_{R'} f(x(u, v, w), y(u, v, w), z(u, v, w)) \left|\frac{\partial(x, y, z)}{\partial(u, v, w)}\right| dV(u, v, w).$$

This is the formula that relates triple integrals that correspond to the change of variables

$$x = x(u, v, w), \quad y = y(u, v, w), \quad z = z(u, v, w).$$

We know that rectangular coordinates $(x, y, z)$ are related to cylindrical coordinates $(r, \theta, z)$ by the equations

$$x = r\cos\theta, \quad y = r\sin\theta, \quad z = z.$$

The Jacobian is

$$\frac{\partial(x, y, z)}{\partial(r, \theta, z)} = \begin{vmatrix} \frac{\partial x}{\partial r} & \frac{\partial x}{\partial \theta} & \frac{\partial x}{\partial z} \\ \frac{\partial y}{\partial r} & \frac{\partial y}{\partial \theta} & \frac{\partial y}{\partial z} \\ \frac{\partial z}{\partial r} & \frac{\partial z}{\partial \theta} & \frac{\partial z}{\partial z} \end{vmatrix} = \begin{vmatrix} \cos\theta & -r\sin\theta & 0 \\ \sin\theta & r\cos\theta & 0 \\ 0 & 0 & 1 \end{vmatrix}.$$

The value of the determinant is $r$. This corresponds to the formula

$$dV = r\,dr\,d\theta\,dz,$$

which we derived in Section 17.8. See Figure 17.10.10.

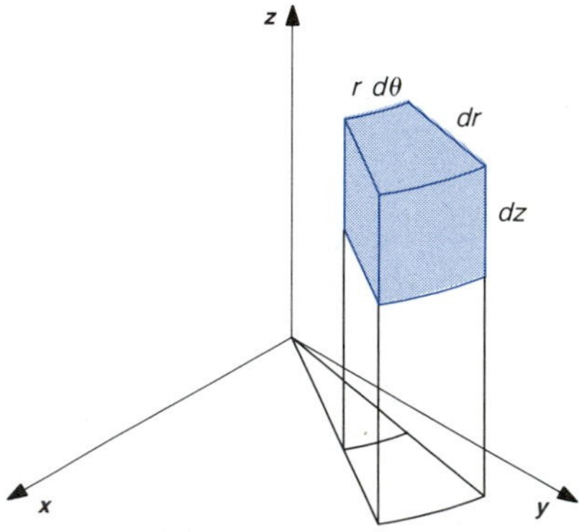

FIGURE 17.10.10

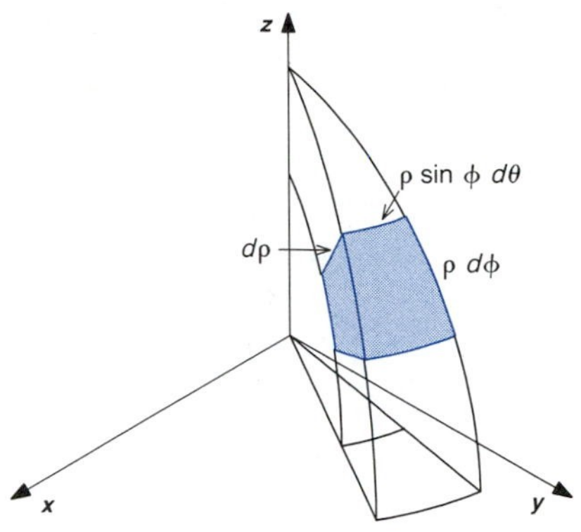


FIGURE 17.10.11

Rectangular coordinates $(x, y, z)$ are related to spherical coordinates $(\rho, \theta, \phi)$ by the equations

$$x = \rho \sin \phi \cos \theta, \quad y = \rho \sin \phi \sin \theta, \quad z = \rho \cos \phi.$$

The Jacobian is

$$\frac{\partial(x, y, z)}{\partial(\rho, \theta, \phi)} = \begin{vmatrix} \dfrac{\partial x}{\partial \rho} & \dfrac{\partial x}{\partial \theta} & \dfrac{\partial x}{\partial \phi} \\ \dfrac{\partial y}{\partial \rho} & \dfrac{\partial y}{\partial \theta} & \dfrac{\partial y}{\partial \phi} \\ \dfrac{\partial z}{\partial \rho} & \dfrac{\partial z}{\partial \theta} & \dfrac{\partial z}{\partial \phi} \end{vmatrix}$$

$$= \begin{vmatrix} \sin \phi \cos \theta & -\rho \sin \phi \sin \theta & \rho \cos \phi \cos \theta \\ \sin \phi \sin \theta & \rho \sin \phi \cos \theta & \rho \cos \phi \sin \theta \\ \cos \phi & 0 & -\rho \sin \phi \end{vmatrix}.$$

As we may expect, the absolute value of the determinant is $\rho^2 \sin \phi$. This corresponds to the formula

$$dV = \rho^2 \sin \phi \, d\rho \, d\phi \, d\theta$$

which we derived in Section 17.8. See Figure 17.10.11.

■ **EXAMPLE 4**

The region $R$ bounded by the planes

$$\sqrt{2}y + z = \sqrt{3}, \qquad \sqrt{2}y + z = 0,$$

$$\sqrt{3}x - y + \sqrt{2}z = \sqrt{6}, \qquad \sqrt{3}x - y + \sqrt{2}z = 0,$$

$$-\sqrt{3}x - y + \sqrt{2}z = \sqrt{6}, \qquad -\sqrt{3}x - y + \sqrt{2}z = 0,$$

is a cube with length of side one, oriented with a diagonal along the $z$-axis. Use the linear transformation

$$x = \sqrt{\frac{1}{2}}v - \sqrt{\frac{1}{2}}w,$$

$$y = \sqrt{\frac{2}{3}}u - \sqrt{\frac{1}{6}}v - \sqrt{\frac{1}{6}}w,$$

$$z = \sqrt{\frac{1}{3}}u + \sqrt{\frac{1}{3}}v + \sqrt{\frac{1}{3}}w,$$

to evaluate $\iiint_R z \, dV$.

FIGURE 17.10.12

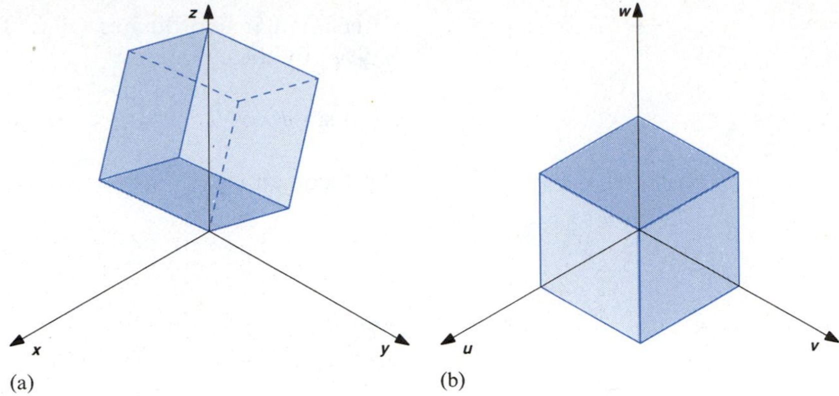


**SOLUTION**

The region $R$ is sketched in Figure 17.10.12a. The surface in $uvw$-space that corresponds to the plane $\sqrt{2}y + z = \sqrt{3}$ is

$$\sqrt{2}\left(\sqrt{\frac{2}{3}}u - \sqrt{\frac{1}{6}}v - \sqrt{\frac{1}{6}}w\right) + \left(\sqrt{\frac{1}{3}}u + \sqrt{\frac{1}{3}}v + \sqrt{\frac{1}{3}}w\right) = \sqrt{3},$$

$$\sqrt{3}u = \sqrt{3},$$

$$u = 1.$$

Substitution into each of the other boundary planes of the region $R$ gives the planes $u = 0$, $v = 1$, $v = 0$, $w = 1$, and $w = 0$. The region in $uvw$-space that corresponds to the region $R$ in $xyz$-space is sketched in Figure 17.10.12b.

The Jacobian of the transformation is

$$\begin{vmatrix} \dfrac{\partial x}{\partial u} & \dfrac{\partial x}{\partial v} & \dfrac{\partial x}{\partial w} \\ \dfrac{\partial y}{\partial u} & \dfrac{\partial y}{\partial v} & \dfrac{\partial y}{\partial w} \\ \dfrac{\partial z}{\partial u} & \dfrac{\partial z}{\partial v} & \dfrac{\partial z}{\partial w} \end{vmatrix} = \begin{vmatrix} 0 & \sqrt{\frac{1}{2}} & -\sqrt{\frac{1}{2}} \\ \sqrt{\frac{2}{3}} & -\sqrt{\frac{1}{6}} & -\sqrt{\frac{1}{6}} \\ \sqrt{\frac{1}{3}} & \sqrt{\frac{1}{3}} & \sqrt{\frac{1}{3}} \end{vmatrix}$$

$$= \begin{vmatrix} -\sqrt{\frac{1}{6}} & -\sqrt{\frac{1}{6}} \\ \sqrt{\frac{1}{3}} & \sqrt{\frac{1}{3}} \end{vmatrix}(0) - \begin{vmatrix} \sqrt{\frac{2}{3}} & -\sqrt{\frac{1}{6}} \\ \sqrt{\frac{1}{3}} & \sqrt{\frac{1}{3}} \end{vmatrix}\left(\sqrt{\frac{1}{2}}\right) + \begin{vmatrix} \sqrt{\frac{2}{3}} & -\sqrt{\frac{1}{6}} \\ \sqrt{\frac{1}{3}} & \sqrt{\frac{1}{3}} \end{vmatrix}\left(-\sqrt{\frac{1}{2}}\right)$$

$$= -\left(\sqrt{\frac{2}{9}} + \sqrt{\frac{1}{18}}\right)\left(\sqrt{\frac{1}{2}}\right) + \left(\sqrt{\frac{2}{9}} + \sqrt{\frac{1}{18}}\right)\left(-\sqrt{\frac{1}{2}}\right)$$

$$= -\frac{1}{3} - \frac{1}{6} - \frac{1}{3} - \frac{1}{6} = -1.$$

It follows that

$$\iiint\limits_{R} z\,dA(x, y, z) = \iiint\limits_{R'} \left(\sqrt{\frac{1}{3}}u + \sqrt{\frac{1}{3}}v + \sqrt{\frac{1}{3}}w\right)|-1|\,dA(u, v, w)$$

$$= \int_0^1 \int_0^1 \int_0^1 \left(\sqrt{\frac{1}{3}}u + \sqrt{\frac{1}{3}}v + \sqrt{\frac{1}{3}}w\right) dw\,dv\,du$$

$$= \int_0^1 \int_0^1 \left(\sqrt{\frac{1}{3}}u + \sqrt{\frac{1}{3}}v + \frac{1}{2}\sqrt{\frac{1}{3}}\right) dv\,du$$

$$= \int_0^1 \left(\sqrt{\frac{1}{3}}u + \frac{1}{2}\sqrt{\frac{1}{3}} + \frac{1}{2}\sqrt{\frac{1}{3}}\right) du$$

$$= \frac{1}{2}\sqrt{\frac{1}{3}} + \frac{1}{2}\sqrt{\frac{1}{3}} + \frac{1}{2}\sqrt{\frac{1}{3}}$$

$$= \frac{3}{2}\sqrt{\frac{1}{3}} = \frac{\sqrt{3}}{2}.$$

■

The change of variables

$$x = au, \quad y = bv, \quad z = cw$$

corresponds to a change of scale of the coordinate axes. This type of transformation changes ellipsoids in *xyz*-space into spheres in *uvw*-space. See Figure 17.10.13. This type of transformation can also be used to change surfaces in *xyz*-space that have elliptical cross sections into surfaces in *uvw*-space that have circular cross sections. We can then use either cylindrical or spherical coordinates in *uvw*-space.

**FIGURE 17.10.13**

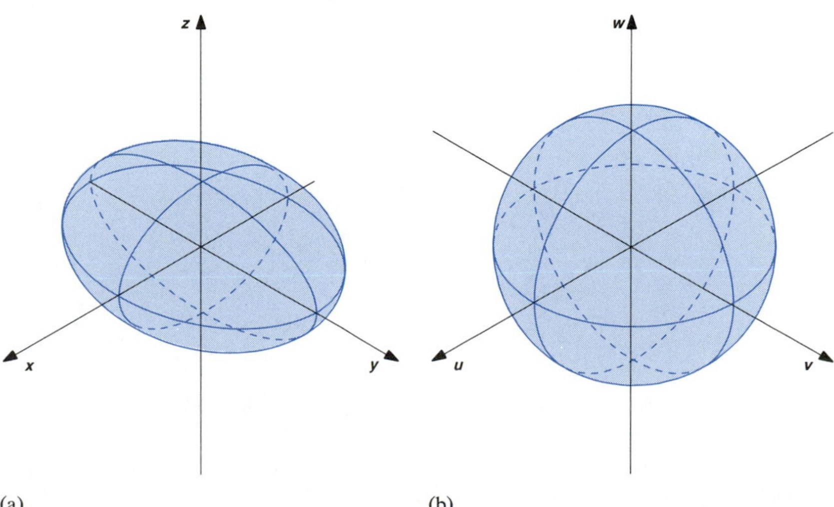

**EXAMPLE 5**

A homogeneous elliptical cam with density $\rho(x, y, z) = 4$ occupies the region $R$ bounded by the elliptical cylinder $4x^2 + y^2 = 4$ and the planes $z = 0$ and $z = 1$. Find the moment of inertia of the cam about the $z$-axis.

**SOLUTION**

The cam is sketched in Figure 17.10.14a. The change of scale transformation

$$x = u, \quad y = 2v, \quad z = w$$

transforms the elliptical cylinder $4x^2 + y^2 = 4$ into the right circular cylinder $u^2 + v^2 = 1$. The planes $z = 1$ and $z = 0$ correspond to $w = 1$ and $w = 0$. The region $R'$ in $uvw$-space that corresponds to the region $R$ in $xyz$-space is sketched in Figure 17.10.14b. The Jacobian of the transformation is

$$\begin{vmatrix} 1 & 0 & 0 \\ 0 & 2 & 0 \\ 0 & 0 & 1 \end{vmatrix} = \begin{vmatrix} 2 & 0 \\ 0 & 1 \end{vmatrix}(1) - \begin{vmatrix} 0 & 0 \\ 0 & 1 \end{vmatrix}(0) + \begin{vmatrix} 0 & 2 \\ 0 & 0 \end{vmatrix}(0) = 2.$$

We then have

$$I_{z\text{-axis}} = \iiint_R (x^2 + y^2)\rho(w, y, z)\, dV(x, y, z)$$

$$= \iiint_{R'} [u^2 + (2v)^2](4)|2|\, dV(u, v, w).$$

Let us use cylindrical coordinates in $uvw$-space to evaluate the latter integral. We set

$$u = r\cos\theta, \quad v = r\sin\theta, \quad w = w.$$

FIGURE 17.10.14

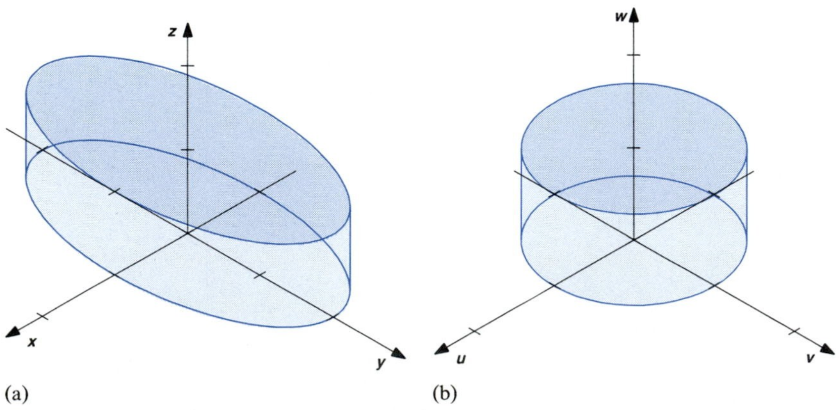


(a) (b)

Then

$$\iiint_{R'} [u^2 + (2v)^2](4)\,|2|\,dV(u, v, w)$$

$$= \int_0^{2\pi}\int_0^1\int_0^1 (r^2\cos^2\theta + 4r^2\sin^2\theta)(8)r\,dw\,dr\,d\theta$$

$$= \int_0^{2\pi}\int_0^1\int_0^1 (1 + 3\sin^2\theta)(8r^3)\,dw\,dr\,d\theta$$

$$= \int_0^{2\pi}\int_0^1 (1 + 3\sin^2\theta)(8r^3)\,dr\,d\theta$$

$$= \int_0^{2\pi} (1 + 3\sin^2\theta)(2)\,d\theta$$

$$= 2\theta - 3\sin\theta\cos\theta + 3\theta\Big]_0^{2\pi} = 10\pi.$$ ■

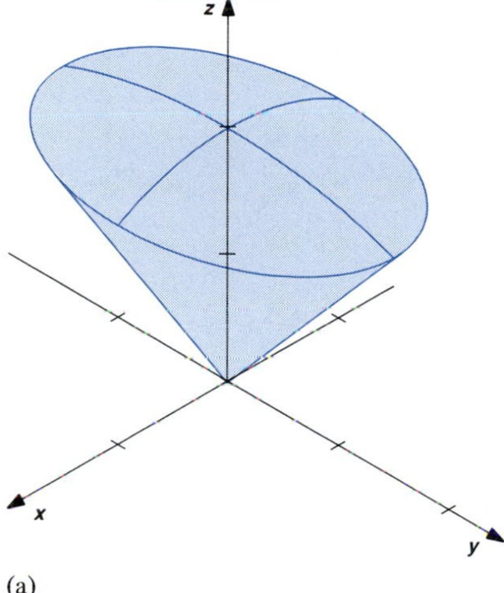


(a)

(b)

**FIGURE 17.10.15**

■ **EXAMPLE 6**

Find the volume of the region $R$ inside both the ellipsoid $9x^2 + 4y^2 + 9z^2 = 36$ and the elliptical cone $\sqrt{3}z = \sqrt{9x^2 + 4y^2}$.

**SOLUTION**

The region $R$ is sketched in Figure 17.10.15a. Let us change scale by using the transformation

$$x = 2u, \quad y = 3v, \quad z = 2w.$$

The ellipsoid then has $uvw$-equation $u^2 + v^2 + w^2 = 1$ and the cone has $uvw$-equation $w = \sqrt{3}\sqrt{u^2 + v^2}$. The corresponding region $R'$ is sketched in Figure 17.10.15b.

The Jacobian of the transformation is

$$\begin{vmatrix} 2 & 0 & 0 \\ 0 & 3 & 0 \\ 0 & 0 & 2 \end{vmatrix} = \begin{vmatrix} 3 & 0 \\ 0 & 2 \end{vmatrix}(2) - \begin{vmatrix} 0 & 0 \\ 0 & 2 \end{vmatrix}(0) + \begin{vmatrix} 0 & 3 \\ 0 & 0 \end{vmatrix}(0) = 12.$$

We then have

$$\text{Volume} = \iiint_R dV(x, y, z)$$

$$= \iiint_{R'} |12|\,dV(u, v, w).$$

Let us use spherical coordinates in *uvw*-space to evaluate the latter integral. We set

$$u = \rho \sin\phi \cos\theta, \quad v = \rho \sin\phi \sin\theta, \quad w = \rho \cos\phi.$$

The sphere given by $u^2 + v^2 + w^2 = 1$ becomes $\rho = 1$ and the cone given by $w = \sqrt{3}\sqrt{u^2 + v^2}$ becomes $\phi = \tan^{-1}(1/\sqrt{3}) = \pi/6$. Then

$$\iiint_{R'} |12|\, dV(u, v, w) = \int_0^{2\pi} \int_0^{\pi/6} \int_0^1 12\rho^2 \sin\phi \, d\rho\, d\phi\, d\theta$$

$$= \int_0^{2\pi} \int_0^{\pi/6} 4 \sin\phi \, d\phi\, d\theta$$

$$= \int_0^{2\pi} -4\cos\phi \Big]_{\phi=0}^{\phi=\pi/6} d\theta$$

$$= \int_0^{2\pi} \left[-4\left(\frac{\sqrt{3}}{2}\right) - 4(-1)\right] d\theta$$

$$= (4 - 2\sqrt{3})(2\pi) \approx 3.3671489. \qquad \blacksquare$$

## EXERCISES 17.10

**1** Evaluate the Jacobian that corresponds to the translation of coordinates $x = u + h$, $y = v + k$, $h$ and $k$ constants.

**2** Evaluate the Jacobian that corresponds to the rotation of coordinates $x = u\cos\theta - v\sin\theta$, $y = u\sin\theta + v\cos\theta$, $\theta$ constant.

**3** Verify that the change from rectangular to cylindrical coordinates given by the equations $x = r\cos\theta$, $y = r\sin\theta$, $z = z$ has Jacobian $r$.

**4** Verify that the change from rectangular to spherical coordinates given by the equations $x = \rho\sin\phi\cos\theta$, $y = \rho\sin\phi\sin\theta$, $z = \rho\cos\phi$ has Jacobian with absolute value $\rho^2\sin\phi$.

**5** Show that a linear transformation $x = au + bv$, $y = cu + dv$ that has nonzero Jacobian transforms a line $Ax + By + C = 0$ in the *xy*-plane into a line in the *uv*-plane.

**6** Show that a linear transformation

$$x = au + bv + cw, \quad y = du + ev + fw, \quad z = gu + hv + iw$$

that has nonzero Jacobian, transforms a plane $Ax + By + Cz + D = 0$ in *xyz*-space into a plane in *uvw*-space.

In Exercises 7–22 express the given integrals as iterated integrals with respect to the indicated coordinates. Do not evaluate the integrals.

**7** $\iint_R xy\, dA$, where $R$ is the region bounded by the lines $y = 2x - 3$, $y = 2x$, $x = 2y - 3$, $x = 2y$; $x = 2u + v$, $y = u + 2v$

**8** $\iint_R (x + y)\, dA$, where $R$ is the region bounded by the lines $y = 2x - 6$, $x = 2y$, $y = x$; $x = 2u + v$, $y = u + 2v$

**9** $\iint_R x\, dA$, where $R$ is the region bounded by the lines $y = 2x$, $x = 2y$, $x + y = 6$; $x = 2u + v$, $y = u + 2v$

**10** $\iint_R x\, dA$, where $R$ is the region bounded by the lines $y = 2x$, $x = 2y$, $x + y = 6$; $x = 2u - v$, $y = u + v$

**11** $\iint_R (2x + 3y)\, dA$, where $R$ is the region bounded by the lines $4y = x - 3$, $4y = x + 2$, $2x + 3y = 6$, $2x + 3y = 17$; $x = 4u - 3v$, $y = u + 2v$

**12** $\iint_R (x + y)\, dA$, where $R$ is the region bounded by the lines $x + y = 1$, $x + y = 2$, $y = x - 1$, $y = x + 1$; $x = -u + v$, $y = u + v$

13 $\iint_R (x + y)\,dA$, where $R$ is the region bounded by the lines $y = 2x$, $x = 2y$, $x + y = 6$; $x = (u - v)/\sqrt{2}$, $y = (u + v)/\sqrt{2}$

14 $\iint_R (x + y)\,dA$, where $R$ is the region bounded by the lines $y = 2x$, $x = 2y$, $x + y = 6$; $x = (u + v)/\sqrt{2}$, $y = (-u + v)/\sqrt{2}$

15 $\iint_R (x^2 - y^2)^2\,dA$, where $R$ is the region bounded by $xy = 1$, $y = x - 1$, $y = x + 1$; $x = (u + v)/\sqrt{2}$, $y = (-u + v)/\sqrt{2}$

16 $\iint_R (x^2 - y^2)^2\,dA$, where $R$ is the region bounded by the parabola $x^2 - 2xy + y^2 + \sqrt{2}x + \sqrt{2}y = 12$ and the line $x + y = 2\sqrt{2}$; $x = (u + v)/\sqrt{2}$, $y = (-u + v)/\sqrt{2}$

17 $\iint_R e^{x^2 - y^2}\,dA$, where $R$ is the region bounded by $x^2 - y^2 = 1$, $x^2 - y^2 = 16$, $x = 2y$, in the first quadrant; $x = r \sec\theta$, $y = r \tan\theta$

18 $\iint_R x\,dA$, where $R$ is the region bounded by $y = x$, $y = -x$, $4x^2 - y^2 = 16$, $4x^2 - y^2 = 4$, in the first quadrant; $x = r\sec\theta$, $y = 2r\tan\theta$

19 $\iiint_R (\sqrt{3}x + y)\,dV$, where $R$ is the region bounded by the planes $y = \sqrt{3}x$, $y = \sqrt{3}x + 2\sqrt{3}x$, $y = -\sqrt{3}x$, $y = -\sqrt{3}x + 2\sqrt{3}$, $z = 0$, $z = 2$; $x = u - v$, $y = \sqrt{3}u + \sqrt{3}v$, $z = 2w$

20 $\iiint_R dV$, where $R$ is the region bounded by the planes $z = 0$, $2z = x$, $x - 2y - z = 0$, $x + 2y + z = 4$; $x = 2u + 2w$, $y = u + 2v + w/2$, $z = w$

21 $\iiint_R dV$, where $R$ is the region bounded by $x = y^2 + z$, $z = x - 4$, $z = 0$, $z = 6$; $x = u + w$, $y = v$, $z = w$

22 $\iiint_R dV$, where $R$ is the region bounded by $x^2 + y^2 + z^2 - 2yz = 1$, $z = 0$, $z = 4$; $x = u$, $y = v + w$, $z = w$

23 Find the area of the region in the first quadrant that is bounded by $x = 2y$, $x^2 + 4y^2 = 4$, and the $x$-axis.

24 Find the center of mass of a homogeneous lamina bounded by $3y = 2\sqrt{9 - x^2}$ and the $x$-axis.

25 Evaluate $\iint_R x\,dA$, where $R$ is the region bounded by the lines $y = x$, $y = x - 4$, $y = -x$, $y = -x + 4$.

26 Evaluate $\iint_R (x + y)\,dA$, where $R$ is the region bounded by the lines $y = x$, $y = x - 1$, $x = 2y$, $x = 2y - 1$.

27 Find the volume of the ellipsoid

$$\frac{x^2}{a^2} + \frac{y^2}{b^2} + \frac{z^2}{c^2} = 1.$$

28 Find the volume of the region bounded by the elliptical paraboloid $2z = 16 - 4x^2 - y^2$ and $z = 0$.

29 Find the volume of the region bounded by

$$\frac{x^2}{2} + \frac{y^2}{8} = z^2 + 1, \qquad z = 1, \quad z = -1.$$

# REVIEW EXERCISES

1 Evaluate $\int_0^1 \int_0^x 6xy^2\,dy\,dx$.

2 Evaluate $\int_0^1 \int_x^1 (2y - 1)\,dy\,dx$.

3 Evaluate $\int_0^{\pi/2} \int_0^{\sin\theta} \cos\theta\, r\,dr\,d\theta$.

4 Evaluate $\int_0^1 \int_0^{\sqrt{1-x^2}} \int_0^{\sqrt{1-x^2-y^2}} xy\,dz\,dy\,dx$.

5 Evaluate $\iint_R x^3y\,dA$, where $R$ is the region bounded by $x = \sqrt{y}$, $y = 4$, and the $y$-axis.

6 Find the area of the region bounded by $y = x/2$, $x + y = 3$, and the $x$-axis.

7 Express

$$\int_0^1 \int_x^1 dy\,dx$$

as an iterated double integral with the opposite order of integration.

8 Express

$$\int_0^1 \int_1^{e^x} dy\,dx$$

as an iterated double integral with the opposite order of integration.

9 Find the surface area of that part of the surface $z = x^2 + y^2$ above the region in the $xy$-plane bounded by $x^2 + y^2 = 2$.

10 Find the surface area of that part of the surface $z = \sqrt{1 - x^2}$ that is inside the cylinder $x^2 + y^2 = 1$.

11 Find the center of mass of the lamina in the first quadrant bounded by $y = x$, $y = 0$, $x^2 + y^2 = 1$, if the density is given by $\rho(x, y) = \sqrt{x^2 + y^2}$.

**12** Find the center of mass of the lamina in the first quadrant, between $x^2 + y^2 = 4$ and $x^2 + y^2 = 1$, if the density is proportional to the distance from the origin.

In Exercises 13–16, use each of rectangular, cylindrical, and spherical coordinates to set up iterated triple integrals that give the volume of the regions. Use one of the integrals to evaluate the volume.

**13** Inside $x^2 + y^2 + z^2 = 4$ and above $z = 1$

**14** Inside both $x^2 + y^2 + z^2 = 4$ and $z = \sqrt{x^2 + y^2}$

**15** Inside $x^2 + y^2 = 1$, between $z = \sqrt{x^2 + y^2}$ and $z = 0$

**16** In the first octant, below $x + y + z = 2$, and above $z = x + y$

In Exercises 17–18, use each of rectangular, cylindrical, and spherical coordinates to set up iterated triple integrals that give the moment of inertia about the $z$-axis of the solids. Do not evaluate the integrals.

**17** The solid bounded by $z = \sqrt{x^2 + y^2}$ and $z = 2$, with density $\rho(x, y, z) = 1 + \sqrt{x^2 + y^2 + z^2}$

**18** The solid bounded by $x^2 + y^2 + z^2 = 4$, with density $\rho(x, y, z) = x^2 + y^2$

In Exercises 19–20, express the volume of the indicated region as a single iterated triple integral with respect to rectangular coordinates for every order of integration for which that is possible. Do not evaluate the integrals.

**19** Bounded by $x + 2y + 3z = 6$, $x = 4$, $x = 0$, $y = 0$, $z = 0$

**20** Bounded by $x + y = 4$, $y = x$, $x + 2z = 4$, $x = 0$, $z = 0$

In Exercises 21–23, express the given integrals as iterated integrals with respect to the indicated coordinates. Do not evaluate the integrals.

**21** $\iint_R y\,dA$, where $R$ is the region bounded by the lines $y = x$, $y = x + 2$, $x + y = 0$, $x + y = 2$; $x = (u - v)/\sqrt{2}$, $y = (u + v)/\sqrt{2}$.

**22** $\iint_R x^2\,dA$, where $R$ is the region bounded by $x^2 - y^2 = 1$ and $x = 2/\sqrt{3}$ in the first quadrant; $x = r\sec\theta$, $y = r\tan\theta$.

**23** $\iiint_R (y - x)\,dV$, where $R$ is the region bounded by the planes $y = x$, $z = y$, $z = y - 2x$, $z = 1$; $x = -v + w$, $y = u - v + w$, $z = u + v + w$

**24** Find the volume of the region bounded by the elliptical cone $z = 4\sqrt{x^2 + 4y^2}$ and the plane $z = 4$.

## CHAPTER EIGHTEEN

# *Topics in vector calculus*

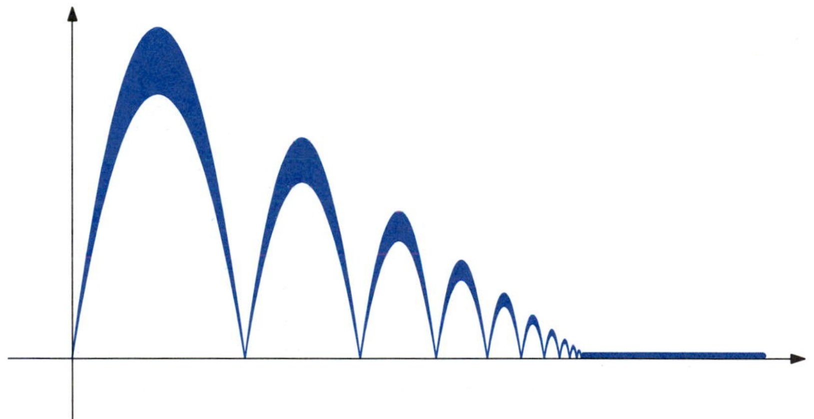

*I*n this chapter we will introduce integrals over curves and surfaces and establish some important results that relate integrals over a region to an integral over the boundary of the region. We will see how integrals of functions that measure components of forces along the unit tangent and normal vectors of curves and along the unit normal vector of surfaces relate to some physical problems.

## 18.1 LINE INTEGRALS

Line integrals are used to measure sums of physical quantities as the values vary along points on a curve. As usual, the definition involves a limit of Riemann sums and applications of the concept depend on the physical interpretation of the terms of the sum.

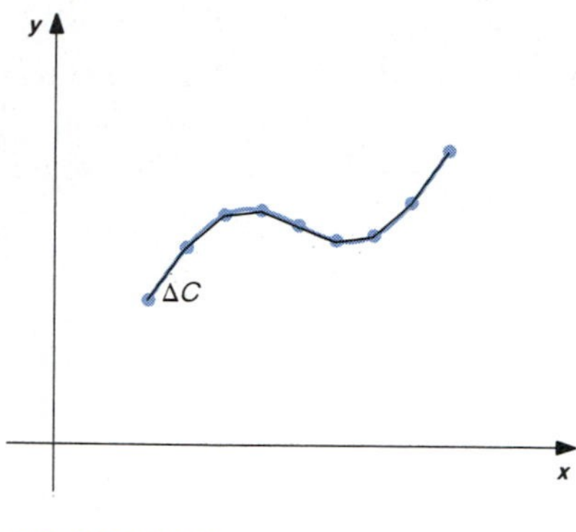


FIGURE 18.1.1

Suppose $C$ is a curve that is in the plane, or in three-dimensional space, and $f$ is defined at each point of $C$. We divide $C$ into small pieces $\Delta C$ of length $\Delta s$, and form the Riemann sum

$$\sum f \, \Delta s,$$

where $f$ is evaluated at a point in $\Delta C$. See Figure 18.1.1. If the curve is "smooth enough" and $f$ is continuous on $C$, theory guarantees that the Riemann sums converge to a single number $\int_C f\,ds$ as the lengths $\Delta s$ approach zero. The number $\int_C f\,ds$ is called the **line integral** of $f$ over the curve $C$. We should expect that

$$\int_C ds \quad \text{is the arc length of } C.$$

If a smooth curve $C$ is given in parametric form, we can express the line integral of a continuous function $f$ over $C$ as an integral with respect to the parameter. That is, let us consider the curve in the plane that is given by the parametric equations

$$C\colon\ x = x(t), \quad y = y(t), \qquad a \le t \le b,$$

where $x(t)$ and $y(t)$ have continuous first partial derivatives and $(dx/dt)^2 + (dy/dt)^2 \neq 0$. If $[a, b]$ is divided into subintervals of length $\Delta t$, the points on the curve that correspond to the points of subdivision divide $C$ into pieces of length $\Delta s$, where

$$\Delta s \approx \sqrt{(\Delta x)^2 + (\Delta y)^2} = \sqrt{\left(\frac{\Delta x}{\Delta t}\right)^2 + \left(\frac{\Delta y}{\Delta t}\right)^2}\,\Delta t.$$

It follows that

$$\sum f\,\Delta s \approx \sum f(x(t), y(t))\sqrt{\left(\frac{\Delta x}{\Delta t}\right)^2 + \left(\frac{\Delta y}{\Delta t}\right)^2}\,\Delta t.$$

It should then seem reasonable that

$$\int_C f\,ds = \int_a^b f(x(t), y(t))\sqrt{\left(\frac{dx}{dt}\right)^2 + \left(\frac{dy}{dt}\right)^2}\,dt. \tag{1}$$

This means that we can evaluate a line integral by substituting parametric values and then evaluating the resulting integral with respect to the real-valued parameter. The formula

$$ds = \sqrt{\left(\frac{dx}{dt}\right)^2 + \left(\frac{dy}{dt}\right)^2}\, dt$$

for the differential arc length of a parametric curve in the plane is the same as that used in Section 15.2. Line integrals over smooth curves in three-dimensional space can also be evaluated by substituting parametric values and using the appropriate formula for $ds$.

**EXAMPLE 1**

(a) $C\colon \quad x = t, \quad y = t^2, \qquad 1 \le t \le 2.$

Express $\int_C 3y/x\, ds$ as a definite integral with respect to the parameter $t$. Evaluate the integral.

(b) $C'\colon \quad x = \sqrt{u}, \quad y = u, \qquad 1 \le u \le 4,$

is a reparametrization of $C$, obtained by using the equation $t = \sqrt{u}$. Express $\int_{C'} 3y/x\, ds$ as a definite integral with respect to the parameter $u$ and then use the change of variables $t = \sqrt{u}$ to express that integral as a definite integral with respect to $t$.

(c) $-C\colon \quad x = 3 - v, \quad y = (3 - v)^2, \qquad 1 \le v \le 2,$

has the same graph as $C$, but opposite direction. Express $\int_{-C} 3y/x\, ds$ as a definite integral with respect to the parameter $v$ and then use the change of variables $t = 3 - v$ to express that integral as a definite integral with respect to $t$.

**SOLUTION**

(a) The equations $x = t$ and $y = t^2$ imply

$$\frac{dx}{dt} = 1 \quad \text{and} \quad \frac{dy}{dt} = 2t.$$

Substitution into formula (1) then gives

$$\begin{aligned}
\int_C \frac{3y}{x}\, ds &= \int_1^2 3\left(\frac{t^2}{t}\right)\sqrt{(1)^2 + (2t)^2}\, dt \\
&= \int_1^2 3t\sqrt{1 + 4t^2}\, dt \qquad [u = 1 + 4t^2,\ du = 8t\, dt] \\
&= \frac{3}{8}\frac{(1 + 4t^2)^{3/2}}{3/2}\Bigg]_1^2 \\
&= \frac{1}{4}(17)^{3/2} - \frac{1}{4}(5)^{3/2}.
\end{aligned}$$

(b) The equations $x = \sqrt{u}$ and $y = u$ imply

$$\frac{dx}{du} = \frac{1}{2\sqrt{u}} \quad \text{and} \quad \frac{dy}{du} = 1.$$

Substitution into formula (1) then gives

$$\begin{aligned}\int_{C'} \frac{3y}{x}\, ds &= \int_1^4 3\left(\frac{u}{\sqrt{u}}\right)\sqrt{\frac{1}{4u} + 1}\, du \\ &= \int_1^4 \frac{3}{2}\sqrt{1 + 4u}\, du.\end{aligned}$$

If $t = \sqrt{u}$, we have $u = t^2$ and $du = 2t\, dt$. To obtain limits of integration in terms of the variable $t$, we note that

$$u = 4 \quad \text{implies} \quad t = 2, \quad \text{and} \quad u = 1 \quad \text{implies} \quad t = 1.$$

Substitution into the above integral then gives

$$\begin{aligned}\int_{C'} \frac{3y}{x}\, ds &= \int_1^2 \frac{3}{2}\sqrt{1 + 4t^2}(2t)\, dt \\ &= \int_1^2 3t\sqrt{1 + 4t^2}\, dt.\end{aligned}$$

This is the same integral as that in (a).

(c) The equations $x = 3 - v$ and $y = (3 - v)^2$ imply

$$\frac{dx}{dv} = -1 \quad \text{and} \quad \frac{dy}{dv} = 2(3 - v)(-1).$$

Substitution into formula (1) then gives

$$\begin{aligned}\int_{-C} \frac{3y}{x}\, ds &= \int_1^2 3\left(\frac{(3 - v)^2}{3 - v}\right)\sqrt{1 + 4(3 - v)^2}\, dv \\ &= \int_1^2 3(3 - v)\sqrt{1 + 4(3 - v)^2}\, dv.\end{aligned}$$

If $t = 3 - v$, we have $v = 3 - t$ and $dv = -dt$.

$$v = 2 \quad \text{implies} \quad t = 1, \quad \text{and} \quad v = 1 \quad \text{implies} \quad t = 2.$$

Substituting into the above integral then gives

$$\begin{aligned}\int_{-C} \frac{3y}{x}\, ds &= \int_2^1 3t\sqrt{1 + 4t^2}(-1)\, dt \\ &= -\int_2^1 3t\sqrt{1 + 4t^2}\, dt \\ &= \int_1^2 3t\sqrt{1 + 4t^2}\, dt.\end{aligned}$$

This is the same integral as that in (a). ■

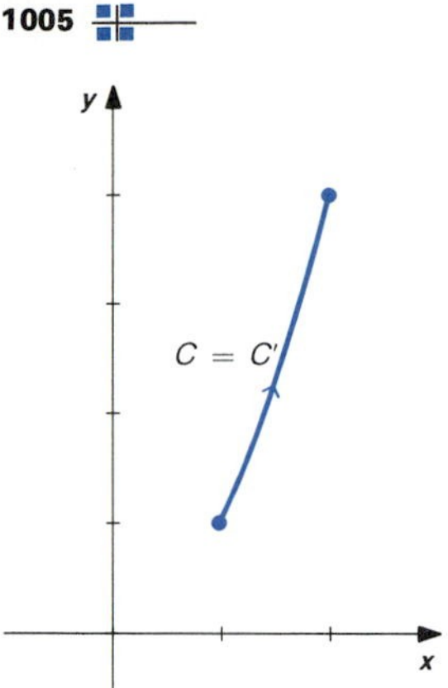

(a)

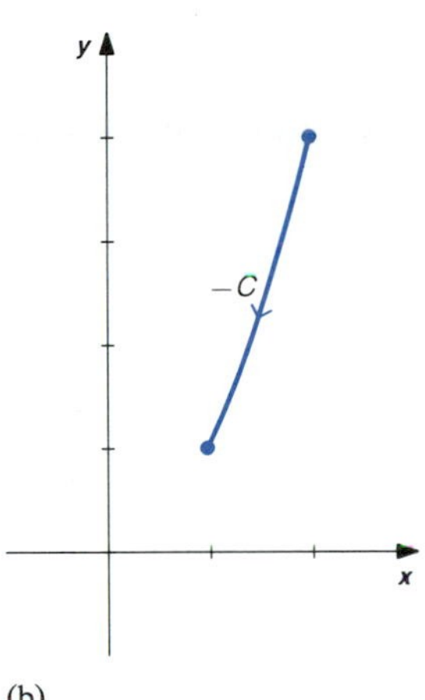

(b)

**FIGURE 18.1.2**



The curves $C$ and $C'$ in Example 1 consist of that portion of the parabola $y = x^2$ from the initial point (1, 1) to the terminal point (2, 4). See Figure 18.1.2a. The curve $-C$ goes along the parabola in the opposite direction, from (2, 4) to (1, 1). See Figure 18.1.2b. From Example 1 we see that

$$\int_C \frac{3y}{x}\,ds = \int_{C'} \frac{3y}{x}\,ds = \int_{-C} \frac{3y}{x}\,ds.$$

We should expect that different smooth parametrizations of $C$ would give the same value of $\int_C f\,ds$. The direction does not matter, because $\Delta s$ represents length in the sum $\sum f\,\Delta s$, and length is nonnegative. The length of a curve is the same in either direction.

We can use the formula for change of variable in a definite integral to verify the following general results.

If $C'$ is a reparametrization of $C$, then

$$\int_{C'} f\,ds = \int_C f\,ds.$$

If $-C$ has the same graph, but opposite direction of $C$, then

$$\int_{-C} f\,ds = \int_C f\,ds.$$

We may choose the most convenient parametric representation of a curve to evaluate a line integral over the curve.

We need to be able to determine parametric equations for curves that are described geometrically or in terms of equations in $x$ and $y$. Let us recall the following rules from Section 15.1.

**The curve along the graph of $y = f(x)$ in the direction of increasing $x$ from $(a, f(a))$ to $(b, f(b))$ can be expressed as** (2)

$$C\colon\ x = t,\quad y = f(t),\qquad a \le t \le b.$$

**The curve along the graph of $x = f(y)$ in the direction of increasing $y$ from $(f(a), a)$ to $(f(b), b)$ can be expressed as** (3)

$$C\colon\ x = f(t),\quad y = t,\qquad a \le t \le b.$$

**The line segment from $(x_0, y_0)$ to $(x_1, y_1)$ is given by** (4)

$$C\colon\ x = x_0 + (x_1 - x_0)t,\quad y = y_0 + (y_1 - y_0)t,\qquad 0 \le t \le 1.$$

**Parametric equations of the curve that goes from the point $(h + r, k)$ counterclockwise once around the circle $(x - h)^2 + (y - k)^2 = r^2$ are** (5)

$$C\colon\ x = h + r\cos t, \quad y = k + r\sin t, \quad 0 \le t \le 2\pi.$$

**Parametric equations of the curve that goes from the point $(h + a, k)$ counterclockwise once around the ellipse**

$$\frac{(x-h)^2}{a^2} + \frac{(y-k)^2}{b^2} = 1 \tag{6}$$

**are**

$$C\colon\ x = h + a\cos t, \quad y = k + b\sin t, \qquad 0 \le t \le 2\pi.$$

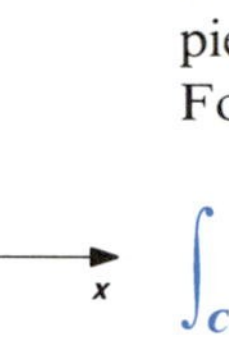

FIGURE 18.1.3

Recall from Section 15.3 that the sum of a finite number of smooth curves is called **piecewise smooth**. We can evaluate line integrals over piecewise smooth curves by adding the integrals over each smooth part. For example, if $C_1$ and $C_2$ are smooth curves, we have

$$\int_{C_1+C_2} f\,ds = \int_{C_1} f\,ds + \int_{C_2} f\,ds.$$

See Figure 18.1.3.

■ **EXAMPLE 2**

Evaluate $\int_C y\,ds$, where $C$ consists of the polygonal path from $(1, 0)$ to $(2, 3)$, to $(4, 3)$.

**SOLUTION**

The curve is sketched in Figure 18.1.4. We need a parametric representation for each of the two smooth parts of the curve.

Formula (4) can be used to find parametric equations of the line segment from $(1, 0)$ to $(2, 3)$. We obtain

$$C_1\colon\ x = 1 + t, \quad y = 3t, \qquad 0 \le t \le 1.$$

$$\frac{dx}{dt} = 1, \quad \frac{dy}{dt} = 3, \quad \text{so} \quad ds = \sqrt{(1)^2 + (3)^2}\,dt = \sqrt{10}\,dt.$$

The horizontal line segment from $(2, 3)$ to $(4, 3)$ can be described by the equation $y = 3$, $2 \le x \le 4$. Formula (2) then gives

$$C_2\colon\ x = t, \quad y = 3, \qquad 2 \le t \le 4.$$

$$\frac{dx}{dt} = 1, \quad \frac{dy}{dt} = 0, \quad \text{so} \quad ds = \sqrt{(1)^2 + (0)^2}\,dt = dt.$$

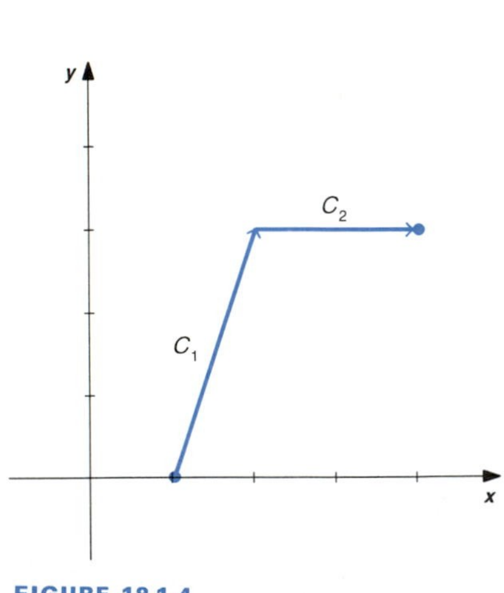

FIGURE 18.1.4

Then

$$\begin{aligned}\int_C y\,ds &= \int_{C_1} y\,ds + \int_{C_2} y\,ds \\ &= \int_0^1 3t\sqrt{10}\,dt + \int_2^4 3\,dt \\ &= \left[3\sqrt{10}\left(\frac{t^2}{2}\right)\right]_0^1 + \left[3t\right]_2^4 \\ &= \left[3\sqrt{10}\left(\frac{1}{2}\right) - (0)\right] + [3(4) - 3(2)] = \frac{3}{2}\sqrt{10} + 6.\end{aligned}$$

■

Either of formulas (2), (3), or (4) could be used to find parametric equations of the line segment from (1, 0) to (2, 3) in Example 2. For example, the equation of the line that contains (1, 0) and (2, 3) is $y = 3x - 3$. Formula (2) then gives

$$C_1\colon\ x = t,\quad y = 3t - 3,\qquad 1 \le t \le 2.$$

Since $y = 3x - 3$ implies $x = (y + 3)/3$, formula (3) gives

$$C_1\colon\ x = \frac{t+3}{3},\quad y = t,\qquad 0 \le t \le 3.$$

The line integral $\int_{C_1} y\,ds$ has the same value for each of these parametrizations.

Let us now see some applications of line integrals.

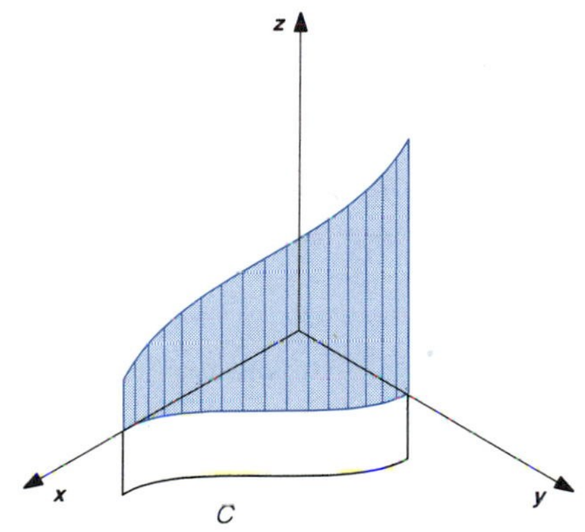

FIGURE 18.1.5

The graph in three-dimensional space of an equation $F(x, y) = 0$ is a generalized cylinder with generating lines parallel to the $z$-axis. The **lateral surface area** of that part of the cylinder between the surfaces $z = h_1(x, y)$ and $z = h_2(x, y)(h_1 \le h_2)$ is

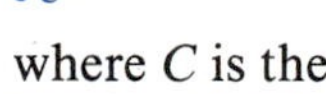

$$\int_C (h_2 - h_1)\,ds,$$

where $C$ is the curve given by the graph of $F(x, y) = 0$ in the $xy$-plane. See Figure 18.1.5.

### ■ EXAMPLE 3

Find the lateral surface area of that part of the cylinder $x^2 + y^2 = 4$ between the planes $z = 0$ and $x + 2y + z = 6$.

#### SOLUTION

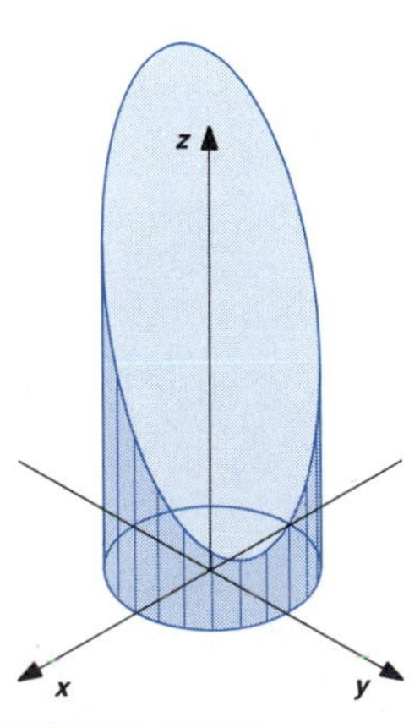

FIGURE 18.1.6

The cylinder is sketched in Figure 18.1.6. We need a parametric representation of the circle $x^2 + y^2 = 4$ in the $xy$-plane. Formula (5) gives

$$C\colon\ x = 2\cos t,\quad y = 2\sin t,\qquad 0 \le t \le 2\pi.$$

Then

$$\frac{dx}{dt} = -2\sin t \quad\text{and}\quad \frac{dy}{dt} = 2\cos t.$$

It follows that

$$ds = \sqrt{(-2\sin t)^2 + (2\cos t)^2}\,dt = \sqrt{4(\sin^2 t + \cos^2 t)}\,dt = 2\,dt.$$

The lateral surface area is

$$S = \int_C (6 - x - 2y)\, ds = \int_0^{2\pi} [6 - 2\cos t - 2(2\sin t)](2)\, dt$$

$$= 12t - 4\sin t + 8\cos t \Big]_0^{2\pi} = 24\pi.$$ ■

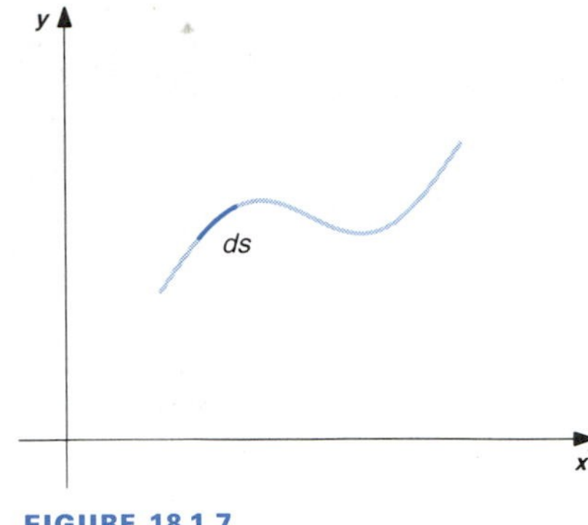


FIGURE 18.1.7

Suppose a wire is bent into the shape of a curve $C$ and the density per unit length of the wire at a point on the curve is given by the function $\rho$. If $\rho$ is continuous, the differential mass of a small piece of the wire is

$$dM = \rho\, ds.$$

See Figure 18.1.7. The total mass of the wire is (the sum)

$$M = \int_C dM = \int_C \rho\, ds.$$

■ **EXAMPLE 4**

A wire is bent in the shape of the curve

$$C\colon\ x = t\cos t, \quad y = t\sin t, \quad z = t, \qquad 0 \le t \le 2\pi.$$

The density per unit length at a point on the curve is given by the distance of the point from the $xy$-plane. Find the mass of the wire.

**SOLUTION**

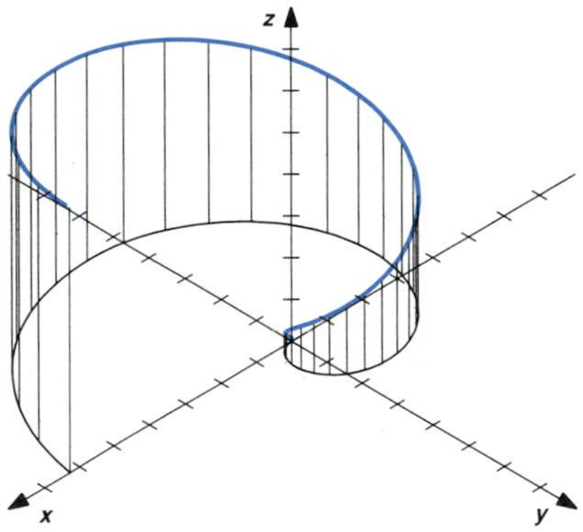

The wire is sketched in Figure 18.1.8. The distance of a point on the curve to the $xy$-plane is the $z$-coordinate of the point, so $\rho = z$.

FIGURE 18.1.8

$$\frac{dx}{dt} = -t\sin t + \cos t, \quad \frac{dy}{dt} = t\cos t + \sin t, \quad \text{and} \quad \frac{dz}{dt} = 1.$$

$$\left(\frac{dx}{dt}\right)^2 + \left(\frac{dy}{dt}\right)^2 + \left(\frac{dz}{dt}\right)^2 = (-t\sin t + \cos t)^2 + (t\cos t + \sin t)^2 + (1)^2$$

$$= t^2\sin^2 t - 2t\sin t\cos t + \cos^2 t$$

$$+ t^2\cos^2 t + 2t\cos t\sin t + \sin^2 t + 1$$

$$= t^2 + 2.$$

The mass is given by

$$M = \int_C \rho\, ds = \int_C z\, ds$$

$$= \int_0^{2\pi} (t)\sqrt{t^2 + 2}\, dt$$

$[u = t^2 + 2,\ du = 2t\, dt]$

$$= \frac{1}{3}(t^2 + 2)^{3/2}\Big]_0^{2\pi} = \frac{(4\pi^2 + 2)^{3/2} - (2)^{3/2}}{3}.$$ ■

## EXERCISES 18.1

1 (a) $C$: $x = \cos t$, $y = \sin t$, $0 \le t \le \pi/2$. Express $\int_C xy\, ds$ as a definite integral with respect to the parameter $t$. Evaluate the integral.

(b) $C'$: $x = \sqrt{1-u^2}$, $y = u$, $0 \le u \le 1$, is a reparametrization of $C$ obtained by using the equation $u = \sin t$. Express $\int_{C'} xy\, ds$ as a definite integral with respect to the parameter $u$ and then use the change of variables $u = \sin t$ to express that integral as a definite integral with respect to $t$.

(c) $-C$: $x = \sin v$, $y = \cos v$, $0 \le v \le \pi/2$, has the same graph as $C$, but opposite direction. Express $\int_{-C} xy\, ds$ as a definite integral with respect to the parameter $v$ and then use the change of variables $t = (\pi/2) - v$ to express that integral as a definite integral with respect to $t$.

2 (a) $C$: $x = t$, $y = t^2/2$, $1 \le t \le 2$. Express $\int_C x\, ds$ as a definite integral with respect to the parameter $t$. Evaluate the integral.

(b) $C'$: $x = \sqrt{u-1}$, $y = (u-1)/2$, $2 \le u \le 5$, is a reparametrization of $C$ obtained by using the equation $u = t^2 + 1$. Express $\int_{C'} x\, ds$ as a definite integral with respect to the parameter $u$ and then use the change of variables $u = t^2 + 1$ to express that integral as a definite integral with respect to $t$.

(c) $-C$: $x = 3 - v$, $y = (3-v)^2/2$, $1 \le v \le 2$, has the same graph as $C$, but opposite direction. Express $\int_{-C} x\, ds$ as a definite integral with respect to the parameter $v$ and then use the change of variables $t = 3 - v$ to express that integral as a definite integral with respect to $t$.

3 (a) $C$: $x = t$, $y = 0$, $0 \le t \le 1$. Express

$$\int_C \frac{x}{x^2+1}\, ds$$

as a definite integral with respect to the parameter $t$. Evaluate the integral.

(b) $C'$: $x = \sqrt{u-1}$, $y = 0$, $1 \le u \le 2$, is a reparametrization of $C$ obtained by using the equation $u = t^2 + 1$. Express

$$\int_{C'} \frac{x}{x^2+1}\, ds$$

as a definite integral with respect to the parameter $u$ and then use the change of variables $u = t^2 + 1$ to express that integral as a definite integral with respect to $t$.

(c) $-C$: $x = 1 - v$, $y = 0$, $0 \le v \le 1$, has the same graph as $C$, but opposite direction. Express

$$\int_{-C} \frac{x}{x^2+1}\, ds$$

as a definite integral with respect to the parameter $v$ and then use the change of variables $t = 1 - v$ to express that integral as a definite integral with respect to $t$.

4 (a) $C$: $x = 2t$, $y = t$, $0 \le t \le 1$. Express $\int_C xe^{y^2}\, ds$ as a definite integral with respect to the parameter $t$. Evaluate the integral.

(b) $C'$: $x = 2\sqrt{u}$, $y = \sqrt{u}$, $0 \le u \le 1$, is a reparametrization of $C$ obtained by using the equation $u = t^2$. Express $\int_{C'} xe^{y^2}\, ds$ as a definite integral with respect to the parameter $u$ and then use the change of variables $u = t^2$ to express that integral as a definite integral with respect to $t$.

(c) $-C$: $x = 2 - 2v$, $y = 1 - v$, $0 \le v \le 1$, has the same graph as $C$, but opposite direction. Express $\int_{-C} xe^{y^2}\, ds$ as a definite integral with respect to the parameter $v$ and then use the change of variables $t = 1 - v$ to express that integral as a definite integral with respect to $t$.

5 Evaluate $\int_C (2x + y)\, ds$, where $C$ consists of the polygonal path from $(0, 0)$ to $(3, 0)$, to $(3, 2)$.

6 Evaluate $\int_C (2x + y)\, ds$, where $C$ consists of the polygonal path from $(0, 0)$ to $(0, 2)$, to $(3, 2)$.

7 Evaluate $\int_C (x + y^2)\, ds$, where $C$ consists of the path from $(2, 0)$, counterclockwise along the circle $x^2 + y^2 = 4$ to $(-2, 0)$, and then along the $x$-axis to $(2, 0)$.

8 Evaluate $\int_C (x + y^2)\, ds$, where $C$ consists of the path from $(2, 0)$, counterclockwise once around the circle $x^2 + y^2 = 4$.

9 Find the lateral surface area of that part of the cylinder $x^2 + y^2 = 4$ between $z = 2 + xy$ and $z = 0$.

10 Find the lateral surface area of that part of the cylinder $x^2 + y^2 = 1$ between $z = x^2$ and $z = 0$.

11 Find the lateral surface area of that part of the vertical generalized cylinder $x = y^2/2$, $-1 \le y \le 1$, between $z = x$ and $z = 0$.

12 Find the lateral surface area of that part of the vertical generalized cylinder $y = \cos x$, $-\pi/2 \le x \le \pi/2$, between $z = y$ and $z = 0$.

In Exercises 13–16 a wire is bent in the shape of the given curve. Find the mass of the wire for the indicated density function.

13 $y = x^2$, $0 \le x \le 2$; $\rho(x, y) = x$

14 $y = x^3$, $0 \le x \le 2$; $\rho(x, y) = y$

15 $C$: $x = \cos t$, $y = \sin t$, $z = t$, $0 \le t \le 2\pi$; $\rho(x, y, z) = z$

16 $C$: $x = t^2$, $y = t$, $z = t$, $0 \le t \le 2$; $\rho(x, y, z) = z$

17 Show that $|\int_C f\, ds| \le$ (Maximum of $|f|$ on $C$)(Arc length of $C$).

## 18.2 APPLICATIONS OF LINE INTEGRALS TO VECTOR FIELDS

The most important applications of line integrals involve vectors. We will introduce some of these applications in this section.

A vector-valued function **F** that is defined on a region $R$ is called a **vector field** on $R$. Each point $P$ of $R$ is associated with a vector $\mathbf{F}(P)$. $R$ may be a region in the plane or in three-dimensional space.

If each vector value of **F** represents a force, the vector field is also called a **force field**. The forces may be due to gravitational or magnetic attraction, or an electric charge, to name a few. A force field due to magnetic attraction is indicated in Figure 18.2.1. Vectors at representative points give the direction and magnitude of the force at those points.

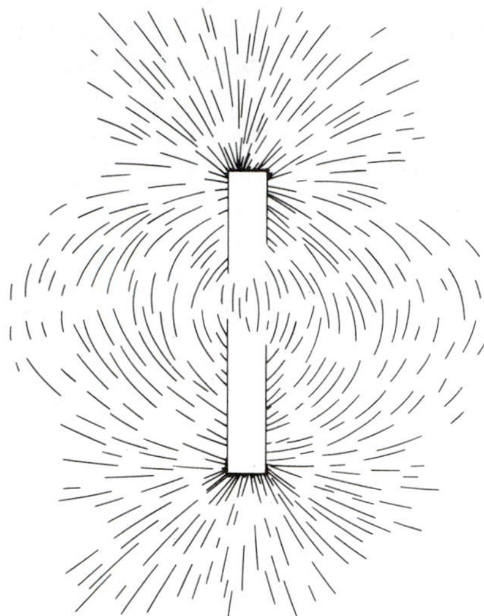

FIGURE 18.2.1

Each vector **F** of a vector field may represent a velocity vector. A vector field that gives the velocity of water at the surface of a stream is indicated in Figure 18.2.2. The direction of the vectors gives the direction of flow and the length of the vectors gives the rate of flow. The sketch shows that the stream is flowing more rapidly near the center than along the banks. Also, the stream appears to flow more rapidly where it is narrower.

FIGURE 18.2.2

If a function $U$ is defined and has first partial derivatives in a region $R$, the vectors $\mathbf{F} = \nabla U$ form a vector field on $R$. Such vector fields are called **gradient fields** or **conservative fields**. Gradient fields are very common in applications.

Many of the applications of line integrals involve integrals of the form

$$\int_C \mathbf{F} \cdot \mathbf{T}\, ds,$$

where **F** is a vector field and **T** is the unit tangent vector of $C$. Let us develop a formula that allows us to evaluate this type of line integral without calculating the unit tangent vector and the differential arc length of $C$ in each case.

Suppose that $\mathbf{F}(x, y) = M(x, y)\mathbf{i} + N(x, y)\mathbf{j}$ and

$$C\colon\ x = x(t), \quad y = y(t), \qquad a \le t \le b.$$

The vector form of the curve is then

$$\mathbf{R}(t) = x(t)\mathbf{i} + y(t)\mathbf{j}, \qquad a \le t \le b.$$

We have

$$\mathbf{R}' = \frac{dx}{dt}\mathbf{i} + \frac{dy}{dt}\mathbf{j} \quad \text{and} \quad |\mathbf{R}'| = \sqrt{\left(\frac{dx}{dt}\right)^2 + \left(\frac{dy}{dt}\right)^2}.$$

The unit tangent vector is

$$\mathbf{T} = \frac{\mathbf{R}'}{|\mathbf{R}'|} = \frac{\dfrac{dx}{dt}\mathbf{i} + \dfrac{dy}{dt}\mathbf{j}}{\sqrt{\left(\dfrac{dx}{dt}\right)^2 + \left(\dfrac{dy}{dt}\right)^2}}.$$

$$\mathbf{F}\cdot\mathbf{T} = (M\mathbf{i} + N\mathbf{j})\cdot\left(\frac{\dfrac{dx}{dt}\mathbf{i} + \dfrac{dy}{dt}\mathbf{j}}{\sqrt{\left(\dfrac{dx}{dt}\right)^2 + \left(\dfrac{dy}{dt}\right)^2}}\right) = \frac{M\dfrac{dx}{dt} + N\dfrac{dy}{dt}}{\sqrt{\left(\dfrac{dx}{dt}\right)^2 + \left(\dfrac{dy}{dt}\right)^2}}.$$

Then formula (1) of Section 18.1 gives

$$\int_C \mathbf{F}\cdot\mathbf{T}\,ds = \int_a^b \frac{M\dfrac{dx}{dt} + N\dfrac{dy}{dt}}{\sqrt{\left(\dfrac{dx}{dt}\right)^2 + \left(\dfrac{dy}{dt}\right)^2}}\sqrt{\left(\frac{dx}{dt}\right)^2 + \left(\frac{dy}{dt}\right)^2}\,dt$$
$$= \int_a^b \left(M\frac{dx}{dt} + N\frac{dy}{dt}\right) dt.$$

Since

$$dx = \frac{dx}{dt}\,dt \quad \text{and} \quad dy = \frac{dy}{dt}\,dt,$$

we can express the latter integral more simply as

$$\int_C M\,dx + N\,dy.$$

Summarizing:

If $C$ is a plane curve and $\mathbf{F} = M\mathbf{i} + N\mathbf{j}$, then

$$\int_C \mathbf{F}\cdot\mathbf{T}\,ds = \int_C M\,dx + N\,dy.$$

Integrals of the form

$$\int_C M\,dx + N\,dy$$

are evaluated by substitution of parametric values. The values of the integrals do not depend on the particular smooth parametrization of $C$ used to evaluate the integrals, *except for direction.* The unit tangent vector of $-C$ is the negative of the unit tangent vector of $C$. See Figure 18.2.3. If the vector $\mathbf{T}$ in the integrand represents the unit tangent of the curve of integration, we have

$$\int_{-C} \mathbf{F}\cdot\mathbf{T}\,ds = -\left(\int_C \mathbf{F}\cdot\mathbf{T}\,ds\right),$$

so

$$\int_{-C} M\,dx + N\,dy = -\left(\int_C M\,dx + N\,dy\right).$$

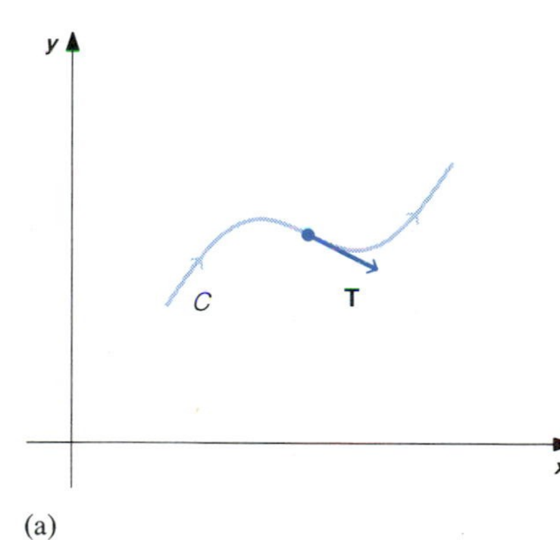

(a)

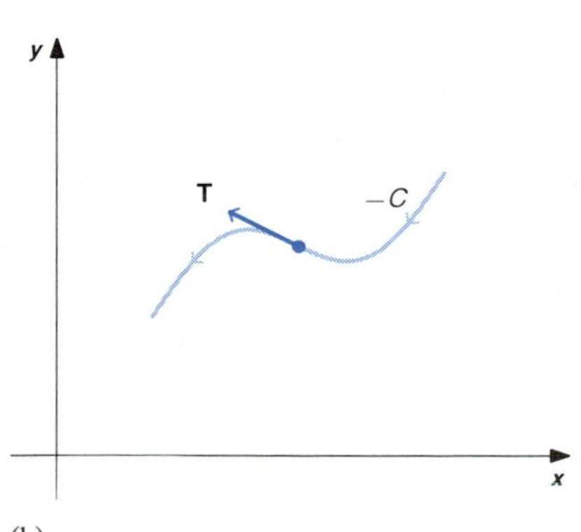

(b)

**FIGURE 18.2.3**

Similarly, if $C$ is a curve in three-dimensional space and $\mathbf{F} = M\mathbf{i} + N\mathbf{j} + P\mathbf{k}$, then

$$\int_C \mathbf{F}\cdot\mathbf{T}\, ds = \int_C M\, dx + N\, dy + P\, dz \quad \text{and}$$

$$\int_{-C} M\, dx + N\, dy + P\, dz = -\left(\int_C M\, dx + N\, dy + P\, dz\right).$$

■ **EXAMPLE 1**

(a) $C$: $x = -\cos t$, $y = \sin t$, $0 \le t \le \pi$. Evaluate $\int_C (2y - x)\, dx - y\, dy$.

(b) $C'$: $x = u$, $y = \sqrt{1 - u^2}$, $-1 \le u \le 1$, is a reparametrization of $C$. Evaluate $\int_{C'} (2y - x)\, dx - y\, dy$.

(c) $-C$: $x = \cos v$, $y = \sin v$, $0 \le v \le \pi$, has the same graph as $C$, but opposite direction. Evaluate $\int_{-C} (2y - x)\, dx - y\, dy$.

**SOLUTION**

The curves are sketched in Figure 18.2.4.

(a) We have $x = -\cos t$, $y = \sin t$, $dx = \sin t\, dt$, and $dy = \cos t\, dt$. Substitution then gives

$$\int_C (2y - x)\, dx - y\, dy$$

$$= \int_0^\pi \{[2(\sin t) - (-\cos t)](\sin t) - (\sin t)(\cos t)\}\, dt$$

$$= \int_0^\pi 2\sin^2 t\, dt$$

[Table of integrals]

$$= -\sin t\cos t + t\Big]_0^\pi = \pi.$$

(b) $x = u$, $y = \sqrt{1 - u^2}$, $dx = du$, $dy = \dfrac{-u}{\sqrt{1 - u^2}}\, du$, so

$$\int_{C'} (2y - x)\, dx - y\, dy = \int_{-1}^1 \left[(2\sqrt{1 - u^2} - u) - (\sqrt{1 - u^2})\left(\frac{-u}{\sqrt{1 - u^2}}\right)\right] du$$

[The factor $-u/\sqrt{1 - u^2}$ becomes unbounded at $\pm 1$, but the product with $\sqrt{1 - u^2}$ remains bounded and can be defined to be continuous at $\pm 1$.]

$$= \int_{-1}^1 2\sqrt{1 - u^2}\, du$$

[Table of integrals]

$$= u\sqrt{1 - u^2} + \sin^{-1} u\Big]_{-1}^1$$

$$= \sin^{-1}(1) - \sin^{-1}(-1) = \frac{\pi}{2} - \left(-\frac{\pi}{2}\right) = \pi.$$

FIGURE 18.2.4

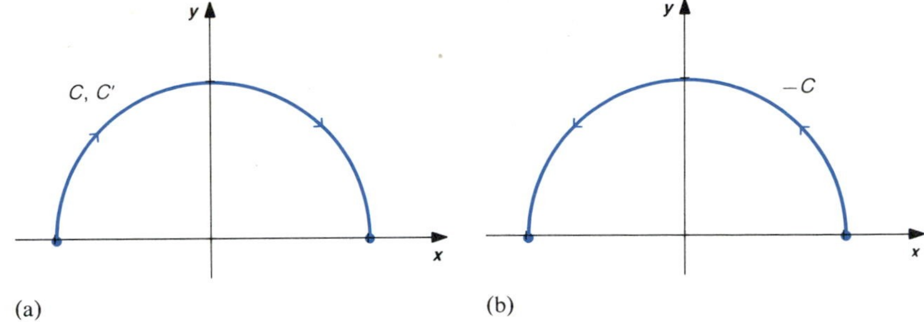


(c) $x = \cos v$, $y = \sin v$, $dx = -\sin v\, dv$, $dy = \cos v\, dv$, so

$$\int_{-C} (2y - x)\, dx - y\, dy$$

$$= \int_0^{\pi} \{[2(\sin v) - (\cos v)](-\sin v) - (\sin v)(\cos v)\}\, dv$$

$$= \int_0^{\pi} -2 \sin^2 v\, dv$$

[Table of integrals]

$$= \sin v \cos v - v \Big]_0^{\pi} = -\pi.$$

We should expect the integrals in (a) and (b) to be equal, since $C$ and $C'$ have the same graph and the same direction. The integral in (c) is the negative of the others, because $-C$ has opposite direction. ■

Let us now see some of the applications of the above type of line integrals.

We begin considering the problem of finding the work associated with a force field **F** as a small object moves along a curve $C$ in the field. The force may vary in both magnitude and direction at different points on the curve. See Figure 18.2.5. Recall from Section 14.1 that the work associated with a

FIGURE 18.2.5

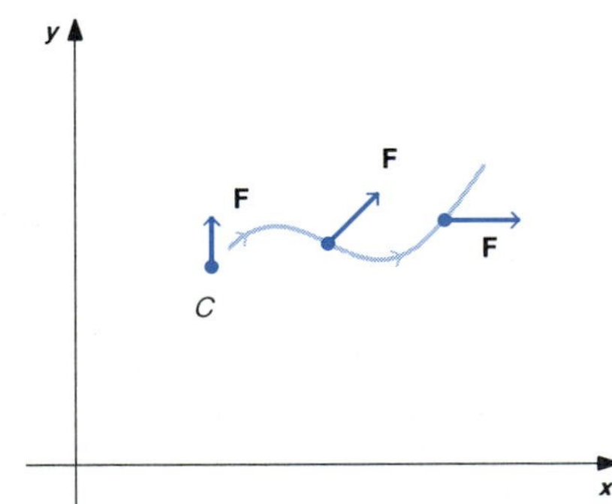

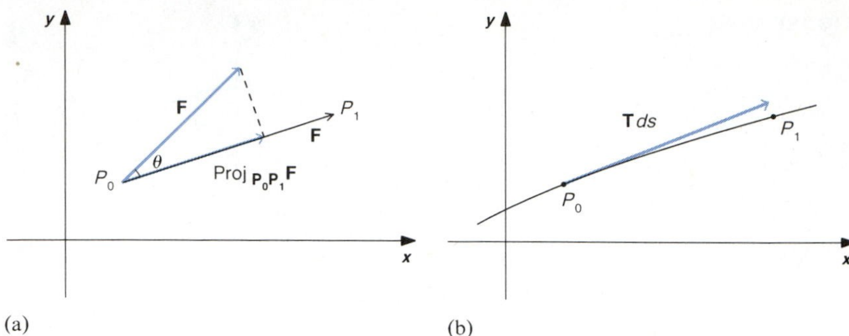


FIGURE 18.2.6

*constant* force **F** as an object is moved in a *straight line* from points $P_0$ to $P_1$ on $C$ is

$$\mathbf{F}\cdot(\mathbf{P}_0\mathbf{P}_1) = |\mathbf{F}||\mathbf{P}_0\mathbf{P}_1| \cos\theta.$$

See Figure 18.2.6a. If the curve is smooth, then small pieces of the curve are nearly straight and we have

$$\mathbf{P}_0\mathbf{P}_1 \approx \mathbf{T}\, ds,$$

where **T** is the unit tangent vector of $C$ at the point $P_0$ in the direction of motion and $ds$ is the arc length of $C$ between $P_0$ and $P_1$. See Figure 18.2.6b. If the force varies continuously, then **F** is nearly constant along small pieces of the curve. The work associated with moving a small distance along the curve is then approximately

$$dW = \mathbf{F}\cdot\mathbf{T}\, ds.$$

It should seem reasonable that the total work is

$$W = \int_C \mathbf{F}\cdot\mathbf{T}\, ds.$$

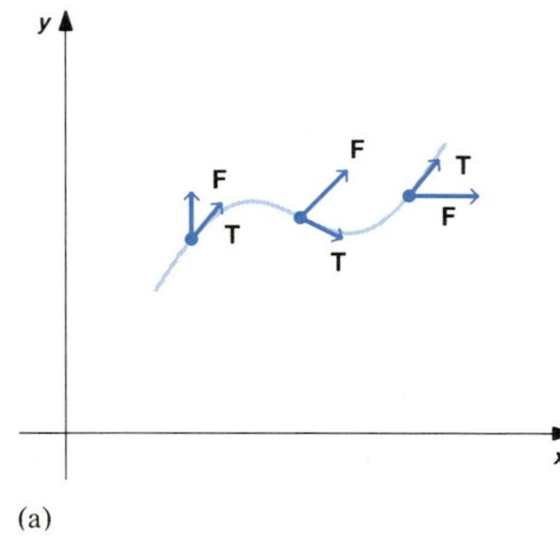


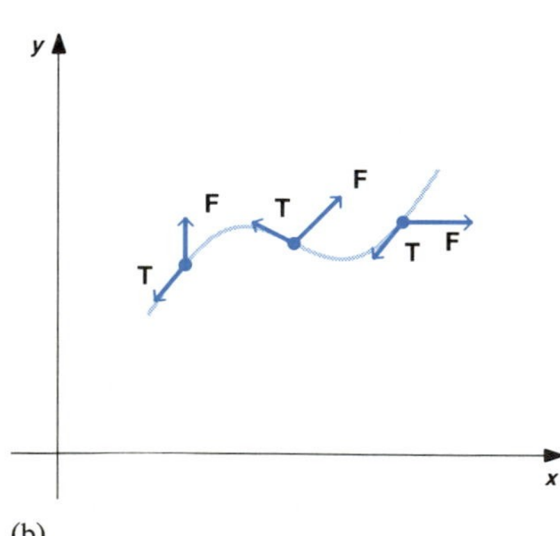


FIGURE 18.2.7

The value of the work associated with a force field as an object moves along a curve depends on the direction of motion along the curve $C$. This corresponds to the formula

$$\int_{-C} \mathbf{F}\cdot\mathbf{T}\, ds = -\int_C \mathbf{F}\cdot\mathbf{T}\, ds.$$

If $\mathbf{F} \neq \mathbf{0}$ and the angle between **F** and **T** is less than $\pi/2$ at all points of $C$, as indicated in Figure 18.2.7a, then $\mathbf{F}\cdot\mathbf{T}$ is positive at every point of $C$ and the work associated with **F** as an object moves along $C$ will be positive. If the angle between **F** and **T** is between $\pi/2$ and $\pi$ at all points of $C$, as indicated in Figure 18.2.7b, the work will be negative. If the angle between **F** and **T** varies between 0 and $\pi$, the net work may be either positive or negative.

FIGURE 18.2.8

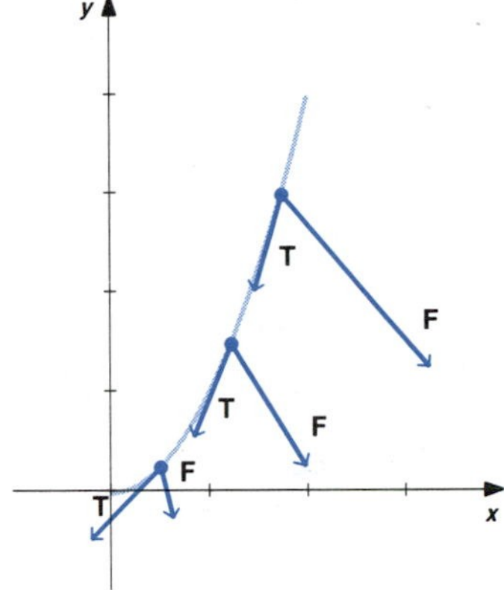


■ **EXAMPLE 2**

Find the work associated with the force field $\mathbf{F} = (y/2)\mathbf{i} - x\mathbf{j}$ as an object moves along the curve $y = x^2$ from $(2, 4)$ to $(0, 0)$. Sketch the curve and representative vectors **F** and **T** at some points along the curve.

**SOLUTION**

The curve and representative vectors **F** and **T** are sketched in Figure 18.2.8. Note that the angle between **F** and **T** appears to be less than $\pi/2$ at all points on the curve, so we should expect the work to be positive.

The work is $W = \int_C y/2\, dx - x\, dy$. We need parametric equations of the curve to evaluate the line integral. Since $C$ is in the direction of decreasing $x$ and decreasing $y$, it is more convenient to parametrize $-C$ than $C$. We can then use the formula

$$\int_C M\, dx + N\, dy = -\left(\int_{-C} M\, dx + N\, dy\right)$$

to evaluate the work. Formula (2) of Section 18.1 gives

$$-C\colon\ x = t, \quad y = t^2, \qquad 0 \le t \le 2.$$

Then $dx = dt$ and $dy = 2t\, dt$. Substitution gives

$$\begin{aligned} W &= \int_C \frac{y}{2}\, dx - x\, dy \\ &= -\int_{-C} \frac{y}{2}\, dx - x\, dy \\ &= -\int_0^2 \left[\frac{t^2}{2} - (t)(2t)\right] dt \\ &= \int_0^2 \frac{3t^2}{2}\, dt = \frac{t^3}{2}\bigg]_0^2 = 4. \end{aligned}$$

As expected, the work is positive. ■

We can use the formula

$$W = \int_C \mathbf{F} \cdot \mathbf{T}\, ds$$

to calculate the work associated with the force **F** as an object moves along the curve $C$, even though there may be other forces acting on the object. In case **F** is an external force that is the only force acting on the object, then **F** determines the motion of the object. In this case, we can establish a relation between the work associated with **F** and the change in kinetic energy of the object. That is, if the position of the object at time $t$ is given by the vector function $\mathbf{R}(t)$, then the velocity is $\mathbf{V} = \mathbf{R}'$ and the acceleration is $\mathbf{A} = \mathbf{V}'$. Newton's Second Law of Motion implies

$$\mathbf{F} = M\mathbf{A},$$

where $M$ is the mass of the object and **A** is the acceleration vector. The unit tangent vector to $C$ is

$$\mathbf{T} = \frac{\mathbf{R}'}{|\mathbf{R}'|} = \frac{\mathbf{V}}{|\mathbf{R}'|}$$

Then

$$\begin{aligned}\mathbf{F} \cdot \mathbf{T} &= (M\mathbf{A}) \cdot \left(\frac{\mathbf{V}}{|\mathbf{R}'|}\right) \\ &= (M\mathbf{V}') \cdot \left(\frac{\mathbf{V}}{|\mathbf{R}'|}\right) \\ &= \frac{M}{|\mathbf{R}'|}(\mathbf{V}' \cdot \mathbf{V})\end{aligned}$$

Since

$$\frac{d}{dt}(|\mathbf{V}|^2) = \frac{d}{dt}(\mathbf{V} \cdot \mathbf{V}) = \mathbf{V} \cdot \mathbf{V}' + \mathbf{V}' \cdot \mathbf{V} = 2\mathbf{V}' \cdot \mathbf{V},$$

we have

$$\mathbf{F} \cdot \mathbf{T} = \frac{M}{|\mathbf{R}'|}\left(\frac{1}{2}\frac{d}{dt}(|\mathbf{V}|^2)\right)$$

Since $ds = |\mathbf{R}'|\, dt$, it follows that

$$\begin{aligned}\int_C \mathbf{F} \cdot \mathbf{T}\, ds &= \int_a^b \frac{M}{|\mathbf{R}'|}\left(\frac{1}{2}\frac{d}{dt}(|\mathbf{V}|^2)\right)|\mathbf{R}'|\, dt \\ &= \int_a^b \frac{M}{2}\frac{d}{dt}(|\mathbf{V}|^2)\, dt \\ &= \frac{M}{2}|\mathbf{V}(b)|^2 - \frac{M}{2}|\mathbf{V}(a)|^2,\end{aligned}$$

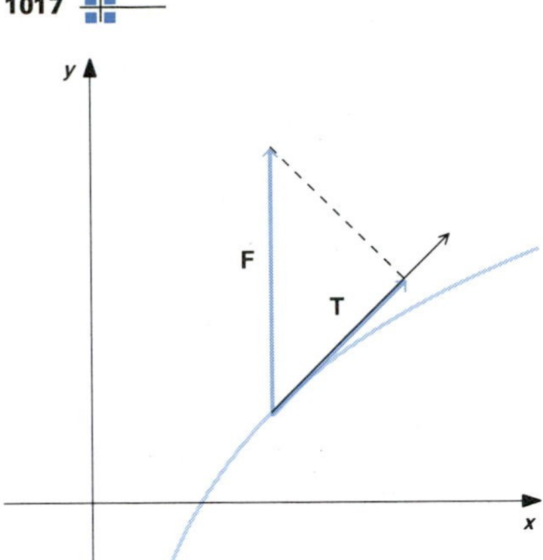

FIGURE 18.2.9

where the last equality is a consequence of the Fundamental Theorem of Calculus. The expression $M|\mathbf{V}|^2/2$ is the **kinetic energy** of the object. We see that, if **F** is an external force that is the only force acting on an object, then the work associated with the force is equal to the change in kinetic energy.

To obtain another interpretation of the line integral, let us consider a thin layer of fluid that is moving on the $xy$-plane and suppose that the velocity of the fluid at a point $(x, y)$ is given by a vector function $\mathbf{F}(x, y)$. If $C$ is a curve with unit tangent vector $\mathbf{T}$, then the inner product $\mathbf{F}\cdot\mathbf{T}$ is the component of the velocity **F** in the direction of **T** and $\mathbf{F}\cdot\mathbf{T}\,ds$ is a measure of the flow of fluid along an arc of the curve of length $ds$. See Figure 18.2.9. The integral $\int_C \mathbf{F}\cdot\mathbf{T}\,ds$ is the net amount of flow along the curve. If $C$ is a *simple closed curve*, the net flow is a measure of the circulation of the fluid around the curve. We may interpret the sign of the line integral as an indication of the net direction of rotation of the fluid along the curve. We say

$\int_C \mathbf{F}\cdot\mathbf{T}\,ds$ is the **circulation** of **F** around the simple closed curve $C$.

We use the term circulation, even though **F** may represent something other than velocity. For example, **F** may represent force due to gravity, an electric charge, or a magnetic force.

■ **EXAMPLE 3**

Evaluate the circulation of $\mathbf{F} = (x - 2y)\mathbf{i} + (2x + y)\mathbf{j}$ around the curve

$$C\colon\ x = \cos t,\quad y = \sin t,\qquad 0 \le t \le 2\pi.$$

**SOLUTION**

The curve and some representative vectors of **F** along the curve are sketched in Figure 18.2.10. In this example the vectors **F** indicate a counterclockwise rotation about the origin. Since the direction of $C$ is also counterclockwise, we should expect the circulation around $C$ to be positive.

The circulation is $\int_C \mathbf{F}\cdot\mathbf{T}\,ds$. We have

$$\mathbf{F} = (x - 2y)\mathbf{i} + (2x + y)\mathbf{j} = M\mathbf{i} + N\mathbf{j},\quad \text{so}$$

$$M = x - 2y \quad \text{and} \quad N = 2x + y.$$

Substitution then gives

$$\int_C \mathbf{F}\cdot\mathbf{T}\,ds = \int_C M\,dx + N\,dy$$

$$= \int_C (x - 2y)\,dx + (2x + y)\,dy.$$

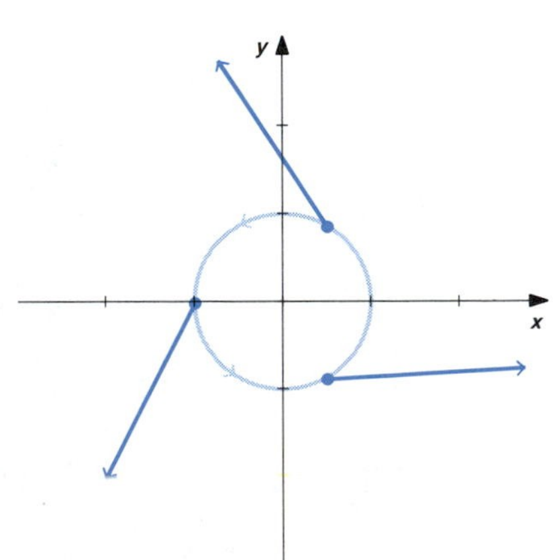

FIGURE 18.2.10

From the parametric equations of $C$, we have

$$x = \cos t,\quad y = \sin t,\quad dx = -\sin t\,dt,\quad dy = \cos t\,dt.$$

The above integral then becomes

$$= \int_0^{2\pi} [(\cos t - 2 \sin t)(-\sin t) + (2 \cos t + \sin t)(\cos t)]\, dt$$

$$= \int_0^{2\pi} 2\, dt = 4\pi. \quad ■$$

Let us return to the model of a fluid flowing along the *xy*-plane with velocity **F**. If *C* is a simple closed curve, the unit tangent vector of *C* is

$$\mathbf{T} = \frac{dx}{ds}\mathbf{i} + \frac{dy}{ds}\mathbf{j}.$$

The vector obtained by a clockwise rotation of **T** by $\pi/2$ is

$$\mathbf{n} = \frac{dy}{ds}\mathbf{i} - \frac{dx}{ds}\mathbf{j}.$$

If the region enclosed by *C* is on the left as *C* is traced, then **n** will point outward; **n** is then called the **outer unit normal** of the region enclosed by *C*. See Figure 18.2.11. The dot product $\mathbf{F}\cdot\mathbf{n}$ is the component of the velocity vector **F** in the direction of **n**. The amount of fluid per unit time that flows in the direction of **n** across a small piece of the curve *C* with length *ds* is $\mathbf{F}\cdot\mathbf{n}\, ds$. The net flow across *C* is the "sum"

$$\int_C \mathbf{F}\cdot\mathbf{n}\, ds.$$

This integral is called the **outward flux** of **F** across *C*. If $\mathbf{F} = M\mathbf{i} + N\mathbf{j}$, then

$$\mathbf{F}\cdot\mathbf{n} = (M\mathbf{i} + N\mathbf{j})\cdot\left(\frac{dy}{ds}\mathbf{i} - \frac{dx}{ds}\mathbf{j}\right) = M\frac{dy}{ds} - N\frac{dx}{ds}, \quad \text{and}$$

$$\int_C \mathbf{F}\cdot\mathbf{n}\, ds = \int_C M\, dy - N\, dx.$$

FIGURE 18.2.11

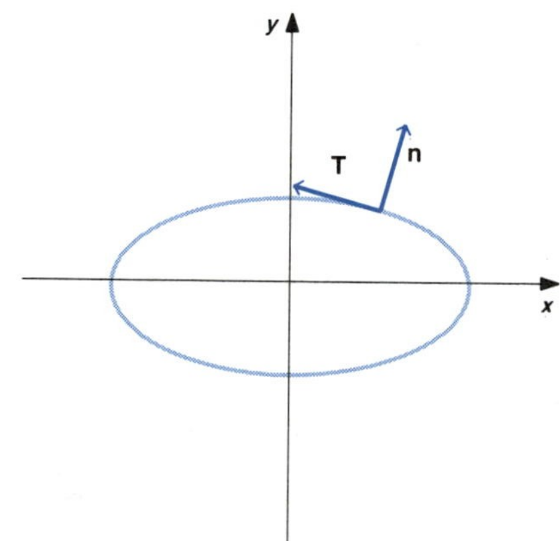

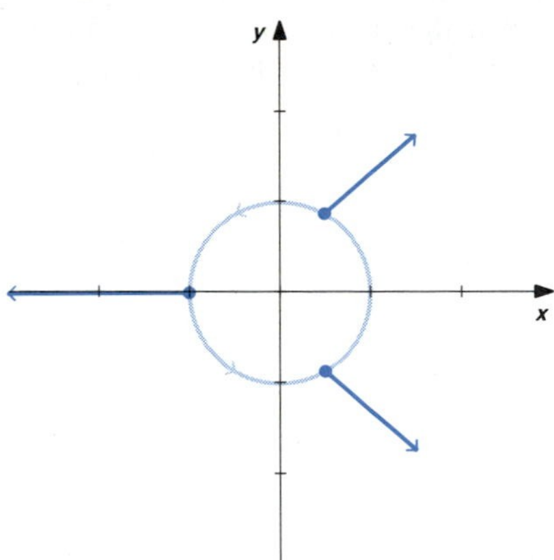


FIGURE 18.2.12

■ **EXAMPLE 4**

Evaluate the outward flux of $\mathbf{F} = 2x\mathbf{i} + y\mathbf{j}$ across the curve

$$C\colon\ x = \cos t, \quad y = \sin t, \qquad 0 \le t \le 2\pi.$$

**SOLUTION**

The sketch in Figure 18.2.12 indicates an outward flow across $C$ at each point of $C$, so we should expect the flux to be positive.

The flux is $\int_C \mathbf{F}\cdot\mathbf{n}\,ds$. We have

$$\mathbf{F} = 2x\mathbf{i} + y\mathbf{j} = M\mathbf{i} + N\mathbf{j}, \quad \text{so} \quad M = 2x \quad \text{and} \quad N = y.$$

Substitution into the formula for the outward flux of $\mathbf{F}$ across $C$ then gives

$$\int_C \mathbf{F}\cdot\mathbf{n}\,ds = \int_C M\,dy - N\,dx$$

$$= \int_C 2x\,dy - y\,dx.$$

From the parametric equations of $C$, we have

$$x = \cos t, \quad y = \sin t, \quad dx = -\sin t\,dt, \quad dy = \cos t\,dt.$$

The integral that gives the flux then becomes

$$\int_0^{2\pi} [(2\cos t)(\cos t) - (\sin t)(-\sin t)]\,dt$$

$$= \int_0^{2\pi} (\cos^2 t + \cos^2 t + \sin^2 t)\,dt$$

$$= \int_0^{2\pi} (\cos^2 t + 1)\,dt$$

[Table of integrals]

$$= \frac{\sin t\cos t}{2} + \frac{t}{2} + t\Big]_0^{2\pi}$$

$$= \frac{2\pi}{2} + 2\pi = 3\pi. \qquad ■$$

# EXERCISES 18.2

Evaluate the line integrals in Exercises 1–6.

1 $\int_C 2x\,dx + xy\,dy$; $C\colon x = 3 - 3t,\ y = 3t,\ 0 \le t \le 1$

2 $\int_C 2x\,dx + xy\,dy$; $C\colon x = 3t,\ y = 3 - 3t,\ 0 \le t \le 1$

3 $\int_C y\,dx - x\,dy$; $C$: $x = 2\cos t$, $y = \sin t$, $0 \le t \le 2\pi$

4 $\int_C x\,dx + y\,dy$; $C$: $x = \cos t$, $y = \sin t$, $0 \le t \le 2\pi$

5 $\int_C yz\,dx + xz\,dy + xy\,dz$; $C$: $x = t$, $y = t^2$, $z = t^3$, $0 \le t \le 2$

6 $\int_C -y\,dx + x\,dy + xy\,dz$; $C$: $x = \cos t$, $y = \sin t$, $z = 2t$, $0 \le t \le 2\pi$

In Exercises 7–12 sketch the given curve and representative vectors **F** and **T** at some points on the curve. Find the work associated with the given force **F** as an object moves along the curve.

7 $C$: $x = 3t$, $y = 2t$, $0 \le t \le 1$; $\mathbf{F} = \mathbf{i} - \mathbf{j}$

8 $C$: $x = 2t$, $y = 3t$, $0 \le t \le 1$; $\mathbf{F} = x\mathbf{i} + y\mathbf{j}$

9 $C$: $x = \cos t$, $y = \sin t$, $0 \le t \le 2\pi$; $\mathbf{F} = y\mathbf{i} - x\mathbf{j}$

10 $C$: $x = \cos t$, $y = \sin t$, $0 \le t \le 2\pi$; $\mathbf{F} = x\mathbf{i} + y\mathbf{j}$

11 $C$: $x = \cos t$, $y = \sin t$, $z = t$, $0 \le t \le 2\pi$; $\mathbf{F} = z\mathbf{k}$

12 $C$: $x = \cos t$, $y = \sin t$, $z = 1/t$, $2\pi \le t \le 6\pi$; $\mathbf{F} = (1/z^2)\mathbf{k}$

13 An object moves in a straight line from (24, 10) to (12, 5) due to an external force

$$\mathbf{F} = \frac{-x}{(x^2 + y^2)^{3/2}}\mathbf{i} + \frac{-y}{(x^2 + y^2)^{3/2}}\mathbf{j}.$$

Find the change in kinetic energy of the object.

14 An object moves in a straight line from (3, 4) to (6, 8) due to an external force

$$\mathbf{F} = \frac{x}{(x^2 + y^2)^{3/2}}\mathbf{i} + \frac{y}{(x^2 + y^2)^{3/2}}\mathbf{j}.$$

Find the change in kinetic energy of the object.

In Exercises 15–16 use the sketch of $C$ and representative vectors **F** and **T** to determine whether the circulation of **F** around $C$ is positive or negative.

15 16

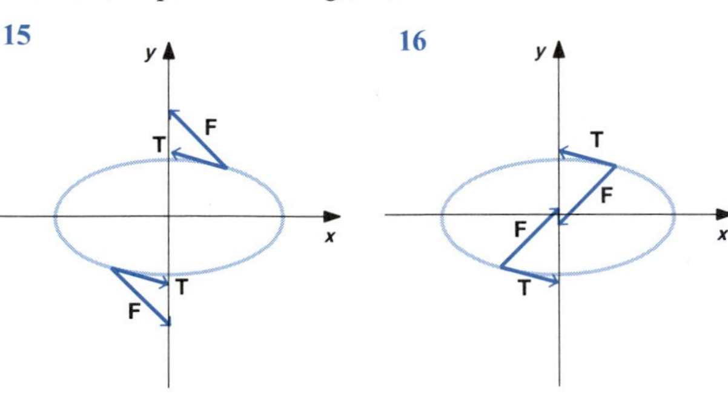


In Exercises 17–18 sketch the curve $C$ and representative vectors **F** and **T**. Find the circulation of **F** around $C$.

17 $\mathbf{F} = x\mathbf{i} + y\mathbf{j}$; $C$: $x = \cos t$, $y = \sin t$, $0 \le t \le 2\pi$

18 $\mathbf{F} = -y\mathbf{i} + x\mathbf{j}$; $C$: $x = \cos t$, $y = \sin t$, $0 \le t \le 2\pi$

In Exercises 19–22 find the circulation of the given vector function **F** around the indicated curve.

19 $\mathbf{F} = (x^2 - y)\mathbf{i} + (x - y^2)\mathbf{j}$; $C$: $x = \cos t$, $y = \sin t$, $0 \le t \le 2\pi$

20 $\mathbf{F} = 2xy\mathbf{i} + (x^2 + y^2)\mathbf{j}$; $C$: $x = \cos t$, $y = \sin t$, $0 \le t \le 2\pi$

21 $\mathbf{F} = xy^2\mathbf{i} + xy^3\mathbf{j}$; $C$ is the polygonal path from (0, 0) to (1, 0), to (0, 1), to (0, 0)

22 $\mathbf{F} = (x - 1)y\mathbf{i} + (x - 1)y^2\mathbf{j}$; $C$ is the polygonal path from (0, 0) to (1, 1), to (1, 0), to (0, 0)

In Exercises 23–24 use the sketch of representative vectors **F** and **T** to determine whether the outward flux of **F** around $C$ is positive or negative.

23 24

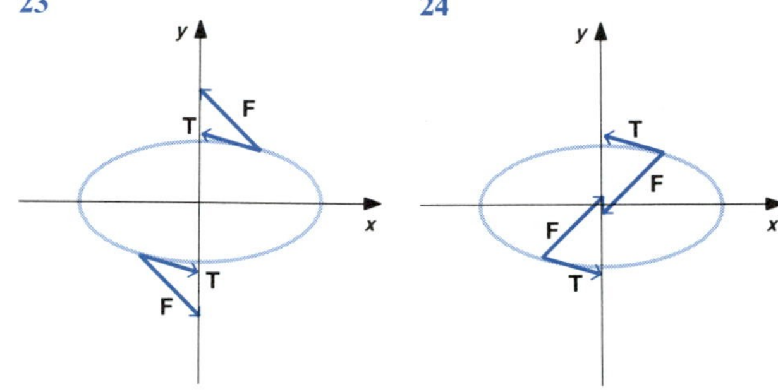


In Exercises 25–26 sketch the curve $C$ and representative vectors **F** and **T**. Find the outward flux of **F** around $C$.

25 $\mathbf{F} = x\mathbf{i} + y\mathbf{j}$; $C$: $x = \cos t$, $y = \sin t$, $0 \le t \le 2\pi$

26 $\mathbf{F} = -y\mathbf{i} + x\mathbf{j}$; $C$: $x = \cos t$, $y = \sin t$, $0 \le t \le 2\pi$

In Exercises 27–30 find the outward flux of the given vector function **F** across the indicated curve.

27 $\mathbf{F} = x\mathbf{i} + y\mathbf{j}$; $C$: $x = h + r\cos t$, $y = k + r\sin t$, $0 \le t \le 2\pi$

28 $\mathbf{F} = -y\mathbf{i} + x\mathbf{j}$; $C$: $x = h + r\cos t$, $y = k + r\sin t$, $0 \le t \le 2\pi$

29 $\mathbf{F} = (x^2 + y^2)\mathbf{i}$; $C$: $x = 2 + \cos t$, $y = \sin t$, $0 \le t \le 2\pi$

30 $\mathbf{F} = x\mathbf{i} + y\mathbf{j}$; $C$ is the polygonal path from (0, 0) to (1, 0), to (0, 1), to (0, 0)

31 Show that

$$\left|\int_C \mathbf{F}\cdot\mathbf{T}\,ds\right| \le (\text{Maximum of } |\mathbf{F}| \text{ on } C)(\text{Length of } C).$$

32 An irrigation canal is 2 m wide. If an $xy$-coordinate system is positioned horizontally with the $x$-axis along the center of the canal, the water flow in $m^3$/min below a point $(x, y)$ on the surface is given by the vector field $\mathbf{F} = (1 - y^4)\mathbf{i}$. Two vertical nets are placed across the canal, one along each of the curves

$$C_1\colon\ x = 2,\quad y = t,\qquad -1 \le t \le 1,\quad \text{and}$$

$$C_2\colon\ x = t,\quad y = t^3,\qquad -1 \le t \le 1.$$

Find the total flow (flux) past each net and compare the answers. Does the total flow depend on the horizontal curve of the net?

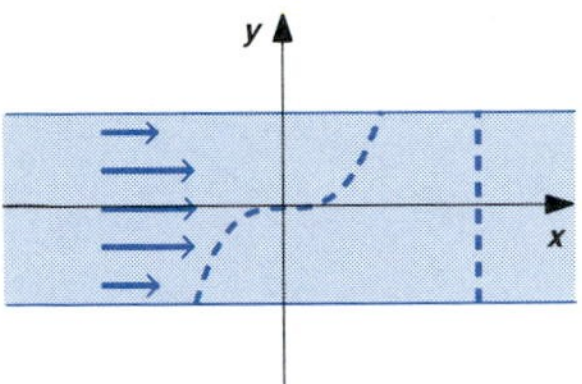


## 18.3 GREEN'S THEOREM

In this section we will establish a relation between a double integral over a region and a line integral over the curve that bounds the region.

Suppose $R$ is a vertically simple region, as illustrated in Figure 18.3.1. The curve $C = C_1 + C_2 + C_3 + C_4$ is the complete boundary of $R$, with direction chosen so the region is on the left as the boundary is traced. We assume that the function $M(x, y)$ and its first partial derivatives are continuous on $R$ and on the boundary of $R$. We then have

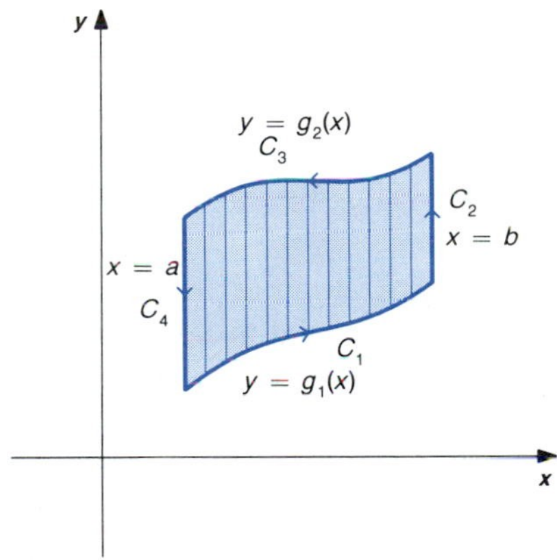


FIGURE 18.3.1

$$\begin{aligned}\int_C M\,dx &= \int_{C_1} M\,dx + \int_{C_2} M\,dx + \int_{C_3} M\,dx + \int_{C_4} M\,dx\\ &= \int_{C_1} M\,dx + \int_{C_2} M\,dx - \int_{-C_3} M\,dx - \int_{-C_4} M\,dx.\end{aligned}$$

We have changed signs to obtain a curve $-C_3$ in the direction of increasing $x$ and a curve $-C_4$ in the direction of increasing $y$. The above integrals can then be evaluated by using the parametric equations

$$C_1\colon\ x = x,\quad y = g_1(x),\qquad a \le x \le b,$$

$$C_2\colon\ x = b,\quad y = y,\qquad g_1(b) \le y \le g_2(b),$$

$$-C_3\colon\ x = x,\quad y = g_2(x),\qquad a \le x \le b,$$

$$-C_4\colon\ x = a,\quad y = y,\qquad g_1(a) \le y \le g_2(a).$$

Note that we have used $x$ as the parameter for $C_1$ and $-C_3$; we have used $y$ as the parameter for $C_2$ and $-C_4$. Since $dx = 0$ for $C_2$ and $-C_4$, we obtain

$$\begin{aligned}\int_C M\,dx &= \int_{C_1} M\,dx - \int_{-C_3} M\,dx\\ &= \int_a^b M(x, g_1(x))\,dx - \int_a^b M(x, g_2(x))\,dx\\ &= \int_a^b [M(x, g_1(x)) - M(x, g_2(x))]\,dx\end{aligned}$$

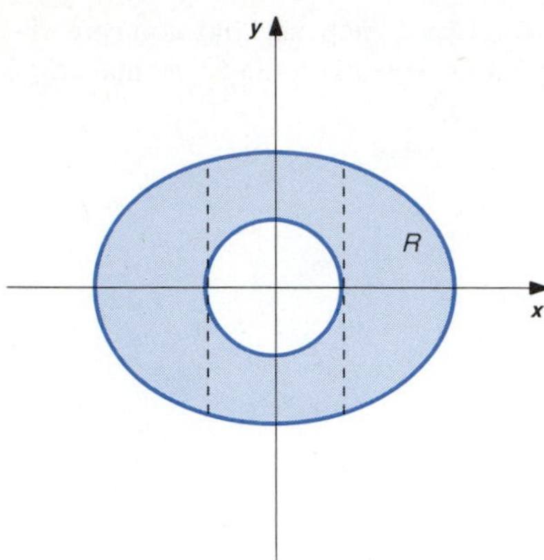


FIGURE 18.3.2

For $x$ fixed, $M(x, y)$ and $M_y(x, y)$ are continuous functions of $y$ for $g_1(x, y) \leq y \leq g_2(x, y)$. Since $M_y(x, y)$ is the derivative of $M(x, y)$ with respect to $y$ with $x$ fixed, the Fundamental Theorem of Calculus tells us

$$M(x, g_1(x)) - M(x, g_2(x)) = \int_{g_1(x)}^{g_2(x)} - M_y(x, y)\, dy.$$

We then obtain

$$\int_C M\, dx = \int_a^b \int_{g_1(x)}^{g_2(x)} - M_y(x, y)\, dy\, dx$$

$$= \iint_R - M_y\, dA.$$

That is,

$$\int_C M\, dx = \iint_R - M_y\, dA. \tag{1}$$

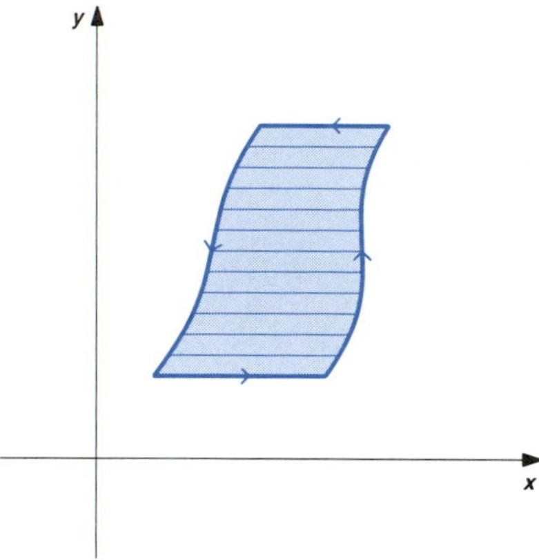


FIGURE 18.3.3

This result was obtained simply by using the Fundamental Theorem to evaluate the inner integral, with respect to $y$, of the iterated integral that corresponds to the double integral.

If a region is divided into subregions by cutting along vertical lines, the total boundary of the subregions consists of the boundary of the original region and the vertical cutting lines. Since the integral of $M\, dx$ over vertical lines is zero, the integral of $M\, dx$ over the original boundary is equal to the sum of the integrals over the boundaries of the subregions. The double integral of $-M_y$ over the region is also equal to the sum of the integrals over the subregions. It follows that (1) holds for regions that can be divided by vertical lines into a finite number of regions that are vertically simply. We call such regions **finitely vertically simple**. For example, the region indicated in Figure 18.3.2 is finitely vertically simple; the two vertical dotted lines divide the region into four vertically simple regions.

If $R$ is a horizontally simple region, as indicated in Figure 18.3.3, and $N(x, y)$ has continuous partial derivatives, we can verify that

$$\int_C N\, dy = \iint_R N_x\, dA.$$

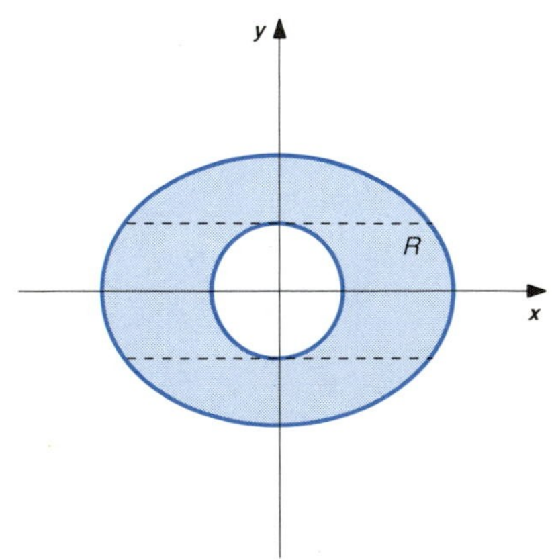


FIGURE 18.3.4

This result holds if the region is **finitely horizontally simple**. That is, if the region can be divided by horizontal lines into a finite number of horizontally simple regions. For example, the region in Figure 18.3.4 is finitely horizontally simple.

Let us say a region in the plane is **finitely simple** if it is both finitely vertically simple and finitely horizontally simple. Then both of the above results are true. We combine the results in the following theorem.

*Green's Theorem*

**Suppose $R$ is a finitely simple region in the plane, $C$ is the complete boundary of $R$ with the direction of each part chosen so $R$ is on the left as the boundary is traced. If $M$, $N$, and their first partial derivatives are continuous on $R$ and its boundary, then**

$$\int_C M\,dx + N\,dy = \iint_R (N_x - M_y)\,dA.$$

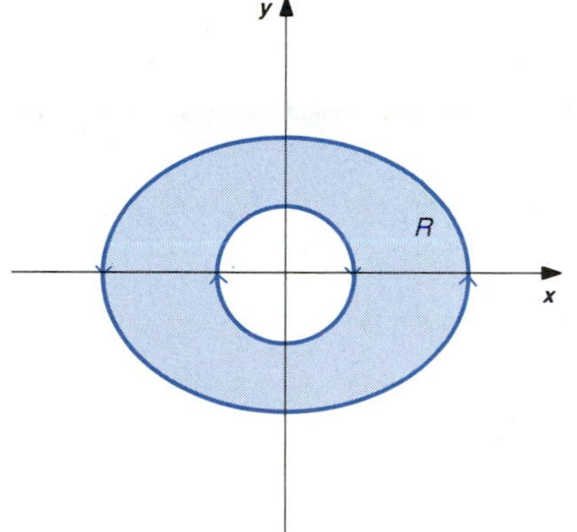

FIGURE 18.3.5

If the region is to be on the left as the boundary is traced, then outside boundary curves are traced counterclockwise, while interior boundary curves are traced clockwise. See Figure 18.3.5.

■ **EXAMPLE 1**

Determine whether Green's Theorem can be used to evaluate

$$\int_C y^2\,dx + 6xy\,dy,$$

where $C$ is the boundary curve of the region bounded by $y = \sqrt{x}$, $y = 0$, and $x = 4$, in the counterclockwise direction. Evaluate the line integral.

**SOLUTION**

The region is sketched in Figure 18.3.6. We have

$$\int_C y^2\,dx + 6xy\,dy = \int_C M\,dx + N\,dy,$$

FIGURE 18.3.6

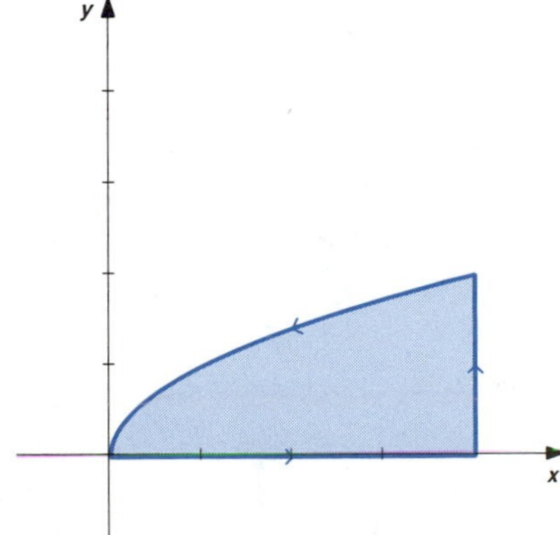

so $M(x, y) = y^2$ and $N(x, y) = 6xy$. We then have $M_y = 2y$ and $N_x = 6y$. We see that $M$, $N$, and their first partial derivatives are continuous at all points of $C$ and at all points of the region enclosed by $C$, so we can use Green's Theorem to evaluate the line integral.

We can either evaluate the line integral as the sum of the line integrals over each of the three smooth boundary curves of the region or use Green's Theorem. Let us use Green's Theorem. We note that the direction of the boundary curve is such that the region is on the left as the boundary curve is traced. We then have

$$\begin{aligned}
\int_C y^2\,dx + 6xy\,dy &= \int_C M\,dx + N\,dy \\
&= \iint_R (N_x - M_y)\,dA \\
&= \iint_R (6y - 2y)\,dA \\
&= \iint_R 4y\,dA \\
&= \int_0^4 \int_0^{\sqrt{x}} 4y\,dy\,dx \\
&= \int_0^4 2y^2 \Big]_{y=0}^{y=\sqrt{x}} dx \\
&= \int_0^4 2x\,dx = x^2 \Big]_0^4 = 16.
\end{aligned}$$

■

■ **EXAMPLE 2**

Determine whether Green's Theorem can be used to evaluate

$$\int_C \frac{-y}{2 - x^2 - y^2}\,dx + \frac{x}{2 - x^2 - y^2}\,dy,$$

and evaluate the integral, where

(a) $C$: $x = \cos t$, $y = \sin t$, $0 \le t \le 2\pi$,
(b) $C$: $x = \sqrt{3}\cos t$, $y = \sqrt{3}\sin t$, $0 \le t \le 2\pi$.

**SOLUTION**

We have

$$M = \frac{-y}{2 - x^2 - y^2} \quad \text{and} \quad N = \frac{x}{2 - x^2 - y^2}.$$

(a)

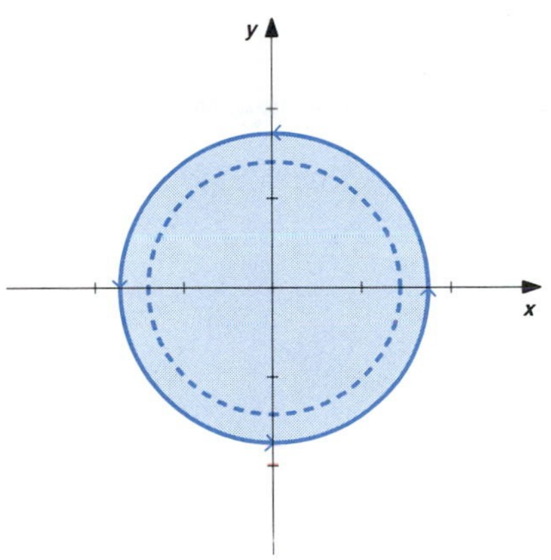


(b)

FIGURE 18.3.7

Using the Quotient Rule to differentiate, we then have

$$M_y = \frac{\partial}{\partial y}\left(\frac{-y}{2 - x^2 - y^2}\right) = \frac{(2 - x^2 - y^2)(-1) - (-y)(-2y)}{(2 - x^2 - y^2)^2}$$

$$= \frac{-2 + x^2 - y^2}{(2 - x^2 - y^2)^2} \quad \text{and}$$

$$N_x = \frac{2 + x^2 - y^2}{(2 - x^2 - y^2)^2}.$$

We see that $M$, $N$, and their first partial derivatives are continuous, except at points on the circle $x^2 + y^2 = 2$.

(a) $C$ encloses points $(x, y)$ with $x^2 + y^2 \leq 1$. See Figure 18.3.7a. Since $M$ and $N$ have continuous first partial derivatives at these points, Green's Theorem may be used to evaluate the line integral. We could also use the parametric equations of $C$ to evaluate the line integral. Let us use Green's Theorem, although that may not necessarily be easier than evaluating the line integral directly. We see that the region is on the left as the boundary curve is traced. We have

$$\int_C \frac{-y}{2 - x^2 - y^2}\,dx + \frac{x}{2 - x^2 - y^2}\,dy$$

$$= \int_C M\,dx + N\,dy$$

$$= \iint_R (N_x - M_y)\,dA$$

$$= \iint_R \left(\frac{2 + x^2 - y^2}{(2 - x^2 - y^2)^2} - \frac{-2 + x^2 - y^2}{(2 - x^2 - y^2)^2}\right) dA$$

$$= \iint_R \frac{4}{(2 - x^2 - y^2)^2}\,dA,$$

which in polar coordinates is

$$= \int_0^{2\pi} \int_0^1 \frac{4}{(2 - r^2)^2}\, r\,dr\,d\theta$$

$$[u = 2 - r^2,\ du = -2r\,dr]$$

$$= \int_0^{2\pi} \frac{2}{2 - r^2}\Bigg]_{r=0}^{r=1} d\theta$$

$$= \int_0^{2\pi} (2 - 1)\,d\theta = 2\pi.$$

(b) $C$ encloses points with $x^2 + y^2 \leq 3$. See Figure 18.3.7b. Since this region includes points where the first partial derivatives of $M$ and $N$ do not exist, we cannot use Green's Theorem to evaluate the line integral. [The integral $\iint_R (N_x - M_y)\,dA$ would be a divergent improper integral in this example.] We can evaluate the line integral by using the parametric equations of $C$. We have

$$x = \sqrt{3}\cos t, \quad y = \sqrt{3}\sin t, \quad dx = -\sqrt{3}\sin t\,dt, \quad dy = \sqrt{3}\cos t\,dt.$$

Substitution then gives

$$\int_C \frac{-y}{2 - x^2 - y^2}\,dx + \frac{x}{2 - x^2 - y^2}\,dy$$

$$= \int_0^{2\pi} \left[\left(\frac{-\sqrt{3}\sin t}{2 - 3}\right)(-\sqrt{3}\sin t) + \left(\frac{\sqrt{3}\cos t}{2 - 3}\right)(\sqrt{3}\cos t)\right] dt$$

$$= \int_0^{2\pi} -3\,dt = -3(2\pi) = -6\pi.$$

■

■ **EXAMPLE 3**

Determine whether Green's Theorem can be used to evaluate

$$\int_C (e^{x^2} - 2y)\,dx + \ln y\,dy,$$

and evaluate the integral, where

$$C\colon\ x = 2\cos t, \quad y = 3 + 2\sin t, \qquad 0 \le t \le 2\pi.$$

**SOLUTION**

We have

$$M = e^{x^2} - 2y \quad \text{and} \quad N = \ln y.$$

The functions $M$ and $N$ are defined and continuous in the region $y > 0$. For $y > 0$, we have

$$M_y = -2 \quad \text{and} \quad N_x = 0.$$

FIGURE 18.3.8

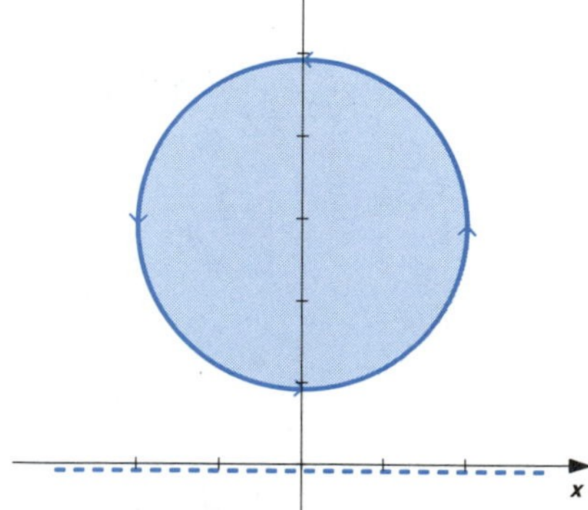

We see that $M$, $N$, and their first partial derivatives are continuous on $C$ and the region enclosed by $C$, so Green's Theorem can be used to evaluate the line integral. See Figure 18.3.8. The region enclosed by $C$ is on the left as $C$ is traced, so

$$\begin{aligned}\int_C (e^{x^2} - 2y)\,dx + \ln y\,dy &= \int_C M\,dx + N\,dy \\ &= \iint_R (N_x - M_y)\,dA \\ &= \iint_R [(0) - (-2)]\,dA \\ &= \iint_R 2\,dA \\ &\qquad\qquad \text{[Since 2 is a } \textit{constant}\text{]} \\ &= 2\left(\iint_R dA\right) \\ &= 2(\text{Area of } R) = 2(\pi 2^2) = 8\pi. \quad ■\end{aligned}$$

Green's Theorem gives us the following formulas for the area of a region in terms of a line integral over the boundary curve of the region.

$$\textbf{Area} = \iint_R dA = \int_C x\,dy = \int_C -y\,dx = \frac{1}{2}\int_C x\,dy - y\,dx.$$

■ **EXAMPLE 4**

Find the area inside the ellipse

$$\frac{x^2}{a^2} + \frac{y^2}{b^2} = 1.$$

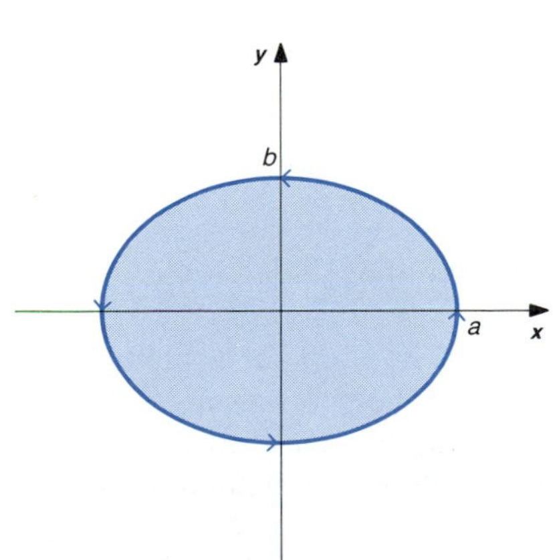

FIGURE 18.3.9

**SOLUTION**

The ellipse is sketched in Figure 18.3.9. A parametric representation of the ellipse with direction so the region enclosed is on the left as the boundary is traced is given by

$$C\colon\ x = a\cos t,\ y = b\sin t,\ 0 \le t \le 2\pi.$$

We use the formula

$$\text{Area} = \frac{1}{2}\int_C x\,dy - y\,dx,$$

which turns out to give a very simple integral. We have

$$\text{Area} = \frac{1}{2}\int_C x\,dy - y\,dx$$
$$= \frac{1}{2}\int_0^{2\pi} [(a\cos t)(b\cos t) - (b\sin t)(-a\sin t)]\,dt$$
$$= \frac{1}{2}\int_0^{2\pi} ab\,dt = ab\pi.$$ ■

We have seen in the previous section that, if $\mathbf{F} = M\mathbf{i} + N\mathbf{j}$, the circulation of $\mathbf{F}$ around a simple closed curve $C$ is

$$\int_C \mathbf{F}\cdot\mathbf{T}\,ds = \int_C M\,dx + N\,dy. \tag{2}$$

If Green's Theorem applies to the function $\mathbf{F}$ and the region $R$ bounded by a simple closed curve $C$, we can express the circulation of $\mathbf{F}$ around $C$ as the integral of $N_x - M_y$ over $R$. That is,

$$\int_C \mathbf{F}\cdot\mathbf{T}\,ds = \int_C M\,dx + N\,dy = \iint_R (N_x - M_y)\,dA. \tag{3}$$

We cannot use Green's Theorem to evaluate the line integral if there is a point inside the region $R$ where either $\mathbf{F}$ or one of its first partial derivatives is discontinuous, even if $\mathbf{F}$ and its first derivatives are continuous on $C$.

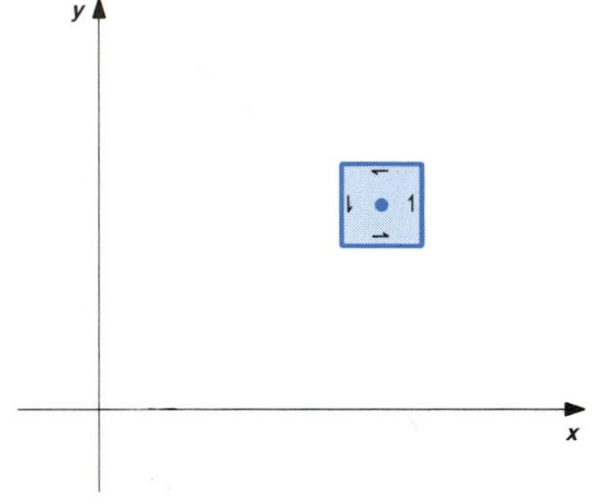

FIGURE 18.3.10

The expression $N_x - M_y$ can be related to the circulation of $\mathbf{F} = M\mathbf{i} + N\mathbf{j}$. To do this, let $C_r$ denote the boundary of a square $R_r$ with center at a fixed point $(x_0, y_0)$ and length of sides $r$, $C_r$ directed counterclockwise around $R_r$. See Figure 18.3.10. Let us assume that $\mathbf{F}$ and its partial derivatives are continuous on some fixed square with center $(x_0, y_0)$. If $r$ is sufficiently small, we then have

$$N_x(x, y) - M_y(x, y) \approx N_x(x_0, y_0) - M_y(x_0, y_0) \qquad \text{for all } (x, y) \text{ on } C_r.$$

Then

$$\int_{C_r} \mathbf{F}\cdot\mathbf{T}\,ds = \iint_{R_r} (N_x - M_y)\,dA$$
$$\approx [N_x(x_0, y_0) - M_y(x_0, y_0)][\text{Area of } R_r].$$

This suggests that we think of the value of $N_x - M_y$ at a point as the circulation per unit area of $\mathbf{F}$ at the point. The expression $N_x - M_y$ is sometimes called the **scalar curl** of $\mathbf{F}$. Green's Theorem then says that the circulation of $\mathbf{F}$ around $C$ is the integral of the curl (circulation per unit area) over the region enclosed by $C$.

It should not be surprising that the circulation of $\mathbf{F}$ around a simple closed curve $C$ involves the circulation at points in the region $R$ enclosed

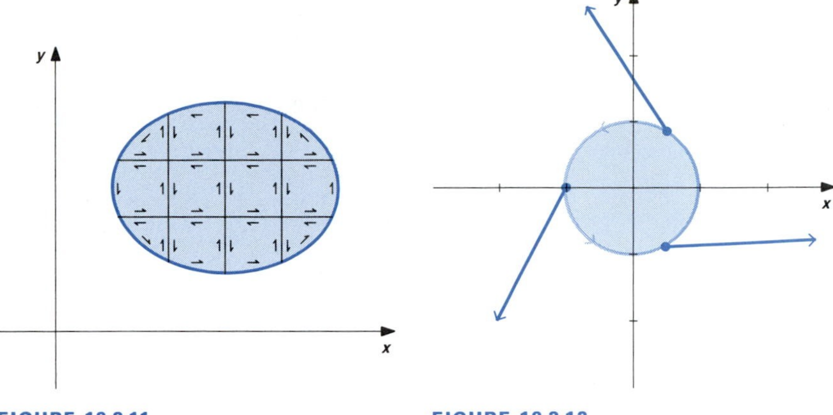


FIGURE 18.3.11

FIGURE 18.3.12

by $C$. That is, let us suppose that $R$ is divided into subregions $R_j$ with simple closed boundary curves $C_j$. We assume that each boundary is directed so the corresponding region is on the left as the boundary is traced. From Figure 18.3.11 we see that the unit tangent vectors of adjacent subregions have opposite direction along their common boundary. It follows that the net contribution of the line integral of $\mathbf{F} \cdot \mathbf{T}$ over the common boundary of two subregions is zero. The only parts of the boundary curves that are not common to two subregions are parts of $C$. It follows that

$$\sum \int_{C_j} \mathbf{F} \cdot \mathbf{T}\, ds = \int_C \mathbf{F} \cdot \mathbf{T}\, ds,$$

where the sum is taken over all subregions. That is, the circulation of $\mathbf{F}$ around $C$ is the sum of the circulation around the boundary curves of the subregions of $R$. In this sense, the circulation around $C$ measures the circulation (curl) at each point in the region enclosed by $C$, as indicated by the interpretation of Green's Theorem given in the preceding paragraph.

### ■ EXAMPLE 5

Find the scalar curl of $\mathbf{F} = (x - 2y)\mathbf{i} + (2x + y)\mathbf{j}$. Evaluate the circulation of $\mathbf{F}$ around the curve

$$C\colon\ x = \cos t,\ y = \sin t,\ 0 \le t \le 2\pi.$$

#### SOLUTION

The curve and some representative vectors $\mathbf{F}$ are sketched in Figure 18.3.12. We have

$$\mathbf{F} = (x - 2y)\mathbf{i} + (2x + y)\mathbf{j} = M\mathbf{i} + N\mathbf{j}, \quad \text{so}$$

$$M = x - 2y \quad \text{and} \quad N = 2x + y.$$

The scalar curl is $N_x - M_y = 2 - (-2) = 4$, a constant function.

Since **F** and its first partial derivatives are continuous at all points, we can use Green's Theorem to evaluate the circulation of **F** around $C$. The region enclosed by $C$ is on the left as $C$ is traced, so

$$\begin{aligned}\int_C \mathbf{F}\cdot\mathbf{T}\,ds &= \int_C M\,dx + N\,dy \\ &= \iint_R (N_x - M_y)\,dA \\ &= \iint_R 4\,dA \\ &= 4\left(\iint_R dA\right) \qquad \text{[Since 4 is a } constant\text{]} \\ &= 4(\text{Area of } R) = 4\pi.\end{aligned}$$

This agrees with the result of Exercise 3 of Section 18.2, where we calculated the circulation by using the parametric equations of $C$ to evaluate the line integral. ■

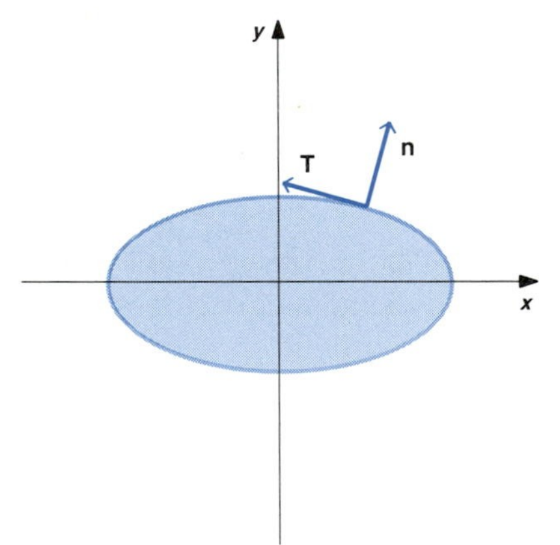

FIGURE 18.3.13

If $C$ is the boundary curve of a region $R$, and if $R$ is on the left as $C$ is traced, then the outer unit normal is

$$\mathbf{n} = \frac{dy}{ds}\mathbf{i} - \frac{dx}{ds}\mathbf{j}.$$

See Figure 18.3.13. The **outward flux** of **F** across $C$ is

$$\int_C \mathbf{F}\cdot\mathbf{n}\,ds = \int_C M\,dy - N\,dx. \tag{4}$$

In cases where Green's Theorem applies, we have

$$\int_C \mathbf{F}\cdot\mathbf{n}\,ds = \int_C M\,dy - N\,dx = \iint_R (M_x + N_y)\,dA. \tag{5}$$

The expression $M_x + N_y$ is called the **divergence** of **F**. We can think of the value of $M_x + N_y$ at a point as the flux per unit area at the point. Positive divergence at a point indicates that there is a net flow from the point, or a source at the point; negative divergence indicates a net flow into the point, or a sink at the point. Green's Theorem in the form of (5) then says that the outward flux of **F** over $C$ is the sum of the flux per unit area (divergence) of **F** over the region enclosed by $C$. That is, the flow over the boundary of $R$ is the net sum of the flow at each point of $R$.

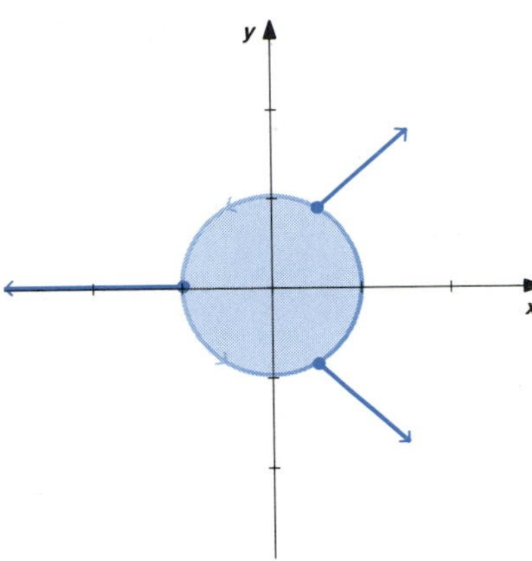


FIGURE 18.3.14

■ **EXAMPLE 6**

Find the divergence of $\mathbf{F} = 2x\mathbf{i} + y\mathbf{j}$. Evaluate the outward flux of $\mathbf{F}$ across the curve

$$C\colon\ x = \cos t,\ y = \sin t,\ 0 \le t \le 2\pi.$$

**SOLUTION**

The curve and some representative vectors $\mathbf{F}$ are sketched in Figure 18.3.14. We have

$$\mathbf{F} = 2x\mathbf{i} + y\mathbf{j} = M\mathbf{i} + N\mathbf{j}, \quad \text{so}$$

$$M = 2x, \quad N = y, \quad M_x = 2, \quad \text{and} \quad N_y = 1.$$

The divergence is $M_x + N_y = 2 + 1 = 3$.

Since $\mathbf{F}$ and its first partial derivatives are continuous at all points, we can use Green's Theorem to evaluate the outward flux of $\mathbf{F}$ across $C$. Formula (5) implies

$$\begin{aligned}\int_C \mathbf{F}\cdot\mathbf{n}\,ds &= \int_C M\,dy - N\,dx \\ &= \iint_R (M_x + N_y)\,dA \\ &= \iint_R 3\,dA \\ &\qquad\qquad [\text{Since 3 is a } constant] \\ &= 3\left(\iint_R dA\right) \\ &= 3(\text{Area of } R) = 3\pi.\end{aligned}$$

This agrees with the result of Exercise 4 of Section 18.2, where we calculated the flux by using the parametric equations of $C$ to evaluate the line integral. ■

■ **EXAMPLE 7**

Find the divergence of

$$\mathbf{F} = \frac{-x}{x^2 + y^2}\mathbf{i} + \frac{-y}{x^2 + y^2}\mathbf{j}.$$

Evaluate the outward flux of $\mathbf{F}$ across the curve

$$C_r\colon\ x = r\cos t,\ y = r\sin t,\ 0 \le t \le 2\pi.$$

FIGURE 18.3.15

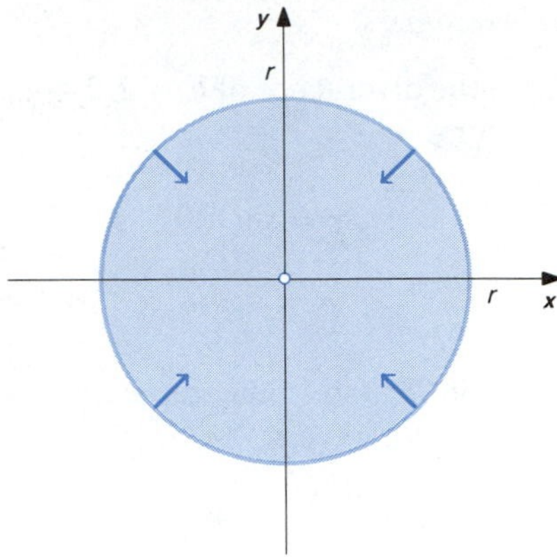


**SOLUTION**

The curve and some representative vectors of **F** are sketched in Figure 18.3.15.

We have

$$\mathbf{F} = \frac{-x}{x^2 + y^2}\,\mathbf{i} + \frac{-y}{x^2 + y^2}\,\mathbf{j} = M\mathbf{i} + N\mathbf{j}, \quad \text{so}$$

$$M = \frac{-x}{x^2 + y^2} \quad \text{and} \quad N = \frac{-y}{x^2 + y^2}.$$

Using the Quotient Rule, we have

$$M_x = \frac{(x^2 + y^2)(-1) - (-x)(2x)}{(x^2 + y^2)^2}$$

$$= \frac{x^2 - y^2}{(x^2 + y^2)^2}.$$

Similarly, we obtain that

$$N_y = \frac{y^2 - x^2}{(x^2 + y^2)^2}.$$

The divergence is then

$$M_x + N_y = 0, \qquad (x, y) \neq (0, 0).$$

Note that the function

$$\mathbf{F} = \frac{-y}{x^2 + y^2}\,\mathbf{i} + \frac{-x}{x^2 + y^2}\,\mathbf{j}$$

cannot be defined at (0, 0) so it is continuous at (0, 0), and that (0, 0) is in the region enclosed by $C_r$. Since Green's Theorem requires that **F** and its first partial derivatives be continuous on $C_r$ and in the region enclosed by $C_r$, we cannot apply formula (5) to evaluate the line integral that gives the flux of $\mathbf{F}$ over the curve $C_r$. We must use formula (4) to evaluate the flux directly as a line integral. We then have

$$\begin{aligned}
\int_{C_r} \mathbf{F}\cdot\mathbf{n}\, ds &= \int_{C_r} M\, dy - N\, dx \\
&= \int_{C_r} \frac{-x}{x^2+y^2}\, dy - \frac{-y}{x^2+y^2}\, dx \\
&= \int_0^{2\pi} \left[\left(\frac{-r\cos t}{(r\cos t)^2+(r\sin t)^2}\right)(r\cos t)\right. \\
&\qquad \left. - \left(\frac{-r\sin t}{(r\cos t)^2+(r\sin t)^2}\right)(-r\sin t)\right] dt \\
&= \int_0^{2\pi} (-1)\, dt = -2\pi.
\end{aligned}$$

We see that the flux is the same for all positive values of $r$. The negative flux indicates that there is a net flow into the region bounded by $C_r$. ■

Even though $M_x + N_y = 0$ at every point except the origin in Example 7, $\int_C \mathbf{F}\cdot\mathbf{n}\, ds \neq \iint_R 0\, dA$ for curves $C$ that enclose the origin. The flux across any simple closed curve that does not enclose the origin is zero.

## EXERCISES 18.3

Determine whether Green's Theorem can be used to evaluate the line integrals in Exercises 1–10. Evaluate the integrals.

1 $\int_C x\, dx + xy\, dy$
$C$ is the boundary curve of the region in the first quadrant bounded by $2x + 3y = 6$, in the counterclockwise direction.

2 $\int_C x^2y\, dx + y\, dy$
$C$ is the boundary curve of the region bounded by $y = x$, $y = 0$, and $x = 2$, in the counterclockwise direction.

3 $\int_C y\, dx + xy\, dy$
$C$ is the boundary curve of the region bounded by $y = x^2$, $y = 0$, and $x = 2$, in the counterclockwise direction.

4 $\int_C xy^2\, dx + x^2y\, dy$
$C$ is the boundary curve of the region bounded by $y = x^2$, $x + y = 2$, and $x = 0$, in the counterclockwise direction.

5 $\int_C (x^2 + y^2)\, dx + 2xy\, dy$
$C$: $x = \cos t$, $y = \sin t$, $0 \le t \le \pi$

6 $\int_C xy\, dx + dy$
$C$: $x = t$, $y = t^2$, $0 \le t \le 2$

7 $\int_C -y \ln(x^2 + y^2)\, dx + x \ln(x^2 + y^2)\, dy$
$C$: $x = 2\cos t$, $y = 2\sin t$, $0 \le t \le 2\pi$.

8 $\int_C \frac{2xy}{(x^2+y^2)^2}\, dx + \frac{y^2 - x^2}{(x^2+y^2)^2}\, dy$
$C$: $x = \cos t$, $y = 2 + \sin t$, $0 \le t \le 2\pi$

9 $\int_C \ln(x^2 + 1)\, dx - x\, dy$
$C$ is the boundary curve of the region bounded by $y = \sqrt{4 - x^2}$ and $y = 0$, in the counterclockwise direction.

10 $\int_C (y + \sin x)\, dx - (x + \cos y)\, dy$
$C$: $x = \cos t$, $y = \sin t$, $0 \le t \le 2\pi$

In Exercises 11 and 12, verify the formulas for the coordinates of the center of mass of a homogeneous lamina in terms of line integrals. $A$ denotes the area of the region occupied by the lamina.

11 $\bar{x} = \dfrac{1}{A} \iint_R x \, dA = \dfrac{1}{2A} \int_C x^2 \, dy$

12 $\bar{y} = \dfrac{1}{A} \iint_R y \, dA = \dfrac{-1}{2A} \int_C y^2 \, dx$

13 (a) Use the formula Area $= \int_C x \, dy$ to find the area of the region bounded by $y = \tan^{-1} x$, $y = -\tan^{-1} x$, and $x = 1$.

(b) Use the line integrals in Exercises 11 and 12 to find the center of mass of a homogeneous lamina that occupies the region in (a).

14 (a) Use the formula Area $= \int_C x \, dy$ to find the area of the region bounded by $y = \ln x$, $y = -\ln x$, and $x = e$.

(b) Use the line integrals in Exercises 11 and 12 to find the center of mass of a homogeneous lamina that occupies the region in (a).

In Exercises 15–20 find the scalar curl of the given vector function **F** and the circulation of **F** around the indicated curve.

15 $\mathbf{F} = x\mathbf{i} + y\mathbf{j}$; $C$: $x = h + r \cos t$, $y = k + r \sin t$, $0 \le t \le 2\pi$

16 $\mathbf{F} = -y\mathbf{i} + x\mathbf{j}$; $C$: $x = h + r \cos t$, $y = k + r \sin t$, $0 \le t \le 2\pi$

17 $\mathbf{F} = [1/(y^2 + 1)]\mathbf{i}$; $C$ is the polygonal path from (0, 0) to (1, 0), to (1, 1), to (0, 1), to (0, 0)

18 $\mathbf{F} = [1/(y^2 + 1)]\mathbf{i}$; $C$ is the polygonal path from $(-1, -1)$ to $(1, -1)$, to $(1, 1)$, to $(-1, 1)$, to $(-1, -1)$

19 $\mathbf{F} = e^x \sin y\mathbf{i} + e^x \cos y\mathbf{j}$; $C$: $x = \cos t$, $y = \sin t$, $0 \le t \le 2\pi$

20 $\mathbf{F}(x, y) = \begin{cases} y^2\mathbf{i}, & y \ge 0, \\ \mathbf{0}, & y < 0; \end{cases}$ $C$: $x = \cos t$, $y = \sin t$, $0 \le t \le 2\pi$

In Exercises 21–26 find the divergence of the given vector function **F** and find the outward flux of **F** across the indicated curve.

21 $\mathbf{F} = xi + yj$; $C$: $x = h + r \cos t$, $y = k + r \sin t$, $0 \le t \le 2\pi$

22 $\mathbf{F} = -y\mathbf{i} + x\mathbf{j}$; $C$: $x = h + r \cos t$, $y = k + r \sin t$, $0 \le t \le 2\pi$

23 $\mathbf{F} = (2x - y)\mathbf{i} + (x + 3y)\mathbf{j}$; $C$: $x = \cos t$, $y = \sin t$, $0 \le t \le 2\pi$

24 $\mathbf{F} = (3x + 2y)\mathbf{i} + (x - 2y)\mathbf{j}$; $C$: $x = \cos t$, $y = \sin t$, $0 \le t \le 2\pi$

25 $\mathbf{F} = \dfrac{x}{(x^2 + y^2)^{3/2}}\mathbf{i} + \dfrac{y}{(x^2 + y^2)^{3/2}}\mathbf{j}$; $C$: $x = 2 \cos t$, $y = 2 \sin t$, $0 \le t \le 2\pi$

26 $\mathbf{F} = \dfrac{x^2 - y^2}{(x^2 + y^2)^2}\mathbf{i} + \dfrac{2xy}{(x^2 + y^2)^2}\mathbf{j}$; $C$: $x = \cos t$, $y = \sin t$, $0 \le t \le 2\pi$

27 Show that the circulation of a constant vector field around any simple closed curve is zero.

28 Show that the outward flux of a constant vector field across any simple closed curve is zero.

29 Show that the outward flux of

$$\mathbf{F} = \frac{-x}{x^2 + y^2}\mathbf{i} + \frac{-y}{x^2 + y^2}\mathbf{j}$$

across any simple closed curve that contains the origin in its interior is $-2\pi$. (*Hint*: Use the result of Example 7. Choose $r$ so small that $C_r$ is inside $C$, and then apply Green's Theorem to the region between $C$ and $C_r$.)

30 Evaluate $\int_C \cos x \, dx + \sin y \, dy$, where $C$: $x = \cos t$, $y = \sin t$, $0 \le t \le \pi$. [*Hint*: Apply Green's Theorem to the region enclosed by $C$ and the line segment from $(-1, 0)$ to $(1, 0)$.]

## 18.4 INDEPENDENCE OF PATH OF LINE INTEGRALS IN THE PLANE

We have seen that applications of line integrals lead to integrals of the form

$$\int_C M \, dx + N \, dy.$$

An integral of this type is evaluated by substitution of parametric values. We know that the value of the integral does not depend on the particular

FIGURE 18.4.1

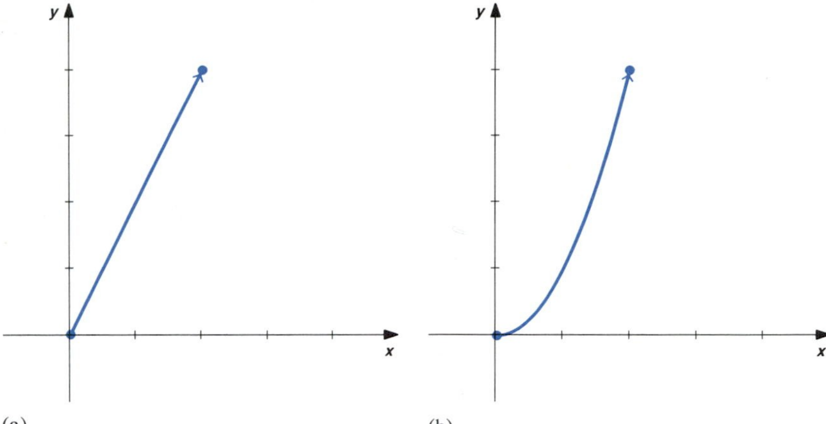


(smooth) parametric representation of $C$, except for the direction in which $C$ is traced as the parameter increases. We will now investigate conditions under which the value of the integral depends only on the initial and terminal points of $C$ and not on the path between these points.

**EXAMPLE 1**

Evaluate $\int_C y\,dx - x\,dy$ for each of the given curves.

(a) $C_1$: $x = t$, $y = 2t$, $0 \le t \le 2$,

(b) $C_2$: $x = t$, $y = t^2$, $0 \le t \le 2$.

**SOLUTION**

The curves are sketched in Figure 18.4.1. The curves follow different paths from (0, 0) to (2, 4).

$$\text{(a)} \quad \int_{C_1} y\,dx - x\,dy = \int_0^2 [(2t) - (t)(2)]\,dt = \int_0^2 0\,dt = 0.$$

$$\text{(b)} \quad \int_{C_2} y\,dx - x\,dy = \int_0^2 [(t^2) - (t)(2t)]\,dt$$

$$= \int_0^2 -t^2\,dt = -\left.\frac{t^3}{3}\right]_0^2 = -\frac{8}{3}.$$

These integrals have unequal values. ■

**EXAMPLE 2**

Evaluate $\int_C -x\,dx + y\,dy$ for each of the given curves.

(a) $C_1$: $x = 1 - t$, $y = t$, $0 \le t \le 1$.

(b) $C_2$: $x = \cos t$, $y = \sin t$, $0 \le t \le \pi/2$,

(c) $C_3$: $x = f(t)$, $y = g(t)$, $a \le t \le b$,

where $f$ and $g$ have continuous derivatives, $(f(a), g(a)) = (1, 0)$, and $(f(b), g(b)) = (0, 1)$.

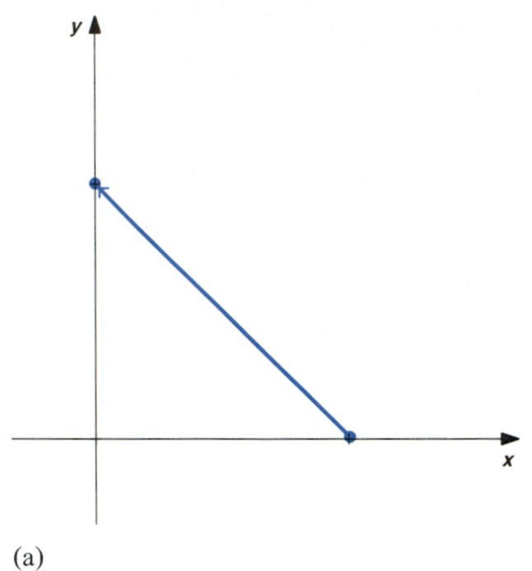


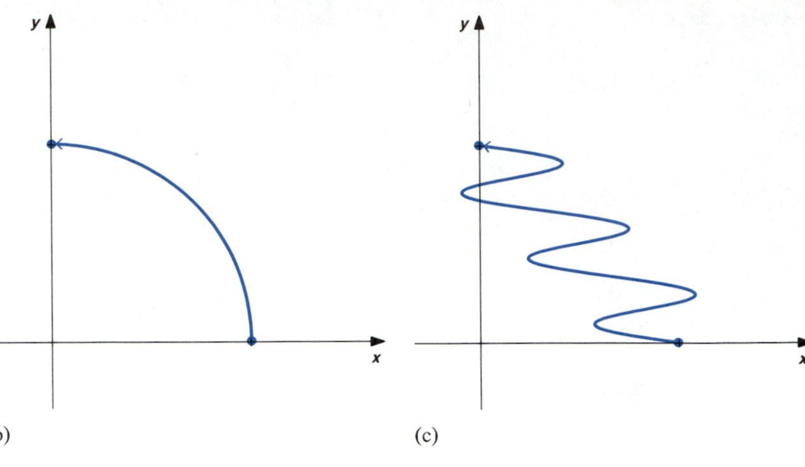


FIGURE 18.4.2

**SOLUTION**

The curves are sketched in Figure 18.4.2. These curves follow different paths from (1, 0) to (0, 1).

(a) $$\int_{C_1} -x\,dx + y\,dy = \int_0^1 [-(1-t)(-1) + (t)]\,dt = \int_0^1 1\,dt = 1.$$

(b) $$\int_{C_2} -x\,dx + y\,dy = \int_0^{\pi/2} [-(\cos t)(-\sin t) + (\sin t)(\cos t)]\,dt$$

$$= \int_0^{\pi/2} 2\sin t\cos t\,dt$$

$[u = \sin t,\ du = \cos t\,dt]$

$$= \sin^2 t\Big]_0^{\pi/2} = 1.$$

(c) $$\int_{C_3} -x\,dx + y\,dy = \int_a^b [-f(t)f'(t) + g(t)g'(t)]\,dt$$

$[u = f(t),\ du = f'(t)\,dt;\ v = g(t),\ dv = g'(t)\,dt]$

$$= -\frac{[f(t)]^2}{2} + \frac{[g(t)]^2}{2}\Big]_a^b$$

$$= \left\{-\frac{[f(b)]^2}{2} + \frac{[g(b)]^2}{2}\right\} - \left\{-\frac{[f(a)]^2}{2} + \frac{[g(a)]^2}{2}\right\}$$

$$= -\frac{0}{2} + \frac{1}{2} + \frac{1}{2} - \frac{0}{2} = 1.$$

Part (c) shows that the value of the integral does not depend on the path of $C$ from (1, 0) to (0, 1). ■

## 18.4 INDEPENDENCE OF PATH OF LINE INTEGRALS IN THE PLANE

For given functions $M$ and $N$ that are defined in a plane region $R$, it is desirable to know if the value of the line integral $\int_C M\,dx + N\,dy$ depends only on the initial and terminal points of $C$ and not on the particular path taken within a region $R$. Such line integrals are said to be **independent of path** in $R$.

*Theorem 1*

**(a) Suppose $C_1$ and $C_2$ are curves that have the same initial and terminal points. If $M$, $N$, and their first partial derivatives are continuous with**

$$M_y = N_x$$

**at all points of $C_1$ and $C_2$ and the region between the curves, then**

$$\int_{C_1} M\,dx + N\,dy = \int_{C_2} M\,dx + N\,dy.$$

**(b) Suppose the plane region $R$ has the property that whenever $C$ is a closed curve in $R$, the region enclosed by $C$ is also in $R$. If $M$, $N$, and their first partial derivatives are continuous with $M_y = N_x$ in $R$, then $\int_C M\,dx + N\,dy$ is independent of path in $R$.**

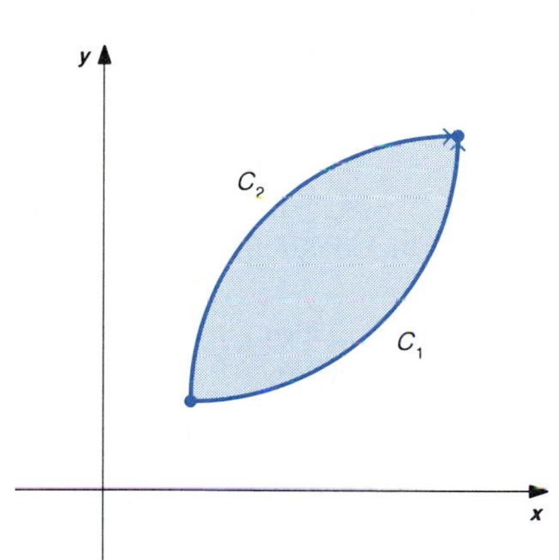


FIGURE 18.4.3

■ *Proof* Statement (a) of Theorem 1 is a consequence of Green's Theorem applied to the region $R$ bounded by $C = C_1 - C_2$, as illustrated in Figure 18.4.3. That is,

$$\int_{C_1} M\,dx + N\,dy - \int_{C_2} M\,dx + N\,dy = \int_C M\,dx + N\,dy$$

$$= \iint_R (N_x - M_y)\,dA = 0.$$

Statement (b) follows from part (a). Note that statement (b) does not apply to regions that have "holes," as illustrated in Figure 18.4.4. ■

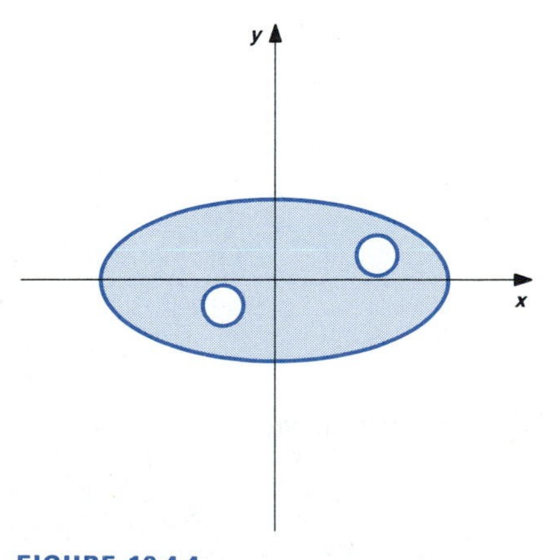


FIGURE 18.4.4

The function $\mathbf{F} = y\mathbf{i} - x\mathbf{j}$ of Example 1 does not satisfy the condition $M_y = N_x$ for any region $R$. We have seen that $\int_C y\,dx - x\,dy$ is not independent of path. The function $\mathbf{F} = -x\mathbf{i} + y\mathbf{j}$ of Example 2 does satisfy $M_y = N_x$ for all $(x, y)$. The integral $\int_C -x\,dx + y\,dy$ is independent of path in the entire plane.

If a line integral is independent of path in a region, it may be easier to evaluate the integral by replacing the given curve with a more convenient path that has the same initial and terminal points. We also have the following result.

**If $\int_C M\,dx + N\,dy$ is independent of path in $R$ and $C$ is a *closed* curve in $R$, then $\int_C M\,dx + N\,dy = 0$.**

FIGURE 18.4.5

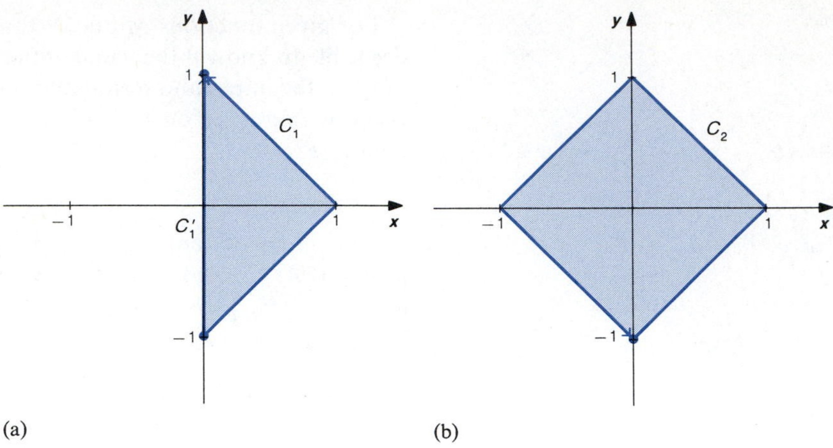


■ **EXAMPLE 3**

Evaluate $\int_C (x - 3y + 2)\,dx - (3x + 2y - 4)\,dy$ over the curves:

(a) $C_1$: The polygonal path from (0, −1) to (1, 0), to (0, 1),

(b) $C_2$: The polygonal path from (0, −1) to (1, 0), to (0, 1), to (−1, 0), to (0, −1).

**SOLUTION**

We have $M = x - 3y + 2$ and $N = -3x - 2y + 4$. This implies $M_y = -3 = N_x$ for all $(x, y)$, so the line integrals are independent of path in the entire $xy$-plane.

We replace $C_1$ by the path $C_1'$ from (0, −1) to (0, 1) along the $y$-axis. See Figure 18.4.5a. This curve has parametric representation

$$C_1'\colon\ x = 0, \quad y = t, \qquad -1 \le t \le 1.$$

Note that $dx = 0$. Then

$$\begin{aligned}\int_{C_1} &(x - 3y + 2)\,dx - (3x + 2y - 4)\,dy \\ &= \int_{C_1'} (x - 3y + 2)\,dx - (3x + 2y - 4)\,dy \\ &= \int_{-1}^{1} -[3(0) + 2t - 4]\,dt \\ &= -t^2 + 4t\Big]_{-1}^{1} = 8.\end{aligned}$$

From Figure 18.4.5b, we see that $C_2$ is a closed curve, so

$$\int_{C_2} (x - 3y + 2)\,dx - (3x + 2y - 4)\,dy = 0. \qquad ■$$

Theorem 1 guarantees that two line integrals over curves that have the same initial and terminal points have equal values only if $M_y = N_x$ for *every* point in the region enclosed by the curves.

**■ EXAMPLE 4**

Evaluate

$$\int_C \frac{-y}{x^2 + y^2}\,dx + \frac{x}{x^2 + y^2}\,dy,$$

where $C$ is

(a) $C_1$: $x = \cos t, \quad y = \sin t, \qquad 0 \le t \le \pi,$

(b) $C_2$: $x = \cos t, \quad y = -\sin t, \qquad 0 \le t \le \pi.$

**SOLUTION**

The curves are sketched in Figure 18.4.6. We see that the curves have the same initial and terminal points. Let us check the condition $M_y = N_x$ for independence of path. Using the Quotient Rule for differentiation, we have

$$M_y = \frac{(x^2 + y^2)(-1) - (-y)(2y)}{(x^2 + y^2)^2} = \frac{y^2 - x^2}{(x^2 + y^2)^2} \quad \text{and}$$

$$N_x = \frac{(x^2 + y^2)(1) - (x)(2x)}{(x^2 + y^2)^2} = \frac{y^2 - x^2}{(x^2 + y^2)^2}.$$

It follows that $M_y = N_x$ for $(x, y) \neq (0, 0)$. (Even if $M$ and $N$ were assigned values at the origin, $M_y$ and $N_x$ would not exist at the origin.) Since the region enclosed by $C_1$ and $C_2$ contains the point where $M_y$ and $N_x$ do not exist, we do not have $M_y = N_x$ for all points between the curves, so Theorem 1 does not guarantee that the values of the line integrals of **F** over the two curves will be the same.

FIGURE 18.4.6

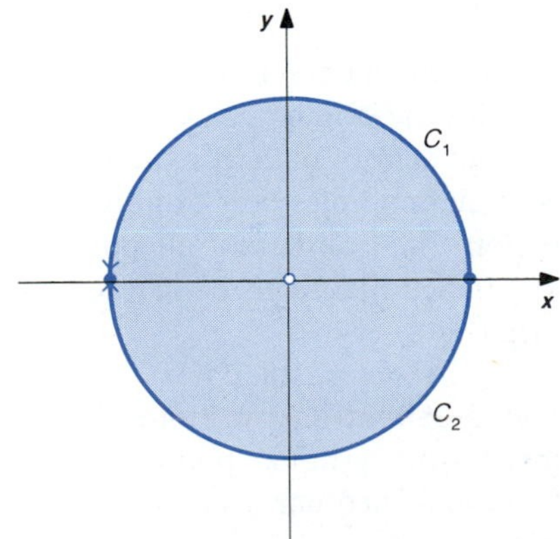

FIGURE 18.4.7

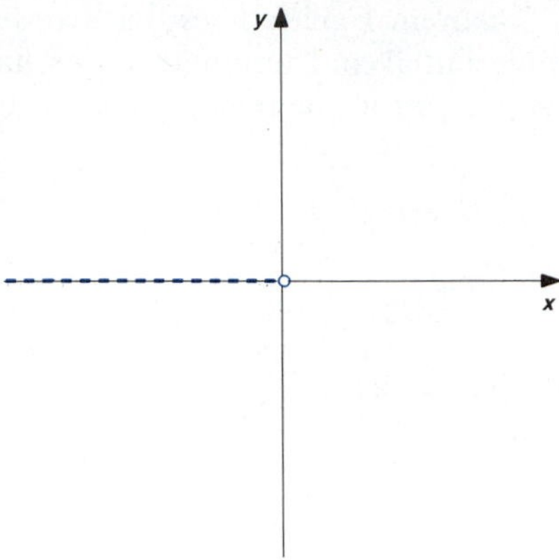


(a) We have

$$\int_{C_1} M\,dx + N\,dy$$

$$= \int_0^{\pi} \left[ \left( \frac{-\sin t}{\cos^2 t + \sin^2 t} \right)(-\sin t) + \left( \frac{\cos t}{\cos^2 t + \sin^2 t} \right)(\cos t) \right] dt$$

$$= \int_0^{\pi} 1\,dt = \pi.$$

(b) Similarly,

$$\int_{C_2} M\,dx + N\,dy$$

$$= \int_0^{\pi} \left[ \left( \frac{-(-\sin t)}{\cos^2 t + \sin^2 t} \right)(-\sin t) + \left( \frac{\cos t}{\cos^2 t + \sin^2 t} \right)(-\cos t) \right] dt$$

$$= \int_0^{\pi} -1\,dt = -\pi.$$ ■

The line integral in Example 4 is not independent of path in regions that contain the origin, a point where $M_y$ and $N_x$ do not exist. The integral is independent of path in regions that do not contain the origin and do not contain any holes. For example, the integral is independent of path in the region that consists of all points in the $xy$-plane except for $(x, 0)$ with $x \leq 0$. See Figure 18.4.7. Closed curves in this region cannot encircle the origin.

We have seen that if $M_y$ and $N_x$ are continuous with $M_y = N_x$ in a region $R$ that has no holes, then $\int_C M\,dx + N\,dy$ depends only on the initial and terminal points of $C$, as long as $C$ is contained in $R$. The following theorem gives a different characterization of independence of path.

■ *Theorem 2*

**Suppose $M$ and $N$ are continuous in a region $R$.**

**(a) If there is a function $U$ such that $\nabla U = M\mathbf{i} + N\mathbf{j}$ in $R$, then $\int_C M\,dx + N\,dy$ is independent of path in $R$. Moreover, if $C$ is a smooth curve in $R$ with initial point $(x(a), y(a))$ and terminal point $(x(b), y(b))$, then**

$$\int_C M\,dx + N\,dy = U(x(b), y(b)) - U(x(a), y(b)).$$

**(b) If $\int_C M\,dx + N\,dy$ is independent of path in $R$, then there is a function $U$ such that $\nabla U = M\mathbf{i} + N\mathbf{j}$ in $R$.**

■ *Proof* To prove part (a) of this theorem, let us consider a smooth curve

$$C\colon\ x = x(t), \quad y = y(t), \qquad a \le t \le b,$$

where $C$ is contained in $R$. Then $U = U(x, y) = U(x(t), y(t))$ is a function of $t$, and the Chain Rule gives

$$\frac{dU}{dt} = \frac{\partial U}{\partial x}\frac{dx}{dt} + \frac{\partial U}{\partial y}\frac{dy}{dt}.$$

Since $\nabla U = (\partial U/\partial x)\mathbf{i} + (\partial U/\partial y)\mathbf{j} = M\mathbf{i} + N\mathbf{j}$, we have $M = \partial U/\partial x$ and $N = \partial U/\partial y$. Then

$$\int_C M\,dx + N\,dy = \int_a^b \left(\frac{\partial U}{\partial x}\frac{dx}{dt} + \frac{\partial U}{\partial y}\frac{dy}{dt}\right) dt$$

$$= \int_a^b \frac{dU}{dt}\,dt = U(x(b), y(b)) - U(x(a), y(a)),$$

where the last equality is a consequence of the Fundamental Theorem of Calculus.

To verify part (b) of the theorem, we assume that $\int_C M\,dx + N\,dy$ is independent of path in $R$. For a fixed point $(x_0, y_0)$, we define a function $U$ by

$$U(x, y) = \int_{C(x,y)} M\,dx + N\,dy,$$

where $C(x, y)$ is any curve in $R$ from $(x_0, y_0)$ to $(x, y)$. We are assuming that the line integral is independent of path, so the value $U(x, y)$ does not depend on the path of $C(x, y)$ from $(x_0, y_0)$ to $(x, y)$, as long as $C(x, y)$ is in $R$. If $R$ has components that cannot be connected to $(x_0, y_0)$ by a path in $R$, then we must treat each such component separately. Each curve in $R$ must remain in a single component of $R$. See Figure 18.4.8.

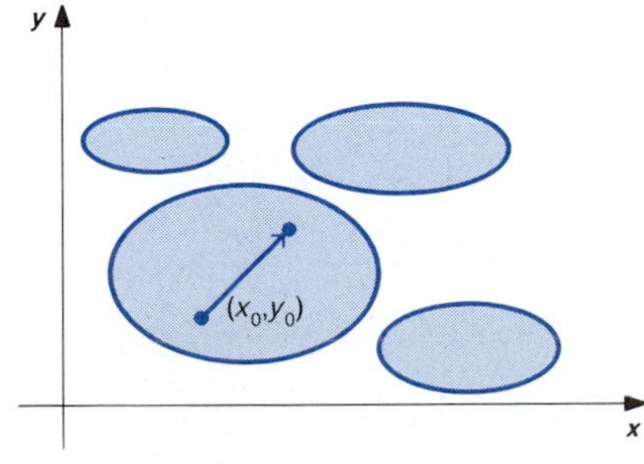

FIGURE 18.4.8

Let us show that $U_x(c, d) = M(c, d)$. The partial derivative $U_x(c, d)$ depends only on the values of $U$ on the horizontal line $y = d$. These values can be determined by choosing $C(x, d)$ to be any path from $(x_0, y_0)$ to a

FIGURE 18.4.9

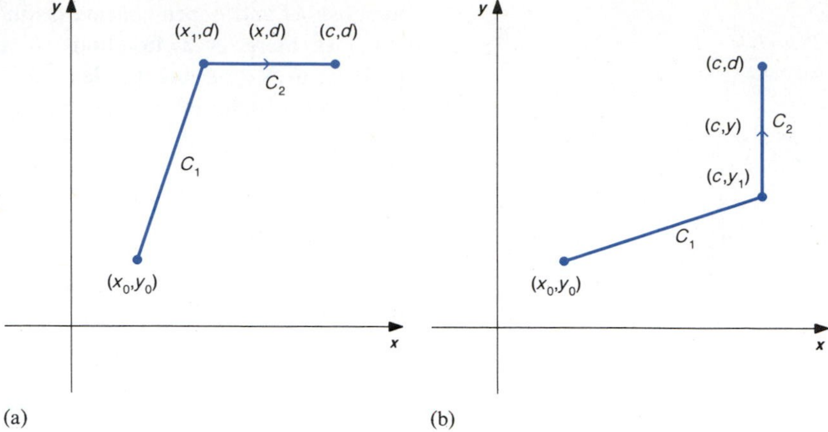


(a) (b)

fixed point $(x_1, d)$, and then along the line $y = d$ from $(x_1, d)$ to $(x, d)$. See Figure 18.4.9a. Then

$$
\begin{aligned}
U(x, d) &= \int_{C(x,d)} M\,dx + N\,dy \\
&= \int_{C_1} M\,dx + N\,dy + \int_{C_2} M\,dx + N\,dy \\
&= \int_{C_1} M\,dx + N\,dy + \int_{x_1}^{x} M(t, d)\,dt.
\end{aligned}
$$

The first term above is a constant. The Fundamental Theorem of Calculus tells us the derivative of the second term with respect to $x$ is $M(x, d)$. It follows that $U_x(x, d) = M(x, d)$, so $U_x(c, d) = M(c, d)$.

We can show that $U_y = N$ similarly, except we use the path indicated in Figure 18.4.9b.

We conclude that $M\mathbf{i} + N\mathbf{j} = U_x\mathbf{i} + U_y\mathbf{j} = \nabla U$. ■

If there is a function $U$ with $\nabla U = \mathbf{F}$, we say that $\mathbf{F}$ is a **conservative** vector field. The negative of $U$ is called the **potential** of $\mathbf{F}$.

We know from Section 18.2 that if $\mathbf{F}$ is an external force that is the only force acting on an object, then the value of $\int_C \mathbf{F}\cdot\mathbf{T}\,ds$ is the change in kinetic energy of the object as it moves along $C$. That is, if $C$ has initial point $A$ and terminal point $B$, then

$$\int_C \mathbf{F}\cdot\mathbf{T}\,ds = \text{K.E.}(B) - \text{K.E.}(A),$$

where K.E. denotes kinetic energy. If $\mathbf{F}$ is *conservative* with potential function $P$, then Theorem 2 implies

$$\int_C \mathbf{F}\cdot\mathbf{T}\,ds = -P(B) + P(A).$$

It follows that the value of line integral is the change of **potential energy** of the object. Combining results, we have

$$\text{K.E.}(B) - K(A) = -P(B) + P(A), \quad \text{so}$$

$$\text{K.E.}(B) + P(B) = \text{K.E.}(A) + P(A).$$

The last equality tells us that the sum of the kinetic energy and the potential energy is constant if **F** is a conservative force field. That is, energy is conserved in a conservative force field. Many electrical, magnetic, and gravitational fields are conservative.

Theorem 2 tells us that $M\mathbf{i} + N\mathbf{j}$ is the gradient of a function $U$ in $R$ if and only if $\int_C M\,dx + N\,dy$ is independent of path in $R$. This raises two questions:

*For a given vector function* **F**, *how do we know when there is a function U with* $\nabla U = \mathbf{F}$?

*If such a function U exists, how do we find it?*

We can combine Theorems 1 and 2 to answer the first question. That is, if $M_y$ and $N_x$ are continuous in a region $R$ that has no holes and $M_y = N_x$ in $R$, then Theorem 1 tells us that $\int_C M\,dx + N\,dy$ is independent of path in $R$. Theorem 2 then implies there is a function $U$ such that $\nabla U = \mathbf{F}$.

We might also note that if $M_y$ and $N_x$ are continuous in $R$, and if $\nabla U = \mathbf{F}$, then $M_y = N_x$. To see this, note that $U_x = M$ implies $U_{xy} = M_y$ and that $U_y = N$ implies $U_{yx} = N_x$. Since $U_{xy}$ and $U_{yx}$ are continuous, $U_{xy} = U_{yx}$, so $M_y = N_x$.

Let us summarize the above remarks as a Theorem.

*Theorem 3*

**If $M_y$ and $N_x$ are continuous in a region $R$ that contains no holes, and $M_y = N_x$ in $R$, then there is a function $U$ such that $\nabla U = \mathbf{F}$ in $R$, so F is conservative in $R$.**

**If $M_y$ and $N_x$ are continuous in $R$ and there is a point of $R$ where $M_y \neq N_x$, then there is no function $U$ such that $\nabla U = F$ in $R$, so $F$ is not conservative in $R$.**

Let us now turn to the second question above. Suppose that the vector function $\mathbf{F} = M\mathbf{i} + N\mathbf{j}$ has continuous first partial derivatives with $M_y = N_x$ in a region $R$ that contains no holes. We then know that there is a function $U$ such that the gradient of $U$ is **F** in $R$. We want to determine a formula for $U(x, y)$. From the proof of Theorem 2b we know that, if $C(x, y)$ is any smooth curve in $R$ with initial point $(x_0, y_0)$ and terminal point $(x, y)$, then we can take $U$ to be

$$U(x, y) = \int_{C(x,y)} M\,dx + N\,dy. \tag{1}$$

This means that we can determine values of $U$ by evaluating a line integral. We will illustrate a different technique for finding a function $U$ with

$\nabla U = \mathbf{F}$. If $\int_C M\,dx + N\,dy$ is not independent of path in $R$, the method we will illustrate does not determine a function $U$; formula (1) would determine a function $U$, but $U$ would not satisfy $\nabla U = \mathbf{F}$ at all points of $R$.

**EXAMPLE 5**

Determine whether there is a function $U$ such that $\nabla U = -x\mathbf{i} + y\mathbf{j}$ for all $(x, y)$. If there is, find such a function $U$.

**SOLUTION**

$\mathbf{F} = -x\mathbf{i} + y\mathbf{j}$ is the function of Example 2. The functions $M_y$ and $N_x$ are continuous with $M_y = N_x = 0$ in the entire $xy$-plane. The entire plane does not have any holes, so Theorem 3 implies that there is a function $U$ such that $\nabla U = \mathbf{F}$ in the entire $xy$-plane. We need a formula for $U(x, y)$.

If $\nabla U = -x\mathbf{i} + y\mathbf{j}$, then $U_x = -x$. It follows that

$$U(x, y) = -\frac{x^2}{2} + h(y) \qquad \text{for some function } h.$$

Note that the partial derivative $U_x$ is found by holding $y$ constant and differentiating with respect to $x$. Thus, we find $U$ from $U_x$ by holding $y$ constant while we integrate $U_x$ with respect to $x$. Also, since $y$ is considered to be a constant, the "constant" of integration can depend on $y$. We have used $h(y)$ to denote the "constant" of integration.

From the above formula for $U$, we have

$$U_y = 0 + h'(y).$$

Also, $\nabla U = -x\mathbf{i} + y\mathbf{j}$ implies that $U_y = y$. Combining results, we obtain

$$h'(y) = y, \quad \text{so} \quad h(y) = \frac{y^2}{2} + C.$$

Substitution of this expression for $h(y)$ in the previous formula for $U$ gives us

$$U(x, y) = -\frac{x^2}{2} + \frac{y^2}{2} + C.$$

This function satisfies $\nabla U = -x\mathbf{i} + y\mathbf{j}$ for any choice of the constant $C$. ■

**EXAMPLE 6**

Determine whether there is a function $U$ such that

$$\nabla U = \frac{-y}{x^2 + y^2}\mathbf{i} + \frac{x}{x^2 + y^2}\mathbf{j} \qquad \text{in the region } y > 0.$$

If there is, find such a function.

**SOLUTION**

$$\mathbf{F} = \frac{-y}{x^2 + y^2}\mathbf{i} + \frac{x}{x^2 + y^2}\mathbf{j}$$

is the function of Example 4. We have already verified that $M_y = N_x$ for $(x, y) \neq (0, 0)$, so we have $M_y = N_x$ for every point in the region $y > 0$. Since this region contains no holes, Theorem 3 implies there is a function $U$ with $\nabla U = \mathbf{F}$. We need a formula for $U(x, y)$.

$$U_x = \frac{-y}{x^2 + y^2} \quad \text{implies} \quad U(x, y) = -\tan^{-1}\left(\frac{x}{y}\right) + h(y), \qquad y > 0.$$

Differentiating the above formula with respect to $y$, we obtain

$$U_y = \left(-\frac{1}{1 + \left(\dfrac{x}{y}\right)^2}\right)\left(-\frac{x}{y^2}\right) + h'(y)$$

$$= \frac{x}{x^2 + y^2} + h'(y).$$

Since we want $U_y = N$, we obtain the equation

$$\frac{x}{x^2 + y^2} + h'(y) = \frac{x}{x^2 + y^2}, \quad \text{so} \quad h'(y) = 0.$$

Then $h(y) = C$, so

$$U(x, y) = -\tan^{-1}\left(\frac{x}{y}\right) + C, \qquad y > 0.$$

This function satisfies

$$\nabla U = \frac{-y}{x^2 + y^2}\mathbf{i} + \frac{x}{x^2 + y^2}\mathbf{j}$$

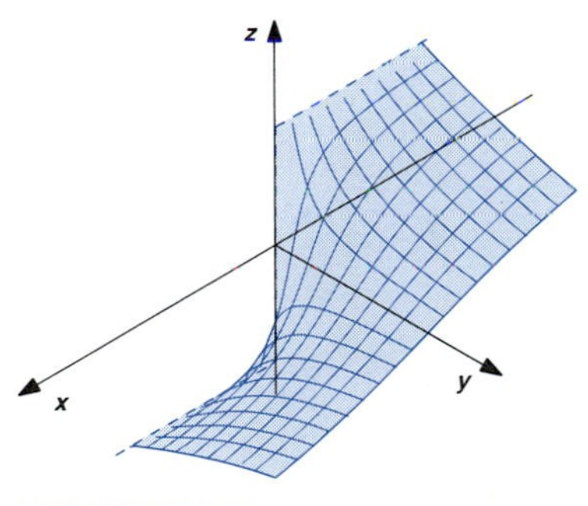

FIGURE 18.4.10

for every choice of the constant $C$. The graph of $U$ with $C = 0$ is sketched in Figure 18.4.10. ■

We cannot find a function $U$ with $\nabla U$ equal to the function $\mathbf{F}$ in Example 6 for all $(x, y) \neq (0, 0)$. If we could, then line integrals of $\mathbf{F}$ around circles with center at the origin would be zero, but they are not. For example, it follows from Example 4 that the line integral of $\mathbf{F}$ counterclockwise around the circle $x^2 + y^2 = 1$ has value $2\pi$.

If $\mathbf{F}$ is conservative in $R$, so $\int_C M\,dx + N\,dy$ is *independent of path* in $R$, we can use the notation

$$\int_A^B M\,dx + N\,dy$$

to denote the integral over any curve $C$ in $R$ that has initial point $A$ and terminal point $B$. The integral can be evaluated either by choosing $C$ to be any convenient curve with initial point $A$ and terminal point $B$, or by finding a function $U$ with $\nabla U = \mathbf{F}$ and using the formula

$$\int_A^B M\,dx + N\,dy = U(B) - U(A).$$

## ■ EXAMPLE 7

Show that the integral $\int_C (2x + y)\,dx + (x - 3y^2)\,dy$ is independent of path in the entire $xy$-plane and evaluate

$$\int_{(0,0)}^{(3,2)} (2x + y)\,dx + (x - 3y^2)\,dy.$$

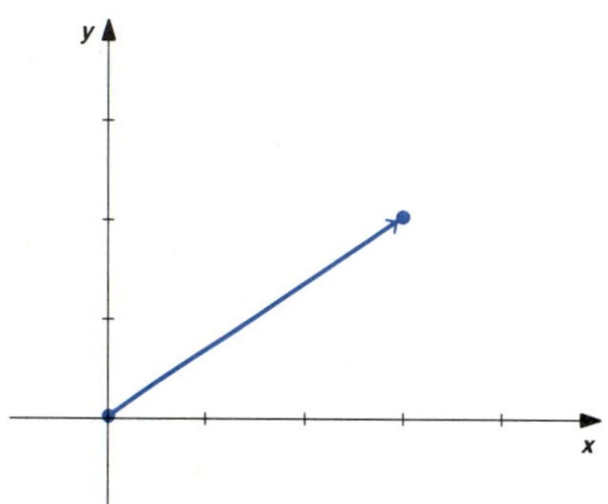


FIGURE 18.4.11

### SOLUTION

We have $M = 2x + y$ and $N = x - 3y^2$, so $M_y = 1 = N_x$ for all $(x, y)$. It follows that the integral is independent of path in the entire plane.

Let us evaluate the line integral by choosing $C$ to be the line segment from (0, 0) to (3, 2). See Figure 18.4.11. Expressing $C$ in parametric form, we have

$$C: \quad x = 3t, \quad y = 2t, \qquad 0 < t < 1.$$

Then

$$\begin{aligned}\int_{(0,0)}^{(3,2)} (2x + y)\,dx + (x - 3y^2)\,dy &= \int_C (2x + y)\,dx + (x - 3y^2)\,dy \\ &= \int_0^1 \{[2(3t) + (2t)](3) + [(3t) - 3(2t)^2](2)\}\,dt \\ &= \int_0^1 (30t - 24t^2)\,dt \\ &= 15t^2 - 8t^3 \Big]_0^1 = 7.\end{aligned}$$

We could also evaluate the line integral by finding a function $U$ such that $\nabla U = (2x + y)\mathbf{i} + (x - 3y^2)\mathbf{j}$. Using the technique of Examples 5 and 6, we have:

$$U_x = 2x + y \quad \text{implies} \quad U = x^2 + xy + h(y).$$

Taking the partial derivative of $U$ with respect to $y$ then gives

$$U_y = x + h'(y).$$

Then $U_y = x - 3y^2$ implies

$$x + h'(y) = x - 3y^2,$$
$$h'(y) = -3y^2,$$
$$h(y) = -y^3 + C.$$

It follows that

$$U(x, y) = x^2 + xy - y^3 + C$$

satisfies $\nabla U = (2x + y)\mathbf{i} + (x - 3y^2)\mathbf{j}$ for any choice of the constant $C$. Let us choose $C = 0$. Then

$$\int_{(0,0)}^{(3,2)} y^2\,dx + 2xy\,dy = U(3, 2) - U(0, 0)$$
$$= [(3)^2 + (3)(2) - (2)^3] - [0] = 7.$$

This agrees with the value obtained earlier. ■

It is worth noting what happens if we try to use the method illustrated above to find $U$ with $\nabla U = \mathbf{F}$, for a vector field $\mathbf{F}$ with $M_y \neq N_x$. Let us try the method on the function $\mathbf{F} = y\mathbf{i} - x\mathbf{j}$ of Example 1. Then

$$U_x = y \quad \text{implies} \quad U(x, y) = yx + h(y), \quad \text{so} \quad U_y = x + h'(y).$$

On the other hand, $U_y = -x$. We conclude that

$$x + h'(y) = -x, \quad \text{so} \quad h'(y) = -2x.$$

The last equality does not make sense, because the left-hand side depends only on $y$ and the right-hand side depends on $x$. The method does not determine a function $U$ if the formula for $h'(y)$ depends on $x$. This indicates that the vector field is not conservative.

## EXERCISES 18.4

Evaluate the line integrals in Exercises 1–14.

1 $\int_C xy\,dx + x^2\,dy$

(a) $C$: $x = t^2$, $y = t$, $0 \le t \le 2$
(b) $C$: $x = 2t$, $y = t$, $0 \le t \le 2$

2 $\int_C y^2\,dx + x^2\,dy$

(a) $C$: $x = t$, $y = t^3$, $0 \le t \le 2$
(b) $C$: $x = t$, $y = 4t$, $0 \le t \le 2$

3 $\int_C x^2\,dx + y^2\,dy$

(a) $C$: $x = \cos t$, $y = \sin t$, $0 \le t \le \pi$
(b) $C$: $x = \cos t$, $y = -\sin t$, $0 \le t \le \pi$

4 $\int_C x\,dx - y\,dy$

(a) $C$: $x = \cos t$, $y = \sin t$, $0 \le t \le \pi$
(b) $C$: $x = \cos t$, $y = -\sin t$, $0 \le t \le \pi$

**5** $\int_C (x^2 + y^2)\,dx + 2xy\,dy$
(a) $C$: $x = \cos t$, $y = \sin t$, $0 \le t \le \pi$
(b) $C$: $x = \cos t$, $y = -\sin t$, $0 \le t \le \pi$

**6** $\int_C (x^2 + y^2)\,dx + 2xy\,dy$
(a) $C$: $x = -2\cos t$, $y = 2\sin t$, $0 \le t \le \pi$
(b) $C$: $x = t$, $y = 4 - t^2$, $-2 \le t \le 2$

**7** $\int_C dx + x\,dy$
(a) $C$: Polygonal path from $(0, 0)$ to $(1, 1)$, to $(0, 2)$
(b) $C$: Along the $y$-axis from $(0, 0)$ to $(0, 2)$

**8** $\int_C y\,dx - x\,dy$
(a) $C$: Polygonal path from $(0, 0)$ to $(1, 1)$, to $(2, 0)$
(b) $C$: Along the $x$-axis from $(0, 0)$ to $(2, 0)$

**9** $\int_C (x^2 - y^2)\,dx + (y^2 - 2xy)\,dy$
(a) $C$: Polygonal path from $(0, 0)$ to $(3, 2)$, to $(3, 0)$
(b) $C$: Polygonal path from $(0, 0)$ to $(3, 2)$, to $(3, 0)$, to $(0, 0)$

**10** $\int_C (x + y)\,dx + (x - y)\,dy$
(a) $C$: Polygonal path from $(0, 0)$ to $(3, 0)$, to $(3, 3)$, to $(0, 3)$
(b) $C$: Polygonal path from $(0, 0)$ to $(3, 0)$, to $(3, 3)$, to $(0, 3)$, to $(0, 0)$

**11** $\int_C (y^2 + x)\,dx + (2xy + y)\,dy$
(a) $C$: Polygonal path from $(0, 0)$ to $(4, 0)$, to $(4, 2)$, to $(0, 2)$
(b) $C$: Polygonal path from $(0, 0)$ to $(4, 0)$, to $(4, 2)$, to $(0, 2)$, to $(0, 0)$

**12** $\int_C (2xy + x^2)\,dx + (x^2 - y^2)\,dy$
(a) $C$: Polygonal path from $(0, 0)$ to $(2, 0)$, to $(2, 2)$
(b) $C$: Polygonal path from $(0, 0)$ to $(2, 0)$, to $(2, 2)$, to $(0, 0)$

**13** $\int_C \frac{-y}{x^2 + y^2}\,dx + \frac{x}{x^2 + y^2}\,dy$
(a) $C$: $x = 3 + \sin t$, $y = -2\cos t$, $0 \le t \le \pi$
(b) $C$: $x = 3 + \sin t$, $y = -2\cos t$, $0 \le t \le 2\pi$

**14** $\int_C \ln y\,dx + x/y\,dy$
(a) $C$: $x = 2\cos t$, $y = 2 + \sin t$, $0 \le t \le \pi$
(b) $C$: $x = 2\cos t$, $y = 2 + \sin t$, $0 \le t \le 2\pi$

In Exercises 15–22 determine whether there is a function $U$ such that $\nabla U$ is equal to the given vector function $\mathbf{F}$. If there is, find such a function $U$.

**15** $\mathbf{F} = 2xy\mathbf{i} + x^2\mathbf{j}$

**16** $\mathbf{F} = 2xy\mathbf{i} + (x^2 - y^2)\mathbf{j}$

**17** $\mathbf{F} = (2x - y)\mathbf{i} + (x - 2y)\mathbf{j}$

**18** $\mathbf{F} = (x - y)\mathbf{i} + (x + y)\mathbf{j}$

**19** $\mathbf{F} = 2e^y\mathbf{i} + (2xe^y + y)\mathbf{j}$

**20** $\mathbf{F} = (x^2 + xy)\mathbf{i} + \left(\frac{x^2}{2} + y^2\right)\mathbf{j}$

**21** $\mathbf{F} = \frac{2x}{x^2 + y^2}\mathbf{i} + \frac{2y}{x^2 + y^2}\mathbf{j}$, $(x, y) \ne (0, 0)$

**22** $\mathbf{F} = 3x\sqrt{x^2 + y^2}\,\mathbf{i} + 3y\sqrt{x^2 + y^2}\,\mathbf{j}$, $(x, y) \ne (0, 0)$

In Exercises 23–28, show that each integral given is independent of path in the entire $xy$-plane and find its value.

**23** $\int_{(0,0)}^{(2,-1)} 2xy^2\,dx + 2x^2y\,dy$

**24** $\int_{(0,0)}^{(2,3)} y\,dx + x\,dy$

**25** $\int_{(-3,-4)}^{(4,3)} 2x\,dx + 2y\,dy$

**26** $\int_A^B 2x\,dx + 2y\,dy$, $A$ and $B$ any points on the circle $x^2 + y^2 = r^2$

**27** $\int_{(1,-3)}^{(3,2)} 3x^2\,dx$

**28** $\int_{(1,b)}^{(3,d)} 3x^2\,dx$, $b$ and $d$ any numbers

**29** Show that the line integral $\int_{(a,b)}^{(c,d)} f(x)\,dx$ is independent of path and that $\int_{(a,b)}^{(c,d)} f(x)\,dx = \int_a^c f(x)\,dx$.

Determine whether the integrals given in Exercises 30–32 are independent of path in the indicated regions.

**30** $\int_C \frac{x}{x^2 + y^2}\,dx + \frac{y}{x^2 + y^2}\,dy$
(a) $R$ consists of all points $(x, y)$ with $x > 0$
(b) $R$ consists of all points $(x, y)$ with $x^2 + y^2 \ne 0$

**31** $\int_C \frac{y}{x^2 - 2x + 1 + y^2}\,dx + \frac{1 - x}{x^2 - 2x + 1 + y^2}\,dy$
(a) $R$ consists of all points $(x, y)$ with $y > 0$
(b) $R$ consists of all points $(x, y)$ with $(x, y) \ne (1, 0)$

**32** $\int_C \frac{2xy}{(x^2 + y^2)^2}\,dx + \frac{y^2 - x^2}{(x^2 + y^2)^2}\,dy$
(a) $R$ consists of all points $(x, y)$ with $y > 0$
(b) $R$ consists of all points $(x, y)$ with $x^2 + y^2 \ne 0$

**33** If $\int_C M\,dx + N\,dy$ is independent of path in a rectangular region $R$, show that

$$\int_{(x_0, y_0)}^{(x, y)} M\,dx + N\,dy = \int_{x_0}^{x} M(u, y)\,du + \int_{y_0}^{y} N(x_0, v)\,dv$$

for $(x, y)$ and $(x_0, y_0)$ in $R$. [*Hint*: Consider the polygonal path from $(x_0, y_0)$ to $(x_0, y)$, to $(x, y)$.]

**34** Suppose $M_y = N_x$ in a rectangular region $R$. Show that formula (1) becomes

$$U(x, y) = \int_{x_0}^{x} M(s, y)\,ds + \int_{y_0}^{y} N(x_0, t)\,dt,$$

if $C(x, y)$ is chosen to be the polygonal path from $(x_0, y_0)$ to $(x_0, y)$, to $(x, y)$.

In Exercises 35–37 use the formula in Exercise 34 to find the function $U$ that corresponds to the given vector function $\mathbf{F} = M\mathbf{i} + N\mathbf{j}$ and the given point $(x_0, y_0)$.

**35** $\mathbf{F} = (2xy - y^2)\mathbf{i} + (x^2 - 2xy)\mathbf{j}$; $(0, 0)$

**36** $\mathbf{F} = (3x^2 - 2xy)\mathbf{i} + (2y - x^2)\mathbf{j}$; $(0, 0)$

**37** $\mathbf{F} = ye^{xy}\mathbf{i} + (xe^{xy} + 1)\mathbf{j}$; $(0, 0)$

## 18.5 SOLUTION OF EXACT DIFFERENTIAL EQUATIONS

In this section we will see how to use line integrals to solve certain differential equations of the form

$$M(x, y) + N(x, y)\frac{dy}{dx} = 0.$$

If $M$ and $N$ have continuous first partial derivatives in an open region $R$, and $M_y = N_x$ in $R$, the differential equation

$$M(x, y) + N(x, y)\frac{dy}{dx} = 0 \tag{1}$$

is said to be **exact** in $R$.

If the differential equation (1) is exact in a region that contains no holes, we know there is a differentiable function $U(x, y)$ such that

$$\nabla U = M\mathbf{i} + N\mathbf{j}, \quad \text{so} \quad U_x = M \quad \text{and} \quad U_y = N.$$

If $y = y(x)$ is differentiable with $(x, y(x))$ in $R$, the Chain Rule then gives

$$\frac{d}{dx}(U(x, y(x))) = U_x(x, y(x)) + U_y(x, y(x))\frac{dy}{dx}$$
$$= M(x, y(x)) + N(x, y(x))\frac{dy}{dx}.$$

It follows that, if $y = y(x)$ is any differentiable function that is defined implicitly by the equation

$$U(x, y) = K,$$

where $K$ is a constant, then we can differentiate the equation with respect to $x$ to obtain

$$M(x, y) + N(x, y)\frac{dy}{dx} = 0,$$

so $y = y(x)$ is a solution of the differential equation (1). On the other hand, if $y = y(x)$ is any solution of (1) with $(x, x(y))$ in $R$, we have

$$M(x, y(x)) + N(x, y(x))\frac{dy}{dx} = 0.$$

Then $U_x = M$ and $U_y = N$ imply

$$U_x(x, y(x)) + U_y(x, y(x))\frac{dy}{dx} = 0, \quad \text{so}$$

$$\frac{d}{dx}(U(x, y(x))) = 0, \quad \text{and}$$

$$U(x, y(x)) = K, \qquad K \text{ a constant.}$$

That is, each solution of (1) is given implicitly by an equation

$$U(x, y) = K.$$

Let us summarize these results.

*Theorem 1*

**If the differential equation**

$$M(x, y) + N(x, y)\frac{dy}{dx} = 0$$

**is exact in a rectangular region $R$, then there is a differentiable function $U(x, y)$ such that $U_x = M$ and $U_y = N$ in $R$. If $y = y(x)$ is defined implicitly by the equation**

$$U(x, y) = K,$$

**then $y = y(x)$ is a solution of the differential equation. If $y = y(x)$ is any solution with $(x, y(x))$ in $R$, then $y$ satisfies the equation**

$$U(x, y) = K$$

**for some constant $K$.**

If $U_x = M$ and $U_y = N$ in a rectangle $R$, then any solution of the exact differential equation

$$M + N\frac{dy}{dx} = 0$$

must satisfy the equation

$$U(x, y) = K, \qquad K \text{ a constant.}$$

We can find a solution that satisfies an **initial condition**

$$y(x_0) = y_0$$

by substituting $(x_0, y_0)$ into the equation $U(x, y) = K$ to obtain $K = U(x_0, y_0)$. There may be points $(x_0, y_0)$ in $R$ for which the equation

$$U(x, y) = U(x_0, y_0)$$

does not define $y$ as a function of $x$. This means that the initial value problem may not have a solution $y = y(x)$ for every $(x_0, y_0)$ in $R$. Even if there is a solution to the initial value problem, there may be a restriction on the values of $x$ for which the solution is defined.

If the differential equation (1) is exact in a rectangle $R$, then $M_y = N_x$ in $R$ and the function $U$ of Theorem 1 satisfies $\nabla U = M\mathbf{i} + N\mathbf{j}$. It follows that we can use the method of Section 18.4 to find $U$. Also, we know that the line integral

$$\int_C M\,dx + N\,dy$$

is independent of path in $R$ and formula (1) of Section 18.4 says that we can choose

$$U(x, y) = \int_{(x_0, y_0)}^{(x, y)} M\,dx + N\,dy, \tag{2}$$

where $(x_0, y_0)$ is any fixed point in $R$. If we choose $C$ to be the polygonal path from $(x_0, y_0)$, to $(x_0, y)$, to $(x, y)$, we have $C = C_1 + C_2$, where

$$C_1\colon\ x = x_0, \quad y = t, \qquad y_0 \le t \le y, \quad \text{and}$$

$$C_2\colon\ x = s, \quad y = y, \qquad x_0 \le s \le x.$$

See Figure 18.5.1. Then

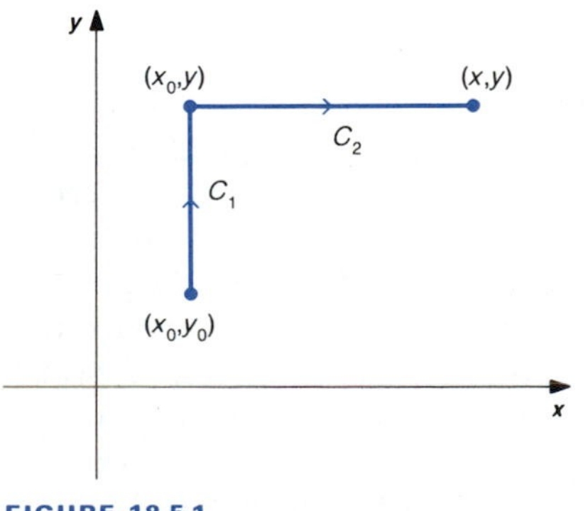

FIGURE 18.5.1

$$\int_{C_1} M\,dx + N\,dy = \int_{y_0}^{y} [M(x_0, t)(0) + N(x_0, t)(1)]\,dt$$
$$= \int_{y_0}^{y} N(x_0, t)\,dt \quad \text{and}$$

$$\int_{C_2} M\,dx + N\,dy = \int_{x_0}^{x} [M(s, y)(1) + N(s, y)(0)]\,ds$$
$$= \int_{x_0}^{x} M(s, y)\,ds.$$

Combining results, we have

$$U(x, y) = \int_{x_0}^{x} M(s, y)\, ds + \int_{y_0}^{y} N(x_0, t)\, dt. \tag{3}$$

We will use formula (3) instead of the method of Section 18.4 to find the function $U$. This is intended to emphasize the role of line integrals in the solution of exact differential equations.

## ■ EXAMPLE 1

Show that the differential equation

$$(2x - 2) + (8y - 16)\frac{dy}{dx} = 0$$

is exact and find the solution.

### SOLUTION

We have $M = 2x - 2$ and $N = 8y - 16$, so $M_y = 0 = N_x$. This means the equation is exact in the entire $xy$-plane. Formula (3) with $(x_0, y_0) = (0, 0)$ gives

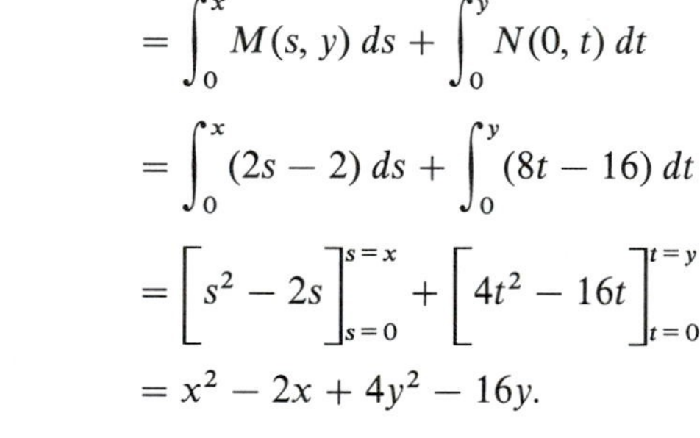

$$\begin{aligned} U(x, y) &= \int_{x_0}^{x} M(s, y)\, ds + \int_{y_0}^{y} N(x_0, t)\, dt \\ &= \int_{0}^{x} M(s, y)\, ds + \int_{0}^{y} N(0, t)\, dt \\ &= \int_{0}^{x} (2s - 2)\, ds + \int_{0}^{y} (8t - 16)\, dt \\ &= \Big[ s^2 - 2s \Big]_{s=0}^{s=x} + \Big[ 4t^2 - 16t \Big]_{t=0}^{t=y} \\ &= x^2 - 2x + 4y^2 - 16y. \end{aligned}$$

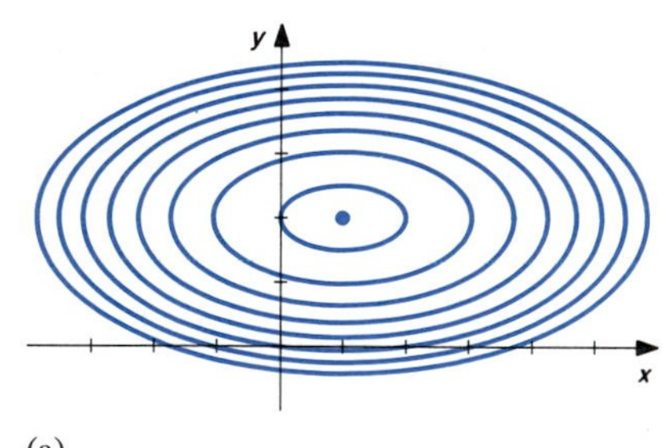


(a)

Each solution of the differential equation is given implicitly by an equation of the form

$$x^2 - 2x + 4y^2 - 16y = K.$$

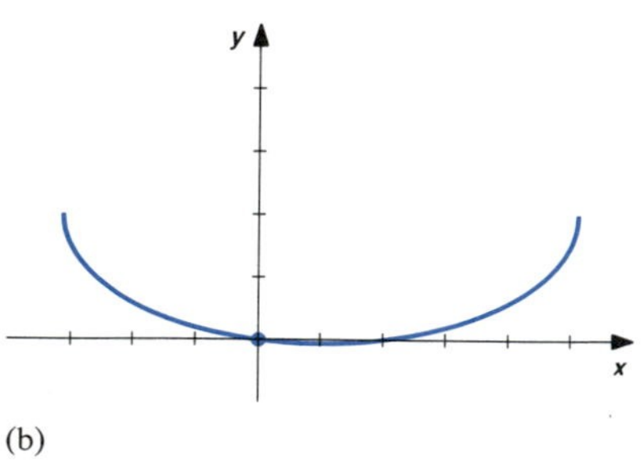


(b)

FIGURE 18.5.2

Graphs of the equation for several choices of the constant $K$ are given in Figure 18.5.2a. The graphs are ellipses with center at the point (1, 2).

None of the ellipses defines $y$ as a function of $x$ in a rectangle with center at a point with $y$-coordinate 2; the differential equation does not have a solution that satisfies an initial condition $y(x_0) = 2$. If $y_0 > 2$, the graph of the solution of the differential equation that satisfies the initial condition $y(x_0) = y_0$ is the upper half of the ellipse that contains the point $(x_0, y_0)$. If $y_0 < 2$, the graph of the solution that satisfies the initial condition $y(x_0) = y_0$ is the lower half of the ellipse that contains the point $(x_0, y_0)$. For example, the graph of the solution that satisfies the initial condition

$y(0) = 0$ is the lower half of the ellipse

$$x^2 - 2x + 4y^2 - 16y = 0.$$

See Figure 18.5.2b. There is no solution of the differential equation that is defined for all $x$. ■

The differential equation in Example 1 is a separable differential equation. That is, it is of the form

$$M(x)\,dx + N(y)\,dy = 0.$$

Separable differentiable equations were introduced in Chapter 9. Separable differential equations are exact, because

$$\frac{\partial}{\partial y}(M(x)) = 0 = \frac{\partial}{\partial x}(N(y)).$$

Let us now consider an exact differential equation that is not separable.

■ **EXAMPLE 2**

Show that the differential equation

$$(4xe^{y/2} + 4) + x^2e^{y/2}\frac{dy}{dx} = 0$$

is exact and find the solution.

**SOLUTION**

We have $M = 4xe^{y/2} + 4$ and $N = x^2e^{y/2}$, so $M_y = 4xe^{y/2}(1/2)$ and $N_x = 2xe^{y/2}$. Since $M_y = N_x$, the equation is exact in the entire $xy$-plane. Let us use $(x_0, y_0) = (0, 0)$ in formula (3). Then

$$\begin{aligned} U(x, y) &= \int_{x_0}^{x} M(s, y)\,ds + \int_{y_0}^{y} N(x_0, t)\,dt \\ &= \int_{0}^{x} M(s, y)\,ds + \int_{0}^{y} N(0, t)\,dt \\ &= \int_{0}^{x} (4se^{y/2} + 4)\,ds + \int_{0}^{y} (0)^2e^{t/2}\,dt \\ &= 2x^2e^{y/2} + 4x. \end{aligned}$$

The solution is given by the equation

$$2x^2e^{y/2} + 4x = K.$$

Graphs of the equation for several choices of the constant $K$ are given in Figure 18.5.3. Note that the equation does not define $y$ as a function of $x$ in any rectangle that has center with $x$-coordinate 0. ■

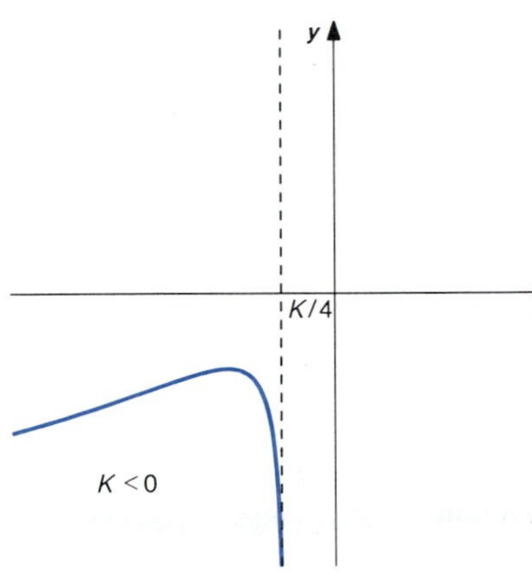


(a)

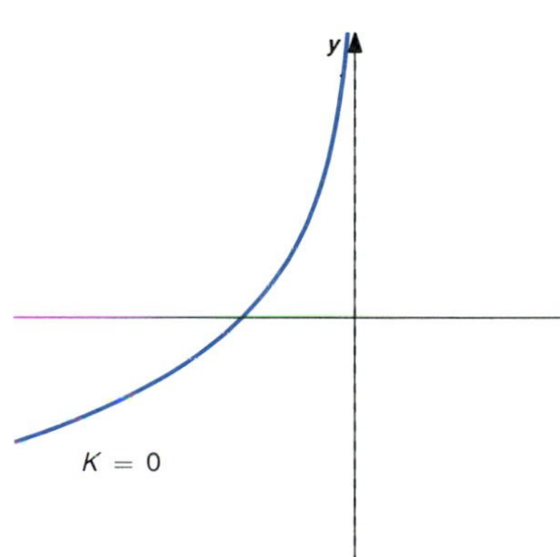


(b)

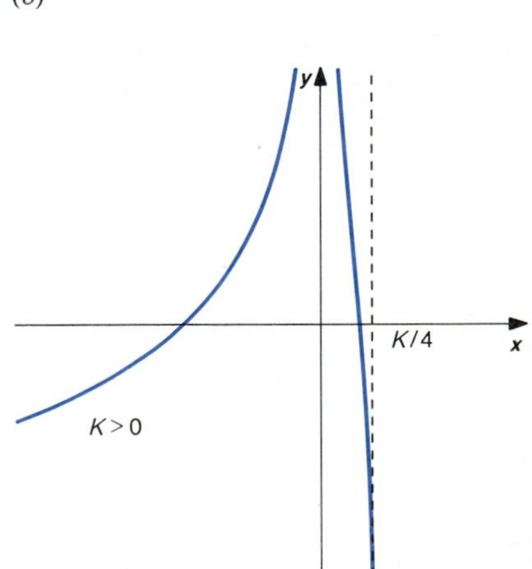


(c)

**FIGURE 18.5.3**

It is common to write the exact differential equation

$$M(x, y) + N(x, y)\frac{dy}{dx} = 0$$

in the form

$$M\,dx + N\,dy = 0.$$

This corresponds to the solution obtained by the line integral

$$\int_{(x_0, y_0)}^{(x, y)} M\,dx + N\,dy.$$

### EXAMPLE 3

Find the solution of the initial value problem

$$\frac{x}{x^2 + y^2}\,dx + \frac{y}{x^2 + y^2}\,dy = 0, \qquad y(3) = 4,$$

and sketch the graph of the solution.

**SOLUTION**

We have

$$M = \frac{x}{x^2 + y^2} \quad \text{and} \quad N = \frac{y}{x^2 + y^2}.$$

For $(x, y) \neq (0, 0)$,

$$M_y = -\frac{2xy}{(x^2 + y^2)^2} = N_x.$$

The differential equation is exact in the rectangular region that consists of, say, $y > 0$. This region contains the initial point (3, 4). Formula (3) with $(x_0, y_0) = (3, 4)$ gives

$$\begin{aligned}
U(x, y) &= \int_{x_0}^{x} M(s, y)\,ds + \int_{y_0}^{y} N(x_0, t)\,dt \\
&= \int_{3}^{x} M(s, y)\,ds + \int_{4}^{y} N(3, t)\,dt \\
&= \int_{3}^{x} \frac{s}{s^2 + y^2}\,ds + \int_{4}^{y} \frac{t}{3^2 + t^2}\,dt \\
&= \frac{1}{2}\ln(x^2 + y^2) - \frac{1}{2}\ln(3^2 + y^2) + \frac{1}{2}\ln(3^2 + y^2) - \frac{1}{2}\ln(3^2 + 4^2) \\
&= \frac{1}{2}\ln(x^2 + y^2) - \frac{1}{2}\ln 25.
\end{aligned}$$

The solution of the differential equation must satisfy the equation

$$\frac{1}{2}\ln(x^2+y^2)-\frac{1}{2}\ln 25=K$$

for some constant $K$. Substitution of the initial values (3, 4) gives $K=0$, so the solution of the initial value problem is given by

$$\frac{1}{2}\ln(x^2+y^2)-\frac{1}{2}\ln 25=0.$$

We can simplify the equation by writing

$$\ln(x^2+y^2)=\ln 25,$$

$$x^2+y^2=25.$$

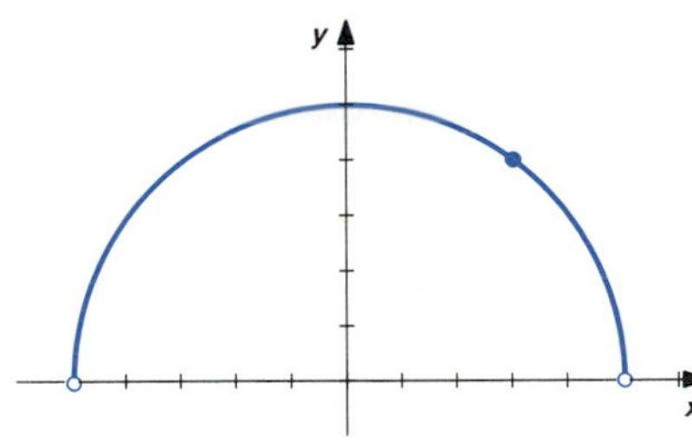

FIGURE 18.5.4

The differentiable function $y=y(x)$ with $y(3)=4$ defined by this equation is

$$y=\sqrt{25-x^2},$$

which is defined and differentiable for $-5<x<5$. The graph of the solution is given in Figure 18.5.4. ■

## EXERCISES 18.5

Show that the differential equations in Exercises 1–12 are exact and find their solutions.

1 $x+y\frac{dy}{dx}=0$

2 $\frac{1}{y}-\frac{x}{y^2}\frac{dy}{dx}=0,\ y\neq 0$

3 $\frac{1}{x}+\frac{1}{y}\frac{dy}{dx}=0,\ x>0,\ y>0$

4 $\frac{y}{x}+\ln x\frac{dy}{dx}=0$

5 $(2x+y-1)+(x-6y+5)\frac{dy}{dx}=0$

6 $(x+2y+3)+(2x-4y+1)\frac{dy}{dx}=0$

7 $(3x^2-y)\,dx+(2y-x)\,dy=0$

8 $(e^x\sin y+x)\,dx+(e^x\cos y+y)\,dy=0$

9 $(ye^x+x)\,dx+(e^x+y)\,dy=0$

10 $(ye^{xy}+1)\,dx+(xe^{xy}+1)\,dy=0$

11 $\cos x\cos^2 y\,dx-2\sin x\sin y\cos y\,dy=0$

12 $\tan y+x\sec^2 y\,dy=0$

Solve the initial value problems in Exercises 13–16 and sketch the graphs of the solutions.

13 $y+x\frac{dy}{dx}=0,\ y(1)=2$

14 $\frac{x}{(x^2+y^2)^{3/2}}+\frac{y}{(x^2+y^2)^{3/2}}\frac{dy}{dx}=0,\ y(0)=2$

15 $\frac{-y}{x^2+y^2}\,dx+\frac{x}{x^2+y^2}\,dy=0,\ y(1)=\sqrt{3}$

16 $(2y-2x)\,dx+2x\,dy=0$; (a) $y(1)=2$, (b) $y(-1)=3$, (c) $y(3)=1$

## 18.6 SURFACE INTEGRALS

Riemann sums of a function $f$ over a surface $S$ in three-dimensional space are obtained by dividing the surface into small pieces with surface areas $\Delta S$. See Figure 18.6.1. We then form the sum

$$\sum f \,\Delta S,$$

FIGURE 18.6.1

where $f$ is evaluated at a point on the corresponding small piece of the surface. The limit of these Riemann sums in the usual sense is called the **surface integral** of $f$ over $S$ and is denoted

$$\iint_S f \, dS,$$

where $dS$ denotes **differential surface area**.

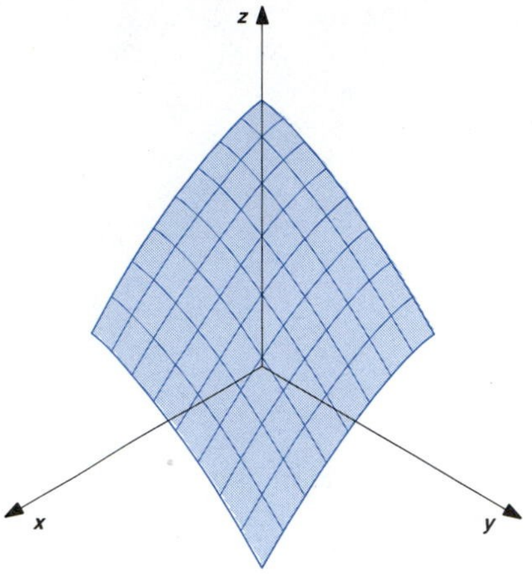

$$\iint_S dS \qquad \text{is the } \textbf{surface area} \text{ of } S.$$

In some cases we can express surface integrals as double integrals over a plane region. We assume the functions involved in the integral are continuous, so the integrals exist.

**If the surface $S$ is given by $z = z(x, y)$, $(x, y)$ in a region $R_{x,y}$ of the $xy$-plane, then**

$$\iint_S f \, dS = \iint_{R_{x,y}} f(x, y, z(x, y))\sqrt{(z_x)^2 + (z_y)^2 + 1} \, dA. \tag{1}$$

**$R_{x,y}$ is the projection of the surface on the $xy$-plane and $dA$ represents differential area in the $xy$-plane. See Figure 18.6.2.**

FIGURE 18.6.2

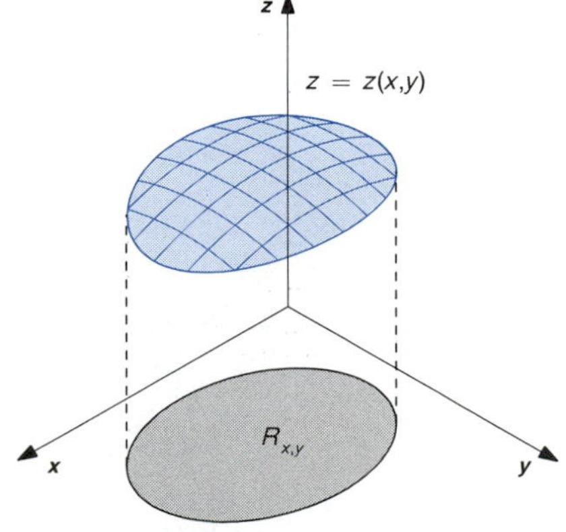

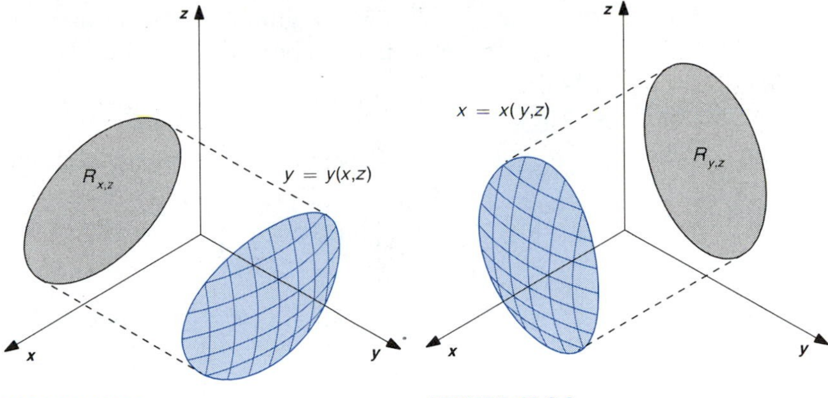


FIGURE 18.6.3

FIGURE 18.6.4

**If the surface $S$ is given by $y = y(x, z)$, $(x, z)$ in a region $R_{x,z}$ of the $xz$-plane, then**

$$\iint_S f\,dS = \iint_{R_{x,z}} f(x, y(x, z), z)\sqrt{(y_x)^2 + 1 + (y_z)^2}\,dA. \tag{2}$$

**$R_{x,z}$ is the projection of the surface on the $xz$-plane and $dA$ represents differential area in the $xz$-plane. See Figure 18.6.3.**

**If the surface $S$ is given by $x = x(y, z)$, $(y, z)$ in a region $R_{y,z}$ of the $yz$-plane, then**

$$\iint_S f\,dS = \iint_{R_{y,z}} f(x(y, z), y, z)\sqrt{1 + (x_y)^2 + (x_z)^2}\,dA. \tag{3}$$

**$R(y, z)$ is the projection of the surface on the $yz$-plane and $dA$ represents differential area in the $yz$-plane. See Figure 18.6.4.**

■ **EXAMPLE 1**

Use each of formulas (1), (2), and (3) to evaluate $\iint_S x\,dS$, where $S$ is that part of the plane $x + 2y + 3z = 6$ that is in the first octant.

**SOLUTION**

The surface is sketched in Figure 18.6.5.

(a) Let us express the surface as

$$z = \frac{6 - x - 2y}{3}, \quad \text{so} \quad z_x = -1/3, \quad z_y = -2/3, \quad \text{and}$$

$$\sqrt{(z_x)^2 + (z_y)^2 + 1} = \sqrt{\frac{1}{9} + \frac{4}{9} + 1} = \sqrt{\frac{14}{9}} = \frac{\sqrt{14}}{3}.$$

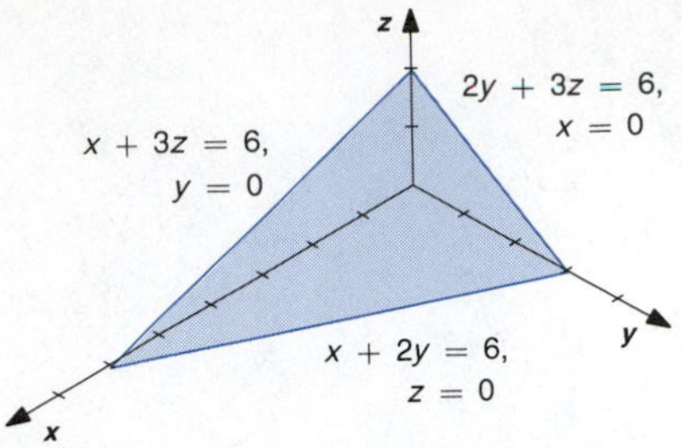

FIGURE 18.6.5

From Figure 18.6.5 we see that we are interested in that part of the surface that is above the region in the $xy$-plane bounded by

$$x + 2y = 6, \quad y = 0, \quad \text{and} \quad x = 0.$$

Substituting into formula (1), we obtain

$$\iint_S x\,dS = \int_0^6 \int_0^{(6-x)/2} x\left(\frac{\sqrt{14}}{3}\right) dy\,dx$$

$$= \int_0^6 \frac{\sqrt{14}}{3} xy \Big]_{y=0}^{y=(6-x)/2} dx$$

$$= \int_0^6 \sqrt{14}\left(x - \frac{x^2}{6}\right) dx$$

$$= \sqrt{14}\left[\frac{x^2}{2} - \frac{x^3}{18}\right]_0^6 = 6\sqrt{14}.$$

(b) Let us express the surface as

$$y = \frac{6 - x - 3z}{2}, \quad \text{so}$$

$$y_x = -1/2, \quad y_z = -3/2, \quad \text{and}$$

$$\sqrt{(y_x)^2 + 1 + (y_z)^2} = \sqrt{\frac{1}{4} + 1 + \frac{9}{4}} = \sqrt{\frac{14}{4}} = \frac{\sqrt{14}}{2}.$$

In this case we integrate over the region in the $xz$-plane bounded by

$$x + 3z = 6, \quad z = 0, \quad \text{and} \quad x = 0.$$

Substituting into formula (2), we obtain

$$\iint_S x\,dS = \int_0^6 \int_0^{(6-x)/3} x\left(\frac{\sqrt{14}}{2}\right) dz\,dx$$

$$= \int_0^6 \frac{\sqrt{14}}{2} xz \Big]_{z=0}^{z=(6-x)/3} dx$$

$$= \int_0^6 \sqrt{14}\left(x - \frac{x^2}{6}\right) dx = 6\sqrt{14}.$$

(c) If we express the surface as

$$x = 6 - 2y - 3z,$$

we have $x_y = -2$, $x_z = -3$, and

$$\sqrt{1 + (x_y)^2 + (x_z)^2} = \sqrt{1 + 4 + 9} = \sqrt{14}.$$

We integrate over the region in the $yz$-plane bounded by

$2y + 3z = 6, \quad y = 0, \quad \text{and} \quad z = 0.$

Substituting into formula (3), we obtain

$$\begin{aligned}\iint_S x\, dS &= \int_0^3 \int_0^{(6-2y)/3} (6 - 2y - 3z)(\sqrt{14})\, dz\, dy \\ &= \int_0^3 \sqrt{14}\left[6z - 2yz - \frac{3z^2}{2}\right]_{z=0}^{z=(6-2y)/3} dx \\ &= \int_0^3 \sqrt{14}\left[2(6-2y) - \frac{2y(6-2y)}{3} - \frac{(6-2y)^2}{6}\right] dy \\ &= \int_0^3 \sqrt{14}\left(12 - 4y - 4y + \frac{4y^2}{3} - 6 + 4y - \frac{2y^2}{3}\right) dy \\ &= \int_0^3 \sqrt{14}\left(6 - 4y + \frac{2y^2}{3}\right) dy \\ &= \sqrt{14}\left[6y - 2y^2 + \frac{2y^3}{9}\right]_0^3 = 6\sqrt{14}.\end{aligned}$$ ■

The three integrals in Example 1 have the same value, since they represent different ways of evaluating the same surface integral.

If a thin surface $S$ has density per unit area at a point $(x, y, z)$ on the surface given by a continuous function $\rho(x, y, z)$, the **differential mass** is

$$dM = \rho\, dS.$$

and the **total mass** is the sum

$$M = \iint_S dM = \iint_S \rho\, dS.$$

The coordinates of the **center of mass** are given by the formulas

$$\bar{x} = \frac{\iint_S x\, dM}{\iint_S dM}, \quad \bar{y} = \frac{\iint_S y\, dM}{\iint_S dM}, \quad \bar{z} = \frac{\iint_S z\, dM}{\iint_S dM}.$$

The **moment of inertia about the $z$-axis** is

$$I_{z\text{-axis}} = \iint_S (x^2 + y^2)\, dM.$$

■ **EXAMPLE 2**

The surface that consists of that part of the cone $z = \sqrt{x^2 + y^2}$ with $x^2 + y^2 \le 1$ has density per unit area given by $\rho(x, y, z) = 1 - z$. Find its mass, center of mass, and its moment of inertia about the $z$-axis.

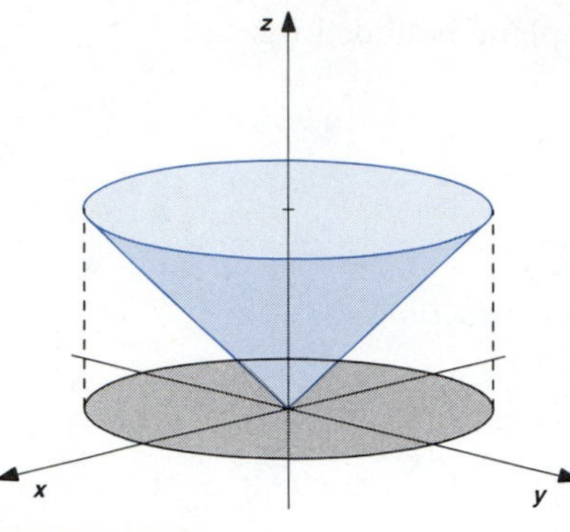


FIGURE 18.6.6

**SOLUTION**

The cone is sketched in Figure 18.6.6. The surface is given by the equation

$z = \sqrt{x^2 + y^2}$, so

$$z_x = \frac{1}{2\sqrt{x^2 + y^2}}(2x) = \frac{x}{\sqrt{x^2 + y^2}} \quad \text{and}$$

$$z_y = \frac{1}{2\sqrt{x^2 + y^2}}(2y) = \frac{y}{\sqrt{x^2 + y^2}}.$$

The partial derivatives of $z$ do not exist for $(x, y) = (0, 0)$. For $(x, y) \neq (0, 0)$, we have

$$(z_x)^2 + (z_y)^2 + 1 = \frac{x^2}{x^2 + y^2} + \frac{y^2}{x^2 + y^2} + 1 = \frac{x^2 + y^2}{x^2 + y^2} + 1 = 2.$$

The region of integration $R_{x,y}$ in the $xy$-plane is bounded by $x^2 + y^2 = 1$. [The formulas for surface integrals that we are using require that the partial derivatives of $z(x, y)$ be continuous in the region $R_{x,y}$. Since the partial derivatives of $z$ do not exist at the origin, we should integrate over the region defined by $a^2 \leq x^2 + y^2 \leq 1$ and then take the limit as $a$ approaches zero from the right. Since the surface area of that part of the cone above $x^2 + y^2 \leq a^2$ approaches zero as $a$ approaches zero, this would give the same result as that obtained by applying the formulas directly to the region $x^2 + y^2 \leq 1$, which we will do.] The mass is

$$M = \iint_S dM = \iint_S \rho \, dS$$

$$= \iint_S (1 - z) \, dS$$

$$= \iint_{R_{x,y}} (1 - z)\sqrt{(z_x)^2 + (z_y)^2 + 1} \, dA$$

$$= \iint_{R_{x,y}} (1 - \sqrt{x^2 + y^2})\sqrt{2} \, dA,$$

which in polar coordinates is

$$= \int_0^{2\pi} \int_0^1 (1 - r)\sqrt{2}\, r \, dr \, d\theta$$

$$= \int_0^{2\pi} \sqrt{2}\left[\frac{r^2}{2} - \frac{r^3}{3}\right]_{r=0}^{r=1} d\theta$$

$$= \sqrt{2}\left(\frac{1}{2} - \frac{1}{3}\right)(2\pi) = \frac{\pi\sqrt{2}}{3}.$$

The first moment about the $xy$-plane is

$$\begin{aligned}\iint_S z\, dM &= \iint_S z\rho\, dS \\ &= \iint_S z(1-z)\, dS \\ &= \int_0^{2\pi}\int_0^1 r(1-r)\sqrt{2}\, r\, dr\, d\theta \\ &= \int_0^{2\pi} \sqrt{2}\left[\frac{r^3}{3}-\frac{r^4}{4}\right]_{r=0}^{r=1} d\theta \\ &= \sqrt{2}\left(\frac{1}{3}-\frac{1}{4}\right)(2\pi) = \frac{\pi\sqrt{2}}{6}.\end{aligned}$$

The $z$-coordinate of the center of mass is then

$$\bar{z} = \frac{\iint_S z\, dM}{\iint_S dM} = \frac{\dfrac{\pi\sqrt{2}}{6}}{\dfrac{\pi\sqrt{2}}{3}} = \left(\frac{\pi\sqrt{2}}{6}\right)\left(\frac{3}{\pi\sqrt{2}}\right) = \frac{1}{2}.$$

By symmetry of the surface and the density function, we have $\bar{x} = \bar{y} = 0$. The center of mass is at the point $(0, 0, 1/2)$. Note that this point is not on the surface. It is inside the region enclosed by the cone, as we should expect it to be.

The moment of inertia about the $z$-axis is

$$\begin{aligned}I_{z\text{-axis}} &= \iint_S (x^2+y^2)\, dM \\ &= \iint_S (x^2+y^2)(1-z)\, dS \\ &= \int_0^{2\pi}\int_0^1 r^2(1-r)\sqrt{2}\, r\, dr\, d\theta \\ &= \int_0^{2\pi} \sqrt{2}\left[\frac{r^4}{4}-\frac{r^5}{5}\right]_{r=0}^{r=1} d\theta \\ &= \sqrt{2}\left(\frac{1}{4}-\frac{1}{5}\right)(2\pi) = \frac{\pi\sqrt{2}}{10}.\end{aligned}$$

■

In Section 18.2 we used line integrals to measure the flux of a two-dimensional vector field across a plane curve. Let us now see how to use surface integrals to measure the flux of a three-dimensional vector field across a surface in three-dimensional space. We will measure the flux of a vector field **F** across a surface $S$ in the direction of a unit normal vector **n**. We assume that the direction of **n** varies continuously over the surface.

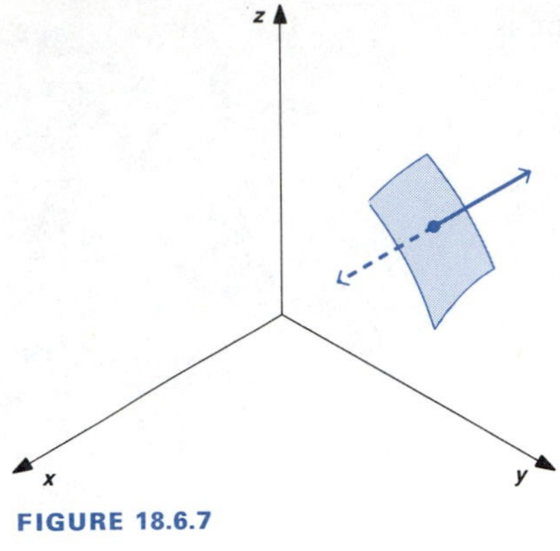


FIGURE 18.6.7

At each point on a surface $S$ there will be two unit normal vectors, one the negative of the other, on opposite sides of the surface. See Figure 18.6.7. For a given surface $S$, we need to establish which of the two unit normal vectors at each point that we are using to measure the flux of **F**. For surfaces that tend to enclose a region, we can describe the direction of the unit normal **n** as either *inward* or *outward*. Spheres, cones, paraboloids, and cylinders are of this type. If the direction of **n** is to vary continuously, then **n** must be chosen inward at each point or outward at each point. See Figure 18.6.8. For surfaces of the form $z = z(x, y)$, $(x, y)$ in a region $R_{x,y}$ of the $xy$-plane, we can describe the direction of **n** as either *upward* or *downward*. The unit normal **n** is upward whenever the $z$-coordinate of **n** is positive; **n** is downward whenever the $z$-coordinate of **n** is negative. If the direction of **n** is to vary continuously, then **n** must be chosen either upward at each point or downward at each point. See Figure 18.6.9.

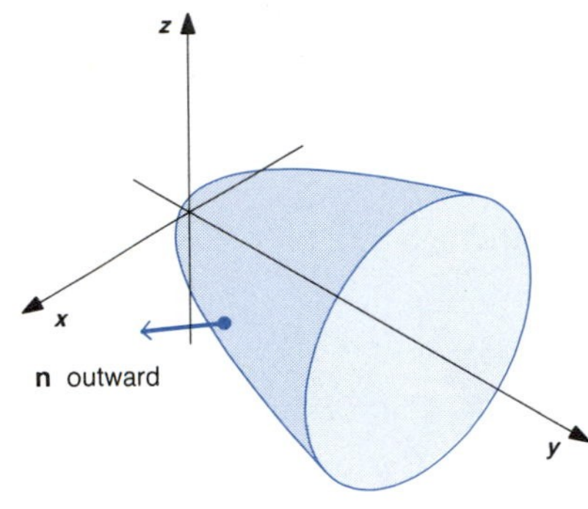


FIGURE 18.6.8

Let $\mathbf{n} = \cos\alpha\,\mathbf{i} + \cos\beta\,\mathbf{j} + \cos\gamma\,\mathbf{k}$ denote a unit vector that is normal to the surface $S$; the components are the direction cosines of **n**. The **flux** of a vector function $\mathbf{F} = M\mathbf{i} + N\mathbf{j} + P\mathbf{k}$ across the surface $S$ in the direction of **n** is

$$\iint_S \mathbf{F}\cdot\mathbf{n}\,dS = \iint_S (M\cos\alpha + N\cos\beta + P\cos\gamma)\,dS. \tag{4}$$

If the surface is appropriate, we can use either formula (1), (2), or (3) to evaluate each of the three terms of the surface integral.

If the surface is given by $z = z(x, y)$, $(x, y)$ in a region $R_{x,y}$ of the $xy$-plane, the upward unit normal to the surface is

$$\mathbf{n} = \frac{-z_x\mathbf{i} - z_y\mathbf{j} + \mathbf{k}}{\sqrt{(z_x)^2 + (z_y)^2 + 1}}.$$

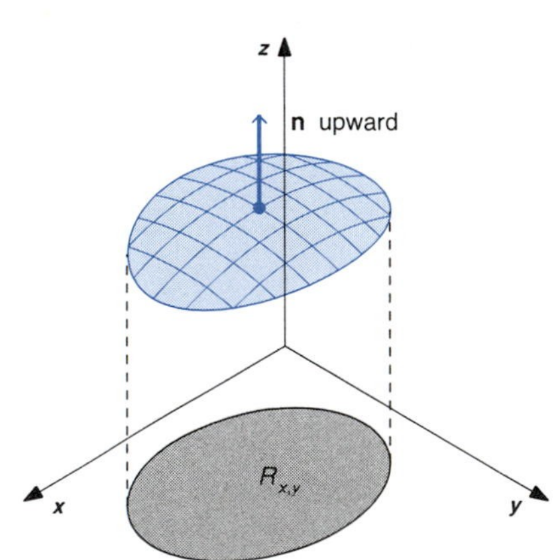


FIGURE 18.6.9

The direction cosines of **n** are

$$\cos\alpha = -\frac{z_x}{\sqrt{(z_x)^2 + (z_y)^2 + 1}},$$

$$\cos\beta = -\frac{z_y}{\sqrt{(z_x)^2 + (z_y)^2 + 1}},$$

$$\cos\gamma = \frac{1}{\sqrt{(z_x)^2 + (z_y)^2 + 1}}.$$

The differential surface area is

$$dS = \sqrt{(z_x)^2 + (z_y)^2 + 1}\,dA = \sec\gamma\,dA$$

and formula (1) gives

$$\iint_S M \cos \alpha \, dS = \iint_{R_{x,y}} -Mz_x \, dA, \tag{5}$$

$$\iint_S N \cos \beta \, dS = \iint_{R_{x,y}} -Nz_y \, dA, \tag{6}$$

$$\iint_S P \cos \gamma \, dS = \iint_{R_{x,y}} P \, dA. \tag{7}$$

The flux of a vector function $\mathbf{F} = M\mathbf{i} + N\mathbf{j} + P\mathbf{k}$ across the surface $z = z(x, y)$, $(x, y)$ in a region $R_{x,y}$ of the $xy$-plane in the direction of the upward unit normal vector $\mathbf{n}$ can then be expressed as

$$\iint_S \mathbf{F} \cdot \mathbf{n} \, dS = \iint_{R_{x,y}} (-Mz_x - Nz_y + P) \, dA. \tag{8}$$

Since the downward unit normal is the negative of the upward unit normal, the flux in the direction of the downward unit vector is given by the negative of the expression in (8).

■ **EXAMPLE 3**

Evaluate $\iint_S \mathbf{F} \cdot \mathbf{n} \, dS$, where $\mathbf{F} = x\mathbf{i} + y\mathbf{j} + z\mathbf{k}$, the surface $S$ is that part of the paraboloid $z = 4 - x^2 - y^2$ with $z \geq 3$, and $\mathbf{n}$ is directed upward.

**SOLUTION**

The paraboloid is sketched in Figure 18.6.10.

Let us use formula (8) to evaluate the integral. We have

$$z = 4 - x^2 - y^2, \quad \text{so}$$

$$z_x = -2x \quad \text{and} \quad z_y = -2y.$$

We have $\mathbf{F} = x\mathbf{i} + y\mathbf{j} + z\mathbf{k}$, so

$$M = x, \quad N = y, \quad \text{and} \quad P = z = 4 - x^2 - y^2.$$

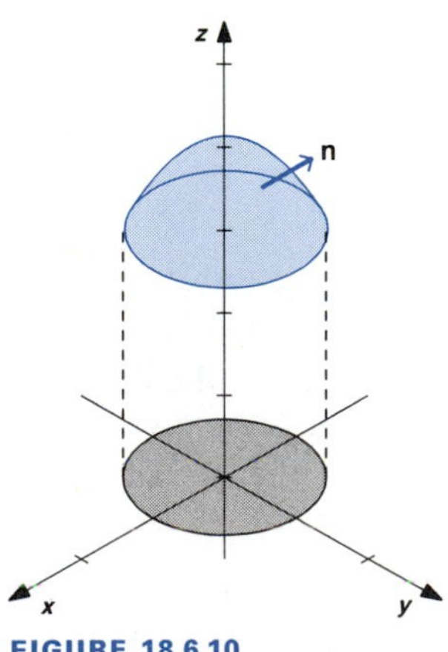

FIGURE 18.6.10

Note that $z$ has been expressed in terms of $x$ and $y$. That part of the parabola $z = 4 - x^2 - y^2$ with $z \geq 3$ satisfies $x^2 + y^2 \leq 1$. This inequality determines the region $R_{x,y}$. The unit normal vector that is given is upward, as in formula (8). Substitution into formula (8) then gives

$$\iint_S \mathbf{F}\cdot\mathbf{n}\,dS = \iint_{R_{x,y}} (-Mz_x - Nz_y + P)\,dA$$

$$= \iint_{R_{x,y}} [-(x)(-2x) - (y)(-2y) + (4 - x^2 - y^2)]\,dA$$

$$= \iint_{R_{x,y}} (4 + x^2 + y^2)\,dA.$$

At this point we notice that the double integral is well suited to polar coordinates. We then have

$$\iint_S \mathbf{F}\cdot\mathbf{n}\,dS = \int_0^{2\pi}\int_0^1 (4 + r^2)r\,dr\,d\theta$$

$$= \int_0^{2\pi}\int_0^1 (4r + r^3)\,dr\,d\theta$$

$$= \int_0^{2\pi}\left[2r^2 + \frac{r^4}{4}\right]_0^1 d\theta$$

$$= \int_0^{2\pi}\left(2 + \frac{1}{4}\right)d\theta = \left(\frac{9}{4}\right)(2\pi) = \frac{9\pi}{2}.$$

■

■ **EXAMPLE 4**

Evaluate the flux of $\mathbf{F} = x\mathbf{i} + y\mathbf{j} + z\mathbf{k}$ across that part of the surface $z = 3$ with $x^2 + y^2 \le 1$, in the direction of the downward unit normal vector.

**SOLUTION**

The surface $S$ is sketched in Figure 18.6.11. The flux of $\mathbf{F}$ across the surface $S$ in the direction of the unit normal $\mathbf{n}$ is

$$\iint_S \mathbf{F}\cdot\mathbf{n}\,dS.$$

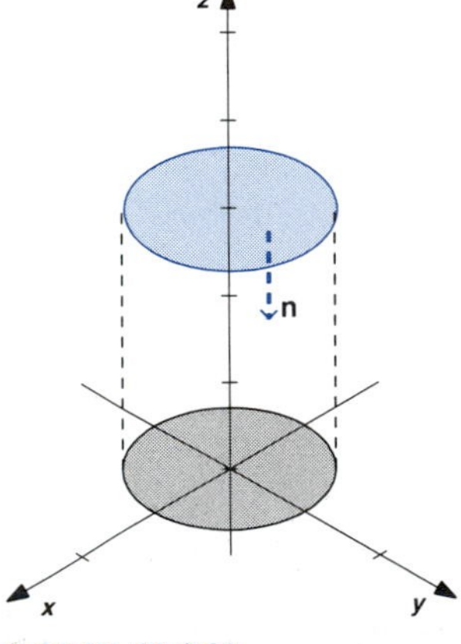

FIGURE 18.6.11

The surface $z = 3$ is of the form $z = z(x, y)$. We are interested in that part of the surface above the region $R_{x,y}$ in the $xy$-plane bounded by $x^2 + y^2 = 1$. We can use formula (8) to evaluate the integral that gives the flux, but since $\mathbf{n}$ is downward, we must change the sign of each term of the integral on the right of the equation.

$z = 3$ implies $z_x = z_y = 0$.

$\mathbf{F} = x\mathbf{i} + y\mathbf{j} + z\mathbf{k}$ implies $M = x$, $N = y$, and $P = z$.

From formula (8), with a change of sign, we have

$$\iint_S \mathbf{F}\cdot\mathbf{n}\,dS = \iint_{R_{x,y}} (Mz_x + Nz_y - P)\,dA$$

$$= \iint_{R_{x,y}} [(x)(0) + (y)(0) - z]\,dA.$$

Since $z = 3$ for points on the surface, the above integral is

$$= \iint_{R_{x,y}} -3\,dA,$$

which, since $-3$ is a *constant*,

$$= -3(\text{Area of } R_{x,y}) = -3\pi.$$ ■

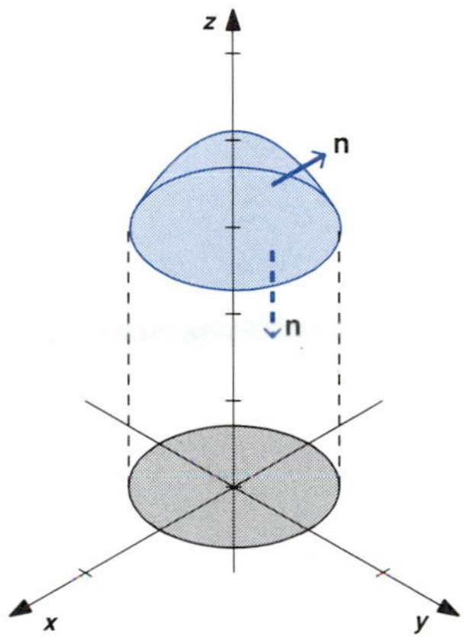

FIGURE 18.6.12

The two surfaces in Examples 3 and 4 form the complete boundary of the region bounded by the paraboloid $z = 4 - x^2 - y^2$ and the plane $z = 3$. The normal is directed outward from the region on each of the boundary surfaces. See Figure 18.6.12. Example 3 tells us that the function $\mathbf{F} = x\mathbf{i} + y\mathbf{j} + z\mathbf{k}$ has an outward flux of $9\pi/2$ from the region across the paraboloid and Example 4 tells us that $\mathbf{F}$ has an outward flux of $-3\pi$ across the plane surface. That is, the flux is out of the region on the paraboloid and into the region on the plane. The net flux out of the region is the algebraic sum of the outward flux across each boundary surface,

$$\left(\frac{9\pi}{2}\right) + (-3\pi) = \frac{3\pi}{2}.$$

■ **EXAMPLE 5**

Find the outward flux of

$$\mathbf{F} = \frac{x\mathbf{i} + y\mathbf{j} + z\mathbf{k}}{(x^2 + y^2 + z^2)^{3/2}}$$

across the sphere $x^2 + y^2 + z^2 = R^2$.

**SOLUTION**

The flux is the integral

$$\iint_S \mathbf{F}\cdot\mathbf{n}\,dS,$$

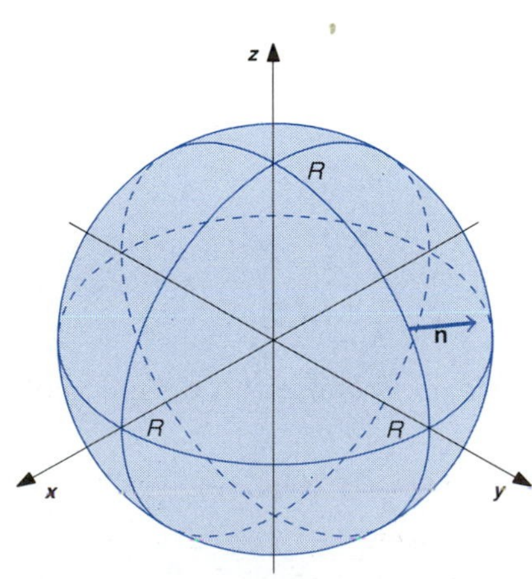

FIGURE 18.6.13

where $\mathbf{n}$ is the outward unit normal to the sphere. See Figure 18.6.13. Let us evaluate the integral by finding $\mathbf{n}$ and then using formula (4).

We write the equation of the sphere in the form $G(x, y, z) = 0$, where $G(x, y, z) = x^2 + y^2 + z^2 - R^2$. We then know that the gradient of $G$ is normal to the surface. We have

$$\nabla G = G_x\mathbf{i} + G_y\mathbf{j} + G_z\mathbf{jk} = 2x\mathbf{i} + 2y\mathbf{j} + 2z\mathbf{k}.$$

The two unit normal vectors that are normal to the sphere are

$$\pm\frac{\nabla G}{|\nabla G|} = \pm\frac{2x\mathbf{i} + 2y\mathbf{j} + 2z\mathbf{k}}{\sqrt{(2x)^2 + (2y)^2 + (2z)^2}} = \pm\frac{x\mathbf{i} + y\mathbf{j} + z\mathbf{k}}{\sqrt{x^2 + y^2 + z^2}}.$$

The outward unit normal to the sphere is

$$\mathbf{n} = \frac{x\mathbf{i} + y\mathbf{j} + z\mathbf{k}}{\sqrt{x^2 + y^2 + z^2}}.$$

We then see that

$$\mathbf{F} = \frac{x\mathbf{i} + y\mathbf{j} + z\mathbf{k}}{(x^2 + y^2 + z^2)^{3/2}} \quad \text{implies}$$

$$\mathbf{F}\cdot\mathbf{n} = \frac{(x)(x) + (y)(y) + (z)(z)}{(x^2 + y^2 + z^2)^{1/2}(x^2 + y^2 + z^2)^{3/2}}$$

$$= \frac{1}{x^2 + y^2 + z^2} = \frac{1}{R^2}$$

for all points $(x, y, z)$ on the surface of the sphere. We then have

$$\iint_S \mathbf{F}\cdot\mathbf{n}\,dS = \iint_S \frac{1}{R^2}\,dS,$$

which, since $R$ is a *constant*,

$$= \frac{1}{R^2}\left(\iint_S dS\right)$$

$$= \frac{1}{R^2}\,(\text{Surface area of a sphere of radius } R)$$

$$= \frac{1}{R^2}\,(4\pi R^2) = 4\pi.$$

■

We could express the surface of the sphere in Example 5 as the union of the two surfaces

$z = \sqrt{R^2 - x^2 - y^2}$ (**n** upward) and

$z = -\sqrt{R^2 - x^2 - y^2}$ (**n** downward).

FIGURE 18.6.14

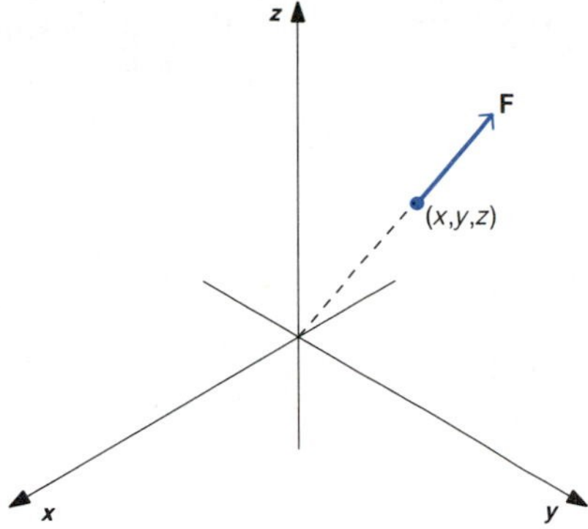


We could then use formula (8) with $R_{x,y}$ the region in the $xy$-plane bounded by $x^2 + y^2 = R^2$ to evaluate the outward flux over each of the two surfaces. [Since the first partial derivatives of the above functions $z$ do not exist for $x^2 + y^2 = R^2$, the integral in formula (8) would be *improper*. However, the improper integrals would be *convergent*. Generally, we can use formula (8) to evaluate surface integrals over hemispheres, even though the resulting integrals are actually improper.]

The vector field $\mathbf{F} = (x\mathbf{i} + y\mathbf{j} + z\mathbf{k})/(x^2 + y^2 + z^2)^{3/2}$ of Example 5 is called an **inverse square field**. The magnitude is inversely proportional to the square of the distance from the origin and the direction of **F** is away from the origin. See Figure 18.6.14. Inverse square fields occur in several physical problems. For example, the force due to gravitational attraction between two objects has magnitude that is inversely proportional to the square of the distance between the objects and is in the direction between the objects. Example 5 shows that the outward flux of the inverse square field across a sphere with center at the origin and any radius is $4\pi$.

## EXERCISES 18.6

Evaluate the surface integrals in Exercises 1–6.

**1** $\iint_S xy\,dS$; $S$ is that part of the plane $z = 2x + 3y$ that is in the first octant and below $z = 6$.

**2** $\iint_S z\,dS$; $S$ is that part of the sphere $x^2 + y^2 + z^2 = 4$ above $z = 1$.

**3** $\iint_S y\,dS$; $S$ is that part of the cylinder $x^2 + y^2 = 4$ between $z = 0$ and $z = 3$ with $y \geq 1$.

**4** $\iint_S x\,dS$; $S$ is that part of the generalized cylinder $y = 1 - x^2$ between $z = 0$ and $z = 3$, in the first octant.

**5** $\iint_S (y + z)\,dS$; $S$ is that part of the cone $x = 2\sqrt{y^2 + z^2}$ with $x \leq 4$, in the first octant.

**6** $\iint_S x\,dS$; $S$ is that part of the cylinder $x = \sin y$ with $0 \leq y \leq \pi$ and $0 \leq z \leq 2$.

Find (a) the mass, (b) the center of mass, and (c) the moment of inertia about the $z$-axis of the surfaces $S$ with density per unit area $\rho$ given in Exercises 7–10.

**7** $S$ is that part of the paraboloid $z = 9 - x^2 - y^2$ with $z \geq 0$; $\rho(x, y, z) = 1/\sqrt{4x^2 + 4y^2 + 1}$.

**8** $S$ is the hemisphere $z = \sqrt{1 - x^2 - y^2}$; $\rho(x, y, z) = z$.

**9** $S$ is that part of the cone $z = \sqrt{x^2 + y^2}$ with $z \leq 2$; $\rho(x, y, z) = x^2 + y^2$.

10 $S$ is that part of the cone $z = \sqrt{x^2 + y^2}$ with $1 \le z \le 2$; $\rho(x, y, z) = 1$.

In Exercises 11–15 evaluate the flux of the given vector function across the given surface for the indicated normal vector.

11 $\mathbf{F} = x\mathbf{i} + y\mathbf{j} + z\mathbf{k}$; $S$ is that part of the plane $x + y + z = 1$ in the first octant; **n** upward.

12 $\mathbf{F} = x\mathbf{i} + y\mathbf{j} + z\mathbf{k}$; $S$ is that part of the paraboloid $z = 4 - x^2 - y^2$ with $0 \le x \le 1$ and $0 \le y \le 1$; **n** outward.

13 $\mathbf{F} = x\mathbf{i} + y\mathbf{j} + z\mathbf{k}$;
(a) $S$ is that part of the cone $z = \sqrt{x^2 + y^2}$ with $z \le 4$; **n** outward.
(b) $S$ is that part of the plane $z = 4$ with $x^2 + y^2 \le 16$; **n** upward.

14 $\mathbf{F} = x\mathbf{i} + y\mathbf{j} + z\mathbf{k}$;
(a) $S$ is that part of the paraboloid $z = 9 - x^2 - y^2$ with $z \ge 0$; **n** outward.
(b) $S$ is that part of the plane $z = 0$ with $x^2 + y^2 \le 9$; **n** downward.

15 $\mathbf{F} = x\mathbf{i} + y\mathbf{j} + z\mathbf{k}$; $S$ is that part of the ellipsoid

$$\frac{x^2}{2^2} + \frac{y^2}{1^2} + \frac{z^2}{1^2} = 1$$

in the first octant; **n** outward. [*Hint*: Use $x(y, z)$.]

16 Find the outward flux of $\mathbf{F} = x\mathbf{i} + y\mathbf{j} + z\mathbf{k}$ across the sphere $x^2 + y^2 + z^2 = R^2$.

17 Find the outward flux of

$$\mathbf{F} = \frac{x\mathbf{i} + y\mathbf{j} + z\mathbf{k}}{x^2 + y^2 + z^2}$$

across the sphere $x^2 + y^2 + z^2 = R^2$.

## 18.7 THE DIVERGENCE THEOREM

Let $D$ be the region in three-dimensional space bounded by the surfaces

$$z = h_2(x, y), \quad z = h_1(x, y), \quad y = g_2(x), \quad y = g_1(x), \quad x = b, \quad \text{and} \quad x = a,$$

and let $R_{x,y}$ be the region in the $xy$-plane bounded by the curves

$$y = g_2(x), \quad y = g_1(x), \quad x = b, \quad \text{and} \quad x = a.$$

Let

$$\mathbf{n} = \cos\alpha\mathbf{i} + \cos\beta\mathbf{j} + \cos\gamma\mathbf{k}$$

denote the outwardly directed unit normal vector to the boundary surfaces of $D$. We have $\cos\gamma = 0$ on the vertical surfaces

$$y = g_2(x), \quad y = g_1(x), \quad x = b, \quad \text{and} \quad x = a.$$

The unit vector **n** is directed upward on the top surface

$$S_2\colon\ z = h_2(x, y), \quad (x, y) \text{ in } R_{x,y}.$$

**n** is directed downward on the bottom surface

$$S_1\colon\ z = h_1(x, y), \quad (x, y) \text{ in } R_{x,y}.$$

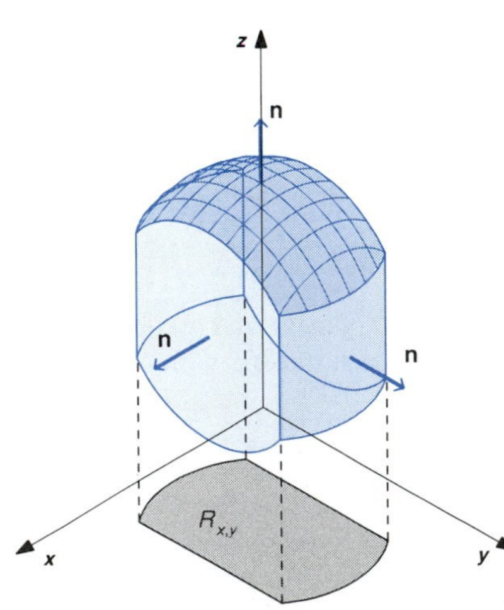

FIGURE 18.7.1

See Figure 18.7.1. If $S$ is the complete boundary of $D$, we can then use formula (7) of Section 18.6 to obtain

$$\begin{aligned}\iint_S P\cos\gamma\,dS &= \iint_{S_2} P\cos\gamma\,dS + \iint_{S_1} P\cos\gamma\,dS\\ &= \iint_{R_{x,y}} P(x,y,h_2(x,y))\,dA - \iint_{R_{x,y}} P(x,y,h_1(x,y))\,dA\\ &= \int_a^b\int_{g_1(x)}^{g_2(x)} [P(x,y,h_2(x,y)) - P(x,y,h_1(x,y))]\,dy\,dx\\ &= \int_a^b\int_{g_1(x)}^{g_2(x)}\left[\int_{h_1(x,y)}^{h_2(x,y)} \frac{\partial P}{\partial z}\,dz\right]dy\,dx\\ &= \iiint_D \frac{\partial P}{\partial z}\,dV.\end{aligned}$$

That is, we have

$$\iint_S P\cos\gamma\,dS = \iiint_D \frac{\partial P}{\partial z}\,dV. \tag{1}$$

If a region is divided into subregions by cutting along a vertical generalized cylinder, the total of the boundaries of the subregions consists of the boundary of the original region and the vertical cutting surface. Since vertical surfaces do not contribute to the integral of $P\cos\gamma$ over a surface, the integral of $P\cos\gamma$ over the boundary of the original region is equal to the sum of the integrals over the boundaries of the subregions. The triple integral of $\partial P/\partial z$ over the region is also equal to the sum of the integrals over the subregions. It follows that formula (1) holds for regions that can be divided by vertical surfaces into a finite number of regions of the type considered above. Let us call such regions **finitely simple in the direction of the $z$-axis**. For example, the region $S$ between the spheres $x^2 + y^2 + z^2 = 4$ and $x^2 + y^2 + z^2 = 1$ can be divided into six regions of the desired type by cutting along the vertical cylinder $x^2 + y^2 = 1$ and along the vertical planes $x = 1$ and $x = -1$, as indicated in Figure 18.7.2.

FIGURE 18.7.2

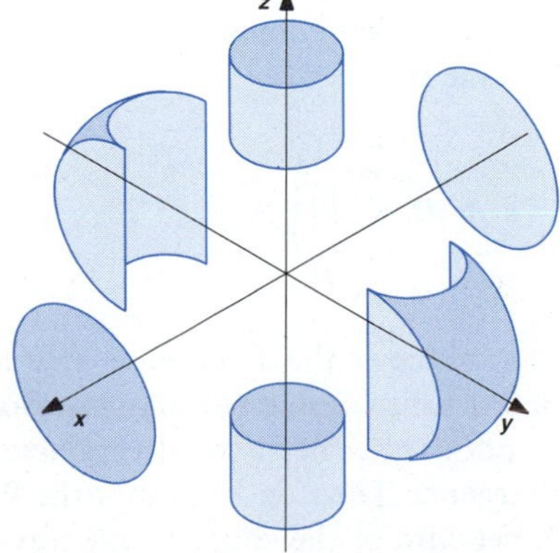

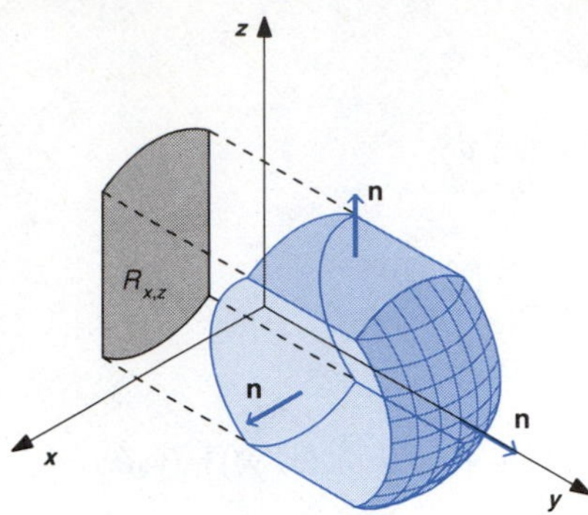
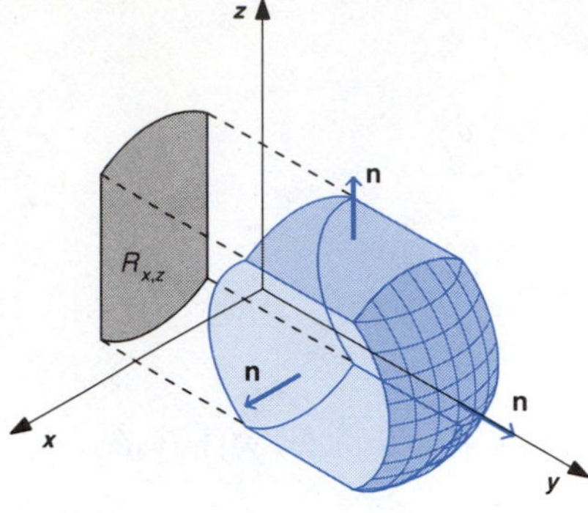


FIGURE 18.7.3

If the region $D$ can be divided into a finite number of regions of the type illustrated in Figure 18.7.3 by cutting along generalized cylinders with generating lines parallel to the $y$-axis, we say the region is **finitely simple in the direction of the $y$-axis**. We then have

$$\iint_S N \cos \beta \, dS = \iiint_D \frac{\partial N}{\partial y} \, dV. \tag{2}$$

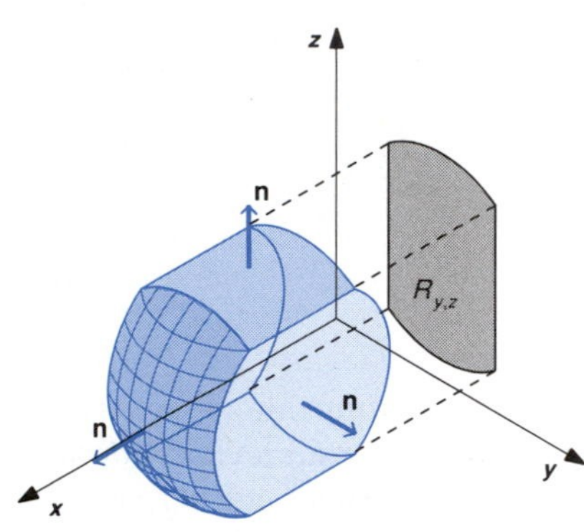


FIGURE 18.7.4

A region that can be divided into a finite number of regions of the type illustrated in Figure 18.7.4 by generalized cylinders with generating lines parallel to the $x$-axis is said to be **finitely simple in the direction of the $x$-axis**. We then have

$$\iint_S M \cos \alpha \, dS = \iiint_D \frac{\partial M}{\partial x} \, dV. \tag{3}$$

Let us say that a region is **finitely simple** if it is finitely simple in the direction of each of the three coordinate axes. We can then combine the formulas (1), (2), and (3). Let us express the result in vector form. We define the symbolic vector **del** to be

$$\nabla = \frac{\partial}{\partial x}\mathbf{i} + \frac{\partial}{\partial y}\mathbf{j} + \frac{\partial}{\partial z}\mathbf{k}.$$

If $\mathbf{F} = M\mathbf{i} + N\mathbf{j} + P\mathbf{k}$, we use the symbolism

$$\nabla \cdot \mathbf{F} = \frac{\partial M}{\partial x} + \frac{\partial N}{\partial y} + \frac{\partial P}{\partial z}.$$

The expression $\nabla \cdot \mathbf{F}$ is called the **divergence** of $\mathbf{F}$. If $\mathbf{n}$ is the outward unit normal to the boundary surface of a region $D$, then $\iint_S \mathbf{F} \cdot \mathbf{n} \, dS$ is the outward **flux** of $\mathbf{F}$ across the boundary of $D$. We then have the following.

*Divergence Theorem*

**Suppose $D$ is a finitely simple region in three-dimensional space, $S$ is the complete boundary of $D$, and n is the outward unit normal to the boundary surface. If F has continuous first partial derivatives in $D$ and on the boundary of $D$, then**

$$\iint_S \mathbf{F} \cdot \mathbf{n} \, dS = \iiint_D \nabla \cdot \mathbf{F} \, dV.$$

The value of the divergence $\nabla \cdot \mathbf{F}$ at a point can be considered to be the flux per unit volume at the point. Positive divergence indicates a source at the point and negative divergence indicates a sink at the point. The Divergence Theorem says that the flux across the boundary of a region is the net sum of the sources and sinks within the region.

FIGURE 18.7.5

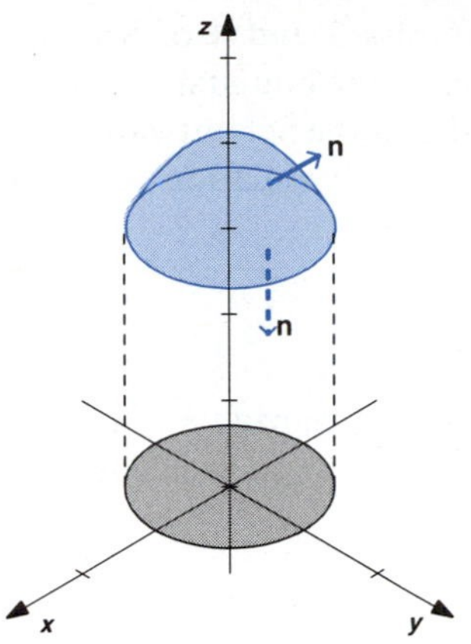


■ **EXAMPLE 1**

Evaluate $\iint_S \mathbf{F}\cdot\mathbf{n}\,dS$, where $\mathbf{F} = x\mathbf{i} + y\mathbf{j} + z\mathbf{k}$, $S$ is the complete boundary of the region $D$ bounded by the paraboloid $z = 4 - x^2 - y^2$ and the plane $z = 3$, and $\mathbf{n}$ is outward.

**SOLUTION**

The region is sketched in Figure 18.7.5. $\mathbf{F}$ has continuous first partial derivatives, so we can use the Divergence Theorem to evaluate the surface integral. $\mathbf{F} = x\mathbf{i} + y\mathbf{j} + z\mathbf{k}$ implies $M = x$, $N = y$, and $P = z$. We then have

$$\iint_S \mathbf{F}\cdot\mathbf{n}\,dS = \iiint_D \nabla\cdot\mathbf{F}\,dV$$

$$= \iiint_D (M_x + N_y + P_z)\,dV$$

$$= \iiint_D (1 + 1 + 1)\,dV$$

which in cylindrical coordinates is

$$= \int_0^{2\pi}\int_0^1\int_3^{4-r^2} 3\,r\,dz\,dr\,d\theta$$

$$= \int_0^{2\pi}\int_0^1 [(4 - r^2) - 3](3r)\,dr\,d\theta$$

$$= \int_0^{2\pi}\int_0^1 (3r - 3r^3)\,dr\,d\theta$$

$$= \int_0^{2\pi}\left(\frac{3}{2} - \frac{3}{4}\right)d\theta = \frac{3}{4}(2\pi) = \frac{3\pi}{2}.$$ ■

In Examples 3 and 4 of Section 18.6 we calculated the outward flux across each of the boundary surfaces of the region in Example 1 above. We also noted that the net outward flux across the two surfaces was $3\pi/2$. This agrees with the calculation using the Divergence Theorem.

■ **EXAMPLE 2**

Evaluate the outward flux of $\mathbf{F} = x\mathbf{i} + z\mathbf{j} + y\mathbf{k}$ across the boundary of the region in the first octant inside the sphere

$$x^2 + y^2 + z^2 = 1.$$

**SOLUTION**

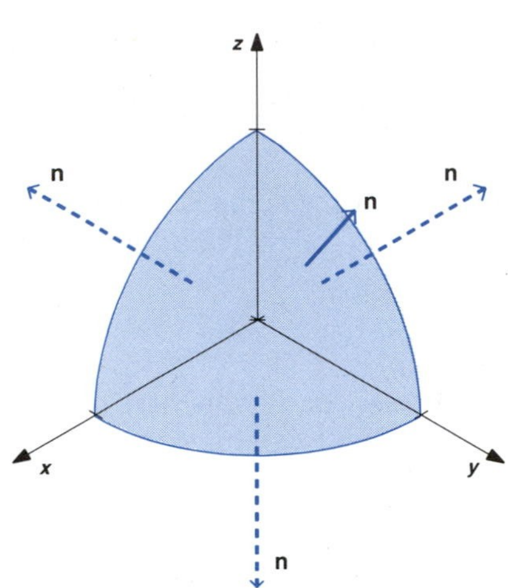


FIGURE 18.7.6

The region is sketched in Figure 18.7.6.

**F** has continuous first partial derivatives, so we can then apply the Divergence Theorem. The flux is

$$\iint_S \mathbf{F}\cdot\mathbf{n}\, dS = \iiint_D \nabla\cdot\mathbf{F}\, dV$$

$$= \iiint_D \left[\frac{\partial}{\partial x}(x) + \frac{\partial}{\partial y}(z) + \frac{\partial}{\partial z}(y)\right] dV$$

$$= \iiint_D 1\, dV$$

$$= \frac{1}{8}(\text{Volume of sphere with radius } 1) = \frac{1}{8}\left(\frac{4\pi(1)^3}{3}\right) = \frac{\pi}{6}. \quad ■$$

■ **EXAMPLE 3**

Evaluate $\iint_S \mathbf{F}\cdot\mathbf{n}\, dS$, where

$$\mathbf{F} = \tan^{-1}(y^2 + z^2)\mathbf{i} + \ln(1 + x^2 + z^2)\mathbf{j} + 2ze^{x^2+y^2}\mathbf{k},$$

$S$ is the complete boundary of the region $D$ that is inside the cylinder $x^2 + y^2 = 1$ and between the planes $z = 0$ and $z = 1$, and **n** is the outward unit normal.

**SOLUTION**

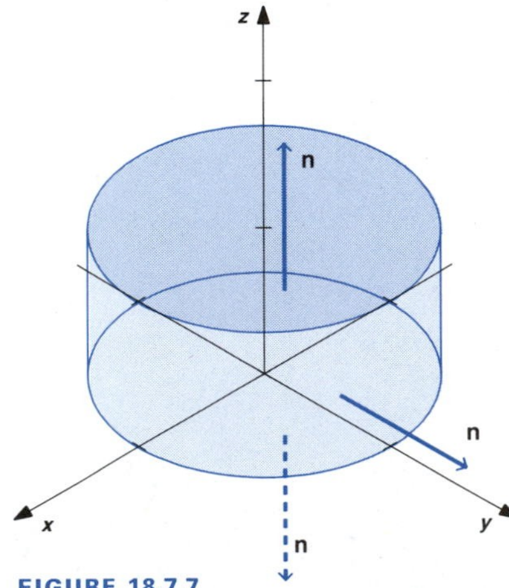


FIGURE 18.7.7

The region is sketched in Figure 18.7.7.

We see **F** has continuous first partial derivatives and

$$\nabla\cdot\mathbf{F} = \frac{\partial}{\partial x}(\tan^{-1}(y^2 + z^2)) + \frac{\partial}{\partial y}(\ln(1 + x^2 + z^2)) + \frac{\partial}{\partial z}(2ze^{x^2+y^2})$$

$$= 2e^{x^2+y^2}.$$

The Divergence Theorem then gives

$$\iint_S \mathbf{F}\cdot\mathbf{n}\,dS = \iiint_D \nabla\cdot\mathbf{F}\,dV$$

$$= \iiint_D 2e^{x^2+y^2}\,dV.$$

We use cylindrical coordinates to evaluate the triple integral over the cylindrical region $D$. See we have

$$\iiint_D 2e^{x^2+y^2}\,dV = \int_0^{2\pi}\int_0^1\int_0^1 2e^{r^2}r\,dz\,dr\,d\theta$$

$$= \int_0^{2\pi}\int_0^1 e^{r^2}2r\,dr\,d\theta \qquad [u = r^2,\ du = 2r\,dr]$$

$$= \int_0^{2\pi} e^{r^2}\Big]_0^1\,d\theta$$

$$= \int_0^{2\pi}(e-1)\,d\theta = 2\pi(e-1).$$ ■

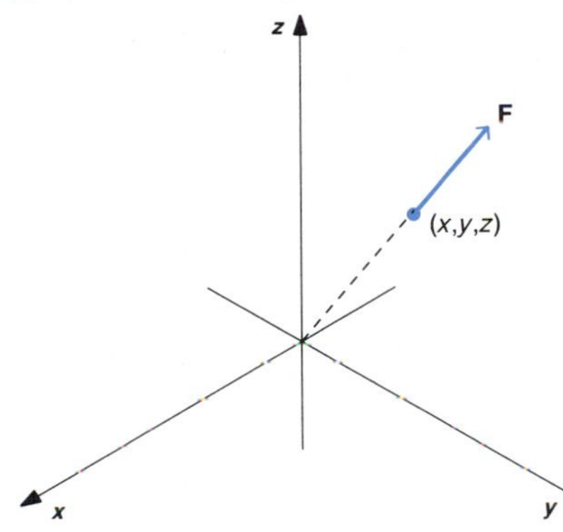


FIGURE 18.7.8

Let us consider the **inverse square field**

$$\mathbf{F} = \frac{x\mathbf{i} + y\mathbf{j} + z\mathbf{k}}{(x^2+y^2+z^2)^{3/2}}.$$

The length of **F** is one over the square of the distance from the origin and the direction is directly away from the origin. See Figure 18.7.8. The divergence of **F** is

$$\nabla\cdot\mathbf{F} = \frac{\partial}{\partial x}\left(\frac{x}{(x^2+y^2+z^2)^{3/2}}\right) + \frac{\partial}{\partial y}\left(\frac{y}{(x^2+y^2+z^2)^{3/2}}\right) + \frac{\partial}{\partial z}\left(\frac{z}{(x^2+y^2+z^2)^{3/2}}\right).$$

Using the Quotient Rule to evaluate the first partial derivative above, we have

$$\frac{\partial}{\partial x}\left(\frac{x}{(x^2+y^2+z^2)^{3/2}}\right) = \frac{(x^2+y^2+z^2)^{3/2}(1) - (x)(3/2)(x^2+y^2+z^2)^{1/2}(2x)}{(x^2+y^2+z^2)^{6/2}}$$

$$= \frac{(x^2+y^2+z^2) - 3x^2}{(x^2+y^2+z^2)^{5/2}}$$

$$= \frac{-2x^2+y^2+z^2}{(x^2+y^2+z^2)^{5/2}}.$$

The other two partial derivatives are evaluated similarly. We obtain

$$\nabla \cdot \mathbf{F} = \frac{-2x^2 + y^2 + z^2}{(x^2 + y^2 + z^2)^{5/2}} + \frac{x^2 - 2y^2 + z^2}{(x^2 + y^2 + z^2)^{5/2}} + \frac{x^2 + y^2 - 2z^2}{(x^2 + y^2 + z^2)^{5/2}}$$

$$= 0, \quad (x, y, z) \neq (0, 0, 0).$$

Let us state the above result as a theorem.

*Theorem 1*

**The inverse square vector field**

$$\mathbf{F} = \frac{x\mathbf{i} + y\mathbf{j} + z\mathbf{k}}{(x^2 + y^2 + z^2)^{3/2}}$$

**has divergence zero at every point of three-dimensional space except the origin.**

If the divergence of **F** is zero at every point of a region $D$ that has complete boundary $S$, the Divergence Theorem implies

$$\iint_S \mathbf{F} \cdot \mathbf{n}\, dS = \iiint_D \nabla \cdot \mathbf{F}\, dV = 0,$$

so the outward flux of **F** across $S$ is zero. Theorem 1 then gives the following.

*Theorem 2*

**If $D$ is any finitely simple region that does not contain the origin either in its interior or on its boundary, the outward flux of the inverse square field**

$$\mathbf{F} = \frac{x\mathbf{i} + y\mathbf{j} + z\mathbf{k}}{(x^2 + y^2 + z^2)^{3/2}}$$

**across the total boundary of $D$ is zero.**

*Theorem 3*

**If $D$ is any finitely simple region that contains the origin in its interior, the outward flux of the inverse square field**

$$\mathbf{F} = \frac{x\mathbf{i} + y\mathbf{j} + z\mathbf{k}}{(x^2 + y^2 + z^2)^{3/2}}$$

**across the total boundary of $D$ is $4\pi$.**

■ *Proof* We cannot apply the Divergence Theorem to the region $D$, because $D$ contains a point where **F** does not have first partial derivatives.

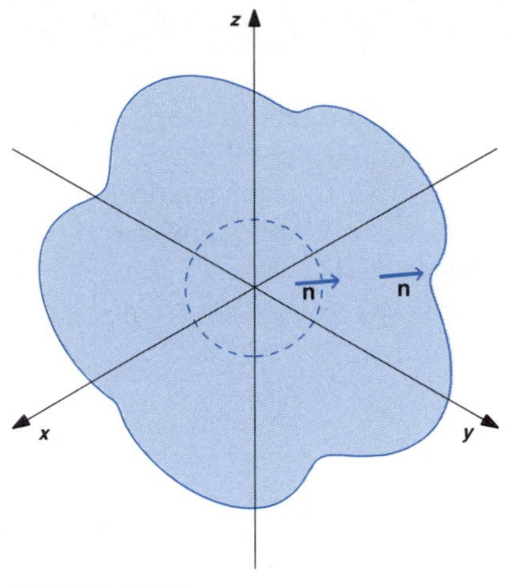


FIGURE 18.7.9

(**F** becomes unbounded at the origin, so it cannot have partial derivatives at that point.) However, we can apply the Divergence Theorem to the region $D - D_r$, where $D_r$ is the sphere bounded by $x^2 + y^3 + z^2 = r^2$ and $r$ is chosen small enough that $D_r$ is contained in $D$. Let $S$ denote the boundary of $D$ and let $S_r$ denote the boundary of $D_r$. Let **n** denote the unit normal that is outward from the region $D$ on $S$ and outward from $D_r$ on $S_r$. The outward flux from $D - D_r$ is then in the direction of $-\mathbf{n}$ on the surface $S_r$. See Figure 18.7.9.

The outward flux of $F$ across the complete boundary of $D - D_r$ is

$$\iint_S \mathbf{F}\cdot\mathbf{n}\,dS - \iint_{S_r} \mathbf{F}\cdot\mathbf{n}\,dS = 0,$$

by Theorem 2. In Example 5 of Section 18.6, we determined that the outward flux of **F** across the surface $S_r$ is $4\pi$. It follows that

$$\iint_S \mathbf{F}\cdot\mathbf{n}\,dS = \iint_{S_r} \mathbf{F}\cdot\mathbf{n}\,dS = 4\pi.$$

■

## EXERCISES 18.7

In Exercises 1–12 find the outward flux of the given vector function across the boundary of the indicated region.

1 $\mathbf{F} = x^2\mathbf{i} + y^2\mathbf{j}$; $D$ bounded by $x = 2$, $y = 3$, $z = 2$, in the first octant

2 $\mathbf{F} = xyz\mathbf{k}$; $D$ bounded by $z = 2 - x - y$, in the first octant

3 $\mathbf{F} = x^2\mathbf{i} + y^2\mathbf{j} + z^2\mathbf{k}$; $D$ bounded by $x^2 + y^2 = 1$, $z = 0$, $z = 4$

4 $\mathbf{F} = -y\mathbf{i} + x\mathbf{j}$; $D$ bounded by $x^2 + y^2 = 4$, $z = 0$, $z = 3$

5 $\mathbf{F} = x^3\mathbf{i} + y^3\mathbf{j} + z^3\mathbf{k}$; $D$ bounded by $z = \sqrt{x^2 + y^2}$ and $z = 1$

6 $\mathbf{F} = x^3\mathbf{i} + y^3\mathbf{j} + z^3\mathbf{k}$; $D$ bounded by $z = \sqrt{1 - x^2 - y^2}$ and $z = 0$

7 $\mathbf{F} = xy\mathbf{i} + yz\mathbf{j} + zx\mathbf{k}$; $D$ bounded by $z = 4 - x^2 - y^2$ and $z = 0$

8 $\mathbf{F} = x^2\mathbf{i} + y^2\mathbf{j} + z^2\mathbf{k}$; $D$ bounded by $z = x^2 + y^2$ and $z = 4$

9 $\mathbf{F} = (x\mathbf{i} + y\mathbf{j} + z\mathbf{k})/(x^2 + y^2 + z^2)$; $D$ the sphere with center at the origin and radius $r$

10 $\mathbf{F} = (x\mathbf{i} + y\mathbf{j} + z\mathbf{k})/(x^2 + y^2 + z^2)^2$; $D$ the sphere with center at the origin and radius $r$

11 $\mathbf{F} = (x\mathbf{i} + y\mathbf{j} + z\mathbf{k})/(x^2 + y^2 + z^2)^{3/2}$; $D$ the sphere with center $(h, k, l)$ and radius $r$, $0 < h^2 + k^2 + l^2 < r^2$

12 $\mathbf{F} = (x\mathbf{i} + y\mathbf{j} + z\mathbf{k})/(x^2 + y^2 + z^2)^{3/2}$; $D$ the sphere with center $(h, k, l)$ and radius $r$, $h^2 + k^2 + l^2 > r^2$

13 The electric field due to a charge at the origin is given by

$$\mathbf{E} = \frac{x\mathbf{i} + y\mathbf{j} + z\mathbf{k}}{(x^2 + y^2 + z^2)^{3/2}}.$$

What is the outward flux across the boundary of a cubic box that has center at the origin and length of each edge 3 cm?

14 What is the outward flux of the electric field of Exercise 13 across the boundary of the ellipsoid $4x^2 + 4y^2 + 9z^2 = 36$?

15 Show that the flux of the electric field of Exercise 13 across that part of the sphere $x^2 + y^2 + z^2 = 25$ with $z \geq 3$ is equal to the flux across that part of the plane $z = 3$ with $x^2 + y^2 \leq 16$.

## 18.8 STOKES' THEOREM

We have seen in Section 18.3 that Green's Theorem relates the circulation of a vector function $\mathbf{F} = M\mathbf{i} + N\mathbf{j}$ around a simple closed curve $C$ in the $xy$-plane to the integral of the scalar curl $N_x - M_y$ over the region $R$ in the $xy$-plane enclosed by $C$. In this section we will investigate a three-dimensional version of this result, called Stokes' Theorem.

Let us define the **curl** of a vector field $\mathbf{F} = M\mathbf{i} + N\mathbf{j} + P\mathbf{k}$ to be

$$\text{curl } \mathbf{F} = \left(\frac{\partial P}{\partial y} - \frac{\partial N}{\partial z}\right)\mathbf{i} + \left(\frac{\partial M}{\partial z} - \frac{\partial P}{\partial x}\right)\mathbf{j} + \left(\frac{\partial N}{\partial x} - \frac{\partial M}{\partial y}\right)\mathbf{k}.$$

If $\mathbf{F} = M(x, y)\mathbf{i} + N(x, y)\mathbf{j}$, so $M$ and $N$ depend only on $x$ and $y$ and $P = 0$, we have $M_z = N_z = P_x = P_y = 0$, and

$$\text{curl } \mathbf{F} = \left(\frac{\partial N}{\partial x} - \frac{\partial M}{\partial y}\right)\mathbf{k};$$

the scalar curl

$$\frac{\partial N}{\partial x} - \frac{\partial M}{\partial y}$$

measures the circulation per unit area about the axis of rotation $\mathbf{k}$. See Figure 18.8.1a. Generally, the direction of the vector curl $\mathbf{F}$ at a point determines an axis of rotation of the circulation of $\mathbf{F}$ and the magnitude of curl $\mathbf{F}$ measures the magnitude of the net circulation about that axis at that point. See Figure 18.8.1b. The expression $(\text{curl } \mathbf{F}) \cdot \mathbf{n}$ measures the magnitude of the net circulation of $\mathbf{F}$ about the axis of rotation determined by the unit vector $\mathbf{n}$. See Figure 18.8.1c. If $\theta$ is the angle between curl $\mathbf{F}$ and the unit vector $\mathbf{n}$, we have

$$(\text{curl } \mathbf{F}) \cdot \mathbf{n} = |\text{curl } \mathbf{F}|\,|\mathbf{n}|\cos\theta = |\text{curl } \mathbf{F}|\cos\theta,$$

FIGURE 18.8.1

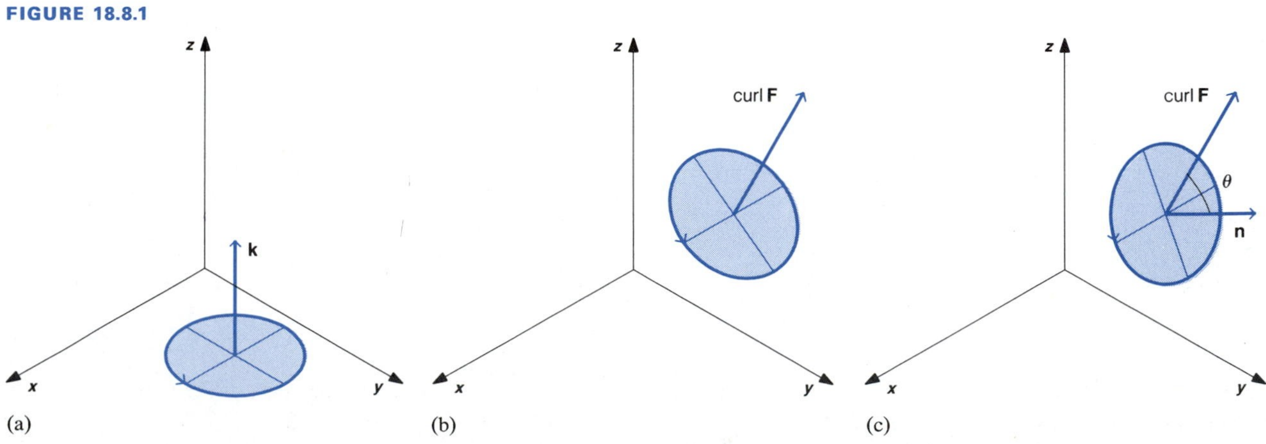

so we see that the circulation of **F** about **n** is greatest with magnitude |curl **F**| when **n** is the unit vector in the direction of curl **F**.

It is convenient to express curl **F** in terms of the symbolic vector del,

$$\nabla = \frac{\partial}{\partial x}\mathbf{i} + \frac{\partial}{\partial y}\mathbf{j} + \frac{\partial}{\partial z}\mathbf{k}.$$

If $\mathbf{F} = M\mathbf{i} + N\mathbf{j} + P\mathbf{k}$, we use the symbolism

$$\nabla \times \mathbf{F} = \begin{vmatrix} \mathbf{i} & \mathbf{j} & \mathbf{k} \\ \dfrac{\partial}{\partial x} & \dfrac{\partial}{\partial y} & \dfrac{\partial}{\partial z} \\ M & N & P \end{vmatrix}$$

$$= \begin{vmatrix} \dfrac{\partial}{\partial y} & \dfrac{\partial}{\partial z} \\ N & P \end{vmatrix}\mathbf{i} - \begin{vmatrix} \dfrac{\partial}{\partial x} & \dfrac{\partial}{\partial z} \\ M & P \end{vmatrix}\mathbf{j} + \begin{vmatrix} \dfrac{\partial}{\partial x} & \dfrac{\partial}{\partial y} \\ M & N \end{vmatrix}\mathbf{k}$$

$$= \left(\frac{\partial P}{\partial y} - \frac{\partial N}{\partial z}\right)\mathbf{i} + \left(\frac{\partial M}{\partial z} - \frac{\partial P}{\partial x}\right)\mathbf{j} + \left(\frac{\partial N}{\partial x} - \frac{\partial M}{\partial y}\right)\mathbf{k},$$

so

$$\text{curl } \mathbf{F} = \nabla \times \mathbf{F}.$$

The cross product form is very convenient for remembering the definition of curl **F**. We can evaluate curl **F** either by substituting into the defining formula or by expressing the cross product as a determinant.

■ **EXAMPLE 1**

$$\mathbf{F} = yz\mathbf{i} - xz\mathbf{j} + xy\mathbf{k}.$$

(a) Find curl **F**.

(b) Evaluate $\iint_S (\text{curl } \mathbf{F})\cdot\mathbf{n}\, dS$, where $S$ is that part of the paraboloid $z = x^2 + y^2$ with $x^2 + y^2 \le 4$ and **n** is directed upward.

(c) Evaluate $\iint_S (\text{curl } \mathbf{F})\cdot\mathbf{n}\, dS$, where $S$ is that part of the plane $z = 4$ with $x^2 + y^2 \le 4$ and **n** is directed upward.

(d) Evaluate $\int_C \mathbf{F}\cdot\mathbf{T}\, ds$, where

$$C\colon\ x = 2\cos t,\quad y = 2\sin t,\quad z = 4,\qquad 0 \le t \le 2\pi.$$

**SOLUTION**

(a) Using the determinant form of curl $F$, we have

$$\nabla \times \mathbf{F} = \begin{vmatrix} \mathbf{i} & \mathbf{j} & \mathbf{k} \\ \dfrac{\partial}{\partial x} & \dfrac{\partial}{\partial y} & \dfrac{\partial}{\partial z} \\ yz & -xz & xy \end{vmatrix}$$

$$= \begin{vmatrix} \dfrac{\partial}{\partial y} & \dfrac{\partial}{\partial z} \\ -xz & xy \end{vmatrix}\mathbf{i} - \begin{vmatrix} \dfrac{\partial}{\partial x} & \dfrac{\partial}{\partial z} \\ yz & xy \end{vmatrix}\mathbf{j} + \begin{vmatrix} \dfrac{\partial}{\partial x} & \dfrac{\partial}{\partial y} \\ yz & -xz \end{vmatrix}\mathbf{k}$$

$$= \left(\frac{\partial}{\partial y}(xy) - \frac{\partial}{\partial z}(-xz)\right)\mathbf{i} - \left(\frac{\partial}{\partial x}(xy) - \frac{\partial}{\partial z}(yz)\right)\mathbf{j}$$

$$+ \left(\frac{\partial}{\partial x}(-xz) - \frac{\partial}{\partial y}(yz)\right)\mathbf{k}$$

$$= (x + x)\mathbf{i} - (y - y)\mathbf{j} + (-z - z)\mathbf{k}$$

$$= 2x\mathbf{i} - 2z\mathbf{k}.$$

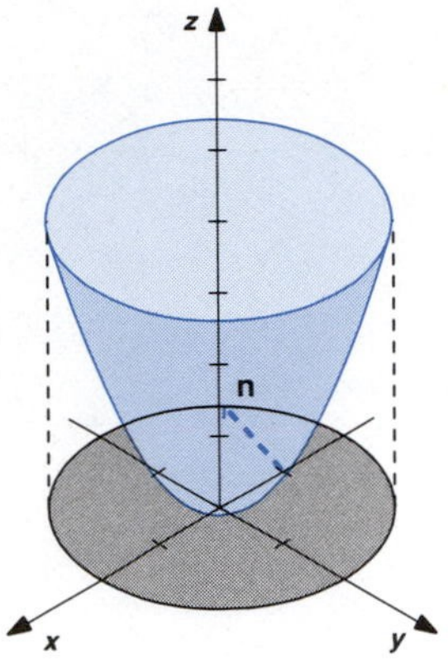

FIGURE 18.8.2

(b) We apply formula (8) of Section 18.6 to the vector field curl $\mathbf{F} = 2x\mathbf{i} - 2z\mathbf{k}$. The surface is given by the formula $z = x^2 + y^2$, for $(x, y)$ in the region $R_{x,y}$ of the $xy$-plane bounded by $x^2 + y^2 = 2^2$. See Figure 18.8.2. The components of curl $\mathbf{F} = \hat{M}\mathbf{i} + \hat{N}\mathbf{j} + \hat{P}\mathbf{k}$ are $\hat{M} = 2x$, $\hat{N} = 0$, and $\hat{P} = -2z = -2(x^2 + y^2)$. Then

$$\iint_S (\text{curl } \mathbf{F}) \cdot \mathbf{n}\, dS = \iint_{R_{x,y}} (-\hat{M}z_x - \hat{N}z_y + \hat{P})\, dA$$

$$= \iint_{R_{x,y}} \{-(2x)(2x) - (0)(2y) + [-2(x^2 + y^2)]\}\, dA,$$

which in polar coordinates is

$$= \int_0^{2\pi}\int_0^2 (-4r^2\cos^2\theta - 2r^2)\, r\, dr\, d\theta$$

$$= \int_0^{2\pi}\int_0^2 (-4r^3\cos^2\theta - 2r^3)\, dr\, d\theta$$

$$= \int_0^{2\pi}\left[-r^4\cos^2\theta - \frac{r^4}{2}\right]_{r=0}^{r=2} d\theta$$

$$= \int_0^{2\pi}(-16\cos^2\theta - 8)\, d\theta$$

[Table of integrals]

$$= -8(\sin\theta\cos\theta + \theta) - 8\theta\Big]_0^{2\pi}$$

$$= -8(2\pi) - 8(2\pi) = -32\pi.$$

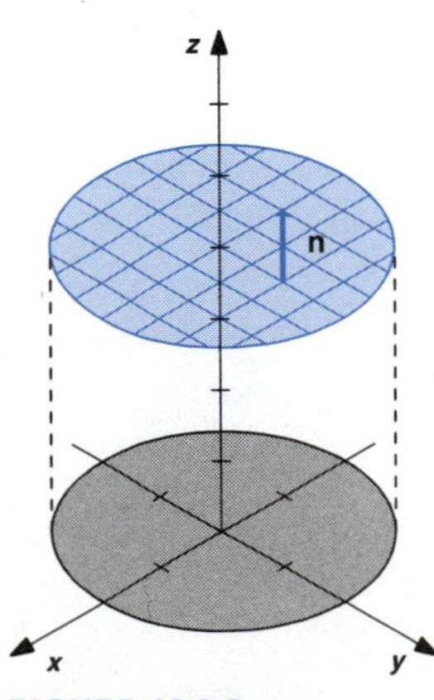


FIGURE 18.8.3

(c) We apply formula (8) found in Section 18.6 to the vector field curl $\mathbf{F} = 2x\mathbf{i} - 2z\mathbf{k}$. The surface is given by $z = 4$, for $(x, y)$ in the region $R_{x,y}$ of the $xy$-plane bounded by $x^2 + y^2 = 2^2$. See Figure 18.8.3. The

FIGURE 18.8.4

components of curl $\mathbf{F} = \hat{M}\mathbf{i} + \hat{N}\mathbf{j} + \hat{P}\mathbf{k}$ are $\hat{M} = 2x$, $\hat{N} = 0$, and $\hat{P} = -2z = -2(4) = -8$. Then

$$\iint_S (\text{curl } \mathbf{F}) \cdot \mathbf{n}\, dS = \iint_{R_{x,y}} (-\hat{M}z_x - \hat{N}z_y + \hat{P})\, dA$$

$$= \iint_{R_{x,y}} [-(2x)(0) - (0)(0) + (-8)]\, dA$$

$$= -8(\text{Area of } R_{x,y}) = -8(\pi 2^2) = -32\pi.$$

(d) The curve is sketched in Figure 18.8.4. Note that $C$ is the boundary curve of the surfaces in parts (b) and (c). We have $\mathbf{F} = yz\mathbf{i} - xz\mathbf{j} + xy\mathbf{k}$, so $M = yz$, $N = -xz$, and $P = xy$. Then

$$\int_C \mathbf{F} \cdot \mathbf{T}\, ds = \int M\, dx + N\, dy + P\, dz$$

$$= \int_C yz\, dx - xz\, dy + xy\, dz$$

$$= \int_0^{2\pi} [(2 \sin t)(4)(-2 \sin t) - (2 \cos t)(4)(2 \cos t)$$

$$+ (2 \cos t)(2 \sin t)(0)]\, dt$$

$$= \int_0^{2\pi} [-16 \sin^2 t - 16 \cos^2 t]\, dt$$

$$= \int_0^{2\pi} -16\, dt = -32\pi.$$ ■

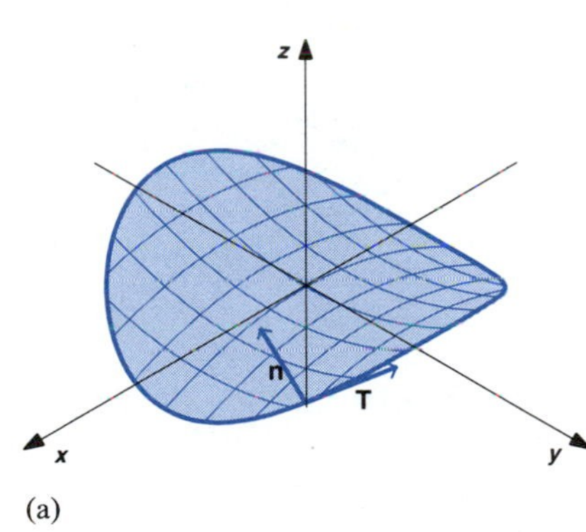

(a)

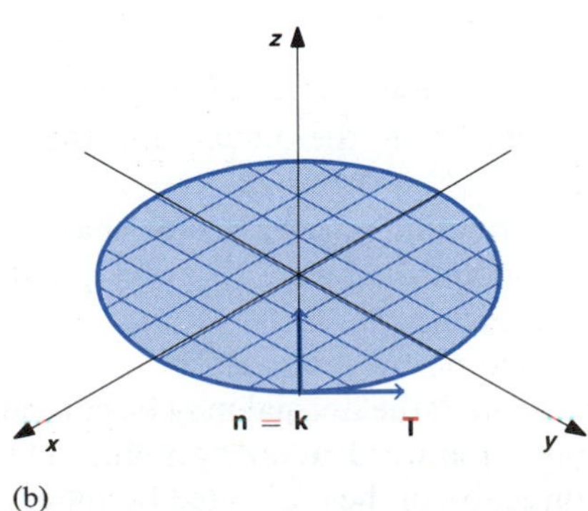

(b)

FIGURE 18.8.5

The equality of the two surface integrals and the line integral in Example 1 are a consequence of a general result called Stokes' Theorem. To state this theorem, we need to establish a relation between the direction of the unit normal vector of a surface and the direction of the boundary curve of the surface.

We will consider surfaces $S$ that are **oriented** in the sense that the surface has a unit normal vector $\mathbf{n}$ at each point and that the direction of $\mathbf{n}$ varies continuously over the surface. The boundary curve $C$ (the edge) of an oriented surface $S$ is said to be **n-directed** if the unit tangent vector $\mathbf{T}$ of $C$ satisfies the condition that $\mathbf{T} \times \mathbf{n}$ points away from the surface. (Recall from Section 14.2 that the vectors $\mathbf{T}$, $\mathbf{n}$, and $\mathbf{T} \times \mathbf{n}$, in that order, form a right-hand system.) Informally, this means if you walk along the boundary of $S$ in the direction of $\mathbf{T}$ with your head in the direction of $\mathbf{n}$, the surface would be on your left. See Figure 18.8.5a. This is consistent with the relation between regions in the plane and the direction of their boundary curves that we used in Green's Theorem, if the unit normal vector $\mathbf{n}$ of the plane surface is chosen to be $\mathbf{k}$. See Figure 18.8.5b.

If $f(x, y)$ has continuous first partial derivatives, the surface

$$z = f(x, y), \qquad (x, y) \text{ in a region of the } xy\text{-plane}$$

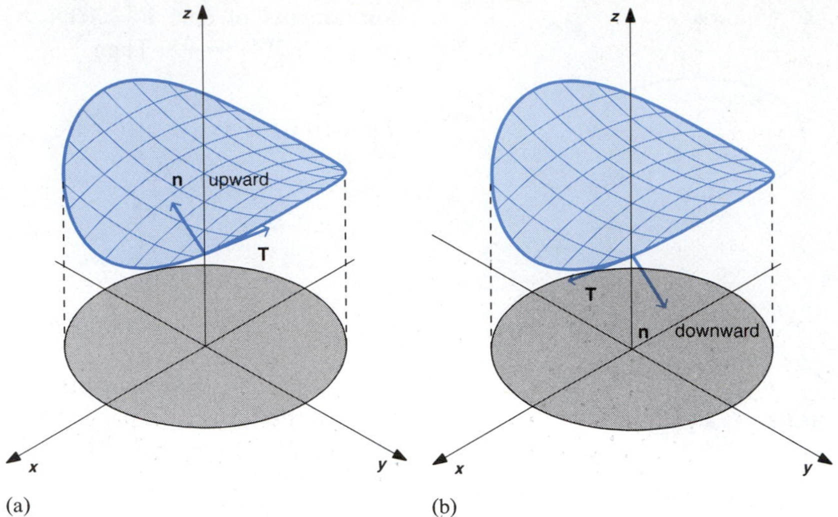


FIGURE 18.8.6

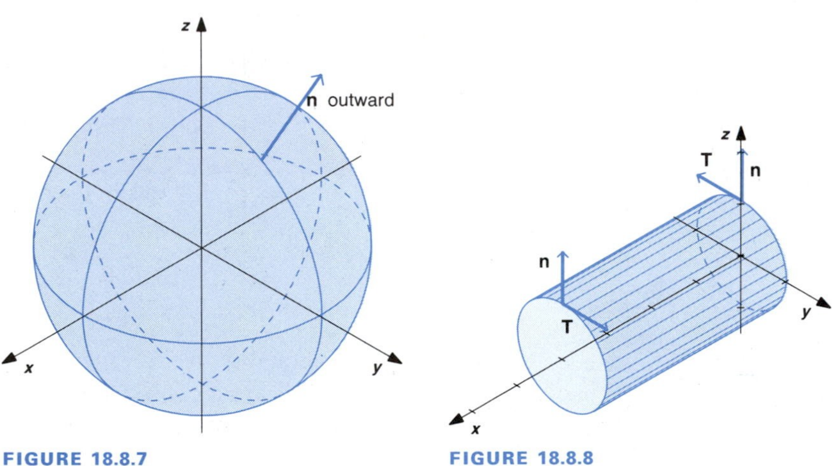


FIGURE 18.8.7

FIGURE 18.8.8

is oriented; **n** may be chosen either to be upward at every point or downward at every point. See Figure 18.8.6. Note the direction of the **n**-directed boundary curve in each case.

The surface of a sphere is oriented. The unit normal vector may be chosen either to point outward at every point or inward at every point. This surface has no boundary curve. See Figure 18.8.7.

The surface that consists of that part of the cylinder $y^2 + z^2 = 1$ between the planes $x = 0$ and $x = 4$ is oriented; the normal may be chosen either to point outward at every point or inward at every point. The boundary consists of two circles. The direction of the **n**-directed boundary curves for **n** outward is indicated in Figure 18.8.8.

FIGURE 18.8.9

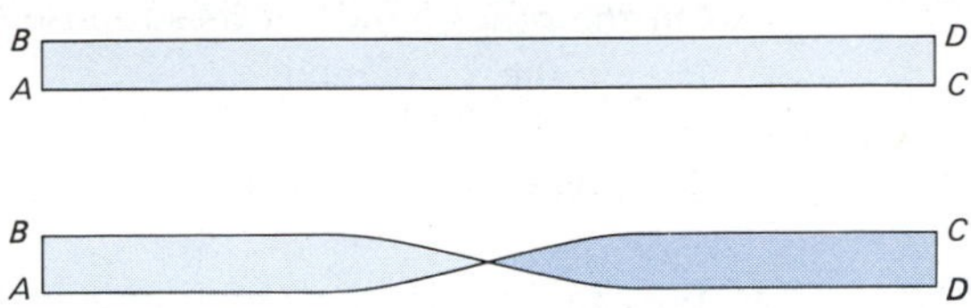

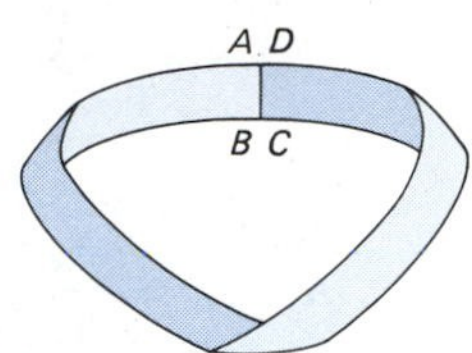


A **Möbius band** is an example of a simple surface that is not oriented. A Möbius band can be constructed from a strip of paper by turning over one end and then joining the ends, as indicated in Figure 18.8.9. There is no choice of unit normal vectors that vary continuously over the surface. That is, if the unit normal vectors varied continuously around the Möbius band, then they would vary continuously over the flat surface from which the band was constructed. This means the normal vectors would all point toward the same side of the flat surface. The unit vectors would then be in the opposite direction on each side of the line where the ends are joined to form the Möbius band. The normal vector **n** cannot vary continuously at points where it abruptly changes direction, so it cannot vary continuously over the entire Möbius band.

■ *Stokes' Theorem*

**Suppose $S$ is an oriented surface with unit normal vector n, $C$ is the n-directed boundary curve of $S$, and T is the unit tangent vector of $C$. If F has continuous first partial derivatives, then**

$$\int_C \mathbf{F}\cdot\mathbf{T}\,ds = \iint_S (\text{curl }\mathbf{F})\cdot\mathbf{n}\,dS.$$

■ *Proof of some special cases* If the surface $S$ consists of a region in the $xy$-plane with $\mathbf{n}=\mathbf{k}$ and $\mathbf{F}=M(x, y)\mathbf{i}+N(x, y)\mathbf{j}$, then curl $\mathbf{F} = (N_x - M_y)\mathbf{k}$ and Stokes' Theorem reduces to Green's Theorem. Let us use Green's Theorem to prove Stokes' Theorem in the case that the surface $S$ is of the form

$$z = z(x, y),\quad (x, y) \text{ in } R_{x,y},\quad \mathbf{n} \text{ upward},$$

where $z$ has continuous second partial derivatives. In this case we can apply formula (8) of Section 18.6 to the function

$$\text{curl }\mathbf{F} = (P_y - N_z)\mathbf{i} + (M_z - P_x)\mathbf{j} + (N_x - M_y)\mathbf{k}$$

to obtain

$$\iint_S (\text{curl }\mathbf{F})\cdot\mathbf{n}\,dS = \iint_{R_{x,y}} (-(P_y - N_z)z_x - (M_z - P_x)z_y + (N_x - M_y))\,dA.$$

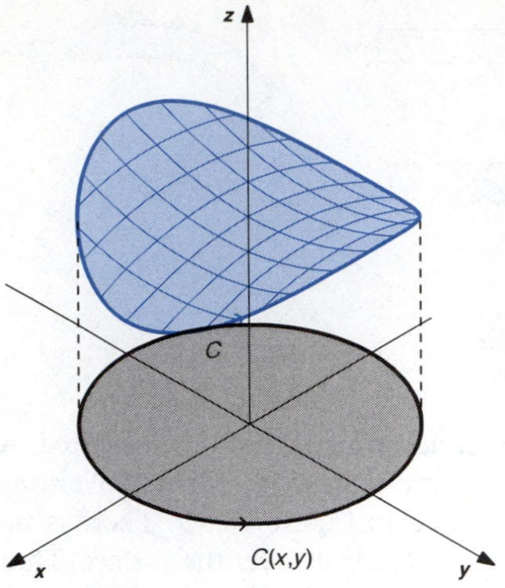

FIGURE 18.8.10

On the other hand, let us suppose that the boundary curve of the region $R_{x,y}$ in the $xy$-plane is

$$C_{x,y}\colon\ x = x(t),\quad y = y(t),\qquad a \le t \le b.$$

We can then use the equation of the surface to express the boundary curve $C$ of $S$ in parametric form as

$$C\colon\ x = x(t),\quad y = y(t),\quad z = z(x(t), y(t)),\qquad a \le t \le b.$$

See Figure 18.8.10. We then have

$$\begin{aligned}\int_C \mathbf{F}\cdot\mathbf{T} &= \int_C M\,dx + N\,dy + P\,dz\\ &= \int_a^b \left(M\frac{dx}{dt} + N\frac{dy}{dt} + P\left(z_x\frac{dx}{dt} + z_y\frac{dy}{dt}\right)\right)dt\\ &= \int_{C_{x,y}} (M + Pz_x)\,dx + (N + Pz_y)\,dy,\end{aligned}$$

which by Green's Theorem is

$$= \iint_{R_{x,y}} ((N + Pz_y)_x - (M + Pz_x)_y)\,dA.$$

Using the Chain Rule to evaluate the partial derivatives, we have

$$\begin{aligned}(N + Pz_y)_x &= \frac{\partial}{\partial x}(N(x, y, z(x, y)) + P(x, y, z(x, y))z_y(x, y))\\ &= (N_x + N_z z_x) + [(P)(z_{yx}) + (z_y)(P_x + P_z z_x)]\\ &= N_x + N_z z_x + Pz_{yx} + P_x z_y + P_z z_x z_y\end{aligned}$$

and, similarly,

$$(M + Pz_x)_y = M_y + M_z z_y + Pz_{xy} + P_y z_x + P_z z_x z_y.$$

Then

$$\begin{aligned}&\iint_{R_{x,y}} ((N + Pz_y)_x - (M + Pz_x)_y)\,dA\\ &\qquad = \iint_{R_{x,y}} ((N_x + N_z z_x + Pz_{yx} + P_x z_y + P_z z_x z_y)\\ &\qquad\qquad - (M_y + M_z z_y + Pz_{xy} + P_y z_x + P_z z_x z_y))\,dA\\ &\qquad = \iint_{R_{x,y}} (-(P_y - N_z)z_x - (M_z - P_x)z_y + (N_x - M_y))\,dA,\end{aligned}$$

FIGURE 18.8.11

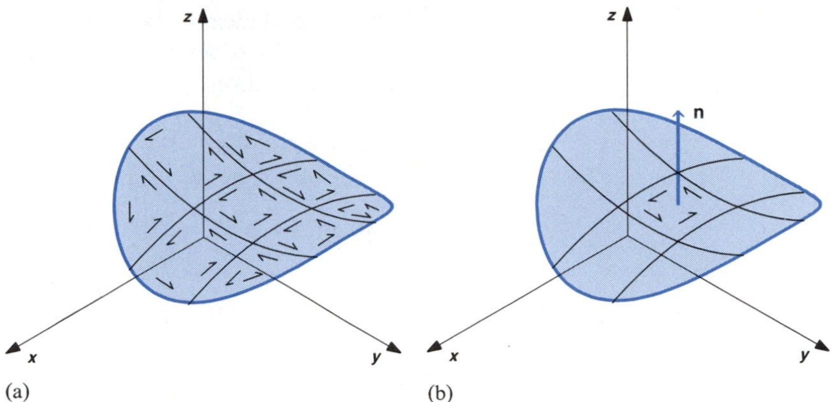


where we have used the fact that the continuity of the second partial derivatives of $z$ implies that $z_{xy} = z_{yx}$. Combining results gives the conclusion of Stokes' Theorem in this special case.

We could prove Stokes' Theorem for surfaces given by either $y = y(x, z)$ or $x = x(y, z)$ in a manner analogous to that used in the case $z = z(x, y)$. We could then obtain Stokes' Theorem for oriented surfaces that can be divided into a finite number of the above type surfaces. ■

Stokes' Theorem expresses the fact that the circulation at each point on a surface $S$ is related to the circulation around the boundary curve $C$ of the surface. If a surface $S$ is divided into parts, the sum of the circulation of **F** around each of the parts is the circulation around the boundary curve $C$ of the surface. This is because of cancellation of the line integrals in opposite directions along the interior dividing curves. See Figure 18.8.11a. If the parts are small, then the circulation of **F** around each part is essentially a measure of the circulation around the axis **n**. See Figure 18.8.11b. If $(\text{curl }\mathbf{F})\cdot\mathbf{n}$ represents the circulation per unit area of **F** around the axis **n**, we should expect that the "sum" $\iint_S (\text{curl }\mathbf{F})\cdot\mathbf{n}\, dS$ to be equal to the circulation of **F** around the boundary of $S$, which is given by $\int_C \mathbf{F}\cdot\mathbf{T}\, dS$.

■ **EXAMPLE 2**

Use Stokes' Theorem to evaluate $\iint_S (\text{curl }\mathbf{F})\cdot\mathbf{n}\, dS$, where $\mathbf{F} = z\mathbf{i} + x\mathbf{j} + y\mathbf{k}$, $S$ is that part of the sphere $z = \sqrt{5 - x^2 - y^2} - 1$ with $z > 0$, and **n** is directed upward.

**SOLUTION**

The boundary of the surface is the intersection of the sphere and the plane $z = 0$. When $z = 0$, $z = \sqrt{5 - x^2 - y^2} - 1$ implies $1 = 5 - x^2 - y^2$, so $x^2 + y^2 = 2^2$. Since the unit normal is upward, we want to choose the direction of the boundary curve to be from the positive $x$-axis toward the positive $y$-axis. See Figure 18.8.12. We choose

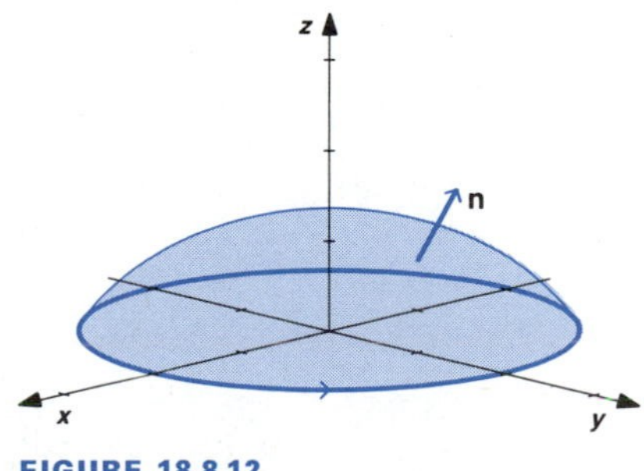


FIGURE 18.8.12

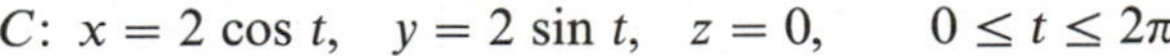

$C$: $x = 2\cos t$, $y = 2\sin t$, $z = 0$, $\quad 0 \le t \le 2\pi$.

(It is a good idea to check the points on the curve that correspond to $t = 0$ and $t = \pi/2$ to determine if the parametric equations of a circle give the correct direction.)

We have $\mathbf{F} = z\mathbf{i} + x\mathbf{j} + y\mathbf{k}$, so $M = z$, $N = x$, and $P = y$. Stokes' Theorem implies

$$\begin{aligned}\iint_S (\text{curl } \mathbf{F})\cdot\mathbf{n}\, dS &= \int_C \mathbf{F}\cdot\mathbf{T}\, dS \\ &= \int_C M\, dx + N\, dy + P\, dz \\ &= \int_C z\, dx + x\, dy + y\, dz \\ &= \int_0^{2\pi} [(0)(-2\sin t) + (2\cos t)(2\cos t) \\ &\quad + (2\sin t)(0)]\, dt \\ &= \int_0^{2\pi} 4\cos^2 t\, dt \qquad \text{[Table of integrals]} \\ &= 2\cos t\sin t + 2t\Big]_0^{2\pi} = 4\pi. \qquad \blacksquare\end{aligned}$$

■ **EXAMPLE 3**

Use Stokes' Theorem to evaluate $\iint_S (\text{curl } \mathbf{F})\cdot\mathbf{n}\, dS$, where $\mathbf{F} = -yz\mathbf{i} + xy\mathbf{k}$, $S$ is that part of the cone $y = 2\sqrt{x^2 + z^2}$ between $y = 2$ and $y = 4$, and $\mathbf{n}$ is outward.

**SOLUTION**

The surface is sketched in Figure 18.8.13. The surface has two boundary curves, one in the plane $y = 2$ and the other in the plane $y = 4$. Since $\mathbf{n}$ is outward, the right-side curve is in the direction from the positive $x$-axis toward the positive $z$-axis; the left-side curve is in the opposite direction.

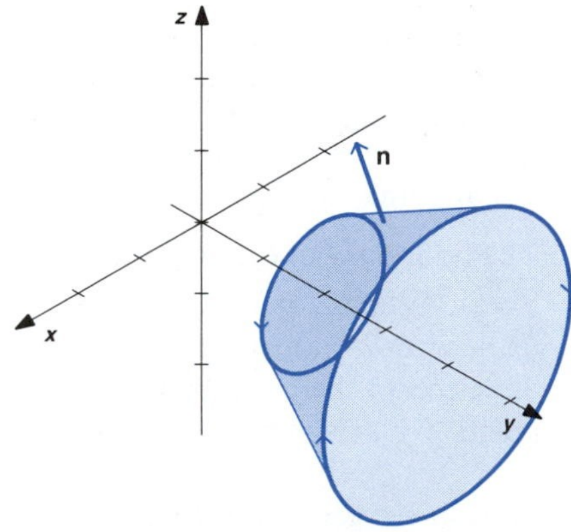


FIGURE 18.8.13

Let us choose

$$C_4\colon\ x = 2\cos t,\quad y = 4,\quad z = 2\sin t,\qquad 0 \le t \le 2\pi,$$

$$C_2\colon\ x = \sin t,\quad y = 2,\quad z = \cos t,\qquad 0 \le t \le 2\pi.$$

(Checking the points on the curves that correspond to $t = 0$ and $t = \pi/2$ shows that these curves are in the correct direction.) $\mathbf{F} = -yz\mathbf{i} + xy\mathbf{k}$ implies $M = -yz$, $N = 0$, and $P = xy$. Stokes' Theorem then gives

$$\begin{aligned}
\iint_S (\text{curl }\mathbf{F})\cdot\mathbf{n}\,dS &= \int_C \mathbf{F}\cdot\mathbf{T}\,ds \\
&= \int_C M\,dx + N\,dy + P\,dz \\
&= \int_{C_4+C_2} -yz\,dx + xy\,dz \\
&= \int_{C_4} -yz\,dx + xy\,dz + \int_{C_2} -yz\,dx + xy\,dz \\
&= \int_0^{2\pi} [-(4)(2\sin t)(-2\sin t) + (2\cos t)(4)(2\cos t)]\,dt \\
&\quad + \int_0^{2\pi} [-(2)(\cos t)(\cos t) + (\sin t)(2)(-\sin t)]\,dt \\
&= \int_0^{2\pi} [16\sin^2 t + 16\cos^2 t]\,dt \\
&\quad + \int_0^{2\pi} [-2\cos^2 t - 2\sin^2 t]\,dt \\
&= \int_0^{2\pi} 16\,dt + \int_0^{2\pi} -2\,dt = 32\pi - 4\pi = 28\pi. \qquad \blacksquare
\end{aligned}$$

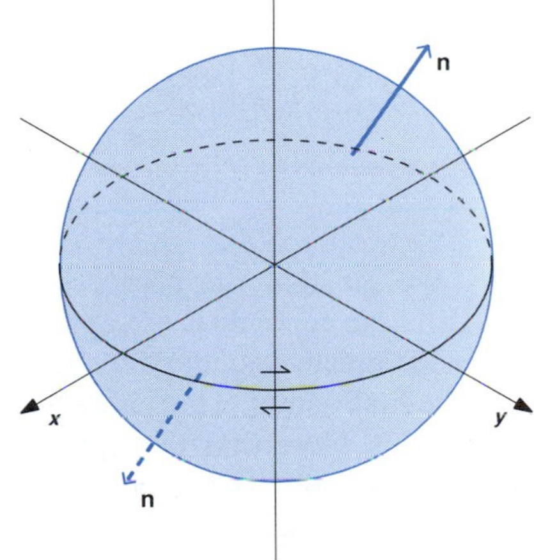


FIGURE 18.8.14

If $S$ is a closed surface that has no boundary curves,

$$\iint_S (\text{curl }\mathbf{F})\cdot\mathbf{n}\,dS = 0.$$

The net circulation of **F** around parts of the surface is zero, because of cancellation. See Figure 18.8.14.

If curl $\mathbf{F} = \mathbf{0}$ for all points in a region $R$ in three-dimensional space, then the circulation of **F** around any smooth closed curve that is the boundary curve of a surface in $R$ is zero. See Figure 18.8.15.

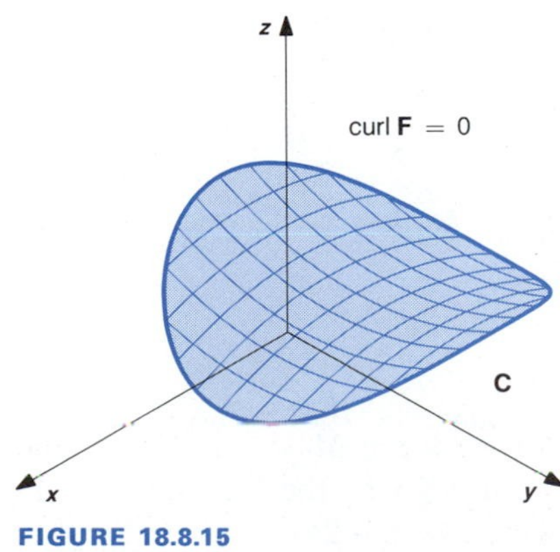


FIGURE 18.8.15

In some cases it may be convenient to evaluate the circulation of **F** around a curve in three-dimensional space by finding a surface $S$ that has $C$ as its complete boundary curve and evaluating the surface integral $\iint_S (\text{curl }\mathbf{F})\cdot\mathbf{n}\,dS$.

FIGURE 18.8.16

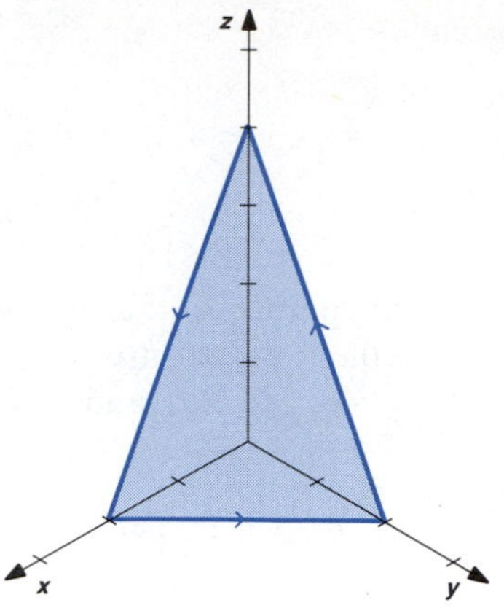


■ **EXAMPLE 4**

Evaluate

$$\int_C [y + \ln(x^2 + 1)]\, dx + [z + \tan^{-1} y]\, dy + e^{z^2}\, dz,$$

where $C$ is the polynomial path from (2, 0, 0), to (0, 2, 0), to (0, 0, 4), and back to (2, 0, 0).

**SOLUTION**

The curve is sketched in Figure 18.8.16. We see that the prospect of evaluating $\mathbf{F} \cdot \mathbf{T}$ along each of the three parts of the curve does not appear attractive. Also, we see that the curve is the complete boundary curve of that part of the plane $2x + 2y + z = 4$ in the first octant. This plane has upward unit normal vector $\mathbf{n} = (2\mathbf{i} + 2\mathbf{j} + \mathbf{k})/3$. Note that the upward normal to the plane corresponds to the given direction of the boundary curve $C$. We have

$$M = y + \ln(x^2 + 1), \quad N = z + \tan^{-1} y, \quad \text{and} \quad P = e^{z^2}, \quad \text{so}$$

$$M_y = 1, \quad M_z = 0, \quad N_x = 0, \quad N_z = 1, \quad \text{and} \quad P_x = P_y = 0.$$

It follows that

$$\begin{aligned} \text{curl}\, \mathbf{F} &= (P_y - N_z)\mathbf{i} + (M_z - P_x)\mathbf{j} + (N_x - M_y)\mathbf{k} \\ &= (0 - 1)\mathbf{i} + (0 - 0)\mathbf{j} + (0 - 1)\mathbf{k} \\ &= -\mathbf{i} - \mathbf{k}, \end{aligned}$$

so the components of curl $\mathbf{F} = \hat{M}\mathbf{i} + \hat{N}\mathbf{j} + \hat{P}\mathbf{k}$ are $\hat{M} = -1$, $\hat{N} = 0$, and $\hat{P} = -1$. We apply formula (8) of Section 18.6 to the function curl $\mathbf{F}$, where the surface $S$ is that part of the plane $z = 4 - 2x - 2y$ with $(x, y)$ in the region $R_{x,y}$ of the $xy$-plane bounded by $x + y = 2$, $x = 0$ and $y = 0$. Stokes' Theorem gives

$$\int_C \mathbf{F} \cdot \mathbf{T}\, ds = \iint_S (\text{curl } \mathbf{F}) \cdot \mathbf{n}\, ds$$

$$= \iint_{R_{x,y}} (-\hat{M}z_x - \hat{N}z_y + \hat{P})\, dA$$

$$= \iint_{R_{x,y}} [-(-1)(-2) - (0)(-2) + (-1)]\, dA$$

$$= \iint_{R_{x,y}} -3\, dA$$

$$= \int_0^2 \int_0^{2-x} -3\, dy\, dx$$

$$= \int_0^2 (-6 + 3x)\, dx$$

$$= -6x + \frac{3x^2}{2}\bigg]_0^2$$

$$= -12 + 6 = -6.$$

■

## EXERCISES 18.8

1 $\mathbf{F} = z\mathbf{i} + x\mathbf{j} + y\mathbf{k}$

(a) Find curl $\mathbf{F}$.

(b) Evaluate $\iint_S (\text{curl } \mathbf{F}) \cdot \mathbf{n}\, dS$, where $S$ is that part of the paraboloid $z = x^2 + y^2$ with $x^2 + y^2 \leq 1$, and $\mathbf{n}$ is downward.

(c) Evaluate $\iint_S (\text{curl } \mathbf{F}) \cdot \mathbf{n}\, dS$, where $S$ is that part of the plane $z = 1$ with $x^2 + y^2 \leq 1$, and $\mathbf{n}$ is downward.

(d) Evaluate $\int_C \mathbf{F} \cdot \mathbf{T}\, ds$, where $C$: $x = \sin t$, $y = \cos t$, $z = 1$, $0 \leq t \leq 2\pi$.

2 $\mathbf{F} = -yz\mathbf{i} + xz\mathbf{j} - xy\mathbf{k}$

(a) Find curl $\mathbf{F}$.

(b) Evaluate $\iint_S (\text{curl } \mathbf{F}) \cdot \mathbf{n}\, dS$, where $S$ is that part of the cone $z = 2\sqrt{x^2 + y^2}$ with $x^2 + y^2 \leq 4$, and $\mathbf{n}$ is upward.

(c) Evaluate $\iint_S (\text{curl } \mathbf{F}) \cdot \mathbf{n}\, dS$, where $S$ is that part of the plane $z = 4$ with $x^2 + y^2 \leq 4$, and $\mathbf{n}$ is upward.

(d) Evaluate $\int_C \mathbf{F} \cdot \mathbf{T}\, ds$, where $C$: $x = 2\cos t$, $y = 2\sin t$, $z = 4$, $0 \leq t \leq 2\pi$.

3 $\mathbf{F} = -y\mathbf{i} + x\mathbf{j} + \sqrt{x^2 + y^2}\mathbf{k}$

(a) Find curl $\mathbf{F}$.

(b) Evaluate $\iint_S (\text{curl } \mathbf{F}) \cdot \mathbf{n}\, dS$, where $S$ is the hemisphere $z = \sqrt{1 - x^2 - y^2}$ and $\mathbf{n}$ is outward.

(c) Evaluate $\iint_S (\text{curl } \mathbf{F}) \cdot \mathbf{n}\, dS$, where $S$ is the hemisphere $z = -\sqrt{1 - x^2 - y^2}$ and $\mathbf{n}$ is outward.

(d) Evaluate $\int_C \mathbf{F} \cdot \mathbf{T}\, ds$, where $C$: $x = \cos t$, $y = \sin t$, $z = 0$, $0 \leq t \leq 2\pi$.

4 $\mathbf{F} = -y^3\mathbf{i} + x^3\mathbf{j} + z^3\mathbf{k}$

(a) Find curl $\mathbf{F}$.

(b) Evaluate $\iint_S (\text{curl } \mathbf{F}) \cdot \mathbf{n}\, dS$, where $S$ is that part of the cone $z = 2\sqrt{x^2 + y^2}$ between $z = 2$ and $z = 4$, and $\mathbf{n}$ is outward.

(c) Evaluate $\int_C \mathbf{F} \cdot \mathbf{T}\, ds$, where $C$: $x = 2\cos t$, $y = 2\sin t$, $z = 4$, $0 \leq t \leq 2\pi$

(d) Evaluate $\int_C \mathbf{F} \cdot \mathbf{T}\, ds$, where $C$: $x = \sin t$, $y = \cos t$, $z = 2$, $0 \leq t \leq 2\pi$

In Exercises 5–14 given below, use Stokes' Theorem in order to evaluate $\iint_S (\text{curl } \mathbf{F}) \cdot \mathbf{n}\, dS$ for the given vector functions $\mathbf{F}$, the given surfaces $S$, and the indicated unit normal vector.

5 $\mathbf{F} = -y\mathbf{i} + x\mathbf{j} + z\mathbf{k}$; $z = 9 - x^2 - y^2$, $x^2 + y^2 \le 9$; $\mathbf{n}$ upward

6 $\mathbf{F} = x\mathbf{i} + y\mathbf{j} + z\mathbf{k}$; $z = 16 - x^2 - y^2$, $z \ge 12$; $\mathbf{n}$ upward

7 $\mathbf{F} = y\mathbf{i} - x\mathbf{j} - z\mathbf{k}$; $z = \sqrt{9 - x^2 - y^2}$; $\mathbf{n}$ upward

8 $\mathbf{F} = y\mathbf{i} - x\mathbf{j} - z\mathbf{k}$; $x^2 + y^2 + z^2 = 9$; $\mathbf{n}$ outward

9 $\mathbf{F} = x\mathbf{i} + y\mathbf{j} + z\mathbf{k}$; $z = \sqrt{x^2 + y^2}$, $z \le 2$; $\mathbf{n}$ outward

10 $\mathbf{F} = x\mathbf{i} + y\mathbf{j} + z\mathbf{k}$; $z = \sqrt{x^2 + y^2}$, $1 \le z \le 2$; $\mathbf{n}$ outward

11 $\mathbf{F} = -yz\mathbf{i} + xz\mathbf{j} - xy\mathbf{k}$; $S$ is that part of the cylinder $x^2 + y^2 = 4$ with $0 \le z \le 2$; $\mathbf{n}$ outward

12 $\mathbf{F} = x^2\mathbf{i} + 2yz\mathbf{j} + z^2\mathbf{k}$; $S$ is that part of the surface $z^2 = x^2 + y^2 - 1$ with $-1 \le z \le 1$; $\mathbf{n}$ outward

13 $\mathbf{F} = y\mathbf{k}$; $x = \sqrt{9 - y^2 - z^2}$; $\mathbf{n}$ outward

14 $\mathbf{F} = yz\mathbf{i} + xz\mathbf{j} - xy\mathbf{k}$; $y = \sqrt{9 - x^2 - z^2}$, $y \ge 2$; $\mathbf{n}$ outward

15 Evaluate $\int_C \sin(x^2)\, dx + ((z + \cos)(y^2))\, dy + 1/(1 + z^4)\, dz$, where $C$ is the polygonal path from $(1, 0, 0)$, to $(0, 1, 0)$, to $(0, 0, 1)$, and back to $(1, 0, 0)$.

16 Evaluate $\int_C [z + (1/(1 + x^2))]\ dx + \ln(1 + y^4)\, dy + e^{2z}\, dz$, where $C$ is the polygonal path from $(1, 0, 0)$, to $(1, 1, 0)$, to $(0, 1, 1)$, to $(0, 0, 1)$, and back to $(1, 0, 0)$.

17 Evaluate $\int_C y^2z^3\, dx + 2xyz^3\, dy + 3xy^2z^2\, dz$, where $C$ is the polygonal path from $(1, 1, 1)$, to $(-1, 1, -1)$, to $(-1, -1, 1)$, to $(1, -1, -1)$, and back to $(1, 1, 1)$.

18 Evaluate $\int_C yz\, dx + xz\, dy + xy\, dz$, where $C$ is the polygonal path from $(1, 0, 1)$, to $(0, 1, -1)$, to $(-1, 0, 1)$, to $(0, -1, -1)$, and back to $(1, 0, 1)$.

19 Evaluate $\int_C \mathbf{F} \cdot \mathbf{T}\, ds$, where $\mathbf{F} = (x\mathbf{i} + y\mathbf{j} + z\mathbf{k})/(x^2 + y^2 + z^2)^{3/2}$ and $C$: $x = \cos t$, $y = \sin t$, $z = z_0$, $0 \le t \le 2\pi$, $z_0$ a constant.

20 Evaluate $\int_C \mathbf{F} \cdot \mathbf{T}\, ds$, where $\mathbf{F} = (x\mathbf{i} + y\mathbf{j} + z\mathbf{k})/(x^2 + y^2 + z^2)^{3/2}$ and $C$: $x = \cos t$, $y = \sin t$, $z = z(t)$, $0 \le t \le 2\pi$, $z(t)$ any function that has a continuous derivative and $z(0) = z(2)$.

21 Evaluate div curl $\mathbf{F}$, the divergence of curl $\mathbf{F}$.

22 Evaluate curl $\nabla U$, the curl of the gradient of $U$. Assume the second partial derivatives are continuous.

## 18.9 INDEPENDENCE OF PATH OF LINE INTEGRALS IN THREE-DIMENSIONAL SPACE

If the region $R$ in the plane has no holes, we know that $\int_C M\, dx + N\, dy$ is independent of path in $R$ whenever $M$ and $N$ are continuous with scalar curl $N_x - M_y = 0$ in $R$. We also know that $\int M\, dx + N\, dy$ is independent of path in a region (with or without holes) if $\mathbf{F} = M\mathbf{i} + N\mathbf{j}$ is the gradient of a function $U$ in the region. Similar results hold for line integrals in three-dimensional space.

If $C_1$ and $C_2$ have the same initial point and the same terminal point, then $C_1 - C_2$ is a closed curve and

$$\int_{C_1} \mathbf{F} \cdot \mathbf{T}\, ds = \int_{C_2} \mathbf{F} \cdot \mathbf{T}\, ds \quad \text{if and only if} \quad \int_{C_1 - C_2} \mathbf{F} \cdot \mathbf{T}\, ds = 0.$$

If $C_1 - C_2$ is the complete boundary of a surface $S$ on which $\mathbf{F}$ has continuous first partial derivatives, then Stokes' Theorem gives

$$\int_{C_1 - C_2} \mathbf{F} \cdot \mathbf{T}\, ds = \iint_S \text{curl } \mathbf{F} \cdot \mathbf{n}\, ds.$$

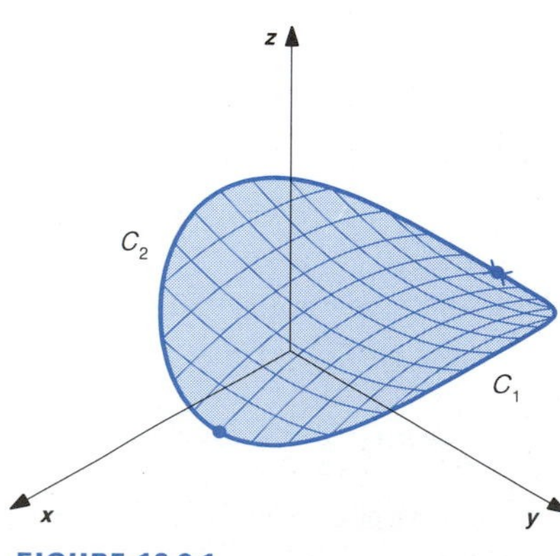

FIGURE 18.9.1

See Figure 18.9.1. If curl $\mathbf{F} = \mathbf{0}$ at all points of $S$, it follows that $\int_{C_1} \mathbf{F} \cdot \mathbf{T}\, ds = \int_{C_2} \mathbf{F} \cdot \mathbf{T}\, ds$. This argument can be used to verify the following theorem.

FIGURE 18.9.2

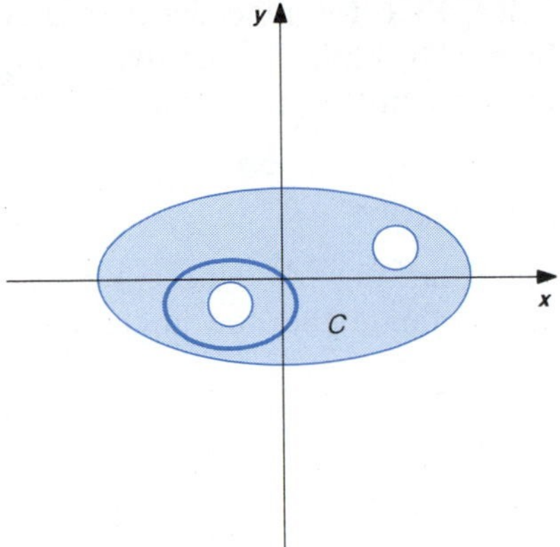


*Theorem 1*

**Suppose an open region $R$ in three-dimensional space has the property that every smooth simple closed curve in $R$ is the complete boundary of a bounded surface $S$ in $R$. If $\mathbf{F}$ has continuous first partial derivatives and curl $F = 0$ at all points of $R$, then $\int_C \mathbf{F} \cdot \mathbf{T}\, ds$ is independent of path in $R$.**

The property that every smooth simple closed curve in the three-dimensional region $R$ be the complete boundary of a bounded surface in $R$ corresponds to the condition that a region in the plane not contain any holes, although the situation is more complicated in three-dimensional space. For example, a region in the plane that has interior holes does not have the property that every smooth simple closed curve in the region is the complete boundary of a surface that is *in the plane region*. See Figure 18.9.2. A curve in three-dimensional space can be the complete boundary of a surface in a three-dimensional region even if the region has interior holes. The surface can go around a hole in three-dimensional space, as indicated in Figure 18.9.3a. However, the region that consists of all of three-dimensional space except for a line does not have the desired property; a curve that encircles the line cannot be the complete boundary of a bounded surface that does not intersect the line. See Figure 18.9.3b.

FIGURE 18.9.3

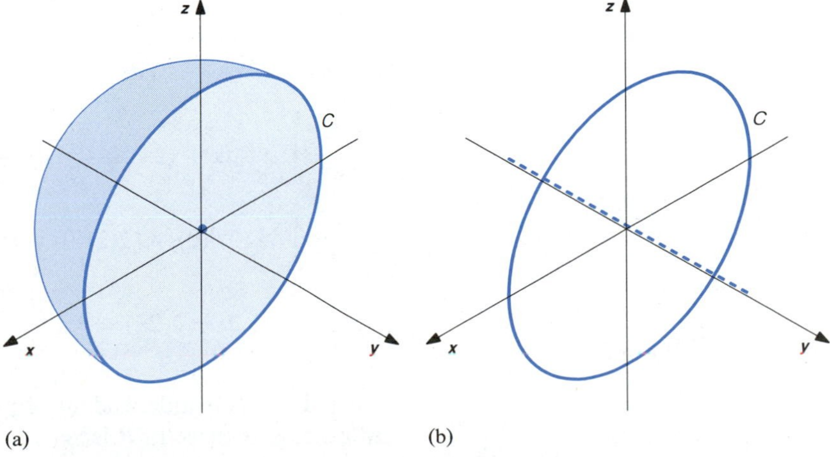

If $\int_C \mathbf{F} \cdot \mathbf{T}\, ds$ is independent of path in $R$, it may be easier to evaluate the integral by replacing the given curve with a more convenient path that has the same initial and terminal points.

**■ EXAMPLE 1**

Show that $\int_C y^2\, dx + (2xy + z^2)\, dy + (2yz + 2)\, dz$ is independent of path and evaluate the line integral for $C$: $x = 2 \cos t$, $y = 2 \sin t$, $z = t$, $0 \leq t \leq 2\pi$.

**SOLUTION**

We have $\mathbf{F} = M\mathbf{i} + N\mathbf{j} + P\mathbf{k}$ with $M = y^2$, $N = 2xy + z^2$, and $P = 2yz + 2$. The vector field $\mathbf{F}$ has continuous first partial derivatives for all $(x, y, z)$. The region $R$ that consists of all of three-dimensional space has the property that every simple closed curve in $R$ is the complete boundary of a surface $S$ in $R$. The line integral is then independent of path if curl $\mathbf{F} = \mathbf{0}$. Using the determinant form to evaluate curl $\mathbf{F}$, we have

$$\text{curl } \mathbf{F} = \begin{vmatrix} \mathbf{i} & \mathbf{j} & \mathbf{k} \\ \dfrac{\partial}{\partial x} & \dfrac{\partial}{\partial y} & \dfrac{\partial}{\partial z} \\ y^2 & 2xy + z^2 & 2yz + 2 \end{vmatrix}$$

$$= \begin{vmatrix} \dfrac{\partial}{\partial y} & \dfrac{\partial}{\partial z} \\ 2xy + z^2 & 2yz + 2 \end{vmatrix} \mathbf{i} - \begin{vmatrix} \dfrac{\partial}{\partial x} & \dfrac{\partial}{\partial z} \\ y^2 & 2yz + 2 \end{vmatrix} \mathbf{j} + \begin{vmatrix} \dfrac{\partial}{\partial x} & \dfrac{\partial}{\partial y} \\ y^2 & 2xy + z^2 \end{vmatrix}$$

$$= (2z - 2z)\mathbf{i} - (0 - 0)\mathbf{j} + (2y - 2y)\mathbf{k} = \mathbf{0}.$$

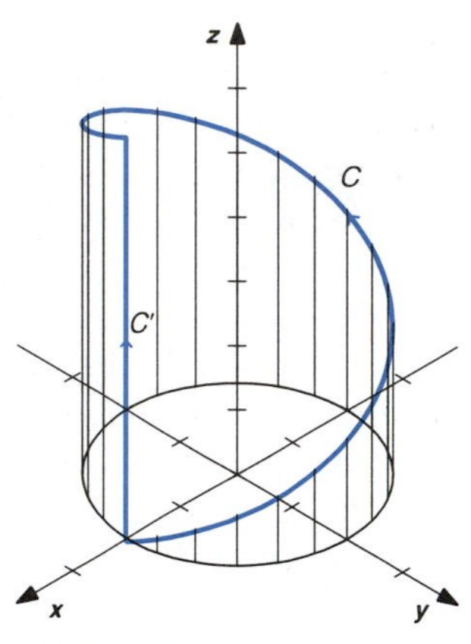


FIGURE 18.9.4

The given curve $C$ has initial point $(2, 0, 0)$ and terminal point $(2, 0, 2\pi)$. Since the integral is independent of path, let us replace $C$ with the line segment between these two points. See Figure 18.9.4. We have

$$C'\colon\ x = 2, \quad y = 0, \quad z = t, \qquad 0 \leq t \leq 2\pi.$$

Then

$$\int_C y^2\, dx + (2xy + z^2)\, dy + (2yz + 2)\, dz$$

$$= \int_{C'} y^2\, dx + (2xy + z^2)\, dy + (2yz + 2)\, dz$$

$$= \int_0^{2\pi} \{(0)^2(0) + [2(2)(0) + (t)^2](0) + [2(0)(t) + 2](1)\}\, dt$$

$$= \int_0^{2\pi} 2\, dt = 2(2\pi) = 4\pi. \quad ■$$

If $\int_C \mathbf{F} \cdot \mathbf{T}\, ds$ is independent of path in $R$, then the line integral of $\mathbf{F} \cdot \mathbf{T}$ over a closed curve in $R$ is zero.

■ **EXAMPLE 2**

If $\mathbf{F}$ is the inverse square field

$$\frac{x\mathbf{i} + y\mathbf{j} + z\mathbf{k}}{(x^2 + y^2 + z^2)^{3/2}},$$

show that $\int_C \mathbf{F} \cdot \mathbf{T}\, ds$ is independent of path in the region $R$ that consists of all of three-dimensional space except the origin, and evaluate the line integral for the curve

$$C\colon\ x = \cos t, \quad y = \sin t, \quad z = \frac{\sin 2t}{2}, \qquad 0 \le t \le 2\pi.$$

**SOLUTION**

We have $M = x(x^2 + y^2 + z^2)^{-3/2}$, $N = y(x^2 + y^2 + z^2)^{-3/2}$, and $P = z(x^2 + y^2 + z^2)^{-3/2}$. For $(x, y, z) \neq (0, 0, 0)$, we then have

$$M_y = x\left(-\frac{3}{2}\right)(x^2 + y^2 + z^2)^{-5/2}(2y) = -3xy(x^2 + y^2 + z^2)^{-5/2},$$

$$M_z = -3xz(x^2 + y^2 + z^2)^{-5/2},$$

$$N_x = -3xy(x^2 + y^2 + z^2)^{-5/2},$$

$$N_z = -3yz(x^2 + y^2 + z^2)^{-5/2},$$

$$P_x = -3xz(x^2 + y^2 + z^2)^{-5/2},$$

$$P_y = -3yz(x^2 + y^2 + z^2)^{-5/2}.$$

It follows that

$$\text{curl } \mathbf{F} = (P_y - N_z)\mathbf{i} + (M_z - P_x)\mathbf{j} + (N_x - M_y)\mathbf{k} = \mathbf{0}$$

for all $(x, y, z)$ in the region $R$ that consists of all of three-dimensional space except the origin. Since every smooth simple closed curve in $R$ is the complete boundary of a bounded surface in $R$, Theorem 1 implies that $\int_C \mathbf{F} \cdot \mathbf{T}\, ds$ is independent of path in $R$.

Since $C$ is a closed path in $R$ and $\int_C \mathbf{F} \cdot \mathbf{T}\, ds$ is independent of path in $R$, we have

$$\int_C \mathbf{F} \cdot \mathbf{T}\, ds = 0.$$

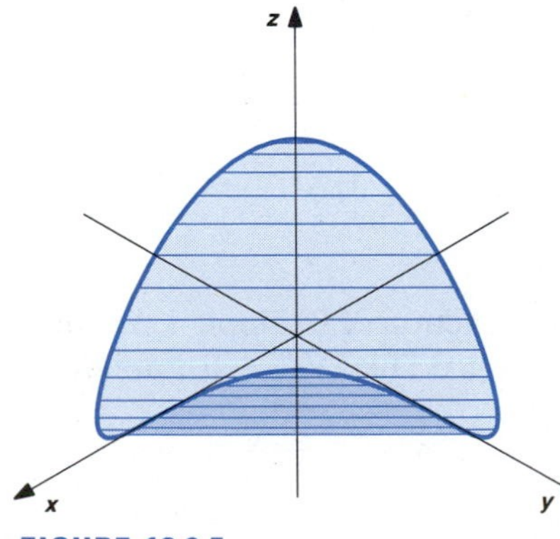

FIGURE 18.9.5

The graph of $C$ and a bounded surface in $R$ with complete boundary $C$ is sketched in Figure 18.9.5. ■

The following theorem can be proved in the same way that the corresponding result in the plane was proved in Section 20.4.

*Theorem 2*

**Suppose $M$, $N$, and $P$ are continuous in a region $R$.**

**(a) If there is a function $U$ such that $\nabla U = M\mathbf{i} + N\mathbf{j} + P\mathbf{k}$ in $R$, then $\int_C M\,dx + N\,dy + P\,dz$ is independent of path in $R$. Moreover, if $C$ is a smooth curve in $R$ with initial point $(x(a), y(a), z(a))$ and terminal point $(x(b), y(b), z(b))$, then**

$$\int_C M\,dx + N\,dy + P\,dz = U(x(b), y(b), z(b)) - U((x(a), y(a), z(a)).$$

**(b) If $\int_C M\,dx + N\,dy + P\,dz$ is independent of path in $R$, then there is a function $U$ such that $\nabla U = N\mathbf{i} + N\mathbf{j} + P\mathbf{k}$ in $R$.**

As in the case of two-dimensional vector fields, a three-dimensional vector field $\mathbf{F}$ is called **conservative** in $R$ if there is a function $U$ such that $\mathbf{F} = \nabla U$ in $R$. The function $P = -U$ is then called the **potential** of $\mathbf{F}$.

The following theorem corresponds to Theorem 3 of Section 18.4.

*Theorem 3*

**If $\mathbf{F}$ has continuous first partial derivatives in a region $R$ that has the property that any smooth simple closed curve in $R$ is the complete boundary of a surface in $R$, and curl $\mathbf{F} = \mathbf{0}$ in $R$, then there is a function $U$ such that $\nabla U = \mathbf{F}$ in $R$, so $\mathbf{F}$ is conservative in $R$.**

**If $\mathbf{F}$ has continuous first derivatives in $R$ and there are points of $R$ where curl $\mathbf{F} \neq \mathbf{0}$, then there is no function $U$ such that $\nabla U = \mathbf{F}$ in $R$, so $\mathbf{F}$ is not conservative in $R$.**

■ *Proof* The first statement of Theorem 3 follows from Theorems 1 and 2. To prove the second statement, we note that $\nabla U = M\mathbf{i} + N\mathbf{j} + P\mathbf{k}$ implies $U_x = M$, $U_y = N$, and $U_z = P$. Then $U_{xy} = M_y$, $U_{xz} = M_z$, $U_{yx} = N_y$, $U_{yz} = N_z$, $U_{zx} = P_x$, and $U_{zy} = P_y$. The continuity of the first partial derivatives of $\mathbf{F}$ implies the continuity of the second partials of $U$, so we have $U_{xy} = U_{yx}$, $U_{xz} = U_{zx}$, and $U_{yz} = U_{zy}$. It follows that curl $\mathbf{F} = (P_y - N_z)\mathbf{i} + (M_z - P_x)\mathbf{j} + (N_x - M_y)\mathbf{k} = \mathbf{0}$. ■

If curl $\mathbf{F} = \mathbf{0}$ in an appropriate region $R$, we can find a function $U$ such that $\nabla U = \mathbf{F}$ by using a method similar to that used in Section 18.3.

■ **EXAMPLE 3**

Find a function $U$ such that $\nabla U = y^2\mathbf{i} + (2xy + z^2)\mathbf{j} + (2yz + 2)\mathbf{k}$.

**SOLUTION**

$\mathbf{F} = y^2\mathbf{i} + (2xy + z^2)\mathbf{j} + (2yz + 2)\mathbf{k}$ is the function of Example 1. We have already verified that curl $\mathbf{F} = \mathbf{0}$ for all $(x, y, z)$. Then $\nabla U = \mathbf{F}$ implies

$$U_x = y^2, \quad U_y = 2xy + z^2, \quad \text{and} \quad U_z = 2yz + 2.$$

$$U_x = y^2 \quad \text{implies} \quad U = xy^2 + h(y, z).$$

Note that we integrated with respect to $x$ while holding $y$ and $z$ constant; the "constant" of integration $h(y, z)$ can depend on $y$ and $z$.

$U = xy^2 + h(y, z)$ implies $U_y = x(2y) + h_y(y, z)$.

$U_y = 2xy + z^2$ implies $x(2y) + h_y(y, z) = 2xy + z^2$, so

$h_y(y, z) = z^2$,

$h(y, z) = yz^2 + k(z)$.

Then $U = xy^2 + yz^2 + k(z)$, so $U_z = 2yz + k'(z)$.

$U_z = 2yz + 2$ implies $2yz + k'(z) = 2yz + 2$, so

$k'(z) = 2$,

$k(z) = 2z + C$.

We conclude that $U = xy^2 + yz^2 + 2z + C$. ■

■ **EXAMPLE 4**

Find a function $U$ such that

$$\nabla U = \frac{x\mathbf{i} + y\mathbf{j} + z\mathbf{k}}{(x^2 + y^2 + z^2)^{3/2}}, \qquad (x, y, z) \neq (0, 0, 0).$$

**SOLUTION**

We have seen in Example 2 that the inverse square field satisfies curl $\mathbf{F} = \mathbf{0}$ for $(x, y, z) \neq (0, 0, 0)$. We want a function $U$ such that

$U_x = x(x^2 + y^2 + z^2)^{-3/2}$, $U_y = y(x^2 + y^2 + z^2)^{-3/2}$, and

$U_z = z(x^2 + y^2 + z^2)^{-3/2}$.

$U_x = x(x^2 + y^2 + z^2)^{-3/2}$ implies

$$\begin{aligned} U &= \int x(x^2 + y^2 + z^2)^{-3/2}\,dx \\ &= -(x^2 + y^2 + z^2)^{-1/2} + h(y, z). \end{aligned} \qquad [u = x^2 + y^2 + z^2,\ du = 2x\,dx]$$

Then

$U_y = -\left(-\dfrac{1}{2}\right)(x^2 + y^2 + z^2)^{-3/2}(2y) + h_y(y, z)$, so

$-\left(-\dfrac{1}{2}\right)(x^2 + y^2 + z^2)^{-3/2}(2y) + h_y(y, z) = y(x^2 + y^2 + z^2)^{-3/2}$,

$h_y(y, z) = 0$,

$h(y, z) = k(z)$.

Then

$$U = -(x^2 + y^2 + z^2)^{-1/2} + k(z),$$

$$U_z = -\left(-\frac{1}{2}\right)(x^2 + y^2 + z^2)^{-3/2}(2z) + k'(z), \quad \text{so}$$

$$-\left(-\frac{1}{2}\right)(x^2 + y^2 + z^2)^{-3/2}(2z) + k'(z) = z(x^2 + y^2 + z^2)^{-3/2},$$

$$k'(z) = 0,$$

$$k(z) = C.$$

We conclude that $U = -(x^2 + y^2 + z^2)^{-1/2} + C$. ■

From Examples 2 and 4 we conclude that the inverse square field **F** is conservative with potential function $P = (x^2 + y^2 + z^2)^{-1/2}$ in the region that consists of all points in three-dimensional space except the origin.

If $\int_C M\,dx + N\,dy + P\,dz$ is independent of path in $R$, we use the notation

$$\int_A^B M\,dx + N\,dy + P\,dz$$

to denote the integral over any curve in $R$ that has initial point $A$ and terminal point $B$. As in the case of line integrals in the plane, the integral can be evaluated either by choosing $C$ to be any convenient curve from $A$ to $B$, or by finding a function $U$ such that $\nabla U = \mathbf{F}$ and using the formula

$$\int_A^B M\,dx + N\,dy + P\,dz = U(B) - U(A).$$

■ **EXAMPLE 5**

Show that the integral

$$\int_C y^2\,dx + (2xy + z^2)\,dy + (2yz + 2)\,dz$$

is independent of path and evaluate

$$\int_{(2,0,0)}^{(2,0,2\pi)} y^2\,dx + (xy + z^2)\,dy + (2yz + 2)\,dz.$$

**SOLUTION**

We showed in Example 1 that $\mathbf{F} = y^2 + (2xy + z^2)\mathbf{j} + (2yz + 2)\mathbf{k}$ satisfies curl $\mathbf{F} = \mathbf{0}$, so the line integral is independent of path. In Example 3 we determined that

$$U(x, y, z) = xy^2 + yz^2 + 2z$$

satisfies $\nabla U = \mathbf{F}$. (We have used $C = 0$ in the formula for $U$ that we found in Example 3.) It follows that

$$\int_{(2,0,0)}^{(2,0,2\pi)} y^2\,dx + (2xy + z^2)\,dy + (2yz + 2)\,dz$$
$$= U(2, 0, 2\pi) - U(2, 0, 0)$$
$$= [(2)(0)^2 + (0)(2\pi)^2 + 2(2\pi)] - [(2)(0)^2 + (0)(0)^2 + 2(0)]$$
$$= 4\pi.$$

This agrees with the result of Example 1, where we evaluated the integral by using parametric equations of the line segment from (2, 0, 0) to (2,0,2π). ■

## EXERCISES 18.9

Show that the line integrals in Exercises 1–4 are independent of path and find their value for the indicated curves.

1 $\int_C 2xyz\,dx + x^2z\,dy + x^2y\,dz$

(a) $C$ is the polygonal path from (0, 0, 0), to (2, 0, 0), to (2, 3, 0), to (2, 3, 4), to (0, 3, 4)

(b) $C$ is the polygonal path from (0, 0, 0), to (2, 0, 0), to (2, 3, 0), to (2, 3, 4), to (0, 3, 4), to (0, 0, 0)

2 $\int_C 2xz^3\,dx + 3x^2z^2\,dz$

(a) $C$: $x = \cos t$, $y = \sin t$, $z = t$, $0 \le t \le 2\pi$

(b) $C$: $x = \cos t$, $y = \sin t$, $z = \sin 2t$, $0 \le t \le 2\pi$

3 $\int_C x\,dx + y\,dy + z\,dz$

(a) $C$: $x = 2\cos t$, $y = 3\sin t$, $z = \cos t$, $0 \le t \le \pi$

(b) $C$: $x = 2\cos t$, $y = 3\sin t$, $z = \cos t$, $0 \le t \le 2\pi$

4 $\int_C (x + y)\,dx + x\,dy + z\,dz$

(a) $C$: $x = t - t^2$, $y = (t - t^2)^2$, $z = t^3$, $0 \le t \le 1$

(b) $C$: $x = t^2 - t$, $y = (t^2 - t)^2$, $z = t^3$, $0 \le t \le 1$

In Exercises 5–8 find a function $U$ such that $\nabla U$ is equal to the given vector function $\mathbf{F}$.

5 $\mathbf{F} = (2xy + z^2)\mathbf{i} + (x^2 + 2yz)\mathbf{j} + (y^2 + 2xz)\mathbf{k}$

6 $\mathbf{F} = (x^2 + y)\mathbf{i} + (x + z)\mathbf{j} + (y + 2z)\mathbf{k}$

7 $\mathbf{F} = \dfrac{x}{\sqrt{x^2 + y^2 + z^2}}\mathbf{i} + \dfrac{y}{\sqrt{x^2 + y^2 + z^2}}\mathbf{j} + \dfrac{z}{\sqrt{x^2 + y^2 + z^2}}\mathbf{k}$, $(x, y, z) \ne (0, 0, 0)$

8 $\mathbf{F} = \dfrac{x}{x^2 + y^2 + z^2}\mathbf{i} + \dfrac{y}{x^2 + y^2 + z^2}\mathbf{j} + \dfrac{z}{x^2 + y^2 + z^2}\mathbf{k}$, $(x, y, z) \ne (0, 0, 0)$

Show that the line integrals in Exercises 9–12 are independent of path and find their value.

9 $\int_{(0,0,0)}^{(1,2,-2)} x^2\,dx + y^2\,dy + z^2\,dz$

10 $\int_{(0,0,0)}^{(1,3,2)} (3x^2 + y^2)\,dx + 2xy\,dy + dz$

11 $\int_{(0,0,0)}^{(1,-1,2)} 3x^2\,dx + z^2\,dy + (2yz + 1)\,dz$

12 $\int_{(0,0,0)}^{(3,\pi/2,4)} yz\cos xy\,dx + xz\cos xy\,dy + \sin xy\,dz$

13 Evaluate

$$\int_C x(x^2 + y^2 + z^2)^{-3/2}\,dx + y(x^2 + y^2 + z^2)^{-3/2}\,dy + z(x^2 + y^2 + z^2)^{-3/2}\,dz,$$

where $C$ is any smooth curve that is contained in the sphere $x^2 + y^2 + z^2 = r^2$.

14 The electrostatic force on a charged particle at the point $(x, y, z)$ is $\mathbf{F} = -5(x\mathbf{i} + y\mathbf{j} + z\mathbf{k})/(x^2 + y^2 + z^2)^{3/2}$. Find the work associated with $\mathbf{F}$ as the particle moves along the polygonal path from (1, 0, 0) to (0, 1, 0), to (0, 1, 1).

## REVIEW EXERCISES

1 Set up and evaluate a line integral that gives the lateral surface area of that part of the cylinder $4x^2 + y^2 = 4$ in the first octant below $z = xy$.

2 Set up and evaluate a line integral that gives the total mass of a wire bent in the shape of the curve

$$x = 2\cos t, \quad y = 3\sin t, \quad z = 4t, \qquad 0 \le t \le \pi/2,$$

if the density at a point on the wire is given by the formula $P(x, y, z) = xy$.

Evaluate the line integrals in Exercises 3–14.

3 $\int_C y\,dx - x\,dy$, $C$: $x = \cos t$, $y = \sin t$, $0 \le t \le 2\pi$

4 $\int_C y\,dx + dy$, $C$ goes from $(0, 0)$, along $y = x^2$ to $(1, 1)$.

5 $\int_C (x^2 - y^2)\,dx - 2xy\,dy$, $C$ goes from $(0, 0)$, along $y = \sin x$ to $(2\pi, 0)$.

6 $\int_C 2xy\,dx + (x^2 + y)\,dy$, $C$ goes along $y = x^2$ from $(0, 0)$ to $(1, 1)$, then along $x + y = 2$ from $(1, 1)$ to $(0, 2)$.

7 $\int_C xy^2\,dx + x^2y\,dy$, $C$ is the polygonal path from $(0, 0)$, to $(1, 0)$, to $(2, 2)$, to $(0, 1)$, to $(0, 0)$.

8 $\int_C 1/2\ln(x^2 + y^2)\,dx + \tan^{-1}(x/y)\,dy$, $C$ goes from $(3, 2)$, counterclockwise once around $(x - 2)^2 + (y - 2)^2 = 1$.

9 $\int_C (y^2 - 2xy)\,dx + 2xy\,dy$, $C$ is the polygonal path from $(0, 0)$, to $(1, 0)$, to $(1, 1)$, to $(0, 0)$.

10 $\int_C (y\sin x - y)\,dx + (y - \cos x)\,dy$, $C$ is the polygonal path from $(0, 0)$, to $(1, 0)$, to $(1, 1)$, and then along $y = x^2$ from $(1, 1)$ to $(0, 0)$.

11 $\int_C x\,dx + y\,dy + z\,dz$; $C$: $x = 4\cos t$, $y = 4\sin t$, $z = \sin 2t$, $0 \le t \le 2\pi$

12 $\int_C y\,dx + x\,dy + z\,dz$; $C$: $x = e^{-t}\cos t$, $y = 2t$, $z = e^{-t}\sin t$, $0 \le t \le 2\pi$

13 $\int_C (z + y)\,dx + (x + z)\,dy + (y + z)\,dz$, $C$ goes from $(1, 0, 0)$, counterclockwise once around the intersection of the sphere $x^2 + y^2 + z^2 = 1$ and the surface $z = xy$.

14 $\int_C x(x^2 + y^2 + z^2)^{-3/2}\,dx + y(x^2 + y^2 + z^2)^{-3/2}\,dy + z(x^2 + y^2 + z^2)^{-3/2}\,dz$; $C$: $x = \cos t$, $y = \sin t$, $z = \sin t + \cos t$, $0 \le t \le \pi/2$.

In Exercises 15–16 an object is to be moved from $(0, 0)$ to $(3, 0)$, subject to the given force **F**. Determine whether the work done is independent of path. If it is, evaluate the work.

15 $\mathbf{F} = x\mathbf{i} - y\mathbf{j}$

16 $\mathbf{F} = y\mathbf{i} - x\mathbf{j}$

Find the scalar curl of **F** and the circulation around $C$ in Exercises 17–18.

17 $\mathbf{F} = y\mathbf{i} - x\mathbf{j}$, $C$ goes from $(2, 0)$, counterclockwise once around $x^2 + y^2 = 4$.

18 $\mathbf{F} = x\mathbf{i} + y\mathbf{j}$, $C$ goes from $(3, 0)$, counterclockwise once around $x^2 + y^2 = 9$.

Find the divergence of **F** and the flux across $C$ in Exercises 19–20.

19 $\mathbf{F} = y\mathbf{i} - x\mathbf{j}$, $C$ goes from $(2, 0)$ counterclockwise once around $x^2 + y^2 = 4$.

20 $\mathbf{F} = x\mathbf{i} + y\mathbf{j}$, $C$ goes from $(3, 0)$, counterclockwise once around $x^2 + y^2 = 9$.

In Exercises 21–24 find a function $U(x, y)$ such that $\nabla U = \mathbf{F}$ for all $(x, y)$ or indicate "not possible."

21 $\mathbf{F} = e^x\cos y\mathbf{i} - (e^x\sin y + 2y)\mathbf{j}$

22 $\mathbf{F} = (1 + 2xy)\mathbf{i} + (x^2 + y)\mathbf{j}$

23 $\mathbf{F} = y\mathbf{i} - x\mathbf{j}$

24 $\mathbf{F} = x\mathbf{i} - y\mathbf{j}$

In Exercises 25–26 find a function $U(x, y, z)$ such that $\nabla U = \mathbf{F}$ for all $(x, y, z)$ or indicate "not possible."

25 $\mathbf{F} = (y^2 - z^2)\mathbf{i} + 2xy\mathbf{j} + (1 - 2xz)\mathbf{k}$

26 $\mathbf{F} = y\cos xy\mathbf{i} + (x\cos xy + z\sin yz)\mathbf{j} + y\sin yz\mathbf{k}$

27 Evaluate $\iint_S z\,dS$, where $S$ is that part of the sphere $x^2 + y^2 + z^2 = 5$ above the plane $z = 1$.

28 Evaluate $\iint_S 1/z\,dS$, where $S$ is that part of the sphere $x^2 + y^2 + z^2 = 5$ above the plane $z = 2$.

Evaluate $\iint_S \mathbf{F}\cdot\mathbf{n}\,dS$ in Exercises 29–34.

29 $\mathbf{F} = x\mathbf{i} + y\mathbf{j} + z\mathbf{k}$; $S$ is that part of the paraboloid $z = x^2 + y^2$ with $z \le 1$; **n** downward.

30 $\mathbf{F} = x\mathbf{i} + y\mathbf{j} + z\mathbf{k}$; $S$ is that part of the plane $z = 1$ with $x^2 + y^2 \le 1$; **n** upward.

31 $\mathbf{F} = x\mathbf{i} + y\mathbf{j} + z\mathbf{k}$; $S$ is the complete boundary of the region bounded by $z = x^2 + y^2$ and $z = 1$; **n** outward.

32 $\mathbf{F} = 2x\mathbf{i} + 2y\mathbf{j} + z\mathbf{k}$; $S$ is the complete boundary of the region bounded by the cone $z = \sqrt{x^2 + y^2}$ and the plane $z = 4$; **n** outward.

33 $\mathbf{F} = 2x\mathbf{i} + 2y\mathbf{j} + z\mathbf{k}$; $S$ is that part of the cone $z = \sqrt{x^2 + y^2}$ with $z \le 4$; **n** downward.

34 $\mathbf{F} = 2x\mathbf{i} + 2y\mathbf{j} + z\mathbf{k}$; $S$ is that part of the plane $z = 4$ with $x^2 + y^2 \le 16$; **n** upward.

35 Evaluate $\iint_S \text{curl}\,\mathbf{F}\cdot\mathbf{n}\,dS$, where $\mathbf{F} = -yz\mathbf{i} + xz\mathbf{j} + xy\mathbf{k}$, $S$ is that part of the surface $z = x^2 + y^2$ below $z = 1$, and **n** is directed downward.

**36** Evaluate $\iint_S \text{curl } \mathbf{F} \cdot \mathbf{n}\, dS$, where $\mathbf{F} = -yz\mathbf{i} + xz\mathbf{j} + xy\mathbf{k}$, $S$ is that part of the sphere $x^2 + y^2 + z^2 = 5$ below $z = 2$, and $\mathbf{n}$ is directed upward.

**37** Evaluate $\iint_S \text{curl } \mathbf{F} \cdot \mathbf{n}\, dS$, where $\mathbf{F} = -yz\mathbf{i} + xz\mathbf{j} + xy\mathbf{k}$, $S$ is that part of the cylinder $x^2 + y^2 = 1$ between $z = 0$, and $z = 4$, and $\mathbf{n}$ is directed outward.

**38** Find a unit vector $\mathbf{n}$ normal to the paraboloid $2y = x^2 + z^2$ and directed outward from the paraboloid.

Show that the differential equations in Exercises 39–40 are exact and find their solution.

**39** $(2x + y) + (x + 2y)\, dy/dx = 0$

**40** $2e^{2x} \cos y - e^{2x} \sin y\, dy/dx = 0$

Solve the initial value problems in Exercises 41–42 and sketch the graphs of the solutions.

**41** $y\, dx + (x - 1)\, dy = 0$, $y(2) = 6$

**42** $(2xy - 6y)\, dx + (x - 3)^2\, dy = 0$: (a) $y(2) = 1$ (b) $y(4) = 1$

# *Answers to odd-numbered exercises*

## CHAPTER ONE

### *Exercises 1.1*

**1** $x < 4$

**3** $x < -5/2$

**5** $1 < x < 2$

**7** $1 < x < 2$

**9** $1 < x < 2$

**11** $2/3 < x < 2$

**13** $1 < x < 2$

**15** 1, 7

**17** 5, 5

**19** 1

**21** 3

**23** $x^{-2}$

**25** $x^5$

**27** $x^{-6}$

**29** $x^{-1/2}$

**31** $x$

**33** 27

**35** 2

**37** 4

**39** 4

**41** 9

**43** 2.8058551

**45** 6.47300784

**47** 0.00781250

**49**

| Interval notation | Inequality notation |
|---|---|
| $(0, 1)$ | $0 < x < 1$ |
| $[1, 3]$ | $1 \le x \le 3$ |
| $(-1, 1]$ | $-1 < x \le 1$ |
| $(0, \infty)$ | $x > 0$ |
| $(-\infty, 1]$ | $x \le 1$ |

## Exercises 1.2

1 $-6x + 5$

3 $(2x + 1)(30x + 1)$

5 $\dfrac{-(x+3)}{2x^{5/2}}$

7 $\dfrac{x(7x+6)}{3(x+1)^{2/3}}$

9 $x + 2,\ x \neq 2$

11 $-\dfrac{x+2}{4x^2},\ x \neq 2$

13 $-1, 6$

15 None

17 $0, 2$

19 $-1/3, -2$

21 (a) 2 (b) 0

23 (a) None (b) $x \leq -1/2$

25 $x < -2,\ x > 1$

27 $-1 < x < 0,\ x > 2$

29 $x > -1$

31 $x < 0,\ x > 1$

33 $x > 0$

35 $x > 1/2$

37 $x < -1,\ 0 < x < 1$

39 $0 < x < 1$

## Exercises 1.3

1 $(1, -1)$

3 $(2, 1), (0, 1)$

5 $(-3, 2)$

7

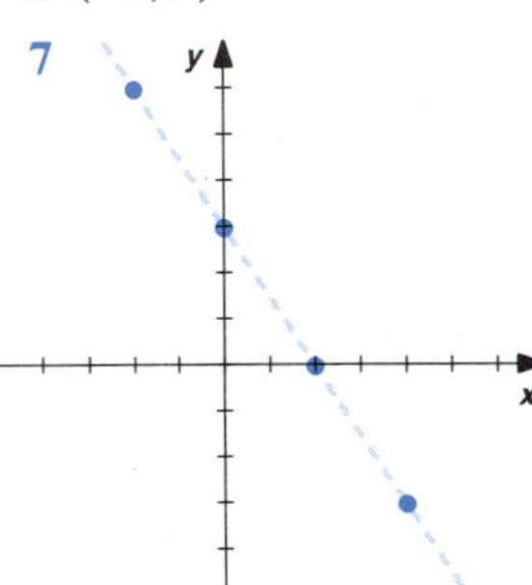


9

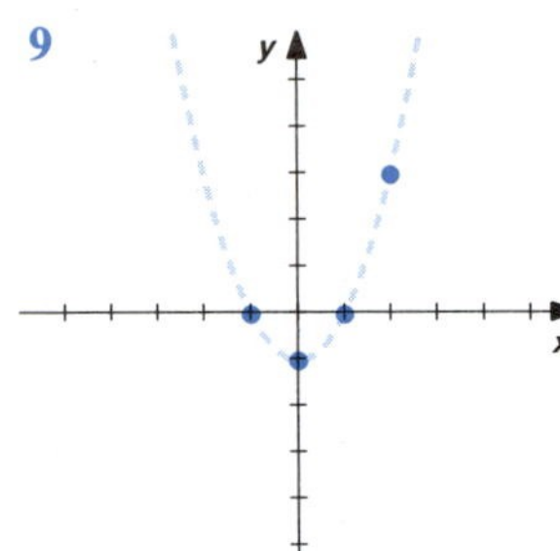


11

13 (a) $(x, 2x)$ (b) $(y/2, y)$

15 (a) $(x, x^3 - 1)$ (b) $((y + 1)^{1/3}, y)$

17 (a) $(x, \sqrt{x+1}), (x, -\sqrt{x+1})$ (b) $(y^2 - 1, y)$

19 5; $-3/4$; $(1, 1/2)$

21 5; undefined; $(1, 1/2)$

23 $\sqrt{5x^2-6x+9}$; $\dfrac{2x}{x-3}$; $\left(\dfrac{x+3}{2}, x\right)$

25 $|x^2 + x - 2|$; undefined; $\left(x, \dfrac{x^2-x}{2}\right)$

27 $|4 - y^2|$; 0, $y \neq \pm 2$; $\left(\dfrac{4-y^2}{2}, y\right)$

29 $-2/3,\ x \neq -6$

31 $x,\ x \neq 0$

33 $x^2 + x + 1,\ x \neq 1$

35 $x - 1,\ x \neq 2$

37 $-1/x,\ x \neq 1$

39 $-\dfrac{2+x}{3(x^2-1)},\ x \neq 2$

## Exercises 1.4

1

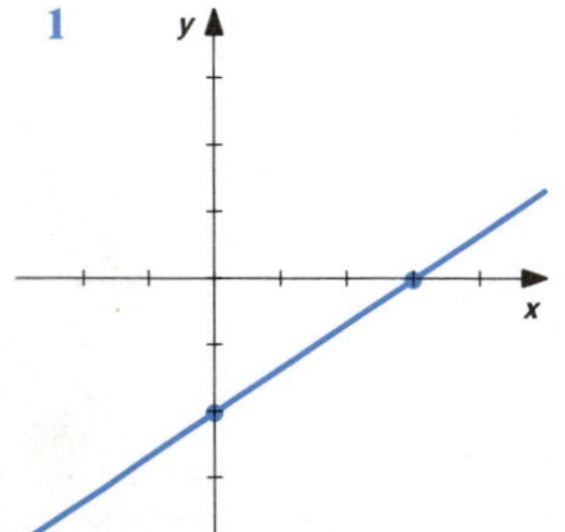


3

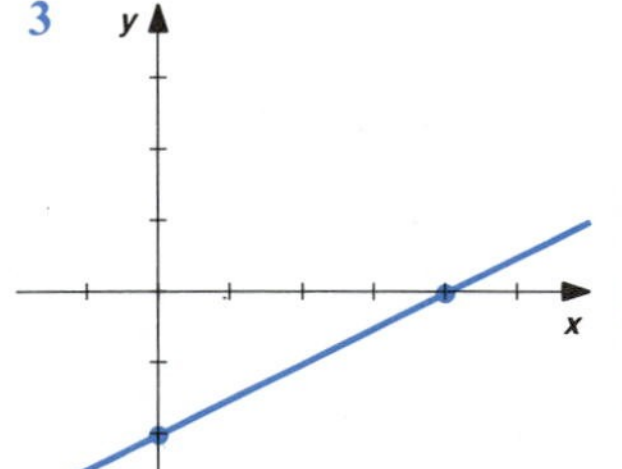


5

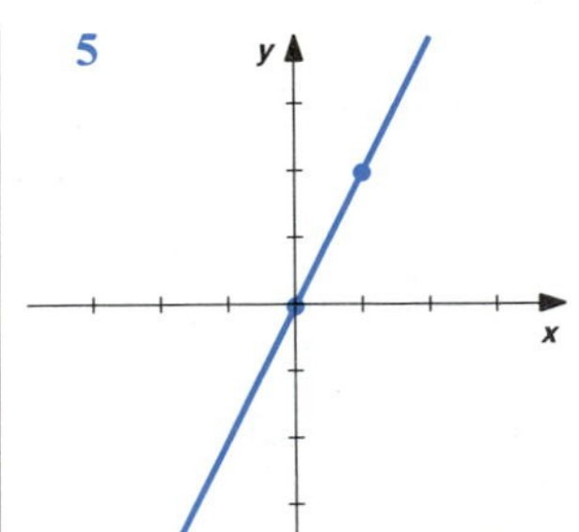


7

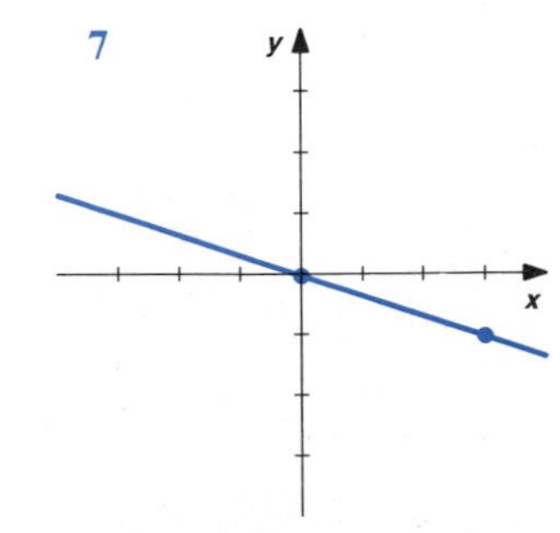

9

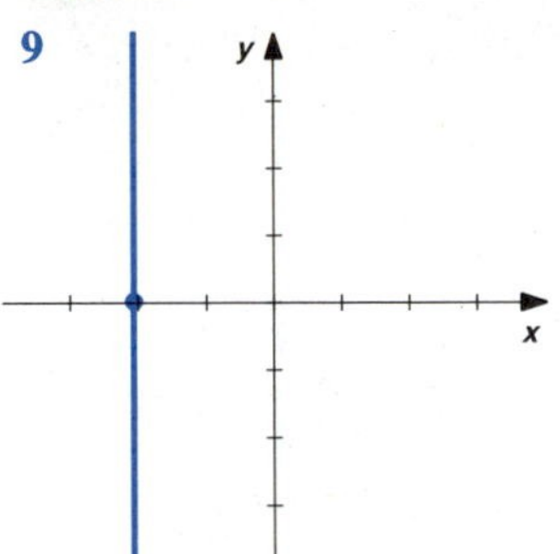


11

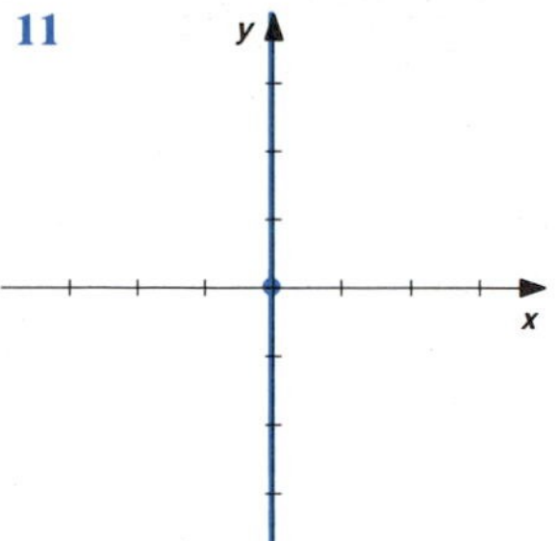


13

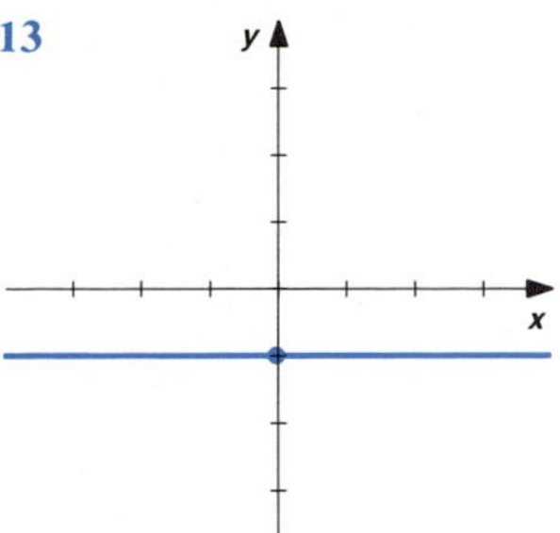


**15** $x + y = 5$ **17** $3y = x + 7$

**19** $x + y = 3$ **21** $2x - y = 1$

**23** $x - y + 2 = 0$ **25** $y = 2$

**27** $x = -1$ **29** $6x + y + 9 = 0$

**31** $y + 1 = 0$ **33** $x + y = 1$

**35** $y = 3x + 2$ **37** $3x + 2y = 6$

**39** $y = 2x$ **41** $4x - 2y = 3$

**43** $y = 1$

## *Exercises 1.5*

1

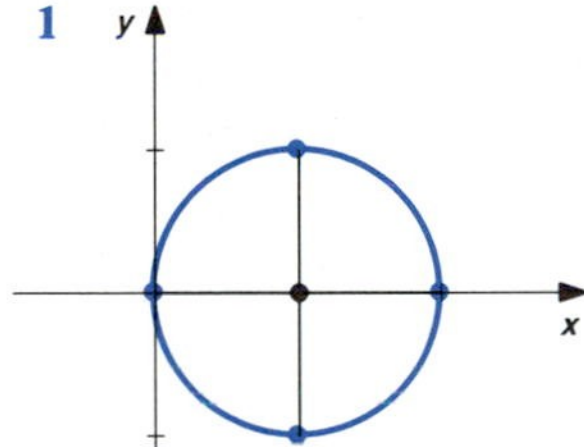


3

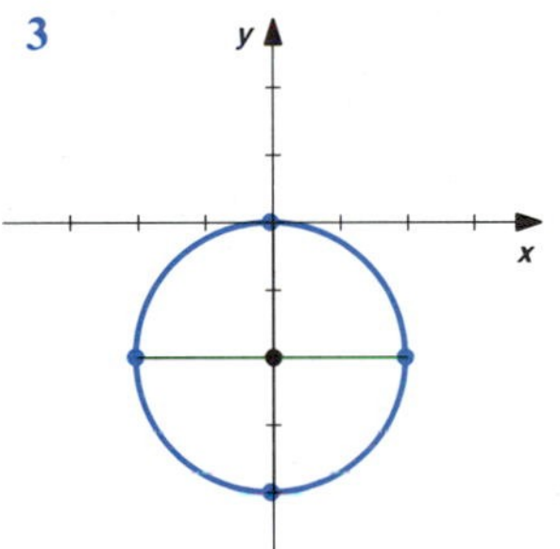


5

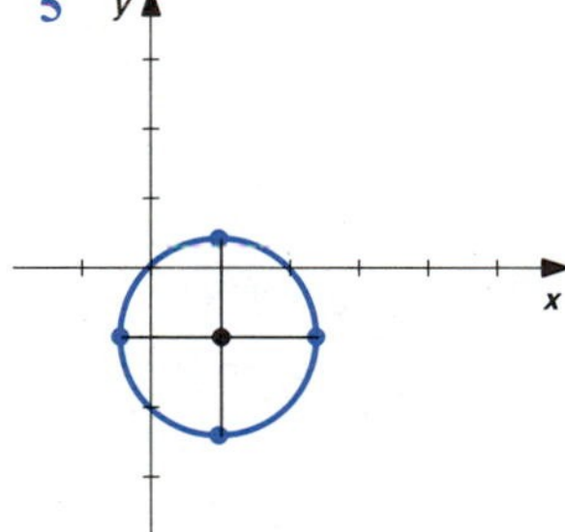


7

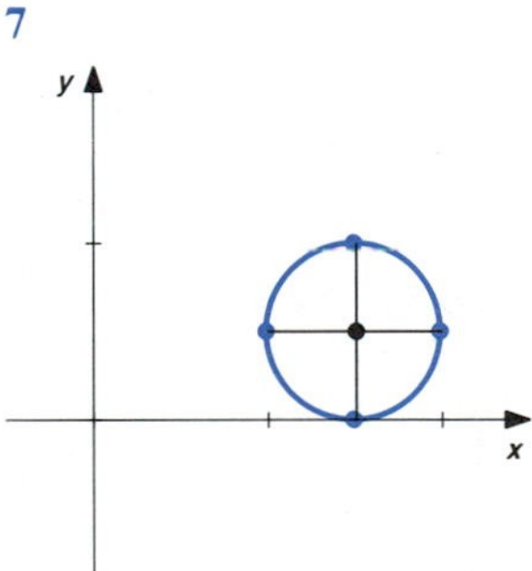


9

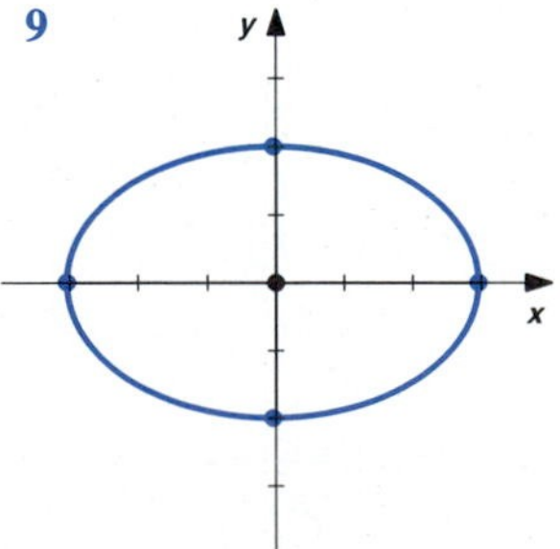


11

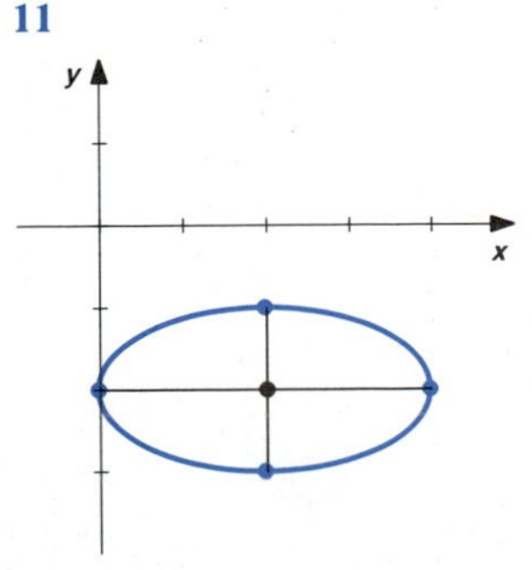


13

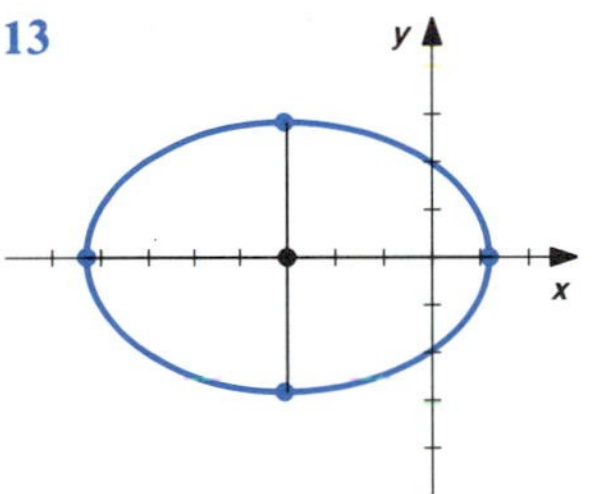


15

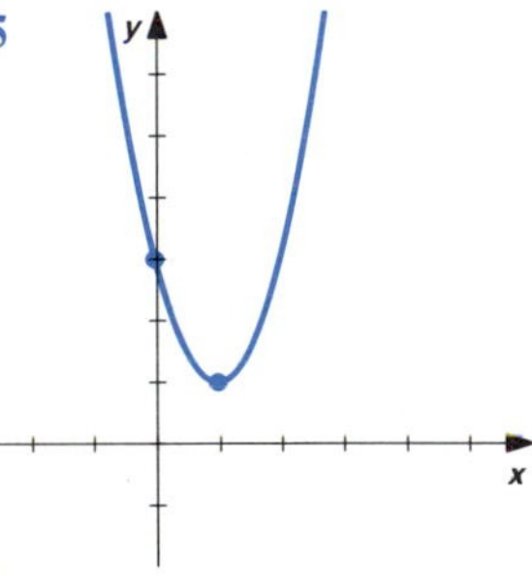


17

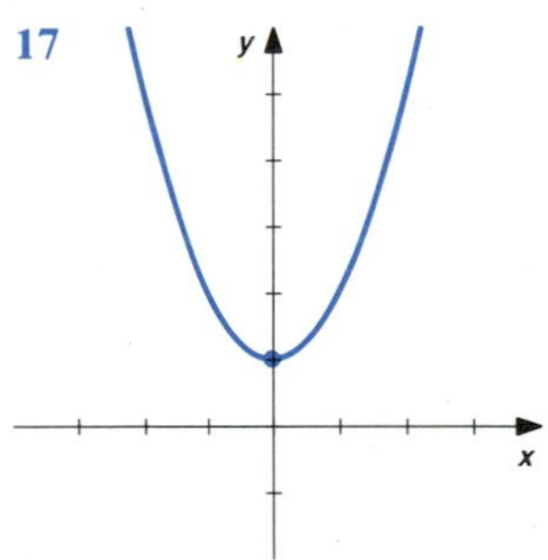


19

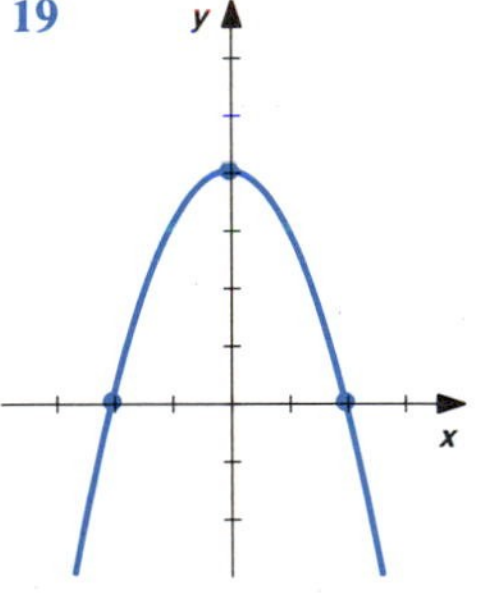


21

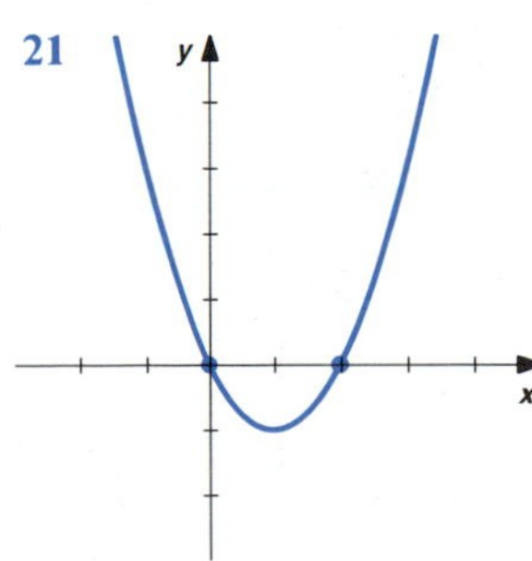


23

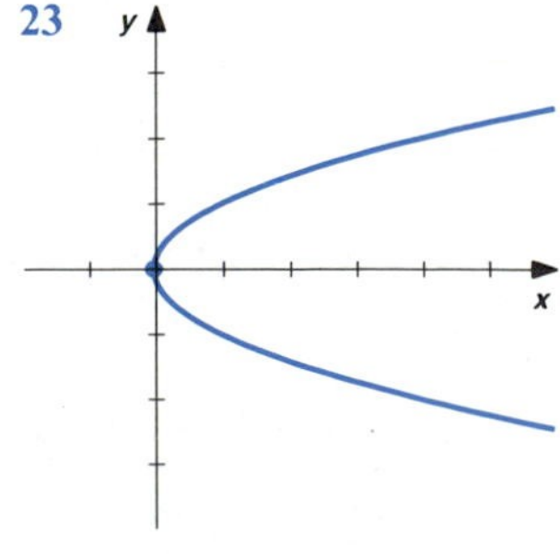

25
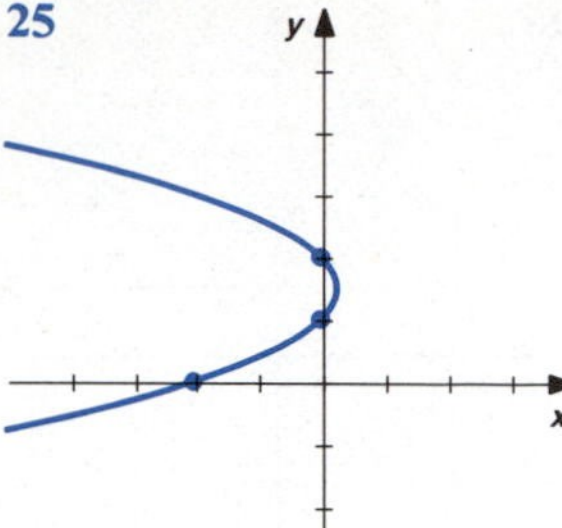


27
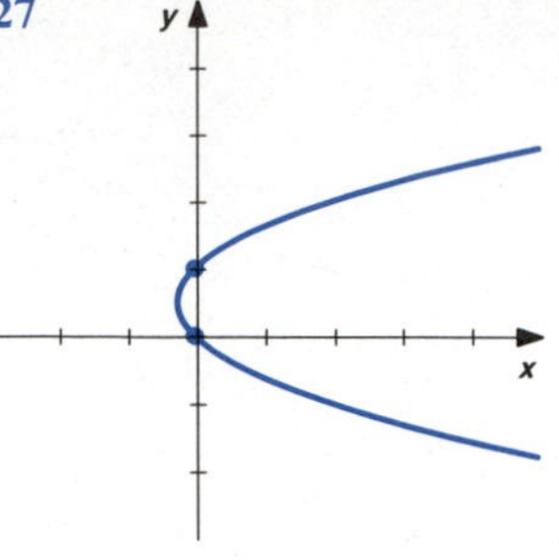


29
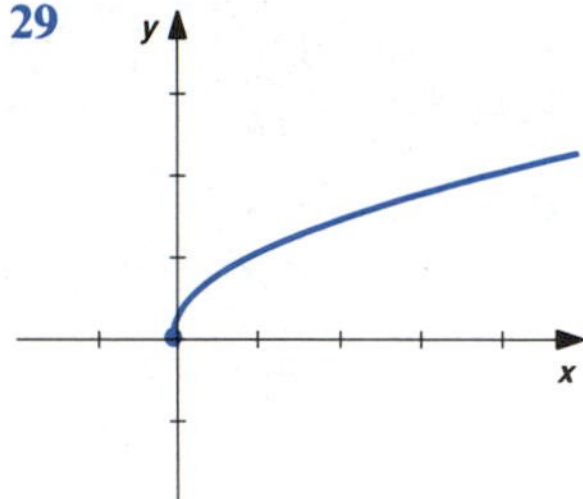


31
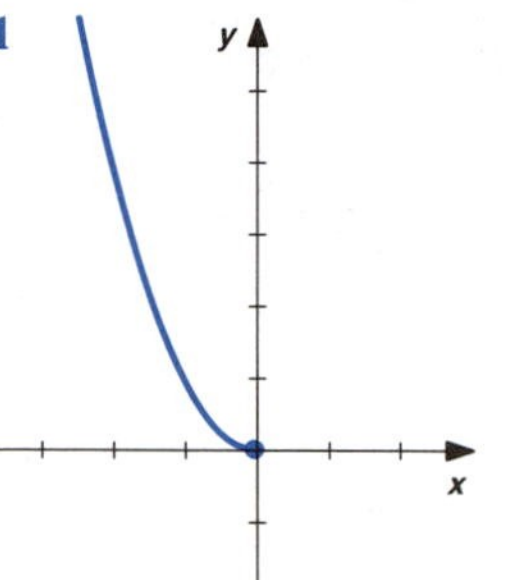


33
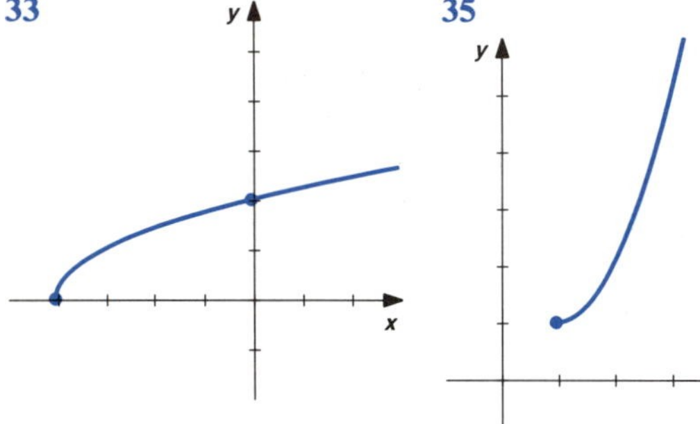


35


37
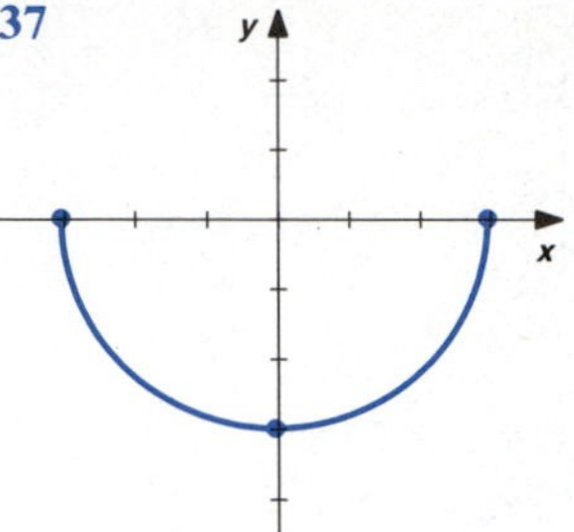


39
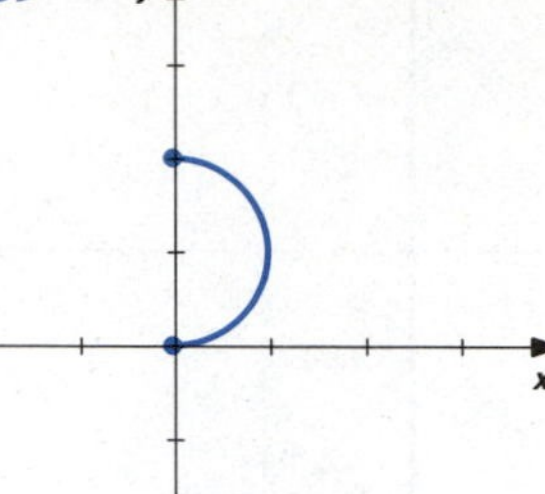


41
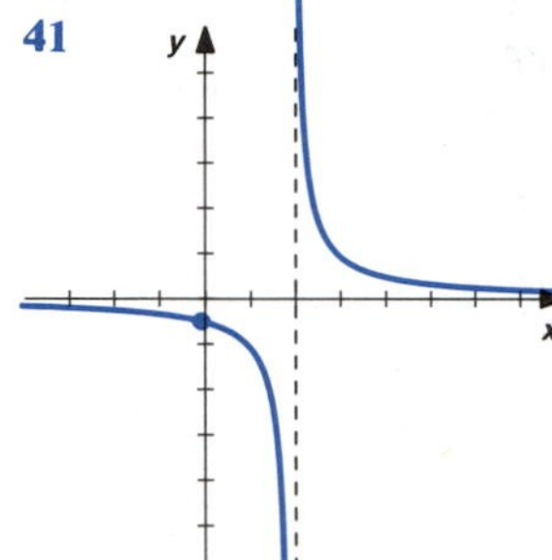


43
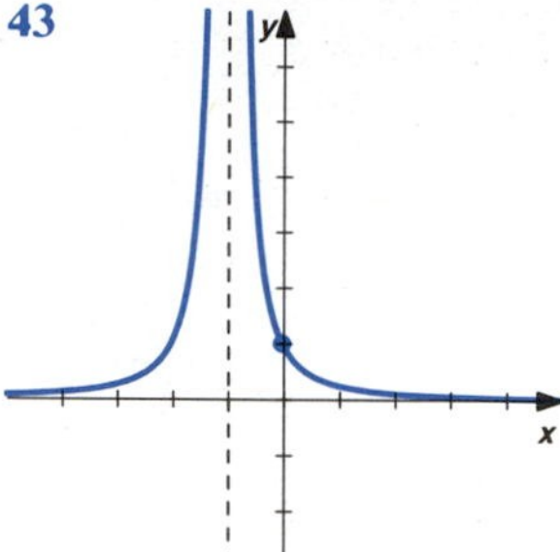


## *Exercises 1.6*

**1** 4/5

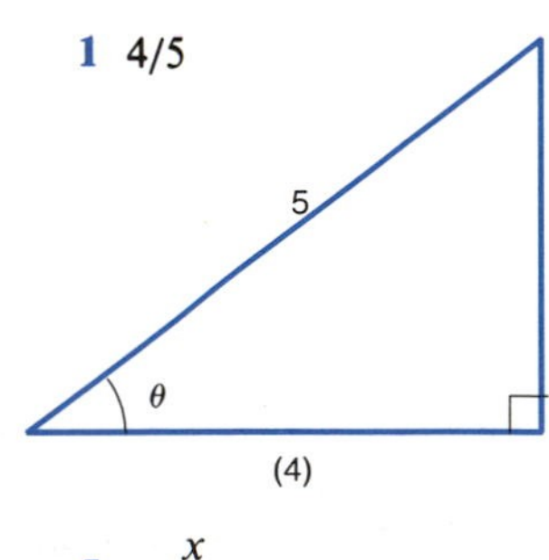


**3** $\dfrac{x}{\sqrt{4-x^2}}$

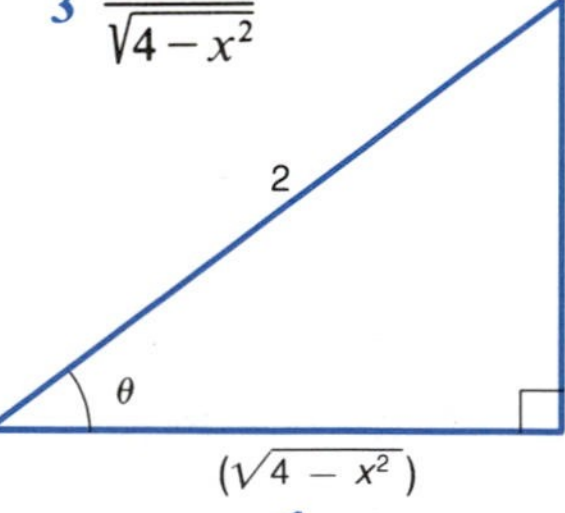


**5** $\dfrac{x}{\sqrt{4+x^2}}$

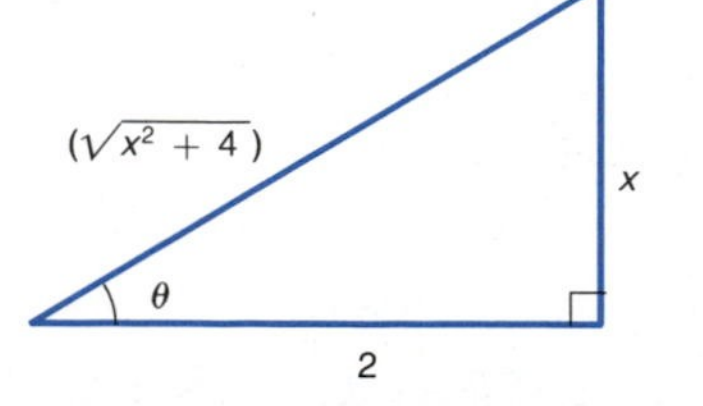


**7** $b = 2 \sin \theta$

**9** $L = 2 \cos \theta$

**11** $h = 5 \cot \theta$

**13** (a) 0 (b) $-1$ (c) 0

**15** (a) 1/2 (b) $-\sqrt{3}/2$ (c) $-1/\sqrt{3}$

**17** (a) $-1/\sqrt{2}$ (b) $1/\sqrt{2}$ (c) $-1$

**19** (a) 1 (b) $-\sqrt{2}$ (c) $-\sqrt{2}$

**21** 4/5

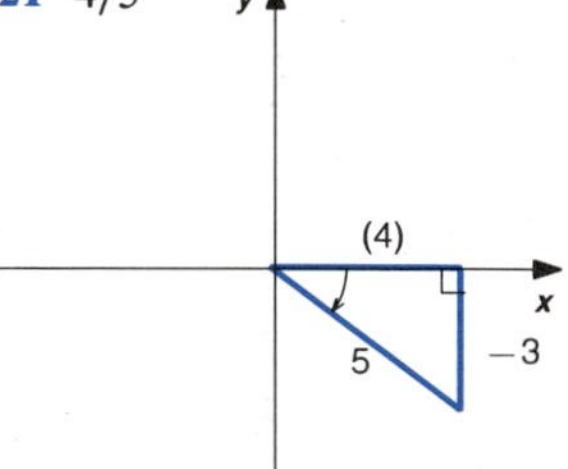


**23** $-2/\sqrt{5}$

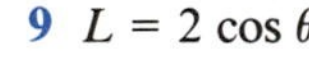

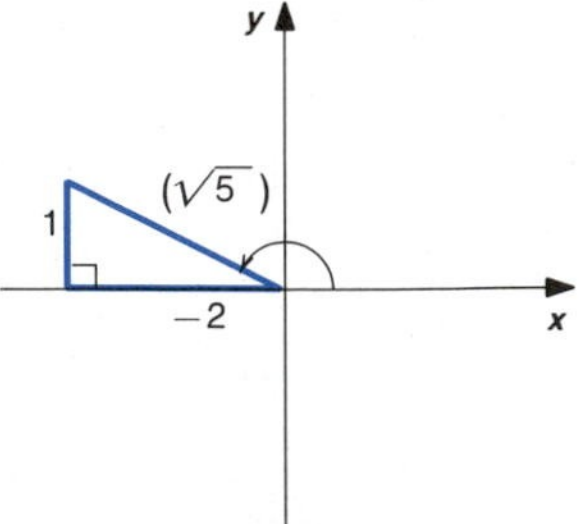

**25** $\sqrt{1-x^2}$

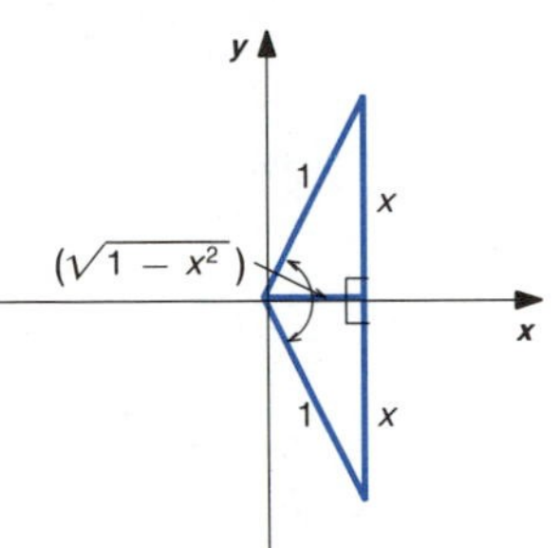


**27** $\pi/6, 5\pi/6$

**29** $\pi/3, 4\pi/3$

**31** $0, \pi/2, \pi, 3\pi/2, 2\pi$

**33** $\pi/6, \pi/2, 5\pi/6, 7\pi/6, 3\pi/2, 11\pi/6$

**35** $L^2 = 164 - 160 \cos \theta$

**37** $b = 1/\sin \theta$

**39** $16 = 5x^2 - 4x^2 \cos \theta$

**41**

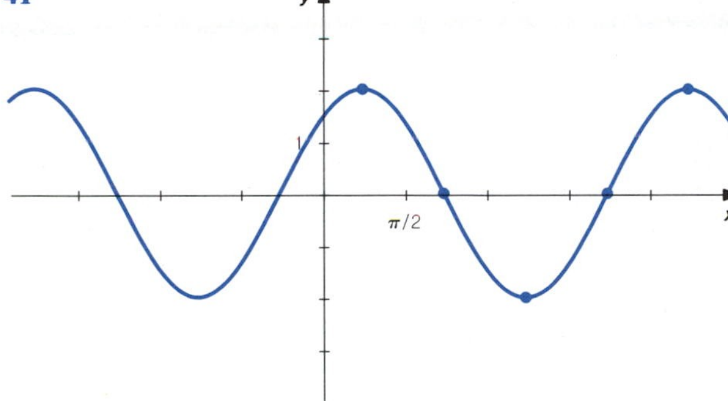

**43**

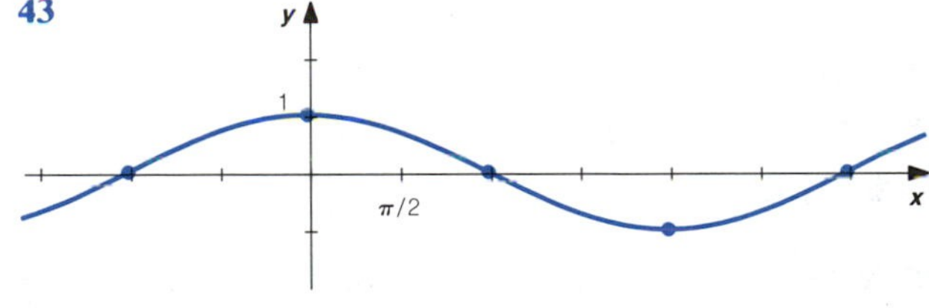


**45**

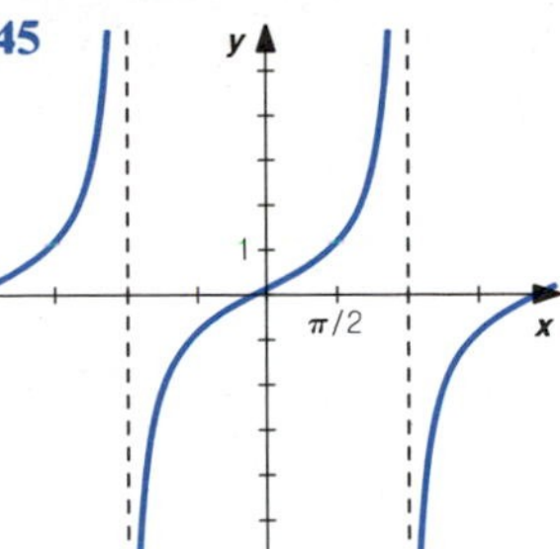


**47**

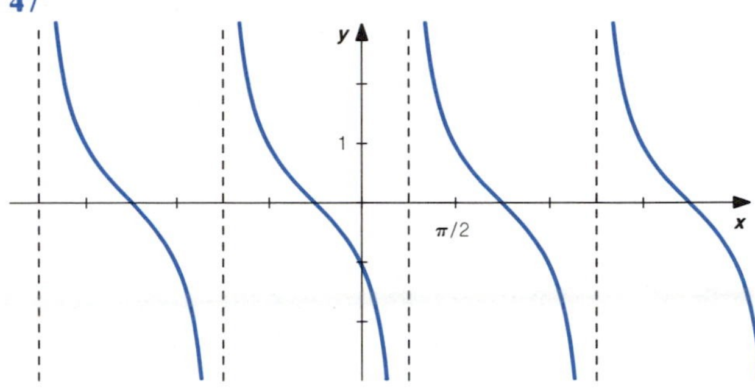


**49**

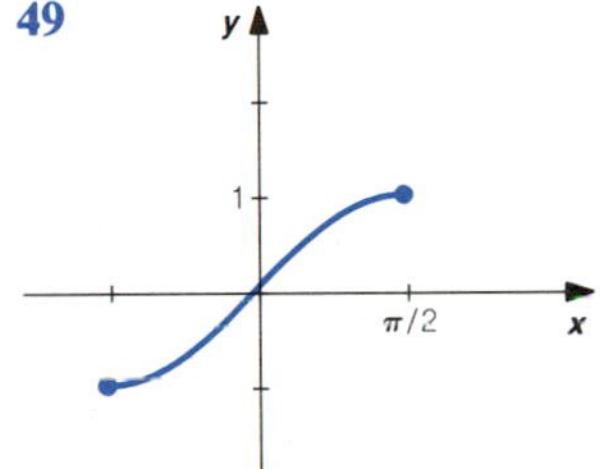


**51**

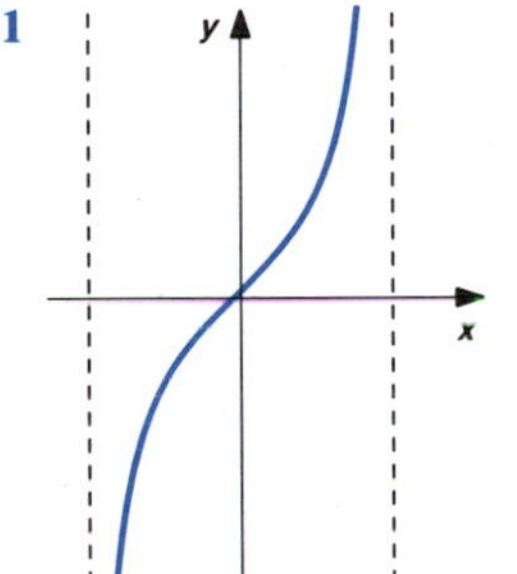


## Review Exercises

**1** $x > 3$

**3** $x^4$

**5** 13/18, 3/2

**7** $\dfrac{3-x^2}{3(x^2+1)^{5/3}}$

**9** $x < -2, x > 1$

**11** $\sqrt{x^2-4x+13}; \dfrac{3}{2-x}; \left(\dfrac{2+x}{2}, \dfrac{3}{2}\right)$

**13**

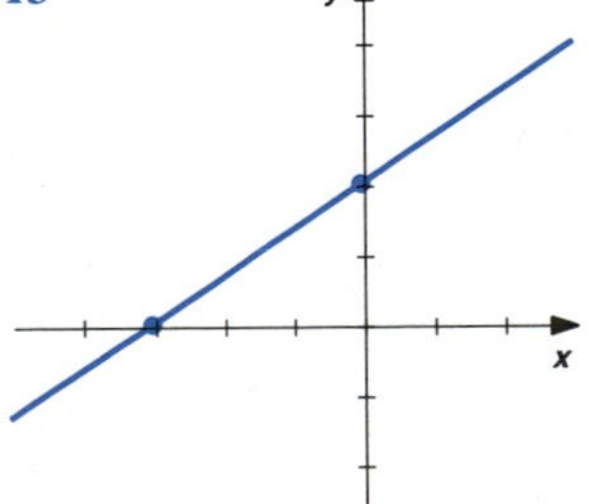


**15**

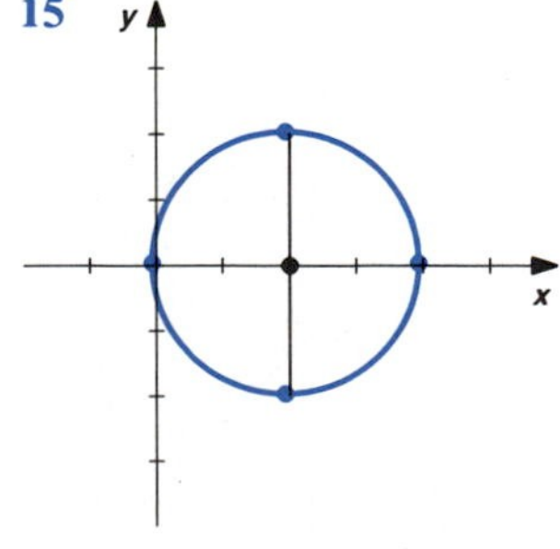

17

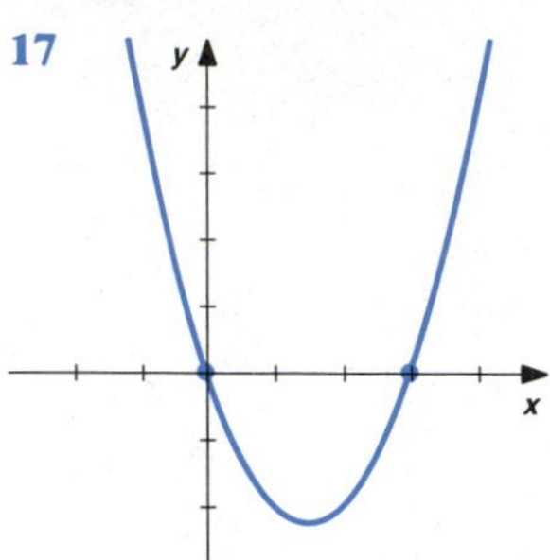


19

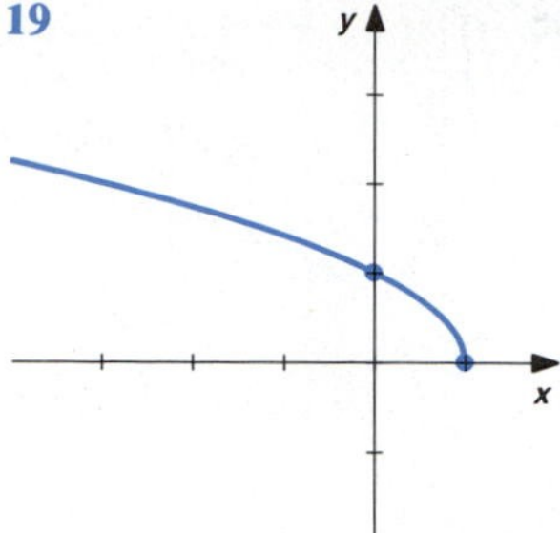


21

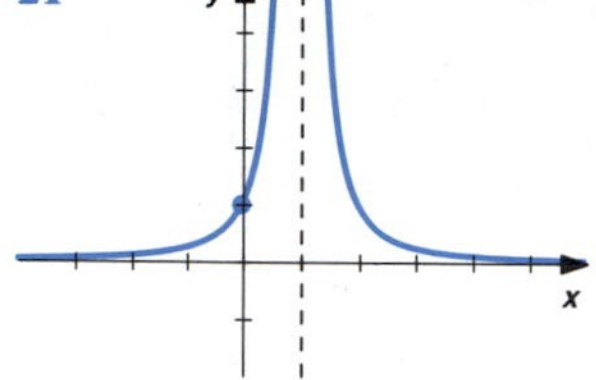


23 $2x - y = 5$

25 $2x + 3y = 6$

27 $x = 2 \sec \theta$

29 $-\sqrt{3}/2$

31 $\pi/4, 3\pi/4, 5\pi/4, 7\pi/4$

33

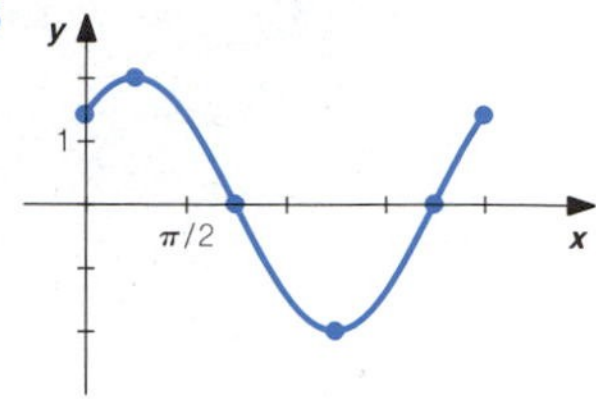


## CHAPTER TWO

### *Exercises 2.1*

1 15; 1

3 $\dfrac{2-x}{x}; \dfrac{1}{2x-1}$

5 $3x^2 - 6x + 1; 3x^2 - 3$

7 $x - 1, x \neq -2$

9 $3x^2; 5x^2 - x$

11 $12x^2 - 16x + 6$

13 Undefined; 0; 4

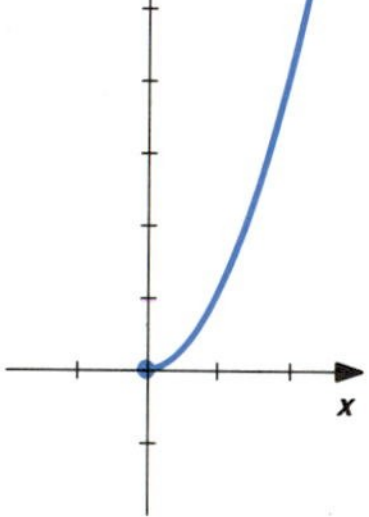


15 Undefined; 0; $-1/2$

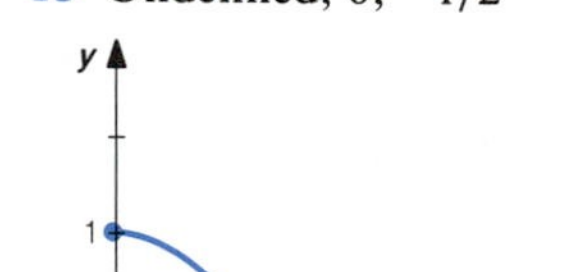


17 1; undefined; $-1$

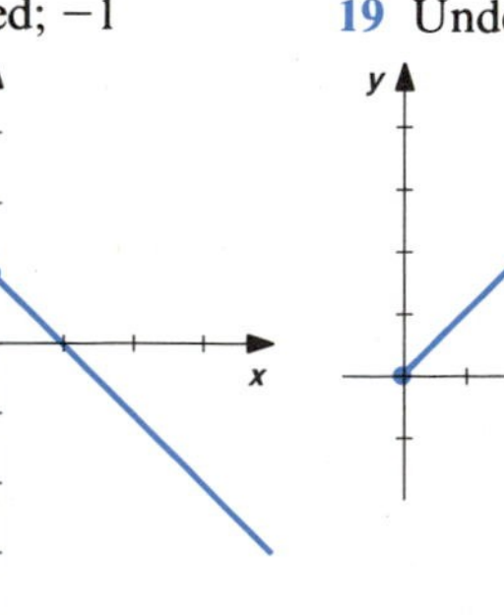


19 Undefined; 2; 2

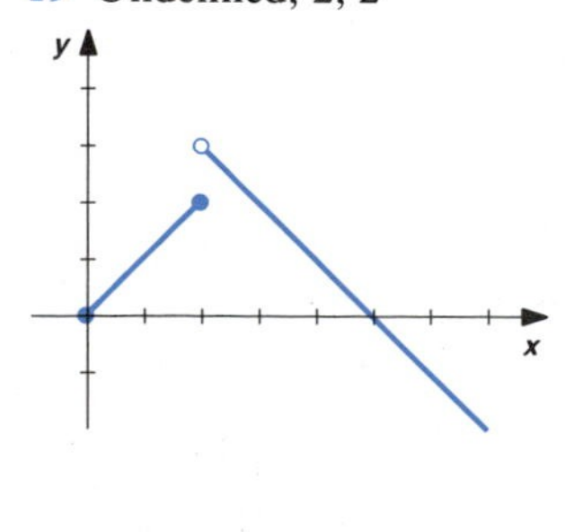


21 $x \neq 1, -1$

23 $x \leq 3/2$

25 $x \neq \dfrac{(2k+1)\pi}{2}$, $k$ an integer

27 $-2, x \neq 1$

29 $x + 1, x \neq 1$

31 $V(x) = x(2 - x)(3 - x), 0 \leq x \leq 2$

33 $S(x) = 20x - 3x^2, 0 \leq x \leq 5$

35 $-4 < x \leq 4; -1 \leq y \leq 2$; undefined; 1; 2

37 $0 \leq x < 4; 1 \leq y < 3$; undefined; 1; 2

39 $f(x) = \dfrac{-2x-6}{3}, -\infty < x < \infty; -\infty < y < \infty$

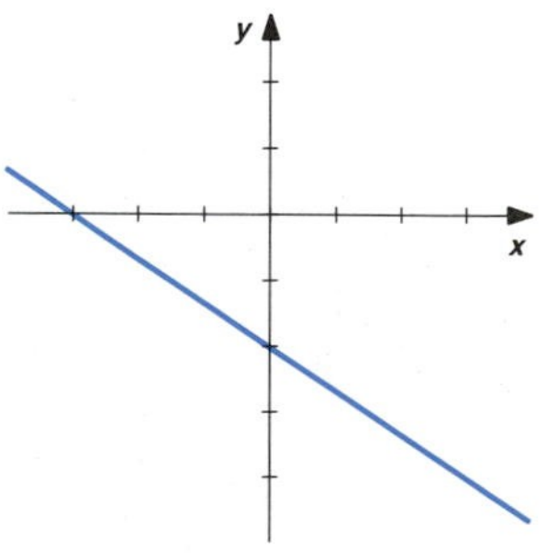

41 $y$ is not a function of $x$

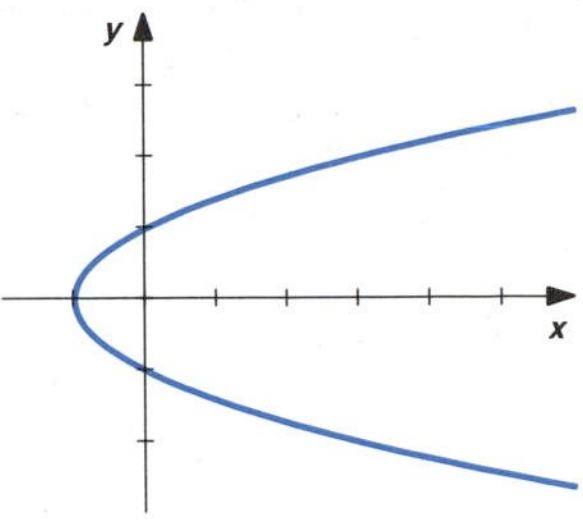


43 $f(x) = -\sqrt{1-x}$, $x \leq 1$; $y \leq 0$

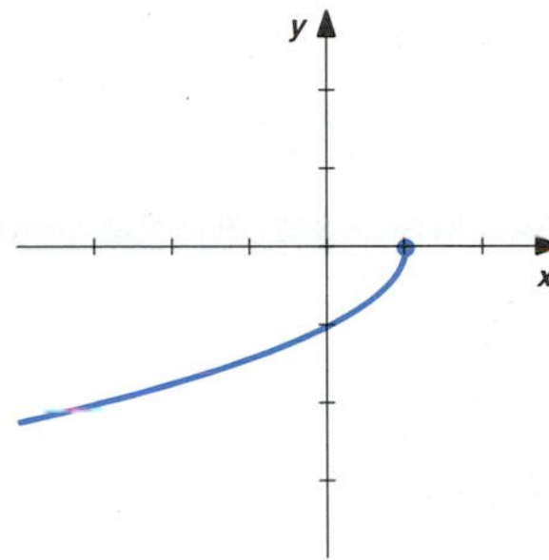


45 $f(x) = 4/x^2$, $x \neq 0$; $y > 0$

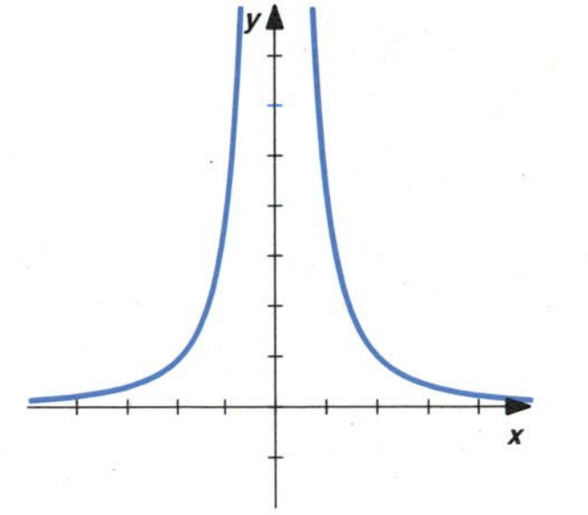


47 $y$ is not a function of $x$

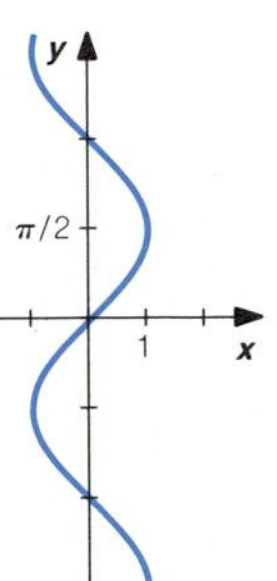


49 Yes; $(2x + y)^2 = 0$ implies $y = -2x$

51 $f(x) = \sqrt{4-x^2}$

53 $f \circ g(x)$ is defined only for $x \geq 0$

*Exercises 2.2*

1 25
3 15
5 Unbounded
7 5
9 2
11 1/5
13 Unbounded
15 1/10
17 2.5
19 1/3
21 1/4
23 1/5
25 3
27 $\pi^2$
29 1
31 3/2
33 1
35 2
37 1/80
39 2
41 4
43 1
45 0.01

*Exercises 2.3*

1 0
3 Does not exist
5 Does not exist
7 2
9 3
11 (a) $-2$ (b) Does not exist (c) 0
13 Does not exist
15 0
17 1
19 0.001 m
21 0.000999 m
23 0.000829
25 0.005 Ω
27 $\varepsilon/2$
29 $\varepsilon^{1/3}$
31 $\varepsilon^3$
33 0.0002
35 0.012
37 $\varepsilon$

## Exercises 2.4

**1** 4
**3** $-10$
**5** 7
**7** 3
**9** 2
**11** 1/2
**13** $\sqrt{3}$
**15** 0
**17** 0
**19** 6
**21** 5
**23** $-1/4$
**25** 1/9
**27** 0, 0, 0
**29** 1, $-1$, does not exist
**31** $-1$, 1, does not exist
**33** Does not exist, 0, does not exist
**35** 0, 0, 0; 1, 1, 1
**37** $-1$, $-1$, $-1$; 1, does not exist, does not exist

## Exercises 2.5

**1** Continuous at $b$, discontinuous at $a$ and $c$
**3** Discontinuous at $a$, $b$, and $c$
**5** 1
**7** Impossible
**9** $-1/2$, 2
**11** 1
**13** 2
**15** 5
**17** 1/2
**19** $\sqrt{2}$
**21** $-1$
**23** 1
**25** Continuous from right and left at zero

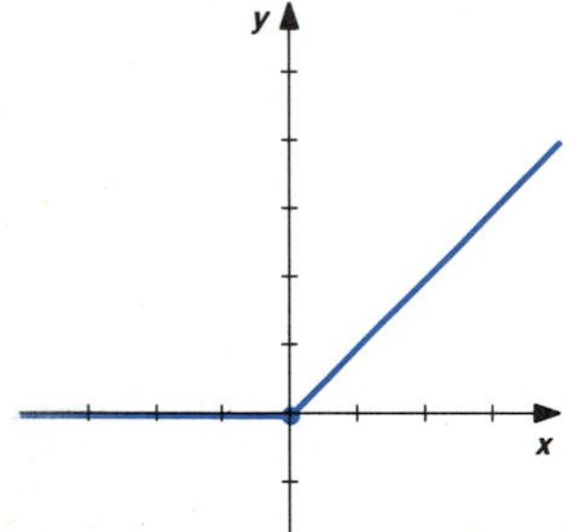


**27** Continuous from left at zero

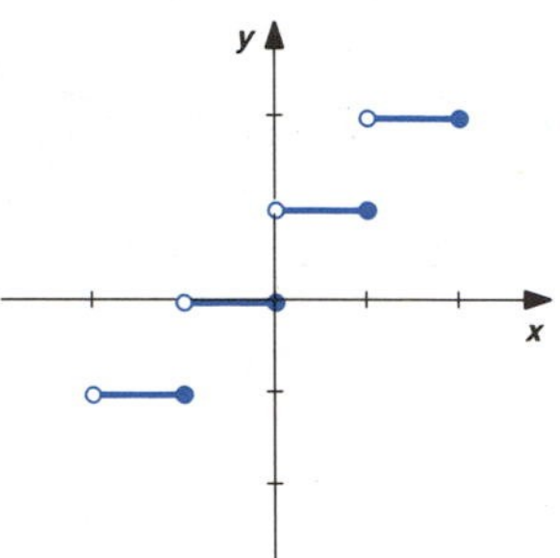


**33**

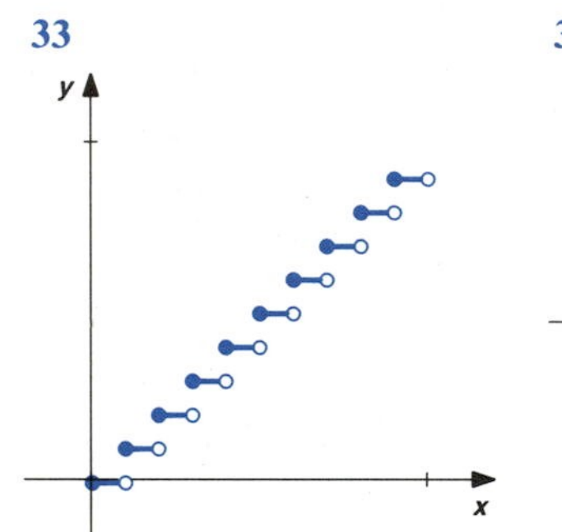


**35**

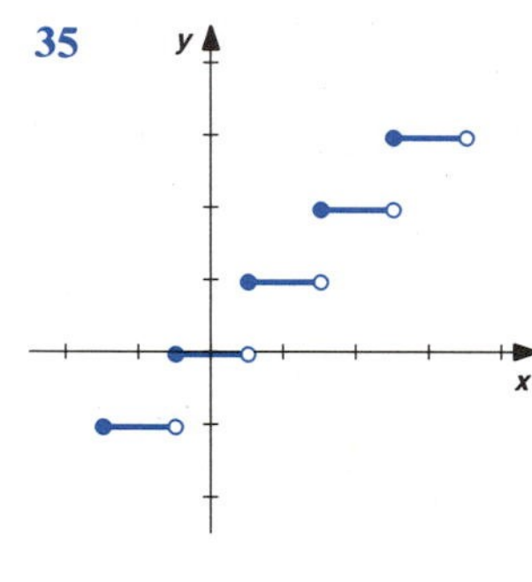


## Exercises 2.6

**1** (a) $x - 2, x \neq 1$ (b) $-1$ (c) $x + y + 1 = 0$
**3** (a) $3, x \neq -3$ (b) 3 (c) $y = 3x + 4$
**5** (a) $x^2 + 2x + 4, x \neq 2$ (b) 12 (c) $y = 12x - 16$
**7** (a) $\dfrac{-1}{4x}, x \neq 4$ (b) $-1/16$ (c) $x + 16y = 8$
**9** (a) $t + 2, t \neq 2$ (b) 4
**11** (a) $-2, t \neq 3$ (b) $-2$
**13** (a) $\dfrac{1}{\sqrt{t+1}+3}, t \neq 8$ (b) 1/6
**15** (a) $\dfrac{1}{4(t+1)}, t \neq 3$ (b) 1/16
**17** $-16t + v_0, t \neq 0; v_0$
**19** $6\pi$
**21** $-4/5$
**23** No
**25** Yes
**27** No
**29** No
**31** Yes
**33** No
**35** Yes
**37** No
**39** 1, 2

*Review Exercises*

**1** $2x^2 - 17x + 17$; $x^2 - 9x + 4$

**3** (a)

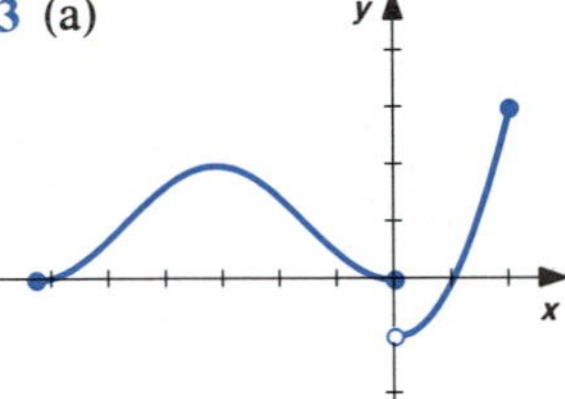


(b) $-2\pi \le x \le 2$; $-1 < y \le 3$

(c) 2, 0, undefined

(d) 0, $-1$ (e) Yes, no, no

**5** $x < 3/2$

**7** $V(x) = x(11 - 2x)(8 - 2x)$, $0 \le x \le 4$

**9** $y$ is not a function of $x$

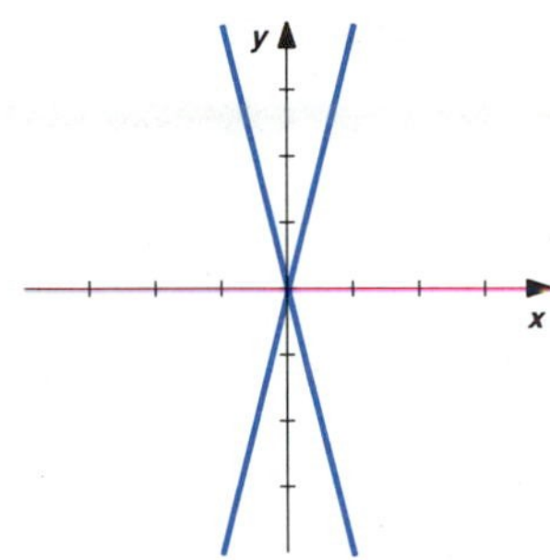


**11** 30

**13** 3/2

**15** 16

**17** Unbounded

**19** $\varepsilon/2$

**21** $\varepsilon^3/27$

**23** 0.025 ft

**25** Does not exist

**27** $x + 3$, $x \ne 3$; $y = 6x - 9$

**29** Slope is positive and increases without bound

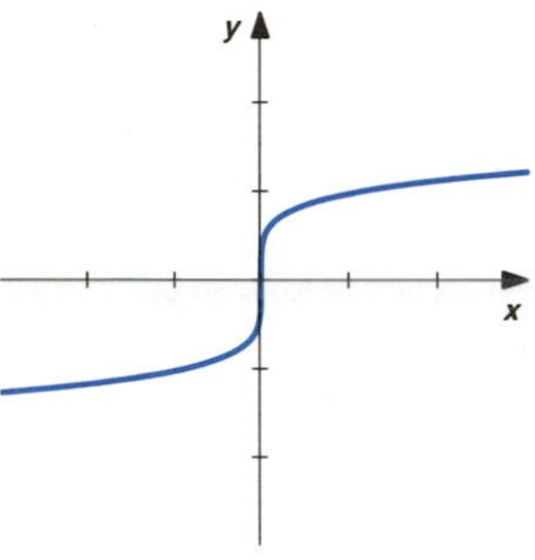


## CHAPTER THREE

*Exercises 3.1*

**1** $2x - 7$

**3** $6x^2 - 1$

**5** $2 - x^{-2}$

**7** $\frac{1}{2}x^{-1/2} - \frac{1}{2}x^{-3/2}$

**9** $\frac{4}{3}x^{1/3} - \frac{4}{3}x^{-2/3}$

**11** $2.6x^{0.3} + 1.5x^{-0.7}$

**13** $3x^{1/2} + \frac{1}{2}x^{-1/2}$

**15** $\frac{1}{4} - \frac{1}{4}x^{-2}$

**17** $12t^3 + 12t$

**19** $-\frac{1}{2}t^{-2} - \frac{1}{4}t^{-3/2}$

**21** $y = 3x + 1$

**23** $y = 4x - 6$

**25** $x + 4y + 4 = 0$

**27** $50 - 32t$; 6; $-46$

**29** $1 + 12t^2$; 5; 13

**31** $t^{-1/2}/2$; 2; 1/4

**33** (a) $(3, -8)$ (b) None

**35** (a) $(2, 4)$, $(-2, -4)$ (b) None

**37** (a) $(-1, -3)$ (b) $(0, 0)$

**39** $(1, 0)$, $(-1, 0)$

**41** $(9, 81)$, $(1, 1)$

**43** $(1, 1)$, 1.2490 rad; $(-2, 4)$, 0.5404 rad

*Exercises 3.2*

**1** $(x^2 + 1)(4x - 3) + (2x^2 - 3x + 1)(2x) = 8x^3 - 9x^2 + 6x - 3$

**3** $(x + x^{-1})(1 + x^{-2}) + (x - x^{-1})(1 - x^{-2}) = 2x + 2x^{-3}$

**5** $\dfrac{(x^2+1)(3)-(3x)(2x)}{(x^2+1)^2} = \dfrac{3-3x^2}{(x^2+1)^2}$

**7** $\dfrac{(4x+1)(2)-(2x-1)(4)}{(4x+1)^2} = \dfrac{6}{(4x+1)^2}$

**9** $x + y = 1$

**11** $5y = 8x - 7$

13 $2x + y = 7$

15 (a) None (b) None

17 (a) $(1, 1/6)$, $(-1, -1/6)$ (b) $(0, 0)$

19 $2ff'$

21 $f'gh + fg'h + fgh'$

23 $-2g'/g^3$

25 13

## Exercises 3.3

1 $\cos x - \sin x$

3 $\sec x \tan x - 2\sec^2 x$

5 $(3x^{1/3})(\cos x) + (\sin x)(x^{-2/3}) = 3x^{1/3}\cos x + x^{-2/3}\sin x$

7 $(2x)(\cos x) + (\sin x)(2) = 2x\cos x + 2\sin x$

9 $(\tan x)(\cos x) + (\sin x)(\sec^2 x) = \sin x + \sin x \sec^2 x$

11 $\dfrac{(x^2+4)(-\sin x) - (\cos x)(2x)}{(x^2+4)^2}$

13 $y = 2x$

15 $42x - 36y = 7\pi - 6\sqrt{3}$

17 $\left(\dfrac{2\pi}{3}, \dfrac{2\pi}{3} + \sqrt{3}\right)$, $\left(\dfrac{4\pi}{3}, \dfrac{4\pi}{3} - \sqrt{3}\right)$

19 $\left(\dfrac{\pi}{3}, 2\right)$, $\left(\dfrac{4\pi}{3}, -2\right)$

21 $0 \le t < \pi/3$, $2\pi/3 < t \le 2\pi$

23 $0 \le t < \pi/4$

29 0

31 2

33 1/2

## Exercises 3.4

1 $5(2x^2 - 3x + 1)^4(4x - 3)$

3 $-\dfrac{2}{3}(2x^3 - 1)^{-5/3}(6x^2) = -4x^2(2x^3 - 1)^{-5/3}$

5 $-(2\sin x + 3)^{-2}(2\cos x)$

7 $(\cos(x^3 - x))(3x^2 - 1)$

9 $(\sec^2\sqrt{x})\left(\dfrac{1}{2\sqrt{x}}\right)$

11 $\left(\sec\dfrac{1}{x}\tan\dfrac{1}{x}\right)\left(-\dfrac{1}{x^2}\right)$

13 $(x)\left(\dfrac{1}{2\sqrt{x^2+1}}\right)(2x) + (\sqrt{x^2+1})(1) = \dfrac{2x^2+1}{\sqrt{x^2+1}}$

15 $\dfrac{(\sqrt{\cos^2 x+1})(\cos x) - (\sin x)\left(\dfrac{1}{2\sqrt{\cos^2 x+1}}\right)(2\cos x)(-\sin x)}{(\sqrt{\cos^2 x+1})^2} = \dfrac{2\cos x}{(\cos^2 x+1)^{3/2}}$

17 $2(\tan 2x)(\sec^2 2x)(2)$

19 $\left(\dfrac{1}{2\sqrt{x^2+\sin^2 x}}\right)(2x + 2(\sin x)(\cos x)) = \dfrac{x + \sin x\cos x}{\sqrt{x^2+\sin^2 x}}$

21 $\dfrac{1}{2}[x + (x + x^2)^{1/2}]^{-1/2}\left[1 + \dfrac{1}{2}(x + x^2)^{-1/2}(1 + 2x)\right]$

23 $16y = 11x - 3$

25 $4x + 125y = 41$

27 $4x + y = \pi + 1$

29 $y = 3x - \pi$

31 (a) $(1, -1)$ (b) $(4/3, 0)$

33 (a) $(-2, -1/2)$ (b) None

35 (a) $(\pi/2, \pi/2 - 1)$ (b) None

37 20

39 2/3

41 0

45 $y - 1/\sqrt{2} = (5/\sqrt{2})(x - \pi/4)$

47 $2/x$

49 $\dfrac{2x}{x^2+1}$

53 $10\pi/9$ ft/s

## Exercises 3.5

1 All points on graph except $(-1, 0)$ and $(1, 0)$

3 All points on graph

5 All points on graph except $(0, 0)$

7 All points on graph except $(0, 0)$, $(-1, 0)$, $(1, 0)$

9 $x/y$

11 $1/(3y^2)$

13 $-\sqrt{y/x}$

15 $\dfrac{1 - \cos(x+y)}{\cos(x+y)}$

17 $\dfrac{2xy - y^2}{2xy - x^2}$

19 $\dfrac{y - x^2}{y^2 - x}$

21 $\dfrac{\cos x\cos y}{1 + \sin x\sin y}$

23 $3x + y = 2$

25 $x + y = 2$

27 $46y = 9x + 120$

29 $32x + 4y = 15$

31 $6\sqrt{3}x + 12y = 3\sqrt{3} + 2\pi$

33 $8x + 4y = 8 + \pi$

37 $2\sqrt{6}/3$

39 $\dfrac{dv}{du} = \dfrac{1}{2 + 6v}$, $\dfrac{du}{dv} = 2 + 6v$, they are reciprocals

### Exercises 3.6

1 $6x - 7; 6; 0$

3 $-\frac{1}{2}(1-x)^{-1/2}; -\frac{1}{4}(1-x)^{-3/2}; -\frac{3}{8}(1-x)^{-5/2}$

5 $\frac{1}{3}x^{-2/3} - \frac{1}{3}x^{-4/3}; -\frac{2}{9}x^{-5/3} + \frac{4}{9}x^{-7/3}; \frac{10}{27}x^{-8/3} - \frac{28}{27}x^{-10/3}$

7 $\sec^2 x$; $2\sec^2 x \tan x$; $2\sec^4 x + 4\sec^2 x \tan^2 x$

9 $-x/y; -4/y^3$

11 $\frac{y}{y-x}; \frac{y^2-2xy}{(y-x)^3}$

13 $\cos^2 y; -2\cos^3 y \sin y$

15 (a) $-32t + 32$ (b) $-32$ (c) $s(1) = 64$ (d) $v(3) = -64$, $v(-1) = 64$

17 (a) $-2\cos t$ (b) $2\sin t$ (c) $s(\pi/2 + 2\pi k) = -1$, $s(3\pi/2 + 2\pi k) = 3$ (d) $v(\pi/6 + 2\pi k) = -\sqrt{3}$, $v(5\pi/6 + 2\pi k) = \sqrt{3}$, $k$ an integer

19 (a) (2, 16) (b) None

21 (a) (1, 2), (−1, −2) (b) None

23 (a) $(\pi/2, \pi/2)$ (b) $(\pi/2, \pi/2)$, $(3\pi/2, 3\pi/2)$

25 $P_1(x) = 2(x-1)$; $P_1(1.2) = 0.4$;
$P_2(x) = 2(x-1) + 2(x-1)^2$; $P_2(1.2) = 0.48$;
$P_3(x) = 2(x-1) + 2(x-1)^2 + (x-1)^3$;
$P_3(1.2) = 0.488$;
$f(1.2) = 0.488$

27 $P_1(x) = -8 + 12x$; $P_1(-0.1) = -9.2$;
$P_2(x) = -8 + 12x - 6x^2$; $P_2(-0.1) = -9.26$;
$P_3(x) = -8 + 12x - 6x^2 + x^3$; $P_3(-0.1) = -9.261$;
$f(-0.1) = -9.261$

29 $P_1(x) = P_2(x) = x$; $P_1(0.2) = P_2(0.2) = 0.2$;
$P_3(x) = x + \frac{x^3}{3}$; $P_3(0.2) = \frac{76}{375} \approx 0.20266667$;
$f(0.2) \approx 0.20271003$

31 $P_1(x) = 1 + \frac{1}{2}(x-1)$; $P_1(0.98) = 0.99$;
$P_2(x) = 1 + \frac{1}{2}(x-1) - \frac{1}{8}(x-1)^2$; $P_2(0.98) = 0.98995$;
$P_3(x) = 1 + \frac{1}{2}(x-1) - \frac{1}{8}(x-1)^2 + \frac{1}{16}(x-1)^3$;
$P_3(0.98) = 0.9899495$; $f(0.98) \approx 0.98994949$

33 $f''g + 2f'g' + fg''$

### Exercises 3.7

1 [1.2, 1.3]

3 [0.7, 0.8]

5 [1.8, 1.9]

7 [0.6, 0.7]

9 [−2, −1], [0, 1], [1, 2]

11 [−1, 0], [0, 1], [2, 3]

15 $f$ is not defined at $x = 0$

### Exercises 3.8

1 1.7321

3 −2.8284

5 2.0801

7 1.6818

9 1.1593

11 1.6180

13 0.2638

15 0.3222

17 0.7391

19 0.8241

21 −1.879, 0.347, 1.532

23 −0.732, 1.000, 2.732

25 −2.422, 0.917, 4.505

### Exercises 3.9

1 $f(3) = 1$; $f(0) = -2$

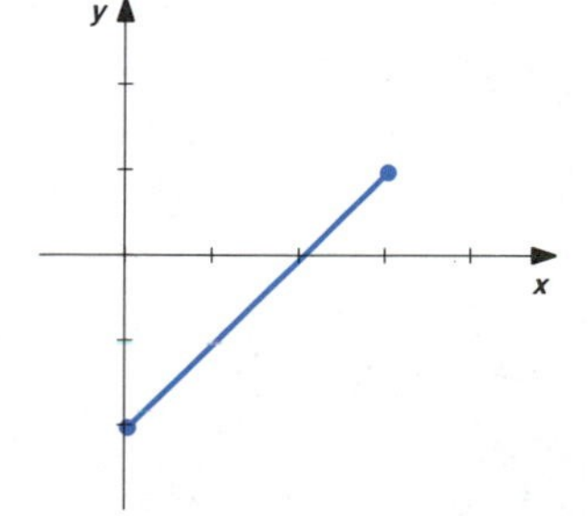


3 $f(0) = 1$; no minimum value

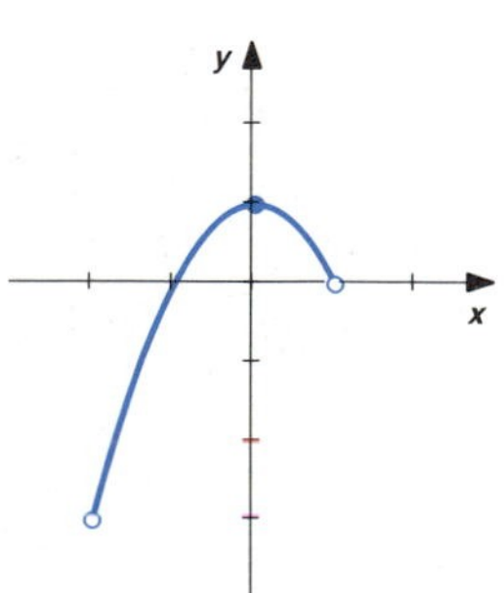

5 No maximum value; no minimum value

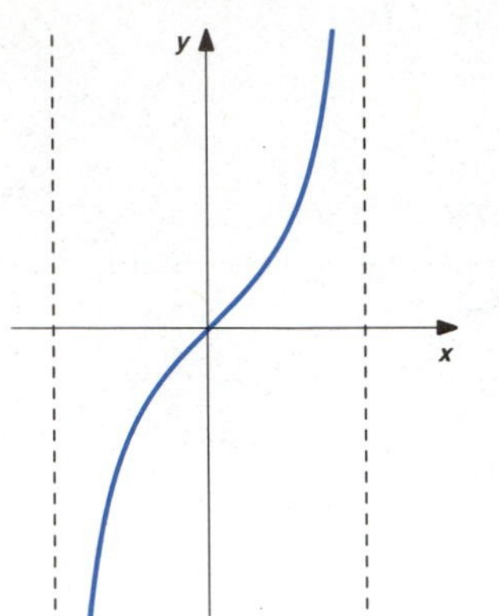


7 No maximum value; $f(0) = -1$

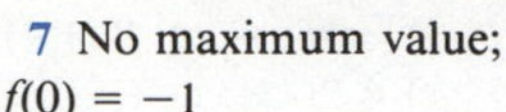

9 $f(0) = 3; f(2) = -5$

11 $f(0) = f(6) = 4; f(4) = -28$

13 $f(8) = 12; f(0) = 0$

15 $f(3/4) = 3/4^{4/3}; f(2) = -2$

17 $f(1) = 1/2; f(0) = 0$

19 $f(5\pi/6) = 1; f(2\pi) = \sin(6\pi/5) \approx -0.58778525$

21 $f(\pi/4) = \sqrt{2}; f(5\pi/4) = -\sqrt{2}$

23 $f(3/2) = 4; f(4) = -1$

25 64 ft

27 490 m

29 9

## *Exercises 3.10*

5 $[4.35, 4.65]$

7 $[2.68, 2.72]$

11 $f'$ does not exist at $x = 0$

13 $y - f(a) = \dfrac{f(b) - f(a)}{b - a}(x - a)$

19 $-1$

21 No

## *Review Exercises*

1 $3x^2 - 4x + 1$

3 $x^2 \sec^2 x + 2x \tan x$

5 $\sec^3 x + \sec x \tan^2 x$

7 $\dfrac{2x \cos x + x^2 \sin x}{\cos^2 x}$

9 $-5(5x - 1)^{-4/3}/3$

11 $\dfrac{-3 \sin 3x}{(1 - \cos 3x)^2}$

13 $2 \sin x \cos x$

15 $\csc(\cot x) \cot(\cot x) \csc^2 x$

17 $\dfrac{6x + 1}{\sqrt{4x + 1}}$

19 $x^2/y^2$

21 $-\csc y$

23 $4y = x + 4$

25 (a) $(-24, -48)$ (b) $(-32, 0)$

27 $0 < t < 1$

29 3

31 0

33 $-9/\sqrt{135}, -18/\sqrt{108}, -27/\sqrt{63}$, increases without bound

35 $1/\sqrt{a^2 - x^2}$

37 $x/y; -3/y^3$

39 $P_4(x) = 1 - 2x^2 + \dfrac{2}{3}x^4$

41 $[1.2, 1.3]$

43 $[-3, -2], [-1, 0], [2, 3]$

45 0.4502

47 $f(0) = 3$, no minimum value

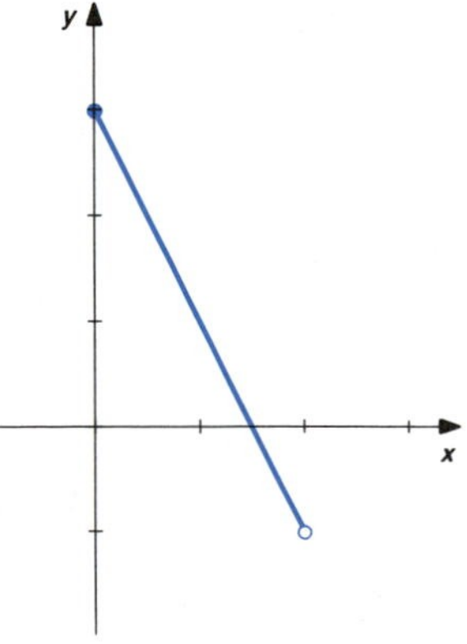


49 $f(0) = 1$, no minimum value

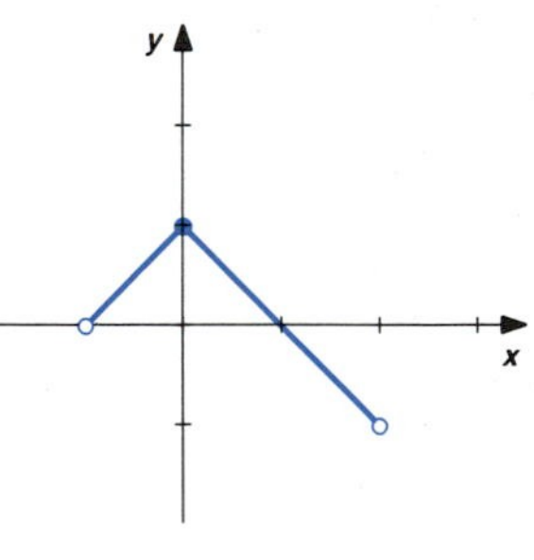


51 $f(2) = 2, f(-1) = -4$

53 $f(3) = 4, f(1/3) = 0$

55 16

## CHAPTER FOUR

### *Exercises 4.1*

1
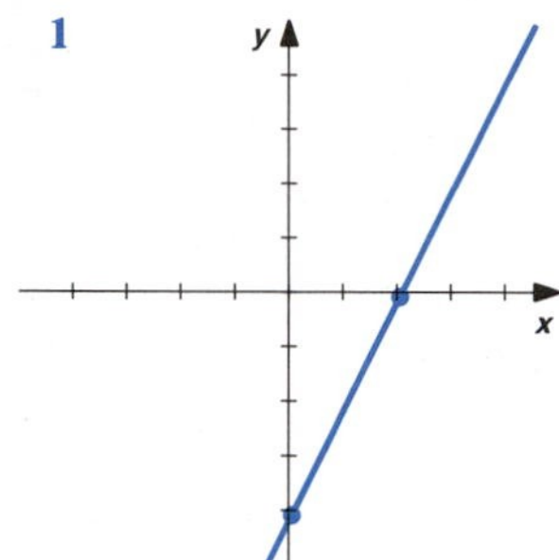

3
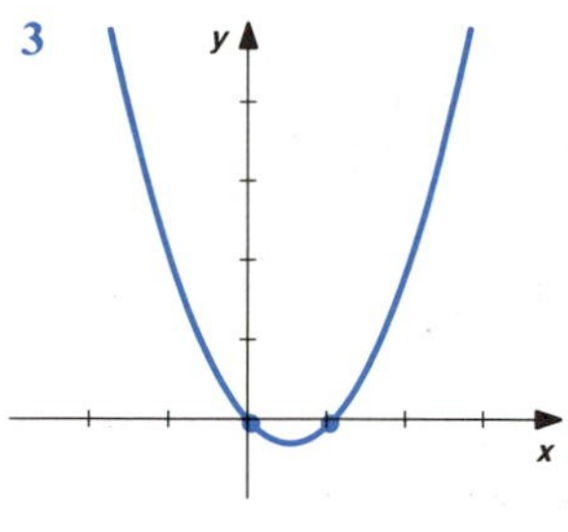

5
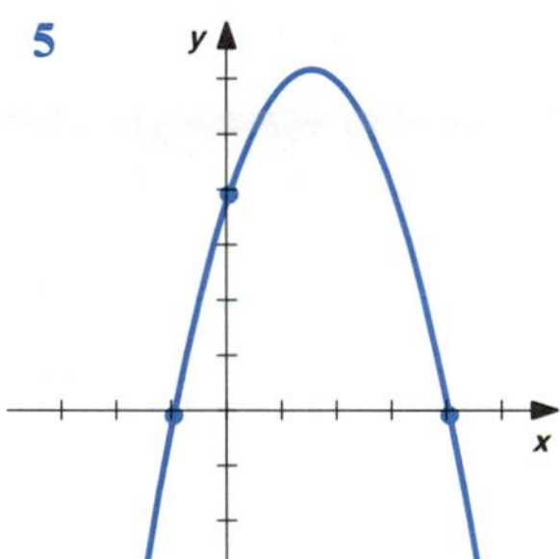

7
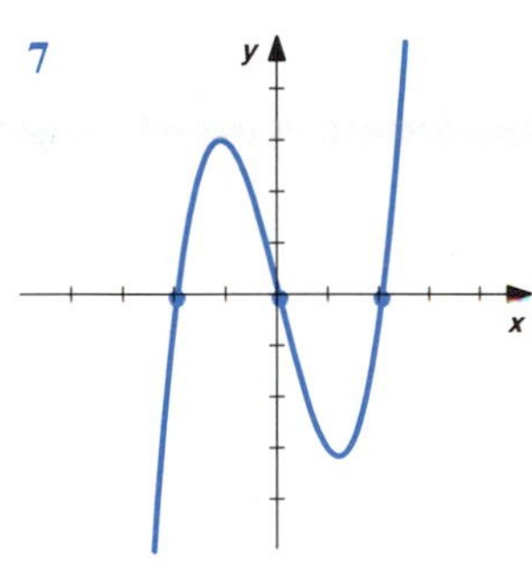

9
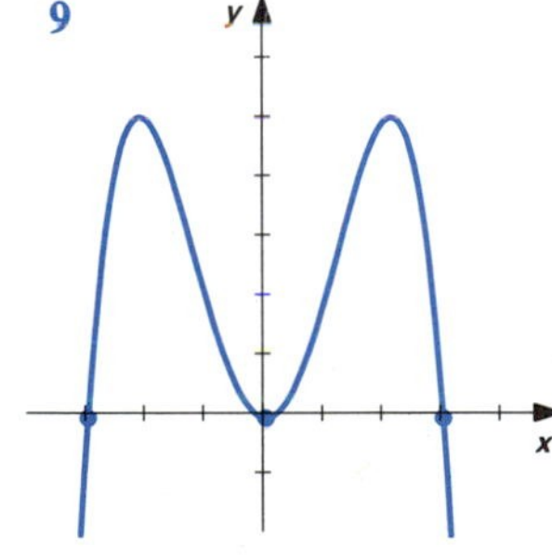

11
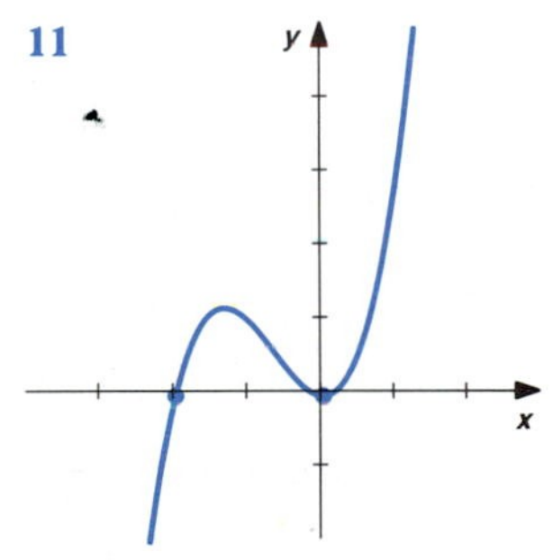

13
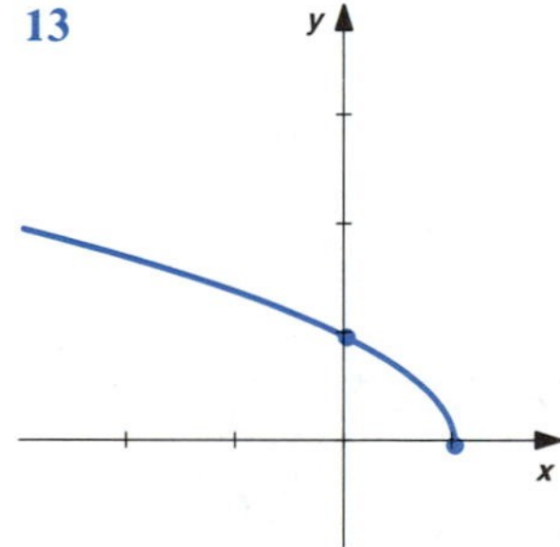

15
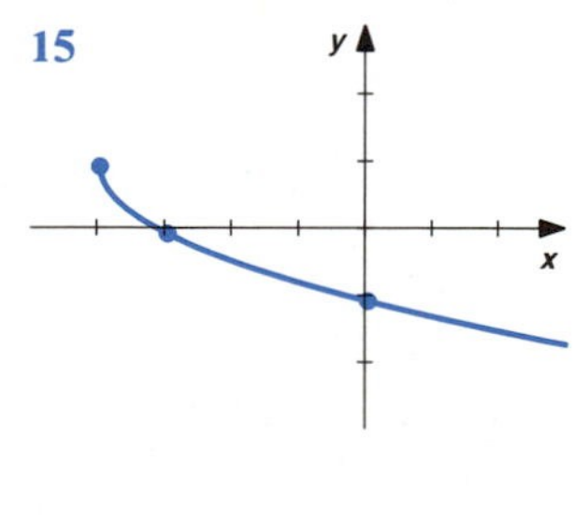

17
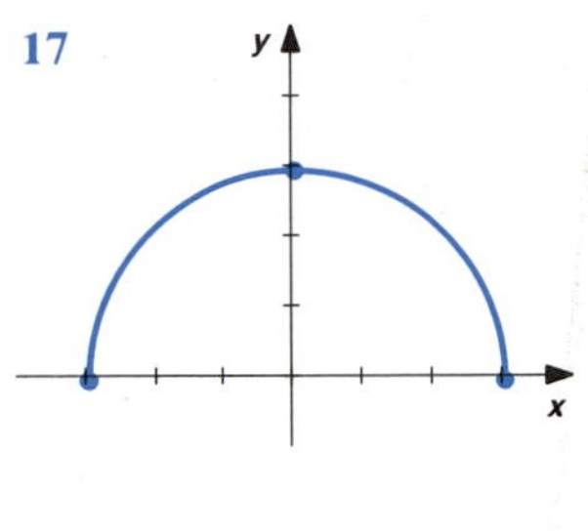

19
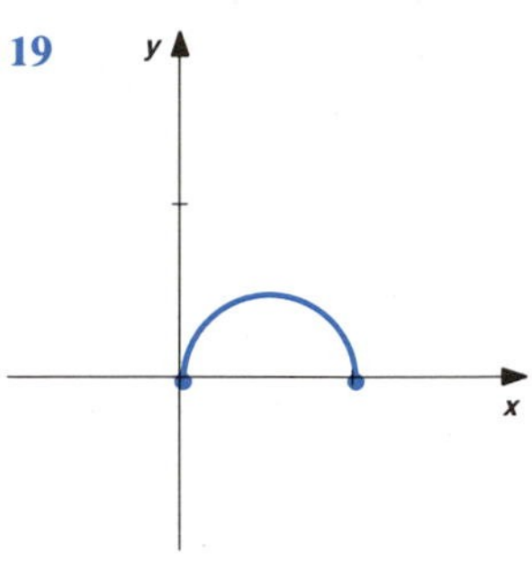

21 $-\infty, \infty$
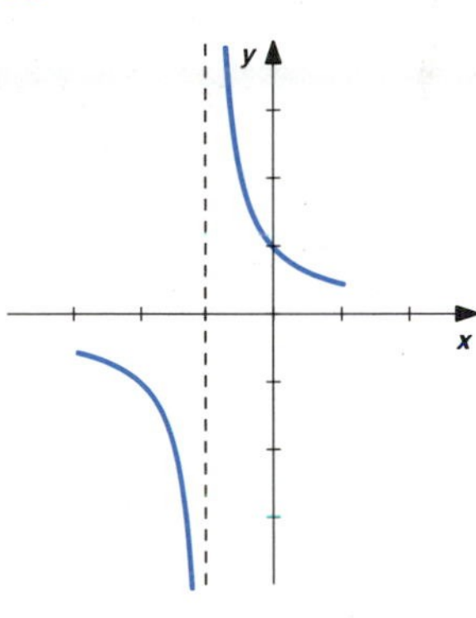

23 $\infty, -\infty$
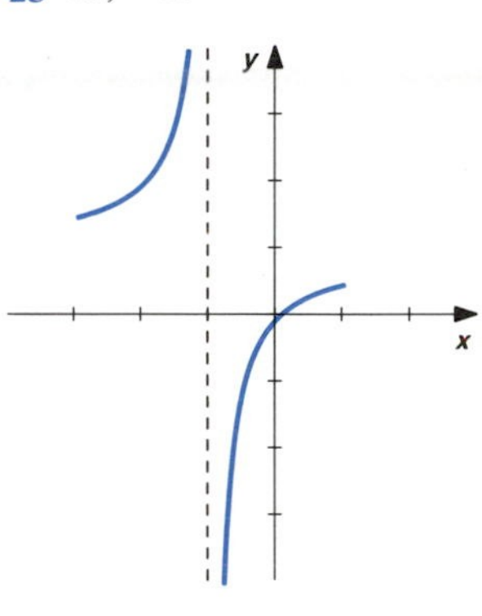

25 $-\infty, \infty$
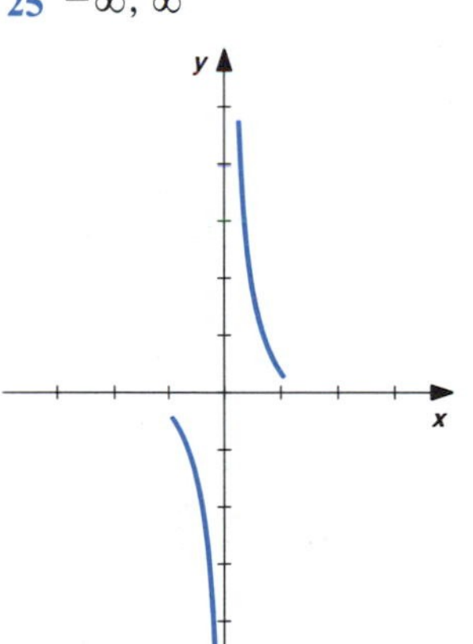

27 $\infty, \infty$
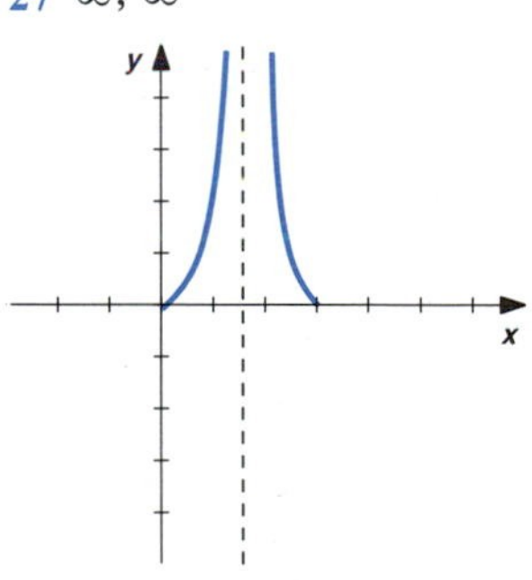

29 $-\infty, -\infty$
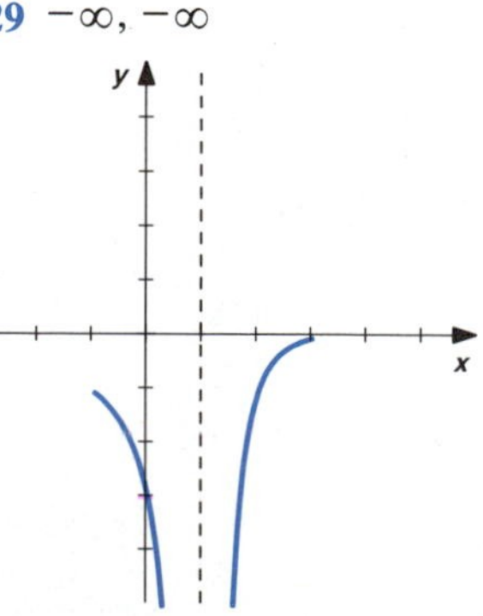

**31** $\infty, -\infty$

**33** $\infty, \infty$

**35** $\infty$

**37** $\infty$

**39** 2

**41** 0

**45** $\lim_{x\to c^+} f(x) = \infty$ if, for each positive number $B$, there is a positive number $\delta$ such that $f(x) > B$ whenever $c < x < c + \delta$.

**51** $x$-axis, $y$-axis, origin

**53** $y$-axis

**55** $x$-axis

**57** Even

**59** Neither

**61** Odd

**63** Even

**65** Even for $n$ even, odd for $n$ odd

**67** Even

**69** Symmetric with respect to the $y$-axis

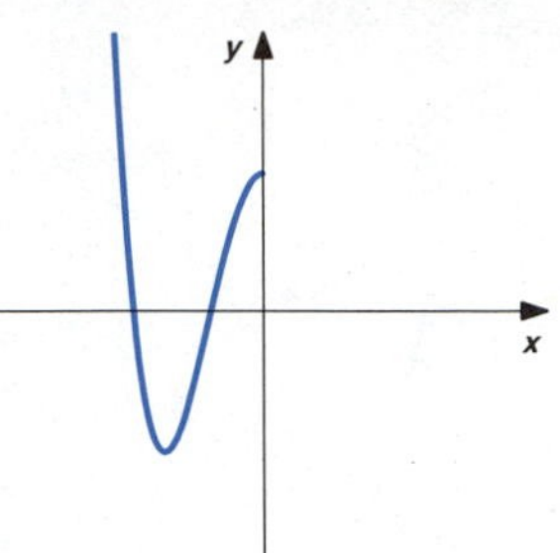


(a)

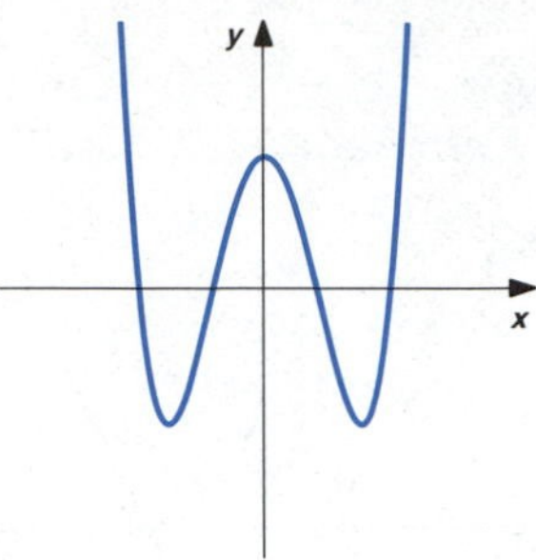


(b)

**71** Symmetric with respect to the $x$-axis

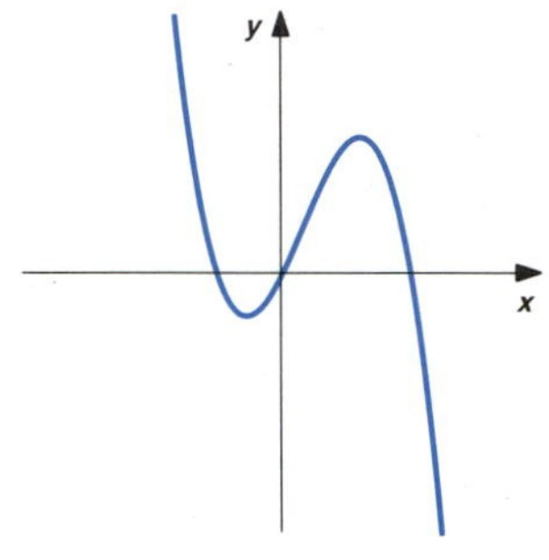


(a)

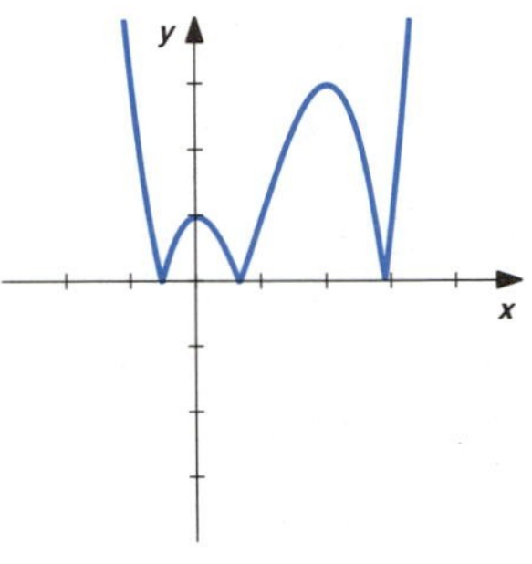


(b)

## *Exercises 4.2*

**1** 0

**3** 1

**5** 1

**7** 1/2

**9** $\infty$

**11** $-\infty$

**13** $-1/2$

**15** 2

**17** $\infty$

**19** $-\infty$

**21** $-\infty$

**23** 0

**25** **27**

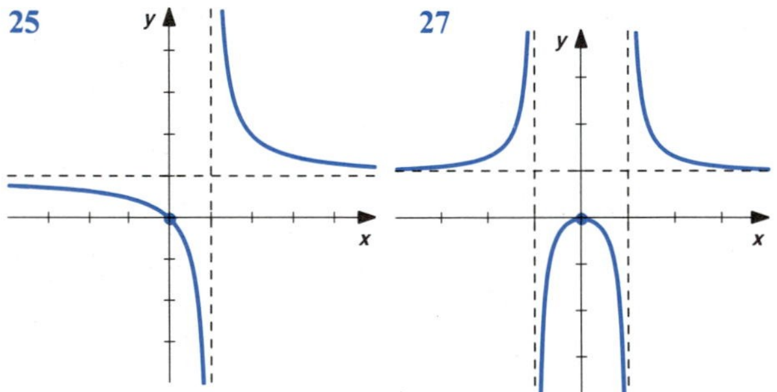


**29**

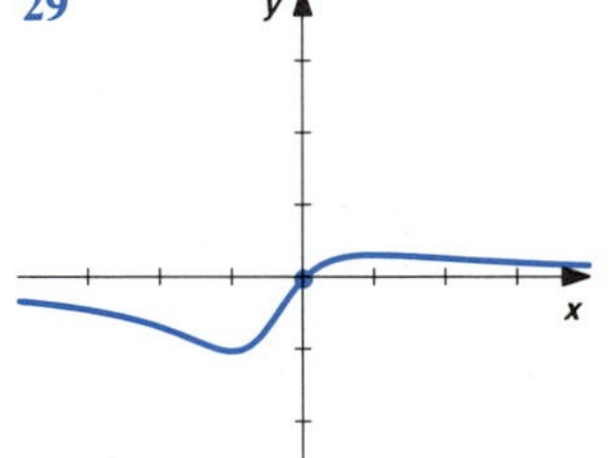


**31**

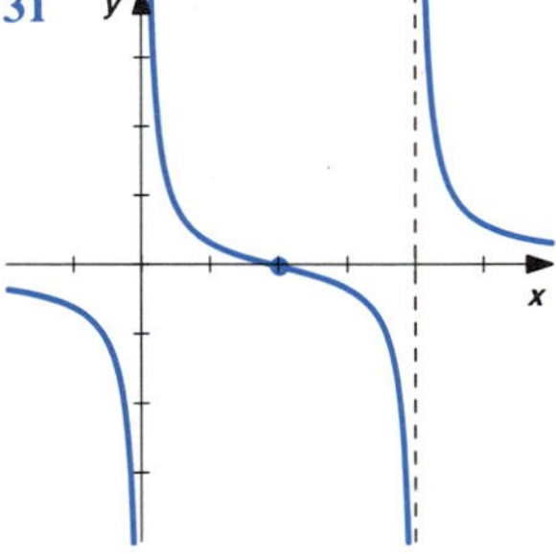


**33**

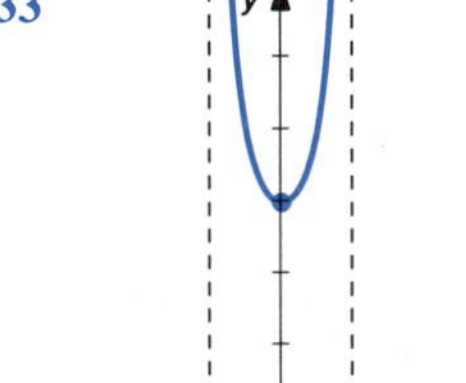

**35**

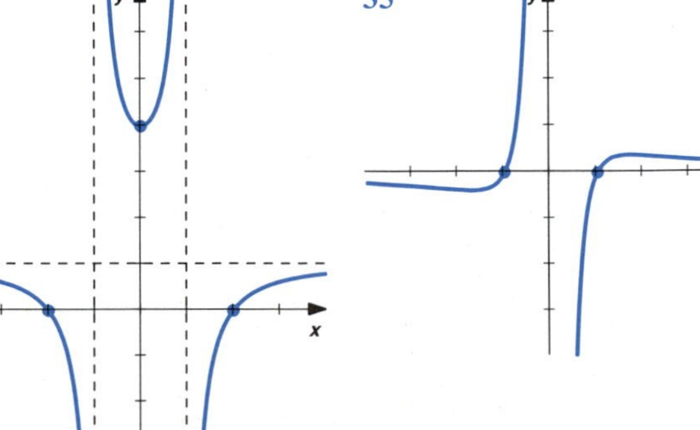

37

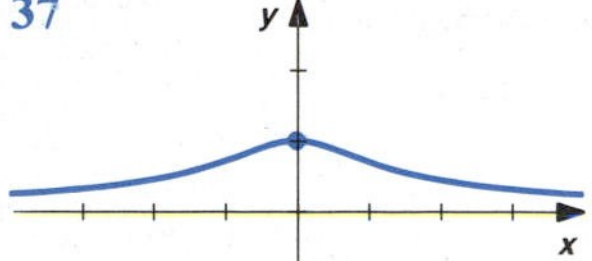


39

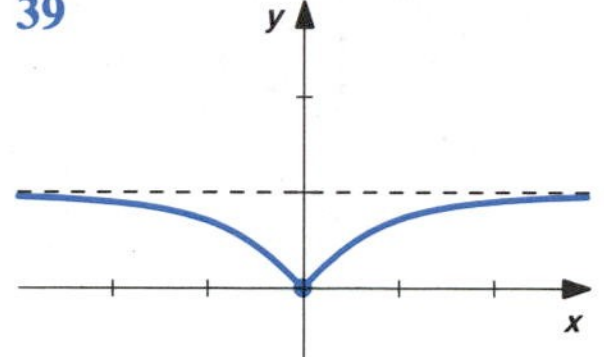


41

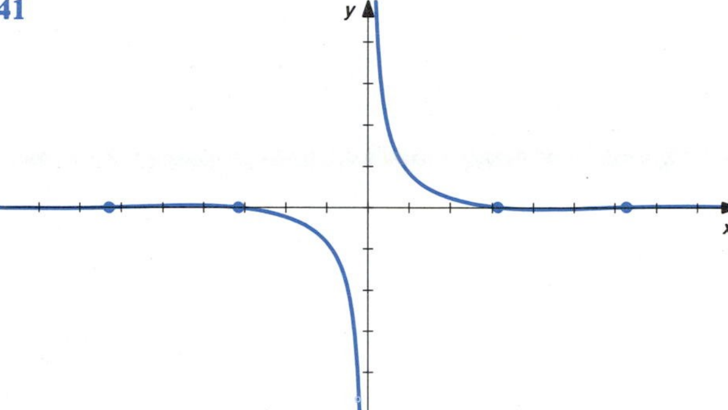


43

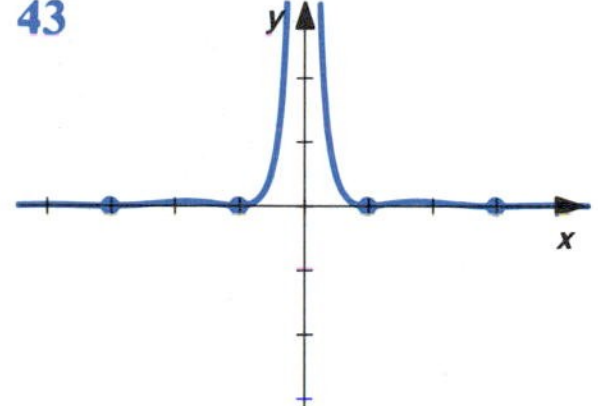


45 $\lim_{x\to\infty} f(x) = \infty$ if, for each positive number $B$, there is a positive number $N$ such that $f(x) > B$ for all $x > N$.

47 $\lim_{x\to-\infty} f(x) = -\infty$ if, for each positive number $B$, there is a positive number $N$ such that $f(x) < -B$ for all $x < -N$.

53 $A(x) = x$

55 $A(x) = -x - 2$

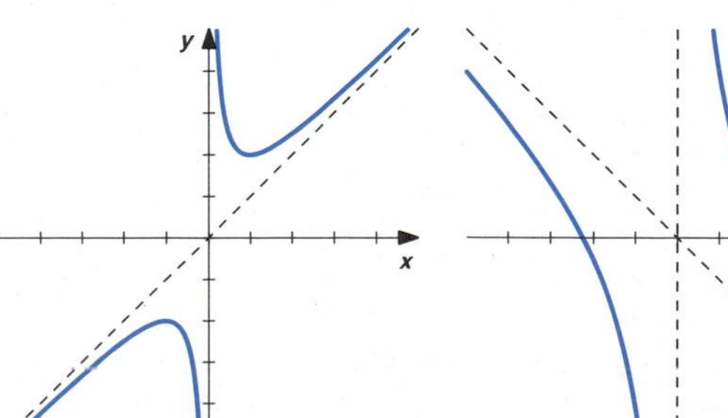


57 $A(x) = x - 1$

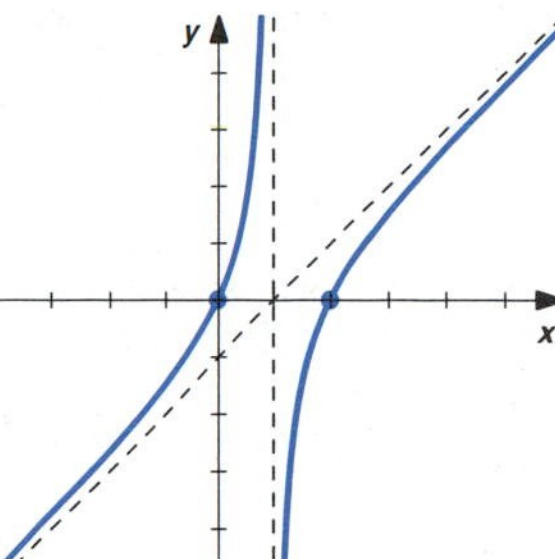


59 $A(x) = x^2$

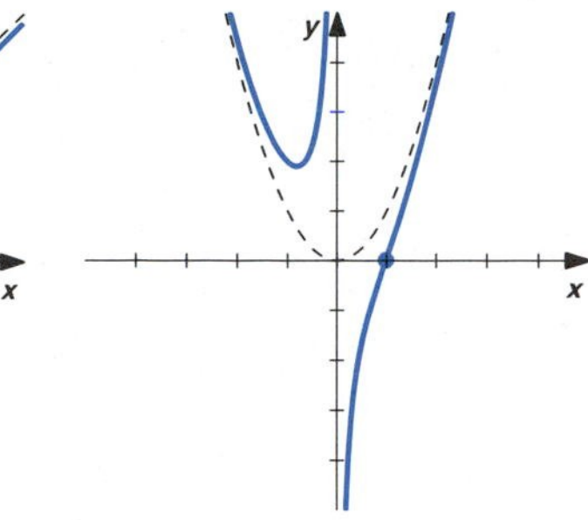

61 $A(x) = x$

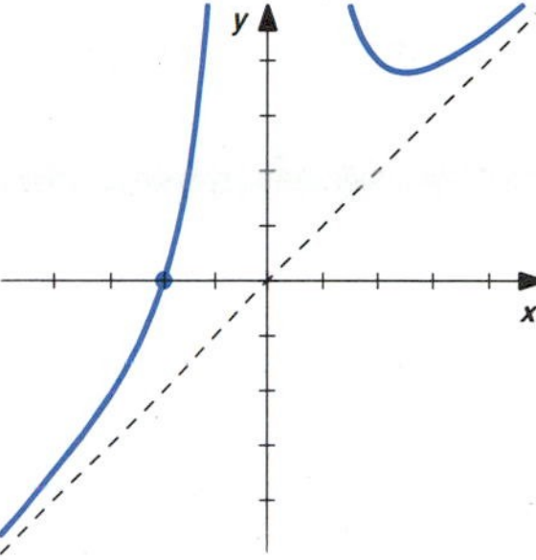


63 $A(x) = x^3$

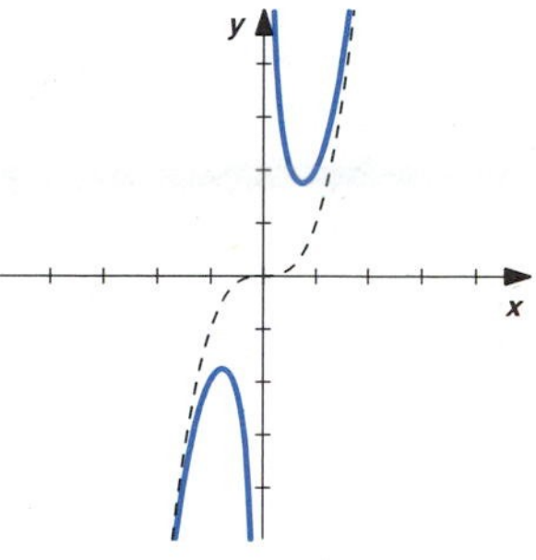


65 $A(x) = x^2$

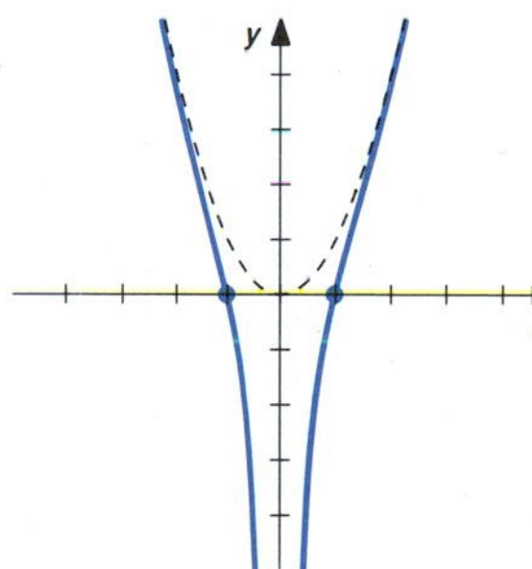


67 $A(x) = x^2 + 1$

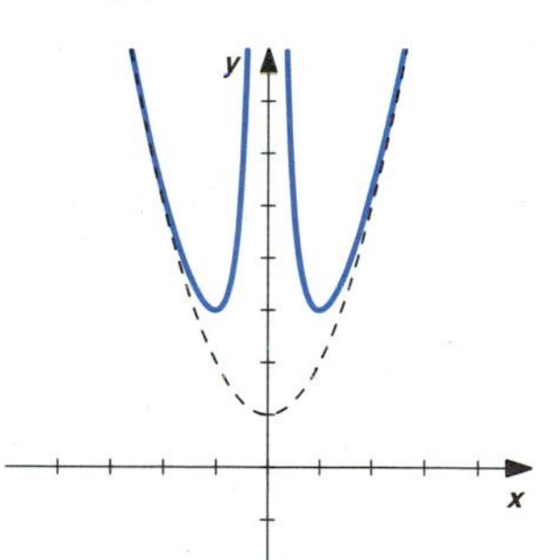


69 $A(t) = 1 + \cos t$

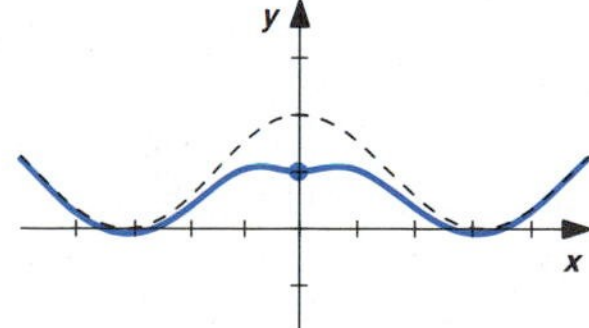

## *Exercises 4.3*

**1** loc min (0, 0)

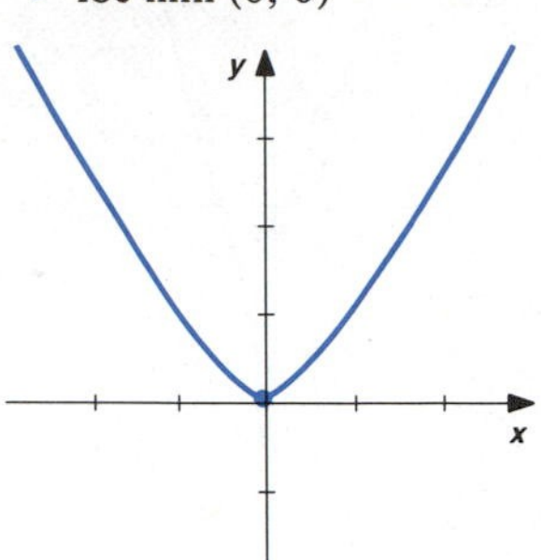

**3**

**5** loc max (3, 19/6)

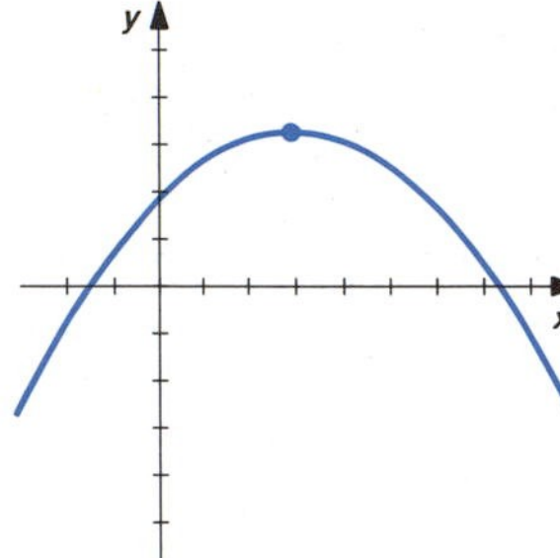

**7** loc max (−2, 2); loc min (2, −2)

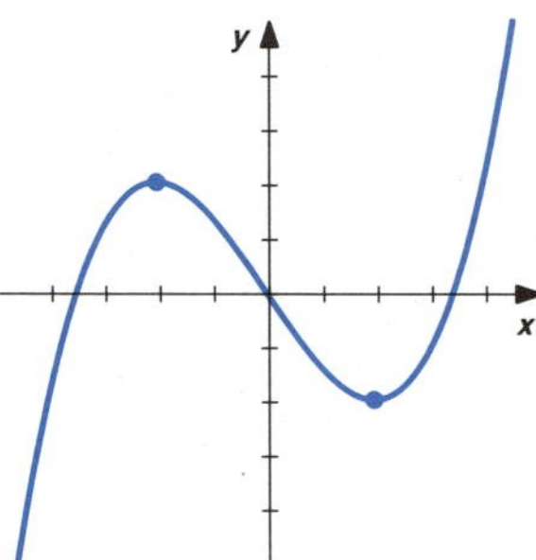

**9** loc max (1, 4); loc min (−1, 0)

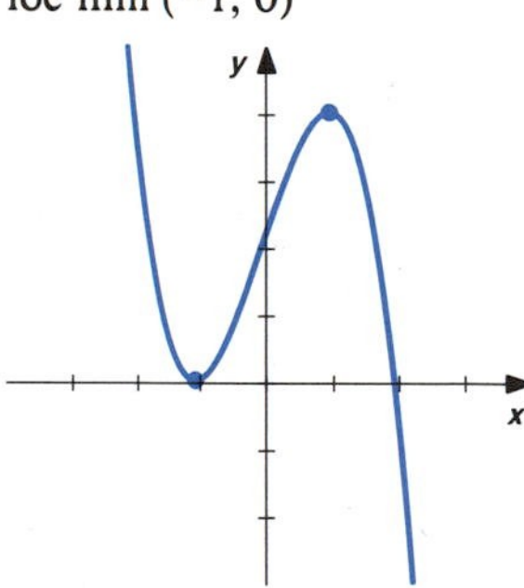

**11**

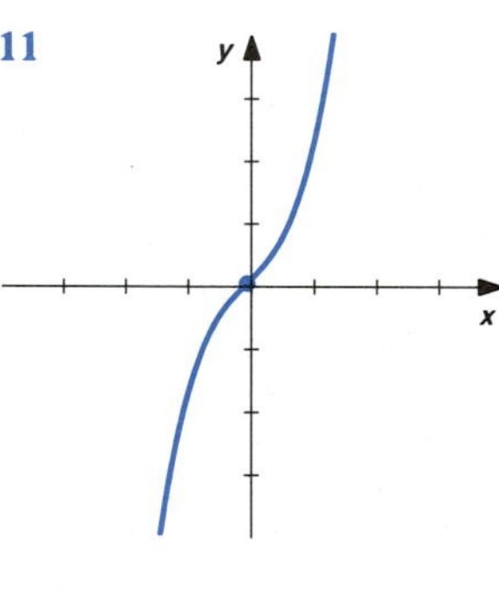

**13** loc max $(-\sqrt{2}, 4)$, $(\sqrt{2}, 4)$; loc min (0, 0)

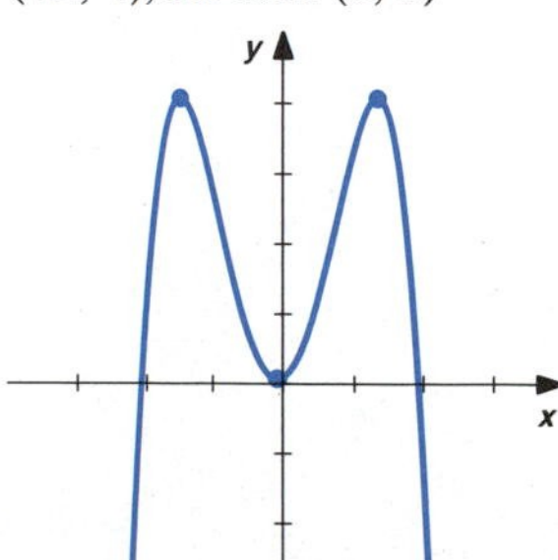

**15** loc min (0, 0)

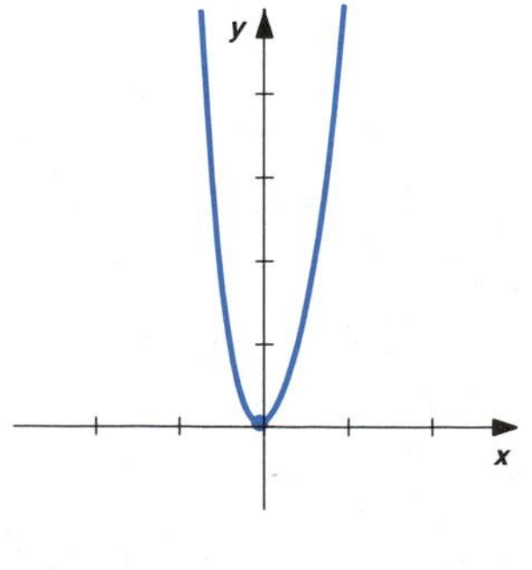

**17** loc max (1, 7/8); loc min (2, −2)

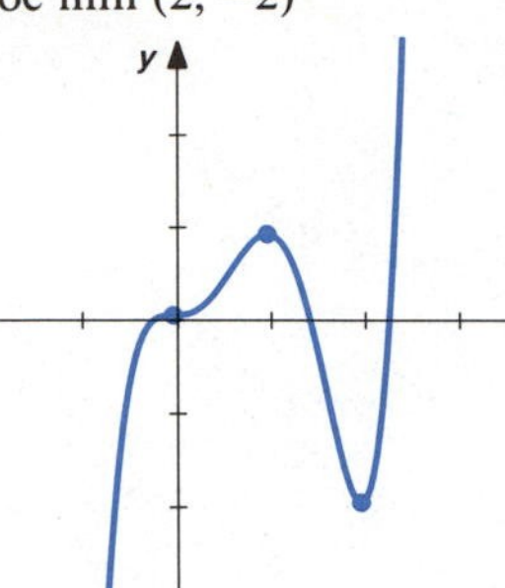

**19** loc max (−1, 6); loc min (1, −6)

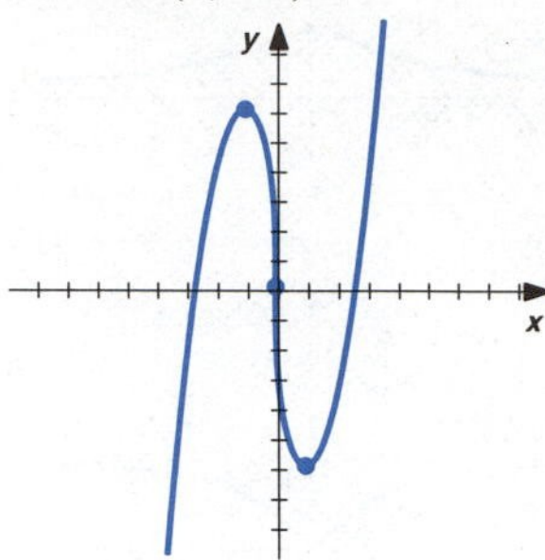

**21** loc min (0, −1)

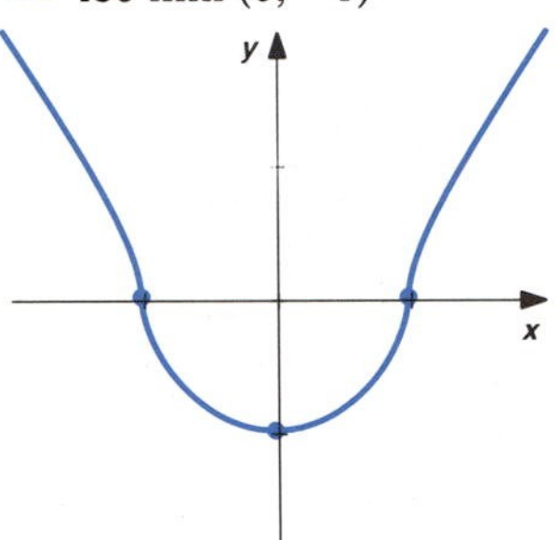

**23** loc min (1, −2)

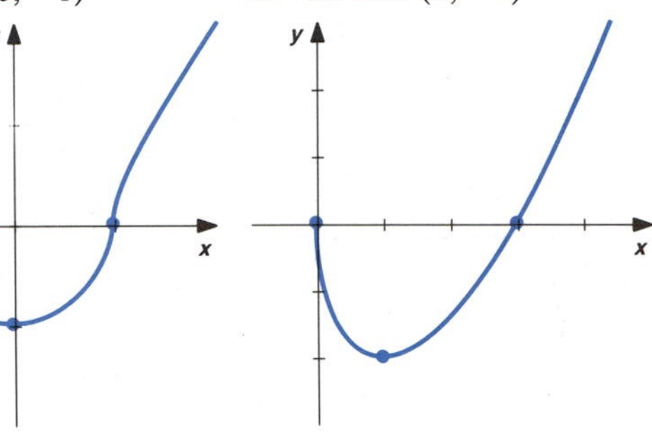

**25** loc max (2, 2); loc min (−2, −2)

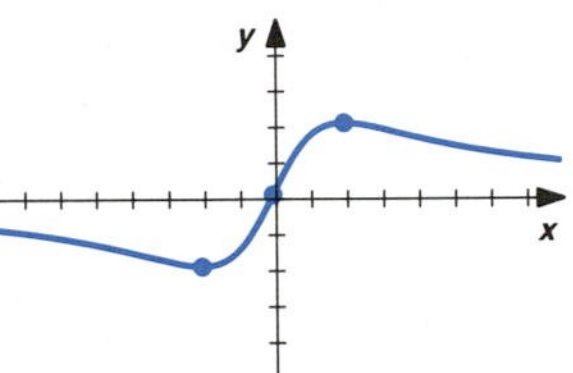

**27** loc min (0, 0)

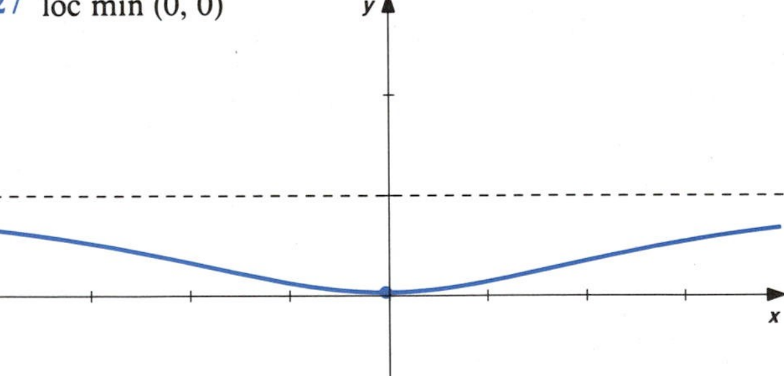

29 loc max (1, 1); loc min (−1, −1)

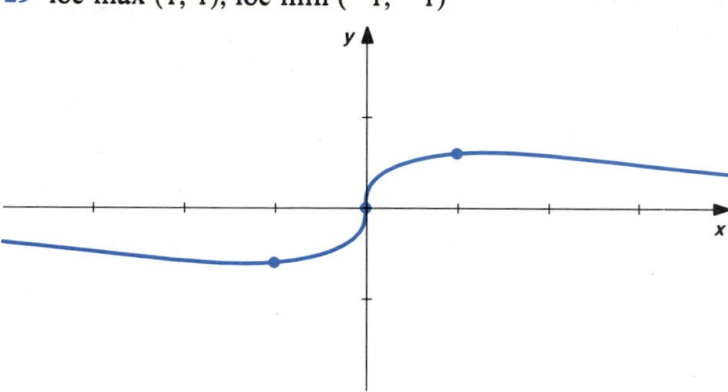


31 loc max (−2, −4);
loc min (2, 4)

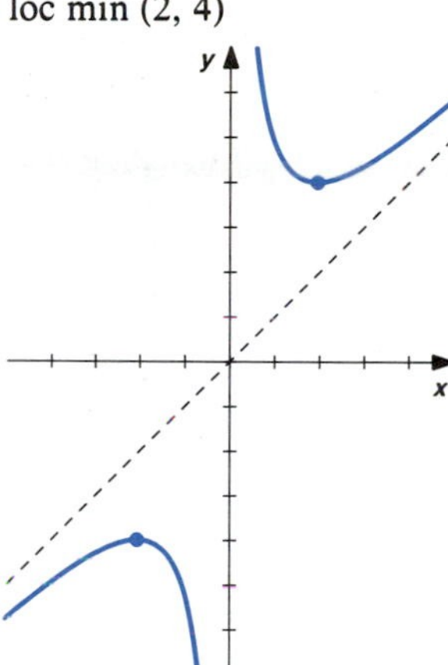


33 loc min (−1, 2), (1, 2)

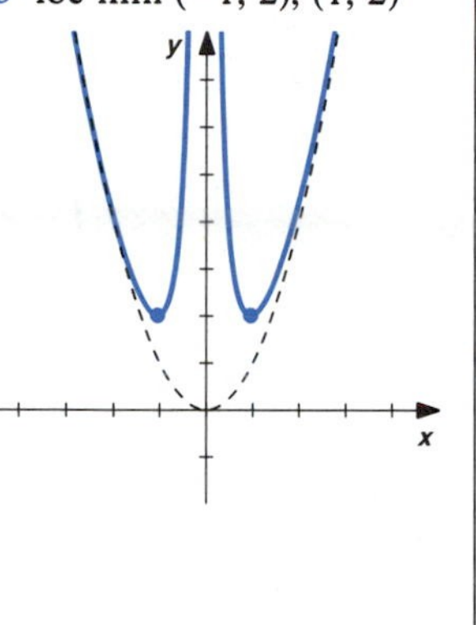


35 loc max (0, −1)

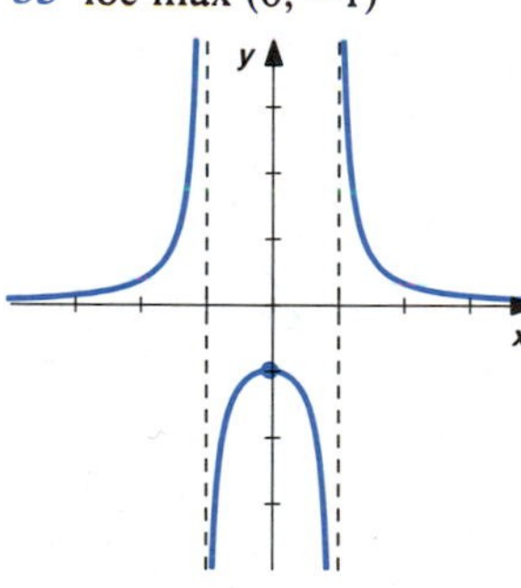


37 loc max (4, 1);
loc min (1, 4)

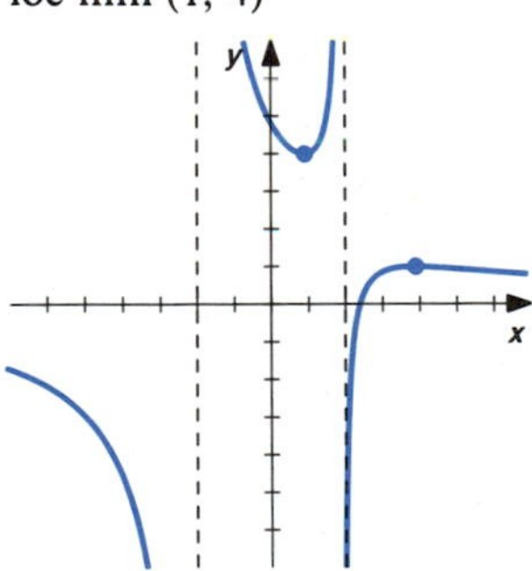


39 loc min (−1, 0), (1, 0)

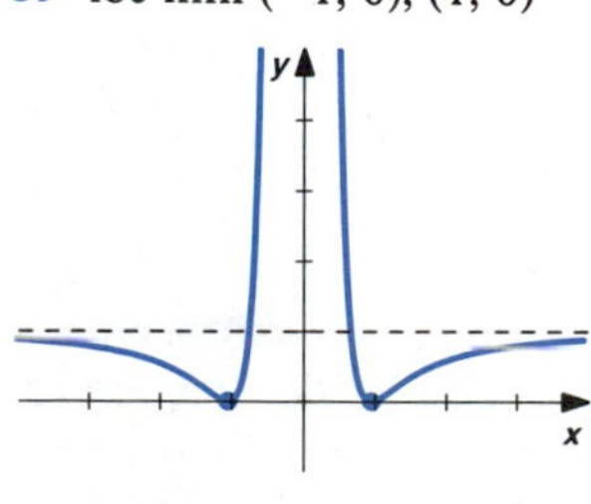


41

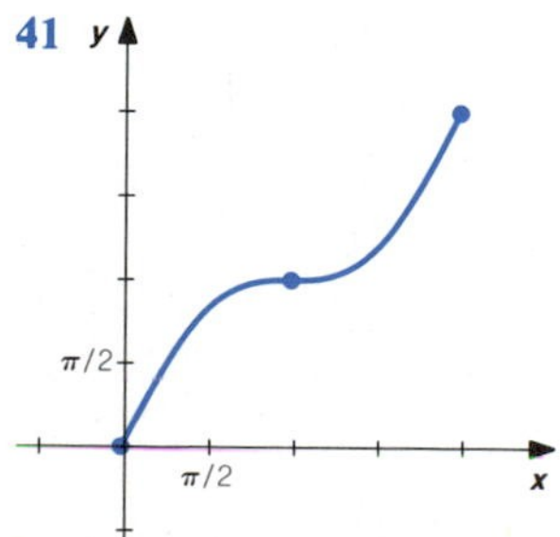


43

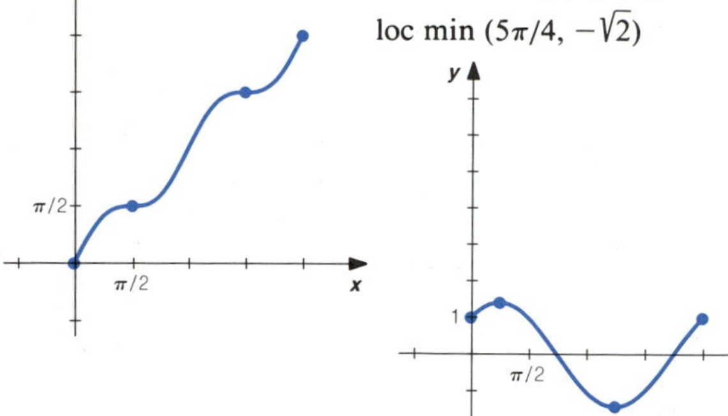


45 loc max ($\pi/4$, $\sqrt{2}$);
loc min ($5\pi/4$, $-\sqrt{2}$)

47

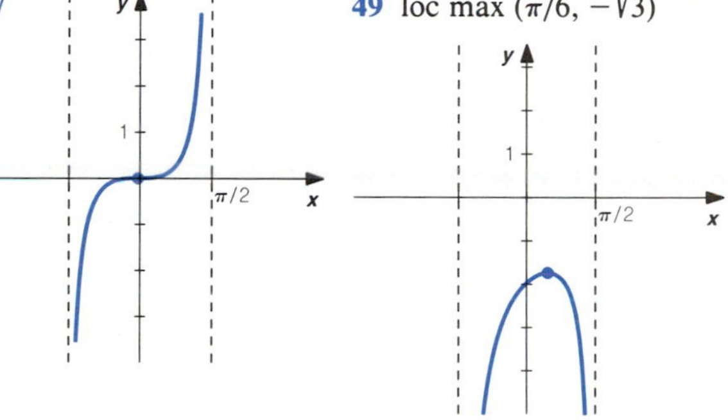


49 loc max ($\pi/6$, $-\sqrt{3}$)

51 loc min (0, 1)

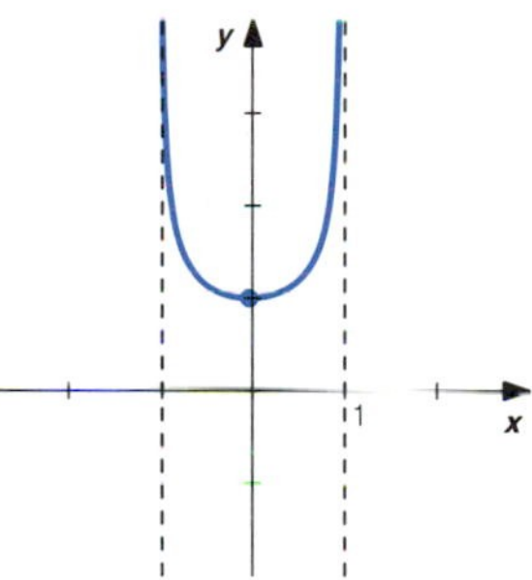


53 loc max (2, 2)

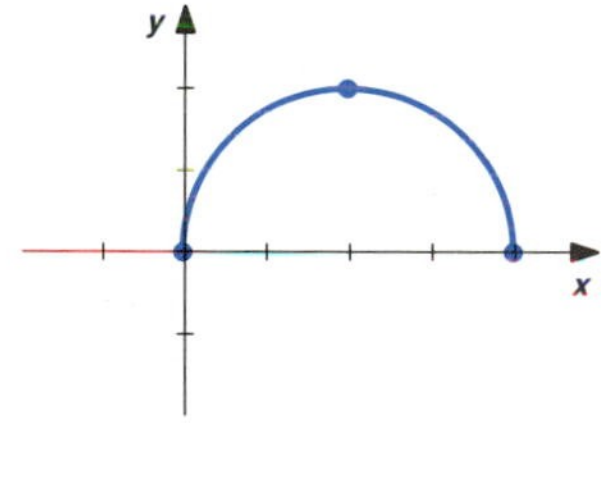


55

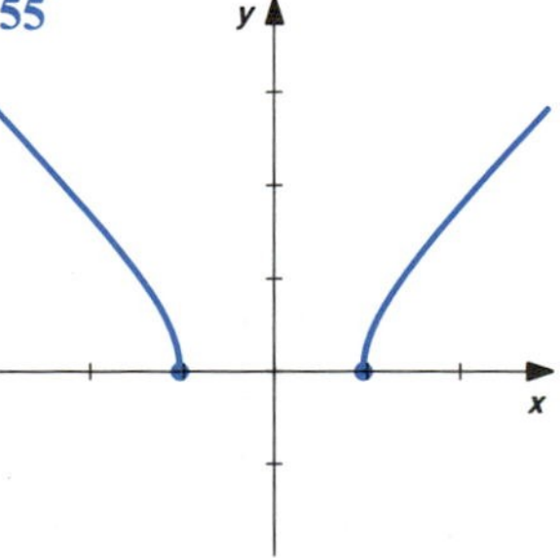


57 $0 \le t < 2$

59 $0 \le t < \pi/2$, $3\pi/2 < t \le 2\pi$

## Exercises 4.4

1

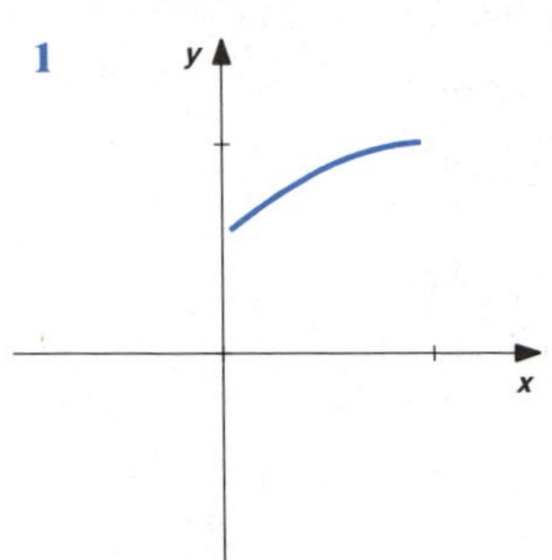

3

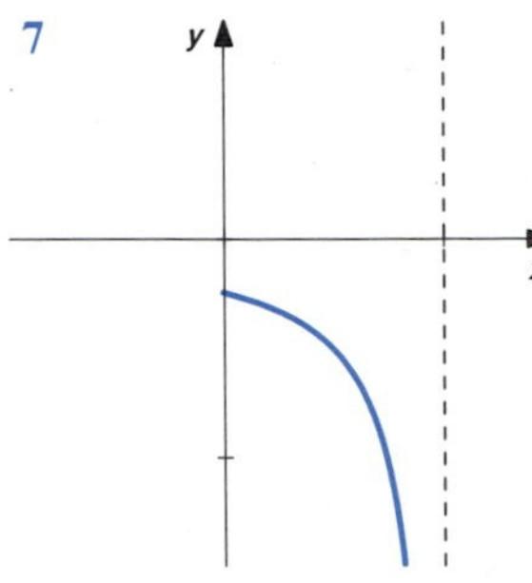

5 Impossible

7

9

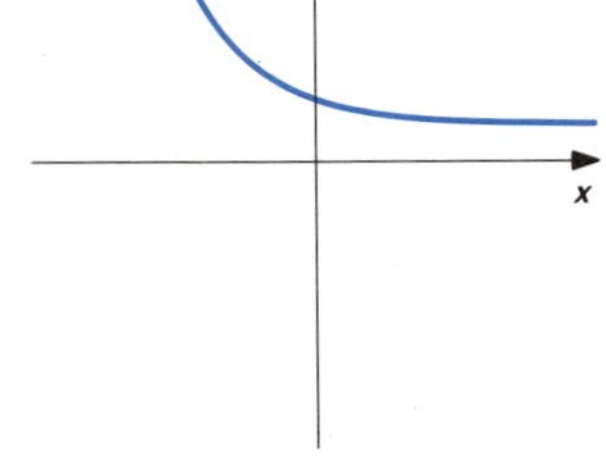

11 Impossible

13

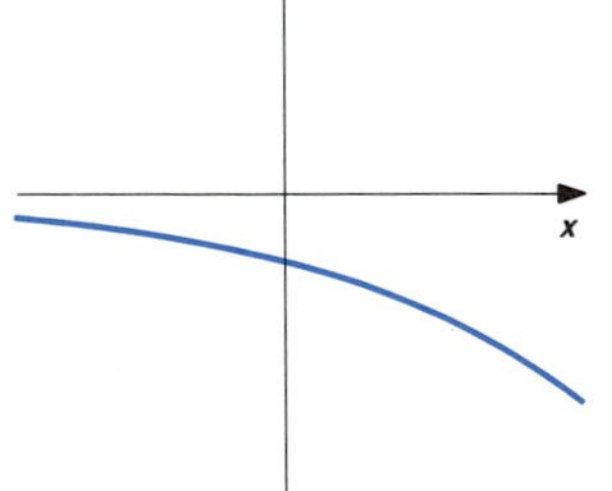

15 Impossible

17 infl pt (0, 1)

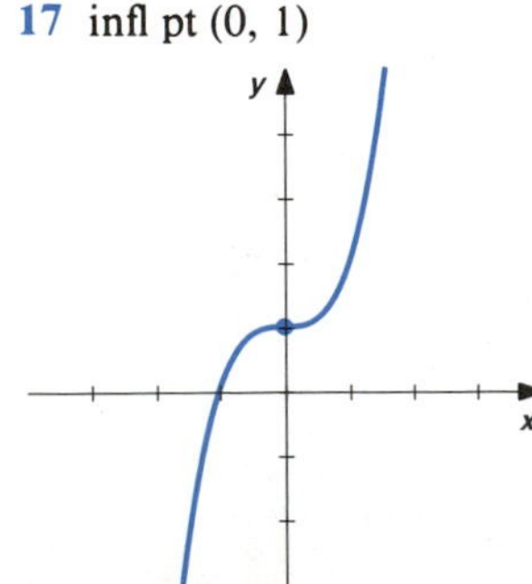

19 loc max (1, 0); loc min (−1, −4); infl pt (0, −2)

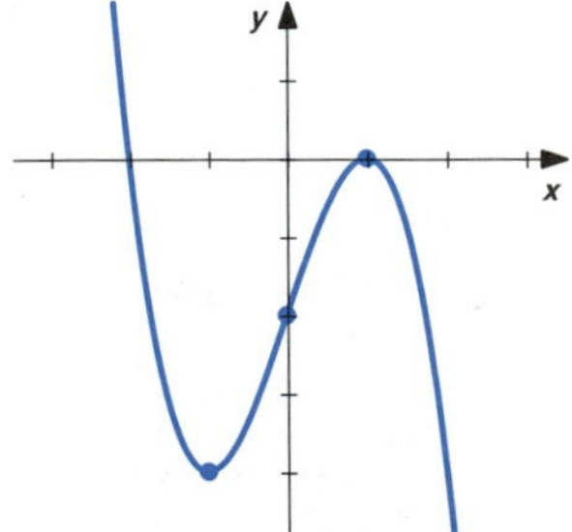

21 infl pt (0, 0)

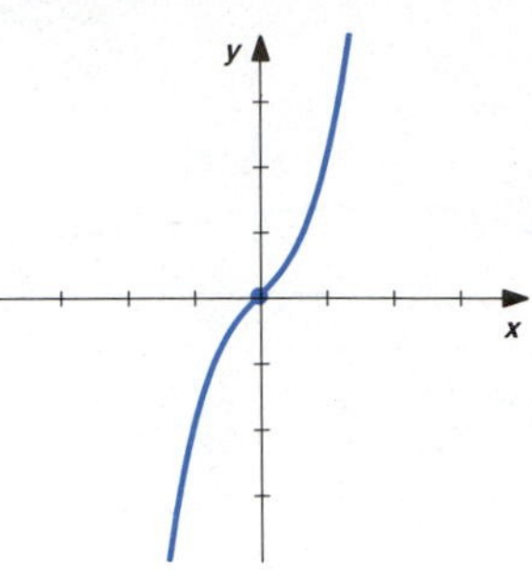

23 loc min (1/8, −1)

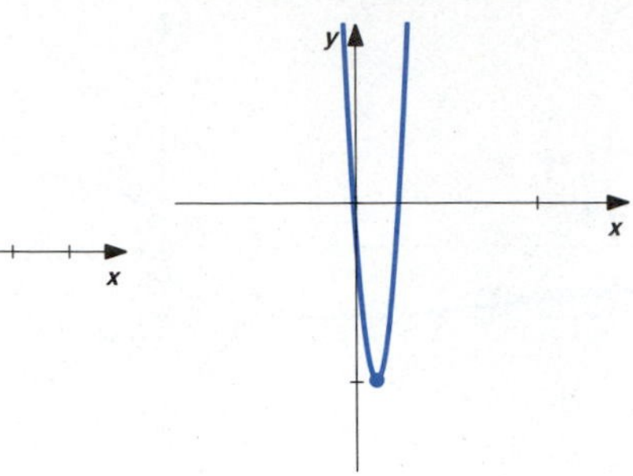

25 loc max (−1, 2); loc min (1, −2); infl pt (0, 0)

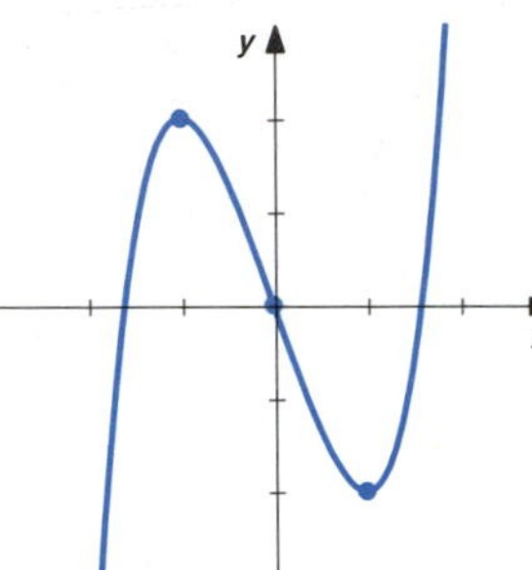

27 loc max (−1, 6); loc min (1, −6); infl pt (0, 0)

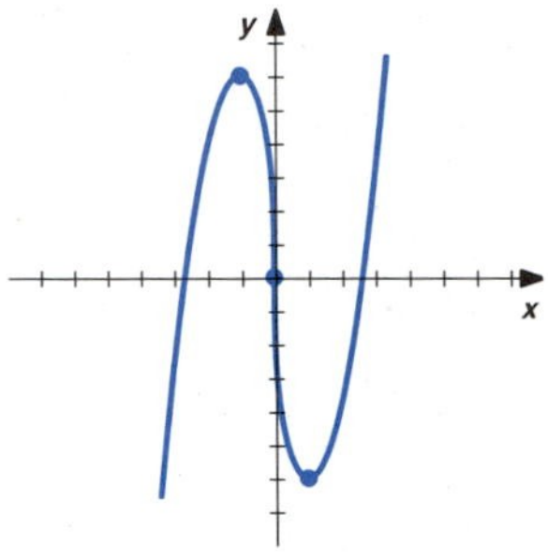

29 loc max $(-16, 24(2)^{2/3})$; loc min (0, 0); infl pt (8, 48)

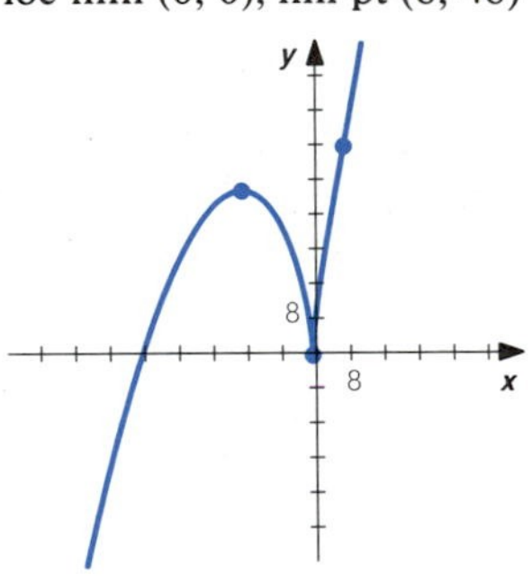


31 loc max (2, 2); loc min (−2, −2); infl pt (0, 0), $(2\sqrt{3}, \sqrt{3})$, $(-2\sqrt{3}, -\sqrt{3})$

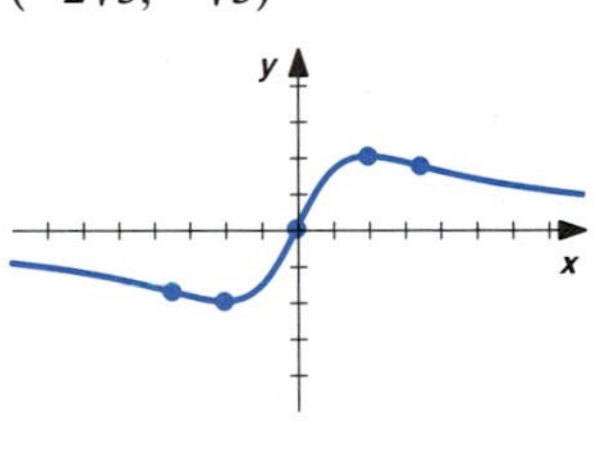

33 infl pt (0, 0)

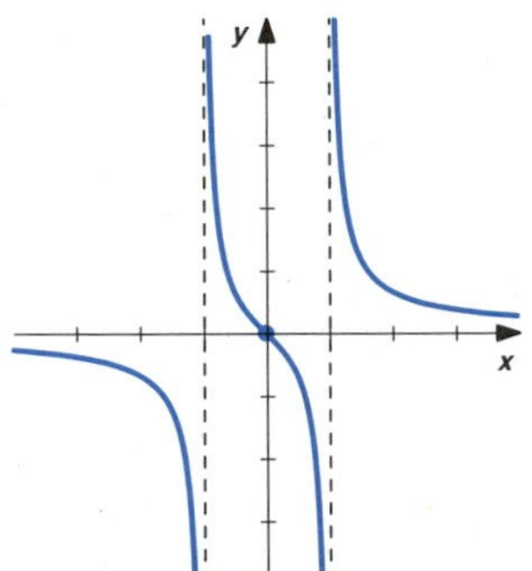

35 loc max (2, 1/4); infl pt (3, 2/9)

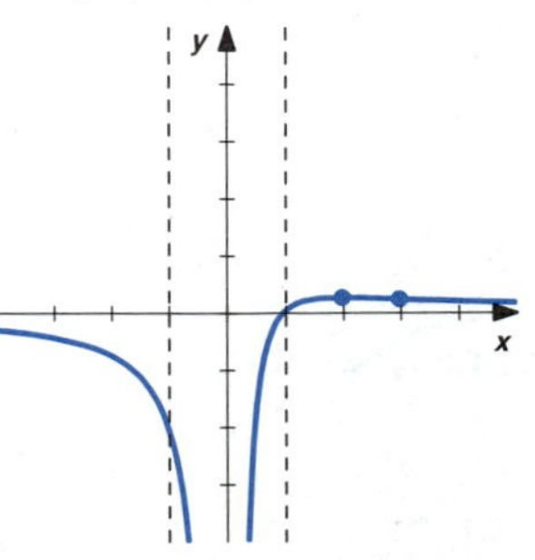

37 loc min $(-1, 0)$, $(1, 0)$; infl pt $(-\sqrt{5/3}, 4/25)$, $(\sqrt{5/3}, 4/25)$

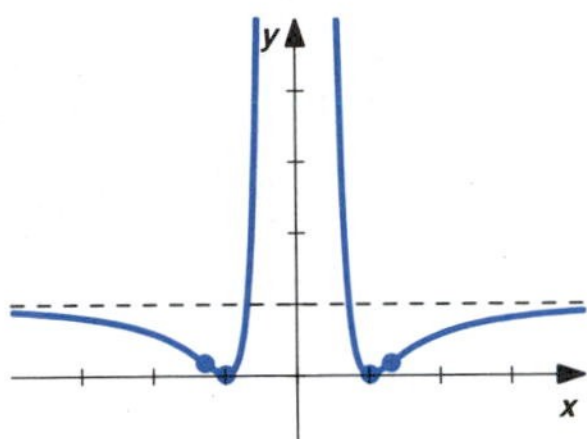

39 loc max $(\pi/3, \pi/3 + \sqrt{3}/2)$, $(4\pi/3, 4\pi/3 + \sqrt{3}/2)$; loc min $(2\pi/3, 2\pi/3 - \sqrt{3}/2)$, $(5\pi/3, 5\pi/3 - \sqrt{3}/2)$, infl pt $(\pi/2, \pi/2)$, $(\pi, \pi)$, $(3\pi/2, 3\pi/2)$

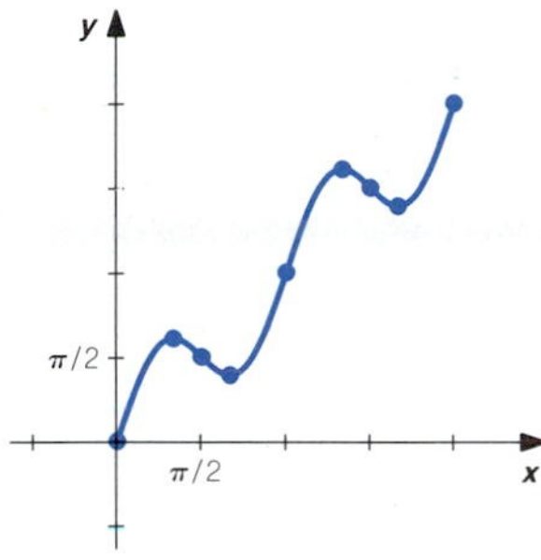

41 loc max $(\pi/4, \sqrt{2})$; loc min $(5\pi/4, -\sqrt{2})$; infl pt $(3\pi/4, 0)$, $(7\pi/4, 0)$

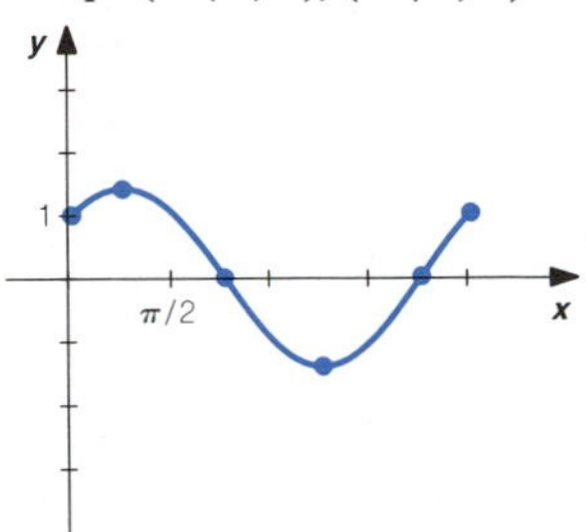

43 45

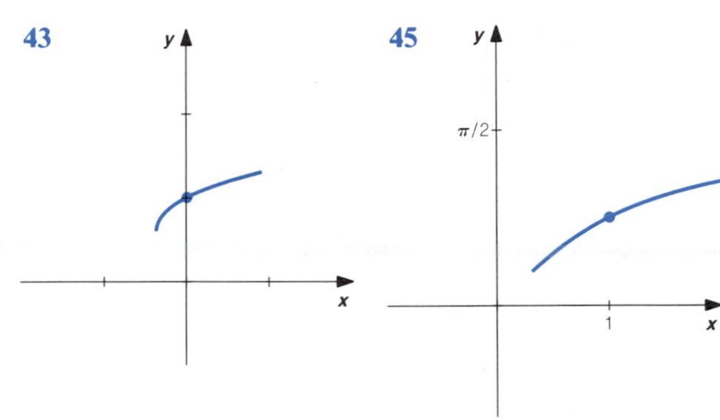

## *Exercises 4.5*

1 (a) $2h + l = 11$ (b) $2h + w = 8.5$ (c) $V = (11 - 2h)(8.5 - 2h)(h)$

3 (a) $l + 2w = 600$ (b) $A = l\left(\frac{600 - l}{2}\right)$

5 (a) $h = \sqrt{3}l/2$ (b) $A = \sqrt{3}l^2/4$

7 (a) $lw = 20{,}000$ (b) $P = 2l + \frac{40{,}000}{l}$

9 (a) $\frac{4}{3}\pi r^3 + \pi r^2 h = 600$ (b) $S = \frac{4}{3}\pi r^2 + \frac{1200}{r}$

11 $s^2 = x^2 + 150^2$

13 $h = 50 \tan\theta$

15 (a) $3l + 4w = 12$ (b) $A = 3l - \frac{3}{4}l^2$

17 $F = x/20$

19 (a) $n = -r/10 + 110$ (b) $R = 110r - r^2/10$

21 $A = 2h\sqrt{16 - h^2}$

23 $V = 9h^2$

25 $V = \frac{\pi h}{3}(a^2 + ab + b^2)$

## *Exercises 4.6*

1 (a) $4\pi$ ft/s (b) $24\pi$ ft$^2$/s

3 (a) 720 cm$^2$/s (b) 3600 cm$^3$/s

5 $2/(9\pi)$ ft/min

7 5/18 ft/min

9 (a) 1/625 m/min (b) 1/1000 m/min

11 (a) 3 ft/s (b) 3 ft/s

13 30/13 ft/s

15 Decreasing at 18 mph

17 $10\sqrt{7}$ mph

19 1/169 rad/s

21 3/80 rad/s

23 $-5$ rad/s

25 0.075 in./s

27 Decreasing at 0.0222 Ω/s

29 Decreasing at 15/16 m$^3$/s

33 $100\sqrt{3}$ m, $10\sqrt{3}$ m/s

35 $\frac{dx}{dt} = \frac{1}{\sqrt{17}}, \frac{dy}{dt} = \frac{4}{\sqrt{17}}$

37 $\approx 88$ mph

39 $\frac{dV}{dt} = -\pi r^3 \sec^2\theta \frac{d\theta}{dt}$

## Exercises 4.7

1 $A(\theta) = \frac{25}{2}\sin\theta\cos\theta, 0 \le \theta \le \pi/2; \pi/4, \pi/4, \pi/2;$ $A(\pi/4) = 25/4$

3 $A(x) = 450x - \frac{3}{2}x^2, 0 \le x \le 300;$ 150 m × 225 m; $A(150) = 33{,}750$ m$^2$

5 (a) $F(x) = x + \frac{30{,}000}{x}, x > 0; 50\sqrt{3}$ m × $100\sqrt{3}$ m; $F(100\sqrt{3}) = 200\sqrt{3}$ m

(b) $F(x) = x + \frac{30{,}000}{x}, 0 < x \le 150;$ 100 m × 150 m; $F(150) = 350$ m

7 (a) Minimum area is $\frac{9}{4(\pi+4)}$ m$^2$ when $\frac{12}{\pi+4}$ m is used for the square and $\frac{3\pi}{\pi+4}$ for the circle.

(b) Use all of wire for circle for maximum area. Maximum area is $\frac{9}{4\pi}$.

9 $V(h) = h(3-2h)(2-2h), 0 \le h \le 1;$ $\left(\frac{5-\sqrt{7}}{6}\text{ft}\right) \times \left(\frac{4+\sqrt{7}}{3}\text{ft}\right) \times \left(\frac{1+\sqrt{7}}{3}\text{ft}\right);$ $V\left(\frac{5-\sqrt{7}}{6}\right) = \frac{10+7\sqrt{7}}{27}$ ft$^3$

11 $C(x) = 8x + \frac{96{,}000}{x}, x > 0; 20\sqrt{30}$ m × $80\sqrt{30}/3$ m; $C(20\sqrt{30}) = \$1752.71$

13 (a) $C(x) = 130\sqrt{x^2+60^2} + 50(100-x), 0 \le x \le 100;$ $C(25) = \$12{,}200$ (b) $C(x) = 130\sqrt{x^2+60^2} + 50(20-x),$ $0 \le x \le 20; C(20) \approx \$8222$

15 $L(x) = \sqrt{x^2+36} + \sqrt{(21-x)^2+64}, 0 \le x \le 21;$ $L(9) \approx 25.24$ ft

17 $R(r) = 60r - r^2/10, 200 \le r \le 600;$ \$300/month

19 $A(r) = r(2-r), 2/(1+\pi) \le r \le 2; \theta = 2; r = 1$

23 $V(h) = \frac{\pi}{3}(36h - h^3), 0 \le h \le 6; 2\pi - \frac{2\pi\sqrt{6}}{3}$

25 (a) $A(\theta) = 36 - 18\csc\theta + 9\cot\theta, \pi/6 \le \theta \le \pi/2;$ $\theta = \pi/3$ (b) $A(\theta) = 66 - 60.5\csc\theta + 30.25\cot\theta,$ $\theta_0 \le \theta \le \pi/2; \theta_0,$ where $\sin\theta_0 = 5.5/6$

27 $I(y) = \frac{ky}{(10{,}000 - 50y + y^2)^{3/2}}, 0 \le y \le 200;$ $\frac{25+\sqrt{80{,}625}}{4} \approx 77.24$ ft

## Exercises 4.8

1 3.99

3 1.51

5 1

7 1

9 4.075

11 0.88539816

13 0.9

15 $3\,dx$

17 $2x\,dx$

19 $(3x^2 + 8x)\,dx$

21 $\frac{1}{2\sqrt{x}}\,dx$

23 $-3\sin 3x\,dx$

25 $\sec^2 x\,dx$

27 $\Delta V = 2\pi rh_0\,dr + \pi h_0(dr)^2, dV = 2\pi rh_0\,dr$

29 $\Delta A = 2s\,ds + (ds)^2, dA = 2s\,ds$

31 $V = 2\pi r^3, dV = \pm 9.0478$ m$^3$, $\pm 2.25\%$

33 $h = 40 \pm 1.396$ ft

## Exercises 4.9

1 0.4; 0.571; 0.171; 0.03078

3 2.0033333; 2.0033278; 0.00000556; $1.5432 \times 10^{-8}$

5 1; 0.9982; 0.0018; 0.000036

7 1.2; 1.24; 0.078125; 0.01953125

9 1.2; 1.22; 0.06; 0.004

11 −0.25; −0.28125; 0.05555556; 0.01234568

13 $P_1(x) = 4 - \frac{3}{4}(x-3); P_1(3.1) = 3.925;$ $P_2(x) = 4 - \frac{3}{4}(x-3) - \frac{25}{128}(x-3)^2;$ $P_2(3.1) = 3.9230469$

15 $P_1(x) = \frac{\pi}{6} + \frac{2}{\sqrt{3}}\left(x - \frac{1}{2}\right); P_1(0.55) \approx 0.5813338;$ $P_2(x) = \frac{\pi}{6} + \frac{2}{\sqrt{3}}\left(x - \frac{1}{2}\right) + \frac{2}{3\sqrt{3}}\left(x - \frac{1}{2}\right)^2;$ $P_2(0.55) \approx 0.582296$

## Exercises 4.10

9 1/3

11 1/3

13 −1

15 1/36

17 $\frac{1}{3}x^3 - x^2 + 3x + C$

19 $\frac{2}{3}x^{3/2} + 2x^{1/2} + C$

21 $-\cos x + \sin x + C$

23 $-\csc x + C$

25 $\frac{1}{3}x^3 - \frac{1}{2}x^2 - 2x + C$

27 $x - \frac{1}{x} + C$

29 $2x^3 + x^2 - 3x + 5$

31 $\frac{2}{3}x^{3/2} + \frac{4}{3}$

33 $3 - \cos x$

35 $2x^3 + 2x + 3$

37 $-\sin x - \cos x + 2x + 2$

39 (a) 16 ft (b) $2s$ (c) −32 ft/s

41 (a) 1 s (b) 80 ft/s

43 (a) 3 s (b) 28.7 m/s

45 (a) 8 m (b) 56 m

---

## Review Exercises

1

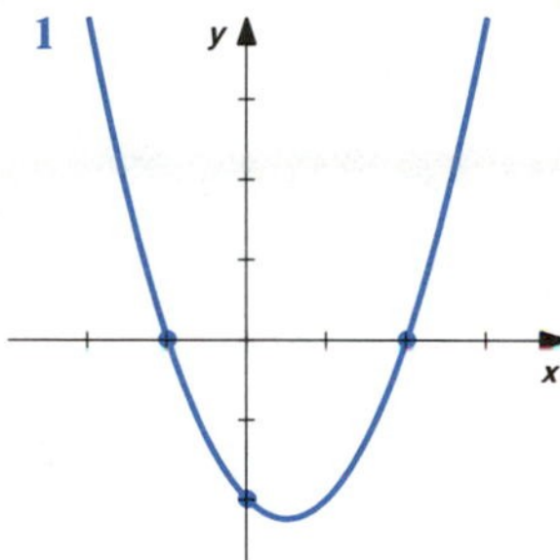


3

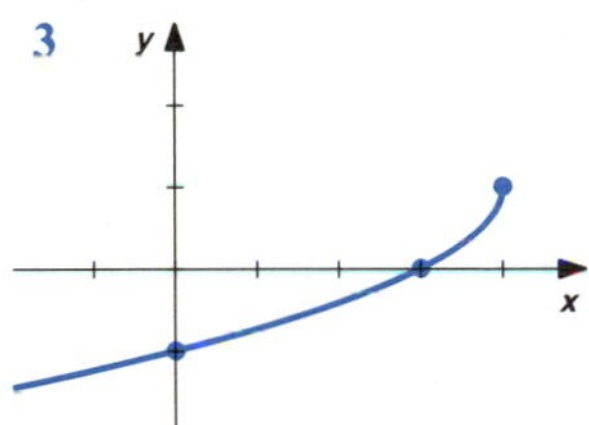


5 ∞

7 −∞

9 3/2

11 −3

13 $A(x) = -2$

15 $A(x) = x + 2$

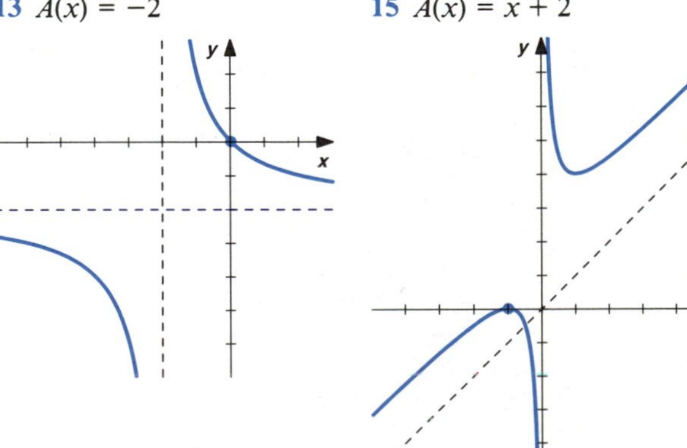


17 $y$-axis

19 $x$-axis

21 Even

23 Neither

25 Symmetric with respect to $y$-axis

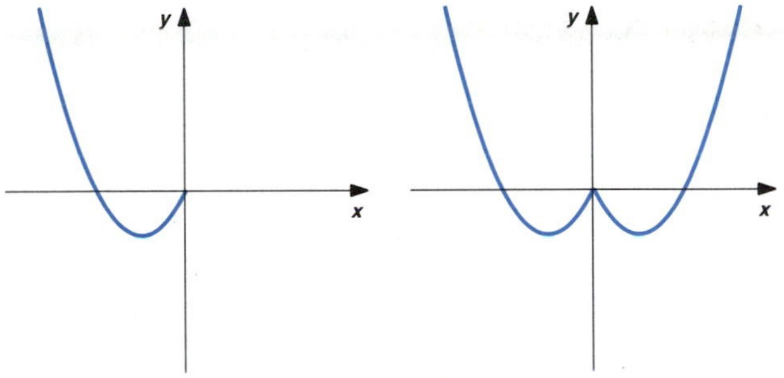

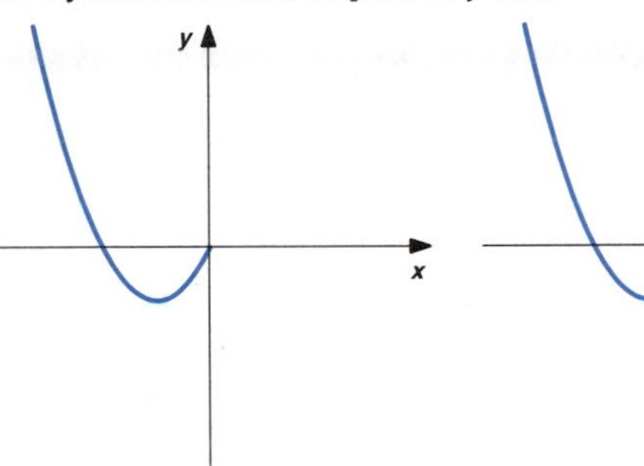


(a) (b)

27

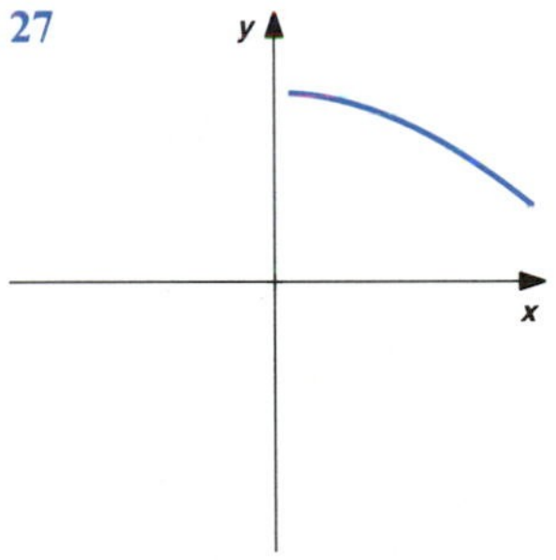

29 Impossible

31 loc max (−1, 2); loc min (1, −2); infl pt (0, 0)

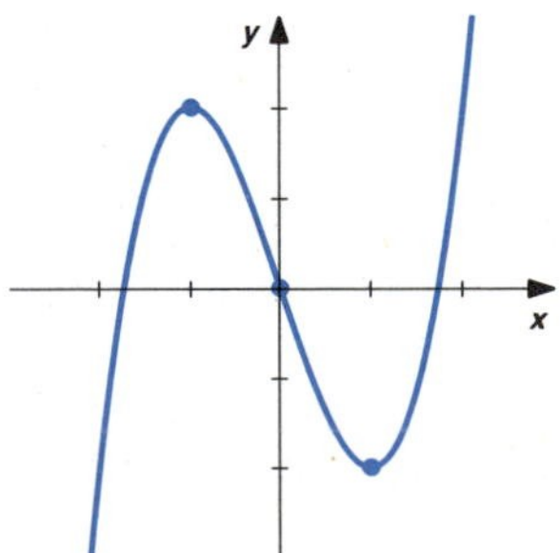

33

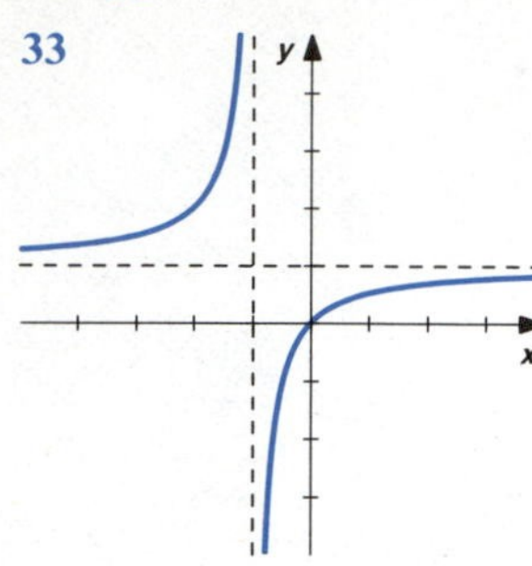


35 loc max $(-2, -3)$

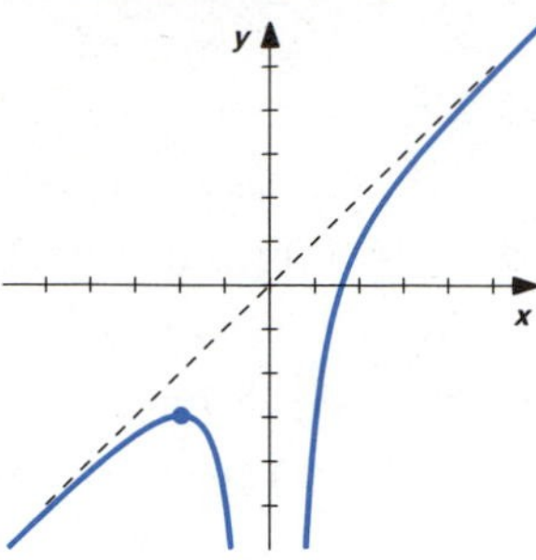


37 loc max $(\pi/3, \sqrt{3} - \pi/3)$; loc min $(5\pi/3, -\sqrt{3} - 5\pi/3)$; infl pt $(\pi, -\pi)$

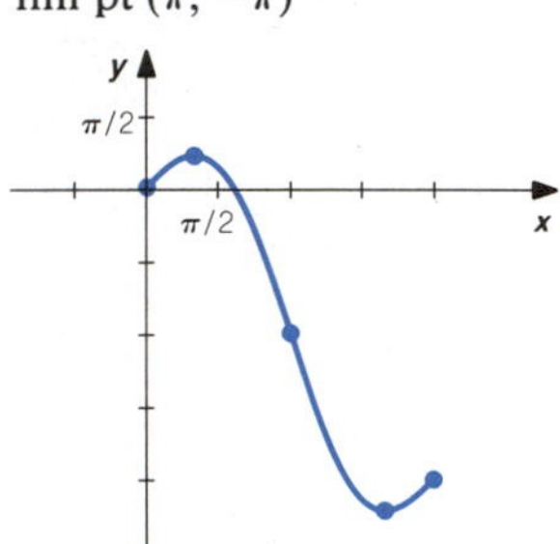


39 loc max $(0, 1)$; loc min $(-1, 0)$, $(1, 0)$; infl pts $(\sqrt{3}, 2^{2/3})$, $(-\sqrt{3}, 2^{2/3})$

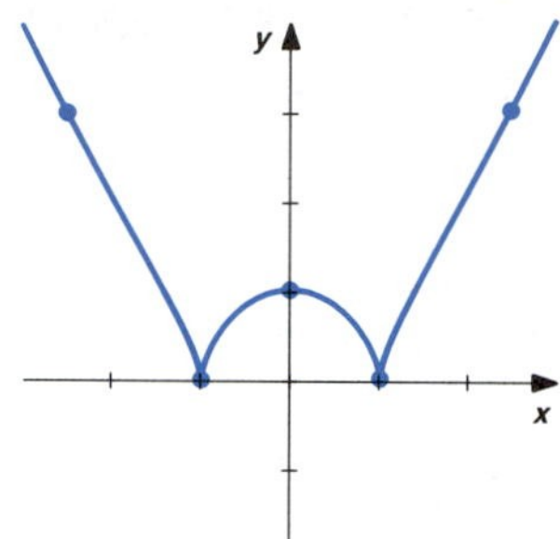


41 Increasing at $144\pi$ in.$^3$/s

43 0.01 rad/s

45 (a) $A(x) = 2000x - 2x^2$, $0 \le x \le 1000$; 500 ft $\times$ 1000 ft; $A(500) = 500{,}000$ ft$^2$ (b) $A(x) = 2000x - 2x^2$, $0 \le x \le 400$; 400 ft $\times$ 1200 ft; $A(400) = 480{,}000$ ft$^2$

47 $A(b) = \dfrac{1}{2}\dfrac{b^2}{b-2}$, $b > 2$, $4 \times 2$, $A(4) = 4$

49 50.5°

51 0.98875; 0.0005625

53 $\pm 0.12$ m$^3$, $\pm 3\%$

55 $-1/2$

57 $-1/3$

59 $\dfrac{1}{4}x^4 - \dfrac{1}{2}x^2 + 2x + C$

61 $-\cos x - \sin x + C$

63 $\dfrac{2}{5}x^{5/2} + \dfrac{2}{3}x^{3/2} + C$

65 $2x^3 + 2x^2 - x - 1$

67 64 ft; 4 s

## CHAPTER FIVE

### *Exercises 5.1*

1 $x^2 + 2$

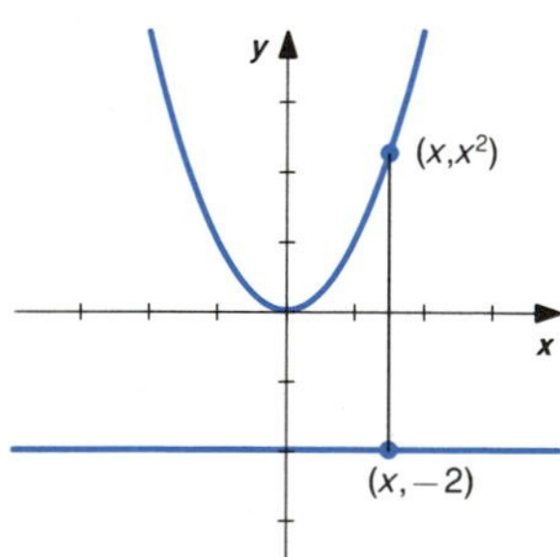


3 $2x^2 + 1$

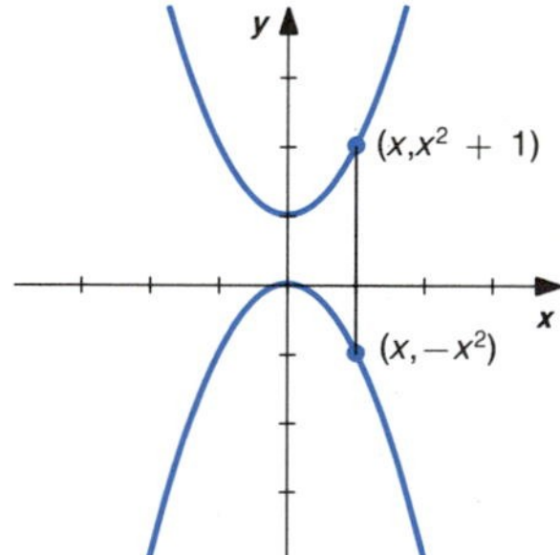


5 $y^2 + y + 1$

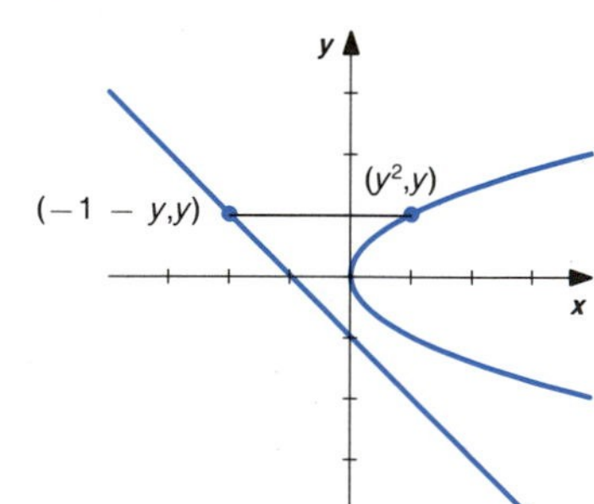

**7** $2\sqrt{4-(x-2)^2}$

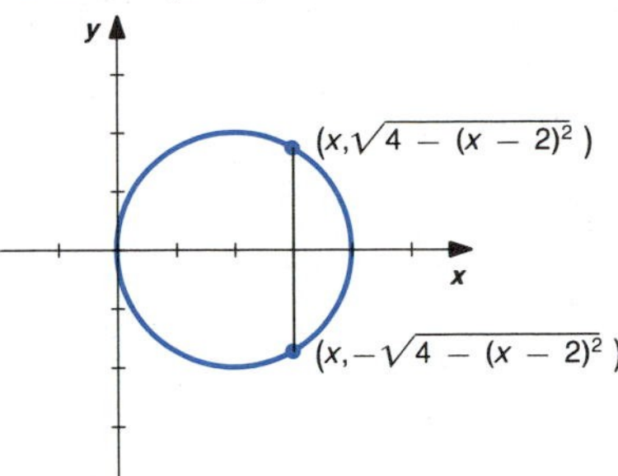


**9** $2\sqrt{y+1}$

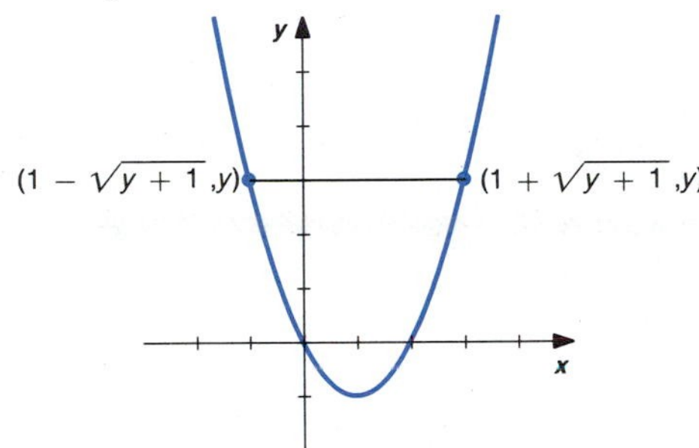


**11** $\begin{cases} 2x, x \geq 0 \\ -2x, x < 0 \end{cases}$

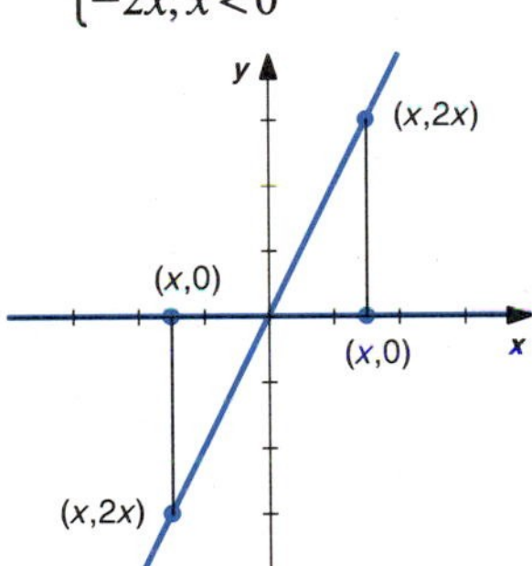


**13** $\begin{cases} 2-x-x^2, -2 \leq x \leq 1 \\ -2+x+x^2, x<-2, x>1 \end{cases}; \quad \left(x, \dfrac{2+x-x^2}{2}\right)$

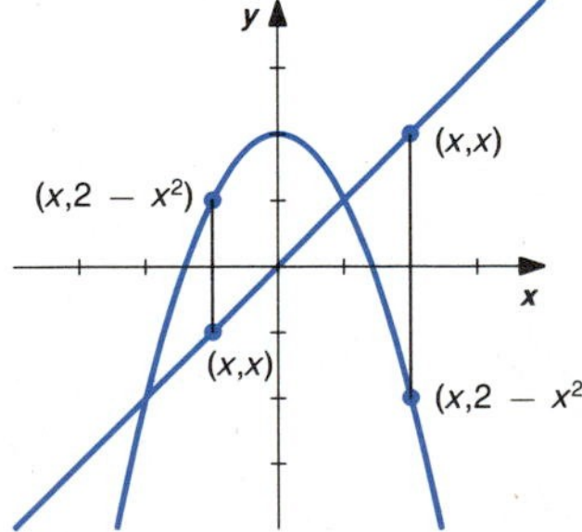


**15** $\begin{cases} (1-y^2)/2, -1 \leq y \leq 1 \\ (y^2-1)/2, y<-1, y>1 \end{cases}; \quad \left(\dfrac{1+3y^2}{4}, y\right)$

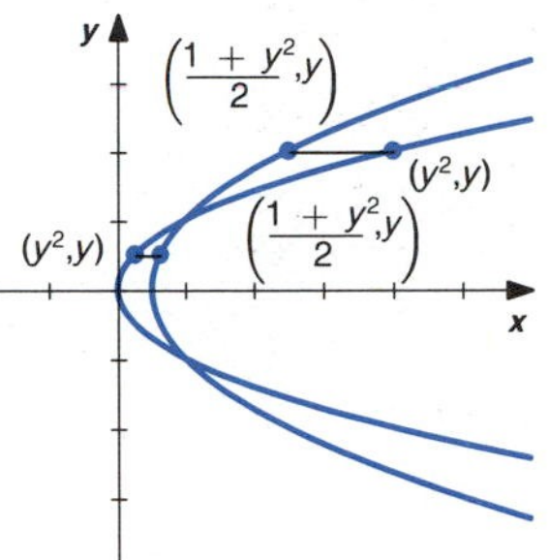


**17** $x-x^2$, $0 \leq x \leq 1$; $\left(x, \dfrac{x+x^2}{2}\right)$

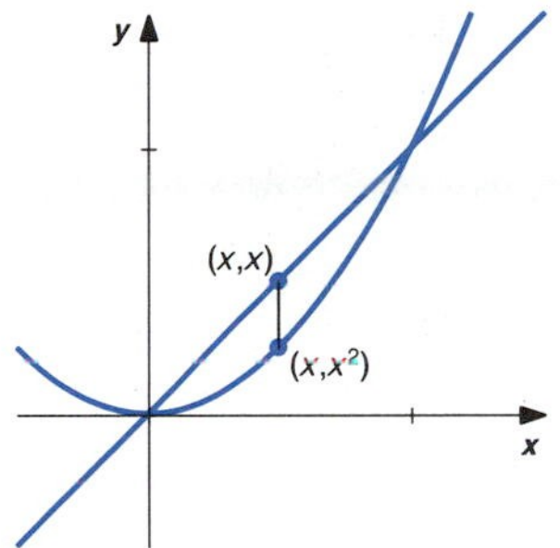


**19** $y^{1/3}-y^{1/2}$, $0 \leq y \leq 1$; $\left(\dfrac{y^{1/2}+y^{1/3}}{2}, y\right)$

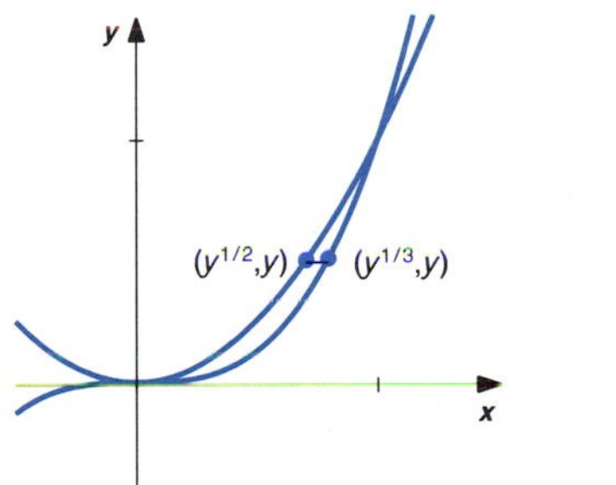


**21** $\pi\left(\dfrac{x}{3}\right)^2$, $0 \leq x \leq 3$

**23** $4\left(1-\dfrac{x}{4}\right)^2$, $0 \leq x \leq 4$

**25** $x^2/8$, $0 \leq x \leq 4$

**27** $y^2/4$, $0 \leq y \leq 4$

**29** $3(3-y)$, $0 \leq y \leq 3$

**31** $\pi(x^2+1)^2$

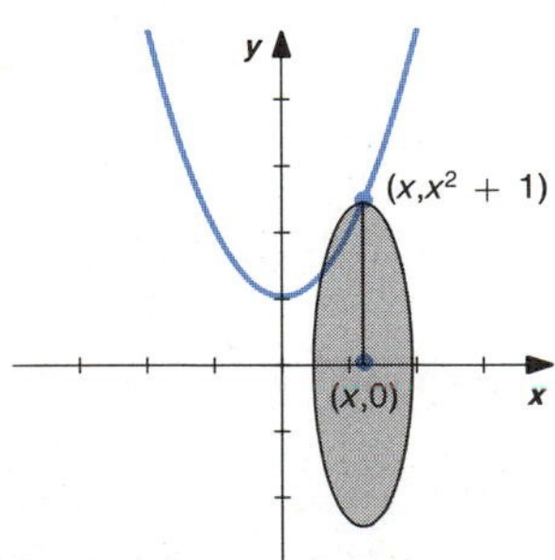

**33** $\pi(x^2+2)^2 - \pi(1)^2$

**35** $\pi\left(\dfrac{rx}{h}\right)^2,\ 0 \le x \le h$

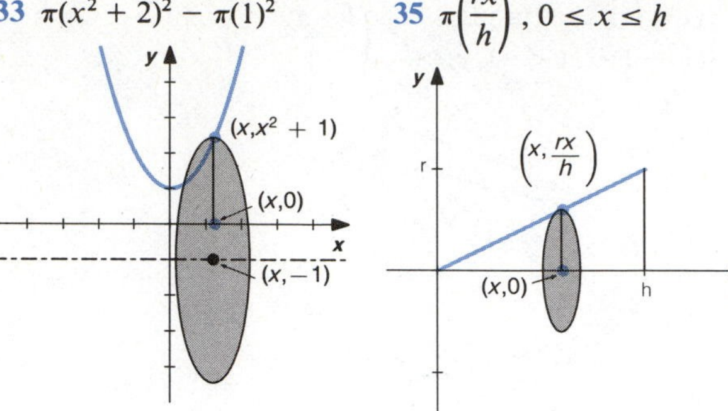


**37** $2\pi x h\left(1-\dfrac{x}{r}\right),\ 0 \le x \le r$

**39** $2\pi y h\left(1-\dfrac{y}{r}\right),\ 0 \le y \le r$

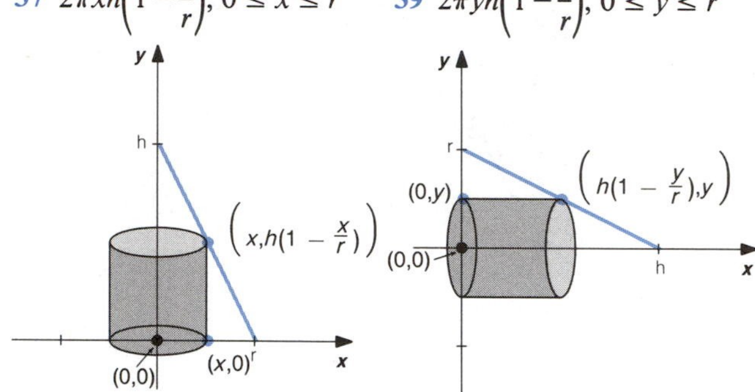


**41** $4\pi y^{3/2}$

**43** $\pi(4-y^2)^2 - \pi(2-y)^2,\ -1 \le y \le 2$

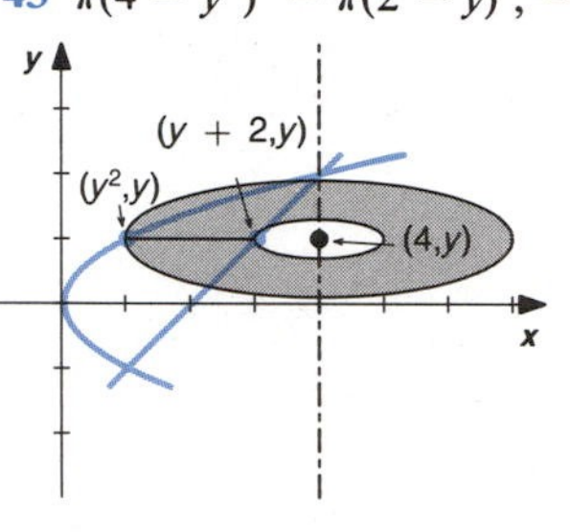


## Exercises 5.2

**1** 0.55

**3** $\dfrac{\pi}{6}(2+\sqrt{3})$

**5** 1.1

**7** 6.6

**9** 0.16

**11** $\pi/2$

**13** 1

**15** $57\pi/100$

**17** 0.385

**19** $123\pi/200$

**21** $0.75 \le s(1) \le 1.25$

**23** $6.1875 \le s(2) \le 7.1875$

**25** $12 \le s(1) \le 20$

**27** $4.001744 \le s(2) \le 4.501744$

**29** 245

**31** $4\pi\sqrt{3}/3$

## Exercises 5.3

**1** 30

**3** −16

**5** 68

**7** 5616

**9** 696,600

**11** (a) $-3 + \dfrac{9}{n}$ (b) $-3$

**13** (a) $-16 + \dfrac{32}{n} + \dfrac{32}{n^2}$ (b) $-16$

**15** (a) $6 + \dfrac{24}{n} + \dfrac{16}{n^2}$ (b) 6

**17** (a) $b$ (b) $b$

**19** (a) $\dfrac{b^3}{6}\left(2+\dfrac{3}{n}+\dfrac{1}{n^2}\right)$ (b) $\dfrac{b^3}{3}$

**21** 36

**23** 8

**25** −4

**27** 7

**29** 30

**31** 17

**37** (a) $1.6 \le \displaystyle\int_0^{1/2} \frac{4}{1+x^2}\,dx \le 2;$

$1 \le \displaystyle\int_{1/2}^{1} \frac{4}{1+x^2}\,dx \le 1.6.$

## Exercises 5.4

1 (a) 0.6015625 (b) 0.6640625 (c) 2/3 (d) 0.125 (e) 0.00260417 (f) 0 (g) $n \geq 10{,}001$ (h) $n \geq 41$ (i) $n \geq 2$

3 (a) 0.68 (b) 0.68 (c) 2/3 (d) 0.4 (e) 0.01333333 (f) 0 (g) $n \geq 40{,}001$ (h) $n \geq 116$ (i) $n \geq 2$

5 (a) 1.8961189 (b) 1.8961189 (c) 2.0045598 (d) 1.2337055 (e) 0.16149102 (f) 0.00664105 (g) $n \geq 49{,}349$ (h) $n \geq 161$ (i) $n \geq 12$

7 (a) 3.0694888 (b) 3.0694888 (c) 3.1429485 (d) 3.4063234 (if $M = 1 + \pi$) (e) 0.3690316 (if $M = 2 + \pi$) (f) 0.0093684 (if $M = 4 + \pi$) (g) $n \geq 204{,}380$ (h) $n \geq 365$ (i) $n \geq 20$

9 (a) 0.69315453 (b) 0.69315453

11 3.1415686

13 11.904271

15 $f^{(4)}(\pm 2)$ do not exist and $f^{(4)}$ is unbounded on the interval $-2 \leq x \leq 2$.

17 1600

19 10.683128 ($n = 6$)

## Exercises 5.5

1 14/3

3 9/2

5 3/8

7 11/6

9 −34/3

11 1

13 1

15 $2\left(1 - \frac{\sqrt{3}}{3}\right)$

17 2

19 0

21 $2\sqrt{3}/3$

23 0

25 (a) −48 (b) 80

27 (a) −2 (b) 2.5

29 (a) 0 (b) 4

31 $\cos(x^2)$

33 $\sec(x^2)$

35 $-(1 - x)^5$

37 $2x \sin(x^2)$

39 $x^4$

41 $1/x$

43 $2x \tan(x^2) - \tan x$

45 −4/25

## Exercises 5.6

1 $-\frac{(1-x)^{11}}{11} + C$

3 $\frac{3}{8}(x^2 + 1)^{4/3} + C$

5 Not of type

7 $\frac{4}{3}(1 + \sqrt{x})^{3/2} + C$

9 $-\frac{1}{2(x^2 + 2x + 3)} + C$

11 $2 \sin \sqrt{x} + C$

13 Not of type

15 $\frac{1}{2}\tan(x^2) + C$

17 $\frac{1}{2}\sec 2x + C$

19 $\frac{2}{3}(1 + \sin x)^{3/2} + C$

21 $\sin^2 x + C$

23 $\frac{\tan^7 x}{7} + C$

25 13/3

27 3

29 1

31 195/4

33 0

35 $\frac{1}{6}(27 - 5^{3/2})$

37 $\int_1^2 \frac{1}{u}\,du$

## Review Exercises

1 $2 + x - x^2, -1 \leq x \leq 2; \left(x, \frac{6 - x - x^2}{2}\right)$

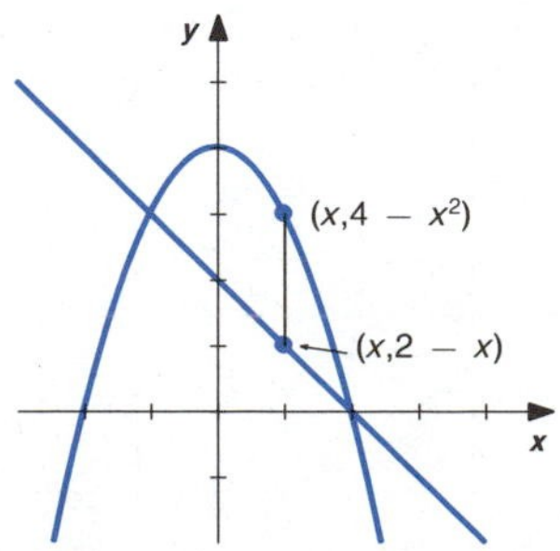


3 $\pi(4 - x)^2/8, 0 \leq x \leq 4$

**5** (a) $\pi \sin^2 x,\ 0 \le x \le \pi$ (b) $2\pi x \sin x,\ 0 \le x \le \pi$ (c) $\pi(\sin x + 1)^2 - \pi(1)^2,\ 0 \le x \le \pi$

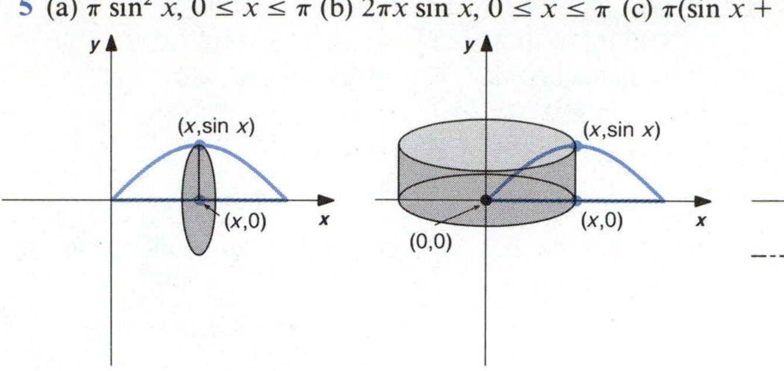


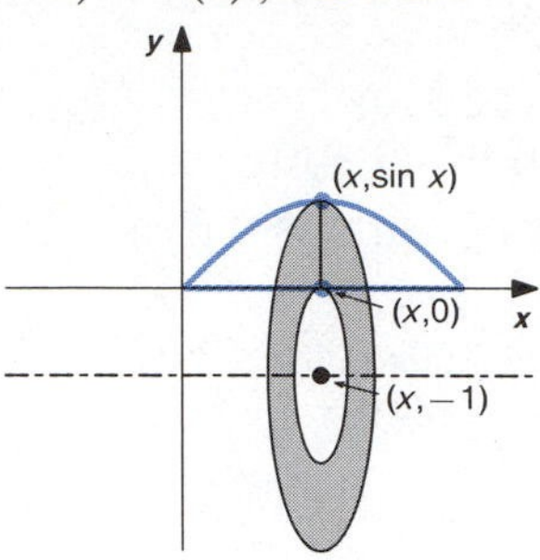


(a) (b) (c)

**7** 3.75

**9** 7.5

**11** 9

**13** 18,768

**15** $\dfrac{4}{n} + \dfrac{4}{n^2}$; 0

**17** $\dfrac{\sqrt{3}+1}{3} \le s(1) - s(0) \le \dfrac{\sqrt{3}+3}{3}$

**19** 7

**23** (a) 1.8961189 (b) 2.0045598 (c) 0.16149102 (d) 0.00664105 (e) $n \ge 23$ (f) $n \ge 6$

**25** $\dfrac{1}{\pi} \sec \pi x + C$

**27** $\dfrac{1}{3}(x^2 + 4)^{3/2} + C$

**29** $\dfrac{1}{3} \tan^3 x + C$

**31** 9

**33** 8/225

**35** $\displaystyle\int_1^{17} \frac{1}{8u}\, du$

**37** $\sin(x^2)$

**39** $-1/x$

## CHAPTER SIX

### *Exercises 6.1*

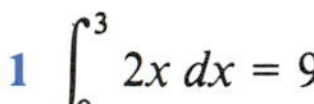

**1** $\displaystyle\int_0^3 2x\, dx = 9$

**3** $\displaystyle\int_0^2 \left(5 - \frac{5}{2}x\right) dx = 5$

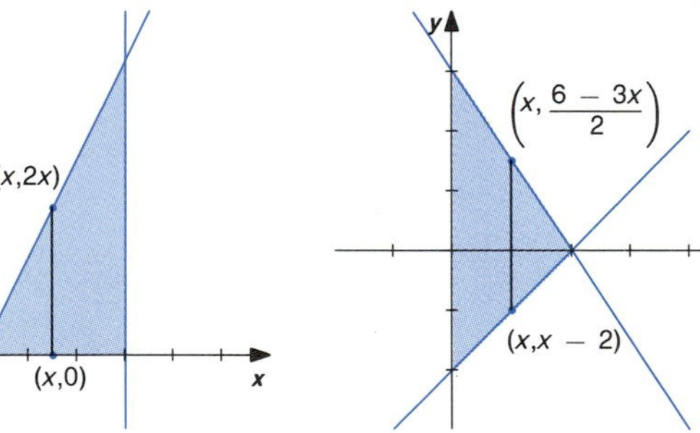


**9** $\displaystyle\int_0^2 (5 - 2y)\, dy = 6$

**11** $\displaystyle\int_0^2 x^2\, dx = 8/3$

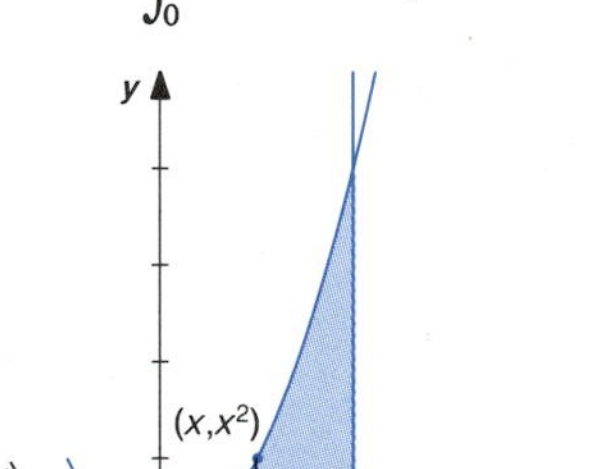


**5** $\displaystyle\int_0^2 (6 - 3y)\, dy = 6$

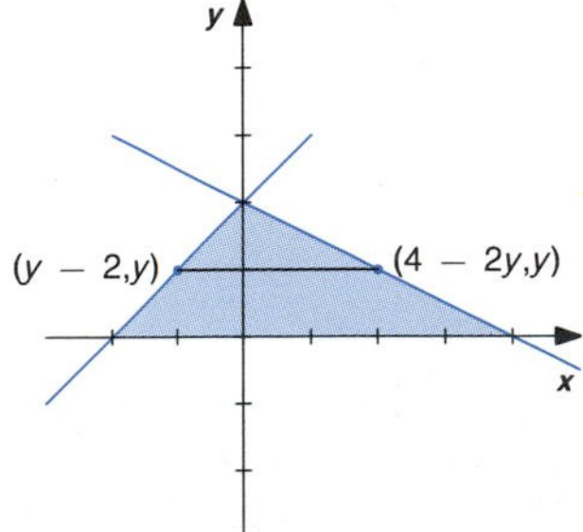


**7** $\displaystyle\int_0^2 3\, dx = 6$

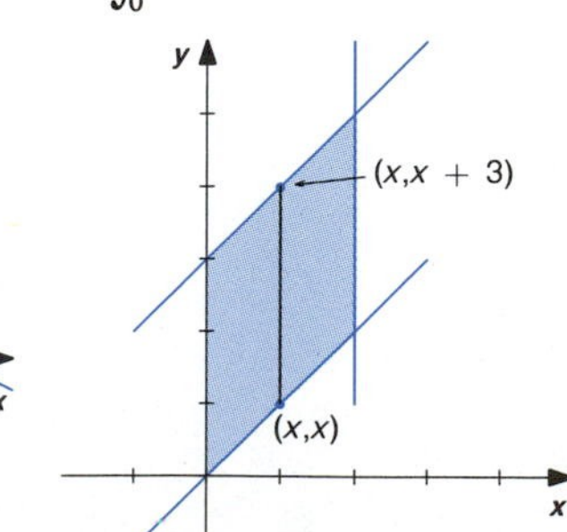
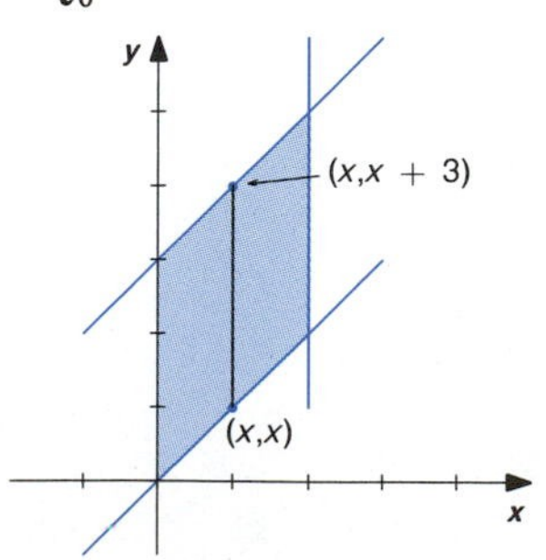


**13** $\displaystyle\int_0^1 (x - x^2)\, dx = 1/6$

**15** $\displaystyle\int_{-\pi/2}^{0} -\sin x\, dx = 1$

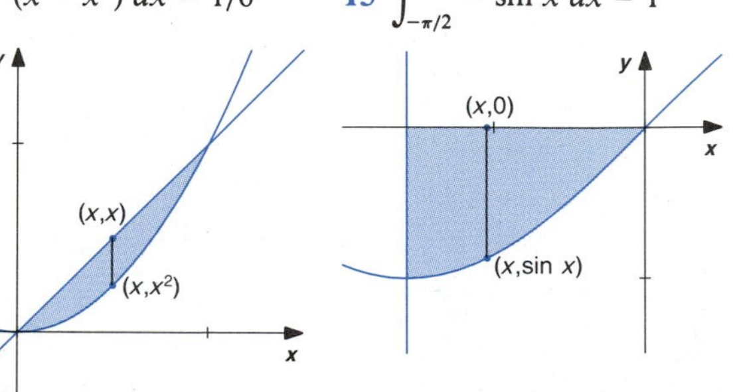

**17** $\int_{-2}^{1} (2 - y - y^2)\, dy = 9/2$

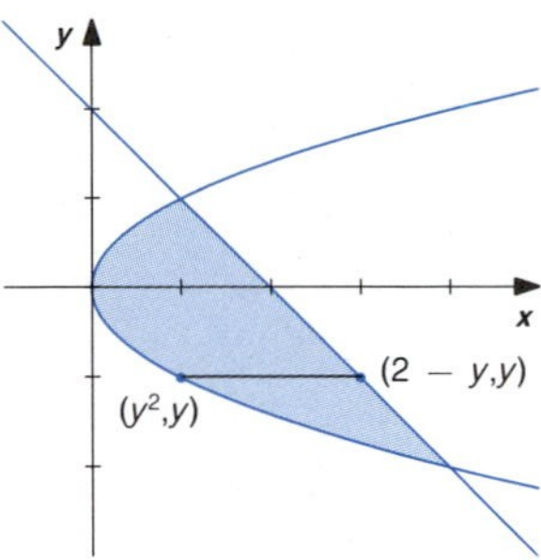


**19** $\int_{-\pi}^{0} - \sin x\, dx + \int_{0}^{\pi} \sin x\, dx = 4$

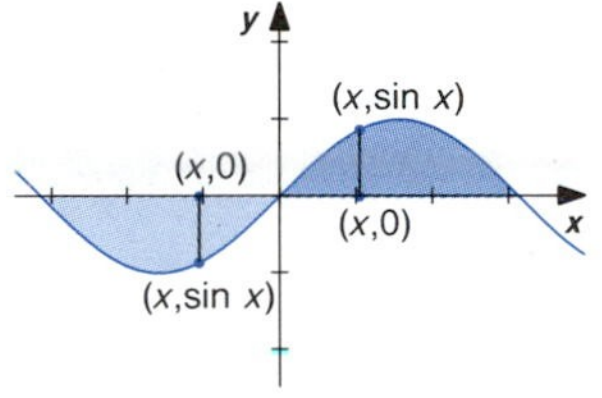


**21** $\int_{0}^{\pi/6} \left(\frac{1}{2} - \sin x\right) dx = \frac{\pi + 6\sqrt{3} - 12}{12}$

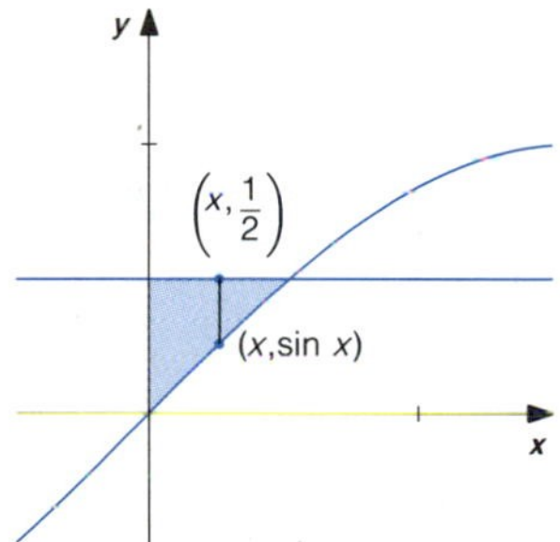


**23** $\int_{0}^{3} \left(x^2 - \frac{x^3}{3}\right) dx = 9/4$

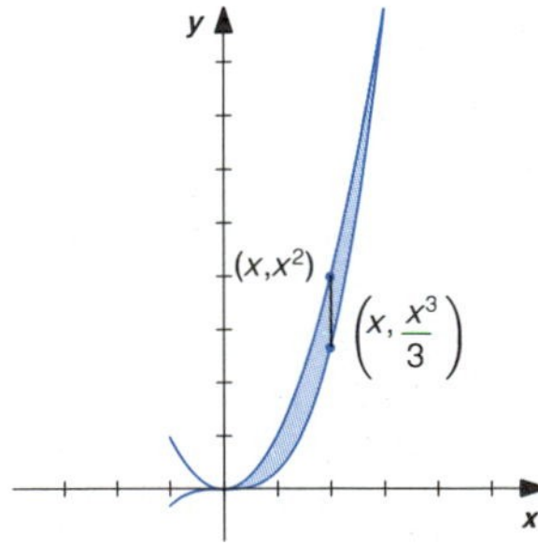


**25** $\int_{0}^{2} [\sqrt{4 - x^2} - (2 - x)]\, dx$

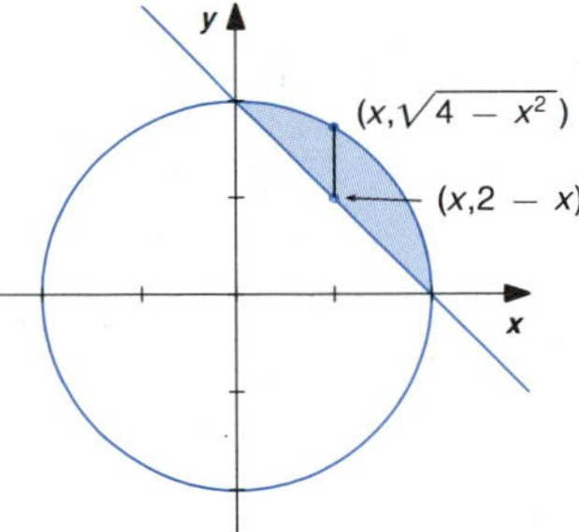


**27** $\int_{-1}^{2} 2\sqrt{4 - y^2}\, dy$

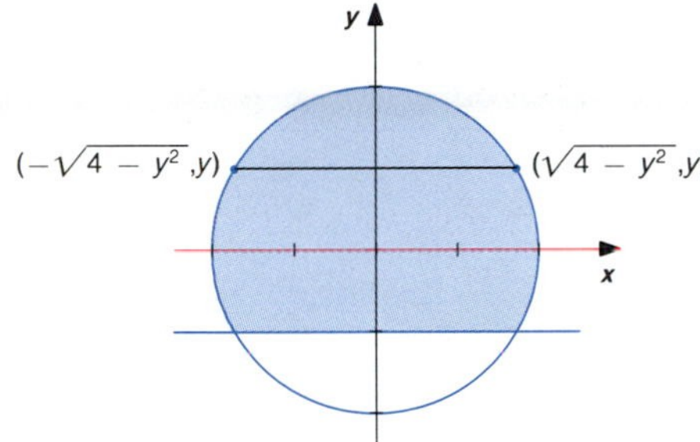


**29** $\int_{0}^{\pi} (x - \sin x)\, dx$

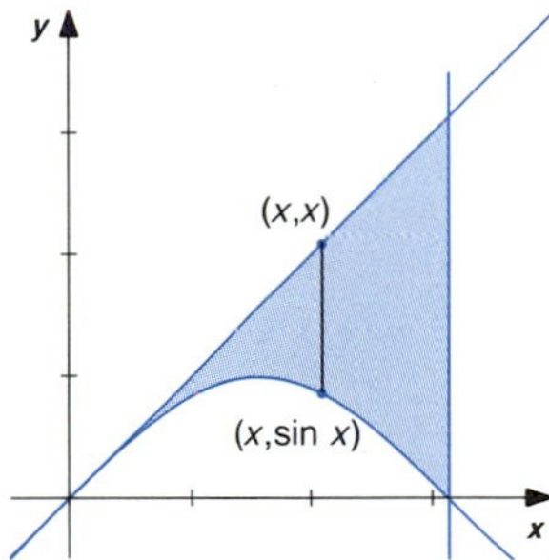


**31**

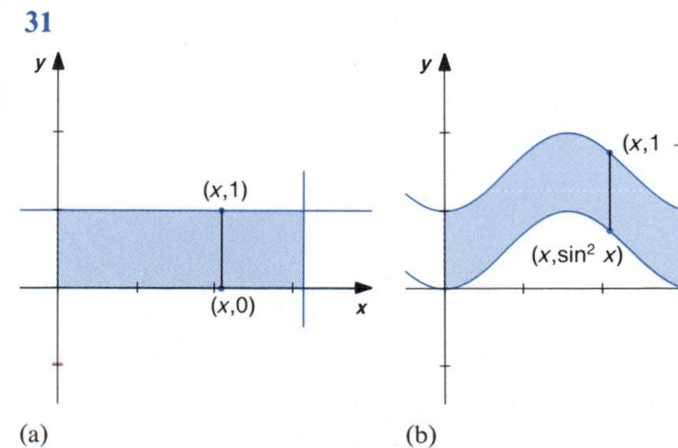


(a) (b)

## Exercises 6.2

**1** $\int_0^3 \left(-\frac{2}{3}x+2\right) dx = 3$ **3** $\int_0^3 \frac{\pi x^2}{9}\, dx = \pi$

**5** $\int_0^h \frac{b^2(h-y)^2}{h^2} dy = \frac{b^2 h}{3}$ **7** $\int_{-1}^1 \pi(1+y^2)\, dy = \frac{8\pi}{3}$

**9** (a) $\int_0^3 \frac{4}{9}(x-3)^2\, dx = 4$ (b) $\int_0^3 \frac{1}{9}(x-3)^2\, dx = 1$
(c) $\int_0^3 \frac{2}{9}(x-3)^2\, dx = 2$

**11** (a) $\int_0^4 \frac{\pi}{2}(4-x)\, dx = 4\pi$ (b) $\int_0^4 \sqrt{3}(4-x)\, dx = 8\sqrt{3}$
(c) $\int_0^4 2(4-x)\, dx = 16$

**13** $\int_0^{2r} \left[\pi\left(2r-\frac{y}{2}\right)^2 - \pi(r)^2\right] dy = \frac{8\pi r^3}{3}$

**15** $\int_{-2}^2 \frac{1}{2\sqrt{3}}(4-x^2)\, dx = \frac{16\sqrt{3}}{9}$ in.$^3$

**17** $\int_0^\pi \pi \sin^2 x\, dx$

**19** $\frac{16r^3}{3\sin\theta}$

**21** $\int_{-r/2}^{r/2} 2r\sqrt{r^2-y^2}\, dy$

**23** The two stacks have equal cross sections at each level, so Cavalieri's Theorem implies they have equal volume.

## Exercises 6.3

**1** (a) $\int_0^1 \pi(x^2)^2\, dx = \pi/5$
(b) $\int_0^1 [\pi(1)^2 - \pi(1-x^2)^2]\, dx = 7\pi/15$
(c) $\int_0^1 2\pi x(x^2)\, dx = \pi/2$
(d) $\int_0^1 2\pi(1-x)(x^2)\, dx = \pi/6$

**3** (a) $\int_0^1 [\pi(x)^2 - \pi(x^2)^2]\, dx = 2\pi/15$
(b) $\int_0^1 [\pi(1-x^2)^2 - \pi(1-x)^2]\, dx = \pi/5$
(c) $\int_0^1 2\pi x(x-x^2)\, dx = \pi/6$
(d) $\int_0^1 2\pi(1-x)(x-x^2)\, dx = \pi/6$

**5** $\int_0^3 [\pi(4x-x^2)^2 - \pi(x)^2]\, dx = \frac{108\pi}{5}$

**7** (a) $\int_0^1 2\pi y(3-3y)\, dy = \pi$
(b) $\int_0^1 [\pi(3-2y)^2 - \pi(y)^2]\, dy = 4\pi$

**9** $\int_0^1 2\pi x(2-x-x^2)\, dx = 5\pi/6$

**11** $\int_{-\sqrt{2}}^{\sqrt{2}} \pi(2-y^2)^2\, dy = 64\pi\sqrt{2}/15$

**13** (a) $\int_1^2 \pi\left(\frac{1}{x}\right)^2 dx = \pi/2$ (b) $\int_1^2 2\pi(x)\left(\frac{1}{x}\right) dx = 2\pi$

**15** (a) $\int_0^1 [\pi(2-x^2)^2 - \pi(2-x)^2]\, dx = \frac{8\pi}{15}$
(b) $\int_0^1 2\pi x(x-x^2)\, dx = \pi/6$

**17** (a) $\int_1^8 2\pi y^{2/3}\, dy = 186\pi/5$ (b) $\int_1^8 \pi y^{-2/3}\, dy = 3\pi$

**19** $\int_0^{\pi/4} [\pi(\sec x)^2 - \pi(1)^2]\, dx = \pi - \frac{\pi^2}{4}$

**21** $\int_0^{\sqrt{\pi}} 2\pi x \sin(x^2)\, dx = 2\pi$

**23** $\int_0^r 2\pi x\left(h - \frac{hx}{r}\right) dx = \pi r^2 h/3$

**25** $\int_{r-h}^r \pi(r^2-y^2)\, dy = \pi r h^2 - \frac{\pi h^3}{3}$

**27** $\int_{-1}^1 [\pi(2+\sqrt{1-y^2})^2 - \pi(2-\sqrt{1-y^2})^2]\, dy$
$= \int_1^3 2\pi x(2\sqrt{1-(x-2)^2})\, dx$

**29** (a) $\int_0^\pi \pi \sin^2 x\, dx$ (b) $\int_0^\pi 2\pi x \sin x\, dx$

**31** (a) $\int_0^{\pi/2} [\pi(1)^2 - \pi(\cos x)^2]\, dx$ (b) $\int_0^{\pi/2} 2\pi x(1-\cos x)\, dx$

**33** (a) $\int_0^{\pi/4} \pi \tan^2 x\, dx$ (b) $\int_0^{\pi/4} 2\pi\left(\frac{\pi}{4}-x\right) \tan x\, dx$

## Exercises 6.4

1 $(17^{3/2} - 5^{3/2})/6$

3 $(52^{3/2} - 125)/27$

5 $\pi(2^{3/2} - 1)/9$

7 $10\pi$

9 (a) $2\sqrt{5}$ (b) $4\pi\sqrt{5}$

11 (a) 17/12 (b) $47\pi/16$

13 (a) 33/8 (b) $225\pi/16$

15 $\pi r\sqrt{r^2 + h^2}$

17 (a) $2\pi r^2 - 2\pi r(r^2 - x_0^2)^{1/2}$ (b) $2\pi r^2$

19 (a) 9.726695 (b) 9.7505254

21 (a) 3.8066080 (b) 3.8194032

23 (a) 0.7848262 (b) 0.785487997

## Exercises 6.5

1 10,800 ft-lb

3 (a) 249.6 ft-lb (b) 1248 ft-lb

5 (a) 5324.8 ft-lb (b) 832 ft-lb

7 (a) 196.04 ft-lb (b) 183.78 ft-lb

9 (a) 40 ft-lb (b) 240 ft-lb

11 (a) 24 ft-lb (b) 56 ft-lb

13 (a) 1.5 ft-lb (b) 6 ft-lb

15 1.44 ft

17 1412 in.-lb

19 1200 ft-lb

## Exercises 6.6

1 4/7

3 1/5

5 (0, −1/2)

7 (1, −2)

9 (11/7, 13/7)

11 (25/18, 19/18)

13 $\left(0, \dfrac{3\pi + 4}{\pi + 4}\right)$

15 (5/8, 17/20)

17 (0, 5/6)

19 (1/2, 8/5)

21 (4/5, 0)

23 (4/5, 1/4)

27 (2/3, 1/2)

29 (5/6, 5/16)

31 (5/12, 5/7)

33 (1/5, 3/5)

37 $18\pi$

## Exercises 6.7

1 (a) $1200\pi$ lb (b) $1200\pi$ lb

3 (a) $\int_0^1 \rho g(1 - y)(y)\, dy$ (b) $\int_0^1 \rho g(2 - y)(y)\, dy$

5 (a) $\int_{-1}^1 \rho g(2 - y)(2\sqrt{1 - y^2})\, dy$
(b) $\int_0^2 \rho g(3 - y)(2\sqrt{2y - y^2})\, dy$

7 $\int_0^{\sqrt{3}} \rho g(\sqrt{3} - y)\left(\dfrac{2}{\sqrt{3}} y\right) dy$

9 $\int_{-2}^{-1} \rho g(-1 - y)(2\sqrt{4 - y^2})\, dy$

11 $45\rho g/2$

13 $81\pi\rho g/8$

15 $40\sqrt{2}\rho g$

17 $37\rho g$

19 $52.5\rho g$

21 $5\sqrt{3}\rho g$

23 $9\sqrt{5}\rho g/4$

25 (a) $-\rho_f g d b^2$ (b) $\rho_f g(d + h)b^2$ (c) $-\rho_b g h b^2$
(d) $hb^2(\rho_f - \rho_b)g$; downward, so it sinks

## Review Exercises

1 $\int_0^2 (8x - x^4)\, dx$

3 $\int_0^3 [y - y(y - 2)]\, dy$

5 $\int_0^1 \left(2x - \dfrac{x}{2}\right) dx + \int_1^2 \left[(3 - x) - \dfrac{x}{2}\right] dx$

7 $\int_0^3 \dfrac{1}{2}\left(\dfrac{2x}{3}\right)^2 dx$

9 $\int_{-1}^1 4(1 - x^2)\, dx$

11 (a) $\int_0^1 (\pi x^2 - \pi x^4)\, dx$ (b) $\int_0^1 2\pi x(x - x^2)\, dx$

**13** (a) $\int_0^{\pi} \sqrt{1+\cos^2 x}\,dx$

(b) $\int_0^{\pi} 2\pi \sin x\sqrt{1+\cos^2 x}\,dx$

(c) $\int_0^{\pi} 2\pi x\sqrt{1+\cos^2 x}\,dx$

**15** 7.5 ft-lb

**17** 218,934 N-m

**19** 4200 ft-lb

**21** (19/15, 14/15)

**23** $(0, 4/(3\pi))$

**25** $256\rho g/15$

**27** $28\pi\rho g$

**29** 2

**31** 2

**33** $k^3h\rho/3$

## CHAPTER SEVEN

### *Exercises 7.1*

**1** 6

**3** $\sqrt{3}/2$

**5** 0

**7** 2

**9** $-1/6$

**11** $\infty$

**13** $-2$

**15** 3/5

**17** 1/3

**19** 0

**21** 1

**23** 1

**25** 1/2

**27** 1

**29** 1

**31** $-2$

**33** 0

### *Exercises 7.2*

**1** 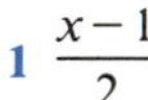 $\frac{x-1}{2}$

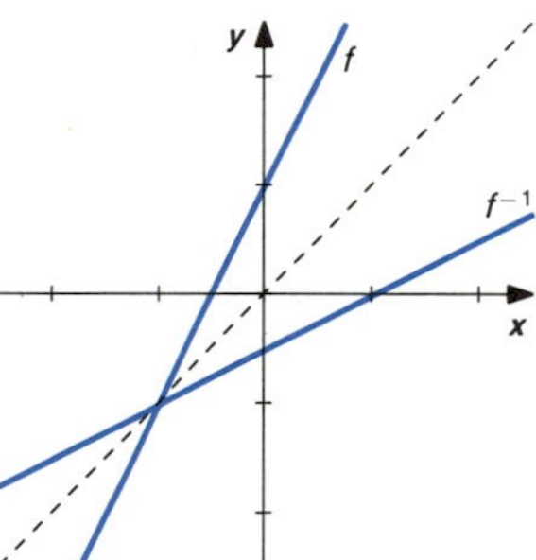


**3** $\frac{x^3+8}{8}$

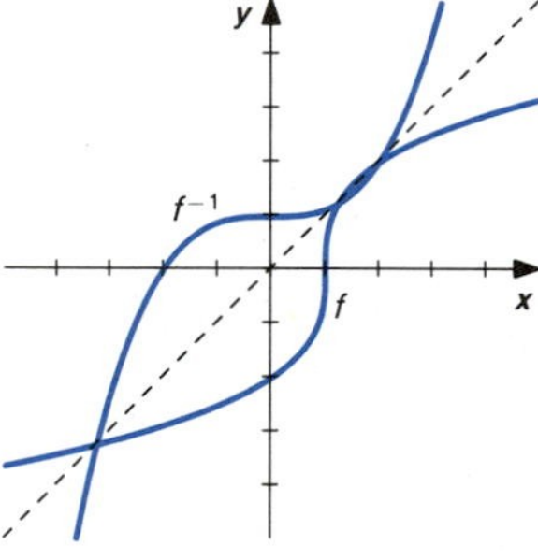


**5** Does not exist

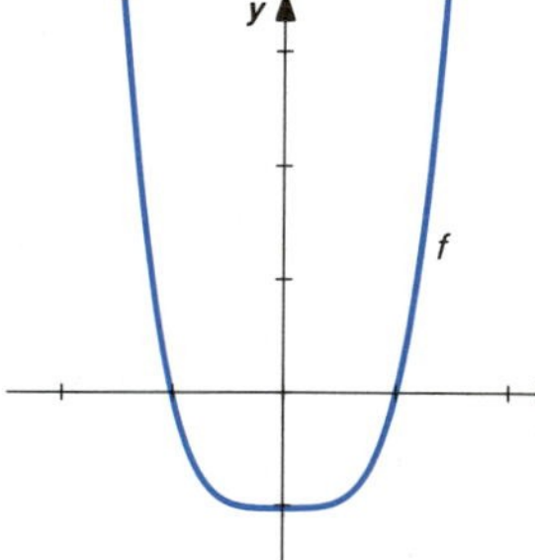


**7** $\sqrt{4-x^2},\ 0 \le x \le 2$

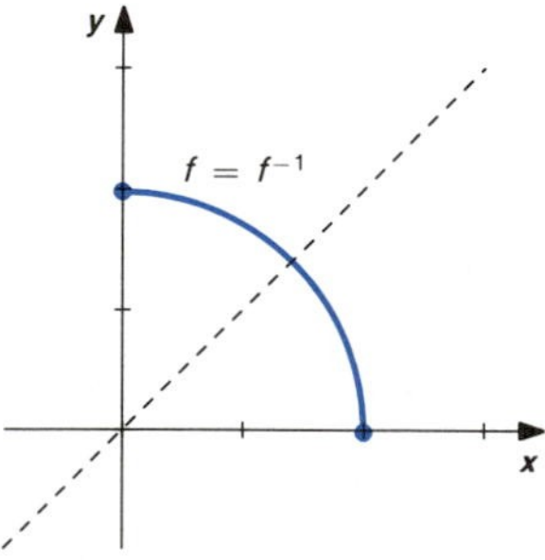


**9** $x^2 - 1,\ x \ge 0$

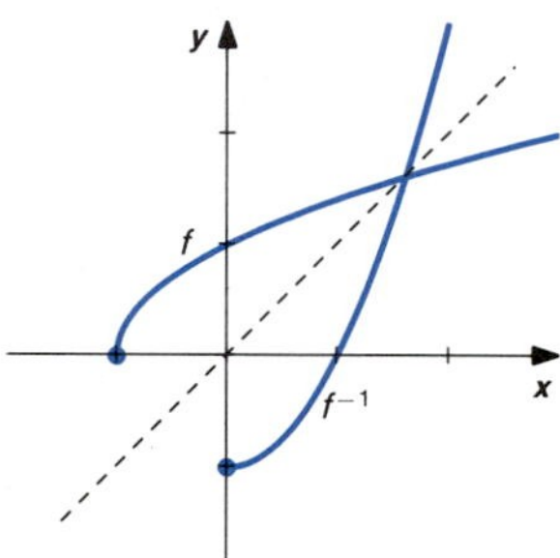


**11** $\frac{1+x}{x}$

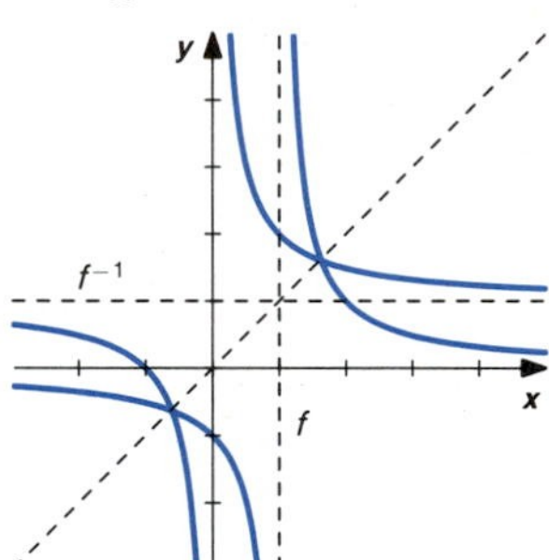


**13** $\pi/6$

**15** $\pi$

**17** $\pi/3$

**19** $\pi/4$

**21** 0.1001674

**23** 1.3694384

**25** 1.2490458

**27** $1/\sqrt{2}$

**29** $\pi/4$

**31** $x/\sqrt{1-x^2}$

**33** $\sqrt{4-x^2}/2$

**35** $\frac{\sqrt{x^2-9}}{|x|}$

**37** 0.1139888

**39** 0.1263401

**41** 1.2309594, 5.0522259

**43** 1.448307, 3.0191033

**47** $f$ cannot be continuous

**53** $\theta(x) = \tan^{-1}\left(\frac{16}{x}\right) - \tan^{-1}\left(\frac{9}{x}\right)$

## Exercises 7.3

1 $\dfrac{1}{\sqrt{9-x^2}}$

3 $\dfrac{-1}{\sqrt{4-x^2}}$

5 $\dfrac{-1}{x^2+1}$

7 $\dfrac{2}{x\sqrt{x^4-1}}$

9 $\dfrac{1}{\sqrt{a^2-x^2}}$

11 $\sin^{-1}\dfrac{x}{3}+C$

13 $\dfrac{1}{6}\tan^{-1}\dfrac{2x}{3}+C$

15 $\sec^{-1}|2x|+C$

17 $\dfrac{\sqrt{x^4-1}}{2}+C$

19 $\dfrac{(\sin^{-1}x)^2}{2}+C$

21 $(\pi+2)x-4y=2$

23 0.02 rad/s

25 $\pi/3$

27 (a) $4\pi^2/3$ (b) $32\pi$

29 $\pi/2$

## Exercises 7.4

1 $\dfrac{2}{2x+1}$

3 $\sec x$

5 $-2/x$

7 $\dfrac{2x^2+1}{x(x^2+1)}$

9 $\dfrac{-13}{(2x+1)(3x-5)}$

11 $\dfrac{3-x^2}{x(x^2+3)}$

13 $\dfrac{2\ln x}{x}$

15 $\dfrac{6x^2-x-3}{3(x-1)^{2/3}(x+1)^{1/3}}$

17 $\dfrac{2x^2+3x-3}{3(x-1)^{2/3}(x+1)^{5/3}}$

19 loc min $(e^{-1/2}, -e^{-1}/2)$; pt of infl $(e^{-3/2}, -3e^{-3}/2)$

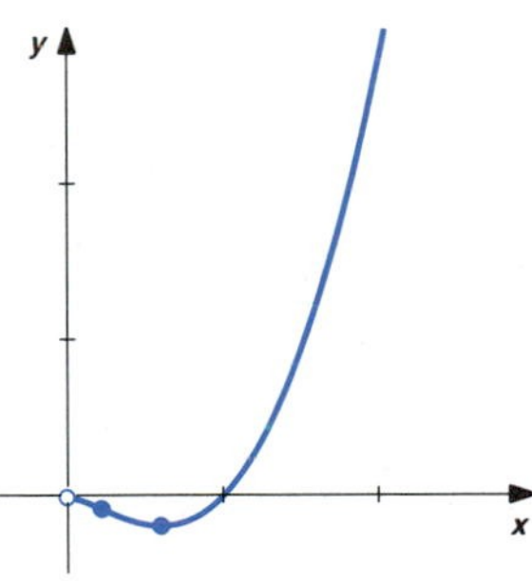


21 loc max $(e, e^{-1})$; pt of infl $(e^{3/2}, 3e^{-3/2}/2)$

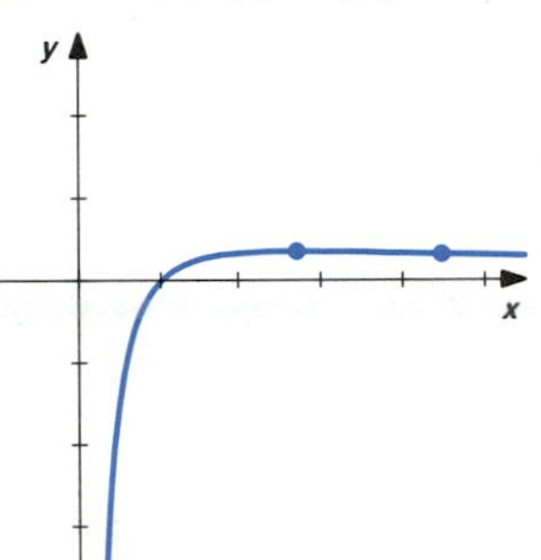


23 loc min (1, 1); pt of infl (2, 0.5 + ln 2)

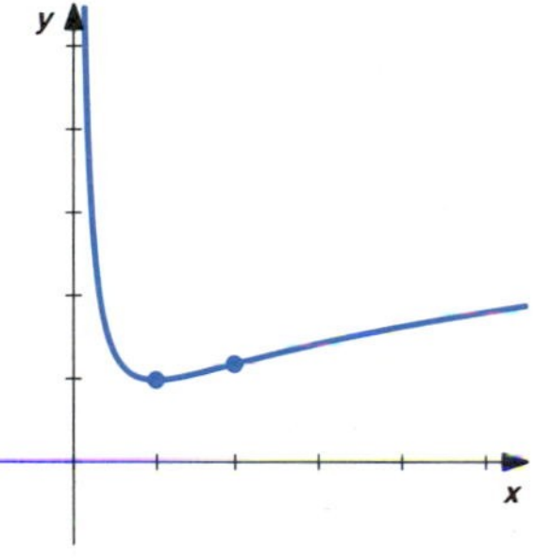


## Exercises 7.5

1 $\dfrac{1}{2}\ln|2x+1|+C$

3 $\dfrac{1}{2}\ln(x^2+1)+C$

5 $x+3\ln|x-3|+C$

7 $-\ln(1+\cos x)+C$

9 $\ln|\sec x+\tan x|+C$

11 $\ln|\ln x|+C$

13 $x-\tan^{-1}x+C$

15 $\dfrac{3}{2}\ln(1+x^{2/3})+C$

17 $\tan^{-1}x+C$

19 $\sin^{-1}\dfrac{x}{2}+C$

21 ln (4/3)

23 (a) $2\pi$ (b) $4\pi(1-\ln 2)$

## Exercises 7.6

1 $-xe^{-x^2/2}$

3 $2e^{2x-1}$

5 $e^{\tan x}\sec^2 x$

7 $e^{-x}(1-x)$

9 $\dfrac{-e^{-x}(x+1)}{x^2}$

11 $\dfrac{e^x}{1+e^x}$

13 $x^{\cos x}\left(\dfrac{\cos x}{x}-\sin x\ln x\right)$

15 $10^x\ln 10$

17 $-e^{-x} + C$

19 $2e^{\sqrt{x}} + C$

21 $\dfrac{2(e^x + 1)^{3/2}}{3} + C$

23 $\tan^{-1} e^x + C$

25 $\dfrac{1}{2} \ln (1 + e^{2x}) + C$

27 $\dfrac{-2^{-x}}{\ln 2} + C$

29 loc min $(-1, -e^{-1})$; pt of infl $(-2, -2e^{-2})$

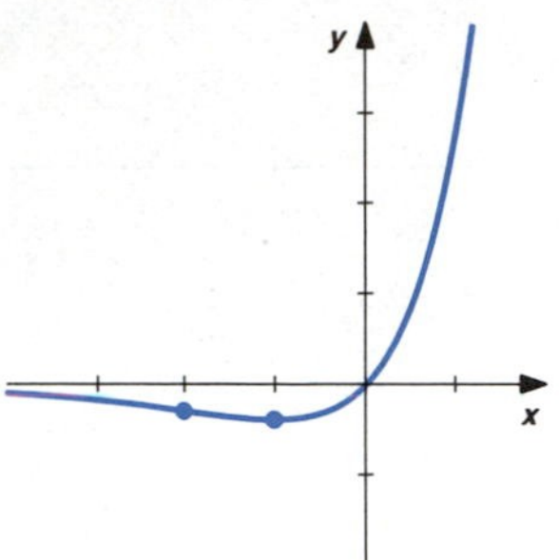


31 pt of infl (0, 0)

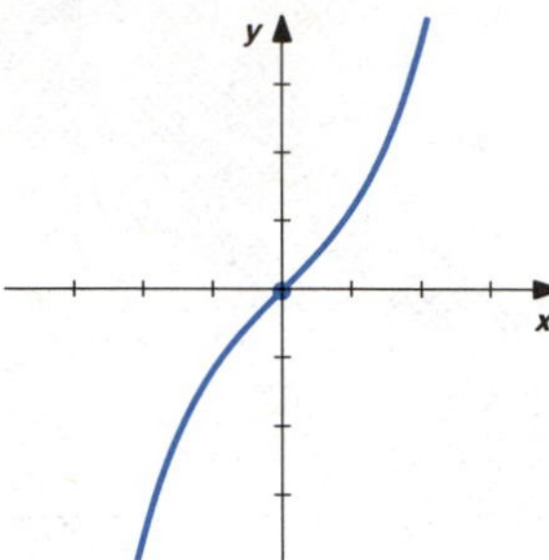


33 loc min $(1, e)$

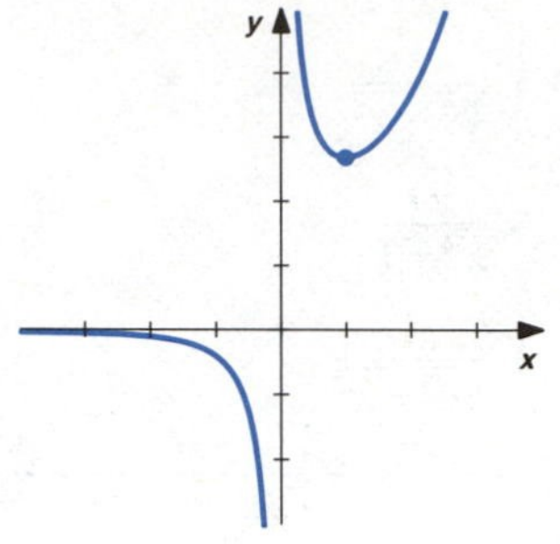


35 $1 - e^{-1}$

37 2

39 1

41 0

43 0

45 1

---

## *Exercises 7.7*

1 800

3 18.8 yr

5 13.51 days

7 7601 yr

9 9.5123 A

11 75°

13 K; the temperature of the object approaches the temperature of its surroundings.

15 $2 - 2e^{-0.06t}$ lb/gal

17 0.495 lb/gal

---

## *Exercises 7.8*

11 $2x \cosh (x^2)$

13 $2 \cosh x \sinh x$

15 $\tanh x$

17 $2/\sqrt{1 + 4x^2}$

19 $\sec x$

21 $\dfrac{1}{3} \cosh 3x + C$

23 $\dfrac{1}{2} \tanh (x^2) + C$

25 $\ln (\cosh x) + C$

27 $\dfrac{1}{2} \sinh^{-1} \left(\dfrac{2x}{3}\right) + C$

29 $\cosh^{-1} (e^x) + C$

31 $\dfrac{\ln 5}{6}$

---

## *Review Exercises*

1 8

3 $e$

5 $\dfrac{x + 2}{3}$

7 $\dfrac{1}{2} \tan^{-1} (x + 1)$

9 $-\pi/3$

11 $\dfrac{\sqrt{4 - x^2}}{2}$

13 $\dfrac{2x}{\sqrt{1 - x^4}}$

15 $\dfrac{-1}{\sqrt{1 - x^2}}$

17 $\dfrac{-\sin x \cos x}{2 - \sin^2 x}$

19 $-x\, e^{-x^2/2}$

21 $\pi/8$

23 $\dfrac{1}{2} \ln (9 + x^2) + C$

25 $-\dfrac{1}{3} \operatorname{sech}^{-1} \left(\dfrac{|x|}{3}\right) + C$

27 $e^{\tan x} + C$

29 $\dfrac{1}{3} (\sin^{-1} x)^3 + C$

31 loc max $(2, 2e^{-1})$; pt of infl $(4, 4e^{-2})$

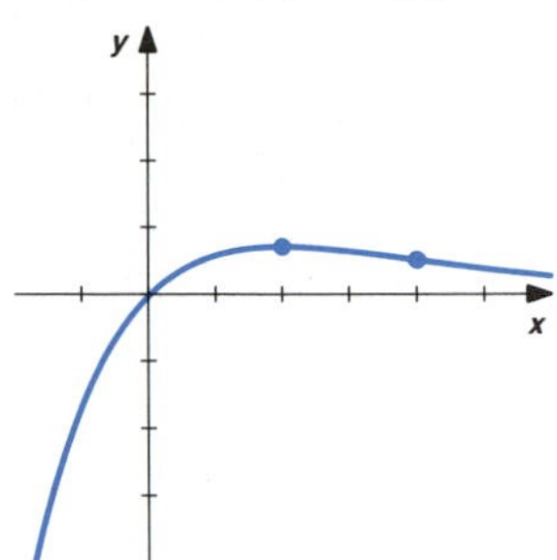

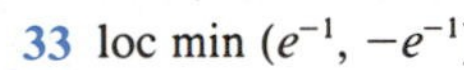

33 loc min $(e^{-1}, -e^{-1})$

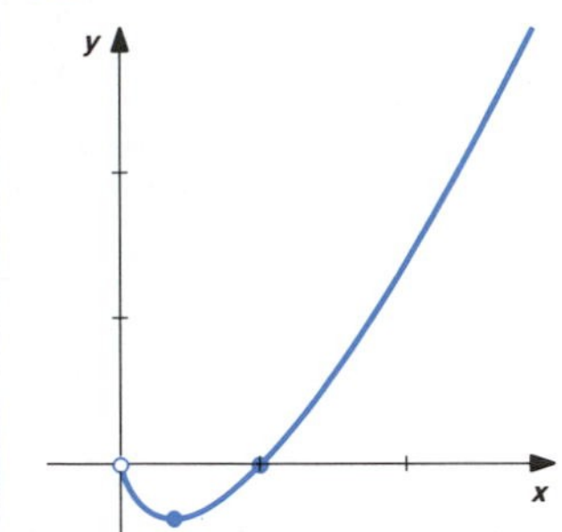

35 1.39 h

37 128.33°F

## CHAPTER EIGHT

### Exercises 8.1

1 $\sqrt{x^2+4} + C$

3 $\frac{1}{3}\sin^{-1} 3x + C$

5 $-\ln(2 - \sin x) + C$

7 $-\frac{1}{2}e^{-x^2} + C$

9 $\ln|\ln x| + C$

11 $-\frac{1}{2}\cos 2x + C$

13 $\frac{3}{2}\sec(x^{2/3}) + C$

15 $\frac{1}{3}e^{3x-1} + C$

17 $-\frac{1}{3}(3x-1)^{-1} + C$

19 $\frac{\sqrt{2}}{6}\tan^{-1}\frac{3x}{\sqrt{2}} + C$

21 $-\frac{1}{2}\cot 2x + C$

23 $\frac{1}{5}\sin^5 x + C$

25 $\frac{1}{6}\tan^6 x + C$

27 $\frac{1}{5}\sec^5 x + C$

29 $\sin^{-1} e^x + C$

31 $\sin^{-1}\frac{x-2}{2} + C$

33 $\frac{1}{2}\tan^{-1}\frac{x+1}{2} + C$

35 $\frac{1}{2}\ln(x^2 - 2x + 5) + \frac{1}{2}\tan^{-1}\frac{x-1}{2} + C$

### Exercises 8.2

1 $\frac{1}{4}e^{2x}(2x-1) + C$

3 $-2x\cos\frac{x}{2} + 4\sin\frac{x}{2} + C$

5 $\frac{2x}{\sqrt{1-x^2}} - 2\sin^{-1} x + C$

7 $-2\sqrt{x}\cos\sqrt{x} + 2\sin\sqrt{x} + C$

9 $-e^{-x}(x^3 + 3x^2 + 6x + 6) + C$

11 $\frac{2}{3}x^{3/2}\ln x - \frac{4}{9}x^{3/2} + C$

13 $x\tan^{-1} x - \frac{1}{2}\ln(1+x^2) + C$

15 $\frac{1}{2}x^2\tan^{-1} x - \frac{1}{2}x + \frac{1}{2}\tan^{-1} x + C$

17 $\frac{e^{2x}}{5}(\sin x + 2\cos x) + C$

19 $\frac{x}{2}[\sin(\ln x) - \cos(\ln x)] + C$

25 $-\frac{\sin x\cos x}{2} + \frac{x}{2} + C$

27 $\frac{1}{6}\cos^2 2x\sin 2x + \frac{1}{3}\sin 2x + C$

29 $-\frac{2}{5}e^{-x}\left(2\sin\frac{x}{2} + \cos\frac{x}{2}\right) + C$

31 $\frac{1}{29}e^{-2x}(5\sin 5x - 2\cos 5x) + C$

33 $x(\ln x)^2 - 2x\ln x + 2x + C$

35 $\frac{1}{4}x^4\ln x - \frac{1}{16}x^4 + C$

37 $\pi^2/2$

## Exercises 8.3

1 $-\frac{1}{3}\cos^3 x + C$

3 $\frac{1}{2}\tan^2 x + C$

5 $\frac{1}{5}\sec^5 x + C$

7 $-\frac{1}{4}\sin 2x\cos 2x + \frac{1}{2}x + C$

9 $\sin x - \frac{1}{3}\sin^3 x + C$

11 $\frac{1}{4}\sec 2x\tan 2x + \frac{1}{4}\ln|\sec 2x + \tan 2x| + C$

13 $\frac{1}{9}\tan^9 x + \frac{1}{7}\tan^7 x + C$

15 $\frac{1}{5}\sin^5 x - \frac{1}{7}\sin^7 x + C$

17 $\sec x + \cos x + C$

19 $-\frac{1}{6}\cos^5 x\sin x + \frac{1}{24}\cos^3 x\sin x + \frac{1}{16}\cos x\sin x + \frac{1}{16}x + C$

21 $\frac{\sin x\cos x}{2} + \frac{x}{2} + C$

23 $\frac{1}{2}\tan^2 x - \ln|\sec x| + C$

25 $\frac{1}{5}\cos^5 x - \frac{1}{3}\cos^3 x + C$

27 $\frac{1}{2}\sin x - \frac{1}{10}\sin 5x + C$

29 $-\frac{1}{10}\cos 5x + \frac{1}{6}\cos 3x + C$

## Exercises 8.4

1 $\frac{x\sqrt{4-x^2}}{2} + 2\sin^{-1}\frac{x}{2} + C$

3 $\frac{x}{2\sqrt{2-x^2}} + C$

5 $\frac{x}{2(1+9x^2)} + \frac{1}{6}\tan^{-1} 3x + C$

7 $\frac{1}{5}(1-x^2)^{5/2} - \frac{1}{3}(1-x^2)^{3/2} + C$

9 $\ln|x + \sqrt{x^2-3}| + C$

11 $\frac{1}{2}x\sqrt{x^2-16} + 8\ln|x + \sqrt{x^2-16}| + C$

13 $-\frac{x(9-x^2)^{3/2}}{4} + \frac{9x\sqrt{9-x^2}}{8} + \frac{81}{8}\sin^{-1}\frac{x}{3} + C$

15 $\frac{1}{2}\ln(4+x^2) + C$

17 $\ln|\sqrt{x^2+4x+5} + x + 2| + C$

19 $\frac{1}{3}\ln|x+3| - \frac{1}{6}\ln|9-x^2| + C$

21 $ab\pi$

23 $\pi\sqrt{2} - \pi\ln(\sqrt{2}+1)$

25 $\sqrt{17} + \frac{1}{4}\ln(\sqrt{17}+4)$

## Exercises 8.5

1 $-\ln|x-2| + \ln|x-3| + C$

3 $2\ln|x| - \ln|x-3| + C$

5 $x^2 + 2\ln|x-2| - 3\ln|x+2| + C$

7 $2\ln|x-1| + \frac{1}{x-1} + C$

9 $\ln|x| + 2\ln|x-2| - 2\ln|x+2| + C$

11 $2\ln|x| - \ln|x-2| + 3\ln|x-1| + C$

13 $-(x-1)^{-1} - \frac{1}{2}(x-1)^{-2} + C$

15 $\frac{x^2}{2} - \frac{2}{x} + \ln|x-1| + C$

17 $\ln|x| - x^{-2} - \ln|x+1| + C$

19 $\ln|x| - x^{-1} - \ln|x-2| - (x-2)^{-1} + C$

21 $x - \frac{\sqrt{3}}{2}\ln|x+\sqrt{3}| + \frac{\sqrt{3}}{2}\ln|x-\sqrt{3}| + C$

23 $2\ln|x| + \frac{1}{2}\ln(x^2+1) + C$

25 $-2\ln|x+1| + 2\ln|x-1| + C$

27 $\tan^{-1} x - \frac{1}{3}\tan^{-1}\frac{x}{3} + C$

29 $x + \frac{7}{6}\ln|x-2| + \frac{\sqrt{3}}{6}\tan^{-1}\frac{x+1}{\sqrt{3}} - \frac{7}{12}\ln(x^2+2x+4) + C$

31 $\frac{1}{6}\tan^{-1}\frac{x}{3} - \frac{x}{2(x^2+9)} + C$

33 $\ln|x| - \frac{1}{2}\ln(x^2+2x+5) - \frac{1}{2}\tan^{-1}\frac{x+1}{2} + C$

35 $\ln|x^2+2x-1| - \frac{1}{\sqrt{2}}\ln|x+1-\sqrt{2}| + \frac{1}{\sqrt{2}}\ln|x+1+\sqrt{2}| + C$

## Exercises 8.6

1 $\frac{1}{6}(2x-1)^{3/2} + \frac{1}{2}(2x-1)^{1/2} + C$

3 $\frac{1}{15}(3x+2)^{5/3} - \frac{1}{3}(3x+2)^{2/3} + C$

5 $2\sqrt{x} - 2\ln(1+\sqrt{x}) + C$

7 $-x + 4\sqrt{x} - 4\ln(1+\sqrt{x}) + C$

9 $2x^{1/2} - 3x^{1/3} + 6x^{1/6} - 6\ln(x^{1/6}+1) + C$

11 $\frac{3}{2}x^{2/3} - \frac{3}{2}\ln(1+x^{2/3}) + C$

13 $-\frac{1}{2}(2x+5)^{-1/2} + \frac{5}{6}(2x+5)^{-3/2} + C$

15 $-\frac{1}{2}(2-x)^{-2} + \frac{2}{3}(2-x)^{-3} + C$

17 $-\ln(e^x+1) + C$

19 $\frac{2}{5}(1+e^x)^{5/2} - \frac{2}{3}(1+e^x)^{3/2} + C$

21 $\frac{2}{1-\tan(x/2)} + C$

23 $\tan\frac{x}{2} + C$

25 $\frac{1}{2}\tan^{-1}\frac{\tan(x/2)}{2} + C$

27 $\frac{1}{5}\ln\left|\tan\frac{x}{2}+\frac{1}{2}\right| - \frac{1}{5}\ln\left|\tan\frac{x}{2}-2\right| + C$

29 $\frac{4}{3}(1+\sqrt{x})^{3/2} - 4(1+\sqrt{x})^{1/2} + C$

31 $2\sqrt{x}e^{\sqrt{x}} - 2e^{\sqrt{x}} + C$

## Exercises 8.7

1 div
3 2
5 div
7 div
9 div
11 1/2
13 div
15 $\pi$
17 div
19 2
21 0
23 $\pi$
25 div
27 4
29 $\ln\sqrt{2}$
31 1/4
33 $p < 1$
37 $\infty$

## Review Exercises

1 $-\frac{1}{2}x\cos 2x + \frac{1}{4}\sin 2x + C$

3 $x\ln x - x + C$

5 $-e^{-x}(x^2+2x+2) + C$

7 $\frac{1}{7}\sin^7 x - \frac{1}{9}\sin^9 x + C$

9 $\frac{1}{4}\cos^3 x\sin x + \frac{3}{8}\cos x\sin x + \frac{3}{8}x + C$

11 $\frac{1}{9}\tan^9 x + \frac{1}{11}\tan^{11} x + C$

13 $\tan x - x + C$

15 $-\frac{1}{3}(1-x^2)^{3/2} + C$

17 $\ln(x+\sqrt{9+x^2}) + C$

19 $\frac{x}{2(x^2+1)} + \frac{1}{2}\tan^{-1}x + C$

21 $\frac{1}{2}x^2 + \frac{1}{2}\ln|x^2-1| + C$

23 $\ln|x-2| - \ln|x-1| + C$

25 $\frac{1}{2}\ln(x^2-4x+8) + \tan^{-1}\frac{x-2}{2} + C$

27 $-\ln|x| - x^{-1} + \ln|x+1| + C$

29 $\frac{2}{5}(x+1)^{5/2} - \frac{2}{3}(x+1)^{3/2} + C$

31 $2\sqrt{x} - 2\ln(1+\sqrt{x}) + C$

33 1/2
35 div
37 1/4

## CHAPTER NINE

### Exercises 9.1

**7** $y = 2x^3 - 2x^2 + 3x + C$ **9** $y^2 = x^2 + C$

**11** $y = Ce^{2x}$ **13** $y = -\frac{1}{2}x^2 + \frac{5}{2}$

**15** $y = 2e^{-3x}$

### Exercises 9.2

**1** $y = 2x^2 - x + 2$ **3** $y = -\ln|1 - x| + 1$

**5** $y^3 = \frac{3}{2}x^2 + \frac{27}{2}$ **7** $y = \tan x$

**9** $y = x^3 + C$; $y = \frac{1}{3}x^{-1} + C$ **11** $y = \tan^{-1} x + C$; $y = -x - \frac{1}{3}x^3 + C$

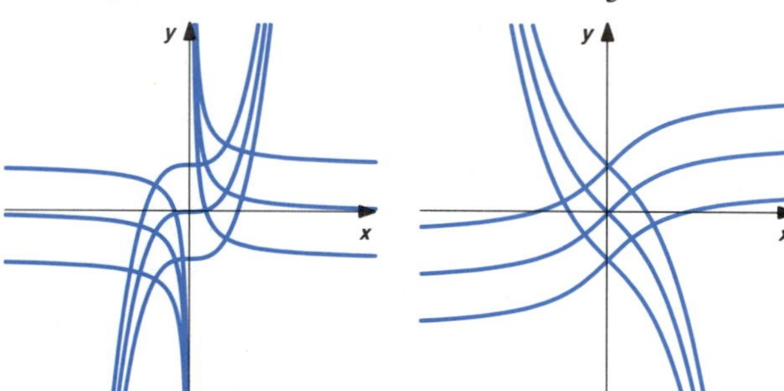


**13** $y = Ce^{2x}$; $y^2 = -x + C$ **15** $y = Cx^3$; $y^2 + \frac{1}{3}x^2 = C$

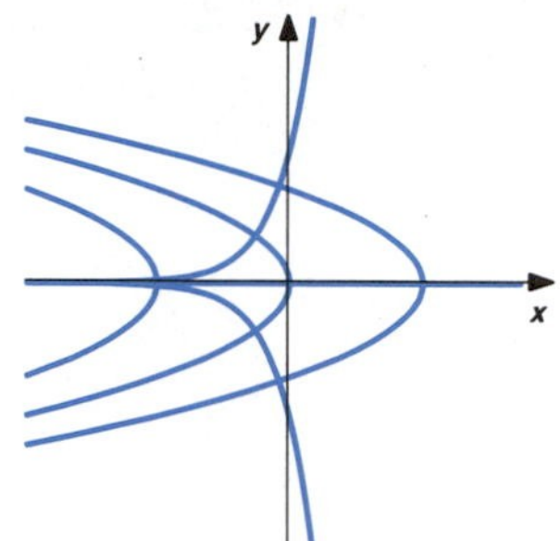


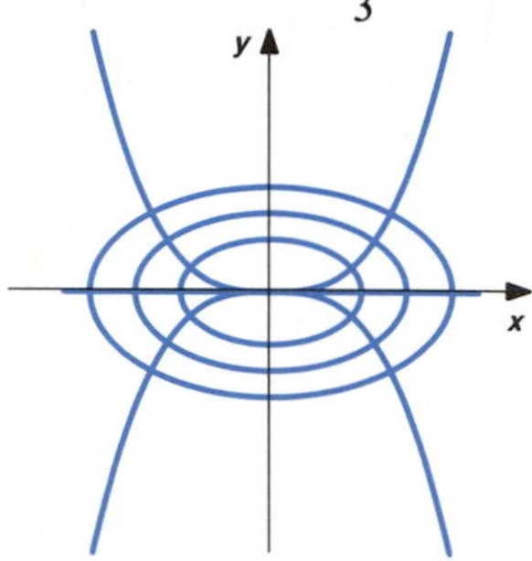


**17** $z = \dfrac{12 - 12e^{kt}}{3 - 4e^{kt}}$

**19** $z = 1 - \dfrac{1}{kt + 1}$

**21** $xy' + y = 0$; $y^2 - x^2 = C$

### Exercises 9.3

**1** 4.78 s **3** 5.42 s

**5** $v(t) = 20e^{-8t/5} - 20$; $s(t) = -12.5e^{-8t/5} - 20t + 3012.5$; 20 ft/s

**7** $v(t) = 40e^{-4t/5} - 40$; $s(t) = -50e^{-4t/5} - 40t + 3050$; 40 ft/s

**9** $v(t) = \dfrac{8 - 8e^{8t}}{1 + e^{8t}}$; 8 ft/s

**11** 8 ft/s; stretched 3/2 ft, compressed 1/2 ft

**13** 2 ft

**15** $\sqrt{80/3}$ ft/s; ±0.5 ft from natural length

**17** $\pm\sqrt{2}$ ft from natural length

**19** 3.94 R

**21** $\sqrt{Rg}$

**23** 313 s

### Exercises 9.4

**1** $y = \dfrac{x^2}{4} + \dfrac{C}{x^2}$ **3** $y = -e^x + Ce^{2x}$

**5** $y = \dfrac{1}{x} + \dfrac{C}{xe^x}$ **7** $y = \dfrac{2x^2 + x^4 + C}{4(1 + x^2)}$

**9** $y = x \ln \dfrac{x}{x+1} + Cx$ **11** $y = 1 + C\cos x$

**13** $y = \dfrac{\ln(\sec^2 x + \sec x \tan x) + C}{\sec x + \tan x}$

**15** $i(t) = 8e^{-5t/6}$

### Review Exercises

1 $y = \frac{1}{5 - x^2}$

3 $y^2 - x^2 = C$; $xy = C$

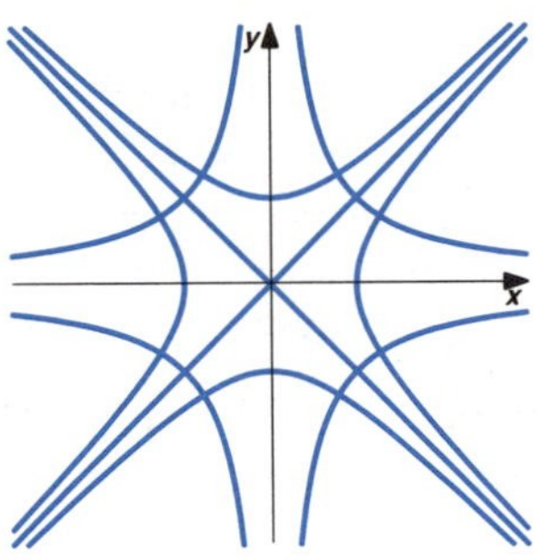


5 $z(t) = \frac{3 - 3e^{2kt}}{1 - 3e^{2kt}}$

7 (a) $v(t) = 32e^{-t} - 32$, 32 ft/s (b) 448 ft/s

9 Compressed 1 ft, stretched 2 ft; $8\sqrt{2}$ ft/s

11 $y = \frac{2}{-x^2 - 2x + C}$

13 $y = Ce^{2x} - 1$

15 $y = x - 1 + Ce^{-x}$

17 $y = 1 + \frac{C}{x}$

19 $y = x \cos x + C \cos x$

## CHAPTER TEN

### Exercises 10.1

1 3/2

3 0

5 4/3

7 1/2

9 div

11 0

13 div

15 div

17 0

19 0

21 0

23 0

### Exercises 10.2

1 0.5, 0.625, 0.66666667, 0.68229167; 0.69314718

3 0.8, 0.97066667, 1.0362027, 1.0661620; 1.0986123

5 1, 3, 5, 6.3333333; 7.3890561

7 0.5, 0.45833333, 0.46458333, 0.46346726; 0.46364761

9 $10[1 - (0.2)^n]$, 10

11 $-5.5[1 - (-1.1)^n]/2.1$, div

13 $6[1 - (0.5)^n]$, 6

15 $\frac{1}{2} - \frac{n+1}{n+2}, -\frac{1}{2}$

17 $\frac{1}{2} - \frac{n+1}{2^{n+1}}, \frac{1}{2}$

19 $\frac{3}{2} - \frac{1}{n+1} - \frac{1}{n+2}, \frac{3}{2}$

21 $\frac{1}{3} - \frac{1}{n+3}, \frac{1}{3}$

23 5/9

25 30 ft

### Exercises 10.3

1 div

3 div

5 conv

7 div

9 conv

11 div

13 div

15 conv

17 conv

19 div

21 conv

23 div

25 conv

27 conv

29 conv

31 conv

33 $S_4 = 1.0787519$

35 1000

37 $p > 1$

*Exercises 10.4*

1 conv
3 conv
5 div
7 conv
9 conv
11 conv
13 conv
15 div
17 conv
19 div
21 conv
23 conv
25 div
27 conv
29 conv
31 $S_4 = -0.625$
33 $S_2 = 0.70465265$

*Exercises 10.5*

1 abs conv
3 div
5 abs conv
7 abs conv
9 cond conv
11 $1, 0 < x < 2$
13 $1, 0 \le x < 2$
15 $2, -2 < x < 2$
17 $1, -1 < x < 1$
19 $2, 0 < x < 4$
21 $3, 0 \le x \le 6$
23 $0, x = 0$
25 $\infty, -\infty < x < \infty$
27 $1/2, -1/2 < x < 1/2$
29 $1/2, -1/2 < x < 1/2$

*Exercises 10.6*

1 (a) $1 + x^2 + x^4 + x^6 + \cdots, |x| < 1$
(b) $x + x^3 + x^5 + x^7 + \cdots, |x| < 1$
(c) $x + 2x^3 + 3x^5 + 4x^7 + \cdots, |x| < 1$

3 (a) $x - \frac{x^2}{2} + \frac{x^3}{3} - \frac{x^4}{4} + \cdots, |x| < 1$
(b) $x^2 - \frac{x^3}{2} + \frac{x^4}{3} - \frac{x^5}{4} + \cdots, |x| < 1$
(c) $2 - 2x + \frac{8x^2}{3} - 4x^3 + \cdots, |x| < 1/2$

5 (a) $-\frac{1}{3} - \frac{2x}{9} - \frac{7x^2}{27} - \frac{20x^3}{81} - \cdots, |x| < 1$
(b) $-\frac{x^2}{3} - \frac{2x^3}{9} - \frac{7x^4}{27} - \frac{20x^5}{81} - \cdots, |x| < 1$

7 (a) $1 + 2x + 2x^2 + 2x^3 + \cdots, |x| < 1$
(b) $-1 - x - 2x^2 - 2x^3 - \cdots, |x| < 1$

9 −0.882352941, −1.111337269, −1.218302266, −1.277786014; −1.3862944

11 0.92307692, 1.185252617, 1.31928800, 1.400864891; 1.6094379

13 0.01866667

15 0.00520833

17 0.11197917

19 8

21 0

27 $1 - 2(x-1) + 3(x-1)^2 - 4(x-1)^3 + \cdots, |x-1| < 1$

*Exercises 10.7*

1 $3 + 6(x-1) + 3(x-1)^2 + (x-1)^3$

3 $-8 + 12x - 6x^2 + x^3$

5 $x + \frac{x^3}{3!} + \frac{x^5}{5!} + \frac{x^7}{7!} + \cdots$

7 $2x - \frac{8x^3}{3!} + \frac{32x^5}{5!} - \frac{128x^7}{7!} + \cdots$

9 $1 - x^2 + \frac{x^4}{2!} - \frac{x^6}{3!} + \frac{x^8}{4!} - \cdots$

11 $1 - x^2 + \frac{x^4}{3} - \frac{2x^6}{45} + \cdots$

13 $\frac{1}{2!} - \frac{x^2}{4!} + \frac{x^4}{6!} - \frac{x^6}{8!} + \cdots$

15 −160
17 0
19 0.125
21 0.60677083
23 0.49305556
25 0.19733333

*Exercises 10.8*

1 $1 + \frac{x}{2} - \frac{x^2}{8} + \frac{x^3}{16} - \cdots$

3 $1 + \frac{x}{3} + \frac{2x^2}{9} + \frac{14x^3}{81} + \cdots$

5 $2 + \frac{x^2}{4} - \frac{x^4}{64} + \frac{x^6}{512} - \cdots$

7 $\frac{1}{8} - \frac{3x}{16} + \frac{3x^2}{16} - \frac{5x^5}{32} + \cdots$

9 $1 + x - \frac{x^2}{2} + \frac{x^3}{2} - \cdots$

11 $1 + \frac{x^2}{6} + \frac{3x^4}{40} + \frac{5x^6}{112} + \cdots$

13 0.150568271

15 0.04112464

17 0.02006748

### *Exercises 10.9*

1 $2 + 2x + x^2 + \frac{x^3}{3}$

3 $1 + x + \frac{x^2}{2} + \frac{x^3}{6}$

5 $x - \frac{x^3}{6} + \frac{x^5}{120} - \frac{x^7}{5040}$

7 $1 + 2x + \frac{x^2}{2} + \frac{x^3}{3}$

9 $2 - x - \frac{7x^2}{2} - \frac{19x^3}{6}$

11 $2 + x - \frac{5x^2}{2} + \frac{x^3}{2}$

13 $1 + x + x^2 + x^3$

15 $3 + x - x^2 + \frac{x^3}{3}$

### *Review Exercises*

1 3/2

3 3

5 0

7 0

9 $\frac{7}{3} - \frac{2n+2}{n+2} - \frac{2n+4}{n+3}; -\frac{5}{3}$

11 $\frac{3}{2} - \frac{1}{n+1} - \frac{1}{n+2}; \frac{3}{2}$

13 $12[1 - (3/4)^n]$; 12

15 div

17 conv

19 conv

21 conv

23 conv

25 div

27 conv

29 cond conv

31 cond conv

33 $0 \le x < 2$

35 $1 < x < 3$

37 $n \ge 23$

39 $n \ge 31$

41 $1 - x + \frac{x^2}{2} + \cdots$

43 $x - x^3 + x^5 - \cdots$

45 $\frac{1}{2\sqrt{2}} + \frac{3x}{8\sqrt{2}} + \frac{15x^2}{64\sqrt{2}} + \cdots$

47 $-2x + 4x^3 - 6x^5 + \cdots$

49 $-\frac{1}{2} - \frac{x}{4} - \frac{x^2}{8} - \cdots$

51 $1 - x^2 + \frac{x^4}{3} - \cdots$

53 0.299757

55 0.19955556

57 $2 + 2x + \frac{3x^2}{2} + \frac{x^3}{2}$

59 $1 + 2x - \frac{x^3}{3} - \frac{x^4}{12}$

## CHAPTER ELEVEN

### *Exercises 11.1*

1 (0, 1); (0, 0); $y = 2$

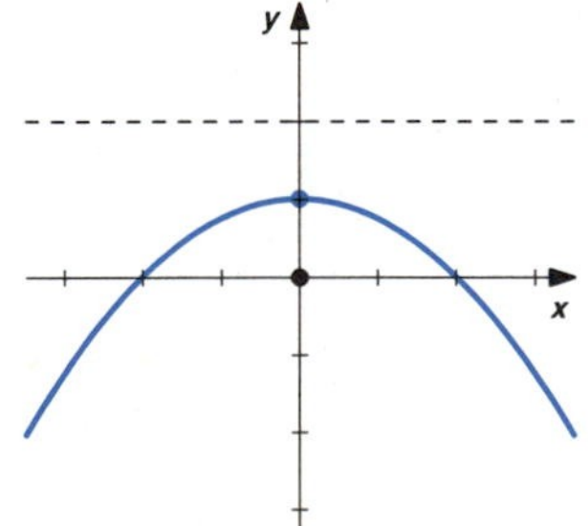


3 (2, 0); (0, 0); $x = 4$

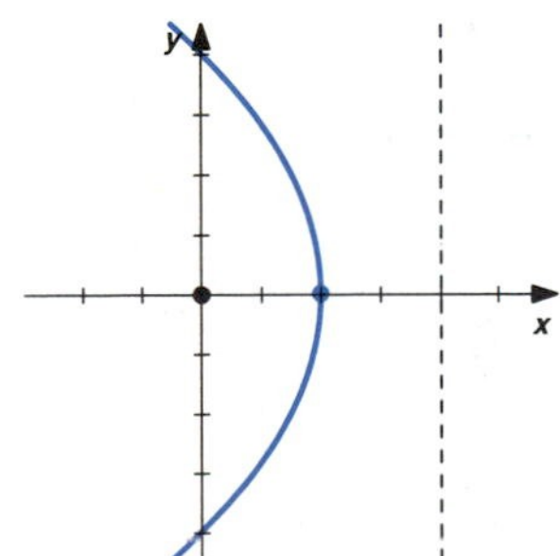


5 (3/2, −9/4); (3/2, −2); $y = -5/2$

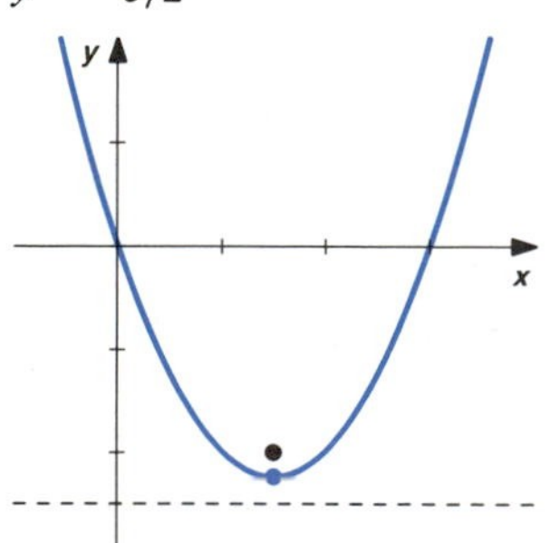

7 $(-1, -1)$; $(-1/3, -1)$; $x = -5/3$

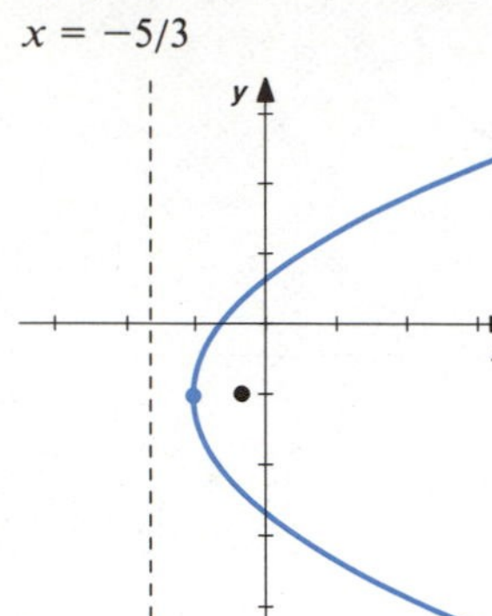


9 $(1, 1)$; $(1, 4/3)$; $y = 2/3$

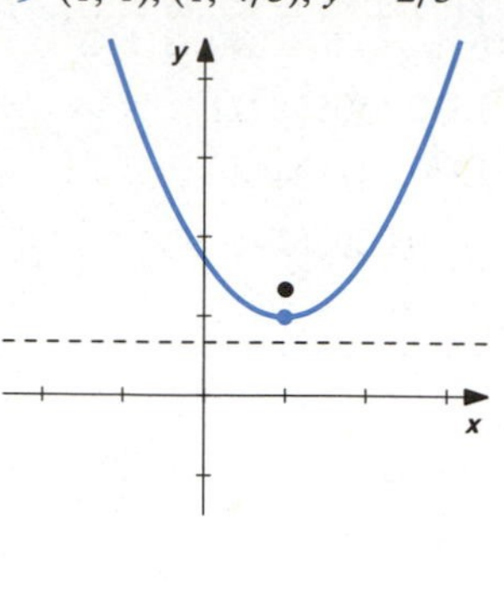


11 $y^2 = 4(x + 1)$

13 $(x - 4)^2 = -8(y - 2)$

15 $x^2 = 8(y + 1)$

17 $(y + 4)^2 = -\frac{8}{3}(x - 6)$

19 $y = -x^2 + 2x + 3$

21 (a) $y = 2x - 5$ (b) $(0, -5)$ (e) $\alpha = \beta$

*Exercises 11.2*

1 $(0, 0)$; $(0, \pm 5)$; $(0, \pm 4)$

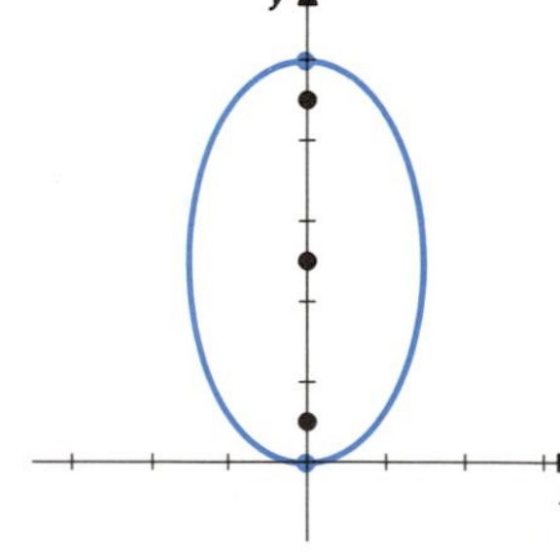


3 $(2, 0)$; $(4, 0)$, $(0, 0)$; $(2 \pm \sqrt{3}, 0)$

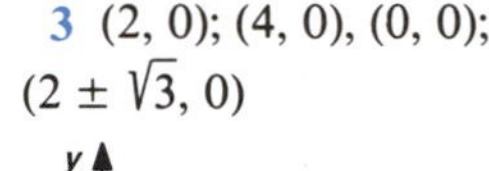

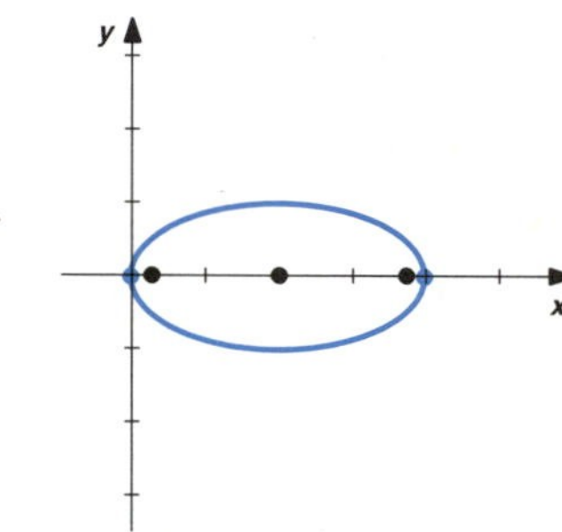


5 $(0, 5/2)$; $(0, 0)$, $(0, 5)$; $(0, 9/2)$, $(0, 1/2)$

7 $(3/2, 1)$; $(3, 1)$, $(0, 1)$; $\left(\frac{3 \pm \sqrt{5}}{2}, 1\right)$

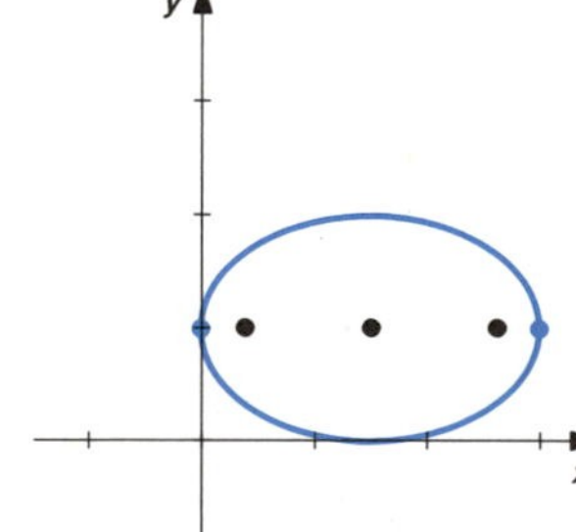


9 $(1, 1)$; $(1, -1)$, $(1, 3)$; $(1, 1 \pm \sqrt{8/3})$

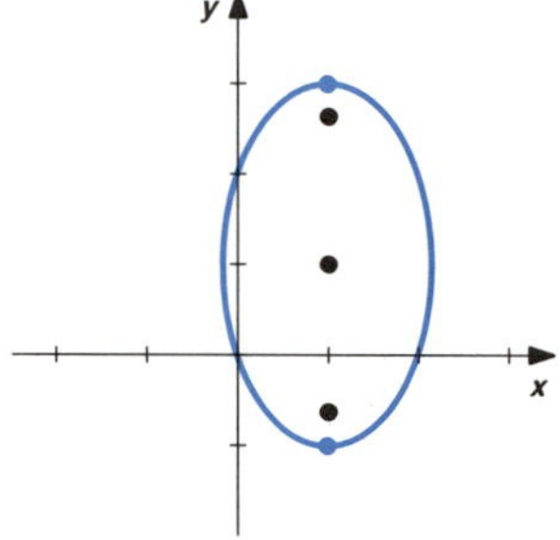


11 $\frac{x^2}{5^2} + \frac{y^2}{13^2} = 1$

13 $\frac{x^2}{5^2} + \frac{(y-3)^2}{3^2} = 1$

15 $\frac{(x-1)^2}{1^2} + \frac{y^2}{4^2} = 1$

17 $\frac{x^2}{4^2} + \frac{y^2}{(2/\sqrt{3})^2} = 1$

19 $\frac{(x+5)^2}{10^2} + \frac{y^2}{(5\sqrt{3})^2} = 1$, $\frac{(x-5/3)^2}{(10/3)^2} + \frac{y^2}{(5\sqrt{3}/3)^2} = 1$

21 (a) $y = 2x - 5$ (b) $(0, -5)$ (e) $\alpha = \beta$

23 $\frac{(x-1)^2}{2^2} + \frac{y^2}{(\sqrt{3})^2} = 1$

25 (a) $-16/15$ (b) $(-3, 0)$, $(3, 0)$ (c) $12/35$, $12/5$ (d) $20/9$, $20/9$ (e) $\alpha = \beta$

## Exercises 11.3

**1** (0, 0); (±12, 0); (±13, 0); $y = \pm 5x/12$

**3** (0, 0); (0, ±3); (0, ±5); $y = \pm 3x/4$

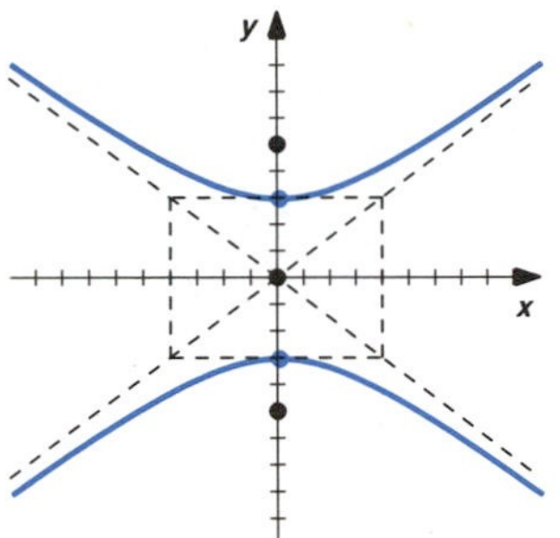

**5** (1, 0); (0, 0), (2, 0); $(1 \pm \sqrt{2}, 0)$; $y = \pm(x - 1)$

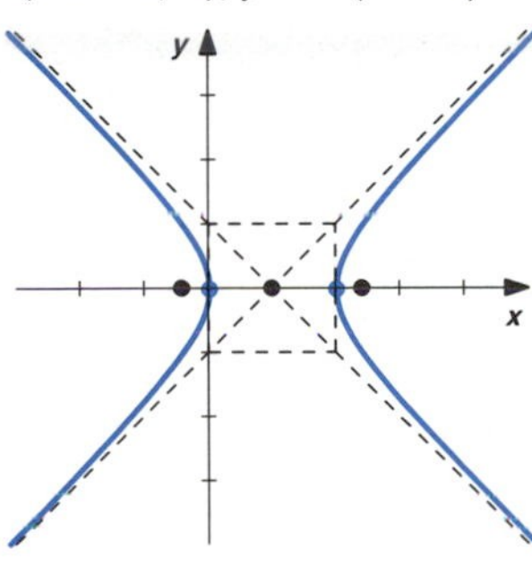

**7** (−1, 2); (−3/2, 2), (−1/2, 2); $(-1 \pm \sqrt{5}/2, 2)$; $x + 1 = \pm(y - 2)/2$

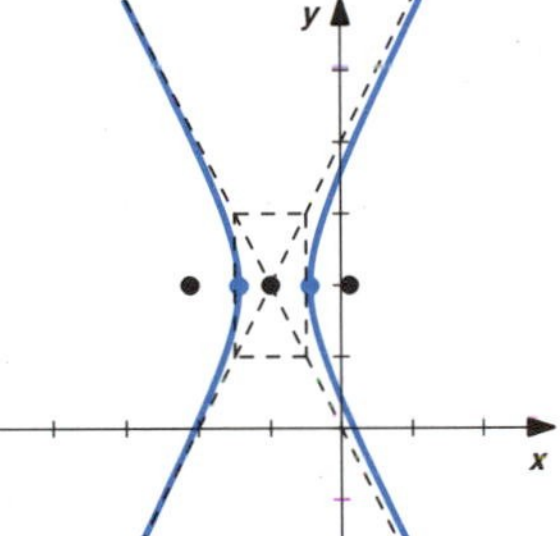

**9** (2, 1); (0, 1), (4, 1); (−1/2, 1), (9/2, 1); $x - 2 = \pm 4(y - 1)/3$

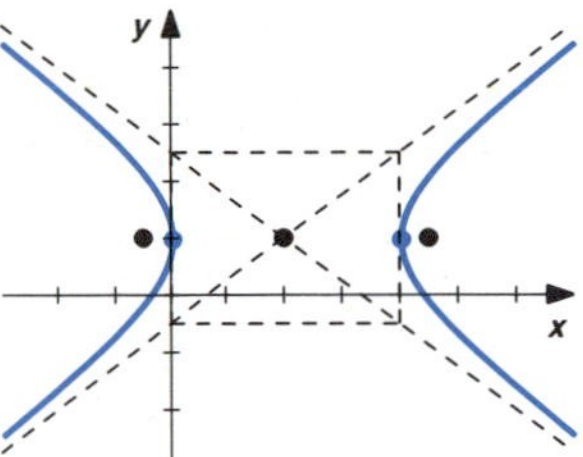

**11** $-\dfrac{x^2}{5^2} + \dfrac{y^2}{12^2} = 1$

**13** $\dfrac{x^2}{4^2} - \dfrac{(y-3)^2}{3^2} = 1$

**15** $-\dfrac{(x-1)^2}{(3/4)^2} + \dfrac{y^2}{3^2} = 1$

**17** $-\dfrac{x^2}{(2/\sqrt{5})^2} + \dfrac{y^2}{4^2} = 1$

**19** $\dfrac{(x-8/3)^2}{(2/3)^2} - \dfrac{y^2}{(\sqrt{60}/3)^2} = 1$, $\dfrac{(x-8/5)^2}{(2/5)^2} - \dfrac{y^2}{(\sqrt{60}/5)^2} = 1$

**21** The coordinates $(x, y)$ must satisfy $y = 0$ and either $x \le -c$ or $x \ge c$, so the set consists of two half-lines.

**23** $\dfrac{(x+4)^2}{2^2} - \dfrac{y^2}{(\sqrt{12})^2} = 1$

**25** (a) 20/9 (b) (±5, 0) (c) −12/5, 12/35 (d) 16/15, 16/15 (e) $\alpha = \beta$

## Exercises 11.4

**1** $x' = y'^2$

**3** $y' = -x'^2$

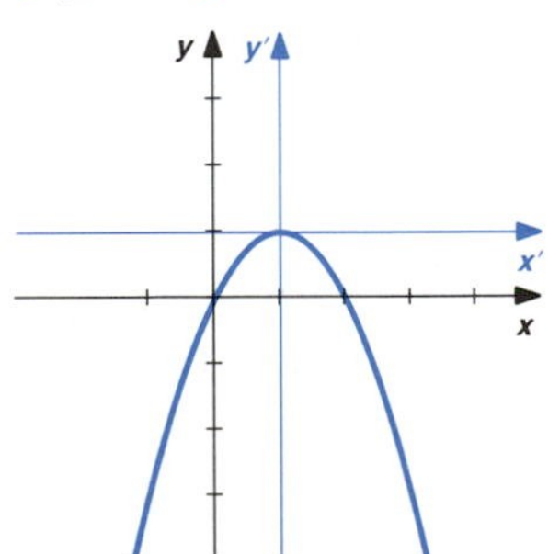

**5** $x = x' + 2$, $y = y' + 1$

**7** $x = x'$, $y = y' + 1$

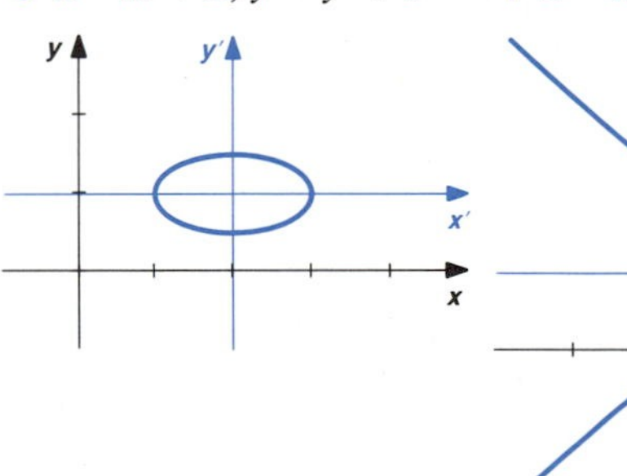

**9** $y' = \sqrt{2}$

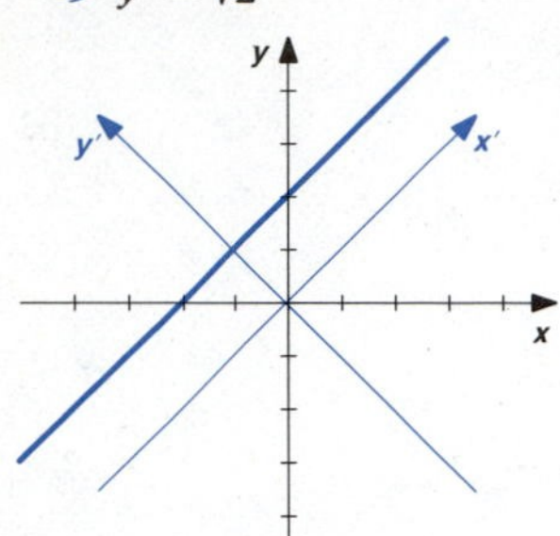


**11** $x' = 2/\sqrt{5}$

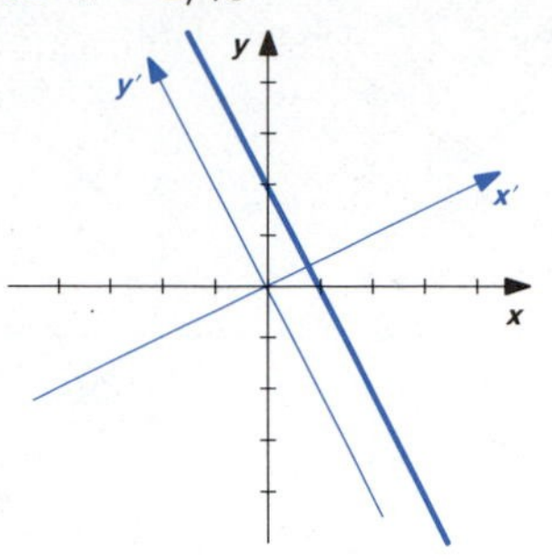


**13** $\frac{\pi}{4}; -\frac{x'^2}{(\sqrt{2})^2} + \frac{y'^2}{(\sqrt{2})^2} = 1$

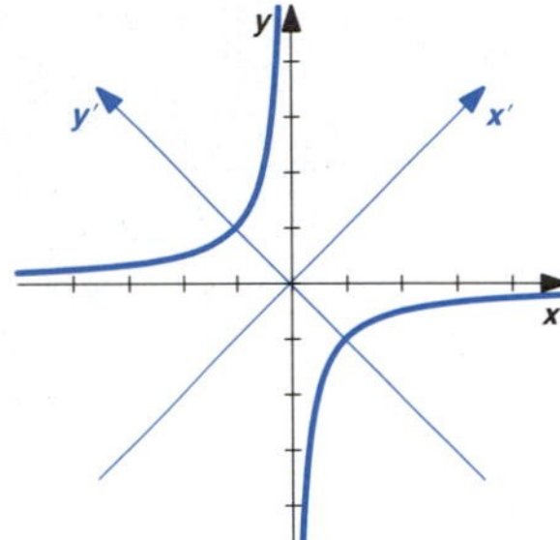


**15** $\frac{\pi}{4}; y'^2 = x'$

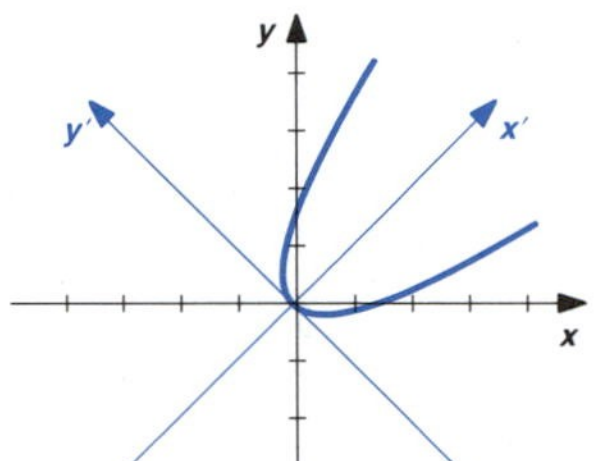


**17** $\frac{\pi}{6}; \frac{x'^2}{2^2} + \frac{y'^2}{1^2} = 1$

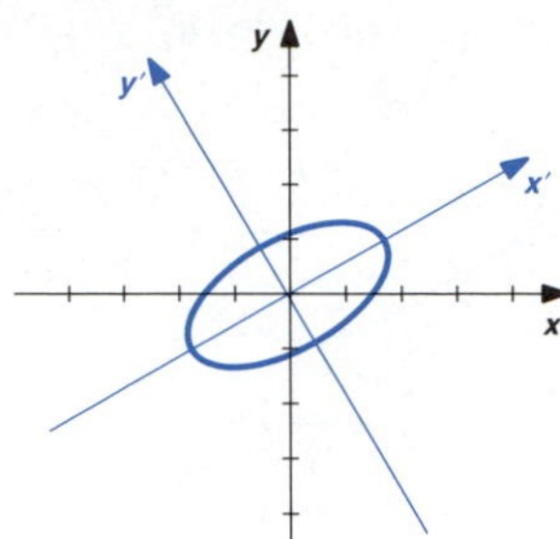


**19** $\tan^{-1}(-4/3)$; $y' = 12/5$

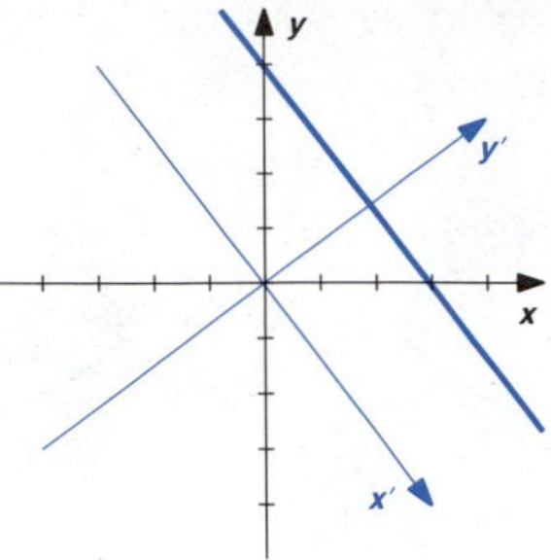


### *Review Exercises*

**1** $\frac{(x-4)^2}{5^2} + \frac{y^2}{3^2} = 1$

**3** $\frac{(x-4)^2}{4^2} - \frac{y^2}{3^2} = 1$

**5** $y^2 = 8x$

**7**

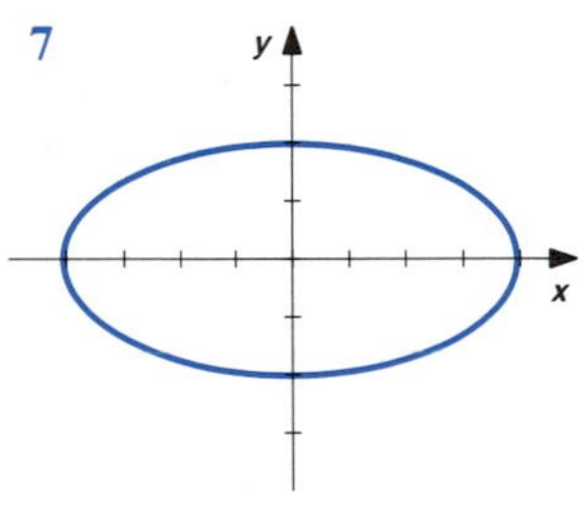


**9**

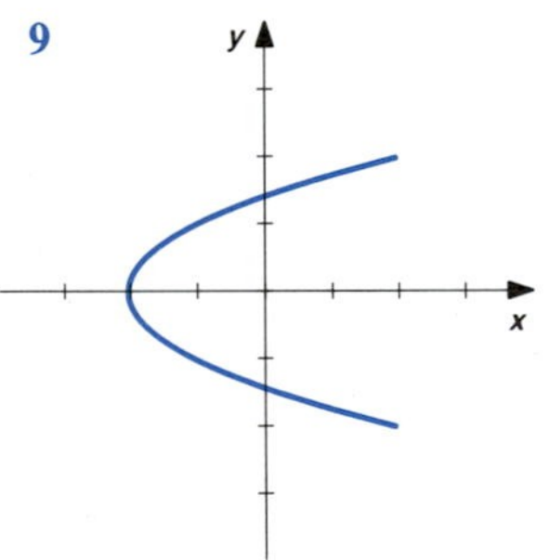


**11**

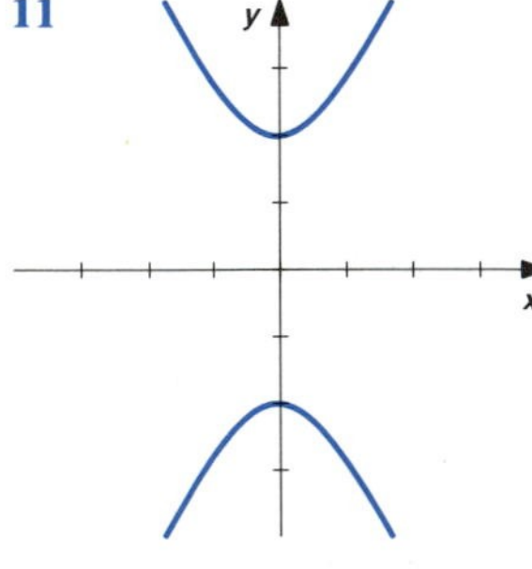


**13** $\pi/3$; $y' = \sqrt{3}/2$

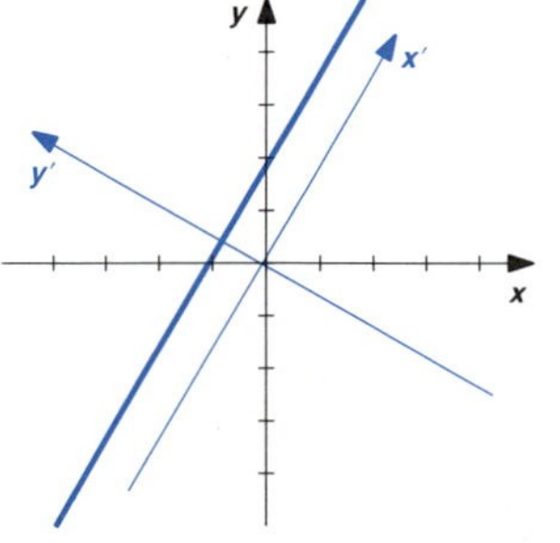


**15** $\pi/4; \frac{(x' - \sqrt{2}/2)^2}{(\sqrt{2})^2} - \frac{(y' + \sqrt{2}/2)^2}{(\sqrt{2})^2} = 1$

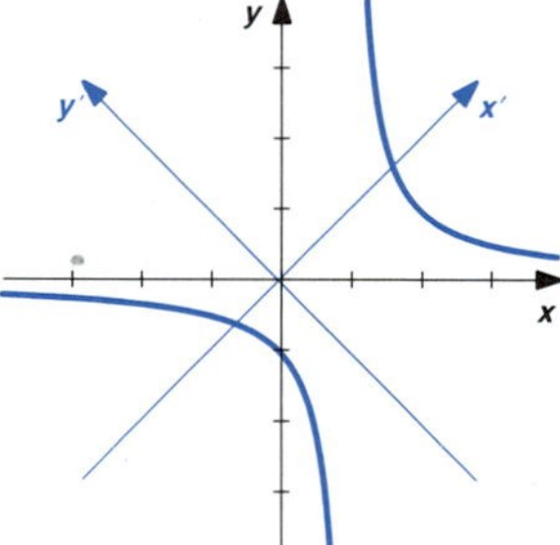


**17** $(-1, -1)$, $(1, 1)$

# CHAPTER TWELVE

## *Exercises 12.1*

**1**

**3** $A(x, y) = (1, 0)$, $B(x, y) = (1, \sqrt{3})$, $C(x, y) = (0, 3)$, $D(x, y) = (-\sqrt{2}, \sqrt{2})$

**5** $A(r, \theta) = (1, 0) = (-1, \pi)$,
$B(r, \theta) = (2, \pi/2) = (-2, -\pi/2)$,
$C(r, \theta) = (2, 2\pi/3) = (-2, -\pi/3)$,
$D(r, \theta) = (2\sqrt{2}, -\pi/4) = (-2\sqrt{2}, 3\pi/4)$

**7** $x^2 + y^2 = 4$

**9** $x = 1$

**11** $x^2 + y^2 = 4x$

**13** $x^2 - 4y = 4$

**15** $r = 3$

**17** $r = \dfrac{3}{\sin\theta}$

**19** $r = -6\cos\theta$

**21** $r = \dfrac{2}{1 - \cos\theta}$

**23** Circle

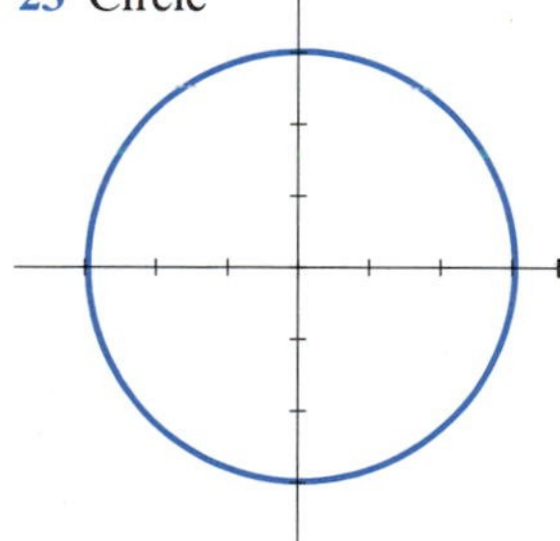

**25** Line

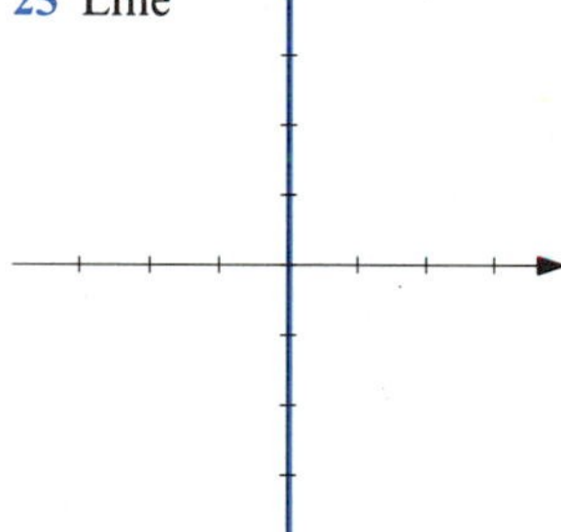

**27** Line

**29** Line

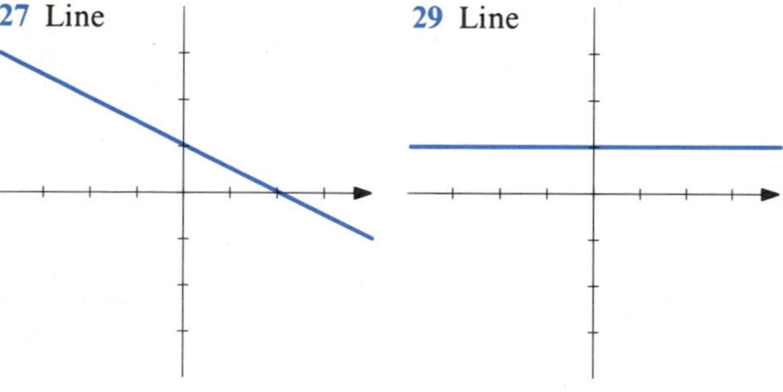

**31** Circle

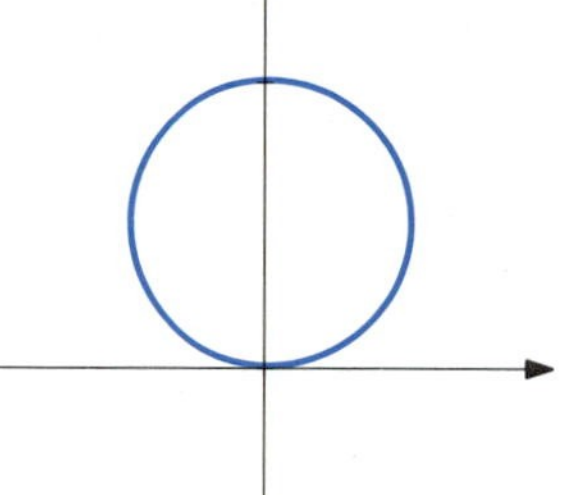

**33** Circle

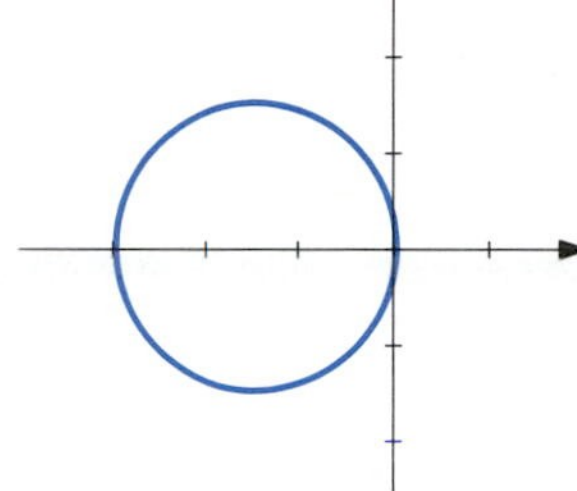

**35** Parabola

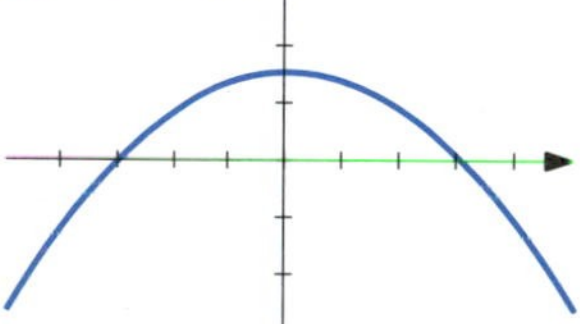

**37** Ellipse

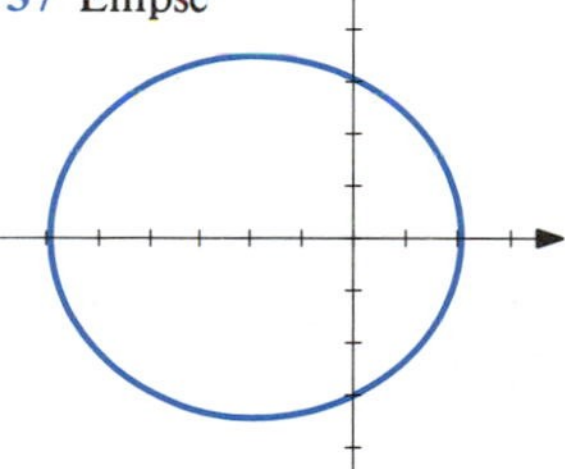

**39** Hyperbola

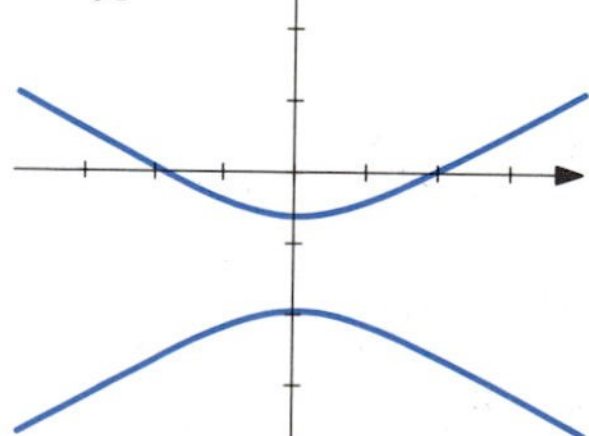

**41** Distance $= \sqrt{r_1^2 + r_0^2 - 2r_1 r_0 \cos(\theta_1 - \theta_0)}$

**43** $r = \dfrac{d}{1 - \cos\theta}$

**45** $r = \dfrac{d}{2 - \cos\theta}$

## *Exercises 12.2*

1 3

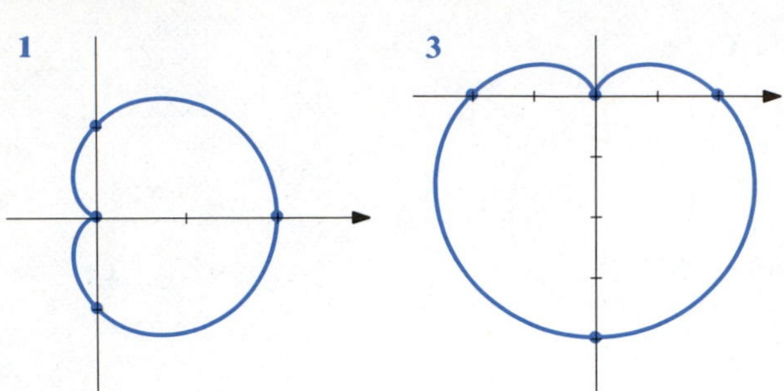

5 7

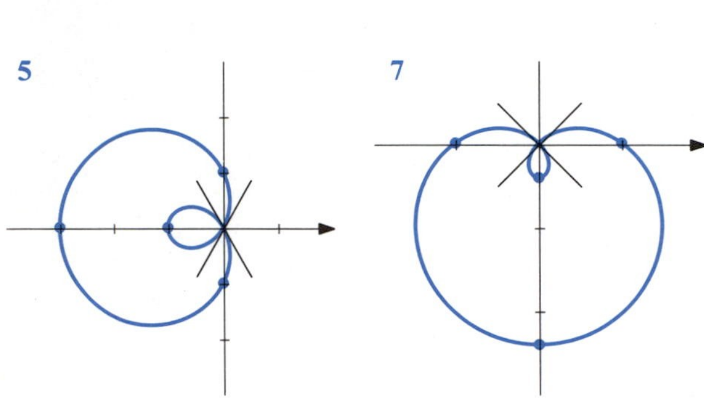

9 11

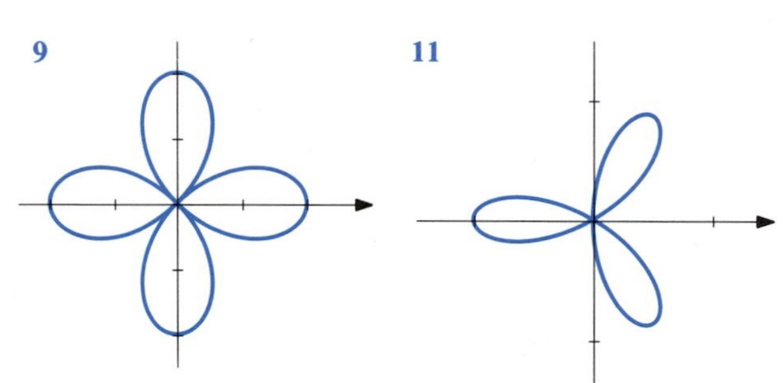

13 15

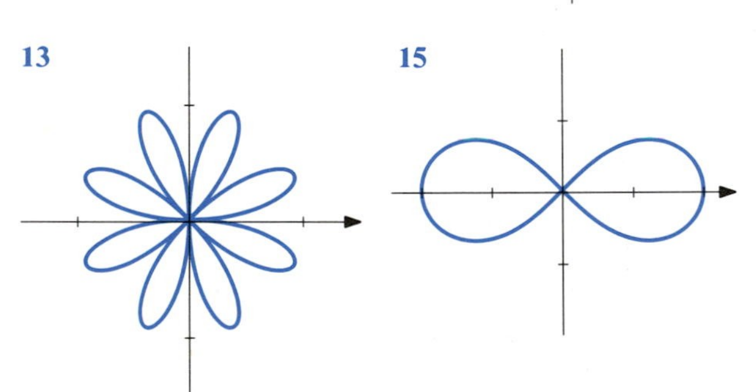

17 19

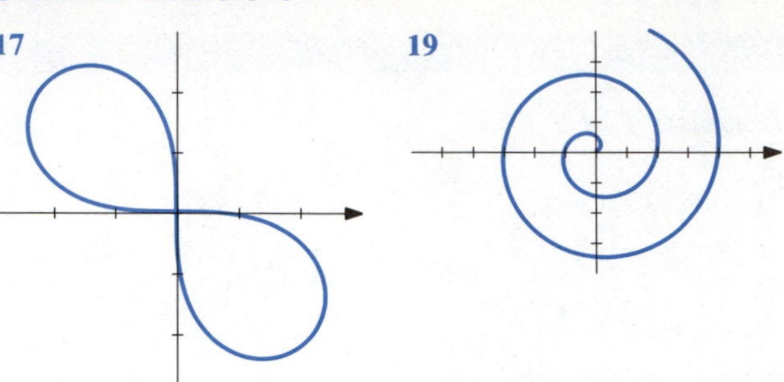

21 23

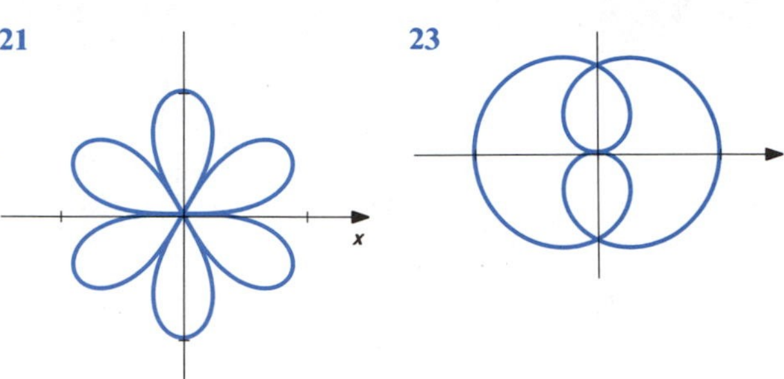


25 27

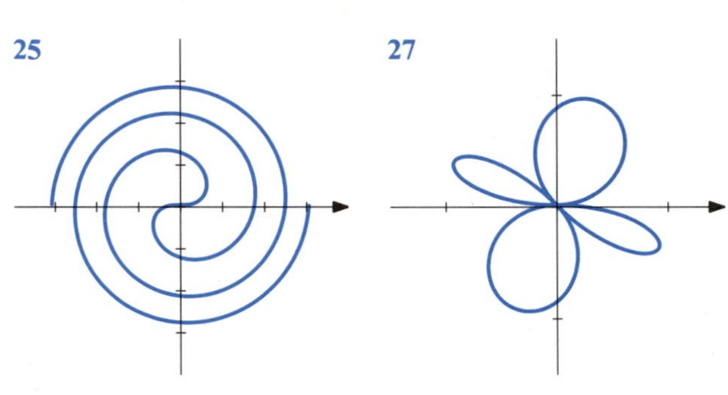

29 $\left(\ln \dfrac{\sqrt{\pi^2+4}-\pi}{2}, \dfrac{\sqrt{\pi^2+4}-\pi}{2}\right)$

*Exercises 12.3*

**1** $\pi/4, \pi/2, -\pi/4, 0, \pi/4$

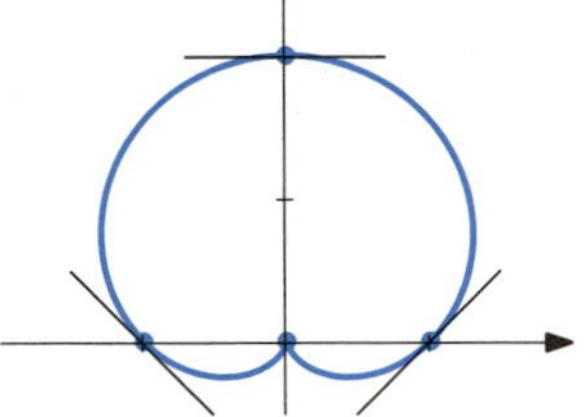

**3** $\tan^{-1}(-1/2), \tan^{-1} 2, \tan^{-1}(-1/2), \tan^{-1} 2, \tan^{-1}(-1/2)$

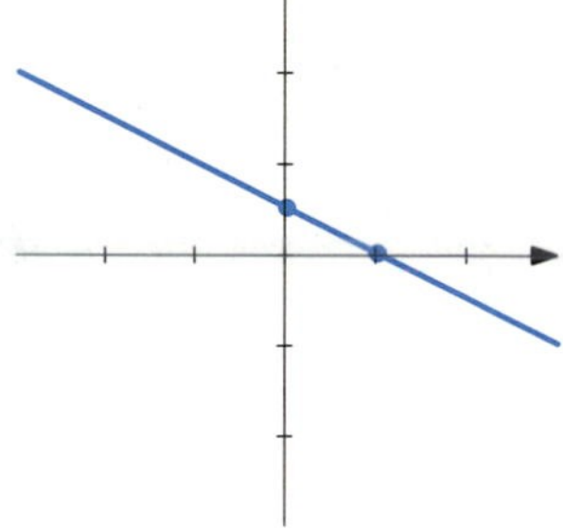

**5** $-\pi/4, \pi/2, \pi/4, \pi/2, -\pi/4$

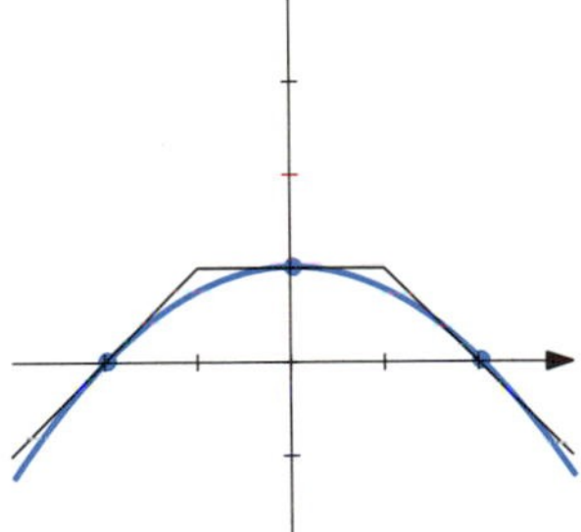

**7** $r = \dfrac{4}{\cos\theta + \sqrt{3}\sin\theta}$

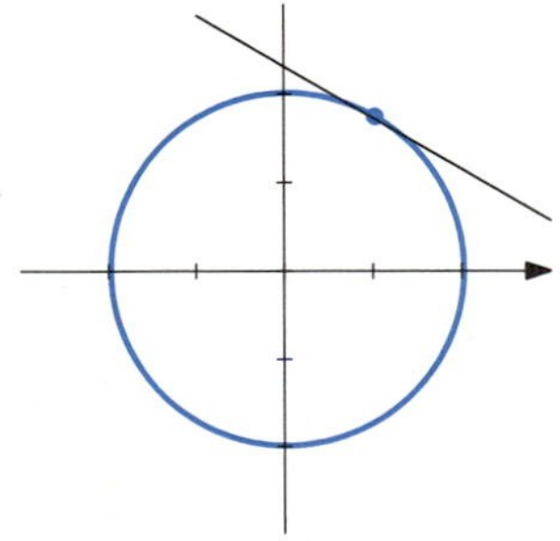

**9** $r = \dfrac{-1}{\sin\theta}$

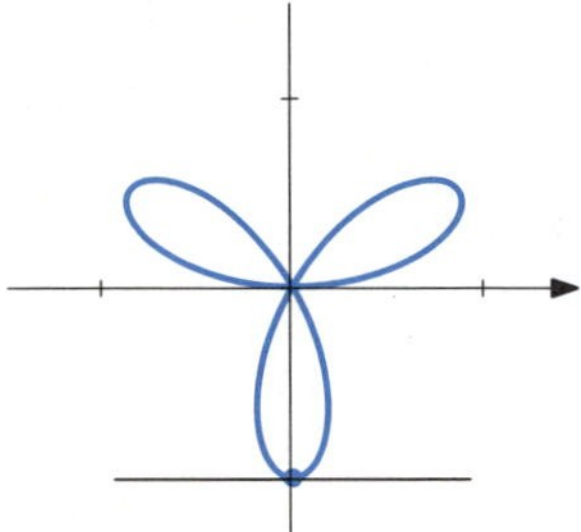

**11** $(2, 0), (1/2, 2\pi/3), (1/2, 4\pi/3)$

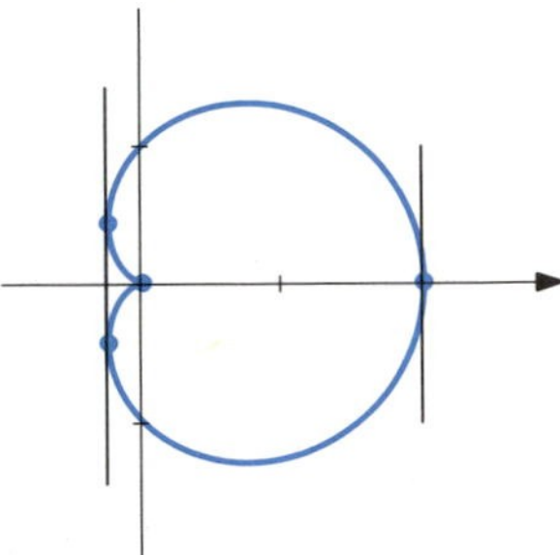

**13** $(3, 0), (1, \pi)$

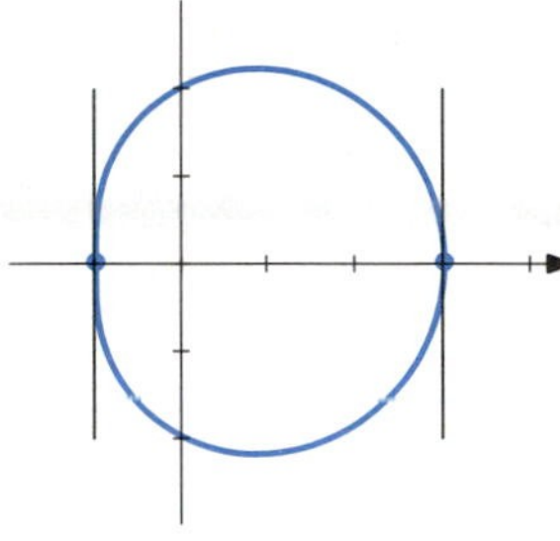

**15** $(3/2, \pi/3), (3/2, 5\pi/3), (0, \pi)$

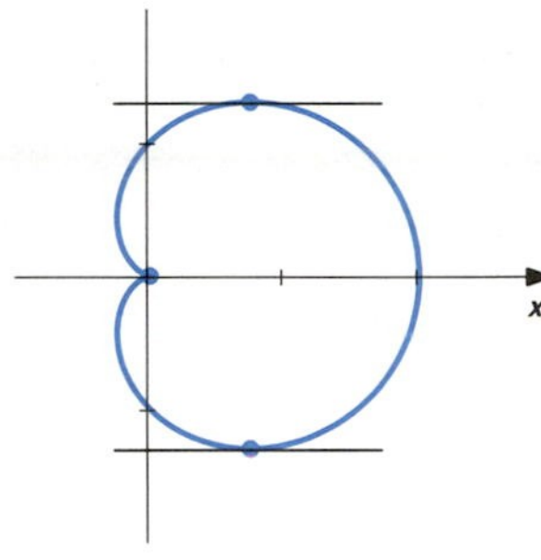


**17** $(2, \pi/2), (6, 3\pi/2)$

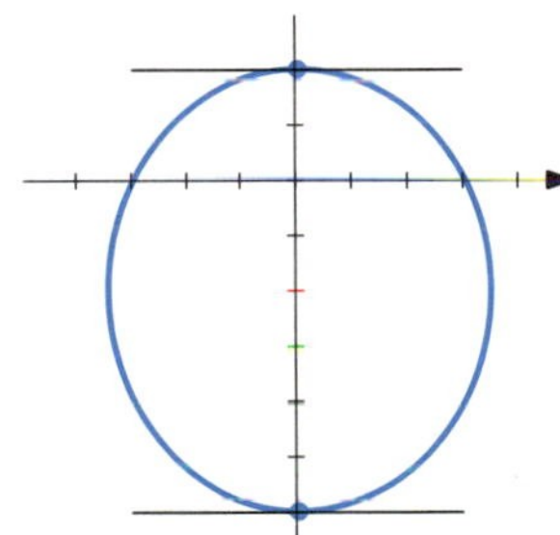

**19**

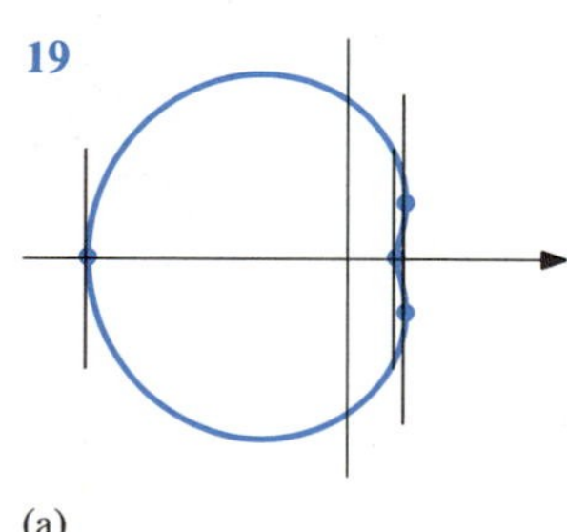

(a)

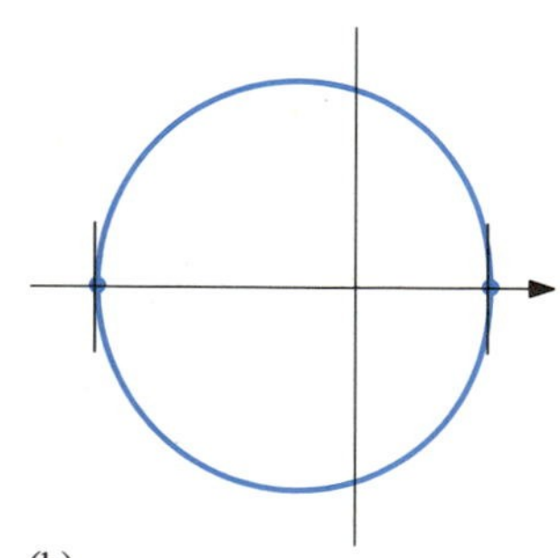

(b)

### *Exercises 12.4*

**1** $2\sqrt{3}$ **3** $\pi^3/6$

**5** $6\pi$ **7** 2

**9** $\pi/2$ **11** $-\dfrac{3\sqrt{3}}{2}+\pi$

**13** $-\sqrt{3}+\dfrac{4\pi}{3}$ **15** $\pi^3/2$

**17** $\dfrac{2\pi}{3}+\sqrt{3}$ **19** $\pi+3\sqrt{3}$

**21** $\dfrac{8\pi}{3}+\sqrt{3}$ **23** $\dfrac{\pi}{6}+\dfrac{\sqrt{3}}{4}$

### *Exercises 12.5*

**1** $4\pi$ **3** $\sqrt{5}(e^{\pi}-1)$

**5** 4 **7** $\dfrac{7^{3/2}-8}{27}$

**9** $\dfrac{8\pi(2\sqrt{2}-1)}{3}$ **11** $4\pi$

**13** $4\pi^2$ **15** $\dfrac{\pi\sqrt{5}(e^{-\pi/2}+e^{\pi/2})}{2}$

### *Review Exercises*

**1**

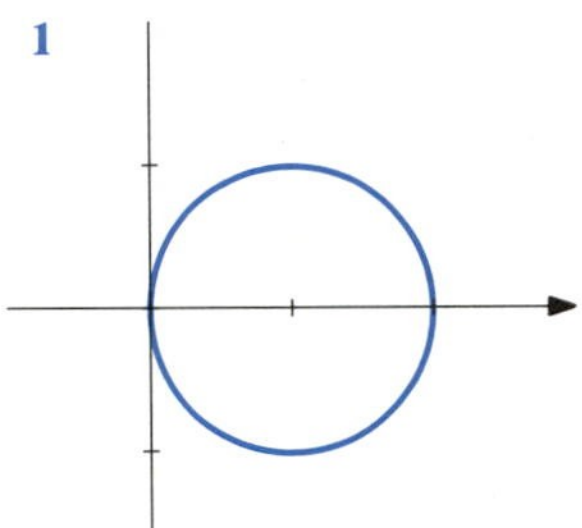

**3**

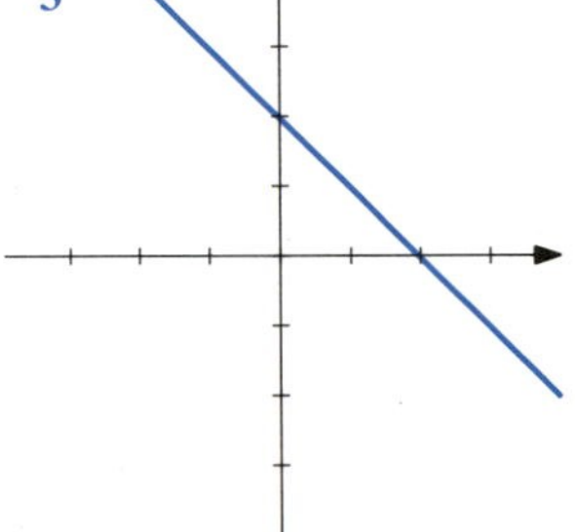

**5**

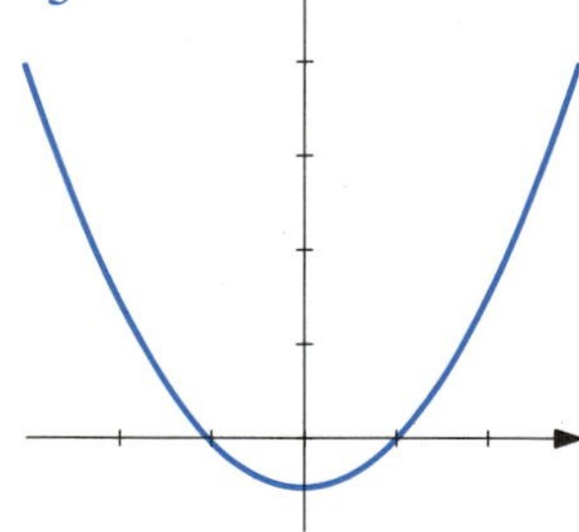

**7**

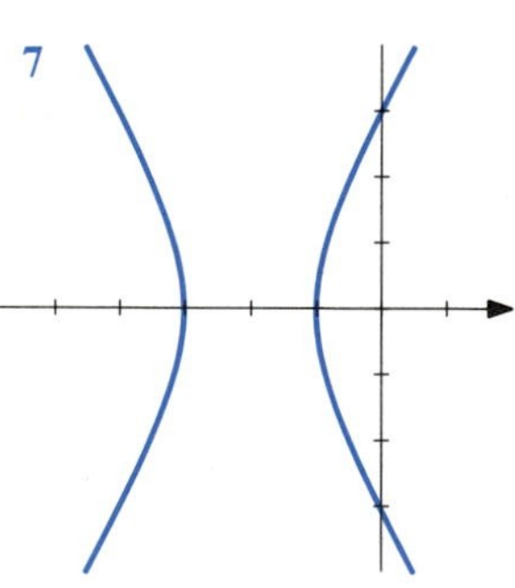

**9**

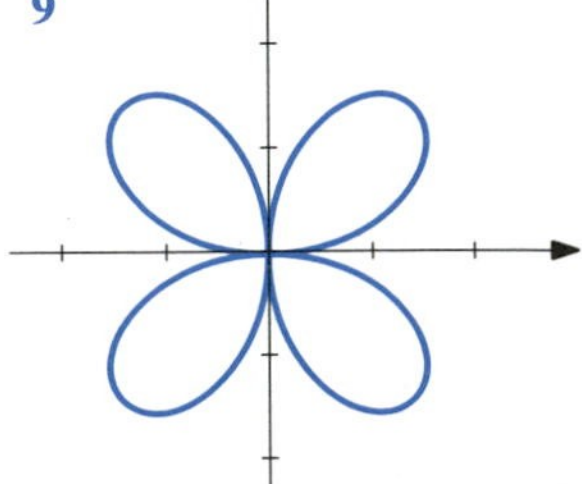

**11**

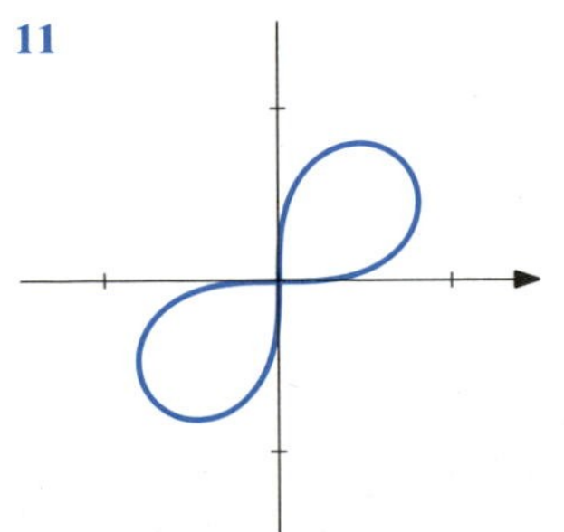

**13**

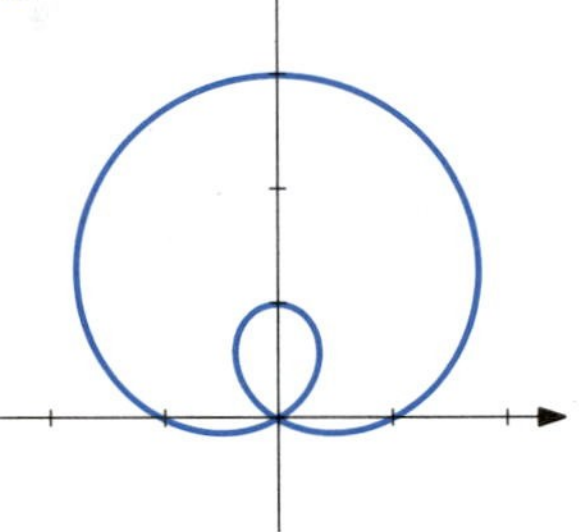

**15**

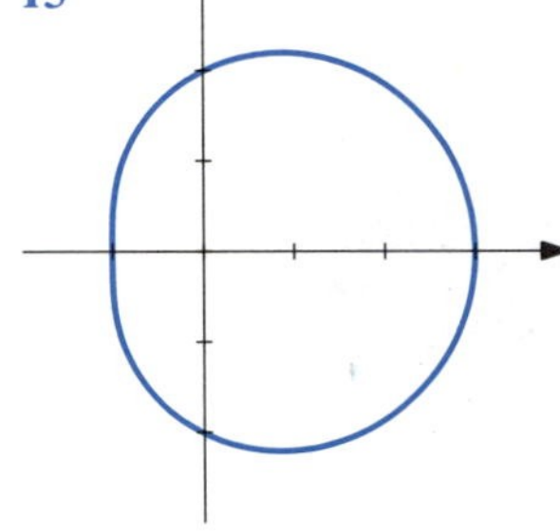

**17** $\tan^{-1}(2/3)$

**19** $\pi/6$

**21** $\displaystyle\int_0^{\pi}(2-2\cos\theta)^2\,d\theta$

**23** $\displaystyle\int_{2\pi/3}^{\pi}(4\cos^2\theta-8\cos\theta-5)\,d\theta$

**25** $\displaystyle\int_0^{2\pi}\sqrt{5-4\sin\theta}\,d\theta$

**27** $\displaystyle\int_0^{\pi/4}4\pi\sin\theta\,d\theta$

## CHAPTER THIRTEEN

### Exercises 13.1

**1** $B$

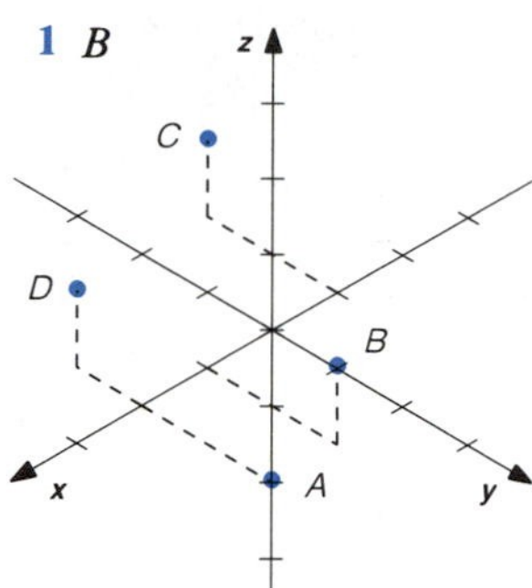


**3** 6, $(-1, 1, -1)$

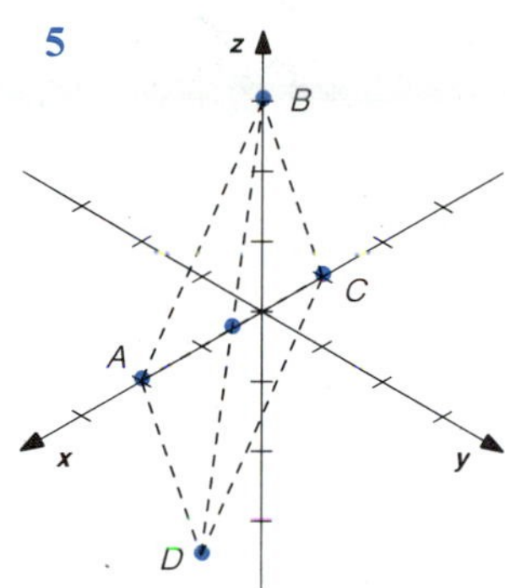


**5**

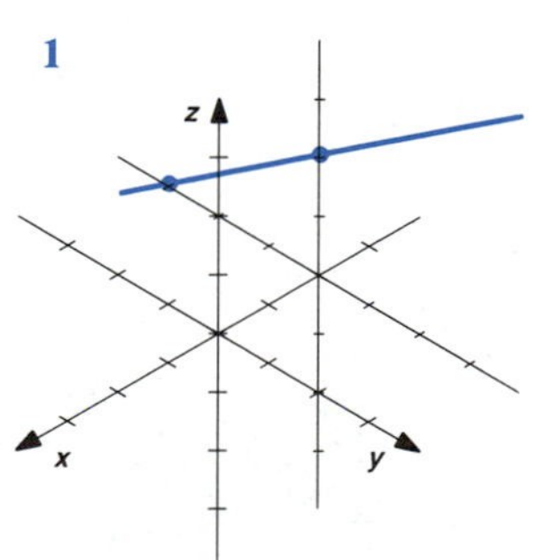


**7** 3

**9** $(x - 1)^2 + (y - 2)^2 + (z - 1)^2 = (\sqrt{6})^2$

**11** $(x - 4)^2 + (y - 3)^2 + (z - 12)^2 = 13^2$

**13** $(x - 1)^2 + (y + 2)^2 + (z - 1)^2 = (\sqrt{5})^2$

**15** $(1, -2, 0)$, $\sqrt{5}$

**17** $\sqrt{13}$

**19** 3

### Exercises 13.2

**1**

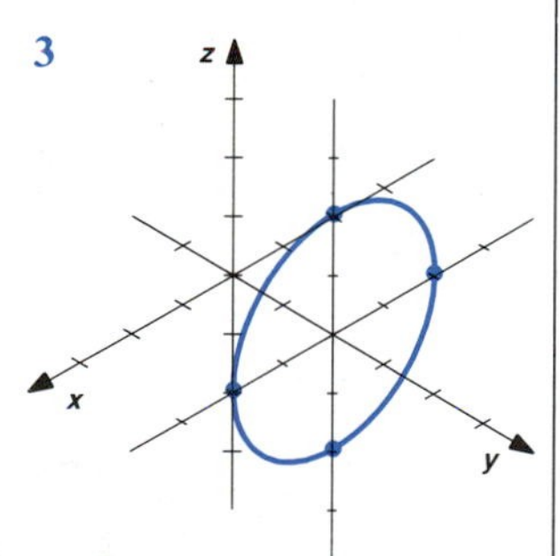


**3**

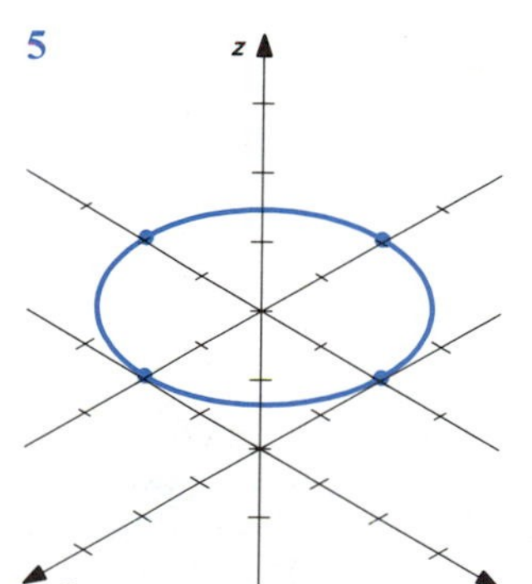


**5**

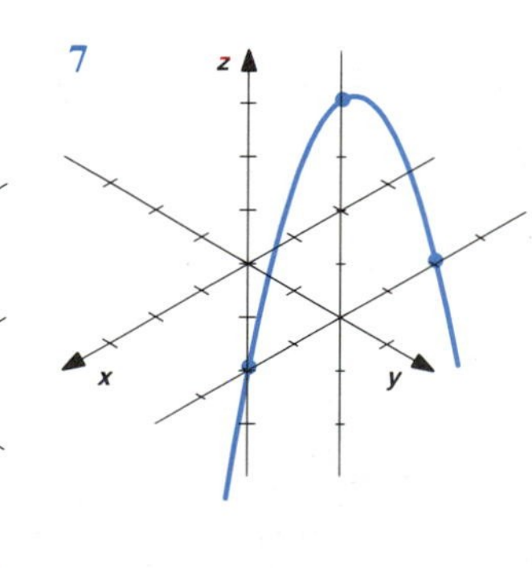


**7**

9

11

13

15

17

19

21

23

25

27

29

*Exercises 13.3*

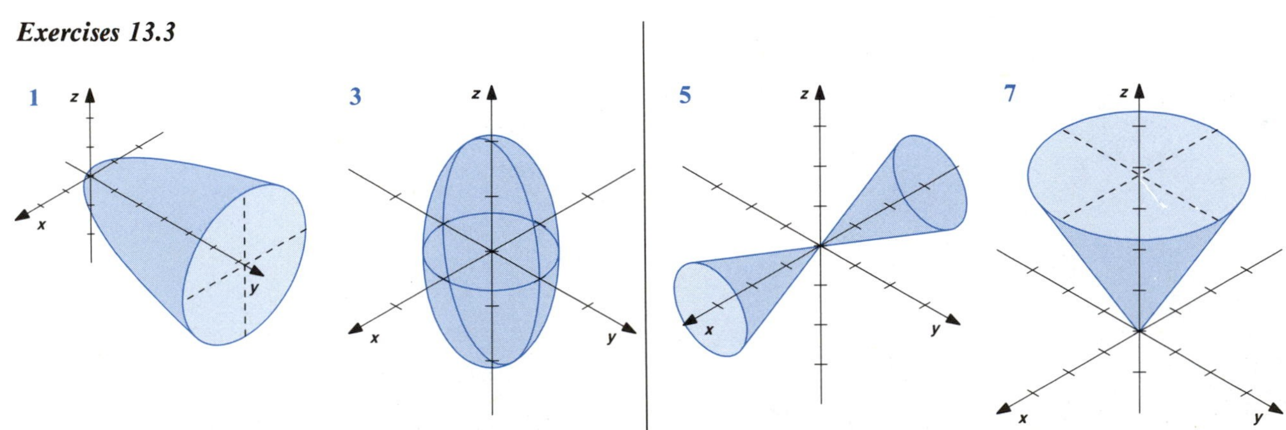

9
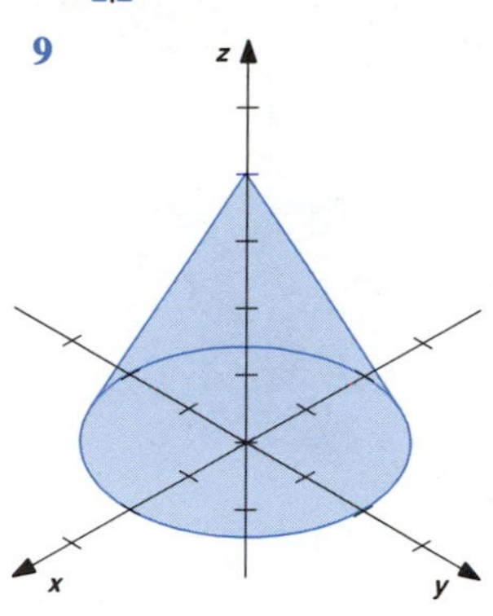


11
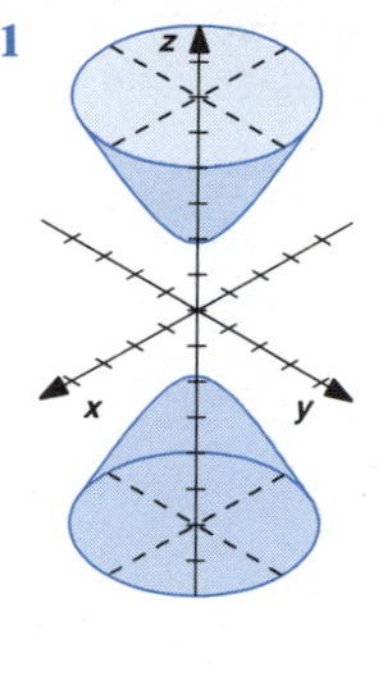


13
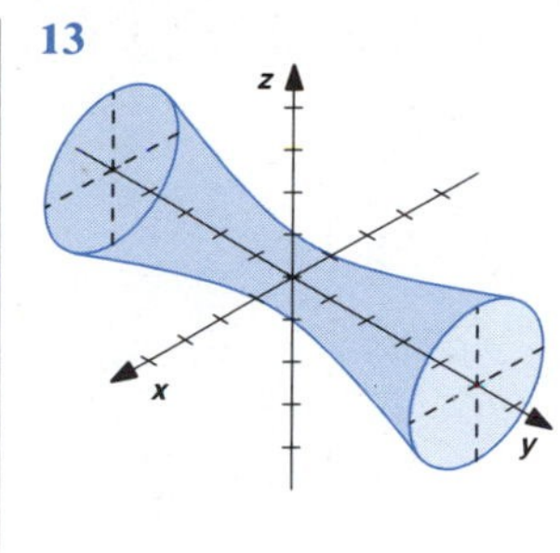


15
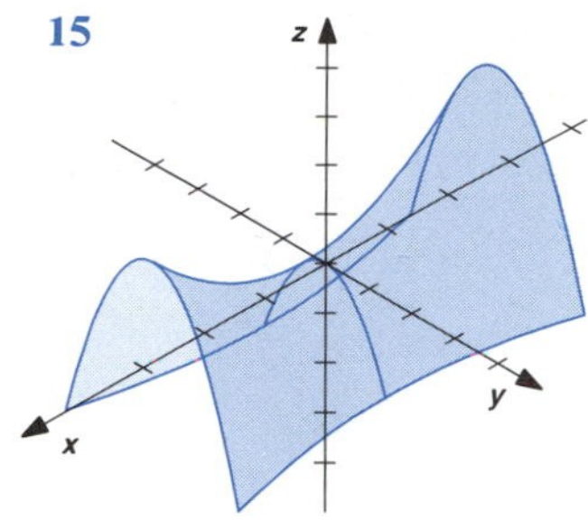


## *Review Exercises*

1 6, (1, 0, 2)

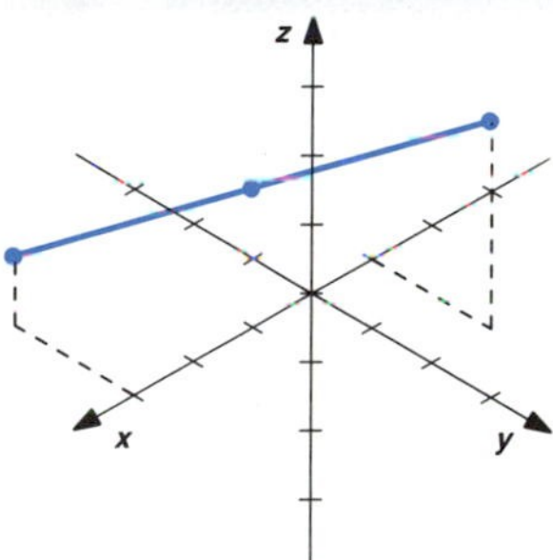


3 $(x-2)^2+y^2+(z-1)^2=3^2$

5 $(-2, 3, 0), \sqrt{13}$

7
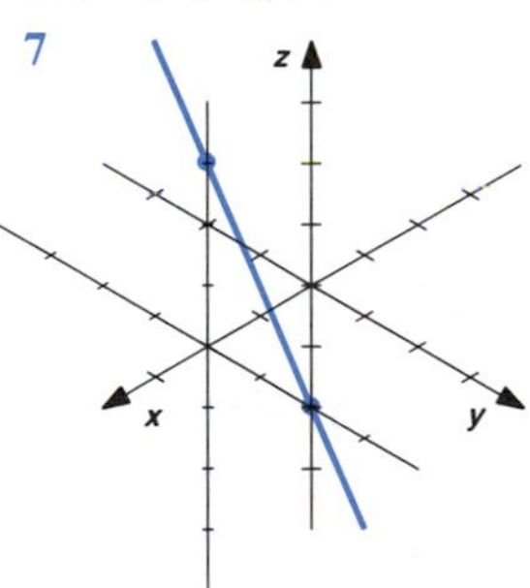


9
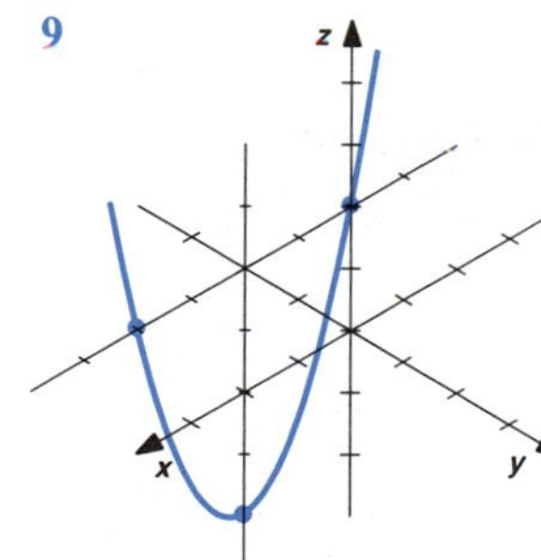


11
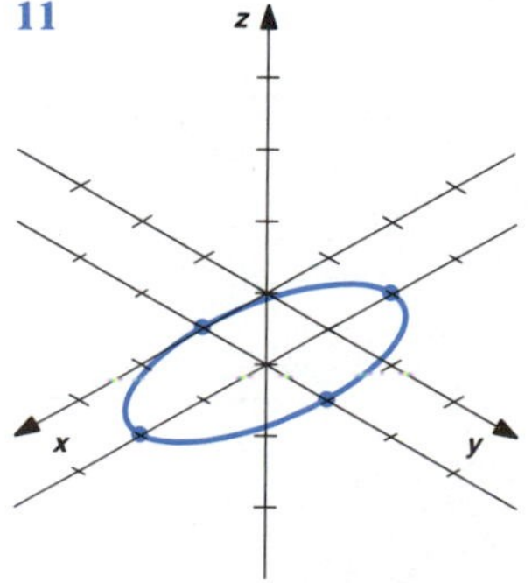


13
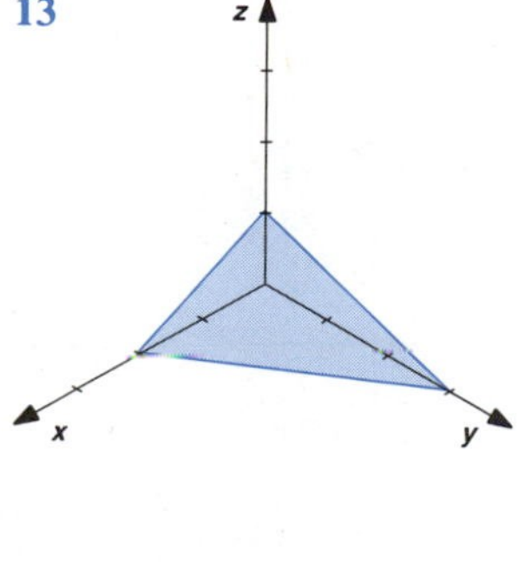


15
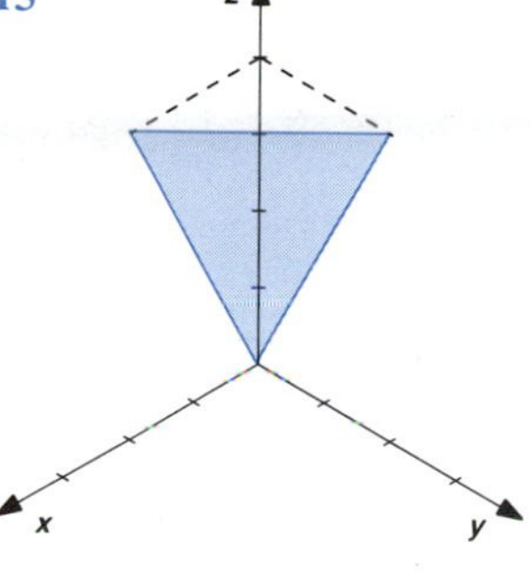


17
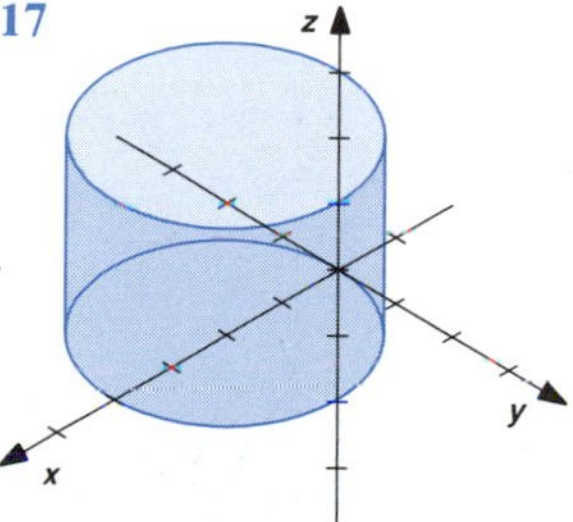


19
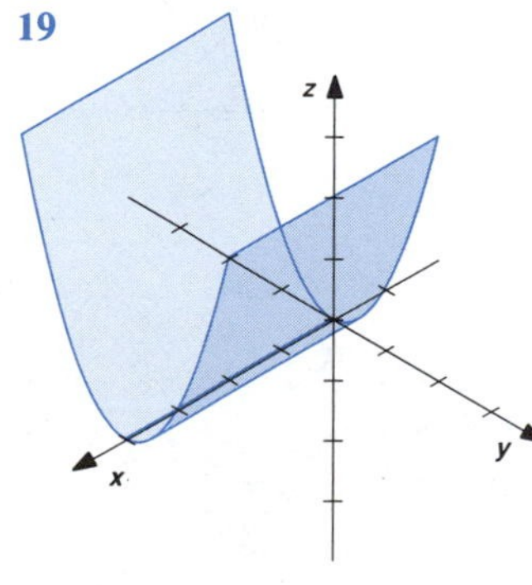

**21**

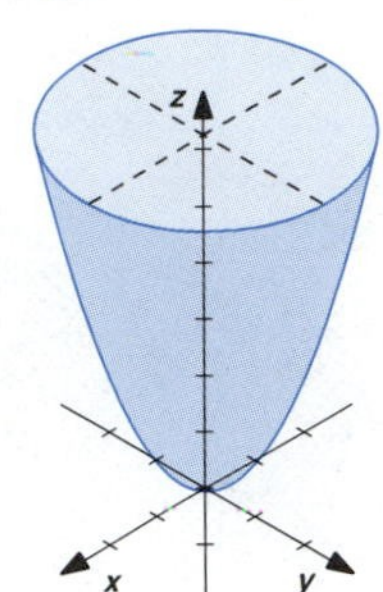


**23**

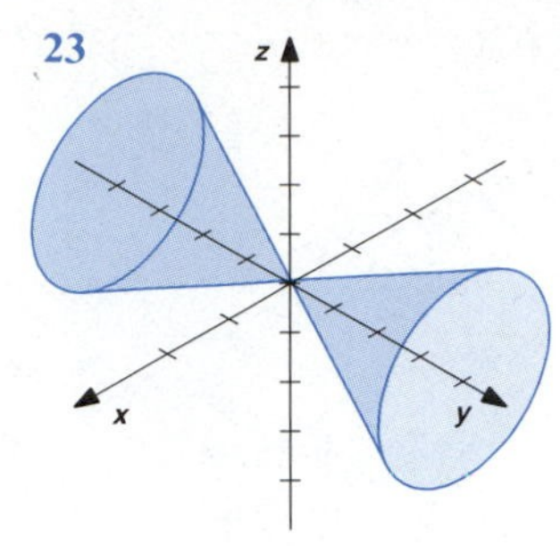


**25** **27**

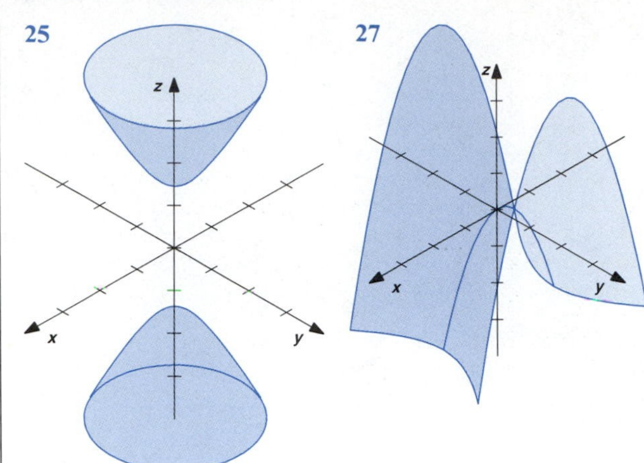


## CHAPTER FOURTEEN

### *Exercises 14.1*

**1** $-2\mathbf{i} + 3\mathbf{j}$

**3** $-4\mathbf{i} + 13\mathbf{j}$

**5** 13

**7** $\sqrt{13}$

**9** 8

**11** 1

**13** $2\mathbf{i}$, $2\mathbf{j}$, $2\mathbf{i} + 2\mathbf{j}$, $-\mathbf{i} + \mathbf{j}$

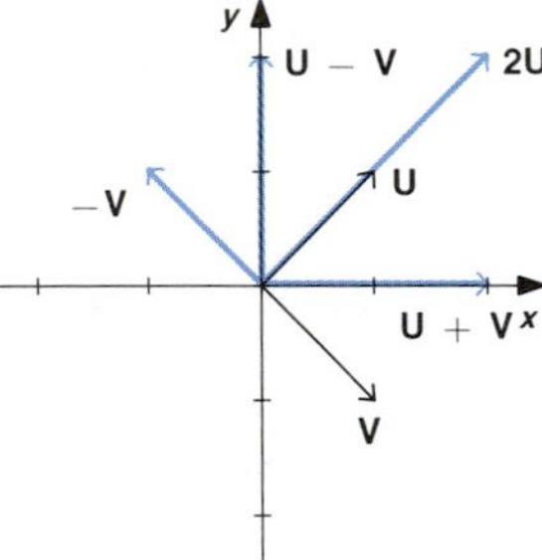


**15** $3\mathbf{i} - \mathbf{j}$, $\mathbf{i} - 5\mathbf{j}$, $4\mathbf{i} - 6\mathbf{j}$, $-\mathbf{i} - 2\mathbf{j}$

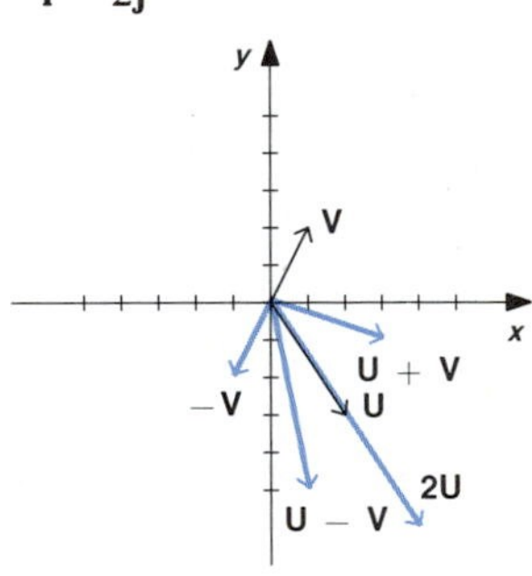


**17** $\pi/2$

**19** 0

**21** $\pi$

**23** $\cos^{-1}[7/(5\sqrt{2})]$

**25** $\pm[(12/13)\mathbf{i} + (5/13)\mathbf{j}]$

**27** $\pm\mathbf{j}$

**29** $(6/25)\mathbf{i} + (8/25)\mathbf{j}$, $(6/25)\mathbf{i} + (8/25)\mathbf{j}$, $(44/25)\mathbf{i} - (33/25)\mathbf{j}$

**31** $(1/2)\mathbf{i} - (1/2)\mathbf{j}$, $(1/2)\mathbf{i} - (1/2)\mathbf{j}$, $(5/2)\mathbf{i} + (5/2)\mathbf{j}$

**33** $\mathbf{i} + \mathbf{j}$

**35** $-3\mathbf{i} - 3\mathbf{j}$

**37** $(-1/5)\mathbf{i} + (2/5)\mathbf{j}$, $[(-1/5)\mathbf{i} + (2/5)\mathbf{j}] + [(6/5)\mathbf{i} + (3/5)\mathbf{j}]$

**39** $720\sqrt{3}$ ft-lb

**41** $4\sqrt{2}$ lb

**43** 14.48°, 38.7 s

### *Exercises 14.2*

**1** $-6\mathbf{i} + 6\mathbf{j} - 2\mathbf{k}$

**3** $2\mathbf{j} - 7\mathbf{k}$

**5** $\sqrt{50}$

**7** 3

**9** 4

**11** 3

**13** $3\mathbf{j} - 2\mathbf{k}$

**15** $9\mathbf{i} - 8\mathbf{j} - \mathbf{k}$

**17** $-\mathbf{j}$

**19** $-1$

**21** $\cos^{-1}(2/\sqrt{6})$, $\mathbf{i} + \mathbf{j}$

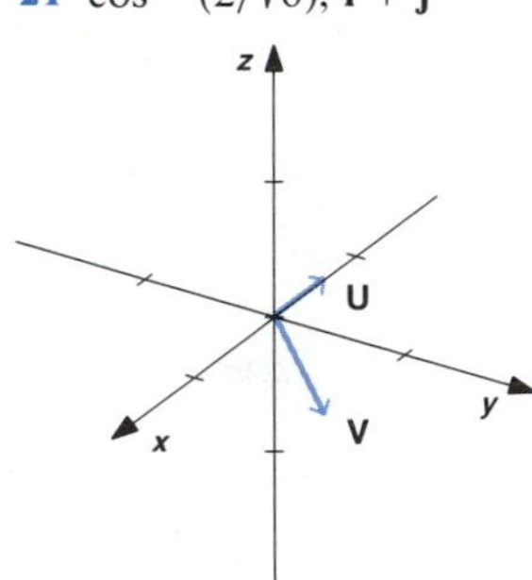


**23** $\cos^{-1}[6/(5\sqrt{13})]$, $(18/25)\mathbf{i} + (24/25)\mathbf{k}$

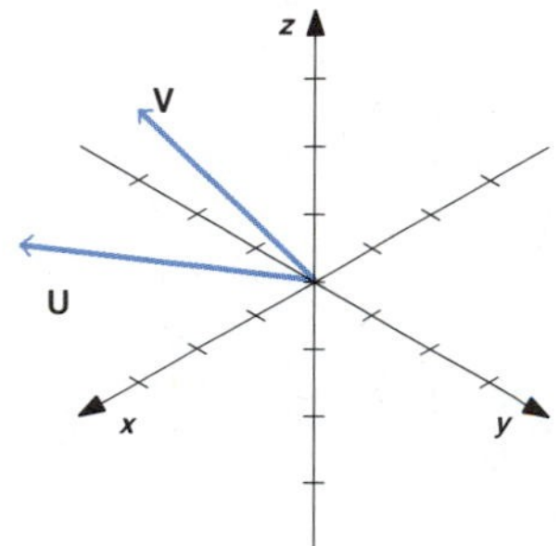

25 $4/\sqrt{41}, 0, -5/\sqrt{41}$

27 $3/\sqrt{22}, 2/\sqrt{22}, 3/\sqrt{22}$

29 $\pm(\mathbf{i} - \mathbf{j} + \mathbf{k})/\sqrt{3}$

31 $\pm(3\mathbf{i} - 3\mathbf{j} - 4\mathbf{k})/\sqrt{34}$

33 $\sqrt{2}$

35 $\sqrt{12}$

37 2

39 12

41 $\sqrt{45}/2$

43 $2\sqrt{34}$

45 $[(2/3)\mathbf{i} + (4/3)\mathbf{j} + (4/3)\mathbf{k}] + [(-2/3)\mathbf{i} + (-4/3)\mathbf{j} + (5/3)\mathbf{k}]$

47 $[2\mathbf{i}] + [-\mathbf{j} + \mathbf{k}]$

## *Exercises 14.3*

1 $4x + 3y - 2z = 15$

3 $x - 3y + 2z = 4$

5 $z = 3$

7 $z = x$

9 $14x - 5y - 3z = 11$

11 $2x - 3y - z = 3$

13 $\pi/2$

15 $\cos^{-1}(8/\sqrt{66})$

17 $6/\sqrt{17}$, (12/17, 18/17, 12/17)

19 $\sqrt{2}$, (0, 2, 2)

21 $4/\sqrt{14}$

23 $3x + y = 3$

25 $x + 3y = 4$

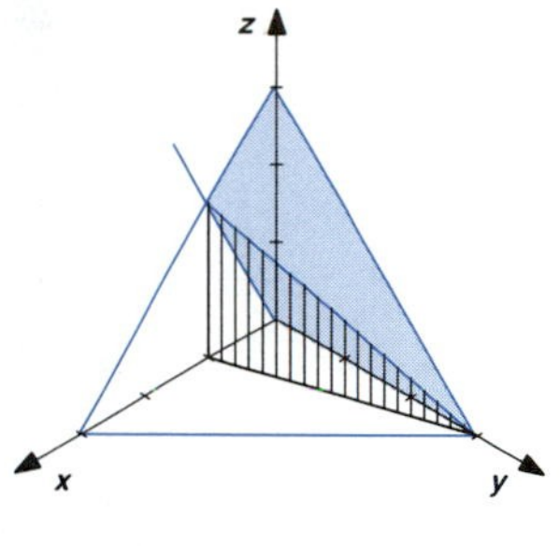


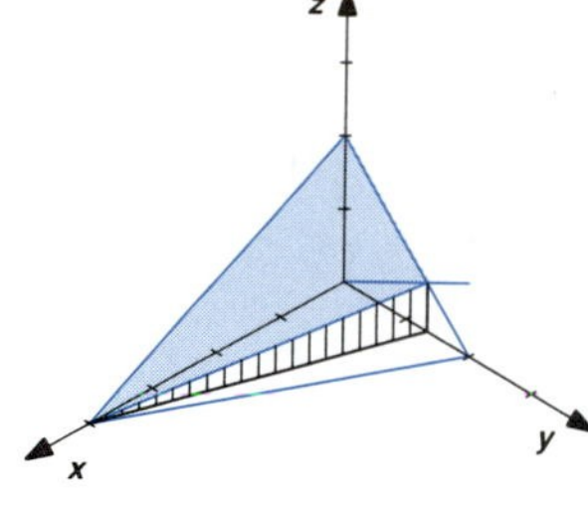


27 $3x + 2y = 3$

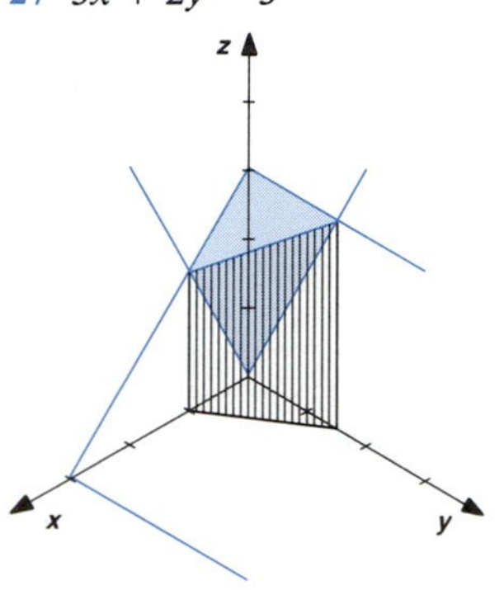


29 $x^2 + y^2 + 3x + 3y = 4$

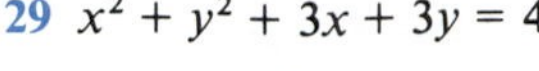

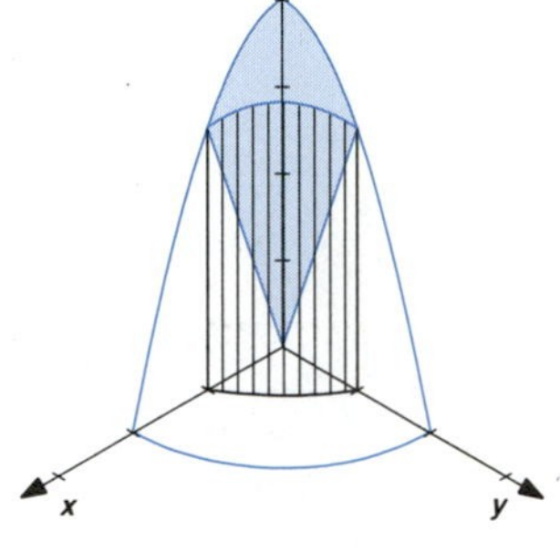


31 $x^2 + y^2 = 4$

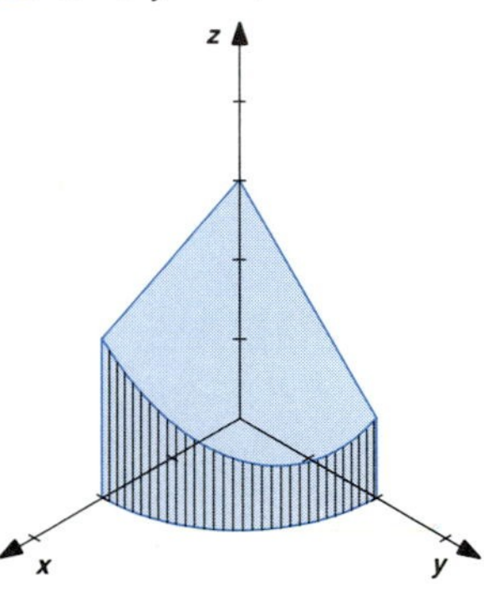


## *Exercises 14.4*

1 $\dfrac{x-1}{2} = \dfrac{y+1}{-2} = \dfrac{z-2}{3}$;
$x = 1 + 2t, y = -1 - 2t, z = 2 + 3t$

3 $\dfrac{x-2}{2} = \dfrac{y-3}{-1} = \dfrac{z+1}{3}$;
$x = 2 + 2t, y = 3 - t, z = -1 + 3t$

5 $y = 4, z = -3; x = 2 + t, y = 4, z = -3$

7 $\dfrac{x-1}{2} = \dfrac{y+2}{-3} = \dfrac{z-3}{1}$;
$x = 1 + 2t, y = -2 - 3t, z = 3 + t$

9 $\dfrac{x-1}{1} = \dfrac{y-1}{1} = \dfrac{z-3}{-2}$; $x = 1 + t, y = 1 + t, z = 3 - 2t$

11 $\dfrac{x-2}{3} = \dfrac{y-4}{-1} = \dfrac{z-3}{4}$;
$x = 2 + 3t, y = 4 - t, z = 3 + 4t$

13 $\dfrac{x}{1} = \dfrac{y-2}{-1}$, $z = 2$; $x = t, y = 2 - t, z = 2$

15 $(-7, -7, 3), \sin^{-1}[2/(3\sqrt{22})]$

17 $(-1, 1, 1), \sin^{-1}[3/(\sqrt{51})]$

19 $(0, 1, 2), \cos^{-1}(4/9)$

21 The lines do not intersect.

23 $\sqrt{117/29}$

25 $\sqrt{8/3}$

27 $4x - 3y + 2z = 5$

29 $4\mathbf{i} - 3\mathbf{j} - 5\mathbf{k}, 7\sqrt{2}/5$

31 There is no such plane.

33 $7x + 6y - 4z = 7$

35 $5x - 3y - 4z = 6$

37 $\sin^{-1}[86/(10\sqrt{110})]$

## Exercises 14.5

1 $4\mathbf{i} - 6\mathbf{j}$

3 $6\mathbf{i} + 2\mathbf{j}$

5 $2t\mathbf{i} - 3\mathbf{j}$

7 $\mathbf{i} + (-e^{-t}\sin t + e^{-t}\cos t)\mathbf{j} + (-e^{-t}\cos t - e^{-t}\sin t)\mathbf{k}$

9 $\sec^2 t\mathbf{i} + \sec t \tan t\mathbf{j}$

11 $t^2\mathbf{i} + 2t^3\mathbf{j} + \mathbf{C}$

13 $-\cos t\mathbf{i} + \sin t\mathbf{j} + \mathbf{C}$

15 $\frac{1}{2}\tan 2t\mathbf{i} + \frac{1}{2}\sec 2t\mathbf{j} + \mathbf{C}$

17 $\frac{1}{2}\ln(t^2+4)\mathbf{i} + \frac{1}{2}\tan^{-1}\frac{t}{2}\mathbf{j} + \mathbf{C}$

19 $\mathbf{i} + 2t\mathbf{j}$, $2\mathbf{j}$, $\sqrt{1+4t^2}$

21 $-2\sin 2t\mathbf{i} + 2\cos 2t\mathbf{j}$, $-4\cos 2t\mathbf{i} - 4\sin 2t\mathbf{j}$, 2

23 $t\mathbf{i} + (2t - 16t^2)\mathbf{j}$

25 $4t\mathbf{i} + (2t^3 + 3)\mathbf{j}$

27 $96\sqrt{3} - 144$ ft, 48 ft

29 2s, 2s

31 $\sqrt{7680}$ ft/s

## Review Exercises

1 (a) $12\mathbf{j}$ (b) $\sqrt{40}$ (c) $\sqrt{10}$ (d) $-16$ (e) $\cos^{-1}(-4/5)$ (f) $(\mathbf{i} - 3\mathbf{j})/\sqrt{10}$ (g) $(-8/5)\mathbf{i} + (24/5)\mathbf{j}$

3 $(69/25)\mathbf{i} + (92/25)\mathbf{j}$

5 3

7 $\sqrt{2}/2$

9 $3x - 4y - 5z + 6 = 0$

11 $3x + 4y - 5z + 10 = 0$

13 $\frac{x-7}{3} = \frac{y-5}{4} = \frac{z-3}{1}$; $x = 7 + 3t$, $y = 5 + 4t$, $z = 3 + t$

15 $\frac{x-1}{2} = \frac{y+1}{-1} = \frac{z-2}{3}$; $x = 1 + 2t$, $y = -1 - t$, $z = 2 + 3t$

17 $\cos^{-1}(4/9)$

19 $(-3, 9/2, -3/2)$, $\sin^{-1}(2/7)$

21 $5/\sqrt{14}$

23 $-\sqrt{3}\mathbf{i} + (3/2)\mathbf{j}$, $\sqrt{21}/2$, $-\mathbf{i} - (3\sqrt{3}/2)\mathbf{j}$

25 $(4t^4 + 6)\mathbf{i} - 5t^3\mathbf{j}$

27 249 ft

# CHAPTER FIFTEEN

## Exercises 15.1

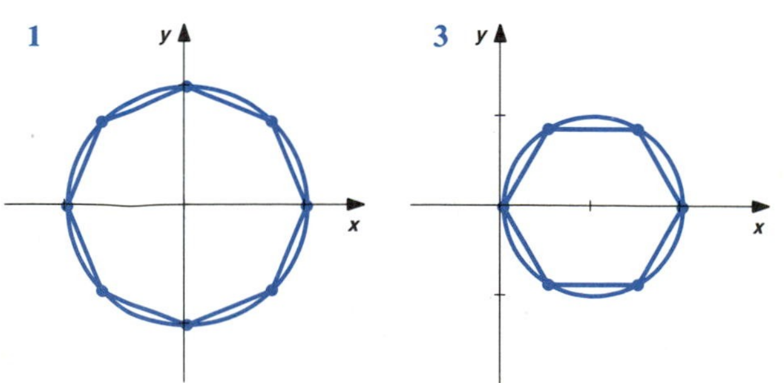


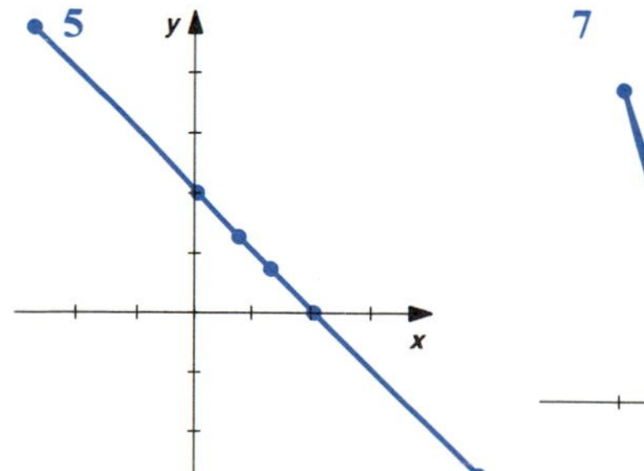


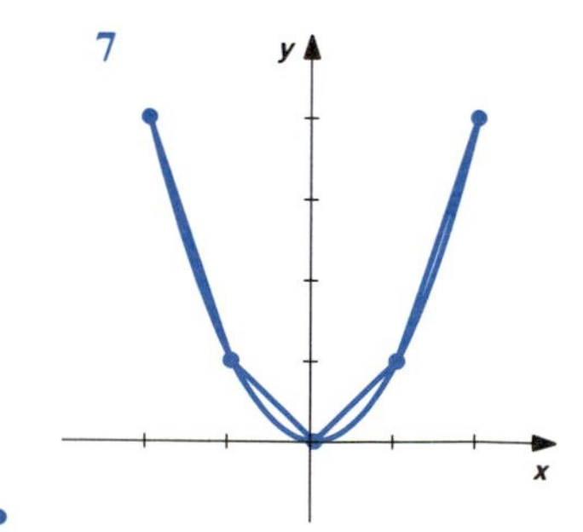

9 Line segment from $(-3, -1)$ to $(3, 3)$

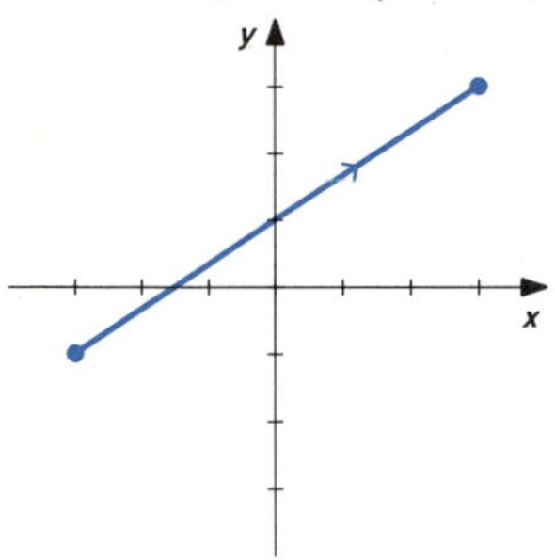

11 Along parabola $y = 2x - x^2$ from $(0, 0)$ to $(2, 0)$

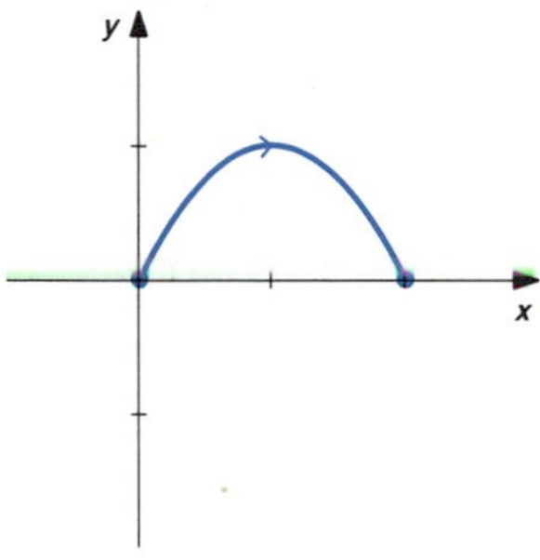

13 Along $y = \cos x$ from $(0, 1)$ to $(2\pi, 1)$

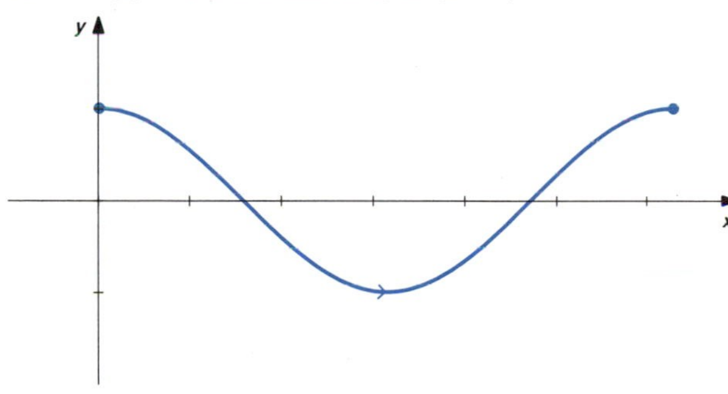

15 Along $y = \sin^{-1} x$ from $(-1, -\pi/2)$ to $(1, \pi/2)$

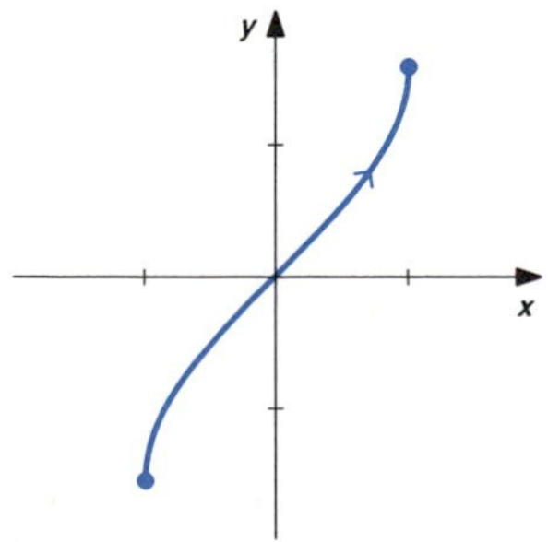

17 Counterclockwise along $x^2 + y^2 = 1$ from $(1, 0)$ to $(-1, 0)$

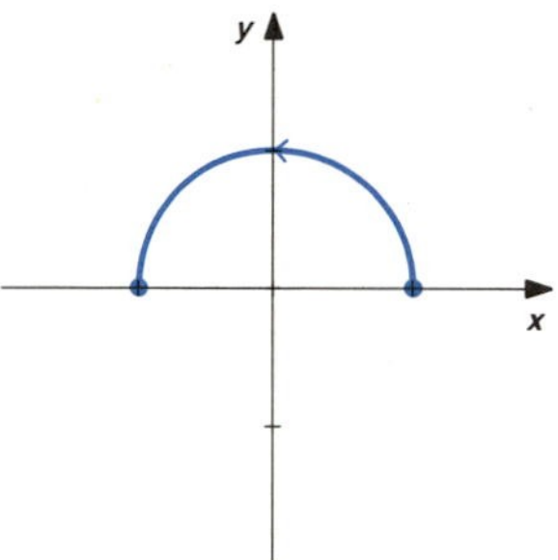

19 From $(0, -1)$, clockwise once around $x^2 + y^2 = 1$, excluding $(0, -1)$

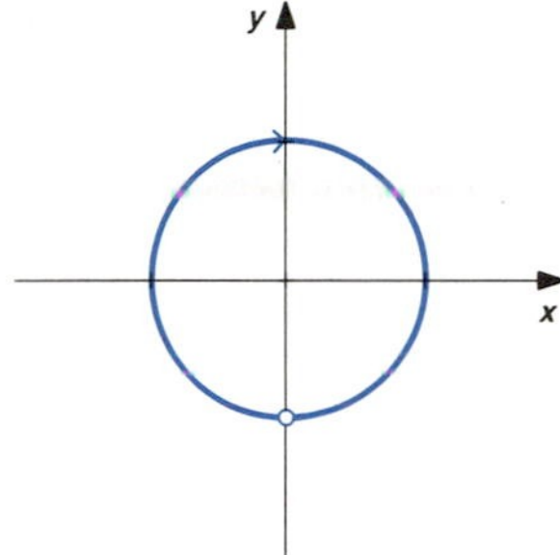

21 $x = t,\ y = t^3,\ -1 \le t \le 2$

23 $x = t^2,\ y = t,\ -1 \le t \le 2$

25 $x = 3t,\ y = 2t,\ 0 \le t \le 1$

27 $x = 2 - 4t,\ y = 1 - t,\ 0 \le t \le 1$

29 $x = 2\cos t,\ y = 2\sin t,\ 0 \le t \le 2\pi$

31 $x = 2\cos t,\ y = 3\sin t,\ 0 \le t \le 2\pi$

33 $x = 2\sec t,\ y = \tan t,\ -\pi/2 < t < \pi/2$

35 $x = 3\tan t,\ y = 2\sec t,\ -\pi/2 < t < \pi/2$

37 $x = -u,\ y = u^2,\ -2 \le u \le 2$

39 $x = \cos u,\ y = -\sin u,\ 0 \le u \le 2\pi$

41 $x = (1 - \cos t)\cos t,\ y = (1 - \cos t)\sin t,\ 0 \le t \le 2\pi$

43 $x = \dfrac{t}{2}\cos t,\ y = \dfrac{t}{2}\sin t,\ 0 \le t \le 2\pi$

45 $x = t - d\sin(t/r),\ y = r - d\cos(t/r)$

47 $x = \dfrac{1}{\sqrt{2}}\cos t - \sin t,\ y = \dfrac{1}{\sqrt{2}}\cos t + \sin t,\ 0 \le t \le 2\pi$

49 (a) $x = -12 + 12\cos\theta - (36 - 12\theta)\sin\theta$, $y = 12\sin\theta + (36 - 12\theta)\cos\theta,\ 0 \le \theta \le 3$ (rad) (b) $\int_0^3 (1/2)(36 - 12\theta)^2\, d\theta$ (c) $648\pi + 1296$

*Exercises 15.2*

**1** $3x - 2y = 1$

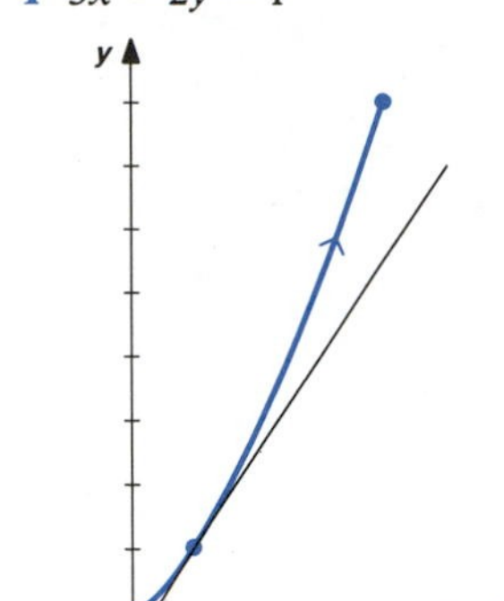


**3** $12x - 6\sqrt{3}y = 6 - \pi\sqrt{3}$

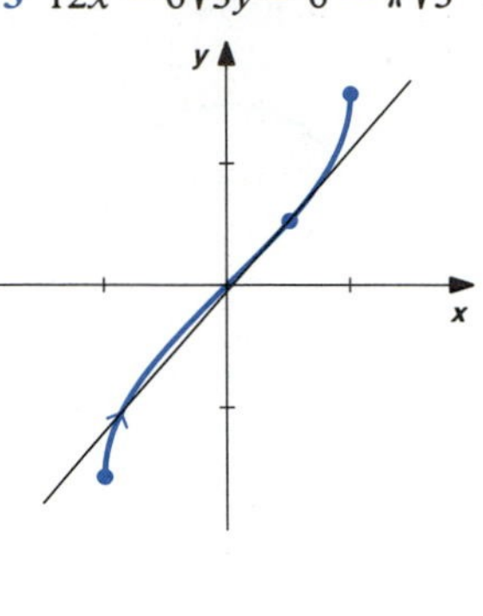


**5** 3/2, 3/4

**7** 4/3, −2/27

**9** 2, 4

**11** 5

**13** $(40^{3/2} - 8)/27$

**15** $2\pi$

**17** $8[(1 + \pi^2)^{3/2} - 1]/3$

**19** 15/4

**21** $2\pi$

**23** $15\pi$

**25** $\pi(17^{3/2} - 1)/6$

**27** $16\pi$

**29** $12\pi^2$

**31** $x = 4s/5,\ y = 3s/5,\ 0 \le s \le 5$

**33** $x = 3\cos(s/3),\ y = 3\sin(s/3),\ 0 \le s \le 6\pi$

**35** $x = \frac{2}{3}\left[\left(\frac{3s}{2}+1\right)^{2/3} - 1\right]^{3/2},\ y = \left(\frac{3s}{2}+1\right)^{2/3} - 1,$

$0 \le s \le \frac{2}{3}(10^{3/2} - 1)$

**37** $x = \left(\frac{s}{\sqrt{2}}+1\right)\cos\left[\ln\left(\frac{s}{\sqrt{2}}+1\right)\right],$

$y = \left(\frac{s}{\sqrt{2}}+1\right)\sin\left[\ln\left(\frac{s}{\sqrt{2}}+1\right)\right],$

$0 \le s \le \sqrt{2}(e^{2\pi} - 1)$

*Exercises 15.3*

**1** Simple, not closed, not smooth, piecewise smooth;
$C_1: x = -(1-u)^{3/2},\ y = 1 - u,\ 0 \le u \le 1;$
$C_2: x = t^{3/2},\ y = t,\ 0 \le t \le 1$

**3** Not simple, closed, not smooth, piecewise smooth;
$C_1: x = t,\ y = t^3,\ -1 \le t \le 1;$
$C_2: x = -u,\ y = -u,\ -1 \le u \le 1$

**5** Not simple, closed, smooth;
$C: x = \cos t,\ y = \sin t,\ 0 \le t \le 4\pi$

**7** Simple, closed, smooth;
$C: x = \cos t,\ y = \sin t,\ 0 \le t \le 2\pi$

**9** $(r, \theta) = (0, 0)$

**11** $-\sin t\mathbf{i} + \cos t\mathbf{j},\ -\cos t\mathbf{i} - \sin t\mathbf{j},\ 1/2$

**13** $\frac{t}{\sqrt{1+t^2}}\mathbf{i} + \frac{1}{\sqrt{1+t^2}}\mathbf{j},\ \frac{1}{\sqrt{1+t^2}}\mathbf{i} - \frac{t}{\sqrt{1+t^2}}\mathbf{j},\ \frac{t}{(1+t^2)^{3/2}}$

**15** $\frac{\cos t - \sin t}{\sqrt{2}}\mathbf{i} + \frac{\cos t + \sin t}{\sqrt{2}}\mathbf{j},$

$\frac{-\cos t - \sin t}{\sqrt{2}}\mathbf{i} + \frac{-\sin t + \cos t}{\sqrt{2}}\mathbf{j},\ \frac{e^{-t}}{\sqrt{2}}$

**17** $\mathbf{i},\ -\mathbf{j},\ 1,\ 1,\ (0, 0)$

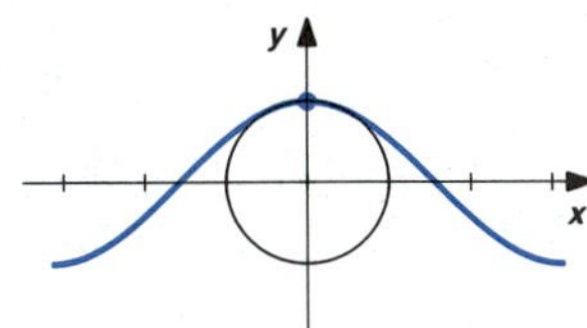

**19** **j**, −**i**, 4/9, 9/4, (7/4, 0); −**i**, −**j**, 3/16, 16/3, (0, −7/3)

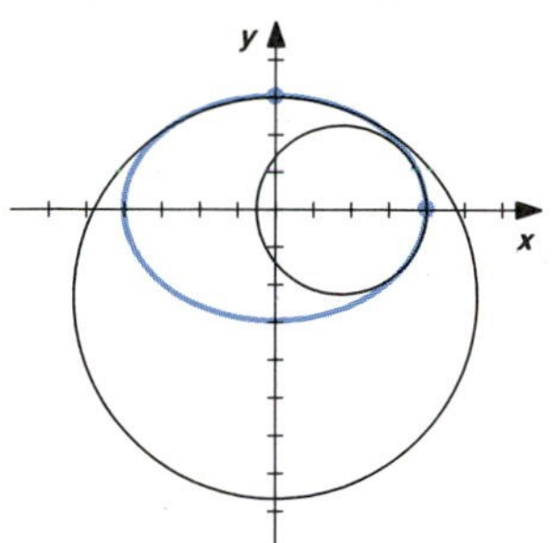


**21** $\dfrac{2x^3}{(1+x^4)^{3/2}}$

**23** $\cos x$

**25** $\dfrac{2}{(4\sin^2 t+\cos^2 t)^{3/2}}$

**27** $(\tan^2 t + \sec^2 t)^{-3/2}$

**29** $\dfrac{2}{3\sqrt{3}}, \left(\dfrac{1}{\sqrt{2}}, \ln\dfrac{1}{\sqrt{2}}\right)$

**31** 1/2, (0, 2)

**33** $\dfrac{6(1-\cos\theta)}{(5-4\cos\theta)^{3/2}}$

**35** $\dfrac{3}{2\sqrt{2-2\sin\theta}}$

**37** 1/2

**39** 1/2

## *Exercises 15.4*

**1**

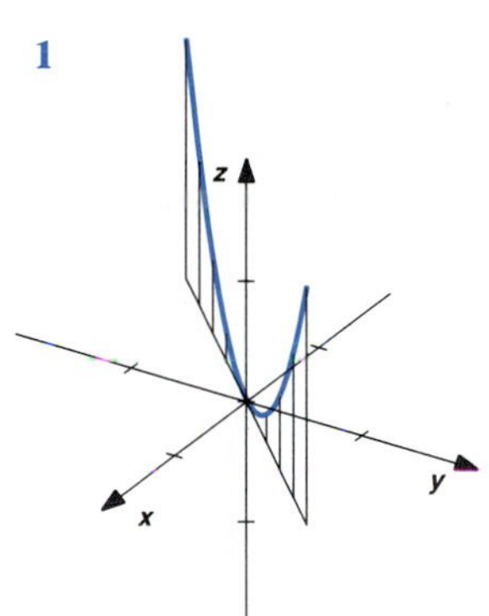


**3**

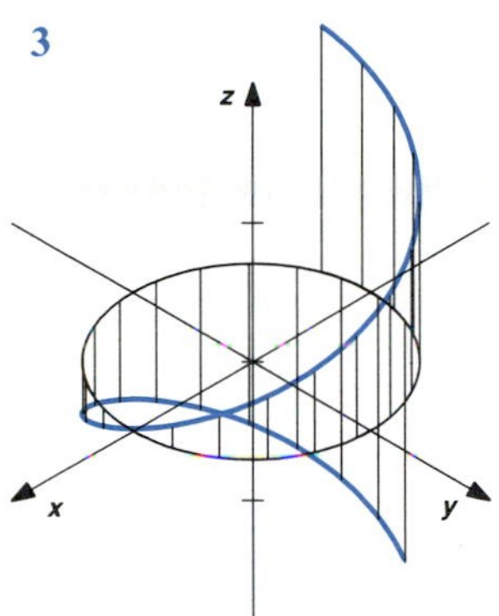


**5**

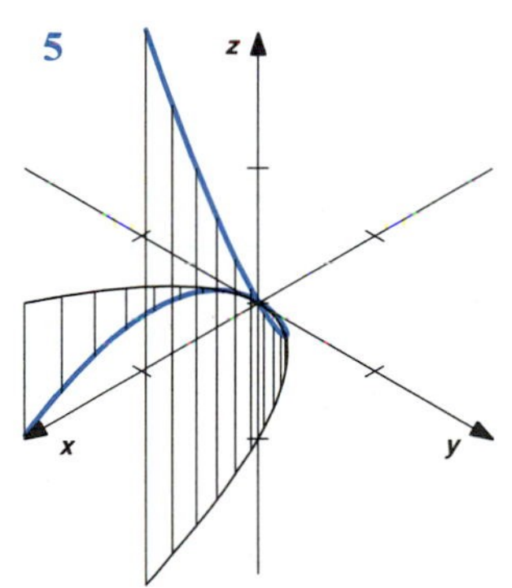


**7**

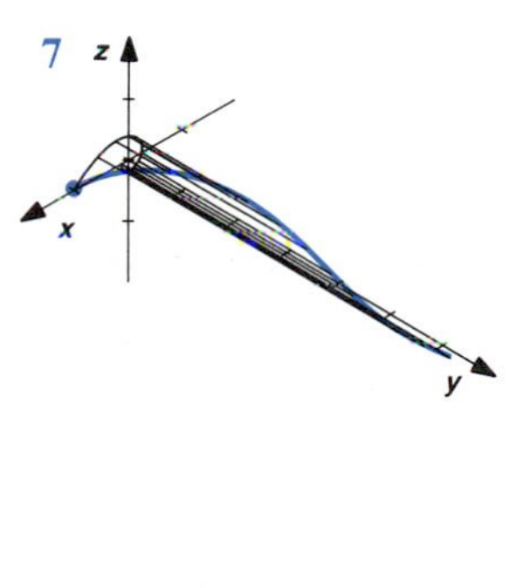


**9** $x = 2 + 3t$, $y = 1 - 3t$, $z = -1 + 4t$, $0 \le t \le 1$

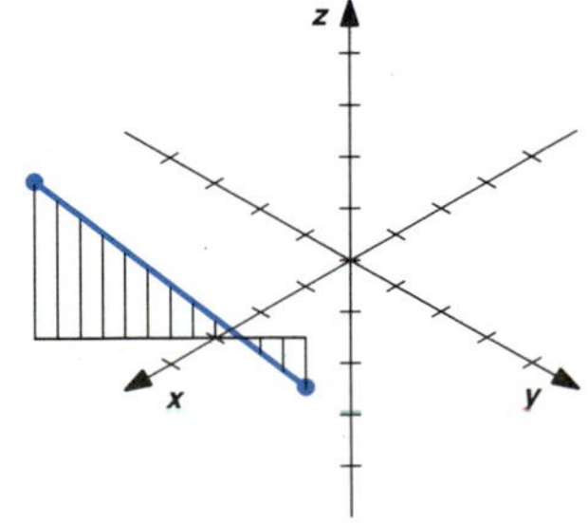


**11** $x = \cos t$,
$y = \sin t$,
$z = 1 - \cos t$,
$0 \le t \le 2\pi$

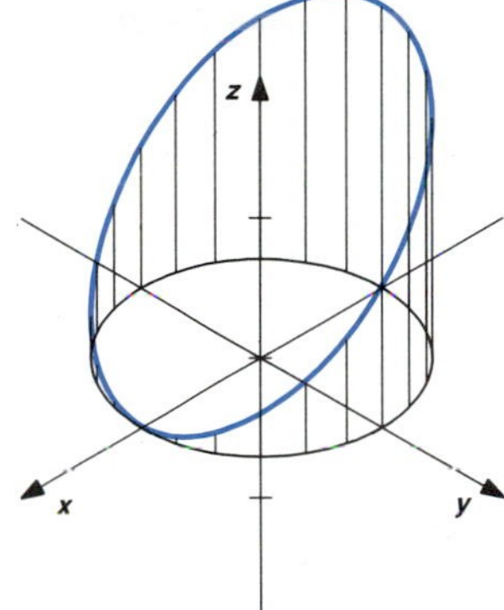


**13** $x = \sqrt{2}\cos t$,
$y = 2\sin t$,
$z = -\sqrt{2}\cos t$,
$0 \le t \le 2\pi$

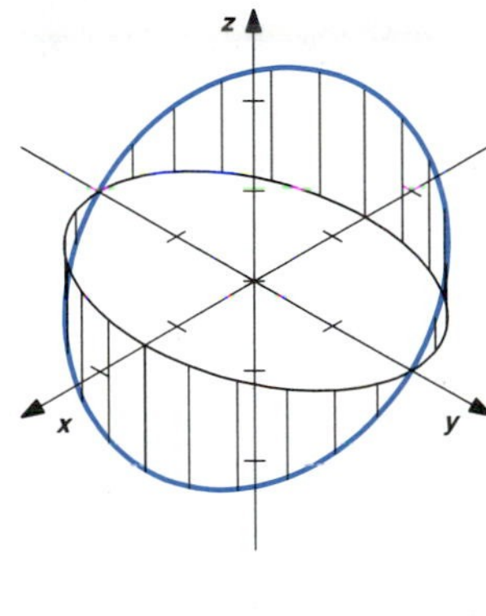


**15** $\dfrac{\cos t-\sin t}{\sqrt{3}}\mathbf{i} + \dfrac{\cos t+\sin t}{\sqrt{3}}\mathbf{j} + \dfrac{1}{\sqrt{3}}\mathbf{k}$,
$\dfrac{-\cos t-\sin t}{\sqrt{2}}\mathbf{i} + \dfrac{\cos t-\sin t}{\sqrt{2}}\mathbf{j}$,
$\dfrac{-\cos t+\sin t}{\sqrt{6}}\mathbf{i} + \dfrac{-\cos t-\sin t}{\sqrt{6}}\mathbf{j} + \dfrac{2}{\sqrt{6}}\mathbf{k}, \dfrac{\sqrt{2}e^{-t}}{3}$

**17** $\dfrac{1}{\sqrt{2}}\mathbf{i} + \dfrac{1}{\sqrt{2}}\mathbf{k}, \mathbf{j}, -\dfrac{1}{\sqrt{2}}\mathbf{i} + \dfrac{1}{\sqrt{2}}\mathbf{k}, \dfrac{1}{2}$

**19** $\dfrac{1}{\sqrt{2}}\mathbf{i} - \dfrac{1}{2}\mathbf{j} + \dfrac{1}{2}\mathbf{k}, \dfrac{1}{\sqrt{2}}\mathbf{j} + \dfrac{1}{\sqrt{2}}\mathbf{k}, -\dfrac{1}{\sqrt{2}}\mathbf{i} - \dfrac{1}{2}\mathbf{j} + \dfrac{1}{2}\mathbf{k}, \dfrac{1}{2\sqrt{2}}$

**21** 0, 12

**23** $\dfrac{4t}{\sqrt{1+4t^2}}, \dfrac{2}{\sqrt{1+4t^2}}$

**25** 0, 1

**27** $-2/\sqrt{3}$, $\sqrt{8}/\sqrt{3}$

**29** $\sqrt{19}/(7\sqrt{14})$

**31** $3x - y - 3z + 1 = 0$

## Review Exercises

**1** $x^2 + y^2 = 1$

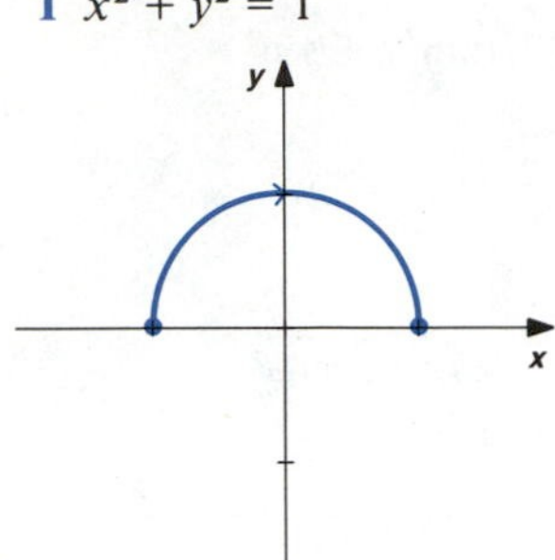


**3** $2x + y = 5$

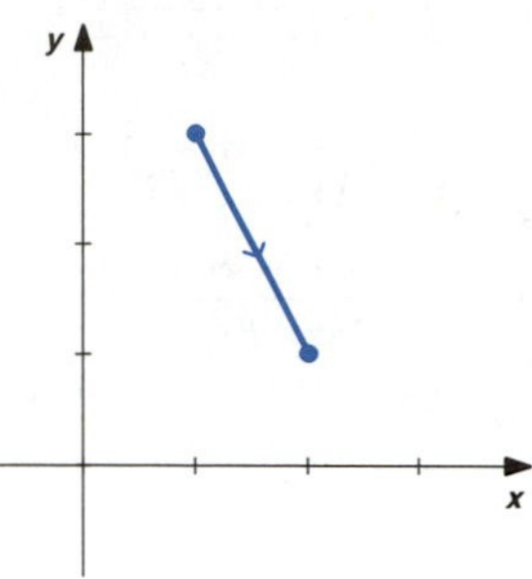


**5**

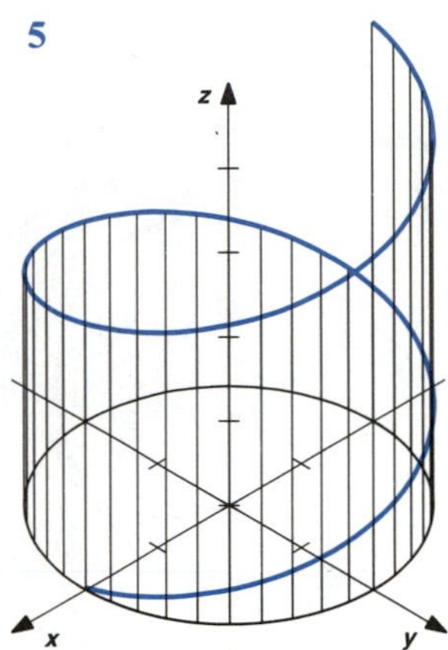


**7** $x = 2 - 3t,\ y = -1 + 4t,\ z = 3 - 2t,\ 0 \le t \le 1$

**9** $x = (2 - u)^2,\ y = 2 - u,\ 0 \le u \le 2$

**11** $x = (1 + \cos t)\cos t,\ y = (1 + \cos t)\sin t,\ -\pi \le t \le \pi$

**13** $x = e + 2et,\ y = e + et$

**15** $-2, 2$

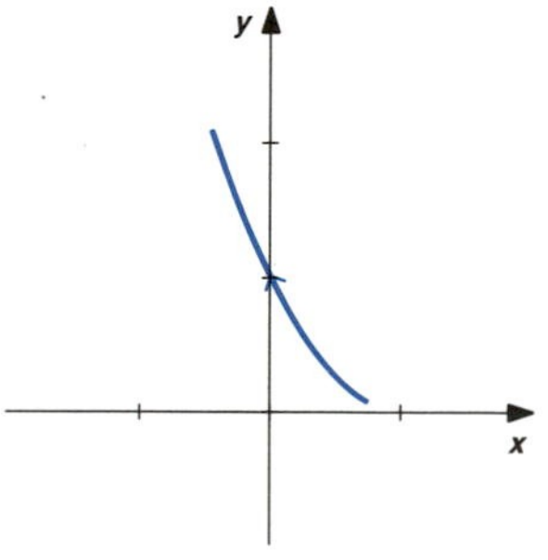


**17** $\dfrac{13^{3/2} - 8}{27}$

**19** $\displaystyle\int_0^{2\pi} 2\pi(1 - \cos t)\sqrt{2 + 2\cos t}\, dt$

**21** $x = 1 - \dfrac{3s}{5},\ y = 3 + \dfrac{4s}{5},\ 0 \le s \le 5$

**23** $x = \dfrac{1}{3}[(3s + 1)^{2/3} - 1]^{3/2},\ y = \dfrac{1}{2}[(3s + 1)^{2/3} - 1],$
$0 \le s \le \dfrac{5^{3/2} - 1}{3}$

**25** Not simple, not closed, not smooth, piecewise smooth;
$C_1$: $x = t,\ y = t^2,\ 0 \le t \le 1$;
$C_2$: $x = -u,\ y = -u,\ -1 \le u \le 1$

**27** **j**, −**i**, 2/9, 9/2, (−5/2, 0)

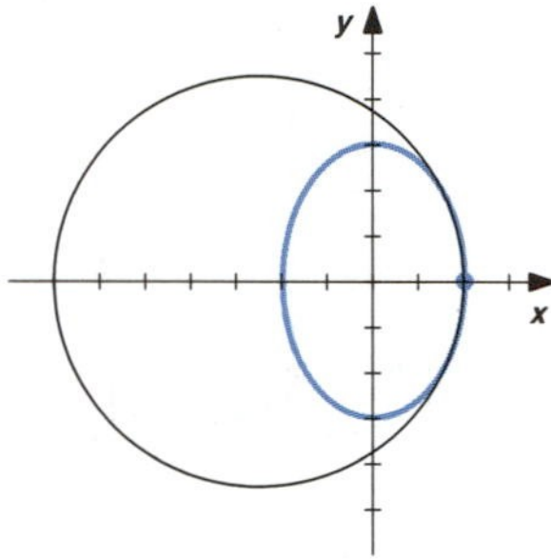


**29** $\dfrac{1}{2}\mathbf{i} + \dfrac{\sqrt{3}}{2}\mathbf{j},\ -\dfrac{\sqrt{3}}{2}\mathbf{i} + \dfrac{1}{2}\mathbf{j},\ \dfrac{1}{16\sqrt{3}}$

**31** $\dfrac{2}{|t|(4 + t^2)^{3/2}}$

**33** $\dfrac{|\cos x|}{(1 + \sin^2 x)^{3/2}}$

**35** $\dfrac{8t}{\sqrt{1 + 8t^2}},\ \dfrac{2\sqrt{2}}{\sqrt{1 + 8t^2}},\ \dfrac{2\sqrt{2}}{(1 + 8t^2)^{3/2}}$

## CHAPTER SIXTEEN

### Exercises 16.1

**1** $x \neq 0$; $0, -1/\pi$

**3** $(x, y) \neq (0, 0)$; $1, 1/2, 1/a$

**5** $x^2 + y^2 \leq 9$; 2, 2

**7** Lines

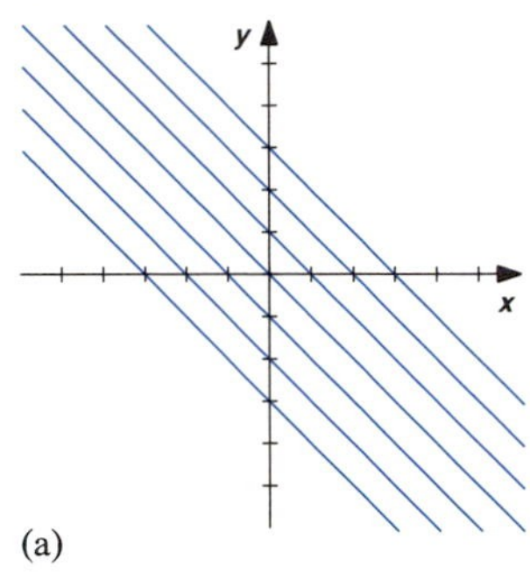

(a)

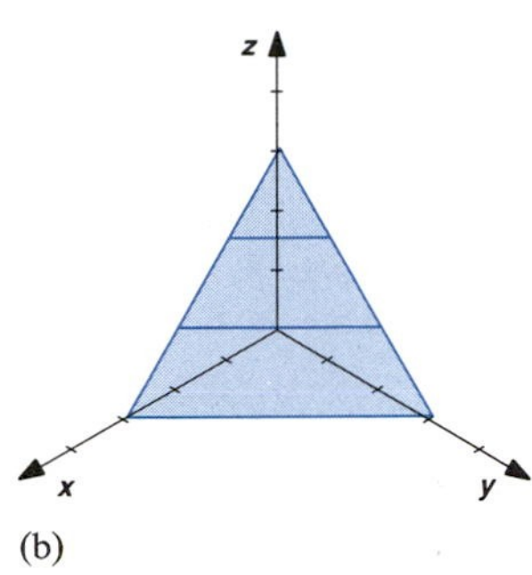

(b)

**9** Circles

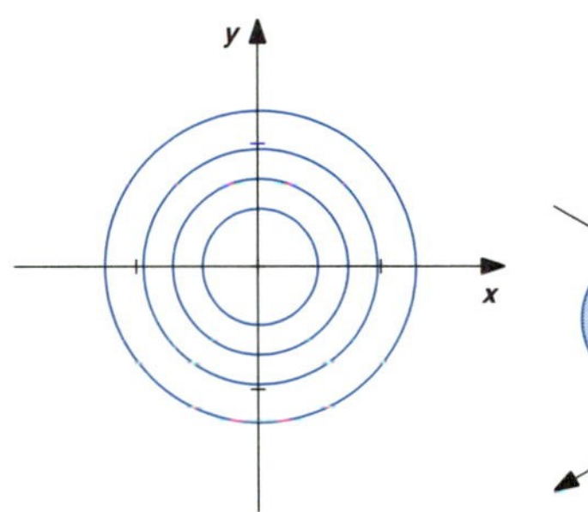

(a)

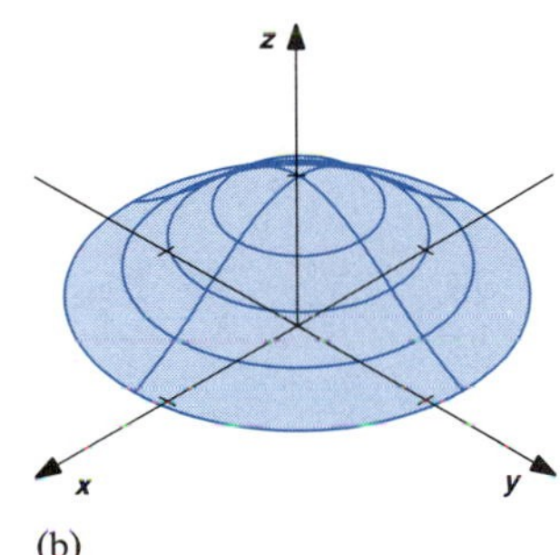

(b)

**11** Circles

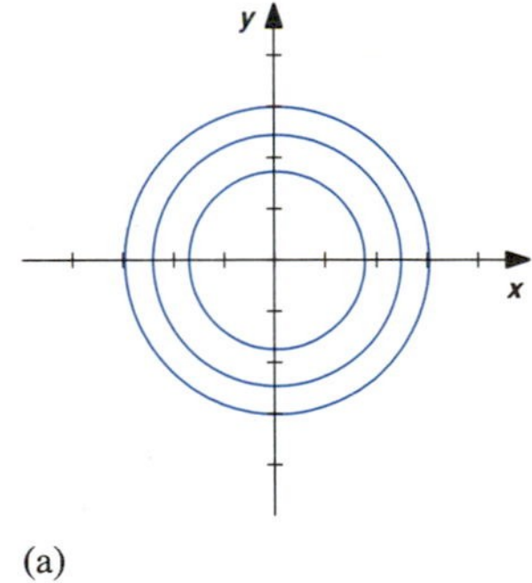

(a)

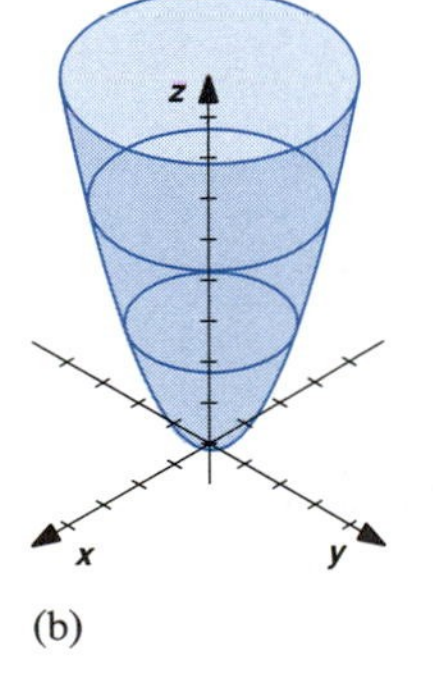

(b)

**13** Hyperbolas

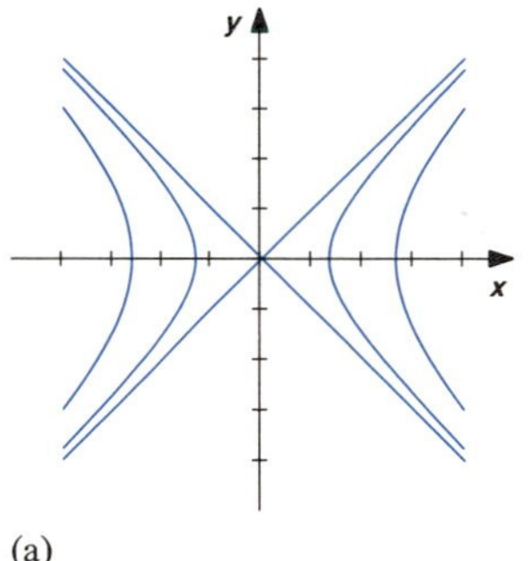

(a)

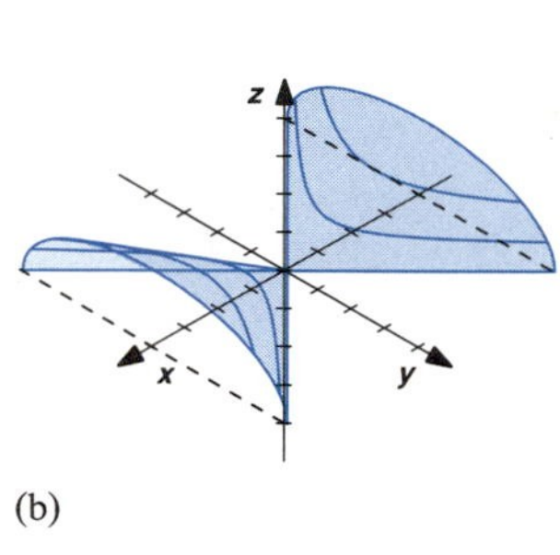

(b)

**15** Planes

**17** Paraboloids

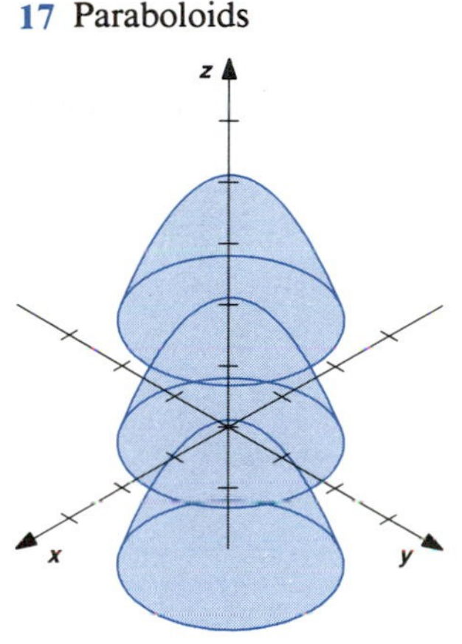


**19** Cones

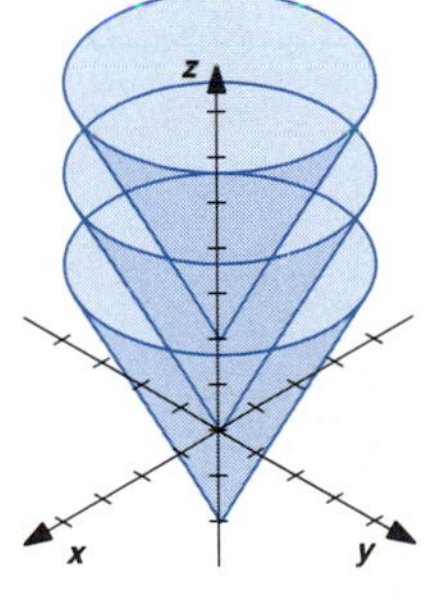


### Exercises 16.2

**1** 7

**3** 1

**5** Does not exist

**7** 1

**9** 4

**11** Does not exist

**13** 0

**15** Does not exist

**17** Does not exist

**19** 0

**21** 0

**23** $x^2 + y^2 \leq 1$; $x^2 + y^2 = 1$; closed

**25** $y \neq x^2$ and $(0, 0)$; $y = x^2$; not open, not closed

**27** $y \neq x^2 + 1$; $y = x^2 + 1$; open

## Exercises 16.3

1 4, −6

3 18, 0

5 $f_x = 4xy - y^3 + 3; f_y = 2x^2 - 3xy^2 - 1$

7 $f_x = \dfrac{y}{(x+y)^2}; f_y = \dfrac{-x}{(x+y)^2}$

9 $f_x = \dfrac{2x^2+y^2}{\sqrt{x^2+y^2}}; f_y = \dfrac{xy}{\sqrt{x^2+y^2}}$

11 $f_x = \dfrac{2x}{x^2+y^2}; f_y = \dfrac{2y}{x^2+y^2}$

13 $f_x = \dfrac{-y}{x^2+y^2}; f_y = \dfrac{x}{x^2+y^2}$

15 $f_x = 2\sec^2(2x - y); f_y = -\sec^2(2x - y)$

17 $f_x = 4xy^2 \sin(x^2y^2)\cos(x^2y^2)$;
$f_y = 4x^2y \sin(x^2y^2)\cos(x^2y^2)$

19 $f_x = y + z; f_y = x + z; f_z = x + y$

21 $f_x = \dfrac{y+z}{(x+z)^2}; f_y = \dfrac{-1}{x+z}; f_z = \dfrac{y-x}{(x+z)^2}$

23 $f_x = 2e^{2x-y}; f_y = -e^{2x-y}; f_z = 0$

25 $f_x = 3x^2 + 2xy - 3y^2; f_y = x^2 - 6xy + 3y^2$;
$f_{xx} = 6x + 2y; f_{yy} = -6x + 6y; f_{xy} = f_{yx} = 2x - 6y$

27 $f_x = \ln y; f_y = x/y; f_{xx} = 0; f_{yy} = -x/y^2; f_{xy} = f_{yx} = 1/y$

29 $f_x = y\cos xy; f_y = x\cos xy; f_{xx} = -y^2\sin xy$;
$f_{yy} = -x^2\sin xy; f_{xy} = f_{yx} = \cos xy - xy\sin xy$

## Exercises 16.4

1 $(6x - 5y)\,dx + (-5x + 2y)\,dy$

3 $2xy^3z\,dx + 3x^2y^2z\,dy + x^2y^3\,dz$

5 5; 4.981967482; −0.018032518; −0.02

7 6; 5.859; −0.141; −0.1

9 $100/\sqrt{3}$ ft; 3.482 ft

11 $20\pi$ m$^2$; $2\pi$ m$^2$

13 9%

15 6%

17 Increased by 1.5 Ω

## Exercises 16.5

1 $4xy^3 + 9x^2y^2$

3 $\left(\dfrac{2x}{x^2+y^2}\right)(-\sin t) + \left(\dfrac{2y}{x^2+y^2}\right)(\cos t)$

5 $\dfrac{\partial z}{\partial r} = (2x)(\cos\theta) + (2y)(\sin\theta)$;
$\dfrac{\partial z}{\partial \theta} = (2x)(-r\sin\theta) + (2y)(r\cos\theta)$

7 $\dfrac{\partial z}{\partial x} = (2r\cos\theta)\left(\dfrac{x}{\sqrt{x^2+y^2}}\right) + (-r^2\sin\theta)\left(\dfrac{-y}{x^2+y^2}\right)$;
$\dfrac{\partial z}{\partial y} = (2r\cos\theta)\left(\dfrac{y}{\sqrt{x^2+y^2}}\right) + (-r^2\sin\theta)\left(\dfrac{x}{x^2+y^2}\right)$

9 $\left(\dfrac{y^2}{z}\right)(1) + \left(\dfrac{2xy}{z}\right)(3) + \left(\dfrac{-xy^2}{z^2}\right)(2)$

11 $\dfrac{\partial w}{\partial x} = \left(\dfrac{1}{v}\right)(2x) + \left(\dfrac{-u}{v^2}\right)(2y)$;
$\dfrac{\partial w}{\partial y} = \left(\dfrac{1}{v}\right)(-2y) + \left(\dfrac{-u}{v^2}\right)(2x)$

13 $-\dfrac{3x^2+y}{x+3y^2}$

15 $-y/x$

17 $\dfrac{\partial z}{\partial x} = -\dfrac{yze^{xyz}-2x}{xye^{xyz}-2z}; \dfrac{\partial z}{\partial y} = -\dfrac{xze^{xyz}-2y}{xye^{xyz}-2z}$

19 $\dfrac{\partial z}{\partial x} = -\dfrac{1+z^2}{y+2xz}; \dfrac{\partial z}{\partial y} = -\dfrac{z}{y+2xz}$;
$\dfrac{\partial^2 z}{\partial x^2} = \dfrac{(1+z^2)(4yz+6xz^2-2x)}{(y+2xz)^3}$

21 31

23 24

27 (a) Decreasing at 1 ft$^3$/s (b) Increasing at 4 ft$^2$/s

## Exercises 16.6

1 52/5

3 2

5 −76/3

7 $-\mathbf{i}$; 4

9 $\dfrac{-4\mathbf{i}-7\mathbf{j}}{\sqrt{65}}$; $\sqrt{65}$

11 $\mathbf{k}$; 6

13 $\dfrac{7\mathbf{i}-8\mathbf{j}+8\mathbf{k}}{\sqrt{177}}$; $\sqrt{177}$

15 $\dfrac{\pm(4\mathbf{i}+3\mathbf{j})}{5}$

17 $\dfrac{\pm(3\mathbf{i}+\mathbf{j})}{\sqrt{10}}$

19 $2\sqrt{2}\mathbf{i} + \sqrt{2}\mathbf{j}$

## Exercises 16.7

1 $-4\mathbf{i} + \mathbf{j}$; 0

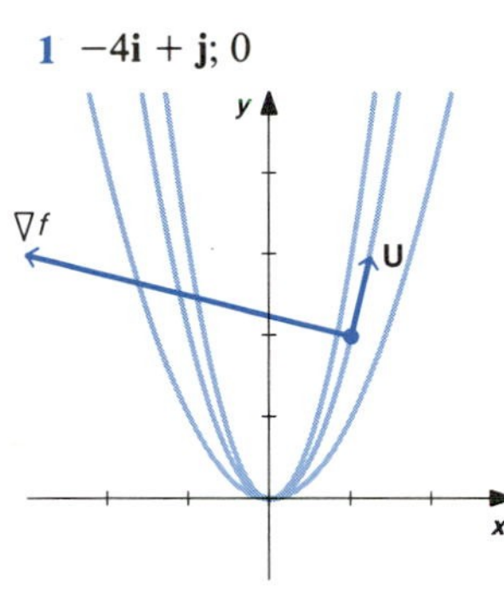


3 $\dfrac{2\mathbf{i}+2\mathbf{j}}{\sqrt{2}}$; 0

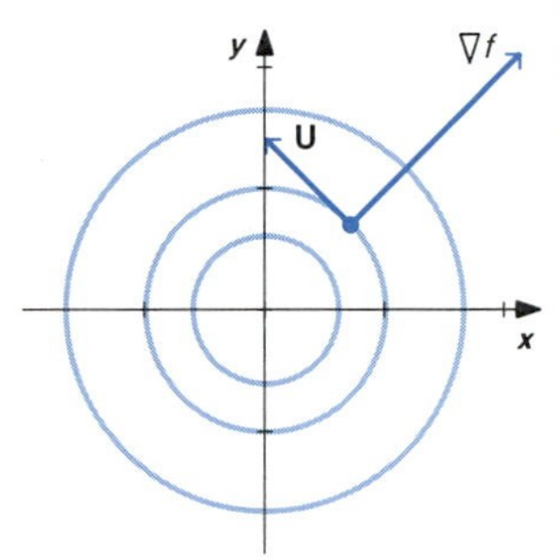


5 $4x + 3y = 0$

7 $2x + y = 5$

9 $3x - 4y + 5z = 0$

11 $2x - 2y - 3ez + 2e = 0$

13 $3x - 2y - z + 6 = 0$

15 $\dfrac{x+1}{2} = \dfrac{y-2}{-4} = \dfrac{z-5}{1}$; $x = -1 + 2t,\ y = 2 - 4t,\ z = 5 + t$

17 $\dfrac{x-2}{-2} = \dfrac{y+3}{-1} = \dfrac{z}{-6}$; $x = 2 - 2t,\ y = -3 - t,\ z = -6t$

19 $\dfrac{x}{-1} = \dfrac{z}{1}$, $y = 0$; $x = -t,\ y = 0,\ z = t$

21 $c(4\mathbf{i} - 5\mathbf{j} + 3\mathbf{k})$

## Exercises 16.8

1 loc min $f(1, 0) = -2$; saddle pt $f(-1, 0) = 2$

3 loc min $f(1, 1) = -4$; loc max $f(-1, -1) = 4$; saddle pts $f(1, -1) = 0$, $f(-1, 1) = 0$

5 loc min $f(1, -1) = -1$; saddle pt $f(0, 0) = 0$

7 loc min $f(0, -1) = -1$

9 loc min $f(0, 0) = 0$

11 max $f(1, 3) = 3$

13 max $f(2, -1) = 3$

15 min $f(1, 0) = -2$; max $f(2, 0) = 2$

17 min $f(0, -3) = -18$; max $f(\pm\sqrt{320}/3, -1/3) = 986/27$

19 min $f(0, 1) = -1$; max $f(3, 3) = 18$

21 $(-6/5, 3/5)$

23 $(1/\sqrt{2}, -1/2)$, $(-1/\sqrt{2}, -1/2)$

25 $1 \times (1/2) \times 1$; $1/2$

27 $\alpha = \dfrac{(\sum x_j)(\sum y_j) - n(\sum x_j y_j)}{(\sum x_j)^2 - n(\sum x_j^2)}$; $\beta = \dfrac{(\sum x_j)(\sum x_j y_j) - (\sum x_j^2)(\sum y_j)}{(\sum x_j)^2 - n(\sum x_j^2)}$

## Exercises 16.9

1 min $f(-3, 4) = -25$; max $f(3, -4) = 25$

3 min $f(-\sqrt{3}, 1) = f(\sqrt{3}, -1) = -\sqrt{3}$; max $f(\sqrt{3}, 1) = f(-\sqrt{3}, -1) = \sqrt{3}$

5 min $f(-1, -2, 2) = -9$; max $f(1, 2, -2) = 9$

7 loc min $f(-2, -2/3, -2/3) = -8/9$; loc max $f(2, 2/3, 2/3) = 8/9$

9 loc min $f(1/2, 2, 1) = 6$

11 min $f(1, 1, -\sqrt{2}) = -\sqrt{2}$; max $f(1, 1, \sqrt{2}) = \sqrt{2}$; $f(2, 0, 0) = f(0, 2, 0) = 0$ not local extrema

13 min $f(-1/\sqrt{2}, 1/\sqrt{2}, -1/\sqrt{2}) = -6/\sqrt{2}$; max $f(1/\sqrt{2}, -1/\sqrt{2}, 1/\sqrt{2}) = 6/\sqrt{2}$; loc min $f\left(-\dfrac{1}{\sqrt{10}}, \dfrac{3}{\sqrt{10}}, \dfrac{1}{\sqrt{10}}\right) = -\sqrt{10}$; loc max $f\left(\dfrac{1}{\sqrt{10}}, -\dfrac{3}{\sqrt{10}}, -\dfrac{1}{\sqrt{10}}\right) = \sqrt{10}$

15 min $f(-4, -2) = -40$; max $f(2, 1) = 5$

17 min $f(2, -1, 3) = -14$; max $f(-4, 2, -6) = 112$

19 min $f(-1/\sqrt{2}, 1/\sqrt{2}, 0) = f(1/\sqrt{2}, -1/\sqrt{2}, 0) = -1$; max $f(1/\sqrt{2}, 1/\sqrt{2}, 0) = f(-1/\sqrt{2}, -1/\sqrt{2}, 0) = 1$

21 $4 \times 4 \times 2$; 48

23 1 ft × 1 ft × (3/2) ft; 3/2 $\text{ft}^3$

27 4 in.; $\pi/3$

## Review Exercises

1 $x \neq y$

3 Parabolas

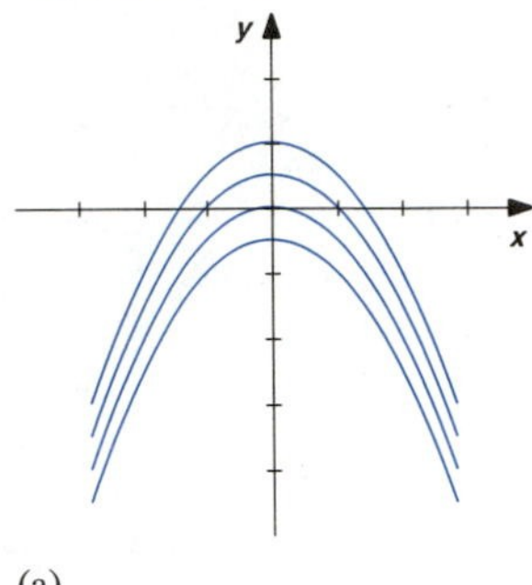


(a)

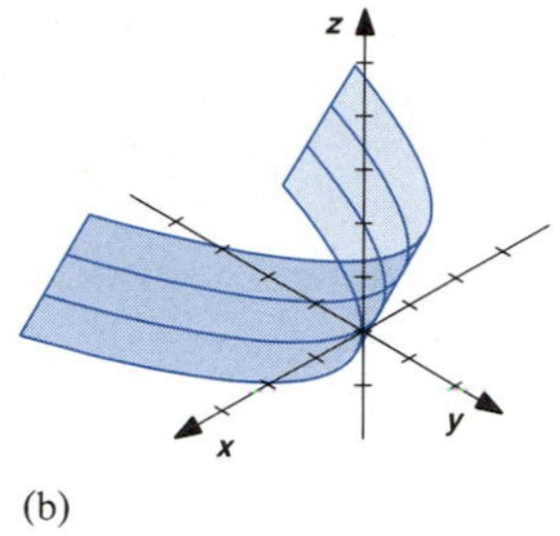


(b)

5 Cones

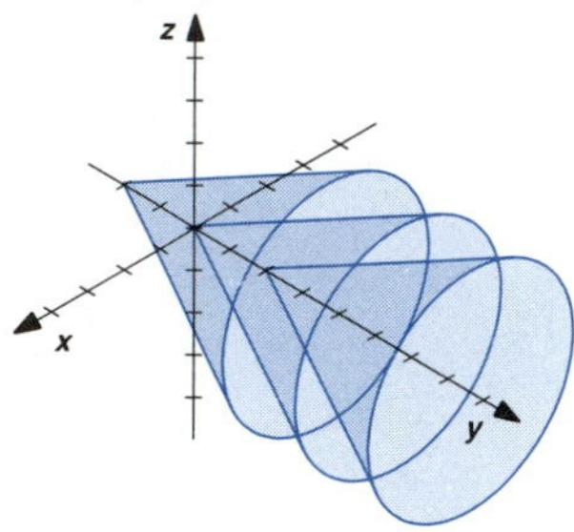


7 Does not exist

9 Does not exist

11 $f_x = 2x - 2y; f_y = -2x + 6y; f_{xx} = 2; f_{yy} = 6;$ $f_{xy} = f_{yx} = -2$

13 $f_x = \dfrac{x}{x^2+y^2}; f_y = \dfrac{y}{x^2+y^2}; f_{xx} = \dfrac{y^2-x^2}{(x^2+y^2)^2};$
$f_{yy} = \dfrac{x^2-y^2}{(x^2+y^2)^2}; f_{xy} = f_{yx} = \dfrac{-2xy}{(x^2+y^2)^2}$

15 3.2%

17 $(2x/y)(\sec^2 t) + (-x^2/y^2)(-2 \cos t \sin t)$

19 $\dfrac{\partial z}{\partial r} = (2x)(\cos \theta) + (-2y)(\sin \theta);$
$\dfrac{\partial z}{\partial \theta} = (2x)(-r \sin \theta) + (-2y)(r \cos \theta)$

21 $\dfrac{\partial z}{\partial x} = -\dfrac{2xy+z^2}{y^2+2xz}; \dfrac{\partial z}{\partial y} = -\dfrac{x^2+2yz}{y^2+2xz}$

23 $\dfrac{\partial^2 z}{\partial y \partial x} = -\dfrac{xy}{z^3}$

25 $4x + 6y - z = 13$

27 $\dfrac{x-5}{-10} = \dfrac{y+4}{-8} = \dfrac{z-3}{6};$
$x = 5 - 10t, y = -4 - 8t; z = 3 + 6t$

29 $\mathbf{i} + \mathbf{j}; \dfrac{\pm(\mathbf{i}-\mathbf{j})}{\sqrt{2}}$

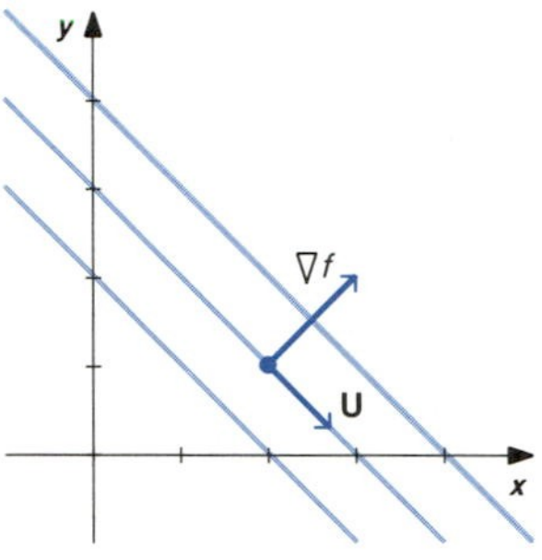


31 $4\mathbf{i} - \mathbf{j}; \pm(\mathbf{i} + 4\mathbf{j})/\sqrt{17}$

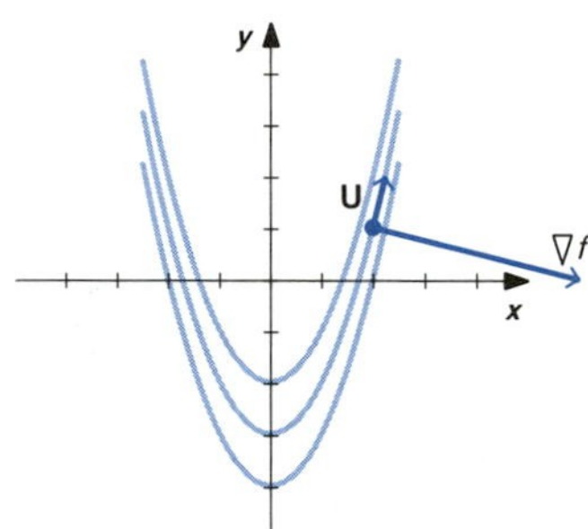


33 $\dfrac{-3\mathbf{i}+8\mathbf{j}}{\sqrt{73}}; \sqrt{73}$

35 loc min $f(0, 0) = 0$

37 min $f(1/3, -1/3, 1/3) = 1/3$

39 (4 ft) × (4 ft) × (2 ft); 48 ft$^2$

## CHAPTER SEVENTEEN

### Exercises 17.1

1 For $x$ fixed, $y$ varies from $y = x$ to $y = x^2 + 1$.
$x$ varies from $x = 0$ to $x = 1$.

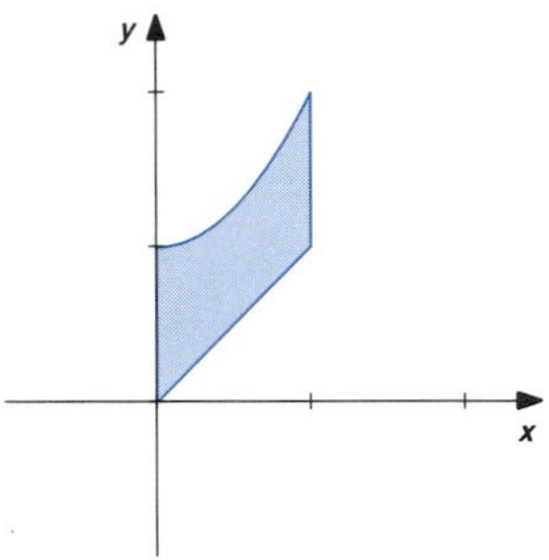

3 For $x$ fixed, $y$ varies from $y = x$ to $y = 1$.
$x$ varies from $x = 0$ to $x = 1$.

For $y$ fixed, $x$ varies from $x = 0$ to $x = y$.
$y$ varies from $y = 0$ to $y = 1$.

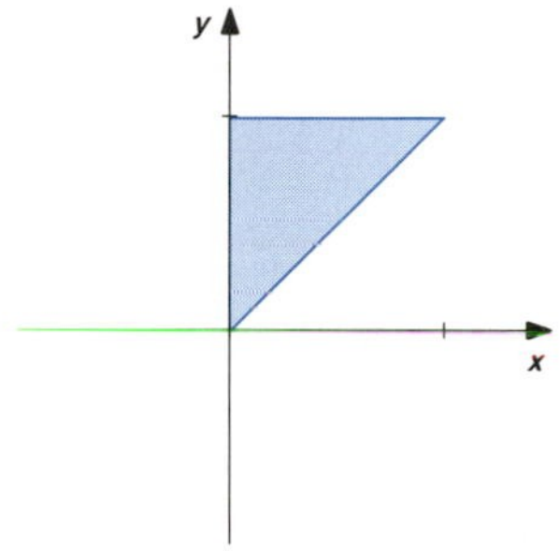

5 For $x$ fixed, $y$ varies from $y = 0$ to $y = (2 - x)/2$.
$x$ varies from $x = 0$ to $x = 2$.

For $y$ fixed, $x$ varies from $x = 0$ to $x = 2 - 2y$.
$y$ varies from $y = 0$ to $y = 1$.

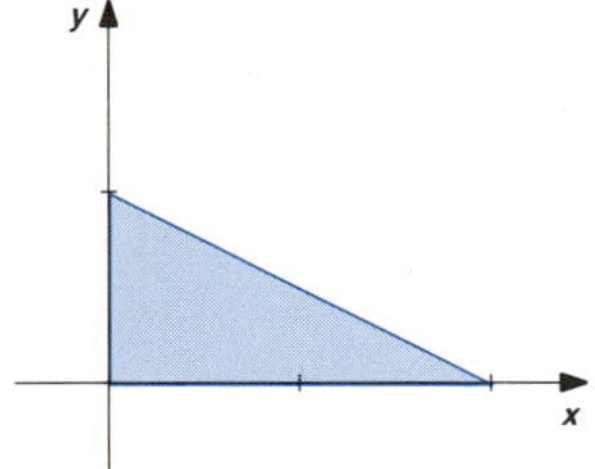

7 For $x$ fixed, $y$ varies from $y = \sin x$ to $y = \cos x$.
$x$ varies from $x = 0$ to $x = \pi/4$.

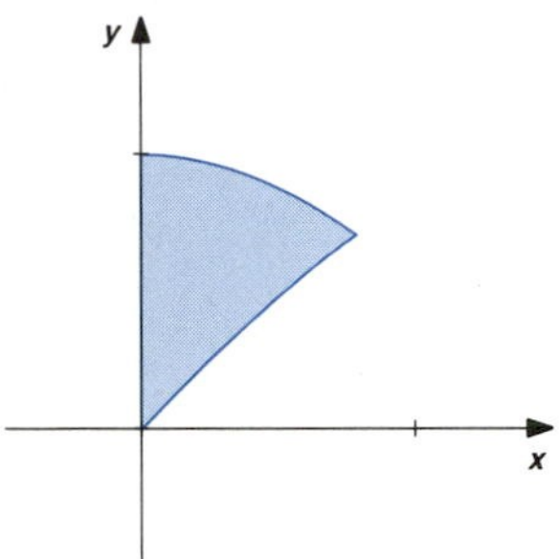

9 For $y$ fixed, $x$ varies from $x = y$ to $x = 2 - y^2$.
$y$ varies from $y = -2$ to $y = 1$.

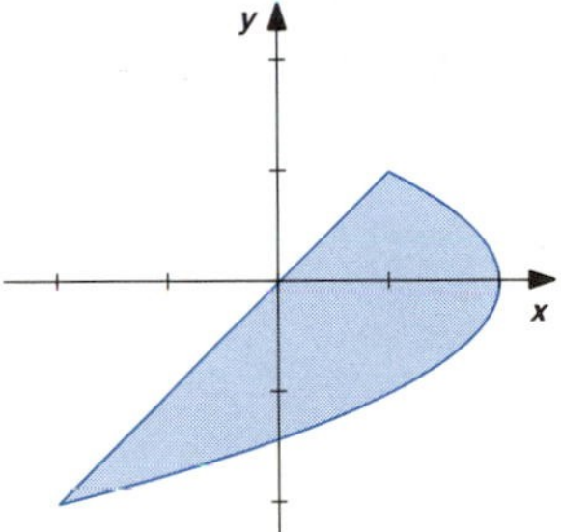

11 For $y$ fixed, $x$ varies from $x = 0$ to $x = e^y$.
$y$ varies from $y = 0$ to $y = 1$.

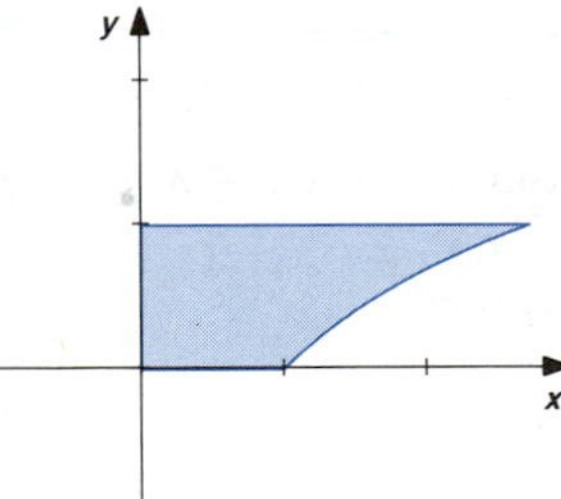

13 For $x$ fixed, $y$ varies from $y = 0$ to $y = \sqrt{4-x^2}$. $x$ varies from $x = -2$ to $x = 2$.

For $y$ fixed, $x$ varies from $x = -\sqrt{4-y^2}$ to $x = \sqrt{4-y^2}$. $y$ varies from $y = 0$ to $y = 2$.

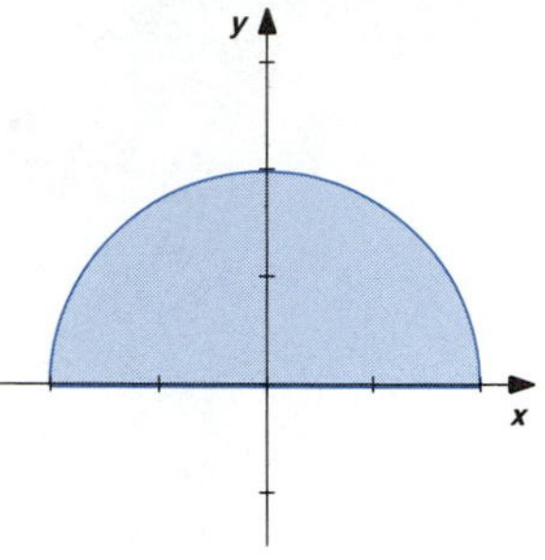


15 For $x$ and $y$ fixed, $z$ varies from $z = x + y$ to $z = 2 - x - y$.

For $x$ fixed, $y$ varies from $y = 0$ to $y = 1 - x$. $x$ varies from $x = 0$ to $x = 1$.

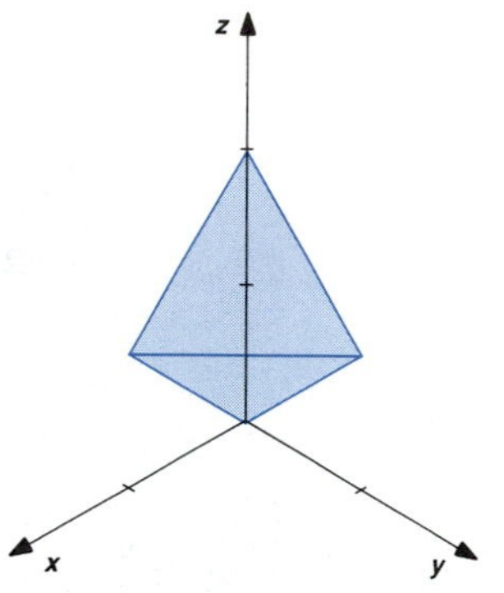


17 For $x$ and $y$ fixed, $z$ varies from $z = x$ to $z = 3 - x - y$.

For $x$ fixed, $y$ varies from $y = 0$ to $y = 3 - 2x$. $x$ varies from $x = 0$ to $x = 3/2$.

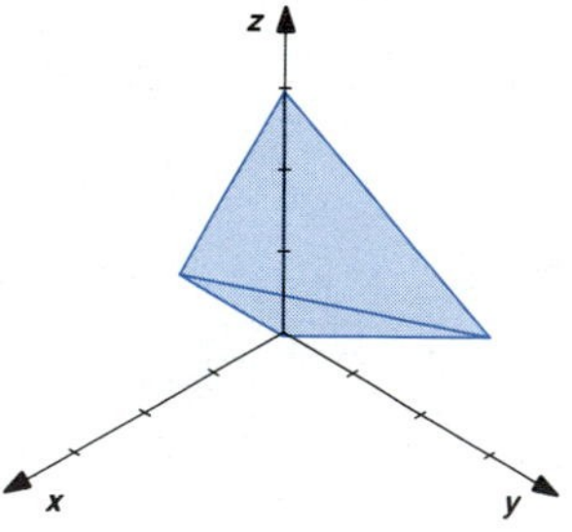


19 For $x$ and $y$ fixed, $z$ varies from $z = 0$ to $z = 2 - x - y$.

For $x$ fixed, $y$ varies from $y = -\sqrt{1-x^2}$ to $y = \sqrt{1-x^2}$.
$x$ varies from $x = -1$ to $x = 1$.

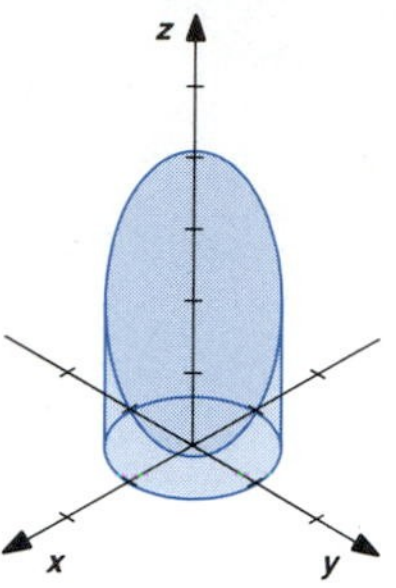


21 For $x$ and $y$ fixed, $z$ varies from $z = 0$ to $z = 3 - y$.

For $x$ fixed, $y$ varies from $y = 0$ to $y = 2 - x$. $x$ varies from $x = 0$ to $x = 2$.

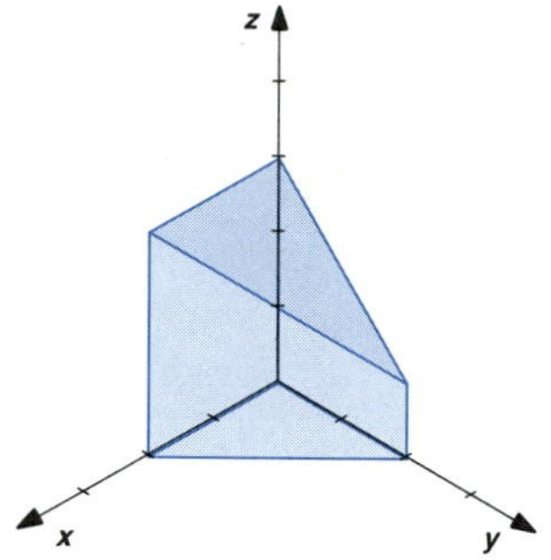


23 For $x$ and $y$ fixed, $z$ varies from $z = 0$ to $z = \sqrt{4-y^2}$.

For $x$ fixed, $y$ varies from $y = 0$ to $y = \sqrt{4-x^2}$. $x$ varies from $x = 0$ to $x = 2$.

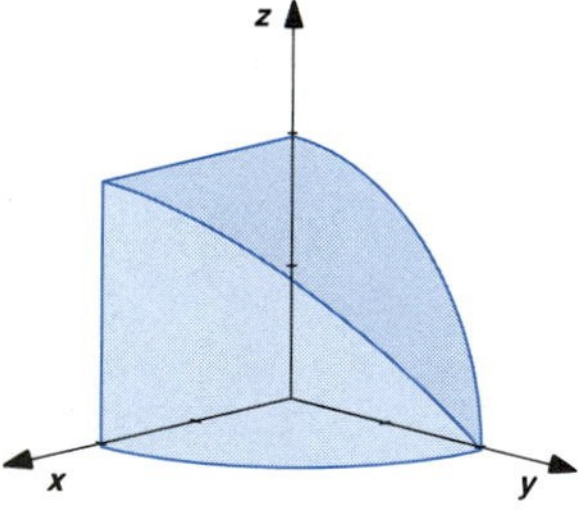

**25** For $x$ and $y$ fixed, $z$ varies from $z = -\sqrt{4x^2+4y^2}$ to $z = \sqrt{4x^2+4y^2}$.

For $x$ fixed, $y$ varies from $y = -\sqrt{1-x^2}$ to $y = \sqrt{1-x^2}$.
$x$ varies from $x = -1$ to $x = 1$.

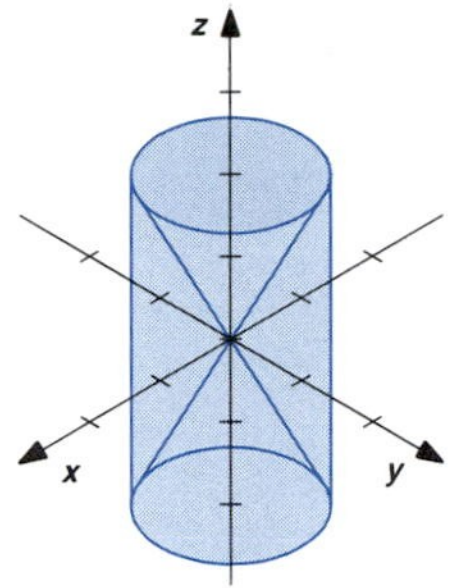


**27** For $x$ and $y$ fixed, $z$ varies from $z = 2x$ to $z = 2$

For $x$ fixed, $y$ varies from $y = 0$ to $y = 3$.
$x$ varies from $x = 0$ to $x = 1$.

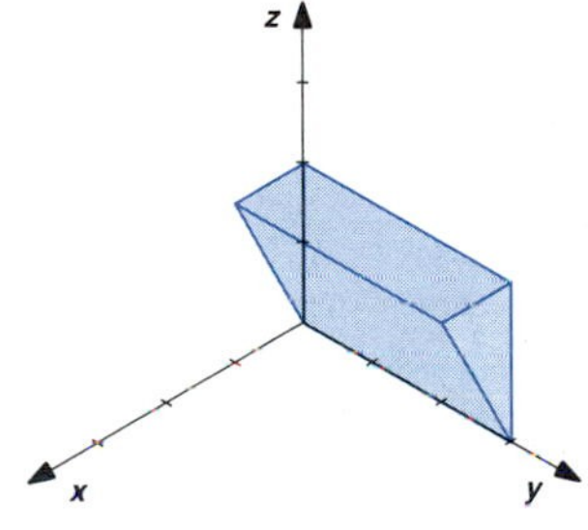


**29** For $x$ and $y$ fixed, $z$ varies from $z = 0$ to $z = 2$.

For $x$ fixed, $y$ varies from $y = 0$ to $y = 2$.
$x$ varies from $x = 0$ to $x = 2$.

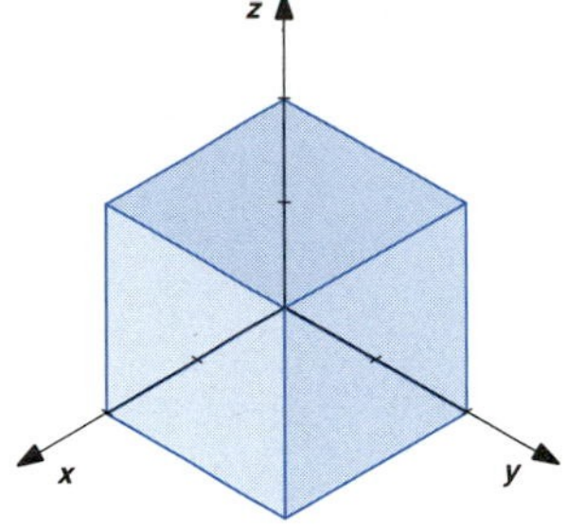


**31** For $x$ and $y$ fixed, $z$ varies from $z = y/2$ to $z = (4 - y)/2$.

For $x$ fixed, $y$ varies from $y = 0$ to $y = 2$.
$x$ varies from $x = 0$ to $x = 2$.

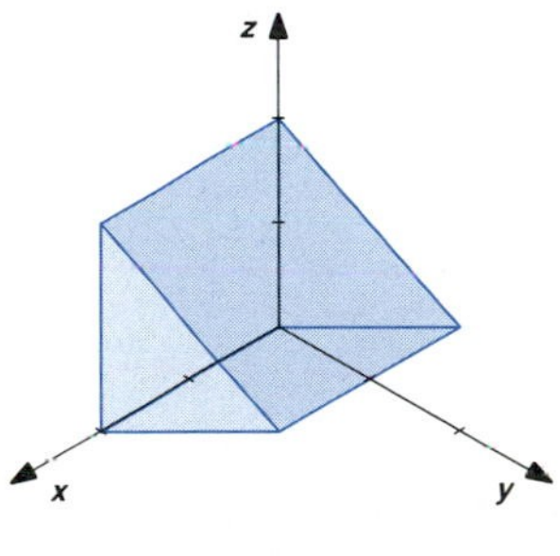


## *Exercises 17.2*

**1** 1/8

**3** 1/4

**5** $\pi/4$

**7** $\ln(\sqrt{2} + 1) - \pi/4$

**9** 20/3

**11** 8

**13** $4e^2 - 4e - 4$

**15** 1

**17** $\sqrt{3} - \pi/3$

**19** 32/3

**21** 3/2

**23** 37/12

**25** $\int_0^4 \int_{y/2}^{\sqrt{y}} xy\,dx\,dy = 8/3$

**27** $\int_0^1 \int_0^{\sin^{-1} y} \cos x\,dx\,dy = 1/2$

**29** $\int_{\pi/6}^{\pi/2} \int_{1/2}^{\sin y} \cot y\,dx\,dy = \frac{1}{2} - \frac{1}{2}\ln 2$

**31** $\int_0^1 \int_1^{e^y} x\,dx\,dy = \frac{e^2}{4} - \frac{3}{4}$

**33** The function $e^{xy}$ is continuous in the closed and bounded region $0 \le x \le 1, 0 \le y \le 1$.

$$\int_0^1 e^{xy}\,dy = \begin{cases} \dfrac{e^x - 1}{x}, & 0 < x \le 1, \\ 1, & x = 0. \end{cases}$$

Yes, it is a continuous function of $x$.

### Exercises 17.3

**1** $27/2$  **3** $2/5$
**5** $8\pi$  **7** $6\sqrt{5}$
**9** $2(3^{3/2} - 2^{3/2})/3$  **11** $\pi rh/2$
**13** $8/3$  **15** $7/3$
**17** $\pi/4$  **19** $(1, 8/3)$
**21** $(3\pi/16, 0)$  **23** $(8(4 - \sqrt{2})/15, 8\sqrt{2}/15)$

### Exercises 17.4

**1** $\pi/2 - 1$  **3** $\frac{9}{2}\tan^{-1}\frac{1}{2}$
**5** $4\pi R^3/3$  **7** $k\pi R^4/2$
**9** $(0, 3\pi/16)$  **11** $(28/(9\pi), 28/(9\pi))$
**13** $R\pi\sqrt{H^2 + R^2}$  **15** $2\pi R^2(1 - \cos\phi_0)$
**17** $4/3$  **19** $\pi \ln 2$
**21** $r_0\,\Delta r\,\Delta\theta + \frac{1}{2}(\Delta r)^2\Delta\theta$

### Exercises 17.5

**1** $9/2$  **3** $1/48$
**5** $\frac{\pi^4}{4} + \frac{\cos(\pi^2)}{2} - \frac{1}{2}$  **7** $(e - 2)/4$
**9** $1/6$  **11** $8/3$
**13** $8/15$  **15** $16r^3/3$

**17** $\int_{-r}^{r}\int_{-\sqrt{r^2-x^2}}^{\sqrt{r^2-x^2}}\int_{-\sqrt{R^2-x^2}}^{\sqrt{R^2-x^2}} dz\,dy\,dx$

**19** $\int_{-1}^{1}\int_{-\sqrt{1-x^2}}^{\sqrt{1-x^2}}\int_{-\sqrt{(y+4)^2-4x^2}/2}^{\sqrt{(y+4)^2-4x^2}/2} dz\,dy\,dx$

**21** $\int_0^{12/7}\int_{z/2}^{6-3z}\int_0^{(6-x-3z)/2} dy\,dx\,dz$,
$\int_0^{12/7}\int_0^{(12-7z)/4}\int_{z/2}^{6-2y-3z} dx\,dy\,dz$,
$\int_0^{3}\int_0^{(12-4y)/7}\int_{z/2}^{6-2y-3z} dx\,dz\,dy$

**23** $\int_0^1\int_0^{4-z}\int_0^{4-x-z} dy\,dx\,dz$, $\int_0^1\int_0^{4-z}\int_0^{4-y-z} dx\,dy\,dz$

**25** $\int_0^{4/3}\int_{2x}^{4-x}\int_0^{6-x-z} dy\,dz\,dx$

### Exercises 17.6

**1** $4$  **3** $1464$
**5** $(7/12, 7/12, 7/6)$  **7** $(12/7, 8/7, 4/7)$
**9** $MR^2/2$  **11** $2MS^2/3$
**13** 3.75 rev/s

**15** $\frac{3m}{\pi r^2 h}\int_{-r}^{r}\int_{-\sqrt{r^2-x^2}}^{\sqrt{r^2-x^2}}\int_{h\sqrt{x^2+y^2}/r}^{h}\frac{(hy-rz)^2 + (r^2+h^2)x^2}{r^2+h^2}\,dz\,dy\,dx$

### Exercises 17.7

**1**  **3**

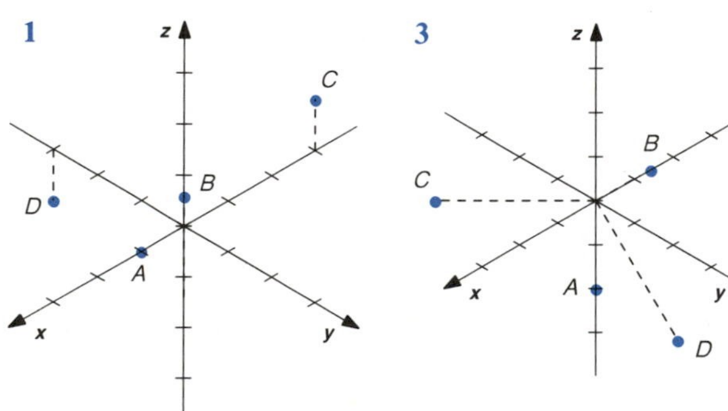


**5** $A(x, y, z) = (0, 0, 2)$, $B(x, y, z) = (\sqrt{2}, \sqrt{2}, -1)$, $C(x, y, z) = (1, \sqrt{3}, 3)$

**7** $A(x, y, z) = (0, 4\sqrt{2}, 4\sqrt{2})$, $B(x, y, z) = (1/\sqrt{2}, 1/\sqrt{2}, 0)$, $C(x, y, z) = (\sqrt{3}/2, 3/2, -1)$

**9** $A(r, \theta, z) = (2\sqrt{2}, \pi/4, 2)$, $B(r, \theta, z) = (0, 0, -1)$, $C(r, \theta, z) = (8, -\pi/3, 1)$

**11** $A(\rho, \theta, \phi) = (2, 0, \pi)$, $B(\rho, \theta, \phi) = (2\sqrt{2}, \pi/4, \pi/2)$, $C(\rho, \theta, \phi) = (2\sqrt{2}, \pi/2, \pi/4)$

13 For $r$ and $\theta$ fixed, $z$ varies from $z = 0$ to $z = 2r\cos\theta + 2r\sin\theta$.
For $\theta$ fixed, $r$ varies from $r = 0$ to $r = 2$.
$\theta$ varies from $\theta = 0$ to $\theta = \pi/2$.

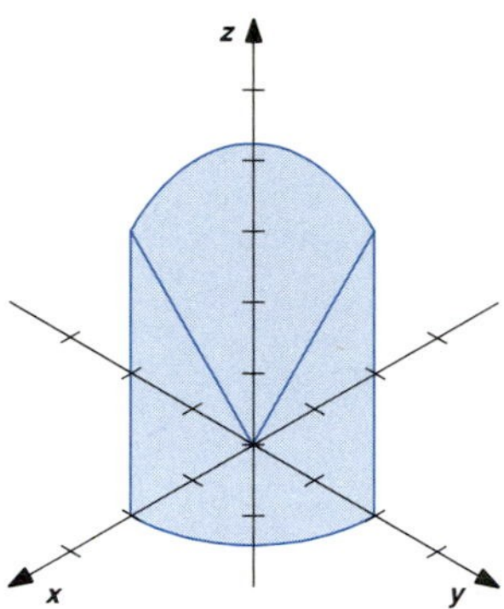


15 For $r$ and $\theta$ fixed, $z$ varies from $z = 0$ to $z = r$.
For $\theta$ fixed, $r$ varies from $r = 0$ to $r = 3$.
$\theta$ varies from $\theta = 0$ to $\theta = 2\pi$.

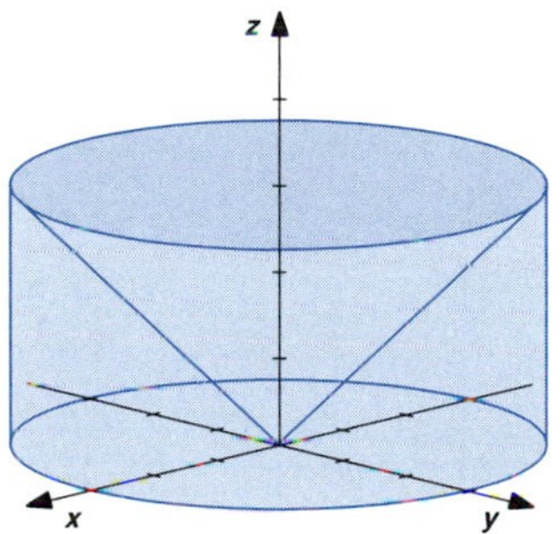


17 For $r$ and $\theta$ fixed, $z$ varies from $z = 0$ to $z = 4 - r^2$.
For $\theta$ fixed, $r$ varies from $r = 0$ to $r = 2$.
$\theta$ varies from $\theta = 0$ to $\theta = 2\pi$.

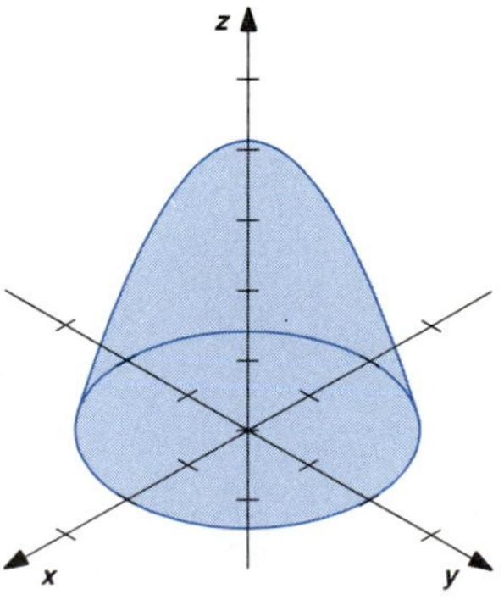


19 For $r$ and $\theta$ fixed, $z$ varies from $z = 0$ to $z = 2 - e^{r^2}$.
For $\theta$ fixed, $r$ varies from $r = 0$ to $r = \sqrt{\ln 2}$.
$\theta$ varies from $\theta = 0$ to $\theta = 2\pi$.

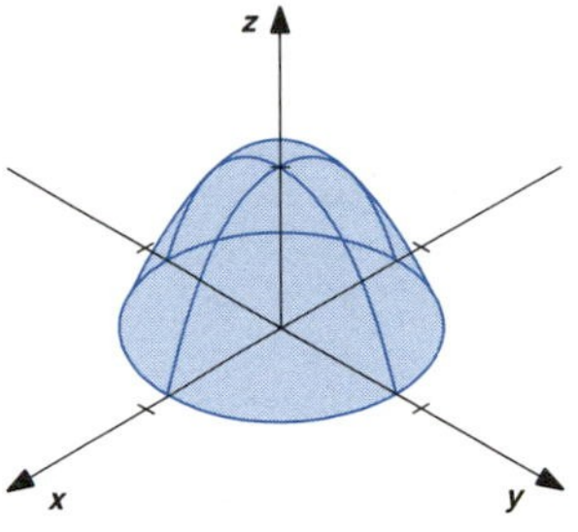


21 For $r$ and $\theta$ fixed, $z$ varies from $z = 0$ to $z = \sqrt{16 - r^2}$.
For $\theta$ fixed, $r$ varies from $r = 0$ to $r = 4/(\cos\theta + \sin\theta)$.
$\theta$ varies from $\theta = 0$ to $\theta = \pi/2$.

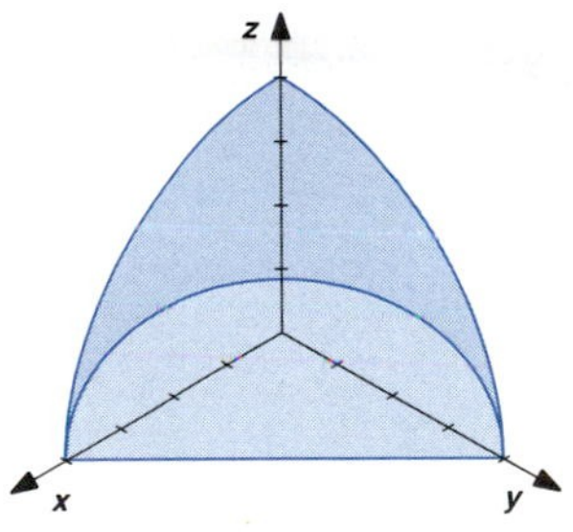


23 For $r$ and $\theta$ fixed, $z$ varies from $z = 0$ to $z = 4$.
For $\theta$ fixed, $r$ varies from $r = 0$ to $r = 2\cos\theta$.
$\theta$ varies from $\theta = -\pi/2$ to $\theta = \pi/2$.

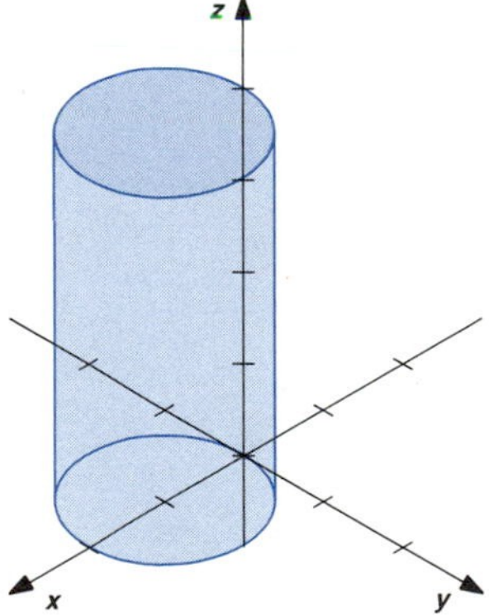

25 For $\theta$ and $\phi$ fixed, $\rho$ varies from $\rho = 0$ to $\rho = 3$.
For $\theta$ fixed, $\phi$ varies from $\phi = 0$ to $\phi = \pi/2$.
$\theta$ varies from $\theta = 0$ to $\theta = 2\pi$.

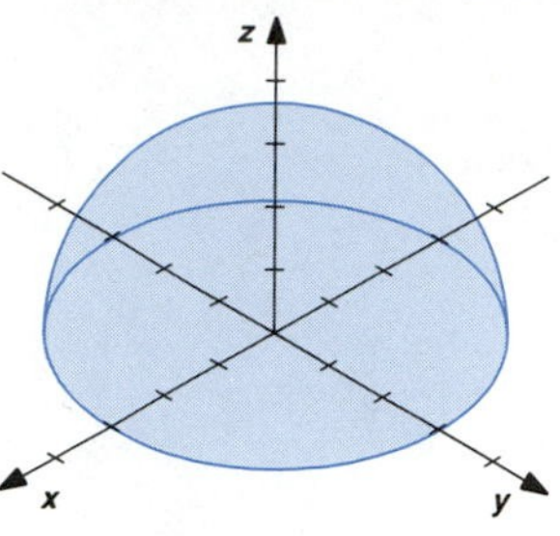


27 For $\theta$ and $\phi$ fixed, $\rho$ varies from $\rho = 0$ to $\rho = 1$.
For $\theta$ fixed, $\phi$ varies from $\phi = 0$ to $\phi = \pi/2$.
$\theta$ varies from $\theta = 0$ to $\theta = \pi/2$.

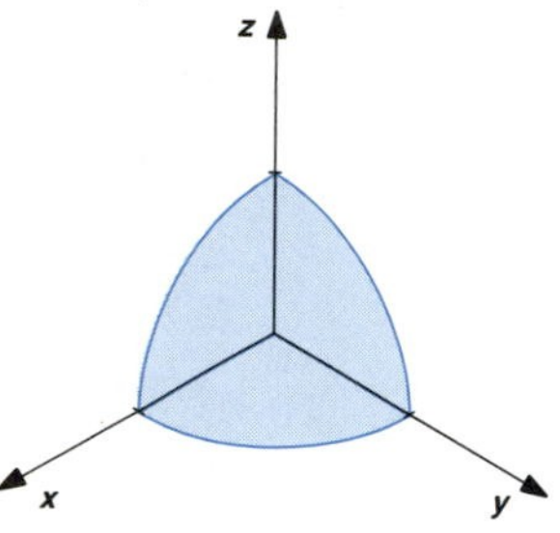


29 For $\theta$ and $\phi$ fixed, $\rho$ varies from $\rho = 0$ to $\rho = 4 \csc \phi$.
For $\theta$ fixed, $\phi$ varies from $\phi = \pi/4$ to $\phi = \pi/2$.
$\theta$ varies from $\theta = 0$ to $\theta = 2\pi$.

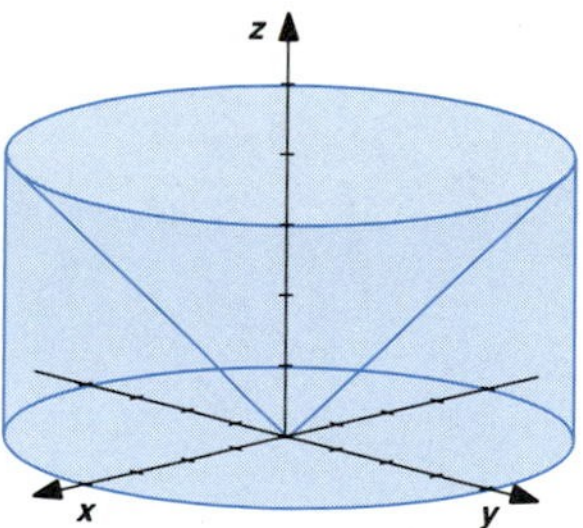


31 For $\theta$ and $\phi$ fixed, $\rho$ varies from $\rho = 3 \sec \phi$ to $\rho = 5$.
For $\theta$ fixed, $\phi$ varies from $\phi = 0$ to $\phi = \cos^{-1}(3/5)$.
$\theta$ varies from $\theta = 0$ to $\theta = 2\pi$.

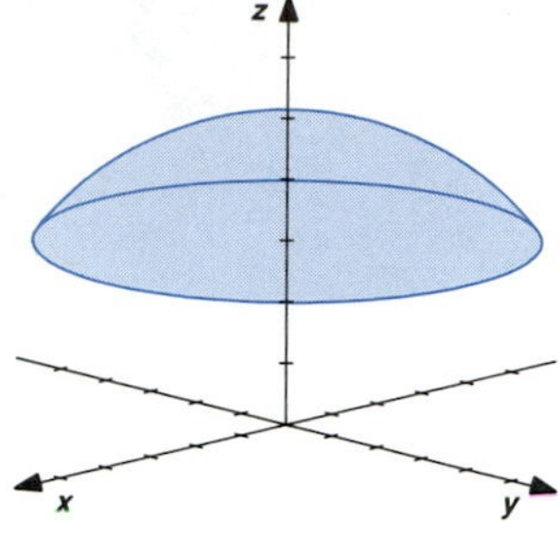


33 For $\theta$ and $\phi$ fixed, $\rho$ varies from $\rho = 0$ to $\rho = 3/(\sin \phi \cos \theta + \sin \phi \sin \theta + \cos \phi)$.
For $\theta$ fixed, $\phi$ varies from $\phi = 0$ to $\phi = \pi/2$.
$\theta$ varies from $\theta = 0$ to $\theta = \pi/2$.

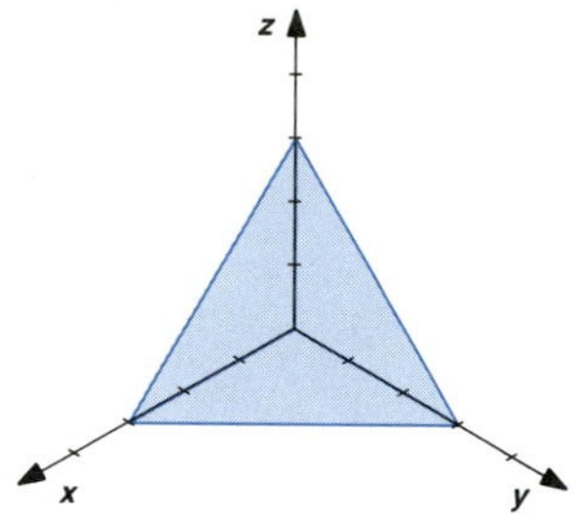


35 For $\theta$ and $\phi$ fixed, $\rho$ varies from $\rho = 0$ to $\rho = 2/(\sin \phi \cos \theta + \cos \phi)$.
For $\theta$ fixed, $\phi$ varies from $\phi = 0$ to $\phi = \tan^{-1}(1/2)$.
$\theta$ varies from $\theta = 0$ to $\theta = 2\pi$.

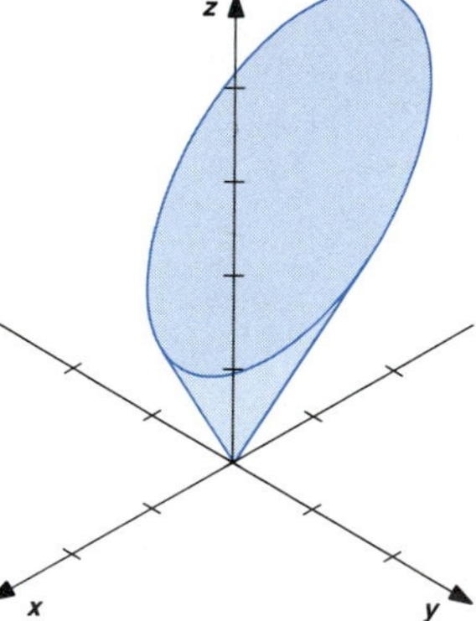

### Exercises 17.8

**1** $\pi R^2H/3$ **3** $32\pi(8 - 3^{3/2})/3$

**5** $2\pi R^3/3$ **7** $5\pi/3$

**9** On axis, $3H/4$ from vertex **11** $\left(0, 0, \dfrac{(9+4HR)H}{6(3+HR)}\right)$

**13** $\dfrac{2M(R^5-r^5)}{5(R^3-r^3)}$ **15** $M/3$

**17** (a) $M\left(\dfrac{R^2}{2}+D^2\right)$ (b) $D = 0$

### Exercises 17.9

**1** (a) $(x, y) = (2, 1)$ (b) $(x, y) = (-1, 2)$ (c) $(u, v) = (2/5, -1/5)$ (d) $(u, v) = (1/5, 2/5)$ (e) $u = (2/5)x + (1/5)y$, $v = (-1/5)x + (2/5)y$

**3** (a) $(x, y, z) = (1, 0, 1)$ (b) $(x, y, z) = (-1, 1, 0)$ (c) $(x, y, z) = (0, -2, 1)$ (d) $(u, v, w) = (1/3, -2/3, -1/3)$ (e) $(u, v, w) = (1/3, 1/3, -1/3)$ (f) $(u, v, w) = (2/3, 2/3, 1/3)$ (g) $u = (x + y + 2z)/3$, $v = (-2x + y + 2z)/3$, $w = (-x - y + z)/3$

**5**

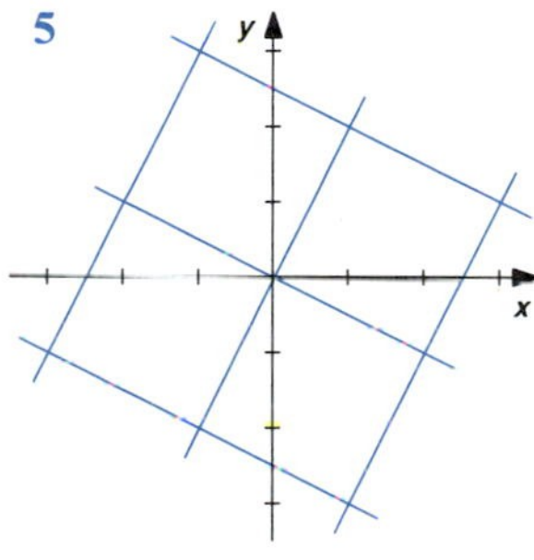


(a)

(b)

**7**

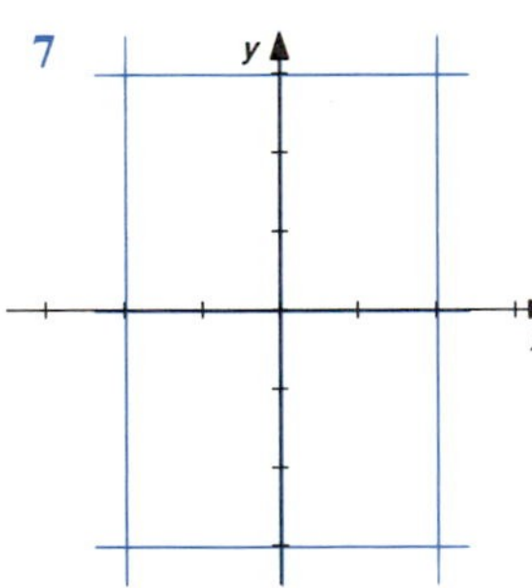


(a)

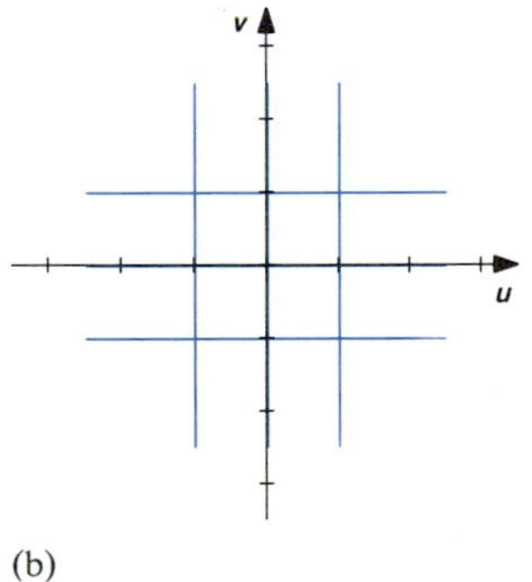


(b)

**9**

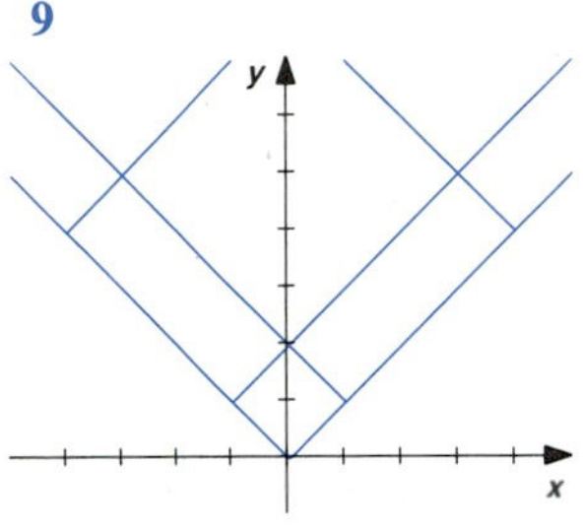


(a)

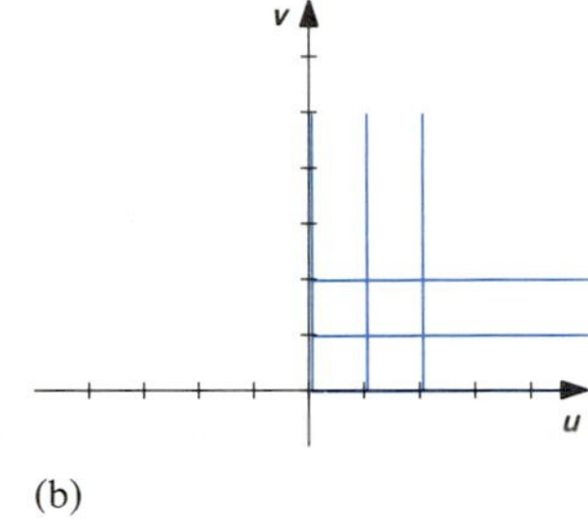


(b)

**11** $u = 2$

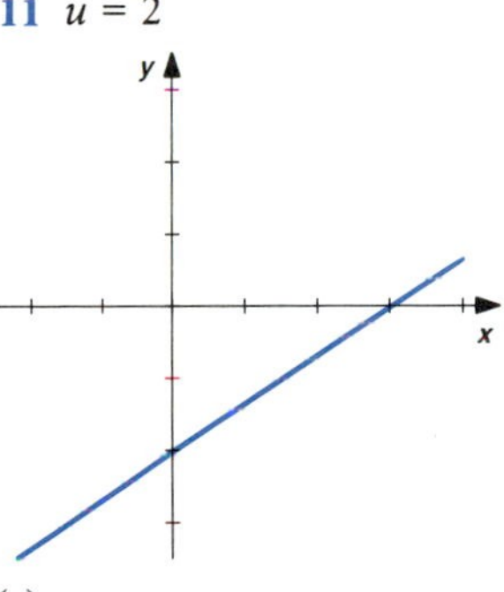


(a)

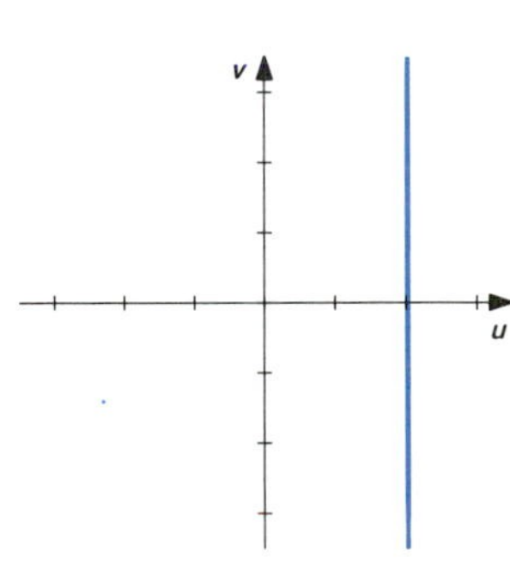


(b)

**13** $u^2 + v^2 = 11$

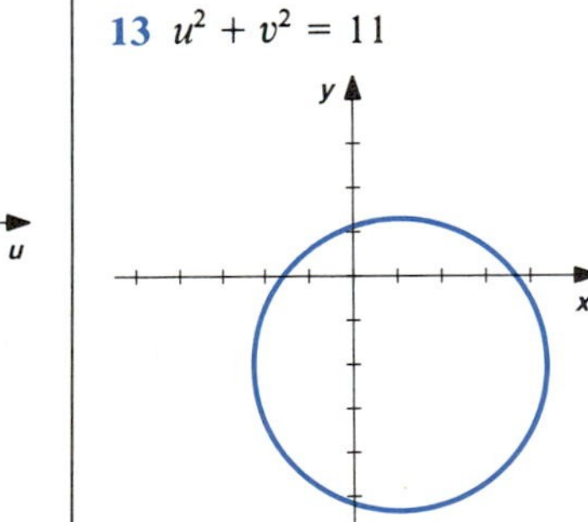


(a)

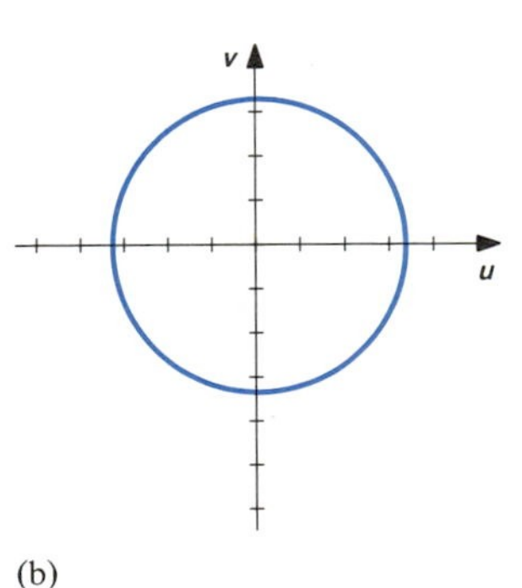


(b)

**15** $u^2 - v^2 = 2$

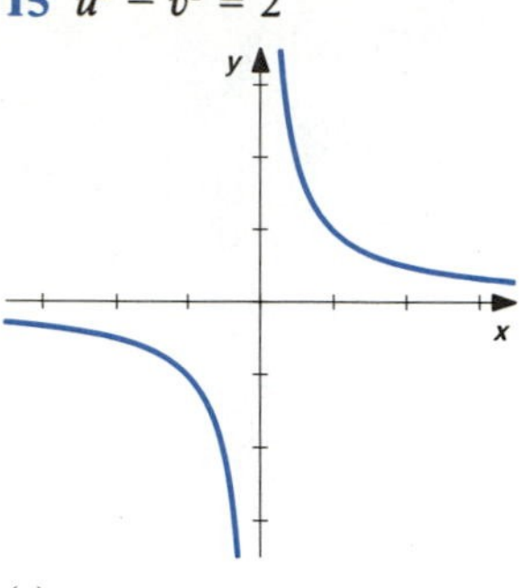


(a) (b)

**17**

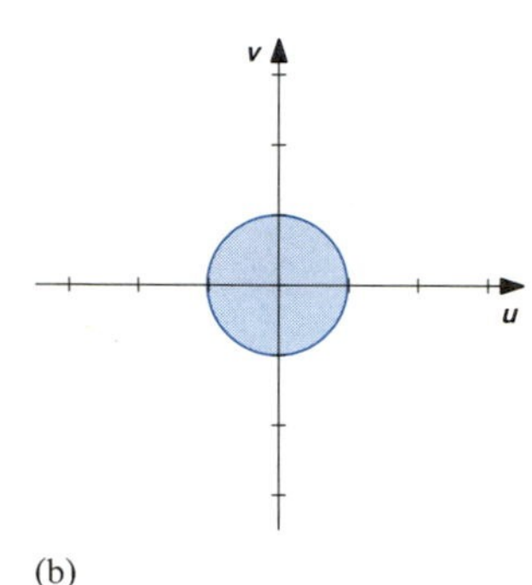


(a) (b)

**19**

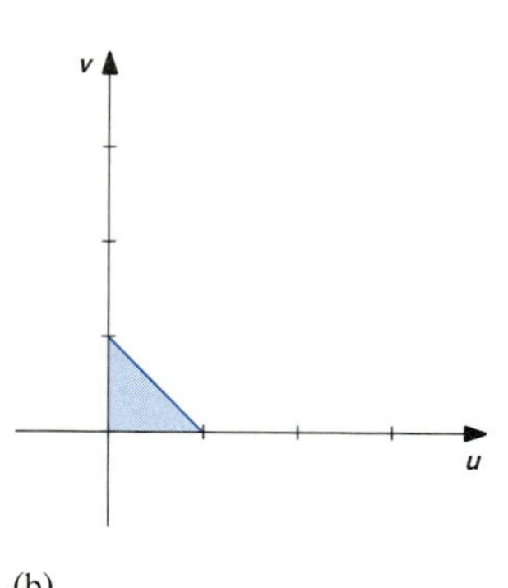


(a) (b)

**21**

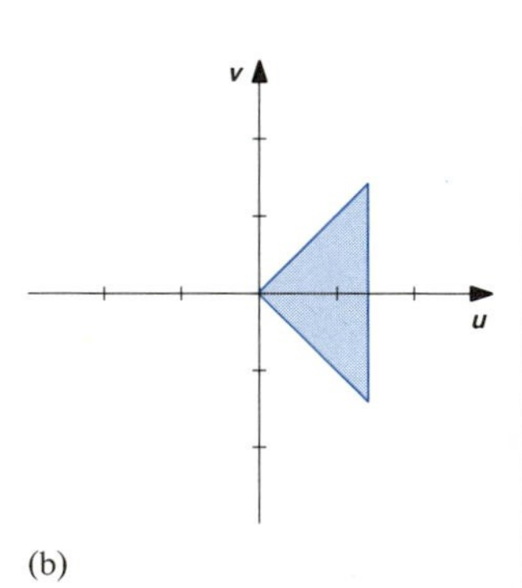


(a) (b)

**23**

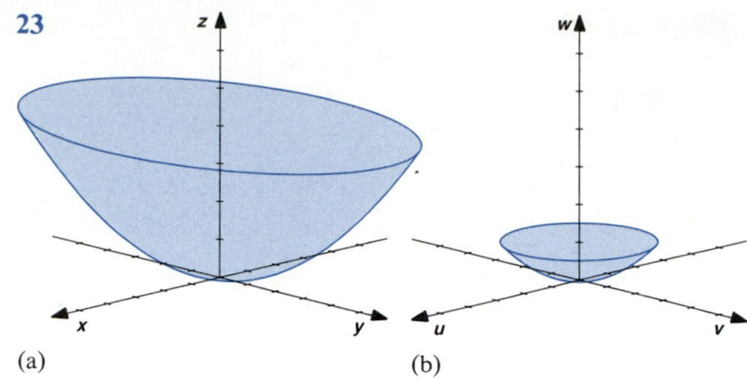


(a) (b)

**25**

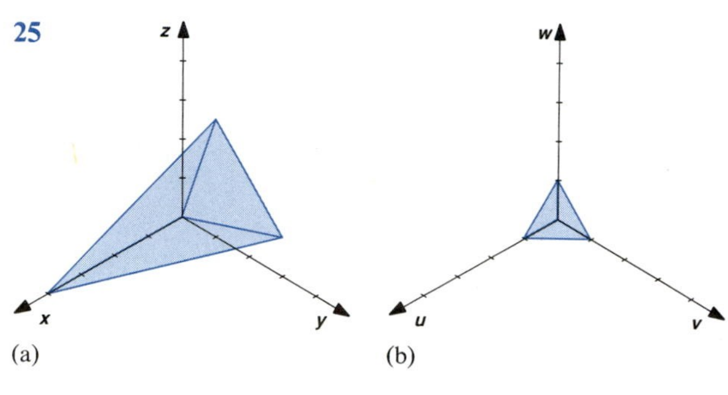


(a) (b)

**27**

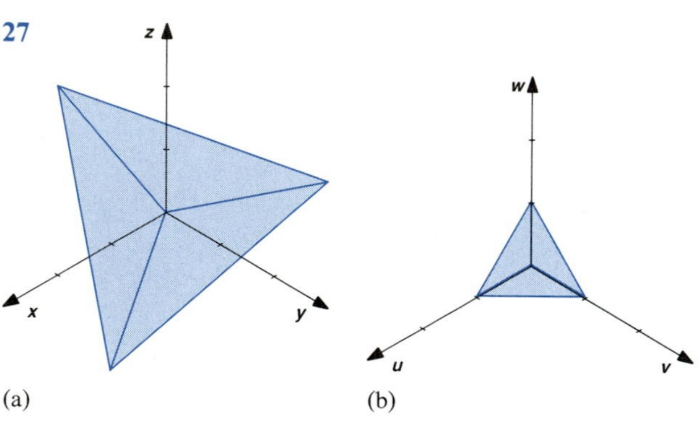


(a) (b)

## Exercises 17.10

**1** 1

**7** $\int_0^1 \int_0^1 (2u + v)(u + 2v)\, 3\, du\, dv$

**9** $\int_0^2 \int_0^{2-u} (2u + v)\, 3\, dv\, du$

**11** $\int_{-3/11}^{2/11} \int_{6/11}^{17/11} 121\, u\, du\, dv$

**13** $\int_0^{3\sqrt{2}} \int_{-u/3}^{u/3} \sqrt{2}\, u\, dv\, du$

**15** $\int_{-1/\sqrt{2}}^{1/\sqrt{2}} \int_{-\sqrt{2+u^2}}^{\sqrt{2+u^2}} 4u^2v^2\, dv\, du$

**17** $\int_0^{\pi/6} \int_1^4 e^{r^2} r \sec\theta\, dr\, d\theta$

**19** $\int_0^1 \int_0^1 \int_0^1 24u\, du\, dv\, dw$

**21** $\int_0^6 \int_{-2}^2 \int_{v^2}^4 du\, dv\, dw$

**23** $\frac{\pi}{4}$

**25** 16

**27** $4abc\pi/3$

**29** $32\pi/3$

## Review Exercises

**1** 2/5

**3** 1/6

**5** 16

**7** $\int_0^1 \int_0^y dx\, dy$

**9** $13\pi/3$

**11** $\left(\frac{3\sqrt{2}}{2\pi}, \frac{3}{\pi} - \frac{3\sqrt{2}}{2\pi}\right)$

**13** $\int_{-\sqrt{3}}^{\sqrt{3}} \int_{-\sqrt{3-x^2}}^{\sqrt{3-x^2}} \int_1^{\sqrt{4-x^2-y^2}} dz\, dy\, dx,$

$\int_0^{2\pi} \int_0^{\sqrt{3}} \int_1^{\sqrt{4-r^2}} r\, dz\, dr\, d\theta,$

$\int_0^{2\pi} \int_0^{\pi/3} \int_{\sec\phi}^2 \rho^2 \sin\phi\, d\rho\, d\phi\, d\theta$, $5\pi/3$

**15** $\int_{-1}^1 \int_{-\sqrt{1-x^2}}^{\sqrt{1-x^2}} \int_0^{\sqrt{x^2+y^2}} dz\, dy\, dx,$

$\int_0^{2\pi} \int_0^1 \int_0^r r\, dz\, dr\, d\theta,$

$\int_0^{2\pi} \int_{\pi/4}^{\pi/2} \int_0^{\csc\phi} \rho^2 \sin\phi\, d\rho\, d\phi\, d\theta$, $2\pi/3$

**17** $\int_{-2}^2 \int_{-\sqrt{4-x^2}}^{\sqrt{4-x^2}} \int_{\sqrt{x^2+y^2}}^2 (x^2 + y^2)(1 + \sqrt{x^2+y^2+z^2})\, dz\, dy\, dx,$

$\int_0^{2\pi} \int_0^2 \int_r^2 r^3(1 + \sqrt{r^2+z^2})\, dz\, dr\, d\theta,$

$\int_0^{2\pi} \int_0^{\pi/4} \int_0^{2\sec\phi} \rho^4 \sin^3\phi(1 + \rho)\, d\rho\, d\phi\, d\theta$

**19** $\int_0^4 \int_0^{(6-x)/2} \int_0^{(6-x-2y)/3} dz\, dy\, dx,$

$\int_0^4 \int_0^{(6-x)/3} \int_0^{(6-x-3z)/2} dy\, dz\, dx$

**21** $\int_0^{\sqrt{2}} \int_0^{\sqrt{2}} \frac{u+v}{\sqrt{2}}\, du\, dv$

**23** $\int_0^1 \int_0^{1-u} \int_0^{1-u-v} 2u\, dw\, dv\, du$

# CHAPTER EIGHTEEN

## Exercises 18.1

**1** (a) $\int_0^{\pi/2} \cos t \sin t\, dt = 1/2$ (b) $\int_0^1 u\, du = \int_0^{\pi/2} \sin t \cos t\, dt$

(c) $\int_0^{\pi/2} \sin v \cos v\, dv = \int_0^{\pi/2} \cos t \sin t\, dt$

**3** (a) $\int_0^1 \frac{t}{t^2+1}\, dt = \frac{1}{2} \ln 2$ (b) $\int_1^2 \frac{1}{2u}\, du = \int_0^1 \frac{t}{t^2+1}\, dt$

(c) $\int_0^1 \frac{1-v}{(1-v)^2+1}\, dv = \int_0^1 \frac{t}{t^2+1}\, dt$

**5** 23

**7** $4\pi$

**9** $8\pi$

**11** $\frac{3\sqrt{2}}{8} + \frac{1}{16} \ln \frac{\sqrt{2}-1}{\sqrt{2}+1}$

**13** $\frac{1}{12}(17\sqrt{17} - 1)$

**15** $2\pi^2\sqrt{2}$

### Exercises 18.2

1 $-9/2$

3 $-4\pi$

5 64

7 1

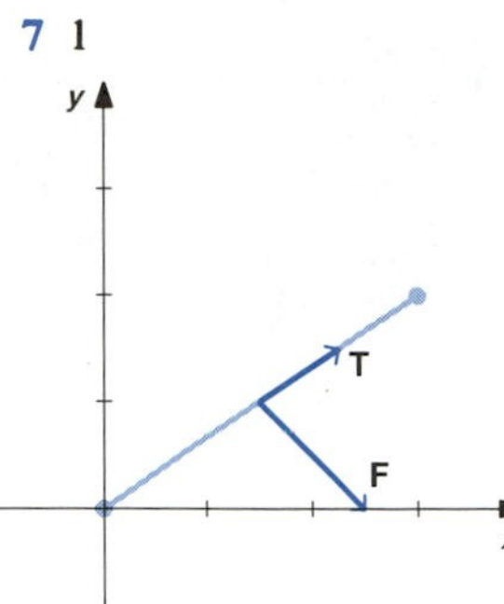


9 $-2\pi$

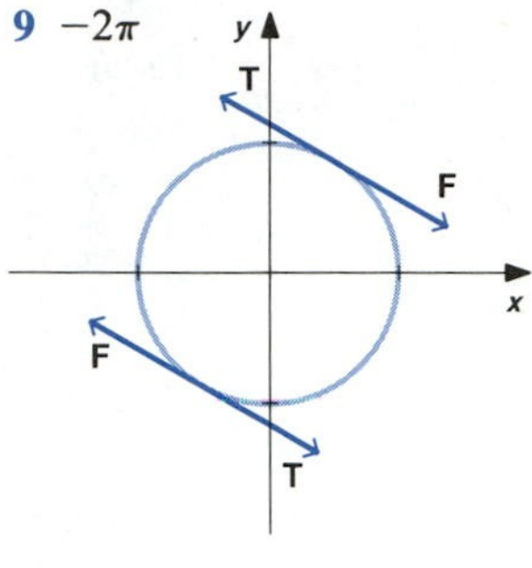


11 $2\pi^2$

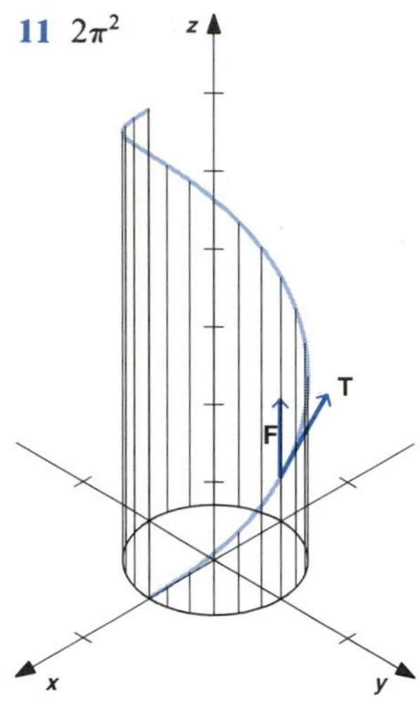


13 1/26

15 Positive

17 0

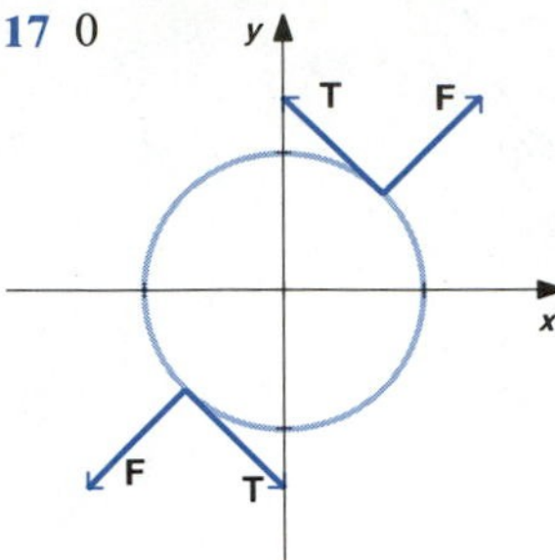


19 $2\pi$

21 $-1/30$

23 Positive

25 $2\pi$

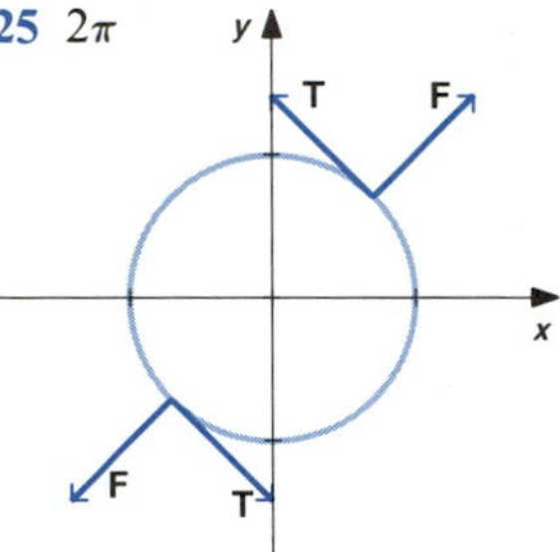


27 $2\pi r^2$

29 $4\pi$

### Exercises 18.3

1 Yes, 2

3 Yes, 8/15

5 Not directly (curve is not closed), $-2/3$

7 No, $16\pi \ln 2$

9 Yes, $-2\pi$

13 (a) $-\ln 2 + \pi/2$ (b) $\left(\dfrac{\pi - 2}{\pi - 2\ln 2}, 0\right)$

15 0, 0

17 $\dfrac{2y}{(y^2+1)^2}, \dfrac{1}{2}$

19 0, 0

21 2, $2\pi r^2$

23 5, $5\pi$

25 $\dfrac{-1}{(x^2+y^2)^{3/2}}, \pi$

### Exercises 18.4

1 (a) 96/5 (b) 64/3

3 (a) $-2/3$ (b) $-2/3$

5 (a) $-2/3$ (b) $-2/3$

7 (a) 1 (b) 0

9 (a) 9 (b) 0

11 (a) 2 (b) 0

13 (a) $2 \tan^{-1}(2/3)$ (b) 0

15 $x^2y + C$

17 No such function

19 $2xe^y + y^2/2 + C$

21 $\ln(x^2 + y^2) + C$

23 4

25 0

27 26

31 (a) Yes (b) No

35 $x^2y - xy^2$

37 $e^{xy} - 1 + y$

## Exercises 18.5

**1** $x^2 + y^2 = K$ **3** $y = K/x$

**5** $x^2 + xy - 3y^2 - x + 5y = K$

**7** $x^3 - xy + y^2 = K$ **9** $x^2 + y^2 + 2ye^x = K$

**11** $\sin x \cos^2 y = K$ **13** $xy = 2, x > 0$

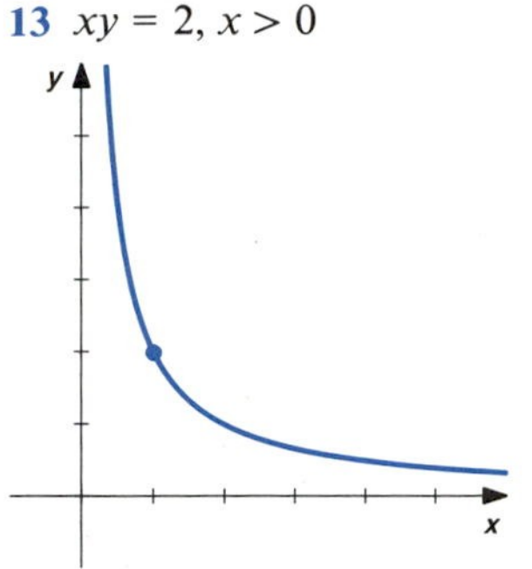


**15** $y = \sqrt{3}x, x > 0$

## Exercises 18.6

**1** $3\sqrt{14}/2$ **3** $12\sqrt{3}$

**5** $16\sqrt{5}/3$

**7** (a) $9\pi$ (b) $(0, 0, 9/2)$ (c) $81\pi/2$

**9** (a) $8\pi\sqrt{2}$ (b) $(0, 0, 8/5)$ (c) $64\pi\sqrt{2}/3$

**11** $1/2$ **13** (a) 0 (b) $64\pi$

**15** $\pi$ **17** $4\pi R$

## Exercises 18.7

**1** 60 **3** $16\pi$

**5** $9\pi/10$ **7** $32\pi/3$

**9** $4\pi r$ **11** $4\pi$

**13** $4\pi$

## Exercises 18.8

**1** (a) $\mathbf{i} + \mathbf{j} + \mathbf{k}$ (b) $-\pi$ (c) $-\pi$ (d) $-\pi$

**3** (a) $\dfrac{y}{\sqrt{x^2+y^2}}\mathbf{i} - \dfrac{x}{\sqrt{x^2+y^2}}\mathbf{j} + 2\mathbf{k}$ (b) $2\pi$ (c) $-2\pi$ (d) $2\pi$

**5** $18\pi$ **7** $-18\pi$

**9** 0 **11** $-16\pi$

**13** $9\pi$ **15** $-1/2$

**17** 0 **19** 0

**21** 0

## Exercises 18.9

**1** (a) 0 (b) 0 **3** (a) 0 (b) 0

**5** $x^2y + y^2z + xz^2 + C$ **7** $(x^2 + y^2 + z^2)^{1/2} + C$

**9** $1/3$ **11** $-1$

**13** 0

*Review Exercises*

**1** $\int_0^{\pi/2} 2 \sin t \cos t\sqrt{\sin^2 t + 4\cos^2 t}\, dt = \dfrac{14}{9}$

**3** $-2\pi$

**5** $\dfrac{8}{3}\pi^3$

**7** 0

**9** 2/3

**11** 0

**13** 0

**15** Yes, 9/2

**17** $-2, -8\pi$

**19** 0, 0

**21** $e^x \cos y - y^2 + C$

**23** Not possible

**25** $xy^2 - xz^2 + z + C$

**27** $4\pi\sqrt{5}$

**29** $\pi/2$

**31** $3\pi/2$

**33** $128\pi/3$

**35** $-2\pi$

**37** $-8\pi$

**39** $x^2 + xy + y^2 = K$

**41** $xy - y = 6$

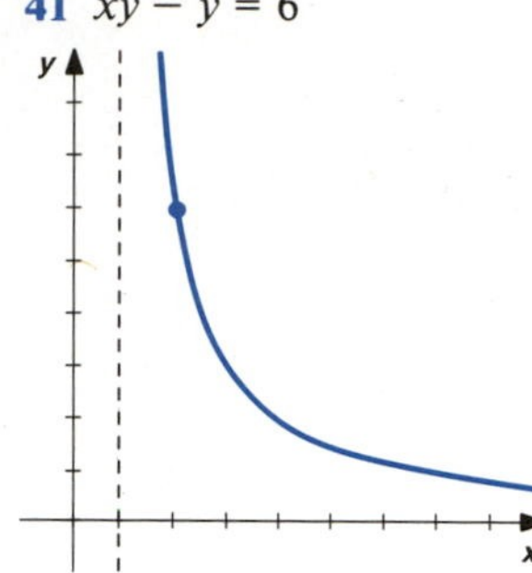

# Index

# TABLE OF INTEGRALS

*Basic integrals*

1 $\int u\,dv = uv - \int v\,du$

2 $\int u^\alpha\,du = \dfrac{u^{\alpha+1}}{\alpha+1} + C, \alpha \neq -1$

3 $\int \dfrac{1}{u}\,du = \ln|u| + C$

4 $\int e^u\,du = e^u + C$

5 $\int \cos u\,du = \sin u + C$

6 $\int \sin u\,du = -\cos u + C$

7 $\int \sec^2 u\,du = \tan u + C$

8 $\int \csc^2 u\,du = -\cot u + C$

9 $\int \sec u \tan u\,du = \sec u + C$

10 $\int \csc u \cot u\,du = -\csc u + C$

11 $\int \dfrac{1}{\sqrt{a^2-u^2}}\,du = \sin^{-1}\dfrac{u}{a} + C$

12 $\int \dfrac{1}{u^2+a^2}\,du = \dfrac{1}{a}\tan^{-1}\dfrac{u}{a} + C$

13 $\int \dfrac{1}{u\sqrt{u^2-a^2}}\,du = \dfrac{1}{a}\sec^{-1}\dfrac{|u|}{a} + C$

*Integrals that involve trigonometric functions*

14 $\int \tan u\,du = \ln|\sec u| + C$

15 $\int \cot u\,du = \ln|\sin u| + C$

16 $\int \sec u\,du = \ln|\sec u + \tan u| + C$

17 $\int \csc u\,du = \ln|\csc u - \cot u| + C$

18 $\int \sin^2 u\,du = -\dfrac{\sin u \cos u}{2} + \dfrac{u}{2} + C$

19 $\int \cos^2 u\,du = \dfrac{\sin u \cos u}{2} + \dfrac{u}{2} + C$

20 $\int \sec^3 u\,du = \dfrac{1}{2}\sec u \tan u + \dfrac{1}{2}\ln|\sec u + \tan u| + C$

21 $\int \sin^n u\,du = -\dfrac{\sin^{n-1} u \cos u}{n} + \dfrac{n-1}{n}\int \sin^{n-2} u\,du$

22 $\int \cos^n u\,du = \dfrac{\sin u \cos^{n-1} u}{n} + \dfrac{n-1}{n}\int \cos^{n-2} u\,du$

23 $\int \sin^m u \cos^n u\,du$

$= -\dfrac{\sin^{m-1} u \cos^{n+1} u}{m+n} + \dfrac{m-1}{m+n}\int \sin^{m-2} u \cos^n u\,du$

24 $\int \sin^m u \cos^n u\,du$

$= \dfrac{\sin^{m+1} u \cos^{n-1} u}{m+n} + \dfrac{n-1}{m+n}\int \sin^m u \cos^{n-2} u\,du$

25 $\int \tan^n u\,du = \dfrac{1}{n-1}\tan^{n-1} u - \int \tan^{n-2} u\,du$

26 $\int \sec^n u\,du = \dfrac{1}{n-1}\sec^{n-2} u \tan u + \dfrac{n-2}{n-1}\int \sec^{n-2} u\,du$

*Integrals that involve $a^2 + u^2$*

27 $\int \dfrac{1}{(u^2+a^2)^n}\,du$

$= \dfrac{u}{a^2(2n-2)(u^2+a^2)^{n-1}} + \dfrac{2n-3}{a^2(2n-2)}\int \dfrac{1}{(u^2+a^2)^{n-1}}\,du$

28 $\int \dfrac{1}{(a^2+u^2)^{3/2}}\,du = \dfrac{u}{a^2\sqrt{a^2+u^2}} + C$

29 $\int \dfrac{1}{u\sqrt{a^2+u^2}}\,du = -\dfrac{1}{a}\ln\left|\dfrac{a+\sqrt{a^2+u^2}}{u}\right| + C$

30 $\int \dfrac{1}{\sqrt{a^2+u^2}}\,du = \ln|u + \sqrt{a^2+u^2}| + C$

31 $\int \sqrt{a^2+u^2}\,du = \dfrac{u}{2}\sqrt{a^2+u^2} + \dfrac{a^2}{2}\ln|u + \sqrt{a^2+u^2}| + C$